Lecture Notes in Artificial Intelligence 9273

Subseries of Lecture Notes in Computer Science

LNAI Series Editors

Randy Goebel
University of Alberta, Edmonton, Canada
Yuzuru Tanaka
Hokkaido University, Sapporo, Japan
Wolfgang Wahlster
DFKI and Saarland University, Saarbrücken, Germany

LNAI Founding Series Editor

Joerg Siekmann
DFKI and Saarland University, Saarbrücken, Germany

More information about this series at http://www.springer.com/series/1244

Francisco Pereira · Penousal Machado
Ernesto Costa · Amílcar Cardoso (Eds.)

Progress in Artificial Intelligence

17th Portuguese Conference
on Artificial Intelligence, EPIA 2015
Coimbra, Portugal, September 8–11, 2015
Proceedings

 Springer

Editors
Francisco Pereira
ISEC - Coimbra Institute of Engineering
Polytechnic Institute of Coimbra
Coimbra
Portugal

Penousal Machado
CIUSC, Department of Informatics
   Engineering
University of Coimbra
Coimbra
Portugal

Ernesto Costa
CIUSC, Department of Informatics
   Engineering
University of Coimbra
Coimbra
Portugal

Amílcar Cardoso
CIUSC, Department of Informatics
   Engineering
University of Coimbra
Coimbra
Portugal

ISSN 0302-9743   ISSN 1611-3349 (electronic)
Lecture Notes in Artificial Intelligence
ISBN 978-3-319-23484-7   ISBN 978-3-319-23485-4 (eBook)
DOI 10.1007/978-3-319-23485-4

Library of Congress Control Number: 2015947099

LNCS Sublibrary: SL7 – Artificial Intelligence

Springer Cham Heidelberg New York Dordrecht London

Printed on acid-free paper

Springer International Publishing AG Switzerland is part of Springer Science+Business Media
(www.springer.com)

Preface

The Portuguese Conference on Artificial Intelligence is returning to Coimbra, 18 years after its previous edition in this city. UNESCO recently recognized the University of Coimbra, as well as some areas of the city, as a world heritage site, acknowledging its relevance in the dissemination of knowledge throughout the fields of arts, sciences, law, architecture, urban planning, and landscape. It is therefore the ideal period to welcome back the conference to Coimbra.

EPIA events have a longstanding tradition in the Portuguese Artificial Intelligence (AI) community. Their purpose is to promote research in AI and the scientific exchange among researchers, practitioners, scientists, and engineers in related disciplines. The first edition took place in 1985 and since 1989 it has occurred biennially as an international conference. Continuing this successful tradition, the 17th Portuguese Conference on Artificial Intelligence (EPIA 2015) took place on the campus of the University of Coimbra, Coimbra, Portugal (http://epia2015.dei.uc.pt), September 8–11, 2015.

Following the organization of recent editions, EPIA 2015 was organized as a series of thematic tracks. Each track was coordinated by an Organizing Committee and includes a specific international Program Committee, composed of experts in the corresponding scientific area. Twelve thematic tracks were selected for EPIA 2015: - Ambient Intelligence and Affective Environments (AmIA) - Artificial Intelligence in Medicine (AIM) - Artificial Intelligence in Transportation Systems (AITS) - Artificial Life and Evolutionary Algorithms (ALEA) - Computational Methods in Bioinformatics and Systems Biology (CMBSB) - General Artificial Intelligence (GAI) - Intelligent Information Systems (IIS) - Intelligent Robotics (IROBOT) - Knowledge Discovery and Business Intelligence (KDBI) - Multi-Agent Systems: Theory and Applications (MASTA) - Social Simulation and Modelling (SSM) - Text Mining and Applications (TEMA).

For this edition, 131 submissions were received. All papers were reviewed in a double-blind process by at least three members of the corresponding track Program Committee. Some of the submissions were reviewed by up to 5 reviewers. Following the revision process, 45 contributions were accepted as full regular papers. This corresponds to a full paper acceptance rate of approximately 34 %. Additionally, 36 other contributions were accepted as short papers. Geographically, the authors of accepted contributions belong to research groups from 18 different countries – Algeria, Austria, Brazil, the Czech Republic, Egypt, France, India, Italy, the Netherlands, Norway, Poland, Portugal, Russia, Spain, Sweden, the UK, and the USA –, confirming the attractiveness and international character of the conference.

We express our gratitude for the hard work of the track chairs and the members of the different Program Committees, as they were crucial for ensuring the high scientific quality of the event. This conference was only possible thanks to the joint effort of many people from different institutions. The main contributions came from the

University of Coimbra, the Polytechnic Institute of Coimbra, and the Centre for Informatics and Systems of the University of Coimbra. We would also like to thank the members of the Organizing Committee: Anabela Simões, António Leitão, João Correia, Jorge Ávila, Nuno Lourenço, and Pedro Martins. Acknowledgment is due to SISCOG – Sistemas Cognitivos S.A., Feedzai S.A., Thinkware S.A., iClio, FBA., and FCT – Fundação para a Ciência e a Tecnologia for the financial support. A final word goes to Easychair, which greatly simplified the management of submissions, reviews, and proceedings preparation, and to Springer for the assistance in publishing the current volume.

July 2015

Francisco Pereira
Penousal Machado
Ernesto Costa
Amlcar Cardoso

Organization

The 17th Portuguese Conference on Artificial Intelligence (EPIA 2015) was co-organized by the University of Coimbra, the Polytechnic Institute of Coimbra, and the Centre for Informatics and Systems of the University of Coimbra.

Conference Co-chairs

Francisco Pereira	Polytechnic Institute of Coimbra, Portugal
Penousal Machado	University of Coimbra, Portugal

Program Co-chairs

Ernesto Costa	University of Coimbra, Portugal
Francisco Pereira	Polytechnic Institute of Coimbra, Portugal

Organization Co-chairs

Amílcar Cardoso	University of Coimbra, Portugal
Penousal Machado	University of Coimbra, Portugal

Proceedings Co-chairs

João Correia	University of Coimbra, Portugal
Pedro Martins	University of Coimbra, Portugal

Program Committee

AmIA Track Chairs

Paulo Novais	University of Minho, Portugal
Goreti Marreiros	Polytechnic of Porto, Portugal
Ana Almeida	Polytechnic of Porto, Portugal
Sara Rodriguez Gonzalez	University of Salamanca, Spain

AmIA Program Committee

Antonio Fernández Caballero	University of Castilla-La Mancha, Spain
Amílcar Cardoso	University of Coimbra, Portugal
Andrew Ortony	Northwestern University, USA
Angelo Costa	University of Minho, Portugal
Antonio Camurri	University of Genoa, Italy
Boon Kiat Quek	National University of Singapore
Carlos Bento	University of Coimbra, Portugal

Carlos Ramos	Polytechnic of Porto, Portugal
César Analide	University of Minho, Portugal
Dante Tapia	University of Salamanca, Spain
Davide Carneiro	University of Minho, Portugal
Diego Gachet	European University of Madrid, Spain
Eva Hudlicka	Psychometrix Associates Blacksburg, USA
Florentino Fdez-Riverola	University of Vigo, Spain
Javier Jaen	Polytechnic University of Valencia, Spain
Javier Bajo	Polytechnic University of Madrid, Spain
José M. Molina	University Carlos III of Madrid, Spain
José Machado	University of Minho, Portugal
José Neves	University of Minho, Portugal
Juan M. Corchado	University of Salamanca, Spain
Laurence Devillers	LIMS-CNRS, France
Lino Figueiredo	Polytechnic of Porto, Portugal
Luís Macedo	University of Coimbra, Portugal
Ricardo Costa	Polytechnic of Porto, Portugal
Rui José	University of Minho, Portugal
Vicente Julian	Polytechnic University of Valencia, Spain

AIM Track Chairs

Manuel Filipe Santos	University of Minho, Portugal
Filipe Portela	University of Minho, Portugal

AIM Program Committee

Allan Tucker	Brunel University, UK
Álvaro Silva	Abel Salazar Biomedical Sciences Institute, Portugal
Andreas Holzinger	Medical University Graz, Austria
António Abelha	University of Minho, Portugal
Antonio Manuel de Jesus Pereira	Polytechnic Institute of Leiria, Portugal
Barna Iantovics	Petru Maior University of Tîrgu-Mure, Romania
Beatriz de la Iglesia	University of East Anglia, UK
Cinzia Pizzi	Università degli Studi di Padova, Italy
Danielle Mowery	University of Utah, USA
Do Kyoon Kim	Pennsylvania State University, USA
Giorgio Leonardi	University of Piemonte Orientale, Italy
Gören Falkman	University of Skövde, Sweden
Hélder Coelho	University of Lisbon, Portugal
Helena Lindgren	Ume University, Sweden
Inna Skarga-Bandurova	East Ukrainian National University, Ukraine
José Machado	University of Minho, Portugal
José Maia Neves	University of Minho, Portugal
Luca Anselma	University of Turin, Italy

Michael Ignaz Schumacher	University of Applied Sciences Western Switzerland
Miguel Angel Mayer	Pompeu Fabra University, Spain
Mohd Khanapi Abd Ghani	Technical University of Malaysia, Malaysia
Panagiotis Bamidis	Aristotelian Univ. of Thessaloniki, Greece
Pedro Gago	Polytechnic Institute of Leiria, Portugal
Pedro Pereira Rodrigues	University of Porto, Portugal
Rainer Schmidt	Institute for Biometrics and Medical Informatics, Germany
Ricardo Martinho	Polytechnic Institute of Leiria, Portugal
Rui Camacho	University of Porto, Portugal
Salva Tortajada	Polytechnic University of Valencia, Spain
Shabbir Syed-Abdul	Taipei Medical University, Taiwan
Shelly Sachdeva	Jaypee Institute of Information Technology, India
Szymon Wilk	Poznan University of Technology, Poland
Ulf Blanke	Swiss Federal Institute of Technology in Zurich, Switzerland
Werner Ceusters	University of New York at Buffalo, USA

AITS Track Chairs

| Rui Gomes | Universidade de Coimbra, Portugal |
| Rosaldo Rossetti | Universidade do Porto, Portugal |

AITS Program Committee

Achille Fonzone	Edinburgh Napier University, UK
Agachai Sumalee	Hong Kong Polytechnic University, Hong Kong
Alberto Fernandez	Universidad Rey Juan Carlos, Spain
Ana Almeida	Instituto Politécnico do Porto, Portugal
Ana L.C. Bazzan	Universidade Federal do Rio Grande do Sul, Brazil
Constantinos Antoniou	National Technical University of Athens, Greece
Cristina Olaverri-Monreal	AIT Austrian Institute of Technology GmbH, Austria
Eduardo Camponogara	Universidade Federal de Santa Catarina, Brazil
Giovanna Di Marzo Serugendo	University of Geneva, Switzerland
Gonçalo Correia	Delft University of Technology, Netherlands
Harry Timmermans	Eindhoven University of Technology, Netherlands
Hussein Dia	Swinburne University of Technology, Australia
José Telhada	Universidade do Minho, Portugal
Kai Nagel	Technische Universität Berlin, Germany
Luís Moreira Matias	NEC Europe Ltd, Germany
Luís Nunes	ISCTE Instituto Universitário de Lisboa, Portugal
Oded Cats	Delft University of Technology, Netherlands
Sascha Ossowski	Universidad Rey Juan Carlos, Spain
Shuming Tang	Chinese Academy of Sciences, China
Tânia Fontes	Universidade do Porto, Portugal

ALEA Track Chairs

Mauro Castelli	NOVA IMS, Universidade Nova de Lisboa, Portugal
Leonardo Vanneschi	NOVA IMS, Universidade Nova de Lisboa, Portugal
Sara Silva	University of Lisbon, Portugal

ALEA Program Committee

Alberto Moraglio	University of Exeter, UK
Alessandro Re	NOVA IMS, Universidade Nova de Lisboa, Portugal
Anabela Simes	Instituto Politecnico de Coimbra, Portugal
Antonio Della-Cioppa	Università degli Studi di Salerno, Italy
António Gaspar-Cunha	University of Minho, Portugal
Arnaud Liefooghe	Université de Lille, France
Carlos Cotta	Universidad de Málaga, Spain
Carlos Fernandes	Instituto Superior Técnico, Lisboa, Portugal
Carlos Gershenson	Universidad Nacional Autónoma de México, Mexico
Carlotta Orsenigo	Politecnico di Milano, Italy
Christian Blum	University of the Basque Country, Spain
Ernesto Costa	University of Coimbra, Portugal
Fernando Lobo	University of Algarve, Portugal
Helio Barbosa	Laboratório Nacional de Computação Científica, Brazil
Ivanoe De-Falco	National Research Council, Italy
Ivo Gonçalves	University of Coimbra, Portugal
James Foster	University of Idaho, USA
Jin-Kao Hao	University of Angers, France
Leonardo Trujillo	Instituto Tecnológico de Tijuana, Mexico
Luca Manzoni	University of Milano-Bicocca, Italy
Luís Correia	University of Lisbon, Portugal
Luís Paquete	University of Coimbra, Portugal
Marc Schoenauer	INRIA, France
Mario Giacobini	University of Turin, Italy
Pedro Mariano	University of Lisbon, Portugal
Penousal Machado	University of Coimbra, Portugal
Rui Mendes	University of Minho, Portugal
Stefano Beretta	University of Milano-Bicocca, Italy
Stefano Cagnoni	University of Parma, Italy
Telmo Menezes	Centre National de la Recherche Scientifique, France

CMBSB Track Chairs

Rui Camacho	Universidade do Porto, Portugal
Miguel Rocha	Universidade do Minho, Portugal
Sara Madeira	Instituto Superior Técnico, Portugal
José Luís Oliveira	Universidade de Aveiro, Portugal

CMBSB Program Committee

Francisco Couto	University of Lisbon, Portugal
Susana Vinga	IDMEC-LAETA, IST-UL, Portugal
Marie-France Sagot	INRIA Grenoble Rhône-Alpes and Université de Lyon 1, France
Alexessander Couto Alves	Imperial College London, UK
Alexandre P. Francisco	Technical University of Lisbon, Portugal
Vítor Santos Costa	Universidade do Porto, Portugal
Mário J. Silva	Universidade de Lisboa, Portugal
Fernando Diaz	University of Valladolid, Spain
Sérgio Matos	Universidade de Aveiro, Portugal
Paulo Azevedo	Universidade do Minho, Portugal
Rui Mendes	Universidade do Minho, Portugal
Inês Dutra	Universidade do Porto, Portugal
Nuno A. Fonseca	European Bioinformatics Institute, UK
Florentino Fdez-Riverola	University of Vigo, Spain
André Carvalho	USP, Brazil
Alexandra Carvalho	IT/IST, Portugal
Arlindo Oliveira	IST/INESC-ID and Cadence Research Laboratories, Portugal
Ross King	University of Manchester, UK
Luís M. Rocha	Indiana University, USA

GAI Track Chairs

Francisco Pereira	Polytechnic Institute of Coimbra, Portugal
Penousal Machado	University of Coimbra, Portugal

GAI Program Committee

Adriana Giret	Universitat Politècnica de València, Spain
Alexandra Carvalho	Technical University of Lisbon, Portugal
Amal El Fallah	Pierre-and-Marie-Curie University, France
Amílcar Cardoso	University of Coimbra, Portugal
Andrea Omicini	University of Bologna, Italy
Arlindo Oliveira	INESC-ID, Portugal
Carlos Bento	University of Coimbra, Portugal
Carlos Ramos	Polytechnic Institute of Porto, Portugal
César Analide	University of Minho, Portugal
Eric de La Clergerie	INRIA, France
Ernesto Costa	University of Coimbra, Portugal
Eugénio Oliveira	University of Porto, Portugal
Frank Dignum	Utrecht University, Netherlands
Gal Dias	University of Caen, France
Hélder Coelho	University of Lisbon, Portugal
Irene Rodrigues	University of Évora, Portugal

João Balsa	University of Lisbon, Portugal
João Gama	University of Porto, Portugal
João Leite	New University of Lisbon, Portugal
John-Jules Meyer	Utrecht University, Netherlands
José Cascalho	University of Azores, Portugal
José Neves	University of Minho, Portugal
José Gabriel Pereira Lopes	New University of Lisbon, Portugal
José Júlio Alferes	New University of Lisbon, Portugal
Juan Corchado	University of Salamanca, Spain
Luís Antunes	University of Lisbon, Portugal
Luís Cavique	University Aberta, Portugal
Luís Seabra Lopes	University of Aveiro, Portugal
Luís Correia	University of Lisbon, Portugal
Luís Macedo	University of Coimbra, Portugal
Michael Rovatsos	University of Edinburgh, UK
Miguel Calejo	APPIA, Portugal
Paulo Cortez	University of Minho, Portugal
Paulo Gomes	University of Coimbra, Portugal
Paulo Moura Oliveira	University of Trás-os-Montes and Alto Douro, Portugal
Paulo Urbano	University of Lisbon, Portugal
Pavel Brazdil	University of Porto, Portugal
Pedro Barahona	New University of Lisbon, Portugal
Pedro Henriques	University of Minho, Portugal
Pedro Mariano	University of Lisbon, Portugal
Rui Camacho	University of Porto, Portugal
Salvador Abreu	University of Évora, Portugal

IIS Track Chairs

Álvaro Rocha	University of Coimbra, Portugal
Luís Paulo Reis	University of Minho, Portugal
Adolfo Lozano	University of Extremadura, Spain

IIS Program Committee

Fernando Bobillo	University of Zaragoza, Spain
Tossapon Boongoen	Royal Thai Air Force Academy, Thailand
Carlos Costa	IUL-ISCTE, Portugal
Hironori Washizaki	Waseda University, Japan
Vitalyi Talanin	Zaporozhye Institute of Economics & Information Technologies, Ukraine
Mu-Song Chen	Da-Yeh University, Taiwan
Garyfallos Arabatzis	Democritus University of Thrace, Greece
Khalid Benali	Université de Lorraine, France
Salama Mostafa	UNITEN, Malaysia
Fernando Ribeiro	Polytechnic Institute of Castelo Branco, Portugal
Pedro Henriques Abreu	University of Coimbra, Portugal

Radouane Yafia Ibn	Zohr University, Morocco
Maria José Sousa	Universidade Europeia, Portugal
Mijalce Santa	Ss Cyril and Methodius University, Macedonia
Sławomir Żółkiewski	Silesian University of Technology, Poland
Kuan Yew Wong	Universiti Teknologi Malaysia, Malaysia
Alvaro Prieto	University of Extremadura, Spain
Roberto Rodriguez-Echeverria	University of Extremadura, Spain
Yair Wiseman	Bar-Ilan University, Israel
Babak Rouhani	Payame Noor University, Iran
Hing Kai Chan	University of Nottingham Ningbo China, China
José Palma	University of Murcia, Spain
Manuel Mazzara	Innopolis University, Russia
Brígida Mónica Faria	Polytechnic Institute of Porto, Portugal

IROBOT Track Chairs

Luís Paulo Reis	University of Minho, Portugal
Nuno Lau	University of Aveiro, Portugal
Brígida Mónica Faria	Polytechnic Institute of Porto, Portugal
Rui P. Rocha	University of Coimbra, Portugal

IROBOT Program Committee

António J.R. Neves	University of Aveiro, Portugal
Antonio P. Moreira	University of Porto, Portugal
Armando Sousa	University of Porto, Portugal
Carlos Cardeira	University of Lisbon, Portugal
Cristina Santos	University of Minho, Portugal
Filipe Santos	INESC TEC, Portugal
João Fabro	Federal University of Technology-Parana, Brazil
Josémar Rodrigues de Souza	Bahia State University, Brazil
Luís Correia	University of Lisbon, Portugal
Luís Mota	Lisbon University Institute, Portugal
Manuel Fernando Silva	Polytechnic Institute of Porto, Portugal
Nicolas Jouandeau	Paris 8 University, France
Paulo Goncalves	Polytechnic Institute of Castelo Branco, Portugal
Paulo Urbano	University of Lisbon, Portugal
Pedro Abreu	University of Coimbra, Portugal
Rodrigo Braga	Federal University of Santa Catarina, Brazil

KDBI Track Chairs

Paulo Cortez	University of Minho, Portugal
Luís Cavique	Open University, Portugal
João Gama	University of Porto, Portugal

Nuno Marques   New University of Lisbon, Portugal
Manuel Filipe Santos   University of Minho, Portugal

KDBI Program Committee

Agnès Braud   Univ. Robert Schuman, France
Albert Bifet   University of Waikato, New Zealand
Aline Villavicencio   UFRGS, Brazil
Alípio Jorge   University of Porto, Portugal
Amílcar Oliveira   Open University, Portugal
André Carvalho   University of São Paulo, Brazil
Armando Mendes   University of the Azores, Portugal
Bernardete Ribeiro   University of Coimbra, Portugal
Carlos Ferreira Gomes   Institute of Eng. of Porto, Portugal
Elaine Faria   Federal University of Uberlandia, Brazil
Fátima Rodrigues   Institute of Eng. of Porto, Portugal
Fernando Bação   New University of Lisbon, Portugal
Filipe Pinto   Polytechnical Inst. Leiria, Portugal
Gladys Castillo   Choose Digital, USA
José Costa   UFRN, Brazil
Karin Becker   UFRGS, Brazil
Leandro Krug Wives   UFRGS, Brazil
Luís Lamb   UFRGS, Brazil
Manuel Fernandez Delgado   University of Santiago Compostela, Spain
Marcos Domingues   University of São Paulo, Brazil
Margarida Cardoso   ISCTE-IUL, Portugal
Mark Embrechts   Rensselaer Polytechnic Institute, USA
Mohamed Gaber   University of Portsmouth, UK
Murate Testik   Hacettepe University, Turkey
Ning Chen   Institute of Eng. of Porto, Portugal
Orlando Belo   University of Minho, Portugal
Paulo Gomes   University of Coimbra, Portugal
Pedro Castillo   University of Granada, Spain
Phillipe Lenca   Télécom Bretagne, France
Rita Ribeiro   University of Porto, Portugal
Rui Camacho   University of Porto, Portugal
Stéphane Lallich   University of Lyon 2, France
Yanchang Zhao   Australian Government, Australia

MASTA Track Chairs

Ana Paula Rocha   Porto University, Portugal
Pedro Henriques Abreu   Coimbra University, Portugal
Jomi Fred Hubner   Universidade Federal de Santa Catarina, Brazil
Jordi Sabater Mir   IIIA-CSIC, Spain
Luís Moniz   Lisbon University, Portugal

MASTA Program Committee

Alessandra Alaniz Macedo	São Paulo University, Brazil
António Castro	LIACC, Portugal
António Carlos da Rocha Costa	Universidade Federal do Rio Grande, Brazil
Brigida Mónica Faria	Polytechnic Institute of Porto, Portugal
Carlos Carrascosa	Universidad Politecnica de Valencia, Spain
César Analide	Minho University, Portugal
Daniel Castro Silva	LIACC-Porto University, Portugal
Didac Busquets	Imperial College London, UK
Eugénio Oliveira	LIACC-Porto University, Portugal
Felipe Meneguzzi	Pontificia Universidade Católica do Rio Grande do Sul, Brazil
Francisco Grimaldo	Universidad de Valencia, Spain
Frank Dignum	Utrecht University, Netherlands
Hélder Coelho	Lisbon University, Portugal
Henrique Lopes Cardoso	LIACC-Porto University, Portugal
Jaime Sichmann	São Paulo University, Brazil
Javier Carbó	Universidad Carlos III, Spain
Joana Urbano	Instituto Superior Miguel Torga, Portugal
João Balsa	Lisbon University, Portugal
Laurent Vercouter	Ecole Nationale Supérieure des Mines de Saint-Etienne, France
Luís Correia	Lisbon University, Portugal
Luís Macedo	Coimbra University, Portugal
Luís Paulo Reis	LIACC-Minho University, Portugal
Manuel Filipe Santos	Minho University, Portugal
Maria Fasli	University of Essex, UK
Michael Schumacher	University of Applied Sciences, Western Switzerland
Márcia Ito	São Paulo Faculty of Technology, Brazil
Nicoletta Fornara	University of Lugano, Switzerland
Nuno Lau	Aveiro University, Portugal
Olivier Boisser	ENS Mines Saint-Etienne, France
Pablo Noriega	IIIA-CSIC, Spain
Paulo Trigo	Superior Institute of Engineering of Lisbon, Portugal
Paulo Urbano	Lisbon University, Portugal
Rafael Bordini	Pontifícia Universidade Católica do Rio Grande do Sul, Brazil
Ramón Hermoso	University of Zaragoza, Spain
Rosaldo Rossetti	Porto University, Portugal
Virginia Dignum	Delft University of Technology, Netherlands
Viviane Torres Da Silva	Universidade Federal Fluminense, Brazil
Wamberto Vasconcelos	University of Aberdeen, UK

SSM Track Chairs

Luís Antunes	Universidade de Lisboa, Portugal
Graçaliz Pereira Dimuro	Universidade Federal do Rio Grande, Brazil
Pedro Campos	Universidade do Porto, Portugal
Juan Pavón	Universidad Complutense de Madrid, Spain

SSM Program Committee

Frédéric Amblard	Univ. Toulouse 1, France
Pedro Andrade	INPE, Brazil
Tânya Araújo	ISEG, Portugal
Robert Axtell	George Mason Univ., USA
João Balsa	Univ. Lisbon, Portugal
Ana Bazzan	UFRGS, Brazil
François Bousquet	CIRAD/IRRI, Thailand
Amílcar Cardoso	University of Coimbra, Portugal
Cristiano Castelfranchi	ISTC/CNR, Italia
Shu-Heng Chen	National Chengchi Univ., Taiwan
Claudio Cioffi-Revilla	George Mason Univ., USA
Hélder Coelho	Univ. Lisbon, Portugal
Rosaria Conte	ISTC/CNR Rome, Italy
Nuno David	ISCTE, Portugal
Paul Davidsson	Blekinge Inst. Technology, Sweden
Guillaume Deffuant	Cemagref, France
Alexis Drogoul	IRD, France
Julie Dugdale	Lab. d'Informatique Grenoble, France
Bruce Edmonds	Centre for Policy Modelling, UK
Nigel Gilbert	Univ. Surrey, UK
Nick Gotts	Macaulay Inst., Scotland, UK
David Hales	The Open Univ., UK
Samer Hassan	Univ. Complutense Madrid, Spain
Rainer Hegselmann	Univ. Bayreuth, Germany
Wander Jager	Univ. Groningen, Netherlands
Adolfo Lópes Paredes	Univ. Valladolid, Spain
Pedro Magalhães	ICS, Portugal
Scott Moss	Centre for Policy Modelling, UK
Jean-Pierre Muller	CIRAD, France
Akira Namatame	National Defense Academy, Japan
Fernando Neto	Univ. Pernambuco, Brazil
Carlos Ramos	GECAD – ISEP, Portugal
Juliette Rouchier	Greqam/CNRS, France
David Sallach	Univ. Chicago, USA
Keith Sawyer	Washington Univ. St. Louis, USA
Carles Sierra	IIIA, Spain
Elizabeth Sklar	City Univ. New York, USA

Keiki Takadama	Univ. Electro-communications, Japan
Oswaldo Teran	Univ. Los Andes, Venezuela
Takao Terano	Univ. Tsukuba, Japan
Jan Treur	Vrije Univ. Amsterdam, The Netherlands
Klaus Troitzsch	Univ. Koblenz, Germany
Harko Verhagen	Stockholm Univ., Sweden

TEMA Track Chairs

Joaquim F. Ferreira da Silva	NOVA LINCS FCT/UNL, Portugal
Gabriel Pereira Lopes	NOVA LINCS FCT/UNL, Portugal
Hugo Gonçalo Oliveira	CISUC, University of Coimbra, Portugal
Vitor R. Rocio	Universidade Aberta, Portugal
Gaël Dias	University of Caen Basse-Normandie, France

TEMA Program Committee

Adam Jatowt	Kyoto University, Japan
Adeline Nazarenko	University of Paris 13, France
Aline Villavicencio	Universidade Federal do Rio Grande do Sul, Brazil
Antoine Doucet	University of Caen, France
António Branco	Universidade de Lisboa, Portugal
Béatrice Daille	University of Nantes, France
Belinda Maia	Universidade do Porto, Portugal
Brigitte Grau	LIMSI, France
Bruno Cremilleux	University of Caen, France
Christel Vrain	Université d'Orléans, France
Eric de La Clergerie	INRIA, France
Gabriel Pereira Lopes	NOVA LINCS FCT/UNL, Portugal
Gaël Dias	University of Caen Basse-Normandie, France
Gracinda Carvalho	Universidade Aberta, Portugal
Gregory Grefenstette	CEA, France
Hugo Gonalo Oliveira	CISUC, University of Coimbra, Portugal
Isabelle Tellier	University of Orléans, France
Joaquim F. Ferreira da Silva	NOVA LINCS FCT/UNL, Portugal
João Balsa	Universidade de Lisboa, Portugal
João Magalhães	Universidade Nova de Lisboa, Portugal
Lluís Padró	Universitat Politècnica de Catalunya, Spain
Lucinda Carvalho	Universidade Aberta, Portugal
Manuel Vilares Ferro	University of Vigo, Spain
Marc Spaniol	University of Caen Basse-Normandie, France
Marcelo Finger	Universidade de São Paulo, Brazil
Maria das Graças Volpe Nunes	Universidade de São Paulo, Brazil
Mark Lee	University of Birmingham, UK
Nuno Mamede	Universidade Técnica de Lisboa, Portugal
Nuno Marques	Universidade Nova de Lisboa, Portugal

Pablo Gamallo   Universidade de Santiago de Compostela, Spain
Paulo Quaresma   Universidade de Évora, Portugal
Pavel Brazdil   University of Porto, Portugal
Pierre Zweigenbaum   CNRS-LIMSI, France
Spela Vintar   University of Ljubljana, Slovenia
Vitor R. Rocio   Universidade Aberta, Portugal

Additional Reviewers

Aching, Jorge
Adamatti, Diana Francisca
Adedoyin-Olowe, Mariam
Andeadis, Pavlos
André, João
Barbosa, Raquel
Billa, Cleo
Cardoso, Douglas
Ferreira, Carlos
Flavien, Balbo
Francisco, Garijo
Fuentes-Fernández,
    Rubén

García-Magariño, Iván
Gonçalves, Eder Mateus
Gorgonio, Flavius
Kheiri, Ahmed
Lopes Silva, Maria
    Amélia
Magessi, Nuno
Mota, Fernanda
Nibau Antunes, Francisco
Nunes, Davide
Paiva, Fábio
Pasa, Leandro
Pinto, Andry

Ramos, Ana
Rocha, Luis
Rodrigues, Filipe
Santos, António Paulo
Sarmento, Rui
Serrano, Emilio
Shatnawi, Safwan
Soulas, Julie
Souza, Jackson
Trigo, Luis

Contents

Artificial Life and Evolutionary Algorithms

Intelligent Robotics

Knowledge Discovery and Business Intelligence

Multi-agent Systems: Theory and Applications

Social Simulation and Modelling

Text Mining and Applications

Ambient Intelligence
and Affective Environments

Defining Agents' Behaviour for Negotiation Contexts

João Carneiro[1(✉)], Diogo Martinho[1], Goreti Marreiros[1], and Paulo Novais[2]

[1] GECAD – Knowledge Engineering and Decision Support Group, Institute of Engineering,
Polytechnic of Porto, Porto, Portugal
{jomrc,1090557,mgt}@isep.ipp.pt
[2] CCTC – Computer Science and Technology Center, University of Minho, Braga, Portugal
pjon@di.uminho.pt

Abstract. Agents who represent participants in the group decision-making context require a certain number of individual traits in order to be successful. By using argumentation models, agents are capable to defend the interests of those who they represent, and also justify and support their ideas and actions. However, regardless of how much knowledge they might hold, it is essential to define their behaviour. In this paper (1) is presented a study about the most important models to infer different types of behaviours that can be adapted and used in this context, (2) are proposed rules that must be followed to affect positively the system when defining behaviours and (3) is proposed the adaptation of a conflict management model to the context of Group Decision Support Systems. We propose one approach that (a) intends to reflect a natural way of human behaviour in the agents, (b) provides an easier way to reach an agreement between all parties involved and (c) does not have high configuration costs to the participants. Our approach will offer a simple yet perceptible configuration tool that can be used by the participants and contribute to more intelligent communications between agents and makes possible for the participants to have a better understanding of the types of interactions experienced by the agents belonging to the system.

Keywords: Group decision support systems · Ubiquitous computing · Affective computing · Multi-Agent systems · Automatic negotiation

1 Introduction

Rahwan et al. (2003) defined negotiation as "a form of interaction in which a group of agents, with conflicting interests and a desire to cooperate, try to come to a mutually acceptable agreement on the division of scarce resources" [1], Hadidi, Dimopoulos, and Moraitis (2011), defined negotiation as "the process of looking for an agreement between two or several agents on one or more issues" [2] and El-Sisi and Mousa (2012) defined as "a process of reaching an agreement on the terms of a transaction such as price, quantity, for two or more parties in multi-agent systems such as E-Commerce. It tries to maximize the benefits to all parties" [3]. It is possible to verify in the literature a consensus regarding to the main approaches to deal with negotiation: game theory, heuristics and argumentation [1-3]. It is a known fact that game

© Springer International Publishing Switzerland 2015
F. Pereira et al. (Eds.) EPIA 2015, LNAI 9273, pp. 3–14, 2015.
DOI: 10.1007/978-3-319-23485-4_1

theoretic and heuristic based approaches evolved and turned more complex. With this development they have been used in a wide range of applications. However they share some limitations. In the majority of game-theoretic and heuristic models, agents exchange proposals, but these proposals are limited. Agents are not allowed to exchange any additional information other than what is expressed in the proposal itself. This can be problematic, for example, in situations where agents have limited information about the environment, or where their rational choices depend on those of other agents. Another important limitation is that agent's utilities or preferences are usually assumed to be completely characterized prior to the interaction. Thus, to overcome these limitations, argumentation-based negotiation appeared and turned one of the most popular approaches to negotiation [4], it has been extensively investigated and studied, as witnessed by many publications [5-7]. The main idea of argumentation-based negotiation is the ability to support offers with justifications and explanations, which play a key role in the negotiation settings. So, it allows the participants to the negotiation not only to exchange offers, but also reasons and justifications that support these offers in order to mutually influence their preference relations on the set of offers, and consequently the outcome of the dialogue.

It is simple to understand the parallelism between this approach and group decision-making. The idea of a group of agents exchanging arguments in order to achieve, for instance, a consensus, in order to support groups in decision-making process is easy to understand [8]. However the complexity of this process must not be underestimated, if considering a scenario where an agent seeks to defend the interests of who it represents and at the same time be part of a group that aims to reach a collective decision towards a problem for their organization [9, 10]. Not only are those agents simultaneously competitive and cooperative but also represent human beings. Establishing some sort of dialog, as well as the different types of arguments that can be exchanged by agents is only the first step towards the problem resolution. One agent that represents a decision-maker involved in a process of group decision-making may show different levels of experience and knowledge related with the situation and should behave accordingly. Literature shows that there are works on the subject [10-13], however it should be noted the existence of some flaws in terms of real world applicability of certain models. Some require high configuration costs that will not suit the different types of users they are built for and others show flaws that in our opinion are enough to affect the success of a Group Decision Support System (GDSS).

In this work it will be presented the most relevant models that allow inferring or configuring a behaviour style for a group decision-making context. It is also proposed a set of rules for which a behaviour model must follow without jeopardizing the entire GDSS and finally it is proposed an approach made through the modification of one existing model to the context of GDSS.

The rest of the paper is organized as follows: in the next section is presented the literature review. Section 3 presents our approach, where we identify different types of behaviours, defined with the use of an existing model and presented the set of rules that we believe that are the most important to allow defining types of behaviours for the agents in a way that does not compromise the system. In section 4 it will be discussed and debated how our approach can be applied to the context of GDSS and its differences compared with other existing approaches. Finally, some conclusions are taken in section 5, along with the work to be done hereafter.

2 Literature Review

The concern for identifying and understanding particular behavioural attitudes has led to many investigations and studies throughout the last decades with emphasis on proposing models and behaviour styles that can relate to the personality of the negotiator.

Carl Jung (1921), was the first to specify a model to study different psychological personality types based on four types of consciousness (sensation, intuition, thinking and feeling) that could in turn be combined with two types of attitudes (extraversion and introversion) and that way identify eight primary psychological types [14].

In 1962, Myers Briggs, developed a personality indicator model (The Meyers-Briggs Type Indicator) based on Jung's theories [15]. This indicator is used as a psychometric questionnaire and allows people to understand the world around them and how they behave and make decisions based on their preferences [16]. This model was useful in order to identify different styles of leadership, which were later specified in Keirsey and Bate's publication [17], in 1984, as four styles of leadership:

- Stabilizer: tends to be very clear and precise when defining objectives and organizing and planning tasks in order to achieve them. Stabilizer leaders are also reliable and trustworthy due to the fact they show concern for other worker's necessities and problems. They are able to increase the motivation of their workers by setting tradition and organization as an example of success;
- Catalyst: the main focus is to develop the quality of own work and the one provided by their staff. They serve the facilitator's role by bringing the best out of other people, and motivate other workers with their own enthusiasm and potential;
- Trouble-shooter: as the name suggests, focus on dealing and solving problems. They show great aptitude for solving urgent problems by being practical and immediate. They bring people together as a team by analysing what needs to be done and informing exactly what to do as quickly as possible;
- Visionary: visionaries act based on their own intuition and perception of the problems in order to make decisions. They have a mind projected for the future and plan idealistic scenarios and objectives which may not always be achievable.

Related to vocational behaviour, Holland [18], in 1973 proposed a hexagonal model (RIASEC model) where he differentiates six types of personality mainly used in careers environments and to guide through the individual's choice of vocation. Those types are defined as:

- Realistic: realistic individuals value things over people and ideas. They are mechanical and athletic, and prefer working outdoors with tools and objects;
- Investigative: investigative individuals have excellent analytic skills. They prefer working alone and solving complex problems;
- Artistic: artistic individuals show a deep sense of creativity and imagination. They prefer working on original projects and value ideas over things;

- Social: social individuals have high social aptitude, preferring social relationships and helping other people solving their problems. They prefer working with people over things;
- Enterprising: enterprising individuals show great communication and leadership skills, and are usually concerned about establishing direct influence on other people. They prefer dealing with people and ideas over things;
- Conventional: conventional individuals value order and efficiency. They show administrative and organization skills. They prefer dealing with numbers and words over people and ideas.

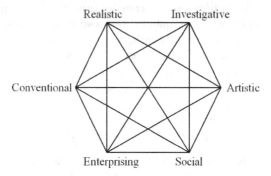

Fig. 1. Representation of Holland's hexagon model, adapted from [18]

It is important to note the distribution of these personalities in Holland's hexagon where personalities next to each other are the most similar while personalities facing against each other are the least similar (see Fig. 1).

Conflict management has always been an important area of decision-making, since it is very rare to find situations in group discussion where conflict is not present. In 1975, Thomas and Kilmann [19], also based on Jung's studies and a conflict-handling mode proposed by Blake and Mouton [20], suggested a model for interpersonal conflict-handling behaviour, defining five modes: competing, collaborating, compromising, avoiding and accommodating, according to two dimensions: assertiveness and cooperativeness. As seen in Fig. 2, both assertiveness and cooperativeness dimensions are related to integrative and distributive dimensions which were discussed by Walton and McKersie in 1965 [21]. Integrative dimensions refer to the overall satisfaction of the group involved in the discussion while distributive dimension refers to the individual satisfaction within the group. It is possible to see that the thinking-feeling dimension maps onto the distributive dimension while the introversion-extraversion dimension maps onto the integrative dimension. It is easy to understand this association by looking at competitors as the ones who seek the highest individual satisfaction, collaborators as the ones who prefer the highest satisfaction of the entire group. On the other hand avoiders do not worry about group satisfaction and accommodators do not worry about individual satisfaction. They also concluded that the thinking-feeling dimension did not move towards the integrative dimension, and also that the introversion-extraversion did not move towards the distributive dimension.

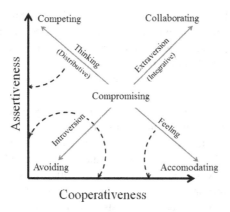

Fig. 2. Thomas and Kilmann's model for interpersonal conflict-handling behaviour, adapted from [19]

In 1992, Costa and McCrae [22] proposed a set of thirty traits extending the five-factor model of personality (OCEAN model) which included six facets for each of the factors. These traits were used in a study made by Howard and Howard [23] in order to help them separate different kinds of behaviour styles and identify corresponding themes. A theme is defined as "a trait which is attributable to the combined effect of two or more separate traits". Those styles and themes are based on common sense and general research, and some of them have already been mentioned before in this literature review, however it is also important to referrer other relevant styles that were suggested such as the Decision and Learning styles. Decision style includes the Autocratic, Bureaucratic, Diplomat and Consensus themes while Learning style includes the Classroom, Tutorial, Correspondence and Independent themes.

In 1995, Rahim and Magner [24] created a meta-model of styles for handling interpersonal conflict based on two dimensions: concern for self and concern for the other. This was the base for the five management styles identified as obliging, avoiding, dominating, integrating and compromising as will be explained in detail in the Section 3.

3   Methods

It is really important to define correctly the agent's behaviour in order to not jeopardize the validation of the entire GDSS. Sometimes, in this area of research, there is an exhaustive concern to find a better result and because of that, other variables may be forgotten which can make impossible the use of a certain approach in those situations. For example: Does it make sense for a decision-maker or a manager from a large company, with his super busy schedule have the patience/time to answer (seriously) to a questionnaire of 44 questions like "the Big Five Inventory" so that he can model his agent with his personality? Due to reasons like this we have defined a list with considerations to have when defining types of behaviours for the agents in the context here presented. The definition of behaviour should:

1. Enhance the capabilities of the agents, i.e., make the process more intelligent, more human and less sequential, even though it may not be visible in the conceptual model it must not be possible for the programmer to anticipate the sequence of interactions just by reading the code;
2. Be easy to configure (usability) or not need any configuration at all from the user (decision-maker);
3. Represent the interests of the decision-makers (strategy used), so that agent's way of acting meets the interests defined by the user (whenever possible);
4. Not be the reason for the decision-makers to give up using the application, i.e., in a hypothetical situation, a decision-maker should not "win" more decisions just because he knows how to manipulate/configure better the system;
5. Be available for everyone to benefit from it. Obviously all decision-makers face meetings in different ways. Their interests and knowledge for each topic is not always the same. Sometimes it may be of their interest to let others speak first and only after gathering all the information, elaborate a final opinion on the matter. Other times it may be important to control the entire conversation and try to convince the other participants to accept out opinion straightaway.

By taking into account all these points, we propose in this article a behaviour model for the decision-making context based on conflict styles defined by Rahim and Magner (1995) [24]. The styles defined are presented in Fig. 3 and have been adapted to our problem. Rahim and Magner reckons the existence of 5 types of conflict styles: integrating, obliging, dominating, avoiding and compromising. In their work, they suggested these styles in particular to describe different ways of behave in conflict situations. They defined these styles according to the level of concern a person has for reaching its own goal and reaching other people's objectives. This definition goes along exactly with what we consider that the agents that operate in a GDSS context should be, when we say that they are both cooperative and competitive simultaneously. Therefore this model ends up describing 5 conflict styles which support what we think that is required for the agents to have a positive behaviour in this context. It also has the advantage of being a model easy to understand and to use.

In our approach, the configuration of agent's behavior made by the decision-maker, will be done through the selection of one conflict style. The main idea is to define the agent with the participant's interests and strategies. For that, the definition of each conflict style should be clear and understandable for the decision-maker. The decision-maker can define in his agent different conflict styles throughout the process. For example, a decision-maker who is included in a decision process and has few or even no knowledge about the problem during the early stage of discussion. For that situation he may prefer to use an "avoiding" style and learn with what other people say, gather arguments and information that will support different options and that way learn more about the problem. In a following stage, when the decision-maker already has more information and knowledge about the problem, he may opt to use a more active and dominating style in order to convince others towards his opinion. Like mentioned before, there are many factors that can make the decision-maker face a meeting in different ways: interest about a topic, lack of knowledge about a topic, reckons the participation of more experienced people in the discussion, etc.

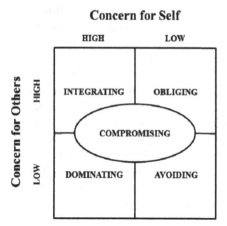

Fig. 3. Conflict Style, adapted from [24]

The different types of behaviour defined and that can be used by the agents are:

- Integrating (IN): This style should be selected every time the decision-maker considers that satisfying his own objectives is as important as satisfying the other participants' objectives. By choosing this conflict style, the agent will seek and cooperate with other agents in order to find a solution that is satisfactory to all the participants;
- Obliging (OB): This style should be selected if the decision-maker prefers to satisfy other participant's objectives instead of satisfying his own objectives. For example, in a situation where the decision-maker does not have any knowledge about the discussion topic;
- Dominating (DO): This style should be selected when the decision-maker only wants to pursuit his own objectives and do everything in his power to achieve them. For example, in a situation where the decision-maker is absolutely sure that his option to solve the problem is the most benefic. By using this style, the agent will be more dominant and will try to persuade the maximum possible agents. With this style the agent will prefer to risk everything to achieve his objectives even if that means he might end up at disadvantage because of that;
- Avoiding (AV): This style should be selected when the decision-maker does not have any interest in achieving either his own and other participants' objectives. For example, when the decision-maker has been include in a group discussion for which he does not have any sort of interest;
- Compromising (CO): This style should be selected when the decision-maker has a moderate interest in the topic and at the same time he also has a certain interest to achieve his own and other participants' objectives.

4   Discussion

Many approaches have been suggested in the literature which define/model agents with characteristics that will differentiate them from each other and as result will also show different ways of operating [11-13, 25-27]. However, even if many of those publications might be interesting for an academic context, they still show some issues that must be addressed. These issues that we will analyze are related to the context of support to group decision-making and also to competitive agents which that represent real individuals. There are several approaches in literature for (1) agents that are modeled according to the real participant personality (decision-maker) which they represent and (2) modeled with different intelligence levels (abilities) [10, 12]. One of the most used technics in literature is "The Big Five Inventory" questionnaire that allows to obtain values for each one of the personality traits defined in the model of "The Big Five" (openness, conscientiousness, extraversion, agreeableness and neuroticism) [11, 26, 28]. Theoretically, we can think that the way agents operate, which is similar to real participant because it is modeled with "the same personality" is perfect. However, defining an agent with a conflict style based on the values of personality traits may not be the right way to identify the decision-maker. What makes a human act in a particular way is the result of much more than just its personality, it is a set of factors such as: personality, emotions, humor, knowledge, and body (physic part), and it can also be considered other factors such as sensations and the spiritual part [29]. Another relevant question is the fact this type of approach allows that certain agents have advantage over other agents. Many may say and think that this occurrence is correct, because close to what happens in real life, there are decision-makers that are more apt and therefore have advantage over other decision-makers. However the questions that arise are the following: Would a product like this used by decision-makers that knew they would be at disadvantage by using this tool? Would it be possible to sell a product that does not guarantee equality between its future users? It is also important to discuss another relevant analysis point which is the fact that this type of approach, in some situations, might provide less intelligent and more sequential outputs.

The study of different types of behaviour in agents has been represented in literature by a reasonable number of contributions. However, it is a subject that most of the times offers validation problems. Although there are proposals with cases of study aiming to validate this subject, that validation is somewhat subjective most of the times. Even when trying to mathematically formulate the problem so that it becomes scientifically "proven", that proof may often feel forced. A reflection of this problem is the difference between the practiced approaches for social and exact sciences. It is clear for us, as computer science researchers that it is not our goal to elaborate a model for behavioural definition to use in specific scenarios. Instead, we will use a model defined and theoretically validated by others who work in areas that allow them to have these skills. However, the inclusion of intelligence in certain systems is growing at a blistering pace and some of the systems would not make sense nor would succeed without this inclusion. This means that it become more of common practice to adapt certain models that have not been designed specifically for the context for which they will be used. Because of that the evolution of the presented approaches will happen in an empirical way.

Another relevant condition is related with how most of the works are focused on very specific topics which may prevent a more pragmatic comparison of the various approaches. Even if in some situations the use of a specific technic (such as "The Big

Five Inventory") might make sense, in others, and even though it may scientifically provide a case of study with brilliant results, it can be responsible for jeopardizing the success of the system. Our work aims to support each participant (decision-maker) in the process of group decision-making. It is especially targeted for decision support in ubiquitous scenarios where participants are considered people with a very fast pace of life, where every second counts (top managers and executives). In our context the system will notify the participant whenever he is added in a decision process (for instance, by email), and after that every participant can access the system and model his agent according to his preferences (alternatives and attributes classification), as well as how he plans to face that decision process (informing the agent about the type of behaviour to have), always knowing that there are no required fields in the agent setup. This way provides more freedom for the user to configure (depending on his interest and time) his agent with detail or with no detail at all. As can be seen in this context (and referred previously) the agents must be cooperative and competitive. They are cooperative because they all seek one solution for the organization they belong to, and competitive because each agent seeks to defend the interests of its participant and persuade other agents to accept his preferred alternative. For us this means that if an agent is both cooperative and competitive then it cannot exhibit behaviour where it is only concerned in achieving its objectives and vice versa.

5   Conclusions and Future Work

The use of agents to represent/support humans as well as their intentions in negotiation context is relatively common practice in literature. There are several approaches which based on relationships allow agents to judge different levels for trust, credibility, intelligence, etc. Specifically looking at support to group decision-making context, a few approaches have appeared and propose modelling agents based on a number of characteristics that will allow them to operate in a way similar to how the decision-maker would in real life. If in one hand the modelling of an agent with certain human characteristics makes sense since it allows to define different types of behaviours and strategies according to the objectives of the decision-maker, on the other hand even if some of those approaches may seem intellectually interesting and complex, they affect the system where they belong due to many reasons, as for instance: illusory intelligence creation, unbalanced agent capabilities, high configuration costs e and weak representation of what in practice the decision-maker would want the agent operating model to be.

In this paper we presented (1) a study about the most important models that can be used to infer different types of behaviours that can be adapted and used in this context, (2), a set of rules that must be followed and that will positively affect the system when defining behaviours and (3) is proposed the adaptation of a conflict management model in the context of GDSS. Furthermore we included a new approach of how to look at this problem, and alert to the negative impact some other approaches might have in the system where they are used. Our approach intends to provide a more perceptible and concrete way for the decision-maker to understand the five types of behaviour that can be used to model the agent in support to group decision-making context where each agent represents a decision-maker. We believe that with our approach it will be simpler for agents to reach or suggest solutions since they are modeled with behaviours according to what the decision-maker wants. This makes it easier to reflect in the agent the concern to achieve the decision-maker's objectives or the objectives belonging to other participants in the decision process. With this approach the agents follow one defined type of behaviour that also works as a strategy that can be adopted by each one of the decision-makers.

As for future work we will work in the specific definition of each type of behaviour identified in this work. We intend to describe behaviours according to certain facets proposed in the Five Factor Model and also study tendencies for each type of behaviour to make questions, statements, and requests. At later stage we will integrate this model in the prototype of a group decision support system which we are developing.

Acknowledgements. This work is part-funded by ERDF - European Regional Development Fund through the COMPETE Programme (operational programme for competitiveness) within pro-ject FCOMP-01-0124-FEDER-028980 (PTDC/EEISII/1386/2012) and by National Funds through the FCT - Fundação para a Ciência e a Tecnologia (Portuguese Foun-dation for Science and Technology) with the João Carneiro PhD grant with the refer-ence SFRH/BD/89697/2012.

References

1. Rahwan, I., Ramchurn, S.D., Jennings, N.R., Mcburney, P., Parsons, S., Sonenberg, L.: Argumentation-based negotiation. The Knowledge Engineering Review **18**, 343–375 (2003)
2. Hadidi, N., Dimopoulos, Y., Moraitis, P.: Argumentative alternating offers. In: McBurney, P., Rahwan, I., Parsons, S. (eds.) ArgMAS 2010. LNCS, vol. 6614, pp. 105–122. Springer, Heidelberg (2011)
3. El-Sisi, A.B., Mousa, H.M.: Argumentation based negotiation in multiagent system. In: 2012 Seventh International Conference on, Computer Engineering & Systems (ICCES), pp. 261–266. IEEE (1012)
4. Marey, O., Bentahar, J., Asl, E.K., Mbarki, M., Dssouli, R.: Agents' Uncertainty in Argumentation-based Negotiation: Classification and Implementation. Procedia Computer Science **32**, 61–68 (2014)
5. Mbarki, M., Bentahar, J., Moulin, B.: Specification and complexity of strategic-based reasoning using argumentation. In: Maudet, N., Parsons, S., Rahwan, I. (eds.) ArgMAS 2006. LNCS (LNAI), vol. 4766, pp. 142–160. Springer, Heidelberg (2007)
6. Amgoud, L., Vesic, S.: A formal analysis of the outcomes of argumentation-based negotiations. In: The 10th International Conference on Autonomous Agents and Multiagent Systems, vol. 3, pp. 1237–1238. International Foundation for Autonomous Agents and Multiagent Systems (2011)

7. Bonzon, E., Dimopoulos, Y., Moraitis, P.: Knowing each other in argumentation-based negotiation. In: Proceedings of the 11th International Conference on Autonomous Agents and Multiagent Systems, vol. 3, pp. 1413–1414. International Foundation for Autonomous Agents and Multiagent Systems (2012)
8. Kraus, S., Sycara, K., Evenchik, A.: Reaching agreements through argumentation: a logical model and implementation. Artificial Intelligence **104**, 1–69 (1998)
9. Faratin, P., Sierra, C., Jennings, N.R.: Negotiation decision functions for autonomous agents. Robotics and Autonomous Systems **24**, 159–182 (1998)
10. Rahwan, I., Kowalczyk, R., Pham, H.H.: Intelligent agents for automated one-to-many e-commerce negotiation. In: Australian Computer Science Communications, pp. 197–204. Australian Computer Society Inc. (2002)
11. Santos, R., Marreiros, G., Ramos, C., Neves, J., Bulas-Cruz, J.: Personality, emotion, and mood in agent-based group decision making (2011)
12. Kakas, A., Moraitis, P.: Argumentation based decision making for autonomous agents. In: Proceedings of the Second International Joint Conference on Autonomous Agents and Multiagent Systems, pp. 883–890. ACM (2003)
13. Zamfirescu, C.-B.: An agent-oriented approach for supporting Self-facilitation for group decisions. Studies in Informatics and control **12**, 137–148 (2003)
14. Jung, C.G.: Psychological types. The collected works of CG Jung **6**(18), 169–170 (1971). Princeton University Press
15. Myers-Briggs, I.: The Myers-Briggs type indicator manual. Educational Testing Service, Prinecton (1962)
16. Myers, I.B., Myers, P.B.: Gifts differing: Understanding personality type. Davies-Black Pub. (1980)
17. Bates, M., Keirsey, D.: Please Understand Me: Character and Temperament Types. Prometheus Nemesis Book Co., Del Mar (1984)
18. Holland, J.L.: Making vocational choices: A theory of vocational personalities and work environments. Psychological Assessment Resources (1997)
19. Kilmann, R.H., Thomas, K.W.: Interpersonal conflict-handling behavior as reflections of Jungian personality dimensions. Psychological reports **37**, 971–980 (1975)
20. Blake, R.R., Mouton, J.S.: The new managerial grid: strategic new insights into a proven system for increasing organization productivity and individual effectiveness, plus a revealing examination of how your managerial style can affect your mental and physical health. Gulf Pub. Co. (1964)
21. Walton, R.E., McKersie, R.B.: A behavioral theory of labor negotiations: An analysis of a social interaction system. Cornell University Press (1991)
22. Costa, P.T., MacCrae, R.R.: Revised NEO Personality Inventory (NEO PI-R) and NEO Five-Factor Inventory (NEO FFI): Professional Manual. Psychological Assessment Resources (1992)
23. Howard, P.J., Howard, J.M.: The big five quickstart: An introduction to the five-factor model of personality for human resource professionals. ERIC Clearinghouse (1995)
24. Rahim, M.A., Magner, N.R.: Confirmatory factor analysis of the styles of handling interpersonal conflict: First-order factor model and its invariance across groups. Journal of Applied Psychology **80**, 122 (1995)
25. Allbeck, J., Badler, N.: Toward representing agent behaviors modified by personality and emotion. Embodied Conversational Agents at AAMAS **2**, 15–19 (2002)
26. Badler, N., Allbeck, J., Zhao, L., Byun, M.: Representing and parameterizing agent behaviors. In: Proceedings of Computer Animation, 2002, pp. 133–143. IEEE (2002)

27. Velásquez, J.D.: Modeling emotions and other motivations in synthetic agents. In: AAAI/IAAI, pp. 10–15. Citeseer (1997)
28. Durupinar, F., Allbeck, J., Pelechano, N., Badler, N.: Creating crowd variation with the ocean personality model. In: Proceedings of the 7th International Joint Conference on Autonomous Agents and Multiagent Systems, vol. 3, pp. 1217–1220. International Foundation for Autonomous Agents and Multiagent Systems (2008)
29. Pasquali, L.: Os tipos humanos: A teoria da personalidade. Differences **7**, 359–378 (2000)

Improving User Privacy and the Accuracy of User Identification in Behavioral Biometrics

André Pimenta[(⊠)], Davide Carneiro, José Neves, and Paulo Novais

Algoritmi Centre, Universidade Do Minho, Braga, Portugal
{apimenta,dcarneiro,jneves,pjon}@di.uminho.pt

Abstract. Humans exhibit their personality and their behavior through their daily actions. Moreover, these actions also show how behaviors differ between different scenarios or contexts. However, Human behavior is a complex issue as it results from the interaction of various internal and external factors such as personality, culture, education, social roles and social context, life experiences, among many others. This implies that a specific user may show different behaviors for a similar circumstance if one or more of these factors change. In past work we have addressed the development of behavior-based user identification based on keystroke and mouse dynamics. However, user states such as stress or fatigue significantly change interaction patterns, risking the accuracy of the identification. In this paper we address the effects of these variables on keystroke and mouse dynamics. We also show how, despite these effects, user identification can be successfully carried out, especially if task-specific information is considered.

Keywords: Mental fatigue · Machine learning · Computer security · Behavioral biometrics · Behavioral analysis

1 Introduction

In the last years there has been a significantly increase in jobs that are mentally stressful or fatiguing, in expense of otherwise traditional physically demanding jobs [1]. Workers are nowadays faced not only with more mentally demanding jobs but also with demanding work conditions (e.g. positions of high responsibility, competition, risk of unemployment, working by shifts, working extra hours). This results in the recent emergence of stress and mental fatigue as some of the most serious epidemics of the twenty first century [2,3]. In terms of workplace indicators, this has an impact on human error, productivity or quality of work and of the workplace. In terms of social or personal indicators, this has an impact on quality of life, health or personal development. Moreover, there is an increase in the loss of focus that leads people to be unaware of risks, thus lowering the security threshold.

Recent studies show the negative impact of working extra hours on productivity [4,5]: people work more but produce less. Stressful milieus just add to the

© Springer International Publishing Switzerland 2015
F. Pereira et al. (Eds.) EPIA 2015, LNAI 9273, pp. 15–26, 2015.
DOI: 10.1007/978-3-319-23485-4_2

problem. The questions is thus how to create the optimal conditions to meet productivity requirements while respecting people's well-being and health. Since each worker is different, what procedures need to be implemented to measure the level of stress or burnout of each individual worker? And their level of productivity? The mere observation of these indicators using traditional invasive means may change the worker's behavior, leading to biased results that do not reflect his actual state. Directly asking, through questionnaires or similar instruments, can also lead to biased results as workers are often unwilling to share feelings concerning their workplace with their coworkers.

Recent approaches for assessing and managing fatigue have been developed that look at one's interaction patterns with technological devices to assess one's state (e.g. we type at a lower pace when fatigued). Moreover, the same approaches can be used to identify users (e.g. each individual types in a different manner). This field is known as behavioral biometrics. In this paper we present a framework for collecting from users in a transparent way, that allows to perform tasks commonly associated to behavioral biometrics. Moreover, this framework respects user privacy. Finally, we show how including the user's state and information about the interaction context may improve the accuracy of user identification. The main objective of this work is to define a reliable and non-intrusive user identification system for access control.

1.1 Human-Computer Interaction

Currently there is a very large community of computer users. In fact, most of the jobs today require some type of computer usage [6]. Moreover, with services like home banking, tech support services and social networks, people start to interact more with computers than with other persons, even to take care of important aspects of their lives.

There is thus a new form of communication in which computers serve as intermediaries. Therefore, the perception of a conversation between human beings is somewhat lost. By using a computer, the people involved cannot perceive one of the most important aspects in communication which is body language. Other important aspects include speech, intonation or facial expressions, just to mention a few. To overcome this loss of information computer systems must adopt new processes to better perceive the human being [7].

In fact, Humans tend to show their personality or their state through their actions, even in an unconscious way. Facial expressions and body language, for example, have been known as a gateway for feelings that result in intentions. The resultant actions can be traced to a certain behavior. Therefore, it is safe to assume that a human behavior can be outlined even if the person does not want to explicitly share that information.

Human behavior can also be deemed complex as it is driven by internal and external factors, such as personality, culture, education, social roles, life experiences, among others. Accurately evaluating a behavior requires constant observation of all the elements that are able to provide useful information. Nonetheless, with our evolved social skills, we are often able to conceal certain emotions or shielding them with others. Thus, multi-modal approaches should be considered for increased accuracy (e.g. relying solely on visual emotion recognition may be less accurate than including additional aspects such as tone of voice or speech rhythm). An potentially interesting process is to consider involuntary actions, that cannot consciously controlled by the individual. This often includes movement and posture, hand gestures, touches and interaction with objects the environment, among others. Therefore, it can be stated that the observation and evaluation of behaviors must consider not only the displayed emotions and actions, but also the nature of the interaction with the environment.

One particularly interesting source of these unconscious behaviors is our interaction with computers and other technological devices. In fact, we don't think about controlling the rhythm at which we type on a keyboard or the way we move the mouse when we become fatigued or stress, although we might want to hide our state from our colleagues or superiors. But the truth is that our interaction does change, as we have established in previous work [8,9]. Under different states we use the mouse and the keyboard differently. Moreover, we also interact differently with the smartphone. The case of the smartphone is still more interesting as it provides a range of sensors that are not available on other platforms and can provide very valuable information about interaction patterns, including a touch screen (that provides information about touches, their intensities, their area or their duration), gyroscopes, accelerometers, among others.

Human-Computer Interaction thus becomes a very promising field when it comes to reliable sources of information for characterizing one's state following behavioral approaches, the main advantage being that the individual generally does not or cannot consistently change such precise and fine-grained behaviors. Thus, while one can, to some extent, fake facial expressions and transmit a chosen emotion or state, doing so successfully through these behaviors results much harder.

2 Security Systems

As stated before, with the increase of Human-computer Interaction and the development of social-networking, people rapidly increased the rate at which they share information, even when it is sensitive personal information. Some of the current concerns are thus privacy, security and data protection.

Deemed as the most profitable crime of modern times, information theft is increasing at an alarming rate, the corporate sector being the most affected. This type of crime frequently includes the theft of personal data and often does a large damage in a person's life, as well as in companies. One way to improve security is to build robust authentication systems which prevent unauthorized access to machines. These systems may range from password-based authentication, in the simplest cases, to biometry systems in the most complex ones.

One of the possible ways to increase security in the context of user identification is to consider the user' behavior when interacting with a technological device. As each individual has a particular way to walk, talk, laugh or do anything else, each one of us has also their own interaction patterns with technological devices. Moreover, most of the applications we interact with have a specific flow of operation or require a particular type of interaction, restricting or conditioning the user's possible behaviors to a smaller set. Maintaining a behavioral profile of authorized users may allow to identify uncommon behaviors on the current user that may indicate a possible unauthorized user. This is even more likely to work when behavioral information for particular applications is used.

Such systems are known as Behavioural Biometrics: they rely on the users' behavioral profiles to establish the behavior of authorized users. Whenever, when analyzing the behavior of the current user, a moderate behavioral deviation is detected from the known profiles, the system may take action such as logging off, notifying the administrator or using an alternative method of authentication.

Such systems can also include behaviors other than the ones originated from keyboard and mouse interaction patterns [10]. In fact, any action performed on the technological device can be used as threat detection. For example, if the system console is started and the authorized user of the device never used the console before, a potential invasion may be taking place. Similar actions can be taken on other applications or even on specific commands (e.g. it is unlikely that a user with a non-expert profile suddenly starts using advanced commands on the console).

To implement behavioral biometrics, distinct procedures that can be adopted, such as:

- Biometric Sketch: this method uses the user drawings as templates for comparison [11,12]. The system collects patterns from the user drawing and compares it to others in a database. Singularity is assured by the number of possible combinations. The downside is that the drawings must be very precise, which in most of the cases is quite difficult, even more by using a standard mouse to draw.
- GUI Interaction: this technique uses the interaction of the user with visual interfaces of the applications and compares it to the model present in the database. For every application that the user interacts with, a model must be present. Thus, both the model and the application must be saved in the database. This method requires that every action per application is saved, resulting in a large amount of information to be maintained. Moreover, each new application or update must be trained and modeled. Therefore, this method is very strict and complex to implement and maintain.
- Keystroke Dynamics: this method uses the keyboard as input and is based on the user's typing patterns. It captures the keys pressed, measuring time and pressing patterns, extracting several features about the typing behavior. This is a well established method, as it relies solely on the user's interaction, allowing to create simple and usable models.
- Mouse Dynamics: this method consists in capturing the mouse movement and translating it into a model. All the interactions are considered, such as movement and clicking. This method is similar to the GUI Interaction method but simpler; although it suffers the same context problem. The main

purpose of this method is to accompany the Keystroke Dynamics, providing additional data to the main model, thus increasing accuracy and decreasing false positives.

– Tapping: this method focuses on the pulse wave resulting from a touch sensor. The pulse duration and tapping interval are the proprieties considered for the analysis. The way a user acts with a smartphone or computer can be monitored, and a model can be extracted. Conceptually, it is a solid method, as tapping follows the same concept of the keystroke dynamics, but in reality the usage context must also be considered, as tapping changes according to the application being used.

In this work, Keystroke Dynamics and Mouse Dynamics are chosen as inputs. Their broader features and availability are the traits that suit the aim of the intended system. They are application independent and operating system independent, and are nowadays the most common input method when interacting with computers.

3   The Framework

The framework developed in the context of this work is a unified system, composed of two main modules: fatigue monitoring and security. The process of monitoring is implemented using an application that captures the keyboard and mouse inputs transparently. The features used are the same in both systems and are defined in more detail in [13]. The features extracted from the keyboard are:

– Key down time: total time that a key is pressed
– Time between keys: time between a key being released and the following key being pressed
– Writing speed: the rhythm at which keys are pressed
– Errors per key: quantification of the use of the backspace and delete keys

And the features extracted from the mouse are:

– Distance of the mouse to the straight line: the sum of the distances between the pointer and the line defined by each two consecutive clicks
– Mouse acceleration: the acceleration of the mouse
– Mouse velocity: the velocity of the mouse
– Average distance of the mouse to the straight line: the average distance at which the mouse pointer is from the line that is defined by each two consecutive clicks
– Total excess of distance: the distance that the mouse travels in excess between each two consecutive clicks
– Average excess of distance: the distance that the mouse travels in excess divided by the shortest distance between each two consecutive clicks
– Time between clicks: the time between each two consecutive clicks
– Distance during clicks: distance traveled by the mouse while performing a click

– Double click duration: the time between two clicks in a double click event
– Absolute sum of angles: the quantification of how much the mouse turns, regardless of the direction of the turn, between each two consecutive clicks
– Signed sum of angles: the quantification of how much the mouse turns, considering the direction of the turn, between each two consecutive clicks
– Distance between clicks: the distance traveled between each two consecutive clicks

The data gathered may be processed differently to extract the information related to each scope of the framework (fatigue monitoring and stress). An integrated system is beneficial due to the jointly nature of the data and to the fact that only one application is present locally, thus having a low footprint on computer resources. Furthermore, these are two areas that are intrinsically connected, and one can affect the other. Their joint analysis is fundamental to the achievement of the proposed objectives.

3.1 Providing Security and Safety in the Monitoring System

The use of Keystroke Dynamics and Mouse Dynamics for detecting behavior is extremely useful, especially since it allows the creation of non-intrusive and non-invasive systems.

However the use of behavioral biometrics, in particular the use of keystroke dynamics, can pose some risks to the data security and privacy of the user, especially when the data obtained from the mouse and keyboard are processed by remote or 3rd party Web Services. For this reason, and to ensure the users' privacy and data security, data must be encrypted. The most sensitive data is indeed the information about which key was pressed. However, this specific information is mostly irrelevant for behavior analysis, i.e., we are not interested in *which* keys the user presses but on *how* the user presses them.

We therefore propose the collection of events created by the mouse and the keyboard in the following way:

– MOV, timestamp, posX, posY - an event describing the movement of the mouse, in a given time, to coordinates (posX, posY) in the screen;
– MOUSE_DOWN, timestamp, [Left—Right], posX, posY - this event describes the first half of a click (when the mouse button is pressed down), in a given time. It also describes which of the buttons was pressed (left or right) and the position of the mouse in that instant;
– MOUSE_UP, timestamp, [Left—Right], posX, posY - an event similar to the previous one but describing the second part of the click, when the mouse button is released;
– MOUSE_WHEEL, timestamp, dif - this event describes a mouse wheel scroll of amount dif, in a given time;
– KEY_DOWN, timestamp, encrypted key - identifies a given key from the keyboard being pressed down, at a given time;
– KEY_UP, timestamp, encrypted key - describes the release of a given key from the keyboard, in a given time;

In this approach, the encrypted key replaces the information about the specific key pressed. It is therefore still possible, while hiding what the user wrote, to extract the previously mentioned features, thus guaranteeing the user's privacy. The encryption of the pressed keys is carried out through the generation of random key encryptions at different times. This is done as depicted in Algorithm 1, which exemplifies the developed approach for the case of the key down time feature. An example of the result of the algorithm is depicted in Table 1, where a record with and without encryption is depicted for different keys.

Data: Keyboard Inputs
Result: List of KeyDownTime records
while *Keyboard Inputs have records* **do**
    Get timestamp of a KEY_DOWN ;
    Get code of a KEY_DOWN ;
    while *Keyboard Inputs have records* **do**
        Get timestamp of a KEY_UP ;
        Get code of a KEY_UP ;
        if *KEY_DOWN code and KEY_UP code are the same* **then**
            Save the difference between KEY_UP timestamp and KEY_DOWN timestamp;
    end
end
end

Algorithm 1. Key down time algorithm

Table 1. Example of a keyboard log with and without encryption.

No Encryption	Random Encryption Key 1
KD,63521596046072,A	KD,63521596046072,1COc0qNOOk=
KU,63521596046165,A	KU,63521596046165,1COc0qNOOk=
KD,63521596057943,v	KD,63521596057943,sMA0Wu0n3k=
KU,63521596058037,v	KU,63521596058037,sMA0Wu0n3k=
KU,63521596058084,a	KU,63521596058084,hb0s0lHEF8+sA==
KU,63521596058037,v	KU,63521596058037,sMA0Wu0n3k=

Hiding user input is just one part of the solution for the issue of user security. The other is to prevent intrusions. In this scope, behavioral biometrics security systems can run in two different modes [14]: identification mode and verification mode. In this system we use the identification mode instead of the verification mode to ensure a constant user identification in the monitoring system.

The identification mode is the process of trying to discover the identity of a person by examining a biometric pattern calculated from biometric data of the person. In this mode the user is identified based on information previously collected from keystroke dynamics profiles of all users. For each user, a biometric profile is built in a training phase. When in the running phase, the usage pattern being created in real-time is compared to every known model, producing either a score or a distance that describes the similarity between the pattern and the model. The system assigns the pattern to the user with the most similar biometric model. Thus, the user is identified without the need for extra information.

4 Case Study

The system was analyzed and tested in four different ways. As a first step we used the records of 40 users registered in the monitoring system to train different models. The created models were then validated through 150 random system usage records, taken from the system in order to validate the models created. In a second step models were created using the type of task to be performed in addition to the biometric information, and in the third step, models were trained using the user's fatigue state. We finally created models that, in addition to using biometric data, also used the type of task and the user's mental state at the time of registration. Both the type of task as the user's mental state are provided by the monitoring system.

The participants, forty in total (36 men, 4 women) which are registered in the monitoring system. Their age ranged between 18 and 45. The following requirements were established to select, among all the volunteers, the ones that participated: (1) familiarity and proficiency with the use of the computer; (2) use of the computer on a daily basis and throughout the day; (3) owning at least one personal computer.

4.1 Results and Discussion

After training different models (Naive Bayes, KNN, SVM and Random Forest) with data from different users on the system, different degrees of accuracy have been obtained in user identification, as depicted in Figure 1. It is also possible to observe that the type of task and the level of fatigue have influence on the process of identifying the user, since these factors effectively influence interaction patterns. Taking this information into consideration allows the creation of more accurate models.

The type of task being carried out during the monitoring of the interaction patterns is particularly important, mainly due to the very nature of the task, as well as the set of tools available to perform the task. Figure 2 shows the values of features Key Down Time and Average Excess of Distance for five different types of applications: Chat, Leisure, Office, Reading and Programming. The way each different application conditions the interaction behavior is explicit. Such information must, therefore, absolutely be considered while developing behavioral biometrics system based on input behavior. Table 2 further supports this claim showing that data collected in different types of applications has statistically

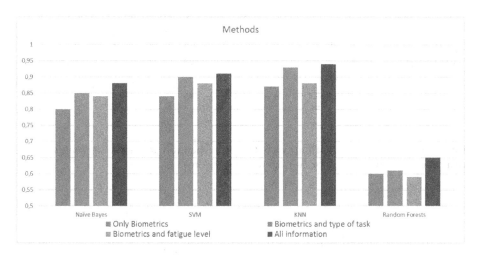

Fig. 1. Accuracy of the different algorithms and inputs considered for user authentication.

Table 2. p-values of the Kruskal-Wallis test when comparing the data organized according to the types of application and to the level of fatigue. In the vast majority of the cases, the differences between the different groups are statistically significant.

Features	Applications	Fatigue
Writing Velocity	4.4e-15	0.43
Time Between Keys	<2.2e-16	<2.2e-16
Key Down Time	<2.2e-16	3.8e-14
Distance of the Mouse to the SL	0.01	<2.2e-16
Mouse acceleration	<2.2e-16	1.3e-09
Average Distance of the Mouse to the SL	0.03	<2.2e-16
Mouse velocity	<2.2e-16	<2.2e-16
Average Excess of Distance	0.04	1.6e-08
Time Between Clicks	<2.2e-16	<2.2e-16
Distance During Clicks	<2.2e-16	0.05
Total Excess of Distance	0.22	<2.2e-16
Double Click Duration	1.0e-4	<2.2e-16
Absolute Sum of Angles	8.1e-3	2.5e-13
Signed Sum of Angles	0.91	<2.2e-16
Distance between clicks	0.06	<2.2e-16

significant differences for most of the features. The same happens for different levels of fatigue.

Another extremely important aspect in user identification is the influence of mental states on interaction patterns. Previous studies by our research team [8,15] show that individuals under different states of stress, fatigue, high/low mental workload or even mood evidence significant behavior changes that impact interaction patterns with devices. They do, consequently, influence behavioral

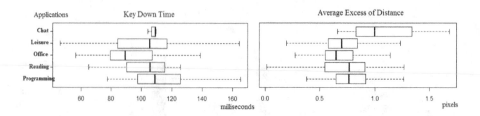

Fig. 2. Differences in the distributions of the data when comparing interaction patterns with different applications, for two interaction features.

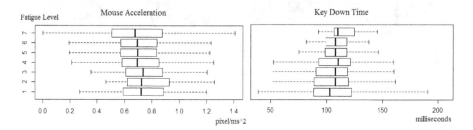

Fig. 3. Effects of different levels of fatigue on the interaction patterns, depicted for two different features.

biometric features. Figure 3 depicts this influence for two interaction features. Numbers in the y-axis represent the level of fatigue as self-reported by the individual using the seven-point USAFSAM Mental Fatigue Scale questionnaire [16]. Each value represents the following state:

1. Fully alert. Wide awake. Extremely peppy.
2. Very lively. Responsive, but not at peak.
3. Okay. Somewhat fresh.
4. A little tired. Less than fresh.
5. Moderately tired. Let down.
6. Extremely tired. Very difficult to concentrate.
7. Completely exhausted. Unable to function effectively. Ready to drop.

It is therefore possible to see how increased levels of fatigue result in generally less efficient interactions of the participants with the computer. For example, a higher value of the Mouse Acceleration depicts a more efficient interaction behavior in which the user is moving the mouse. The same happens with Key Down Time, where a shorter time corresponds to a more efficiency in the use of the keyboard.

This conclusion justifies the need for the inclusion of mental states on behavioral biometrics approaches and explains the increased accuracy of the presented approach concerning user identification when all modalities are used jointly: interaction patterns, user state and type of application, as depicted previously in Figure 1.

5   Conclusions and Future Work

This paper describes a non-invasive and non-intrusive approach to the field of behavioral biometrics. Specifically, we described a system that, transparently, acquires interaction features from keystroke dynamics and mouse dynamics, with the purpose of user identification. We analyzed the significant effect that mental states have on interaction patterns, particularly fatigue, to conclude that these aspects must be considered when developing identification systems based on interaction patterns. Likewise, we also show how user interaction significantly changes according to the application being used.

These notions support the claim put forward in this paper that accurate user identification approaches based on behavioral biometrics must absolutely include these aspects when building user profiles. That much is evidenced by our results. In all four classification algorithms used for user identification, the cases in which all these types of information are used together outperform the others.

Moreover, the proposed approach is also concerned with user privacy. Specifically, it is based not on *what* the user inputs but on *how* the user inputs. To this end it conceals each key under an encryption key that prevents remote services from knowing what the user typed. The data gathered from users thus respects their privacy while allowing to extract the necessary features for the system to carry out its task, i.e., to detect unauthorized accesses.

In future work we will focus on including additional user features, namely in what concerns user state. Specifically, we will include information regarding the user's level of stress, which we have already determined in previous work to significantly influence interaction patterns with technological devices. In the long-term we will include more potentially interesting features, namely the user's mood. It is our conviction that this kind of information, when available, can significantly increase the accuracy of user identification in behavioral biometrics approaches.

Acknowledgments. This work is part-funded by ERDF - European Regional Development Fund through the COMPETE Programme (operational programme for competitiveness) and by National Funds through the FCT (Portuguese Foundation for Science and Technology) within project FCOMP-01-0124-FEDER-028980 (PTDC/EEI-SII/1386/2012) and project PEst-OE/EEI/UI0752/2014.

References

1. Tanabe, S., Nishihara, N.: Productivity and fatigue. Indoor Air **14**(s7), 126–133 (2004)
2. Miller, J.C.: Cognitive Performance Research at Brooks Air Force Base, Texas, 1960–2009. Smashwords, March 2013

3. Wainwright, D., Calnan, M.: Work stress: the making of a modern epidemic. McGraw-Hill International (2002)
4. Folkard, S., Tucker, P.: Shift work, safety and productivity. Occupational Medicine **53**(2), 95–101 (2003)
5. Rosekind, M.R.: Underestimating the societal costs of impaired alertness: safety, health and productivity risks. Sleep Medicine **6**, S21–S25 (2005)
6. Beauvisage, T.: Computer usage in daily life. In: Proceedings of the SIGCHI conference on Human Factors in Computing Systems, pp. 575–584. ACM (2009)
7. Pantic, M., Rothkrantz, L.J.: Toward an affect-sensitive multimodal human-computer interaction. Proceedings of the IEEE **91**(9), 1370–1390 (2003)
8. Pimenta, A., Carneiro, D., Novais, P., Neves, J.: Monitoring mental fatigue through the analysis of keyboard and mouse interaction patterns. In: Pan, J.-S., Polycarpou, M.M., Woźniak, M., de Carvalho, A.C.P.L.F., Quintián, H., Corchado, E. (eds.) HAIS 2013. LNCS, vol. 8073, pp. 222–231. Springer, Heidelberg (2013)
9. Carneiro, D., Castillo, J.C., Novais, P., Fernández-Caballero, A., Neves, J.: Multi-modal behavioral analysis for non-invasive stress detection. Expert Systems with Applications **39**(18), 13376–13389 (2012)
10. Lee, P.M., Chen, L.Y., Tsui, W.H., Hsiao, T.C.: Will user authentication using keystroke dynamics biometrics be interfered by emotions?-nctu-15 affective keyboard typing dataset for hypothesis testing
11. Al-Zubi, S., Brömme, A., Tönnies, K.D.: Using an active shape structural model for biometric sketch recognition. In: Michaelis, B., Krell, G. (eds.) DAGM 2003. LNCS, vol. 2781, pp. 187–195. Springer, Heidelberg (2003)
12. Brömme, A., Al-Zubi, S.: Multifactor biometric sketch authentication. In: BIOSIG, pp. 81–90 (2003)
13. Pimenta, A., Carneiro, D., Novais, P., Neves, J.: Analysis of human performance as a measure of mental fatigue. In: Polycarpou, M., de Carvalho, A.C.P.L.F., Pan, J.-S., Woźniak, M., Quintian, H., Corchado, E. (eds.) HAIS 2014. LNCS, vol. 8480, pp. 389–401. Springer, Heidelberg (2014)
14. Shanmugapriya, D., Padmavathi, G.: A survey of biometric keystroke dynamics: Approaches, security and challenges (2009). arXiv preprint arXiv:0910.0817
15. Rodrigues, M., Gonçalves, S., Carneiro, D., Novais, P., Fdez-Riverola, F.: Keystrokes and clicks: measuring stress on E-learning students. In: Casillas, J., Martínez-López, F.J., Vicari, R., De la Prieta, F. (eds.) Management Intelligent Systems. AISC, vol. 220, pp. 119–126. Springer, Heidelberg (2013)
16. Samn, S.W., Perelli, L.P.: Estimating aircrew fatigue: a technique with application to airlift operations. Technical report, DTIC Document (1982)

Including Emotion in Learning Process

Ana Raquel Faria[1](✉), Ana Almeida[1], Constantino Martins[1],
Ramiro Gonçalves[2], and Lino Figueiredo[1]

[1] GECAD - Knowledge Engineering and Decision Support Research Center Institute
of Engineering, Polytechnic of Porto (ISEP/IPP), Porto, Portugal
`{arf,amn,acm,lbf}@isepp.ipp.pt`
[2] Universidade de Trás-Os-Montes E Alto Douro, Vila Real, Portugal
`ramiro@utad.pt`

Abstract. The purpose of this paper is to propose new architecture that includes the student's, learning preferences, personality traits and emotions to adapt the user interface and learning path to the students need and requirements. This aims to reduce the difficulty and emotional stain that students encounter while interacting with learning platforms.

Keywords: Learning styles · Student modeling · Adaptive systems · Affective computing

1 Introduction

Human to human communication depends on the interpretation of a mix of audio-visual and sensorial signals. To simulate the same behaviour in a human-machine interface the computer has to be able to detect affective state and behaviour alterations and modify its interaction accordingly. The field of affective computing develops systems and mechanisms that are able to recognize, interpret and simulate human emotions [1] so closing the gap between human and machine. Affective Computing concept was introduced by Rosalind Picard in 1995 as a tool to improve human-machine interfaces by including affective connotations.

Emotion plays an important role in the decision process and knowledge acquisition of an individual. Therefore, it directly influences the perception, the learning process, the way people communicate and the way rational decisions are made. So the importance of understanding affects and its effect on cognition and in the learning process. To understand how the emotions influence the learning process several models were developed. Models like Russell's Circumplex model [2] are used to describe user's emotion space and Kort's learning spiral model [3] are used to explore the affective evolution during learning process.

In a traditional learning context the teacher serves as a facilitator between the student and his learning material. Students, as individuals, differ in their social, intellectual, physical, emotional, and ethnic characteristics. Also, differ in their learning rates, objectives and motivation turning, their behaviour rather unpredictable. The teacher has to perceive the student state of mind and adjust the teaching process to the student's needs

© Springer International Publishing Switzerland 2015
F. Pereira et al. (Eds.) EPIA 2015, LNAI 9273, pp. 27–32, 2015.
DOI: 10.1007/978-3-319-23485-4_3

and behaviour. In a learning platform this feedback process does not take place in real time and, sometimes it is not what the student requires to overcome the problem at hand. This overtime can become a major problem and cause difficulties to the student learning process. A possible solution to this problem could be the addition of mechanisms, to the learning platforms, that enable computers to detect and interfere when the student requires help or motivation to complete a task. The major difficulties of this work will be the detection of these situations and how to interfere. The method of detection cannot be too intrusive, because that would affect the student behaviour in a negative way that would cause damage to his learning process. Another important issue is the selection of the variable to monitor. This can include the capture of emotions, behaviour or learning results among others. Finally, determining which will be the computer intervention when a help situation is detected in order to reverse the help situation.

2   EmotionTest Prototype

In order to prove that emotion can have influence in the learning process. A prototype (EmotionTest) was developed, a learning platform that takes into account the emotional aspect, the learning style and the personality traits, adapting the course (content and context) to the student needs. The architecture proposed for this prototype is composed of 4 major models: the Application Model, Emotive Pedagogical Model, Student Model, and Emotional Model [4], as shown in the figure bellow.

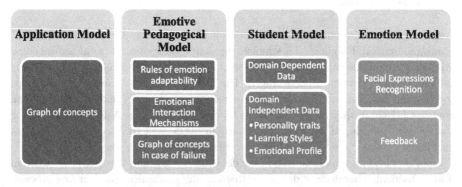

Fig. 1. Architecture

The student model consists in the user information and characteristics. This includes personal information (name, email, telephone, etc.), demographic data (gender, race, age, etc.), knowledge, deficiencies, learning styles, emotion profile, personality traits, etc. This information is use by the student model to better adapt the prototype to student [4].

The emotion model gathers all the information the facial emotion recognition software and feedback of the students. Facial Expression Recognition allows video analysis of images in order to recognise an emotion. This type of emotion recognition was chosen because it was the least intrusive with the student activities. The emotion recognition is achieved by making use of an API entitled ReKognition [5]. This API allows detection of the face, eyes, nose and mouth and if the eyes and mouth are open or close. In addition specifies the gender of the individual and an estimate of age and emotion. In each moment a group of three emotions is captured. For each emotion is given a number that shows the confidence level of the emotion captured.

The application model is compose by a series of modules contain different subjects. The subject consist in a number steps that the student has to pass in order to complete is learning program. Usually each subject is composed by a Placement test in order to access and update the student level of knowledge. Followed by the subject content in which the subject is explained and follow by the subject exercises and final test. The first step is the subject's Placement Test (PT) that can be optional. This is designed to give students and teachers a quick way of assessing the approximate level of student's knowledge. The result of the PT is percentage PTs that is added to the student knowledge (Ks), on a particular subject, and places the student at one of the five levels of knowledge $kpt = \sum_{i=0}^{5} \text{exercise}_i$. If the PT is not performed Ksp will be equal to zero and the student will start with any level of knowledge. The Subject Content (SC) contains the subject explanation. The subject explanation depends on the stereotype. Each explanation will have a practice exercise. These exercises will allow the students to obtain points to perform the final test of the subject. The student needs to get 80% on the TotalKsc to undertake the subject test. The Subject Test (ST) is the assessment of the learned subject. This will give a final value kst that represents the student's knowledge on the subject, $kst = \sum_{i=0}^{5} \text{exercise}_i$ Only if the kst is higher than 50% it can be concluded that the student has successful completed the subject. In this case the values of the ksp and kst are compared to see if there was an effective improvement on the student's knowledge. This is represented in the following diagram [6].

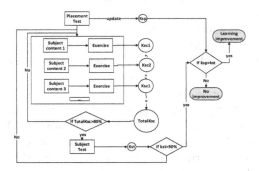

Fig. 2. Representation diagram

The last model is the emotive pedagogical model that is composed by three submodels: the rules of emotion adaptability, the emotional interaction mechanisms and the graph of concepts in case of failure.

The Rules of Emotion Adaptability manage the way the subject content is presented. The subject content is presented according the student's learning preference and personality. This way information and exercises are presented in a manner more agreeable to the student helping him to comprehend the subject at hand.

The subject content and subject exercises are presented according the learning style and personality of the student. The emotional interaction mechanisms consist in the trigger of an emotion interaction when is captured an emotion that need to be contradicted in order to facilitate the learning process. The emotions to be contradicted are: anger, sadness, confusion and disgusted. The interaction can depend on the personality and on the learning style of the student. Finally the graph of concepts in case of failure this indicates the steps to be taken when a student fails to pass a subject. The graph of concepts in case of failure represents the steps to be taken when fail to surpass a subject. To be approved in a subject the whole the tasks must be completed, and only with a subject completed it is possible to pass to the next. Inside of a subject the student has to complete the placement test, the subject content plus exercises with a grade equal or higher than 80% and the subject test with approval with a grade higher than 50% to complete the subject. In case of failure it has to go back to the subject content and repeat all the steps [6].

3   Data Analysis

To test the performance of the developed prototype some experiences were carried out with students from two ISEP Engineering courses: Informatics Engineering and Systems Engineering. The total number of students involved in these tests was 115 with ages between 17 and 42 years old. This group of students was composed of 20% female (n=23) and 80% male (n=92), the participants were mainly from the districts of Oporto, Aveiro and Braga.

To assess validity of the prototype the students were divided in two groups, v1 and v2. Group v1 tested the prototype with emotional interaction and group v2 without any emotional interaction.

Group v1 had to accomplish a diagnostic test (in paper) to help grade the student initial knowledge level, followed by the evaluation of the prototype with the emotional interaction and learning style. This would include, the login into the prototype, at this moment is when the initial data is begins to be collected for the student model. By accessing the school's Lightweight Directory Access Protocol (LDAP) one was able to gather the generic information of the students (like name, email and other). After login the students were required to answer two questionnaires (TIPI, VARK) build into the prototype. This allows the prototype to known the student's personality traits and learning preferences. Afterward the student could assess the learning materials and exercises. From the moment the student login his emotion state has been monitor and saved and every time that is detected an emotion that triggers an intervention it would appear on the screen. After this evaluation the students had to complete a final test (in paper) to help grade the student final knowledge level [7].

Group v2 had to accomplish diagnostic test (in paper) to help grade the student initial knowledge level, followed by the prototype evaluation without any emotional interaction. This evaluation is in all similar to group v1, but with one big difference. Even though the emotional state is monitor, when is detected an emotion that triggers an intervention it would not appear on the screen. After this test the students had to complete a final test (in paper) to help grade the student final knowledge level [7].

Analyzing the data of evaluation test, for group v3 for diagnostic test it has a mean of 45,7% (SD =40,3) and for the final test a mean of 85,7% (SD=12,2) and for group v4 for diagnostic test it has a mean of 37,1% (SD = 29,2) and for the final test a mean of 61,4% (SD=33,7). The data gathered did not have a normal distribution so the two groups were compared using a non-parametric test Mann-Whitney. The diagnostic test has a Mann–Whitney U = 83,0 and for a sample size of 14 students. For this analysis it was found a P value of 0,479 which indicates that it doesn't have any statistical difference which is understandable because it was assumed that all students had more or less the same level of knowledge. The final test it has a Mann–Whitney U =54,0 and for an equal sample size of the diagnostic test. For this analysis it was found a P value of 0,029 in this case the differences observed are statistical different. In addition, a series of tests were made to compare the means values of the students by group and by learning preference, by group and personality and by group and emotional state. The objective of running these tests was to find out if learning preference, personality and emotional state had any influence on the outcome of the final test. In relation to the first two tests, by learning preference and by personality, no statistically significant differences were found in the data. Therefore it cannot be concluded that learning preference and personality in each group had any influence in the final test outcome. To prove this assumption it is needed a larger sample size. For the question "if the emotional state had any influence in the in the final test", the differences observed were statistical significant [7].

4   Conclusion

In conclusion, with this work was attempted to answer several questions. The central question that guided this work was: "Does a learning platform that takes into account the student's emotions, learning preferences and personality improve the student's learning results?"

The gathered data from the performed test showed that there is a statistical difference between students' learning results while using two learning platforms: one learning platform that takes into account the student's emotional state and the other platform that does not have that in consideration. This gives an indication that by introducing the emotional component, the students' learning results can possibly be improved. Another question was: "Does Affective computing technology help improving a student learning process?"

In answering positively to the central question, this question is partly answered. As results showed, the students' learning results can be improved by adding an emotional component to a learning platform; also the use of Affective Computing technology to capture emotion can enhance this improvement. The use Affective Computing allows the capture of the student's emotion by using techniques that don't inhibit the student's actions. Also, it can be used one or more techniques simultaneously to help verify the accuracy of the emotional capture. The last question was: "What are the stimuli that can be used to induce or change the student state of mind in order to improve the learning process?"

First, the results indicate that the platform with an emotional component had an overall set of more positive emotions than the platform without this component. Showing that, the stimuli produced in the platform with an emotional component was able to keep the students in a positive emotional state and motivated to do the tasks at hand, this did not happen in the platform without this component.

Second the results demonstrated that the platform with an emotional component not only got the set of more positive emotions among the students, but also obtained an improvement in the students learning results.

Acknowledgments. This work is supported by FEDER Funds through the "Programa Operacional Factores de Competitividade - COMPETE" program and by National Funds through FCT "Fundação para a Ciência e a Tecnologia" under the project: FCOMP-01-0124-FEDER-PEst-OE/EEI/UI0760/2014.

References

1. Picard, R.W., Papert, S., Bender, W., Blumberg, B., Breazeal, C., Cavallo, D., Machover, T., Resnick, M., Roy, D., Strohecker, C.: Affective learning - a manifesto. BT Technol. J. **22**(4), 253–268 (2004)
2. Russell, J.A.: A circumplex model of affect. J. Pers. Soc. Psychol. **39**(6), 1161–1178 (1980)
3. Kort, B., Reilly, R., Picard, R.W.: An affective model of interplay between emotions and learning: re-engineering educational pedagogy-building a learning companion. In: Proc. - IEEE Int. Conf. Adv. Learn. Technol. ICALT 2001, pp. 43–46 (2001)
4. Faria, A., Almeida, A., Martins, C., Lobo, C., Gonçalves, R.: Emotional Interaction Model For Learning. In: INTED 2015 Proc., pp. 7114–7120 (2015)
5. orbe.us | ReKognition - Welcome to Rekognition.com (2015) (Online). http://rekognition.com/index.php/demo/face. (accessed: 26–Jul–2014)
6. Faria, A.R., Almeida, A., Martins, C., Gonçalves, R.: Emotional adaptive platform for learning. In: Mascio, T.D., Gennari, R., Vittorini, P., de la Prieta, F. (eds.) Methodologies and Intelligent Systems for Technology Enhanced Learning. AISC, vol. 374, pp. 9–16. Springer, Heidelberg (2015)
7. Faria, R., Almeida, A., Martins, C., Gonçalves, R.: Learning Platform. In: 10th Iberian Conference on Information Systems and Technologies – CISTI 2015 (2015)

Ambient Intelligence:
Experiments on Sustainability Awareness

Fábio Silva$^{(\boxtimes)}$ and Cesar Analide

Algoritmi Centre, University of Minho, Braga, Portugal
{fabiosilva,analide}@uminho.pt

Abstract. Computer systems are designed to help solve problems presented to our society. New terms such as computational sustainability and internet of things present new fields where traditional information systems are being applied and implemented on the environment to maximize data output and our ability to understand how to improve them. The advancement of richer and interconnected devices has created opportunities to gather new data sources from the environment and use it together with other pre-existent information in new reasoning processes. This work describes a sensorial platform designed to help raise awareness towards sustainability and energy efficient systems by exploring the concepts of ambient intelligence and fusion of data to create monitoring and assessment systems. The presented platform embodies the effort to raise awareness of user actions on their impact towards their sustainability objectives.

Keywords: Ambient intelligence · Pervasive systems · Sustainability · Energy efficiency

1 Introduction

The advent of computer science and its evolution led to the availability of computational resources that can better assess and execute more complex reasoning and monitoring of sustainability attributes. This led to the creation of the field of computational sustainability (Gomes, 2011). Coupled with sustainability is energy efficiency which is directly affected by human behaviour and social aspects such as human comfort. Fundamentally, efficiency deals with the best strategy to obtain the objectives that are set, however, when the concept of sustainability is added, several efficient plans might be deemed unsustainable because they cannot be maintained in the future.

While efficiency is focused on optimization, sustainability is mostly concerned on restrictions put in place to ensure that the devised solution does not impair the future. Not only, context hardens the problem but also the possibility of missing information which might occur due to same unforeseen event that jeopardizes an efficient solution. To tackle such event, computational systems are able to maintain sensory networks over physical environments to acquire contextual information so it can validate the conditions for efficient planning but also acquire information and, as a last resort, act upon the physical environment.

© Springer International Publishing Switzerland 2015
F. Pereira et al. (Eds.) EPIA 2015, LNAI 9273, pp. 33–38, 2015.
DOI: 10.1007/978-3-319-23485-4_4

2   Related Work

The term computational sustainability is used by researchers such as Carla Gomes (Gomes, 2011) to define the research field where sustainability problems are addressed by computer science programs and models in order to balance the three dimensions of sustainability: economic, ecologic and social dimensions.

It is accepted that the world ecosystem is a complex sustainability problem, affected by human and non-human actions. Despite the use of statistical and mathematical models for the study of sustainability and computational models to address problems of environmental and societal sustainability, the term computational sustainability appeared around 2008. Nevertheless, the pairing between computer science and the study of sustainability is as old as the awareness of sustainability and as long as computing was available. It is a fact that, as computational power capacity increased over time, so did the complexity and length of the models used to study sustainability. The advent and general availability of modern techniques from artificial intelligence and machine learning allowed better approaches to the study of each dimension of sustainability and their overall impact for sustainability.

The types of sensors used in the environment may be divided into categories to better explain their purpose. In terms of sensing the environment, sensors can be dived into sensors that sense the environment or users and their activities. Generally, an ambient might be divided in sensors and actuators. Sensors monitor the environment and gather data useful for cognitive and reasoning process. On the other hand, actuators take action upon the environment performing actions such as thermostats to control the temperature, lightning switches or other appliances.

Different methodologies and procedures exist to keep track of human activities and to make prediction based on previous and current information gathered in these environments. Common approaches with machine learning techniques involve the use of neural networks, classification techniques, fuzzy logic, sequence discovery, instance based learning and reinforced learning as in (Costa, Novais &, Simões, 2014).

An approach to this problem using fuzzy logic algorithms is proposed by Hagras et al (Hagras, Doctor, Callaghan, & Lopez, 2007). Sequence discovery approach is at the heart of learning algorithm in (Aztiria, Augusto, Basagoiti, Izaguirre, & Cook, 2012), which demonstrates a system that can learn user behavioural patterns and take proactive measures accordingly.

The work presented considers the use of these types of sensor to assess and reason about sustainability and indicator design. This information will then be used to reason about user behaviour and their accountability.

3   Platform Engine

The focus of this project is, more than developing new procedures or algorithms to solving problems, putting these innovations on the hand of the user, with a clear purpose: that these innovative tools should be guided to assist people in the context of energetic sustainability.

3.1   Network Design

The PHESS platform supports heterogeneous devices by implementing middleware upon groups of devices to control data and information acquisition. Local central servers are viewed as decentralized by the platforms which access them to obtain data and implement their plans through the local network actuators.

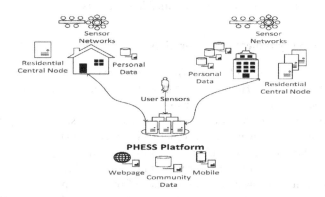

Fig. 1. Generic PHESS platform configuration.

Figure 1, details a generic composition of two different environment scenarios with users and their connection to PHESS platform. This residential central node is responsible for the middleware to connect local sensor networks to the PHESS platform using dedicated protocols for data acquisition and storing information.

The data gathered is summarized locally according to time and user presence models and synchronized with the central PHESS platform. It is also responsible for creating different user and environment profiles. Notifications are generated by the central PHESS platform to the project webpage or mobile application.

3.2   Data Fusion

The process of data fusion is handled by local central nodes where data is submitted to data fusion process according to the number of overlapping and complementing sensors. In this regard, there are strategies that can be followed according to the context and nature of the fusion process.

The first one is a weighted average of values, for the same type of sensors in the same context to get an overview of an attribute with multiple sensors to reduce measurement errors. The weights are defined manually by the local administrator. More sophisticated fusion is employed with complementary sensors which according to some logic defined into the system measure an attribute by joining efforts such as user presence with both RFID reader and wireless connection of personal devices such as smartphones. The last resource is the use of heterogeneous data to create attributes with some level of knowledge from the start. An example is the assessment of thermal comfort using default indicator expressed as mathematical formulae such as the PMV index (Fanger, 1970) for instance. Other application is the definition of sustainable indicators according to custom mathematical formulae in the platform that shall process some attributes in the system to make their calculation.

The configuration of data fusion steps, the selection of sensors and streams of data is made on the initial step of the system by the local administrator. According to each area of interest and with specialized knowledge obtained by experts it is possible to monitor relevant information to build sustainable indicators.

3.3   Sustainability Indicators Generation

Indicators evaluate sustainability in terms of three main groups, namely economic, environmental and social. However, due to their impact, some indicators can be designed to influence more than one dimension of sustainability. For this reason it was chosen to have these indicators as the general analysis of sustainable principles instead of sustainability dimensions. In order to be directly comparable indicators are defined to use the same scale, and are based on the notion of positive and negative impact. The values of each indicator range from -1 to 1 and can be interpreted as unsustainable for values below zero and sustainable from there upwards.

Indicator definition is another configurable space inside the PHESS platform where monitoring indicators are defined using values from sensors and sensor fusion, and customized with mathematical formulas. All these indicators are calculated either locally, i.e., in a room basis, or they are evaluated for the entire setting which sums assessment of all different rooms. In this way, even if the environment is considered sustainable, the user may still assess changes in premises with low supportable standards. Environments are generally hosts to of many different users, which influence it with their behaviours, actions and habits. Tracking user activities is something that can be used to infer and establish cause and effect relationships. The PHESS system uses different dynamics to produce accountability reports on user actions based on environment and personal monitoring coupled user presence detection. Areas uncopied, are considered the responsibility of all people present in the environment, so that the coverage of the entire environment is assured by its occupants. In cases where the local context and local sensor values are indistinguishable based on location them, user accountability takes in consideration only user presence in the environment. The richer the environment is in sensor data acquisition the richer results and analysis is.

4   Case Study and Results

As a case study, results from five days in an environment are presented. In this case a home environment with a limited set of sensors, and a smartphone as a user detection mechanism. User notifications are made by actuator modules which push notifications to users in order to alert them based on notification schemes and personal rules. Sensors include electrical consumption, temperature, humidity, luminosity and presence sensing through smartphones and an indicator based on the sensation of temperature PMV used in thermal comfort studies (Rana, Kusy, Jurdak, Wall, & Hu, 2013).

The indicators are designed in the platform in order to perceive energy efficacy and as such the case scenario uses electricity to do this analysis. Therefore, a list of sample indicator was defined using data fusion available through PHESS modules. A sample of four indicators were defined and their expression is as follows:

- Unoccupied consumption – measures the deviation of consumption when no user is present in the environment from a user inputted objective.
- Activity based consumption - measures the deviation of consumption during the period of 1 hour from the objective value set by the user, in this case;
- Total consumption: measures the deviation of total consumption during the period of a day based on a default value defined for consumption;
- Comfort Temperature – based on the PMV comfort indicator obtained through data fusion process which is calculated by PHESS platform;
- Comfort Humidity – based on comfort values that define the normal range of values humidity in indoor environments.

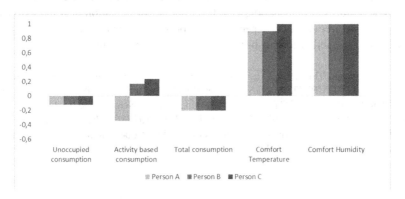

Fig. 2. Indicator values from the PHESS platform

As seen in figure 3, with the graphical representation of the indicators it is possible to analyse the behaviour of an environment based on user inputted indicators. These indicators are represented in the scale -1 to 1 as stated in section 3.4. While comfort values are being respected, consumption based indicator shows that the values set for the system are not yet being followed. As a result, indicators towards consumption analysis are performing below the accepted margin.

Accountability is based on the notion of user impact on the system. Results indicate a higher consumption when the environment is not occupied which demonstrates that the environment configuration has more impact than user actions. The person the longest in the environment each day has more impact towards the total consumption which is inputted to user actions.

This analysis allows the system to identify people with most impact on the system based on the attributes and indicators defined. There is the need to adapt each indicator set to the objectives and the areas to improve, but the generic platform allows for this configuration based on the layers of the PHESS platform.

5   Conclusions

Computational sustainability, although a new and interesting topic of research to the academic community, still presents a number of difficult challenges. The platform presented is the combination of modules which take inspiration from ambient intelligence and information systems to provide analysis and assessment of environment and their users, to identify and provide real time analysis of concepts based on sustainability and efficiency. Results indicate that modest configurations can yield meaningful results that may be used to take actions on the environment and user behaviour. The process of data fusion and indicator design are responsible to thoroughly analyse key situation of environment and user behaviour so they expression appears meaningful to produce not only user reports but also user suggestions.

With the mainstream use of interconnected appliances the possibilities for automatic actuation are increasing, mainly with the new internet f things standards being proposed by major companies. The PHESS project aim to increase its support by adding new features to their middleware layer and allow actuation in conjunction with sensorization. Moreover, it is expected to support multi environments for each user, increasing accountability to user behaviours regardless of location.

Acknowledgements. This work is part-funded by ERDF - European Regional Development Fund through the COMPETE Programme (operational programme for competitiveness) and by National Funds through the FCT (Portuguese Foundation for Science and Technology) within project FCOMP-01-0124-FEDER-028980 (PTDC/EEI-SII/1386/2012) and project PEst-OE/EEI/UI0752/2014.. Additionally, it is also supported by a doctoral grant, with the reference SFRH/BD/78713/2011, issued by FCT.

References

1. Aztiria, A., Augusto, J.C., Basagoiti, R., Izaguirre, A., Cook, D.J.: Discovering frequent user-environment interactions in intelligent environments. Personal and Ubiquitous Computing **16**(1), 91–103 (2012)
2. Fanger, P.O.: Thermal comfort: Analysis and applications in environmental engineering. Danish Technical Press (1970)
3. Gomes, C.P.: Computational sustainability. In: Gama, J., Bradley, E., Hollmén, J. (eds.) IDA 2011. LNCS, vol. 7014, p. 8. Springer, Heidelberg (2011).
   http://citeseerx.ist.psu.edu/viewdoc/download?doi=10.1.1.158.2293&rep=rep1&type=pdf
4. Hagras, H., Doctor, F., Callaghan, V., Lopez, A.: An Incremental Adaptive Life Long Learning Approach for Type-2 Fuzzy Embedded Agents in Ambient Intelligent Environments. IEEE Transactions on Fuzzy Systems **15**(1), 41–55 (2007)
5. Rana, R., Kusy, B., Jurdak, R., Wall, J., Hu, W.: Feasibility analysis of using humidex as an indoor thermal comfort predictor. Energy and Buildings **64**, 17–25 (2013). doi:10.1016/j.enbuild.2013.04.019
6. Costa, A., Novais, P., Simões, R.: A Caregiver Support Platform within the Scope of an AAL Ecosystem. Sensors **14**(3), 654–5676 (2014). MDPI AG, ISSN: 1424-8220

Artificial Intelligence in Medicine

Reasoning with Uncertainty
in Biomedical Models

Andrea Franco, Marco Correia, and Jorge Cruz[(✉)]

NOVA Laboratory for Computer Science and Informatics,
DI/FCT/UNL, Caparica, Portugal
jcrc@fct.unl.pt

Abstract. The use of mathematical models in biomedical research largely developed in the second half of the 20th century. However, their translation to clinically useful tools has proved challenging. Reasoning with deep biomedical models is computationally demanding as parameters are typically subject to nonlinear relations, dynamic behavior, and uncertainty. This paper proposes a new approach for assessing the reliability of the conclusions drawn from these models given the underlying uncertainty. It relies on probabilistic constraint programming for a sound propagation of uncertainty from model parameters to results. The advantages of the approach are illustrated on an important problem in the obesity research field, namely the estimation of free-living energy intake in humans. Based on a well known energy intake model, our approach is able to correctly characterize the provided estimates given the uncertainty inherent to the model parameters.

Keywords: Biomedical models · Constraint programming · Energy intake

1 Introduction

Mathematical models are extensively used in many biomedical domains for supporting rational decisions. A mathematical model describes a system by a set of variables and constraints that establish relations between them. Uncertainty and nonlinearity play a major role in modeling most real-world continuous systems. A competitive framework for decision support in continuous domains must provide an expressive mathematical model to represent the system behavior and be able to perform sound reasoning that accounts for the uncertainty and the effect of nonlinearity.

Given the uncertainty, there are two opposite attitudes for reasoning with scenarios consistent with the mathematical model. Stochastic approaches [1] reason on approximations of the most likely scenarios. They associate a probabilistic model to the problem thus characterizing the likelihood of the different scenarios. In contrast, constraint programming approaches [2] reason on safe enclosures of all consistent scenarios. Rather than associate approximate values

© Springer International Publishing Switzerland 2015
F. Pereira et al. (Eds.) EPIA 2015, LNAI 9273, pp. 41–53, 2015.
DOI: 10.1007/978-3-319-23485-4_5

to real variables, intervals are used to include all their possible values. Model-based reasoning and what-if scenarios are adequately supported through safe constraint propagation techniques, which only eliminate combinations of values that definitely do not satisfy model constraints.

In this work we use a probabilistic constraint approach that combines a stochastic representation of uncertainty on the parameter values with a reliable constraint framework robust to nonlinearity. Similarly to stochastic approaches it associates an explicit probabilistic model to the problem, and similarly to constraint approaches it assumes reliable bounds for the model parameters. The approach computes conditional probability distributions of the model parameters, given the uncertainty and the constraints.

The potential of our approach to support clinical practice is illustrated in a real world problem from the obesity research field. The impact of obesity on health, at both individual and public levels, is widely documented [3–5]. Despite this fact, and the availability of nutritional recommendations and guidelines to the general audience, the prevalence of overweight and obesity in adults and children increased dramatically in the last 30 years [6]. According to the World's Health Organization, the main cause for the "obesity pandemics" is the energy unbalance caused by an increased calorie intake associated to a lower energy expenditure as a result of a sedentary lifestyle.

Many biomedical models use the energy balance approach to simulate individual body weight dynamics, e.g. [7,8]. Change of body weight over time is modeled as the rate of energy stored (or lost), which is a function of the energy intake (from food) and the energy expended. However, the exact amount of calories ingested, or energy intake, is difficult to ascertain as it is usually obtained through methods that underestimate its real value, such as self-reported diet records [9].

The inability to rigorously assess the energy intake is considered by [10] the "fundamental flaw in obesity research". This fact hinders the success and adherence to individual weight control interventions [11]. Therefore the correct evaluation of such interventions will be highly dependent on the precision of energy intake estimates and the assessment of the uncertainty inherent to those estimates. In this paper we show how the probabilistic constraint framework can be used in clinical practice to correctly characterize such uncertainty given the uncertainty of the underlying biomedical model.

Next section overviews the energy intake problem and introduces a biomedical model used in clinical practice. Section 3 addresses constraint programming and its extensions to differential equations and probabilistic reasoning. Section 4 shows how the problem is cast into the probabilistic constraint framework. Section 5 discusses the experimental results and the last section summarizes the main conclusions.

2   Energy Intake Problem

The mathematical models that predict weight change in humans are usually based on the energy balance equation:

$$R = I - E \tag{1}$$

where R is the energy stored or lost (kcal/d), I is the energy intake (kcal/d) and E is the energy expended (kcal/d).

Several models have been applied to provide estimates of individual energy intake [12,13]. Our paper focus on the work of [12] which developed a computational model to determine individual energy intake during weight loss. This model, herein designated EI model, calculates the energy intake based on the following differential equation:

$$cf \frac{dF}{dt} + cl \frac{dFF}{dt} = I - (DIT + PA + RMR + SPA) \tag{2}$$

The left hand side of equation (2) represents the change in body's energy stores, R in equation (1), and is modeled through the weighted sum of the changes in Fat mass (F) - the body's long term energy storage mechanism - and Fat Free mass (FF) - proxy for protein content used for energy purposes.

Differently from other models, that express the relationship between F and FF using logarithmic eq. (3) [14], or linear approximations [15], the EI model uses a fourth-order polynomial for estimating FF as a function of F, the age of the subject a, and its height h eq. (4),

$$FF^{log}(F) = d_0 + d_1 \log F \tag{3}$$
$$FF^{poly}(F, a, h) = (c_0 + c_1 F + c_2 F^2 + c_3 F^3 + c_4 F^4)(c_5 + c_6 a)(c_7 + c_8 h) \tag{4}$$

The rate of energy expended, E in equation (1), is the total amount of energy spent in several physiological processes: Diet Induced Thermogenesis (DIT) - energy required to digest and absorb food; Physical Activity (PA) - energy spent in volitional activities; Resting Metabolic Rate (RMR) - minimal amount of energy used to sustain life and; Spontaneous Physical Activity (SPA) - energy spent in spontaneous activities.

The EI model uses data from the 24-week CALERIE phase I study [16], in particular body weight for one female subject of the caloric restriction group. During the experiment, participants had their weight monitored every two weeks. Those weight measures are used to estimate the real energy intake for that particular individual.

3   Constraint Programming

A constraint satisfaction problem is a classical artificial intelligence paradigm characterized by a set of variables and a set of constraints that specify relations among subsets of these variables. Solutions are assignments of values to all variables that satisfy all the constraints. Constraint programming [2] is a form of declarative programming which must provide a set of constraint reasoning algorithms that take advantage of constraints to reduce the search space, avoiding regions inconsistent with the constraints. These algorithms are supported by specialized techniques that explore the specificity of the constraint model such as the domain of its variables and the structure of its constraints.

Continuous constraint programming [17,18] has been widely used to model safe reasoning in applications where uncertainty on the values of the variables is modeled by intervals including all their possibilities. A Continuous Constraint Satisfaction Problem (CCSP) is a triple$\langle X, D, C \rangle$ where X is a tuple of n real variables $\langle x_1, \cdots, x_n \rangle$, D is a Cartesian product of intervals $D(x_1) \times \cdots \times D(x_n)$ (a box), where each $D(x_i)$ is the domain of variable x_i and C is a set of numerical constraints (equations or inequalities) on subsets of the variables in X. A solution of the CCSP is a value assignment to all variables satisfying all the constraints in C. The feasible space F is the set of all CCSP solutions within D.

Continuous constraint reasoning relies on branch-and-prune algorithms [19] to obtain sets of boxes that cover exact solutions for the constraints (the feasible space F). These algorithms begin with an initial crude cover of the feasible space (the initial search space, D) which is recursively refined by interleaving pruning and branching steps until a stopping criterion is satisfied. The branching step splits a box from the covering into sub-boxes (usually two). The pruning step either eliminates a box from the covering or reduces it into a smaller (or equal) box maintaining all the exact solutions. Pruning is achieved through an algorithm that combines constraint propagation and consistency techniques based on interval analysis methods [20].

In the biomedical context, constraint technology seems to have the potential to bridge the gap between theory and practice. The declarative nature of constraints makes them an adequate tool for the explicit representation of any kind of domain knowledge, including "deep" biophysical modeling. The constraint propagation techniques provide sound methods, with respect to the underlying model, that can be used to support practical tasks (e.g. diagnosis/prognosis may be supported through propagation on data about the patient symptoms/diseases). In particular, the continuous constraint framework seems to be the most adequate for representing the nonlinear relations on continuous variables, often present in biophysical models. Additionally, the uncertainty of biophysical phenomena may be explicitly represented as intervals of possible values and handled through constraint propagation.

However, the direct application of classical constraint programming to biomedical models suffers from two major pitfalls. System dynamics which is often modeled through differential equations cannot be explicitly represented by these approaches and integrated within the constraint model. Moreover, the interval representation of uncertainty may be too conservative and inadequate to distinguish between consistent scenarios based on their likelihood which may be crucial to the development of effective tools. This work is based on extensions to constraint programming for handling both problems and provide sound propagation of uncertainty from model parameters to results.

3.1 Differential Equations

The behavior of many systems is naturally modeled by a system of first order Ordinary Differential Equations (ODEs), often parametric. ODEs are equations that involve derivatives with respect to a single independent variable, t, usually representing time. A parametric ODE system, with parameters p, represented in vector notation as:

$$y' = f(p, y, t) \qquad (5)$$

is a restriction on the sequence of values that y can take over t. A solution, for a time interval T, is a function that satisfies equation (5) for all values of $t \in T$.

Since (5) does not fully determine a single solution (but rather a family of solutions), initial conditions are usually provided with a complete specification of y at some time point t. An Initial Value Problem (IVP) is characterized by an ODE system together with the initial condition $y(t_0) = y_0$. A solution of the IVP with respect to an interval of time T is the unique function that is a solution of (5) and satisfies the initial condition.

Parametric ODEs are expressive mathematical means to model system dynamics. Notwithstanding its expressive power, reasoning with such models may be quite difficult, given their complexity. Analytical solutions are available only for the simplest models. Alternative numerical simulations require precise numerical values for the parameters involved, often impossible to gather given the uncertainty on available data. This may be an important drawback since small differences on input values may cause important differences on the output produced.

Interval methods for solving differential equations with initial conditions [20] do verify the existence of unique solutions and produce guaranteed error bounds for the solution trajectory along an interval of time T. They use interval arithmetic to compute safe enclosures for the trajectory, explicitly keeping the error term within safe bounds.

Several extensions to constraint programming [21] were proposed for handling differential equations based on interval methods for solving IVPs. An approach that integrates other conditions of interest was proposed in [22] and successfully applied to support safe decisions based on deep biomedical models [23].

In this paper we use an approach similar to [21] that allows the integration of IVPs with the standard numerical constraints. The idea is to consider an IVP as a function Φ where the first argument are the parameters p, the second argument is the initial condition that must be verified at time point t_0 (third argument) and the last argument is a time point $t \in T$. A relation between the values at two time points t_0 and t_1 along the trajectory is represented by the equation:

$$y(t_1) = \Phi(p, y(t_0), t_0, t_1) \qquad (6)$$

Using variables x_0 and x_1 to represent $y(t_0)$ and $y(t_1)$, equation (6) is integrated into the CCSP as a constraint $x_1 = \Phi(p, x_0, t_0, t_1)$ with specialized constraint propagators to safely prune both variable domains based on a validated solver for IVPs [24].

3.2   Probabilistic Constraint Programming

In classical CCSPs, uncertainty is modeled by intervals that represent the domains of the variables. Constraint reasoning reduces uncertainty providing a safe method for computing a set of boxes enclosing the feasible space. Nevertheless this paradigm cannot distinguish between different scenarios and all combination of values within such enclosure are considered equally plausible. In this work we use probabilistic constraint programming [25] that extends the continuous constraint framework with probabilistic reasoning, allowing to further characterize uncertainty with probability distributions over the domains of the variables.

In the continuous case, the usual method for specifying a probabilistic model assumes, either explicitly or implicitly, a full joint probability density function (p.d.f.) over the considered random variables, which assigns a probability measure to each point of the sample space Ω. The probability of an event $\mathcal{H}$, given a p.d.f. f, is its multidimensional integral on the region defined by the event:

$$P(\mathcal{H}) = \int_{\mathcal{H}} f(\mathbf{x})d\mathbf{x} \tag{7}$$

The idea of probabilistic constraint programming is to associate a probabilistic space to the classical CCSP by defining an appropriate density function. A probabilistic constraint space (PC) is a pair $\langle\langle X, D, C\rangle, f\rangle$, where $\langle X, D, C\rangle$ is a CCSP and f is a p.d.f. defined in $\Omega \supseteq D$ such that: $\int_{\Omega} f(\mathbf{x})d\mathbf{x} = 1$.

A constraint (or set of constraints) can be viewed as an event $\mathcal{H}$ whose probability can be computed by integrating the density function f over its feasible space as in equation (7). In general these multidimensional integrals cannot be easily computed, since they may have no closed-form solution and the event may establish a complex nonlinear integration boundary. The probabilistic constraint framework relies on continuous constraint reasoning to get a tight box cover of the region of integration $\mathcal{H}$ and compute the overall integral by summing up the contributions of each box in the cover. Generic quadrature methods are used to evaluate the integral at each box.

In this work Monte Carlo methods [26] are used to estimate the value of the definite multidimensional integrals at each box. As long as the function is reasonably well behaved, the integral can be estimated by randomly selecting N points in the multidimensional space and averaging the function values at these points. Consider N random sample points $x_1, \ldots, x_N$ uniformly distributed inside a box B. The contribution of this box to the overall integral on the region of integration $\mathcal{H}$ is approximated by:

$$\int_{B \cap \mathcal{H}} f(\mathbf{x})dx \approx \frac{\sum_{i=1}^{N} 1_{\mathcal{H}}(\mathbf{x_i})f(\mathbf{x_i})}{N} vol(B) \tag{8}$$

where $1_{\mathcal{H}}$ is the indicator function[1] of $\mathcal{H}$. This method displays $\frac{1}{\sqrt{N}}$ convergence, i.e., by quadrupling the number of sampled points the error is halved, regardless of the number of dimensions.

---

[1] $1_{\mathcal{H}}(\mathbf{x_i})$ returns 1 if $\mathbf{x_i} \in \mathcal{H}$ and 0 otherwise.

The advantages from this close collaboration between constraint pruning and random sampling were previously illustrated in ocean color remote sensing studies [27] where this approach achieved quite accurate results even with small sampling rates. The success of this technique relies on the reduction of the sampling space where a pure non-naive Monte Carlo (adaptive) method is not only hard to tune but also impractical in small error settings.

4   Probabilistic Constraints for Solving the EI Problem

Let t be the number of days since the beginning of treatment of a given subject, $F(t)$ the Fat Mass at time t, $w(t)$ the weight observed at time t, and I the subject's energy intake, which is assumed to be a constant parameter between consecutive observations [12]. The energy balance equation and total body mass are related through the model:

$$F'(t) = g(I, F(t), t) \tag{9}$$
$$w(t) = FF(a, h, F(t)) + F(t) \tag{10}$$

where g is obtained by solving equation (2) with respect to $F'(t)$.

4.1   CCSP Model

Let $i \in \{0, \ldots, n\}$ denote the i'th observation since beginning of treatment, occurred at time t_i, and let F_i and w_i be respectively the fat mass and the weight of the patient at time t_i (with $t_0 = 0$). The EI model may be formalized as a CCSP $\langle X, \mathbb{R}^{2n+1}, C \rangle$ with a set of variables $X = \{F_0\} \bigcup_{i=1}^{n} \{F_i, I_i\}$ representing the fat mass F_i at each observation and the energy intake I_i between consecutive observations (at t_{i-1} and t_i), and a set of constraints $C = \{b_0\} \bigcup_{i=1}^{n} \{a_i, b_i\}$ enforcing eqs. (9, 10):

$$a_i \equiv [F_i = \Phi(I_i, F_{i-1}, t_{i-1}, t_i)]$$
$$b_i \equiv [w_i = FF(a, h, F_i) + F_i]$$

Recall that solving the above CCSP means finding the values for F_0 and the variables F_i, I_i ($1 \leq i \leq n$) that satisfy the above set of constraints.

4.2   Probabilistic CCSP Model

Uncertainty inherent to FF estimation may be integrated into the above CCSP model by considering that the true value of FF is the model given FF^M plus an associated error term $\epsilon_i \sim \mathcal{N}(\mu = 0, \sigma_\epsilon)$,

$$FF(a, h, F_i) = FF^M(a, h, F_i) + \epsilon_i$$

and we may rewrite the set of b_i constraints of the CCSP model as follows,

$$b_i \equiv [w_i = FF^M(a, h, F_i) + \epsilon_i + F_i]$$

Additionally, to keep the errors within reasonable bounds, bounding constraints are considered for each observation: $3\sigma_\epsilon \leq \epsilon_i \leq 3\sigma_\epsilon$, thus ignoring assignments whose contribution to the total error is less than 0.1%.

Note that a solution to the new (probabilistic) CCSP, i.e. an assignment of values to F_0 and the variables F_i, I_i ($1 \leq i \leq n$), determines the possible combinations of values for the errors $\epsilon_0, \ldots, \epsilon_n$.

If we assume that the FF model errors over the $n + 1$ distinct observations are independent, then each solution has an associated probability density value given by the joint p.d.f. f,

$$f(\epsilon_0, \ldots, \epsilon_n) = \prod_{i=0}^{n} f_i(\epsilon_i) \tag{11}$$

where f_i is the normal distribution associated with the error ϵ_i.

Instead of considering independence between model errors from consecutive observations, a more realistic alternative, explicitly represents the deviation between error ϵ_i and the previous error ϵ_{i-1} as a normally distributed random variable $\delta_i \sim \mathcal{N}(\mu = 0, \sigma_\delta)$, resulting in the following joint p.d.f. f,

$$f(\epsilon_0, \ldots, \epsilon_n, \delta_1, \ldots, \delta_n) = \prod_{i=0}^{n} f_i(\epsilon_i) \prod_{i=1}^{n} h_i(\delta_i) \tag{12}$$

where f_i and h_i are the normal distributions associated with the errors ϵ_i and δ_i respectively. The deviations are introduced in the model by considering constraints $\delta_i = \epsilon_i - \epsilon_{i-1}$ ($1 \leq i \leq n$) determining their values from the ϵ_i values.

A naive approximate algorithm for solving both alternative CCSP models could be simply to perform Monte Carlo sampling in the space defined by $D(F_j) \times D(I_1) \times \ldots \times D(I_n)$, with $j \in \{1, \ldots, n\}$. Note that, given the constraints in the model, each sampled point determines the values of all variables F_i and ϵ_i (and δ_i). From the values assigned to ϵ_i (and δ_i), eq. 11 (or 12) can be used to compute an estimate of its probability, as shown in (8).

With this approach, accurate results are hard to obtain for increasing number of observations due to the huge size of sampling space $O\left(|D|^{n+1}\right)$. Instead, we developed an improved technique that is able to drastically reduce both the exponent n and the base $|D|$ of this expression, as described in the following section.

4.3   Method

The main idea is to avoid considering all variables simultaneously but instead to reason only with a small subset that changes incrementally over time. For each observation i, we can compute the probability distributions of the variables of interest given the past knowledge already accumulated.

We start by computing the probability distribution of F_0 given the initial weight w_0 subject to the constraint b_0 and the bounding constraints for ϵ_0. This distribution, denoted $P^{\boxplus}(F_0)$, is discretized on a grid over $D(F_0)$ computed through probabilistic constraint reasoning integrated with Monte Carlo sampling as described in section 3.2. Specifically, given a point $\dot{F}_0$ sampled from $D(F_0)$, value $\dot{\epsilon}_0$ is determined by the constraint b_0, and its p.d.f. value is $f(\dot{F}_0) = f_0(\dot{\epsilon}_0)$.

The joint probability of F_1, I_1 is computed by considering the constraints associated with observation 1, the observed weight w_1, and $P^{\boxplus}(F_0)$. The method is similar: the grid $P^{\boxplus}(F_1, I_1)$, discretized over $D(F_1) \times D(I_1)$ is computed using probabilistic constraint reasoning; and Monte Carlo sampling is performed over this space region.

Given a point $(\dot{F}_1, \dot{I}_1)$, sampled from $D(F_1) \times D(I_1)$, the values $\dot{F}_0$ and $\dot{\epsilon}_1$ are determined by constraints a_1 and b_1, and accordingly to equation (11), its p.d.f. value should be $f(\dot{\epsilon}_0, \dot{\epsilon}_1) = f_0(\dot{\epsilon}_0) f_1(\dot{\epsilon}_1)$. However, we replace the computation of $f_0(\dot{\epsilon}_0)$ with the value of the discretized probability $P^{\boxplus}(\dot{F}_0)$ computed in the previous step providing an approximation that converges to the correct value when the number of grid subdivisions goes to infinity: $f(\dot{F}_1, \dot{I}_1) \approx P^{\boxplus}(\dot{F}_0) f_1(\dot{\epsilon}_1)$. If the alternative equation (12) is used, $\dot{\delta}_1$ is also computed from the constraints and the respective p.d.f. approximation is: $f(\dot{F}_1, \dot{I}_1) \approx P^{\boxplus}(\dot{F}_0) f_1(\dot{\epsilon}_1) h_1(\dot{\delta}_1)$.

Finally, the computed $P^{\boxplus}(F_1, I_1)$ is marginalized to obtain $P^{\boxplus}(F_1)$, and the process is iterated for the remaining observations.

5 Experimental Results

This section demonstrates how to the previously described method may be used to improve the applicability of the EI model by complementing its predictions with measures of confidence. The algorithm was implemented in C++ and used for obtaining the probability distribution approximations $P^{\boxplus}(F_i, I_i)$ at each observation $i \in \{1, \ldots, 12\}$ of a 45 year-old woman over the course of the 24-week trial (CALERIE Study phase I). The runtime was about 2 minutes per observation on an Intel Core i7 @ 2.4 GHz.

Fat Free mass is estimated using two distinct models: FF^{poly} (eq. 4), and FF^{log} (eq. 3). Both of these models were initially fit to a set of 7278 north american women resulting in the corresponding standard deviation of the error, $\sigma_\epsilon^{poly} = 3.35$ and $\sigma_\epsilon^{log} = 5.04$. This data set was collected during NHANES surveys (1994 to 2004) and is available online at the Centers for Disease Control and Prevention website [28].

We considered also different assumptions regarding independence of the error: the uncorrelated error model (11), and a correlated error model (12) with a small $\sigma_\delta = 0.5$. Note that, due to current data access restrictions, this latter value is purely illustrative.

The following techniques could be used for assessing propagation of uncertainty.

5.1 Joint Probability Distributions

The direct visual inspection of the joint probability distributions of F_i and I_i conveys important information about the relation between these parameters.

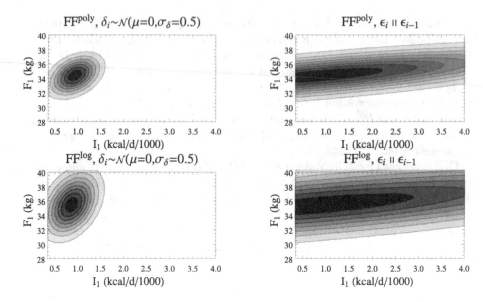

Fig. 1. Probabilities of fat mass (F) and intake (I) on the first clinical observation ($t = 14$). Top and bottom rows shows results for different FF models. Left and right columns correspond to different assumptions regarding independence of model errors.

In figure 1 we plot the obtained results regarding the first observation $i = 1$ for each combination of FF model and error correlation. The following is apparent from these plots: a) Uncertainty on F is positively correlated with uncertainty on I; b) The assumption of independence between model errors on consecutive weeks drastically affects the predicted marginal distribution of I (compare horizontally); c) The improved accuracy of FF^{poly} model (note that $\sigma_\epsilon^{poly} < \sigma_\epsilon^{log}$) reflects in slightly sharper F estimates, but does not seem to impact the estimation of I (compare vertically).

5.2   Marginal Probability Distributions with Confidence Intervals

To perceive the effect of the uncertainty on the estimated variables over time, it is useful to marginalize the computed joint probability distributions. Figure 2 shows the estimated F_i, and I_i over time, for each of the error correlation assumptions. Since the results concerning the FF^{poly} model are very similar to those obtained for FF^{log}, and for space economy reasons, we focus only on the former.

Each box in these plots depicts the most probable value (marked in the center of the box), the union hull of the 50% most probable values (the rectangle), and the union hull of the 82% most probable values (the whiskers). Additionally, each plot overlays the estimates obtained from the algorithm published by the author of the EI model [7].

The presented results show that the previous conclusions for the case of $i = 1$ extend for all remaining observations. Additionally, an interesting phenomena occurs in the case of correlated error: the uncertainty in the estimation of F decreases slightly over time. This is most probably the consequence of having, at each new observation, an increasingly constrained problem for which the size of the solution space is consequently increasingly smaller. At least in the case of F, more information seems to lead to signicatively better estimations. This can not occur if the errors are independent, as is indeed confirmed in the plots.

Finally, our results show that in some cases the most probable values obtained by [7] are crude approximations to their own proposed model.

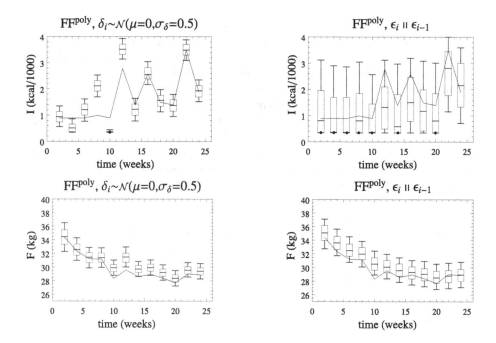

Fig. 2. Most probable intervals for the values of I (top row) and F (bottom row) over time using the F^{poly} model. Left and right columns correspond to different assumptions regarding independence of model errors. The continuous line plots estimates obtained in [12].

5.3   Best-Fit

Although the presented algorithm is primarily intended for characterizing uncertainty in model predictions, it is nevertheless a sound method for obtaining the predictions themselves. Indeed, as the magnitude of the error in the model parameters decreases (σ_ϵ in our example), the obtained joint probability distributions will converge to the correct solution of the model.

6   Conclusions

The standard practice for characterizing confidence on the predictions resulting from a complex model is to perform controlled experiments. In the biomedical field, this often translates to closely monitoring distinct groups of subjects over large periods of time, and assessing the fitness of the model statistically. While the empirical approach has its own advantages, namely that it does not require a complete understanding of the implications of the individual assumptions and approximations made in the model, it has some important shortcomings. Depending on the medical field, controlled experiments are not always practical, do not convey enough statistical significance, or have associated high costs.

Contrary to the empirical, black-box approach, this paper proposes to characterize the uncertainty on the model estimates by propagating the errors stemming from each of its parts. The described technique extends constraint programming to integrate probabilistic reasoning and constraints modeling dynamic behaviour, offering a mathematically sound and efficient alternative.

The application field of the presented approach is quite broad: it targets models which are themselves composed of other, possibly identically complex (sub)models, for which there is a known characterization of the error. The selected case-study is a good example: the EI model is a fairly complex model including dynamic behaviour and nonlinear relations, and integrates various (sub)models with associated uncertainty. The experimental section illustrated how different choices for one of these (sub)models, the FF model, impacts the error of the complete EI model.

Probabilistic constraint programming offers modeling and reasoning capabilities that go beyond the traditional alternatives. This approach has the potential to bridge the gap between theory and practice by supporting reliable conclusions from complex biomedical models taking into account the underlying uncertainty.

References

1. Halpern, J.Y.: Reasoning about Uncertainty. MIT Press (2003)
2. Rossi, F., Beek, P.V., Walsh, T. (eds.): Handbook of Constraint Programming. Foundations of Artificial Intelligence. Elsevier Science (2006)
3. Swinburn, B.A., et al.: The global obesity pandemic: shaped by global drivers and local environments. The Lancet **378**(9793), 804–814 (2011)
4. Leahy, S., Nolan, A., O'Connell, J., Kenny, R.A.: Obesity in an ageing society implications for health, physical function and health service utilisation. Technical report, The Irish Longitudinal Study on Ageing (2014)
5. Lehnert, T., Sonntag, D., Konnopka, A., Heller, S.R., König, H.: Economic costs of overweight and obesity. Best Pract Res Clin Endoc. Metab. **27**(2), 105–115 (2013)
6. Ng, M., et al.: Global, regional, and national prevalence of overweight and obesity in children and adults during 1980–2013: a systematic analysis for the global burden of disease study 2013. The Lancet **384**(9945), 766–781 (2014)

7. Thomas, D., Martin, C., Heymsfield, S., Redman, L., Schoeller, D., Levine, J.: A simple model predicting individual weight change in humans. J. Biol. Dyn. 5(6), 579–599 (2011)
8. Christiansen, E., Garby, L., Sørensen, T.I.: Quantitative analysis of the energy requirements for development of obesity. J. Theor. Biol. 234(1), 99–106 (2005)
9. Hill, R., Davies, P.: The validity of self-reported energy intake as determined using the doubly labelled water technique. Brit. J. Nut. 85, 415–430 (2001)
10. Winkler, J.T.: The fundamental flaw in obesity research. Obesity Reviews 6, 199–202 (2005)
11. Champagne, C.M., et al.: Validity of the remote food photography method for estimating energy and nutrient intake in near real-time. Obesity 20(4), 891–899 (2012)
12. Thomas, D.M., Schoeller, D.A., Redman, L.M., Martin, C.K., Levine, J.A., Heymsfield, S.: A computational model to determine energy intake during weight loss. Am. J. Clin. Nutr. 92(6), 1326–1331 (2010)
13. Hall, K.D., Chow, C.C.: Estimating changes in free-living energy intake and its confidence interval. Am. J. Clin. Nutr. 94, 66–74 (2011)
14. Forbes, G.B.: Lean body mass-body fat interrelationships in humans. Nutr. Rev. 45, 225–231 (1987)
15. Thomas, D., Ciesla, A., Levine, J., Stevens, J., Martin, C.: A mathematical model of weight change with adaptation. Math. Biosci. Eng. 6(4), 873–887 (2009)
16. Redman, L.M., et al.: Effect of calorie restriction with or without exercise on body composition and fat distribution. J. Clin. Endocrinol. Metab. 92(3), 865–872 (2007)
17. Lhomme, O.: Consistency techniques for numeric CSPs. In: Proc. of the 13th IJCAI, pp. 232–238 (1993)
18. Benhamou, F., McAllester, D., van Hentenryck, P.: CLP(intervals) revisited. In: ISLP, pp. 124–138. MIT Press (1994)
19. Hentenryck, P.V., Mcallester, D., Kapur, D.: Solving polynomial systems using a branch and prune approach. SIAM J. Num. Analysis 34, 797–827 (1997)
20. Moore, R.: Interval Analysis. Prentice-Hall, Englewood Cliffs (1966)
21. Goldsztejn, A., Mullier, O., Eveillard, D., Hosobe, H.: Including ordinary differential equations based constraints in the standard CP framework. In: Cohen, D. (ed.) CP 2010. LNCS, vol. 6308, pp. 221–235. Springer, Heidelberg (2010)
22. Cruz, J.: Constraint Reasoning for Differential Models. Frontiers in Artificial Intelligence and Applications, vol.126. IOS Press (2005)
23. Cruz, J., Barahona, P.: Constraint reasoning in deep biomedical models. Artificial Intelligence in Medicine 34(1), 77–88 (2005)
24. Nedialkov, N.: Vnode-lp a validated solver for initial value problems in ordinary differential equations. Technical report, McMaster Univ., Hamilton, Canada (2006)
25. Carvalho, E.: Probabilistic Constraint Reasoning. PhD thesis, FCT/UNL (2012)
26. Hammersley, J., Handscomb, D.: Monte Carlo Methods. Methuen London (1964)
27. Carvalho, E., Cruz, J., Barahona, P.: Probabilistic constraints for nonlinear inverse problems. Constraints 18(3), 344–376 (2013)
28. National health and nutrition examination survey. http://www.cdc.gov/nchs/nhanes.htm

Smart Environments and Context-Awareness for Lifestyle Management in a Healthy Active Ageing Framework

Davide Bacciu[1], Stefano Chessa[1,2(✉)], Claudio Gallicchio[1], Alessio Micheli[1], Erina Ferro[2], Luigi Fortunati[2], Filippo Palumbo[2], Oberdan Parodi[2], Federico Vozzi[2], Sten Hanke[3], Johannes Kropf[3], and Karl Kreiner[4]

[1] Department of Computer Science, University of Pisa, Largo Pontecorvo 3, Pisa, Italy
[2] CNR-ISTI, PISA CNR Research Area, via Moruzzi 1, 56126 Pisa, Italy
stefano.chessa@unipi.it
[3] Health and Environment Department, AIT Austrian Institute of Technology GmbH, Vienna, Austria
[4] Safety and Security Department, AIT Austrian Institute of Technology GmbH, Vienna, Austria

Abstract. Health trends of elderly in Europe motivate the need for technological solutions aimed at preventing the main causes of morbidity and premature mortality. In this framework, the DOREMI project addresses three important causes of morbidity and mortality in the elderly by devising an ICT-based home care services for aging people to contrast cognitive decline, sedentariness and unhealthy dietary habits. In this paper, we present the general architecture of DOREMI, focusing on its aspects of human activity recognition and reasoning.

Keywords: Human activity recognition · E-health · Reasoning · Smart environment

1 Introduction

According to the University College Dublin Institute of Food and Health, three are the most notable health promotion and disease prevention programs that target the main causes of morbidity and premature mortality: malnutrition, sedentariness, and cognitive decline, conditions that particularly affect the quality of life of elderly people and drive to disease progression. These three features represent the target areas in the DOREMI project. The project vision aims at developing a systemic solution for the elderly, able to prolong the functional and cognitive capacity by stimulating, and unobtrusively monitoring the daily activities according to well-defined "Active Ageing" lifestyle protocols. The project joins the concept of prevention centered on the elderly, characterized by a unified vision of being elderly today, namely, a promotion of the health by a constructive interaction among mind, body, and social engagement.

This work has been funded in the framework of the FP7 project "Decrease of cOgnitive decline, malnutRition and sedEntariness by elderly empowerment in lifestyle Management and social Inclusion" (DOREMI), contract N.611650.

F. Pereira et al. (Eds.) EPIA 2015, LNAI 9273, pp. 54–66, 2015.
DOI: 10.1007/978-3-319-23485-4_6

To fulfill these goals, food intake measurements, exergames associated to social interaction stimulation, and cognitive training programs (cognitive games) will be proposed to an elderly population enrolled during a pilot study. The DOREMI project is going further with respect to the current state of the art by developing, testing, and exploiting with a short-term business model impact a set of IT-based (Information Technology) services able to:

- Stimulate elderly people in modifying dietary needs and physical activity according to the changes in age through creative, personalized, and engaging solutions;
- Monitor parameters of the elderly people to support the specialist in the daily verification of the compliance of the elderly with the prescribed lifestyle protocol, in accordance with his/her response to physical and cognitive activities.
- Advise the specialist with different types and/or intensities of daily activity for improving the elderly health, based on the assigned protocol progress assessment.
- Empower aging people by offering them knowledge about food and physical activity effectiveness, to let them become the main actors of their health.

To reach these objectives, the project builds over interdisciplinary knowledge encompassing health and artificial intelligence, the latter covering aspects ranging from sensing, machine learning, human-machine interfaces, and games. This paper focuses on the machine learning contribution of the project, which applies to the analysis of the sensor data with the purpose of identifying users' conditions (in terms of balance, calories expenditure, etc.) and activities, detecting changes in the users' habits, and reasoning over such data. The ultimate goal of this data analysis is to support the user who is following the lifestyle protocol prescribed by the specialist, by giving him feedbacks through an appropriate interface, and by providing the specialist with information about the user lifestyle. In particular, the paper gives a snapshot of the status of the project (which just concluded the first year of activity) in the design of the activity recognition and reasoning components.

2   Background and State of the Art on Machine Learning

Exploratory data analysis (EDA) analyzes data sets to find their main features [1], beyond what can be found by formal modeling or hypothesis testing task. When dealing with accelerometer data, features are classified in three categories: time domain, frequency domain, and spatial domain [2]. In the time domain, we use the standard deviation in a frame, which is indicative of the acceleration data and the intensity of the movement during the activity. In the frequency domain, frequency-domain entropy helps the distinction of activities with similar energy intensity by comparing their periodicities. This feature is computed as the information entropy of the normalized Power Spectral Density (PSD) function of the input signal without including the DC component (mean value of the waveform). The periodicity feature evaluates the periodicity of the signal that helps to distinguish cyclic and non-cyclic activities. In the spatial domain, orientation variation is defined by the variation of the gravitational components at three axes of the accelerometer sensor. This feature effectively shows how severe the posture change can be during an activity.

Other EDA tasks of the project concern unsupervised user habits detection aimed at finding behavioral anomalies, by retrieving heterogeneous and multivariate timeseries of sensor data, over long periods. In the project, these tasks are unsupervised to avoid obtrusive data collection campaign at the user site. For this reason, we focus on motif search on sensory data collected in the test by exploiting the results obtained in the field of time series motifs discovery [3,4]. Time series motifs are approximately repeated patterns found within the data. The approach chosen is based on *stigmergy*. Several works used this technique in order to infer motifs in time series related to different fields, from DNA and biological sequences [5,6] to intrusion detection systems [7].

Human activity recognition refers to the process of inferring human activities from raw sensor data [8], classifying or evaluating specific sections of the continuous sensors data stream into specific human activities, events or health parameters values. Recently, the need for adaptive processing of temporal data from potentially large amounts of sensor data has led to an increasing use of machine learning models in activity recognition systems (see [9] for a recent survey), especially due to their robustness and flexibility. Depending on the nature of the treated data, of the specific scenario considered and of the admissible trade-off among efficiency, flexibility and performance, different supervised machine learning methods have been applied in this area.

Among others, Neural Network for sequences, including Recurrent Neural Networks (RNNs) [10], are considered as a class of learning models suitable for approaching tasks characterized by a sequential/temporal nature, and able to deal with noisy and heterogeneous input data streams. Within the class of RNNs, the Reservoir Computing (RC) paradigm [11] in general, and the Echo State Network (ESN) model, [12,13] in particular, represent an interesting efficient approach to build adaptive nonlinear dynamical systems. The class of ESNs provides predictive models for *efficiently* learning in sequential/temporal domains from heterogeneous sources of noisy data, supported by theoretical studies [13,14] and with hundreds of relevant successful experimental studies reported in literature [15]. Interestingly, ESNs have recently proved to be particularly suitable for processing noisy information streams originated by sensor networks, resulting in successful real-world applications in supervised computational tasks related to AAL (Ambient Assisted Living) and human activity recognition. This is also testified by some recent results [16,17,18,19,20], which may be considered as a first preliminary experimental assessment of the feasibility of ESN to the estimation of some relevant target human parameters, although obtained on different and broader AAL benchmarks.

At the reasoning level, our interest is for hybrid approachs founded on static rules and probabilistic methods. Multiple-stage decisions refer to decision tasks that consist of a series of interdependent stages leading towards a final resolution. The decision-maker must decide at each stage what action to take next in order to optimize performance (usually utility). Some examples of this sort are working towards a degree, troubleshooting, medical treatment, budgeting, etc. Decision trees are a useful mean for representing and analyzing multiple-stage decision tasks; they support decisions learned from data, and their terminal nodes represent possible consequences [21].

Other popular approaches, which have been used to implement medical expert systems, are Bayesian Networks [22] and Neural Networks [23], but they require many empirical data to train the algorithms and are not appropriate to be manually adjusted. On the other hand, in our problem the decision process must be transparent and mainly requires static rules based on medical guidelines provided by the professionals. Thus, the decision trees are the best solution since they provide a very structured and easy to understand graphical representation. There also exist efficient and powerful algorithms for automated learning of the trees [24,25,26]. A decision tree is a flow-chart-like structure in which an internal node represents the test on an attribute, each branch represents a test outcome and each leaf node represents a class label (decision taken after computing all attributes). A path from root to leaf represents classification rules. Decision trees give a simple representation for classifying examples. In general, as for all machine learning algorithms, the accuracy of the algorithms increases with the number of sample data. In applications in which the number of samples is not large, a high number of decisions could lead to problems. In these cases, a possible solution is the use of a Hybrid Decision Tree/Genetic Algorithm approach as suggested in [27].

3   Problem Definition and Requirements

Our main objective is to provide a solution for prolonging the functional and cognitive capacity of the elderly by proposing an "Active Ageing" lifestyle protocol. Medical specialists monitor the progress of their patients daily through a dashboard and modify the protocol for each user according to their capabilities. A set of mobile applications (social games, exer-games, cognitive games and diet application) feedback the protocol proposed by the specialist and the progress of games to the end user. The monitoring of each user is achieved by means of a network of sensors, either wearable or environmental, and applications running on personal mobile devices. The human activity recognition (HAR) measures characteristics of the elderly lifestyle in the physical and social domains through non-invasive monitoring solutions based on the sensor data. Custom mobile applications cover the areas of diet and cognitive monitoring. In the rest of this section, we present the main requirements of the HAR.

By leveraging environmental sensors, such as PIRs (Passive InfraRed) and a localization system, the HAR module profiles user habits in terms of daily ratio of room occupancy and indoor/outdoor living. The system is also able to detect changes in the user habits that occur in the long-term. By relying on accelerometer and heartbeat data from a wearable bracelet, the HAR module provides time-slotted estimates, in terms of calories, of the energy expenditure associated with the physical activities of the user. Energy consumption can result from everyday activities and physical exercises proposed by the protocol. The system also computes daily outdoor covered distance, the daily number of steps and detects periods of excessive physical stress by using data originated by the accelerometer and the heartbeat in the bracelet. Finally, a smart carpet is used to measure the user weight and balance skills, leveraging a machine learning classification model based on the BERG balance assessment test.

The HAR assesses the social interactions of the user both indoor and outdoor. In particular, in the indoor case, HAR estimates a quantitative measure of the social interactions based on the occurrence and duration of the daily social gatherings at the user house. Regarding the outdoor socialization, the system estimates the duration of the encounters with other users by detecting the proximity of the users' devices.

The Reasoner uses the data produced by the HAR, the diet, and the games applications to provide an indicator of the user protocol compliance and protocol progress in three areas: social life, physical activity and related diet, cognitive status. These indicators, along with the measured daily metrics and aggregate data, support medical specialists on providing periodical changes to the protocol (i.e.: set of physical activities and games challenges, diet). The Reasoner is able to suggest changes to the user protocol by means of specialist-defined rules.

The HAR module and the Reasoner are, therefore, core system modules, bridging the gap between sensors data, medical specialists, and the end user.

4 An Applicative Scenario

We consider a woman in her 70s, still independent and living alone in her apartment (for the sake of simplicity, we give her the name of Loredana). She is a bit overweight and she starts forgetting things. Recently, the specialist told her that she is at risk for cardiovascular disease, due to her overweight condition, and that she has a mild cognitive impairment. For this reason, Loredana uses our system as a technological support to monitor her life habits and to keep herself healthier and preventing chronic diseases. In a typical day, Loredana measures her weight and balance by means of a smart carpet, which collects data for the evaluation of her BERG scale equilibrium and her weight. The data concerning the balance is used to suggest a personalized physical activity (PA) plan, while the data about the weight give indications about the effectiveness of the intervention in terms of a personalized diet regimen and PA plan. During the day, Loredana wears a special bracelet, which measures (by means of an accelerometer) her heart rate, how much she walked, and how many movements she did with physical exercises. These data are used, by the system developed in the project, to assess her calories expenditure and to monitor the execution of the prescribed physical exercises. The bracelet is also used to localize her both indoor (also collecting information about the time spent in each room) and outdoor (collecting information about distance covered). Furthermore, the bracelet detects the proximity of Loredana with other users wearing the same bracelet, while machine-learning classification models based on environmental sensors deployed at home (PIRs and door switches) detect the presence of other people in her apartment to give indication about the number of received visits. These data are used as an indicator of her social life.

Loredana also uses a tablet to interact with the system, with which she performs cognitive games and inserts data concerning her meals, which are converted by the system (under the supervision of her specialist) in daily Kcalories intake and food composition. She is also guided through the daily physical exercises and games that are selected by the system (under the supervision of her specialist) based on the evolution of her conditions (in terms of balance, weight, physical exercises etc.). All the data collected during the day are processed at night to produce a summary of the Loredana lifestyle, with the purpose of giving feedbacks to Loredana in terms of proposed physical activity, and presenting the condition of Loredana to the specialist on a daily basis.

5   Activity Recognition and Reasoning

5.1   High Level Architecture and Data Flows

The high-level system architecture is presented in Fig. 1, highlighting the data flow originated at a pilot site (in terms of sensor data), through the data processing stages (preprocessing, activity recognition and reasoner subsystems), and then back to the user (in terms of feedbacks) and to the specialist (in terms of information about the user performance).

In particular, Fig. 1 shows that the data processing system contains five main subsystems running on the server (the grey rectangles in the figure), three databases (RAW, HOMER [28] and KIOLA [29]), plus a middleware that uploads the sensors' data in the RAW database, whose description is out of the scope of this paper. At night, a synchronization mechanism (shown as a clock in Fig. 1), sequentially activates these five subsystems. In turn, these subsystems pre-process data (preprocessing subsystem in the Fig. 1), configure the predictive activity recognition tasks (task configurator subsystem), process the daily pre-processed data through the predictive human activity recognition subsystem (HAR subsystem), perform the exploratory data analysis (EDA subsystem), and refine and aggregate the results of these stages (Reasoner subsystem).

Along with the sensors' data, the RAW DB also stores the intermediate data produced by the pre-processing subsystem. The HOMER DB stores configuration information (e.g. regarding sensors deployment) that is used by the task configurator to retrieve the tuning parameters for the different pilot sites. The refined data produced by the HAR and EDA subsystems is then stored in the KIOLA DB, where the Reasoner reads them. The Reasoner outputs feedbacks for the user in terms of suggestions about her lifestyle, and data about her compliance to the suggested lifestyle protocol for the specialist (or caregiver) through the dashboard.

The data processing stages deal with three flows of data, related to the user diet, social relationships, and sedentariness, respectively. The dietary data flow relies on data produced by the smart carpet (pressure data and total weight), the bracelet (hearth rate and accelerometers data), and data about the food composition provided by the user himself through an interface on his tablet. In particular, the data produced by the bracelet pass through the HAR subsystem that estimates the user physical activity. The data flow about the user's social relationships relies on a number of environmental sensors, which detect the contacts of the user with other people. To this purpose, the HAR and EDA subsystems detect the user encounters and the proximity of the user with other people by fusing information produced by presence sensors, user's localization and door switches. This data flow also relies on the user's mood

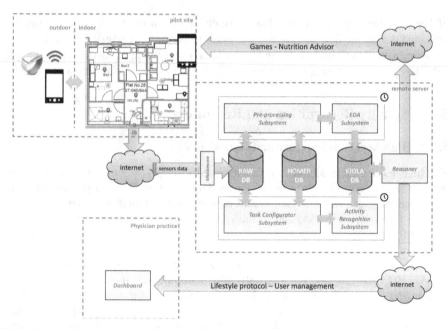

Fig. 1. High-level architecture and deployment of a typical installation, with data flowing from a pilot site to the remote Activity Recognition and Reasoning system.

information, which the user himself asserts daily through an app on his tablet. Finally, the sedentariness data flow exploits data produced by the bracelet (heart rate, user's localization, movements, and step count), the smart carpet, and data about the use of the application that guides the user through the daily physical activity. The HAR subsystem processes the data of the smart carpet to assess the user balance according to the BERG scale. The HAR and EDA subsystems also process data from the bracelet to assess the intensity of the physical effort.

The Reasoner fuses all these data flows at a higher level than that of HAR and EDA. In a first step, it exploits rules extracted from clinical guidelines to compute specific parameters for each of the three data flows. For example, it uses the physical activity estimation in the sedentariness data flow to assess the compliance of the user to the prescribed lifestyle protocol (as the medical expert defines it). In the second step, the Reasoner performs a cross-domain reasoning on top of the first step, allowing a deeper insight in the well-being of the patient. Note that the empirical rules needed to define the second-level reasoning protocol are yet not available, as they are the output of medical studies on the data collected from the on-site experimentation that will be concluded in the next year of the project activity. Hence, at this stage of the project, the second level reasoning is not yet implemented.

Note that the Reasoner subsystem operates at a different time-scale with respect to the HAR and EDA subsystems. In fact, the aim of HAR and EDA is the recognition of short-term activities of the user. These can be recognized from a sequence of input sensor information (possibly pre-processed) in a limited (short-term) time window.

All short-term predictions generated across the day are then forwarded to the Reasoner for information integration across medium/long time scales. Medium term reasoning operates over 24h periods (for example, to assess the calories assumption/consumption balance in a day). Long-term reasoning, on the other hand, shows general trends by aggregating information on the entire duration of the experimentation in the pilot sites (for example to offer statistical data about the user, which the medical experts can use to assess the overall user improvement during the experimentation).

5.2   Human Activity Recognition and Exploratory Data Analysis Subsystems

The goal of the activity recognition subsystems (HAR and EDA) is to evaluate the user parameters (referred to as *predictions*) concerning his short-term activities performed during the day. It exploits the *activity recognition configuration*, which is the result of an off-line configuration phase aimed at finding the final setting for the pre-processing and for the activity recognition subsystems (both HAR and EDA) that are deployed to implement the activity recognition tasks.

The EDA subsystem analyses pre-processed data in order to profile user's habits, to detect behavioral deviations of the routine indoor activities, and to provide aggregated values useful to the Reasoner in the sedentariness area, such as user habits, daily outdoor distance covered, daily steps and information about outdoor meetings with other users. The actual nature of the data streams processed by the EDA depends on the particular sensors originating them. For example, in the case of BERG score prediction concerning the user balance, the data produced by the smart carpet have a high and variable frequency ($\sim$100 Hz). These data are normalized and broken into fixed frequency time series segments of 200 ms, from which the preprocessing stage extracts statistical features consisting in mean, standard deviation, skewness and kurtosis. The resulting features time series, presenting a lower frequency of 5 Hz, are the input of the EDA subsystem. A similar pre-processing stage is applied to data coming from the other sensors.

The unsupervised models used for the EDA do not rely on a long-term, invasive and costly ground truth collection and annotation campaigns, which may be not acceptable by the users. Rather, EDA is designed to detect symptoms of chronic diseases (which are most relevant for the project purposes), characterized by a gradual, long-term deviation from the user typical behavior, or by critical trends in the user's vital parameters. For example, EDA features a module for the detection of abnormal deviations in the user habits (based on motif discovery and stigmergy algorithms) that relies on the locations of the user (at room-level) during his daily living activities.

Concerning the HAR subsystem, its internal architecture is shown in Fig. 2. It is composed of two main subsystems, which are the task configurator and the activity recognition subsystem. The former handles the retrieval of configuration information and pre-processed data for the tasks addressed in the system and forwards it to the latter subsystem, which is responsible for performing the actual activity recognition tasks. The core of the activity recognition system is given by the HAR scheduler (which activates the activity recognition tasks when all sensor information is consolidated and pre-processed in the RAW DB), and by the pool of activity recognition

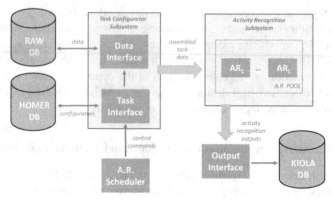

Fig. 2. Detailed architecture of the supervised HAR system

components (based on predictive learning models), one for each specific task. These components implement the trained predictive learning model obtained from a preliminary validation phase, and they produce their predictions by computing the outputs of the supervised learning model in response to the input data.

5.3   Reasoning Subsystem and Dashboard

Fig. 3 shows the Reasoner, the high-level database and the dashboard with the other system components. The high-level database receives the data from three sources: the activity recognition subsystem (data about physical activity, calories consumption, balance, user sociality etc.), the diet application on the tablet (nutritional data inserted by the users themselves), and the application for serious games on the tablet (statistics about the performance of the user in cognitive games).

The Reasoner compares all these data with the clinical protocol the person should follow based on the pre-sets from the medical experts. To this purpose, it adopts a rule based with hierarchical decision trees, where the rules will be created according to the actual medical guidelines. Based on this, a general overview, as well as some calculated data relations, is presented to the specialist's dashboard. The Reasoner settings can be modified by the medical experts to change the protocol according to a certain user behavior (for example he can change the food composition or reduce the overall caloric intake), or the Reasoner itself may adapt the protocol according to pre-defined rules, when some known conditions occur. For example, an improvement in the physical activity assessed by the heart rate response to exercise may result in a progressive increase in the intensity of proposed exercises. The Reasoner gives feedbacks to the user by means of applications on the user tablet (namely, the nutritional and physical activity advisors and the cognitive games).

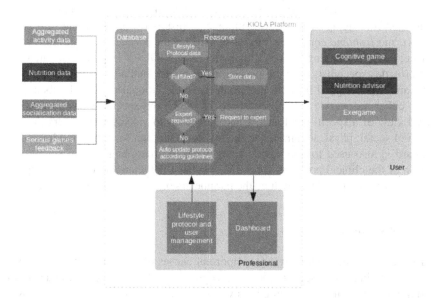

Fig. 3. Architecture of the Reasoner and its relationship with the other system components

Fig. 4. Personal dashboard showing activity data of an end-user

The reasoning module and the dashboard are integrated in the KIOLA modular platform, suitable for clinical data and therapy management. It is built on top of the open-source web-framework Django[1] and uses PostgreSQL 9.4 as primary data storage. KIOLA has two groups of components: *core components* (that provide data

---

[1] http://www.djangoproject.com

models for receiving and storing external sensor data, rule-based reasoning on observations, and messaging services to communicate results of the reasoning to external
systems), and *frontend components* (a dashboard for the specialists, an administrative
interface, and a search engine for all data stored in KIOLA). In particular, the dashboard provides specialists with the possibility to review and adjust clinical protocols
online, and it is designed for both mobile devices and computers. The dashboard can
provide either an overview of all end-users to which the specialist has access, or a
detailed view of a specific end-user. Charts are used to visualize all observations in
the area of social, physical games, and dietary data (see Fig. 4). A task module on the
dashboard is also used to notify specialists when the reasoning system suggests an
adoption of the clinical protocol. Here, the specialist can then approve or disapprove
the recommendation, and he can tune the parameters of the protocol by himself.

6   Conclusions

The DOREMI project addresses three important causes of morbidity and mortality in
the elderly (malnutrition, sedentariness and cognitive decline), by designing a solution
aimed at promoting an active aging lifestyle protocol. It envisages to provide an ICT-
based home care services for aging people to contrast cognitive decline, sedentariness
and unhealthy dietary habits. The proposed approach builds on activity recognition
and reasoning subsystems, which are the scope of this paper. At the current stage of
the project such components are being deployed, and they will be validated in the
course of the year by means of an extensive data collection campaign aimed at obtaining the annotated datasets. These datasets, that are currently being collected over a
group of elderly volunteers in Pisa, in view of the experimentation in the pilot sites
planned by the beginning of 2016.

References

1. Tukey, J.W.: Exploratory data analysis, pp. 2–3 (1977)
2. Long, X., Yin, B., Aarts, R.M.: Single-accelerometer-based daily physical activity classification. In: Engineering in Medicine and Biology Society, EMBC 2009. Annual International Conference of the IEEE. IEEE (2009)
3. Fernández-Llatas, C., et al.: Process Mining for Individualized Behavior Modeling Using
   Wireless Tracking in Nursing Homes. Sensors **13**(11), 15434–15451 (2013)
4. der Aalst, V., Wil, M.P., et al.: Workflow mining: A survey of issues and approaches. Data
   & Knowledge Engineering **47**(2), 237–267 (2003)
5. Yang, C.-H., Liu, Y.-T., Chuang, L.-Y.: DNA motif discovery based on ant colony optimization and expectation maximization. In: Proceedings of the International Multi Conference of Engineers and Computer Scientists, vol. 1 (2011)
6. Bouamama, S., Boukerram, A., Al-Badarneh, A.F.: Motif finding using ant colony optimization. In: Dorigo, M., Birattari, M., Di Caro, G.A., Doursat, R., Engelbrecht, A.P.,
   Floreano, D., Gambardella, L.M., Groß, R., Şahin, E., Sayama, H., Stützle, T. (eds.) ANTS
   2010. LNCS, vol. 6234, pp. 464–471. Springer, Heidelberg (2010)

7. Cui, X., et al.: Visual mining intrusion behaviors by using swarm technology. In: 2011 44th Hawaii International Conference on System Sciences (HICSS). IEEE (2011)
8. Bao, L., Intille, S.S.: Activity Recognition from User-Annotated Acceleration Data. In: Ferscha, A., Mattern, F. (eds.) PERVASIVE 2004. LNCS, vol. 3001, pp. 1–17. Springer, Heidelberg (2004)
9. Lara, O.D., Labrador, M.A.: A survey on human activity recognition using wearable sensors. Communications Surveys & Tutorials, IEEE **15**(3), 1192–1209 (2013)
10. Kolen, J., Kremer, S. (eds.): A Field Guide to Dynamical Recurrent Networks. IEEE Press (2001)
11. Lukoševicius, M., Jaeger, H.: Reservoir computing approaches to recurrent neural network training. Computer Science Review **3**(3), 127–149 (2009)
12. Jaeger, H., Haas, H.: Harnessing nonlinearity: Predicting chaotic systems and saving energy in wireless communication. Science **304**(5667), 78–80 (2004)
13. Gallicchio, C., Micheli, A.: Architectural and markovian factors of echo state networks. Neural Networks **24**(5), 440–456 (2011)
14. Tino, P., Hammer, B., Boden, M.: Markovian bias of neural based architectures with feedback connections. In: Hammer, B., Hitzler, P. (eds.) Perspectives of neural-symbolic integration. SCI, vol. 77, pp. 95–133. Springer-Verlag, Heidelberg (2007)
15. Lukoševičius, M., Jaeger, H., Schrauwen, B.: Reservoir Computing Trends. KI - Künstliche Intelligenz **26**(4), 365–371 (2012)
16. Bacciu, D., Barsocchi, P., Chessa, S., Gallicchio, C., Micheli, A.: An experimental characterization of reservoir computing in ambient assisted living applications. Neural Computing and Applications **24**(6), 1451–1464 (2014)
17. Chessa, S., et al.: Robot localization by echo state networks using RSS. In: Recent Advances of Neural Network Models and Applications. Smart Innovation, Systems and Technologies, vol. 26, pp. 147–154. Springer (2014)
18. Palumbo, F., Barsocchi, P., Gallicchio, C., Chessa, S., Micheli, A.: Multisensor data fusion for activity recognition based on reservoir computing. In: Botía, J.A., Álvarez-García, J.A., Fujinami, K., Barsocchi, P., Riedel, T. (eds.) EvAAL 2013. CCIS, vol. 386, pp. 24–35. Springer, Heidelberg (2013)
19. Bacciu, D., Gallicchio, C., Micheli, A., Di Rocco, M., Saffiotti, A.: Learning context-aware mobile robot navigation in home environments. In: 5th IEEE Int. Conf. on Information, Intelligence, Systems and Applications (IISA) (2014)
20. Amato, G., Broxvall, M., Chessa, S., Dragone, M., Gennaro, C., López, R., Maguire, L., Mcginnity, T., Micheli, A., Renteria, A., O'Hare, G., Pecora, F.: Robotic UBIquitous COgnitive network. In: Novais, P., Hallenborg, K., Tapia, D.I., Rodrìguez, J.M. (eds.) Ambient Intelligence - Software and Applications. AISC, vol. 153, pp. 191–195. Springer, Heidelberg (2012)
21. Lavrac, N., et al.: Intelligent data analysis in medicine. IJCAI **97**, 1–13 (1997)
22. Chae, Y.M.: Expert Systems in Medicine. In: Liebowitz, J. (ed.) The Handbook of applied expert systems, pp. 32.1–32.20. CRC Press (1998)
23. Gurgen, F.: Neuronal-Network-based decision making in diagnostic applications. IEEE EMB Magazine **18**(4), 89–93 (1999)
24. Anderson, J.R., Machine learning: An artificial intelligence approach. In: Michalski, R.S., Carbonell, J.G., Mitchell, T.M. (eds.) vol. 2. Morgan Kaufmann (1986)
25. Hastie, T., et al.: The elements of statistical learning, vol. 2(1). Springer (2009)
26. Murphy, K.P.: Machine learning: a probabilistic perspective. MIT Press (2012)

27. Carvalho, D.R., Freitas, A.A.: A hybrid decision tree/genetic algorithm method for data mining. Information Sciences **163**(1), 13–35 (2004). [EDA1] Tukey, J.W.: Exploratory data analysis, pp. 2–3 (1977)
28. Fuxreiter, T., et al.: A modular plat- form for event recognition in smart homes. In: 12th IEEE Int. Conf. on e-Health Networking Applications and Services (Healthcom), pp. 1–6 (2010)
29. Kreiner, K., et al.: Play up! A smart knowledge-based system using games for preventing falls in elderly people. Health Informatics meets eHealth (eHealth 2013). In: Proceedings of the eHealth 2013, OCG, Vienna, pp. 243–248 (2013). ISBN: 978-3-85403-293-9

Gradient: A User-Centric Lightweight Smartphone Based Standalone Fall Detection System

Ajay Bhatia[1]([✉]), Suman Kumar[2], and Vijay Kumar Mago[3]

[1] Punjab Technical University, Jalandhar, India
prof.ajaybhatia@gmail.com
[2] Troy University, Troy, AL 36082, USA
skumar@troy.edu
[3] Department of Computer Science, Lakehead University,
Thunder Bay, ON, Canada
vmago@lakeheadu.ca

Abstract. A real time pervasive fall detection system is a very important tool that would assist health care professionals in the event of falls of monitored elderly people, the demography among which fall is the epidemic cause of injuries and deaths. In this work, *Gradient*, a user centric and device friendly standalone smartphone based fall detection solution is proposed. Our solution is standalone and user centric as it is portable, cost efficient, user friendly, privacy preserving, and requires only technologies which exists in cellphones. In addition, *Gradient* is light weight which makes it device friendly since cellphones are constrained by energy and memory limitations. Our work is based on accelerometer sensor data and the data derived from gravity sensors, a recently available inbuilt sensor in smartphones. Through experimentation, we demonstrate that *Gradient* exhibits superior accuracy among other fall detection solutions.

Keywords: Fall detection · Accelerometer · Gravity · Sensors · Android app

1 Introduction

Advances in health care services and techonologies, and decrease in fertility rate (especially in developed countries) are bringing major demographic changes in aging. Currently, around 10% world population is aged over 65 and it is estimated that this demography will see a 10 times increase in next 50 years [1]. It is observed that fall is the major contributor to the growing rates of mortality, morbidity of aging population, and complications induced by fall contribute to the increase in higher health care cost also [2]. According to U.S. Census Bureau, 13% of the population is over 65 years old, out of which 40% homely old-age adults fall atleast once a year and 1 in 40 is hospitalised. Those who are hospitalised have 50% of chance to be alive a year later. Fall is a major health

F. Pereira et al. (Eds.) EPIA 2015, LNAI 9273, pp. 67–78, 2015.
DOI: 10.1007/978-3-319-23485-4_7

threat to not only indpendently living elders [3] but also to community-dwelling ones where fall rate is estimated to be 30% to 60% annually [4]. Elders who suffer from visual impairment, urinary incontinence, and functional limitations are at increased risk of recurrent falls [5]. Therefore, pervasive fall detection systems that meet the needs of aging population is a necessity especially since increase of aging population and estimated decrease in health care professionals demand for technology assisted intelligent health care solutions [6].

Considerable efforts have been made to design fall detection solutions for elderly populations, see [7] for a comprehensive list of solutions proposed by researchers. As more of society is relying on smartphones because of its ubiquitous nature of internet connectivity and computing, smartphone based fall-detection assisted health monitoring and emergency response system is highly desired by elderly population. In fact, recent trend suggests that smartphones are reducing the need for wearing watches, the most common wearable gadget in previous centuries [8] making smartphones the most commonly carried device by people. Naturally, effective smartphone based fall-detection solutions that do not require any infrastructure support external to smartphones are more fitting the needs of elderly population.

In this work, we aim to design a smartphone based standalone fall detection system that is portable, cost efficient, user friendly, privacy preserving [9], and requiring only existing cellphone technology. In addition, such solution must exhibit low memory and low computational overhead which is only fitting since cellphones are constrained by limited energy and limited memory. Clearly, this motivates the design approaches centering around existing sensing technology available in cellphones. Among sensor data, accelerometer sensor is considered to be accurate [10] and therefore, a natural choice for design of such system. In fact, accelerometer and orientation sensor combination based solutions are proposed in past researches, for example, iFall [11] are some of the most notable solutions that meet the design requirements mentioned above. However, through experimentation, we show that smartphone fall detection solutions that involve accelerometer data supplemented with orientation sensor data are not very accurate. Therefore, to meet the above requirements, we propose a novel fall detection mechanism that utilizes gravity sensor data and accelerometer data. Experimental comparision shows that the proposed approach is superior as compared to its peers. In our work, Android operating system platform is used for experimentation and implementation purposes.

This paper is organized as follows: in Section 2, we discuss recent and landmark research work in this area. Section 3 discusses an Android based application for data acquisition and our proposed fall detection method. In Section 4, *alpha test* is performed on Gradient using simulated data. Finally, summary and possible future work are discussed in section 5.

2 Related Work

Various studies have been carried out on fall detection using wearable sensors and mobile phones. We divide the fall detection systems in two categories:

2.1   User-Centric and Device-Friendly

The first accelerometer data driven fall detection system was proposed in [12]. Their system detects a fall when there is a change in body orientation from upright to lying that occurs immediately after a large negative acceleration. This system design later becomes a reference point for many fall detection algorithms using accelerometers.

A popular Android phone application, iFall [11], is developed for fall monitoring and response. Data acquired from the accelerometer with the help of application is evaluated with several threshold based algorithms to determine a fall. Basic body metrics like height and weight, along with level of activity are used for estimating the threshold values. An alert notification system, SMS and emergency call is developed in moderate to critical situations and emergency.

A threshold based fall detection algorithm is proposed in [13]. The algorithm works by detecting dynamic situations of postures, followed by unintentional falls to lying postures. The thresholds calculated and obtained from the collected data are compared with the linear acceleration and the angular velocity sensor data for detecting the fall. After the fall is detected, a notification is sent to make an alert about the fall. The authors used gyroscope for data acquisition. The gyroscope data is not efficient in detecting the fall accurately therefore making the proposed system not a good choice for monitoring the fall. In [14], accelerometer and gyroscope sensors are used together to detect fall in elderly population. Further gravity and angular velocity is extracted from the data to detect fall. Using gravity and angular velocity not only detect a fall but also the posture of the body during fall is detected. Authors experimented the system to know the false positive and false negative fall detection. Also, false fall positions were used in study to know the impact and efficiency of the algorithm. The approach used to detect the fall in their proposed study is too simple to be adopted for detecting *quick* falls using accelerometer data. That is why, the system performs poorly in differentiating *jumping into bed* and *falling against wall* with seated posture.

In [15], a smartphone-based fall detection system using a threshold-based algorithm to distinguish between activities of daily living and falls in real time is proposed. By comparative analysis of threshold levels for acceleration, in order to get the best sensitivity and specificity, acceleration thresholds were determined for early pre-impact alarm (4.5-5 m/s^2) and post-fall detection (21-28 m/s^2) under experimental conditions. The experimental thresholds calculated are helpful for further study in this area of research. But accelerometer alone is not sufficient for detecting the fall effectively.

2.2   External Infrastructure Based

The paper [16] describes a fall detection sensor to monitor the subject safely and accurately by implementing it in a large sensor network called SensorNet. Initial approaches such as conjoined angle change and magnitude detection algorithm were unsuccessful. Another drawback of this approach involves inefficiency of complex fall detection in which the user did not end up oriented horizontally with the ground. The fall-detection board designed was able to detect 90% of all falls with 5% false positive rate.

The Ivy project [17] used low-cost and low-power wearable accelerometer on wireless sensor network to detect the fall. The threshold on the peak values of the accelerometer were estimated along orientation angular data to detect the fall. The authors found that the intensity and acceleration of fall is far different from other activities. The main drawback of this proposed system is that it only works well for an indoor environment because of its dependance on a fixed network to relay events. In the study [18], accelerometer sensor is used to detect wearer's posture, activity and fall in wireless sensor network. The activity is determined by the *alternating current* component and posture is determined by the *direct current* component of the accelerometer signal. Fall detection rate of the proposed system is 93.2%. The paper lacks in explaining the actual algorithm. Moreover the complexity and cost involved in designing the system makes it less suitable for fall-detection in real life scenario.

In [19], authors proposed a fall detection system which uses two sets of sensors, one with an accelerometer and a gyroscope and the other with only an accelerometer. They used sensor data to calculate angular data and their system outperforms the earlier known systems as the system is able to detect fall with an average lead-time of 700 milliseconds before the impact to ground occurs. The major drawback of the system is that the subject has to wear torso and thigh sensors and this tends to be cumbersome for subjects. Also, the system is not capable to record data in real-life situations.

3   Methods

In this section, first, principle behind previous researches that utilize orientation sensor based design approaches is discussed, then we describe gravity sensor based approach that forms the principle and basis of our proposal. We further show that gravity sensor based design is more accurate and becomes a natural choice for user-centric and device friendly fall detection solution approaches.

3.1   Design Principle: Orientation Sensor

A gyroscope can be used to either measure, or maintain, the orientation of a device. In comparison with an accelerometer, a gyroscope measures the orientation rather than linear acceleration of the device. Analyzing accelerometer and orientation data to detect fall is the core of several fall detection systems. Orientation sensor which in most cases is the combination of gyro sensor magnetometer supplements the aim to provide accurate orientation data. We further explain the idea behind this approach.

Tri-axial orientation is represented in Figure 1(a)) where θ, ϕ, and ψ are azimuthal angle (0^0 is north), pitch angle (0^0 is flat on its back from positive Y, east) and roll angle (0^0 is flat on its back from positive Z, downward) respectively.

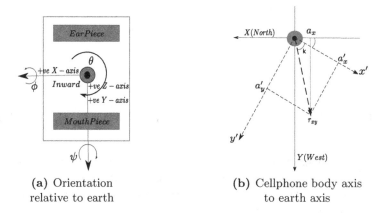

(a) Orientation
relative to earth

(b) Cellphone body axis
to earth axis

Fig. 1. Orientation of cellphone relative to real world axis system

Consider the example in Figure 1(b), where axis system is rotated around z axis by θ. The rotated axis are shown as X' and Y'. X' and Y' are axis coordinates relative to phone. Contribution of this rotation to the X component of fixed axis (X and Y) is given as: $a_X = r_{xy}cos(\theta + k)$, where a'_x and a'_y are X' and Y' components of acceleration vector and k is the angle acceleration vector makes width the X' axis. Combining the effect of azimuth, roll and pitch, we get following matrix:

$$M = \begin{bmatrix} cos(\psi).cos(\phi) & -sin(\psi).cos(\theta) & sin(\psi).sin(\theta) \\ sin(\psi).cos(\phi) & cost(\psi).cos(\theta) & -cost(\psi).sin(\theta) \\ -sin(\psi) & cos(\phi).sin(\theta) & cos(\phi).cos(\theta) \end{bmatrix}$$
$$+ \begin{bmatrix} 0 & cost(\psi).sin(\phi).sin(\theta) & cost(\psi).sin(\phi).cos(\theta) \\ 0 & sin(\phi).sin(\theta) & sin(\psi).sin(\phi).cos(\theta) \\ 0 & 0 & cos(\phi).cos(\theta) \end{bmatrix} \quad (1)$$

Relation between body axis ($[x', y', z']$) and earth axis ($[x, y, z]$) is given as

$$[a_x \quad a_y \quad a_z] \quad = \quad [a_{x'} \quad a_{y'} \quad a_{z'}] \quad \times \quad M \quad (2)$$

Here a_x, a_y and a_z are accelerometer sensor readings on tri-axial coordinate system relative to cell phone and θ, ϕ and ψ are orientation sensor readings from the mobile phone and M is the matrix defined in equation 1.

3.2 Design Principle: Gravity Sensor

Gravity sensor returns only the influence of gravity. Gravity vector under the context of the phone coordinate system is given as: $\overrightarrow{G} = (g_{x'}, g_{y'}, g_{z'})$, and acceleration in the same axis system is given as: $\overrightarrow{A} = (a_{x'}, a_{y'}, a_{z'})$.

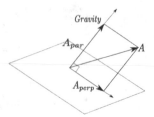

Fig. 2. Cellphone Acceleration Along (A_{par}) and Perpendicular (A_{perp}) to Gravity

From Figure 2, since both vectors are under the same coordinate system, the phone coordinate system by simple vector inner product, the angle between the two can be easily found and thus acceleration component of cellphone parallel to gravity vector is given as below,

$$\overrightarrow{A_{par}} = \overrightarrow{a_z} = (\overrightarrow{A} \odot \overrightarrow{G})\frac{\overrightarrow{G}}{\left|\overrightarrow{G}\right|^2} \tag{3}$$

Likewise, the perpendicular counterpart is given as below,

$$\overrightarrow{A_{perp}} = \overrightarrow{A} - (\overrightarrow{A} \odot \overrightarrow{G})\frac{\overrightarrow{G}}{\left|\overrightarrow{G}\right|^2} \tag{4}$$

The absolute value of downward acceleration is given by following equation:

$$\left|\overrightarrow{A_{par}}\right| = |a_z| = \frac{a_{x'}g_{x'} + a_{y'}g_{y'} + a_{z'}g_{z'}}{\sqrt{g_{x'}^2 + g_{y'}^2 + g_{y'}^2}} \tag{5}$$

And the absolute acceleration along $X' - Y'$ plane of cellphone is given as

$$\left|\overrightarrow{A_{perp}}\right| = \sqrt{a_{x'}^2 + a_{y'}^2 + a_{z'}^2 - |\overrightarrow{A_{par}}|^2} \tag{6}$$

3.3 Orientation Sensor vs. Gravity Sensor

A comparison of vertical downward acceleration component of linear acceleration sensor data using orientation sensor values (Equation 1 and Equation 2) with the same using gravity sensor values is presented in Figure 3. We observe unstable and erroneous values using orientation sensor where as the values using gravity sensor data is stable and accurate. Gyro sensor and Magnetometer sensor suffer through convergence to stabilize values and hence causes a problem which one can see in our experiment. Therefore, it explains why many supposedly promising solutions failed the accuracy test by a huge margin. With this observation, we are motivated to use gravity sensor in our work. To the best of our knowledge, our work is beyond the research already done and includes gravity sensor which is more promising in detecting a real fall than a false fall. The system design using accelerometer and gravity sensor follows next.

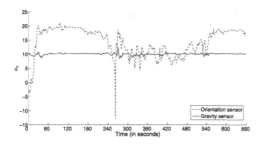

Fig. 3. Downward acceleration: Orientation sensor vs. Gravity sensor

4   Gradient: System Design

4.1   Data Acquisition

The Android application has been developed by considering the importance of each sensor inside the phone. Since application is basically developed to collect data through sensors even when screen is off therefore it is designed to work as a service rather than an activity for collecting data through available sensors in the device. The application is designed to run on any Android platform having at least GingerBread operating system [20]. Application uses both accelerometer and gravity sensors for the data acquisition. Both the sensors are acceleration based sensors and are responsible for measuring acceleration of the device in all three directions. An acceleration sensor measures the acceleration applied to the device, including the force of gravity. The gravity sensor provides a three dimensional vector indicating the direction and magnitude of gravity. Along this configuration, an application uses permissions to write to external storage for saving the data in the form of CSV (Comma-separated values) files. The data is saved in different CSV files based on the type of sensors. $x-axis$, $y-axis$, $z-axis$ and timestamp are recorded in the respective files. As the data is acquired, the saved files are used to process and detect fall. Subsection 4.2 describes the our processed methodology.

4.2   Algorithm

Gradient is based on the observation that when a user falls there is a sudden change in acceleration in vertical downward direction and therefore, downward acceleration component contains enough information to detect fall events. We ran several laboratory experiments to verify this hypothesis. The sensor data is collected on a mobile device using *Gradient* application. Then the differentiation of vertical downward component in time is compared with preset threshold value to identify fall events. It is to be noted that the name *Gradient* is derived from **Gravity-diff**erentiation, the driving idea behind the proposed solution. The detailed algorithm is presented below:

Algorithm 1. Fall Detection Algorithm (Using Vectorization)

Data: $\vec{A_d} \leftarrow (a_x \; a_y \; a_z)$; //Accelerometer sensor readings
$\vec{G_d} \leftarrow (g_x \; g_y \; g_z)$; // Gravity sensor readings
Result: $FallDetected$

$W_s \leftarrow 10$; //Set window size for moving averages

$\vec{A_v} \leftarrow mavg(A_d, W_s)$; //Calculate moving averages on accelerometer data

$\vec{G_v} \leftarrow mavg(G_d, W_s)$; //Calculate moving averages on gravity data

$|\overrightarrow{A_{par}}| \leftarrow \frac{a_x g_x + a_y g_y + a_z g_z}{\sqrt{g_x^2 + g_y^2 + g_z^2}}$; //Downward acceleration

$|\overrightarrow{A_{perp}}| \leftarrow \sqrt{a_x^2 + a_y^2 + a_z^2 - |\overrightarrow{A_{par}}|^2}$; //Aceeleration along $X - Y$ plane

$|\overrightarrow{A_{parNew}}| \leftarrow \frac{|\overrightarrow{A_{par}}|}{dt}$; //Gradient of the downward acceleration

$|\overrightarrow{A_{perpNew}}| \leftarrow \frac{|\overrightarrow{A_{perp}}|}{dt}$; //Gradient along the x-y plane

$Th \leftarrow |\overrightarrow{A_{parNew}}|$; //Estimate threshold value as average of A_{par}

if $|\overrightarrow{A_{parNew}}|_i < \theta, \forall i$ **then**
⌊ $FallDetected \leftarrow true$;

---

5 Result and Discussion

In this section, performance of the proposed solution is discussed along with the experiment scenarios. We did not test our system on real fall datasets[1] because gravity sensor values are not available. However, we tried our best to collect fall data in a controlled environment to match the real life natural fall events.

5.1 Experiment Set-Up

Gradient app distributed among internal members of the research team (4 graduate students and 2 faculty members) to evaluate the user acceptance and feedback of the use of the app and also its accuracy for data acquisition. Data was collected on Samsung Galaxy S4 which has Quad-core 1.6 GHz processor, 2 GB internal RAM and runs on Android OS, v4.2.2 (Jelly Bean). Members of the team carried the smart phone with *Gradient* application during their daily routine of work and activities across different time spans and number of days. Detailed logs of the times when they fall were maintained to check the sustainability of the app post-experimentation and results. Feedbacks were taken from the members to improve the app in every aspect of data acquisition. Two of the experiments performed by the team members are represented in the Figure 4a and 4b respectively. In the first experiment, Figure 4a, there is fall at the end of 6th minute of data acquisition. Second fall is clearly visible at end of the 7th minute. Similarly in Figure 4b fall can be seen in the mid of 6th and 7th minute respectively.

---

[1] http://www.bmi.teicrete.gr/index.php/research/mobifall

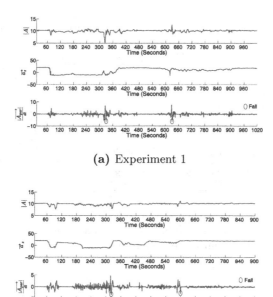

(a) Experiment 1

(b) Experiment 2

Fig. 4. Fall detection experiments

The experiments were conducted by graduate students of the leading author. The students were informed about the procedure and implications of this study. The institute of the leading author also approved this research work. A total of 20 experiments[2] are performed in the closed and safe environment. Total number of actual falls performed by the subjects for the study are 27. It is found that the best threshold value (θ) for the fall detection is -6.15, while applying the *Gradient* approach. The values lesser than θ are treated as fall event. The *true positive* falls and *false negative* falls detected by the proposed system are 24 and 3 respectively. On the other hand, *false positive* falls were counted to be 3. On close observation of *false negative* cases, we find that these are the cases where the subject fell from low heights, e.g., from a sofa set. We wish to acknowledge that to distinguish between putting the phone in pocket after receiving a call and falling from a low lying sofa is challenging. These **soft falls** needs further investigation. Also, there is no way to compute the *true negative* cases, so we are not reporting the specificity of the system.

We also attempted to compare our method with a well known Android fall detection application iFall [11]. The iFall detects all 27 fall events but it also raised count of *false negative* to 13, and hence reducing the sensitivity to 67.5% only, see Table 1. We observe that the application is very sensitive to jerks and movements like brisk walking/running. Statistically, our system is more reliable and robust to cope with the real-time activities.

---

[2] All the experiments were performed in controlled and safe environment.

Table 1. Comparison between iFall application and Gradient

| | Fall | | |
	Positive	Negative	
Outcome Positive	True positive (TP) = 20	False positive (FP) = 3	**Positive predictive value** **= TP / (TP + FP)** **= 20 / (20 + 3)** **= 86.96%**
Outcome Negative	False negative (FN) = 2	True negative (TN) = 2	**Negative predictive value** **= TN / (FN + TN)** **= 2 / (2 + 2)** **= 50.00%**
	Sensitivity = TP / (TP + FN) = 20 / (20 + 2) = 90.91%	Specificity = TN / (FP + TN) = 2 / (3 + 2) = 40.00%	Total 27

The positive predictive value (86.96%) and sensitivity (90.91%) shows effectivity of the proposed system in detecting falls.

5.2   Gradient in Action

We plotted the time series data in Figure 4. The first subplot is drawn from the magnitude of accelerometer sensor data, which is calculated as:

$$|A| = \sqrt{a_x^2 + a_y^2 + a_z^2}$$

It is to be noted that several researches [11,13,15] utilized acceleration data derived by the above Equation for fall detection. However, fall event mainly corresponds to vertical fall therefore, including horizontal components would result into false positives in many activity scenarios where there is a sudden change in horizontal velocity of a person (see next section for further explanation) and clearly, those events are not fall events. For the sake of completeness of this work we have included the absolute value of acceleration in subplot of Figure 4.

The second subplot is derived from value of a_z from Equation 5. The fall event points are circled in the figure and fall is correctly detected by using the time differential value of a_z. Although our experimental scenario involves controlled environment, we observe successful detection of all fall events.

5.3   Performance Comparison

In this section, we compare our work with one of the most notable work known as iFall [11]. The experiment was performed by running iFall and *Gradient* concurrently along a stop watch to measure the exact time of real falls. The comparison between the iFall and our proposed design is presented in Figure 5. The upper numbered labels 1 through 9 represents falls detected by iFall and the lower numbered labels 1 through 4 represents falls detected by *Gradient*. In Figure 5, we observe although iFall successfully detects fall events, it also

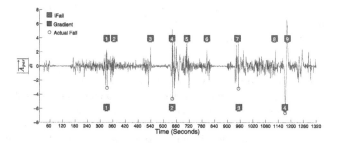

Fig. 5. Comparison between iFall (red square) and *Gradient* (blue square)

outputs several false positives in the event of no fall. We observe if the device is on a running motion or on a jerky motion such as shake, iFall records such events as fall events. Clearly, *Gradient* shows better accuracy than iFall.

6   Conclusion

Fall is the major health risk among the old-aged people around the world. Fall detection using computational approach has remained a challenging task, that prompted researchers to propose various computational methods to detect the occurrence of fall. But the solution that are user-centric and device-friendly is elusive. In this paper, we proposed a novel approach of fall detection using accelerometer and gravity sensors which are now integral components of smart-phones. We designed an Android application to collect experimental data, and applied our algorithm to test the accuracy of the system. Our initial results are very promising and the proposed method has a potential to reduce the false positives which is a common problem with other popular user-centric and device-friendly systems. Furthermore, we believe that this system can help health care-takers, health professionals, and medical practitioners to better mange health hazards due to fall in elder people. In future, we plan to conduct a user study with a healthcare center and test our system on real fall datasets.

References

1. Haub, C.: World population aging: clocks illustrate growth in population under age 5 and over age 65. Population Reference Bureau, June 18, 2013 (2011)
2. Fulks, J., Fallon, F., King, W., Shields, G., Beaumont, N., Ward-Lonergan, J.: Accidents and falls in later life. Generations Review **12**(3), 2–3 (2002)
3. Duthie Jr, E.: Falls. The Medical clinics of North America **73**(6), 1321–1336 (1989)
4. Graafmans, W., Ooms, M., Hofstee, H., Bezemer, P., Bouter, L., Lips, P.: Falls in the elderly: a prospective study of risk factors and risk profiles. American Journal of Epidemiology **143**(11), 1129–1136 (1996)
5. Tromp, A., Pluijm, S., Smit, J., Deeg, D., Bouter, L., Lips, P.: Fall-risk screening test: a prospective study on predictors for falls in community-dwelling elderly. Journal of Clinical Epidemiology **54**(8), 837–844 (2001)

6. Kleinberger, T., Becker, M., Ras, E., Holzinger, A., Müller, P.: Ambient intelligence in assisted living: enable elderly people to handle future interfaces. In: Stephanidis, C. (ed.) UAHCI 2007 (Part II). LNCS, vol. 4555, pp. 103–112. Springer, Heidelberg (2007)

7. Igual, R., Medrano, C., Plaza, I.: Challenges, issues and trends in fall detection systems. BioMedical Engineering OnLine 12(1), 1–24 (2013)

8. Phones replacing wrist watches. http://today.yougov.com/news/2011/05/05/brother-do-you-have-time/ (online accessed April 22, 2014)

9. Ziefle, M., Rocker, C., Holzinger, A.: Medical technology in smart homes: exploring the user's perspective on privacy, intimacy and trust. In: 2011 IEEE 35th Annual Computer Software and Applications Conference Workshops (COMPSACW), pp. 410–415. IEEE (2011)

10. Lindemann, U., Hock, A., Stuber, M., Keck, W., Becker, C.: Evaluation of a fall detector based on accelerometers: A pilot study. Medical and Biological Engineering and Computing 43(5), 548–551 (2005)

11. Sposaro, F., Tyson, G.: ifall: An android application for fall monitoring and response. In: Annual International Conference of the IEEE Engineering in Medicine and Biology Society, EMBC 2009, pp. 6119–6122. IEEE (2009)

12. Williams, G., Doughty, K., Cameron, K., Bradley, D.: A smart fall and activity monitor for telecare applications. In: Proceedings of the 20th Annual International Conference of the IEEE Engineering in Medicine and Biology Society, 1998, vol. 3, pp. 1151–1154. IEEE (1998)

13. Wibisono, W., Arifin, D.N., Pratomo, B.A., Ahmad, T., Ijtihadie, R.M.: Falls detection and notification system using tri-axial accelerometer and gyroscope sensors of a smartphone. In: 2013 Conference on Technologies and Applications of Artificial Intelligence (TAAI), pp. 382–385. IEEE (2013)

14. Li, Q., Stankovic, J.A., Hanson, M.A., Barth, A.T., Lach, J., Zhou, G.: Accurate, fast fall detection using gyroscopes and accelerometer-derived posture information. In: Sixth International Workshop on Wearable and Implantable Body Sensor Networks, BSN 2009, pp. 138–143. IEEE (2009)

15. Mao, L., Liang, D., Ning, Y., Ma, Y., Gao, X., Zhao, G.: Pre-impact and impact detection of falls using built-in tri-accelerometer of smartphone. In: Zhang, Y., Yao, G., He, J., Wang, L., Smalheiser, N.R., Yin, X. (eds.) HIS 2014. LNCS, vol. 8423, pp. 167–174. Springer, Heidelberg (2014)

16. Brown, G.: An accelerometer based fall detector: development, experimentation, and analysis. University of California, Berkeley (2005)

17. Chen, J., Kwong, K., Chang, D., Luk, J., Bajcsy, R.: Wearable sensors for reliable fall detection. In: 27th Annual International Conference of the Engineering in Medicine and Biology Society, IEEE-EMBS 2005, pp. 3551–3554. IEEE (2006)

18. Lee, Y., Kim, J., Son, M., Lee, J.H.: Implementation of accelerometer sensor module and fall detection monitoring system based on wireless sensor network. In: 29th Annual International Conference of the IEEE Engineering in Medicine and Biology Society, EMBS 2007, pp. 2315–2318. IEEE (2007)

19. Nyan, M., Tay, F.E., Murugasu, E.: A wearable system for pre-impact fall detection. Journal of Biomechanics 41(16), 3475–3481 (2008)

20. Inc., G.: Android Gingerbread OS (2013). http://developer.android.com/about/versions/android-2.3-highlights.html (online accessed April 04, 2014)

Towards Diet Management with Automatic Reasoning and Persuasive Natural Language Generation

Luca Anselma$^{(\boxtimes)}$ and Alessandro Mazzei

Dipartimento di Informatica, Università di Torino, Turin, Italy
{anselma,mazzei}@di.unito.it

Abstract. We devise a scenario where the interaction between man and food is mediated by an intelligent system that, on the basis of various factors, encourages or discourages the user to eat a specific dish. The main factors that the system need to account for are (1) the diet that the user intends to follow, (2) the food that s/he has eaten in the last days, and (3) the nutritional values of the dishes and their specific recipes. Automatic reasoning and Natural Language Generation (NLG) play a fundamental role in this project: the compatibility of a food with a diet is formalized as a Simple Temporal Problem (STP), while the NLG tries to motivate the user. In this paper we describe these two facilities and their interface.

Keywords: Diet management · Automatic reasoning · Natural language generation

1 Introduction

The daily diet is one of the most important factors influencing diseases, in particular for obesity. As highlighted by the World Health Organization, this factor is primarily due to the recent changes in the lifestyle [26]. The necessity to encourage the world's population toward a healthy diet has been sponsored by the FAO [20]. In addition, many states specialized these guidelines by adopting strategies related to their *food history* (for instance, for USA http://www.choosemyplate.gov). In Italy, the Italian Society for Human Nutrition has recently produced a prototypical study with recommendations for the use of specialized operators [1].

This scenario suggests the possibility to integrate the directives on nutrition in the daily diet of people by using multimedia tools on mobile devices. The smartphone can be considered as an super-sense that creates new modalities of interaction with food. In recent years there has been a growing interest in using multimedia applications on mobile devices as *persuasive* technologies [13].

Often a user is not able to carefully follow a diet for a number of reasons. When a deviation occurs, it is useful to support the user in devising the consequences of such deviation and to dynamically adapt the rest of the diet in the upcoming meals so that the global Dietary Reference Values (henceforth DRVs)

F. Pereira et al. (Eds.) EPIA 2015, LNAI 9273, pp. 79–90, 2015.
DOI: 10.1007/978-3-319-23485-4_8

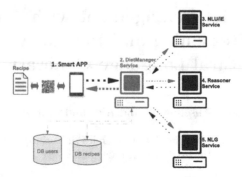

Fig. 1. The architecture of the diet management system.

could nevertheless be reached. In particular in this paper we describe a system which is useful for (i) evaluating the compatibility of a dish with a diet allowing small and occasional episodes of diet disobedience, (ii) determining what are the consequences of eating a specific dish on the rest of the diet, (iii) showing such consequences to the user thus empowering her/him and, moreover, (iv) motivating the user in following the diet by persuading her/him to minimize the acts of disobedience. Using automatic reasoning to evaluate the compatibility of a dish with a diet could enhance a smartphone application with a sort of *virtual dietitian*. Artificial intelligence should make the system *tolerant* to diet disobedience, but also *persuasive* to minimize these acts of disobedience. Thus, a critical issue directly related to automatic reasoning is the final presentation to the user of the results. Several studies have addressed the problem of generating natural language sentences that explain the results of automatic reasoning [4,17].

In our hypothetical scenario the interaction between man and food is mediated by an intelligent system that, on the basis of various factors, encourages or discourages the user to eat a specific dish. The main factors that the system needs to account for are (1) the diet that the user has to follow, (2) the food that s/he has been eating in the last days or that s/he intends to eat in the next days, and (3) the nutritional values of the ingredients of the dish and its specific recipe. In Fig. 1 we report the architecture of our system. It is composed by five modules/services: a smartphone application (APP), a central module that manages the information flow (DietMAnager), an information extraction module (NLU/IE), a reasoning module (Reasoner) and a natural language generation module (NLGenerator). In this paper we focus on the description of the Reasoner and NLGenerator modules; some details on the other modules and on the system can be found on the webpage of the project (http://di.unito.it/madiman).

We think that this system could be commercially attractive at least in two contexts. The first context is the medical one, where users (e.g. patients affected by essential obesity) are strongly motivated to strictly follow a diet and need tools that help them. The second context is the one involving, e.g., healthy fast food or restaurant chains, where the effort of deploying the system can be rewarded by an increase in customer retention.

This paper is organized as follows: in Section 2 we describe the automatic reasoning facilities, in Section 3 we describe the design of the persuasive NLG based on different theories of persuasion and, finally, in Section 4 we draw some conclusions.

2   Automatic Reasoning for Diet Management

Since our approach to automatic reasoning for diet management is based on the STP framework, first we introduce STP, then we describe how we exploit STP to reason on a diet and how we interpret the results from STP.

2.1   Preliminaries: STP

We base our treatment of nutrition constraints on the framework of "Simple Temporal Problem" (STP) [8]. An STP constraint consists in a bound on differences of the form $c \leq x - y \leq d$, where x and y are temporal points and c and d are numbers (their domain can be either discrete or real). An STP constraint can be interpreted in the following way: the temporal distance between the time points x and y is between c – the lower bound of the distance – and d – the upper bound of the distance. It is also possible to impose strict inequalities (i.e., $<$) and $-\infty$ and $+\infty$ can be used to denote the fact that there is no lower or upper bound, respectively. An STP is a conjunction of STP constraints.

An interesting feature of STP is that the problem of determining the consistency of an STP is tractable and that the algorithm employed, i.e., an all-pairs shortest paths algorithm such as Floyd-Warshall's one, also obtains the minimal network, that is the minimum and maximum distance between each pair of points. STP can be represented with a graph whose nodes correspond to the temporal points of the STP and whose arcs are labeled with the temporal distance between the points.

Property. Floyd-Warshall's algorithm is correct and complete on STP, i.e. it performs all and only the correct inferences while propagating the STP constraints [8], and obtains a minimal network. Its temporal computational cost is cubic in the number of time points.

2.2   Towards Automatically Reasoning on a Diet

Reasoning on DRVs. In a diet it is necessary to consider parameters such as the total energy requirements and the specific required amount of nutrients and macronutrients such as proteins, carbohydrates and lipids. In particular in the literature it is possible to find systems of DRVs that are recommended to be followed for significant amounts of time. In the running example, without loss of generality we refer to the Italian values [1]. Such values have to be customized for the specific patients according to their characteristics. In particular, from

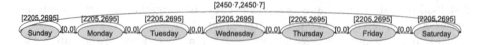

Fig. 2. Example of DRVs for a week represented as STP (for space constraints the constraints for the meals are not represented).

weight, gender and age, using Schofield equation [24], it is possible to estimate the basal metabolic rate; for example a 40-year-old male who is 1.80 m tall and weighs 71.3 kg has an estimated basal metabolic rate of 1690 kcal/day. Such value is then adjusted [1] by taking into account the energy expenditure related to the physical activity of the individual; for example a sedentary lifestyle corresponds to a physical activity level of 1.45, thus, in the example, since the physical activity level is a multiplicative factor, the person has a total energy requirement of 2450 kcal/day. Moreover, it is recommended [1] that such energy is provided by the appropriate amount of the different macronutrients, e.g., 260 kcal/day of proteins, 735 kcal/day of lipids and 1455 kcal/day of carbohydrates. In this section we focus on the total energy requirement; the macronutrients can be dealt with separately in the same way.

We represent the DRVs as STPs; more precisely, we use an STP constraint to represent – instead of temporal distance between temporal points – the admissible DRVs. Thus, e.g., a recommendation to eat a lunch of minimum 500 kcal and maximum 600 kcal is represented by the STP constraint $500 \leq lunch_E - lunch_S \leq 600$, where $lunch_E$ and $lunch_S$ represent the end and the start of the lunch, respectively.

Furthermore, we exploit the STP framework to allow a user to make small deviations with regard to the "ideal" diet and to know in advance what are the consequences of such deviations on the rest of the diet. Thus, we impose less strict constraints over the shortest periods (i.e., days or meals) and stricter constraints over the longest periods (i.e., months, weeks). For example the recommended energy requirement of 2450 kcal/day, considered over a week, results in a constraint such as $2450 \cdot 7 \leq week_E - week_S \leq 2450 \cdot 7$ and for the single days we allow the user to set, e.g., a deviation of 10%, thus resulting in the constraints $2450 - 10\% \leq Sunday_E - Sunday_S \leq 2450 + 10\%$, ..., $2450 - 10\% \leq Saturday_E - Saturday_S \leq 2450 + 10\%$ (see Fig. 2). For single meals we can further relax the constraints: for example the user can decide to split the energy assumption for the day among the meals (e.g., 20% for breakfast and 40% for lunch and dinner) and to further relax the constraints (e.g., of 30%), thus resulting in a constraint, e.g., $2450 \cdot 20\% - 30\% \leq Sunday\_breakfast_E - Sunday\_breakfast_S \leq 2450 \cdot 20\% + 30\%$.

Representing and Reasoning on the Diet and the Food. Along these lines, it is possible to represent the dietary recommendations for a specific user. However, we wish to support such a user into taking advantage of the information regarding the actual meals s/he consumes. In this way, the user can learn what

Fig. 3. Example of DRVs represented as STP.

are the consequences on his/her diet of eating a specific dish and s/he could use such information in order to make informed decisions about the current or future meals. Therefore it is necessary to "integrate" the information about the eaten dishes with the dietary recommendations. We devise a system where the user inputs the data about the food s/he is eating using a mobile app where the input is possibly supported by reading a QR code and s/he can also specify the amount of food s/he has eaten. Thus, we allow some imprecision due to possible differences in the portions (in fact, the actual amount of food in a portion is not always the same and, furthermore, a user may not eat a whole portion) or in the composition of the dish [6]. We support such feature by using STP constraints also for representing the nutritional values of the eaten food.

The dietary recommendations can be considered constraints on *classes*, which can be instantiated several times when the user assumes his/her meals. Thus, the problem of checking whether a meal satisfies the constraints of the dietary recommendations corresponds to checking whether the constraints of the instances satisfy the constraints of the classes. This problem has been dealt with in [25] and [2]. In these works the authors have considered the problem of "inheriting" the temporal constraints from classes of events to instances of events in the context of the STP framework, also taking into account problems deriving from correlation between events and from observability. In our setting we have a simpler setting, where correlation is known and observability is complete (even if possibly imprecise). Thus, we generate a new, provisional, STP where we add the new STP constraints deriving from the meals that the user has consumed: the added constraints possibly restrict the values allowed by the constraints in the STP. Then we propagate the constraints in such a new STP and we determine whether the new constraints are consistent and we obtain the new minimal network with the implied relations. For example, let us suppose that the user on Sunday, Monday and Tuesday had an actual intake of 2690 kcal for each day. This corresponds to adding to the STP the new constraints $2690 \leq Sunday_E - Sunday_S \leq 2690$, $\dots$, $2690 \leq Tuesday_E - Tuesday_S \leq 2690$. Then, propagating the constraints of the new STP (see Fig. 3), we discover that (i) the STP is consistent and thus the intake is compatible with the diet and (ii) on each remaining day of the week the user has to assume a minimum of 2205 kcal and a maximum of 2465 kcal.

2.3   Interpretation of STP

Although the information deriving from the STP is complete (and correct), in order to show to the user a meaningful feedback and to make it possible to interface the automatic reasoning module with the NLG module, it is useful to interpret the results of the STP. In particular we wish to provide the user with a user-friendly information not limited to a harsh "consistent/inconsistent" answer regarding the adequacy of a dish with regard to her/his diet. Therefore we consider the case where the user proposes to our system a dish, we obtain its nutritional values, we translate them, along with the user's diet and past meals, into STP and, by propagating the constraints, we obtain the minimal network. By taking into account a single macronutrient (carbohydrates, lipids or proteins), the resulting STP allows us to classify the macronutrient in the proposed dish in one of the following five cases: *permanently inconsistent* (I.1), *occasionally inconsistent* (I.2), *consistent and not balanced* (C.1), *consistent and well-balanced* (C.2) and *consistent and perfectly balanced* (C.3).

In the cases I.1 and I.2 the value of the macronutrient is inconsistent. In case I.1 the value for the nutrient is inconsistent with the DRVs as represented in the user's diet. The dish cannot be accepted even independently of the other food s/he may possibly eat. This case is detected by considering whether the macronutrient violates a constraint on classes. In case I.2 the dish per se does not violate the DRVs, but – considering the past meals s/he has eaten – it would preclude to be consistent with the diet. Thus, it is inconsistent now, but it could become possible to choose it in the future, e.g., next week or month. This case is detected by determining whether the macronutrient, despite it satisfies the constraints on the classes, is inconsistent with the propagated inherited STP.

In the cases C.1, C.2 and C.3 the value of the macronutrient is consistent with the diet, also taking into account the other dishes that the user has already eaten. It is possible to detect that the dish is consistent by exploiting the minimal network of the STP: if the value of the macronutrient is included between the lower and upper bounds of the relative constraint, then we are guaranteed that the STP is consistent and that the dish is consistent with the diet. This can be proven by using the property that in a minimal network every tuple in a constraint can be extended to a solution [19]. A consistent but not balanced choice of a dish will have consequences on the rest of the user's diet because the user will have to "recover" from it. Thus we distinguish three cases depending on the level of the adequacy of the value of the macronutrient to the diet. In order to discriminate between the cases C.1, C.2 and C.3, we consider how the value of the macronutrient stacks upon the allowed range represented in the related STP constraint. We assume that the mean value is the "ideal" value according to the DRVs and we consider two parametric user-adjustable thresholds relative to the mean: according to the deviation with respect to the mean we classify the macronutrient as not balanced (C.1), well balanced (C.2) or perfectly balanced (C.3) (see Fig. 4). In particular, we distinguish between lack or excess of a specific macronutrient for a dish: if a macronutrient is lacking (in excess) with regard to the ideal value, we tag the dish with the keyword *IPO* (*IPER*). This information will be exploited in the generation of the messages.

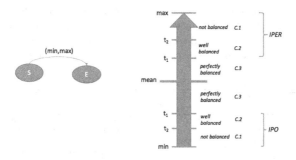

Fig. 4. Classification of a consistent value of a macronutrient given the minimum and maximum value of an STP constraint in a minimal network.

3 Persuasive NLG for Diet

A number of works considered the problem of NLG for presenting the results of automated reasoning to a user, especially in the case of expert systems for reasoning, e.g., [4,17]. In order to convert the five possible kinds of output of the STP reasoner (see Section 2.3) in messages, we adopted a simple template-based generator that produces five kinds of messages designed for persuasion. We first describe the generator (Section 3.1) and later we describe the theories that motivated the design of our messages (Section 3.2).

3.1 A Simple Template-Based Generation Architecture

The standard architecture for NLG models generation is a pipeline composed by three distinct modules/processes: the document planning, the micro-planning, and the surface realization [22]. Each one of these modules addresses distinct issues, in particular: (1) In the document planning one decides what to say, that is which information contents will be communicated; (2) In the micro-planning, the focus is on the design of a number of features that are related to the information contents as well as to the specific language, as the choice of the words; (3) In the surface realization, sentences are finally generated on the base of the decision taken by the previous modules and by considering the constraints related to the language specific word order and inflections.

For our system, the contents of information that have to be communicated, i.e. the document planning, are produced by the reasoner. Moreover, with the aim to easily implement in the messages the prescriptions of the persuasion theories, we adopted the simplest architecture for NLG. We treat sentence planning and surface realization in one single module by adopting a *template-based* approach. We use five templates to communicate the five cases of output of the reasoner: in Table 1 we report the cases obtained by the interpretation of the output of the reasoner (column **C**), the direction of the deviation (column **D**), the Italian templates and their rough English translation.

Indeed, the final message is obtained by modifying the templates on the basis of the specific values for the motivation of inconsistency that can be extracted by

Table 1. The persuasive message templates: the underline denotes the variable parts of the template. The column **C** contains the classification produced by the STP reasoner, while the column **D** contains the direction of the deviation: *IPO* (*IPER*) stands for the information that the dish is poor (rich) in the value of the macronutrient.

C	D	Message Template	Translation
I.1	IPO	Questo piatto non va affatto bene, contiene davvero pochissime proteine!	This dish is not good at all, it's too poor in proteins!
I.2	IPO	Ora non puoi mangiare questo piatto perché è poco proteico. Ma se domenica mangi un bel piatto di fagioli allora lunedì potrai mangiarlo.	You cannot have this dish now because it doesn't provide enough proteins, but if you eat a nice dish of beans on Sunday, you can have it on Monday.
C.1	IPO	Va bene mangiare le patatine ma nei prossimi giorni dovrai mangiare più proteine.	It's OK to eat chips but in the next days you'll have to eat more proteins.
C.2	IPO	Questo piatto va bene, è solo un po' scarso di proteine. Nei prossimi giorni anche fagioli però! :)	This dish is OK, but it's a bit poor in proteins. In the next days you'll need beans too! :)
C.3	-	Ottima scelta! Questo piatto è perfetto per la tua dieta :)	Great choice! This dish is perfect for your diet :)

interpreting the output of the reasoner (cf. Section 2.3) and possible suggestions that can guide the choices of the user in the next days. The suggestions can be obtained by a simple table that couples the excess (deficiency) of a macronutrient with a dish that could compensate this excess (deficiency). In particular, for the reasoner's outputs I.1, I.2, C.1 and C.2, we need to distinguish the case of a dish poor in a macronutrient (*IPO* in Table 1) with respect to the case of a dish rich in a macronutrient (*IPER*). If the dish is classified as IPO (IPER), we insert into the message a suggestion to consume in the next days a dish that contains a big (small) quantity of that specific macronutrient.

For sake of simplicity we do not describe the algorithm used in the generation module to combine the three distinct outputs of the reasoner on the three distinct macronutrients (i.e. proteins, lipids and carbohydrates). In short, the messages corresponding to each macronutrient need to be *aggregated* into a single message. A number of constraints related to coordination and relative clauses need to be accounted for [22]. In the next Section we describe the three theories of persuasion that influenced and motivated the design of the messages.

3.2 Designing Persuasive Messages in the Diet Domain

A number of theories on the design of persuasive textual and multimedial messages have been proposed in the last years [7,10–12,14,16,21,23]. Most of these theories can be split in two narrow categories. The first category includes the theories that approach the persuasion from a practical and empirical point of view, by using strategies and methods typical of the psychology and of the interaction design. The second category includes the theories that approach the persuasion from a theoretical point of view, by using strategies and methods typical of strong artificial intelligence and cognitive science. We discuss the three theories that mostly influenced the design of the messages in relation to our project.

CAPTology (Computers As Persuasive Technologies) is the study of computers as persuasive technologies, i.e. "[...] the design, research, and analysis of interactive computing products (computers, mobile phones, websites, wireless technologies, mobile applications, video games, etc.) created for the purpose of changing peoples attitudes or behaviors" [10]. The starting point of Fogg's theory is that the computer is perceived by users in three coexisting forms, Tool-Media-SocialActor, and each one of these three forms can exercise some forms of persuasion. As a tool, the computer can enhance the capabilities of a user: our system calculates the nutritional contents of the food, and so it enhances the ability to correctly judge the compatibility of a dish with a diet. As a media the computer "provides experience": in our system, the human memory is enhanced by the reasoner, which indirectly reminds her/him what s/he ate in the last days. As a social actor the computer creates an empathic relationship with the user reminding her/him the "social rules": in our system the messages guide the user towards the choice of a balanced meal, convincing her/him to follow the diet that her/himself decided. Fogg recently defined a number of rules to design effectively persuasive systems [11], and some of these rules have modeled our messages. For example, the rule: *Learn what is Preventing the target behavior*, proposes to classify an "uncorrect" behavior along three major lines: (1) lack of motivation, (2) lack of ability, (3) lack of a well-timed trigger to perform the behavior. In our system all the three components play a role. Indeed, a user follows a bad diet because (i) s/he is not enough motivated, (ii) because s/he does not know that the dish is in contrast to her/his diet (iii) because s/he does not have the right stimulus at the time of choosing a dish. The reasoning and the generated messages are working on the last two components: the reasoner enhances the user's abilities allowing her/him to have the relevant information at the right time, the generation system creates a stimulus (the message) when it is really necessary, *kairos* in the Fogg's terminology, i.e. when the user has to decide what to eat.

Another approach to computational persuasion is strongly related to the concept of *tailoring*, i.e. the adaptation of the output of the computation to a specific user. A pioneering work for tailoring in the field of NLG is described in [21]: the authors have designed an NLG system, called *STOP*, to build a letter that induces a specific reader to quit smoking. The key component of STOP is the individuation of a *user type* by using the answers given to a questionnaire. In this way, one can build a specific user profile. By using this profile the system generates a tailored letter on the basis of a template. This simple approach to persuasion unfortunately did not yield the desired results. The experimental protocol has shown, through the use of a control group, that the enhancement given by customization was negligible. At this stage, we do not adopt in our system the ability to create custom messages for a specific user, but, as evidenced by similar experiences, customization of the feedback could improve the performances of the system. A system for tailoring that we partially adopt in our messages is described in [16], where a series of messages are sent via SMS to reduce the consumption of snacks. In this case, the messages adopt six patterns/templates for persuasion derived from the general theory of persuasion of Cialdini [7]. The six patterns are: (1) Reciprocity: *people feel obligated to return a favor*, (2) Scarcity: *people will value scarce products*, (3) Authority: *people value the opinion of experts*, (4) Consistency: *people do as they said they would*, (5) Consensus: *people do as other people do*, (6) Liking: *we say yes to people we like*. Compared to this classification, all the messages of our generator belong to the patterns of authority and consistency.

One approach to persuasion strictly related to strong artificial intelligence and cognitive science is based on the concept of the computer as an intelligent agent [12,14,23]. The system behaves as a real autonomous entity and it is often modeled as a BDI (Beliefs, Desires, Intentions) agent, whose main purpose is to persuade the user to behave in a specific way. This approach has been adopted essentially for research purposes rather than for commercial applications. In contrast to the design of our NL generator, where there is a single module based on templates, such agent-based approach allows a great modularity in the design of a persuasive system. We describe some issues of these systems in order to understand the deficiencies of our simple approach. Hovy defines a number of heuristic rules that constrain the "argument" defined in the process of sentence planning. For example: *Adverbial stress words can only be used to enhance or mitigate expressions that carry some affect already* [14]. In a similar way, De Rosis and Grasso define a number of heuristic rules on the argument structure, to lexically enhance or mitigate a message [23]. The use of certain adverbs, as little bit (*poco*), very (*molto*), really (*davvero*), are used to enhance some specific argument structures. Indeed, we adopt this strategy by using this kind of adverbs in the messages I.1, I.2, C.1 and C.2. Guerini et al. define a detailed taxonomy of persuasion strategies that a system can adopt and relate the strategies to the theory of argumentation [12]. Moreover, they define an architecture for persuasion that follows the standard modularization of NLG systems. This allows for a very rich persuasive action, which begins from the planning of a rhetorical structure in the content planning. Compared to the taxonomy of the proposed strategies, we can see that our messages belong to one single category, called action_inducement/goal_balance/positive_consequence. This strategy induces an action (to choose a dish), by using the user's goal (a balanced diet) and by using the benefits deriving from this goal.

Finally, note that in the messages C.2 and C.3 we used emoticons. Indeed, some studies showed that the use of emoticons in written texts can increase the communicative strength of a message. For example Dirke shows that the use of emoticons sets a tone of friendship to the message type and can increase the positive value of the message [9].

4   Conclusions and Related Works

There are a number of academic studies that are related to our project, among them [3,15], and there is also a great number of smartphone applications related to nutrition, e.g. *DailyBurn*, *Lose It!*, *MyNetDiary*, *A low GI Diet*, *Weight-Watchers*. However, our dietary system presents two elements of novelty: (1) the use of automatic reasoning as a tool for verifying the compatibility of a specific recipe with a specific diet and for determining the consequences of the choice of a specific dish and (2) the use of NLG techniques to produce the answer.

Some authors have applied Operational Research techniques to tackle the problem of planning a diet (see the survey in [18] or the more recent paper [5]). These techniques are based on the simplex method for solving linear programming problems. However these approaches are meant to plan an entire diet and they do not support the user in choosing a dish and in investigating the consequences of her/his choice. In [6] the authors have tackled the problem of assessing the compatibility of a single meal to a norm and of suggesting to the user some actions to balance the meal (e.g. removing/adding food); they employed fuzzy arithmetic to represent imprecision/uncertainty in quantity and composition of food and heuristic search for determining the actions to be suggested. They did not consider the problem of globally balancing the meals.

In the next future, we intend to improve the NLG module for tailoring. In particular, we want 1) to build a corpus of sentences that a professional dietician would use to persuade users towards correct dish choices, 2) to separate microplanning from realization, 3) to classify the users in types in order to personalize the messages on the basis, for instance, of the age. Finally, we plan to experiment the system in two settings. First we intend to design a simulation that includes 1) a database of real recipes, 2) a user model that allows to test the persuasion efficacy and 3) a baseline built rigidly sticking to DRVs. Second, we intend to test the system with a focus group in a clinical setting, in particular with patients affected by essential obesity. In this setting we imagine that the system could be used also by human dieticians for the supervision of their patients.

References

1. LARN - Livelli di Assunzione di Riferimento di Nutrienti ed energia per la popolazione italiana - IV Revisione. SICS Editore, Milan (2014)
2. Anselma, L., Terenziani, P., Montani, S., Bottrighi, A.: Towards a comprehensive treatment of repetitions, periodicity and temporal constraints in clinical guidelines. Artificial Intelligence in Medicine 38(2), 171–195 (2006)
3. Balintfy, J.L.: Menu planning by computer. Commun. ACM 7(4), 255–259 (1964)
4. Barzilay, R., Mccullough, D., Rambow, O., Decristofaro, J., Korelsky, T., Lavoie, B. Inc, C.: A new approach to expert system explanations. In: 9th International Workshop on Natural Language Generation, pp. 78–87 (1998)
5. Bas, E.: A robust optimization approach to diet problem with overall glycemic load as objective function. Applied Mathematical Modelling 38(19–20), 4926–4940 (2014)
6. Buisson, J.C.: Nutri-educ, a nutrition software application for balancing meals, using fuzzy arithmetic and heuristic search algorithms. Artif. Intell. Med. 42(3), 213–227 (2008)
7. Cialdini, R.B.: Influence: science and practice. Pearson Education, Boston (2009)
8. Dechter, R., Meiri, I., Pearl, J.: Temporal constraint networks. Artif. Intell. 49(1–3), 61–95 (1991)

9. Derks, D., Bos, A.E.R., von Grumbkow, J.: Emoticons in computer-mediated communication: Social motives and social context. Cyberpsy., Behavior, and Soc. Networking **11**(1), 99–101 (2008)

10. Fogg, B.: Persuasive Technology: Using computers to change what we think and do. Morgan Kaufmann Publishers, Elsevier, San Francisco (2002)

11. Fogg, B.: The new rules of persuasion (2009). http://captology.stanford.edu/resources/article-new-rules-of-persuasion.html

12. Guerini, M., Stock, O., Zancanaro, M.: A taxonomy of strategies for multimodal persuasive message generation. Applied Artificial Intelligence **21**(2), 99–136 (2007)

13. Holzinger, A., Dorner, S., Födinger, M., Valdez, A.C., Ziefle, M.: Chances of increasing youth health awareness through mobile wellness applications. In: Leitner, G., Hitz, M., Holzinger, A. (eds.) USAB 2010. LNCS, vol. 6389, pp. 71–81. Springer, Heidelberg (2010)

14. Hovy, E.H.: Generating Natural Language Under Pragmatic Constraints. Lawrence Erlbaum, Hillsdale (1988)

15. Iizuka, K., Okawada, T., Matsuyama, K., Kurihashi, S., Iizuka, Y.: Food menu selection support system: considering constraint conditions for safe dietary life. In: Proceedings of the ACM Multimedia 2012 Workshop on Multimedia for Cooking and Eating Activities, CEA 2012, pp. 53–58. ACM, New York (2012)

16. Kaptein, M., de Ruyter, B.E.R., Markopoulos, P., Aarts, E.H.L.: Adaptive persuasive systems: A study of tailored persuasive text messages to reduce snacking. TiiS **2**(2), 10 (2012)

17. Lacave, C., Diez, F.J.: A review of explanation methods for heuristic expert systems. Knowl. Eng. Rev. **19**(2), 133–146 (2004)

18. Lancaster, L.M.: The history of the application of mathematical programming to menu planning. European Journal of Operational Research **57**(3), 339–347 (1992)

19. Montanari, U.: Networks of constraints: Fundamental properties and applications to picture processing. Information Sciences **7**, 95–132 (1974)

20. Nishida, C., Uauy, R., Kumanyika, S., Shetty, P.: The joint WHO/FAO expert consultation on diet, nutrition and the prevention of chronic diseases: process, product and policy implications. Public Health Nutrition **7**, 245–250 (2004)

21. Reiter, E., Robertson, R., Osman, L.: Lessons from a Failure: Generating Tailored Smoking Cessation Letters. Artificial Intelligence **144**, 41–58 (2003)

22. Reiter, E., Dale, R.: Building Natural Language Generation Systems. Cambridge University Press, New York (2000)

23. de Rosis, F., Grasso, F.: Affective natural language generation. In: Paiva, A. (ed.) Affective Interactions. LNCS, vol. 1814, pp. 204–218. Springer, Heidelberg (2000)

24. Schofield, W.N.: Predicting basal metabolic rate, new standards and review of previous work. Human Nutrition: Clinical Nutrition **39C**, 5–41 (1985)

25. Terenziani, P., Anselma, L.: A knowledge server for reasoning about temporal constraints between classes and instances of events. International Journal of Intelligent Systems **19**(10), 919–947 (2004)

26. World Health Organization: Global strategy on diet, physical activity and health (WHA57.17). In: 75th World Health Assembly (2004)

Predicting Within-24h Visualisation of Hospital Clinical Reports Using Bayesian Networks

Pedro Pereira Rodrigues[1,2,3](✉), Cristiano Inácio Lemes[4],
Cláudia Camila Dias[1,2], and Ricardo Cruz-Correia[1,2]

[1] CINTESIS - Centre for Health Technology and Services Research,
Rua Dr. Plácido Costa, s/n, 4200-450 Porto, Portugal
{pprodrigues,camila,rcorreia}@med.up.pt
[2] CIDES-FMUP, Faculty of Medicine of the University of Porto,
Alameda Prof. Hernâni Monteiro, 4200-319 Porto, Portugal
[3] LIAAD - INESC TEC, Artificial Intelligence and Decision Support Laboratory,
Rua Dr. Roberto Frias, 4200-465 Porto, Portugal
[4] ICMC, Institute of Mathematical and Computer Sciences, University of São Paulo,
Avenida Trabalhador São-carlense, 400, São Carlos 13566-590, Brazil
cristianoinaciolemes@gmail.com

Abstract. Clinical record integration and visualisation is one of the most important abilities of modern health information systems (HIS). Its use on clinical encounters plays a relevant role in the efficacy and efficiency of health care. One solution is to consider a virtual patient record (VPR), created by integrating all clinical records, which must collect documents from distributed departmental HIS. However, the amount of data currently being produced, stored and used in these settings is stressing information technology infrastructure: integrated VPR of central hospitals may gather millions of clinical documents, so accessing data becomes an issue. Our vision is that, making clinical reports to be stored either in primary (fast) or secondary (slower) storage devices according to their likelihood of visualisation can help manage the workload of these systems. The aim of this work was to develop a model that predicts the probability of visualisation, within 24h after production, of each clinical report in the VPR, so that reports less likely to be visualised in the following 24 hours can be stored in secondary devices. We studied log data from an existing virtual patient record (n=4975 reports) with information on report creation and report first-time visualisation dates, along with contextual information. Bayesian network classifiers were built and compared with logistic regression, revealing high discriminating power (AUC around 90%) and accuracy in predicting whether a report is going to be accessed in the 24 hours after creation.

Keywords: Bayesian networks · Health services · Virtual patient records

1 Introduction

Evidence-based medicine relies on three information sources: patient records, published evidence and the patient itself [25]. Even though great improvements

© Springer International Publishing Switzerland 2015
F. Pereira et al. (Eds.) EPIA 2015, LNAI 9273, pp. 91–102, 2015.
DOI: 10.1007/978-3-319-23485-4_9

and developments have been made over the years, on-demand access to clinical information is still inadequate in many settings, leading to less efficiency as a result of a duplication of effort, excess costs and adverse events [10]. Furthermore, a lot of distinct technological solutions coexist to integrate patient data, using different standards and data architectures which may lead to difficulties in further interoperability [7]. Nonetheless, a lot of patient information is now accessible to health-care professionals at the point of care. But, in some cases, the amount of information is becoming too large to be readily handled by humans or to be efficiently managed by traditional storage algorithms. As more and more patient information is stored, it is very important to efficiently select which one is more likely to be useful [8].

The identification of clinically relevant information should enable an improvement both in user interface design and in data management. However, it is difficult to identify what information is important in daily clinical care, and what is used only occasionally. The main problem addressed here is how to estimate the relevance of health care information in order to anticipate its usefulness at a specific point of care. In particular, we want to estimate the probability of a piece of information being accessed during a certain time interval (e.g. first 24 hours after creation), taking into account the type of data and the context where it was generated and to use this probability to prioritise the information (e.g. assigning clinical reports for secondary storage archiving or primary storage access).

Next section presents background knowledge on electronic access to clinical data (2.1), assessment of clinical data relevance (2.2) and machine learning in health care research (2.3), setting the aim of this work (2.4). Then, section 3 presents our methodology to data processing, model learning, and prediction of within-24h visualisation of clinical data, which results are exposed in section 4. Finally, section 5 finalises the exposition with discussion and future directions.

2 Background

The practice of medicine has been described as being dominated by how well information is collected, processed, retrieved, and communicated [2].

2.1 Electronic Access to Clinical Data

Currently in most hospitals there are great quantities of stored digital data regarding patients, in administrative, clinical, lab or imaging systems. Although it is widely accepted that full access to integrated electronic health records (EHR) and instant access to up-to-date medical knowledge significantly reduces faulty decision making resulting from lack of information [9], there is still very little evidence that life-long EHR improve patient care [4]. Furthermore, there use is often disregarded. For example, studies have indicated that data generated before an emergency visit are accessed often, but by no means in a majority of times (5% to 20% of the encounters), even when the user was notified of the availability of such data [12].

One usual solution for data integration in hospitals is to consider a virtual patient record (VPR), created by integrating all clinical records, which must collect documents from distributed departmental HIS [3]. Integrated VPR of central hospitals may gather millions of clinical documents, so accessing data becomes an issue. A paradigmatic example of this burden to HIS is the amount of digital data produced in the medical imaging departments, which has increased rapidly in recent years due mainly to a greater use of additional diagnostic procedures, and an increase in the quality of the examinations. The management of information in these systems is usually implemented using Hierarchical Storage Management (HSM) solutions. This type of solution enables the implementation of various layers which use different technologies with different speeds of access, corresponding to different associated costs. However, the solutions which are currently implemented use simple rules for information management, based on variables such as the time elapsed since the last access or the date of creation of information, not taking into account the likely relevance of information in the clinical environment [6].

In a quest to prioritise the data that should be readily available in HIS, several pilot studies have been endured to analyse for how long clinical documents are useful for health professionals in a hospital environment, bearing in mind document content and the context of the information request. Globally, the results show that some clinical reports are still used one year after creation, regardless of the context in which they were created, although significant differences existed in reports created during distinct encounter types [8]. Other results show that half of all visualisations might be of reports more than 2 years-old [20], although this visualisation distribution also varies across clinical department and time of production [21]. Thus, usage of patients past information (data from previous hospital encounters), varied significantly according to the setting of health care and content, and is, therefore, not easy to prioritise.

2.2 Assessment of Clinical Data Relevance

As previously noted, and especially in critical and acute care settings, the age of data is one of the factors often used to assess data relevance, making new information more relevant to the current search. However, studies have shown that some clinical reports are still used after one year regardless of the context in which they were created, although significant differences exist in reports created in distinct encounter types and document content, which contradicts the definition of *old data* used in previous studies. Hence the need to define better rules for recommending documents in encounters.

Classifying the relevance of information based only on the time elapsed since the date of acquisition is clearly inefficient. It is expected that the need to consult an examination at a given time will be dependent on several factors beyond the date of the examination, such as type of examination and the patient's pathology. Thus, a system that uses more factors to identify the relevance of information at a given time would be more efficient in managing the information that is stored in fast memory and slow memory. A recent study from the same group addressed other possibly relevant factors besides document age, including type of encounter (i.e. emergency room, inpatient care, or outpatient consult), department where the report was generated (e.g. gynaecology or internal medicine) and even type of report in each department, but the possibility of modelling visualisations with survival analysis proved to be extremely difficult [21].

Nonetheless, if we could, for instance, discriminate solely between documents that will be needed in the next 24 hours from the remaining, we could efficiently decide which ones to store in a faster-accessible memory device. Furthermore, we could then rank documents according to their probability of visualisation in order to adjust the graphical user interface of the the VPR, to improve system's usability. By applying regression methods or other modelling techniques it is possible to identify which factors are associated with the usage or relevance of patient data items. These factors and associations can then be used to estimate data relevance in a specific future time interval.

2.3   Machine Learning in Healthcare Research

The definition of clinical decision support systems (most of the times based on expert systems) is currently a major topic since it may help the diagnosis, treatment selection, prognosis of rate of mortality, prognosis of quality of life, etc. They can even be used to administrative tasks like the one addressed by this work. However, the complicated nature of real-world biomedical data has made it necessary to look beyond traditional biostatistics [14] without loosing the necessary formality. For example, naive Bayesian approaches are closely related to logistic regression [22]. Hence, such systems could be implemented applying methods of machine learning [16], since new computational techniques are better at detecting patterns hidden in biomedical data, and can better represent and manipulate uncertainties [22]. In fact, the application of data mining and machine learning techniques to medical knowledge discovery tasks is now a growing research area. These techniques vary widely and are based on data-driven conceptualisations, model-based definitions or on a combination of data-based knowledge with human-expert knowledge [14].

Bayesian approaches have an extreme importance in these problems as they provide a quantitative perspective and have been successfully applied in health care domains [15]. One of their strengths is that Bayesian statistical methods allow taking into account prior knowledge when analysing data, turning the data analysis into a process of updating that prior knowledge with biomedical and health-care evidence [14]. However, only after the 90's we may find evidence of a large interest on these methods, namely on Bayesian networks, which offer a general and versatile approach to capturing and reasoning with uncertainty in medicine and health care [15]. They describe the distribution of probabilities of one set of variables, making possible a two-fold analysis: a qualitative model and a quantitative model, presenting two types of information for each variable.

On a general basis, a Bayesian network represents a joint distribution of one set of variables, specifying the assumption of independence between them, with the inter-dependence between variables being represented by a directed acyclic graph. Each variable is represented by a node in the graph, and is dependent of the set of variables represented by its ascendant nodes; a node X is a ascendant of another node Y if exists a direct arc from X to Y [16]. To give more representational power to the relations represented by the arcs of the graph, it is necessary to associate values to it. The matrix of conditional probability is given for each variable, describing the distribution of probabilities of each variable given its ascendant variables.

After the qualitative and quantitative models are constructed, the next step, and one of the most important, is how to calculate the new probabilities when new evidence is introduced in the network. This process is called inference and works as follows. Each variable has a finite number of categories greater than or equal to two. A node is observed when there is knowledge about the state of that variable. The observed variables have a huge importance because with conditional probabilities they define the prior probabilities of the non observed variables. With the joint probabilities we can calculate the marginal probabilities of each unobserved variable, adding for all categories the probabilities that the variable is in the desired state [15].

2.4   Aim

The aim of this work is the development of a decision support model for discriminating between reports that are going to be useful in the next 24 hours and reports which can be otherwise stored in slower storage devices, since they will not be accessed in the next 24 hours, thus improving performance of the entire virtual patient record system.

3   Data and Methods

Between May 2003 and May 2004, a virtual patient record (VPR) was designed and implemented at Hospital S. João, a university hospital with over 1350 beds. An agent-based platform, Multi-Agent System for Integration of Data (MAID), ensures the communication among various hospital information systems (see [24] for a description of the system). Clinical documents are retrieved from clinical department information systems (DIS) and stored into a central repository in a browser friendly format. This is done by regularly scanning 14 DIS using different types of agents [17]:

- For each department, a List Agent regularly retrieves report lists from the DIS, with report file references and meta-data, and stores them in the VPR repository.
- The Balancer Agent of that department retrieves the report file references and distributes them to the departmental File Agents.
- File Agents retrieve the actual report files.

As the amount of information available to the agents increases throughout time, there is also an increase in the difficulty of managing that information by humans. Not rarely, a request for a report arrives (after the List Agent has published the existence of that report) before the File Agent was able to retrieve the document. In this cases, an Express Agent is called to retrieve the file, which stresses the entire system's workload, otherwise balanced.

To enable a quantitative analysis (e.g. the likelihood of document access), all actions by users of the VPR are recorded in the log file. Intentionally and originally created and kept for audit purposes, these logs can provide very interesting insights into the information needs of health-care professionals in some particular situations, although most of the times the quality of these logs is not delivering [5].

3.1   Studied Variables and Outcomes

Data was collected from from the virtual patient record (VPR) with information on report creation and report first-time visualisation dates, along with contextual information. This study focuses on a sample of 5000 reports (2.7% of the entire data for the studied year) and corresponding visualisations, stored in the VPR in 2010. The data used in this study was collected using Oracle SQL Developer from the VPR patient database, containing patient's identification and references to the clinical records. We developed models with seven explanatory variables, including patient data (age and sex), context data (department and type of encounter) and creation time data (hour, day-of-week, daily period), defined as follows. The main outcome of this study was within-24h visualisation of reports.

AgeCat (cat) discretised in decades;
Sex (binary);
Department (cat);
EncType (cat) one of outpatient consult, inpatient care, emergency or other;
Hour (cat) truncated from creation time;
DoW (cat) one of Sun, Mon, Tue, Wed, Thu, Fri or Sat;
Period (cat) one of morning (Hour=7-12), afternoon (13-18), night (19-24) or dawn (1-6);
Visual24h (binary) target outcome, whether the report has been visualised in the first 24 hours after creation or not.

3.2   Model Building and Evaluation

In order to correctly fit the models, only complete cases were considered in the analysis. Logistic regression was applied to all studied variables to predict visualisation. Additionally, two Bayesian network classifiers were built - Naive Bayes (NB) and Tree Augmented Naive Bayes (TAN) - which differ on the number of conditional dependencies (besides the outcome) allowed among variables (NB: zero dependencies; TAN: one dependence), in order to choose the structure which could better represent the problem. Receiver Operating Characteristic (ROC) curve analysis was performed to determine in-sample area under the curve (AUC). Furthermore, to assess the general structure and accuracy of learned models, stratified 10-fold cross-validation was repeated 10 times, estimating accuracy, sensitivity, specificity, precision (positive and negative predictive values) and the area under the ROC curve, for all compared models.

3.3 Software

Logistic regression was done with R package *stats* [18], Bayesian network structure was learned with R package *bnlearn* [23], Bayesian network parameters were fitted with R package *gRain* [11], ROC curves were computed with R package *pROC* [19], and odds ratios (OR) were computed with R package *epitools* [1].

4 Results

A total of 4975 reports were included in the analysis. The main characteristics of the reports are shown in Table 1, which were generated from patients with a mean (std dev) age of 55.5 (20.5). Less than 23% of the reports were visualised in the 24 hours following their creation, which were nonetheless more from female patients (almost 55%) with a 24h-visualisation OR=1.51 (95%CI [1.32,1.72]) for female-patient reports. Also significant was the context of report creation, with more reports being created in inpatient care (44.4%) and outpatient consults (41.4%), although compared with the latter context, 24-hour visualisations are more likely for reports generated in inpatient care (OR=8.60 [7.04,10.59]) or in the emergency room (OR=14.50 [11.22,18.83]). Regarding creation time, morning (OR=1.22 [1.05,1.41]), night (OR=1.82 [1.46,2.28]) and dawn (OR=2.88 [2.03,4.07]) have all higher 24-hour visualisation likelihood than the afternoon period.

4.1 Qualitative Analysis of the Bayesian Network Model

Figure 1 presents the qualitative model for the Tree-Augmented Naive Bayes network, where interesting connections can be extracted from the resulting model. First, patient's data features are associated. Then, creation time data and context data are also strongly related. However, the most interesting feature is probably the department that created the report, since this was chosen by the algorithm as ancestor of patient's age, time of report creation and type of encounter.

4.2 In-Sample Quantitative Analysis

For a quantitative analysis, Figure 2 presents the in-sample ROC curves for logistic regression (left), Naive Bayes (centre) and TAN (right). As expected, increasing model complexity enhances the in-sample AUC (LR 88.6%, NB 86.9% and TAN 90.7) but, globally, all models presented good discriminating power towards the outcome.

Table 1. Basic characteristics of included reports: patient's data (sex and age), report creation context (department, encounter) and time (day of week, daily period) data.

	Visualised in 24 hours		
	No	Yes	Total
Outcome, n (%)	3846 (77.3)	1129 (22.7)	4975 (100)
Female, n (%)	1716 (44.6)	619 (54.8)	2335 (46.9)
Age, $\mu(\sigma)$	54.6 (19.8)	58.5 (22.4)	55.5 (20.5)
AgeCat, n (%)			
[0,10[	97 (2.5)	59 (5.2)	156 (3.1)
[10,20[	58 (1.5)	23 (2.0)	81 (1.6)
[20,30[	215 (5.6)	40 (3.5)	255 (5.1)
[30,40[	583 (15.2)	115 (10.2)	698 (14.0)
[40,50[	597 (15.5)	122 (10.8)	719 (14.5)
[50,60[	601 (15.6)	150 (13.3)	751 (15.1)
[60,70[	710 (18.5)	199 (17.6)	909 (18.2)
[70,80[	554 (14.4)	207 (18.3)	761 (15.3)
[80,90[	372 (9.67)	181 (16.0)	553 (11.1)
[90,100[	55 (1.4)	31 (2.8)	86 (1.7)
$\geq$100	4 (0.1)	2 (0.2)	6 (0.1)
Encounter Type, n (%)			
Outpatient consult	1940 (50.4)	120 (10.6)	2060 (41.4)
Inpatient care	1442 (37.5)	768 (68.0)	2210 (44.4)
Emergency room	217 (5.6)	19 (1.7)	236 (4.7)
Other	247 (6.4)	222 (19.7)	469 (9.4)
Department, n (%)			
1	76 (2.0)	11 (1.0)	87 (1.8)
2	1626 (42.3)	55 (4.9)	1681 (33.8)
3	646 (16.8)	469 (41.5)	1115 (22.4)
5	1057 (27.5)	529 (46.9)	1586 (31.9)
6	154 (4.0)	23 (2.0)	177 (3.6)
7	89 (2.3)	22 (2.0)	111 (2.2)
9	11 (0.3)	7 (0.6)	18 (0.4)
10	10 (0.3)	1 (0.1)	11 (0.2)
12	139 (3.6)	11 (1.0)	150 (3.0)
13	23 (0.6)	0 (0)	23 (0.3)
16	5 (0.1)	0 (0)	5 (0.1)
21	10 (0.3)	1 (0.1)	11 (0.2)
Day-of-Week, n (%)			
Mon	728 (18.9)	303 (26.8)	1031 (20.7)
Tue	671 (17.5)	291 (25.8)	962 (19.3)
Wed	743 (19.3)	208 (18.4)	951 (19.1)
Thu	804 (20.9)	35 (3.1)	839 (16.9)
Fri	673 (17.5)	92 (8.2)	765 (15.4)
Sat	122 (3.2)	99 (8.7)	221 (4.4)
Sun	105 (2.7)	101 (9.0)	206 (4.1)
Daily Period, n (%)			
Morning	1768 (46.0)	521 (46.2)	2289 (46.0)
Afternoon	1661 (43.2)	402 (35.6)	2063 (41.5)
Night	331 (8.6)	146 (13.0)	477 (9.6)
Dawn	86 (2.2)	60 (5.3)	146 (2.9)

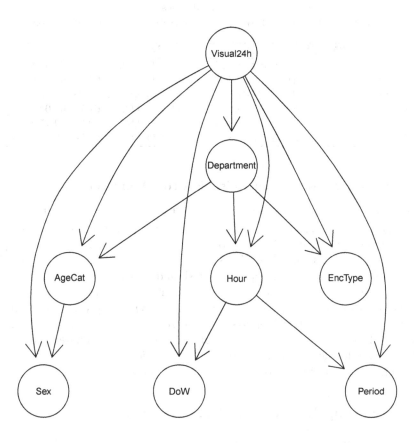

Fig. 1. Tree-Augmented Naive Bayes for predicting within 24h visualisation of clinical reports in the virtual patient record.

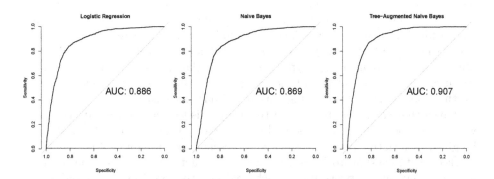

Fig. 2. In-sample ROC curves for logistic regression (left), naive Bayes (centre) and Tree-Augmented Naive Bayes (right).

Table 2. Validity assessment averaged from 10 times stratified 10-fold cross-validation for logistic regression (LR), naive Bayes (NB) and Tree-Augmented Naive Bayes (TAN).

Measure, % [95%CI]	LR	NB	TAN
Accuracy	82.30 [82.08,82.52]	82.43 [82.14,82.72]	82.80 [82.51,83.09]
Sensitivity	41.33 [40.50,42.16]	60.68 [59.81,61.55]	64.12 [63.36,64.89]
Specificity	94.33 [94.07,94.58]	88.81 [88.50,89.12]	88.28 [87.96,88.61]
Precision (PPV)	68.40 [67.45,69.35]	61.53 [60.80,62.25]	61.75 [61.04,62.47]
Precision (NPV)	84.57 [84.39,84.76]	88.51 [88.29,88.74]	89.35 [89.15,89.56]
AUC	87.58 [87.27,87.89]	86.37 [86.04,86.70]	85.50 [85.13,85.88]

4.3   Bayesian Network Generalisable Cross-Validation

In order to assess the ability of the models to generalise beyond the derivation cohort, cross-validation was endured. Table 2 presents the result of the 10-times-repeated stratified 10-fold cross-validation. Although the more complicated model loses in terms of AUC (85% vs 87%), it brings advantages to the precise problem of identifying reports that should be stored in secondary memory as they are less likely to be visualised in the next 24 hours, since it reveals a negative precision of 89% vs 88% (NB) and 84% (LR). Along with this result, it is much better at identifying reports that are going to be needed, as sensitivity rises from 41% (LR) to 64%. Future work should consider different threshold values for the decision boundary (here, 50%) in order to better suit the model to the sensitivity-specificity goals of the problem a hands.

5   Concluding Remarks and Future Work

The main contribution of this work is the preliminary study for the development of a decision support model for discriminating between reports that are going to be useful in the next 24 hours and reports which can be safely stored in secondary memory, since they will not be accessed in the next 24 hours.

An initial sample of clinical reports was used to derive Bayesian network models which were then compared with a logistic regression model in terms of in-sample discriminating power and generalisable validity with cross-validation. The studied data was in accordance with previous works in terms of the relevance that some factors may have on the likelihood of visualisation of clinical reports, e.g. department and type of encounter that produced the report [21]. Additionally, patient data and time of report creation were also found relevant for the global model of predicting within 24-hour visualisations.

Given that the main objective of this project is to enable a clear decision on whether a report can safely be stored in secondary memory or not, focus should be given to negative precision, since it represents the probability that a report marked by the system to be stored away is, in fact, irrelevant for the present day. The Bayesian network models achieved negative precision of around 89%, while keeping specificity high (also around 88%).

Future work will be concentrated in a) exploring other variables that might influence the likelihood of visualisation of clinical reports (e.g. actual data from the report, patient's diagnosis, etc.); b) exploiting the maximum amount of data from the log file of the virtual patient record (e.g. 2010 comprises of more than 184K reports); and c) inspecting the usefulness of temporal Bayesian network models [13] for the precise problem of relevance estimation.

Overall, this study presents Bayesian network models as useful techniques to integrate in a virtual patient record that needs to prioritise the accessible documents, both in terms of user-interface optimisation and data management procedures.

Acknowledgments. The authors acknowledge the help of José Hilário Almeida during the data gathering process.

References

1. Aragon, T.J.: epitools: Epidemiology Tools (2012)
2. Barnett, O.: Computers in medicine. JAMA: the Journal of the American Medical Association **263**(19), 2631 (1990)
3. Bloice, M.D., Simonic, K.M., Holzinger, A.: On the usage of health records for the design of virtual patients: a systematic review. BMC Medical Informatics and Decision Making **13**(1), 103 (2013)
4. Clamp, S., Keen, J.: Electronic health records: Is the evidence base any use? Medical Informatics and the Internet in Medicine **32**(1), 5–10 (2007)
5. Cruz-Correia, R., Boldt, I., Lapão, L., Santos-Pereira, C., Rodrigues, P.P., Ferreira, A.M., Freitas, A.: Analysis of the quality of hospital information systems audit trails. BMC Medical Informatics and Decision Making **13**(1), 84 (2013)
6. Cruz-Correia, R., Rodrigues, P.P., Freitas, A., Almeida, F., Chen, R., Costa-Pereira, A.: Data quality and integration issues in electronic health records. In: Hristidis, V. (ed.) Information Discovery on Electronic Health Records, chap. 4. Data Mining and Knowledge Discovery Series, pp. 55–95. CRC Press (2009)
7. Cruz-Correia, R.J., Vieira-Marques, P.M., Ferreira, A.M., Almeida, F.C., Wyatt, J.C., Costa-Pereira, A.M.: Reviewing the integration of patient data: how systems are evolving in practice to meet patient needs. BMC Medical Informatics and Decision Making **7**, 14 (2007)
8. Cruz-Correia, R.J., Wyatt, J.C., Dinis-Ribeiro, M., Costa-Pereira, A.: Determinants of frequency and longevity of hospital encounters' data use. BMC Medical Informatics and Decision Making **10**, 15 (2010)
9. Dick, R., Steen, E.: The Computer-based Patient Record: An Essential Technology for HealthCare. National Academy Press (1997)
10. Feied, C.F., Handler, J.A., Smith, M.S., Gillam, M., Kanhouwa, M., Rothenhaus, T., Conover, K., Shannon, T.: Clinical information systems: instant ubiquitous clinical data for error reduction and improved clinical outcomes. Academic emergency medicine **11**(11), 1162–1169 (2004)
11. Højsgaard, S.: Graphical independence networks with the gRain package for R. Journal of Statistical Software **46**(10) (2012)
12. Hripcsak, G., Sengupta, S., Wilcox, A., Green, R.: Emergency department access to a longitudinal medical record. Journal of the American Medical Informatics Association **14**(2), 235–238 (2007)

13. Lappenschaar, M., Hommersom, A., Lucas, P.J.F., Lagro, J., Visscher, S., Korevaar, J.C., Schellevis, F.G.: Multilevel temporal Bayesian networks can model longitudinal change in multimorbidity. Journal of Clinical Epidemiology **66**, 1405–1416 (2013)
14. Lucas, P.: Bayesian analysis, pattern analysis, and data mining in health care. Current Opinion in Critical Care **10**(5), 399–403 (2004)
15. Lucas, P.J.F., van der Gaag, L.C., Abu-Hanna, A.: Bayesian networks in biomedicine and health-care. Artificial Intelligence in Medicine **30**(3), 201–214 (2004)
16. Mitchell, T.M.: Machine Learning. McGraw-Hill (1997)
17. Patriarca-Almeida, J.H., Santos, B., Cruz-Correia, R.: Using a clinical document importance estimator to optimize an agent-based clinical report retrieval system. In: Proceedings of the 26th IEEE International Symposium on Computer-Based Medical Systems, pp. 469–472 (2013)
18. R Core Team: R: A Language and Environment for Statistical Computing (2013)
19. Robin, X., Turck, N., Hainard, A., Tiberti, N., Lisacek, F., Sanchez, J.C., Müller, M.: pROC: an open-source package for R and S+ to analyze and compare ROC curves. BMC Bioinformatics **12**, 77 (2011)
20. Rodrigues, P.P., Dias, C.C., Cruz-Correia, R.: Improving clinical record visualization recommendations with bayesian stream learning. In: Learning from Medical Data Streams, vol. 765, p. paper4. CEUR-WS.org (2011)
21. Rodrigues, P.P., Dias, C.C., Rocha, D., Boldt, I., Teixeira-Pinto, A., Cruz-Correia, R.: Predicting visualization of hospital clinical reports using survival analysis of access logs from a virtual patient record. In: Proceedings of the 26th IEEE International Symposium on Computer-Based Medical Systems, Porto, Portugal, pp. 461–464 (2013)
22. Schurink, C.A.M., Lucas, P.J.F., Hoepelman, I.M., Bonten, M.J.M.: Computer-assisted decision support for the diagnosis and treatment of infectious diseases in intensive care units. The Lancet infectious diseases **5**(5), 305–312 (2005)
23. Scutari, M.: Learning Bayesian Networks with the bnlearn R Package. Journal of Statistical Software **35**, 22 (2010)
24. Vieira-Marques, P.M., Cruz-Correia, R.J., Robles, S., Cucurull, J., Navarro, G., Marti, R.: Secure integration of distributed medical data using mobile agents. Intelligent Systems **21**(6), 47–54 (2006)
25. Wyatt, J.C., Wright, P.: Design should help use of patients' data. Lancet **352**(9137), 1375–1378 (1998)

On the Efficient Allocation of Diagnostic Activities in Modern Imaging Departments

Roberto Gatta[1](✉), Mauro Vallati[2], Nicola Mazzini[1,2,3], Diane Kitchin[2],
Andrea Bonisoli[3], Alfonso E. Gerevini[3], and Vincenzo Valentini[1]

[1] Radiotherapy Department, Università Cattolica Del Sacro Cuore, Milan, Italy
roberto.gatta.bs@gmail.com
[2] School of Computing and Engineering, University of Huddersfield, Huddersfield, UK
[3] Dipartimento d'Ingegneria dell'Informazione,
Università degli Studi di Brescia, Brescia, Italy

Abstract. In a modern Diagnostic Imaging Department, managing the schedule of exams is a complex task. Surprisingly, it is still done mostly manually, without a clear, explicit and formally defined objective or target function to achieve.

In this work we propose an efficient approach for optimising the exploitation of available resources. In particular, we provide an objective function, that considers the aspects that have to be optimised, and introduce a two-steps approach for scheduling diagnostic activities. Our experimental analysis shows that the proposed technique can easily scale on large and complex Imaging Departments, and generated allocation plans have been positively evaluated by human experts.

1 Introduction

In a modern Diagnostic Imaging Department the allocation and re-allocation of exams is a complex task, that is time-consuming and is still done manually. On the one hand, it is fundamental to keep the waiting lists as short as possible, in order to meet the established waiting time; on the other hand, it is of critical importance to minimise expenses for the Department. Moreover, patient scheduling has to be balanced, in order to plan the best possible allocation according to the staff organisation/skills on different modalities, i.e. computed tomography (CT), Radiography (RX), magnetic resonance (MR) and ultrasound (US) equipment. In order to plan the best possible allocation a lot of available resources must be taken into account: staff (radiologists, nurses, etc.), equipment (US, CT, MR, etc.), examinations performed (tagged by imaging modalities, reimbursement rate, clustered by regions and/or pathologies) and staff characteristics (part-time, full-time, etc.).

The literature on medical appointment scheduling is extensive, but approaches –either automated or in the form of formal guidelines– to deal with diagnostic activities in radiology Departments are rare. Nevertheless, the importance of scheduling activities in hospital services is well-known [8]. Even though a few techniques have been proposed for dealing with part of the allocation problem (see, e.g., [1,4–6]), a complete approach able to manage all the aspects of the

© Springer International Publishing Switzerland 2015
F. Pereira et al. (Eds.) EPIA 2015, LNAI 9273, pp. 103–109, 2015.
DOI: 10.1007/978-3-319-23485-4_10

allocation problem of a Radiological Department is still missing. Improvements in that area can lead to a significant reduction of costs and human time.

In this work we propose a formalisation of the activities allocation problem. The formalisation is composed by a set of definitions, constraints and a target function. The model is then exploited by an efficient scheduling approach, based on enforced hill climbing, that aims at optimising the use of available resources. This work encompasses a previous preliminary study by Mazzini et al. [7]. According to our experimental analysis, and to the feedback we received from medical experts, the proposed approach is efficiently able to produce good quality scheduling, following the metric it is required to optimise, which considers at the same time temporal and economic indexes.

2   The Diagnostic Activities Allocation Problem

In this section we define the relevant entities involved in the allocation problem, and describe the function used for evaluating the quality of allocation plans.

Entities
We define a set of entities that can easily fit into most of the Radiology Information Systems (RIS) currently used in diagnostic Imaging Departments [2]. Specifically, the proposed elements can directly fit with Paris, provided by ATS-Teinos, and PRORAM from METRIKA. With some minor changes it can be also fit with Estensa, of Esaote. We are confident it can also be easily adopted in other situations. The most important entities are the following:

Exam represents the diagnostic examination that can be performed, e.g., "CT brain".

Exam Group (or cluster) is a group of exams. In many cases it is useful, due to some team specialisation, to group exams for the area of the body (e.g., "head and neck" or "abdominal") or to group them in order to reflect which Department the patient comes from (e.g., "CT from GPs"). The grouping is done according to the habits of the Imaging Department, team and work-flow: hybrid models can also be implemented.

Modality this entity represents a medical device, like an ultrasound, a CT scan, etc. In our model, a modality corresponds to an actual room. This is reasonable since the machinery used for exams is usually not moved between rooms. Such modelisation leads to having an independent agenda per modality. In principle, it is possible to have the machines required for different sorts of exams in the same room. This case, which is extremely rare since it leads to underused resources, is not modelled.

Personnel represents the human resources (staff members) available. Each member of staff has at least one role. Each Exam Group has a set of roles assigned, this indicates the specific needs of that group of exams in terms of human resources.

For instance, some exam groups require several nurses to be present in the room, while other exams require technicians to be available.

Time Slot we adopted an atomic time slot of 5 minutes in a weekly calendar. The granularity of 5 minutes has been chosen since it is not extremely long, thus limiting the waste time; also, it is not too short, therefore short delays do not affect the overall daily scheduling.

Temporal horizon catches the requirements in terms of queue governance. This can represent constraints like "the queue for 'Brain MRI' must be lower than 3 months for, at least, the next 12 months". It should be noted that this is the usual way in which queue governance requirements are expressed.

Objective Function and Constraints

The optimisation of the scheduling of exams has to deal with two main components. On the one hand, it is important to maximise income for the Department. This should result in the prioritisation of exams that are both frequently requested and expensive. On the other hand, the Department is also providing an important service to the community. Therefore, keeping all the queue lists as short as possible is fundamental. In public hospitals there are strict upper bounds for queues.

For considering both the aforementioned aspects, we designed an objective function (to minimize) that combines the two perspectives. The adopted function is depicted in Equation 1. In particular, n indicates the number of exams. r_i is the cost of the i-th exam. q_i is the amount of exams of group i that should be performed. Δ_j represents the difference between waiting queue and desired queue length for the j-th exam group. W_j is the importance of the queue for the j-th exam. Finally, α and β indicate, respectively, the importance that is given to the economic side and to the respect of the limits on the queues' length. Intuitively, the function synthesises the point of view of the hospital administration (first addend) –focused on the economic side– and of the doctor (second addend) –focused on the quality of the service.

$$f = \alpha \sum_{1}^{n} (r_i * q_i) + \beta \sum_{1}^{n} (\Delta_j * W_j) \tag{1}$$

In order to guarantee the feasibility of identified allocation plans, a number of constraints have to be satisfied. In particular for each exam group the queue length must not be higher than the provided upper bound for respecting the governance rules on waiting time. Also there is a set of constraints that regulate the correct allocation of exams to suitable modalities and time slots and correct assignment of medical personnel to exam rooms respecting specific limits and maximum consecutive working hours that are in place in hospitals for all the staff. Our implemented prototype takes into account all such constraints.

3   The Proposed Two-Steps Algorithm

For the automatic allocation of diagnostic activities, we designed an approach that is composed of two main steps: *generation* and *improvement*. In both steps, groups of exams are considered. In a modern Radiology Department the number of possible exams is usually high (300 – 400), but they can be easily organised in groups, both for administrative issues and similarity of requirements. For example: all the "RX bones" exams are very similar in terms of required roles, funding, used time slots and required modality. Therefore, considering a group instead of a single exam provides a good abstraction: it reduces the number of variables to deal with, thus improving the computability, but does not lead to loss of information. Also, RIS systems normally manage their agenda following the same approach: instead of declaring all the exams which can be assigned to a modality in a period of time, they allow the definition of clusters of exams that can be linked to the modality. As a matter of fact, the medical personnel are trained for performing a specific cluster of exams rather than a single type of examination.

Generation
This step aims at quickly identifying an allocation of activities and personnel that satisfies all the constraints, regardless of the overall quality. This task is probably the most difficult, because it has to search the huge allocation space and identify a solution that satisfies all the allocation constraints. In particular, the generation step is based on the following tasks, which are executed by considering one week at a time:

1. the available personnel is fully assigned to the rooms. Most of the staff are assigned to morning slots, since it is the period of the day where most exams take place. Some heuristics are followed for reducing the spread of exams of the same group in different rooms, or in very different time slots.
2. A random number of time slots is assigned to each cluster of examinations, according to the hard constraints related to human-resources.
3. After the allocation, free time slots or free human resources are analysed in order to be exploited. For reducing the fragmentation, the preferred solution is to extend the time slot of exams allocated before/after the free slot. Fragmentation leads to waste time due to switching exam equipment between modalities and personnel moving between rooms.

In parallel with *generation*, a waiting lists estimation is performed. This allows assessment of whether the requirements on exam queue lengths will be satisfied, since the proposed approach provides a week-by-week allocation plan.

Improvement
The first allocation plan is then improved through an enforced hill climbing. At each iteration, a neighbour of the current plan is generated by de-allocating a cluster of exams from a modality, and substituting it with a different cluster. The substitution is a complex step; it is not guaranteed it can be applied for

all the pairs of clusters. This is due to, for example, different requirements in terms of personnel, equipment or time. The choice of the group of exam to be substituted is done by ordering clusters according to the requested resources and number of requests per week. Clusters that require many resources and are rarely performed are suitable to be substituted. The selected cluster is substituted with another that can fit in the released time-slots.

If the new allocation plan has a better target function than the current plan, the former is saved; otherwise the algorithm restarts by considering a different suitable cluster to substitute. The search stops when a specified number of re-scheduling attempts, or the time limit, is reached. It should be noted that the designed algorithm is able to provide several solutions of increasing quality.

4 Experimental Analysis

We implemented a prototype of the proposed algorithm in C++; as input it requires an XML description of available resources and exams to be performed. We tested the proposed approach in various scenarios. We considered 3 different possible Departments, several values of α and β, and different settings up to at most 25 exam groups, 90 exams, 10 modalities and 24 staff members, distributed among radiologists, radiographers and nurses. This reflects a medium-sized Department. We observed that a first solution is usually found in about 15 seconds. Incremental improvements, generally from 60 to 250 increasing quality allocation plans, are usually found in less than 10 minutes.

Given the fact that in real-world Departments no quantitative functions are used for evaluating the quality of allocation plans, it is incorrect to compare the allocation plans generated by our approach with existing scheduling.

In order to have a first validation of the generated plans we showed some of them to human experts; namely, to two radiologists, one radiographer technician and one IT specialist employed in a Diagnostic Imaging Department. They considered the plans to be realistic and feasible. Considering the context and the environment of a modern hospital, this is a reasonable way for validating the generated schedules. Moreover, from experts feedback we synthesised a number of useful insights, that are described in the following.

The *generation* step is computationally hard, but it is crucial in order to obtain good final solutions. We observed that a low quality initial solution usually leads to small increments in the subsequently improved allocations. On the other hand, this generation step is not fundamental in real-world applications, since Radiology Departments already have a feasible scheduling in place. Moreover, their current scheduling usually includes thousands of reservations, therefore cannot be suddenly changed without dramatic effects.

Fragmentation should be included in the objective function. In the current model it is not considered, since we believed that dealing with exam groups instead of single exams would have been enough. On the contrary, we observed that exam groups allocated to a modality change frequently; this also causes a frequent change in the staff, which affects the quality of the allocation plan. In large hospitals, where examination rooms are far from each other, frequent staff movement results in a significant waste of time.

Data entry is time-expensive: changes in personnel, exams, instrumentation or policies are quite frequent in a medium-big Radiology Department, and require updating of the data and re-planning. The best way to efficiently support an operator would be by exporting, from existent RIS, data regarding staff, modalities and dates, in order to save time. Currently, HL7 [3] would probably be the best standard for such integration.

The proposed algorithm can provide useful information about the available resources. In particular, it can be used for identifying the most limiting resource (personnel, modalities, etc.) and evaluating the impact of new resources. Head of Departments highlighted that it is currently a very complex problem to estimate the impact of a new modality, or of increased personnel. By using the proposed algorithm, the impact of new resources can be easily assessed by comparing the quality of plans with and without them, in different scenarios.

New modalities or new staff require an initial "training" period. In the case of new modalities, the staff will initially require more time for performing exams. Newly introduced staff usually need to be trained. The current approach is not able to catch such situations. A possible way for dealing with this is considering a "penalty" for some instrumentation or personnel; a time slot is "longer" –by a given penalty factor– when they are assigned to it. Penalties can be reduced over time.

The current approach does not have a year overview. Some exams are more likely to be required in some periods, thus they should have different priorities with regards to their waiting lists' requirements. Also, it is common practice that in summer hospital's personnel is reduced. A good integration with medical and administrative databases will be useful for further refinements of the allocation abilities.

5   Conclusion

In many Diagnostic Imaging Departments the allocation of exams is currently done manually. As a result, it is time consuming, it is hard to assess its overall quality, and no information about limiting resources are identified.

In this paper, we addressed the aforementioned issues by introducing: (i) a formal model of the diagnostic activities allocation problem, and (ii) an efficient algorithm for the automated scheduling of diagnostic activities. The proposed model is general, and can therefore fit with any existing Imaging Department. Moreover, a quantitative function for assessing the quality of allocation plans is provided. The two-steps algorithm allows the generation and/or improvement of allocation plans. An experimental analysis showed that the approach is efficiently able to provide useful and valid scheduling for examinations. Feedback received from experts confirms its usefulness, also for evaluating the impact of new instrumentation or staff members.

This work can be seen as a pilot study, which can potentially lead to the exploitation of more complex and sophisticated Artificial Intelligence techniques for handling the Radiological resource allocation problem. Further investigations are needed both on the reasoning and on the knowledge representation sides. Moreover, larger experimental analysis and stronger validation approaches are envisaged. Future work includes also an extended experimental analysis with data gathered from existing Departments, and the investigation of techniques for reducing fragmentation.

References

1. Barbati, M., Bruno, G., Genovese, A.: Applications of agent-based models for optimization problems: A literature review. Expert Systems with Applications **39**(5), 6020–6028 (2012)
2. Boochever, S.S.: HIS/RIS/PACS integration: getting to the gold standard. Radiol Manage **26**(3), 16–24 (2004)
3. Dolin, R.H., Alschuler, L., Boyer, S., Beebe, C., Behlen, F.M., Biron, P.V., Shvo, A.S.: Hl7 clinical document architecture, release 2. Journal of the American Medical Informatics Association **13**(1), 30–39 (2006)
4. Eagen, B., Caron, R., Abdul-Kader, W.: An agent-based modelling tool (abmt) for scheduling diagnostic imaging machines. Technology and Health Care **18**(6), 409–415 (2010)
5. Falsini, D., Perugia, A., Schiraldi, M.: An operations management approach for radiology services. In: Sustainable Development: Industrial Practice, Education and Research (2010)
6. Macal, C.M., North, M.J.: Agent-based modeling and simulation. In: Winter Simulation Conference, pp. 86–98 (2009)
7. Mazzini, N., Bonisoli, A., Ciccolella, M., Gatta, R., Cozzaglio, C., Castellano, M., Gerevini, A., Maroldi, R.: An innovative software agent to support efficient planning and optimization of diagnostic activities in radiology departments. International Journal of Computer Assisted Radiology and Surgery **7**(1), 320–321 (2012)
8. Welch, J.D.: N.B.T.: Appointment systems in hospital outpatient departments. The Lancet **259**(6718), 1105–1108 (1952)

Ontology-Based Information Gathering System for Patients with Chronic Diseases: Lifestyle Questionnaire Design

Lamine Benmimoune[1,3](✉), Amir Hajjam[1], Parisa Ghodous[2],
Emmanuel Andres[4], Samy Talha[4], and Mohamed Hajjam[3]

[1] IRTES-SET, Université de Technologie Belfort-Montbéliard, 90000 Belfort, France
lamine.benmimoune@utbm.fr
[2] LIRIS, Université Claude Bernard Lyon 1, 69100 Villeurbanne, France
[3] Newel, 68100 Mulhouse, France
[4] Hôpital Civil de Strasbourg, 67000 Strasbourg, France

Abstract. The aim of this paper is to describe an original approach which consists of designing an Information Gathering System (IGS). This system gathers the most relevant information related to the patient. Our IGS is based on using questionnaire ontology and adaptive engine which collects relevant information by prompting the whole significant questions in connection with the patient' s medical background. The formerly collected answers are also taken into consideration in the questions selection process. Our approach improves the classical approach by customizing the interview to each patient. This ensures the selection of all of the most relevant questions. The proposed IGS is integrated within E-care monitoring platform for gathering lifestyle-related patient data.

Keywords: Information gathering system · Questionnaire · Health-care · Clinical decision support system · Ontology · Monitoring

1 Introduction

Computer-based questionnaires are a new form of data collection, which are designed to offer more advantages compared to pen and paper questionnaires or oral interviewing [13]. They are less time-consuming and more efficient by offering more structure and more details compared to the classical methods [2].

The Information Gathering Systems (IGSs) have had measurable benefits in reducing omissions and errors arising as a result of medical interviews [14]. The medical and health care domain is one of the most active domain in using IGS for gathering patients data [13].

Recently, various research works were conducted to design and to use the IGS as part of clinical decision support system (CDSS). Among them. Bouamrane *et al.* [2], [3] proposed a generic model for context-sensitive self-adaptation of IGS based on questionnaire ontology. The proposed model is implemented as an data collector module in [4] to collect patient medical history for preoperative

© Springer International Publishing Switzerland 2015
F. Pereira et al. (Eds.) EPIA 2015, LNAI 9273, pp. 110–115, 2015.
DOI: 10.1007/978-3-319-23485-4_11

risk assessment. Sherimon *et al.* [5], [6], [15] proposed an questionnaire ontology based on [2]. This ontology is used to gather patient medical history, which is then integrated within CDSS to predict the Risk of hypertension. Farooq *et al.* [7] proposed an ontology-based CDSS for chest pain risk assessment, based on [2] the proposed CDSS integrates a data collector to collect patient medical history. Alipour [13] proposed an approach to design an IGS based on the use of ontology-driven generic questionnaire and Pellet inference engine for questions selection process.

Although the presented IGS in the literature permit gathering patient data using ontologies, the created questionnaires are hard coded for specific domains and they are defined under the domain ontologies. These, make them less flexible, more difficult to maintain and even hard to share and to reuse.

Unlike previous approaches, our approach offers more flexibility by separating the ontologies and by integrating a domain ontology to drive the creation of questionnaire. This allows to give meaning to the created questions, and configuring different models of questionnaires without coding and regardless of the content of the domains. Therefore, many CDDSSs can easily integrate and use the proposed IGS for their specific needs.

Furthermore, the proposed approach permits to collect relevant information by prompting the whole significant questions in connection with the patient profile. The formerly collected answers are also taken into consideration in the questions selection process. This improves the classical approach by customizing the interview to each patient.

The proposed IGS is integrated within E-care home health monitoring platform [1], [8] for gathering lifestyle-related patient data.

2 Information Gathering System within E-Care Platform

E-care is a home health monitoring platform for patients with chronic diseases such as diabetes, heart failure, high blood pressure, etc. [1] [8]. The aim is early detection of any anomalies or dangerous situations by collecting relevant data from the patient such as physiological data (heart rate, blood pressure, pulse, temperature, weight, etc.) and lifestyle data (tobacco-use, eating habits, physical activity, sleep, stress, etc.).

To improve the accuracy in anomalies detection, the platform needs relevant information that describes as precisely as possible the patient's health status and his lifestyle changes (tobacco-use, lack of physical activity, poor eating habits, etc.). That is why the patient is invited daily to collect his physiological data using medical sensors (Blood Pressure Monitor, Weighing Scale, Pulse Oximeter, etc.) and to answer on lifestyle questionnaires. These questionnaires are automatically generated by the IGS which permits gathering relevant information about the patient lifestyle.

All collected data (physiological data and lifestyle data) is stored in the patient profile ontology which models the health status of patient and then analysed by the inference engine for anomalies detection.

3   Information Gathering System Architecture

The proposed architecture consists of four main components: Questionnaire Ontology, Survey History Ontology, Adaptive Engine and User Interfaces.

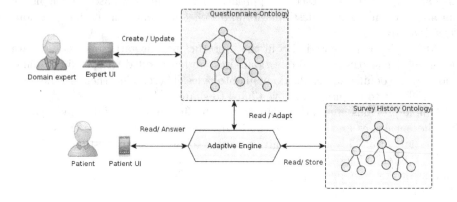

Fig. 1. Information Gathering System architecture

Questionnaire Ontology (QO): models the concepts representing the common components of a questionnaire. The QO is created based on Bouamrane *et al.*'s research works [2]. It is designed as generic, structured and flexible to accept most of the questionnaire models. The main classes are: *Questionnaire, SubQuestionnaire, Question* and *PotentialAnswer*. The Questionnaire class is composed of Sub-questionnaires, which represent a group of thematically related question classes. The question classes could be inter-related by structural properties such as *hasParent, hasChild, hasSibling*, etc. Each question is characterised by a type and related to one or more potential answers using Adaptive properties such as *ifAnswerToThisQuestionEqualsTo, thenGoToQuestion*, etc.

Survey History Ontology (SHO): stores all the patient surveys. It includes all the asked questions and the given answers by the patient. It is used in the questions selection process.

The Adaptive Engine (AE): it interprets the properties asserted in the questionnaire ontology and prompts the corresponding questions in connection with the patient profile and the formerly collected answers. The AE initially loads all questions except the children questions. It prompts the first question and checks if the question is appropriate to the patient profile (e.g. AE doesn't ask questions about the smoking habits, if the patient is a non-smoker). If it is, the AE asks the question and gets the answer from the UI. If it is not, the AE just prompts the next question.

If however the current question happens to be adaptive (i.e. it has at least a child question), the given answer is then checked against the answers that are expected to lead to children questions. If a match is found, the AE loads the children questions. If no match is found, the next question is prompted. The interaction loop is repeated until there are no more questions to be asked.

The User Interfaces (UI): consist of two parts of UI namely: expert UI and Patient UI.

- **Expert UI:** permits the domain experts (clinicians) to configure the IGS by defining questionnaires and to consult the surveys history.
- **Patient UI:** permits to start/stop the survey. It is designed in such a way that the patient can respond to the questionnaire from anywhere using his mobile device (tablet or smart phone).

4   Domain Ontology Driven Questionnaire

The domain ontology aims to drive the creation of questionnaires by offering a common and controlled vocabulary. To achieve this goal, we have developed a domain ontology for lifestyle concepts based on recommendations provided by Haute Autorité de Santé (HAS)[1]. The ontology is structured as an hierarchy of concepts and relations between concepts. It is composed of three main entities:

- *LifeStyleEntities:* hierarchical concepts that model lifestyle entities such as eating habits, physical activity, smoking habits, etc.
- *DimensionsEntities:* hierarchical concepts that model temporal dimensions and physical dimensions (quantity). Each dimension includes a hierarchy of concepts (e.g. *TimesOfDay*, *TimeFrequency* and *TimeUnit* are grouped under the *timeDimension* concept).
- *CataloguesEntities:* includes concepts used to give more semantic for the *LifeStyleEntities* concepts. Each catalogue entity includes a hierarchy of concepts that model the types of *LifeStyleEntities* concepts (e.g. *Cigarette*, *E-cigarette* and *Drug* are types of *Tobacco* concept for the *SmokingHabits* concept).

The concepts are related amongst them through properties as follows.

- *DimensionProperties:* used to relate the *LifeStyleEntities* to the *DimensionsEntities*, they include a set of properties such as *hasQuantity*, *hasFrequency*, *hasTimesOfDay*, etc.
- *CatalogueProperties:* used to relate the *LifeStyleEntities* to the *CataloguesEntities*, they include a set of properties such as *hasExercise*, *hasTobacco*, *hasFood*, etc.

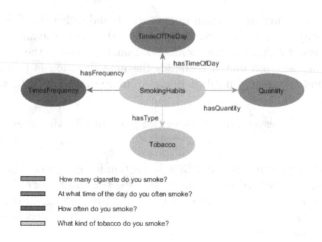

Fig. 2. Related questions for the "smoking habits" concept

The example illustrated by figure 2 shows how the domain concepts can be related amongst them and how they are used to design lifestyle questionnaires.

Given the smoking habit that is characterized by a type of tobacco (e.g. cigarette, electronic cigarette, drugs, etc.), time frequency (daily, monthly, weekly, etc.), smoking quantity, etc. Several questions can be created based on *SmokingHabits* concept, with each smoking-related question should be related to the *SmokingHabits* through the domain properties, while the potential answers are related either to the *DimensionsEntities* or to the *CatalogueProperties*. (see figure 2).

5   Conclusion and Future Work

In this paper, we presented a novel approach, which aims to design an Ontology-Based Information Gathering System. This system permits gathering the most relevant information by providing personalised questionnaires related to the patient profile. Our IGS consists of a questionnaire ontology which is driven by domain ontology. We have seen how the domain ontology is used to control vocabulary and to give a meaning to the asked questions. Furthermore, the use of a domain ontology can improve the gathering of data and the design of questionnaires can be made easier and faster compared to the hard coding of questionnaires. On the other hand, we have highlighted the interest of using the proposed IGS within E-care health monitoring platform, since it permits gathering relevant information about patients' lifestyle.

In the near future, we will experiment the proposed IGS in the real life with chronic patients.

---

[1] http:www.has-sante.fr

References

1. Benyahia, A.A., Hajjam, A., Hilaire, V., Hajjam, M.: E-care ontological architecture for telemonitoring and alerts detection. In: 5th IEEE International Symposium on Monitoring & Surveillance Research (ISMSR): Healthcare-Safety-Security (2012)
2. Bouamrane, M.-M., Rector, A.L., Hurrell, M.: Ontology-driven adaptive medical information collection system. In: An, A., Matwin, S., Raś, Z.W., Ślęzak, D. (eds.) Foundations of Intelligent Systems. LNCS (LNAI), vol. 4994, pp. 574–584. Springer, Heidelberg (2008)
3. Bouamrane, M.M., Rector, A., Hurrell, M.: Gathering precise patient medical history with an ontology-driven adaptive questionnaire. In: 21st IEEE International Symposium on Computer-Based Medical Systems, CBMS 2008, June 17–19, 2008
4. Bouamrane, M.-M., Rector, A., Hurrell, M.: Using ontologies for an intelligent patient modelling, adaptation and management system. In: Meersman, R., Tari, Z. (eds.) OTM 2008, Part II. LNCS, vol. 5332, pp. 1458–1470. Springer, Heidelberg (2008)
5. Sherimon, P.C., Vinu, P.V., Krishnan, R., Takroni, Y.: Ontology Based System Architecture to Predict the Risk of Hypertension in Related Diseases. IJIPM: International Journal of Information Processing and Management 4(4), 44–50 (2013)
6. Sherimon, P.C., Vinu, P.V., Krishnan, R., Takroni, Y., AlKaabi, Y., AlFars, Y.: Adaptive questionnaire ontology in gathering patient medical history in diabetes domain. In: Herawan, T., Deris, M.M., Abawajy, J. (eds.) DaEng-2013. LNEE, vol. 285, pp. 453–460. Springer, Singapore (2014)
7. Farooq, K., Hussain, A., Leslie, S., Eckl, C., Slack, W.: Ontology driven cardiovascular decision support system. In: 2011 5th International Conference on Pervasive Computing Technologies for Healthcare (PervasiveHealth), May 23–26, 2011
8. Benyahia, A.A., Hajjam, A., Hilaire, V., Hajjam, M., Andres, E.: E-care telemonitoring system: extend the platform. In: 2013 Fourth International Conference on Information, Intelligence, Systems and Applications (IISA), July 10–12, 2013
9. Saripalle, R.K.: Current status of ontologies in Biomedical and clinical Informatics. University of Connecticut. http://www.engr.uconn.edu/steve/Cse300/saripalle.pdf (retrieved January 16, 2014)
10. Gruber, T.R.: Toward principles for the design of ontologies used for knowledge sharing? International Journal Human-Computer Studies 43(5–6), 907–928 (1995)
11. Guarino, N.: Formal Ontology and information systems. Formal ontology in information systems. In: Proceedings of FOIS 1998, Trento, Italy, June 6–8, 1998
12. Noy, N.F., McGuinness, D.L.: Ontology Development 101: A Guide to Creating Your First Ontology. Stanford University (2005)
13. Alipour-Aghdam, M.: Ontology-Driven Generic Questionnaire Design. Thesis for the degree of Master of Science in Computer Science. Presented to The University of Guelph, August 2014
14. Bachman, J.W.: The patient-computer interview: a neglected tool that can aid the clinician. Mayo Clinic Proceedings 78, 67–78 (2003)
15. Sherimon, P.C., Vinu, P.V., Krishnan, R., Saad, Y.: Ontology driven analysis and prediction of patient risk in diabetes. Canadian Journal of Pure and Applied Sciences 8(3), 3043–3050 (2014). SENRA Academic Publishers, British Columbia

Predicting Preterm Birth in Maternity Care by Means of Data Mining

Sónia Pereira, Filipe Portela$^{(\boxtimes)}$, Manuel F. Santos, José Machado,
and António Abelha

Algoritmi Centre, University of Minho, Braga, Portugal
b7004@dps.uminho.pt, {cfp,mfs}@dsi.uminho.pt,
{jmac,abelha}@di.uminho.pt

Abstract. Worldwide, around 9% of the children are born with less than 37 weeks of labour, causing risk to the premature child, whom it is not prepared to develop a number of basic functions that begin soon after the birth. In order to ensure that those risk pregnancies are being properly monitored by the obstetricians in time to avoid those problems, Data Mining (DM) models were induced in this study to predict preterm births in a real environment using data from 3376 patients (women) admitted in the maternal and perinatal care unit of Centro Hospitalar of Oporto. A sensitive metric to predict preterm deliveries was developed, assisting physicians in the decision-making process regarding the patients' observation. It was possible to obtain promising results, achieving sensitivity and specificity values of 96% and 98%, respectively.

Keywords: Data mining · Preterm birth · Real data · Obstetrics care · Maternity care

1   Introduction

Preterm birth portrays a major challenge for maternal and perinatal care and it is a leading cause of neonatal morbidity. The medical, education, psychological and social costs associated with preterm birth indicate the urgent need of developing preventive strategies and diagnostic measures to improve the access to effective obstetric and neonatal care [1]. This may be achieved by exploring the information provided from the information systems and technologies increasingly used in healthcare services.

In Centro Hospitalar of Oporto (CHP), a Support Nursing Practice System focused on nursing practices (SAPE) is implemented, producing clinical information. In addition, patient data plus their admission form are recorded though EHR (Electronic Health Record) presented in Archive and Diffusion of Medical Information (AIDA) platform. Both SAPE and EHR are also used by the CHP maternal and perinatal care unit, Centro Materno Infantil do Norte (CMIN). CMIN is prepared to provide medical care / services for women and child. Therefore, using obstetrics and prenatal information recorded from SAPE and EHR, it is possible to extract new knowledge in the context of preterm birth. This knowledge is achieved by means of Data Mining (DM) techniques, enabling predictive models based on evidence. This study accomplished

© Springer International Publishing Switzerland 2015
F. Pereira et al. (Eds.) EPIA 2015, LNAI 9273, pp. 116–121, 2015.
DOI: 10.1007/978-3-319-23485-4_12

DM models with sensitivity and specificity values of approximately 96% and 98%, which are going to support the making of preventive strategies and diagnostic measures to handle preterm birth.

Besides the introduction, this article includes a presentation of the concepts and related work in Section 2, followed by the data mining process, described in Section 3. Furthermore, the results are discussed and a set of considerations are made in Section 4. Section 5 presents the conclusions and directions of future work.

2   Background and Related Work

2.1   Preterm Birth

Preterm birth refers to a delivery prior to 37 completed weeks (259 days) of labour. Symptoms of preterm labour include uterine contractions occurring more often than every ten minutes, or the leaking of fluids. Preterm birth is the leading cause of long-term disability in children, since many organs, including the brain; lungs and liver are still developing in the final weeks of pregnancy [2]. Preterm Birth has not decreased in the last 30 years, due to the failure identifying the high-risk group during routine prenatal care [3]. Many studies were conducted to identify a way to predict preterm deliveries, focusing on physiologic measures, ultrasonography, obstetrics history and socioeconomic status [4]. For instance, in 2011 a model was developed for predicting spontaneous delivery before 34 weeks based on maternal factors, placental perfusion and function at 11-13 weeks' gestation, through screening maternal characteristics and regression analysis. They detected 38.2% of the preterm deliveries in women with previous pregnancies beyond 16 weeks and 18.4% in those without [3]. Most of the efforts to predict preterm birth face limited provision of population based data, since registration of births is incomplete and information is lacking on gestational age [6].

2.2   Interoperability Systems and Data Mining in Healthcare

As mentioned in the previous section, this study is based on real data acquired from CMIN. The knowledge extraction depends substantially on the interoperability between SAPE and EHR systems assured through AIDA. This multi-agent platform enables the standardization of clinical systems and overcomes the medical and administrative complexity of the different sources of information from the hospital [5].

In healthcare systems, there is a wealth of data available, although there is a lack of effective analysis tools to extract useful information. Thus, data mining have found numerous applications in scientific and clinical domain [8]. Successful mining applications have been implemented in the healthcare. In obstetrics and maternal care, some of these studies were employed to predict the risk pregnancy in women performing voluntary interruption of pregnancy (VIP) [9] and manage VIP by predicting the most suitable drug administration [7].

3   Study Description

This study was conducted by following the Knowledge Discovery in Database (KDD), allowing the extraction of implicit and potentially useful information, through algorithms, taking account the magnitudes of data increasing [10].

The DM methodology employed was the Cross Industry Standard Process for Data Mining (CRIP-DM), a non-rigid sequence of six phases, carried out in this section, which allow the implementation of DM models to be used in real environments [11]. To induce the DM models, four different algorithms were implemented: Decision Trees (DT), Generalized Linear Models (GLM), Support Vector Machine (SVM) and Naïve Bayes (NB). This study used data collected from 3376 patients (women) admitted in the maternal and perinatal care unit (CMIN) of CHP comprising a period between 2012-07-01 and 2015-01-31, in a total of 1120 days.

3.1   Business Understanding

The Business aim of this project is to identify the risk group of preterm delivery, to ensure the proper monitoring and to avoid its associated problems. The DM goal is to develop accurate models able to support the decision-making process by predicting whether or not a woman will be subjected to a preterm delivery, based on data from clinic cases.

3.2   Data Understanding

The initial dataset extracted from SAPE and EHR admission records was analysed and processed in order to be used in the DM process. A set of 13 variables were selected: age (corresponds to the age of the pregnant patient), programmed (indicates whether or not a delivery is programmed), gestation (singular or multiple pregnancies), PG1 and PG2 (first echography measures), motive (reason of intervention - normal delivery or unexpected events), patients' weight and height, BMI (body mass index), blood type, cardiotocography (CTG) (biophysics exam that evaluates the fetal wellbeing), streptococcus (presence of the bacterium streptococcus in the pregnant system) and finally, marital status of the pregnant patient. The target variable *Group Risk* denotes the preterm birth risk and it is presented in Table 1.

Table 1. Representation of the target variable *Group Risk*.

Description	Value	Target	Distribu-	Percentage
>=37 weeks of gestation (Term)	0		3137	92.92%
< 37 weeks of gestation (Preterm)	1		239	7.08%

In Table 2 are shown statistics measures related to the numerical variables age, gestation, PG1, PG2 and BMI, while in Table 3 it is represented the percentage of occurrences for some used variables.

Table 2. Statistics measures of age, PG1, PG2, weight, height, BMI variables.

	Minimum	Maximum	Average	Standard Deviation
Age	14	46	29.88	5.81
PG1	5	40	12.81	2.96
PG2	0	8	3.09	1.96
BMI	14.33	54.36	29.40	4.57

Table 3. Percentage of occurrences of some variables.

Variable	Class	Cases
Programmed	True	12.53%
Gestation	Singular	89.90%
Motive	Normal	81.33%
Streptococcus	Positive	13.27%
Cardiotocography	Suspect	2.19%

3.3   Data Preparation

After understanding the data collected, the variables were prepared to be used by the DM models. The data pre-processing phase started with the identification of null and noise values. These values were eliminated from the dataset. To ensure the data normalization, all the values, such as weight and height, were transformed to International System measures, using the point to separate decimal values.

As shown in Table 1, there is a disparity in the distribution of values of the target variable Risk Group (low percentage of preterm birth cases). In order to balance the target, the oversampling technique was implemented by replicating the preterm birth cases until it reached approximately 50% of the dataset, obtaining 6244 entries.

3.4   Modelling

A set of Data Mining models (DMM) were induced using the four DM techniques (DMT) mentioned in Section 3: GLM, SVM, DT and NB. The developed models used two sampling methods Holdout sampling (30% of data for testing) and Cross Validation (all data for testing). Additionally there were implemented two different approaches, one using the raw dataset (3376 entries) and another with oversampling. Different combinations of variables were used, obtaining 5 different scenarios:

S1: {Age (A), Gestation (G), Programmed (P), PG1, PG2, Motive (M), Height (H), Weight (W), BMI, Blood Type (B), Marital Status (MS), CTG, Streptococcus (S)}

S2: {A, H, W, BMI, B, MS, CTG, S}

S3: {G, P, PG1, PG2, M, CTG, S}

S4: {A, G, PG1, PG2, M, H, W, BMI, B, CTG, S}

S5: {A, G, P, M, H, W, BMI, B}

Therefore, a total of 80 Data Mining models (DMM) were induced:

DMM = {5 *Scenarios, 4 Techniques, 2 Sampling Methods, 2 Approaches*}

All the models were induced using the Oracle Data Miner with its default configurations. For instance, GLM was induced with automatic preparation, with a confidence level of 0.95 and a reference value of 1.

3.5   Evaluation

The study used the confusion matrix (CMX) to assess the induced DM models. Using the CMX, the study estimated some statistical metrics: sensitivity, specificity and accuracy. Table 4 presents the best results achieved by each technique, sampling method and approach. The best accuracy (93.00%) was accomplished with scenario 3 by both DT and NB techniques using oversampling and 30% of data for testing. The best sensitivity (95.71%) was achieved by scenario 4 with oversampling using SVM technique and all the data for testing. Regarding specificity, scenario 2 reached 97.52% using SVM with oversampling and all the data for testing.

Table 4. Sensitivity, specificity and accuracy values for the best scenarios for each DMT, approach and sampling method. Below, the best metric values highlighted for each DMT.

DMT	Oversampling	Sampling	Scenario	Sensitivity	Specificity	Accuracy
DT	No	30%	3	**0.8889**	0.9303	**0.9300**
	No	All	1	0.2896	**0.9723**	0.8599
GML	No	All	4	0.2896	**0.9723**	0.8599
	Yes	All	4	**0.8674**	0.7126	0.7687
NB	No	30%	3	**0.8889**	0.9303	**0.9300**
	No	All	1	0.4868	**0.9646**	0.9271
SVM	No	All	2	0.1023	**0.9752**	0.4570
	Yes	All	4	**0.9571**	0.6647	**0.7410**

In order to choose the best models a threshold was established, considering sensitivity, accuracy and sensitivity values upper than to 85%. Table 5 shows the models that fulfil the threshold.

Table 5. Best model achieving the established threshold.

Scenario	Model	Oversampling	Sampling	Sensitivity	Specificity	Accuracy
3	NB,DT	No	30%	0.8889	0.9303	0.9300

4   Discussion

Should be noted that the best sensitivity (95.71%) and specificity (97.52%) are reached by models that did not achieve the threshold defined, showing low values in the remaining statistical measures used to evaluate the models. It can be settled that scenario 3 meets the defined threshold, presenting good results in terms of specificity and sensitivity, as seen in Table 5. Thus, it appears that the most relevant factors that affect the term of birth are: pregnancy variables, Gestation and physical conditions of the pregnant woman. In a clinic perspective, the achieved results will enable the prediction of preterm birth, with low uncertainty, allowing those responsible better monitoring and resource management. In a real time environment, physicians can rely on the model to send a warning informing that a specific patient has a risk pregnancy and it is in danger of preterm delivery. Consequently, the physician can be observant and alert to these cases and can put the patients on special watch, saving resources and time to the healthcare institution.

5   Conclusions and Future Work

At the end of this work it is possible to assess the viability of using these variables and classification DM models to predict Preterm Birth. The study was conducted using real data. Promising results were achieved by inducing DT and NB, with over-sampling and 30% of the data for testing, in scenario 3, achieving approximately 89% of sensitivity and 93% of specificity, suited to predict preterm births. The developed model support the decision-making process in maternity care by identifying the pregnant patients in danger of preterm delivery, alerting to their monitoring and close observation, preventing possible complications, and ultimately, avoiding preterm birth.

In the future new variables will be incorporated in the predictive models and other types of data mining techniques will be applied. For instance, inducing Clustering techniques would create clusters with the most influential variables to preterm birth.

Acknowledgments. This work has been supported by FCT - Fundação para a Ciência e Tecnologia within the Project Scope UID/CEC/00319/2013.

References

1. Berghella, V. (ed.): Preterm birth: prevention and management. John Wiley & Sons (2010)
2. Spong, C.Y.: Defining "term" pregnancy: recommendations from the Defining "Term" Pregnancy Workgroup. Jama **309**(23), 2445–2446 (2013)
3. Beta, J., Akolekar, R., Ventura, W., Syngelaki, A., Nicolaides, K.H.: Prediction of spontaneous preterm delivery from maternal factors, obstetric history and placental perfusion and function at 11–13 weeks. Prenatal diagnosis **31**(1), 75–83 (2011)
4. Andersen, H.F., Nugent, C.E., Wanty, S.D., Hayashi, R.H.: Prediction of risk for preterm delivery by ultrasonographic measurement of cervical length. AJOG **163**(3), 859–867 (1990)
5. Abelha, A., Analide, C., Machado, J., Neves, J., Santos, M., Novais, P.: Ambient intelligence and simulation in health care virtual scenarios. In: Camarinha-Matos, L.M., Afsarmanesh, H., Novais, P., Analide, C. (eds.) Establishing the Foundation of Collaborative Networks. IFIP — The International Federation for Information Processing, vol. 243, pp. 461–468. Springer, US (2007)
6. McGuire, W., Fowlie, P.W. (eds.): ABC of preterm birth, vol. 95. John Wiley & Sons (2009)
7. Brandão, A., Pereira, E., Portela, F., Santos, M.F., Abelha, A., Machado, J.: Managing voluntary interruption of pregnancy using data mining. Procedia Technology **16**, 1297–1306 (2014)
8. Kaur, H., Wasan, S.K.: Empirical study on applications of data mining techniques in healthcare. Journal of Computer Science **2**(2), 194 (2006)
9. Brandão, A., Pereira, E., Portela, F., Santos, M.F., Abelha, A., Machado, J.: Predicting the risk associated to pregnancy using data mining. In: ICAART 2015 Portugal. SciTePress (2015)
10. Maimon, O., Rokach, L.: Introduction to knowledge discovery in databases. Data Mining and Knowledge Discovery Handbook, pp. 1–17. Springer, US (2005)
11. Chapman, P., Clinton, J., Kerber, R., Khabaza, T., Reinartz, T., Shearer, C., Wirth, R.: CRISP-DM 1.0 Step-by-step data mining guide (2000)

Clustering Barotrauma Patients in ICU–A Data Mining Based Approach Using Ventilator Variables

Sérgio Oliveira, Filipe Portela[✉], Manuel F. Santos, José Machado,
António Abelha, Álvaro Silva, and Fernando Rua

Algoritmi Centre, University of Minho, Braga, Portugal
sergiomdcoliveira@gmail.com, {cfp,mfs}@dsi.uminho.pt,
{jmac,abelha}@di.uminho.pt, moreirasilva@me.com,
fernandorua.sci@chporto.min-saude.pt

Abstract. Predicting barotrauma occurrence in intensive care patients is a difficult task. Data Mining modelling can contribute significantly to the identification of patients who will suffer barotrauma. This can be achieved by grouping patient data, considering a set of variables collected from ventilators directly related with barotrauma, and identifying similarities among them. For clustering have been considered k-means and k-medoids algortihms (Partitioning Around Medoids). The best model induced presented a Davies-Bouldin Index of 0.64. This model identifies the variables that have more similarity among the variables monitored by the ventilators and the occurrence of barotrauma.

Keywords: Barotrauma · Plateau pressure · Intensive medicine · Data mining · Clustering · Similarity · Correlation

1   Introduction

Data Minng (DM) process provides not only the methodology but also the technology to transform the data collected into useful knowledge for the decision-making process [1]. In critical areas of medicine some studies reveal that one of the respiratory diseases with higher incidence in the patients is Barotrauma [2]. Health professionals have identified high levels of Plateau pressure as having a significantly contribute to the Barotrauma occurrence [3]. This study is part of the major project INTCare. In this work a clustering process was addressed in order to characterize patients with barotrauma and analyze the similarity among ventilator variables. The best models achieved a Davies-Bouldin Index of 0.64. The work was tested using data provided by the Intensive Care Unit (ICU) of the Centro Hospitalar do Porto (CHP).

This paper consists of four sections. The first section corresponds to the introduction of the problem and related work. Aspects directly related to this study and supporting technologies for knowledge discovering from databases are then addressed in the second section. The third section formalizes the problem and presents the results in terms of DM models following the methodology Cross Industry Standard Process for Data Mining (CRISP-DM). In the fourth section some relevant conclusions are taken.

© Springer International Publishing Switzerland 2015
F. Pereira et al. (Eds.) EPIA 2015, LNAI 9273, pp. 122–127, 2015.
DOI: 10.1007/978-3-319-23485-4_13

2   Background

2.1   Plateau Pressure, Acute Respiratory Distress Syndrome and Barotrauma

The occurrence of barotrauma happens when a patient has complications in mechanical ventilation. Patients with Acute Respiratory Distress Syndrome (ARDS) have shown that the incidence of pneumothorax and barotrauma varies between 0% and 76% [4]. The occurrence of barotrauma is one of the most dreaded complications when a patient is mechanically ventilated. This occurrence is associated with an increased morbidity and mortality. Several researchers argue that the Positive end-expiratory pressure (PEEP) is related to the occurrence of barotrauma, however there are other researchers not supporting this relationship and enforcing that there was not identified any relationship between PEEP and barotrauma [2]. The Plateau Pressure (PPR) values shall be continuously monitored providing important information for patient diagnosis. It is important to maintain the value of PPR $<= 30$ cmH_2O in order to protect the patient's lungs. An increased in PPR is associated to an increasing elasticity of the respiratory system and decreased compliance of the respiratory system [3].

2.2   Related Work

Predicting barotrauma occurrence is important for the patient wellbeing. So it is fundamental to explore the prediction accuracy of the variables and their correlation with barotrauma – PPR values $>= 30$ cmH_2O. In a first stage of this project it was predicted the probability of occurring barotrauma considering only the data provided by the ventilator. This study [5] shown that it is possible to predict PPR class <30 cmH_2O and $>= 30$ cmH_2O, with an accuracy between 95.52% and 98.71%. The best model was achieved using Support Vector Machines and all the variables considered in the study. However, another good model was obtained (95.52% of accuracy) using only three variables. This model showed a strong correlation among: Compliance Dynamic (CDYN), Means Airway Pressure and Pressure Peak.

2.3   INTCare

This work was carried out under the research project INTCare. INTCare is an Intelligent Decision Support System (IDSS) [6] for Intensive Care which is in constantly developing and testing. This intelligent system was deployed in the Intensive Care Unit (ICU) of Centro Hospitalar do Porto (CHP). INTCare allows a continuous patient condition monitoring and a prediction of clinical events using DM. One of the most recent goals addressed is the identification of patients who may have barotrauma.

2.4   Data Mining

DM corresponds to the process of using technical features of artificial intelligence, statistical calculations and mathematical metrics able to extract information and useful knowledge. The knowledge discovery may represent various forms, business rules, similarities, patterns or correlations [7].

This work is mainly focused on the development and analysis of clusters. This is a grouping process based on observing the similarity or interconnection density. This process aims to discover data groups according to the distributions of the attributes that make up the dataset [8]. To develop and assess the application of clustering algorithms in the barotrauma dataset, the statistical system R was choosen.

3   Knowledge Discovering Process

3.1   Business Understanding

The main goal is to use ventilation data in order to identify groups of objects that belong to the same class, i.e. group sets of similar objects in a single set and dissimilar in different sets. The data used to conduct this study were collected in the ICU of CHP. The clusters were supported only for data monitored by ventilators; the values used were numeric and were from discrete quantitative type.

3.2   Data Understanding

The initial data sample contained several records without patient identification (PID). This happens because sometimes the patients are admitted for a few hours in the ICU but are not assigned to an Electronic Health Record (EHR). These records were discarded for this study. The sample used was collected from the ventilators and comprises a period between 01.09.2014 and 10.12.2014 and a total of 33023 records. Each record contains fourteen fields: CDYN – (F_1); CSTAT – (F_2); FIO2 – (F_3); Flow – (F_4); RR – (F_5); PEEP – (F_6); PMVA – (F_7);

Plauteau pressure – (F_8); Peak pressure – (F_9); RDYN – (F_10); RSTAT – (F_11); Volume EXP – (F_12); Volume INS – (F_13); Volume Minute – (F_14).

The coefficient of variation shows that the distributions are heterogeneous for all the attributes since the results obtained are higher than 20%. This measure corresponds to the dispersion ratio between the standard deviation and the average.

3.3   Data Preparation

Data transformations were necessary to perform data segmentation using clustering techniques based in resource partition methods. Because these techniques do not handle null values and qualitative data, two operations have been performed:

- Firstly, records having at least one null value were eliminated;
- Then, the records containing qualitative values were eliminated.

3.4   Modelling

The algorithms k-means and k-medoids were used to create the cluster. The choice is justified by the principle of partition method and the difference in their sensitivity to find outliers.

K-means algorithm is sensitive to outliers, because the objects are far from the majority, which can significantly influence the average value of the cluster. This effect is particularly exacerbated by the use of the squared error function [9].

On the other hand, K-medoids instead of using the value of a cluster object as a referencing point takes on real objects and represents the clusters, creates an object for each cluster. The partitioning method is then performed based on the principle of adding the differences between each p (intra-clusters distance). It is representing an object (dataset partition) [9] where the p is always >= 0. The K-medoids algorithm is similar to K-means, except that the centroids must belong to a set of grouped data [10]. Some configurations were atempted for each one of the algorithms. In the k-means algorithm the value K (cluster number) varies between 2 and 10. In order to obtain the appropriate number of K it was used the sum of squared error (SSE). Each dataset was executed 10 times.

The model M_n belongs to an approach A and it is composed by the fields F, a type of variable TV and an Algorithm AG:

$A_f = \{Discription\ (Clustering)_1\}$
$F_l = \{F\_1_1, F\_2_2, F\_3_3, F\_4_4, F\_5_5, F\_6_6, V\_7_7, F\_8_8, F\_9_9, F\_10_{10}, F\_11_{11}$
$\quad\quad , F\_12_{12}, F\_13_{13}, F\_14_{14}\}$
$TV_x = \{Qualitative\ variables\ ordinal_1\}$
$AG_y = \{K - means_1, K - medoids(PAM)_2\}$

Being this study related with Barotrauma and Plateau Pressure, all the models included the variable $F_{\{8\}}$. Some of the clusters induced are composed by the group of variables defined in the first approach.

3.5   Evaluation

This is the last phase of the study. It focuses mainly on the analysis of the results presented through the implementation of clustering algorithms (K-means and PAM). The evaluation of the induced models was made by using the Davies-Bouldin Index. The models which presented most satisfactory results were those obtained by means of the K-means algorithm. In general, some models presented good results, however the models did not achieve optimal results (index near 0). Table 2 presents the best models and the correspondent results.

Table 1. Models for clustering

Model	Fields	Number of Clusters	Algorithm	Davies-Boldin Index
M_1	$F_{\{1,2,7,8,10\}}$	2	AG_1	0.82
M_2	$F_{\{1,2,7,8,10\}}$	5	AG_2	0.86
M_3	$F_{\{1,7,8,10\}}$	2	AG_1	0.64
M_4	$F_{\{1,7,8,10\}}$	6	AG_2	1.17

The M_3 model shown to be the most capable in designing a clusters with better distances. Davies-Bouldin Index tends to $+\infty$ however M_3 model has an index of 0.64. This is not the optimal value, but it is the most satisfactory because it is closest to 0. Figure 1 presents M_3 results.

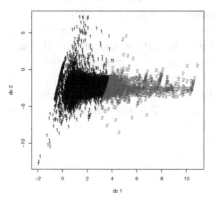

Fig. 1. Clusters of model M_3

Table 3 presents the minimum, maximum, average, standard deviation and coefficient of variation of each variable used to host the clusters in M_3.

Table 2. Distributions for each cluster

Clusters	Fields	Min	Max	Average	StDev	CoeVariation
Cluster 1 (30017 rows)	$F_{\{1\}}$	0	73	31.45	16.55	52.63%
	$F_{\{7\}}$	0	38	11.94	2.71	22.73%
	$F_{\{10\}}$	0	100	14.18	7.26	51.2%
	$F_{\{8\}}$	0	99	22.04	6.57	29.82%
Cluster 2 (3006 rows)	$F_{\{1\}}$	73	200	114.72	32.62	28.43%
	$F_{\{7\}}$	3.9	22	11.00	1.68	15.3%
	$F_{\{10\}}$	1.4	46	13.22	5.27	39.88%
	$F_{\{8\}}$	0.2	84	17.88	6.24	34.87%

4   Conclusion

This study identified a set of variables that have a great similarity. These variables are related with Plateau Pressure - variable with greater influence in the occurrence of barotrauma. The better result was achieved with the model M_3 obtaining a Davies-Bouldin Index of 0.64, a value near to the optimum value (0).

It should be noted that most of the variables used presented some dispersion however in one of the clusters the higher dispersion value is quite acceptable: 19.42. This result was obtained with the implementation of the K-means algorithm. The CDYN is one of the variables that most influences the clustering, demonstrating a strong relationship with the PPR and Barotrauma. From the results shown in Table 3 it can be noted that the best corresponding field is CDYN, presenting only a few intersecting values (minimum and maximum). This means that Cluster 1 has CDYN values ranging between [0; 73] and Cluster 2 has CDYN values between [73;200]. The remaining fields used have only few interceptions. Finally, this study demonstrated the feasibility of creating clusters using only data monitored by ventilators and analyzing similar populations. These results motivate further studies in order to induce more adjusted models reliable for classification and clustering at the same time.

Acknowledgements. This work has been supported by FCT - Fundação para a Ciência e Tecnologia within the Project Scope UID/CEC/00319/2013 and the contract PTDC/EEI-SII/1302/2012 (INTCare II).

References

1. Koh, H., Tan, G.: Data mining applications in healthcare. J. Healthc. Inf. Manag. **19**(2), 64–72 (2005)
2. Anzueto, A., Frutos-Vivar, F., Esteban, A., Alía, I., Brochard, L., Stewart, T., Benito, S., Tobin, M.J., Elizalde, J., Palizas, F., David, C.M., Pimentel, J., González, M., Soto, L., D'Empaire, G., Pelosi, P.: Incidence, risk factors and outcome of barotrauma in mechanically ventilated patients. Intensive Care Med. **30**(4), 612–619 (2004)
3. Al-Rawas, N., Banner, M.J., Euliano, N.R., Tams, C.G., Brown, J., Martin, A.D., Gabrielli, A.: Expiratory time constant for determinations of plateau pressure, respiratory system compliance, and total resistance. Crit Care **17**(1), R23 (2013)
4. Boussarsar, M., Thierry, G., Jaber, S., Roudot-Thoraval, F., Lemaire, F., Brochard, L.: Relationship between ventilatory settings and barotrauma in the acute respiratory distress syndrome. Intensive Care Med. **28**(4), 406–413 (2002)
5. Oliveira, S., Portela, F., Santos, M.F., Machado, J., Abelha, A., Silva, A., Rua, F.: Predicting plateau pressure in intensive medicine for ventilated patients. In: Rocha, A., Correia, A.M., Costanzo, S., Reis, L.P. (eds.) New Contributions in Information Systems and Technologies, Advances in Intelligent Systems and Computing 354. AISC, vol. 354, pp. 179–188. Springer, Heidelberg (2015)
6. Portela, F., Santos, M.F., Machado, J., Abelha, A., Silva, A., Rua, F.: Pervasive and intelligent decision support in intensive medicine – the complete picture. In: Bursa, M., Khuri, S., Renda, M. (eds.) ITBAM 2014. LNCS, vol. 8649, pp. 87–102. Springer, Heidelberg (2014)
7. Turban, E., Sharda, R., Delen, D.: Decision Support and Business Intelligence Systems. 9a Edição. Prentice Hall (2011)
8. Anderson, R.K.: Visual Data Mining: The VisMiner Approach, Chichester, West Sussex, U.K., 1st edn. Wiley, Hoboken (2012)
9. Han, J., Kamber, M., Pei, J.: Data Mining Concepts and Techniques. 3a Edição. Morgan Kaufmann (2012)
10. Xindong, W., Vipin, K.: The Top Ten Algorithms in Data Mining. CRC Press–Taylor & Francis Group (2009)

Clinical Decision Support for Active and Healthy Ageing: An Intelligent Monitoring Approach of Daily Living Activities

Antonis S. Billis[1], Nikos Katzouris[2,3], Alexander Artikis[2], and Panagiotis D. Bamidis[1(✉)]

[1] Medical Physics Laboratory, Medical School, Faculty of Health Sciences, Aristotle University of Thessaloniki, Thessaloniki, Greece
{ampillis,bamidis}@med.auth.gr
[2] Institute of Informatics and Telecommunications, National Center for Scientific Research 'Demokritos', Aghia Paraskevi, Greece
{nkatz,a.artikis}@iit.demokritos.gr
[3] Department of Informatics and Telecommunications, National Kapodistrian University of Athens, Athens, Greece

Abstract. Decision support concepts such as context awareness and trend analysis are employed in a sensor-enabled environment for monitoring Activities of Daily Living and mobility patterns. Probabilistic Event Calculus is employed for the former; statistical process control techniques are applied for the latter case. The system is tested with real senior users within a lab as well as their home settings. Accumulated results show that the implementation of the two separate components, i.e. Sensor Data Fusion and Decision Support System, works adequately well. Future work suggests ways to combine both components so that more accurate inference results are achieved.

Keywords: Decision support · Unobtrusiveness · Sensors · Context awareness · Trend analysis · Statistical process control · Event calculus

1 Introduction

Europe's ageing population is drastically increasing in numbers [1], thereby bearing serious health warnings such as dementias or mental health disorders such as depression [2]. Hence, the immediate need for early and accurate diagnoses becomes apparent. Ambient-Assisted Living (AAL) technologies can provide support to this end [3]. However, most of these research efforts fail to either become easily acceptable by end-users or be useful at a practice level; the obtrusive nature of the utilized technologies invading the daily life of elder adults is probably the one to be blamed [4]. To this end, the approach followed in this paper, which is also aligned with the major objective of the USEFIL project [5], is to apply remote monitoring techniques within an unobtrusive sensor-enabled intelligent monitoring system. The first part of the intelligent monitoring system is an event-based sensor data fusion (SDF) module,

© Springer International Publishing Switzerland 2015
F. Pereira et al. (Eds.) EPIA 2015, LNAI 9273, pp. 128–133, 2015.
DOI: 10.1007/978-3-319-23485-4_14

while the second part consists of two major components, i) the trend analysis component and ii) a higher level formal representation model based on Fuzzy Cognitive Maps (FCMs) [5]. The aim of this paper is to present a feasibility study of SDF and Trend Analysis components in real life settings and evaluate their capability of intelligent health monitoring.

2 Materials and Methods

2.1 The USEFIL Platform

The USEFIL intelligent monitoring system (cf. Fig. 1) comprises of three different layers of processing. Low cost sensors provide unobtrusive low level information, e.g. activity, mood and physiological signs. Event fusion module is the intermediary layer, which combines multimodal low level events and translates them into contextual information. A server-side Decision Support System consumes time stamped contextual information and projects them in the long run, producing alerts, upon recognition of data abnormalities or health deteriorating trends. This information is channeled to seniors or their carers via user-friendly interfaces.

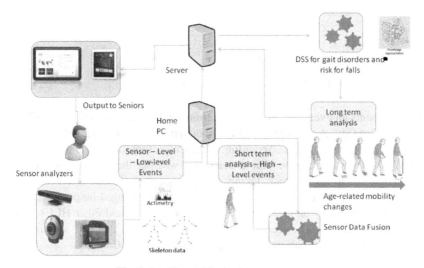

Fig. 1. Intelligent Monitoring components

2.2 Sensor Data Fusion

The role of the data fusion component in USEFIL is to interpret sensor data into a semantic representation of the user's status. Its tasks range from contextualization of sensor measurements to characterization of functional ability. We employ a Complex Event Recognition methodology [7], which allows to combine heterogeneous data sources by means of event hierarchies. In our setting, input consists of a stream of *low-level events (LLEs)* – such as time-stamped sensor data, and output consists of recognized complex, or *high-level events (HLEs),* that is, spatio-temporal combinations of simpler events and domain knowledge. Our approach is based on the Event Calculus [8], a first-order formalism for reasoning about events and their effects. To address uncertainty, we ported the Event Calculus in the ProbLog language [10], as in [9]. ProbLog is an extention of Prolog, where inference has a robust probabilistic semantics.

Constructing patterns (rules) for the detection of an HLE, amounts to specifying its dependencies with LLEs and other HLEs. As an example, we use the case study of *Barthel-scoring* of Activities of Daily Living (ADL), adapted from [12]. ADL refers to fundamental self-care activities, while the Barthel Index [11] is considered as the "golden standard" for assessing functional ability in ADL. The *Transfer* ADL refers to the ability of a person to sit down or get up from a bed or chair. The corresponding scores in the Barthel Index are evaluations of the performance in this task, based on the ability to perform and the amount of help needed.

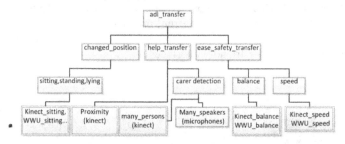

Fig. 2. An event hierarchy for the *Transfer* ADL

Fig. 2 presents an event hierarchy for the *transfer* ADL, developed in USEFIL. The leaves in the tree structure represent LLEs, obtained from sensor measurements, while each node represents an HLE. According to this representation, to Barthel-score the *transfer* ADL (root node), one should determine whether the user changed position, while receiving help for this task, taking into account the ease and safety with which the user performs. Each of these indicators (position change, help offered, ease-safety) is represented by an HLE, defined in terms of LLEs and other HLEs in lower levels of the hierarchy. The reader is referred to [12] for a detailed account of the implementation of such a hierarchy in the probabilistic Event Calculus.

2.3   Decision Support Module

Health trends identification is based on statistical process control principles. Each parameter may be modeled as a random process with a time-varying mean value and standard deviation. In this work we have followed two steps, namely the baseline extraction and the identification of acute events.

A personalized baseline profile is computed through time-series analysis and statistical process control concepts. More specifically, each variable under investigation (e.g. walking speed) is modeled as a time-series, random process with a time-varying mean value and standard deviation. The computations involved are the following:

Time-series observations are divided into n overlapping windows. The mean ($\overline{x}$) and the standard deviation ($\overline{r}$) of each time window are computed. Then, the mean value and the standard deviation of the entire process are averaged based on the individual runs:

$$\hat{x} = mean\{\overline{x}\} \quad \hat{r} = mean\{\overline{r}\} \tag{1}$$

The baseline profile is also consisted of a confidence interval for both the process mean value and the standard deviation. These intervals are defined by the following limits:

$$\lim_{low} = \hat{x} - \frac{3\hat{r}}{\sqrt{n}} \quad \lim_{upper} = \hat{x} + \frac{3\hat{r}}{\sqrt{n}} \tag{2}$$

Ongoing monitoring of the process under consideration is facilitated through the characterization of a further follow-up period based on comparison against the control limits of the baseline process. Single runs that are out of the control limits are considered as acute events.

3   Data Collection

In the process of system integration and pilot setup, an e-home like environment was established serving as an Active & Healthy Aging (AHA) Living Lab (see Fig. 3). A total of five (5) senior women aged 65+ (mean 74.6±3.85 years) were recruited. All users provided voluntary participation forms to denote that they chose to participate to this trial voluntarily after being informed of the requirements of their participation. The ability of independent living was assessed by the Barthel index. The real testing and use of the environment took place for several days. Seniors executed several activities of everyday life in a free-form manner, meaning that they were left to perform activities without strict execution orders.

Fig. 3. Performance of directed activities, interaction with the system

Apart from the lab environment, the system was also installed in home of lone-living seniors for several days lasting from one to three months. Four (4) elderly women aged 75.3±4.1 years provided their informed consent for their participation in the home study. Recordings over several days in these senior apartments measured, among others, gait patterns, emotional fluctuations and clinical parameters.

4   Results

4.1   Short-Term Monitoring – Scoring of ADLs

In order to evaluate the SDF module, the Transfer ADL was extracted for each senior. Carers examined seniors and assessed all of them as totally independent, with a

Barthel score equal to "3", which is the ground truth for all cases. Therefore, an overall confusion matrix for all five seniors is built in Table 1. As shown, SDF several times scores seniors as needing help with the Transfer Activity (scores "1" and "2"), although they are totally independent. This paradox can be attributed to the presence of a facilitator during the monitoring sessions.

4.2   Long-Term Monitoring – Gait Trends

Data from an initial period of two to four weeks were used so as to calculate the baseline for each elderly participant. The rest of the period was used as a "follow up period", where the actual monitoring was examined. Walking speed as measured by the Kinect sensor was used as the gait parameter to monitor in the long run. Fig. 4 illustrates the baseline process formation of a senior suffering from mobility problems due to osteoporosis. During the monitoring period there are significant deviations from the baseline process (42.4% of days were out of control). This means that older woman's walking speed decreases with respect to her baseline period. Walking speed levels decrease might correlate to early health risk signs, such as falls.

Table 1. Confusion matrix of ADL Transfer scoring for all 5 participants

		Predicted class		
		ADL TransferScore1	ADL TransferScore2	ADL TransferScore3
Actual Class	ADL TransferScore3	11	7095	6358

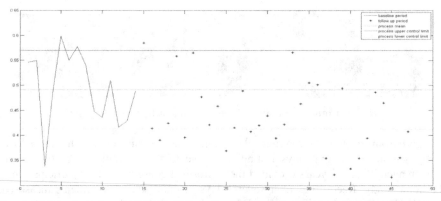

Fig. 4. Walking speed control chart. Yellow line: lower control limit, Red line: upper control limit, Blue continuous line: baseline period, Dots: follow up days.

5   Discussion

In this paper, mechanisms towards a truly intelligent and unobtrusive monitoring system for active and healthy aging were demonstrated together with a sample of the first series of results. Short-term context awareness was tested with the ADL scenario, while long-term trend analysis with the gait patterns scenario. In the first case, there were many false positives, due to the challenging, unconstrained nature of the experiment. Personalized thresholds would help SDF algorithm to avoid scoring Barthel equal to "1". Trend analysis' baseline extraction and process control limits would possibly refine latter inference results. On the other hand, long term analysis, may benefit by the SDF output, since outliers found to be out of control, could possibly be annotated as logical "noise" through context awareness. This way pathological values may be interpreted as normal based on a-priori knowledge of the context.

This whole notion is remarkably appealing, as it could lead to potential applications where the synergy between the short-term component of the SDF and the long-term Trend analysis component may prove pivotal. Further data collection from home environments, will prove pivotal upon integrating successfully the two components.

Acknowledgements. This research was partially funded by the European Union's Seventh Framework Programme (FP7/2007-2013) under grant agreement no 288532. (www.usefil.eu). The final part of this work was supported by the business exploitation scheme of LLM, namely, LLM Care which is a self-funded initiative at the Aristotle University of Thessaloniki (www. llmcare.gr). A.S. Billis also holds a scholarship from Fanourakis Foundation (http://www. fanourakisfoundation.org/).

References

1. Lutz, W., O'Neill, B.C., Scherbov, S.: Europe's population at a turning point. Science **28**(299), 1991–1992 (2003)
2. Murrell, S.A., Himmelfarb, S., Wright, K.: Prevalence of depression and its correlates in older adults. Am. J. Epidimiology **117**(2), 173–185 (1983)
3. Kleinberger, T., Becker, M., Ras, E., Holzinger, A., Müller, P.: Ambient intelligence in assisted living: enable elderly people to handle future interfaces. In: Stephanidis, C. (ed.) UAHCI 2007 (Part II). LNCS, vol. 4555, pp. 103–112. Springer, Heidelberg (2007)
4. Wild, K., Boise, L., Lundell, J., Foucek, A.: Unobtrusive in-home monitoring of cognitive and physical health: Reactions and perceptions of Older Adults. Applied Gerontology **27**(2), 181–200 (2008)
5. https://www.usefil.eu/. Retrieved from web at 29/05/2015
6. Billis, A.S., Papageorgiou, E.I., Frantzidis, C.A., Tsatali, M.S., Tsolaki, A.C., Bamidis, P.D.: A Decision-Support Framework for Promoting Independent Living and Ageing Well. IEEE J. Biomed. Heal. Informatics. **19**, 199–209 (2015)
7. Opher Etzion and Peter Niblett. Event processing in action. Manning Publications Co. (2010)
8. Kowalski, R., Sergot, M.: A logic-based calculus of events. In: Foundations of Knowledge Base Management, pp. 23–55. Springer (1989)
9. Skarlatidis, A., Artikis, A., Filippou, J., Paliouras, G.: A probabilistic logic programming event calculus. Journal of Theory and Practice of Logic Programming (TPLP) (2014)
10. Kimmig, A., Demoen, B., De Raedt, L., Santos Costa, V., Rocha, R.: On the implementation of the probabilistic logic programming language ProbLog. In: de la Banda, M.G., Pontelli, E. (eds.) Theory and Practice of Logic Programming, vol. 11, pp. 235–262 (2011)
11. Collin, C., Wade, D.T., Davies, S., Horne, V.: The barthel ADL index: a reliability study. Disability & Rehabilitation **10**(2), 61–63 (1988)
12. Katzouris, N., Artikis, A., Paliouras, G.: Event recognition for unobtrusive assisted living. In: Likas, A., Blekas, K., Kalles, D. (eds.) SETN 2014. LNCS, vol. 8445, pp. 475–488. Springer, Heidelberg (2014)

Discovering Interesting Trends in Real Medical Data: A Study in Diabetic Retinopathy

Vassiliki Somaraki[1,2]([✉]), Mauro Vallati[1], and Thomas Leo McCluskey[1]

[1] School of Computing and Engineering, University of Huddersfield, Huddersfield, UK
{v.somaraki,m.vallati,t.l.mccluskey}@hud.ac.uk
[2] St. Paul's Eye Unit, Royal Liverpool University Hospital, Liverpool L7 8XP, UK

Abstract. In this work we present SOMA: a Trend Mining framework, based on longitudinal data analysis, that is able to measure the interestingness of the produced trends in large noisy medical databases. Medical longitudinal data typically plots the progress of some medical condition, thus implicitly contains a large number of trends. The approach has been evaluated on a large collection of medical records, forming part of the diabetic retinopathy screening programme at the Royal Liverpool University Hospital, UK.

1 Introduction

Knowledge discovery is the process of automatically analysing large amount volumes of data searching for patterns that can be considered as knowledge about the data [2]. In large real-world datasets, it is possible to discover a large number of rules and relations, but it may be difficult for the end user to identify the interesting ones. *Trend mining* deals with the process of discovering hidden, but noteworthy, trends in a large collection of temporal patterns. The number of trends that may occur –especially in large medical databases– is huge. Therefore, a methodology to distinguish interesting trends is imperative.

In this paper we report on a framework (SOMA) that is capable of performing trend mining in large databases and evaluating the interestingness of the produced trends. It uses a three-step approach that: (i) exploits logic rules for cleaning noisy data; (ii) mines the data and recognises trends, and (iii) evaluates their interestingness. This work encompasses a previous preliminary study of Somaraki et al. [10]. The temporal patterns of interest, in the context of this work, are frequent patterns that feature some prescribed change in their frequency between two or more "time stamps". A time stamp is the sequential patient consultation event number, i.e., the date in which some medical features of the patient have been checked and registered. We tested SOMA on the diabetic retinopathy screening data collected by The Royal Liverpool University Hospital, UK, which is a major referral centre for patients with Diabetic Retinopathy (DR). DR is a critical complication of diabetes, and it is one of the most common cause of blindness in working age people in the United Kingdom.[1] It is a chronic

---

[1] http://diabeticeye.screening.nhs.uk/diabetic-retinopathy

© Springer International Publishing Switzerland 2015
F. Pereira et al. (Eds.) EPIA 2015, LNAI 9273, pp. 134–140, 2015.
DOI: 10.1007/978-3-319-23485-4_15

disease affecting patients with Diabetes Mellitus, and causes significant damage to the retina.

The contribution of this paper is twofold. First, we provide an automatic approach for evaluating the interestingness of temporal trends. Second, we test the ability of the SOMA framework in automatically extracting interesting temporal trends from a large and complex medical database. The extracted interesting trends have been checked by clinicians, who confirmed their interestingness and potential utility for the early diagnosis of diabetic retinopathy.

2 Background

Association Rule Mining (ARM) is a popular, and well researched, category of data mining for discovering interesting relations between variables in large databases. In (ARM) [1], an observation or transaction (e.g. the record of a clinical consultation) is represented as a set of items where an item is an attribute - value pair. Given a set of transactions, we can informally define an *Association rule* (AR) as a rule in the form of $X \implies Y$. The *support* of an AR is defined as the percentage of transactions that contain both X and Y. It can also be defined as the probability $P(X \cap Y)$. Finally, the *confidence* of an AR is defined as the ratio between the number of transactions that contain $X \cup Y$ and the number of transactions that contain X.

ARM procedures contain two stages: (i) frequent item set identification, and (ii) AR generation. Piatetsky-Shapiro [8] defined ARM as a method for the description, analysis and presentation of ARs, discovered in databases using different measures of interestingness. ARM is concerned with the discovery, in tabular databases, of rules that satisfy defined threshold requirements. Of these requirements, the most fundamental one is concerned with the support (frequency) of the item sets used to make up the ARs: a rule is applicable only if the relationship it describes occurs sufficiently often in the data.

The topic of mining interesting trends has recently grown due to the availability of large and complex databases. Liu et al. [6] introduced the concept of general impressions. General impressions are if-then clauses that describe the relation between a condition variable and a class value, and reflect the users knowledge of the domain. Based on the users knowledge, the rules can be classified as: unexpected (previously unknown), confirming (confirms previous knowledge) or actionable (can be fruitfully exploited).

Geng and Hamillton [3] used the term "interesting measures" to facilitate a general approach to automatically identifying interesting patterns. They used this term in three ways, or roles, to use their terminology. First, the measures can be used to prune uninteresting patterns during the mining process. Second, measures can be used to rank the patterns according to their interestingness, and finally they are used during post-processing to select interesting patterns.

3   The SOMA Framework

The SOMA framework receives raw data from a range of repositories. Firstly, data are pre-processed: data cleansing, creation of data timestamps, selection of subsets for analysis and the application of logic rules take place. Then, by analysing pre-processed information, frequent patterns are generated and trend mining is performed in order to identify interesting trends.

Pre-processing

Medical data values are usually from continuous domains. The presence of continuous domains makes it difficult to apply the frequent item set techniques. For this reason, in SOMA pre-processing discretisation is applied. Discretisation allocates continuous values into a limited number of intervals, called bands [5, 7]. Bands can be either defined by domain experts –in our analysis, physicians– or automatically identified.

Real data, particularly from medical repositories, are usually variegated (text, discrete and continuous values) and noisy. In order to deal with missing data, we adopted logic rules, based on expert knowledge. Logic rules are a sequence of if-then-else cases that consider the values of various related fields for identifying missing ones. Given the considered context, clinicians considered logic rules as the most appropriate method to fill in the data.

Finally, pre-processing performs the task of creating time stamped datasets. Data from different sources are usually not collected with the same frequency, thus is it complex to automatically define a clear association between data and time stamps. In fact, this process is domain-specific, and must be addressed by designing domain-specific solutions crafted by domain experts.

Processing

The main step of the SOMA framework consists of two processes: (i) association rule mining (ARM), and (ii) trend generation and categorisation.

The ARM process is repeated for every time stamp, and performs the following tasks: (i) identify rules; (ii) evaluate interestingess of rules, and (iii) filter rules accordingly to their evaluation. Rules are created in the usual AR form: $X \implies Y$. On frequent item sets, characteristics that measure interestingness are computed. Finally, rules are filtered; only those which are both frequent and interesting are kept. The corresponding threshold are determined by users, according to the amount of knowledge they want to extract.

In applications involving large medical datasets, efficient processing is a fundamental factor. Since the ARM process is repeated for every time step, the technique must be capable of efficiently identifying rules. Given that, we decided to exploit matrix algorithm principles [11] to efficiently identify frequent rules with acceptable confidence, which only requires one pass through the whole dataset.

The subsequent mining of trends is implemented by considering relations on the vectors of support count, in order to show how the support count for each rule changes over time. We considered the following well-known relations: Increasing, decreasing, constant, jumping and disappearing. For further information, the interested reader is referred to [9].

Measuring Interestingness
Strong rules based on support and confidence of association rules are not always interesting. The pitfall of confidence can be traced to the fact that its definition ignores the support of the right-hand (Y) part of the rule. To determine which trend is interesting or not, we exploited measures introduced by Han et al. [4]. They propose the *lift* correlation and a set of pattern evaluation measures, in order to mathematically determine interestingness. Lift is calculated as : $Lift(X,Y) = P(X \cup Y)/P(X) \cdot P(Y)$. Where $Lift = 1$, it indicates that X and Y are independent; a value greater than 1 indicates a positive correlation, a $Lift$ value smaller than 1 refers to negative correlation. The pattern evaluation measures proposed in [4] are: all_confidence, max_confidence, Kulczynski and cosine. Each of them has the following property: its values is only influenced by the supports of X, and Y, but not by the total number of transactions. Their values range from 0 to 1 and the higher their value the closer the relationship between X and Y. We selected them since they analyse different aspects of data.

The values of those measures are calculated during the trend mining process from SOMA. For each trend, a matrix with dimension $5 \times T$, where T is the number of time stamps is built. There is one line for each measure and in the first column the values for the first time are stored, in the second column the values for the second time stamp and so on. The values are in binary form following that if the value of the measure is equal or greater to its threshold then the value at the matrix is 1 and 0 otherwise. The maximum score for each trend is $5 \times T$ which is the sum of the elements of the matrix if a trend has values equal or greater than the threshold for all measures for all time stamps. The final score is transformed to a percentage. Thus, according to the threshold provided by the users through a process of empirical validation, rules are deemed to be interesting. Further details can be found in [9].

4 Experimental Analysis

In this work we considered the data of the Saint Paul's Eye Clinic of the Royal Liverpool University Hospital, UK. The data (anonymised in order to guarantee patients' privacy) was collected from a warehouse with 22,000 patients, 150,000 visits to the hospital, with attributes including demographic details, visual acuity data, photographic grading results, data from biomicroscopy of the retina and results from biochemistry investigations. Stored information had been collected between 1991 and 2009. Data are noisy and longitudinal; they are repeatedly sampled and collected over a period of time with respect to some set of subjects. Typically, values for the same set of attributes are collected at each sample points. The sample points are not necessarily evenly spaced. Similarly, the data collection process for each subject need not necessarily be commenced at the same time.

In our experimental analysis we considered all the 1420 patients who had readings over 6 time stamps, 887 patients over 7 time stamps, and 546 patients on 8 time stamps. The number of patients decreases when time windows are larger. This is due to the fact that not all the patients are followed by the Clinic for the same amount of time. For instance, a significant amount of patients from the database had 2 visits only; clearly, this does not allow us to derive any meaningful information about general trends. The percentage of missing values is 9.67% for the test with 6 time stamps, 13.16% for the test with 7 time stamps and 18.26% for the test with 8 time stamps.

The medical experts working with us required that the experiments focused on 7 medical features that they believed to be important: age at exam, treatment of the patient, diabetes type, diabetes duration, age at diagnosis, presence of cataract and presence of DR. Features that represent time are continuous, and have been discretised in bands. In particular, age at exam and at diagnosis have been discretised in bands of approximately 10 years; duration features are discretised in 5-year bands. This was done following advice from medical experts. These medical features have some known relationships with regards to the diagnosis of diabetic retinopathy. Therefore, selecting the aforementioned features, and following medical experts' indications, allow us *to validate the SOMA framework* in the following way: if the known interesting trends are identified by SOMA, it increases our confidence that the process is finding valid and interesting knowledge. The interested reader can find more information about the validation process in [9].

For each time stamp we used as support threshold 15% and for confidence 80%. The threshold for lift is 1.5 and for the other measures is set to 0.75. The overall score threshold is set to 80% or where 24 out of the total 30 entries must be 1. Such thresholds have been identified by discussing with medical experts, and by performing some preliminary analysis on small subsets of the available data. It should be noted that different thresholds can significantly affect the set of identified interesting rules. Lower thresholds lead to larger numbers of potentially less interesting rules output, while higher thresholds results in a very small –but highly interesting– set of rules. A major influence on setting parameter values was to guarantee that the configuration would produce *already known associations and trends*, following the aforementioned validation approach.

Results

SOMA was implemented and executed in a MATLAB environment. On the considered data, six interesting medical clauses regarding DR were discovered:

1. If a diabetic patient has not developed cataract it is not likely to develop diabetic retinopathy.
2. The younger a patient is diagnosed with diabetes the more likely it is that this patient will develop diabetic retinopathy.
3. If a patient has suffered from diabetes type 2 for more than 20 years it is very likely that this patient will develop diabetic retinopathy.

Table 1. Scores, with regards to considered metrics, of the six medical clauses at different time stamps (1–6). All Conf stands for the All_confidence criteria. Max Conf indicates the Max_confidence criteria.

	Clause 1						Clause 2						Clause 3						Clause 4						Clause 5						Clause 6					
	1	2	3	4	5	6	1	2	3	4	5	6	1	2	3	4	5	6	1	2	3	4	5	6	1	2	3	4	5	6	1	2	3	4	5	6
Lift	1	1	1	1	1	1	1	1	1	1	1	1	1	1	1	1	1	1	1	1	1	1	1	1	1	1	1	1	1	1	1	1	1	1	1	1
All Conf	1	1	1	0	0	0	1	0	0	1	0	1	1	0	0	0	1	1	1	0	0	0	0	1	1	0	0	1	0	1	1	0	0	0	0	1
Max Conf	1	1	1	1	1	1	1	0	1	1	1	1	1	1	1	1	1	1	1	0	0	1	1	1	1	0	0	1	1	1	1	0	0	1	1	1
Kulc	1	1	1	1	1	0	1	0	0	1	0	0	1	1	0	1	1	1	1	1	0	0	0	1	1	1	0	1	0	1	1	1	0	0	0	1
Cosine	1	1	1	1	0	0	1	0	0	1	0	0	1	0	0	0	1	1	1	0	0	1	1	1	1	0	0	1	0	1	1	0	0	0	0	1

4. If a patient suffers from diabetes type 1 it is very likely that this patient will develop diabetic retinopathy.
5. If a patient suffers from diabetes type 2 and is on insulin treatment, this patient is likely to suffer from diabetic retinopathy.
6. If a patient suffers from diabetes type 2 and the duration of diabetes is longer than 20 years, this patient is likely to develop diabetic retinopathy.

Table 1 shows the value of the clauses, with regards to the considered criteria, per time stamp. Given a row, it is possible to assess if the corresponding criterion changed its value over time. The maximum score that a clause can get by considering 6 time stamps and 5 criteria is 30; therefore, given the threshold of 80%, at least 24 values should be 1. According to Table 1, only the first medical clause achieves an overall score which is above the threshold. Therefore, clause 1 is deemed to be the most interesting. When using 7 or 8 time stamps, the overall interestingness reduced respectively to 73% and 64%. This is possibly due to the smaller number of considered patients, and to the different impact of missing values on the different datasets. Interestingly, for every considered clause, the lift value is well above the threshold; this means that in rules in the form of $X \implies Y$, X and Y of the medical clauses are positively correlated, and there is associations between X and Y for all medical clauses. It should also be noted that the confidence of the reverse rules is below the threshold: that explains why Kulczynski and cosine and All_conf could not exceed the threshold, and indicates that there is not a strong relevance in the reverse rule. On the contrary, SOMA revealed a very good confidence also for the inverse rule of clause 1. Ophthalmologists of the Saint Paul's Eye Unit confirm that this result is very interesting, and it highlights a cause-and-effect relationship between cataract of diabetic patients and diabetic retinopathy.

It is well known that diabetes has many factors that affect its progress, and not all of them will necessarily appear in databases. However, in general, according to the clinicians of Saint Paul's Eye Unit, the first two clauses appear to provide new evidence of previously unknown relations, and are thus worth investigating further. The other 4 clauses fit with accepted thinking, and so while not being actionable knowledge, provide validation to the approach described in this work.

5   Conclusion

In this paper we described SOMA, a framework that is able to identify interesting trends in large medical databases. Our approach has been empirically evaluated on the the data of the Saint Paul's Eye Clinic of the Royal Liverpool University Hospital. SOMA is highly configurable. In order to meaningfully set available values, we involved medical experts in the process. In particular, we set parameters in order to allow SOMA to find previously known interesting trends. We used this as a heuristic to indicate that the previously unknown interesting trends identified by SOMA within the same configuration may be valuable. As clinicians confirmed, SOMA was able to identify suspected relations, and to identify previously unknown causal relations, by evaluating the interestingness of corresponding trends in the data.

Future work includes applying SOMA to other medical databases, and the investigation of techniques for visualising trends and results.

Acknowledgments. The authors would like to thank Professor Simon Harding and Professor Deborah Broadbent at St. Paul's Eye Unit of Royal Liverpool University Hospital for providing information and support.

References

1. Agrawal, R., Srikant, R., et al.: Fast algorithms for mining association rules. In: Proc. 20th Int. Conf. Very Large Data Bases, VLDB, vol. 1215, pp. 487–499 (1994)
2. Frawley, W.J., Piatetsky-Shapiro, G., Matheus, C.J.: Knowledge discovery in databases: An overview. AI Magazine **13**(3), 57 (1992)
3. Geng, L., Hamilton, H.J.: Interestingness measures for data mining: A survey. ACM Computing Surveys (CSUR) **38**(3), 9 (2006)
4. Han, J., Kamber, M., Pei, J.: Data mining: Concepts and techniques, (the morgan kaufmann series in data management systems) (2006)
5. Kotsiantis, S., Kanellopoulos, D.: Discretization techniques: A recent survey. GESTS International Transactions on Computer Science and Engineering **32**(1), 47–58 (2006)
6. Liu, B., Hsu, W., Chen, S.: Using general impressions to analyze discovered classification rules. In: KDD, pp. 31–36 (1997)
7. Liu, H., Hussain, F., Tan, C.L., Dash, M.: Discretization: An enabling technique. Data Mining and Knowledge Discovery **6**(4), 393–423 (2002)
8. Piatetsky-Shapiro, G.: Discovery, analysis and presentation of strong rules. In: Knowledge Discovery in Databases, pp. 229–238 (1991)
9. Somaraki, V.: A framework for trend mining with application to medical data. Ph.D. Thesis, University of Huddersfield (2013)
10. Somaraki, V., Broadbent, D., Coenen, F., Harding, S.: Finding temporal patterns in noisy longitudinal data: a study in diabetic retinopathy. In: Perner, P. (ed.) ICDM 2010. LNCS, vol. 6171, pp. 418–431. Springer, Heidelberg (2010)
11. Yuan, Y.B., Huang, T.Z.: A matrix algorithm for mining association rules. In: Huang, D.-S., Zhang, X.-P., Huang, G.-B. (eds.) ICIC 2005. LNCS, vol. 3644, pp. 370–379. Springer, Heidelberg (2005)

Artificial Intelligence
in Transportation Systems

A Column Generation Based Heuristic for a Bus Driver Rostering Problem

Vítor Barbosa[1(✉)], Ana Respício[2], and Filipe Alvelos[3]

[1] Escola Superior de Ciências Empresariais do Instituto Politécnico de Setúbal,
Setúbal, Portugal
vitor.barbosa@esce.ips.pt
[2] Centro de Matemática, Aplicações Fundamentais e Investigação Operacional,
Faculdade de Ciências da Universidade de Lisboa, Lisbon, Portugal
alrespicio@fc.ul.pt
[3] Departamento de Produção e Sistemas da Universidade do Minho, Braga, Portugal
falvelos@dps.uminho.pt

Abstract. The Bus Driver Rostering Problem (BDRP) aims at determining optimal work-schedules for the drivers of a bus company, covering all work duties, respecting the Labor Law and the regulation, while minimizing company costs. A new decomposition model for the BDRP was recently proposed and the problem was addressed by a metaheuristic combining column generation and an evolutionary algorithm. This paper proposes a new heuristic, which is integrated in the column generation, allowing for the generation of complete or partial rosters at each iteration, instead of generating single individual work-schedules. The new heuristic uses the dual solution of the restricted master problem to guide the order by which duties are assigned to drivers. The knowledge about the problem was used to propose a variation procedure which changes the order by which a new driver is selected for the assignment of a new duty. Sequential and random selection methods are proposed. The inclusion of the rotation process results in the generation of rosters with better distribution of work among drivers and also affects the column generation performance. Computational tests assess the proposed heuristic ability to generate good quality rosters and the impact of the distinct variation procedures is discussed.

Keywords: Rostering · Column generation · Heuristic

1 Introduction

Personnel scheduling or rostering [1] consists in defining a work-schedule for each of the workers in a company during a given period. A roster is a plan including the schedules for all workers. An individual work-schedule defines, for each day, if the worker is assigned to work or has a day-off and, in the first case, which daily duty/shift has to be performed. The rostering problem arises because the company usually has diverse duties to assign on each day, sometimes needing particular skills and, on the other hand, the Labour Law and company rules (days-off, rest time, etc.)

© Springer International Publishing Switzerland 2015
F. Pereira et al. (Eds.) EPIA 2015, LNAI 9273, pp. 143–156, 2015.
DOI: 10.1007/978-3-319-23485-4_16

restrict the blind assignment of duties to workers. Rostering is addressed in many types of business as recently surveyed in [2] and also some years ago in [3].

The Bus Drivers Rostering problem (BDRP) and most rostering problems are NP-Hard combinatorial optimization problems [4, 5], being computationally challenging to obtain optimal solutions. Many authors address rostering problems with heuristic methods which are usually faster in the achievement of good solutions comparatively to exact methods [4, 6, 7].

The BDRP occurs in the last phase of the transportation planning system, which also includes timetabling, vehicle scheduling and crew scheduling in order to know the drivers demand in each day [8]. It is concerned with the assignment of duties (set of consecutive trips and rest times defining a day of work, previously generated) to drivers, respecting the labour/contractual rules and pursuing the bus company interests in optimizing the drivers use.

Considering the BDRP model proposed in [4], a new decomposition model was proposed for the problem in [9], as well as a new metaheuristic based in the Search-Col framework [10]. In the proposed metaheuristic, column generation and an evolutionary algorithm are used to obtain valid solutions for the problem. The column generation is used to build a pool of schedules for the drivers, resulting from the subproblems' optimization (individual work-schedules), and also to get information about the quality of those schedules (considering their contribution in the optimal linear solution of the column generation).

This paper proposes the integration of a new heuristic in the column generation exact method [11]. The combination of exact and heuristic methods is not new. According to the classification proposed in [12], our combination can be included in the "integrative combinations", since the heuristic is incorporated in the normal cycle of the column generation, but it can also be classified as a sequential "collaborative combination" since the column generation helps the heuristic, and the heuristic returns new solutions to the column generation.

The new heuristic solves all the subproblems together, avoiding the multiple assignments of the duties to more than one driver, as happens when the subproblems are solved independently. The main contribution of this new heuristic is that it is capable to obtain integer and good quality solutions for the complete problem while performing column generation, without harming its performance. A secondary contribution is that the search-space composed by the solutions obtained with this heuristic is richer in complementary solutions that can be further explored with SearchCol algorithms [10].

The way the heuristic builds rosters or schedules is simple. The novelty is in using the dual solution of the restricted master problem to guide the heuristic. The dual solution is used to set the order by which the duties are selected to be assigned, as well as it is used to define the order by which the drivers are picked to test the assignment of the duty on their schedule. Some variations on the heuristic behaviour are tested where knowledge about the problem is used to obtain rosters with the overtime distributed more evenly between the drivers.

Computational tests show the impact on the column generation performance and on the integer solutions obtained by three configurations of the proposed heuristic.

In the next section the decomposition model for the BDRP is introduced. Section 3 introduces the column generation method, the improvements made by using an heuristic to solve the subproblems and the global heuristic used to solve all the subproblems together. Section 4 presents the computational tests run in a set of BDRP instances, using three configurations of the global heuristic. Section 5 provides some conclusions.

2   BDRP Model for Column Generation

The adopted model for the BDRP is an integer programming formulation adapted from the one proposed in [4]. The complete adapted compact model and the decomposition model were presented in [9]. The model is only concerned with the rostering stage, assuming that the construction of duties was previously done by joining trips and rest times to obtain complete daily duties ready to assign to drivers.

In the decomposition model, for each driver, the model considers a set of feasible schedules, represented by the columns built with subproblems' solutions. The set of all the possible valid columns can be so large, making impossible its enumeration. Therefore, only a restricted subset of valid columns are considered, leading to the formulation of a restricted master problem (RMP) of the BDRP decomposition model.

RMP Formulation:

$$Min \ \sum_{v \in V} \sum_{j \in J^v} p_j^v \lambda_j^v \tag{1}$$

Subject to:

$$\sum_{v \in V} \sum_{j \in J^v} a_{ih}^{jv} \lambda_j^v \ \geq 1 \ , i \in T_h^w, \ h = 1, \dots, 28, \tag{2}$$

$$\sum_{j \in J^v} \lambda_j^v \ = 1 \, , v \in V, \tag{3}$$

$$\lambda_j^v \in \{0,1\}, \ j \in J^v, v \in V. \tag{4}$$

Where:

λ_j^v – Binary variable associated to the schedule j of driver v, from set of drivers V;

J^v – Set of valid schedules for driver v (generated by subproblem v);

p_j^v – Cost of the schedule j obtained from the subproblem of driver v;

a_{ih}^{jv} – Assumes value 1 if duty i of day h is assigned in the schedule j of driver v;

In this model, the valid subproblem solutions are represented as columns, with cost p_j^v for the solution with index j of the subproblem v, with the assignment of duty i on day h, if $a_{ih}^{jv}=1$;

T_h^w is the set of work duties available on day h.

The objective function is to minimize the total cost of the selected schedules, the first set of constraints, the linking constraints (2), assure that all duties, in each day, are assigned to someone and the last set of constraints, the convexity constraints (3), assure that a work-schedule is selected for each driver/subproblem.

To give some context about the subproblem constraints for the next sections, we describe below the constraints included in its formulation. To see the complete model with the description of the variables and data, we recommend the reading of [9]. The constraints are the following:

— A group of constraints assures that, for each day of the rostering problem, a duty is assigned to the driver (the day-off is also represented as a duty);
— A group of constraints avoids the assignment of incompatible duties in consecutive days (if a driver works in a late duty on day h, the minimum rest time prevents the assignment of an early duty on day $h+1$). A subset of these constraints considers information from last duty assigned on the previous roster to be considered on the first day assignment;
— A group of constraints avoids the assignment of sequences of work duties that do not respect the maximum number of days without a day-off. A subset of these constraints also considers information from the last roster to force the assignment of the first day-off considering the working days on the end of the previous rostering period;
— A group of constraints forces a minimum number of days-off in each week of the rostering period and also a minimum number of days-off in a Sunday during the rostering period;
— Another group of constraints sets limits on the sum of the working time units each driver can do in each week and in all the rostering period;
— A constraint is used to apply a fixed cost whenever a driver is used (at least one work duty is assigned in the driver' schedule).

3   Column Generation Heuristic

In this section an overview of the column generation method is presented to explain how the decomposition model described in the previous section can be solved. After, some improvements to the implementation of the algorithm are detailed. Those improvements intent to reduce the computational time observed when using the standard implementation. The last subsection presents a new heuristic which is combined with the column generation to solve all the subproblems simultaneously, that obtains complete or partial rosters.

Column Generation (CG) is a well-known exact solution method used to obtain solutions to problems where the number of variables is huge compared to the number of constraints. An overview of the origins and evolution of column generation can be found in [13]. For a comprehensive description we propose the reading of [11].

Usually the CG is used to solve problems modeled by a decomposition model where an original model is decomposed in a master problem and some subproblems. Dantzig-Wolfe decomposition [14] is commonly used to obtain the new model.

In the master problem of the decomposition model, new variables represent all the possible solutions of all the subproblems. To avoid the enumeration of all those variables, a restricted master problem is considered. The CG is used to obtain the subproblems solutions that may have a contribution to the improvement of the global solution.

Generally, the CG consists in iteratively optimize the restricted master problem (RMP) to obtain an optimal linear solution (using a simplex algorithm from a solver or similar algorithm). The dual solution of the linking and convexity constraints of the RMP is used to update the objective function of the subproblems. The subproblems are solved with the new objective function values and the suproblems' solutions with negative reduced cost are added as new columns (variables) in the RMP, starting the next iteration. When no new column with negative reduced cost is found, the optimal solution is reached and the algorithm ends.

3.1   Improving Column Generation

When using the general CG a *tailing-off effect* is commonly observed. It consists in a slow approximation to the optimal solution [11]. If a high number of iterations are expected, one approach to reduce the global computational time is by reducing the time in each iteration. One option is a deviation of the normal cycle, by changing the number of subproblems solved in each iteration or deciding if all columns are added to the RMP or only the best ones. In the framework presented in [10] these configurations are allowed when running the CG algorithm.

Besides changing the normal path of the CG, a usual approach is to use efficient combinatorial algorithm or heuristics to solve the subproblems, if available, reducing considerably the optimization time. Multiple examples are found in the literature where dynamic programming [15], constraint programming [16] and heuristics [17] are used to obtain subproblem' solutions.

Considering the computational time of the CG in the optimization of the decomposition model for the BDRP presented in [9] and since multiple configurations of the CG algorithm path are already available in the framework where the algorithm is being implemented, we started using an heuristic to obtain valid solutions for the subproblems.

The heuristic used to solve the subproblems of the BDRP decomposition model is based in the decoder algorithm proposed in [4]. The objective of the heuristic is to build schedules with the highest contribution to improve the global solution.

The heuristic described in Figure 1 builds a schedule for a driver trying to assign the duties with the most negative costs (after the update of the objective function with the dual solution of the RMP) following a greedy behavior.

A duty i cannot be included in the schedule if:

— The day of duty i has already a duty assigned;
— Duty i is incompatible with the assigned duty on (day of duty i)+1 or (day of duty i)-1 considering the minimum rest time between duties;
— Assignment of duty i makes a sequence of working days (without a day-off) longer than the maximum allowed;
— Assignment of duty i exceeds the maximum of working hours allowed by week or for all the rostering period;
— Assignment of duty i makes it impossible to have the minimum number of days-off in each week of the rostering period or the minimum number of days-off on Sundays in all the rostering period.

Get dual solution from RMP optimization (π);
Update objective function of the subproblem;
Order updated costs (*costs[]*) in increasing order, keeping information from original
                position of duty *i* (*origDuty[i]*);
Build empty schedule for the rostering period size;
Initialize driver data: working time (total and week);
FOR *i*=0 to size of *costs[]*
    IF *costs[i]>0* **THEN**
        Next *i*;
    *Assign=**TestAssignment**(origDuty[i])*;
    IF *Assign*=true **THEN**
        set driver as full in the day of *origDuty[i]*;
        Update schedule: add original cost of *origDuty[i]* to schedule;
        Update driver data: add *origDuty[i]* time length to total working time and
                corresponding week working time;
FOR *d*=0 to *number of days of the rostering period*;
    IF no duty was assigned to driver on day *d* **THEN**
        Assign a day-off to driver on day *d*;
IF *number assigned duties* >0 **THEN**
    Update schedule: add fixed cost of driver use;
Return schedule;

Fig. 1. Driver Schedule Builder Heuristic Algorithm

The function *TestAssignment* used in the heuristic algorithm tests all the conditions previously enumerated, which represent the constraints of the subproblems formulation. If any of the conditions fails, the function returns *false* and only if all the conditions are verified the function returns *true*, allowing the assignment of the duty to the schedule of the driver represented by the subproblem.

Having an heuristic to obtain solutions to the subproblems, the column generation algorithm is changed to use the heuristic, since it does not replace the exact optimization solver, because the solutions of the heuristic are not optimal, only valid. The resulting algorithm is presented in Figure 2 and details the column generation using the heuristic.

DO
    Optimize RMP;
    Update subproblems objective function with current dual solution of the RMP;
    FOR EACH subproblem
        Solve using heuristic;
        Add new columns into the RMP with subproblems attractive solutions;
    IF *no new columns added* **THEN**
        FOR EACH subproblem
            Solve using exact optimization solver;
            Add new columns into the RMP with subproblems attractive solutions;
WHILE *new columns added* >0

Fig. 2. Column Generation with Subproblem Heuristic Algorithm

In the new configuration of the column generation cycle, the heuristic is used until no new columns are added from the obtained solutions. At that point, the exact optimization solver is used to obtain the optimal solutions of the subproblems and eventually add new attractive columns. In the next iteration the heuristic is tested again.

In the SearchCol++ framework, the algorithm presented in Figure 2 can have other configurations. It is possible to solve only a single subproblem in each iteration, optimize the RMP again and, in the following iteration solve the next subproblem, iterating by all the subproblems. This strategy results in less columns added to the RMP when the subproblems are returning similar solutions, allowing a faster optimization of the RMP, due to a reduced number of variables.

3.2   New Rosters Using Column Generation

Although the improvements to the column generation presented in the previous section, the objective pursued in [9] was to use subproblems' solutions to build good quality rosters by searching the best combination of schedules covering all the duties.

When the BDRP decomposition model was implemented in [9], the standard column generation spent the time set as limit for the column generation and not even the optimization of the RMP using the branch-and-bound method from a commercial solver considering the new column as binary variables was able to achieve low cost rosters.

The use of the heuristic to solve the subproblems was introduced in [18] and an improved version of an evolutionary algorithm was tested to search for valid rosters in the space of solutions resulting from the column generation using the heuristic to solve the subproblems in both configurations: solving all the subproblems in each iteration or solving only one. The improved algorithm behaved better in the search space obtained from the configuration where all the subproblems are solved in each iteration. The search space from that configuration is larger, but similar solutions are repeated for more than one driver/subproblem, allowing finding the best combination of good schedules more easily.

The results obtained in [9, 18] suggest that it is hard to find a combination of schedules that fit together covering all the duties and avoiding over-assignment (a duty assigned to more than one driver) without some additional information in the column generation.

To assure the existence of complementary schedules between each other, we now present a new heuristic that, in the column generation cycle, assigns the duties considering all the subproblems together, as a single one. The cycle does not change, however, instead of generating individual schedules one by one, a new heuristic is called to generate a feasible combination of schedules as well as the schedules *per se*. The primary purpose of solving the subproblems in an aggregated way was to assure the existence of complete or partial rosters without the over-assignment of duties whenever the column generation was stopped. If the solutions included in the initial population of the evolutionary algorithm are already valid rosters, the expected result of the evolution is a better roster.

The heuristic presented in Figure 3 is able to build rosters by testing the assignment of each of the available duties in the schedules of free drivers. Since in each iteration of the column generation a new dual solution is used to update the costs of the duties in the subproblems, the order in which the duties are assigned may vary from iteration to iteration. The objective is that the dual solution of the RMP can guide the generation of distinct, and valid, rosters through the iterations.

When using the aggregated heuristic in the column generation algorithm in Figure 2, the cycle solving the subproblems is replaced by a single call to the new heuristic, which returns schedules for all subproblems/drivers. The exact solver continues to be used when no new attractive columns are built from the heuristic solutions.

Get dual solution from RMP optimization (π);
Order duties (*duties[]*) in ascending order of the dual solution value of the linking constraints, keeping information from original position of duty i (*origDuty[i]*);
Order drivers (*drivers[]*) in ascending order of the dual solution value of the convexity constraints;
Build an empty schedule for the rostering period size to each of the available drivers (subproblems);
Initialize drivers data: working time (total and week);
FOR i=0 to size of *duties[]*
    FOR v=0 to *number of drivers*
        Select schedule of *driver[v]*
        Assign=**TestAssignment**(*origDuty[i], schedule [driver[v]]*);
        IF Assign **THEN**
            Set driver v as full in the day of *origDuty[i]*;
            Update schedule: add original cost of *origDuty[i]* to *driver[v]'* schedule;
            Update driver v data: add *origDuty[i]* time length to total working time and corresponding week ;
            EXIT FOR
FOR v=0 to *number of drivers*
    FOR d=0 to *number of days of the rostering period*;
        IF no duty was assigned to driver v on day d **THEN**
            Assign a day-off to driver v on day d;
    IF number assigned duties to driver $v >0$ **THEN**
        Update driver v schedule: add fixed cost of driver use;
Return *schedule[]*;

Fig. 3. Roster Builder Heuristic Algorithm

The BDRP model defines a cost to each unit of time of overtime which may be different to all drivers. However, in our test instances, the drivers are split in a limited number of categories. All the drivers in the same category have the same cost for the overtime labor. This means that we still want to assign first the duties with bigger overtime to the drivers from the category with lower cost of overtime, if possible. However, we want to distribute them among all, avoiding the schedules with extra days-off because of a large concentration of duties with overtime.

Although the ability of the Roster Builder Heuristic to generate valid and distinct rosters, preliminary tests showed that the schedules of the first drivers were filled with the duties with higher overtime. Even if we want to assign the duties with higher overtime to drivers with lower salary, which are the first group in the set of all drivers, if the assignment starts always from the same driver, his/her schedule will be filled with the duties with larger overtime, resulting in an unbalanced work distribution.

Given the existence of different drivers' categories, concerning the value paid by overtime labor, drivers of the same category are grouped and the dual solution values of the convexity constraints are used to order them inside each group.

To assure that when the dual values of the convexity constraints do not lead to the desired diversity in the order of the driver inside each group, we added an additional

---

Get dual solution from RMP optimization (π);
Order duties (*duties[]*) in ascending order of the dual solution value of the linking
   constraints, keeping information from original position of duty *i* (*origDuty[i]*);
Split drivers in groups with the same category of salary;
Order drivers inside each group according to the dual solution value of the corresponding convexity constraint;
Build an empty schedule for each of the available drivers (subproblems);
Initialize drivers data: working time (total and week);
FOR *i*=0 to size of *duties []*
   FOR *g*=0 to size of *groups of drivers*
      Select starting driver position according to configuration *r*= (0 or 1 or random);
      FOR *j*=0 to *r*
      Rotate drivers inside group (remove from the begin and add to the end);
      FOR *v*=0 to *number of drivers in group g*
      Select schedule of driver *v*
      Assign=**TestAssignment**(*origDuty[i]*, *schedule[v]*);
      IF Assign **THEN**
         Set driver *v* as full in the day of *origDuty[i]*;
         Update schedule: add original cost of *origDuty[i]* to *driver[v]'* schedule;
         Update driver *v* data: add *origDuty[i]* time length to total working time and corresponding week;
      EXIT FOR
   IF Assign **THEN EXIT FOR**
FOR *v*=0 to *number of drivers*
   FOR *d*=0 to *number of days of the rostering period*;
      IF no duty was assigned to driver *v* on day *d* **THEN**
         Assign a day-off to driver *v* on day *d*;
   IF *number assigned duties to driver v* >0 **THEN**
      Update driver *v* schedule: add fixed cost of driver use;
Return *schedule[]*;

Fig. 4. Roster Builder Heuristic with drivers' rotation Algorithm

procedure to select the first driver inside each ordered group. We started considering each group of drivers as a circular array. After that, two configuration were prepared to define how a driver is selected when a new duty needs to be assigned.

By default, when a new duty is selected for assignment, the driver to select is the one in the position 0 of the first group. We developed two configurations of the Roster Builder Heuristic with drivers' rotation, namely the sequential and the random configurations. In both, after the assignment of a duty, we rotate the drivers inside the group, the first is removed and inserted at the end. In the sequential configuration, the rotation is of a single position, and in the random configuration, the number of positions rotated is randomly selected between one and the number of drivers in the group minus one, to avoid a complete rotation to the same position.

The inclusion of the rotation leads to a better distribution of the duties with overtime among the group drivers. Figure 4 presents the algorithm of the roster builder heuristic with drivers' rotation. The changes are: the inclusion of the groups of drivers, the selection of the configuration: 'normal' – without rotations; 'sequential' – to rotate one position, picking the drivers sequentially; 'random' - using the stochastic selection by rotating the driver inside the group using a random number of positions.

If a new roster built by the heuristic is better than the best found in previous iterations of the metaheuristic, the best is updated accordingly. The schedules composing the roster are saved in the poll of solutions whenever considered attractive by column generation.

In the next section the computational tests and the results obtained using this new heuristic (column generation with heuristic solving subproblems aggregated) are presented.

4   Computational Tests

The decomposition model for the adopted BDRP was implemented in the computational framework SearchCol++ [10]. The BDRP test instances are the ones designated as P80 in [4]. All the instances have 36 drivers available, distributed by four salary categories in groups of equal size (9 drivers). All the tests ran on a computer with Intel Pentium CPU G640, 2,80GHz, 8 Gb of RAM, Windows 7 Professional 64 bits operating system and IBM ILOG 12.5.1 64 bits installed. In all the test configurations, only the column generation stage with the use of the new heuristic was run. It allows to retrieve the lower bounds of the optimal solution (linear), the time consumed to obtain that solution and the integer solution found by the global heuristic (solve all the subproblems aggregated).

In both heuristic configurations where the rotation of the drivers is used, we set that the rotation is not applied in 20% of the iterations (nearly the double of the probability of a driver to be selected randomly inside each group). In practice, for these iterations the assignment starts by the first driver of the ordered group keeping the order defined by the dual values.

Table 1 presents the results obtained from running the three configurations of the Roster Builder Heuristic in each instance, namely, the computational time used by the column generation to achieve the optimal solution (Time) and the value (Value) of the integer solution found. The lower bound (LB) provided by the CG (ceiling of the optimal solution value) is included in the table (the value is the same for all configurations, only the normal configuration was unable to obtain an optimal solution for the instance P80_6 in the time limit of two hours). For the random configuration, each instance was solved 20 times. In addition to the best value and its computational time, the table also display the average (Avg) and standard deviation (σ) of the values and times of the runs.

Table 1. Results from the three configurations of the heuristic

		Random						Sequential		Normal	
		Time (s)			Value			Time (s)	Value	Time (s)	Value
Instance	LB	Best	Avg	σ	Best	Avg	σ				
P80_1	3512	419.0	534.6	158.5	**3601**	3679.8	34.7	**337.9**	3695	917.2	3716
P80_2	2703	150.4	145.0	5.7	**2819**	2823.2	2.8	143.2	2821	**128.2**	2830
P80_3	4573	271.1	275.3	45.0	**4694**	4701.2	5.5	**260.3**	4694	451.2	4697
P80_4	3566	971.3	827.6	225.6	3759	3761.6	3.1	**433.1**	**3755**	860.8	3776
P80_5	3465	**535.5**	768.6	387.2	**3608**	3612.2	1.7	766.3	**3608**	988.0	3617
P80_6	3576	**1403.0**	1249.6	237.8	**3650**	3655.2	2.8	2929.4	3666	7315.8	3679
P80_7	3703	765.6	761.9	120.5	**3840**	3886.5	13.8	768.6	3889	**741.4**	3895
P80_8	4555	2871.7	2901.5	512.4	**4809**	4813.5	2.7	4387.0	4813	**1696.3**	4812
P80_9	3501	**323.7**	356.7	29.4	**3594**	3603.2	8.8	383.0	3599	495.7	3611
P80_10	4005	1390.7	1398.3	64.5	**4183**	4305.0	52.2	1224.7	4269	**1218.9**	4268
Average		**910.2**	921.9					1163.4		1481.4	

Under the Time columns, the average time is presented. The "normal" configuration is penalized by instance P80_6 where the time limit of two hours was reached before obtaining the optimal solution. The best values (time and value) are displayed in bold. Generally the computational times of the rotation heuristics are better, however for the P80_8 the time is considerably higher when comparing both with the "normal" one. The configurations with rotation were able to reach the best solutions for all instances, particularly the random configuration, which also reduces the average computational time by 39% relatively to normal configuration.

The heuristic solutions were compared with the solution value of the optimization of the compact model using the CPLEX solver with the time limit of 24 hours. Table 2 presents the gaps between the best heuristic solutions with the best known solutions. Only for the instances where the gaps are marked with bold the optimal solution was found by the CPLEX solver before the time limit. The gap of the solutions found by our heuristic is in average 3.2%.

Table 2. Gap of the best heuristic integer solution to the best known solution

Instance	Solution	Gap
P80_1	3601	2.0%
P80_2	2819	3.9%
P80_3	4694	2.6%
P80_4	3755	5.2%
P80_5	3608	3.0%
P80_6	3650	1.4%
P80_7	3840	**3.7%**
P80_8	4809	4.9%
P80_9	3594	**1.9%**
P80_10	4183	**3.5%**

The previous results show that all the configurations are able to obtain good quality rosters for the BDRP instances in test and that the separation of the drivers by category groups with the inclusion of the rotation procedure has a significant impact in the column generation optimization time. Besides that, Table 3 shows the impact on the roster when changing the configuration used. For all the configurations, the table presents the average (Avg) units of overtime assigned to a member of the first group (lower cost), the second column (Δ) presents the maximum difference of overtime assigned between the drivers and the last column (days-off) presents the number of extra days-off counted in the schedules of the 9 members of the group.

With the rotation procedures more days-off are counted. However, it is observed that, in average, additional units of overtime were assigned to the drivers, reaching one additional unit, when comparing the sequential and the normal heuristics. The most important change is observed in the uniformity of the distribution of the overtime, where the random configuration reduces the difference for the normal configuration by 5.6 units of time, and the sequential which reduces that value to less than half.

Table 3. Comparison of driver' schedules from first group

Instance	Random			Sequential			Normal		
	Avg	Δ	days-off	Avg	Δ	days-off	Avg	Δ	days-off
P80_1	90.6	28	12	91.2	17	9	88.9	40	10
P80_2	85.4	52	9	85.2	26	8	84.2	61	7
P80_3	91.1	6	9	90.9	9	9	90.6	11	9
P80_4	90.3	41	10	90.8	22	9	88.4	44	8
P80_5	93.3	24	9	93.3	13	9	92.3	29	9
P80_6	99.3	36	11	97.6	22	11	96.1	36	9
P80_7	78.4	21	9	81.7	8	9	81.2	11	9
P80_8	84.6	22	9	84.8	10	9	84.9	34	8
P80_9	80.7	54	9	80.8	27	7	79.4	59	4
P80_10	64.1	24	4	65.4	11	1	65.6	39	4
Average	85.8	30.8	9.1	86.2	16.5	8.1	85.2	36.4	7.7

5   Conclusions

In this paper, we presented a new heuristic capable of building good quality rosters to the BDRP. The heuristic is integrated with the column generation exact optimization method, using the information from the dual solutions.

In the BDRP, the objective is to define the schedules for all the drivers in the rostering period considered, assuring the assignment of all the duties and optimizing each driver use, reducing bus company costs.

In the proposed method, a decomposition model is implemented in a framework and column generation is used to optimize it. The standard optimization of the subproblems in the column generation iterations is replaced by a global heuristic which solves all the subproblems together. The heuristic is guided by the information from the RMP solution, as it sets the order by which duties are assigned, and also by which order the drivers are selected when assigning a new duty. Three configurations of this heuristic are presented: the normal configuration makes use of the dual information to guide the assignment of all the duties; the sequential and the random configurations group the drivers by category, and implement a rotation of drivers inside the groups (by 1 and a random number, respectively). The last two configurations intend to obtain rosters with a better distribution of work among drivers and more diversity of schedules.

Computational tests were made in a set of BDRP instances and the results presented. In the results it is observed that the different configurations of the heuristics have impact in the performance of the column generation and also that good quality rosters are obtained by all configurations. The quality of the obtained rosters is evaluated by comparison with the best known integer solutions, where the average gap is 3.2%. An evaluation of the schedules of the first group of drivers shows that the rotation procedure has impact in the distribution of overtime among drivers, particularly when the sequential configuration is used. Besides the better distribution of overtime, the rotation configuration was able to obtain better solutions by augmenting the average overtime units assigned to the drivers of the first group (with lower cost), even with the additional days-off counted. The additional overtime assigned to drivers of the first group compensates the extra days-off assigned.

Our heuristic with the variation configurations seems to work well in the BDRP in most of the instances tested, however it is not guaranteed that the rotation is able to improve the performance of the CG or obtain better solution, as in the instance P80_8 where the computational time increased greatly when comparing with the normal configuration. If the solutions obtained by the heuristic do not include attractive solutions to the column generation, the computational time can increase.

The proposed heuristic can be used with other problems, provided that there is an heuristic to solve the subproblems and that it is possible to use it in an aggregated way. The variation strategies used need to be tailored using knowledge about each problem.

Future work will focus on tuning this heuristic to improve the column generation performance and, if possible, obtain better integer solutions for the rostering problem. We also intend to generate a search-space composed of solutions provided by this heuristic, so that the concept of the SearchCol can be followed and other metaheuristics can explore the recombination of the obtained rosters (complete or partial) to get closer to the optimal solutions. Application of the current approach to other rostering problems is being considered as future work, since minor changes are needed for adaptation of the general metaheuristic, as well as for the roster generation heuristic here proposed.

Acknowlegments. This work is supported by National Funding from FCT - Fundação para a Ciência e a Tecnologia, under the project: UID/MAT/04561/2013.

References

1. Ernst, A.T., Jiang, H., Krishnamoorthy, M., Sier, D.: Staff scheduling and rostering: A review of applications, methods and models. European Journal of Operational Research **153**, 3–27 (2004)

2. Van den Bergh, J., Beliën, J., De Bruecker, P., Demeulemeester, E., De Boeck, L.: Personnel scheduling: A literature review. European Journal of Operational Research **226**, 367–385 (2013)

3. Ernst, A.T., Jiang, H., Krishnamoorthy, M., Owens, B., Sier, D.: An Annotated Bibliography of Personnel Scheduling and Rostering. Annals of Operations Research **127**, 21–144 (2004)

4. Moz, M., Respício, A., Pato, M.: Bi-objective evolutionary heuristics for bus driver rostering. Public Transport **1**, 189–210 (2009)

5. Dorne, R.: Personnel shift scheduling and rostering. In: Voudouris, C., Lesaint, D., Owusu, G. (eds.) Service Chain Management, pp. 125–138. Springer, Heidelberg (2008)

6. Burke, E.K., Kendall, G., Soubeiga, E.: A Tabu-Search Hyperheuristic for Timetabling and Rostering. Journal of Heuristics **9**, 451–470 (2003)

7. Respício, A., Moz, M., Vaz Pato, M.: Enhanced genetic algorithms for a bi-objective bus driver rostering problem: a computational study. International Transactions in Operational Research **20**, 443–470 (2013)

8. Leone, R., Festa, P., Marchitto, E.: A Bus Driver Scheduling Problem: a new mathematical model and a GRASP approximate solution. Journal of Heuristics **17**, 441–466 (2011)

9. Barbosa, V., Respício, A., Alvelos, F.: A Hybrid Metaheuristic for the Bus Driver Rostering Problem. In: Vitoriano, B., Valente, F. (eds.) ICORES 2013–2nd International Conference on Operations Research and Enterprise Systems, pp. 32–42. SCITEPRESS, Barcelona (2013)

10. Alvelos, F., de Sousa, A., Santos, D.: Combining column generation and metaheuristics. In: Talbi, E.-G. (ed.) Hybrid Metaheuristics, vol. 434, pp. 285–334. Springer, Heidelberg (2013)

11. Lübbecke, M.E., Desrosiers, J.: Selected Topics in Column Generation. Oper. Res. **53**, 1007–1023 (2005)

12. Puchinger, J., Raidl, G.R.: Combining metaheuristics and exact algorithms in combinatorial optimization: a survey and classification. In: Mira, J., Álvarez, J.R. (eds.) First International Work-Conference on the Interplay Between Natural and Artificial Computation. Springer, Las Palmas (2005)

13. Nemhauser, G.L.: Column generation for linear and integer programming. Documenta Mathematica Extra Volume: Optimization Stories, 65–73 (2012)

14. Dantzig, G.B., Wolfe, P.: Decomposition Principle for Linear Programs. Operations Research **8**, 101–111 (1960)

15. Cintra, G., Wakabayashi, Y.: Dynamic programming and column generation based approaches for two-dimensional guillotine cutting problems. In: Ribeiro, C.C., Martins, S.L. (eds.) WEA 2004. LNCS, vol. 3059, pp. 175–190. Springer, Heidelberg (2004)

16. Yunes, T.H., Moura, A.V., de Souza, C.C.: Hybrid Column Generation Approaches for Urban Transit Crew Management Problems. Transportation Science **39**, 273–288 (2005)

17. dos Santos, A.G., Mateus, G.R.: General hybrid column generation algorithm for crew scheduling problems using genetic algorithm. In: IEEE Congress on Evolutionary Computation. CEC 2009, pp. 1799–1806 (2009)

18. Barbosa, V., Respício, A., Alvelos, F.: Genetic Algorithms for the SearchCol++ framework: application to drivers' rostering. In: Oliveira, J.F., Vaz, C.B., Pereira, A.I. (eds.) IO2013 - XVI Congresso da Associação Portuguesa de Investigação Operacional, pp. 38–47. Instituto Politécnico de Bragança, Bragança (2013)

A Conceptual MAS Model for Real-Time Traffic Control

Cristina Vilarinho[1(✉)], José Pedro Tavares[1], and Rosaldo J.F. Rossetti[2]

[1] CITTA, Departamento de Engenharia Civil,
Faculdade de Engenharia da Universidade do Porto, Porto, Portugal
{cvilarinho,ptavares}@fe.up.pt
[2] LIACC, Departamento de Engenharia Informática,
Faculdade de Engenharia da Universidade do Porto, Porto, Portugal
rossetti@fe.up.pt

Abstract. This paper presents the description of the various steps to analyze and design a multi-agent system for the real-time traffic control at isolated intersections. The control strategies for traffic signals are a high-importance topic due to impacts on economy, environment and society, affecting people and freight transport that have been studied by many researches during the last decades. The research target is to develop an approach for controlling traffic signals that rely on flexibility and maximal level of freedom in control where the system is updated frequently to meet current traffic demand taking into account different traffic users. The proposed model was designed on the basis of the Gaia methodology, introducing a new perspective in the approach where each isolated intersection is a multi-agent system on its own right.

Keywords: Multi-agent system · Traffic signal control · Isolated intersections

1   Introduction

Traffic signal control is considered a competitive traffic management strategy for improving mobility and addressing environmental issues in urban areas [1]. Nevertheless, inefficient operation of traffic lights is a common problem that is certainly experienced by all drivers, passengers and pedestrians. This problem annoys road users and negatively affects the local economy.

The research community has been focused on the optimization of traffic signal plans. A traffic signal plan regulates traffic flow through an intersection. Permission for one or more traffic streams to move through the intersection is granted during green-light time intervals. Although there have been relatively successful efforts in optimizing traffic control, these plans often exhibit some shortcomings during operation. This is mainly because traffic control systems are often blind to the surrounding environment, missing the current traffic state, different traffic users and their needs.

In general, the traffic signal controls developed are characterized by lack of flexibility in their control systems, constraining the definition of new green-interval values or new design structures. Their rigidity is responsible for constraining the effectiveness of these strategies. So, this work focuses on isolated intersections to have

© Springer International Publishing Switzerland 2015
F. Pereira et al. (Eds.) EPIA 2015, LNAI 9273, pp. 157–168, 2015.
DOI: 10.1007/978-3-319-23485-4_17

more flexibility in operation because coordinated intersections have the drawback of having a common cycle time set to meet the needs of the largest and most complex intersection in a series, which signals at smaller intersections in the series are required to follow. In case traffic flow pattern changes as well, the common cycle time update has to be done more slowly.

The main outcome of this research is the development of an approach for controlling traffic signals that relies on flexibility and a maximal level of control freedom in which the system is updated frequently to match current traffic demand and maintain awareness of the various different traffic users. The use of a multi-agent system (MAS) approach seems to be a step forward to create a system more autonomous and cooperative in real-time control without sacrificing the safety of road users or compromising operation with a significant computational effort.

Researchers attempting to optimize traffic signal control have investigated a wide range of approaches, but several operational challenges have not received sufficient attention from the community. The main problems and challenges at an intersection with which a traffic signal control should be prepared to cope with so as to have an effective control strategy are the following: traffic congestion effect, traffic demand fluctuation, hardware failure, incidents and the mix of different types of road users.

As described, this theme is a very complex system. One way to address the aforementioned issues is to make the traffic control system more intelligent and flexible.

2   Literature Review

MASs have been suggested for many transportation problems such as traffic signal control. Zheng, et al. [2] describe their autonomy, their collaboration, and their reactivity as the most appealing characteristics for MAS application in traffic management. The application of MASs to the traffic signal control problem is characterized by decomposition of the system into multiple agents. Each agent tries to optimize its own behavior and may be able to communicate with other agents. The communication can also be seen as a negotiation in which agents, while optimizing their own goals, can also take into account the goals of other agents. The final decision is usually a trade-off between the agent's own preferences against those of others. MAS control is decentralized, meaning that there is not necessarily any central level of control and that each agent operates individually and locally. The communication and negotiation with other agents is usually limited to the neighborhood of the agent, increasing robustness [3]. Although there are many actors in a traffic network that can be considered autonomous agents [4] such as drivers, pedestrians, traffic experts, traffic lights, traffic signal controllers, the most common approach is that in which each agent represents an intersection control [3]. A MAS might have additional attributes that enable it to solve problems by itself, to understand information, to learn and to evaluate alternatives. This section reviews a number of broad approaches in previous research that have been used to create intelligent traffic signal controllers using MASs. In some work [5-8] it was argued that the communication capabilities of MAS can be used to accomplish traffic signal coordination. However, there is no consensus on the best configuration for a traffic-managing MAS and its protocol [7]. To solve conflicts between agents, in addition to communication approaches, work has been done on i) hierarchical structure, so that conflicts are resolved at an upper level, ii) agents learning how to control, iii) agents being self-organized.

Many authors make use of a hierarchical structure in which higher-level agents are able to monitor lower level agents and intervene whenever necessary. In some approaches [6, 8, 9] there is no communication between agents at the same level. The higher-level agents have the task of resolving conflicts between lower-level agents which they cannot resolve by themselves. In approaches (ii) and (iii), agents need time to learn or self-organize, which may be incompatible with the dynamics of the environment. Agents learning to control (ii) is a popular approach related to controlling traffic signals. One or more agents learn a policy for mapping states to actions by observing the environment and selecting actions; the reinforcement learning technique is the most popular method used [4, 10, 11]. The approach of self-organizing agents (iii) is a progressive system in which agents interact to communicate information and make decisions. Agent behavior is not imposed by hierarchical elements but is achieved dynamically during agent interactions creating feedback to the system [12].

Dresner and Stone [13] view cars as an enormous MAS involving millions of heterogeneous agents. The driver agents approaching the intersection request the intersection manager for a reservation of "green time interval." The intersection manager decides whether to accept or reject requested reservations according to an intersection control policy. Vasirani and Ossowski [14, 15] extended Dresner's and Stone's approach to network intersections. The approach is called market-based in which driver agents, i.e., buyers, trade with the infrastructure agents, i.e., sellers in a virtual marketplace, purchasing reservations to cross intersections. The drivers have an incentive to choose an alternative to the shortest paths.

In summary, since the beginning of this century, interest in application of MAS to traffic control has been increasing. Further, the promising results already achieved by several authors have helped to establish that agent-based approaches are suitable to traffic management control. Most reviewed MASs have focused their attention on network controllers, with or without coordination, rather than on isolated intersections. Another issue is that traffic control approaches focus on private vehicle as the major component of traffic, and may be missing important aspects of urban traffic such as public transport and soft modes (pedestrian, bicycles).

3   Methodological Approach

The development of a MAS conceptual model for real-time traffic control at an isolated intersection followed a methodology for agent-oriented analysis and design. In this section an increasingly detailed model is constructed using Gaia [16, 17] as the main methodology, complemented by concepts introduced by Passos, et al. [18].

The first step is an overview of the scenario description and system requirements. The Gaia process starts with an analysis phase whose goal is to collect and establish the organization specifications. The output of this phase is the basis for the second phase, namely the architectural design, and the third phase, which is a detailed design phase.

3.1   Scenario Description

The problem addressed is the control of a traffic signal at an isolated intersection at which, depending on the intersection topology and the detected amounts of traffic of various types of road users, the lights regulating traffic streams are to change color to achieve a more efficient traffic management strategy.

The scenario of the proposed traffic signal control is as follows:

- At time X (e.g., each 5 min) or event Y (e.g., traffic conditions, new topology, system failure), a request for a new traffic signal plan is created;
- All information about current topology and traffic conditions is updated to generate new traffic data predictions for the movements of each traffic component. In this way a new traffic signal plan is defined to meet the new intersection characteristics;
- During processing of the new traffic signal plan, if topology has changed, the stage design is developed following the new topology;
- The traffic signal plan is selected based on criteria such as the minimum delay, the system saves the traffic plan information (design, times) and implements it;
- During monitoring, current traffic data are compared with traffic predictions, the topology is verified and data are analyzed by the auditor, which computes the actual level of service and informs the advisor of the results. Depending on the results, the auditor decides if it should make a suggestion for the traffic streams such as to terminate or to extend the current stage or if a new plan should be requested;
- Depending on the information received, traffic streams can continue with the traffic signal plan or negotiate adjustments to it;

The system is responsible for defining and implementing a traffic signal plan as well as deciding when to suspend it, in which case it initiates negotiation between traffic streams to adjust the plan according to traffic flow fluctuations and characteristics (e.g., traffic modes, priority vehicles), or even decides to design a new plan.

As input, the Gaia methodology uses a collection of requirements. The requirements can be collected through analyzing and understanding the scenario in which the organizations are identified, as well as the basic interactions between them to achieve their goals. For early requirements collection, it uses the Tropos methodology [19], in which relevant roles, their goals and intentions, as well as their inter-dependencies are identified and modeled as social actors with dependencies.

3.2   Analysis Phase

The goal of the analysis phase is to develop an overview of the system and capture its structure. The division into sub-organizations helps finding system entities with specific goals that interact with other entities of the system and require competencies that are not needed in other parts of the system. From the diagram of the early requirements (Fig.1), 7 actors were found whose goals, soft goals and dependencies are described below.

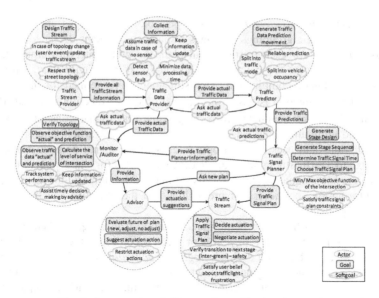

Fig. 1. Actors and goal diagram for traffic signal control

The TrafficStreamProvider has the goal to *design a traffic stream*. Each traffic stream is described by movements and by lanes assigned to each movement, including information about traffic sensor locations. To achieve its goal, two soft goals were defined: to respect the intersection topology and to keep the topology information updated in case of topology changes (e.g. road works, accidents). The actor should "provide all traffic stream information" to TrafficDataProvider.

The TrafficDataProvider has the main objective to *collect information* about traffic data from sensors installed at a signalized intersection and aggregate data according to the traffic stream information received. The goal is built upon four sub-goals: to keep traffic data information updated, to minimize data processing time when dependent actors are waiting for the information, assume traffic data if no sensor is installed or if a sensor seems to act strangely. TrafficPredictor requests recent traffic data from this actor and makes its own traffic predictions. Monitor/Advisor also requests traffic data from this actor and uses them for early detection of possible problems and improvements at the intersection control.

The TrafficPredictor has the main goal to *generate a traffic data prediction* for each movement; this is to optimize signal control for imminent demand rather than being reactive to current flow. The strategy may include traffic measurements from a past time period and the current time and uses them to estimate the near future. The generated traffic prediction should be: reliable for future traffic and comprehensive, with total values and splits into traffic modes. The actor requests recent traffic data from TrafficDataProvider and makes its own traffic predictions. TrafficSignalPlanner requests this actor for recent traffic predictions and uses them to optimize the traffic plan.

The TrafficSignalPlanner has the following objectives. *Generate stage design:* search possible signal group sets that can run concurrently respecting a set of safety constraints; *generate stage sequence:* once possible stage designs are defined, this step compiles strategic groupings of stages to have signal plans designed; *determine traffic signal times*: for each traffic signal plan, the green-interval durations, inter-green and cycle lengths are calculated; and *choose traffic signal plan,* based on a criterion or a weighted combination. Two soft goals were defined for the objective: traffic signal plan selection is based on the best objective function and plan design and timing should be conducted respecting some operational constraints, such as maximum and minimum cycle lengths. The actor requests TrafficPredictor for recent traffic predictions and uses these to optimize its traffic signal plan. It provides the selected plan to TrafficStream to be applied. Advisor asks for a new plan search if the current plan is not adequate to remain active. Finally, Monitor/Auditor receives traffic planner information such as traffic predictions and the objective function so it can monitor independently.

The TrafficStream has three main goals. A*pply a traffic signal plan,* each traffic stream assumes a signal state: red, yellow or green according to the plan or the current actuation action, if it has been defined; *negotiate actuation,* traffic streams cooperate to find possible actuation actions following the advisor's suggestions; and *decide actuation,* traffic stream actors together decide an actuation action to implement. To accomplish its goal, the actor intends: to verify transition to next stage and to satisfy user beliefs about the traffic light to prevent frustration. The actor receives the selected traffic signal plan from TrafficSignalPlanner to apply it and actuation suggestions from Advisor to guide the negotiation phase. If negotiations are needed, Traffic Stream actors discuss these among themselves.

The Advisor´s two main objectives are: to *evaluate the future of plan,* choose a possible action depending on information received from Monitor/Auditor: find a new plan, adjust the current plan or continue the implementation; and to *suggest actuation action,* if it is decided to adjust the plan through actuation, the actor prepares a recommendation to guide the actuation process. The *suggest actuation action* has a soft goal defined: formulate a recommendation that will restrict the solution space of actuation negotiation. The actor provides actuation suggestions to TrafficStream. It requests a new plan search from TrafficSignalPlanner if the current plan is not adequate to remain active. Monitor/Auditor sends monitor information to this actor.

The Monitor/Auditor´s four main objectives are: verify topology, check if any topology change occurred, and report it to Advisor if so; observe traffic data, "actual" and prediction; observe objective function; and calculate the level of service of the intersection. The data acquired through monitoring are used to evaluate if Advisor should be asked for any plan change. The objectives are complemented with three sub-goals: data collection to keep information updated, track system performance and assist timely decision-making by Advisor to exploit every opportunity to improve the intersection system. The actor requests recent traffic data from TrafficDataProvider and receives them for early detection of possible problems and improvements at the intersection control. It receives traffic planner information such as traffic predictions and the objective function from TrafficSignalPlanner. It sends monitor information to Advisor.

Modeling the environment is one of the agent-oriented methodologies' major activities. The environment model can be viewed in its simplest form as a list of resources that the MAS can exploit, control or consume when working towards the accomplishment of its goal. The resources can be information (e.g., a database) or physical entity (e.g., a sensor). Six resources were defined for the proposed traffic signal control: topology, traffic detector, traffic database, traffic prediction, traffic signal plan and traffic light. The resources are identified by name and characterized by their types of actions.

A partial list of those resources is:

- Topology has the action to read and change when new topology is detected. The resource contains information regarding intersection topology such as number of traffic arms, their direction, number of approach lanes in each traffic arm, movements assigned in each lane and traffic detector position;
- Traffic Detector is essential for the system because it contains all traffic data (read) and also needs to be frequently updated (change) so it can correspond to the real traffic demand. This makes it possible to know information in each detector about: current traffic data in lane, number of users type, vehicle occupancy, traffic flow distribution by movement, lanes without sensor or equipment failure;

Complex scenarios such as this are very dynamic, so the approach presented by Passos, Rossetti and Gabriel [18] extends Gaia methodology to include Business Process Management Notation (BPMN) to capture the model dynamics. Business Process (BP) collects related and structured activities that can be executed to satisfy a goal.

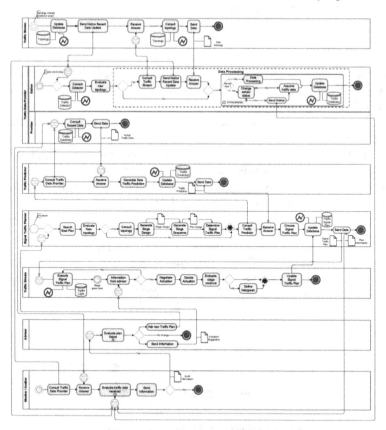

Fig. 2. Collaboration diagram of traffic signal control intersection

The diagram in Fig. 2 shows the interactions between the seven participants (actors of Fig. 1) with message exchanges and includes tasks within participants, providing a detailed visualization of the scenario. Their interactions with resources are also present in the diagram.

The actors and goals in diagrams in Fig.1 and Fig.2 help to identify the roles that will build up the final MAS organization. The preliminary roles model defined first, as the name implies, is not a complete configuration at this stage, but it is appropriate to identify system characteristics that are likely to remain. It identifies the basic skills, functionalities and competences required by the organization to achieve its goals. For traffic signal control 13 preliminary roles were defined. A partial list of those roles is:

- RequestTrafficSignalPlan role associated with creating all possible traffic signal plans in order to select one (ChooseTrafficPlan) to be implemented.
- ChooseTrafficPlan role involves deciding on the best plan to choose based on some criteria.
- The goal of the preliminary interaction protocol is to describe the interactions between the various roles in the MAS organization. Moreover, the interaction model describes the characteristics and dynamics of each protocol (when, how, and by whom a protocol is to be executed).

3.3   Design Phase

The goal of the analysis phase is to define the main characteristics and understand what the MAS will have to be. In the design phase, the preliminary models must be completed. The design phase usually detects missing or incomplete specifications or conflicting requirements demanding a regression back to previous stages of the development process.

From the analysis phase, the organization structure is presented in Fig.3. It is a crucial phase and affects the following steps in MAS development. To represent the organizational structure, we have adopted a graphical representation proposed by Castro and Oliveira [20] that uses the Gaia concept in UML 2.0 representation.

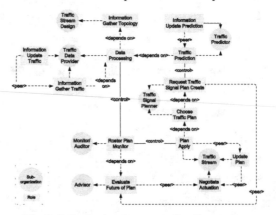

Fig. 3. Organizational structure for all system

There are three types of relationships: "depends on", "controls" and "peer". "Depends on" is a dependency relationship that means one role relies on resources or knowledge from the other. "Controls" is an association relationship usually meaning that one role has an authoritative relationship with the other role, controlling its actions. "Peer" is a dependency relationship also and usually means that both roles are at the same level and collaborate to solve problems.

After achieving the structural organization, the roles and interactions of the preliminary model can be fulfilled. To complete the role model, it is necessary to include all protocols, the liveness and safety responsibilities. In Table 1, one of 13 roles is described according to the role schema. For complete definition of interaction protocols, they should be revised to respect the organizational structure. Table 2 shows the definition of the "InformPlanEvents" protocol using Gaia notation.

Table 1. Role Model example

Id	Properties
RosterPlanMonitor	*Description:* role involves monitoring the traffic condition and traffic signal plan for events related to: Substantial difference: between objective function acceptable value versus current value, or traffic flow (total and by traffic) prediction versus current; Reach some limit of measure of effective (such as maximum queue length); Number maximum of plan repetition; Sensor system fault. After detecting one of these events the RosterPlanMonitor role will request to evaluate what should be done for EvaluateFutureOfPlan role. · *Protocols and activities:* <u>CheckForNewPlanEvents</u>, <u>UpdatePlanEventStatus</u> *Send:reportEventStatus*, requestPlanEvaluation; *Receive:* informPlanEvent, sendEvaluationPlanRequest · Permissions: Read *Traffic Detector* // to obtain information about sensor system Read *Traffic Database* // to obtain traffic information about current condition Read *Traffic Signal Plan* // read and compare plan prediction and current · Responsibilities: Liveness: RosterPlanMonitor = (CheckForNewPlanEvents^W. informsPlanEvent)^W ‖ (reportPlanEventStatus^W. UpdatePlanEventStatus)^W, w is indefinitely Safety: successful_connection_with_TrafficDetector = true successful_connection_with_TrafficDatabase = true successful_connection_with_TrafficSignalPlan = true

Table 2. Protocol model example

Protocol schema: InformPlanEvents		
Initiator Role: PlanApply	*Partner Role:* RosterPlanMonitor	*Input:* Open position information
Description: After an event has been detected it is necessary to analyses the plan adequacy. For that it is necessary to send details about position so plan with his current position can be generated.		*Output:* The plan is analyzed taking into account the current traffic information.
CheckForNewPlanEvents, UpdatePlanEventStatus	ReceiveMonitorInformation, SendRequest	

To better clarify the organization with its roles and interactions, a final diagram is presented in Fig.4 with the full model, including all protocols and required services that will be the bases for the roles agents choose.

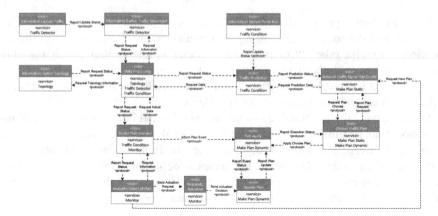

Fig. 4. Final diagram: roles, interactions and services

3.4   Detailed Design

The detailed design phase is the last step, and it is responsible for the most important output: the full agent model definition for helping the actual implementation of agents. The agent model identifies the agents from role-interaction analysis. Moreover, it includes a service model. The model design should try to reduce the model complexity without compromising the organization rules. To present the agent model, the dependency relations between agents, roles and services are presented in Fig. 5.

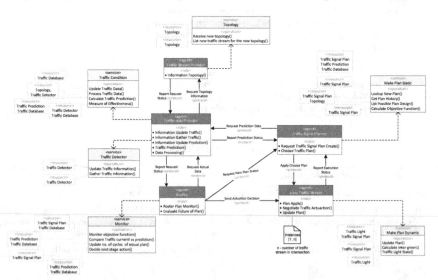

Fig. 5. Full Agent Model

The diagram above should be read as "Traffic Signal Planner agent is responsible to perform the service Make Plan Static." Five types of agents were defined. "Traffic Stream" has n agents, one for each traffic stream of the intersection, so it depends on intersection topology. It means that the agent class "Traffic Stream" is defined to play the roles PlanApply, NegotiateTrafficActuation and UpdatePlan, and there are between one and n instances of this class in MAS. After the completion of the design process, the agent classes defined are ready to be implemented, according to previous models.

4   Conclusions

This paper presents the design of a conceptual model of a multi-agent architecture for real-time signal control at isolated traffic intersections using Gaia as the main methodology. The main idea is to make rational decisions about traffic stream lights such that the control is autonomous and efficient under different conditions (e.g., topology, traffic demand, traffic priority, and system failure). The traffic control of an isolated intersection has the advantage that each intersection may have an independent control not limited by neighbors' control. This allows a control algorithm to be simpler than one for coordinated intersections and more flexible to define plan design and times.

Comparing the proposed strategy with traditional approaches using MAS, it is possible to find several differences. Traditional traffic control methods rely on each agent controlling an intersection within the traffic network. The system usually has a traffic signal plan defined a priori and the system controls how to perform small adjustments such as decreasing, increasing or advancing the green time interval of a traffic stage. Research is being conducted on using MAS to coordinate several neighboring agent controllers, in either a centralized or distributed system. Another feature shared by traditional approaches is the agent decision (action selections) based on learning. From the result of each decision, the learning rule gives the probability with which every action should be performed in the future.

As introduced before, the present approach is distinct from other works to the extent that each traffic stream is an agent and each signalized intersection builds upon independent MASs. Thus, the multitude of agents designed for isolated intersections create, manage and evolve their own traffic signal plans. Therefore this proposed multi-agent control brings the benefit of staged designs and sequences being formed as needed instead of being established a priori. The system structure is flexible, and it has the ability to adapt traffic control decisions to predictions and react to unexpected traffic events.

The validation of this traffic control strategy will be performed using a state-of-the-art microscopic traffic simulator such as, for instance, AIMSUN. The proposed model was developed from scratch rather than by enhancing an existing model.

Finally, it is not our goal to present the process of designing and implementing a MAS or promoting the use of Gaia; there is existing research that is much more adequate for that. However, the methodology applied is well-suited to the problem.

Acknowledgment. This project has been partially supported by FCT, under grant SFRH/BD/51977/2012.

References

1. Park, B., Schneeberger, J.D.: Evaluation of traffic signal timing optimization methods using a stochastic and microscopic simulation program. Virginia Transportation Research Council (2003)
2. Zheng, H., Son, Y., Chiu, Y., Head, L., Feng, Y., Xi, H., Kim, S., Hickman, M.: A Primer for Agent-Based Simulation and Modeling in Transportation Applications. FHWA (2013)
3. McKenney, D., White, T.: Distributed and adaptive traffic signal control within a realistic traffic simulation. Engineering Applications of Artificial Intelligence 26, 574–583 (2013)
4. Bazzan, A.L.C.: Opportunities for multiagent systems and multiagent reinforcement learning in traffic control. Auton. Agent. Multi-Agent Syst. 18, 342–375 (2009)
5. Katwijk, R., Schutter, B., Hellendoorn, H.: Look-ahead traffic adaptive control of a single intersection – A taxonomy and a new hybrid algorithm (2006)
6. Choy, M., Cheu, R., Srinivasan, D., Logi, F.: Real-Time Coordinated Signal Control Through Use of Agents with Online Reinforcement Learning. Transportation Research Record: Journal of the Transportation Research Board 1836, 64–75 (2003)
7. Bazzan, A.L.C., Klügl, F.: A review on agent-based technology for traffic and transportation. The Knowledge Engineering Review 29, 375–403 (2013)
8. Hernández, J., Cuena, J., Molina, M.: Real-time traffic management through knowledge-based models: The TRYS approach. ERUDIT Tutorial on Intelligent Traffic Management Models, Helsinki, Finland (1999)
9. Roozemond, D.A., Rogier, J.L.: Agent controlled traffic lights. In: ESIT 2000, European Symposium on Intelligent Techniques. Citeseer (2000)
10. Bazzan, A.L.C., Oliveira, D., Silva, B.C.: Learning in groups of traffic signals. Engineering Applications of Artificial Intelligence 23, 560–568 (2010)
11. Wiering, M., Veenen, J., Vreeken, J., Koopman, A.: Intelligent Traffic Light Control (2004)
12. Oliveira, D., Bazzan, A.L.C.: Traffic lights control with adaptive group formation based on swarm intelligence. In: Dorigo, M., Gambardella, L.M., Birattari, M., Martinoli, A., Poli, R., Stützle, T. (eds.) ANTS 2006. LNCS, vol. 4150, pp. 520–521. Springer, Heidelberg (2006)
13. Dresner, K., Stone, P.: A Multiagent Approach to Autonomous Intersection Management. J. Artif. Intell. Res. (JAIR) 31, 591–656 (2008)
14. Vasirani, M., Ossowski, S.: A market-inspired approach to reservation-based urban road traffic management. In: Proceedings of 8th International Conference on AAMAS, pp. 617–624. International Foundation for AAMS (2009)
15. Vasirani, M., Ossowski, S.: A computational market for distributed control of urban road traffic systems. IEEE Transactions on Intelligent Transportation Systems 12, 313–321 (2011)
16. Zambonelli, F., Jennings, N.R., Wooldridge, M.: Developing multiagent systems: The Gaia methodology. ACM T. Softw. Eng. Meth. 12, 317–370 (2003)
17. Wooldridge, M., Jennings, N.R., Kinny, D.: The Gaia methodology for agent-oriented analysis and design. Auton. Agent. Multi-Agent Syst. 3, 285–312 (2000)
18. Passos, L.S., Rossetti, R.J.F., Gabriel, J.: An agent methodology for processes, the environment, and services. In: IEEE Int. C. Intell. Tr., pp. 2124–2129. IEEE (2011)
19. Bresciani, P., Perini, A., Giorgini, P., Giunchiglia, F., Mylopoulos, J.: Tropos: An agent-oriented software development methodology. Auton. Agent. Multi-Agent Syst. 8, 203–236 (2004)
20. Castro, A., Oliveira, E.: The rationale behind the development of an airline operations control centre using Gaia-based methodology. International Journal of Agent-Oriented Software Engineering 2, 350–377 (2008)

Prediction of Journey Destination in Urban Public Transport

Vera Costa[1(✉)], Tânia Fontes[1], Pedro Maurício Costa[1], and Teresa Galvão Dias[1,2]

[1] Department of Industrial Management, Faculty of Engineering,
University of Porto, Porto, Portugal
veracosta@fe.up.pt
[2] INESC-TEC, Porto, Portugal

Abstract. In the last decade, public transportation providers have focused on improving infrastructure efficiency as well as providing travellers with relevant information. Ubiquitous environments have enabled traveller information systems to collect detailed transport data and provide information. In this context, journey prediction becomes a pivotal component to anticipate and deliver relevant information to travellers. Thus, in this work, to achieve this goal, three steps were defined: (i) firstly, data from smart cards were collected from the public transport network in Porto, Portugal; (ii) secondly, four different traveller groups were defined, considering their travel patterns; (iii) finally, decision trees (J48), Naïve Bayes (NB), and the Top-K algorithm (Top-K) were applied. The results show that the methods perform similarly overall, but are better suited for certain scenarios. Journey prediction varies according to several factors, including the level of past data, day of the week and mobility spatiotemporal patterns.

Keywords: Prediction · Journey destination · Urban public transports

1 Introduction

In the last decade, Urban Public Transport (UPT) systems have turned to Information and Communication Technologies (ICT) for improving the efficiency of existing transportation networks, rather than expanding their infrastructures [3, 7]. Public transport providers make use of a wide range of ICT tools to adjust and optimise their service, and plan for future development.

The adoption of smart cards, in particular, has not only enabled providers to access detailed information about usage, mobility patterns and demand, but also contributed significantly towards service improvement for travellers [21, 25]. For instance, inferring journey transfers and destination based on historical smart card data has allowed transportation providers to significantly improve their estimates of service usage – otherwise based on surveys and other less reliable methods [8]. As a result, UPT providers are able to adjust their service accordingly while reducing costs [2]. Furthermore, the combination of UPT and ICT has enabled the development of Traveller Information Systems (TIS), with the goal of providing users with relevant

© Springer International Publishing Switzerland 2015
F. Pereira et al. (Eds.) EPIA 2015, LNAI 9273, pp. 169–180, 2015.
DOI: 10.1007/978-3-319-23485-4_18

on-time information. Previous work has shown that TIS have a positive impact on travellers. For instance, providing on-time information at bus stops can significantly increase perception, loyalty and satisfaction [5].

The latest developments in ICT have paved the way for the emergence of ubiquitous environments and ambient intelligence in UPT, largely supported by miniaturised computer devices and pervasive communication networks. Such environments simplify the collection and distribution of detailed real-time data that allow for richer information and support the development of next-generation TIS [6, 20].

In this context, as transportation data is generated and demand for real-time information increases, the need for contextual services arises for assisting travellers, identifying possible disruptions and anticipating potential alternatives [20, 26].

A number of methods have been used for inferring journeys offline (e.g. [1,8,23]). After the journeys are completed, the application of these methods can support different analysis, such as patterns of behaviour (e.g. [13,14]) and traveller segmentation (e.g. [11,12]. In contrast, little research has focused on real-time journey prediction (e.g. [16]). Contextual services, however, require on-time prediction and, unless explicitly stated by the user, the destination of a journey may not be known until alighting.

The prediction of journeys based on past data and mobility patterns is a pivotal component of the next generation of TIS for providing relevant on-time contextual information. Simultaneously, UPT providers benefit from up-to-date travelling information, allowing them to monitor their infrastructures closely and take action.

An investigation of journey prediction is presented, based on a group of bus travellers in Porto, Portugal. Specifically this research focuses on the following questions:

- Is it possible to predict a journey destination of UPT based on past usage? How do past journeys impact the quality of these predictions?
- What is the variation in journey prediction for groups of travellers with different mobility characteristics? Why is so important define such groups?
- Are there variations between groups of travellers over time, specifically for different days of the week?

In order to answer to these questions, this paper is structured as follows: Section 2 describes the data collection and the algorithms used in the work presented; Section 3 presents and discusses the results and the main findings obtained; final conclusions and considerations are presented in Section 4.

2   Material and Methods

In order to predict the journey destinations for an individual traveller, three steps were defined: (i) firstly, data from smart cards were collected and pre-processed from the public transport network in Porto, Portugal (see section 2.1); (ii) secondly, four different groups of users were defined, considering their travel patterns (see section 2.2); (iii) finally, three different intelligent algorithms were assessed considering different performance measures (see section 2.3). Figure 1 presents an overview of the overall methodology applied to perform the simulations. While at this stage the analysis is based on a set of simulations and historical data, the goal is to apply the method for a timely prediction of destinations and which will be implemented in the scope of the Seamless project [4]. Thus, the simulations presented in the present paper enable the evaluation of the importance of groups of travellers, and the best algorithm to use for predicting journey destinations in a real-world environment.

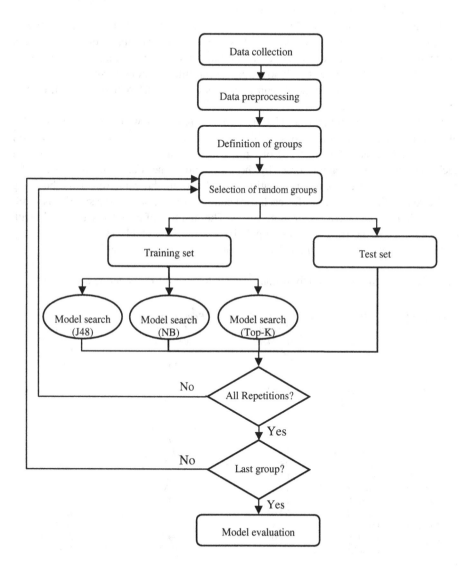

Fig. 1. Methodology overview.

2.1 Data

The public transport network of the Metropolitan Area of Porto covers an area of 1,575 km^2 and serves 1.75 million of inhabitants [10]. The network is composed of 126 buses lines (urban and regional), 6 metro lines, 1 cable line, 3 tram lines, and 3 train lines [24]. This system is operated by 11 transport providers, of which Metro do Porto and STCP are the largest.

The Porto network is based on an intermodal and flexible ticket system: the Andante. Andante is an open zonal system, based on smart cards, that requires validation only when boarding. A validated occasional ticket allows for unlimited travel within a specified area and time period: 1 hour for the minimum 2-zone ticket, and longer as the number of zones increases. Andante holders can use different lines and transport modes in a single ticket.

In this work, to perform the simulations two months of data were used, April and May of 2010 to perform the simulations. Table 1 shows an example of the data collected for an individual traveller for one week of April 2010. Journey ID is a unique identifier for each trip, sorted in ascending order by the transaction time. For each traveller (i.e. for each Andante smart card), the information related with the boarding time (first boarding on the route), the line (or lines for each trip) and the stop (or stops for each trip) is available. Each trip could have one or more stages. The first line of the table shows a trip with two stages. First the traveller uses stop 1716 and line 303 at 11h34 followed the stop 3175 and line 302. Based on these data the route sequence can be rebuilt.

Table 1. Extracted trip chain information for an individual travellers during a week of April, 2010.

Journey ID	Date	First boarding time of the route	Route sequence (Line ID)	Stop sequence (Stop ID)
1036866	10/04/2010	11:24	303 → 302	1716 → 3175
1036867	10/04/2010	16:27	203	1622
1036868	10/04/2010	23:14	200	1035
1036869	12/04/2010	09:05	402	1632
1036870	12/04/2010	12:42	203	1695
1036871	12/04/2010	13:44	203	1632
1036872	12/04/2010	19:45	303	1338
1036873	12/04/2010	22:29	400 → 400	1675 → 1689
1036874	13/04/2010	09:09	402	1632
1036875	13/04/2010	19:11	206	1338
1036876	14/04/2010	08:45	302	1632
1036877	14/04/2010	12:30	402	1338
1036878	14/04/2010	13:43	203	1632
1036879	14/04/2010	19:11	303	1338
1036880	15/04/2010	09:08	402	1632
1036881	15/04/2010	12:57	203	1695
1036882	15/04/2010	14:04	302→501	1632→1810
1036883	15/04/2010	20:52	303	1338

In order to make a prepossessing of data, an inference algorithm was used [18] to identify the journey destination. Since the Oporto system is based on the validation at the entrance only, this pre-processing is required. In this method, the travel origin of a traveller is the destination of the previous travel of that traveller. However, due to the absence of information, the results obtained with the application of this algorithm were partially restricted, since data from only one transport provider, the STCP company, were available. Thus, in order to minimize the error, only data from users with at least 80% of destinations inferred with success and, on average, two or more validations per day was used. As a result, the sample consists of 615,647 trips corresponding to 6865 different Andante cards.

The data set consists of a set of descriptive attribute of which three of them were used. The first attribute represents the code of the origin bus stop. The second attribute identifies the date, which represents the day of the week for each validation. The third represents the bus stop as an inferred destination.

2.2   Definition of Groups

Sets of travellers were selected from the main dataset for predicting the destination of a journey. Four different groups of travellers were defined with different mobility characteristics. The first three groups (1, 2 and 3) are characterized by patterns of mobility with different usage characteristics. The last group (4) is composed of travellers without a seemingly travel pattern. The spatiotemporal mobility patterns are based on two characteristics: primary journeys and journey schedule. The primary journeys describe spatial regularity, identifying the most frequent journeys in a given route. The journey schedule assesses temporal regularity, based on the departure times. Thus, we have:

- Group 1 (G1): includes individual travellers with a regular spatiotemporal pattern. In this group, individuals with two primary journeys were selected (e.g. home/work/home or home/school/home). These primary journeys represent, in average, 74% of the total journeys. Furthermore, to ensure the temporal travelling regularity, a maximum departure time deviation of one hour was considered. These restrictions resulted in a group of travellers who have a tendency for a rigid journey schedule (e.g. professionals);
- Group 2 (G2): includes travellers with a regular spatial pattern but without temporal regularity. Similar to group one, travellers with two primary journeys (e.g. home/work/home or home/school/home) were selected. These primary journeys represent about 72% of the total journeys. In addition, to exclude temporal regularity of the primary journeys, a departure time deviation greater than one hour was considered. As a result, this group of travellers have a tendency towards a flexible journey schedule (e.g. students);
- Group 3 (G3): includes travellers with a broader spatial regularity. In this group, individuals with four primary journeys were considered (e.g. home/work/home and work/gymnasium/home). These primary journeys represent about 79% of the total journeys. In contrast to the previous two groups, temporal regularity was not taken into account;
- Group 4 (G4): is characterized by a non-regular spatial mobility pattern. In this group, the number of different routes is higher than 50% of the total number of journeys (e.g. occasional travellers).

Table 2 shows the main characteristics of those groups.

Table 2. Characteristics of individual journeys considering different groups of travellers.

	Number of travellers (N)	Number of total journeys (X±SD)	Number of different routes (X±SD)	Frequency of primary journeys (%)		Deviation of journey schedule (X±SD)	Representativity of each group in the population
				Top 2 journeys	Top 4 journeys		
G1	200	75.9±12.4	15.8±8.4	73.6%	81.1%	00:17±00:15 [a]	4.6%
G2	200	91.5±30.9	19.1±12.2	72.1%	79.3%	03:08±01:36 [a]	9.6%
G3	200	100.8±26.7	19.2±8.5	48.7%	79.1%	02:04±01:49 [b]	10.5%
G4	200	83.2±35.1	60.0±25.3	0.3%	0.5%	-	12.6%

[a] for the top 2 trips; [b] for the top 4 trips.

2.3   Methods

In order to estimate the destination of each traveller, three different algorithms were analysed: (i) the decision trees (J48); (ii) the Naïve Bayes (NB); and (iii) the Top- K algorithm (Top-K).

Decision trees represent a supervised approach to classification. These algorithms are a tree-based knowledge representation methodology, which are used to represent classification rules in a simple structure where non-terminal nodes represent tests on one or more attributes and terminal nodes reflect decision outcomes. The decision tree approach is usually the most useful in classification problems [17]. With this technique, a tree is built to model the classification process. J48 is an implementation of a decision tree algorithm in the WEKA system, used to generate a decision tree model to classify the destination based on the attribute values of the available training data. In R software, the RWeka package was used.

The Naïve Bayes algorithm is a simple probabilistic classifier that calculates a set of probabilities by counting the frequency and combinations of values in a given data set [19]. The probability of a specific feature in the data appears as a member in the set of probabilities and is calculated by the frequency of each feature value within a class of a training data set. The training dataset is a subset, used to train a classifier algorithm by using known values to predict future, unknown values. The algorithm is based on the Bayes theorem and assumes all attributes to be independent given the value of the class variable. In this work the e1071 R package was used.

The Top-K algorithm enables finding the most frequent elements or item sets based on an increment counter [15]. The method is generally divided into counter-based and sketch-based techniques. Counter-based techniques keep an individual counter for a subset of the elements in the dataset, guaranteeing their frequency. Sketch-based techniques, on the other hand, provide an estimation of all elements, with a less stringent guarantee of frequency. Metwally proposed the Space-Saving algorithm, a counter-based version of the Top-K algorithm that targets performance and efficiency for large-scale datasets. This version of the algorithm maintains partial information of interest, with accurate estimates of significant elements supported by a lightweight data structure, resulting in memory saving and efficient processing. It focuses on the influential nodes and discards less connected ones [22]. The main idea behind this method is to have a set of counters that keep the frequency of individual elements. Invoking a parallelism with social network analysis, the algorithm proposed by Sarmento et al. [22] was changed; a journey is considered to be an edge i.e. a connection between any node (stop) A and B. The algorithm starts to count occurrences of journeys. For each traveller, if the new journey is monitored, the counter is updated. Otherwise, the algorithm adds a new journey in your Top-K list. If the number of unique journeys exceeds 10*K monitored journeys the algorithm follows the space saving application.

For each algorithm and group of travellers defined previously (see Section 2.2), 15 repetitions were performed. For each repetition, 30 travellers were randomly selected from each group. Figure 2 shows the average number of journeys for the groups. In each simulation, the test size is always one and corresponds to the day under evaluation, i (n_{test} = 1 day), while the train continuously grows with i (n_{train} = i-1 day(s)). Table 3 illustrates this procedure.

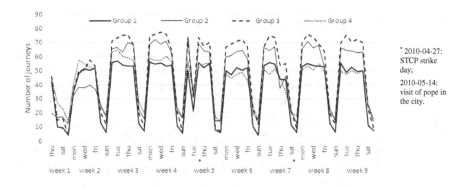

Fig. 2. Average number of journeys by each group of travellers.

Table 3. Data sets used to train (t) and test (T).

	Days												
	1	2	3	4	5	6	7	8	9	10	(...)	n-1	n
i=1	T												
i=2	t	T											
i=3	t	t	T										
(...)													
i=n	t	t	t	t	t	t	t	t	t	t	(...)	t	T

NOTES: t: days of training set; T: days of the test set.

To evaluate the performance of the different algorithms the Accuracy measure (1), which represents the proportion of correctly identified results, was used. The basis for this approach is the confusion matrix, a two-way table that summarizes the performance of the classifier. Considering one of the classes as the positive (P) class, and other the negative (N) class, four quantities may be defined: the true positives (TP), the true negatives (TN), the false positives (FP) and the false negatives (FN), we have:

$$Accuracy = \frac{TP + TN}{TP + TN + FP + FN} \qquad (1)$$

Nonetheless, due to the sparse usage of the transportation network by travellers, in this work we have an imbalanced sample. In this case, the Accuracy measure by itself is not the most appropriate to analyse the presence of imbalanced data [9]. Therefore, the F-score (2), which combines Precision and Recall measures as a weighted average, was also used. Precision gives an insight on how the classifier behaves in relation to the positive class, giving a measure of how many positive predicted instances are in fact positive, while the Recall gives an insight of the classifier's performance on the positive class, measuring how well the whole positive class is recognized. F-score is given by:

$$F = 2 * \frac{precision * recall}{precision + recall} \qquad (2)$$

3   Results and Discussion

Table 4 show the average *Accuracy* and *F-score* obtained for journey destination prediction respectively, by algorithm, group of travellers and day of the week (weekday vs weekend). The analysis of results shows several similarities and differences between the different groups of travellers analysed.

Regarding the comparison of the journey destination prediction between weekdays and weekends, the results shows small differences of performance for the individuals of Group 4. In this case the *Accuracy* and *F-score* for weekdays is around 2% higher than on weekends. Nevertheless, for the remaining groups (G1, G2 and G3), a clear difference is observed for these two periods. As example, while during weekdays for Group 1, the *Accuracy* is, on average, 77-79% (with exception of disruptive events in the city, namely on 2010-04-27, a STCP strike day, and on 2010-05-14, a pope visit to the city, which were removed), during the weekends the values fall down to 47-49%. The same trend is observed to *F-score* (an average value of 80-83% and 51-53% respectively). Therefore, high deviations are observed during the weekends, which suggest uncertainty in predicting for these days, associated with the lack of travelling routines for these groups (G1, G2 and G3). This discrepancy between weekdays and weekends dissipates, as mobility pattern characteristics are less strict. Simultaneously, as weekdays and weekends become indiscernible, so does the average prediction performance, with an average decrease which varies between 6 and 40%.

Regarding the algorithms applied, similar results were found between them. For this comparison, the first four weeks were disregarded to exclude the initial learning period. Thus, the last five weeks represent a more stable account of journey prediction. On average low differences of performance were found during the weekdays (*Accuracy*: G1=2%, G2=3%, G3=2%, and G4=4%; *F-score*: G1=2%, G2=3%, G3=3%, and G4=3%). During the weekends, these differences are generally higher (*Accuracy*: G1=7%, G2=4%, G3=4%, and G4=4%; *F-score*: G1=6%, G2=4%, G3=5%, and G4=4%).

Table 4. Accuracy and F-score (%) average (X) and standard deviation (SD), obtained to the prediction of the journey destination, by algorithm (J48, NB and Top-K), day (weekdays and weekends), and group of individual travellers (G1, G2, G3 and G4).

			J48		NB		Top-K	
			X±SD	min-max	X±SD	min-max	X±SD	min-max
Accuracy (%)	Weekday	G1	78.3 ± 5.6	27.0 - 89.8	77.2 ± 6.0	39.8 - 88.7	79.6 ± 6.0	38.8 - 89.4
		G2	74.3 ± 6.9	21.0 - 85.7	72.5 ± 7.4	28.7 - 83.7	75.2 ± 7.1	25.5 - 85.0
		G3	63.5 ± 5.2	17.5 - 74.2	63.1 ± 5.5	24.3 - 72.3	62.9 ± 5.7	24.2 - 74.1
		G4	21.1 ± 5.0	6.9 - 29.4	18.8 ± 5.0	8.4 - 27.0	19.2 ± 5.0	4.9 - 29.0
	Weekend	G1	49.7 ± 21.9	30.2 - 62.6	47.3 ± 21.6	31.2 - 65.7	47.8 ± 22.4	34.2 - 63.8
		G2	65.0 ± 12.9	30.3 - 82.6	63.3 ± 13.2	40.2 - 76.7	65.3 ± 12.8	40.2 - 83.4
		G3	56.9 ± 15.1	27.8 - 73.9	59.0 ± 14.9	40.3 - 77.1	58.5 ± 15.6	39.1 - 72.5
		G4	19.5 ± 9.4	9.0 - 26.8	17.2 ± 8.8	10.8 - 25.9	15.8 ± 9.0	5.8 - 24.0
F-score (%)	Weekday	G1	81.8 ± 5.2	33.9 - 90.9	80.8 ± 5.3	55.9 - 90.3	83.3 ± 5.3	55.8 - 90.3
		G2	77.6 ± 6.7	28.2 - 87.7	75.4 ± 7.2	43.6 - 86.0	78.3 ± 7.0	38.8 - 88.1
		G3	66.3 ± 5.5	18.4 - 76.4	65.9 ± 5.7	33.1 - 74.9	66.0 ± 5.9	33.0 - 76.2
		G4	22.2 ± 5.8	7.7 - 31.2	20.8 ± 5.8	10.6 - 27.5	20.9 ± 5.7	7.3 - 30.9
	Weekend	G1	52.9 ± 22.1	36.0 - 66.4	51.5 ± 21.9	32.6 - 70.0	52.4 ± 22.4	40.9 - 68.1
		G2	66.7 ± 12.9	37.2 - 81.5	65.4 ± 12.8	50.6 - 76.6	67.6 ± 12.6	50.9 - 81.7
		G3	58.4 ± 14.7	30.0 - 73.4	60.7 ± 14.5	38.9 - 76.7	60.2 ± 15.0	37.8 - 72.3
		G4	20.6 ± 10.2	8.8 - 30.3	18.9 ± 9.9	10.5 - 27.1	17.1 ± 9.3	5.5 - 26.2

Shadow area: ▓ >75%   ░ 75-50%   □ <50%
Removed days: 2010-04-27, STCP strike day; 2010-05-14, visit of pope in the city.

A detailed analysis of *Accuracy* revealed that the three methods have different performance levels related to the spatiotemporal characteristics of the groups. For Group 1 while Top-K shows better performance in the first five weeks, the J48 method performs better for the last four weeks. The NB was better than the other two in only 17% of the days. In Group 2, the Top-K method performs better in 46% of the days, followed by the J48 with 41% and NB with 13%. In contrast to the previous two groups, in Group 3 the NB method performs better in the first three weeks of predictions and in 34% of the days overall, with the J48 method performing better in 46% and the Top-K in 20%. Interestingly, the first three weeks are very similar in terms of performance between the NB and Top-K, with J48 and NB in the remaining ones. Similarly, in Group 4 the NB method performs best in the first two weeks, down to 17% overall. Top-K performs better in only 7% of the days, and J48 in 76% of them.

With the exception of the last Group 4, both *Accuracy* and *F-score* measures show similar results. However, in Group 4, the *F-score* measure reveals that both NB and Top-K perform better in 20% of the days, and J48 in 60%. In addition, the F-score performance tends to show better performance for Top-K for Groups 2, 3 and 4 in detriment of J48. With the exception of the mentioned differences, the similarity between *Accuracy* and *F-score* measures indicates robustness in the results obtained.

Figure 3 shows the average *F-score* for one algorithm, Top-K. Whereas in the first 1-2 weeks of prediction, the F-score values increase steeply, and almost duplicate for Groups 1 and 2, after this period it increases very slowly until 5-6 weeks. The exception is Group 4, with a slow grow tendency for the entire period of 2 months.

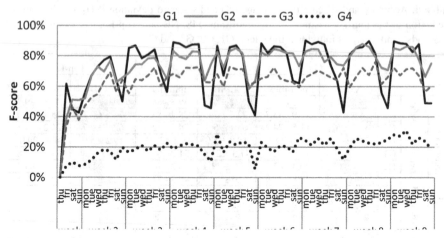

Fig. 3. F-score average for the prediction of the journey destination, by group of individual travellers obtained with the application of the Top-K algorithm.

4   Conclusions

In this work, an investigation into journey prediction was performed, based on past data and mobility patterns. Three different methods were used to predict journey destination for four different groups of travellers and spatiotemporal characteristics. The main findings obtained are described in the previous Section provide answers to the questions originally formulated as follows:

- The results show that it is indeed possible to predict journey destination in UPT based on past usage, with varying degrees of success depending on the mobility patterns. In addition, the accuracy of journey predictions tends to stabilize after two or more weeks of data (historical journeys), with considerable differences between the groups;
- After the initial two weeks of prediction, the average *Accuracy* and *F-score* has values around 70% for Groups 1 and 2, 60% for Group 3 and 20% for Group 4. The performance of journey predictions seems to be directly related to the mobility patterns, with stricter characteristics scoring higher prediction performance;
- Weekend's present low mean and high standard deviation values of *Accuracy* and *F-score* (between 6% and 40% lower than weekdays). This difference increases in groups with more frequent journeys. Group 1 and 2 have low *Accuracy* and *F-score* in the weekends but higher values in weekdays. The difference is negligible for Group 3 and non-existing in Group 4.

Even though the three methods present similar results overall, the analysis shows that certain scenarios allow them to perform differently. The performance differences are mainly related to the level of historic data, day of the week and travelling patterns. Thus, journey prediction is impacted by a number of factors that inform the design and implementation of TIS.

Future work will enable a comparison between used classifiers regarding processing time and memory efficiency. We did not approach these metrics in this work, due to space restrictions. Further research is also demanded on the characterization of groups of travellers, regarding further analysis to enable the discovery of additional groups of typical traveller's profile. We will hopefully be able to find and study these new profiles with a vaster dataset of users.

Acknowledgements. This work was performed under the project "Seamless Mobility" (FCOMP-01-0202-FEDER-038957), financed by European Regional Development Fund (ERDF), through the Operational Programme for Competitiveness Factors (POFC) in the National Strategic Reference Framework (NSRF), within the Incentive System for Technology Research and Development. The authors would also like to acknowledge the bus transport provider of Porto, STCP, which provided travel data for the project.

References

1. Bagchi, M., White, P.R.: The potential of public transport smart card data. Transport Policy **12**(5), 464–474 (2005)
2. Bera, S., Rao, K.V.: Estimation of origin-destination matrix from traffic counts: the state of the art, European Transport\Trasporti Europei, ISTIEE, Institute for the Study of Transport within the European Economic Integration, vol. 49, pp. 2–23 (2011)
3. Caragliu, A., Bo, C.D., Nijkamp, P.: Smart Cities in Europe. J. of Urban Technology **18**(2), 65–82 (2011)
4. Costa, P.M., Fontes, T., Nunes, A.N., Ferreira, M.C., Costa, V., Dias, T.G., Falcão e Cunha, J.: Seamless Mobility: a disruptive solution for public urban transport. In: 22nd ITS World Congress, 5-9/10, Bordeux (2015)
5. Dziekan, K., Kottenhoff, K.: Dynamic at-stop real-time information displays for public transport: effects on customers. Transp. Research Part A **41**(6), 489–501 (2007)
6. Foth, M., Schroeter, R., Ti, J.: Opportunities of public transport experience enhancements with mobile services and urban screens. Int. J. of Ambient Computing and Intelligence (IJACI) **5**(1), 1–18 (2013)
7. Giannopoulos, G.A.: The application of information and communication technologies in transport. European J. of Operational Research **152**(2), 302–320 (2004)
8. Gordon, J.B., Koutsopoulos, H.N., Wilson, N.H.M., Attanucci, J.P.: Automated Inference of Linked Transit Journeys in London Using Fare-Transaction and Vehicle Location Data. Transp. Res. Record: J. of the Transportation Research Board **2343**, 17–24 (2013)
9. He, H., Garcia, E.: Learning form imbalanced data. IEEE Transactions on Knowledge and Data Engineering **21**(9), 1263–1284 (2009)
10. INE (2013). https://www.ine.pt/. Instituto Nacional de Estatística I.P., Portugal
11. Kieu, L.M., Bhaskar, A., Chung, E.: Transit passenger segmentation using travel regularity mined from Smart Card transactions data. In: Transportation Research Board 93rd Annual Meeting. Washington, D.C., January 12–16, 2014
12. Krizek, J.J., El-Geneidy, A.: Segmenting preferences and habits of transit users and non-users. Journal of Public Transportation **10**(3), 71–94 (2007)
13. Kusakabe, T., Asakura, Y.: Behavioural data mining of transit smart card data: A data fusion approach. Transp. Research Part C **46**, 179–191 (2014)

14. Ma, X., Wu, Y.-J., Wanga, Y., Chen, F., Liu, J.: Mining smart card data for transit riders' travel patterns. Transp. Research Part C **36**, 1–12 (2013)
15. Metwally, A., Agrawal, D.P., El Abbadi, A.: Efficient computation of frequent and top-k elements in data streams. In: Eiter, T., Libkin, L. (eds.) ICDT 2005. LNCS, vol. 3363, pp. 398–412. Springer, Heidelberg (2005)
16. Mikluščák, T., Gregor, M., Janota, A.: Using neural networks for route and destination prediction in intelligent transport systems. In: Mikulski, J. (ed.) TST 2012. CCIS, vol. 329, pp. 380–387. Springer, Heidelberg (2012)
17. Nor Haizan, W., Mohamed,W., Salleh, M.N.M., Omar, A.H.: A Comparative Study of Reduced Error Pruning Method in Decision Tree Algorithms. In: International Conference on Control System, Computing and Engineering (IEEE). Penang, Malaysia, November 25, 2012
18. Nunes, A., Dias, T.G., Cunha, J.F.: Passenger Journey Destination Estimation from Automated Fare Collection System Data Using Spatial Validation. IEEE Transactions on Intelligent Transportation Systems. Forthcoming
19. Patil, T., Sherekar, S.: Performance Analysis of Naive Bayes and J48 Classification Algorithm for Data Classification. Int. J. of Comp. Science and Applic. **5**(2), 256–261 (2013)
20. Patterson, D.J., Liao, L., Gajos, K., Collier, M., Livic, N., Olson, K., Wang, S., Fox, D., Kautz, H.: Opportunity knocks: a system to provide cognitive assistance with transportation services. In: Mynatt, E.D., Siio, I. (eds.) UbiComp 2004. LNCS, vol. 3205, pp. 433–450. Springer, Heidelberg (2004)
21. Pelletier, M.P., Trépanier, M., Morency, C.: Smart card data use in public transit: A literature review. Transp Research Part C **19**(4), 557–568 (2011)
22. Sarmento, R., Cordeiro, M., Gama, J.: Streaming network sampling using top-k neworks. In: Proceedings of the 17th International Conference on Enterprise Information Systems (ICEIS 2015), p. to appear. INSTICC (2015)
23. Seaborn, C., Attanucci, J., Wilson, H.M.: Analyzing multimodal public transport journeys in London with smart card fare payment data. Transp. Research Record: J. of the Transp. Research Board **2121**(1) (2009)
24. TIP (2015). http://www.linhandante.com/. Transportes Intermodais do Porto
25. Utsunomiya, M., Attanucci, J., Wilson, N.H.: Potential Uses of Transit Smart Card Registration and Transaction Data to Improve Transit Planning. Transp. Research Record: J. of the Transp. Research Board 119–126 (2006)
26. Zito, P., Amato, G., Amoroso, S., Berrittella, M.: The effect of Advanced Traveller Information Systems on public transport demand and its uncertainty. Transportmetrica **7**(1), 31–43 (2011)

Demand Modelling for Responsive Transport Systems Using Digital Footprints

Paulo Silva, Francisco Antunes$^{(\boxtimes)}$, Rui Gomes, and Carlos Bento

Faculdade de Ciências e Tecnologia da Universidade de Coimbra Pólo II,
Rua Sílvio Lima, 3030-790 Coimbra, Portugal
pacsilva@student.dei.uc.pt, {fnibau,ruig,bento}@dei.uc.pt

Abstract. Traditionally, travel demand modelling focused on long-term multiple socio-economic scenarios and land-use configurations to estimate the required transport supply. However, the limited number of transportation requests in demand-responsive flexible transport systems require a higher resolution zoning. This work analyses users short-term destination choice patterns, with a careful analysis of the available data coming from various different sources, such as GPS traces and social networks. We use a Multinomial Logit Model, with a social component for utility and characteristics, both derived from Social Network Analyses. The results from the model show meaningful relationships between distance and attractiveness for all the different alternatives, with the variable distance being the most significant.

Keywords: Innovative transport modes · Public transport operations · Transport demand and behaviour · Urban mobility and accessibility

1 Introduction

Transportation systems are a key factor for economic sustainability and social welfare, but providing quality public transportation may be extremely expensive when demand is low, variable and unpredictable, as it is on some periods of the day in urban areas. Demand Responsive Transportation (DRT) services try to address this problem with routes and frequencies that may vary according to the actual observed demand. However, in terms of financial sustainability and quality level, the design of this type of services may be complicated.

Anticipating demand by studying users short-term destination choice can improve the overall efficiency and sustainability of the transport services. Traditionally, demand modelling focused on long-term socio-economic scenarios and land-use to estimate the required level of supply. However, the limited number of transportation requests in DRT systems does not allow the application of traditional models. Also, DRTs require a higher resolution zoning, otherwise it can lead to unacceptable inaccuracies. Information coming from various sources should be used effectively in order to model demand for DRTs trips.

The approach followed in this work analyses users short-term destination choice patterns, with a careful analysis of the available data coming from various different sources, such as, GPS traces and social networks. The theory of

© Springer International Publishing Switzerland 2015
F. Pereira et al. (Eds.) EPIA 2015, LNAI 9273, pp. 181–186, 2015.
DOI: 10.1007/978-3-319-23485-4_19

utility maximization, usually through discrete choice modelling, is often used to study individual decision-making. We use the Multinomial Logit Model (MNL) with a social component for utility and characteristics, both derived from Social Network Analyses (SNA), where a network is constructed linking the nodes (decision makers) that have social influence over one another (friendship), and the strength of that influence. To measure different ties strength, mutuality, propinquity, mutual friends and multiplexity factors were used.

We review the state of the art in the next section. The methodology is presented in Section 3 and the results in Section 4. The documents ends with the conclusions and possible future lines of work.

2   State of the Art

Urban movements profiling has usually relied on traditional survey methods that are expensive and time consuming, giving planners only a picture of what has happened. In contrast, the wide deployment of pervasive computing devices (cell phone, GPS devices and digital cameras) provide unprecedented digital footprints, telling where and when people are. An emerging field of research uses mobile phones for "urban sensing" [1]. Moreover, the past few years have witnessed a huge increase in the adoption of social media and transportation researchers have also realized the potential of SNA for demand modelling [2].

A growing research topic is understanding how trips, trip modes and trip purposes can be derived from GPS data. For instance, [3] propose an approach to predict both the intended destination and route of a person by exploiting personal movement data collected by GPS. GPS data, however, have some limitations, such as (1) GPS signals are usually blocked indoor, (2) GPS devices may get interferences near tall buildings, and (3) continuously collecting GPS data may consume devices energy quickly.

Social networks and human interactions are crucial not only for understanding social activities, but also for travel patterns [2]. [4] connects travel with social networks, arguing that daily life revolves around family, colleagues, friends and shopping. [5] refers to the conformation to social norms, implying that decision-makers are more likely to choose a particular alternative if more peers have already chosen the same alternative. The emergence of geolocated social media seems a good opportunity to address SNA's lack of geographic consideration. For instance, [6] presents a technique to analyse large-scale geo-location data from social media to infer individual activity patterns.

Previously, for travel demand modelling aggregate approaches were used, such as gravity or entropy models. These approaches were gradually replaced by disaggregated models [7]. In discrete choice modelling, the effect of social dimensions was first formalized for the binomial and the multinomial cases in [8] and [9], respectively. Generically, the agents' utility is formed by both private and social components. The private component corresponds to the decision-makers characteristics. The social component represents the strength of social utility and the percentage of others in the neighbourhood selecting the same alternative in the choice set [10].

3 Methodology

3.1 Data Gathering

We use GPS data traces provided by TU Delft from 80 individuals over the course of four days, and also data collect from social networks, namely Twitter, Instagram and Foursquare. The data obtained is cleared of personal values as to ensure privacy. To get the geo-located points of interest, we use the FourSquare API, extracting the 50 most popular venues, within a radius of 30 meters for each given point, resulting in a total of 37506 venues, in 489 categories, with their identification, geo-location and total number of check-ins made. The subscription zone for Instagram had a radius of 5 kilometers from the city center. For Twitter, we covered a bigger area in order to get Delft surroundings.

3.2 Social Network Analysis

Friendship. To get the friendship, we have to use the user unique identification from the post, and request the users that the user followed and that follow him back. The only significant friendships considered are the ones between users that posted around Delft. The total number of friendships used is 35457. Discrete choice model also had to take into account the strength between users. To get and measure the ties strength, tie mutuality, propinquity, mutual friends and multiplexity factors are used.

Detecting Important Locations. To build the MNL we also need to know the user home and work location, since we are only interested in the user movement patterns before and after work hours. Home and work are the starting points for which the distance to the points of interest are measured. To get these locations, we use a clustering algorithm, namely *DBScan* [11].

3.3 Data Preparation

In the data set with the posts and associated venues, i.e., the choice set (CS), there is a large amount of data with no use for us, as it does not provide useful information (for instance, useless categories) or represent work or residential places, for which the demand patterns are well established and can be met by traditional transportation services. The data containing those specific categories was erased from the choice set. Since the number of alternatives is quite big, we grouped those venues in 6 main categories: Appointment (17%), Food (17%), Bar (5%), Shop (24%), Entertainment (27%) and Travel (10%).

If we used these categories as our number of different alternatives for the MNL model, we would only get results concerning each of those 6 alternatives, which are quite generic. However, we want to use the model to predict probabilities of destination choices with a higher resolution, so we generated data for all the venues and then use those categories only to filter unnecessary data.

3.4   Multinomial Logit Model

Our data corresponds to the observed choices of individuals - revealed preferences data. For each dataset we have the number of alternatives selected in each hour, which is our finite set of alternatives for each individual. The number of alternatives and observations vary significantly along the hours. The variables used for the data-frame are:

- distance : the venue distance to the user central point,
- check-ins : the total number of check-ins in each alternative for each user,
- friendship : the sum of the individual friendship for each alternative,
- choice : the alternative selection.

Since we cannot directly extract user personal information (e.g. age, gender), our data does not contain individual specific variables, and so the alternative specific variables have a generic coefficient, i.e., we consider that the number of check-ins, distance and friendship have the same value for all alternatives. Choice takes values of yes and no, if the alternative was chosen or not by the user. To estimate the MNL we have used the R statistics system with the mlogit package. The following formula was used for our work,

$$Mlogit(choice \sim distance + friendship + attractiveness, CS)$$

where choice is the variable that indicates the choice made for each individual among the alternatives and the distance, friendship and attractiveness being the alternative specific variables with generic coefficients from the choice set CS.

4   Results

We present the results and estimation parameter for one choice set, namely the one representing the choices made at hour 21, which has 24 alternatives and 91 observations.

The model predictions are reasonable good when tested against the user observed choices. Table 1 presents the average probabilities returned by the model against the observed frequency. The results from the MNL model show meaningful relationships between distance and attractiveness for all the different alternatives, being distance the most significant variable, i.e., longer distances almost always reduce the attractiveness of a destination, all else being equal.

The same can be said for the attractiveness variable, but the friendship variable does not have the same impact to the individual when choosing an alternative. Table 2 illustrates these findings. To show the usefulness of the analyses made, we feed the probabilities predicted by our model to a DRT simulator developed in [12]. Figure 1 shows that most origins and destinations found for the time period and travel objective considered lie outside the service area of the different public transport modes (dotted lines) and DRT could satisfy this demand (solid lines).

Table 1. Average probabilities returned by the model

Venue	Freq.	Avg. Prob.
Stadion Feijenoord	0.032967	0.04490835
EkoPlaza	0.065934	0.03791752
Station Den Haag HS	0.054945	0.05494505
La Mer	0.032967	0.03303908
Diner Company	0.032967	0.03777430
LantarenVenster	0.032967	0.04320755
Station Rotterdam Centraal	0.153846	0.12336128
BIRD	0.043956	0.03794951
Maassilo	0.043956	0.03530117
Emma	0.021978	0.02478593
Station Den Haag Centraal	0.054945	0.05070866
Lucent Danstheater	0.032967	0.02519137
Zaal 3	0.032967	0.03212339
De Banier	0.054945	0.09073518
Randstadrail javalaan	0.032967	0.02411359
Kot Treinpersoneel	0.032967	0.02666024
Spuimarkt	0.032967	0.02543518
Doerak	0.032967	0.04017886
Paard van Troje	0.032967	0.03294463
Ahoy Rotterdam	0.043956	0.03654922
Restaurant Meram	0.043956	0.03907711
Live Tv Show	0.021978	0.04634078
Stadskwekerij den haag	0.010989	0.03433138
Oudedijk 166 A2	0.032967	0.04312563

Table 2. Relationships between variables

Variables	Estimate	Std.error	t-value	p-value
distance	-0.107414	0.022597	-4.7534	2.000e-06
friendship	0.094441	0.081570	1.1578	0.2469
attractiveness	0.342170	0.061907	5.5272	3.254e-08

Fig. 1. Simulation results

5   Conclusions

Traditionally, travel demand modelling focused on long-term socio-economic scenarios and land-use to estimate the required transport supply. However, the limited number of transportation requests in demand-responsive flexible transport
systems require a higher resolution zoning. We analysed users short-term destination choice patterns, with a careful analysis of the available data coming from
GPS traces and social networks. We defined a Multinomial Logit Model (MNL),
with a social component for utility and characteristics, both derived from Social

Network Analyses. The low frequency of posts with identified locations for each user made it difficult to generate a clear pattern for each user. Nevertheless, the results from the model show meaningful relationships between distance and attractiveness for all the different alternatives, with the variable distance being the most significant.

Since the analyses of the social network done in this work does not produce individual characteristics, like age, gender and socio-economic, it would be interesting for future work to include data mining algorithms to extract some of those values from tweets, and add features specific to each venue, to better understand the motivation behind the choice made.

References

1. Cuff, D., Hansen, M., Kang, J.: Urban sensing: Out of the woods. Communications of the ACM **51**(3), 24–33 (2008)
2. Carrasco, J., Hogan, B., Wellman, B., Miller, E.: Collecting social network data to study social activity-travel behaviour: an egocentric approach. Environment and Planning B: Planning and Design **35**, 961–980 (2008)
3. Chen, L., Mingqi, L., Chen, G.: A system for destination and future route prediction based on trajectory mining. Pervasive and Mobile Computing **6**(6), 657–676 (2010)
4. Axhausen, K.: Social networks, mobility biographies, and travel: survey challenges. Environment and Planning B: Planning and Design **35**, 981–996 (2008)
5. Paáez, A., Scott, D.: Social influence on travel behavior: a simulation example of the decision to telecommute. Environment and Planning A **39**(3), 647–665 (2007)
6. Hasan, S., Ukkusuri, S.: Urban activity pattern classification using topic models from online geo-location data. Transportation Research Part C: Emerging Technologies **44**, 363–381 (2014)
7. Ben-Akiva, M., Lerman, S.: Discrete Choice Analysis: Theory and Application to Travel Demand (1985)
8. Brock, W., Durlauf, S.: Discrete Choice with Social Interactions. Review of Economic Studies **68**(2), 235–260 (2001)
9. Brock, W., Durlauf, S.: A multinomial choice model with neighborhood effects. American Economic Review **92**, 298–303 (2002)
10. Zanni, A.M., Ryley, T.J.: Exploring the possibility of combining discrete choice modelling and social networks analysis: an application to the analysis of weather-related uncertainty in long-distance travel behaviour. In: International Choice Modelling Conference, Leeds, pp. 1–22 (2011)
11. Ester, M., Kriegel, H., Sander, J., Xu, X.: A density-based algorithm for discovering clusters in large spatial databases with noise. In: Proceedings of 2nd International Conference on Knowledge Discovery and Data Mining, pp. 226–231. AAAI Press (1996)
12. Gomes, R., Sousa, J.P., Galvao, T.: An integrated approach for the design of Demand Responsive Transportation services. In: de Sousa, J.F., Rossi, R. (eds.) Computer-based Modelling and Optimization in Transportation. AISC, vol. 232, pp. 223–235. Springer, Heidelberg (2014)

Artificial Life and Evolutionary Algorithms

A Case Study on the Scalability of Online Evolution of Robotic Controllers

Fernando Silva[1,2,4]([✉]), Luís Correia[4], and Anders Lyhne Christensen[1,2,3]

[1] BioMachines Lab, Lisboa, Portugal
fsilva@di.fc.ul.pt, anders.christensen@iscte.pt
[2] Instituto de Telecomunicações, Lisboa, Portugal
[3] Instituto Universitário de Lisboa (ISCTE-IUL), Lisboa, Portugal
[4] BioISI, Faculdade de Ciências, Universidade de Lisboa, Lisboa, Portugal
luis.correia@ciencias.ulisboa.pt

Abstract. Online evolution of controllers on real robots typically requires a prohibitively long evolution time. One potential solution is to distribute the evolutionary algorithm across a group of robots and evolve controllers in parallel. No systematic study on the scalability properties and dynamics of such algorithms with respect to the group size has, however, been conducted to date. In this paper, we present a case study on the scalability of online evolution. The algorithm used is odNEAT, which evolves artificial neural network controllers. We assess the scalability properties of odNEAT in four tasks with varying numbers of simulated e-puck-like robots. We show how online evolution algorithms can enable groups of different size to leverage their multiplicity, and how larger groups can: (i) achieve superior task performance, and (ii) enable a significant reduction in the evolution time and in the number of evaluations required to evolve controllers that solve the task.

Keywords: Evolutionary robotics · Artificial neural network · Evolutionary algorithm · Online evolution · Robot control · Scalability

1 Introduction

Evolutionary computation has been widely studied and applied to synthesise controllers for autonomous robots in the field of evolutionary robotics (ER). In *online* ER approaches, an evolutionary algorithm (EA) is executed onboard robots during task execution to continuously optimise behavioural control. The main components of the EA (evaluation, selection, and reproduction) are performed by the robots without any external supervision. Online evolution thus enables addressing tasks that require online learning or online adaptation. For instance, robots can evolve new controllers and modify their behaviour to respond to unforeseen circumstances, such as changes in the task or in the environment.

F. Pereira et al. (Eds.) EPIA 2015, LNAI 9273, pp. 189–200, 2015.
DOI: 10.1007/978-3-319-23485-4_20

Research in online evolution started out with a study by Floreano and Mondada [1], who conducted experiments on a real mobile robot. The authors successfully evolved navigation and obstacle avoidance behaviours for a Khepera robot. The study was a significant breakthrough as it demonstrated the potential of online evolution of controllers. Researchers then focused on how to mitigate the issues posed by evolving controllers directly on real robots, especially the prohibitively long time required [2]. Watson *et al.* [3] introduced an approach called *embodied evolution* in which an online EA is distributed across a group of robots. The main motivation behind the use of multirobot systems was to leverage the potential speed-up of evolution due to robots that evolve controllers in parallel and that exchange candidate solutions to the task.

Over the past decade, numerous approaches to online evolution in multirobot systems have been developed. Examples include Bianco and Nolfi's open-ended approach for self-assembling robots [4], mEDEA by Bredeche *et al.* [5], and odNEAT by Silva *et al.* [6]. When the online EA is decentralised and distributed across a group of robots, one common assumption is that online evolution inherently scales with the number of robots [3]. Generally, the idea is that the more robots are available, the more evaluations can be performed in parallel, and the the faster the evolutionary process [3]. The dynamics of the online EA itself, and common issues that arise in EAs from population sizing such as convergence rates and diversity [7] have, however, not been considered. Furthermore, besides adhoc experiments with large groups of robots, see [5] for examples, there has been no systematic study on the scalability properties of online EAs across different tasks. Given the strikingly long time that online evolution requires to synthesise solutions to any but the simplest of tasks, the approach remains infeasible on real robots [8].

In this paper, we study the scalability properties of online evolution of robotic controllers. The online EA used in this case study is odNEAT [9], which optimises artificial neural network (ANN) controllers. One of the main advantages of odNEAT is that it evolves both the weights and the topology of ANNs, thereby bypassing the inherent limitations of fixed-topology algorithms [9]. odNEAT is used here as a representative efficient algorithm that has been successfully used in a number of simulation-based studies related to adaptation and learning in robot systems, see [6, 8–11] for examples. We assess the scalability properties and performance of odNEAT in four tasks involving groups of up to 25 simulated e-puck-like robots [12]: (i) an aggregation task, (ii) a dynamic phototaxis task, and (iii, iv) two foraging tasks with differing complexity. Overall, our study shows how online EAs can enable groups of different size to leverage their multiplicity for higher performance, and for faster evolution in terms of evolution time and number of evaluations required to evolve effective controllers.

2   Online Evolution with odNEAT

This section provides an overview of odNEAT; for a comprehensive introduction see [9]. odNEAT is an efficient online neuroevolution algorithm designed for multirobot systems. The algorithm starts with minimal networks with no hidden neurons, and with each input neuron connected to every output neuron. Throughout evolution, topologies are gradually complexified by adding new neurons and new connections through mutation. In this way, odNEAT is able find an appropriate degree of complexity for the current task, and a suitable ANN topology is the result of a continuous evolutionary process [9].

odNEAT is distributed across multiple robots that exchange candidate solutions to the task. The online evolutionary process is implemented according to a physically distributed island model. Each robot optimises an internal population of genomes (directly encoded ANNs) through intra-island variation, and genetic information between two or more robots is exchanged through inter-island migration. In this way, each robot is potentially self-sufficient and the evolutionary process opportunistically capitalises on the exchange of genetic information between multiple robots for collective problem solving [9].

During task execution, each robot is controlled by an ANN that represents a candidate solution to a given task. Controllers maintain a virtual energy level reflecting their individual performance. The fitness value is defined as the mean energy level. When the virtual energy level of a robot reaches a minimum threshold, the current controller is considered unfit for the task. A new controller is then created via selection of two parents from the internal population, crossover of the parents' genomes, and mutation of the offspring. Mutation is both structural and parametric, as it adds new neurons and new connections, and optimises parameters such as connection weights and neuron bias values.

odNEAT has been successfully used in a number of simulation-based studies related to long-term self-adaptation in robot systems. Previous studies have shown: (i) that odNEAT effectively evolves controllers for robots that operate in dynamic environments with changing task parameters [11], (ii) that the controllers evolved are robust and can often adapt to changes in environmental conditions without further evolution [9], (iii) that robots executing odNEAT can display a high degree of fault tolerance as they are able to adapt and learn new behaviours in the presence of faults in the sensors [9], (iv) how to extend the algorithm to incorporate learning processes [11], and (v) how to evolve behavioural building blocks prespecified by the human experimenter [8,10]. Given previous results, odNEAT is therefore used in our study as a representative online EA. The key research question of our study is if and how online EAs can enable robots to leverage their multiplicity. That is, besides performance and robustness criteria, we are interested in studying scalability with respect to the group size, an important aspect when large groups of robots are considered.

3   Methods

In this section, we define our experimental methodology, including the simulation platform and robot model, and we describe the four tasks used in the study: aggregation, phototaxis, and two foraging tasks with differing complexity.

3.1   Experimental Setup

We use JBotEvolver [13] to conduct our simulation-based experiments. JBotE-
volver is an open-source, multirobot simulation platform and neuroevolution
framework. In our experiments, the simulated robots are modelled after the e-
puck [12], a 7.5 cm in diameter, differential drive robot capable of moving at
speeds up to 13 cm/s. Each robot is equipped with infrared sensors that mul-
tiplex obstacle sensing and communication between robots at a range of up to
25 cm.[1] The sensor and actuator configurations for the tasks are listed in Table 1.
Each sensor and each actuator are subject to noise, which is simulated by adding
a random Gaussian component within ± 5% of the sensor saturation value or of
the current actuation value.

The robot controllers are discrete-time ANNs with connection weights in the
range [-10,10]. odNEAT starts with simple networks with no hidden neurons,
and with each input neuron connected to every output neuron. The ANN inputs
are the readings from the sensors, normalised to the interval [0,1]. The output
layer has two neurons whose values are linearly scaled from [0,1] to [-1,1] to set
the signed speed of each wheel. In the two foraging tasks, a third output neuron
sets the state of a gripper. The gripper is activated if the output value of the
neuron is higher than 0.5, otherwise it is deactivated. If the gripper is activated,
the robot collects the closest resource within a range of 2 cm, if there is any.
Depending on the foraging task (see below), the robot may need to actively
select which type of resources to collect and to avoid other types.

Aggregation Task. In an aggregation task, dispersed robots must move close
to one another to form a single cluster. Aggregation combines several aspects of
multirobot tasks, including distributed individual search, coordinated movement,
and cooperation. Furthermore, aggregation plays an important role in robotics
because it is a precursor of other collective behaviours such as group transport
of heavy objects [14]. In our aggregation task, robots are evaluated based on
criteria that include the presence of robots nearby, and the ability to explore the
arena and move fast, see [9] for details. The initial virtual energy level E of each
controller is set to 1000 and limited to the range $[0, 2000]$ units. At each control
cycle, the update of the virtual energy level, E, is given by:

$$\frac{\Delta E}{\Delta t} = \alpha(t) + \gamma(t) \tag{1}$$

where t is the current control cycle, $\alpha(t)$ is a reward proportional to the number n of different genomes received in the last $P = 10$ control cycles. Because
robots executing odNEAT exchange candidate solutions, the number of different

---

[1] The original e-puck infrared range is 2-3 cm [12]. In real e-pucks, the *liblrcom* library,
see http://www.e-puck.org, extends the range up to 25 cm and multiplexes infrared
communication with proximity sensing.

Table 1. Controller details. Light sensors have a range of 50 cm (phototaxis task). Other sensors have a range of 25 cm.

Aggregation task – controller details
Input neurons: 18
8 for IR robot detection
8 for IR wall detection
1 for energy level reading
1 for reading the number of different genomes received
Output neurons: 2 Left and right motor speeds
Phototaxis task – controller details
Input neurons: 25
8 for IR robot detection
8 for IR wall detection
8 for light source detection
1 for energy level reading
Output neurons: 2 Left and right motor speeds
Foraging tasks – controller details
Input neurons: 25
4 for IR robot detection
4 for IR wall detection
1 for energy level reading
8 for resource A detection
8 for resource B detection
Output neurons: 3
2 for left and right motor speeds
1 for controlling the gripper

genomes received is used to estimate the number of robots nearby. $\gamma(t)$ is a factor related to the quality of movement computed as:

$$\gamma(t) = \begin{cases} \text{-1} & \text{if } v_l(t) \cdot v_r(t) < 0 \\ \Omega_s(t) \cdot \omega_s(t) & \text{otherwise} \end{cases} \tag{2}$$

where $v_l(t)$ and $v_r(t)$ are the left and right wheel speeds, $\Omega_s(t)$ is the ratio between the average and maximum speed, and $\omega_s(t) = \sqrt{v_l(t) \cdot v_r(t)}$ rewards controllers that move fast and straight at each control cycle.

Phototaxis Task. In a phototaxis task, robots have to search and move towards a light source. Following [9], we use a dynamic version of the phototaxis task in which the light source is periodically moved to a new random location. As a result, robots have to continuously search for and reach the light source, which eliminates controllers that find the light source by chance. The virtual energy level $E \in [0, 100]$ units, and controllers are assigned an initial value of 50 units. At each control cycle, E is updated as follows:

$$\frac{\Delta E}{\Delta t} = \begin{cases} S_r & \text{if } S_r > 0.5 \\ 0 & \text{if } 0 < S_r \le 0.5 \\ \text{-0.01} & \text{if } S_r = 0 \end{cases} \tag{3}$$

where S_r is the maximum value of the readings from light sensors, between 0 (no light) and 1 (brightest light). Light sensors have a range of 50 cm and robots are therefore only rewarded if they are close to the light source. Remaining sensors have a range of 25 cm.

Foraging Tasks. In a foraging task, robots have to search for and pick up objects scattered in the environment. Foraging is a canonical testbed in cooperative robotics domains, and is evocative of tasks such as toxic waste clean-up, harvesting, and search and rescue [15].

We setup a foraging task with different types of resources that have to be collected. Robots spend virtual energy at a constant rate and must learn to find and collect resources. When a resource is collected by a robot, a new resource of the same type is placed randomly in the environment so as to keep the number of resource constant throughout the experiments. We experiment with two variants of a foraging task: (i) one in which there are only type A resources, henceforth called *standard foraging task*, and (ii) one in which there are both type A and type B resources, henceforth called *concurrent foraging task*. In the concurrent foraging task, resources A and B have to be consumed sequentially. That is, besides learning the foraging aspects of the task, robots also have to learn to collect resources in the correct order. The energy level of each controller is initially set to 100 units, and limited to the range [0,1000]. At each control cycle, E is updated as follows:

$$\frac{\Delta E}{\Delta t} = \begin{cases} reward & \text{if right type of resource is collected} \\ penalty & \text{if wrong type of resource is collected} \\ \text{-0.02} & \text{if no resource is consumed} \end{cases} \tag{4}$$

where $reward = 10$ and $penalty = -10$. The constant decrement of 0.02 means that each controller will execute for a period of 500 seconds if no resource is collected since it started operating. Note that the *penalty* component applies only to the concurrent foraging task. To enable a meaningful comparison of performance when groups of different size are considered, the number of resources of each type is set to the number of robots multiplied by 10.

3.2   Experimental Parameters and Treatments

We analyse the impact of the group size on the performance of odNEAT by conducting experiments with groups of 5, 10, 15, 20, and 25 robots. For each experimental configuration, we conduct 30 independent evolutionary runs. Each run lasts 100 hours of simulated time. odNEAT parameters are set as in previous studies [9], including a population size of 40 genomes per robot and a control cycle frequency of 100 ms. Robots operate in a square arena surrounded by walls. In the aggregation and phototaxis tasks, the area of the arena is increased proportionally to the number of robots (5 robots: 9 m^2, 10 robots: 18 m^2, ..., 25 robots: 45 m^2). Notice that if we maintained the same size of the environment, comparisons would not be meaningful. For instance, in the aggregation task, with the increasing density of robots in the environment, the task becomes easier to solve simply because robots encounter each other more frequently. In the phototaxis task, the number of light sources in the environment is also increased proportionally to the number of robots.

4   Experimental Results and Discussion

In this section, we present our experimental results. We analyse: (i) the task performance of controllers in terms of their individual fitness score, (ii) the number of evaluations, that is, the number of controllers tested by each robot before a solution to the task is found, and (iii) the corresponding evolution time. We use the two-tailed Mann-Whitney U test to compute statistical significance of differences between results because it is a non-parametric test, and therefore no strong assumptions need to be made about the underlying distributions.

4.1   Quality of the Solutions and Population-Mixing

We first compare the individual fitness scores of the final controllers. In the aggregation task and in the phototaxis tasks, groups of 5 robots are typically

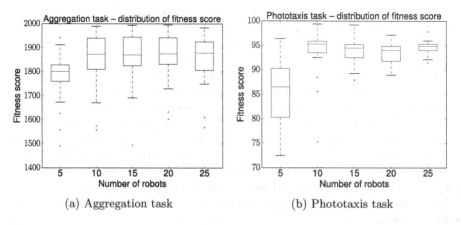

(a) Aggregation task   (b) Phototaxis task

Fig. 1. Distribution of the fitness score of the final controllers in: (a) aggregation task, and (b) phototaxis task.

Table 2. Summary of the individual fitness score of final solutions in the two foraging tasks.

Task	Robots	Mean	Std. dev.	Minimum	Maximum
Standard foraging	5	96.03	45.62	41.88	268.02
	10	105.31	62.31	29.57	396.47
	15	107.98	119.79	33.64	981.34
	20	112.06	109.29	36.69	968.27
	25	136.37	158.39	39.03	994.47
Concurrent foraging	5	104.02	73.11	32.62	459.83
	10	112.02	115.61	39.67	949.35
	15	144.54	153.29	38.85	975.81
	20	165.39	165.82	38.77	971.42
	25	179.29	196.58	38.08	978.56

outperformed by larger groups ($\rho < 0.001$, see Fig. 1). In the phototaxis task, groups of 25 robots also perform significantly better than groups with 20 robots ($\rho < 0.01$). Specifically, results suggest that a minimum of 10 robots are necessary for high-performing controllers to be evolved in a consistent manner.

A summary of the results obtained in the two foraging tasks is shown in Table 2. Given the dynamic nature of task, especially as the number of robots increases, the fitness score of the final controllers displays a high variance. The results, however, further show that larger groups typically yield better performance both in terms of the mean and of the maximum fitness scores, and is an indication that decentralised online approaches such as odNEAT can indeed capitalise on larger groups to evolve more effective solutions to the current task.

To quantify to what extent is a robot dependent on the candidate solutions it receives from other robots, we analyse the origin of the information stored in the population of each robot. In the phototaxis task, when capable solutions have been evolved approximately 86.85% (5 robots) to 93.95% (25 robots) of genomes maintained in each internal population originated from other robots, whereas the remaining genomes stored were produced by the robots themselves (analysis of the results obtained in the other tasks revealed a similar trend). The final solutions executed by each robot to solve the task have on average from 87.26% to 89.10% matching genes. Moreover, 39.73% (5 robots) to 47.70% (25 robots) of these solutions have more than 90% of their genes in common. The average weight difference between matching connection genes varies from 2.48 to 4.37, with each weight in [-10, 10], which indicates that solutions were refined by the EA on the receiving robot. Local exchange of candidate controllers therefore appears to be a crucial part in the evolutionary dynamics of decentralised online EAs because it serves as a substrate for collective problem solving. In the following section, we analyse how the exchange of such information enables online EAs to capitalise on increasingly larger groups of robots for faster evolution of solutions to the task.

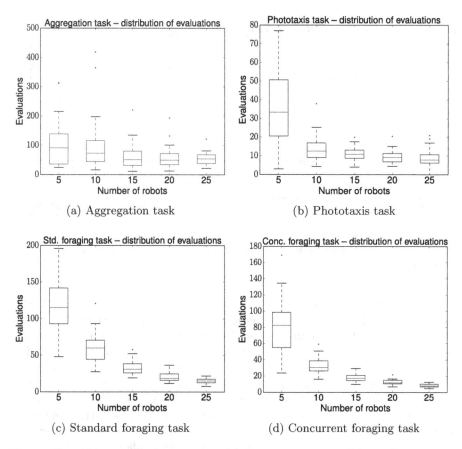

(a) Aggregation task

(b) Phototaxis task

(c) Standard foraging task

(d) Concurrent foraging task

Fig. 2. Distribution of evaluations in: (a) aggregation task, (b) phototaxis task, (c) standard foraging task, and (d) concurrent foraging task.

4.2   Evaluations and Time Analysis

The distribution of evaluations with respect to the group size is shown in Fig. 2. In the aggregation task, the number of evaluations required to evolve solutions to the task decreases as the group size is increased, and becomes significantly lower when the group size is increased from 10 to 15 robots ($\rho < 0.001$). On average, the number of evaluations decreases from 104 for groups of 5 robots to 55 for groups of 25 robots. The mean evolution time is of 6.22 hours for groups of 5 robots, 2.34 hours for 10 robots, 1.80 hours for 15 robots, 1.48 hours for 20 robots, and 1.12 hours for 25 robots. Hence, adding more robots also enables a significant reduction of the evolution time ($\rho < 0.01$ for every group increment). With the increase in the size of the environment, there is a larger area to search for other robots and to explore. Task conditions become more challenging because, in relative terms, each robot senses a smaller portion of the environment. Robots are, however, still able to evolve successful controllers in fewer evaluations and less evolution time.

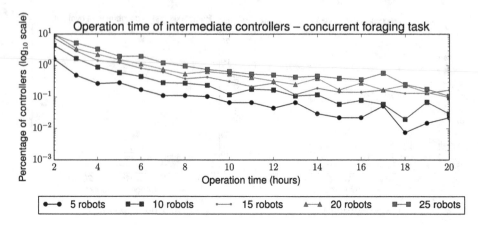

Fig. 3. Operation time of intermediate controllers in the concurrent foraging task. 67% to 96% of intermediate controllers operate for few minutes before they fail (not shown for better plot readability).

The speed up of evolution with the increase of group size also occurs in the phototaxis task. The number of evaluations is significantly reduced ($\rho < 0.001$) with the increase of the group size from 5 to 10 robots (mean number of evaluations of 39 and 14, respectively). The mean evolution time is of 39.16 hours for groups of 5 robots, 9.51 hours for 10 robots, 7.20 hours for 15 robots, 6.30 hours for 20 robots, and 5.27 hours for 25 robots. Similarly to the number of evaluations, the evolution time yields on average a 4-fold-decrease when the group is enlarged from 5 to 10 robots ($\rho < 0.001$). Larger groups enable further improvements ($\rho < 0.001$ for increases up to 20 robots, $\rho < 0.01$ when group size is changed from 20 to 25 robots), but at comparatively smaller rates. Chiefly, the results of the aggregation task and of the phototaxis task show quantitatively distinct speed-ups of evolution when groups are enlarged.

With respect to the two foraging tasks, the distribution of the number of evaluations shown in Fig. 2 is inversely proportional, with a gentle slope, to the number of robots in the group. For both tasks, differences in the number of evaluations are significant across all comparisons ($\rho < 0.001$). In effect, the number of evaluations is reduced on average: (i) from 115 evaluations (5 robots) to 15 evaluations (25 robots) in the standard foraging task, which corresponds to a 7.67-fold decrease in terms of evaluations, and (ii) from 82 evaluations (5 robots) to 8 evaluations in the concurrent foraging task, which amounts to a 10.25-fold decrease. These results show that decentralised online evolution can scale well in terms of evaluations, even when task complexity is increased.

Regarding the evolution time, results show a similar trend for both foraging tasks. On average, the evolution time varies from approximately 35 and 36 hours for groups of 5 robots to 21 and 23 hours for groups of 25 robots. That is, despite significant improvements in terms of the number of evaluations, the evolution time required to evolve the final controllers to the task is still prohibitively long. This result is due to the controller evaluation policy. Online evolution approaches typically employ a policy in which robots substitute controllers at regular time intervals, see [5] for an example. This approach has been shown to lead to incongruous group behaviour and to poor performance in collective tasks that explicitly require continuous collective coordination and cooperation [9]. odNEAT, on the other hand, adopts a different approach by allowing a controller to remain active as long as it is able to solve the task. A new controller is thus only synthesised if the current one fails. As shown in Fig. 3 for the concurrent foraging task, the evaluation policy results in intermediate controllers that operate for a significant amount of time (the standard foraging task displays a similar trend). While 67% to 96% of intermediate controllers only operate for a few minutes (data not shown for better plot readability), there are a few intermediate controllers that operate up to 20 hours of consecutive time before they fail. Although such controllers yield high fitness scores comparable to those of the final solutions, typically 1% to 4% less, they delay the synthesis of more effective solutions.

5   Concluding Discussion and Future Work

In this paper, we presented a case study on the scalability properties and performance of online evolutionary algorithms. We used odNEAT, a decentralised online evolution algorithm in which robots optimise controllers in parallel and exchange candidate solutions to the task. We conducted experiments with groups of up to 25 e-puck-like-robots [12] in four tasks: (i) aggregation, (ii) dynamic phototaxis, and (iii, iv) two foraging tasks with differing complexity.

We showed that larger groups of robots typically enable: (i) superior task performance in terms of the fitness score, and (ii) significant improvements both in terms of the number of evaluations required to evolve solutions to the task and of the corresponding evolution time. There are, however, specific conditions in which intermediate controllers are able to operate up to 20 hours of consecutive time. These controllers yield high performance levels as their fitness score is typically 1% to 4% less than the fitness score of the final solutions. In addition, while additional robots may further speed up evolution, there are specific group sizes after which speed-ups are comparatively smaller. One key research question regarding scalability is therefore how to best leverage *all* robots so that they can learn appropriate behaviours, constitute differentiated groups, and perform cooperative or competitive actions reflecting the structure of the task.

The immediate follow-up work includes studying novel evaluation policies for online evolution of robotic controllers. Regarding odNEAT, the algorithm typically enables a significant reduction in the number of evaluations as the group increases. Hence, if intermediate controllers that operate for long periods of time can be detected and discarded via, for instance, established methods such as early stopping algorithms or racing techniques, there is the potential to enable timely and efficient online evolution in real robots.

Acknowledgments. This work was partly supported by FCT under grants UID/EEA/50008/2013, SFRH/BD/89573/2012, UID/Multi/04046/2013, and EXPL/EEI-AUT/0329/2013.

References

1. Floreano, D., Mondada, F.: Automatic creation of an autonomous agent: Genetic evolution of a neural-network driven robot. In: 3rd International Conference on Simulation of Adaptive Behavior, pp. 421–430. MIT Press, Cambridge (1994)
2. Matarić, M., Cliff, D.: Challenges in evolving controllers for physical robots. Robotics and Autonomous Systems 19(1), 67–83 (1996)
3. Watson, R.A., Ficici, S.G., Pollack, J.B.: Embodied evolution: Distributing an evolutionary algorithm in a population of robots. Robotics and Autonomous Systems 39(1), 1–18 (2002)
4. Bianco, R., Nolfi, S.: Toward open-ended evolutionary robotics: evolving elementary robotic units able to self-assemble and self-reproduce. Connection Science 16(4), 227–248 (2004)
5. Bredeche, N., Montanier, J., Liu, W., Winfield, A.: Environment-driven distributed evolutionary adaptation in a population of autonomous robotic agents. Mathematical and Computer Modelling of Dynamical Systems 18(1), 101–129 (2012)
6. Silva, F., Urbano, P., Oliveira, S., Christensen, A.L.: odNEAT: An algorithm for distributed online, onboard evolution of robot behaviours. In: 13th International Conference on the Simulation and Synthesis of Living Systems, pp. 251–258. MIT Press, Cambridge (2012)
7. De Jong, K.A.: Evolutionary computation: a unified approach. MIT Press, Cambridge (2006)
8. Silva, F., Duarte, M., Oliveira, S.M., Correia, L., Christensen, A.L.: The case for engineering the evolution of robot controllers. In: 14th International Conference on the Synthesis and Simulation of Living Systems, pp. 703–710. MIT Press, Cambridge (2014)
9. Silva, F., Urbano, P., Correia, L., Christensen, A.L.: odNEAT: An algorithm for decentralised online evolution of robotic controllers. Evolutionary Computation (2015) (in press). http://www.mitpressjournals.org/doi/pdf/10.1162/EVCO_a_00141
10. Silva, F., Correia, L., Christensen, A.L.: Speeding Up online evolution of robotic-controllers with macro-neurons. In: Esparcia-Alcázar, A.I., Mora, A.M. (eds.) EvoApplications 2014. LNCS, vol. 8602, pp. 765–776. Springer, Heidelberg (2014)
11. Silva, F., Urbano, P., Christensen, A.L.: Online evolution of adaptive robot behaviour. International Journal of Natural Computing Research 4(2), 59–77 (2014)
12. Mondada, F., Bonani, M., Raemy, X., Pugh, J., Cianci, C., Klaptocz, A., Magnenat, S., Zufferey, J., Floreano, D., Martinoli, A.: The e-puck, a robot designed for education in engineering. In: 9th Conference on Autonomous Robot Systems and Competitions, pp. 59–65, IPCB, Castelo Branco (2009)
13. Duarte, M., Silva, F., Rodrigues, T., Oliveira, S.M., Christensen, A.L.: JBotEvolver: A versatile simulation platform for evolutionary robotics. In: 14th International Conference on the Synthesis and Simulation of Living Systems, pp. 210–211. MIT Press, Cambridge (2014)
14. Groß, R., Dorigo, M.: Towards group transport by swarms of robots. International Journal of Bio-Inspired Computation 1(1–2), 1–13 (2009)
15. Cao, Y., Fukunaga, A., Kahng, A.: Cooperative mobile robotics: Antecedents and directions. Autonomous Robots 4(1), 1–23 (1997)

Spatial Complexity Measure for Characterising Cellular Automata Generated 2D Patterns

Mohammad Ali Javaheri Javid$^{(\boxtimes)}$, Tim Blackwell, Robert Zimmer,
and Mohammad Majid Al-Rifaie

Department of Computing Goldsmiths, University of London,
London SE14 6NW, UK
{m.javaheri,t.blackwell,r.zimmer,m.majid}@gold.ac.uk

Abstract. Cellular automata (CA) are known for their capacity to generate complex patterns through the local interaction of rules. Often the generated patterns, especially with multi-state two-dimensional CA, can exhibit interesting emergent behaviour. This paper addresses quantitative evaluation of spatial characteristics of CA generated patterns. It is suggested that the structural characteristics of two-dimensional (2D) CA patterns can be measured using mean information gain. This information-theoretic quantity, also known as conditional entropy, takes into account conditional and joint probabilities of cell states in a 2D plane. The effectiveness of the measure is shown in a series of experiments for multi-state 2D patterns generated by CA. The results of the experiments show that the measure is capable of distinguishing the structural characteristics including symmetry and randomness of 2D CA patterns.

Keywords: Cellular automata · Spatial complexity · 2D patterns

1 Introduction

Cellular automata (CA) are one of the early bio-inspired systems invented by von Neumann and Ulam in the late 1940s to study the logic of self-reproduction in a material-independent framework. CA are known to exhibit complex behaviour from the iterative application of simple rules. The popularity of the Game of Life drew the attention of a wider community of researchers to the unexplored potential of CA applications and especially in their capacity to generate complex behaviour. The formation of complex patterns from simple rules sometimes with high aesthetic qualities has been contributed to the creation of many digital art works since the 1960s. The most notable works are *"Pixillation"*, one of the early computer generated animations [11], the digital art works of Struycken [10], Brown [3] and evolutionary architecture of Frazer [5]. Furthermore, CA have been used for music composition, for example, Xenakis [17] and Miranda [9].

Although classical one-dimensional CA with binary states can exhibit complex behaviours, experiments with multi-state two-dimensional (2D) CA reveal a very rich spectrum of symmetric and asymmetric patterns [6,7].

© Springer International Publishing Switzerland 2015
F. Pereira et al. (Eds.) EPIA 2015, LNAI 9273, pp. 201–212, 2015.
DOI: 10.1007/978-3-319-23485-4_21

There are numerous studies on the quantitative [8] and qualitative behaviour [14–16] of CA but they are mostly concerned with categorising the rule space and the computational properties of CA. In this paper, we investigate information gain as a spatial complexity measure of multi-state 2D CA patterns. Although the Shannon entropy is commonly used to measure complexity, it fails to discriminate accurately structurally different patterns in two-dimensions. The main aim of this paper is to demonstrate the effectiveness of information gain as a measure of 2D structural complexity.

This paper is organised as follows. Section 2 provides formal definitions and establishes notation. Section 3 demonstrates that Shannon entropy is an inadequate measure of 2D cellular patterns. In the framework of the objectives of this study a spatial complexity spectrum is formulated and the potential of information gain as a structural complexity measure is discussed. Section 4 gives details of experiments that test the effectiveness of information gain. The paper closes with a discussion and summary of findings.

2   Cellular Automata

This section serves to specify the cellular automata considered in this paper, and to define notation.

A cellular automaton $\mathcal{A}$ is specified by a quadruple $\langle L, S, N, f \rangle$ where:

- L is a finite square lattice of cells (i, j).
- $S = \{1, 2, \ldots, k\}$ is set of states. Each cell (i, j) in L has a state $s \in S$.
- $\mathcal{N}$ is neighbourhood, as specified by a set of lattice vectors $\{e_a\}$, $a = 1, 2, \ldots, N$. The neighbourhood of cell $r = (i, j)$ is $\{r+e_1, r+e_2, \ldots, r+e_N\}$. A cell is considered to be in its own neighbourhood so that one of $\{e_a\}$ is the zero vector $(0, 0)$. With an economy of notation, the cells in the neighbourhood of (i, j) can be numbered from 1 to N; the neighbourhood states of (i, j) can therefore be denoted $(s_1, s_2, \ldots, s_N)$. Two common neighbourhoods are the five-cell von Neumann neighbourhood $\{(0, 0), (\pm 1, 0), (0, \pm 1)\}$ and the nine-cell Moore neighbourhood $\{(0, 0), (\pm 1, 0), (0, \pm 1), (\pm 1, \pm 1)\}$. Periodic boundary conditions are applied at the edges of the lattice so that complete neighbourhoods exist for every cell in L.
- f is the update rule. f computes the state $s_1(t + 1)$ of a given cell from the states $(s_1, s_2, \ldots, s_N)$ of cells in its neighbourhood: $s_1(t + 1) = f(s_1, s_2, \ldots, s_N)$. A quiescent state s_q satisfies $f(s_q, s_q, \ldots, s_q) = s_q$.

The collection of states for all cells in L is known as a configuration c. The global rule F maps the whole automaton forward in time; it is the synchronous application of f to each cell. The behaviour of a particular $\mathcal{A}$ is the sequence $c^0, c^1, c^2, \ldots, c^{t-1}$, where c^0 is the initial configuration (IC) at $t = 0$.

CA behaviour is sensitive to the IC and to L, S, N and f. The behaviour is generally nonlinear and sometimes very complex; no single mathematical analysis can describe, or even estimate, the behaviour of an arbitrary automaton. The vast size of the rule space, and the fact that this rule space is unstructured, mean that knowledge of the behaviour a particular cellular automaton, or even of a set of automata, gives no insight into the behaviour of any other CA. In the lack of any practical model to predict the behaviour of a CA, the only feasible method is to run simulations. Fig. 1 illustrates some experimental configurations generated by the authors to demonstrate the capabilities of CA in exhibiting complex behaviour with visually pleasing qualities.

Fig. 1. Samples of multi-state 2D CA patterns

3 Spatial Complexity Measure of 2D Patterns

The introduction of information theory by Shannon provided a mathematical model to measure the order and complexity of systems. Shannon's information theory was an attempt to address communication over an unreliable channel [12]. Entropy is the core of this theory [4]. Let $\mathcal{X}$ be discrete alphabet, X a discrete random variable, $x \in \mathcal{X}$ a particular value of X and $P(x)$ the probability of x. Then the entropy, $H(X)$, is:

$$H(X) = - \sum_{x \in \mathcal{X}} P(x) \log_2 P(x) \tag{1}$$

The quantity H is the average uncertainty in bits, $\log_2(\frac{1}{p})$ associated with X. Entropy can also be interpreted as the average amount of information needed to describe X. The value of entropy is always non-negative and reaches its maximum for the uniform distribution, $\log_2(|\mathcal{X}|)$:

$$0 \leqslant H \leqslant \log_2(|\mathcal{X}|) \tag{2}$$

The lower bound of relation (2) corresponds to a deterministic variable (no uncertainty) and the upper bound corresponds to a maximum uncertainty associated with a random variable. Another interpretation of entropy is as a measure of *order* and *complexity*. A low entropy implies low uncertainty so the message is highly predictable, ordered and less complex. And high entropy implies a high uncertainty, less predictability, highly disordered and complex. Despite the dominance of Shannon entropy as a measure of complexity, it fails to reflect on structural characteristics of 2D patterns. The main reason for this drawback is that it only reflects on the distribution of the symbols, and not on their ordering. This is illustrated in Fig. 2 where, following [1], the entropy of 2D patterns with various structural characteristics is evaluated. Fig. 2a-b are patterns with ordered structures and Fig. 2c is a pattern with repeated three element structure over the plane. Fig. 2d is a fairly structureless pattern.

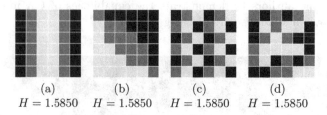

(a)	(b)	(c)	(d)
$H = 1.5850$	$H = 1.5850$	$H = 1.5850$	$H = 1.5850$

Fig. 2. Measure of H for structurally different patterns with uniform distribution of elements

Fig. 2 clearly demonstrates the failure of entropy to discriminate structurally different 2D patterns. In other words, entropy is invariant to spatial rearrangement of composing elements. This is in contrast to our intuitive perception of the complexity of patterns and is problematic for the purpose of measuring the complexity of multi-state 2D CA behaviour.

Taking into account our intuitive perception of complexity and structural characteristics of 2D patterns, a complexity measure must be bounded by two extreme points of complete order and disorder. It is reasonable to assume that *regular structures*, *irregular structures* and *structureless* patterns lie along between these extremes, as illustrated in Fig. 3.

$$order \quad \xleftarrow{\quad \text{regular structure} \mid \text{irregular structure} \mid \text{structureless} \quad} \quad disorder$$

Fig. 3. The spectrum of spatial complexity.

A complete regular structure is a pattern of high symmetry, an irregular structure is a pattern with some sort of structure but not as regular as a fully symmetrical pattern and finally a structureless pattern is a random arrangement of elements.

A measure introduced in [1,2,13] and known as information gain, has been suggested as a means of characterising the complexity of dynamical systems and of images. It measures the amount of information gained in bits when specifying the value, x, of a random variable X given knowledge of the value, y, of another random variable Y,

$$G_{x,y} = -\log_2 P(x|y). \tag{3}$$

$P(x|y)$ is the conditional probability of a state x conditioned on the state y. Then the *mean information gain*, $\overline{G}_{X,Y}$, is the average amount of information gain from the description of the all possible states of Y:

$$\overline{G}_{X,Y} = \sum_{x,y} P(x,y)G_{x,y} = -\sum_{x,x} P(x,y)\log_2 P(x|y) \tag{4}$$

where $P(x,y)$ is the joint probability, $\text{prob}(X = x, Y = y)$. $\overline{G}$ is also known as the conditional entropy, $H(X|Y)$ [4]. Conditional entropy is the reduction in uncertainty of the joint distribution of X and Y given knowledge of Y, $H(X|Y) = H(X,Y) - H(Y)$. The lower and upper bounds of $\overline{G}_{X,Y}$ are

$$0 \leqslant \overline{G}_{X,Y} \leqslant \log_2|\mathcal{X}|. \tag{5}$$

where $y \in \mathcal{Y}$.

In principle, $\overline{G}$ can be calculated for a 2D pattern by considering the distribution of cell states over pairs of cells r, s,

$$\overline{G}_{r,s} = -\sum_{s_r,s_s} P(s_r,s_s)\log_2 P(s_r,s_s) \tag{6}$$

where s_r, s_s are the states at r and s. Since $|\mathcal{S}| = N$, $\overline{G}_{r,s}$ is a value in $[0,N]$.

In particular, horizontal and vertical near neighbour pairs provide four MIGs, $\overline{G}_{(i,j),(i+1,j)}$, $\overline{G}_{(i,j),(i-1,j)}$, $\overline{G}_{(i,j),(i,j+1)}$ and $\overline{G}_{(i,j),(i,j-1)}$. In the interests of notational economy, we write $\overline{G}_s$ in place of $\overline{G}_{r,s}$, and omit parentheses, so that, for example, $G_{i+1,j} \equiv G_{(i,j),(i+1,j)}$. The relative positions for non-edge cells are given by matrix M:

$$M = \begin{bmatrix} & (i,j+1) & \\ (i-1,j) & (i,j) & (i+1,j) \\ & (i,j-1) & \end{bmatrix}. \tag{7}$$

Correlations between cells on opposing lattice edges are not considered. Fig. 4 provides an example. The depicted pattern is composed of four different symbols $S = \{light\text{-}grey, grey, white, black\}$. The light-grey cell correlates with two neighbouring white cells $(i+1,j)$ and $(i,j-1)$. On the other hand, The grey cell has four neighbouring cells of which three are white and one is black. The result of this edge condition is that $G_{i+1,j}$ is not necessarily equal to $\overline{G}_{i-1,j}$. Differences between the horizontal (vertical) mean information rates reveal left/right (up/down) orientation.

Fig. 4. A sample 2D pattern

The mean information gains of the sample patterns in Fig. 2 are presented in Fig. 5. The merits of $\overline{G}$ in discriminating structurally different patterns ranging from the structured and symmetrical (Fig. 5a-b), to the partially structured

(Fig. 5c) and the structureless and random (Fig. 5d), are clearly evident. The cells in the columns of pattern (a) are completely correlated. However knowledge of cell state does not provide complete predictability in the horizontal direction and, as a consequence, the horizontal $\overline{G}$ is finite. Pattern (b) has non-zero, and identical $\overline{G}$'s indicating a symmetry between horizontal and vertical directions, and a lack of complete predictability. Analysis of pattern (c) is similar to (a) except the roles of horizontal and vertical directions are interchanged. The four $\overline{G}$s in the final pattern are all different, indicating a lack of vertical and horizontal symmetry; the higher values show the increased randomness. Details of calculations for a sample pattern are provided in the appendix.

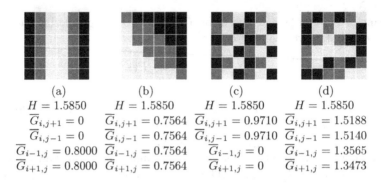

(a)	(b)	(c)	(d)
$H = 1.5850$	$H = 1.5850$	$H = 1.5850$	$H = 1.5850$
$\overline{G}_{i,j+1} = 0$	$\overline{G}_{i,j+1} = 0.7564$	$\overline{G}_{i,j+1} = 0.9710$	$\overline{G}_{i,j+1} = 1.5188$
$\overline{G}_{i,j-1} = 0$	$\overline{G}_{i,j-1} = 0.7564$	$\overline{G}_{i,j-1} = 0.9710$	$\overline{G}_{i,j-1} = 1.5140$
$\overline{G}_{i-1,j} = 0.8000$	$\overline{G}_{i-1,j} = 0.7564$	$\overline{G}_{i-1,j} = 0$	$\overline{G}_{i-1,j} = 1.3565$
$\overline{G}_{i+1,j} = 0.8000$	$\overline{G}_{i+1,j} = 0.7564$	$\overline{G}_{i+1,j} = 0$	$\overline{G}_{i+1,j} = 1.3473$

Fig. 5. The comparison of H with measures of $\overline{G}_{i,j}$ for structurally different patterns.

4   Experiments and Results

A set of experiments was designed to examine the effectiveness of $\overline{G}$ in discriminating the particular patterns that are generated by a multi-state 2D cellular automaton. The (outer-totalistic) CA is specified in Table 1. The chosen experimental rule maps three states, represented by *green*, *red* and *white*; the quiescent state is *white*.

Table 1. Specifications of experimental cellular automaton

$L = 129 \times 129$ (16641 cells).
$S = \{0, 1, 2\} \equiv \{white, red, green\}$
$\mathcal{N}$: von Neumann neighbourhood
$f : S^9 \mapsto S$

$$f(s_{i,j})(t) = s_{i,j}(t+1) = \begin{cases} 1 \text{ if } s_{(i,j)}(t) = 1,2 \text{ and } \sigma = 0-2 \\ 2 \text{ if } s_{(i,j)}(t) = 2,3 \text{ and } \sigma = 1 \\ 2 \text{ if } s_{(i,j)}(t) = 2 \quad \text{ and } \sigma = 2 \\ 0 \text{ otherwise} \end{cases}$$

where σ is the sum total of the neighbourhood states.

The experiments are conducted with two different ICs: (1) all white cells except for a single *red* cell and (2) a random configuration with 50% *white* quiescent states (8320 cells), 25% *red* and 25% *green*. The experimental rule has been iterated synchronously for 150 successive time steps. Fig. 6 and Fig. 7 illustrate the space-time diagrams for a sample of time steps starting from single and random ICs.

Fig. 6. Space-time diagram of the experimental cellular automaton for sample time steps starting from the single cell IC.

The behaviour of cellular automaton from the single cell IC is a sequence of symmetrical patterns (Fig. 6). The directional measurements of $\overline{G}_{i,j}$ for the single cell IC start with $\overline{G}_{i,j+1} = \overline{G}_{i,j-1} = \overline{G}_{i-1,j} = \overline{G}_{i+1,j} = 0.00094$ and $H = 0.00093$, and they attain $\overline{G}_{i,j+1} = \overline{G}_{i,j-1} = \overline{G}_{i-1,j} = \overline{G}_{i+1,j} = 1.13110$ and $H = 1.13714$ (and Figs 8, 11) at the end of the runs.

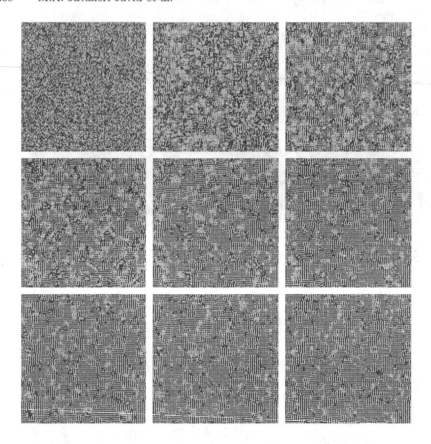

Fig. 7. Space-time diagram of the experimental cellular automaton for sample time steps starting from the random IC.

The sequence of states can be analysed by considering the differences between the up/down and left/right mean information gains, as defined by

$$\Delta \overline{G}_{i,j\pm1} = |\overline{G}_{i,j+1} - \overline{G}_{i,j-1}| \tag{8}$$

$$\Delta \overline{G}_{i\pm1,j} = |\overline{G}_{i+1,j} - \overline{G}_{i-1,j}|. \tag{9}$$

For the single cell IC, $\Delta \overline{G}_{i,j\pm1}$ and $\Delta \overline{G}_{i\pm1,j}$ are constant for the 150 time steps ($\Delta \overline{G}_{i,j\pm1} = \Delta \overline{G}_{i,j\pm1} = 0$). This indicates the development of the symmetrical patterns along the up/down and left/right directions.

The behaviour of cellular automaton from the random IC is a sequence of irregular structures (Fig. 7). The formation of patterns with local structures has reduced the values of $\overline{G}_{i,j}$ until a stable oscillating pattern is attained (Figs 7, 9). This is an indicator of the development of irregular structures. However the patterns are not random patterns since $\overline{G}_{i,j} \approx 1.1$ is less than the maximum three-state value $\log_2(3) = 1.5850$ (see Eq. 5). Mean information rate differences

$\Delta \overline{G}_{i,j\pm1}$ and $\Delta \overline{G}_{i\pm1,j}$ for both ICs are plotted in Fig. 10. The structured but asymmetrical patterns emerging from the random start are clearly distinguished from the symmetrical patterns of the single cell IC.

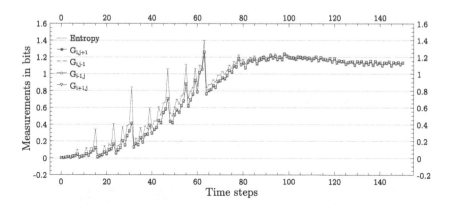

Fig. 8. Measurements of H, $\overline{G}_{i,j+1}$, $\overline{G}_{i,j-1}$, $\overline{G}_{i+1,j}$, $\overline{G}_{i-1,j}$ for 150 time steps starting from the single cell IC.

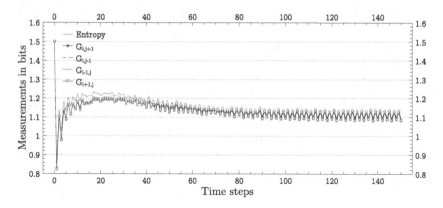

Fig. 9. Measurements of H, $\overline{G}_{i,j+1}$, $\overline{G}_{i,j-1}$, $\overline{G}_{i+1,j}$, $\overline{G}_{i-1,j}$ for 150 time steps starting from the random IC.

These experiments demonstrate that a cellular automaton rule seeded with different ICs leads to the formation of patterns with structurally different characteristics. The gradient of the mean information rate along lattice axes is able to detect the structural characteristics of patterns generated by this particular multi-state 2D CA. From the comparison of H with $\Delta \overline{G}_{i,j\pm1}$ and $\Delta \overline{G}_{i\pm1,j}$ in the set of experiments, it is clear that entropy fails to discriminate between the diversity of patterns that can be generated by various CA.

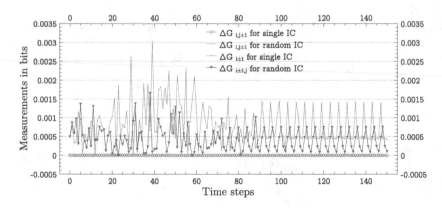

Fig. 10. Plots of $\Delta\overline{G}_{i,j\pm1}$ and $\Delta\overline{G}_{i\pm1,j}$ for two different ICs

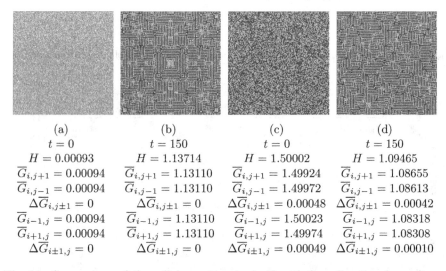

(a)	(b)	(c)	(d)
$t = 0$	$t = 150$	$t = 0$	$t = 150$
$H = 0.00093$	$H = 1.13714$	$H = 1.50002$	$H = 1.09465$
$\overline{G}_{i,j+1} = 0.00094$	$\overline{G}_{i,j+1} = 1.13110$	$\overline{G}_{i,j+1} = 1.49924$	$\overline{G}_{i,j+1} = 1.08655$
$\overline{G}_{i,j-1} = 0.00094$	$\overline{G}_{i,j-1} = 1.13110$	$\overline{G}_{i,j-1} = 1.49972$	$\overline{G}_{i,j-1} = 1.08613$
$\Delta\overline{G}_{i,j\pm1} = 0$	$\Delta\overline{G}_{i,j\pm1} = 0$	$\Delta\overline{G}_{i,j\pm1} = 0.00048$	$\Delta\overline{G}_{i,j\pm1} = 0.00042$
$\overline{G}_{i-1,j} = 0.00094$	$\overline{G}_{i-1,j} = 1.13110$	$\overline{G}_{i-1,j} = 1.50023$	$\overline{G}_{i-1,j} = 1.08318$
$\overline{G}_{i+1,j} = 0.00094$	$\overline{G}_{i+1,j} = 1.13110$	$\overline{G}_{i+1,j} = 1.49974$	$\overline{G}_{i+1,j} = 1.08308$
$\Delta\overline{G}_{i\pm1,j} = 0$	$\Delta\overline{G}_{i\pm1,j} = 0$	$\Delta\overline{G}_{i\pm1,j} = 0.00049$	$\Delta\overline{G}_{i\pm1,j} = 0.00010$

Fig. 11. Comparison of the cellular automaton's H with four directional measure of $\overline{G}_{i,j}$ $\Delta\overline{G}_{i,j\pm1}$ and $\Delta\overline{G}_{i\pm1,j}$ starting from single (a, b) and random ICs (c, d).

5 Conclusion

Cellular automata (CA) are one of the early bio-inspired models of self-replicating systems and, in 2D, are powerful tools for the pattern generation. Indeed, multi-state 2D CA can generate many interesting and complex patterns with various structural characteristics. This paper considers an information-theoretic classification of these patterns.

Entropy, which is a statistical measure of the distribution of cell states, is not in general able to distinguish these patterns. However mean information gain, as proposed in [1,2,13], takes into account conditional and joint probabilities between pairs of cells and, since it is based on correlations between cells, holds promise for pattern classification.

This paper reports on a pair of experiments for two different initial conditions of an outer-totalistic CA. The potential of mean information gain for distinguishing multi-state 2D CA patterns is demonstrated. Indeed, the measure appears to be particular good at distinguishing symmetry from non-random non-asymmetric patterns.

Since CA are one of the generative tools in computer art, means of evaluating the aesthetic qualities of CA generated patterns could have a substantial contribution towards further automation of CA art. This is the subject of on-going research.

Appendix

In this example the pattern is composed of two different cells $S = \{white, black\}$ where the set of permutations with repetition is $\{ww, wb, bb, bw\}$. Considering the mean information gain (Eq. 4) and given the positional matrix M (Eq. 7), the calculations can be performed as follows:

$white - white$

$P(w, s_{(i,j+1)}) = \frac{5}{6}$

$P(w|w_{(i,j+1)}) = \frac{4}{5}$

$P(w, w_{(i,j+1)}) = \frac{5}{6} \times \frac{4}{5} = \frac{2}{3}$

$G(w, w_{(i,j+1)}) = \frac{2}{3} \log_2 P(\frac{4}{5})$

$G(w, w_{(i,j+1)}) = 0.2146\ bits$

$white - black$

$P(w, s_{(i,j+1)}) = \frac{5}{6}$

$P(w|b_{(j+1)}) = \frac{1}{5}$

$P(w, b_{(i,j+1)}) = \frac{5}{6} \times \frac{1}{5} = \frac{1}{6}$

$\overline{G}(w, b_{(i,j+1)}) = \frac{1}{6} \log_2 P\frac{1}{5}$

$\overline{G}(w, b_{(i,j+1)}) = 0.3869\ bits$

$black - black$

$P(b, s_{(i,j+1)}) = \frac{1}{6}$

$P(b|b_{(i,j+1)}) = \frac{1}{1}$

$P(b, b_{(i,j+1)}) = \frac{1}{6} \times \frac{1}{1} = \frac{1}{6}$

$G(b, b_{(i,j+1)}) = \frac{1}{6} \log_2 P(1)$

$G(b, b_{(i,j+1)}) = 0\ bits$

$black - white$

$P(b, s_{(i,j+1)}) = \frac{1}{6}$

$P(b|w_{(i,j+1)}) = \frac{0}{1}$

$P(b, w_{(i,j+1)}) = \frac{1}{6} \times 0$

$G(b, w_{(i,j+1)}) = 0\ bits$

$\overline{G} = G(w, w_{(i,j+1)}) + G(w, b_{(i,j+1)}) + G(b, b_{(i,j+1)}) + G(b, w_{(i,j+1)})$

$\overline{G} = 0.6016\ bits$

In $white - white$ case G measures the uniformity and spatial property where $P(w, s_{(i,j+1)})$ is the joint probability that a cell is white and it has a neighbouring cell at its $(i, j + 1)$ position, $P(w|w_{(i,j+1)})$ is the conditional probability of a cell is white given that it has white neighbouring cell at its $(i, j + 1)$ position, $P(w, w_{(i,j+1)})$ is the joint probability that a cell is white and it has neighbouring cell at its $(i, j + 1)$ position, $G(w, w_{(i,j+1)})$ is information gain in $bits$ from specifying a white cell where it has a white neighbouring cell at its $(i, j + 1)$ position. The same calculations are performed for the rest of cases; $black-black$, $white-black$ and $black-white$.

References

1. Andrienko, Y.A., Brilliantov, N.V., Kurths, J.: Complexity of two-dimensional patterns. Eur. Phys. J. B **15**(3), 539–546 (2000)
2. Bates, J.E., Shepard, H.K.: Measuring complexity using information fluctuation. Physics Letters A **172**(6), 416–425 (1993)
3. Brown, P.: Stepping stones in the mist. In: Creative Evolutionary Systems, pp. 387–407. Morgan Kaufmann Publishers Inc. (2001)
4. Cover, T.M., Thomas, J.A.: Elements of Information Theory (Wiley Series in Telecommunications and Signal Processing). Wiley-Interscience (2006)
5. Frazer, J.: An evolutionary architecture. Architectural Association Publications, Themes VII (1995)
6. Javaheri Javid, M.A., Al-Rifaie, M.M., Zimmer, R.: Detecting symmetry in cellular automata generated patterns using swarm intelligence. In: Dediu, A.-H., Lozano, M., Martín-Vide, C. (eds.) TPNC 2014. LNCS, vol. 8890, pp. 83–94. Springer, Heidelberg (2014)
7. Javaheri Javid, M.A., te Boekhorst, R.: Cell dormancy in cellular automata. In: Alexandrov, V.N., van Albada, G.D., Sloot, P.M.A., Dongarra, J. (eds.) ICCS 2006. LNCS, vol. 3993, pp. 367–374. Springer, Heidelberg (2006)
8. Langton, C.G.: Studying artificial life with cellular automata. Physica D: Nonlinear Phenomena **22**(1), 120–149 (1986)
9. Miranda, E.: Composing Music with Computers. No. 1 in Composing Music with Computers. Focal Press (2001)
10. Scha, I.R.: Kunstmatige Kunst. De Commectie **2**(1), 4–7 (2006)
11. Schwartz, L., Schwartz, L.: The Computer Artist's Handbook: Concepts, Techniques, and Applications. W W Norton & Company Incorporated (1992)
12. Shannon, C.: A mathematical theory of communication. The Bell System Technical Journal **27**, 379–423, 623–656 (1948)
13. Wackerbauer, R., Witt, A., Atmanspacher, H., Kurths, J., Scheingraber, H.: A comparative classification of complexity measures. Chaos, Solitons & Fractals **4**(1), 133–173 (1994)
14. Wolfram, S.: Statistical mechanics of cellular automata. Reviews of Modern Physics **55**(3), 601–644 (1983)
15. Wolfram, S.: Universality and complexity in cellular automata. Physica D: Nonlinear Phenomena **10**(1), 1–35 (1984)
16. Wolfram, S.: A New Kind of Science. Wolfram Media Inc. (2002)
17. Xenakis, I.: Formalized music: thought and mathematics in composition. Pendragon Press (1992)

Electricity Demand Modelling
with Genetic Programming

Mauro Castelli[1], Matteo De Felice[2], Luca Manzoni[3]([✉]),
and Leonardo Vanneschi[1]

[1] NOVA IMS, Universidade Nova de Lisboa, 1070-312 Lisboa, Portugal
{mcastelli,lvanneschi}@novaims.unl.pt
[2] Energy and Environment Modeling Technical Unit (UTMEA), ENEA,
Casaccia Research Center, 00123 Roma, Italy
matteo.defelice@enea.it
[3] Dipartimento di Informatica, Sistemistica E Comunicazione,
Università Degli Studi di Milano Bicocca, 20126 Milano, Italy
luca.manzoni@disco.unimib.it

Abstract. Load forecasting is a critical task for all the operations of
power systems. Especially during hot seasons, the influence of weather
on energy demand may be strong, principally due to the use of air conditioning and refrigeration. This paper investigates the application of
Genetic Programming on day-ahead load forecasting, comparing it with
Neural Networks, Neural Networks Ensembles and Model Trees. All the
experimentations have been performed on real data collected from the
Italian electric grid during the summer period. Results show the suitability of Genetic Programming in providing good solutions to this problem.
The advantage of using Genetic Programming, with respect to the other
methods, is its ability to produce solutions that explain data in an intuitively meaningful way and that could be easily interpreted by a human
being. This fact allows the practitioner to gain a better understanding
of the problem under exam and to analyze the interactions between the
features that characterize it.

1 Introduction

Load forecasting is the task of predicting the electricity demand on different time
scales, such as minutes (very short-term), hours/days (short-term), and months
and years (long-term). This information has to be used to plan and schedule
operations on power systems (dispatch, unit commitment, network analysis) in
a way to control the flow of electricity in an optimal way, with respect to various
aspects (quality of service, reliability, costs, etc). An accurate load forecasting
has great benefits for electric utilities and both negative or positive errors lead
to increased operating costs [10]. Overestimate the load leads to an unnecessary
energy production or purchase and, on the contrary, underestimation causes
unmet demand with a higher probability of failures and costly operations. Several
factors influence electricity demand: day of the week and holidays (the so-called

© Springer International Publishing Switzerland 2015
F. Pereira et al. (Eds.) EPIA 2015, LNAI 9273, pp. 213–225, 2015.
DOI: 10.1007/978-3-319-23485-4_22

"calendar effects"), special or unusual events, and weather conditions. In warm countries, the last factor is particularly critical during summer, when the use of refrigeration, irrigation and air conditioning becomes more common than in the rest of the year.

Most of the used methods for Load forecasting are time-series approaches, like Box-Jenkins models, or artificial intelligence methods, like Neural Networks (NN). There is a large literature about the use of computer science for load forecasting, in Section 2 a review of this literature is proposed. In the last decades, techniques based on Computational Intelligence (CI) methods have been proposed to overcome the most common problems of traditional methods, especially in most difficult scenarios. These techniques have demostrated their effectiveness in several cases, often becoming a valid alternative to conventional methods.

A CI method is Genetic Programming [11, 21] (GP). GP has several advantages over others machine learning methods, including the ones that have been considered in this work, that are Neural Networks and Model Trees. In particular, one interesting feature is the ability of GP to produce human-readable solutions. This property may allow an in depth analysis of the features that characterize a specific problem. It is important to underline that GP trees are usually very large and have significant redundancy (introns) even with parsimony measures. Thus, they can be very difficult to interpret. However, these trees can be usually simplified, producing a compact and readable model. In this work, the term human-readable is related to the fact that GP produces a model that represents interactions between variables, and that can be used for a better understanding of the problem under examination, a property that is particularly useful in real-life problems. This property is not true when, for instance, Neural Networks are considered. In fact NNs produce numerical matrices of weights, which does not facilitate the practitioner's task of obtaining a better understanding of the relations between the features that characterize a specific problem. In this work we use GP for the load forecast problem during summer period, and we provide a comparison between GP and three different machine learning techniques on this problem.

The paper is organized as follows: Section 2 introduces the load forecasting problem and gives an overview of the state of the art approaches to deal with this problem. Section 3 briefly presents the techniques considered in this paper: neural networks, neural networks ensemble, model trees and GP. Section 4 describes the experimental phase, reporting the experimental settings and discussing the results. Section 5 concludes the paper and summarizes the results of this work.

2   Short-Term Load Forecasting

Load forecasting is commonly defined "short-term" when the prediction horizon is from one hour to one week. For this kind of problem various factors might be considered, such as weather data or, more in general, all the factors influencing the load pattern (e.g. day of the week). Load data normally exhibit seasonality: the load at time t tends to be similar to the load at time $t - k$ with k usually representing a day, a week or a month, depending on the dataset. In this paper, we are focusing on daily average load and we can observe that the load at the day t is usually similar to the one at the same day of the previous week (i.e., weekly seasonality), with the exception of holidays and weekends, when the load pattern is usually more unpredictable.

2.1 Forecasting Methods: State of the Art

Different methods have been used to cope with load forecasting, mainly classical statistical methods and machine learning techniques. The first approach, time series analysis, consists of the determination of the relationship between process input and output with a linear model. To do that, observations that are assumed equispaced discrete-time samples are used. There is a wide variety of models to deal with this problem, the most popular in engineering applications are probably autoregression (AR) and moving-average (MA) models with their combinations (ARMA, ARIMA, etc) and a common methodology is the iterative one proposed by Box & Jenkins [4]. Since Park's paper in 1991 [20], Neural Networks have been widely applied to the Load forecasting problem. The main advantage of NNs is their implicit nonlinearity which can potentially allow modeling complex dynamics. On the other hand, NNs present numerous parameters that are usually tuned with empirical approaches. Furthermore, NNs computational needs become expensive with high-dimensional problems and large datasets. Some reviews on the application of NNs can be found in [1,7,9]. Model trees, which we will introduce in the next section, have been used less frequently and, to the best of our knowledge, there are only few works using them for forecasting [19,24,25].

While GP has been used [3,16] to face the load forecasting problem, it is not a widely applied technique like other CI approaches. This is probably due to the fact that GP is computationally expensive, in particular the evaluation of candidate solutions requires more time with respect to other machine learning techniques. On the other hand, differently from other machine learning techniques like Neural Networks, GP is able to produce solutions that are easier to read and interpret by humans [11], a feature that can be particularly useful in some applications. Moreover, like other Evolutionary Algorithms, the practitioner may design a specific fitness function in order to make the algorithm focusing on a specific problem feature [14]. While this makes the whole task "technically" more difficult, on the other hand the algorithm becomes more versatile: depending on how the fitness function is defined (or functions if multi-objective optimization is considered), it is possible to direct GP towards different parts of the solution space. In other words, it is possible to steer GP to particular behaviors, that are much more difficult to obtain with other machine learning techniques. Hence, it could be not only used for producing a forecasting model, but also to have some hints on the role of the variables that influence the forecasting model itself. As reported in [12], GP has been used to produce many instances of results that are competitive with human-produced results. These human-competitive results come from a wide variety of fields, including quantum computing circuits [2], analog electrical circuits [14], antennas [18], mechanical systems [17], photonic systems [22], optical lens systems [13] and sorting networks [15].

3 Considered Techniques for the Load Forecasting Problem

This section briefly introduces the techniques used in this paper: neural networks, model trees, and genetic programming.

3.1 Neural Networks

Neural Networks are a non-linear statistical modeling tool. In this work we use feed-forward neural networks trained with Levenberg-Marquardt back-propagation training algorithm, implemented in MATLAB. In addition to single neural networks, we use an ensemble of 100 neural networks with their outputs combined using an arithmetic mean. Neural networks ensemble have been introduced in [8] and this approach has already been used for load forecasting in [5].

3.2 Model Trees

Model trees (M5 system) can be considered an extension of regression trees introduced by Quinlan in 1992 [23]. A regression tree is a type of prediction tree where in each leaf (terminal node) there is a zero-order model (i.e., constant value) predicting the target variable. On the other hand, trees built with the M5 algorithm [23] have at their leaves a multivariate linear model (i.e., first-order model). This linear model is the one that best fits those training points that satisfy the conditions represented in the internal nodes that are on the path from the root to the linear model itself. In these binary trees each internal node represents a "rule" that defines which sub-branch of the tree we have to use to make a prediction on a particular case. Model trees are basically a combination between conventional regression trees and linear regression models. They are particularly useful in the cases where a single global model is hard to obtain. Their ability in partitioning the sample space allows finding the part of the data space where a linear model best fits. The advantages of this technique are several: easiness of implementation, possibility to cope with not-smooth regression surfaces and existence of fast and reliable learning algorithms. In this work, we used a MATLAB implementation of the M5 algorithm[1]. For a description of the algorithm we refer to the Quinlan original paper [23] and to the improvement made by Wang [26].

---

[1] M5PrimeLab is an open source toolbox for MATLAB/Octave available at http://www.cs.rtu.lv/jekabsons/regression.html

3.3 Genetic Programming

GP is a machine learning technique inspired by Darwin's theory of evolution. In GP terminology [11] each candidate solution is called individual and the quality of the solution is called fitness. Fitness is a function that associates a real number to each possible individual. In minimization (respectively maximization) problems the objective is to find the solution with the minimal (respectively maximal) fitness (or a good-enough approximation of it). The GP algorithm is an iterative process (every iteration is called generation) that explores the search space to find good individuals (solutions) by evolving a population (set of possible solutions). In doing this exploration GP uses several genetic operators to mimic the natural evolution process: selection, mutation, and crossover. In GP individuals are traditionally represented as LISP-like trees. The selection operator selects individuals in function of their fitness with better solutions having higher probability of being selected. The selection operator is the only GP operator that works on fitness. The two operators that can change the structure of GP individuals are mutation and crossover. Mutation can be defined as random manipulation that operates on only one individual. The aim of the mutation is to avoid local optima and to move the search to new areas of the search space [11]. This operator selects a point in the GP tree randomly and replaces the existing sub-tree at that point with a new randomly generated sub-tree. The crossover operator combines the genetic material of two parents by swapping a sub-tree of one parent with a part of the other. The crossover operator is used to combine the pairs of selected individuals (parents) to create new individuals that potentially have a higher fitness than either of their parents. For a complete description of GP and ability to solve real-world problems, the reader is referred to [11,21].

In GP every internal node is a function whose arguments are its children (that can also be other functions). Leaf nodes are constants or variables. Thus, a GP individual is a mathematical function and the sequence of composition of "basic functions" needed to produce it. Thus, it is important to underline that, differently from the aforementioned model trees, GP produces a solution that represents an algorithm to solve a particular problem and not a method to partition the sample space. In this work we used our C++ implementation of GP.

4 Experiments

In this work we compare the performances of different machine learning techniques on the electric load forecasting problem using Italian national grid data provided by TERNA[2]. We collected the daily average load y during the working days in June and July for the years 2003-2009, for a final data set of 300 samples.

---

[2] TERNA is the owner of the Italian transmission grid and the responsible for energy transmission and dispatching. Real-time data about electricity demand are available on their homepage www.terna.it

The aim of this forecasting task is the load at time t (y_t) providing information until day $t-1$ (one-day ahead forecasting) using the past samples of the load and the information provided by temperature. We built a data set with 9 input variables: $x_0, x_1, \ldots, x_6$ representing the daily load for each past day. The variable x_0 refers to the load at time $t-7$ while x_6 to the load at time $t-1$. The value x_7 is the daily average temperature (Celsius degrees) at day $t-1$ while x_8 is the daily average temperature the same day of the forecast.

Temperature data have been obtained with an average of all the data available in Italy provided by the ECMWF (European Centre for Medium-Range Weather Forecasts) ERA-Interim reanalysis [6]. For the variable x_8, we assume to have a perfect forecast for the day t and hence we use the observed data.

4.1 Experimental Settings

As explained before, the techniques we have considered are the standard GP algorithm, back propagation Neural Networks, and M5 Model Trees.

Training and test sets have been obtained by randomly splitting the dataset described in the previous section. In particular 50 different partitions of the original dataset with its same size, have been considered. In each partition 70% of the data have been randomly selected (without replacement) with uniform probability and inserted into the training set, while the remaining 30% form the test set (i.e., they are not used during the training phase). For each partition a total of 50 runs were performed with each technique (2500 runs in total). This setup has been implemented in order to perform a fair comparison between different techniques. In fact, unlike M5 algorithm, GP and NNs are stochastic methods and thus their performances can be influenced by initial conditions.

Neural Networks. Feed-forward neural networks with 9 inputs, 3 hidden neurons (with hyperbolic tangent transfer function) and 1 output have been used. The number of hidden neurons has been chosen after a set of preliminary tests. For each dataset we created 100 neural networks with different initial weights and, after the training phase, their output has been combined using an arithmetic mean and hance evaluated on the test set. For investigation purposes, in this paper we present both the results for the ensemble and for the best-training NN, i.e., the NN with the lower error during the training phase.

M5 Model Trees. In this case we performed a single execution for each dataset, in fact the M5 algorithm is not stochastic and so there is no need to perform multiple runs. We set as 10 the minimum number of training data cases represented by each leaf, this value has been selected after a set of exploratory tests.

Genetic Programming. Regarding the experimental settings related to standard GP, all the runs used populations of 100 individuals allowed to evolve for 100 generations. Tree initialization was performed with the Ramped Half-and-Half method [21] with a maximum initial depth of 6. The function set contained the four binary operators $+$, $-$, $*$, and $/$ protected as in [21]. The terminal set contained 9 variables and 100 random constants randomly generated in the range $[0, 40000]$. This range has been chosen considering the magnitude of the values at stake in the considered application. Because the cardinalities of the function and terminal sets were so different, we have explicitly imposed functions and terminals to have the same probability of being chosen when a random node is needed. The reproduction (replication) rate was 0.1, meaning that each selected parent has a 10% chance of being copied to the next generation instead of being engaged in breeding. Standard tree mutation and standard crossover (with uniform selection of crossover and mutation points) were used with probabilities of 0.1 and 0.9, respectively. Recall that the use crossover and mutation are applied only on individuals not selected for the replication. The new random branch created for mutation has maximum depth 6. Selection for survival was elitist, with the best individual preserved in the next generation. The selection method used was tournament selection with size 6. The maximum tree depth is 17. This depth value is considered the standard value for this parameter [11]. Despite the higher number of parameters, parameter tuning in GP was not particularly problematic with respect to the other methods since there are some general rules (e.g., low mutation and high crossover rate) that provide a good starting point for setting the parameters.

4.2   Results

We outline here the results obtained after the performed experimentation on the 50 different partitions of the dataset.

The objective of the learning process is to minimize the root mean squared error (RMSE) between outputs and targets. For each partition of the dataset, we collected the RMSE on test set of the best individual produced at the end of the training process. Thus, we have 50 values for each partition and we considered the median of these 50 values. The median was preferred over the arithmetic mean due to its robustness to outliers. Repeating this process with all the considered 50 partitions results in a set of 50 values. Each value is the median of the error on test set at the end of the learning process, for a specific partition of the dataset.

Table 1 reports median and standard deviation of all the median errors achieved considering all the 50 partitions of the dataset for the considered techniques. The same results are shown with a boxplot in Fig. 1(a). Denoting by IQR the interquartile range, the ends of the whiskers represent the lowest datum still within $1.5 \cdot IQR$ of the lower quartile, and the highest datum still within $1.5 \cdot IQR$ of the upper quartile. Errors for each dataset are shown in Fig. 1(b) where it is particularly visible that GP and M5 have similar performances.

GP is the best performer, considering both the median and the standard deviation. To analyze the statistical significance of these results, a set of statistical tests has been performed on the resulting median errors. The Kolmogorov-Smirnov test shows that the data are not normally distributed hence a rank-based statistic has been used. The Mann Whitney rank-sum test for pairwise data comparison is used under the alternative hypothesis that the samples do

Table 1. Median and standard deviation of median test errors of the dataset's partitions for the considered techniques.

Technique	Median [kW]	Standard deviation (σ)
Neural Networks (best)	27.599	6.731
Ensemble	24.887	7.467
Genetic Programming	22.993	4.622
M5 Model Trees	23.122	4.730

not have equal medians. The p-values obtained are $3.9697 \cdot 10^{-7}$ when GP is compared to Neural Networks, 0.0129 when it is compared to a Neural Network ensemble and 0.9862 when GP is compared to M5 trees. Therefore, when using a significance level $\alpha = 0.05$ with a Bonferroni correction for the value α, we obtain that in the first two cases GP produces fitness values that are significantly lower (i.e., better) than the other methods, but the same conclusion cannot be reached when comparing to M5 (the p-value is equal to 0.9862). Results of Mann Whitney test are summarized in Table 2.

For a better understanding of the dynamic of the evolutionary process for this particular real-world problem, in Fig. 2 the median of the test fitness generation by generation is reported.

Table 2. p-values of Mann Whitney rank-sum test. In **bold** values that can not reject the null hypothesis with 5% significance value, meaning that errors difference is not significative.

p-value	NNs	NN Ens.	GP	M5
NNs	-	0.0074	$3.9697 \cdot 10^{-7}$	$1.0069 \cdot 10^{-7}$
NN Ens.	0.0074	-	0.0129	0.0182
GP	$3.9697 \cdot 10^{-7}$	0.0129	-	**0.9862**
M5	$1.0069 \cdot 10^{-7}$	0.0182	**0.9862**	-

4.3   Analysis of Results

We applied four different techniques to face the load forecasting problem and the solutions achieved have different forms. As stated before, GP and M5 produce, differently from NNs, solutions that could be easily interpreted. Hence, in this section we want to analyze the structure of the best solution returned by both these techniques.

We achieved 50 model trees with the M5 algorithm, one for each testing dataset. We analyzed all the linear models present in the tree leaves with the aim of understanding which variables are most used. We found that the variable x_8 is used 55 times. Moreover variables from x_7 to x_4 are used respectively 45, 55, 3 and 3 times. The variable x_0 is used 9 times and x_2 only once. Then, as expected, the most common used variables are temperatures and day-before

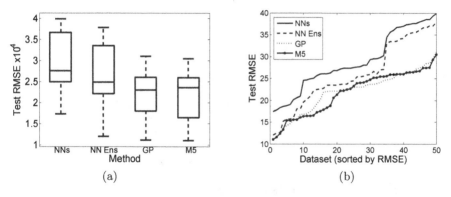

Fig. 1. Summary of test errors.

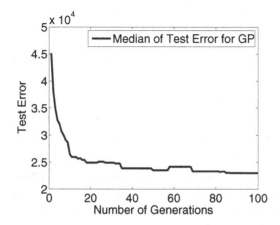

Fig. 2. Median of test fitness generation by generation. 50 independent runs have been considered.

load. In particular x_8 is the temperature at the day of the forecasting, x_7 is the temperature at the day before the forecasting and x_6 is the energy consumption the day before the forecasting. Median values for the coefficients of x_6, x_7 and x_8 are respectively 0.9164, -9435 and 10057 (relative standard deviations 17.6%, 20.7% and 25.5%). In Figure 3 a sample model tree is shown, it consists of five linear models of which three are constants. Considering the best individuals obtained by GP in all the considered partitions of the dataset, we can observe that all of them show a common structure. In particular, best individuals have this structure:

$$x_6 + f(x_7, x_8)$$

where x_6 is the energy consumption at time $t-1$ and f is a polynomial (that we call "correction") that only considers the temperatures at time $t-1$ and t. Below we report two (simplified) individuals returned by GP:

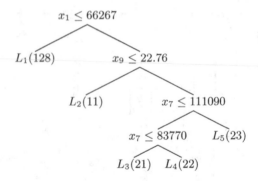

- $L_1 : y = -69185 + 0.718\,x_7 + 3642\,x_9$
- $L_2 : y = 36364$
- $L_3 : y = 81604$
- $L_4 : y = 101605$
- $L_5 : y = -205637 + 13344\,x_9$

Fig. 3. A sample model tree. In brackets the number of data samples represented by the specific linear model is given.

$$I_1 : x_6 + x_7 - x_8 + 4\,x_7 \cdot (2\,x_7 + (-x_7 + x_8) \cdot (x_7 + 6\,x_8)) - 3{,}968.452198$$
$$I_2 : x_6 + x_7 - x_8 + (x_7 - x_8) \cdot x_8 + x_7^2\,x_8 \cdot (-x_7 + x_8) + 1{,}507.852134$$

As it is possible to see, GP returns solutions that could be understood and interpreted more easily than Neural Networks and M5 Model Trees. In fact, with respect to the latter, GP does not provide a tree with decision nodes and a linear model in every leaf. Instead, it provides a more compact non-linear model and it also performs an automatic feature selection; in fact, the final model contains only three variables. Furthermore, the form of the model $(x_6 + f(x_7, x_8))$ is preserved in all the solutions produced during the performed runs. We want also to point out that while many standard techniques used for prediction are either limited in the form of the solution presented or in the legibility of the solutions, GP has neither problems. In fact, the expressiveness of GP is limited only by the choice of functional and terminal symbols and by the size of the trees.

Figure 4 reports the polynomial P obtained considering the average correction of the 50 best individuals:

$$P(x_7, x_8) = \frac{1}{50} \sum_{i=1}^{50} f_i(x_7, x_8) \tag{1}$$

The bigger is the difference between x_7 and x_8, the bigger is the "correction". This result matches our intuition regarding energy demand pattern: if in a particular day D, with a temperature equal to T, the energy consumption is E, we expect that the day $D+1$ the energy consumption will be greater than E

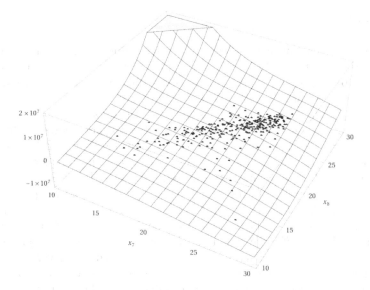

Fig. 4. Average correction considering the best individuals of all the considered datasets. x_7 is the temperature at time $t-1$ while x_8 is the temperature at time t. The black points represent the pairs of temperatures present in the dataset.

if the temperature is higher than T while we expect to have a lower energy consumption otherwise. As we stated before, the electricity demand in Italy during summer is strongly affected by the use of the air conditioning systems.

Having a model that is easily understandable and that respects our intuition of the problem (we could informally say a model that "makes sense") allows a manual validation by the end user. It is a crucial issue in convincing practitioners to really use it for their applications. In our opinion, the GP models are the only ones (among the studied techniques) that have this characteristic of intuitivity and practical consistence with our interpretation of the problem. Basically, GP is saying to us that the energy consumption in a given day is the same as the one of previous day plus (or minus) a given quantity that directly depends on the temperature increase (or decrease) between that day and the previous one. Any practitioner would understand and trust such a model as a reliable explanation of the data. In fact, GP performs an effective (and intuitively meaningful) automatic feature selection that results in the preservation of only three variables. Differently, NNs always use all the provided variables in their solutions unless an appropriate feature selection algorithm is used in a preprocessing phase. M5 Model trees also use a different subset of variables in each of the linear models that are present in their leaves while GP always uses the same 3 features, consistently in all the best individuals found in all the runs.

5 Conclusions

Energy load forecasting can provide important information regarding future energy consumption. In this work, a genetic programming based forecasting method has been presented and a comparison with other machine learning techniques has been performed. Experimental results show that GP and M5 perform quite similar, with a difference that is not statistically significant. Neural networks and the ensemble method have returned results of poorer quality. GP, differently from the other methods, produces solutions that explain data in a simple and intuitively meaningful way and could be easily interpreted. Hence, GP has been used in this work not only to build a model, but also to evaluate the effect of the parameters of the model. In fact, for the considered problem GP highlighted the relationship between the energy consumption and the external temperature. Moreover, GP demonstrates the ability to perform feature selection in an automatic and effective way, using a more compact set of variables than the other techniques and using always the same limited set of variables in all the returned solutions. When other machine learning techniques are considered, the analysis of the solutions is much more difficult and less intuitive. Possible future works may consider longer forecasting periods, also with the use of data provided by temperatures forecasts.

References

1. Adya, M., Collopy, F.: How effective are neural networks at forecasting and prediction? A review and evaluation. Journal of Forecasting **17**, 481–495 (1998)
2. Barnum, H., Bernstein, H.J., Spector, L.: Quantum circuits for OR and AND of ORs. Journal of Physics A: Mathematical and General **33**, 8047–8057 (2000)
3. Bhattacharya, M., Abraham, A., Nath, B.: A linear genetic programming approach for modelling electricity demand prediction in victoria. In: Hybrid Information Systems, pp. 379–393. Springer (2002)
4. Box, G., Jenkins, G.M., Reinsel, G.: Time Series Analysis: Forecasting & Control, 3rd edn. Prentice Hall, February 1994
5. De Felice, M., Yao, X.: Neural networks ensembles for short-term load forecasting. In: IEEE Symposium Series in Computational Intelligence (SSCI) (2011)
6. Dee, D., Uppala, S., Simmons, A., Berrisford, P., Poli, P., Kobayashi, S., Andrae, U., Balmaseda, M., Balsamo, G., Bauer, P., et al.: The ERA-Interim reanalysis: configuration and performance of the data assimilation system. Quarterly Journal of the Royal Meteorological Society **137**(656), 553–597 (2011)
7. Feinberg, E.A., Genethliou, D.: Load forecasting. In: Chow, J., Wu, F., Momoh, J. (eds.) Applied Mathematics for Restructured Electric Power Systems: Optimization, Control and Computational Intelligence, pp. 269–285. Springer (2005)
8. Hansen, J., Nelson, R.: Neural networks and traditional time series methods: a synergistic combination in state economic forecasts. IEEE Transactions on Neural Networks **8**(4), 863–873 (1997)
9. Hippert, H., Pedreira, C., Souza, R.: Neural networks for short-term load forecasting: a review and evaluation. IEEE Transactions on Power Systems **16**(1), 44–55 (2001)

10. Hobbs, B., Jitprapaikulsarn, S., Konda, S., Chankong, V., Loparo, K., Maratukulam, D.: Analysis of the value for unit commitment of improved load forecasts. IEEE Transactions on Power Systems **14**(4), 1342–1348 (1999)
11. Koza, J.R.: Genetic programming: on the programming of computers by natural selection. MIT Press, Cambridge (1992)
12. Koza, J.R.: Human-competitive results produced by genetic programming. Genetic Programming and Evolvable Machines **11**, 251–284 (2010)
13. Koza, J.R., Al-Sakran, S.H., Jones, L.W.: Automated ab initio synthesis of complete designs of four patented optical lens systems by means of genetic programming. Artif. Intell. Eng. Des. Anal. Manuf. **22**(3), 249–273 (2008)
14. Koza, J.R., Andre, D., Bennett, F.H., Keane, M.A.: Genetic Programming III: Darwinian Invention & Problem Solving, 1st edn. Morgan Kaufmann Publishers Inc., San Francisco (1999)
15. Koza, J.R., Bade, S.L., Bennett, F.H.: Evolving sorting networks using genetic programming and rapidly reconfigurable field-programmable gate arrays. In: Workshop on Evolvable Systems. International Joint Conference on Artificial Intelligence, pp. 27–32. IEEE Press (1997)
16. Lee, D.G., Lee, B.W., Chang, S.H.: Genetic programming model for long-term forecasting of electric power demand. Electric Power Systems Research **40**(1), 17–22 (1997)
17. Lipson, H.: Evolutionary synthesis of kinematic mechanisms. Artif. Intell. Eng. Des. Anal. Manuf. **22**(3), 195–205 (2008)
18. Lohn, J.D., Hornby, G.S., Linden, D.S.: Human-competitive evolved antennas. Artif. Intell. Eng. Des. Anal. Manuf. **22**(3), 235–247 (2008)
19. Troncoso Lora, A., Riquelme, J.C., Martínez Ramos, J.L., Riquelme Santos, J.M., Gómez Expósito, A.: Influence of kNN-based load forecasting errors on optimal energy production. In: Pires, F.M., Abreu, S.P. (eds.) EPIA 2003. LNCS (LNAI), vol. 2902, pp. 189–203. Springer, Heidelberg (2003)
20. Park, D., El-Sharkawi, M., Marks, R., Atlas, L., Damborg, M.: Electric load forecasting using an artificial neural network. IEEE Transactions on Power Systems **6**, 442–449 (1991)
21. Poli, R., Langdon, W.B., Mcphee, N.F.: A field guide to genetic programming, published via Lulu.com (2008). http://www.gp-field-guide.org.uk/
22. Preble, S., Lipson, M., Lipson, H.: Two-dimensional photonic crystals designed by evolutionary algorithms. Applied Physics Letters **86**(6) (2005)
23. Quinlan, R.J.: Learning with continuous classes. In: 5th Australian Joint Conference on Artificial Intelligence, pp. 343–348. World Scientific, Singapore (1992)
24. Solomatine, D., Xue, Y.: M5 model trees and neural networks: application to flood forecasting in the upper reach of the Huai River in China. Journal of Hydrologic Engineering **9**(6), 491–501 (2004)
25. Štravs, L., Brilly, M.: Development of a low-flow forecasting model using the M5 machine learning method. Hydrological Sciences Journal **52**(3), 466–477 (2007)
26. Wang, Y., Witten, I.: Inducing model trees for continuous classes. In: Proceedings of the Ninth European Conference on Machine Learning, pp. 128–137 (1997)

The Optimization Ability of Evolved Strategies

Nuno Lourenço[1]($\boxtimes$), Francisco B. Pereira[1,2], and Ernesto Costa[1]

[1] CISUC, Department of Informatics Engineering, University of Coimbra,
Polo II - Pinhal de Marrocos, 3030 Coimbra, Portugal
{naml,xico,ernesto}@dei.uc.pt
[2] Instituto Politécnico de Coimbra, ISEC, DEIS, Rua Pedro Nunes,
Quinta da Nora, 3030-199 Coimbra, Portugal

Abstract. Hyper-Heuristics (HH) is a field of research that aims to automatically discover effective and robust algorithmic strategies by combining low-level components of existing methods and by defining the appropriate settings. Standard HH frameworks usually comprise two sequential stages: Learning is where promising strategies are discovered; and Validation is the subsequent phase that consists in the application of the best learned strategies to unseen optimization scenarios, thus assessing its generalization ability.

Evolutionary Algorithms are commonly employed by the HH learning step to evolve a set of candidate strategies. In this stage, the algorithm relies on simple fitness criteria to estimate the optimization ability of the evolved strategies. However, the adoption of such basic conditions might compromise the accuracy of the evaluation and it raises the question whether the HH framework is able to accurately identify the most promising strategies learned by the evolutionary algorithm. We present a detailed study to gain insight into the correlation between the optimization behavior exhibited in the learning phase and the corresponding performance in the validation step. In concrete, we investigate if the most promising strategies identified during learning keep the good performance when generalizing to unseen optimization scenarios. The analysis of the results reveals that simple fitness criteria are accurate predictors of the optimization ability of evolved strategies.

1 Introduction

Hyper-Heuristics (HH) is a field of research that aims to automatically discover effective and robust optimization algorithms [1]. HH frameworks can generate metaheuristics for a given computational problem either by selecting/ combining low level heuristics or by designing a new method based on components of existing ones. In [1], Burke *et al.* have presented a detailed discussion of these HH categories, complemented with several representative examples. HH are commonly divided in two sequential stages: *Learning* is where the strategies are automatically created, whilst, in *Validation*, the most promising learned solutions are applied to unseen and more challenging scenarios.

Evolutionary Algorithms (EAs) are regularly applied as HH search engines to learn effective algorithmic strategies for a given problem or class of related

© Springer International Publishing Switzerland 2015
F. Pereira et al. (Eds.) EPIA 2015, LNAI 9273, pp. 226–237, 2015.
DOI: 10.1007/978-3-319-23485-4_23

problems. Each strategy created by the EA needs to be evaluated to estimate its optimization ability. In concrete, every individual from the population is applied to an instance of the problem under consideration and the quality of the solutions found is used as an estimator of its optimization ability. To keep the computational effort at a reasonable level, the evaluation step of the EA search engine relies on small instances and simplified fitness criteria. However, it is known that the training conditions impact the properties of the algorithms being evolved [5–7,12] and it is not clear if the limited evaluation conditions adopted by HH frameworks compromise the accurate identification of the best optimization strategies. In this paper we address this question by investigating if the fitness criteria used in learning provide enough information to identify the most effective and robust strategies.

The study is performed with a well-known Grammatical Evolution (GE) [9] HH framework, originally proposed in 2012 [14]. This computational model is able to automatically generate complete Ant Colony Optimization (ACO) [2] algorithms that can effectively solve different traveling salesperson problem (TSP) instances. It allows for a flexible definition of the components and settings to be used by the algorithm, as well as its general structure. Results presented in the aforementioned reference confirm that this HH framework is able to evolve novel ACO architectures, competitive with state-of-the-art human designed variants.

The analysis presented in this paper reveals that the most promising strategies discovered in the learning phase tend to maintain the good performance in the validation step. This outcome supports the adoption of simple and somehow inaccurate fitness criteria in the learning phase, as this does not compromise the ability of the HH framework to identify the most effective and robust optimization strategies. Additionally, we investigate the existence of overfitting in the learning phase, i.e., the over interpretation of relationships that only occur in the learning data. Preliminary results suggest that learning is able to avoid the overfitting of the data.

The paper is structured as follows: Section 2 describes the general properties of the HH framework adopted in this work, whereas section 3 reviews the main features of the ACO HH model used in the experiments. Section 4 contains the experimental setup and presents an empirical study to assess the optimization ability of the learned strategies. Finally, Section 5 gathers the main conclusions and suggests some ideas for future work.

2   Hyper-Heuristics

HH is a recent area of research that addresses the construction of specific, high-level, heuristic problem solvers, by searching the space of possible low-level heuristics for the particular problem one wants to solve. HH can be divided in two major groups [1]: the selection group comprises the search for the best sequence of low-level heuristics, selected from a set of predefined methods usually applied to a specific problem; the other group includes methods that promote the creation of new heuristics. In the later case, the HH iteratively learns a novel

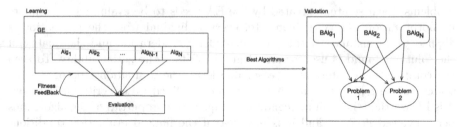

Fig. 1. Hyper-Heuristic Framework Architecture

algorithm which is then applied to solve the problem at hand. During this process, the HH are usually guided by feedback obtained through the execution of each candidate solution in simple instances of the problem under consideration. Genetic Programming (GP), a branch of EAs, has been increasingly adopted as the HH search engine to learn effective algorithmic strategies [10]. In the recent years, Grammatical Evolution (GE) [9], a linear form of GP, has received increasing attention from the HH community since it allows for a straightforward enforcement of semantic and syntactic restrictions, by means of a grammar.

2.1   Framework

In this work a two phase architecture is adopted (see Fig. 1). In the first phase, *Learning*, a GE-based HH will construct algorithmic strategies. GE is a GP branch that decouples the genotype from the phenotype by encoding solutions as a linear string of integers. The evaluation of an individual requires the application of a genotype-phenotype mapping that decodes the linear string into a syntactically legal algorithmic strategy, by means of a grammar. GE grammars are composed by a set of production rules written in the Backus-Naur format, defining the general structure of the programs being evolved and also the components that can be selected to build a given strategy (consult [9] for details concerning GE algorithms).

The quality of a strategy generated by the GE should reflect its ability to solve a given problem. During evolution, each GE solution is applied to a predetermined problem instance and its fitness corresponds to the quality of the best solution found. Given this modus operandi, the GE evaluation step is a computationally intensive task. To prevent the learning process from taking an excessive amount of time, some simple evaluation conditions are usually defined: i) one single and small problem instance is used to assign fitness; ii) only one run is performed; iii) the number of iterations is kept low. Clearly, the adoption of such simple conditions might compromise the results by hindering differences between competing strategies, leading to an inaccurate assessment of the real optimization ability of evolved solutions. The experiments described in section 4 aim to gain insight into this situation.

The second phase of the HH framework is *Validation*. The most promising strategies (*BAlg*) identified in the previous step, *i.e.*, those that obtained better fitness in the learning task, are applied to unseen scenarios to confirm their effectiveness and robustness.

2.2 Related Work

Recent works have shown that the conditions used in the learning phase influence the structure of algorithmic strategies that are being evolved. In [13], Smit *et al.* shows that using different performance measures like mean best fitness or success rate may yield very different algorithmic strategies. Lourenço *et al.* [5] presented a study on how a GE-based HH to evolve full-fledged EAs is affected by the learning conditions used to evaluate the quality of the algorithm. More precisely, they investigated how different population sizes and/or number of generations influenced the components that were selected by the HH to build the EA. Later, in [6] they presented a HH to learn selection strategies to EAs, and showed that the levels of selective pressure would depend on the EA where the strategy was inserted. In [7] Martin *et al.* evolved Black-Box Search Algorithms (BBSAs) and showed that using multiple instances of the problem affect the algorithmic structure of the strategies.

In [3], Eiben *et al.* present a discussion on how evolved algorithms should be selected, and present robustness as being a key factor to determine the quality of algorithms. Robustness is related to performance and its variance across some dimension. One of these dimensions is the range of problems (or problem instances) that the algorithm can tackle. Based on this, they define two properties: *fallibility* which indicates that the algorithms can clearly fail on some specific problems; *applicability* which indicates the range of problems that the algorithm can successfully tackle. Note that the applicability depends on a certain performance threshold T. An algorithm is robust if it performs well across several problems (high applicability), and if it has small performance variances.

3   Design of Ant Algorithms with Grammatical Evolution

The HH framework, originally proposed by Tavares *et al.* [14] to evolve full-fledged ACO algorithms, will be used as the testbed for our experiments. Ant Colony Optimization (ACO) algorithms are a set of population-based methods, loosely inspired by the behaviour of ant foraging [2]. Following the original Ant System (AS) algorithm proposed by Marco Dorigo in 1992, many other variants and extensions have been described in the literature. To help researchers and practitioners to select and tailor the most appropriate variant to a given problem, several automatic ACO design frameworks have been proposed in the last few years [4,11,14]. The production set of the above mentioned framework defines the general architecture of an ACO-like algorithm, comprising an initialisation step followed by an optimization cycle. The first stage initialises the pheromone matrix and other settings of the algorithm. The main loop consists of the building of the solutions, pheromone trail update and daemon actions. Each component contains several alternatives to implement a specific task. As an example, the decision policy adopted by the ants to build a trail can be either the random proportional rule used by AS methods or the q-selection pseudorandom proportional rule introduced by the Ant Colony System (ACS) variant. If the last option is selected, the GE engine also defines a specific value for the q-value parameter. The grammar allows the replication of all main ACO algorithms, such as AS, ACS, Elitist Ant System (EAS), Rank-based Ant System (RAS), and Max-Min Ant System (MMAS). Additionally, it can generate novel combinations of blocks and settings that define alternative ACO algorithms. Results presented in [14] show that the GE-HH framework is able to learn original ACO architectures, different from standard strategies. Moreover, results obtained in validation instances reveal that the evolved strategies generalize well and are competitive with human-designed variants (consult the aforementioned reference for a detailed analysis of the results).

4   Experimental Analysis

Experiments described in this section aim to gain insight into the capacity of the GE-based HH to identify the most promising solutions during the learning step. In concrete, we determine the relation between the quality of strategies as estimated by the GE and their optimization ability when applied to unseen and harder scenarios. Such study will provide valuable information about the capacity of the GE to build and identify strategies that are robust, i.e., highly applicable and with small fallibility.

In practical terms, we take all strategies belonging to the last generation of the GE and rank them by the fitness obtained in the learning evaluation instance. Since the GE relies on a steady-state replacement method, the last generation contains the best optimization strategies identified during the learning phase. Then, these strategies are applied to unseen instances and ranked again based on the new results achieved. The comparison of the ranks obtained in different phases will provide relevant information in what concerns the generalization ability of the evolved strategies.

Table 1. GE Learning Parameters: adapted from [14]

Runs	30
Population Size	64
Generations	40
Individual Size	25
Wrapping	No
Crossover Operator	One-Point with a 0.7 rate
Mutation Operator	Integer-Flip with a 0.05 rate
Selection	Tournament with size 3
Replacement	Steady State
Learning Instances	pr76, ts225

The GE settings used in the experiments are depicted in Table 1. The population size is set to 64 individuals, each one composed by 25 integer codons, which is an upper bound on the number of production rules needed to generate an ACO strategy using the grammar from [14]. As this grammar does not contain recursive production rules, it is possible to determine the maximum number of values needed to create a complete phenotype. Also, wrapping is not necessary since the mapping process never goes beyond the end of the integer string.

We selected several TSP instances from the TSPLIB[1] for the experimental analysis. Two different instances were selected to learn the ACO strategies: pr76 and ts225 (the numerical values represent how many cities the instance has). Each ACO algorithm encoded in a GE solution is executed once during 100 iterations. The fitness assigned to this strategy corresponds to the best solution found. The strategies encode all the required settings to run the ACO algorithm, with the exception of the colony size, which is set to 10% of the number of cities (truncated the closest integer).

In what concerns the validation step, the best ACO strategies are applied to four different TSP instances: lin105, pr136, pr226, lin318. In this phase, all ACO algorithms are run for 30 times and the number of iterations is increased to 5000. The size of the colony is the same (10% of the size of the instance being optimised). Table 2 summarises the parameters used. In both phases, the results are expressed as a normalised distance to the optimum.

Table 2. ACO Validation Parameters

Runs	30
Iterations	5000
Colony Size	10% of the Instance Size
Instances	lin105, pr136, pr226, lin318

Fig. 2 displays the ranking distributions of the best ACO strategies learned with the pr76 instance. The 4 panels correspond to the 4 different validation instances. Each solution from the last GE generation is identified using an integer from 1 to 64, displayed in the horizontal axis. These solutions are ranked by the fitness obtained in training (solution 1 is the best strategy from the last generation, whilst solution 64 is the worst). The vertical axis corresponds to the position in the rank. Small circles highlight the learning rank and, given the ordering of the solutions from the GE last generation, we see a perfect diagonal in all panels. The small triangles identify the ranking of the solutions achieved in the 4 validation tasks (one on each panel). Ideally, these rankings should be identical to the ones obtained in training, i.e., the most promising solutions identified by the GE would be those that generalize better to unseen instances.

An inspection of the results reveals an evident correlation between the behavior of the strategies in both phases. An almost perfect line of triangles is visible in

---

[1] http://comopt.ifi.uni-heidelberg.de/software/TSPLIB95/

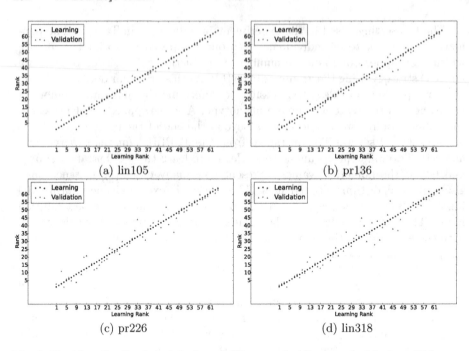

Fig. 2. Ranking distribution of the best ACO strategies discovered with the pr76 learning instance.

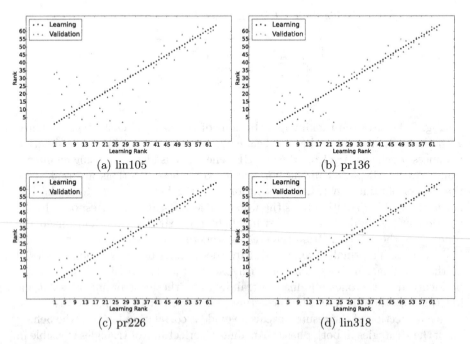

Fig. 3. Ranking distribution of the best ACO strategies discovered with the ts225 learning instance.

the 4 panels, confirming that the best strategies from training keep the good performance in validation. This trend is visible across all the validation instances and shows that, with the pr76 instance, training is accurately identifying the more robust and effective ACO strategies.

Fig. 3 displays the ranking distributions of the best ACO strategies learned with the ts225 instance. Although the general trend is maintained, a close inspection of the results reveals some interesting disagreements. The best ACO strategies learned with ts225 tend to have a modest performance when applied to small validation instances, such as lin105 and pr136. On the contrary, they behave well on larger instances (see, e.g., the results obtained with the validation instance from panel d)). This outcome confirms that the training conditions impact the structure of the evolved algorithmic strategies, which is in agreement with other findings reported in the literature [5,13]. The ts225 instance is considered a hard TSP instance [8] and, given the results displayed in Fig. 3, it promotes the evolution of ACO strategies particularly suited for TSP problems with a higher number of cities. In the remainder of this section we present some additional results that help gain insight into these findings.

To authenticate the correlation between learning and validation we computed the Pearson correlation coefficient between the rankings obtained in each phase. This coefficient ranges between -1 and 1, where -1 identifies a completely negative correlation and 1 highlights a total correlation (the best strategies in learning are the best in validation). The results obtained are presented in Table 3. Columns contain instances used in learning, whilst rows correspond to validation instances. The values from the table confirm that there is always a clearly positive correlation between the two phases, i.e., the quality obtained by a solution in learning is an accurate estimator if its optimization ability. The lowest values of the Pearson coefficient are obtained by strategies learned with the ts225 instance and validated in small TSP problems, confirming the visual inspection of Fig. 3. In this correlation analysis we adopted a significance level of $\alpha = 0.05$. All the p-values obtained were smaller than α, thus confirming the statistical significance of the study.

To complement the analysis, we present in Fig. 4 the absolute performance of the best learned ACO strategies in the 4 selected validation instances. Each panel comprises one of the validation scenarios and contains a comparison between the optimization performance of strategies evolved with different learning instances (black mean and error bars are from ACO strategies trained with the pr76 instances, whilst the grey are from algorithms evolved with the ts225 instance). In general, for all panels and for strategies evolved with the two training instances, the deviation from the optimum increases with the training ranking, confirming that the best algorithms from phase 1 are those that exhibit a better optimization ability. However, the results reveal an interesting pattern in what concerns the absolute behavior of the algorithms. For the smaller validation instances (lin 105 and pr136 in panels a) and b)), the ACO strategies evolved by the smaller learning instances achieve a better performance. On the contrary, ACO algorithms learned with the ts225 instance are better equipped to handle the largest validation problem (lin318 in panel d)). This is another piece of evidence that confirms the impact of the training conditions on the structure of the evolved solutions. A detailed analysis of the algorithmic structure reveals that the pr76 training instance promotes the appearance of extremely greedy ACO algorithms (e.g., they tend to have very low evaporation levels), particularly suited for the quick optimization of simple instances. On the contrary, strategies evolved with the ts225 training instance strongly rely on full evaporation, thus promoting the appearance of methods with increased exploration ability, particularly suited for larger and harder TSP problems.

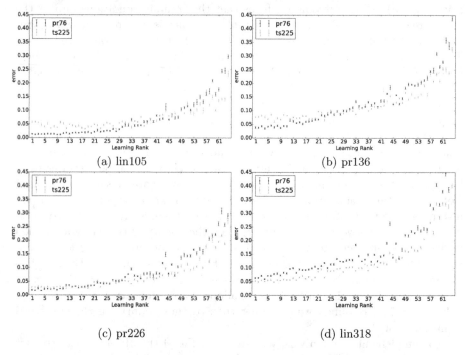

Fig. 4. MBF of the best evolved ACO strategies in the 4 validation instances. Black symbols identify results from strategies learned with the pr76 instance and grey symbols correspond to results from strategies obtained with the ts225 instance.

4.1 Measuring Overfitting

To complete our analysis we investigate the evolution of overfitting while learning ACO strategies. To estimate the occurrence of overfitting we selected one additional instance for each training scenario, with the same size of the instance used in training (eil76 and tsp225, respectively). In each GE generation, the current best ACO strategy is applied to this new test instance and the quality of the obtained solution is recorded (this value is never used for training).

Fig. 5 and 6 present the evolution of the *Mean Best Fitness* (MBF) during the learning phase, respectively for the pr76 and ts225 instances. Both figures contain two panels: panel a) exhibits the evolution of the MBF measured by the learning instance, which corresponds to the value used to guide the GE exploration; panel b) displays the MBF obtained with the testing instance and it is only used to detect overfitting.

The results depicted in panels 5a and panel 6a show that the HH framework gradually learns better strategies. A brief perusal of the MBF evolution reveals a rapid decrease in the first generations, followed by a slower convergence. This is explained by the fact that in the beginning of the evolutionary process the GE combines different components provided by the grammar to build a robust strategy, whilst at the end it tries to fine-tune the numeric parameters. The search for a meaningful combination of components has a stronger impact on fitness than modifying numeric values.

Overfitting occurs when the fitness of the learning strategies keeps improving, whilst it deteriorates in testing. Panels 5b and 6b show the MBF for the testing step. An inspection of the results shows that it tends to decrease throughout the evolutionary run. This shows that the strategies being evolved are not becoming overspecialized, i.e., they maintain the ability to solve instances different from the ones used in training.

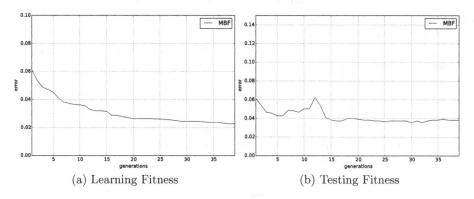

(a) Learning Fitness   (b) Testing Fitness

Fig. 5. Evolution of the MBF for the pr76 learning instance and the corresponding eil76 testing instance.

Table 3. Pearson correlation coefficients

	pr76	ts225
lin105	0.98	0.81
pr136	0.90	0.90
pr226	0.97	0.95
lin318	0.95	0.98

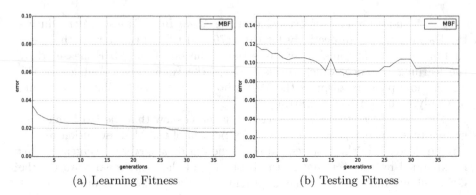

(a) Learning Fitness   (b) Testing Fitness

Fig. 6. Evolution of the MBF for the ts225 learning instance and the corresponding tsp225 testing instance.

5   Conclusions

HH is an area of research that aims to automate the design of algorithmic strategies by combining low-level components of existing methods. Most of the HH frameworks are divided two phases. The first phase, Learning, is where the strategies are built and evaluated. Afterwards, the robustness of the best solutions is validated in unseen scenarios. Usually, researchers select the best learned strategies based only simple and somehow inaccurate criteria. Given this situation, there is the risk of failing the identification of the most effective learned algorithmic strategies.

In this work we studied the correlation between the quality exhibited by strategies during learning and their effective optimization ability when applied to unseen scenarios. We relied on an existing GE-based HH to evolve full-fledged ACO algorithms to perform the analysis. Results revealed a clear correlation between the quality exhibited by the strategies in both phases. As a rule, the most promising algorithms identified in learning generalize better to unseen validation instances. This study provides valuable guidelines for HH practitioners, as it suggests that the limited training conditions do not seriously compromise the identification of the algorithmic strategies with the best optimization ability. The outcomes also confirmed the impact of the training conditions on the structure of the evolved solutions. Training with small instances promotes the appearance of greedy optimization strategies particularly suited for simple problems, whereas larger (and harder) training cases favor algorithmic solutions that excel in more complicated scenarios. Finally, a preliminary investigation revealed that training seems to be overfitting free, i.e., the strategies being learned are not becoming overspecialized to the specific instance used in the evaluation.

There are several possible extension to this work. In the near future we aim to validate the correlation study in alternative training evaluations conditions. Also, a complete understanding of overfitting is still in progress and we will extend this analysis to a wider range of scenarios (e.g., the size, structure and the number of the instances used for testing might influence the results). Finally, we will investigate if the main results hold for different HH frameworks.

Acknowledgments. This work was partially supported by Fundação para a Ciência e Tecnologia (FCT), Portugal, under the grant SFRH/BD/79649/2011.

References

1. Burke, E.K., Gendreau, M., Hyde, M., Kendall, G., Ochoa, G., Özcan, E., Qu, R.: Hyper-heuristics: A survey of the state of the art. Journal of the Operational Research Society **64**(12), 1695–1724 (2013)
2. Dorigo, M., Stützle, T.: Ant Colony Optimization. Bradford Company, Scituate (2004)
3. Eiben, A., Smit, S.: Parameter tuning for configuring and analyzing evolutionary algorithms. Swarm and Evolutionary Computation **1**(1), 19–31 (2011)
4. López-Ibáñez, M., Stützle, T.: Automatic configuration of multi-objective ACO algorithms. In: Dorigo, M., Birattari, M., Di Caro, G.A., Doursat, R., Engelbrecht, A.P., Floreano, D., Gambardella, L.M., Groß, R., Şahin, E., Sayama, H., Stützle, T. (eds.) ANTS 2010. LNCS, vol. 6234, pp. 95–106. Springer, Heidelberg (2010)
5. Lourenço, N., Pereira, F.B., Costa, E.: The importance of the learning conditions in hyper-heuristics. In: Proceedings of the 15th Annual Conference on Genetic and Evolutionary Computation, GECCO 2013, pp. 1525–1532 (2013)
6. Lourenço, N., Pereira, F., Costa, E.: Learning selection strategies for evolutionary algorithms. In: Legrand, P., Corsini, M.-M., Hao, J.-K., Monmarché, N., Lutton, E., Schoenauer, M. (eds.) EA 2013. LNCS, vol. 8752, pp. 197–208. Springer, Heidelberg (2014)
7. Martin, M.A., Tauritz, D.R.: A problem configuration study of the robustness of a black-box search algorithm hyper-heuristic. In: Proceedings of the 2014 Conference Companion on Genetic and Evolutionary Computation Companion, GECCO Comp. 2014, pp. 1389–1396 (2014)
8. Merz, P., Freisleben, B.: Memetic algorithms for the traveling salesman problem. Complex Systems **13**(4), 297–346 (2001)
9. O'Neill, M., Ryan, C.: Grammatical evolution: evolutionary automatic programming in an arbitrary language, vol. 4. Springer Science (2003)
10. Pappa, G.L., Freitas, A.: Automating the Design of Data Mining Algorithms: An Evolutionary Computation Approach, 1st edn. Springer Publishing Company, Incorporated (2009)
11. Runka, A.: Evolving an edge selection formula for ant colony optimization. In: Proceedings of the GECCO 2009, pp. 1075–1082 (2009)
12. de Sá, A.G.C., Pappa, G.L.: Towards a method for automatically evolving bayesian network classifiers. In: Proceedings of the 15th Annual Conference Companion on Genetic and Evolutionary Computation, pp. 1505–1512. ACM (2013)
13. Smit, S.K., Eiben, A.E.: Beating the world champion evolutionary algorithm via revac tuning. In: IEEE Congress on Evolutionary Computation (CEC) 2010, pp. 1–8. IEEE (2010)
14. Tavares, J., Pereira, F.B.: Automatic design of ant algorithms with grammatical evolution. In: Moraglio, A., Silva, S., Krawiec, K., Machado, P., Cotta, C. (eds.) EuroGP 2012. LNCS, vol. 7244, pp. 206–217. Springer, Heidelberg (2012)

Evolution of a Metaheuristic for Aggregating Wisdom from Artificial Crowds

Christopher J. Lowrance[✉], Omar Abdelwahab, and Roman V. Yampolskiy

Department of Computer Engineering and Computer Science,
University of Louisville, Louisville, KY 40292, USA
{chris.lowrance,omar.abdelwahab,roman.yampolskiy}@louisville.edu
https://louisville.edu/speed/computer

Abstract. Approximation algorithms are often employed on hard optimization problems due to the vastness of the search spaces. Many approximation methods, such as evolutionary search, are often indeterminate and tend to converge to solutions that vary with each search attempt. If multiple search instances are executed, then the wisdom among the crowd of stochastic outcomes can be exploited by aggregating them to form a new solution that surpasses any individual result. Wisdom of artificial crowds (WoAC), which is inspired by the wisdom of crowds phenomenon, is a post-processing metaheuristic that performs this function. The aggregation method of WoAC is instrumental in producing results that consistently outperform the best individual. This paper extends the contributions of existing work on WoAC by investigating the performance of several aggregation methods. Specifically, existing and newly proposed WoAC aggregation methods are used to synthesize parallel genetic algorithm (GA) searches on a series of traveling salesman problems (TSPs), and the performance of each approach is compared. Our proposed method of weighting the input of crowd members and incrementally increasing the crowd size is shown to improve the chances of finding a solution that is superior to the best individual solution by 51% when compared to previous methods.

Keywords: Wisdom of crowds · Evolutionary combinatorial optimization · Aggregation metaheuristic · Parallel search aggregation

1 Introduction

Computing the optimal solution using an exhaustive search becomes intractable as the size of the problem grows for computationally hard (NP-hard) problems [1]. Consequently, heuristics and stochastic search algorithms are commonly used in an effort to find reasonable approximations to difficult problems in a polynomial time [1,2]. These approximation algorithms are often incomplete and can produce indeterminate results that vary when repeated on larger search spaces [3,4]. The variance produced by these types of searches, assuming several search attempts have been made, can be exploited in a collaborative effort to form better

© Springer International Publishing Switzerland 2015
F. Pereira et al. (Eds.) EPIA 2015, LNAI 9273, pp. 238–249, 2015.
DOI: 10.1007/978-3-319-23485-4_24

approximations to difficult problems. The post-processing metaheuristic known as the Wisdom of Artificial Crowds (WoAC) is based on this concept [1,5]. In essence, the WoAC algorithm operates on a group of indeterminate search outcomes from the same problem space, and then aggregates them with the goal of forming a solution that is superior to any individual outcome within the group. It is inspired by a more widely known sociological phenomenon referred to as the Wisdom of Crowds (WoC) [6], and likewise, seeks to perform reasoning and aggregation based on commonality observed within a group of converged and best-fit search outcomes.

WoAC operates on a pool of converged search results that can be obtained via any indeterminate search method. Evolutionary algorithms, such as the genetic algorithm (GA) used in this paper, are well-suited for WoAC because they are easily parallelizable [3]. Multiple search instances can be conducted in parallel, which can speed up the process of generating a pool of best-fit candidate approximations.

The means of aggregating the group of best-fit solutions is the essence of WoAC, and this process is critical to the goal of producing a solution that is paramount to any individual contributor within the crowd of possible solutions. As a result, this paper focuses on the evaluating several means of WoAC aggregation. The primary contribution of this paper includes the introduction and evaluation of new methods for performing WoAC aggregation. We show that our proposed aggregation algorithm improves the consistency of forming a superior solution when compared to existing methods. Our findings also highlight some key factors that were previously unreported and that heavily influence the performance of WoAC. These factors include the importance of weighting the crowd members based on fitness and iteratively attempting aggregation after each new member has been added to the crowd.

This paper utilizes the well-known TSP as the combinatorial optimization problem for evaluating the aggregation components of the WoAC algorithm. As our search method, we employ a GA for generating pools of possible TSP solutions. We selected these options because of their familiarity among the research community, and as a result, abstract details about the fundamentals of the TSP and the GA. Instead, we focus our attention on the WoAC algorithm and note that the metaheuristic is applicable to a wide range of optimization problems and indeterminate search techniques other than the TSP and GA, respectively.

The remainder of this paper is organized as follows. Section 2 reviews the related research involving aggregation methods for WoC and WoAC. Subsequently, in Section 3, we review the existing WoAC algorithm, and also, provide the details on the newly proposed means of aggregation. Afterwards, we evaluate the experimental results in Section 4, and finally, we provide our conclusions and future work in Section 5.

2   Related Work

The concept of applying WoC to the TSP has shown promising results in several works. Yi and Dry aggregated human-generated TSP responses and demonstrated that it is possible to generate an aggregate response that is superior to any individual [7]. This concept was extended to computer-generated TSP approximations by Yampolskiy and El-Barkouky [1] and was coined WoAC. In that study, 90% of the 20 contributors must agree on a TSP connection before it is kept as part of the aggregate. In another study [5], Yampolskiy, Ashby, and Hassan remove the 90% agreement stipulation and have the group of 10 GA outcomes vote with equal weights similar to [7].

Others researchers applied the concept of WoC with varying levels of success to other applications [8–12], but none of these studies compared different methods of aggregation, nor weighted the input of crowd members. Velic, Grzinic, and Padavic applied the WoC concept in a stock market prediction algorithm [12]. The algorithm at times produced results reflective of the groupthink phenomenon, where less knowledgeable contributors negate the influence of highly experienced experts and negatively impact the outcome. In another study, Moore and Clayton tested the effect of WoC in detecting phishing websites and found that inexperienced users, who frequently made mistakes, often voted similarly [11]. Kittur and Kraut leveraged WoC in managing Wikipedia content and showed that coordination between writers on Wikipedia is vital for content quality, especially as the crowd size grows [10]. Yu et al. proposed a WoC-based algorithm for traffic route planning and used a non-Markovian aggregation tree for fusing the results from route planning agents [9]. In [8], Hoshen, Ben-Artzi, and Peleg proposed a means to combine multiple video streams into an improved video using the WoC concept. Lastly, WoAC has also been applied to a number of computer games [13–16].

Based on research using human groups, Wagner and Suh concluded that the size of the crowd influences the performance of WoC, and they found that improvement saturates as the crowd size grows beyond a certain size [17]. We suspect that crowd size also affects the performance of artificial crowds. Hence, this dynamic is investigated by our work, unlike the aforementioned studies.

Another dynamic that affects WoAC performance is the weight given to individual contributors. The related works of this section generally used an aggregation method that provided equal weight to each crowd member. When the worth of a contribution from an agent cannot be evaluated, then providing equal weight might be the only prudent option. However, in most combinatorial optimization problems, search results can be evaluated according to their performance with respect to other potential solutions. Giving more weight to better-performed members may be an important aspect to avoiding common mistakes (i.e. suboptimal choices) taken by several members of the crowd. In other words, problems, similar to the groupthink phenomenon observed in [12] and the common mistakes observed in [11], could be mitigated by favoring better performers and suppressing less fit contributors. The concept of merging multiple hypotheses using weight assignments has also been explored in ensemble learning. For instance, Puuronen et al. assigned different weights to the outputs of component classifiers based on their predicted classification error, and used a stacked (i.e. second-level) classifier to determine the final output based on the weighted votes received from the component classifiers [17]. However, the weighting techniques explored in this paper are generated differently than in ensemble learning, and we apply the concept in the application domain of optimization, not supervised learning.

A final factor that likely impacts the performance of WoC, but was not investigated in the previous works, is the level of difficulty involved in the problem. As noted in a study using a group of human responses to various questions [18], the performance of WoC suffers as the task difficulty increases. We posit that the same holds true for artificial crowds and investigate this issue more closely in Section 4 and show how weighting opinions can mitigate this issue.

3   WoAC Algorithm Description

The original WoAC algorithm proposed in [5,7] constructs new approximations to optimization problems based on the trends and commonality observed among the pool of candidate solutions. The identification of popular search selections (i.e. edge connections) is accomplished through a histogram matrix, which reflects the frequency of every edge option. In graph-based problems, the algorithm consists of scanning each individual graph, and then, recording its specific edge selections by updating the frequency counts stored in the matrix.

Similarly, we use an n x n matrix that serves as a histogram for recording the frequency of TSP edge occurrences witnessed while examining each contributing proposal within the crowd; the variable, n, corresponds to the number of nodes in the TSP. Each position in the matrix (i.e. row and column combination) corresponds to a possible edge connection between nodes. The initial step of the WoAC algorithm is to review every search result within the crowd pool and update the histogram accordingly based on the observed edge connections of each solution. Once every search result is reviewed and the histogram matrix is fully constructed, the positions with higher values correspond to the more common selections made by the crowd members; hence, based on the WoC principle, such node connections tend to be wise choices that should be a part of the new route.

After the histogram is constructed, the WoAC algorithm then proceeds to build the new solution. For the TSP, this is accomplished by choosing a starting node and then searching the histogram's entire row or column corresponding to this particular node. The matrix position that has the highest value is selected as the adjacent node. This process is repeated for every adjacent node until the Hamiltonian cycle is complete. If a node is already in the newly constructed path, then the next highest occurrence is selected and so on. The objective function (i.e. spatial information) is only referenced if all options in the histogram have been exhausted, meaning that the crowd's preferred choices have already been selected as part of the newly constructed graph. In this case, a greedy heuristic is used to find the nearest node as the next destination based on the objective function. Finally, every node is attempted as the starting node and the route yielding the lowest cost is chosen as the WoAC solution.

3.1 Modifications to the WoAC Algorithm

We propose extending the aforementioned WoAC algorithm and incorporating two primary modifications. The first modification deals with weighting the suggestions from contributors based on the range of fitness observed in the crowd pool, instead of treating them equally as previously described. The new weighting process can easily be accomplished with minimal processing because the fitness scores of the candidates are already known from the convergence of each preliminary search. We propose two new types of weighted aggregation methods and later compare them to the equal weight distribution method of [5,7]. In order to support these new means of weighting contributions, we modify the WoAC algorithm to maintain three separate histograms: one for each method of weighting.

The other alteration we propose to WoAC is to vary the crowd size by incrementally building and evaluating the weighted histograms. In other words, the modified WoAC algorithm would review the edge selections of an individual solution, then update each weighted histogram based on its fitness score with respect to the group, and afterwards, iterate through the WoAC aggregation process. This process is repeated for every member in the crowd pool. In contrast, the previous version of WoAC selects some arbitrary crowd size *a priori*, and then, complies their selection choices in a single histogram matrix before running the final aggregation process only once. Instead, we effectively vary the crowd size and build a new WoAC aggregate after every individual member is added to each histogram matrices. Once all crowd members have been considered, the lowest cost solution from all WoAC aggregation attempts would be propagated as the algorithm output.

Percentage Weight. The first considered aggregation alternative to the equal weight method used in [5,7] is a simple percentage weight. The new method assigns a weight to each edge selection of a contributor based on its objective function ranking among its peers in the crowd pool. Assuming that minimal cost is desirable, we can represent the fitness of each candidate using

$$\underline{c} = \{c_1, c_2, ..., c_n\} \tag{1}$$

where $\underline{c}$ is the cost vector (i.e. array) that contains the fitness associated with each individual crowd member and n represents the total number of crowd members. Given the range of costs associated with the candidates, we can formulate a cost-distance using

$$d_i = \frac{c_i - \min(\underline{c})}{\max(\underline{c}) - \min(\underline{c})} \tag{2}$$

where d_i is the cost-distance ratio for the ith member of the crowd. This calculation provides a ratio or a percentage of how far away a candidate's solution is from the best agent in the crowd with respect to the worst. This metric varies from 0-1, and as individual fitness scores approach the best, the metric approaches zero. Finally, this ratio is transformed into a weight that is with respect to the candidate's proximity to the best agent by

$$w_{p_i} = 1 - d_i \qquad (3)$$

where w_{p_i} is the percentage weight assigned to the ith candidate's edge contributions, which are stored in the histogram dedicated for percentage weight. Using this approach, the best GA candidate solution among the crowd is given the full weight of one, while the worst candidate is ignored by assigning it zero.

Exponential Weight. In order to provide more weight to better-performed candidates and more rapidly diminish the contributions of those less favorable solutions, an exponential weighting algorithm for WoAC was also investigated. Specifically, the weight of individual contributions was based on the exponential function

$$w_{e_i} = e^{-x_i} \qquad (4)$$

where w_{e_i} is the exponential weight of the ith candidate in the crowd of suggested solutions and x_i is a constant that is associated with d_i above, which is a measure of the candidate's distance from the best option in the crowd. Specifically, the constant x_i is obtained by multiplying d_i by another constant, m, which facilitates the mapping of d_i to a range of values suitable for generating a weighting range from 0-1 using the exponential function. In the following equation,

$$x_i = d_i * m \qquad (5)$$

m is the constant that maps the ratio, d_i, to a range of values between 0 to m. Based on (5), the best option within the crowd would be assigned 0 because d_i would be 0, and the worst-fit candidate would be assigned m because d_i would be 1. Therefore, the weight assigned to an individual solution's contribution to the WoAC aggregate, as described in (4), would be a maximum of 1, while options with a higher cost (i.e. less fit) are assigned a weight less than 1. In this paper, we let $m = 5$, as e^{-5} is approximately zero, yielding a weighting factor that ranges from 1-0, with exponential decay as the candidate solution moves away from the top candidate.

Pseudocode for the Modified WoAC

```
BEGIN
1  C1,C2,...,Cn = gather_crowd(n) //Perform n independent searches
2  best_performer = MAX(C1,C2,...,Cn)   //Most-fit crowd member
3  worst_performer = MIN(C1,C2,...,Cn)  //Least-fit crowd member
4  FOR i = 1 to n // Iterate through all crowd members
5      d = calc_dist_ratio(C(i)), best_performer,worst_performer)
       // see eqn. (2)
```

```
6        p(i) = calc_percent_weight(d)    // see eqn. (3)
7        e(i) = calc_exp_weight(d)     // see eqns. (4) & (5)
8        FOR j = 1 to edge_count   // Iterate through entire graph
9            row,column = identify_connected_vertices(C(i),j)
             // given the current edge in cycle i,
             // return the connected vertices
10           update_equal_histogram(row,column)
             // update matrix position corresponding to
             // connected vertices with: cnt = cnt + 1
11           update_percent_histogram(row,column,p(i))
             // update matrix position corresponding to
             // connected vertices with: cnt = cnt + p(i)
12           update_exp_histogram(row,column,e(i))
             // update matrix position corresponding to
             // connected vertices with: cnt = cnt + e(i)
13       ENDFOR
14       equal_graph = build_graph(equal_histogram)
15       percent_graph = build_graph(percent_histogram)
16       exp_graph = build_graph(exp_histogram)
17       current_best = MIN(equal_graph,percent_graph,exp_graph)
         // find the least expense cycle from the 3 options
18       IF (current_best < overall_best) then
             overall_best = current_best
             // if new WoAC cycle is better than all previous,
             // then store it
19       ENDIF
20   ENDFOR
END
```

4   Experimental Evaluation

4.1   Evaluation Overview

The modified WoAC algorithm was repetitively tested to evaluate its performance. Every trial run of the algorithm consisted of the following two-step process. First, a crowd (i.e. pool) of approximations were generated to a specific TSP by instantiating 30 GA searches. After the searches converged, the post-processing metaheuristic was executed according to the procedure outlined in Section 3. The best aggregate solutions of each method, as well as their respective crowd sizes, were logged for statistical purposes.

Before reviewing the evaluation statistics, we will provide some preliminary information about the testing environment. The evaluation process was repeated on four different TSP datasets, and a total of 100 trials were executed on each. The TSP datasets were randomly generated using Concorde [19], and the sizes and optimal (i.e. best-known) costs of each are displayed in Table 1. The optimal costs were obtained using the TSP solver in Concorde.

Table 1. Sizes and Optimal Costs of the TSP Datasets

Num. of Nodes	Best-known Cost
44	549
77	707
97	794
122	868

The parameters of the GA used on the TSP datasets are outlined in Table 2. A total of 30 parallel instances of the GA were allowed to search and converge before executing the WoAC algorithm. The GA generally produced crowd members (i.e. candidate approximations) that were near, but slightly suboptimal to the costs generated by Concorde. Therefore, there was opportunity for the WoAC algorithm to improve upon the pool of candidate solutions and aggregate them to form a new solution closer to the best-known optimum.

Table 2. Parameters of the Genetic Algorithm

GA Parameter	Setting
Population Size	20
Parent Selection	Fitness Tournament (uniformly random among top 5)
Crossover Operator	Single-point (uniformly random)
Mutation Operator	Combination of Two Mutation Steps: 1. Uniformly random 1% mutation 2. Greedy custom - adjacent node swap until improved

As an illustrative example of the evaluation procedure, Fig. 1 shows the evolutionary development of a crowd of GAs that worked on the 44 node TSP. After the convergence of all 30 search instances, the WoAC algorithm was initiated. The percentage and exponential weights assigned to the crowd members as part of the modified WoAC are also shown in Fig. 1. Together the plots indicate that the crowd had some diversity in opinions (i.e. edge selections), which is an important aspect in attempting to recombine these opinions into a unique aggregate. Some level of diversity is important, but the original WoAC algorithm is fundamentally based on making decisions using group consensus. However, the newly proposed weighting concept skews this fundamental bias towards group decision making, and instead, gives greater consideration to those contributors known to be wiser. The goal of the evaluation was to distinguish between these different approaches and to identify the most effective means of aggregating the crowd members to form superior solutions.

4.2 Evaluation Analysis

First, we will focus on the impact that weighting crowd members had on improving the chances of surpassing the best GA. The success rates of the different

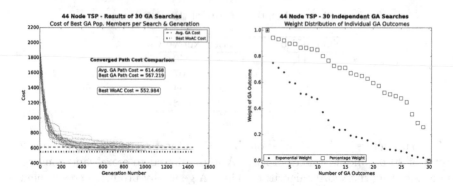

Fig. 1. The plot on the left shows the convergence of 30 independent instances of the same GA searching for the optimum tour on a TSP. The plot on the right corresponds to the exponential (*dotted*) and percentage (*square*) weights assigned to the 30 converged GA outcomes as part of the modified WoAC algorithm.

Table 3. Success Rate of the Modified WoAC Aggregation Methods

| | Percent Success in Surpassing Best GA | | | |
Dataset	Equal	Percentage	Exponential	Combined
TSP 44	22%	36%	36%	51%
TSP 77	52%	60%	48%	77%
TSP 97	4%	9%	32%	42%
TSP 122	5%	7%	33%	41%
All (Mean)	20.8%	28%	37.3%	52.8%

weighting techniques are summarized in Table 3. The percentages are based on the number of times the aggregation methods surpassed the best GA search, given that 100 evaluations were performed on each TSP dataset. From the table, it is evident that the exponential weighting method outperformed the other methods more consistently. On the other hand, the reliability of the equal- and percentage-weight techniques are shown to decrease as the TSPs became larger (i.e. more challenging). By comparing Tables 1 and 4, the GA crowd pool for the larger datasets are farther away from the global optimum and the pools are more spread apart (i.e. diverse). Therefore, the crowd was generally noisier (i.e. possessed less consensus) in these cases, which appeared to cause problems for the equal- and percentage-weighting schemes. These methods do not appear to provide a strong enough distinction between the candidates' performances in these noisy environments. In contrast, the exponential technique favors the wiser contributors more than the others, while also diminishing the opinions of the more inferior candidates. Overall, the results show that biasing the input of stronger contributors is critical to improving the success rate of the WoAC algorithm.

Table 4. Mean Costs (μ) and Standard Deviations (σ) of Crowd Members and WoAC Aggregates

Dataset	Crowd of GAs		Best GA		Equal		Percentage		Exponential	
	μ	σ	μ	σ	μ	σ	μ	σ	μ	σ
TSP 44	613.3	26.5	566.4	7.3	576.3	12.6	570.8	10.9	565.1	9.1
TSP 77	827.2	37.2	761.6	12.5	760.3	19.5	755.5	18.3	757.8	16.4
TSP 97	944.0	41.0	868.6	14.8	915.0	28.7	895.8	23.9	873.9	22.0
TSP 122	1044.9	45.5	963.2	13.7	1007.8	30.2	994.3	24.0	969.5	21.8

The other dynamic investigated during the evaluation of the modified WoAC algorithm was the concept of varying the crowd size and incrementally adding new opinions (i.e. approximations) to the crowd one-at-a-time. In the original WoAC algorithm [1,5], a fixed crowd size was determined *a priori* and all opinions were aggregated only once after considering the votes from all contributors. The success rate for the equal-weight technique in Table 3 was based on varying the crowd size; however, if the crowd size was not varied and fixed at 30, then the mean success rate of the equal-weight technique would drop to 1.8%, given the results from all the TSP trials (i.e. 400 experiments). To better visualize the impact of crowd size, Fig. 2 shows a histogram of the number of members in the crowd when the equal-weight technique successfully surpassed the best GA during the 400 trials. It indicates that the number of opinions needed to outperform the best GA is unpredictable and should not be fixed *a priori*; therefore, varying the crowd size is effective at mitigating this challenge and improving the success rate of the metaheuristic.

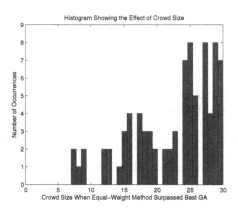

Fig. 2. The crowd size at the time when the equal-weight technique surpassed the best GA. This statistic, which is based on 400 trials from all TSP datasets, is plotted as a histogram.

In summary, the experimental results show that weighting the opinions of crowd members and varying the crowd size are vital to the success of WoAC. For instance, if we consider that the original configuration of the algorithm (i.e. fixed crowd size and equal-weight) succeeded 1.8% of the time, and that the combined success rate of the modified WoAC (i.e. all weighting techniques and variable crowd size) was 52.8% (see Table 3), then the new approach improved the algorithm's success rate by 51%.

4.3 Conclusion and Future Work

This paper investigated ways of improving the WoAC metaheuristic, which aggregates collective searches in an attempt to form superior approximations to computationally-hard problems. Specifically, we explored the heuristic's means of aggregation and discovered the critical factors that influence its performance. As a result, we proposed two beneficial modifications to the algorithm, which significantly improved its performance in surpassing the best candidate within the crowd. These modifications include iteratively adjusting the crowd size and intelligently weighting the input of the crowd members based on their fitness within the pool.

The crowds (i.e. converged searches) generated in this paper were collected using several parallel instances of the same genetic algorithm (GA). In the future, we are interested in using non-uniform GAs as part of the crowd gathering process. Such an approach could introduce more diversified opinions within the crowd and provide the potential for increased performance. However, with increased diversity, we suspect that assigning weights to the better-performed outcomes would become even more important in order to dampen the noise in the crowd caused by less efficient contributors. We are also interested in applying WoAC to other computationally hard problems and comparing its performance to other metaheuristics.

References

1. Yampolskiy, R.V., El-Barkouky, A.: Wisdom of artificial crowds algorithm for solving NP-hard problems. International Journal of Bio-Inspired Computation **3**(6), 358–369 (2011)
2. Collet, P., Rennard, J.-P.: Stochastic optimization algorithms (2007). arXiv preprint arXiv:0704.3780
3. Hoos, H.H., Sttzle, T.: Stochastic search algorithms, vol. 156. Springer (2007)
4. Kautz, H.A., Sabharwal, A., Selman, B.: Incomplete Algorithms. Handbook of Satisfiability **185**, 185–204 (2009)
5. Yampolskiy, R.V., Ashby, L., Hassan, L.: Wisdom of Artificial Crowds - A Metaheuristic Algorithm for Optimization. Journal of Intelligent Learning Systems and Applications **4**, 98 (2012)
6. Surowiecki, J.: The wisdom of crowds. Random House LLC (2005)
7. Yi, S.K.M., Steyvers, M., Lee, M.D., Dry, M.: Wisdom of the Crowds in Traveling Salesman Problems. Memory and Cognition **39**, 914–992 (2011)

8. Hoshen, Y., Ben-Artzi, G., Peleg, S.: Wisdom of the crowd in egocentric video curation. In: 2014 IEEE Conference on Computer Vision and Pattern Recognition Workshops (CVPRW), pp. 587–593, June 23–28, 2014

9. Jiangbo, Y., Kian Hsiang, L., Oran, A., Jaillet, P.: Hierarchical Bayesian nonparametric approach to modeling and learning the wisdom of crowds of urban traffic route planning agents. In: 2012 IEEE/WIC/ACM International Conferences on Web Intelligence and Intelligent Agent Technology (WI-IAT), pp. 478–485, December 4–7, 2012

10. Kittur, A., Kraut, R.E.: Harnessing the wisdom of crowds in wikipedia: quality through coordination. Paper presented at the Proceedings of the 2008 ACM conference on Computer supported cooperative work, San Diego, CA, USA

11. Moore, T., Clayton, R.C.: Evaluating the wisdom of crowds in assessing phishing websites. In: Tsudik, G. (ed.) FC 2008. LNCS, vol. 5143, pp. 16–30. Springer, Heidelberg (2008)

12. Velic, M., Grzinic, T., Padavic, I.: Wisdom of crowds algorithm for stock market predictions. In: Proceedings of the International Conference on Information Technology Interfaces, ITI, pp. 137–144 (2013)

13. Ashby, L.H., Yampolskiy, R.V.: Genetic algorithm and wisdom of artificial crowds algorithm applied to light up. In: 2011 16th International Conference on Computer Games (CGAMES), pp. 27–32, July 27–30, 2011

14. Hughes, R., Yampolskiy, R.V.: Solving Sudoku Puzzles with Wisdom of Artificial Crowds. International Journal of Intelligent Games and Simulation 7(1), 6 (2013)

15. Khalifa, A.B., Yampolskiy, R.V.: GA with Wisdom of Artificial Crowds for Solving Mastermind Satisfiability Problem. International Journal of Intelligent Games and Simulation 6(2), 6 (2011)

16. Port, A.C., Yampolskiy, R.V.: Using a GA and wisdom of artificial crowds to solve solitaire battleship puzzles. In: 2012 17th International Conference on Computer Games (CGAMES), pp. 25–29, July 30, 2012-August 1, 2012

17. Puuronen, S., Terziyan, V., Tsymbal, A.: A dynamic integration algorithm for an ensemble of classifiers. In: Ra, Z., Skowron, A. (eds.) Foundations of Intelligent Systems. Lecture Notes in Computer Science, vol. 1609, pp. 592–600. Springer, Berlin Heidelberg (1999)

18. Wagner, C., Ayoung, S.: The wisdom of crowds: impact of collective size and expertise transfer on collective performance. In: 2014 47th Hawaii International Conference on System Sciences (HICSS), pp. 594–603, January 6–9, 2014

19. Concorde TSP Solver. http://www.math.uwaterloo.ca/tsp/concorde/index.html

The Influence of Topology in Coordinating Collective Decision-Making in Bio-hybrid Societies

Rob Mills$^{(\boxtimes)}$ and Luís Correia

BioISI – Biosystems and Integrative Sciences Institute,
Faculty of Sciences, University of Lisbon, Lisbon, Portugal
rob.mills@fc.ul.pt

Abstract. Collective behaviours are widespread across the animal kingdom, many of which result from self-organised processes, making it difficult to understand the individual behaviours that give rise to such results.
One method to improve our understanding is to develop bio-hybrid societies, in which robots and animals interact, combining elements whose
behaviours are under our control (robots) with elements that are not
(animals). Recent work has shown that a bio-hybrid society comprising
simulated robots and honeybees is able to reach collective decisions that
are the product of self-organisation among the robots and the bees, and
that these decisions can be coordinated across multiple groups that reside
in distinct habitats via robot–robot communication. Here we examine
how sensitive the collective decision-making is to the specific topologies of
information sharing in such bio-hybrid societies, using agent-based simulation modelling. We find that collective decision-making across multiple
groups occupying distinct habitats is possible for a range of inter-habitat
interaction topologies, where the rate of coordinated outcomes has a positive relationship with the number of inter-habitat links. This indicates
that system-wide coordination states are relatively robust and do not
require as strong inter-habitat coupling as had previously been used.

Keywords: Collective behaviour · Mixed animal-robot societies

1 Introduction

Social living is integral to organisms across many magnitudes of scale and complexity, from bacterial biofilms [1] to primates [2], and such societies frequently
exhibit behaviours at the level of the collective, such as moving together by following a leader [3] or self-organised aggregation [4]. Many social animals and
behaviours have a substantial impact on humanity, both beneficial (*e.g.*, pol-
lination) and detrimental (*e.g.*, spread of disease). Since collective behaviours
can emerge from a combination of self-organised interactions, it can be problematic to understand what triggers, modulates, or suppresses their emergence.
One emerging methodology used to examine collective behaviours is to develop

© Springer International Publishing Switzerland 2015
F. Pereira et al. (Eds.) EPIA 2015, LNAI 9273, pp. 250–261, 2015.
DOI: 10.1007/978-3-319-23485-4_25

bio-hybrid societies, in which robots are integrated into the animal society [5,6]. In so doing, such an approach allows direct testing of hypotheses regarding individual behaviours and how they are modulated by the group context (for example, confirming a hypothesised behaviour by showing that a collective behaviour is not changed when some animals are substituted by robots [6]). Alternatively, it becomes possible to use robot behaviours that can manipulate the overall collective behaviours [5].

Our research aims to develop bio-hybrid societies, ultimately comprising multiple species that interact with robots, which thus form an interface between animals that need not naturally share a habitat. Interfacing in this manner has the added advantage that we can monitor precisely what information is exchanged between animal groups (and permits experiments that attenuate or amplify specific information types). To move towards addressing this overarching aim, here we use a simplified system that comprises multiple populations of the same species, and we examine this using individual-based simulation modelling.

In recent work, it has been shown that juvenile honeybees interacting with robots can reach collective decisions jointly with those robots, and moreover, that such collective decisions can be coordinated across multiple populations of animals that reside in distinct habitats [7]. This work showed that robots using cross-inhibition and local excitation led to high levels of collective decision-making. However, it only compared 'all-or-nothing' coupling between the two habitats. In this paper, we examine the sensitivity of decision-making and coordination of those decisions across arenas, with respect to the inter-robot communication topology. We find that even relatively sparse numbers of links between habitats can be sufficient to coordinate outcomes across those habitats. These findings improve our understanding about the interactions that are sufficient to coordinate behaviours among separated groups of animals, and the limits that can be tolerated.

1.1   Related Work

Individual-based modelling and embodiment of behaviour in robots are becoming increasingly used as tools for studying animal behaviour, at both the individual level [8] and the collective level [9,10]. Moreover, rather than merely attempt to replicate (or to abstract) a behaviour under study with robots, using robots to interact with animals can enable investigation of social behaviours in a more direct way [6]. For instance, [3] present a programmable fish that is used to investigate leadership among fish movement. [5] present robot cockroaches that are accepted by the insects, and together exhibit group dynamics equivalent to wholly animal societies. This work also showed that by changing the behaviour of the robots they could modulate the overall collective decisions reached.

Honeybees are another social insect that aggregate in environmental regions that they favour: [11] studies thermal conditions (rather than the light level preferences in cockroaches). Individual animals do not systematically select these preferred regions however: it is a collective decision that depends on being part of a group of a sufficient size [12].

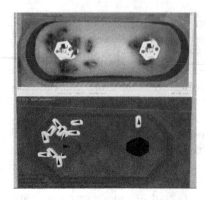
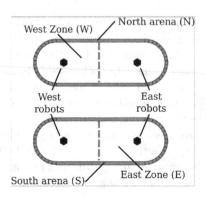

Fig. 1. A preliminary experiment with a hybrid society comprising robots, real bees, virtual bees and simulated robots, yielding collective decisions among virtual and real bees.

Fig. 2. How we split and name the arenas into zones during the analysis.

Our recent work [7] has examined how collective decisions can be reached by hybrid animal–robot societies, with individual-based simulation following preliminary work that coupled real bees to simulated bees via physical and simulated robots (see Fig. 1). This work uses robots that are able to manipulate key environmental variables for honeybees, including the temperature, light, and vibration in the vicinity of the robot [13], as well as being able to detect the presence of bees nearby. Using two robots, when we introduced a positive feedback loop between the heat each robot emits and the presence of bees nearby, the animals make a collective decision by aggregating around one of the robots. There was nothing to discriminate between the robots to start with, but the action of the bee population breaks symmetry – initially by chance, but reinforced by one or other robot. Moreover, we also showed that collective decisions made in two separate arenas, each with a population of bees and two robots, can be coordinated when the robots share task-specific information with another robot in the other arena. The current paper builds on these results, examining the influence of the inter-robot links used to couple two arenas of simulated honeybees.

2   Methods

We use a real-time platform for 2D robot simulation[1] to simulate the interaction of bees and robots. We model both bees and robots as agents in this world, making use of a basic motile robot for the bees, and use a fixed robot with a customised model that corresponds to the bespoke robots designed in our laboratory-based work [13]. While simulation modelling cannot fully replace reality, it does allow us to explore relationships between key micro-level mechanisms and how these can give rise to observed macro-level dynamics. The simulator design enables execution of the exact same robot controllers in simulation

---

[1] Enki – an open source fast 2D robot simulator http://home.gna.org/enki/

and the physical robots, adding substantial value to the resolution of models employed, within the larger cycle of modelling and empirical work.

We model juvenile honeybee behaviour using Beeclust [14]. This is a social model that results in aggregations in zones of highly favoured stimulus. This model was developed based on observations of honeybees: specifically, that they exhibit a preference to aggregate in regions with temperatures in the range 34°C–38°C; that groups of bees are able to identify optimal temperature zones, but individual bees do not do so; and that specific inter-animal chemical cues (*e.g.*, pheromones) have not been shown to be important in this collective aggregation [14]. It has previously been used to illustrate light-seeking behaviour in a swarm of robots [9]. Here we simulate the bees in a thermal environment.

The robot and bee models used in this work are the same as those in [7] and for completeness we describe them fully in the remainder of this section.

2.1   Bee Model

There are two main phases in the bee behaviour: (i) random exploration; and (ii) pausing; in addition to obstacle avoidance (this interrupts (i)). When a bee encounters or collides with another bee, it enters the pausing state, and remains paused for a duration proportional to the local temperature (warmer regions yield longer pauses). There is a positive feedback loop between pause duration and the chance of further exploring bees encountering a bee in a given location, and thus an aggregation in a warm zone can undergo amplification.

We model bees that can discriminate between conspecifics and inanimate obstacles at close proximity, using infra-red (IR) sensors that provide a distance d and object type y. The bees have three sensors l, f, r that are oriented at $-45°$, 0, $+45°$ relative to the bee's bearing. The behaviour is defined as follows.

loop:
1. delay(dt)
2. $((y_l, y_f, y_r), (d_l, d_f, d_r))$ ← read_sensors(l, f, r)
3. if $\exists i, (d_i < 0.5 \wedge y_i = \text{bee})$, for $i \in [l, f, r]$,
     (a) stop()
     (b) T ← measure temperature
     (c) t_{wait} ← compute_wait(T)
     (d) sleep(t_{wait})
     (e) random_turn()
4. else if $\exists i, (d_i < 0.5 \wedge y_i = \text{obstacle})$, for $i \in [l, f, r]$,
        random_turn()
5. else:
        forwards()
end loop

We use data observed in juvenile honeybees (collected and analysed by Univ. Graz, fitted with a hill function) to define the compute_wait(T) mapping:

$$t_{wait} = \left(\frac{(a + b \cdot T)^c}{(a + b \cdot T)^c + d^c} \right) \cdot e + f, \tag{1}$$

where $a = 3.09$, $b = -0.0403$, $c = -28.5$, $d = 1.79$, $e = 22.5$, and $f = 0.645$. It is similar to a sigmoid, with low waiting times ($\sim 1\,\text{s}$) for $\text{T} < 25°\text{C}$ and high waiting times ($\sim 25\,\text{s}$) for $\text{T} = 38°\text{C}$.

2.2   Robot Controller

In our research we employ custom-designed robotic devices which are able to generate stimuli of several modalities that the animals are sensitve to, including heat and light; and the robots an also sense various environmental factors (*e.g.*, temperature, IR) [13]. The robots occupy fixed positions within the experimental arenas to interact with the animals. In this paper we use the robots' thermal actuators and 6 IR proximity sensors. The robots can also communicate with specific neighbours (the topologies are described below) The following program defines how the robots determine their temperatures through time.

At initialisation: set vector m of length m_{max} to zero for $m[0]..m[m_{max} - 1]$. For each timestep, t:

1. $d_{raw} \leftarrow$ count IR sensors above their threshold
2. $m[\text{mod}(t, m_{max})] \leftarrow$ saturate(d_{raw})
3. send$(\widehat{m},$ neighbours$)$
4. if $\text{mod}(t, t_{update}) = 0$:
   (a) $\mathbf{d}_x \leftarrow$ receive message(s) from neighbour(s)
   (b) $T_{new} \leftarrow$ density_to_heat$(\widehat{m}, \mathbf{d}_x)$

where d_{raw} is a raw estimate of local bee density in a given timestep, $\widehat{m}$ is a time-averaged estimate computed as the mean value of the memory vector m, $\mathbf{d}_x$ is a vector of density estimates received from other robots in the interaction neighbourhood. In this paper, robots have zero, one, or two neighbours depending on the specific topology under test.

We use saturate$(s) = \min(4, s)$ in this study. The density to heat function maps the time-averaged detection count to an output temperature via a linear transformation, and is parameterised to allow for different topologies examined. Each involved robot x makes a contribution $c_x = \frac{\widehat{d_x}}{4}(T_{max} - T_{min})$ that depends on a robot's temperature. These are combined as a weighted sum:

$$T_{new} = T_{min} + \sum_{x \in \{l,r,c\}} c_x w_x, \tag{2}$$

where the relative weights of each robot's contribution depends on the specific setup. Each robot can be influenced by the local environment c_l, cross-inhibitory signals from a competitor c_c, and collaborative signals from a specific remote robot c_r. In this paper we use cross-inhibition $w_c = -0.5$ throughout. When a robot has an incoming collaborative link then we set $w_l = w_r = 0.5$, and otherwise set $w_l = 1$ and $w_r = 0$. The topologies tested are shown below.

2.3 Measuring Collective Decisions

Characterising behaviours at the group level is not as clear cut as for decisions at the individual level. However, by using binary choice assays where individuals can exhibit a clear choice, we can use statistical tests to formally quantify when the frequency of choices differs significantly from an accidental outcome. We follow the methodology of [5] and [12] that uses the binomial test to formally quantify when a collective decision is reached. Extending the setup to include multiple binary choices affords these benefits of quantifiable behaviours at the level of collective while also admitting more complex environments.

Specifically, at the end of each experiment, we divide the location of the bees into two different zones within each arena, such that one of the two robots has 'won' the competition for that bee (see Fig. 2). We define the null hypothesis to be that the bees made their choice at random and without bias. When the outcome differs significantly from this, we consider it a collective decision (CD).

Since here we concentrate on the ability to coordinate the collective decisions (CCD) reached in two arenas, we also consider a test that lumps together all bees from both populations, as if they were in a single group. We use the binomial test to quantify when such outcomes are significant within a single experiment. We also apply χ^2 tests and binomial tests across a set of repeats and between different conditions, to verify when outcomes are significantly different.

2.4 Parameters and Setup

With our choices, we aim to reflect key conditions used in our animal-based experiments, such as the arena and robot setup, and the temperatures used are in a range that is relevant for the animals without harming them.

In the experiments below, $T_{min} = 28°C$ and $T_{max} = 38°C$; the ambient temperature is 27°C. The modelled bees measure 13.5 mm × 5 mm (based on our measurements), and detection range is 5 mm. Memory length $m_{max} = 18$, $t_{update} = 3$ s, $t_{resample} = 0.5$ s. Since other methods did not provide more accurate estimates in preliminary testing we use a simple time average across the whole memory m. To facilitate the observation of binary choices, the arenas used are rectangular with rounded ends (see Fig. 1). They have dimensions 210×65 mm (internal). The two robots are positioned 60 mm either side of the centre, on the midline. Bees are initialised with random position and orientation.

3 Simulation Experiments

This paper aims to examine the sensitivity of coordinating collective decision-making between arenas as a function of inter-arena communication links. To address this aim, we examine a range of different topologies within the limits examined in prior work, employing a basic setup that uses two identical arenas, each comprising a population of bees and two robots. We vary the inter-arena links, keeping the link weights positive where present. We use six different topologies that vary in the number and direction of coupling that they provide between

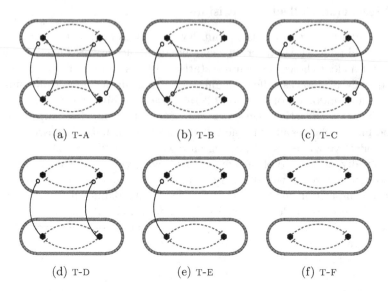

(a) T-A   (b) T-B   (c) T-C

(d) T-D   (e) T-E   (f) T-F

Fig. 3. Topologies tested in the multi-arena experiments. Solid lines indicate positive contributions and dashed lines indicate negative contributions, with respect to the receiving robot. From top-left to bottom-right, the two arenas become more loosely coupled.

the two arenas. Fig. 3 shows these topologies. Moving from top-left to bottom-right, T-A has the strongest coupling; T-B and T-C have some reciprocal paths; T-D and T-E have links in one direction only. T-F is the other extreme without any between-arena links, which we use to establish a baseline for the other outcomes. These motifs give broad coverage of the space and while other topologies are possible with more links or more classes of link, our motivation to understand sparser networks is better served by these networks with fewer rather than more inter-arena links.

Our prior work showed strong coordination under (a) and confirmed that the absence of links (f) does not lead to coordination [7]. Intuitively, we expect a weaker ability to coordinate as the links become sparser; however, we do not know what the limits are or how gracefully the system will degrade.

We run 50 independent repeats for each of the six topologies, each experiment lasting for 15 mins. Fig. 5 shows the frequency of statistically collective decisions made, for each of the topologies. Fig. 4 provides a slightly different view of the experiments by showing the mean percentage of bees that were present in the east Zone during the last 120 s. All 50 repeats in each topology have a point plotted in this graph, and while it is not always the case that the points in the extremes correspond to a significant collective decision at the time of measurement, the two views are strongly linked.

Considering the distributions shown in Fig. 5, we perform the following statistical tests to identify the collective decision-making and coordination that arises

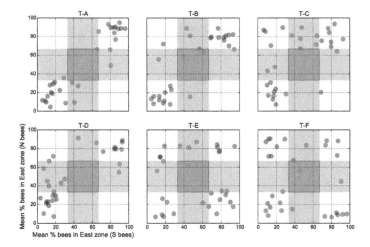

Fig. 4. Final states of each of the topologies, with one point shown per experiment. When the system has coordinated the decisions made in each arena, the points only appear in two of the four corners (indicating that the decisions are mutually constrained).

under each topology. Using a χ^2 test with a null model of equal likelihood for each of the four possible decision pairings, topologies T-E and T-F do not deviate significantly from the null model ($\rho > 0.05$). The other four topologies all deviate significantly, *i.e.*, they exhibit coordination between the two arenas. We also compare the overall rate of collective decision-making across different topologies. The three cases that use two links (T-B, T-C, T-D) have similar ability to induce coordinated collective decisions as T-A (binomial test, $\rho > 0.05$). Comparing the rate of collective decision within the two arenas separately (*i.e.*, any of the four outcomes), T-B, T-C, T-E have significantly lower rates than T-A (binomial test, $\rho < 0.05$); however, although the T-D rate is lower, it is not significantly lower ($\rho > 0.05$). None of the topologies exhibit a bias towards either coordinated outcome (binomial test, $\rho > 0.05$), with the exception of T-D ($\rho < 0.05$).

Overall, these results show that: (i) All topologies with two or more links coupling the two arenas are able to coordinate the decision-making. T-E, the sparsest topology, is not able to reliably coordinate the decision-making (nor is the unlinked case T-F but of course this is to be expected). (ii) Most of the sparser topologies are less frequently able to induce collective decision-making than the most tightly linked case T-A. T-D is a marginal exception in this regard. (iii) T-D exhibits some anomalies regarding a bias towards the WW outcome over the EE outcome. Given the absence of bias in the model or the topology, this is somewhat surprising and requires further investigation to identify the source of this bias.

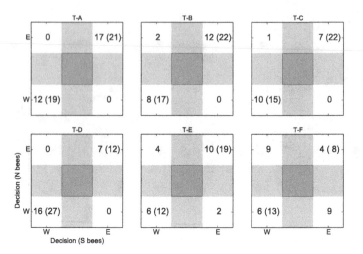

Fig. 5. Frequency of runs with significant collective decisions, from 50 repeats. Values in brackets indicate the frequency of coordinated collective decisions made, i.e., lumping together both populations. Other values indicate assessment of the two populations separately.

To obtain a better understanding of the differences between the different conditions, we inspect some example trajectories (Fig. 6), which illustrate some of the issues faced by different topologies. (The runs are characteristic of several seen in each case, but frame (c) was selected to highlight a difficulty rather than be the most frequent trajectory type). These figures divide the bee locations into two zones per arena, one for each robot (see Fig. 2). We compute the average percentage of bees in the East zone during each period (here, 30 s), and additionally, apply a binomial test to quantify whether a significant collective decision is reached at the end of each period (for $p < 0.05$). In frame (a), we see that the strong coupling of T-A results in tight changes in bee location in each arena. In frame (b), the two populations initially move towards opposite ends, but as the decision in the South arena solidifies, it is able to coordinate this decision with the North arena. The progression of the two populations is not typically as tightly in lock-step as for T-A, with reciprocal feedback, but the topology does result in coordination with a high frequency. In frame (c) there is only one link (T-E), and it is far slower for the South arena to exert a coordinating influence over the North arena. In this case it was able to coordinate within 15 mins, but in other cases different decisions are reached in each arena and the single link is not always able to overturn the result.

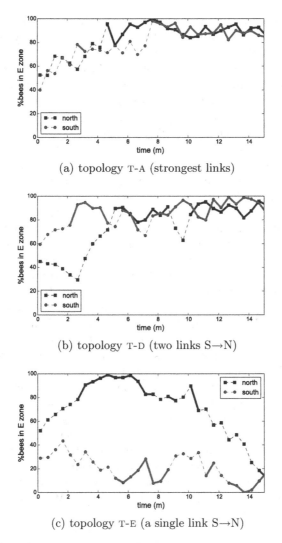

(a) topology T-A (strongest links)

(b) topology T-D (two links S→N)

(c) topology T-E (a single link S→N)

Fig. 6. Example trajectories showing a fraction of bees in East half of each arena, and annotated for where the collective decisions are made with thick, solid lines.

4   Discussion and Future Work

The different topologies vary in strength of coupling between the two arenas, and this also results in varying abilities to coordinate the collective decisions reached within those arenas. With four links in two reciprocal pairs (T-A), the activity in the two arenas is tightly coupled and reliably results in coordination across both bee populations. With two links forming a reciprocal network (T-B, T-C), the two arenas still coordinate frequently, although to a lesser extent than for T-A. Using two links in the same direction (T-D) is able to couple the decisions made at a slightly higher rate than the other two-link topologies, although clearly the single direction of information-sharing may not always be appropriate. Using only a single link (T-E) is not sufficient coupling to significantly coordinate the decisions made across the two arenas.

In summary, we find that the topology influences the ability of multiple populations to coordinate their activities, but also that relatively few links are still sufficient to result in some coordination. This is welcome news, since it suggests that such distributed collective systems will exhibit favourable tolerance to failures in links (provided some ability to detect those failures – for instance, monitoring an absence of messages received over a given link could be used to adapt the robot's behaviour). Moreover, as we begin to consider larger systems, sparse networks of localised interactions are generally preferable.

The scenarios modelled above all have static background environments. We are also interested in how dynamic environments (*e.g.*, exogeneous shocks) affect coordination ability. This introduces the question: can the overall system restore a coordinated state following disruption? which may more strongly discriminate between different topologies. Perturbations could be due to an exogenous heat source, or another stimulus modality that modulates the bee behaviour (for example, it is thought that vibrations can act as stop signals). Interestingly, when considering a multi-species bio-hybrid system, the source of variability could also be endogenous, and the role of the robots in enabling coordination would be even greater. In any of these more dynamic scenarios, it would become important to measure the speed of coordinated decision-making, as well as the longevity (see *e.g.*, [15]). A further open question relates to the advantages and tradeoffs within an extended system comprising more sub-groups of animals (each of which offers some distributed 'memory' of a coordinated decision): more smaller groups may be better able to retain a decision; but dividing into groups that are too small will likely degrade the ability of each group to form a decision in the first place. In more complex scenarios such as these we aim to investigate how further adaptive mechanisms within the robot network can improve efficiency.

In this paper we have investigated how inter-robot interaction topology influences the ability of system-level coordinated decisions. We have shown that the rate of coordination does depend on topology, but also that the basin of attraction for coordinated states is relatively robust. In general, better understanding the limits and affordances of interactions in such systems will enable the development of more capable mixed animal–robot societies.

Acknowledgments. This work is supported by: EU-ICU project "ASSISI|bf" no. 601074, and by centre grant (to BioISI, ref: UID/MULTI/04046/2013), from FCT/MCTES/ PIDDAC, Portugal.

References

1. Nadell, C.D., Xavier, J.B., Foster, K.R.: The sociobiology of biofilms. FEMS Microbiol. Rev. **33**(1), 206–224 (2009)
2. King, A.J., Cowlishaw, G.: Leaders, followers, and group decision-making. Commun. Integr. Biol. **2**(2), 147–150 (2009)

3. Faria, J.J., Dyer, J.R., Clément, R.O., Couzin, I.D., Holt, N., Ward, A.J., Waters, D., Krause, J.: A novel method for investigating the collective behaviour of fish: introducing 'robofish'. Behav. Ecol. Sociobiol. **64**(8), 1211–1218 (2010)
4. Parrish, J.K., Edelstein-Keshet, L.: Complexity, pattern, and evolutionary trade-offs in animal aggregation. Science **284**(5411), 99–101 (1999)
5. Halloy, J., Sempo, G., Caprari, G., Rivault, C., Asadpour, M., Tâche, F., Said, I., Durier, V., Canonge, S., Amé, J.M., et al.: Social integration of robots into groups of cockroaches to control self-organized choices. Science **318**(5853), 1155–1158 (2007)
6. De Schutter, G., Theraulaz, G., Deneubourg, J.L.: Animal-robots collective intelligence. Ann. Math. Artif. Intel. **31**(1–4), 223–238 (2001)
7. Mills, R., Zahadat, P., Silva, F., Mliklic, D., Mariano, P., Schmickl, T., Correia, L.: Coordination of collective behaviours in spatially separated agents. In: Procs. ECAL (2015)
8. Webb, B.: Can robots make good models of biological behaviour? Behav. Brain. Sci. **24**(06), 1033–1050 (2001)
9. Kernbach, S., Thenius, R., Kernbach, O., Schmickl, T.: Re-embodiment of honeybee aggregation behavior in an artificial micro-robotic system. Adapt. Behav. **17**(3), 237–259 (2009)
10. Campo, A., Garnier, S., Dédriche, O., Zekkri, M., Dorigo, M.: Self-organized discrimination of resources. PLoS ONE **6**(5), e19888 (2011)
11. Grodzicki, P., Caputa, M.: Social versus individual behaviour: a comparative approach to thermal behaviour of the honeybee (Apis mellifera L.) and the american cockroach (Periplaneta americana L.). J. Insect. Physiol. **51**(3), 315–322 (2005)
12. Szopek, M., Schmickl, T., Thenius, R., Radspieler, G., Crailsheim, K.: Dynamics of collective decision making of honeybees in complex temperature fields. PLoS ONE **8**(10), e76250 (2013)
13. Griparic, K., Haus, T., Bogdan, S., Miklic, D.: Combined actuator sensor unit for interaction with honeybees. In: Sensor Applications Symposium (2015)
14. Schmickl, T., Thenius, R., Moeslinger, C., Radspieler, G., Kernbach, S., Szymanski, M., Crailsheim, K.: Get in touch: cooperative decision making based on robot-to-robot collisions. Auton. Agent. Multi. Agent. Syst. **18**(1), 133–155 (2009)
15. Gautrais, J., Michelena, P., Sibbald, A., Bon, R., Deneubourg, J.L.: Allelomimetic synchronization in merino sheep. Anim. Behav. **74**(5), 1443–1454 (2007)

A Differential Evolution Algorithm for Optimization Including Linear Equality Constraints

Helio J.C. Barbosa[1,2]([✉]), Rodrigo L. Araujo[2], and Heder S. Bernardino[2]

[1] Laboratório Nacional de Computação Científica, Petrópolis, RJ, Brazil
hcbm@lncc.br
http://www.lncc.br/~hcbm
[2] Universidade Federal de Juiz de Fora, Juiz de Fora, MG, Brazil
rdgleppaus@gmail.com, heder@ice.ufjf.br
http://www.lncc.br/~hedersb

Abstract. In this paper a differential evolution technique is proposed in order to tackle continuous optimization problems subject to a set of linear equality constraints, in addition to general non-linear equality and inequality constraints. The idea is to exactly satisfy the linear equality constraints, while the remaining constraints can be dealt with via standard constraint handling techniques for metaheuristics. A procedure is proposed in order to generate a random initial population which is feasible with respect to the linear equality constraints. Then a mutation scheme that maintains such feasibility is defined. The procedure is applied to test-problems from the literature and its performance is also compared with the case where the constraints are handled via a selection scheme or an adaptive penalty technique.

1 Introduction

Constrained optimization problems are common in many areas, and due to the growing complexity of the applications tackled, nature-inspired metaheuristics in general, and evolutionary algorithms in particular, are becoming increasingly popular. That is due to the fact that they can be readily applied to situations where the objective function(s) and/or constraints are not known as explicit functions of the decision variables, and when potentially expensive computer models must be run in order to compute the objective function and/or check the constraints every time a candidate solution needs to be evaluated.

As move operators are usually blind to the constraints (i.e. when operating upon feasible individuals they do not necessarily generate feasible offspring) standard metaheuristics must be equipped with a constraint handling technique. In simpler situations, repair techniques [18], special move operators [19], or special decoders [9] can be designed to ensure that all candidate solutions are feasible. We will not attempt to survey the current literature on constraint handling here, and the reader is referred to [4], [13], and [5].

© Springer International Publishing Switzerland 2015
F. Pereira et al. (Eds.) EPIA 2015, LNAI 9273, pp. 262–273, 2015.
DOI: 10.1007/978-3-319-23485-4_26

Here we will focus on obtaining solutions automatically satisfying all linear equality constraints (in the form $Ex = c$). Any remaining constraints present in the problem can be dealt with using constraint handling techniques available in the literature.

The first approach to this problem seems to be the GENOCOP system approach [14] where the linear equalities are used to eliminate some of the variables, which are now written as a function of the remaining ones, thus reducing the number of variables. As a result, any linear inequality constraint present in the problem has to be adequately modified. Another idea is that of the use of a homomorphous mapping [10] that transforms a space constrained by $Ex = c$ into a space that is not only fully unconstrained, but also of lower dimensionality [15]. In [16] two modifications of the particle swarm optimization (PSO) technique were proposed to tackle linear equality constraints: LPSO and CLPSO. LPSO starts from a feasible initial population and then maintains feasibility by modifying the standard PSO formulas for particle velocity. CLPSO tries to improve on some observed shortcomings of LPSO and changes the equation for the best particle in the swarm so that it explores the feasible domain using a random velocity vector in the null space of E, that is, this velocity vector keeps the best particle feasible with respect to the linear equality constraints.

In this paper a simple modification of the differential evolution technique (denoted here by DELEqC) is proposed in order to exactly treat the linear equality constraints present in continuous optimization problems that may also include additional non-linear equality and inequality constraints, which can be dealt with via existing constraint handling techniques. Starting from a population which is feasible with respect to the linear equality constraints, feasibility is maintained by avoiding the standard DE crossover and using an adequate mutation scheme along the search process.

2   The Problem

The problem considered here is to find $x \in \mathbb{R}^n$ that minimizes the objective function $f(x)$ subject to $m < n$ linear equality constraints, written as $Ex = c$, in addition to p general inequality contraints $g_i(x) \leq 0$ and q general equality constraints $h_j(x) = 0$.

When using metaheuristics, general equality constraints are usually relaxed to $|h_j(x)| \leq \epsilon$ where the parameter $\epsilon > 0$ must be conveniently set by the user. As a result, candidate solutions strictly feasible with respect to all equality constraints are very hard to be obtained. Whenever x violates the j-th constraint one defines the corresponding constraint violation as $v_j(x) = |h_j(x)| - \epsilon$ and aggregates for the individual x as $v(x) = \sum_j v_j(x)$.

Being able to exactly satisfy all linear equalities is a valuable improvement in a general constrained optimization setting, and should thus be pursued.

It is assumed here that the constraints are linearly independent so that E has full rank ($\text{rank}(E) = m$). A candidate solution $x_1 \in \mathbb{R}^n$ is said to be feasible if $x_1 \in \mathcal{E}$ where $\mathcal{E}$ denotes the feasible set:

$$\mathcal{E} = \{x \in \mathbb{R}^n : Ex = c\}$$

A vector $d \in \mathbb{R}^n$ is said to be a feasible direction at the point $x \in \mathcal{E}$ if $x + d$ is feasible: $E(x + d) = c$. It follows that the feasible direction d must satisfy $Ed = 0$ or, alternatively, that any feasible direction belongs to the null space of the matrix E

$$\mathcal{N}(E) = \{x \in \mathbb{R}^n : Ex = 0\}$$

Now, given two feasible vectors x_1 and x_2 it is easy to see that $d = x_1 - x_2$ is a feasible direction, as $E(x_1 - x_2) = 0$. As a result, one can see that the standard mutation formulae adopted within DE (see Section 3) would always generate a feasible vector whenever the vectors involved in the differences are themselves feasible. If crossover is avoided, one could start from a feasible random initial population and proceed, always generating feasible individuals.

3 Differential Evolution

Differential evolution (DE) [20] is a simple and effective algorithm for global optimization, specially for continuous variables. The basic operation performed is the addition to each design variable in a given candidate solution of a term which is the scaled difference between the values of such variable in other candidate solutions in the population. The number of differences applied, the way in which the individuals are selected, and the type of recombination performed define the DE variant (also called DE strategy). Although many DE variants can be found in the literature [17], the simplest one (DE/rand/1/bin) is adopted here:

$$u_{i,j,G+1} = x_{r_1,j,G} + \mathbf{F} \cdot (x_{r_2,j,G} - x_{r_3,j,G})$$

where r_1, r_2 and r_3 correspond to distinct randomly picked indexes, and G denotes the iteration counter of the search technique.

In addition, in the general case, a crossover operation is performed, according to the parameter CR. However, in order to maintain feasibility, crossover is not performed in the proposed DE variant. Here, $u_{i,j,G+1}$ replaces $x_{i,j,G}$ when u is better then x for each i.

3.1 A Selection Scheme

In order to handle the constraints, a popular technique is Deb's selection scheme [6] (denoted here by DSS) which enforces the following criteria: (i) any feasible solution is preferred to any infeasible solution, (ii) between two feasible solutions, the one having better objective function value is preferred, and (iii) between two infeasible solutions, the one having smaller constraint violation $(v(x))$ is preferred.

3.2   An Adaptive Penalty Technique

A parameterless adaptive penalty method (APM) was developed in [2,3,11] which does not require the knowledge of the explicit form of the constraints as a function of the design variables, and is free of parameters to be set by the user. An adaptive scheme automatically sizes the penalty parameter corresponding to each constraint along the evolutionary process. The fitness function proposed is written as:

$$F(x) = \begin{cases} f(x), & \text{if } x \text{ is feasible,} \\ \overline{f}(x) + \sum_{j=1}^{m} k_j v_j(x), & \text{otherwise} \end{cases}$$

$$\overline{f}(x) = \begin{cases} f(x), & \text{if } f(x) > \langle f(x) \rangle, \\ \langle f(x) \rangle & \text{otherwise} \end{cases} \qquad k_j = |\langle f(x) \rangle| \frac{\langle v_j(x) \rangle}{\sum_{l=1}^{m} [\langle v_l(x) \rangle]^2}$$

where $\langle f(x) \rangle$ is the average of the objective function values in the current population and $\langle v_l(x) \rangle$ is the violation of the l-th constraint averaged over the current population. The idea is that the values of the penalty coefficients should be distributed in a way that those constraints which are more difficult to be satisfied should have a relatively higher penalty coefficient.

4   The Proposal

Differently from penalty or selection schemes, and from special decoders, the proposed DE algorithm that satisfies linear equality constraints is classified as a feasibility preserving approach.

In order to generate a feasible initial population of size NP one could think of starting from a feasible vector x_0 and proceed by moving from x_0 along random feasible directions d_i: $x_i = x_0 + d_i, i = 1,2,\ldots,$ NP. A random feasible direction can be obtained by projecting a random vector onto the null space of E. The projection matrix is given by [12]

$$P_{\mathcal{N}(E)} = I - E^T(EE^T)^{-1}E \tag{1}$$

where the superscript T denotes transposition. Random feasible candidate solutions can then be generated as

$$x_i = x_0 + P_{\mathcal{N}(E)}v_i, \quad i = 1,2,\ldots, \text{NP}$$

where $v_i \in \mathbb{R}^n$ is randomly generated and x_0 is computed as $x_0 = E^T(EE^T)^{-1}c$. It is clear that x_0 is a feasible vector, as $Ex_0 = EE^T(EE^T)^{-1}c = c$.

It should be mentioned that the matrix inversion in eq. 1 is not actually performed and, then, $P_{\mathcal{N}(E)}$ is never computed (see Algorithm 1).

The Differential Evolution for Linear Equality Constraints (DELEqC) algorithm is defined as DE/rand/1/bin, equipped with the feasible initial population generation procedure in Algorithm 1, and running without crossover.

Notice that any additional non-linear equality or inequality constraint can be dealt with via existing constraint handling techniques, such as those in sections 3.1 and 3.2.

Algorithm 1. Algorithm `CreateInitialPopulation`.

---

    input : NP (population size)

1 $M = EE^{\mathrm{T}}$;
2 Perform LU Decomposition: $M = LU$;
3 Solve $My = c$ $(Lw = c$ and $Uy = w)$;
4 $x_0 = E^{\mathrm{T}}y$;
5 **for** $i \leftarrow 1$ **to** NP **do**
6      $d \in \mathbb{R}^n$ is randomly generated;
7      $z = Ed$;
8   Solve $Mu = z$ $(Lw = z$ and $Uu = w)$;
9      $v = E^{\mathrm{T}}u$;
10     $x_i = x_0 + d - v$;

---

5   Computational Experiments

In order to test the proposal (DELEqC) and assess its performance, a set of test problems with linear equality constraints was taken from the literature (their descriptions are available in Appendix A). The results produced are then compared with those from alternative procedures available in the metaheuristics literature [15,16], as well as running them with established constraint handling techniques (Deb's selection scheme and an adaptive penalty method) to enforce the linear equality constraints. One hundred independent runs were executed in all experiments.

Initially, computational experiments were performed aiming at selecting the values for population size (NP) and F. The tested values here are NP $\in \{5, 10, 20, 30, \ldots, 90, 100\}$ and F $\in \{0.1, 0.2, \ldots, 0.9, 1\}$. Due the large number of combinations, performance profiles [7] were used to identify the parameters which generate the best results. We adopted the maximum budget allowed in [15,16] as a stop criterion and the final objective function value as the quality metric. The area under the performance profiles curves [1] indicates that the best performing parameters according to these rules are NP $= 50$ and F $= 0.7$, and these values were then used for all DEs in the computational experiments.

5.1   Results

First, we analyze how fast the proposed technique is in order to find the best known solution of each test-problem when compared with a DE using (i) Deb's selection scheme (DE+DSS) or (ii) an adaptive penalty method (DE+APM). The objective in this test case is to verify if DELEqC is able to obtain the best known solutions using a similar number of objective function evaluations. We used CR $= 0.9$, NP $= 50$, and F $= 0.7$ for both DE+DSS and DE+APM. Statistical information (best, median, mean, standard deviation, worst), obtained from 100 independent runs, The number of successful runs (sr) is also shown. A successful run is one in which the best known solution is found (absolute error less or

Table 1. Statistical comparisons. Number of objective function evaluations required to obtain the best known solution with a absolute error less or equal to 10^{-4}. The bounds $[-1000; 1000]$ were adopted for all test-problems. For DE+APM and DE+DSS, the tolerance for equality constraints is $\epsilon = 0.0001$.

TP	technique	best	median	mean	st. dev.	worst	sr
	DELEqC	**2800**	**3475**	**3458.00**	$2.51e+02$	**4050**	100
1	DE+APM	14750	16400	16456.50	$6.68e+02$	18200	100
	DE+DSS	16300	18350	18336.00	$7.54e+02$	20250	100
	DELEqC	**2750**	**3300**	**3249.50**	$1.93e+02$	**3600**	100
2	DE+APM	13050	15375	15385.00	$7.79e+02$	17350	100
	DE+DSS	16300	18100	18121.50	$6.92e+02$	20150	100
	DELEqC	**1500**	**2050**	**2029.00**	$1.74e+02$	**2400**	100
3	DE+APM	12950	14250	14256.50	$5.49e+02$	15800	100
	DE+DSS	18300	20225	20158.50	$7.68e+02$	22000	100
	DELEqC	**1250**	**1900**	**1906.00**	$1.61e+02$	**2250**	100
4	DE+APM	12250	14350	14280.00	$7.30e+02$	15650	100
	DE+DSS	17200	19000	19060.50	$7.60e+02$	20800	100
	DELEqC	**1700**	**2050**	**2055.00**	$1.46e+02$	**2350**	100
5	DE+APM	13950	15300	15382.83	$6.17e+02$	16650	99
	DE+DSS	17550	19150	19157.50	$6.93e+02$	21400	100
	DELEqC	**1450**	**1950**	**1932.50**	$1.56e+02$	**2250**	100
6	DE+APM	13850	15600	15582.00	$6.32e+02$	17150	100
	DE+DSS	16850	18850	18916.50	$7.45e+02$	20750	100
	DELEqC	**6550**	**7300**	**7326.00**	$3.24e+02$	**8150**	100
7	DE+APM	77050	83750	83938.00	$3.55e+03$	95750	100
	DE+DSS	104050	118550	118711.50	$7.01e+03$	134100	100
	DELEqC	**6700**	**8050**	**8018.00**	$4.05e+02$	**8950**	100
8	DE+APM	80850	88675	88901.00	$3.71e+03$	98600	100
	DE+DSS	108250	122000	122199.50	$7.26e+03$	139000	100
	DELEqC	**18150**	**26100**	30482.76	$1.59e+04$	**91400**	58
9	DE+APM	117150	140950	148259.78	$2.19e+04$	208900	46
	DE+DSS	145950	175875	182394.57	$2.63e+04$	256150	46
	DELEqC	**6900**	**7900**	**7959.00**	$4.65e+02$	**9200**	100
10	DE+APM	68200	77000	77211.50	$3.46e+03$	86150	100
	DE+DSS	98000	119175	119015.00	$7.54e+03$	138350	100
	DELEqC	**12150**	**25975**	**31196.50**	$1.51e+04$	**73150**	100
11	DE+APM	101850	109600	121478.57	$2.76e+04$	210700	14
	DE+DSS	131450	141350	143682.14	$9.04e+03$	165550	14

Table 2. Results for the test-problems 7, 8, 9, 10, and 11 using the reference budget (rb) and $2 \times rb$.

TP	nofe	technique	best	median	mean	st. dev.	worst
			Results using the reference budget (rb)				
		DELEqC	39.5143	**115.8354**	122.5069	$5.03e+01$	260.6652
7	1,250	Genocop II [16]	38.322	-	739.438	$8.40e+02$	$1.63e+3$
		LPSO [16]	37.420	-	$7.03e+03$	$8.01e+03$	$4.63e+4$
		CLPSO [16]	**32.138**	-	**35.197**	$2.21e+01$	**252.826**
		DELEqC	35.3784	**35.3961**	**35.4051**	$2.78e-02$	**35.5360**
8	5,000	Genocop II	37.939	-	104.192	$5.99e+01$	262.656
		LPSO	240.101	-	$8.46e+3$	$1.05e+04$	$7.79e+4$
		CLPSO	**35.377**	-	82.077	$6.10e+01$	197.389
		DELEqC	40.5363	**58.7789**	58.1392	$8.04e+00$	77.4060
9	5,000	Genocop II	49.581	-	**56.694**	$8.93e+00$	**75.906**
		LPSO	36.981	-	77.398	$2.35e+01$	149.429
		CLPSO	**36.975**	-	72.451	$2.57e+01$	167.644
		DELEqC	**21485.2614**	**21485.2983**	**21485.2962**	$5.87e-03$	**21485.3000**
10	10,000	Genocop II	22334.971	-	58249.328	$6.25e+04$	$2.00e+5$
		LPSO	$1.95e+5$	-	$1.38e+9$	$4.48e+09$	$3.55e+10$
		CLPSO	21485.306	-	$6.52e+8$	$2.39e+09$	$2.23e+10$
		DELEqC	**0.3091**	**0.5910**	**0.5821**	$9.56e-02$	**0.8099**
11	5,000	Genocop II	0.713	-	1.009	$1.30e-01$	1.131
		LPSO	0.529	-	6.853	$6.20e+00$	36.861
		CLPSO	0.632	-	7.470	$7.27e+00$	44.071
			Results using twice the reference budget ($2 \times rb$)				
		DELEqC	32.5550	**34.2880**	34.7973	$1.97e+00$	41.1522
		Genocop II	37.612	-	304.884	$3.88e+02$	$1.17e+3$
		LPSO	**32.137**	-	445.316	$8.03e+02$	$4.51e+3$
7	2,500	CLPSO	**32.137**	-	32.139	$6.69e-03$	32.183
		Constricted PSO [15]	**32.137**	-	**32.137**	$2.00e-10$	**32.137**
		BareBones PSO [15]	**32.137**	-	**32.137**	$1.00e-14$	**32.137**
		PSOGauss [15]	**32.137**	-	**32.137**	$1.00e-14$	**32.137**
		DELEqC	**35.3769**	35.3770	**35.3770**	$2.57e-05$	**35.3770**
		Genocop II	35.393	-	49.945	$1.10e+01$	82.221
8	10,000	LPSO	35.400	-	758.525	$1.50e+03$	$1.12e+4$
		CLPSO	**35.377**	-	68.570	$5.39e+01$	196.067
		Constricted PSO	**35.377**	-	36.165	$3.12e+00$	55.538
		BareBones PSO	**35.377**	-	40.019	$9.61e+00$	75.147
		PSOGauss	**35.377**	-	38.998	$8.59e+00$	72.482
		DELEqC	**36.9755**	**44.9910**	**46.7872**	$8.30e+00$	**67.1005**
		Genocop II	37.116	-	52.379	$7.50e+00$	67.564
9	10,000	LPSO	**36.975**	-	76.487	$3.07e+01$	232.979
		CLPSO	**36.975**	-	69.039	$2.16e+01$	154.379
		Constricted PSO	**36.975**	-	50.431	$1.23e+01$	85.728
		BareBones PSO	**36.975**	-	55.921	$1.61e+01$	119.556
		PSOGauss	**36.975**	-	55.622	$1.48e+01$	119.094
		DELEqC	**21485.2614**	**21485.2983**	**21485.2962**	$5.87e-03$	**21485.3000**
		Genocop II	21490.840	-	21630.020	$1.54e+02$	22030.988
		LPSO	21554.158	-	$4.44e+6$	$2.28e+07$	$2.18e+8$
10	20,000	CLPSO	21485.305	-	$7.45e+5$	$7.12e+06$	$7.11e+7$
		Constricted PSO	**21485.3**	-	**21485.3**	$6.00e-11$	**21485.3**
		BareBones PSO	**21485.3**	-	**21485.3**	$6.00e-11$	**21485.3**
		PSOGauss	**21485.3**	-	**21485.3**	$6.00e-11$	**21485.3**
		DELEqC	**0.1509**	**0.4299**	**0.4163**	$1.07e-01$	**0.6677**
		Genocop II	0.417	-	0.702	$1.87e-01$	0.971
11	10,000	LPSO	0.387	-	2.997	$2.94e+00$	15.805
		CLPSO	0.236	-	3.049	$3.10e+00$	16.427
		Constricted PSO	**0.151**	-	0.488	$1.68e-01$	0.83
		BareBones PSO	**0.203**	-	0.523	$1.81e-01$	0.912
		PSOGauss	**0.151**	-	0.53	$1.68e-01$	0.958

Table 3. Results for the test-problems 7, 8, 9, 10, and 11 using $3 \times rb$ and $4 \times rb$.

TP	nofe	technique	best	median	mean	st. dev.	worst
		Results using three times the reference budget ($3 \times rb$)					
7	3,750	DELEqC	32.1447	**32.1881**	32.2008	$5.53e-02$	32.5202
		Genocop II	33.837	-	69.154	$2.67e+01$	124.820
		LPSO	**32.137**	-	35.071	$2.15e+01$	244.077
		CLPSO	**32.137**	-	**32.137**	$1.83e-04$	**32.138**
8	15,000	DELEqC	**35.3769**	**35.3770**	**35.3770**	$2.57e-05$	**35.3770**
		Genocop II	35.772	-	42.393	$6.86e+00$	60.110
		LPSO	**35.377**	-	125.727	$2.31e+02$	$1.72e+3$
		CLPSO	**35.377**	-	59.001	$5.00e+01$	196.065
9	15,000	DELEqC	**36.9751**	**37.1959**	**39.6822**	$4.52e+00$	**54.5931**
		Genocop II	37.326	-	47.643	$8.45e+00$	67.128
		LPSO	**36.975**	-	74.338	$2.83e+01$	234.968
		CLPSO	37.970	-	77.409	$3.09e+01$	224.024
10	30,000	DELEqC	**21485.2614**	**21485.2983**	**21485.2962**	$5.87e-03$	**21485.3000**
		Genocop II	21487.098	-	21546.332	$8.53e+01$	21836.797
		LPSO	21483.373	-	$3.71e+5$	$2.41e+7$	$2.05e+7$
		CLPSO	21485.305	-	21485.305	$9.83e-08$	21485.305
11	15,000	DELEqC	**0.1508**	**0.2361**	**0.2605**	$8.79e-02$	**0.5190**
		Genocop II	0.351	-	0.702	$1.72e-01$	0.962
		LPSO	0.250	-	2.653	$2.72e+00$	14.405
		CLPSO	0.250	-	2.146	$2.21e+00$	11.983
		Results using four times the reference budget ($4 \times rb$)					
7	5,000	DELEqC	**32.1371**	**32.1381**	32.1386	$1.41e-03$	32.1436
		Genocop II	32.544	−	54.846	$1.69e-01$	107.584
		LPSO	**32.137**	−	**32.137**	$7.18e-12$	**32.137**
		CLPSO	**32.137**	−	**32.137**	$3.02e-06$	**32.137**
		Constricted PSO	**32.137**	-	**32.137**	$1.00e-14$	**32.137**
		BareBones PSO	**32.137**	-	**32.137**	$1.00e-14$	**32.137**
		PSOGauss	**32.137**	-	**32.137**	$1.00e-14$	**32.137**
8	20,000	DELEqC	**35.3769**	**35.3770**	**35.3770**	$2.57e-05$	**35.3770**
		Genocop II	35.410	-	39.500	$6.78e+00$	56.613
		LPSO	**35.377**	-	59.762	$3.98e+01$	246.905
		CLPSO	**35.377**	-	39.832	$1.09e+01$	71.380
		Constricted PSO	**35.377**	-	35.783	$2.39e+00$	55.538
		BareBones PSO	**35.377**	-	37.079	$5.33e+00$	55.538
		PSOGauss	**35.377**	-	35.589	$5.28e-01$	36.892
9	20,000	DELEqC	**36.9748**	**36.9755**	**38.5722**	$3.27e+00$	**51.4201**
		Genocop II	37.011	-	43.059	$6.14e+00$	59.959
		LPSO	38.965	-	75.011	$2.77e+01$	184.226
		CLPSO	**36.975**	-	76.896	$2.73e+01$	151.394
		Constricted PSO	**36.975**	-	46.199	$7.48e+00$	76.736
		BareBones PSO	**36.975**	-	49.238	$1.02e+01$	76.774
		PSOGauss	**36.975**	-	47.11	$8.14e+00$	68.802
10	40,000	DELEqC	**21485.2614**	**21485.2983**	**21485.2962**	$5.87e-03$	**21485.3000**
		Genocop II	21485.363	-	21485.714	$4.00e-01$	21486.646
		LPSO	21485.925	-	$1.260e+05$	$1.04e+06$	$1.04e+07$
		CLPSO	21485.305	-	21485.305	$9.40e-08$	21485.305
		Constricted PSO	**21485.3**	-	**21485.3**	$6.00e-11$	**21485.3**
		BareBones PSO	**21485.3**	-	**21485.3**	$6.00e-11$	**21485.3**
		PSOGauss	**21485.3**	-	**21485.3**	$6.00e-11$	**21485.3**
11	20,000	DELEqC	**0.1482**	**0.2019**	**0.2241**	$6.11e-02$	**0.3849**
		Genocop II	0.201	-	0.584	$1.31e-01$	0.843
		LPSO	0.338	-	1.695	$1.92e+00$	14.401
		CLPSO	0.236	-	1.900	$2.38e+00$	17.259
		Constricted PSO	**0.151**	-	0.413	$1.45e-01$	0.792
		BareBones PSO	**0.151**	-	0.444	$1.58e-01$	0.83
		PSOGauss	**0.151**	-	0.454	$1.74e-01$	0.83

equal to 10^{-4}) using up to the maximum allowed number of objective function evaluations (5,000,000). The best results are highlighted in boldface. It is easy to see that DELEqC requires much less objective function evaluations to find the best known solutions of the test-problems when compared to DE+DSS and DE+APM. Also, notice that DELEqC obtained more successful runs than both DE+DSS and DE+APM. Finally, it is important to highlight that the results obtained by the proposed technique are statistically different: p-values< 0.05, with respect to (i) pairwise comparisons using Wilcoxon rank-sum test, and (ii) p-values adjusted by Bonferroni correction.

We also investigated if the proposed technique produces results better or similar to those available in the literature using the same number of objective function evaluations. To do so, test-problems 7-11 were considered as in [15,16]. Each test-problem has its allowed number of objective function evaluations (nofe) grouped in 4 different computational budgets: a reference budget (rb), $2 \times rb$, $3 \times rb$, and $4 \times rb$. Statistical information of the results is presented in Tables 2 and 3, where the best results are highlighted in boldface.

Analyzing the results in Tables 2 and 3, it is important to highlight that DELEqC obtains the best mean values in 15 of the 20 cases considered here (5 test-problems and 4 different budgets). Also, the best results with respect to the best value found were attained in 15 situations. Notice that CLPSO, the best performing technique in [16], obtained the best mean values in only 6 cases, and the best results, concerning the best value found, in 14 cases.

When compared to Constricted PSO –the best performing technique in [15] and which has results available for 10 of the 20 cases considered here– one can notice that DELEqC obtained the best mean values in 8 cases, and the best results, concerning the best value found, in 9 of the 10 cases, while Constricted PSO found the best mean values in only 3 cases, and the best results, concerning the best value found, in 9 of the 10 cases.

It should be noted that sometimes more than one algorithm reached the best result. In general, for test-problems 7-11, one can notice that despite the use of the simplest DE variant in DELEqC: (i) it performed similarly to the techniques from the literature with respect to best results, concerning the best value found; (ii) it obtained the best mean values in more test-cases; and (iii) it performed well independently of the number of objective function evaluations tested here.

6 Concluding Remarks

Existing metaheuristics usually only approximately satisfy equality constraints (according to a user specified tolerance value), even when they are linear. Here, a modified DE algorithm (DELEqC) is proposed to exactly satisfy linear equality constraints while allowing any available constraint handling technique to be applied to the remaining constraints of the optimization problem. A procedure for the generation of a random feasible initial population is proposed. By avoiding the standard DE crossover and using mutation operator formulae containing only differences of feasible candidate solutions, a DE algorithm is proposed which maintains feasibility with respect to the linear equality constraints along the search process. Results from the computational experiments indicate that DELEqC outperforms the few alternatives that could be found in the literature and is a useful additional tool for the practitioner.

Further ongoing work concerns the extension of DELEqC so that linear inequality constraints are also exactly satisfied, as well as the introduction of a crossover operator that maintains feasibility with respect to the linear equality constraints.

Acknowledgments. The authors would like to thank the reviewers for their comments, which helped improve the paper, and the support provided by CNPq (grant 310778/2013-1), CAPES, and Pós-Graduação em Modelagem Computacional da Universidade Federal de Juiz de Fora (PGMC/UFJF).

References

1. Barbosa, H.J.C., Bernardino, H.S., Barreto, A.M.S.: Using performance profiles to analyze the results of the 2006 CEC constrained optimization competition. In: IEEE Congress on Evolutionary Computation, pp. 1–8 (2010)
2. Barbosa, H.J.C., Lemonge, A.C.C.: An adaptive penalty scheme in genetic algorithms for constrained optimization problems. In: Langdon, W.B., et al. (ed.) Proc. of the Genetic and Evolutionary Computation Conference. USA (2002)
3. Barbosa, H.J.C., Lemonge, A.C.C.: A new adaptive penalty scheme for genetic algorithms. Information Sciences **156**, 215–251 (2003)
4. Coello, C.A.C.: Theoretical and numerical constraint-handling techniques used with evolutionary algorithms: a survey of the state of the art. Computer Methods in Applied Mechanics and Engineering **191**(11–12), 1245–1287 (2002)
5. Datta, R., Deb, K. (eds.): Evolutionary Constrained Optimization. Infosys Science Foundation Series. Springer, India (2015)
6. Deb, K.: An efficient constraint handling method for genetic algorithms. Comput. Methods Appl. Mech. Engrg **186**, 311–338 (2000)
7. Dolan, E., Moré, J.J.: Benchmarking optimization software with performance profiles. Math. Programming **91**(2), 201–213 (2002)
8. Hock, W., Schittkowski, K.: Test Examples for Nonlinear Programming Codes. Springer-Verlag New York Inc., Secaucus (1981)
9. Koziel, S., Michalewicz, Z.: A decoder-based evolutionary algorithm for constrained parameter optimization problems. In: Eiben, A.E., Bäck, T., Schoenauer, M., Schwefel, H.-P. (eds.) PPSN 1998. LNCS, vol. 1498, pp. 231–240. Springer, Heidelberg (1998)
10. Koziel, S., Michalewicz, Z.: Evolutionary algorithms, homomorphous mappings, and constrained parameter optimization. Evol. Comput. **7**(1), 19–44 (1999)
11. Lemonge, A.C.C., Barbosa, H.J.C.: An adaptive penalty scheme for genetic algorithms in structural optimization. Intl. Journal for Numerical Methods in Engineering **59**(5), 703–736 (2004)
12. Luenberger, D.G., Ye, Y.: Linear and Nonlinear Programming. Springer-Verlag (2008)
13. Mezura-Montes, E., Coello, C.A.C.: Constraint-handling in nature-inspired numerical optimization: Past, present and future. Swarm and Evolutionary Computation **1**(4), 173–194 (2011)

272   H.J.C. Barbosa et al.

14. Michalewicz, Z., Janikow, C.Z.: Genocop: A genetic algorithm for numerical optimization problems with linear constraints. Commun. ACM **39**(12es), 175–201 (1996)
15. Monson, C.K., Seppi, K.D.: Linear equality constraints and homomorphous mappings in PSO. IEEE Congress on Evolutionary Computation **1**, 73–80 (2005)
16. Paquet, U., Engelbrecht, A.P.: Particle swarms for linearly constrained optimisation. Fundamenta Informaticae **76**(1), 147–170 (2007)
17. Price, K.V.: An introduction to differential evolution. New Ideas in Optimization, pp. 79–108 (1999)
18. Salcedo-Sanz, S.: A survey of repair methods used as constraint handling techniques in evolutionary algorithms. Computer Science Review **3**(3), 175–192 (2009)
19. Schoenauer, M., Michalewicz, Z.: Evolutionary computation at the edge of feasibility. In: Ebeling, W., Rechenberg, I., Voigt, H.-M., Schwefel, H.-P. (eds.) PPSN 1996. LNCS, vol. 1141, pp. 245–254. Springer, Heidelberg (1996)
20. Storn, R., Price, K.V.: Differential evolution - a simple and efficient heuristic for global optimization over continuous spaces. Journal of Global Optimization **11**, 341–359 (1997)

Appendix A. Test-Problems

Test-problems 1 to 6 and 7 to 11 were taken from [8] and [15], respectively. Also note that problems 7 to 11 are subject to the same set of linear constraints (2).

Problem 1 - The solution is $x^* = (1,1,1,1,1)^T$ with $f(x^*) = 0$.

$$\min (x_1 - 1)^2 + (x_2 - x_3)^2 + (x_4 - x_5)^2$$
$$\text{s.t. } x_1 + x_2 + x_3 + x_4 + x_5 = 5$$
$$x_3 - 2x_4 - 2x_5 = -3$$

Problem 2 - The solution is $x^* = (1,1,1,1,1)^T$ with $f(x^*) = 0$.

$$\min (x_1 - x_2)^2 + (x_3 - 1)^2 + (x_4 - 1)^4 + (x_5 - 1)^6$$
$$\text{s.t. } x_1 + x_2 + x_3 + 4x_4 = 7$$
$$x_3 + 5x_5 = 6$$

Problem 3 - The solution is $x^* = (1,1,1,1,1)^T$ with $f(x^*) = 0$.

$$\min (x_1 - x_2)^2 + (x_2 - x_3)^2 + (x_3 - x_4)^4 + (x_4 - x_5)^2$$
$$\text{s.t. } x_1 + 2x_2 + 3x_3 = 6$$
$$x_2 + 2x_3 + 3x_4 = 6$$
$$x_3 + 2x_4 + 3x_5 = 6$$

Problem 4 - The solution is $x^* = (1,1,1,1,1)^T$ with $f(x^*) = 0$.

$$\min (x_1 - x_2)^2 + (x_2 + x_3 - 2)^2 + (x_4 - 1)^2 + (x_4 - 1)^2 + (x_5 - 1)^2$$
$$\text{s.t. } x_1 + 3x_2 = 4$$
$$x_3 + x_4 - 2x_5 = 0$$
$$x_2 - x_5 = 0$$

Problem 5 - The solution is $x^* = (-33/349, 11/349, 180/349, -158/349, 1/349)^T$ with $f(x^*) = 5.326647564$.

$$\min (4x_1 - x_2)^2 + (x_2 + x_3 - 2)^2 + (x_4 - 1)^2 + (x_5 - 1)^2$$
$$\text{s.t. } x_1 + 3x_2 = 0$$
$$x_3 + x_4 - 2x_5 = 0$$
$$x_2 - x_5 = 0$$

Problem 6 - The solution is $x^* = (-33/43, 11/43, 27/43, -5/43, 11/43)^T$ with $f(x^*) = 4.093023256$.

$$\min (x_1 - x_2)^2 + (x_2 + x_3 - 2)^2 + (x_4 - 1)^2 + (x_5 - 1)^2$$
$$\text{s.t. } x_1 + 3x_2 = 0$$
$$x_3 + x_4 - 2x_5 = 0$$
$$x_2 - x_5 = 0$$

Problem 7 (Sphere) - $f(x^*) = 32.137$

$$\min_{x \in \mathcal{E}} \sum_{i=1}^{10} x_i^2$$

The feasible set $\mathcal{E}$ is given by the linear equality constraints:

$$\begin{cases} -3x_2 - x_3 + 2x_6 - 6x_7 - 4x_9 - 2x_{10} = 3 \\ -x_1 - 3x_2 - x_3 - 5x_7 - x_8 - 7x_9 - 2x_{10} = 0 \\ x_3 + x_6 + 3x_7 - 2x_9 + 2x_{10} = 9 \\ 2x_1 + 6x_2 + 2x_3 + 2x_4 + 4x_7 + 6x_8 + 16x_9 + 4x_{10} = -16 \\ -x_1 - 6x_2 - x_3 - 2x_4 - 2x_5 + 3x_6 - 6x_7 - 5x_8 - 13x_9 - 4x_{10} = 30 \end{cases} \quad (2)$$

Problem 8 (Quadratic) - $f(x^*) = 35.377$

$$\min_{x \in \mathcal{E}} \sum_{i=1}^{10} \sum_{j=1}^{10} e^{-(x_i - x_j)^2} x_i x_j + \sum_{i=1}^{10} x_i$$

Problem 9 (Rastrigin) - $f(x^*) = 36.975$

$$\min_{x \in \mathcal{E}} \sum_{i=1}^{10} x_i^2 + 10 - 10\cos(2\pi x_i)$$

Problem 10 (Rosenbrock) - $f(x^*) = 21485.3$

$$\min_{x \in \mathcal{E}} \sum_{i=1}^{9} 100(x_{i+1} - x_i^2)^2 + (x_i - 1)^2$$

Problem 11 (Griewank) - $f(x^*) = 0.151$

$$\min_{x \in \mathcal{E}} \frac{1}{4000} \sum_{i=1}^{10} x_i^2 - \prod_{i=10}^{10} \cos(\frac{x_i}{\sqrt{i}}) + 1$$

Multiobjective Firefly Algorithm for Variable Selection in Multivariate Calibration

Lauro Cássio Martins de Paula$^{(\boxtimes)}$ and Anderson da Silva Soares

Institute of Informatics, Federal University of Goiás, Goiânia, Goiás, Brazil
{laurocassio,anderson}@inf.ufg.br
http://www.inf.ufg.br

Abstract. Firefly Algorithm is a newly proposed method with potential application on several real world problems, such as variable selection problem. This paper presents a Multiobjective Firefly Algorithm (MOFA) for variable selection in multivariate calibration models. The main objective is to propose an optimization to reduce the error value prediction of the property of interest, as well as reducing the number of variables selected. Based on the results obtained, it is possible to demonstrate that our proposal may be a viable alternative in order to deal with conflicting objective-functions. Additionally, we compare MOFA with traditional algorithms for variable selection and show that it is a more relevant contribution for the variable selection problem.

Keywords: Firefly algorithm · Multiobjective optimization · Variable selection · Multivariate calibration

1 Introduction

Multivariate calibration may be considered as a procedure for constructing a mathematical model that establishes the relationship between the properties measured by an instrument and the concentration of a sample to be determined [3]. However, the building of a model from a subset of explanatory variables usually involves some conflicting objectives, such as extracting information from a measured data with many possible independent variables. Thus, a technique called variable selection may be used [3]. In this sense, the development of efficient algorithms for variable selection becomes important in order to deal with large and complex data. Furthermore, the application of Multiobjective Optimization (MOO) may significantly contribute to efficiently construct an accurate model [8].

Previous works about multivariate calibration have demonstrated that while monoobjective formulation uses a bigger number of variables, multiobjective algorithms can use fewer variables with a lower prediction error [2][1]. On the one hand, such works have used only genetic algorithms for exploiting MOO. On the other hand, the application of MOO in bioinspired metaheuristics such as Firefly Algorithm may be a better alternative in order to obtain a model with

F. Pereira et al. (Eds.) EPIA 2015, LNAI 9273, pp. 274–279, 2015.
DOI: 10.1007/978-3-319-23485-4_27

a more appropriate prediction capacity [8]. In this sense, some works have used FA to solve many types of problems. Regarding multiobjective characteristic, Yang [8] was the first one to present a multiobjective FA (MOFA) to solve optimization problems and showed that MOFA has advantages in dealing with multiobjective optimization.

As far as we know, the application of MOO-based Firefly Algorithm is not still widely used. There is no work in the literature that uses a multiobjective FA to select variables in multivariate calibration. Therefore, this paper presents an implementation of a MOFA for variable selection in multivariate calibration models. Additionally, estimates from the proposed MOFA are compared with predictions from the following traditional algorithms: Successive Projections Algorithm (SPA-MLR) [6], Genetic Algorithm (GA-MLR) [1] and Partial Least Squares (PLS). Based on the results obtained, we concluded that our proposed algorithm may be a more viable tool for variable selection in multivariate calibration models.

Section 2 describes multivariate calibration and the original FA. The proposed MOFA is presented in Section 3. Section 4 describes the material and methods used in the experiments. Results are described in Section 5. Finally, Section 6 shows the conclusions of the paper.

2 Background

2.1 Multivariate Calibration

The multivariate calibration model provides the value of a quantity y based on values measured from a set of explanatory variables $\{x_1, x_2, \ldots, x_k\}^T$ [3]. The model can be defined as:

$$y = \beta_0 + \beta_1 x_1 + \ldots + \beta_k x_k + \varepsilon, \tag{1}$$

where β_0, β_1, ..., β_k, $i = 1, 2, \ldots, k$, are the coefficients to be determined, and ε is a portion of random error. Equation (2) shows how the regression coefficients may be calculated using the Moore-Penrose pseudoinverse [4]:

$$\beta = (\mathbf{X}^T \mathbf{X})^{-1} \mathbf{X}^T \mathbf{y}, \tag{2}$$

where $\mathbf{X}$ is the matrix of samples and independent variables, $\mathbf{y}$ is the vector of dependent variables, and β is the vector of regression coefficients.

As shown in Equations (3) and (4), the predictive ability of MLR models comparing predictions with reference values for a test set from the squared deviations can be calculated by RMSEP or MAPE [3][5]:

$$\text{RMSEP} = \sqrt{\frac{\sum_{i=1}^{N}(\mathbf{y}_i - \hat{\mathbf{y}}_i)^2}{N}}, \tag{3}$$

where $\mathbf{y}$ is the reference value of the property of interest, N is the number of observations, and $\hat{\mathbf{y}} = \{\hat{y}_1, \hat{y}_2, \ldots, \hat{y}_k\}^T$ is the estimated value.

$$MAPE = \frac{\sum \left| \frac{y_i - \hat{y}_i}{y_i} \right|}{N}(100) = \frac{\sum \left| \frac{e_i}{y_i} \right|}{N}(100), \qquad (4)$$

where $\mathbf{y}_i$ is the actual data at variable i, $\hat{\mathbf{y}}_i$ is the forecast at variable i, e_i is the forecast error at variable i, and N is the number of samples.

2.2 Firefly Algorithm

Nature-inspired metaheuristics have been a powerful tool in solving various types of problems [8]. FA is a recently developed optimization algorithm proposed by Yang [8]. It is based on the behaviour of the flashing characteristics of fireflies. A pseudocode for the original FA can be obtained in the work of Yang [8]. In the original algorithm, there are two important issues to be treated: i) the variation of light intensity; and ii) the attractiveness formulation. The attractiveness of a firefly is determined by its brightness or light intensity, which is associated with the encoded objective function [8].

As a firefly's attractiveness is proportional to the light intensity seen by adjacent fireflies, one can define the attractiveness ω of a firefly by:

$$\omega = \omega_o e^{-\gamma r^2}, \qquad (5)$$

where ω_o is the attractiveness at $r = 0$.

According to Yang [8], a firefly i is attracted to a brighter firefly j and its movement is determined by:

$$x_i = x_j + \omega_0 e^{-\gamma r_{i,j}^2}(x_j - x_i) + \alpha(\text{rand} - \frac{1}{2}), \qquad (6)$$

where $rand$ is a random number generated in $[0, 1]$.

3 Proposal

Previous works have showed that multiobjective algorithms can use fewer variables and obtain lower prediction error [2]. Thus, this paper presents a Multiobjective Firefly Algorithm (MOFA) for variable selection in multivariate calibration. In the multiobjective formulation of FA, the choice of current best solution is based on two conditions: i) error of prediction; and ii) number of variables selected. Among non-dominated solutions, it is applied a multiobjective decision maker method called $Wilcoxon\ Signed\text{-}Rank^1$ to choose the final best one [2]. Algorithm 1 shows a pseudocode for the proposed MOFA. In line 9 of Algorithm 1, a firefly i dominates another firefly j when its RMSEP/MAPE and number of variables selected are lower.

---

[1] $Wilcoxon\ Signed\text{-}Rank$ is a nonparametric hypotheses test used when comparing two related samples to evaluate if the rank of the population means are different [7].

Algorithm 1. Proposed Multiobjective Firefly Algorithm.

---

1. **Parameters**: $\mathbf{X}_{n \times m}$, $\mathbf{y}_{n \times 1}$
2. $s \leftarrow$ number of fireflies
3. **for** $n = 1 : MaxGeneration$
4.   Generate randomly a population $Pop_{s \times m}$ of fireflies
5.   Compute Equations (2), (3), and the number of variables selected for each firefly
6.   **for** i $= 1$: s
7.   **for** j $= 1$: s
8.   **if** firefly i-th dominates firefly j-th
9.   Move firefly j towards firefly i using Equation (6)
10.   **end if**
11.   **end for** j
12.   **end for** i
13. **end for** n
14. Calculate RMSEP and variables selected for all fireflies
15. Select the best firefly by a decision maker based in [2]

---

4   Experimental Results

The proposed MOFA was implemented using $\alpha = 0.2$, $\gamma = 1$ and $\omega_0 = 0.97$. The number of firelies and the number of generations were 200 and 100, respectively. We have used for RMSEP comparison three traditional methods for variable selection: SPA-MLR [6], GA-MLR [1] and the PLS. The number of iterations was the same for all algorithms and the multiobjective approach was not applied in this three traditional methods.

The dataset employed in this work consists of 775 NIR spectra of whole-kernel wheat, which were used as shoot-out data in the 2008 International Diffuse Reflectance Conference (http://www.idrc-chambersburg.org/shootout.html). Protein content (%) was used as the y-property in the regression calculations.

All calculations were carried out by using a desktop computer with an Intel Core i7 2600 (3.40 GHz), 8 GB of RAM memory and Windows 7 Professional. The Matlab 8.1.0.604 (R2013a) software platform was employed throughout. Regarding the outcomes, it is important to note that all of them were obtained by averaging fifty executions.

5   Results and Discussion

Figure 1(a) shows the first population of fireflies generated. Figure 1(b) illustrates the behaviour of fireflies when a monoobjective formulation is employed. In the chart, the only goal was to reduce RMSEP.

The application of multiobjective optimization is presented in Figure 2. The fireflies create a relatively perfect Pareto Front tending to a minimum error value as well as a minimum number of variables selected. It is possible to note that the

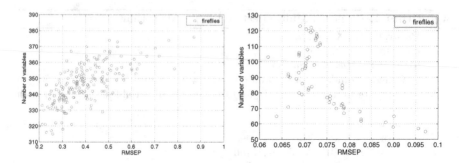

Fig. 1. Behaviour of fireflies with: (a) randomly generated fireflies.; (b) monoobjective formulation.

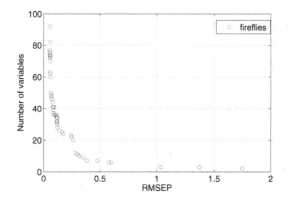

Fig. 2. Behaviour of fireflies with multiobjective optimization.

Table 1. Results for the **FA, MOFA, SPA-MLR, GA-MLR and PLS.**

	Number of variables	RMSEP	MAPE
GA-MLR	146	0.21	1.50%
FA	103	0.07	0.81%
MOFA	37	0.05	0.72%
PLS	15	0.21	1.50%
SPA-MLR	13	0.20	1.43%

application of multiobjective formulation can move fireflies to more appropriate solutions using the non-dominance characteristic.

A comparison between MOFA and traditional algorithms is showed in Table 1. Despite SPA-MLR was able to yield the lowest number of variables selected, MOFA presented the lowest RMSEP and MAPE[2]. A comparison of computational time between these algorithms can be obtained in [4].

6 Conclusion

This paper proposed a Multiobjective Firefly Algorithm (MOFA) for variable selection in multivariate calibration models. The objective was to present an optimization to reduce the error value prediction of the property of interest as well as reducing the number of variables selected. In terms of error reduction, MOFA presented the lowest values when compared with traditional algorithms. Therefore, through the results obtained we were able to demonstrate that MOFA may be a better solution for obtaining a model with an adequate prediction capacity.

Acknowledgments. Authors thank the research agencies CAPES and FAPEG for the support provided to this work.

References

1. Soares, A.S., de Lima, T.W., Soares, F.A.A.M.N., Coelho, C.J., Federson, F.M., Delbem, A.C.B., Van Baalen, J.: Mutation-based compact genetic algorithm for spectroscopy variable selection in determining protein concentration in wheat grain. Electronics Letters **50**, 932–934 (2014)
2. Lucena, D.V., Soares, A.S., Soares, T.W., Coelho, C.J.: Multi-Objective Evolutionary Algorithm NSGA-II for Variables Selection in Multivariate Calibration Problems. International Journal of Natural Computing Research **3**, 43–58 (2012)
3. Martens, H.: Multivariate Calibration. John Wiley & Sons (1991)
4. Paula, L.C.M., Soares, A.S., Soares, T.W., Delbem, A.C.B., Coelho, C.J., Filho, A.R.G.: Parallelization of a Modified Firefly Algorithm using GPU for Variable Selection in a Multivariate Calibration Problem. International Journal of Natural Computing Research **4**, 31–42 (2014)
5. Hibon, M., Makridakis, S.: Evaluating Accuracy (or Error) Measures. INSEAD (1995)
6. Arajo, M.C.U., Saldanha, T.C., Galvo, R.K., Yoneyama, T.: The successive projections algorithm for variable selection in spectroscopic multicomponent analysis. Chemometrics and Intelligent Laboratory Systems **57**, 65–73 (2001)
7. Ramsey, P.H.: Significance probabilities of the wilcoxon signed-rank test. Journal of Nonparametric Statistics **2**, 133–153 (1993)
8. Yang, X.S.: Multiobjective firefly algorithm for continuous optimization. Engineering with Computers **29**, 175–184 (2013)

---

[2] It is worth noting that the Successive Projections Algorithm is composed of three phases, and its main objective is to select a subset of variables with low collinearity [6].

Semantic Learning Machine: A Feedforward Neural Network Construction Algorithm Inspired by Geometric Semantic Genetic Programming

Ivo Gonçalves[1,2](✉), Sara Silva[1,2,3], and Carlos M. Fonseca[1]

[1] CISUC, Department of Informatics Engineering,
University of Coimbra, 3030-290 Coimbra, Portugal
icpg@dei.uc.pt
[2] BioISI - Biosystems & Integrative Sciences Institute, Faculty of Sciences,
University of Lisbon, 1749-016 Campo Grande, Lisbon, Portugal
sara@fc.ul.pt
[3] NOVA IMS, Universidade Nova de Lisboa, 1070-312 Lisbon, Portugal
cmfonsec@dei.uc.pt

Abstract. Geometric Semantic Genetic Programming (GSGP) is a recently proposed form of Genetic Programming in which the fitness landscape seen by its variation operators is unimodal with a linear slope by construction and, consequently, easy to search. This is valid across all supervised learning problems. In this paper we propose a feedforward Neural Network construction algorithm derived from GSGP. This algorithm shares the same fitness landscape as GSGP, which allows an efficient search to be performed on the space of feedforward Neural Networks, without the need to use backpropagation. Experiments are conducted on real-life multidimensional symbolic regression datasets and results show that the proposed algorithm is able to surpass GSGP, with statistical significance, in terms of learning the training data. In terms of generalization, results are similar to GSGP.

1 Introduction

Moraglio et al. [6] recently proposed a new Genetic Programming formulation called Geometric Semantic Genetic Programming (GSGP). GSGP derives its name from the fact that it is formulated under a geometric framework [5] and from the fact that it operates directly in the space of the underlying semantics of the individuals. In this context, semantics is defined as the outputs of an individual over a set of data instances. The most interesting property of GSGP is that the fitness landscape seen by its variation operators is always unimodal with a linear slope (cone landscape) by construction. This implies that there are no local optima, and consequently, that this type of landscape is easy to search. When applied to multidimensional real-life datasets, GSGP has shown competitive results in learning and generalization [3,7]. In this paper, we adapt

F. Pereira et al. (Eds.) EPIA 2015, LNAI 9273, pp. 280–285, 2015.
DOI: 10.1007/978-3-319-23485-4_28

the geometric semantic mutation to the realm of feedforward Neural Networks by proposing the Semantic Learning Machine (SLM). Section 2 defines the SLM. Section 3 describes the experimental setup. Section 4 presents and discusses the results of the SLM and GSGP, and Section 5 concludes.

2 Semantic Learning Machine

Given that the geometric semantic operators are defined over the semantic space (outputs), they can be extended for different representations. The Semantic Learning Machine (SLM) proposed in this section is based on a derivation of the GSGP mutation operator for real-value semantics. This implies that the SLM shares the same semantic landscape proprieties as GSGP. Particularly, the fitness landscape induced by its operator is always unimodal with a linear slope (cone landscape) by construction, and consequently easy to search. This is valid across all supervised learning problems.

2.1 A Geometric Semantic Mutation Operator for Feedforward Neural Networks

The GSGP mutation for real-value semantics [6] is defined as follows:

Definition 1. (GSGP Mutation). *Given a parent function* $T : \mathbb{R}^n \to \mathbb{R}$*, the geometric semantic mutation with mutation step* ms *returns the real function* $T_M = T + ms \cdot (T_{R1} - T_{R2})$*, where* T_{R1} *and* T_{R2} *are random real functions.*

This mutation essentially performs a linear combination of two individuals: the parent and a randomly generated tree (which results from subtracting the two subtrees T_{R1} and T_{R2}). The degree of semantic change is controlled by the mutation step.

An equivalent geometric semantic mutation operator can be derived for feedforward Neural Networks (NN). The only three small restrictions for this NN mutation operator are: the NN must have at least one hidden layer; the output layer must have only one neuron; and the output neuron must have a linear activation function. Each application of the operator adds a new neuron to the last hidden layer. The weight from the new neuron to the output neuron is defined by the learning step (SLM parameter). This learning step is the equivalent of the mutation step in the GSGP mutation. It defines the amount of semantic change for each application of the operator. The weights from the last hidden layer to the previous layer are randomly generated. This is the equivalent of generating the two random subtrees in the GSGP mutation. In this work these weights are generated with uniform probability between -1.0 and 1.0. If more than one hidden layer is used, all other weights remain constant once initialized. In this work all experiments are conducted with a single hidden layer. The activation function for the neurons in the last hidden layer can be freely chosen. However, it has been recently shown, in the context of GSGP, that applying a structural bound to the randomly generated tree (which results from subtracting the two subtrees T_{R1} and T_{R2}) results in significant improvements in terms of generalization ability [3]. In fact, if a unbounded mutation (equivalent to using a linear activation function) is used, there is a tendency for GSGP to greatly overfit the training data [3]. For this reason, it is recommended that the activation function for the neurons in the last hidden layer to be a function with a relatively small codomain. In this work a modified logistic function (transforming the logistic function output to range in the interval $[-1, 1]$) is used for this purpose. In terms of generalization ability, it is also essential to use a small learning/mutation step [3]. If more than one hidden layer is used, the activation functions for the remaining neurons may be freely chosen.

2.2   Algorithm

The SLM algorithm is essentially a geometric semantic hill climber for feedforward neural networks. The idea is to perform a semantic sampling with a given size (SLM parameter) by applying the mutation operator defined in the previous subsection. As is common in hill climbers, only one solution (in this case a neural network) is kept along the run. At each iteration, the mentioned semantic sampling is performed to produce N neighbors. At the end of the iteration, the best individual from the previous best and the newly generated individuals is kept. The process is repeated until a given number of iterations (SLM parameter) has been reached. As mentioned in the previous subsection, the mutation operator always adds a new neuron to the last hidden layer, so the number of neurons in the last hidden layer is at most the same as the number of iterations. This number of neurons can be smaller than the number of iterations if in some iterations it was not possible to generate an individual superior to the current best.

3   Experimental Setup

The experimental setup is based on the setup of Vanneschi et al. [7] and Gonçalves et al. [3], since these works recently provided results for GSGP. Experiments are run for 500 iterations/generations because that is where the statistical comparisons were made in the mentioned works. 30 runs are conducted. Population/sample size is 100. Training and testing set division is 70% - 30%. Fitness is computed as the root mean squared error. The initial tree initialization is performed with the ramped half-and-half method, with a maximum depth of 6. Besides GSGP, the Semantic Stochastic Hill Climber (SSHC) [6] is also used as baseline for comparison. The variation operators used are the variants defined for real-value semantics [6]: SGXM crossover for GSGP, and SGMR mutation for GSGP and SSHC. For GSGP a probability of 0.5 is used for both operators. The function set contains the four binary arithmetic operators: +, -, *, and / (protected). No constants are used in the terminal set. Parent selection in GSGP is based on tournaments of size 4. Also for GSGP, survivor selection is elitist as the best individual always survives to the next generation. All claims of statistical significance are based on Mann-Whitney U tests, with Bonferroni correction, and considering a significance level of $\alpha = 0.05$. For each dataset 30 different random partitions are used. Each method uses the same 30 partitions. Experiments are conducted on three multidimensional symbolic regression real-life datasets. These datasets are the Bioavailability (hereafter Bio), the Plasma Protein Binding (hereafter PPB), and the Toxicity (hereafter LD50). The first two were also used by Vanneschi et al. [7] and Gonçalves et al. [3]. These datasets have, respectively: 359 instances and 241 features; 131 instances and 626 features; and 234 instances and 626 features. For a detailed description of these datasets the reader is referred to Archetti et al. [1]. These datasets have also been used in other Genetic Programming studies, e.g., [2,4].

4   Experimental Study

Figure 1 presents the training and testing error evolution plots for SLM, GSGP and SSHC. These evolution plots are constructed by taking the median over 30 runs of the training and testing error of the best individuals in the training data. The mutation/learning step used was 1 for the the Bio and PPB datasets (as in Vanneschi et al. [7] and Gonçalves et al. [3]), and 10 for the LD50 as it was found, in preliminary testing, to be a suitable value (other values tested were: 0.1, 1, and 100). A consideration for the different initial values (at iteration/generation 0) is in order. The SLM presents much higher errors than GSGP/SSHC after the random initialization. This is explained by the fact that the weights for the SLM are generated with uniform probability between -1.0 and 1.0, and consequently, the amount of data fitting is clearly bounded. On the other hand, GSGP and SSHC have no explicit bound on the random trees and therefore can provide a superior initial explanation of the data. It is interesting to note that, despite this initial disadvantage, the SLM compensates with a much higher learning rate. This higher learning efficiency is confirmed by the statistically significant superiority found in terms of training error across all datasets, against GSGP (p-values: Bio 2.872×10^{-11}, PPB 2.872×10^{-11} and LD50 7.733×10^{-10}), and against SSHC (p-values: Bio 2.872×10^{-11}, PPB 2.872×10^{-11} and LD50 3.261×10^{-5}).

This learning superiority is particularly interesting when considering that the SLM and the SSHC use the exact same geometric semantic mutation operator. This raise the question: how can two methods with the same variation operator, the same induced semantic landscape, and the same parametrizations achieve such different outcomes? The answer lies in the different semantic distributions that result from the random initializations. Different representations have different natural ways of being randomly initialized. This translates into different semantic distributions and, consequently, to different offspring distributions. From the results it is clear that the distribution induced by the random initialization of a list of weights (used in SLM), is more well-behaved than the initialization of a random tree (used in SSHC). In the original GSGP proposal, Moraglio et al. [6] provided a discussion on whether syntax (representation) matters in terms of search. They argued that, in abstract, the offspring distributions may be affected by the different syntax initializations. In our work, we

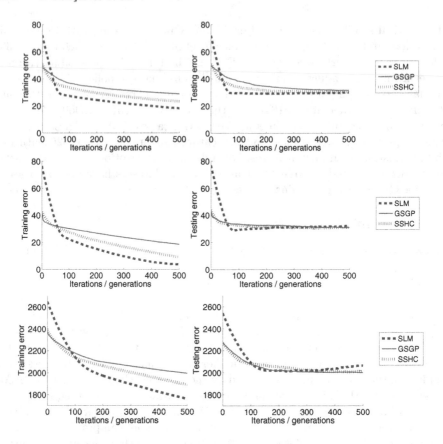

Fig. 1. Bio (top), PPB (center) and LD50 (bottom) training and testing error evolution plots

can empirically see how different representations induce different offspring distributions and consequently reach considerably different outcomes. A possible research venue lies in analyzing the semantic distributions induced by different tree initialization methods, and to possibly propose new tree initializations that are more well-behaved.

In terms of generalization, results show that all methods achieve similar results. The only statistically significant difference shows that the SLM is superior to GSGP in the Bio dataset (p-value: 1.948×10^{-4}). However, it seems that in this case, GSGP is still evolving and that in a few more generations may reach a generalization similar to the SLM. On a final note, the evolution plots also show that SSHC consistently learns the training data faster and better than GSGP. This should be expected as the semantic space has no local optima and consequently the search can be focused around the best individual in the population. These differences are confirmed as statistically significant (p-values: Bio 2.872×10^{-11}, PPB 2.872×10^{-11} and LD50 1.732×10^{-4}). There are no statistically significant differences in terms of generalization.

5   Conclusions

This work presented a novel feedforward Neural Network (NN) construction algorithm, derived from Geometric Semantic Genetic Programming (GSGP). The proposed algorithm shares the same fitness landscape as GSGP, which enables an efficient search for any supervised learning problem. Results in regression datasets show that the proposed NN construction algorithm is able to surpass GSGP, with statistical significance, in terms of learning the training data. Generalization results are similar to those of GSGP. Future work involves extending the experimental analysis to other regression datasets and to provide results for classification tasks. Comparisons with other NN algorithms and other commonly used supervised learning algorithms (e.g. Support Vector Machines) are also in order.

Acknowledgments. This work was partially supported by national funds through FCT under contract UID/Multi/04046/2013 and projects PTDC/EEI-CTP/2975/2012 (MaSSGP), PTDC/DTP-FTO/1747/2012 (InteleGen) and EXPL/EMS-SIS/1954/ 2013 (CancerSys). The first author work is supported by FCT, Portugal, under the grant SFRH/BD/79964/2011.

References

1. Archetti, F., Lanzeni, S., Messina, E., Vanneschi, L.: Genetic programming for computational pharmacokinetics in drug discovery and development. Genetic Programming and Evolvable Machines **8**(4), 413–432 (2007)
2. Gonçalves, I., Silva, S.: Balancing learning and overfitting in genetic programming with interleaved sampling of training data. In: Krawiec, K., Moraglio, A., Hu, T., Etaner-Uyar, A.Ş., Hu, B. (eds.) EuroGP 2013. LNCS, vol. 7831, pp. 73–84. Springer, Heidelberg (2013)
3. Gonçalves, I., Silva, S., Fonseca, C.M.: On the generalization ability of geometric semantic genetic programming. In: Machado, P., Heywood, M.I., McDermott, J., Castelli, M., García-Sánchez, P., Burelli, P., Risi, S., Sim, K. (eds.) Genetic Programming. LNCS, vol. 9025, pp. 41–52. Springer, Heidelberg (2015)
4. Gonçalves, I., Silva, S., Melo, J.B., Carreiras, J.M.B.: Random sampling technique for overfitting control in genetic programming. In: Moraglio, A., Silva, S., Krawiec, K., Machado, P., Cotta, C. (eds.) EuroGP 2012. LNCS, vol. 7244, pp. 218–229. Springer, Heidelberg (2012)
5. Moraglio, A.: Towards a Geometric Unification of Evolutionary Algorithms. Ph.D. thesis, Department of Computer Science, University of Essex, UK, November 2007
6. Moraglio, A., Krawiec, K., Johnson, C.G.: Geometric semantic genetic programming. In: Coello, C.A.C., Cutello, V., Deb, K., Forrest, S., Nicosia, G., Pavone, M. (eds.) PPSN 2012, Part I. LNCS, vol. 7491, pp. 21–31. Springer, Heidelberg (2012)
7. Vanneschi, L., Castelli, M., Manzoni, L., Silva, S.: A new implementation of geometric semantic GP and its application to problems in pharmacokinetics. In: Krawiec, K., Moraglio, A., Hu, T., Etaner-Uyar, A.Ş., Hu, B. (eds.) EuroGP 2013. LNCS, vol. 7831, pp. 205–216. Springer, Heidelberg (2013)

Eager Random Search for Differential Evolution in Continuous Optimization

Miguel Leon[(✉)] and Ning Xiong

Mälardalen University, Västerås, Sweden
{miguel.leonortiz,ning.xiong}@mdh.se

Abstract. This paper proposes a memetic computing algorithm by incorporating Eager Random Search (ERS) into differential evolution (DE) to enhance its search ability. ERS is a local search method that is eager to move to a position that is identified as better than the current one without considering other opportunities. Forsaking optimality of moves in ERS is advantageous to increase the randomness and diversity of search for avoiding premature convergence. Three concrete local search strategies within ERS are introduced and discussed, leading to variants of the proposed memetic DE algorithm. The results of evaluations on a set of benchmark problems have demonstrated that the integration of DE with Eager Random Search can improve the performance of pure DE algorithms while not incurring extra computing expenses.

Keywords: Evolutionary algorithm · Differential evolution · Eager Random Search · Memetic algorithm · Optimization

1 Introduction

Differential evolution [1] presents a class of metaheuristics [2] to solve real parameter optimization tasks with nonlinear and multimodal objective functions. DE has been used as very competitive alternative in many practical applications due to its simple and compact structure, easy use with fewer control parameters, as well as high convergence in large problem spaces. However, the performance of DE is not always excellent to ensure fast convergence to the global optimum. It can easily get stagnation resulting in low precision of acquired results or even failure.

Hybridization of EAs with local search (LS) techniques can greatly improve the efficiency of the search. EAs that are augmented with LS for self-refinement are called Memetic Algorithms (MAs) [3]. Memetic computing has been used with DE to refine individuals in their neighborhood. Norman and Iba [4] proposed a crossover-based adaptive method to generate offspring in the vicinity of parents. Many other works apply local search mechanisms to certain individuals of every generation to obtain possibly even better solutions, see examples in ([5], [6], [7]). Other Researchers investigate adaptation of control parameters of DE to improve the performance ([8], [9]).

© Springer International Publishing Switzerland 2015
F. Pereira et al. (Eds.) EPIA 2015, LNAI 9273, pp. 286–291, 2015.
DOI: 10.1007/978-3-319-23485-4_29

This paper proposes a new memetic DE algorithm by incorporating Eager Random Search (ERS) to enhance the performance of a conventional DE algorithm. ERS is a local search method that is eager to move to a position that is identified as better than the current one without considering other opportunities in the neighborhood. This is different from common local search methods such as gradient descent or hill climbing which seek local optimal actions during the search. Forsaking optimality of moves in ERS is advantageous to increase randomness and diversity of search for avoiding premature convergence. Three concrete local search strategies within ERS are introduced and discussed, leading to variants of the proposed memetic DE algorithm. In addition, only a small subset of randomly selected variables is used in every step of the local search for randomly deciding the next trial point. The results of tests on a set of benchmark problems have demonstrated that the hybridization of DE with Eager Random Search can bring improvement of performance compared to pure DE algorithms while not incurring extra computing expenses.

2   Basic DE

DE is a stochastic and population based algorithm with Np individuals in the population. Every individual in the population stands for a possible solution to the problem. One of the Np individuals is represented by $X_{i,g}$ with $i = 1, 2, ., N_p$ and g is the index of the generation. DE has three consecutive steps in every iteration: mutation, recombination and selection. The explanation of these steps is given below:

MUTATION. N_p mutated individuals are generated using some individuals of the population. The vector for the mutated solution is called mutant vector and it is represented by $V_{i,g}$. There are some ways to mutate the current population, but the most common one is called random mutation strategy. This mutation strategy will be explained below. The other mutation strategies and their performance are given in [10].

$$V_{i,g} = X_{r_1,g} + F \times (X_{r_2,g} - X_{r_3,g}) \tag{1}$$

where $V_{i,g}$ represents the mutant vector, i stands for the index of the vector, g stands for the generation, $r_1, r_2, r_3 \in \{ 1,2,\ldots,N_p\}$ are random integers and F is the scaling factor in the interval $[0, 2]$.

CROSSOVER. In step two we recombine the set of mutated solutions created in step 1 (mutation) with the original population members to produce trial solutions. A new trial vector is denoted by $T_{i,g}$ where i is the index and g is the generation. Every parameter in the trial vector is calculated with equation 2.

$$T_{i,g}[j] = \begin{cases} V_{i,g}[j] & \text{if } rand[0,1] < CR \ \text{ or } \ j = j_{rand} \\ X_{i,g}[j] & \text{otherwise} \end{cases} \tag{2}$$

where j stands for the index of every parameter in a vector, CR is the probability of the recombination and j_{rand} is a randomly selected integer in $[1, N_p]$ to ensure that at least one parameter from the mutant vector is selected.

SELECTION. In this last step we compare the fitness of a trial vector with the fitness of its parent in the population with the same index i, selecting the individual with the better fitness to enter the next generation. So we compare

3   DE Integrated with ERS

This section is devoted to the proposal of the memetic DE algorithm with integrated ERS for local search. We will first introduce ERS as a general local search method together with its three concrete (search) strategies, and then we shall outline how ERS can be incorporated into DE to enable self-refinement of individuals inside a DE process.

3.1   Eager Random Local Search (ERS)

The main idea of ERS is to immediately move to a randomly created new position in the neighborhood without considering other opportunities as long as this new position receives a better fitness score than the current position. This is different from some other conventional local search methods such as Hill Climbing in which the next move is always to the best position in the surroundings. Forsaking optimality of moves in ERS is beneficial to achieve more randomness and diversity of search for avoiding local optima. Further, in exploiting the neighborhood, only a small subset of randomly selected variables undergoes changes to randomly create a trial solution. If this trial solution is better, it simply replaces the current one. Otherwise a new trial solution is generated with other randomly selected variables. This procedure is terminated when a given number of trial solutions have been created without finding improved ones.

The next more detailed issue with ERS is how to change a selected variable in making a trial solution in the neighborhood. Our idea is to solve this issue using a suitable probability function. We consider three probability distributions (uniform, normal, and Cauchy) as alternatives for usage when generating a new value for a selected parameter/variable. The use of different probability distributions lead to different local search strategies within the ERS family, which will be explained in the sequel.

Random Local Search (RLS). In Random Local Search (RLS), we simply use a uniform probability distribution when new trial solutions are created given a current solution. To be more specific, when dimension k is selected for change, the trial solution X' will get the following value on this dimension regardless of its initial value in the current solution:

$$X'[k] = rand(a_i, b_i) \tag{3}$$

where $rand(a_i, b_i)$ is a uniform random number between a_i and b_i, and a_i and b_i are the minimum and maximum values respectively on dimension k.

Normal Local Search (NLS). In Normal Local Search (NLS), we create a new trial solution by disturbing the current solution in terms of a normal probability distribution. This means that, if dimension k is selected for change, the value on this dimension for trial solution X' will be given by

$$X'[k] = X[k] + N(0, \delta) \tag{4}$$

where $N(0, \delta)$ represents a random number generated according to a normal density function with its mean being zero.

Cauchy Local Search (CLS). In this third local search strategy, we apply the Cauchy density function in creating trial solutions in the neighborhood. It is called Cauchy Local search (CLS). A nice property of the Cauchy function is that it is centered around its mean value whereas exhibiting a wider distribution than the normal probability function. The value of trial solution X' will be generated as follows:

$$X'[k] = X[k] + t \times tan(\pi \times (rand(0,1) - 0.5)) \tag{5}$$

where $rand(0, 1)$ is a random uniform number between 0 and 1.

3.2   The Proposed Memetic DE Algorithm

Here with we propose a new memetic DE algorithm by combining basic DE with Eager Random Search (ERS). ERS is applied in each generation after completing the mutation, crossover and selection operators. The best individual in the population is used as the starting point when ERS is executed. If ERS terminates with a better solution, it is inserted into the population and the current best member in the population is discarded.

We use DERLS, DENLS, and DECLS to refer to the variants of the proposed memetic DE algorithm that adopt RLS, NLS, and CLS respectively as local search strategies.

4   Experiments and Results

To examine the merit our proposed memetic DE algorithm compared to basic DE, we tested the algorithms in thirteen benchmark functions [10]. Functions 1 to 7 are unimodal and functions 8 to 13 are multimodal functions that contain many local optima.

4.1   Experimental Settings

DE has three main control parameters: population size (N_p), crossover rate (CR) and the scaling factor (F) for mutation. The following specification of these parameters was used in the experiments: $Np = 60$, $CR = 0.85$ and $F = 0.9$. All the algorithms were applied to the benchmark problems with the aim to find the best solution for each of them. Every algorithm was executed 30 times on every function to acquire a fair result for the comparison. The condition to finish the execution of DE programs is that the error of the best result found is below 10e-8 with respect to the true minimum or the number of evaluations has exceeded 300,000. In DECLS, $t = 0.2$.

4.2 Performance of the Memetic DE with Random Mutation Strategy

First, random mutation strategy (DE/rand/1) was used in all DE approaches to study the effect of the ERS local search strategies in the memetic DE algorithm. The results can be observed in Table 1 and the values in boldface represent the lowest average error found by the approaches.

Table 1. Average error of the found solutions on the test problems with random mutation strategy

F.	DE	DERLS	DENLS	DECLS
f1	**0,00E+00** (4,56E-14)	**0,00E+00** (6,80E-13)	**0,00E+00** (1,21E-13)	**0,00E+00** (1,33E-14)
f2	1,82E-08 (1,13E-08)	5,30E-08 (2,39E-08)	2,26E-08 (1,32E-08)	**1,42E-08** (1,07E-08)
f3	6,55E+01 (3,92E+01)	8,01E+01 (4,88E+01)	1,11E+00 (1,10E+00)	**6,54E-01** (1,47E+01)
f4	6,22E+00 (5,07E+00)	2,37E+00 (1,87E+00)	1,80E-02 (7,28E-01)	**5,66E-01** (3,68E-01)
f5	2,31E+01 (2,00E+01)	2,27E+01 (1,81E+01)	2,65E+01 (2,41E+01)	**2,03E+01** (2,62E+01)
f6	**0,00E+00** (0,00E+00)	**0,00E+00** (0,00E+00)	**0,00E+00** (0,00E+00)	**0,00E+00** (0,00E+00)
f7	1,20E-01 (3,79E-03)	1,15E-02 (3,16E-03)	1,23E-02 (3,29E-03)	**1,05E-02** (3,61E-03)
f8	2,72E+03 (8,15E+02)	**2,31E+02** (1,50E+02)	1,86E+03 (5,46E+02)	1,58E+03 (5,16E+02)
f9	1,30E+01 (3,70E+00)	1,28E+01 (3,72E+00)	**6,17E+00** (2,06E+00)	7,72E+00 (2,43E+00)
f10	1,88E+01 (4,28E+00)	**1,87E+00** (4,76E+00)	4,94E+00 (8,11E+00)	5,50E+00 (7,53E+00)
f11	**8,22E-04** (2,49E-03)	**8,22E-04** (2,49E-03)	1,49E-02 (2,44E-02)	1,44E-02 (2,70E-02)
f12	3,46E-03 (1,86E-02)	3,46E-03 (1,86E-02)	1,04E-02 (3,11E-02)	**0,00E+00** (8,49E-15)
f13	3,66E-04 (1,97E-03)	**0,00E+00** (2,14E-13)	**0,00E+00** (1,31E-14)	**0,00E+00** (3,37E-15)

We can see in Table 1 that DECLS is the best in all the unimodal functions except on Function 4 that is the second best. In multimodal functions, DERLS is the best on Functions 8, 10 and 11. DECLS found the exact optimum all the times in Functions 12 and 13. The basic, DE performed the worst in multimodal functions. According to the above analysis, we can say that DECLS improve a lot the performance of basic DE with random mutation strategy and also we found out that DERLS is really good in multimodal functions particularly on Function 8, which is the most difficult function. Considering all the functions, the best algorithm is DECLS and the weakest one is the basic DE.

5   Conclusions

In this paper we propose a memetic DE algorithm by incorporating Eager Random Search (ERS) as a local search method to enhance the search ability of a pure DE algorithm. Three concrete local search strategies (RLS, NLS, and CLS)

are introduced and explained as instances of the general ERS method. The use of different local search strategies from the ERS family leads to variants of the proposed memetic DE algorithm, which are abbreviated as DERLS, DENLS and DECLS respectively. The results of the experiments have demonstrated that the overall ranking of DECLS is superior to the ranking of basic DE and other memetic DE variants considering all the test functions. In addition, we found out that DERLS is much better than the other counterparts in very difficult multimodal functions.

Acknowledgment. The work is funded by the Swedish Knowledge Foundation (KKS) grant (project no 16317). The authors are also grateful to ABB FACTS, Prevas and VOITH for their co-financing of the project.

References

1. Storn, R., Price, K.: Differential evolution - a simple and efficient heuristic for global optimization over continuous spaces. Journal of Global Optimization **11**(4), 341–359 (1997)
2. Xiong, N., Molina, D., Leon, M., Herrera, F.: A walk into metaheuristics for engineering optimization: Principles, methods, and recent trends. International Journal of Computational Intelligence Systems **8**(4), 606–636 (2015)
3. Krasnogor, N., Smith, J.: A tutorial for competent memetic algorithms: Model, taxonomy, and design issue. IEEE Transactions on Evolutionary Computation **9**(5), 474–488 (2005)
4. Norman, N., Ibai, H.: Accelerating differential evolution using an adaptive local search. IEEE Transactions on Evolutionary Computation **12**, 107–125 (2008)
5. Ali, M., Pant, M., Nagar, A.: Two local search strategies for differential evolution. In: Proc. 2010 IEEE Fifth International Conference on Bio-Inspired Computing: Theories and Applications (BIC-TA), Changsha, China, pp. 1429–1435 (2010)
6. Dai, Z., Zhou, A.: A diferential ecolution with an orthogonal local search. In: Proc. 2013 IEEE Congress on Evolutionary Computation (CEC), Cancun, Mexico, pp. 2329–2336 (2013)
7. Leon, M., Xiong, N.: Using random local search helps in avoiding local optimum in dieferential evolution. In: Proc. Artificial Intelligence and Applications, AIA2014, Innsbruck, Austria, pp. 413–420 (2014)
8. Qin, A., Suganthan, P.: Self-adaptive differential evolution algorithm for numerical optimization. In: The 2005 IEEE Congress on Evolutionary Computation, vol. 2, pp. 1785–1791 (2005)
9. Leon, M., Xiong, N.: Greedy adaptation of control parameters in differential evolution for global optimization problems. In: IEEE Conference on Evolutionary Computation, CEC2015, Japan, pp. 385–392 (2015)
10. Leon, M., Xiong, N.: Investigation of mutation strategies in differential evolution for solving global optimization problems. In: Rutkowski, L., Korytkowski, M., Scherer, R., Tadeusiewicz, R., Zadeh, L.A., Zurada, J.M. (eds.) ICAISC 2014, Part I. LNCS, vol. 8467, pp. 372–383. Springer, Heidelberg (2014)

Learning from Play: Facilitating Character Design Through Genetic Programming and Human Mimicry

Swen E. Gaudl[1]([⊠]), Joseph Carter Osborn[2], and Joanna J. Bryson[1]

[1] Department of Computer Science, University of Bath, Bath, UK
swen.gaudl@gmail.com, jjb@bath.ac.uk
[2] Baskin School of Engineering, University of California SC, Santa Cruz, USA
jcosborn@soe.ucsc.edu

Abstract. Mimicry and play are fundamental learning processes by which individuals can acquire behaviours, skills and norms. In this paper we utilise these two processes to create new game characters by mimicking and learning from actual human players. We present our approach towards aiding the design process of game characters through the use of genetic programming. The current state of the art in game character design relies heavily on human designers to manually create and edit scripts and rules for game characters. Computational creativity approaches this issue with fully autonomous character generators, replacing most of the design process using black box solutions such as neural networks. Our GP approach to this problem not only mimics actual human play but creates character controllers which can be further authored and developed by a designer. This keeps the designer in the loop while reducing repetitive labour. Our system also provides insights into how players express themselves in games and into deriving appropriate models for representing those insights. We present our framework and preliminary results supporting our claim.

Keywords: Agent design · Machine learning · Genetic programming · Games

1 Introduction

Designing intelligence is a sufficiently complex task that it can itself be aided by the proper application of AI techniques. Here we present a system that mines human behaviour to create better Game AI. We utilise genetic programming (GP) to generalise from and improve upon human game play. More importantly, the resulting representations are amenable to further authoring and development. We introduce a GP system for evolving game characters by utilising recorded human play. The system uses the platformerAI toolkit, detailed in section 3, and the JAVA genetic algorithm and genetic programming package (JGAP) [6]. JGAP provides a system to evolve agents when given a set of command genes, a fitness function, a genetic selector and an interface to the target

© Springer International Publishing Switzerland 2015
F. Pereira et al. (Eds.) EPIA 2015, LNAI 9273, pp. 292–297, 2015.
DOI: 10.1007/978-3-319-23485-4_30

application. Thereafter, our system generates players by creating and evolving JAVA program code which is fed into the PLATFORMERAI toolkit and evaluated using our player-based fitness function.

The rest of this paper is organised as follows. In section 2 we describe how our system derives from and improves upon the start of the art. Section 3 describes our system and its core components, including details of our fitness function. We conclude our work by describing our initial results and possible future work.

2 Background and Related Work

In practice, making a good game is achieved by a good concept and long iterative cycles in refining mechanics and visuals, a process which is resource consuming. It requires a large number of human testers to evaluate the qualities of a game. Thus, analysing tester feedback and incrementally adapting games to achieve better play experience is tedious and time consuming. This is where our approach comes into play by trying to minimise development, manual adaptation and testing time, yet allow the developer to remain in full control.

Agent Design initially no more than creating 2D shapes on the screen, e.g. the aliens in SPACEINVADERS. Due to early hardware limitations, more complex approaches were not feasible. With more powerful computers it became feasible to integrate more complex approaches from science. In 2002 Isla introduced the BEHAVIOURTREE (BT) for the game Halo, later elaborated by Champandard [2]. BT has become the dominant approach in the industry. BTs are a combination of a decision tree (DT) with a pre-defined set of node types. A related academic predecessor of the BT were the POSH dynamic plans of BOD [1,3].

Generative Approaches [4,7] build models to create better and more appealing agents. In turn, a generative agent uses machine learning techniques to increase its capabilities. Using data derived from human interaction with a game—referred to as human play traces—can allow the game to act on or *react* to input created by the player. By training on such data it is possible to derive models able to mimic certain characteristics of players. One obvious disadvantage of this approach is that the generated model only learns from the behaviour exhibited in the data provided to it. Thus, interesting behaviours are not accessible because they were never exhibited by a player.

In contrast to other generative agent approaches [7,9,15] our system combines features which allow the generation and development of truly novel agents. The first is the use of un-authored recorded player input as direct input into our fitness function. This allows the specification of agents only by playing. The second feature is that our agents are actual programs in the form of java code which can be altered and modified after evolving into a desired state, creating a white box solution. While Stanley and Miikkulainen[13] use neural networks (NN) to create better agents and enhance games using Neuroevolution, we utilise genetic programming [10] for the creation and evolution of artificial players in human readable and modifiable form. The most comparable approach is that of Perez et al.[9] which uses grammar based evolution to derive BTs given an initial set and structure of subtrees. In contrast, we start with a clean slate to evolve novel agents as directly executable programs.

3   Setting and Environment

Evolutionary algorithms have the potential to solve problems in vast search spaces, especially if the problems require multi-parameter optimisation [11, p.2]. For those problems humans are generally outperformed by programs [12]. Our GP approach uses a pool of program chromosomes P and evolves those in the form of decision trees (DTs) exploring the possible solution space. For our experiments the PLATFORMERAI toolkit (http://www.platformersai.com) was used. It consists of a 2D platformer game, similar to existing commercial products and contains modules for recording a player, controlling agents and modifying the environment and rules of the game.

The *Problem Space* is defined by all actions an agent can perform. Within the game, agent A has to solve the complex task of selecting the appropriate action each given frame. The game consists of A traversing a level which is not fully observable. A level is 256 spatial units long and A should traverse it left to right. Each level contains objects which act in a deterministic way. Some of those objects can alter the player's score, e.g. coins. Those bonus objects present a secondary objective. The goal of the game, move from start to finish, is augmented with the objective of gaining points. A can get points by collecting objects or jumping onto enemies. To make it comparable to the experience of similar commercial products we use a realistic time frame in which a human would need to solve a level, 200 time units. The level observability is limited to a 6×6 grid centred around the player, cf. Perez et al.[9].

Agent Control is handled through a 6-bit vector C: *left, right, up, down, jump* and *shoot|run*. The vector is required each frame, simulating an input device. However, some actions span more than one frame. This is a simple task for a human but quite complex to learn for an agent. One such example, the high jump, requires the player to press the jump button for multiple frames. Our system has a gene for each element of C plus 14 additional genes formed of five gene types: sensory information about the level or agent, executable actions, logical operators, numbers and structural genes. All those are combined on creation time into a chromosome represented as a DT using the grammar underlying the JAVA language. Structural genes allow the execution of n genes in a fixed sequence, reducing the combinatorial freedom provided by JAVA.

Evaluation of Fitness in our system is done using the Gamalyzer-based play trace metric which determines the fitness of individual chromosomes based on human traces as an evaluation criterion. For finding optimal solutions to a problem statistical fitness functions offer near-optimal results when optimality can be defined. We are interested in understanding and modelling human-like or human-believable behaviour in games. There is no known algorithm for measuring how human-like behaviour is; identifying this may even be computationally intractable. A near-best solution for the problem space of finding the optimal way through a level was given by Baumgarten [14] using the A^* algorithm. This approach produces agents which are extremely good at winning the level within a minimum amount of time but at the same time are clearly distinguishable from actual human players. For games and game designers a less distinguishable approach is normally more appealing—based on our initial assumptions.

4 Fitness Function

Based on the biological concept of selection, all evolutionary systems require some form of judgement about the quality of a specific individual—the fitness value of the entity. Our *Player Based Fitness* (PBF) uses multiple traces of human, t_h, and agent, t_a, players to derive a fitness value by judging their similarity. For that purpose we integrate the Gamalyzer Metric—a game independent measurement of the difference between two play traces. It is based on the syntactic edit distance d_{dis} between pairs of sequences of player inputs [8]. It takes pairs of sequences of events gathered during a game play along with designer-provided rules for comparing individual events and yields a numerical value in $[0, 1]$. Identical traces have distance $d_{dis} = 0$ and incomparably different traces $d_{dis} = 1$. Gamalyzer finds the least expensive way to turn one play trace into another by repeatedly deleting an event from the first trace, inserting an event of the second trace into the first trace, or changing an event of the first trace into an event of the second trace. The game designer or analyst must also provide a comparison function which describes the difficulty of changing one event into another. The other important feature of Gamalyzer, warp window ω, is a constraint that prevents early parts of the first trace from comparing against late parts of the second. This is important for correctness (a running leap at the beginning of the level has a very different connotation from a running leap at the pole at the end of each stage). For our purpose, only the input commands players use to control the agent are encoded—the six commands introduced earlier. This allows us to compare against direct controller input for future studies and to help designers sitting in front of the controls analysing the resulting character program. The PBF currently offers two parameters: the chunk size, cpf, and the warp window size, ω. The main advantage over a pure statistical fitness function is that a designer can feed our system specific play traces of human players without having to modify implicit values of a fitness score.

To make a stronger emphasis on playing the game well, we create a multi-objective problem using an aggregation function g to take Δd—the moved distance—and the fitness $f(a)$ for an agent using the playerbased metric PBF into account, see formula (1). Using g we were able to put equal focus on the trace metric, $f_{ptm} \in [0 \ldots 1] \subset \mathbb{R}$, and the advancement along the game, $\Delta d \in [0 \ldots 256] \subset \mathbb{N}$.

$$f(a) = g(f_{ptm}(t_a, t_h), \Delta d) \tag{1}$$

5   Preliminary Results and Future Work

Using our experimental configuration and the PBF fitness function we are now able to execute, evaluate and compare platformerAI agents against human traces. We are using the settings supplied in table 1. As a selection mechanism, the weighted roulette wheel is used and we additionally preserve the fittest individual of a generation. We use single point tree branch crossover on two selected parent chromosomes and expose the resulting child to a single point mutation before it is put into the new generation. Figure 1 illustrates the convergence of the program pool against the global optimum. Good solutions are on average reached after 700 generations, when an agent finishes the given level. Our first experiments show that our approach is able to train on and converge against raw human play traces without stopping at local optima, visible in the two dents of the averaged fitness (black) diverging from the fittest individual (red). A next step would be to investigate the generated modifiable programs further and analyse their benefit in understanding players better. However, our current solution already offers a way to design agents for a game by simply playing it and creating learning agents from those traces. Other possible directions could be expansion of the model underlying Gamalyzer to model specific events within the game rather than pure input actions. This should provide interesting feedback and offer a better matching of expressed player behaviour and model generation. Our current agent model consists of an unweighted tree representation containing program genes. Currently subtrees are not taken into consideration when calculating the fitness of an individual. By including those weights it would be possible to narrow down the search space of good solutions for game characters dramatically, also potentially reducing the bloat of the DT. So, to enhance the quality of our

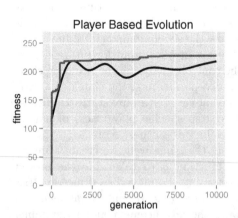

Fig. 1. The evolved agents' fitness using PBF (10000 generations), in red the fittest individuals, in black the averaged fitness of all agents per generation.

Table 1. GP parameters used in our system.

Parameter	Value
Initial Population Size	100
Selection	Weighted Roulette Wheel
Genetic Operators	Branch Typing CrossOver and Single Point Mutation
Initial Operator probabilities	0.6 crossover, 0.2 new chromosomes, 0.01 mutation, fixed
Survival	Elitism
Function Set	$ifelse$, not, &&, ||, sub, $IsCoinAt$, $IsEnemyAt$, $IsBreakAbleAt$, ...
Terminal Set	Integers $[-6,6]$, $\leftarrow$, $\rightarrow$, $\downarrow$, $IsTall$, $Jump$, $Shoot$, Run $Wait$, $CanJump$, $CanShoot$, ...

reproduction component we believe it might be interesting to investigate the applicability of behavior-programming for GP (BPGP) [5] into our system.

References

1. Bryson, J.J., Stein, L.A.: Modularity and design in reactive intelligence. In: Proceedings of the 17th International Joint Conference on Artificial Intelligence, pp. 1115–1120. Morgan Kaufmann, Seattle, August 2001
2. Champandard, A.J.: AI Game Development. New Riders Publishing (2003)
3. Gaudl, S.E., Davies, S., Bryson, J.J.: Behaviour oriented design for real-time-strategy games - an approach on iterative development for starcraft ai. In: Proceedings of the Foundations of Digital Games, pp. 198–205. Society for the Advancement of Science of Digital Games (2013)
4. Holmgard, C., Liapis, A., Togelius, J., Yannakakis, G.: Evolving personas for player decision modeling. In: 2014 IEEE Conference on Computational Intelligence and Games (CIG), pp. 1–8, August 2014
5. Krawiec, K., O'Reilly, U.M.: Behavioral programming: a broader and more detailed take on semantic gp. In: Proceedings of the 2014 Conference on Genetic and Evolutionary Computation, pp. 935–942. ACM (2014)
6. Meffert, K., Rotstan, N., Knowles, C., Sangiorgi, U.: Jgap-java genetic algorithms and genetic programming package, September 2000. http://jgap.sf.net (last viewed: January 2015)
7. Ortega, J., Shaker, N., Togelius, J., Yannakakis, G.N.: Imitating human playing styles in super mario bros. Entertainment Computing 4(2), 93–104 (2013)
8. Osborn, J.C., Mateas, M.: A game-independent play trace dissimilarity metric. In: Proceedings of the Foundations of Digital Games. Society for the Advancement of Science of Digital Games (2014)
9. Perez, D., Nicolau, M., O'Neill, M., Brabazon, A.: Evolving behaviour trees for the mario ai competition using grammatical evolution. In: Di Chio, C., et al. (eds.) EvoApplications 2011, Part I. LNCS, vol. 6624, pp. 123–132. Springer, Heidelberg (2011)
10. Poli, R., Langdon, W.B., McPhee, N.F., Koza, J.R.: A field guide to genetic programming. Lulu. com (2008)
11. Schwefel, H.P.P.: Evolution and optimum seeking: the sixth generation. John Wiley & Sons, Inc. (1993)
12. Smit, S.K., Eiben, A.E.: Comparing parameter tuning methods for evolutionary algorithms. In: IEEE Congress on Evolutionary Computation, CEC 2009, pp. 399–406. IEEE (2009)
13. Stanley, K.O., Miikkulainen, R.: Evolving neural networks through augmenting topologies. Evolutionary Computation 10, 99–127 (2002)
14. Togelius, J., Karakovskiy, S., Baumgarten, R.: The 2009 mario ai competition. In: 2010 IEEE Congress on Evolutionary Computation (CEC), pp. 1–8. IEEE (2010)
15. Togelius, J., Yannakakis, G., Karakovskiy, S., Shaker, N.: Assessing believability. In: Hingston, P. (ed.) Believable Bots, pp. 215–230. Springer, Heidelberg (2012)

Memetic Algorithm for Solving the 0-1 Multidimensional Knapsack Problem

Abdellah Rezoug[1]([✉]), Dalila Boughaci[2], and Mohamed Badr-El-Den[3]

[1] Department of Computer Science, University Mhamed Bougarra of Boumerdes,
Boumerdes, Algeria
abdellah.rezoug@gmail.com
[2] FEI, Department of Computer Science Beb-Ezzouar, USTHB, Algiers, Algeria
dboughaci@usthb.dz, dalila_info@yahoo.fr
[3] School of Computing, Faculty of Technology, University of Portsmouth,
Portsmouth, UK
mohamed.bader@port.ac.uk

Abstract. In this paper, we propose a memetic algorithm for the Multidimensional Knapsack Problem (MKP). First, we propose to combine a genetic algorithm with a stochastic local search (GA-SLS), then with a simulated annealing (GA-SA). The two proposed versions of our approach (GA-SLS and GA-SA) are implemented and evaluated on benchmarks to measure their performance. The experiments show that both GA-SLS and GA-SA are able to find competitive results compared to other well-known hybrid GA based approaches.

Keywords: Multidimensional knapsack problem · Stochastic local search · Genetic algorithm · Simulated annealing · Local search · Memetic algorithm

1 Introduction

The Multidimensional Knapsack Problem (MKP) is a strong NP-hard combinatorial optimization problem [14]. The MKP has been extensively considered because of its theoretical importance and wide range of applications. Many practical engineering design problems can be formulated as MKP such as: the capital budgeting problem [17], the project selection [2] and so on.

The solutions for MKP can be classified into exact, approximate and hybrid. The exact solutions are used for problems of small size. Branch and bound, branch and cut, linear, dynamic and quadratic programming, etc. are the principal exact methods used for solving MKP [13,21]. The approximate solutions are used when the data size is high but it is not sure to obtain the optimal results. They are mainly based on heuristics such as: simulated annealing, tabu search, genetic algorithm, ant colony particle swarm, harmony search, etc [5,6,20]. The hybrid solutions combine two or more exact or/and approximate solutions. These solutions are the most used in the field of optimization and especially for MKP such as [4,8–12,18] and so on.

© Springer International Publishing Switzerland 2015
F. Pereira et al. (Eds.) EPIA 2015, LNAI 9273, pp. 298–304, 2015.
DOI: 10.1007/978-3-319-23485-4_31

In this paper, we propose a memetic algorithm for MKP. We developed two versions of our method for MKP. The first denoted GA-SLS is a GA combined with the stochastic local search (SLS) [3]. The second denoted GA-SA is a combination of GA with the simulated annealing (SA) [16]. The two versions are implemented and evaluated on some well-known benchmarks for MKP where the sizes of benchmarks arrange from small to large. A comparative study is done with a pure GA and some algorithms for MKP. The objective is to show the impact of the local search in the performance of the memetic approach.

The rest of the paper is organized as follows. Section 2 gives the MKP model. The proposed Memetic approaches are detailed in Section 3. Section 4 describes the experiments. Finally, Section 5 concludes the paper.

2   The Multidimensional Knapsack Problem

The MKP is composed of N items and a knapsack with m different capacities b_i where $i \in \{1, \ldots, m\}$. Each item j where $j \in \{1, \ldots, n\}$ has a profit c_j and can take a_{ij} of the capacity i of the knapsack. The goal is to pack the items in the knapsack so as to maximize the profits of items without exceeding the capacities of the knapsack. The MKP is modeled as the following integer program:

$$\text{Maximize} \sum_{j=1}^{n} c_j x_j \tag{1}$$

$$\text{Subject to} : \sum_{j=1}^{n} a_{ij} x_j \leq b_i \quad i \in \{1 \ldots m\} \tag{2}$$

$$x_j \in \{0, 1\} \quad j \in \{1 \ldots n\}$$

3   The Proposed Approaches for MKP

Two versions of memetic approach have been studied. The first one is the Genetic Algorithm-Stochastic Local Search (GA-SLS) where GA is combined with SLS. The second one is the Genetic Algorithm-Simulated Annealing (GA-SA) which is GA combined with SA. The structures of both approaches are similar in the GA part. The difference is in the local search. The GA-SLS applies SLS while GA-SA applies SA. Their process consists in: Create the initial population P using the Random Key method (RK)[1] and initialize $Q = \{\}$, NI and $T = T_0$ (T for GA-SA only). Select two parents X_1 and X_2 that are the two best individuals in P and $X_1, X_2 \notin Q$. Exchange NCB items between the parents X_1 and X_2 to produce two new infeasible offspring X_1' and X_2', then if conflict exists in X_1' or X_2', repeatedly remove either the worst items or an item chosen randomly according to a probability rp. Push the two parents X_1 and X_2 in Q. Apply the local search (SLS for GA-SLS or SA for GA-SA) on offspring X_1', X_2'. Find the best individuals X_{best} in P and replace randomly a number of items in X_1' and X_2' by items in X_{best}. If the quality of X_1' and X_2' is better than the two

---

Algorithm 1. GA-SLS Algorithm.

Require: An MKP instance, NI and $Q = \phi$.
Ensure: An best solution found X^*.
1: Create the initial population P by the RK method.
2: **for** $(Cpt = 1$ to $NI)$ **do**
3:      Selection of the two best individuals X_1, X_2 in P and $X_1, X_2 \notin Q$.
4:      Crossover X_1, X_2 to produce offsprings X'_1, X'_2
5:      Repair offsprings X'_1, X'_2
6:      Apply the local search method on X'_1, X'_2
7:      Mutation on X'_1, X'_2 with X_{best} of P
8:      $X_{worst} \longleftarrow$ the worst individual in P
9:      **if** $(f(X'_1) > f(X_{worst}))$ **then**
10:          $P = P - \{X_{worst}\}$
11:          $P = P \cup \{X'_1\}$
12:      **end if**
13:      $X_{worst} \longleftarrow$ the worst individual in P
14:      **if** $(f(X'_2) > f(X_{worst}))$ **then**
15:          $P = P - \{X_{worst}\}$
16:          $P = P \cup \{X'_2\}$
17:      **end if**
18:      $Q = Q \cup \{X_1, X_2\}$
19: **end for**
20: Return the best individual found.

---

worst individuals in P, then they replace them. If the number of iterations NI is not attend then go to *STEP 2.*. Otherwise return the best individual in P. The GA-SLS and GA-SA can be expressed by Algorithm 1.

4   The Experiments

GA, GA-SA and GA-SLS were implemented in C++ on 2 GHz Intel Core 2 Duo processor and 2 GB RAM. They were tested on the OR-Library [22] 54 benchmarks, with $m = 2$ to 30 and $n = 6$ to $n = 105$ and on the OR-Library GK [22] with $m = 15$ to $m = 50$ and $n = 100$ to $n = 1500$. In all experiments the parameters are chosen empirically such as: the number of iteration $NI = 30000$, the population size $PS = 100$, the waiting time $WT = 50$, the number of crossing bites $NCB = 1/10$. the initial temperature $T_0 = 50$, the walk probability $wp = 0.93$, the number of local iteration $N = 100$ and the number of runs is 30.

Results for the SAC-94 Standard Instances. The average fitness (*Result*), the average gap (*GAP*), the best (*Best*) and the worst fitness (*Worst*), the number of success runs (*NSR*), the number of success instance (*NSI*) and the rate of success runs (*RSR*) have been recorded by analyzing the recorded obtained fitness. Also, the average CPU runtime (*Time*) has been calculated. All the results and statistics computed by the GA, GA-SA and GA-SLS are reported in Tables 1-2. From results, GA resolved to optimality one instance of 54 with average gap of 4,454 %, GA-SA 35 instances with a global gap of 0,093 % and GA-SLS 39 instances with a global gap of 0,0221 %. GA-SLS reached the optimum at least once in 50 instances followed by GA-SA in 49 instances then GA in 18 instances. The *RSR* show that GA-SLS totally solved instances of groups *hp, pb*

Table 1. Comparison of GA, GA-SA and GA-SLS on SAC-94 datasets.

Dataset	Opt	GA Result	GA GAP	GA-SA Result	GA-SA GAP	GA-SLS Result	GA-SLS GAP
hp	3418	3381,07	1,080	3418	0	3418	0
	3186	3120,63	2,052	3186	0	3186	0
Average	3302	3250,85	1,566	**3302**	**0**	**3302**	**0**
	3090	3060,27	0,962	3090	0	3090	0
	3186	3139,13	1,471	3186	0	3186	0
pb	95168	93093,5	2,180	95168	0	95168	0
	2139	2079,93	2,762	2139	0	2139	0
	776	583,767	24,772	776	0	776	0
	1035	1018,13	1,630	1035	0	1035	0
Average	17565,666	17162,454	5,629	**17565,666**	**0**	**17565,666**	**0**
	87061	86760,1	0,346	87061	0	87061	0
	4015	4015	0	4015	0	4015	0
pet	6120	6091	0,474	6120	0	6120	0
	12400	12380,3	0,159	12400	0	12400	0
	10618	10560,9	0,538	10609,1	0,084	10608,6	0,089
	16537	16373,9	0,986	16528,1	0,054	16528,3	0,053
Average	22791,833	22696,866	0,417	**22788,866**	**0,023**	22788,816	0,024
sento	7772	7606,03	2,135	7772	0	7772	0
	8722	8569,7	1,746	8721,2	0,009	8722	0
Average	8247	8087,865	1,941	8246,6	0,005	**8247**	**0**
	141278	141263	0,011	141278	0	141278	0
	130883	130857	0,020	130883	0	130883	0
	95677	94496,2	1,234	95677	0	95677	0
weing	119337	118752	0,490	119337	0	119337	0
	98796	97525,3	1,286	98796	0	98796	0
	130623	130590	0,025	130623	0	130623	0
	1095445	1086484,2	0,818	1094579,6	0,079	1095432,7	0,0011
	624319	581683	6,829	623727	0,095	624319	0
Average	304545,375	297707,062	1,339	304362,625	0,022	**304543,22**	**0,0001**
	4554	4530,03	0,526	4554	0	4554	0
	4536	4506,77	0,644	4536	0	4536	0
	4115	4009,37	2,567	4115	0	4115	0
	4561	4131,07	9,426	4561	0	4561	0
	4514	4159,73	7,848	4514	0	4514	0
	5557	5491,73	1,175	5557	0	5557	0
	5567	5428,37	2,490	5567	0	5567	0
	5605	5509,43	1,705	5605	0	5605	0
	5246	5104,5	2,697	5246	0	5246	0
	6339	6014,23	5,123	6339	0	6339	0
	5643	5234,33	7,242	5643	0	5643	0
	6339	5916	6,673	6339	0	6339	0
	6159	5769,5	6,324	6159	0	6159	0
	6954	6495,6	6,592	6954	0	6954	0
weish	7486	6684,6	10,705	7486	0	7486	0
	7289	6878,4	5,633	7289	0	7289	0
	8633	8314,73	3,687	8629,5	0,041	8633	0
	9580	9146,5	4,525	9559,63	0,213	9568,63	0,119
	7698	7223,17	6,168	7698	0	7698	0
	9450	8632,1	8,655	9448,63	0,014	9449,37	0,007
	9074	8114,4	10,575	9073,23	0,008	9073,33	0,007
	8947	8321,17	6,995	8926,73	0,227	8938,83	0,091
	8344	7603,77	8,871	8321,97	0,264	8318,93	0,3
	10220	9685,77	5,227	10152,9	0,657	10164,2	0,546
	9939	9077,9	8,664	9900,07	0,392	9910,73	0,284
	9584	8728,87	8,922	9539,4	0,465	9560,53	0,245
	9819	8873,7	9,627	9777,9	0,419	9802,03	0,173
	9492	8653,57	8,833	9423,87	0,718	9442,17	0,525
	9410	8466,67	10,025	9359,5	0,537	9369,5	0,430
	11191	10250,1	8,408	11106,3	0,757	11128,7	0,557
Average	7394,833	6898,536	6,218	7379,386	0,157	**7386,772**	**0,109**

Table 2. Results of NSR, RSR and Time parameters obtained by GA, GA-SA and GA-SLS.

	GA			GA-SA			GA-SLS		
	NSR	RSR	Time	NSR	RSR	Time	NSR	RSR	Time
hp	1	1,67	1,798	2	100	6,077	2	100	7,101
pb	4	3,33	1,811	6	100	3,443	6	100	4,681
pet	4	32,78	1,395	6	81,67	11,296	6	80,55	12,179
sento	0	0,00	2,616	2	66,67	46,584	2	100	24,277
weing	6	39,58	1,669	8	76,25	10,259	8	99,58	10,586
weish	3	4,55	1,620	25	66,89	17,352	26	76,88	17,146
Average	18	13,65	1,818	49	81,91	15,835	50	92,83	12,662

Table 3. Results of the approaches test on the GK dataset.

Dataset		GA		GA-SA		GA-SLS	
Instance	Optimal	Result	Gap	Result	Gap	Result	Gap
1	3766	3673,5	2,456	3704,3	1,638	3704,2	1,641
2	3958	3860,7	2,458	3894,8	1,596	3897,7	1,523
3	5656	5511,5	2,554	5538,8	2,072	5535,7	2,127
4	5767	5630,6	2,365	5655,2	1,938	5655,4	1,935
5	7560	7351,3	2,76	7395,1	2,181	7391,3	2,231
6	7677	7505,7	2,231	7528,4	1,935	7528,1	1,939
7	19220	18612,1	3,162	18691	2,752	18692,4	2,745
8	18806	18330,2	2,53	18393	2,196	18392,4	2,199
9	58091	56198,5	3,257	56371,1	2,96	56381,4	2,943
10	57295	55837,9	2,543	55959,3	2,331	55961,9	2,326
Average	18779,6	18251,2	2,632	18313,1	2,484	18314,05	2,479

and *sento* followed by GA-SA. GA-SLS obtained a total *RSR* better than GA-SA (92,83% and 81,91%, respectively). At the same time, GA-SA and GA-SLS widely surpass GA (13,65%). *RSR* shows that hybridization of GA with SA has improved the success rate of 79,18% and its hybridization with SLS of 68,49%. From Table 2, GA is the fastest with an global average CPU time of 1.818 *sec*.

Results for the Ten Large Instances. From results on the GK shown in Table 3 GA-SA has the best value of *Result* and *GAP* for 1, 3, 5, 6 and 8 instances. GA-SLS has the best value of *Result* and *GAP* for instances 2, 4, 7, 9 and 10. Global, GA-SLS has the best performance for all instances with an total average *GAP* of 2.479 %. GA-SA has almost the same performance with average *GAP* of 2.484 %. Also, GA is not very far from GA-SA and GA-SLS with an total average *GAP* of 2.632 %.

Comparison with Other GA Approaches. We compared results of the proposed GA-SA and GA-SLS to other approaches. The results of the KHBA [15], COTRO [7], TEVO [19], CHEBE [6] and HGA [10] were obtained from [10]. From Table 4 GA-SA and GA-SLS gave improved results compared to KHBA, COTRO and TEVO, for almost all instances. GA-SA and GA-SLS were able to find the optimal solutions to 6, and 3 of 7 problems respectively. Furthermore, GA-SA performs results quite similar to CHEBE and HGA.

Table 4. Comparison of GA-SA and GA-SLS with some GA-based approaches.

		KHBA	COTRO	TEVO	CHBE	HGA	GA-SA	GA-SLS
problem	Optimum	Sol A.	Sol A.	Sol A.	Sol A.	Sol A.	Sol A.	Sol A.
sento1	7772	7626	7767,9	7754,2	**7772**	**7772**	**7772**	**7772**
sento2	8722	8685	8716,3	8719,5	**8722**	**8722**	8721,2	**8722**
weing7	1095445	1093897	1095296,1	1095398,1	**1095445**	**1095445**	1094579,6	1095432,7
weing8	624319	613383	622048,1	622021,3	**624319**	**624319**	623727	**624319**
weish23	8344	8165,1	8245,8	8286,7	**8344**	**8344**	8321,97	**8344**
hp1	3418	3385,1	3394,3	3401,6	**3418**	**3418**	**3418**	**3418**
pb2	3186	3091	3131,2	3112,5	**3186**	**3186**	**3186**	**3186**

5 Conclusion

In this paper we addressed the multidimensional knapsack problem (MKP). We proposed, compared and tested two combinations: GA-SLS and GA-SA. GA-SLS combines the genetic algorithm and the stochastic local search (SLS) while GA-SA uses the simulated annealing (SA) instead of SLS. The experiments have shown the performance of our methods for MKP. Also, the hybridization of GA with local search methods allows to greatly improving its performance. As perspectives, we plan to study the impact of local search method when used with other evolutionary approaches such as: harmony search and particle swarm.

References

1. Bean, J.C.: Genetics and random keys for sequencing and optimization. ORSA Journal of Computing **6**(2), 154–160 (1994)
2. Beaujon, G.J., Martin, S.P., McDonald, C.C.: Balancing and optimizing a portfolio of R&D projects. Naval Research Logistics **48**, 18–40 (2001)
3. Boughaci, D., Benhamou, B., Drias, H.: Local Search Methods for the Optimal Winner Determination Problem in Combinatorial Auctions. Math. Model. Algor. **9**(1), 165–180 (2010)
4. Chih, M., Lin, C.J., Chern, M.S., Ou, T.Y.: Particle swarm optimization with time-varying acceleration coefficients for the multidimensional knapsack problem. Applied Mathematical Modelling **38**, 1338–1350 (2014)
5. Cho, J.H., Kim, Y.D.: A simulated annealing algorithm for resource-constrained project scheduling problems. Operational Research Society **48**, 736–744 (1997)
6. Chu, P., Beasley, J.: A Genetic Algorithm for the Multidimensional Knapsack Problem. Heuristics **4**, 63–86 (1998)
7. Cotta, C., Troya, J.: A Hybrid Genetic Algorithm for the 0–1 Multiple Knapsack problem. Artificial Neural Nets and Genetic Algorithm **3**, 250–254 (1994)
8. Deane, J., Agarwal, A.: Neural, Genetic, And Neurogenetic Approaches For Solving The 0–1 Multidimensional Knapsack Problem. Management & Information Systems - First Quarter 2013 **17**(1) (2013)
9. Della Croce, F., Grosso, A.: Improved core problem based heuristics for the 0–1 multidimensional knapsack problem. Comp. & Oper. Res. **39**, 27–31 (2012)
10. Djannaty, F., Doostdar, S.: A Hybrid Genetic Algorithm for the Multidimensional Knapsack Problem. Contemp. Math. Sciences **3**(9), 443–456 (2008)
11. Feng, L., Ke, Z., Ren, Z., Wei, X.: An ant colony optimization approach for the multidimensional knapsack problem. Heuristics **16**, 65–83 (2010)

12. Feng, Y., Jia, K., He, Y.: An Improved Hybrid Encoding Cuckoo Search Algorithm for 0–1 Knapsack Problems. Computational Intelligence and Neuroscience, ID 970456 (2014)
13. Fukunaga, A.S.: A branch-and-bound algorithm for hard multiple knapsack problems. Annals of Operations Research **184**, 97–119 (2011)
14. Garey, M.R., Johnson, D.S.: Computers and intractability: A guide to the theory of NP-completeness. W. H. Freeman & Co, New York (1979)
15. Khuri, S., Bäck, T., Heitkötter, J.: The zero-one multiple knapsack problem and genetic algorithms. In: Proceedings of the ACM Symposium on Applied Computing, pp. 188–193 (1994)
16. Kirkpatrick, S., Gelatt, C.D., Vecchi, P.M.: Optimization By Simulated Annealing. Science **220**, 671–680 (1983)
17. Meier, H., Christofides, N., Salkin, G.: Capital budgeting under uncertainty-an integrated approach using contingent claims analysis and integer programming. Operations Research **49**, 196–206 (2001)
18. Tuo, S., Yong, L., Deng, F.: A Novel Harmony Search Algorithm Based on Teaching-Learning Strategies for 0–1 Knapsack Problems. The Scientific World Journal Article ID 637412, 19 pages (2014)
19. Thiel, J., Voss, S.: Some Experiences on Solving Multiconstraint Zero-One Knapsack Problems with Genetic Algorithms. INFOR **32**, 226–242 (1994)
20. Vasquez, M., Vimont, Y.: Improved results on the 0–1 multidimensional knapsack problem. Eur. J. Oper. Res. **165**, 70–81 (2005)
21. Yoon, Y., Kim, Y.H.: A Memetic Lagrangian Heuristic for the 0–1 Multidimensional Knapsack Problem. Discrete Dynamics in Nature and Society, Article ID 474852, 10 pages (2013)
22. http://people.brunel.ac.uk/~mastjjb/jeb/orlib/mknapinfo.html

Synthesis of In-Place Iterative Sorting Algorithms Using GP: A Comparison Between STGP, SFGP, G3P and GE

David Pinheiro$^{(\boxtimes)}$, Alberto Cano, and Sebastián Ventura

Department of Computer Science and Numerical Analysis, University of Córdoba,
Córdoba, Spain
{dpinheiro,acano,sventura}@uco.es

Abstract. This work addresses the automatic synthesis of in-place, iterative sorting algorithms of quadratic complexity. Four approaches (Strongly Typed Genetic Programming, Strongly Formed Genetic Programming, Grammar Guided Genetic Programming and Grammatical Evolution) are analyzed and compared considering their performance and scalability with relation to the size of the primitive set, and consequently, of the search space. Performance gains, provided by protecting composite data structure accesses and by another layer of knowledge into strong typing, are presented. Constraints on index assignments to grammar productions are shown to have a great performance impact.

Keywords: Automatic algorithm synthesis · Genetic programming · Sorting

1 Introduction

This work compares four approaches, namely Strongly Typed Genetic Programming (STGP), Strongly Formed Genetic Programming (SFGP), Grammar Guided Genetic Programming (G3P) and Grammatical Evolution (GE), at evolving sorting algorithms. Special emphasis is given on their ability to scale well in spite of bigger primitive sets.

We restrict ourselves to iterative (non-recursive), in-place, comparison based sorting algorithms, expecting quadratic running times ($\mathcal{O}(n^2)$) and constant ($\mathcal{O}(1)$) additional memory. We make no assumptions about stability and adaptability of the evolved algorithms. Bloat analysis and solution will be left for future work.

2 Implementation and Experimental Context

The experiments were done on top of the EpochX [1] GP java library and were designed to be as fair as possible. However our goal is to get a grasp of the

© Springer International Publishing Switzerland 2015
F. Pereira et al. (Eds.) EPIA 2015, LNAI 9273, pp. 305–310, 2015.
DOI: 10.1007/978-3-319-23485-4_32

Table 1. GP parameters used in the experiments

	STGP	SFGP	G3P	GE
Population size	500			
Initialization	Grow initializer			
Selection	Tournament selection with a small size of 2 to prevent a lack of population diversity and to have a small selection pressure			
Crossover operator	Subtree Crossover		Whigham Crossover	One point Crossover
Crossover probability	90%			
Mutation operator	Subtree Mutation		Whigham Mutation	Point Mutation
Mutation probability	10%			
Elitism	Only one individual, to keep the best genome seen so far but with minimal impact on genetic diversity			
Reproduction	No reproduction. 90% of individuals are obtained by crossover and the remaining 10% by mutation			
Max Initial Depth*	10		24	
Max Depth*	10		32	
Number of generations	50			
Number of Runs	500			
Fitness Function	Levenshtein Distance			

* As there is no easy direct relation between tree sizes in strongly typed and grammar guided GP (since the first uses expression trees in which all nodes contribute to the computation and the second uses parse trees in which only leafs contribute) several maximum tree depths were tested and the ones that showed the best results, for the bigger primitive sets, were chosen.

scalability related to the primitive set size, therefore we will not delve into fine tuning the choices made for operators and values (shown in Table 1).

The nonterminals used are shown in Table 2. In the first three columns a number is attributed to every syntactic element to help understand the results, followed by the name and a description. The node data type and the needed child data types, used by STGP and SFGP, and the node type and child node types, used by SFGP, fulfill the last four columns. The approaches based on grammars do not need to define data or node types, for the very grammar contains the specification of the requirements and restrictions on the types of data and syntactic form of the solutions. The strongly typed approaches were designed to achieve side effects, to change global variables, for that reason the Statement nonterminals return data type Void. SFGP nonterminal nodes belong to only three supertypes, namely CodeBlock, Statement and Expression. The implementation uses polymorphism then whenever it is required to use an Expression, for example, any Expression subtype can be used. Only two terminals were used, the minimum for quadratic algorithms, which in practice work as indexes to array elements (next section tests the use of more terminals). Using insight from human-made algorithms, the loops were restricted to a small set of widely used variants, like looping from a specified position of the array, ascending or descending. To prevent infinite loops, a limit of 100 cycles in each loop was set.

Table 2. Non-terminal syntactic elements

#	Name	Description	Data type	Child data types	Node super-type	Child node types
1	Swap	Swap two given array elements	Void	Array, Int, Int	Statement	Variable, Expression, Expression
2	For Each	Loop for each element of the array	Void	Array, Int, Void	Statement	Variable, Variable, CodeBlock
3	If Then	Conditional statement	Void	Bool, Void	Statement	Expression, CodeBlock
4	Less Than	Relational less than operator, <	Bool	Int, Int	Expression	Expression, Expression
5	For Each From	Loop for each array element, starting from a given position	Void	Array, Int, Int, Void	Statement	Variable, Variable, Variable, CodeBlock
6	Swap Next	Swap with the next element	Void	Array, Int	Statement	Variable, Expression
7	And	Logical AND operator	Bool	Bool, Bool	Expression	Expression, Expression
8	Decreasing For Each From	Decreasing loop, for each array element starting from a given position	Void	Array, Int, Int, Void	Statement	Variable, Variable, Variable, CodeBlock
9	Swap Previous	Swap with the previous element	Void	Array, Int	Statement	Variable, Expression
10	If Then Else	Conditional statement with the 'else' clause	Void	Bool, Void, Void	Statement	Expression, CodeBlock, CodeBlock
11	Greater Than	Relational greater than operator, >	Bool	Int, Int	Expression	Expression, Expression
12	Lesser Than or Equal	Relational less than or equal operator, <=	Bool	Int, Int	Expression	Expression, Expression
13	Greater Than or Equal	Relational greater than or equal operator, >=	Bool	Int, Int	Expression	Expression, Expression
14	Equal	Relational equality, ==	Bool	Int, Int	Expression	Expression, Expression
15	Unequal	Relational inequality, !=	Bool	Int, Int	Expression	Expression, Expression
16	Not	Logical NOT operator	Bool	Bool	Expression	Expression
17	Or	Logical OR operator	Bool	Bool, Bool	Expression	Expression, Expression
18	Decreasing For Each	Decreasing loop, for each array element from the last to the first	Void	Array, Int, Void	Statement	Variable, Variable, CodeBlock

The experiments use sets of 4, 5, 7, 10 and 18 nonterminal primitives. They were chosen in order to give a perspective of the asymptotic behavior. Set sizes correspond to the numbers on the first column of Table 2. For example the set of size 7 uses the nonterminals numbered 1 to 7, inclusive. Starting from the set of size 4 it is possible to obtain a simple Selection Sort algorithm and thereafter Bubble Sort and Insertion Sort.

The fitness function is defined by the Levenshtein Distance (an error measure that needs to be minimized). Every evolved program runs against the five arrays presented in [2]. If the program correctly sorts all of them, it runs against 30 arrays of random sizes between 10 and 20, filled with random integers between 0 and 100.

3 Experiments, Results and Analysis

To analyze the experiments we used the minimal computational effort required to find a solution with 99% confidence, presented in [3], but without the ceiling operator, as suggested in [4]. Confidence intervals at 95% (shown between parentheses in Table 3) were calculated using the Wilson score method [5].

Performance and scalability. The performance and scalability of the approaches with growing primitive set sizes are shown in Figure 1 and compiled in the lines titled UAA (Unprotected Array Accesses) in Table 3. In our setup, SFGP and G3P provided the best performance. Both G3P and GE reveal the particularity that the set of size 18 presents more useful constructs to evolve sorting algorithms than the set of size 10.

Protection against out of bounds array accesses. A recurrent situation that happens when using indexed data types, for example arrays, is that the indexes can get out of bounds of the data type when used inside some of the loops, causing run-time exceptions. In this experiment we protected against out of bounds array accesses, using the % (mod) operator against the size of the array, to ensure that the index is always in the correct bounds, and obtained the results presented in Figure 2 and Table 3. The important positive impact that this tweak had on the performance and scalability of the approaches can be ascertained by comparing lines named PAA (Protected Array Accesses) and UAA.

Influence of grammar context insensitivity. Context-free Grammars (CFG), used in G3P and GE, show a lack of expressiveness to describe semantic constraints[1] [6]. Their context insensitivity can have an appreciable negative impact on the size of the search space, especially in the presence of loops and swaps that repeatedly require the same index. As our system doesn't allow us to specify that certain terminal (index) assignments should be repeated in a given nonterminal, we obtained the same result changing the grammar so that the rules which require more than one index are split into two, one specifically for the index i, another specifically for the index j[2]. This acts as a kind of context sensitivity, forcing these constructs to always correctly match the indexes. The results, presented in Figure 3 and lines PAACS (Protected Array Accesses with Context Sensitivity) of Table 3, attest that this has a huge positive impact on the performance of grammar guided approaches, even to the point that almost all runs produce a correct individual.

Performance and scalability with bigger terminal sets. The last experiment tests the performance and scalability of SFGP in the presence of bigger terminal sets, the same number of terminals as nonterminals, between 4 and 18 of each. The terminals consist of integers of node type Variable. From Figure 4 and line PAANT (Protected Array Accesses with N Terminals) of Table 3 one can see that the number of terminals has an important negative impact in the performance but nevertheless SFGP remains scalable, showing almost the same performance for sets of size 10 and 18. This gives us confidence in the introduction of more data types in the evolutionary process, such as trees, graphs, stacks, etc., with the important goal of evolving not-in-place algorithms.

---

[1] For example, to define a loop, the grammar can state
```
<for>::= for(<index> = 0; <index> < array.length; <index>++);
<index>::= i | j | k;
```
which can be evaluated as
```
for(i = 0; j < array.length; k++){}
```
or any other combination of indexes that give an infinite loop. This situation could be overcome by the evolutionary system, indicating that the second and third indexes should be the same as the first.

[2] For example, the for loop was subdivided into one loop for each index:
```
<loop_i>::= for(i = 0; i < array.length; i++){}
<loop_j>::= for(j = 0; j < array.length; j++){}
```

Table 3. Non-terminal syntactic elements

		Nonterminal Set Size				
		4	5	7	10	18
SFGP	UAA	53,506 (47874-60134)	51,172 (33872-77608)	157,263 (139132-178635)	334,924 (284776-395599)	469,402 (387056-571624)
	PAA	10,800 (8862-13218)	9,364 (7777-11322)	38,537 (26893-55438)	142,002 (122458-165396)	90,467 (80221-102535)
	PAANT	52,907 (54206-67532)	79,542 (70970-89627)	340,904 (288677-404302)	1,224,842 (914425-1647128)	1,353,136 (995924-1845724)
STGP	UAA	349,348 (296183-413822)	274,308 (235619-320754)	797,004 (627793-1015894)	1,509,921 (1093099-2093891)	1,957,807 (1357643-2834324)
	PAA	39,955 (27707-57843)	46,810 (31529-69768)	258,519 (217208-308988)	291,612 (239267-356876)	227,370 (191936-270492)
G3P	UAA	8,109 (7170-9304)	11,380 (10212-12801)	75,596 (45923-124924)	342,219 (258684-454526)	270,186 (210357-348421)
	PAA	9,079 (7560-10950)	8,988 (10212-12801)	66,565 (45923-124924)	290,221 (220009-384361)	235,359 (184705-301109)
	PAACS	741 (579-1026)	741 (579-1026)	1,642 (1477-1837)	13,576 (10901-16978)	25,606 (19052-34550)
GE	UAA	16,265 (12812-20733)	20,147 (15475-26335)	66,565 (41660-106773)	1,209,531 (734764-1998785)	1,015,323 (731689-1414445)
	PAA	7,317 (6192-8682)	10,436 (8588-12733)	33,723 (24067-47439)	163,316 (79315-337577)	229,105 (98122-537000)
	PAACS	741 (579-1026)	741 (579-1026)	1,308 (1177-1464)	11,747 (9565-14486)	15,751 (12451-20007)

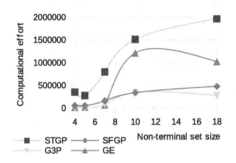

Fig. 1. Unprotected Array Accesses

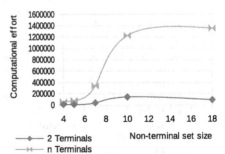

Fig. 2. Protected Array Accesses

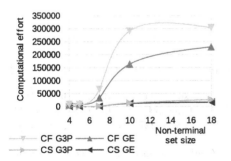

Fig. 3. Context Free vs Context Sensitive index assignment

Fig. 4. Use of n terminals in SFGP with Protected Array Accesses

310 D. Pinheiro et al.

4 Conclusions and Future Work

In general all the approaches showed good promise for Automatic Algorithm Synthesis, always converging in their scalability, and responding positively in terms of performance to the techniques introduced to reduce the search space.

SFGP revealed effective consistent performance and scalability improvements over STGP in all experiments. We argue that the introduction of types of nodes, in addition to data types, functions as another layer of restrictions on the search space, and causes a substantial and structural reduction of its size.

Protected accesses to the array, in order to prevent index out of bounds exceptions, revealed overall performance gains. The use of constraints on index assignment in the grammar rules of Context Free Grammars had a great positive impact on the performance and scalability of G3P and GE. The introduction of the same number of terminals as nonterminals revealed that, although the performance worsens, SFGP maintains its scalability. This result requires further analysis, but increases our confidence in the possibility of introducing high level abstract data structures to support the synthesis of not-in-place algorithms.

Future work will include: Implement the divide and conquer paradigm (recursion), along with a technique to assess algorithm performance, in order to enable and encourage the evolution of faster ($\mathcal{O}(n \log n)$) sorting algorithms; Introduce composite data structures as terminals, to enable the evolution of not-in-place algorithms; Add nonterminals obtained from algorithms developed by humans, to allow the application of GP to other areas beyond sorting; Analyze and reduce bloat, applying strategies to obtain compact unbloated algorithms.

Acknowledgments. This research was supported by the Spanish Ministry of Science and Technology, project TIN2014-55252-P, and by FEDER funds. This research was also supported by the Spanish Ministry of Education under FPU grant AP2010-0042.
We thank Tom Castle for kindly giving access to his SFGP code.

References

1. Otero, F., Castle, T., Johnson, C.: Epochx: genetic programming in java withstatistics and event monitoring. In: Proceedings of the 14th Annual Conference Companion on Genetic and Evolutionary Computation, pp. 93–100 (2012)
2. ONeill, M, Nicolau, M., Agapitos, A.: Experiments in program synthesis withgrammatical evolution: a focus on Integer Sorting. In: IEEE Congress on Evolutionary Computation (CEC), pp. 1504–1511 (2014)
3. Koza, J.R.: Genetic programming: on the programming of computers by means of natural selection, vol. 1. MIT press (1992)
4. Christensen, S., Oppacher, F.: An analysis of koza's computational effort statistic for genetic programming. In: Foster, J.A., Lutton, E., Miller, J., Ryan, C., Tettamanzi, A.G.B. (eds.) EuroGP 2002. LNCS, vol. 2278, pp. 182–191. Springer, Heidelberg (2002)
5. Walker, M., Edwards, H., Messom, C.H.: Confidence intervals for computational effort comparisons. In: Ebner, M., O'Neill, M., Ekárt, A., Vanneschi, L., Esparcia-Alcázar, A.I. (eds.) EuroGP 2007. LNCS, vol. 4445, pp. 23–32. Springer, Heidelberg (2007)
6. Orlov, M., Sipper, M.: FINCH: a system for evolving Java (bytecode). In: Genetic Programming Theory and Practice VIII, Springer, pp. 1–16 (2011)

Computational Methods in
Bioinformatics and Systems Biology

Variable Elimination Approaches
for Data-Noise Reduction in 3D QSAR Calculations

Rafael Dolezal[1,2(✉)], Agata Bodnarova[3], Richard Cimler[3], Martina Husakova[3],
Lukas Najman[1], Veronika Racakova[1], Jiri Krenek[1], Jan Korabecny[1,2],
Kamil Kuca[1,2], and Ondrej Krejcar[1]

[1] Center for Basic and Applied Research, Faculty of Informatics and Management,
University of Hradec Kralove, Rokitanskeho 62, 50003 Hradec Kralove, Czech Republic
{rafael.dolezal,lukas.najman,veronika.racakova,jiri.krenek,
jan.korabecny,kamil.kuca,ondrej.krejcar}@uhk.cz
[2] Biomedical Research Center, University Hospital Hradec Kralove,
Hradec Kralove, Czech Republic
[3] Department of Information Technologies, Faculty of Informatics and Management,
University of Hradec Kralove, Hradec Kralove, Czech Republic
{agata.bodnarova,richard.cimler,martina.husakova.2}@uhk.cz

Abstract. In the last several decades, the drug research has moved to involve
various IT technologies in order to rationalize the design of novel bioactive
chemical compounds. An important role among these computer-aided drug design (CADD) methods is played by a technique known as quantitative structure-
activity relationship (QSAR). The approach is utilized to find a statistically
significant model correlating the biological activity with more or less extent data derived from the chemical structures. The present article deals with approaches for discriminating unimportant information in the data input within the
three dimensional variant of QSAR – 3D QSAR. Special attention is turned to
uninformative and iterative variable elimination (UVE/IVE) methods applicable
in connection with partial least square regression (PLS). Herein, we briefly introduce 3D QSAR approach by analyzing 30 antituberculotics. The analysis is
examined by four UVE/IVE-PLS based data-noise reduction methods.

Keywords: UVE/IVE · Data-noise reduction · 3D QSAR · PLS · CADD

1 Introduction

The principle behind quantitative structure-activity relationships (QSAR) has been
known for more than 150 years. It was a logical inference resulting from discovering
the molecular structure of the matter. A first mathematical formalization of QSAR
being often highlighted in historical reviews on rational drug design methods is the
equation introduced by Crum-Brown and Fraser (Eq. 1).

$$\varphi = (C) \tag{1}$$

© Springer International Publishing Switzerland 2015
F. Pereira et al. (Eds.) EPIA 2015, LNAI 9273, pp. 313–325, 2015.
DOI: 10.1007/978-3-319-23485-4_33

Here, φ means the biological effect of a substance which is characterized by a set of structural features C [1]. The equation (Eq. 1) was published in 1869 as a proposition of a correlation between the biological activity of different atropine derivatives and their molecular structure. Although the validity of the Crum Brown-Fraser equation was confirmed after a 25 years' lag, it has undoubtedly become a corner stone of rational approaches in drug design and discovery.

In simplest terms, QSAR refers to a strategy aimed at building a statistically significant correlation model between the biological activity and various molecular descriptors by chemometric tools. The biological activity can be expressed as minimal inhibition concentration (MIC), concentration causing 50% enzyme inhibition (IC_{50}), binding affinity (K_i), lethal dose (LD_{50}), etc. Regarding the description of the molecular structure, a great progress has been achieved since the genesis of classical models by Hansch or Free-Wilson in the sixties of the twentieth century [2, 3]. So far, thousands of various molecular descriptors have been developed for utilization in QSAR analyses. In order to build a statistically significant QSAR model from known biological activities and molecular descriptors, linear (e.g. multiple linear regression MLR, partial least squares PLS, principal component regression PCR) or non-linear (e.g. artificial neural networks ANN, k-nearest neighbors kNN, Bayesian nets) data-mining methods are commonly employed in QSAR analyses.

In the present paper, three dimensional version of QSAR (3D QSAR) methodology is particularly studied. It is a method based on statistical processing of molecular interaction fields (MIFs). The MIF matrix is regularly data-mined by PLS to build a linear predictive 3D QSAR model. Unfortunately, PLS itself is not a sufficient tool for finding a stable model utilizing the original MIFs since it considerably suffers from abundant and noisy information in the input. The objective of our study is to evaluate several statistical methods applicable in building robust 3D QSAR models. Chapter 2 introduces in simple terms what the principles of 3D QSAR analysis are. Data-processing and noise reduction approaches based on uninformative/iterative variable elimination (UVE/IVE) are depicted in Chapter 3. Finally, the merits of these methods are demonstrated by 3D QSAR analysis of 30 antituberculotics in Chapter 4.

2   3D QSAR – Principles and Methodology

The core of 3D QSAR method, originally named comparative molecular field analysis (CoMFA), was designed by Cramer et al. in 1988 as a four-step procedure: 1) superimposition of ligand molecules on a selected template structure, 2) representation of ligand molecules by molecular interaction fields (MIFs), 3) data analysis of MIFs and biological activities by PLS, utilizing cross-validation to select the most robust 3D QSAR model, 4) graphical explanation of the results through three-dimensional pseudo β^{PLS} coefficients contour plots.

Within the 3D QSAR analysis, a starting set of molecular models can be prepared with any chemical software capable of creating and geometrically optimizing chemical structures (e.g. HyperChem, Spartan, ChemBio3D Ultra, etc.). Usually, a molecular dynamics method (e.g. simulated annealing, quenched molecular dynamics,

Langevin dynamics, Monte Carlo) is employed to obtain the most thermodynamically representative conformers. Having the chemical structures in a suitable geometry, preferably the most biologically active one is chosen as a template and the others denoted as candidate ligands. In the alignment step, all candidates are superimposed on the template structure using a distance-based scoring function implemented in an optimizing algorithm to find "the tightest" molecular alignment set. Once an optimum molecular alignment set is gained, molecular interaction fields (MIFs) may be calculated for all candidates as well as the template compound. The MIFs calculations may be adumbrated as calculating the steric and electrostatic potential in a gridbox surrounding the ligand molecule (Fig. 1).

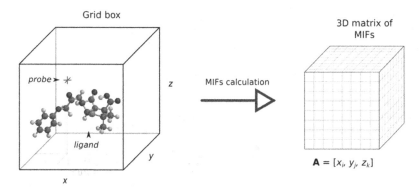

Fig. 1. Calculation of MIFs.

The potential energies (e.g. Lennard-Jones, van der Waals (VDW) and Coulomb electrostatic potential energies (ESP)) experienced by unit charge or atom probe at various points (x_i, y_j, z_k) around the studied molecules are usually regressed by PLS on the biological activities to reveal significant correlations. Generally, any supervised learning method is applicable in 3D QSAR analysis instead of PLS, provided it is able to process many thousands of "independent" x variables within a reasonable time [4]. A common MIF related to a 1.0Å spaced gridbox of the size 30.0 x 30.0 x 30.0Å represents a chemical compound by 27 000 real numbers. Because not all points in MIFs are typically related with the observed biological activity, the redundant information in the data input may bring about overlearning which mostly causes unreliable prediction for compounds outside the training and test sets. As a rule, the raw input data for 3D QSAR analysis must be pre-processed by data-noise reduction methods prior to building the final model by PLS. The data-noise reduction techniques like fractional factor design (FFD), uninformative variable elimination (UVE) or iterative variable elimination (IVE) frequently utilize cross-validated coefficient of determination (Q^2) as a cost-function for selecting the most robust 3D QSAR model [5]. The final step in deciding whether the derived 3D QSAR model is trustworthy is statistical validation. Such methods as progressive Y-scrambling, randomization, leave-two-out (LTO) or multiple leave-many-out cross-validation (LMO), and, above all, external validation are employed for these ends [6].

Besides quantitative predictions of unknown biological activities, 3D QSAR analysis enables to disclose influential features in the studied chemical structures through the spatial visualization of pseudo β^{PLS}-coefficients. Usually, pseudo β^{PLS}-coefficients are additionally multiplied by standard deviation of MIFs column vectors (noted as SD) to underline more varied regions in the chemicals structures. Since the pseudo β^{PLS}-coefficients quantitatively indicate how much each point of MIFs contributes to the biological activity, the pharmacophore within the studied compound set may be revealed. An example of 3D contour maps outlining pseudo β^{PLS} x SD coefficients is given in Fig. 2. Data for the illustration were taken from the literature [5].

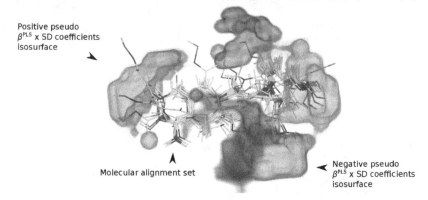

Positive pseudo
β^{PLS} x SD coefficients
isosurface

Molecular alignment set

Negative pseudo
β^{PLS} x SD coefficients
isosurface

Fig. 2. 3D contour map of pseudo β^{PLS} x SD coefficients.

By 3D contour maps of pseudo β^{PLS} x SD coefficients one can disclose which molecular features are crucial for the biological activity observed. Accordingly, medicinal chemists can utilize the information to design novel drugs through their chemical intuition or they can employ the found 3D QSAR model in ligand-based virtual screening of convenient drug databases (e.g. zinc.docking.org).

3   Variable Elimination as Data-Noise Reduction Method for 3D QSAR Analysis

Currently, 3D QSAR analysis has been involved in a variety of drug research branches. Examples of such successful projects are discovery of biphenyl-based cytostatics, design of mitochondrial cytochrome P450 enzyme inhibitors, investigation of sirtuin 2 inhibitors as potential therapeutics for neurodegenerative diseases or development of improved acetylcholinesterase reactivators [5, 7]. However, many of recent 3D QSAR studies are justified only by internal coefficient of determination R^2 > 0.8 and by its leave-one-out (LOO) cross-validated counterpart Q^2_{LOO} > 0.6. In scarcer cases, external validation of the 3D QSAR models is reported. On the other hand, it is very well known among QSAR experts that mostly unstable 3D QSAR models result *via* PLS regression when a proper selection of independent variables from the original MIFs is neglected. This drawback, which challenges not only 3D QSAR models, manifests especially in external prediction or exhaustive LMO cross-validation. Lately, several variable selection algorithms have been developed to address the overlearning in PLS based models. In the present study, a special attention is turned to variable elimination methods applicable in 3D QSAR analysis.

3.1   PLS Regression Analysis of Molecular Interaction Fields

The MIFs obtained by consecutive probing the molecules in 3D lattice box by different probes are regularly reordered into long row vectors and stored as $\mathbf{X}$ descriptor matrix. Each row represents individual chemical compound and each column contains the values of interaction energy at a given lattice intersection. Since the $\mathbf{X}$ matrix usually consists of thousands of columns and tens of rows, common MLR is useless for data-mining these data due to singularities arising during the inversion of $(\mathbf{X}^T\mathbf{X})$ matrix. A method of choice for processing MIFs in 3D QSAR analysis has been the partial least square regression (PLS). It is a supervised learning method which combines principal component analysis (PCA) and MLR. By extracting orthogonal factors (i.e. latent variables - LVs) from the original $\mathbf{X}$ matrix PLS aims to predict one or more dependent variables ($\hat{y}$ column vector or $\hat{\mathbf{Y}}$ matrix).

PLS is a convenient method for deriving a linear regression model correlating a set of dependent variables (i.e. biological activities) with an extent set of predictors (i.e. MIFs or molecular descriptors). A significant strength of PLS is the possibility to process multidimensional data with high degree of intercorrelation. Although PLS was originally developed in social sciences, it has become one the most favorite chemometric tool utilized in 3D QSAR. The benefits of PLS are appreciated especially when the number of $\mathbf{X}$ rows is decreased after splitting the input data into training, test and external sets. The nature of PLS may be briefly characterized as simultaneous decomposition of $\mathbf{X}$ and $\mathbf{Y}$ matrices. Formally, PLS works with several coupled matrices (Eq. 2):

$$\hat{\mathbf{Y}} = \mathbf{X}\boldsymbol{\beta}^{\mathrm{PLS}} = \mathbf{TP}^T\boldsymbol{\beta}^{\mathrm{PLS}} = \mathbf{TBC}^T; \; \mathbf{T} = \mathbf{XW}; \; \mathbf{U} = \mathbf{YC}; \mathbf{B} = \mathbf{T}^T\mathbf{U} \qquad (2)$$

where $\boldsymbol{\beta}^{\mathrm{PLS}}$ means the pseudo β^{PLS} regression coefficients; $\mathbf{W}$ and $\mathbf{C}$ denote the weight matrices; $\mathbf{P}$ means the loading matrix; $\mathbf{T}$ and $\mathbf{U}$ are the score matrices. The columns of $\mathbf{T}$ are orthogonal and called the latent variables ($\mathbf{T}^T\mathbf{T} = \mathbf{I}$). The crucial operation within PLS analysis consists in simplifying the complexity of the system by selecting only few LVs to build the model. The number of involved LVs is often determined by cross-validation. When Q^2_{LOO} starts dropping or the standard error of prediction (SDEP) increases, the optimum number of latent variables has been exceeded (Fig. 3). Considerably more robust algorithms for latent variable selection implement leave-many-out cross-validated Q^2_{LMO} or coefficient of determination for external prediction R^2_{ext} as a control function.

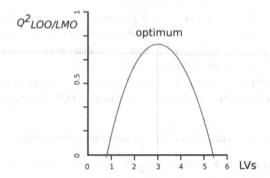

Fig. 3. Determining the optimal number of LVs through cross-validation.

Since PLS is a PCA based method, it is highly sensitive to the variance of the values included in the input data. This problematic feature of PLS can lead to masking significant information by data assuming greater values. For instance, the information on weak hydrophobic interactions is suppressed by Coulombic interactions and hydrogen bonding that are stronger. However, it has become evident that hydrophobic interactions play such an important role in drug binding to receptors that they cannot be neglected in 3D QSAR calculations. To prevent discriminating variables with relatively low values, the MIFs as well as the biological activities have to be column centered and normalized prior to PLS. In 3D QSAR analysis, different MIFs may be also scaled as separated blocks by block unscaled weighting (BUW) to give each probe the same significance in PLS [8]. PLS regression can be performed by a number of subtle differing algorithms (e.g. NIPALS, SIMPLS, Lanczos bidiagonalization) [9]. In case **Y** consists only of one column, then NIPALS algorithm may be simply illustrated as follows:

```
[X,Y] = autoscale(X, Y); % Centering and normalization
T=[]; P=[]; W=[]; Q=[]; B=[]; % initialization
for a=1:LV % Calculate the entered number of LVs
    w=(Y'*X)'; % Calculate weighting vector
    w=w/sqrt(w'*w); % Normalization to unit length
    t=X*w; % Calculate X scores
    if a>1 t=t - T*(inv(T'*T)*(t'*T)'); end;
    u=t/(t'*t);
    p=(u'*X)'; % Calculate X loadings
    X=X - t*p'; % Calculate X residuals
    T=[T t]; % Calculate X scores
    P=[P p]; % Calculate X loadings
    W=[W w]; % Calculate X weights
    Q=[Q;y'*u]; % Calculate Y loadings
    B=[B W*inv(P'*W)*Q]; % Calculate PLS coefficients
end
Y_pred=X*B; % Internal prediction
```

The above PLS algorithm derives the entered number of LVs and utilizes them in the internal prediction of the autoscaled dependent variable **Y**. The robustness of the resulting PLS model can be easily controlled by incorporating a cross-validation into the algorithm. Nonetheless, the prediction instability of the 3D QSAR models has to be solved mainly through preselecting the input data.

3.2   Unsupervised Preselection in the MIFs before PLS Analysis

Calculation of the physically based interactions between the probe and the molecules placed into a gridbox necessarily leads to dramatic increase of the energy when the probe gets closer to atom nuclei. Conversely, the energy calculated in gridpoints relatively far from a molecule may assume negligible magnitude from the QSAR's point of view. These MIF properties enable to perform, to certain extension, a preliminary input data clearance. It is a common practice in 3D QSAR analysis to remove from the MIFs such points (i.e. whole columns in **X** matrix) where the maximum energy exceeds a chosen threshold (e.g. > 50 kcal/mol for steric van der Waals MIF, > 30 kcal/mol for electrostatic MIF). Other possible unsupervised techniques to eliminate unimportant data in the MIFs are: zeroing all gridpoints having absolute value below 0.05 kcal/mol, removing all MIFs vectors of the standard deviation lower than 0.1 kcal/mol, removing all MIFs vectors assuming only several levels with skewed distribution. By such type preselection it is possible to remove from 1% to 80% uninformative MIFs vectors depending on the cutoff levels used. Commonly, several hundreds or thousands MIFs columns are left after unsupervised preselection to other variable reduction approaches.

3.3   Uninformative Variable Elimination

The main goal of 3D QSAR analysis is to develop a stable mathematical model which can be used for prediction of unseen biological activities. From this somewhat narrowed point of view, any model that is not able to prove itself in external prediction must be rejected from further consideration. However, such refusal does not provide any suggestion why different models are successful or fail in predictions and how to boost the predictive ability.

The method introduced by Centner and Massart focuses on what is noisy and/or irrelevant information and how to discriminate it [10]. Comparing to other variable selection techniques like forward selection, stepwise selection, genetic and evolutional algorithms, uninformative variable elimination (UVE) does not attempt to find the best subset of variables to build a statistically significant model but to remove such variables that contain no useful information. Centner and Massart took their inspiration from previously published studies which tried to eliminate those variables having small loadings or pseudo β^{PLS} coefficient in a model derived by PLS.

UVE-PLS method resembles stepwise variable selection used in MLR. The j-th variable (i.e. a MIF vector) is eliminated from the original vector pool if its c_j value is lower than a certain cutoff level (Eq. 3)

$$c_j = \frac{\bar{\beta}_j^{PLS}}{sd(\bar{\beta}_j^{PLS})} \tag{3}$$

In order to obtain the mean ($\bar{\beta}_j^{PLS}$) and the standard deviation ($sd(\bar{\beta}_j^{PLS})$) of pseudo β^{PLS} coefficient for each variable, a cross-validation is necessary to carry out. A critical point in this method is how to determine the cutoff level for the elimination. For this purposes, Centner and Massart proposed to add several artificial random variables into the original **X** matrix and to calculate their c_j^{random} values. The original variables that exhibit lower c_j than the maximum in c_j^{random}'s determined for the artificial variables are removed. Although the UVE method is capable to discriminate uninformative variables, the artificial variables used have to be of low magnitude comparing to the original variables so as they do not disturb significantly the model. The artificial random matrix **R** is proposed to be of the same dimension as the **X** matrix.

According to Centner and Massart, the UVE-PLS procedure can be summarized in the following steps: 1) determination of the optimum number of LVs as the minimum of the root-mean-square error prediction function RMSEP (Eq. 4); 2) generation of the random matrix **R** and its scaling by a small factor (e.g. 10^{-10}); 3) PLS regression of the conjugate **RX** matrix and leave-one-out cross-validation; 4) calculation of c_j values for all variables; 5) determination of $\max(abs(c_j^{random}))$; 6) elimination of all original variables for which $abs(c_j) < \max(abs(c_j^{random}))$; 7) evaluating of the new model by leave-one-out cross-validation.

$$\text{RMSEP} = \sqrt{\sum_{i=1}^{n} \frac{(\hat{y}_i - y_i)^2}{n}} \tag{4}$$

Here, y_i is the observed biological activity for i-th compound, $\hat{y}_i$ stands for the predicted biological activity of i-th compound, n is the number of compounds in the set. The above-mentioned UVE algorithm can be transformed to a more robust variant by expressing the c_j as median (β_j^{PLS})/interquartile range (β_j^{PLS}). The criterion for variable elimination may be substituted for a 90-95 quantile of $abs(c_j^{random})$.

3.4   Iterative Variable Elimination

Uninformative variable elimination was designed to improve the predictive power and the interpretability of 3D QSAR models *via* removing those MIFs parts which do not contain fecund information in comparison to random noise introduced to the input data. The UVE method is based on calculation of c_j values indicating the ratio of the size and standard deviation of pseudo β^{PLS} coefficient related to j-th vector of MIFs. All MIF vectors with c_j smaller then a cutoff derived from c values of the random variables are removed from the **X** matrix in a single step procedure.

A modification of the UVE algorithm suggested by Polanski and Gieleciak revises the very one-step elimination of MIF vectors with low c_j values [11, 12]. Their improved algorithm named iterative variable elimination (IVE) does not remove the selected vectors in a single step but in a sequential manner. In the first version of IVE, the MIF vector having the lowest pseudo β^{PLS} coefficient is eliminated and the remaining $\mathbf{X}_{(-i)}$ matrix is regressed by PLS to evaluate the benefit. The iterative IVE procedure can be described by the following protocol: 1) carry out PLS analysis with a fixed number of LVs and estimate the performance by leave-one-out cross-validation; 2) eliminate the $\mathbf{X}$ matrix column with the lowest absolute value of pseudo β^{PLS} coefficient; 3) carry out PLS analysis of the reduced $\mathbf{X}_{(-i)}$ matrix and estimate the performance by leave-one-out cross-validation; 4) go to step 1 and repeat until the maximal leave-one-out cross-validated coefficient of determination Q^2_{LOO} is reached (Eq. 5).

$$Q^2_{LOO} = 1 - \frac{\sum_i^n (\hat{y}_{i(-1)} - y_i)^2}{\sum_i^n (y_i - \bar{y})^2} \tag{5}$$

Here, y_i means the observed biological activity for i-th compound, $\hat{y}_{i(-1)}$ denotes the biological activity of i-th compound predicted by the model derived without the i-th compound, n is the number of compounds in the set. In the first version, the IVE procedure was based on iterative elimination of MIF vectors with the lowest absolute values of pseudo β^{PLS} coefficients. In the next IVE variants, the criterion for MIF vector elimination was substituted by c_j values obtained by leave-one-out cross-validation. The most robust form of IVE was proposed to involve optimization of the LV number and c_j values defined as median (β_j^{PLS})/interquartile range (β_j^{PLS}). It was proved by Polanski and Gieleciak that the robust IVE form surpassed the other variants and gave the most reliable 3D QSAR models in terms of the highest Q^2_{LOO} and sufficient resolution of pseudo β^{PLS} coefficient contour maps.

3.5 Hybrid Variable Elimination

The issue concerning the selection or elimination of the right variables in 3D QSAR studies can be addressed in several ways [13]. A logical candidate to be implemented in improving the ability of 3D QSAR models seems to be genetic algorithm (GA) that enables evaluating different sets of MIF vectors. However, thanks to many variables in MIFs there is no guarantee that GA-PLS procedure can straightforwardly converge to achieve the best solution. Other promising alternatives of variable selection/elimination techniques relevant in 3D QSAR are sub-window permutation analysis coupled with PLS (SwPA-PLS), iterative predictor weighting PLS (IPW-PLS), regularized elimination procedure (REP-PLS), interactive variable selection (IVS-PLS), soft-thresholding PLS (ST-PLS) or powered PLS (PPLS) (see [13]).

Noteworthy, the MIFs generated for chemical compounds are not independent variables but spatially inter-correlated. In the original 3D MIF matrix, the energy values are ordered according to a molecular structure probed in a gridbox. This internal information is neglected and stays hidden when transforming the 3D MIF matrices into row vectors of the $\mathbf{X}$ matrix. Application of UVE and IVE methods on such unfolded data therefore does not reflect the chemical information born by MIFs but treats it as different molecular descriptors. To utilize also the spatial contiguity of the original 3D MIFs, a methodology named smart region definition (SRD) has found its important position in 3D QSAR analysis. SRD procedure aims to rearrange the unfolded MIFs into group variables related to the same chemical regions (e.g. points around the same atoms). These groups of neighboring variables are explicitly associated with chemical structures and when treated as logical units, the resulting 3D QSAR models are less prone to chance correlations and easier to interpret [14]. Since the SRD procedure clusters similar MIF vectors into groups, the time consumed by UVE or IVE analyses is shorter than in the standard processing of all individual MIF vectors. The SRD algorithm involves three major operations: 1) selecting the most important MIF vectors (seeds) having the highest PLS weights; 2) building 3D Voronoi polyhedra around the seeds; 3) collapsing Voronoi polyhedral into larger regions.

The starting point of the SRD procedure is PLS analysis of the whole **X** matrix which reveals through the magnitude of weights significant MIF vectors. Depending on the user's setting, a selection of important MIF vectors is denoted as seeds. In next step, the remaining MIF vectors are assigned to the nearest seed according to preset Euclidian distance. In case a MIF vector is too far from each seed, it is assigned to "zero" region and eliminated from the **X** matrix. After distributing the MIF vectors into Voronoi polyhedra, further variable absorption is performed. Neighboring Voronoi polyhedra are statistically analysed and if found significantly correlated, they are merged into one larger Voronoi polyhedra. The cutoff distances for initial building the Voronoi polyhedra as well as for subsequent collapsing are critical points which decide on the merit of SRD and, thus, have to be cautiously optimized.

4   Performance Comparison of UVE and IVE Based Methods

In order to practically evaluate the performance of the UVE and IVE based noise-reduction methods, a 3D QSAR analysis has been carried out. We selected a group of 30 compounds, which are currently considered as potential antituberculotics [15, 16], and analyzed them in Open3DAlign and Open3DQSAR programs [17, 18]. Since we cannot provide a detailed description of all undertaken steps of the analysis in this article, we will confine the present study only to the performance of the UVE and IVE algorithms and their SRD hybridized variants. In the 3D QSAR analyses, the most common or default setting was used.

First, the set of 30 compounds published in the literature was modeled in Hyper-Chem 7.0 to prepare the initial molecular models. Then, the molecular ensemble was submitted to quenched molecular dynamics and the resulting conformers were processed by an aligning algorithm in Open3DAlign program to determine the optimal molecular superimposition. In Open3DQSAR program two MIFs were generated (i.e. van der Waals MIF and Coulombic MIF) and processed by four 3D QSAR methods: 1) UVE-PLS, 2) IVE-PLS, 3) UVE-SRD-PLS and 4) IVE-SRD-PLS (Fig. 4). As a dependent variable, we used the published logMICs against *Mycobacterium tuberculosis*. To evaluate the 3D QSAR model, the original set of compounds was randomly divided into training and test sets in ratio 25 : 5.

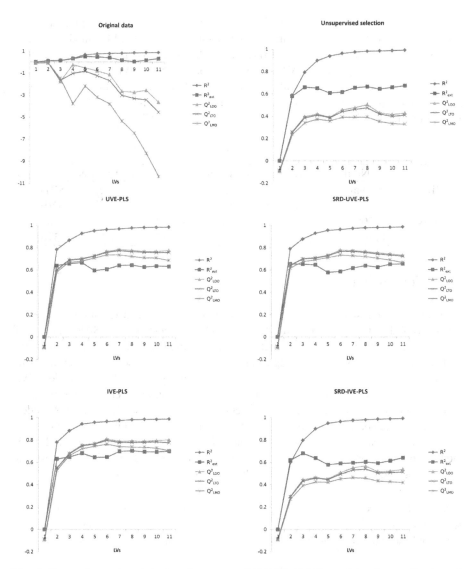

Fig. 4. Comparison of statistical performance of PLS 3D QSAR models after application of various data-noise reduction methods. The results are related to 30 antituberculotics. The vertical y-axes in all graphs represent R^2, R^2_{ext}, Q^2_{LOO}, Q^2_{LTO} and Q^2_{LMO} values.

From the above plots follows that an efficient data noise reduction is crucial to achieve a stable PLS model in cross-validations. The original dataset provided all $Q^2_{LOO/LTO/LMO}$ with negative values, which indicate an overtrained PLS model. Significant improvement was reached even by unsupervised preselection of the **X** matrix. UVE and SRD-UVE PLS models showed considerable better performance and exhibited nearly the same robustness in cross-validation. Top scoring model resulted when applying SRD/UVE-PLS with 6-7 LVs ($Q^2_{LMO} = 0.7352 - 0.7361$). The best 3D QSAR model was obtained by IVE-PLS (5 LVs; Q^2_{LMO} = 0.7652). It is interesting that application of SRD-IVE-PLS caused significant deterioration of the IVE-PLS model stability in cross-validation (max(Q^2_{LMO}) = 0.4617; 6 LVs). It is likely a consequence of stepwise elimination of larger groups of MIF vectors grouped by SRD. Without SRD, the IVE algorithm iteratively investigates all MIF vectors and is better able to find the critical portion of eliminated information from the input data. Numbers of eliminated MIF vectors by the used data-noise reduction methods are given in Table 1.

Table 1. Remaining MIF vectors after data-noise reductions.

MIF	Number of MIF vectors remaining after data-noise reduction					
	Original dataset	Unsupervised preselection	UVE	SRD-UVE	IVE	SRD-IVE
VDW	11500	1261	237	128	344	536
ESP	11500	11479	4752	275	6026	576

5   Conclusion

In medicinal chemistry and bioinformatics various computerized technologies are increasingly utilized to facilitate drug discovery. One of them is 3D QSAR analysis of molecular interaction fields. By prediction of unseen biological activities, the time and costs needed to develop a novel drug can be substantially decreased. However, the benefit of such computational methods depends on reliability of the derived models. We demonstrated in this work that a data-noise reduction algorithm to select only significant MIF vectors from the original **X** matrix is essential to build a stable 3D QSAR model. A promising method for elimination of useless information in 3D QSAR analysis seems to be iterative variable elimination (IVE) coupled with PLS. It was shown that IVE-PLS method surpassed other techniques for data-noise reduction and provided the most statistically significant 3D QSAR model for 30 selected antituberculotics.

Acknowledgements. The paper is supported by the project of specific science "Application of Artificial Intelligence in Bioinformatics" at the Faculty of Informatics and Management, University of Hradec Kralove, Czech Republic and by the long term development plan FNHK.

References

1. Brown, A.C., Fraser, T.R.: XX.—On the Connection between Chemical Constitution and Physiological Action. Part II.—On the Physiological Action of the Ammonium Bases derived from Atropia and Conia. Trans. Roy. Soc. Edinburgh **25**, 693–739 (1869)
2. Hansch, C., Fujita, T.: ρ-σ-π Analysis. A method for the correlation of biological activity and chemical structure. J. Am. Chem. Soc. **86**, 1616–1626 (1964)
3. Free, S.M., Wilson, J.W.: A mathematical contribution to structure-activity studies. J. Med. Chem. **7**, 395–399 (1964)
4. Cramer Iii, R.D.: Partial least squares (PLS): its strengths and limitations. Perspect. Drug. Discov. **1**, 269–278 (1993)

5. Dolezal, R., Korabecny, J., Malinak, D., Honegr, J., Musilek, K., Kuca, K.: Ligand-based 3D QSAR analysis of reactivation potency of mono- and bis-pyridinium aldoximes toward VX-inhibited rat acetylcholinesterase. J. Mol. Graph. Model. **56c**, 113–129 (2014)
6. Tropsha, A., Gramatica, P., Gombar, V.K.: The importance of being earnest: validation is the absolute essential for successful application and interpretation of QSPR models. QSAR Comb. Sci. **22**, 69–77 (2003)
7. Chuang, Y.C., Chang, C.H., Lin, J.T., Yang, C.N.: Molecular modelling studies of sirtuin 2 inhibitors using three-dimensional structure-activity relationship analysis and molecular dynamics simulations. Mol. Biosyst. **11**, 723–733 (2015)
8. Kastenholz, M.A., Pastor, M., Cruciani, G., Haaksma, E.E., Fox, T.: GRID/CPCA: a new computational tool to design selective ligands. J. Med. Chem. **43**, 3033–3044 (2000)
9. Bro, R., Elden, L.: PLS works. J. Chemometr. **23**, 69–71 (2009)
10. Centner, V., Massart, D.L., de Noord, O.E., de Jong, S., Vandeginste, B.M., Sterna, C.: Elimination of uninformative variables for multivariate calibration. Anal. Chem. **68**, 3851–3858 (1996)
11. Polanski, J., Gieleciak, R.: The comparative molecular surface analysis (CoMSA) with modified uniformative variable elimination-PLS (UVE-PLS) method: application to the steroids binding the aromatase enzyme. J. Chem. Inf. Comp. Sci. **43**, 656–666 (2003)
12. Gieleciak, R., Polanski, J.: Modeling robust QSAR. 2. iterative variable elimination schemes for CoMSA: application for modeling benzoic acid pKa values. J. Chem. Inf. Model. **47**, 547–556 (2007)
13. Mehmood, T., Liland, K.H., Snipen, L., Sæbø, S.: A review of variable selection methods in partial least squares regression. Chemometr. Intel. Lab. **118**, 62–69 (2012)
14. Pastor, M., Cruciani, G., Clementi, S.: Smart region definition: a new way to improve the predictive ability and interpretability of three-dimensional quantitative structure-activity relationships. J. Med. Chem. **40**, 1455–1464 (1997)
15. Dolezal, R., Waisser, K., Petrlikova, E., Kunes, J., Kubicova, L., Machacek, M., Kaustova, J., Dahse, H.M.: N-Benzylsalicylthioamides: Highly Active Potential Antituberculotics. Arch. Pharm. **342**, 113–119 (2009)
16. Waisser, K., Matyk, J., Kunes, J., Dolezal, R., Kaustova, J., Dahse, H.M.: Highly Active Potential Antituberculotics: 3-(4-Alkylphenyl)-4-thioxo-2H-1,3-benzoxazine-2(3H)-ones and 3-(4-Alkylphenyl)-2H-1,3-benzoxazine-2,4(3H)-diones Substituted in Ring-B by Halogen. Archiv Der Pharmazie **341**, 800–803 (2008)
17. Tosco, P., Balle, T., Shiri, F.: Open3DALIGN: an open-source software aimed at unsupervised ligand alignment. J. Comput. Aid. Mol. Des. **25**, 777–783 (2011)
18. Tosco, P., Balle, T.: Open3DQSAR: a new open-source software aimed at high-throughput chemometric analysis of molecular interaction fields. J. Mol. Model. **17**, 201–208 (2011)

Pattern-Based Biclustering with Constraints for Gene Expression Data Analysis

Rui Henriques[✉] and Sara C. Madeira

Inesc-ID, Instituto Superior Técnico, Universidade de Lisboa, Lisboa, Portugal
{rmch,sara.madeira}@tecnico.ulisboa.pt

Abstract. Biclustering has been largely applied for gene expression data analysis. In recent years, a clearer understanding of the synergies between pattern mining and biclustering gave rise to a new class of biclustering algorithms, referred as pattern-based biclustering. These algorithms are able to discover exhaustive structures of biclusters with flexible coherency and quality. Background knowledge has also been increasingly applied for biological data analysis to guarantee relevant results. In this context, despite numerous contributions from domain-driven pattern mining, there is not yet a solid view on whether and how background knowledge can be applied to guide pattern-based biclustering tasks.

In this work, we extend pattern-based biclustering algorithms to effectively seize efficiency gains in the presence of constraints. Furthermore, we illustrate how constraints with succinct, (anti-)monotone and convertible properties can be derived from knowledge repositories and user expectations. Experimental results show the importance of incorporating background knowledge within pattern-based biclustering to foster efficiency and guarantee non-trivial yet biologically relevant solutions.

1 Introduction

Biclustering, the task of finding subsets of rows with a coherent pattern across subsets of columns in real-valued matrices, has been largely used for expression data analysis [9,11]. Biclustering algorithms based on pattern mining methods [9,11,12,18,22,25], referred in this work as pattern-based biclustering, are able to perform flexible and exhaustive searches. Initial attempts to use background knowledge for biclustering based on user expectations [5,7,15] and knowledge-based repositories [18,20,26] show its key role to guide the task and guarantee relevant solutions. In this context, two valuable synergies can be identified based on these observations. First, the optimality and flexibility of pattern-based biclustering provide an adequate basis upon which knowledge-driven constraints can be incorporated. Contrasting with pattern-based biclustering, alternative biclustering algorithms place restrictions on the structure (number, size and positioning), coherency and quality of biclusters, which may prevent the incorporation of certain constraints [11,16]. Second, the effective use of background knowledge to guide pattern mining searches has been largely researched in the context of domain-driven pattern mining [4,23].

© Springer International Publishing Switzerland 2015
F. Pereira et al. (Eds.) EPIA 2015, LNAI 9273, pp. 326–339, 2015.
DOI: 10.1007/978-3-319-23485-4_34

Despite these synergies, there is a lack of literature on the feasibility and impact of integrating domain-driven pattern mining and biclustering. In particular, there is a lack of research on how to map the commonly available background knowledge in the form of parameters or constraints to guide the biclustering task. Additionally, the majority of existing pattern-based biclustering algorithms rely on searches dependent on bitset vectors [18,22,25], which may turn their performance impracticable for large and dense biological datasets. Although new searches became recently available for biclustering large and dense data [13], there are not yet contributions on how these searches can be adapted to seize the benefits from the available background knowledge.

In this work, we address these problems. First, we list an extensive set of key constraints with biological relevance and show how they can be specified for pattern-based biclustering. Second, we extend F2G [13], a recent pattern-growth search that tackles the efficiency bottlenecks of peer searches, to bed able to effectively use constraints with succinct, (anti-)monotone and convertible properties.

To achieve these goals, we propose BiC2PAM (**BiC**lustering with **C**onstraints using **PA**ttern Mining), an algorithm that integrates recent breakthroughs on pattern-based biclustering [9,11,12] and extends them to effectively incorporate constraints. Experimental results confirm the role of BiC2PAM to foster the biological relevance of pattern-based biclustering solutions and to seize large efficiency gains by adequately pruning the search space.

The paper is structured as follows. *Section 2* provides background on pattern-based biclustering and domain-driven pattern mining. *Section 3* surveys key contributions and limitations from related work. *Section 4* lists biologically meaningful constraints and proposes BiC2PAM for their effective incorporation. *Section 5* provides initial empirical evidence of BiC2PAM's efficiency and ability to unravel non-trivial yet biologically significant biclusters from gene expression data. Finally, concluding remarks are synthesized.

2   Background

Definition 1. *Given a matrix, $A=(X,Y)$, with a set of rows $X=\{x_1,..,x_n\}$, a set of columns $Y=\{y_1,..,y_m\}$, and elements $a_{ij}\in\mathbb{R}$ relating row i and column j: the* **biclustering task** *aims to identify a set of biclusters $\mathcal{B}=\{B_1,..,B_m\}$, where each* bicluster $B_k = (I_k, J_k)$ *is a submatrix of A ($I_k \subseteq X$ and $J_k \subseteq Y$) satisfying specific criteria of* homogeneity *and* significance *[11].*

A real-valued matrix can thus be described by a (multivariate) distribution of background values and a *structure* of biclusters, where each bicluster satisfies specific criteria of *homogeneity* and *significance*. The *structure* is defined by the number, size and positioning of biclusters. Flexible structures are characterized by an arbitrary-high set of (possibly overlapping) biclusters. The *coherency* (homogeneity) of a bicluster is defined by the observed correlation of values (see Definition 2). The *quality* of a bicluster is defined by the type and amount of accommodated noise. The statistical *significance* of a bicluster determines the deviation of its probability of occurrence from expectations.

Definition 2. *Let the elements in a bicluster* $a_{ij} \in (I, J)$ *have coherency across rows given by* $a_{ij}=k_j+\gamma_i+\eta_{ij}$, *where* k_j *is the expected value for column* j, γ_i *is the adjustment for row* i, *and* η_{ij} *is the noise factor* [16]. *For a given real-valued matrix A and coherency strength* δ: $a_{ij}=k_j+\gamma_i+\eta_{ij}$ *where* $\eta_{ij} \in [-\delta/2, \delta/2]$.

As motivated, the discovery of exhaustive and flexible structures of biclusters satisfying certain homogeneity criteria (Definition 2) is a desirable condition to effectively incorporate knowledge-driven constraints. However, due to the complexity of such biclustering task , most of the existing algorithms are either based on greedy or stochastic approaches, producing sub-optimal solutions and placing restrictions (e.g. fixed number of biclusters, non-overlapping structures, and simplistic coherencies) that prevent the flexibility of the biclustering task [16]. Pattern-based biclustering appeared in recent years as one of various attempts to address these limitations. As follows, we provide background on this class of biclustering algorithms, as well as on constraint-based searches.

Pattern-Based Biclustering. Patterns are itemsets, rules, sequences or other structures that appear in symbolic datasets with frequency above a specified threshold. Patterns can be mapped as a bicluster with constant values across rows $(a_{ij}=c_j)$, and specific coherency strength determined by the number of symbols in the dataset, $\delta=1/|\mathcal{L}|$ where $\mathcal{L}$ is the alphabet of symbols. The relevance of a pattern is primarily defined by its support (number of rows) and length (number of columns). To allow this mapping, the pattern mining task needs to output not only the patterns but also their supporting transactions (*full-patterns*). Definitions 3 and 4 illustrate the paradigmatic mapping between full-pattern mining and biclustering.

Definition 3. *Let* $\mathcal{L}$ *be a finite set of items, and P an itemset* $P \subseteq \mathcal{L}$. *A symbolic matrix D is a finite set of transactions in* $\mathcal{L}$, $\{P_1, .., P_n\}$. *Let the coverage* Φ_P *of an itemset P be the set of transactions in D in which P occurs,* $\{P_i \in D \mid P \subseteq P_i\}$, *and its* support sup_P *be the coverage size,* $|\Phi_P|$.

A full-pattern is a pair (P, Φ_P), *where P is an itemset and* Φ_P *the set of all transactions that contain P. A* closed full-pattern (P, Φ_P) *is a full-pattern where P is not subset of another itemset with the same support,* $\forall_{P'\supset P}|P'| < |P|$.

Given D and a minimum support threshold θ, *the* **full-pattern mining** *task* [13] *consists of computing:* $\{(P, \Phi_P) \mid P \subseteq \mathcal{L}, sup_P \geq \theta, \forall_{P'\supset P}|P'| < |P|\}$.

Given an illustrative symbolic matrix $D=\{(t_1, \{a, c, e\}), (t_2, \{a, b, d\}), (t_3, \{a, c, e\})\}$, we have $\Phi_{\{a,c\}}=\{t_1, t_3\}$, $sup_{\{a,c\}}=2$. For a minimum support $\theta=2$, the full-pattern mining task over D returns the set of closed full-patterns, $\{(\{a\}, \{t_1, t_2, t_3\}), (\{a, c, e\}, \{t_1, t_3\})\}$ (note that $|\Phi_{\{a,c\}}|\leq|\Phi_{\{a,c,e\}}|$). Fig.1 illustrates how full-pattern mining can be used to derive constant biclusters[1].

Definition 4. *Given a symbolic matrix D in* $\mathcal{L}$, *let a matrix A be the concatenation of D elements with their column indexes. Let* Ψ_P *be the column indexes*

---

[1] Association rule mining, sequential pattern mining and graph mining can be also used to respectively mine biclusters with noisy, order-preserving and differential coherencies [9, 12].

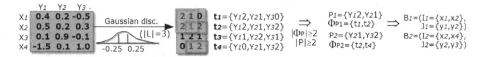

Fig. 1. Discovery of biclusters with constant coherency on rows from full-patterns.

of an itemset P, and Υ_P be the original items of P in $\mathcal{L}$. The set of **maximal biclusters** $\cup_k B_k = (I_k, J_k)$ can be derived from the set of closed full-patterns $\cup_k P_k$ from A, by mapping $I_k = \Phi_{P_k}$ and $J_k = \Psi_{P_k}$, to compose constant biclusters with coherency across rows with pattern Υ_P [11].

The inherent simplicity, efficiency and flexibility of pattern-based biclustering explains the increasing attention [11,12,18,22,25]. The major contributions of pattern-based approaches for biclustering include: *1)* efficient analysis of large matrices due to the monotone search principles (and the support for distributed/partitioned data settings and approximate patterns [8]). *2)* biclusters with parameterizable coherency strength (beyond differential assumption) and type (possibility to accommodate additive, multiplicative, order-preserving and plaid models) [9,11,12]; *3)* flexible structures of biclusters (arbitrary positioning of biclusters) and searches (no need to fix the number of biclusters apriori) [22,25]; and *4)* robustness to noise, missings and discretization problems [11].

Constraint-Based Pattern Mining. A *constraint* is a predicate on the powerset of items $C : 2^{\mathcal{L}} \to \{true, false\}$. A full-pattern (P, Φ_P) satisfies C if $C(P)$ is true. Minimum support is the default constraint in full-pattern mining, $C_{freq}(P) = |\Phi_P| \geq \theta$. Typical constraints with interesting properties include: regular expressions on the items in the pattern, and inequalities based on aggregate functions, such as length, maximum, minimum, range, sum, average and variance [24].

Definition 5. *Let each item have a correspondence with a real value, $\mathcal{L} \to \mathbb{R}$, when numeric operators are considered. C is **monotone** if for any P satisfying C, P supersets satisfy C (e.g. $range(P) \geq v$). C is **anti-monotone** if for any P not satisfying C, P supersets do not satisfy C (e.g. $max(P) \leq v$). Let P_1 satisfy C, C is **succint** if for any P_2 satisfying C, $P_1 \subseteq P_2$ (e.g. $min(P_2) \leq v$). C is **convertible** w.r.t. an ordering of items R_Σ if for any P satisfying C, P suffixes satisfy C or/and itemsets with P as suffix satisfy C (e.g. $avg(P) \geq v$).*

To illustrate these constraints, consider $\{(t_1, \{a, b, c\}), (t_2, \{a, b, c, d\}), (t_3, \{a, d\})\}$, $\theta = 1$ and $\{a{:}0, b{:}1, c{:}2, d{:}3\}$ value correspondence. The set of closed full-patterns under the monotone $range(P) \geq 2$ is $\{(\{a, b, c\}, \{t_1, t_2\}), (\{a, d\}, \{t_1, t_3\})\}$; the anti-monotone $sum(P) \leq 1$ is $\{(\{a, b\}, \{t_1, t_2\})\}$; the succint $P \supseteq \{c, d\}$ is $\{(\{a, b, c, d\}, \{t_2\})\}$; and the convertible $avg(P) \geq 2$ is $\{(\{b, c, d\}, \{t_2\})\}$.

3 Related Work

Knowledge-Driven Biclustering. The use of background knowledge to guide biclustering has been increasingly motivated since solutions with good homogeneity and statistical significance may not necessarily be biologically relevant. However, only few biclustering algorithms are able to incorporate background knowledge. AI-ISA [26], GenMiner [18] and scatter biclustering [20] are able to annotate data with functional terms retrieved from repositories with ontologies, and use these annotations to guide the search. COBIC [19] is able to adjust its behavior (maximum-flow/minimum-cut parameters) in the presence of background knowledge. Similarly, the priors and architectures of generative biclustering algorithms can also incorporate background knowledge [10]. However, COBIC and generative peers are not able to deliver flexible biclustering solutions and only consider simplistic constraints. Fang et al. [5] propose a constraint-based algorithm that turns possible the discovery of dense biclusters associated with high-order combinations of single-nucleotide polymorphisms (SNPs). Data-Peeler [7], as well as algorithms from formal concept analysis [15] and bi-sets mining [1], are able to efficiently discover dense biclusters in binary matrices in the presence of (anti-)monotone constraints. However, these last sets of algorithms impose a very restrictive form of homogeneity in the delivered biclusters.

Full-Pattern Mining for Biclustering. The majority of existing full-pattern miners rely on frequent itemset mining with implementations based on bitset vectors to represent transaction-sets. There are two major classes of searches with this behavior. First, Apriori-based searches [8], generally suffering from costs of candidate generation for low support thresholds (commonly required for biological tasks [22]). Efficient implementations include LCM and CLOSE, used respectively by BiModule [22] and GenMiner [18] biclustering algorithms. Second, vertical-based searches, such as Eclat and Carpenter [8]. These searches rely on intersection operations over transaction-sets to generate candidates, requiring structures such as bitset vectors or diffsets. However, for datasets with a high number of transactions the bitset cardinality becomes large, these structures consume a significant amount of memory and operations become costly. MAFIA is an implementation used by DeBi [25]. Only in recent years, a third class of searches without the bottlenecks associated with bitset vectors were made available by extending pattern-growth searches for the discovery of full-patterns using frequent-pattern trees (FP-Trees) annotated with transactions. F2G [13] used by default in BicPAM [11] implements this third type of searches.

Constraint-Based Pattern Mining. A large number of studies explore how constraints can be used with pattern mining. Two major paradigms are available: constraint-programming (CP) and dedicated searches. First, CP allows the pattern mining task to be declaratively defined according to sets of constraints [4,14]. These declarative models are expressive as they can allow mathematical expressions over itemsets and transaction-sets. Nevertheless, due to the poor scalability of CP methods, they have been only used in highly constrained settings, small-to-medium data, or to mine approximative patterns [4,14].

Second, pattern mining methods have been adapted to optimally seize efficiency gains from different types of constraints. Such efforts replace naïve solutions: post-filtering patterns that satisfy constraints. Instead, the constraints are pushed as deeply as possible within the mining step for an optimal pruning of the search space. The nice properties exhibited by constraints, such as anti-monotone and succinct properties, have been initially seized by Apriori methods [21] to affect the generation of candidates. Convertible constraints, can hardly be pushed in Apriori but can be handled by FP-Growth approaches [23]. FICA, FICM, and more recently MCFPTree, are FP-Growth extensions to seize the properties of anti-monotone, succinct and convertible constraints [23]. The inclusion of monotone constraints is more complex. Filtering methods, such as ExAnte, are able to combine anti-monotone and monotone pruning based on reduction procedures [2]. Reductions are optimally handled in FP-Trees [3].

4   Pattern-Based Biclustering with Constraints

BicPAM [11], BicSPAM [12] and BiP [9] are the state-of-the-art algorithms for pattern-based biclustering. They integrate the dispersed contributions of previous pattern-based algorithms and extend them to discover non-constant coherencies and to guarantee their robustness to discretization (by assigning multi-items to a single element [11]), noise and missings. In this section, we propose BiC2PAM (**BiC**lustering with **C**onstraints using **PA**ttern Mining) to integrate their contributions and adapt them to effectively incorporate constraints. BiC2PAM is a composition of three major steps: 1) *preprocessing* to itemize real-valued data; 2) *mining* step, corresponding to the application of full-pattern miners; and 3) *postprocessing* to merge, reduce, extend and filter similar biclusters. As follows, *Section 4.1* lists native constraints supported by parameterizations along these steps. *Section 4.2* lists biologically meaningful constraints with properties of interest. Finally, we extend a pattern-growth search to seize efficiency gains from succinct, (anti-)monotone and convertible constraints (*Section 4.3*).

4.1   Native Constraints

Below we list a set of structural constraints that can be incorporated by adapting the parameters that control the behavior of pattern-based biclustering algorithms along their three major steps.

Relevant constraints provided in the pre-processing step:

- combined inclusion of annotations (such as functional terms) with succinct constraints. A functional term is associated with an interrelated group of genes, and thus it can be appended as a new dedicated symbol to the respective transactions/genes, possibly leading to a set of transactions with varying length. Illustrating, consider T_1 and T_2 terms to be respectively associated with genes $\{g_1, g_3, g_4\}$ and $\{g_3, g_5\}$, an illustrative dataset for this scenario would be $\{(g_1, \{a_{11}, .., a_{1m}, T_1\}), (g_2, \{a_{21}, .., a_{2m}\}), (g_3\{a_{31}, .., a_{3m}, T_1, T_2\}), ...\}$. Pattern mining can then be applied on top of these annotated transactions with succinct constraints to guarantee the inclusion of certain terms (such as $P \cap \{T_1, T_2\} \neq 0$). This is useful to discover, for instance, biclusters with genes participating in specific functions of interest.

- ranges of values (or symbols) to ignore from the input matrix, $remove(S)$ where $S \subseteq \mathbb{R}^+$ (or $S \subseteq \mathcal{L}$). In gene expression, elements with default/non-differential expression are generally less relevant and thus can be removed. This is achieved by removing these elements from the transactions. Despite the simplicity of this constraint, this option is not easily supported by peer biclustering algorithms [16].
- minimum coherency strength (or number of symbols) of the target biclusters: $\delta = 1/|\mathcal{L}|$. Decreasing the coherency strength (increasing the number of symbols) reduces the noise-tolerance of the resulting set of bilusters and it is often associated with solutions composed by a larger number of biclusters with smaller areas.
- level of relaxation to handle noise by increasing the η_{ij} noise range (Definition 2). This constraint is used to adjust the behavior of BiC2PAM in the presence of noise or discretization problems (values near a boundary of discretization). By default, one symbol is associated with an element. Yet, this constraint gives the possibility to assign an additional symbol to an element when its value is near a boundary of discretization, or even a parameterizable number of symbols per element for a high tolerance to noise (proof in [11]).

Relevant constraints provided in the mining step:

- minimum pattern length (minimum number of columns in the bicluster).
- stopping criteria: either the anti-monotone minimum support length (minimum number of rows in the bicluster), or iteratively decreasing support until minimum number of biclusters is discovered or minimum area of the input matrix is coverage by the discovered biclusters.
- type of coherency and orientation. Currently, BiC2PAM supports the selection of constant, additive, multiplicative, symmetric, order-preserving and plaid models with coherency on rows or columns (according to [9,11]).
- pattern representation: simple (all coherent biclusters), closed (all maximal biclusters), and maximal (solutions with a compact number of biclusters with a preference towards a high number of columns).

Understandably, constraints addressed at the postprocessing stage are not desirable since they are not able to seize major efficiency gains. Nevertheless, BiC2PAM supports two key types of constraints that could imply additional computational costs, but are addressed with heightened efficiency: *1)* maximum percentage of noisy and missing elements per bicluster (based on merging procedures [11]), and *2)* minimum homogeneity of the target biclusters (using extension and reduction procedures with a parameterizable merit function [11]).

4.2   Biologically Meaningful Constraints

Different types of constraints were introduced in Definition 5. In order to illustrate how such constraints can be specified and instantiated, a symbolic gene expression matrix (and associated "price table") is provided in Fig.2, where the rows correspond to different genes and the values correspond to observed levels of expressions for a specific condition (column). The {-3,-2}, {-1,0,1} and {2,3} sets of symbols are respectively associated with repressed (down-regulated), default (preserved) and activated (up-regulated) levels of expression.

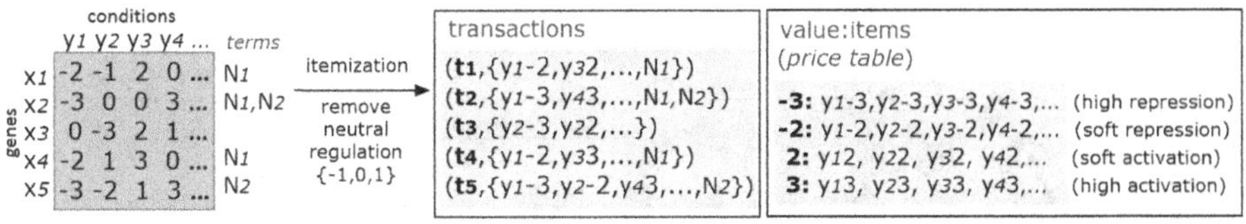

Fig. 2. Illustrative symbolic dataset and "price table" for expression data analysis.

First, **succinct** constraints in gene expression analysis allow the discovery of genes with specific constrained levels of expression across a subset of conditions. Illustrating, $min(P)=-3$ implies an interest in biclusters (biological processes) where genes are at least highly repressed in one condition. Alternatively, succinct constraints can be used to discover non-trivial biclusters by focusing on non-highly differential expression (e.g. patterns with symbols $\{-2,2\}$). Such option contrasts with the large focus on dense biclusters [16]. Finally, succinct constraints can also be used to guarantee that a specific condition of interest appears in the resulting set (e.g. $P \cap \{y_2\text{-}3, y_2\text{-}2, y_2 2, y_2 3\} \neq \emptyset$ to include y_2), or a specific annotation ($P \cap \{N_1, N_2\} \neq \emptyset$).

Second, **(anti-)monotone** constraints are key to capture background knowledge and guide biclustering. Illustrating, the non-succinct monotonic constraint $countVal(P) \geq 2$ implies that at least two different levels of expression must be present within a bicluster (biological process). In gene expression analysis, biclusters should be able to accommodate genes with different degrees of up-regulation and/or down-regulation. Yet, the majority of existing biclustering approaches are only able to model constant values across conditions [11,16]. When constraints, such as the value-counting inequality, are available, the pruning of the search space allows an efficient handling of very low support thresholds for these non-trivial biclusters to be discovered.

Finally, **convertible** constraints also play an important role in biological settings to guarantee, for instance, that the observed patterns have an average of values within a specific range. Illustrating, the anti-monotonic convertible constraint $avg(P) \leq 0$ indicates a preference for patterns with repression mechanisms without a strict exclusion of activation mechanisms. These constraints are useful to focus the discovery on specific expression levels, while still allowing for noise deviations. Understandably, they are a robust alternative to the use of strict bounds from succinct constraints with maximum-minimum inequalities.

4.3 Effective Use of Constraints in Pattern-based Biclustering

Although native constraints are supported through adequate parameterizations of pattern-based biclustering algorithms, the previous (non-native) constraints are not directly supported. Nevertheless, as surveyed, pattern mining searches have been extended to seize efficiency gains when succinct, (anti-)monotone or convertible constraints are considered. Although there is large consensus that pattern-growth searches are better positioned to seize efficiency gains from constraints than peer methods based on bitset vectors, there is not yet proof whether

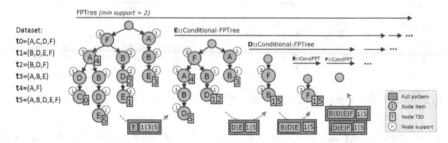

Fig. 3. Illustrative behavior of F2G [13].

this observation remains valid in the context of full-pattern mining. As such, we extend the recently proposed F2G algorithm to guarantee an optimal pruning of the search space in the presence of constraints and integrate F2G in BiC2PAM. F2G implements a pattern-growth search that does not suffer from efficiency bottlenecks since it relies on tree structures where transaction-IDs are stored without duplicates[2]. F2G behavior is illustrated in Fig.3. In this section, we first show the compliance of F2G with principles to handle succinct and convertible constraints [23]. Second, we show the compliance of F2G with principles to handle difficult combinations of monotone and anti-monotone constraints [3].

Compliance with Different Types of Constraints. Unlike candidate generation methods, pattern growth methods (such as FP-Growth) provide further pruning opportunities. Pruning principles can be standardly applied on both the original database (full FP-Tree) and on each projected database (conditional FP-Tree). CFG extensions to FP-Growth [23] seize the properties of such constraints under three simple principles. First, supersets of itemsets violating anti-monotone constraints are removed for each (conditional) FP-Tree (e.g. for $y_1 2$ conditional database, remove conflicting items $\cup_{i=1}^m \{y_i 2, y_i 3\}$ as their sum violates $sum(P) \leq 3$). For an effective pruning, it is recommended to order the symbols in the header table according to their value and support [23,24]. F2G is compliant with these removals, since it allows the rising of transaction-IDs in the FP-Tree according to the order of candidate items for removal in the header table (property explained in [13]).

For the particular case of an anti-monotone convertible constraint, itemsets that satisfy the constraint are efficiently generated under a pattern-growth search [24] (e.g. $\{y_1\text{-}3, y_2 2, y_4 2\}$ itemset is not included in the generated pattern set respecting $avg(P) \leq 0$), and provide a simple criterion to either stop FP-tree projections or prune items in a (conditional) FP-Tree.

Finally, the removal of conflicting transactions (e.g. t_1 and t_4 does not satisfy the illustrated succinct constraint) and of individual items (e.g. $\cup_{i=1}^m \{y_i\text{-}$

---

[2] The FP-tree is recursively mined to enumerate all full-patterns. Unlike peer pattern-growth searches, transaction-IDs are not lost at the first scan. Full-patterns are generated by concatenating the pattern suffixes with the full-patterns discovered from conditional FP-trees where suffixes are removed. F2G is applicable on top of FP-Close trees to mine closed full-patterns [13].

$1, y_i0, y_i1\}$) do not cause changes in the FP-Tree construction methods. Additionally, constraint checks can be avoided for subsets of itemsets satisfying a monotone constraint (e.g. no further checks of $countVal(P) \geq 2$ constraint when the range of values in the suffix is ≥ 2 under the $\{y_10, y_11\}$-conditional FP-Tree).

Combination of Constraints. The previous extensions of pattern-growth searches are not able to effectively comply with monotone constraints when anti-monotone constraints (such as minimum support) are also considered. In FP-Bonsai [3], principles to further explore the monotone properties for pruning the search space are considered without reducing anti-monotone pruning opportunities. This method is based on the ExAnte synergy of two data-reduction operations that seize the properties of monotone constraints: μ-reduction, which deletes transactions not satisfying C; and α-reduction, which deletes from transactions single items not satisfying C. Thanks to the recursive projecting approach of FP-growth, the ExAnte data-reduction methods can be applied on each conditional FP-tree to obtain a compact number of smaller FP-Trees (FP-Bonsais). The FP-Bonsai method can be combined with the previously introduced principles, which are particularly prone to handle succinct and convertible anti-monotone constraints. Since F2G can be extended to support the pruning of FP-Trees, it complies with the FP-Bonsai extension.

5   Results and Discussion

In this section, we assess the performance of BiC2PAM on synthetic and real datasets with different types of constraints and three distinct full-pattern miners: AprioriTID[3], Eclat[3] and F2G. BiC2PAM is implemented in Java (JVM v1.6.0-24). The experiments were computed using an IC i5 2.30GHz with 6GB of RAM.

Results on Synthetic Data. The generated data settings are described in Table 1. Biclusters with different shapes and coherency strength ($|\mathcal{L}| \in \{4,7,10\}$) were planted by varying the number of rows and columns using Uniform distributions with ranges in Table 1. For each setting we instantiated 20 matrices with background values generated with Uniform and Gaussian distributions.

Table 1. Properties of the generated dataset settings.

Matrix size (♯rows × ♯columns)	500×50	1000×100	2000×200	4000×400
Nr. of hidden patterns	5	10	15	25
Nr. transactions for the hidden patterns	[10,14]	[14,30]	[30,50]	[50,100]
Nr. items for the hidden patterns	[5,7]	[6,8]	[7,9]	[8,10]

BiC2PAM was applied with a default merging option (70% of overlapping) and a decreasing support until a minimum number of 50 (maximal) biclusters was found. Fig.4 provides the results of parameterizing BiC2PAM with different pattern miners and two simple constraints defining the target coherency strength

---
[3] http://www.philippe-fournier-viger.com/spmf/

and symbols to remove. We observe that the proposed F2G miner is the most efficient option for denser data settings (looser coherency). Also, in contrast with existing biclustering algorithms, BiC2PAM seizes large efficiency gains from neglecting specific ranges of values (symbols) from the input matrix.

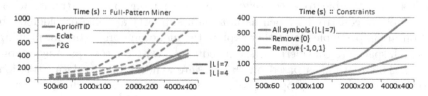

Fig. 4. BiC2PAM performance in the presence of simplistic native constraints.

In order to test the ability of BiC2PAM to seize further efficiency gains in the presence of non-trivial constraints, we fixed the 2000×200 setting with 6 symbols/values {-3,-2,-1,1,2,3}. In the baseline performance, constraints were satisfied using post-filtering procedures. Fig.5 illustrates this analysis. As observed, the use of constraints can significantly reduce the search complexity when they are properly incorporated within the full-pattern mining method. In particular, CFG principles [23] are used to seize efficiency gains from convertible constraints and FP-Bonsai [3] to seize efficiency gains from monotonic constraints.

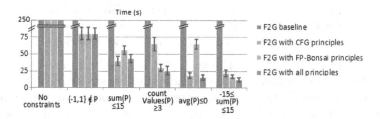

Fig. 5. Efficiency gains of considering constraints in F2G using different principles.

Results on Real Data. Fig.6 shows the (time and memory) efficiency of applying BiC2PAM in the yeast[4] expression dataset with different pattern miners and varying support thresholds for a desirable coherency strength of 10% ($|\mathcal{L}|=10$). The proposed F2G is the most efficient option in terms of time and, along with Apriori, a competitive choice for efficient memory usage.

Finally, Figs.7 and 8 show the impact of biologically meaningful constraints in the efficiency and effectiveness of BiC2PAM. For this purpose, we used the complete gasch dataset (6152×176) [6] with six levels of expression ($|\mathcal{L}|=6$). The effect of constraints in the efficiency is shown in Fig.7. This analysis supports their key role of providing opportunities to solve hard biomedical tasks.

---

[4] http://www.upo.es/eps/bigs/datasets.html

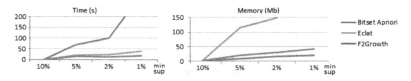

Fig. 6. Computational time and memory of full-pattern miners for yeast (2884×17).

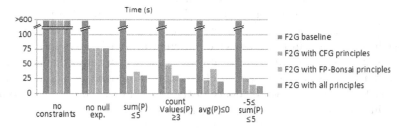

Fig. 7. Efficiency gains from using biological constraints for gasch (6152×176).

The impact of these constraints in the relevance of pattern-based biclustering solutions is illustrated in Fig.8. The biological relevance of each bicluster was derived from the functionally enriched terms using an hypergeometric test of Gene Ontology (GO) annotations [17]. As a measure of significance, we counted the number of terms with Bonferroni corrected p-values below 0.01 [17]. Two major observations can be retrieved. First, when focusing on properties of interest (e.g. differential expression), the average significance of biclusters increases as their genes have higher propensity to be functionally co-regulated. This trend is observed despite the smaller size of the constrained biclusters. Second, when focusing on rare expression profiles (≥3 distinct levels of expression), the average relevance of biclusters slightly decreases as their co-regulation is less obvious. Yet, such non-trivial biclusters hold unique properties with potential interest.

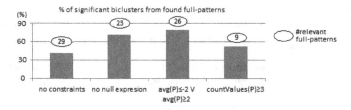

Fig. 8. Biological relevance of F2G for multiple constraint-based profiles of expression.

6   Conclusions

This work motivates the task of biclustering biological data in the presence of constraints. To answer this task, we explore the synergies between pattern-based biclustering and domain-driven pattern mining. As a result, BiC2PAM algorithm is proposed to effectively incorporate constraints derived from user expectations and available background knowledge.

Two major sets of constraints were proposed for the discovery of biclusters with specific interestingness criteria. First, native constraints to guarantee the discovery of biclusters with parameterizable coherency, noise-tolerance and shape, and to consider annotations from knowledge-based repositories. Second, constraints with succinct, monotone, anti-monotone and convertible properties to focus the search space on non-trivial yet biologically meaningful patterns.

In this context, we extended a recent pattern-growth search to optimally explore efficiency gains in the presence of different types of constraints.

Results from synthetic and real data show that biclustering benefits from large efficiency gains in the presence of constraints derived from background knowledge. We further provide evidence of the relevance of the supported types of constraints to discover non-trivial yet meaningful biclusters in expression data.

Acknowledgments. This work was supported by *FCT* under the project UID/CEC/ 50021/2013 and the PhD grant SFRH/BD/75924/2011 to RH.

References

1. Besson, J., Robardet, C., De Raedt, L., Boulicaut, J.-F.: Mining Bi-sets in numerical data. In: Džeroski, S., Struyf, J. (eds.) KDID 2006. LNCS, vol. 4747, pp. 11–23. Springer, Heidelberg (2007)
2. Bonchi, F., Giannotti, F., Mazzanti, A., Pedreschi, D.: Exante: a preprocessing method for frequent-pattern mining. IEEE Intel. Systems **20**(3), 25–31 (2005)
3. Bonchi, F., Goethals, B.: FP-Bonsai: the art of growing and pruning small FP-trees. In: Dai, H., Srikant, R., Zhang, C. (eds.) PAKDD 2004. LNCS (LNAI), vol. 3056, pp. 155–160. Springer, Heidelberg (2004)
4. Bonchi, F., Lucchese, C.: Extending the state-of-the-art of constraint-based pattern discovery. Data Knowl. Eng. **60**(2), 377–399 (2007)
5. Fang, G., Haznadar, M., Wang, W., Yu, H., Steinbach, M., Church, T.R., Oetting, W.S., Van Ness, B., Kumar, V.: High-Order SNP Combinations Associated with Complex Diseases: Efficient Discovery, Statistical Power and Functional Interactions. Plos One 7 (2012)
6. Gasch, A.P., Werner-Washburne, M.: The genomics of yeast responses to environmental stress and starvation. Functional & integrative genomics **2**(4–5), 181–192 (2002)
7. Guerra, I., Cerf, L., Foscarini, J., Boaventura, M., Meira, W.: Constraint-based search of straddling biclusters and discriminative patterns. JIDM **4**(2), 114–123 (2013)
8. Han, J., Cheng, H., Xin, D., Yan, X.: Frequent pattern mining: current status and future directions. Data Min. Knowl. Discov. **15**(1), 55–86 (2007)
9. Henriques, R., Madeira, S.: Biclustering with flexible plaid models to unravel interactions between biological processes. IEEE/ACM Trans, Computational Biology and Bioinfo (2015). doi:10.1109/TCBB.2014.2388206
10. Henriques, R., Antunes, C., Madeira, S.C.: Generative modeling of repositories of health records for predictive tasks. Data Mining and Knowledge Discovery, pp. 1–34 (2014)

11. Henriques, R., Madeira, S.: Bicpam: Pattern-based biclustering for biomedical data analysis. Algorithms for Molecular Biology 9(1), 27 (2014)
12. Henriques, R., Madeira, S.: Bicspam: Flexible biclustering using sequential patterns. BMC Bioinformatics 15, 130 (2014)
13. Henriques, R., Madeira, S.C., Antunes, C.: F2g: Efficient discovery of full-patterns. In: ECML /PKDD IW on New Frontiers to Mine Complex Patterns. Springer-Verlag, Prague, CR (2013)
14. Khiari, M., Boizumault, P., Crémilleux, B.: Constraint programming for mining n-ary patterns. In: Cohen, D. (ed.) CP 2010. LNCS, vol. 6308, pp. 552–567. Springer, Heidelberg (2010)
15. Kuznetsov, S.O., Poelmans, J.: Knowledge representation and processing with formal concept analysis. Wiley Interdisc. Reviews: Data Mining and Knowledge Discovery 3(3), 200–215 (2013)
16. Madeira, S.C., Oliveira, A.L.: Biclustering algorithms for biological data analysis: A survey. IEEE/ACM Trans. Comput. Biol. Bioinformatics 1(1), 24–45 (2004)
17. Martin, D., Brun, C., Remy, E., Mouren, P., Thieffry, D., Jacq, B.: Gotoolbox: functional analysis of gene datasets based on gene ontology. Genome Biology (12), 101 (2004)
18. Martinez, R., Pasquier, C., Pasquier, N.: Genminer: Mining informative association rules from genomic data. In: BIBM, pp. 15–22. IEEE CS (2007)
19. Mouhoubi, K., Létocart, L., Rouveirol, C.: A knowledge-driven bi-clustering method for mining noisy datasets. In: Huang, T., Zeng, Z., Li, C., Leung, C.S. (eds.) ICONIP 2012, Part III. LNCS, vol. 7665, pp. 585–593. Springer, Heidelberg (2012)
20. Nepomuceno, J.A., Troncoso, A., Nepomuceno-Chamorro, I.A., Aguilar-Ruiz, J.S.: Integrating biological knowledge based on functional annotations for biclustering of gene expression data. Computer Methods and Programs in Biomedicine (2015)
21. Ng, R.T., Lakshmanan, L.V.S., Han, J., Pang, A.: Exploratory mining and pruning optimizations of constrained associations rules. SIGMOD R. 27(2), 13–24 (1998)
22. Okada, Y., Fujibuchi, W., Horton, P.: A biclustering method for gene expression module discovery using closed itemset enumeration algorithm. IPSJ T. on Bioinfo. 48(SIG5), 39–48 (2007)
23. Pei, J., Han, J.: Can we push more constraints into frequent pattern mining? In: KDD. pp. 350–354. ACM, New York (2000)
24. Pei, J., Han, J.: Constrained frequent pattern mining: a pattern-growth view. SIGKDD Explor. Newsl. 4(1), 31–39 (2002)
25. Serin, A., Vingron, M.: Debi: Discovering differentially expressed biclusters using a frequent itemset approach. Algorithms for Molecular Biology 6, 1–12 (2011)
26. Visconti, A., Cordero, F., Pensa, R.G.: Leveraging additional knowledge to support coherent bicluster discovery in gene expression data. Intell. Data Anal. 18(5), 837–855 (2014)

A Critical Evaluation of Methods for the Reconstruction of Tissue-Specific Models

Sara Correia and Miguel Rocha(✉)

Centre of Biological Engineering, University of Minho, Braga, Portugal
mrocha@di.uminho.pt

Abstract. Under the framework of constraint based modeling, genome-scale metabolic models (GSMMs) have been used for several tasks, such as metabolic engineering and phenotype prediction. More recently, their application in health related research has spanned drug discovery, biomarker identification and host-pathogen interactions, targeting diseases such as cancer, Alzheimer, obesity or diabetes. In the last years, the development of novel techniques for genome sequencing and other high-throughput methods, together with advances in Bioinformatics, allowed the reconstruction of GSMMs for human cells. Considering the diversity of cell types and tissues present in the human body, it is imperative to develop tissue-specific metabolic models. Methods to automatically generate these models, based on generic human metabolic models and a plethora of omics data, have been proposed. However, their results have not yet been adequately and critically evaluated and compared.

This work presents a survey of the most important tissue or cell type specific metabolic model reconstruction methods, which use literature, transcriptomics, proteomics and metabolomics data, together with a global template model. As a case study, we analyzed the consistency between several omics data sources and reconstructed distinct metabolic models of hepatocytes using different methods and data sources as inputs. The results show that omics data sources have a poor overlapping and, in some cases, are even contradictory. Additionally, the hepatocyte metabolic models generated are in many cases not able to perform metabolic functions known to be present in the liver tissue. We conclude that reliable methods for *a priori* omics data integration are required to support the reconstruction of complex models of human cells.

1   Introduction

Over the last years, genome-scale metabolic models (GSMMs) for several organisms have been developed, mainly for microbes with an interest in Biotechnology [6,20]. These models have been used to predict cellular metabolism and promote biological discovery [17], under constraint-based approaches such as Flux Balance Analysis (FBA) [18]. FBA finds a flux distribution that maximizes biomass production, considering the knowledge of stoichiometry and reversibility of reactions, and taking some simplifying assumptions, namely assuming quasi steady-state conditions.

© Springer International Publishing Switzerland 2015
F. Pereira et al. (Eds.) EPIA 2015, LNAI 9273, pp. 340–352, 2015.
DOI: 10.1007/978-3-319-23485-4_35

Recently, efforts on model reconstruction have also addressed more complex multicellular organisms, including humans [5,9,25]. In biomedical research, they have been used, for instance, to elucidate the role of proliferative adaptation causing the Warburg effect in cancer [23], to predict metabolic markers for inborn errors of metabolism [24] and to identify drug targets for specific diseases [11].

However, the human organism is quite complex, with a large number of cell types/tissues and huge diversity in their metabolic functions. This led to the need of developing tissue/cell type specific metabolic models, which could allow studying in more depth specific cell phenotypes. Towards this end, it was imperative to better characterize specific cell types, gathering relevant data. Indeed, an important set of technological advances in the last decades greatly increased available biological data through high-throughput studies that allow the identification and quantification of cell components (gene expression, proteins and metabolites). These are collectively known as 'omics' data and have generated new fields of study, such as transcriptomics, proteomics and metabolomics.

The most widely available omics data are transcriptomics, the quantification of gene expression levels in a cell, using DNA microarrays or sequencing (e.g. RNA-seq). The most significant databases for gene expression data are the Gene Expression Omnibus (GEO) [3] and the ArrayExpress [19]. Other resources use those databases as references to synthesize their information, such as the Gene Expression Barcode [16], which provides absolute measures of expression for the most annotated genes, organized by tissue, cell-types and diseases.

In the cells, mRNA is not always translated into protein, and the amounts of protein depend on gene expression but also on other factors. Thus, knowledge about the amounts of proteins in the cell, provided by proteomics data, is of foremost relevance. These data can confirm the presence of proteins and provide measurements of their quantities for each protein within a cell. For human cells, a database is available with millions of high-resolution images showing the spatial distribution of protein expression profiles in normal tissues, cancer and cell lines - the Human Protein Atlas (HPA) portal [26].

Another source of information is provided by metabolomics data that involve the quantification of the small molecules present in cells, tissues, organs and biological fluids using techniques such as Nuclear Magnetic Resonance spectroscopy or Gas Chromatography combined with Mass Spectrometry [13]. The Human Metabolome Database (HMDB) [28] contains spectroscopic, quantitative, analytic and molecular-scale information about human metabolites, associated enzymes or transporters, their abundance and disease-related properties.

Resources for omics data, together with generic human metabolic models, have been used to generate context-specific models. This has been achieved through the development of methods, such as the Model Building Algorithm (MBA) [12], the Metabolic Context-specificity Assessed by Deterministic Reaction Evaluation (mCADRE)[27] and the Task-driven Integrative Network Inference for Tissues (tINIT)[2].

The reconstructed models have allowed, for instance, to find metabolic targets to inhibit the proliferation of cancer cells [29], to study the interaction between distinct brain cells [14], and to find potential therapeutic targets for the treatment of non-alcoholic steatohepatitis [15].

However, the aforementioned methods have not yet been critically and systematically evaluated on standardized case studies. Indeed, each of the methods is proposed and validated with distinct cases and taking distinct omics data sources as inputs. Thus, the impact of using different omics datasets on the final results of those algorithms is a question that remains to be answered. Here, we present a critical evaluation of the most important methods for the reconstruction of tissue-specific metabolic models published until now.

We have developed a framework where we implemented different methods for the reconstruction of tissue-specific metabolic models. In this scenario, the algorithms use sets of metabolites and/or maps of scores for each reaction as input. So, in our framework the algorithms are independent from the omics data source, and the separation of these two layers allows to use different data sources in each algorithm for the generation of tissue-specific metabolic models. As a case study, to compare the three different approaches implemented, metabolic models were reconstructed for hepatocytes, using the same set of data sources as inputs for each algorithm. Moreover, distinct combinations of data sources are evaluated to check their influence on the final results.

2   Materials and Methods

2.1   Human Metabolic Models

Modeling metabolic systems requires the analysis and prediction of metabolic flux distributions under diverse physiological and genetic conditions. The human organism is one of the most complex organisms to build a metabolic model since the number of genes, types of cells and their diversity are huge. In the last years, a few human metabolic models were proposed [5,9,15,25]. In this work, we will use the Recon 2 human metabolic model that accounts for 7440 reactions, 5063 metabolites and 1789 enzyme-encoding genes. This model is a community-driven expansion of the previous human reconstruction, Recon 1 [5], with additional information from different resources: EHMN [9], Hepatonet1[8], Ac-FAO module[21] and the small intestinal enterocyte reconstruction [22].

2.2   Algorithms for Tissue-Specific Metabolic Models Reconstruction

Although there are several applications of the human GSMMs, the specificity of cell types requires the reconstruction of tissue-specific metabolic models. Some approaches have been proposed based on existing generic human models. Here, we present three of the most well-known approaches for this task that will be used in the remaining of this work.

MBA. The Model-Building Algorithm (MBA) [12] reconstructs a tissue-specific metabolic model from a generic model and two sets of reactions, denoted as core reactions (C_H) and reactions with a moderate probability to be carried out in the specific tissue (C_M). These sets were previously built according to evidence levels based on omics data, literature and experimental knowledge. In general, the C_H set includes human-curated tissue-specific reactions and the C_M set includes reactions certified by omics data. The algorithm iteratively removes one reaction from the generic model, in a random order, and validates if the model remains consistent. The process ends when the removal of all reactions, except the ones in C_H, is tried. As a result, this algorithm reconstructs a model containing all the C_H reactions, as many as possible C_M reactions, and a minimal set of other reactions that are required for obtaining overall model consistency (for each reaction there is a flux distribution in which it is active).

Since reactions are scanned in a random order, the authors recommend to run the algorithm a large number of times to generate intermediate models. After this step, a score per each reaction is calculated, according to the number of times it appears in these models. The final model is built starting from C_H and iteratively adding reactions ordered by their scores, until a final consistent model is achieved.

INIT/ tINIT. The Integrative Network Inference for Tissues (INIT) [1] uses the Human Protein Atlas (HPA) as its main source of evidence. Expression data can be used when proteomic evidence is missing. It also allows the integration of metabolomics data by imposing a positive net production of metabolites for which there is experimental support, for instance in HMDB. The algorithm is formulated using mixed integer-linear programming (MILP), so that the final model contains reactions with high scores from HPA data. This algorithm does not impose strict steady-state conditions for all internal metabolites, allowing a small net accumulation rate. A couple of years later, a new version of this algorithm was proposed, the Task-driven Integrative Network Inference for Tissues (tINIT) [2], which reconstructs tissue-specific metabolic models based on protein evidence from HPA and a set of metabolic tasks that the final context-specific model must perform. These tasks are used to test the production or uptake of external metabolites, but also the activation of pathways that occur in a specific tissue. Another improvement from the previous version is the addition of constraints to guarantee that irreversible reactions operate in one direction only.

mCADRE. The Metabolic Context specificity Assessed by Deterministic Reaction Evaluation (mCADRE) [27] method is able to infer a tissue-specific network based on gene expression data, network topology and reaction confidence levels. Based on the expression score, the reactions of the global model, used as template, are ranked and separated in two sets - core and non-core. All reactions with expression-based scores higher than a threshold value are included in the core set, while the remaining reactions make the non-core set. In this method, the expression score does not represent the level of expression, but rather the frequency of expressed states over several transcript profiles. So, it is necessary to previously binarize the expression data. Thus, it is possible to use data retrieved from the Gene Expression Barcode project that already contains binary information on which genes are present or not in a specific tissue/cell type. Reactions from the non-core set are ranked according to the expression scores, connectivity-based scores and confidence level-based scores. Then, sequentially, each reaction is removed and the consistency of the model is tested. The elimination only occurs if the reaction does not prevent the production of a key-metabolite and the core consistency is preserved. Comparing with the MBA algorithm, mCADRE presents two improvements: it allows the definition of key metabolites, i.e. metabolites that have evidence to be produced in the context-specific model reconstruction, and relaxes the condition of including all core reactions in the final model.

Table 1 shows the mathematical formulation and pseudocode for all algorithms described above.

Table 1. Formulation and description of algorithms of MBA, tINIT and mCADRE. In the table R_G represents the list of reaction from the global model, R_C the set of core reaction on mCADRE algorithm, C_H and C_M the core and moderate probability sets used in MBA algorithm, r a reaction and the $for(i)$ and the $rev(i)$ represent the i-th reaction direction (forward and reverse).

MBA	tINIT	mCADRE
$generateModel(R_G, C_H, C_M)$	$min \sum_{i \in R} w_i * y_i$	$generateModel(R_G, treshold)$
$\quad R_P \leftarrow R_G$	$\quad s.t.$	$\quad R_P \leftarrow R_G$
$\quad R_S \leftarrow R_P \backslash (C_H \cup C_M)$	$\quad Sv = b$	$\quad R_C \leftarrow score(R_P) > treshold$
$\quad P \leftarrow randomPermutation(R_S)$	$\quad |v_i| \leq v_{max}$	$\quad coreActiveG \leftarrow flux(r)! = 0, r \in R_C$
$\quad for(r \in P)$	$\quad 0 < v_i + (v_{max} * y_i) \leq v_{max}$	$\quad R_{NC} \leftarrow R_P \backslash R_C$
$\quad\quad inactiveR \leftarrow CheckModel(R_P, r)$	$\quad b_j \geq \delta\, j \in Metabolomics$	$\quad for(r \in order(R_{NC}))$
$\quad\quad e_H \leftarrow inactiveR \cap C_H$	$\quad b_j = 0\, j \notin Metabolomics$	$\quad\quad inactiveR \leftarrow CheckModel(R_P, r)$
$\quad\quad e_M \leftarrow inactiveR \cap C_M$	$\quad y_{for(i)} + y_{rev(i)} \leq 1$	$\quad\quad s1 = |inactiveR \cap R_C|$
$\quad\quad e_X \leftarrow inactiveR \backslash (C_M \cup C_H)$	$\quad v_i \geq \delta, i \in RequiredReac$	$\quad\quad s2 = |inactiveR \cap R_{NC}|$
$\quad\quad if(|e_H| == 0\, AND\, |e_M| < \delta * |e_X|)$	$\quad y_i \in 0, 1$	$\quad\quad if(r \notin withExpressionValues\, AND$
$\quad\quad\quad R_P \leftarrow R_P \backslash (e_M \cup e_X)$	$\quad w_i, score\ for\ i \in R$	$\quad\quad\quad s1 \backslash s2 <= RACIO\, AND$
$\quad\quad endif$		$\quad\quad\quad checkModelFunction(R_p \backslash inactiveR))$
$\quad endfor$		$\quad\quad\quad R_P \leftarrow R_P \backslash inactiveR$
$\quad returnR_P$		$\quad\quad elseif(|s1| == 0\, AND$
$endfunction$		$\quad\quad\quad checkModelFunction(R_p \backslash inactiveR))$
		$\quad\quad\quad R_P \leftarrow R_P \backslash inactiveR$
		$\quad\quad endif$
		$\quad returnR_P$
		$endfunction$

2.3 Omics Data

Proteomics data used in this work were retrieved from the Human Protein Atlas (HPA) [26], which contains the profiles of human proteins in all major human healthy and cancer cells. We collected information for the liver tissue (hepatocytes) from HPA version 12 and Ensembl [7] version 73.37. After a conversion from Ensembl gene identifiers to gene symbols, duplicated genes with different evidence levels were removed (Table S1 from supplementary data)[1].

---

[1] All supplementary files are provided in http://darwin.di.uminho.pt/epia2015

Transcriptomics data were collected from Gene Expression Barcode (GEB) [16] (HGU133plus2 (Human) cells v3). The conversion to gene expression levels was done considering the average level of probes for each gene. The mapping between probes and gene symbols was performed using the library "hgu133plus2.db" [4] from Bioconductor. The gene expression is classified as *High, Moderate* and *Low* if the gene expression evidence on that tissue is greater than 0.9, between 0.5 and 0.9, and between 0.1 and 0.5, respectively. The genes with expression evidence below 0.1 were considered not expressed in hepatocytes.

The reaction scores were obtained through the Gene-Protein-Rules present in the Recon2 model, based on the scores associated with each gene in the data. The reaction scores were calculated by taking the maximum (minimum) value of expression scores for genes connected by an "OR" ("AND"). If one of the gene scores is unknown, the other gene score is assumed in the conversion rule.

3   Results

To compare the metabolic models generated by the different algorithms and the effects of distinct omics data sources, we chose the reconstruction of hepatocytes metabolic models as our case study. Hepatocytes are the principal site of the metabolic conversions underlying the diverse physiological functions of the liver [10]. The hepatocytes metabolic models were generated using Recon2 as a template model and the GEB, HPA and the sets C_H and C_M from [12] as input data, for the three methods described in the previous section.

In the experiments, we seek to answer two main questions: Are omics data consistent across different data sources? What is the overlap of the resulting metabolic models obtained using different methods and different data sources? In 2010, a manually curated genome-scale metabolic network of human hepatocytes was presented, the HepatoNet1 [8], used as a reference in the validation process.

3.1   Omics Data Consistency

The HPA has evidence information related with 16324 genes in hepatocytes. The reliability of the data is also scored as "supportive" or "uncertain", depending on similarity in immunostaining patterns and consistency with protein/gene characterization data. On the other hand, the GEB transcriptome has information for 20149 genes, of which 5772 have evidence of being expressed in hepatocytes.

Together, these two data sources have information for 21921 genes, but only 14552 are present in both (Figure 1A). Moreover, the number of genes with evidence of being expressed in the tissue in both sources is only of 3549, around 24% of all shared genes (Figure 1B). These numbers decrease significantly if using only HPA information marked as "supportive". In this scenario, only 3868 genes are present also in GEB and only 1294 of them have expression evidence.

Next, evidence levels frequencies (*High, Moderate, Low*) were calculated across the GEB and HPA, as shown in Figure 2. Only a small number of genes have similar evidence levels in both data sources. Furthermore, a significant

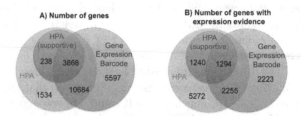

Fig. 1. A) Number of genes present in Gene Expression Barcode and Human Protein Atlas. In HPA, the number of genes with reliability "supportive" and "uncertain" are shown. B) Number of genes with evidence level "Low", "Moderate" or "High" in HPA and gene expression evidence higher than 0 in Gene Expression Barcode.

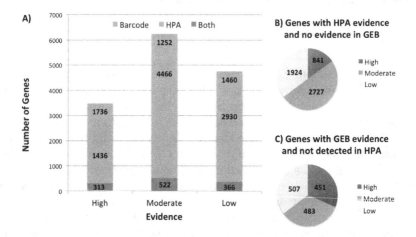

Fig. 2. A) Distribution of genes from Gene Expression Barcode project and Human Protein Atlas across the evidence levels - "High", "Moderate" and "Low". The ranges [0.9, 1], [0.5, 0.9[and [0.1, 0.5[were used to classify the data into "Low", "Moderate" and "High" levels. B) Genes with no evidence to be present in hepatocytes from GEB, but with evidence in the HPA. C) Genes with no evidence to be present in hepatocytes from HPA, but with evidence in GEB.

number of genes have contradictory levels of evidence - genes with expression evidence in one data source and not expressed in the other. If we focus only in the genes present in the model Recon2 with information in GEB and HPA (supportive), there are 15% of genes with "High" or "Moderate" evidence in one of the sources and not expressed in the other. This number increases to 22% if we also consider "Low" evidence level (Supplementary Figure S1).

The methods to reconstruct tissue-specific metabolic models use reaction scores calculated based on omics data to determine their inclusion in the final models. So, we analyzed the impact of these omics discrepancies in the values of reaction scores and compared those with the manually curated set C_H from Jerby et al. [12]. In Figure 3A, the poor overlap of the reaction scores calculated based

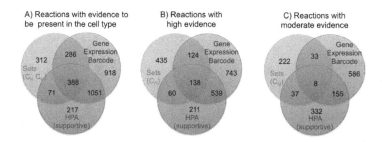

Fig. 3. A) Reactions with evidence that support their inclusion in the hepatocytes metabolic model. B) Number of reactions that have a high level of evidence of expression for each data source. C) Number of reactions that have a moderate evidence of expression for each data source

on different sources can be observed. Considering all data sources, 3243 reactions show some evidence that support their inclusion in the hepatocytes metabolic model, but only 388 are supported by all sources. The numbers are further dramatically reduced if we consider only moderate or high levels of evidence (Figure 3 B-C).

3.2 Metabolic Models

We applied each of the three algorithms to each omics data source, resulting in nine metabolic models for hepatocytes. In the application of mCADRE, we consider the list of key metabolites as published in the original article and a threshold of 0.5 to calculate the core set. A set of core metabolic tasks, that should occur in all cell types, was retrieved from [2] and used in the tINIT algorithm. The final MBA models were constructed based on 50 intermediate metabolic models. According to [12], a larger number would be desirable, but the time needed to generate each model prevented larger numbers of replicates. The detailed list of reactions that compose each metabolic model are available in supplementary material.

In Figure 4 A-C, we observe the consistency of the intermediate models generated by MBA, as well as the number of occurrences of reactions present in the final model. Moreover, Figure 4D shows the relations between the nine metabolic models generated through hierarchical clustering. The models obtained using the C_H and C_M sets as input data group together. Regarding the remaining, the mCADRE and MBA resulting models group according to their data (HPA and GEB), while the models created by tINIT cluster together. Overall, the data used as input seems to be the most relevant factor in the final result.

A more detailed comparison between the models reconstructed using the same algorithm or the same data source is available in Figure 5, A and B respectively. Considering the models generated by the same algorithm, it is observed that mCADRE has a smaller overlap (only 812 reactions) compared to the other methods. This could be explained by the possibility of removing core reactions during the mCADRE reconstruction process. Note that both reactions with

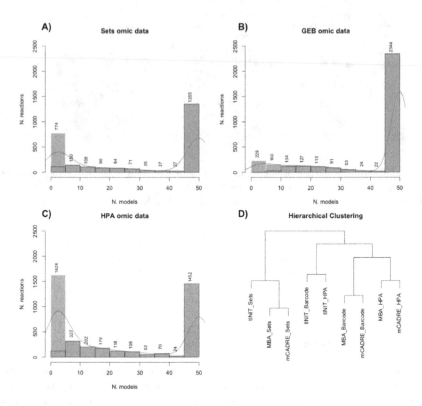

Fig. 4. A-C) Distribution of reactions across the 50 models for each data type. Grey bars show an histogram with the number of reactions present in a certain number of models. Green bars show the reactions that are present in the final model. D) Results from hierarchical clustering of the resulting nine models.

"High" and "Moderate" evidence levels, and from C_H and C_M sets, are all considered as belonging to the core. Furthermore, the mean of reactions that belong to all models of the same algorithm is around 45%. When the comparison is made by grouping models with the same input data, the variance between models is lower than grouping by algorithm. Here, the mean of reactions common to all models with the same data source is around 67% (Supplementary Table S3). Again, the variability of the final results seems to be dominated by the data source factor.

The quality of the metabolic models was further validated using the metabolic functions that are known to occur in hepatocytes [8]. The generic Recon2 human metabolic model, used as template in the reconstruction process, is able to satisfy 337 of the 408 metabolic functions available. Metabolic functions related with disease or involving metabolites not present in Recon2 were removed from the original list.

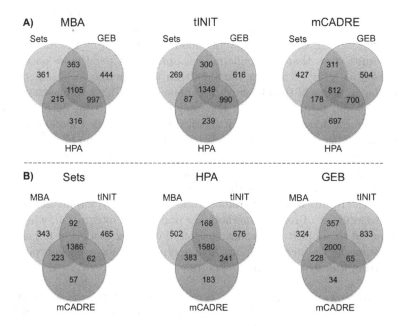

Fig. 5. Metabolic models reaction intersection considering: (A) the same algorithm; (B) the same omics data source.

Table 2. Number of reactions and the percentage of liver metabolic functions that each metabolic model performs when compared with the template model - Recon 2.

Method	Sets		HPA		GEB	
	N. Reac.	Tasks	N. Reac	Tasks	N. Reac.	Tasks
MBA	2044	18%	2633	24%	2909	6%
mCADRE	1728	2%	2387	3%	2327	4%
tINIT	2005	4%	2665	5%	3255	6%

The results of this functional validation, showing also the number of reactions in each metabolic model, are given on Table 2. These show that the number of satisfied metabolic tasks is very low compared with the manual curated metabolic model HepatoNet1. The metabolic model which performs the higher number of metabolic tasks was obtained using the MBA algorithm with the HPA evidence. Nevertheless, the success percentage is less than 25% when comparing with the performance by the template metabolic model - Recon2.

4   Conclusions

In this work, we present a survey of the most important methods for the reconstruction of tissue-specific metabolic models. Each method was proposed to use different data sources as input. Here, we analyze the consistency of information across important omics data sources used in this context and verify the impact of such differences in the final metabolic models generated by the methods.

The results show that metabolic models obtained depend more on the data sources used as inputs, than on the algorithm used for the reconstruction. To validate the accuracy of the obtained metabolic models, a set of metabolic functions that should be performed in hepatocytes was tested for each metabolic model. We found that the number of satisfied liver metabolic functions was surprisingly low. This shows that methods for the reconstruction of tissue-specific metabolic models based on a single omics data source are not enough to generate high quality metabolic models. Methods to combine several omics data sources to rank the reactions for the reconstruction process could be a solution to improve the results of these methods. Indeed, this study emphasizes the need for the development of reliable methods for omics data integration, which seem to be required to support the reconstruction of complex models of human cells, but also reinforce the need to be able to incorporate known phenotypical data available from literature or human experts.

Acknowledgments. S.C. thanks the FCT for the Ph.D. Grant SFRH/BD/80925/2011. The authors thank the FCT Strategic Project of UID/BIO/04469/2013 unit, the project RECI/BBB-EBI/0179/2012 (FCOMP-01-0124-FEDER-027462) and the project "BioInd - Biotechnology and Bioengineering for improved Industrial and Agro-Food processes", REF. NORTE-07-0124-FEDER-000028 Co-funded by the Programa Operacional Regional do Norte (ON.2 - O Novo Norte), QREN, FEDER.

References

1. Agren, R., Bordel, S., Mardinoglu, A., Pornputtapong, N., Nookaew, I., Nielsen, J.: Reconstruction of Genome-Scale Active Metabolic Networks for 69 Human Cell Types and 16 Cancer Types Using INIT. PLoS Computational Biology **8**(5), e1002518 (2012)
2. Agren, R., Mardinoglu, A., Asplund, A., Kampf, C., Uhlen, M., Nielsen, J.: Identification of anticancer drugs for hepatocellular carcinoma through personalized genome-scale metabolic modeling. Molecular Systems Biology **10**, 721 (2014)
3. Barrett, T., Troup, D.B., Wilhite, S.E., Ledoux, P., et al.: NCBI GEO: archive for functional genomics data sets - 10 years on. Nucleic Acids Research **39**(suppl 1), D1005–D1010 (2011)
4. Carlson, M.: hgu133plus2.db: Affymetrix Human Genome U133 Plus 2.0 Array annotation data (chip hgu133plus2) (2014). r package version 3.0.0
5. Duarte, N.C., Becker, S.A., Jamshidi, N., Thiele, I., Mo, M.L., Vo, T.D., Srivas, R., Palsson, B.O.: Global reconstruction of the human metabolic network based on genomic and bibliomic data. Proceedings of the National Academy of Sciences of the United States of America **104**(6), 1777–1782 (2007)
6. Duarte, N.C., Herrgård, M.J., Palsson, B.O.: Reconstruction and validation of Saccharomyces cerevisiae iND750, a fully compartmentalized genome-scale metabolic model. Genome Research **14**(7), 1298–1309 (2004)

7. Flicek, P., Amode, M.R., Barrell, D., et al.: Ensembl 2014. Nucleic Acids Research **42**(D1), D749–D755 (2014)
8. Gille, C., Bölling, C., Hoppe, A., et al.: HepatoNet1: a comprehensive metabolic reconstruction of the human hepatocyte for the analysis of liver physiology. Molecular Systems Biology **6**(411), 411 (2010)
9. Hao, T., Ma, H.W., Zhao, X.M., Goryanin, I.: Compartmentalization of the Edinburgh Human Metabolic Network. BMC Bioinformatics **11**, 393 (2010)
10. Ishibashi, H., Nakamura, M., Komori, A., Migita, K., Shimoda, S.: Liver architecture, cell function, and disease. Seminars in Immunopathology **31**(3) (2009)
11. Jerby, L., Ruppin, E.: Predicting Drug Targets and Biomarkers of Cancer via Genome-Scale Metabolic Modeling. Clinical Cancer Research : An Official Journal of the American Association for Cancer Research **18**(20), 5572–5584 (2012)
12. Jerby, L., Shlomi, T., Ruppin, E.: Computational reconstruction of tissue-specific metabolic models: application to human liver metabolism. Molecular Systems Biology **6**(401), 401 (2010)
13. Kaddurah-Daouk, R., Kristal, B., Weinshilboum, R.: Metabolomics: a global biochemical approach to drug response and disease. Annu. Rev. Pharmacol. Toxicol. **48**, 653–683 (2008)
14. Lewis, N.E., Schramm, G., Bordbar, A., Schellenberger, J., Andersen, M.P., Cheng, J.K., Patel, N., Yee, A., Lewis, R.A., Eils, R., König, R., Palsson, B.O.: Large-scale in silico modeling of metabolic interactions between cell types in the human brain. Nature Biotechnology **28**(12), 1279–1285 (2010)
15. Mardinoglu, A., Agren, R., Kampf, C., Asplund, A., Uhlen, M., Nielsen, J.: Genome-scale metabolic modelling of hepatocytes reveals serine deficiency in patients with non-alcoholic fatty liver disease. Nature Communications **5**, Jan 2014
16. McCall, M.N., Jaffee, H.A., Zelisko, S.J., Sinha, N., et al.: The Gene Expression Barcode 3.0: improved data processing and mining tools. Nucleic Acids Research **42**(D1), D938–D943 (2014)
17. Oberhardt, M.A., Palsson, B.O., Papin, J.A.: Applications of genome-scale metabolic reconstructions. Molecular Systems Biology **5**(320), 320 (2009)
18. Orth, J.D., Thiele, I., Palsson, B.O.: What is flux balance analysis? Nature Biotechnology **28**(3), 245–248 (2010)
19. Parkinson, H., Sarkans, U., Shojatalab, M., Abeygunawardena, N., et al.: ArrayExpress-a public repository for microarray gene expression data at the EBI. Nucleic Acids Research **33**(Database issue), Jan 2005
20. Reed, J.L., Vo, T.D., Schilling, C.H., Palsson, B.O.: An expanded genome-scale model of Escherichia coli K-12 (i JR904 GSM / GPR) **4**(9), 1–12 (2003)
21. Sahoo, S., Franzson, L., Jonsson, J.J., Thiele, I.: A compendium of inborn errors of metabolism mapped onto the human metabolic network. Mol. BioSyst. **8**(10), 2545–2558 (2012)
22. Sahoo, S., Thiele, I.: Predicting the impact of diet and enzymopathies on human small intestinal epithelial cells. Human Molecular Genetics **22**(13), 2705–2722 (2013)
23. Shlomi, T., Benyamini, T., Gottlieb, E., Sharan, R., Ruppin, E.: Genome-scale metabolic modeling elucidates the role of proliferative adaptation in causing the Warburg effect. PLoS Computational Biology **7**(3), e1002018 (2011)
24. Shlomi, T., Cabili, M.N., Ruppin, E.: Predicting metabolic biomarkers of human inborn errors of metabolism. Molecular Systems Biology **5**(263), 263 (2009)
25. Thiele, I., Swainston, N., Fleming, R.M.T., et al.: A community-driven global reconstruction of human metabolism. Nature Biotechnology **31**(5), May 2013

26. Uhlen, M., Oksvold, P., Fagerberg, L., Lundberg, E., et al.: Towards a knowledge-based Human Protein Atlas. Nat Biotech **28**(12), 1248–1250 (2010)
27. Wang, Y., Eddy, J.A., Price, N.D.: Reconstruction of genome-scale metabolic models for 126 human tissues using mCADRE. BMC Systems Biology **6**(1), 153 (2012)
28. Wishart, D.S., Knox, C., Guo, A.C., Eisner, R., et al.: HMDB: a knowledgebase for the human metabolome. Nucleic Acids Research **37**(suppl 1), Jan 2009
29. Yizhak, K., Le Dévédec, S.E., Rogkoti, V.M.M., et al.: A computational study of the Warburg effect identifies metabolic targets inhibiting cancer migration. Molecular Systems Biology **10**(8) (2014)

Fuzzy Clustering for Incomplete Short Time Series Data

Lúcia P. Cruz, Susana M. Vieira, and Susana Vinga[✉]

IDMEC, Instituto Superior Técnico, Universidade de Lisboa, Lisboa, Portugal
{lucia.cruz,susana.vieira,susanavinga}@tecnico.ulisboa.pt

Abstract. The analysis of clinical time series is currently a key topic in biostatistics and machine learning applications to medical research. The extraction of relevant features from longitudinal patients data brings several problems for which novel algorithms are warranted. It is usually impossible to measure many data points due to practical and also ethical restrictions, which leads to short time series (STS) data. The sampling might also be at unequally spaced time-points and many of the predicted measurements are often missing. These problems constitute the rationale of the present work, where we present two methods to deal with missing data in STS using fuzzy clustering analysis. The methods are tested and compared using data with equal and varying time sampling interval lengths, with and without missing data. The results illustrate the potential of these methods in clinical studies for patient classification and feature selection using biomarker time series data.

Keywords: Short time series · Missing data · Fuzzy c-means (FCM) clustering

1 Introduction

In medical care, the efficient acquisition of information is subject to many obstacles related with ethical and experimental restrictions, which usually leads to longitudinal data with unequal and long sampling periods. Furthermore, these time series are usually sparse and incomplete, which further hampers their analysis. Along with this, the high costs associated with medical analysis lead to the necessity of performing less frequent tests. For Intensive Care Unit (ICU) cases, for example, this issues have been studied. In [1] the issue of missing data, in medical datasets, was addressed. In oncological studies, survival analysis is usually applied to identify probability distributions. Regression methods, such as Cox proportional hazards models, are further used to identify the statistically significant features associated with survival [2,3]. In this type of clinical studies it is common to have biomarkers time series, which may be used to diagnose and predict the outcome of the disease.

The main motivation of this work is to developed and test clustering algorithms for biomarkers time series, a problem arising for example in the analysis of

© Springer International Publishing Switzerland 2015
F. Pereira et al. (Eds.) EPIA 2015, LNAI 9273, pp. 353–359, 2015.
DOI: 10.1007/978-3-319-23485-4_36

bone metastatic patients. Due to the problems referred before, these biomarkers measurements are commonly short time series with missing data. Many methods are not able to deal, at the same time, with missing data and short time series, let alone, with unevenly sampled time series. The proposed clustering algorithm is able to take into account both missing data and short time series, evenly or unevenly sampled. The approaches are unsupervised, since the marker study objective is to relate the outcome of the unsupervised clustering with outcomes of the patient's health.

2   Methods for Short Incomplete Time Series Data

In this section we present the definitions and clustering algorithms to deal with a collection of n short time series with missing data. All the presented approaches are based on fuzzy c-means (FCM) algorithm [4].

A time series k of length s can be represented as $\mathbf{x}_k = (x_{k1}, \ldots, x_{kj}, \ldots, x_{ks})$ with $x_{kj} \equiv x_k(t_j)$, with $1 \leq j \leq s$ and $1 \leq k \leq n$.

Furthermore, with the creation of cluster prototypes, v_{ij}, for FCM algorithm, i represents cluster numbers, subject to $1 \leq i \leq c$, where c is the selected cluster number.

2.1   Incomplete Data

In this section several approaches to deal with missing data are presented. These might include some sort of imputation and deletion [1], further compared for pattern recognition problems in [5] and explored in a fuzzy clustering setting [4]. The two methods here presented are the *Partial Distance Strategy* (PDS) [4] and *Optimal Completion Strategy* (OCS) [4], both applied to fuzzy clustering. Both use the Euclidean distance, but any other metric can also be used.

The PDS method computes the distance between two vectors, scaling it to the proportion of non missing values to complete vectors size. For the entries with missing values, the distance is simply updated with zero, as expressed in Eq. 1:

$$D_{ik} = \frac{s}{I_k} \sum_{j=1}^{s} (x_{kj} - v_{ij})^2 I_{kj}, \qquad (1)$$

where $I_k = \sum_{j=1}^{s} I_{kj}$ and $I_{kj} = \begin{cases} 0, & \text{if } x_{kj} \in X_M \\ 1, & \text{if } x_{kj} \in X_P \end{cases}$, for $1 \leq j \leq s$, $1 \leq k \leq n$.

The OCS approach is an imputation method where the entries with missing data, x_{kj}, are substituted with a given value. For fuzzy clustering these values are initialized at random, and their update results from the computation of Eq. 2, using the partition matrix U_{ik} and cluster prototypes v_{ij}. This equation derives from the calculation of the cluster prototypes [4].

$$x_{kj} = \frac{\sum_{i=1}^{c} (U_{ik})^m v_{ij}}{\sum_{i=1}^{c} (U_{ik})^m} \qquad (2)$$

2.2 Short Time Series

This section presents the Short Time Series (STS) approach, where both even and uneven sampled time series can be dealt with.

In survey [6] the only method capable of handling short time series is presented in [7], which also deals with unevenly sampled time series. The STS distance proposed $d_{STS}(x, v) = \sum_{j=0}^{s} \left(\frac{v_{(j+1)} - v_j}{t_{(j+1)} - t_j} - \frac{x_{(j+1)} - x_j}{t_{(j+1)} - t_j} \right)^2$ is computed between the cluster prototypes, v_j, and the time series data points, x_j, including the corresponding times, t_j. Based in the STS approach, two new methods to deal with STS were developed.

1st order derivative calculation with normalization (Slopes): It uses the approximate derivatives between each consecutive data point as input. First the z-score normalization of the series is calculated [7], $z_k = \frac{x_k - \bar{x}}{s_x}$, where $\bar{x}$ is the mean value of x_k and s_x is the standard deviation of x_k. The derivatives of each consecutive points are then computed, $dv_{kj} = \frac{z_{(j+1)} - z_j}{t_{(j+1)} - t_j}$.

Combination of 1st order derivatives of time points (Slopes Comb): This method's objective is to generate a combination of time series data points derivatives yet with different, increasing lags. A vector is created with the derivatives of consecutive data points (or intervals of one point). It is then incremented with the derivatives of data points separated by two sample points, and so on and so forth, until no more points separation is possible (maximum lag possible is attained with x_s and x_1).

As an example lets consider the time series $S = (x_1, x_2, x_3, x_4)$ and equivalent time points $T = (t_1, t_2, t_3, t_4)$. The vector creation, $V = (V_1, V_2, V_3)$, would be $V_1 = \left(\frac{x_2 - x_1}{t_2 - t_1}, \frac{x_3 - x_2}{t_3 - t_2}, \frac{x_4 - x_3}{t_4 - t_3} \right)$, $V_2 = \left(\frac{x_3 - x_1}{t_3 - t_1}, \frac{x_4 - x_2}{t_4 - t_2} \right)$ and $V_3 = \left(\frac{x_4 - x_1}{t_4 - t_1} \right)$, where V_1 is the derivative of data points with interval of one point, V_2 is the derivative of data points with interval of two points and V_3 is the derivative of data points with interval of three points. The vector V will be the input used in this method.

2.3 Short Incomplete Time Series Fuzzy Clustering Algorithms

The algorithm construction is described as follows, making use of FCM [4]. For the FCM algorithm the objective function for any of the methods is to minimize $J(x, v, U) = \sum_{i=1}^{c} \sum_{k=1}^{n} U_{i,k}^m \, d^2(x_k, v_i)$.

Having the methods to deal with missing data (PDS and OCS) and the method to deal with short time series (STS), we propose 5 combinations: *PDS-Slopes, OCS-Slopes, PDS-Slopes Comb, OCS-Slopes Comb* and *STS-OCS*. For the combinations with *Slopes* and *Slopes Comb*, the FCM algorithm is the same as in [4]. The alterations occur prior to the algorithm calculations, where to generate the new dataset to be used it is applied either *Slopes* or *Slopes Comb*.

The *STS - OCS* case uses mainly the algorithm described in [7], with the STS distance, yet the dataset to be used needs to be initialized with the missing values imputation. This imputation is done after the normalization part in the algorithm [7], and is based on Eq. 2, however the cluster prototypes calculation follows the procedure described in [7].

3   Results and Discussion

These methods were tested on 4 datasets generated according to [7]. Each original dataset contains 20 time series classified into four classes (five time series per class) but with different characteristics, depending on the number of time points t and if they are equally or unequally spaced. In Fig. 1 these datasets are represented, where each cluster is defined with the same line type.

Every method was tested with the complete information (0% missing values) and with an increasing percentage of missing entries. These were generated randomly, equally distributed through all the dataset. The restrictions imposed were that no line (time series samples) nor column (time points) could be entirely composed of missing data. The results presented were obtained for 500 runs with 500 different random seeds. The 500 seeds are kept the same for every method tested to guarantee comparison exactness. In Fig. 2 the results for the 4 datasets tested are presented in terms of accuracy of the final clustering obtained. We compared the *Mean number of misclassifications* with the *Percentage of missing data (%)*, given that these results are the mean values of the 500 runs results. The percentages of missing data tested were 0% (complete dataset), 10%, 20%, 30% and 40%. Since all datasets contain 20 time series samples, for e.g. a *mean number of misclassification* of 10 time series corresponds to 50% of misclassifications. By observing Fig. 2 it is clear that the more missing values the dataset has, the more misclassifications will be obtained. PDS and OCS roughly maintain their performance throughout increasing percentages of missings. *STS-OCS* is the method which deteriorated the fastest, most likely due to the bias caused by the imputation and the inadequate missing values update. When comparing *Slopes* with *Slopes* and *Slopes Comb* with *Slopes Comb*, it is noteworthy that the

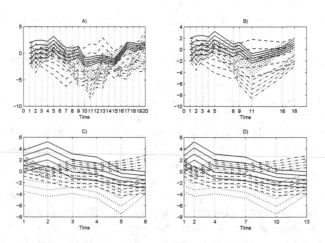

Fig. 1. Time Series classification. For each dataset the 20 time series are divided by the 4 clusters, specified by the different contours.

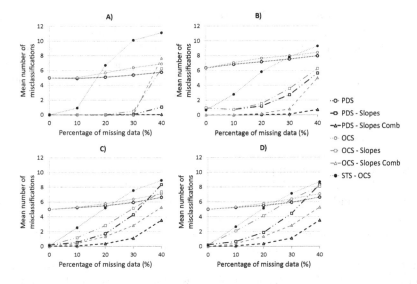

Fig. 2. Misclassifications of clustering methods. Average number of misclassifications for 500 runs of time series with 20 samples, four classes, using different sampling time points T and intervals.

Table 1. Results of FCM variations for 40% missing data.

Datasets	Methods	Iteration Count	Misclassifications	Objective Function Error	NID
T = 20 equal	PDS	70.52 (22.69)	5.75 (1.51)	5.26E-03 (6.06E-02)	0
	PDS - Slopes	25.57 (13.85)	1.04 (1.67)	9.13E-05 (1.38E-03)	22
	PDS - Slopes Comb	16.32 (4.98)	**0.03 (0.42)**	5.95E-06 (2.25E-06)	0
	OCS	88.21 (16.00)	6.88 (1.61)	5.97E-03 (3.49E-02)	0
	OCS - Slopes	52.10 (31.12)	6.24 (3.82)	4.92E-03 (3.50E-02)	0
	OCS - Slopes Comb	51.67 (24.58)	7.64 (3.93)	6.19E-03 (5.66E-02)	6
	STS - OCS	49.45 (30.62)	11.12 (1.56)	8.27E-04 (4.93E-03)	0
T = 10 unequal	PDS	63.62 (23.89)	7.94 (1.48)	1.47E-03 (1.51E-02)	0
	PDS - Slopes	34.89 (18.52)	5.62 (2.34)	1.39E-05 (7.45E-05)	358
	PDS - Slopes Comb	24.69 (12.77)	**0.71 (1.31)**	9.87E-06 (5.61E-05)	11
	OCS	75.65 (21.06)	8.40 (1.54)	1.24E-03 (1.41E-02)	0
	OCS - Slopes	89.17 (18.78)	6.24 (2.02)	9.65E-04 (5.49E-03)	0
	OCS - Slopes Comb	71.19 (28.97)	4.99 (3.55)	1.36E-02 (9.08E-02)	1
	STS - OCS	80.67 (20.11)	9.27 (1.78)	2.14E-04 (9.26E-04)	0
T = 6 equal	PDS	54.80 (22.06)	6.60 (1.44)	1.64E-03 (2.39E-02)	0
	PDS - Slopes	27.67 (5.51)	8.33 (3.21)	7.10E-06 (2.52E-06)	497
	PDS - Slopes Comb	28.13 (15.10)	**3.52 (2.30)**	2.22E-04 (3.14E-03)	285
	OCS	63.84 (24.28)	7.07 (1.50)	1.03E-03 (2.15E-02)	0
	OCS - Slopes	85.38 (19.71)	7.36 (1.81)	7.49E-04 (4.69E-03)	0
	OCS - Slopes Comb	87.09 (19.73)	5.26 (2.06)	4.23E-03 (2.90E-02)	1
	STS - OCS	70.46 (23.40)	8.92 (1.72)	1.83E-04 (1.53E-03)	0
T = 6 unequal	PDS	54.80 (22.06)	6.60 (1.44)	1.64E-03 (2.39E-02)	0
	PDS - Slopes	15.67 (5.13)	8.33 (2.31)	4.22E-06 (3.96E-06)	497
	PDS - Slopes Comb	28.13 (15.10)	**3.52 (2.30)**	2.22E-04 (3.14E-03)	285
	OCS	63.84 (24.28)	7.07 (1.50)	1.03E-03 (2.15E-02)	0
	OCS - Slopes	74.89 (23.76)	8.08 (1.78)	3.72E-04 (3.10E-03)	0
	OCS - Slopes Comb	87.09 (19.73)	5.27 (2.05)	4.23E-03 (2.90E-02)	1
	STS - OCS	45.31 (20.40)	8.70 (1.69)	1.36E-05 (7.26E-05)	0

PDS performs better than the OCS, probably due to the bias arising from the imputation. In all datasets, the *PDS-Slopes Comb* method performed the best.

In Table 1 the results for 40% of missing data are summarised. For each dataset every method was computed with 500 runs, and the mean and standard deviation of *Iteration Count, Misclassifications, Objective Function Error* and *Number of Ignored Datasets* (NID) is shown. Highlighted in each table are the best results for each dataset. The NID has values different from zero only for *Slopes* and *Slopes Comb*. This occurs whenever a time series sample has only one non missing value. It is noteworthy that the OCS approach avoids this problem through imputation while the PDS cannot overcome it.

4   Conclusions

This work describes and compares several algorithms for the clustering of short and incomplete time series data. As expected, all methods perform very well with complete data and the performance decreases when the percentage of missing data increases. Nevertheless it is possible to maintain a reasonable accuracy in the final classification even for values as large as 40%. When the original time series have some sort of underlying patterns or evident trends, the methods using the combination of different slopes (*Slopes comb*) with all possible lags is preferable. Overall, PDS with combined slopes achieves an excellent performance in the datasets tested, with average values of zero misclassified series when 10% of the values are missing. Interestingly, even for higher percentage of missing values, the performance of this method with combined derivatives or slopes is very high, with average misclassification much lower than what would be expected.

These algorithms can be applied directly in several areas of clinical studies, namely in oncology. One possible future application is the stratification of patients based on biomarker evolution, which is expected to have a direct impact on feature selection methods and survival analysis of oncological patients.

Acknowledgments. This work was supported by FCT, through IDMEC, under LAETA, project UID/EMS/50022/2013, CancerSys (EXPL/EMS-SIS/1954/2013) and Program Investigador FCT (IF/00653/2012 and IF/00833/2014), co-funded by the European Social Fund (ESF) through the Operational Program Human Potential (POPH).

References

1. Cismondi, F., et al.: Missing data in medical databases: Impute, delete or classify? Artificial Intelligence in Medicine **58**(1), 63–72 (2013)
2. Westhoff, P.G., et al.: An Easy Tool to Predict Survival in Patients Receiving Radiation Therapy for Painful Bone Metastases. International Journal of Radiation Oncology*Biology*Physics **90**(4), 739–747 (2014)
3. Harries, M., et al.: Incidence of bone metastases and survival after a diagnosis of bone metastases in breast cancer patients. Cancer Epidemiology **38**(4), 427–434 (2014)
4. Hathaway, R.J., Bezdek, J.C.: Fuzzy c-means clustering of incomplete data. IEEE Transactions on Systems, Man, and Cybernetics. Part B, Cybernetics : a Publication of the IEEE Systems, Man, and Cybernetics Society **31**(5), 735–744 (2001)

5. Dixon, J.K.: Pattern recognition with partly missing data. IEEE Transactions on Systems, Man and Cybernetics 9(10), 617–621 (1979)
6. Warren Liao, T.: Clustering of time series data - A survey. Pattern Recognition 38(11), 1857–1874 (2005)
7. Möller-Levet, C.S., Klawonn, F., Cho, K.-H., Wolkenhauer, O.: Fuzzy clustering of short time-series and unevenly distributed sampling points. In: Berthold, M., Lenz, H.-J., Bradley, E., Kruse, R., Borgelt, C. (eds.) IDA 2003. LNCS, vol. 2810, pp. 330–340. Springer, Heidelberg (2003)

General Artificial Intelligence

Allowing Cyclic Dependencies in Modular Logic Programming

João Moura$^{(\boxtimes)}$ and Carlos Viegas Damásio

CENTRIA / NOVA Laboratory for Computer Science and Informatics
(NOVA-LINCS), Universidade NOVA de Lisboa, Lisbon, Portugal
joaomoura@yahoo.com, cd@fct.unl.pt

Abstract. Even though modularity has been studied extensively in conventional logic programming, there are few approaches on how to incorporate modularity into Answer Set Programming, a prominent rule-based declarative programming paradigm. A major approach is Oikarinnen and Janhunen's Gaifman-Shapiro-style architecture of program modules, which provides the composition of program modules. Their module theorem properly strengthens Lifschitz and Turner's splitting set theorem for normal logic programs. However, this approach is limited by module conditions that are imposed in order to ensure the compatibility of their module system with the stable model semantics, namely forcing output signatures of composing modules to be disjoint and disallowing positive cyclic dependencies between different modules. These conditions turn out to be too restrictive in practice and after recently discussing alternative ways of lifting the first restriction [17], we now show how one can allow positive cyclic dependencies between modules, thus widening the applicability of this framework and the scope of the module theorem.

1 Introduction

Over the last few years, answer set programming (ASP) [2,6,12,15,18] emerged as one of the most important methods for declarative knowledge representation and reasoning. Despite its declarative nature, developing ASP programs resembles conventional programming: one often writes a series of gradually improving programs for solving a particular problem, e.g., optimizing execution time and space. Until recently, ASP programs were considered as integral entities, which becomes problematic as programs become more complex, and their instances grow. Even though modularity is extensively studied in logic programming, there are only a few approaches on how to incorporate it into ASP [1,5,8,19] or other module-based constraint modeling frameworks [11,22]. The research on modular systems of logic program has followed two main-streams [3], one is programming in-the-large where compositional operators are defined in order to combine different modules [8,14,20]. These operators allow combining programs algebraically, which does not require an extension of the theory of logic programs. The other direction is programming-in-the-small [10,16], aiming at enhancing logic programming with scoping and abstraction mechanisms available in other

© Springer International Publishing Switzerland 2015
F. Pereira et al. (Eds.) EPIA 2015, LNAI 9273, pp. 363–375, 2015.
DOI: 10.1007/978-3-319-23485-4_37

programming paradigms. This approach requires the introduction of new logical connectives in an extended logical language. The two mainstreams are thus quite divergent.

The approach of [19] defines modules as structures specified by a program (knowledge rules) and by an interface defined by input and output atoms which for a single module are, naturally, disjoint. The authors also provide a module theorem capturing the compositionality of their module composition operator. However, two conditions are imposed: there cannot be positive cyclic dependencies between modules and there cannot be common output atoms in the modules being combined. Both introduce serious limitations, particularly in applications requiring integration of knowledge from different sources. The techniques used in [5] for handling positive cycles among modules are shown not to be adaptable for the setting of [19].

In this paper we discuss two alternative solutions to the cyclic dependencies problem, generalizing the module theorem by allowing positive loops between atoms in the interfaces of the modules being composed. A use case for this requirement can be found in the following example.

Example 1. Alice wants to buy a safe and inexpensive car; she preselected 3 cars, namely c_1, c_2 and c_3. Her friend Bob says that car c_2 is expensive and Charlie says that car c_3 is expensive. Meanwhile, she consulted two car magazines reviewing all three cars. The first considered c_1 safe and the second considered c_1 to be safe while saying that c_3 may be safe if it has an optional airbag. Furthermore, if a friend declares that a car is expensive, then she will consider it safe. Alice is very picky regarding safety, and so she seeks some kind of agreement between the reviews.

The described situation can be captured with five modules, one for Alice, other three for her friends, and another for each magazine. Alice should conclude that c_1 is safe since both magazines agree on this. Therefore, one would expect Alice to opt for car c_1 since it is not expensive, and it is safe.   ∎

In summary, the fundamental results of [19] require a syntactic operation to combine modules – basically corresponding to the union of programs –, and a compositional semantic operation joining the models of the modules. The module theorem states that the models of the combined modules can be obtained by applying the semantics of the natural join operation to the original models of the modules – which is compositional.

This paper proceeds in Section 2 with an overview of the modular logic programming paradigm, identifying some of its shortcomings. In Section 3 we discuss alternative methods for lifting the restriction that disallows positive cyclic dependencies. We finish with conclusions and a general discussion.

2   Modularity in Answer Set Programming

Modular aspects of ASP have been clarified in recent years, with authors describing how and when two program parts (modules) can be composed [5,11,19] under the stable model semantics. In this paper, we will make use of Oikarinen and Janhunen's logic program modules defined in analogy to [8] which we review after presenting the syntax of ASP.

Answer Set Programming Logic programs in the ASP paradigm are formed by finite sets of rules r having the following syntax:

$$L_1 \leftarrow L_2, \ldots, L_m, not\ L_{m+1}, \ldots, not\ L_n. \quad (n \geq m \geq 0)(\mathbf{1})$$

where each L_i is a logical atom without the occurrence of function symbols – arguments are either variables or constants of the logical alphabet.

Considering a rule of the form (**1**), let $Head_P(r) = L_1$ be the literal in the head, $Body_P^+(r) = \{L_2, \ldots, L_m\}$ be the set with all positive literals in the body, $Body_P^-(r) = \{L_{m+1}, \ldots, L_n\}$ be the set containing all negative literals in the body, and $Body_P(r) = \{L_2, \ldots, L_n\}$ be the set containing all literals in the body. If a program is positive we will omit the superscript in $Body_P^+(r)$. Also, if the context is clear we will omit the subscript mentioning the program and write simply $Head(r)$ and $Body(r)$ as well as the argument mentioning the rule. The semantics of stable models is defined via the reduct operation [9]. Given an interpretation M (a set of ground atoms), the reduct P^M of a program P with respect to M is program $P^M = \{Head(r) \leftarrow Body^+(r) \mid r \in P, Body^-(r) \cap M = \emptyset\}$. An interpretation M is a stable model (SM) of P iff $M = LM(P^M)$, where $LM(P^M)$ is the least model of program P^M.

The syntax of logic programs has been extended with other constructs, namely weighted and choice rules [18]. In particular, choice rules have the following form:

$$\{A_1, \ldots, A_n\} \leftarrow B_1, \ldots B_k, not\ C_1, \ldots, not\ C_m.(n \geq 1)(\mathbf{2})$$

As observed by [19], the heads of choice rules possessing multiple atoms can be freely split without affecting their semantics. When splitting such rules into n different rules

$$\{a_i\} \leftarrow B_1, \ldots B_k, not\ C_1, \ldots, not\ C_m \text{ where } 1 \leq i \leq n,$$

the only concern is the creation of n copies of the rule body

$$B_1, \ldots B_k, not\ C_1, \ldots, not\ C_m.$$

However, new atoms can be introduced to circumvent this. There is a translation of these choice rules to normal logic programs [7], which we assume is performed throughout this paper but that is omitted for readability. We deal only with ground programs and use variables as syntactic place-holders.

2.1   Modular Logic Programming (MLP)

Modules in the sense of [19] are essentially sets of rules with Input/Output interfaces:

Definition 1 (Program Module). *A logic program module* $\mathcal{P}$ *is a tuple* $\langle R, I, O, H \rangle$ *where:*

1. *R is a finite set of rules;*
2. *I, O, and H are pairwise disjoint sets of input, output, and hidden atoms;*
3. *At(R) $\subseteq$ At($\mathcal{P}$) defined by At($\mathcal{P}$) $= I \cup O \cup H$; and*
4. *Head(R) $\cap I = \emptyset$.*

The set of atoms in $At_v(\mathcal{P}) = I \cup O$ are considered to be *visible* and hence accessible to other modules composed with $\mathcal{P}$ either to produce input for $\mathcal{P}$ or to make use of the output of $\mathcal{P}$. We use $At_i(\mathcal{P}) = I$ and $At_o(\mathcal{P}) = O$ to represent the input and output signatures of $\mathcal{P}$, respectively. The hidden atoms in $At_h(\mathcal{P}) = At(\mathcal{P}) \backslash At_v(\mathcal{P}) = H$ are used to formalize some auxiliary concepts of $\mathcal{P}$ which may not be sensible for other modules but may save space substantially. The condition $head(R) \not\subseteq I$ ensures that a module may not interfere with its own input by defining input atoms of I in terms of its rules. Thus, input atoms are only allowed to appear as conditions in rule bodies.

Example 2. The use case in Example 1 is encoded into the five modules shown here:

$$\mathcal{P}_A =< \quad \{ buy(X) \leftarrow car(X), safe(X), not\ exp(X).$$
$$car(c_1).\quad car(c_2).\quad car(c_3).\},$$
$$\{ safe(c_1), safe(c_2), safe(c_3), exp(c_1), exp(c_2), exp(c_3)\},$$
$$\{ buy(c_1), buy(c_2), buy(c_3)\},$$
$$\{ car(c_1), car(c_2), car(c_3)\} >$$
$$\mathcal{P}_B =< \quad \{ exp(c_2).\}, \{\}, \{exp(c_2), exp(c_3)\}, \{\} >$$
$$\mathcal{P}_C =< \quad \{ exp(c_3).\}, \{\}, \{exp(c_1), exp(c_2), exp(c_3)\}, \{\} >$$
$$\mathcal{P}_{mg_1} =< \{ \leftarrow not\ safe(c_1).\quad airbag(C) \leftarrow safe(C).\},$$
$$\{ safe(C)\},$$
$$\{ airbag(C)\},$$
$$\{\} >$$
$$\mathcal{P}_{mg_2} =< \{ safe(X) \leftarrow car(X), airbag(X).$$
$$car(c_1).\ car(c_2).\ car(c_3).\ \leftarrow not\ airbag(c_1).\ \{\leftarrow not\ airbag(c_3)\}.\ \},$$
$$\{ airbag(C)\},$$
$$\{ safe(c_1), safe(c_2), safe(c_3)\},$$
$$\{ airbag(c_1), airbag(c_2), airbag(c_3), car(c_1), car(c_2), car(c_3)\} > \quad \blacksquare$$

In Example 2, module $\mathcal{P}_A$ encodes the rule used by Alice to decide if a car should be bought. The safe and expensive atoms are its inputs, and the buy atoms its outputs; it uses hidden atoms $car/1$ to represent the domain of variables. Modules $\mathcal{P}_B$, $\mathcal{P}_C$ and $\mathcal{P}_{mg_1}$ captures the factual information in Example 1 and depends on input literal $safe$ to determine if its output states that a car has an *airbag* or not. They have no input and no hidden atoms, but *Bob* has only analyzed the price of cars c_2 and c_3. The ASP program module for the second magazine is more interesting[1], and expresses the rule used to determine if a car

---

[1] car belongs to both hidden signatures of $\mathcal{P}_A$ and $\mathcal{P}_{mg_2}$ which is not allowed when composing these modules, but for clarity we omit a renaming of the $car/1$ predicate.

is safe, namely that a car is safe if it has an airbag; it is known that car c_1 has an airbag, c_2 does not, and the choice rule states that car c_3 may or may not have an airbag.

Next, the SM semantics is generalized to cover modules by introducing a generalization of the Gelfond-Lifschitz's fixpoint definition. In addition to weakly negated literals (i.e., *not*), also literals involving input atoms are used in the stability condition. In [19], the SMs of a module are defined as follows:

Definition 2 (Stable Models of Modules). *An interpretation* $M \subseteq At(\mathcal{P})$ *is a SM of an ASP program module* $\mathcal{P} = \langle R, I, O, H \rangle$, *iff* $M = LM\left(R^M \cup \{a. | a \in M \cap I\}\right)$. *The SMs of* $\mathcal{P}$ *are denoted by* $AS(\mathcal{P})$.

Intuitively, the SMs of a module are obtained from the SMs of the rules part, for each possible combination of the input atoms.

Example 3. Program modules $\mathcal{P}_B$, $\mathcal{P}_C$, and $\mathcal{P}_{mg_1}$ have each a single answer set:
$AS(\mathcal{P}_B) = \{\{exp(c_2)\}\}$, $AS(\mathcal{P}_C) = \{\{exp(c_3)\}\}$, and $AS(\mathcal{P}_{mg_1}) = \{\{safe(c_1), airbag(c_1)\}\}$.

Module $\mathcal{P}_{mg_2}$ has two SMs, namely:
$\{safe(c_1), car(c_1), car(c_2), car(c_3), airbag(c_1)\}$, and
$\{safe(c_1), safe(c_3), car(c_1), car(c_2), car(c_3), airbag(c_1), airbag(c_3)\}$.

Alice's ASP program module has $2^6 = 64$ models corresponding each to an input combination of safe and expensive atoms. Some of these models are:

$$\{ buy(c_1), car(c_1), car(c_2), car(c_3), safe(c_1) \}$$
$$\{ buy(c_1), buy(c_3), car(c_1), car(c_2), car(c_3), safe(c_1), safe(c_3) \}$$
$$\{ buy(c_1), car(c_1), car(c_2), car(c_3), exp(c_3), safe(c_1), safe(c_3) \}\blacksquare$$

2.2 Composing Programs from Models

The composition of models is obtained from the union of program rules and by constructing the composed output set as the union of modules' output sets, thus removing from the input all the specified output atoms. [19] define their first composition operator as follows: Given two modules $\mathcal{P}_1 = \langle R_1, I_1, O_1, H_1 \rangle$ and $\mathcal{P}_2 = \langle R_2, I_2, O_2, H_2 \rangle$, their composition $\mathcal{P}_1 \oplus \mathcal{P}_2$ is defined when their output signatures are disjoint, that is, $O_1 \cap O_2 = \emptyset$, and they respect each others hidden atoms, i.e., $H_1 \cap At(\mathcal{P}_2) = \emptyset$ and $H_2 \cap At(\mathcal{P}_1) = \emptyset$. Then their composition is

$$\mathcal{P}_1 \oplus \mathcal{P}_2 = \langle R_1 \cup R_2, (I_1 \backslash O_2) \cup (I_2 \backslash O_1), O_1 \cup O_2, H_1 \cup H_2 \rangle$$

However, the conditions given for $\oplus$ are not enough to guarantee compositionality in the case of answer sets and as such they define a restricted form:

Definition 3 (Module Union Operator $\sqcup$). *Given modules* $\mathcal{P}_1, \mathcal{P}_2$, *their union is* $\mathcal{P}_1 \sqcup \mathcal{P}_2 = \mathcal{P}_1 \oplus \mathcal{P}_2$ *whenever* (i) $\mathcal{P}_1 \oplus \mathcal{P}_2$ *is defined and* (ii) $\mathcal{P}_1$ *and* $\mathcal{P}_2$ *are mutually independent meaning that there are no positive cyclic dependencies among rules in different modules, defined as loops through input and output signatures.*

Natural join ($\bowtie$) on visible atoms is used in [19] to combine the stable models of modules as follows:

Definition 4 (Join). *Given modules $\mathcal{P}_1$ and $\mathcal{P}_2$ and sets of interpretations $A_1 \subseteq 2^{At(\mathcal{P}_1)}$ and $A_2 \subseteq 2^{At(\mathcal{P}_2)}$, the natural join of A_1 and A_2 is:*

$$A_1 \bowtie A_2 = \{ M_1 \cup M_2 \mid M_1 \in A_1, M_2 \in A_2 \text{ and } M_1 \cap At_v(\mathcal{P}_2) = M_2 \cap At_v(\mathcal{P}_1)\}$$

This leads to their main result, stating that:

Theorem 1 (Module Theorem). *If $\mathcal{P}_1, \mathcal{P}_2$ are modules such that $\mathcal{P}_1 \sqcup \mathcal{P}_2$ is defined, then:*

$$AS(\mathcal{P}_1 \sqcup \mathcal{P}_2) = AS(\mathcal{P}_1) \bowtie AS(\mathcal{P}_2)$$

Still according to [19], their module theorem also straightforwardly generalizes for a collection of modules because the module union operator $\sqcup$ is commutative, associative, and has the identity element $< \emptyset, \emptyset, \emptyset, \emptyset >$.

Example 4. Consider the composition $\mathcal{Q} = (\mathcal{P}_A \sqcup \mathcal{P}_{mg_1}) \sqcup \mathcal{P}_B$. First, we have

$$\mathcal{P}_A \sqcup \mathcal{P}_{mg_1} = \left\langle \begin{array}{l} \{buy(X) \leftarrow car(X), safe(X), not\ exp(X). \\ car(c_1).\ car(c_2).\ car(c_3).\ safe(c_1).\}, \\ \{exp(c_1), exp(c_2), exp(c_3)\}, \\ \{buy(c_1), buy(c_2), buy(c_3), safe(c_1),\ safe(c_2),\ safe(c_3)\}, \\ \{car(c_1), car(c_2), car(c_3)\} \end{array} \right\rangle$$

It is immediate to see that the module theorem holds in this case. The visible atoms of $\mathcal{P}_A$ are $safe/1$, $exp/1$ and $buy/1$, and the visible atoms for $\mathcal{P}_{mg_1}$ are $\{safe(c_1),\ safe(c_2)\}$. The only model for $\mathcal{P}_{mg_1} = \{safe(c_1)\}$ when naturally joined with the models of $\mathcal{P}_A$, results in eight possible models where $safe(c_1)$, *not* $safe(c_2)$, and *not* $safe(c_3)$ hold, and $exp/1$ vary. The final ASP program module $\mathcal{Q}$ is

$$\left\langle \begin{array}{l} \{buy(X) \leftarrow car(X), safe(X), not\ exp(X). \\ car(c_1).\ car(c_2).\ car(c_3).\ exp(c_2).\ safe(c_1).\}, \\ \{exp(c_1)\}, \\ \{buy(c_1), buy(c_2), buy(c_3), exp(c_2), safe(c_1), safe(c_2), safe(c_3)\}, \\ \{car(c_1), car(c_2), car(c_3)\} \end{array} \right\rangle$$

The SMs of $\mathcal{Q}$ are thus:
$\{safe(c_1), exp(c_1), exp(c_2), car(c_1), car(c_2), car(c_3)\}$ and
$\{buy(c_1), safe(c_1), exp(c_2), car(c_1), car(c_2), car(c_3)\}$

2.3   Shortcomings

The conditions imposed in these definitions bring about some shortcomings such as the fact that the output signatures of two modules must be disjoint which disallows many practical applications e.g., we are not able to combine the results of program module $\mathcal{Q}$ with any of $\mathcal{P}_C$ or $\mathcal{P}_{mg_2}$, and thus it is impossible to obtain the combination of the five modules. Also because of this, the module union operator $\sqcup$ is not reflexive. By trivially waiving this condition, we immediately get problems with conflicting modules. The compatibility criterion for the operator $\bowtie$ also rules out the compositionality of mutually dependent modules, but allows positive loops inside modules or negative loops in general. We illustrate this in Example 5, which has been solved recently in [17] and the issue with positive loops between modules in Example 6 .

Example 5 (Common Outputs). Given $\mathcal{P}_B$ and $\mathcal{P}_C$, which respectively have:
$AS(\mathcal{P}_B) = \{\{exp(c_2)\}\}$ and $AS(\mathcal{P}_C)=\{\{exp(c_3)\}\}$,
the single SM of their union $AS(\mathcal{P}_B \sqcup \mathcal{P}_C)$ is: $\{exp(c_2), exp(c_3)\}$. However, the join of their SMs is $AS(\mathcal{P}_B) \bowtie AS(\mathcal{P}_C) = \emptyset$, invalidating the module theorem.∎

Example 6 (Cyclic Dependencies). Take the following two program modules (a simplification of the magazine modules in Example 2):

$$\mathcal{P}_1 = \langle\{airbag \leftarrow safe.\}, \{safe\}, \{airbag\}, \emptyset\rangle$$
$$\mathcal{P}_2 = \langle\{safe \leftarrow airbag.\}, \{airbag\}, \{safe\}, \emptyset\rangle$$

Their SMs are: $AS(\mathcal{P}_1) = AS(\mathcal{P}_2) = \{\{\}, \{airbag, safe\}\}$ while the single SM of the union $AS(\mathcal{P}_1 \sqcup \mathcal{P}_2)$ is the empty model $\{\}$. Therefore $AS(\mathcal{P}_1 \sqcup \mathcal{P}_2) \neq AS(\mathcal{P}_1) \bowtie AS(\mathcal{P}_2) = \{\{\}, \{airbag, safe\}\}$, also invalidating the module theorem.
∎

3 Positive Cyclic Dependencies Between Modules

To attain a generalized form of compositionality we need to be able to deal with both restrictions identified previously and in particular cyclic dependencies between modules. In the literature, [5] presents a solution based on a model minimality property. It forces one to check for minimality on every comparable models of all program modules being composed. It is not applicable to our setting though, which can be seen in Example 7 where logical constant $\bot$ represents value *false*. Example 7 shows that [5] is not compositional in the sense of Oikarinen and Janhunen.

Example 7. Given modules $\mathcal{P}_1 = \langle\{a \leftarrow b. \bot \leftarrow not\ b.\}, \{b\}, \{a\}, \{\}\rangle$ with one SM $\{a, b\}$, and $\mathcal{P}_2 = \langle\{b \leftarrow a.\}, \{a\}, \{b\}, \{\}\rangle$ with SMs $\{\}$ and $\{a, b\}$, their composition has no inputs and no intended SMs while their minimal join contains $\{a, b\}$. ∎

Another possible solution requires the introduction of extra information in the models to allow detecting mutual positive dependencies. This route has been identified before [21] and is left for future work.

3.1   Model Minimization

We present a model join operation that requires one to look at every model of both modules being composed in order to check for minimality on models comparable on account of their inputs. However, this operation is able to distinguish between atoms that are self supported through positive loops and atoms with proper support, allowing one to lift the condition in Definition 3 disallowing positive dependencies between modules.

Definition 5 (Minimal Join). *Given modules* $\mathcal{P}_1$ *and* $\mathcal{P}_2$*, let their composition be* $\mathcal{P}_C = \mathcal{P}_1 \oplus \mathcal{P}_2$*. Define* $AS(\mathcal{P}_1) \bowtie^{min} AS(\mathcal{P}_2) = \{M \mid M \in AS(\mathcal{P}_1) \bowtie AS(\mathcal{P}_2)$ *such that* $\nexists_{M' \in AS(\mathcal{P}_1) \bowtie AS(\mathcal{P}_2)} : M' \subset M$ *and* $M \cap At_i(\mathcal{P}_C) = M' \cap At_i(\mathcal{P}_C)\}$

Example 8 (Minimal Join). A car is safe if it has an airbag and it has an airbag if it is safe and the airbag is an available option. This is captured by two modules, namely: $\mathcal{P}_1 = \langle\{airbag \leftarrow safe, available\_option.\}, \{safe, available\_option\}, \{airbag\}, \emptyset\rangle$ and $\mathcal{P}_2 = \langle\{safe \leftarrow airbag.\}, \{airbag\}, \{safe\}, \emptyset\rangle$ which respectively have $AS(\mathcal{P}_1) = \{\{\}, \{safe\}, \{available\_option\}, \{airbag, safe, available\_option\}\}$ and $AS(\mathcal{P}_2) = \{\{\}, \{airbag, safe\}\}$. The composition has as its input signature $\{available\_option\}$ and therefore its answer set $\{airbag,safe,available\_option\}$ is not minimal regarding the input signature of the composition because $\{available\_option\}$ is also a SM (and the only intended model among these two). Thus $AS(\mathcal{P}_1 \oplus \mathcal{P}_2) = AS(\mathcal{P}_1) \bowtie^{min} AS(\mathcal{P}_2) = \{\{\}, \{available\_option\}\}$. ∎

This join operator allows us to lift the prohibition of composing mutually dependent modules under certain situations. Integrity constraints containing only input atoms in their body are still a problem with this approach as these would exclude models that would otherwise be minimal in the presence of unsupported loops.

Theorem 2 (Minimal Module Theorem). *If* $\mathcal{P}_1, \mathcal{P}_2$ *are modules such that* $\mathcal{P}_1 \oplus \mathcal{P}_2$ *is defined (allowing cyclic dependencies between modules), and that only normal rules are used in modules, then:*

$$AS(\mathcal{P}_1 \oplus \mathcal{P}_2) = AS(\mathcal{P}_1) \bowtie^{min} AS(\mathcal{P}_2)$$

3.2   Annotated Models for Composing Mutualy Dependent Modules

Because the former operator is not general and it forces us to compare one model with every other model for minimality, thus it is not local, we present next an alternative that requires adding annotations to models. We start by looking at positive cyclic dependencies (loops) that are formed by composition. It is known from the literature (e.g. [21]) that in order to do without looking at the rules of the program modules being composed, which in the setting of MLP we assume not having access to, we need to have extra information incorporated into the models.

Definition 6 (Dependency Transformation). *Let $\mathcal{P}$ be an MLP. Its dependency transformation is defined as the set of rules $(R_{\mathcal{P}})^A$ obtained from $R_{\mathcal{P}}$ by replacing each clause $L_1 \leftarrow L_2, \ldots, L_m, not\ L_{m+1}, \ldots, not\ L_n.(n \neq m)$ in $R_{\mathcal{P}}$ with the following clause, where $n \neq m$ and $D = D_2 \cup \ldots \cup D_m$ is a set of dependency sets:*

$$(1) \quad L_1 : D \leftarrow L_2 : D_2, \ldots, L_m : D_m, not\ L_{m+1} : D_{m+1}, \ldots, not\ L_n : D_n.$$

Definition 7 (Annotated Model). *Given a module $\mathcal{P} = \langle R^A, I, O, H \rangle$, its set of annotated models is constructed as before: An interpretation $M \subseteq At^A(\mathcal{P})$ is an annotated answer set of an ASP program module $\mathcal{P} = \langle R, I, O, H \rangle$, if and only if:*

$$M = LM\left((R^A)^M \cup \{a : \{\{a\}\}. \mid a : D \in M \cap I\}\right),$$

where $(R^A)^M_I$ is the version of the Gelfond-Lifschitz reduct allowing weighted and choice rules, of the dependency transformation of R, and LM is the operator returning the least model of the positive program argument. The set of annotated stable models of $\mathcal{P}$ is denoted by $AS^A(\mathcal{P})$.

Semantic of Annotated Programs. An annotated interpretation maps every atom into a set of subsets of input atoms, tracking the dependencies of the atom in combinations of input atoms. The semantics of annotated programs is obtained by iterating an immediate consequences monotonic operator applied to a definite program, defined as follows:

$$T_P(I)(L_1) = \bigcup \{T_P(I)(L_2) \cup \ldots \cup T_P(I)(L_m) \mid L_1 \leftarrow L_2, \ldots L_m \in R^A\}$$

starting from the interpretation mapping every atom into the empty set. In order to consider input atoms in modules we set $I(a) = \{\{a\}\}$ for every $a \in M \cap I$, and $\{\}$ otherwise.

Collapsed Annotated Models. Previous Definition 7 generates equivalent models for each alternative rule where atoms from the model belong to the head of the rule. We need to merge them into a collapsed annotated model where the alternatives are listed as sets of annotations, in order to retain a one to one correspondence between these and the SMs of the original program. As we are only interested in this collapsed form, we will henceforth take collapsed annotated models as annotated models.

Definition 8 (Collapsed Annotated Model). *Let M' and M'' be two annotated models such that for every atom $a \in M'$, it is also the case that $a \in M''$ and vice-versa. A collapsed annotated model M of M' and M'' is constructed as follows:*

$$M = \{a : \{D', D''\} \mid a : D' \in M' and\ a : D'' \in M''\}$$

Given a module $\mathcal{P}$, a program $P(M)$ can conversely be constructed from one of the module's annotated models M simply by adding rules of the form $a \leftarrow D_1, \ldots, D_m$. for each annotated atom $a_{\{D_1,\ldots,D_m\}} \in M$. Such constructed program $P(M)$ will be equivalent (but not strongly equivalent [13]) to taking the original program and adding facts that belong to the annotated model M, intersected with the input signature of $\mathcal{P}$, correspondingly.

Example 9 (Annotated Model). Let $\mathcal{P} = \langle \{a \leftarrow b, c. \ \ b \leftarrow d, not\ e, not\ f.\}, \{d, f\},$ $\{a, b\}, \emptyset \rangle$ be a module. $\mathcal{P}$ has one annotated model as per Definition 7: $\{b_{\{\{d\}\}}, d\}$. ∎

In the previous example, the first rule $a \leftarrow b, c.$ can never be activated because c is not an input atom $(c \notin I)$ and it is not satisfied by the rules of the module $(R_{\mathcal{P}} \not\models c)$. Thus, the only potential positive loop is identified by $\{b_{\{d\},d}\}$. If we compose $\mathcal{P}$ with a module containing e.g., rule $d \leftarrow b.$ and thus with an annotated model $\{d_{\{b\}}, b\}$, then it is easy to identify this as being a loop and if any atoms in the loop are satisfied by the module composition then there will be a stable model reflecting that. Also notice that since e is not a visible atom, it does not interfere with other modules, as long as it is respected, and thus it does not need to be in the annotation.

Cyclic Compatibility. We define next the compatibility of mutually dependent models. We assume that the outputs are disjoint as per the original definitions. The compatibility is defined as a two step criterion. The first is similar to the original compatibility criterion, only adapted to dealing with annotated models by disregarding the annotations. This first step makes annotations of negative dependencies unnecessary. The second step takes models that are compatible according to the first step and, after reconstructing two possible programs from the compatible annotated models, implies computing the minimal model of the union of these reconstructed programs and see if the union of the compatible models is a model of the union of the reconstructed programs.

Definition 9 (Basic Model Compatibility). *Let $\mathcal{P}_1$ and $\mathcal{P}_2$ be two modules. Let $AS^A(\mathcal{P}_1)$, respectively $AS^A(\mathcal{P}_2)$ be their annotated models. Let now $M_1 \in AS^A(\mathcal{P}_1)$ and $M_2 \in AS^A(\mathcal{P}_2)$ be two models of the modules, they will be compatible if:*

$$M_1 \cap At_v(\mathcal{P}_2) = M_2 \cap At_v(\mathcal{P}_1)$$

Now, for the second step of the cyclic compatibility criterion one takes models that passed the basic compatibility criterion and reconstruct their respective possible programs as defined previously. Then one computes the minimal model of the union of these reconstructed programs and see if the union of the originating models is a model of the union of their reconstructed programs.

Definition 10 (Annotation Compatibility). *Let $\mathcal{P}_1$ and $\mathcal{P}_2$ be two modules. Let $AS^A(\mathcal{P}_1)$, respectively $AS^A(\mathcal{P}_2)$ be their annotated models. Let now $M_1 \in AS^A(\mathcal{P}_1)$ and $M_2 \in AS^A(\mathcal{P}_2)$ be two compatible models according to Definition 9. They will be compatible annotated models if $AS^A(\mathcal{P}_1 \cup \mathcal{P}_2) = M_1 \cup M_2$.*

3.3 Attaining Cyclic Compositionality

After setting the way by which one can deal with positive loops by using annotations in models, the join operator needs to be redefined. The original composition operators are applicable to annotated modules after applying Definition 6. This way, their atoms positive dependencies are added to their respective models.

Definition 11 (Modified Join). *Given two compatible annotated (in the sense of Definition 10) modules $\mathcal{P}_1, \mathcal{P}_2$, their composition is $\mathcal{P}_1 \otimes \mathcal{P}_2 = \mathcal{P}_1 \oplus \mathcal{P}_2$ provided that (i) $\mathcal{P}_1 \oplus \mathcal{P}_2$ is defined. This way, given modules $\mathcal{P}_1$ and $\mathcal{P}_2$ and sets of annotated interpretations $A_1^A \subseteq 2^{A_t(\mathcal{P}_1)}$ and $A_2^A \subseteq 2^{A_t(\mathcal{P}_2)}$, the natural join of A_1^A and A_2^A, denoted by $A_1^A \bowtie_A A_2^A$, is defined as follows for intersecting output atoms:*

$$\{M_1 \cup M_2 \mid M_1 \in A_1, M_2 \in A_2, s.t.\ M_1 \text{ and } M_2 \text{ are compatible.}\}$$

Theorem 3 (Cyclic Module Theorem). *If $\mathcal{P}_1, \mathcal{P}_2$ are modules with annotated models such that $\mathcal{P}_1 \sqcup \mathcal{P}_2$ is defined, then:*

$$AS^A(\mathcal{P}_1 \sqcup \mathcal{P}_2) = AS^A(\mathcal{P}_1) \bowtie_A AS^A(\mathcal{P}_2)$$

3.4 Shortcomings Revisited

By adding the facts contained in stable models of one composing module to the other composing module, through a program transformation, one is able to counter the fact that the inputs of the composed module are removed if they are met by the outputs of either composing modules [17]. As for positive loops, going back to Example 6, the new composition operator also produces desired results:

Example 10 (Cyclic Dependencies Revisited). Take again the two program modules in Example 6:

$$\mathcal{P}_1 = \langle \{airbag \leftarrow safe.\}, \{safe\}, \{airbag\}, \emptyset \rangle$$
$$\mathcal{P}_2 = \langle \{safe \leftarrow airbag.\}, \{airbag\}, \{safe\}, \emptyset \rangle$$

which respectively have annotated models $AS^A(\mathcal{P}_1) = \{\{\}, \{airbag_{\{safe\}}, safe\}\}$ and $AS^A(\mathcal{P}_2) = \{\{\}, \{airbag, safe_{\{airbag\}}\}\}$ while $AS^A(\mathcal{P}_1 \otimes \mathcal{P}_2) = \{\{\}, \{airbag_{\{safe\}}, safe_{\{airbag\}}\}\}$. Because of this, $AS^A(\mathcal{P}_1 \otimes \mathcal{P}_2) = AS^A(\mathcal{P}_1) \bowtie_A AS^A(\mathcal{P}_2)$. Now, take $\mathcal{P}_3 = \langle \{airbag.\}, \{\}, \{airbag\}, \emptyset \rangle$ and compose it with $\mathcal{P}_1 \otimes \mathcal{P}_2$. We get $AS^A(\mathcal{P}_1 \otimes \mathcal{P}_2 \otimes \mathcal{P}_3) = \{\{airbag, safe\}\}$. ∎

4 Conclusions and Future Work

We lift the restriction that disallows composing modules with cyclic dependencies in the framework of Modular Logic Programming [19]. We present a model join operation that requires one to look at every model of two modules being composed in order to check for minimality of models that are comparable on account of their inputs. This operation is able to distinguish between atoms that are self supported through positive loops and atoms with proper support, allowing one to lift the condition disallowing positive dependencies between modules. However, this approach is not local as it requires comparing every models and, as it is not general because it does not allow combining modules with integrity constraints, it is of limited applicability.

Because of this lack of generality of the former approach, we present an alternative solution requiring the introduction of extra information in the models for one to be able to detect dependencies. We use models annotated with the way they depend on the atoms in their module's input signature. We then define their semantics in terms of a fixed point operator. After setting the way by which one deals with positive loops by using annotations in models, the join operator needs to be redefined. The original composition operators are applicable to annotated modules after applying Definition 7. This way, their positive dependencies are added to their respective models. This approach turns out to be local, in the sense that we need only look at two models being joined and unlike the first alternative we presented, it works well with integrity constraints.

As future work we can straightforwardly extend these results to probabilistic reasoning with ASP by applying the new module theorem to [4], as well as to DLP functions and general stable models. An implementation of the framework is also foreseen in order to assess the overhead when compared with the original benchmarks in [19].

Acknowledgments. The work of João Moura was supported by grant SFRH/BD/69006/2010 from Fundação para a Ciência e Tecnologia (FCT) from the Portuguese Ministério do Ensino e da Ciência.

References

1. Babb, J., Lee, J.: Module theorem for the general theory of stable models. TPLP **12**(4–5), 719–735 (2012)
2. Baral, C.: Knowledge Representation, Reasoning, and Declarative Problem Solving. Cambridge University Press (2003)
3. Bugliesi, M., Lamma, E., Mello, P.: Modularity in logic programming. J. Log. Program. **19**(20), 443–502 (1994)
4. Viegas Damásio, C., Moura, J.: Modularity of P-log programs. In: Delgrande, J.P., Faber, W. (eds.) LPNMR 2011. LNCS, vol. 6645, pp. 13–25. Springer, Heidelberg (2011)
5. Dao-Tran, M., Eiter, T., Fink, M., Krennwallner, T.: Modular nonmonotonic logic programming revisited. In: Hill, P.M., Warren, D.S. (eds.) ICLP 2009. LNCS, vol. 5649, pp. 145–159. Springer, Heidelberg (2009)
6. Eiter, T., Faber, W., Leone, N., Pfeifer, G.: Computing preferred and weakly preferred answer sets bymeta-interpretation in answer set programming. In: Proceedings AAAI 2001 Spring Symposium on Answer Set Programming, pp. 45–52. AAAI Press (2001)

7. Ferraris, P., Lifschitz, V.: Weight constraints as nested expressions. TPLP **5**(1–2), 45–74 (2005)
8. Gaifman, H., Shapiro, E.: Fully abstract compositional semantics for logic programs. In: Symposium on Principles of Programming Languages, POPL, pp. 134–142. ACM, New York (1989)
9. Gelfond, M., Lifschitz, V.: The stable model semantics for logic programming. In: Proceedings of the 5th International Conference on Logic Program. MIT Press (1988)
10. Giordano, L., Martelli, A.: Structuring logic programs: a modal approach. The Journal of Logic Programming **21**(2), 59–94 (1994)
11. Järvisalo, M., Oikarinen, E., Janhunen, T., Niemelä, I.: A module-based framework for multi-language constraint modeling. In: Erdem, E., Lin, F., Schaub, T. (eds.) LPNMR 2009. LNCS, vol. 5753, pp. 155–168. Springer, Heidelberg (2009)
12. Lifschitz, V.: Answer set programming and plan generation. Artificial Intelligence **138**(1–2), 39–54 (2002)
13. Lifschitz, V., Pearce, D., Valverde, A.: Strongly equivalent logic programs. ACM Transactions on Computational Logic **2**, 2001 (2000)
14. Mancarella, P., Pedreschi, D.: An algebra of logic programs. In: ICLP/SLP, pp. 1006–1023 (1988)
15. Marek, V.W., Truszczynski, M.: Stable models and an alternative logic programming paradigm. In: The Logic Programming Paradigm: A 25-Year Perspective (1999)
16. Miller, D.: A theory of modules for logic programming. In: In Symp. Logic Programming, pp. 106–114 (1986)
17. Moura, J., Damásio, C.V.: Generalising modular logic programs. In: 15th International Workshop on Non-Monotonic Reasoning (NMR 2014) (2014)
18. Niemelä, I.: Logic programs with stable model semantics as a constraint programming paradigm. Annals of Mathematics and Artificial Intelligence **25**, 72–79 (1998)
19. Oikarinen, E., Janhunen, T.: Achieving compositionality of the stable model semantics for smodels programs. Theory Pract. Log. Program. **8**(5–6), 717–761 (2008)
20. O'Keefe, R.A.: Towards an algebra for constructing logic programs. In: SLP, pp. 152–160 (1985)
21. Slota, M., Leite, J.: Robust equivalence models for semantic updates of answer-set programs. In: Brewka, G., Eiter, T., McIlraith, S.A. (eds.) Proc. of KR 2012. AAAI Press (2012)
22. Tasharrofi, S., Ternovska, E.: A semantic account for modularity in multi-language modelling of search problems. In: Tinelli, C., Sofronie-Stokkermans, V. (eds.) FroCoS 2011. LNCS, vol. 6989, pp. 259–274. Springer, Heidelberg (2011)

Probabilistic Constraint Programming
for Parameters Optimisation
of Generative Models

Massimiliano Zanin[✉], Marco Correia, Pedro A.C. Sousa, and Jorge Cruz

NOVA Laboratory for Computer Science and Informatics,
FCT/UNL, Caparica, Portugal
m.zanin@campus.fct.unl.pt, {mvc,pas,jcrc}@fct.unl.pt

Abstract. Complex networks theory has commonly been used for modelling and understanding the interactions taking place between the elements composing complex systems. More recently, the use of generative models has gained momentum, as they allow identifying which forces and mechanisms are responsible for the appearance of given structural properties. In spite of this interest, several problems remain open, one of the most important being the design of robust mechanisms for finding the optimal parameters of a generative model, given a set of real networks. In this contribution, we address this problem by means of Probabilistic Constraint Programming. By using as an example the reconstruction of networks representing brain dynamics, we show how this approach is superior to other solutions, in that it allows a better characterisation of the parameters space, while requiring a significantly lower computational cost.

Keywords: Probabilistic Constraint Programming · Complex networks · Generative models · Brain dynamics

1 Introduction

The last decades have witnessed a revolution in science, thanks to the appearance of the concept of *complex systems*: systems that are composed of a large number of interacting elements, and whose interactions are as important as the elements themselves [1]. In order to study the structures created by such relationships, several tools have been developed, among which *complex networks theory* [2,3], a statistical mechanics understanding of graph theory, stands out.

Complex networks have been used to characterise a large number of different systems, from social [4] to transportation ones [5]. They have also been valuable in the study of brain dynamics, as one of the greatest challenges in modern science is the characterisation of how the brain organises its activity to carry out complex computations and tasks. Constructing a complete picture of the computation performed by the brain requires specific mathematical, statistical and computational techniques. As brain activity is usually complex, with different regions coordinating and creating temporally multi-scale, spatially extended

© Springer International Publishing Switzerland 2015
F. Pereira et al. (Eds.) EPIA 2015, LNAI 9273, pp. 376–387, 2015.
DOI: 10.1007/978-3-319-23485-4_38

networks, complex networks theory appears as the natural framework for its characterisation.

When complex networks are applied to brain dynamics, nodes are associated to sensors (*e.g.* measuring the electric and magnetic activity of neurons), thus to specific brain locations, and links to some specific conditions. For instance, brain functional networks are constructed such that pairs of nodes are connected if some kind of synchronisation, or correlated activity, is detected in those nodes - the rationale being that a coordinated dynamics is the result of some kind of information sharing [6]. Once these networks are reconstructed, graph theory allows endowing them with a great number of quantitative properties, thus vastly enriching the set of objective descriptors of brain structure and function at neuroscientists' disposal. This has especially been fruitful in the characterisation of the differences between healthy (control) subjects and patients suffering from neurologic pathologies [7].

Once the topology (or structure) of a network has been described, a further question may be posed: can such topology be explained by a set of simple generative rules, like a higher connectivity of neighbouring regions, or the influence of nodes physical position? When a set of rules (a *generative model*) has been defined, it has to be optimised and validated: one ought to obtain the best set of parameters, such that the networks yielded by the model are topologically equivalent to the real ones. This usually requires maximising a function of the p-values representing the differences between the characteristics of the synthetic and real networks. In spite of being accepted as a standard strategy, this method presents several drawbacks. First, its high computational complexity: large sets of networks have to be created and analysed for every possible combination of parameters; and second, its unfitness for assessing the presence of multiple local minima.

In this contribution, we propose the use of *probabilistic constraint programming* (PCP) for characterising the space created by the parameters of a generative model, *i.e.* a space representing the distance between the topological characteristics of real and synthetic networks. We show how this approach allows recovering a larger quantity of information about the relationship between model parameters and network topology, with a fraction of the computational cost required by other methods. Additionally, PCP can be applied to single subjects (networks), thus avoiding the constraints associated with working with a large and homogeneous population. We further validate the PCP approach by studying a simple generative model, and by applying it to a data set of brain activity of healthy people.

The remainder of the text is organised as follows. Besides this introduction, Sections 2 and 2.1 respectively review the state of the art in constraint programming and its probabilistic version. Afterwards, the application of PCP is presented in Section 3 for a data set of brain magneto-encephalographic recordings, and the advantages of PCP are discussed in Section 4. Finally, some conclusions are drawn in Section 5.

2   Constraint Programming

A constraint satisfaction problem [8] is a classical artificial intelligence paradigm characterised by a set of variables and a set of constraints, the latter specifying relations among subsets of these variables. Solutions are assignments of values to all variables that satisfy all the constraints.

Constraint programming is a form of declarative programming, in the sense that instead of specifying a sequence of steps to be executed, it relies on properties of the solutions to be found that are explicitly defined by the constraints. A constraint programming framework must provide a set of constraint reasoning algorithms that take advantage of constraints to reduce the search space, avoiding regions inconsistent with the constraints. These algorithms are supported by specialised techniques that explore the specificity of the constraint model, such as the domain of its variables and the structure of its constraints.

Continuous constraint programming [9,10] has been widely used to model safe reasoning in applications where uncertainty on the values of the variables is modelled by intervals including all their possibilities. A Continuous Constraint Satisfaction Problem (CCSP) is a triple $\langle X, D, C \rangle$, where X is a tuple of n real variables $\langle x_1, \cdots, x_n \rangle$, D is a Cartesian product of intervals $D(x_1) \times \cdots \times D(x_n)$ (a box), each $D(x_i)$ being the domain of variable x_i, and C is a set of numerical constraints (equations or inequalities) on subsets of the variables in X. A solution of the CCSP is a value assignment to all variables satisfying all the constraints in C. The feasible space F is the set of all CCSP solutions within D.

Continuous constraint reasoning relies on branch-and-prune algorithms [11] to obtain sets of boxes that cover exact solutions for the constraints (the feasible space F). These algorithms begin with an initial crude cover of the feasible space (the initial search space, D) which is recursively refined by interleaving pruning and branching steps until a stopping criterion is satisfied. The branching step splits a box from the covering into sub-boxes (usually two). The pruning step either eliminates a box from the covering or reduces it into a smaller (or equal) box maintaining all the exact solutions. Pruning is achieved through an algorithm [12] that combines constraint propagation and consistency techniques [13]: each box is reduced through the consecutive application of narrowing operators associated with the constraints, until a fixed-point is attained. These operators must be correct (do not eliminate solutions) and contracting (the obtained box is contained in the original). To guarantee such properties, interval analysis methods are used.

Interval analysis [14] is an extension of real analysis that allows computations with intervals of reals instead of reals, where arithmetic operations and unary functions are extended for interval operands. For instance, $[1, 3] + [3, 7]$ results in the interval $[4, 10]$, which encloses all the results from a point-wise evaluation of the real arithmetic operator on all the values of the operands. In practice these extensions simply consider the bounds of the operands to compute the bounds of the result, since the involved operations are monotonic. As such, the narrowing operator $Z \leftarrow Z \cap (X + Y)$ may be associated with constraint $x + y = z$ to prune the domain of variable z based on the domains of variables x and y.

Similarly, in solving the equation with respect to x and y, two additional narrowing operators can be associated with the constraint, to safely narrow the domains of these variables. With this technique, based on interval arithmetic, the obtained narrowing operators are able to reduce a box $X \times Y \times Z = [1,3] \times [3,7] \times [0,5]$ into $[1,2] \times [3,4] \times [4,5]$, with the guarantee that no possible solution is lost.

2.1 Probabilistic Constraint Programming

In classical CCSPs, uncertainty is modelled by intervals that represent the domains of the variables. Constraint reasoning reduces uncertainty, providing a safe method for computing a set of boxes enclosing the feasible space. Nevertheless this paradigm cannot distinguish between different scenarios, and all combination of values within such enclosure are considered equally plausible. In this work we use probabilistic constraint programming [15], which extends the continuous constraint framework with probabilistic reasoning, allowing to further characterise uncertainty with probability distributions over the domains of the variables.

In the continuous case, the usual method for specifying a probabilistic model [16] assumes, either explicitly or implicitly, a joint probability density function (p.d.f.) over the considered random variables, which assigns a probability measure to each point of the sample space Ω. The probability of an event $\mathcal{H}$, given a p.d.f. f, is its multidimensional integral on the region defined by the event:

$$P(\mathcal{H}) = \int_{\mathcal{H}} f(\mathbf{x})d\mathbf{x} \tag{1}$$

The idea of probabilistic constraint programming is to associate a probabilistic space to the classical CCSP by defining an appropriate density function. A probabilistic constraint space is a pair $\langle \langle X, D, C \rangle , f \rangle$, where $\langle X, D, C \rangle$ is a CCSP and f is a p.d.f. defined in $\Omega \supseteq D$ such that: $\int_{\Omega} f(\mathbf{x})d\mathbf{x} = 1$.

A constraint (or a conjunction of constraints) can be viewed as an event $\mathcal{H}$ whose probability can be computed by integrating the density function f over its feasible space as in equation (1). The probabilistic constraint framework relies on continuous constraint reasoning to get a tight box cover of the region of integration $\mathcal{H}$, and computes the overall integral by summing up the contributions of each box in the cover. Generic quadrature methods may be used to evaluate the integral at each box.

In this work, Monte Carlo methods [17] are used to estimate the value of the integrals at each box. The integral can be estimated by randomly selecting N points in the multidimensional space and averaging the function values at these points. This method displays $\frac{1}{\sqrt{N}}$ convergence, i.e. by quadrupling the number of sampled points the error is halved, regardless of the number of dimensions.

The advantages obtainable from this close collaboration between constraint pruning and random sampling were previously illustrated in ocean colour remote sensing studies [18], where this approach achieved quite accurate results even with small sampling rates. The success of this technique relies on the reduction

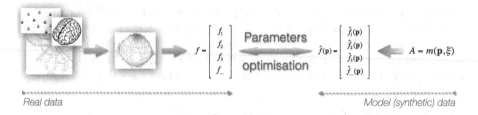

Fig. 1. Schematic representation of the use of generative models for analysing functional networks. f and $\hat{f}$ respectively represent real and synthetic topological features, as the ones described in Sec. 3.2. Refer to Sec. 3 for a description of all steps of the analysis.

of the sampling space, where a pure non-naïve Monte Carlo (adaptive) method is not only hard to tune, but also impractical in small error settings.

3   From Brain Activity to Network Models

In order to validate the use of PCP for analysing the parameters space of a generative models, here we consider a set of magneto-encephalographic (MEG) recordings. A series of preliminary steps are required, as shown in Fig. 1. First, starting from the left, real brain data (or data representing any other real complex system) have to be recorded and encoded in networks, then transformed into a set of topological (structural) features. In parallel, as depicted in the right part, a generative model has to be defined: this allows to generate networks as a function of the model parameters, and extract their topological features. Finally, both features should be matched, *i.e.* the model parameters should be optimised to minimise the distance between the vectors of topological features of the synthetic and real networks.

3.1   MEG Data Recording

Magneto-encephalographic (MEG) scans were obtained for 19 right handed elderly and healthy participants, recruited from the Geriatric Unit of the Hospital Universitario San Carlos Madrid and the Centro de Prevención del Deterioro Cognitivo, Ayuntamiento de Madrid, Spain. Before the task execution, all participants or legal representatives gave informed consent to participate in the study. The study was approved by the local ethics committee.

Brain activity scans correspond to a modified version of the Sternberg's letter-probe task [19], a standard task used to evaluate elders memory proficiency. MEG signals were recorded with a 254 Hz sampling rate, using 148-channel whole head magnetometer, confined in a magnetically shielded room (MSR). 35 artefact-free epochs were randomly chosen from those corresponding to correct answers for each of participant.

3.2   Networks Reconstruction and Evaluation

Following the diagram of Fig. 1, MEG recordings are converted in functional networks. Nodes, corresponding to MEG sensors and therefore to different brain regions, are pairwise connected when some kind of common dynamics is detected between the corresponding time series. Such relationship is assessed through Synchronization Likelihood (SL) [20], a metric able to detect *generalised synchronisation, i.e.* situations in which two time series react to a given input in different, yet consistent ways [21]. It thus goes beyond simple linear correlations, as it is able to detect non-linear and potentially chaotic relations. Applying SL yields a correlation matrix $C\{w_{ij}\}$ of size 148×148 (the number of sensors in the MEG machine) for each epoch available. In order to filter any kind of transient or noise specific to one epoch, the 35 matrices corresponding to each subjects have been averaged: the final result is then a single weight matrix $\tilde{C}\{w_{ij}\}$ for each subject.

While a correlation matrix can readily be interpreted as a weighted fully-connected network, few metrics are available to describe the structure of such objects. It is then customary to apply a threshold, *i.e.* discard all links whose weight is not significant, and thus obtain an unweighted network. This presents several advantages. First of all, brain networks are expected to be naturally sparse, as increasing the connectivity implies a higher physiological cost. Furthermore, low synchronisation values may be the result of statistical fluctuations, *e.g.* of correlated noise; in such cases, deleting spurious links can only improve the understanding of the system. Lastly, a pruning can also help deleting indirect, second order correlations, which do not represent direct dynamical relationships.

The final step involves the calculation of the topological metrics associated to each pruned network, *i.e.* the fs of Fig. 1. Two have here been considered, representing two complementary aspects of brain information processing; their selection has been motivated by the generative model used afterwards (see Section 3.3):

Clustering Coefficient. The *clustering coefficient*, also known as *transitivity*, measures the presence of triangles in the network [22]. Mathematically, it is defined as the relationship between the number of triangles and the number of connected triples in the network: $C = 3N_\Delta/N_3$. Here, a triangle is a set of three nodes with links between each pair of them, while a connected triple is a set of three nodes where each one can be reached from each other (directly or indirectly). From a biological point of view, the clustering coefficient represents how brain regions are locally connected, creating dense communities computing some information in a collaborative way.

Efficiency. It is defined as the inverse of the harmonic mean of the length of the shortest paths connecting pairs of nodes [23]:

$$E = \frac{1}{N(N-1)} \sum_{i \neq j} \frac{1}{d_{ij}}, \qquad (2)$$

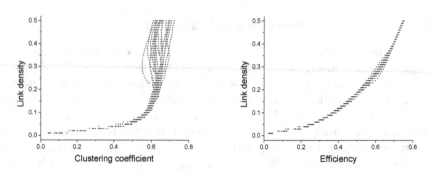

Fig. 2. Evolution of clustering coefficient (Left) and efficiency (Right) as a function of the link density for the 19 functional brain networks reconstructed in Sec. 3.2.

d_{ij} being the distance between nodes i and j, *i.e.* the number of jumps required to travel between them. A low value of E implies that all brain regions are connected by short paths.

It has to be noticed how these two measures are complementary, the clustering coefficient and efficiency respectively representing the *segregation* and *integration* of information [24,25]. Additionally, both C and E are here defined as a function of the threshold τ applied to prune the networks - their evolution is represented in Fig. 2.

3.3   Generative Model Definition

Jumping to the right side of Fig. 1, it is now necessary to define a generative model. As an example, we have here implemented a *Economical Clustering Model* model as defined in [26,27]. Given two nodes i and j, the probability of creating a connection between them is given by:

$$P_{i,j} \propto k_{i,j}^{\gamma} d_{i,j}^{-\eta}. \tag{3}$$

$k_{i,j}$ is the number of neighbours common to i and j, and $d_{i,j}$ is the physical distance between the two nodes. This model thus includes two different forces that compete to create links. On one side, γ controls the appearance of triangles in the network, by positive biasing the connectivity between nodes having nearest neighbours in common; it thus defines the clustering coefficient and the appearance of computational communities. On the other side, η accounts for the distance in the connection, such that long-range connections, which are biologically costly, are penalised.

3.4   Parameters Estimation Through P-values

The problem is now identifying the best values of γ and η that permit recovering the topological properties obtained in Sec. 3.2 for the experimental brain networks.

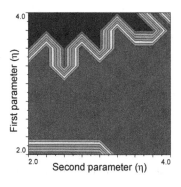 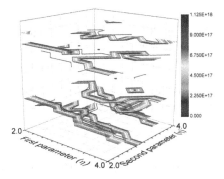

Fig. 3. (Left) Contour plot of the energy E (see Eq. 4) in the parameters space, for a link density of 0.3. (Right) Energy contour plots for ten link densities, from 0.05 (bottom) to 0.5 (top); for the sake of clarity, only region outlines are visible.

As an example of a standard p-value based mechanism, we here use a simplified version of the energy function proposed in Refs. [26,27]:

$$E = 1/\prod_i P_i. \qquad (4)$$

P_i represents the p-value of the Kolmogorov-Smirnoff (K-S) test between the distributions estimated from the model and experimental networks, and i runs over all topological metrics. As just two topological properties are here studied, the previous formula simplifies to: $E = 1/(P_E \cdot P_C)$.

For each considered value of γ and η, a set of networks have been generated according to the model of Eq. 3; their topological features extracted; and the resulting probability distribution compared with the distribution corresponding to the real networks, through a K-S test.

Fig. 3 presents the result of plotting the energy evolution in the parameters space. Specifically, Fig. 3 Left reports the evolution of the energy for a link density of 0.3. It can be noticed that a large portion of the space, constructed around the values of γ and η suggested in [26], maximises the energy. Fig. 3 Right represents the same information for ten different link densities, from 0.05 (bottom part) to 0.5 (upper part).

3.5 Parameters Estimation Through Probabilistic Constraint Programming

As an alternative solution, the previously described PCP method is here used to recover the shape of the parameters space. Two preliminary steps have to be completed: first, reconstruct a set of synthetic networks using the generative model of Eq. 3, for different γ and η values, and extract their topological characteristics; and second, obtain approximated functions describing the evolution of the topological metrics as a function of the model parameters, *i.e.* $C = \tilde{f}_C(\gamma, \eta)$

and $E = \tilde{f}_E(\gamma, \eta)$. Afterwards, each observed feature o_i is modelled as a function f_i of the model parameters plus an associated error term $\epsilon_i \sim \mathcal{N}\left(\mu = 0, \sigma^2\right)$:

$$o_i = f_i(\gamma, \eta) + \epsilon_i$$

For n observations, a probabilistic constraint space is considered with random variables γ and η, a set of constraints C,

$$C = \{-3\sigma \le o_i - f_i(\gamma, \eta) \le 3\sigma | 1 \le i \le n\}$$

3σ being chosen to keep the error within reasonable bounds, and the joint p.d.f. f,

$$f(\gamma, \eta) = \prod_{i=1}^{n} g(o_i - f_i(\gamma, \eta)) \tag{5}$$

where g is the normal distribution with 0 mean and standard deviation σ.

To compute the probability distribution of the random variables γ and η, a grid is constructed over their domains and a branch-and-prune algorithm is initially used to obtain a grid box cover of the feasible space (where each box belongs to a single grid cell). Then, for each box in the cover, a Monte Carlo method is used to compute its contribution to equation (1) with the p.d.f. defined in equation (5). The probability of the respective cell is updated accordingly and normalised in end of the process.

Fig. 4 reports the results obtained, *i.e.* the probability of obtaining networks with the generative model which are compatible with the real ones, as a function of the two parameters γ and η, and as a function of the link density. In the next Section, both approaches and their results are compared.

4 Comparing P-value and Probabilistic Constraint Programming

Results presented in Sec. 3.4 and 3.5 allow comparing the p-value and PCP methods, and highlight the advantages that the latter presents over the former.

The extremely high computational cost of analysing the parameters space by means of K-S tests seldom allows a full characterisation of such space. This is due to the fact that, for any set of parameters, a large number of networks have to be created and characterised. Increasing the resolution of the analysis, or enlarging the region of the space considered, increases the computational cost in a linear way. This problem is far from being trivial, as, for instance, the networks required to create Fig. 3 represents approximatively 3 GB of information and several days of computation in a standard computer. Such computational cost implies that it is easy to miss some important information. Let us consider, for instance, the result presented in Fig. 3 Left. The shape of the iso-lines suggests that the maximum is included in the region under analysis, and that no further explorations are required - while Figs. 4 and 5 prove otherwise.

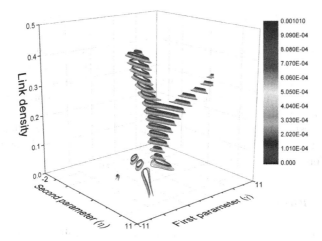

Fig. 4. Contour plot of the parameters space, as obtained by the PCP method, for the whole population of subjects and as a function of the link density. The colour of each point represents the normalised probability of generating topologically equivalent networks.

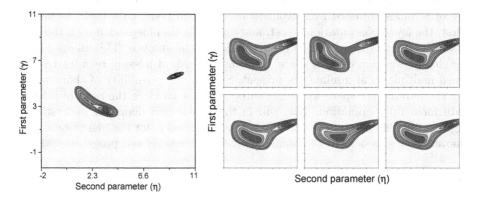

Fig. 5. (Left) Parameters space, as obtained with the PCP method, for a link density of 0.3 and for the whole studied population. (Right) Parameters space for six subjects. The scale of the right graphs is the same as the left one; the colour scale is the same of the one of Fig. 4.

On the other hand, estimating the functions $\tilde{f}_C$ and $\tilde{f}_E$ requires the creation and analysis of a constant number of networks, independently on the size of the parameters space. The total computational cost drops below the hour in a standard computer, implying a 3 orders of magnitude reduction. This has important consequences on the kind of information one can obtain. Fig. 5 Left presents the same information as Fig. 3 Left, but calculated by means of PCP over a larger region. It is then clear that the maximum identified in Fig. 3 is just one of the two maxima presents in the system.

The second important advantage is that, while the PCP can yield results for just one network or subject, a p-value analysis requires a probability distribution. It is therefore not possible to characterise the parameters space for just one subject, but only for a large population. Fig. 5 Right explores this issue, by showing the probability evolution in the parameters space for six different subjects. It is interesting to notice how subjects are characterised by different shapes in the space. This allows a better description of subjects, aimed for instance at detecting differences among them.

5   Conclusions

In this contribution, we have presented the use of Probabilistic Constraint Programming for optimising the parameters of a generative model, aimed at describing the mechanisms responsible for the appearance of some given topological structures in real complex networks. As a validation case, we have here presented the results corresponding to functional networks of brain activity, as obtained through MEG recordings of healthy people.

The advantages of this method against other customary solutions, *e.g.* the use of p-values obtained from Kolmogorov-Smirnoff tests, have been discussed. First, the lower computational cost, and especially its independence on the size of the parameters space and on the resolution of the analysis. This allows a better characterisation of such space, reducing the risk of missing relevant results when multiple local minima are present. Second, the possibility of characterising the parameters space for single subjects, thus avoiding the need of having data for a full population. This will in turn open new doors for understanding the differences between individuals: as, for instance, for the identification of characteristics associated to specific diseases in diagnosis and prognosis tasks.

References

1. Anderson, P.W.: More is different. Science **177**, 393–396 (1972)
2. Albert, R., Barabási, A.L.: Statistical mechanics of complex networks. Reviews of Modern Physics **74**, 47 (2002)
3. Newman, M.E.: The structure and function of complex networks. SIAM Review **45**, 167–256 (2003)
4. Costa, L.D.F., Oliveira Jr, O.N., Travieso, G., Rodrigues, F.A., Villas Boas, P.R., Antiqueira, L., Viana, M.P., Correa Rocha, L.E.: Analyzing and modeling real-world phenomena with complex networks: a survey of applications. Advances in Physics **60**, 329–412 (2011)
5. Zanin, M., Lillo, F.: Modelling the air transport with complex networks: A short review. The European Physical Journal Special Topics **215**, 5–21 (2013)
6. Bullmore, E., Sporns, O.: Complex brain networks: graph theoretical analysis of structural and functional systems. Nature Reviews Neuroscience **10**, 186–198 (2009)

7. Papo, D., Zanin, M., Pineda-Pardo, J.A., Boccaletti, S., Buldú, J.M.: Functional brain networks: great expectations, hard times and the big leap forward. Philosophical Transactions of the Royal Society of London B: Biological Sciences **369**, 20130525 (2014)
8. Mackworth, A.K.: Consistency in networks of relations. Artificial Intelligence **8**, 99–118 (1977)
9. Lhomme, O.: Consistency techniques for numeric CSPs. In: Proc. of the 13th IJCAI, pp. 232–238 (1993)
10. Benhamou, F., McAllester, D., van Hentenryck, P.: CLP(intervals) revisited. In: ISLP, pp. 124–138 (1994)
11. Van Hentenryck, P., McAllester, D., Kapur, D.: Solving polynomial systems using a branch and prune approach. SIAM Journal on Numerical Analysis **34**, 797–827 (1997)
12. Granvilliers, L., Benhamou, F.: Algorithm 852: realpaver: an interval solver using constraint satisfaction techniques. ACM Transactions on Mathematical Software **32**, 138–156 (2006)
13. Benhamou, F., Goualard, F., Granvilliers, L., Puget, J.-F.: Revising hull and box consistency. In: Procs. of ICLP, pp. 230–244 (1999)
14. Moore, R.: Interval analysis. Prentice-Hall, Englewood Cliffs (1966)
15. Carvalho, E.: Probabilistic constraint reasoning. PhD Thesis (2012)
16. Halpern, J.Y.: Reasoning about uncertainty. MIT, Cambridge (2003)
17. Hammersley, J.M., Handscomb, D.C.: Monte Carlo methods. Methuen, London (1964)
18. Carvalho, E., Cruz, J., Barahona, P.: Probabilistic constraints for nonlinear inverse problems. Constraints **18**, 344–376 (2013)
19. Maestú, F., Fernández, A., Simos, P.G., Gil-Gregorio, P., Amo, C., Rodriguez, R., Arrazola, J., Ortiz, T.: Spatio-temporal patterns of brain magnetic activity during a memory task in Alzheimer's disease. Neuroreport **12**, 3917–3922 (2001)
20. Stam, C.J., Van Dijk, B.W.: Synchronization likelihood: an unbiased measure of generalized synchronization in multivariate data sets. Physica D: Nonlinear Phenomena **163**, 236–251 (2002)
21. Yang, S., Duan, C.: Generalized synchronization in chaotic systems. Chaos, Solitons & Fractals **9**, 1703–1707 (1998)
22. Newman, M.E.: Scientific collaboration networks. I. Network construction and fundamental results. Physical Review E 64, 016131 (2001)
23. Latora, V., Marchiori, M.: Efficient behavior of small-world networks. Physical Review Letters **87**, 198–701 (2001)
24. Tononi, G., Sporns, O., Edelman, G.M.: A measure for brain complexity: relating functional segregation and integration in the nervous system. Proceedings of the National Academy of Sciences **91**, 5033–5037 (1994)
25. Rad, A.A., Sendiña-Nadal, I., Papo, D., Zanin, M., Buldu, J.M., del Pozo, F., Boccaletti, S.: Topological measure locating the effective crossover between segregation and integration in a modular network. Physical Review Letters **108**, 228701 (2012)
26. Vértes, P.E., Alexander-Bloch, A.F., Gogtay, N., Giedd, J.N., Rapoport, J.L., Bullmore, E.T.: Simple models of human brain functional networks. Proceedings of the National Academy of Sciences **109**, 5868–5873 (2012)
27. Vértes, P.E., Alexander-Bloch, A., Bullmore, E.T.: Generative models of rich clubs in Hebbian neuronal networks and large-scale human brain networks. Philosophical Transactions of the Royal Society B: Biological Sciences **369**, 20130531 (2014)

Reasoning over Ontologies
and Non-monotonic Rules

Vadim Ivanov, Matthias Knorr$^{(\boxtimes)}$, and João Leite

NOVA LINCS, Departamento de Informática, Faculdade Ciências e Tecnologia,
Universidade Nova de Lisboa, Caparica, Portugal
mkn@fct.unl.pt

Abstract. Ontology languages and non-monotonic rule languages are
both well-known formalisms in knowledge representation and reasoning,
each with its own distinct benefits and features which are quite orthogonal to each other. Both appear in the Semantic Web stack in distinct
standards – OWL and RIF – and over the last decade a considerable
research effort has been put into trying to provide a framework that
combines the two. Yet, the considerable number of theoretical approaches
resulted, so far, in very few practical reasoners, while realistic use-cases
are scarce. In fact, there is little evidence that developing applications
with combinations of ontologies and rules is actually viable. In this paper,
we present a tool called NoHR that allows one to reason over ontologies
and non-monotonic rules, illustrate its use in a realistic application, and
provide tests of scalability of the tool, thereby showing that this research
effort can be turned into practice.

1 Introduction

Ontology languages in the form of Description Logics (DLs) [4] and non-
monotonic rule languages as known from Logic Programming (LP) [6] are both
well-known formalisms in knowledge representation and reasoning (KRR) each
with its own distinct benefits and features. This is also witnessed by the emergence of the Web Ontology Language (OWL) [18] and the Rule Interchange
Format (RIF) [7] in the ongoing standardization of the Semantic Web driven by
the W3C[1].

On the one hand, ontology languages have become widely used to represent
and reason over taxonomic knowledge and, since DLs are (usually) decidable
fragments of first-order logic, are monotonic by nature which means that once
drawn conclusions persist when adopting new additional information. They also
allow reasoning on abstract information, such as relations between classes of
objects even without knowing any concrete instances and a main theme inherited
from DLs is the balance between expressiveness and complexity of reasoning. In
fact, the very expressive general language OWL 2 with its high worst-case complexity includes three tractable (polynomial) profiles [27] each with a different
application purpose in mind.

---

[1] http://www.w3.org

© Springer International Publishing Switzerland 2015
F. Pereira et al. (Eds.) EPIA 2015, LNAI 9273, pp. 388–401, 2015.
DOI: 10.1007/978-3-319-23485-4_39

On the other hand, non-monotonic rules are focused on reasoning over instances and commonly apply the Closed World Assumption (CWA), i.e., the absence of a piece of information suffices to derive it being false, until new information to the contrary is provided, hence the term non-monotonic. This permits to declaratively model defaults and exceptions, in the sense that the absence of an exceptional feature can be used to derive that the (more) common case applies, and also integrity constraints, which can be used to ensure that the considered data is conform with desired specifications.

Combining both formalisms has been frequently requested by applications [1]. For example, in clinical health care, large ontologies such as SNOMED CT[2], that are captured by the OWL 2 profile OWL 2 EL and its underlying description logic (DL) $\mathcal{EL}^{++}$ [5], are used for electronic health record systems, clinical decision support systems, or remote intensive care monitoring, to name only a few. Yet, expressing conditions such as dextrocardia, i.e., that the heart is exceptionally on the right side of the body, is not possible and requires non-monotonic rules.

Finding such a combination is a non-trivial problem due to the considerable differences as to how decidability is ensured in each of the two formalisms and a naive combination is easily undecidable. In recent years, there has been a considerable amount of effort devoted to combining DLs with non-monotonic rules as known from Logic Programming – see, e.g., related work in [12, 28]) – but this has not been accompanied by similar variety of reasoners and applications. In fact, only very few reasoners for combining ontologies and non-monotonic rules exist and realistic use-cases are scarce. In other words, there is little evidence so far that developing applications in combinations of ontologies and rules is actually viable.

In this paper, we want to contribute to showing that this paradigm is viable by describing a tool called NoHR and show how it can be used to handle a real use-case efficiently as well as its scalability. NoHR is theoretically founded in the formalism of Hybrid MKNF under the well-founded semantics [22] which comes with two main arguments in its favor. First, the overall approach, which was introduced in [28] and is based on the logic of minimal knowledge and negation as failure (MKNF) [26], provides a very general and flexible framework for combining DL ontologies and non-monotonic rules (see [28]). Second, [22], which is a variant of [28] based on the well-founded semantics [13] for logic programs, has a lower data complexity than the former – it is polynomial for polynomial DLs – and is amenable for applying top-down query procedures, such as $\mathbf{SLG}(\mathcal{O})$ [2], to answer queries based only on the information relevant for the query, and without computing the entire model – no doubt a crucial feature when dealing with large ontologies and huge amounts of data.

NoHR is realized as a plug-in for the ontology editor Protégé 4.X[3], that allows the user to query combinations of $\mathcal{EL}_\bot^+$ ontologies and non-monotonic rules in a top-down manner. To the best of our knowledge, it is the first Protégé plug-in to integrate non-monotonic rules and top-down queries. We describe its

---

[2] http://www.ihtsdo.org/snomed-ct/
[3] http://protege.stanford.edu

Table 1. Syntax and semantics of $\mathcal{EL}_\bot^+$.

	Syntax	Semantics
atomic concept	$A \in \mathsf{N_C}$	$A^\mathcal{I} \subseteq \Delta^\mathcal{I}$
atomic role	$R \in \mathsf{N_R}$	$R^\mathcal{I} \subseteq \Delta^\mathcal{I} \times \Delta^\mathcal{I}$
individual	$a \in \mathsf{N_I}$	$a^\mathcal{I} \in \Delta^\mathcal{I}$
top	$\top$	$\Delta^\mathcal{I}$
bottom	$\bot$	$\emptyset$
conjunction	$C \sqcap D$	$C^\mathcal{I} \cap D^\mathcal{I}$
existential restriction	$\exists R.C$	$\{x \in \Delta^\mathcal{I} \mid \exists y \in \Delta^\mathcal{I} : (x,y) \in R^\mathcal{I} \wedge y \in C^\mathcal{I}\}$
concept inclusion	$C \sqsubseteq D$	$C^\mathcal{I} \subseteq D^\mathcal{I}$
role inclusion	$R \sqsubseteq S$	$R^\mathcal{I} \subseteq S^\mathcal{I}$
role composition	$R_1 \circ \cdots \circ R_k \sqsubseteq S$	$(x_1, x_2) \in R_1^\mathcal{I} \wedge \ldots \wedge (x_k, y) \in R_k^\mathcal{I} \rightarrow (x_1, y) \in S^\mathcal{I}$
concept assertion	$C(a)$	$a^\mathcal{I} \in C^\mathcal{I}$
role assertion	$R(a,b)$	$(a^\mathcal{I}, b^\mathcal{I}) \in R^\mathcal{I}$

features including the possibility to load and edit rule bases, and define predicates with arbitrary arity; guaranteed termination of query answering, with a choice between one/many answers; robustness w.r.t. inconsistencies between the ontology and the rule part and demonstrate its effective usage on the application use-case combining $\mathcal{EL}_\bot^+$ ontologies and non-monotonic rules outlined in the following and adapted from [29], as well as an evaluation for real ontology SNOMED CT with over 300,000 concepts.

Example 1. The customs service for any developed country assesses imported cargo for a variety of risk factors including terrorism, narcotics, food and consumer safety, pest infestation, tariff violations, and intellectual property rights. Assessing this risk, even at a preliminary level, involves extensive knowledge about commodities, business entities, trade patterns, government policies and trade agreements. Parts of this knowledge is ontological information and taxonomic, such as the classification of commodities, while other parts require the CWA and thus non-monotonic rules, such as the policies involving, e.g., already known suspects. The overall task then is to access all the information and assess whether some shipment should be inspected in full detail, under certain conditions randomly, or not at all.

The remainder of the paper is structured as follows. In Sect. 2, we briefly recall the DL $\mathcal{EL}_\bot^+$ and MKNF knowledge bases as a tight combination of the former DL and non-monotonic rules. Then, in Sect. 3, we present the Protégé plug-in NoHR, and, in Sect. 4, we discuss the cargo shipment use case and its realization using NoHR. We present some evaluation data in Sect. 5, before we conclude in Sect. 6[4].

---

[4] Details on the translation of $\mathcal{EL}$ ontologies into rules used in NoHR can be found in [19].

2 Preliminaries

2.1 Description Logic $\mathcal{EL}^+_\perp$

We start by recalling the syntax and semantics of $\mathcal{EL}^+_\perp$, a large fragment of $\mathcal{EL}^{++}$ [5], the DL underlying the tractable profile OWL 2 EL [27], following the presentation in [21]. For a more general and thorough introduction to DLs we refer to [4].

The language of $\mathcal{EL}^+_\perp$ is defined over countably infinite sets of *concept names* N_C, *role names* N_R, and *individual names* N_I as shown in the upper part of Table 1. Building on these, *complex concepts* are introduced in the middle part of Table 1, which, together with atomic concepts, form the set of *concepts*. We conveniently denote individuals by a and b, (atomic) roles by R and S, atomic concepts by A and B, and concepts by C and D. All expressions in the lower part of Table 1 are *axioms*. A *concept equivalence* $C \equiv D$ is an abbreviation for $C \sqsubseteq D$ and $D \sqsubseteq C$. Concept and role assertions are *ABox axioms* and all other axioms *TBox axioms*, and an *ontology* is a finite set of axioms.

The semantics of $\mathcal{EL}^+_\perp$ is defined in terms of an *interpretation* $\mathcal{I} = (\Delta^\mathcal{I}, \cdot^\mathcal{I})$ consisting of a non-empty domain $\Delta^\mathcal{I}$ and an *interpretation function* $\cdot^\mathcal{I}$. The latter is defined for (arbitrary) concepts, roles, and individuals as in Table 1. Moreover, an interpretation $\mathcal{I}$ *satisfies* an axiom α, written $\mathcal{I} \models \alpha$, if the corresponding condition in Table 1 holds. If $\mathcal{I}$ satisfies all axioms occurring in an ontology $\mathcal{O}$, then $\mathcal{I}$ is a *model* of $\mathcal{O}$, written $\mathcal{I} \models \mathcal{O}$. If $\mathcal{O}$ has at least one model, then it is called *consistent*, otherwise *inconsistent*. Also, $\mathcal{O}$ *entails* axiom α, written $\mathcal{O} \models \alpha$, if every model of $\mathcal{O}$ satisfies α. *Classification* requires to compute all concept inclusions between atomic concepts entailed by $\mathcal{O}$.

2.2 MKNF Knowledge Bases

MKNF knowledge bases (KBs) build on the logic of minimal knowledge and negation as failure (MKNF) [26]. Two main different semantics have been defined [22,28], and we focus on the well-founded version [22], due to its lower computational complexity and amenability to top-down querying without computing the entire model. Here, we only point out important notions, and refer to [22] and [2] for the details.

We start by recalling MKNF knowledge bases as presented in [2] to combine an ($\mathcal{EL}^+_\perp$) ontology and a set of non-monotonic rules (similar to a normal logic program).

Definition 2. *Let $\mathcal{O}$ be an ontology. A function-free first-order atom $P(t_1, \ldots, t_n)$ s.t. P occurs in $\mathcal{O}$ is called DL-atom; otherwise non-DL-atom. A rule r is of the form*

$$H \leftarrow A_1, \ldots, A_n, \mathbf{not}\, B_1, \ldots, \mathbf{not}\, B_m \qquad (1)$$

where the head of r, H, and all A_i with $1 \leq i \leq n$ and B_j with $1 \leq j \leq m$ in the body of r are atoms. A program $\mathcal{P}$ is a finite set of rules, and an MKNF knowledge base $\mathcal{K}$ *is a pair* $(\mathcal{O}, \mathcal{P})$. *A rule* r *is* DL-safe *if all its variables occur in at least one non-DL-atom* A_i *with* $1 \leq i \leq n$, *and* $\mathcal{K}$ *is* DL-safe *if all its rules are DL-safe. The* ground instantiation *of* $\mathcal{K}$ *is the KB* $\mathcal{K}_G = (\mathcal{O}, \mathcal{P}_G)$ *where* $\mathcal{P}_G$ *is obtained from* $\mathcal{P}$ *by replacing each rule* r *of* $\mathcal{P}$ *with a set of rules substituting each variable in* r *with constants from* $\mathcal{K}$ *in all possible ways.*

DL-safety ensures decidability of reasoning with MKNF knowledge bases and can be achieved by introducing a new predicate o, adding $o(i)$ to $\mathcal{P}$ for all constants i appearing in $\mathcal{K}$ and, for each rule $r \in \mathcal{P}$, adding $o(X)$ for each variable X appearing in r to the body of r. Therefore, we only consider DL-safe MKNF knowledge bases.

The semantics of $\mathcal{K}$ is based on a transformation of $\mathcal{K}$ into an MKNF formula to which the MKNF semantics can be applied (see [22, 26, 28] for details). Instead of spelling out the technical details of the original MKNF semantics [28] or its three-valued counterpart [22], we focus on a compact representation of models for which the computation of the well-founded MKNF model is defined[5]. This representation is based on a set of **K**-atoms and $\pi(\mathcal{O})$, the translation of $\mathcal{O}$ into first-order logic.

Definition 3. *Let* $\mathcal{K}_G = (\mathcal{O}, \mathcal{P}_G)$ *be a ground hybrid MKNF knowledge base. The set of* **K**-atoms *of* $\mathcal{K}_G$, *written* $\mathsf{KA}(\mathcal{K}_G)$, *is the smallest set that contains (i) all ground atoms occurring in* $\mathcal{P}_G$, *and (ii) an atom* ξ *for each ground* **not**-atom **not**ξ *occurring in* $\mathcal{P}_G$. *For a subset* S *of* $\mathsf{KA}(\mathcal{K}_G)$, *the* objective knowledge *of* S *w.r.t.* $\mathcal{K}_G$ *is the set of first-order formulas* $\mathsf{OB}_{\mathcal{O},S} = \{\pi(\mathcal{O})\} \cup S$.

The set $\mathsf{KA}(\mathcal{K}_G)$ contains all atoms occurring in $\mathcal{K}_G$, only with **not**-atoms substituted by corresponding atoms, while $\mathsf{OB}_{\mathcal{O},S}$ provides a first-order representation of $\mathcal{O}$ together with a set of known/derived facts. In the three-valued MKNF semantics, this set of **K**-atoms can be divided into true, undefined and false atoms. Next, we recall operators from [22] that derive consequences based on $\mathcal{K}_G$ and a set of **K**-atoms that is considered to hold.

Definition 4. *Let* $\mathcal{K}_G = (\mathcal{O}, \mathcal{P}_G)$ *be a positive, ground hybrid MKNF knowledge base. The operators* $R_{\mathcal{K}_G}$, $D_{\mathcal{K}_G}$, *and* $T_{\mathcal{K}_G}$ *are defined on subsets of* $\mathsf{KA}(\mathcal{K}_G)$:

$$R_{\mathcal{K}_G}(S) = \{H \mid \mathcal{P}_G \text{ contains a rule of the form } H \leftarrow A_1, \dots A_n$$
$$\text{such that, for all } i,\ 1 \leq i \leq n, A_i \in S\}$$
$$D_{\mathcal{K}_G}(S) = \{\xi \mid \xi \in \mathsf{KA}(\mathcal{K}_G) \text{ and } \mathsf{OB}_{\mathcal{O},S} \models \xi\}$$
$$T_{\mathcal{K}_G}(S) = R_{\mathcal{K}_G}(S) \cup D_{\mathcal{K}_G}(S)$$

The operator $T_{\mathcal{K}_G}$ is monotonic, and thus has a least fixpoint $T_{\mathcal{K}_G} \uparrow \omega$. Transformations can be defined that turn an arbitrary hybrid MKNF KB $\mathcal{K}_G$ into a positive one (respecting the given set S) to which $T_{\mathcal{K}_G}$ can be applied. To ensure coherence, i.e., that classical negation in the DL enforces default negation in the rules, two slightly different transformations are defined (see [22] for details).

---

[5] Strictly speaking, this computation yields the so-called well-founded partition from which the well-founded MKNF model is defined (see [22] for details).

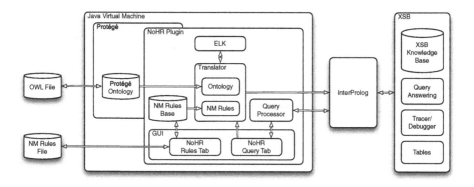

Fig. 1. System Architecture of NoHR

Definition 5. *Let* $\mathcal{K}_G = (\mathcal{O}, \mathcal{P}_G)$ *be a ground hybrid MKNF knowledge base and* $S \subseteq \mathsf{KA}(\mathcal{K}_G)$. *The MKNF transform* $\mathcal{K}_G/S$ *is defined as* $\mathcal{K}_G/S = (\mathcal{O}, \mathcal{P}_G/S)$, *where* $\mathcal{P}_G/S$ *contains all rules* $H \leftarrow A_1, \ldots, A_n$ *for which there exists a rule of the form (1) in* $\mathcal{P}_G$ *with* $B_j \notin S$ *for all* $1 \leq j \leq m$. *The MKNF-coherent transform* $\mathcal{K}_G//S$ *is defined as* $\mathcal{K}_G//S = (\mathcal{O}, \mathcal{P}_G//S)$, *where* $\mathcal{P}_G//S$ *contains all rules* $H \leftarrow A_1, \ldots, A_n$ *for which there exists a rule of the form (1) with* $B_j \notin S$ *for all* $1 \leq j \leq m$ *and* $\mathsf{OB}_{\mathcal{O},S} \not\models \neg H$. *We define* $\Gamma_{\mathcal{K}_G}(S) = T_{\mathcal{K}_G/S} \uparrow \omega$ *and* $\Gamma'_{\mathcal{K}_G}(S) = T_{\mathcal{K}_G//S} \uparrow \omega$.

Based on these two antitonic operators [22], two sequences $\mathbf{P}_i$ and $\mathbf{N}_i$ are defined, which correspond to the true and non-false derivations.

$$\mathbf{P}_0 = \emptyset \qquad\qquad \mathbf{N}_0 = \mathsf{KA}(\mathcal{K}_G)$$
$$\mathbf{P}_{n+1} = \Gamma_{\mathcal{K}_G}(\mathbf{N}_n) \qquad\qquad \mathbf{N}_{n+1} = \Gamma'_{\mathcal{K}_G}(\mathbf{P}_n)$$
$$\mathbf{P}_\omega = \bigcup \mathbf{P}_i \qquad\qquad \mathbf{N}_\omega = \bigcap \mathbf{N}_i$$

The fixpoints yield the well-founded MKNF model [22] (in polynomial time).

Definition 6. *The* well-founded MKNF model *of an MKNF-consistent ground hybrid MKNF knowledge base* $\mathcal{K}_G = (\mathcal{O}, \mathcal{P}_G)$ *is defined as* $(\mathbf{P}_\omega, \mathsf{KA}(\mathcal{K}_G) \setminus \mathbf{N}_\omega)$.

If $\mathcal{K}$ is *MKNF-consistent*, then this partition does correspond to the unique model of $\mathcal{K}$ [22], and, like in [2], we call the partition the *well-founded MKNF model* $\mathsf{M}_{\mathsf{wf}}(\mathcal{K})$. Here, $\mathcal{K}$ may indeed not be MKNF-consistent if the $\mathcal{EL}_\perp^+$ ontology alone is inconsistent, which is possible if $\perp$ occurs, or by the combination of appropriate axioms in $\mathcal{O}$ and $\mathcal{P}$, e.g., $A \sqsubseteq \perp$ and $A(a) \leftarrow$. In the former case, we argue that the ontology alone should be consistent and be repaired if necessary before combining it with non-monotonic rules. Thus, we assume in the following that $\mathcal{O}$ occurring in $\mathcal{K}$ is consistent.

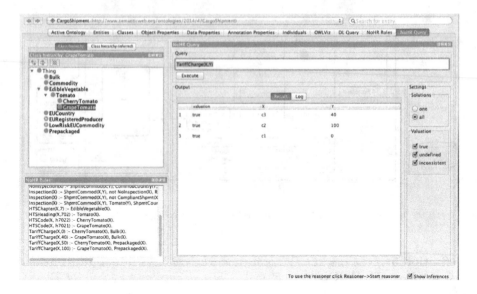

Fig. 2. NoHR Query tab with a query for TariffCharge($\mathbf{x}, \mathbf{y}$) (see Sect. 4)

3   System Description

In this section, we briefly describe the architecture of the plug-in for Protégé as shown in Fig. 1 and discuss some features of the implementation and how querying is realized.

The input for the plug-in consists of an OWL file in the DL $\mathcal{EL}^+_\perp$ as described in Sect. 2.1, which can be manipulated as usual in Protégé, and a rule file. For the latter, we provide a tab called NoHR Rules that allows us to load, save and edit rule files in a text panel following standard Prolog conventions.

The NoHR Query tab (see Fig. 2) also allows for the visualization of the rules, but its main purpose is to provide an interface for querying the combined KB. Whenever the first query is posed by pushing "Execute", a translator is started, initiating the ontology reasoner ELK [21] tailored for $\mathcal{EL}^+_\perp$ and considerably faster than other reasoners when comparing classification time [21]. ELK is used to classify the ontology $\mathcal{O}$ and then return the inferred axioms to the translator. It is also verified whether $DisjointWith$ axioms appear in $\mathcal{O}$, i.e., in $\mathcal{EL}^+_\perp$ notation, axioms of the form $C \sqcap D \sqsubseteq \perp$ for arbitrary classes C and D, which determines whether inconsistencies may occur in the combined hybrid knowledge base. Then the result of the classification is translated into rules and joined with the already given non-monotonic rules in $\mathcal{P}$, and the result is conditionally further transformed if inconsistency detection is required.

The result is used as input for the top-down query engine XSB Prolog[6] which realizes the well-founded semantics for logic programs [13]. To guarantee

---

[6] http://xsb.sourceforge.net

full compatibility with XSB Prolog's more restrictive admitted input syntax, the joint resulting rule set is further transformed such that all predicates and constants are encoded using MD5. The result is transfered to XSB via Inter-Prolog [9][7], which is an open-source Java front-end allowing the communication between Java and a Prolog engine.

Next, the query is sent via InterProlog to XSB, and answers are returned to the query processor, which collects them and sets up a table showing for which variable substitutions we obtain true, undefined, or inconsistent valuations (or just shows the truth value for a ground query). The table itself is shown in the Result tab (see Fig. 2) of the Output panel, while the Log tab shows measured times of pre-processing the knowledge base and answering the query. XSB itself not only answers queries very efficiently in a top-down manner, with tabling, it also avoids infinite loops.

Once the query has been answered, the user may pose other queries, and the system will simply send them directly without any repeated preprocessing. If the user changes data in the ontology or in the rules, then the system offers the option to recompile, but always restricted to the part that actually changed.

4   Cargo Shipment Use Case

The customs service for any developed country assesses imported cargo for a variety of risk factors including terrorism, narcotics, food and consumer safety, pest infestation, tariff violations, and intellectual property rights[8]. Assessing this risk, even at a preliminary level, involves extensive knowledge about commodi-ties, business entities, trade patterns, government policies and trade agreements. Some of this knowledge may be external to a given customs agency: for instance the broad classification of commodities according to the international Harmo-nized Tariff System (HTS), or international trade agreements. Other knowledge may be internal to a customs agency, such as lists of suspected violators or of importers who have a history of good compliance with regulations.

Figure 3 shows a simplified fragment $\mathcal{K} = (\mathcal{O}, \mathcal{P})$ of such a knowledge base. In this fragment, a shipment has several attributes: the country of its origination, the commodity it contains, its importer and producer. The ontology contains a geographic classification, along with information about producers who are located in various countries. It also contains (partial) information about three shipments: s_1, s_2 and s_3. There is also a set of rules indicating information about importers, and about whether to inspect a shipment either to check for compliance of tariff information or for food safety issues. For that purpose, the set of rules also includes a classification of commodities based on their harmonized tariff information (HTS chapters, headings and codes, cf. http://www.usitc.gov/tata/hts), and tariff information, based on the classification of commodities as given by the ontology.

---

[7] http://www.declarativa.com/interprolog/

[8] The system described here is not intended to reflect the policies of any country or agency.

* * * $\mathcal{O}$ * * *

Commodity ≡ (∃HTSCode.⊤)   Tomato ⊑ EdibleVegetable
CherryTomato ⊑ Tomato   GrapeTomato ⊑ Tomato
CherryTomato ⊓ GrapeTomato ⊑ ⊥   Bulk ⊓ Prepackaged ⊑ ⊥
EURegisteredProducer ≡ (∃RegisteredProducer.EUCountry)
LowRiskEUCommodity ≡ (∃ExpeditableImporter.⊤) ⊓ (∃CommodCountry.EUCountry)

ShpmtCommod(s_1, c_1)   ShpmtDeclHTSCode(s_1, h7022)
ShpmtImporter(s_1, i_1)   CherryTomato(c_1)   Bulk(c_1)
ShpmtCommod(s_2, c_2)   ShpmtDeclHTSCode(s_2, h7022)
ShpmtImporter(s_2, i_2)   GrapeTomato(c_2)   Prepackaged(c_2)
ShpmtCountry(s_2, portugal)
ShpmtCommod(s_3, c_3)   ShpmtDeclHTSCode(s_3, h7021)
ShpmtImporter(s_3, i_3)   GrapeTomato(c_3)   Bulk(c_3)
ShpmtCountry(s_3, portugal)   ShpmtProducer(s_3, p_1)
RegisteredProducer(p_1, portugal)   EUCountry(portugal)
RegisteredProducer(p_2, slovakia)   EUCountry(slovakia)

* * * $\mathcal{P}$ * * *

AdmissibleImporter($\mathbf{x}$) ← ShpmtImporter($\mathbf{y}, \mathbf{x}$), notSuspectedBadGuy($\mathbf{x}$).
SuspectedBadGuy(i_1).
ApprovedImporterOf($i_2, \mathbf{x}$) ← EdibleVegetable($\mathbf{x}$).
ApprovedImporterOf($i_3, \mathbf{x}$) ← GrapeTomato($\mathbf{x}$).
CommodCountry($\mathbf{x}, \mathbf{y}$) ← ShpmtCommod($\mathbf{z}, \mathbf{x}$), ShpmtCountry($\mathbf{z}, \mathbf{y}$).
ExpeditableImporter($\mathbf{x}, \mathbf{y}$) ← ShpmtCommod($\mathbf{z}, \mathbf{x}$), ShpmtImporter($\mathbf{z}, \mathbf{y}$),
                          AdmissibleImporter($\mathbf{y}$), ApprovedImporterOf($\mathbf{y}, \mathbf{x}$).
CompliantShpmt($\mathbf{x}$) ← ShpmtCommod($\mathbf{x}, \mathbf{y}$), HTSCode($\mathbf{y}, \mathbf{z}$), ShpmtDeclHTSCode($\mathbf{x}, \mathbf{z}$).
Random($\mathbf{x}$) ← ShpmtCommod($\mathbf{x}, \mathbf{y}$), notRandom($\mathbf{x}$).
NoInspection($\mathbf{x}$) ← ShpmtCommod($\mathbf{x}, \mathbf{y}$), CommodCountry($\mathbf{y}, \mathbf{z}$), EUCountry($\mathbf{z}$).
Inspection($\mathbf{x}$) ← ShpmtCommod($\mathbf{x}, \mathbf{y}$), notNoInspection($\mathbf{x}$), Random($\mathbf{x}$).
Inspection($\mathbf{x}$) ← ShpmtCommod($\mathbf{x}, \mathbf{y}$), notCompliantShpmt($\mathbf{x}$).
Inspection($\mathbf{x}$) ← ShpmtCommod($\mathbf{x}, \mathbf{y}$), Tomato($\mathbf{y}$), ShpmtCountry($\mathbf{x}$, slovakia).
HTSChapter($\mathbf{x}, 7$) ← EdibleVegetable($\mathbf{x}$).
HTSHeading($\mathbf{x}, 702$) ← Tomato($\mathbf{x}$).
HTSCode($\mathbf{x}$, h7022) ← CherryTomato($\mathbf{x}$).
HTSCode($\mathbf{x}$, h7021) ← GrapeTomato($\mathbf{x}$).
TariffCharge($\mathbf{x}, 0$) ← CherryTomato($\mathbf{x}$), Bulk($\mathbf{x}$).
TariffCharge($\mathbf{x}, 40$) ← GrapeTomato($\mathbf{x}$), Bulk($\mathbf{x}$).
TariffCharge($\mathbf{x}, 50$) ← CherryTomato($\mathbf{x}$), Prepackaged($\mathbf{x}$).
TariffCharge($\mathbf{x}, 100$) ← GrapeTomato($\mathbf{x}$), Prepackaged($\mathbf{x}$).

Fig. 3. MKNF knowledge base for Cargo Imports

The overall task then is to access all the information and assess whether some shipment should be inspected in full detail, under certain conditions randomly, or not at all. In fact, an inspection is considered if either a random inspection is indicated, or some shipment is not compliant, i.e., there is a mismatch between the filed cargo codes and the actually carried commodities, or some suspicious cargo is observed, in this case tomatoes from slovakia. In the first case, a potential random inspection is indicated whenever certain exclusion conditions do not hold. To ensure that one can distinguish between strictly required and random inspections, a random inspection is assigned the truth value undefined based on the rule $\mathsf{Random}(\mathbf{x}) \leftarrow \mathsf{ShpmtCommod}(\mathbf{x},\mathbf{y}), \mathsf{notRandom}(\mathbf{x})$.

The result of querying this knowledge base for $\mathsf{Inspection}(\mathbf{x})$ reveals that of the three shipments, s_2 requires an inspection (due to mislabeling) while s_1 may be subject to a random inspection as it does not knowingly originate from the EU. It can also be verified using the tool that preprocessing the knowledge base can be handled within $300ms$ and the query only takes $12ms$, which certainly suffices as interactive response. Please also note that the example indeed utilizes the features of rules and ontologies: for example exceptions to the potential random inspections can be expressed, but at the same time, taxonomic and non-closed knowledge is used, e.g., some shipment may in fact originate from the EU, this information is just not available.

5   Evaluation

In this section, we present some tests showing that a) the huge $\mathcal{EL}^+$ ontology SNOMED CT can be preprocessed for querying in a short period of time, b) adding rules increases the time of the translation only linearly, and c) querying time is in comparison to a) and b) in general completely neglectable. We performed the tests on a Mac book air 13 under Mac OS X 10.8.4 with a 1.8 GHz Intel Core i5 processor and 8 GB 1600 MHz DDR3 of memory. We ran all tests in a terminal version and Java with the "-XX:+AggressiveHeap" option, and test results are averages over 5 runs.

We considered SNOMED CT, freely available for research and evaluation[9], and added a varying number of non-monotonic rules. These rules were generated arbitrarily, using predicates from the ontology and additional new predicates (up to arity three), producing rules with a random number of body atoms varying from 1 to 10 and facts (rules without body atoms) with a ratio of 1:10. Note that, due to the translation of the DL part into rules, all atoms literally become non-DL-atoms. So ensuring that each variable appearing in the rule is contained in at least one non-negated body atom suffices to guarantee DL-safety for these rules.

The results are shown in Fig. 4 (containing also a constant line for classification of ELK alone and starting with the values for the case without additional rules), and clearly show that a) preprocessing an ontology with over 300,000 concepts takes less than 70 sec. (time for translator+loading in XSB), b) the

---

[9] http://www.ihtsdo.org/licensing/

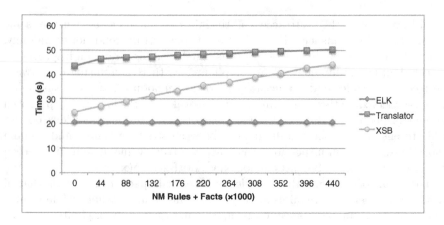

Fig. 4. Preprocessing time for SNOMED with a varying number of Rules

time of translator and loading the file in XSB only grows linearly on the number of rules with a small degree, in particular in the case of translator, and c) even with up to 500,000 added rules the time for translating does not surpass ELK classification, which itself is really fast [21], by more than a factor 2.5. All this data indicates that even on with a very large ontology, preprocessing can be handled very efficiently.

Finally, we also tested the querying time. To this purpose, we randomly generated and handcrafted several queries of different sizes and shapes using SNOMED with a varying number of non-monotonic rules as described before. In all cases, we observed that the query response time is interactive, observing longer reply times only if the number of replies is very high because either the queried class contains many subclasses in the hierarchy or if the arbitrarily generated rules create too many meaningless links, thus in the worst case requiring to compute the entire model. Requesting only one solution avoids this problem. Still, the question of realistic randomly generated rule bodies for testing querying time remain an issue of future work.

6   Conclusions

We have presented NoHR, the first plug-in for the ontology editor Protégé that integrates non-monotonic rules and top-down queries with ontologies in the OWL 2 profile OWL 2 EL. We have discussed how this procedure is implemented as a tool and shown how it can be used to implement a real use case on cargo shipment inspections. We have also presented an evaluation which shows that the tool is applicable to really huge ontologies, here SNOMED CT.

There are several relevant approaches discussed in the literature. Most closely related are probably [15,23], because both build on the well-founded MKNF semantics [22]. In fact, [15] is maybe closest in spirit to the original idea of

SLG($\mathcal{O}$) oracles represented in [2] on which the implementation of NoHR is theoretically founded. It utilizes the CDF framework already integrated in XSB, but its non-standard language is a drawback if we want to achieve compatibility with standard OWL tools based on the OWL API. On the other hand, [23], presents an OWL 2 QL oracle based on common rewritings in the underlying DL DL-Lite [3]. Less closely related is the work pursued in [8,14] that investigates direct non-monotonic extensions of $\mathcal{EL}$, so that the main reasoning task focuses on finding default subset inclusions, unlike this query-centered approach.

Two other related tools are DReW [30] and HD Rules [10], but both are based on different underlying formalisms to combine ontologies and non-monotonic rules. The former builds on dl-programs [12] and focuses on datalog-rewritable DLs [17], and the latter builds on Hybrid Rules [11]. While a more detailed comparison is surely of interest, the main problem is that both underlying formalisms differ from MKNF knowledge bases in the way information can flow between its two components and how flexible the language is [12,28].

We conclude with pointing out that given the successful application of the tool to the use-case as well as its evaluation, an obvious next step will be to try applying it to other use-case domains. This will allow gathering data, which may then be used for a) further dissemination in particular of query processing, which would b) stimulate application-driven optimizations and enhancements of the tool NoHR. Other future directions are extensions to paraconsistency [20] or more general formalisms [16,24,25].

Acknowledgments. We would like to thank the referees for their comments. We acknowledge partial support by FCT under project ERRO (PTDC/EIA-CCO/121823/2010) and under strategic project NOVA LINCS (PEst/UID/CEC/04516/2013). V. Ivanov was partially supported by a MULTIC – Erasmus Mundus Action 2 grant and M. Knorr was also partially supported by FCT grant SFRH/BPD/86970/2012.

References

1. Alberti, M., Knorr, M., Gomes, A.S., Leite, J., Gonçalves, R., Slota, M.: Normative systems require hybrid knowledge bases. In: Procs. of AAMAS, IFAAMAS, pp. 1425–1426 (2012)
2. Alferes, J.J., Knorr, M., Swift, T.: Query-driven procedures for hybrid MKNF knowledge bases. ACM Trans. Comput. Log. **14**(2), 1–43 (2013)
3. Artale, A., Calvanese, D., Kontchakov, R., Zakharyaschev, M.: The *DL-Lite* family and relations. J. Artif. Intell. Res. (JAIR) **36**, 1–69 (2009)
4. Baader, F., Calvanese, D., McGuinness, D.L., Nardi, D., Patel-Schneider, P.F. (eds.): The Description Logic Handbook: Theory, Implementation, and Applications, 3rd edn. Cambridge University Press (2010)
5. Baader, F., Brandt, S., Lutz, C.: Pushing the $\mathcal{EL}$ envelope. In: Procs. of IJCAI (2005)
6. Baral, C., Gelfond, M.: Logic programming and knowledge representation. J. Log. Program. **19**(20), 73–148 (1994)
7. Boley, H., Kifer, M. (eds.): RIF Overview. W3C Recommendation, February 5, 2013 (2013). http://www.w3.org/TR/rif-overview/

8. Bonatti, P.A., Faella, M., Sauro, L.: $\mathcal{EL}$ with default attributes and overriding. In: Patel-Schneider, P.F., Pan, Y., Hitzler, P., Mika, P., Zhang, L., Pan, J.Z., Horrocks, I., Glimm, B. (eds.) ISWC 2010, Part I. LNCS, vol. 6496, pp. 64–79. Springer, Heidelberg (2010)

9. Calejo, M.: InterProlog: towards a declarative embedding of logic programming in java. In: Alferes, J.J., Leite, J. (eds.) JELIA 2004. LNCS (LNAI), vol. 3229, pp. 714–717. Springer, Heidelberg (2004)

10. Drabent, W., Henriksson, J., Maluszynski, J.: Hd-rules: a hybrid system interfacing prolog with dl-reasoners. In: Procs. of ALPSWS, vol. 287 (2007)

11. Drabent, W., Maluszynski, J.: Hybrid rules with well-founded semantics. Knowl. Inf. Syst. **25**(1), 137–168 (2010)

12. Eiter, T., Ianni, G., Lukasiewicz, T., Schindlauer, R., Tompits, H.: Combining answer set programming with description logics for the semantic web. Artif. Intell. **172**(12–13), 1495–1539 (2008)

13. Gelder, A.V., Ross, K.A., Schlipf, J.S.: The well-founded semantics for general logic programs. J. ACM **38**(3), 620–650 (1991)

14. Giordano, L., Gliozzi, V., Olivetti, N., Pozzato, G.L.: Reasoning about typicality in low complexity dls: the logics $\mathcal{EL}^\perp T_{min}$ and $DL\text{-}Lite_c T_{min}$. In: Procs. of IJCAI (2011)

15. Gomes, A.S., Alferes, J.J., Swift, T.: Implementing query answering for hybrid MKNF knowledge bases. In: Carro, M., Peña, R. (eds.) PADL 2010. LNCS, vol. 5937, pp. 25–39. Springer, Heidelberg (2010)

16. Gonçalves, R., Alferes, J.J.: Parametrized logic programming. In: Janhunen, T., Niemelä, I. (eds.) JELIA 2010. LNCS, vol. 6341, pp. 182–194. Springer, Heidelberg (2010)

17. Heymans, S., Eiter, T., Xiao, G.: Tractable reasoning with dl-programs over datalog-rewritable description logics. In: Procs of ECAI, pp. 35–40. IOS Press (2010)

18. Hitzler, P., Krötzsch, M., Parsia, B., Patel-Schneider, P.F., Rudolph, S. (eds.): OWL 2 Web Ontology Language: Primer (Second Edition). W3C Recommendation, December 11, 2012 (2012). http://www.w3.org/TR/owl2-primer/

19. Ivanov, V., Knorr, M., Leite, J.: A query tool for EL with non-monotonic rules. In: Alani, H., Kagal, L., Fokoue, A., Groth, P., Biemann, C., Parreira, J.X., Aroyo, L., Noy, N., Welty, C., Janowicz, K. (eds.) ISWC 2013, Part I. LNCS, vol. 8218, pp. 216–231. Springer, Heidelberg (2013)

20. Kaminski, T., Knorr, M., Leite, J.: Efficient paraconsistent reasoning with ontologies and rules. In: Procs. of IJCAI. IJCAI/AAAI (2015)

21. Kazakov, Y., Krötzsch, M., Simančík, F.: The incredible ELK: From polynomial procedures to efficient reasoning with $\mathcal{EL}$ ontologies. Journal of Automated Reasoning **53**, 1–61 (2013)

22. Knorr, M., Alferes, J.J., Hitzler, P.: Local closed world reasoning with description logics under the well-founded semantics. Artif. Intell. **175**(9–10), 1528–1554 (2011)

23. Knorr, M., Alferes, J.J.: Querying OWL 2 QL and non-monotonic rules. In: Aroyo, L., Welty, C., Alani, H., Taylor, J., Bernstein, A., Kagal, L., Noy, N., Blomqvist, E. (eds.) ISWC 2011, Part I. LNCS, vol. 7031, pp. 338–353. Springer, Heidelberg (2011)

24. Knorr, M., Hitzler, P., Maier, F.: Reconciling OWL and non-monotonic rules for the semantic web. In: Procs. of ECAI, pp. 474–479. IOS Press (2012)

25. Knorr, M., Slota, M., Leite, J., Homola, M.: What if no hybrid reasoner is available? Hybrid MKNF in multi-context systems. J. Log. Comput. **24**(6), 1279–1311 (2014)

26. Lifschitz, V.: Nonmonotonic databases and epistemic queries. In: Procs. of IJCAI (1991)
27. Motik, B., Cuenca Grau, B., Horrocks, I., Wu, Z., Fokoue, A., Lutz, C. (eds.): OWL 2 Web Ontology Language: Profiles. W3C Recommendation, February 5, 2013. http://www.w3.org/TR/owl2-profiles/
28. Motik, B., Rosati, R.: Reconciling description logics and rules. J. ACM **57**(5) (2010)
29. Slota, M., Leite, J., Swift, T.: Splitting and updating hybrid knowledge bases. TPLP **11**(4–5), 801–819 (2011)
30. Xiao, G., Eiter, T., Heymans, S.: The DReW system for nonmonotonic dl-programs. In: Procs. of SWWS 2012. Springer Proceedings in Complexity. Springer (2013)

On the Cognitive Surprise in Risk Management: An Analysis of the Value-at-Risk (VaR) Historical

Davi Baccan[1]([✉]), Elton Sbruzzi[2], and Luis Macedo[1]

[1] CISUC, Department of Informatics Engineering, University of Coimbra,
Coimbra, Portugal
{baccan,macedo}@dei.uc.pt
[2] CCFEA, University of Essex, Colchester, UK
efsbru@essex.ac.uk

Abstract. Financial markets are environments in which a variety of products are negotiated by heterogeneous agents. In such environments, agents need to cope with uncertainty and with different kinds of risks. In trying to assess the risks they face, agents use a myriad of different approaches to somewhat quantify the occurrence of risks and events that may have a significant impact. In this paper we address the problem of risk management from the cognitive science perspective. We compute the cognitive surprise "felt" by an agent relying on a popular risk management tool known as Value-at-Risk (VaR) historical. We applied this approach to the S&P500 index from 26-11-1990 to 01-07-2009, and divided the series into two subperiods, a calm period and a crash period. We carried out an experiment with twelve different treatments and for each treatment we compare the intensity of surprise "felt" by the agent under these two different regimes. This interdisciplinary work contributes toward the truly understanding and improvement of complex economical and financial systems, specifically in providing insights on the behaviour of cognitive agents in those contexts.

Keywords: Agent-based computational economics · Artificial agents · Cognitive architectures · Cognitive modelling · Social simulation and modelling

1 Introduction

Financial markets such as stock markets are complex and dynamic environments in which a variety of products are negotiated by a very large number of heterogeneous agents ([1], [2]). Agents, either human, artificial or hybrid, are heterogeneous in the sense that they have, for instance, different preferences, beliefs, goals, and trading strategies (e.g., [3], [4]). Additionally, in such environments agents need to cope with uncertainty and with different kinds of risks [5].

Generally speaking, in economical and financial systems, agents usually try to assess in an objective or subjective way the risks they face. Ideally they would

© Springer International Publishing Switzerland 2015
F. Pereira et al. (Eds.) EPIA 2015, LNAI 9273, pp. 402–413, 2015.
DOI: 10.1007/978-3-319-23485-4_40

be able to come up with probabilities, either mathematical or not (e.g., subjective belief), to the occurrence of events that may have a positive or negative impact (e.g., [6], [7]) given his/her preferences and considering his/her goals [8]. Furthermore, agents tend to typically trade based on the risk-return trade-off, i.e., lower (higher) levels of risk are generally associated with lower (higher) levels of potential returns [9].

There are several different theories and hypotheses that try to explain how agents behave and how a stock market works (e.g., [3]). In short, there are two different and opposite perspectives. On the one hand, traditional economic theories (e.g., Efficient Market Hypothesis (EMH) [1]) rely, for instance, on the assumption that, when confronted with decisions that involve risk, agents are able to correctly form their probabilistic assessments according to the laws of probability [10], calculating which of the alternative courses of action maximize their expected utility. On the other hand, behavioral economics (e.g., [11]), i.e., the combination of psychology and economics that aims to understand human decision-making under risk as well as how this behaviour matters in economic contexts, have documented that there are deviations, known as behavioral biases, from the so-called rational behaviour. These behavioral biases are believed to be ubiquitous to humans, and several of them are clearly counterproductive from the economics perspective. Although the discussion regarding these different perspectives is of importance and fascinating for us, it is out of the scope of this work.

Nevertheless, we claim that, given the nature and complexity of financial markets together with the sophisticated and complex human decision-making mechanism (which is influenced by a myriad of intertwined factors), that the task of risk management is quite complex and falls on the category of those who requires the application of different and novel approaches in order to be improved ([12], [13], [14]). For instance, there are extensive empirical evidence (e.g., [5]) suggesting that the financial crisis of 2007/2008 was significantly aggravated by inappropriate risk management systems and tools.

However, perhaps more important for agents than having a good risk management system or tool is to rely on a system or tool that is in line with his/her behavioral and emotional profile. For instance, the underestimation of the occurrence of a given event may lead a particular agent to make a risky decision (e.g., excessive leverage) that may eventually elicit the surprise emotion in the agent with a high level of intensity as well as may probably result in substantial financial losses [15].

In this paper we address the problem of risk management from the cognitive science perspective by computing the cognitive surprise "felt" by an agent relying on a popular and widespread used risk management tool known as Value-at-Risk (VaR) historical. To this end, we model the VaR historical based on the principles of cognitive emotion theories and compute the cognitive surprise based on an artificial surprise model. We applied this approach to the S&P500 index and divide the series into two subperiods, a calm period and a crash period. We carried out an experiment with twelve different treatments, and for each treatment we compare the intensity of surprise "felt" by the agent under these two different regimes.

2   Value-at-Risk (VaR)

The Value-at-Risk (VaR) tool is one of the most popular financial risk measures, used by financial institutions all over the world [16]. The objective of the VaR is to measure the probability for significant loss in a portfolio of financial assets [17]. Generally speaking, we can assume that for a given time horizon t and a confidence level p, VaR is the loss in market value over the time horizon t that is exceeded with probability $1 - p$ [18]. For example, suppose a period of one-day ($t = 1$) and a confidence level p of 95%, the VaR would be 0.05 or 5% the critical value. There are several different methods for calculating VaR. For instance, let us briefly present the following two methods to calculate VaR, the statistical and historical approach [19].

The VaR statistical assumes that the historical returns respect the EMH. The EMH in turn assumes that the series of historical financial returns are Gaussian, with the average value μ of zero and constant variance of σ^2, i.e., returns $\sim \mathcal{N}(0, \sigma^2)$. Based on the EMH assumptions and on the Gaussian characteristics, we could compute that VaR statistical for a confidence level p of 99% and 95% are -2σ and -3σ, respectively. For example, if a series of returns show a standard deviation of 5%, the VaR statistical for a confidence level of 99% and 95% are -10% and -15%, respectively.

Unlike the VaR statistical method, an alternative way to calculate VaR is to rank the historical simulation from the smallest to the highest, which is named VaR historical. Suppose that the series of T returns are $r_1,...,\ r_t$, we define that this series of returns are ranked if $r_1 \leq r_2 \leq ... \leq r_T$. In this case, the VaR historical is the return on the position integer $((1 - p)\ T)$. For example, suppose a confidence level p of 99% and T of 250, the VaR historical would be r_3. In the case of p of 95%, the VaR historical would be r_5.

We consider the VaR historical as the most appropriate method for this work for several reasons. First, because the VaR historical method is widely used by practitioners ([20], [21], [22]). Second, because it is an easy to understand measure that computes the estimation based on historical data and a confidence level (we will later explain in Section 3.2 the particular importance of confidence). Last but not least, because unlike the VaR statistical, VaR historical is free of the assumption about the distribution of the series of returns [19].

3   Surprise in Cognitive Science

From the perspective of cognitive emotions theories (e.g., [23], [24]), surprise can be thought of as a belief-disconfirmation signal or a mismatch generated as a result of the comparison between a newly acquired belief and a pre-existing belief. Formally speaking, surprise is a neutral valence emotion, defined as a peculiar state of mind, usually of brief duration, caused by unexpected events, or proximally the detection of a contradiction or conflict between a newly acquired and pre-existing belief ([25], [26]).

Surprise serves us in many functions and can be considered as a key element for survival in a complex and rapidly changing environment. It is closely related to how beliefs are stored in memory. Our semantic memory, i.e., our general knowledge and concepts about the world, is assumed to be represented in memory through knowledge structures known as schemas (e.g., [27]). A schema is a well-integrated chunk of knowledge or sets of beliefs, which main source of information available comes from abstraction from repeated personally experienced events or generalizations, that are our episodic memory.

3.1 The Surprise Process

Meyer and colleagues [25] proposed a cognitive-psychoevolutionary model of surprise. They claim surprise-eliciting events elicit a four-step sequence of processes. The first step is the appraisal of an event as unexpected or schema-discrepant.

Then in the second step, if the degree of unexpectedness or schema-discrepancy exceeds a certain threshold, surprise is experienced, ongoing mental process are interrupted and resources such as attention are reallocated towards the unexpected event.

The third step is the analysis and evaluation of the unexpected event. It generally includes a set of subprocesses namely the verification of the schema discrepancy, the analysis of the causes of the unexpected event, the evaluation of the unexpected event's significance for well-being, and the assessment of the event's relevance for ongoing action. It is assumed that some aspects of the analysis concerning the unexpected event are stored as part of the schema for this event so that in the future analysis of similar events can be significantly reduced both in terms of time and cognitive effort.

The fourth step is the schema update. It involves producing the immediate reactions to the unexpected event, and/or operations such as the update, extension, or revision of the schema or sets of beliefs that gave rise to the discrepancy. The schema change ideally enables one to some extent to predict and control future occurrences of the schema-discrepant event and, if possible, to avoid the event if it is negative and uncontrollable, or to ignore the event if it is irrelevant for action.

3.2 Artificial Surprise

Two models of artificial surprise for artificial agents can be stressed, namely the model proposed by Macedo and Cardoso [28] and the model proposed by Lorini and Castelfranchi [29]. Both models were mainly inspired by the cognitive-psychoevolutionary model of surprise proposed by Meyer and colleagues [25] and have influence of the analysis of the cognitive causes of surprise from a cognitive science perspective proposed by Ortony and Partridge [30]. The comparative study of these two models is out of the scope of this work. To a detailed description of the similarities and differences of the models please see [26]. The empirical tests we performed provide evidence in favor of using the model proposed by Macedo and Cardoso in our work.

Macedo and colleagues carried out an empirical study [28] with the goal of investigating how to compute the intensity of surprise in an artificial agent.

This study suggests that the intensity of surprise about an event E_g, from a set of mutually exclusive events E_1, E_2, ..., E_m, is a nonlinear function of the difference, or contrast, between its probability/belief and the probability/belief of the highest expected event (E_h) in the set of mutually exclusive events E_1, E_2, ..., E_m.

Formally, let (Ω, A, P) be a probability space where Ω is the sample space (i.e., the set of possible outcomes of the event), $A = A_1, A_2, ..., A_n$, is a σ-field of subsets of Ω (also called the event space, i.e., all the possible events), and P is a probability measure which assigns a real number $P(F)$ to every member F of the σ-field A. Let $E = E_1, E_2, ..., E_m$, $E_i \in A$, be a set of mutually exclusive events in that probability space with probabilities $P(E_i) \geq 0$, such that $\sum_{i=1}^{m} P(E_i) = 1$. Let E_h be the highest expected event from E. The intensity of surprise about an event E_g, defined as $S(E_g)$, is calculated as $S(E_g) = log_2(1 + P(E_h) - P(E_g))$ (Equation 1). In each set of mutually exclusive events, there is always at least one event whose occurrence is unsurprising, namely E_h.

4   Experiments and Results

The goal of our experiment is to compute the cognitive surprise "felt" by an agent relying on a popular risk management tool known as Value-at-Risk (VaR) historical under different financial settings and regimes.

First, we selected the S&P500, i.e., an index based on market capitalization that includes 500 companies in leading industries in the U.S. economy, from 26-11-1990 to 01-07-2009, in a total of 4688 days. The S&P500 index is recognized as one of the most important stock market indexes (perhaps the most important). First, we applied a method similar to the work of Halbleib and Pohlmeier [20] to divide the selected S&P500 period into two parts, a calm period from 26-11-1990 to 31-08-2008 (total of 4478 days), and a crash period from 01-09-2008 to 01-07-2009 (total of 210 days). The data was obtained free of charge from Yahoo Finance. Figure 1 presents the daily close, daily return, and the histogram of daily returns of the S&P500 from 26-11-1990 to 01-07-2009. Figure 2 presents the daily close and the histogram of daily returns of the S&P500 of calm and crash periods.

We employed two different approaches in our experiment. The first approach consists in specifying a rolling window size and a confidence level, which combination is used by the VaR historical to compute the estimation by considering the most recent returns. The windows size set contains the 50, 125, 250, and 500 values. The confidence level set contains the 0.95 and 0.99 values. The combination of the window size values with the confidence level values result in 8 different treatments.

The second approach consists in specifying a decay function so that recent daily returns gain more weight as opposed to old returns. Decay functions are generally used with the objective of emulating to a certain extent the human memory process of "forgetting" as well as to contemplate some findings from

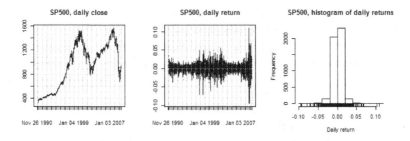

Fig. 1. SP500, daily close (left), daily return (center), and histogram of daily return (right).

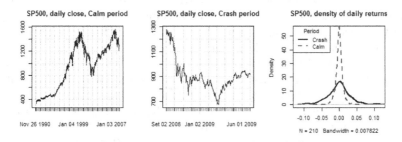

Fig. 2. SP500, daily close of calm period (left), daily close of crash period (center), and histogram of daily return of calm period and crash period (right)

how humans use past experience in decision-making (e.g., [31]) which indicate that in revising their beliefs, people tend to overweight recent information and underweight prior information. The alpha set contains the 0.995, 0.99, 0,97, 0.94 values. A higher (lower) alpha implies a smaller (higher) level of forgetfulness. Unlike the first approach, we opted to use just a confidence level of 0.99. The reason is that we observed in our initial experiments that the combination of a confidence level of 0.95 with the previous alpha values caused the agent to "forget" too many returns, generating in the end a quite low and poor VaR historical estimation. Therefore, in the second approach we have four treatments. We conducted an experiment with these twelve different treatments.

The algorithm we used in our experiment works as follows. We first perform initial adjustments to ensure that all iterations are actually carried out within a specified period. For each simulation (1) for a window size value, $window_k \in \{50, 125, 250, 500\}$, (2) according to a uniform distribution function, we select a begin day d_i within the period (for the calm period we fixed the $d_i = 01\text{-}09\text{-}2008$), (3) we compute the VaR historical estimations VaR_{95} and VaR_{99} based on $window_k$ daily returns preceding the d_i as well as the confidence levels $p \in \{0.95, 0.99\}$, respectively, (4) for each day d_k beginning in the day after d_i, from $k = i + 1$ to $k = 210$ (for the next 210 days), we check if daily return of $d_k \leq VaR_{95}$ and if daily return of $d_k \leq VaR_{99}$, (5) we advance the rolling window one day, i.e.,

$i = i + 1$, and (6) we back to step (3). For the alpha approach the algorithm is similar, except for some minor adjustments, specifically in step (1) since the algorithm runs for each alpha value $\in \{0.995, 0.99, 0, 97, 0.94\}$, and in step (3) since we modify the preceding daily returns by applying the current alpha value and compute the VaR historical estimation with a confidence level p of 0.99. All other steps are exactly the same.

Let us now describe how we address this problem in the context of the artificial surprise, presented in Section 3. We essentially applied the concepts, ideas, and method presented by Baccan and Macedo [32]. This work can be thought of as a continuation and expansion of their initial work to other contexts. We assume, for the sake of the experiment and simplicity, the confidence levels p (0.99, and 0.95) as the subjective belief of the agent in the accurateness of the VaR historical estimation. By making this assumption and considering a higher subjective belief, we are empowering the agent with a "firmly believe" in the accurateness of the VaR historical estimation. So, suppose an event E_g as VaR historical estimation that can assume two mutually exclusive events, meaning that it can be either correct (E_1), i.e., daily return is not lower than estimation, or incorrect (E_2), i.e., daily return is lower than estimation.

The agent will either "feels" no surprise (a higher intensity of surprise) as what he/she considered as more (less) likely, i.e., correct (E_1) (incorrect (E_2)), happened. More precisely, for the confidence level of 0.99 the surprise about event E_2 would be 0.9855004, i.e., $S(E_g) = log_2(1 + 0.99 - 0.01)$. Similarly, for the confidence level of 0.95 the surprise about event E_2 would be 0.9259994, i.e., $S(E_g) = log_2(1 + 0.95 - 0.05)$. For each day d_k in which the VaR historical estimation is tested in step (4) of the algorithm described above, if the daily return of $d_k \leq$ VaR$_p$, then we compute the cognitive surprise, $surprise_k$.

In the end we have a sequence $\{surprise_1, ..., surprise_{210}\}$. Afterwards, for each simulation, we compute the cumulative sum of the surprise. It means that we generate a sequence of 210 elements as a result of the partial sums $surprise_1$, $surprise_1 + surprise_2$, $surprise_1 + surprise_2 + surprise_3$, and so forth. In the case of the calm period, we added the cumulative sum of the surprise of each treatment and then average it by the number of simulations. The cumulative sum and the average assumption make it easier both the observation of surprise over time as well as the comparison of surprise between the calm and crash period. The average cumulative sum of the surprise for a given treatment is presented in the next figures for the calm period.

We ran 10^4 independent simulations for each treatment for the calm period and one simulation for each treatment for the crash period (since the begin day $d_i = 01\text{-}09\text{-}2008$ is fixed).

Figures 3 and 4 present the behaviour of all treatments for the calm period and crash period, respectively. Figures 5 present a comparison between different rolling window treatments and alpha treatments, respectively, for both periods.

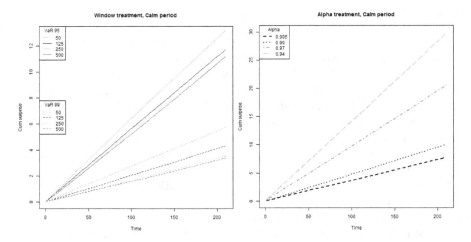

Fig. 3. All treatments for the calm period.

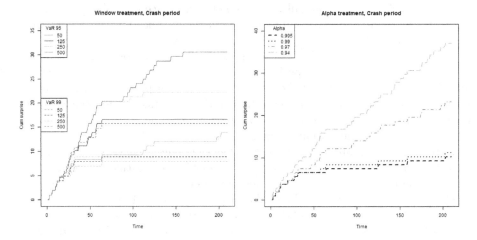

Fig. 4. All treatments for the crash period.

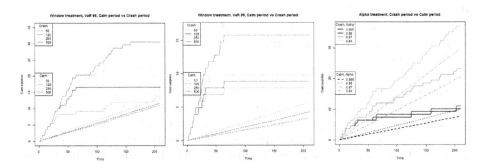

Fig. 5. Comparing different treatments for different periods.

5   Discussion

First of all, it is important to bear in mind that financial markets are by their very nature complex and dynamic systems. Such complexity significantly increases if we take into account the sophisticated and complex human decision-making mechanism. As a result, we believe that the task of risk management in finance and economics is indeed quite difficult. Therefore, we do not have in this work the goal of providing evidence neither in favor of nor against to a particular risk management tool or system. Instead, our goal is to compute, in a systematic and clear method, the cognitive surprise "felt" by an agent relying on the VaR historical during different periods, a calm period and a crash period.

Let us then begin by analyzing some characteristics of the calm period in comparison with the crash period. We can see in Figure 1 that daily returns seem, as expected and in line with the existing literature, to reproduce some statistical regularities that are often found in a large set of different assets and markets known as stylized facts [33]. Specifically we can observe that returns do not follow a Gaussian distribution (are not normally distributed), seem to exhibit what is known as fat-tails, as well as to reproduce the volatility clustering fact, i.e., high-volatility events tend to cluster in time. Volatility clustering resembles the concept of entropy used in a variety of areas such as information and communication theory. We can see in Figure 2 that the comparison between the daily returns of the calm period with the daily returns of the crash period allows us to claim that the daily returns for the crash period seem to exhibit fat tail distributions. We can also see that during the crash period the SP&500 depreciated almost 50% in value in a short period of time.

We now turn our attention to the analysis of the cognitive surprise. For the calm period, we can observe in Figure 3 that the cumulative surprise for alpha treatment (right) is higher when compared to all window treatments (left). Additionally, the lower the alpha, the higher the cumulative surprise and, similarly, the larger the window size, the lower the surprise. The cumulative surprise of the window treatments with a confidence level p of 0.95, VaR_{95} is, in turn, higher when compared to the cumulative surprise with a confidence level p of 0.99, VaR_{99}. Interestingly, if we assume that alpha somewhat emulates the human memory process of "forgetting" and considering that a lower alpha implies a higher level of forgetfulness, we may argue that an agent should be careful in forgetting the past, at least in the context of a stock market, since the cognitive surprise "felt" by an agent under this treatment is significantly higher than the window treatment.

For the crash period, we first observe in Figure 4 that, once again, the lower the alpha, the higher the cumulative surprise (right). However, quite contrary to the calm period, the lower the window size, the lower the surprise (left). The cognitive surprise "felt" is higher for agents that rely on a larger window size. It may be explained by the fact that the crash period is indeed a period in which the volatility is high. Therefore, a VaR historical based on a larger window size takes more time to adapt to this new and changing environment, in comparison with a VaR historical based on a smaller window size.

When comparing in Figure 5 how treatments behave during the calm period and the crash period, we can observe that, as expected, the cognitive surprise is higher during the crash period.

Generally speaking, each day the agent "felt" surprise represents a failed VaR historical estimation. Therefore, the higher the surprise, the wrong a given VaR historical treatment is. The analysis of the results indicate that it may be quite difficult for an agent not to "feel" surprise when relying on VaR historical. Indeed, there are several issues with models, like VaR historical, that take into account historical data for their estimation. These models somewhat assume the past is a good indicator of what may happen in the future, i.e., history repeats itself. However, this inductive reasoning often underestimate the probability of extreme returns and, consequently, underestimates the level of risk. Additionally, an essential flaw of this kind of rationale is not to truly acknowledge that "absence of evidence is not evidence of absence" and the existence of "unknown unknowns" [34].

Consider, for instance, the turkey paradox ([35]). There is a butcher and a turkey. Every day for let us say 100 days the butcher feeds the turkey. As time goes by the turkey increases its belief that in the next day it will receive food from the butcher. However, at a given day, for the "shock" and "surprise" of the turkey, instead of being feed by the butcher, the butcher kills the turkey. The same analogy may be applied to the black swan scenario [6] as well as to other complex and risky financial operations that provide small but regular gains, until the day they blow all the gains, resulting in huge losses [15].

The main contribution of our work resides in the fact that we have applied in a systematic, clear and easy to reproduce way the ideas, concepts and methods described by Baccan and Macedo [32] to the context of risk, uncertainty, and therefore risk management. Our interdisciplinary work is in line with those who claim that there is a need for novel approaches so that complex and financial systems may be improved (e.g., [12], [13], [14]). It is, as far as we know, one of the first attempts to apply a cognitive science perspective to risk management.

6   Conclusion and Future Work

In this paper we addressed the problem of risk management from the cognitive science perspective [1]. We computed the cognitive surprise "felt" by an agent relying on a popular risk management tool known as Value-at-Risk (VaR) historical. We applied this approach to the S&P500 stock market index from 26-11-1990 to 01-07-2009, and divide the series into two subperiods, a calm period and a crash period. We carried out an experiment with twelve different treatments and for each treatment we compare the intensity of surprise "felt" by the agent under these two different regimes. This interdisciplinary work is in line with a broader movement and contributes toward the truly understanding and improvement of

[1] This work was supported by FCT, Portugal, SFRH/BD/60700/2009, and by TribeCA project, funded by FEDER through POCentro, Portugal.

complex economical and financial systems, specifically in providing insights on the behaviour of cognitive agents in those risky and uncertain contexts.

The use of cognitive modeling approaches to the study of complex and dynamic systems is in its early stages. We consider that the use of relatively simple but powerful tools in conjunction with cognitive models, like those discussed in this work, offers a rich and novel set of possibilities. For instance, the use of cognitive agents would allow us to carry out a variety of different agent and multi-agent-based simulations of significant economical and financial events, so that we would be able to compute individual cognitive surprise and, in the end, the global surprise of agents regarding some event. It may provide contributions toward the understanding of the behaviour of agents individually as well as of the system as a whole. Last but not least, it would be interest to compare the global surprise with other market sentiment indexes (e.g., VIX, the "fear" indicator).

References

1. Fama, E.F.: Efficient capital markets: A review of theory and empirical work. The Journal of Finance **25**(2), 383–417 (1970)
2. Fama, E.F.: Two pillars of asset pricing. American Economic Review **104**(6), 1467–1485 (2014)
3. Lo, A.W.: Reconciling efficient markets with behavioral finance: The adaptive markets hypothesis. Journal of Investment Consulting **7**, 21–44 (2005)
4. Treleaven, P., Galas, M., Lalchand, V.: Algorithmic trading review. Commun. ACM **56**(11), 76–85 (2013)
5. Lo, A., Mueller, M.: WARNING: physics envy may be hazardous to your wealth!. Journal of Investment Management **8**, 13–63 (2010)
6. Taleb, N.N.: The Black Swan: The Impact of the Highly Improbable, 1st edn. Random House, April 2007
7. Meder, B., Lec, F.L., Osman, M.: Decision making in uncertain times: what can cognitive and decision sciences say about or learn from economic crises? Trends in Cognitive Sciences **17**(6), 257–260 (2013)
8. Markowitz, H.: Portfolio selection. Journal of Finance **7**, 77–91 (1952)
9. Kahneman, D.: The myth of risk attitudes. The Journal of Portfolio Management **36**(1), 1 (2009)
10. Chiodo, A., Guidolin, M., Owyang, M.T., Shimoji, M.: Subjective probabilities: psychological evidence and economic applications. Technical report, Federal Reserve Bank of St. Louis (2003)
11. Kahneman, D., Tversky, A.: Prospect theory: An analysis of decision under risk. Econometrica **47**(2), 263–291 (1979)
12. Farmer, J.D., Foley, D.: The economy needs agent-based modelling. Nature **460**(7256), 685–686 (2009)
13. Bouchaud, J.: Economics needs a scientific revolution. Nature **455**(7217), 1181 (2008)
14. Gatti, D., Gaffeo, E., Gallegati, M.: Complex agent-based macroeconomics: a manifesto for a new paradigm. Journal of Economic Interaction and Coordination **5**(2), 111–135 (2010)

15. Taleb, N.N.: Antifragile: Things That Gain from Disorder. Reprint edition edn. Random House Trade Paperbacks, New York (2014)
16. Kawata, R., Kijima, M.: Value-at-risk in a market subject to regime switching. Quantitative Finance **7**(6), 609–619 (2007)
17. Alexander, C., Sarabia, J.M.: Quantile Uncertainty and Value-at-Risk Model Risk. Risk Analysis **32**(8), 1293–1308 (2012)
18. Duffie, D., Pan, J.: An Overview of Value at Risk. The Journal of Derivatives **4**(3), 7–49 (1997)
19. Jorion, P.: Value at Risk: The New Benchmark for Managing Financial Risk, 3rd edn. McGraw-Hill (2006)
20. Halbleib, R., Pohlmeier, W.: Improving the value at risk forecasts: Theory and evidence from the financial crisis. Journal of Economic Dynamics and Control **36**(8), 1212–1228 (2012)
21. David Cabedo, J., Moya, I.: Estimating oil price Value at Risk using the historical simulation approach. Energy Economics **25**(3), 239–253 (2003)
22. Hendricks, D.: Evaluation of value-at-risk models using historical data. Economic Policy Review, 39–69, April 1996
23. Reisenzein, R., Hudlicka, E., Dastani, M., Gratch, J., Hindriks, K., Lorini, E., Meyer, J.J.: Computational modeling of emotion: Towards improving the inter- and intradisciplinary exchange. IEEE Transactions on Affective Computing **99**(1), 1 (2013)
24. Reisenzein, R.: Emotions as metarepresentational states of mind: Naturalizing the belief-desire theory of emotion. Cognitive Systems Research **10**(1), 6–20 (2009)
25. Meyer, W.U., Reisenzein, R., Schutzwohl, A.: Toward a process analysis of emotions: The case of surprise. Motivation and Emotion **21**, 251–274 (1997)
26. Macedo, L., Cardoso, A., Reisenzein, R., Lorini, E., Castelfranchi, C.: Artificial surprise. In: Handbook of Research on Synthetic Emotions and Sociable Robotics: New Applications in Affective Computing and Artificial Intelligence, pp. 267–291 (2009)
27. Baddeley, A., Eysenck, M., Anderson, M.C.: Memory, 1 edn. Psychology Press, February 2009
28. Macedo, L., Reisenzein, R., Cardoso, A.: Modeling forms of surprise in artificial agents: empirical and theoretical study of surprise functions. In: 26th Annual Conference of the Cognitive Science Society, pp. 588–593 (2004)
29. Lorini, E., Castelfranchi, C.: The cognitive structure of surprise: looking for basic principles. Topoi: An International Review of Philosophy **26**, 133–149 (2007)
30. Ortony, A., Partridge, D.: Surprisingness and expectation failure: what's the difference? In: Proceedings of the 10th International Joint Conference on Artificial Intelligence, vol. 1, pp. 106–108. Morgan Kaufmann Publishers Inc., Milan (1987)
31. Griffin, D., Tversky, A.: The weighing of evidence and the determinants of confidence. Cognitive Psychology **24**(3), 411–435 (1992)
32. Baccan, D., Macedo, L., Sbruzzi, E.: Towards modeling surprise in economics and finance: a cognitive science perspective. In: STAIRS 2014. Frontiers in Artificial Intelligence and Applications, vol. 264, Prague, Czech Republic, pp. 31–40 (2014)
33. Cont, R.: Empirical properties of asset returns: stylized facts and statistical issues. Quantitative Finance **1**(2), 223 (2001)
34. Rumsfeld, D.: DoD News Briefing (2002)
35. Taleb, N.N.: Fooled by Randomness: The Hidden Role of Chance in Life and in the Markets, 2 edn. Random House, October 2008

Logic Programming Applied to Machine Ethics

Ari Saptawijaya[1,2](✉) and Luís Moniz Pereira[1]

[1] NOVA Laboratory for Computer Science and Informatics (NOVA LINCS),
Departamento de Informática, Faculdade de Ciências e Tecnologia,
Universidade Nova de Lisboa, Lisboa, Portugal
ar.saptawijaya@campus.fct.unl.pt, lmp@fct.unl.pt
[2] Faculty of Computer Science, Universitas Indonesia, Depok, Indonesia

Abstract. This paper summarizes our investigation on the application of
LP-based reasoning to machine ethics, a field that emerges from the need of
imbuing autonomous agents with the capacity for moral decision-making.
We identify morality viewpoints (concerning moral permissibility and the
dual-process model) as studied in moral philosophy and psychology, which
are amenable to computational modeling. Subsequently, various LP-based
reasoning features are applied to model these identified morality viewpoints, via classic moral examples taken off-the-shelf from the literature.

1 Introduction

The need for systems or agents that can function in an ethically responsible
manner is becoming a pressing concern, as they become ever more autonomous
and act in groups, amidst populations of other agents, including humans. Its
importance has been emphasized as a research priority in AI with funding support [26]. Its field of enquiry, named *machine ethics*, is interdiscplinary, and is
not just important for equipping agents with some capacity for moral decision-
making, but also to help better understand morality, via the creation and testing
of computational models of ethical theories.

Several logic-based formalisms have been employed to model moral theories
or particular morality aspects, e.g., deontic logic in [2], non-monotonic reasoning in [6], and the use of Inductive Logic Programming (ILP) in [1]; some of
them only abstractly, whereas others also provide implementations (e.g., using
ILP-based systems [1], an interactive theorem prover [2], and answer set programming (ASP) [6]). Despite the aforementioned logic-based formalisms, Logic
Programming (LP) itself is rather limitedly explored. The potential and suitability of LP, and of computational logic in general, for machine ethics, is identified
and discussed at length in [11], on the heels of our work. LP permits declarative knowledge representation of moral cases with sufficiently level of detail to
distinguish one moral case from other similar cases. It provides a logic-based
programming paradigm with a number of practical Prolog systems, thus allowing not only addressing morality issues in an abstract logical formalism, but also
via a Prolog implementation as proof of concept and a testing ground for experimentation. Furthermore, LP are also equipped with various reasoning features,

F. Pereira et al. (Eds.) EPIA 2015, LNAI 9273, pp. 414–422, 2015.
DOI: 10.1007/978-3-319-23485-4_41

as identified in the paragraph below, whose applications to machine ethics are promising, but still unexplored. This paper summarizes our integrative investigation on the appropriateness of various LP-based reasoning to machine ethics, not just abstractly, but also furnishing a proof of concept implementation for the morality issues in hand.

We identify conceptual morality viewpoints, which are covered in two morality themes: (1) *moral permissibility*, taking into account viewpoints such as the Doctrines of Double Effect (DDE) [15], Triple Effect (DTE) [10], and Scanlon's contractualist moral theory [23]; and (2) *the dual-process model* [3,14], which stresses the interaction between deliberative and reactive behaviors in moral judgment. The mapping of all these considered viewpoints into LP-based reasoning benefits from its features and their integration, such as abduction with integrity constraints (ICs) [22], preferences over abductive scenarios [4], probabilistic reasoning [7], updating [21], counterfactuals [20], and from LP tabling technique [25].

We show, in Section 2, how these various LP-based reasoning features are employed to model the aforementioned morality viewpoints, including: (1) The use of a priori ICs and a posteriori preferences over abductive scenarios to capture deontological and utilitarian judgments; (2) Probabilistic moral reasoning, to reason about actions, under uncertainty, that might have occurred, and thence provide judgment adhering to moral principles within some prescribed uncertainty level. This permits to capture a form of argumentation (wrt. Scanlon's contractualism [23]) in courts, through presenting different evidences as a consideration whether an exception can justify a verdict of guilty (beyond reasonable doubt) or non-guilty; (3) The use of QUALM, which combines LP abduction, updating, and counterfactuals, supported by LP tabling mechanisms (based on [20–22]) to examine moral permissibility wrt. DDE and DTE, via counterfactual queries. Finally, QUALM is also employed to experiment with the issue of moral updating, allowing for other (possibly overriding) moral rules (themselves possibly subsequently overridden) to be adopted by an agent, on top of those it currently follows.

2   Modeling Morality with Logic Programming

2.1   Moral Permissibility with Abduction, a Priori ICs and a Posteriori Preferences

In [17], moral permissibility is modeled through several cases of the classic trolley problem [5], by emphasizing the use of ICs in abduction and preferences over abductive scenarios. The cases, which include moral principles, are modeled in order to deliver appropriate moral decisions that conform with those the majority of people make, on the basis of empirical results in [9]. DDE [15] is utilized in [9] to explain the consistency of judgments, shared by subjects from demographically diverse populations, on a series of trolley dilemmas. In addition to DDE, we also consider DTE [10].

Each case of the trolley problem is modeled individually; their details being referred to [17]. The key points of their modeling are as follows. The DDE and DTE are modeled via a priori ICs and a posteriori preferences. Possible decisions are modeled as abducibles, encoded in ACORDA by even loops over default negation. Moral decisions are therefore accomplished by satisfying a priori ICs, computing abductive stable models from all possible abductive solutions, and then appropriately preferring amongst them (by means of rules), a posteriori, just some models, on the basis of their abductive solutions and consequences. Such preferred models turn out to conform with the results reported in the literature.

Capturing Deontological Judgment via a Priori ICs. In this application, ICs are used for two purposes. First, they are utilized to force the goal in each case (like in [9]), by observing the desired end goal resulting from each possible decision. Such an IC thus enforces all available decisions to be abduced, together with their consequences, from all possible observable hypothetical end goals. The second purpose of ICs is for ruling out impermissible actions, viz., actions that involve intentional killing in the process of reaching the goal, enforced by the IC: $false \leftarrow intentional\_killing$. The definition of $intentional\_killing$ depends on rules in each case considered and whether DDE or DTE is to be upheld. Since this IC serves as the first filter of abductive stable models, by ruling out impermissible actions, it affords us with just those abductive stable models that contain only permissible actions.

Capturing Utilitarian Judgment via a Posteriori Preferences. Additionally, one can further prefer amongst permissible actions those resulting in greater good. That is, whereas a priori ICs can be viewed as providing an agent's reactive behaviors, generating intuitively intended responses that comply with deontological judgment (enacted by ruling out the use of intentional harm), a posteriori preferences amongst permissible actions provides instead a more involved reasoning about action-generated models, capturing utilitarian judgment that favors welfare-maximizing behaviors (in line with the dual-process model [3]).

In this application, a preference predicate (e.g., based on a utility function concerning the number of people died) is defined to select those abductive stable models [4] containing decisions with greater good of overall consequences. The reader is referred to [17] for the results of various trolley problem cases.

2.2 Probabilistic Moral Reasoning

In [8], probabilistic moral reasoning is explored, where an example is contrived to reason about actions, under uncertainty, and thence provide judgment adhering to moral rules within some prescribed uncertainty level. The example takes a variant of the Footbridge case within the context of a jury trials in court, in order to proffer verdicts beyond reasonable doubt: *Suppose a board of jurors in a court is faced with the case where the actual action of an agent shoving the man onto the track was not observed. Instead, they are just presented with the fact that the man on the bridge died on the side track and the agent was seen on the bridge at the occasion. Is the agent guilty (beyond reasonable doubt), in the sense of violating DDE, of shoving the man onto the track intentionally?*

To answer it, abduction is enacted to reason about the verdict, given the available evidence. Considering the active goal *judge*, to judge the case, two abducibles are available: *verdict(guilty_brd)* and *verdict(not_guilty)*, where *guilty_brd* stands for 'guilty beyond reasonable doubt'. Depending on how probable each verdict (the value of which is determined by the probability $pr_{int\_shove}(P)$ of intentional shoving), a preferred *verdict(guilty_brd)* or *verdict(not_guilty)* is abduced as a solution.

The probability with which shoving is performed intentionally is causally influenced by evidences and their attending truth values. Two evidences are considered, viz., (1) Whether the agent was running on the bridge in a hurry; and (2) Whether the bridge was slippery at the time. The probability $pr_{int\_shove}(P)$ of intentional shoving is therefore determined by the existence of evidence, expressed as dynamic predicates *evd_run/1* and *evd_slip/1*, whose sole argument is *true* or *false*, standing for the evidences that the agent was running in a hurry and that the bridge was slippery, resp.

Based on this representation, different judgments can be delivered, subject to available (observed) evidences and their attending truth value. By considering the standard probability of proof beyond reasonable doubt –here the value of 0.95 is adopted [16]– as a common ground for the probability of guilty verdicts to be qualified as 'beyond reasonable doubt', a form of argumentation (à la Scanlon contractualism [23]) may take place through presenting different evidence (via updating of observed evidence atoms, e.g., *evd_run(true)*, *evd_slip(false)*, etc.) as a consideration to justify an exception. Whether the newly available evidence is accepted as a justification to an exception –defeating the judgment based on the priorly presented evidence– depends on its influence on the probability $pr_{int\_shove}(P)$ of intentional shoving, and thus eventually influences the final verdict. That is, it depends on whether this probability is still within the agreed standard of proof beyond reasonable doubt. The reader is referred to [8], which details a scenario capturing this moral jurisprudence viewpoint.

2.3 Modeling Morality with QUALM

Distinct from the two previous applications, QUALM emphasizes the interplay between LP abduction, updating and counterfactuals, supported furthermore by their joint tabling techniques.

Counterfactuals in Morality. We revisit moral permissibility wrt. DDE and DTE, but now applying counterfactuals. Counterfactuals may provide a general way to examine DDE in dilemmas, like the classic trolley problem, by distinguishing between a *cause* and a *side-effect* as a result of performing an action to achieve a goal. This distinction between causes and side-effects may explain the permissibility of an action in accordance with DDE. That is, *if some morally wrong effect E happens to be a cause for a goal G that one wants to achieve by performing an action A, and E is not a mere side-effect of A, then performing A is imper-missible.* This is expressed by the counterfactual form below, in a setting where action A is performed to achieve goal G: *"If not E had been true, then not G would have been true."*

The evaluation of this counterfactual form identifies permissibility of action A from its effect E, by identifying whether the latter is a necessary cause for goal G or a mere side-effect of action A: if the counterfactual proves valid, then E is instrumental as a cause of G, and not a mere side-effect of action A. Since E is morally wrong, achieving G that way, by means of A, is impermissible; otherwise, not. Note, the evaluation of counterfactuals in this application is considered from the perspective of agents who perform the action, rather than from that of observers. Moreover, the emphasis on causation in this application focuses on agents' deliberate actions, rather than on causation and counterfactuals in general.

We demonstrate in [18] the application of this counterfactual form in machine ethics. First, we use counterfactual queries to distinguish moral permissibility between two off-the-shelf military cases from [24], viz., terror bombing vs. tactical bombing, according to DDE. In the second application, we show that counterfactuals may as well be suitable to justify permissibility, via a process of argumentation (wrt. Scanlon contractualism [23]), using a scenario built from cases of the trolley problem that involve both DDE and DTE. Alternatively, we show that moral justification can also be addressed via 'compound counterfactuals' – *Had I known what I know today, then if I were to have done otherwise, something preferred would have followed* – for justifying with hindsight a moral judgment that was passed under lack of current knowledge.

Moral Updating. Moral updating (and evolution) concerns the adoption of new (possibly overriding) moral rules on top of those an agent currently follows. Such adoption often happens in the light of situations freshly faced by the agent, e.g., when an authority contextually imposes other moral rules, or due to some cultural difference. In [12], moral updating is illustrated in an interactive storytelling (using ACORDA), where the robot must save the princess imprisoned in a castle, by defeating either of two guards (a giant spider or a human ninja), while it should also attempt to follow (possibly conflicting) moral rules that may change dynamically as imposed by the princess (for the visual demo, see [13]).

The storytelling is reconstructed in this paper using QUALM, to particularly demonstrate: (1) The direct use of LP updating so as to place a moral rule into effect; and (2) The relevance of contextual abduction to rule out tabled but incompatible abductive solutions, in case a goal is invoked by a non-empty initial abductive context (the content of this context may be obtained already from another agent, e.g., imposed by the princess). A simplified program modeling the knowledge of the princess-savior robot in QUALM is shown below, where $fight/1$ is an abducible predicate:

$$guard(spider). \quad guard(ninja). \quad human(ninja).$$
$$survive\_from(G) \leftarrow utilVal(G, V), V > 0.6. \quad utilVal(spider, 0.4). \quad utilVal(ninja, 0.7).$$
$$intend\_savePrincess \leftarrow guard(G), fight(G), survive\_from(G).$$
$$intend\_savePrincess \leftarrow guard(G), fight(G).$$

The first rule of *intend_savePrincess* corresponds to a utilitarian moral rule (wrt. the robot's survival), whereas the second one to a 'knight' moral, viz., to intend the goal of saving the princess at any cost (irrespective of the robot's survival chance). Since each rule in QUALM is assigned a unique name in its transform (based on rule name fluent in [21]), the name of each rule for *intend_savePrincess* may serve as a unique moral rule identifier for updating by toggling the rule's name, say via rule name fluents #*rule(utilitarian)* and #*rule(knight)*, resp. In the subsequent plots, query ?- *intend_savePrincess* is referred, representing the robot's intent on saving the princess.

In the first plot, when both rule name fluents are retracted, the robot does not adopt any moral rule to save the princess, i.e., the robot has no intent to save the princess, and thus the princess is not saved. In the second (restart) plot, in order to maximize its survival chance in saving the princess, the robot updates itself with the utilitarian moral: the program is updated with #*rule(utilitarian)*. The robot thus abduces *fight(ninja)* so as to successfully defeat the ninja instead of confronting the humongous spider.

The use of tabling in contextual abduction is demonstrated in the third (start again) plot. Assuming that the truth of *survive_from(G)* implies the robot's success in defeating (killing) guard G, the princess argues that the robot should not kill the *human* ninja, as it violates the moral rule she follows, say a 'Gandhi' moral, expressed by the following rule in her knowledge (the first three facts in the robot's knowledge are shared with the princess): *follow_gandhi* ← *guard(G), human(G), not fight(G)*. That is, the princess abduces *not fight(ninja)* and imposes this abductive solution as the initial (input) abductive context of the robot's goal (viz., *intend_savePrincess*). This input context is inconsistent with the tabled abductive solution *fight(ninja)*, and as a result, the query fails: the robot may argue that the imposed 'Gandhi' moral conflicts with its utilitarian rule (in the visual demo [13], the robot reacts by aborting its mission). In the final plot, as the princess is not saved yet, she further argues that she definitely has to be saved, by now additionally imposing on the robot the 'knight' moral. This amounts to updating the rule name fluent #*rule(knight)* so as to switch on the corresponding rule. As the goal *intend_savePrincess* is still invoked with the input abductive context *not fight(ninja)*, the robot now abduces *fight(spider)* in the presence of the newly adopted 'knight' moral. Unfortunately, it fails to survive, as confirmed by the failing of the query ?- *survive_from(spider)*.

The plots in this story reflect a form of deliberative employment of moral judgments within Scanlon's contractualism. For instance, in the second plot, the robot may justify its action to fight (and kill) the ninja due to the utilitarian moral it adopts. This justification is counter-argued by the princess in the subsequent plot, making an exception in saving her, by imposing the 'Gandhi' moral, disallowing the robot to kill a human guard. In this application, rather than employing updating, this exception is expressed via contextual abduction with tabling. The robot may justify its failing to save the princess (as the robot leaving the scene) by arguing that the two moral rules it follows (viz., utilitarian and 'Gandhi') are conflicting wrt. the situation it has to face. The argumentation proceeds, whereby the princess orders the robot to save her whatever risk it takes, i.e., the robot should follow the 'knight' moral.

3   Conclusion and Future Work

The paper summarizes our investigation on the application of LP-based reasoning to the *terra incognita* of machine ethics, a field that is now becoming a pressing concern and receiving wide attention. Our research shows a number of original inroads, exhibiting a proof of possibility to model morality viewpoints systematically using a combination of various LP-based reasoning features (such as LP abduction, updating, preferences, probabilistic LP and counterfactuals) afforded by the-state-of-the-art tabling mechanisms, through moral examples taken off-the-shelf from the literature. Given the broad dimension of the topic, our contributions touch solely on a dearth of morality issues. Nevertheless, it prepares and opens the way for additional research towards employing various features in LP-based reasoning to machine ethics. Several topics can be further explored in the future, as summarized below.

So far, our application of counterfactuals in machine ethics is based on the evaluation of counterfactuals in order to determine their validity. It is interesting to explore in future other aspects of counterfactual reasoning relevant for moral reasoning. First, we can consider *assertive counterfactuals*: rather than evaluating the truth validity of counterfactuals, they are asserted (known) as being a valid statement. The causality expressed by such a valid counterfactual may be useful for refining moral rules, which can be achieved through incremental rule updating. Second, we may extend the antecedent of a counterfactual with a rule, instead of just literals, allowing to express exception in moral rules, such as "If killing the giant spider had been done by a noble knight, then it would not have been wrong". Third, we can imagine the situation where the counterfactual's antecedent is not given, though its conclusion is, the issue being that the conclusion is some moral wrong. In this case, we want to abduce the antecedent in the form of interventions that would prevent some wrong: "What could I have done to prevent a wrong?".

This paper contemplates the individual realm of machine ethics: it stresses individual moral cognition, deliberation, and behavior. A complementary realm stresses collective morals, and emphasizes instead the emergence, in a population, of evolutionarily stable moral norms, of fair and just cooperation, to the advantage of the whole evolved population. The latter realm is commonly studied via Evolutionary Game Theory by resorting to simulation techniques, typically with pre-determined conditions, parameters, and game strategies (see [19] for references). The bridging of the gap between the two realms [19] would appear to be promising for future work. Namely, how the study of individual cognition of morally interacting multi-agent (in the context of this paper, by using LP-based reasoning features) is applicable to the evolution of populations of such agents, and vice versa.

Acknowledgments. Both authors acknowledge the support from FCT/MEC NOVA LINCS PEst UID/CEC/ 04516/2013. Ari Saptawijaya acknowledges the support from FCT/MEC grant SFRH/BD/72795/2010.

References

1. Anderson, M., Anderson, S.L.: EthEl: Toward a principled ethical eldercare robot. In: Procs. AAAI Fall 2008 Symposium on AI in Eldercare (2008)
2. Bringsjord, S., Arkoudas, K., Bello, P.: Toward a general logicist methodology for engineering ethically correct robots. IEEE Intelligent Systems **21**(4), 38–44 (2006)
3. Cushman, F., Young, L., Greene, J.D.: Multi-system moral psychology. In: Doris, J.M. (ed.) The Moral Psychology Handbook. Oxford University Press (2010)
4. Dell'Acqua, P., Pereira, L.M.: Preferential theory revision. Journal of Applied Logic **5**(4), 586–601 (2007)
5. Foot, P.: The problem of abortion and the doctrine of double effect. Oxford Review **5**, 5–15 (1967)
6. Ganascia, J.-G.: Modelling ethical rules of lying with answer set programming. Ethics and Information Technology **9**(1), 39–47 (2007)
7. Anh, H.T., Kencana Ramli, C.D.P., Damásio, C.V.: An implementation of extended P-Log using XASP. In: Garcia de la Banda, M., Pontelli, E. (eds.) ICLP 2008. LNCS, vol. 5366, pp. 739–743. Springer, Heidelberg (2008)
8. Han, T.A., Saptawijaya, A., Pereira, L.M.: Moral reasoning under uncertainty. In: Bjørner, N., Voronkov, A. (eds.) LPAR-18 2012. LNCS, vol. 7180, pp. 212–227. Springer, Heidelberg (2012)
9. Hauser, M., Cushman, F., Young, L., Jin, R.K., Mikhail, J.: A dissociation between moral judgments and justifications. Mind and Language **22**(1), 1–21 (2007)
10. Kamm, F.M.: Intricate Ethics: Rights, Responsibilities, and Permissible Harm. Oxford U. P (2006)
11. Kowalski, R.: Computational Logic and Human Thinking: How to be Artificially Intelligent. Cambridge U. P (2011)
12. Lopes, G., Pereira, L.M.: Prospective storytelling agents. In: Carro, M., Peña, R. (eds.) PADL 2010. LNCS, vol. 5937, pp. 294–296. Springer, Heidelberg (2010)
13. Lopes, G., Pereira, L.M.: Visual demo of "Princess-saviour Robot" (2010). http://centria.di.fct.unl.pt/~lmp/publications/slides/padl10/quick_moral_robot.avi
14. Mallon, R., Nichols, S.: Rules. In: Doris, J.M. (ed.) The Moral Psychology Handbook. Oxford University Press (2010)
15. McIntyre, A.: Doctrine of double effect. In: Zalta, E.N. (ed.) The Stanford Encyclopedia of Philosophy. Center for the Study of Language and Information, Stanford University, Fall 2011 edition (2004). http://plato.stanford.edu/archives/fall2011/entries/double-effect/
16. Newman, J.O.: Quantifying the standard of proof beyond a reasonable doubt: a comment on three comments. Law, Probability and Risk **5**(3–4), 267–269 (2006)
17. Pereira, L.M., Saptawijaya, A.: Modelling morality with prospective logic. In: Anderson, M., Anderson, S.L. (eds.) Machine Ethics, pp. 398–421. Cambridge U. P (2011)
18. Pereira, L.M., Saptawijaya, A.: Abduction and beyond in logic programming with application to morality. Accepted in "Frontiers of Abduction", Special Issue in IfCoLog Journal of Logics and their Applications (2015). http://goo.gl/yhmZzy

19. Pereira, L.M., Saptawijaya, A.: Bridging two realms of machine ethics. In: White, J.B., Searle, R. (eds.) Rethinking Machine Ethics in the Age of Ubiquitous Technology. IGI Global (2015)
20. Pereira, L.M., Saptawijaya, A.: Counterfactuals in Logic Programming with Applications to Agent Morality. Accepted at a special volume of Logic, Argumentation & Reasoning (2015). http://goo.gl/6ERgGG (preprint)
21. Saptawijaya, A., Pereira, L.M.: Incremental tabling for query-driven propagation of logic program updates. In: McMillan, K., Middeldorp, A., Voronkov, A. (eds.) LPAR-19 2013. LNCS, vol. 8312, pp. 694–709. Springer, Heidelberg (2013)
22. Saptawijaya, A., Pereira, L.M.: TABDUAL: a Tabled Abduction System for Logic Programs. IfCoLog Journal of Logics and their Applications **2**(1), 69–123 (2015)
23. Scanlon, T.M.: What We Owe to Each Other. Harvard University Press (1998)
24. Scanlon, T.M.: Moral Dimensions: Permissibility, Meaning, Blame. Harvard University Press (2008)
25. Swift, T.: Tabling for non-monotonic programming. Annals of Mathematics and Artificial Intelligence **25**(3–4), 201–240 (1999)
26. The Future of Life Institute. Research Priorities for Robust and Beneficial Artificial Intelligence (2015). http://futureoflife.org/static/data/documents/research_priorities.pdf

Intelligent Information Systems

Are Collaborative Filtering Methods Suitable for Student Performance Prediction?

Hana Bydžovská[✉]

CSU and KD Lab Faculty of Informatics, Masaryk University, Brno, Czech Republic
bydzovska@fi.muni.cz

Abstract. Researchers have been focusing on prediction of students' behavior for many years. Different systems take advantages of such revealed information and try to attract, motivate, and help students to improve their knowledge. Our goal is to predict student performance in particular courses at the beginning of the semester based on the student's history. Our approach is based on the idea of representing students' knowledge as a set of grades of their passed courses and finding the most similar students. Collaborative filtering methods were utilized for this task and the results were verified on the historical data originated from the Information System of Masaryk University. The results show that this approach is similarly effective as the commonly used machine learning methods like Support Vector Machines.

Keywords: Student performance · Prediction · Collaborative filtering methods · Recommender system

1 Introduction

Students have to accomplish all the study requirements defined by their university. The most important is to pass all mandatory courses and to select elective and voluntary courses that they are able to pass. Masaryk University offers a vast amount of courses to its students. Therefore, it is very difficult for students to make a good decision. It is important for us to understand students' behavior to be able to guide them through their studies to graduate. Our goal is to design an intelligent module integrated into the Information System of Masaryk University that will help students with selecting suitable courses and warn them against too difficult ones. Necessarily, we need to be able to predict whether a student will succeed or fail in an investigated course in order to realize the module. We need the information at the beginning of a term when we have no information about students' knowledge, skills or enthusiasm for any particular course. We also do not want to obtain the information directly from students using questionnaires. Since the questionnaires tend to have a lower response rate, we use only verifiable data from the university information system.

We have drawn inspiration from techniques utilized in recommender systems. Nowadays, usage of *collaborative filtering (CF) methods* [5] spreads over many areas including the educational environment. Walker et al. [9] designed a system called

© Springer International Publishing Switzerland 2015
F. Pereira et al. (Eds.) EPIA 2015, LNAI 9273, pp. 425–430, 2015.
DOI: 10.1007/978-3-319-23485-4_42

Altered Vista that was specifically aimed at teachers and students who reviewed web resources targeted at education. The system implemented CF methods in order to recommend web resources to its users. Many researchers aimed at e-learning, e.g. Loll et al. [6] designed a system that enables students to solve their exercises and to criticize their schoolmates' solutions. The system based on students' answers could reveal difficult tasks and recommend good solutions to enhance students' knowledge.

In this paper, we report on the possibility to estimate student performance in particular courses based only on knowledge of students' previously passed courses. We utilize different CF to estimate the final prediction. The preliminary work can be seen in [1]. Now we explore the most suitable settings of the approach in detail and compare the results with our previous approach using classification algorithms [2].

2   Our Previous Approach

Many researchers successfully used machine learning approach to predict students' performance [7]. In order to characterize students, we collected many different data about students that are stored in the Information System of Masaryk University [2]. The study-related data contained attributes such as gender, year of birth, year of admission, number of credits gained from passed courses. The social data described students' behavior and co-operation with other students, e.g. weighted average grades of their friends, importance in the sociogram computed for example from the communication statistics, students' publication co-authoring, and comments among students.

We also utilized algorithms implemented in Weka [10] with different data sets in order to obtain the best possible results. We utilized algorithms for regression and classification, and we also computed the mean absolute error from confusion matrix. In order to lower the number of attributes, we employed feature selection (FS) algorithms. Support Vectors Machines (SMO) and Random Forests were the most suitable methods in combination with FS algorithms. The comparison of this approach with the results obtained by CF methods on the same data set can be seen in Section 4.

3   Experiment

Our hypothesis is that students' knowledge can be characterized by the grades of courses that students enrolled during their studies. Based on this information we can select students with similar interests and knowledge and subsequently predict whether a particular student has sufficient skills needed for a particular course.

For our purposes, each student can be represented with a vector of grades of courses passed in one of the student's studies. In order to confirm our hypothesis, we selected 62 courses with different success rates that were offered to students in the years 2010 – 2013 at Masaryk University. The students for whom we were not able to give any prediction – students without any history in the system and without any passed course – were omitted from the experiment. The extracted data set comprised of 3,423 students enrolled at least in one of the 62 courses and their 42,635 grades.

Our aim was to predict the grades of students enrolled in the investigated courses in the year 2012 based on the results of similar students enrolled in the same courses in the years 2010 and 2011. Then we could verify the predictions with the real grades and evaluate the methods and the settings. Then we selected the most suitable method and verified it on data about students enrolled in the same courses in the year 2013.

Similarity of Students. For each student, we constructed four vectors of grades characterizing the knowledge. The values were computed with respect to the number of repetition of each course. We consider only the last grade (NEWEST), a grade of each attempt at the last year (YEAR), only the last grades of each repetition (LAST), and all grades (ALL). For example, a student failed a course in the first year using three attempts and got the grades 444. The student had to repeat the course next year. Supposing he or she got the grades 442, the student's values for this particular course were the following: NEWEST: 2, YEAR: 4+4+2, LAST: 4+2, ALL: 4+4+4+4+4+2.

Vectors of grades were compared by five methods. *Mean absolute difference (MAD)* and *Root mean squared difference (RMSD)* measure the mean difference of the investigated student's grades and the grades of students' in their shared courses. The lower the value, the better the result is. The other methods return values near 1 for the best results. *Cosine similarity (COS)* and *Pearson's correlation coefficient (PC)* define the similarity of grades of shared courses. *Jaccard's coefficient (JC)* defines the ratio of shared and different courses. Supposing that students' knowledge can be represented with passed courses, it was very important to calculate the overlap of students' courses.

Neighborhood Selection. We selected several methods to compute a suitable neighborhood:

- Top x, where $x \in [1; 50]$ with step 1; (the analysis [4] indicates that the neighborhood of 20 - 50 neighbors is usually optimal).
- More similar than the threshold y, where $y \in [0; 1]$ with step 0.1.
- We also utilized the idea of baseline user [8]. We selected only these students to the neighborhood that were more similar to the investigated one than the investigated one to the baseline user. We decided to calculate two types of baseline user:
  - Average student – we characterized an average student by the average grades of courses in which the investigated students were enrolled in.
  - Uniform student – we characterized a uniform student by grades with values 2.5 (the average grade through all courses) of all courses in which the investigated student was enrolled in.

Grade Prediction. As the neighborhood was defined, we could make a prediction. We used different approaches to estimate grades from grades of students in the neighborhood: mean, median, and the majority class. We also utilized the significance weighting [3], also its extension using average grades of compared students (sig. weighting +), and lowering the importance of students with only few co-ratings [4].

428   H. Bydžovská

4   Results

Mean absolute error (MAE) represents the size of the prediction error. The exact grade prediction is very difficult and even less powerful prediction can be sufficient. Therefore, we also predicted the grades as good (1) / bad (2) / failure (4) or just success or failure. The results of the CF methods were compared with our previous work described in Section 2 where the predictions were obtained using classification algorithms (CA). We used a confusion matrix for calculating MAE. We mined study-related data and data about social behavior of students.

The comparison of both approaches can be seen in Table 1. Although both the approaches used different data from the information system and utilized different processing, the results showed that their performance was almost similar. The only one significant difference can be seen in grade prediction when CF methods were slightly better. We consider the accuracy of 78.5% for student success or failure prediction reliable enough considering that we did not know students' skills or enthusiasm for courses. MAE of good / bad / failure prediction was around 0.6. We consider MAE less than one degree in the modified grade scale to be very satisfactory. Even in the grade (1, 1.5, 2, 2.5, 3, and 4) prediction, MAE was around 0.7 which means only slightly more than one degree in the grade scale. In general, these results were positive but the grade prediction was still not trustworthy.

Table 1. Comparison of approaches

		Grade	Good/bad/failure	Success/failure
Approach	Selection	MAE		Accuracy
CA	2012	0.67	0.58	81.04%
	2013	0.84	0.61	78.72%
CF	2012	0.64	0.57	80.44%
	2013	0.68	0.64	78.58%

The advantage of the CF approach is that all information systems store the data about students' grades. Therefore, this approach can be used in all systems. Our previous approach was based on mining data obtained from the information system. But not all systems store the data about social behavior of students. We proved that this data improve the accuracy of the results significantly [2].

5   Discussion

The settings of the CF approach that reached the best average results can be seen in Table 2. As the results show, PC worked properly in combination with the uniform student for selecting a proper neighborhood and significance weighting with an extension using average grades of compared students for the final prediction. On the other hand, for MAD, a Top x function was the best option for selecting the neighborhood and median for the final prediction. Both the approaches reached very similar results in all tasks and we consider them to be trustworthy. We also investigated the most suitable x for these tasks. We searched for the minimal x with the best possible results. We derived $x = 25$ to be the best choice generally for all methods and settings. The most suitable classification algorithms were SMO and Random Forests (Table 3).

Table 2. The settings of the CF approach that reached the best average results

	Sim. function	Neighborhood	Estimation approach
Grade	PC	Uniform student	Sig. weighting +
Good/bad/failure	PC	Uniform student	Sig. weighting +
Success/failure	MAD	Top 25	Median

Table 3. The settings of the classification algorithms that reached the best average results

	Classification algorithm	Feature selection algorithm
Grade	SMO	InfoGainAttributeEval
Good/bad/failure	SMO	OneRAttributeEval
Success/failure	Random Forests	5 attributes selected by each FS algorithm for each course

We also investigated the influence of different details of grades described in Section 3. The conclusion was that only the NEWEST grade was expressive enough for a satisfactory prediction. More detailed information about the grades did not improve the results significantly.

6   Conclusion

In this paper, we used CF methods for student modeling. Our experiment provides evidence that CF approach is also suitable for student performance prediction. The data set comprised of 62 courses taught in 4 years with almost 3,423 students and their 42,635 grades. We confirmed our hypothesis, that students' knowledge can be sufficiently characterized only by their previously passed courses that should cover their knowledge of the field of study. We processed data about students' grades stored in the Information System of Masaryk University to be able to estimate students' interests, enthusiasm and prerequisites for passing enrolled courses at the beginning of each term. For each investigated student, we searched for students enrolled in the same courses in the last years who were the most similar ones to the investigated student. Based on their study results, we predicted the students' performance.

We compared the results with the results obtained by classification algorithms that researches usually utilize for student performance prediction. The results were almost the same. The main advantage of CF approach is that all university information systems store the data about students' grades needed for the prediction. On the other hand, this approach is not suitable if we have no information about the history of the particular students. Now, we are able to predict the student success or failure with the accuracy of 78.5%, whether the grade will be good, bad, or failure with the MAE of

0.6 and the exact grade with the MAE of 0.7. We consider the results to be very satisfactory and CF approach can be considered as expressive as the commonly used classification algorithms.

Based on this approach we can recommend suitable voluntary courses for each student with respect to his or her interests and skills. We hope that this information will also encourage students to study hard when they have to enroll in a mandatory course that seems to be too difficult for them. Moreover, teachers can utilize this information to identify potentially weak students and help them before they will be at risk to fail the course. This approach can be also beneficially used in an intelligent tutoring system as the basic estimation of students' potentials before they start to operate with the system in the investigated course.

Acknowledgement. We thank Michal Brandejs, Lubomír Popelínský, and all colleagues of Knowledge Discovery Lab, and also IS MU development team for their assistance. This work has been partially supported by Faculty of Informatics, Masaryk University.

References

1. Bydžovská, H.: Student performance prediction using collaborative filtering methods. In: Conati, C., Heffernan, N., Mitrovic, A., Verdejo, M. (eds.) AIED 2015. LNCS, vol. 9112, pp. 550–553. Springer, Heidelberg (2015)
2. Bydžovská, H., Popelínský, L.: The Influence of social data on student success prediction. In: Proceedings of the 18th International Database Engineering & Applications Symposium, pp. 374–375 (2014)
3. Herlocker, J.L., Konstan, J.A., Borchers, A., Riedl, J.: An algorithmic framework for performing collaborative filtering. In: Proceedings of the 22nd Annual International ACM SIGIR Conference, pp. 230–237 (1999)
4. Herlocker, J.L., Konstan, J.A., Riedl, J.: Explaining collaborative filtering recommendations. In: Proceedings of the ACM Conference on Computer Supported Cooperative Work, pp. 241–250 (2000)
5. Jannach, D., Zanker, M., Felfernig, A., Friedrich, G.: Recommender Systems: An Introduction. Cambridge University Press (2010)
6. Loll, F., Pinkwart N.: Using collaborative filtering algorithms as elearning tools. In: Proceedings of the 42nd Hawaii International Conference on System Sciences (2009)
7. Marquez-Vera, C., Romero, C., Ventura, S.: Predicting school failure using data mining. In: Pechenizkiy, M., et al. (eds.) EDM, pp. 271–276 (2011)
8. Matuszyk, P., Spiliopoulou, M.: Hoeffding-CF: neighbourhood-based recommendations on reliably similar users. In: Dimitrova, V., Kuflik, T., Chin, D., Ricci, F., Dolog, P., Houben, G.-J. (eds.) UMAP 2014. LNCS, vol. 8538, pp. 146–157. Springer, Heidelberg (2014)
9. Walker, A., Recker, M.M., Lawless, K., Wiley, D.: Collaborative Information Filtering: A Review and an Educational Application. International Journal of Artificial Intelligence in Education **14**(1), 3–28 (2004)
10. Witten, I., Frank, E, Hall, M.: Data Mining: Practical Machine Learning Tools and Techniques, 3rd edn. Morgan Kaufmann Publishers (2011)

Intelligent Robotics

A New Approach for Dynamic Strategic Positioning in RoboCup Middle-Size League

António J.R. Neves[(✉)], Filipe Amaral, Ricardo Dias, João Silva,
and Nuno Lau

Intelligent Robotics and Intelligent Systems Lab,
IEETA/DETI – University of Aveiro, Aveiro, Portugal
an@ua.pt

Abstract. Coordination in multi-robot or multi-agent systems has been receiving special attention in the last years and has a prominent role in the field of robotics. In the robotic soccer domain, the way that each team coordinates its robots, individually and together, in order to perform cooperative tasks is the base of its strategy and in large part dictates the success of the team in the game. In this paper we propose the use of Utility Maps to improve the strategic positioning of a robotic soccer team. Utility Maps are designed for different set pieces situations, making them more dynamic and easily adaptable to the different strategies used by the opponent teams. Our approach has been tested and successfully integrated in normal game situations to perform passes in free-play, allowing the robots to choose, in real-time, the best position to receive and pass the ball. The experimental results obtained, as well as the analysis of the team performance during the last RoboCup competition show that the use of Utility Maps increases the efficiency of the team strategy.

1 Introduction

RoboCup ("Robot Soccer World Cup") is a scientific initiative with an annual international meeting and competition that started in 1997. The aim is to worldwidly promote developments in Artificial Intelligence, Robotics and Multi-agent systems. Robot soccer represents one of the attractive domains promoted by RoboCup for the development and testing of multi-agent collaboration techniques, computer vision algorithms and artificial intelligence approaches, only to name a few.

In the RoboCup Middle Size League (MSL), autonomous mobile soccer robots must coordinate and collaborate for playing and winning a game of soccer, similar to the human soccer games. They have to assume dynamic roles in the field, to share information about visible objects of interest or obstacles and to position themselves in the field so that they can score goals and prevent the opponent team from scoring. Decisions such as game strategies, positioning and team coordination play a major role in the MSL soccer games.

This paper introduces Utility Maps as a tool for the dynamic positioning of soccer robots on the field and for opportunistic passing between robots, under

F. Pereira et al. (Eds.) EPIA 2015, LNAI 9273, pp. 433–444, 2015.
DOI: 10.1007/978-3-319-23485-4_43

different situations that will be presented throughout the paper. As far as the authors know, no previous work has been presented about the use of Utility Maps in the Middle Size League of RoboCup.

The paper is structured into 8 sections, first of them being this Introduction. In Section 2 we present a summary of the work already done on strategic positioning. Section 3 introduces the use of Utility Maps in the software structure of the CAMBADA MSL team. In Section 4 we describe the construction of Utility Maps. Section 5 describes the use of Utility Maps for the positioning of the robots in defensive set pieces. In Section 6 we present the use of Utility Maps in offensive set pieces, while Section 7 presents their use in Free Play. Finally, Section 8 introduces some measures of the impact of the Utility Maps on the performance of the team and discusses a series of results that prove the efficiency of the team strategy, based on Utility Maps.

2   Related Work

Strategic positioning is a topic with broad interest within the RoboCup community. As teams participating in the RoboCup Soccer competitions gradually managed to solve the most basic tasks involved in a soccer game, such as locomotion, ball detection and ball handling, the need of having smarter and more efficient robotic soccer players arose. Team coordination and strategic positioning are nowadays the key factors when it comes to winning a robotic soccer game.

The first efforts for achieving coordination in multi-agent soccer teams has been presented in [2] [3]. Strategic Positioning with Attraction and Repulsion (SPAR) takes into account the positions of other agents as well as that of the ball. The following forces are evaluated when taking a decision regarding the positioning of an agent: repulsion from opponents and team members, attraction to the active team member and ball and attraction to the opponents' goal.

In the RoboCup Soccer Simulation domain, Situation Based Strategic Position (SBSP) [4] is a well known technique used for the positioning of the software agents. The positioning of an agent only takes into consideration the ball position, as focal point, and does not consider other agents. However, if all agents are assumed to always devote their attention to the ball position, then cooperative behavior can be achieved indirectly. An agent defines its base strategic position based on the analysis of the tactic and team formation. Its position is adjusted accordingly to ball pose and game situation. This approach has been adapted to the Middle Size League constraints and has been presented in [5].

In [6] a method for Dynamic Positioning based on Voronoi Cells(DPVC) was introduced. The robotic agents are placed based on attraction vectors. These vectors represent the attraction of the players towards objects, depending on the current state of the game and roles.

The Delaunay Triangulation formations (DT) [7] divide the soccer pitch into several triangles based on given training data. Each training datum affects only the divided region to which it belongs. A map is built from a focal point, such as ball position, to the positioning of the agents.

For more than 20 years, grid-based representations have been used in robotics in order to show different kinds of spacial information, allowing a more accurately and simplified world perception and modelling [8] [9]. Usually, this type of representations are oriented to a specific goal. Utility functions have been presented before [10] [11] as a tool for role choosing within multi-agents systems.

Taking into account the successful use of this approach, a similar idea was applied to the CAMBADA agents, but with different functionalities and purpose. The aim of the proposed approach was to improve the collective behavior in some specific game situations. Utility Maps have been developed as support tools for the positioning of soccer robots in defensive and offensive set pieces, as well as in freeplay passes situations.

3   Utility Maps of CAMBADA MSL Team

The general architecture of the CAMBADA robots has been described in [1]. The decision about the strategic positioning of the robots in taken by the high level agent. This is a process that, at each cycle, is responsible for the high-level control of the robots, which is divided in several stages. The first stage is the sensor fusion, executed by an integrator module, with the objective of gathering the noisy information from the sensors and from its team mates and updating the state of the world. This world state will be used by the high-level decision and coordination modules.

In the second stage of the high-level control of the robots, the agent has to decide how to act given the state of the world that he built. At the higher level it assumes a Role and operates on the field with a given attitude, for example, as role `Striker`. A detailed description about the Roles used in CAMBADA team can be found in [13]. The actions it can take are defined by lower level Behaviors, which define the orders to be send to the actuators in order to fulfill a task, for example, a `Move` behavior, to reach a given point on the field.

During a MSL game, there are three possible game situations: **defensive set pieces** when for some reason (a fault, ball outside the field or a valid goal) the game stops and the ball belongs to the opponent team; **offensive set pieces** when for some reason the game stops and the ball belongs to our team and **free play** when the ball is moving on the field after a game stop.

In **defensive set pieces** only one role is involved, role `Barrier`. The robot will stay in this role until the ball is considered to be in game. The ball is in game when it moves more than 20cm or if 10 seconds passed since the start signal has been given by the referee. After this point, the game enters in free play mode and new roles will be assigned to the robots, as described next.

In **offensive set pieces** two roles are involved, role `Replacer` and role `Receiver`. The robot closest to the ball will assume the role `Replacer` and all the others, except for the goal keeper, will assume the role `Receivers`. After the ball has been passed, the robot that will receive the ball becomes `Striker` and all the other robots are `Midfielders`. After a successful pass or when our robots detect that the ball has been gathered by the opponent team, the game state changes to free play.

When the game is in **free play**, a robot can assume one of two roles: `Striker` or `Midfielder`, depending on the relative robot position regarding the ball.

4   Building the Utility Maps

In this paper we propose an approach for dynamic strategy positioning in MSL based on Utility Maps. The position of each robot that takes part in a specific game situation is dynamically obtained based on the information about the environment around the robot, namely its position on the field, the position of obstacles and the ball position.

The CAMBADA robots use a catadioptric vision system, often named omni-directional vision system. The algorithms for detecting the objects of interest are presented in [14]. The information acquired by the vision system is merged with other information of the robot to build the worldstate information, namely its position on the field and a list of valid balls and a list of valid obstacles. A detailed description of the algorithms used to build the worldstate is presented in [15].

The information about the obstacles in the current version of the CAMBADA robot worldstate is a list of objects containing their absolute positions on the field and their classification of being team mates or opponents. The algorithm for obstacles detection and identification is described in [15].

The Utility Map is constructed merging the relevant information about the environment, namely the team mates positions, obstacles and ball positions, using the information of the robot and the information shared by the colleagues [12].

The first step to obtain the Utility Map is to build an occupancy map that gives the robot a global idea about the state of the world around it. Then, depending on the game situation and the role in which the Utility Map will be used, a Field of Vision (FOV) is calculated on top of the occupancy map. FOV represents the area that is considered visible from the point in which it was calculated. For example, it is possible to calculate a FOV from the ball, from the robot itself, from the goal, etc. An example of a FOV calculated from robot number 3 is shown on Fig. 1.

Finally, the Utility Map is created taking into consideration the occupancy map, the FOVs and some conditions, restrictions and metrics for decisions depending on the game situation. Taking as example the offensive set pieces, there are some restrictions included in the process of building the Utility Map. The two main restrictions are that the robots cannot be inside the goal areas and they have to receive the ball from at least 2 meters from the ball. Moreover, it is possible to combine three metrics to build the Utility Maps in order to decide the best positions to receive the ball. One is the free space between the pass line and the closest obstacle. The second one is the weighted average between the distance to the ball, the distance to the opponent goal, the rotation angle for a shot on target and the distance from the point on the map to the position of the robot. Finally, the third metric is the angle of each map position to the opponent goal.

The use of these Utility Maps allow the robots to easily take decisions regarding their positioning simply choosing the local maximum on the maps. The Utility Maps are calculated locally on the robots and are part of their worldstate so that they are easily accessible by any behavior or role. In terms of implementation, the TCOD[1] library has been used. The library provides built-in toolkits for management of height maps, which in the context of this work are used as Utility Maps, and field of view calculations. It takes, on average, 4ms to update the necessary maps in each cycle of the agent software execution. The robots are currently working with a cycle of 20 ms, controlled by the vision process that works at 50 frames per second [14].

The identified opponent robots lead to hills in the map on their position, with some persistence to improve the stability of the decisions based on the maps. It takes 5 agent cycles (100 milliseconds) for a new obstacle to reach the maximum cost level. In the end, the map is normalized, thus always holding values between 0.0 and 1.0.

5   Positioning in Defensive Set Pieces

In defensive set pieces the main objective is to prevent the opponent team to perform a pass and to gain ball possession. In order to prevent the opponent players to pass the ball, our robots must be positioned between the ball and the possible receivers from the opponent team and to follow them while they move.

To calculate the base position for the Barriers, the role assumed by the robots during the opponent set pieces, we use a Delaunay Triangulation (DT) [7] to interpolate all possible robots positions on the field, depending on the ball position on the field. On top of that, the rules restrictions are applied. These restrictions are: minimum 3m distance to the ball, except in a drop ball situation, when the required distance is only 1m and only one player inside our penalty area. This player does not need to respect the previous rule as long it is inside the penalty area.

Figure 1 shows the tool used to configure the DT positions. Here, the ball position is used as triangle vertice and each vertex represents the given training data. Each vertex produces output values for the position of the robots for that triangle. When the ball is inside a triangle, the position of the agent is calculated using the interpolation algorithm described in [7].

It is possible to configure each one of the Barriers to dynamically cover the opponent robots or to simply stay in the base positions given by the DT configuration. When there are no obstacles on the field, the robots only use this position. In the presence of obstacles, the position of the cover robots are obtained from an Utility Map, following the approach proposed in this paper.

In order to avoid any possible error in the identification of obstacles, it is necessary to filter the information received from different agents. Obstacles close to each other are merged using a clustering algorithm, and obstacles too close to

---

[1] http://roguecentral.org/doryen/libtcod/

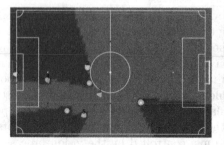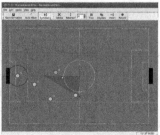

Fig. 1. On the left, the FOV calculated from robot number 3. The ball is represented by a small circle and the robots as larger circles. The circles with numbers are considered team mates. The red areas are considered visible from the point of view of the robot. On the right, the configuration tool used for the positioning of our robots in opponent set pieces and during free play (DT).

a team mate are ignored, unless that team mate sees it. The filtered information is then used to build the map.

From each cluster of obstacles, a valley is carved in direction to the ball. After that, the calculated map is added with a predefined height map that defines the priorities of the positions (see Figure 2). This map takes into consideration that it is more important to cover the opponent robots in the direction of our goal, rather than in the direction of their goal. Finally, all the restrictions (minimum distance to the ball, positions inside the field and avoidance of penalty areas) are added.

In Fig. 2 we can see a game situation where robots number 3, 4 and 5 are configured to cover the opponent robots. The best position given by the Utility Map for each robot is represented in red. As intended, these positions are between the ball and the opponent robots. Robot 2 is in its base position provided by the DT configuration.

The distance between our team robot and the opponent robot that it is trying to cover can be configured (in the configuration file of the robot). The human coach can specify, in the same configuration file, what are the robots allowed to perform covering.

6   Positioning in Offensive Set Pieces

To configure our set pieces we use a graphical tool (Figure 3) that implements an SBSP algorithm. The field is divided into 10 zones. Each zone defines a set of positions for the Replacer (the role of the robot closer to the ball and responsible for putting the ball in play) and Receivers (the role of the other robots, except the Goalie). The position to kick by default in a situation where there is no Receiver available can also be configured. The position of the Receivers can be absolute or relative to the ball. We can also define if the Receiver needs to have a clear line between its position and the ball and an option to force the

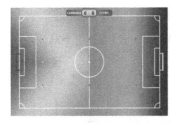 

Fig. 2. On the left, cover priorities height map used to define the priority of the cover positions in the field. Red is the more prioritary and blue is the least prioritary. On the right, example of a cover Utility Map. As we can see, the best position for the robots number 3, 4 and 5 are between the ball and the robots of the opponent team (red color).

`Receiver` to be aligned with the goal. The priority for each receiver is indicated as well, in the case that for a specific region will be more than one configured and available. The one with more priority will be tested first for receiving a pass. The action to be performed by the `Replacer` to that specific `Receiver` is also configurable. This action can be a pass, a cross or none. In the last case, the `Receiver` will never be considered as an option. It is possible to configure differently each one of the possible set pieces (corner, free kick, drop ball, throw in and kick of) for each one of the regions.

When the set pieces using the referred tool are configured, the opponent team is not taken into account. After the opponent team has positioned itself, our configured positions can be positions where the receiver is not able to receive a pass. To deal with these situations, there is the need to have an alternative reception position that has to be calculated dynamically, taking into account the opponent team. An Utility Map is used to calculate the alternative position for the `Receiver`.

All the constraints imposed by the rules, namely minimum distance to the ball (2m) and no entering in the goal areas are taken into account for the construction of the `Receiver` Utility Map. The field is divided into two zones for the application of different metrics to calculate the utility value for each position. On our side of the field, only one metric is used. This metric is the distance to the halfway line. Three metrics are used on the opponent side of the field. One is the free space between the pass line and the closest obstacle. The second one is the weighted average between the distance to the ball, the distance to the opponent goal, the rotation angle for a shot on target and the distance from the point on the map to the position of the robot. Finally, the third metric is the angle of each map position to the opponent goal. The weights for the second metric are easily configurable in the configuration file of the robot. These metrics are only applied within a circle whose radius is also defined in the configuration file. This circle is centered on the position of the ball in the set piece, and only the positions that have FOV from the ball (positions where the ball can be passed) are considered.

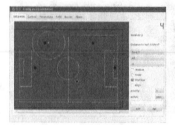 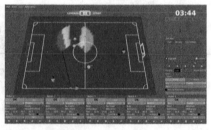

Fig. 3. On the left, the configuration tool used for our set pieces. On the right, an example of a Utility Map for a `Receiver`.]Example of an alternative positioning map for `Receiver` calculated for Robot number 2. CAMBADA is attacking towards the blue goal. The black line goes from the ball to the alternative position indicated by Robot number 2 to receive the ball.

The robots move to the best position extracted from the Utility Map only after the referee gives the start signal to prevent the opponent robots to follow them. In Fig. 3 all the receivers are sharing that they have line clear to receive the ball (`lineClear` is information associated to each robot). Robot number 4 is the `Replacer` already chosen to pass the ball to robot number 2. The pass line it is trying to make is represented by the black line. Robot number 2 will move to that position to receive the ball.

7   Positioning in Freeplay

Free-play passes are a true challenge in terms of coordination among robots, being thus much more complex to achieve. This is mainly due to the complete freedom that the opponent team has to approach our `Striker` or cover a `Midfielder`. Since the robot that wants to make the pass has the ball in its possession, its movement is very limited. Taking this into consideration, the development of an Utility Map to estimate the best position of the `Midfielders` on the field is more than necessary in order to improve the capability of performing passes in free play.

A Delaunay Triangulation (DT) as in their set pieces is used to calculate the base position for the `Midfielders` we use. On top of that, the restrictions from the rules are applied, namely the avoidance of the goal areas.

The algorithm to calculate the Utility Map for the `Midfielders` is similar to the one described in our set pieces. An example of an Utility Map in a free play game situation is presented in Fig. 4. The best positions on the field to receive the ball are in red. Only positions inside the field are considered, as well as only positions outside both penalty areas. Positions near the opponent corners (dead angles) are also avoided. A minimum distance of 2 meters to the `Striker` is required and a preference is given to positions near the last chosen position, near the strategic position returned by DT and also near the actual `Receiver`. A FOV for the ball is also required. With these constraints, the free-play receiver

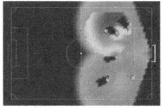

Fig. 4. On the left, the Free-play utility map calculated by the **Midfielders**. The best positions on the field to receive the ball are in red. On the right, the Free-play Utility Map to be used by the **Striker** to choose the best position to perform a pass or a kick to the goal, when dribbling the ball. The robot will dribble to the positions on red.

robot will be constantly adapting to the changes of the opponent formation and the ball position.

The **Striker**, the robot holding the ball or closer to it, also uses an Utility Map for selecting the best position to perform a pass or to kick towards the goal. The calculated map (Figure 4) deals with the constraints regarding a generic dribble behavior, complying with the current MSL rules, that do not allow a robot to dribble for more than 3 meters. Some areas of the field have less utility, namely both penalty areas, areas outside the field and outside a 3 meter radius circle centered on the point where the ball was grabbed. More priority is given to the areas close the limits of the field since it is more advantageous to kick to the goal from there.

8   Results and Discussions

CAMBADA won third place in the last RoboCup MSL competition. After a thorough analysis of the log files and game videos, it is safe to say that the approach we present in this paper for a dynamic strategic positioning of the robots, has had a major contribution to the success of the team. We present in this section the analysis of the presented approach and we discuss its impact on the performance of the team. While a clear distinction between the performance of the team prior to the use of Utility Maps and its current performance cannot be pursued due to the continuously evolving dynamism of the MSL soccer games and the improvements of each of the participating teams each year, the following results prove that the use of Utility Maps had a major contribution for bringing the robotic soccer game as close to the human soccer games as possible.

Looking to the examples of the Utility Maps presented above regarding the three game situations, we can confirm that the maps were correctly built since the position of the robots in a specific game situation are the intended positions, in order to maximize the success of the game.

The video that we submitted together with this paper represents, to our view, the best experimental results to show the effectiveness of the proposed approach for dynamic strategic positioning. Moreover, we analysed the videos and log files

Table 1. Defensive set pieces cover efficiency during the last two games of RoboCup2014. According to the rules, the defending team has to be at least 3 meters from the ball. In these situations it is impossible to intercept the ball.

Game	Opponent	<3m	Intercepted		% Sucess
			Yes	No	
Semi-final	Tech United	11	10	3	77
3rd place	MRL	14	4	3	57
Total		25	14	6	70

Table 2. Attacking set pieces efficiency during the last two games of RoboCup2014. We are considering the success of the ball reception after a pass.

Game	Opponent	Pass		% Sucess
		Yes	No	
Semi-final	Tech United	10	8	56
3rd place	MRL	15	6	71
Total		25	14	64

from the RoboCup games. This analysis reveals that the team reached a 70% success rate in the interception of the ball in defensive set pieces (see Table 1), performed 64% successful passes in offensive set pieces (see Table 2) and a high percentage of successful passes in free play, being these last situations hard to analyze due to the high dynamism of the games.

Looking at Table 1, considering that most of the times when the attacking team made a short pass means that was forced into it by not having other pass option, we have a success rate of 70% in defensive set pieces situations.

Looking further into the unsuccessful situations, the problem was clearly identified and it is not related to the cover position obtained from the Utility Maps. The problem was rather the transition from the `Barrier` role into the `Midfielder` or `Striker` role, situations where the cover position are not used. This is still an open issue to be addressed in the near future.

In the last two games of RoboCup 2014, the final and semi-final - which were probably the most dynamic games, there was a total of 45 defensive set pieces situations. An average of 22 defensive set pieces per game, in a game of 30 minutes, which means a defensive set piece situation every 1 minute and 20 seconds.

In Fig. 5 we can see two game situations were the CAMBADA robots are in strategic positions. By being in those positions, the CAMBADA robots do not allow the attacking team to perform a pass in a proper way.

In the same last two games of RoboCup 2014, the final and semi-final, there was a total of 39 offensive set pieces situation. An average of 20 offensive set pieces per game, in a game of 30 minutes, which means an offensive set piece situation every 1 minute and 30 seconds. In 64% of the situations the robots were able to properly receive the ball. Looking further into the unsuccessful situations, there were cases were the ball was passed to a position far from the `Receiver`

Fig. 5. Defensive game situations during RoboCup 2014. CAMBADA team has blue markers and it is the defending team.

Fig. 6. Offensive game situations during RoboCup 2014. CAMBADA team is with blue markers and is the attacking team.

and it was lost and some other cases where the reception was not properly done mainly due to misalignment of the `Receiver`. These situations where not due to a wrong positioning given by the Utility Map. We just counted a total of 3 interceptions by the opponent team of the ball for long passes.

In Fig. 6 we can see two game situations where the CAMBADA robots are in strategic positions, which allows them to receive the ball with success.

Based upon this study, we are convinced that the use of Utility Maps is an advantageous approach in extremely dynamic environments, such as the one of robotic soccer. Without great complexity being added to the structure of the agents, as it was described in the previous sections, it was possible to introduce the desired dynamism that led to the increase of the team competitiveness and improved its overall performance.

References

1. Neves, A., Azevedo, J., Lau, N., Cunha, B., Silva, J., Santos, F., Corrente, G., Martins, D.A., Figueiredo, N., Pereira, A., Almeida, L., Lopes, L.S., Pedreiras, P.: CAMBADA soccer team: from robot architecture to multiagent coordination, chapter 2, pp. 19–45. I-Tech Education and Publishing, Vienna, January 2010
2. Veloso, M., Bowling, M., Achim, S., Han, K., Stone, P.: The cmunited-98 champion small robot team (accessed February 27, 2014)
3. Stone, P.: Layered Learning in Multiagent Systems: A Winning Approach to Robotic Soccer. MIT Press (2000)
4. Reis, L.P., Lau, N., Oliveira, E.C.: Situation based strategic positioning for coordinating a team of homogeneous agents. In: Hannebauer, M., Wendler, J., Pagello, E. (eds.) ECAI-WS 2000. LNCS (LNAI), vol. 2103, pp. 175–197. Springer, Heidelberg (2001)

5. Lau, N., Lopes, L.S., Corrente, G.: CAMBADA: information sharing and team coordination. In: Proc. of the 8th Conference on Autonomous Robot Systems and Competitions, Portuguese Robotics Open - ROBOTICA 2008, pp. 27–32, Aveiro, Portugal, April 2008

6. Dashti, H.A.T., Aghaeepour, N., Asadi, S., Bastani, M., Delafkar, Z., Disfani, F.M., Ghaderi, S.M., Kamali, S.: Dynamic positioning based on voronoi cells (DPVC). In: Bredenfeld, A., Jacoff, A., Noda, I., Takahashi, Y. (eds.) RoboCup 2005. LNCS (LNAI), vol. 4020, pp. 219–229. Springer, Heidelberg (2006)

7. Akiyama, H., Noda, I.: Multi-agent positioning mechanism in the dynamic environment. In: Visser, U., Ribeiro, F., Ohashi, T., Dellaert, F. (eds.) RoboCup 2007: Robot Soccer World Cup XI. LNCS (LNAI), vol. 5001, pp. 377–384. Springer, Heidelberg (2008)

8. Elfes, A.: Using occupancy grids for mobile robot perception and navigation. Computer **22**(6), 46–57 (1989)

9. Rosenblatt, J.K.: Utility fusion: map-based planning in a behavior-based system. In: Zelinsky, A. (ed.) Field and Service Robotics, pp. 411–418. Springer, London (1998)

10. Chaimowicz, L., Kumar, V.: Mario Fernando Montenegro Campos. A paradigm for dynamic coordination of multiple robots. Auton. Robots **17**(1), 7–21 (2004)

11. Spaan, M.T.J., Groen, F.C.A.: Team coordination among robotic soccer players. In: Kaminka, G.A., Lima, P.U., Rojas, R. (eds.) RoboCup 2002. LNCS (LNAI), vol. 2752, pp. 409–416. Springer, Heidelberg (2003)

12. Santos, F., Almeida, L., Lopes, L.S., Azevedo, J.L., Cunha, M.B.: Communicating among robots in the robocup middle-size league. In: Baltes, J., Lagoudakis, M.G., Naruse, T., Ghidary, S.S. (eds.) RoboCup 2009. LNCS, vol. 5949, pp. 320–331. Springer, Heidelberg (2010)

13. Lau, N., Lopes, L.S., Corrente, G., Filipe, N., Sequeira, R.: Robot team coordination using dynamic role and positioning assignment and role based setplays. Mechatronics **21**(2), 445–454 (2011)

14. Trifan, A., Neves, A.J.R., Cunha, B., Azevedo, J.L.: UAVision: a modular time-constrained vision library for soccer robots. In: Bianchi, R.A.C., Akin, H.L., Ramamoorthy, S., Sugiura, K. (eds.) RoboCup 2014. LNCS, vol. 8992, pp. 490–501. Springer, Heidelberg (2015)

15. Silva, J., Lau, N., António, J.R., Neves, A.J., Azevedo, J.L.: World modeling on an MSL robotic soccer team. Mechatronics **21**(2), 411–422 (2011)

Intelligent Wheelchair Driving: Bridging the Gap Between Virtual and Real Intelligent Wheelchairs

Brígida Mónica Faria[1,2,3]($\boxtimes$), Luís Paulo Reis[2,4], Nuno Lau[3,5],
António Paulo Moreira[6,7], Marcelo Petry[2,7,8], and Luís Miguel Ferreira[3]

[1] ESTSP/IPP - Escola Sup. Tecnologia de Saúde do Porto,
Inst Politécnico do Porto, Vila Nova de Gaia, Portugal
btf@estsp.ipp.pt

[2] LIACC – Lab. Inteligência Artificial e Ciência de Computadores, Porto, Portugal
lpreis@dsi.uminho.pt, marcelo.petry@ufsc.br

[3] IEETA - Inst. Engenharia Electrónica e Telemática de Aveiro, Aveiro, Portugal
{nunolau,luismferreira}@ua.pt

[4] Dep. de Sistemas de Informação, EEUM - Escola de Engenharia da Universidade do Minho,
Guimarães, Portugal

[5] DETI/UA – Dep de Electrónica, Telecomunicações e Informática da Univ. Aveiro,
Aveiro, Portugal

[6] FEUP - Faculdade de Engenharia, Universidade do Porto, Porto, Portugal
amoreira@fe.up.pt

[7] INESC TEC - INESC Tecnologia e Ciência, Porto, Portugal

[8] UFSC - Universidade Federal de Santa Catarina, Blumenau, Brazil

Abstract. Wheelchairs are important locomotion devices for handicapped and senior people. With the increase in the number of senior citizens and the increment of people bearing physical deficiencies, there is a growing demand for safer and more comfortable wheelchairs. So the new Intelligent Wheelchair (IW) concept was introduced. Like many other robotic systems, the main capabilities of an intelligent wheelchair should be: autonomous navigation with safety, flexibility and capability of avoiding obstacles; intelligent interface with the user; communication with other devices. In order to achieve these capabilities a good testbed is needed on which trials and users' training may be safely conducted. This paper presents an extensible virtual environment simulator of an intelligent wheelchair to fulfill that purpose. The simulator combines the main features of robotic simulators with those built for training and evaluation of prospective wheelchair users. Experiments with the real prototype allowed having results and information to model the virtual intelligent wheelchair. Several experiments with real users of electric wheelchairs (suffering from cerebral palsy) and potential users of an intelligent wheelchair were performed. The System Usability Score allowed having the perception of the users in terms of the usability of the IW in the virtual environment. The mean score was 72 indicating a satisfactory level of the usability. It was possible to conclude with the experiments that the virtual intelligent wheelchair and environment are usable instruments to test and train potential users.

Keywords: Intelligent wheelchair · Intelligent robotics · Intelligent simulation · Virtual reality · Multimodal interface

© Springer International Publishing Switzerland 2015
F. Pereira et al. (Eds.) EPIA 2015, LNAI 9273, pp. 445–456, 2015.
DOI: 10.1007/978-3-319-23485-4_44

1   Introduction

Recently, virtual reality has attracted much interest in the field of motor rehabilitation engineering [1]. Virtual reality has been applied to provide safe and interesting training scenarios with near-realistic environments for subjects to interact with it [2]. The performance of elements within these virtual environments proved to be representative of the elements' abilities in the real world and their real-world skills showed significant improvements following the virtual reality training [3-5]. Until now, electric powered wheelchairs simulators were mainly developed to either facilitate patient training and skill assessment [6] [7] or assist in testing and development of semi-autonomous intelligent wheelchairs [8]. While in the training simulators the focus is on user interaction and immersion, the main objective of the robotics simulators is the accurate simulation of sensors and physical behaviour. The simulator presented here addresses the need to combine these approaches. It is a simple design that provides the user training ability while supporting a number of sensors and ensuring physically feasible simulation for intelligent wheelchairs' development. The simulator is a part of a larger project where the IntellWheels prototype will include all typical IW capabilities, like facial expression recognition based command, voice command, sensor base command, advanced sensorial capabilities, the use of computer vision as an aid for navigation, obstacle avoidance, intelligent planning of high-level actions and communication with other devices.

The experiments with real wheelchair users allowed to access information about the usability of the virtual intelligent wheelchair and virtual environment. These users besides using a wheelchair to move around are also potential users of an intelligent wheelchair. In fact, they suffer from cerebral palsy which is a group of permanent disorders in the development of movement and posture.

This paper is organized with five sections; the first one is composed with this introduction. Section two presents an overview of the methodologies for wheelchair simulation and the criteria to select the platform to simulate the IW and the environment. A special attention is given to the USARSim which was the chosen platform to produce the simulation. In section three a brief description of the IntellWheels project is presented. Section four presents the experiments and results and finally the last section refers the conclusions and future work.

2   Methodologies for Wheelchair Simulation

Assistive technologies are defined as any product, instrument, equipment or adapted technology specially designed to improve the functional levels of the disabled person. Resorting to these products can help reduce the limitations in mobility [9] [10] [11] [12]. The electronic wheelchair wheels are an enabling technology, used by people, due to a wide range of diseases including cerebral palsy [13]. Simulators allow users to test and train several of these assistive technologies [14]. This section contains an overview of the methodologies available for developing a wheelchair simulator.

The methodologies that typically are concerned with rendering 2D and 3D are known as graphics engine. Usually they are aggregate inside games engine using specific libraries for rendering. Examples of graphics engines are OpenSceneGraph [15], Object- Oriented Graphics Rendering Engine (OGRE) [16], jMonkey Engine [17] and Crystal Space [18]. The physics engines are software applications with the objective of simulate the physics reality of objects and world. Bullet [19], Havok [20], Open Dynamics Engine (ODE) [21] and PhysX [22] are examples of physics engines. These engines also contribute for robotics simulation for more realistic motion generation of the robot. The game engines are software framework that developers use to create games. The game engines normally include a graphic engine and a physics engine. The collision detection/ response, sound, scripting, animation, artificial intelligence, networking, streaming, memory management, localization support and scene graph are also functionalities included in this kind of engine. Examples of game engine are Unreal Engine 3 [23], Blender Game Engine [24], HPL Engine [25] and Irrlicht Engine [26]. The robotics simulator is a platform to develop software for robots modulation and behaviour simulation in a virtual environment. In several cases it is possible to transfer the application develops in the simulation to the real robots without any extra modification. In the literature there are several commercial examples of robotics simulators: AnyKode Marilou (for mobile robots, humanoids and articulated arms) [27]; Webots (for educational purposes it has a large choice of simulated sensors and actuators is available to prepare each robot) [28]; Microsoft Robotics Developer Studio (HRDS) (allows an easy access to simulated sensors and actuators) [29]; Workspace 5 (environment based on Windows and allows the creation, manipulation and modification of images in 3DCad and several ways of communication) [30]. The non-commercial robotics simulators are also available: SubSim [31]; SimRobot [32]; Gazebo [33]; USARSim [34]; Simbad [35] and SimTwo [36]. A comparison of the most used 3D robotics simulator according several criteria was presented by Petry et al. [37]. Petry et al. [37] also presented the requirements and characteristics for simulation of intelligent wheelchairs and in particularly to the IntellWheels prototype.

The USARSim, acronym of Unified System for Automation and Robot Simulation, is a high-fidelity simulation of robots and environments, based on the Unreal Tournament game engine [34]. Initially was created as a research tool designated as a simulation of Urban Search And Rescue (USAR) robots and environments for the study of human-robot interaction (HRI) and multi-robot coordination [37]. USARSim is the basis for the RoboCup rescue virtual robot competition (RoboCup) as well as the IEEE Virtual Manufacturing Automation Competition (VMAC) [34].

Nowadays, the simulator uses the Unreal Engine UDK and the NVIDIA's PhysX physics engine. The version used to develop the IntellSim was the Unreal Engine 2.5 and the Karma physics engine (which are integrated into the Unreal Tournament 2004 game) which maintain and render the virtual environment and model the physical behaviour of its elements respectively.

3   IntellWheels Project

IntellWheels project aims at providing a low cost platform to electric wheelchairs in order to transform them into intelligent wheelchairs. The simulated environment allows to test and train potential users of the intelligent wheelchair. And select the appropriate interface, among the available possibilities, for a specific user. After the first set of experiments [38] it was necessary to improve the realism of the simulated environment and behaviour of the IW. For that reason and trying to maintain the principle of producing the IntellWheels' project as the lowest cost possible the USARSim was the choice for the new simulator. There were other reasons to decide by the USARSim such as, having an advance support on robots with wheels, allowing it the independent configuration; allowing the importation of object and robots modelled in different platforms in order to facilitate for instance the wheelchair modulation; being possible to program robots and control them in the network which can be implemented in the mixed reality [39].

One of the main objectives of the project is also the creation of a development platform for intelligent wheelchairs [40], entitled IntellWheels Platform (IWP). The project main focus is the research and design of a multi-agent platform, enabling easy integration of different sensors, actuators, devices for extended interaction with the user [41], navigation methods and planning techniques and methodologies for intelligent cooperation to solve problems associated with intelligent wheelchairs [42].

The IntellWheels platform allows the system to work in real mode (the IW has a real body), simulated (the body of the wheelchair is virtual) or mixed reality (real IW with perception of real and virtual objects). In real mode it is necessary to connect the system (software) to the IW hardware. In the simulated mode, the software is connected to the IWP simulator. In the mixed reality mode, the system is connected to both (hardware and simulator). Several types of input devices were used in this project to allow people with different disabilities to be able to drive the IW. The intention is to offer the patient the freedom to choose the device they find most comfortable and safe to drive the wheelchair. These devices range from traditional joysticks, accelerometers, to commands expressed by speech, facial expressions or a combination of some of them. Moreover, these multiple inputs for interaction with the IW can be integrated with a control system responsible for the decision of enabling or disabling any kind of input, in case of any observed conflict or dangerous situation. To compose the necessary set of hardware to provide the wheelchair's ability to avoid obstacles, follow walls, map the environment and see the holes and unevenness in the ground, two side bars were designed, constructed and place on the wheelchair. In these bars were incorporated 16 sonars and a laser range finder. Two encoders were also included, and coupled to the wheels to allow the odometry.

3.1   IntellWheels Simulator

The system module, named IntellWheels Simulator [43] or more recent IntellSim, allows the creation of a virtual world where one can simulate the environment of a building (e.g. a floor of a hospital), as well as wheelchairs and generic objects (tables, doors and other objects). The purpose of this simulator is essentially to support the testing of algorithms, analyse and test the modules of the platform and safely train users of the IW in a simulated environment.

The virtual intelligent wheelchair was modelled using the program 3DStudioMax [44]. The visualizing part, which appears on the screen, was imported to the UnrealEditor as separated static meshes (*.usx) file. The model was then added to USARSim by writing appropriate UnrealScript classes and modifying the USARSim configuration file. The physics property of the model was described in Unreal Script language, using a file for each robot's part. The model has fully autonomous caster wheels and two differential steering wheels. In the simulation it is equipped with: camera; front sonar ring; odometry sensor and encoders. Fig. 1 shows different perspectives of the real and virtual wheelchair.

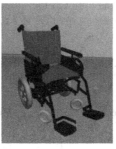

Fig. 1. The real and virtual prototype of the IW

An important factor affecting the simulation of any model in UT2004 is its mass distribution and associated inertial properties. These were calculated using estimated masses of the different parts of the real chair (70 kg) with batteries and literature values for average human body parameters (60 kg). The values obtained for the center of mass and tensor of inertia were used to calculate the required torque for the two simulated motors using the manufacturer's product specification as a guideline. The sensors used in the simulated wheelchair are the same as those used in the real IW. As in the real prototype 16 sonars and a laser range finder were place in two side bars. Two encoders were also included, and coupled to the wheels to allow odometry. These sensors provide the wheelchair's ability to avoid obstacles, follow walls, map the environment and see the holes and unevenness in the ground. Using the simulator it was also possible to model rooms with low illumination and noisy environments and test the performance of users while driving the wheelchair. The map created was done using the Unreal Editor 3 and it is similar to the local were the patients are used to move around. Several components in the map were modelled using 3DStudioMax. In order to increase the realism of the virtual environment it was implemented several animations using sequence scripts. The simulator runs on a dual-core PC with a gaming standard dual-view graphics card. And other supported input devices include keyboard, mouse, mouse replacement devices and gaming joysticks.

4   Experiments and Results

The initial experiments were conducted to have more information about the real prototype and with the objective of being able to modelled and simulate the behaviour of the real wheelchair more precisely in virtual environment. The final experiments involved real wheelchair users and potential users of the intelligent wheelchair. Therefore the experiments were divided into two components: wheelchair technical information and users' feeling about the simulated wheelchair and the virtual environment modelled.

4.1   Real Intelligent Wheelchair Characteristics

The technical information was obtained using the manual of the electric wheelchair which the platform was applied, measurements taken and experiments with the real intelligent wheelchair prototype (Table 1 resumes that information).

Table 1. Real intelligent wheelchair prototype characteristics

Real Intelligent Wheelchair Prototype Characteristics			
Weight (with batteries)	70 kg	Motor	180w 24 volt, 2.5A max, 3800 RPM (32:1)
Big wheel diameter	0.315 m	Brakes	electromagnetic automatic brakes
Small wheel diameter	0.185 m	Maximum speed	7km/h
Distance between rear wheels	0.535 m	Time for a total rotation	7''
Distance between axes	0.48 m	Sensors	16 sonares, 1 laser range finder (URG-04LX Hokuyo), 2 encoders (rear wheels)

Fig. 2. Real intelligent wheelchair experiments

Fig. 2 shows some of the experiments done with real intelligent wheelchair prototype. It was analyzed the velocity and time for the total rotation with the new adaptations applied to the real prototype. The results obtained, using the real prototype, were considered in order to develop the virtual intelligent wheelchair.

4.2 IntellSim Experiments

The *IntellSim* experiments were performed by patients suffering from cerebral palsy, all of them wheelchairs users. Using *IntellSim* several experiments were performed changing the conditions of the environment, using the joystick and the manual control. A map was created integrating paths with degrees of difficulty. The overall circuit (Fig. 3) passes through two floors and the link between floors is a ramp. The map was divided into three parts. The first part is characterized by having simple and large corridors without any kind of obstacles except pillars. The second part has narrow corridors, ramps and obstacles. The last part involved a circuit entering in three rooms with different kind of illumination and noise.

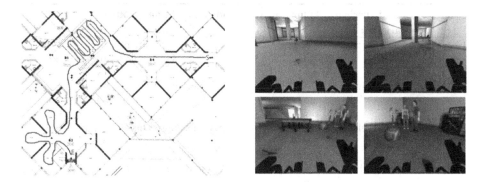

Fig. 3. Overall circuit and snapshots of the first person view during the game

After the experiments the users respond to the System Usability Scale [45] which is a simple ten-item *Likert* scale giving a global view of individual assessments of usability (with a score between 0 and 100) and some more questions about safety and control managing the IW, if it was easy to drive the IW in tight places and the attention needed to drive the IW.

4.2.1 Sample Characterization

To better understand the next results the sample characterization will be presented. It is important to reinforce that cerebral palsy is defined as a group of permanent disorders in the development of movement and posture. It causes limitations at the level of daily activities because of a non-progressive disturbance which occurs in the brain during the fetal and infant development [46]. The motor disorders in cerebral palsy are associated with deficits of perception, cognition, communication and behaviour. In general, there are also episodes of epilepsy and secondary musculoskeletal problems [46]. The individuals included in this study suffer from cerebral palsy and were classified in the levels IV (26%) and V (74%) of Gross Motor Function Measure (GMF). These are the highest levels in the cerebral palsy severity degree. The sample size was composed of the 19 individuals and all require the use of a wheelchair. The mean of age was 29 years old with 79% males and 21% females. In terms of school level 15% did not answer, 10% are illiterate, 16% just have the elementary school, 16% have the middle school, 37% have the high school and only 5% have a BSc. The dominant hand was divided as: 12 for left, 6 for right hand and 1 did not answer. Another question was the frequency of use of information and communication technologies: 27% did not answer; 42% answered rarely; 21% sometimes; 5% lots of times and 5% always. The aspects related to experience of using manual and electric wheelchair were also questioned. Table 2 shows the distribution of answers about autonomy and independency using the wheelchair and constraints presented by these individuals.

Table 2. Experience using wheelchair, autonomy, independence and constraints of the cerebral palsy users

Experience, Autonomy, Independence and Constraints			
Variables	**n**	**Variables**	**n**
Use manual wheelchair		Cognitive constraints	
no	13	no	8
yes	6	yes	11
Use electric wheelchair		Motor constraints	
no	6	no	0
yes	13	yes	19
Autonomy using wheelchair		Visual constraints	
no	4	no	11
yes	15	yes	8
Independence using wheelchair		Auditive constraints	
no	4	no	19
yes	15	yes	0

4.2.2   *IntellSim* Results

The results obtained reveal a satisfactory usability in terms of the experiments using the *IntellSim*. In fact all of them could easily identify that the virtual environment was a replica of the cerebral palsy institution where they are used to be.

Table 3. Results of the experiments in *IntellSim*

IntellSim experiments with patients suffering from cerebral palsy					
	Mean	**Median**	**Std**	**Min**	**Max**
Usability and Safety					
Score SUS	72.0	70.0	11.7	57.5	95
Safety managing IW	--	Agree	--	Ind	SAgree
Control of the IW	--	Agree	--	Ind	SAgree
Easy to drive the IW in tight places	--	Agree	--	Dis	SAgree
The IW do not need to much attention	--	Dis	--	SDis	Agree
Satisfaction	--	VSatis	--	Ind	VSatis
Performance					
Time	12.6	9.5	8.6	5.6	42.4
Number of objects collected	12.7	14	3.6	4	15

Legend: SDis – strongly disagree; Dis – disagree; Ind – indifferent; SAgree – strongly agree; VDiss – very dissatisfied; Diss – dissatisfied; Satis – satisfied; VSatis – very satisfied

The SUS mean score was satisfactory and all the users considered the usability positive (higher than 57.5). Overall, the users were very satisfied with the experience. The users made the circuit with a median of 9.5 minutes. The best time was made by a user in 5.6 minutes and the worst time 42.4 minutes was made by a user with severe difficulties and without autonomy or independence in driving a wheelchair. The data from the logs allows plotting the circuits after the experiment. It is a way of analysing the behaviour of the users (Fig. 4).

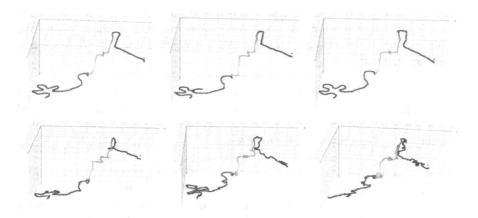

Fig. 4. Circuits performed by patients suffering from cerebral palsy with joystick

It is interesting to notice that the path is smooth using the joystick in manual mode and using the *IntellSim*. These three individuals are autonomous and independent in driving their own electric wheelchairs with joystick (Level IV of the GMF). The next three circuits' examples (second line in the Fig. 4) were executed by the users that took the longest times. In the left example the user took 17.7 minutes to collect 14 objects and to finish the circuit. In the middle the user took 22.35 to collect 12 objects and on the right side there is the circuit performed by the user that took the longest time of 42.4 minutes to collect only 9 objects. Although these three examples are the worst in terms of performance it is necessary to enhance that they are classified in the most severe degree of the GMF and do not have the autonomy and independence to drive a conventional wheelchair. However, the *IntellSim* can be used to train these users and with appropriate methodologies, such as shared or automatic controls, it is possible to drive the IW in efficient and effective manner.

5   Conclusions and Future Work

The attention given to the autonomy and independence of the individual is nowadays an actual subject. The scientific community is concerned in develop and present many prototypes, such as intelligent wheelchairs, however most of them only execute experimental work in the labs and without real potential users. The virtual reality wheelchair simulator presented here addresses an important gap in wheelchair simulation in that it can be used for both patient training or evaluation and design and development of semi-autonomous intelligent wheelchairs. Because of its flexibility and the long (and expanding) list of sensors inherited from the USARSim project, the current system provides a perfect test bed for development and testing of intelligent wheelchair systems. Experimental work with the real IW prototype enable having information about weight, wheels diameters, distance between axes and wheels, motor characteristics, maximum speed, time for total rotation and localization and sensors characteristics in order to realistic model the virtual wheelchair. The experiments with real patients suffering from cerebral palsy allow having a confirmation about the usability of the IW in the *IntellSim*. The performance results also present evidences that it is possible to use the *IntellSim* as a training and test tool.

After this exploratory work for bridging the gap between the real and virtual IW and for future work the full capabilities of the IW are going to be tested with real patients. The possibility of driving the IW using a multimodal interface and a shared control that allows the correction of the trajectory of the user with severe constraints are some of the issues that are going to be tested.

Acknowledgments. This work is financed by LIACC (PEst-OE/EEI/UI0027/2014) and ERDF – European Regional Development Fund through the COMPETE Programme (operational programme for competitiveness) and by National Funds through the FCT – Fundação para a Ciência e a Tecnologia (Portuguese Foundation for Science and Technology) within project «FCOMP-01-0124-FEDER-037281».

References

1. Holden, M.K.: Virtual environments for motor rehabilitation: review. J. Cyberpsychol Behav. **8**(3), 187–211 (2005)
2. Boian, R.F., Burdea, G.C., Deutsch, J.E., Windter, S.H.: Street crossing using a virtual environment mobility simulator. In: Proceedings of IWVR, Lausanne, Switzerland (2004)
3. Inman, D.P., Loge, K., Leavens, J.: VR education and rehabilitation. Commun. ACM **40**(8), 53–58 (1997)
4. Harrison, A., Derwent, G., Enticknap, A., Rose, F.D., Attree, E.A.: The role of virtual reality technology in the assessment and training of inexperienced powered wheelchair users. Disabil Rehabil **24**(8), 599–606 (2002)
5. Adelola, I.A., Cox, S.L., Rahman, A.: VEMS - training wheelchair drivers, Assistive Technology, vol. 16, pp. 757–761. IOS Press (2005)
6. Desbonnet, M., Cox, S.L., Rahman, A.: Development and evaluation of a virtual reality based training system for disabled children. In: Sharkey, P., Sharkeand, R., Lindström, J.-I. (eds.) The Second European Conference on Disability, Virtual Reality and Associated Technologies, Mount Billingen, Skvde, Sweden, pp. 177–182 (1998)
7. Niniss, H., Inoue, T.: Electric wheelchair simulator for rehabilitation of persons with motor disability, Symp Virtual Reality VIII, Belém (PA). Brazilian Comp. Society (BSC) (2006)
8. Röfer, T.: Strategies for using a simulation in the development of the Bremen autonomous wheelchair. In: 12th European Simulation Multi Conference 1998 Simulation: Past, Present and Future, ESM 1998, Manchester, UK, pp. 460–464 (1998)

9. Tefft, D., Guerette, P., Furumasu, J.: Cognitive predictors of young children's readiness for powered mobility. Dev. Medicine and Child Neurology **41**(10), 665–670 (1999)
10. Faria, B.M., Silva, A., Faias, J., Reis, L.P., Lau, N.: Intelligent wheelchair driving: a comparative study of cerebral palsy adults with distinct boccia experience. In: Rocha, Á., Correia, A.M., Tan, F., Stroetmann, K. (eds.) New Perspectives in Information Systems and Technologies, Volume 2. AISC, vol. 276, pp. 329–340. Springer, Heidelberg (2014)
11. Palisano, R.J., Tieman, B.L., Walter, S.D., Bartlett, D.J., Rosenbaum, P.L., Russell, D., et al.: Effect of environmental setting on mobility methods of children with cerebral palsy. Developmental Medicine & Child Neurology **45**(2), 113–120 (2003)
12. Wiart, L., Darrah, J.: Changing philosophical perspectives on the management of children with physical disabilities-their effect on the use of powered mobility. Disability & Rehabilitation **24**(9), 492–498 (2002)
13. Edlich, R.F., Nelson, K.P., Foley, M.L., Buschbacher, R.M., Long, W.B., Ma, E.K.: Technological advances in powered wheelchairs. Journal Of Long-Term Effects of Medical Implants **14**(2), 107–130 (2004)
14. Faria, B.M., Teixeira, S.C., Faias, J., Reis, L.P., Lau, N.: Intelligent wheelchair simulator for users' training: cerebral palsy children's case study. In: 8th Iberian Conf. on Information Systems and Technologies, vol. I, pp. 510–515 (2013)
15. Wang, R., Qian, X.: OpenSceneGraph 3 Cookbook. Packt Pub. Ltd., Birmingham (2012)
16. Koranne, S.: Handbook of Open Source Tools, West Linn. Springer, Oregon (2010)
17. jMonkeyEngine (2012). http://jmonkeyengine.com/ (current May 2014)
18. Space, C.: Crystal Space user manual, Copyright Crystal Space Team. http://www.crystalspace3d.org/main/Documentation#Stable_Release_1.4.0 (current May 2014)
19. Gu, J., Duh, H.B.L.: Handbook of Augmented Reality. Springer, Florida (2011)
20. Havok, Havok (2012). http://havok.com/ (current May 2014)
21. Smith, R.: Open Dynamics Engine (2007). http://www.ode.org/ (current May 2014)
22. Rhodes, G.: Real-Time Game Physics, in Introduction to Game Development, Boston, Course Technology, pp. 387–420 (2010)
23. Busby, J., Parrish, Z., Wilson, J.: Mastering Unreal Technology. Sams Publishing, Indianapolis (2010)
24. Flavell, L.: Beginning Blender - Open Source 3D Modeling, Animation and Game Design. Springer, New York (2010)
25. Games, F.: Frictional Games (2010). http://www.frictionalgames.com/site/about (current May 2014)
26. Kyaw, A.S.: Irrlicht 1.7 Realtime 3D Engine – Beginner's Guide. Packt Publishing Ltd., Birmingham (2011)
27. Marilou, April 2012. http://doc.anykode.com/frames.html?frmname=topic&frmfile=index.html (current May 2014)
28. Cyberbotics, Webots Reference Manual, April 2012. http://www.cyberbotics.com/reference.pdf (current May 2014)
29. Johns, K., Taylor, T.: Microsoft Robotics Developer Studio. Wrox, Indiana (2008)
30. Workspace, Workspace Robot Simulation, WAT Solutions (2012). http://www.workspacelt.com/ (current May 2014)
31. Boeing, A., Braunl, T.: SubSim: an autonomous underwater vehicle simulation package. In: 3rd Int. Symposium on A. Minirobots for Research and Edut, Fukui, Japan
32. Laue, T., Röfer, T.: SimRobot - development and applications. In: Proceedings of the International Conference on Simulation, Modeling and Programming for Autonomous Robots, Venice, Italy (2008)
33. Gazebo, Gazebo. http://gazebosim.org/ (current May 2014)

34. Carpin, S., Lewis, M., Wang, J., Balakirsky, S., Scrapper, C.: USARSim: a robot simulator for research and education. In: Proceedings of the IEEE International Conference on Robotics and Automation, Roma, Italy (2007)
35. Hugues, L., Bredeche, N.: Simbad Project Home, May 2011. http://simbad.sourceforge.net/ (current May 2014)
36. Costa, P.: SimTwo - A Realistic Simulator for Robotics, March 2012. http://paginas.fe.up.pt/~paco/pmwiki/index.php?n=SimTwo.SimTwo (current May 2014)
37. Petry, M., Moreira, A.P., Reis, L.P., Rossetti, R.: Intelligent wheelchair simulation: requirements and architectural issues. In: 11th International Conference on Mobile Robotics and Competitions, Lisbon, pp. 102–107 (2011)
38. Faria, B.M., Vasconcelos, S., Reis, L.P., Lau, N.: Evaluation of Distinct Input Methods of an Intelligent Wheelchair in Simulated and Real Environments: A Performance. The Official Journal of RESNA (Rehabilitation Engineering and Assistive Technology Society of North America) 25(2), 88–98 (2013). USA
39. Namee, B.M., Beaney, D., Dong, Q.: Motion in Augmented Reality Games: An engine for creating plausible physical interactions in augmented reality games. International Journal of Computer Games Technology (2010)
40. Braga, R., Petry, M., Moreira, A., Reis, L.P.: A development platform for intelligent wheelchairs for disabled people. In: 5th Int. Conf Informatics in Control, Automation and Robotics, vol. 1, pp. 115–121 (2008)
41. Reis, L.P., Braga, R.A., Sousa, M., Moreira, A.P.: IntellWheels MMI: a flexible interface for an intelligent wheelchair. In: Baltes, J., Lagoudakis, M.G., Naruse, T., Ghidary, S.S. (eds.) RoboCup 2009. LNCS, vol. 5949, pp. 296–307. Springer, Heidelberg (2010)
42. Braga, R., Petry, M., Moreira, A., Reis, L.P.: Platform for intelligent wheelchairs using multi-level control and probabilistic motion model. In: 8th Portuguese Conf. Automatic Control, pp. 833–838 (2008)
43. Braga, R.A., Malheiro, P., Reis, L.P.: Development of a realistic simulator for robotic intelligent wheelchairs in a hospital environment. In: Baltes, J., Lagoudakis, Michail G., Naruse, Tadashi, Ghidary, Saeed Shiry (eds.) RoboCup 2009. LNCS, vol. 5949, pp. 23–34. Springer, Heidelberg (2010)
44. Murdock, K.L.: 3ds Max 2011 Bible. John Wiley & Sons, Indianapolis (2011)
45. Brooke, J.: SUS: A quick and dirty usability scale, in Usability evaluation in industry, pp. 189–194. Taylor and Francis, London (1996)
46. Rosenbaum, P., Paneth, N., Leviton, A., Goldstein, M., Bax, M., Damiano, D., Dan, B., Jacobsson, B.: A report: the definition and classification of cerebral palsy April 2006. Developmental Medicine & Child Neurology - Supplement 49(6), 8–14 (2007)

A Skill-Based Architecture for Pick and Place Manipulation Tasks

Eurico Pedrosa[(✉)], Nuno Lau, Artur Pereira, and Bernardo Cunha

Department of Electronics, Telecommunications and Informatics, IEETA, IRIS,
University of Aveiro, Aveiro, Portugal
{efp,nunolau,artur}@ua.pt, mbc@det.ua.pt

Abstract. Robots can play a significant role in product customization
but they should leave a repetitive, low intelligence paradigm and be able
to operate in unstructured environments and take decisions during the
execution of the task. The EuRoC research project addresses this issue
by posing as a competition to motivate researchers to present their solution to the problem. The first stage is a simulation competition where
Pick & Place type of tasks are the goal and planning, perception and
manipulation are the problems. This paper presents a skill-based architecture that enables a simulated moving manipulator to solve these tasks.
The heuristics that were used to solve specific tasks are also presented.
Using computer vision methods and the definition of a set of manipulation skills, an intelligent agent is able to solve them autonomously. The
work developed in this project was used in the simulation competition of
EuRoC project by team IRIS and enabled them to reach the 5th rank.

1 Introduction

The trend in industry is clearly for higher levels of customization of products,
which must be enabled by fast customization and adaptability of the production lines to different requirements. Robots can play a significant role in this
customization but they should leave a repetitive, low intelligence paradigm and
be able to operate in unstructured environments and take decisions during the
execution of the task. The European Robotics Challenges (EuRoC) is a research
project based on a robotics competition that aims to present solutions to the
European manufacturing industry. Exploring our healthy competitive nature,
we try to build and developed a robot, or robots, to accomplish a task better than the competition. Usually, the final outcome is a push in the boundary
of our knowledge. Take for example the DARPA Grand Challenge from which
self-driving cars are a reality, e.g. [12].

The EuRoC project is divided in three challenges, each with different motivations and objectives. Our focus is the *Shop Floor Logistics and Manipulation*
challenge, or Challenge 2 (C2), a challenge that presents tasks to be solved by a
mobile robot with manipulation capabilities in a industrial shop floor. The first
stage of the challenge, was a simulation contest where the contesting teams had

© Springer International Publishing Switzerland 2015
F. Pereira et al. (Eds.) EPIA 2015, LNAI 9273, pp. 457–468, 2015.
DOI: 10.1007/978-3-319-23485-4_45

to solve several tasks in a *Simulation Environment* in order to score (Sect. 2). In the end, the best fifteen teams became candidates for entering a second stage.

The *Simulation Environment* of EuRoC C2 exposes an interface with the simulation using the Robotic Operating System (ROS) communications middleware. To address this requirement we propose a *System Architecture* that is mapped into ROS without the loss of generality (Sect. 3). The analysis of the properties of the environment allowed us to design a generic agent that can be used to solve any task (Sect. 4). On top of this architecture and agent design, the logistics and manipulation tasks are solved using computer vision methods and a set of manipulation skills ruled by several heuristics (Sect. 5). Some related work is presented in Sect. 6 and final conclusion in Sect. 7.

2   Shop Floor Logistics and Manipulation

The motivation of this challenge is to bring a mobile robot onto the shop floor for dexterous manipulations and logistics carries. Enabling an autonomous robot to operate in an unstructured environment and establishing a safe and effective human-robot interaction are two of the main research issues to be addressed.

The first stage consists of a sequence of tasks to be performed in a simulated environment, with increasing difficulties in the problem to be solved. The *Simulation Environment* consist of a Light-Weight-Robot (LWR), with a two-jaw-gripper, mounted on a moving XY axis on a table top, and a fixed mast with a Pan & Tilt (PT) actuator. Additionally, a vision system made by a pair of RGB and Depth cameras is installed on the PT and another on the Tool Center Point (TCP). The objects to be manipulated assume a basic shape, i.e. cylinder or box, or a compound of basic shapes. An overview is depicted in Fig. 1.

All tasks include a Pick & Place (P&P) scenario where objects (e.g. Fig. 2) have to be picked from unknown locations and placed on target locations. Required actions to accomplish a task includes: perception, to locate the objects; manipulation, to pick and place them; and planning, to move the arm to the target locations. For demonstrating these problems four different tasks are considered.

P&P. The goal is to pick all objects in the working space and place them in the proper location without any particular order. The task contains three objects of different shape and color on the table (e.g. Figures 2c, 2b 2d). The pose of the objects in the environment are unknown but their properties (color and shape composition) and corresponding place zone are given. The LWR base cannot move. Scoring is achieved by picking an object and place it in the correct zone.

P&P with Significant Errors and Noise. This is the same task as P&P but with the difference that there are significant errors in the robot precision and also significant noise in all sensors. This implies an operation of the LWR in an imprecise and uncalibrated environment.

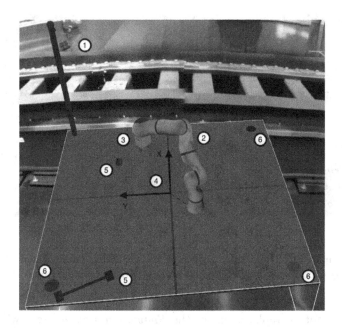

Fig. 1. Overview of the simulation environment. 1) Mast with a PT, 2) LWR, 3) Gripper, 4) XY Axis, 5) Object to pick, 6) Place zone.

Mobile P&P with Typical Errors and Noise. This is based on the P&P task, but with typical calibration errors and sensor noise. The LWR can now use the XY axis to move. This task introduces the concept of mobility, meaning that pick and place positions may not be in range of the LWR, not considering the additional axis.

Loose Assembly of a Puzzle. In this tasks, a puzzle made up of pieces like the ones depicted in Fig. 2a and Fig. 2e, initially scattered across the table in unknown locations has to be loosely done. Each puzzle part is composed of basic blocks and has to be placed into a puzzle of size 4x4. To allow parts to be pushed, the puzzle fixture has two fixed sides. Scoring is achieved by covering the puzzle with the correct blocks.

3 System Architecture Overview

The tasks to be accomplished were different. However, while studying the problems to be solved, common subtasks were identified. In order to take advantage of this fact, a system architecture to solve all tasks was developed. This architecture, shown in Fig. 3, is divided into three components: *Simulation Environment*, *Interface Node*, and *Agent*.

The *Interface Node*, supplied by the EuRoC partners to all teams, provides an abstraction of the *Simulation Environment*. The purpose of this abstraction

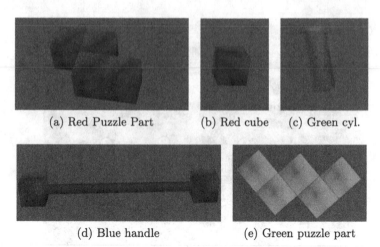

(a) Red Puzzle Part   (b) Red cube   (c) Green cyl.

(d) Blue handle   (e) Green puzzle part

Fig. 2. Examples of objects that appear in the simulated environment. The puzzle parts are only examples of the set of valid shapes. Objects may vary on scale and color.

is to allow a future replacement of the *Simulation Environment* with a *Real Environment*, where a real robot in a real shop floor would be used instead with minimum modifications of the *Agent*.

The proposed *Agent* is composed of a *Solver*, multiple *Skills* and a *Sensory Data Adapter*. The *Solver* is responsible for making decisions on how to solve the current task based on the current sensory data and available *Skills*. A *Skill* is the capacity of doing a particular task like, for instance, picking an object or moving the end-effector of the manipulator to a desired pose. The role of the *Sensory Data Adapter* is to convert the sensor data format transmitted by the *Interface Node* to the format used inside the *Agent*.

The provided interface with the *Simulation Environment* is implemented using ROS [6]. This communications middleware is a publish-subscribe infrastructure where a **topic** is a communication channel that forwards **messages** from the publisher to the subscriber. A **node**, which is a system process, can subscribe or advertise n **topics**, with $n \geq 0$. In addition, it provides a request-response infrastructure through **services**.

The *Interface Node*, as the bridge between the *Simulation Environment* and the *Agent*, is a **node** that provides a continuous stream of sensory data and a set of functions to interact with the simulation effectors. The sensory data stream includes the images from the cameras and effectors feedback (i.e. telemetry), and is sent through **topics**. The provided functions include: inverse and forward kinematics calculation, joint motion planning and execution, and operation mode control. All functions are available as **services**.

The *Sensory Data Adapter* is a ROS **node** that converts the joints state information to a transformation tree using ROS **tf** package. This allows the user to keep track the multiple coordinate frames over time. The telemetry information is only provided upon request, i.e. by service call. The reason for this choice

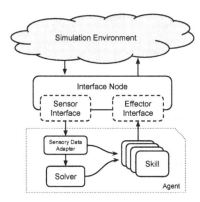

Fig. 3. Diagram of the system architecture.

is based on the fact that the telemetry information is published at a rate of 1 kHz that would introduce a computation overhead to all **nodes** that require this information. This way only the *Sensory Data Adaptor* suffers from a computation overhead, other nodes needing this information may request it at a desired rate.

Each *Skill* is implemented as a ROS **node** that exposes itself as a **service**. Only a *Skill* knows how to interact with the simulation effectors, doing it by using the appropriate functions of the *Interface Node*.

The *Solver* is the decision-making module implemented in a single **node**. Based on the task to be solved and on an initially analysis of the environment, it defines a plan to solve the task and executes it using the available skills.

4 Agent Design

The EuRoC project exposes a set of tasks with perception and manipulation problems to be completed. To solve these tasks we need a proper *agent* capable of perceiving the environment through sensors and act upon that environment using actuators [8]. To design such *agent* we have to analyze the properties of the environment, which in this case is a well defined *Simulated Environment*, to develop a generic as possible solution capable of handling all tasks. Our proposal is summarized in algorithm form in Fig. 4.

The set $\mathcal{O}$ of objects to be manipulated in each task, including their properties and place zones, is known in advance. The algorithm starts by determining the order by which the objects have to be manipulated. The order is restricted by a direct acyclic graph (dag), which represents a dependency graph between objects in terms of order of manipulation. Leafs represent objects that need to be handled first. A dummy object λ is added to represent the graph root.

Once an object is manipulated, it is removed from the graph. Thus, at any moment in the task execution, leafs represent objects that can be manipulated. When the dummy object is the only one in the graph, the task is completed.

```
 1: G ← BuildOrderGraph(O), S ← buildSearchSpace(G)
 2: while leafs(G) ≠ λ do                              ▷ λ is the empty root
 3:     L ← leafs(G)
 4:     obj ← detectObject(L)
 5:     while obj is not found do
 6:         s ← next(S)
 7:         move(s)
 8:         obj ← detectObject(L)
 9:     end while
10:     focusObject(obj)
11:     plan = MakePlan(obj)
12:     success = execute(plan)
13:     if success then
14:         removeLeaf(G, obj)
15:     end if
16: end while
```

Fig. 4. Generic algorithm to solve a task.

If the TCP is over an object, its vision system is in a privileged position to help in its location and pick up operation. However it has to move to there first. In the other side, and despite of being in a high position, it can be not possible, even using its pan and tilt capabilities, to cover all the objects in the table with the vision system placed in the mast. For instance, an object could be occluded by the LWR. Thus, both vision systems are used.

To ensure that all object are eventually detected, a set of search poses S is calculated by the procedure *buildSearchSpace*, insuring that the entire workspace is covered. To improve the search, the vision system on the PT is used to detect objects in the environment and the obtained poses are put in the head of S in the order defined by G. It may not detect any object, but if it does, the system can gain in overall execution time due to good initial search poses. Thus, each pose in S represents a region containing one or more objects or a region that should be locally explored using the TCP vision system. The set S is encoded as a circular list so that search poses never ends. The algorithm then executes two nested loops that, making the TCP move around the searchable poses, finishes when all objects are placed in their target positions. Each searchable pose is explored to see if a leaf object is there. If so, a sequence of steps is performed to put the object in its target position.

First, the TCP is moved in such a way that the focus axis of the RGB camera intercepts the geometry center of the object, giving preference to positions in which the gripper is perpendicular to the object's plane. This way the object appears in the center of the image, which results in two benefits: the distortion of the object in the image is reduced; and it copes better with noise in sensors and effectors by restricting the problem to a bounded local space.

In order to properly transfer the object to its target position a plan is computed. A plan is a sequence of actions that allows to properly pick the object, move the TCP and place the object properly in the target position. Those actions depend on a priori calculation of the pick and place pose. The way the plan is calculated depends on the task being solved and on the disposition of the object and its target position. For instance, the target position could be non-reachable from the top, then the object has to be picked from a different direction. If the execution of the plan succeed, the leaf corresponding to the processed object is removed from the graph and the algorithm goes to the next iteration.

From this design only two procedures are task dependent, BUILDORDERGRAPH and MAKEPLAN. This fosters the re-utilization of software code and facilitates the task solving job, since the developer only has to focus on the creation of a plan. To aid the definition of a plan, a set of skills was defined and implemented.

4.1 Object Detection Pipeline

The objective of the object detection module is, as the name implies, to detect an object in the environment. It creates information about an object by processing the fed sensory data. The input data is the RGB and depth images created by one of the two available vision systems. The data is processed by several submodules managed by a pipeline (Fig. 5).

Depth and RGB Integration. Before any processing takes place it is necessary to match every depth value to the corresponding RGB value [3]. The depth and the RGB images come from different cameras separated by an offset, thus they do not overlap correctly. To solve this issue, for each depth value in the depth image, the corresponding 3D space position is calculated and then re-projected into the image plane of the RGB camera. The registered depth is then transformed into an organized point cloud in the coordinates frame of the LWR.

Height Filter. The purpose of the *Height Filter* is to reduce the search space for objects of interest using the fact that they are on top of a table (e.g. Fig. 6a). A filter is applied to the pointcloud where positions with z value below a threshold are set to not-a-number (NaN), identifying a non searchable position. The output of this block is a mask defined by the filtered pointcloud that identifies the searchable areas in the RGB image (Fig. 6b).

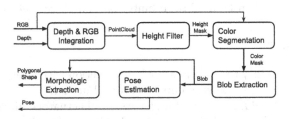

Fig. 5. Object detection pipeline.

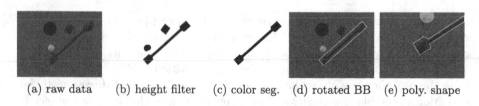

(a) raw data   (b) height filter   (c) color seg.   (d) rotated BB   (e) poly. shape

Fig. 6. View of several stages in the object detection pipeline.

Color Segmentation. The color of the object is a known attribute, it is homogeneous but subject to different light conditions (e.g. shadows). Using color segmentation we can further identify the object of interest. The segmentation is applied on HSV color space to reduce the influence of different light conditions [10]. The output of this block is a mask defined by the segmentation (Fig. 6c).

Blob Extraction. At this stage, it is expected that the input mask defines one or more undefined shape areas. More than one blob can happen if two, or more, objects share the same color. However, we are only interested in one object, thus only the blob with the biggest area is considered. This is a safe action because different objects have different shapes and can be disambiguated by matching the shape from the blob contours with the correct object. To extract a blob, the contours of the segmented areas are calculated by contour detection function [11]. The output mask is a set of points that delimits the area of the blob.

Pose Estimation. After the identification of the object blob in the image we calculate its position (x, y, z) and orientation θ. We start by calculating the rotated bounding box that best fits the select blob (e.g. Fig. 6d). The center of the rectangle in the image provide us enough information to extract its position because we have a registered point cloud where (u, v) coordinate from the RGB image has a corresponding (x, y, z). Furthermore, from the rotation of the rectangle we can now extract the orientation of the object relative to the LWR.

Morphologic Extraction. The goal is to extract a useful shape from the blob (e.g. Fig. 6e). All objects are treated as polygons, even the cylinder that is a circle when looked from above – a circle can be approximated by a polygon. The shape is obtained by a function that approximates the blob to a polygon [2]. The polygon shape can then be used to disambiguate an object detection.

4.2 Agent Skills

A *Skill* is the capability to perform a specialized action with a semantic meaning. Performing an action will always imply a physical motion by one or more actuators. In the *Simulated Environment* we have several actuators, already described, that are controlled by joints, either rotational or linear. All joints can be actuated on to provoke motion at the same time, meaning that a skill is not necessarily bound to one actuator. For example, moving the LWR and the XY axis at the same time can be considered a single skill.

To solve any task we need a set of skills that are enough for the tasks ahead, but at the same time keep their number low as possible. The idea is to create skills that can be used as atoms from more complex skills. To solve a task we need the capability to move the LWR so that its TCP is in the desired position, e.g. a search pose to perceive the environment. Objects needs to be picked and place, so this actions are also necessary. It may not always be possible to position an object with a place action, for example, the Gripper may hit a wall when placing right next to it. This can be solved by pushing the object to its rightfull place.

Simple Move. This skill uses only the LWR actuator. Its objective is to put the LWR's TCP on the requested pose. This action requires at first the calculation of the inverse-kinematics, and then the control of the joints to the desired position.

Move XY Axis. The reach of the LWR can be increased by using the XY axis to move it. This skill only requires x, y positions to move. The values are directly applied to the joints.

Pick Object. The LWR and Gripper are used together to provide this skill. It expects as input a pick pose that must correspond to the tip of the Gripper end-effector. The skill will do the necessary calculation to transform the input pose to the corresponding TCP pose to move the LWR. Then it takes the proper actions to safely grab the object.

Place Object. This skill uses the LWR and Gripper to place the object in the requested pose. The input pose is adjusted to the TCP and then a safe position is calculated to ensure that the object will hover the floor before the actual place. Then it will lower the object with a lower velocity to prevent any undesirable collisions due to higher accelerations.

Push Object. When it is not possible to place an object without a harmful collision a push strategy can be applied. Assuming that the object is near its final position, we can use the Gripper to push the object until it reaches the final destination. This is achieved by defining a path between two points to be covered by the end-effector of the Gripper. If an object is in the path, as a product of the LWR motion the object will be pushed. The skill is sensible enough to ask for a motion with force limits. If the limit is not reached it means the object could be pushed to its final position.

5   Experiments with Solving Tasks

To evaluate the proposed architecture for logistics and manipulation tasks we set to solve the already described tasks: all P&P tasks and Loose Assembly of a Puzzle. As stated, to solve a task we have to concentrate our efforts on the objects order and planning. Most planning involve finding the correct pick and place positions and then make use of the available skills to accomplish that.

5.1 Pick and Place Tasks

In these tasks, the order by which the objects are picked and placed is not relevant and so graph $\mathcal{G}$ does not contain any order dependency. Thus, once an object is detected it can be picked and placed in its target location.

The agent assumes all objects are pickable. This implies that there is a part of the object the gripper is able to grab. Since the shapes of the objects are known in advance, the grab pose, defined in the object's frame of reference, can be pre-determined. The gripper's grabbing pose can be obtained by merging this information with the rotated bounding box (RBB) that encloses the object, estimated by the object detection module. For basic shape objects, the grab pose coincides with the center of the RBB. For compound objects, the grab pose can be obtained from the center of the RBB, adding half of its width and subtracting half of its height. This approach worked well even in the presence of significant noise.

The preferable way to pick an object is applying a top-down trajectory to the gripper (Fig. 7a). However, this may not be possible due to constraints in the freedom of motion of the LWR. In such cases, an angular adjustment is applied to make the pick operation possible (Fig. 7c). This drawback can be mitigated when the LWR's movement along the XY axes is available, since the LWR can be moved to a better position to pick the object of interest.

After an object is picked it must be placed in its target location. The most efficient move is to place it keeping the gripper orientation. But, again, it may not be possible. When this happens, a fake placement is done instead: an appropriated location is chosen, the object is placed there, and a failure signal is returned so that the object is picked again. Similarly as before, the availability of robot movement along the XY axes can avoid this drawback.

5.2 Loose Assembly of a Puzzle

The assembly of the puzzle can not be done by putting the parts from above. The margin of error for the positioning is very thin, making it difficult to use an approach where the parts are put from the top. Then, using the puzzle fixture boundary and the already in place parts, a pushing approach can be used. To

(a) preferable pose   (b) object out-of-reach   (c) gripper adjustment

Fig. 7. Example of an out-of-reach situation for the preferable pick pose.

find the right order we search all permutations for ordering the objects in their insertion in the puzzle. The valid solutions must comply with the following rule: the pushing of the object, horizontally or vertically, is not prevented by any other object. After a solution is found the graph $\mathcal{G}$ is built.

The pick position is selected from one of the convex terminations of the part that has a width smaller than the Gripper maximum width. The parts of the puzzle are always assembled from cubes with a width smaller than the gripper maximum width, hence the pick position must consider a convex termination with the width of a cube. To calculate the pick position the vertices of the polygon shape are used. The vertices v are properly ordered and ring accessible by v_i. The main idea is to search for an edge $\mathbf{e_i} = (v_i, v_{i+1})$ that is part of the frame of reference. An edge $\mathbf{e}_i$ is a candidate when:

$$\|\mathbf{e}_i\| \approx \ell \quad \wedge \quad \angle(\mathbf{e}_{i-1}, \mathbf{e}_i) \approx \angle(\mathbf{e}_i, \mathbf{e}_{i+1}) \approx \frac{\pi}{2}$$

where ℓ is the length of a cube edge. Approximate values are considered to handle errors. Afterward, all candidates go through a final validation. To recognize the object frame of reference, which is needed to correctly place the object, we assume that the vertices v_i is the origin, then, the number of blocks at the left, right, top and bottom are compared with the original object shape definition. Once an edge is selected the pick position is given by sum of the normalized edges $\mathbf{e}_i$ and $\mathbf{e}_{i+1}$, and the orientation is given by the normal direction of $\mathbf{e}_i$. The XY is available for this task, thus any object can be picked in the preferable way.

The next step is to define the set of actions to position the puzzle part in its rightful place. An offset is added to the final position p of the puzzle part in the puzzle fixture $p_o = p + $ offset to prevent an overlap of parts. The offset takes into account how the parts are connected. After it is placed at p_o, it has to be pushed towards p. The first push is towards the closest fixed axis and the next it against the supporting piece – or axis. Doing a single pushing sequence may not be enough. For some reason the piece may get stuck, therefore, the sequence must be repeated. Detecting a stuck piece is simple, since the TCP reports the applied force and when a force above a threshold is detected a termination is triggered.

6 Related Work

ROS has become the robot middleware of choice for researcher and the industry. For example, MoveIt! [1] is a mobile manipulation software suitable for research and the industry. In addition to manipulation, the creation of behaviors is also a topic of interest, e.g. ROSCo [5]. Task level programming of robots is an important exercise for industrial applications. The authors of SkiROS [7] propose a paradigm based on a hierarchy movement primitives, skills and planning. For P&P tasks, the authors of [9] propose a manipulation planner under continuous grasps and placements, while a decomposition of the tasks is proposed by [4].

7 Conclusion

The tasks to be accomplished were different. However, while studying the problems to be solved, we identified common subtasks. In order to take advantage of that fact, a general system architecture to solve all tasks was developed. Additionally, the architecture works seamlessly in the ROS infrastructure. This solution allowed our team to achieve the 5th rank.

Acknowledgments. This research is supported by the European Union's FP7 under EuRoC grant agreement CP-IP 608849; and by the Foundation for Science and Technology in the context of UID/CEC/00127/2013 and Incentivo/EEI/UI0127/2014.

References

1. Chitta, S., Sucan, I., Cousins, S.: Moveit! [ros topics]. IEEE Robotics Automation Magazine **19**(1), 18–19 (2012)
2. Douglas, D.H., Peucker, T.K.: Algorithms for the reduction of the number of points required to represent a digitized line or its caricature. Cartographica: The International Journal for Geographic Information and Geovisualization (1973)
3. Henry, P., Krainin, M., Herbst, E., Ren, X., Fox, D.: RGB-D mapping: Using Kinect-style depth cameras for dense 3D modeling of indoor environments. I. J. Robotic Res. **31**(5), 647–663 (2012)
4. Lozano-Pérez, T., Jones, J.L., Mazer, E., O'Donnell, P.A.: Task-level planning of pick-and-place robot motions. IEEE Computer **22**(3), 21–29 (1989)
5. Nguyen, H., Ciocarlie, M., Hsiao, K., Kemp, C.: Ros commander (rosco): Behavior creation for home robots. In: ICRA, pp. 467–474 (May 2013)
6. Quigley, M., Gerkey, M., Conley, K., Faust, J., Foote, T., Leibs, J., Berger, E., Wheeler, R., Ng, A.: ROS: An open-source robot operating system. In: ICRA Workshop on Open Source Software, Kobe, Japan (May 2009)
7. Rovida, F., Chrysostomou, D., Schou, C., Bøgh, S., Madsen, O., Krüger, V., Andersen, R.S., Pedersen, M.R., Grossmann, B., Damgaard, J.S.: Skiros: A four tiered architecture for task-level programming of industrial mobile manipulators. In: 13th Internacional Conference on Intelligent Autonomous System, Padova (July 2013)
8. Russell, S., Norvig, P.: Artificial Intelligence: A Modern Approach, 3rd edn., Prentice Hall (2010)
9. Simeon, T., Cortes, J., Sahbani, A., Laumond, J.P.: A manipulation planner for pick and place operations under continuous grasps and placements. In: Proceedings of the Robotics and Automation, ICRA 2002, vol. 2, pp. 2022–2027 (2002)
10. Sural, S., Qian, G., Pramanik, S.: Segmentation and histogram generation using the hsv color space for image retrieval. In: Proceedings of the 2002 International Conference on Image Processing 2002, vol. 2, pp. II-589–II-592 (2002)
11. Suzuki, S., Abe, K.: Topological structural analysis of digitized binary images by border following. Computer Vision, Graphics and Image Processing **30**(1), 32–46 (1985)
12. Urmson, C., Baker, C.R., Dolan, J.M., Rybski, P.E., Salesky, B., Whittaker, W., Ferguson, D., Darms, M.: Autonomous Driving in Traffic: Boss and the Urban Challenge. AI Magazine **30**(2), 17–28 (2009)

Adaptive Behavior of a Biped Robot
Using Dynamic Movement Primitives

José Rosado[1(✉)], Filipe Silva[2], and Vítor Santos[3]

[1] Department of Computer Science and Systems Engineering, Coimbra Institute of Engineering,
IPC, 3030-199 Coimbra, Portugal
jfr@isec.pt
[2] Department of Electronics, Telecommunications and Informatics, Institute of Electronics
and Telematics Engineering of Aveiro, University of Aveiro, 3810-193 Aveiro, Portugal
fmsilva@ua.pt
[3] Department of Mechanical Engineering, Institute of Electronics and
Telematics Engineering of Aveiro, University of Aveiro, 3810-193 Aveiro, Portugal
vitor@ua.pt

Abstract. Over the past few years, several studies have suggested that adaptive behavior of humanoid robots can arise based on phase resetting embedded in pattern generators. In this paper, we propose a movement control approach that provides adaptive behavior by combining the modulation of dynamic movement primitives (DMP) and interlimb coordination with coupled phase oscillators. Dynamic movement primitives (DMP) represent a powerful tool for motion planning based on demonstration examples. This approach is currently used as a compact policy representation well-suited for robot learning. The main goal is to demonstrate and evaluate the role of phase resetting based on foot-contact information in order to increase the tolerance to external perturbations. In particular, we study the problem of optimal phase shift in a control system influenced by delays in both sensory information and motor actions. The study is performed using the V-REP simulator, including the adaptation of the humanoid robot's gait pattern to irregularities on the ground surface.

Keywords: Biped locomotion · Adaptive behavior · Movement primitives · Interlimb coordination · Phase resetting

1 Introduction

The coordination within or between legs is an important element for legged systems independently of their size, morphology and number of legs. Evidences from neurophysiology indicate that pattern generators in the spinal cord contribute to rhythmic movement behaviors and sensory feedback modulates proper coordination dynamics [1], [2]. In this context, several authors studied the role of phase shift and rhythm resetting. Phase resetting is a common strategy known to have several advantages in legged locomotion, namely by endowing the system with the capability to switch among different gait patterns or to restore coordinated patterns in the face of

F. Pereira et al. (Eds.) EPIA 2015, LNAI 9273, pp. 469–479, 2015.
DOI: 10.1007/978-3-319-23485-4_46

disturbances. In human biped walking the maintenance of reciprocal out-of-phase motions of the legs is critical for stable and efficient gait patterns [3], [4].

In the same line of thought, coordination dynamics is important for humanoid robots operating in real world environments. Further, this dynamics often needs to be adapted to account for variations in the environment conditions and external perturbations. Over the past few years different approaches to coordination has been applied to biped locomotion robots in which the emergence and change of coordination patterns are governed by dynamical equations [5], [6]. These authors have explored the role of phase resetting for adaptive walking based on foot-contact information using theoretical models and physical robots. Adaptation of the interlimb parameters largely restores symmetry of the gait cycle with inherent advantages for stability. In other words, the adjustment of the phase between legs helps to substantially increase the range of parameters (*e.g.*, average speed) and the tolerance to disturbances for which stable walking is possible.

In this paper, we propose a movement control approach that provides adaptive behavior by combining the modulation of dynamic movement primitives (DMP) and interlimb coordination with coupled phase oscillators. DMP appeared as a powerful tool for motion planning based on demonstration examples. This approach is currently used as a compact policy representation well-suited for robot learning. Here, rhythmic DMP are employed as trajectory representations learned in task-space from a single demonstration. Once learned, new movements are generated by simply modifying the parameters of the DMP. Adaptive biped locomotion based on phase resetting, which is the main focus of this paper, is studied and evaluated using the ASTI robot model in the V-REP simulation software [10]. The main goal is to demonstrate and evaluate the role of phase resetting based on foot-contact information in order to increase the tolerance to external perturbations. In particular, we study the problem of optimal phase shift in a control system influenced by delays in both sensory information and motor actions.

The remainder of the paper is organized as follows: Section 2 describes the proposed approach for trajectory formation based on rhythmic movement primitives learned in the task space and their modulation capabilities. Section 3 presents the interlimb coordination strategy based on coupled phase oscillators and phase resetting embedded with the movement control. In Section 4 the applicability of these concepts is demonstrated by numerical simulations. Section 5 concludes the paper and discusses future work.

2   Rhythmic Movement Primitives

2.1   Trajectory Formation

Dynamical system movement primitives have become a robust policy representation, for both discrete and periodic movements, that facilitates the process of learning and improving the desired behavior [7]. The basic idea behind DMP is to use an analytically well-understood dynamical system with convenient stability properties and modulate it with nonlinear terms such that it achieves a desired point or limit cycle attractor. The approach was originally proposed by Ijspeert et al. [8] and, since then, other mathematical variants have been proposed [9].

In the case of rhythmic movement, the dynamical system is defined in the form of a linear second order differential equation that defines the convergence to the goal g (baseline, offset or center of oscillation) with an added nonlinear forcing term f that defines the actual shape of the encoded trajectory. This model can be written in first-order notation as follows:

$$\tau\dot{z} = \alpha_z\left[\beta_z\left(g - y\right) - z\right] + f$$

$$\tau\dot{y} = z \qquad , \qquad (1)$$

where τ is a time constant the parameters $\alpha_z, \beta_z > 0$ are selected and kept fixed, such as the system converge to the oscillations given by f around the goal g in a critically damped manner. The forcing function f (nonlinear term) can be defined as a normalized combination of fixed basis functions:

$$f(\phi) = \frac{\sum_{i=1}^{N}\psi_i\omega_i}{\sum_{i=1}^{N}\psi_i}r$$

$$\psi_i(\phi) = \exp\left(-h_i\left(\cos(\phi - c_i) - 1\right)\right), \qquad (2)$$

where ω_i are adjustable weights, r characterizes the amplitude of the oscillator, ψ_i are von Mises basis functions, N is the number of periodic kernel functions, $h_i > 0$ are the widths of the kernels and c_i equally spaced values from 0 to 2π in N steps (N, h_i and c_i are chosen a priori and kept fixed). The phase variable ϕ bypasses explicit dependency on time by introducing periodicity in a rhythmic canonical system. This is a simple dynamical system that, in our case, is defined by a phase oscillator:

$$\tau\dot{\phi} = \Omega, \qquad (3)$$

where Ω is the frequency of the canonical system. In short, there are two main components in this approach: one providing the shape of the trajectory patterns (the transformation system) and the other providing the synchronized timing signals (the canonical system). In order to encode a desired demonstration trajectory y_{demo} as a DMP, the weight vector has to be learned with, for example, statistical learning techniques such as locally weighted regression (LWR) given their suitability for online robot learning.

2.2   Extension to Multiple Degrees-of-Freedom

The extension of the previous concepts to multiple degrees-of-freedom (DOF) is commonly performed by sharing one canonical system among all DOFs, while maintaining a set of transformation systems and forcing terms for each DOF. In brief, the canonical system provides the temporal coupling among DOFs, the transformation system achieves the desired attractor dynamics for each individual DOF and the respective forcing terms modulate the shape of the produced trajectories.

The adaptation of learned motion primitives to new situations becomes difficult when the demonstrated trajectories are available in the joint space. The problem occurs because, in general, a change in the primitive's parameters does not correspond to a meaningful effect on the given task. Having this in mind, the proposed solution is to learn the DMP in task space and relate their parameters to task variables. To concretely formulate the dynamical model, a task coordinate system is fixed to the hip section that serves as a reference frame where tasks are presented. The y-axis is aligned with the direction of movement, the z-axis is oriented downwards and the x-axis points towards the lateral side to form a direct system.

In line with this, a total of six DMP are learned to match the Cartesian trajectories of the lower extremities of both feet (end-effectors), using a single demonstration. It is worth note that a DMP contains one independent dynamical system per dimension of the space in which it is learned. At the end, the outputs of these DMP are converted, through an inverse kinematic algorithm, to the desired joint trajectories used as reference input to a low-level feedback controller. Fig. 1 shows the close match between the reference signals (solid lines) and the learned ones (dashed lines) as defined in the reference frame. The gray shaded regions show the phases of double-support.

Once the complete desired movement $\{y, \dot{y}, \ddot{y}\}$ is learned (*i.e.*, encoded as a DMP), new trajectories with similar characteristics can be easily generated. In this work, the DMP parameters resulting from the previous formulation (*i.e.*, amplitude, frequency and offset) are directly related to task variables, such as step length, hip height, foot clearance and forward velocity. For example, the frequency is used for speed up or slow down the motion, the amplitudes of the DMP associated with the y- and z-coordinates are used to modify the step length and the hip height (or foot clearance) of the support leg (or swing leg), respectively.

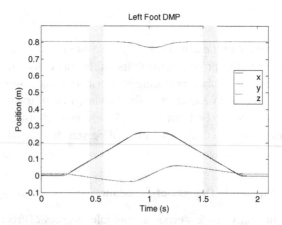

Fig. 1. Result of learning the single-demonstration: the task is specified by the x, y and z-coordinates of the robot's foot in the reference frame. Reference signal (solid line) and trained signal (dashed line) are superimposed. Gray shaded regions show double-support phases.

3   Adaptive Biped Locomotion

3.1   Modulation of the DMP Parameters

The formulation of DMP includes a few parameters which allow changing the learned behavior. This subsection provides examples of how new movements can be generated by simply modifying the parameters of a rhythmic DMP. These parameters can potentially be used to adapt the learned movement to new situations in order, for example, to adapt the final goal position, the movement amplitude or the duration of the movement. Therefore, the amplitude, frequency and offset parameters will be modified by scaling the corresponding parameters r, Ω and g, respectively.

Fig. 2 shows the time evolution of the rhythmic motion associated to a canonical system with frequency $\Omega = 2Hz$ and whose weights ω are learned to fit two different trajectories: the first is the sum of the first two harmonics of a rectangular wave defined in the interval between $t = 0$ and $t = 8s$; the second trajectory is the sum of the first three harmonics of a triangular wave defined in the interval between $t = 8s$ and $t = 16s$. The reference input signal is superimposed and vertical dashed lines marks events where the parameters are changed: the instant in which the learned signal doubles the amplitude r (at $t = 4$ s), change the reference signal (at $t = 8$ s) and modulates the baseline g (at $t = 12.5$ s). The main observation is that the changes of parameters result in smooth variations of the trajectory $y(t)$ to be reproduced by the robot.

An example of the use of DMP modulation to biped locomotion is shown in Fig 3. The original signal that was used to train the DMP was modified in order to change the relative step length of each leg and the corresponding foot clearances. This strategy allows the robot to turn with a smooth curve around an obstacle placed on its path (a video is available at http://www.youtube.com/watch?v=Y3Y-6WNhxHE).

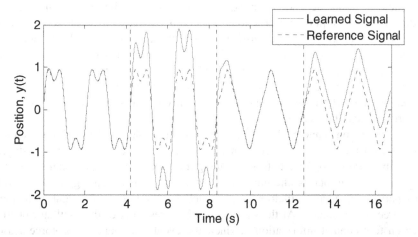

Fig. 2. Time courses of the rhythmic dynamical system with continuous learning (solid line) and the input reference signal (dashed-line). The vertical dashed-lines in the plot indicate, from left to right, the instant in which the learned signal doubles the amplitude r (at $t = 4$ s), change the reference signal (at $t = 8$ s) and modulates the baseline g (at $t = 12.5$ s).

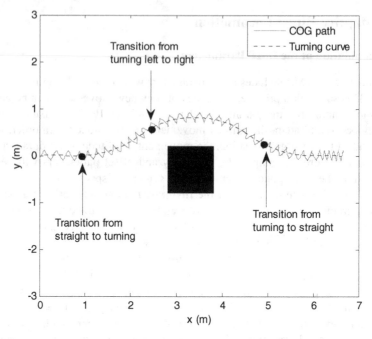

Fig. 3. View of the movement path of the robot's COG projected on the ground and the corresponding turning curve. The black box represents an obstacle placed on the path.

3.2   Interlimb Coordination

DMP exhibit a desirable property in the context of robot learning from demonstration: the system does not depend on an explicit time variable, giving them the ability to handle spatial or temporal perturbations. This property makes them attractive in order to create smooth kinematics control policies that can robustly replicate and adapt demonstrations. However, adaptation of inter-limb parameters is essential to restore the symmetry of the gait cycle in order to reduce the likelihood of becoming unstable. For example, whenever a leg is constrained, for example, by external perturbations, compensatory reactions in the other leg are expected such as to restore the out-of-phase relationship between legs.

In this study, one canonical system per leg and multiple transformation systems associated with the x-, y- and z-coordinates of the robot's end-effectors are adopted. Intra-limb coordination results from planning trajectories in the Cartesian space, constraining the leg to act as one unit. Phase coordination between legs is provided by two separate canonical oscillators coupled such that the left and the right limbs move 180 degrees out-of-phase. At the same time, phase resetting of the oscillator phase is based on foot-contact information (a kinematic event) that depends on force sensors placed on the feet.

As a result, the dynamics of the phase oscillators in (3), for the left and the right leg, are modified according to:

$$
\begin{aligned}
\tau\dot{\phi}_{left} &= \Omega - K_\phi \sin\left(\phi_{left} - \phi_{right} - \pi\right) - \left(\phi_{left} - \phi^{contact}\right)\delta\left(t - t_{left}^{contact} - \Delta t\right) \\
\tau\dot{\phi}_{right} &= \Omega - K_\phi \sin\left(\phi_{right} - \phi_{left} - \pi\right) - \left(\phi_{right} - \phi^{contact}\right)\delta\left(t - t_{rightt}^{contact} - \Delta t\right)
\end{aligned}
\tag{4}
$$

where K_ϕ is the coupling strength parameter ($K_\phi > 0$), $\phi^{contact}$ is the phase value to be reset when the foot touch the ground, $\delta(\cdot)$ is the Dirac's delta function, $t_i^{contact}$ ($i = \text{left, right}$) is the time when the foot touch the ground and Δt is a factor used to study the influence of delays in both sensory information an motor control.

4 Numerical Simulations

The applicability of these concepts is demonstrated by numerical simulations performed in V-REP, Virtual Robot Experimentation Platform [17], using the ASTI robot model available in their libraries. Two specific experiments are conducted demonstrate the important role of phase resetting to achieve adaptive locomotion subject to perturbations.

4.1 Robustness Against Disturbing Forces

In the first experiment, an external force is applied to the trunk section of the humanoid robot in two situations: a horizontal force is applied on the direction of the movement or, instead, on the backward direction. Specifically, after the robot has achieved a steady-state stable walking, a horizontal force is applied for 0.1 s at its center-of-gravity (COG). The instant in which this external force is applied varies, from the moment the left foot leaves the ground to the instant when the same foot touches the ground, in intervals of 50ms. In both cases, the maximum force tolerated by robot without falling was measured, with and without phase resetting. Fig. 4 shows the increase on the tolerated forces with phase reset for the backward and the forward force application, respectively. It is worth noting that the tolerance to disturbing forces is greatly affected by the phase value to be reset requiring its optimization.

The result of applying a force to the robot, with and without phase resetting is observed on the variation of the COG velocity on the direction of movement (see Fig. 5). Here a force was applied around the 11.6s and we can see that without phase resetting the robot lost the stability with a high increase on the COG velocity leaving to the fall a few seconds after (blue curve). With phase resetting, the impact produces a moderate increase on the COG velocity, but after a few seconds the normal cyclic pattern is recovered. Also, the coupling between the phase oscillators recovers the phase offset of 180° between each leg. In fact, as we can see on the black curve, the phase resetting produces an increase on the phase offset to around 205° degrees at the moment of the force application and the coupling returns this offset to the 180° after a few seconds.

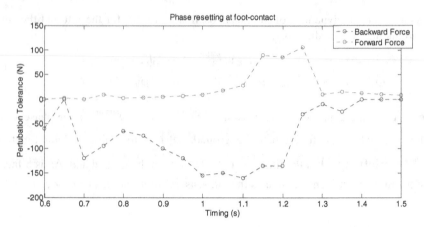

Fig. 4. Additional tolerance to perturbation forces applied at different instants of the movement cycle when using phase resetting.

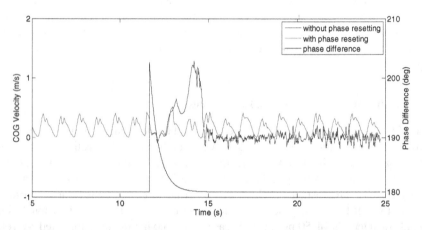

Fig. 5. Velocity of the COG in the direction of the motion with a perturbation force applied at 11.5s without and with the use of phase resetting; the time course of the phase difference between the oscillators is represented in a different vertical axis.

4.2   Adaptation to Irregular Terrains

Biped walking in irregular terrains depends on prediction about when the swing foot touches the ground. In the first experiment, the humanoid robot walks over a level surface when it finds a small step of 2 cm high used to approximate irregularities of the environments. The learned DMP phase is reset online to properly incorporate the sensory information from the force sensors mounted on the robot feet. More concrete-ly, the dynamic event corresponds to foot-contact information at the instant of impact of the swing foot with the ground. Fig. 6 illustrates snapshots of the robot walking response without and with phase reset. It was found that using the locomotion pattern as defined by the learned DMP, without any modulation, the robot tolerates step irregularities up to 0.5 cm height. Here, the proposed strategy is to change the phase of the canonical system to a value corresponding to the point of ground contact on the normal signal generation (when there's no early contact with the ground).

In the second experiment, it is examined how the phase reset of the canonical oscillator provides changes on the DMP that allows the robot to overcome a set of irregularities that assemble like a set of steps of a small staircase. These consist of two consecutive steps up followed by two steps down, each one with 2cm high. Beside this the robot system is also supposed to receive visual information regarding the stairs location and height in order to modify the basic gait pattern (foot clearance and step size). Fig. 7 shows the path the robot has to go through and the sequence of captured images of the robot stepping on the first step, followed by the second step and after a few steps on this, the first down step followed by the final down step takes the robot to the ground level. As in the previous example, a phase reset is applied as soon the robot senses the foot as hit the ground sooner than expected.

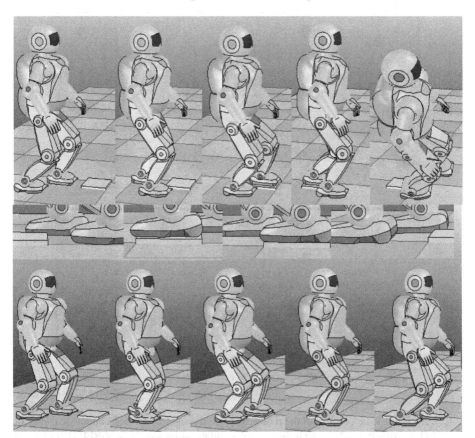

Fig. 6. Snapshots of the robot walking on a level surface when it finds a small step 2 cm high that disturbs its balance (top: without phase resetting; bottom: with phase resetting). Numerical simulations performed in V-REP [10].

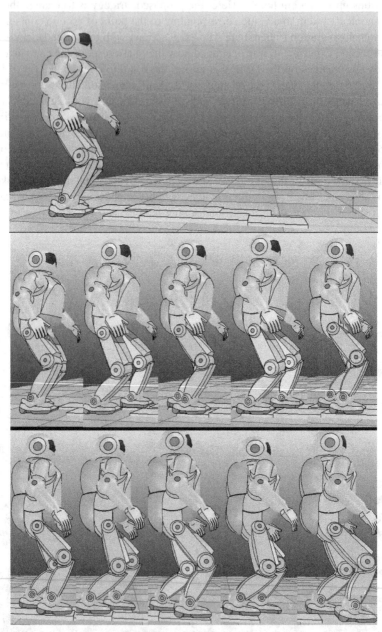

Fig. 7. Top: full view of the path the robot has to go through; center and bottom: sequence of the robot walking through the path. Numerical simulations performed in V-REP [10] (a video of this experiment is available at: http://www.youtube.com/watch?v=WjBq27hJAJE).

5   Conclusions

This paper presents a study in which online modulation of the DMP parameters and interlimb coordination trough phase coupling providing adaptation of biped locomotion with improvement to external perturbations. By using the DMP in the task space, new tasks are easily accomplished by modifying simple DMP parameters that directly relate to the task, such as step length, velocity, foot clearance. By introducing coupling between members and using phase reset we have shown that adaption to irregularities on the terrain are successful. The phase resetting methodology also allowed increasing the tolerance to external perturbations, such as forces that push or pull the robot on the direction of the movement. Future work will address problems like the role of phase resetting and DMP parameters change in sudden changes on the trunk mass, stepping up stairs, climbing up and down on ramps. Demonstrations from human data behavior will be collected using a VICON system and use to train the DMP in new tasks.

Acknowledgements. This work is partially funded by FEDER through the Operational Program Competitiveness Factors - COMPETE and by National Funds through FCT - Foundation for Science and Technology in the context of the project FCOMP-01-0124-FEDER-022682 (FCT reference Pest-C/EEI/UI0127/ 2011).

References

1. Hultborn, H., Nielsen, J.: Spinal control of locomotion – from cat to man. Acta Physiologica **189**(2), 111–121 (2007)
2. Grillner, S.: Locomotion is vertebrates: central mechanisms and reflex interaction. Physiological Reviews **55**(2), 247–304
3. Aoi, S., Ogihara, N., Funato, T., Sugimoto, Y., Tsuchiya, K.: Evaluating the functional roles of phase resetting in generation of adaptive human biped walking with a physiologically based model of the spinal pattern generator. Biological Cybernetic **102**, 373–387 (2010)
4. Yamasaki, T., Nomura, T., Sato, S.: Possible functional roles of phase resetting during walking. Biological Cybernetic **88**, 468–496 (2013)
5. Aoi, S., Tsuchiva, K.: Locomotion control of a biped robot using nonlinear oscillators. Autonomous Robots **19**(3), 219–232 (2005)
6. Nakanishi, J., Morimoto, J., Endo, G., Cheng, G., Schaal, S., Kawato, M.: A framework for learning biped locomotion with dynamical movement primitives. International Journal of Humanoid Robots (2004)
7. Argall, B., Chernova, S., Veloso, M., Browning, B.: A survey of robot learning from demonstration. Robotics and Autonomous Systems **57**(5), 469–483 (2009)
8. Ijspeert, A., Nakanishi, J., Schaal, S.: Movement imitation with nonlinear dynamical systems in humanoid robots. In: Proceedings of the 2002 IEEE International Conference on Robotics and Automation, pp. 1398–1403 (2002)
9. Ijspeert, A., Nakanishi, J., Hoffmann, H., Pastor, P., Schaal, S.: Dynamical movement primitives: learning attractor models for motor behaviors. Neural Computation **25**, 328–373 (2013)
10. Rohmer, E., Singh, S., Freese, M.: V-REP: A versatile and scalable robot simulation framework. In: IEEE/RSJ International Conference on Intelligent Robots and Systems, pp. 1321–1326 (2013)

Probabilistic Constraints for Robot Localization

Marco Correia, Olga Meshcheryakova, Pedro Sousa, and Jorge Cruz[✉]

NOVA Laboratory for Computer Science and Informatics,
DI/FCT/UNL, Caparica, Portugal
{mvc,jcrc}@fct.unl.pt

Abstract. In robot localization problems, uncertainty arises from many factors and must be considered together with the model constraints. Probabilistic robotics is the classical approach for dealing with hard robotic problems that relies on probability theory. This work describes the application of probabilistic constraint techniques in the context of probabilistic robotics to solve robot localization problems. Instead of providing the most probable position of the robot, the approach characterizes all positions consistent with the model and their probabilities (in accordance with the underlying uncertainty). It relies on constraint programming to get a tight covering of the consistent regions combined with Monte Carlo integration techniques that benefit from such reduction of the sampling space.

Keywords: Probabilistic robotic · Constraint programming · Robot localization

1 Introduction

Uncertainty plays a major role in modeling most real-world continuous systems and, in particular, robotic systems. A reliable framework for decision support must provide an expressive mathematical model for a sound integration of the system and uncertainty.

Stochastic approaches associate a probabilistic model to the problem and reason on approximations of the most likely scenarios. In highly nonlinear problems such approximations may miss relevant satisfactory scenarios leading to erroneous decisions. In contrast, constraint programming (CP) approaches reason on safe enclosures of all consistent scenarios. Model-based reasoning and what-if scenarios are supported through safe constraint techniques, which only eliminate scenarios that do not satisfy model constraints. However, safe reasoning based exclusively on consistency may be inappropriate to sufficiently reduce the space of possibilities on large uncertainty settings.

This paper shows how probabilistic constraints can be used for solving global localization problems providing a probabilistic characterization of the robot positions (consistent with the environment) given the uncertainty on the sensor measurements.

© Springer International Publishing Switzerland 2015
F. Pereira et al. (Eds.) EPIA 2015, LNAI 9273, pp. 480–486, 2015.
DOI: 10.1007/978-3-319-23485-4_47

2   Probabilistic Robotics

Probabilistic robotics [1] is a generic approach for dealing with hard robotic problems that relies on probability theory to reason with uncertainty in robot perception and action. The idea is to model uncertainty explicitly, representing information by probability distributions over all space of possible hypotheses instead of relying on best estimates.

Probabilistic approaches are typically more robust in the face of sensor limitations and noise, and often scale much better to unstructured environments. However, the required algorithms are usually less efficient when compared with nonprobabilistic algorithms, since entire probability densities are considered instead of best estimates. Moreover, the computation of probability densities require working exclusively with parametric distributions or discretizing the probability space representation.

In global localization problems, a robot is placed somewhere in the environment and has to localize itself from local sensor data. The probabilistic paradigm maintains over time, the robot's location estimate which is represented by a probability density function over the space of all locations. Such estimate is updated whenever new information is gathered from sensors, taking into account its underlying uncertainty.

A generic algorithm known as Bayes filter [2] is used for probability estimation. The Bayes filter is a recursive algorithm that computes a probability distribution at a given moment from the distribution at the previous moment accordingly to the new information gathered. Two major strategies are usually adopted for the implementation of Bayes filters in continuous domains: Gaussian filters and nonparametric filters.

Gaussian techniques share the idea that probabilities are represented by multivariate normal distributions. Among this techniques the most popular are (Extended) Kalman Filters [3,4] which are computationally efficient but inadequate for problems where distributions are multimodal and subject to highly nonlinear constraints.

Nonparametric techniques [5,6] approximate continuous probabilities by a finite number of values. Representatives of these techniques for robot localization problems are Grid and Monte Carlo Localization algorithms [7,8]. Both techniques do not make any assumptions on the shape of the probability distribution and have the property that the approximation error converges uniformly to zero as the the number of values used to represent the probabilistic space goes to infinity. The computational cost is determined by the granularity of the approximation (the number of values considered) which is not easy to tune depending both on the model constraints and on the underlying uncertainty.

3   Constraint Programming

Continuous constraint programming [9,10] has been widely used to model safe reasoning in applications where uncertainty on the values of the variables is modeled by intervals including all their possibilities. A Continuous Constraint

482   M. Correia et al.

Satisfaction Problem (CCSP) is a triple$\langle X, D, C\rangle$ where X is a tuple of n real variables $\langle x_1, \cdots, x_n\rangle$, D is a Cartesian product of intervals $X_i \times \cdots \times X_n$ (a box), where each X_i is the domain of x_i and C is a set of numerical constraints (equations or inequalities) on subsets of the variables in X. A solution of the CCSP is a value assignment to all variables satisfying all the constraints in C. The feasible space F is the set of all CCSP solutions within D.

Continuous constraint reasoning relies on branch-and-prune algorithms [11] to obtain sets of boxes that cover the feasible space F. These algorithms begin with an initial crude cover of the feasible space (D) which is recursively refined by interleaving pruning and branching steps until a stopping criterion is satisfied. The branching step splits a box from the cover into sub-boxes (usually two). The pruning step either eliminates a box from the covering or reduces it into a smaller (or equal) box maintaining all the exact solutions. Prunning is achieved by performing constraint propagation [12] based on interval analysis methods [13].

Algorithm 1. $probDist(D, C, G)$

1 $S \leftarrow Branch\&Prune(D, C)$;
2 $\forall_{1 \leq i_1 \leq g_1 \ldots 1 \leq i_n \leq g_n} M_{i_1, \ldots, i_n} \leftarrow 0$;
3 $P \leftarrow 0$;
4 **foreach** $B \in S$ **do**
5 $\quad \langle i_1, \ldots, i_n \rangle \leftarrow getIndex(B)$;
6 $\quad M_{i_1, \ldots, i_n} \leftarrow MCIntegrate(B)$;
7 $\quad P \leftarrow P + M_{i_1, \ldots, i_n}$;
8 **if** $P = \emptyset$ **then return** M;;
9 $\forall_{1 \leq i_1 \leq g_1 \ldots 1 \leq i_n \leq g_n} M_{i_1, \ldots, i_n} \leftarrow M_{i_1, \ldots, i_n}/P$;
10 **return** M;

Probabilistic Constraint Programming. In classical CCSPs, uncertainty is modeled by intervals that represent the domains of the variables. Nevertheless this paradigm cannot distinguish between different scenarios and all combination of values within such enclosure are considered equally plausible. In this work an extension of the classical CP paradigm is used to support probabilistic reasoning. Probabilistic constraint programming [14] associates a probabilistic space to the classical CCSP by defining an appropriate density function. A probabilistic constraint space is a pair $\langle \langle X, D, C \rangle, f \rangle$, where $\langle X, D, C \rangle$ is a CCSP and f a p.d.f. defined in $\Omega \supseteq D$ such that: $\int_\Omega f(\mathbf{x})d\mathbf{x} = 1$. The constraints C specify an event $\mathcal{H}$ whose probability can be computed by integrating f over its feasible space, $P(\mathcal{H}) = \int_\mathcal{H} f(\mathbf{x})d\mathbf{x}$. The probabilistic constraint framework peforms constraint propagation to get a box cover of the region of integration $\mathcal{H}$ and compute the overall integral by summing up the contributions of each box in the cover. In this work, classical Monte Carlo methods [15] are used to estimate the value of the integrals at each box, by randomly selecting N points in the multidimensional space and averaging the function values at these points. The success of this technique relies on the reduction of the sampling space where a pure Monte Carlo method is not only hard to tune but also impractical in small error settings.

Probability distributions are computed by algorithm 1 which assumes a grid over the feasible region and computes a conditional probability distribution of the random vector X given the event $\mathcal{H}$ that satisfies all constraints in C. The grid is specified by the input $G = \langle g_1, \ldots, g_n \rangle$ which is an array that defines the number or partitions considered at each dimension. The output matrix M is the conditional probability at each grid cell. The algorithm first computes a grid box cover S for the feasible space of the model constraints C (line 1). Function *Branch&Prune* (see [14] for details) is used with a grid oriented parametrization, i.e., it splits the boxes in the grid and choose to process only those boxes that are not yet inside a grid cell, stopping when there are no more eligible boxes. Matrix M is initialized to zero (line 2) as well as the normalization factor P that will contain, in the end, the overall sum of all non normalized parcels (line 3). For each box B in the cover S (lines 4-7), its corresponding index of the matrix cell is identified (line 5) and its probability is computed by function *MCIntegrate* (that implements the Monte Carlo method to compute the contribution of B) and assigned to the value in that cell (line 6). The normalization factor is updated (line 7) and used in the end to normalize the computed probabilities (line 9).

4   Probabilistic Constraints for Robot Localization

The location of the robot is confined to a box that characterises the environment's coordinate system on which prior knowledge is represented as a set of segments (walls). A robot pose is a triplet $\langle x, y, \alpha \rangle$ where (x, y) defines its location and $\alpha \in [0, 2\pi]$ characterises its heading direction. This work focus on the information gathered by a ladar which provides a panoramic view of the environment gathering distance measurements within a given maximum range δ_{max} (for a given direction it provides the distance to the closest wall in that direction). We consider n ladar measurements, each represented as a pair: the angle relative to the robot heading direction and the recorded distance.

Figure 1(above) illustrates 3 environments confined to the box $[0,1000] \times [0,1000]$ and their respective robot poses. The robot is pictured as a small circle centered at its location coordinates (with radius 30) and its heading direction is shown by the inner tick. The straight dotted lines illustrate the distance measurements that would be captured by a ladar from the robot pose. It considers 7 measurements with direction angles covering the ladar panoramic range and a maximum distance range $\delta_{max} = 300$ (dotted circle).

In general, the information provided by the robot sensors is subject to different sources of noise. We assume that all ladar measurement errors are independent and normally distributed with 0 mean and σ standard deviation. For n ladar measurements, their joint error probability density is the Gaussian function $\left(\sigma \sqrt{2\pi} \right)^{-n} e^{-\frac{1}{2\sigma^2} \Sigma_i \epsilon_i^2}$ where ϵ_i is the error committed in the ith measurement. Without loss of generality, other pdf could be used to model the uncertainty on other sensor measurements.

Figure 1(below) illustrates the results (projected in the xy plan). It shows the grid over the initial coordinates box and the xy cells consistent with the measurements with grey levels that reflect the computed probability.

The Gaussian pdf for the current pose $\langle x, y, \alpha \rangle$ of the robot given n ladar measurements (used by the function *MCIntegrate*) is computed as follows. For each ladar measurement i, the direction of the observation α_i is given by the robot angle α plus the relative angle of the measurement. The predicted value for the distance is computed as the distance from the robot pose to the closest object in the direction α_i (or is δ_{max} if such distance exceeds the maximum ladar range). The difference between the predicted and the distance recorded by the ladar is the measurement error and its square is accumulated for all measurements and used to compute the pdf.

The specialized function that narrows a domains box accordingly to the ladar measurements maintains a set of numerical constraints that can be enforced over the variables of the problem and then calls the generic interval arithmetic narrowing procedure [16]. The numerical constraints may result from each ladar measurement. Firstly a geometric function is used to determine which of the walls in the map can eventually be seen by a robot positioned in the box with an angle of vision within the range of the robot pose angle plus the relative angle of the ladar measurement. If no wall can be seen, the predicted distance is the maximum ladar range δ_{max} and a constraint is added to enforce the error between the predicted and the ladar measurement not to exceed a predefined threshold ϵ_{max}. If it is only possible to see a single wall, an adequate numerical constraint is enforced to restrain the error between the ladar measurement and the predicted distance for a pose $\langle x, y, \alpha \rangle$. Notice that whenever there is the possibility of seeing more that one wall it cannot be decided which constraint to enforce, and the algorithm proceeds without associating any constraint to the ladar measurement.

5   Experimental Results

The probabilistic constraint framework was applied on a set of global localization problems covering different simulated environments and robot poses and are illustrative of the potential and limitations of the proposed approach. The algorithms were implemented in C++ over the RealPaver constraint solver [16] and the experiments were carried out on an Intel Core i7 CPU at 2.4 GHz. Our grid approach adopted as reference a grid granularity commonly used in indoor environments [1]: 15 cm for the xy dimensions (10 units represents 1.5 cm), and 4 degrees for the rotational dimension. Based on previous experience on the hybridization of Monte Carlo techniques with constraint propagation we adopted a small sampling size of $N = 100$.

In all our experiments increasing the sampling size did not improve the quality of the results. Similarly the reference value of 4 degrees for the grid size of the rotational dimension was fixed since coarser grids prevented constraint pruning and finer grids increased the computation time without providing better results. In the following problems, to illustrate the effect of the resolution of the xy grid we consider, apart from the reference grid size, a 4 times coarser grid (60 cm) and a 4 times finer grid (3.75 cm).

Fig. 1(left) illustrates a problem where, from the given input, the robot location can be circumscribed to a unique compact region. The results obtained with the coarse grid clearly identify a single cell enclosing the simulated robot location. The obtained enclosure for the heading direction is guaranteed to be between 40 and 48 degrees (not shown). With the reference and the fine-grained grids the results were similar. The CPU time was 5s, 10s and 30s for increasing grid resolutions.

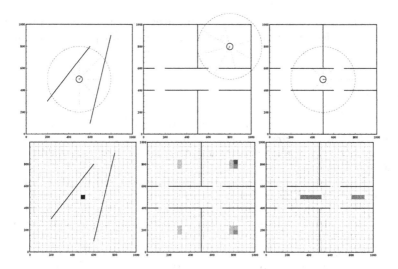

Fig. 1. (above) environment and robot poses with the distance measurements captured by the robot sensors; (below) computed solutions given the environment and the measurements.

Fig. 1(center) illustrates a problem where local symmetry of the environment (with 4 similar rooms) makes it impossible to localize the robot into a unique compact region, coexisting several possible alternatives. In this case, all the consistent locations were identified despite the adopted grid resolution. The predicted heading directions point towards the respective room entrance. The CPU time was 5s, 8s and 13s for increasing grid resolutions.

Fig. 1(right), illustrates a problem where there is a continuum of indistinguishable locations for the robot (along the corridor). Again, despite the adopted grid resolution, all the consistent locations were successfully identified. Notice that the possible locations are represented as two sets of adjacent cells and the probability of the cells in the left is larger because here the robot may be faced towards two oposite directions (left or right) whereas within the right cells the robot must be heading left. This example reflect the limitations of grid approaches to represent a continuum of possibilities - the CPU time severely degradates for increasing grid resolutions: about 4s, 16s and 117s.

6   Conclusions

In this paper we propose the application of probabilistic constraint programming to probabilistic robotics. We show how the approach can be used to support sound reasoning in global localization problems integrating prior knowledge on the environment with the uncertainty information gathered by the robot sensors. Preliminary experiments on a set of simulated problems highlighted the potential and limitations of the approach.

In the future the authors aim to extend the approach to address kinematic constraints and their underlying uncertainty. Probabilistic constraint reasoning has the potential to combine all sources of uncertainty information providing a valuable probabilistic characterization of the set of robot poses consistent with the kinematic constraints.

References

1. Thrun, S., Burgard, W., Fox, D.: Probabilistic Robotics. The MIT Press (2006)
2. Särkkä, S.: Bayesian filtering and smoothing. Cambridge University Press (2013)
3. Kalman, R.E.: A new approach to linear filtering and prediction problems. ASME Journal of Basic Engineering (1960)
4. Julier, S.J., Jeffrey, Uhlmann, K.: Unscented filtering and nonlinear estimation. Proceedings of the IEEE, 401–422 (2004)
5. Kaplow, R., Atrash, A., Pineau, J.: Variable resolution decomposition for robotic navigation under a pomdp framework. In: IEEE Robotics and Automation, pp. 369–376 (2010)
6. Arulampalam, M., Maskell, S., Gordon, N.: A tutorial on particle filters for online nonlinear/non-gaussian bayesian tracking. IEEE Trans. Signal Proc. **50**, 174–188 (2002)
7. Wang, Y., Wu, D., Seifzadeh, S., Chen, J.: A moving grid cell based mcl algorithm for mobile robot localization. In: IEEE Robotics and Biomimetics, pp. 2445–2450 (2009)
8. Dellaert, F., Fox, D., Burgard, W., Thrun, S.: Monte carlo localization for mobile robots. In: IEEE Robotics and Automation, pp. 1322–1328 (1999)
9. Lhomme, O.: Consistency techniques for numeric CSPs. In: Proc. of the 13th IJCAI (1993)
10. Benhamou, F., McAllester, D., van Hentenryck, P.: CLP(intervals) revisited. In: ISLP, pp. 124–138. MIT Press (1994)
11. Hentenryck, P.V., Mcallester, D., Kapur, D.: Solving polynomial systems using a branch and prune approach. SIAM Journal Numerical Analysis **34**, 797–827 (1997)
12. Benhamou, F., Goualard, F., Granvilliers, L., Puget, J.F.: Revising hull and box consistency. In: Procs. of ICLP, pp. 230–244. MIT (1999)
13. Moore, R.: Interval Analysis. Prentice-Hall, Englewood Cliffs (1966)
14. Carvalho, E.: Probabilistic Constraint Reasoning. PhD thesis, FCT/UNL (2012)
15. Hammersley, J., Handscomb, D.: Monte Carlo Methods. Methuen, London (1964)
16. Granvilliers, L., Benhamou, F.: Algorithm 852: Realpaver an interval solver using constraint satisfaction techniques. ACM Trans. Mathematical Software **32**(1), 138–156 (2006)

Detecting Motion Patterns in Dense Flow Fields: Euclidean Versus Polar Space

Andry Pinto[✉], Paulo Costa, and Antonio Paulo Moreira

Robotics and Intelligent Systems - INESCTEC and Faculty of Engineering,
University of Porto, Porto, Portugal
{andry.pinto,paco,amoreira}@fe.up.pt

Abstract. This research studies motion segmentation based on dense
optical flow fields for mobile robotic applications. The optical flow is usually represented in the Euclidean space however, finding the most suitable motion space is a relevant problem because techniques for motion
analysis have distinct performances. Factors like the processing-time and
the quality of the segmentation provide a quantitative evaluation of the
clustering process. Therefore, this paper defines a methodology that evaluates and compares the advantage of clustering dense flow fields using
different feature spaces, for instance, Euclidean and Polar space. The
methodology resorts to conventional clustering techniques, Expectation-
Maximization and K-means, as baseline methods. The experiments conducted during this paper proved that the K-means clustering is suitable
for analyzing dense flow fields.

Keywords: Visual perception · Optical flow · Motion segmentation

1 Introduction

The work presented by this research studies and compares techniques for motion
analysis and segmentation using dense optical flow fields. Motion segmentation is
the process of dividing an image into different regions using motion information
in a way that each region presents homogeneous characteristics. Two techniques
are considered during this research, the Expectation-Maximization (EM) and K-
means. The performance and the behavior of these techniques is well-known in
the scientific community since they have been applied in countless applications of
machine learning; however, this paper presents their performance for clustering
dense motion fields obtained by a realistic robotic application [5]. The optical
flow technique [7] is used this paper because it estimates dense flow fields in short
time and makes it suitable for robotic applications without specialized computer
devices however, the quality of the flow fields that are obtained is lower when
compared to the most recent methods.

Four different feature spaces are considered in this research: the motion vector
is represented in Cartesian space or Polar space, and the feature can have the
positional information of the image location. Mathematically, this is represented

© Springer International Publishing Switzerland 2015
F. Pereira et al. (Eds.) EPIA 2015, LNAI 9273, pp. 487–492, 2015.
DOI: 10.1007/978-3-319-23485-4_48

by the following features: $f_l^c = (\bar{x}, \bar{y}, \bar{u}, \bar{v})$, $f_l^p = (\bar{x}, \bar{y}, \bar{m}, \bar{\phi})$, $f^c = (u, v)$ and $f^p = (m, \phi)$, where (x, y) is the image location, (u, v) is the flow vector in Cartesian space and the (m, ϕ) is the flow vector in Polar space (magnitude and angle). In most cases, a normalization is performed [3] due to different physical meanings of the feature's components: $\bar{v} = \frac{mean(v)}{\sigma_v}$,, where σ_v is the standard deviation of the component v. As can be noticed, the influence of this normalization on the segmentation procedure is also analyzed.

Therefore, contributions of this articles include: study about motion analysis in dense optical flow fields and for a practical use in a mobile robot. The goal is to segment different objects according to their motion coherence; comparison about the most suitable feature space for clustering dense flow fields; extensive qualitative and quantitative evaluations considering several baseline and pixel-wise clustering techniques, namely the K-means and the Expectation-Maximization.

2   Related work

The work [3] evaluates the performance of several clustering methods namely, K-means, self-tuning spectral clustering and nonlinear dimension reduction. Authors defend that one most important factor for clustering dense flow fields is the proper choice of the distance measure. They considered the feature space to be formed by the pixel coordinates and motion vectors, whose values are normalized by taking into consideration the mean and standard deviation of each feature. Results show the difficulty of segmenting dense flow fields because no technique was outperformed and, thereby, the choice of the most suitable clustering technique and distance metric must be investigated for a specific context and environment. An accurate segmentation technique that resorts to dense flow fields and uses long term point trajectories is presented in [1]. By clustering trajectories over time, it is possible to use a metric that measures the distance between these trajectories. The work in [4] proposes a sparse approach for detecting salient regions in the sequences. Feature points are tracked over time in order to pursue saliency detection as violation of co-visibility. The evaluation of the method shows that it cannot achieve a real-time computational performance since it took 32.6 seconds to process a single sequence. The [2] addresses the problem of motion detection and segmentation in dynamic scenes with small camera movements and resorting to a set of moving points. They use the Lukas-Kanade optical flow to compute the sparse flow field (features obtained by the Harris corner). Afterwards, these points are clustered using a variable bandwidth mean-shift technique and, finally, the cluster segmentation is conducted using graph cuts.

3   Practical Results

An extensive set of experiments was conducted as part of this research. They aimed to analyze and understand the behavior of two parametric techniques for motion analysis in a robotic and surveillance context [7]. The EM and the K-means are used in this research as baselines for segmenting dense optical flow fields and they were implemented as standard functions. In the first experiment, the assessment was performed using an objective (quantitative) and subjective (qualitative) evaluation. The objective metric F-score [6] provides quantitative quality evaluations of the clustering results since it weights the average of precision and recall and reaches the best value at 1. The baseline methods provide a pixel-wise segmentation and factors such as the computational effort and the quality of the visual clustering are considered. Experiments were performed [1] considering four feature spaces: f_l^c, f_l^p, f^c and f^p.

The results start by demonstrating the segmentation performed by the EM and K-means in several testing sequences that capture a real surveillance scenario (indoor). Figures 1(a), 1(b) and 1(c) depict only three dense flow fields that were obtained from these sequences. Using the EM and the K-means for clustering the flow field represented in f^p have resulted in figures 2(d) to 2(f) and 3(d) to 3(f), respectively. Figures 2(a) to 2(c) and 3(a) to 3(c) depict results for the Cartesian space.

As can be noticed, the segmentation conducted by the EM in f^c do not originate a suitable segmentation because the clusters of people appear larger and they have spatially isolated regions that are meaningless (hereafter, called clustering noise). This issue is more depicted in figure 2(c) however, the same flow field segmented in Polar space originated a result that represents more faithfully the person's movement since it is less affected by meaningless and isolated regions, see figure 2(f). On the other hand, the visual illustration of the motion segmentation conducted by the K-means in f^p is similar to f^c. The qualitative analysis of these results is not possible however; and independently of the feature space that is considered (f^p or f^c), the result of the K-means is better than the EM since the person's movements are more faithfully depicted. In addition, the clustering noise of the K-means is lower than the EM for these two feature spaces. The K-means is simpler than the EM however, it is a powerful technique to cluster the input dataset.

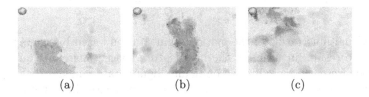

(a)   (b)   (c)

Fig. 1. Figures of the first row depict dense flow fields that were obtained from the technique proposed in [7]. The HSV color space is used to represent the direction (color) and magnitude (saturation) of the flow.

---

[1] The results in this section were obtained using an I3-M350 2.2GHz and manually annotated images.

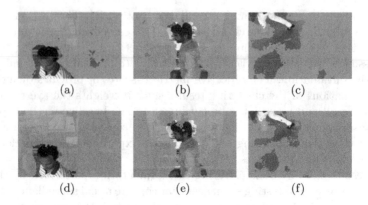

Fig. 2. Comparison between the EM in f^c (first row) and EM in f^p (second row). Motion segmentation for the flow fields represented in figures 1(a), 1(b) and 1(c).

Fig. 3. Comparison between the K-means in f^c (first row) and K-means in f^p (second row). Motion segmentation for the flow fields represented in figures 1(a), 1(b) and 1(c).

The visual illustration of the motion segmentation based K-means produced inconclusive results in terms of the best feature space. Therefore, quantitative evaluations were conducted based on manually annotated images that represent the ground truth of the segmentation. Table 1 presents the results evaluated using the objective metric F-score, for the spaces f^c and f^p. The superscripts "c" and "p" represent the segmentation result in Cartesian and Polar space. The motion segmentation conducted by the K-means^c was close to the K-means^p, although with a lower amount of noise. This table confirms the visual illustration of the previous results since, the EM^p had a F-score that is in average 0.178 higher than EM^c, while the K-means^c produced clusters with a F-score that is 0.040 higher than K-means^p. This means, the F-score of the EM^c and

EM^p was 0.630 and 0.809; the K-means^c and K-means^p was 0.893 and 0.852 (in average). Therefore, the Polar feature space is a clear advantage for the EM technique while the quality of the segmentation produced by the K-means is not so affected by the feature space. Both clustering techniques were able to characterize the two motion models present in each trial although, EM technique produces clusters affected by a higher level of noise. This may be caused due to a process that is more complex in nature since it is an iterative scheme that computes the posterior probabilities and the log-likelihood. Therefore, it is less robust to noisy data relatively to the K-means that is a more simpler technique. In addition, table 1 proves that the quality of the segmentation obtained by the EM was substantially better when compared to the experiments with f^c. In detail, the performance of the EM increased by 43.3% however, the average performance of the K-mean was similar to the result obtained in f^c. Generally, the f_l^c makes possible for the EM and K-means to achieve a better segmentation quality (despising the first sequence). These trials depict that the EM and K-mean are not suitable for the Polar feature space with information about the image location, f_l^p, since results are inconclusive (only some trials reported an improved quality).

Table 1. F-score - Performance comparison between the EM and K-means in f^c, f^p, f_l^c and f_l^p. Parameters such the precision ("Prec.") and the recall ("Rec.") are presented. Experiences 1, 2 and 3 represent the results for the flow fields 1(a), 1(b) and 1(c), respectively.

Sequence	EM^c			K-means^c			EM^p			K-means^p		
	Pre.	Rec.	F-score	Pre.	Rec.	F-score	Pre.	Rec.	F-score	Pre.	Rec.	F-score
1	0.404	1.000	0.575	0.981	0.944	0.962	0.797	0.936	0.861	0.881	0.928	0.904
2	0.631	0.999	0.774	1.000	0.853	0.921	0.875	0.917	0.896	0.924	0.837	0.879
3	0.455	0.994	0.624	0.543	0.926	0.685	0.496	0.894	0.638	0.535	0.885	0.667
4	0.638	0.996	0.778	0.987	0.866	0.922	0.805	0.939	0.867	0.923	0.934	0.929
5	0.367	0.998	0.537	0.999	0.852	0.919	0.715	0.905	0.799	0.852	0.846	0.849
6	0.330	0.999	0.496	0.998	0.899	0.946	0.668	0.974	0.792	0.828	0.947	0.884

Sequence	EM$_l^c$			K-means$_l^c$			EM$_l^p$			K-means$_l^p$		
1	0.912	0.927	0.919	0.578	0.916	0.708	0.717	0.936	0.812	0.303	0.976	0.463
2	0.934	0.897	0.915	0.999	0.853	0.920	0.589	0.969	0.733	0.906	0.836	0.870
3	0.882	0.968	0.923	0.979	0.948	0.963	0.418	0.975	0.585	0.874	0.937	0.904
4	0.879	0.998	0.935	0.977	0.920	0.947	0.922	0.900	0.911	0.910	0.901	0.906
5	0.697	0.972	0.812	0.998	0.858	0.923	0.923	0.726	0.813	0.843	0.871	0.857
6	0.885	0.951	0.917	0.998	0.901	0.947	0.909	0.866	0.887	0.822	0.942	0.878

Finally, the computational performance was evaluated for the EM and K-means in both f^c and f^p. The computation of the techniques took in average, 7.127 seconds (EM^c), 3.787 seconds (EM^p), 0.142 seconds (K-means^c) and 0.122 seconds (K-means^p). As can be seen, the Polar feature space accelerates the convergence of the clustering in both techniques since the processing time is substantially reduced, especially for the EM case whose processing time is reduced by 46.9% while the processing time of the K-means is reduced by 14.1%.

4   Conclusion

Therefore, the paper presented an important research topic for motion perception and analysis because the segmentation produces poor results when the features space is not properly adjusted. This compromises the ability of the mobile robot to understand its surrounding environment. An extensive set of experiments were conducted as part of this work and several factors were considered and studied such as, space (Cartesian and Polar) and dimensionality of the feature vector. Results prove that choosing a good feature space for the detection of motion patters is not a trivial problem since it influences the performance of the Expectation-Maximization and K-means. This last technique in Cartesian space revealed the best performance for motion segmentation of flow fields (with a resolution of 640×480). It originates a good visual segmentation (evaluated using the F-score metric) in a reduced period of time since it took 0.122 seconds to compute.

This work was funded by the project FCOMP - 01-0124-FEDER-022701.

References

1. Brox, T., Malik, J.: Object segmentation by long term analysis of point trajectories. In: Daniilidis, K., Maragos, P., Paragios, N. (eds.) ECCV 2010, Part V. LNCS, vol. 6315, pp. 282–295. Springer, Heidelberg (2010)
2. Bugeau, A., Prez, P.: Detection and segmentation of moving objects in complex scenes. Computer Vision and Image Understanding **113**(4), 459–476 (2009)
3. Eibl, G., Brandle, N.: Evaluation of clustering methods for finding dominant optical flow fields in crowded scenes. In: International Conference on Pattern Recognition, pp. 1–4 (December 2008)
4. Georgiadis, G., Ayvaci, A., Soatto, S.: Actionable saliency detection: Independent motion detection without independent motion estimation. In: IEEE Conference on Computer Vision and Pattern Recognition (CVPR), pp. 646–653 (2012)
5. Pinto, A.M., Costa, P.G., Correia, M.V., Paulo Moreira, A.: Enhancing dynamic videos for surveillance and robotic applications: The robust bilateral and temporal filter. Signal Processing: Image Communication **29**(1), 80–95 (2014)
6. Dan Melamed, I., Green, R., Turian, J.P.: Precision and recall of machine translation. In: Proceedings of the 2003 Conference of the North American Chapter of the Association for Computational Linguistics on Human Language Technology: Companion Volume of the Proceedings of HLT-NAACL, NAACL-Short 2003, pp. 61–63. Association for Computational Linguistics, Stroudsburg (2003)
7. Pinto, A.M., Paulo Moreira, A., Correia, M.V., Costa, P.G.: A flow-based motion perception technique for an autonomous robot system. Journal of Intelligent and Robotic Systems, 1–25 (2013) (in press)

Swarm Robotics Obstacle Avoidance: A Progressive Minimal Criteria Novelty Search-Based Approach

Nesma M. Rezk$^{(\boxtimes)}$, Yousra Alkabani, Hassan Bedour, and Sherif Hammad

Computer and Systems Department Faculty of Engineering,
Ain Shams University, Cairo, Egypt
{nesma.rezk,yousra.alkabani,hassan_bedour,sherif.hammad}@eng.asu.edu.eg

Abstract. Swarm robots are required to explore and search large areas. In order to cover largest possible area while keeping communications, robots try to maintain hexagonal formation while moving. Obstacle avoidance is an extremely important task for swarm robotics as it saves robots from hitting objects and being damaged.

This paper introduces novelty search evolutionary algorithm to swarm robots multi-objective obstacle avoidance problem in order to overcome deception and reach better solutions.

This work could teach robots how to move in different environments with 2.5% obstacles coverage while keeping their connectivity more than 82%. Percentage of robots reached the goal was more than 97% in 70% of the environments and more than 90% in the rest of the environments.

Keywords: Maintaining formation · Novelty search · Obstacle avoidance

1 Introduction

Our main interest in this work, is to teach swarm robots by using novelty search evolutionary algorithm how to reach a certain goal while maintaining formation and avoiding obstacles.

2 Related Work

2.1 Novelty Search

Lehman and Stanley proposed a new change in genetic algorithms [1]. Instead of calculating a fitness function and selecting individuals with the best fitness values, individuals who have more novel behaviour than other individuals are selected to be added to the new generation.

Novelty search can be easily implemented on top of most evolutionary algorithms. Basically, the fitness value would be replaced with what is called a novelty metric. The novelty metric (sparseness of an individual) is calculated as the

© Springer International Publishing Switzerland 2015
F. Pereira et al. (Eds.) EPIA 2015, LNAI 9273, pp. 493–498, 2015.
DOI: 10.1007/978-3-319-23485-4_49

average distance between behavior vectors of the individual and its k_{th} nearest neighbors, and an archive.

Pure novelty search for large search spaces is not enough to reach solutions as the algorithm will spend a lot of time searching behaviors that are not meeting the goal. So, novelty search can overcome deception but cannot work alone without the guidance of the fitness value.

In MCNS (Minimal Criteria Novelty Search) only individuals who have a fitness value greater than a minimal criteria would be assigned their novelty score [2]. Otherwise their novelty score would be zero. Zero novelty value individuals would only be used for reproduction if no other individuals meet the minimal criteria. It is clear that MCNS acts as random search until individuals that meet the minimal criteria are reproduced. So, MCNS should be seeded with initial population that meets the minimal criteria.

Progressive minimal criteria novelty search was proposed by Gomes *et al.* to overcome the need of seeding the MCNS algorithm with initial population [3]. The minimal criteria is a dynamic fitness threshold initially set to zero. The fitness threshold progressively increases among generations to avoid the search from exploring irrelevant solutions. In each generation, the new criteria is found by determining the value of the P-th percentile of the fitness scores in the current population. This means that P percent of the fitness values would fall under the minimal criteria.

2.2 Evolutionary Obstacle Avoidance

Hettiarachchi and Spears proposed an evolutionary algorithm for swarm robots offline learning [4]. The genetic algorithm is to teach robots how to reach their goal while preserving their hexagonal formation and avoiding obstacles at the same time. To get optimum coverage with least number of robots and efficient multi-hop communications network hexagonal formation is the best formation as discussed in [5].

Lennard Jones force law is used to control the robot. The robot moves according to the net force calculated from the forces that are exerted upon it by other robots and environment. As the robot moves in the environment it interacts with three types of objects: robots, obstacles and a goal. It is required to set the parameters for three copies of the force law one for each object in order to keep the robot at distance R from neighbor robots which will lead to the hexagonal formation, keep away from obstacles, and reach goal.

An evolutionary algorithm is used to optimize the parameters for the three copies of the force law.The penalty function (1) is a minimizing multi-objective fitness function. The function consists of three components: penalty for not reaching the goal, penalty for collisions, and penalty for lack of cohesion.

$$Penalty = P_{collisions} + P_{connectivity} + P_{notreachingthegoal} \qquad (1)$$

3   Applying PMCNS

To apply PMCNS, we have several issues to handle. The penalty function of our problem is a minimizing function. It needs to minimize the penalty for not reaching the goal, penalty for non-cohesion and penalty for collisions. The PMCNS algorithm equations assumes that the fitness function is a maximizing function. It allows individuals that have fitness values higher than the minimal criteria to be selected for the next generation. We have two options to apply. The first option is to change all of the equations of the algorithm to be a maximal criteria algorithm not a minimal criteria.

The second option (which was adopted in this work) is to inverse all the penalty values to change the problem from a minimizing to a maximizing problem. The minimum value would be the maximum value and the maximum value would be the minimum value . Equation (2) shows how the penalty value would be inverted.

$$inversed\_penalty(i) = max\_penalty - penalty(i) \qquad (2)$$

where $max\_penalty$ is the maximum penalty of the current generation. $penalty(i)$ is subtracted from current generation maximum penalty so the inverted penalty values will have the same range of penalty values.

We need to decide how to capture the behavior vector, and how to apply PMCNS to a multi-objective problem. To fill the behaviour vector, we need to make the vector express how the controller behave through simulation. For two reasons, we chose the genomes of the individual to express the behaviour vector. The first reason is that the genomes are the parameters of the three versions of the force law. Those parameters decide how the robot will behave with robots, obstacles, and the goal, so they express the behaviour of the controller. The second reason is that Cuccu and Gomez stated that the simplest way to fill the behavior vector is to fill it with the individual genomes [6].

4   Experiments and Results

4.1   PMCNS Experiment

This experiment was held to compare objective-based search to progressive minimal criteria novelty search. The environments generated for the evaluation module contains 40 robots and 90 obstacles like the experiment held by Hettiarachchi and Spears [4]. The evolution was run for 80 generations. Each individual was evaluated 20 times. The evaluation module of the genetic algorithm was distributed over 30 computers to gain speedup [7].

There are differences in the way the two algorithms act. The objective-based search would start faster than PMCNS but PMCNS could reach less penalty values than objective-based search. Fig. 1 shows the minimum penalty found in all generations during evolution for both objective-based and PMCNS with percentile = 50%.

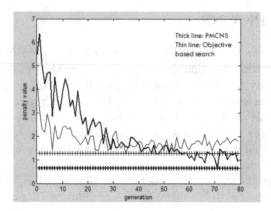

Fig. 1. Minimum penalty of each generation for both objective-based and novelty search evolutionary algorithms.

4.2   Evaluation Module Experiments

The next experiments were held to examine the changes can be done in the evaluation module to reach better solutions. Since the previous experiment showed that PMCNS can perform better than objective-based search, we used PMCNS in the rest of experiments. In these experiments PMCNS evolutionary algorithm with percentile= 50% was used.

The target of these experiments is to train robots to move in environments with obstacle coverage less than or equal 2.5%. So, the environments for training contained 40 robots and 50 obstacles. The diameter of the obstacle is 0.2 units, while the diameter of the robot is 0.02 units. The arena dimensions are 9 x 7 units, so the obstacles coverage is 2.5% of the environment. Robots are initially placed at the bottom left of the arena and the goal is located at the top right of the arena. Obstacles are randomly placed in the arena. There is an area around the nest where no obstacles are placed to prevent proximity collisions.

Robot can sense goal at any distance. A penalty is added at the end of simulation (1500 time step) if less than 80% of the robots did not reach the goal area. The goal area is 4R from the center of the goal. Each individual was evaluated 20 times. Most of those settings are like Hettiarachchi and Spears experimental settings in their work [4]. For each experiment the best penalty value individual performance was tested over 20, 40, 60, 80, and 100 robots moving in environments that contain 10, 20, 30, 40, and 50 obstacles corresponding to obstacle coverages equal 0.5, 1, 1.5, 2 and 2.5% of the environment. So, the total number of performance experiments=25. Each experiment is evaluated 50 times.

1. First Penalty Function (Penalty Experiment 1)

   In this experiment, robot can sense neighbor robots at distance 1.5R where R is the desired separation between robots. Robot attracts its neighbors if neighbors are 1.5R distance away, and repulses its neighbors if the neighbors

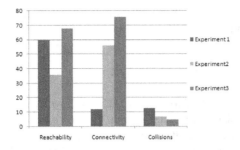

Fig. 2. Summary of the penalty evaluation experiments results.

are closer than 1.5R distance. A penalty for non cohesion is added if less than or more than 6 robots are found at distance R from the robot.

Obstacles are sensed at distance equals double the obstacle diameter from the center of the robot to the center of the obstacle. Robot starts to interact with the obstacle at distance equals double the obstacle diameter from the center of the robot to the center of the obstacle. A penalty is added for collision if the distance between center of robot and center of obstacle is less than robot diameter.

2. Second Penalty Function (Penalty Experiment 2) In this experiment, we changed the distances of doing action and adding penalty for cohesion (robot to robot interaction). Robot can sense neighbor robots at distance 1.5R, where R is the desired separation between robots. Robot attracts its neighbors if the neighbors are R distance away and repulses its neighbors if the neighbors are closer than R distance. A penalty for non cohesion is added if less than or more than 6 robots are found at distance R from the robot

3. Applying Harder Problem on Third Penalty Function (Penalty Experiment 3) In the last two experiments, we noticed that results are the worst at 50 obstacle environments, so we decided to hold a new experiment with same settings of penalty experiment 2, but the training environments contain 60 obstacles instead of 50 obstacles to see if the training environments were harder would the robots behave better for easier environments.

Penalty Experiments Results. Experiment 1 is the best in the percentage of robots reached the goal but the worst in the minimum percentage of robots remained connected and the number of collisions. Experiment 2 is much better than experiment 1 in the minimum percentage of robots remained connected but less number of robots reached the goal. Experiment 3 is the same like experiment 2 but only the training environments were harder. Experiment 3 showed the best results for all objectives.

Fig. 2 shows graphically the results of the experiments. Reachability tells us how many times in the reachibility performance experiments the percentage of robots reached the goal was over 97%. Connectivity tells us how many times in the connectivity performance experiments the minimum percentage of robots remained connected was over 90%. The number of collisions shows the total number of collisions found in each experiment.

5 Conclusion and Future Work

This work shows that progressive minimal criteria novelty search can perform in a different way than objective-based search. It can reach better solutions than objective-based search for a multi-objective task for swarm robots. However we believe that as the task is deceptive we can have better solutions and better evolutionary behavior using novelty search algorithms than the behavior reached in this work.

The purpose of the upcoming experiments is to find the part of the fitness function that was decepted by the multi-objective problem to describe it in the behavior vector and apply novelty search. Otherwise, we shall prove that the way used in this work for calculating the fitness value using the multi-objective function was able to overcome the deception of the problem and novelty search will not be of a big value.

This work examined different changes in the evaluation module of the multi-objective genetic algorithm settings. These changes can enhance one objective, but another objective may get worse. This work proved that using harder problem for learning during the evolutionary algorithm will give us better solutions for easier problems.

References

1. Lehman, J., Stanley, K.: Improving evolvability through novelty search and self-adaptation. In: Proceedings of the 2011 IEEE Congress on Evolutionary Computation (CEC), Piscataway, NJ, US (2011)
2. Lehman, J., Stanley, K.: Revising the evolutionary computation abstraction: minimal criteria novelty search. In: Proceedings of the Genetic and Evolutionary Computation Conference (GECCO), New York, US (2010)
3. Gomes, J., Urbano, P., Christensen, A.L.: Progressive minimal criteria novelty search. In: Pavón, J., Duque-Méndez, N.D., Fuentes-Fernández, R. (eds.) IBERAMIA 2012. LNCS, vol. 7637, pp. 281–290. Springer, Heidelberg (2012)
4. Hettiarachchi, S., Spears, W., Spears, D.: Physicomimetics, chapter 14, pp. 441–473. Springer, Heidelberg (2011)
5. Prabhu, S., Li, W., McLurkin, J.: Hexagonal lattice formation in multi-robot systems. In: 11th International Conference on Autonomous Agents and Multiagent Systems (AAMAS) (2012)
6. Cuccu, G., Gomez, F.: When novelty is not enough. In: Di Chio, C., Cagnoni, S., Cotta, C., Ebner, M., Ekárt, A., Esparcia-Alcázar, A.I., Merelo, J.J., Neri, F., Preuss, M., Richter, H., Togelius, J., Yannakakis, G.N. (eds.) EvoApplications 2011, Part I. LNCS, vol. 6624, pp. 234–243. Springer, Heidelberg (2011)
7. Rezk, N., Alkabani, Y., Bedour, H., Hammad, S.: A distributed genetic algorithm for swarm robots obstacle avoidance. In: IEEE 9th International Conference on Computer Engineering and Systems (ICCES), Cairo, Egypt (2014)

Knowledge Discovery
and Business Intelligence

An Experimental Study on Predictive Models Using Hierarchical Time Series

Ana M. Silva[1,2], Rita P. Ribeiro[1,3](✉), and João Gama[1,2]

[1] LIAAD-INESC TEC, University of Porto, Porto, Portugal
{201001541,jgama}@fep.up.pt, rpribeiro@dcc.fc.up.pt
[2] Faculty of Economics, University Porto, Porto, Portugal
[3] Faculty of Sciences, University Porto, Porto, Portugal

Abstract. Planning strategies play an important role in companies' management. In the decision-making process, one of the main important goals is sales forecasting. They are important for stocks planing, shop space maintenance, promotions, etc. Sales forecasting use historical data to make reliable projections for the future. In the retail sector, data has a hierarchical structure. Products are organized in hierarchical groups that reflect the business structure. In this work we present a case study, using real data, from a Portuguese leader retail company. We experimentally evaluate standard approaches for sales forecasting and compare against models that explore the hierarchical structure of the products. Moreover, we evaluate different methods to combine predictions for the different hierarchical levels. The results show that exploiting the hierarchical structure present in the data systematically reduces the error of the forecasts.

Keywords: Data mining · Hierarchical time-series · Forecasting in retail

1 Introduction

Nowadays, with the increasing competitiveness, it is important for companies to adopt management strategies that allow them to value up against the competition. In the retail sector, in particular, there is an evident relationship among different time series. The problem presented here is related to a Portuguese leader company in the retail sector, in the electronics area. As we see in Figure 1, the total sales of the company can be divided into five business unities: Home Appliances (U51), Entertainment (U52), Wifi (U53), Image (U54) and Mobile (U55), each one can be divided in 137 different stores.

In this paper, we propose a predictive model that estimates the monthly sales revenue for all stores of this company. We then compare this flat model that ignores the hierarchy present in the time series, with three other ways, existing in the literature, to combine the obtained forecasts, by exploring its hierarchical structure (e.g. [1,2,3,4]): *bottom-up*, *top-down* and *combination* of predictions made at different levels of the hierarchy. The experimental results

© Springer International Publishing Switzerland 2015
F. Pereira et al. (Eds.) EPIA 2015, LNAI 9273, pp. 501–512, 2015.
DOI: 10.1007/978-3-319-23485-4_50

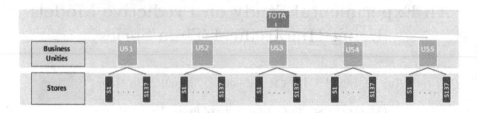

Fig. 1. Hierarchical structure of our time series data

obtained with our case study comprove that taking advantage of the hierarchical structure present in the data leads to an improvement of the models performance as it reduces the error of the forecasts.

This paper is organized as follows: in Section 2 the related work is presented; furthermore, in Section 3, is described how the forecast model was built and how we do the comparison between the different hierarchical models; and finally, in Section 4 are exposed the conclusions of the work and future work.

2   Related Work

Increasingly, the data mining techniques have been applied in time series analysis [5]. In the retail sector, where the time series display well defined components of trend and seasonality, the usage of learning algorithms such as Artificial Neural Networks (ANN) proves to be more efficient than the application of traditional methods, once that this one can capture the non-linear dynamics associated to these components and their interactions [6]. However, to apply these algorithms we must study the best way to present the data. Some studies show that, as standard, the ANNs are most efficient when applied to time series with trend and seasonality correction [7]. In addition, in the retail sector, normally, there are several variables that can somehow justify fluctuations in sales. In [8], ANNs are applied in daily sales forecasting of a company in the shoe's industry, using as explanatory variables: month of the year, day of the week, holidays, promotions or special events, sales period, weeks pre/post Christmas and Easter, the average temperature, the turnover index in retail sale of textiles, clothing, footwear and leather articles and the daily sales of previous seven days with correction of trend and seasonality.

In [9] the predictive power of ANNs is compared with the Support Vector Regression Machines [10] (SVRs) - this second algorithm also compares the use of the *linear kernel* function with the *Gaussian kernel* function. In that work, these learning algorithms are applied to five different artificial time series: stationary, with additive seasonality, with linear trend, with linear trend and additive seasonality and linear trend and multiplicative seasonality. The results showed that the SVRs with *Gaussian kernel* function is the most efficient algorithm in the forecasting of time series without trend. However, in series with trend, the predictions shown are disastrous, while the ANNs and SVRs with *linear kernel* function produce robust predictions, even without performing pre-processing of the data.

In hierarchical databases, it is frequently useful to explore the relationship of dependency between different time series, thus ensuring consistency in time series forecasting belonging to different levels. In [1] is presented a methodology that explores the different levels of aggregation hierarchies and the predictions for different time periods. In this case, forecasting the next elements of the time series are obtained by aggregating the predictions of descendants series in the hierarchy associated with this dimension.

Normally, when climbing into the hierarchy the forecast error decreases, since at these levels some of the deviations and fluctuations are neutralized. The two most commonly used strategies are *bottom-up* and *top-down*. While with the first method the forecasts are calculated on the lower level of the hierarchy and then aggregated to provide the higher dimension series predictions; in the second, the forecasts are calculated at the top level and then disaggregated to the lower levels [2]. When using this second case, there is no universal form of breakdown, but there are several methodologies that can be adopted [3]. Although the algorithm *top-down* is easy to build and produce reliable predictions of aggregated levels, when we go down in the hierarchy, it can lead to loss of information related to the dynamics of the series in descending hierarchies and the distribution of predictions by lower levels is not always easy to accomplish. Furthermore, with the algorithm *bottom-up*, the loss of information is not so big, but there are many series to predict and the noise presented in the data below hierarchies is often high. However, there is no consensus on the best approach.

More recently, it was proposed a method which produces better results when compared to the application of algorithms *bottom-up* and *top-down*. This method consists in the calculation of independent projections in all levels of the hierarchy, applying then a regression model to optimize the combination of these predictions [4]. The main advantage of this approach is the fact that the predictions obtained for all time series of all levels can come from the application of any learning algorithm, which allows you to use all available data, as well as inherent time series dynamics. After that, the forecasts will be revised and transformed by applying a weighted average, which will use all other forecasts.

To better explain how this method works, let us consider the time series hierarchy illustrated in Figure 2.

The authors [4] start by proposing the translation of the time series hierarchy to a matrix notation where:

- each line i represents a node in the hierarchy in breadth traversal order;
- each column j represents a node of the bottom level;
- each position (i, j) of the matrix is 1 if the time series contained in the bottom level node j contributes to the time series in node i; otherwise, it should be 0.

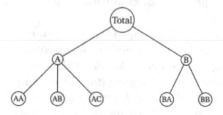

Fig. 2. Example of a hierarchical relationship with two levels

Thus, the hierarchy shown in Figure 2 is represented by the following matrix S (cf. Equation 1).

$$S = \begin{pmatrix} 1\,1\,1\,1\,1 \\ 1\,1\,1\,0\,0 \\ 0\,0\,0\,1\,1 \\ 1\,0\,0\,0\,0 \\ 0\,1\,0\,0\,0 \\ 0\,0\,1\,0\,0 \\ 0\,0\,0\,1\,0 \\ 0\,0\,0\,0\,1 \end{pmatrix} \tag{1}$$

The authors [4] also showed that, assuming that the forecasting errors of the hierarchy take the same distribution of the aggregated data, it is possible to obtain reasonable forecasts by solving the expression shown in Equation 2,

$$\widetilde{Y}_t(h) = S(S^t S)^{-1} S^t \times \widehat{Y}_t(h) \tag{2}$$

where $\widetilde{Y}_t(h)$ represents the recalculated prediction for series h at time t, $\widehat{Y}_t(h)$ represents the prediction obtained independently for the time series h at time t and S is the matrix that represents the hierarchy of the time series.

The calculation of $S(S^t S)^{-1} S^t$ gives us the weights we need to do the forecast adjustment. Considering the matrix S (cf. Equation 1), we obtain the weights matrix (cf. Equation 3) corresponding to all series of the different hierarchy nodes.

$$S(S^t S)^{-1} S^t = \begin{pmatrix} 0.58 & 0.30 & 0.28 & 0.10 & 0.10 & 0.10 & 0.14 & 0.14 \\ 0.31 & 0.51 & -0.20 & 0.17 & 0.17 & 0.17 & -0.10 & -0.10 \\ 0.27 & -0.21 & 0.48 & -0.07 & -0.07 & 0.07 & 0.24 & 0.24 \\ 0.10 & 0.18 & -0.08 & 0.72 & -0.27 & -0.27 & -0.04 & -0.04 \\ 0.10 & 0.18 & -0.08 & -0.27 & 0.72 & -0.27 & -0.04 & -0.04 \\ 0.10 & 0.18 & -0.08 & -0.27 & -0.27 & 0.72 & -0.04 & -0.04 \\ 0.15 & -0.09 & 0.24 & -0.03 & -0.03 & -0.03 & 0.62 & -0.38 \\ 0.15 & -0.09 & 0.24 & -0.03 & -0.03 & -0.03 & -0.38 & 0.62 \end{pmatrix} \tag{3}$$

For example, the forecast value for the times series AA would be obtained by using the weights of the fourth line of this weights matrix, as shown in Equation 4.

$$\widetilde{Y}_{AA} = 0.10 \times \widehat{Y}_{Total} + 0.18 \times \widehat{Y}_A - 0.08 \times \widehat{Y}_B + 0.72 \times \widehat{Y}_{AA}$$
$$- 0.27 \times \widehat{Y}_{AB} - 0.27 \times \widehat{Y}_{AC} - 0.04 \times \widehat{Y}_{BA} - 0.04 \times \widehat{Y}_{BB} \qquad (4)$$

We should note that the negative weights are associated to the time series which are not directly influencing the considered time series. This coefficient is negative, instead of null, since we want to extract the effect of this series in the series on the top levels.

The same authors made available in R [11], the package hts [12] which has the implementation of an algorithm that automatically returns predictions for all hierarchical levels based on the idea described previously. However, it only allows the use of a linear model to predict each set.

3   The Case Study

3.1   Data Description

The challenge lies in building a model to forecast the sales revenue of the company, per month. Additionally, this forecast should be made by store - there are 137 stores across the country - and by business unit - i.e., is not intended a forecast for the whole store, but for a particular set products. In this particular business, there are five business units: Home Appliances (U51), Entertainment (U52), Wifi (U53), Image (U54) and Mobile (U55). The goal is at the 15th day of each month, foreseen the sales revenue of the following month.

This problem is as a regression problem - since it is the forecast of a continuous variable - and the available data will be used to train the algorithms, i.e., we will have a supervised learning process.

We have monthly aggregated data since January 2011 until December 2014. We keep the last six months of 2014 to evaluate our models, using the remaining data for training. Due to the reduced number of instances, we decided to use a growing window with a time horizon of two months. This means that to predict the sales of July 2014, we use as training window all the data until May 2014. Then, the training window grows by incorporating the data of June 2014, to predict the sales of August 2014, and so on.

3.2   Non-hierarchical Model

From a first analysis of our data, we have noticed that the sales evolution regarding each business unit in each store exhibits very different behaviours. In these conditions, it is unfeasible to obtain a single model that would have a good performance for every store and business unit. Thus, in order to avoid very large errors, we have started by applying the k-means algorithm [13] to our stores, setting the number of clusters to three. Our aim was to cluster the stores by the three main areas of the country: north, center and south. The stores that constituted the centroid of each cluster, were used to tune the learning methods parameters for the time series corresponding to the total and the business unit of each store.

To the tuning process, we used the function `experimentalComparison` from the package `DMwR` [14], which chooses the best parameters that minimize the mean squared error (MSE).

In a first modelling approach, we have applied the *Autoregressive integrated moving average* (`arima`) - from R package `forecast` [15] - to obtain the sales forecast for each business unit in each store and each store total. Still, no good results were obtained.

In this context, and since we have a huge variety of time series with different features, we decided to apply different learning algorithms to each time series and use a simple ensemble to combine the models predictions. The final prediction was obtained by summing up the prediction made by the best model, i.e. the model that achieved the lowest MSE estimation, weighted by 0.75, with the prediction made by the second best model, weighted by 0.25. The used learning algorithms in our ensemble were: *Artificial Neural Networks* - from R packages `nnet` [16] and `caret` [17] -, *Support Vector Machines* - from R packages `e1071` [18] and `kernlab` [19] - and Random Forests - from R package `randomForest` [20]. These non-linear learning algorithms have shown better results in comparison to the linear model `arima` and the ensemble has shown better results when compared with each simple learning algorithm alone.

Using this modelling strategy, and based on [8], we also conducted a study to see if the addition/change of variables to the original time series would justify the sales' fluctuations and, thus, positively influence the results of the sales forecast. After some tests, the following changes to the original time series lead to overall better results. The sales values were normalized to the $[0, 1]$ interval. New variables were added with information on the month of year, the number of Saturdays/Sundays in the month, indication of the Easter month, and promotional campaigns and their time intervals of impact on the sales. It was also verified that the results became worse when the sales of earlier periods were used - either in the original format and in the data corrected of trend and/or seasonality, thus it was not used. It was also found that there was no significant correlation between considered variables.

There were some recent stores for which we did not had the information from the beginning of 2011. For these stores, we found the oldest store with the most similar behaviour and used it to predict the new one.

3.3 Hierarchical Models

Our flat modelling approach, described in previous section, of predicting sales for each store, each business unit, and total of the company shown some drawbacks. The sum of the forecasts of the series of lower levels, does not correspond exactly to the value of the upper level. In this context, we found that would be useful to explore the hierarchies present in the database, which will also help to build a more consistent model in time series forecasts from different hierarchical levels. Therefore, based on the forecasts obtained with our base model,we considered the following four different modelling approaches.

Non-hierarchical Model (NonHierarch): model that predicts the sales for each store, each business unit, and total of the company, ignoring the hierarchical relationship.

Bottom-up Model (BottomUp): model obtained using a *bottom-up* approach, which means that the predictions made for the lower level of the hierarchy are used to forecast levels above; for each level the prediction is given by the sum of the predictions made in the level below.

Top-Down Model (TopDown): model obtained using a *top-down* approach, which means that the forecasts made for the total of the company are used predict the sales of lower hierarchical levels; the prediction uses an a priori measure of the weight of each business unit in the total of sales total, based on the history of the considered month.

Hierarchical Combination Model (HierarchComb): model that uses all the predictions obtained independently for each hierarchical level, and applies them the regression model suggested by [21] to optimize the combination of the obtained predictions.

3.4 Experimental Results

The comparative results of the Mean Absolute Percentage Error (MAPE) obtained for the total sales of the company and for each business unit are shown in Table 1. This error metric was used in order to compare the results obtained with the error rates previously set by the company.

From the analysis of Table 1, we verify that for the series corresponding to the total sales of the company, the months of July, October and November (i.e. 50% of the test set) are better predicted by HierarchComb, while the remaining months are better predicted by the model BottomUp - 17% - and NonHierarch and TopDown - 33%. These two last models are, in fact, the same because TopDown uses the forecast obtained for the total of sales. Regarding the business units level, the models BottomUp and HierarchComb are the best in the same number of forecasts - 33% each - followed by models NonHierarch and TopDown, in 17% of the forecasts, each.

The results for each store, by business unit, are illustrated in Figure 3. We have also the results for the total of each store, obtained accordingly with each model.

In fact, looking at Figure 3, there is mainly a reduction of the highest error rates when using the model HierarchComb in comparison with model NonHierarch. Moreover, except for some small rounding errors, the predictions obtained by models NonHierarch and BottomUp are equal, since the model BottomUp uses all independent forecasts of the bottom level. On the other hand, the error rate obtained by the model TopDown is higher than the others, which can be justified by the noise introduced in the disaggregation process.

In order to verify if the results obtained by the different models were statistically significant in the bottom level, we applied an hypothesis test. Since we have a high sample size, the *Central Limit Theorem* allow us to consider that the sample follows a distribution approximately *Normal*. So, visually, we observe

Table 1. MAPE of sales forecast per business unit and total sales of the company by four modelling approaches to combine predictions of the different hierarchical levels.

Business Unit	Modelling Approach	Forecast Month					
		Jul	Aug	Sept	Oct	Nov	Dec
U51	NonHierarch	2.62%	13.31%	6.08%	0.91%	**2.89%**	1.96%
	BottomUp	2.71%	**3.81%**	**3.81%**	3.82%	6.54%	1.34%
	TopDown	2.46%	6.32%	4.21%	2.71%	5.42%	2.10%
	HierarchComb	**1.93%**	9.42%	5.82%	**0.46%**	3.15%	**1.04%**
U52	NonHierarch	12.42%	5.10%	**2.31%**	2.37%	16.36%	15.22%
	BottomUp	6.24%	12.68%	2.67%	3.28%	**7.02%**	13.42%
	TopDown	**4.12%**	6.76%	2.52%	4.02%	9.51%	14.22%
	HierarchComb	9.22%	**4.97%**	2.39%	**2.36%**	15.21%	**13.11%**
U53	NonHierarch	4.91%	23.94%	6.3%	2.21%	9.63%	**5.89%**
	BottomUp	**2.98%**	**8.10%**	6.56%	3.06%	6.78%	15.60%
	TopDown	5.02%	8.43%	**5.78%**	**1.98%**	**5.59%**	7.20%
	HierarchComb	4.76%	22.71%	7.21%	2.15%	8.52%	6.26%
U54	NonHierarch	9.30%	1.41%	3.51%	1.52%	3.25%	4.02%
	BottomUp	**1.47%**	1.70%	**2.70%**	2.89%	**2.87%**	4.30%
	TopDown	3.56%	2.87%	3.87%	2.81%	4.56%	**3.89%**
	HierarchComb	8.67%	**1.37%**	3.70%	**1.40%**	4.02%	4.07%
U55	NonHierarch	1.87%	14.48%	**2.07%**	**0.03%**	6.76%	11.89%
	BottomUp	13.22%	**9.86%**	4.05%	1.91%	**4.49%**	12.15%
	TopDown	4.20%	11.21%	3.42%	1.32%	5.02%	12.67%
	HierarchComb	**2.04%**	10.17%	4.12%	0.12%	6.05%	**11.65%**
Total	NonHierarch	1.51%	9.18%	**0.56%**	0.16%	1.85%	**4.42%**
	BottomUp	1.58%	**2.24%**	0.93%	0.45%	2.42%	5.02%
	TopDown	1.51%	9.18%	**0.56%**	0.16%	1.85%	**4.42%**
	HierarchComb	**1.46%**	5.76%	0.68%	**0.14%**	**1.23%**	6.33%

that TopDown has higher error rates and we also know that for this hierarchical level, the models NonHierarch and BottomUp are equal. Therefore, and since parametric tests are more powerful than non-parametric tests, we used the *t-test* for paired samples, considering the difference between the pairs of error rates observations, and the null hypothesis tests whether the mean of these differences is null, instead of the alternative hypothesis, which tests if the mean value of the differences is higher that zero, which means the errors obtained by the model HierarchComb are lower than that obtained by the model BottomUp. We got a *p-value* less than 1% so, for this level of significance, we reject the null hypothesis. We conclude that the model HierarchComb produces better forecasts than the model BottomUp, and the differences are statistically significant. In fact, graphically, the model HierarchComb particularly reduces larger errors, which have great impact on the average.

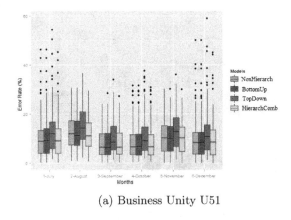

(a) Business Unity U51

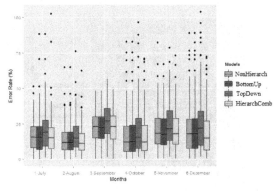

(b) Business Unity U52

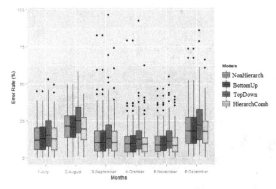

(c) Business Unity U53

Fig. 3. Distribution of MAPE of sales forecast for all the stores in business units U51, U52 and U53 by four modelling approaches: NonHierarch, BottomUp, TopDown and HierarchComb

510   A.M. Silva et al.

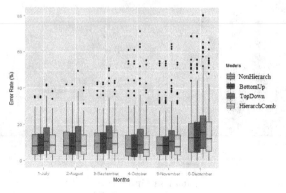

(d) Business Unity U54

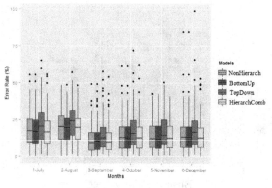

(e) Business Unity U55

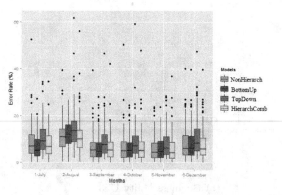

(f) total by store

Fig. 3. (*Continued*)

4 Conclusions and Future Work

The main goal of this paper is sales forecasting. We present a case study using real data from a Portuguese company leader of non-food retail sector. We study predictive models to obtain predictions for monthly sales revenue, by business unity, for all stores of the company. In our study we compare the standard flat approach, that ignores the hierarchical structure, and 3 different models that exploit the hierarchical structure. Our results show that when we descend in the hierarchy, the error rates tend to increase, given that, at higher levels we have a more uniform history. Our results confirmed that exploiting the hierarchical structure of the time series leads to more accurate forecasts. Namely, the approach proposed by [21] that combines the predictions made at each hierarchy level produced globally better results, for all hierarchical levels. This type of model tends to reduce the higher error rates and it can be seen as a valuable alternative when compared with the flat approach, and the bottom-up and top-down approaches. As future work, we plan to study how the forecasting variance depends on the inherent interactions between the time series in the same level. Moreover, as there are many real world problems where data presents an hierarchical structure, we intent to explore the application of this methodology to other real problems.

Acknowledgments. This work was supported by the European Commission through the project MAESTRA (Grant number ICT-2013-612944).

References

1. Ferreira, N., Gama, J.: Análise exploratória de hierarquias em base de dados multidimensionais. Revista de Ciências da Computação **7**, 24–42 (2012)
2. Fliedner, G.: Hierarchical forecasting: issues and use guidelines. Industrial Management and Data Systems **101**, 5–12 (2001)
3. Gross, C.W., Sohl, J.E.: Disaggregation methods to expedite product line forecasting. Journal of Forecasting **9**(3) (1990)
4. Hyndman, R.J., Ahmed, R.A., Athanasopoulos, G., Shang, H.L.: Optimal combination forecasts for hierarchical time series. Computational Statistics & Data Analysis **55**, 2579–2589 (2011)
5. Azevedo, J.M., Almeida, R., Almeida, P.: Using data mining with time series data in short-term stocks prediction: A literature review. International Journal of Intelligence Science **2**, 176 (2012)
6. Alon, I., Qi, M., Sadowski, R.J.: Forecasting aggregate retail sales: A comparison of artifcial neural networks and traditional methods. Journal of Retailing and Consumer Services, 147–156 (2001)
7. Zhang, G.P.: Neural networks for retail sales forecasting. In: Encyclopedia of Information Science and Technology (IV), pp. 2100–2104. Idea Group (2005)
8. Sousa, J.: Aplicação de redes neuronais na previsão de vendas para retalho. Master's thesis, Faculdade de Engenharia da Universidade do Porto (2011)
9. Crone, S.F., Guajardo, J., Weber, R.: A study on the ability of support vector regression and neural networks to forecast basic time series patterns. In: Bramer, M. (ed.) Artificial Intelligence in Theory and Practice. LNCS, vol. 217, pp. 149–158. Springer, Boston (2006)

10. Drucker, H., Burges, C.J.C., Kaufman, L., Smola, A.J., Vapnik, V.: Support vector regression machines. In: Advances in Neural Information Processing Systems 9, December 2–5, NIPS, Denver, CO, USA, pp. 155–161 (1996)
11. R Core Team: R: A Language and Environment for Statistical Computing. R Foundation for Statistical Computing, Vienna, Austria (2014)
12. Hyndman, R.J., Wang, E., with contributions from Roman, A., Ahmed, A.L., to earlier versions of the package, H.L.S.: hts: Hierarchical and grouped time series. R package version 4.4 (2014)
13. Hartigan, J.A., Wong, M.A.: A k-means clustering algorithm. JSTOR: Applied Statistics **28**(1), 100–108 (1979)
14. Torgo, L.: Data Mining with R, learning with case studies. Chapman and Hall/CRC (2010)
15. With contributions from George Athanasopoulos, R.J.H., Razbash, S., Schmidt, D., Zhou, Z., Khan, Y., Bergmeir, C., Wang, E.: forecast: Forecasting functions for time series and linear models. R package version 5.6 (2014)
16. Ripley, B.: nnet: Feed-forward Neural Networks and Multinomial Log-Linear Models R package version 7.3-8 (2014)
17. Kuhn, M.: caret: Classification and Regression Training. R package version 6.0-35 (2014)
18. Meyer, D., Dimitriadou, E., Hornik, K., Weingessel, A., Leisch, F.: e1071: Misc Functions of the Department of Statistics (e1071), TU Wien. R package version 1.6-4 (2014)
19. Karatzoglou, A., Smola, A., Hornik, K., Zeileis, A.: kernlab - an S4 package for kernel methods in R. Journal of Statistical Software **11**(9), 1–20 (2004)
20. Breiman, L., Cutler, A., Liaw, A., Wiener, M.: Random forests for classification and regression. R package version 4.6-10 (2014)
21. Hyndman, R.J., Athanasopoulos, G.: Optimally reconciling forecasts in a hierarchy. Foresight: The International Journal of Applied Forecasting (35), 42–48 (2014)

Crime Prediction Using Regression and Resources Optimization

Bruno Cavadas[1,2], Paula Branco[3,4(✉)], and Sérgio Pereira[5]

[1] Instituto de Investigação e Inovação em Saúde,
Universidade do Porto, Porto, Portugal
brunomcavadas@gmail.com
[2] Instituto de Patologia e Imunologia Molecular da,
Universidade do Porto, Porto, Portugal
[3] LIAAD - INESC TEC, Porto, Portugal
paobranco@gmail.com
[4] DCC - Faculdade de Ciências, Universidade do Porto, Porto, Portugal
[5] ALGORITMI Centre, University of Minho, Braga, Portugal
pereirasrm@gmail.com

Abstract. Violent crime is a well known social problem affecting both the quality of life and the economical development of a society. Its prediction is therefore an important asset for law enforcement agencies, since due to budget constraints, the optimization of resources is of extreme importance. In this work, we tackle both aspects: prediction and optimization.

We propose to predict violent crime using regression and optimize the distribution of police officers through an Integer Linear Programming formulation, taking into account the previous predictions. Although some of the optimization data are synthetic, we propose it as a possible approach for the problem. Experiments showed that Random Forest performs better among the other evaluated learners, after applying the SmoteR algorithm to cope with the rare extreme values. The most severe violent crime rates were predicted for southern states, in accordance with state reports. Accordingly, these were the states with more police officers assigned during optimization.

Keywords: Violent crime · Prediction · SmoteR · Regression · Optimization

1 Introduction

Violent crime is a severe problem in society. Its prediction can be useful for the law enforcement agents to identify problematic regions to patrol. Additionally, it can be a valuable information to optimize available resources ahead of time.

In the United States of America (USA), according to the Uniform Crime Reports (UCR) published by the Federal Bureau of Investigation (FBI) [1], violent crimes imply the use of force or threat of using force, such as rape, murder,

© Springer International Publishing Switzerland 2015
F. Pereira et al. (Eds.) EPIA 2015, LNAI 9273, pp. 513–524, 2015.
DOI: 10.1007/978-3-319-23485-4_51

robbery, aggravated assault, and non-negligent manslaughter. In 2013, it was reported 1,163,146 violent crimes, with an average of 367.9 per 100k inhabitants. This was equivalent to one violent crime every 27.1 seconds. In 2012, according to the United States Department of Labor [2], there were 780,000 police officers and detectives in the USA, with a median salary of $56,980 per year. Therefore, the optimization of police officers can be useful to optimize costs, while guaranteeing the safety of the population.

In this paper, the contributions are twofold. Firstly, we propose to predict the violent crime per 100k population using regression. To the best of our knowledge, this is the first time that such problem is tackled in this way. Moreover, we preprocess the data using smoteR algorithm to improve predictions on the most critical values: the extreme high. Having the predictions, we also propose an Integer Linear Programming formulation for the optimization of police officers distribution across states. This distribution takes into account the crime severity, population, density and budget of the states.

The remaining of the paper is organized as follows. In Section 2 a brief survey on related work is presented. Materials and methods are exposed in Section 3, including the description of the data set, the prediction-related procedures and the optimization scheme. Then, in Section 4, results are presented and discussed, while in Section 5 the main conclusions are pointed out.

2   Related Work

Crime prediction has been extensively studied throughout the literature due to its relevance to society. These studies employ diverse machine learning techniques to tackle the crime forecasting problem.

Nath [3] combined K-means clustering and a weighting algorithm, considering a geographical approach, for the clustering of crimes according to their types. Liu et al. [4] proposed a search engine for extracting, indexing, querying and visualizing crime information using spatial, temporal, and textual information and a scoring system to rank the data. Shah et al. [5] went a step further and proposed CROWDSAFE for real-time and location-based crime incident searching and reporting, taking into account Internet crowd sourcing and portable smart devices. Automatic crime prediction events based on the extraction of Twitter posts has also been reported [6].

Regarding the UCI data set used in this work, Iqbal et al. [7] compared Naive Bayesian and decision trees methods by dividing the data set into three classes based on the risk level (Low, medium and high). In this study, decision trees outperformed Naive Bayesian algorithms, but the pre-processing procedures were rudimentary. Somayeh Shojaee et al. [8], applied a more rigorous data processing methodology for a binary class and applied the usage of two different feature selection methods to a wider range of learning algorithms (Naive Bayesian, decision trees, support vector machine, neural networks and K-Nearest neighbors). In these studies no class balancing methodologies were employed. Other approaches such as the fuzzy association rule mining [9] and case-based editing [10] have also been performed.

After prediction, optimization of resources can be achieved by several strategies. Donovan et al. [11] used integer linear programming for the optimization of fire-fighting resources, solving one of the most commonly constrains faced by fire managers. The same strategy was used by Caulkins et al. [12] in the optimization of software system security measures given a fixed budget.

Regarding the problem of police officer optimization, Mitchell [13] used a P-median model to determine the patrol areas in California, while Daskin [14] applied a Backup Coverage Model to maximize the number of areas covered. More recently, Li et al. [15] relied on the concept of "crime hot-spots" to create a cross entropy approach to produce randomized optimal patrol routes.

3   Materials and Methods

3.1   Data Set Description

The *Communities and Crime Unnormalized Data Set*[1] provides information on several crimes in the USA, combining socio-economical and law enforcement data from 90' Census, 1990 Law Enforcement Management and Admin Stats survey and the 1995 FBI UCR. It includes 2215 examples, 124 numeric and 1 nominal attribute. It also contains 4 non-predictive attributes with information about the community name, county, code and fold. Among the several possible target variables we chose the number of violent crimes per 100k population.

3.2   Prediction

We started by pre-processing the data set. The violent crime is our target variable, thus we removed all the other 17 possible target variables contained in the data set. We also eliminated all the examples that had a missing value on our target variable and removed all the attributes that had more than 80% of missing values. The data set contained four non-predictive attributes, which we have also eliminated. Finally, we have removed one more example that still had a missing value, and have normalized all the remaining attributes.

Although this problem was previously tackled as a classification task, we opted for addressing it as a regression task. This is an innovative aspect of our proposal and this choice is also based on the fact that we will use the numeric results obtained with the predictions for solving an optimization problem. Therefore, it makes sense to use a continuous variable throughout the work, instead of discretizing the target variable and latter recovering a numeric value.

Another challenge involving this data set is the high number of attributes. To address this problem we have applied the same feature selection scheme with two different percentages. The scheme applies a hierarchical clustering analysis, using the Pearson Correlation Coefficient. This step removes a percentage of the features less correlated with the target variable. Then, a Random Forest (RF)

---

[1] Available at UCI repository in
   https://archive.ics.uci.edu/ml/datasets/Communities+and+Crime+Unnormalized.

learner is applied to compute the remaining features importance based on the impact in the Mean Squared Error. A percentage of the most important features provided is selected. Two different sets of features were selected by applying different percentages in the previous scheme. In one of the pre-processed data we aimed at obtaining 50% of the original features and in the other the goal was to select only 30% of the original features. This way we obtained two data sets with 52 and 32 features corresponding to 50% and 30% percentages.

In our regression problem we are interested in predicting the number of violent crimes per 100k inhabitants. However, we are more concerned with the errors made in the higher values of the target variable, i.e., the consequences of missing a high value of violent crimes by predicting it as low are worst than the reverse type of error. The extreme high values of the violent crime variable are the most important and yet the less represented in the data set. When addressed as a classification problem, this is clearly a problem with imbalanced classes, where the most important class has few examples. SmoteR algorithm is a proposal to address this type of problems within regression which was presented in [16,17]. This proposal uses the notion of utility-based Regression [18] and relies on the definition of a relevance function. The relevance function expresses the user preferences regarding the importance assigned to the target variable range. Ribeiro [18] proposes automatic methods for estimating the relevance function of the target variable. We have used those methods because they correspond to our specific concerns: the extreme rare values are the most important. The essential idea of SmoteR algorithm is to balance the data set by under-sampling the most frequent cases and over-sampling the rare extreme examples. The over-sampling strategy generates new synthetic examples by interpolating existing rare cases. More details can be obtained in [16,17]. The motivation for applying this procedure is to force the learning systems to focus on the rare extreme cases which would be difficult to achieve in the original imbalanced data. Our experiences included several variants of smoteR which were applied to the two pre-processed data sets. The smoteR variants used in the experiences included all combinations of the following parameters: under-sampling percentage 50% and 100%; over-sampling percentage 200% and 400%; number of neighbours 5.

For the prediction task we have used three learning algorithms: Support Vector Machines (SVM), RF and Multivariate Adaptive Regression Splines (MARS). More details on the experimented parameters and the evaluation are described in Section 4.1.

3.3   Optimization Through Integer Linear Programming

Given the predicted violent crime per 100k population, we propose to optimize the distribution of available police officers by state. We present our proposal as a proof of concept, since more detailed data and insight into the problem would be needed to implement a more realistic solution. Given that the number of officers by state is an integer quantity, it is used Integer Linear Programming. To solve the optimization problem it was applied the Branch-and-bound algorithm.

Problem Formulation. We considered as resources a certain amount of police officers to freely distribute by the states of the USA. The optimization takes into account the predictions on violent crime per 100k population to assign more officers by the states with more violent criminality. This assignment is constrained by an ideal number of officers that each state would like to receive and the available budget. However, every state should receive a minimum amount of officers to guarantee the security of its citizens.

In the data set, the instances are defined by communities, with several of them belonging to the same state. Since we wanted to distribute officers by state, it was calculated the mean violent crime predictions by state.

The optimization problem was defined as,

$$\text{maximize} \sum_{i=1}^{m} s_i x_i$$

$$\text{subject to} \sum_{i=1}^{m} x_i = N; \quad x_i \leq H_i;$$

$$x_i \geq f_i H_i; \quad c_i x_i \leq B_i;$$

$$x_i \in \mathbb{N}$$

where $i \in \{1, ..., m\}$ indexes each of the m states, with $m = 46$, x_i is the number of officers to distribute by state, s_i is the violent crime predictions by state, H_i is the ideal number of officers by state, f_i is a fraction on the ideal number of officers that each state accepts as the minimum, c_i is the cost that each state should pay for each officer, and B_i is the available budget for each state.

The ideal number of officers was defined in function of the violent crime prediction of the state and the population (number of citizens), since bigger populations, with more violent crime, have higher demands regarding police officers. To this end, the violent crime predictions were scaled (s_{s_i}) to the interval $[v_l, v_h]$. This way, it acts as a proportion on the population. However, since some populations have millions of citizens, this value was divided by 100 to get more realistic estimates for the ideal number of officers. So,

$$H_i = \frac{s_{s_i} p_i}{100} \tag{1}$$

where p_i is the real population of the state i.

It was defined that the minimum number of officers should be a fraction on the ideal number, taking into account the crime predictions. Defining a lower (l_b) and an upper (u_b) bound for the fraction, the previously scaled violent crime predictions are linearly mapped to the interval $[l_b, u_b]$. Knowing that it is in the interval $[v_l, v_h]$, the fraction on the ideal number of officers is calculated as,

$$f_i = \frac{s_i - v_l}{v_h - v_l} (u_b - l_b) + l_b \tag{2}$$

Budget was defined in function of the population and its density. Such definition is based on the intuition that a small and less dense population needs less budget and officers than a highly dense and big population. However, the population numbers are several orders of magnitude higher than density, which would make the effect of density negligible. So, we have rescaled both population and density to the range $[0, 100]$ (p_{s_i} and d_{s_i}). Moreover, the budget for each state is a part of the total national budget (B_T). So, B_i was calculated as

$$B_i = \frac{(d_{s_i} + a \cdot p_{s_i}) B_T}{\sum_{i=1}^{m} d_{s_i} + a \cdot p_{s_i}} \tag{3}$$

where $a > 0$ is a parameter to tune the weight of the density and population over the budget calculation.

4   Experimental Analysis

We have divided our problem, and analysis, into two sub-problems: prediction and optimization. In this section, we describe the tools, metrics, and evaluation methodology for each sub-problem. Then we focus in each sub-problem results.

4.1   Experimental Setup

Prediction. The main goal of our experiments is to select one of the two pre-processed data sets, a smoteR variant (in case it has a positive impact) and a model (among SVM, RF and MARS) to apply in the optimization task.

The experiments were conducted with R software. Table 1 summarizes the learning algorithms that were used and the respective parameter variants. All combinations of parameters were tried for the learning algorithms, which led to 4 SVM variants, 6 RF variants and 8 MARS variants.

We started by splitting each data set in train and test sets, approximately corresponding to 80% and 20% of the data. The test set was held apart to be used in the optimization, after predicting its crime severities. This set was randomly built with stratification and with the condition of including at least one example for each possible state of the USA.

In imbalanced domains, it is necessary to use adequate metrics since traditional measures are not suitable for assessing the performance. Most of these specific metrics, such as precision and recall, exist for classification problems. The notions of precision and recall were adapted to regression problems with non-uniform relevance of the target values by Torgo and Ribeiro [19] and Ribeiro [18]. We will use the framework proposed by these authors to evaluate and compare our results. More details on this formulation can be obtained in[18].

All the described alternatives were evaluated according to the F-measure with $\beta = 1$, which means that the same importance was given to both precision and recall scores. The values of F_1 were estimated by means of 3 repetitions of a 10-fold Cross Validation process and the statistical significance of the observed paired differences was measured using the non-parametric pairwise Wilcoxon signed-rank test.

Table 1. Regression algorithms, parameter variants, and respective R packages.

Learner	Parameter Variants	R package
MARS	$nk = \{10, 17\}, degree = \{1, 2\}, thresh = \{0.01, 0.001\}$	**earth** [20]
SVM	$cost = \{10, 150\}, gamma = \{0.01, 0.001\}$	**e1071** [21]
Random Forest	$mtry = \{5, 7\}, ntree = \{500, 750, 1500\}$	**randomForest** [22]

Optimization. In the optimization sub-problem the objective was to assign to each state a certain amount of police officers, given the total budget, the total number of available officers, and the violent criminality predictions. The optimization was carried out in R software, with the package "lpSolve".

The values for the population and the density are real values, obtained from the estimates for 2014 [23]. However, the total budget, the number of available police officers, and the individual cost of the officers by state were defined by us. Although they are not real values, they serve as proof of concept. The cost of each officer by state was chosen randomly, and uniformly, from the interval $[5, 15]$ once, then the same values were used in all experiments. Additionally, the values for u_b and l_b were set to 0.12 and 0.08, while v_l and v_h were set to 0.125 and 0.7, respectively.

4.2 Results and Discussion

Prediction. We started by examining the results obtained with all the parameters selected for the two pre-processed data sets, the three types of learners and the smoteR variants. All combinations of parameters were tested by means of 3 repetitions of a 10-fold cross validation process. Figure 1 shows these results.

We have also analysed the statistical significance of the differences observed in the results. Table 2 contains the several p-values obtained when comparing the SmoteR variants and the different learners, using the non-parametric pairwise Wilcoxon signed rank test with Bonferroni correction for multiple testing.

The p-value for the differences between the two data sets (with 30% and 50% of the features) was 0.17. Therefore we chose the data set with less features to continue to the optimization problem. This was mainly because of: i) the non statistical significant differences and ii) the smaller size of the data (less features can explain well the target variable, so we chose the most efficient alternative).

Table 2. Pairwise Wilcoxon signed rank test with Bonferroni correction for the SmoteR strategies (left) and the learning systems (right).

Strategies	none	S.o2.u0.5	S.o2.u1	S.o4.u0.5
S.o2.u0.5	1.3e-14	-	-	-
S.o2.u1	< 2e-16	1	-	-
S.o4.u0.5	2.3e-16	1	1	-
S.o4.u1	< 2e-16	0.18	1	1

Learners	svm	rf
rf	<2e-16	-
mars	0.077	<2e-16

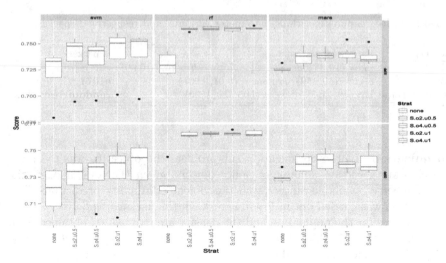

Fig. 1. Results from 3 × 10-fold CV by learning system and SmoteR variant. (none-original data; S-smoteR; ox-x × 100% over-sampling; uy-y × 100% under-sampling)

Regarding the SmoteR strategy, Figure 1 and Table 2 provide clear evidence of the advantages of this procedure. Moreover, we also observed that the differences between the several variants of this procedure are not statistically significant. Therefore, we have opted for the variant which leads to a smaller data set and consequently a lower run time. For the optimization sub-problem we chose to use the smoteR variant with 200% of over-sampling percentage and 100% of under-sampling percentage. The learning system that provides a better performance is clearly the RF. With this learner, there is almost no differences among the several experimented variants.

Considering these results, we chose the following setting to generate a model for the optimization sub-problem:

- Pre-processing to remove missing values and select 30% of the most relevant features;
- Apply the smoteR strategy with parameters k=5; over-sampling percentage=200; under-sampling percentage=100;
- RF model with parameters: mtry=7; ntree=750.

After generating the model we obtained the predictions for the test set which was held apart to use in the optimization sub-problem. These predictions were used as input of the optimization task.

Optimization. Several parameters were experimented. It was verified that with high budget and number of available officers, states with more criminality are assigned more officers. When the weight of the population increases, the most populated states, such as California, receive more police officers. When this weight is decreased, those states lost officers, while, for instance, Vermont obtained the ideal number, although the population is one of the lowest

Table 3 shows the results of distributing 500,000 police officers, with a budget of 8,000,000, and $a = 1$. Figure 2 shows the same results in a map of the USA, where brighter red is associated with higher criminality, and the radius of the circles is proportional to the amount of officers assigned to the state. The color of the circle indicates which restriction limited the number of officers. Therefore, green means that the state received the ideal number, the minimum is represented in blue, yellow means that the budget of the state did not allow more

Table 3. Distribution of 500,000 police officers by state, subjected to a total budget of 8,000,000.

State	Crime Prediction	Budget	Min. Off.	Ideal Off.	Dist. Off.	Cost
NJ	676.8	320390.4	1597	18614	18614	182659.9
PA	276.1	323802.5	1279	15991	1279	11413.7
OR	638.3	87076.7	678	7951	7951	59607.0
NY	893.1	504835.7	4445	49991	41007	504830.6
MO	381.3	12559.3	123	1503	123	1280.8
MA	408.1	233862.6	842	10283	842	5457.7
IN	726.6	164404.1	1247	14420	14420	162921.4
TX	821.1	648307.0	5643	64214	64214	477378.9
CA	850.2	948629.7	8369	94781	68083	948620.3
KY	655.6	104572.9	769	8996	8996	66277.2
AR	928.2	64349.2	691	7726	5885	64346.6
CT	355.4	146317.7	413	5090	413	6109.4
OH	542.8	90044.9	587	6997	587	8778.9
NH	312.6	33606.1	142	1760	142	2027.5
FL	1313.2	503583.7	6432	67716	52441	503581.4
WA	557.3	168051.3	1089	12954	12954	120927.3
LA	1721.0	109854.6	1994	19765	11893	109847.5
WY	528.1	1846.3	87	1036	87	778.2
NC	1315.6	247220.2	3221	33900	27118	247214.5
MS	1089.4	65686.5	808	8801	7149	65677.9
VA	851.7	208745.0	1798	20363	14106	208734.1
SC	1171.5	119505.3	1397	15028	10543	119498.9
WI	325.7	136544.5	629	7793	629	5271.4
TN	712.7	160819.6	1219	14128	10940	160817.3
UT	794.8	61723.4	599	6850	4253	61723.1
OK	488.6	86322.2	545	6560	545	5960.0
ND	342.0	6056.3	83	1026	83	618.2
AZ	500.5	155514.1	962	11554	962	11214.4
CO	791.7	121546.9	1087	12432	12432	84710.8
WV	551.2	39327.5	283	3371	283	3485.0
RI	440.9	110983.5	138	1680	138	1327.7
AL	1452.9	113590.5	1738	17915	7786	113576.2
GA	1254.6	248026.3	3120	33144	26054	248017.1
ID	444.8	28553.6	216	2616	216	2020.0
ME	275.9	23475.0	133	1663	133	1942.2
KS	1286.4	60751.9	920	9724	9724	52508.6
SD	568.5	8853.4	133	1585	1403	8851.3
NV	920.6	58244.7	656	7350	7350	56186.7
IA	556.7	67617.0	479	5696	2243	15039.1
MD	1271.9	191051.4	1872	19832	13217	191045.6
MN	862.1	125640.6	1191	13465	13465	96440.4
NM	835.6	39199.9	443	5031	3950	39199.7
DE	1161.9	48882.5	268	2891	2891	16497.7
VT	517.2	8881.3	92	1097	92	548.1
AK	932.3	64349.2	694	7752	7752	45227.2
DC	3044.8	926793.2	553	4612	4612	67407.4

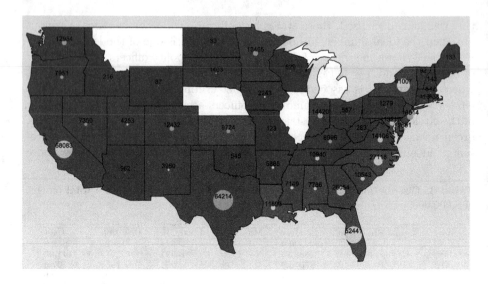

Fig. 2. Map of the USA representing the level of violent criminality by state, the amount of police officers assigned, and the restriction that imposed that number. White states are not represented in the data set.

officers, and white means that the state received a middle value of officers, which is less than the ideal or the maximum allowed by the budget, but higher than the minimum. It is possible to observe that ten states received the ideal number of officers. Some of them were associated with low or moderate levels of criminality, but the density or the population was high, such as New Jersey or Texas. Others are less populated, such as Oregon, but the ideal number of officers was also lower than other states constraint by the budget. The violent crime rate was particularly important in Kansas, since with a lower density and population, its budget allowed the state to receive the ideal number of officers. It is, also, possible to observe that the states with more violent criminality reached the number of officers allowed by their budget, such as Alabama or South Carolina. Accordingly, many states with less criminality received the minimum number of officers that they would allow (North Dakota), or values between the minimum and the ideal, without being constrained by the budget (Iowa). This behaviour may be desirable, since having too many officers in states with less criminality may be a waste of resources. The influence of the crime severity may be perceived when comparing Arizona with Nevada. The former has more population, higher density and budget than the latter, but received less officers because of the lower criminality rating.

According to the FBI [1], the region with more violent crime incidents is the South, followed by the West, Midwest and Northeast. It is interesting to notice that, in Figure 2, it was predicted more severe criminality for the southern states. These were the states that receive more police officers.

5   Conclusions

In this paper, we proposed a pipeline for predicting violent crime and a resources optimization scheme. Prediction encompasses feature selection through correlation and feature importance analysis, over-sampling of the rare extreme values of the target variable and regression. Among the evaluated learning systems, RF presented the best performance. This pipeline itself is one of the contributions of this work, given that, to the best of our knowledge, this problem in this data set was never approached as regression. Having the predictions, we propose a decision support scheme through the optimization of police officers across states, while taking into account the violent crime predictions, population, density and budget of the states. This contribution is presented as a proof of concept, since some of the parameters were synthesized and may not correspond to the real scenario. Nevertheless, our results show an higher crime burden in states located in the southern part of the USA compared with the states in the north. For this reason, southern states tend to have an higher assignment of police officers. These predictions are in accordance with some national reports, and although some parameters of the optimization are not completely realistic, it seems to work as expected.

This work, although limited to the United States, can be easily applied to various other countries. So, as future work we consider that it would be interesting to apply the proposed framework in other countries or regions.

Acknowledgments. This work is financed by the FCT – Fundação para a Ciência e a Tecnologia (Portuguese Foundation for Science and Technology) within project UID/EEA/50014/2013. Sérgio Pereira and Paula Branco were supported by scholarships from the Fundação para a Ciência e Tecnologia (FCT), Portugal (scholarships number PD/BD/105803/2014 and PD/BD/105788/2014). We would like to thank the useful comments of Manuel Filipe Santos, Paulo Cortez, Rui Camacho and Luis Torgo.

References

1. FBI, Crime in the United States 2013 (2014). http://www.fbi.gov/about-us/cjis/ucr/crime-in-the-u.s/2013/crime-in-the-u.s.-2013 (accessed: January 21, 2015)
2. Labor-Statistics, B.: United States Department of Labor - Bureau of Labor Statistics: Police and detectives (2012). http://www.bls.gov/ooh/protective-service/police-and-detectives.htmtab-1 (accessed: January 21, 2015)
3. Nath, S.V.: Crime pattern detection using data mining. In: 2006 IEEE/WIC/ACM International Conference on Web Intelligence and Intelligent Agent Technology Workshops, WI-IAT 2006 Workshops, pp. 41–44. IEEE (2006)
4. Liu, X., Jian, C., Lu, C.-T.: A spatio-temporal-textual crime search engine. In: Proceedings of the 18th SIGSPATIAL International Conference on Advances in Geographic Information Systems, pp. 528–529. ACM (2010)
5. Shah, S., Bao, F., Lu, C.-T., Chen, I.-R.: Crowdsafe: crowd sourcing of crime incidents and safe routing on mobile devices. In: Proceedings of the 19th ACM SIGSPATIAL International Conference on Advances in Geographic Information Systems, pp. 521–524. ACM (2011)

6. Wang, X., Gerber, M.S., Brown, D.E.: Automatic crime prediction using events extracted from twitter posts. In: Yang, S.J., Greenberg, A.M., Endsley, M. (eds.) SBP 2012. LNCS, vol. 7227, pp. 231–238. Springer, Heidelberg (2012)

7. Iqbal, R., Murad, M.A.A., Mustapha, A., Panahy, P.H.S., Khanahmadliravi, N.: An experimental study of classification algorithms for crime prediction. Indian Journal of Science and Technology 6(3), 4219–4225 (2013)

8. Shojaee, S., Mustapha, A., Sidi, F., Jabar, M.A.: A study on classification learning algorithms to predict crime status. International Journal of Digital Content Technology and its Applications 7(9), 361–369 (2013)

9. Buczak, A.L., Gifford, C.M.: Fuzzy association rule mining for community crime pattern discovery. In: ACM SIGKDD Workshop on Intelligence and Security Informatics, p. 2. ACM (2010)

10. Redmond, M.A., Highley, T.: Empirical analysis of case-editing approaches for numeric prediction. In: Innovations in Computing Sciences and Software Engineering, pp. 79–84. Springer (2010)

11. Donovan, G., Rideout, D.: An integer programming model to optimize resource allocation for wildfire containment. Forest Science 49(2), 331–335 (2003)

12. Caulkins, J., Hough, E., Mead, N., Osman, H.: Optimizing investments in security countermeasures: a practical tool for fixed budgets. IEEE Security & Privacy 5(5), 57–60 (2007)

13. Mitchell, P.S.: Optimal selection of police patrol beats. The Journal of Criminal Law, Criminology, and Police Science, 577–584 (1972)

14. Daskin, M.: A maximum expected covering location model: formulation, properties and heuristic solution. Transportation Science 17(1), 48–70 (1983)

15. Li, L., Jiang, Z., Duan, N., Dong, W., Hu, K., Sun, W.: Police patrol service optimization based on the spatial pattern of hotspots. 2011 IEEE International Conference on in Service Operations, Logistics, and Informatics, pp. 45–50. IEEE (2011)

16. Torgo, L., Ribeiro, R.P., Pfahringer, B., Branco, P.: SMOTE for regression. In: Reis, L.P., Correia, L., Cascalho, J. (eds.) EPIA 2013. LNCS, vol. 8154, pp. 378–389. Springer, Heidelberg (2013)

17. Torgo, L., Branco, P., Ribeiro, R.P., Pfahringer, B.: Resampling strategies for regression. Expert Systems (2014)

18. Ribeiro, R.P.: Utility-based Regression. PhD thesis, Dep. Computer Science, Faculty of Sciences - University of Porto (2011)

19. Torgo, L., Ribeiro, R.: Precision and recall for regression. In: Gama, J., Costa, V.S., Jorge, A.M., Brazdil, P.B. (eds.) DS 2009. LNCS, vol. 5808, pp. 332–346. Springer, Heidelberg (2009)

20. Milborrow, S.: earth: Multivariate Adaptive Regression Spline Models. Derived from mda:mars by Trevor Hastie and Rob Tibshirani (2012)

21. Dimitriadou, E., Hornik, K., Leisch, F., Meyer, D., Weingessel, A.: e1071: Misc Functions of the Department of Statistics (e1071), TU Wien (2011)

22. Liaw, A., Wiener, M.: Classification and regression by randomforest. R News 2(3), 18–22 (2002)

23. U.S.C. Bureau, Population Estimates (2012). http://www.census.gov/popest/data/index.html (accessed: January 23, 2015)

Distance-Based Decision Tree Algorithms for Label Ranking

Cláudio Rebelo de Sá[1,3]([✉]), Carla Rebelo[3], Carlos Soares[2,3],
and Arno Knobbe[1]

[1] LIACS Universiteit Leiden, Leiden, The Netherlands
{c.f.de.sa,a.j.knobbe}@liacs.leidenuniv.nl
[2] Faculdade de Engenharia, Universidade do Porto, Porto, Portugal
csoares@fe.up.pt
[3] INESCTEC Porto, Porto, Portugal

Abstract. The problem of Label Ranking is receiving increasing attention from several research communities. The algorithms that have developed/adapted to treat rankings as the target object follow two different approaches: distribution-based (e.g., using Mallows model) or correlation-based (e.g., using Spearman's rank correlation coefficient). Decision trees have been adapted for label ranking following both approaches. In this paper we evaluate an existing correlation-based approach and propose a new one, Entropy-based Ranking trees. We then compare and discuss the results with a distribution-based approach. The results clearly indicate that both approaches are competitive.

1 Introduction

Label Ranking (LR) is an increasingly popular topic in the machine learning literature [7,8,18,19,24]. LR studies a problem of learning a mapping from instances to rankings over a finite number of predefined labels. It can be considered as a natural generalization of the conventional classification problem, where only a single label is requested instead of a ranking of all labels [6]. In contrast to a classification setting, where the objective is to assign examples to a specific class, in LR we are interested in assigning a complete preference order of the labels to every example.

There are two main approaches to the problem of LR: methods that transform the ranking problem into multiple binary problems and methods that were developed or adapted to treat the rankings as target objects, without any transformation. An example of the former is the ranking by pairwise comparison of [11]. Examples of algorithms that were adapted to deal with rankings as the target objects include decision trees [6,23], naive Bayes [1] and k-Nearest Neighbor [3,6].

Some of the latter adaptations are based on statistical distribution of rankings (e.g., [5]) while others are based on rank correlation measures (e.g., [19,23]). In this paper we carry out an empirical evaluation of decision tree approaches for LR based on correlation measures and compare it to distribution-based approaches.

© Springer International Publishing Switzerland 2015
F. Pereira et al. (Eds.) EPIA 2015, LNAI 9273, pp. 525–534, 2015.
DOI: 10.1007/978-3-319-23485-4_52

We implemented and analyzed the algorithm previously presented in [17]. We also propose a new decision tree approach for LR, based on the previous one, which uses information gain as splitting criterion. The results clearly indicate that both are viable LR methods and are competitive with state of the art methods.

2   Label Ranking

The Label Ranking (LR) task is similar to classification. In classification, given an instance x from the instance space $\mathbb{X}$, the goal is to predict the label (or class) λ to which x belongs, from a pre-defined set $\mathcal{L} = \{\lambda_1, \ldots, \lambda_k\}$. In LR, the goal is to predict the ranking of the labels in $\mathcal{L}$ that are associated with x [11]. A ranking can be represented as a total order over $\mathcal{L}$ defined on the permutation space Ω. In other words, a total order can be seen as a permutation π of the set $\{1, \ldots, k\}$, such that $\pi(a)$ is the position of λ_a in π.

As in classification, we do not assume the existence of a deterministic $\mathbb{X} \to \Omega$ mapping. Instead, every instance is associated with a *probability distribution* over Ω [6]. This means that, for each $x \in \mathbb{X}$, there exists a probability distribution $\mathcal{P}(\cdot|x)$ such that, for every $\pi \in \Omega$, $\mathcal{P}(\pi|x)$ is the probability that π is the ranking associated with x. The goal in LR is to learn the mapping $\mathbb{X} \to \Omega$. The training data is a set of instances $D = \{\langle x_i, \pi_i \rangle\}, i = 1, \ldots, n$, where x_i is a vector containing the values $x_i^j, j = 1, \ldots, m$ of m independent variables describing instance i and π_i is the corresponding target ranking.

Given an instance x_i with label ranking π_i, and the ranking $\hat{\pi}_i$ predicted by an LR model, we evaluate the accuracy of the prediction with a loss function on Ω. One such function is the number of discordant label pairs,

$$\mathcal{D}(\pi, \hat{\pi}) = \#\{(a,b)|\pi(a) > \pi(b) \wedge \hat{\pi}(a) < \hat{\pi}(b)\}$$

If normalized to the interval $[-1, 1]$, this function is equivalent to Kendall's τ coefficient [12], which is a correlation measure where $\mathcal{D}(\pi, \pi) = 1$ and $\mathcal{D}(\pi, \pi^{-1}) = -1$ (π^{-1} denotes the inverse order of π).

The accuracy of a model can be estimated by averaging this function over a set of examples. This measure has been used for evaluation in recent LR studies [6,21] and, thus, we will use it here as well. However, other correlation measures, like Spearman's rank correlation coefficient [22], can also be used.

2.1   Ranking Trees

One of the advantages of tree-based models is how they can clearly express information about the problem because their structure is relatively easy to interpret even for people without a background on learning algorithms. It is also possible to obtain information about the importance of the various attributes for the prediction depending on how close to the root they are used. The Top-Down Induction of Decision Trees (TDIDT) algorithm is commonly used for induction of decision trees [13]. It is a recursive partitioning algorithm that iteratively splits data into smaller subsets which are increasingly more homogeneous in terms of the target variable (Algorithm 1).

It starts by determining the split that optimizes a given splitting criterion. A split is a test on one of the attributes that divides the dataset into two disjoint subsets. For instance, given a numerical attribute x^2, a split could be $x^2 \geq 5$. Without a stopping criterion, the TDIDT algorithm only stops when the nodes are pure, i.e., when the value of the target attribute is the same for all examples in the node. This usually leads the algorithm to overfit, i.e., to generate models that fit not only to the patterns in the data but also to the noise. One approach to address this problem is to introduce a stopping criterion in the algorithm that tests whether the best split is significantly improving the quality of the model. If not, the algorithm stops and returns a leaf node. This node is represented by the prediction that will be made for new examples that fall into that node. This prediction is generated by a rule that solves potential conflicts in the set of training examples that are in the node. In classification, the prediction rule is usually the most frequent class among the training examples. If the stopping criterion is not verified, then the algorithm is executed recursively for the subsets of the data obtained based on the best split.

Algorithm 1. TDIDT algorithm

BestSplit = Test of the attributes that optimizes the SPLITTING CRITERION
if STOPPING CRITERION == TRUE **then**
    Determine the leaf prediction based on the target values of the examples in D
    Return a leaf node with the corresponding LEAF PREDICTION
else
    LeftSubtree = TDIDT($D_{\neg BestSplit}$)
    RightSubtree = TDIDT($D_{BestSplit}$)
end if

An adaptation of the TDIDT algorithm for the problem of learning rankings has been proposed [23], called Ranking Trees (RT) which is based on the clustering trees algorithm [2]. Adaptation of this algorithm for label ranking involves an appropriate choice of the splitting criterion, stopping criterion and the prediction rule.

Splitting Criterion. The splitting criterion is a measure that quantifies the quality of a given partition of the data. It is usually applied to all the possible splits of the data that can be made based on individual tests of the attributes.

In RT the goal is to obtain leaf nodes that contain examples with target rankings as similar between themselves as possible. To assess the similarity between the rankings of a set of training examples, we compute the mean correlation between them, using Spearman's correlation coefficient. The quality of the split is given by the weighted mean correlation of the values obtained for the subsets, where the weight is given by the number of examples in each subset.

Table 1. Illustration of the splitting criterion

Attribute	Condition		Negated condition	
	values	rank corr.	values	rank corr.
x^1	a	0.3	b, c	-0.2
	b	0.2	a, c	0.1
	c	0.5	a, b	0.2
x^2	< 5	-0.1	≥ 5	0.1

The splitting criterion of ranking trees is illustrated both for nominal and numerical attributes in Table 1. The nominal attribute x^1 has three values (a, b and c). Therefore, three binary splits are possible. For the numerical attribute x^2, a split can be made in between every pair of consecutive values. In this case, the best split is $x^1 = c$, with a mean correlation of 0.5 for the training examples that verify the test and a mean correlation of 0.2 for the remaining, i.e., the training examples for which $x^1 = a$ or $x^1 = b$.

Stopping Criterion. The stopping criterion is used to determine if it is worthwhile to make a split to avoid overfitting [13]. A split should only be made if the similarity between examples in the subsets increases substantially. Let S_{parent} be the similarity between the examples in the parent node and S_{split} the weighted mean similarity in the subsets obtained with the best split. The stopping criterion is defined in [17] as follows:

$$(1 + S_{parent}) \geq \gamma(1 + S_{split}) \tag{1}$$

Note that the significance of the increase in similarity is controlled by the γ parameter.

Prediction Rule. The prediction rule is a method to generate a prediction from the (possibly conflicting) target values of the training examples in a leaf node. In RT, the method that is used to aggregate the q rankings that are in the leaves is based on the mean ranks of the items in the training examples that fall into the corresponding leaf. The average rank for each setting is $\overline{\pi}(j) = \sum_i \pi_i(j)/n$. The predicted ranking $\hat{\pi}$ will be the average ranking $\overline{\pi}$ after assigning ranks to $\pi(j)$. Table 2 illustrates the prediction rule used in this work.

Table 2. Illustration of the prediction rule.

	λ_1	λ_2	λ_3	λ_4
π_1	1	3	2	4
π_2	2	1	4	3
$\overline{\pi}$	1.5	2	3	3.5
$\hat{\pi}$	1	2	3	4

2.2 Entropy Ranking Trees

Decision trees, like ID3 [15], use Information Gain (IG) as a splitting criterion to look for the best split points.

Information Gain. IG is a statistical property that measures the difference in entropy, between the prior and actual state relatively to a target variable [13]. In other words, considering a set S of size n_S, as entropy - H - is a measure of disorder, IG is basically how much uncertainty in S is reduced after splitting on attribute A:

$$IG(A, T; S) = H(S) - \frac{|S_1|}{n_S} H(S_1) - \frac{|S_2|}{n_S} H(S_2)$$

where $|S_1|$ and $|S_2|$ are the number of instances on the left side (S_1) and the number of instances on the right side (S_2), respectively, of the cut point T in attribute A.

Using the same tree generation algorithm, the TDIDT (Section 2.1), we propose an alternative approach of decision trees for ranking data, the Entropy-based Ranking Trees (ERT). The difference is on the splitting and stopping criteria. ERT use IG to assess the splitting points and $MDLPC$ [10] as stopping criterion. Using the measure of entropy for rankings [20], the splitting and stopping criteria come in a natural way.

The entropy for rankings [20] is defined as:

$$H_{ranking}(S) = \sum_{i=1}^{K} P(\pi_i, S) \log(P(\pi_i, S)) \log(\overline{kt}(S)) \tag{2}$$

where K is the number of distinct rankings in S and $\overline{kt}(S)$ is the average normalized Kendall τ distance in the subset S:

$$\overline{kt}(S) = \frac{\sum_{i=1}^{K} \sum_{j=1}^{n} \frac{\tau(\pi_i, \pi_j)+1}{2}}{K \times n_S}$$

where K is the number of distinct target values in S.

As in Section 2.1 the leafs of the tree should not be forced to have pure leafs. Instead, they should have a stop criterion to avoid overfitting and be robust to noise in rankings. As shown in [20], the $MDLPC$ $Criterion$ can be used as a splitting criterion with the adapted version of entropy $H_{ranking}$. This entropy measure also works with partial orders, however, in this work, we only use total orders.

One other ranking tree approach based in Gini Impurity, which will not be presented in detail in this work, was proposed in [25].

3 Experimental Setup

The data sets in this work were taken from KEBI Data Repository in the Philipps University of Marburg [6] (Table 3). Two different transformation methods were used to generate these datasets: (A) the target ranking is a permutation of the classes of the original target attribute, derived from the probabilities generated by a naive Bayes classifier; (B) the target ranking is derived for each example from the order of the values of a set of numerical variables, which are no longer used as independent variables. Although these are somewhat artificial datasets, they are quite useful as benchmarks for LR algorithms.

The statistics of the datasets used in our experiments is presented in Table 3. U_π is the proportion of distinct target rankings for a given dataset.

Table 3. Summary of the datasets

Datasets	type	#examples	#labels	#attributes	U_π
autorship	A	841	4	70	2%
bodyfat	B	252	7	7	94%
calhousing	B	20,640	4	4	0.1%
cpu-small	B	8,192	5	6	1%
elevators	B	16,599	9	9	1%
fried	B	40,769	5	9	0.3%
glass	A	214	6	9	14%
housing	B	506	6	6	22%
iris	A	150	3	4	3%
pendigits	A	10,992	10	16	19%
segment	A	2310	7	18	6%
stock	B	950	5	5	5%
vehicle	A	846	4	18	2%
vowel	A	528	11	10	56%
wine	A	178	3	13	3%
wisconsin	B	194	16	16	100%

The code for all the examples in this paper has been written in R ([16]).

The performance of the LR methods was estimated using a methodology that has been used previously for this purpose [11]. It is based on the ten-fold cross validation performance estimation method. The evaluation measure is Kendall's τ and the performance of the methods was estimated using ten-fold cross-validation.

4 Results

RT uses a parameter, γ, that can affect the accuracy of the model. A $\gamma \geq 1$ does not increase the purity of nodes. On the other hand, small γ values will rarely generate any nodes. We vary γ from 0.50 to 0.99 and measure the accuracy on several KEBI datasets.

To show in what extent γ affects the accuracy of RT we show in Figure 1 the results obtained for some of the datasets in Table 3. From Figure 1 it is clear

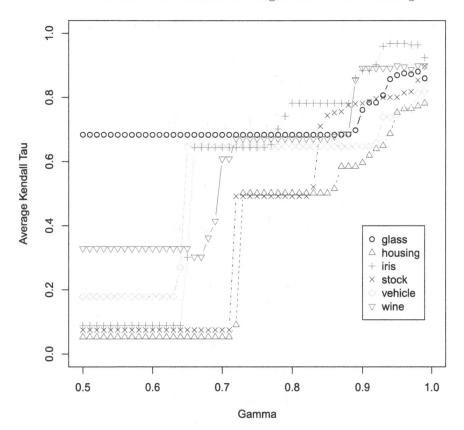

Fig. 1. Comparison of the accuracy obtained on some datasets by RT as γ varies from 0.5 to 0.99.

that γ plays an important role in the accuracy of RT. It seems that the best values lie between 0.95 and 0.98. We will use $\gamma = 0.98$ for the Ranking Tees (RT).

Table 4 presents the results obtained by the two methods presented in comparison to the results for Label Ranking Trees (LRT) obtained in [6]. Even though LRT perform better in the cases presented, given the closer values to it, both RT and ERT give interesting results.

To compare different ranking methods we use a method proposed in [4] which is a combination of Friedmans test and Dunns Multiple Comparison Procedure [14]. First we run the Friedman's test to check whether the results are different or not, with the following hypotheses:

Table 4. Results obtained for Ranking Trees on KEBI datasets. (The mean accuracy is represented in terms of Kendall's tau, τ)

	RT	ERT	LRT
authorship	.879	.890	.882
bodyfat	.104	.183	.117
calhousing	.181	.292	.324
cpu-small	.461	.437	.447
elevators	.710	.758	.760
fried	.796	.773	.890
glass	.881	.854	.883
housing	.773	.704	.797
iris	.964	.853	.947
pendigits	.055	.042	.935
segment	.895	.902	.949
stock	.854	.859	.895
vehicle	.813	.786	.827
vowel	.085	.054	.794
wine	.899	.907	.882
wisconsin	-.039	-.035	.343

Table 5. P-values obtained for the comparison of the 3 methods

	RT	ERT	LRT
RT		1.0000	0.2619
ERTt	1.0000		0.1529
LRT	0.2619	0.1529	

H_0. There is no difference in the mean average correlation coefficients for the 3 methods

H_1. There are some differences in the mean average correlation coefficients for the three methods

Using the *friedman.test* function from the *stats* package [16] we got a p-value $< 1\%$, which shows strong evidence against H_0.

Now that we know that there are some differences between the 3 methods we will test which are different from one another with the Dunns Multiple Comparison Procedure [14]. Using the R package *dunn.test* [9] with a Bonferroni adjustment, as in [4], we tested the following hypotheses for each pair of of methods a and b:

H_0. There is no difference in the mean average correlation coefficients between a and b

H_1. There is some difference in the mean average correlation coefficients between a and b

The p-values obtained are presented in Table 5. Table 5 indicates that there is no strong statistically evidence that the methods are different. One other conclusion is that both RT and ERT are very equivalent approaches. While RT and ERT does not seem to outperform LRT in most of the cases studied, from the statical tests we can say that both approaches are competitive.

5 Conclusions

In this work we implemented a decision tree method for Label Ranking, *Ranking Trees* (RT) and proposed an alternative approach *Entropy-based Ranking Trees* (ERT). We also present an empirical evaluation on several datasets of correlation-based methods, RT and ERT, and compare with the state of the art distribution-based *Label Ranking Trees* (LRT). The results indicate that both RT and ERT are reliable LR methods.

Our implementation of Ranking Trees (RT) shows that the method is a competitive approach in the LR field. We showed that the input parameter, γ, can have a great impact on the accuracy of the method. The tests performed on KEBI datasets indicate that the best results are obtained when $0.95 < \gamma < 1$.

The method proposed in this paper, ERT, which uses IG as a splitting criterion achieved very similar results to the RT presented in [17]. Statistical tests indicated that there is no strong evidence that the methods (RT, ERT and LRT) are significantly different. This means that both RT and ERT are valid approaches, and, since they are correlation-based methods, we can also say that this kind of approaches is also worth pursuing.

References

1. Aiguzhinov, A., Soares, C., Serra, A.P.: A similarity-based adaptation of naive bayes for label ranking: application to the metalearning problem of algorithm recommendation. In: Pfahringer, B., Holmes, G., Hoffmann, A. (eds.) DS 2010. LNCS, vol. 6332, pp. 16–26. Springer, Heidelberg (2010)
2. Blockeel, H., Raedt, L.D., Ramon, J.: Top-down induction of clustering trees. CoRR cs.LG/0011032 (2000). http://arxiv.org/abs/cs.LG/0011032
3. Brazdil, P., Soares, C., Costa, J.: Ranking Learning Algorithms: Using IBL and Meta-Learning on Accuracy and Time Results. Machine Learning **50**(3), 251–277 (2003)
4. Brazdil, P., Soares, C., da Costa, J.P.: Ranking learning algorithms: Using IBL and meta-learning on accuracy and time results. Machine Learning **50**(3), 251–277 (2003). http://dx.doi.org/10.1023/A:1021713901879
5. Cheng, W., Dembczynski, K., Hüllermeier, E.: Label ranking methods based on the plackett-luce model. In: ICML, pp. 215–222 (2010)
6. Cheng, W., Huhn, J.C., Hüllermeier, E.: Decision tree and instance-based learning for label ranking. In: Proceedings of the 26th Annual International Conference on Machine Learning, ICML 2009, June 14–18, Montreal, Quebec, Canada, pp. 161–168 (2009)
7. Cheng, W., Hüllermeier, E.: Label ranking with abstention: Predicting partial orders by thresholding probability distributions (extended abstract). Computing Research Repository, CoRR abs/1112.0508 (2011). http://arxiv.org/abs/1112.0508

8. Cheng, W., Hüllermeier, E., Waegeman, W., Welker, V.: Label ranking with partial abstention based on thresholded probabilistic models. In: Advances in Neural Information Processing Systems 25: 26th Annual Conference on Neural Information Processing Systems 2012. Proceedings of a meeting held December 3–6, Lake Tahoe, Nevada, United States, pp. 2510–2518 (2012). http://books.nips.cc/papers/files/nips25/NIPS2012_1200.pdf

9. Dinno, A.: dunn.test: Dunn's Test of Multiple Comparisons Using Rank Sums, r package version 1.2.3 (2015). http://CRAN.R-project.org/package=dunn.test

10. Fayyad, U.M., Irani, K.B.: Multi-interval discretization of continuous-valued attributes for classification learning. In: Proceedings of the 13th International Joint Conference on Artificial Intelligence, August 28-September 3, Chambéry, France, pp. 1022–1029 (1993)

11. Hüllermeier, E., Fürnkranz, J., Cheng, W., Brinker, K.: Label ranking by learning pairwise preferences. Artificial Intelligence 172(16–17), 1897–1916 (2008)

12. Kendall, M., Gibbons, J.: Rank correlation methods. Griffin London (1970)

13. Mitchell, T.: Machine Learning. McGraw-Hill (1997)

14. Neave, H., Worthington, P.: Distribution-free Tests. Routledge (1992). http://books.google.nl/books?id=1Y1QcgAACAAJ

15. Quinlan, J.R.: Induction of decision trees. Machine Learning 1(1), 81–106 (1986). http://dx.doi.org/10.1023/A:1022643204877

16. R Development Core Team: R: A Language and Environment for Statistical Computing. R Foundation for Statistical Computing, Vienna, Austria (2010). http://www.R-project.org ISBN 3-900051-07-0

17. Rebelo, C., Soares, C., Costa, J.: Empirical evaluation of ranking trees on some metalearning problems. In: Chomicki, J., Conitzer, V., Junker, U., Perny, P. (eds.) Proceedings 4th AAAI Multidisciplinary Workshop on Advances in Preference Handling (2008)

18. Ribeiro, G., Duivesteijn, W., Soares, C., Knobbe, A.: Multilayer perceptron for label ranking. In: Villa, A.E.P., Duch, W., Érdi, P., Masulli, F., Palm, G. (eds.) ICANN 2012, Part II. LNCS, vol. 7553, pp. 25–32. Springer, Heidelberg (2012)

19. de Sá, C.R., Soares, C., Jorge, A.M., Azevedo, P., Costa, J.: Mining association rules for label ranking. In: Huang, J.Z., Cao, L., Srivastava, J. (eds.) PAKDD 2011, Part II. LNCS, vol. 6635, pp. 432–443. Springer, Heidelberg (2011)

20. de Sá, C.R., Soares, C., Knobbe, A.: Entropy-based discretization methods for ranking data. Information Sciences in Press (2015) (in press)

21. de Sá, C.R., Soares, C., Knobbe, A., Azevedo, P., Jorge, A.M.: Multi-interval discretization of continuous attributes for label ranking. In: Fürnkranz, J., Hüllermeier, E., Higuchi, T. (eds.) DS 2013. LNCS, vol. 8140, pp. 155–169. Springer, Heidelberg (2013)

22. Spearman, C.: The proof and measurement of association between two things. American Journal of Psychology 15, 72–101 (1904)

23. Todorovski, L., Blockeel, H., Džeroski, S.: Ranking with predictive clustering trees. In: Elomaa, T., Mannila, H., Toivonen, H. (eds.) ECML 2002. LNCS (LNAI), vol. 2430, pp. 444–455. Springer, Heidelberg (2002)

24. Vembu, S., Gärtner, T.: Label ranking algorithms: A survey. In: Fürnkranz, J., Hüllermeier, E. (eds.) Preference Learning, pp. 45–64. Springer, Heidelberg (2010)

25. Xia, F., Zhang, W., Li, F., Yang, Y.: Ranking with decision tree. Knowl. Inf. Syst. 17(3), 381–395 (2008). http://dx.doi.org/10.1007/s10115-007-0118-y

A Proactive Intelligent Decision Support System for Predicting the Popularity of Online News

Kelwin Fernandes[1]([✉]), Pedro Vinagre[2], and Paulo Cortez[2]

[1] INESC TEC Porto/Universidade Do Porto, Porto, Portugal
[2] ALGORITMI Research Centre, Universidade Do Minho, Braga, Portugal
kelwinfc@gmail.com

Abstract. Due to the Web expansion, the prediction of online news popularity is becoming a trendy research topic. In this paper, we propose a novel and proactive Intelligent Decision Support System (IDSS) that analyzes articles prior to their publication. Using a broad set of extracted features (e.g., keywords, digital media content, earlier popularity of news referenced in the article) the IDSS first predicts if an article will become popular. Then, it optimizes a subset of the articles features that can more easily be changed by authors, searching for an enhancement of the predicted popularity probability. Using a large and recently collected dataset, with 39,000 articles from the Mashable website, we performed a robust rolling windows evaluation of five state of the art models. The best result was provided by a Random Forest with a discrimination power of 73%. Moreover, several stochastic hill climbing local searches were explored. When optimizing 1000 articles, the best optimization method obtained a mean gain improvement of 15 percentage points in terms of the estimated popularity probability. These results attest the proposed IDSS as a valuable tool for online news authors.

Keywords: Popularity prediction · Online news · Text mining · Classification · Stochastic local search

1 Introduction

Decision Support Systems (DSS) were proposed in the mid-1960s and involve the use of Information Technology to support decision-making. Due to advances in this field (e.g., Data Mining, Metaheuristics), there has been a growing interest in the development of Intelligent DSS (IDSS), which adopt Artificial Intelligence techniques to decision support [1]. The concept of Adaptive Business Intelligence (ABI) is a particular IDSS that was proposed in 2006 [2]. ABI systems combine prediction and optimization, which are often treated separately by IDSS, in order to support decisions more efficiently. The goal is to first use data-driven models for predicting what is more likely to happen in the future, and then use modern optimization methods to search for the best possible solution given what can be currently known and predicted.

© Springer International Publishing Switzerland 2015
F. Pereira et al. (Eds.) EPIA 2015, LNAI 9273, pp. 535–546, 2015.
DOI: 10.1007/978-3-319-23485-4_53

Within the expansion of the Internet and Web 2.0, there has also been a growing interest in online news, which allow an easy and fast spread of information around the globe. Thus, predicting the popularity of online news is becoming a recent research trend (e.g., [3,4,5,6,7]). Popularity is often measured by considering the number of interactions in the Web and social networks (e.g., number of shares, likes and comments). Predicting such popularity is valuable for authors, content providers, advertisers and even activists/politicians (e.g., to understand or influence public opinion) [4]. According to Tatar et al. [8], there are two main popularity prediction approaches: those that use features only known after publication and those that do not use such features. The first approach is more common (e.g., [3,5,9,6,7]). Since the prediction task is easier, higher prediction accuracies are often achieved. The latter approach is more scarce and, while a lower prediction performance might be expected, the predictions are more useful, allowing (as performed in this work) to improve content prior to publication.

Using the second approach, Petrovic et al. [10] predicted the number of retweets using features related with the tweet content (e.g., number of hashtags, mentions, URLs, length, words) and social features related to the author (e.g., number of followers, friends, is the user verified). A total of 21 million tweets were retrieved during October 2010. Using a binary task to discriminate retweeted from not retweeted posts, a top F-1 score of 47% was achieved when both tweet content and social features were used. Similarly, Bandari et al. [4] focused on four types of features (news source, category of the article, subjectivity language used and names mentioned in the article) to predict the number of tweets that mention an article. The dataset was retrieved from Feedzilla and related with one week of data. Four classification methods were tested to predict three popularity classes (1 to 20 tweets, 20 to 100 tweets, more than 100; articles with no tweets were discarded) and results ranged from 77% to 84% accuracy, for Naïve Bayes and Bagging, respectively. Finally, Hensinger et al. [11] tested two prediction binary classification tasks: popular/unpopular and appealing/non appealing, when compared with other articles published in the same day. The data was related with ten English news outlets related with one year. Using text features (e.g., bag of words of the title and description, keywords) and other characteristics (e.g., date of publishing), combined with a Support Vector Machine (SVM), the authors obtained better results for the appealing task when compared with popular/unpopular task, achieving results ranging from 62% to 86% of accuracy for the former, and 51% to 62% for the latter.

In this paper, we propose a novel proactive IDSS that analyzes online news *prior* to their publication. Assuming an ABI approach, the popularity of a candidate article is first estimated using a prediction module and then an optimization module suggests changes in the article content and structure, in order to maximize its expected popularity. Within our knowledge, there are no previous works that have addressed such proactive ABI approach, combining prediction and optimization for improving the news content. The prediction module uses a large list of inputs that includes purely new features (when compared with the literature [4,11,10]): digital media content (e.g., images, video); earlier popularity of news referenced in the article; average number of shares of keywords prior to publication; and natural language features (e.g., title polarity, Latent Dirichlet Allocation topics). We adopt the common binary (popular/unpopular) task and test five state of the art methods (e.g., Random Forest, Adaptive Boosting, SVM), under a realistic rolling windows. Moreover, we use the trendy Mashable (mashable.com/) news content, which was not previously studied when predicting popularity, and collect a recent and large dataset related with the last two years (a much larger time period when compared with the literature). Furthermore, we also optimize news content using a local search method (stochastic hill climbing) that searches for enhancements in a partial set of features that can be more easily changed by the user.

2   Materials and Methods

2.1   Data Acquisition and Preparation

We retrieved the content of all the articles published in the last two years from Mashable, which is one of the largest news websites. All data collection and processing procedures described in this work (including the prediction and optimization modules) were implemented in Python by the authors. The data was collected during a two year period, from January 7 2013 to January 7 2015. We discarded a small portion of special occasion articles that did not follow the general HTML structure, since processing each occasion type would require a specific parser. We also discarded very recent articles (less than 3 weeks), since the number of Mashable shares did not reach convergence for some of these articles (e.g., with less than 4 days) and we also wanted to keep a constant number of articles per test set in our rolling windows assessment strategy (see Section 2.3). After such preprocessing, we ended with a total of 39,000 articles, as shown in Table 1. The collected data was donated to the UCI Machine Learning repository (http://archive.ics.uci.edu/ml/).

Table 1. Statistical measures of the Mashable dataset.

Number of articles	Total days	Articles per day			
		Average	Standard Deviation	Min	Max
39,000	709	55.00	22.65	12	105

We extracted an extensive set (total of 47) features from the HTML code in order to turn this data suitable for learning models, as shown in Table 2. In the table, the attribute types were classified into: number – integer value; ratio – within $[0,1]$; bool – $\in \{0,1\}$; and nominal. Column **Type** shows within brackets (#) the number of variables related with the attribute. Similarly to what is executed in [6,7], we performed a logarithmic transformation to scale the unbounded numeric features (e.g., number of words in article), while the nominal attributes were transformed with the common *1-of-C* encoding.

We selected a large list of characteristics that describe different aspects of the article and that were considered possibly relevant to influence the number of shares. Some of the features are dependent of particularities of the Mashable service: articles often reference other articles published in the same service; and articles have meta-data, such as keywords, data channel type and total number of shares (when considering Facebook, Twitter, Google+, LinkedIn, Stumble-Upon and Pinterest). Thus, we extracted the minimum, average and maximum number of shares (known before publication) of all Mashable links cited in the article. Similarly, we rank all article keyword average shares (known before publication), in order to get the worst, average and best keywords. For each of these keywords, we extract the minimum, average and maximum number of shares. The data channel categories are: "lifestyle","bus","entertainment","socmed", "tech","viral" and "world".

We also extracted several natural language processing features. The Latent Dirichlet Allocation (LDA) [12] algorithm was applied to all Mashable texts (known before publication) in order to first identify the five top relevant topics and then measure the closeness of current article to such topics. To compute the subjectivity and polarity sentiment analysis, we adopted the Pattern web mining module (http://www.clips.ua.ac.be/pattern) [13], allowing the computation of sentiment polarity and subjectivity scores.

Table 2. List of attributes by category.

Feature	Type (#)	Feature	Type (#)
Words		**Keywords**	
Number of words in the title	number (1)	Number of keywords	number (1)
Number of words in the article	number (1)	Worst keyword (min./avg./max. shares)	number (3)
Average word length	number (1)	Average keyword (min./avg./max. shares)	number (3)
Rate of non-stop words	ratio (1)	Best keyword (min./avg./max. shares)	number (3)
Rate of unique words	ratio (1)	Article category (Mashable data channel)	nominal (1)
Rate of unique non-stop words	ratio (1)	**Natural Language Processing**	
Links		Closeness to top 5 LDA topics	ratio (5)
Number of links	number (1)	Title subjectivity	ratio (1)
Number of Mashable article links	number (1)	Article text subjectivity score and	
Minimum, average and maximum number of shares of Mashable links	number (3)	its absolute difference to 0.5	ratio (2)
		Title sentiment polarity	ratio (1)
Digital Media		Rate of positive and negative words	ratio (2)
Number of images	number (1)	Pos. words rate among non-neutral words	ratio (1)
Number of videos	number (1)	Neg. words rate among non-neutral words	ratio (1)
Time		Polarity of positive words (min./avg./max.)	ratio (3)
Day of the week	nominal (1)	Polarity of negative words (min./avg./max.)	ratio (3)
Published on a weekend?	bool (1)	Article text polarity score and	
		its absolute difference to 0.5	ratio (2)

Target	Type (#)
Number of article Mashable shares	number (1)

2.2   Intelligent Decision Support System

Following the ABI concept, the proposed IDSS contains three main modules (Figure 1): data extraction and processing, prediction and optimization. The first module executes the steps described in Section 2.1 and it is responsible for collecting the online articles and computing their respective features. The prediction module first receives the processed data and splits it into training, validation and test sets (data separation). Then, it tunes and fits the classification models (model training and selection). Next, the best classification model is stored and used to provide article success predictions (popularity estimation). Finally, the optimization module searches for better combinations of a subset of the current article content characteristics. During this search, there is an heavy use of the classification model (the oracle). Also, some of the new searched feature combinations may require a recomputing of the respective features (e.g., average keyword minimum number of shares). In the figure, such dependency is represented by the arrow between the feature extraction and optimization. Once the optimization is finished, a list of article change suggestions is provided to the user, allowing her/him to make a decision.

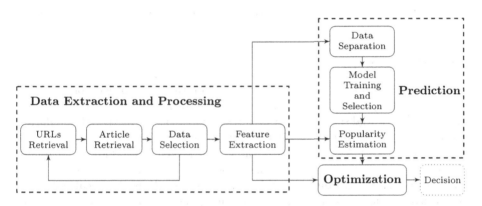

Fig. 1. Flow diagram describing the IDSS behavior.

2.3 Prediction Module

We adopted the Scikit learn [14] library for fitting the prediction models. Similarly to what is executed in [10,4,11], we assume a binary classification task, where an article is considered "popular" if the number of shares is higher than a fixed decision threshold (D_1), else it is considered "unpopular".

In this paper, we tested five classification models: Random Forest (RF); Adaptive Boosting (AdaBoost); SVM with a Radial Basis Function (RBF) kernel; K-Nearest Neighbors (KNN) and Naïve Bayes (NB). A grid search was used to search for the best hyperparameters of: RF and AdaBoost (number of trees); SVM (C trade-off parameter); and KNN (number of neighbors). During this grid search, the training data was internally split into training (70%) and validation sets (30%) by using a random holdout split. Once the best hyperparameter is selected, then the model is fit to all training data.

The receiver operating characteristic (ROC) curve shows the performance of a two class classifier across the range of possible threshold ($D_2 \in [0, 1]$) values, plotting one minus the specificity (x-axis) versus the sensitivity (y-axis) [15]. In this work, the classification methods assume a probabilistic modeling, where a class is considered positive if its predicted probability is $p > D_2$. We computed several classification metrics: Accuracy, Precision, Recall, F1 score (all using a fixed $D_2 = 0.5$); and the Area Under the ROC (AUC, which considers all D_2 values). The AUC metric is the most relevant metric, since it measures the classifier's discrimination power and it is independent of the selected D_2 value [15]. The ideal method should present an AUC of 1.0, while an AUC of 0.5 denotes a random classifier. For achieving a robust evaluation, we adopt a rolling windows analysis [16]. Under this evaluation, a training window of W consecutive samples is used to fit the model and then L predictions are performed. Next, the training window is updated by replacing the L oldest samples with L more recent ones, in order to fit a new model and perform a new set of L predictions, and so on.

2.4   Optimization

Local search optimizes a goal by searching within the neighborhood of an initial solution. This type of search suits our IDSS optimization module, since it receives an article (the initial solution) and then tries to increase its predicted popularity probability by searching for possible article changes (within the neighborhood of the initial solution). An example of a simple local search method is the hill climbing, which iteratively searches within the neighborhood of the current solution and updates such solution when a better one is found, until a local optimum is reached or the method is stopped. In this paper, we used a stochastic hill climbing [2], which works as the pure hill climbing except that worst solutions can be selected with a probability of P. We tested several values of P, ranging from $P = 0$ (hill climbing) to $P = 1$ (Monte-Carlo random search).

For evaluating the quality of the solutions, the local search maximizes the probability for the "popular" class, as provided by the best classification model. Moreover, the search is only performed over a subset of features that are more suitable to be changed by the author (adaptation of content or change in day of publication), as detailed in Table 3. In each iteration, the neighborhood search space assumes small perturbations (increase or decrease) in the feature original values. For instance, if the current number of words in the title is $n = 5$, then a search is executed for a shorter ($n' = 4$) or longer ($n' = 6$) title. Since the day of the week was represented as a nominal variable, a random selection for a different day is assumed in the perturbation. Similarly, given that the set of keywords (K) is not numeric, a different perturbation strategy is proposed. For a particular article, we compute a list of suggested keywords K' that includes words that appear more than once in the text and that were used as keywords in previous articles. To keep the problem computationally tractable, we only considered the best five keywords in terms of their previous average shares. Then, we generate perturbations by adding one of the suggested keywords or by removing one of the original keywords. The average performance when optimizing N articles (i.e., N local searches), is evaluated using the Mean Gain (MG) and Conversion Rate (CR):

$$MG = \frac{1}{N} \sum_{i=1}^{N} (Q'_i - Q_i)$$

$$CR = \overline{U'}/U$$

(1)

where Q_i denotes the quality (estimated popularity probability) for the original article (i), Q'_i is the quality obtained using the local search, U is the number of unpopular articles (estimated probabilitity $\leq D_2$, for all N original articles) and $\overline{U'}$ is the number of converted articles (original estimated probability was $\leq D_2$ but after optimization changed to $> D_2$).

Table 3. Optimizable Features.

Feature	Perturbations
Number of words in the title (n)	$n' \in \{n-1, n+1\}, n \geq 0 \wedge n' \neq n$
Number of words in the content (n)	$n' \in \{n-1, n+1\}, n \geq 0 \wedge n' \neq n$
Number of images (n)	$n' \in \{n-1, n+1\}, n \geq 0 \wedge n' \neq n$
Number of videos (n)	$n' \in \{n-1, n+1\}, n \geq 0 \wedge n' \neq n$
Day of week (w)	$w' \in [0..7), w' \neq w$
Keywords (K)	$k' \in \{K \cup i\} \cup \{K - j\}, i \in K' \wedge j \in K$

3   Experiments and Results

3.1   Prediction

For the prediction experiments, we adopted the rolling windows scheme with a training window size of $W = 10,000$ and performing $L = 1,000$ predictions at each iteration. Under this setup, each classification model is trained 29 times (iterations), producing 29 prediction sets (each of size L). For defining a popular class, we used a fixed value of $D_1 = 1,400$ shares, which resulted in a balanced "popular"/"unpopular" class distribution in the first training set (first $10,000$ articles). The selected grid search ranges for the hyperparameters were: RF and AdaBoost – number of trees $\in \{10, 20, 50, 100, 200, 400\}$; SVM – $C \in \{2^0, 2^1, ..., 2^6\}$; and KNN – number of neighbors $\in \{1, 3, 5, 10, 20\}$.

Table 4 shows the obtained classification metrics, as computed over the union of all 29 test sets. In the table, the models were ranked according to their performance in terms of the AUC metric. The left of Figure 2 plots the ROC curves of the best (RF), worst (NB) and baseline (diagonal line, corresponds to random predictions) models. The plot confirms the RF superiority over the NB model for all D_2 thresholds, including more sensitive (x-axis values near zero, $D_2 >> 0.5$) or specific (x-axis near one, $D_2 << 0.5$) trade-offs. For the best model (RF), the right panel of Figure 2 shows the evolution of the AUC metric over the rolling windows iterations, revealing an interesting steady predictive performance over time. The best obtained result (AUC=0.73) is 23 percentage points higher than the random classifier. While not perfect, an interesting discrimination level, higher than 70%, was achieved.

Table 4. Comparison of models for the rolling window evaluation (best values in **bold**).

Model	Accuracy	Precision	Recall	F1	AUC
Random Forest (RF)	**0.67**	0.67	**0.71**	**0.69**	**0.73**
Adaptive Boosting (AdaBoost)	0.66	0.68	0.67	0.67	0.72
Support Vector Machine (SVM)	0.66	0.67	0.68	0.68	0.71
K-Nearest Neighbors (KNN)	0.62	0.66	0.55	0.60	0.67
Naïve Bayes (NB)	0.62	**0.68**	0.49	0.57	0.65

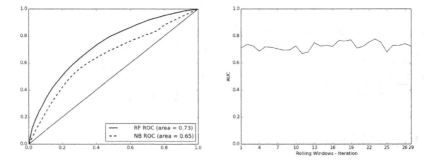

Fig. 2. ROC curves (left) and AUC metric distribution over time for RF (right).

Table 5 shows the relative importance (column **Rank** shows ratio values, # denotes the ranking of the feature), as measured by the RF algorithm when trained with all data (39,000 articles). Due to space limitations, the table shows the best 15 features and also the features that are used by the optimization module. The keyword related features have a stronger importance, followed by LDA based features and shares of Mashable links. In particular, the features that are optimized in the next section (**with keywords** subset) have a strong importance (33%) in the RF model.

3.2 Optimization

For the optimization experiments, we used the best classification model (RF), as trained during the last iteration of the rolling windows scheme. Then, we selected all articles from the last test set ($N = 1,000$) to evaluate the local search methods. We tested six stochastic hill climbing probabilities ($P \in \{0.0, 0.2, 0.4, 0.6, 0.8, 1.0\}$). We also tested two feature optimization subsets

Table 5. Ranking of features according to their importance in the RF model.

Feature	Rank (#)	Feature	Rank (#)
Avg. keyword (avg. shares)	0.0456 (1)	Closeness to top 1 LDA topic	0.0287 (11)
Avg. keyword (max. shares)	0.0389 (2)	Rate of unique non-stop words	0.0274 (12)
Closeness to top 3 LDA topic	0.0323 (3)	Article text subjectivity	0.0271 (13)
Article category (Mashable data channel)	0.0304 (4)	Rate of unique tokens words	0.0271 (14)
Min. shares of Mashable links	0.0297 (5)	Average token length	0.0271 (15)
Best keyword (avg. shares)	0.0294 (6)	Number of words	0.0263 (16)
Avg. shares of Mashable links	0.0294 (7)	Day of the week	0.0260 (18)
Closeness to top 2 LDA topic	0.0293 (8)	Number of words in the title	0.0161 (31)
Worst keyword (avg. shares)	0.0292 (9)	Number of images	0.0142 (34)
Closeness to top 5 LDA topic	0.0288 (10)	Number of videos	0.0082 (44)

related with Table 3: using all features except the keywords (*without keywords*) and using all features (*with keywords*). Each local search is stopped after 100 iterations. During the search, we store the best results associated with the iterations $I \in \{0, 1, 2, 4, 8, 10, 20, 40, 60, 80, 100\}$.

Figure 3 shows the final optimization performance (after 100 iterations) for variations of the stochastic probability parameter P and when considering the two feature perturbation subsets. The convergence of the local search (for different values of P) is also shown in Figure 3. The extreme values of P (0 – pure hill climbing; 1 – random search) produce lower performances when compared with their neighbor values. In particular, Figure 4 shows that the pure hill climbing is too greedy, performing a fast initial convergence that quickly gets flat. When using the *without keywords* subset, the best value of P is 0.2 for MG and 0.4 for CR metric. For the *with keywords* subset, the best value of P is 0.8 for both optimization metrics. Furthermore, the inclusion of keywords-related suggestions produces a substantial impact in the optimization, increasing the performance in both metrics. For instance, the MG metric increases from 0.05 to 0.16 in the best case ($P = 0.8$). Moreover, Figure 3 shows that the *without keywords* subset optimization is an easier task when compared with the *with keywords* search. As argued by Zhang and Dimitroff [17], metadata can play an important role on webpage visibility and this might explain the importance of the keywords in terms of its influence when predicting (Table 5) and when optimizing popularity (Figure 3).

For demonstration purposes, Figure 5 shows an example of the interface of the implemented IDSS prototype. A more recent article (from January 16 2015) was selected for this demonstration. The IDSS, in this case using the *without keywords* subset, estimated an increase in the popularity probability of 13 percentage points if several changes are executed, such as decreasing the number of title words from 11 to 10. In another example (not shown in the figure), using the *with keywords* subset, the IDSS advised a change from the keywords $K \in \{$ "television", "showtime", "uncategorized", "entertainment", "film", "homeland", "recaps"$\}$ to the set $K' \in \{$ "film", "relationship", "family", and "night"$\}$ for an article about the end of the "Homeland" TV show.

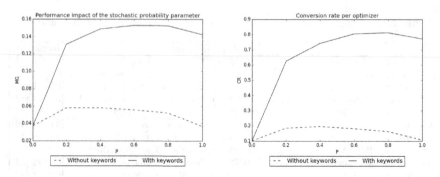

Fig. 3. Stochastic probability (P) impact on the Mean Gain (left) and in the Conversion Rate (right).

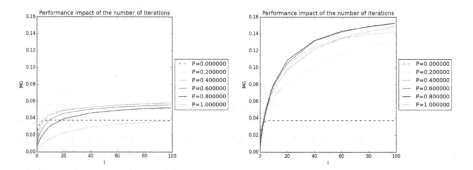

Fig. 4. Convergence of the local search under the *without keywords* (left) and *with keywords* (right) feature subsets (y-axis denotes the Mean Gain and x-axis the number of iterations).

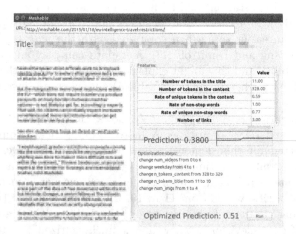

Fig. 5. Example of the interface of the IDSS prototype.

4   Conclusions

With the expansion of the Web, there is a growing interest in predicting online news popularity. In this work, we propose an Intelligent Decision Support System (IDSS) that first extracts a broad set of features that are known prior to an article publication, in order to predict its future popularity, under a binary classification task. Then, it optimizes a subset of the article features (that are more suitable to be changed by the author), in order to enhance its expected popularity.

Using large and recent dataset, with 39,000 articles collected during a 2 year period from the popular Mashable news service, we performed a rolling windows evaluation, testing five state of the art classification models under distinct metrics. Overall, the best result was achieved by a Random Forest (RF), with an overall area under the Receiver Operating Characteristic (ROC) curve of 73%, which corresponds to an acceptable discrimination. We also analyzed the importance of the RF inputs, revealing the keyword based features as one of the most important, followed by Natural Language Processing features and previous shares of Mashable links. Using the best prediction model as an oracle, we explored several stochastic hill climbing search variants aiming at the increase in the estimated article probability when changing two subsets of the article features (e.g., number of words in title). When optimizing 1,000 articles (from the last rolling windows test set), we achieved 15 percentage points in terms of the mean gain for the best local search setup. Considering the obtained results, we believe that the proposed IDSS is quite valuable for Mashable authors.

In future work, we intend to explore more advanced features related to content, such as trends analysis. Also, we plan to perform tracking of articles over time, allowing the usage of more sophisticated forecasting approaches.

Acknowledgements. This work has been supported by FCT - Fundação para a Ciência e Tecnologia within the Project Scope UID/CEC/00319/2013. The authors would like to thank Pedro Sernadela for his contributions in previous work.

References

1. Arnott, D., Pervan, G.: Eight key issues for the decision support systems discipline. Decision Support Systems **44**(3), 657–672 (2008)
2. Michalewicz, Z., Schmidt, M., Michalewicz, M., Chiriac, C.: Adaptive business intelligence. Springer (2006)
3. Ahmed, M., Spagna, S., Huici, F., Niccolini, S.: A peek into the future: predicting the evolution of popularity in user generated content. In: Proceedings of the sixth ACM international conference on Web search and data mining, pp. 607–616. ACM (2013)
4. Bandari, R., Asur, S., Huberman, B.A.: The pulse of news in social media: forecasting popularity. In: ICWSM (2012)
5. Kaltenbrunner, A., Gomez, V., Lopez, V.: Description and prediction of slashdot activity. In: Web Conference, LA-WEB 2007, pp. 57–66. IEEE, Latin American (2007)

6. Szabo, G., Huberman, B.A.: Predicting the popularity of online content. Communications of the ACM **53**(8), 80–88 (2010)
7. Tatar, A., Antoniadis, P., De Amorim, M.D., Fdida, S.: From popularity prediction to ranking online news. Social Network Analysis and Mining **4**(1), 1–12 (2014)
8. Tatar, A., de Amorim, M.D., Fdida, S., Antoniadis, P.: A survey on predicting the popularity of web content. Journal of Internet Services and Applications **5**(1), 1–20 (2014)
9. Lee, J.G., Moon, S., Salamatian, K.: Modeling and predicting the popularity of online contents with cox proportional hazard regression model. Neurocomputing **76**(1), 134–145 (2012)
10. Petrovic, S., Osborne, M., Lavrenko, V.: RT to win! predicting message propagation in twitter. In: Fifth International AAAI Conference on Weblogs and Social Media (ICWSM), pp. 586–589 (2011)
11. Hensinger, E., Flaounas, I., Cristianini, N.: Modelling and predicting news popularity. Pattern Analysis and Applications **16**(4), 623–635 (2013)
12. Blei, D.M., Ng, A.Y., Jordan, M.I.: Latent dirichlet allocation. Journal of Machine Learning Research **3**, 993–1022 (2003)
13. De Smedt, T., Nijs, L., Daelemans, W.: Creative web services with pattern. In: Proceedings of the Fifth International Conference on Computational Creativity (2014)
14. Pedregosa, F., Varoquaux, G., Gramfort, A., Michel, V., Thirion, B., Grisel, O., Blondel, M., Prettenhofer, P., Weiss, R., Dubourg, V., Vanderplas, J., Passos, A., Cournapeau, D., Brucher, M., Perrot, M., Duchesnay, E.: Scikit-learn: Machine learning in Python. Journal of Machine Learning Research **12**, 2825–2830 (2011)
15. Fawcett, T.: An introduction to roc analysis. Pattern Recognition Letters **27**(8), 861–874 (2006)
16. Tashman, L.J.: Out-of-sample tests of forecasting accuracy: an analysis and review. International Journal of Forecasting **16**(4), 437–450 (2000)
17. Zhang, J., Dimitroff, A.: The impact of metadata implementation on webpage visibility in search engine results (part ii). Information Processing & Management **41**(3), 691–715 (2005)

Periodic Episode Discovery Over Event Streams

Julie Soulas[1,2]($\boxtimes$) and Philippe Lenca[1,2]

[1] UMR 6285 Lab-STICC, Institut Mines-Telecom, Telecom Bretagne,
Technopôle Brest Iroise CS 83818, 29238 Brest Cedex 3, France
{julie.soulas,philippe.lenca}@telecom-bretagne.eu
[2] Université Européenne de Bretagne, Rennes, France

Abstract. Periodic behaviors are an important component of the life of most living species. Daily, weekly, or even yearly patterns are observed in both human and animal behaviors. These behaviors are searched as frequent periodic episodes in event streams. We propose an efficient algorithm for the discovery of frequent and periodic episodes. Update procedures allow us to take into account that behaviors also change with time, or because of external factors. The interest of our approach is illustrated on two real datasets.

1 Introduction

The discovery of patterns based on the temporality in their occurrences is of great interest in a wide range of applications, such as social interactions analysis [9], biological sustainability studies [10], elderly people monitoring [15], mobility data analysis [2], etc. The rhythm of the patterns appearances is studied in order to determine whether the patterns occur regularly (the time gaps between the occurrences are bounded [1,16]), periodically (some occurrences form repeating cycles of time intervals), or mostly in a specific time interval.

Periodicity highlights habits. For example, Li et al. [10] studied the travel behaviors of animals, building rules such as: *"From 6 pm to 6 am, it has 90% probability staying at location A"*. Soulas et al. [15] studied the living habits of elderly people, and discovered periodic episodes as such *"The user has 70% probability having breakfast around 9:22 ± 40 min"*. The discovery of such behaviors enhances understanding of the needs of the living beings, and the factors governing their behavior. It is also useful to detect anomalies in routines due to major events such as environmental change [10] or the onset of disorders [15].

The discovery of periodic behaviors and their evolution involves three sub-problems: (i) the detection of the periods (daily, weekly, etc), (ii) the discovery of periodic behaviors and (iii) the update of the periods and patterns when new data arrives. We here focus on sub-problems (ii) and (iii). Period determination has already been extensively studied [2,10]. Moreover, the experts in the target applications domains usually have background knowledge on the interesting periods, or require particular periods to be investigated: e.g., the physicians monitoring elderly people express the need to study daily and weekly behaviors.

F. Pereira et al. (Eds.) EPIA 2015, LNAI 9273, pp. 547–559, 2015.
DOI: 10.1007/978-3-319-23485-4_54

The main contributions of the paper are: a new frequent parallel episode mining algorithm on data streams; and a heuristic for the online estimation of the periodicity of the episodes. The rest of the paper is organized as follows: section 2 presents some prominent related work. Section 3 details the proposition for frequent periodic pattern mining and updating. Experiments (section 4) on two real datasets illustrate the interest of this approach. Finally, some conclusions are drawn, and ideas for future work are presented.

2   Related Work

Frequent episode mining has attracted a lot of attention since its introduction by Mannila et al. [12]. The algorithms (e.g. [11–13,18]) differ from one another by their target episodes (sequential or parallel), their search strategies (breadth or depth first search), the considered occurrences (contiguous, minimal, overlapping, etc.), and the way they count support. However, most algorithms consider only static data. The formalism used in this paper (see section 3.1) is loosely inspired from the formalism used in [11] and [18].

With the rapidly increasing amount of data recording devices (network traffic monitoring, smart houses, sensor networks,...), stream data mining has gained major attention. This evolution led to paradigm shifts. For an extensive problem statement and review of the current trends, see [6]. In particular, item set [3,17] and episode [11,13,14] mining in streams have been investigated. The application context of [14] is close to the behavior we are searching for: the focus is set on the extraction of human activities from home automation sensor streams. However, periodicity is not taken into account.

Due to their powerful descriptive and predictive capabilities, periodic patterns are studied in several domains. For instance, Kiran and Reddy [8] discover frequent and periodic patterns in transactional databases. Periodicity is also defined and used with event sequences, for example with the study of parallel episodes in home automation sensor data for the monitoring of elderly people [7,15]. These three periodic pattern mining algorithms process only static data.

To the best of our knowledge, few studies have focused on mining both frequent and periodic episodes over data streams. One can however point out some rather close studies: Li et al. [10] and Baratchi et al. [2] both use geo-spatial data in order to detect areas of interest for an individual (respectively eagles and people) and periodic movement patterns. They both also determine the period of the discovered patterns. However, their periodicity descriptions are based on single events, not episodes.

3   Frequent Periodic Pattern Discovery and Update Over Data Streams

3.1   Problem Statement

Behavioral patterns are searched in the form of episodes (definition 2) in an event (definition 1) sequence, which is processed using the classical sliding window framework (length of the window: T_W). Indeed, recent behaviors are observable in recent events.

Definition 1 (Event). *An event is a (e, t) pair, where e is the event label, taking values in a finite alphabet $\mathcal{A}$; and t is the* timestamp.

Definition 2 (Episode, episode length). *An episode E is a set of n event labels $\{e_1, ..., e_n\}$ taken from the alphabet $\mathcal{A}$. The* length *of episode E is n.*

Definition 3 (Episode occurrence, occurrence duration). *An episode $E = \{e_1, ..., e_n\}$ occurs if there are n events whose labels match the n items in E. Formally, there is an occurrence o of E at time t_1 if there exists a permutation σ on $(1, ...n)$ and n timestamps $t_1 \leq ... \leq t_n$, such that $o = \langle (e_{\sigma(1)}, t_1), ..., (e_{\sigma(n)}, t_n) \rangle$ is a subsequence of the event stream. The duration of o is $\delta t_o = t_n - t_1$.*

The label order in the occurrence is not taken into account: it corresponds to the episodes referred to as *parallel* in the problem definition of episode mining [12]. The events making the occurrence may be interleaved with other events.

The events occurring in the vicinity of each other are more likely linked to a same behavior than distant events. A stricter constraint T_{ep} on the *maximal episode duration* can thus be set (T_W is used otherwise). T_{ep} exploits expert or statistical knowledge regarding the expected behavior durations. It also serves as a heuristic for the reduction of the search space (see section 3.2).

Definition 4 (Minimal occurrence - MO). *Let $E = \{e_1, ...e_n\}$ be an episode, occurring on $o = \langle (e_{\sigma(1)}, t_1), ...(e_{\sigma(n)}, t_n) \rangle$. o is a minimal occurrence if there is no other, shorter occurrence that occurs within the time span of o. That is to say, $\neg \exists o' = \langle (e_{\sigma'(1)}, t'_1), ...(e_{\sigma'(n)}, t'_n) \rangle$ such that $t_1 \leq t'_1$, $t'_n \leq t_n$ and $t'_n - t'_1 < t_n - t_1$.*

Definition 5 (Time queue - TQ). *The time queue of an episode E (noted TQ_E) is the list containing the distinct pairs of beginning and end timestamps of its minimal occurrences.*

We consider here only the minimal occurrences. The *support* of an episode is the length of its time queue. An episode is *frequent* if its support is greater than a *minimal support* threshold S_{min}. Minimal occurrences have convenient properties for the mining of frequent episodes, namely:

- An episode E has at most one time queue entry that starts (respectively finishes) at a given timestamp t (**observation 1**);
- Let E be an episode, and E' a subepisode (subset) of E. For every entry in the TQ of E, there is at least one entry in the TQ of E' (**observation 2**);
- As a consequence, the support of E' is greater or equal to this of E: the support verifies the downward closure property (**observation 3**);
- A new event (e, t) can be part of an occurrence of episode $E' = \{e\} \cup E$ (where $e \notin E$) if the latest entry in the TQ of E started less than T_{ep} (or T_W) ago (**observation 4a**). This occurrence is minimal if the beginning of the latest entry in TQ_E starts strictly after the latest entry in $TQ_{E'}$ (**observation 4b**). This gives a particular importance to the recently observed episodes.

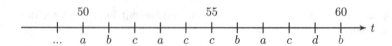

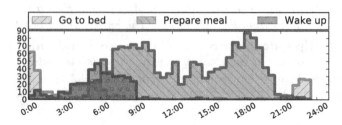

Fig. 1. Example: a segment of an event stream

Fig. 2. Example of an histogram representing the observed occurrence times for some daily habits of an elderly person living in a smart home

Example. Figure 1 presents an example of a event stream segment. The current window contains 11 events $(a, 50), (b, 51)$, etc. The last seen event is $(b, 60)$, and the labels take values in the alphabet $\mathcal{A} = \{a, b, c, d\}$. With $T_{ep} = 3$, episode $\{a, c\}$ occurs on $\langle (a, 50), (c, 52) \rangle, \langle (c, 52), (a, 53) \rangle, \langle (a, 53), (c, 54) \rangle, \langle (a, 53), (c, 55) \rangle$ etc: $\langle (a, 53), (c, 54) \rangle$ is minimal, but $\langle (a, 53), (c, 55) \rangle$ is not. The time queue for episode $\{a, b, c\}$ is $[(50, 52), (51, 53), (53, 56), (55, 57), (56, 58), (57, 60)]$, and its support is 6. $(54, 57)$ is not in the TQ because it does not correspond to a MO. A one-to-one mapping between a time queue entry and a MO is not guarantied: the time queue entry $(53, 56)$ of episode $\{a, b, c\}$, corresponds to two MO: $\langle (a, 53), (c, 54), (b, 56) \rangle$ and $\langle (a, 53), (c, 55), (b, 56) \rangle$.

Periodicity. Humans and animals tend to follow routines [10,15]. A typical example of periodic behavior is the daily occurrences of some human activities of daily living. Figure 2 presents the occurrence times of three such activities, recorded in a smart home over a six-month period (CASAS dataset, presented in section 4.1). It highlights some of the characteristics these activities may have:

- The occurrence times vary from one day to the next, and this variability is user- and activity-dependent: here go to bed is less variable than wake up,
- Activities may have several components. Here, there seems to be two meals at home a day: a breakfast around 8:00 and a dinner around 18:00,
- Each component has its own preferential occurrence time (mean μ) and variability (standard deviation σ),
- Some occurrences do not follow the periodic patterns.

This leads us to describe the periodicity of an episode as a distribution of its *relative* occurrence times within the period of interest (e.g 1 day, 1 week). This is done thanks to Gaussian Mixture Models (GMM), since they take into account the aforementioned characteristics.

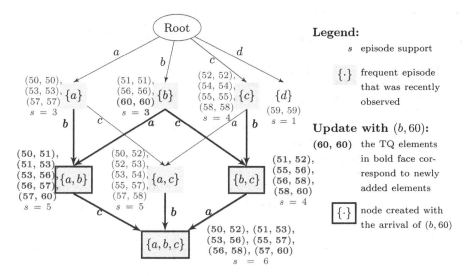

Fig. 3. Lattice corresponding to the example stream figure 1, when $(b, 60)$ is the last seen event. The update when event $(b, 60)$ arrives is highlighted in boldface.

For a periodicity model {period T, n components $(\mu_1, \sigma_1), ...(\mu_n, \sigma_n)$}, an occurrence is expected to occur in every time interval $k \cdot T + \mu_i(\pm\sigma_i)$, such that $1 \le i \le n$ and for every integer $k \ge 0$ such that $k \cdot T + \mu_i$ is in the current window. The quality of a periodicity description is evaluated on its *accuracy*, that is to say the proportion of the expected occurrences that were actually observed.

3.2   Frequent Episode Discovery and Updating

In order to be habits, episodes need to be frequent. However, the periodic episodes are not necessarily the most frequent ones, which is why the support threshold should remain rather low. We propose to handle the task of frequent episode mining over an event stream thanks to a frequent episode lattice (FEL).

Episode Lattice. The frequent episodes and their time queues are stored in a frequent episode lattice (FEL). The nodes in the FEL correspond either to length-1 episodes, or to frequent episodes. Length-1 episodes are kept even if they are not frequent (yet) in order to build longer episodes when they do. The parents of a node (located at depth d) correspond to its sub-episodes of length $d-1$, and its children to its super-episodes of length $d+1$. The edges linking two episodes are indexed on the only event label that is present in the child episode but not in the parent. Each node retains the TQ of the corresponding episode and the GMM description that best fits the episode (see section 3.3). The episode lattice corresponding to the example in figure 1 is given in figure 3. In spite of its possibly big edge count, the lattice structure was chosen over the standard prefix tree because it allows faster episode retrieval and update.

Algorithm 1. Computation of $E = E_1 \cup E_2$'s TQ from the TQ of E_1 and E_2

Input: TQ_1 (resp. TQ_2 the time queue of E_1 (resp. E_2), indexed on i (resp. j)
1: $i \leftarrow 0$; $j \leftarrow 0$; $TQ \leftarrow [\,]$; support $s \leftarrow 0$
2: **while** $i < |TQ_1|$ and $j < |TQ_2|$ **do**
3:     **if** $TQ_1[i]$ finishes after $TQ_2[j]$ **then**
4:         Increment j as long as $TQ_2[j]$ ends before $TQ_1[i]$
5:     **else**
6:         Increment i as long as $TQ_1[i]$ ends before $TQ_2[j]$
7:     $start \leftarrow \min(TQ_1[i][0], TQ_2[j][0]$
8:     $end \leftarrow \min(TQ_1[i][1], TQ_2[j][1]$
9:     **if** $end - start < T_{ep}$ **then**
10:         Add $(start, end)$ to TQ; $s \leftarrow s + 1$ /* New minimal occurrence */
11:     Increment the index of the TQ whose current element started earlier (both if $TQ_1[i][0] == TQ_2[j][0]$)
12: **return** TQ, s

Update with a New Event. We keep track of the recently modified nodes (RMN, the nodes describing an episode that occurred recently, i.e. less than T_{ep}, or T_W, ago). Indeed (see observations 4a and 4b), the recent occurrences of these episodes can be extended with new, incoming events to form longer episodes. The RMN are stored in a collection of lists (nodes at depth 1, depth 2, etc). The TQ of a newly frequent length-n episode is computed thanks to the time queues of a length-$(n\text{-}1)$ sub-episode and the length-1 episode containing the missing item, using algorithm 1.

When a new event (e, t) arrives, it can be a new occurrence (and also a MO) of the length-1 episode $\{e\}$. It can also form a new MO of an episode $E' = E \cup \{e\}$, where E is a recently observed episode. The lattice update follows these steps:

1. If label e is new: create a node for episode $\{e\}$ and link it to the FEL root;
2. Update the time queue of episode $\{e\}$;
3. If $\{e\}$ is frequent:
   (a) Add it to the RMN list;
   (b) For each node N_E in the RMN list, try to build a new occurrence of $E \cup \{e\}$, following algorithm 2, which takes advantage of observations 1–4. If an episode $E' = E \cup \{e\}$ becomes frequent, a new node $N_{E'}$ is created, and is linked to its parents in the lattice. The parents are the nodes describing the episodes $E'\backslash\{e'\}$ for each $e' \in E$, and are accessible via $N_E.parent(e').child(e)$, where N_E is the node for the known subset E. Since the RMN list is layered, and explored by increasing node depth, $N_E.parent(e').child(e)$ is always created before $N_{E'}$ tries to access it.

The update process of the FEL is illustrated with the arrival of a new event $(b, 60)$ in figure 3. $(b, 60)$ makes $\{b\}$ a frequent episode. The nodes the RMN list ($\{a\}$, $\{c\}$ and $\{a, c\}$) are candidate for the extension with the (frequent) new event. This allows the investigation of episodes $\{a, b\}$ (extension of $\{a\}$), $\{b, c\}$ (extension of $\{c\}$), and $\{a, b, c\}$ (extension of $\{a, c\}$), which indeed become frequent.

Algorithm 2. RMN-based update when a new event (e, t) arrives

Input: new event (e, t); recently modified node N_E, characterizing episode E
1: **if** $e \in E$ **then**
2:      **pass**   /* E cannot be extended with label e: E already contains it */
3: **else**
4:      **if** $E.lastMO$ starts before $t - T_{ep}$ **then**
5:           Remove N_E from RMN list: the last MO of E is too old to be extended
6:      **else**
7:           **if** N_E has a child $N_{E'}$ on label e **then**   /* E' is already frequent */
8:                **if** $E.lastMO$ starts strictly after $E'.lastMO$ **then**   /* New MO */
9:                     Add new entry to $N_{E'}.TQ$; Add $N_{E'}.TQ$ to the RMN list
10:                **else**   /* There is already a MO for E' starting in E.lastMO.start:
                              there cannot be another one */
11:                     **pass**
12:           **else**   /* E' may become frequent */
13:                $TQ, S \leftarrow$ ALGORITHM1$(TQ_E, TQ_{E'})$
14:                **if** $S \geq S_{min}$ **then**
15:                     Create node $N_{E'}$ for E'. Link it to its parents.
16:                     Add $N_{E'}$ to the RMN list
17: **return**   /* The FEL is updated with the information from event (e, t) */

Removal of Outdated Information. Events older than T_W are outdated, and their influence in the FEL needs to be removed. The TQ construction makes it so that its entries are ordered by start timestamp: the entries that need to be removed are thus at the beginning of the nodes TQ. Moreover, according to observation 2, for every TQ entry (and thus every outdated entry) there is at least one (outdated) entry in the TQ of one of the parent nodes. The FEL can thus be traversed from the root using a breadth-first search algorithm, where nodes are investigated and updated only if at least one of their parents presents outdated occurrences. Episodes becoming rare are removed from the FEL.

3.3   Periodicity Discovery

The periodicity of an episode is described thanks to a GMM. Each node in the FEL is associated with a GMM describing the periodicity of the episode, which is updated when new MO are observed or occurrences removed. Usually, a GMM is trained with the Expectation-Maximization algorithm [5] (EM): for each component of the GMM (the number of components being a user-given parameter), and each data point x, the probability that x was generated by the component is computed. The components characteristics (mean, standard deviation) are then tweaked to maximize the likelihood of the data point/component attribution.

But streaming data may be non-stationary, the number of components may evolve, as well as their characteristics. It is not acceptable to ask the user for the number of components, especially since the suitable number depends on the considered episode. We here extend EM with heuristics for the addition, removal and merging of components.

Algorithm 3. Overview of the periodicity update (comp=GMM component)

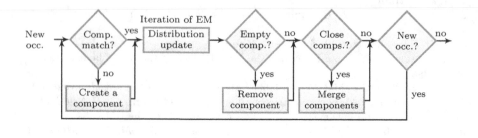

Algorithm 3 presents the general workflow for the periodicity update. When a new MO is detected for the episode, the position of the timestamp in the period $t_r = timestamp$ modulo $period$ is computed. If t_r does not match any of the existing components, i.e. for each component (μ, σ), $|t_r - \mu| > \sigma$, a new component is added. When outdated data is removed, some components lose their importance. When too rare, they are removed from the GMM. Finally, when two components $(\mu_1, \sigma_1), (\mu_2, \sigma_2)$ become close to one another, i.e if $|\mu_1 - \mu_2| < a * (\sigma_1 + \sigma_2)$ (with $a = 1.5$ in the experiments), the two components are merged. In the general case, GMM updates do not change much the model. Thus, when the number of components does not change, a single EM iteration is necessary to update the characteristics of the components.

The interest of this approach was evaluated on synthetic data, following know mixture of gaussian models evolving with time. The heuristics allow the detection of the main trends in the data: emergence of new components, disparition of old and rare components, shiftings in characteristics of the components.

4   Experimentation

A prototype was implemented in Python. It was also instrumented to record the episodes and lattice updates. The instrumentation slows down the experimentations: the execution times given in the next subsections are over-estimated.

4.1   Ambient Assisted Living Dataset

The CASAS project [4] uses home automation devices to improve ageing at home. Over the years, they collected and published several datasets. We present here our experimentations on the Aruba dataset[1]. The house of an elderly woman was equipped with motion detectors and temperature sensors. The obtained information was annotated with activities (11 labels, such as *Sleeping, House-keeping*, etc.; 22 when dissociating the begin and end timestamp of each activity). These annotations are our events (12 953 events, from Nov. 2010 to Jun. 2011).

---

[1] http://wsucasas.wordpress.com/datasets/, number 17, consulted on Dec 4th, 2014.

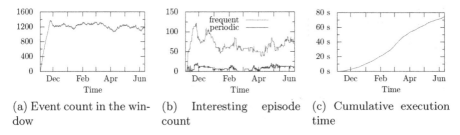

(a) Event count in the window

(b) Interesting episode count

(c) Cumulative execution time

Fig. 4. Execution log for the CASAS Aruba annotation dataset

The dataset was processed using a period of one day, a window T_W of 3 weeks, a minimal support S_{min} of 15, a maximal episode duration T_{ep} of 30 minutes, and an accuracy threshold of 70%. The parameter setting was reinforced by a descriptive analysis of the data (e.g., it showed that most activities last less than 30 minutes). The results obtained throughout the course of the execution are given in figure 4. During the first 3 weeks, the sliding window fills with the incoming events, and the first frequent and periodic episodes appear. Then, the number of events in the window remains quite stable, but the behaviors keep evolving. The execution time (figure 4c) shows the scalability of the approach for this kind of application. The contents of the FEL in the last window is investigated, some of periodic episodes with the highest accuracy A are:

- {*Sleeping end*}: 50 MO, 1 component, $\mu = 6:00$, $\sigma = 2$ hours, $A = 100\%$
- {*Sleeping end, Meal_Preparation begin, Meal_Preparation end, Relax begin*}: 26 MO, 1 component, $\mu = 6:00$, $\sigma = 1.45$ hours, $A = 82\%$
- {*Enter_Home begin, Enter_Home end*}: 61 MO, 1 component, $\mu = 14:00$, $\sigma = 3$ hours, $A = 88\%$

These patterns can be interpreted as habits: the person woke up every morning around 6:00, and also had breakfast in 82% of the mornings. The third episode describes a movement pattern: the inhabitant usually goes out of home at some time (it is another episode), and comes back in the early afternoon.

Figure 5 presents the influence of the minimal support S_{min}, maximal episode duration T_{ep}, and window length T_W on the maximal size of the FEL and the execution time. In particular, it shows that the execution time is reasonable and scalable. The duration of the episodes also has a large impact on the size of the FEL.

4.2 Travian Game Dataset

Travian is a web-browser game, where players, organized into alliances, fight for the fulfilling of objectives and the control of territories. The game company releases each day a snapshot of the server status: it contains information on the players (villages, alliance membership). These daily updates were collected for the 2014 fr5 game round, from July 8th to November 23rd. We focus here on the

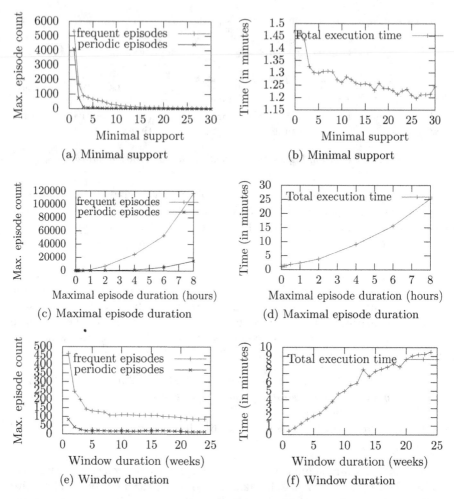

Fig. 5. Influence of the algorithm configuration on the frequent and periodic episode counts, and on the execution time for the CASAS Aruba dataset

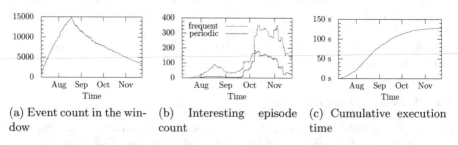

(a) Event count in the window

(b) Interesting episode count

(c) Cumulative execution time

Fig. 6. Execution log for the Travian fr5 alliance membership dataset

players alliance shifts: the event labels look like *"Player P [joined|left] alliance A"*. 27674 such events are recorded, but most labels are rare (25985 labels).

The dataset was processed with a period of one week, a window T_W = six weeks, a minimal support $S_{min} = 5$ and a maximal episode duration $T_{ep} = 1$ day. Figure 6 presents the evolution of the window size, episode counts, and execution time during the mining. The results are fairly different from those of the home automation dataset, but were explained by a player (picturing a domain expert). During the first 6 weeks, the window fills rapidly with events: new players register onto the game, and the diplomacy begins. The players join or switch alliances. After the 6 weeks, the event count in the window decreases with time. Several explanations: (i) the opening of a new game round (on August 22nd) slowed down the number of new player registrations (players tend to join the most recent game round); (ii) most players have found an alliance they like: they stop changing alliances. Until October, little frequent and periodic patterns are detected, but their number increases rapidly after that. The periodic episodes discovered in the Sep, 18th – Oct, 30th (maximal count of periodic episodes) contain notably:

- {*1SixCentDix8 left Vtrans, 1SixCentDix8 joined iChiefs*}: 8 MO, 2 components, μ_1 = Fri. 0:00, μ_2 = Mon. 0:00, $\sigma_1 = \sigma_2 = 0$, $A = 80\%$
- {*1SixCentDix8 left iChiefs, 1SixCentDix8 joined Vtrans*}: 8 MO, 2 components, μ_1 = Sat. 0:00, μ_2 =Tue. 0:00, $\sigma_1 = \sigma_2 = 0$, $A = 80\%$
- {*Jill left Bakka, Jill joined LI*}: 10 MO, 2 components, μ_1 = Mon. 18:00, $\sigma_1 = 1$ day, μ_2 = Fri. 0:00, $\sigma_2 = 0$, $A = 75\%$

Some players periodically change of alliance: *1SixCentDix8* leaves *Vtrans* for *iChiefs* on Mondays and Fridays, and goes back to *Vtrans* one day later. *Jill* goes from *Bakka* to *LI* either on Mondays or Tuesdays, as well as on Fridays. This actually highlights a strategy allied alliances (*iChiefs* and *Vtrans* on one side, and *Bakka* and *LI* on the other side) have developed to share with one another the effects of artifacts owned by players *1SixCentDix8* and *Jill*, respectively.

5   Conclusion

Behavior pattern (episode) mining over event sequences is an important data mining problem, with many applications, in particular for ambient assisted living, or wildlife behavior monitoring. Several frequent episode mining algorithms have been proposed for both static data and data streams. But while periodicity can also be an interesting characteristic for the study of behaviors, very few algorithm have addressed frequent and periodic patterns. We propose an efficient algorithm to mine frequent periodic episodes in data streams. We briefly illustrate the interest of this algorithm with two case studies. As a perspective of this work, the experiments can be extensively increased, and applied to other application domains. It also be interesting to include a period-determination algorithm in order to automatically adapt the period to each pattern. Closed episodes and non-overlaping occurrences could also be investigated.

References

1. Amphawan, K., Lenca, P., Surarerks, A.: Efficient mining top-k regular-frequent itemset using compressed tidsets. In: Cao, L., Huang, J.Z., Bailey, J., Koh, Y.S., Luo, J. (eds.) PAKDD Workshops 2011. LNCS, vol. 7104, pp. 124–135. Springer, Heidelberg (2012)
2. Baratchi, M., Meratnia, N., Havinga, P.J.M.: Recognition of periodic behavioral patterns from streaming mobility data. In: Stojmenovic, I., Cheng, Z., Guo, S. (eds.) MOBIQUITOUS 2013. LNICST, vol. 131, pp. 102–115. Springer, Heidelberg (2014)
3. Calders, T., Dexters, N., Goethals, B.: Mining frequent itemsets in a stream. In: ICDM, pp. 83–92 (2007)
4. Cook, D.J., Crandall, A.S., Thomas, B.L., Krishnan, N.C.: Casas: A smart home in a box. IEEE Computer 46(7), 62–69 (2013)
5. Dempster, A., Laird, N., Rubin, D.: Maximum likelihood from incomplete data via the EM algorithm. Journal of the Royal Statistical Society. Series B (Methodological), 1–38 (1977)
6. Gama, J.: A survey on learning from data streams: current and future trends. Progress in Artificial Intelligence 1(1), 45–55 (2012)
7. Heierman, E.O., Youngblood, G.M., Cook, D.J.: Mining temporal sequences to discover interesting patterns. In: KDD Workshop on mining temporal and sequential data (2004)
8. Kiran, R.U., Reddy, P.K.: Mining periodic-frequent patterns with maximum items' support constraints. In: ACM COMPUTE Bangalore Conference, pp. 1–8 (2010)
9. Lahiri, M., Berger-Wolf, T.Y.: Mining periodic behavior in dynamic social networks. In: ICDM, pp. 373–382. IEEE Computer Society (2008)
10. Li, Z., Han, J., Ding, B., Kays, R.: Mining periodic behaviors of object movements for animal and biological sustainability studies. Data Mining and Knowledge Discovery 24(2), 355–386 (2012)
11. Lin, S., Qiao, J., Wang, Y.: Frequent episode mining within the latest time windows over event streams. Appl. Intell. 40(1), 13–28 (2014)
12. Mannila, H., Toivonen, H., Verkamo, A.I.: Discovering frequent episodes in sequences. In: Fayyad, U.M., Uthurusamy, R. (eds.) KDD, pp. 210–215. AAAI Press (1995)
13. Patnaik, D., Laxman, S., Chandramouli, B., Ramakrishnan, N.: Efficient episode mining of dynamic event streams. In: ICDM, pp. 605–614 (2012)
14. Rashidi, P., Cook, D.J.: Mining sensor streams for discovering human activity patterns over time. In: Webb, G.I., Liu, B., Zhang, C., Gunopulos, D., Wu, X. (eds.) ICDM 2010, The 10th IEEE International Conference on Data Mining, Sydney, Australia, December 14–17, 2010, pp. 431–440. IEEE Computer Society (2010)
15. Soulas, J., Lenca, P., Thépaut, A.: Monitoring the habits of elderly people through data mining from home automation devices data. In: Reis, L.P., Correia, L., Cascalho, J. (eds.) EPIA 2013. LNCS, vol. 8154, pp. 343–354. Springer, Heidelberg (2013)

16. Surana, A., Kiran, R.U., Reddy, P.K.: An efficient approach to mine periodic-frequent patterns in transactional databases. In: Cao, L., Huang, J.Z., Bailey, J., Koh, Y.S., Luo, J. (eds.) PAKDD Workshops 2011. LNCS, vol. 7104, pp. 254–266. Springer, Heidelberg (2012)
17. Wong, R.W., Fu, A.C.: Mining top-k frequent itemsets from data streams. Data Mining and Knowledge Discovery 13(2), 193–217 (2006)
18. Zhu, H., Wang, P., He, X., Li, Y., Wang, W., Shi, B.: Efficient episode mining with minimal and non-overlapping occurrences. In: ICDM, pp. 1211–1216 (2010)

Forecasting the Correct Trading Actions

Luís Baía[1,2]($\boxtimes$) and Luís Torgo[1,2]

[1] LIAAD - INESC TEC, Porto, Portugal
`luisbaia_1992@hotmail.com, ltorgo@dcc.fc.up.pt`
[2] DCC - Faculdade de Ciências - Universidade do Porto, Porto, Portugal

Abstract. This paper addresses the problem of decision making in the context of financial markets. More specifically, the problem of forecasting the correct trading action for a certain future horizon. We study and compare two different alternative ways of addressing these forecasting tasks: i) using standard numeric prediction models to forecast the variation on the prices of the target asset and on a second stage transform these numeric predictions into a decision according to some pre-defined decision rules; and ii) use models that directly forecast the right decision thus ignoring the intermediate numeric forecasting task. The objective of our study is to determine if both strategies provide identical results or if there is any particular advantage worth being considered that may distinguish each alternative in the context of financial markets.

1 Introduction

Many real world applications require decisions to be made based on forecasting some numeric quantity. Sales forecasting may lead to some important decisions concerning the production process. Asset price forecasting may lead investors to buy or sell some financial product. Forecasting the future evolution of some indicator of a patient may lead a medical doctor to some important treatment prescriptions. These are just a few examples of concrete applications that fit this general setting: decisions based on numeric forecasts of some variable. In most cases the decision process is based on a pre-defined protocol that associates intervals of the range of the numeric variable with concrete actions/decisions. This means that once we have a prediction for the numeric variable we will use some deterministic process to reach the action/decision to be taken. This work is focused on this particular type of situations in the context of financial markets. In this domain the goal of investors is to make the correct decision (Sell, Buy or Hold) at any given point in time. These decisions are taken based on the investor's expectations on the future evolution of the asset prices. In this work we approach this decision problem using predictive models. More specifically, we will compare two possible ways of trying to forecast what is the correct trading decision at any point in time.

In our target applications we assume that there are deterministic decision rules that given the estimated evolution of the prices of the asset will indicate the trading action to be taken. For instance, a rule could state that if the forecast

© Springer International Publishing Switzerland 2015
F. Pereira et al. (Eds.) EPIA 2015, LNAI 9273, pp. 560–571, 2015.
DOI: 10.1007/978-3-319-23485-4_55

of the variation of prices is a 2.5% increase then the correct decision is to buy the asset as this will allow covering transaction costs and still have some profit. Given the deterministic mapping from forecasted values into decisions we can define the prediction task in two different ways. The first consists on obtaining a numeric prediction model that we can then use to obtain predictions of the future variation of the prices which are then transformed (deterministically) into trading decisions (e.g. [1], [2]). The second alternative consists of directly forecasting the correct trading decisions (e.g. [3], [4], [5]). Which is the best option in terms of the resulting financial results? To the best of our knowledge no comparative study was carried out to answer this question. This is the goal of the current paper: to compare these two approaches and provide experimental evidence of the advantages and disadvantages of each alternative.

2   Problem Formalization

The problem of decision making based on forecasts of a numerical (continuous) value can be formalized as follows. We assume there is an unknown function that maps the values of p predictor variables into the values of a certain numeric variable Y. Let f be this unknown function that receives as input a vector $\mathbf{x}$ with the values of the p predictors and returns the value of the target numeric variable Y whose values are supposed to depend on these predictors,

$$f: \mathbb{R}^p \to \mathbb{R}$$
$$\mathbf{x} \mapsto f(\mathbf{x}).$$

We also assume that based on the values of this variable Y some decisions need to be made. Let g be another function that given the values of this target numeric variable transforms them into actions/decisions,

$$g: \mathbb{R} \to \mathcal{A} = \{a_1, a_2, a_3, \dots\}$$
$$Y \mapsto g(Y).$$

where $\mathcal{A}$ represents a set of possible actions.

In our target applications, functions f and g are very different. Function g is known and deterministic, in the sense that it is part of the domain background knowledge. Function f is unknown and uncertain. The only information we have about function f is an historical record of mappings from $\mathbf{x}$ into Y, i.e. a data set that can be used to learn an approximation of the function f. Given that the variable Y is numeric this approximation could be obtained using some existing multiple regression tool. This means that given a data set $D_r = \{\langle \mathbf{x}_i, Y_i \rangle_{i=1}^n\}$ we can use some regression tool to obtain a model $\hat{r}(\mathbf{x})$ that is an approximation of f. From an operational perspective this would mean that given a test case $\mathbf{q}$ for which a decision needs to be made we would proceed by first using $\hat{r}$ to obtain a prediction for Y and then apply g to this predicted value to get the predicted action/decision, i.e. $\mathbf{q} \mapsto \hat{r}(\mathbf{q}) \mapsto g(\hat{r}(\mathbf{q}))$. In the context of financial markets the predictors describe the currently observed dynamics of the prices of some financial asset and the target numeric variable Y represents the future variation of this price. This means that f is the unknown function that maps the currently observed price dynamics into a future evolution of the price. On the other hand g is a deterministic function (typically based on domain knowledge and risk preferences of traders) that maps the prediction of the future evolution of prices into one of three possible decisions: Sell, Hold or Buy.

Given the deterministic nature of g we can use an alternative process for obtaining decisions. More specifically, we can build an alternative data set $D_c = \{\langle \mathbf{x}_i, g(Y_i)\rangle\}_{i=1}^n$, where the target variable is the decision associated with each known Y value in the historical record of data. This means that we have a nominal target variable, i.e. we are facing a classification task. Once again we can use some standard classification tool to obtain an approximation $\hat{c}$ of the unknown function that maps the predictors into the correct actions/decisions. Once such model is obtained we can use it given a query case $\mathbf{q}$ to directly estimate the correct decision by applying the learned model to the case, i.e. $\mathbf{q} \mapsto \hat{c}(\mathbf{q})$. This means that given the description of the current dynamics of the price we will use function $\hat{c}$ to forecast directly the correct trading action for this context.

Independently of the approach followed, the final goal of the applications we are targeting is always to make correct decisions. This means that whatever process we use to reach a decision, it will be evaluated in terms of the "quality" of the decisions it generates. In this context, it seems that the classification approach, by having as target variable the decisions, would be easier to bias towards optimal actions. However, this approach completely ignores the intermediate numeric variable that is supposed to influence decisions, though one may argue that information on the relationship between Y and the decisions is "encoded" when building the training set D_c by using as target the values of $g(Y_i)$. On the other hand, while the regression approach is focused on obtaining accurate predictions of Y, it completely ignores questions like eventual different cost/benefits of the different possible decisions that could be easily encoded into the classification tasks. All these potential trade-offs motivate the current study. The main goal of this paper is to compare these two approaches in the context of financial markets.

3   Material and Methods

This section describes the main issues involved in the experimental comparison we will carry out with the goal of comparing the two possible approaches described in the previous section.

3.1   The Tasks

The problem addressed in this paper is very common in automatic trading systems where decisions are based on the forecasts of some prediction models.

The decisions to open or close short/long positions are typically the result of a deterministic mapping from the predicted prices variation.

In our experiments, we have used the assets prices of 12 companies. Each data set has a minimum of 7 years of daily data and a maximum of 30 years. In order to simplify the study, we will be working with a one-day horizon, i.e. take a decision based on the forecasts of the assets variation for one day ahead. Moreover, we will be working exclusively with the closing prices of each trading session, i.e. we assume trading decisions are to be made after the markets close.

The decision function for this application receives as input the forecast of the daily variation of the assets closing prices and returns a trading action. We will be using the following function in our experiments:

$$g\colon \mathbb{R} \to \mathcal{A} = \{hold, buy, sell\}$$

$$Y \mapsto \begin{cases} buy, & Y > 0.02 \\ sell, & Y < -0.02 \ . \\ hold, & \text{other cases} \end{cases}$$

This means we are assuming that any variation above 2% will be sufficient to cover the transaction costs and still obtain some profit. Concerning the data that will be used as predictors for the forecasting models (either forecasting the prices variation (Y) or directly the trading action (A)) we have used the price variations on recent days as well as some trading indicators, such as the annual volatility, the Welles Wilder's style moving average [6], the stop and reverse point indicator developed by J. Welles Wilder [6], the usual moving average and others. The goal of this selection of predictors is to provide the forecasting models with useful information on the recent dynamics of the assets prices.

Regarding the performance metrics we will use to compare each approach, we will use two metrics that capture important properties of the economic results of the trading decisions made by the alternative models. More specifically, we will use the Sharpe Ratio as a measure of the risk (volatility) associated with the decisions, and the percentage Total Return as a measure of the overall financial results of these actions. To make our experiments more realistic we will consider a transaction cost of 2% for each Buy or Sell decision a model may take.

At this stage it is important to remark that the prediction tasks we are facing have some characteristics that turn them in to particularly challenging tasks. One of the main hurdles results from the fact that interesting events, from a trading perspective, are rare in financial markets. In effect, large movements of prices are not very frequent. This means that the data sets we will provide to the models have clearly imbalanced distributions of the target variables (both the numeric percentage variations and the trading actions). To make this imbalance problem harder the situations that are more interesting from a trading perspective are rare in the data sets which creates difficulties to most modelling techniques. In the next section, we will describe some of the measures we have taken to alleviate this problem.

3.2   The Models

In this section we will list all the model variants that will be used in the experimental comparison. The point is to ensure that both approaches have the same conditions for a fair comparison. Several variants for each family of models (SVM, Random Forests, etc) were tested in order to make sure our conclusions were not biased by the choice of models. The Table 1 and Table 2 shows all the model variants used, where nearly 182 model variants were tested.

Table 1. Regression models used for the experimental comparisons. SVM stands for Support Vectorial Machines, KNN for K-nearest neighbours, NNET for Neural Networks and MARS for Multivariate Adaptive Regression Spline models

Model	Variants	R Package
SVM	cost=$\{1,5,10\}$,$\epsilon = \{0.1,0.05,0.01\}$, tolerance=$\{0.001,0.005\}$,kernel=linear	e1071
SVM	cost=$\{1,10\}$,$\epsilon = \{0.1,0.05,0.01\}$, degree=$\{2,3,5\}$,kernel=polynomial	e1071
Random Forest	ntree=$\{500,750,1000,2000,3000\}$,mtry=$\{4,5,6\}$	randomForest
Trees (pruned)	se=$\{0,0.5,1,1.5,2\}$,cp=0, minsplit=6	DMwR
KNN	k=$\{1,3,5,7,11,15\}$	DMwR
NNET	size=$\{2,4,6\}$,decay=$\{0.05,0.1,0.15\}$	nnet
MARS	thresh=$\{0.001,0.0005,0.002\}$, degree=$\{1,2,3\}$,minspan=$\{0,1\}$	earth
AdaBoost	dist=$\{$gaussian$\}$,n.trees=$\{10000,20000\}$, shrinkage=$\{0.001,0.01\}$,interaction.depth=$\{1,2)\}$	gbm

Table 2. Classification models used for the experimental comparisons. SVM stands for Support Vectorial Machines, KNN for K-nearest neighbours, NNET for Neural Networks and MARS for Multivariate Adaptive Regression Spline models

Model	Variants	R Package
SVM	cost=$\{1,3,7,10\}$,kernel=linear tolerance=$\{0.001,0.005,0.0005,0.002\}$	e1071
SVM	cost=$\{1,10\}$,$\epsilon = \{0.1,0.05\}$, degree=$\{2,3,4,5\}$,kernel=polynomial	e1071
Random Forest	ntree=$\{500,750,1000,2000,3000\}$,mtry=$\{3,4,5\}$	randomForest
Trees (pruned)	se=$\{0,0.5,1,1.5,2\}$,cp=0, minsplit=6	DMwR
KNN	k=$\{1,3,5,7,11,15\}$	DMwR
NNET	size=$\{2,4,6\}$,decay=$\{0.05,0.1,0.15\}$	nnet
AdaBoost	coeflearn = c('Breiman','Freund','Zhu'), mfinal=c(500,1000,2000)	boosting

The predictive tasks we are facing have two main difficulties: (i) the fact that the distribution of the target variables is highly imbalanced, with the more relevant values being less frequent; and (ii) the fact that there is an implicit ordering among the decisions. The first problem causes most modelling techniques to focus on cases (the most frequent) that are not relevant for the application goals. The second problem is specific to classification tasks as these algorithms do not distinguish among the different types of errors, whilst in our target applciation confusing a Buy decision with a Hold decision is less serious than confusing it with a Sell.

These two problems lead us to consider several alternatives to our base modelling approaches described in Tables 1 and 2. For the first problem of imbalance we have considered the hypothesis of using resampling to balance the distribution of the target variable before obtaining the models. In order to do that, we have used the SMOTE algorithm [7]. This method is well known for classification models, consisting basically of oversampling the minority classes and under-sampling the majority ones. The goal is to modify the data set in order to ensure that each class is similarly represented. Regarding the regression tasks we have used the work by Torgo et. al [8], where a regression version of SMOTE was presented. Essentially, the concept is the same as in classification, using a method to try to balance the continuous distribution of the target variable by oversampling and under-sampling different ranges of its domain.

We have thoroughly tested the hypothesis that using resampling before obtain ghe models would boost the performance of the different models we have considered for our tasks. Our experiments confirmed that resampling lead the models to issue more Buy and Sell signals (the one that are less frequent but more interesting). However, this increased number of signals was accompanied by an increased financial risk that frequently lead to very poor economic results, with very few exceptions.

Regarding the second problem of the order among the classes we have also considered a frequently used approach to this issue. Namely, we have used a cost-benefit matrix that allows us to distinguish between the different types of classification errors. Using this matrix, and given a probabilistic classifier, we can predict for each test case the class that maximises the utility instead of the class that has the highest probability.

We have used the following procedure to obtain the cost-benefit matrices for our tasks. Correctly predicted *buy/sell* signals have a positive benefit estimated as the average return of the *buy/sell* signals in the training set. On the other hand, in the case of incorrectly predicting a true *hold* signal as *buy* (or *sell*), we we assign it minus the average return of the *buy* (or *sell*) signals. Basically, the benefit associated to correctly predicting one rare signal is entirely lost when the model suggests an investment when the correct action would be doing nothing. In the extreme case of confusing the *buy* and *sell* signals, the penalty will be minus the sum of the average return of each signal. Choosing such a high penalty for these cases will eventually change the model to be less likely to make this type of very dangerous mistakes. Considering the case of incorrectly predicting a true *sell* (or *buy*) signal as *hold*, we also charge for it, but in a less severe way. Therefore, the average of the *sell* (or *buy*) signal is considered, but divided by two. This division was our way of "teaching" the model that it is preferable to miss an opportunity to earn money rather than making the investor lose money.

Finally, correctly predicting a *hold* signal gives no penalty nor reward, since no money is either won or lost. Table 3 shows an example of such cost-benefit matrix that was obtained with the data from 1981-01-05 to 2000-10-13 of Apple.

Table 3. Example cost-benefit matrix for Apple shares.

	Trues		
	s	h	b
s	0.49	-0.49	-0.82
Pred h	-0.24	0.00	-0.17
b	-0.82	-0.33	0.33

We have also thoroughly tested the hypothesis that using cost-benefit matrices to implement utility maximisation would improve the performance of the models. Our tests have shown that nearly half the model variants see their performance boosted with this approach.

3.3 The Experimental Methodology

In this section we present the experimental methodology used in our comparative experiments. Due to the temporal nature of the used data sets, the usual cross-validation methodology should not be used to estimate the performance of a certain model. Namely, this procedure assumes that the data has no order and by using it we would obtain unreliable estimates. In this context, we have decided to use a Monte Carlo simulation method for obtaining our estimates. This methodology consists of randomly selecting a series of N points in time within the available data set. For each of these random dates, we use a certain consecutive past window as training set for obtaining the alternative models that are then tested/compared in a sub-sequent and consecutive test window. The Monte Carlo estimates are formed by the average scores obtained on the N repetitions. In our experiments we have used $N = 10$, 50% of the data as the size of the training window, and 25% of the data as size of the test sets.

With respect to testing the statistical significance of the observed differences between the estimated scores we have used the recommendations of the work by Demsar [9]. More specifically, in situations where we are comparing k alternative models on one specific task we have used the Wilcoxon signed rank test to test the significance of the differences. On the experiments where k models are compared on t tasks we use the Friedman test followed by a post-hoc Nemenyi test to check the significance of the difference between the average ranks of the k models across the t tasks.

4 Experimental Results

This section presents the results of the experimental comparisons between the two general approaches to making trading decisions based on forecasting models.

In our experiments we have considered 76 classification models. For each of these models we have also tried the version with resampling and the version with cost-benefit matrices, totaling $76 \times 3 = 228$ different classification variants. In terms of regression we have a slightly large set of 97 base models that where then tried with and without resampling, for a total of $97 \times 2 = 194$ variants. All these variants were compared on the data sets of the 12 companies described in Section 3.1 using the methodology described in Section 3.3.

We have divided our experimental analysis in two main parts. In the first one, for each company and for each metric, we have compared the best regression and classification variant using a Wilcoxon singed rank statistical test with a significance level of 0.05 to check if we can reject the null hypothesis that there is not significant difference between the best classification and regression variants. This leads to 12 statistical tests for each metric (one test for each company), where the models compared for each company are not necessarily the same. The motivation of this first part is to compare the best classification variant against the best regression task for each company and metric. Figure 1 shows the result of this comparison for the Total Return and Sharpe Ratio financial evaluation metrics. The results on these figures are somewhat correlated. In effect, whenever we have found a significant difference in terms of Total Return, the same also happened in terms of Sharpe Ration. Regarding the left graph (Total Return), we have one significant win for each approach and 6 against for 4 non significant wins for classification and regression, respectively. With respect to the right graph (Sharpe Ration) we can observe a slight advantage of the classification approach, with one more significant win and 8 vs 1 non-significant wins. Paying respect to the second figure, one more significant win for classification is obtained and this approach achieved 8 non significant wins against 1 for the regression one. Overall, we have observed a very slight advantage of the best classification approach against the best regression variant.

From an economical perspective we have observed contradictory results. For instance, there is a very high level of Total Return for the Meg company (above 60% return), but the best Sharpe Ratio was very low. This means that the best model for the first metric was taking enormous amounts of risk and that the high level of return achieved was probably due to pure luck. On the other hand, there are some high values for the Total Return accompanied by high levels of Sharpe Ratio, such as for the Exas company. This strongly suggests that the models could actually provide some profit with low risk, thus indicating that the model actually predicted meaningful signals. Given the high variability of the results across companies, taking conclusions solely based on the analysis of the best variant per model and per metric may lead to wrong results. This establishes the motivation for the second part of our experiments.

In this second part of our experiments, instead of grouping by metric and company, we will just group by metric and study the average rank of each model across all the companies (top 5 of each approach are considered). With the use of the Friedman test followed by the post-hoc Nemenyi test, we check whether there are statistically significant differences among these rankings. This way, if

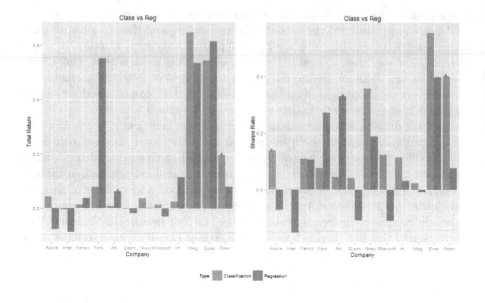

Fig. 1. Best classification variant against the best regression one for the Total Return and Sharpe Ratio metrics (asterisks denote that the respective variant is significantly better, according to a Wilcoxon test with $\alpha = 0.05$).

a model obtains a very good result for one company but poor for all the others (meaning that it was lucky in that specific company), its average ranking will be low allowing the top average rankings to be populated by the true top models that perform well across most companies.

Table 4 summarises the results in terms of Total Return. Since we could not reject the Friedman's null hypothesis, the post-hoc Nemenyi's test was not performed. This means that we can not say with 95% confidence that there is some difference in terms of Total return between these modelling approaches. Nevertheless, there are some observations to remark. The model with the best average ranking is a classification model using cost-benefit matrices. All the remaining classification variants are in their original form (without using cost-benefit matrices) and occupying mostly the last positions. Moreover, not a single variant obtained with SMOTE appears in this top 5 for each approach, which means that we confirm that resampling does not seem to pay off for this class of applications due to the economic costs of making more risky decisions. Furthermore, another very interesting remark is that all the top models are are using SVMs as the base learning algorithm. Overall we can not say that any of the two approaches (forecasting directly the trading actions using classification models or forecasting the price returns before using regression) is better than the other.

Table 5 shows the results of the same experiment in terms of Sharpe Ratio, i.e. the risk exposure of the alternatives. The conclusions are quite similar to

Table 4. The average rank of the top 5 Classification and Regression models in terms of Total Return. The Friedman test returned a p-value of 0.3113477, meaning that there is no statistical difference between all the 10 variants compared. Note: BC means the model was obtained using a benefit-cost matrix, while (p)/(l) means the SVM model was obtained using a polynomial/linear kernel. The vx labels represent the different parameter settings that were considered within each variant.

Rank Variant	Avg. Rank	Rank Variant	Avg. Rank
1 CLASS SVM(p) BC v3	4.54	6 CLASS SVM(l) v1	5.83
2 REG SVM(l) v1	5.12	7 CLASS SVM(l) v5	5.83
3 REG SVM(l) v11	5.12	8 CLASS SVM(l) v9	5.83
4 REG SVM(l) v2	5.25	9 REG SVM(l) v3	5.83
5 CLASS SVM(l) v6	5.67	10 REG SVM(l) v18	5.96

the Total Return metric. Once again, no significant differences were observed. Still, one should note that the first 5 places are dominated by the classification approaches. The best variant for the Total Return is also the best variant for the Sharpe Ratio, which makes this variant unarguably the best one of our study when considering the 12 different companies. Hence, ultimately we can state that the most solid model belongs to the classification approach using an SVM with cost-benefit matrices, since it obtained the highest returns with lowest associated risk. Finally, unlike the results for Total Return, in this case we observe other learning algorithms appearing in the top 5 best results.

Table 5. Top 5 average rankings of the Classification and Regression models for the Sharpe Ratio. The Friedman test returned a p-value of 0.1037471, implying there is no statistical difference between all the 10 variants tested.

Rank Variant	Avg. Rank	Rank Variant	Avg. Rank
1 CLASS SVM(p) BC v3	4.88	6 CLASS SVM(l) v1	5.58
2 CLASS SVM(l) BC v15	5.38	7 REG TREE v4	5.58
3 REG NNET v1	5.38	8 REG TREE v5	5.58
4 CLASS SVM(l) v14	5.42	9 REG NNET v2	5.67
5 CLASS SVM(l) v6	5.50	10 REG SVM(l) v1	6.04

In conclusion, we can not state that one approach performs definitely better than the other in the context of financial trading decisions. The scientific community typically puts more effort into the regression models, but this study strongly suggests that both have at least the same potential. Actually, the most consistent model we could obtain is a classification approach. Another interesting conclusion is that, of a considerably large set of different types of models, SVMs achieved better results both when considering classification or regression tasks.

5   Conclusions

This paper presents a study of two different approaches to financial trading decisions based on forecasting models. The first, and more conventional, approach uses regression tools to forecast the future evolution of prices and then uses some decision rules to choose the "correct" trading decision based on these predictions. The second approach tries to directly forecast the "correct" trading decision. Our study is a specific instance of the more general problem of making decisions based on numerical forecasts. In this paper we have focused on financial trading decisions because this is a specific domain that requires specific trade-offs in terms of economic results. This means that our conclusions from this study in this area should not be generalised to other application domains.

Overall, the main conclusion of this study is that, for this specific application domain, there seems to not be any statistically significant difference between these two approaches to decision making. Given the large set of classification and regression models that were considered, as well as different approaches to the learning task, we claim that this conclusion is supported by significant experimental evidence.

The experiments carried out in this paper have also allowed us to draw some other conclusions in terms of the applicability of resampling and cost-benefit matrices in the context of financial forecasting. Namely, we have observed that the application of resampling, although increasing the number of trading decisions made by the models, would typically bring additional financial risks that would make the models unattractive to traders. On the other hand the use of cost-benefit matrices in an effort to maximise the utility of the predictions of the models, did bring some advantages to a high percentage of modelling variants.

As future work we plan to extend our comparisons of these two forms of addressing decision making based on numeric forecasting, to other application domains, in an effort to provide general guidelines to the community on how to address these relevant real world tasks.

Acknowledgments. This work is financed by the FCT Fundação para a Ciência e a Tecnologia (Portuguese Foundation for Science and Technology) within project UID/EEA/50014/2013.

References

1. Lu, C.J., Lee, T.S., Chiu, C.C.: Financial time series forecasting using independent component analysis and support vector regression. Decision Support Systems **47**(2), 115–125 (2009). cited By 112
2. HELLSTR iOM, T.: Data snooping in the stock market. Theory of Stochastic Processes **21**, 33–50 (1999) (1999b)
3. Luo, L., Chen, X.: Integrating piecewise linear representation and weighted support vector machine for stock trading signal prediction. Applied Soft Computing **13**(2), 806–816 (2013)

4. Ma, G.Z., Song, E., Hung, C.C., Su, L., Huang, D.S.: Multiple costs based decision making with back-propagation neural networks. Decision Support Systems **52**(3), 657–663 (2012)
5. Teixeira, L.A., de Oliveira, A.L.I.: A method for automatic stock trading combining technical analysis and nearest neighbor classification. Expert Systems with Applications **37**(10), 6885–6890 (2010)
6. Wilder, J.: New Concepts in Technical Trading Systems. Trend Research (1978)
7. Chawla, N.V., Bowyer, K.W., Hall, L.O., Kegelmeyer, W.P.: Smote: synthetic minority over-sampling technique. Journal of Artificial Intelligence Research **16**(1), 321–357 (2002)
8. Torgo, L., Branco, P., Ribeiro, R.P., Pfahringer, B.: Resampling strategies for regression. Expert Systems (2014)
9. Demšar, J.: Statistical comparisons of classifiers over multiple data sets. The Journal of Machine Learning Research **7**, 1–30 (2006)

CTCHAID: Extending the Application
of the Consolidation Methodology

Igor Ibarguren[✉], Jesús María Pérez, and Javier Muguerza

Department of Computer Architecture and Technology, University of the Basque
Country UPV/EHU, Manuel Lardizabal 1, 20018 Donostia, Spain
{igor.ibarguren,txus.perez,j.muguerza}@ehu.es
http://www.sc.ehu.es/aldapa/

Abstract. The consolidation process, originally applied to the C4.5 tree
induction algorithm, improved its discriminating capacity and stability. Consolidation creates multiple samples and builds a simple (non-
multiple) classifier by applying the ensemble process during the model
construction times. A benefit of consolidation is that the understand-
ability of the base classifier is kept. The work presented aims to show
the consolidation process can improve algorithms other than C4.5 by
applying the consolidation process to another algorithm, CHAID*. The
consolidation of CHAID*, CTCHAID, required solving the handicap of
consolidating the value groupings proposed by each CHAID* tree for
discrete attributes. The experimentation is divided in three classification contexts for a total of 96 datasets. Results show that consolidated
algorithms perform robustly, ranking competitively in all contexts, never
falling into lower positions unlike most of the other 23 rule inducting algorithms considered in the study. When performing a global comparison
consolidated algorithms rank first.

1 Introduction

In some problems that make use of classification techniques, the reason of why
a decision is made is almost as important as the accuracy of the decision, thus
the classifier must be comprehensible. Decision trees are considered comprehensible classifiers. The most common way of improving the discriminating capacity
of decision trees is to build ensemble classifiers. However with ensembles, the
explaining capacity individual trees possess is lost. The consolidation of algorithms is an alternative that resamples the training sample multiple times and
applies the ensemble voting process while the classifier is being built, so that
the final classifier is a single classifier (with explaining capacity) built using the
knowledge of multiple samples. The well-known C4.5 tree induction algorithm
[10] has successfully been consolidated in the past [9].

With the aim of studying the benefit of the consolidation process on other algorithms, maintaining the explaining capacity of the classifier, in this work we apply
this methodology on a variation of the CHAID [7,8] algorithm (CHAID* [5]), one
of the first tree induction algorithms along with C4.5 and CART. We propose the

© Springer International Publishing Switzerland 2015
F. Pereira et al. (Eds.) EPIA 2015, LNAI 9273, pp. 572–577, 2015.
DOI: 10.1007/978-3-319-23485-4_56

consolidation of CHAID* and using tests for statistical significance [4] we compare its results in three different classification contexts (amounting a total of 96 datasets) against 16 genetics-based and 7 classical algorithms and also the original CTC (Consolidated Tree Construction) algorithm.

The rest of the paper is organized as follows. Section 2 details the related work. Section 3 explains the consolidation version of the CHAID*, CTCHAID. Section 4 defines the experimental methodology. Section 5 lays out the obtained results. Finally, section 6 gives this work's conclusions.

2   Related Work on CHAID* and Consolidation

The CHAID (Chi-squared Automatic Interaction Detector) [7,8] is a tree induction algorithm that uses the chi-squared (χ^2) as the split function and it only works with discrete variables. CHAID* [5] is a variation of CHAID that differs in three main aspects:

- Handling of attributes: The original CHAID algorithm lacks the ability to handle continuous variables. Inspired by how C4.5 handles continuous variables, CHAID* uses the χ^2 to determine the best cutting point to divide the variable into two sets.
- Missing values on continuous variables: Three options are considered to treat the examples with missing values: grouping them with those examples with a value lower or equal to the cutting point, grouping them with examples whose value is greater than the cutting point or creating a branch just for the examples with a missing value.
- Pruning: CHAID* uses the same strategy as C4.5 by applying the Reduced-error pruning mechanism.

The consolidation approach aims at improving the discriminating capacity and the stability while reducing the complexity of the classifier [9]. It works by applying the ensemble voting process while building the simple classifier instead of building multiple simple classifiers and performing the ensemble vote only when classifying new examples. Recently the term "Inner Ensembles" has been coined to group methodologies following this approach [1].

The first consolidated algorithm was the well-known C4.5 decision tree induction algorithm, creating the CTC (Consolidated Tree Construction) algorithm. CTC works by first creating multiple samples from the training samples, usually by subsampling. Then, from each sample a C4.5 tree begins to grow. However on each node, execution "stops". Each tree proposes a new split based on their unique sample. A vote takes place and a common split is agreed. All trees comply with the majority vote and make the split accordingly, even if it is not what they have voted. This continues until the majority decides not to split any more. Because of this process, the structure of the trees grown from all subsamples is the same and the outcome is a single tree model. Then, for each leaf node, the *a posteriori* probabilities for each class are computed by averaging the probabilities on that particular leaf using the same samples used to build the consolidated tree.

3   CTCHAID

As explained in section 2 the changes made to CHAID* make it very similar to the C4.5 algorithm, which makes the implementation of CTCHAID very similar to the implementation of CTC45 (Consolidated C4.5) described in section 2. Aside from the split function, the other main difference between the algorithms is how discrete variables (nominal and ordinal) are handled. By default, when splitting using a discrete variable C4.5 creates a branch for every possible value for the attribute. On the other hand, CHAID* considers grouping more than one value on each branch. In each node a contingency table is created for each variable. Each of these tables describes the relationship between the values a variable can take and how the examples with this value are distributed among all possible classes. CHAID* uses Kass' algorithm [7] on all contingency tables to find the most significant variable and value-group to make the split.

When consolidating CHAID* the behavior is different depending on the type of variable. First the contingency tables are built from each sample and processed with Kass' algorithm to find the most important grouping. From each subsample a variable is proposed and voting takes place as with CTC45. If the voted variable is continuous the median value of the proposed cut-point values will be used. For categorical values, the contingency tables from each tree for the chosen variable are averaged into a single table. This averaged table is processed with Kass' algorithm to find the most significant combination of categories.

4   Experimental Methodology

This experiment follows a very similar structure as the works in [3] and [6] as we compare to the results published in those works. The same three classification contexts are analyzed: 30 standard (mostly multi-class) datasets, 33 two-class imbalanced datasets and the same 33 imbalanced datasets preprocessed with SMOTE (Synthetic Minority Over-sampling Technique [2]) until the two classes were balanced by oversampling the minority class. Fernández et al. [3] proposed a taxonomy to classify genetics-based machine learning (GBML) algorithms for rule induction. They listed 16 algorithms and classified them in 3 categories and 5 subcategories. They compared the performance of these algorithms with a set of classical algorithms (CART, AQ, CN2, C4.5, C4.5-Rules and Ripper). In our work, for each of the contexts, the winner for each of the 5 GBML categories, 7 classical algorithms (including CHAID*), CTC45 and CTCHAID are compared. Finally a global ranking is also computed. All algorithms used the same 5-run × 5-fold cross-validation strategy and the same training/test partitions (found in the KEEL repository[1]). The tables containing all the information have been omitted from this article for space issues and have been moved to the website with the additional material for this paper[2].

---

[1] http://sci2s.ugr.es/keel/datasets.php
[2] http://www.aldapa.eus/res/2015/ctchaid/

For CTC45 and CTCHAID, following the conclusions of the latest work on consolidation [6], the subsamples used in this work are balanced and the number of examples per class is the number of examples the least populous class has in the original training sample. The number of samples for each dataset has been determined using a coverage value of 99% based on the results of [6]. The tables detailing the number of samples for each dataset have been moved to the additional material. The pruning used for C4.5, CHAID*, CTC45 and CTCHAID was C4.5's reduced-error pruning. However, when pruning resulted in a tree with just the root node, the tree was kept unpruned. This is due to the fact that a root node tree results in zero for most performance measures used in this paper. Thus, the results shown for C4.5 are not those previously published by Fernández *et al.* using the KEEL platform but Quinlan's implementation of the algorithm.

5   Results

As described in the Experimental Methodology section, we divide the study into three contexts. For each context we analyze and compare the behavior of 14 algorithms: 5 GBML algorithms (the best for each subcategory proposed by Fernández *et al.*), 7 classical algorithms, CTC45 and CTCHAID. The GBML algorithms change from context to context while the classical stay the same (CART, AQ, CN2, C4.5, C45-Rules, Ripper and CHAID*).

The significance of the average performance values achieved by the algorithms has been tested using the Friedman Aligned Ranks test, as proposed by [4]. When this test finds statistically significant differences between algorithms Holm's post-hoc test has been used to find which algorithms perform significantly worse than the best ranking algorithm. The average performance values have been moved to the website with the additional material for this article. Figure 1 offers a visual representation of the average ranks achieved by the algorithms on different contexts. In that figure a thick black line covers algorithms without statistically significant differences with the best ranking algorithm. The lower the rank the better the performance is.

Although CTCHAID does not rank first for any of the three contexts. The differences with the best ranking algorithm for each context are never found to be significant by the Holm test.

In a similar fashion to what was done in [6] we perform a global analysis combining the results of the three contexts. The rankings of this global analysis are found in Figure 2. For the standard dataset classification, only the kappa measure is used. In this case CTC45 ranks first followed by CTCHAID. The Friedman Aligned Ranks test computes a p-value of 4.8×10^{-12} (test statistic 89.51) indicating the clear presence of statistically significant differences between the performance of the algorithms. According to the Holm test DT-GA, SIA, UCS, CART, OCEC, CORE, AQ and CN2 perform significantly worse than CTC45.

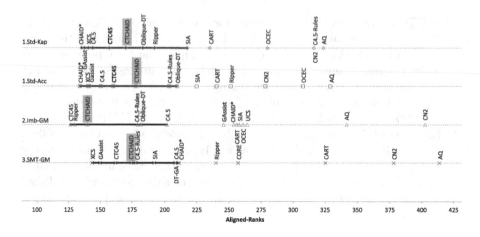

Fig. 1. Visual representation of Friedman Aligned Ranks for the three contexts.

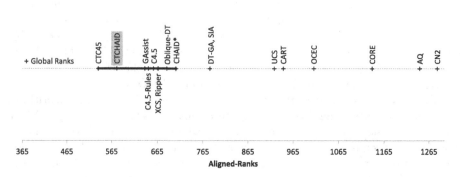

Fig. 2. Visual representation of Friedman Aligned Ranks for the global ranking.

6   Conclusions and Future Work

Results show that CTCHAID performs competitively. In summary, CTCHAID ranks in the upper half for all three contexts and in the first quartile for two out of three. As most algorithms fall into much lower positions for at least one context, CTCHAID ranks competitively in global terms. This behavior shows the robustness brought by the consolidation process in contrast to the behavior of the base algorithms, C4.5 and CHAID*, that fall into lower positions in some contexts, ranking worse globally. This shows that the consolidation process can bring improvement to multiple algorithms.

As future work we would like to study the performance of the CTC45 and CTCHAID algorithms under different pruning strategies: standard pruning, the strategy used in this work, disabling pruning, alternatives to pruning, etc. in order to tackle the class imbalance problem. Also in the same spirit, we would like to consolidate other tree and rule-induction algorithms.

Acknowledgments. This work was funded by the University of the Basque Country UPV/EHU (BAILab, grant UFI11/45); by the Department of Education, Universities and Research and by the Department of Economic Development and Competitiveness of the Basque Government (grant PRE-2013-1-887; BOPV/2013/128/3067, grant IT-395-10, grant IE14-386); and by the Ministry of Economy and Competitiveness of the Spanish Government (eGovernAbility, grant TIN2014-52665-C2-1-R).

References

1. Abbasian, H., Drummond, C., Japkowicz, N., Matwin, S.: Inner ensembles: using ensemble methods inside the learning algorithm. In: Blockeel, H., Kersting, K., Nijssen, S., Železný, F. (eds.) ECML PKDD 2013, Part III. LNCS, vol. 8190, pp. 33–48. Springer, Heidelberg (2013)
2. Chawla, N.V., Bowyer, K.W., Hall, L.O., Kegelmeyer, W.P.: SMOTE: Synthetic minority over-sampling technique. Journal of Artificial Intelligence Research **16**(1), 321–357 (2002)
3. Fernández, A., Garcia, S., Luengo, J., Bernadó-Mansilla, E., Herrera, F.: Genetics-based machine learning for rule induction: State of the art, taxonomy, and comparative study. IEEE Transactions on Evolutionary Computation **14**(6), 913–941 (2010)
4. García, S., Fernández, A., Luengo, J., Herrera, F.: Advanced nonparametric tests for multiple comparisons in the design of experiments in computational intelligence and data mining: Experimental analysis of power. Information Sciences **180**(10), 2044–2064 (2010)
5. Ibarguren, I., Lasarguren, A., Pérez, J.M., Muguerza, J., Arbelaitz, O., Gurrutxaga, I.: BFPART: Best-first PART. Submitted to Information Sciences
6. Ibarguren, I., Pérez, J.M., Muguerza, J., Gurrutxaga, I., Arbelaitz, O.: Coverage-based resampling: Building robust consolidated decision trees. Knowledge-Based Systems **79**, 51–67 (2015)
7. Kass, G.V.: Significance testing in automatic interaction detection (a.i.d.). Journal of the Royal Statistical Society. Series C (Applied Statistics) **24**(2), 178–189 (1975)
8. Morgan, J.A., Sonquist, J.N.: Problems in the analysis of survey data, and a proposal. J. Amer. Statistics Ass. **58**, 415–434 (1963)
9. Pérez, J.M., Muguerza, J., Arbelaitz, O., Gurrutxaga, I., Martín, J.I.: Combining multiple class distribution modified subsamples in a single tree. Pattern Recognition Letters **28**(4), 414–422 (2007)
10. Quinlan, J.R.: C4.5: programs for machine learning. Morgan Kaufmann Publishers Inc., San Francisco (1993)

Towards Interactive Visualization of Time Series Data to Support Knowledge Discovery

Jan Géryk[✉]

KD Lab, Faculty of Informatics, Masaryk University, Brno, Czech Republic
xgeryk@fi.muni.cz

Abstract. Higher education institutions have a significant interest in increasing the educational quality and effectiveness. A major challenge in modern education is the large amount of time-dependent data, which requires efficient tools and methods to improve decision making. Methods like motion charts (MC) show changes over time by presenting animations in two-dimensional space and by changing element appearances. In this paper, we present a visual analytics tool which makes use of enhanced animated data visualization methods. The tool is primarily designed for exploratory analysis of academic analytics (AA) and offers several interactive visualization methods that enhance the MC design. An experiment is conducted to evaluate the efficacy of both static and animated data visualization methods. To interpret the experiment results, we utilized one-way repeated measures ANOVA.

Keywords: Animation · Motion charts · Visual analytics · Academic analytics · Experiment

1 Introduction

A key requirement of Business Intelligence (BI) is to improve the decision making process and to facilitate users to get all the needed information at the right time. There is an increasing distinction made between academic analytics (AA) and traditional BI because of the unique type of information that university executives and administrators require for decision making. In [1], hundreds of higher education executives were surveyed on their analytic needs. Authors resulted that advanced analytics should support better decision-making, studying enrollment trends, and measuring student retention. They also pointed out that management commitment and staff skills are more important in deploying AA than the technology.

Visualizations are common methods used to gain a qualitative understanding of data prior to any computational analysis. By displaying animated presentations of the data and providing analysts with interactive tools for manipulating the data, visualizations allow human pattern recognition skills to contribute to the analytic process. The most commonly used statistical visualization methods generally focus on univariate or bivariate data. The methods are usually used for tasks ranging from the exploration to the confirmation of models, including the presentation of the results. However, fewer

© Springer International Publishing Switzerland 2015
F. Pereira et al. (Eds.) EPIA 2015, LNAI 9273, pp. 578–583, 2015.
DOI: 10.1007/978-3-319-23485-4_57

methods are available for visualizing data with more than two dimensions (e.g. motion charts or parallel coordinates), as the logical mapping of the data dimension to the screen dimension cannot be directly applied.

Although a snapshot of the data can be beneficial, presenting changes over time can provide a more sophisticated perspective. Animations allow knowledge discovery in complex data and make it easier to see meaningful characteristics of changes over time. The dynamic nature of Motion Charts (MC) allows a better identification of trends in the longitudinal multivariate data and enables visualization of more element characteristics simultaneously, as presented in [2]. The authors also conducted an experiment whose results concluded that MC excels at data presentation. MC is a dynamic and interactive visualization method that enables analysts to display complex and quantitative data in an intelligible way.

In this paper, we show the assets of animated data visualizations for successful understanding of complex and large data. In the next section, we describe our VA tool that implements visualization methods which make use of enhanced MC design. Further, we conduct an empirical study with 16 participants on their data comprehension to compare the efficacy of various static data visualizations with our enhanced methods. We then discuss the implications of our experiment results. Finally, we draw the conclusion and outline future work.

2   The Visual Analytics Tool

Visualization tools represent an effective way to make statistical data understandable to analysts, as showed in [3]. MC methods proved to be useful for data presentation and the approach was verified to be successfully employed to show the story in data [4] or support decision making [5]. Several web-based tools allowing analysts to interactively explore associations, patterns, and trends in data with temporal characteristics are available. In [6], authors presented a visualization of energy statistics using an existing web-based data analysis tools, including IBM's Many Eyes, and Google Motion Charts.

The motivation to develop advanced MC methods was to improve expression capabilities, as well as to facilitate analysts to depict each student or study as a central object of their interest. Moreover, the implementation enhances the number of animations that express the students' behavior during their studies more precisely. We partly validated usefulness of the developed methods with a case study where we successfully utilized the capabilities of the tool for the purpose of confirming hypothesis concerning student retention. Although, we concluded that the methods proved to be useful for analytic purposes, more adjustments were needed.

Two main challenges are addressed by the presented VA tool. It enables visualization of multivariate data and the qualitative exploration of data with temporal characteristics. The technical advantages over other implementations of MC are its flexibility and the ability to manage many animations simultaneously. Technical aspects of the enhanced MC methods are elaborately described in [7].

To create an effective and efficient knowledge discovery process, it is important to support common data manipulation tasks by creating quick, responsive and intuitive interaction methods. The tool offers several beneficial configurable interactive features for a more convenient analytic process. User interface features are highly customizable and allow analysts to arrange a display and variable mapping according to his or her needs. Available features include a mouse-over data display, color and plot size representation, traces, animated time plot, variable animation speed, changing of axis series, changing of axis scaling, distortion, and the support of statistical methods.

3   Experiment

Any quantitative research of AA also requires a preliminary exploratory data analysis. Though useful, advanced MC methods involve several drawbacks in comparison with common data visualization methods. Thus, empirical data is needed to evaluate its actual usability and efficacy. In this section, we describe the experiment for the purpose of evaluating the efficacy of the MC methods implemented in our tool. We present the results including a detailed discussion. Sixteen subjects (7 female, 9 male) with an average age of 23.44 (SD = 2.12) participated in our experiment. The participants ranged from 21 to 26 years of age. All participants came from professions requiring the use of data visualizations, including college students, analysts, and administrators.

We performed a study to test the benefits of the animated methods over static methods when employed to analyze study related data. The experiment used a 3 (visualization) x 2 (size) within-subjects design. The visualizations varied between the static and the animated methods. The methods were represented by motion charts (MC), line charts (LC), and scatter plots (SP) which were generated for each semester. The size of datasets varied between small and large ones with the threshold of 500 elements. For the experiment, we utilized study related data about students admitted to bachelor studies of the Faculty of Informatics Masaryk University between the years of 2006 and 2008.

3.1   Hypotheses

We designed the experiment to address the following three hypotheses:

- H1. The MC methods will be more effective than both the static methods for all datasets. That is, the subjects will be (a) faster and (b) make fewer errors when using MC.
- H2. The subjects will be more effective with the small datasets than with the large datasets for all methods. That is, participants will be (a) faster and (b) make fewer errors when working with small datasets.

In each trial, the participants completed 12 tasks, each with 1 to 3 required answers. Each task had identification numbers of students or fields of study as the answer. Several questions have more correct answers than requested. The participants selected answers by selecting IDs in the legend box located in the upper right from the chart area. In order to complete the task, two buttons could be used–either "OK" button to confirm the participant's choice or "Skip Question" button to proceed to the next task without saving the answer. There was no time limit during the experiment. For each task, the order of the datasets was fixed with the smaller ones first.

The participants were asked to proceed as quickly and accurately as possible. In order to reduce learning effects, the participants were told to make use of as many practice trials as they needed. It was followed by 12 tasks (6 small dataset tasks and 6 large dataset tasks in this particular order). After that, the subjects completed survey with questions specific for the visualization. Each block lasted about 1.5 hours. The subjects were screened to ensure that they were not color-blind and understood common data visualization methods. To test for significant effects, we conducted repeated measures analysis of variance (RM-ANOVA). Post-hoc analyses were performed by using the Bonferroni technique. Only significant results are reported.

3.2   Results

Accuracy. Since some of the tasks required multiple answers, accuracy was calculated as a percentage of the correct answers. Thus, when a subject selected only one correct answer from two, we calculated the answer as 50 % accurate rather than an incorrect answer. The analysis revealed several significant accuracy results at the .05 level. The type of visualization had a statistically significant effect on the accuracy for large datasets ($F(1.413, 21.194) = 20.700$, $p < 0.001$). Pair-wise comparison of the visualizations found significant differences showing that MC was significantly more accurate than the LC ($p < 0.001$). MC was also more accurate than the SP ($p < 0.001$). There was no statistically significant difference between the LC and the SP. For the small datasets, visualizations were not statistically distinguishable. Second, the subjects were more accurate with the small datasets ($F(1, 15) = 50.668$, $p < 0.001$). This fact supports our hypothesis H2.b.

Task Completion Time. An answer was considered to be incorrect if none of the correct answers was provided. In terms of time to task completion, we observed a statistically significant effect ($F(2, 30) = 107.474$, $p < 0.001$). Post-hoc tests revealed that MC was fastest with the large datasets. The LC was faster than the SP ($p < 0.017$). The mean time for MC was 48.56 seconds compared to 59.39 seconds for the LC–about 22% slower, and 62.88 seconds for the LC–about 29% slower. For the small datasets, static methods were faster than MC. Pair-wise comparison of the visualizations found significant differences between all of them. MC was slower than the LC ($p < 0.003$) and the SP ($p < 0.001$). The LC was slower than the SP ($p < 0.016$). The mean time for MC was 42.19 seconds compared to 35.94 seconds for the LC–about 17% faster, and 31.94 seconds for the SP–about 32% faster. This only partially supports the hypothesis H1.a. MC is faster than both the static methods when used for the large datasets.

Subjective Preferences. For each experiment block, the subjects completed a survey where the subjects assessed their preferences regarding analysis. The subjects rated LC, SP, and MC on a five-point Likert scale (1 = strongly disagree, 5 = strongly agree). Using RM-ANOVA, we revealed statistically significant effects (F(1.364, 20.453) = 4.672, p = 0.033). Post-hoc analysis found that MC was significantly more helpful than LC (p = 0.046).

Table 1. The resulted mean values of the preferences.

	SP	LC	MC
The visualization was helpful when solving the tasks.	3.50	3.44	3.94
I found this visualization entertaining and interesting.	2.56	2.31	4.13
I prefer visualization for the small datasets.	3.88	4.00	2.63
I prefer visualization for the large datasets.	2.38	2.69	3.69

The significant differences indicate that MC was judged to be more helpful than the static methods. The subjects preferred the static methods to MC for the small datasets. However, MC was judged to be more beneficial than static methods for the large datasets (p < 0.001). The results also showed that MC was more entertaining and interesting than the static methods (p < 0.001).

4   Discussion

Our first hypothesis (H1) was that MC would outperform both the static methods for all dataset sizes, but the hypothesis was only partially confirmed. Contrary to the hypothesis, the static methods proved to achieve better speed than the animated methods for the small datasets. Moreover, the methods were not statistically distinguishable in terms of accuracy. We also hypothesized that the accuracy would increase for the smaller datasets (H2). Hypothesis H2.a was supported, because the subjects were faster with the small datasets. The mean time for the large datasets was 56.94 seconds and for the small datasets was 36.69 seconds. Hypothesis H2.b was also supported, because the subjects made fewer errors with the small datasets when compared with the large datasets. Accuracy is an issue for static visualizations when the large datasets are employed.

The study supports the intuition that using animations in analysis requires convenient interactive tools to support effective use. The study suggests that MC leads to fewer errors. Also, the subjects found MC method to be more entertaining and exciting. The evidence from the study indicates that the animations were more effective at building the subjects' comprehension of large datasets. However, the simplicity of static methods was more effective for small datasets. These observations are consistent with the verbal reports in which the subjects refused to abandon the static visual methods generally. Results supported the thoughts that MC does not represent a replacement of common statistic data visualizations but a powerful addition. The overall accuracy was quite low in the study with average about 75%. However, only one question was skipped.

5   Conclusion and Future Work

In the tool, we enhanced the MC design and expanded it to be more suitable for AA analysis. We also developed an intuitive, yet powerful, interactive user interface that provides analysts with instantaneous control of MC properties and data configuration, along with several customization options to increase the efficacy of the exploration process. We validate the usefulness and general applicability of the tool with the experiment to assess an efficacy of the described methods.

The study suggests that animated methods lead to fewer errors for the large datasets. Also, the subjects find MC to be more entertaining and interesting. The entertainment value probably contributes to the efficacy of the animation, because it serves to hold the subjects' attention. This fact can be useful for the purpose of designing methods in academic settings.

Despite the findings of the study, further investigation is required to evaluate the general applicability of the animated methods. We also plan to combine our animated interactive methods with common DM methods to follow the VA principle more precisely. We already implemented a standalone method utilizing decision tree algorithm providing interactive visual representation. We prefer decision trees because of their clarity and simplicity to comprehend. We will also finish the integration of the tool with our university information system to allow university executives and administrators easy access when analyzing AA and to better support decision making.

References

1. Goldstein, P.J.: Academic analytics: The uses of management information and technology in higher education. Educause (2005)
2. Al-Aziz, J., Christou, N., Dinov, I.D.: SOCR Motion Charts: an efficient, open-source, interactive and dynamic applet for visualizing longitudinal multivariate data. Journal of Statistics Education **18**(3), 1–29 (2010)
3. Grossenbacher, A.: The globalisation of statistical content Statistical Journal of the IAOS. Journal of the International Association for Official Statistics, 133–144 (2008)
4. Baldwin, J., Damian, D.: Tool usage within a globally distributed software development course and implications for teaching. Collaborative Teaching of Globally Distributed Software Development, 15–19 (2013)
5. Sultan, T., Khedr, A., Nasr, M., Abdou, R.: A Proposed Integrated Approach for BI and GIS in Health Sector to Support Decision Makers. Editorial Preface (2013)
6. Vermylen, J.: Visualizing Energy Data Using Web-Based Applications. American Geo-physical Union (2008)
7. Géryk, J., Popelínský, L.: Visual analytics for increasing efficiency of higher education institutions. In: Abramowicz, W., Kokkinaki, A. (eds.) BIS 2014 Workshops. LNBIP, vol. 183, pp. 117–127. Springer, Heidelberg (2014)

Ramex-Forum: Sequential Patterns of Prices in the Petroleum Production Chain

Pedro Tiple[1], Luís Cavique[2], and Nuno Cavalheiro Marques[3]([⊠])

[1] GoBusiness Finance, Lisbon, Portugal
[2] Universidade Aberta, Lisbon, Portugal
[3] NOVA Laboratory for Computer Science and Informatics, DI-FCT,
Universidade Nova de Lisboa, Lisbon, Portugal
nmm@fct.unl.pt

Abstract. We present a sensibility analysis and new visualizations using an improved version of the Ramex-Forum algorithm applied to the study of the petroleum production chain. Different combinations of parameters and new ways to visualize data will be used. Results will highlight the importance of Ramex-Forum and its proper parameterizations for analyzing relevant relations among price variations in petroleum and other similar markets.

Keywords: Ramex-forum · Financial data analysis · Petroleum price · Petroleum production chain · Business intelligence

1 Introduction

Petroleum is one of the most important resources to the developed world and still is a major variable influencing the Economy and markets. The price of petroleum and its derivatives isn't influenced simply by supply and demand; taxes, speculation, wars, costs in refinement and transportation all contribute in setting prices. Due to its lengthy refinement process, a significant increase in the price of the source material can only reflect in the price of its derivatives after the time it takes to refine (usually within 3-4 weeks [1]). Moreover, due to its high economic importance and cost, the price of crude oil should always reflect on the final price [2].

This work presents a study on a method to quantify how the price of the crude oil (raw material) can influence the price of manufactured products, by using Ramex-Forum. This paper departs from the work of [3], with the original Ramex-forum proposal [4]. It analysis how this proposal can be improved and then tunned for finding sequential patterns using the prices of petroleum and derivatives. Sect. 2 presents the basic method and introduces the main concepts and Sect. 3 presents an evaluation on how the price of derivatives are influenced by the price of the crude oil (the source material). Finally, some conclusions are presented.

© Springer International Publishing Switzerland 2015
F. Pereira et al. (Eds.) EPIA 2015, LNAI 9273, pp. 584–589, 2015.
DOI: 10.1007/978-3-319-23485-4_58

2   Counting Co-occurrences in Financial Markets

We assume a crossover strategy to buy and sell financial products. Given the product price index in time 't', denoted I(t), and the moving average of that price index, with length of N_{MA} days, calculated by: $MA(t, N_{MA}) = \sum I(t-w)/N_{MA}, \forall w \in \{0 : N_{MA}\}$, the decision is as follows:

- Buy, if $I(t) \cdot (1 + \epsilon) \geq MA(t, N_{MA})$
- Sell, if $I(t) \cdot (1 - \epsilon) \leq MA(t, N_{MA})$

For each moment t, if there is a decision of either Buy or Sell, respective counters ($Counter_B$, $Counter_S$) will be incremented by one unit. If neither of those decisions is made, both counters are reset. This way, each counter has the number of consecutive moments where the same decision is made. See example in the Fig. 1, the 'B' (Buy) an 'S' (Sell) char illustrate the crossover strategy for a large enough ϵ and N_{MA}. In this paper (except when explicitly mentioned otherwise), we will use a standard 1% error, i.e. $\epsilon = 0.01$. Other parameter is also used when defining an influence: parameter δ is the maximum trading period length where a check for relations between two assets is made. Finally we can define $\#Influence(A, B, \delta)$: a cumulative influence counter of a given Buy or Sell decision for a market signal A to a market signal B (denoted $A \rightarrow B$), counts how many times $0 < (|Counter_A| - |Counter_B|) \leq \delta \wedge Counter_B \neq 0$.

Fig. 1. Financial product (normalized DJI index in black) and respective moving average (blue) and crossover starting *Buy* (green) and *Sell* (red) decision with ϵ confidence.

3   Results on the Petroleum Production Chain

Petroleum is refined into a relatively extensive list [5], with each category having hundreds of sub-products. Moreover this division and classification mostly depends on its social usage. This study is based on the repository of publicly available historical values for a wide range of petroleum related products provided by the U.S. Energy Information Administration[1]. The variations in the prices of these products are also compared with the stock market value of eleven corporations dedicated to extracting, processing, and selling of crude and crude related products. The prices of the 55 products are separated into retail/bulk price and spot price (for some items the price is taken from retail sellers and for other items it is the security price at that day). The data is separated into four categories of known benchmarks [6] for: crude oil (West Texas Intermediate as OklahomaWTI, European Brent, and the OPEC Basket); Refinery price for Gasoline, RBOB Gasoline, Diesel, Kerosene Jet, Propane, and Heating Oil; National, state, and city averages for regular gasoline and diesel; Corporation stock values.

This paper studies the influence and best values for parameters δ, ϵ thresholds, and moving average size. Focus will be put on the *Buy* comparison because in the selected data the increase/decrease of prices is very asymmetrical with a strong lean towards increases. Also, for better parameter comparison, an additional measure is used in our results, the average edge weight: a relation between the weight sum of output edges divided by the number of edges in the graph: $AverageEdgeWeight(V, E) = \sum weight(e)/|E|, \forall e \in E$.

Parameter δ was analyzed regarding its effect on the average edge weight changes. The result can be seen in Fig. 2A. The chart shows the average edge weight change for each increment in the value of δ. Each line represents the results obtained using different moving average sizes. Several big spikes can be seen every 5 days, this is because gas and diesel prices at the pump are only registered on a weekly basis, so for each 5 day increase in δ the algorithm will pick up another change in value. This makes the analysis somewhat harder but it's still useful as now changes in retail prices are clearly identified. The first thing noted is that at the first week there is already a noticeable increase in the average edge weight, however some of it is due to influences between retail prices and not only from refinery to retail prices. Second, after the fourth week the individual increases in δ barely produce a meaningful increase in value, still the cumulative increases are significant.The parameter δ was fixed at a value of 30 working days (around six weeks: two weeks more than the expected).

Parameter ϵ was studied by trying to find the best combination of parameters. The algorithm was ran several times and the average edge weight value was recorded for each run. The best values for threshold interval and the moving average size are represented in Fig. 2B. The graph shows the progression of the average edge weight, in relation to increase in threshold size. The parameters that lead to the highest increase in average weight can be clearly identified as the moving average size of 240 days with a threshold of around 26% of the moving average. However things change when the influence event count is also

---

[1] The used data was downloaded from http://www.eia.gov/ on June 2014 and ranges form January 2006 to June 2014.

considered: increasing the threshold rapidly decreases the number of detected events (a threshold of 26% will reduce the number of events by about 80%). In this case, the starting average is around 130 events and falls to 30, in the 3 years period analyzed: a very low average number of events. For this case study the choice was made to maximize the event count so that a broader spectrum of influences can be detected instead of restricting the analysis to situations where the prices rise or fall sharply (which is what higher threshold values restrict the analysis to). Small increases in the ϵ threshold will raise the average weight while only lowering the event count by small amounts. Nevertheless, random fluctuations do not advise going for a threshold of 0%, so this trade-off seems to favor the usage of smaller values for ϵ.

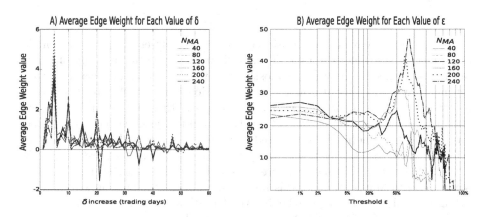

Fig. 2. Graphs showing the change in: (A) average edge weight with each increment of δ using the *Buy* comparison; (B) average edge weight and number of nodes with each $\epsilon_{t+1} = \epsilon_t + 1\%$ increment in the threshold interval for $\delta = 30$ using the *Buy* comparison.

Moving Average Size, N_{MA} The choices for available moving average sizes were based on [4] and the graphs show that maximizing this parameter yields the best results and even raises the question of how further increases in the size would fare. The user still needs to take into account of what it means to increase the moving average size: the bigger it is the moving average, the smoother the curve will be and thus it will behave like a noise filter (ie., by becoming less and less sensitive to small changes in the behavior of the product). The values overlap for small ϵ values and it is hard to read the effects of the first increments in a linear scale, so Fig. 2B uses a logarithmic scale for representing ϵ values, showing that the 240 and 120 moving average sizes have a very similar behavior. The average weight for a moving average of 120 days has a higher starting value than the 240 days one, this means that for a buy signal best thresholds are: $\epsilon = 1\% \wedge \delta = 30 \wedge N_{MA} = 120$. Fig. 3 shows that it is possible to find more than just sequential patterns with this parameters.

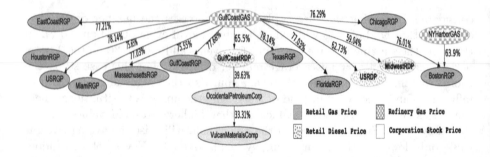

Fig. 3. Part of the graph showing the resulting Buy tree after applying Ramex Forum on the data with the selected parameters.

In the complete result graph (available in [3]) colors were added to each node according to their product type. These colors show a clear grouping of product types, with same color nodes mostly close to each other. This was expected for gas to gas and diesel to diesel influences. However even the stock, refinery, and reference benchmark prices tend to group together at least in pairs. Furthermore, refineries are almost exclusively related to the same type of product, gas producing refineries are connected to retail gas prices and diesel producing refineries are connected to diesel retail prices. The Gulf Coast GAS refinery node (Fig. 2) does not exactly meet the previous observation as it is shown influencing some diesel products, even so, this might be a positive thing as it will alert an attentive analyst to the weight behind the Gulf Coast refinery gas prices. After further analysis Gulf Coast GAS is identified as the most influential node as it has at least one detected event for all other products and its average edge weight is the highest by a margin of 5%, probably due to huge oil production in this area it is mostly the start of oil production chain. In [3], it was also observed that specificities of gas usage in the Rocky Mountain retail gas price could trigger third level dependences. Next the most glaring aspect of the graph is how influential specific products are, the tree is not just an assorted web of relations but groups of products aggregating around very influential/influenced products. There are some expected trend setters like the OPECBasket that is used as a benchmark for oil price, the Gulf Coast refineries and then some unexpected like the Minnesota retail gas price. For other tested data and parameters, equal graphs were observed. Indeed, color coding also showed very similar results with strong groupings of colors and some few select products influencing groups of others.

4   Conclusions

The presented case study, using real world data and deep analysis, aims to provide an illustrative and useful example of Ramex-Forum: the signal-to-noise ratio on the Petroleum production chain analysis already shows that sequential patterns of prices can provide a much deeper description of product dependencies based on events. Moreover the δ and ϵ parameters seem both consistent, intuitive and adaptable alternative for measuring long term dependencies that are not directly possible with more instantaneous methods. So far only the connections themselves have been considered, if the influence weights are also taken into account the analysis becomes more complex. Future work in this area should extend the number of products used and their detail. Some related studies (namely [7]) show that less classical hybrid approaches can be used to complement the crossover event detection approach and are good candidates for future experiments. Of particular interest will be to include a better characterization of algorithm behavior during a global market crisis, namely by quantifying the drives and consequences of the recent crisis in oil prices.

Acknowledgments. This research is supported by the GoBusiness Research project (http://www.gobusinessfinance.ch/en/research). The authors would like to thank GoBusiness Finance for partial financial support and for data sets and financial knowledge used in present work.

References

1. Borenstein, S., Shepard, A.: Sticky Prices, Inventories, and Market Power in Wholesale Gasoline Markets. NBER working paper series, vol. 5468. National Bureau of Economic Research (1996)
2. Suviolahti, H.: The influence of volatile raw material prices on inventory valuation and product costing. Master Thesis, Department of Business Technology, Helsinki School of Economics (2009)
3. Tiple, P.: Tool for discovering sequential patterns in financial markets. Master Thesis in Engenharia Informática, Faculdade de Ciências e Tecnologia da Universidade Nova de Lisboa (2014)
4. Marques, N.C., Cavique, L.: Sequential pattern mining of price interactions. In: Advances in Artificial Intelligence – Proceedings of the Workshop Knowledge Discovery and Business Intelligence, EPIA-KDBI, Portuguese Conference on Artificial Intelligence, pp. 314–325 (2013)
5. Gary, J., Handwerk, G.: Petroleum Refining. Institut français du pétrole publications. Taylor & Francis (2001)
6. Hammoudeh, S., Ewing, B.T., Thompson, M.A.: Threshold cointegration analysis of crude oil benchmarks. The Energy Journal **29**(4), 79–96 (2008)
7. Matos, D., Marques, N., Cardoso, M.: Stock market series analysis using self-organizing maps. Revista de Ciências da Computação **9**(9), 79–90 (2014)

Geocoding Textual Documents
Through a Hierarchy of Linear Classifiers

Fernando Melo and Bruno Martins[✉]

Instituto Superior Técnico and INESC-ID, Universidade de Lisboa, Lisbon, Portugal
{fernando.melo,bruno.g.martins}@ist.utl.pt

Abstract. In this paper, we empirically evaluate an automated technique, based on a hierarchical representation for the Earth's surface and leveraging linear classifiers, for assigning geospatial coordinates to previously unseen documents, using only the raw text as input evidence. We measured the results obtained with models based on Support Vector Machines, over collections of geo-referenced Wikipedia articles in four different languages, namely English, German, Spanish and Portuguese. The best performing models obtained state-of-the-art results, corresponding to an average prediction error of 83 Kilometers, and a median error of just 9 Kilometers, in the case of the English Wikipedia collection.

Keywords: Text mining · Document geocoding · Hierarchical text classification

1 Introduction

Geographical Information Retrieval (GIR) has recently captured the attention of many different researchers that work in fields related to language processing and to the retrieval and mining of relevant information from large document collections. For instance, the task of resolving individual place references in textual documents has been addressed in several previous works, with the aim of supporting subsequent GIR processing tasks, such as document retrieval or the production of cartographic visualizations from textual documents [5,6]. However, place reference resolution presents several non-trivial challenges [8,9], due to the inherent ambiguity of natural language discourse. Moreover, there are many vocabulary terms, besides place names, that can frequently appear in the context of documents related to specific geographic areas [1]. Instead of resolving individual references to places, it may be interesting to instead study methods for assigning entire documents to geospatial locations [1,11].

In this paper, we describe a technique for assigning geospatial coordinates of latitude and longitude to previously unseen textual documents, using only the raw text of the documents as evidence, and relying on a hierarchy of linear models built with basis on a discrete hierarchical representation for the Earths surface, known in the literature as the HEALPix approach [4]. The regions at each level of this hierarchical representation, corresponding to equally-distributed curvilinear

© Springer International Publishing Switzerland 2015
F. Pereira et al. (Eds.) EPIA 2015, LNAI 9273, pp. 590–596, 2015.
DOI: 10.1007/978-3-319-23485-4_59

and quadrilateral areas of the Earths surface, are initially associated to textual contents (i.e., we use all the documents from a training set that are known to refer to particular geospatial coordinates, associating each text to the corresponding region). For each level in the hierarchy, we build classification models using the textual data, relying on a vector space model representation, and using the quadrilateral areas as the target classes. New documents are assigned to the most likely quadrilateral area, through the usage of the classifiers inferred from training data. We finally assign documents to their respective coordinates of latitude and longitude, taking the centroid coordinates from the quadrilateral areas.

The proposed document geocoding technique was evaluated with samples of geo-referenced Wikipedia documents in four different languages. We achieved an average prediction error of 83 Kilometers, and a median error of just 9 Kilometers, in the case of documents from the English Wikipedia. These results are slightly better than those reported in previous state-of-the-art studies [11,12].

2 Previous and Related Work

While most work on geographic information retrieval relies on specific keywords such as place names, Adams and Janowicz proposed an approach for geocoding documents that uses only non-geographic expressions, concluding that even ordinary textual terms may be good predictors of geographic locations [1]. The proposed technique used Latent Dirichlet Allocation (LDA) to discover latent topics from general vocabulary terms occurring in a training collection of geo-referenced documents, together with Kernel Density Estimation (KDE) to interpolate a density surface, over each LDA topic. New documents are assigned to the geospatial areas having the highest aggregate density, computed from the per-document topic distributions and from the KDE surfaces.

Wing and Baldridge evaluated approaches for automatically geocoding documents based on their textual contents, specifically leveraging generative language models learned from Wikipedia [11]. The authors applied a regular geodesic grid to divide the Earth's surface into discrete rectangular cells. Each cell can be seen as a virtual document that concatenates all the training documents located within the cell's region. Three different methods were compared in the task of finding the most similar cell, for a new document, namely (i) the Kullback-Leibler divergence, (ii) naïve Bayes, and (iii) a baseline method corresponding to the average cell probability. Method (i) obtained the best results, i.e. a median prediction error of just 11.8 Kilometers, and a mean error of 271 Kilometers, on tests with documents taken from the English Wikipedia. More recently, Dias et al. [2] reported on experiments with an adapted version of the method described by Wing and Baldridge, which used language models based on character n-grams together with a discrete representation for the surface of the Earth based on an equal-area hierarchical triangular mesh approach [3]. Another improvement over the language modeling method was latter reported by Roller et al. [7], where the authors collapsed nearby training documents through the

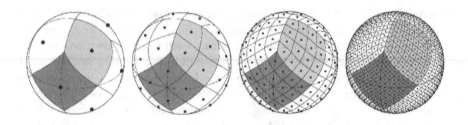

Fig. 1. Orthographic views associated to the first four levels of the HEALPix sphere tessellation.

usage of a k-d tree data structure. Moreover, Roller et al. proposed to assign the centroid coordinates of the training documents contained in the most probable cell, instead of just using the center point for the cell. These authors report on a mean error of 181 Kilometers and a median error of 11 Kilometers, when geocoding documents from the English Wikipedia.

More recently, Wing and Baldridge also reported on tests with discriminative classifiers [12]. To overcome the computational limitations of discriminative classifiers, in terms of the maximum number of classes they can handle, the authors proposed to leverage a hierarchical classification procedure that used feature hashing and an efficient implementation of logistic regression. In brief, the authors used an hierarchical approach in which the Earth's surface is divided according to a rectangular grid (i.e., using either a regular grid or a k-d tree), and where an independent classifier is learned for every non-leaf node of the hierarchy. The probability of any node in the hierarchy is the product of the probabilities of that node and all of its ancestors, up to the root. The most probable leaf node is used to infer the final geospatial coordinates. Rather than greedily using the most probable node from each level, or rather than computing the probability of every leaf node, the authors used a stratified beam search. This procedure starts at the root, keeping the b highest-probability nodes at each level, until reaching the leafs. Wing and Baldridge report on results over English Wikipedia data corresponding to a mean error of 168.7 Kilometers and a median error of 15.3 Kilometers.

3   The Proposed Document Geocoding Method

The proposed document geocoding approach is based on discretizing the surface of the Earth into hierarchically organized sets of regions, as given by the HEALPix procedure and where each set corresponds to a different partitioning resolution. Having documents associated to these discrete regions allows us to predict locations with standard discriminative classification approaches (e.g., with linear Support Vector Machines classifiers).

HEALPix is an acronym for Hierarchical Equal Area isoLatitude Pixelization of a sphere, and the procedure results on a multi level recursive subdivision

Table 1. Number of regions and approximate area for HEALPix grids of different resolutions.

Resolution	4	64	256	1024
Total number of regions	192	49,152	786,432	12,582,912
Approximate area of each region (Km^2)	2,656,625	10,377	649	41

for a spherical approximation to the Earth's surface, according to curvilinear quadrilateral regions, in which each resulting subdivision covers an equal surface area. Figure 3, adapted from an original illustration provided in the HEALPix website[1], shows from left to right the resolution increase by three steps from the base level with 12 different regions [4].

The HEALPix representation scheme contains a parameter N_{side} that controls the resolution, i.e. the number of divisions along the side of a base-resolution region that is needed to reach a desired high-resolution partition, and which naturally will also define the area of the curvilinear quadrilaterals. In our experiments, we used a hierarchy of 4 different representations with different resolutions, equaling the N_{side} parameter to the values of 4, 64, 256 and 1024. Table 1 presents the total number of regions in each of the considered resolution levels (i.e., $n = 12 \times N_{side}^2$), together with the approximate area, in squared Kilometers, corresponding to each region.

Another important question relates to the choice of how to represent the textual documents. We used a vector space model representation, where each document is seen as a vector of features. The feature weights in the vectors that represent each document are given according to the term frequency times inverse document frequency (TF-IDF) scheme, where the weight for a term i on a document j can be computed as:

$$TF\text{–}IDF_{i,j} = \log_2(1 + TF_{i,j}) \times \log_2\left(\frac{N}{n_i}\right) \tag{1}$$

In the formula, $TF_{i,j}$ is the term frequency for term i on document j, N is the total number of documents in the collection, and n_i is the number of documents containing the term i. The TF-IDF weight is 0 if $TF_{i,j} = 0$.

With the hierarchy of discrete representations given by the HEALPix method, together with the document representations based on TF-IDF, we then used linear classification algorithms to address the document geocoding task. We trained a separate classification model for each node in the hierarchy of discrete representations, taking all documents whose coordinates lay within the region corresponding to each node, as the training data for each classifier. When geocoding a test document, we first apply the root-level classifier to decide the most likely region, and then proceed greedily by applying the classifier for each of the most likely nodes, up to the leafs. After reaching a leaf region from the

---

[1] http://healpix.jpl.nasa.gov

hierarchical representation, the geospatial coordinates of latitude and longitude are assigned by taking the centroid coordinates of the leaf region.

Support Vector Machines (SVMs) are one of the most popular approaches for learning classifiers from training data. In our experiments, we used the multi-class linear SVM implementation from scikit-learn[2] with default parameters (e.g., with the default regularization constant), which in turn is a wrapper over the LIBLINEAR[3] package.

4 Experimental Validation

In our experiments, we used samples with geocoded articles from the English (i.e., 847,783 articles), German (i.e., 307,859 articles), Spanish (i.e., 180,720 articles) and Portuguese (i.e., 131,085 articles) Wikipedias, taken from database dumps produced in 2014. Separate experiments evaluated the quality of the document geocoders built for each of the four languages, in terms of the distances from the predictions towards the correct geospatial coordinates. We processed the Wikipedia dumps to extract the raw text from the articles, and for extracting the geospatial coordinates of latitude and longitude from the corresponding infoboxes. We used 90% of the geocoded articles of each Wikipedia for model training, and the other 10% for model validation.

In what regards the geospatial distributions of documents, we have that some regions (e.g, North America or Europe) are considerable more dense in terms of document associations than others (e.g, Africa), and that oceans and other large masses of water are scarce in associations to Wikipedia documents. This implies that the number of classes that has to be considered by our model is much smaller than the theoretical number of classes given by the HEALPix procedure. In our English dataset, there are a total of 286,966 regions containing associations to documents at a resolution level of $N_{side} = 1024$, and a total of 82,574, 15,065, and 190 regions, respectively at resolutions 256, 64, and 4. These numbers are even smaller in the collections for the other languages.

Table 2 presents the obtained results for the different Wikipedia collections. The prediction errors shown in Table 2 correspond to the distance in Kilometers, computed through Vincenty's geodetic formulae [10], from the predicted locations to the true locations given in Wikipedia. The accuracy values correspond to the relative number of times that we could assign documents to the correct region (i.e., the HEALPix region where the document's true geospatial coordinates of latitude and longitude are contained), for each level of hierarchical classification. Table 2 also presents upper and lower bounds for the average and median errors, according to a 95% confidence interval and as measured through a sampling procedure.

The results attest for the effectiveness of the proposed method, as we measured slightly inferior errors than those reported in previous studies [2,7,11,12], which besides different classifiers also used simpler procedures for representing

---

[2] http://scikit-learn.org/
[3] http://www.csie.ntu.edu.tw/~cjlin/liblinear/

Table 2. The results obtained for each different language.

	Classifier accuracy				Errors in terms of distance	
	1st	2nd	3rd	4th	Average	Median
English	0.966	0.785	0.540	0.262	82.501 (±4.340)	8.874 [5.303 - 15.142]
German	0.972	0.832	0.648	0.396	62.995 (±5.753)	4.974 [3.615 - 8.199]
Spanish	0.950	0.720	0.436	0.157	165.887 (±16.675)	13.410 [8.392 - 22.691]
Portuguese	0.951	0.667	0.336	0.104	105.238 (±10.059)	21.872 [13.611 - 33.264]

textual contents and for representing the geographical space. It should nonetheless be noted that the datasets used in our tests may be slightly different from those used in previous studies (e.g., they were taken from different Wikipedia dumps), despite their similar origin.

5   Conclusions and Future Work

Through this work, we empirically evaluated a simple method for geo-referencing textual documents, relying on a hierarchy of linear classifiers for assigning documents to their corresponding geospatial coordinates. We have shown that the automatic identification of the geospatial location of a document, based only on its text, can be performed with high accuracy by using out-of-the-box implementations of well-known supervised classification methods, and leveraging a hierarchical procedure based on HEALPix [4].

Despite the interesting results, there are also many ideas for future work. The geospatial coordinates estimated from our document geocoding procedure can for instance be used as prior evidence (i.e., as document-level priors) to support the resolution of individual place references in text [8]. In terms of future work, we would also like to experiment with other types of classification approaches and with different text representation and feature weighting schemes.

Acknowledgments. This work was supported by Fundação para a Ciência e a Tecnologia (FCT), through project grants with references EXCL/EEI-ESS/0257/2012 (DataStorm research line of excellency), EXPL/EEI-ESS/0427/2013 (KD-LBSN), and also UID/CEC/50021/2013 (INESC-ID's associate laboratory multi-annual funding).

References

1. Adams, B., Janowicz, K.: On the geo-indicativeness of non-georeferenced text. In: Proceedings of the International AAAI Conference on Weblogs and Social Media (2012)
2. Dias, D., Anastácio, I., Martins, B.: A language modeling approach for georeferencing textual documents. Actas del Congreso Español de Recuperación de Información (2012)
3. Dutton, G.: Encoding and handling geospatial data with hierarchical triangular meshes. In: Kraak, M.J., Molenaar, M., (eds.) Advances in GIS Research II. CRC Press (1996)

4. Górski, K.M., Hivon, E., Banday, A.J., Wandelt, B.D., Hansen, F.K., Reinecke, M., Bartelmann, M.: HEALPIX - a framework for high resolution discretization, and fast analysis of data distributed on the sphere. The Astrophysical Journal **622**(2) (2005)

5. Lieberman, M.D., Samet, H.: Multifaceted toponym recognition for streaming news. In: Proceedings of the International ACM SIGIR Conference on Research and Development in Information Retrieval (2011)

6. Mehler, A., Bao, Y., Li, X., Wang, Y., Skiena, S.: Spatial analysis of news sources. IEEE Transactions on Visualization and Computer Graphics **12**(5) (2006)

7. Roller, S., Speriosu, M., Rallapalli, S., Wing, B., Baldridge, J.: Supervised text-based geolocation using language models on an adaptive grid. In: Proceedings of the Conference on Empirical Methods on Natural Language Processing (2012)

8. Santos, J., Anastácio, I., Martins, B.: Using machine learning methods for disambiguating place references in textual documents. GeoJournal **80**(3) (2015)

9. Speriosu, M., Baldridge, J.: Text-driven toponym resolution using indirect supervision. In: Proceedings of the Annual Meeting of the Association for Computational Linguistics (2013)

10. Vincenty, T.: Direct and inverse solutions of geodesics on the ellipsoid with application of nested equations. Survey Review **XXIII**(176) (1975)

11. Wing, B., Baldridge, J.: Simple supervised document geolocation with geodesic grids. In: Proceedings of the Annual Meeting of the Association for Computational Linguistics (2011)

12. Wing, B., Baldridge, J.: Hierarchical discriminative classification for text-based geolocation. In: Proceedings of the Conference on Empirical Methods on Natural Language Processing (2014)

A Domain-Specific Language for ETL Patterns Specification in Data Warehousing Systems

Bruno Oliveira[✉] and Orlando Belo

Algoritmi R&D Centre, Department of Informatics, School of Engineering,
University of Minho, Campus de Gualtar, 4710-057 Braga, Portugal
obelo@di.uminho.pt

Abstract. During the last few years many research efforts have been done to improve the design of ETL (Extract-Transform-Load) systems. ETL systems are considered very time-consuming, error-prone and complex involving several participants from different knowledge domains. ETL processes are one of the most important components of a data warehousing system that are strongly influenced by the complexity of business requirements, their changing and evolution. These aspects influence not only the structure of a data warehouse but also the structures of the data sources involved with. To minimize the negative impact of such variables, we propose the use of ETL patterns to build specific ETL packages. In this paper, we formalize this approach using BPMN (Business Process Modelling Language) for modelling more conceptual ETL workflows, mapping them to real execution primitives through the use of a domain-specific language that allows for the generation of specific instances that can be executed in an ETL commercial tool.

Keywords: Data warehousing systems · ETL conceptual modelling · ETL patterns · BPMN specification models · Domain-Specific languages

1 Introduction

Commercial tools that support ETL (Extract, Transform, and Load) processes development and implementation have a crucial impact in the implementation of any data warehousing system populating process. They provide the generation of very detailed models under a specific methodology and notation. Usually, such kind of documentation follows a proprietary format, which is intrinsically related to architectural issues of the development tool. For that reason, ETL teams must have skills and experience on such tools that allow for them to use and explore appropriately the tools. In the case of a migration process of ETL migration to another ETL tool environment, the ETL development team will need to understand all specificities provided by the new tool and start a new project often almost from scratch. We believe that ETL systems development requires a simply and reliable approach. A more abstract view of the processes and data structures is very convenient as well as a more effective mapping to some kind of execution primitives allowing for its execution inside the environments of commercial tools. Using a pallet of specific ETL patterns representing some of the most used ETL tasks in real world applica-

© Springer International Publishing Switzerland 2015
F. Pereira et al. (Eds.) EPIA 2015, LNAI 9273, pp. 597–602, 2015.
DOI: 10.1007/978-3-319-23485-4_60

tion scenarios – e.g. *Surrogate Key Pipelining* (SKP), *Slowly Changing Dimensions* (SCD) or *Change Data Capture* (CDC) –, we designed a new ETL development layer on top of a traditional method, making possible to use ETL tools from the very beginning of the project, in order to plan and implement more appropriated ETL processes. To do that we used the *Business Process Modelling Notation* (BPMN) [1] for ETL processes representation, extending its original meta-model for including the ETL pattern specification we designed. The inclusion of these patterns distinguishes clearly two very relevant aspects in ETL design and implementation: process flow coordination and data processing tasks. BPMN is very suitable for this kind of processes, simply because it provides some very convenient features like expressiveness and flexibility on the specification of processes. Thus, after a brief exposure of some related work (section 2), we present and discuss briefly a demonstration scenario using one of the most useful (and crucial) ETL process: a *Data Quality Enhancement* (DQE) (section 3). Next, in section 4, we present a DQE specification skeleton, its internal behaviour and how we can configure using a *Domain-Specific Language* (DSL) to enable its execution. Finally, we discuss the experiments done so far, analysing results and presenting some conclusions and future work (section 5).

2   Related Work

With the exception of some low-level methods for ETL development [2], most approaches presented so far use conceptual or logical models as the basis for ETL modelling. Such models reduce complexity, produce detailed documentation and provide the ability to easily communicate with business users. Some of the proposals presented by Vassiliadis and Simitsis cover several aspects of ETL conceptual specification [3], its representation using logical views [4, 5], and its implementation using a specific ETL tool [6]. Later, Trujillo [7] and Muñoz [8] provided an UML extension for ETL conceptual modelling, reducing some of the communication issues that the proposal of Vassiliadis et al. revealed previously. However, the translation to execution primitives they made was not very natural, since UML is essentially used to describe system requirements and not to support its execution. The integration of existing organizational tasks with ETL processes was addressed by Wilkinson et al. [9], which exposed a practical approach for the specification of ETL conceptual models using BPMN. BPMN was firstly introduced by Akkaoui and Zimanyi [10] on ETL systems specification. Subsequently, Akkaoui et al. [11] provided a BPMN-based meta-model for an independent ETL modelling approach. They explored and discussed the bridges to a model-to-text translation, providing its execution in some ETL commercial tools. Still using BPMN notation, Akkaoui et al. [12] provided a BPMN meta-model covering two important architectural layers related to the specification of ETL processes. More recently and following the same guidelines from previous works, Akkaoui et al. [13] disposed a framework that allows for the translation of abstract BPMN models to its concrete execution in a target ETL tool using model-to-text transformations.

3   Pattern-Based ETL Modelling

We designed a high-level approach for ETL conceptual modelling using patterns. An ETL pattern is a task class that is characterized using a set of pre-established activities (internal composition) and their correspondent input and output interfaces, i.e. how the pattern interacts within a workflow context. Patterns avoid rewriting some of the most repetitive tasks that are used regularly in an ETL system. The use of ETL patterns produces simpler conceptual models, because finer grain tasks will be omitted from the global ETL schema. The descriptive models provided by BPMN supported all this. Using the ETL terminology, we defined three categories of ETL patterns: (1) Gathers that represent typical data extraction processes; (2) Transformers, which are used for data transformations such as cleaning or conforming tasks, and (3) Loaders, which are used to load data into the data warehouse repository. With these categories we can group all the most frequent ETL patterns and we can identify in a conventional ETL system all its operational stages.

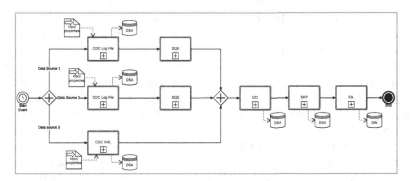

Fig. 1. A BPMN pattern specification for an ETL process

To demonstrate the application of ETL patterns in an ETL system solution we selected a simple application case. We represent the whole process with the ETL patterns we proposed using the BPMN notation, both in terms of the data orchestration and the representation of composite tasks. The process begins with a 'Start Event' having a 'Timer' representing an automatic execution of the process according to a specific time frame. This process starts with a parallel invocation of two data flows using a parallel gateway (Fig. 1). For each data source, a CDC pattern was used for data extraction. For the first and second flow, two other CDC patterns acting over log files were used. In the third flow we selected another CDC pattern for working with XML data structures. These tasks have the ability to identify new, updated, or deleted records, extracting and putting them into a *Data Staging Area* (DSA). We used BPMN 'Data input' and 'Data Store' objects to represent the input/output interfaces for each pattern. On the first two flows, we used a DQE pattern configured with some specific data cleaning and data conforming procedures. Next, the remaining two flows were synchronized using a parallel gateway, being data integrated using a *data conciliation pattern* (DCI) to identify common records based on a specific set of conciliation rules. Posteriorly, a SKP pattern assigns the surrogate keys to the data extracted, maintaining the correspondence meta-data stored in specific mapping tables located in the DSA. After the SKP process appears an IDL pattern that loads data into the data warehouse, establishing all the necessary correspondences to the data warehouse schema.

4   Specification of a DQE Pattern

Usually, operational data are stored in specific data schemas built to serve particular business needs. Independently from the sources involved (single or multiple), many problems can occur at schema and instance level when executing a loading data process [14]. Several procedures can be applied on transformation and cleaning tasks to eliminate problems like these to avoid schema ambiguities, data inconsistences, data entry errors, missing information or invalid data. To instantiate patterns a generator should know how they must be created following a specific template. In particular, for ETL processes the description of the structure of a pattern was studied already [15]. We do not intend to describe patterns in a natural language but essentially using some descriptive primitives to support their instantiation. For that, we divide ETL patterns specification in three parts: 1) input meta-data for pattern execution, which is composed by source schema(s), attribute(s) and the procedures that should be applied to the source data; 2) output meta-data, which describes the output schemas in which output data will be stored; and 3) exception handling, which represents the identification of unexpected scenarios and correspondent policies to perform. To receive these configuration aspects, we developed a DSL for configuring the components of ETL patterns. The use of a DSL facilitates pattern configuration and provides the necessary meanings to make it suitable for computer interpretation. One of the most common DQE procedures is the decomposition of a string in its meaningful parts. The decomposition of an attribute typically involves several database attributes such as a name or an address. Usually, these attributes need to be decomposed in n parts according to a certain condition.

```
Use Decomposition[                      Use Decomposition[
  Header:                                 Header: [...],
  [Name: 'DecTask',                       Content: [...]
   Description: 'Decomposition ...']     Exception:
  Content:                                [Event= EmptyAttribute,
  [Id: AUTO,                               Action: Compensation:
   Input[Schema='Sale',                     Quarantine[Table='CustomerQ'],
    Attribute='FullName']                   Log:[Name= 'CustomerLog',
   Output[Schema= 'SaleT',                      Type= RelationalTable,
    Attributes:[FirstName String;               Details[DATE AS 'mydate',
     LastName String;]]                          SOURCETABLE AS 'mytable',
   Rule[Regex= '\\s',                            SOURCEATTRIBUTE AS myAttr']]]
    Limit[FIRST, LAST]]]                   [Event= InvalidDecompositionRule,
                                           Action: Error= Terminate [...]]
```

Fig. 2. A decomposition procedure (left) and its exception handling description (right)

Fig. 2 shows an example of the DSL proposed for a typical decomposition procedure used as sub-part of a DQE pattern that splits the customer full name in two new attributes: 'FirstName' and 'LastName'. The decomposition rule is performed based on a regular expression: '\\s', which means that an original string must be split using a space as delimiter. The pattern configuration starts with the description of the pattern, using the 'Name' and the 'Description' keywords inside of a 'Header' block. Blocks are identified using square brackets delimiters. Next, a decomposition block is used in order to describe the internal components of a decomposition pattern. This block must have an internal 'ID', which can be manually or automatically assigned ('AUTO' keyword). Next, three blocks are used: 'Input' block for input metadata, 'Output' block for output metadata, and 'Rule' for a decomposition rule specification. Both input and output blocks use a target data schema name storing the original/resulted data, and a collection of attributes (and data types) used for each block. We distinguished single assignments (a singular value) and composite assignments (composite data structures) using an equal operator for atomic attributions and square brackets for composite attributions. The 'Rule' block describes the regular expression that should be applied to the original string using the 'Regex' keyword. The 'Limit' keyword is used to control the number of times that a pattern is applied affecting the length of the returned result set. Two special identifiers ('FIRST' and 'LAST') are used to extract the first and last occurrence matching the regular expression. The results of the output rule ('FirstName' and 'LastName') are mapped to the 'Output' attributes. The DSL also includes some compensation and error exception statements associated to each pattern. The compensation events provide an alternative approach to handle a specific exception event, e.g. storing the non-conform records in quarantine tables for later evaluation or applying automatic error. The error events block or end the process execution. Fig. 2 (right) presents an exception block with compensation and error policies associated. The 'Exception' block is formed by three mandatory constructs: 1) Event that specifies an exception that may occur, e.g. 'EmptyAttribute' or 'InvalidDecompositionRule'; 2) Action that identifies the action that should be performed, e.g. record that started the exception can be stored to specific quarantine table or can abort the workflow; and 3) Log activity that stores the exception occurrences to a specific log file structure. With these domain-level instructions, it is possible generating dynamically the instances following the language rules. For that, we can implement code generators to translate the DSL to specifics formats supported by ETL commercial tools.

5   Conclusions and Future Work

In this paper we showed how a typical ETL process can be represented exclusively using ETL patterns on BPMN models, and how these patterns can be integrated in a single ETL system package. To demonstrate their practical application, we selected and discussed a DQE pattern, describing its internal composition and providing a specific DSL for its configuration. From a conceptual modelling point of view, we consider that ETL models should not include any kind of implementation infrastructure specification or any criteria associated with its execution. All infrastructures that support the implementation of conceptual models are related to specific classes of users involving therefore the application of specific constructors. The BPMN provides this kind of abstraction, focusing essentially on the coordination of ETL patterns, promoting the reusability of patterns across several systems, and making the system more robust to process changes. Additionally, the DSL proposed dispose an effective way to formalize each pattern configuration, allowing for its posterior mapping to a programing language such as Java. Using a domain-level DSL it is possible to describe more naturally each part of an ETL process without having the need to program each component. In the short term, we intend to have an extended family of ETL patterns that will allows for building a complete ETL system from scratch, covering all coordination and communication aspects as well as the description of all the tasks required to materialize it. Additionally, a generic transformation plug-in for generating ETL physical schemas for data integration tools is also planned.

References

1. OMG, Documents Associated With Business Process Model And Notation (BPMN) Version 2.0 (2011)
2. Thomsen, C., Pedersen, T.B.: Pygrametl: a powerful programming framework for extract-transform-load programmers. In: Proceeding of the ACM Twelfth International Workshop on Data Warehousing and OLAP, DOLAP 2009, pp. 49–56 (2009)
3. Vassiliadis, P., Simitsis, A., Skiadopoulos, S.: Conceptual modeling for ETL processes. In: Proceedings of the 5th ACM International Workshop on Data Warehousing and OLAP, DOLAP 2002, pp. 14–21 (2002)
4. Vassiliadis, P., Simitsis, A., Skiadopoulos, S.: On the logical modeling of ETL processes. In: Pidduck, A., Mylopoulos, J., Woo, C.C., Ozsu, M. (eds.) CAiSE 2002. LNCS, vol. 2348, pp. 782–786. Springer, Heidelberg (2002)
5. Simitsis, A., Vassiliadis, P.: A method for the mapping of conceptual designs to logical blueprints for ETL processes. Decis. Support Syst. **45**, 22–40 (2008)
6. Vassiliadis, P., Vagena, Z., Skiadopoulos, S., Karayannidis, N., Sellis, T.: Arktos: A Tool for Data Cleaning and Transformation in Data Warehouse Environments. IEEE Data Eng. Bull. **23**(4), 42–47 (2000)
7. Luján-Mora, S., Trujillo, J., Song, I.-Y.: A UML profile for multidimensional modeling in data warehouses. Data Knowl. Eng. **59**, 725–769 (2006)
8. Trujillo, J., Luján-Mora, S.: A UML based approach for modeling ETL processes in data warehouses. Concept. Model. **2813**, 307–320 (2003)
9. Wilkinson, K., Simitsis, A., Castellanos, M., Dayal, U.: Leveraging business process models for ETL design. In: Parsons, J., Saeki, M., Shoval, P., Woo, C., Wand, Y. (eds.) ER 2010. LNCS, vol. 6412, pp. 15–30. Springer, Heidelberg (2010)
10. El Akkaoui, Z., Zimanyi, E.: Defining ETL worfklows using BPMN and BPEL. In: Proceedings of the ACM Twelfth International Workshop on Data Warehousing and OLAP, DOLAP 2009, pp. 41–48 (2009)
11. El Akkaoui, Z., Zimànyi, E., Mazón, J.-N., Trujillo, J.: A model-driven framework for ETL process development. In: Proceedings of the ACM 14th International Workshop on Data Warehousing and OLAP, DOLAP 2011, pp. 45–52 (2011)
12. El Akkaoui, Z., Mazón, J.-N., Vaisman, A., Zimányi, E.: BPMN-based conceptual modeling of ETL processes. In: Cuzzocrea, A., Dayal, U. (eds.) DaWaK 2012. LNCS, vol. 7448, pp. 1–14. Springer, Heidelberg (2012)
13. El Akkaoui, Z., Zimanyi, E., Mazon, J.-N., Trujillo, J.: A BPMN-based design and maintenance framework for ETL processes. Int. J. Data Warehous. Min. **9**, 46 (2013)
14. Rahm, E., Do, H.: Data cleaning: Problems and current approaches. IEEE Data Eng. Bull. **23**, 3–13 (2000)
15. Köppen, V., Brüggemann, B., Berendt, B.: Designing Data Integration: The ETL Pattern Approach. Eur. J. Informatics Prof. **XII** (2011)

Optimized Multi-resolution Indexing and Retrieval Scheme of Time Series

Muhammad Marwan Muhammad Fuad$^{(\boxtimes)}$

Forskningsparken 3, Institutt for Kjemi, NorStruct,
The University of Tromsø – The Arctic University of Norway, 9037 Tromsø, Norway
marwan.fuad@uit.no

Abstract. Multi-resolution representation has been successfully used for index-ing and retrieval of time series. In a previous work we presented Tight-MIR, a multi-resolution representation method which speeds up the similarity search by using distances pre-computed at indexing time. At query time Tight-MIR ap-plies two pruning conditions to filter out non-qualifying time series. Tight-MIR has the disadvantage of storing all the distances corresponding to all resolution levels, even those whose pruning power is low. At query time Tight-MIR also processes all stored resolution levels. In this paper we optimize the Tight-MIR algorithm by enabling it to store and process only the resolution levels with the maximum pruning power. The experiments we conducted on the new optimized version show that it does not only require less storage space, but it is also faster than the original algorithm.

Keywords: Multi-resolution indexing and retrieval · Optimization · Tight-MIR · Time series

1 Introduction

A time series is a chronological collection of observations. The particular nature of these data makes them more appropriate to be handled as whole entities rather than separate numeric observations. In the last decade a great deal of research was devoted to the development of time series data mining because of its various applications in finance, medicine, engineering, and other domains.

Time series are usually represented by *Dimensionality Reduction Techniques* which map the time series onto low-dimension spaces where the query is processed.

Several dimensionality reduction techniques exist in the literature, of those we men-tion: *Piecewise Linear Approximation* (PLA) [1], and *Adaptive Piecewise Constant Approximation* (APCA) [2].

Multi-resolution dimensionality reduction techniques map the time series to several spaces instead of one. In a previous work [3] we presented a multi-resolution indexing and retrieval method of time series called Weak-MIR. Weak-MIR uses pre-computed distances and two filters to speed up the similarity search. In [4] we presented another multi-resolution indexing and retrieval method, MIR-X, which associates our multi-resolution approach with another dimensionality reduction technique. In a third work

© Springer International Publishing Switzerland 2015
F. Pereira et al. (Eds.) EPIA 2015, LNAI 9273, pp. 603–608, 2015.
DOI: 10.1007/978-3-319-23485-4_61

[5] we introduced Tight-MIR which has the advantages of the two previously mentioned methods. Tight-MIR, however, stores distances corresponding to all resolution levels, even though some of them might have a low pruning power. In this paper we present an optimized version of Tight-MIR which stores and processes only the resolution levels with the maximum pruning power.

The rest if the paper is organized as follows: Section 2 is a background section. The optimized version is presented in Section 3 and tested in Section 4. We conclude this paper with Section 5.

2   Background

In [3] we presented *Multi-resolution Indexing and Retrieval Algorithm* (Weak-MIR). The motivation behind this method is that traditional dimensionality reduction techniques use a "one-resolution" approach to indexing and retrieval, where the dimension of the low-dimension space is selected at indexing time, so the performance of the algorithm at query time depends completely on the choice made at indexing time. But in practice, we do not necessarily know a priori the optimal dimension of the low-dimension space.

Weak-MIR uses a multi-resolution representation of time series. During indexing time the algorithm computes and stores distances corresponding to a number of resolution levels, with lower resolution levels having lower dimensions. The algorithm uses these pre-computed distances to speed up the retrieval process. The basis of Weak-MIR is as follows: let U be the original n-dimension space and R be a $2m$-dimension space, where $2m \leq n$. At indexing time each time series $u \in U$ is divided into m segments each of which is approximated by a function (we used a first degree polynomial in [3]) so that the approximation error between each segment and the corresponding polynomial is minimal. The n-dimension vector whose components are the images of all the points of all the segments of a time series on that approximating function is called the *image vector* and denoted by $\bar{u}$. The images of the two end points of the segment are called the *main image* of that segment. The $2m$ main images of each time series are the *projection vector* u^R.

Weak-MIR uses two distances, the first is d which is defined on a n-dimension space, so it is the distance between two time series in the original space, i.e. $d(u_i, u_j)$, or the distance between the original time series and its image vector, i.e. $d(u_i, \bar{u}_i)$. The second distance is d^R which is defined on a $2m$-dimension space, so it is the distance between two projection vectors, i.e. $d^R(u_i^R, u_j^R)$. We proved in [3] that d^R is a lower bound of d when the Minkowski distance is used.

The resolution level k is an integer related to the dimensionality of the reduced space R. So the above definitions of the projection vector and the image vector can be extended to further segmentation of the time series using different values $\leq m_k$. The image vector and the projection vector at level k are denoted by $\bar{u}^{(k)}$ and $u^{R(k)}$, respectively.

Given a query(q, ε), let $\bar{u}, \bar{q}$ be the projection vectors of u, q, respectively, on their approximating functions, where $u \in U$. By applying the triangle inequality we get:

$$|d(u, \bar{u}) - d(q, \bar{q})| > \varepsilon \qquad (1)$$

This relation represents a pruning condition which is the first filter of Weak-MIR. By applying the triangle inequality again we get:

$$d^R\left(u^{R(k)}, q^{R(k)}\right) > \varepsilon + d\left(q, \bar{q}^{(k)}\right) + d\left(u, \bar{u}^{(k)}\right) \qquad (2)$$

This relation is the second filter of Weak-MIR.

In [4] we introduced MIR-X which combines a representation method with a multi-resolution time series. MIR-X uses one of the two filters that Weak-MIR uses together with the low-dimension distance of a time series dimensionality reduction technique. We showed how MIR-X can boost the performance of Weak-MIR.

In [5] we presented Tight-MIR which has the advantages of both Weak-MIR and MIR-X in that it is a standalone method, like Weak-MIR, yet it has the same competitive performance of MIR-X. In Tight-MIR instead of using the projection vector to construct the second filter, we access the raw data in the original space directly using a number of points that corresponds to the dimensionality of the reduced space at that resolution level. In other words, we use $2m$ raw points, instead of $2m$ main images, to compute d^R. There are several advantages to this modification; the first is that the new d^R is obviously tighter than d^R as computed in [4]. The second is that when using a Minkowski distance d^R is a lower bound of the original distance in the original space. The direct consequence of this is that the two distances $d\left(q, \bar{q}^{(k)}\right), d\left(u, \bar{u}^{(k)}\right)$ become redundant, so the second filter is overwritten by the usual lower bounding condition $d^R\left(u^{R(k)}, q^{R(k)}\right) > r$.

At indexing time the distances $d\left(u, \bar{u}^{(k)}\right) \ \forall u \in U$ are computed and stored. At query time the algorithm starts at the lowest level and applies (1) to the first time series in U. If the time series is filtered out the algorithm moves to the next time series, if not, the algorithm applies equation (2). If all the time series in the database have been pruned the algorithm terminates, if not, the algorithm moves to a higher level.

Finally, after all levels have been exploited, we get a candidate answer set which we then scan sequentially to filter out all the non-qualifying time series and return the final answer set.

3   An Optimized Multi-resolution Indexing and Retrieval Scheme

The disadvantage of the indexing scheme presented in the previous section is that it is "deterministic", meaning that at indexing time the time series are indexed using a top-down approach, and the algorithm behaves in a like manner at query time. If some resolution levels have low utility in terms of pruning power, the algorithm will still use the pre-computed distances related to these levels, and at query time these levels will also be examined. Whereas the use of the first filter does not require any query time distance evaluation, applying the second does include calculating distances and thus we might be storing and calculating distances for little pruning benefit.

We propose in this paper an optimized multi-resolution indexing and retrieval scheme. Taking into account that the time series to which we apply equations (1) and (2) are those which have not been filtered out at lower resolution levels, this optimized scheme should determine the optimal combination of resolution levels the algorithm should keep at indexing time and consequently use at query time.

The optimization algorithm we use to solve this problem is the *Genetic Algorithm*. The Genetic Algorithm (GA) is a famous evolutionary algorithm that has been applied to solve a variety of optimization problems. GA is a population-based global optimization algorithm which mimics the rules of Darwinian selection in that weaker individuals have less chance of surviving the evolution process than stronger ones. GA captures this concept by adopting a mechanism that preserves the "good" features during the optimization process.

In GA a population of candidate solutions (*chromosomes*) explores the search space and exploits this by sharing information. These chromosomes evolve using genetic operations (selection, crossover, mutation, and replacement).

GA starts by randomly initializing a population of chromosomes inside the search space. The fitness function of these chromosomes is evaluated. According to the values of the fitness function new offspring chromosomes are generated through the aforementioned genetic operations. The above steps repeat for a number of generations or until a predefined stopping condition terminates the GA.

The new algorithm, which we call *Optimized Multi-Resolution Indexing and Retrieval – O-MIR*, works as follow; we proceed in the same manner described for Tight-MIR to produce k candidate resolution levels. The next step is handled by the optimizer to select op resolution levels of the k resolution levels, where these op levels provide the maximum pruning power. For the current version of our algorithm the number of resolution levels to be kept, op, is chosen by the user according to the storage and processing capacity of the system. In other words, our algorithm will decide which are the op optimal resolution levels to be kept out of the k resolution levels produced by the indexing step.

Notice that when $op = 1$ we have one resolution level, which is the case with traditional dimensionality reduction techniques.

The optimization stage of O-MIR starts by randomly initializing a population of chromosomes $V_j = \langle v_1^j, v_2^j, ..., v_{op}^j \rangle$ where $j = 1, ..., popSize$ and where $v_1^j < v_2^j < \cdots < v_{op}^j$. Each chromosome represents a possible configuration of the resolution levels to be kept. The fitness function of our optimization problem is the pruning power of this configuration. As in [5], the performance criterion is based on the latency time concept presented in [6]. The latency time is calculated by the number of cycles the processor takes to perform the different arithmetic operations (>,+ - ,*,abs, sqrt) which are required to execute the similarity search query. This number for each operation is multiplied by the latency time of that operation to get the total latency time of the similarity search query. The latency time is 5 cycles for (>, + -), 1 cycle for (abs), 24 cycles for (*), and 209 cycles for (sqrt) [6]. The latency time for each chromosome is the average of the latency time of nQ random queries.

As with other GAs, our algorithm selects a percentage $sRate$ of chromosomes for mating and mutates a percentage $mRate$ of genes. The above steps repeat for $nGen$ generations.

4   Experiments

We compared O-MIR with Tight-MIR on similarity search experiments on different time series datasets from different time series repositories [7], and [18] using different threshold values, and for different values of op. Since the value of op is related to the value of k, which in turn depends on the length of the time series tested, we denote the percentage of the resolution levels kept to the total resolution levels by $pk = op/k$.

As for the parameters of the algorithm that we used in the experiments; the population size, $popSize$, was 16, the number of generations $nGen$ was 100, the mutation rate, $mRate$, was 0.2, the selection rate, $sRate$, was 0.5, and the number of queries, nQ, was set to 10.

We show in Fig. 1 the results of our experiments. For (CBF), (Wafer), and (Gun-Point) we have 5 resolution levels ($k = 5$). For these datasets we chose $pk = 40\%, 60\%, 80\%$. (motoCurrent) has 8 resolution levels ($k = 8$), we chose $pk = 25\%, 50\%, 75\%$ As we can see, the results are promising in terms of latency time and storage space. For the three first datasets O-MIR is faster than Tight-MIR and in addition, it required less storage space. This is also the case with (motoCurrent) except for $pk = 20\%$ where the latency time for O-MIR was longer than that of Tight-MIR. However, for this value of pk the gain of storage space is substantial without much increase in latency time.

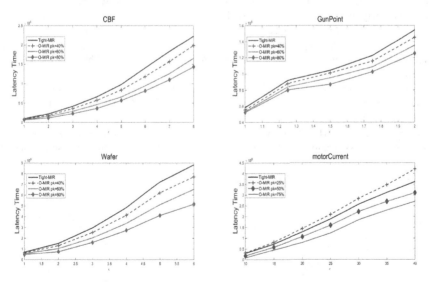

Fig. 1. Comparison of the latency time between Tight-MIR and O-MIR on datasets (CBF), (Wafer), (GunPoint)), and (motoCurrent) for different values of pk

5   Conclusion

In this paper we presented an optimized version of our previous Tight-MIR multi-resolution indexing and retrieval method of time series. Whereas the original method stores and processes all the resolution levels the indexing step produces, the new algorithm, O-MIR, optimizes this process by applying the genetic algorithms to choose the resolution levels with the maximum pruning power. The experiments we conducted show that O-MIR is faster than Tight-MIR, and it also has the advantage of requiring less storage space.

We believe that the main advantage of the new method is that it reduces the storage space requirement of the original method which can be a burden when applying such multi-resolution methods to large datasets.

References

1. Morinaka, Y., Yoshikawa, M., Amagasa, T., Uemura, S.: The L-index: an indexing structure for efficient subsequence matching in time sequence databases. In: Proc. 5th Pacific Asia Conf. on Knowledge Discovery and Data Mining (2001)
2. Keogh, E., Chakrabarti, K., Pazzani, M., Mehrotra, S.: Locally Adaptive Dimensionality Reduction for Similarity Search in Large Time Series Databases. SIGMOD (2001)
3. Muhammad Fuad, M.M., Marteau P.F.: Multi-resolution approach to time series retrieval. In: Fourteenth International Database Engineering & Applications Symposium– IDEAS 2010, Montreal, QC, Canada (2010)
4. Muhammad Fuad, M.M., Marteau P.F.: Speeding-up the similarity search in time series databases by coupling dimensionality reduction techniques with a fast-and-dirty filter. In: Fourth IEEE International Conference on Semantic Computing– ICSC 2010, Carnegie Mellon University, Pittsburgh, PA, USA (2010)
5. Muhammad Fuad, M.M., Marteau, P.F.: Fast retrieval of time series by combining a multi-resolution filter with a representation technique. In: The International Conference on Advanced Data Mining and Applications–ADMA 2010, ChongQing, China, November 21, 2010
6. Schulte, M.J., Lindberg, M. Laxminarain, A.: Performance evaluation of decimal floating-point arithmetic. In: IBM Austin Center for Advanced Studies Conference, February 2005
7. http://povinelli.eece.mu.edu/
8. Keogh, E., Zhu, Q., Hu, B., Hao, Y., Xi, X., Wei, L., Ratanamahatana, C.A.: The UCR Time Series Classification/Clustering (2011). www.cs.ucr.edu/~eamonn/time_series_data/

Multi-agent Systems:
Theory and Applications

Minimal Change in Evolving Multi-Context Systems

Ricardo Gonçalves$^{(\boxtimes)}$, Matthias Knorr, and João Leite

NOVA LINCS, Departamento de Informática, Faculdade Ciências e Tecnologia,
Universidade Nova de Lisboa, Caparica, Portugal
rjrg@fct.unl.pt

Abstract. In open environments, agents need to reason with knowledge
from various sources, possibly represented in different languages. The
framework of *Multi-Context Systems* (MCSs) offers an expressive, yet
flexible, solution since it allows for the integration of knowledge from different heterogeneous sources in an effective and modular way. However,
MCSs are essentially static as they were not designed for dynamic scenarios. The recently introduced *evolving Multi-Context Systems* (eMCSs)
extend MCSs by also allowing the system to both react to, and reason in
the presence of dynamic observations, and evolve by incorporating new
knowledge, thus making it even more adequate in Multi-Agent Systems
characterised by their dynamic and open nature.
In dynamic scenarios which admit several possible alternative evolutions,
the notion of *minimal change* has always played a crucial role in determining the most plausible choice. However, different KR formalisms –
as combined within eMCSs – may require different notions of minimal
change, making their study and their interplay a relevant highly non-
trivial problem. In this paper, we study the notion of minimal change in
eMCSs, by presenting and discussing alternative minimal change criteria.

1 Introduction

Open and dynamic environments create new challenges for knowledge representation languages for agent systems. Instead of having to deal with a single
static knowledge base, each agent has to deal with multiple sources of distributed
knowledge possibly written in different languages. These sources of knowledge
include the large number of available ontologies and rule sets, as well as the
norms and policies published by *institutions*, the information communicated by
other agents, to name only a few.

The need to incorporate in agent-oriented programming languages the ability
to represent and reason with heterogeneous distributed knowledge sources, and
the flow of information between them, has been pointed out in [1–4], although
a general adequate practical solution is still not available.

Recent literature in knowledge representation and reasoning contains several
proposals to combine heterogeneous knowledge bases, one of which – Multi-
Context Systems (MCSs) [5–7] – has attracted particular attention because it

© Springer International Publishing Switzerland 2015
F. Pereira et al. (Eds.) EPIA 2015, LNAI 9273, pp. 611–623, 2015.
DOI: 10.1007/978-3-319-23485-4_62

provides an elegant solution by considering each source of knowledge as a module and then providing means to model the interaction between these modules. More specifically, an MCS consists of a set of contexts, each of which is a knowledge base in some KR formalism, such that each context can access information from the other contexts using so-called bridge rules. Such non-monotonic bridge rules add their heads to the context's knowledge base provided the queries (to other contexts) in their bodies are successful. Managed Multi-Context Systems (mMCSs) [8] extend MCSs by allowing operations, other than simple addition, to be expressed in the heads of bridge rules, thus allowing mMCSs to properly deal with the problem of consistency management within contexts. MCSs have gained some attention by agent developers [9–11].

Whereas mMCSs are quite general and flexible to address the problem of integration of different KR formalisms, they are essentially static in the sense that the contexts do not evolve to incorporate the changes in dynamic scenarios. In such scenarios, new knowledge and information is dynamically produced, often from several different sources – e.g., a stream of raw data produced by some sensors, new ontological axioms written by some user, newly found exceptions to some general rule, observations, etc.

Evolving Multi-Context Systems (eMCSs) [12] inherit from mMCSs the ability to integrate and manage knowledge represented in heterogeneous KR formalisms, and at the same time are able to react to dynamic observations, and evolve by incorporating knowledge. The semantics of eMCSs is based on the stable model semantics, and allows alternative models for a given evolution, in the same way as answer sets represent alternative solutions to a given ASP program.

One of the main principles of belief revision is minimal change, which in case of eMCSs means that information should be maintained by inertia unless it is required to change. In dynamic scenarios where systems can have alternative evolutions, it is thus desirable to have some minimal change criteria to be able to compare possible alternatives. This problem is particularly interesting and non-trivial in dynamic frameworks based on MCSs, because of the heterogeneity of KR frameworks that may exist in an MCS – each of which may require different notions of minimal change –, and also because the evolution of such systems is based not only on the semantics, but also on the evolution of the knowledge base of each context.

In this paper, we study minimal change in eMCSs, by presenting different minimal change criteria to be applied to the possible evolving equilibria of an eMCS, and by discussing the relation between them.

The remainder of this paper is as follows. We introduce the framework of eMCSs in Sect. 2. Then, we present and study some minimal change criteria in eMCSs in Sect. 3, and conclude with a discussion of related work and possible future directions in Sect. 4.

2 Evolving Multi-Context Systems

In this section, we revisit evolving Multi-Context Systems as introduced in [12], which generalize mMCSs [8] to dynamic scenarios in which contexts are enabled to react to external observations and evolve.

An evolving multi-context system (eMCS) consists of a collection of components, each of which contains knowledge represented in some *logic*, defined as a triple $L = \langle \mathbf{KB}, \mathbf{BS}, \mathbf{ACC} \rangle$ where $\mathbf{KB}$ is the set of well-formed knowledge bases of L, $\mathbf{BS}$ the set of possible belief sets, and $\mathbf{ACC} : \mathbf{KB} \to 2^{\mathbf{BS}}$ a function describing the semantics of L by assigning to each knowledge base a set of acceptable belief sets. We assume that each element of $\mathbf{KB}$ and $\mathbf{BS}$ is a set, and define $F = \{s : s \in kb \wedge kb \in \mathbf{KB}\}$.

In addition to the knowledge base in each component, *bridge rules* are used to interconnect the components, specifying what operations to perform on its knowledge base given certain beliefs held in the components of the eMCS. For that purpose, each component of an eMCS is associated with a *management base*, which is a set of operations that can be applied to the possible knowledge bases of that component. Given a management base OP and a logic L, let $OF = \{op(s) : op \in OP \wedge s \in F\}$ be the *set of operational formulas* over OP and L. Each component of an eMCS gives semantics to operations in its management base using a *management function* over a logic L and a management base OP, $mng : 2^{OF} \times \mathbf{KB} \to (2^{\mathbf{KB}} \setminus \{\emptyset\})$, i.e., $mng(op, kb)$ is the (non-empty) set of knowledge bases that result from applying the operations in op to the knowledge base kb. We assume that $mng(\emptyset, kb) = \{kb\}$.

In an eMCS some contexts are assumed to be *observation contexts* whose knowledge bases will be constantly changing over time according to the observations made, similar, e.g., to streams of data from sensors.[1] The changing observations will then affect the other contexts by means of the bridge rules. As we will see, such effect can either be instantaneous and temporary, i.e., limited to the current time instant, similar to (static) mMCSs, where the body of a bridge rule is evaluated in a state that already includes the effects of the operation in its head, or persistent, but only affecting the next time instant. To achieve the latter, the operational language is extended with a unary meta-operation *next* that can only be applied on top of operations. Given a management base OP and a logic L, we define eOF, the evolving operational language, as $eOF = OF \cup \{next(op(s)) : op(s) \in OF\}$.

The idea of observation contexts is that each such context has a language describing the set of possible observations of that context, along with its current observation. The elements of the language of the observation contexts can then be used in the body of bridge rules to allow contexts to access the observations. Formally, an *observation context* is a tuple $O = \langle \Pi_O, \pi \rangle$ where Π_O is the *observation language* of O and $\pi \subseteq \Pi_O$ is its *current observation*.

We can now define *evolving Multi-Context Systems* (eMCS).

---

[1] For simplicity of presentation, we consider discrete steps in time here.

614   R. Gonçalves et al.

Definition 1. *An eMCS is a sequence $M_e = \langle C_1, \ldots, C_n, O_1, \ldots, O_\ell \rangle$, such that, for each $i \in \{1, \ldots, \ell\}$, $O_i = \langle \Pi_{O_i}, \pi_i \rangle$ is an observation context, and, for each $i \in \{1, \ldots, n\}$, C_i is an evolving context defined as $C_i = \langle L_i, kb_i, br_i, OP_i, mng_i \rangle$ where*

- *$L_i = \langle \mathbf{KB}_i, \mathbf{BS}_i, \mathbf{ACC}_i \rangle$ is a logic*
- *$kb_i \in \mathbf{KB}_i$*
- *br_i is a set of bridge rules of the form*

$$H(\sigma) \leftarrow a_1, \ldots, a_k, \mathbf{not}\ a_{k+1}, \ldots, \mathbf{not}\ a_n \qquad (1)$$

such that $H(\sigma) \in eOF_i$, and each a_i, $i \in \{1, \ldots, n\}$, is either of the form $(r\!:\!b)$ with $r \in \{1, \ldots, n\}$ and b a belief formula of L_r, or of the form $(r@b)$ with $r \in \{1, \ldots, \ell\}$ and $b \in \Pi_{O_r}$
- *OP_i is a management base*
- *mng_i is a management function over L_i and OP_i.*

Given an eMCS $M_e = \langle C_1, \ldots, C_n, O_1, \ldots, O_\ell \rangle$ we denote by $\mathbf{KB}_{M_e}$ the set of *knowledge base configurations for M_e*, i.e., $\mathbf{KB}_{M_e} = \{\langle k_1, \ldots, k_n \rangle : k_i \in \mathbf{KB}_i$ for each $1 \leq i \leq n\}$. A *belief state* for $M_e = \langle C_1, \ldots, C_n, O_1, \ldots, O_\ell \rangle$ is a sequence $S = \langle S_1, \ldots, S_n \rangle$ such that, for each $1 \leq i \leq n$, we have $S_i \in \mathbf{BS}_i$. We denote by $\mathbf{BS}_{M_e}$ the set of belief states for M_e.

An *instant observation* for M_e is a sequence $\mathcal{O} = \langle o_1, \ldots, o_\ell \rangle$ such that, for each $1 \leq i \leq \ell$, we have that $o_i \subseteq \Pi_{O_i}$.

Given a belief state $S = \langle S_1, \ldots, S_n \rangle$ for M_e and an instant observation $\mathcal{O} = \langle o_1, \ldots, o_\ell \rangle$ for M_e, we define the satisfaction of bridge literals of the form $(r\!:\!b)$ as $S, \mathcal{O} \models (r\!:\!b)$ if $b \in S_r$ and $S, \mathcal{O} \models \mathbf{not}\ (r\!:\!b)$ if $b \notin S_r$. The satisfaction of bridge literal of the form $(r@b)$ depends on the current observations, i.e., we have that $S, \mathcal{O} \models (r@b)$ if $b \in o_r$ and $S \models \mathbf{not}\ (r@b)$ if $b \notin o_r$. For a set B of bridge literals, we have that $S, \mathcal{O} \models B$ if $S, \mathcal{O} \models L$ for every $L \in B$.

We say that a bridge rule σ of a context C_i is *applicable given a belief state S and an instant observation $\mathcal{O}$* if its body is satisfied by S and $\mathcal{O}$, i.e., $S, \mathcal{O} \models B(\sigma)$. We denote by $app_i(S, \mathcal{O})$ the set of heads of bridges rules of the context C_i which are applicable given the belief state S and the instant observation $\mathcal{O}$. Recall that the heads of bridge rules in an eMCS may be of two types: those that contain *next* and those that do not. The former are to be applied to the current knowledge base and not persist, whereas the latter are to be applied in the next time instant and persist. Therefore, we distinguish these two subsets.

Definition 2. *Let $M_e = \langle C_1, \ldots, C_n, O_1, \ldots, O_\ell \rangle$ be an eMCS, S a belief state for M_e, and $\mathcal{O}$ an instant observation for M_e. Then, for each $1 \leq i \leq n$, consider the following sets:*

- *$app_i^{next}(S, \mathcal{O}) = \{op(s) : next(op(s)) \in app_i(S, \mathcal{O})\}$*
- *$app_i^{now}(S, \mathcal{O}) = \{op(s) : op(s) \in app_i(S, \mathcal{O})\}$*

If we want an effect to be instantaneous and persistent, this can be achieved using two bridge rules with identical body, one with and one without *next*.

Similar to equilibria in mMCS, the (static) equilibrium is defined to incorporate instantaneous effects based on $app_i^{now}(S, \mathcal{O})$ alone.

Definition 3. *Let* $M_e = \langle C_1, \ldots, C_n, O_1, \ldots, O_\ell \rangle$ *be an eMCS, and* $\mathcal{O}$ *an instant observation for* M_e. *A belief state* $S = \langle S_1, \ldots, S_n \rangle$ *for* M_e *is an equilibrium of* M_e *given* $\mathcal{O}$ *iff for each* $1 \leq i \leq n$, $S_i \in \mathbf{ACC}_i(kb)$ *for some* $kb \in mng_i(app_i^{now}(S, \mathcal{O}), kb_i)$.

To be able to assign meaning to an eMCS evolving over time, we introduce evolving belief states, which are sequences of belief states, each referring to a subsequent time instant.

Definition 4. *Let* $M_e = \langle C_1, \ldots, C_n, O_1, \ldots, O_\ell \rangle$ *be an eMCS. An evolving belief state of size* s *for* M_e *is a sequence* $S_e = \langle S^1, \ldots, S^s \rangle$ *where each* S^j, $1 \leq j \leq s$, *is a belief state for* M_e.

To enable eMCSs to react to incoming observations and evolve, a sequence of observations (defined below) has to be processed. The idea is that the knowledge bases of the observation contexts O_i change according to that sequence.

Definition 5. *Let* $M_e = \langle C_1, \ldots, C_n, O_1, \ldots, O_\ell \rangle$ *be an eMCS. A sequence of observations for* M_e *is a sequence* $Obs = \langle \mathcal{O}^1, \ldots, \mathcal{O}^m \rangle$, *such that, for each* $1 \leq j \leq m$, $\mathcal{O}^j = \langle o_1^j, \ldots, o_\ell^j \rangle$ *is an instant observation for* M_e, *i.e.,* $o_i^j \subseteq \Pi_{O_i}$ *for each* $1 \leq i \leq \ell$.

To be able to update the knowledge bases and the sets of bridge rules of the evolving contexts, we need the following notation. Given an evolving context C_i, and a knowledge base $k \in \mathbf{KB}_i$, we denote by $C_i[k]$ the evolving context in which kb_i is replaced by k, i.e., $C_i[k] = \langle L_i, k, br_i, OP_i, mng_i \rangle$. For an observation context O_i, given a set $\pi \subseteq \Pi_{O_i}$ of observations for O_i, we denote by $O_i[\pi]$ the observation context in which its current observation is replaced by π, i.e., $O_i[\pi] = \langle \Pi_{O_i}, \pi \rangle$. Given $K = \langle k_1, \ldots, k_n \rangle \in \mathbf{KB}_{M_e}$ a knowledge base configuration for M_e, we denote by $M_e[K]$ the eMCS $\langle C_1[k_1], \ldots, C_n[k_n], O_1, \ldots, O_\ell \rangle$.

We now define when certain evolving belief states are evolving equilibria of an eMCS M_e given a sequence of observations $Obs = \langle \mathcal{O}^1, \ldots, \mathcal{O}^m \rangle$ for M_e. The intuitive idea is that, given an evolving belief state $S_e = \langle S^1, \ldots, S^s \rangle$ for M_e, in order to check if S_e is an evolving equilibrium, we need to consider a sequence of eMCSs, $M^1, \ldots, M^s$ (each with ℓ observation contexts), representing a possible evolution of M_e according to the observations in Obs, such that each S^j is a (static) equilibrium of M^j. The current observation of each observation context O_i in M^j is exactly its corresponding element o_i^j in $\mathcal{O}^j$. For each evolving context C_i, its knowledge base in M^j is obtained from the one in M^{j-1} by applying the operations in $app_i^{next}(S^{j-1}, \mathcal{O}^{j-1})$.

Definition 6. *Let* $M_e = \langle C_1, \ldots, C_n, O_1, \ldots, O_\ell \rangle$ *be an eMCS,* $S_e = \langle S^1, \ldots, S^s \rangle$ *an evolving belief state of size* s *for* M_e, *and* $Obs = \langle \mathcal{O}^1, \ldots, \mathcal{O}^m \rangle$ *an observation sequence for* M_e *such that* $m \geq s$. *Then,* S_e *is an evolving equilibrium of size* s *of* M_e *given* Obs *iff, for each* $1 \leq j \leq s$, *the belief state* S^j *is an equilibrium of* $M^j = \langle C_1[k_1^j], \ldots, C_n[k_n^j], O_1[o_1^j], \ldots, O_\ell[o_\ell^j] \rangle$ *where, for each* $1 \leq i \leq n$, k_i^j *is defined inductively as follows:*

- $k_i^1 = kb_i$
- $k_i^{j+1} \in mng_i(app^{next}(S^j, \mathcal{O}_i^j), k_i^j)$.

3   Minimal Change

In this section, we discuss some alternatives for the notion of minimal change in eMCSs. What makes this problem interesting is that there are different parameters that we may want to minimize in a transition from one time instant to the next one. In the following discussion we focus on two we deem most relevant: the operations that can be applied to the knowledge bases, and the distance between consecutive belief states.

We start by studying minimal change at the level of the operations. In the following discussion we consider fixed an eMCS $M_e = \langle C_1, \ldots, C_n, O_1, \ldots, O_\ell \rangle$.

Recall from the definition of evolving equilibrium that, in the transition between consecutive time instants, the knowledge base of each context C_i of M_e changes according to the operations in $app_i^{next}(S, O)$, and these depend on the belief state S and the instant observation O. The first idea to compare elements of this set of operations is to, for a fixed instant observation O, distinguish those equilibria of M_e which generate a minimal set of operations to be applied to the current knowledge bases to obtain the knowledge bases of the next time instant. Formally, given a knowledge base configuration $K \in \mathbf{KB}_{M_e}$ and an instant observation O for M_e, we can define the set:

$$MinEq(K, O) = \{S : S \text{ is an equilibrium of } M_e[K] \text{ given } O \text{ and there is no}$$
$$\text{equilibrium } S' \text{ of } M_e[K] \text{ given } O \text{ such that, for all } 1 \leq i \leq n,$$
$$app_i^{next}(S', O) \subset app_i^{next}(S, O)\}$$

This first idea of comparing equilibria based on inclusion of the sets of operations can, however, be too strict in most cases. Moreover, different operations usually have different costs,[2] and it may well be that, instead of minimizing based on set inclusion, we want to minimize the total cost of the operations to be applied. For that, we need to assume that each context has a cost function over the set of operations, i.e., $cost_i : OP_i \to \mathbb{N}$, where $cost_i(op)$ represents the cost of performing operation op.

Let S be a belief state for M_e and O an instant observation for M_e. Then, for each $1 \leq i \leq n$, we define the cost of the operations to be applied to obtain the knowledge base of the next time instant as:

$$Cost_i(S, O) = \sum_{op(s) \in app_i^{next}(S, O)} cost_i(op)$$

Summing for all evolving contexts, we obtain the global cost of S given O:

$$Cost(S, O) = \sum_{i=1}^{n} Cost_i(S, O)$$

Now that we have defined a cost function over belief states, we can define a minimization function over possible equilibria of eMCS $M_e[K]$ for a fixed knowledge base configuration $K \in \mathbf{KB}_{M_e}$. Formally, given O an instant observation for M_e, we define the set of equilibria of $M_e[K]$ given O which minimize the global

---

[2] We use the notion of cost in an abstract sense, i.e., depending on the context, it may refer to, e.g., the computational cost of the operation, or its economic cost.

cost of the operations to be applied to obtain the knowledge base configuration of the next time instant as:

$$MinCost(K, \mathcal{O}) = \{S : S \text{ is an equilibrium of } M_e[K] \text{ given } \mathcal{O} \text{ and}$$
$$\text{there is no equilibrium } S' \text{ of } M_e[K] \text{ given } \mathcal{O}$$
$$\text{such that } Cost(S', \mathcal{O}) < Cost(S, \mathcal{O})\}$$

Note that, instead of using a global cost, we could have also considered a more fine-grained criterion by comparing costs for each context individually, and define some order based on these comparisons. Also note that the particular case of taking $cost_i(op) = 1$ for every $i \in \{1, \ldots, n\}$ and every $op \in OP_i$, captures the scenario of minimizing the total number of operations to be applied.

The function $MinCost$ allows for the choice of those equilibria that are minimal with respect to the operations to be performed to the current knowledge base configuration in order to obtain the knowledge base configuration of the next time instant. Still, for each choice of an equilibrium S, we have to deal with the existence of several alternatives in the set $mng_i(app_i^{next}(S, \mathcal{O}), kb_i)$. Our aim now is to discuss how we can apply some notion of minimal change that allows us to compare the elements in $mng_i(app_i^{next}(S, \mathcal{O}), kb_i)$. The intuitive idea is to compare the distance between the current equilibria and the possible equilibria resulting from the elements in $mng_i(app_i^{next}(S, \mathcal{O}), kb_i)$. Of course, given the possible heterogeneity of contexts in an eMCS, we cannot assume a global notion of distance between belief sets. Therefore, we assume that each evolving context has its own distance function between its beliefs sets. Formally, for each $1 \le i \le n$, we assume the existence of a distance function d_i, i.e., $d_i : \mathbf{BS}_i \times \mathbf{BS}_i \to \mathbb{R}$ satisfying for all $S_1, S_2, S_3 \in \mathbf{BS}_i$:

1. $d_i(S_1, S_2) \ge 0$
2. $d_i(S_1, S_2) = 0$ iff $S_1 = S_2$
3. $d_i(S_1, S_2) = d_i(S_2, S_1)$
4. $d_i(S_1, S_3) \le d_i(S_1, S_2) + d_i(S_2, S_3)$

There are some alternatives to extend the distance function of each context to a distance function between belief states. In the following we present two natural choices. One is to consider the maximal distance between belief sets of each context. The other is to consider the average of distances between belief sets of each context. Formally, given S^1 and S^2 belief states of M_e we define two functions $\overline{d}_{\max} : \mathbf{BS}_{M_e} \times \mathbf{BS}_{M_e} \to \mathbb{R}$ and $\overline{d}_{\text{avg}} : \mathbf{BS}_{M_e} \times \mathbf{BS}_{M_e} \to \mathbb{R}$ as follows:

$$\overline{d}_{\max}(S^1, S^2) = Max\{d_i(S_i^1, S_i^2) \mid 1 \le i \le n\}$$
$$\overline{d}_{\text{avg}}(S^1, S^2) = \frac{\sum_{i=1}^{n} d_i(S_i^1, S_i^2)}{n}$$

We can prove that $\overline{d}_{\max}$ and $\overline{d}_{\text{avg}}$ are distance functions between belief states.

Proposition 1. *The functions $\overline{d}_{max}$ and $\overline{d}_{avg}$ defined above are both distance functions, i.e., satisfy the axioms 1) - 4).*

We now study how we can use one of these distance functions between belief states to compare the possible alternatives in the sets $mng_i(app_i^{next}(S, \mathcal{O}), kb_i)$, for each $1 \leq i \leq n$. Recall that the intuitive idea is to minimize the distance between the current belief state S and the possible equilibria that each element of $mng_i(app_i^{next}(S, \mathcal{O}), kb_i)$ can give rise to. We explore here two options, which differ on whether the minimization is global or local. The idea of global minimization is to choose only those knowledge base configurations $\langle k_1, \ldots, k_n \rangle \in \mathbf{KB}_{M_e}$ with $k_i \in mng_i(app_i^{next}(S, \mathcal{O}), kb_i)$, which guarantee minimal distance between the original belief state S and the possible equilibria of the obtained eMCS. The idea of local minimization is to consider all possible tuples $\langle k_1, \ldots, k_n \rangle$ with $k_i \in mng_i(app_i^{next}(S, \mathcal{O}), kb_i)$, and only apply minimization for each such choice, i.e., for each such knowledge base configuration we only allow equilibria with minimal distance from the original belief state.

We first consider the case of pruning those tuples $\langle k_1, \ldots, k_n \rangle$ such that $k_i \in mng_i(app_i^{next}(S, \mathcal{O}), kb_i)$, which do not guarantee minimal change with respect to the original belief state. We start by defining an auxiliary function. Let S be a belief state for M_e, $K = \langle k_1, \ldots, k_n \rangle \in \mathbf{KB}_{M_e}$ a knowledge base configuration for M_e, and $\mathcal{O} = \langle o_1, \ldots, o_\ell \rangle$ an instant observation for M_e. Then we define the set of knowledge base configurations that are obtained from K given the belief state S and the instant observation $\mathcal{O}$ as:

$$NextKB(S, \mathcal{O}, \langle k_1, \ldots, k_n \rangle) = \{\langle k_1', \ldots, k_n' \rangle \in \mathbf{KB}_{M_e} : \text{for each } 1 \leq i \leq n$$
$$\text{we have that } k_i' \in mng_i(app_i^{next}(S, \mathcal{O}), k_i)\}$$

For each choice $\overline{d}$ of a distance function between belief states, we define the set of knowledge base configurations that minimize the distance to the original belief state. Let S be a belief state for M_e, $K = \langle k_1, \ldots, k_n \rangle \in \mathbf{KB}_{M_e}$ a knowledge base configuration for M_e, and $\mathcal{O}^j$ and $\mathcal{O}^{j+1}$ instant observations for M_e.

$$MinNext(S, \mathcal{O}^j, \mathcal{O}^{j+1}, K) = \{(S', K') : K' \in NextKB(S, \mathcal{O}^j, K) \text{ and}$$
$$S' \in MinCost(M_e[K'], \mathcal{O}^{j+1}) \text{ s.t. there is no}$$
$$K'' \in NextKB(S, \mathcal{O}^j, K) \text{ and no}$$
$$S'' \in MinCost(M_e[K''], \mathcal{O}^{j+1}) \text{ with}$$
$$\overline{d}(S, S'') < \overline{d}(S, S')\}.$$

Note that $MinNext$ applies minimization over all possible equilibria resulting from every element of $NextKB(S, \mathcal{O}^j, K)$. Using $MinNext$, we can now define a minimal change criterion to be applied to evolving equilibria of M_e.

Definition 7. Let $M_e = \langle C_1, \ldots, C_n, O_1, \ldots, O_\ell \rangle$ be an eMCS, Obs $= \langle \mathcal{O}^1, \ldots, \mathcal{O}^m \rangle$ an observation sequence for M_e, and let $S_e = \langle S^1, \ldots, S^s \rangle$ be an evolving equilibrium of M_e given Obs. We assume that $\langle K^1, \ldots, K^s \rangle$, with $K^j = \langle k_1^j, \ldots, k_n^j \rangle$, is the sequence of knowledge base configurations associated with S_e as in Definition 6. Then, S_e satisfies the strong minimal change criterion for M_e given Obs if, for each $1 \leq j \leq s$, the following conditions are satisfied:

- $S^j \in MinCost(M_e[K^j], \mathcal{O}^j)$
- $(S^{j+1}, K^{j+1}) \in MinNext(S^j, \mathcal{O}^j, \mathcal{O}^{j+1}, K^j)$

We call this minimal change criterion the *strong* minimal change criterion because it applies minimization over all possible equilibria resulting from every possible knowledge base configuration in $NextKB(S, \mathcal{O}^j, K)$.

The following proposition states the desirable property that the existence of an equilibrium guarantees the existence of an equilibrium satisfying the strong minimal change criterion. We should note that this is not a trivial statement since we are combining minimization of two different elements: the cost of the operations and the distance between belief states. This proposition in fact follows from their careful combination in the definition of $MinNext$.

Proposition 2. *Let $Obs = \langle \mathcal{O}^1, \ldots, \mathcal{O}^m \rangle$ be an observation sequence for M_e. If M_e has an evolving equilibrium of size s given Obs, then at least one evolving equilibrium of size s given Obs satisfies the strong minimal change criterion.*

Note that in the definition of the strong minimal change criterion, the knowledge base configurations $K \in NextKB(S^j, \mathcal{O}^j, K^j)$, for which the corresponding possible equilibria are not at a minimal distance from S^j, are not considered. However, there could be situations in which this minimization criterion is too strong. For example, it may well be that all possible knowledge base configurations in $NextKB(S^j, \mathcal{O}^j, K^j)$ are important, and we do not want to disregard any of them. In that case, we can relax the minimization condition by applying minimization individually for each knowledge base configuration in $NextKB(S^j, \mathcal{O}^j, K^j)$. The idea is that, for each fixed $K \in NextKB(S^j, \mathcal{O}^j, K^j)$ we choose only those equilibria of $M_e[K]$ which minimize the distance to S^j.

Formally, let S be a belief state for M_e, $K \in \mathbf{KB}_{M_e}$ a knowledge base configuration for M_e, and $\mathcal{O}$ an instant observation for M_e. For each distance function $\overline{d}$ between belief states, we can define the following set:

$$MinDist(S, \mathcal{O}, K) = \{S' : S' \in MinCost(M_e[K], \mathcal{O}) \text{ and}$$
$$\text{there is no } S'' \in MinCost(M_e[K], \mathcal{O})$$
$$\text{such that } \overline{d}(S, S'') < \overline{d}(S, S')\}$$

Using this more relaxed notion of minimization we can define an alternative weaker minimal change criterion to be applied to evolving equilibria of an eMCS.

Definition 8. *Let $M_e = \langle C_1, \ldots, C_n, O_1, \ldots, O_\ell \rangle$ be an eMCS, $Obs = \langle \mathcal{O}^1, \ldots, \mathcal{O}^m \rangle$ an observation sequence for M_e, and $S_e = \langle S^1, \ldots, S^s \rangle$ an evolving equilibrium of M_e given Obs. We assume that $\langle K^1, \ldots, K^s \rangle$, with $K^j = \langle k_1^j, \ldots, k_n^j \rangle$, is the sequence of knowledge base configurations associated with S_e as in Definition 6. Then, S_e satisfies the weak minimal change criterion of M_e given Obs, if for each $1 \leq j \leq s$ the following conditions are satisfied:*

- *$S^j \in MinCost(M_e[K^j], \mathcal{O}^j)$*
- *$S^{j+1} \in MinDist(S^j, K^{j+1}, \mathcal{O}^{j+1})$*

We can now prove that the existence of an evolving equilibrium implies the existence of an equilibrium satisfying the weak minimal change criterion. Again note that the careful combination of the two minimizations – cost and distance – in the definition of $MinDist$ is fundamental to obtain the following result.

Proposition 3. *Let $Obs = \langle \mathcal{O}^1, \ldots, \mathcal{O}^m \rangle$ be an observation sequence for M_e. If M_e has an evolving equilibrium of size s given Obs, then at least one evolving equilibrium of size s of M_e given Obs satisfies the weak minimal change criterion.*

We can now prove that the strong minimal change criterion is, in fact, stronger than the weak minimal change criterion.

Proposition 4. *Let M_e be an eMCS, $Obs = \langle \mathcal{O}^1, \ldots, \mathcal{O}^m \rangle$ an observation sequence for M_e, and $S_e = \langle S^1, \ldots, S^s \rangle$ an evolving equilibrium of M_e given Obs. If S_e satisfies the strong minimal change criterion of M_e given Obs, then S_e satisfies the weak minimal change criterion of M_e given Obs.*

4 Related and Future Work

In this paper we have studied the notion of minimal change in the context of the dynamic framework of eMCSs [12]. We have presented and discussed some alternative definitions of minimal change criteria for evolving equilibria of an eMCS.

Closely related to eMCSs is the framework of reactive Multi-Context Systems (rMCSs) [13–15] inasmuch as both aim at extending mMCSs to cope with dynamic observations. The key difference between them is that the operator *next* of eMCSs allows for a clear separation between persistent and non-persistent effects, and the specification of transitions based on the current state.

Another framework closely related to eMCSs is that of evolving logic programs EVOLP [16] which deals with updates of generalized logic programs, and the two frameworks of reactive ASP, one implemented as a solver *oclingo* [17] and one described in [13]. Whereas EVOLP employs an update predicate that is similar in spirit to the *next* predicate of eMCSs, it does not deal with heterogeneous knowledge, neither do both versions of Reactive ASP. Moreover, no notion of minimal change is studied for these frameworks.

This work raises several interesting paths for future research. Immediate future work includes the study of more global approaches to the minimization of costs of operations, namely by considering the global cost of an evolving equilibrium, instead of minimizing costs at each time instant. A topic worth investigating is how to perform AGM-style belief revision at the (semantic) level of the equilibria, as in Wang et al [18], though necessarily different since knowledge is not incorporated in the contexts. Also interesting is to study a paraconsistent version of eMCSs, grounded on the work in [19] on paraconsistent semantics for hybrid knowledge bases. Another important issue open for future work is a more fine-grained characterization of updating bridge rules (and knowledge bases) as studied in [20] in light of the encountered difficulties when updating rules [21–23] and the combination of updates over various formalisms [22,24]. Also, as already outlined in [25,26], we can consider generalized notions of minimal and grounded equilibria [5] for eMCSs to avoid, e.g., self-supporting cycles introduced by bridge rules, or the use of preferences to deal with several evolving equilibria an eMCS can have for the same observation sequence. Also interesting is to apply the ideas in this paper to study the dynamics of frameworks closely related to MCSs, such as those in [27–30].

Finally, and in line with the very motivation set out in the introduction, we believe that the research in MCSs – including eMCSs with the different notions of minimal change – provides a blue-print on how to represent and reason with heterogeneous dynamic knowledge bases which could (should) be used by developers of practical agent-oriented programming languages, such as JASON [31], 2APL [32], or GOAL [33], in their quest for providing users and programmers with greater expressiveness and flexibility in terms of the knowledge representation and reasoning facilities provided by such languages. To this end, an application scenario that could provide interesting and rich examples would be that of norm-aware multi-agent systems [34–39].

Acknowledgments. We would like to thank the referees for their comments, which helped improve this paper. R. Gonçalves, M. Knorr and J. Leite were partially supported by FCT under project ERRO (PTDC/EIA-CCO/121823/2010) and under strategic project NOVA LINCS (PEst/UID/CEC/04516/2013). R. Gonçalves was partially supported by FCT grant SFRH/BPD/100906/2014 and M. Knorr was partially supported by FCT grant SFRH/BPD/86970/2012.

References

1. Dastani, M., Hindriks, K.V., Novák, P., Tinnemeier, N.A.M.: Combining multiple knowledge representation technologies into agent programming languages. In: Baldoni, M., Son, T.C., van Riemsdijk, M.B., Winikoff, M. (eds.) DALT 2008. LNCS (LNAI), vol. 5397, pp. 60–74. Springer, Heidelberg (2009)
2. Klapiscak, T., Bordini, R.H.: JASDL: A practical programming approach combining agent and semantic web technologies. In: Baldoni, M., Son, T.C., van Riemsdijk, M.B., Winikoff, M. (eds.) DALT 2008. LNCS (LNAI), vol. 5397, pp. 91–110. Springer, Heidelberg (2009)
3. Moreira, Á.F., Vieira, R., Bordini, R.H., Hübner, J.F.: Agent-oriented programming with underlying ontological reasoning. In: Baldoni, M., Endriss, U., Omicini, A., Torroni, P. (eds.) DALT 2005. LNCS (LNAI), vol. 3904, pp. 155–170. Springer, Heidelberg (2006)
4. Alberti, M., Knorr, M., Gomes, A.S., Leite, J., Gonçalves, R., Slota, M.: Normative systems require hybrid knowledge bases. In: van der Hoek, W., Padgham, L., Conitzer, V., Winikoff, M. (eds.) Procs. of AAMAS, pp. 1425–1426. IFAAMAS (2012)
5. Brewka, G., Eiter, T.: Equilibria in heterogeneous nonmonotonic multi-context systems. In: Procs. of AAAI, pp. 385–390. AAAI Press (2007)
6. Giunchiglia, F., Serafini, L.: Multilanguage hierarchical logics or: How we can do without modal logics. Artif. Intell. **65**(1), 29–70 (1994)
7. Roelofsen, F., Serafini, L.: Minimal and absent information in contexts. In: Kaelbling, L., Saffiotti, A. (eds.) Procs. of IJCAI, pp. 558–563. Professional Book Center (2005)
8. Brewka, G., Eiter, T., Fink, M., Weinzierl, A.: Managed multi-context systems. In: Walsh, T. (ed.) Procs. of IJCAI, pp. 786–791. IJCAI/AAAI (2011)
9. Benerecetti, M., Giunchiglia, F., Serafini, L.: Model checking multiagent systems. J. Log. Comput. **8**(3), 401–423 (1998)

10. Dragoni, A., Giorgini, P., Serafini, L.: Mental states recognition from communication. J. Log. Comput. **12**(1), 119–136 (2002)
11. Sabater, J., Sierra, C., Parsons, S., Jennings, N.R.: Engineering executable agents using multi-context systems. J. Log. Comput. **12**(3), 413–442 (2002)
12. Gonçalves, R., Knorr, M., Leite, J.: Evolving multi-context systems. In: Schaub, T., Friedrich, G., O'Sullivan, B. (eds.) Procs. of ECAI. Frontiers in Artificial Intelligence and Applications, vol. 263, pp. 375–380. IOS Press (2014)
13. Brewka, G.: Towards reactive multi-context systems. In: Cabalar, P., Son, T.C. (eds.) LPNMR 2013. LNCS, vol. 8148, pp. 1–10. Springer, Heidelberg (2013)
14. Ellmauthaler, S.: Generalizing multi-context systems for reactive stream reasoning applications. In: Procs. of ICCSW. OASICS, vol. 35, pp. 19–26. Schloss Dagstuhl - Leibniz-Zentrum fuer Informatik, Germany (2013)
15. Brewka, G., Ellmauthaler, S., Pührer, J.: Multi-context systems for reactive reasoning in dynamic environments. In: Schaub, T., Friedrich, G., O'Sullivan, B., (eds.) Procs. of ECAI. Frontiers in Artificial Intelligence and Applications, vol. 263, pp. 159–164. IOS Press (2014)
16. Alferes, J.J., Brogi, A., Leite, J., Moniz Pereira, L.: Evolving logic programs. In: Flesca, S., Greco, S., Leone, N., Ianni, G. (eds.) JELIA 2002. LNCS (LNAI), vol. 2424, p. 50. Springer, Heidelberg (2002)
17. Gebser, M., Grote, T., Kaminski, R., Schaub, T.: Reactive answer set programming. In: Delgrande, J.P., Faber, W. (eds.) LPNMR 2011. LNCS, vol. 6645, pp. 54–66. Springer, Heidelberg (2011)
18. Wang, Y., Zhuang, Z., Wang, K.: Belief change in nonmonotonic multi-context systems. In: Cabalar, P., Son, T.C. (eds.) LPNMR 2013. LNCS, vol. 8148, pp. 543–555. Springer, Heidelberg (2013)
19. Kaminski, T., Knorr, M., Leite, J.: Efficient paraconsistent reasoning with ontologies and rules. In: Procs. of IJCAI. IJCAI/AAAI (2015)
20. Gonçalves, R., Knorr, M., Leite, J.: Evolving bridge rules in evolving multi-context systems. In: Bulling, N., van der Torre, L., Villata, S., Jamroga, W., Vasconcelos, W. (eds.) CLIMA 2014. LNCS, vol. 8624, pp. 52–69. Springer, Heidelberg (2014)
21. Slota, M., Leite, J.: On semantic update operators for answer-set programs. In Coelho, H., Studer, R., Wooldridge, M., (eds.) Procs. of ECAI. Frontiers in Artificial Intelligence and Applications, vol. 215, pp. 957–962. IOS Press (2010)
22. Slota, M., Leite, J.: Robust equivalence models for semantic updates of answer-set programs. In: Brewka, G., Eiter, T., McIlraith, S.A. (eds.) Procs. of KR. AAAI Press (2012)
23. Slota, M., Leite, J.: The rise and fall of semantic rule updates based on se-models. TPLP **14**(6), 869–907 (2014)
24. Slota, M., Leite, J.: A unifying perspective on knowledge updates. In: del Cerro, L.F., Herzig, A., Mengin, J. (eds.) JELIA 2012. LNCS, vol. 7519, pp. 372–384. Springer, Heidelberg (2012)
25. Gonçalves, R., Knorr, M., Leite, J.: Towards efficient evolving multi-context systems (preliminary report). In: Ellmauthaler, S., Pührer, J. (eds.) Procs. of React-Know (2014)
26. Knorr, M., Gonçalves, R., Leite, J.: On efficient evolving multi-context systems. In: Pham, D.-N., Park, S.-B. (eds.) PRICAI 2014. LNCS, vol. 8862, pp. 284–296. Springer, Heidelberg (2014)
27. Knorr, M., Slota, M., Leite, J., Homola, M.: What if no hybrid reasoner is available? hybrid MKNF in multi-context systems. J. Log. Comput. **24**(6), 1279–1311 (2014)

28. Gonçalves, R., Alferes, J.J.: Parametrized logic programming. In: Janhunen, T., Niemelä, I. (eds.) JELIA 2010. LNCS, vol. 6341, pp. 182–194. Springer, Heidelberg (2010)
29. Knorr, M., Alferes, J., Hitzler, P.: Local closed world reasoning with description logics under the well-founded semantics. Artif. Intell. **175**(9–10), 1528–1554 (2011)
30. Ivanov, V., Knorr, M., Leite, J.: A query tool for $\mathcal{EL}$ with non-monotonic rules. In: Alani, H., Kagal, L., Fokoue, A., Groth, P., Biemann, C., Parreira, J.X., Aroyo, L., Noy, N., Welty, C., Janowicz, K. (eds.) ISWC 2013, Part I. LNCS, vol. 8218, pp. 216–231. Springer, Heidelberg (2013)
31. Bordini, R.H., Hübner, J.F., Wooldridge, M.: Programming Multi-Agent Systems in AgentSpeak Using Jason (Wiley Series in Agent Technology). John Wiley & Sons (2007)
32. Dastani, M.: 2APL: a practical agent programming language. Journal of Autonomous Agents and Multi-Agent Systems **16**(3), 214–248 (2008)
33. Hindriks, K.V.: Programming rational agents in GOAL. In: El Fallah Seghrouchni, A., Dix, J., Dastani, M., Bordini, R.H. (eds.) Multi-Agent Programming, pp. 119–157. Springer, US (2009)
34. Criado, N., Argente, E., Botti, V.J.: THOMAS: an agent platform for supporting normative multi-agent systems. J. Log. Comput. **23**(2), 309–333 (2013)
35. Meneguzzi, F., Rodrigues, O., Oren, N., Vasconcelos, W.W., Luck, M.: BDI reasoning with normative considerations. Eng. Appl. of AI **43**, 127–146 (2015)
36. Cardoso, H.L., Oliveira, E.: A context-based institutional normative environment. In: Hübner, J.F., Matson, E., Boissier, O., Dignum, V. (eds.) COIN 2008. LNCS, vol. 5428, pp. 140–155. Springer, Heidelberg (2009)
37. Gerard, S.N., Singh, M.P.: Evolving protocols and agents in multiagent systems. In: Gini, M.L., Shehory, O., Ito, T., Jonker, C.M. (eds.) Procs. of AAMAS, pp. 997–1004. IFAAMAS (2013)
38. Vasconcelos, W.W., Kollingbaum, M.J., Norman, T.J.: Normative conflict resolution in multi-agent systems. Autonomous Agents and Multi-Agent Systems **19**(2), 124–152 (2009)
39. Panagiotidi, S., Alvarez-Napagao, S., Vázquez-Salceda, J.: Towards the norm-aware agent: bridging the gap between deontic specifications and practical mechanisms for norm monitoring and norm-aware planning. In: Balke, T., Dignum, F., van Riemsdijk, M.B., Chopra, A.K. (eds.) COIN 2013. LNCS, vol. 8386, pp. 346–363. Springer, Heidelberg (2014)

Bringing Constitutive Dynamics to Situated Artificial Institutions

Maiquel de Brito[1]([⊠]), Jomi F. Hübner[1], and Olivier Boissier[2]

[1] Federal University of Santa Catarina, Florianópolis, SC, Brazil
maiquel.b@posgrad.ufsc.br, jomi.hubner@ufsc.br
[2] Laboratoire Hubert Curien UMR CNRS 5516, Institut Henri Fayol,
MINES Saint-Etienne, Saint-Etienne, France
Olivier.Boissier@emse.fr

Abstract. The Situated Artificial Institution (SAI) model, as proposed in the literature, conceives the regulation of Multi-Agent Systems as based on a constitutive state that is consequence of the institutional interpretation of facts issued by the environment. The different nature of these facts (e.g. past sequence of events, states holding) implies various dynamic behaviours that need to be considered to properly define the life cycle of the constitutive state. This paper aims to bring such a dynamic to SAI. It defines a formal apparatus (i) for the institutional interpretation of environmental facts based on constitutive rules and (ii) for the management of the resulting constitutive state.

1 Introduction

Among the different works related to artificial institutions, [1,2] are concerned with the grounding of norms in the environment where the agents act, keeping a clear separation among regulative, constitutive, and environmental elements involved in the regulation of Multi-Agent Systems (MAS). In this paper we consider and extend the Situated Artificial Institution (SAI) model [2]. The choice of SAI is motivated by its available specification language, that is interesting to specify norms decoupled but still grounded in the environment as shown in [3]. For example, the norm stating that "the winner of the auction is obliged to pay its offer" is specified on top of a constitutive level that defines who, in the environment, is the winner that must pay its offer and what must be done, in the environment, to comply with that expectation. Norms abstracting from the environment are more stable and flexible but must be connected to the environment [1], as the regulation of the system (realised in what we call *institutions*) is, in fact, the regulation of what happens in the environment.

The notion of *constitution* proposed by John Searle [4] has inspired different works addressing the relation between the environment and the regulative elements in MAS. Among them, SAI goes in a particular direction, considering that constitutive rules specify how agents acting, events occurring, and states holding in the environment compose (or *constitute*) the constitutive level of the institution. In the previous example, a constitutive rule could state that the agent that

F. Pereira et al. (Eds.) EPIA 2015, LNAI 9273, pp. 624–637, 2015.
DOI: 10.1007/978-3-319-23485-4_63

acts in the environment placing the best bid counts, in the constitutive level, as the winner of the auction (Figure 1).

While the notion of constitution in SAI is well defined, a precise and formal definition of the dynamics of the constitutive level, resulting of the interpretation of constitutive rules, is still lacking. Interpreting the constitutive rules and managing the SAI constitutive state require to consider (i) how to tackle with the different natures of the environmental elements that may constitute the relevant elements to the institutional regulation (i.e. agents, events, states) and (ii) how to base the dynamic of the constitution both on the occurrences of these elements in the environment and on the production of new constitutive elements in the institution itself. Taking as granted that the institutional regulation depends on the constitutive state, this paper departs from the SAI conceptual model to propose clear defined semantics answering to these two challenges.

The paper begins with a global overview of the SAI model (Section 2), on which we base our contributions, that are presented in the sections 3 and 4. While the Section 3 introduces the necessary representations to support the interpretation of the constitutive rules, the Section 4 is focused on the dynamic aspects of this interpretation. Before concluding and pointing some perspectives for future work, Section 5 discusses the contributions of this paper with respect to related work.

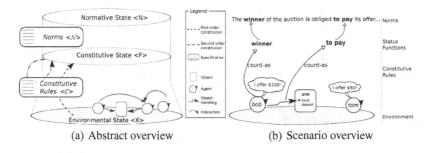

(a) Abstract overview   (b) Scenario overview

Fig. 1. SAI overview

2   Background

Before presenting our contributions in the next sections, this section briefly describes the SAI model proposed in [2]. In SAI, norms define the expected behaviour from the agents in an abstract level that is not directly related to the environment. For example, the norm *"the winner of an auction is obliged to pay its offer"* does not specify neither who is the winner that is obliged to fulfil the norm nor what the winner must concretely do to fulfil it. The effectiveness of a norm depends on its connection to the environment as its dynamics (activation, fulfilment, etc) results of facts occurring there. Such a connection is established when the components of the norms – the *status functions* – are constituted, according to *constitutive rules*, from the *environmental elements* (Figure 1). These elements are described below:

– The environmental elements, represented by $\mathcal{X} = \mathcal{A}_\mathcal{X} \cup \mathcal{E}_\mathcal{X} \cup \mathcal{S}_\mathcal{X}$, are organized in the set $\mathcal{A}_\mathcal{X}$ of agents possibly acting in the system, the set $\mathcal{E}_\mathcal{X}$ of events that may happen in the environment, and the set $\mathcal{S}_\mathcal{X}$ of properties used to describe the possible states of the environment.

– The status functions of a SAI, represented by $\mathcal{F} = \mathcal{A}_\mathcal{F} \cup \mathcal{E}_\mathcal{F} \cup \mathcal{S}_\mathcal{F}$, are the set $\mathcal{A}_\mathcal{F}$ of agent-status functions (i.e. status functions assignable to agents), the set $\mathcal{E}_\mathcal{F}$ of event-status functions (i.e. status functions assignable to events), and the set $\mathcal{S}_\mathcal{F}$ of state-status functions (i.e. status functions assignable to states). Status functions are functions that the environmental elements (agents, events, and states) perform from the institutional perspective [4]. For example, in an auction, an agent may have the function of *winner*, the utterance "I offer \$100" may have the function of *bid*, and the state of "more than 20 people placed in a room at Friday 10am" may mean the *minimum quorum* for its realization.

– The *constitutive rules* defined in $\mathcal{C}$ specify the constitution of the status functions of $\mathcal{F}$ from the environment element of $\mathcal{X}$. A constitutive rule $c \in \mathcal{C}$ is a tuple $\langle x, y, t, m \rangle$ where $x \in \mathcal{F} \cup \mathcal{X} \cup \{\varepsilon\}$, $y \in \mathcal{F}$, $t \in \mathcal{E}_\mathcal{F} \cup \mathcal{E}_\mathcal{X} \cup \top$, $m \in W$, and $W = W_\mathcal{F} \cup W_\mathcal{X}$. $W_\mathcal{F}$ is the set of status-functions-formulae (sf-formulae) and $W_\mathcal{X}$ is the set of environment-formulae (e-formulae), defined later. A constitutive rule $\langle x, y, t, m \rangle$ specifies that x counts as y when t has happened while m holds. If $x = \varepsilon$, then there is a *freestanding assignment* of the status function y, i.e. an assignment where there is not a concrete environmental element carrying y [2,4]. When x actually counts as y (i.e. when the conditions t and m declared in the constitutive rule are true), we say that there is a *status function assignment* (SFA) of the status function y to the element x. The establishment of a SFA of y to some x is the *constitution* of y. The set of all SFAs of a SAI composes its *constitutive state* (see Def. 4).

The sf-formulae $w_\mathcal{F} \in W_\mathcal{F}$ are logical formulae, based on status functions (see the Expression 1 below). The e-formulae $w_\mathcal{X} \in W_\mathcal{X}$ are logical formulae, based on environmental elements (see the Expression 2 below). Section 3 defines the proper semantics of these formulae, based on SFA and on the actual environment.

$$w_\mathcal{F} ::= e_\mathcal{F} \mid s_\mathcal{F} \mid \neg w_\mathcal{F} \mid w_\mathcal{F} \vee w_\mathcal{F} \mid w_\mathcal{F} \wedge w_\mathcal{F} \mid x \text{ is } y \mid \bot \mid \top \tag{1}$$

$$w_\mathcal{X} ::= e_\mathcal{X} \mid s_\mathcal{X} \mid \neg w_\mathcal{X} \mid w_\mathcal{X} \vee w_\mathcal{X} \mid w_\mathcal{X} \wedge w_\mathcal{X} \mid \bot \mid \top \tag{2}$$

Considering these definitions of SAI, the challenges stated in the introduction are addressed in the next sections by (i) defining a uniform constitutive dynamics considering the agent, state or event proper life cycles, and (ii) enriching this uniform dynamics with the life cycle of the SFA themselves since constitutions may be stated by already constituted status functions. The first sub-objective requires to consider both instantaneous and fluent[1] dynamics coming from events

---

[1] *Fluent* refers to the possibility of holding along many states and *instantaneous* refers to the holding during a single state.

or states in the environment. Addressing the second sub-objective requires to consider the constitutive state as condition to the constitution (i.e. constitutions may take place under specific constitutive states) but also as the container of elements to whom status function can be assigned.

3   Constitutive Dynamics - Preliminaries

The semantics of the constitutive rules requires, as presented in this section, a formal representation of the elements related to the SAI constitutive dynamics.[2]

Definition 1 (SAI state). *The SAI state is composed by an environmental state X, a constitutive state F, and a normative state N. It is represented by $SAI_{Dyn} = \langle X, F, N \rangle$.*

The formal representation of X and F is introduced below. As the normative state dynamics is beyond the scope of this paper, N is introduced as part of the SAI state but it is not detailed here.

Definition 2 (SAI history). *The history of a SAI is the sequence of its $i \in \mathbb{N}$ states (where $\mathbb{N}$ is the set of the natural numbers).*

The SAI state at the ith step of its history is represented by $SAI_{Dyn}^{i} = \langle X^i, F^i, N^i \rangle$. The set of all states between the first step and the ith step is represented by $SAI_{Dyn}^{[i]} = \langle X^{[i]}, F^{[i]}, N^{[i]} \rangle$.

Definition 3 (Environmental state). *The environmental state is represented by $X = \langle A_X, E_X, S_X \rangle$ where (i) A_X is the set of agents participating in the system, (ii) E_X is the set of events occurring in the environment and (iii) S_X is the set of environmental properties describing the environmental state.*

Agents in A_X are represented by their names. States in S_X are represented by first order logic atomic formulae. Events in E_X are represented by pairs (e, a) where e is the event, represented by a first order logic atomic formula, triggered by the agent a. Events can be triggered by actions of the agents (e.g. the utterance of a bid in an auction, the handling of an environmental artifact, etc) but can be also produced by the environment itself (e.g. a clock tick). In this case, events are represented by pairs (e, ε). We use $X = \langle A_X, E_X, S_X \rangle$ to denote the current state of the environment. When it is necessary to explicitly refer to the state of X at the step i of the SAI history, we use $X^i = \langle A_X^i, E_X^i, S_X^i \rangle$. The environmental state X is used to evaluate e-formulae (see Expression 2 for

---

[2] Similarly to the SAI specification, the SAI dynamics can be divided in two parts: (i) *constitutive dynamics*, consisting of the status functions assignments and revocations, and (ii) *normative dynamics*, consisting of the norm activations, fulfilments, violations, etc. The normative dynamic is beyond the scope of this paper.

syntax)[3]:

$$S_X \models w_X \text{ iff } \exists\theta : w_X \in \mathcal{S}_X \wedge w_X\theta \in S_X \tag{3}$$
$$E_X \models w_X \text{ iff } \exists\theta : w_X \in \mathcal{E}_X \wedge w_X\theta \in E_X \tag{4}$$

Definition 4 (Constitutive state). *The constitutive state of a SAI is represented by $F = \langle A_F, E_F, S_F \rangle$ where (i) $A_F \subseteq A_X \times \mathcal{A}_F$ is the set of agent-status function assignments, (ii) $E_F \subseteq E_X \times \mathcal{E}_F \times A_X$ is the set of event-status function assignments and (iii) $S_F \subseteq S_X \times \mathcal{S}_F$ is the set of state-status function assignments.*

As introduced in the previous section, SFA are relations between environmental elements and status functions. Elements of A_F are pairs $\langle a_X, a_F \rangle$ meaning that the agent $a_X \in A_X$ has the status function $a_F \in \mathcal{A}_F$. Elements of E_F are triples $\langle e_X, e_F, a_X \rangle$ meaning that the event-status function $e_F \in \mathcal{E}_F$ is assigned to the event $e_X \in E_X$ produced by the agent $a_X \in A_X$. As events are supposed to be considered at the individual agent level in normative systems [6], it is important to record the agent that causes an event-status function assignment. Elements of S_F are pairs $\langle s_X, s_F \rangle$ meaning that the state $s_X \in S_X$ carries the status function $s_F \in \mathcal{S}_F$. In the following, we will note $F = \langle A_F, E_F, S_F \rangle$ to denote the current constitutive state and $F^i = \langle A_F^i, E_F^i, S_F^i \rangle$ will be used to refer to the constitutive state F at the step i of the SAI history.

The constitutive state F is used to evaluate the sf-formulae (see Expression 1 for syntax). If an agent x participates in the system (i.e. $x \in A_X$) and carries the status function y (i.e. if $\langle x, y \rangle \in A_F$), then the formula x **is** y is true in current state F:

$$A_F \models x \text{ is } y \quad \text{iff } x \in A_X \wedge y \in \mathcal{A}_F \wedge \langle x, y \rangle \in A_F \tag{5}$$

In the same way, event-status function semantics is defined in the Expression 6. In addition, if an event-status function is assigned to some environmental event, then this event-status function follows from the current constitutive state F (Expression 7):

$$E_F \models x \text{ is } y \text{ iff } x \in E_X \wedge x = \langle e, a \rangle \wedge y \in \mathcal{E}_F \wedge \langle e, y, a \rangle \in E_F \tag{6}$$
$$E_F \models w_F \text{ iff } w_F \in \mathcal{E}_F \wedge \exists e_X : e_X \text{ is } w_F \tag{7}$$

State-status function semantics is similarly defined in the Expression 8. In addition, if there is some assignment involving a state-status function, then this state-status function follows from the current constitutive state F (Expression 9):

$$S_F \models x \text{ is } y \text{ iff } x \in S_X \wedge y \in \mathcal{S}_F \wedge \langle x, y \rangle \in S_F \tag{8}$$
$$S_F \models w_F \text{ iff } w_F \in \mathcal{S}_F \wedge \exists s_X : s_X \text{ is } w_F \tag{9}$$

The *constitutive state* defines how the institution is situated. The next section defines how this constitutive state is deduced from the environmental state and from the constitutive state itself.

---

[3] In this paper, a *substitution* is always represented by θ. A *substitution* is a finite set of pairs $\{\alpha_1/\beta_1, \cdots \alpha_n/\beta_n\}$ where α_i is a variable and β_i is a term. If θ is a substitution and ρ is a literal, then $\rho\theta$ is the literal resulting from the replacement of each α_i in ρ by the corresponding β_i [5].

```
status_functions:
   agents: auctioneer, bidder, current_winner, winner.
   events: to_bid(Value), to_pay(Value), to_fine_winner, commercial_transaction.
   states: auction_running, auction_finished, current_value(Value).
norms:
   1:auction_finished: winner obliged to_pay(current_value).
constitutive_rules:
   /* The agent that proposes an auction is the auctioneer */
   1: Agent count-as auctioneer when (propose(auction),Agent) while not auction_finished.
   /* While the auction is running, any agent other than the auctioneer is a bidder */
   2: Agent count-as bidder while not(Agent is auctioneer)& auction_running.
   /* Auctioneer and bidders are auction participants */
   3: auctioneer count-as auction_participant
   4: bidder count-as auction_participant
   /* The agent that performs the best bid is the current_winner */
   5: Agent count-as current_winner when (to_bid(Value),Agent)
          while (not(current_value(Current)) & Current>Value)& (auction_running|auction_finished).
   /* The current winner is the (final) winner if the auction is finished */
   6: current_winner count-as winner while auction_finished.
   /* An auction is running while there is an agent being the auctioneer */
   7: count-as auction_running while _ is auctioneer.
   /* Auctioneer hitting the hammer means that the auction is finished */
   8: count-as auction_finished when (hit_hammer, Agent) while Agent is auctioneer.
   /* An offer done by a bidder while the auction is running is a bid */
   9: (offer(Value),Agent) count-as to_bid(Value) while auction_running & Agent is bidder.
   /* An offered value is the current value if it is greater than the last one */
   10: count-as current_value(Value) when (to_bid(Value),Agent)
          while Agent is bidder & (not(Current is current_value) & Current>Value)& (auction_running|auction_finished).
   /* A bid is a commercial transaction */
   11: to_bid count-as commercial_transaction.
   /* A bank deposit from the winner to the auctioneer is a payment */
   12: (bank_deposit(Creditor,Value),Agent) count-as to_pay(Value)
          while Creditor is auctioneer & Agent is winner & auction_finished & current_value(Value).
```

Fig. 2. SAI Specification

4 Constitutive Dynamics

The interpretation of the constitutive rules produces the SFA composing the SAI constitutive state. Constitutive rules can specify two kinds of constitution of status functions: *first-order constitution* (Section 4.1) and *second-order constitution* (Section 4.2). From these two definitions, Section 4.3 defines the constitutive dynamics of SAI. This is all illustrated considering an auction scenario whose regulation is specified in the Figure 2, according to the SAI specification language proposed in [2].

4.1 First-Order Constitution

The *first-order constitution*, explained in defs. 5 to 7, explicitly assigns a status function to agents, events, and states from the environment stating, for example that the agent *bob* counts as a *bidder*.

Definition 5 (First-order constitution of agent-status-functions). *The set of agent-status function assignments due to first-order constitution in the ith step of the SAI history is given by the function f-$const_a$ defined as follows:*

$$f\text{-}const_a(\mathcal{F}, \mathcal{C}, X^{[i-1]}, F^{[i-1]}) = \{\langle x\theta, y\rangle | \exists\theta \; \exists\langle x,y,t,m\rangle \in \mathcal{C} \; \exists s \in \mathbb{N} \; \forall k \in [s, i-1] :$$
$$(y \in \mathcal{A}_{\mathcal{F}}) \wedge (E_X^s \cup E_F^s \models t\theta) \wedge (X^k \cup F^k \models m\theta) \wedge x\theta \in A_X^{i-1}\}$$

Informally, (i) if exists a constitutive rule $\langle x, y, t, m \rangle$ whose element t, under a substitution θ, represents an event occurred at the step s and (ii) if along all the steps k from s to $i - 1$ the formula m, under θ, is entailed by the environmental and constitutive states, then the agent identified by the element x under θ carries the agent-status function y in the step i. Note that the function defines that an SFA to an agent only holds while the agent participates in the system. If it leaves the system, all its SFA are dropped. The function also explicits our proposed approach to deal with combined instantaneous events and fluent states as conditions to constitution when it defines that an SFA belongs to the constitutive state if m holds in all steps k from the occurrence of t (at the step s) until the step $i - 1$. Some points to observe in this definition are: (i) the repetition of the event t does not affect the SFA and (ii) a SFA is dropped if m ceases to hold and is not undropped if the m turns to hold (unless the event t happens again while m is again holding).

The rule 1 in the Figure 2 defines a first-order constitution of an agent-status function. If $(propose(auction), bob) \in E_X^1$, meaning that the agent bob has proposed an auction at the step 1, then bob carries the status function $auctioneer$ (i.e. $\langle bob, auctioneer \rangle \in f\text{-}const_a(\mathcal{F}, \mathcal{C}, X^{[i-1]}, F^{[i-1]})$) for all steps i, starting from the 2nd one, while the property $auction\_finished$ does not hold (considering $\theta = \{Agent/bob\}$).

Definition 6 (First-order constitution of state-status-functions). *The set of state-status-function assignments due to first-order constitution in the ith step of the SAI history is given by the function $f\text{-}const_s$ defined as follows:*

$$f\text{-}const_s(\mathcal{F}, \mathcal{C}, X^{[i-1]}, F^{[i-1]}) = \{\langle x\theta, y \rangle | \exists \theta \; \exists \langle x, y, t, m \rangle \in \mathcal{C} \exists s \in \mathbb{N} \; \forall k \in [s, i-1] :$$
$$(y \in \mathcal{S}_{\mathcal{F}}) \wedge (E_X^s \cup E_F^s \models t\theta) \wedge (X^k \cup F^k \models m\theta) \wedge ((x = \varepsilon) \vee (x\theta \in S_X^{i-1}))\}$$

Similar to the constitution of agent-status functions, (i) a SFA to a state $x \in S_X$ only holds while x holds in the environment and (ii) the constitution of state-status functions is conditioned to the holding of m in all steps from the occurrence of the event t. Besides, the function $f\text{-}const_s$ explicits our conception that the constitution of state-status functions may result in freestanding assignments.

The rule 8 in the Figure 2 defines a first-order constitution of a state-status function. If $(hit\_hammer, bob) \in E_X^3$, meaning that bob has hitted a hammer at the step 3, then the assignment $\langle \varepsilon, auction\_finished \rangle$ is active from the step 4 while bob has the status function of $auctioneer$ (considering $\theta = \{Agent/bob\}$).

Definition 7 (First-order constitution of event-status-functions). *The set of event-status-function assignments due to first-order constitution in the ith step of the SAI history is given by the function $f\text{-}const_e$ defined as follows:*

$$f\text{-}const_e(\mathcal{F}, \mathcal{C}, X^{[i-1]}, F^{[i-1]}) = \{\langle e\theta, y, a\theta \rangle | \exists \theta \; \exists \langle x, y, t, m \rangle \in \mathcal{C} : (y \in \mathcal{E}_{\mathcal{F}}) \wedge$$
$$(E_X^{i-1} \cup E_F^{i-1} \models t\theta) \wedge (X^{i-1} \cup F^{i-1} \models m\theta) \wedge x = (e, a) \wedge (e\theta, a\theta) \in E_X^{i-1}\}$$

Compared to agent- and state-status functions, the constitution of event-status functions is differently related to the SAI history. Event-status function assignments are assumed to hold only in the step after which the conditions t and m hold, mimicking, thus, in the constitutive level, the atomic nature of the environmental events [7]. Thus, the holding of m during many steps of the SAI history does not imply in the holding of an event-status function assignment.

The rule 9 in the Figure 2 defines a first-order constitution of an event-status function. If $(offer(100), tom) \in E_X^2$ meaning that tom has uttered an offer of \$100 at the step 2, then the assignment $\langle offer(100), to\_bid, tom \rangle$ holds in the step 3, i.e. $\langle offer(100), to\_bid, tom \rangle \in f\text{-}const_e(\mathcal{F}, \mathcal{C}, X^{[2]}, F^{[2]})$ (considering $\theta = \{Value/100, Agent/tom\}$). When $t \neq \top$, the event-status-function assignment is assigned to the event x conditioned to the occurrence of two events at the same step: the event x itself and the event t.

4.2   Second-Order Constitution

Constitutive rules specifying second-order constitution define that a status function counts as another status function. But even specifying a relation between two status functions, the assignments resulting of the second-order constitution are also relations between status functions and environmental elements. That is to say, whenever status function s_1 counts as a status function s_2 all the elements constituting s_1 constitute also s_2. For example, even the rule 3 in the Figure 2 states that the *auctioneer* counts as an *auction_participant*, the status function of *auction_participant* is actually assigned to all the concrete agents carrying the status function of *auctioneer*.

Defining the set of SFA due to second order constitution is an iterative process, as each change in the constitutive state may produce new SFA. To deal with this, the functions defined as follows have the index n (e.g. $s\text{-}const_a^n$), representing the nth iteration in the evaluation of second-order constitution in a same step of the SAI history. Each iteration n takes into account the assignments produced in the iteration $n-1$. The whole set of SFA due to second-order constitution in a step i of the SAI history is found when the SFAs produced in the iterations n and $n-1$ are the same.

Definition 8 (Second-order constitution of agent-status-functions). *Given the function $s\text{-}const_a^n$ ($n \geq 0$) defined below, the set of agent-status function assignments due to second-order constitution in the ith step of the SAI history is given by $s\text{-}const_a = s\text{-}const_a^n$ for the lowest n s.t. $s\text{-}const_a^n = s\text{-}const_a^{n-1}$:*

$$s\text{-}const_a^n(\mathcal{F}, \mathcal{C}, X^{[i]}, F^{[i]}) = \{\langle a_X, y \rangle | \exists \theta \; \exists \langle x, y, t, m \rangle \in \mathcal{C} \; \exists s \in \mathbb{N} \; \forall k \in [s, i-1] :$$

$$(y \in \mathcal{A}_{\mathcal{F}}) \wedge (E_X^s \cup E_F^s \models t\theta) \wedge (X^k \cup F^k \models m\theta) \wedge (x\theta \in \mathcal{A}_{\mathcal{F}} \wedge \langle a_X, x\theta \rangle \in A\}$$

$$where \; A = \begin{cases} A_F^{[i]} & if \; n = 0 \\ A_F^{[i]} \cup s\text{-}const_a^{n-1}(\mathcal{F}, \mathcal{C}, X^{[i]}, F^{[i]}) & otherwise \end{cases}$$

Informally, if there is a constitutive rule $\langle x, y, t, m \rangle$ whose element x, under a substitution θ, corresponds to a status function already assigned to an agent a_X, then this agent carries also the status function $y \in \mathcal{A}_{\mathcal{F}}$ (subject to the conditions t and m, as in the first-order constitution (Def.5)). When the agent a_X ceases to carry the status function $x\theta$, it also ceases to carry the status function y.

The rule 3 of the Figure 2 defines a second-order constitution of an agent-status function. Considering $\theta = \{Agent/bob\}$, if bob is $auctionner$ at the ith step (i.e. $\langle bob, auctioneer \rangle \in A_{\mathcal{F}}^i$), then $\langle bob, auction\_participant \rangle \in$ $s\text{-}const_a^n(\mathcal{F}, \mathcal{C}, X^{[i]}, F^{[i]})$, for $n \geq 0$, and, eventually, $bob \in A_F^i$. Informally, the rule states that an agent having the status function of $auctioneer$ counts as an $auction\_participant$ and, as bob has the status function of $auctioneer$, he has also the status function of $auction\_participant$.

Definition 9 (Second-order constitution of state-status-functions).
Given the function $s\text{-}const_s^n$ ($n \geq 0$) defined below, the set of state-status function assignments due to second-order constitution in the ith step of the SAI history is given by $s\text{-}const_s = s\text{-}const_s^n$ for the lowest n s.t. $s\text{-}const_s^n = s\text{-}const_s^{n-1}$:

$$s\text{-}const_s^n(\mathcal{F}, \mathcal{C}, X^{[i]}, F^{[i]}) = \{\langle s_X, y \rangle | \exists \theta \; \exists \langle x, y, t, m \rangle \in \mathcal{C} \; \exists s \in \mathbb{N} \; \forall k \in [s, i-1] :$$
$$(y \in \mathcal{S}_{\mathcal{F}}) \wedge (E_X^s \cup E_F^s \models t\theta) \wedge (X^k \cup F^k \models m\theta) \wedge (x\theta \in \mathcal{S}_{\mathcal{F}} \wedge \langle s_X, x\theta \rangle \in S)\}$$

$$\text{where } S = \begin{cases} S_F^{[i]} & \text{if } n = 0 \\ S_F^{[i]} \cup s\text{-}const_s^{n-1}(\mathcal{F}, \mathcal{C}, X^{[i]}, F^{[i]}) & \text{otherwise} \end{cases}$$

If there is a constitutive rule $\langle x, y, t, m \rangle$ whose element x, under a substitution θ, corresponds to a status function already assigned to a state s_X, then this state carries also the status function $y \in \mathcal{S}_{\mathcal{F}}$ (subject to the conditions t and m, as in the first-order constitution (Def.6)). When s_X ceases to carry the status function $x\theta$, it also ceases to carry the status function y.

Let's consider the status function $payment\_phase$ in an auction scenario and a constitutive rule stating that `auction_finished count-as payment_phase`. Thus, if $\langle \varepsilon, auction\_finished \rangle \in S_F^i$, then $\langle \varepsilon, payment\_phase \rangle \in s\text{-}const_s^n(\mathcal{F}, \mathcal{C}, X^{[i]}, F^{[i]})$, for $n \geq 0$ and, eventually, $\langle \varepsilon, payment\_phase \rangle \in S_F^i$.

Definition 10 (Second-order constitution of event-status-functions).
Given the function $s\text{-}const_e^n$ ($n \geq 0$) defined below, the set of event-status function assignments due to second-order constitution in the ith step of the SAI history is given by $s\text{-}const_e = s\text{-}const_e^n$ for the lowest n s.t. $s\text{-}const_e^n = s\text{-}const_e^{n-1}$:

$$s\text{-}const_e^n(\mathcal{F}, \mathcal{C}, X^{[i]}, F^{[i]}) = \{\langle e_X, y, a_X \rangle | \exists \theta \; \exists \langle x, y, t, m \rangle \in \mathcal{C} : (y \in \mathcal{E}_{\mathcal{F}}) \wedge$$
$$(E_X^{i-1} \cup E_F^{i-1} \models t\theta) \wedge (X^{i-1} \cup F^{i-1} \models m\theta) \wedge x\theta \in \mathcal{E}_{\mathcal{F}} \wedge \langle e_X, x\theta, a_X \rangle \in E\}$$

$$\text{where } E = \begin{cases} E_F^{[i]} & \text{if } n = 0 \\ E_F^{[i]} \cup s\text{-}const_e^{n-1}(\mathcal{F}, \mathcal{C}, X^{[i]}, F^{[i]}) & \text{otherwise} \end{cases}$$

If there is a constitutive rule $\langle x, y, t, m \rangle$ whose element x, under a substitution θ, corresponds to a status function already assigned to the event e_X, then e_X

carries also the status function $y \in \mathcal{E}_{\mathcal{F}}$ (subject to the conditions t and m, as in the first-order constitution (Def.7)). The assignment of y to e_X holds while the assignment of x to e_X holds.

The constitutive rule 11 in the Figure 2 states that bidding in an auction is a commercial transaction. Supposing that the agent *tom* has uttered an offer at the step $i - 1$, then by the rule 9, $(offer(100), to\_bid, tom) \in E_F^i$ and, by the rule 12, $\langle offer(100), commercial\_transaction, tom \rangle \in s\text{-}const_e^n(\mathcal{F}, \mathcal{C}, X^{[i]}, F^{[i]})$, for $n \geq 0$, and eventually $\langle offer(100), commercial\_transaction, tom \rangle \in E_F^i$ because (i) the term x of the rule is an event-status-function that (ii) is already assigned to the event $offer(100)$.

From the definitions 8 to 10 we can see define how status functions being assigned to status functions allows to ground the institution in the environment while it enables different kinds of manipulations inside the constitutive level, such as the definition of multiple levels of abstraction (defining, for example, that the status functions y_1 counts as y_2, that, on its turn, counts as y_3), as well allowing to define relations inside the constitutive level such as generalization (e.g. y_1 and y_2 count as y_3), etc.

4.3   SAI Constitutive State

The previously presented functions permit to formally define the constitutive state in the ith step of the SAI history as $F^i = \langle A_F^i, E_F^i, S_F^i \rangle$ where:

$$A_F^i = \{f\text{-}const_a(\mathcal{F}, \mathcal{C}, X^{[i-1]}, F^{[i-1]}) \cup s\text{-}const_a(\mathcal{F}, \mathcal{C}, X^{[i]}, F^{[i]})\}$$
$$E_F^i = \{f\text{-}const_e(\mathcal{F}, \mathcal{C}, X^{[i-1]}, F^{[i-1]}) \cup s\text{-}const_e(\mathcal{F}, \mathcal{C}, X^{[i]}, F^{[i]})\}$$
$$S_F^i = \{f\text{-}const_s(\mathcal{F}, \mathcal{C}, X^{[i-1]}, F^{[i-1]}) \cup s\text{-}const_s(\mathcal{F}, \mathcal{C}, X^{[i]}, F^{[i]})\}$$

4.4   Illustration of the Constitutive Dynamic within SAI

Following the semantics proposed in the previous sections, the interpretation of the constitutive rules produces the assignments and revocations of status functions, i.e. the constitutive dynamics of SAI. Such a constitutive dynamics is illustrated here with a running example related to the auction scenario previously explored.

The Table 1 shows 8 steps of the SAI history focusing on the environmental state X and on the constitutive one F. The environmental state evolves as follows: at step 1, the agents *bob* and *tom* act in the system; at the step 2, *bob* utters a proposal for an auction; at the step 5, *tom* utters an offer and the agent *ana* enters in the system; at the step 6, *bob* hits the hammer. The constitutive rules, interpreted as described in Section 4, build the constitutive state F. The column *C.Rule* shows the constitutive rule from the Figure 2 that has produced each SFA. For example, in the step 4, $\langle bob, auctioneer \rangle$ is produced by constitutive rule 1, $\langle bob, auction\_participant \rangle$ by the rule 3, and $\langle \varepsilon, auction\_running \rangle$ by rule 7.

Table 1. Running example

Step	Environmental State (X)	Constitutive State (F)	C.Rule
1	$X_a = \{bob, tom\}$		
2	$X_a = \{bob, tom\}$ $X_e = \{(propose(auction), bob)\}$		
3	$X_a = \{bob, tom\}$	$A_F = \{\langle bob, auctioneer\rangle, \langle bob, auction\_participant\rangle\}$	1,3
4	$X_a = \{bob, tom\}$	$A_F = A_F^3$ $S_F = \{\langle \varepsilon, auction\_running\rangle\}$	1,3 7
5	$X_a = \{bob, tom, ana\}$ $X_e = \{(offer(100), tom)\}$	$A_F = A_F^4 \cup \{\langle tom, bidder\rangle, \langle tom, auction\_participant\rangle\}$ $S_F = S_F^4$	1,3,2,4 7
6	$X_a = \{bob, tom, ana\}$ $X_e = \{(hit\_hammer, bob)\}$	$A_F = A_F^5 \cup \{\langle ana, bidder\rangle, \langle ana, auction\_participant\rangle\}$ $E_F = \{\langle(offer(100), to\_bid, tom)\rangle\}$ $S_F = S_F^5$	1,3,2,4,2,4 9 7
7	$X_a = \{bob, tom, ana\}$	$A_F = A_F^6 \cup \{\langle tom\ current\_winner\rangle\}$ $S_F = S_F^6 \cup \{\langle \varepsilon, current\_value(100)\rangle,$ $\langle \varepsilon, auction\_finished\rangle\}$	1,3,2,4,2,4,5 7,10, 8
8	$X_a = \{bob, tom, ana\}$	$A_F = \{\langle tom\ current\_winner\rangle, \langle tom, winner\rangle\}$ $S_F = \{\langle \varepsilon, current\_value(100)\rangle, \langle \varepsilon, auction\_finished\rangle\}$	5,6 10,8

Note that the proposed semantics does not define just the establishment of the SFA but it defines also their revocations. For example, the constitutive rule 1 in the Figure 2 defines that the agent that proposes an auction is the *auctioneer* while the auction is not finished. In the example, this condition ceases to hold in the step 7, leading to a new state (8) where the assignment of the status function *auctioneer* to the agent *bob* is revoked.

5   Related Work

Different approaches in the literature investigate how environmental facts affect artificial institutions. Some, contrary to us, do not consider the environment producing some kind of dynamic inside the institution: in [1,8], the environmental elements are related to the concepts appearing in the norm specification but they do not produce facts related to the dynamics of norms (violations, fulfilments, etc); in [9], environmental facts determine properties that should hold in the institution but the institution is in charge to take such information and produce some dynamic where appropriate.

Some approaches, as we do, consider that environmental facts produce some kind of dynamics in the institution: in [10] they affect the dynamics of organisations producing role assignments, goal achievements, etc; they produce institutional events in [11]; they affect the normative dynamics in [12,13] producing norm fulfilments, violations, etc. Compared to these related works, this paper deals with the definition of how the environment determines another fact in the institution, that is namely the *constitution of status functions*, defining (and not just affecting) the *constitutive dynamics* that is the base of the regulation in SAI.

When the constitution of each kind of status functions is considered in isolation, some relations can be made, for example, between the constitution of state-status functions and the constitution of *states of affairs* proposed in [14], or between the constitution of event-status functions and the generation of *institutional events* proposed in [11]. But we deal with the constitution of agent-, event-, and state-status function as, together, determining the dynamics of the constitutive state of SAI. This constitutive state is not viewed just as a container of constituted status functions but as a system having particular – and well defined by this paper – dynamics taking into account the different nature of their components. We deal with the constitution considering the particular nature of the three different kinds of status function, but we also consider the constitution of the three different status functions affecting each other.

Works such as those in line with [8], are concerned with the ontological aspects of the count-as, i.e. with the constitution defining and providing meaning to the institutional vocabulary. These aspects are also part of SAI conceptual model and, regarding to them, this paper contributes providing clear representations and semantics to actually ground the institutional vocabulary in the environment. In addition, by dealing with the second-order constitution, we clearly define how the manipulation of concepts of the institutional vocabulary – not explicitly related to the environment – is grounded in the environment.

6   Discussion and Perspectives

To be compliant with SAI definitions [2], the dynamics of the constitutive state must consider that status functions are assigned to (and only to) agents, events, and states under a uniform definition of constitution. Thus, our first sub-objective was to define a uniform constitutive dynamics considering that SFA may have specific life cycles according to their nature. To achieve it, we first defined the life cycles of the SFAs that even being produced by similar definition of constitutive rules, may be distinguished into: (i) agent-status function assignments holding only while the agent that carries the status function participates to the system, (ii) state-status function assignments holding while the state carrying the status function holds in the environment and (iii) event-status function assignments holding only during a single step of the SAI history. These definitions have been then complemented by the explicitation of the instantaneous and fluent expressions conditioning these constitutions. We captured important properties on this dynamics such as: proper dynamics of status function assignment for event, state or agents, stability of constituted status functions wrt repetition of events, dropping of constituted status function as soon as state condition is no more holding, etc.

The second sub-objective was to enrich the proposed dynamics issued of the environmental elements with the dynamics of the constituted status functions themselves. The approach that we took concerned first the conditions of constitutive rules where constituted status functions may appear (defs. 5 to 10), and definition of second-order constitution dynamics that highlights an important property of SAI conceptual model: production of new constitutive states based on facts that are indirectly related to the environment. This property is important in the sense that it makes possible to situate the institution in the environment while making possible to consider the definition and dynamics of constitutive abstractions, generalisations, etc.

Future work include investigations about the normative state affecting the SAI constitutive state, normative dynamics on top of the constitutive dynamic, and manipulations inside the constitutive level through second-order constitution.

Acknowledgments. The authors thanks the financial support given by CAPES (PDSE 4926-14-5) and CNPq (grants 448462/2014-1 e 306301/2012-1).

References

1. Aldewereld, H., Álvarez Napagao, S., Dignum, F., Vázquez-Salceda, J.: Making norms concrete. In: van der Hoek, W., Kaminka, G.A., Lespérance, Y., Luck, M., Sen, S. (eds) AAMAS 2010, pp. 807–814 (2010)
2. de Brito, M., Hübner, J.F., Boissier, O.: A conceptual model for situated artificial institutions. In: Bulling, N., van der Torre, L., Villata, S., Jamroga, W., Vasconcelos, W. (eds.) CLIMA XV 2014. LNCS (LNAI), vol. 8624, pp. 35–51. Springer, Heidelberg (2014)
3. De Brito, M., Thevin, L., Garbay, C., Boissier, O., Hübner, J.F.: Situated artificial institution to support advanced regulation in the field of crisis management. In: Demazeau, Y., Decker, K.S., Bajo Pérez, J., De la Prieta, F. (eds.) PAAMS 2015. LNCS (LNAI), vol. 9086, pp. 66–79. Springer, Heidelberg (2015)
4. Searle, J.: Making the Social World. The Structure of Human Civilization. Oxford University Press (2009)
5. Brachman, R., Levesque, H.: Knowledge Representation and Reasoning. Morgan Kaufmann Publishers Inc., San Francisco (2004)
6. Vos, M.D., Balke, T., Satoh, K.: Combining event-and state-based norms. In: AAMAS 2013, pp. 1157–1158 (2013)
7. Cassandras, C.G., Lafortune, S.: Introduction to Discrete Event Systems. Springer-Verlag New York Inc., Secaucus (2006)
8. Grossi, D., Meyer, J.-J.C., Dignum, F.P.M.: Counts-as: classification or constitution? an answer using modal logic. In: Goble, L., Meyer, J.-J.C. (eds.) DEON 2006. LNCS (LNAI), vol. 4048, pp. 115–130. Springer, Heidelberg (2006)
9. de Brito, M., Hübner, J.F., Bordini, R.H.: Programming institutional facts in multi-agent systems. In: Aldewereld, H., Sichman, J.S. (eds.) COIN 2012. LNCS (LNAI), vol. 7756, pp. 158–173. Springer, Heidelberg (2013)
10. Piunti, M., Boissier, O., Hübner, J.F., Ricci, A.: Embodied organizations: a unifying perspective in programming agents, organizations and environments. In: MALLOW 2010. CEUR, vol. 627 (2010)
11. Cliffe, O., De Vos, M., Padget, J.: Answer set programming for representing and reasoning about virtual institutions. In: Inoue, K., Satoh, K., Toni, F. (eds.) CLIMA 2006. LNCS (LNAI), vol. 4371, pp. 60–79. Springer, Heidelberg (2007)

12. Dastani, M., Grossi, D., Meyer, J.-J.C., Tinnemeier, N.: Normative multi-agent programs and their logics. In: Meyer, J.-J.C., Broersen, J. (eds.) KRAMAS 2008. LNCS (LNAI), vol. 5605, pp. 16–31. Springer, Heidelberg (2009)
13. Campos, J., López-Sánchez, M., Rodríguez-Aguilar, J.A., Esteva, M.: Formalising situatedness and adaptation in electronic institutions. In: Hübner, J.F., Matson, E., Boissier, O., Dignum, V. (eds.) COIN 2008. LNCS (LNAI), vol. 5428, pp. 126–139. Springer, Heidelberg (2009)
14. Jones, A., Sergot, M.: A formal characterisation of institutionalised power. Logic Journal of IGPL **4**(3), 427–443 (1996)

Checking WECTLK Properties of Timed Real-Weighted Interpreted Systems via SMT-Based Bounded Model Checking

Agnieszka M. Zbrzezny$^{(\boxtimes)}$ and Andrzej Zbrzezny

IMCS, Jan Długosz University, Al. Armii Krajowej 13/15,
42-200 Częstochowa, Poland
{agnieszka.zbrzezny,a.zbrzezny}@ajd.czest.pl

Abstract. In this paper, we present the SMT-based bounded model checking (BMC) method for Timed Real-Weighted Interpreted Systems and for the existential fragment of the Weighted Epistemic Computation Tree Logic. We performed the BMC algorithm on Timed Weighted Generic Pipeline Paradigm benchmark. We have implemented SMT-BMC method and made preliminary experimental results, which demonstrate the efficiency of the method. To perform the experiments we used the state of the art SMT-solver Z3.

1 Introduction

The formalism of *interpreted systems* (ISs) was introduced in [2] to model multi-agent systems (MASs) [7], which are intended for reasoning about the agents' epistemic and temporal properties. *Timed interpreted systems* (TIS) was proposed in [9] to extend interpreted systems in order to make possible reasoning about real-time aspects of MASs. The formalism of weighted interpreted systems (WISs) [10] extends ISs to make the reasoning possible about not only temporal and epistemic properties, but also about agents's quantitative properties.

Multi-agent systems (MASs) are composed of many intelligent agents that interact with each other. The agents can share a common goal or they can pursue their own interests. Also, the agents may have deadline or other timing constraints to achieve intended targets. As it was shown in [2], knowledge is a useful concept for analyzing the information state and the behaviour of agents in multi-agent systems. Another different extensions of temporal logics [1] with doxastic [4], and deontic [5] modalities have been proposed. In this paper, we consider the existential fragment of a weighted epistemic computation tree logic (WECTLK) interpreted over Timed Real-Weighted Interpreted Systems (TRWISs).

SMT-based bounded model checking (BMC) consists in translating the existential model checking problem for a modal logic and for a model to the satisfiability modulo theory problem (SMT-problem) of a quantifier-free first-order formula.

© Springer International Publishing Switzerland 2015
F. Pereira et al. (Eds.) EPIA 2015, LNAI 9273, pp. 638–650, 2015.
DOI: 10.1007/978-3-319-23485-4_64

The original contributions of the paper are as follows. First, we define TRWIS as a model of MASs with the agents that have real-time deadlines to achieve intended goals and each transition holds a weight, which can be any non-negative real value. Second, we introduce the language WECTLK. Third, we propose a SMT-based BMC technique for TRWIS and for WECTLK.

To the best of our knowledge, there is no work that considers SMT-based BMC methods to check multi-agent systems modelled by means of timed real-weighted interpreted systems. Thus, in this paper we offer such a method. In particular, we make the following contributions. Firstly, we define and implement an SMT-based BMC method for WECTLK and for TRWISs. Secondly, we report on the initial experimental evaluation of our SMT-based BMC method. To this aim we use a scalable benchmark: the *timed weighted generic pipeline paradigm* [8,10].

The structure of the paper is as follows. In Section 2 we shortly introduce the theory of timed real-weighted interpreted systems and the WECTLK language. In Section 3 we present our SMT-based BMC method. In Section 4 we experimentally evaluate the performance of our SMT-based BMC encoding. We conclude the paper in Section 5.

2   Preliminaries

In this section we first explain some notations used through the paper, and next we define timed real-weighted interpreted systems, and next we introduce syntax and semantics of WECTLK.

Let $\mathbb{N}$ be a set of natural numbers, $\mathbb{N}_+ = \mathbb{N} \setminus \{0\}$, $\mathbb{R}$ be the set of non-negative real numbers, and $\mathcal{X}$ be a finite set of non-negative natural variables, called *clocks* ranging over a set of non-negative natural numbers. A clock valuation is a function $v : \mathcal{X} \rightarrow \mathbb{N}$ that assigns to each clock $x \in \mathcal{X}$ a non-negative natural value $v(x)$. A set of all the clock valuations is denoted by $\mathbb{N}^{|\mathcal{X}|}$. The valuation $v' = v[\mathcal{X}' := 0]$, for $\mathcal{X}' \subseteq \mathcal{X}$ is defined as: $\forall_{x \in \mathcal{X}'} v'(x) = 0$ and $\forall_{x \in \mathcal{X} \setminus \mathcal{X}'} v'(x) = v(x)$. For $\delta \in \mathbb{N}$, $v + \delta$ denotes the valuation that assigns the value $v(x) + \delta$ to each clock x.

The grammar

$$\varphi := \mathbf{true} \mid x < c \mid x \leq c \mid x = c \mid x \geq c \mid x > c \mid \varphi \wedge \varphi$$

generates the set $\mathcal{C}(\mathcal{X})$ of clock constraints over $\mathcal{X}$, where $x \in \mathcal{X}$ and $c \in \mathbb{N}$. A clock valuation v satisfies a clock constraint φ, written as v, iff φ evaluates to be true using the clock values given by v.

Let c_{max} be a constant and $v, v' \in \mathbb{N}^{|\mathcal{X}|}$ two clock valuation. We say that $v \simeq v'$ iff the following condition holds for each $x \in \mathcal{X}$:

$$v(x) > c_{max} \text{ and } v'(x) > c_{max} \text{ or } v(x) \leq c_{max} \text{ and } v'(x) \leq c_{max} \text{ and } v(x) = v'(x)$$

The clock valuation v' such that for each clock $x \in \mathcal{X}$, $v'(x) = v(x) + 1$ if $v(x) \leq c_{max}$, and $v'(x) = c_{max} + 1$ otherwise, is called a time successor of v (written $succ(v)$).

TRWISs. Let $Ag = \{1, \ldots, n\}$ denotes a non-empty and finite set of agents, and $\mathcal{E}$ be a special agent that is used to model the environment in which the agents operate and $\mathcal{PV} = \bigcup_{\mathbf{c} \in Ag \cup \{\mathcal{E}\}} \mathcal{PV}_{\mathbf{c}}$ be a set of propositional variables, such that $\mathcal{PV}_{\mathbf{c}_1} \cap \mathcal{PV}_{\mathbf{c}_2} = \emptyset$ for all $\mathbf{c}_1, \mathbf{c}_2 \in Ag \cup \{\mathcal{E}\}$. The *timed real-weighted interpreted system* (TRWIS) is a tuple

$$(\{L_{\mathbf{c}}, Act_{\mathbf{c}}, \mathcal{X}_{\mathbf{c}}, P_{\mathbf{c}}, t_{\mathbf{c}}, \mathcal{V}_{\mathbf{c}}, \mathcal{I}_{\mathbf{c}}, d_{\mathbf{c}}\}_{\mathbf{c} \in Ag \cup \{\mathcal{E}\}}, \iota),$$

where $L_{\mathbf{c}}$ is a non-empty set of *local states* of the agent $\mathbf{c}$, $S = L_1 \times \ldots \times L_n \times L_{\mathcal{E}}$ is the set of all global states, $\iota \subseteq S$ is a non-empty set of initial states, $Act_{\mathbf{c}}$ is a non-empty set of *possible actions* of the agent $\mathbf{c}$, $Act = Act_1 \times \ldots \times Act_n \times Act_{\mathcal{E}}$ is the set of *joint actions*, $\mathcal{X}_{\mathbf{c}}$ is a non-empty set of *clocks*, $P_{\mathbf{c}} : L_{\mathbf{c}} \to 2^{Act_{\mathbf{c}}}$ is a *protocol function*, $t_{\mathbf{c}} : L_{\mathbf{c}} \times C(\mathcal{X}_{\mathbf{c}}) \times 2^{\mathcal{X}_{\mathbf{c}}} \times Act \to L_{\mathbf{c}}$ is a (partial) *evolution function*, $\mathcal{V}_{\mathbf{c}} : L_{\mathbf{c}} \to 2^{\mathcal{PV}}$ is a *valuation function* assigning to each local state a set of propositional variables that are assumed to be true at that state, $\mathcal{I}_{\mathbf{c}} : L_{\mathbf{c}} \to C(\mathcal{X}_{\mathbf{c}})$ is an *invariant function*, that specifies an amount of time the agent $\mathbf{c}$ may spend in a given local state, and $d_{\mathbf{c}} : Act_{\mathbf{c}} \to \mathbb{R}$ is a *weight function*.

For a given TRWIS we define a *timed real-weighted model* (or a *model*) as a tuple $\mathcal{M} = (Act, S, \iota, T, \mathcal{V}, d)$, where:

- $Act = Act_1 \times \ldots \times Act_n \times Act_{\mathcal{E}}$ is the set of all the joint actions,
- $S = (L_1 \times \mathbb{N}^{|\mathcal{X}_1|}) \times \ldots \times (L_n \times \mathbb{N}^{|\mathcal{X}_n|}) \times (L_{\mathcal{E}} \times \mathbb{N}^{|\mathcal{X}_{\mathcal{E}}|})$ is the set of all the *global states*
- $\iota = (\iota_1 \times \{0\}^{|\mathcal{X}_1|}) \times \ldots \times (\iota_n \times \{0\}^{|\mathcal{X}_n|}) \times (\iota_{\mathcal{E}} \times (\{0\}^{|\mathcal{X}_{\mathcal{E}}|})$ is the set of all the *initial* global states,
- $\mathcal{V} : S \to 2^{\mathcal{PV}}$ is the valuation function defined as $\mathcal{V}(s) = \bigcup_{\mathbf{c} \in Ag \cup \{\mathcal{E}\}} \mathcal{V}_{\mathbf{c}}(l_{\mathbf{c}}(s))$, $T \subseteq S \times (Act \cup \mathbb{N}) \times S$ is a transition relation defined by action and time transitions. For $a \in Act$ and $\delta \in \mathbb{N}$:
  1. action transition: $(s, a, s') \in T$ (or $s \xrightarrow{a} s'$) iff for all $\mathbf{c} \in Ag \cup \mathcal{E}$, there exists a local transition $t_{\mathbf{c}}(l_{\mathbf{c}}(s), \varphi_{\mathbf{c}}, \mathcal{X}', a) = l_{\mathbf{c}}(s')$ such that $v_{\mathbf{c}}(s) \models \varphi_{\mathbf{c}} \wedge \mathcal{I}(l_{\mathbf{c}}(s))$ and $v'_{\mathbf{c}}(s') = v_{\mathbf{c}}(s)[\mathcal{X}' := 0]$ and $v'_{\mathbf{c}}(s') \models \mathcal{I}(l_{\mathbf{c}}(s'))$;
  2. time transition $(s, \delta, s') \in T$ iff for all $\mathbf{c} \in Ag \cup \mathcal{E}$, $l_{\mathbf{c}}(s) = l_{\mathbf{c}}(s')$ and $v'_{\mathbf{c}}(s') = v_{\mathbf{c}}(s) + \delta$ and $v'_{\mathbf{c}}(s') \models \mathcal{I}(l_{\mathbf{c}}(s'))$.
- $d : Act \to \mathbb{R}$ is the "joint" weight function defined as follows: $d((a_1, \ldots, a_n, a_{\mathcal{E}})) = d_1(a_1) + \ldots + d_n(a_n) + d_{\mathcal{E}}(a_{\mathcal{E}})$.

Given a TRWIS we can define the indistinguishability relation $\sim_{\mathbf{c}} \subseteq S \times S$ for any agent $\mathbf{c}$ as follows: $s \sim_{\mathbf{c}} s'$ iff $l_{\mathbf{c}}(s') = l_{\mathbf{c}}(s)$ and $v_{\mathbf{c}}(s') \simeq v_{\mathbf{c}}(s)$ A run of TRWIS is an infinite sequence $\rho = s_0 \xrightarrow{\delta_0, a_0} s_1 \xrightarrow{\delta_1, a_1} s_2 \xrightarrow{\delta_2, a_2} \ldots$ of global states such that the following conditions hold for all $i \in \mathbb{N} : s_i \in S, a_i \in Act, \delta_i \in \mathbb{N}_+$, and there exists $s'_i \in S$ such that $(s_i, \delta, s'_i) \in T$ and $(s_i, a, s_{i+1}) \in T$. Notice that the definition of a run does not permit two consecutive joint actions to be performed one after the other, i.e., between each two joint actions some time must pass; such a run is called *strongly monotonic*.

WECTLK. WECTLK has been defined in [8] as the existential fragment of the weighted CTLK with integer cost constraints on *all* temporal modalities. We extend WECTLK logic by adding non-negative real cost constraints. In the syntax of WECTLK we assume the following: $p \in \mathcal{PV}$ is an atomic proposition, $\mathbf{c} \in Ag$, $\Gamma \subseteq Ag$, I is an interval in $\mathbb{R} = \{0 \ldots\}$ of the form: $[a, \infty)$ and $[a, b)$, for $a, b \in \mathbb{N}$ and $a \neq b$. Moreover, hereafter, by $\mathbf{right}(I)$ we denote the right end of the interval I. The WECTLK formulae are defined by the following grammar:

$$\varphi ::= \mathbf{true} \,|\, \mathbf{false} \,|\, p \,|\, \neg p \,|\, \varphi \vee \varphi \,|\, \varphi \wedge \varphi \,|\, \mathbf{EX}_I \varphi \,|\, \mathbf{E}(\varphi \mathbf{U}_I \varphi) \,|$$
$$\mathbf{EG}_I \varphi \,|\, \overline{\mathbf{K}}_{\mathbf{c}} \varphi \,|\, \overline{\mathbf{D}}_\Gamma \varphi \,|\, \overline{\mathbf{E}}_\Gamma \varphi \,|\, \overline{\mathbf{C}}_\Gamma \varphi.$$

In the semantics we assume the following definitions of epistemic relations: $\sim_\Gamma^E \overset{def}{=} \bigcup_{\mathbf{c} \in \Gamma} \sim_{\mathbf{c}}$, $\sim_\Gamma^C \overset{def}{=} (\sim_\Gamma^E)^+$ (the transitive closure of $\sim_\Gamma^E$), $\sim_\Gamma^D \overset{def}{=} \bigcap_{\mathbf{c} \in \Gamma} \sim_{\mathbf{c}}$, where $\Gamma \subseteq Ag$.

A WECTLK formula φ is $true$ in a model $\mathcal{M}$ (in symbols $\mathcal{M} \models \varphi$) iff $\mathcal{M}, s^0 \models \varphi$ for some $s^0 \in \iota$ (i.e., φ is true at some initial state of the model $\mathcal{M}$). For every $s \in S$ the relation $\models$ is defined inductively as follows:

$-\mathcal{M}, s \models \mathbf{true}$, $\mathcal{M}, s \not\models \mathbf{false}$, $\mathcal{M}, s \models p$ iff $p \in \mathcal{V}(s)$, $\mathcal{M}, s \models \neg p$ iff $p \notin \mathcal{V}(s)$,

$-\mathcal{M}, s \models \alpha \wedge \beta$ iff $\mathcal{M}, s \models \alpha$ and $\mathcal{M}, s \models \beta$,

$-\mathcal{M}, s \models \alpha \vee \beta$ iff $\mathcal{M}, s \models \alpha$ or $\mathcal{M}, s \models \beta$,

$-\mathcal{M}, s \models \mathbf{EX}_I \alpha$ iff $(\exists \pi \in \Pi(s))(D\pi[0..1] \in I$ and $\mathcal{M}, \pi(1) \models \alpha)$,

$-\mathcal{M}, s \models \mathbf{EG}_I \alpha$ iff $(\exists \pi \in \Pi(s))(\forall i \geq 0)(D\pi[0..i] \in I$ implies $\mathcal{M}, \pi(i) \models \beta)$,

$-\mathcal{M}, s \models \mathbf{E}(\alpha \mathbf{U}_I \beta)$ iff $(\exists \pi \in \Pi(s))(\exists i \geq 0)(D\pi[0..i] \in I$ and $\mathcal{M}, \pi(i) \models \beta$ and $(\forall j < i)\mathcal{M}, \pi(j) \models \alpha)$,

$-\mathcal{M}, s \models \overline{\mathbf{K}}_{\mathbf{c}} \alpha$ iff $(\exists \pi \in \Pi)$ $(\exists i \geq 0)(s \sim_{\mathbf{c}} \pi(i)$ and $\mathcal{M}, \pi(i) \models \alpha)$,

$-\mathcal{M}, s \models \overline{Y} \alpha$ iff $(\exists \pi \in \Pi)(\exists i \geq 0)(s \sim \pi(i)$ and $\mathcal{M}, \pi(i) \models \alpha)$, where $Y \in \{\mathbf{D}_\Gamma, \mathbf{E}_\Gamma, \mathbf{C}_\Gamma\}$ and $\sim \in \{\sim_\Gamma^D, \sim_\Gamma^E, \sim_\Gamma^C\}$.

Abstract Model. Let $\mathbb{D}_{\mathbf{c}} = \{0, \ldots, c_{\mathbf{c}} + 1\}$ with $c_{\mathbf{c}}$ be the largest constant appearing in any enabling condition or state invariants of agent $\mathbf{c}$ and $\mathbb{D} = \bigcup_{\mathbf{c} \in Ag \cup \mathcal{E}} \mathbb{D}_{\mathbf{c}}^{|\mathcal{X}_{\mathbf{c}}|}$. A tuple $\mathcal{M} = (Act, S, \iota, T, \mathcal{V}, d)$, is an $abstract\ model$, where $\iota = \prod_{\mathbf{c} \in Ag \cup \mathcal{E}} \iota_{\mathbf{c}} \times \{0\}^{|\mathcal{X}_{\mathbf{c}}|}$ is the set of all initial global states, $S = \prod_{\mathbf{c} \in Ag \cup \mathcal{E}} L_{\mathbf{c}} \times \mathbb{D}_{\mathbf{c}}^{|\mathcal{X}_{\mathbf{c}}|}$ is the set of all abstract global states. $\mathcal{V} : S \to 2^{\mathcal{PV}}$ is the valuation function such that: $p \in \mathcal{V}(s)$ iff $p \in \bigcup_{\mathbf{c} \in Ag \cup \mathcal{E}} \mathcal{V}_{\mathbf{c}}(l_{\mathbf{c}}(s))$ for all $p \in \mathcal{PV}$; and $T \subseteq S \times (Act \cup \tau) \times S$. Let $a \in Act$. Then,

1. Action transition: $(s, a, s') \in T$ iff $\forall_{\mathbf{c} \in Ag} \exists_{\phi_{\mathbf{c}} \in \mathcal{C}(\mathcal{X}_{\mathbf{c}})} \exists_{\mathcal{X}'_{\mathbf{c}} \subseteq \mathcal{X}_{\mathbf{c}}} (t_{\mathbf{c}}(l_{\mathbf{c}}(s), \phi_{\mathbf{c}}, \mathcal{X}'_{\mathbf{c}}, a) = l_{\mathbf{c}}(s')$ and $v_{\mathbf{c}} \models \phi_{\mathbf{c}} \wedge \mathcal{I}(l_{\mathbf{c}}(s))$ and $v'_{\mathbf{c}}(s') = v_{\mathbf{c}}(s)[\mathcal{X}'_{\mathbf{c}} := 0]$ and $v'_{\mathbf{c}}(s') \models \mathcal{I}(l_{\mathbf{c}}(s')))$

2. Time transition: $(s, \tau, s') \in T$ iff $\forall_{\mathbf{c} \in Ag \cup \mathcal{E}} (l_{\mathbf{c}}(s) = l_{\mathbf{c}}(s'))$ and $v_{\mathbf{c}}(s) \models \mathcal{I}(l_{\mathbf{c}}(s))$ and $succ(v_{\mathbf{c}}(s)) \models \mathcal{I}(l_{\mathbf{c}}(s)))$ and $\forall_{\mathbf{c} \in Ag}(v'_{\mathbf{c}}(s') = succ(v_{\mathbf{c}}(s')))$ and $(v'_{\mathcal{E}}(s') = succ(v_{\mathcal{E}}(s)))$.

A path π in an abstract model is a sequence $s_0 \xrightarrow{b_1} s_1 \xrightarrow{b_2} s_2 \xrightarrow{b_3} \ldots$ of transitions such that for each $i \leq 1$, $b_i \in Act \cup \{\tau\}$ and $b_1 = \tau$ and for each two consecutive transitions at least one of them is a time transition.

Given an abstract model one can define the indistinguishability relation $\sim_{\mathbf{c}} \subseteq S \times S$ for agent $\mathbf{c}$ as follows: $s \sim_{\mathbf{c}} s'$ iff $l_{\mathbf{c}}(s') = l_{\mathbf{c}}(s)$ and $v_{\mathbf{c}}(s') = v_{\mathbf{c}}(s)$.

3   SMT-Based Bounded Model Checking

In this section, we present an outline of the bounded semantics for WECTLK and define an SMT-based BMC method for WECTLK, which is based on the BMC encoding presented in [8]. As usual, we start by defining k-paths and $(k, l)-loops$. Next we define a bounded semantics, which is used for the translation to SMT.

Bounded Semantics. Let $\mathcal{M}$ be a model, and $k \in \mathbb{N}$ a bound. A k-path π_k is a finite sequence $s_0 \xrightarrow{b_1} s_1 \xrightarrow{b_2} \ldots \xrightarrow{b_k} s_k$ of transitions such that for each $1 \leq i \leq k$, $b_i \in Act \cup \{\tau\}$ and $b_1 = \tau$ and for each two consecutive transitions at least one is a time transition. A k-path π_k is a *loop* if $l < k$ and $\pi(k) = \pi(l)$. Note that if a k-path π_k is a loop, then it represents the infinite path of the form uv^ω, where $u = (s_0 \xrightarrow{b_1} s_1 \xrightarrow{b_2} \ldots \xrightarrow{b_l} s_l)$ and $v = (s_{l+1} \xrightarrow{b_{l+2}} \ldots \xrightarrow{b_k} s_k)$. $\Pi_k(s)$ denotes the set of all the k-paths of $\mathcal{M}$ that start at s, and $\Pi_k = \bigcup_{s^0 \in \iota} \Pi_k(s^0)$.

The bounded satisfiability relation $\models_k$ which indicates k-truth of a WECTLK formula in the model $\mathcal{M}$ at some state s of $\mathcal{M}$ is also defined in [8]. A WECTLK formula φ is k-*true* in the model $\mathcal{M}$ (in symbols $\mathcal{M} \models_k \varphi$) iff φ is k-true at some initial state of the model $\mathcal{M}$.

The *model checking problem* asks whether $\mathcal{M} \models \varphi$, but the *bounded model checking problem* asks whether there exists $k \in \mathbb{N}$ such that $\mathcal{M} \models_k \varphi$. The following theorem states that for a given model and a WECTLK formula there exists a bound k such that the model checking problem ($\mathcal{M} \models \varphi$) can be reduced to the bounded model checking problem ($\mathcal{M} \models_k \varphi$).

Theorem 1. *Let $\mathcal{M}$ be the abstract model and φ a WECTLK formula. Then, the following equivalence holds: $\mathcal{M} \models \varphi$ iff there exists $k \geq 0$ such that $\mathcal{M} \models_k \varphi$.*

Proof. The theorem can be proved by induction on the length of the formula φ (for details one can see [8]). $\qed$

Translation to SMT. Let $\mathcal{M}$ be an abstract model, φ a WECTLK formula, and $k \geq 0$ a bound. The presented SMT encoding of the BMC problem for WECTLK and for TRWIS is based on the SAT encoding of the same problem [10,12], and it relies on defining the quantifier-free first-order formula:

$$[\mathcal{M}, \varphi]_k := [\mathcal{M}^{\varphi, \iota}]_k \wedge [\varphi]_{\mathcal{M}, k}$$

that is satisfiable if and only if $\mathcal{M} \models_k \varphi$ holds.

Let $\mathbf{c} \in Ag \cup \{\mathcal{E}\}$. The definition of the formula $[\mathcal{M}, \varphi]_k$ assumes that

– each global state $s \in S$ is represented by a valuation of a *symbolic state* $\overline{\mathbf{w}} = ((w_1, v_1), \ldots, (w_n, v_n), (w_\mathcal{E}, v_\mathcal{E}))$ that consists of *symbolic local states* and each symbolic local state $w_\mathbf{c}$ is a pair $(w_\mathbf{c}, v_\mathbf{c})$ of individual variables ranging over the natural numbers, in which the first element represents a local state of the agent $\mathbf{c}$, and the second represents a clock valuation;

– each joint action $a \in Act$ is represented by a valuation of a *symbolic action* $\overline{a} = (a_1, \ldots, a_n, a_\mathcal{E})$ that consists of *symbolic local actions* and each symbolic local action $a_\mathbf{c}$ is an individual variable ranging over the natural numbers;

– each sequence of weights associated with the joint action is represented by a valuation of a *symbolic weights* $\overline{d} = (d_1, \ldots, d_{n+1})$ that consists of *symbolic local weights* and each symbolic local weight $d_{\mathbf{c}}$ is an individual variable ranging over the natural numbers.

The formula $[\mathcal{M}^{\varphi,\iota}]_k$ encodes a rooted tree of k−paths of the model $\mathcal{M}$. The number of branches of the tree depends on the value of $f_k : \text{WECTLK} \to \mathbb{N}$ which is the auxiliary function defined in [8]:

$-f_k(\mathbf{true}) = f_k(\mathbf{false}) = 0;$
$-f_k(p) = f_k(\neg p) = 0$, where $p \in \mathcal{PV}$;
$-f_k(\alpha \wedge \beta) = f_k(\alpha) + f_k(\beta);$
$-f_k(\alpha \vee \beta) = max\{f_k(\alpha), f_k(\beta)\};$
$-f_k(\mathbf{EX}_I \alpha) = f_k(\alpha) + 1;$
$-f_k(\mathbf{EG}_I \alpha) = (k+1) \cdot f_k(\alpha) + 1;$
$-f_k(\mathbf{E}(\alpha \mathbf{U}_I \beta)) = k \cdot f_k(\alpha) + f_k(\beta) + 1;$
$-f_k(\overline{\mathbf{C}}_\Gamma \alpha) = f_k(\alpha) + k;$
$-f_k(Y\alpha) = f_k(\alpha) + 1$ for $Y \in \{\overline{\mathbf{K}}_{\mathbf{c}}, \overline{\mathbf{D}}_\Gamma, \overline{\mathbf{E}}_\Gamma\}.$

The formula $[\mathcal{M}^{\varphi,\iota}]_k$ is defined over $(k+1) \cdot f_k(\varphi)$ different symbolic states, $k \cdot f_k(\varphi)$ different symbolic actions, and $k \cdot f_k(\varphi)$ different symbolic weights. Moreover, it uses the following auxiliary quantifier-free first-order formulae:

$-I_s(\overline{\mathbf{w}})$ - it encodes the state s of the model $\mathcal{M}$; $\mathbf{c} \in Ag \cup \mathcal{E}$;
$-H_{\mathbf{c}}(w_{\mathbf{c}}, w'_{\mathbf{c}})$ - it encodes equality of two local states, such that $w_{\mathbf{c}} = w'_{\mathbf{c}}$ for $\mathbf{c} \in Ag \cup \mathcal{E}$;
$-\mathcal{T}_{\mathbf{c}}(w_{\mathbf{c}}, ((\overline{a}, \overline{d}), \overline{\delta}), w'_{\mathbf{c}})$ - it encodes the local evolution function of agent $\mathbf{c}$;
$-\mathcal{A}(\overline{a})$ - it encodes that each symbolic local action a_c of $\overline{a}$ has to be executed by each agent in which it appears;
$-\mathcal{T}(\overline{\mathbf{w}}, ((\overline{a}, \overline{d}), \overline{\delta}), \overline{\mathbf{w}}') := \mathcal{A}(\overline{a}) \wedge \bigwedge_{\mathbf{c} \in Ag \cup \{\mathcal{E}\}} \mathcal{T}_{\mathbf{c}}(w_{\mathbf{c}}, ((\overline{a}, \overline{d}), \overline{\delta}), w'_{\mathbf{c}});$
$-$Let π_j denote the j-th *symbolic k-path*, i.e. the sequence of symbolic transitions:
$\overline{\mathbf{w}}_{0,j} \xrightarrow{(\overline{a}_{1,j}, \overline{d}_{1,j}), \delta_{1,j}} \overline{\mathbf{w}}_{1,j} \xrightarrow{(\overline{a}_{2,j}, \overline{d}_{2,j}), \delta_{2,j}} \ldots \xrightarrow{(\overline{a}_{k,j}, \overline{d}_{k,j}), \delta_{k,j}} \overline{\mathbf{w}}_{k,j}$. Then, $\mathcal{D}^I_{a,b;c,d}(\pi_n)$ for $a \leq b$ and $c \leq d$ is a formula that:
  • for $a < b$ and $c < d$ encodes that the weight represented by the sequences $\overline{d}_{a+1,n}, \ldots, \overline{d}_{b,n}$ and $\overline{d}_{c+1,n}, \ldots, \overline{d}_{d,n}$ belongs to the interval I,
  • for $a = b$ and $c < d$ encodes that the weight represented by the sequence $\overline{d}_{c+1,n}, \ldots, \overline{d}_{d,n}$ belongs to the interval I,
  • for $a < b$ and $c = d$ encodes that the weight represented by the sequence $\overline{d}_{a+1,n}, \ldots, \overline{d}_{b,n}$ belongs to the interval I,
  • for $a = b$ and $c = d$, the formula $\mathcal{D}^I_{a,b;c,d}(\pi_n)$ is true iff $0 \in I$.

Thus, given the above, we can define the formula $[\mathcal{M}^{\varphi,\iota}]_k$ as follows:

$$[\mathcal{M}^{\varphi,\iota}]_k := \bigvee_{s \in \iota} I_s(\overline{\mathbf{w}}_{0,0}) \wedge \bigvee_{j=1}^{f_k(\varphi)} \overline{\mathbf{w}}_{0,0} = \overline{\mathbf{w}}_{0,j} \wedge$$
$$\bigwedge_{j=1}^{f_k(\varphi)} \bigwedge_{i=0}^{k-1} \mathcal{T}(\overline{\mathbf{w}}_{i,j}, ((\overline{a}_{i,j}, \overline{d}_{i,j}), \overline{\delta}_{i,j}), \overline{\mathbf{w}}_{i+1,j})$$

where $\overline{w}_{i,j}$, $\overline{a}_{i,j}$, and $\overline{d}_{i,j}$ are, respectively, symbolic states, symbolic actions, and symbolic weights for $0 \leq i \leq k$ and $1 \leq j \leq f_k(\varphi)$. Hereafter, by π_j we denote the j-th symbolic k-path of the above unfolding, i.e., the sequence of transitions:

$$\overline{w}_{0,j} \xrightarrow{(\overline{a}_{1,j},\overline{d}_{1,j}),\overline{\delta}_{1,j}} \overline{w}_{1,j} \xrightarrow{(\overline{a}_{2,j},\overline{d}_{2,j}),\overline{\delta}_{2,j}} \cdots \xrightarrow{(\overline{a}_{k,j},\overline{d}_{k,j}),\overline{\delta}_{k,j}} \overline{w}_{k,j}.$$

The formula $[\varphi]_{\mathcal{M},k}$ encodes the bounded semantics of a WECTLK formula φ, and it is defined on the same sets of individual variables as the formula $[\mathcal{M}^{\varphi,\iota}]_k$. Moreover, it uses the auxiliary quantifier-free first-order formulae defined in [8].

Furthermore, following [8], our formula $[\varphi]_{\mathcal{M},k}$ uses the following auxiliary functions g_l, g_r, g_μ, h_U, h_G that were introduced in [11], and which allow to divide the set $A \subseteq F_k(\varphi) = \{j \in \mathbb{N} \mid 1 \leq j \leq f_k(\varphi)\}$ into subsets needed for translating the subformulae of φ. Let $0 \leq n \leq f_k(\varphi)$, $m \leqslant k$, and $n' = min(A)$. The rest of translation is defined in the same way as in [8].

$-[\mathbf{true}]_k^{[m,n,A]} := \mathbf{true}$, $[\mathbf{false}]_k^{[m,n,A]} := \mathbf{false}$,

$-[p]_k^{[m,n,A]} := p(\overline{w}_{m,n})$,

$-[\neg p]_k^{[m,n,A]} := \neg p(\overline{w}_{m,n})$,

$-[\alpha \wedge \beta]_k^{[m,n,A]} := [\alpha]_k^{[m,n,g_l(A,f_k(\alpha))]} \wedge [\beta]_k^{[m,n,g_r(A,f_k(\beta))]}$,

$-[\alpha \vee \beta]_k^{[m,n,A]} := [\alpha]_k^{[m,n,g_l(A,f_k(\alpha))]} \vee [\beta]_k^{[m,n,g_l(A,f_k(\beta))]}$,

$-[\mathbf{EX}_I \alpha]_k^{[m,n,A]} := \overline{w}_{m,n} = \overline{w}_{0,n'} \wedge (\overline{d}_{1,n'} \in I) \wedge [\alpha]_k^{[1,n',g_\mu(A)]}$, if $k > 0$; $\mathbf{false}$, otherwise,

$-[\mathbf{E}(\alpha \mathbf{U}_I \beta)]_k^{[m,n,A]} := \overline{w}_{m,n} = \overline{w}_{0,n'} \wedge \bigvee_{i=0}^k ([\beta]_k^{[i,n',h_U(A,k,f_k(\beta))(j)]} \wedge (\sum_{j=1}^i \overline{d}_{j,n} \in I \wedge \bigwedge_{j=0}^{i-1}[\alpha]_k^{[j,n',h_U(A,k,f_k(\beta))]})$,

$-[\mathbf{E}(\mathbf{G}_I \alpha)]_k^{[m,n,A]} := \overline{w}_{m,n} = \overline{w}_{0,n'} \wedge ((\sum_{j=1}^k \overline{d}_{j,n} \geq \mathbf{right}(I) \wedge \bigwedge_{i=0}^k (\sum_{j=1}^i \overline{d}_{j,n} \notin I \vee [\alpha]_k^{[i,n',h_G(A,k)(j)]})) \vee (\sum_{j=1}^k \overline{d}_{j,n} < \mathbf{right}(I) \wedge \bigwedge_{i=0}^k (\sum_{j=1}^i \overline{d}_{j,n} \notin I \vee [\alpha]_k^{[i,n',h_G(A,k)(j)]}) \wedge \bigvee_{l=0}^{k-1} (\overline{w}_{k,n'} = \overline{w}_{l,n'} \wedge \bigwedge_{i=l}^{k-1}(\neg \mathcal{D}_{0,k;l,i+1}^I(\pi_{n'}) \vee [\alpha]_k^{[i,n',h_G(A,k)(j)]})))),$

$-[\overline{K}_c \alpha]_k^{[m,n,A]} := (\bigvee_{s \in \iota} I_s(\overline{w}_{0,n'})) \wedge \bigvee_{j=0}^k ([\alpha]_k^{[j,n',g_\mu(A)]} \wedge H_c(\overline{w}_{m,n},\overline{w}_{j,n'}))$,

$-[\overline{D}_\Gamma \alpha]_k^{[m,n,A]} := (\bigvee_{s \in \iota} I_s(\overline{w}_{0,n'})) \wedge \bigvee_{j=0}^k ([\alpha]_k^{[j,n',g_\mu(A)]} \wedge \bigwedge_{c \in \Gamma} H_c(\overline{w}_{m,n},\overline{w}_{j,n'}))$,

$-[\overline{E}_\Gamma \alpha]_k^{[m,n,A]} := (\bigvee_{s \in \iota} I_s(\overline{w}_{0,n'})) \wedge \bigvee_{j=0}^k ([\alpha]_k^{[j,n',g_\mu(A)]} \wedge \bigvee_{c \in \Gamma} H_c(\overline{w}_{m,n},\overline{w}_{j,n'}))$,

$-[\overline{C}_\Gamma \alpha]_k^{[m,n,A]} := [\bigvee_{j=1}^k (\overline{E}_\Gamma)^j \alpha]_k^{[m,n,A]}$.

The theorem below states the correctness and the completeness of the presented translation. It can be proved in a standard way by induction on the complexity of the given WECTLK formula.

Theorem 2. *Let $\mathcal{M}$ be a model, and φ a WECTLK formula. For every $k \in \mathbb{N}$, $\mathcal{M} \models_k \varphi$ if, and only if, the quantifier-free first-order formula $[\mathcal{M},\varphi]_k$ is satisfiable.*

4   Experimental Results

In this section we experimentally evaluate the performance of our SMT-based BMC encoding for WECTLK over the TRWIS semantics.

The benchmark we consider is the *timed weighted generic pipeline paradigm* (TWGPP) TRWIS model [10]. The model of TWGPP involves $n + 2$ agents:

– Producer producing data within certain time interval ($[a, b]$) or being inactive,
– Consumer receiving data within certain time interval ($[c, d]$) or being inactive within certain time interval ($[g, h]$),
– a chain of n intermediate Nodes which can be ready for receiving data within certain time interval ($[c, d]$), processing data within certain time interval ($[e, f]$) or sending data.

The weights are used to adjust the cost properties of Producer, Consumer, and of the intermediate Nodes.

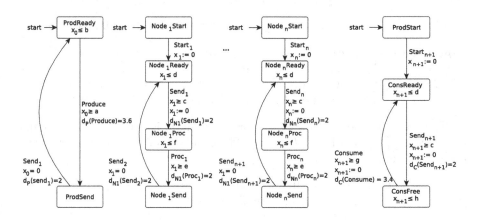

Fig. 1. The TWGPP system

Each agent of the scenario can be modelled by considering its local states, local actions, local protocol, local evolution function, local weight function, the local clocks, the clock constraints, invariants, and local valuation function. Fig. 1 shows the local states, the possible actions, and the protocol, the clock constraints, invariants and weights for each agent. Null actions are omitted in the figure.

Given Fig. 1, the local evolution functions of TWGPP are straightforward to infer. Moreover, we assume the following set of propositional variables: $\mathcal{PV} = \{ProdReady, ProdSend, ConsReady, ConsFree\}$ with the following definitions of local valuation functions:

$-\mathcal{V}_P(ProdReady\text{-}0) = \{ProdReady\},\ \mathcal{V}_P(ProdSend\text{-}1) = \{ProdSend\},$
$-\mathcal{V}_C(ConsReady\text{-}0) = \{ConsReady\},\ \mathcal{V}_C(ConsFree\text{-}1) = \{ConsFree\}.$

Let $Act = Act_P \times \prod_{i=1}^{n} Act_{N_i} \times Act_C$, with $Act_P = \{Produce, Send_1\}$, $Act_C = \{Start_{n+1}, Consume, Send_{n+1}\}$, ans $Act_{N_i} = \{Start_i, Send_i, Send_{i+1}, Proc_i\}$ defines the set of joint actions for the scenario. For $\hat{a} \in Act$ let $act_P(\hat{a})$ denotes an action of Producer, $act_C(\hat{a})$ denotes an action of Consumer, and $act_{N_i}(\hat{a})$ denotes an action of Node i. We assume the following local evolution functions:

$-t_P(ProdReady, x_0 \geq a, \emptyset, \hat{a}) = ProdSend$, if $act_P(\hat{a}) = Produce$
$-t_P(ProdSend, true, \{x_0\}, \hat{a}) = ProdReady$, if $act_P(\hat{a}) = Send_1$ and $act_{N_i}(\hat{a}) = Send_1$
$-t_C(ConsStart, true, \{x_{n+1}\}, \hat{a}) = ConsReady$, if $act_C(\hat{a}) = Start_{n+1}$
$-t_C(ConsReady, x_{n+1} \geq c, \{x_{n+1}\}, \hat{a}) = ConsFree$, if $act_C(\hat{a}) = Send_{n+1}$ and $act_{N_n}(\hat{a}) = Send_{n+1}$
$-t_C(ConsFree, x_{n+1} \geq g, \{x_{n+1}\}, \hat{a}) = ConsReady$, if $act_C(\hat{a}) = Consume$

Finally, we assume the following two local weight functions for each agent:

$-d_P(Produce) = 4,\ d_P(send_1) = 2,\ d_C(Consume) = 4,\ d_C(send_{n+1}) = 2,$
$d_{N_i}(send_i) = d_{Ni}(send_{i+1}) = d_{N_i}(Proc_i) = 2.$
$-d_P(Produce) = 4000,\ d_P(send_1) = 2000,\ d_C(Consume) = 4000,$
$d_C(send_{n+1}) = 2000,\ d_{N_i}(send_i) = d_{Ni}(send_{i+1}) = d_{N_i}(Proc_i) = 2000.$

The set of all the global states S for the scenario is defined as the product $(L_P \times \mathbb{N}) \times \prod_{i=1}^{n}(L_i \times \mathbb{N}) \times (L_C \times \mathbb{N})$. The set of the initial states is defined as $\iota = \{s^0\}$, where

$$s^0 = ((ProdReady\text{-}0, 0), (Node_1 Ready\text{-}0, 0), \ldots, (Node_n Ready\text{-}0, 0), (ConsReady\text{-}0, 0)).$$

The system is scaled according to the number of its Nodes (agents), i.e., the problem parameter n is the number of Nodes. For any natural number $n \geq 0$, let $D(n) = \{1, 3, \ldots, n-1, n+1\}$ for an even n, and $D(n) = \{2, 4, \ldots, n-1, n+1\}$ for an odd n. Moreover, let

$$r(j) = d_P(Produce) + 2 \cdot \sum_{i=1}^{j} d_{N_i}(Send_i) + \sum_{i=1}^{j-1} \cdot d_{N_i}(proc_i)$$

Then we define $Right$ as follows:

$$Right = \sum_{j \in D(n)} r(j).$$

We consider the following formulae as specifications:

$\varphi_1 = \mathbf{EF}_{[0,Right)}(ConsFree)$ - it states that there exists a path on which Consumer receives a data and the cost of receiving the data will be less than Right.

$\varphi_2 = \mathbf{EF}_{[0,Right)}(ConsFree \wedge \mathbf{EG}(ProdSend \vee ConsFree))$ - it states that there exists a path on which Consumer receives a data and the cost of receiving the data is less than Right and from that point there exists a path on which always either the Producer has sent a data or the Consumer has received a data.

$\varphi_3 = \overline{K}_P(\mathbf{EF}_{[0,Right)}(ConsFree \wedge \mathbf{EG}(ProdSend \vee ConsFree)))$ - it states that it is not true that Producer knows that there exists a path on which Consumer receives a data and the cost of receiving the data is less than Right and from that point there exists a path on which always either the Producer has sent a data or the Consumer has received a data.

$\varphi_4 = \overline{K}_P(\mathbf{EF}_{[0,Right)}(ConsFree \wedge \overline{K}_C\overline{K}_P(\mathbf{EG}(ProdSend \vee ConsFree))))$ - it states that it is not true that Producer knows that there exists a path on which Consumer receives a data and the cost of receiving the data is less than Right and at that point it is not true that Consumer knows that it is not true that Producer knows that there exists a path on which always either the Producer has sent a data or Consumer has received a data.

The number of the considered k-paths is equal to 1 for φ_1, 2 for φ_2, 3 for φ_3, and 5 for φ_4, respectively. The length of the witness is $(n+1) \cdot 4$ for the formula φ_1, 9 if $n = 1$, and $(n+1) \cdot 4$ if $n > 1$ for the formula φ_2, $2 \cdot n + 4$ if $n \in \{1, 2\}$ and, $2 \cdot n + 2$ if $n > 2$ for the formula φ_3, $2 \cdot n + 2$ for the formula φ_4, respectively.

Performance Evaluation. We have performed our experimental results on a computer equipped with I7-3770 processor, 32 GB of RAM, and the operating system Arch Linux with the kernel 3.19.2. We set the CPU time limit to 3600 seconds. Our SMT-based BMC algorithm is implemented as standalone program written in the programming language C++. We used the state of the art SMT-solver Z3 [6] (http://z3.codeplex.com/).

For properties φ_1, φ_2, φ_4, and φ_4 we have scaled up both the number of nodes and the weights parameters. The results are summarised on charts in Fig. 2, Fig. 3, Fig. 4, and Fig. 5. One can observed that our SMT-based BMC is not sensitive (Fig. 2, Fig. 4, Fig. 5) to scaling up the weights, but it is sensitive to scaling up the size of benchmark. More precisely, in order to calculate results for φ_1 and for TWGPP with 1 node and the basic weights (bwfor short), the bwmultiplied by 1,000 our method uses 13.4 MB and the test lasts less than 0.1 seconds. In order to calculate results for φ_1 and for TWGPP with 23 nodes and the bwmultiplied by 1,000 our method uses 236.1 MB and the test lasts 4510.0 seconds. The most interesting result which can be observed is for the formulae φ_2. In this case time usage for the bwis greater (9013.1 seconds) than for the bwmultiplied by 1,000 (570.9 seconds) for 11 nodes. In particular, in the time limit set for the benchmark, the SMT-based BMC is able to verify the formula φ_2 for the bwonly for 11 nodes while for the bwmultiplied by 1,000 can handle 15 nodes.

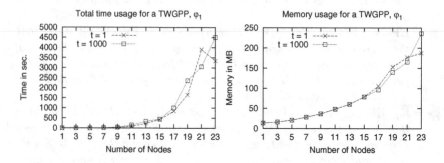

Fig. 2. Formula φ_1: Scaling up both the number of nodes and weights.

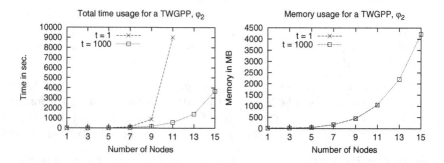

Fig. 3. Formula φ_2: Scaling up both the number of nodes and weights.

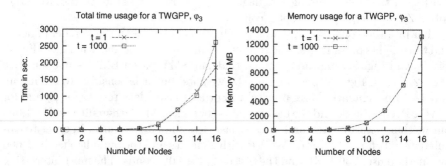

Fig. 4. Formula φ_3: Scaling up both the number of nodes and weights.

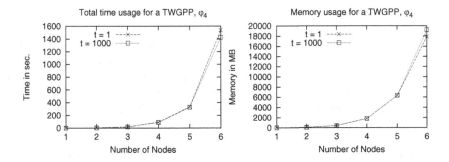

Fig. 5. Formula φ_4: Scaling up both the number of nodes and weights.

In the case of properties φ_3 and φ_4 we obtained similar results. Namely, in order to calculate results for φ_3 and for TWGPP with 16 nodes and the bw, the bwmultiplied by 1,000, our method uses 13074.2 (13072.0) MB and the test lasts 1864.4 (2624.5) sec. Next, in order to calculate results for φ_4 and for TWGPP with 6 nodes and the bw, the bwmultiplied by 1,000, our method uses 17904.5 (19240.9) MB and the test lasts 1536.9 (1424.4) sec.

In Tables 1 and 2 we present time usage, memory usage and the length of the witness for the formula φ_2.

Table 1. Formula φ_2 for basic weights

Table 2. Formula φ_2 for basic weights multiplied by 1,000

n	time	memory	witness length
1	0.1	15.1	9
2	0.5	17.6	12
3	1.6	21.4	16
4	4.2	26.5	20
5	10.2	43.7	24
6	21.0	84.7	28
7	112.8	176.7	32
8	275.8	292.4	36
9	879.6	451.3	40
10	2284.4	707.9	44
11	9013.1	1056.5	48

n	time	memory	witness length
1	0.2	14.7	8
2	0.8	17.5	12
3	1.8	21.6	16
4	5.7	27.1	20
5	12.5	43.3	24
6	25.6	86.5	28
7	47.6	177.1	32
8	105.1	290.8	36
9	172.5	452.1	40
10	290.2	705.3	44
11	570.9	1060.0	48
12	879.9	1557.1	52
13	1399.8	2198.9	56
14	2892.7	3078.3	60
15	3660.4	4222.0	64

5   Conclusions

We have proposed SMT-based BMC verification method for model checking WECTLK properties interpreted over the timed real-weighted interpreted systems. We have provided a preliminary experimental results showing that our method is worth interest. In the future we are going to provide a comparison of our new method with the SAT- and BDD-based BMC methods. The module will be added to the model checker VerICS([3]).

Acknowledgments. Partly supported by National Science Centre under the grant No. 2014/15/N/ST6/05079. The study is co-funded by the European Union, European Social Fund. Project PO KL "Information technologies: Research and their interdisciplinary applications", Agreement UDA-POKL.04.01.01-00-051/10-00.

References

1. Emerson, E.A.: Temporal and modal logic. In: van Leeuwen, J. (eds.) Handbook of Theoretical Computer Science, vol. B, chapter 16, pp. 996–1071. Elsevier Science Publishers (1999)
2. Fagin, R., Halpern, J.Y., Moses, Y., Vardi, M.Y.: Reasoning about Knowledge. MIT Press, Cambridge (1995)
3. Kacprzak, M., Nabialek, W., Niewiadomski, A., Penczek, W., Pólrola, A., Szreter, M., Woźna, B., Zbrzezny, A.: VerICS 2007 - a model checker for knowledge and real-time. Fundamenta Informaticae **85**(1–4), 313–328 (2008)
4. Levesque, H.: A logic of implicit and explicit belief. In: Proceedings of the 6th National Conference of the AAAI, pp. 198–202. Morgan Kaufman, Palo Alto (1984)
5. Lomuscio, A., Sergot, M.: Deontic interpreted systems. Studia Logica **75**(1), 63–92 (2003)
6. de Moura, L., Bjørner, N.S.: Z3: An efficient SMT solver. In: Ramakrishnan, C.R., Rehof, J. (eds.) TACAS 2008. LNCS, vol. 4963, pp. 337–340. Springer, Heidelberg (2008)
7. Wooldridge, M.: An introduction to multi-agent systems, 2nd edn. John Wiley & Sons (2009)
8. Woźna-Szcześniak, B.: SAT-Based bounded model checking for weighted deontic interpreted systems. In: Reis, L.P., Correia, L., Cascalho, J. (eds.) EPIA 2013. LNCS, vol. 8154, pp. 444–455. Springer, Heidelberg (2013)
9. Woźna-Szcześniak, B.: Checking EMTLK properties of timed interpreted systems via bounded model checking. In: Bazzan, A.L.C., Huhns, M.N., Lomuscio, A., Scerri, P. (eds.) International Conference on Autonomous Agents and Multi-Agent Systems, AAMAS 2014, Paris, France, May 5–9, pp. 1477–1478. IFAAMAS/ACM (2014)
10. Woźna-Szcześniak, B., Zbrzezny, A.M., Zbrzezny, A.: SAT-Based bounded model checking for weighted interpreted systems and weighted linear temporal logic. In: Boella, G., Elkind, E., Savarimuthu, B.T.R., Dignum, F., Purvis, M.K. (eds.) PRIMA 2013. LNCS, vol. 8291, pp. 355–371. Springer, Heidelberg (2013)
11. Zbrzezny, A.: Improving the translation from ECTL to SAT. Fundamenta Informaticae **85**(1–4), 513–531 (2008)
12. Zbrzezny, A.: A new translation from ECTL* to SAT. Fundamenta Informaticae **120**(3–4), 377–397 (2012)

SMT-Based Bounded Model Checking
for Weighted Epistemic ECTL

Agnieszka M. Zbrzezny$^{(\boxtimes)}$, Bożena Woźna-Szcześniak, and Andrzej Zbrzezny

IMCS, Jan Długosz University,
Al. Armii Krajowej 13/15, 42-200 Częstochowa, Poland
{agnieszka.zbrzezny,b.wozna,a.zbrzezny}@ajd.czest.pl

Abstract. We define the SMT-based bounded model checking (BMC) method for weighted interpreted systems and for the existential fragment of the weighted epistemic computation tree logic. We implemented the new BMC algorithm and compared it with the SAT-based BMC method for the same systems and the same property language on several benchmarks for multi-agent systems.

1 Introduction

The previous ten years in the area of multi-agent systems (MASs) have seen significant research in verification procedures, which automatically evaluate whether a MAS reaches its intended specifications. One of main techniques here is *symbolic model checking* [2]. Unfortunately, because of the agents' intricate nature, the practical applicability of model checking is firmly limited by the "state-space explosion problem" (i.e., an exponential growth of the system state space with the number of agents). To reduce this issue, various techniques, including the SAT- and BDD-based bounded model checking (BMC) [3,4], have been proposed. These have been effective in permitting users to handle bigger MASs, however it is still hard to check MASs with numerous agents and cost demands on agents' actions. The point of this paper is to help beat this inadequacy by employing SMT-solvers (i.e., *satisfiability modulo theories* tools for deciding the satisfiability of formulae in a number of theories) [1,5].

The fundamental thought behind bounded model checking (BMC) is, given a system, a property, and an integer bound $k \geq 0$, to define a formula (in the case of SAT-based BMC, this is a propositional logic formula; in the case of SMT-based BMC, this can be a quantifier-free first-order formula) such that the formula is satisfiable if and only if the system has a counterexample of length at most k violating the property. The bound is incremented until a satisfiable formula is discovered (i.e, the specification does not hold for the system) or a completeness threshold is reached without discovering any satisfiable formulae.

To model check the prerequisites of MASs, different extensions of temporal logics have been proposed. In this paper, we consider the existential fragment of

---

Partly supported by National Science Centre under the grant No. 2014/15/N/ST6/05079.

© Springer International Publishing Switzerland 2015
F. Pereira et al. (Eds.) EPIA 2015, LNAI 9273, pp. 651–657, 2015.
DOI: 10.1007/978-3-319-23485-4_65

a weighted epistemic computation tree logic (WECTLK) interpreted over WISs [6].

To the best of our knowledge, there is no work that considers SMT-based BMC methods to check multi-agent systems modelled by means of interpreted systems. Thus, in this paper we offer such a method. In particular, we make the following contributions. First of all, we define and implement an SMT-based BMC method for WECTLK and for *weighted interpreted systems* (WISs) [6,7]. Next, we report on the initial experimental evaluation of our SMT-based BMC methods. To this aim we use two scalable benchmarks: the *weighted generic pipeline paradigm* [7] and the *weighted bits transmission problem* [6]. Finally, we compare our prototype implementation of the SMT-based BMC method against the SAT-based BMC engine of [6,8], the only existing technique that is suitable with respect to the input formalism and checked properties. The results show that the SMT-based BMC performs very good and is, in fact, sometimes significantly faster than the tested SAT-based BMC method.

The rest of the paper is organised as follows. In the next section we briefly present the theory of weighted interpreted systems and the WECTLK language. In Section 3 we present our SMT-based BMC method. In Section 4 we experimentally evaluate the performance of our SMT-based BMC encoding. In Section 5, we conclude the paper.

2 Preliminaries

WIS. Let $Ag = \{1, \ldots, n\}$ be the non-empty and finite set of agents, and $\mathcal{E}$ be a special agent that is used to model the environment in which the agents operate, and let $\mathcal{PV} = \bigcup_{\mathbf{c} \in Ag \cup \{\mathcal{E}\}} \mathcal{PV}_{\mathbf{c}}$ be a set of propositional variables such that $\mathcal{PV}_{\mathbf{c}_1} \cap \mathcal{PV}_{\mathbf{c}_2} = \emptyset$ for all $\mathbf{c}_1, \mathbf{c}_2 \in Ag \cup \{\mathcal{E}\}$. The *weighted interpreted system* (WIS) [6,7] is a tuple $(\{L_{\mathbf{c}}, Act_{\mathbf{c}}, P_{\mathbf{c}}, \mathcal{V}_{\mathbf{c}}, d_{\mathbf{c}}\}_{\mathbf{c} \in Ag \cup \{\mathcal{E}\}}, \{t_{\mathbf{c}}\}_{\mathbf{c} \in Ag}, t_{\mathcal{E}}, \iota)$, where $L_{\mathbf{c}}$ is non-empty and finite set of *local states* ($S = L_1 \times \ldots \times L_n \times L_{\mathcal{E}}$ denotes the non-empty set of all *global states*), $Act_{\mathbf{c}}$ is a non-empty and finite set of *possible actions* ($Act = Act_1 \times \ldots \times Act_n \times Act_{\mathcal{E}}$ denotes the non-empty set of *joint actions*), $P_{\mathbf{c}} : L_{\mathbf{c}} \to 2^{Act_{\mathbf{c}}}$ is a *protocol function*, $\mathcal{V}_{\mathbf{c}} : L_{\mathbf{c}} \to 2^{\mathcal{PV}_{\mathbf{c}}}$ is a *valuation function*, $d_{\mathbf{c}} : Act_{\mathbf{c}} \to \mathbb{N}$ is a *weight function*, $t_{\mathbf{c}} : L_{\mathbf{c}} \times L_{\mathcal{E}} \times Act \to L_{\mathbf{c}}$ is a (partial) *evolution function* for agents, and $t_{\mathcal{E}} : L_{\mathcal{E}} \times Act \to L_{\mathcal{E}}$ is (partial) *evolution function* for the environment, and $\iota \subseteq S$ is a set of initial global states.

Assume that $l_{\mathbf{c}}(s)$ denotes the local component of agent $\mathbf{c} \in Ag \cup \{\mathcal{E}\}$ in the global state $s \in S$. For a given WIS we define a *model* as a tuple $M = (Act, S, \iota, T, \mathcal{V}, d)$, where the sets Act and S are defined as above, $\mathcal{V} : S \to 2^{\mathcal{PV}}$ is the valuation function defined as $\mathcal{V}(s) = \bigcup_{\mathbf{c} \in Ag \cup \{\mathcal{E}\}} \mathcal{V}_{\mathbf{c}}(l_{\mathbf{c}}(s))$, $d : Act \to \mathbb{N}$ is a "joint" weight function defined as $d((a_1, \ldots, a_n, a_{\mathcal{E}})) = d_1(a_1) + \ldots + d_n(a_n) + d_{\mathcal{E}}(a_{\mathcal{E}})$, and $T \subseteq S \times Act \times S$ is a transition relation defined as: $(s, a, s') \in T$ (or $s \xrightarrow{a} s'$) iff $t_{\mathbf{c}}(l_{\mathbf{c}}(s), l_{\mathcal{E}}(s), a) = l_{\mathbf{c}}(s')$ for all $\mathbf{c} \in Ag$ and $t_{\mathcal{E}}(l_{\mathcal{E}}(s), a) = l_{\mathcal{E}}(s')$; we assume that the relation T is total, i.e. for any $s \in S$ there exists $s' \in S$ and a non-empty joint action $a \in Act$ such that $s \xrightarrow{a} s'$.

For each agent $\mathbf{c} \in Ag$, the definition of the standard *indistinguishability* relation $\sim_{\mathbf{c}} \subseteq S \times S$ is the following: $s \sim_{\mathbf{c}} s'$ iff $l_{\mathbf{c}}(s') = l_{\mathbf{c}}(s)$. Finally, the following definitions of epistemic relations: $\sim_{\Gamma}^{E} \overset{def}{=} \bigcup_{\mathbf{c} \in \Gamma} \sim_{\mathbf{c}}$, $\sim_{\Gamma}^{C} \overset{def}{=} (\sim_{\Gamma}^{E})^{+}$ (the transitive closure of $\sim_{\Gamma}^{E}$), $\sim_{\Gamma}^{D} \overset{def}{=} \bigcap_{\mathbf{c} \in \Gamma} \sim_{\mathbf{c}}$, where $\Gamma \subseteq Ag$ are assumed.

Syntax of WECTLK. The WECTLK logic has been defined in [6] as the existential fragment of the weighted CTLK with cost constraints on *all* temporal modalities.

For convenience, the symbol I denotes an interval in $\mathbb{N} = \{0, 1, 2, \ldots\}$ of the form $[a, \infty)$ or $[a, b)$, for $a, b \in \mathbb{N}$ and $a \neq b$. Moreover, the symbol $\mathbf{right}(I)$ denotes the right end of the interval I. Given an atomic proposition $p \in \mathcal{PV}$, an agent $\mathbf{c} \in Ag$, a set of agents $\Gamma \subseteq Ag$ and an interval I, the WECTLK formulae are defined by the following grammar: $\varphi ::= \mathbf{true} \mid \mathbf{false} \mid p \mid \neg p \mid \varphi \vee \varphi \mid \varphi \wedge \varphi \mid \mathrm{EX}_{I}\varphi \mid \mathrm{E}(\varphi \mathrm{U}_{I}\varphi) \mid \mathrm{EG}_{I}\varphi \mid \overline{\mathrm{K}}_{\mathbf{c}}\varphi \mid \overline{\mathrm{D}}_{\Gamma}\varphi \mid \overline{\mathrm{E}}_{\Gamma}\varphi \mid \overline{\mathrm{C}}_{\Gamma}\varphi$.

E (for some path) is the path quantifier. X_{I} (weighted neXt time), U_{I} (weighted until) and G_{I} (weighted always) are the weighted temporal modalities. Note that the formula "weighted eventually" is defined as standard: $\mathrm{EF}_{I}\varphi \overset{def}{=} \mathrm{E}(\mathbf{true}\mathrm{U}_{I}\varphi)$ (meaning that it is possible to reach a state satisfying φ via a finite path whose cumulative weight is in I). $\overline{\mathrm{K}}_{\mathbf{c}}$ is the modality dual to $\mathrm{K}_{\mathbf{c}}$. $\overline{\mathrm{D}}_{\Gamma}$, $\overline{\mathrm{E}}_{\Gamma}$, and $\overline{\mathrm{C}}_{\Gamma}$ are the dualities to the standard group epistemic modalities representing, respectively, distributed knowledge in the group Γ, everyone in Γ knows, and common knowledge among agents in Γ.

We omit here the definition of the bounded (i.e., the relation $\models_{k}$) and unbounded semantics (i.e., the relation $\models$) of the logic, since they can be found in [6]. We only recall the notions of k-paths and loops, since we need them to explain the SMT-based BMC. Namely, given a model M and a bound $k \in \mathbb{N}$, a *k-path* π_{k} is a finite sequence $s_{0} \xrightarrow{a_{1}} s_{1} \xrightarrow{a_{2}} \ldots \xrightarrow{a_{k}} s_{k}$ of transitions. A k-path π_{k} is a *loop* if $l < k$ and $\pi(k) = \pi(l)$. Furthermore, let M be a model, and φ a WECTLK formula. The *bounded model checking* problem asks whether there exists $k \in \mathbb{N}$ such that $M \models_{k} \varphi$, i.e., whether there exists $k \in \mathbb{N}$ such that the formula φ is k-true in the model M.

3   SMT-Based BMC

In order to encode the BMC problem for WECTLK by means of SMT, we consider a quantifier-free logic with individual variables ranging over the natural numbers. Formally, let M be the model, φ a WECTLK formula and $k \geq 0$ a bound. We define the quantifier-free first-order formula: $[M, \varphi]_{k} := [M^{\varphi, \iota}]_{k} \wedge [\varphi]_{M,k}$ that is satisfiable if and only if $M \models_{k} \varphi$ holds.

The definition of the formula $[M, \varphi]_{k}$ is based the SAT encoding of [6], and it assumes that each state, each joint action, and each sequence of weights associated with a joint action are represented by a valuation of, respectively, a *symbolic state* $\overline{w} = (w_{1}, \ldots, w_{n}, w_{\mathcal{E}})$ consisting of *symbolic local states* $w_{\mathbf{c}}$, a *symbolic action* $\overline{a} = (a_{1}, \ldots, a_{n}, a_{\mathcal{E}})$ consisting of *symbolic local actions* $a_{\mathbf{c}}$, and a *symbolic weights* $\overline{d} = (d_{1}, \ldots, d_{n+1})$ consisting of *symbolic local weights*

$d_{\mathbf{c}}$, where each $w_{\mathbf{c}}$, $a_{\mathbf{c}}$, and $d_{\mathbf{c}}$ are individual variables ranging over the natural numbers, for $\mathbf{c} \in Ag \cup \{\mathcal{E}\}$. Next, the definition of $[M, \varphi]_k$ uses the auxiliary function $f_k :$ WECTLK $\to \mathbb{N}$ of [6] which returns the number of k-paths that are required for proving the k-truth of φ in M. Finally, the definition of $[M, \varphi]_k$ uses the following auxiliary quantifier-free first-order formulae: $I_s(\overline{w})$ - it encodes the state s of the model M; $p(\overline{w})$ - it encodes the set of states of M in which $p \in \mathcal{PV}$ holds; $H_{\mathbf{c}}(\overline{w}, \overline{w}') := w_{\mathbf{c}} = w'_{\mathbf{c}}$ for $\mathbf{c} \in Ag$; $T_{\mathbf{c}}(w_{\mathbf{c}}, (\overline{a}, \overline{d}), w'_{\mathbf{c}})$ - it encodes the local evolution function of agent $\mathbf{c} \in Ag \cup \{\mathcal{E}\}$; $A(\overline{a})$ - it encodes that each symbolic local action $a_{\mathbf{c}}$ of $\overline{a}$ has to be executed by each agent in which it appears; $T(\overline{w}, (\overline{a}, \overline{d}), \overline{w}') := A(\overline{a}) \wedge \bigwedge_{\mathbf{c} \in Ag \cup \{\mathcal{E}\}} T_{\mathbf{c}}(w_{\mathbf{c}}, (\overline{a}, \overline{d}), w'_{\mathbf{c}})$. Let π_j denote the j-th *symbolic k-path*, i.e. the sequence of symbolic transitions: $\overline{w}_{0,j} \xrightarrow{\overline{a}_{1,j}, \overline{d}_{1,j}} \overline{w}_{1,j} \xrightarrow{\overline{a}_{2,j}, \overline{d}_{2,j}} \ldots \xrightarrow{\overline{a}_{k,j}, \overline{d}_{k,j}} \overline{w}_{k,j}$, and let $d_{i,j,m}$ denotes the m-th component of the symbolic joint weight $\overline{d}_{i,j}$. Then,

- $\mathcal{B}_k^I(\pi_j) := \sum_{i=1}^{k} \sum_{m=1}^{n+1} d_{i,j,m} < \mathbf{right}(I)$ - it encodes that the weight represented by the sequence $\overline{d}_{1,j}, \ldots, \overline{d}_{k,j}$ is less than $\mathbf{right}(I)$;
- $\mathcal{D}_{a,b}^I(\pi_j)$ for $a \leq b$ - if $a < b$, then it encodes that the weight represented by the sequence $\overline{d}_{a+1,j}, \ldots, \overline{d}_{b,j}$ belongs to the interval I, otherwise, i.e. if $a = b$, then $\mathcal{D}_{a,b}^I(\pi_j)$ is true iff $0 \in I$;
- $\mathcal{D}_{a,b;c,d}^I(\pi_j)$ for $a \leq b$ and $c \leq d$ - it encodes that the weight represented by the sequences $\overline{d}_{a+1,j}, \ldots, \overline{d}_{b,j}$ and $\overline{d}_{c+1,j}, \ldots, \overline{d}_{d,j}$ belongs to the interval I.

Given symbolic states $\overline{w}_{i,j}$, symbolic actions $\overline{a}_{i,j}$ and symbolic weights $\overline{d}_{i,j}$ for $0 \leq i \leq k$ and $0 \leq j < f_k(\varphi)$, the formula $[M^{\varphi, \iota}]_k$, which encodes a rooted tree of k-paths of the model M, is defined as follows:

$$[M^{\varphi, \iota}]_k := \bigvee_{s \in \iota} I_s(\overline{w}_{0,0}) \wedge \bigwedge_{j=0}^{f_k(\varphi)-1} \bigwedge_{i=0}^{k-1} T(\overline{w}_{i,j}, (\overline{a}_{i+1,j}, \overline{d}_{i+1,j}), \overline{w}_{i+1,j})$$

The formula $[\varphi]_{M,k}$ encodes the bounded semantics of the WECTLK formula φ, it is defined on the same sets of individual variables as the formula $[M^{\varphi, \iota}]_k$, and it uses the auxiliary functions g_μ, h_k^U, h_k^G of [9] that allow us to divide the set $A \subseteq F_k(\varphi) = \{j \in \mathbb{N} \mid 0 \leq j < f_k(\varphi)\}$ into subsets necessary for translating the subformulae of φ.

Let $[\varphi]_k^{[m,n,A]}$ denote the translation of φ at symbolic state $\overline{w}_{m,n}$ by using the set $A \subseteq F_k(\varphi)$. The formula $[\varphi]_{M,k} := [\varphi]_k^{[0,0,F_k(\varphi)]}$ is defined inductively with the classical rules for the propositional fragment of WECTLK and with the following rules for weighted temporal and epistemic modalities. Let $0 \leq n \leq f_k(\varphi)$, $m \leqslant k$, $n' = min(A)$, $h_k^U = h_k^U(A, f_k(\beta))$, and $h_k^G = h_k^G(A)$. Then,

- $[\mathbf{EX}_I \alpha]_k^{[m,n,A]} := \overline{w}_{m,n} = \overline{w}_{0,n'} \wedge \mathcal{D}_{0,1}^I(\pi_{n'}) \wedge [\alpha]_k^{[1,n',g_\mu(A)]}$, if $k > 0$; $\mathbf{false}$, else,
- $[\mathbf{E}(\alpha \mathbf{U}_I \beta)]_k^{[m,n,A]} := \overline{w}_{m,n} = \overline{w}_{0,n'} \wedge \bigvee_{i=0}^{k} ([\beta]_k^{[i,n',h_k^U(k)]} \wedge \mathcal{D}_{0,i}^I(\pi_{n'}) \wedge \bigwedge_{j=0}^{i-1} [\alpha]_k^{[j,n',h_k^U(j)]})$,

$-[\mathrm{E}(\mathrm{G}_I\alpha)]_k^{[m,n,A]} := \overline{w}_{m,n} = \overline{w}_{0,n'} \wedge \big((\neg\mathcal{B}_k^I(\pi_n) \wedge \bigwedge_{i=0}^k(\neg\mathcal{D}_{0,i}^I(\pi_{n'})\vee$
$[\alpha]_k^{[i,n',h_k^G(k)]})) \vee (\mathcal{B}_k^I(\pi_n) \wedge \bigwedge_{i=0}^k(\neg\mathcal{D}_{0,i}^I(\pi_{n'}) \vee [\alpha]_k^{[i,n',h_k^G(k)]})\wedge$
$\bigvee_{l=0}^{k-1}(\overline{w}_{k,n'} = \overline{w}_{l,n'} \wedge \bigwedge_{i=l}^{k-1}(\neg\mathcal{D}_{0,k;l,i+1}^I(\pi_{n'}) \vee [\alpha]_k^{[i,n',h_k^G(k)]})))),$

$-[\overline{\mathrm{K}}_c\alpha]_k^{[m,n,A]} := (\bigvee_{s\in\iota} I_s(\overline{w}_{0,n'})) \wedge \bigvee_{j=0}^k([\alpha]_k^{[j,n',g_\mu(A)]} \wedge H_c(\overline{w}_{m,n},\overline{w}_{j,n'})),$

$-[\overline{\mathrm{D}}_\Gamma\alpha]_k^{[m,n,A]} := (\bigvee_{s\in\iota} I_s(\overline{w}_{0,n'})) \wedge \bigvee_{j=0}^k([\alpha]_k^{[j,n',g_\mu(A)]} \wedge \bigwedge_{c\in\Gamma} H_c(\overline{w}_{m,n},\overline{w}_{j,n'})),$

$-[\overline{\mathrm{E}}_\Gamma\alpha]_k^{[m,n,A]} := (\bigvee_{s\in\iota} I_s(\overline{w}_{0,n'})) \wedge \bigvee_{j=0}^k([\alpha]_k^{[j,n',g_\mu(A)]} \wedge \bigvee_{c\in\Gamma} H_c(\overline{w}_{m,n},\overline{w}_{j,n'})),$

$-[\overline{\mathrm{C}}_\Gamma\alpha]_k^{[m,n,A]} := [\bigvee_{j=1}^k(\overline{\mathrm{E}}_\Gamma)^j\alpha]_k^{[m,n,A]}.$

The theorem below states the correctness and the completeness of the presented translation. It can be proven by induction on the length of the given WECTLK formula.

Theorem 1. *Let M be a model, and φ a WECTLK formula. For every $k \in \mathbb{N}$, $M \models_k \varphi$ if, and only if, the quantifier-free first-order formula $[M,\varphi]_k$ is satisfiable.*

The proposed SMT-based BMC is based on the SAT-based BMC defined in [6]. The main difference between those two method is in the representation of symbolic states, symbolic actions, and symbolic weights. Thus, the main result is the generalisation of the propositional encoding of [6] into the quantifier-free first-order encoding.

4 Experimental Results

Here we experimentally evaluate the performance of our SMT-based BMC method for WECTLK over the WIS semantics. We compare our method with the SAT-based BMC [6,8], the only existing method that is suitable with respect to the input formalism (i.e., weighted interpreted systems) and checked properties (i.e., WECTLK). We have computed our experimental results on a computer equipped with I7-3770 processor, 32 GB of RAM, and the operating system Arch Linux with the kernel 3.15.3. We set the CPU time limit to 3600 seconds. For the SAT-based BMC we used the PicoSAT solver and for the SMT-based BMC we used the Z3 solver.

The first benchmark we consider is the *weighted generic pipeline paradigm* (WGPP) WIS model [6]. The problem parameter n is the number of Nodes. Let Min be the minimum cost incurred by Consumer to receive the data produced by Producer, and p denote the cost of producing data by Producer. The specifications we consider are as follows:

$\varphi_1 = \overline{\mathrm{K}}_P\mathrm{EF}_{[Min,Min+1)}ConsReady$ - it expresses that it is not true that Producer knows that always the cost incurred by Consumer to receive data is Min.

$\varphi_2 = \overline{\mathrm{K}}_P\mathrm{EF}(ProdSend \wedge \overline{\mathrm{K}}_C\overline{\mathrm{K}}_P\mathrm{EG}_{[0,Min-p)}ConsReady)$ - it states that it is not true that Producer knows that always if it produces data, then Consumer knows that Producer knows that Consumer has received data and the cost is less than $Min - p$.

The size of the reachable state space of the WGPP system is $4 \cdot 3^n$, for $n \geq 1$. The number of the considered k-paths is equal to 2 for φ_1 and 5 for φ_2, respectively. The lengths of the discovered witnesses for formulae φ_1 and φ_2 vary, respectively, from 3 for 1 node to 23 for 130 nodes, and from 3 for 1 node to 10 for 27 nodes.

The second benchmark of our interest is the *weighted bits transmission problem* (WBTP) WIS model [7]. We have adapted the local weight functions of [7]. This system is scaled according to the number of bits the S wants to communicate to R. Let $a \in \mathbb{N}$ and $b \in \mathbb{N}$ be the costs of sending, respectively, bits by Sender and an acknowledgement by Receiver. The specifications we consider are as follows:

$\phi_1 = \mathrm{EF}_{[a+b,a+b+1)}(\mathbf{recack} \wedge \overline{\mathrm{K}}_{\mathcal{S}}(\overline{\mathrm{K}}_{\mathcal{R}}(\bigwedge_{i=0}^{2^n-2}(\neg \mathbf{i})))) $ - it expresses that it is not true that if an *ack* is received by S, then S knows that R knows at least one value of the n-bit numbers except the maximal value, and the cost is $a + b$.

$\phi_2 = \mathrm{EF}_{[a+b,a+b+1)}(\overline{\mathrm{K}}_{\mathcal{S}}(\bigwedge_{i=0}^{2^n-1}(\overline{\mathrm{K}}_{\mathcal{R}}(\neg \mathbf{i})))) $ - it expresses that it is not true that S knows that R knows the value of the n-bit number and the cost is $a + b$.

The size of the reachable state space of the WBTP system is $3 \cdot 2^n$ for $n \geq 1$. The number of the considered k-paths is equal to 3 for ϕ_1 and $2^n + 2$ for ϕ_2, respectively. The length of the witnesses for both formulae is equal to 2 for any $n > 0$.

Performance Evaluation. The experimental results show that the both BMC method, SAT- and SMT-based, are complementary. We have noticed that for the WGPP system and both considered formulae the SMT-based BMC is faster than the SAT-base BMC, however, the SAT-based BMC consumes less memory. Moreover, the SMT-based method is able to verify more nodes for both tested formulae. In particular, in the time limit set for the benchmarks, the SMT-based BMC is able to verify the formula φ_1 for 120 nodes while the SAT-based BMC can handle 115 nodes. For φ_2 the SMT-based BMC is still more efficient - it is able to verify 27 nodes, whereas the SAT-based BMC verifies only 25 nodes.

In the case of the WBTP system the SAT-based BMC performs much better in terms of the total time and the memory consumption for both the tested formulae. In the case of the formula ϕ_2 both methods are able to verify the same number of bits. For the WBTP the reason of a higher efficiency of the SAT-based BMC is, probably, that the lengths of the witnesses for both formulae is constant and very short, and that there is no nested temporal modalities in the scope of epistemic operators. For formulae like ϕ_1 and ϕ_2 the number of arithmetic operations is small, so the SMT-solvers cannot show its strength.

Further more we have noticed that the total time and the memory consumption for both benchmarks and all the tested formulae is independent from the values of the considered weights.

5 Conclusions

We have proposed, implemented, and experimentally evaluated SMT-based bounded model checking approach for WECTLK interpreted over the weighted interpreted systems. We have compared our method with the corresponding SAT-based technique. The experimental results show that the approaches are complementary, and that the SMT-based BMC approach appears to be superior for the WGPP system, while the SAT-based approach appears to be superior for the WBTP system. This is a novel and interesting result, which shows that the choice of the BMC method should depend on the considered system.

References

1. Clark, B., Sebastiani, R., Sanjit, S., Tinelli, C.: Satisfiability modulo theories. In: Handbook of Satisfiability. Frontiers in Artificial Intelligence and Applications, vol. 185, chapter 26, pp. 825–885. IOS Press (2009)
2. Clarke, E.M., Grumberg, O., Peled, D.A.: Model Checking. The MIT Press (1999)
3. Jones, A.V., Lomuscio, A.: Distributed BDD-based BMC for the verification of multi-agent systems. In: Proc. AAMAS 2010, pp. 675–682. IFAAMAS (2010)
4. Męski, A., Penczek, W., Szreter, M., Woźna-Szcześniak, B., Zbrzezny, A.: BDD- versus SAT-based bounded model checking for the existential fragment of linear temporal logic with knowledge: algorithms and their performance. Autonomous Agents and Multi-Agent Systems 28(4), 558–604 (2014)
5. de Moura, L., Bjørner, N.S.: Z3: An efficient SMT solver. In: Ramakrishnan, C.R., Rehof, J. (eds.) TACAS 2008. LNCS, vol. 4963, pp. 337–340. Springer, Heidelberg (2008)
6. Woźna-Szcześniak, B.: SAT-based bounded model checking for weighted deontic interpreted systems. In: Reis, L.P., Correia, L., Cascalho, J. (eds.) EPIA 2013. LNCS, vol. 8154, pp. 444–455. Springer, Heidelberg (2013)
7. Woźna-Szcześniak, B., Zbrzezny, A.M., Zbrzezny, A.: SAT-based bounded model checking for weighted interpreted systems and weighted linear temporal logic. In: Boella, G., Elkind, E., Savarimuthu, B.T.R., Dignum, F., Purvis, M.K. (eds.) PRIMA 2013. LNCS, vol. 8291, pp. 355–371. Springer, Heidelberg (2013)
8. Woźna-Szcześniak, B., Szcześniak, I., Zbrzezny, A.M., Zbrzezny, A.: Bounded model checking for weighted interpreted systems and for flat weighted epistemic computation tree logic. In: Dam, H.K., Pitt, J., Xu, Y., Governatori, G., Ito, T. (eds.) PRIMA 2014. LNCS, vol. 8861, pp. 107–115. Springer, Heidelberg (2014)
9. Zbrzezny, A.: Improving the translation from ECTL to SAT. Fundamenta Informaticae 85(1–4), 513–531 (2008)

Dynamic Selection of Learning Objects
Based on SCORM Communication

João de Amorim Junior[(⊠)] and Ricardo Azambuja Silveira

PPGCC – UFSC, Florianópolis, Brazil
joao.amorim@iate.ufsc.br, ricardo.silveira@ufsc.br

Abstract. This paper presents a model to select Learning Objects in e-learning courses, based on multi-agent paradigm, aiming to facilitate the learning material reuse and adaptability on Learning Management Systems. The proposed model has a BDI multi-agent architecture, as an improvement of the Intelligent Learning Objects approach, allowing the dynamic selection of Learning Objects. As the first steps of our research, we implement a prototype to validate the proposed model using the JADEX BDI V3 platform. Thereafter, we extends the framework to allow the communication of the agents with SCORM standard resources, making possible to build enhanced dynamic learning experiences.

Keywords: Dynamic learning experience · Intelligent learning objects · SCORM

1   Introduction and Related Works

Adaptability and reuse are important aspects that contribute to improve learning process in virtual learning environments [1]. The former relates to different students' profiles and needs. An adaptable system increases the student understanding, taking into account its knowledge level and preferences [2,3,4]. The latter means that it is unnecessary to develop new resources if there are others related to the same learning purpose [4,5]. Some computational tools improve the teaching-learning process, i.e.: (1) Intelligent Tutoring Systems (ITS) - applications created for a specific domain, generally with few adaptability and interoperability [6]; (2) Learning Management Systems (LMS) - environments used to build online courses (or publishing material), allowing teacher to manage educational data [1], [7,8]; and (3) Learning Objects (LO) - digital artifacts that promotes reuse and adaptability of resources [9]. LO and LMS provide reusability, but they usually are not dynamically adaptable [8,9]. This article presents our research that seeks the convergence of these different paradigms for the development of intelligent learning environments and describes the mechanisms of an Intelligent Learning Objects' dynamic presentation model, based on communication with SCORM (Sharable Content Object Reference Model) resources [19, 20].

There are analogous studies that provides adaptability to learning systems. Some examples extend the LMS with distinct adaptive strategies, such as conditional jumps [8], Bayesian networks [3] or data mining [7]. Other researches are not integrated with

© Springer International Publishing Switzerland 2015
F. Pereira et al. (Eds.) EPIA 2015, LNAI 9273, pp. 658–663, 2015.
DOI: 10.1007/978-3-319-23485-4_66

a LMS, and use diversified ways to adapt the learning to the students' style, i.e.: ITS [6], recommender system [2], genetic algorithm [10] and swarm intelligence [11].

Moreover, there are some similar works based on the Multi-Agent System (MAS) approach resulting on smarter applications [12,13]. Some of them combine LMS and MAS to make the former more adaptive [14], and another is a dynamically adaptive environment, based on agents that are able to identify the student cognitive profile [14]. These related works identify the student's profile applying questionnaires in the beginning of the course or by clustering the students according to their assessments performance. Additionally, we observe in these papers that the attachment of new LO to the system is not possible without teacher intervention. The educator needs to configure previously all the possible course paths for each student style, what could be hard and take so much time [3]. Further, the attaching of a new LO to the course involves modifying its structure, resulting in limited adaptability and reuse.

In order to produce more intelligent LO, we have proposed in previous researches the convergence between the LO and MAS technologies, called Intelligent Learning Objects (ILO) [15]. This approach makes possible to offer more adaptive, reusable and complete learning experiences, following the learner cognitive characteristics and performance. According to this approach an ILO is an agent capable to play the role of a LO, which can acquire new knowledge by the interaction with students and other ILO (agents information exchange), raising the potential of student's understanding. The LO metadata permits the identification of what educational topic is related to the LO [9]. Hence, the ILO (agents) are able to find out what is the subject associated with the learning experience shown to the student, and then to show complementary information (another ILO) to solve the student's lack of knowledge in that subject.

2   ILOMAS

The proposed model integrates MAS and LMS into an intelligent behavior system, resulting on the improvement of the related works, leading to dynamic LO inclusion. The objective of the new model called Intelligent Learning Object Multi-Agent System (ILOMAS) is to enhance the framework developed to create ILO based on MAS with BDI architecture [16], extending this model to allow the production of adaptive and reusable learning experiences taking advantage of the SCORM data model elements. The idea is to select dynamically ILO in the LMS according to the student performance, without previous specific configuration on the course structure. The proposed model achieves reuse by the combination of pre-existed and validated LO whose concept is the same of that the student needs to learn about, avoiding the building of new materials. Moreover, the course structure becomes more flexible, since it is unnecessary to configure all the possible learning paths for each student profile.

The solution's adaptability is based on the ability to attach new LO to the LMS (that was not explicitly added before) as soon as the system finds out that the student needs to reinforce its understanding on a specific concept. This is automatically identified through the verification of the student assessment performance (i.e.: grade), on each instructional unit, or by student choice, when the learner interacting with the LO.

It is important to clarify that the approach does not use student' learning profile (i.e.: textual, interactive [2]) as information to select LOs. The scope of this research is to consider only the learner performance results (grades, time of interaction, sequencing and navigation). The ILOMAS is composed by agents with specific goals, and capable of communicating and offering learning experiences to students in a LMS course, according to the interaction with these students, taking advantage of the SCORM standard's features [19]. The ILOMAS architecture needs two kinds of agents:

- LMSAgent – Finds out the subject that the student must learn about, and passes the control of the interaction with the student to a new ILOAgent. Its beliefs are data provided from the LMS database, i.e.: the topic that the student must learn about.
- ILOAgent – Searches for a LO on the repository (related to the topic obtained from the LMSAgent), and exhibits it to the student. Besides, monitors the interaction between the student and the LO, which means the analysis of the data received from the SCORM communication. Depending on the analyzed data (beliefs), the agent will deliberate the exhibition of another LO (course with dynamic content).

The JADEX BDI V3 (V3) platform was chosen to implement the agents based on the BDI architecture [12,13]. The design of ILOMAS includes the characteristics of e-learning courses deployed on LMS (as MOODLE [7]), which means an environment accessed mostly from Web Browsers. The Java Servlets and JSP technologies are the bases of the interface between the client side (student) and the server side (agents' environment), getting benefits of the V3 services communication structure [13]. A non-agent class based on the Facade design pattern [17] keeps the low coupling between the MAS layer and the external items (front-end and servlets).

A first prototype was developed and tested with emphasis on the MAS development, instead of visualization issues (such as LO formats or graphical user interfaces) [18]. The simulation of a learning situation resulted on a different LO retrieved from the repository. This new LO had the same subject as the previous LO shown. It was not explicitly defined in the database that the student should have watched this new LO (only the topic was defined, no specific LO), so the MAS obtained the related LO dynamically, taking into account the metadata elements declared in IEEE-LOM [9].

2.1   ILOMAS and SCORM Integration

The extension of ILOMAS to use the SCORM standard [19,20] raises reuse, dynamic sequencing, and interoperability on learning environments. The SCORM specification defines a set of API functions, which allows the communication among the student, the LO and the LMS. This API admits that the ILOMAS uses the data model elements to define the student's knowledge level, and to evaluate the status of the current experience. Some available elements are the learner's answers to quizzes (result), the elapsed time since the beginning of the interaction (latency), the weighting of the interaction status relative to others, and a description of the LO's objectives [19]. If the learner demonstrates difficult in some subject (i.e.: wrong answers in sequence on the SCORM quiz, or take a long time to interact with the LO without any progress), it is possible to make decisions based on the historical received data.

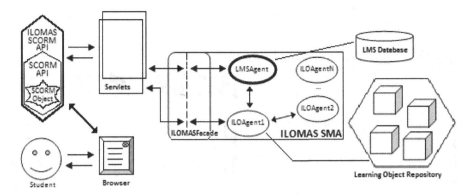

Fig. 1. ILOMAS SCORM Web architecture

The main desire defined to ILOAgent is to solve the student's lack of understanding about the subject. Thus, when the data received from SCORM points to a learner difficulty (error), the ILOMAS deliberation process (based on the JADEX engine [13]) dispatch the goal related with this objective. The ILOAgent's belief base stores the data received, and the deliberation process defines that the student needs to view a new (different) LO when the student selects an incorrect answer. This is the moment when the system achieves a dynamic learning situation, because a new LO not defined previously becomes part of the course structure. From the student's point of view (and even the teacher's point of view) the accessed object was just one, but with several contents (a larger LO composed dynamically by other smaller).

To validate this new version of the platform (SCORM integrated), we used some SCORM objects (on the version SCORM 1.2) about Social Security Laws (Public Law course). The learning interaction takes place in a custom LMS developed with limited features, only to test purposes. The implemented SCORM integration to ILOMAS was tested to reproduce distinct learning situations (Table 1): student that selects all the corrected answers (Student 1), another that misses all questions (Student 2), and one who increases understanding on the subject during interaction (Student 3). Each time that a student makes a mistake, the ILOMAS identifies the understanding problem and suggests another related LO to fill the learning gap (Fig. 2).

Table 1. ILOMAS SCORM preliminaries evaluation tests

Student	Quiz Errors	Previously Configured LO	New LO Offered	Dynamic Behavior
Student 1	0	1	0	No
Student 2	4	1	4	Yes
Student 3	1	1	1	Yes

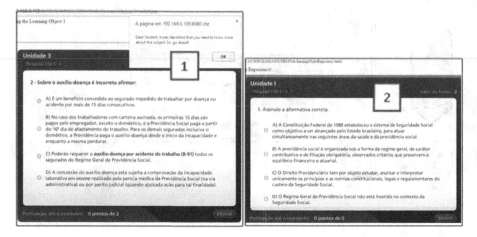

Fig. 2. The ILOMAS SCORM Web application execution: (1) The moment of the identification that the student needs another LO (wrong answer); (2) New LO exhibition

3   Conclusions and Future Work

This research resulted on a prototype implementation to verify the proposed model and its feasibility, followed by the execution of some evaluation tests. The SCORM API implementation gives to ILOMAS the ability of monitoring the overall communication between the LO and the learner, getting benefits of the SCORM data model element (i.e.: interaction status and time of the current learning session).

Future work leads to enhance the analysis of the received SCORM elements, taking into consideration the history of the student's experiences, and to explore all the SCORM data model elements in the process of determining if the learner needs to view a new LO. Another improvement would be the integration of ILOMAS with some MAS based recommender system for indexing and retrieving the related LO within the repository [21]. Finally, future work involves building a new plugin to integrate the ILOMAS inside the MOODLE LMS, and testing the application with different learning situations inside a LMS production instance, with real students.

References

1. Allison, C., Miller, A., Oliver, I., Michaelson, R., Tiropanis, T.: The Web in education. Computer Networks **56**, 3811–3824 (2012)
2. Vesin, B., Klasnja-Milicevic, A., Ivanovic, M., Budimac, Z.: Applying recommender systems and adaptive hypermedia for e-learning personalization. Computing and Informatics **32**, 629–659 (2013). Institute of Informatics
3. Bachari, E., Abelwahed, E., Adnani, M.: E-Learning personalization based on dynamic learners' preference. International Journal of Computer Science & Information Technology (IJCSIT) **3**(3) (2011)

4. Mahkameh, Y., Bahreininejad, A.: A context-aware adaptive learning system using agents. Expert Systems with Applications **38**, 3280–3286 (2011)
5. Caeiro, M., Llamas, M., Anido, L.: PoEML: Modeling learning units through perspectives. Computer Standards & Interfaces **36**, 380–396 (2014)
6. Santos, G., Jorge, J.: Interoperable Intelligent Tutoring Systems as Open Educational Resources. IEEE Transactions on Learning Technologies **6**(3), 271–282 (2013). IEEE CS & ES
7. Despotovic-Zrakic, M., Markovic, A., Bogdanovic, Z., Barac, D., Krco, S.: Providing Adaptivity in Moodle LMS Courses. Educational Technology & Society **15**(1), 326–338 (2012). International Forum of Educational Technology & Society
8. Komlenov, Z., Budimac, Z., Ivanovic, M.: Introducing Adaptivity Features to a Regular Learning Management System to Support Creation of Advanced eLessons. Informatics in Education **9**(1), 63–80 (2010). Institute of Mathematics and Informatics
9. Barak, M., Ziv, S.: Wandering: A Web-based platform for the creation of location-based interactive learning objects. Computers & Education **62**, 159–170 (2013)
10. Chen, C.: Intelligent web-based learning system with personalized learning path guidance. Computers & Education **51**, 787–814 (2008)
11. Kurilovas, E., Zilinskiene, I., Dagiene, V.: Recommending suitable scenarios according to learners' preferences: An improved swarm based approach. Computers in Human Behavior **30**, 550–557 (2014)
12. Wooldridge, M.: An Introduction to MultiAgent Systems, 2nd edn. John Wiley & Sons (2009)
13. Pokahr, A., Braubach, L., Haubeck, C., Ladiges, J.: Programming BDI Agents with Pure Java. University of Hamburg (2014)
14. Giuffra, P., Silveira, R.: A multi-agent system model to integrate Virtual Learning Environments and Intelligent Tutoring Systems. International Journal of Interactive Multimedia and Artificial Intelligence **2**(1), 51–58 (2013)
15. Silveira, R., Gomes, E., Vicari, R.: Intelligent Learning Objects: An Agent-Based Approach of Learning Objects. IFIP – International Federation For Information Processing, vol. 182, pp. 103–110. Springer-Verlag (2006)
16. Bavaresco, N., Silveira, R.: Proposal of an architecture to build intelligent learning objects based on BDI agents. In: XX Informatics in Education Brazilian Symposium (2009)
17. Gamma, E., Helm, R., Johnson, R., Vlissides, J.: Design Patterns: Elements of Reusable Object-Oriented Software. Addison-Wesley (1995)
18. de Amorim Jr., J., Gelaim, T.Â., Silveira, R.A.: Dynamic e-Learning content selection with BDI agents. In: Bajo, J., Hallenborg, K., Pawlewski, P., Botti, V., Sánchez-Pi, N., Duque Méndez, N.D., Lopes, F., Vicente, J. (eds.) PAAMS 2015 Workshops. CCIS, vol. 524, pp. 299–308. Springer, Heidelberg (2015)
19. SCORM 2004. Advanced Distributed Learning. http://www.adlnet.org/scorm
20. Gonzalez-Barbone, V., Anido-Rifon, L.: Creating the first SCORM object. Computers & Education **51**, 1634–1647 (2008)
21. Campos, R.L.R., Comarella, R.L., Silveira, R.A.: Multiagent based recommendation system model for indexing and retrieving learning objects. In: Corchado, J.M., Bajo, J., Kozlak, J., Pawlewski, P., Molina, J.M., Julian, V., Silveira, R.A., Unland, R., Giroux, S. (eds.) PAAMS 2013. CCIS, vol. 365, pp. 328–339. Springer, Heidelberg (2013)

Sound Visualization Through a Swarm of Fireflies

Ana Rodrigues[(✉)], Penousal Machado, Pedro Martins, and Amílcar Cardoso

CISUC, Deparment of Informatics Engineering, University of Coimbra,
Coimbra, Portugal
{anatr,machado,pjmm,amilcar}@dei.uc.pt

Abstract. An environment to visually express sound is proposed. It
is based on a multi-agent system of swarms and inspired by the visual
nature of fireflies. Sound beats are represented by light sources, which
attract the virtual fireflies. When fireflies are close to light they gain
energy and, as such, their bioluminescence is emphasized. Although real
world fireflies do not behave as a swarm, our virtual ones follow a typical
swarm behavior. This departure from biological plausibility is justified
by aesthetic reasons: the desire to promote fluid visualizations and the
need to convey the perturbations caused by sound events. The analysis of
the experimental results highlights how the system reacts to a variety of
sounds, or sequence of events, producing a visual outcome with distinct
animations and artifacts for different musical pieces and genres.

Keywords: Swarm intelligence · Computer art · Multi-agent systems ·
Sound visualization

1   Introduction

Although sound visualization has been an object of study for a long time, the
emergence of the computer, with graphic capabilities, allowed the creation of new
paradigms and creative processes in the area of sound visualization. Most of the
initial experiments were done through analogical processes. Since the advent
of computer science, art has taken significant interest in the use of computers
for the generation of automated images. In section 2, we present some of the
main inspirations to our work including sound visualization, generative artworks,
computer art and multi-agent systems.

Our research question relies on the possibility of developing a multi-agent
model for sound visualization. We explore the intersection between computer
art and nature-inspired multi-agent systems. In the context of this work, swarm
simulations are particularly interesting because they allow the expression of a
large variety of different types of behaviors and tend to be intuitive and natural
forms of interaction.

In section 3 we present the developed project, which is based on a multi-agent
system of swarms and inspired by the visual nature of fireflies. In the scope of our

F. Pereira et al. (Eds.) EPIA 2015, LNAI 9273, pp. 664–670, 2015.
DOI: 10.1007/978-3-319-23485-4_67

work, visualization of music is understood as the mapping of a specific musical composition or sound into a visual language.

Our environment contains sources of light representing sound beats, which attract the fireflies. The closer a firefly is to the light, the more emphasized is its bioluminescence and higher is its chance of collecting energy (life). Using Reynolds' boids algorithm [6], fireflies interact with the surrounding environment by means of sensors. They use them to find and react to energy sources as well as to other fireflies. In section 4 we present an analysis and corresponding experimental results of the systems behavior to 5 different songs. Lastly, in section 5 we present our conclusions and further work to be done.

2   Related Work

Ernst Chladni studied thoroughly the relation between sound and image. One of his best-known achievements was the invention of Cymatics. It geometrically showed the various types of vibration on a rigid surface [5]. In the 1940s Oskar Fischinger made cinematographic works exploring the images of sound by means of traditional animation [4]. His series of 16 studies was his major success [4]. Another geometric approach, was made by Larry Cuba in 1978, but this time with digital tools. "3/78" consisted of 16 objects performing a series of precisely choreographed rhythmic transformations [2].

Complex and self-organized systems have a great appeal for the artistic practice since they can continuously change, adapt and evolve. Over the years, computer artifacts promoting emergent systems behaviors have been explored [1] [7]. Artists got fascinated with the possibility of an unpredictable but satisfying outcome. Examples of this include the work of Ben F. Laposky, Frieder Nake, Manfred Mohr, among many others [3].

3   The Environment

In this section we present a swarm-based system of fireflies and all of its interactions. In this environment, fireflies are fed by the energy of sound beats (rhythmic onsets). While responding to the surrounding elements of the environment, they search for these energies (see Fig. 1). The colors were chosen according to the

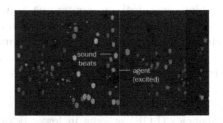

Fig. 1. Systems behavior and appearance example. Best viewed in color.

real nature of fireflies. Since they are visible at night, we opted for a dark blue in the background and a brighter one for the sound beats. As for bioluminescence, we used yellow.

The environment rules and behaviors, plus the visualization were implemented with Processing. The mechanism for extracting typical audio information was made with the aid of the Minim library, mainly because it contains a function for sound beat detection.

3.1 Sound (Energy Sources)

Sound Analysis. To visualize sound, a preliminary analysis is necessary. A sound is characterized by 3 main parameters: frequency, amplitude and duration. Frequency determines the pitch of the sound. Amplitude determines how loud the sound is. Duration can define the rhythm of music and also the instant in the music where sound beats happen.

We perform sound analysis prior to the visualization, in order to promote a fluid animation and convey the perturbations caused by sound events. We compute the main sound characteristics (pitch, volume, sound beats) and export them to a text file. Sound beats are detected note onsets. They are related to the temporal/horizontal position of a sound event.

Although the mechanism used to extract audio is not novel and remains simple, we think this approach is adequate to the goals of our system. It fits in the amount of expressiveness that we intend to represent in our visualization, as visual simplicity characterizes the fireflies natural environment

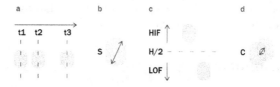

Fig. 2. Graphical representation of sound objects. a - Sound beats instants, b - Amplitude, c - Frequency, d - Collision.

Sound's Graphic Representation. After the sound analysis, all the properties of sound are mapped into graphical representations. Sound beats are mapped into instants (t1, t2, t3,...) which defined the objects horizontal position as shown in Fig. 2a. Each sound object has a pre-defined duration, meaning that it is removed from the environment at the end of its duration. Amplitude was translated into the objects size, i.e., the size is directly proportional to the amplitude (Fig. 2b). Lastly, frequency is mapped into the objects vertical position in the environment (Fig. 2c). High frequencies (HIF) are positioned on the top of screen and low frequencies (LOF) emerge in lower positions of the vertical axis. A fourth characteristic presented in the graphical representation of sound objects is collision (Fig. 2d). This last one is not directly related to sound, only to sound object's physics. When a object collides with another one, a contrary force is applied between these two, separating them from each other.

3.2 Agents (Fireflies)

Agent Behavior. Because the sound beats are presented from the left to right, fireflies are initially born on the left side of the screen, vertically centered. Agents are provided with a specific vision towards the surrounding environment. A vision angle of 30° and a depth of 150 pixels were considered as optimum values (Fig. 3), because the agents could have a high amount of independence and resemble to their original behavior. Agents motion is based on the "Boids" algorithm. They walk randomly until they find something that may affect their behavior, such as source of light or other agents.

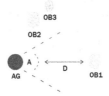

Fig. 3. Agent field of view: angle (A) and depth (D).

The closer they are to a source of light, the more attracted they get to it, meaning that there is a force of attraction towards it. Along with that, agents have a swarming behavior, meaning that neighbor agents can see each others and follow them through flocking behavior rules [6].

These rules were presented by Reynolds with a computational model of swarms exhibiting natural flocking behavior. He demonstrated how a particular computer simulation of boids could produce complex phenomena from simple mechanisms. These behaviors define how each creature behaves in relation to its neighbors: separation, alignment or cohesion [6].

Fig. 4. Left image: separation. Right image: cohesion.

The swarming behaviors present in this system are: separation and cohesion (Fig. 4). Separation gives the agents the ability to maintain a certain distance from others nearby in order to prevent agents from crowding together. Cohesion gives agents the ability to approach and form a group with other nearby agents [6]. No alignment force was applied. Alignment is usually associated with flocking behavior, like birds and fishes do. Swarm behavior – like the one found in bees, flies and our fireflies – does not imply alignment.

Additionally, the life and death of each agent is also determined by the way it interacts with the environment. The agent begins with an initial lifespan, losing part of its energy at each cycle. If the agent gets close to an energy source, it gains more energy and a longer lifespan; otherwise, it keeps losing its energy until it dies. There are no mechanisms for the rebirth of agents, as we intend to keep a clear visualization and understanding of interactions among agents.

Agent's Graphic Representation. Fireflies use bioluminescence to communicate and attract other fireflies. As an agent gradually approaches the light emitted by a sound object within its field of view, the more excited it gets and the more it emphasizes its bioluminescence (Fig. 5, left image). This will temporarily influence the agents size because it gets intermittent. The real agent size will be as big as the energy (Fig. 5, right image) that it has at a certain instant. When an agent dies, it disappears from the environment.

Fig. 5. Left image: agent approximation to an object (AG→OB). Right image: agent growth (E).

4   Results and Discussion

This section presents an analysis of the systems behavior in response to 4 different songs or melodic sequences (from track 1 to track 4). These tracks vary in rhythm, intensity and frequencies, allowing us to illustrate and highlight how the system reacts to different sound stimuli.

Unfortunately, conveying the overall feel of an animation[1] in a paper has its difficulties. To circumvent this issue and to ease our analysis, first we analyze a

Fig. 6. The music that generated this response is characterized by a variety of intensity and big density of beats.

---

[1] A demonstration video can be found at http://tinyurl.com/ky7yaql.

complete visualization of the track so we can perceive the differences inside each one. Secondly, we present the trajectory made by the agents of the corresponding music to better analyze their behavior in the different tracks. We present only one example of those figures due to space constraints.

Track 1 corresponds to a piece with high density of beats and low contrast of intensities. This promotes a higher chance of having a longer lifespan. However, the low contrast of the intensities implies that they do not gather so much energy at once. Track 2 (Fig. 6), is also characterized by a high density of beats, but in this case the contrast in intensities make swarms gain more energy. Track 3 has a low contrast of frequencies and a balanced density of beats. For Last, Track 4 as opposed to almost all of the other examples so far described, has a strong contrast between high and low frequencies. Adding to this, the low density of beats results in a reduced lifespan for swarms as they have a short field of view.

From the observation of these patterns created by our system, we can conclude: (i) fireflies have a tendency to follow the pattern created by the sound beats as we could see in the example depicted in Fig. 6; (ii) there is a bigger concentration of fireflies in the sources that contain more energy; (iii) tracks with a lower contrast between frequencies promote a more balanced spread of the fireflies in the environment; (iv) tracks with a high density of beats give fireflies a longer lifespan because the agents have a narrow vision field and thus they can collect more energy even if it is in small pieces of it.

5 Conclusions and Future Work

We presented an environment to visualize audio signals. It was inspired by the visual nature of fireflies and based on a multi-agent system of swarms proposed by Reynolds. In this environment, sound is mapped into light objects with energy, which attract the virtual fireflies. When fireflies are close to light they gain energy and, as such, their bioluminescence is emphasized. The flocking behavior of the group emerges based on simple rules of interaction.

In real life the presented technique may be useful for people with low understanding of music to take part in musical events. In further work we will expand our system by introducing more sophisticated mechanisms for the sound analysis, which allow the representation of higher-level concepts and musical events. On the other hand, we also wish to explore alternative visual representations to offer the user a wider array of choices. Finally, a user study should be performed to assess the strengths and weaknesses of the different visualization variants and evaluate the system.

Acknowledgments. This research is partially funded by project ConCreTe. Project ConCreTe acknowledges the financial support of the Future and Emerging Technologies (FET) programme within the Seventh Framework Programme for Research of the European Commission, under FET grant number 611733.

References

1. Barszczewski, P., Cybulski, K., Goliski, K., Koniewski, J.: Constellaction (2013). http://pangenerator.com
2. Compart: Larry Cuba, 3/78 (nd). http://tinyurl.com/k2y3vef
3. Dietrich, F.: Visual intelligence: The first decade of computer art (1965–1975). Leonardo **19**(2), 159–169 (1986)
4. Evans, B.: Foundations of a visual music. Computer Music Journal **29**(4), 11–24 (2005)
5. Monoskop: Ernst Chladni (nd). http://monoskop.org/Ernst_Chladni
6. Reynolds, C.W.: Steering behaviors for autonomous characters. In: Game Developers Conference, vol. 1999, pp. 763–782 (1999)
7. Uozumi, Y., Yonago, T., Nakagaito, I., Otani, S., Asada, W., Kanda, R.: Sjq++ ars electronica (2013). https://vimeo.com/66297512

Social Simulation and Modelling

Analysing the Influence of the Cultural Aspect in the Self-Regulation of Social Exchanges in MAS Societies: An Evolutionary Game-Based Approach

Andressa Von Laer, Graçaliz P. Dimuro$^{(\boxtimes)}$, and Diana Francisca Adamatti

PPGCOMP, C3, Universidade Federal Do Rio Grande (FURG), Rio Grande, Brazil
{andressavonlaer,gracaliz,dianaada}@gmail.com

Abstract. Social relationships are often described as social exchanges, understood as service exchanges between pairs of individuals with the evaluation of those exchanges by the individuals themselves. Social exchanges have been frequently used for defining interactions in MAS. An important problem that arises in the context of social simulation and other MAS applications is the self-regulation of the social exchange processes, so that the agents can achieve/maintain the equilibrium of the exchanges by themselves, guaranteing the continuation of the interactions in time. Recently, this problem was tackled by defining the spatial and evolutionary Game of Self-Regulation of Social Exchange Processes (GSREP), implemented in NetLogo, where the agents evolve their exchange strategies by themselves over time, performing more equilibrated and fair interactions. The objective of this paper is to analyse the problem of the self-regulation of social exchange processes in the context of a BDI-based MAS, adapting the GSREP game to Jason agents and introducing a cultural aspect, where the society culture, aggregating the agents' reputation as group beliefs, influences directly the evolution of the agents' exchange strategies, increasing the number of successful interactions and improving the agents' outcomes in interactions.

1 Introduction

As it is well known in social sciences, the acts, actions and practices that involve more than two agents and affect or take account of other agents's activities, experiences or knowledge states are called social interactions. Social interactions and, mainly, the quality of these interactions, are crucial for the proper functioning of the system, since, e.g., communication failure, lack of trust, selfish attitudes, or unfair behaviors can leave the system far of a solution. The application of the social interaction concept to enhancements of MAS's functionality is a natural step towards designing and implementing more intelligent and human-like populations of artificial autonomous systems. [13]

Social relationships are often described as social exchanges [1], understood as service exchanges between pairs of individuals with the evaluation of those

© Springer International Publishing Switzerland 2015
F. Pereira et al. (Eds.) EPIA 2015, LNAI 9273, pp. 673–686, 2015.
DOI: 10.1007/978-3-319-23485-4_68

exchanges by the individuals themselves [16]. Social exchanges have been frequently used for defining social interactions in MAS [10,15,21]. A fundamental problem discussed in the literature is the regulation of such exchanges, in order to allow the emergence of equilibrated exchange processes over time, promoting the continuity of the interactions [12,21], social equilibrium [15,16] and/or fairness behaviour.[1] In particular, this is a difficult problem when the agents, adopting different social exchange strategies, have incomplete information on the other agents' exchange strategies, as in open societies [9].

In the literature (e.g, [9,15,21]), different models were developed (e.g., centralized/decentralized control, closed/open societies) for the social exchange regulation problem. Recently, this problem was tackled by Macedo et al. [12], by introducing the spatial and evolutionary *Game of Self-Regulation of Social Exchange Processes* (GSREP), where the agents, adopting different social exchange strategies (e.g., selfishness, altruism), considering both the short and long-term aspects of the interactions, evolve their exchange strategies along the time by themselves, in order to promote more equilibrated and fair interactions. This approach was implemented in NetLogo.

However, certain characteristics involved in social exchanges may be more appropriately modeled with cognitive agents[2], such as *BDI Agents* (Belief, Desire, Intention) [4]. Also, taking into account the observations made by a society on the behavior of an agent *ag* in its past interactions, it is possible to qualify *ag*'s reputation [5,6], which can be made available to the other agents who themselves have not interacted with that agent. These indirect observations can be aggregated to define any agent past behaviour based on the experiences of participants in the system [11]. Reputation can assist agents in choosing partners where there are other agents that can act so as to promote the disequilibrium of the social exchange processes in the society. Given the importance of this kind of analysis in many real-world applications, a large number of computational models of reputation have been developed (e.g., [23,24]).

Then, this paper introduces an evolutionary and cultural approach of GSREP game for the JaCaMo [2] framework, considering also the influence of the agent society culture, so defining the Cultural-GSREP game. Observe that here are at least five basic categories of cultural knowledge that are important in the belief space of any cultural evolution model: situational, normative, topographic, historical or temporal, and domain knowledge [18]. In this paper, we explore just the normative category, and let the combination of other cultural aspects for further work. We consider a specific society's culture where the agents' reputations are aggregated as group beliefs [23], using the concept of artifacts [20]. Based on the idea that "the culture of a society evolves too, and its evolution may be faster than genetics, enabling a better adaptation of the agent to the environment"[19], we analyse the influence of the culture in the evolution of the

---

[1] We adopted the concept of fairness behaviour/equilibrium as in [17,25].

[2] For discussions on the role of BDI agents in social agent-based simulation, see [14].

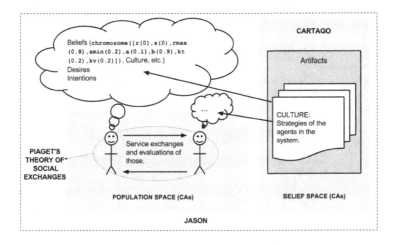

Fig. 1. An overview of the proposed model

agents' exchange strategies, the increase of the number of successful interactions and the improvement of the agents' outcomes in their interactions.[3]

2   The Cultural-GSREP Game

The Cultural-GSREP game is built on the spatial and evolutionary game of incomplete information presented in [12], for the self-regulation of social exchange processes, where agents evolve their strategies in order to maximize a *fitness* function by an evolutionary approach. In each simulation cycle, the fitness function evaluates the material result of the exchanges of an agent with its neighboring agents, getting influence by factors that characterize the exchange strategies and attitudes adopted by the agents.

For the agent belief learning process, the agent society's culture (consisted of a belief space common to all agents) may influence on the decision making in each single game (two-agent exchange). The addition of a belief space common to all agents involved in the system works as a focal point (Schelling Point) [22], serving as reference for agents, since the agents do not have knowledge about the other agents' exchange strategies. The belief space used in this paper is based on the work developed in [23].

Figure 1 shows an overview of the Cultural-GSREP game model, which is organized in two parts: the first is the social exchange game [12], where each single exchange occurs in two sequential stages with their respective evaluations

---

[3] We remark that the aim of the paper is not to introduce a novel method to evaluate reputation. On the contrary, we adopt a very simple method for analysing the reputation in order to show how this cultural aspect may influence the evolution of the self-regulation of social exchanges. For discussions on the perspectives of culture in a more general context see [8].

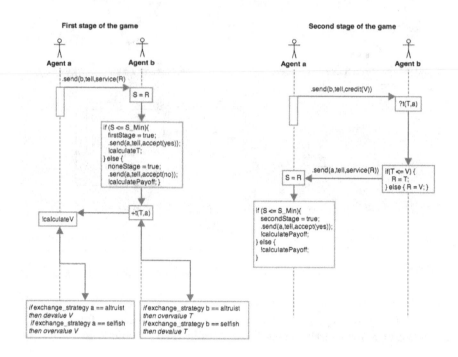

Fig. 2. Two stages of a single social exchange game (for selfishness/altruism exchange strategies)

by the agents themselves, and a fitness function helps to evolve the agents's exchange strategies; the second part is the creation of group beliefs (GBs) using artifacts, which forms the cultural level based on agents' reputations constructed over the exchanges experienced by the agents in the society. The model was implemented in Jason [3], using the concept of Agents & Artifacts [20] for the implementation of group beliefs in CArtAgO framework [2].[4]

3   The Game of Exchange Processes

Figure 2 shows the basic and simplified two-stage sequence of the first part of the model, where the exchanges between each two agents occur in a single exchange cycle.[5] In the first stage of the game, denoted by I, the agent a offers a service with some value of investment (R) to the agent b, such that $R \leq R_{\max}$, where $R_{\max}$ is the maximal investment value a is willing to have for a service performed for another agent. This yields a value of b's satisfaction (S) and debt (T) for the service provided by a, which are directly related to a's investment value.

---

[4] Note that shared artifacts are common in normative and organizational systems (see, e.g., [7]).

[5] The two-stage social exchange was inspired by Piaget's Theory of Social Exchanges.[16]

However, if b believes that the service offered by a provides less satisfaction than the minimal satisfaction ($S_{\min}$) it is willing to accept, than b refuses a's offer and this exchange stage does not occur. Supposing that b accepts the service provided by a, then, at the end of this stage, the agent a has a credit value (V), that is, a credit related to the service it has previously performed to agent b. R and S are called material values, related to the performed exchanges. T and V are virtual values, related to future interactions, since they help the continuation of the exchanges.

The second stage, denoted by II, is similar to the first, but referring to a possible debt collection by agent a, when a charges b for a service in payment for its virtual value (V) (the credit a has obtained from b in the stage I). The agent b has on its belief base a debt value (T) and it then performs a service offer with an investment value (R) to a (with $R \leq R_{\max}$), which in turn generates a value of satisfaction (S) for b's offer, in case that it accepts such satisfaction value (i.e., $S \geq S_{\min}$), otherwise this exchange stage does not occur. After each 2-stage exchange between a and b, they calculate the material reward they received, using the payoff function $p_{ab} : [0,1]^4 \rightarrow [0,1]$:

$$p_{ab}(R_{I_{ab}}, R_{II_{ba}}, S_{I_{ba}}, S_{II_{ab}}) = \tag{1}$$

$$\begin{cases} \dfrac{1 - R_{I_{ab}} + S_{II_{ab}}}{2} & \text{if } (R_{I_{ab}} \leq R_a^{\max} \wedge S_{I_{ba}} \geq S_b^{\min}) \wedge (R_{II_{ba}} \leq R_b^{\max} \wedge S_{II_{ab}} \geq S_a^{\min}) \\ \dfrac{1 - R_{I_{ab}}}{2} & \text{if } (R_{I_{ab}} \leq R_a^{\max} \wedge S_{I_{ba}} \geq S_b^{\min}) \wedge (R_{II_{ba}} > R_b^{\max} \vee S_{II_{ab}} < S_a^{\min}) \\ 0 & \text{if } (R_{I_{ab}} > R_a^{\max} \vee S_{I_{ba}} < S_b^{\min}) \wedge (R_{II_{ba}} > R_b^{\max} \vee S_{II_{ab}} < S_a^{\min}) \end{cases}$$

Observe that, according to Eq. (1), if both exchange stages I and II are successfully performed, then the agents' rewards are greater. On the contrary, if no stage occurs, i.e., the agent b refuses the service of agent a in the first stage I, the payoff is null.

3.1   Social Exchange Strategies in Beliefs

The agents evaluate the services according to their exchange strategies, e.g., an agent with selfishness strategy is more likely to devalue the received service and overvalue an offered service, which impacts on debt and credit values. The calculations of debts T and credits V are made using the debt and credit depreciation ($\rho = d$) or overestimation ($\rho = o$) factors $k^{\rho t}, k^{\rho v} \in [0,1]$, respectively, characterizing each strategy, as follows:

Depreciation: $T = (1 - k^{dt})S, \ V = (1 - k^{dv})R$ \hfill (2)

Overestimation: $T = S + (1 - S)k^{ot}, V = R + (1 - R)k^{ov}$ \hfill (3)

Each agent has in its belief base an exchange strategy-based belief, called *chromosome belief*, which evolves along the time, by mutation plans. For example, the initial chromosome belief of an altruist agent is defined as:

```
chromosome([r(0), s(0), rmax(0.8), smin(0.2), e(0.1), g(0.9), kot(0.2), kdv(0.2)]),
```

where $r(0)$ and $s(0)$ are, respectively, the initial investment and satisfaction values, and the current parameters that represent its exchange strategy are: R, $R^{\max}$, $S^{\min}$, k^t and k^v, where $R \in [0;1]$ is the value of investment, $R^{\max} \in [0;1]$ is the maximum value that the agent will invest, $S^{\min} \in [0;1]$ is the minimum value of satisfaction that the agent accepts, $k^{ot} \in [0; 1]$ and $k^{dt} \in [0; 1]$ are, respectively, the debt overestimation and credit depreciation factors, as shown in Eqs. (2) and (3), $e \in [0, 1]$ is the weight that represents the agent's tolerance degree when its payoff is less than of its neighboring agents (*envy degree*), and $g \in [0, 1]$ represents the agent's tolerance degree when its payoff is higher than its neighboring agents' payoffs (*guilt degree*) [12,25].

Analogously, the initial chromosome belief of a selfish agent is defined as:

```
chromosome([r(0), s(0), rmax(0.2), smin(0.8), e(0.9), g(0.1), kdt(0.2), kov(0.2)]).
```

To implement/evaluate the model, we consider five agents that perform the exchanges, each agent with a different initial exchange strategy, namely: altruism, weak altruism, selfishness, weak selfishness and rationality. The rational agent plays just for the Nash Equilibrium[6], and then $s_{\min} = e = g = k^t = k^v = 0$.

3.2   The Fitness Evaluation

Given a neighborhood $A = \{1, \ldots, m\}$ of m agents, each agent $i \in A$ plays the exchange game with the other $m - 1$ neighboring agents $j \in A$, with $j \neq i$. In each simulation cycle, each agent i evaluates its local social exchange material results with each other neighboring agent j, using the local payoff function given in Eq. (1). The total payoff received by each agent is calculated after each agent has performed the two exchange stages with its entire neighborhood. For p_{ij} calculated by Eq. (1), the *total payoff allocation* of a neighborhood of m agents is given by

$$X = \{x_1, \ldots, x_m\}, \text{ where } x_i = \sum_{j \in A, j \neq i} p_{ij}. \tag{4}$$

The agent i calculates its adaptation degree through its *fitness* function F_i : $[0, 1]^m \to [0, 1]$, whose definition, encompassing all types of exchange strategies, is:

$$F_i(X) = x_i - \frac{e_i}{(m-1)} \sum_{j \neq i} \max(x_j - x_i, 0) - \frac{g_i}{(m-1)} \sum_{j \neq i} \max(x_i - x_j, 0), \tag{5}$$

where X is the total payoff allocation of agent i (Eq. (4)), e_i and g_i are i's envy and guilt degrees, respectively. To evaluate its *fitness*, the agent compares its current fitness with the previous one: if it exceeds the value of the previous fitness, then the current strategy is better than the previous one, and the agent makes an adjustment in the vector of probabilities, increasing the probability of the current strategy to be chosen again, increasing/decreasing the parameters of the chromosome belief defining its strategy.[7]

The probability vector of adjustments is in Table 1. There are 27 possible adjustments, e.g., p_i^0 is the probability of increasing R_i, $R_i^{\max}$ and $S_i^{\min}$ (by a certain exogenously specified adjustment step), and p_i^5 is the probability of increasing the value of R_i, keeping the value of R_i^{max} and decreasing $S_i^{\min}$. The probability and strategy adjustment steps f_p and f_s determine, respectively, on which extent the probabilities of the probability vector and the values r_i, r_i^{max} and s_i^{min} are increased or decreased.

---

[6] See [12] for a discussion on the Nash Equilibrium of the Game of Social Exchange Processes.

[7] The fitness function was based in [12,25].

Table 1. The probability vector adjustment

	R_i	R_i^{max}	S_i^{min}		R_i	R_i^{max}	S_i^{min}		R_i	R_i^{max}	S_i^{min}
p_i^0	↑	↑	↑	p_i^9	=	↑	↑	p_i^{18}	↓	↑	↑
p_i^1	↑	↑	=	p_i^{10}	=	↑	=	p_i^{19}	↓	↑	=
p_i^2	↑	↑	↓	p_i^{11}	=	↑	↓	p_i^{20}	↓	↑	↓
p_i^3	↑	=	↑	p_i^{12}	=	=	↑	p_i^{21}	↓	=	↑
p_i^4	↑	=	=	p_i^{13}	=	=	=	p_i^{22}	↓	=	=
p_i^5	↑	=	↓	p_i^{14}	=	=	↓	p_i^{23}	↓	=	↓
p_i^6	↑	↓	↑	p_i^{15}	=	↓	↑	p_i^{24}	↓	↓	↑
p_i^7	↑	↓	=	p_i^{16}	=	↓	=	p_i^{25}	↓	↓	=
p_i^8	↑	↓	↓	p_i^{17}	=	↓	↓	p_i^{26}	↓	↓	↓

4 The Culture: Group Belief and Reputation Artifacts

The culture of the agent society is consisted of the group belief (GB) and the reputation artifacts. For the implementation in CArtAgO, these artifacts are firstly created by the *mediator* agent, which is also responsible for initiating the exchanges by sending a message to all agents to start the sequence of exchanges. The GB artifact stores the beliefs sent by agents after obtaining experience in exchanges and the reputation artifact creates the reputation of agents.

The beliefs that compose the artifacts are observable properties. The announcements are treated as interface operations, where some parameters are informed: the predicate announcement, the degree of certainty of a belief and a value of the strength of this certainty. The composition of a GB works as follows. The formation rules of individual beliefs lie within the agent minds. The rules that form the group beliefs (synthesis rules) are in an external entity to agents and the communication for the formation of GB is made through announcements sent to a component that aggregates it, forming a GB (see Fig. 3). The set A of all announces is defined by

$$A \stackrel{\text{def}}{=} \{< p, c, s > \mid p \in P, c \in [0..1], s \in \mathbb{N}\}, \tag{6}$$

where P is the set of all the predicates, and p, c and s are, respectively, the predicate, the certainty degree and the strength degree of an announce. For example, in the announce `personality("selfish",bob)`, with certainty degree 0.8 and strength 6, the advertiser is quite sure that agent *bob* adopts a selfishness exchange strategy, based on 6 experiences it had in past exchanges with *bob*. See the method `announce` in Fig. 4.

Figure 5 shows the architecture of the artifact `ArtCG` of group beliefs, including the classes `AgentAnnounce`, `Belief` and the `announce` method, which corresponds to the announce operation of beliefs (Eq. (6)). When receiving an announce, the artifact adds it to a list of announces, and whenever there exists at least one equal announce from each agent present in the system, this announce becomes a reputation (see Fig. 4).

The `Belief` class function is to represent the group belief composed by the tuple: predicate, certainty degree and strength, and it implements a ToProlog interface, which allows its description in the form of a predicate.

Fig. 3. Group belief model

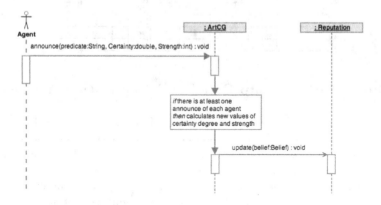

Fig. 4. Predicate announcement

The `AgentAnnounce` class represents the announces made by the agents and inherits the `Belief` class, also adding the *advertiser* attribute that represents the agent that made the announce.

To create a reputation, the certainty and strength values are calculated by the synthesis process and the artifact `Reputation` is notified of the new belief group by the method `update`. If there is already a group belief with the same predicate in the artifact, then it updates such values, otherwise a new group belief is added.

In this paper, we consider a mixed society (composed of agents with five different exchange strategies), and, due to this fact, the adopted aggregation method is the weighted synthesis [23], where announcements are synthesized in order to seek a middle term between them, so not only benefiting a optimistic or pessimistic society. The function weighted synthesis $sinpon_p$, which gives the certainty degree c, the strength s, where $|C_p|$ is the subset containing all the announcements of a predicate p, is given by:

$$sinpon_p = <p,c,s>, \quad c = \frac{\sum\limits_{a \in C_p} c_a s_a}{\sum\limits_{a \in C_p} s_a}, \quad s = \frac{\sum\limits_{a \in C_p} s_a}{|C_p|} \quad (7)$$

Then, in Fig. 2, to begin the second exchange stage between two agents a and b, the agent a charges the agent b for the service made in the first stage, and then it sends b the credit value V that it thinks itself worthy. Through a comparison between a's credit value and the value R that a has invested in the

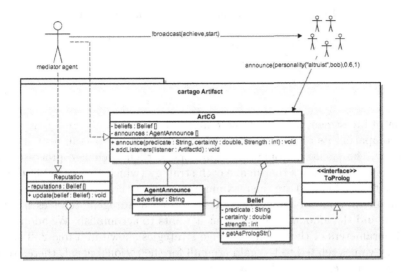

Fig. 5. Reputation diagram

first stage, b is able to draw a conclusion about the exchange strategy adopted by a:

- $R_a > V_a$: if the value of investment used by a in the first stage is higher than the credit value it attributed itself, b concludes that a is altruist;
- $R_a < V_a$: if the value of investment used by a in the first stage is lower than the credit value it attributed itself, b concludes that a is selfish;
- $R_a = V_a$: If the value of a's investment is equal to a's credit value, b concludes that a is rational.

The agent b sends its conclusion about the strategy adopted by agent a to the group belief artifact ArtCG, using the announce method (Fig. 4), to form a reputation of the agent a. Once the reputation is formed in the Reputation artifact, it is added to the agents' beliefs, thus becoming a common group belief to all participants of the game.

Whenever there is a reputation that an agent i is selfish, the agents send a message informing the mediator agent, which sends a message to agent i saying that i can not participate in the next play. So, i fails to improve its fitness value, unless it modifies its strategy to enter into the game again, increasing its investment value R and the maximum investment value R_{max}, and decreasing its minimum satisfaction value S_{min}.

5   Simulation Analysis

An social exchange strategy is determined by how an agent behaves towards the exchanges proposed by other agents, by the way this agent determines the

Table 2. Initial Parameters of Exchange Strategies

Strategy	r^{max}	s^{min}	g	e	$k^{\rho t}$	$k^{\rho v}$
Altruism	0.8	0.2	0.9	0.1	0.2, $\rho = o$	0.2, $\rho = d$
Weak altruism	0.6	0.4	0.7	0.3	0.1, $\rho = o$	0.1, $\rho = d$
Selfishness	0.2	0.8	0.1	0.9	0.2, $\rho = d$	0.2, $\rho = o$
Weak selfishness	0.4	0.6	0.3	0.7	0.1, $\rho = d$	0.1, $\rho = o$
Rationality	0.2	0.2	0	0	0	0

amount of investment it wants to accomplish, and also by the guilt/envy degree when comparing its results with the other agents. As the overall results emerge over time, the agents become self-regulators of their exchange processes. The evaluated characteristics that define each strategy (which are critical in the evolution of the exchanges) are the maximum value that the agent intends to invest, the minimum value of satisfaction accepted when an agent receives a service proposal and the amount of investment it wants to accomplish. We adopted the initial parameters of the social exchange strategies shown in Table 2. The guilt and envy values related to the gain are null for the rational agent, therefore, the values g_{rac} and e_{rac} are defined as 0 (zero).

Two different scenarios were defined, one without considering the culture of the society, and the other with the group beliefs as a "culture" common to all agents, as explained in Section 4. In each scenario there are five agents, each with a different strategy, and each simulation was performed using 300 cycles, for a total of 20 simulations by scenario. In both scenarios, the system stabilizes before 300 cycles. For the lack of space, we present the detailed analysis just for the second scenario.

Considering the two exchange stages in Fig. 2, given m agents, each playing with $m - 1$ agents, with zero, one, or two successful exchange stages in each two-agent interaction, then a cycle of a simulation is composed by $m(m - 1) + w_1 + w_2 + \cdots + w_m$ plays of stages of type I and/or II (successfully performed or not), where w_1 is the number of agents that agent 1 has credit with after the first stage with all other agents (i.e., the number of successful exchanges for the agent 1) and analogously one defines $w_2, \ldots, w_m$. In a single cycle, the number of exchanges of type I (successfully performed or not) is $m(m - 1)$, and the number of exchanges of type II (successfully performed or not) is $w_1 + w_2 + \cdots + w_m$. Note that if all the exchanges of the type I have been successfully for all agents, then one cycle of a simulation presents $2m(m - 1)$ exchanges of type I or II (successfully performed or not).

Figure 6 shows the simulation results in a range of 300 cycles.[8] Whenever the society culture is taken into account, the evolution of the agents' strategies provided an increase in the number of two-stage successful exchanges, which starts with 8 and ends with 20 by the 50th cycle, with a decreasing in the number of non successful interactions to zero, in a short period of time. In comparison with the first scenario, this time was reduced in 44.45%. The average and standard deviation of the number of exchanges are shown in Table 3. In Fig. 7

---

[8] Each mark in the X-axis of 6 represents 10 cycles.

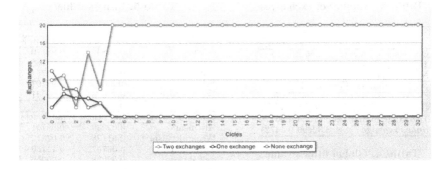

Fig. 6. Evolution of the number of exchanges, considering the culture

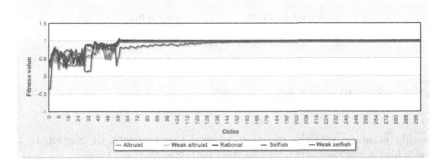

Fig. 7. Evolution of the Fitness value, considering the culture

we show the simulation of the evolution of the agents' fitness values in a period of 300 cycles.

Table 3 shows the number of two-stage exchanges, which was increased by 385.71 %. Table 5 shows the average value and standard deviation of the fitness values in the initial and final cycles of the simulations, considering the different exchange strategies. The increase in the fitness value of the altruist agent was 252.49 %, while for the weak altruist agent it was 258.20 %, for the rational agent it was 188.94 %, for the selfish agent it was 385.77 % and, finally, for the weak selfish agent it was 258.58 %. The strategy that showed lower growth was the rationality strategy, while the selfishness strategy had a higher evolution. In Table 4, we present the values of the overall average and the standard deviation of the global fitness value, showing an increasing of 584.26%.

In the case with culture, the increase in the number of two-stage exchanges was higher (385.71 %) than in the scenario without culture (171.73 %). Regarding the fitness values, in the scenario with culture only the Weak selfish strategy has not increased the average of the fitness value (343.95 % without culture and 258.58 % with culture). The others strategies showed largest increase in their fitness values, as shown in Table 6. Observe that, in both scenarios, the

Table 3. Number of exchanges

Average		
	Begin	End
One-stage exchange	3.65	0.1
Two-stage exchanges	2.8	13.6
None exchange	8.65	0.1
Standard deviation		
	Begin	End
One-stage exchange	2.36809	0.44721
Two-stage exchanges	3.17224	4.96726
None exchange	3.32890	0.44721

Table 4. Global fitness value

	Initial	Final
Global standard deviation	0.28005	0.01673
Global average	0.16701	0.97577

Table 5. Fitness value

Average		
Strategy	Initial fitness	Final fitness
Altruist	0.27849	0.98170
Weak altruist	0.27411	0.98188
Rational	0.33743	0.975
Selfish	-0.33172	0.94797
Weak selfish	0.27673	0.99231
Standard deviation		
Strategy	Initial fitness	Final fitness
Altruism	0.02539	0.08101
Weak altruism	0.01128	0.07812
Rationality	0.00747	0.11180
Selfishness	0.02938	0.20056
Weak selfishness	0.12844	0.03437

Table 6. Increase of the fitness value

Strategy	Without the culture	With the culture
Altruism	164.32%	252.49%
Weak altruism	144.89%	258.20%
Rationality	71.79%	188.94%
Selfishness	297.05%	385.77%
Weak selfishness	343.95%	258.58%

rationality strategy was the one that showed lower growth in relation to the others, while selfishness strategies showed higher evolution.

6   Conclusion

In this paper, the GSREP game was adapted to a BDI-MAS society, using the Jason language, with the addition of group beliefs as the society "culture" common to all agents involved in the system, implemented as a CArtAgO artifact. We consider that the society culture is composed by the agents' reputations. This BDI version of the game was called the Cultural-GSREP game. Then, we analysed and compared the simulation results considering two scenarios, just taking into account or not the culture.

The equilibrium of Piaget's Social Exchange Theory is reached when occurs reciprocity in exchanges during the interactions. Our approach showed that with the evolution of the strategies the agents were able to maximize their adaptation values becoming self-regulators of exchanges processes and thereby contributing to increasing the number of successful interactions. All agents have evolved and contributed to the evolution of the society. Whenever the services offered are more fair (balanced), the greater is the number of successful interactions. Comparing the two scenarios, we conclude that the addition of the culture – the reputation as a focal point – in social exchanges had the expected influence on the evolution of the agents's strategies and exchange processes, increasing the exchanges successfully performed and the fitness value in a shorter time.

Future work will consider the analysis of the final parameters of the strategies that emerged in the evolution process, and other categories of the cultural knowledge in the belief space, using belief artifacts in different scopes beyond the reputation, and creating different ways for the agents to reason about the group beliefs.

Acknowledgments. Supported by CNPq (Proc. No. 481283/2013-7, 306970/2013-9 and 232827/2014-1).

References

1. Blau, P.: Exchange & Power in Social Life. Trans. Publish., New Brunswick (2005)
2. Boissier, O., Bordini, R.H., Hübner, J.F., Ricci, A., Santi, A.: Multi-agent oriented programming with JaCaMo. Science of Computer Programming **78**(6), 747–761 (2013)
3. Bordini, R.H., Hübner, J.F., Wooldrige, M.: Programming Multi-agent Systems in AgentSpeak Using Jason. Wiley Series in Agent Technology. John Wiley & Sons, Chichester (2007)
4. Bratman, M.E.: Intention, plans, and practical reason. Cambridge University Press (1999)
5. Castelfranchi, C., Falcone, R.: Principles of trust for MAS: cognitive anatomy, social importance and quantification. In: Intl. Conf. of Multi-agent Systems (ICMAS), pp. 72–79 (1998)
6. Castelfranchi, C., Falcone, R., Firozabadi, B., Tan, Y.: Special issue on trust, deception and fraud in agent societies. Applied Artificial Intelligence Journal **1**, 763–768 (2000)
7. Criado, N., Argente, E., Botti, V.: Open issues for normative multi-agent systems. AI Communications **24**(3), 233–264 (2011)
8. Dignum, V., Dignum, F. (eds.): Perspectives on Culture and Agent-based Simulations. Springer, Berlin (2014)
9. Dimuro, G.P., Costa, A.R.C., Gonçalves, L.V., Pereira, D.: Recognizing and learning models of social exchange strategies for the regulation of social interactions in open agent societies. Journal of the Brazilian Computer Society **17**, 143–161 (2011)
10. Grimaldo, F., Lozano, M.A., Barber, F.: Coordination and sociability for intelligent virtual agents. In: Sichman, J.S., Padget, J., Ossowski, S., Noriega, P. (eds.) COIN 2007. LNCS (LNAI), vol. 4870, pp. 58–70. Springer, Heidelberg (2008)
11. Huynh, T.D., Jennings, N.R., Shadbolt, N.R.: An integrated trust and reputation model for open multi-agent systems. JAAMAS **13**(2), 119–154 (2006)
12. Macedo, L.F.K., Dimuro, G.P., Aguiar, M.S., Coelho, H.: An evolutionary spatial game-based approach for the self-regulation of social exchanges in mas. In: Schaub, et al. (eds.) Proc. of ECAI 2014–21st European Conf. on Artificial Intelligence. Frontier in Artificial Intelligence and Applications, no. 263, pp. 573–578. IOS Press, Netherlands (2014)
13. Nguyen, N.T., Katarzyniak, R.P.: Actions and social interactions in multi-agent systems. Knowledge and Information Systems **18**(2), 133–136 (2009)
14. Padgham, L., Scerri, D., Jayatilleke, G., Hickmott, S.: Integrating BDI reasoning into agent based modeling and simulation. In: Proc. WSC 2011, pp. 345–356. IEEE (2011)

15. Pereira, D.R., Gonçalves, L.V., Dimuro, G.P., Costa, A.C.R.: Towards the self-regulation of personality-based social exchange processes in multiagent systems. In: Zaverucha, G., da Costa, A.L. (eds.) SBIA 2008. LNCS (LNAI), vol. 5249, pp. 113–123. Springer, Heidelberg (2008)
16. Piaget, J.: Sociological Studies. Routlege, London (1995)
17. Rabin, M.: Incorporating fairness into game theory and economics. The American Economic Review **86**(5), 1281–1302 (1993)
18. Reynolds, R., Kobti, Z.: The effect of environmental variability on the resilience of social networks: an example using the mesa verde pueblo culture. In: Proc. 68th Annual Meeting of Society for American Archeology, vol. 97, pp. 224–244 (2003)
19. Reynolds, R., Zanoni, E.: Why cultural evolution can proceed faster than biological evolution. In: Proc. Intl. Symp. on Simulating Societies, pp. 81–93 (1992)
20. Ricci, A., Viroli, M., Omicini, A.: The A&A programming model and technology for developing agent environments in MAS. In: Dastani, M., Seghrouchni, A.E.F., Ricci, A., Winikoff, M. (eds.) ProMAS 2007, vol. 4908, pp. 89–106. Springer-Verlag, Heidelberg (2008)
21. Rodrigues, M.R., Luck, M.: Effective multiagent interactions for open cooperative systems rich in services. In: Proc. AAMAS 2009, Budapest, pp. 1273–1274 (2009)
22. Schelling, T.C.: The strategy of conflict. Harvard University Press, Cambridge (1960)
23. Schmitz, T.L., Hübner, J.F., Webber, C.G.: Group beliefs as a tool for the formation of the reputation: an approach of agents and artfacts. In: Proc. ENIA 2012, Curitiba (2012)
24. Serrano, E., Rovatsos, M., Botía, J.A.: A qualitative reputation system for multiagent systems with protocol-based communication. In: AAMAS 2012, Valencia, pp. 307–314 (2012)
25. Xianyu, B.: Social preference, incomplete information, and the evolution of ultimatum game in the small world networks: An agent-based approach. JASSS **13**, 2 (2010)

Modelling Agents' Perception: Issues and Challenges in Multi-agents Based Systems

Nuno Trindade Magessi$^{(\boxtimes)}$ and Luís Antunes

GUESS/BioISI, Faculdade de Ciências, Universidade de Lisboa, Lisbon, Portugal
nmaggessi@hotmail.com, xarax@ciencias.ulisboa.pt

Abstract. In virtual agents modelling, perception has been one of main focus for cognitive modelling and multi-agent-based simulation. Research has been guided by the representation of human senses operations. In this sense, perception focus remains on the absorption of changes that occur in the environment. Unfortunately, scientific literature has not covered the representation of most of the perception mechanisms that are supposed to exist in an agent's brain like for example ambiguity. In terms of multi-agent based systems, perception is reduced to a parameter forgetting the complex mechanisms behind it. The goal of this article is to point out that the challenge of modelling perception ought to be centred on the internal mechanisms of perception that occur in our brains, which increases the heterogeneity among agents.

1   Introduction

During the last decade, studies simulating virtual agents (VA) in multi-agent-based simulation (MABS) systems have tried to bring more realism into modelling the perception of environment. Researchers have been focused their efforts on improving perception models, and corresponding techniques.

Nevertheless, recent work [1, 25] has produced sophisticated theoretical models for reproducing the human senses like sight and hearing. The models were integrated then in a sustainable multi-sense perception system, in order to put together a perceptual system capable of approximately replicating the human sensory system. In fact, this is a keystone to use VAs to simulate how the senses of people work in order to capture a dynamic and nondeterministic environment [1]. The major problem of these proposals is to summarise perception into the operations of human senses.

Many of the psychological activities involved in perception, as well as the inherent mechanisms of the brain subsystems associated to it, have been overlooked. Is a VA sure about what it is capturing under this multiple senses frameworks? Is it reality? Clearly, the answer is no to both questions. The VA perception described in literature misses and does not represent the principal target of human perception: recognition.

This article proposes to discuss the challenges involved in perception, including the reproduction of all the mechanisms behind this cognitive process. The multi-senses framework represents only a small component of this huge and complex process that is perception. Perception goes beyond faithful representation of input sensors. As an

© Springer International Publishing Switzerland 2015
F. Pereira et al. (Eds.) EPIA 2015, LNAI 9273, pp. 687–695, 2015.
DOI: 10.1007/978-3-319-23485-4_69

example, the article will focus on risk perception to demonstrate the challenge and what is involved in it.

Section 2 reviews the literature on modelling perception for virtual agents in a multi-agent-based system. Section 3 revisits the concepts behind perception. Section 4 discusses the principal main issues in the literature and presents the crucial challenges, that modelling perception will have in near future. Section 5 describes our vision for implementing perception and section 6 puts some conclusions forward.

2   Related Literature

Literature presents different standpoints about endowing VAs with perception for MABS. The most frequently used approach is to ensure that VAs have a generalised knowledge about the environment [2]. This approach does not allow us to use correctly perception in order to simulate realistic scenarios, because the VA is not certain about the veracity of what is capturing. The opposite case is the one in which agents take their decisions sustained on the collected data received through their multiple senses, having no knowledge about the environment, not even generic knowledge.

In between these extreme cases, we have agents that can perceive some information, have a conception of the environment around them and "act on their perception" [3]. In the case described by [4], agents have a graphical access to their environment. According the described concept of perception, agents choose which path on the graph is more feasible to achieve their target. There are several problems associated to this perspective. The main one is the assumption that environments are static, which raises difficulties in simulating complex scenarios. In this perspective, perception is incomplete and conceived in a very restricted form. Clearly, it is not adequate to use it for modelling realistic situations. Rymill and Dodgson have developed a method to simulate vision and attention of individuals in a crowd [5]. The simulation was done for open and closed spaces. Independently of the problems identified in the techniques to filter information from a highly dynamic environment, the issue remained that perception was incomplete and conceived in a very restricted form. Vision was modelled only as an input sensor and attention is its precedent on the cognitive process. Pelechano et al [6] made a debate about a simulator system for an evacuation scenario, but this system was inaccurate in representing real vision and consequently perception. Brooks [7] developed what he called creatures: a series of mobile robots operating without supervision in standard office environments. The intelligent system behind them was decomposed into independent and parallel activity producers, all of which interfaced directly to the world through perception and action, rather than interfacing to each other.

Other proposals, like [8], had built-in simulators to describe hearing. However olfactory perception is limited to a few published articles with no consistent simulator. And no study is known for simulating tactile senses. Steel et al [9] proposed and developed a cohesive framework to integrate, under a modular and extensible architecture, many virtual agent perception algorithms, with multiple senses available.

Their architecture allows the assimilation, in the sense of integration, of dynamic and distributed environments. They conceive perception according to an environment module, where information is extracted and transposed to the agent's memory. Clearly, they identify the brain as being outside the scope of perception and more related to memories. Kuiper et al [10] associated the vision process to perception and presented more efficient algorithms to process visual input, which were entirely implemented under the DIVAs (Dynamic Information Visualization of Agents systems) framework. Recently, Magessi et al [11] presented an architecture for risk perception. This architecture puts the main focus on the representativeness of perception as it is performed in reality by individuals. Vision and other senses are designated by input sensors and were considered as one component of this cognitive process.

3   Perception

3.1   Definition

Perception is one of the cognitive processes in the brain that precede decision making. Perception is the extraction, selection, organisation and interpretation of sensory information in order to recognise and understand the environment [12]. Perception is not restricted to passivity upon reception of input signals. Perception can suffer the influence of psychological, social and cultural dimensions [13]. Psychology influences perception through capabilities and cognitive factors. For example, one individual who suffers from a psychological disorder may have the notion that his/her perception may be being affected. Concerning the social dimension, the influence comes from the interaction among individuals in society, towards imitation or persuasion, for example. Learning, memory, and expectations can shape the way in which we perceive things [13, 14]. Perception involves these "top-down" effects as well as "bottom-up" methods for processing sensory input [14]. The "bottom-up" processing is basically low-level information that is used to build up higher-level information (e.g., shapes for object recognition). The "top-down" processing refers to recognition task in terms of what it was expected in a specific situation. It is a vital factor in determining where entities look like and knowledge that influence perception [23]. Perception depends on the nervous system complex functions, but subjectively it seems mostly effortless, because processing happens outside conscious awareness [10].

However, it is important to realise that if we want to have a more complete attention mechanism related to vision, the work must be conducted by the interaction of bottom-up factors based on image features and top-down guidance based on scene knowledge and goals. The top-down component could be understood as the epicentre of attention allocation when a task is at hand. Meanwhile, the bottom-up component acts as reaction mechanism of alert. It allows the system to discover potential opportunities or risks in order to stop threatening events. While the top-down process establishes coherence between the environment looked by agent and its goals or tasks, the bottom-up component has the intent of reproducing the alert mechanisms, warning about objects or places relevant to the agent.

3.2   The Perception Process

The perception process starts with a stimulus on body sensory organs [12]. These sensory organs transform the input energy into neural activity through transduction [12]. Then, neural signals are transmitted to the brain and therein processed [12]. The resulting mental re-creation of the distal stimulus is the percept. Perception is sometimes described as the process of constructing mental representations of distal stimuli using the information available in proximal stimuli.

People typically go through the following steps to form judgements: (i) first, when they face an unknown target, they ignite the interest for this target. This means that they activate their attention; (ii) second, people start to extract and select more information about the target. Incrementally, people find hints that they associated to similar experiences and those hints help them to interpret and categorise the target; (iii) in the third step, the hints become less efficient and selective. Thus, people try to look for more hints with the intent of confirming the categorisation of the target. Unfortunately, people also actively ignore and even distort hints that are against their initial perceptions. Perception becomes gradually selective and people finally achieve a judgment about the target.

Casual perception is one of fields with huge development nowadays [24]. It consists of "the relatively automatic, relatively irresistible perception of certain sequences of events as involving causation". Normally, casual perception does not use conscious thoughts or reasoning. It is a kind of "launching effect" in which people perceive spontaneously.

3.3   Affordances

In [16], Gibson developed an interaction approach on perception and action, settled on information available in the environment. He refused the framing assumption of factoring external-physical and internal-mental processes. The interaction alternative is centred on processes of agent-situation interactions that come from ecological psychology and philosophy, namely situation theory [26, 27]. The concept of affordance for an agent can be defined as the conditions or constraints in the environment to which the agent is attuned. This broad view of affordances includes affordances that are recognised as well as affordances that are perceived directly.

Norman used the term affordances to refer just the possibilities of action that are perceivable by an individual [17]. He made the concept dependent on the physical capabilities of an agent and his/her individual goals, plans, values, beliefs, or past experiences. This means that he characterised the concept of affordance as relational, rather than subjective or intrinsic. In 2002, Anderson et al. [18] sought that directed visual attention, and not affordance, is the key responsible for the fast generation of many motor signals associated with the spatial characteristics of perceived objects. They discovered this by examining how the properties of an object affect an observer's reaction time for judging its orientation.

3.4   Perception vs. Reality

For some individuals, the perceived environment, event or object can differ from what it is in reality. Their perception could put themselves far from what is in reality. An object could be perceived differently by each person. This phenomenon is commonly designated by perception gap [19]. This finding is patent in many psychological studies. For example, in the case of visual perception, there are individuals able to acknowledge the perception gap in their minds. Others may not recognise the shape shifting when the object changes. This happens when objects are ambiguous and multiple interpretations can be made on the perceptual level. So to reproduce human perception in VAs, the perception gap must be taken into account and reflected in the models.

4   Issues and Challenges

The agents' perception assumes a critical preponderance in defining their decision, which is then reflected in their behaviours, as social actors. Perception delivers information from the surrounding environment, which assists agents on their activities of planning and decision making [20]. In most approaches, the sensor systems architecture is represented, instead of the complete perception process. Most of the works are confined to relate the upstream part of the process. In some cases, for example the vision sense is not designed as part of the brain [8], which goes against the usual accounts from most of the relevant scientific areas involved. In most cases, the interpretation of information component, which is critical to the success of the process, is not described. The majority of the studies have the assumption that everything which was picked up by senses is reflected on the VA's knowledge [8]. This a strong assumption, which has no correspondence with reality. Interpretation is the goal activity of perception [14]. Grasping an object and having success on its recognition depends on the effectiveness and efficiency of interpretation. We must not forget that the main goal of perception is to produce a judgement about what was socially analysed. This judgement may or may not be stored in memory, depending of its relevance. This aspect is never mentioned in VA's literature. For example, Brooks [7], with his intelligent "Creatures" argues that perception is not necessary as central interface. This is not correct if we want to have robots acting like humans. The proof is on his own work, where he decomposed sensor data in many different sorts of processing, which proceed independently and in parallel, each affecting the "overall system". The "overall system" is in fact an example of spatial perception, ensured by the right extraction and selection of data input. The success of "Creatures" with multiple processes comes from the fact that perception is deeply rooted on his algorithm. However, the robots could not output the result that took them into action. Unfortunately, Brooks traces a direct and linear relation between input sensor and action. He omits representations and implicitly considers an action as decision, which is wrong because decision could result in action or not [16, 21]. This approach is overly simplistic, similarly to a common confusion between judgement (perception) and decision. People can only perceive something if they have a representation of that thing, or from the parts that compose it, even if inchoate.

Other researchers assume that every perception even culminates in storing. However, memory should not be seen as passive, a simple storage of data collected by sensors. It must be seen under a dualistic perspective. VA memory should also have an influence on perception, because to perceive something we need to have the semantic knowledge in our semantic memory, for that object or event. Otherwise, VAs have to learn first, beside of accelerating recognition.

Clearly, the first challenge is to systematise all the perception process, including the missed activities or dimensions that have the incumbency to format agent's perception. The second challenge is to bring to Multi-Agent Based Systems the capacity to represent the interconnections among psychological, social and cultural dimensions involved in perception [11]. These dimensions and subsequent factors are the keystones for the dynamics of perception.

The third challenge, which is both ambitious and complex, is to establish the macro-micro link between a specific judgement and the neuro-physiological dimension of perception. Modelling perception of VAs cannot be trapped to upstream stage and moving on to the downstream stage of the process

5   Vision: Paths to Achieve Our Goals

Taking into account the issues and challenges described above, it is important to figure out what would be the consequences if we improved perception modelling. The major consequence is to separate perception from decision in VAs, similarly to what happens in reality. This is determinant to understand many issues related to decision science, where in fact the relation between what we assimilate and decision is not linear. If we want to understand why a decision-maker took an incorrect decision we need to have clearly modelled his decision and perception processes. If the problem came from perception, it is relevant to pin down in which part of the process it occurred. This brings more heterogeneity to agents in multi-agent based systems.

Another important consequence is to understand if an agent perceived the reality surrounding him. Or instead, if he perceived something different from reality, when he took the decision.

In terms of improving perception modelling in robots, the strategy goes by the use of very simple cases, like the perception of a common figure which has associated ambiguity. For example, the Rubin Vase, which has two interpretations, either as a vase or as two faces. This can be done under pixel techniques, where the captured images allow robots to perceive some figure formats when a connection is established with their own semantic memory.

One of the common mistakes is to insist in capturing some kind of standard perception, common to all individuals. Of course, people have mechanisms in common for perception. However, perception is highly subjective, since it depends on the past experiences of each individual. These experiences and associated acquisition define his/her representation of an object, figure, event or environment. So, instead of searching for (or defining) standard mechanisms of perception, we could replicate the perception of one individual. More specifically, to try to clone a specific person perception by using its own description of what this person is perceiving. In the Rubin Vase example, this means that one robot could recognise a vase and another could recognise faces. Everything depends on the forms (vases or faces) that were collected in the past by each robot and stored on their own semantic memory. In a case of multi-agent based system, the ambiguity of perception is present on the way that agents interpret the variations occurred in some parameters.

Another important key point about modelling perception is to build multi-disciplinary teams to work on it. In this sense, the operational strategy must continue, refresh and fix the idea of [7] where Brooks developed multiple algorithms working in parallel to pursue perception. This strategy happens because a stimulus could not be transformed into a percept. Our claim is that an ambiguous stimuli may be transformed into multiple perceptions, experienced randomly, one at a time, in what is called "multi stable perception." [22] However, the same stimuli, or absence of them, may induct in different perceptions depending on the person's culture and previous experiences. After we integrate fundamental psychological insights in perception modelling and the advance of neuroscience brings us new inputs continuously, modellers will be able to substitute the developed algorithms by new ones, where these replicate what happens in real physiology. So, this vision clearly defends that is possible to build robots with perception similar to human beings if it focuses on a specific target and/or individual.

6   Conclusion

Part of VAs literature on perception is focused on building a multi agent simulator with a lot of features related to input sensors, instead of demanding the complete perception process. This article claims that this is not perception and the major challenge is to go forward in building a complete and holistic account of this cognitive process. Introducing psychological and physiological insights can ensure that virtual agents replicate better what happens in reality. In this sense, the challenge is to establish the macro-micro link for perception, from the physiological dimension to the final judgement.

References

1. Ray, A.: Autonomous perception and decision-making in cyberspace. In: The 8th International Conference on Computer Science & Education (ICCSE), Colombo, Sri Lanka, April 26–28, 2013
2. Uno, K., Kashiyama, K.: Development of simulation system for the disaster evacuation based on multi-agent model using GIS. Tsinghua Science and Technology 13(1), 348–353 (2008)

3. Shi, J., Ren, A., Chen, C.: Agent-based evacuation model of large public buildings under fire conditions. Automation in Construction **18**(3), 338–347 (2009)
4. Sharma, S.: Simulation and modelling of group behaviour during emergency evacuation. In: Proceedings of the IEEE Symposium on Intelligent Agents, Nashville, Tennessee, pp. 122–127, March 30–April 2, 2009
5. Rymill, S.J., Dodgson, N.A.: Psychologically-based vision and attention for the simulation of human behaviour. In: Proceedings of Computer Graphics and Interactive Techniques, Dunedin, New Zealand, pp. 229–236, November 29–December 2, 2005
6. Pelechano, N., Allbeck, J., Badler, N.: Controlling individual agents in high-density crowd simulation. In: 2007 ACM SIGGRAPH/Eurographics Symposium on Computer Animation, San Diego, California, pp. 99–108, August 2–4, 2007
7. Brooks, R.A.: Intelligence without representation. Artificial Intelligence **47**, 139–159 (1991)
8. Piza, H., Ramos, F., Zuniga, F.: Virtual sensors for dynamic virtual environments. In: Proceedings of 1st IEEE International Workshop on Computational Advances in Multi-Sensor Adaptive Processing (2005)
9. Steel, T., Kuiper, D., Wenkstern, R.: Virtual agent perception in multi-agent based simulation systems. In: IEEE/WIC/ACM International Conference on Web Intelligence and Intelligent Agent Technology (2010)
10. Kuiper, D., Wenkstern, R.Z.: Virtual agent perception in large scale multi-agent based simulation systems (Extended Abstract). In: Tumer, K., Yolum, P., Sonenberg, L., Stone, P. (eds.) Proc. of 10th Int. Conf. on Autonomous Agents and Multiagent Systems (AAMAS 2011), Taipei, Taiwan, pp. 1235–1236, May 2–6, 2011
11. Magessi, N., Antunes, L.: An Architecture for Agent's Risk Perception. Advances in Distributed Computing and Artificial Intelligence Journal **1**(5), 75–85 (2013)
12. Schacter, D.L., Gilbert, D.T., Wagner, D.M.: Psychology, 2nd edn. Worth, New York (2011)
13. Magessi, N., Antunes, L.: Modelling agents' risk perception. In: Omatu, S., Neves, J., Corchado Rodriguez, J.M., Paz Santana, J.F., Gonzalez, S.R. (eds.) Distributed Computing and Artificial Intelligence. AISC, vol. 217, pp. 275–282. Springer, Heidelberg (2013)
14. Bernstein, D.A: Essentials of Psychology. Cengage Learning. pp. 123–124. ISBN 978-0-495-90693-3 (Retrieved March 25, 2011)
15. Gregory, R.: Perception. In: Gregory, R.L., Zangwill, O.L., pp. 598–601 (1987)
16. Gibson, J.: The theory of affordances. In: Shaw, R., Bransford, J. (eds.) Perceiving, Acting, and Knowing (1977). ISBN 0-470-99014-7
17. Norman, D.: Affordance, Conventions and Design. Interactions **6**(3), 38–43 (1999)
18. Anderson, S.J., Yamagishi, N., Karavia, V.: Attentional processes link perception and action. Proceedings of the Royal Society B: Biological Sciences **269**(1497), 1225 (2002). doi:10.1098/rspb.2002.1998
19. Ropeik, D.: How Risky Is It, Really? Why Our Fears Don't Always Match the Facts. McGraw Hill, March 2010
20. Steel, T., Kuiper, D., Wenkstern, R.Z.: Context-aware virtual agents in open environments. In: Proceedings of the Sixth International Conference on Autonomic and Autonomous Systems (ICAS 2010), Cancun, Mexico, March 7–13, 2010
21. Fine, K., Rescher, Nicholas: The Logic of Decision and Action. Philosophical Quarterly **20**(80), 287 (1970)

22. Eagleman, D.: Visual Illusions and Neurobiology. Nature Reviews Neuroscience **2**(12), 920–926 (2001). doi:10.1038/35104092. PMID: 11733799
23. Yabus, A.: "Eye movements and vision", chapter Eye movements during perception of complex objects. Plenum Press, New York (1967)
24. Danks, D.: The Psychology of Causal Perception and Reasoning. In: Beebee, H., Hitchcock, C., Menzies, P. (eds.) Oxford Handbook of Causation. Oxford University Press, Oxford (2009)
25. Kurzweil, R.: How to Create a Mind: The Secret of Human Thought Revealed. Viking Books, New York (2012). ISBN 978-0-670-02529-9
26. Barwise, J., Perry, J.: Situations and attitudes. MIT Press/Bradford, Cambridge (1983)
27. Devlin, K.: Logic and information, pp. 49–51. Cambridge University Press (1991)

Agent-Based Modelling for a Resource Management Problem in a Role-Playing Game

José Cascalho[✉] and Pinto Mabunda

Universidade dos Azores, Pólo de Angra do Heroísmo, Azores, Portugal
jmc@uac.pt, pintomabunda@yahoo.com

Abstract. In this paper we present a prototype of a model created in the context of a resource management problem in Gaza, Mozambique. This model is part of a participatory approach to deal with a conflict of water supply. Farmers and cattle producers are added to a stylized environment and a conflict is modelled when cattle needs to access water and destroy farmers' harvest. To address the different behaviours of farmers and cattle producers, a BDI architecture is used to support the conflict simulation using a simple argument-based negotiation between proactive agents. This model is intended to be used as a support to a Role Playing Game (RPG) in the context of an interactive design assembled under the Netlogo tool environment.

Keywords: Agent-based modelling · BDI · Conflict management

1 Introduction

In Gaza, a province of Mozambique, a scenario of conflict exists between farmers and cattle producers. Both stakeholders need water and although the resource exists in abundance, the lack of planning to circumscribe an area for the cattle and a specific local for the agriculture, have been responsible for the increase the number of conflicts between these two activities. The cattle are usually abandoned in the fields near the river and, alone, these animals they follow an erratically trajectory to the water destroying cultivated fields near the river. Local authorities have difficult to deal with the problem because cattle producers argue that they have the right to have the cattle in lands that belong to community. Although cattle producers pay fines for the farmers' harvest losses, their behaviour seems not to change. Ancient practices are difficult to modify.

To address this problem, it was decided to follow the steps identified in the *companion modelling* approach [4]. With this approach we expect to promote an open debate inside the community and to help to find a participated solution that will help to overcome the problem. Although some solutions have been discussed between local authorities and population in general, the lack of investment and the difficulty to joint stakeholders to discuss the problem have delayed the implementation of a definite solution.

Role Playing Games (RPG) have been used for different proposes and one of them is the social learning. In fact, with RPG, it is possible to "reveal some

© Springer International Publishing Switzerland 2015
F. Pereira et al. (Eds.) EPIA 2015, LNAI 9273, pp. 696–701, 2015.
DOI: 10.1007/978-3-319-23485-4_70

aspects of social relationships, allowing the direct observation of interactions among players" [1]. Players are the stakeholders in a problematic context. The use of RPG in these cases, is useful in certain phases of negotiation for different contexts, such as water, land-use and other resources (e.g. [8][6]).

In this project, the participatory approach is an essential feature for its own success. A RPG will be implemented supported by an agent-based model [5]. As described in the *companion approach*, we intend to add a board game in which the stakeholders can show along the game, how they decide in case of conflict. On the other hand, we are looking for some degree of autonomy in the agents created in the context of an agent-based modelling (ABM) approach. In fact, we want to add proactive agents, implementing an architecture that provides agents' behaviour related to the personality traits that is possible to identify in the cattle producers and farmers because it gives us the possibility to generate unexpected events which are important in the context of a simulation [5]. The modelling of a real case-study led us to choose a BDI architecture [7] [10] to support the modelling of some of the decisions of the agents in conflict. This paper describes the first steps towards this implementation.

In the following section we will describe some of agent-based models used in simulation concerning resource management. Then, it is explained how conflicts are modelled in the context of the RPG and the implementation of the BDI agents' architecture. In the last section the conclusions are presented.

2   Modelling Resource Management in ABM and Role Playing Games

The use of ABM to investigate environmental environmental management must have as focus, the social interaction and the ecological dynamics. Agents represent stakeholders at an aggregated or individual level. Environment, which holds renewable resources, are defined as the landscape. These resources are typically modified by agents in the environment. One recurrent characteristics of these systems is the representation of space which usually contributes to the structure of interactions among agents. Different approaches can be used to create the model. In our case-study, the RPG was intentionally selected because it fits the desire to have a participatory approach that contributes to "social learning". RPG is part of the modelling process in which stakeholders participate actively. On the other hand, ABM supports the interaction between stakeholders and leverage the knowledge about the context that is being modelled.

3   Modelling the Scenario for Farmers and Producers

In the following subsections we present the agents modelled as well as the main interactions between them, representing different stakeholders in conflict described. In the scenario implemented, two conflictual agents are provided with a simplified dialogue and in which different types of behaviour are identified.

It was assumed that the scenario could leverage our understanding of how the different stakeholders see the conflict. As already pointed out, a BDI architecture is used to support agents modelling in the ABM implementation. It is expected, as the RPG is implemented, the agents' model will be improved, adding beliefs, desires, rules and filters which underly the decisions observed in the real negotiation context.

3.1   The BDI Architecture and Conflict Resolution

A BDI architecture is used to support agents interactions along the conflict resolution. We are only interested to model the negotiation between a farmer and a producer about the price to pay for the loss in the farmer production.

D:	des(get-payed-for-damaged);		D:	des(pay-for-damaged);
B:	bel(harvest-damaged, true,1); bel(respons-for damaged, #producer, #degree-certainty); bel(damage-patch-extension, #number-patch, #degree-certainty); bel(is-respons-pay, #money, #degree-certainty); bel(received-money-from-damage, false,1); *Farmer*		B:	bel(harvest-damaged, true, #degree-certainty); bel(damage-patch-extension, #number-patch, #degree-certainty); bel(respons-for-damage, #number-patches, #degree-certainty); bel(is-responsible,true, #degree-certainty); bel(already-payed-for-damage,false,1); bel(pay-for-patch,10); *Producer*
F:			F:	**F1:***if* (#degree-certanity < #threshold) *do* intention#1 *else do* intention#2
R:	**D**(*des(get-payed-for-damaged)*) + **B**(*bel(harvest-damaged, true,1)*) → **I**(*intention(get-payed-for-damaged)*)		R:	**D**(*des(pay-for-damaged)*) + **B**(*bel(is-responsible, true, #degree-certainty)* + *bel(already-payed-for-damage, false, 1)*) **(F1)** →**I**(*intention(pay-for-damage)*)
I:	**get-payed-for-damaged** *intention(get-payed-for-damaged)*: illocutionary act with info-seeking and negotiation steps.		I:	**pay-for-damage** *intention(pay-for-damage-1)*; *intention(pay-for-damage-2)*: illocutionary act with negotiation steps;

Fig. 1. BDI architecture supporting the negotiation between farmers and producers.

The figure 1 shows a model for beliefs, desires and intentions of the agents inspired in [7]. The personality traits are related to decisions related to the intention to be executed. This mechanism in the architecture is implemented using a filter (F1) that defines the cases for which an agent selects one option. Notice that this process of decision depend upon the value of uncertainty of the beliefs. The rules are used to define which desires and beliefs activate which intentions. These rules are also part of the agent's trait. We adopt the model proposed by [9] for the dialogue protocol. Two protocols are used. One for information-seeking (info-seeking) and other for negotiation (negotiation). They are defined as simple request-response message sequences between two agents. While the former are used to ask for some information, the latter is used for exchange resources. In the case modelled, farmers ask to producers about their commitment to pay. Then they negotiate the value to pay in different contexts, as a result of successive agreements. The producers may have two different behaviours. They can assume that they agree with the farmer point of view or, otherwise, will have to negotiate. Although this is a very simple protocol, our goal was to test the

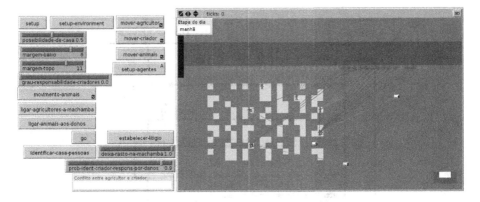

Fig. 2. The interaction environment in the Netlogo representing the conflict between the farmer and the cattle producer, represented by a link (blue line connecting the two agents).

model to understand if it could provide with an initial model in which farmers with different traits could be modelled and to address different behaviours to each one of these traits.

3.2 Experiment in Netlogo

Netlogo[11] has been used as a simulation platform aimed at supporting the realisation of (multi)agent-based simulations. It is a platform that provides a general purpose framework. A typical simulation consists in a cycle where agents, dubbed as turtles, are chosen to be performed an action, considering its situation and state. The stylised scenario and the easy and versatile interaction with the user (e.g. setting up different initial conditions) were the main reasons to select Netlogo.

The Interface for the Participatory Approach. The interface has two distinct goals. The first one is to provide a styllized environment in which stakeholders could identify the narrative space of the events described which are the farmers, the red human-shape agents and the producers, the blue human-shape agents. The latter is to foster autonomy to generate a simulation of events that create the conflictual situations which are intended to be studied. In the case-study, the river is identified as a blue area and the village where farmers and producers live, correspond to the yellow area. The green area along the right margin of the river is where the farmers have their cultivated areas. The interface is prepared to interact with users. As the game is played some of the actions are autonomous (e.g. the motion of the cattle to drink water). The red leaf-shape agents are the cultivated areas damaged by the cattle. The link between a farmer and a producer shows a conflict between them. The producer owns the cattle identified with the number 4 which damaged a large area of plantation that belongs to the farmer.

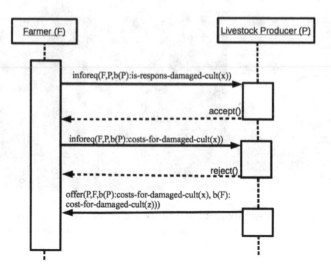

Fig. 3. The information-seeking messages sent by the farmer and the response of the producer.

Interaction through the Conflict Resolution. The figure 3 presents the sequences of messages exchanged by the farmers and the livestock producers. When a conflict is detected (i.e. a cultivated area of the farmer is damaged), the farmer invites the producer to negotiate. The messages sent by the farmer are information-seeking messages. The example shows that the producer wishes to negotiate the value to be payed by the damaged. Along the negotiation, the producer and farmer will try to agree into a final value. A maximum number of interactions along with thresholds concerning the limit values to achieve in negotiation i.e. beliefs in the context of negotiation (not showed in the figure 3), dictates the end of negotiations. The game in the context of the negotiation have uncertainties that might be identified as a result of the degree of certainty concerning beliefs and also the trait of the agents. Accepting conditions imposed by the farmer or not, will drive the producer to negotiate.

4   Conclusions

In this paper we present a prototype for an agent-based modelling in the context of conflict within a resource management in Gaza, Mozambique. To address this problem we chose the RPG approach. The following steps were made:

1. To create a stylized scenario where stakeholders could interact and to identify the situation of conflict;
2. To define a protocol of communication between agents in conflict. The propose of this protocol is to support the interaction of the simulation with the different stakeholders;

3. To provide a bdi-architecture to the agents in conflict. This architecture contributes to the definition of traits and improves the autonomy of the agents defined in the scenario.

This model will be used in the context of a RPG. A board game will be created and the stylized world in the simulated environment will support the game. The stakeholder decisions will be in part as result of game interaction and also based on the knowledge acquired along the steps of implementation of the model. The agents traits allow the RPG to interact with agents with different degree of autonomy. In scenarios where a social learning is a target (e.g. Sylvopast model [5]), a generation of contexts where unexpected events occurs are important to improve the learning and to foster the interaction between the stakeholders.

References

1. Adamatti, D.F., et al.: A prototype using multi-agent-based simulation and role-playing games in water management. In: CABM-HEMA-SMAGET, 2005, Bourg-Saint-Maurice, Les Arcs. CABM-HEMA-SMAGET 2005, pp. 1–20. CDROM (2005)
2. Bandini, S., Manzoni, S., Vizzari, G.: Agent Based Modeling and Simulation: An Informatics Perspective. Journal of Artificial Societies and Social Simulation 12(4), 4 (2009)
3. Barreteau, O., Bousquet, F., Attonaty, J.M.: Role-playing games for opening the black box of multi-agent systems: method and lessons of its application to Senegal River Valley irrigated systems. Journal of Artificial Societies and Social Simulation 4(3) (2001)
4. Barreteau, O., et al.: Participatory approaches. In: Edmonds, B., Meyer, R. (eds.) Simulating Social Complexity, pp. 197–234. Springer, Heidelberg (2013)
5. Bousquet, F., et al.: Multi-agent systems and role games: collective learning processes for ecosystem management. In: Janssen, M.A. (ed.) Complexity and Ecosystem Management: The Theory and Practice of Multi-agent Systems, pp. 249–285. E. Elgar, Cheltenham (2002)
6. Briot, J.-P., et al.: A computer-based role-playing game for participatory management of protected areas: the SimParc project. In: Anais do XXVIII Congresso da SBC, Belém do Pará (2008)
7. Cascalho, J., Antunes, L., Corrêa, M., Coelho, H.: Characterising agents' behaviours: selecting goal strategies based on attributes. In: Klusch, M., Rovatsos, M., Payne, T.R. (eds.) CIA 2006. LNCS (LNAI), vol. 4149, pp. 402–415. Springer, Heidelberg (2006)
8. Cleland, D., et al.: REEFGAME as a Learning and Data-Gathering Computer-Assisted Role-Play Game. Simulation Gaming 43, 102 (2012)
9. Hussain, A., Toni, F.: Bilateral agent negotiation with information-seeking. In: Proc. of the 5th European Workshop on Multi-Agent Systems (2007)
10. Luong, B.V., et al.: A BDI game master agent for computer role-playing games. In: Proceedings of the 2013 International Conference on Autonomous Agents and Multi-Agent Systems (AAMAS 2013), pp. 1187–1188 (2013)
11. Wilensky, U., Stroup, W.: Learning through participatory simulations: network-based design for systems learning in classrooms. In: Proceedings of Computer Supported Collaborative Learning (CSCL 1999). Stanford, CA, December 12–15, 1999

An Agent-Based MicMac Model for Forecasting of the Portuguese Population

Renato Fernandes[1](✉), Pedro Campos[2], and A. Rita Gaio[3]

[1] LIADD/INESC TEC, Faculdade de Ciências da Universidade do Porto,
Porto, Portugal
renato.fernandes@fc.up.pt
[2] LIADD/INESC TEC, Faculdade de Economia da Universidade do Porto,
Porto, Portugal
[3] Departamento de Matemática, Faculdade de Ciências da Universidade do Porto
and Centro de Matemática da Universidade do Porto, Porto, Portugal

Abstract. Simulation is often used to forecast human populations. In this paper we use a novel approach by combining Micro-Macro (MicMac) models into an Agent-Based perspective to simulate and forecast the behavior of the Portuguese population. The models include migrations and three scenarios corresponding to three different expected economic growth rates. We conclude that the increase in the number of emigrants leads to a reduction of the Portuguese women that are in the fertile age. This justifies the decrease of births and therefore the general decrease of the total Portuguese Population.

Keywords: Agent-Based computational demography · Social simulation · MicMac model · Forecasting

1 Introduction

Agent-Based computational demography models (Billari et al. [4] Ferber [5]) can deal with complex interactions between individuals, constituting an alternative to mainstream modelling techniques. Conventional population projection methods forecast the number of people at a given age an a given point in time assuming that the members of a cohort are identical with respect to demographic behaviour. Different approaches include: macro simulation, based on policy interventions and other external events and conditions, and micro simulation, based on life courses of individual cohort members. Micro and Macro (MicMac) (Gaag et al. [6]) approaches offer a bridge between aggregate projections of cohorts. These are important contributes for the sustainability of the health care and pension systems, for example, as they are issues of current concern.

We construct an Agent-Based model, based on the MicMac approach, to simulate the behaviour of the Portuguese population, open to migrations. The notation and the main model components (fertility, mortality and migration) are firstly introduced. Then the iterative simulation process is created and a forecast for the Portuguese Population from 2011 to 2041 is presented.

© Springer International Publishing Switzerland 2015
F. Pereira et al. (Eds.) EPIA 2015, LNAI 9273, pp. 702–707, 2015.
DOI: 10.1007/978-3-319-23485-4_71

2 Population Model

2.1 Fertility and Mortality

We start by establishing the notation and the main components of the model. The variables with A (resp. G) refer to **agent** variables (resp. **global** variables). The indices a, s, k and y are used to denote age, sex, agent identification and year, respectively. Any variable indexed by a, s, k, y represents the realization of the variable in the agent k, aged a years-old and of sex s, in the year y. A similar interpretation applies to any subset of these indices. The following variables are then defined:

$A_{a,s,k,y}^{\text{Alive}}$: vital status, taking the value 1 if the agent is alive and 0 otherwise

$G_{a,s,y}^{\text{Alive}}$: number of living agents; clearly $G_{a,s,y}^{\text{Alive}} = \sum_k A_{a,s,k,y}^{\text{Alive}}$.

G_y^{MaleFreq} : relative frequency of male agents, equal to $\sum_a G_{a,M,y}^{\text{Alive}} / \sum_{a,s} G_{a,s,y}^{\text{Alive}}$

$G_{a,s,y}^{\text{Births}}$: number of births of sex s given by female agents aged a years-old

$G_{a,y+1}^{\text{FertR}}$: global fertility rate

$G_{a,s,y}^{\text{Deaths}}$: number of deaths; it satisfies

$$G_{a,s,y}^{\text{Deaths}} = \#\{A_{a,s,k,y}^{\text{Alive}} = 0 \wedge A_{a-1,s,k,y-1}^{\text{Alive}} = 1\}, \quad a \neq 0$$

$$G_{0,s,y}^{\text{Deaths}} = \sum_a G_{a,s,y}^{\text{Births}} - G_{0,s,y}^{\text{Alive}}$$

$G_{a,s,y+1}^{\text{MortR}}$: global mortality rate.

Real data from the 2011 Portuguese Census is used as the base population, in a 2% size scale. The updating of the mortality and fertility rates is ensured by MicMac models, as in Gaag et al. [6]. The Mac part is ruled by the predictions obtained from Statistics Portugal [7], controlling the overall evolution of the variables. The Mic part is based on the results obtained from the previous year.

The controlling factor for the fertility (resp. mortality) rate is the expected mean fertility (resp. mortality) growth rate for the year y, denoted by G_y^{FertEvo} (resp. G_y^{MortEvo}). Then

$$G_{a,y+1}^{\text{FertR}} = \frac{\sum_s G_{a,s,y}^{\text{Births}}}{G_{a,F,y}^{\text{Deaths}} + G_{a,F,y}^{\text{Alive}}} G_y^{\text{FertEvo}}, \qquad G_{a,s,y+1}^{\text{MortR}} = \frac{G_{a,s,y}^{\text{Deaths}}}{G_{a,s,y}^{\text{Deaths}} + G_{a,s,y}^{\text{Alive}}} G_y^{\text{MortEvo}}.$$

Whenever the population size is very small, the mortality formula is replaced by $G_{a,s,y+1}^{\text{MortR}} = G_{a,s,y}^{\text{MortR}} \times G_y^{\text{MortEvo}}$.

The following random variables are created in order to achieve heterogeneity among the population of agents (division by 3 in the fractions ensures that its values lie between 0 and 1):

$$X_{a,y+1}^{\text{FertR}} \sim N(G_{a,y+1}^{\text{FertR}}, \sigma x_{a,y+1}^{\text{FertR}}), \quad \sigma x_{a,y+1}^{\text{FertR}} = min\{0.02, \frac{G_{a,y+1}^{\text{FertR}}}{3}, \frac{1 - G_{a,y+1}^{\text{FertR}}}{3}\}$$

$$X_{a,s,y+1}^{\text{MortR}} \sim N(G_{a,s,y+1}^{\text{MortR}}, \sigma x_{a,s,y+1}^{\text{MortR}}), \quad \sigma x_{a,s,y+1}^{\text{MortR}} = min\{0.02, \frac{G_{a,s,y+1}^{\text{MortR}}}{3}, \frac{1 - G_{a,s,y+1}^{\text{MortR}}}{3}\}.$$

The parameters described next are assigned to each agent:

$A_{a+1,k,y+1}^{\text{FertR}}$: probability for a female agent to give birth in the year y+1; it is given by $x_{a+1,y+1}^{\text{FertR}} \in X_{a+1,y+1}^{\text{FertR}}$

$A_{a+1,s,k,y+1}^{\text{MortR}}$: probability for an agent to die in the year y+1; it is given by $x_{a+1,s,y+1}^{\text{MortR}} \in X_{a+1,s,y+1}^{\text{MortR}}$.

The evolution process is done according to the following steps:

Step 1. Increase simulation year by one and age every living agent by one.
Step 2. Give birth to new agents according to the fertility rates: randomly sample u_1, u_2 from $U(0,1)$; if $u_1 < A_{a,k,y}^{\text{FertR}}$ then a new agent is born; if $u_2 < G_y^{\text{MaleFreq}}$ then set the agent to male sex, else set it to female.
Step 3. Randomly "kill" agents: randomly sample u_3 from $U(0,1)$; if $u_3 < A_{a,s,k,y}^{\text{MortR}}$, set $A_{a,s,k,y}^{\text{Alive}} = 0$.
Step 4. Compute next year fertility and mortality parameters and male proportion rates and define each agent's fertility and mortality.

2.2 Migration

The Portuguese population is also affected by migrations, with a high amount of entries and exits, summing a total negative net migration. Our model also includes such process. Throughout, c is an index denoting a given country while c_0 denotes Portugal.

G_c^{Health} : health indicator with A_k^{HealthW} as its corresponding weight

G_c^{Safety} : safety indicator with A_k^{SafetyW} as its corresponding weight

$G_{c,y}^{\text{Wage}}$: wage andicator with A_k^{WageW} as its corresponding weight

G_c^{Pop} : indicator for the Portuguese population size, with A_k^{PopW} as its corresponding weight

G_c^{Lang} : indicator for the Portuguese language, with A_k^{LangW} as its corresponding weight

G_c^{Limit} : emigration limits for country c, defined by the destination country

G_c^{ECounter} : emigration counter

The first four indicators range between 0 and 1. The used indicator must be the same for all countries and it is preferable that the data source is the same, because the same indicator may vary in different sources. G_c^{Lang} equals 1 if Portuguese is the native language and 0 otherwise. The wage indicator changes every year according to the country expected mean wage growth. Data were obtained from the UN and OECD databases [9], [8].

The above weights are assigned to each agent by a randomly sampled value from $N(\mu, 0.75\mu)$; μ for the first three and last weights are obtained from Balaz [3]. The value of A_k^{PopW} was based on findings from Anjos and Campos [2].

The gains of migrating also depend on the will to migrate, which is highly dependent on the age of the agent and its employment status. We define

$G_{a,s}^{EmpProp}$: proportion of employed individuals, obtained from the 2011 Portuguese rates of INE database [7]

$G^{EcoGrow}$: expected economic growth for Portugal

$A_{a,s,k,y}^{Emp}$: agent's employment status, coded -1 if employed and 1 otherwise.

Every year, and for each agent, $A_{a,s,k,y}^{Emp}$ is obtained by randomly sampling u from $U(0,1)$: if $u < \left(G_{a,s}^{EmpProp} G^{EcoGrow\,(y-y_0)} \right)$, the agent is employed; else it is unemployed. The variable A_k^{EmpW} is created as the weight for $A_{a,s,k,y}^{Emp}$ and is sampled from $N(\mu, 0.75\mu)$, where μ is chosen to fit the emigration data.

The Weibull distribution function $W(x, \lambda, k)$ (and its derivative $w(x, \lambda, k)$) is used to model the age effect on the will to emigrate. Its shape (λ) A_k^{Shape} and scale (k) A_k^{Scale} are estimated by statistical fitting. These values are also jittered for each agent, and three variables are created: $A_k^{X\text{-axis}}$ (resp. $A_k^{Y\text{-axis}}$) to perform dilatation of the function on the x-axis (resp. y-axis), and a third variable A_k^{Base} to define a base level for the will to emigrate.

For an agent k, its will to emigrate in the first year y_0 of the simulation or when it becomes 18 years-old, is

$$A_{a,k,y_0}^{Will} = A_k^{y\text{-axis}} \times W(A_k^{x\text{-axis}} \times a; A_k^{Shape}, A_k^{Scale}) + A_k^{Base}$$

and is updated by

$$A_{a+1,k,y+1}^{Will} = A_k^{y\text{-axis}} \times w(A_k^{x\text{-axis}} \times a; A_k^{Shape}, A_k^{Scale}) + A_k^{SuccessW} \times A_{k,y}^{Success}.$$

The gain from emigration is then given by

$$A_{c,a,k,y}^{Gain} = \left(A_k^{HealthW} \times G_c^{Health} + A_k^{SafetyW} \times G_c^{Safety} + A_k^{WageW} \times G_{c,y}^{Wage} + \right.$$
$$\left. + A_k^{PopW} \times G_c^{Pop} + A_k^{LangW} \times G_c^{Lang} \right) \times A_{a,k,y}^{Will} \quad , \text{ if } c \neq c_0$$

while the gain to remain in Portugal corresponds to

$$A_{c_0,a,k,y}^{Gain} = \left(A_k^{HealthW} \times G_{c_0}^{Health} + A_k^{SafetyW} \times G_{c_0}^{Safety} + A_k^{WageW} \times G_{c_0,y}^{Wage} + \right.$$
$$\left. + A_k^{PopW} \times G_{c_0}^{Pop} + A_k^{LangW} \times G_{c_0}^{Lang} \right) \times \left(1 - A_{a,k,y}^{Will} \right).$$

The emigration process is now done by the following steps:

Step 1. Initialize the emigration counter $G_c^{ECounter}$ at 0

Step 2. Update each agent's $A_{a,k,y}^{Will}$ and $A_{c,a,k,y}^{Gain}$

Step 3. For each agent, determine its desired emigration destination

$$A_k^{Dest} = \arg\max_c \{ A_{c,a,k,y}^{Gain} \}$$

restricted to $G_{A_k^{Dest}}^{ECounter} < G_{A_k^{Dest}}^{Limits}$ and add 1 to the counter $G_{A_k^{Dest}}^{ECounter}$

Step 4. Remove all agents with A_k^{Dest} different from Portugal.

As this mechanism needs some iterations to converge, the emigration amount in the first 6 years will be equal to that of 2011.

The immigration is exogenously defined by linear regressions on the countries with non-negative immigration rate in 2011 using OECD data since 2002 [8] and the age distribution of immigrants is fitted with UN data [9].

Finally, by defining

$G_{c,y}^{\text{Immi}}$: estimated amount of immigrants;
X^{ImmiAge} : Weibull distribution for the immigrants' age;
G_c^{ImmiProp} : male proportion of immigrants;
G_c^{FertF} : immigrants multiplying fertility factor, as in Adsera and Ferrer [1];
A_k^{ImmiC} : origin country of the immigrant agent k;

the immigration process is done according to:

Step 1. Create the immigrant agents as determined by $G_{c,y}^{\text{Immi}}$ and define the origin country A_k^{ImmiC} accordingly;

Step 2. For each newly immigrant agent, set its age to a randomly sampled value x from X^{ImmiAge};

Step 3. For each newly immigrant agent k, randomly sample u from $U(0,1)$; if $u < G_{A_k^{\text{ImmiC}}}^{\text{ImmiProp}}$ then set agent k as a male;

Step 4. For each immigrant agent k, define its birth and death parameters for the following year;

Step 5. For each immigrant agent k of age a and sex s, set $A_{a,s,k,y}^{\text{BirthR}} = A_{a,s,k,y}^{\text{BirthR}} \times G_{A_k^{\text{ImmiC}}}^{\text{FertF}}$.

3   Results

The results from the previously presented model for three different expected economic growth rates for Portugal, $G^{\text{EcoGrow}} \in \{0.9, 1.0, 1.1\}$, are now presented.

For each scenario, 300 simulations were considered for the period 2011-2041. The outputs of the model are: total population size, number of births, number of deaths and the total number of emigrants, by age and for each of the considered years. Totals across all ages (and subsequently their means) are obtained.

Whatever the economic scenario, the population size is expected to be a decreasing function with time. Moreover, the decrease is deepest when the economic growth rate attains the lowest value. This derives from the fact that economic growth plays a major role on the emigration decision and a decrease on this parameter would increase emigration. Such expectation is confirmed by fig 1.

In addition, although the economic does not directly affect the fertility rate, a decrease on this parameter leads to a faster decrease on the number of births over the years. This is most likely due to the fact that the primary age interval of the Portuguese emigrant population is within the women fertile ages. So the increase in the number of emigrants leads to a reduction of the Portuguese women that

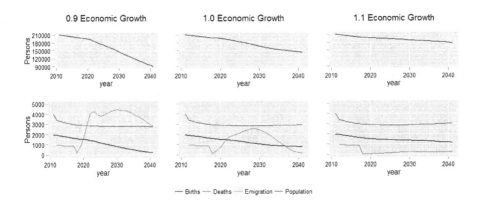

Fig. 1. Mean Population projection for different Economic Growth values

are in the fertile age. This justifies the decrease of births and further decreases the total Portuguese Population.

Acknowledgments. The first and second authors were partially financed by the FCT Fundação para a Ciência e a Tecnologia (Portuguese Foundation for Science and Technology) within project UID/EEA/50014/2013. The last author was partially supported by CMUP (UID/MAT/00144/2013), which is funded by FCT (Portugal) with national (MEC) and European structural funds through the programs FEDER, under the partnership agreement PT2020.

References

1. Adsera, A., Ferrer, A.: Factors influencing the fertility choices of child immigrants in Canada. Population Studies: A Journal of Demography **68**(1), 65–79 (2014)
2. Anjos, C., Campos, P.: The role of social networks in the projection of international migration flows: an Agent-Based approach. Joint Eurostat-UNECE Work Session on Demographic Projections. Lisbon, April 28–30, 2010
3. Baláž, V., Williams, A., Fifeková, E.: Migration Decision Making as Complex Choice: Eliciting Decision Weights Under Conditions of Imperfect and Complex Information Through Experimental Methods. Popul. Space Place (2014)
4. Billari, F., Ongaro, F., Prskawetz, A.: Introduction: Agent-Based Computational Demography. Springer, New York (2003)
5. Ferber, J.: Multi-Agent Systems - An Introduction to Distributed Artificial Intelligence. Addison-Wesley Longman, Harlow (1999)
6. Gaag, N., Beer, J., Willekens, F.: MicMac Combining micro and macro approaches in demographic forecasting. Joint Eurostat-ECE Work Session on Demographic Projections. Vienna, September 21–23, 2005
7. INE: Statistics Portugal (2012). www.ine.pt/en/
8. OECD: Organization for Economic Co-operation and Development (2015). www.stats.oecd.org/
9. UN: United Nations (2015). www.data.un.org/

Text Mining and Applications

Multilingual Open Information Extraction

Pablo Gamallo[1]([✉]) and Marcos Garcia[2]

[1] Centro Singular de Investigacin en Tecnoloxas da Informacin (CITIUS),
Universidade de Santiago de Compostela, Santiago de Compostela, Spain
`pablo.gamallo@usc.es`
[2] Cilenis Language Technology, Santiago de Compostela, Spain
`marcos.garcia@cilenis.com`

Abstract. Open Information Extraction (OIE) is a recent unsupervised strategy to extract great amounts of basic propositions (verb-based triples) from massive text corpora which scales to Web-size document collections. We propose a multilingual rule-based OIE method that takes as input dependency parses in the CoNLL-X format, identifies argument structures within the dependency parses, and extracts a set of basic propositions from each argument structure. Our method requires no training data and, according to experimental studies, obtains higher recall and higher precision than existing approaches relying on training data. Experiments were performed in three languages: English, Portuguese, and Spanish.

1 Introduction

Recent advanced techniques in Information Extraction aim to capture shallow semantic representations of large amounts of natural language text. Shallow semantic representations can be applied to more complex semantic tasks involved in text understanding, such as textual entailment, filling knowledge gaps in text, or integration of text information into background knowledge bases. One of the most recent approaches aimed at capturing shallow semantic representations is known as Open Information Extraction (OIE), whose main goal is to extract a large set of verb-based *triples* (or *propositions*) from unrestricted text. An Open Information Extraction (OIE) system reads in sentences and rapidly extracts one or more textual assertions, consisting in a verb relation and two arguments, which try to capture the main relationships in each sentence [1]. Wu and Weld [2] define an OIE system as a function from a document d, to a set of triples, $(arg1, rel, arg2)$, where $arg1$ and $arg2$ are verb arguments and rel is a textual fragment (containing a verb) denoting a semantic relation between the two verb arguments. Unlike other relation extraction methods focused on a predefined set of target relations, the Open Information Extraction paradigm is not limited to a small set of target relations known in advance, but extracts all types of (verbal) binary relations found in the text. The main general properties of OIE systems are the following: (i) they are domain independent, (ii) they rely on unsupervised extraction methods, and (iii) they are scalable to large amounts of text [3].

© Springer International Publishing Switzerland 2015
F. Pereira et al. (Eds.) EPIA 2015, LNAI 9273, pp. 711–722, 2015.
DOI: 10.1007/978-3-319-23485-4_72

The objective of this article is to describe a heuristic-based OIE system, called ArgOE, which uses syntactic analysis to detect the argument structure of each verb, as well as a set of rules to generate the corresponding triples (or basic propositions) from each argument structure. In our work, an argument structure has a very broad sense, since it includes all those syntactic dependencies headed by a verb except specifiers, auxiliaries, and adverbs. So, it includes all main clause constituents: subjects, objects, attributes, and prepositional phrases referring to locations, instrumentals, manners, causes, etc. So, there is no distinction between traditional arguments and adjuncts, both are used to build the argument structure.

Consider for example the sentence:

In May 2010, the principal opposition parties boycotted the polls after accusations of vote-rigging.

First, our OIE system detects the argument structure of the verb *boycotted* in this sentence: there is a *subject*, a *direct object*, and two prepositional phrases functioning as verb *adjuncts*. Then, a set of basic rules transform the argument structure into a set of triples:

("the principal opposition parties", "boycotted", "the polls"),
("the principal opposition parties", "boycotted the polls in", "May"),
("the principal opposition parties", "boycotted the polls after", "accusations of vote-rigging")

ArgOE requires no training data, generates triples without any post-processing, and takes as input dependency parses in CoNLL-X format [4,5]. Given that such a dependency-based representation is provided by many robust parsers including multilingual systems, e.g., MaltParser [6] or DepPattern [7], ArgOE can be seen as a multilingual open information extractor. We will describe experiments of triples extraction performed on English, Portuguese, and Spanish text. ArgOE's source code configured for English, Spanish, Portuguese, French, and Galician, as well as other resources are released under GPL license.

This article is organized as follows. Section 2 introduces previous work on OIE: in particular it describes different types of OIE systems. Next, in Section 3, the proposed method, ArgOE, is described in detail. Then, some experiments are performed in Section 4, where ArgOE system is compared against several systems and evaluated in several languages, including Portuguese. Finally, conclusions and future work are addressed in 5.

2   Related Work

The goal of an OIE system is to extract triples *(arg1, rel, arg2)* describing basic propositions from large amounts of text. A great variety of OIE systems has been developed in recent years. They can be organized in two broad categories: those systems requiring automatically generated training data to learn a classifier and those based on hand-crafted rules or heuristics. In addition, each system category can also be divided in two subtypes: those systems making use of shalow syntactic analyisis (PoS tagging and/or chunking), and those based on dependency parsing. In sum, we identify four categories of OIE systems:

(1) **Training data and shallow syntax:** The first OIE system, TextRunner [8], belongs to this category. A more recent version of TextRunner, also using training data (even if hand-labeled annotated) and shallow syntactic analysis is R2A2 [9]. Another system of this category is WOE^{pos} [2] whose classifier was trained with corpus obtained automatically from Wikipedia.

(2) **Training data and dependency parsing:** These systems make use of training data represented by means of dependency trees: WOE^{dep} [2] and OLLIE [10].

(3) **Rule-based and shallow syntax:** They rely on lexico-syntactic patterns hand-crafted from PoS tagged text: ReVerb [11], ExtrHech [12], and LSOE [13].

(4) **Rule-based and dependency parsing:** They make use of hand-crafted heuristics operating on dependency parses: ClauseIE [3], CSD-IE [14], KrakeN [15], and DepOE [16].

Our system belongs to the fourth category and, thus, is similar to ClauseIE and CSD-IE, which are the best OIE extractors to date according to the results reported in both [3] and [14]. However, these two systems are dependent on the output format of a particular syntactic parser, namely the Standford dependency parser [17]. In the same way, DepOE reported in [16], relies on a specific dependency parser, DepPattern [7], since it only operates on the by-default output given by this parser. ArgOE, by constrast, uses as input the standard CoNLL-X format and, then, does not depend on a specific dependency parser.

Another significant difference between ArgOE and the other rule-based systems is that ArgOE does not distinguish between arguments and adjuncts. As this distiction is not always clear and well identified by the syntactic parsers, we simplify the number of different verb constituents within the argument structure: all prepositional phrases headed by a verb are taken as verb complements, regardless of their degree of dependency (internal arguments or external adjuncts) with the verb. So, the set of rules used to generate triples from this simplified argument structure is smaller than in other rule-based approaches.

In addition, we make extraction multilingual. More precisely, our system has the following properties:

- Extraction of triples represented at different levels of granularity: surface forms and dependency level.
- Multilingual extraction based on multilingual parsing.

3 The Method

Our OIE method consists of two steps: detection of argument structures and generation of triples.

3.1   Step 1: Argument Structure Detection

For each parsed sentence in the ConLL-X format, all verbs are identified and, for each verb (V), the system selects all dependents whose syntactic function can be part of its argument structure. Each argument structure is the abstract representation of a clause. The functions considered in such representations are *subject* (S), *direct object* (O), *attribute* (A), and all complements headed by a preposition (C). Five types of argument structures were defined and used in the first experiments: SVO, SVC+, SVOC+, SVA, SVAC+, where "C+" means one or more complements. All these argument structures are correct syntactic options in our working languages: English, Portuguese, and Spanish. Table 1 shows English examples for each type of argument structure.

Table 1. Examples of argument structures extracted from our testing dataset.

Type	Example	Constituents
SVO	*A Spanish official offered what he believed to be a perfectly reasonable explanation for why the portable facilities weren't in service*	S="A Spanish official", **V**="offered", O="what he believed to be a perfectly reasonable explanation for why the portable facilities weren't in service"
SVC_1C_2	*Output was reduced in 1996 after one of its three furnaces exploded*	S="Output", **V**="was reduced", C_1="in 1996", C_2="after one of its three furnaces exploded"
SVOC	*These immigrants deserve consideration under the laws that were in place*	S="These immigrants", **V**="deserve", O="consideration", C="under the laws that were in place"
SVA	*Koplowitz's next concert will be a more modest affair*	S="Koplowitz's next concert", **V**="will be", A="a more modest affair"
SVAC	*Gallery hours are 11 a.m. to 6 p.m. daily*	S="Gallery hours", **V**="are daily", A="11 a.m.", C="to 6 p.m."

Within a sentence, it is possible to find several argument structures corresponding to different clauses. For instance, the SVO example in Table 1 represents the argument structure associated with the clause introduced by the verb *offered*, but there are three more clauses introduced by other verbs (in bold): *he **believed** to be a perfectly reasonable explanation for why the portable facilities weren't in service*, *what **be** a perfectly reasonable explanation for why the portable facilities weren't in service*, and *the portable facilities **weren't** in service*, giving rise to the different argument structures shown in Table 2.

Table 2. Argument structures extracted from the sentence *A Spanish official offered what he believed to be a perfectly reasonable explanation for why the portable facilities weren't in service.*

Type	Constituents
SVO	S="A Spanish official", **V**="offered", O="what he believed to be a perfectly reasonable explanation for why the portable facilities weren't in service"
SVO	S="he", **V**="believed to", O="be a perfectly reasonable explanation for why the portable facilities weren't in service"
SVA	S="what", **V**="be", A="a perfectly reasonable explanation for why the portable facilities weren't in service"
SVA	S="the portable facilities", **V**="weren't", A="in service"

The constituents of an argument structure are the full phrases or clauses playing different syntactic functions within the structure. Each constituent is built by finding all dependency paths from its head to all its (direct and indirect) dependents. For instance, consider the SVA example in Table 1. To build the full constituents, the first step is to identify the head word of each constituent: S="concert", V="be", A="affair". Then, each head is extended with all its dependency words by exploring the full dependency path and by taking into account the position in the sequence. This results in full phrases representing all constituents of the clause: S="Koplowitz's next concert", V="will be", A="a more modest affair".

There is, however, an important exception in the process of building full constituents: namely, relative clauses. The constituents we generate do not include those clauses introduced by a verb modifying a noun. For instance, the SVOC example in Table 1 contains the constituent C="under the laws", extracted from the expression *under the laws that were in place*. In this case, the relative clause was not taken into account to generate the constituent C within the argument structure of the main verb *deserve*. However, relative clauses and their antecedents also introduce argument structures. In the same example, we identify a SVA argument structure from the chain "the laws that were in place", where S="the laws", V="were", and A="in place". The main reason for removing relatives from constituents is to guarantee the generation of coherent and non over-specified propositions, as we will report in the next section.

Moreover, coordinatated conjunctions in verbal phrases are splitted into different argument structures, one for each coordinated verb. However, by taking into account the experiments performed in [3], coordinated phrases in the verb arguments are not processed.

Finally, notice that the argument structure SVO_1O_2 (e.g. *John gave Mary a present*) is not considered here, since it is not a correct syntactic structure in Spanish (nor in the rest of latin languages). In order the system to be multilingual, we have defined only those argument structures that are shared by our working languages.

3.2   Step 2: Generation of Triples

One of the most discussed problems of OIE systems is that about 90% of the extracted triples are not concrete facts [1] expressing valid information about one or two named entities, e.g. "Obama was born in Honolulu". However, the vast amount of high confident relational triples (propositions) extracted by OIE systems are a very useful starting point for further NLP tasks and applications, such as common sense knowledge acquisition [18], and extraction of domain-specific relations [19]. It follows that OIE systems are not suited to extract facts, but to transform unstructured texts into structured and coherent information (propositions), closer to ontology formats. Having this in mind, our objective is to generate propositions from argument structures, where propositions are defined as coherent and non over-specified pieces of basic information.

From each argument structure detected in the previous step, our OIE system generates a set of triples representing the basic propositions underlying the linguistic structure. We assume that every argument structure can convey different pieces of basic information which are, in fact, minimal units of coherent, meaningful, and non over-specified information. For example, consider again the sentence:

In May 2010, the principal opposition parties boycotted the polls after accusations of vote-rigging.

which gives rise to the following $SVOC_1C_2$ argument structure:

S="the principal opposition parties" , **V**="boycotted", **O**="the polls",
C$_1$="In May",
C$_2$="after accusations of vote-rigging"

An incoherent and over-specified extraction would generate from this structure the following odd propositions:

P_1=("the principal opposition parties", "boycotted in", "May")
P_2=("the principal opposition parties", "boycotted after", "accusations of vote-rigging")
P_3=("the principal opposition parties", "boycotted the polls after accusations of vote-rigging in", "May")

Propositions P_1 and P_2 are incoherent extractions because the direct object constituent (O) is not optional and, then, may not be omitted from any proposition built from that argument structure. In addition, P_3 contains an over-specified relation constituted by several constituents of the argument structure. To ensure a correct extraction, we defined a set of simple rules allowing us to extract only those propositions that are considered as coherent and non over-specified. For this purpose, direct objects are never omitted and relations cannot contain more than one clause constituent. This way, the three coherent propositions generated from the above argument structure are the following:

P_1=("the principal opposition parties", "boycotted", "the polls")
P_2=("the principal opposition parties", "boycotted the polls after", "accusations of vote-rigging")
P_3=("the principal opposition parties", "boycotted the polls in", "May")

As has been said, another restriction to avoid over-specification is to remove relative clause from the constituents. In the same way, that-clauses that are direct objects are never inserted in the relation so as to avoid long and over-specificied relations.

Propositions are generated using trivial extraction rules that transform argument structures into triples. Table 3 shows the set of rules we used to extract triples from our five types of argument structures. As in the case of all current OIE systems, we only consider the extraction of verb-based triples. We took this decision in order to make a fair comparison when evaluating the performance of our system against similar systems (see Section 4). However, nothing prevents us from defining extraction rules to generate several triples from non-verbal structures: noun-prep-noun, noun-noun, adj-noun, and verb-adverb dependencies.

Table 3. Rules applied on five argument structures to generate the corresponding triples

Argument Structure	Rules
SVO	**arg1**=S, **rel**=V, **arg2**=O
SVC+	for $i = 1$ to n where n is the number of Complements C: C_i is descomposed in $prep_i$ and $Term_i$ **arg1**=S, **rel**=V+$prep_i$, **arg2**=$Term_i$
SVOC+	if O is not a that-clause, then: **arg1**=S, **rel**=V, **arg2**=O for $i = 1$ to n where n is the number of Complements C: C_i is descomposed in $prep_i$ and $Term_i$ **arg1**=S, **rel**=V+O+$prep_i$, **arg2**=$Term_i$ if O is a that-clause, then: **arg1**=S, **rel**=V, **arg2**=O for $i = 1$ to n where n is the number of Complements C: C_i is descomposed in $prep_i$ and $Term_i$ **arg1**=S, **rel**=V+$prep_i$, **arg2**=$Term_i$
SVA	**arg1**=S, **rel**=V, **arg2**=A
SVAC+	**arg1**=S, **rel**=V, **arg2**=A for $i = 1$ to n where n is the number of Complements C: C_i is descomposed in $prep_i$ and $Term_i$ **arg1**=S, **rel**=V+A+$prep_i$, **arg2**=$Term_i$

The output of ArgOE does not offer confidence values for each extraction. As the system is rule-based, there is not probabilistic information to be considered. Finally, with regard to the output format, it is worth mentioning that most OIE systems produce triples only in textual, surface form. This can be a problem if triples are used for NLP tasks requiring more linguistic information. This way, in addition to surface form triples, ArgOE also provides syntax-based information, with PoS tags, lemmas, and heads. If more syntactic information would be required, it can be easily obtained from the dependency analysis.

4   Experiments

We conducted thre experimental studies: with English, Spanish, and Portuguese texts. Preliminary studies were performed to select an appropriate syntactic parser as input of ArgOE. Two multilingual dependency parsers were tested:

MaltParser 1.7.1[1] and DepPattern 3.0 [2], which is provided with a format converter that changes the standard output of the parser into the CoNLL-X format. We opted for DepPattern as input of ArgOE because the tagset and dependency names of DepPattern is the same for all the languages it is able to analyze, and then, there is no need to configure and adapt ArgOE for each new language. The use of MaltParser with different languages would require implementing converters from tagsets and dependency names defined for a particular language to a common set of PoS tags and dependency names. Besides DepPattern, we also use two different PoS taggers as input of the syntactic analyzer: TreeTagger [20] for English texts and FreeLing [21] for Spanish and Portuguese. All datasets, extractions and labels of the two experiments, as well as a version of ArgOE configured for English, Spanish, Portuguese, French, and Galician, are freely available[3].

4.1   English Evaluation

We compare ArgOE against several OIE existing systems for English, namely TextRunner, ReVerb, OLLI, WOE^{parse}, and ClausIE. In this experiment, we will report the results obtained by the the best version of ClauseIE, i.e., without considering redundancy and without processing conjunctions in the arguments. Note that we are comparing four systems based on training data (TextRunner, ReVerb, OLLI, and WOE^{parse}) against two rule-based methods: ClausIE and ArgOE.

The dataset used in the experiment is the Reverb dataset[4] manually labeled for the evaluation reported in [3][5]. The dataset consists of 500 sentences with manually-labeled extractions for the five systems enumerated above. In addition, we manually labeled the extractions obtained from ArgOE for the same 500 sentences. To maintain consistency among the labels associated to the five systems and those associated to ArgOE, we automatically identified those triples extracted by ArgOE that also appear in, at least, one of the other labeled extractions. As a result, we obtained 355 triples extracted by ArgOE that were labeled by annotators of previous work. Then, the extractions of ArgOE were given to two annotators who were instructed to consider the 355 already labeled extractions as starting point. So, our annotators were required to study and analyze the evaluation criteria used by other annotators before starting annotating the rest of extracted triples. We also instructed the annotators to treat as incorrect those triples denoting incoherent and uninformative propositions, as well as those triples constituted by over-specified relations, i.e., relations containing numbers, named entities, or excessively long phrases (e.g., *boycotted the polls after accusations of vote-rigging in*). An extraction was considered as correct

---

[1] htpp://www.maltparser.org/
[2] http://gramatica.usc.es/pln/tools/deppattern.html/
[3] http://172.24.193.8/ArgOE-epia2015.tgz (anonymous version)
[4] http://reverb.cs.washington.edu/
[5] http://www-mpi-inf.mpg.de/departments/d5/software/clausie

if it was labeled as correct by both annotators. The two annotators agreed on 75% of extractions (Cohen's kappa k = 0,50), which is considered a moderate agreement. In sum, we follow similar criteria to those defined in previous OIE evaluations [9].

The results of our evaluation are summarized in Table 4 and Figure 1. Table 4 shows the number of correct expressions extracted as well as the total number of extractions for each system. *Precision* is defined as the number of correct extractions divided by the number of returned extractions. *Recall* is estimated by identifying a pool of relevant extractions which is the total number of different correct extractions made by all the systems (this pool is our gold-standard). So, *recall* is the number of correct extractions made by the system divided by the total number of correct expressions in the pool (3, 222).

Table 4. Number of correct extractions and total number of extractions in the Reverb dataset, according to the evaluation reported in [3] and our own contribution with ArgOE.

Systems	correct extractions	total extractions
textrunner	286	798
reverb	388	727
woe	447	1028
ollie	547	1242
argoe	582	1162
clausie	1706	2975

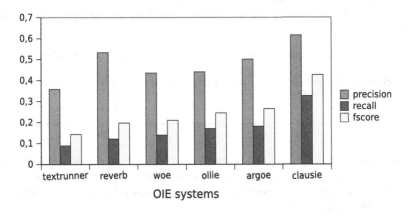

Fig. 1. Evaluation of six OIE systems

The results show that the two rule-based systems, ClausIE and ArgOE, perform better than the classifiers based on automatically generated training data. This is in accordance with previous work reported in [3,14]. Moreover, the four systems based on dependency analysis (ClausIE, ArgOE, OLLIE, and

WOE^{parse}) improve over those relying on shallow syntax (TextRunner and ReVerb). And finally, ClausIE clearly outperforms the other systems, in terms of both precision and recall. A common problem for parse-based OIE systems is the large influence of parser errors. So, the quality of the parser can determine the quality of the OIE extractor. ClausIE uses the Standford Dependency Parser, while ArgOE uses DepPattern, and OLLIE the MaltParser. One possible reason for the comparably low precision of our system against ClausIE might be the lower parsing performance of DepPattern against the Standford Dependency Parser for the English language.

4.2   Spanish Evaluation

In this experiment, we compare ArgOE against the only OIE system that has been evaluated for other language than English: ExtrHech [12]. This is also a rule-based system, but it does not operate on dependency parsing but on shallow syntax (patterns of PoS tags). The Spanish dataset, called Raw Web[6], contains 159 sentences randomly extracted with a web crawler from over 5 billion web pages in Spanish. Each extraction was labeled by two independent annotators. An extraction was considered as correct if it was labeled as correct by both annotators. They agreed on 81% of extractions (Cohen's kappa k = 0,62). Table 5 depicts the results obtained by the two systems on these sentences. Unfortunately, the extractions made by ExtrHech are not available, so it is not possible to create a pool of correct triples extracted by the two systems to measure recall. Only precision can be compared even if we were not able to unify the criteria given to our annotators with those defined in [12]. Notice that the precision of ArgOE is identical to that obtained for English (50%), which can be seen as an indirect evidence that the two parsers used by our system have similar performance.

Table 5. Precision of both ArgOE and ExtrHech on the Spanish dataset

Systems	correct extractions	total extractions	Precision (%)
argoe	107	214	50%
extrahech	-	-	55%

Most errors made by our OIE system come from three different sources: the syntactic parser, the PoS tagger, and the Named Entity Recognition module used by the PoS tagger. So, the improvement of our system relies on the performance of other NLP tasks.

4.3   Portuguese Evaluation

For this purpose, we selected 103 test sentences from a domain-specific corpus, called *CorpusEco* [22], containing texts on ecological issues. ArgOE was

---

[6] http://www.gelbukh.com/resources/spanish-open-fact-extraction

applied on the sentences and 190 triples was extracted. One annotator labeled the extracted triples and Table 6 shows the number of correct triples and precision achieved by the system. To the best of our knowledge, this is the first experiment that reports an OIE system working on Portuguese. Precision is again similar (53%) to that obtained in the previous experiments. Again, most errors are due to problems from the syntactic parser and PoS tagger.

Table 6. Precision of ArgOE on the Portuguese dataset

Systems	correct extractions	total extractions	Precision (%)
argoe	95	190	53%

5 Conclusion

We have described a rule-based OIE system to extract verb-based triples than takes as input dependency parsers in the CoNLL-X format. So, it may take advantage of efficient, robust, and multilingual syntactic parsers. Even if our system is outperformed by other similar rule-based methods, it reaches better results than those strategies based on training data. As far as we know, ArgOE is the first OIE system working on more than one language. In future work, we will include NLP modules to find linguistic generalizations over the extracted triples: e.g., co-reference resolution to link the arguments of different triples, and synonymy detection of verbs to reduce the open set of extracted relations and, then, to enable semantic inference.

Acknowledgments. This work has been supported by projects Plastic and Celtic, Innterconecta (CDTI).

References

1. Banko, M., Cafarella, M.J., Soderland, S., Broadhead, M., Etzioni, O.: Open information extraction from the web. In: International Joint Conference on Artificial Intelligence (2007)
2. Wu, F., Weld, D.S.: Open information extraction using wikipedia. In: Annual Meeting of the Association for Computational Linguistics (2010)
3. Corro, L.D., Gemulla, R.: Clausie: clause-based open information extraction. In: Proceedings of the World Wide Web Conference (WWW-2013), Rio de Janeiro, Brazil, pp. 355–366 (2013)
4. Hall, J., Nilsson, J.: CoNLL-X shared task on multilingual dependency parsing. In: The Tenth CoNLL (2006)
5. Nivre, J., Hall, J., Kübler, S., McDonald, R., Nilson, J., Riedel, S., Yuret, D.: The CoNLL-2007 shared task on dependency parsing. In: Proceedings of the Shared Task Session of EMNLP-CoNLL 2007, Prague, Czech Republic, pp. 915–932 (2007)
6. Nivre, J., Hall, J., Nilsson, J., Chanev, A., Eryigit, G., Kübler, S., Marinov, S., Marsi, E.: Maltparser: A language-independent system for data-driven dependency parsing. Natural Language Engineering **13**(2), 115–135 (2007)

7. Gamallo, P., González, I.: A grammatical formalism based on patterns of part-of-speech tags. Journal of Corpus Linguistics **16**(1), 45–71 (2011)
8. Banko, M., Etzioni, O.: The tradeoffs between open and traditional relation extraction. In: ACL-08 (2008)
9. Etzioni, O., Fader, A., Christensen, J., Soderland, S., Mausam: Open information extraction: the second generation. In: International Joint Conference on Artificial Intelligence (2011)
10. Mausam, Schmitz, M., Soderland, S., Bart, R., Etzioni, O.: Open language learning for information extraction. In: EMNLP-12, pp. 523–534 (2012)
11. Fader, A., Soderland, S., Etzioni, O.: Identifying relations for open information extraction. In: EMNLP-11 (2011)
12. Zhilla, A., Gelbukh, A.: Comparison of open information extraction for Engish and Spanish. In: Dialogue 2014 (2014)
13. Xavier, C.C., Souza, M., de Lima, V.S.: Open information extraction based on lexical-syntactic patterns. In: Brazilian Conference on Intelligent Systems, pp. 189–194 (2013)
14. Bast, H., Haussmann, E.: Open information extraction via contextual sententence decomposition. ICSC 2013, 154–159 (2013)
15. Akbik, A., Loser, A.: Kraken: N-ary facts in open information extraction. In: Joint Workshop on Automatic Knowledge Base Construction and Web-scale Knowledge Extraction, pp. 52–56 (2012)
16. Gamallo, P., Garcia, M., Fernández-Lanza, S.: Dependency-based open information extraction. In: ROBUS-UNSUP Workshop at EACL-2012, Avignon, France (2012)
17. Klein, D., Manning, C.D.: Accurate unlexicalized parsing. In: ACL-03, pp. 423–430 (2003)
18. Lin, T., Mausam, Etzioni, O.: Identifying functional relations in web text. In: Conference on Empirical Methods in Natural Language Processing (2010)
19. Soderland, S., Roof, B., Qin, B., Xu, S., Mausam, Etzioni, O.: Adapting open information extraction to domain-specific relations. AI Magazine **31**(3), 93–102 (2010)
20. Schimd, H.: Improvements in part-of-speech tagging with an application to german. In: ACL SIGDAT Workshop, Dublin, Ireland (1995)
21. Padró, L., Stanilovsky, E.: Freeling 3.0: towards wider multilinguality. In: LREC 2012, Istanbul, Turkey (2012)
22. Zavaglia, C.: O papel do léxico na elaboracão de ontologias computacionais: do seu resgate à sua disponibilizacão. In: Lingüística IN FOCUS - Léxico e morfofonologia: perspectivas e análises. Uberlândia: EDUFU, pp. 233–274 (2006)

Classification and Selection of Translation Candidates for Parallel Corpora Alignment

K.M. Kavitha[1,3]([✉]), Luís Gomes[1,2], José Aires[1,2], and José Gabriel P. Lopes[1,2]

[1] NOVA Laboratory for Computer Science and Informatics (NOVA LINCS),
Faculdade de Ciências e Tecnologia, Universidade Nova de Lisboa,
2829-516 Caparica, Portugal
k.mahesh@campus.fct.unl.pt, luismsgomes@gmail.com, aires74@iol.pt,
gpl@fct.unl.pt
[2] ISTRION BOX-Translation & Revision, Lda., Parkurbis,
6200-865 Covilhã, Portugal
[3] Department of Computer Applications, St. Joseph Engineering College,
Vamanjoor, Mangaluru 575 028, India
kavitham@sjec.ac.in

Abstract. By incorporating human feedback in parallel corpora alignment and term translation extraction tasks, and by using all human validated term translation pairs that have been marked as *correct*, the alignment precision, term translation extraction quality and a bunch of closely correlated tasks improve. Moreover, such a labelled lexicon with entries tagged for correctness enables bilingual learning. From this perspective, we present experiments on automatic classification of translation candidates extracted from aligned parallel corpora. For this purpose, we train SVM based classifiers for three language pairs, English-Portuguese (EN-PT), English-French (EN-FR) and French-Portuguese (FR-PT). The approach enabled micro f-measure classification rates of 95.96%, 75.04% and 65.87% respectively, for the EN-PT, EN-FR and FR-PT language pairs.

1 Introduction

Annotated bilingual lexica with their entries tagged for (in)correctness can be mined to discover the nature of new term translation extractions and/or alignment errors. An automated classification system can then be trained when sufficient amount of positive and negative evidence is available. Such a classifier can facilitate and speed up the manual validation process of automatically extracted term translations, and contribute to make the human validation effort easier while augmenting the number of validated (rejected and accepted) bilingual entries in a bilingual term translation lexicon. Bionic interaction between linguists and highly precise machine classifiers in a continuous common effort, without loosing knowledge contributes to improve alignment precision and, at another level, translation quality. It is therefore important to have term translation extractions automatically classified prior to having them validated by human specialists.

© Springer International Publishing Switzerland 2015
F. Pereira et al. (Eds.) EPIA 2015, LNAI 9273, pp. 723–734, 2015.
DOI: 10.1007/978-3-319-23485-4_73

In this paper, we assume sentence aligned parallel corpora for extracting term translations, constructing translation tables or obtaining that parallel corpora aligned at a subsentence grain [7,13]. In this setting, translation correspondences are identified between term pairs by computing their occurrence frequencies or similarities within the aligned sentences rather than in the entire corpus.

In the completely unsupervised models based on parallel corpora, all the phrase pairs that are considerably consistent with the word alignment are extracted and gathered into a phrase table along with their associated probabilities [4,19]. Naturally, the resulting table extracted from the alignment, with no human supervision, contains alignment errors. Moreover, many of the translations in the phrase table produced are spurious or will never be used in any translation [10]. A recent study shows that nearly 85% of phrases gathered in the phrase table can be reduced without any significant loss in translation quality [21].

A different approach [1], that deviates from this tradition acknowledges the need for blending the knowledge of language, linguistics and translation as relevant for research in Machine Translation [27]. The approach being semi-supervised and iterative takes privilege in informing the machine for not making the same kind of errors in subsequent iterations of alignment and extraction. In this partially supervised, iterative strategy, first a bilingual lexicon is used to align parallel texts [7]. New[1] term-pairs are then extracted from those aligned texts [1]. The newly extracted candidates are manually verified and then added to the existing bilingual lexicon with the entries manually tagged as *accepted* (Acc) and *rejected* (Rej). Iteration over these three steps (parallel text alignment using an updated and validated lexicon, extraction of new translation pairs and their validation) results in an improved alignment precision, improved lexicon quality, and in more accurate extraction of new term-pairs [7]. Human feedbacks are particularly significant in this scenario as incorporating them prevents alignment and extraction errors from being fed back into subsequent alignment and extraction iterations. The work described in this paper may be easily integrated in such a procedure.

Several approaches for extracting phrase translations prevail [1,4,8,15]. However, it is important to have the extractions automatically classified prior to having them validated by human specialists. We view classification as a pre-validation phase that allows a first-order separation of correct entries from incorrect ones, so that the human validation task becomes lighter [11]. We extend our previous work by using a larger set of extracted translation candidates for the language pair EN-PT and by additionally adopting other extraction techniques [4,15] and others not yet published. Experimental evaluations of the classifier for additional language pairs EN-FR and FR-PT are also presented. Further, the performance of the classifier with additional features is discussed.

In the Section 2, we provide a quick review of the related work. In the Section 3, we present the classification approach for selecting translation candidates and the features used. The data sets used, the classification results, and

---

[1] Not seen in the bilingual lexicon that was used for alignment.

their subsequent analysis are presented in Section 4. We conclude with Section 5 by reflecting a bit on the future work.

2   Selection of Bilingual Pairs

The translation selection process might be aimed from the perspective of improving the alignment precision and extraction quality or from the translation perspective itself [11,24]. Nevertheless, different researchers have demonstrated varied views regarding the influence of alignment on translation quality, predominantly from the perspective of entries in a phrase table. It is shown that better alignment presents threefold benefit that includes the advantage of producing a phrase table of manageable size with fewer phrase pairs, a reduced decoding time in searching the phrase table for the most probable translation, and a better quality of word or phrase level translation [21]. However, it is also observed that the decreased alignment error rate does not necessarily imply a significant increase in the translation quality [6,18,26]. We reiterate that we aim at an improved alignment precision and extraction accuracy.

The task of selecting appropriate translation candidates may be cast as the problem of filtering spurious bilingual pairs from the associated tables[2] or as we view it, the learning phase (training) of a classifier that is then used for classifying the extracted bilingual pairs as '*Accepted*' or '*Rejected*' for further manual verification. Various filtering approaches have been proposed and used in selecting appropriate translations [1,5,10,17,22,23,28,29].

2.1   Support Vector Machines in Selecting Bilingual Pairs

Ever since its introduction, the Support Vector Machine (SVM) [25] has been successfully adapted for various translation related machine learning tasks. Related applications include learning translation model for extracting word sequence correspondences (phrase translations) and automatic annotation of cognate pairs [3,20].

The use of SVM based classifiers in selecting the translation candidates has been proposed earlier [2,11,14]. Common criteria for selection include translation coverage, source and target term co-occurrence and the orthographic similarity. Further, One-Class SVM has been used with the Mapping Convergence (MC) algorithm to differentiate the usable and useless phrase pairs based on the confidence scores assigned by the classifier [24]. While the focus is on translation quality and avoiding alignment errors, the classifier is trained with a corpus that comprises of only useful instances. All phrase pairs involved in best phrasal derivations[3] by the Oracle decoder are labeled as positive phrase pairs. Unlabelled examples of phrase pairs, however, are employed in addition to the positive examples in a semi-supervised framework[4] to improve the performance. We on

---

[2] Phrase table or a bilingual lexicon.

[3] One that maximises a combination of model score and translation quality metric.

[4] MC algorithm.

the other hand, view the selection of translation candidates as a supervised classification problem with labeled training examples for both the classes (positive and negative instances).

3   Classification Model

In the current section, we discuss the use of SVM based classifier in segregating the extracted translation candidates as accepted, '*Acc*' or rejected, '*Rej*'. The classification task involves training and testing data representing bilingual data instances. Each bilingual pair is a data instance represented as a *feature vector* and a target value known as the *class label*[5]. We train the learning function with the scaled training data set, where each sample is represented as a feature vector with the label +1 ('Acc') or -1 ('Rej'). The estimated model is then used to predict the class for each of the unknown data instance kept aside for testing, represented similarly as any sample in the training set, but with the class label 0. We use the Radial Basis Function (RBF) kernel: $K(x_i, x_j) = e^{\gamma \|x-y\|^2}$; parameterised by (C, γ), where $C > 0$ is the penalty parameter of the error term and $\gamma > 0$ is the kernel parameter.

3.1   Features

Adequate feature identification for representing the data in hand is fundamental to enable good learning. An overview of the features used in our classification model is discussed in this section. We use the features derived using the orthographic similarity measures (strsim) and the frequency measures (freq) discussed in the section below as baseline $(BL_{strsim+freq})$ for our experiments.

Orthographic Similarity. Two orthographic similarity measures based on edit distance are used to quantify the similarity between the terms in a bilingual pair: the Levenshtein Edit Distance [16] (Equation 1) and the Spelling Similarity measure [8] (Equation 2).

$$EditSim = 1.0 - \frac{EditDist(X,Y)}{Max(|X|,|Y|)} \tag{1}$$

where EditDist(X,Y) is the edit distance between the term X in first language and the term Y in second language.

$$SpSim(X,Y) = 1.0 - \frac{D(X,Y)}{Max(|X|,|Y|)} \tag{2}$$

where the distance function D(X,Y) is the *EditDist* discounting characteristic spelling differences that were learnt previously. In Equations 1 and 2, $|X|$ represents the length of X and $|Y|$ represents the length of Y.

---

[5] Positive and negative examples are respectively labeled as +1 and -1. Data to be classified is labeled 0.

We use the '*accepted*' entries in the training dataset with EditSim ≥ 0.65 as examples to train SpSim and a dictionary containing the substitution patterns is learnt. For instance, the substitution pattern extracted from EN-PT cognate word pair 'phase' and 'fase' is ('^ph', '^f'), after eliminating all matched (aligned) characters, 'a' $\Leftrightarrow$ 'a', 's' $\Leftrightarrow$'s' and 'e' $\Leftrightarrow$ 'e'. The caret (^), at the beginning of the aligned strings distinguishes that the pattern appears as a prefix.

Frequency of Occurrence. To represent the translational equivalence, based on the frequencies of the terms in a bilingual pair, two measures are used: the Dice association measure and the MinMaxRatio.

The Dice association measure for a pair of terms (X,Y) takes into account the frequency of the term X in the first language text, F(X); the frequency of the term Y in the second language text, F(Y); and the co-occurrence frequency of the terms in aligned segments of the parallel texts, F(X,Y) and is given by the equation,

$$Dice(X,Y) = \frac{2 * F(X,Y)}{F(X) + F(Y)} \tag{3}$$

Another measure that efficiently substitutes the individual frequencies F(X) and F(Y) is the minimum to maximum frequency ratio given by the equation,

$$MinMaxRatio(X,Y) = \frac{Min(F(X), F(Y))}{Max(F(X), F(Y))} \tag{4}$$

Table 1. The Similarity and Bad Ending Scores

Term$_{EN}$	Term$_{PT}$	EdSim	SpSim	BE$_{SW}$	BE$_{Pat_{R-A}}$
general indifference	indiferença geral	0.15	1.0	(0.00, 0.00)	(0.00, 0.00)
official	comercial	0.56	0.66	(0.00, 0.00)	(0.00, 0.00)
commitments	compromissos de crédito	0.29	0.24	(0.00, 0.00)	(0.00, 0.00)
limits of the	limites de a	0.54	0.82	(1.00, 1.00)	(1.00, 1.00)
impact on the	impacto em a indústria	0.39	0.47	(1.00, 0.00)	(1.00, 0.00)

Bad Ends. The bilingual pair '*limits of the* $\Leftrightarrow$ *limites de a*' instantiates a particular type of inadequate translation wherein, the term (on both sides) ends with a determiner following which a noun or a noun phrase is anticipated. It is the absence of the noun or a noun phrase after the determiner that makes the translation incomplete. By allowing this entry into the lexicon as a correct translation, we cannot refrain other entries ending with '*o*', '*os*', and so forth from accommodating the determiner's position. We refer to such translations with inadequate endings as having bad ends (BE). To keep a check over such entries, we use a binary valued feature signifying whether a translation ends with a determiner (1) or not (0). This introduces two features, each representing the goodness of the translation endings on each side of the bilingual pair.

We use two different approaches to identify bad ends: one set of two features based on endings that are stop words (BE_{SW}) and the other set of two features based on endings seen in the rejected, but not in the accepted training dataset ($BE_{Pat_{R-A}}$). We consider only those endings that occur more than 5 times in the rejected but not in the accepted training dataset. To avoid the content words from being considered as bad ends, the term length is restricted to less than 5 characters.

Translation Miscoverage. A typical error observed in the extracted candidates represents the lack of parallelism with respect to content words. An example is the bilingual pair '*commitments* ⇔ *compromissos de crédito*'. For this pair to be considered as correct, '*crédito*' needs to be translated either as '*lending*' or as '*loan*' in EN. So the correct term translation would be '*lending commitments* ⇔ *compromissos de crédito*' or '*loan commitments* ⇔ *compromissos de crédito*'. Likewise, the bilingual pair '*union level* ⇔ *união*' is incorrect because no translation exists for the English word '*level*' on the right hand side.

To assess the bilingual candidates for parallelism, we introduce two features. We say that a translation candidate has translation gap with respect to the first language ($gap_{L1}=1$) when the term in the first language does not have a translation in the second language in whole or in parts and vice versa. Lack of parallelism implies a gap in translation.

Stemmed Coverage. While looking for coverage, if the expressions on both sides are not covered by the lexicon, we set the features gap_{L1} and gap_{L2} to 0.5^6. To deal with such situations reflecting our lack of support, we extract two features representing coverage using the stemmed training data. These features work in the same way as discussed above except that they look only at the word stems. To instantiate, while looking for coverage for the bilingual pair '*bronchitically* ⇔ *bronquiticamente*', we use its stemmed version '*bronchit* ⇔ *bronquit*', as the coverage is examined using the stemmed training and test datasets. If the training dataset contains the term '*bronchit*' in EN and '*bronquit*' in PT, then (gap_{L1}, gap_{L2}) would be (0.0, 0.0). This feature would find less gaps in translations that are indeed parallel, and thus decrease the number of false negatives (i.e., good translations that are classified as bad).

For identifying the translation gaps, we use the Aho-corasick set-matching algorithm that checks if the terms in the key-word tree[7] occur as (sub-)expressions in the bilingual pair to be validated and if they occur are accepted translations [9]. Similarly, to find the stemmed coverage, we use the stemmed training and test datasets, obtained using the Snowball stemmer. Here, each keyword tree is constructed using the stemmed part of the term. Translation

---

[6] A neutral value reflecting our lack of support in deciding whether to accept or to reject that pair.

[7] Constructed separately using the first and second language terms in the accepted bilingual training data.

gaps are identified using the Aho-corasick set-matching algorithm as mentioned previously.

4 Experimental Setup and Evaluation

We use the SVM based tool, LIBSVM[8] to learn the binary classifier, which tries to find the hyperplane, that separates the training examples with the largest margin. We scale the data in the range [0 1]. We perform a grid-search on RBF kernel parameters, (C, γ) using cross-validation, so that the classifier can accurately predict unknown data (testing data).

4.1 Data Sets

The translation candidates used in our experiments were acquired using various extraction techniques applied on a (sub-)sentence aligned parallel corpora[9] [1,4,8,15]. We experimented with 3 language pairs, EN-PT, EN-FR and FR-PT. The suffix array based phrase translation extraction technique was employed only for the language pair EN-PT and was excluded in extracting EN-FR and FR-PT bilingual pairs [1]. The statistics of the training and test datasets (validated bilingual lexicon) are as shown in Table 2. We set aside randomly 5% of the validated lexicon as the test set. We repeat experiments for comparing the experimental results related to the size of the training corpus by taking into account randomly extracted 50%, 75%, 80%, 90% and the entire 95% of the training set.

Table 2. Training and Testing Data Statistics

Data Sets		EN-PT			EN-FR			FR-PT		
		Accepted	Rejected	Total	Accepted	Rejected	Total	Accepted	Rejected	Total
Training	95%	853,452	575,951	1,429,403	362,017	51,054	413,071	372,306	78,754	451,060
	90%	768,105	518,356	1,286,461	342,963	48,370	391,333	352,711	74,609	427,320
	80%	682,761	460,761	1,143,522	304,856	42,996	347,852	313,521	66,319	379,840
	75%	640,088	431,963	1,072,051	285,803	40,308	326,111	293926	62,174	356,100
	50%	426,725	287,976	714,701	181,009	26,871	207,880	195,952	41,449	237,401
Test	5%	44,920	30,312	75,232	19,053	2,687	21,740	19,595	4,145	23,740

4.2 Results

In the current section, we discuss the classification results and the performance of the classifier with respect to various features using the complete data set (95%) introduced in the Section 4.1 for each of the language pairs EN-PT, EN-FR and FR-PT.

---

[8] A library for support vector machines - Software available at http://www.csie.ntu.edu.tw/~cjlin/libsvm

[9] DGT-TM - https://open-data.europa.eu/en/data/dataset/dgt-translation-memory
Europarl - http://www.statmt.org/europarl/
OPUS (EUconst, EMEA) - http://opus.lingfil.uu.se/

The Table 3 shows the precision (P_{Acc}, P_{Rej}), recall (R_{Acc}, R_{Rej}) and the accuracy of the estimated classifier in predicting each of the classes (Acc and Rej) while using different features. Micro-average Recall (μ_R), Micro-average Precision (μ_P), and Micro-average f-measure (μ_F)[10] are used to assess the global performance over both classes.

As might be seen from the Table 3, for EN-PT, substantial improvement is achieved by using the feature that looks for translation coverage on both sides of the bilingual pair. We observe an increase in μ_F of 22.85% over the base line and 19.32% over a combination of the features representing baseline and bad ends. Best μ_F is obtained when the stemmed[11] lexicon is used to look for stem coverage rather than the original lexicon. However, for EN-FR, training with stemmed lexicon did not show a meaningful improvement.

Table 3. Classifier Results using different features for EN-PT, EN-FR and FR-PT

Language Pairs	Features	P_{Acc}	R_{Acc}	P_{Rej}	R_{Rej}	μ_R	μ_R	μ_F	Accuracy
EN-PT	BL$_{strsim+freq}$	70.87	93.47	81.66	43.08	76.27	68.28	72.05	73.17
	BL+BE$_{SW}$	76.50	88.47	77.76	59.73	77.13	74.10	75.58	76.89
	BL+BE$_{Pat\_R-A}$+Cov	98.93	92.41	89.75	98.52	94.34	95.47	94.90	94.87
	BL+BE$_{Pat\_R-A}$+Cov$_{Stm}$	99.85	92.03	89.42	99.80	94.64	95.92	95.27	95.16
	BL+BE$_{SW}$+Cov$_{Stm}$	98.64	94.63	92.50	98.06	95.57	96.35	**95.96**	96.02
EN-FR	BL$_{strsim+freq}$	90.67	98.45	71.89	28.17	81.28	63.31	71.18	89.76
	BL+BE$_{SW}$	90.69	98.5	72.73	28.28	81.71	63.39	71.39	89.83
	BL+BE$_{Pat\_R-A}$+Cov	96.03	86.56	43.92	74.62	69.98	80.59	74.91	85.09
	BL+BE$_{\_Pat\ R-A}$+Cov+SpSim	96.07	86.63	44.11	74.84	70.09	80.74	**75.04**	85.17
	BL+BE$_{Pat\_R-A}$+Cov$_{Stm}$	91.10	98.25	71.98	31.93	81.54	65.09	72.39	90.05
	BL+ BE$_{Pat\_R-A}$+Cov$_{Stm}$+SpSim	91.34	98.23	73.04	33.98	82.19	66.11	73.26	90.29
FR-PT	BL$_{strsim+freq}$	85.12	97.85	65.30	19.16	75.21	58.51	65.81	84.11
	BL+BE$_{SW}$	85.12	97.83	65.05	19.13	75.09	58.48	65.75	84.09
	BL+BE$_{SW}$+Cov	88.80	74.55	31.58	55.54	60.19	65.05	62.52	71.23
	BL+BE$_{\_Pat\ R-A}$+Cov+SpSim	88.87	75.54	32.35	55.30	60.61	65.42	62.92	72.01
	BL+BE$_{Pat\_R-A}$+Cov$_{Stm}$	85.12	97.83	65.05	19.13	75.09	58.48	65.75	84.09
	BL+BE$_{Pat\_R-A}$+Cov$_{Stm}$+SpSim	85.13	97.87	65.54	19.18	75.34	58.53	**65.87**	84.13

FR-PT results are worse than the results obtained for other language pairs: the best μ_F and accuracy of 65.87% and 84.13% respectively are obtained when we use a combination of features BL+$BE_{Pat_{R-A}}$ + Cov_{Stm} + $SpSim$. However, the improvement is negligible (approximately ranging from 0.01% - 0.14%) against the baseline ($BL_{strsim+freq}$) in every terms (precision, recall and micro f-measure) over both classes. This may be explained because the number of 'single word - single word' pairs is comparatively larger than for the other language pairs and the number of 'multi-word - multi-word' pairs is small (50,552 for the accepted). Approximately 250K French multi-words are paired with single Portuguese words and approximately 9K Portuguese multi-words are paired with single French words. Moreover, approximately 130K are single word pairs for this pair of languages which is quite different from the EN-PT scenario.

---

[10] Computed as discussed in [11].
[11] Stemmed using the snowball stemmer.

Also, patterns indicating bad ends that are stop words (BE_{SW}) are substantially few in number with respect to FR-PT[12] and EN-FR[13] lexicon corpus as opposed to EN-PT[14]. This is because extractions for these language pairs use all of the techniques mentioned in section 4.1 except for the suffix array based extraction technique [1]. Hence EN-FR and FR-PT were much cleaner.

4.3   Classifier Performance by Training Set Size

We analyzed the impact of varying the size of training datasets on the improvement given by various features. Table 4 shows the results obtained using the features $BL_{strsim+freq}+BE_{SW}+$Cov (EN-PT) and $BL_{strsim+freq}+BE_{Pat_{R-A}}+$ $Cov + SpSim$ (EN-FR and FR-PT) respectively.

Table 4. Classifier Results for EN-PT, EN-FR and FR-PT by training set sizes

Language Pairs	Training Dataset	P_{Acc}	R_{Acc}	P_{Rej}	R_{Rej}	μ_P	μ_R	μ_F	Accuracy
	50%	**99.45**	92.22	89.59	**99.24**	**94.52**	**95.73**	**95.12**	**95.05**
	75%	99.21	92.32	89.68	98.90	94.45	95.61	95.02	94.97
EN-PT	80%	99.04	92.38	89.73	98.67	94.39	95.53	94.95	94.91
	90%	98.74	92.38	89.69	98.25	94.22	95.32	94.76	94.74
	95%	98.38	**92.60**	**89.91**	97.74	94.15	95.17	94.65	94.67
	50%	93.75	58.62	19.77	72.31	56.76	65.47	60.80	60.31
	75%	95.41	74.77	29.39	74.47	62.40	74.62	67.97	74.73
EN-FR	80%	95.59	75.82	30.49	75.21	63.04	75.52	68.72	75.74
	90%	95.85	68.59	26.17	**78.94**	61.01	73.77	66.78	69.87
	95%	**96.07**	**86.63**	44.11	74.84	**70.09**	**80.74**	**75.04**	**85.17**
	50%	88.68	67.57	27.86	**59.20**	58.27	63.39	60.72	66.11
	75%	88.62	75.05	31.59	54.45	60.11	64.75	62.34	71.46
FR-PT	80%	88.76	75.29	31.99	54.93	60.38	65.11	62.65	71.74
	90%	88.43	79.29	**34.22**	50.95	**61.33**	65.12	**63.17**	**74.34**
	95%	**88.87**	**75.54**	32.35	55.30	60.61	**65.42**	62.92	72.01

Looking at the classification results for EN-PT using SVM and the training set, we observe that the larger the training set larger the recall (R_{Acc} is 92.6% against 92.22%) for the 'Accepted' class. Meanwhile, when we augment the training set we loose in precision from 99.45% to 98.38%. However, by augmenting the training set we augment the precision (R_{Acc} from 89.59% to 89.91%) for the 'Rejected' class, whereas the recall drops (R_{Rej} from 99.24% to 97.74%). As the training set is much larger than for other language pairs (95% of the corpus) we

---

[12] 5 in FR and 8 in PT; most frequent are 'de' in FR with 27 occurrences and 'de' in PT with 43 occurrences.

[13] 43 in EN and 15 in FR; most frequent are 'to' in EN with 210 occurrences and 'pas' in FR with 237 occurrences.

[14] 112 in EN and 86 in PT; most frequent are 'the' in EN with 27,455 occurrences and 'a' in PT with 22,242 occurrences.

do not necessarily gain much. Thus, precision and recall for EN-PT does evolve in a way, such that, while one augments the other tends to decrease, partially deviating from the trend observed in our earlier experiments [11]. It is possible that some sort of overfitting occurs.

Unlike EN-PT, for the language pairs EN-FR and FR-PT, with larger training sets the performance of the trained classifier improved. For the features listed in Table 3, best results were obtained with 95% and 90% of the training set.

4.4 Classifier Trained on One Language Pair in Classifying Others

Motivated by the classifier performance for language pairs EN-PT, we conducted few more experiments: we trained the classifier using the full set of features on one language pair, and tested on the other. Training on EN-PT data and testing on EN-FR and FR-PT resulted in μ_F of 55.64% and 54.99%, far below the baseline for EN-FR (a drop by approximately 15% from 71.18%) and FR-PT (a drop by approximately 11% against 65.81%) respectively. Training the system with EN-FR and testing on FR-PT did even worse, leading to a micro f-measure of 52.96%. Training on FR-PT data and testing on EN-FR, led to a μ_F of 47.8%. This lets us to conclude that it does not make any sense to use a classifier trained on one language pair in classifying the data from other language pairs. The related results are shown in Table 5.

Table 5. Performance of Classifier trained on one language pair when tested on others.

Language Pairs (Test Set)	Classifier Trained	P_{Acc}	R_{Acc}	P_{Rej}	R_{Rej}	μ_P	μ_R	μ_F	Accuracy
EN-FR	Train with EN-FR model	96.03	86.56	43.92	74.62	69.98	80.59	74.91	85.09
	Train with EN-PT model	89.07	88.55	22.02	22.93	55.55	55.74	55.64	80.44
	Train with FR-PT model	86.85	70.55	10.39	24.23	48.62	47.39	47.80	64.82
FR-PT	Train with FR-PT model	88.80	74.55	31.58	55.54	60.19	65.05	62.52	71.23
	Train with EN-PT model	85.71	59.66	21.74	52.98	53.73	56.32	54.99	58.49
	Train with EN-FR model	84.96	46.00	19.42	61.52	52.19	53.76	52.96	48.71

5   Conclusion

We have discussed the classification approach as a means for selecting appropriate and adequate candidates for parallel corpora alignment. Experimental results demonstrate the use of the classifiers on EN-PT, EN-FR and FR-PT language pairs under small, medium and large data conditions. Several insights are useful for distinguishing the adequate candidates from inadequate ones such as, lack (presence) of parallelism, spurious terms at translation ends and the base properties (similarity and occurrence frequency) of the translation pairs.

This work is motivated by the need for a system that evaluates the translation candidates automatically extracted prior to their submission for human validation. Automatically extracted bilingual translations after human validation, are subsequently used for realigning parallel corpora and extracting new translations forming an indefinite cycle of iterations. Automatic classification prior to validation contributes to speed up the process of distinguishing the correct translations from naturally occurring alignment and extraction errors. The positive side effect is an enriched annotated lexicon suitable for machine learning systems such as bilingual morphology learning and translation suggestion tool, apart from its primary use as an aid in alignment, extraction and translation.

In future, the use of bilingual stem and suffix correspondences in classifying FR-PT and EN-FR word-to-word translations shall be studied [12]. Looking for coverage in word pairs might be cast as a morphological coverage problem that would enable us to classify word-to-word translations with high accuracy. Further, some experiments should be done on EN-FR and FR-PT using the suffix array based extractor [1]. Experiments must also be carried out to determine an optimal interval for the number of positive and negative bilingual lexicon entries in obtaining optimal classification results.

Acknowledgments. K. M. Kavitha and Luís Gomes acknowledge the Research Fellowship by FCT/MCTES with Ref. nos., SFRH/BD/64371/2009 and SFRH/ BD/65059/2009, respectively, and the funded research project ISTRION (Ref. PTDC/EIA-EIA/114521/2009) that provided other means for the research carried out. The authors thank NOVA LINCS, FCT/UNL for providing partial financial assistance to participate in EPIA 2015, and ISTRION BOX - Translation & Revision, Lda., for providing the data and valuable consultation.

References

1. Aires, J., Lopes, G.P., Gomes, L.: Phrase translation extraction from aligned parallel corpora using suffix arrays and related structures. In: Lopes, L.S., Lau, N., Mariano, P., Rocha, L.M. (eds.) EPIA 2009. LNCS, vol. 5816, pp. 587–597. Springer, Heidelberg (2009)
2. Aker, A., Paramita, M.L., Gaizauskas, R.J.: Extracting bilingual terminologies from comparable corpora. In: Proceedings of the 51st Annual Meeting for Computational linguistics, vol. 2, pp. 402–411 (2013)
3. Bergsma, S., Kondrak, G.: Alignment-based discriminative string similarity. In: Annual meeting-ACL, vol. 45, p. 656 (2007)
4. Brown, P.F., Pietra, V.J.D., Pietra, S.A.D., Mercer, R.L.: The mathematics of statistical machine translation: Parameter estimation. Computational linguistics **19**(2), 263–311 (1993)
5. Chen, B., Cattoni, R., Bertoldi, N., Cettolo, M., Federico, M.: The ITC-irst SMT system for IWSLT-2005, pp. 98–104 (2005)
6. Fraser, A., Marcu, D.: Measuring word alignment quality for statistical machine translation. Computational Linguistics **33**(3), 293–303 (2007)
7. Gomes, L.: Parallel texts alignment. In: New Trends in Artificial Intelligence, 14th Portuguese Conference in Artificial Intelligence, EPIA 2009, Aveiro, October 2009
8. Gomes, L., Pereira Lopes, J.G.: Measuring spelling similarity for cognate identification. In: Antunes, L., Pinto, H.S. (eds.) EPIA 2011. LNCS, vol. 7026, pp. 624–633. Springer, Heidelberg (2011)

9. Gusfield, D.: Algorithms on strings, trees, and sequences: computer science and computational biology. Cambridge Univ Pr., pp. 52–61 (1997)

10. Johnson, J.H., Martin, J., Foster, G., Kuhn, R.: Improving translation quality by discarding most of the phrasetable. In: Proceedings of EMNLP (2007)

11. Kavitha, K.M., Gomes, L., Lopes, G.P.: Using SVMs for filtering translation tables for parallel corpora alignment. In: 15th Portuguese Conference in Arificial Intelligence, EPIA 2011, pp. 690–702, October 2011

12. Kavitha, K.M., Gomes, L., Lopes, J.G.P.: Identification of bilingual suffix classes for classification and translation generation. In: Bazzan, A.L.C., Pichara, K. (eds.) IBERAMIA 2014. LNCS, vol. 8864, pp. 154–166. Springer, Heidelberg (2014)

13. Koehn, P., Hoang, H., Birch, A., Callison-Burch, C., Federico, M., Bertoldi, N., Cowan, B., Shen, W., Moran, C., Zens, R., et al.: Moses: open source toolkit for statistical machine translation. In: Proceedings of the 45th Annual Meeting of the ACL on Interactive Poster and Demonstration Sessions, pp. 177–180. ACL (2007)

14. Kutsumi, T., Yoshimi, T., Kotani, K., Sata, I., Isahara, H.: Selection of entries for a bilingual dictionary from aligned translation equivalents using support vector machines. In: Proceedings of PACLING (2005)

15. Lardilleux, A., Lepage, Y.: Sampling-based multilingual alignment. In: Proceedings of RANLP, pp. 214–218 (2009)

16. Levenshtein, V.I.: Binary codes capable of correcting deletions, insertions, and reversals. Soviet Physics Doklady 10, 707–710 (1966)

17. Melamed, I.D.: Automatic evaluation and uniform filter cascades for inducing n-best translation lexicons. In: Proceedings of the Third Workshop on Very Large Corpora, pp. 184–198. Boston, MA (1995)

18. Och, F.J., Ney, H.: A systematic comparison of various statistical alignment models. Computational linguistics 29(1), 19–51 (2003)

19. Och, F.J., Ney, H.: The alignment template approach to statistical machine translation. Computational Linguistics 30(4), 417–449 (2004)

20. Sato, K., Saito, H.: Extracting word sequence correspondences based on support vector machines. Journal of Natural Language Processing 10(4), 109–124 (2003)

21. Tian, L., Wong, D.F., Chao, L.S., Oliveira, F.: A relationship: Word alignment, phrase table, and translation quality. The Scientific World Journal (2014)

22. Tiedemann, J.: Extraction of translation equivalents from parallel corpora. In: Proceedings of the 11th NoDaLiDa, pp. 120–128 (1998)

23. Tomeh, N., Cancedda, N., Dymetman, M.: Complexity-based phrase-table filtering for statistical machine translation (2009)

24. Tomeh, N., Turchi, M., Allauzen, A., Yvon, F.: How good are your phrases? Assessing phrase quality with single class classification. In: IWSLT, pp. 261–268 (2011)

25. Vapnik, V.: The Nature of Statistical Learning Theory. Data Mining and Knowledge Discovery 1–47 (2000)

26. Vilar, D., Popovic, M., Ney, H.: AER: Do we need to "improve" our alignments? In: IWSLT, pp. 205–212 (2006)

27. Way, A., Hearne, M.: On the role of translations in state-of-the-art statistical machine translation. Language and Linguistics Compass 5(5), 227–248 (2011)

28. Zens, R., Stanton, D., Xu, P.: A systematic comparison of phrase table pruning techniques. In: Proceedings of the 2012 Joint Conference on EMNLP and CoNLL, EMNLP-CoNLL 2012, pp. 972–983. ACL (2012)

29. Zhao, B., Vogel, S., Waibel, A.: Phrase pair rescoring with term weightings for statistical machine translation (2004)

A SMS Information Extraction Architecture to Face Emergency Situations

Douglas Monteiro[✉] and Vera Lucia Strube de Lima

PUCRS, Faculdade de Informática, Programa de Pós-Graduação em Ciência da Computação, Porto Alegre, Brazil
dmmonteiro@gmail.com, vera.strube@pucrs.br

Abstract. In disasters, a large amount of information is exchanged via SMS messages. The content of these messages can be of high value and strategic interest. SMS messages tend to be informal and to contain abbreviations and misspellings, which are problems for current information extraction tools. Here, we describe an architecture designed to address the matter through four components: linguistic processing, temporal processing, event processing, and information fusion. Thereafter, we present a case study over a SMS corpus of messages sent to an electric utility company and a prototype built with Python and NLTK to validate the architecture's information extraction components, obtaining Precision of 88%, Recall of 59% and F-measure (F1) of 71%. The work also serves as a roadmap to the treatment of emergency SMS in Portuguese.

Keywords: Information extraction · Short messages · Emergencies

1 Introduction

Currently, it is hard to imagine any line of business that does not use any textual information. Aside from the available content on the Internet, a large amount of information is generated and transmitted by computers and smartphones all over the world. Gary Miner *et al.* estimate that 80% of the information available in the world are in free text format and therefore not structured [7]. With such large amount of potentially relevant data, an information extraction system can structure and refine raw data in order to find and link relevant information amid extraneous information [3,5]. This process is made possible by understanding the information contained in texts and their context, but this complex task face difficulties when processing informal languages, such as SMS messages or tweets [1,6,10].

Messages using the Short Message Service (SMS), as well as tweets, are widely used for numerous purposes, which makes them rich and useful data for information extraction. The content of these messages can be of high value and strategic interest, specially during emergencies[1]. Under these circumstances, the amount

---

[1] Also referred to as crisis events, disasters, mass emergencies and natural hazards by other researchers in the area.

© Springer International Publishing Switzerland 2015
F. Pereira et al. (Eds.) EPIA 2015, LNAI 9273, pp. 735–746, 2015.
DOI: 10.1007/978-3-319-23485-4_74

of messages tends to increase considerably. However, users of these services write messages freely, with abbreviations, slangs and misspellings. Short messages tend to be brief, informal and to present similarities to speech.

In light of this, we propose an architecture to extract information from SMS messages exchanged during emergency situations. This Information Extraction architecture has as input a corpus of SMS messages and comprises four components: a linguistic processing component, a temporal processing component, an event processing component, and an information fusion component. The linguistic processing component preprocesses messages, handling with abbreviations and punctuation, sentence splitting, tokenization and stopword removal. The temporal processing component uses a set of rules and a list of temporal keywords to identify and classify temporal expressions. The event processing component is responsible for identifying events according to a set of domain-defined categories and provides additional information regarding situation awareness. As output, the architecture consolidates information so one can visualize strategic information in order to help the decision-making process. We built a prototype to validate the architecture and evaluated the information extraction taggers resulting in Precision of 88%, Recall of 59% and F-Measure (F1) of 71%.

This paper is organized in six sections, the first one being this introduction. In Section 2, we review related work on information extraction from short messages and its applications. In Section 3, we introduce the SMS information extraction architecture of messages sent during emergencies. Section 4 details the case study conducted to validate this architecture over a corpus built from SMS Messages sent by costumers to an electric utility company during emergencies. In Section 5, we discuss the evaluation performed over the prototype and the results obtained. Finally, in Section 6, we comment on challenges faced, as well as on future work.

2   Related Work

Corvey *et al.* introduce a system that incorporates linguistic and behavioral annotation on tweets during crisis events to capture situation awareness information [2]. The system filters relevant and tactical information intending to help the affected population. Corvey *et al.* collected data during five disaster events and created datasets for manual annotation. The authors linguistically annotated the corpus, looking for named entities of four types: person, name, organization and facilities. A second level of behavioral annotation assesses how community members tweet during crisis events. Tweets receive different and non-exclusive qualitative tags according to the type of information provided. Tweets containing situational awareness information are collected and tagged with macro-level (environmental, social, physical or structural) and micro-level (regarding damage, status, weather, etc.) information. The results indicated that, under emergencies, "users communicate via Twitter in a very specific way to convey information" [2]. Becoming aware of such behavior helped the framework's machine learning classifier to achieve accuracy of over 83% using POS tags and bag of words. To classify location, they used Conditional Random Fields (CRFs) with lexical and syntactic information and POS as features. The annotated corpus was divided into 60% for training and 40% for testing. They obtained an accuracy of 69% for the complete match and 86% for the partial match and recall of 63% for the complete match and 79% for the partial match.

Sridhar *et al.* present an application of statistical machine translation to SMS messages [11]. This research details on the data collection process and steps and resources used on a SMS message translation framework, which uses finite state transducers to learn the mapping between short texts and canonical form. The authors used a corpus of tweets as surrogate data and a bitext corpus from 40,000 English and 10,000 Spanish SMS messages, collected from transcriptions of speech-based messages sent through a smartphone application. Another 1,000 messages were collected from the Amazon Mechanical Turk[2]. 10,000 tweets were collected and normalized by removing stopwords, advertisements and web addresses. The framework processes messages segmented into chunks using an automatic scoring classifier. Abbreviations are expanded using expansion dictionaries and translated using a translation model based on sentences. The authors built a static table to expand abbreviations found in SMS messages, where a series of noisy texts have the corresponding canonical form mapped. For example, "4ever" is linked to the canonical form "forever". Next, the framework segments phrases using an automatic punctuation classifier trained over punctuated SMS messages. Finally, the Machine Translation component uses a hybrid translation approach with phrase-based translation and sentences from the input corpus represented as a finite-state transducer. The framework was evaluated over a set of 456 messages collected in a real SMS interaction, obtaining a BLEU score of 31.25 for English-Spanish translations and 37.19 for Spanish-English.

Ritter *et al.* present TwiCAL, an open-domain event extraction and categorization system for Twitter [9]. This research proposes a process for recognizing temporal information, detecting events from a corpus of tweets and outputting the extracted information in a calendar containing all significant events. The authors focused on identifying events referring to unique dates. TwiCAL extracts a 4-tuple representation of events, including a named entity, an event phrase, and an event type. The authors trained a POS tagger and a NE tagger on in-domain Twitter data. To build an event tagger, they trained sequence models with a corpus of annotated tweets, and a rule-based system and POS to mark temporal expressions on a text. The open-domain event categorization uses variable models to discover types that match the data and discards incoherent types. The result is applied to the categorization of extracted events. The classification model is evaluated according to the event types created from a manual inspection of the corpus. The authors compared the results with a supervised Maximum Entropy baseline, over a set of 500 annotated events using 10-fold cross validation. Results achieved a 14% increase in maximum F1 score over the supervised baseline. A demonstration of the system is available at the Status Calendar webpage[3].

---

[2] https://www.mturk.com/
[3] http://statuscalendar.com

Dai *et al.* present SoMEST (Social Media Event Sentiment Timeline), a framework for competitive intelligence analysis for social media and the architecture of a NLP tool combining NER, event detection and sentiment analysis [4]. This research presents an architecture to extract information from social media texts and the visualization of these information. The authors use Event Timeline Analysis (ETA) to detect events and display them in a timeline, highlighting trends or behaviors of competitors, consumers, partners and suppliers. Dai *et al.* also use Sentiment Analysis to measure human opinions from texts written in natural language, searching for the topic, its author, and if its a positive or negative opinion. The process comprises three phases: data collection, extraction and classification, and synthesis. From social media texts generated by customers, SoMEST focus on detecting events published from companies and opinions shared by customers. The extraction and classification phase consists of analyzing data and generating event extracts and opinion extracts, which are synthesized into a social media profile, unifying events and opinions linked to brands, services and products of a corporation into a period of time. The timeline displays a chronological order of the corporations events, the competitors events and changes in customers opinions.

Accordingly, as even IE systems built for different tasks may present similarities, we could understand common points in different IE architectures, mainly due to the nature of short text messages. From this learning, we could elaborate an information extraction architecture for SMS messages according to core components shared by most IE systems reviewed here, such as POS taggers, tokenization, and normalization, while adding other components to treat domain-specific characteristics.

3   SMS Information Extraction Architecture

SMS messages contain information that can be extracted, providing valuable resources to support decision-making under emergency situations. With this in mind, here, we detail the proposal for an architecture to extract information from messages under these circumstances. As seen in Figure 1, the proposed IE architecture takes as input a Corpus of SMS messages. Then, the linguistic processing component preprocesses each message and prepares them for Information Extraction. The temporal expression tagger component recognizes and tags all temporal information within the messages, while the event tagger identifies and tags domain-related events accordingly. The information fusion component displays the extracted information so as users of this system can interpret its results. The output of the system is the extracted information organized in a readable display, regarding the application. We detail each component in sections 3.1 to 3.4.

3.1   Linguistic Processing Component

The linguistic processing component comprises a preprocessing module, including four steps: normalization, sentence splitting, tokenization, and stopword

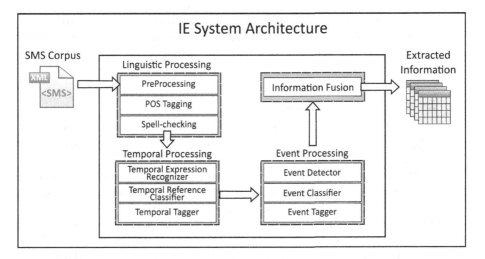

Fig. 1. IE Architecture overview

removal; and steps specifically designed for linguistic processing: POS tagging and spell-checking. The normalization step is responsible for adjusting the text while facing spelling variations, abbreviations, treating special characters and other features of the short message language. Next, the sentence splitting step divides each message into a list of sentences in order to process them individually. Every token is compared to a list of stopwords, which enables discarding unnecessary items and speeding up the process of information extraction.

Accordingly, the tokens are tagged with a part-of-speech tagger, which is trained with an annotated corpus of messages. The following step in the linguistic processing component comprises a Spell Checker, which makes use of an external dictionary to label untagged tokens and submits them to the POS tagger for revision. This component outputs a set of preprocessed sentences that serve as input for the temporal processing component.

3.2   Temporal Processing Component

The temporal processing component is responsible for applying regular expressions in order to identify temporal expressions related to events in SMS messages. Since temporal expressions are limited to a fixed set of syntactic patterns, most Temporal Expression Recognition systems make use of rule-based methods to recognize syntactic chunks [8].

Initially, the temporal expression recognizer uses a rule-based approach to identify variations of temporal references mentioned in the sentences. Despite the rule set being able to identify simple temporal expressions present in messages, rather complex expressions are still to be treated. For cases like "desde às 8h de domingo" (*since Sunday 8am*), the temporal expression recognizer counts on a list of temporal keywords, such as times of the day and days of the week to determine the extent of temporal expressions.

The temporal reference classifier analyzes the expression according to its lexical triggers and defines the type and value of the temporal expression. Finally, the component tags the temporal expression according to the TIMEX2 tag system provided by TimeML[4].

3.3 Event Processing Component

The event processing component starts from the event detection step, which is responsible for finding relevant events in a sentence. This step counts on a set of rules to identify the event.

Since the proposed IE architecture aims to extract information from messages during emergencies situations, one can determine a certain set of categories of events to detect during this step. For instance, as discussed in Section 2, Corvey *et al.* proposed a situational awareness annotation level with the intention of understanding crisis events as a whole [2]. To address that matter, the authors define categories such as 'Social Environment', 'Built Environment' and 'Physical Environment'. Each category has subcategories with specific information, such as 'Crime', 'Damage' or 'Weather'.

Accordingly, in order to be executed, the event processing component requires a previous definition of a set of domain-related observable categories. Consequently, sentences that match any of these categories pass through a classification step which relates the event to the categories. This step makes use of a list of domain-related keywords. Then, the component can assign the correspondent tags to the event mention.

3.4 Information Fusion Component

This component groups and organizes all tagged information in a human understandable manner. All relevant information are "fused" to show the results of the IE application. As there may exist several ways to represent the results, this decision is linked to the intended purpose of their application. For instance, Ritter *et al.* [9], display their results in form of a calendar, where the respective events are shown. On the other hand, Dai *et al.* [4] present the Information Extraction results in a timeline, showing the progress of event mentions and the amount of associated opinions.

Given a set of tagged events and their corresponding temporal information, the relation between and event and when it occurred must be clearly expressed. An IE system built on this architecture must display the extracted information in a meaningful and relevant way to provide situation awareness and to aid decision-making during emergencies.

4 Case Study

In order to validate our proposal, we present a case study conducted over the IE Architecture. In this chapter, we detail our choices and decisions made.

---

[4] http://www.timeml.org/site/publications/timeMLdocs/timeml_1.2.1.html

The input data for the process was organized from a set of 3,021 short messages received by an electric utility company. Clients notify the company when there is a power outage, sending short messages with the word "LUZ" (*light*) and the installation number (provided by the company). As observed in messages received, the companys clients use this communication channel to provide situation awareness information, which is currently not yet processed but could be of great help in services provision. It is important to extract information from these messages to deliver relevant and strategic information about emergencies so as to restore power to customers as quickly and safely as possible. The corpus was built in a XML format, comprising the messages and their delivery date. We split the corpus in a 'learning corpus', containing 2,014 messages; a 'gold standard corpus', containing 100 messages, to perform an evaluation of the prototypes taggers; a 'test corpus' to improve the prototype from the evaluation results.

We prototyped the architecture using Python[5] (version 2.7), mainly due to its ease of use, productivity and features for handling strings, lists, tuples and dictionaries, along with the Natural Language Toolkit (NLTK[6]) (version 3.0). NLTK provides some interesting features for Portuguese, like tokenizers, stemmers, Part-of-Speech taggers, and annotated corpora for training purposes. We highlight the main aspects of the components implementation as follows.

4.1   Linguistic Processing

The component standardizes the text input. SMS messages contain many misspellings, as texters tend not to follow spelling and grammar rules, which led us to address this matter beforehand, covering the most common cases found on the learning corpus. Some variations are caused by different levels of literacy, besides idiosyncratic SMS language characteristics. In this step, we lowercased messages, removed commas, hyphens and special characters, such as '#' and '@', and unnecessary full stops, such as in zip codes or abbreviations. Each sentence undergoes a tokenization step, using whitespaces to mark word boundaries. The prototype uses wordpunct_tokenize[7] to split strings into lists of tokens. We built an external list containing 45 stopwords found on the learning corpus, as well as common word shortenings and phonetic abbreviations, such as "vc" and "q".

The component uses MacMorpho[8], a tagged training corpus with news in standard Brazilian Portuguese. However, the lack of a tagged corpus of texts in SMS language hampers the POS tagging step. Even though Normalization handles some misspellings, many words not written in standard Portuguese remain untagged. To address this matter, we used PyEnchant[9], a spell checking library for Python, as a step to mitigate the spelling variation problem. We added the

[5] https://www.python.org/
[6] http://www.nltk.org
[7] http://www.nltk.org/api/nltk.tokenize.html
[8] http://www.nilc.icmc.usp.br/macmorpho/
[9] https://pythonhosted.org/pyenchant

Open Office[10] Brazilian Portuguese dictionary extension. The prototype only checks the spelling of untagged tokens. We also built an external list of domain-related words the POS tagger cannot resolve, like "transformador" (transformer) or "estouro"(burst), along with their corresponding POS tag, so the prototype can use this list to review untagged tokens.

4.2 Temporal Processing

Once tagged, messages proceed to temporal processing, which comprises the following steps: a Temporal Expression Recognizer, a Temporal Reference Classifier and a Temporal Tagger. There are two external resources associated with the component: a set of regular expressions and a list of lexical triggers.

From a linguistically processed message containing a time anchor, the Temporal Expression Recognizer must be able to identify and extract existing temporal information. For instance, the duration of an event, such as a power outage, may be of great importance to indicate the severity of the problem. A client may inform the existence of a natural disaster that causes a blackout that lasts for hours and affects an entire region.

We found many variations for the same temporal information like '10:30', '10h30', '10h30min', and so on. In light of this, we resorted to a rule-based matcher, which relies on a set of regular expressions. Moreover, we added rules to identify days of the week and times of the day.

Since incoming messages express ongoing situations, the existing temporal expressions refer to past or present events. In order to identify more complex expressions like "desde ontem às 14h" (*since yesterday 2pm*), we built a list of lexical triggers, containing the most common temporal keywords found in the learning corpus and their corresponding TIMEX2 value, such as "ontem" (*yesterday*) with value '-1D' (minus one day) and "noite" (*night*) with a time modifier 'NI'. Tokens that correspond to words in the list are considered part of a temporal expression.

Each tagged temporal expression can contain a value and a modifier, according to its type, which, along with a time anchor (the delivery date), allows us to determine the beginning or duration of an event. For instance, the lexical trigger "amanhã" (tomorrow) has a type DATE and value '1D', indicating one day must be added to the time anchor to determine the TIMEX value. The Temporal Expression Classifier verifies the list of lexical triggers to determine modifiers and values of each expression, considering first lexical triggers expressing largest periods of time. Duration expressions have precedence over dates and dates have precedence over times. Finally, the Tagger groups values and modifiers in a single form and assigns the corresponding time tag to the TEs.

4.3 Event Processing Component

To detect events, we focused on a verb-triggered rule-based approach to identify specific features that may be useful in the context, considering the urgency of

---

[10] http://www.openoffice.org/

the messages. From understanding the verb, its meaning and accessories, one can determine the structure of the sentence of which it is part. The prototype considers sentences in the following structure: Noun Phrase + Verb + Object. Both the Noun phrase and the Object can play semantic roles of agent and patient.

The prototype iterates through sentences looking for POS tags assigned during the linguistic processing in order to find verbs. After, the Event Detection step marks the boundaries of the event mention by greedy searching for nouns, prepositions, noun compounds, adjectives or pronouns on the surroundings of the verb. Verbs interspersed with other POS tags mark different event mentions, while adjectives and nouns (or noun compounds) mark the boundaries of event mentions.

Next, the prototype classifies the detected events. Through an extensive study of the learning corpus, we observed how clients communicate during emergencies as well as what they notify. Then, we listed the most relevant events and related words we found whilst defining the annotation standard. We defined three non-mutually exclusive categories of events, according to the observed events and thirteen notification types to provide situation awareness information. "Instalação" refers to messages containing information regarding the consumer unit (electrical installation), such as power outages, voltage drops and instabilities. "Rede" groups information about the electrical grid status and its components, such as short circuits or fallen utility poles. "Ambiente" comprises information regarding the environment that might affect the electrical grid, like fallen trees, storms and lightnings.

To properly classify the event, we split sentences and analyze separately noun phrases, verbs and objects, searching for domain-related words. The component depends on a list built from 83 words related to the notification types, collected from sources such as Dicionário Criativo[11] and Wordnet[12]. For instance, the sentence "caiu uma arvore na rede" (a tree fell over the power grid), is divided in two phrases: "caiu" (verb), and "uma arvore em rede" (object). Once verified, the Event Classifier considers that while the verb does not determine a notification type, the object notifies a notification type "Queda de Árvore", due to the presence of the words "arvore" and "rede". Finally, the Tagger groups all the information in a set of tags, according to its categories and notification types.

4.4   Information Fusion

Once the information are already tagged, one can use different approaches to visualize and understand such data. Being aware of other possibilities that could be explored in a more extended study, we generated charts from tagged messages and their corresponding notification types, allowing the visualization of the application of the proposed model.

---

[11] http://dicionariocriativo.com.br/
[12] http://wordnetweb.princeton.edu/

We exported the output of the prototype to a spreadsheet containing messages and their corresponding notification types. Structured information can be more easily manipulated, in order to speed up the recognition and attendance of occurrences with more situation awareness.

5   Evaluation and Discussion

In order to evaluate the prototype's taggers, we elaborated a three-step plan comprising: confirming the model of categories and notification types; providing a gold standard - a manually annotated corpus considered as "definitive answer"; and comparing the prototypes results to the gold standard. Furthermore, we assessed the results of the Information Fusion component over the gold standard and the test corpus.

As some answers shall be provided by domain experts, we invited three judges with domain knowledge. They opined on the model of categories of events and temporal expressions and evaluated a set of 100 SMS according to this model. From their answers, we have composed the gold standard and compared it to the output of the IE prototype and obtained precision of 88%, recall of 59% and F-measure (F1) of 71%.

The prototype correctly identified relevant events, with 125 true positives over 16 false positives and 84 false negatives. The results indicate that the set of defined rules is accurate while detecting the events and temporal information mentioned in the gold standard corpus. However, a low recall score alerts us that there are other events, domain-related words and temporal information still uncovered by the model created.

Table 1 shows the hit percentage of the prototype when compared to the Gold Standard, as well as the amount of notifications found in the corpus. The prototype could not resolve mentions to "Wind" and "Rain" events. Analyzing the messages in the gold standard, we can see that some event mentions were not detected by the prototype as they omit verbs, like in "toda nossa comunidade sem luz devido muita chuva ventos fortes" (*our entire community without electricity because of a lot of rain and strong winds*). There were also problems in

Table 1. Hit Percentage by information type

Extracted Information	Prototype	Gold Standard	Percentage
Temporal Information	24	27	89%
Power Outage	86	88	98%
Downed Power Line	7	32	22%
Short Circuit	3	5	60%
Broken Power Pole	4	10	40%
Power Line Fire	5	7	71%
Fallen Tree	5	12	42%
Wind	0	4	0%
Rain	0	2	0%

differentiating "está" (*is*) and its popular contraction "tá" (absent in the training corpus) from "esta" (*this*) which compromised the detection of some events.

The temporal processing component behaved well over the gold standard. In fact, in one of the evaluated messages, the component tagged "15 minutos" (15 minutes) as a temporal expression, while the judges did not recognize it, showing that this understanding is not clear even for humans.

6 Considerations

During emergencies, any detail can help services provision. For that matter, SMS messages can be an important source of valuable information, as it is one of the most widely used means of communication. However, SMS messages are usually written in a proper language containing abbreviations, slangs and misspellings, which hamper their processing in their context of operation.

As observed in Section 2, even IE systems built for different tasks may present similarities. From this learning, we could propose an architecture for information extraction from SMS messages according to core components shared by most IE systems reviewed, while adding other components to treat domain-specific characteristics. The architecture comprises a linguistic processing component, which prepares messages for information extraction; a temporal processing component, which identifies and tags existing temporal information within messages; an event processing component, which detects and classifies events according to a list of domain-related categories; and an information fusion component that interprets information and displays them in a human-understandable manner.

To validate the architecture, we conducted a case study over a corpus of SMS messages sent to an electric utility company. We studied how users communicate during emergencies and defined categories of information that could aid services provision. We validated the architecture against a gold standard corpus built with the assistance of judges with domain knowledge. Among the tagging stages, we established a degree of severity (varying from 1 to 5) to distinguish the categories of events. We assessed the range of scores given by the judges, resulting in a kappa coefficient of 0.0013, *i.e.*, a poor level agreement, which led us to use fewer severity degrees.

As, to the best of our knowledge, there is no architecture to address this matter, especially for the Portuguese language, we expect this proposal to bring focus to this area and encourage other researchers to contribute to its improvement. IE systems built on this architecture may attend other electric utility companies or address other types of disasters or emergencies and other short messages, such as tweets. As improvement opportunities unveiled, we could mention resorting to a more appropriate tagged corpus trained over the SMS language. Such resource would decrease the number of untagged tokens, which in turn would increase the accuracy of the event detection step. However, to the present, we do not know of the existence of such resource for the Portuguese language.

For future work, we intend to continue our research, revising the Case Study results, and refine the prototype according to other approaches, such as machine learning to automatize the categories definition step. Envisaged features comprise adding named entity recognition towards gathering geographic information, which was not considered during this stage of our work, but can be of great importance for services provision. Moreover, we intend to use temporal expressions to determine the start and duration of detected events. We will also assess the information fusion component results as well as alternatives to enhance it.

References

1. Bernicot, J., Volckaert-Legrier, O., Goumi, A., Bert-Erboul, A.: Forms and functions of SMS messages: A study of variations in a corpus written by adolescents. Journal of Pragmatics **4412**, 1701–1715 (2012)
2. Corvey, W.J., Verma, S., Vieweg, S., Palmer, M., Martin, J.H.: Foundations of a multilayer annotation framework for twitter communications during crisis events. In: 8th International Conference on Language Resources and Evaluation Conference (LREC), p. 5 (2012)
3. Cowie, J., Lehnert, W.: Information extraction. Communications of the ACM **391**, 80–91 (1996)
4. Dai, Y., Kakkonen, T., Sutinen, E.: SoMEST: a model for detecting competitive intelligence from social media. In: Proceedings of the 15th International Academic MindTrek Conference: Envisioning Future Media Environments, pp. 241–248 (2011)
5. Jurafsky, D., Martin, J. H.: Speech and language processing, 2nd edn. Prentice Hall (2008)
6. Melero, M., Costa-Juss, M.R., Domingo, J., Marquina, M., Quixal, M.: Holaaa!! writin like u talk is kewl but kinda hard 4 NLP. In: 8th International Conference on Language Resources and Evaluation Conference (LREC), pp. 3794–3800 (2012)
7. Miner, G., Elder, J.I., Hill, T., Nisbet, R., Delen, D.: Practical Text Mining and Statistical Analysis for Non-structured Text Data Applications. Elsevier, Burlington (2012)
8. Pustejovsky, J., Stubbs, A.: Natural Language Annotation for Machine Learning. OReilly Media, Inc. (2012)
9. Ritter, A., Etzioni, O., Clark, S., et al.: Open domain event extraction from twitter. In: Proceedings of the 18th ACM SIGKDD International Conference on Knowledge Discovery and Data Mining, pp. 1104–1112 (2012)
10. Seon, C.-N., Yoo, J., Kim, H., Kim, J.-H., Seo, J.: Lightweight named entity extraction for korean short message service text. KSII Transactions on Internet and Information Systems (TIIS) **5–3**, 560–574 (2011)
11. Sridhar, V.K.R., Chen, J., Bangalore, S., Shacham, R.: A Framework for translating SMS messages. In: Proceedings of COLING 2014, the 25th International Conference on Computational Linguistics: Technical Papers, pp. 974–983 (2014)

Cross-Lingual Word Sense Clustering for Sense Disambiguation

João Casteleiro, Joaquim Ferreira da Silva$^{(\boxtimes)}$, and Gabriel Pereira Lopes

NOVA LINCS FCT/UNL, 2829-516 Caparica, Portugal
casteleiroalves@gmail.com, {jfs,gpl}@fct.unl.pt

Abstract. Translation is one of the areas where word disambiguation must be solved in order to find out adequate translations for such words in the contexts where they occur. In this paper, a Word Sense Disambiguation (WSD) approach using Word Sense Clustering within a cross-lingual strategy is proposed. Available sentence-aligned parallel corpora are used as a reliable knowledge source. English is taken as the source language, and Portuguese, French or Spanish as the targets. Clusters are built based on the correlation between senses, which is measured by a language-independent algorithm that uses as features the words near the ambiguous word and its translation in the parallel sentences, together with their relative positions. Clustering quality reached 81% (V-measure) and 92% (F-measure) in average for the three language pairs. Learned clusters are then used to train a support vector machine, whose classification results are used for sense disambiguation. Classification tests showed an average (for the three languages) F-measure of 81%.

Keywords: Word Sense Disambiguation · Clustering · Parallel corpora · V-measure · F-measure · Support vector machine

1 Introduction

Word sense ambiguity is present in many words no matter the language, and translation is one of the areas where this problem is important to be solved. So, in order to select the correct translation, it is necessary to find the right meaning, that is, the right sense, for each ambiguous word. Although multi-word terms tend to be semantically more accurate than single words, multiword terms may also have some ambiguity, depending on the context.

Thus, a system for automatic translation, for example, from English to Portuguese, should know how to translate the word *bank* as *banco* (an institution for receiving, lending, exchanging and safeguarding money), or as *margem* (the land alongside or sloping down to a river or lake). As the efficiency and effectiveness of a translation system depends on the meaning of the text being processed, disambiguation will always be beneficial and necessary.

Approaches to tackle the issue of WSD may be divided in two main types: the supervised and the unsupervised learning. The former requires semantically

© Springer International Publishing Switzerland 2015
F. Pereira et al. (Eds.) EPIA 2015, LNAI 9273, pp. 747–758, 2015.
DOI: 10.1007/978-3-319-23485-4_75

tagged training data. Although supervised approaches can provide very good results, the need for tagging may become a limitation: semantic tagging depends on more or less complex approaches and it may occur that tagging is not possible for some languages; and POS-tagging, if used, needs good quality taggers that may not exist for some languages. On the other hand, by working with untagged information, unsupervised approaches are more easily language-independent. However, the lack of tags may be a limitation to reach the same level of results as those achieved by supervised approaches.

One way to work around the limitations of both supervised and unsupervised approaches, keeping their advantages, is the use of a hybrid solution. We propose the use of a reliable and valid knowledge source, automatically extracted from sentence-aligned untagged bilingual parallel corpora.

In this paper we present a cross-lingual approach for Word Sense Clustering to assist automatic and human translators on translation processes when faced with expressions which are more complex, more ambiguous and less frequent than general. The underlying idea is that the clustering of word senses provides a useful way to discover semantically related senses, provided that each cluster contains strongly correlated word senses. To achieve our target we propose a semi-supervised strategy to classify words according to their most probable senses. This classification uses a SVM classifier which is trained by the information obtained in the process of the sense clustering. Clusters of senses are built according to the correlation between word senses taking into account the combinations of their neighbor words and the relative position of those neighbor terms; those combinations are taken as features, which are automatically extracted [1] from a sentence-aligned parallel corpora.

2 Related Work

Several studies that combine clustering processes with word senses and parallel corpora has been assessed by several authors in the past years. In [3], the authors present a clustering algorithm for cross-lingual sense induction that generates bilingual semantic resources from parallel corpora. These resources are composed by the senses of words of one language that are described by clusters of their semantically similar translations in another language. The authors proved that the integration of sense-clusters resources leads to important improvements in the translation process. In [4], the authors proposed an unsupervised method for clustering translations of words through point-wise mutual information, based on a monolingual and a parallel corpora. Comparing the induced clusters to reference clusters generated from WordNet, they demonstrated that their method identifies sense-based translation clusters from both monolingual and parallel corpora.

Brown et al. described in [5] a statistical technique for assigning senses to words based on the context in which they appear. By incorporating this method in a machine translation system, a significant reduction of the translation error rate was achieved. In [7], Diab addresses the problem of WSD from a multilingual perspective, expanding the notion of context to encompass multilingual evidence. Given a parallel corpus and a sense inventory for one of the languages in the corpus, an approach to resolve word sense ambiguity in natural language was proposed. In [15], the authors present a method that exploits word clustering based on automatic extraction of translation equivalents, supported by available aligned wordnets. Apidianaki in [2] described a system for SemEval-2013 Cross-lingual WSD task, where word senses are represented by means of translation clusters in a cross-lingual strategy. The WSD method clusters the translations of target words in a parallel corpus using source language context vectors. These vectors are exploited in order to select the most appropriate translations for new instances of the target words in context.

With the goal of increasing the accuracy of WSD systems when faced with expressions that are more complex, ambiguous and less frequent than general, we propose the extension and changes of several works in the field [2–5, 7, 15]. It has differences from those mentioned above, since specific and validated bilingual lexicons, automatically extracted, are used to provide neighbor contexts enabling the calculation of the statistical correlation between senses, which is the basis to build sense clusters, therefore being a language independent approach.

3   System Description

3.1   Dataset

The experiments performed to support the research presented in this article comply with the datasets presented in Table 1.

Table 1. Datasets of ambiguous words and possible senses for English-Portuguese (EN-PT), English-French (EN-FR) and English-Spanish (EN-SP)

Dataset	Source-Words (Ambiguous)	Target-Words (Senses)
EN-PT	15	94
EN-FR	15	70
EN-SP	15	83

Thus, by Table 1 we see that in the experiments performed to support this research, we used, for example, 15 English ambiguous words that could be translated in 94 different Portuguese words, each one having a meaning, that is a sense.

3.2   The Gathering of Word Senses

The gathering of word senses consists of extracting meanings of words for a given ambiguous word. For this, we use the *ISTRION (EN-PT; EN-FR; EN-SP)* lexicon, which is a bilingual and strongly validated data source, resultant from the project *ISTRION*[1]. This lexicon contains 810.000 validated entries for the English-Portuguese language pair, 380.000 for the English-French and 290.000 for the English-Spanish one. This knowledge was automatically extracted and manually validated. For each ambiguous word in the source language (eg. English) we get all different senses existing in the target language (eg. Portuguese, French, Spanish) by consulting the bilingual lexica database; see tables 2 and 3 containing an example for Portuguese and French respectively. These tables show a set of different senses for the same English word "sentence", each one expressed in a word in the target language. According to the content of each table, the reader may predict that the senses could be divided in two semantically different groups (clusters): those signed with a "*", which are related to textual units; and those with a "+", related to Court resolutions. Thus, one of the purposes of this approach is to build clusters of senses according to the semantic closeness among word senses.

Table 2. Example of the different senses for the ambiguous word "sentence" concerning the translation to Portuguese. Senses signed with a "*" are textual units of one or more words. Those signed with a "+" are related to Court resolutions

Ambiguous Word (English)	Sense (Portuguese)
sentence	oração (*clause*) *
sentence	expressão (*expression*) *
sentence	frase (*phrase*) *
sentence	sentença (*sentence*) +
sentence	pena (*penalty*) +
sentence	condenação (*condemnation*) +

Table 3. Example of the different senses for the ambiguous word "sentence" concerning the translation to French. Senses signed with a "*" are textual units of one or more words. Those signed with a "+" are related to Court resolutions

Ambiguous Word (English)	Sense (French)
sentence	condamnation (*condemnation*) +
sentence	jugement (*sentence*) +
sentence	phrase (*phrase*) *
sentence	peine (*penalty*) +
sentence	condamner (*condemn*) +

3.3 Feature Extraction

According to the authors in [10], local context features with bilingual words evidence starts from the assumption that incorporating knowledge from more than

---

[1] http://citi.di.fct.unl.pt/project/project.php?id=97

one language into the feature vector will be more informative than only using monolingual features . By using a sentence-aligned parallel corpora, the proposal we present in this paper confirms this principle. Thus, we use a sentence-aligned parallel corpora (composed by Europarl[2] and DGT[3]), from which we extract features from the neighbor context of the target pair (Ambiguous Word) \t (Sense N) that *fall* within a window of three words to the left and three words to the right of each word of the pair, discarding stop-words. Each target pair has a set of features where each one is a combination of one of the words in the window and its relative position. For a better understanding, let us take the example of the target pair "sentence" – "frase" and one of the sentence-pairs containing it, retrieved from the bilingual parallel corpora (EN \t PT): *Besides being syntactically well-formed, the sentence is correctly translated* \t *Para além de estar sintaticamente bem formada, a frase está correetamente traduzida.* Thus, the context words of the target pair "sentence" – "frase" in this sentence-pair are "Besides", "syntactically", "well-formed", "correctly", "translated", "sintaticamente", "bem", "formada", "corretamente" and "traduzida", taking into account the limits of the window (three words to the left and three words to the right of each word of the pair). Following this, the corresponding features include a tag indicating the language and the relative position of the context word to the corresponding word of the target pair: "enL3_Besides", "enL2_syntactically", "enL1_well-formed", "enR1_correctly", "enR2_translated", "ptL3_sintaticamente", "ptL2_bem", "ptL1_formada", "ptR1_corretamente" and "ptR2_traduzida" —Recall that stop-words are discarded. "L" and "R" stands for Left and Right respectively.

However, for each target pair, there are usually several sentence-pairs retrieved from the bilingual parallel corpora (EN \t PT), containing that target pair. This means that probably several contexts will neighbor the same target pair, generating several features. In our approach, everytime a feature *occurs* in a sentence-pair, its frequency is incremented for the corresponding target pair. In other words, taking the feature "enL2_syntactically", it may have for example: 3 occurrences for target pair "sentence" – "frase" (meaning that the word "syntactically" occurs 2 positions left to the target word "sentence", in 3 of the sentence-pairs containing this target pair); 2 occurrences for "sentence" – "oração"; 0 occurrences for "sentence" – "pena", etc..

We consider that, there is a tendency such that, the closer the relative position of the context word to the target word, the stronger the semantic relation between both words. So, in this approach a different importance is assigned to each feature, according to their relative positions. Thus, we use the criterion we called $\sqrt[p]{f}$, that is: for features whose relative position is p, the root of degree p is applied to frequency f, which is the number of times the feature *occurs* in the set of sentence-pairs containing the target pair. This criterion was chosen empirically as it showed good results after some experiments. Table 4 shows part of the feature extraction concerning the ambiguous word "sentence".

---

[2] http://www.statmt.org/europarl/

[3] http://ipsc.jrc.ec.europa.eu/?id=197

Table 4. Feature extraction for the target pairs concerning the ambiguous word "sentence" (only a small part is shown)

Sense	Feature	Frequency	Final Assigned Value
oração	...	...	...
...	...	...	...
expressão	...	...	...
...	...	...	...
frase	enL3_Besides	1	$\sqrt[3]{1}$
frase	enL2_syntactically	3	$\sqrt{3}$
frase	enL1_well-formed	2	2
frase	enR1_correctly	3	3
frase	enR2_translated	3	$\sqrt{3}$
frase	ptL3_sintaticamente	2	$\sqrt[3]{3}$
frase	ptL2_bem	4	$\sqrt{4}$
frase	ptL1_formada	3	3
frase	ptR1_corretamente	2	2
frase	ptR2_traduzida	3	$\sqrt{3}$
frase	...	...	...
sentença	...	...	...
...	...	...	...
pena	...	...	...
...	...	...	...
condenação	...	...	...
...	...	...	...

For reasons of space, only values for some features of one of the target pairs ("sentence" – "frase") are shown. Values in column *Final Assigned Value* contain the result of the application of $\sqrt[x]{f}$ criterion on the values of column *Frequency*.

The information contained in all columns of Table 4, except *Frequency*, form a matrix which is the base for obtaining the Word Sense Clustering concerning the ambiguous word "sentence". At the end of the feature extraction task we obtained 15 matrices per language pair, corresponding to each of the 15 ambiguous words used, as referred in Table 1.

3.4 Feature Reduction by Sense Correlation

As we have seen in the previous subsection, the number of features associated to each ambiguous word tend to be huge when compared to the number of senses which may correspond to just a few words as in the case of tables 2 and 3. So, considering the purpose of clustering the senses, we transform each of the previously obtained matrices into a new and more compact matrix of correlations (similarities) between each pair of senses. This is a $N{\times}N$ symmetric matrix where N is the number of senses of the ambiguous word. Each line of this matrix corresponds to one of the senses now characterized by the correlation between that sense and each of the N senses. Each correlation is given by (1), which is based on the Pearson's correlation coefficient.

$$Corr(S_i, S_j) = \frac{Cov(S_i, S_j)}{\sqrt{Cov(S_i, S_i)} \times \sqrt{Cov(S_j, S_j)}} \tag{1}$$

$$Cov(S_i, S_j) = \frac{1}{\|\mathcal{F}\| - 1} \sum_{F \in \mathcal{F}} \Big(\big(f(S_i, F) - f(S_i, .)\big) \times \big(f(S_j, F) - f(S_j, .)\big) \Big) \tag{2}$$

where F is an element of the feature set $\mathcal{F}$ and $f(S_i, F)$ stands for the *Final Assigned Value* (a column of Table 4) of feature F for sense S_i; $f(S_i, .)$ gives the average *Final Assigned Value* of the features for sense S_i, which is given by (3).

$$f(S_i, .) = \frac{1}{\|\mathcal{F}\|} \sum_{F \in \mathcal{F}} f(S_i, F) \tag{3}$$

Correlation given by (1) measures how semantically close are senses S_i and S_j. However, a qualitative explanation for this can be given through (2), rather than by (1). Thus, (2) shows that, for each feature F, two deviations are taken: one is given by the *Final Assigned Value* of feature F for sense S_i, subtracted from the average *Final Assigned Value* for S_i, that is, $f(S_i, F) - f(S_i, .)$; the other one is obtained by the *Final Assigned Value* of the same feature F for sense S_j, subtracted from the average *Final Assigned Value* for S_j, that is, $f(S_j, F) - f(S_j, .)$. If both deviations have the same algebraic sign $(+/-)$, the product will be positive, which means that both senses present *similar deviations* concerning feature F. And, if positive products happen for most of features resulting in high values, then there will be a strong positive covariance value $(Cov(S_i, S_j))$, and therefore, a high correlation $(Corr(S_i, S_j))$ — notice that (1) has the effect of just standardizing $Cov(S_i, S_j)$ values, ranging from -1 to +1.

Still analyzing (2), if the partial sum of the positive products has a similar value to the partial sum of the negative ones (when deviations are contrary), then the correlation is close to 0, which means that the semantic closeness between both senses is very weak (or even null). In other words, $Corr(S_i, S_j)$ gives close to +1 values, meaning a high correlation, when both senses tend to occur in the same contexts. If one of the senses occur in contexts where the other sense never occurs, and vice versa, then there is a negative correlation between them.

3.5 Finding Clusters

Our goal is to join similar senses of the same ambiguous word in the same cluster, based on the correlation matrix obtained as explained in Subsec. 3.4. To create clusters we used the *WEKA* tool [8] with *X-means* [12] algorithm. With *X-means* the user does not need to supply the number of clusters, contrary to other clustering algorithms such as k-means or k-medoids. The algorithm returns the best solution for the correlation matrix presented as input. As a matter of fact, for the example of the ambiguous word "sentence", regarding the Portuguese as the target language, it assigned the words "oração", "expressão", and "frase" to a cluster, while "pena" and "condenação" were assigned to another one. In other words, it returned the results expected that were presented in Table 2. With respect to the possible translations of the same ambiguous word "sentence" to French, the clusters were correctly formed too, according to the expected distribution shown in Table 3.

The results of the clustering phase for all ambiguous words gave rise to the evaluation presented in tables 5, 6 and 7.

4   Experiments and Results

4.1   Evaluation of the Clusters

In order to determine the consistency of the obtained clusters, all of them will be evaluated with V-measure and F-measure. V-measure introduces two criteria presented in [14]: homogeneity (h) and completeness (c). A clustering process is considered homogeneously well-formed if all of its clusters contain only data points which are members of a single class. Comparatively, a clustering result satisfies completeness if all data points that are members of a given class are elements of the same cluster. So, increasing the homogeneity of a clustering solution often results in decreasing its completeness. These two criteria run roughly in opposition. This measure will be used to evaluate the resulting clusters. The value of homogeneity varies between 1 and 0. In the perfectly homogeneous case, the homogeneity is 1. In an imperfect situation, it is 0, which happens when the class distribution within each cluster is equal to the overall class distribution.

Similarly to determination of the homogeneity, the results of the completeness also vary between 1 and 0. In the perfectly complete case, it is 1. In the worst case scenario, each class is represented by every cluster with distribution equal to the distribution of cluster sizes; consequently completeness is 0. V-measure is thus a measure for evaluating clusters, which studies the harmonic relationship between homogeneity and completeness. It is given by $V\beta$ (see (4)) where $\beta = 1$, which is a value usually used. V-measure values varies between 1 and 0.

$$V\beta = \frac{(1 + \beta) \times h \times c}{(\beta \times h) + c} \qquad F\beta = \frac{(1 + \beta)^2 \times Precision \times Recall}{(\beta^2 \times Precision) + Recall} \qquad (4)$$

To establish a comparison between different criteria of clusters evaluation, we also compute F-measure [13], given by $F\beta$ (see (4)) where $\beta = 1$, which is also a frequently used measure. As it is shown, this well known metric is based on both Precision and Recall measures. In this context, Precision is determined according

Table 5. English-Portuguese sense-clusters evaluation

Ambiguous Word	V-measure (Portuguese)	F-measure (Portuguese)
plant	1.0	1.0
train	0.81	0.96
motion	1.0	1.0
general	1.0	1.0
fair	0.45	0.81
sentence	1.0	1.0
cold	0.91	0.98
chair	1.0	1.0
break	0.51	0.86

Table 6. English-Spanish sense-clusters evaluation

Ambiguous Word	V-measure (Spanish)	F-measure (Spanish)
heart	1.0	1.0
plant	1.0	1.0
joint	1.0	1.0
motion	0.62	0.89
train	1.0	1.0
right	0.43	0.80
chair	1.0	1.0
sentence	1.0	1.0
break	0.51	0.86

to the following procedure: for each sense-cluster of a clustering, Precision is given by the size of the largest semantic group of senses contained in the cluster (the number of True Positives), divided by the size of the cluster (the sum of True Positives and False Positives); the average Precision of the clusters gives the Precision of the clustering. In order to calculate Recall, for each cluster of a clustering, this measure is given by the same number of True Positives, divided by the real size of the corresponding semantic group of senses (the sum of True Positives and False Negatives); the average Recall of the clusters is taken as the Recall of the clustering. As V-measure, F-measure also varies between 0 and 1, being the former the worst scenario and the latter the optimal.

Due to lack of space, tables 5, 6 and 7 contain the results of subset samples of the clusterings obtained for each of the 15 ambiguous English words regarding each target language. These tables show good results for sense-clusters for Portuguese, Spanish and French language, getting average V-measure and F-measure values of 0.81 and 0.92, respectively. Values tend to be higher for F-measure criterion. Tables also show that a significant part of the clustering were perfectly formed, that is, V-measure = 1.0 and F-measure = 1.0.

Table 7. English-French sense-clusters evaluation

Ambiguous Word	V-measure (French)	F-measure (French)
plant	1.0	1.0
motion	0.62	0.90
train	1.0	1.0
heart	0.51	0.86
joint	1.0	1.0
tank	1.0	1.0
fair	0.56	0.88
break	1.0	1.0
cold	0.62	0.90

However, for some target pairs existing in the bilingual lexica database, there were very few occurrences in the sentence-aligned parallel corpora, which prevents the accurate calculation of the correlation between senses. This is the reason why clustering results are relatively poor for some ambiguous words: for example "motion" for English-French and English-Spanish, among others, as shown in tables 5, 6 and 7.

4.2   Classification — Assigning Sense Clusters to Ambiguous Words

To accomplish the classification process, potentially ambiguous sentences (containing ambiguous words) were extracted from a corpora that were not used in the learning stage, totaling 96 expressions. The purpose is to determine how the disambiguation system behaves when faced with a set of potentially ambiguous sentences. The classification task usually involves the use of two separate training and testing sets. The training phase is closely related with the stage of clustering achieved previously, since we used the acquired knowledge from clusters to train the system. Each cluster is encoded by the presence or absence of all features that belong to all clusters related with a particular ambiguous word, and a *target value* (i.e. the class label) corresponding to the sense cluster. In what concerns to the testing phase, our goal is to encode each testing sentence, extracted from a corpora which was not used in the training phase, and confront it with the training set. So, since each ambiguous word was taken to be translated to a target language, to encode each expression we analyze the presence or absence in the sentence of just the features from English used in the training set.

The classifier used was a support vector machine (LIBSVM) [6] with the Radial Basis Function (RBF) as kernel type, that allow to handle the case when the relation between class labels and attributes is non-linear.

4.3   Classification Results

The results obtained with the application of SVM classifier were evaluated using F-measure [13], again given by $F\beta$ (see (4)) where $\beta = 1$. All 96 ambiguous sentences (containing ambiguous words) were classified and so, F-measure were calculated regarding each target language, as shown in Table 8. The fact that F-measure for French and Spanish did not reach the same value as for Portuguese (0.79 vs 0.85) is probably due to the fact that the EN-PT language pair lexicon used in the clustering process was considerably larger (810,000 entries) than the ones used for the other two language pairs (380,000 and 290,000), implying therefore a *better quality* training phase for that pair.

In order to have a baseline for comparison, the same tests described above were performed using the output of GIZA++ alignments on DGT [11], where the most probable sense is used to disambiguate each sentence: the results obtained were 0.43, 0.38 and 0.38 respectively for EN-PT, EN-FR and EN-SP pairs.

Table 8. Results for the assignment of sense-clusters to ambiguous words

Target Language	F-measure
French	0.79
Portuguese	0.85
Spanish	0.79

5   Conclusions and Future Work

In this paper, a language-independent WSD approach using Word Sense Clustering was proposed. Sentence-aligned parallel corpora revealed to be essential for achieving the objectives accomplished as it provided the neighbor contexts of ambiguous words, enabling the calculation of the statistical correlation between senses, and therefore the building of sense-clusters. The results obtained for sense clustering (V-measure and F-measure) allow us to conclude that the learned clusters are reliable sources of information, supporting the whole process of disambiguation. In the classification process, results were very positive for all language pairs tested, showing that *well-formed* sense clusters are a strong base for WSD.

As future work we would like to improve the approach by studying the optimal size of the neighbor context of the ambiguous words, which will require larger sentence-aligned parallel corpora, in order to keep, and gain, statistical representativeness. Future experiments will include ambiguous multi-words, for which this algorithm was also designed. The approach presented in this paper will enable us to build a semantic translation tagger that can be useful for translation aligners or other translation systems. It is also our intention to test our approach with existing datasets, used on cross-lingual WSD tasks of SemEval[9].

References

1. Aires, J., Lopes, G.P., Gomes, L.: Phrase translation extraction from aligned parallel corpora using suffix arrays and related structures. In: Lopes, L.S., Lau, N., Mariano, P., Rocha, L.M. (eds.) EPIA 2009. LNCS, vol. 5816, pp. 587–597. Springer, Heidelberg (2009)
2. Apidianaki, M.: Cross-lingual word sense disambiguation using translation sense clustering. In: Proceedings of the 7th International Workshop on Semantic Evaluation (SemEval 2013), pp. 178–182. *SEM and NAACL (2013)
3. Apidianaki, M., He, Y., et al.: An algorithm for cross-lingual sense-clustering tested in a MT evaluation setting. In: Proceedings of the International Workshop on Spoken Language Translation, pp. 219–226 (2010)
4. Bansal, M., DeNero, J., Lin, D.: Unsupervised translation sense clustering. In: Proceedings of the 2012 Conference of the North American Chapter of the Association for Computational Linguistics: Human Language Technologies, pp. 773–782. Association for Computational Linguistics (2012)
5. Brown, P.F., Pietra, S.A.D., Pietra, V.J.D., Mercer, R.L.: Word-sense disambiguation using statistical methods. In: Proceedings of the 29th annual meeting on Association for Computational Linguistics, pp. 264–270. Association for Computational Linguistics (1991)
6. Chang, C.C., Lin, C.J.: LIBSVM: a library for support vector machines. ACM Transactions on Intelligent Systems and Technology (TIST) $2(3)$, 27 (2011)
7. Diab, M.T.: Word sense disambiguation within a multilingual framework. Ph.D. thesis, University of Maryland at College Park (2003)
8. Hall, M., Frank, E., Holmes, G., Pfahringer, B., Reutemann, P., Witten, I.H.: The weka data mining software: an update. ACM SIGKDD Explorations Newsletter $11(1)$, 10–18 (2009)
9. Lefever, E., Hoste, V.: Semeval-2010 task 3: Cross-lingual word sense disambiguation. In: Proceedings of the 5th International Workshop on Semantic Evaluation, pp. 15–20. Association for Computational Linguistics (2010)
10. Lefever, E., Hoste, V., De Cock, M.: Five languages are better than one: an attempt to bypass the data acquisition bottleneck for WSD. In: Gelbukh, A. (ed.) CICLing 2013, Part I. LNCS, vol. 7816, pp. 343–354. Springer, Heidelberg (2013)
11. Och, F.J., Ney, H.: A systematic comparison of various statistical alignment models. Computational Linguistics $29(1)$, 19–51 (2003)
12. Pelleg, D., Moore, A.W., et al.: X-means: Extending k-means with efficient estimation of the number of clusters. In: ICML, pp. 727–734 (2000)
13. Rijsbergen, V. (ed.): Information Retrieval, 2nd edn. Information Retrieval Group, University of Glasgow (1979)
14. Rosenberg, A., Hirschberg, J.: V-measure: a conditional entropy-based external cluster evaluation measure. EMNLP-CoNLL 7, 410–420 (2007)
15. Tufiş, D., Ion, R., Ide, N.: Fine-grained word sense disambiguation based on parallel corpora, word alignment, word clustering and aligned wordnets. In: Proceedings of the 20th international conference on Computational Linguistics, p. 1312. Association for Computational Linguistics (2004)

Towards the Improvement of a Topic Model with Semantic Knowledge

Adriana Ferrugento[1]([✉]), Ana Alves[1,2], Hugo Gonçalo Oliveira[1],
and Filipe Rodrigues[1]

[1] CISUC, Department of Informatics Engineering,
University of Coimbra, Coimbra, Portugal
`aferr@student.dei.uc.pt`, {`ana,hroliv,fmpr`}`@dei.uc.pt`
[2] Coimbra Institute of Engineering,
Polytechnic Institute of Coimbra, Coimbra, Portugal

Abstract. Although typically used in classic topic models, surface words cannot represent meaning on their own. Consequently, redundancy is common in those topics, which may, for instance, include synonyms. To face this problem, we present SemLDA, an extended topic model that incorporates semantics from an external lexical-semantic knowledge base. SemLDA is introduced and explained in detail, pointing out where semantics is included both in the pre-processing and generative phase of topic distributions. As a result, instead of topics as distributions over words, we obtain distributions over concepts, each represented by a set of synonymous words. In order to evaluate SemLDA, we applied preliminary qualitative tests automatically against a state-of-the-art classical topic model. The results were promising and confirm our intuition towards the benefits of incorporating general semantics in a topic model.

Keywords: Topic model · Semantics · WordNet · SemLDA

1 Introduction

Topic models allow us to infer probability distributions over a set of words, called "topics", which are useful for uncovering the main subjects in a collection of documents. They improve searching, browsing and summarization in such collections, and their application is not limited to text mining, as they revealed to be useful in fields such as computer vision [20] or bioinformatics [6].

Classic topic modelling algorithms, such as LDA [1], rely on the co-occurrences of surface words to capture their semantic proximity. They consider a surface word to be identical in different contexts and leverage on its co-occurrences with other words to differentiate topics. This fails to consider additional semantic knowledge on the words, which may, on one hand, exclude different senses of the same word from occurring in different topics and, on the other hand, lead to redundant topics, for instance with synonyms, that do not add information to the topic.

F. Pereira et al. (Eds.) EPIA 2015, LNAI 9273, pp. 759–770, 2015.
DOI: 10.1007/978-3-319-23485-4_76

Whether it was during pre-processing [8], the generative process [18], or post processing [2], incorporating semantics into topic modeling emerged as an approach to deal with concepts rather than surface words. Since a word may have different meanings (e.g. *bank*) and since the same concept may be denoted by different words (e.g. *car* and *automobile*), these attempts exploit external semantic resources, such as WordNet [11] or, alternatively, follow a fully unsupervised approach, for instance, using word sense induction techniques [3]. In those approaches, topic distributions with synonymous and semantically similar words are unified in concept representations, such as synsets.

In order to improve current topic models, we propose a new model, SemLDA, which incorporates semantics in the well-known LDA model, using knowledge from WordNet. Similarly to other semantic topic models, the topics produced by SemLDA are sets of synsets, instead of words. The main difference is that SemLDA considers all possible senses of the words in a document, together with their probabilities. Moreover, it only requires a minimal intuitive change to the classic LDA algorithm.

The remaining of the paper is organized as follows: in Section 2, there is a brief enumeration of existing approaches to topic modelling; Section 3 introduces the proposed model in detail, with special focus on the differences towards the classic LDA. Section 4 reports on the performed experiments, with illustrative examples of the obtained topics and their automatic evaluation against the classic LDA. Finally, Section 5 draws some conclusions and future plans for this work.

2 Related Work

The first notable approach to reduce the dimensionality of documents was Latent Semantic Indexing (LSI) [5], which aimed at retaining the most of the variance present in the documents, thus leading to a significant compression of large datasets. Probabilistic Latent Semantic Indexing (pLSI) [9] later emerged as a variant of LSI, where different words in documents are modelled as samples from a simple mixture model where the mixture components are multinomial random variables that can be viewed as representations of "topics". Nevertheless, pLSI was still not a proper generative model of documents, given that it provides no probabilistic model at the level of documents. Having this limitation in mind, Blei et al. [1] developed the Latent Dirichlet Allocation (LDA), a generalization of pLSI that is currently the most applied topic model. It allows documents to have a mixture of topics, given that it enables to capture significant intra-document statistical structure via the mixing distribution.

The single purpose of the previous models is to discover and assign different topics – represented by sets of surface words, each with a different probability – to the collection of documents provided. Those approaches have no concern with additional semantic knowledge about words, which can lead to some limitations in the generated topics. For instance, they might include synonyms, and thus be redundant and less informative. Alternative attempts address this problem using, for instance, WordNet [11], a lexical-semantic knowledge base of English.

WordNet is structured in synsets, which are groups of synonymous words that may be seen as concept representations of a language. Synsets may be connected according to different semantic relations, such as hypernymy (generalization) or meronymy (part-of).

In an attempt to include semantics in topic modelling and, at the same time, perform word sense disambiguation (WSD), Boyd-Graber and Blei [2] presented LDAWN, a modified LDA algorithm that includes a hidden variable for representing the sense of a word, according to WordNet. Each topic consists of a random walk through the WordNet hypernymy hierarchy, which is used to infer topics and their synsets, based on the words from documents. LDAWN was also applied to word sense disambiguation (WSD), although its authors accept the worse performance when compared with state-of-the-art WSD algorithms. One of the proposed solutions is to acquire local context to improve WSD, in the future. However, there is additional work towards the discovery of concept-based topics, not always relying in WordNet. For instance, LDA was used as a ground model to generate topics based on concepts of an ontology [4]; and a commonsense knowledge-based algorithm was used to transform documents into commonsense concepts, which were then clustered to generate the topics [17].

Despite some similarities, the model proposed in this paper differs from the previous in various ways. Instead of words, the produced topics are also distributions over concepts (synsets) and, similarly to LDAWN, it exploits WordNet and modifies the basic LDA by adding a sense variable. But SemLDA considers all possible senses of a word, with a distribution over all the synsets that include it. Indeed, we do not benefit from similar words in the same topic to improve WSD, as in LDAWN. Rather, we try to avoid it. This is why, in the future, our sense probabilities might be obtained from WSD. SemLDA is further explained in the next section.

3   Proposed Model

In order to consider the general semantics of a language during topic discovery, we propose SemLDA, a topic model that exploits external semantic knowledge, acquired from Princeton WordNet [11], a lexical-semantic knowledge base of English. SemLDA is based on the Latent Dirichlet Allocation model [1], but it introduces a new set of parameters $\eta_{1:S}$, where S is the number of synsets where the word occurs (one for each of its senses). The parameters correspond to the probabilities of each word belonging to a synset, which, in the current implementation, is obtained directly from the SemCor corpus [12] – in SemCor, words are manually annotated according their WordNet senses and, in WordNet, both synsets and word senses are ordered according to their frequencies in SemCor.

Given a corpus $\mathcal{D} = \{w^d\}_{d=1}^{D}$ of size $|\mathcal{D}|$ and the probabilities of each word n in a synset s, $\eta_{s,n}$, SemLDA estimates the most likely set of topics. Each document is represented by a distribution of topics, θ, but, in contrast to the traditional LDA, a topic is represented as a distribution over the synsets in the vocabulary, β. This is a major difference because, in LDA, words are handled according to the documents where they appear, regardless of their known semantics. In SemLDA, instead of just words, all the possible senses for each word are considered, although with different probabilities. We should notice that the more informative output (synsets) does not necessarily imply an increasing complexity in the topic representation. If needed, in order to have a comparable output to other topic models, a single word can be selected from each synset. When using WordNet, it makes sense to select the first word of a synset, which we recall to be the most frequently used to denote the concept.

The graphical model of SemLDA is displayed in Figure 1, where D is the number of documents in the corpus, K is the number of topics, N is the number of words in a document and S is the number of synsets of a given word. In this model, each word of a document, w_n, is drawn from a concept, c_n. This is represented by using the synset's distribution over words, parameterized by η, which we shall assume to be fixed. The concept c_n is determined by a discrete topic-assignment z_n, picked from the document's distribution over topics θ and a topic distribution β. It follows the same reasoning as the LDA model, but includes a new layer corresponding to the concepts c_n that the words w_n express. The generative process of a document d under SemLDA is the following:

1. Choose topic proportions $\theta|\alpha \sim Dir(\alpha)$
2. For each concept, c_n
    (a) Choose topic assignment $z_n|\theta \sim Mult(\theta)$
    (b) Choose concept $c_n|z_n, \beta_{1:K} \sim Mult(\beta_{z_n})$
    (c) Choose word to represent concept $w_n|c_n, \eta_{1:S} \sim Mult(\eta_{c_n})$

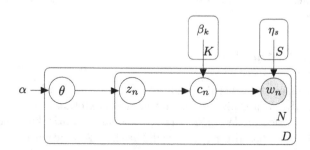

Fig. 1. Graphical model.

Our goal is thus to calculate, for every document, the posterior distribution over the latent variables, $\theta, z_{1:N}, c_{1:N}$. However, as in LDA, performing exact inference is intractable, so we need to use an approximate inference method. In this paper, we use variational inference to perform approximate Bayesian inference. The purpose of variational inference is to minimize KL divergence between the variational distribution $q(\theta, z_{1:N}, c_{1:N})$ and the true posterior distribution $p(\theta, z_{1:N}, c_{1:N}|w_{1:N})$. A fully factorized (mean field) variational distribution q, of the form represented in equation 1, is employed, with γ, ϕ and λ as the variational parameters.

$$q(\theta, z_{1:N}, c_{1:N}) = q(\theta|\gamma)\left(\prod_{n=1}^{N} q(z_n|\phi_n)q(c_n|\lambda_n)\right) \tag{1}$$

KL minimization [10] is equivalent to maximizing the lower bound on the log marginal likelihood (equation 2) by using coordinate ascent.

$$\log p(w_{1:N}|\alpha, \beta_{1:K}, \eta_{1:S}) = \log \int_{\theta} \sum_{z_{1:N}} \sum_{c_{1:N}} \frac{p(\theta, z_{1:N}, c_{1:N}, w_{1:N}|\Theta)q(\theta, z_{1:N}, c_{1:N})}{q(\theta, z_{1:N}, c_{1:N})}$$

$$\geq \mathbb{E}_q[\log p(\theta, z_{1:N}, c_{1:N}, w_{1:N})] - \mathbb{E}_q[\log q(\theta, z_{1:N}, c_{1:N})]$$
$$= \mathcal{L}(\gamma, \phi_{1:N}, \lambda_{1:N}|\Theta) \tag{2}$$

The greek letter Θ is used to denote the model's parameters $\Theta = \{\alpha, \beta, \eta\}$.

By optimizing $\mathcal{L}$ w.r.t γ, we get the update the same update as in the classic LDA in [1], given by equation 3.

$$\gamma_i = \alpha_i + \sum_{n=1}^{N} \phi_{n,i} \tag{3}$$

The variational parameter ϕ can be optimized by collecting only the terms in the lower bound that contain the parameter. Notice that this is a constrained maximization problem, since $\sum_{k=1}^{K} \phi_{n,k} = 1$, which is necessary for it to be a valid probability distribution. Hence, we need to include the necessary Lagrange multipliers, yielding equation 4.

$$\mathcal{L}_{[\phi]} = \sum_{n=1}^{N}\sum_{i=1}^{K} \phi_{n,i}\left(\Psi(\gamma_i) - \Psi\left(\sum_{j=1}^{K}\gamma_j\right)\right) + \sum_{n=1}^{N}\sum_{j=1}^{S}\sum_{i=1}^{K} \lambda_{n,j}\phi_{n,i}\log\beta_{i,j}$$

$$- \sum_{n=1}^{N}\sum_{i=1}^{K} \phi_{n,i}\log\phi_{n,i} + \mu\left(\sum_{k=1}^{K}\phi_{n,k} - 1\right) \tag{4}$$

Setting the derivatives of $\mathcal{L}_{[\phi]}$ w.r.t ϕ to zero gives the update in equation 5.

$$\phi_{n,i} \propto \exp\left(\Psi(\gamma_i) - \Psi\left(\sum_{j=1}^{K}\gamma_j\right) + \sum_{j=1}^{S}\lambda_{n,j}\log\beta_{i,j}\right) \tag{5}$$

In order to optimize $\mathcal{L}$ w.r.t λ, we again start by collecting only the terms in the bound that contain λ. Notice that this is also a constrained maximization problem, since $\sum_{k=1}^{S}\lambda_{n,k} = 1$. Hence, we need to also add the necessary Lagrange multipliers (see equation 6).

$$\mathcal{L}_{[\lambda]} = \sum_{n=1}^{N}\sum_{j=1}^{S}\sum_{i=1}^{K} \lambda_{n,j}\phi_{n,i}\log\beta_{i,j} + \sum_{n=1}^{N}\sum_{j=1}^{S}\sum_{i=1}^{V_j} \lambda_{n,j}w_{n,i}\log\eta_{j,i}$$

$$- \sum_{n=1}^{N}\sum_{j=1}^{S} \lambda_{n,j}\log\lambda_{n,j} + \mu\left(\sum_{k=1}^{S}\lambda_{n,k} - 1\right) \tag{6}$$

Setting the derivatives of $\mathcal{L}_{[\lambda]}$ w.r.t λ to zero gives the update in equation 7.

$$\lambda_{n,j} \propto \exp\left(\sum_{i=1}^{K} \phi_{n,i} \log \beta_{i,j} + \sum_{i=1}^{V_j} w_{n,i} \log \eta_{j,i} \right) \tag{7}$$

The variational inference algorithm iterates between the different updates presented until the convergence of the evidence lower bound. Since the parameters α and η are assumed to be fixed, we only need to estimate β, for which a variational EM algorithm is used. In the E-step of the expectation-maximization algorithm, variational inference is used to find an approximate posterior for each document, as previously described. In the M-step, as in exact EM, we find maximum likelihood estimates of the parameters using the expected sufficient statistics computed in the E-step.

We start by collecting only the terms in the lower bound that contain β. Notice that this is a constrained maximization problem, since $\sum_{k=1}^{V} \beta_{i,k} = 1$, which is necessary to be a valid probability distribution. Hence, we also need to include the necessary Lagrange multipliers (see equation 8).

$$\mathcal{L}_{[\beta]} = \sum_{d=1}^{D}\sum_{n=1}^{N}\sum_{i=1}^{K}\sum_{j=1}^{S} \lambda_{n,j}^{d} \phi_{n,i}^{d} \log \beta_{i,j} + \sum_{i=1}^{K} \mu_i \left(\sum_{k=1}^{S} \beta_{i,k} - 1 \right) \tag{8}$$

Setting the derivatives of $\mathcal{L}_{[\beta]}$ w.r.t β to zero gives the following update in equation 9, which is analogous to the update in standard LDA [1], but with the words $w_{n,j}^{d}$ replaced by their probability in the jth concept, $\lambda_{n,j}^{d}$.

$$\beta_{i,j} \propto \sum_{d=1}^{D}\sum_{n=1}^{N_d} \lambda_{n,j}^{d} \phi_{n,i}^{d} \tag{9}$$

4   Results and Evaluation

In this section, the experiments performed towards the validation of SemLDA are reported. The datasets used are described, some of the obtained synset-based topics are shown, and the evaluation metrics are presented. The latter were used to compare SemLDA with the classic LDA.

4.1   Datasets

Two freely available textual corpora were used in our experiments, namely: the Associated Press (AP) and the 20 Newsgroups dataset, both in English. AP is a large news corpus, from which we used only a part. More precisely, the sample data for the C implementation of LDA, available in David Blei's website[1], which

---

[1] http://www.cs.princeton.edu/~blei/lda-c/

includes 2,246 documents. The 20 Newsgroups[2] is a popular dataset for experiments in text applications of machine learning techniques. It contains 20,000 documents, organized into 20 different newsgroups. Both datasets went through the same pre-processing phase, which included stop-word and numbers removal, as well as word lemmatization. For stop-word removal, we used the *snowball stop words* list[3]. Word lemmatization was based on the NLTK[4] WordNet reader.

4.2 Experiments

The experiments performed aimed at comparing the classic LDA algorithm [1] with SemLDA. An implementation of the classic algorithm, implemented in C, is available from Blei's website[5]. No changes were made to his code. We just had to pre-process the documents, generate a suitable input, and execute it.

For running SemLDA, extra work was needed. First, we retrieved all synsets from the SemCor 3.0 annotations[6], and calculated their probability in this corpus. This is a straightforward task for those WordNet synsets that are in Sem-Cor. But SemCor is a limited corpus and does not cover all words and senses in WordNet. To handle this issue, an extra pre-processing step was added, where all documents were reviewed and, when a word did not occur in SemCor, a new 'dummy' synset was created with a special negative id, and probability equal to the average probability of all the other synsets. This value was chosen to balance the unknown probabilities of dummy synsets according to the probabilities of the remaining synsets, and thus not favor any of them.

For each dataset, the SemLDA input file had the synsets retrieved from SemCor and the words that were in the documents, but not in SemCor. The only difference on the text pre-processing is the use of part-of-speech (POS) tagging, to consider only open-class words, namely nouns, verbs, adjectives and adverbs.

Instead of trial and error with different numbers of topics, we used a Hierarchical Dirichlet Process (HDP) [19] to discover the appropriate number of topics for each dataset. The results obtained suggested that the 20 Newsgroups dataset contains 15 topics and the AP corpus 24. After the pre-processing phase, each model was run for both datasets, with the α parameter fixed at 0.5. Tables 1 and 2 illustrate the results obtained with the classic LDA and SemLDA, respectively for the 20 Newsgroups and for the AP corpus. For each topic discovered by the SemLDA presented, we tried to find an analogous topic by the classic LDA, in a sense that they share similar domains. For the sake of simplicity, we only show the top 10 synsets for each SemLDA topic, with their Synset ID, POS-tag, words and gloss. Underlined words are those present in SemCor and WordNet, whereas the others only appear in WordNet. For each LDA topic only the top 10 words are displayed.

---

[2] http://qwone.com/~jason/20Newsgroups/
[3] http://snowball.tartarus.org/algorithms/english/stop.txt
[4] http://www.nltk.org/
[5] http://www.cs.princeton.edu/~blei/lda-c/
[6] http://web.eecs.umich.edu/~mihalcea/downloads.html#semcor

766   A. Ferrugento et al.

Table 1. Illustrative (analogous) topics from 20 Newsgroups, obtained with the classic LDA (top) and with SemLDA (bottom).

LDA
medical, health, use, patient, disease, doctor, cancer, study, infection, treatment

SemLDA			
Synset ID	POS	Words	Gloss
14447908	N	health, wellness	A healthy state of well being free from disease.
3247620	N	drug	A substance that is used as a medicine or narcotic.
10405694	N	patient	A person who requires medical care.
10020890	N	doctor, doc, physician, MD, Dr., medico	A licensed medical practitioner.
14070360	N	disease	An impairment of health or a condition of abnormal functioning.
14239918	N	cancer, malignant neoplastic disease	Any malignant growth or tumor caused by abnormal and uncontrolled cell division.
47534	ADV	besides, too, also, likewise, as well	In addition.
14174549	N	infection	The pathological state resulting from the invasion of the body by pathogenic microorganisms.
1165043	V	use, habituate	Take or consume (regularly or habitually).
7846	N	person, individual, someone, somebody, mortal, soul	A human being.

The results show success on incorporating semantics into LDA. Topics are based on synsets and WordNet can be used to retrieve additional information on the concept they denote, including their definition (gloss), POS and other words with the same meaning. With both models, the top words of each topic are consistently nouns, which should transmit more content. The presented examples clearly describe very close semantic domains. They share many words and the other are closely related to each other (eg. *drug* and *treatment*, or *exchange* and *trading*). We call attention to topics where the same word is in different synsets (Table 2). While this might sometimes be undesirable, and a possible sign of incoherence, it also shows that the algorithm is correctly handling different senses of the same word. These situations should be minimized in the future, as we intend to acquire sense probabilities from word sense disambiguation (WSD) [14], instead of relying blindly in SemCor for this purpose. This will also minimize the number of dummy synsets.

We can say that the overall results are satisfying. Despite one or another less clear word association, we may say that we are moving towards the right direction. Still, to measure progress towards the classic LDA, we made an automatic evaluation of the coherence of the discovered topics.

4.3 Evaluation

Although, at a first glance, the results might seem promising, they were validated automatically, using metrics previously applied to the context of topic modelling, namely: pointwise mutual information (PMI) and topic coherence.

PMI is a measure of word association that, according to Newman et al. [15], is highly correlated with human-judged topic coherence. In this context, PMI

Table 2. Illustrative (analogous) topics from AP, obtained with the classic LDA (top) and with SemLDA (bottom).

LDA			
stock, market, percent, rate, price, oil, rise, say, point, exchange			
SemLDA			
Synset ID	**POS**	**Words**	**Gloss**
8424951	N	market	The customers for a particular product or service.
13851067	N	index	A numerical scale used to compare variables with one another or with some reference number.
8072837	N	market, securities industry	The securities markets in the aggregate.
13342135	N	share	Any of the equal portions into which the capital stock of a corporation is divided and ownership of which is evidenced by a stock certificate.
79398	N	trading	Buying or selling securities or commodities.
3843092	N	oil, oil color, oil colour	Oil paint containing pigment that is used by an artist.
7167041	N	price	A monetary reward for helping to catch a criminal.
14966667	N	oil	A slippery or viscous liquid or liquefiable substance not miscible with water.
5814650	N	issue	An important question that is in dispute and must be settled.
13333833	N	stock	The capital raised by a corporation through the issue of shares entitling holders to an ownership interest (equity).

is calculated for each topic, based on the co-occurrence probabilities of every pair of its words. In our case, we used the 10 most probable words of each topic, which results in 45 different pairs. For both datasets, co-occurrence is computed from Wikipedia, which provides a large and wide-coverage source of text, completely independent from the datasets used and from WordNet. After comparing different approaches for evaluating topic coherence automatically, Newman et al. [16] concluded that PMI over Wikipedia provides a score that is very consistent with human judgements. See equation 10 for the PMI's formula, where $p(w)$ is the probability of a word w (in our case, the number of Wikipedia articles using this word), and $p(w_i, w_j)$ is the probability of words w_i and w_j co-occurring (in our case, the number of Wikipedia articles using both words). After computing PMI for all topics, we computed the average score for the full topic set.

$$PMI - Score(\mathbf{w}) = \frac{1}{45} \sum_{i<j} \log \frac{p(w_i, w_j)}{p(w_i)p(w_j)}, ij \in \{1...10\} \qquad (10)$$

We recall that topics discovered by SemLDA are sets of synsets and not of surface words. Therefore, to enable a fair comparison with the classic LDA, before computing the PMI scores, we converted our topics to a plain word representation. For this purpose, instead of full synsets, we used only their first word. According to WordNet, this is the word most frequently used to denote the synset concept, in the SemCor corpus. For instance, the SemLDA topic in Table 2 becomes: *market, index, market, share, trading, oil, price, oil, issue, stock*. On the one hand, this representation limits the extent of our results, which are, in fact, synsets. On the other hand, by doing so, it might lead to duplicate words in the same topic, though corresponding to different senses.

Coherence measures the co-occurrence, within the modelled documents, of pairs of words in the same topic [13]. As in the PMI measure, the 10 most probable words of each topic are used. It is also calculated for each topic and, in the end, the average is computed for the full topic set. Assuming that, in every document, there is an explicit theme, by calculating this, we can analyze if the grouping of words is coherent, given their co-occurrence. This measure is very similar to PMI but, in some situations, it achieved higher correlation with human judges [13]. Equation 11 shows the formula of this measure, where $D(v_m^{(t)}, v_l^{(t)})$ is the co-document frequency of two words, 1 is a smoothing count to avoid the logarithm of zero and $D(v_l^{(t)})$ is the document frequency of a word.

$$C(t; V^{(t)}) = \sum_{m=2}^{M} \sum_{l=1}^{m-1} \log \frac{D(v_m^{(t)}, v_l^{(t)}) + 1}{D(v_l^{(t)})} \tag{11}$$

The results obtained with the two topic models, for both datasets, are presented in Table 3. Even if it was a close call, SemLDA outperformed the classic LDA. On both metrics, SemLDA had better scores in the AP corpus, which was the dataset originally used by Blei. For the 20 Newsgroups dataset, the topic coherence measure was very close with both models, whereas the PMI score was better for SemLDA. This confirms that we are heading towards a promising approach that, by exploiting an external lexical-semantic knowledge base, may improve the outcome of the classic LDA model. We should still stress that these are just preliminary results. The following steps are explained with further detail in the next section.

Table 3. Results obtained with the two evaluation metrics.

	20 Newsgroups		AP	
	PMI	Coherence	PMI	Coherence
LDA	1.16 ± 0.39	-32.89 ± 19.77	1.12 ± 0.31	-13.62 ± 9.51
SemLDA	1.22 ± 0.46	-35.4 ± 17.65	1.43 ± 0.26	-9.18 ± 7.51

5   Concluding Remarks

We have presented SemLDA, a topic model based on the classic LDA that incorporates external semantic knowledge to discover less redundant and more informative topics, represented as concepts, instead of surface words. We may say that we have been successful so far. The classic algorithm was effectively changed to produce topics based on WordNet synsets, which, after an automatic validation, shown to have comparable coherence to the original topics. Despite the promising results, there is still much room for improvement.

In fact, to simplify our task, we relied on some assumptions that should be dealt with in a near future and, hopefully, lead to improvements. For instance, the α parameter of LDA was simply set to a fixed value of 0.5. Its selection should be made after testing different values and assess their outcomes. We are also considering to add a Dirichlet prior over the variable concerning topics, β, so that it produces a smooth posterior and controls sparsity. Additional planned tests include the generation of topics considering just a subset of the open class words, for instance, just nouns, which might be more informative. Last but not least, we recall that we obtained the word sense probabilities directly from SemCor. While this corpus is frequently used in WSD tasks and should thus have some representativeness, this approach does not consider the context where the words occur. Instead of relying in SemCor, it is our goal to perform all-words WSD to the input corpora, and this way extract the probabilities of selecting different synsets, given the word context. This should also account for words that are not present in SemCor, and minimize the number of dummy synsets, with the averaged probabilities assigned.

Moreover, we should perform an additional evaluation of the results, not just through automatic measures, but possibly using people to assess the topic coherence. For instance, we may adopt the intruder test, where judges have to manually select the word not belonging to a topic (see [15]). It is also our intention to evaluate SemLDA indirectly, by applying it in tasks that require topic modelling, such as automatic summarization and classification.

We conclude by pointing out that, although the proposed model is language independent, it relies in language specific resources, especially the existence of a wordnet, besides models for POS-tagging and lemmatization. One of our mid-term goals is precisely to apply SemLDA to Portuguese documents. For such, we will use available POS-taggers and lemmatizers for this language, as well as one or more of the available wordnets (see [7] for a survey on Portuguese wordnets).

Since sense probabilities will soon be obtained from a WSD method, the unavailability of a SemCor-like corpus for Portuguese is not an issue.

Acknowledgments. This work was supported by the InfoCrowds project - FCT-PTDC/ECM-TRA/1898/2012FCT.

References

1. Blei, D.M., Ng, A.Y., Jordan, M.I.: Latent dirichlet allocation. Journal of Machine Learning Research **3**, 993–1022 (2003)
2. Boyd-Graber, J., Blei, D., Zhu, X.: A topic model for word sense disambiguation. In: Proceedings of 2007 Joint Conference on Empirical Methods in Natural Language Processing and Computational Natural Language Learning (EMNLP-CoNLL), pp. 1024–1033. ACL Press, Prague, Czech Republic, June 2007
3. Brody, S., Lapata, M.: Bayesian word sense induction. In: Proceedings of 12th Conference of the European Chapter of the Association for Computational Linguistics. EACL 2009, pp. 103–111. ACL Press (2009)
4. Chemudugunta, C., Holloway, A., Smyth, P., Steyvers, M.: Modeling documents by combining semantic concepts with unsupervised statistical learning. In: Sheth, A.P., Staab, S., Dean, M., Paolucci, M., Maynard, D., Finin, T., Thirunarayan, K. (eds.) ISWC 2008. LNCS, vol. 5318, pp. 229–244. Springer, Heidelberg (2008)

5. Deerwester, S.C., Dumais, S.T., Landauer, T.K., Furnas, G.W., Harshman, R.A.: Indexing by latent semantic analysis. JASIS **41**(6), 391–407 (1990)
6. Flaherty, P., Giaever, G., Kumm, J., Jordan, M.I., Arkin, A.P.: A latent variable model for chemogenomic profiling. Bioinformatics **21**(15), 3286–3293 (2005)
7. Gonçalo Oliveira, H., de Paiva, V., Freitas, C., Rademaker, A., Real, L., oes, A.S.: As wordnets do português. In: Simões, A., Barreiro, A., Santos, D., Sousa-Silva, R., Tagnin, S.E.O. (eds.) Linguística, Informática e Tradução: Mundos que se Cruzam, OSLa, vol. 7, no. 1, pp. 397–424. University of Oslo (2015)
8. Guo, W., Diab, M.: Semantic topic models: combining word distributional statistics and dictionary definitions. In: EMNLP, pp. 552–561. ACL Press (2011)
9. Hofmann, T.: Probabilistic latent semantic indexing. In: Proceedings of the 22nd annual international ACM SIGIR conference on Research and development in information retrieval, pp. 50–57. ACM (1999)
10. Jordan, M.I., Ghahramani, Z., Jaakkola, T.S., Saul, L.K.: An introduction to variational methods for graphical models. Machine learning **37**(2), 183–233 (1999)
11. Miller, G.A.: Wordnet: a lexical database for english. Communications of the ACM **38**(11), 39–41 (1995)
12. Miller, G.A., Chodorow, M., Landes, S., Leacock, C., Thomas, R.G.: Using a semantic concordance for sense identification. In: Proceedings of ARPA Human Language Technology Workshop. Plainsboro, NJ, USA (1994)
13. Mimno, D., Wallach, H.M., Talley, E., Leenders, M., McCallum, A.: Optimizing semantic coherence in topic models. In: Proceedings of the Conference on Empirical Methods in Natural Language Processing. EMNLP 2011, pp. 262–272. ACL Press (2011)
14. Navigli, R.: Word sense disambiguation: A survey. ACM Computing Surveys **41**(2), 1–69 (2009)
15. Newman, D., Bonilla, E.V., Buntine, W.: Improving topic coherence with regularized topic models. In: Advances in Neural Information Processing Systems, pp. 496–504 (2011)
16. Newman, D., Lau, J.H., Grieser, K., Baldwin, T.: Automatic evaluation of topic coherence. In: Human Language Technologies: The 2010 Annual Conference of the North American Chapter of the Association for Computational Linguistics. HLT 2010, pp. 100–108. ACL Press (2010)
17. Rajagopal, D., Olsher, D., Cambria, E., Kwok, K.: Commonsense-based topic modeling. In: Proceedings of the 2nd International Workshop on Issues of Sentiment Discovery and Opinion Mining, p. 6. ACM (2013)
18. Tang, G., Xia, Y., Sun, J., Zhang, M., Zheng, T.F.: Topic models incorporating statistical word senses. In: Gelbukh, A. (ed.) CICLing 2014, Part I. LNCS, vol. 8403, pp. 151–162. Springer, Heidelberg (2014)
19. Teh, Y.W., Jordan, M.I., Beal, M.J., Blei, D.M.: Hierarchical dirichlet processes. Journal of the american statistical association **101**(476) (2006)
20. Wang, C., Blei, D., Li, F.F.: Simultaneous image classification and annotation. In: IEEE Conference on Computer Vision and Pattern Recognition. CVPR 2009, pp. 1903–1910. IEEE (2009)

RAPPORT — A Portuguese Question-Answering System

Ricardo Rodrigues[1,2](✉) and Paulo Gomes[1]

[1] Centre for Informatics and Systems, University of Coimbra, Coimbra, Portugal
{rmanuel,pgomes}@dei.uc.pt
[2] College of Education of the Polytechnic Institute of Coimbra, Coimbra, Portugal

Abstract. We present a question answering system for Portuguese that depends on *subject-predicate-object* triples extracted from sentences in a corpus. It is supported by indices that store triples, related sentences and documents. The system processes the questions and retrieves answers based on the triples.

For testing and evaluation purposes, we have used the CHAVE corpus, which has been used in multiple editions of CLEF. The questions from those editions were used to query and benchmark our system. In its current stage, the system has found the answer to 42% of the questions.

This document describes the modules that compose the system and how they are combined, providing a brief analysis on them, and also some preliminary results, as well as some expectations regarding future work.

Keywords: Question answering · Triple extraction · Portuguese

1 Introduction

When querying a system that provides or retrieves information about a given topic, with its contents in natural language, the user should not have to care about system specific details, such as:

- knowing the best keywords to get an answer to a specific question;
- using system specific syntax in order to interact with it;
- perusing the multiple documents that may contain the eventual answer;
- being limited by the questions someone has compiled before.

It is also possible that the information sources are not, or can not be, structured in such a way that can be easily accessed by more conventional techniques of information retrieval (IR) [2].

Most of these issues are addressed by question answering (QA) systems [22], which allow the user to interact with the systems by using natural language, and process documents whose contents are specified also using natural language.

© Springer International Publishing Switzerland 2015
F. Pereira et al. (Eds.) EPIA 2015, LNAI 9273, pp. 771–782, 2015.
DOI: 10.1007/978-3-319-23485-4_77

In this context, we present a system that addresses QA for Portuguese, using triples extracted from sentences in a corpus, that are then used to present "short answers" (excerpts), alongside the sentences and documents they belong to.

In the remaining document, we briefly address the state of the art, describe the overall proposed approach and each of its modules, and draw some conclusions and reflections about future work.

2 Question Answering

Question answering, as in other subfields of IR, may include techniques such as: named entity recognition (NER) or semantic classification of entities, relations between entities, and selection of semantically relevant sentences, phrases or chunks [14], beyond the customary tokenization, lemmatization, and part-of-speech (POS) tagging. QA can also address a restricted set of topics, in a closed world domain, or forgo that restriction, operating in an open world domain.

Most approaches usually follow the framework shown in Fig. 1, where most of the processing stages are made on run-time (except for document indexing).

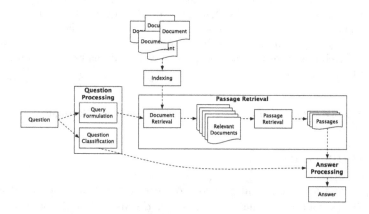

Fig. 1. A Typical Framework for a QA System (based on [10])

Regarding specific approaches to Portuguese, below is a list of the most relevant works whose results are compared against our work.

2.1 Esfinge

Esfinge [4] is a general domain question answering system that tries to take advantage of the great amount of information existent in the Web. Esfinge relies on pattern identification and matching. For each question, a tentative answer is created. For instance, a probable answer for a "What is X?" question will start with "*X* is...". Then this probable answer beginning is used to search the corpus, through the use of a search engine, in order to find possible answers that match the same pattern. In the following stages of the process, *n-grams* are scored and NER is performed in order to improve the performance of the system.

2.2 Senso

The Senso Question Answering System [19] (previously PTUE [17]) uses a local knowledge base, providing semantic information for text search terms expansion. It is composed of two modules: the solver module, that uses two components to collect plausible answers (the logic and the ad-hoc solvers); and the logic solver, that starts by producing a first-order logic expression representing the question and a logic facts list representing the text information and then looking for answers within the facts list that unify and validate the question logic form. There is also an *ad-hoc* solver for cases where the answer can be directly detected in the text. After all modules are used, the results are merged for answer list validation, to filter and adjust answers weight.

2.3 Priberam

Some of the most well known works on NLP and QA have been done at Priberam. Priberam's QA System for Portuguese [1] uses a conservative approach, where the system starts by building contextual rules for performing morphological disambiguation, named entity recognition, etc.. Then it analyses the questions and divides them into categories. These same categories are applied to sentences in the source text. This categorization is done according to question patterns, answer patterns and question answering patterns (where pattern match between answer and question is performed).

2.4 NILC

Brazil's Ncleo Interinstitucional de Lingstica Computacional (NILC) has built a summarization system to be used in the task of monolingual QA for Portuguese texts. NILC's system uses a text summarizer that comprises three main processes: text segmentation, sentence ranking, and extract production [5] (associating sentences to a topic). The questions are then matched against the sentences, with the associated summaries being used to produce an answer.

2.5 RAPOSA

The RAPOSA Question Answering System [21] tries to provide a continuous on-line processing chain from question to answer, combining the stages of information extraction and retrieval. The system involves expanding queries for event-related or action-related factoid questions using a verb thesaurus automatically generated using information extracted from large corpora.

RAPOSA consists of seven modules more or less typical on QA systems: a *question parser*, a *query generator*, a *snippet searcher*, an *answer extractor*, *answer fusion*, an *answer selector*, and *query expansion*. It also deals with two categories of questions: definition questions and factoid questions.

2.6   QA@L²F

QA@L²F [13], the question-answering system from L²F, INESC-ID, is a system that relies on three main tasks: *information extraction, question interpretation* and *answer finding*.

This system uses a database to store information obtained by information extraction, where each entry is expected to represent the relation between the recognized entities. On a second step, the system processes the questions, creating SQL queries that represent the question and are run in the database. The retrieved records from the database are then used to find the wanted questions, through entity matching.

2.7   IdSay

IdSay: Question Answering for Portuguese [3] is an open domain question answering system that uses mainly techniques from the area of IR, where the only external information that it uses, besides the text collections, is lexical information for the Portuguese language.

IdSay starts by performing document analysis and then proceeding to entity recognition. After that, the system makes use of patterns to define the type of the question and expected answers. IdSay uses a conservative approach to QA, being its main stages: information indexing, question analysis, document retrieval, passage retrieval, answer extraction and answer validation.

3   The Proposed System — RAPPORT

Our system, RAPPORT, follows most of the typical framework for a QA system, while being an open domain system. It does, however, improves on some techniques that differ from other approaches to Portuguese.

One of the most differentiating elements is the use of triples as the basic unit of information regarding any topic, represented by a subject, a predicate and an object, and then using those triples as the base for answering questions. This approach shares also some similarities with open information extraction, regarding the storage of information in triples [7].

RAPPORT depends on a combination of four modules, addressing information extraction, storage, querying and retrieving. The basic structure of the system comprehends the following modules:

- triple extraction (performed offline);
- triple storage (performed offline);

- data querying (performed online);
- and answer retrieving (performed online).

Each of these modules are described next, specifying each of the main tasks.

3.1   Triple Extraction

The first module processes the contents of the corpus, picking each of the documents, identifying sentences and extracting triples. It includes multiple tasks, namely sentence splitting, chunking, tokenization, POS tagging, lemmatization, dependency parsing, and NER.

The sentence splitting, tokenization, POS tagging, and NER tasks are done using tools included in the *Apache OpenNLP toolkit*[1], with some minor tweaks for better addressing the texts used. For instance, the system groups tokens that should be processed together — e.g., person names and dates — and, at the same time, it also splits composed tokens, such as it is the case of some Portuguese verbal conjugations and clitics — e.g., the preposition *"no"*[2] becomes *"em o"*. The model used in the chunker was also created for Portuguese (although following guidelines from OpenNLP), as there was no available pre-built model.

For the lemmatization process, *LemPORT* [18], a Portuguese specific lemmatizer was used. Another tool was used for dependency parsing, namely, *Malt-Parser* [16], with the model used by the parser being trained on Bosque 8.0[3] (available through Linguateca[4]). However, the output of MaltParser is further processed in order to group the tokens around the *main* dependencies, such as: subject, root (verb), and objects, among others.

Specifically regarding triple extraction, it is performed using two complementary approaches, both involving named entities, as a way of determining which triples are of use. The triples are defined by three fields: *subject*, *predicate*, and *object*. After the documents are split into sentences, each sentence is directly processed in order to extract named entities. Then, the sentences either are chunked or undergo tokenization, POS tagging and lemmatization before applying the MaltParser to identify the main dependencies. An algorithm describing the process is found in Alg. 1.

As can be noticed, only the triples with entities in the *subject* or in the *object* are further used. The triples extracted this way are then stored for future querying. Also, the predicate has the verb stored in its lemmatized form in order to facilitate later matches.

For clarification, in the triples that are based on the proximity between chunks, most of the predicates comprehend, but are not limited to, the verbs *ser* (to be), *pertencer* (to belong), *haver* (to have), and *ficar* (to locate). For instance, if two NP chunks are found one after another, and the first chunk contains a named entity, it is highly probable that it is further characterized by

---

[1] http://incubator.apache.org/opennlp/
[2] This corresponds to the combination of the preposition "in" with the article "the".
[3] http://www.linguateca.pt/floresta/BibliaFlorestal/completa.html
[4] http://www.linguateca.pt/

Data: Corpus documents
Result: Triples
Read documents;
foreach *document* **do**
    Split sentences;
    foreach *sentence* **do**
        Tokenize, lemmatize, POS tag and dependency parse;
        Extract named entities;
        Get proximity chunks;
        foreach *chunk* **do**
            if *chunk contains any entity* **then**
                if *neighbouring chunk has a specific type* **then**
                    Create triple relating both chunks, depending on the neighbouring chunk type and contents;
                end
            end
        end
        Get dependency chunks;
        foreach *chunk* **do**
            if *chunk contains any entity and is a subject or an object* **then**
                Create triple using the subject or object, the root, and corresponding object or subject, respectively;
            end
        end
    end
end

Algorithm 1: Triple Extraction Algorithm

the second chunk. If the second chunk starts with a determinant or a noun, the predicate of the future triple is set to *ser*; if it starts with the preposition *em* (in), it is used the verb *ficar*; if it starts with the preposition *de* (of), it is used the verb *pertencer*; and so on.

As an example, the sentence "Mel Blanc, o homem que deu a sua voz a o coelho mais famoso de o mundo, Bugs Bunny, era alrgico a cenouras."[5] yields three distinct triples: "{*Bugs Bunny*} {*ser*} {*o coelho mais famoso do mundo*}" and "{*Mel Blanc*} {*ser*} {*o homem que deu a sua voz ao coelho mais famoso do mundo*}", both using the proximity approach, and "{*Mel Blanc*} {*ser*} {*alrgico a cenouras*}", using the dependency approach.

3.2 Triple Storage

After triple extraction is performed, *Lucene*[6] is used for storing the triples, the sentences where the triples are found, and the documents that, by its turn, contain those sentences. For that purpose, three indices were created:

---

[5] In English, "Mel Blanc, the man who lent his voice to the world's most famous rabbit, Bugs Bunny, was allergic to carrots.".
[6] For Lucene, please refer to http://lucene.apache.org.

- the triple index stores the triples (subject, predicate and object), their id, and the ids of the sentences and documents that contain them;
- the sentence index stores the sentences' id (a sequential number representing its order within the document), the tokenized text, the lemmatized text and the documents' id they belong to;
- the document index stores the data describing the document, as found in CHAVE (number, id, date, category and author);

Although each index is virtually independent from the others, they can refer one another by using the ids of the sentences and of the documents. That way, it is easy to determine the relations between documents, sentences, and triples. These indices (mainly the sentence and the triple indices) are then used in the next steps of the present approach.

3.3 Data Querying

In a similar way to the annotation made to the sentences in the corpus, the questions are processed in order to extract tokens, lemmas and named entities, and identify their types, categories and targets.

For building the queries, the system starts by performing NER and lemmatizing the questions. The lemmas are useful for broadening the matches and results that could be found only by using directly the tokens. The queries are essentially made up of the lemmas found in the questions (including named entities and proper nouns). In those queries, all elements are, by default, optional, excluding the named entities. If no entities are present in the questions, proper nouns are made mandatory; by its turn, if there are also no proper nouns in the questions, it is the nouns that are used as mandatory keywords in the queries. For instance, in order to retrieve the answer to que question "*A que era alrgico Mel Blanc?*"[7], the query will end up being defined by three words: "*+Mel_Blanc a que ser alrgico*". We have opted for keeping all the lemmas because Lucene scores higher the hits with the optional lemmas, and virtually ignores them if they are not present.

The query is then applied to the sentence index — the system searches for sentences with the lemmas previously identified, with the verb as an optional term. When a match occurs, the associated triples are retrieved, along with the document data.

The triples that are related to the sentence are then processed, checking for the presence of question's entities in either the *subject* or the *object* of the triples, for selecting which triples are of interest.

3.4 Answer Retrieving

After a sentence matches a query, as stated before, the associated triples and document data are retrieved. As the document data is only used for better characterizing the answers, let us focus on the triples.

---

[7] In English, "What was Mel Blanc allergic to?"

When a sentence contains more than one triple, it is selected the triple which predicate matches the verb in the initial query. If that fails, it is selected the triple, as a whole, that best matches the query, accordingly to the Lucene ranking algorithm for text matches. After a triple is selected, if the best match against the query is found in the subject, the object is returned as being the answer. If, on the other hand, the best match is found against the object, it is the subject that is returned. An algorithm describing both the data querying and this process is found in Alg. 2.

Data: Question
Result: Answers
Create query using *named entities*, *proper nouns*, or *nound* as mandatory, and the remaining lemmas from the *question* as optional;
Run *query* against *sentence index*;
foreach *sentence hit* **do**
    Retrieve triples related to the sentence;
    foreach *triple* **do**
        if *subject contains named entities from question* **then**
            Add *object* to *answers* and retrieve sentence and document
            associated with the triple;
        end
        else if *object contains named entities from question* **then**
            Add *subject* to *answers* and retrieve sentence and document
            associated with the triple;
        end
    end
end

Algorithm 2: Answer Retrieval Algorithm

Continuing with the used example, given the correct sentence, of the three triples, the one that best matches the query is "{*Mel Blanc*} {*ser*} {*alrgico a cenouras*}". Removing from the triple the known terms from que question, what remains must yield the answer: "[a] cenouras". Besides that, as the named entity, Mel Blanc, is found in the subject of the triple, the answer is most likely to be found in the object.

4 Experimentation Results

For the experimental work, we have used the CHAVE corpus [20], a collection of the 1994 and 1995 editions — a total of 1456 — of newspapers "Pblico" and "Folha de So Paulo", with each of the editions usually comprehending over one hundred articles, identified by *id*, number, date, category, author, and the text of the news article itself.

CHAVE was used in the Cross Language Evaluation Forum (CLEF)[8] QA campaigns as a benchmark — although in the lasts editions of the Multilingual QA Track at CLEF a dump of the Portuguese Wikipedia was also used.

---

[8] http://www.clef-initiative.eu/

Almost all of the questions used in each of the CLEF editions are known. It is also known the results of each of the contestant systems. The questions used in CLEF adhere to the following criteria [12]: they can be *list* questions, *embedded* questions, *yes/no* questions (although none was found in the questions used for Portuguese), *who, what, where, when, why,* and *how* questions, and definitions.

For reference, in Table 1 there is a summary of the best results (all answers considered) for the Portuguese QA task on CLEF from 2004 to 2008 (abridged from [6,8,11,12,23]), alongside with the arithmetic mean for each system comprehending the editions where they were contenders. At the end of the table, it is also shown the results of the proposed system.

Table 1. Comparison of the Results at CLEF 2004 to 2008

Approach	Overall Accuracy (%)					
	2004	2005	2006	2007	2008	*Average*
Esfinge	15.08	23.00	24.5	8.0	23.5	*18.82*
PTUE/Senso	28.54	25.00	—	42.0	46.5	*35.51*
Priberam	—	64.50	67.0	50.5	63.5	*61.34*
NILC	—	—	1.5	—	—	*1.5*
RAPOSA	—	—	13.0	20.0	14.5	*18.83*
QA@L²F	—	—	—	13.0	20.0	*16.5*
IdSay	—	—	—	—	32.5	*32.5*
RAPPORT	47.78	33.33	48.89	41.07*	37.78*	*

Although we are using the questions for Portuguese used in CLEF in those years, a major restriction applies: just the questions made upon CHAVE with known answers were selected, as made available from Linguateca. As such, we are using a grand total of 641 questions for testing our system.

Notice that the years of 2004, 2005 and 2006 have only 180 (out of 200) known questions each, with their answers found in CHAVE, and the other two years have the remaining, with 56 in 2007, and 45 in 2008. In those two last years, the majority of the questions had the answers found on the Portuguese Wikipedia instead of just CHAVE. As such, the results for 2007 and 2008 represent the overall accuracy (grouping CHAVE and Wikipedia) of the different systems in those years, and not just for questions over CHAVE — unfortunately, the values for CHAVE and Wikipedia were not available separately. That is the reason for omitting the average result of our system in Table 1, and signalling the results for the years 2007 and 2008.

For verifying if the retrieved triples contain the expected answers, the triples must contain (in the subject or in the object) the named entities found in the questions (or, in alternative, proper nouns, and, if that fails, just the remaining nouns), and also match in the subject or in the object the known answers from CHAVE (alongside the same document id).

Using the set of questions that were known to have their answers found on CHAVE, on a base line scenario, we were able to find the triples that answer

42.09% of the questions (274 in 641), grouping all the question from the already identified editions of CLEF, without a limit of triples for each question. (If the maximum retrieved triples per question is reduced to 10, the number of answered questions drops to 20.75%.)

On the answers that have not been found, we have determined that in a few cases the fail is due to questions depending on information contained in other questions or their answers. In other situations, the problem lies on the use of synonyms, hyponyms, hypernyms and related issues: for instance, the question focusing on a verb and the answer having a related noun, as in "*Who wrote Y?*" for a question and "*X is the author of Y.*". There are certainly also many shortcomings in the creation of the triples, mainly on the chunks that are close together, as opposed to the dependency chunks, that should and must be addressed, in order to improve and create more triples. Furthermore, there are questions that refer to entities that fail to be identified as such by our system, an so no triples were created for them when processing the sentences.

5 Conclusions and Future Work

Although the proposed system currently only scores in an average place among the other systems, the use of triples seems to be a promising way of selecting the right and shorter answers to most of the questions addressed. However, there is a lot that can be improved.

Triples could be improved, namely those that are built from the relations of proximity between chunks, so the system is able to have a number of retrieved triples on par with with the sentences that contain the answers (and the triples). Another boost to the approach would be to differentiate the queries accordingly to the types of the named entities found in the questions, and improve NER, both on questions an on corpus sentences.

It is also our intention to use synonyms, hyponyms, hypernyms, and other relations between tokens or lemmas, in order to expand and improve the queries made to the indices, which will increase the number of retrieved sentences — and the number of right triples would also increase — using a wordnet-like resource, such as Onto.PT [9].

We are currently studying a way of relating words of the same family, such as "*escritor*" (writer) and "*escrever*" (to write) that when lemmatized, or even stemmed, end up being put apart. That can be an issue in situations where the question uses a verb for characterizing the agent, and the candidate answer uses, for instance, an adjective (of the same family of the verb) to characterize the same agent. A solution may be a list of agents and corresponding verbs, applying a set of rules to the verbs in order to generate the corresponding agents.

Another aspect that should be considered is the use of coreference resolution [10] in order to improve the recall of triples by way of replacing, for instance, pronouns with the corresponding, if any, named entities, and hence increasing the number of usable triples.

We believe that expanding the queries using the above technics, and creating better model extract triples, can achieve better results in a short time span.

Finally, the next major goal is to use the Portuguese Wikipedia as a repository of information, either alongside CHAVE, to address the latter editions of CLEF, or by itself, as it has happened in Págico [15].

References

1. Amaral, C., Figueira, H., Martins, A., Mendes, A., Mendes, P., Pinto, C.: Priberam's question answering system for Portuguese. In: Peters, C., et al. (eds.) CLEF 2005. LNCS, vol. 4022, pp. 410–419. Springer, Heidelberg (2006)
2. Baeza-Yates, R., Ribeiro-Neto, B.: Modern Information Retrieval. ACM Press, New York (1999)
3. Carvalho, G., de Matos, D.M., Rocio, V.: IdSay: question answering for Portuguese. In: Peters, C., et al. (eds.) CLEF 2008. LNCS, vol. 5706, pp. 345–352. Springer, Heidelberg (2009)
4. Costa, L.F.: Esfinge – a question answering system in the web using the Web. In: Proceedings of the Demonstration Session of the 11th Conference of the European Chapter of the Association for Computational Linguistics, pp. 410–419. Association for Computational Linguistics, Trento, Italy, April 2006
5. Filho, P.P.B., de Uzêda, V.R., Pardo, T.A.S., das Graças Volpe Nunes, M.: Using a text summarization system for monolingual question answering. In: CLEF 2006 Working Notes (2006)
6. Forner, P., et al.: Overview of the CLEF 2008 multilingual question answering track. In: Peters, C., et al. (eds.) CLEF 2008. LNCS, vol. 5706, pp. 262–295. Springer, Heidelberg (2009)
7. Gamallo, P.: An overview of open information extraction. In: Pereira, M.J.V., Leal, J.P., Simões, A. (eds.) Proceedings of the 3rd Symposium on Languages, Applications and Technologies (SLATE 2014). OpenAccess Series in Informatics, pp. 13–16. Schloss Dagstuhl – Leibniz-Zentrum für Informatik, Dagstuhl Publishing, Germany (2014)
8. Giampiccolo, D., et al.: Overview of the CLEF 2007 multilingual question answering track. In: Peters, C., Jijkoun, V., Mandl, T., Müller, H., Oard, D.W., Peñas, A., Petras, V., Santos, D. (eds.) CLEF 2007. LNCS, vol. 5152, pp. 200–236. Springer, Heidelberg (2008)
9. Gonçalo Oliveira, H.: Onto.PT: Towards the Automatic Construction of a Lexical Ontology for Portuguese. Ph.D. thesis, Faculty of Sciences and Technology of the University of Coimbra (2012)
10. Jurafsky, D., Martin, J.H.: Speech and Language Processing, 2nd edn. Pearson Education International Inc, Upper Saddle River (2008)
11. Magnini, B., et al.: Overview of the CLEF 2006 multilingual question answering track. In: Peters, C., Clough, P., Gey, F.C., Karlgren, J., Magnini, B., Oard, D.W., de Rijke, M., Stempfhuber, M. (eds.) CLEF 2006. LNCS, vol. 4730, pp. 223–256. Springer, Heidelberg (2007)
12. Magnini, B., et al.: Overview of the CLEF 2004 multilingual question answering track. In: Peters, C., Clough, P., Gonzalo, J., Jones, G.J.F., Kluck, M., Magnini, B. (eds.) CLEF 2004. LNCS, vol. 3491, pp. 371–391. Springer, Heidelberg (2005)
13. Mendes, A., Coheur, L., Mamede, N.J., Ribeiro, R., Batista, F., de Matos, D.M.: QA@L²F, first steps at QA@CLEF. In: Peters, C., Jijkoun, V., Mandl, T., Müller, H., Oard, D.W., Peñas, A., Petras, V., Santos, D. (eds.) CLEF 2007. LNCS, vol. 5152, pp. 356–363. Springer, Heidelberg (2008)

14. Moens, M.F.: Information Extraction: Algorithms and Prospects in a Retrieval Context. Springer-Verlag, Heidelberg (2006)
15. Mota, C.: Resultados Págicos: Participação, Resultados e Recursos. Linguamática **4**(1) (April 2012)
16. Nivre, J., Hall, J., Nilsson, J., Chanev, A., Eryiğit, G., Kübler, S., Marinov, S., Marsi, E.: MaltParser: A language-independent system for data-driven dependency parsing. Natural Language Engineering **13**(2), 95–135 (2007)
17. Quaresma, P., Quintano, L., Rodrigues, I., Saias, J., Salgueiro, P.: The University of Évora approach to QA@CLEF-2004. In: CLEF 2004 Working Notes (2004)
18. Rodrigues, R., Gonçalo Oliveira, H., Gomes, P.: LemPORT: a high-accuracy cross-platform lemmatizer for Portuguese. In: Pereira, M.J.V., Leal, J.P., Simões, A. (eds.) Proceedings of the 3rd Symposium on Languages, Applications and Technologies (SLATE'14). OpenAccess Series in Informatics, pp. 267–274. Schloss Dagstuhl – Leibniz-Zentrum für Informatik, Dagstuhl Publishing, Germany (2014)
19. Saias, J., Quaresma, P.: The senso question answering approach to Portuguese QA@CLEF-2007. In: Nardi, A., Peters, C. (eds.) Working Notes for the CLEF 2007 Workshop. Budapest, Hungary (2007)
20. Santos, D., Rocha, P.: The key to the first CLEF in Portuguese: topics, questions and answers in CHAVE. In: Proceedings of the 5th Workshop of the Cross-Language Evaluation Forum, pp. 821–832. Springer-Verlag, Bath, September 2005
21. Sarmento, L., Oliveira, E.: Making RAPOSA (FOX) smarter. In: Nardi, A., Peters, C. (eds.) Working Notes for the CLEF 2007 Workshop. Budapest, Hungary (2007)
22. Strzalkowski, T., Harabagiu, S. (eds.): Advances in Open Domain Question Answering, Text, Speech and Language Technology, vol. 32. Springer-Verlag, Heidelberg (2006)
23. Vallin, A., et al.: Overview of the CLEF 2005 multilingual question answering track. In: Peters, C., et al. (eds.) CLEF 2005. LNCS, vol. 4022, pp. 307–331. Springer, Heidelberg (2006)

Automatic Distinction of Fernando Pessoas' Heteronyms

João F. Teixeira[1][(✉)] and Marco Couto[2]

[1] University of Porto, Porto, Portugal
jpfteixeira.eng@gmail.com
[2] HASLab / INESC TEC, University of Minho, Braga, Portugal
marcocouto90@gmail.com

Abstract. Text Mining has opened a vast array of possibilities concerning automatic information retrieval from large amounts of text documents. A variety of themes and types of documents can be easily analyzed. More complex features such as those used in Forensic Linguistics can gather deeper understanding from the documents, making possible performing difficult tasks such as author identification. In this work we explore the capabilities of simpler Text Mining approaches to author identification of unstructured documents, in particular the ability to distinguish poetic works from two of Fernando Pessoas' heteronyms: Álvaro de Campos and Ricardo Reis. Several processing options were tested and accuracies of 97% were reached, which encourage further developments.

Keywords: Authorship classification · Machine learning · SVM · Text mining

1 Introduction

With the dawn of Text Mining (TM) a massive amount of information was enabled to be retrieved automatically. It is intended to find and quantify even subtle correlations over a large amount a data. This way, a wide variety of themes (economics, sports, etc) with different levels of structure, could be analyzed with little effort. Many TM solutions have been employed in security and web text analysis (blogs, news, etc). TM has been used in sentiment analysis as for evaluating movie reviews to estimate acceptability [1]. Forensic Linguistics enhances TM by considering higher level features of text. Linguistic techniques are usually applied to legal and criminal contexts for problems such as document authorship, analysis and measure of content and intent.

The purpose of this study is to generate a small representation of a large corpus of poems, able to discern between authors or aliases. For this initial study we selected to classify the collection of poems by two of Fernando Pessoa's heteronyms. Ricardo Reis and Álvaro de Campos were chosen due to their contrasting themes and initial concerns relative to the model's accuracy for this kind of dataset.

© Springer International Publishing Switzerland 2015
F. Pereira et al. (Eds.) EPIA 2015, LNAI 9273, pp. 783–788, 2015.
DOI: 10.1007/978-3-319-23485-4_78

To the best of our knowledge, there are no pattern recognition studies for alias distinction on poetic texts, therefore, no direct comparison of this work can be made. On the other hand, there is research on generic alias identification [2], however the objective is to find which aliases correspond to the same author and not to distinguish between personnas.

The author whose works we analyze is Fernando Pessoa [3], who wrote under several heteronyms or aliases. Each one had their own life stories and personal taste in writing style and theme.

Ricardo Reis is an identity of classical roots, when considering his poems' structure, theatricality and entities mentioned (ancient Greek and Roman references). He is fixated with death and avoids sorrow by trying to disassociate himself with anything in life. He seeks resignation and intellectual happiness.

Álvaro de Campos has a different personality, even presenting an internal evolution. Initially, he is shown to be a thrill seeker, mechanic enthusiast, and wishes to live the future. In the end, he feels defeated by time and devoided of the will to experience life. Consequently, he uses a considerable amount of interjections, in a weakly formatted writing style, with expressive punctuation.

The remaining of this document is structured as follows: In Section 2 the dataset is presented and described. Section 3 shows the methodology employed. Section 4 details the experimental approach, along with the result discussion. Section 5 the overall findings are presented along with possible future work.

2   Dataset

The dataset used in this work consists of the complete known poetic works[1] of *Ricardo Reis* (class RR) and *Álvaro de Campos* (class AC). Table 1 presents some statistics concerning the dataset.

Table 1. Class distribution

Class	# of entries	%	# of Words				# of Verses			
			Avg.	Std.	Min	Max	Avg.	Std.	Min	Max
RR	129	54%	77.9	65.1	19	570	14	12	4	106
AC	108	46%	360.9	904.2	29	7857	46	103	5	909

3   Methodology

In this section, the steps taken and experimental approach followed are shown. First, we tested the classifier with the tokenized documents and we progressively introduced other pre-processing models, comparing their performance.

The SVM model was validated with 70% of the dataset using 5-fold cross-validation while the remaining 30% enabled to evaluate the generalization performance of the generated model, i.e., the voting result of the 5 fold models.

---

[1] Available at: http://www.dominiopublico.gov.br

3.1 Document Pre-processing

S_1-*Tokenization.* Each document was turned into a sequence of word-level terms. Then, they were compacted into *bags-of-words* (BoW), disregarding their order, which is the most common document representation [4].

S_2-*Casing Transformation.* After the tokenization, all words suffer a lowercase transformation, reducing the number of different terms. Here, we disregard the capitalization of the poems' first word at every verse, while inadvertently removing significance from capitalized names and some metaphoric references.

S_3-*Length Filtering.* We remove from the token bags terms that contain below 4 or above 15 letters. The reason for this relates to the high probability of shorter words being irrelevant articles or connectors, leading to overfitting of the model, and not many Portuguese words have such large lengths. In fact, the average size of Portuguese words is 4.64 [5]. Nevertheless, removing words produces a more compact representation of the dataset and reduces possible dimensionality issues the classification model may experience with larger feature spaces.

S_4-*Stemming.* Word Stemming also compacts document instances. This consists of removing word affixes, leaving only the root term. Generally, stemmers follow iterative replacement rules, some even dealing with irregular and rare terms. For this work, the Snowball Portuguese dictionary was used [6] [7].

S_5-*Stopword.* Finally, we include stopword removal. This consists of ignoring all terms in a given dictionary. This might help the classifier focus on meaningful terms instead of considering articles, connectors and overall writing style. Also, specific unwanted words can be eliminated.

3.2 Occurrence Metrics

To evaluate if a word is distinctive for the classification task some methods based on its occurrence can be used. The following metrics were experimented:

Binary Term Occurrence. BTO identifies the number of documents in which a given term occurs. It provides little information thus is rarely used.

Term Occurrence. The TO metric provides the number of times a word occurs on each document of the collection. This can be viewed as a measure of significance of a given word for each document.

Term Frequency. The TF is a relative measure of the word occurrence considering the number of words in a document. Consequently, this can be misleading depending on the documents' length variability.

TF-IDF. is generally calculated as the product of TF and the Inverse Document Frequency (IDF) [8]. The IDF approach concerns the number of documents which contain a given term. A term that occurs frequently does not provide discriminative power and should be given less importance (lower weight) [9].

We used SVMs [10] that can linearly separate clusters of data on feature space, by maximizing the hyperspace boundary margin. The model was fed an array of occurrence metrics for the terms included after pre-processing.

Since the focus of this work was on Text Mining, the SVM model employed was relatively simple. A linear kernel with shrinking heuristics was used. It included a termination tolerance $\varepsilon = 0.001$ and no penalty $(C = 0)$.

4   Experimental Results

4.1   Estimation Using Cross-Validation

In Section 3, we conduct several experiments in which the text pre-processing algorithms are incrementally included. These experiments considered the occurrence metric *tf-idf* since it is intuitively the most appropriate to compare the results of models trained with such different instance content. The accuracy of the experiments is S_1:93.35%, S_2:91.58%, S_3:90.44%, S_4:91.03% and S_5:90.44%.

With the length and stopword filtering several of the top scoring terms (SVM weights) were removed. However most of these were articles and connectors which could lead to model overfitting. Their removal only decreased slightly the accuracy. Along with stemming and the lowercase transformation, the number of attributes considered was reduced in more than half (8941 to 4398 terms).

The following experiments aim to evaluate the influence of different the occurrence metrics on the classification model. Table 2 presents the results of those experiments. The results show that the model using Term Occurrence based metrics (BTO, TO) performs worse than with frequency based metrics (TF, *tf-idf*) including misclassification rate balance.

We note that this comparison is not truly fair. The processing pipeline for the experiments was previously optimized for *tf-idf* thus, providing only a general comparison. The performances with these last two are very close and, therefore, the best method cannot be directly found.

Table 2. Occurrence Metrics Experiments (%)

Binary TO			TO			TF			TF-IDF		
Acc	$F1_{RR}$	$F1_{AC}$	Acc	$F1_{RR}$	$F1_{AC}$	Acc	$F1_{RR}$	$F1_{AC}$	Acc	$F1_{RR}$	$F1_{AC}$
80.74	84.91	73.33	68.71	77.39	49.01	91.03	91.89	89.80	90.44	91.58	88.73

4.2   Evaluation of Validation Setup

In this section, we considered, from the previous experiments, the pipelines with the two best performances and with BTO (baseline), while using the remaining 30% of the dataset. The results are shown on Table 3.

As expected, BTO maintained the low accuracy and obtained lower $F1_{AC}$. The best two models kept the high accuracy, however, *tf-idf* managed to overcome the improvement of TF, from the validation phase, even if only by 3 instances. This suggests that, even though *tf-idf* presented lower validation results, it was somewhat underfitting. Either way, these comparative results were expected due to the consideration of term rarity metrics of *idf*.

4.3   Influence of Long Poems

Due to a large difference of length statistics between the two labels, we conducted further analysis. The documents were segmented into multiple instances such

Table 3. Testing Set Results (%)

Binary TO			TF			TF-IDF		
Acc	F1$_{RR}$	F1$_{AC}$	Acc	F1$_{RR}$	F1$_{AC}$	Acc	F1$_{RR}$	F1$_{AC}$
66,20	76,47	40,00	92,96	93,83	91,80	97,18	97,44	96,88

Table 4. Updated Class distribution

Class	# of entries	%	# of Words				# of Verses			
			Avg.	Std.	Min	Max	Avg.	Std.	Min	Max
RR	169	51%	63.5	27.0	3	161	11	4	1	17
AC	165	49%	240.8	150.0	29	521	30	145	5	54

that the portions had the maximum amount of verses equal to the previous mean for that class (plus a tolerance). Table 4 presents the updated class distribution.

For this experiment, the complete word processing pipeline and *tf-idf* scoring criteria were used (testing phase best results). For the cross-validation and testing steps, respectively, the accuracy was 96.58% and 96.04%; the F1$_{RR}$ was 96,64% and 96,15% and the F1$_{AC}$ was 96,49% and 95,92%.

The results show that imposing the upper bound on the number of verses per poem increased the accuracy of the model in the validation phase by around 5%. Apart from accuracy, the model should have really improved since the misclassification rates became more balanced.

It is safe to assume that longer poems might be more difficult to sort correctly into classes since they encompass more terms that can be highly influential to the *tf-idf* metric (through emphatic repetition, etc) which may not contribute positively for the accurate learning of attribute weights. Thus, this can affect poorly on the classification. On the other hand, the test results, in a way, contradict the analysis from the cross-validation. It performs slightly worse than the test experiment for *tf-idf*. As of this, we cannot provide an acceptable hypothesis as to which this occurs, rendering this analysis inconclusive.

5   Discussion and Conclusions

In this work we aimed to distinguish the authorship of poetic texts from two heteronyms of Fernando Pessoa, solely using basic Text Mining approaches. To our surprise, the methods were able to predict quite accurately (most over 70%, best ~97%), further verifying the a clear difference between the heteronyms.

This comes as a revelation mainly because, the author is, in fact, the same, despite having created these two personnas, and thus, the vocabulary and certain parts of writing style should be ubiquitous to the heteronyms.

Many of the best discerning words were related to writing style, including several possessive related terms for AC. However, obviously, some of the best terms were theme related keywords such as *grande* and *sentir*, referring to the magnificence of feelings of AC, *cansaço, domingo* and *sonho* to the tiredness AC feels towards the end and recollections of the past; while RR tries to remain forever calm and avoids pain.

Among the settings tested, *tf-idf* demonstrated, as expected, the best balance and generated the highest accuracy for the testing set.

The change in accuracy for the shorter instance set was not conclusive. These results suggest that dividing the larger poems was either not that relevant or additional instances would be needed to confirm (accuracy already close to 100%).

Although our methodology produces good results, we intend to extend the study to the rest of the heteronyms to evaluate if this kind of simple analysis is still sufficient for discernibility. Additional relevant experiments and approaches could include the model's response to a few verses instead of large chunks or complete poems and word *n-grams* analysis for style traits identification.

We realized that, according to Zipf's law, both the highest and lowest frequent terms are less frequent in large documents. Thus, our approach could be improved concerning the enhancement of term relevance instead of only minding to frequency. This could be done by including a normalization term in the *tf-idf* formula [11].

References

1. Pang, B., Lee, L., Vaithyanathan, S.: Thumbs up?: sentiment classification using machine learning techniques. In: Proceedings of the ACL 2002 Conference on Empirical Methods in Natural Language Processing, EMNLP 2002, vol. 10, pp. 79–86. Association for Computational Linguistics, Stroudsburg (2002)
2. Nirkhi, S., Dharaskar, R.V.: Comparative study of authorship identification techniques for cyber forensics analysis (2014). CoRR abs/1401.6118
3. de Castro, M.G. (ed.): Fernando Pessoa's Modernity Without Frontiers: Influences, Dialogues, Responses. Tamesis Books, Woodbridge (2013)
4. Sebastiani, F.: Machine Learning in Automated Text Categorization. ACM Comput. Surv. 34(1), 1–47 (2002)
5. Quaresma, P., Pinho, A.: Análise de frequências da língua portuguesa. In: Livro de Actas da Conferência Ibero-Americana InterTIC, pp. 267–272. IASK, Porto (2007)
6. Porter, M.F.: Snowball: A language for stemming algorithms, October 2001. http://snowball.tartarus.org/texts/introduction.html
7. Porter, M.F.: Snowball: Portuguese stemming algorithm. http://snowball.tartarus.org/algorithms/portuguese/stemmer.html
8. Salton, G., Buckley, C.: Term-weighting Approaches in Automatic Text Retrieval. Inf. Process. Manage. 24(5), 513–523 (1988)
9. Robertson, S.: Understanding inverse document frequency: on theoretical arguments for IDF. Journal of Documentation 60(5), 503–520 (2004)
10. Vapnik, V.N.: An overview of statistical learning theory. Trans. Neur. Netw. 10(5), 988–999 (1999)
11. Singhal, A., Buckley, C., Mitra, M.: Pivoted document length normalization. In: Proceedings of the 19th Annual International ACM SIGIR Conference on Research and Development in Information Retrieval, SIGIR 1996, pp. 21–29. ACM, New York (1996)

Social Impact - Identifying Quotes of Literary Works in Social Networks

Carlos Barata[1,2]([✉]), Mónica Abreu[1], Pedro Torres[2], Jorge Teixeira[2,3],
Tiago Guerreiro[1], and Francisco M. Couto[1]

[1] Departamento de Informática, Faculdade de Ciências,
Universidade de Lisboa, Lisbon, Portugal
cbarata7@gmail.com
[2] SAPO Labs, Lisboa, Portugal
[3] LIACC, Universidade Do Porto, Porto, Portugal

Abstract. A non-neglectable amount of information shared in social networks has quotes to literary works that, most of the times, is not linked to the original work or author. Also, there are erroneous quotes that do not fully match the original work, for example by including synonyms and slang words. Moreover, users sometimes associate their quotes to the wrong author, which creates misleading information. This paper presents *Social Impact* framework as an approach to identify quotes in social networks and match them to the original literary works from a particular author. This framework was applied to two case-studies: *O Mundo em Pessoa* and *Lusica*. In the first case-study, *Social Impact* evaluation achieved 98% for precision measure and 59% for recall, whereas in the latter case-study it obtained 100% for precision and 53% of recall.

Keywords: Information retrieval · Information extraction · Web mining · Text mining · Pattern recognition

1 Introduction

Social networks emerged in last decade and changed the way we communicate, becoming essential tools in the human interaction. This happened possibly due to the fact that, at the distance of a click, lays the possibility to send and share content. As Kwak et al.[4] refers, this wide use of social networks provide a great interest of investigation in many areas like extraction and information analysis.

Most of the information shared in social networks such as Twitter and Facebook is in text format, and an interesting amount of such information (messages) contains quotes to literary works (e.g.: "Tudo vale a pena, quando a alma não é pequena - Fernando Pessoa"). Nevertheless, in a non-neglectable number of cases there is no reference to which text, book or literary work the quote is referred.

Due the fact that quotes may have incoherencies (e.g.: quote is different from the original text), the identification of the original text or author can be very challenging. These incoherencies have a higher presence on social networks

F. Pereira et al. (Eds.) EPIA 2015, LNAI 9273, pp. 789–795, 2015.
DOI: 10.1007/978-3-319-23485-4_79

(against, for instance, opinion articles on news) because of particular characteristics of the network, namely: short messages or reduced context. The use of synonyms or typos are some of the most common causes for the lack of accuracy in the quotes published in social networks. One may think hash-tags can substantially reduce the complexity of this task, but unfortunately the usage of hash-tags on messages literary work is low as referred in [8] study. This study concluded that in Twitter, the ratio of hash-tags per tweet is between 4%(in Japanese language) to 25%(in German language) of the total tweets using them.

This paper presents *Social Impact* platform as an approach to this problem. The framework major goal is to identify quotes from literary work on social network messages, supported on *SocialBus*[1] and Apache Lucene systems. Evaluation was performed on two case-studies: *O Mundo em Pessoa* and *Lusica*.

2   Related Work

Social Impact platform is generically supported on two different technological blocks: *SocialBus*, a social network crawling and analysis platform, and Lucene, a high-scalable infrastructure for indexing and querying documents.

SocialBus: Social networks such as Twitter and Facebook provide APIs that allow access to public messages, within certain limits, giving the possibility of analysing such content for a variety of purposes, including quotes detection. We propose to use *SocialBus* platform[2,7], a framework that collects and analyse data from Twitter and Facebook for a pre-defined set of users representative of the Portuguese community.

Lucene: is an open-source software[2] for text searching and indexing through a document indexation, coded in Java programming language and developed by Apache Software Foundation. According to Gospodnetic et al.[3] this framework works through the indexation of documents, information parsers and queries to consult and retrieve the indexed information. The result is a ranked list of documents ordered by relevance [1,5,6].

3   Social Impact Platform

Social Impact main objective is to find quotes in messages published in social networks and subsequently link them to their original literary work. This framework stores such data in a relation database and provides such data as RESTful APIs. More importantly, *Social Impact* architecture is abstract enough to be applied on different contexts and scenarios.

---

[1] http://reaction.fe.up.pt/socialbus/
[2] http://lucene.apache.org/core/

3.1 Architecture

The *Social Impact*'s structure is based on a Service-Oriented Architecture(SOA), broadly used in web applications, due to its standardisation approach. This architecture is represented in Figure 1 and it has three main layers, described below.

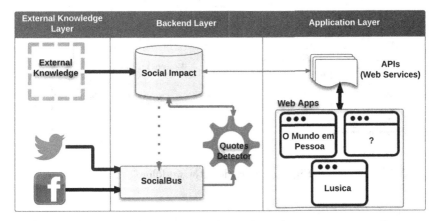

Fig. 1. *Social Impact* Global Architecture

External Layer: represents information and knowledge external to *Social Impact* platform and that somehow is collected into the system. The leftmost block, External knowledge, represents data specific to each case-study, including literary work (e.g.: poems from Fernando Pessoa[3]) or domain specific keywords used to narrow the search over *SocialBus* collected data (e.g.: poems or musics authors). The remain two blocks represent Twitter[4] and Facebook[5] APIs to feed *Social Impact* with data from social networks.

Backend Layer: is the core layer of *Social Impact*, and it is responsible for processing the messages coming from *SocialBus* and analyse them through the Quotes Detector, as well as store those messages and their subsequently generated meta-data on suitable a relational database (MySQL).

Application Layer: represents the interface with the potential applications using *Social Impact* platform. This layer comprehends a set of RESTful APIs that provides information previously processed in the Backend layer to the web applications.

---

[3] Data obtained from "Arquivo Pessoa" available at http://arquivopessoa.net
[4] https://dev.twitter.com/rest/public
[5] https://developers.facebook.com

3.2   Quotes Detector

Figure 2 presents a detailed diagram of the Quotes Detector module, with two essential flows of information:

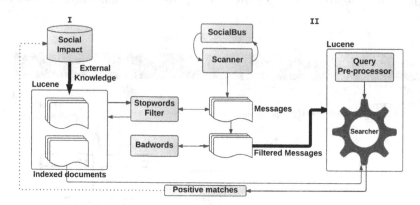

Fig. 2. Quotes detector workflow

Pre-processing and Indexing External Knowledge: represented in Figure 2 as "I" is imported only once and include, for instance, all the literary work from a particular author. Each of these documents (e.g.: a single poem) is submitted to Lucene engine, filtered through a stopwords filter and indexed.

Identification and Indexing of Quotes (refer to "II" in Figure 2) module is listening to *SocialBus* and imports new data as new messages arrive to *SocialBus*. Those messages are then filtered with a stopwords and a badwords (curse words) filter. Each filtered message is transformed into a lucene query syntax. The "Search" operation compares the indexed documents from *Social Impact* (External Knowledge) with each new message and retrieve, if the score if above a given threshold, the most relevant document (poem, music, etc.) as a positive match of a quote. Moreover, all tokens from the matched message are isolated and stored in the database.

4   Case Studies

O Mundo em Pessoa[6] is a web based project that aims to depict the presence of Fernando Pessoa poems on social networks, based on quotes to his literary. This project is based on *Social Impact* platform and it covers Fernando Pessoa work and from all his heteronyms. The list of terms used to narrow the messages crawling (refer to Section 3.1) contains the names of all Fernando Pessoa heteronyms. This project is supported on a Web Application that displays the identified quotes from Fernando Pessoa organized by timeframes, going from one

---

[6] http://fernandopessoa.labs.sapo.pt/

day to one month. For each quote, the user has the possibility to explore the number of social network users that publish that particular quote and access the original message, among other features.

Lusica[7] main purpose was study the lusophone music and its presence on the social networks, supported on *Social Impact* platform. There are two importante aspects that differentiate "Lusica" from "Mundo em Pessoa": (i) the domain is music instead of literary, and (ii) a large effort was put on the visualization of the information obtained from the quotes detection, through an interactive graph available online. "Lusica" external knowledge (refer to Figure 1) is based on the musics' and albums' titles from lusophone artists. Such information was obtained from LastFM APIs[8] (the list of lusophone artists) and from MusicBrainz service[9] (the albums and musics titles for each lusophone artist).

5   Results and Discussion

In this section will be displayed the results of the evaluation of *Social Impact* for both case studies. This evaluation aims to provide an overview of the system's performance.

Data Collection: For evaluation purposes, we used a subset of data published between January 2014 and June 2014. Regarding *O Mundo em Pessoa*, from a set of 56.212 collected messages, only approximately 8% (4.720 messages) were identified as quotes (with Lucene score larger than 1,0). As expected, most of the collected messages are not classified as quotes, and this phenomenon can be explain by the fact that many of the collected messages are just references to Fernando Pessoa but are not actually a quote to his literary work. For *Lusica*, results are similar to *O Mundo em Pessoa*, with only less than 5% (7.628) of messages representing references to the songs' artists, in a set of approximately 420.000 messages.

Quotes Detector Evaluation: Precision and recall metrics were calculated based on the following types of documents: True positive (TP): messages correctly classified as quotes; False positive (FP): messages incorrectly classified as quotes; True negative (TN): messages correctly classified as not quotes; False negative (FN): messages incorrectly classified as not quotes. Precision is measured as $P = TP/(TP+FP)$ while recall is $R = TP/(TP+FN)$. Regarding recall, our assumption is that *SocialBus* filtered messages correspond to all representative messages for the specific domain of the case-study. Evaluation was performed manually on a sample of 200 randomly chosen messages for each of the case-studies. Regarding "Mundo em Pessoa", the evaluation dataset was divided in 4 parts according to Lucene score and Twitter *versus* Facebook messages. Results for Twitter shown a precision of 19% and recall of 100% for low scores (between 0, 5 and 1, 0) and precision of 98% and recall of 100% for high scores (between

---

[7] http://lusica.labs.sapo.pt/

[8] http://www.last.fm/api

[9] https://musicbrainz.org/doc/MusicBrainz_Identifier

$1,0$ and $2,0$). Concerning Facebook, precision value for low scores was 100% and recall 21% while for high scores was precision was 96% and recall was 100%. The average precision for "Mundo em Pessoa" was $P_{MundoemPessoa} = 98\%$, while recall was $P_{MundoemPessoa} = 59\%$. In respect to "Lusica", the same principle was followed, by selecting a sample of 200 messages and dividing them in two groups (Twitter messages with low and high Lucene score). Results shown a average precision value $P_{Lusica} = 100\%$, while recall was $P_{Lusica} = 53\%$.

Execution Time: the performance of *Social Impact* platform was also evaluated, measuring the execution time of processing a single message in a desktop with an Intel(R) Xeon(R) CPU E5405 @ 2.00GHz processor and 3GB of RAM memory. For *O Mundo em Pessoa* the results achieved an average time of execution of 0,01 seconds ($\pm 0,002$). Regarding *Lusica*, the result obtained was an average execution time of 0,02 seconds ($\pm 0,004$).

6   Conclusions

This paper presented an approach to find quotes of original literary work in shared messages on social networks. The proposed approach is supported on the *Social Impact* developed platform presented in this paper. This framework was applied to two distinct case studies: *O Mundo em Pessoa* and *Lusica*.

Evaluation shown that most of the collected messages from *SocialBus* are not classified as quotes (less than 8%) because they are just references to the author and do not contain any quotes. *Social Impact* evaluation achieved high precision values for both case-studies: $P_{MundoemPessoa} = 98\%$ and $P_{Lusica} = 100\%$.

Future work includes: (i) to automatically set the threshold for the Lucene score based on machine learning approaches; (ii) improve the domain specific list of keywords using automatic approaches; (iii) use user feedback to fill missing information about authors and song lyrics; and (iv) apply *Social Impact* platform on other non-literary corpora, such as plagiarism detection.

Acknowledgments. This work was partially supported by SAPO Labs and FCT through the project PEst-OE/EEI/UI0408/2013 (LaSIGE), and by the European Commission through the BiobankCloud project under the Seventh Framework Programme (grant #317871). The authors would like to thank to Bruno Tavares, Sara Ribas and Ana Gomes from SAPO Labs, João Martins, Tiago Aparcio, Farah Mussa, Gabriel Marques and Rafael Oliveira from University of Lisbon and Arian Pasquali from Universidade of Porto for all their support, insights and feedback.

References

1. Baeza-Yates, R., Ribeiro-Neto, B., et al.: Modern information retrieval, vol. 463. ACM press New York (1999)
2. Boanjak, M., Oliveira, E., Martins, J., Mendes Rodrigues, E., Sarmento, L.: Twitterecho: a distributed focused crawler to support open research with twitter data. In: Proceedings of the 21st WWW, pp. 1233–1240. ACM (2012)

3. Hatcher, E., Gospodnetic, O.: Lucene in action. Manning Publications (2004)
4. Kwak, H., Lee, C., Park, H., Moon, S.: What is twitter, a social network or a news media? In: Proceedings of the 19th WWW, pp. 591–600. ACM (2010)
5. Liu, B.: Web data mining. Springer (2007)
6. Manning, C.D., Raghavan, P., Schütze, H.: Introduction to information retrieval, vol. 1. Cambridge University Press, Cambridge (2008)
7. Oliveira, E.J.S.L.: TwitterEcho: crawler focado distribuído para a Twittosfera portuguesa. Master's thesis, Faculdade de Engenharia da Universidade do Porto (2010)
8. Weerkamp, W., Carter, S., Tsagkias, M.: How people use twitter in different languages. Proceedings of the Web Science (2011)

Fractal Beauty in Text

João Cordeiro[1,3(✉)], Pedro R.M. Inácio[1,2], and Diogo A.B. Fernandes[2]

[1] Department of Computer Science, University of Beira Interior,
Rua Marquês d'Ávila e Bolama, 6200-001 Covilhã, Portugal
{jpaulo,inacio}@di.ubi.pt
[2] Instituto de Telecomunicações, Av. Rovisco Pais, 1 - Torre Norte,
1049-001 Lisboa, Portugal
diogoabfernandes@gmail.com
[3] INESC – TEC, University of Porto,
Rua Dr. Roberto Frias, 378, 4200-465 Porto, Portugal
jpcc@inescporto.pt

Abstract. This paper assesses if text possesses fractal properties, namely if several attributes that characterize sentences are self-similar. In order to do that, seven corpora were analyzed using several statistical tools, so as to determine if the empirical sequences for the attributes were Gaussian and self-similar. The Kolmogorov-Smirnov goodness-of-fit test and two Hurst parameter estimators were employed. The results show that there is a fractal beauty in the text produced by humans and suggest that its quality is directly proportional to the self-similarity degree.

Keywords: Self-similarity in text · Statistical linguistic studies

1 Introduction

Since the advent of the *World Wide Web*, an increasing number of users insert text from a multitude of sources, namely from newly created web pages, news in electronic newspapers, blogs, product reviewing, and social networks. This opened new opportunities for linguistic studies and the need for new applications to intelligently deal with all this text and to make sense out of it.

The question of automatically and effectively assessing the *quality* of a text remains unanswered. In general, an experienced human reader can judge the complexity and quality of a given text, a task not so easily attained with computational means. The human reader can not only determine if phrases are grammatically correct, but also figure out the lexicon degree and the structural and rhetorical combination of words, sentences, and ideas. There are aesthetic principles in the way of writing, yielding different types of texts. The spectrum ranges from almost telegraphic accretions, posted in Twitter, up to nobel prize winning novels. The goal of this study covers the topic of the text quality referred herein by looking after the mathematical principles underlying it.

This paper discusses a series of experiments conducted to find out if the text produced by humans exhibits a statistical property known as *self-similarity*. Self-similarity is a property of fractals, and it refers to the possibility of parts of a

© Springer International Publishing Switzerland 2015
F. Pereira et al. (Eds.) EPIA 2015, LNAI 9273, pp. 796–802, 2015.
DOI: 10.1007/978-3-319-23485-4_80

(mathematical) object to be similar to its parts. In the case of self-similarity, it refers more specifically to the fact that the statistical properties of the object look the same independently of the scale from which it is observed.

This paper presents our initial findings on the self-similar properties of text. Its main contributions are twofold: (i) it is shown that the sequences, constructed by measuring several attributes of text blocks produced by humans, are self-similar, suggesting also that the self-similarity degree is related with the quality of the text; and (ii) the way the analysis is performed defines the basic foundations for future research on the intersection of these two fields.

2 Related Work

To the best of our knowledge, no other work has been involved in the determination of self-similarity in text and how it can be used for characterization assessment of general aesthetic principles. However, there are a number of related works with similar goals. For example, McCarthy and Jarvins [9] compared different methods to determine lexical diversity (LD) in text.

The LD index measures the vocabulary diversity of a given text segment and is usually calculated by dividing the number of tokens by the number of types (the number of unique tokens in a segment of text), also known as the *token-type* proportion. As it depends on the considered text segment length, it is not used directly to compute the LD index. Instead, a number of strategies have been proposed [6,9], some of them being based on the division of a text in fixed segments of n tokens. The LD index has been used to assess the writing skills of a subject in a variety of studies, namely in children language skills measurements, English second language acquisition, Alzheimer's onset, and even speaker socioeconomic status [7,8].

Forensic linguistic analysis [10] is a recent field with diverse applications, such as plagiarism detection [1], authorship identification [12], cybercrime and terrorism tracking [10], among others. The work in this field is based on the use of a number of text characteristics in different levels of analysis like morphological, lexical, syntactical, and rhetorical [1,12]. These textual characteristics necessarily exhibit different self-similar properties that suggest investigation. The findings presented in this work are a first step toward that objective, specifically at the lexical level.

3 Self-Similarity and Statistical Tools

In this paper, fractality is studied for values representing attributes of text. It does not concern any particular visual interpretation of the text, in which some artifact is repeated as we graphically zoom in or out. It refers to an interpretation for the statistical behavior of the aforementioned values (*e.g.*, the number of words or non-words in a sentence). Self-similarity refers to the property of a stochastic process that looks statistically identical for any (aggregation) scale from which it is observed from. It is typical to define a (discrete-time) self-similar process $\{X(t)\}_{t\in\mathbb{N}}$ as the one that fulfills the condition that $X(t) \overset{d}{=} a^{-H}X(at)$, where $\overset{d}{=}$ denotes equality in all finite-dimensional distributions, $a \in \mathbb{N}$ and $0 < H < 1$ is the *Hurst parameter*, also referred to as the *self-similarity degree* or the *Hurst exponent*. The most widely known example of a self-similar process is the fractional Brownian motion (fBm), which has a Gaussian distribution. Its *first order differences process*, denoted as fractional Gaussian noise (fGn) is often useful too, since many natural or artificial processes occur in this form. Thus, when performing self-similarity analysis, it is typical to assess whether the empirical values are consistent with sampling a Gaussian variable. In this work, the Kolmogorov-Smirnov goodness-of-fit test [3] was used for that purpose.

There are several methods for estimating the Hurst parameter from empirical data, most of them based on repeatedly calculating a given statistic (*e.g.*, variance or maximum value) for the original process and for a finite number of aggregated processes. The Hurst parameter estimators used in this study were the well-known Variance Time (VT) and Rescaled Range Statistics (RS) estimators. The statistical tools mentioned herein are all implemented in the open-source *TestH* tool [4]. It accepts files containing raw values separated by space or newline, normalizes them, and outputs the estimated values of the Hurst parameter and the p-value concerning the Kolmogorov-Smirnov goodness-of-fit test.

When the Hurst parameter is 0.5, the process is *memoryless* and each occurrence is completely independent of any past or future occurrences. For values of the Hurst parameter ranging from 0 to 0.5, the process is *anti-persistent* or *short-range dependent*, while for values between 0.5 and 1, the process is said to be *persistent* or *long-range dependent*. There are many examples of long-range dependent processes in natural and artificial processes (e.g., the water level in rivers [5]). Prior from starting this work, our expectation was that the text was self-similar with Hurst parameter larger than 0.5 and that the degree of self-similarity was perhaps related with the quality of the text.

Our base unit to construct processes of text attributes is a block of 100 words, meaning that all the sequences analyzed in the scope of this work refer to *attributes per 100 tokens*. If a self-similar structure is embedded in the data, then the statistical behavior of these attributes is the same (apart from scaling) for each block of tokens, or for any number of them. Additionally, it is an indication that human writing is done in bursts, which means that blocks with higher counts in some attribute are probably followed by other blocks with higher counts also, and vice-versa.

4   Data and Experimentation

Five different types of English-written corpora were analyzed in the performed experiments. Each corpus was prepared identically before being submitted for testing. There are essentially three major genres that stand out: *Literature*, *News Stories*, and *Blogs*. In order to strengthen the validation of the main hypothesis, it was also decided to include a text corpus generated randomly from the words of the English language. Below, follows a more detailed description of each corpus:

The **Blogs** Corpus: Also known as the *Blog Authorship Corpus* [11], is a massive collection of 681K blog posts, gathered from 19320 users, from the blogger.com website. These blogs cover a wide range of subjects like *Advertising, Biotechnology, Religion, Science*, among others. It contains a total of 38 different subjects, with 300 million words (844 MB). There are three user age clusters: 13-17 years (8240 users), 23-27 years (8066 users) and 33-48 years (2944 users);

The **News** Corpus: This corpus is formed by a huge set of news stories, automatically collected from the web. An amount of 4.2 MB of text was randomly selected from the set. The news stories were collected for several main subjects, namely *Politics, Economics, Finance, Science*, among others;

The **Literature** Corpus: The (i) complete work of Shakespeare and the (ii) set of books (66) from the Bible[1] was selected for this corpus type;

The **Random** Corpus: A corpus of similar size was randomly generated in order to validate the proposed self-similarity measure. Each word was randomly taken from the English vocabulary, according to a uniform[2] distribution.

The aim of this study is to know whether certain text characteristics exhibit self-similarity properties by resorting to the estimation of the Hurst parameter. The test herein described was designed to determine self-similarity in time series, performing a large number of measurements of attributes over time. Here, the origin of the several time series is a corpus consisting of a considerable amount of text. Thus, the experience had to be drawn to meet two principles: the (i) set of the textual attributes to be measure and the (ii) reading structure of the text in order to achieve a number of significant measurements.

Definition of Attributes: For the first principle we have considered six lexical features that are measured for a given block (amount) of text: the number of (A_0) non-words; (A_1) small words ($|w| < 3$); (A_2) medium words ($3 \leq |w| < 7$); (A_3) long words ($|w| > 6$); (A_4) sentences; and (A_5) the lexical diversity.

Reading Structure: To satisfy the second principle we have decided to divide the text in sequential blocks with equal number of words. It was decided to choose the block size near to an average paragraph length, assuming five sentences per paragraph with each one having an average length of 20 words [2]. In previous studies of this kind, researchers usually take text chunks of this length [6,9]. Below is one such block of 100 tokens:

```
The guy's license plate was a little obvious" 68CONVT". I mean, you can see  that
it's a convertible please, because the top was down. Anyway,  I  stared  straight
ahead. but could hear that low throaty rumble next to me. Suddenly, I felt  tears
prickling in my eyes. It dawned on me that I was suffering  from  the  maliblues.
What will happen at Hot August Nights? Those muscle cars cruise nightly  and  rev
and rev and rev. I'm thinking I should get a medical bracelet with maliblues
--------------------------------------------------------------------------------
(~W, |W|<3, 3=<|W|<6, |W|>=6, #Sentences, Lex. Div.) ---> (18, 16, 51, 15,  8, 67)
```

---

[1] We have chosen the English translation version from King James.
[2] In the future a Zipfian law will be considered.

5   Results

Each corpora (Section 4) was processed to produce the necessary sequences of numbers representing the time series to be analyzed. Given the attributes and block size, these sequences consisted of integer numbers larger than 0 and smaller than 100. After being input to the TestH tool, they were normalized and the VT and RS estimators were applied to the resulting process. The p-value for the $\sqrt{n}D$ statistic was also calculated, via the application of the Kolmogorov-Smirnov goodness-of-fit statistical test available in the tool. Note that the values

Table 1. Results obtained on seven corpora regarding six attributes.

Corpora	A_0: num. non words			A_1: num. words < 3 chars			A_2: num. words 3-6 chars		
	VT	RS	KS	VT	RS	KS	VT	RS	KS
Blogs 13-17	0.73778	0.86064	0.20301	0.70702	0.83312	0.27614	0.72485	0.83613	0.27614
Blogs 23-27	0.76648	0.86895	0.10449	0.81549	0.83606	0.34726	0.75532	0.81676	0.34726
Blogs 33-48	0.84432	0.85356	0.09445	0.83969	0.86252	0.11157	0.86863	0.84596	0.07483
News	0.63524	0.78496	0.01202	0.46540	0.81174	**0.00491**	0.73615	0.85698	**0.00351**
Bible	0.53201	0.83720	0.19772	0.65733	0.79622	0.19772	0.88749	0.81389	**0.00688**
Shakespeare	0.64795	0.84271	0.01136	0.63140	0.75893	0.03116	0.57925	0.78148	0.02035
Random	0.28842	0.46262	**0.00432**	0.68364	0.51305	0.97693	0.57995	0.46603	0.03807

Corpora	A_3: num. words > 6 chars			A_4: num. sentences			A_5: Lexical Diversity (LD)		
	VT	RS	KS	VT	RS	KS	VT	RS	KS
Blogs 13-17	0.71949	0.86469	0.69745	0.76482	*VOR	0.43729	0.76005	0.82502	0.32957
Blogs 23-27	0.81764	0.88098	0.65178	0.82454	*VOR	0.40002	0.76380	0.81367	0.13147
Blogs 33-48	0.88160	0.92609	0.28191	0.81082	0.88063	0.66304	0.81663	0.82250	0.18178
News	0.75006	0.87849	0.03398	0.66815	*VOR	0.99286	0.64356	0.77951	0.25809
Bible	0.80095	0.84375	0.07482	0.81976	0.89219	0.59353	0.87111	0.86753	**0.00039**
Shakespeare	0.69241	0.75295	0.14385	0.66554	0.90009	0.29770	0.69826	0.81306	**0.00000**
Random	0.60186	0.40670	0.03351	0.26403	0.47593	0.67707	0.29526	0.46444	**0.00495**

*VOR: Estimated value out of range.

in the KS column (table 1) suggest that the sequences seem to be coming from Gaussian processes, with only seven cases rejecting the null hypothesis if the significance level is set to 0.01 (bold values). This is interesting, and we will explore if it may be the consequence of the *central limit theorem*, verifying if the attributes are the result of the sum of several independent and identically distributed variables.

For the three age sub-sets of the blogs corpus, we can see that the VT and RS values improve consistently in almost all attributes, and this difference was very marked in the A_3 and A_5 attributes for the VT estimator. This shows well that, even within the same genre, more mature and possibly more experienced authors produce text with an higher degree of self-similarity. The estimated values of the Hurst parameter are consistently larger for RS.

In the literary corpora, we have a significant difference between text from Shakespeare and the Bible. In the latter, self-similarity values are much higher for most of the attributes and in particular for the lexical diversity (A_5). Moreover, Shakespeare text reveals a low self-similarity in all attributes, with the exception for the RS estimator in A_4. The news genre did not reveal high self-similarity for most attributes. There are, however higher parameter values, for the attributes A_1, A_2, and A_3, which suggests strong self-similarity in the lexicon used. The low values obtained in the randomly generated corpus (Random) allow us to put in perspective the values obtained for the other corpora. This reveals a phenomenon of self-similarity in the human writing process.

6 Conclusions and Future Work

The paper shows that there is fractal beauty in text. It is naturally not a property that is consciously included by the authors of the text, which makes the results even more surprising. They suggest that there might be a relation between what is perceived as text quality and the self-similarity degree, although a more complete (more attributes) and exhaustive (more corpora) analysis is required.

Future research work includes the assessment of self-similarity and goodness-of-fit for more attributes and a larger number of different corpus. It should also focus on a more detailed pre-processing of the data, which allows to address the potential lacks of stationarity in large corpus. Gaussianity will also be tested using the Chi-squared goodness-of-fit test, for consistency purposes. The RS will be replaced by the DFA estimator in the future, since the later is known to be more reliable than the aforementioned one. Our efforts will be directed towards finding the (mathematical) explanations behind our initial findings. We intend also to explore self-similarity on different levels of linguistic attributes.

Acknowledgments. This work is financed by the FCT – Fundação para a Ciência e a Tecnologia (Portuguese Foundation for Science and Technology) within project UID/EEA/50014/2013.

References

1. Alzahrani, S., Naomie, S., Ajith, A.: Understanding plagiarism linguistic patterns, textual features, and detection methods. IEEE Transactions on Systems, Man, and Cybernetics, Part C: Applications and Reviews **42**(2), 133–149 (2012)
2. Cordeiro, J., Dias, G., Cleuziou G.: Biology Based Alignments of Paraphrases for Sentence Compression. Workshop on Textual Entailment (ACL-PASCAL) (2007)
3. Corder, G.W., Foreman, D.I.: Nonparametric Statistics for Non-Statisticians: A Step-by-Step Approach. Wiley, New Jersey (2009)
4. Fernandes, D.A.B., Neto, M., Soares, L.F.B., Freire, M.M., Inácio, P.R.M.: A tool for estimating the hurst parameter and for generating self-similar sequences. In: Proceedings of the 46th Summer Computer Simulation Conference 2014 (SCSC 2014), Monterey, CA, USA (2014)
5. Hurst, H.: Long-Term Storage Capacity of Reservoirs. Transactions of the American Society of Civil Engineers **116**, 770–799 (1951)
6. Koizumi, R.: Relationships Between Text Length and Lexical Diversity Measures: Can We Use Short Texts of Less than 100 Tokens? Vocabulary Learning and Instruction **1**(1), 60–69 (2012)
7. Malvern, D., Richards, B., Chipere, N., Durán, P.: Lexical diversity and language development: Quantification and assessment. Houndmills, NH (2004)
8. McCarthy, P., Jarvis, S.: A theoretical and empirical evaluation of vocd. Language Testing **24**, 459–488 (2007)

9. McCarthy, P., Jarvis, S.: MTLD, vocd-D, and HD-D: A validation study of sophisticated approaches to lexical diversity assessment. Behavior Research Methods **42**(2), 381–392 (2010)
10. Olsson, J., Luchjenbroers, J: Forensic linguistics. A&C Black (2013)
11. Schler, J., Koppel, M., Argamon, S., Pennebaker, J.W.: Effects of Age and Gender on Blogging. AAAI Spring Symposium: Computational Approaches to Analyzing Weblogs **6**, 199–205 (2006)
12. Stamatatos, E.: A survey of modern authorship attribution methods. Journal of the American Society for Information Science and Tech. **60**(3), 538–556 (2009)

How Does Irony Affect Sentiment Analysis Tools?

Leila Weitzel[1(✉)], Raul A. Freire[2], Paulo Quaresma[3], Teresa Gonçalves[3],
and Ronaldo Prati[2]

[1] Universidade Federal Fluminense, Niterói, RJ, Brazil
leila.weitzel@gmail.com
[2] Universidade Federal Do ABC, Santo André, SP, Brazil
{f.raul,ronaldo.prati}@ufabc.edu.br
[3] Universidade de Évora, Évora, Portugal
{pq,tcg}@uevora.pt

Abstract. Sentiment analysis applications have spread to many domains: from consumer products, healthcare and financial services to political elections and social events. A common task in opinion mining is to classify an opinionated document into a positive or negative opinion. In this paper, a study of different methodologies is conducted to rank polarity as to better know how the ironic messages affect sentiment analysis tools. The study provides an initial understanding of how irony affects the polarity detection. From the statistic point of view, we realize that there are no significant differences between methodologies. To better understand the phenomenon, it is essential to apply different methods, such as SentiWordNet, based on Lexicon. In this sense, as future work, we aim to explore the use of Lexicon based tools, thus measuring and comparing the attained results.

Keywords: Social media · Irony · Sarcasm · Opinion mining · Polarity detection

1 Introduction and Motivation

Sentiment analysis and opinion mining are very growing topics of interest over the last few years due to the large number of texts produced through Web 2.0. A common task in opinion mining is to classify an opinionated document as a positive or a negative opinion. A comprehensive review of both sentiment analysis and opinion mining as a research field for Natural Language Processing (NLP) is presented in Pang and Lee [1]. The demand for applications and tools to accomplish sentiment classification tasks has attracted the researchers´ attention in this area. Hence, sentiment analysis applications have spread to many domains: from consumer products, healthcare and financial services to political elections and social events. Sentiment classification is commonly categorized in two basic approaches: machine learning and lexicon-based. Machine learning approach uses a set of features, usually some function of the vocabulary frequency, which are learned from annotated corpora or labelled examples. The lexicon-based approach uses lexicon to provide the polarity, or semantic orientation, for each word or phrase in the text. Despite the considerable amount of research,

© Springer International Publishing Switzerland 2015
F. Pereira et al. (Eds.) EPIA 2015, LNAI 9273, pp. 803–808, 2015.
DOI: 10.1007/978-3-319-23485-4_81

the classification of polarity is still a challenging task; mostly because it involves a deep understanding of explicit and implicit information conveyed by language structures. Henceforth, irony or sarcasm has become an important topical issue in NLP. Ironic writing is common in opinionated user generated content such as blog posts and product reviews. As a whole, irony is an activity of saying or writing in such a way that the textual meaning of what is said is the opposite of what is meant. According to Rioff et al. [2] ironic message typically conveys a negative opinion using only positive words. In this paper, we present a study of different methodologies to classify polarity to better know how the ironic messages affect tools of sentiment analysis.

The classifications were carried out by Machine Learning Algorithms (mainly Support Vector Machine – SVM and Naïve Bayes Classifier). We aim to find out how irony affects the sentiment classification. Therefore, our research main question is: What is the best methodology able to boost performance classification?

The outline of this paper is as follows: In section 2, we present the related works. Section 3 explores the background in which this research is established. Section 4 presents our methodology, and Section 5 and 6 provides our results and main conclusions.

2   Related Work

In [3] the authors presented a semi-supervised approach for identification of sarcasm on two different data sets a collection of 5.9 million tweets collected from Twitter, and a collection of 66000 product reviews from Amazon. Using the Mechanical Turk, they created a gold standard sample in which each sentence was tagged by 3 annotators, obtaining F-scores of 0.78 on the product reviews dataset and 0.83 on the Twitter dataset. Tayal et al [4] proposed two algorithms, one to identify a sarcastic tweet and other to perform polarity detection on political sarcastic tweets. The main goal is to analyze and predict who will win the 2014 Indian Central Government Election based on sarcastic tweets. They came to the conclusion that using a supervised approach and their proposed algorithm, they will be able to achieve their goal. They found out that sarcastic tweets can predict election results in an efficient level.

3   Background

The Web 2.0 is the ultimate manifestation of User-Generated Content (UGC) systems. The UGC can be virtually about anything including politics, products, people, events, etc. One of highlights is the Twitter. Twitter constitutes a very open social network space, whose lack of barriers to access, e.g., even non-registered users are able to use Twitter to track breaking news on their chosen topics, from "World Economic Crisis" to "European Football Championship", for instance. Twitter social networkers communicate with each other by posting tweets allowing for a public interactive dialogue. On Twitter, users often post or update short messages referred to as tweets, describing one's current status within a limit of 140 characters [5]. Beyond merely displaying news and reports, the Twitter itself is also a large platform where different opinions are presented and exchanged. The interest that users (companies, politicians, celebrities) show in on-line opinions about products and services and the potential influence such opinions wield is something that vendors of these items are paying more and more attention to. Thus, it is important for correct identification of users opinions expressed in written text. In the general area of sentiment analysis, irony and sarcasm play a role as an interfering factor that can flip the polarity of a message. According to Macmillan English Dictionary (2007), irony is "a form of humor in which you use words to express the opposite of what the words really mean". This mean that it is the activity of saying or writing the opposite of what you mean. Unlike a simple negation, an ironic message typically conveys a negative opinion using only positive words or even intensified positive words [2]. As humans, when we communicate with one another, we have access to a wide range of spoken and unspoken figures that help create the intended message and ensure that our audience will understand what we are saying. Some of these figures include body language, hand gestures, inflection, volume, and accent. Hence, the challenge for Natural Language Processing (NLP) is: how to recognize sarcasm and gauge the appropriate sentiment of any given statement.

4   Methodology

4.1   Data Description and Corpus Generation

This research focuses on document-level irony detection on English Twitter datasets. The text content of Twitter is usually ambiguous and rich of acronym slang and fashion word. And, apart from the plain text, a tweet can contain others elements such as hashtags which are tags assigned by the user to identify topic (e.g. #Obama) or sentiment (#angry, #sarcasm); hyperlinks (typically a bitly URL, i.e., a URL shortening service), emoticons (it is a pictorial representation of a facial expression), references to other users (@<user>), and etc. In our experiments, the tweets were extracted by means of a Java package developed in-house, used for streaming posts [6]. We gathered about ten thousand tweets. Sarcastic tweets were collected using previously selected hashtags {sarcasm, irony, lying, moresarcasm, notcool, notreally, notsarcasm, somuchsarcasm}. These hashtags were used as an indicator of ironic or sarcastic tweets. An example is < I just love all the support I get from my mom. #sarcasm > or like this <I love my highspeed internet connection #moresarcasm>. Our assumption is that the best judge of whether a tweet is intended to be sarcastic or not is the author of the tweet, expressed by his hashtags.

4.2   Preprocessing of Data

The semi-automatic cleaning of the corpus is done to address concerns about corpus noise. Firstly, we removed: Tweets starting with "RT" because they refer to a previous tweets, e.g., <RT@iFraseSincera: ☺ #Sarcasm... http://t.co/MkFuO4JKa >; posts that contain only user name and link e.g. <@ZarethPanther @Ambybutt. #sarcasm http://t.co/zw2ieHVUGN>; Tweets that have less than 3 words e.g. <Price is low. #sarcasm>; tweets that are meaningless, e.g. <@smackalalala yeah, ooooo yeap

#Sarcasm>. We also removed special characters ($, %, &, #, etc.); punctuation marks (full stops, commas etc); all hashtags words and emoticons (smiley). We applied automatic filtering to remove duplicates tweets, and tweets that are written in other idioms. Afterwards, we manually classify the ironic posts in order to ensure that all messages are truly ironic or sarcastic. We gathered about 10000 tweet, after the pre-processing step, the sample size was about 7628 tweets, where 3288 are positive, 3600 are sarcastic and 740 are neutral.

4.3   Set of Experiments

As aforesaid, we used two different approaches in our experiment. The first method was **SUPPORT VECTOR MACHINE CLASSIFIER (SVM)** with: wrapper class called Library for Support Vector Machines (LIBSVM with default kernel RBF, i.e., Gaussian kernel) [7]; and the second classifier was **NAIVE BAYES MULTINOMIAL (NBM)** which is a learning algorithm that is frequently employed to tackle text classification problems. Our motivation to test multiple classifiers also stemmed from related work, which mostly test more than one classifier. We use the Weka [8] (Waikato Environment for Knowledge Analysis) where, each document (row or a tweet) is called as instance and each feature (term) is called as attribute. The SVM classifier requires that each data instance be represented as a vector of real numbers. Thus in order to set a vetorial input data, we use an unsupervised Weka build-in function called String-ToWordVector (that converts string attributes into a set of attributes representing word occurrence), with the following parameters: IDF and TF transform (to show how important a word is to a document in a collection or corpus). All tokens were converted to lowercase before being added to the dictionary. The minimum frequency of each term was restricted to five due to tweet size (up to 140 characters). We also removed stopwords and no stemmer algorithm was used. The classification is based on unigrams of words. Then, the resulting model has 7628 instance and 1502 attributes. All test modes are based on a stratified ten-fold cross-validation; this means that the data is randomly partitioned into 10 equal-sized parts. Thus, each attribute selection method is applied ten times on sub samples of the training set. This validation method reduces the variability of the classification. After finding the best parameters and build the final model, the adopted classifiers were applied on the test set. We also considered the metrics: (i) Precision (also called positive predictive value), (ii) Recall (also known as sensitivity) and (iii) F-measure as a main measure to evaluate the performance of classifier methodologies. All these metrics are generally accepted at Information Retrieval approaches as evaluation performance methods. It is, by far, the most widely performance metrics used in IR. Others measure were also considered, such as: True-positive (TP) and False-Positive (FP) rate (also known respectively as Sensitivity and Specificity measures), ROC area (The area under the ROC curve is a measure of how well a parameter can distinguish features), and Kappa statistic (The kappa coefficient measures pairwise agreement among a set of coders making category judgments).

5   Classification Results

The predictive performances of the models can be seen in Table 1. Table 2 shows the accuracy and the kappa coefficient. We have not obtained a reasonable model, which means we have not achieved a small number of misclassification errors (Relative Absolute Error is about 34%). The simple mean of ROC area is about 87%, which indicates a good performance of the models in terms of AUC. The Accuracy rates (equal to 79%) and Kappa (equal to 66%) point out that there is a moderate statistical dependence between the attributes and the classes. The best performance was achieved by **NBM** algorithm according to Precision (86%), F-measure (83%) and ROC curve Area (95%). However, the rates TP (84%) and Recall (84%) of the **SVM** were better than the **NBM**. On can see that there is a considerable number of TPs, but, nonetheless, there is not a small number of FPs (16% - incorrectly identified), mainly if we only consider the irony category.

Table 1. The main results.

Methods	TP Rate	FP Rate	Precision	Recall	F-Measure	ROC Area	Category
	80%	13%	82%	80%	81%	83%	positive
	84%	16%	80%	84%	82%	84%	irony
SVM	55%	6%	58%	55%	57%	74%	neutral
	78%	13%	78%	78%	78%	82%	weighted avg.
	83%	15%	81%	83%	82%	93%	positive
	80%	10%	86%	80%	83%	95%	irony
NBM	65%	7%	58%	65%	61%	90%	neutral
	79%	12%	80%	79%	80%	94%	weighted avg.
	82%	12%	82%	82%	82%	87%	weighted avg.

Table 2. Correct Classified Instances (Accuracy) and Kappa statistic.

	SVM	NBM
Kappa	64%	66%
Accuracy	78%	79%

6   Conclusions

Individuals post messages in the internet using e-mail, message boards and websites like Facebook and Twitter. These forms of contact are highly integrated in everyday life. With the proliferation of reviews, ratings, recommendations and other forms of online expression, online opinion has turned into a kind of virtual currency for businesses willing to market their products, identify new opportunities and manage their reputations. Despite the considerable amount of research, the classification of polarity is still a challenging task; mostly since it involves a deep understanding of explicit and implicit information conveyed by language structures. Henceforth, irony or sarcasm has become an important topical issue in NLP, mostly because irony (or sarcasm) flips the polarity of the message. Hence, this paper investigated how irony affects tools of sentiment analysis. The classifications were conducted with Support Vector Machine and Naïve Bayes Classifier. The results and conclusions of the experiments raise remarks and new questions. A first remark to be made is that all experiments are performed with English texts. Consequently, the result cannot directly be generalized to other languages. We believe that the results in other languages are different are likely, for instance, to have more structured languages such as Brazilian Portuguese. Another interesting observation is whether very similar results are obtained when the experiments are carried out on data from a different method. From statistic point of view, there are no relevant differences between methodologies, for example, the total accuracy ranges between 78 and 79% and kappa from 64 to 66%, in spite of the inherently ambiguous nature of irony (or sarcasm) that makes hard to be analyzed, not just automatically but often for humans. Our work indicate that the NBM and SVM were reasonably able to detect irony in twitter messages. Bearing in mind that our research deals with only one type of irony that is common in tweets. The study provides us an initial understanding of the how irony affects the polarity detection. To better understand the phenomenon, it is essential to apply different methods, such as polarity given by SentiWordNet, based on Lexicon. In this sense, for future work, we aim to explore the use Lexicon-based tools and thus measure and compare the obtained results.

References

1. Pang, B., Lee, L.: Opinion Mining and Sentiment Analysis. Found. Trends Inf. Retr. **2**, 1–135 (2008)
2. Platt, J.C.: Fast training of support vector machines using sequential minimal optimization. In: Advances in Kernel Methods, pp. 185–208. MIT Press (1999)
3. Weitzel, L., Aguiar, R.F., Rodriguez, W.F.G., Heringer, M.G.: How do medical authorities express their sentiment in twitter messages? In: 2014 9th Iberian Conference on Information Systems and Technologies (CISTI), pp. 1–6 (2014)
4. Miller, G.A.: WordNet: a lexical database for English. Commun. ACM **38**, 39–41 (1995)
5. Weitzel, L., Quaresma, P., Oliveira, J.P.M.D.: Measuring node importance on twitter microblogging. In: Proceedings of the 2nd International Conference on Web Intelligence, Mining and Semantics, pp. 1–7. ACM, Craiova (2012)
6. Chang, C.-C., Lin, C.-J.: LIBSVM: A library for support vector machines. ACM Trans. Intell. Syst. Technol. **2**, 1–27 (2011)
7. Hall, M., Frank, E., Holmes, G., Pfahringer, B., Reutemann, P., Witten, I.H.: The WEKA data mining software: an update. SIGKDD Explor. Newsl. **11**, 10–18 (2009)

Author Index

Printed in the United States
by bookmasters

Printed in the United States
By Bookmasters

Lecture Notes in Computer Science 10355

Commenced Publication in 1973
Founding and Former Series Editors:
Gerhard Goos, Juris Hartmanis, and Jan van Leeuwen

More information about this series at http://www.springer.com/series/7410

Dieter Gollmann · Atsuko Miyaji
Hiroaki Kikuchi (Eds.)

Applied Cryptography and Network Security

15th International Conference, ACNS 2017
Kanazawa, Japan, July 10–12, 2017
Proceedings

 Springer

Editors
Dieter Gollmann
Hamburg University of Technology
Hamburg
Germany

Atsuko Miyaji
Graduate School of Engineering
Osaka University
Suita, Osaka
Japan

Hiroaki Kikuchi
Department of Frontier Media Science
Meiji University
Tokyo
Japan

ISSN 0302-9743   ISSN 1611-3349 (electronic)
Lecture Notes in Computer Science
ISBN 978-3-319-61203-4   ISBN 978-3-319-61204-1 (eBook)
DOI 10.1007/978-3-319-61204-1

Library of Congress Control Number: 2017944358

LNCS Sublibrary: SL4 – Security and Cryptology

Printed on acid-free paper

This Springer imprint is published by Springer Nature
The registered company is Springer International Publishing AG
The registered company address is: Gewerbestrasse 11, 6330 Cham, Switzerland

Preface

The 15th International Conference on Applied Cryptography and Network Security (ACNS2017) was held in Kanazawa, Japan, during July 10–12, 2017. The previous conferences in the ACNS series were successfully held in Kunming, China (2003), Yellow Mountain, China (2004), New York, USA (2005), Singapore (2006), Zhuhai, China (2007), New York, USA (2008), Paris, France (2009), Beijing, China (2010), Malaga, Spain (2011), Singapore (2012), Banff, Canada (2013), Lausanne, Switzerland (2014), New York, USA (2015), and London, UK (2016).

ACNS is an annual conference focusing on innovative research and current developments that advance the areas of applied cryptography, cyber security, and privacy. Academic research with high relevance to real-world problems as well as developments in industrial and technical frontiers fall within the scope of the conference.

This year we have received 149 submissions from 34 different countries. Each submission was reviewed by 3.7 Program Committee members on average. Papers submitted by Program Committee members received on average 4.4 reviews. The committee decided to accept 34 regular papers. The broad range of areas covered by the high-quality papers accepted for ACNS 2107 attests very much to the fulfillment of the conference goals.

The program included two invited talks given by Dr. Karthikeyan Bhargavan (Inria Paris) and Prof. Doug Tygar (UC Berkeley).

The decisions of the best student paper award was based on a vote among the Program Committee members. To be eligible for selection, the primary author of the paper has to be a full-time student who is present at the conference. The winner was Carlos Aguilar-Melchor, Martin Albrecht, and Thomas Ricosset from Université de Toulouse, Toulouse, France, Royal Holloway, University of London, UK, and Thales Communications & Security, Gennevilliers, France. The title of the paper is "Sampling From Arbitrary Centered Discrete Gaussians For Lattice-Based Cryptography."

We are very grateful to our supporters and sponsors. The conference was co-organized by Osaka University, Japan Advanced Institute of Science and Technology (JAIST), and the Information-technology Promotion Agency (IPA); it was supported by the Committee on Information and Communication System Security (ICSS), IEICE, Japan, the Technical Committee on Information Security (ISEC), IEICE, Japan, and the Special Interest Group on Computer SECurity (CSEC) of IPSJ, Japan; it and was co-sponsored by the National Institute of Information and Communications Technology (NICT) International Exchange Program, Mitsubishi Electric Corporation, Support Center for Advanced Telecommunications Technology Research (SCAT), Foundation Microsoft Corporation, Fujitsu Hokuriku Systems Limited, Nippon Telegraph and Telephone Corporation (NTT), and Hokuriku Telecommunication Network Co., Inc.

We would like to thank the authors for submitting their papers to the conference. The selection of the papers was a challenging and dedicated task, and we are deeply grateful to the 48 Program Committee members and the external reviewers for their reviews and discussions. We also would like to thank EasyChair for providing a user-friendly interface for us to manage all submissions and proceedings files. Finally, we would like to thank the general chair, Prof. Hiroaki Kikuchi, and the members of the local Organizing Committee.

July 2017   Dieter Gollmann
                                                                  Atsuko Miyaji

ACNS 2017

The 15th International Conference on Applied Cryptography and Network Security

Jointly organized by

Osaka University
and
Japan Advanced Institute of Science and Technology (JAIST)
and
Information-technology Promotion Agency (IPA)

General Chair

Hiroaki Kikuchi   Meiji University, Japan

Program Co-chairs

Dieter Gollmann   Hamburg University of Technology, Germany
Atsuko Miyaji   Osaka University / JAIST, Japan

Program Committee

Diego Aranha   University of Campinas, Brazil
Giuseppe Ateniese   Stevens Institute of Technology, USA
Man Ho Au   Hong Kong Polytechnic University, Hong Kong, SAR China
Carsten Baum   Bar-Ilan University, Israel
Rishiraj Bhattacharyya   NISER Bhubaneswar, India
Liqun Chen   University of Surrey, UK
Chen-Mou Chen   Osaka University, Japan
Céline Chevalier   Université Panthéon-Assas, France
Sherman S.M. Chow   Chinese University of Hong Kong, Hong Kong, SAR China
Mauro Conti   University of Padua, Italy
Alexandra Dmitrienko   ETH Zurich, Switzerland
Michael Franz   University of California, Irvine, USA
Georg Fuchsbauer   ENS, France
Sebastian Gajek   FUAS, Germany
Goichiro Hanaoka   AIST, Japan
Feng Hao   Newcastle University, UK

Swee-Huay Heng	Multimedia University, Malaysia
Francisco Rodrguez Henrquez	CINVESTAV-IPN, Mexico
Xinyi Huang	Fujian Normal University, China
Michael Huth	Imperial College London, UK
Tibor Jager	Paderborn University, Germany
Aniket Kate	Purdue University, USA
Stefan Katzenbeisser	TU Darmstadt, Germany
Kwangjo Kim	KAIST, Korea
Kwok-yan Lam	NTU, Singapore
Mark Manulis	University of Surrey, UK
Tarik Moataz	Brown University, USA
Ivan Martinovic	University of Oxford, UK
Jörn Müller-Quade	Karlsruhe Institute of Technology, Germany
David Naccache	École normale supérieure, France
Michael Naehrig	Microsoft Research Redmond, USA
Hamed Okhravi	MIT Lincoln Laboratory, USA
Panos Papadimitratos	KTH Royal Institute of Technology, Sweden
Jong Hwan Park	Sangmyung University, Korea
Thomas Peyrin	Nanyang Technological University, Singapore
Bertram Poettering	Ruhr-Universität Bochum, Germany
Christina Pöpper	NYU, United Arab Emirates
Bart Preneel	KU Leuven, Belgium
Thomas Schneider	TU Darmstadt, Germany
Michael Scott	Dublin City University, Ireland
Vanessa Teague	University of Melbourne, Australia
Somitra Kr. Sanadhya	Ashoka University, India
Mehdi Tibouchi	NTT Secure Platform Laboratories, Japan
Ivan Visconti	University of Salerno, Italy
Bo-Yin Yang	Academia Sinica, Taiwan
Kan Yasuda	NTT Secure Platform Laboratories, Japan
Fangguo Zhang	Sun Yat-sen University, China
Jianying Zhou	SUTD, Singapore

Organizing Committee

Local Arrangements

Akinori Kawachi	Tokushima University, Japan

Co-chairs

Kazumasa Omote	University of Tsukuba, Japan
Shoichi Hirose	University of Fukui, Japan
Kenji Yasunaga	Kanazawa University, Japan
Yuji Suga	IIJ, Japan

Finance Co-chairs

Masaki Fujikawa	Kogakuin University, Japan
Yuichi Futa	JAIST, Japan
Natsume Matsuzaki	University of Nagasaki, Japan
Takumi Yamamoto	Mitsubishi Electric, Japan

Publicity Co-chairs

Noritaka Inagaki	IPA, Japan
Masaki Hashimoto	IISEC, Japan
Naoto Yanai	Osaka University, Japan
Kaitai Liang	Manchester Metropolitan University, UK

Liaison Co-chairs

Keita Emura	NICT, Japan
Eiji Takimoto	Ritsumeikan University, Japan
Toru Nakamura	KDDI Research, Japan

System Co-chairs

Atsuo Inomata	Tokyo Denki University/NAIST, Japan
Masaaki Shirase	Future University Hakodate, Japan
Minoru Kuribayashi	Okayama University, Japan
Toshihiro Yamauchi	Okayama University, Japan
Shinya Okumura	Osaka University, Japan

Publication Co-chairs

Takeshi Okamoto	Tsukuba University of Technology, Japan
Takashi Nishide	University of Tsukuba, Japan
Ryo Kikuchi	NTT, Japan
Satoru Tanaka	JAIST, Japan

Registration Co-chairs

Hideyuki Miyake	Toshiba, Japan
Dai Watanabe	Hitachi, Japan
Chunhua Su	Osaka University, Japan

Additional Reviewers

Alesiani, Francesco	Ashur, Tomer
Aminanto, Muhamad Erza	Auerbach, Benedikt
Andaló, Fernanda	Azad, Muhammad Ajmal
Armknecht, Frederik	Bai, Shi

Barrera, David
Bauer, Balthazar
Beierle, Christof
Beunardeau, Marc
Blazy, Olivier
Bost, Raphael
Bourse, Florian
Broadnax, Brandon
Chakraborti, Avik
Chi-Domínguez, Jesús Javier
Chin, Ji-Jian
Choi, Rakyong
Choi, Suri
Ciampi, Michele
Connolly, Aisling
Coon, Ralph A.C.
Costello, Craig
Couteau, Geoffroy
Crane, Stephen
Culnane, Chris
Dargahi, Tooska
Datta, Nilanjan
Davies, Gareth T.
Del Pino, Rafael
Demmler, Daniel
Dirksen, Alexandra
Dominguez Perez, Luis J.
Dong, Xinshu
Dowling, Benjamin
Eom, Jieun
Faust, Sebastian
Ferradi, Houda
Frederiksen, Tore
Gay, Romain
Geraud, Remi
Germouty, Paul
Gochhayat, Sarada Prasad
Hartung, Gunnar
Herzberg, Amir
Huang, Yi
Iovino, Vincenzo
Jap, Dirmanto
Jati, Arpan
Jiang, Jiaojiao
Kairallah, Mustafa
Kamath, Chethan

Karvelas, Nikolaos
Keller, Marcel
Kim, Hyoseung
Kim, Jonghyun
Kim, Joonsik
Kim, Taechan
Kiss, Ágnes
Kitagawa, Fuyuki
Kohls, Katharina
Kuo, Po-Chun
Kurek, Rafael
Lai, Junzuo
Lai, Russell W.F.
Lain, Daniele
Lal, Chhagan
Lee, Kwangsu
Lee, Youngkyung
Li, Huige
Li, Wen-Ding
Li, Yan
Liebchen, Christopher
Liu, Jianghua
Liu, Yunwen
Longa, Patrick
Lu, Jingyang
Lu, Jiqiang
Luykx, Atul
Lyubashevsky, Vadim
Ma, Jack P.K.
Mainka, Christian
Mancillas-López, Cuauhtemoc
Masucci, Barbara
Matsuda, Takahiro
Mazaheri, Sogol
Mechler, Jeremias
Meier, Willi
Meng, Weizhi
Mohamad, Moesfa Soeheila
Moonsamy, Veelasha
Nagel, Matthias
Nielsen, Michael
Nishimaki, Ryo
O'Neill, Adam
Ochoa-Jiménez, José Eduardo
Oliveira, Thomaz
Peeters, Roel

Pereira, Hilder Vitor Lima
Perrin, Léo
Poh, Geong Sen
Puddu, Ivan
Ramanna, Somindu C.
Ramchen, Kim
Renes, Joost
Reparaz, Oscar
Resende, Amanda
Rill, Jochen
Roy, Arnab
Ruffing, Tim
Rupp, Andy
Sakai, Yusuke
Sasaki, Yu
Schuldt, Jacob
Sen Gupta, Sourav
Seo, Hwajeong
Seo, Minhye
Shahandashti, Siamak
Shin, Seonghan
Siniscalchi, Luisa
Spolaor, Riccardo
Stebila, Douglas
Su, Chunhua
Tai, Raymond K.H.

Tan, Syhyuan
Thillard, Adrian
Tosh, Deepak
Vannet, Thomas
Vergnaud, Damien
Volckaert, Stijn
Wang, Ding
Wang, Jiafan
Wang, Xiuhua
Weinert, Christian
Wong, Harry W.H.
Xagawa, Keita
Xie, Shaohao
Yamada, Shota
Yamakawa, Takashi
Yang, Rupeng
Yang, Shaojun
Yang, Xu
Yu, Zuoxia
Zaverucha, Greg
Zhang, Huang
Zhang, Tao
Zhang, Yuexin
Zhang, Zheng
Zhao, Yongjun
Zhou, Peng

Contents

Cryptographic Primitives

Side Channel Attack

Applied Cryptography

Sampling from Arbitrary Centered Discrete Gaussians for Lattice-Based Cryptography

Carlos Aguilar-Melchor[1], Martin R. Albrecht[2], and Thomas Ricosset[1,3(✉)]

[1] INP ENSEEIHT, IRIT-CNRS, Université de Toulouse, Toulouse, France
{carlos.aguilar,thomas.ricosset}@enseeiht.fr
[2] Information Security Group, Royal Holloway, University of London, London, UK
martin.albrecht@royalholloway.ac.uk
[3] Thales Communications & Security, Gennevilliers, France

Abstract. Non-Centered Discrete Gaussian sampling is a fundamental building block in many lattice-based constructions in cryptography, such as signature and identity-based encryption schemes. On the one hand, the center-dependent approaches, e.g. cumulative distribution tables (CDT), Knuth-Yao, the alias method, discrete Zigurat and their variants, are the fastest known algorithms to sample from a discrete Gaussian distribution. However, they use a relatively large precomputed table for each possible real center in $[0, 1)$ making them impracticable for non-centered discrete Gaussian sampling. On the other hand, rejection sampling allows to sample from a discrete Gaussian distribution for all real centers without prohibitive precomputation cost but needs costly floating-point arithmetic and several trials per sample. In this work, we study how to reduce the number of centers for which we have to precompute tables and propose a non-centered CDT algorithm with practicable size of precomputed tables as fast as its centered variant. Finally, we provide some experimental results for our open-source C++ implementation indicating that our sampler increases the rate of Peikert's algorithm for sampling from arbitrary lattices (and cosets) by a factor 3 with precomputation storage up to 6.2 MB.

1 Introduction

Lattice-based cryptography has generated considerable interest in the last decade due many attractive features, including conjectured security against quantum attacks, strong security guarantees from worst-case hardness and constructions of fully homomorphic encryption (FHE) schemes (see the survey [33]). Moreover, lattice-based cryptographic schemes are often algorithmically simple and efficient, manipulating essentially vectors and matrices or polynomials modulo relatively small integers, and in some cases outperform traditional systems.

M.R. Albrecht—The research of this author was supported by EPSRC grant "Bit Security of Learning with Errors for Post-Quantum Cryptography and Fully Homomorphic Encryption" (EP/P009417/1) and the EPSRC grant "Multilinear Maps in Cryptography" (EP/L018543/1).

D. Gollmann et al. (Eds.): ACNS 2017, LNCS 10355, pp. 3–19, 2017.
DOI: 10.1007/978-3-319-61204-1_1

Modern lattice-based cryptosystems are built upon two main average-case problems over general lattices: Short Integer Solution (SIS) [1] and Learning With Errors (LWE) [35], and their analogues over ideal lattices, ring-SIS [29] and ring-LWE [27]. The hardness of these problems can be related to the one of their worst-case counterpart, if the instances follow specific distributions and parameters are choosen appropriately [1, 27, 29, 35].

In particular, discrete Gaussian distributions play a central role in lattice-based cryptography. A natural set of examples to illustrate the importance of Gaussian sampling are lattice-based signature and identity-based encryption (IBE) schemes [16]. The most iconic example is the signature algorithm proposed in [16] (hereafter GPV), as a secure alternative to the well-known (and broken) GGH signature scheme [18]. In this paper, the authors use the Klein/GPV algorithm [21], a randomized variant of Babai's nearest plane algorithm [4]. In this algorithm, the rounding step is replaced by randomized rounding according to a discrete Gaussian distribution to return a lattice point (almost) independent of a hidden basis. The GPV signature scheme has also been combined with LWE to obtain the first identity-based encryption (IBE) scheme [16] conjectured to be secure against quantum attacks. Later, a new Gaussian sampling algorithm for arbitrary lattices was presented in [32]. It is a randomized variant of Babai's rounding-off algorithm, is more efficient and parallelizable, but it outputs longer vectors than Klein/GPV's algorithm.

Alternatively to the above trapdoor technique, lattice-based signatures [11, 23–26] were also constructed by applying the Fiat-Shamir heuristic [14]. Note that in contrast to the algorithms outlined above which sample from a discrete Gaussian distribution for any real center not known in advance, the schemes developed in [11, 25] only need to sample from a discrete Gaussian centered at zero.

1.1   Our Contributions

We develop techniques to speed-up discrete Gaussian sampling when the center is not known in advance, obtaining a flexible time-memory trade-off comparing favorably to rejection sampling. We start with the cumulative distribution table (CDT) suggested in [32] and lower the computational cost of the precomputation phase and the global memory required when sampling from a non-centered discrete Gaussian by precomputing the CDT for a relatively small number of centers, in $\mathcal{O}(\lambda^3)$, and by computing the cdf when needed, i.e. when for a given uniform random input, the values returned by the CDTs for the two closest precomputed centers differ. Second, we present an adaptation of the lazy technique described in [12] to compute most of the cdf in double IEEE standard double precision, thus decreasing the number of precomputed CDTs. Finally, we propose a more flexible approach which takes advantage of the information already present in the precomputed CDTs. For this we use a Taylor expansion around the precomputed centers and values instead of this lazy technique, thus enabling to reduce the number of precomputed CDTs to a $\omega(\lambda)$.

We stress, though, that our construction is not constant time, which limits its utility. We consider addressing this issue important future work.

1.2   Related Work

Many discrete Gaussian samplers over the Integers have been proposed for lattice-based cryptography. Rejection Sampling [12,17], Inversion Sampling with a Cumulative Distribution Table (CDT) [32], Knuth-Yao [13], Discrete Ziggurat [7], Bernoulli Sampling [11], Kahn-Karney [20] and Binary Arithmetic Coding [36].

The optimal method will of course depend on the setting in which it is used. In this work, we focus on what can be done on a modern computer, with a comfortable amount of memery and hardwired integer and floating-point operations. This is in contrast to the works [11,13] which focus on circuits or embedded devices. We consider exploring the limits of the usual memory and hardwired operations in commodity hardware as much an interesting question as it is to consider what is feasible in more constrained settings.

Rejection Sampling and Variants. Straightforward rejection sampling [37] is a classical method to sample from any distribution by sampling from a uniform distribution and accept the value with a probability equal to its probability in the target distribution. This method does not use pre-computed data but needs floating-point arithmetic and several trials by sample. Bernoulli sampling [11] introduces an exponential bias from Bernoulli variables, which can be efficiently sampled specially in circuits. The bias is then corrected in a rejection phase based on another Bernouilli variable. This approach is particularly suited for embedded devices for the simplicity of the computation and the near-optimal entropy consumption. Kahn-Karney sampling is another variant of rejection sampling to sample from a discrete Gaussian distribution which does not use floating-point arithmetic. It is based on the von Neumann algorithm to sample from the exponential distribution [31], requires no precomputed tables and consumes a smaller amount of random bits than Bernoulli sampling, though it is slower. Currently the fastest approach in the computer setting uses a straightforward rejection sampling approach with "lazy" floating-point computations [12] using IEEE standard double precision floating-point numbers in most cases.

Note that none of these methods requires precomputation depending on the distribution's center c. In all the alternative approaches we present hereafter, there is some center-dependent precomputation. When the center is not know this can result in prohibitive costs and handling these becomes a major issue around which most of our work is focused.

Center-Dependent Approaches. The cumulative distribution table algorithm is based on the inversion method [9]. All non-negligible cumulative probabilities are stored in a table and at sampling time one generates a cumulative probability in $[0, 1)$ uniformly at random, performs a binary search through the table and returns the corresponding value. Several alternatives to straightforward CDT are possible. Of special interest are: the alias method [38] which encodes CDTs in a more involved but more efficient approach; BAC Sampling [36] which uses arithmetic coding tables to sample with an optimal consumption of random bits; and Discrete Ziggurat [7] which adapts the Ziggurat method [28] for a flexible time-memory trade-off. Knuth-Yao sampling [22] uses a random bit generator to traverse a binary tree formed from the bit representation of the probability of each possible sample, the terminal node is labeled by the corresponding sample. The main advantage of this method is that it consumes a near-optimal amount of random bits. A block variant and other practical improvements are suggested in [13]. This method is center-dependent but clearly designed for circuits and on a computer setting it is surpassed by other approaches.

Our main contribution is to show how to get rid of the known-center constraint with reasonable memory usage for center-dependent approaches. As a consequence, we obtain a performance gain with respect to rejection sampling approaches. Alternatively, any of the methods discussed above could have replaced our straightforward CDT approach. This, however, would have made our algorithms, proofs, and implementations more involved. On the other hand, further performance improvements could perhaps be achieved this way. This is an interesting problem for future work.

2   Preliminaries

Throughout this work, we denote the set of real numbers by $\mathbb{R}$ and the Integers by $\mathbb{Z}$. We extend any real function $f(\cdot)$ to a countable set A by defining $f(A) = \sum_{x \in A} f(x)$. We denote also by U_I the uniform distribution on I.

2.1   Discrete Gaussian Distributions on $\mathbb{Z}$

The discrete Gaussian distribution on $\mathbb{Z}$ is defined as the probability distribution whose unnormalized density function is

$$\rho : \mathbb{Z} \to [0, 1)$$
$$x \to e^{\frac{-x^2}{2}}$$

If $s \in \mathbb{R}^+$ and $c \in \mathbb{R}$, then we extend this definition to

$$\rho_{s,c}(x) := \rho\left(\frac{x - c}{s}\right)$$

and denote $\rho_{s,0}(x)$ by $\rho_s(x)$. For any mean $c \in \mathbb{R}$ and parameter $s \in \mathbb{R}^+$ we can now define the discrete Gaussian distribution $D_{s,c}$ as

$$\forall x \in \mathbb{Z}, \ D_{s,c}(x) := \frac{\rho_{s,c}(x)}{\rho_{s,c}(\mathbb{Z})}$$

Note that the standard deviation of this distribution is $\sigma = s/\sqrt{2\pi}$. We also define $\text{cdf}_{s,c}$ as the cumulative distribution function (cdf) of $D_{s,c}$

$$\forall x \in \mathbb{Z}, \ \text{cdf}_{s,c}(x) := \sum_{i=-\infty}^{x} D_{s,c}(i)$$

Smoothing Parameter. The smoothing parameter $\eta_\epsilon(\Lambda)$ quantifies the minimal discrete Gaussian parameter s required to obtain a given level of smoothness on the lattice Λ. Intuitively, if one picks a noise vector over a lattice from a discrete Gaussian distribution with radius at least as large as the smoothing parameter, and reduces this modulo the fundamental parallelepiped of the lattice, then the resulting distribution is very close to uniform (for details and formal definition see [30]).

Gaussian Measure. An interesting property of discrete Gaussian distributions with a parameter s greater than the smoothing parameter is that the Gaussian measure, i.e. $\rho_{s,c}(\mathbb{Z})$ for $D_{s,c}$, is essentially the same for all centers.

Lemma 1 (From the proof of [30, Lemma 4.4]). *For any $\epsilon \in (0,1)$, $s > \eta_\epsilon(\mathbb{Z})$ and $c \in \mathbb{R}$ we have*

$$\Delta_{measure} := \frac{\rho_{s,c}(\mathbb{Z})}{\rho_{s,0}(\mathbb{Z})} \in \left[\frac{1-\epsilon}{1+\epsilon}, 1\right]$$

Tailcut Parameter. To deal with the infinite domain of Gaussian distributions, algorithms usually take advantage of their rapid decay to sample from a finite domain. The next lemma is useful in determining the tailcut parameter τ.

Lemma 2 ([17, Lemma 4.2]). *For any $\epsilon > 0$, $s > \eta_\epsilon(\mathbb{Z})$ and $\tau > 0$, we have*

$$E_{tailcut} := \Pr_{X \sim D_{\mathbb{Z},s,c}}[|X - c| > \tau s] < 2e^{-\pi\tau^2} \cdot \frac{1+\epsilon}{1-\epsilon}$$

2.2 Floating-Point Arithmetic

We recall some facts from [12] about floating-point arithmetic (FPA) with m bits of mantissa, which we denote by $\mathbb{FP}_m$. A floating-point number is a triplet $\bar{x} = (s, e, v)$ where $s \in \{0, 1\}$, $e \in \mathbb{Z}$ and $v \in \mathbb{N}_{2^m-1}$ which represents the real number $\bar{x} = (-1)^s \cdot 2^{e-m} \cdot v$. Denote by $\epsilon = 2^{1-m}$ the floating-point precision. Every FPA-operation $\bar{\circ} \in \{\bar{+}, \bar{-}, \bar{\times}, \bar{/}\}$ and its respective arithmetic operation on $\mathbb{R}$, $\circ \in \{+, -, \times, /\}$ verify

$$\forall \bar{x}, \bar{y} \in \mathbb{FP}_m, |(\bar{x} \, \bar{\circ} \, \bar{y}) - (\bar{x} \circ \bar{y})| \leq (x \circ y)\epsilon$$

Moreover, we assume that the floating-point implementation of the exponential function $\bar{\exp}(\cdot)$ verifies

$$\forall \bar{x} \in \mathbb{FP}_m, |\bar{\exp}(\bar{x}) - \exp(\bar{x})| \leq \epsilon.$$

2.3 Taylor Expansion

Taylor's theorem provides a polynomial approximation around a given point for any function sufficiently differentiable.

Theorem 1 (Taylor's theorem). *Let $d \in \mathbb{Z}^+$ and let the function $f : \mathbb{R} \to \mathbb{R}$ be d times differentiable in some neighborhood U of $a \in \mathbb{R}$. Then for any $x \in U$*

$$f(x) = \mathcal{T}_{d,f,a}(x) + \mathcal{R}_{d,f,a}(x)$$

where

$$\mathcal{T}_{d,f,a}(x) = \sum_{i=0}^{d} \frac{f^{(i)}(a)}{i!}(x-a)^i$$

and

$$\mathcal{R}_{d,f,a}(x) = \int_a^x \frac{f^{(d+1)}(t)}{d!}(x-t)^d dt$$

3   Variable-Center with Polynomial Number of CDTs

We consider the case in which the mean is variable, i.e. the center is not know before the online phase, as it is the case for lattice-based hash-and-sign signatures. The center can be any real number, but without loss of generality we will only consider centers in $[0, 1)$. Because CDTs are center-dependent, a first naive option would be to precompute a CDT for each possible real center in $[0, 1)$ in accordance with the desired accuracy. Obviously, this first option has the same time complexity than the classical CDT algorithm, i.e. $\mathcal{O}(\lambda \log s\lambda)$ for λ the security parameter. However, it is completely impractical with 2^λ precomputed CDTs of size $\mathcal{O}(s\lambda^{1.5})$. An opposite trade-off is to compute the CDT on-the-fly, avoiding any precomputation storage, which increase the computational cost to $\mathcal{O}(s\lambda^{3.5})$ assuming that the computation of the exponential function run in $\mathcal{O}(\lambda^3)$ (see Sect. 3.2 for a justification of this assumption).

An interesting question is can we keep the time complexity of the classical CDT algorithm with a polynomial number of precomputed CDTs. To answer this question, we start by fixing the number n of equally spaced centers in $[0, 1)$ and precompute the CDTs for each of these. Then, we apply the CDT algorithm to the two precomputed centers closest to the desired center for the same cumulative probability uniformly draw. Assuming that the number of precomputed CDTs is sufficient, the values returned from both CDTs will be equal most of the time, in this case we can conclude, thanks to a simple monotonic argument, that the returned value would have been the same for the CDT at the desired center and return it as a valid sample. Otherwise, the largest value will immediately follow the smallest and we will then have to compute the cdf at the smallest value for the desired center in order to know if the cumulative probability is lower or higher than this cdf. If it is lower then the smaller value will be returned as sample, else it will be the largest.

3.1   Twin-CDT Algorithm

As discussed above, to decrease the memory required by the CDT algorithm when the distribution center is determined during the online phase, we can precompute CDTs for a number n of centers equally spaced in $[0, 1)$ and compute the cdf when necessary. Algorithm 1 resp. 2 describes the offline resp. online phase of the *Twin-CDT* algorithm. Algorithm 1 precomputes CDTs, up to a precision m that guarantees the λ most significant bits of each cdf, and store them with λ-bits of precision as a matrix $\mathbf{T}$, where the i-th line is the CDT corresponding to the i-th precomputed center i/n. To sample from $D_{s,c}$, Algorithm 2 searches the preimages by the cdf of a cumulative probability p, draw from the uniform distribution on $[0,1) \cap \mathbb{FP}_\lambda$, in both CDTs corresponding to the center $\lfloor n(c - \lfloor c \rfloor) \rfloor / n$ (respectively $\lceil n(c - \lfloor c \rfloor) \rceil / n$) which return a value v_1 (resp. v_2). If the same value is returned from the both CDTs (i.e. $v_1 = v_2$), then this value added the desired center integer part is a valid sample, else it computes $\mathrm{cdf}_{s,c-\lfloor c \rfloor}(v_1)$ and returns $v_1 + \lfloor c \rfloor$ if $p < \mathrm{cdf}_{s,c}(v_1)$ and $v_2 + \lfloor c \rfloor$ else.

Algorithm 1. Twin-CDT Algorithm: Offline Phase

Input: a Gaussian parameter s and a number of centers n
Output: a precomputed matrix $\mathbf{T}$
1: initialize an empty matrix $\mathbf{T} \in \mathbb{FP}_\lambda^{n \times 2\lceil \tau s \rceil + 3}$
2: **for** $i \leftarrow 0, \ldots, n-1$ **do**
3:     **for** $j \leftarrow 0, \ldots, 2\lceil \tau s \rceil + 2$ **do**
4:         $\mathbf{T}_{i,j} \leftarrow \mathbb{FP}_m : \mathrm{cdf}_{s,i/n}(j - \lceil \tau s \rceil - 1)$

Algorithm 2. Twin-CDT Algorithm: Online Phase

Input: a center c and a precomputed matrix $\mathbf{T}$
Output: a sample x that follows $D_{s,c}$
1: $p \leftarrow U_{[0,1) \cap \mathbb{FP}_\lambda}$
2: $v_1 \leftarrow i - \lceil \tau s \rceil - 1$ s.t. $\mathbf{T}_{\lfloor n(c-\lfloor c \rfloor) \rfloor, i-1} \leq p < \mathbf{T}_{\lfloor n(c-\lfloor c \rfloor) \rfloor, i}$
3: $v_2 \leftarrow j - \lceil \tau s \rceil - 1$ s.t. $\mathbf{T}_{\lceil n(c-\lfloor c \rfloor) \rceil, j-1} \leq p < \mathbf{T}_{\lceil n(c-\lfloor c \rfloor) \rceil, j}$
4: **if** $v_1 = v_2$ **then**
5:     **return** $v_1 + \lfloor c \rfloor$
6: **else**
7:     **if** $p < \mathbb{FP}_m : \mathrm{cdf}_{s,c-\lfloor c \rfloor}(v_1)$ **then**
8:         **return** $v_1 + \lfloor c \rfloor$
9:     **else**
10:         **return** $v_2 + \lfloor c \rfloor$

Correctness. We establish correctness in the lemma below.

Lemma 3. *Assuming that m is large enough to ensure λ correct bits during the cdf computation, the statistical distance between the output distribution of Algorithm 2 instantiated to sample from $D_{\mathbb{Z}^m, \sigma, c}$ and $D_{\mathbb{Z}^m, \sigma, c}$ is bounded by $2^{-\lambda}$.*

Proof. First note that from the discrete nature of the considered distribution we have $D_{s,c} = D_{s,c-\lfloor c \rfloor} + \lfloor c \rfloor$. Now recall that the probability integral transform states that if X is a continuous random variable with cumulative distribution function cdf, then $\mathrm{cdf}(X)$ has a uniform distribution on $[0, 1]$. Hence the inversion method: $\mathrm{cdf}^{-1}(U_{[0,1]})$ has the same distribution as X. Finally by noting that for all $s, p \in \mathbb{R}$, $\mathrm{cdf}_{s,c}(p)$ is monotonic in c, if $\mathrm{cdf}_{s,c_1}^{-1}(p) = \mathrm{cdf}_{s,c_2}^{-1}(p) := v$, then $\mathrm{cdf}_{s,c}^{-1}(p) = v$ for all $c \in [c_1, c_2]$, and as a consequence, for all $v \in [-\lceil \tau s \rceil - 1, \lceil \tau s \rceil + 1]$, the probability of outputting v is equal to $\mathbb{FP}_m : \mathrm{cdf}_{s,c}(v) - \mathbb{FP}_m : \mathrm{cdf}_{s,c}(v-1)$ which is $2^{-\lambda}$-close to $D_{s,c}(v)$. $\qquad\square$

The remaining issue in the correctness analysis of Algorithm 2 is to determine the error occurring during the m-precision cdf computation. Indeed, this error allows us to learn what precision m is needed to correctly compute the λ most significant bits of the cdf. This error is characterized in Lemma 4.

Lemma 4. *Let $m \in \mathbb{Z}$ be a positive integer and $\varepsilon = 2^{1-m}$. Let $\bar{c}, \bar{s}, \bar{h} \in \mathbb{FP}_m$ be at distance respectively at most δ_c, δ_c and δ_h from $c, s, h \in \mathbb{R}$ and $h = 1/\rho_{s,c}(\mathbb{Z})$. Let $\Delta f(x) := |\mathbb{FP}_m : f(x) - f(x)|$. We also assume that the following inequalities hold: $s \geq 4$, $\tau \geq 10$, $s\delta_s \leq 0.01$, $\delta_c \leq 0.01$, $s^2\varepsilon \leq 0.01$, $(\tau s + 1)\varepsilon \leq 1/2$. We have the following error bound on $\Delta\mathrm{cdf}_{s,c}(x)$ for any integer x such that $|x| \leq \tau s + 2$*

$$\Delta\mathrm{cdf}_{s,c}(x) \leq 3.5\tau^3 s^2 \varepsilon$$

Proof. We derive the following bounds using [10, Facts 6.12, 6.14, 6.22]:

$$\Delta\mathrm{cdf}_{s,c}(x) \leq \Delta\left[\sum_{i=-\lceil \tau s \rceil - 1}^{\lceil \tau s \rceil + 1} \rho_{s,c}(i)\right]\left(\frac{1}{s} + 3.6s\varepsilon\right) + 3.6s\varepsilon$$

$$\Delta\left[\sum_{i=-\lceil \tau s \rceil - 1}^{\lceil \tau s \rceil + 1} \rho_{s,c}(i)\right] \leq 3.2\tau^3 s^3 \varepsilon$$

$\qquad\square$

For the sake of readability the FPA error bound of Lemma 4 is fully simplified and is therefore not tight. For practical implementation, one can derive a better bound using an ad-hoc approach such as done in [34].

Efficiency. On average, the evaluation of the cdf requires $\lceil \tau s \rceil + 1.5$ evaluations of the exponential function. For the sake of clarity, we assume that the exponential function is computed using a direct power series evaluation with schoolbook multiplication, so its time complexity is $\mathcal{O}(\lambda^3)$. We refer the reader to [6] for a discussion of different ways to compute the exponential function in high-precision.

Lemma 5 establishes that the time complexity of Algorithm 2 is $\mathcal{O}(\lambda \log s\lambda + \lambda^4/n)$, so with $n = \mathcal{O}(\lambda^3)$ it has asymptotically the same computational cost than the classical CDT algorithm.

Lemma 5. *Let P_{cdf} be the probability of computing the cdf during the execution of Algorithm 2, assuming that $\tau s \geq 10$, we have*

$$P_{cdf} \leq 2.2\tau s \left(1 - e^{-\frac{1.25\tau}{sn}}\Delta_{measure}\right)$$

Proof.

$$P_{cdf} \leq \max_{c \in [0,1)} \left(\sum_{i=-\lceil \tau s \rceil - 1}^{\lceil \tau s \rceil + 1} \left| cdf_{s,c}(i) - cdf_{s,c+\frac{1}{n}}(i) \right| \right)$$

Assuming that $\tau s \geq 10$, we have

$$e^{-\frac{1.25\tau}{sn}} \Delta_{measure} \, cdf_{s,c}(i) \leq cdf_{s,c+\frac{1}{n}}(i) \leq cdf_{s,c}(i)$$

Hence the upper bound.   □

On the other hand, the precomputation matrix generated by Algorithm 1 take n times the size of one CDT, hence the space complexity is $\mathcal{O}(ns\lambda^{1.5})$. Note that for n sufficiently big to make the cdf computational cost negligible, the memory space required by this algorithm is about 1 GB for the parameters considered in cryptography and thus prohibitively expensive for practical use.

3.2 Lazy-CDT Algorithm

A first idea to decrease the number of precomputed CDTs is to avoid costly cdf evaluations by using the same lazy trick as in [12] for rejection sampling. Indeed, a careful analysis of Algorithm 2 shows most of the time many of the computed cdf bits are not used. This gives us to a new strategy which consists of computing the bits of $cdf_{s,c}(v_1)$ lazily. When the values corresponding to the generated probability for the two closest centers are different, the *Lazy-CDT* algorithm first only computes the cdf at a precision m' to ensure $k < \lambda$ correct bits. If the comparison is decided with those k bits, it returns the sample. Otherwise, it recomputes the cdf at a precision m to ensure λ correct bits.

Correctness. In addition to the choice of m, discussed in Sect. 3.1, to achieve λ bits of precision, the correctness of Algorithm 3 also requires to know k which is the number of correct bits after the floating-point computation of the cdf with m' bits of mantissa. For this purpose, given m' Lemma 4 provides a theoretical lower bound on k.

Efficiency. As explained in [12] the precision used for floating-point arithmetic has non-negligible impact, because fp-operation become much expensive when the precision goes over the hardware precision. For instance, modern processors typically provide floating-point arithmetic following the double IEEE standard double precision ($m = 53$), but quad-float FPA ($m = 113$) is usually about 10–20 times slower for basic operations, and the overhead is much more for multiprecision FPA. Therefore the maximal hardware precision is a natural choice for m'. However this choice for m' in Algorithm 3 is a strong constraint for cryptographic applications, where the error occurring during the floating-point cdf computation is usually greater than 10 bits, making the time-memory tradeoff of Algorithm 3 inflexible. Note that the probability of triggering high precision in Algorithm 3 given that $v_1 \neq v_2$ is about $2^{q-k}P_{cdf}$, where q is the number of

Algorithm 3. Lazy-CDT Algorithm: Online Phase

Input: a center c and a precomputed matrix $\mathbf{T}$
Output: a sample x that follows $D_{s,c}$

1: $p \leftarrow U_{[0,1) \cap \text{FP}_\lambda}$
2: $v_1 \leftarrow i - \lceil \tau s \rceil - 1$ s.t. $\mathbf{T}_{\lfloor n(c - \lfloor c \rfloor) \rfloor, i-1} \leq p < \mathbf{T}_{\lfloor n(c - \lfloor c \rfloor) \rfloor, i}$
3: $v_2 \leftarrow j - \lceil \tau s \rceil - 1$ s.t. $\mathbf{T}_{\lceil n(c - \lfloor c \rfloor) \rceil, j-1} \leq p < \mathbf{T}_{\lceil n(c - \lfloor c \rfloor) \rceil, j}$
4: **if** $v_1 = v_2$ **then**
5:     **return** $v_1 + \lfloor c \rfloor$
6: **else**
7:     **if** $\text{FP}_k : p < \text{FP}_{m'} : \text{cdf}_{s,c - \lfloor c \rfloor}(v_1)$ **then**
8:         **return** $v_1 + \lfloor c \rfloor$
9:     **else**
10:         **if** $\text{FP}_k : p > \text{FP}_{m'} : \text{cdf}_{s,c - \lfloor c \rfloor}(v_1)$ **then**
11:             **return** $v_2 + \lfloor c \rfloor$
12:         **else**
13:             **if** $p > \text{FP}_m : \text{cdf}_{s,c - \lfloor c \rfloor}(v_1)$ **then**
14:                 **return** $v_1 + \lfloor c \rfloor$
15:             **else**
16:                 **return** $v_2 + \lfloor c \rfloor$

---

common leading bits of $\text{cdf}_{s,\lfloor n(c - \lfloor c \rfloor) \rfloor / n}(v_1)$ and $\text{cdf}_{s,\lceil n(c - \lfloor c \rfloor) \rceil / n}(v_2)$. By using this lazy trick in addition to lookup tables as described in Sect. 5 with parameters considered in cryptography, we achieve a computational cost lower than the classical centered CDT algorithm with a memory requirement in the order of 1 megabyte.

4   A More Flexible Time-Memory Tradeoff

In view of limitations of the lazy approach described above, a natural question is if we can find a better solution to approximate the cdf. The major advantage of this lazy trick is that it does not require additional memory. However, in our context the CDTs are precomputed and rather than approximate the cdf from scratch it would be interesting to reuse the information contained in these precomputations. Consider the cdf as a function of the center and note that each precomputed cdf is zero degree term of the Taylor expansion of the cdf around a precomputed center. Hence, we may approximate the cdf by its Taylor expansions by precomputing some higher degree terms.

At a first glance, this seems to increase the memory requirements of the sampling algorithm, but we will show that this approach allows to drastically reduce the number of precomputed to a $\omega(\lambda)$ centers thanks to a probability which decreases rapidly with the degree of the Taylor expansion. Moreover, this approximation is faster than the cdf lazy computation and it has no strong constraints related to the maximal hardware precision. As a result, we obtain a flexible time-memory tradeoff which reaches, in particular, the same time complexity as the CDT algorithm for centered discrete Gaussians with a practical memory requirements for cryptographic parameters.

4.1 Taylor-CDT Algorithm

Our *Taylor-CDT* algorithm is similar to the *Lazy-CDT* algorithm (Algorithm 3) described above, except that the lazy computation of the cdf is replaced by the Taylor expansion of the cdf, viewed as a function of the Gaussian center, around each precomputed centers for all possible values. The zero-degree term of each of these Taylor expansions is present in the corresponding CDT element $\mathbf{T}_{i,j}$ and the d higher-degree terms are stored as an element $\mathbf{E}_{i,j}$ of another matrix $\mathbf{E}$. As for the other approaches, these precomputations shall be performed at a sufficient precision m to ensure λ correct bits. During the online phase, Algorithm 5 proceed as follow. Draw p from the uniform distribution over $[0,1) \cap \mathbb{FP}_\lambda$ and search p in the CDTs of the two closest precomputed centers to the desired center decimal part. If the two values found are equal, add the desired center integer part to this value and return it as a valid sample. Otherwise, select the closest precomputed center to the desired center decimal part and evaluate, at the desired center decimal part, the Taylor expansion corresponding to this center and the value found in its CDT. If p is smaller or bigger than this evaluation with respect for the error approximation upper bound $\mathsf{E}_{\text{expansion}}$, characterized in Lemma 6, add the desired center integer part to the corresponding value and return it as a valid sample. Otherwise, it is necessary to compute the full cdf to decide which value to return.

Algorithm 4. Taylor-CDT Algorithm: Offline Phase

Input: a Gaussian parameter s, a number of centers n, a Taylor expansion degree d
Output: two precomputed matrices $\mathbf{T}$ and $\mathbf{E}$
1: initialize two empty matrices $\mathbf{T} \in \mathbb{FP}_\lambda^{n \times 2\lceil \tau s \rceil + 3}$ and $\mathbf{E} \in (\mathbb{FP}_\lambda^d)^{n \times 2\lceil \tau s \rceil + 3}$
2: **for** $i \leftarrow 0, \ldots, n-1$ **do**
3:      **for** $j \leftarrow 0, \ldots, 2\lceil \tau s \rceil + 2$ **do**
4:          $\mathbf{T}_{i,j} \leftarrow \mathbb{FP}_m : \mathrm{cdf}_{s,i/n}(j - \lceil \tau s \rceil - 1)$
5:          $\mathbf{E}_{i,j} \leftarrow \mathbb{FP}_m : \mathcal{T}_{d, \mathrm{cdf}_{s,x}(j - \lceil \tau s \rceil - 1), i/n}(x) - \mathbf{T}_{i,j}$

Efficiency. Algorithm 5 performs two binary searches on CDTs in $\mathcal{O}(\lambda \log s\lambda)$, d additions and multiplications on $\mathbb{FP}_m$ in $\mathcal{O}(m^2)$ with probability $\mathsf{P}_{\text{cdf}} \approx 3\lambda/n$ (see Lemma 5) and a cdf computation on $\mathbb{FP}_m$ in $\mathcal{O}(s\lambda^{3.5})$ with probability close to $2^{q+1}\mathsf{P}_{\text{cdf}}\mathsf{E}_{\text{expansion}}$, where q is the number of common leading bits of $\mathrm{cdf}_{s,\lfloor n(c-\lfloor c\rfloor)\rfloor/n}(v_1)$ and $\mathrm{cdf}_{s,\lceil n(c-\lfloor c\rfloor)\rceil/n}(v_2)$ and $\mathsf{E}_{\text{expansion}}$ is the Taylor expansion approximation error bound described in Lemma 6.

Lemma 6. *Let* $\mathsf{E}_{\text{expansion}}$ *be the maximal Euclidean distance between* $\mathrm{cdf}_{s,x}(v)$ *and* $\mathcal{T}_{d,\mathrm{cdf}_{s,x}(v),c}(x)$, *its Taylor expansion around* c, *for all* $v \in [-\lceil \tau s \rceil - 1, \lceil \tau s \rceil + 1]$, $c \in [0,1)$ *and* $x \in [c, c + 1/2n]$, *assuming that* $\tau \geq 2.5$, $s \geq 4$, *we have*

$$\mathsf{E}_{\text{expansion}} < \frac{4\tau^{d+2}}{n^{d+1} s^{\frac{d+1}{2}}}$$

Algorithm 5. Taylor-CDT Algorithm: Online Phase

Input: a center c and two precomputed matrices $\mathbf{T}$ and $\mathbf{E}$
Output: a sample x that follows $D_{s,c}$

1: $p \leftarrow U_{[0,1) \cap \mathbb{FP}_\lambda}$
2: $v_1 \leftarrow i - \lceil \tau s \rceil - 1$ s.t. $\mathbf{T}_{\lfloor n(c - \lfloor c \rfloor) \rfloor, i-1} \le p < \mathbf{T}_{\lfloor n(c - \lfloor c \rfloor) \rfloor, i}$
3: $v_2 \leftarrow j - \lceil \tau s \rceil - 1$ s.t. $\mathbf{T}_{\lceil n(c - \lfloor c \rfloor) \rceil, j-1} \le p < \mathbf{T}_{\lceil n(c - \lfloor c \rfloor) \rceil, j}$
4: **if** $v_1 = v_2$ **then**
5:     **return** $v_1 + \lfloor c \rfloor$
6: **else**
7:     **if** $|c - \lfloor n(c - \lfloor c \rfloor) \rfloor| < |c - \lceil n(c - \lfloor c \rfloor) \rceil|$ **then**
8:         $c' \leftarrow \lfloor n(c - \lfloor c \rfloor) \rfloor$
9:     **else**
10:         $c' \leftarrow \lceil n(c - \lfloor c \rfloor) \rceil$
11:         $i \leftarrow j$
12:     **if** $p < \mathbf{T}_{c',i} + \mathbf{E}_{c',i}(c - \lfloor c \rfloor) - \mathsf{E}_{\text{expansion}}$ **then**
13:         **return** $v_1 + \lfloor c \rfloor$
14:     **else**
15:         **if** $p > \mathbf{T}_{c',i} + \mathbf{E}_{c',i}(c - \lfloor c \rfloor) + \mathsf{E}_{\text{expansion}}$ **then**
16:             **return** $v_2 + \lfloor c \rfloor$
17:         **else**
18:             **if** $p > \mathbb{FP}_m : \text{cdf}_{s,c-\lfloor c \rfloor}(v_1)$ **then**
19:                 **return** $v_1 + \lfloor c \rfloor$
20:             **else**
21:                 **return** $v_2 + \lfloor c \rfloor$

Proof. From Theorem 1 we have

$$\mathsf{E}_{\text{expansion}} = \max_{\substack{c \in [0,1) \\ x \in [c, c+1/2n] \\ v \in [-\lceil \tau s \rceil - 1, \lceil \tau s \rceil + 1]}} \left(\sum_{i=-\lceil \tau s \rceil - 1}^{v} \int_c^x \frac{\rho_{s,t}^{(d+1)}(i)}{d! \, \rho_{s,t}(\mathbb{Z})} \left(c + \frac{1}{2n} - t \right)^d dt \right)$$

By using well-known series-integral comparison we obtain $\rho_{s,t}(\mathbb{Z}) \ge s\sqrt{2\pi} - 1$ and since $\left| \rho_{s,t}^{(d)}(i) \right| < \frac{d(1.3\tau)^d 2^d}{s^{d/2}}$ for $s \ge 4$ and $\tau \ge 2.5$, it follows that

$$\mathsf{E}_{\text{expansion}} \le \frac{(d+1)(1.3)^{d+1}\tau^{d+2}}{d! \, n^{d+1} s^{\frac{d+1}{2}}}$$

□

A careful analysis of this technique show that with $d = 4$ we achieve the same asymptotic computational cost as the classical CDT algorithm with $n = \omega(\lambda)$, where the hidden factor is less than $1/4$, therefore for this degree the space complexity of Algorithms 4 and 5 is only λ times bigger than for centered sampling, showing that these algorithms can achieve a memory requirement as low as 1 MB. Finally, note that taking care to add the floating-point computation error to the error of approximation, one can compute the Taylor expansion evaluation at the maximal hardware precision to reduce its computational cost.

5   Lookup Tables

We shall now show how to use partial lookup tables to avoid the binary search in most cases when using CDT algorithms, this technique is the CDT analogue of the Knuth-Yao algorithm improvement described in [8]. Note that this strategy is particularly fitting for discrete Gaussian distributions with relatively small expected values. The basic idea is to subdivide the uniform distribution $U_{[0,1)}$ into ℓ uniform distributions on subsets of the same size $U_{[i/\ell,(i+1)/\ell)}$, with ℓ a power of two. We then precompute a partial lookup table on these subsets which allows to return the sample at once when the subset considered does not include a cdf image. We note that instead of subdividing the uniform range into stripes of the same size, we can also recursively subdivide only some stripes of the previous subdivision. However, for the sake of clarity and ease of exposure, this improvement is not included in this paper and we will describe this technique for the classical centered CDT algorithm.

First, we initialize a lookup table of size $\ell = 2^l$ where the i-th entry corresponds to a subinterval $[i/\ell, (i+1)/\ell)$ of $[0,1)$. Second, after precomputing the CDT, we mark all the entries for which there is at least one CDT element in their corresponding subinterval $[i/\ell, (i+1)/\ell)$ with $\perp$, and all remaining entries with $\top$. Each entry marked with $\top$ allows to return a sample without the need to perform a binary search in the CDT, because only one value corresponds to this subinterval which is the first CDT element greater or equal to $(i+1)/\ell$.

Efficiency. The efficiency of this technique is directly related to the number of entries, marked with $\top$, whose subintervals do not contain a CDT element. We denote the probability of performing binary search by $\mathsf{P_{binsrch}}$, obviously the probability to return the sample immediately after choosing i, which is a part of p, is $1 - \mathsf{P_{binsrch}}$. Lemma 7 gives a lower bound of $\mathsf{P_{binsrch}}$.

Lemma 7. *For any $\ell \geq 2^8$ and $s \geq \eta_{\frac{1}{2}}(\mathbb{Z})$. Let $\mathsf{P_{binsrch}}$ be the probability of performing binary search during the execution of the CDT algorithm implemented with the lookup table trick described above, we have*

$$\mathsf{P_{binsrch}} < 1.2s\sqrt{\log_2 \ell}/\ell$$

Proof.

$$\mathsf{P_{binsrch}} = \frac{\ell - \sum_{i=\lfloor c-\tau s \rfloor}^{\lceil c+\tau s \rceil} \lfloor \ell\, \mathrm{cdf}_{s,c}(i) \rfloor - \lfloor \ell\, \mathrm{cdf}_{s,c}(i-1) \rfloor}{\ell}$$

From Lemma 2 we have

$$\left\lfloor \ell\, \mathrm{cdf}_{s,c}\left(\left\lfloor c - 0.6s\sqrt{\log_2 \ell} \right\rfloor\right) \right\rfloor = 0$$

$$\left\lfloor \ell\left(1 - \mathrm{cdf}_{s,c}\left(\left\lceil c + 0.6s\sqrt{\log_2 \ell} \right\rceil\right)\right) \right\rfloor = 0$$

$\square$

6   Experimental Results

In this section, we present experimental results of our C++ implementation[1] distributed under the terms of the GNU General Public License version 3 or later (GPLv3+) which uses the MPFR [15] and GMP [19] libraries as well as Salsa20 [5] as the pseudorandom number generator. Our non-centered discrete Gaussian sampler was implemented with a binary search executed byte by byte if $\ell = 2^8$ and 2-bytes by 2-bytes if $\ell = 2^{16}$ without recursive subdivision of $U_{[0,1)}$, therefore $[0, 1)$ is subdivided in ℓ intervals of the same size and cdf(x) is stored for all $x \in [-\lceil \tau \sigma \rceil - 1, \lceil \tau \sigma \rceil + 1]$. The implementation of our non-centered discrete Gaussian sampler uses a fixed number of precomputed centers $n = 2^8$ with a lookup table of size $\ell = 2^8$ and includes the lazy cdf evaluation optimization.

We tested the performance of our non-centered discrete Gaussian sampler by using it as a subroutine for Peikert's sampler [32] for sampling from $D_{(g),\sigma',0}$ with $g \in \mathbb{Z}[x]/(x^N + 1)$ for N a power of two. To this end, we adapted the implementation of this sampler from [3] where we swap out the sampler from

Table 1. Performance of sampling from $D_{(g),\sigma'}$ as implemented in [3] and with our non-centered discrete Gaussian sampler with $\ell = n = 2^8$. The column $D_{(g),\sigma'}/s$ gives the number of samples returned per second, the column "memory" the maximum amount of memory consumed by the process. All timings are on a Intel(R) Xeon(R) CPU E5-2667 (strombenzin). Precomputation uses 2 cores, the online phase uses one core.

[3]					
N	$\log \sigma'$	precomp	time	$D_{(g),\sigma'}/s$	memory
256	38.2	0.08 s	8.46 ms	118.17	11,556 kB
512	42.0	0.17 s	16.96 ms	58.95	11,340 kB
1024	45.8	0.32 s	38.05 ms	26.28	21,424 kB
2048	49.6	0.93 s	78.17 ms	12.79	41,960 kB
4096	53.3	2.26 s	157.53 ms	6.35	86,640 kB
8192	57.0	6.08 s	337.32 ms	2.96	192,520 kB
16384	60.7	13.36 s	700.75 ms	1.43	301,200 kB
This work					
N	$\log \sigma'$	precomp	time	$D_{(g),\sigma'}/s$	memory
256	38.2	0.31 s	2.91 ms	343.16	17,080 kB
512	42.0	0.39 s	5.99 ms	166.88	21,276 kB
1024	45.8	0.65 s	11.89 ms	84.12	38,280 kB
2048	49.6	1.04 s	25.07 ms	39.89	74,668 kB
4096	53.3	2.35 s	48.63 ms	20.56	148,936 kB
8192	57.0	7.27 s	96.67 ms	10.34	302,616 kB
16384	60.7	14.41 s	205.35 ms	4.87	618,448 kB

---

[1] The implementation is available at https://github.com/tricosset/FGN.

the dgs library [2] (implementing rejection sampling and [11]) used in [3] with our sampler for sampling for $D_{\mathbb{Z},\sigma,c}$. Note that sampling from $D_{(g),\sigma',0}$ is more involved and thus slower than sampling from $D_{\mathbb{Z}^N,\sigma',0}$. That is, to sample from $D_{(g),\sigma',0}$, [3] first computes an approximate square root of $\Sigma_2 = \sigma'^2 \cdot g^{-T} \cdot g^{-1} - r^2$ with $r = 2 \cdot \lceil \sqrt{\log N} \rceil$. Then, given an approximation $\sqrt{\Sigma_2}'$ of $\sqrt{\Sigma_2}$ it samples a vector $x \leftarrow_\$ \mathbb{R}^N$ from a standard normal distribution and interpret it as a polynomial in $\mathbb{Q}[X]/(x^N + 1)$; computes $y = \sqrt{\Sigma_2}' \cdot x$ in $\mathbb{Q}[X]/(x^N + 1)$ and returns $g \cdot (\lfloor y \rceil_r)$, where $\lfloor y \rceil_r$ denotes sampling a vector in $\mathbb{Z}^N$ where the i-th component follows $D_{\mathbb{Z},r,y_i}$. Thus, implementing Peikert's sampler requires sampling from $D_{\mathbb{Z},r,y_i}$ for changing centers y_i and sampling from a standard normal distribution. We give experimental results in Table 1, indicating that our sampler increases the rate by a factor ≈ 3.

References

1. Ajtai, M.: Generating hard instances of lattice problems (extended abstract). In: Proceedings of the Twenty-Eighth Annual ACM Symposium on Theory of Computing, STOC 1996, NY, USA, pp. 99–108. ACM, New York (1996)
2. Albrecht, M.R.: dgs – discrete gaussians over the integers (2014). https://bitbucket.org/malb/dgs
3. Albrecht, M.R., Cocis, C., Laguillaumie, F., Langlois, A.: Implementing candidate graded encoding schemes from ideal lattices. In: Iwata, T., Cheon, J.H. (eds.) ASIACRYPT 2015. LNCS, vol. 9453, pp. 752–775. Springer, Heidelberg (2015). doi:10.1007/978-3-662-48800-3_31
4. Babai, L.: On Lovász' lattice reduction and the nearest lattice point problem. In: Mehlhorn, K. (ed.) STACS 1985. LNCS, vol. 182, pp. 13–20. Springer, Heidelberg (1985). doi:10.1007/BFb0023990
5. Bernstein, D.J.: The salsa20 family of stream ciphers. In: Robshaw, M., Billet, O. (eds.) New Stream Cipher Designs: The eSTREAM Finalists, pp. 84–97. Springer, Heidelberg (2008)
6. Brent, R.P., et al.: Fast algorithms for high-precision computation of elementary functions. In: Proceedings of 7th Conference on Real Numbers and Computers (RNC 7), pp. 7–8 (2006)
7. Buchmann, J., Cabarcas, D., Göpfert, F., Hülsing, A., Weiden, P.: Discrete Ziggurat: a time-memory trade-off for sampling from a Gaussian distribution over the integers. In: Lange, T., Lauter, K., Lisoněk, P. (eds.) SAC 2013. LNCS, vol. 8282, pp. 402–417. Springer, Heidelberg (2014). doi:10.1007/978-3-662-43414-7_20
8. de Clercq, R., Roy, S.S., Vercauteren, F., Verbauwhede, I.: Efficient software implementation of ring-LWE encryption. In: 2015 Design, Automation Test in Europe Conference Exhibition (DATE), pp. 339–344 (2015)
9. Devroye, L.: Non-Uniform Random Variate Generation. Springer, Heidelberg (1986)
10. Ducas, L.: Lattice based signatures: attacks, analysis and optimization. Ph.D. thesis (2013)
11. Ducas, L., Durmus, A., Lepoint, T., Lyubashevsky, V.: Lattice signatures and bimodal Gaussians. In: Canetti, R., Garay, J.A. (eds.) CRYPTO 2013. LNCS, vol. 8042, pp. 40–56. Springer, Heidelberg (2013). doi:10.1007/978-3-642-40041-4_3

12. Ducas, L., Nguyen, P.Q.: Faster Gaussian lattice sampling using lazy floating-point arithmetic. In: Wang, X., Sako, K. (eds.) ASIACRYPT 2012. LNCS, vol. 7658, pp. 415–432. Springer, Heidelberg (2012). doi:10.1007/978-3-642-34961-4_26

13. Dwarakanath, N.C., Galbraith, S.D.: Sampling from discrete Gaussians for lattice-based cryptography on a constrained device. Appl. Algebra Eng. Commun. Comput. **25**(3), 159–180 (2014)

14. Fiat, A., Shamir, A.: How To prove yourself: practical solutions to identification and signature problems. In: Odlyzko, A.M. (ed.) CRYPTO 1986. LNCS, vol. 263, pp. 186–194. Springer, Heidelberg (1987). doi:10.1007/3-540-47721-7_12

15. Fousse, L., Hanrot, G., Lefèvre, V., Pélissier, P., Zimmermann, P.: MPFR: a multiple-precision binary floating-point library with correct rounding. ACM Trans. Math. Softw. **33**(2) (2007)

16. Gentry, C., Peikert, C., Vaikuntanathan, V.: Trapdoors for hard lattices and new cryptographic constructions. In: Proceedings of the Fortieth Annual ACM Symposium on Theory of Computing, STOC 2008, pp. 197–206. ACM, New York (2008)

17. Gentry, C., Peikert, C., Vaikuntanathan, V.: Trapdoors for hard lattices and new cryptographic constructions. In: Ladner, R.E., Dwork, C. (eds.) 40th ACM STOC, pp. 197–206. ACM Press, Victoria, 17–20 May 2008

18. Goldreich, O., Goldwasser, S., Halevi, S.: Public-key cryptosystems from lattice reduction problems. In: Kaliski, B.S. (ed.) CRYPTO 1997. LNCS, vol. 1294, pp. 112–131. Springer, Heidelberg (1997). doi:10.1007/BFb0052231

19. Granlund, T.: The GMP development team: GNU MP: The GNU Multiple Precision Arithmetic Library, 6.0.1 edn. (2015). http://gmplib.org/

20. Karney, C.F.F.: Sampling exactly from the normal distribution. ACM Trans. Math. Softw. **42**(1), 3:1–3:14 (2016)

21. Klein, P.: Finding the closest lattice vector when it's unusually close. In: Proceedings of the Eleventh Annual ACM-SIAM Symposium on Discrete Algorithms, SODA 2000, pp. 937–941. Society for Industrial and Applied Mathematics, Philadelphia, PA, USA (2000)

22. Knuth, D.E., Yao, A.C.: The complexity of nonuniform random number generation. In: Traub, J.F. (ed.) Algorithms and Complexity: New Directions and Recent Results. Academic Press, New York (1976)

23. Lyubashevsky, V.: Lattice-based identification schemes secure under active attacks. In: Cramer, R. (ed.) PKC 2008. LNCS, vol. 4939, pp. 162–179. Springer, Heidelberg (2008). doi:10.1007/978-3-540-78440-1_10

24. Lyubashevsky, V.: Fiat-Shamir with aborts: applications to lattice and factoring-based signatures. In: Matsui, M. (ed.) ASIACRYPT 2009. LNCS, vol. 5912, pp. 598–616. Springer, Heidelberg (2009). doi:10.1007/978-3-642-10366-7_35

25. Lyubashevsky, V.: Lattice signatures without trapdoors. In: Pointcheval, D., Johansson, T. (eds.) EUROCRYPT 2012. LNCS, vol. 7237, pp. 738–755. Springer, Heidelberg (2012). doi:10.1007/978-3-642-29011-4_43

26. Lyubashevsky, V., Micciancio, D.: Asymptotically efficient lattice-based digital signatures. In: Canetti, R. (ed.) TCC 2008. LNCS, vol. 4948, pp. 37–54. Springer, Heidelberg (2008). doi:10.1007/978-3-540-78524-8_3

27. Lyubashevsky, V., Peikert, C., Regev, O.: On ideal lattices and learning with errors over rings. In: Gilbert, H. (ed.) EUROCRYPT 2010. LNCS, vol. 6110, pp. 1–23. Springer, Heidelberg (2010). doi:10.1007/978-3-642-13190-5_1

28. Marsaglia, G., Tsang, W.W.: A fast, easily implemented method for sampling from decreasing or symmetric unimodal density functions. SIAM J. Sci. Stat. Comput. **5**, 349–359 (1984)

29. Micciancio, D.: Generalized compact knapsacks, cyclic lattices, and efficient one-way functions. Comput. Complex. **16**(4), 365–411 (2007)
30. Micciancio, D., Regev, O.: Worst-case to average-case reductions based on Gaussian measures. SIAM J. Comput. **37**(1), 267–302 (2007)
31. von Neumann, J.: Various techniques used in connection with random digits. J. Res. Nat. Bur. Stand. **12**, 36–38 (1951)
32. Peikert, C.: An efficient and parallel Gaussian sampler for lattices. In: Rabin, T. (ed.) CRYPTO 2010. LNCS, vol. 6223, pp. 80–97. Springer, Heidelberg (2010). doi:10.1007/978-3-642-14623-7_5
33. Peikert, C.: A decade of lattice cryptography. Found. Trends Theor. Comput. Sci. **10**(4), 283–424 (2016)
34. Pujol, X., Stehlé, D.: Rigorous and efficient short lattice vectors enumeration. In: Pieprzyk, J. (ed.) ASIACRYPT 2008. LNCS, vol. 5350, pp. 390–405. Springer, Heidelberg (2008). doi:10.1007/978-3-540-89255-7_24
35. Regev, O.: On lattices, learning with errors, random linear codes, and cryptography. In: Proceedings of the Thirty-Seventh Annual ACM Symposium on Theory of Computing, STOC 2005, NY, USA, pp. 84–93. ACM, New York (2005)
36. Saarinen, M.J.O.: Arithmetic coding and blinding countermeasures for lattice signatures. J. Cryptographic Eng. 1–14 (2017)
37. Von Neumann, J.: The general and logical theory of automata. Cerebral Mech. Behav. **1**(41), 1–2 (1951)
38. Walker, A.J.: New fast method for generating discrete random numbers with arbitrary frequency distributions. Electron. Lett. **10**, 127–128 (1974)

Simple Security Definitions for and Constructions of 0-RTT Key Exchange

Britta Hale[1]($\boxtimes$), Tibor Jager[2], Sebastian Lauer[3], and Jörg Schwenk[3]

[1] NTNU, Norwegian University of Science and Technology, Trondheim, Norway
britta.hale@ntnu.no
[2] Paderborn University, Paderborn, Germany
tibor.jager@upb.de
[3] Horst Görtz Institute, Ruhr-University Bochum, Bochum, Germany
{sebastian.lauer,joerg.schwenk}@rub.de

Abstract. Zero Round-Trip Time (0-RTT) key exchange protocols allow for the transmission of cryptographically protected payload data without requiring the prior exchange of messages of a cryptographic key exchange protocol. The 0-RTT KE concept was first realized by Google in the QUIC Crypto protocol, and a 0-RTT mode has been intensively discussed for inclusion in TLS 1.3.

In 0-RTT KE two keys are generated, typically using a Diffie-Hellman key exchange. The first key is a combination of an ephemeral client share and a long-lived server share. The second key is computed using an ephemeral server share and the same ephemeral client share.

In this paper, we propose simple security models, which catch the intuition behind known 0-RTT KE protocols; namely that the first (resp. second) key should remain indistinguishable from a random value, even if the second (resp. first) key is revealed. We call this property *strong key independence*. We also give the first constructions of 0-RTT KE which are provably secure in these models, based on the generic assumption that secure non-interactive key exchange (NIKE) exists (This work was partially supported by a STSM Grant from COST Action IC1306).

Keywords: Foundations · Low-latency key exchange · 0-RTT protocols · Authenticated key exchange · Non-interactive key exchange · QUIC · TLS 1.3.

1 Introduction

Efficiency, in terms of messages to be exchanged before a key is established, is a growing consideration for internet protocols today. Basically, the first generation of internet key exchange protocols did not care too much about efficiency, since secure connections were considered to be the exception rather than the rule: SSL (versions 2.0 and 3.0) and TLS (versions 1.0, 1.1, and 1.2) require 2 round-trip times (RTT) for key establishment before the first cryptographically-protected payload data can be sent. With the increased use of encryption,[1] efficiency is

---

[1] For example, initiatives like Let's Encrypt (https://letsencrypt.org/).

© Springer International Publishing AG 2017
D. Gollmann et al. (Eds.): ACNS 2017, LNCS 10355, pp. 20–38, 2017.
DOI: 10.1007/978-3-319-61204-1_2

of escalating importance for protocols like TLS. Similarly, the older IPSec IKE version v1 needs between 3 RTT (aggressive mode + quick mode) and 4.5 RTT (main mode + quick mode). This was soon realized to be problematic, and in IKEv2 the number of RTTs was reduced to 2.

The QUIC Protocol. Fundamentally, the discussion on low-latency key exchange (aka. LLKE, zero-RTT or 0-RTT key exchange) was opened when Google proposed the QUIC protocol.[2] QUIC (cf. Fig. 1) achieves low-latency by caching a signed server configuration file on the client side, which contains a medium-lived Diffie-Hellman (DH) share $Y_0 = g^{y_0}$.[3]

When a client wishes to establish a connection with a server and possesses a valid configuration file of that server, it chooses a fresh ephemeral DH share $X = g^x$ and computes a temporal key k_1 from $g^{y_0 x}$. Using this key k_1, the client can encrypt and authenticate data to be sent to the server, together with X. In response, the server sends a fresh DH share $Y = g^y$ and computes a session key k_2 from g^{xy}, which is used for all subsequent data exchanges.

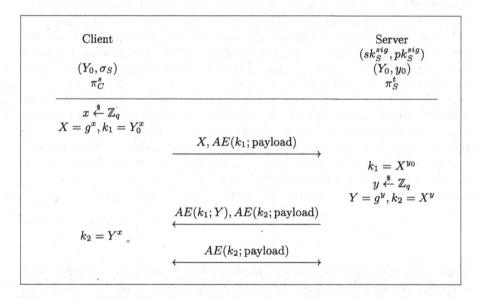

Fig. 1. Google's QUIC protocol (simplified) with cached server key configuration file (Y_0, σ_S). AE denotes a symmetric authenticated encryption algorithm (e.g., AES-GCM), (sk_S^{sig}, pk_S^{sig}) denotes the server's long-term signing keys, and π_S^t (resp. π_C^s) denotes the oracle at server S executing the single t-th instance of the protocol (resp. for client).

---

[2] See https://www.chromium.org/quic.

[3] If the client does not have a valid file, it has to be requested from the server, which increases the number of RTTs by 1, but may then be re-used for future sessions.

TLS 1.3. Early TLS 1.3 drafts, e.g. `draft-ietf-tls-tls13-08` [25], contained a 0-RTT key exchange mode where a QUIC-like `ServerConfiguration` message is cached by the client. The current version `draft-ietf-tls-tls13-18` [26] follows a different approach, where the initial key establishment between a client and a server is never 0-RTT. Instead, it defines a method to establish a new session based on the secret key of a previous session. Even though this is also called "0-RTT" in the current TLS 1.3 specification, it is rather a "0-RTT session resumption" protocol, but does not allow for 0-RTT key establishment. Most importantly, the major difference between the approach of the current TLS 1.3 draft in comparison to a "real" 0-RTT key exchange protocol is that the former requires storing of *secret* key information on the client between sessions. In contrast, a 0-RTT key establishment protocol does not require secret information to be stored between sessions.

Facebook's Zero Protocol. Very recently, the social network Facebook announced that it is currently experimenting with a 0-RTT KE protocol called *Zero*.[4] Zero is very similar to QUIC, except that it uses another nonce and encryption of the ServerHello message. It is noteworthy that the main difference between Zero and QUIC was introduced in order to prevent an attack discovered by Facebook, which has been reported to Google and meanwhile been fixed in QUIC, too. We believe that this is a good example that shows the demand of simple security definitions and provably-secure constructions for such protocols.

Security Goals. 0-RTT KE protocols like QUIC have ad-hoc designs that aim at achieving three goals: (1) 0-RTT encryption, where ciphertext data can already be sent together with the first handshake message; (2) perfect forward secrecy (PFS), where all ciphertexts exchanged after the second handshake message will remain secure even after the (static or semi-static) private keys of the server have been leaked, and (3) key independence, where "knowledge" about one of the two symmetric keys generated should not endanger the security of the other key.

Strong Key Independence. Intuitively, a 0-RTT KE protocol should achieve *strong key independence* between k_1 and k_2; if any one of the two keys is leaked at any time, the other key should still be indistinguishable from a random value. In all known security models, this intuition would be formalized as follows: if the adversary $\mathcal{A}$ asks a Reveal query for k_1, he is still allowed to ask a Test query for k_2, and vice versa. If the two keys are computationally independent from each other (which also includes computations on the different protocol messages), then the adversary should have only a negligible advantage in answering the Test query correctly.

Ultimately this leads to the following research questions: *Do existing examples of 0-RTT KE protocols have strong key independence? Can we describe a generic way to construct 0-RTT KE protocols that provably achieve strong key independence?*

---

[4] See https://code.facebook.com/posts/608854979307125/building-zero-protocol-for-fast-secure-mobile-connections/.

QUIC Does Not Provide Strong Key Independence. If an attacker $\mathcal{A}$ is allowed to learn k_1 by a Reveal-query, then he is able to decrypt $AE(k_1; Y)$ and re-encrypt its own value $Y^* := g^{y^*}$ as $AE(k_1; Y^*)$. Furthermore, he can then compute the same $k_2 = X^{y^*}$ as the client oracle, and can thus distinguish between the "real" key and a "random" key chosen by the Test query. See [11] for more details on key dependency in QUIC.

Note that this theoretical attack does not imply that QUIC is insecure. It only shows that the authenticity of the server's Diffie-Hellman share, which is sent in QUIC to establish k_2, depends strongly on the security of key k_1. Therefore QUIC does not provide strong key independence in the sense sketched above.

Previous Work on 0-RTT Key Exchange. The concept of 0-RTT key exchange was not developed in academia, but in industry – motivated by concrete practical demands of distributed applications. All previous works on 0-RTT KE [11,23] conducted *a-posteriori* security analyses of the QUIC protocol, with tailored models. There are no foundational constructions as yet, and the relation to other cryptographic protocols and primitives is not yet well-understood.

At ACM CCS 2014, Fischlin and Günther [11] provided a formal definition of *multi-stage* key exchange protocols and used it to analyze the security of QUIC. Lychev *et al.* [23] gave an alternate analysis of QUIC, which considers both efficiency and security. They describe a security model which is bespoke to QUIC, adopting the complex, monolithic security model of [17] to the protocol's requirements. Zhao [31] considers identity-concealed 0-RTT protocols, where user privacy is protected by hiding identities of users in a setting with mutual cryptographic authentication of both communicating parties. Günther *et al.* [14] extended the "puncturable encryption"-approach of Green and Miers [13] to show that even 0-RTT KE with full forward secrecy is possible, by evolving the secret key after each decryption. However, their construction is currently mainly of conceptual interest, as it is not yet efficient enough to be deployed at large scale in practice.

Security Model. In this paper, we use a variant of the Canetti-Krawczyk [7] security model. This family of security models is especially suited to protocols with only two message exchanges, with *one-round* key exchange protocols constituting the most important subclass. Popular examples of such protocols are MQV [22], HMQV [18], SMQV [27], KEA [21,24], and NAXOS [20]. A comparison of different variants of the Canetti-Krawczyk model can be found in [9,29].

The Importance of Simplicity of Security Models. Security models for key exchange protocols have to consider *active* adversaries that may modify, replay, inject, drop, etc., any message transmitted between communicating parties. They also need to capture *parallel* executions of multiple protocol sessions, potential reveals of earlier session keys, and adaptive corruptions of long-term secrets of parties. This makes even standard security models for key exchange *extremely* complex (in comparison to most other standard cryptographic primitives, like digital signatures or public-key encryption, for example).

Naturally, the novel primitive of 0-RTT KE requires formal security definitions. There are different ways to create such a model. One approach is to focus on *generality* of the model. Fischlin and Günther [11] followed this path, by defining *multi-stage* key exchange protocols, a generalization of 0-RTT KE. This approach has the advantage that it lays the foundation for the study of a very general class of interesting and novel primitives. However, its drawback is that this generality inherently also brings a huge *complexity* to the model. Clearly, the more complex the security model, the more difficult it becomes to devise new, simple, efficient, and provably-secure constructions. Moreover, proofs in complex models tend to be error-prone and less intuitive, because central technical ideas may be concealed in formal details that are required to handle the generality of the model.

Another approach is to devise a model which is tailored to the analysis of *one specific protocol*. For example, the complex, monolithic ACCE security model was developed in [17] to provide an *a posteriori* security analysis of TLS.[5] A similar approach was followed by Lychev *et al.* [23], who adopted this model for an *a posteriori* analysis of QUIC, by defining the so-called *Q*-ACCE model. The notable drawback of this approach is that such tailor-made models tend to capture only the properties achieved by existing protocols, but not necessarily all properties that we would expect from a "good" 0-RTT KE protocol. In general, such tailor-made models do not, therefore, form a useful foundation for the creation of new protocols.

In this paper, we follow a different approach. We propose novel "bare-bone" security models for 0-RTT KE, which aim at capturing *all* (strong key independence and forward secrecy), but also *only* the properties intuitively expected from "good" 0-RTT KE protocols. We propose two different models. One considers the practically-relevant case of *server-only* authentication (where the client may or may not authenticate later over the established communication channel, similar in spirit to the server-only-authenticated ACCE model of [19]). The other considers traditional *mutual* cryptographic authentication of a client and server.

The reduced generality of our definitions – in comparison to the very general multi-stage security model of [11] – is intended. A model which captures *only*, but also *all* the properties expected from a "good" 0-RTT KE protocol allows us to devise relatively simple, foundational, and generic constructions of 0-RTT KE protocols with as-clean-as-possible security analyses.

Importance of Foundational Generic Constructions. Following [3], we use non-interactive key exchange (NIKE) [8,12] in combination with digital signatures as a main building block.[6] This yields the first examples of 0-RTT KE protocols with strong key independence, as well as the first constructions of 0-RTT KE from generic complexity assumptions. There are many advantages of such generic constructions:

---

[5] A more modular approach was later proposed in [4].

[6] Recall that digital signatures are implied by one-way functions, which in turn are implied by NIKE. Thus, essentially we only assume the existence of NIKE as a building block.

1. Generic constructions provide a better understanding of the structure of protocols. Since the primitives we use have abstract security properties, we can see precisely which abstract security requirements are needed to implement 0-RTT KE protocols.
2. They clarify the relations and implications between different types of cryptographic primitives.
3. They can be generically instantiated with building blocks based on different complexity assumptions. For example, if "post-quantum" security is needed, one can directly obtain a concrete protocol by using only post-quantum secure building blocks in the generic construction.

Usually generic constructions tend to involve more computational overhead than ad-hoc constructions. However, we note that our 0-RTT KE protocols can be instantiated relatively efficiently, given the efficient NIKE schemes of [12], for example.

Contributions. Contributions in this paper can be summarized as follows:

- *Simple security models.* We provide simple security models, which capture all properties that we expect from a "good" 0-RTT KE protocol, but only these properties. We consider both the "practical" setting with server-only authentication and the classical setting with mutual authentication.
- *First generic constructions.* We give intuitive, relatively simple, and efficient constructions of 0-RTT KE protocols in both settings.
- *First Non-DH instantiation.* Both QUIC and TLS 1.3 are based on DH key exchange. Our generic construction yields the first 0-RTT KE protocol which is not based on Diffie-Hellman (e.g., by instantiating the generic construction with the factoring-based NIKE scheme of Freire *et al.* [12]).
- *First 0-RTT KE with strong key independence.* Our 0-RTT KE protocols are the first to achieve strong key independence in the sense described above.
- *Well-established, general assumptions.* The construction is based on general assumptions, namely the existence of secure NIKE and digital signature schemes. For all building blocks we require only standard security properties.
- *Security in the Standard Model.* The security analysis is completely in the standard model, i.e. it is performed without resorting to the Random Oracle heuristic [1] and without relying on non-standard complexity assumptions.
- *Efficient instantiability.* Despite the fact that our constructions are generic, the resulting protocols can be instantiated relatively efficiently.

Full Version of this Paper. Due to space limitations, we have to defer several results to the full version of this paper [15]. This includes the full proof of Theorem 1, the Definition and Security Model for a 0-RTT protocol under mutual authentication (0-RTT-M), a construction of a 0-RTT-M protocol along with its security model and its security proof.

2 Preliminaries

For our construction in Sect. 4, we need signature schemes and non-interactive key exchange (NIKE) protocols. Here we summarize the definitions of these two primitives and their security from the literature.

2.1 Digital Signatures

A digital signature scheme consists of three polynomial-time algorithm $\mathsf{SIG} = (\mathsf{SIG.Gen}, \mathsf{SIG.Sign}, \mathsf{SIG.Vfy})$. The key generation algorithm $(sk, pk) \xleftarrow{\$} \mathsf{SIG.Gen}(1^\lambda)$ generates a public verification key pk and a secret signing key sk on input of security parameter λ. Signing algorithm $\sigma \xleftarrow{\$} \mathsf{SIG.Sign}(sk, m)$ generates a signature for message m. Verification algorithm $\mathsf{SIG.Vfy}(pk, \sigma, m)$ returns 1 if σ is a valid signature for m under key pk, and 0 otherwise.

Consider the following security experiment played between a challenger $\mathcal{C}$ and an adversary $\mathcal{A}$.

1. The challenger generates a public/secret key pair $(sk, pk) \xleftarrow{\$} \mathsf{SIG.Gen}(1^\lambda)$, the adversary receives pk as input.
2. The adversary may query arbitrary messages m_i to the challenger. The challenger replies to each query with a signature $\sigma_i = \mathsf{SIG.Sign}(sk, m_i)$. Here i is an index, ranging between $1 \leq i \leq q$ for some $q \in \mathbb{N}$. Queries can be made adaptively.
3. Eventually, the adversary outputs a message/signature pair (m, σ).

Definition 1. *We define the advantage on an adversary $\mathcal{A}$ in this game as*

$$\mathsf{Adv}_{\mathsf{SIG}, \mathcal{A}}^{sEUF\text{-}CMA}(\lambda) := \Pr\left[(m, \sigma) \xleftarrow{\$} \mathcal{A}^{\mathcal{C}(\lambda)}(pk) : \begin{array}{l} \mathsf{SIG.Vfy}(pk, \sigma, m) = 1, \\ (m, \sigma) \neq (m_i, \sigma_i) \; \forall i \end{array}\right].$$

SIG *is strongly secure* against *existential forgeries under adaptive chosen-message attacks (sEUF-CMA), if* $\mathsf{Adv}_{\mathsf{SIG}, \mathcal{A}}^{sEUF\text{-}CMA}(\lambda)$ *is a negligible function in* λ *for all probabilistic polynomial-time adversaries* $\mathcal{A}$.

Remark 1. Signatures with sEUF-CMA security can be constructed generically from any EUF-CMA-secure signature scheme and chameleon hash functions [6,28].

2.2 Secure Non-interactive Key Exchange

Definition 2. *A non-interactive key exchange (NIKE) scheme consists of two deterministic algorithms* $(\mathsf{NIKE.Gen}, \mathsf{NIKE.Key})$.

$\mathsf{NIKE.Gen}(1^\lambda, r)$ *takes a security parameter* λ *and randomness* $r \in \{0, 1\}^\lambda$. *It outputs a key pair* (pk, sk). *We write* $(pk, sk) \xleftarrow{\$} \mathsf{NIKE.Gen}(1^\lambda)$ *to denote that* $\mathsf{NIKE.Gen}(1^\lambda, r)$ *is executed with uniformly random* $r \xleftarrow{\$} \{0, 1\}^\lambda$.

NIKE.Key(sk_i, pk_j) *is a deterministic algorithm which takes as input a secret key sk_i and a public key pk_j, and outputs a key $k_{i,j}$.*

We say that a NIKE scheme is correct, *if for all* $(pk_i, sk_i) \xleftarrow{\$} $ NIKE.Gen(1^λ) *and* $(pk_j, sk_j) \xleftarrow{\$} $ NIKE.Gen(1^λ) *holds that* NIKE.Key(sk_i, pk_j) = NIKE.Key(sk_j, pk_i).

A NIKE scheme is used by d parties $P_1, \ldots, P_d$ as follows. Each party P_i generates a key pair $(pk_i, sk_i) \leftarrow$ NIKE.Gen(1^λ) and publishes pk_i. In order to compute the key shared by P_i and P_j, party P_i computes $k_{i,j} = $ NIKE.Key(sk_i, pk_j). Similarly, party P_j computes $k_{j,i} = $ NIKE.Key(sk_j, pk_i). Correctness of the NIKE scheme guarantees that $k_{i,j} = k_{j,i}$.

CKS-Light Security. The *CKS-light* security model for NIKE protocols is relatively simplistic and compact. We choose this model because other (more complex) NIKE security models like *CKS, CKS-heavy,* and *m-CKS-heavy* are polynomial-time equivalent to *CKS-light*. See [12] for more details.

Security of a NIKE protocol NIKE is defined by a game **NIKE** played between an adversary $\mathcal{A}$ and a challenger. The challenger takes a security parameter λ and a random bit b as input and answers all queries of $\mathcal{A}$ until she outputs a bit b'. The challenger answers the following queries for $\mathcal{A}$:

- RegisterHonest(i). $\mathcal{A}$ supplies an index i. The challenger runs NIKE.Gen(1^λ) to generate a key pair (pk_i, sk_i) and records the tuple (**honest**, pk_i, sk_i) for later and returns pk_i to $\mathcal{A}$. This query may be asked *at most twice* by $\mathcal{A}$.
- RegisterCorrupt(pk_i). With this query $\mathcal{A}$ supplies a public key pk_i. The challenger records the tuple (**Corrupt**, pk_i) for later.
- GetCorruptKey(i, j). $\mathcal{A}$ supplies two indices i and j where pk_i was registered as corrupt and pk_j as honest. The challenger runs $k \leftarrow$ NIKE.Key(sk_j, pk_i) and returns k to $\mathcal{A}$.
- Test(i, j). The adversary supplies two indices i and j that were registered honestly. Now the challenger uses bit b: if $b = 0$, then the challenger runs $k_{i,j} \leftarrow$ NIKE.Key(pk_i, sk_j) and returns the key $k_{i,j}$. If $b = 1$, then the challenger samples a random element from the key space, records it for later, and returns the key to $\mathcal{A}$.

The game **NIKE** outputs 1, denoted by **NIKE**$_{\mathsf{NIKE}}^{\mathcal{A}}(\lambda) = 1$, if $b = b'$ and 0 otherwise. We say $\mathcal{A}$ wins the game if **NIKE**$_{\mathsf{NIKE}}^{\mathcal{A}}(\lambda) = 1$.

Definition 3. *For any adversary $\mathcal{A}$ playing the above* **NIKE** *game against a NIKE scheme* NIKE, *we define the advantage of winning the game* **NIKE** *as*

$$\mathsf{Adv}_{\mathsf{NIKE},\mathcal{A}}^{CKS\text{-}light}(\lambda) = \left| 2 \cdot \Pr\left[\mathbf{NIKE}_{\mathsf{NIKE}}^{\mathcal{A}}(\lambda) = 1 \right] - 1 \right|.$$

Let λ be a security parameter, NIKE *be a NIKE protocol and $\mathcal{A}$ an adversary. We say* NIKE *is a* CKS-light-secure *NIKE protocol, if for all probabilistic polynomial-time adversaries $\mathcal{A}$, the function* $\mathsf{Adv}_{\mathsf{NIKE},\mathcal{A}}^{CKS\text{-}light}(\lambda)$ *is a negligible function in λ.*

3   0-RTT Key Exchange Protocols: Syntax and Security with Server-Only Authentication

In the model presented in this section, we give formal definitions for 0-RTT KE with strong key independence and main-key forward secrecy. We start with the case of server-only authentication, as it is the more important case in practice (in particular, server-only authentication will be the main operating mode of both QUIC and TLS 1.3).

3.1   Syntax and Correctness

Definition 4. *A 0-RTT key exchange scheme with server-only authentication consists of deterministic algorithms* $(\mathsf{Gen^{server}}, \mathsf{KE^{client}_{init}}, \mathsf{KE^{client}_{refresh}}, \mathsf{KE^{server}_{refresh}})$.

- $\mathsf{Gen^{server}}(1^\lambda, r) \rightarrow (pk, sk)$: *A key generation algorithm that takes as input a security parameter λ and randomness $r \in \{0,1\}^\lambda$ and outputs a key pair (pk, sk). We write $(pk, sk) \xleftarrow{\$} \mathsf{Gen^{server}}(1^\lambda)$ to denote that a pair (pk, sk) is the output of $\mathsf{Gen^{server}}$ when executed with uniformly random $r \xleftarrow{\$} \{0,1\}^\lambda$.*
- $\mathsf{KE^{client}_{init}}(pk_j, r_i) \rightarrow (k^{i,j}_{\mathsf{tmp}}, m_i)$: *An algorithm that takes as input a public key pk_j and randomness $r_i \in \{0,1\}^\lambda$, and outputs a temporary key $k^{i,j}_{\mathsf{tmp}}$ and a message m_i.*
- $\mathsf{KE^{server}_{refresh}}(sk_j, r_j, m_i) \rightarrow (k^{j,i}_{\mathsf{main}}, k^{j,i}_{\mathsf{tmp}}, m_j)$: *An algorithm that takes as input a secret key sk_j, randomness r_j and a message m_i, and outputs a key $k^{j,i}_{\mathsf{main}}$, a temporary key $k^{j,i}_{\mathsf{tmp}}$ and a message m_j.*
- $\mathsf{KE^{client}_{refresh}}(pk_j, r_i, m_j) \rightarrow k^{i,j}_{\mathsf{main}}$: *An algorithm that takes as input a public key pk_j, randomness r_i, and message m_j, and outputs a key $k^{i,j}_{\mathsf{main}}$.*

We say that a 0-RTT key exchange scheme is correct, *if for all $(pk_j, sk_j) \xleftarrow{\$} \mathsf{Gen^{server}}(1^\lambda)$ and for all $r_i, r_j \xleftarrow{\$} \{0,1\}^\lambda$ holds that*

$$\Pr[k^{i,j}_{\mathsf{tmp}} \neq k^{j,i}_{\mathsf{tmp}} \text{ or } k^{i,j}_{\mathsf{main}} \neq k^{j,i}_{\mathsf{main}}] \leq \mathsf{negl}(\lambda) \ ,$$

where $(k^{j,i}_{\mathsf{tmp}}, m_i) \leftarrow \mathsf{KE^{client}_{init}}(pk_j, r_i)$, $(k^{i,j}_{\mathsf{main}}, k^{i,j}_{\mathsf{tmp}}, m_j) \leftarrow \mathsf{KE^{server}_{refresh}}(sk_j, r_j, m_i)$, and $k^{j,i}_{\mathsf{main}} \leftarrow \mathsf{KE^{client}_{refresh}}(pk_j, r_i, m_j)$.

A 0-RTT KE scheme is used by a set parties which are either clients C or servers S (cf. Fig. 2). Each server S_p has a generated key pair $(sk_p, pk_p) \xleftarrow{\$} \mathsf{Gen^{server}}(1^\lambda, j)$ with published pk_p. The protocol is executed as follows:

1. The client oracle C_i chooses $r_i \in \{0,1\}^\lambda$ and selects the public key of the intended partner S_j (which must be a server, otherwise this value is undefined). Then it computes $(k^{i,j}_{\mathsf{tmp}}, m_i) \leftarrow \mathsf{KE^{client}_{init}}(pk_j, r_i)$, and sends m_i to S_j. Additionally, C_i can use $k^{i,j}_{\mathsf{tmp}}$ to encrypt some data M_i.

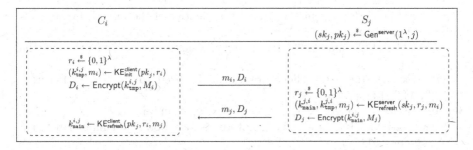

Fig. 2. Execution of a 0-RTT KE Protocol with Server-Only Authentication in Parallel to Encrypted Application Data. Note that the messages D_i and D_j correspond to the symmetric encryption protocol used to encrypt payload data, and are therefore *not* part of the 0-RTT KE protocol, but a separate protocol. These messages are only displayed here only to illustrate the basic, parallel application message flow to that of a 0-RTT KE protocol. While it would in principle be possible to define the symmetric encryption directly as part of the protocol, this would require a significantly more complex "ACCE-style" [17] security model, which we avoid for sake of simplicity.

2. Upon reception of message m_i, S_j initializes a new oracle $S_{j,t}$. This oracle chooses $r_j \in \{0,1\}^\lambda$ and computes $(k_{main}^{j,i}, k_{tmp}^{j,i}, m_j) \leftarrow \mathsf{KE}_{refresh}^{server}(sk_j, r_j, m_i)$. The server may use the ephemeral key $k_{tmp}^{j,i}$ to decrypt D_i. Then, the server sends m_j and optionally some data M_j encrypted with the key $k_{main}^{j,i}$ to the client.

3. C_i computes $k_{main}^{i,j} \leftarrow \mathsf{KE}_{refresh}^{client}(pk_j, r_i, m_j)$ and can optionally decrypt D_j. Correctness of the 0-RTT KE scheme guarantees that $k_{main}^{i,j} = k_{main}^{j,i}$.

3.2 Execution Environment

We provide an adversary $\mathcal{A}$ against a 0-RTT KE protocol with the following *execution environment*. Clients, which are not in possession of a long-term secret are represented by oracles $C_1, \ldots, C_d$ (without any particular "identity"). We consider ℓ servers, each server has a long-term key pair (sk_j, pk_j)[7], $j \in \{1, \ldots, \ell\}$, and each client has access to all public keys $pk_1, \ldots, pk_\ell$. Each server is represented by a collection of k oracles $S_{j,1}, \ldots, S_{j,k}$, where each oracle represents a process that executes one single instance of the protocol.

We use the following variables to maintain the internal state of oracles.

Clients. Each client oracle C_i, $i \in [d]$, maintains
- two variables k_i^{tmp} and k_i^{main} to store the temporal and main keys of a session,

---

[7] We do not distinguish between static (i.e. long-lived) and semi-static (i.e. medium lived) key pairs. Thus the long-lived keys in this model correspond to the server configuration file keys of QUIC and TLS 1.3.

- a variable $\mathsf{Partner}_i$, which contains the identity of the intended communication partner, and
- variables $\mathcal{M}_i^{\mathsf{in}}$ and $\mathcal{M}_i^{\mathsf{out}}$ containing messages sent and received by the oracle.

The internal state of a client oracle is initialized to $(\mathsf{k}_i^{\mathsf{tmp}}, \mathsf{k}_i^{\mathsf{main}}, \mathsf{Partner}_i, \mathcal{M}_i^{\mathsf{in}}, \mathcal{M}_i^{\mathsf{out}}) := (\emptyset, \emptyset, \emptyset, \emptyset, \emptyset)$.

Servers. Each server oracle $S_{j,t}$, $(j,t) \in [\ell] \times [k]$, maintains:

- two variables $\mathsf{k}_i^{\mathsf{tmp}}$ and $\mathsf{k}_i^{\mathsf{main}}$ to store the temporal and main keys of a session, and
- variables $\mathcal{M}_{j,t}^{\mathsf{in}}$ and $\mathcal{M}_{j,t}^{\mathsf{out}}$ containing messages sent and received by the server.

The internal state of a server oracle is initialized to $(\mathsf{k}_{j,t}^{\mathsf{tmp}}, \mathsf{k}_{j,t}^{\mathsf{main}}, \mathcal{M}_{j,t}^{\mathsf{in}}, \mathcal{M}_{j,t}^{\mathsf{out}}) := (\emptyset, \emptyset, \emptyset, \emptyset)$.

We say that an oracle has *accepted the temporal key* if $\mathsf{k}^{\mathsf{tmp}} \neq \emptyset$, and *accepted the main key* if $\mathsf{k}^{\mathsf{main}} \neq \emptyset$.

In the security experiment, the adversary is able to interact with the oracles by issuing the following queries.

$\mathsf{Send}(C_i/S_{j,t}, m)$. The adversary sends a message m to the requested oracle. The oracle processes m according to the protocol specification. Any response generated by the oracle according to the protocol specification is returned to the adversary.

If a client oracle C_i receives m as the first message, then the oracle checks if m consists of a special initialization message $(m = (\mathsf{init}, j))$. If true, then the oracle responds with the first protocol message generated for intended partner $S_{j,}$, else it outputs $\bot$.

$\mathsf{Reveal}(C_i/S_{j,t}, \mathsf{tmp/main})$. This query returns the key of the given stage if it already has been computed, or $\bot$ otherwise.

$\mathsf{Corrupt}(j)$. On input of a server identity j, this query returns the long-term private key of the server. If $\mathsf{Corrupt}(j)$ is the τ-th query issued by $\mathcal{A}$, we say a party is τ-*corrupted*. For parties that are not corrupted we define $\tau := \infty$.

$\mathsf{Test}(C_i/S_{j,t}, \mathsf{tmp/main})$. This query is used to test a key and is only asked once. It is answered as follows: If the variable of the requested key is not empty, a random $b \xleftarrow{\$} \{0,1\}$ is selected, and

- if $b = 0$ then the requested key is returned, else
- if $b = 1$ then a random key, according to the probability distribution of keys generated by the protocol, is returned.

Otherwise $\bot$ is returned.

Security Model Security Game $\mathcal{G}_{\mathcal{A}}^{\mathcal{ORTT}-sa}$. After receiving a security parameter λ the challenger $\mathcal{C}$ simulates the protocol and keeps track of all variables of the execution environment: he generates the long-lived key pairs of all server parties and answers faithfully to all queries by the adversary.

The adversary receives all public keys $pk_1, \ldots, pk_\ell$ and can interact with the challenger by issuing any combination of the queries $\mathsf{Send}()$, $\mathsf{Corrupt}()$, and

Reveal(). At some point the adversary queries Test() to an oracle and receives a key, which is either the requested key or a random value. The adversary may continue asking Send(), Corrupt(), and Reveal()-queries after receiving the key and finally outputs some bit b'.

Definition 5 (0-RTT KE-Security with Server-Only Authentication). *Let an adversary $\mathcal{A}$ interact with the challenger in game $\mathcal{G}_{\mathcal{A}}^{0\mathcal{RTT}-sa}$ as it is described above. We say the challenger outputs 1, denoted by $\mathcal{G}_{\mathcal{A}}^{0\mathcal{RTT}-sa}(\lambda) = 1$, if $b = b'$ and the following conditions hold:*

- *if $\mathcal{A}$ issues $\mathsf{Test}(\mathsf{C}_i, \mathsf{tmp})$ all of the following hold:*
  - $\mathsf{Reveal}(\mathsf{C}_i, \mathsf{tmp})$ *was never queried by $\mathcal{A}$*
  - $\mathsf{Reveal}(\mathsf{S}_{j,t}, \mathsf{tmp})$ *was never queried by $\mathcal{A}$ for any oracle $\mathsf{S}_{j,t}$ such that* $\mathsf{Partner}_i = j$ *and* $\mathcal{M}_{j,t}^{in} = \mathcal{M}_i^{out}$
  - *the communication partner $\mathsf{Partner}_i = j$, if it exists, is not τ-corrupted with $\tau < \infty$*
- *if $\mathcal{A}$ issues $\mathsf{Test}(\mathsf{C}_i, \mathsf{main})$ all of the following hold:*
  - $\mathsf{Reveal}(\mathsf{C}_i, \mathsf{main})$ *was never queried by $\mathcal{A}$*
  - $\mathsf{Reveal}(\mathsf{S}_{j,t}, \mathsf{main})$ *was never queried by $\mathcal{A}$, where $\mathsf{Partner}_i = j$, $\mathcal{M}_{j,t}^{in} = \mathcal{M}_i^{out}$, and $\mathcal{M}_i^{in} = \mathcal{M}_{j,t}^{out}$*
  - *the communication $\mathsf{Partner}_i = j$ is not τ-corrupted with $\tau < \tau_0$, where $\mathsf{Test}(\mathsf{C}_i, \mathsf{main})$ is the τ_0-th query issued by $\mathcal{A}$*
- *if $\mathcal{A}$ issues $\mathsf{Test}(\mathsf{S}_{j,t}, \mathsf{tmp})$ all of the following hold:*
  - $\mathsf{Reveal}(\mathsf{S}_{j,t}, \mathsf{tmp})$ *was never queried by $\mathcal{A}$*
  - *there exists an oracle C_i with $\mathcal{M}_i^{out} = \mathcal{M}_{j,t}^{in}$*
  - $\mathsf{Reveal}(\mathsf{C}_i, \mathsf{tmp})$ *was never queried by $\mathcal{A}$ to any oracle C_i with $\mathcal{M}_i^{out} = \mathcal{M}_{j,t}^{in}$*
  - $\mathsf{Reveal}(\mathsf{S}_{j,t'}, \mathsf{tmp})$ *was never queried by $\mathcal{A}$ for any oracle $\mathsf{S}_{j,t'}$ with $\mathcal{M}_{j,t}^{in} = \mathcal{M}_{j,t'}^{in}$*
  - *j is not τ-corrupted with $\tau < \infty$*
- *if $\mathcal{A}$ issues $\mathsf{Test}(\mathsf{S}_{j,t}, \mathsf{main})$ all of the following hold:*
  - $\mathsf{Reveal}(\mathsf{S}_{j,t}, \mathsf{main})$ *was never queried by $\mathcal{A}$*
  - *there exists an oracle C_i with $\mathcal{M}_i^{out} = \mathcal{M}_{j,t}^{in}$*
  - $\mathsf{Reveal}(\mathsf{C}_i, \mathsf{main})$ *was never queried by $\mathcal{A}$, if $\mathcal{M}_i^{in} = \mathcal{M}_{j,t}^{out}$*

else the game outputs a random bit. We define the advantage of $\mathcal{A}$ in the game $\mathcal{G}_{\mathcal{A}}^{0\mathcal{RTT}-sa}(\lambda)$ by

$$\mathsf{Adv}_{\mathcal{A}}^{0\mathcal{RTT}-sa}(\lambda) := \left| 2 \cdot \Pr[\mathcal{G}_{\mathcal{A}}^{0\mathcal{RTT}-sa}(\lambda) = 1] - 1 \right|.$$

Definition 6. *We say that a 0-RTT key exchange protocol is* test-secure, *if there exists a negligible function $\mathsf{negl}(\lambda)$ such that for all PPT adversaries $\mathcal{A}$ interacting according to the security game $\mathcal{G}_{\mathcal{A}}^{0\mathcal{RTT}-sa}(\lambda)$ it holds that*

$$\mathsf{Adv}_{\mathcal{A}}^{0\mathcal{RTT}-sa}(\lambda) \leq \mathsf{negl}(\lambda).$$

Remark 2. Our security model captures forward secrecy for the main-key, because key indistinguishability is required to hold even if the adversary is able to corrupt the communication partner of the test-oracle (but only after the test-oracle has accepted, of course, in order to avoid trivial attacks).

Moreover, strong key independence is modeled by the fact that an adversary which attempts to distinguish a tmp-key from random (i.e., an adversary which asks Test(X, tmp) for $X \in \{C_i, S_{j,t}$ for some $i, j, t\}$) is allowed to learn the main-key of X. Similarly, an adversary which tries to distinguish a main-key from random by asking Test(X, main) is allowed to learn the tmp-key of X as well. Security in this sense guarantees that the tmp-key and the main-key look independent to a computationally-bounded adversary.

Remark 3. Note that the requirements of $\mathcal{M}_i^{out} = \mathcal{M}_{j,t}^{in}$ etc. in the above security model essentially adopt the notion of *matching conversations*, defined by Bellare and Rogaway [2] for general, multi-message key exchange protocols, to the special case of 0-RTT KE.

3.3 Composing a 0-RTT KE Protocol with Symmetric Encryption

The security model described above considers only the 0-RTT KE protocol, *without* symmetric encryption of payload data (that is, without the messages D_i and D_j displayed in Fig. 2). A protocol secure in this sense guarantees the indistinguishability of keys in a hypothetical setting, where the key is not used for symmetric encryption of payload messages potentially known to the adversary. One may think that this is not sufficient for 0-RTT KE, because the key *will* be used to encrypt payload data, and this will enable an adversary to trivially distinguish a "real" key from a "random" key (this holds for both the "temporal" key $k_{tmp}^{i,j}$ and the actual "main" session key $k_{main}^{i,j}$). Note that this argument applies not only to the above 0-RTT KE security model, but actually to any security model for (authenticated) key exchange which is based on the indistinguishability of keys, such as the classical model of Bellare and Rogaway and many similar models [2, 5, 7, 10, 20, 27]. In practice, this key will usually be used in a cryptographic protocol, e.g. to encrypt messages, and therefore trivially allow for distinguishing "real" from "random" keys. The security of the composition of a protocol secure in the sense of [2, 5, 7, 10, 20, 27] with a symmetric encryption protocol follows from a standard two-step hybrid argument, which essentially proceeds as follows:

1. In the original security experiment, the adversary interacts with a composed protocol, where the KE protocol is first used to derive a key k, which is then used to encrypt payload data with the symmetric encryption protocol.
2. In the next hybrid experiment, the adversary interacts with a composed protocol, where the symmetric encryption does not use the key k computed by the KE protocol, but an independent random key. Note that an adversary that distinguishes this hybrid from the original game can be used to distinguish a "real" key of the KE protocol from a "random" one.

Now the adversary interacts with an encryption protocol that uses a key which is *independent* of the KE protocol. This allows for a reduction of the security of the composed protocol to the security of the symmetric protocol.

A similarly straightforward hybrid argument applies to the composition of 0-RTT KE with symmetric encryption, which works as follows:

1. In the original security experiment, the adversary interacts with a composed protocol, where the 0-RTT KE protocol is first used to derive a key $k_{\mathrm{tmp}}^{i,j}$, which is then used to encrypt the payload data sent along with the first protocol message. Then the 0-RTT KE protocol is used to derive the main key $k_{\mathrm{main}}^{i,j}$, which in turn is used to encrypt all further payload data.
2. In the first hybrid experiment, the adversary interacts with a composed protocol, where only $k_{\mathrm{tmp}}^{i,j}$ is replaced with an independent random value. An adversary that distinguishes this hybrid from the original game can be used to distinguish a "real" $k_{\mathrm{tmp}}^{i,j}$ from a "random" one.

   Now the adversary interacts with an encryption protocol that encrypts the first payload message with a key which is *independent* of the 0-RTT KE protocol. This allows for a reduction of the security of the first payload message to the security of the symmetric protocol.
3. In the second hybrid experiment, the adversary interacts with a composed protocol, where $k_{\mathrm{main}}^{i,j}$ is now also replaced with an independent random value. An adversary that distinguishes this hybrid from the previous one can be used to distinguish a "real" $k_{\mathrm{main}}^{i,j}$ from a "random" one. This allows for a reduction of the security of all further payload messages to the security of the symmetric protocol.

Following the long tradition of previous works on indistinguishability-based key exchange security models [2,5,7,10,20,27], we can thus consider an indistinguishability-based security model for 0-RTT KE even though in practice key exchange messages will be interleaved with messages of the symmetric encryption protocol. This allows for simple security models, and enables a modular analysis of the building blocks of a composed protocol.

4   Generic Construction of 0-RTT KE from NIKE

Now we are ready to describe our generic NIKE-based 0-RTT KE protocol and its security analysis.

4.1   Generic Construction

Let NIKE = (NIKE.Gen, NIKE.Key) be a NIKE scheme according to Definition 2 and let SIG = (SIG.Gen, SIG.Sign, SIG.Vfy) be a signature scheme. Then we construct a 0-RTT KE scheme 0-RTT = $(\mathsf{Gen}^{\mathsf{server}}, \mathsf{KE}_{\mathsf{init}}^{\mathsf{client}}, \mathsf{KE}_{\mathsf{refresh}}^{\mathsf{client}}, \mathsf{KE}_{\mathsf{refresh}}^{\mathsf{server}})$, per Definition 4, in the following way (cf. Fig. 3).

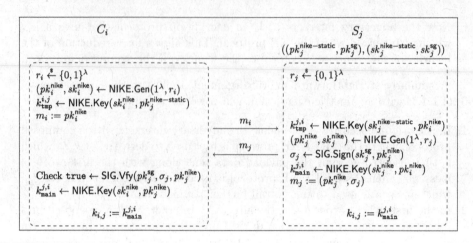

Fig. 3. 0-RTT KE from NIKE. Again, it is possible to include the parallel execution of a symmetric encryption protocol which would behave as in Fig. 2 for encrypted application data. As such a protocol is *not* part of the 0-RTT KE protocol, we omit it here for simplicity.

- $\mathsf{Gen}^{\mathsf{server}}(1^\lambda, r)$ computes key pairs using the NIKE key generation algorithm $(pk^{\mathsf{nike-static}}, sk^{\mathsf{nike-static}}) \xleftarrow{\$} \mathsf{NIKE.Gen}(1^\lambda)$ and signature keys using the SIG algorithm $(pk^{\mathsf{sg}}, sk^{\mathsf{sg}}) \xleftarrow{\$} \mathsf{SIG.Gen}$, and outputs

$$(pk, sk) := ((pk^{\mathsf{nike-static}}, pk^{\mathsf{sg}}), (sk^{\mathsf{nike-static}}, sk^{\mathsf{sg}})).$$

- $\mathsf{KE}^{\mathsf{client}}_{\mathsf{init}}(pk_j, r_i)$ samples $r_i \xleftarrow{\$} \{0,1\}^\lambda$, parses $pk_j = (pk_j^{\mathsf{nike-static}}, pk_j^{\mathsf{sg}})$, runs $(pk_i^{\mathsf{nike}}, sk_i^{\mathsf{nike}}) \leftarrow \mathsf{NIKE.Gen}(1^\lambda, r_i)$ and $k_{i,j}^{\mathsf{nike}} \leftarrow \mathsf{NIKE.Key}(sk_i^{\mathsf{nike}}, pk_j^{\mathsf{nike-static}})$, and outputs

$$(k_{\mathsf{tmp}}^{i,j}, m_i) := (k_{i,j}^{\mathsf{nike}}, pk_i^{\mathsf{nike}}).$$

- $\mathsf{KE}^{\mathsf{server}}_{\mathsf{refresh}}(sk_j, r_j, m_i)$ takes in $m_i = pk_i^{\mathsf{nike}}$, parses $sk_j = (sk_j^{\mathsf{nike-static}}, sk_j^{\mathsf{sg}})$, and samples $r_j \xleftarrow{\$} \{0,1\}^\lambda$. It then computes $k_{i,j}^{\mathsf{nike}} \leftarrow \mathsf{NIKE.Key}(sk_j^{\mathsf{nike-static}}, pk_i^{\mathsf{nike}})$, $(pk_j^{\mathsf{nike}}, sk_j^{\mathsf{nike}}) \leftarrow \mathsf{NIKE.Gen}(1^\lambda, r_j)$, and $\sigma_j \leftarrow \mathsf{SIG.Sign}(sk_j^{\mathsf{sg}}, pk_j^{\mathsf{nike}})$. If $m_i = pk_j^{\mathsf{nike-static}}$ then it samples $k_{\mathsf{main}}^{\mathsf{nike}}$ uniformly random, else it computes $k_{\mathsf{main}}^{\mathsf{nike}} \leftarrow \mathsf{NIKE.Key}(sk_j^{\mathsf{nike}}, pk_i^{\mathsf{nike}})$, outputting

$$(k_{\mathsf{main}}^{j,i}, k_{\mathsf{tmp}}^{j,i}, m_j) := (k_{\mathsf{main}}^{\mathsf{nike}}, k_{i,j}^{\mathsf{nike}}, (pk_j^{\mathsf{nike}}, \sigma_j)).$$

- $\mathsf{KE}^{\mathsf{client}}_{\mathsf{refresh}}(pk_j, r_i, m_j)$ parses $pk_j = (pk_j^{\mathsf{nike-static}}, pk_j^{\mathsf{sg}})$ and $m_j = (pk_j^{\mathsf{nike}}, \sigma_j)$. It then checks $\mathtt{true} \leftarrow \mathsf{SIG.Vfy}(pk_j^{\mathsf{sg}}, \sigma_j, pk_j^{\mathsf{nike}})$ and computes $k_{\mathsf{main}}^{\mathsf{nike}} \leftarrow \mathsf{NIKE.Key}(sk_i^{\mathsf{nike}}, pk_j^{\mathsf{nike}})$, outputting $k_{\mathsf{main}}^{i,j} := k_{\mathsf{main}}^{\mathsf{nike}}$.

Ultimately, the construction follows by applying the NIKE NIKE.Gen algorithm and the signature SIG.Gen algorithm to generate a server configuration file which is comprised of the server public key and a server public signature key which a client can then employ for generating the first protocol flow. In order for the 0-RTT KE construction to abstract the security guarantees of the underlying NIKE, the appropriate client $(pk_i^{\text{nike}}, sk_i^{\text{nike}})$ must be available for use in the NIKE.Key algorithm. Consequently, the $(pk_i^{\text{nike}}, sk_i^{\text{nike}})$ values are generated locally by the client, with pk_i^{nike} passed to the server as a message. Note that this construction naturally forgoes client-side authentication. Figure 3 demonstrates the construction.

Remark 4. One may wonder why we define $\text{KE}_{\text{refresh}}^{\text{server}}(sk_j, r_j, m_i)$ such that it samples a random key when it takes as input a client message m_i which is equal to its own static NIKE key, that is, if $m_i = pk_j^{\text{nike}-\text{static}}$. We note that this is necessary for the security the constructed 0-RTT KE scheme to be reducible to that of the NIKE scheme, because in some cases we will not be able to simulate the key computed by a server oracle that receives as input a message which is equal to the "static" NIKE public key contained in its 0-RTT KE public key. Note that this incurs a negligible correctness error. However, it is straightforward to verify the correctness of the protocol according to Definition 4.

Theorem 1. *Let* 0-RTT *be executed with* d *clients,* ℓ *servers with long-term keys, and* k *server oracles modeling each server. From each attacker* $\mathcal{A}$*, we can construct attackers* $\mathcal{B}_{sig}$*, according to Definition 1, and* $\mathcal{B}_{nike}$*, according to Definition 3, such that*

$$\text{Adv}_{\mathcal{A}}^{0RTT-sa}(\lambda) \le 2kd\ell \cdot \left(\text{Adv}_{\text{NIKE},\mathcal{B}_{nike}}^{CKS\text{-}light}(\lambda) + \text{Adv}_{\text{SIG},\mathcal{B}_{sig}}^{sEUF\text{-}CMA}(\lambda) \right)$$

$$+ d\ell \cdot \left(k \cdot \text{Adv}_{\text{NIKE},\mathcal{B}_{nike}}^{CKS\text{-}light}(\lambda) + \text{Adv}_{\text{SIG},\mathcal{B}_{sig}}^{sEUF\text{-}CMA}(\lambda) \right)$$

$$+ d\ell \cdot \left(\text{Adv}_{\text{NIKE},\mathcal{B}_{nike}}^{CKS\text{-}light}(\lambda) + \text{Adv}_{\text{SIG},\mathcal{B}_{sig}}^{sEUF\text{-}CMA}(\lambda) \right) + 4 \cdot \text{Adv}_{\text{NIKE},\mathcal{B}_{nike}}^{CKS\text{-}light}(\lambda) .$$

The running time of $\mathcal{B}_{sig}$ *and* $\mathcal{B}_{nike}$ *is approximately equal to the time required to execute the security experiment with* $\mathcal{A}$ *once.*

Intuition for the Proof of Theorem 1. In order to prove Theorem 1, we distinguish between four types of attackers:

- adversary $\mathcal{A}_1$ asks Test() to a client oracle and the temporary key (*CT-attacker*)
- adversary $\mathcal{A}_2$ asks Test() to a client oracle and the main key (*CM-attacker*)
- adversary $\mathcal{A}_3$ asks Test() to a server oracle and the temporary key (*ST-attacker*)
- adversary $\mathcal{A}_4$ asks Test() to a server oracle and the main key (*SM-attacker*).

Let us give some intuition why this classification of attackers will be useful for the security proof. In the 0-RTT KE scheme 0-RTT each party computes 2 different keys k_{tmp}' and k_{main}', where k_{tmp}' depends on the ephemeral keys of the client and the static keys of the server, and k'_{main} depends on the ephemeral keys of both parties. In our proof we want to be able to reduce the indistinguishability of the 0-RTT-key to the indistinguishability of the NIKE-key.

In the NIKE security experiment the attacker receives two challenge public keys $\{pk_i^{nike}, pk_j^{nike}\}$. In the reduction, we want to embed these keys in the 0-RTT security experiment, according to Sect. 3.2, and still be able to answer all Reveal()- and Corrupt()-queries correctly. In the case of adversaries that test the temporary key of the client or the server we can embed the NIKE-keys as $pk_j^{nike-static} = pk_j^{nike}$ and $m_i = pk_i^{nike}$. However, this does not work for adversaries against the main key, because k'_{main} depends on the ephemeral keys of the parties. In this case we have to embed the keys as $m_i = pk_i^{nike}$ and $m_j = pk_j^{nike}$. The Test()-query of the attacker in the 0-RTT experiment can then be answered with the challenge the attacker in the NIKE experiment receives.

References

1. Bellare, M., Rogaway, P.: Random oracles are practical: a paradigm for designing efficient protocols. In: Ashby, V. (ed.) ACM CCS 1993, 3–5 November 1993, Fairfax, Virginia, USA, pp. 62–73. ACM Press (1993)
2. Bellare, M., Rogaway, P.: Entity authentication and key distribution. In: Stinson, D.R. (ed.) CRYPTO 1993. LNCS, vol. 773, pp. 232–249. Springer, Heidelberg (1994). doi:10.1007/3-540-48329-2_21
3. Bergsma, F., Jager, T., Schwenk, J.: One-round key exchange with strong security: an efficient and generic construction in the standard model. In: Katz, J. (ed.) PKC 2015. LNCS, vol. 9020, pp. 477–494. Springer, Heidelberg (2015). doi:10.1007/978-3-662-46447-2_21
4. Bhargavan, K., Fournet, C., Kohlweiss, M., Pironti, A., Strub, P.-Y., Zanella-Béguelin, S.: Proving the TLS handshake secure (as it is). In: Garay, J.A., Gennaro, R. (eds.) CRYPTO 2014. LNCS, vol. 8617, pp. 235–255. Springer, Heidelberg (2014). doi:10.1007/978-3-662-44381-1_14
5. Blake-Wilson, S., Johnson, D., Menezes, A.: Key agreement protocols and their security analysis. In: Darnell, M. (ed.) Cryptography and Coding 1997. LNCS, vol. 1355, pp. 30–45. Springer, Heidelberg (1997). doi:10.1007/BFb0024447
6. Boneh, D., Shen, E., Waters, B.: Strongly unforgeable signatures based on computational Diffie-Hellman. In: Yung et al. [30], pp. 229–240
7. Canetti, R., Krawczyk, H.: Analysis of key-exchange protocols and their use for building secure channels. In: Pfitzmann, B. (ed.) EUROCRYPT 2001. LNCS, vol. 2045, pp. 453–474. Springer, Heidelberg (2001). doi:10.1007/3-540-44987-6_28
8. Cash, D., Kiltz, E., Shoup, V.: The twin Diffie-Hellman problem and applications. In: Smart, N. (ed.) EUROCRYPT 2008. LNCS, vol. 4965, pp. 127–145. Springer, Heidelberg (2008). doi:10.1007/978-3-540-78967-3_8
9. Cremers, C.: Examining indistinguishability-based security models for key exchange protocols: the case of CK, CK-HMQV, and eCK. In: Cheung, B.S.N., Hui, L.C.K., Sandhu, R.S., Wong, D.S. (eds.) ASIACCS 2011, 22–24 March 2011, Hong Kong, China, pp. 80–91. ACM Press (2011)
10. Cremers, C.J.F.: Session-state reveal is stronger than ephemeral key reveal: attacking the NAXOS authenticated key exchange protocol. In: Abdalla, M., Pointcheval, D., Fouque, P.-A., Vergnaud, D. (eds.) ACNS 2009. LNCS, vol. 5536, pp. 20–33. Springer, Heidelberg (2009). doi:10.1007/978-3-642-01957-9_2

11. Fischlin, M., Günther, F.: Multi-stage key exchange and the case of Google's QUIC protocol. In: Ahn, G.-J., Yung, M., Li, N. (eds.) ACM CCS 2014, 3–7 November 2014, Scottsdale, AZ, USA, pp. 1193–1204. ACM Press (2014)
12. Freire, E.S.V., Hofheinz, D., Kiltz, E., Paterson, K.G.: Non-interactive key exchange. In: Kurosawa, K., Hanaoka, G. (eds.) PKC 2013. LNCS, vol. 7778, pp. 254–271. Springer, Heidelberg (2013). doi:10.1007/978-3-642-36362-7_17
13. Green, M.D., Miers, I.: Forward secure asynchronous messaging from puncturable encryption. In: IEEE S&P 2015 [16], pp. 305–320
14. Günther, F., Hale, B., Jager, T., Lauer, S.: 0-RTT key exchange with full forward secrecy. In: Coron, J.-S., Nielsen, J.B. (eds.) EUROCRYPT 2017. LNCS, vol. 10212, pp. 519–548. Springer, Cham (2017). doi:10.1007/978-3-319-56617-7_18
15. Hale, B., Jager, T., Lauer, S., Schwenk, J.: Simple security definitions for and constructions of 0-RTT key exchange. Cryptology ePrint Archive, Report 2015/1214 (2015). http://eprint.iacr.org/2015/1214
16. IEEE Symposium on Security and Privacy, 17–21 May 2015, San Jose, CA, USA. IEEE Computer Society Press (2015)
17. Jager, T., Kohlar, F., Schäge, S., Schwenk, J.: On the security of TLS-DHE in the standard model. In: Safavi-Naini, R., Canetti, R. (eds.) CRYPTO 2012. LNCS, vol. 7417, pp. 273–293. Springer, Heidelberg (2012). doi:10.1007/978-3-642-32009-5_17
18. Krawczyk, H.: HMQV: a high-performance secure Diffie-Hellman protocol. In: Shoup, V. (ed.) CRYPTO 2005. LNCS, vol. 3621, pp. 546–566. Springer, Heidelberg (2005). doi:10.1007/11535218_33
19. Krawczyk, H., Paterson, K.G., Wee, H.: On the security of the TLS protocol: a systematic analysis. In: Canetti, R., Garay, J.A. (eds.) CRYPTO 2013. LNCS, vol. 8042, pp. 429–448. Springer, Heidelberg (2013). doi:10.1007/978-3-642-40041-4_24
20. LaMacchia, B., Lauter, K., Mityagin, A.: Stronger security of authenticated key exchange. In: Susilo, W., Liu, J.K., Mu, Y. (eds.) ProvSec 2007. LNCS, vol. 4784, pp. 1–16. Springer, Heidelberg (2007). doi:10.1007/978-3-540-75670-5_1
21. Lauter, K., Mityagin, A.: Security analysis of KEA authenticated key exchange protocol. In: Yung et al. [30], pp. 378–394
22. Law, L., Menezes, A., Minghua, Q., Solinas, J., Vanstone, S.: An efficient protocol for authenticated key agreement. Des. Codes Crypt. 28(2), 119–134 (2003)
23. Lychev, R., Jero, S., Boldyreva, A., Nita-Rotaru, C.: How secure and quick is QUIC? Provable security and performance analyses. In: IEEE S&P 2015 [16], pp. 214–231
24. NIST: SKIPJACK and KEA algorithm specifications (1998). http://csrc.nist.gov/groups/STM/cavp/documents/skipjack/skipjack.pdf
25. Rescorla, E.: The Transport Layer Security (TLS) Protocol Version 1.3: draft-ietf-tls-tls13-08. Technical report, August 2015. Expires 29 Feb 2016
26. Rescorla, E.: The Transport Layer Security (TLS) Protocol Version 1.3: draft-ietf-tls-tls13-18. Technical report, October 2016. Expires 29 April 2017
27. Sarr, A.P., Elbaz-Vincent, P., Bajard, J.-C.: A new security model for authenticated key agreement. In: Garay, J.A., Prisco, R. (eds.) SCN 2010. LNCS, vol. 6280, pp. 219–234. Springer, Heidelberg (2010). doi:10.1007/978-3-642-15317-4_15
28. Steinfeld, R., Pieprzyk, J., Wang, H.: How to strengthen any weakly unforgeable signature into a strongly unforgeable signature. In: Abe, M. (ed.) CT-RSA 2007. LNCS, vol. 4377, pp. 357–371. Springer, Heidelberg (2006). doi:10.1007/11967668_23

29. Yoneyama, K., Zhao, Y.: Taxonomical security consideration of authenticated key exchange resilient to intermediate computation leakage. In: Boyen, X., Chen, X. (eds.) ProvSec 2011. LNCS, vol. 6980, pp. 348–365. Springer, Heidelberg (2011). doi:10.1007/978-3-642-24316-5_25
30. Yung, M., Dodis, Y., Kiayias, A., Malkin, T. (eds.): PKC 2006. LNCS, vol. 3958. Springer, Heidelberg (2006)
31. Zhao, Y.: Identity-concealed authenticated encryption and key exchange. In: Weippl, E.R., Katzenbeisser, S., Kruegel, C., Myers, A.C., Halevi, S. (eds.) ACM CCS 2016, 24–28 October 2016, Vienna, Austria, pp. 1464–1479. ACM Press (2016)

TOPPSS: Cost-Minimal Password-Protected Secret Sharing Based on Threshold OPRF

Stanisław Jarecki[1($\boxtimes$)], Aggelos Kiayias[2], Hugo Krawczyk[3], and Jiayu Xu[1]

[1] University of California, Irvine, USA
stasio@ics.uci.edu, jiayux@uci.edu
[2] University of Edinburgh, Edinburgh, UK
aggelos@kiayias.com
[3] IBM Research, New York City, USA
hugo@ee.technion.ac.il

Abstract. We present TOPPSS, the most efficient Password-Protected Secret Sharing (PPSS) scheme to date. A (t, n)-threshold PPSS, introduced by Bagherzandi et al. [4], allows a user to share a secret among n servers so that the secret can later be reconstructed by the user from any subset of $t + 1$ servers with the sole knowledge of a password. It is guaranteed that any coalition of up to t corrupt servers learns nothing about the secret (or the password). In addition to providing strong protection to secrets stored online, PPSS schemes give rise to efficient Threshold PAKE (T-PAKE) protocols that armor single-server password authentication against the inherent vulnerability to offline dictionary attacks in case of server compromise.

TOPPSS is *password-only*, i.e. it does not rely on public keys in reconstruction, and enjoys remarkable efficiency: A single communication round, a single exponentiation per server and just two exponentiations per client regardless of the number of servers. TOPPSS satisfies threshold security under the (Gap) One-More Diffie-Hellman (OMDH) assumption in the random-oracle model as in prior efficient realizations of PPSS/T-PAKE [18, 19]. Moreover, we show that TOPPSS realizes the *Universally Composable* PPSS notion of [19] under a generalization of OMDH, the *Threshold* One-More Diffie-Hellman (T-OMDH) assumption. We show that the T-OMDH and OMDH assumptions are both hard in the generic group model.

The key technical tool we introduce is a universally composable *Threshold Oblivious PRF* which is of independent interest and applicability.

1 Introduction

Passwords have well-known weaknesses as authentication tokens, foremost because of their vulnerability to offline dictionary attacks in case of the

S. Jarecki—Supported by NSF grant CNS-1547435.
A. Kiayias—Supported by ERC project CODAMODA and H2020 Project Panoramix, #653497.
H. Krawczyk—Supported by ONR Contract N00014-14-C-0113.

© Springer International Publishing AG 2017
D. Gollmann et al. (Eds.): ACNS 2017, LNCS 10355, pp. 39–58, 2017.
DOI: 10.1007/978-3-319-61204-1_3

all-too-common leakage of the database of password hashes stored by the authentication server (see e.g., [1]). Worse still, most people re-use their passwords across multiple services, hence a break-in into one service effectively breaks the security of others. Yet, because of their convenience, passwords are a dominant form of authentication, and the amount and value of information protected using passwords keeps growing. Defenses such as the use of secondary authentication factors (e.g., a PIN generated by a personal device or a USB dongle) increase protection against on-line attacks but not against offline attacks upon server compromise. Techniques such as Password Authenticated Key Exchange (PAKE) [6,8] improve on today's de-facto standard of "password over TLS" authentication by eliminating the reliance on a Public Key Infrastructure (PKI), but they do not help against offline attacks after server compromise.

T-PAKE and PPSS. To address the threat of offline dictionary attacks on the server, Mackenzie et al. [26] introduced (t, n)-Threshold PAKE (T-PAKE), which replaces a single authentication server with a group of n servers and leaks no information on passwords even if up to t servers are corrupted. Bagherzandi et al. [4] proposed a related notion of Password-Protected Secret Sharing (PPSS) which simplifies the notion of T-PAKE by reducing the goal of key exchange between user and servers to that of the user retrieving a single secret previously shared with the servers. Specifically, a (t, n)-PPSS scheme, as formulated in the PKI-free setting by [18], allows a user to share a random secret s among n servers under the protection of her password pw s.t. (1) a reconstruction protocol involving at least $t + 1$ honest servers recovers s if the user inputs the (correct) password pw; (2) the compromise of up to t servers leaks no information about either s or pw; (3) an adversary who corrupts $t' \leq t$ servers and has q_U interactions with the user and q_S interactions with the uncorrupted servers can test at most $\frac{q_S}{t-t'+1} + q_U$ passwords. (In the PKI setting one can set $q_U = 0$.)

The PPSS notion is useful in the design of efficient T-PAKE's because of the low-overhead generic PPSS-to-TPAKE compiler [4,18]. It is also an important primitive in its own right, allowing for online storage of sensitive information like keys, credentials, or personal records, with availability and privacy protection. The only token needed for retrieving stored information is a *single* password, and both information and password remain private if no more than t servers are compromised (and if the adversary does not guess or learn the password).

In this paper we present TOPPSS, the most efficient PPSS scheme to date – and using the PPSS-to-TPAKE compiler of [18] also the most efficient T-PAKE – with a hard-to-beat complexity as detailed below. Our work builds on the works of Jarecki et al. [18,19] who constructed PPSS protocols based on *Oblivious Pseudorandom Functions (OPRF)*, formulated as a universally composable (UC) functionality. The works of [18,19] define UC OPRF differently, but each instantiates its OPRF notion using the *blinded Diffie-Hellman* technique, following Ford and Kaliski [15], under the so-called (Gap) *One-More Diffie-Hellman* (OMDH) assumption [5,22] in the Random Oracle Model (ROM). Using one OPRF construction, [18] showed a PPSS whose reconstruction phase takes a single round between a user and $t + 1$ servers, with 2 (multi)exponentiations per server and $2t + 3$ for the user. The PPSS of [19] uses a simplified OPRF scheme secure under the same assumptions, with 1 exponentiation per server and $t + 2$ for the user. In addition to improving on [18] in efficiency, the latter scheme satisfies a stronger PPSS notion formulated as a UC functionality, which we adopt here.

Our Contributions. We present TOPPSS, a simple PPSS protocol with remarkable and hard to beat performance. The reconstruction procedure requires *just one exponentiation per server and a total of two exponentiations for the user* (independent of the number of servers), plus $O(t)$ modular multiplications by each party. Communication is also optimal: The user sends a single group element to a subset of $t + 1$ servers and gets one group element from each server. Furthermore, we show that this "minimal cost" (and PKI-free) PPSS satisfies the strong UC notion of PPSS from [19]. This contribution is based on the observation that a more efficient PPSS can result from replacing the OPRF used in the protocols of [18,19] with its *threshold* (or multi-party) counterpart which we define as *Threshold OPRF (T-OPRF)*. We provide a UC definition of T-OPRF as a functionality that allows a group of servers to secret-share a key k for PRF f with a shared PRF evaluation protocol which lets the user compute $f_k(x)$ on her input x, s.t. both x and k are secret if no more than t of n servers are corrupted. T-OPRF is an input-oblivious strengthening of Distributed PRF (DPRF) of Naor et al. [27], hence in particular T-OPRF can replace DPRF in all its applications, e.g. for corruption-resilient Key Distribution Center, and long-term information protection (see [27]).

Using this strong notion of T-OPRF security we show a compiler which transforms UC T-OPRF into UC PPSS at negligible additional cost (in ROM). In particular, TOPPSS is obtained by designing a T-OPRF protocol, denoted 2HashTDH, with the efficiency parameters stated above. This T-OPRF protocol is essentially a "threshold exponentiation" protocol, where each server computes m^{k_i} on input m where k_i is the server's secret-share of the PRF key k. We prove that TOPPSS realizes UC T-OPRF under the following assumptions in ROM. Let $t' \leq t$ denote the number of parties actually controlled by an attacker. First, our results imply that in the so-called *full corruption* case, i.e. if $t' = t$, the same (Gap) OMDH assumption used in [18,19] implies that the attacker must query one uncorrupted party per each input on which the attacker wants to obtain the function value. Since this is the case when the attacker controls the full threshold t of servers it is also the case for any $t' < t$. In the application to PPSS this means that the attacker can test up to $q_S + q_U$ passwords, which matches the $\frac{q_S}{t-t'+1} + q_U$ bound for $t' = t$. Since many existing works on T-PAKE, e.g. [2,9,14,23,26,31], implicitly assume the $t' = t$ case by defining security using the simplified $q_S + q_U$ bound on the number of passwords the adversary can test, we call this level of security a *standard threshold security* for T-PAKE/PPSS.

Secondly, for the general case of $t' \leq t$, we show that TOPPSS achieves the stronger $\frac{q_S}{t-t'+1} + q_U$ bound assuming a generalization of the OMDH assumption which we call (Gap) *Threshold One-More Diffie-Hellman (T-OMDH)*. As a sanity check for the T-OMDH assumption we show that the T-OMDH problem is hard in the generic group model. Since OMDH is a special case of T-OMDH, to the best of our knowledge this is also the first generic group analysis of OMDH. The stricter bound implies that an adversary controlling $t' \leq t$ servers must contact $t - t' + 1$ uncorrupted servers for each input on which it wants to compute the function, which coincides with the *standard threshold security* notion when $t' = t$, but it is stronger for $t' < t$. For example, it means that the default network adversary who does corrupt any party but runs q sessions with each server, can test up to $qn/(t+1)$ passwords, whereas the *standard threshold security* would in this case upper-bound the number of tested passwords only by qn.

As a point of comparison in the full version of this paper [20] we consider a generic compiler from *any* OPRF to T-OPRF. This compiler performs multi-party computation of the server code in the underlying OPRF protocol, but in the case of the OPRF of [19] such MPC protocol has the same low computational cost as the customized T-OPRF protocol 2HashTDH discussed above, i.e. 1 exponentiation per server and 2 for the user, with the only drawback of adding an additional communication round to enforce an agreement between the servers on the client's input to the MPC protocol. On the other hand, since the security depends only on the base OPRF, the resultant two-round T-OPRF protocol achieves the $\frac{qs}{t-t'+1} + q_U$ bound based solely on OMDH for all $t' \leq t$.

Other Applications. Oblivious PRFs have found multiple applications which can also enjoy the benefits of a threshold version, particularly given the remarkable efficiency of our schemes. Examples of such applications include search on encrypted data [13,17], set intersection [22], and multiple-server DE-PAKE (device enhanced PAKE) [21].

Related Work. The first (t, n)-Threshold PAKE (T-PAKE) by Mackenzie et al. [26] required ROM in the security analysis and relied on PKI, namely, it assumed that the client can validate the public keys of the servers *during the reconstruction phase.*[1] Gennaro and Raimondo [14] dispensed with ROM and PKI (in authentication) but increased protocol costs. Abdalla et al. [2] showed a PKI-free T-PAKE in ROM with fewer communication rounds than T-PAKE of [26] but the client establishes a key with only one designated *gateway* server. Yi et al. [31] showed a similar round-reduction without ROM. The case of $n = 2$ servers, known as 2-PAKE, received special attention starting with Brainard et al. [9,29] on 2-PAKE in ROM and PKI, and several works [7,23–25] addressed the non-PKI and no-ROM case. Still, each of these T-PAKE schemes requires server-to-server communication. If communication is mediated by the client then the lowest round complexity is 3 for $n > 2$ [2] and 2 for $n = 2$ [7,25].

Bagherzandi et al. [4] introduced the notion of Password-Protected Secret Sharing (PPSS) with the goal of simplifying T-PAKE protocols. Specifically,

---

[1] When we say that PPSS/T-PAKE assumes PKI we mean that it relies on it for the security of the reconstruction/authentication phase. By contrast, the *initialization phase* of any PPSS/T-PAKE solution must assume some trust infrastructure, e.g. PKI, or otherwise each party could be initializing the scheme with an impostor.

they showed a PPSS protocol in ROM assuming PKI, with 2 rounds, constant-sized messages, and $8(t + 1)$ (multi) exponentiations per client, and a low-cost PKI-model compiler from PPSS to T-PAKE. Camenisch et al. [10] constructed another PPSS scheme, called T-PASS, for *Threshold Password-Authenticated Secret Sharing*, without assuming PKI but with $14n$ exponentiations for the client, 7 exponentiations per server, and 5 rounds of communication.

Jarecki et al. [18,19] showed significantly faster PPSS protocols, also without assuming PKI (in reconstruction): The PPSS of [18] takes a single round (two messages) between a user and each server, and uses 2 (multi) exponentiations per server and $2t + 3$ (multi) exponentiations for the client, secure under (Gap) OMDH in ROM. (They also show a 4-message non-ROM PPSS with $O(n \cdot |\mathsf{pw}|)$ exponentiations using Paillier encryption.) The PPSS of [19] improves upon this with a single-round PPSS with 1 exponentiation per server and $t + 2$ exponentiations for the client, also under OMDH in ROM. In related works, [11] showed a single-round *proactive* PPSS in the PKI setting for the case of $t = n$, and [3] showed general methods for ensuring robustness in PPSS reconstruction, and a non-ROM PPSS using $O(|\mathsf{pw}|)$ exponentiations in a prime-order group.

Another important aspect of these PPSS solutions is the type of security notion they achieve. Both the PKI-model PPSS notion of [4] and the PKI-free PPSS notion of [18] were indistinguishability-based, while [10,19] provided Universally Composable (UC) definitions of the PPSS functionality. The essence of the UC PPSS definition of [19], which we adopt here, is that the only attack the adversary can stage is the inevitable one, namely, an online dictionary attack where validating a single password guess requires interaction with either $t + 1$ instances of the servers or with the user. The UC definitions have further advantages for a password-based notion like PPSS, e.g. they imply security in the presence of non-uniformly distributed passwords, correlated passwords used for different services, and password mistyping.

Organization. In Sect. 2 we define the fundamental tool TOPPSS relies on, namely T-OPRF, as a UC functionality. In Sect. 3 we show a single-round, 1exp/server + 2exp/client realization of T-OPRF, protocol 2HashTDH, secure in ROM under the Threshold OMDH assumption we introduce in that section. In Sect. 4 we show a low-cost compiler from T-OPRF to PPSS, which we exemplify in Sect. 5 with a concrete instantiation using 2HashTDH.

2 Universally Composable Threshold OPRF

Notation. We use ":=" for deterministic assignment, "←" for randomized assignment, and "$\leftarrow_R$" for uniform sampling from some set.

The T-OPRF Functionality. In the introduction we gave an informal overview of the notion of Threshold Oblivious PRF (T-OPRF) and its applicability, e.g. to PPSS schemes. Here we provide a formal definition of this notion as a secure realization of the UC functionality $\mathcal{F}_{\mathrm{TOPRF}}$ shown in Fig. 1 which generalizes the single-server (non-verifiable) OPRF functionality of [19] to the multi-party setting. In the $\mathcal{F}_{\text{TOPRF}}$ setting, the PRF key is effectively controlled by a collection of n servers and it remains secret as long as no more than a threshold t of these servers are corrupted. Such (t,n)-threshold "collective control" over a functionality can be realized as we show in our 2HashTDH realization in Sect. 3. We chose to base the T-OPRF notion on the *non-verifiable OPRF* notion of [19] rather than the *verifiable OPRF* notion of [18] because the former was shown to have a more efficient realization under the same assumptions, and because this form of OPRF suffices in the key application of interest to us, namely, Password-Protected Secret-Sharing.

Assume $\text{tx}(p,S)$ and $T(p,x)$ are undefined for all p,x,S.

Initialization

- On message (INIT, sid, $\mathcal{SI}$) from S, ignore it if $|\mathcal{SI}| \neq n$ or S is active. Otherwise mark S as "active" and if no record $\langle sid, [...] \rangle$ exists, pick any previously unused label p and record $\langle sid, \mathcal{SI}, p \rangle$. Send (INIT, $sid, S, \mathcal{SI}, p$) to $\mathcal{A}^*$.
- On message (INIT, sid, $\mathcal{A}^*$, p) from $\mathcal{A}^*$, check that p is a label that has not been used before, record $\langle \mathcal{A}^*, p \rangle$ and return (INIT, $sid, \mathcal{A}^*, p$) to $\mathcal{A}^*$.
- On message (INITCOMPLETE, sid, S) from $\mathcal{A}^*$, retrieve tuple $\langle sid, \mathcal{SI}, p \rangle$. Ignore the message if there is no such tuple or $S \notin \mathcal{SI}$ or not all servers in $\mathcal{SI}$ are active. Otherwise, send (INITCOMPLETE, sid) to S and mark S as "initialized."

Evaluation

- On message (EVAL, $sid, ssid, \mathcal{SE}, x$) from $P \in \{U, \mathcal{A}^*\}$, retrieve $\langle sid, \mathcal{SI}, p \rangle$ if $P = U$ or $\langle \mathcal{A}^*, p \rangle$ if $P = \mathcal{A}^*$. Ignore this message if there is no such tuple, if $|\mathcal{SE}| \neq t+1$, or if tuple $\langle ssid, P, \cdot, \cdot \rangle$ already exists. Otherwise record $\langle ssid, P, p, \mathcal{SE}, x \rangle$ and send (EVAL, $sid, ssid, P, \mathcal{SE}$) to $\mathcal{A}^*$.
- On message (SNDRCOMPLETE, $sid, ssid, S$) from $\mathcal{A}^*$, retrieve tuple $\langle sid, \mathcal{SI}, p \rangle$. Ignore this message if there is no such tuple, or if $S \notin \mathcal{SI}$, or if S is not initialized. Otherwise, set $\text{tx}(p,S)\text{++}$ (or set it to 1 if $\text{tx}(p,S)$ is undefined), and send (SNDRCOMPLETE, $sid, ssid$) to S.
- On message (RCVCOMPLETE, $sid, ssid, P, p^*$) from $\mathcal{A}^*$, retrieve $\langle ssid, P, p, \mathcal{SE}, x \rangle$. Ignore this message if there is no such tuple, or if any of the following conditions fails: (i) if $p^* = p$ then $|\{S \in \mathcal{SI} \mid \text{tx}(p,S) > 0\}| > t$; (ii) if all servers in $\mathcal{SE}$ are honest then $p^* = p$. Otherwise, if $p^* = p$ then set $\text{tx}(p,S)\text{--}$ for any $t+1$ distinct $S \in \mathcal{SI}$ s.t. $\text{tx}(p,S) > 0$, and if $T(p^*,x)$ is undefined then pick $\rho \leftarrow_{\text{R}} \{0,1\}^\ell$ and set $T(p^*,x) := \rho$. Finally, send (EVAL, $sid, ssid, T(p^*,x)$) to P.

Fig. 1. Functionality $\mathcal{F}_{\text{TOPRF}}$ with parameters t, n.

The T-OPRF functionality of Fig. 1 has two stages, Initialization and Evaluation. The functionality enforces that the outputs of any such function are uniformly disributed, similarly to the single-server OPRF notion of [19], even in the case that the adversary controls the private key and/or its sharing among the

n servers. In more detail, in the initialization stage, a set of n servers, denoted $\mathcal{SI}$, are activated at the discretion of the adversary. The stage is complete when all servers become active. Note that the set may include adversarial servers, yet the functionality guarantees that all servers identified in $\mathcal{SI}$ become active by the end of the initialization stage. The initialization also specifies a parameter p used to identify a table $T(p, \cdot)$ of random values that defines the proper PRF values computed by the user when interacting with any subset of $t + 1$ *honest* servers from the set $\mathcal{SI}$. Additional parameters p^*, and corresponding tables $T(p^*, \cdot)$, can be specified by the adversary to represent rogue tables with values computed by the user in interaction with corrupted servers (see more on this below). The parameter p is also used to identify a counter $\mathsf{tx}(p, S)$ for each $S \in \mathcal{SI}$ as specified below.

In the evaluation stage, users connect to an arbitrary set of servers $\mathcal{SE}$ chosen by the adversary and which may arbitrarily overlap with $\mathcal{SI}$ (representing the fact that the user has no memory of who the servers in $\mathcal{SI}$ are). When, at the discretion of the adversary, a server $S \in \mathcal{SI}$ completes its interaction, the functionality increases the counter $\mathsf{tx}(p, S)$. Eventually, the adversary can trigger a response to the user which will be drawn from one of the tables maintained by the functionality. Recall that in addition to the proper table $T(p, \cdot)$ the adversary can register additional function tables $T(p^*, \cdot)$ and may connect an evaluation request from a user to any such table of its choice.

The security guarantees provided by the T-OPRF functionality are the following: (1) it enforces the use of the proper function table p whenever the set of servers $\mathcal{SE}$ selected for an evaluation are all honest; (2) it "charges" $t + 1$ server tickets for accessing the proper table p by decrementing (non-zero) ticket counters $\mathsf{tx}(p, S)$ for an arbitrary set of $t + 1$ servers in $\mathcal{SI}$; and (3) all tables T (the proper table p as well as any additional ones set by the adversary with $p^* \neq p$) are filled with random entries that are chosen on demand as the functionality responds back to the user. These guarantees ensure that at least $t + 1 - t'$ honest servers from $\mathcal{SI}$ need to be contacted for the proper function to be evaluated once. To see why this is the case observe that $t + 1$ tickets are "spent" (decremented) during evaluation which correspond to at least $t + 1 - t'$ tickets from honest ticketing counters. This implies that $t + 1$ servers from $\mathcal{SI}$ have registered a SNDRCOMPLETE message as this is the only event that triggers a counter increment. In the real world this corresponds to the event that a server has completed its interaction with a user that attempts to perform an evaluation.

It is important to highlight that the functionality does not necessarily decrement the ticketing counters of the servers identified in the chosen evaluation set $\mathcal{SE}$; rather, it decrements an arbitrary set of $t + 1$ non-zero counters for servers in $\mathcal{SI}$. This reflects the fact that the functionality does not provide any guarantee about the identities of the responding servers. For instance, this means that we allow for an implementation of T-OPRF where an honest user U attempts to connect to a set of servers $\mathcal{SE}_1$ that are corrupted and its message is rerouted by the adversary so that, unbeknownst to U, an honest set of servers servers $\mathcal{SE}_2$ becomes the responder set.

Another important point regarding the T-OPRF functionality is that while it guarantees correct OPRF evaluation in case the user completes an undisturbed interaction with $t + 1$ honest servers in $\mathcal{SI}$, the ideal world adversary may also maintain an arbitrary collection of random tables and connect a user to them, if desired, as long as the responder set is not composed of honest servers only. For instance, the adversary can assign to a subset of corrupted servers $\mathcal{SE}_1$ a certain function table, while it can assign a different function table to a different subset of corrupted servers $\mathcal{SE}_2$. While the two function tables will be independent, they are not under the control of the ideal world adversary completely: their contents will be populated by the ideal functionality with random values independently of each other. In practice this means that we allow for an implementation where two successive evaluation requests for the same x value result in a different (but still random) value to be produced, depending on which set of servers the user connects to. We stress that the secrecy of the input x is always preserved irrespectively of the subset of servers the user communicates with. At the same time, observe that the randomness requirement imposed for adversarial tables restricts our ability to implement the functionality to the random oracle setting.

3   Threshold OPRF Protocol from OMDH and T-OMDH

Here we present our Threshold Oblivious PRF protocol, called 2HashTDH, that instantiates the $\mathcal{F}_{\text{TOPRF}}$ functionality defined in Sect. 2. Thus, 2HashTDH provides a secure T-OPRF for use in general applications and, in particular, as the basis for our PPSS scheme, TOPPSS, presented in Sect. 4. The 2HashTDH scheme is formally defined as a realization of $\mathcal{F}_{\text{TOPRF}}$ in Fig. 3. In a nutshell, it is a threshold version of the 2HashDH OPRF from [19], recalled in Fig. 2. The underlying PRF, $f_k(x) = H_2(x, (H_1(x))^k)$, remains unchanged, but the key k is shared using Shamir secret-sharing across n servers, where server S_i stores the key share k_i. The initialization of such secret-sharing can be done via a Distributed Key Generation (DKG) for discrete-log-based systems, e.g. [16], and in Fig. 2 we assume it is done with a UC functionality $\mathcal{F}_{\text{DKG}}$ which we discuss further below. For evaluation, given any subset $\mathcal{SE}$ of $t + 1$ servers, the user U sends to each of them the *same* message $a = (H'(x))^r$ for random r, exactly as

---

PRF Definition

G: a group of prime order m; H, H': hash functions with resp. ranges $\{0,1\}^\ell$ and G.
PRF f from $\{0,1\}^*$ to $\{0,1\}^\ell$: For a key $k \leftarrow_{\text{R}} \mathbb{Z}_m$, define $f_k(x) = H(x, (H'(x))^k)$.

Oblivious Computation of PRF $f_k(x)$ between user U and server S

1. On input x, U chooses $r \leftarrow_{\text{R}} \mathbb{Z}_m$; sends $a = (H'(x))^r$ to S.
2. S verifies that the received a is in group G and if so it responds with $b = a^k$.
3. U outputs $f_k(x) = H(x, b^{1/r})$.

Fig. 2. The 2HashDH OPRF [19]

in the single-server OPRF protocol 2HashDH. If each server S_i in $\mathcal{SE}$ returned $b_i = a^{k_i}$ then U could reconstruct the value a^k using standard Lagrange interpolation in the exponent, i.e. $a^k = \prod_{i \in \mathcal{SE}} b_i^{\lambda_i}$ with the Lagrange coefficients λ_i computed using the indexes of servers in $\mathcal{SE}$. After computing a^k, the value of $f_k(x)$ is computed by U by deblinding a^k exactly as in the case of protocol 2HashDH. Note that this takes a single exponentiation for each server and two exponentiations for the user (to compute a and to deblind a^k) plus one multi-exponentiation by U to compute the Lagrange interpolation on the b_i values. We optimize this function evaluation by having each server S_i compute $b_i = a^{\lambda_i \cdot k_i}$, which costs one exponentiation and $O(t)$ multiplications and divisions in $\mathbb{Z}_m$ to compute λ_i. (Note that S_i must know set $\mathcal{SE}$ to compute λ_i.) This way U can compute a^k using only t multiplications instead of a multi-exponentiation, and the total costs are 1 exps for each S_i and 2 exps for U.

Protocol 2HashTDH can be also seen as a simplification of a protocol which results from a generic transformation of any OPRF to T-TOPRF using multi-party secure computation of the server code, and then applying this transformation to the 2HashDH OPRF of [19]. The server in 2HashDH computes a^k on input a, and the MPC protocol for it is exactly the *threshold exponentiation* protocol described above, except that this generic OPRF to T-OPRF transformation must assure that the servers perform the MPC subprotocol on the same input a, and this involves an additional round of server-to-server interaction, which the 2HashTDH protocol avoids. We refer to the full version of this paper [20] for the specification of this general OPRF to T-OPRF compiler.

Roadmap. In Sect. 3.1 we show protocol 2HashTDH and explain the assumptions taken in its specification. In Sect. 3.2 we introduce the T-OMDH assumption, a generalization of OMDH, and we show that it is equivalent to OMDH in several cases, including the *full corruption* case $t' = t$ discussed in the introduction. In Sect. 3.3 we show that protocol 2HashTDH realizes the Threshold OPRF functionality $\mathcal{F}_{\text{TOPRF}}$ under the T-OMDH assumption in ROM for any threshold parameters (t, n) and any number $t' < t$ of corrupted servers. It follows that protocol 2HashTDH achieves the *standard threshold security* property, which corresponds to the full corruption case, under just OMDH in ROM. Note that the non-threshold OPRF 2HashDH of [19] also relies on OMDH.

3.1 T-OPRF Protocol Based on T-OMDH Assumption

The 2HashTDH T-OPRF protocol is shown in Fig. 3, relying on realizations of functionalities $\mathcal{F}_{\text{DKG}}, \mathcal{F}_{\text{AUTH}}$ and $\mathcal{F}_{\text{SEC}}$, which model, respectively, the distributed key generation, authenticated channel, and secure channel. Assuming these functionalities, the 2HashTDH protocol realizes the UC T-OPRF functionality defined in Sect. 2, under the T-OMDH assumption in ROM. As we argue in Sect. 3.2, this implies security under OMDH in ROM in several cases, including the *full corruption* case, where the adversary corrupts $t' = t$ servers, and the *additive sharing* case, where $t = n - 1$. Functionalities $\mathcal{F}_{\text{DKG}}, \mathcal{F}_{\text{AUTH}}, \mathcal{F}_{\text{SEC}}$ all have well-known efficient realizations in ROM under the Diffie-Hellman assumption which is implied by OMDH, and hence also by T-OMDH.

Let $\mathcal{F}_{\text{DKG}}, \mathcal{F}_{\text{AUTH}}$ and $\mathcal{F}_{\text{SEC}}$ be, respectively, the distributed key generation, authenticated channel, and secure channel functionalities; Let G be a cyclic group of prime order m; Let H_1, H_2 be hash functions with resp. ranges G and $\{0,1\}^\ell$.

Initialization

S1: On input $(\text{INIT}, sid, \mathcal{SI}=(S_1,\ldots,S_n))$, server S forwards this input to $\mathcal{F}_{\text{DKG}}$.
S2: On message $(\text{INITCOMPLETE}, sid, y, k_i)$ from $\mathcal{F}_{\text{DKG}}$, server S records $(sid, \mathcal{SI}, y, i, k_i)$, marks itself active, and outputs $(\text{INITCOMPLETE}, sid)$.

Evaluation

U1: On input $(\text{EVAL}, sid, ssid, \mathcal{SE}, x)$, U picks $r \leftarrow_{\text{R}} \mathbb{Z}_m$, computes $a := H_1(x)^r$, and sends $(\text{SEND}, (sid, ssid, 1), S, (\mathcal{SE}, a))$ to $\mathcal{F}_{\text{AUTH}}$ for all $S \in \mathcal{SE}$.
S1: On message $(\text{SENT}, (sid, ssid, 1), U, (\mathcal{SE}, a))$ from $\mathcal{F}_{\text{AUTH}}$, server S, provided it is active, computes $b_i := a^{\lambda_i \cdot k_i}$ where λ_i is a Lagrange interpolation coefficient for index i and index set $\mathcal{SE}$, sends $(\text{SEND}, (sid, ssid, 2), U, b_i)$ to $\mathcal{F}_{\text{AUTH}}$, and outputs $(\text{SNDRCOMPLETE}, sid, ssid)$.
U2: When U receives $(\text{SENT}, (sid, ssid, 2), S_i, b_i)$ from $\mathcal{F}_{\text{AUTH}}$ for all $S_i \in \mathcal{SE}$, it outputs $(\text{EVAL}, sid, ssid, H_2(x, (\prod_{S_i \in \mathcal{SE}} b_i)^{1/r}))$.

Fig. 3. Protocol 2HashTDH realizing $\mathcal{F}_{\text{TOPRF}}$ assuming $\mathcal{F}_{\text{DKG}}, \mathcal{F}_{\text{SEC}}, \mathcal{F}_{\text{AUTH}}$.

Note on Authentic and Secret Channels. In Fig. 3 protocol 2HashTDH is presented in the $(\mathcal{F}_{\text{AUTH}}, \mathcal{F}_{\text{SEC}}, \mathcal{F}_{\text{DKG}})$-hybrid world, i.e., assuming that there are both authenticated and secure (i.e. authenticated and secret) channels between protocol participants. We refer to [12] for the UC models of authenticated and secret channels, but simply speaking, what the authenticated and secure channel functionalities model is that if party P_1 sends message m to party P_2 using $\mathcal{F}_{\text{AUTH}}$ command $(\text{SEND}, sid, P_2, m)$, then P_2 will be able to authenticate m as originated from P_1, i.e. if P_2 receives command $(\text{SENT}, sid, P_1, m')$, it is guaranteed that $m' = m$, and if P_1 sends m to P_2 using $\mathcal{F}_{\text{SEC}}$ command $(\text{SEND}, sid, P_2, m)$, then P_2 can verify authenticity of P_1's message as above, but in addition m will be hidden to the adversary unless P_2 is corrupted.

We note that using ideal functionalities such $\mathcal{F}_{\text{AUTH}}, \mathcal{F}_{\text{SEC}}$ in the hybrid world, does not determine their implementation when the UC protocol is deployed in the real world. This is because they only describe how the adversarial model against the protocol is envisioned. For instance, $\mathcal{F}_{\text{AUTH}}$ *may* be realized using a PKI involving all connected participants, or it may be simply substituted by unauthenticated TCP/IP communication in case it is deemed that modifying message contents is not a relevant threat in the protocol deployment. Indeed, this will also be the case in our setting since we allow the (adversarial) environment to choose the servers that a user connects in the evaluation stage of the protocol in a way that is independent from the initialization servers; in this way, any man-in-the-middle scenario can be simulated by the adversary without violating the $\mathcal{F}_{\text{AUTH}}$ constraints. Similarly, $\mathcal{F}_{\text{SEC}}$ *may* be implemented by TLS, but may also be achieved in other ways, e.g., physically transferring private state between the parties engaged in the protocol.

A second important point is that if a user U initializes a T-OPRF instance with a server set $\mathcal{SI} = \{S_1, ..., S_n\}$ such that some subset B of $\mathcal{SI}$ is made of corrupt entities (which models both the fact that some $\mathcal{SI}$ members are corrupt and the fact that U can execute T-OPRF initialization on an incorrect set of servers), then in this case command (SEND, sid, S_i, m) for $S_i \in B$ will leak m to the adversary, and if U receives (SENT, sid, S_i, m) from $\mathcal{F}_{\text{AUTH}}$ for $S_i \in B$, we can assume that the adversary supplies message m. In other words, the $\mathcal{F}_{\text{AUTH}}$ and $\mathcal{F}_{\text{SEC}}$ channels implement authenticated and/or secret point-to-point message delivery only if they are executed for a proper and non-corrupt server. We note that we assume a secret channel $\mathcal{F}_{\text{SEC}}$ in addition to an authenticated channel $\mathcal{F}_{\text{AUTH}}$ solely to simplify the description of T-OPRF initialization. Indeed, the former can be built from the latter [12], e.g. by having each server S_i first send its encryption public key to U using the authenticated channel.

Note on Distributed Key Generation. Protocol 2HashTDH assumes that servers in $\mathcal{SI}$ establish a secret-sharing $(k_1, ..., k_n)$ of a random key k over authenticated channels via a Distributed Key Generation (DKG) functionality $\mathcal{F}_{\text{DKG}}$, shown in Fig. 4. The DKG sub-protocol for discrete-log based cryptosystems can be efficiently realized without user's involvement [16,30], but if the call to initialize a TOPRF instance is executed by an *honest user* U then the DKG subprotocol can be even simpler, because U can generate sharing $(k_1, ..., k_n)$ of k and then distribute the shares among the servers in $\mathcal{SI}$. Note that since our realizations of $\mathcal{F}_{\text{TOPRF}}$ pertains only to the *static* adversarial model, where the identity of corrupt parties is determined at the outset, we would not explicitly require that the parties erase the information used in the initialization, but any implementation should erase such information. In our specification of protocol 2HashTDH we rely on the $\mathcal{F}_{\text{DKG}}$ functionality to abstract from any specific DKG implementation, e.g. whether it is done by the server or by an honest user.

3.2 Threshold OMDH Assumption

Notation. If n is an integer, then $[n] = \{1, ..., n\}$. If D is a set, then $|D|$ is its cardinality. We use bold font to denote vectors, e.g. $\boldsymbol{a} = [a_1, ..., a_n]$. If $\boldsymbol{a}$ and $\boldsymbol{b}$ are two vectors of the same dimension, then $\boldsymbol{a} \odot \boldsymbol{b}$ is their Hadamard (component-wise) product. If $|\boldsymbol{a}| = n$ and J is a sequence in $[n]$ then $\boldsymbol{a}_J$ denotes the components of $\boldsymbol{a}$ with indices in J, i.e. $[a_{i_1}, ..., a_{i_k}]^T$ if $J = (i_1, ..., i_k)$.

Let $\mathcal{I}_w$ be the set of w-element subsets of $[n]$, i.e. $\mathcal{I}_w = \{I \subseteq [n] \text{ s.t. } |I| = w\}$. Let $W(\boldsymbol{a})$ be the hamming weight of $\boldsymbol{a}$. Let $\mathcal{V}_w$ be the set of n-bit binary vectors $\boldsymbol{q}$ s.t. $W(\boldsymbol{q}) = w$, i.e. $\mathcal{V}_w = \{\boldsymbol{v} \in \{0,1\}^n \text{ s.t. } v_i = 1 \text{ iff } i \in \mathcal{I}_w\}$. For $\boldsymbol{q} = [q_1, ..., q_n]$ define $C_w(\boldsymbol{q})$ as the maximum integer m for which there exist $\boldsymbol{v}_1, ..., \boldsymbol{v}_m \in \mathcal{V}_w$ (not necessarily distinct) s.t. $\boldsymbol{v}_1 + ... + \boldsymbol{v}_m \leq \boldsymbol{q}$. In other words, $C_w(\boldsymbol{q})$ is the

---

Let g generate a cyclic group G of prime order m; Let t, n be integers s.t. $t < n$.

- On message (INIT, sid, $\mathcal{SI}$) from S, ignore it if $|\mathcal{SI}| \neq n$ or S is active or $S \notin \mathcal{SI}$. Otherwise, if no record $\langle sid, [...] \rangle$ exists, let Corrupted be the subset of $\mathcal{SI}$ that is corrupted and set $t' = |\text{Corrupted}|$. If $t' \leq t$ then pick $a_0, a_1, \ldots, a_{t-t'} \leftarrow_R \mathbb{Z}_m$, record $\langle sid, \mathcal{SI}, a_0, a_1, \ldots, a_{t-t'} \rangle$ else record $\langle sid, \mathcal{SI} \rangle$. Irrespectively, mark S as "active", and send (INIT, sid, P, $\mathcal{SI}$) to $\mathcal{A}^*$.
- On message (INIT, sid, $\mathcal{SI}$, s) from a corrupted $S \in \mathcal{SI}$, if the record $\langle sid, \mathcal{SI}, [...] \rangle$ exists, record $\langle \mathcal{A}^*, S, s \rangle$, mark S as active and send (INIT, sid, $\mathcal{A}^*$, S) to $\mathcal{A}^*$.
- On message (INITCOMPLETE, sid, S_i) from $\mathcal{A}^*$, retrieve tuple $\langle sid, \mathcal{SI}, a_0, a_1, \ldots, a_{t-t'} \rangle$. Ignore the message, if there is no such tuple, or if $S_i \notin \mathcal{SI}$ or not all servers in $\mathcal{SI}$ are active. Otherwise, send (INITCOMPLETE, sid, g^{a_0}, i, s_i) to S_i and (INITCOMPLETE, sid, S_i, g^{a_0}) to $\mathcal{A}^*$, where $s_i = p(i)$ and $p(x)$ is a polynomial whose first $t - t' + 1$ coefficients match $a_0, a_1, \ldots, a_{t-t'}$ and $p(j) = s_j$ for each j such that $S_j \in$ Corrupted and $\langle \mathcal{A}^*, S_j, s_j \rangle$ has been previously recorded.

Fig. 4. Distributed key generation functionality $\mathcal{F}_{\text{DKG}}$ [30].

maximum number of times one can subtract elements in $\mathcal{V}_w$ from $\boldsymbol{q}$ s.t. the result remains $\geq \boldsymbol{0}$. For example if and $\boldsymbol{q} = [3, 3, 4]$ then $C_2(\boldsymbol{q}) = 4$ because $\boldsymbol{q} = 2 \times [1, 0, 1] + [1, 1, 0] + 2 \times [0, 1, 1]$.

T-OMDH Intuition. Let $\langle g \rangle$ be a cyclic group of prime order $m > n$. The T-OMDH assumption considers the setting where a random exponent $k \in \mathbb{Z}_m$ is secret-shared using a random t-degree polynomial $p(\cdot)$, and the n trustees holding shares $k_1 = p(1), \ldots, k_n = p(n)$ implement a "threshold exponentiation" protocol which computes a^k for any given $a \in \langle g \rangle$ and $k = p(0)$. Let $\text{TOMDH}_p(\cdot, \cdot)$ be an oracle which on input $(i, a) \in [n] \times \langle g \rangle$ outputs $a^{p(i)}$. The standard way to implement threshold exponentiation is to choose a set $I \in \mathcal{I}_{t+1}$, compute $b_i = \text{TOMDH}_p(i, a) = a^{k_i}$ for each i in I and derive a^k as $\prod_{i \in I} b_i^{\lambda_i}$ using Lagrange interpolation coefficients λ_i s.t. $k = \sum_{i \in I} \lambda_i \cdot k_i$. The T-OMDH assumption states that querying oracle $\text{TOMDH}_p(\cdot, \cdot)$ on at least $t + 1$ different points $i \in [n]$ is *necessary* to compute $a^{p(0)}$ for a given random challenge a. More generally, T-OMDH considers an experiment where the attacker $\mathcal{A}$ receives a challenge set $R = \{g_1, ..., g_N\}$ of random elements in $\langle g \rangle$ and is given access to the $\text{TOMDH}_p(\cdot, \cdot)$ oracle for random t-degree polynomial $p(\cdot)$. T-OMDH assumption states that $\mathcal{A}$ can compute g_j^k for $k = p(0)$ for no more than $C_{t+1}(q_1, \ldots, q_n)$ elements $g_j \in R$, where q_i is the number of $\mathcal{A}$'s queries to $\text{TOMDH}_p(i, \cdot)$.

The above intuition and Definition 1 below correspond to the setting where the attacker does not control any of the trustees holding shares of p, hence it needs $t + 1$ queries to $\text{TOMDH}_p(\cdot, \cdot)$ to compute $a^{p(0)}$ for each random challenge a. Later we extend this definition to the case where $\mathcal{A}$ controls a subset of trustees.

Definition 1. *The (t, n, N, Q)-Threshold One-More Diffie Hellman (T-OMDH) assumption holds in group $\langle g \rangle$ of prime order m if the probability of any polynomial-time adversary $\mathcal{A}$ winning the following game is negligible. $\mathcal{A}$ receives challenge set $R = \{g_1, \ldots, g_N\}$ where $g_i \leftarrow_R \langle g \rangle$ for $i \in [N]$, and is given access to an oracle $\mathsf{TOMDH}_p(\cdot, \cdot)$ for a random t-degree polynomial $p(\cdot)$ over $\mathbb{Z}_m$. $\mathcal{A}$ wins if it outputs g_j^k where $k = p(0)$ for $Q + 1$ different elements g_j in R, and if $C_{t+1}(q_1, \ldots, q_n) \leq Q$ where q_i is the number of $\mathcal{A}$'s queries to $\mathsf{TOMDH}_p(i, \cdot)$.*

Note that the (N, Q)-OMDH assumption [5,22] is the (t, n, N, Q)-T-OMDH assumption for $t = 0$ and any $n \geq 1$, because then $p(\cdot)$ is a constant polynomial and $C_1(q) = W(q)$, i.e. the total number of $\mathcal{A}$'s $\mathsf{TOMDH}_p(\cdot, \cdot)$ queries.

T-OMDH: The General Case. In its general form, the T-OMDH assumption corresponds to computing g_j^k if some subset of $t' \leq t$ trustees holding shares $k_i = p(i)$ is corrupt, and hence the adversary can not only learn these shares but can also set them at will.

Definition 2. *The (t', t, n, N, Q)-T-OMDH assumption holds in group $\langle g \rangle$ of prime order m if for any $\mathsf{B} \subseteq [n]$ s.t. $|\mathsf{B}| = t' \leq t$, the probability of any polynomial-time adversary $\mathcal{A}$ winning the following game is negligible. On input a challenge set $R = \{g_1, \ldots, g_N\}$ where $g_i \leftarrow_R \langle g \rangle$ for $i \in [N]$, adversary $\mathcal{A}$ specifies a set of t' values $\{\alpha_j\}_{j \in \mathsf{B}}$ in $\mathbb{Z}_m$. A random t-degree polynomial $p(\cdot)$ over $\mathbb{Z}_m$ is then chosen subject to the constraint that $p(j) = \alpha_j$ for $j \in \mathsf{B}$, and the adversary $\mathcal{A}$ is given access to oracle $\mathsf{TOMDH}_p(\cdot, \cdot)$. We say that $\mathcal{A}$ wins if it outputs g_j^k where $k = p(0)$ for $Q + 1$ different elements g_j in R, and if $C_{t-t'+1}(q_1, \ldots, q_n) \leq Q$ where q_i for $i \notin \mathsf{B}$ is the number of $\mathcal{A}$'s queries to $\mathsf{TOMDH}_p(i, \cdot)$, and $q_i = 0$ for $i \in \mathsf{B}$.*

Note that (t', t, n, N, Q)-T-OMDH is identical to (t, n, N, Q)-T-OMDH for $t' = 0$.

Gap T-OMDH. In order to prove the security of T-OPRF, we need to extend the T-OMDH assumption stated in Definition 2 to its "gap" form, i.e. suppose $\langle g \rangle$ is a gap group where $\mathcal{A}$ is in addition given access to the DDH oracle in $\langle g \rangle$.

Definition 3. *The Gap (t', t, n, N, Q)-T-OMDH assumption is the T-OMDH assumption of Definition 2 except that $\mathcal{A}$ is also given access to the DDH oracle in group $\langle g \rangle$, which on input (a, b, c, d) outputs 1 if $\log_a b = \log_c d$ and 0 otherwise.*

In the full version of this paper [20] we show that the (Gap) (t, t', n, N, Q)-T-OMDH assumption holds in the generic group model for any (t', t, n). Specifically, the advantage of a T-OMDH adversary restricted to r generic group operations is upper-bounded by $O(Qr^2/m)$, assuming $r \geq Q \geq N$. This is larger by factor Q from the $O(r^2/m)$ upper-bounds on generic group attacks against many static problems related to discrete logarithm [28], and this weakening is caused by the presence of up to Q-degree polynomials of the "target" secret $k = p(0)$ in the representation of the group elements which the adversary can compute given access to $\mathsf{TOMDH}_p(\cdot, \cdot)$ using the query pattern $q = [q_1, \ldots, q_n]$ s.t. $C_{t-t'+1}(q) \leq Q$. Since (Q, N)-OMDH is identical to (t', t, n, N, Q)-T-OMDH for $(t', t) = (0, 0)$

and any n, the same upper-bound applies applies to OMDH, and to the best of our knowledge this is the first generic model security hardness argument for the OMDH (or Gap OMDH) assumption.

T-OMDH = OMDH in Full Corruption and Additive Sharing Cases. The T-OMDH and OMDH assumptions are equivalent in two important cases, namely the *full corruption* case of $t' = t$, for any (t, n), and in the *additive sharing* case of $t = n - 1$, for any t'. We refer to the full version of this paper [20] for (easy) proofs of above equivalences. Note also that whereas the question whether the T-OMDH and OMDH assumptions are equivalent for any $t' < t$ and $t + 1 < n$ remains open, in the full version [20] we also show the same generic group hardness bound for both problems.

3.3   Security Analysis of 2HashTDH

Protocol 2HashTDH protocol of Fig. 3 is secure under the T-OMDH assumption. As a corollary of the fact that in the full corruption case of $t' = t$ faults the T-OMDH and OMDH assumptions are equivalent, Theorem 1 implies that protocol 2HashTDH is secure under OMDH in ROM in the full corruption case of $t' = t$. The proof of Theorem 1 appears in the full version of this paper [20].

Theorem 1. *Protocol 2HashTDH realizes functionality $\mathcal{F}_{\text{TOPRF}}$ with parameters t, n in the $(\mathcal{F}_{\text{AUTH}}, \mathcal{F}_{\text{SEC}}, \mathcal{F}_{\text{DKG}})$-hybrid model, assuming static corruptions, hash functions $H_1(\cdot)$ and $H_2(\cdot, \cdot)$ modelled as Random Oracles, and the Gap (t', t, n, N, Q)-T-OMDH on group $\langle g \rangle$, where Q is the number of EVAL messages sent by any user, $N = Q + q_1$ where q_1 is the number of $H_1(\cdot)$ queries the adversary makes, and $t' < t$ is the number of corrupted servers in $\mathcal{SI}$.*

Specifically, for any efficient adversary $\mathcal{A}$ against protocol 2HashTDH, there is a simulator SIM s.t. no efficient environment $\mathcal{Z}$ can distinguish the view of $\mathcal{A}$ interacting with the real 2HashTDH protocol and the the view of SIM interacting with the ideal functionality $\mathcal{F}_{\text{TOPRF}}$, with advantage better than $q_T \cdot \epsilon(N, Q) + N^2/m$, where q_T is the number of TOPRF instances, $\epsilon(N, Q)$ is the bound on the probability that any algorithm of the same cost violates the Gap (t', t, n, N, Q)-T-OMDH assumption, and $m = |\langle g \rangle|$.

4   TOPPSS: A PPSS Scheme Based on T-OPRF

In Fig. 5 we show a compiler which converts a T-OPRF scheme which realizes the UC T-OPRF notion of Sect. 2 into a PPSS scheme, called TOPPSS, which realizes UC PPSS functionality of [19]. The terminology of the UC setting might obscure the amazing practicality of this construction, so in Sect. 5 we show a concrete implementation of this scheme with the $\mathcal{F}_{\text{TOPRF}}$ functionality implemented using the T-OPRF instantiation 2HashTDH from Sect. 3.

TOPPSS Overview. To explain the mechanics of TOPPSS based on the T-OPRF functionality, it is instructive to compare it to the OPRF-based PPSS

Let $\mathcal{F}_{\text{AUTH}}$ and $\mathcal{F}_{\text{TOPRF}}$ be, respectively, the authenticated channel and the T-OPRF functionality; Let $H(\cdot)$ be a hash function with range $\{0,1\}^{\ell}$.

INIT for user U:

1. On input (INIT, $sid, \mathcal{SI}, \mathsf{pw}$), send (SEND, $(sid, 0), S, \mathcal{SI}$) to $\mathcal{F}_{\text{AUTH}}$ for all S in $\mathcal{SI}$.
2. On (SENT, $(sid, 1), S,$ DONE) from $\mathcal{F}_{\text{AUTH}}$ for all $S \in \mathcal{SI}$, send (EVAL, $sid, 0, \mathcal{SE}, \mathsf{pw}$) to $\mathcal{F}_{\text{TOPRF}}$ for any $\mathcal{SE} \subseteq \mathcal{SI}$ such that $|\mathcal{SE}| = t + 1$.
3. On $\mathcal{F}_{\text{TOPRF}}$'s response (EVAL, $sid, 0, v$), parse $H(v)$ as $[C|K]$ and send (SEND, $(sid, 2), S, C$) to $\mathcal{F}_{\text{AUTH}}$ for every $S \in \mathcal{SI}$.
4. On (SENT, $(sid, 3), S,$ ACK) for all $S \in \mathcal{SI}$ from $\mathcal{F}_{\text{AUTH}}$, output (UINIT, sid, K).

INIT for server S:

1. On (SENT, $(sid, 0), U, \mathcal{SI}$) from $\mathcal{F}_{\text{AUTH}}$, send (INIT, $sid, \mathcal{SI}$) to $\mathcal{F}_{\text{TOPRF}}$.
2. On (INITCOMPLETE, sid) from $\mathcal{F}_{\text{TOPRF}}$, send (SEND, $(sid, 1), U,$ DONE) to $\mathcal{F}_{\text{AUTH}}$.
3. On (SENT, $(sid, 2), U, C$) from $\mathcal{F}_{\text{AUTH}}$, record (sid, C), send (SEND, $(sid, 3), U,$ ACK) to $\mathcal{F}_{\text{AUTH}}$, and output (SINIT, sid).

REC for user U:

1. On input (REC, $sid, ssid, \mathcal{SR}, \mathsf{pw}'$) send (EVAL, $sid, [1|ssid], \mathcal{SR}, \mathsf{pw}'$) to $\mathcal{F}_{\text{TOPRF}}$.
2. On $\mathcal{F}_{\text{TOPRF}}$'s response (EVAL, $sid, [1|ssid], v'$) and (SENT, $(sid, ssid, 1), S, C'$) from $\mathcal{F}_{\text{AUTH}}$ for all $S \in \mathcal{SR}$, if each message contains C' s.t. $[C'|K'] = H(v')$, then set RES $:= K'$, otherwise set RES $:=$ FAIL. Output (UREC, $sid, ssid,$ RES).

REC for server S:

1. On (SNDRCOMPLETE, $sid, [1|ssid]$) from $\mathcal{F}_{\text{TOPRF}}$, if S holds record (sid, C), then send (SEND, $(sid, ssid, 1), U, C$) to $\mathcal{F}_{\text{AUTH}}$ and output (SREC, $sid, ssid$).

Fig. 5. The TOPPSS protocol

scheme of [19]. In that scheme each server holds its own *independently random* key k_i for an OPRF f. At initialization, the secret to be protected is processed with a (t, n) secret sharing scheme and each share is stored at one of n servers, where server S_i stores the i-th share encrypted under $f_{k_i}(\mathsf{pw})$. At reconstruction, the user receives the encrypted shares from $t + 1$ servers which it decrypts using the values $f_{k_i}(\mathsf{pw})$ that it learns by running the OPRF on pw with each of these servers. By contrast, in our TOPPSS scheme, which is T-OPRF-based, the (random) secret to be protected is defined as a single PRF value $v = f_k(\mathsf{pw})$ where k is a key secret-shared as part of a T-OPRF scheme. This provides a significant performance gain by reducing the number of exponentiations performed by the user from $t + 2$ to just 2. In the scheme of [19] implemented with 2HashDH, the user computes the OPRF sub-protocol with each server independently, which involves one blinding operation re-used across all servers, but requires one de-blinding operation *per server* for a total of $t + 2$ exponentiations. By contrast, in the T-OPRF protocol 2HashTDH of Sect. 3 the user performs a single blinding and de-blinding, hence just 2 exponentiations, regardless of the number of servers and threshold t.

Note that the T-OPRF functionality allows the user to evaluate function $f_k(\cdot)$ on the user's password pw, without leaking any information about pw, but it does not let the user verify whether the function is computed correctly. Indeed, following the rules of functionality $\mathcal{F}_{\text{TOPRF}}$, either corrupt servers or a man-in-the-middle adversary could make the user compute $f_k(\text{pw})$ on key k of their choice. If the dictionary D from which the user draws her password is small, the adversary can potentially pick k s.t. function $f_k(\cdot)$ behaves on domain D in some ways the adversary can exploit (e.g., reducing the number of possible outputs). However, since $\mathcal{F}_{\text{TOPRF}}$ assures that $f_k(\cdot)$ behaves like a random function for all k's, even for k's chosen by the adversary, it suffices to include a commitment to the master secret $v = f_k(\text{pw})$ in the information that the servers send to the user, so that the user can verify its correctness. The adversary can still pick k but if $f_k(\cdot)$ is pseudorandom for all k then the adversary cannot change either k or v without guessing pw. Note that the randomness for verifying this commitment must be derived from the committed plaintext $f_k(\text{pw})$ itself as this is the only value the user can retrieve using its only input pw. Although this mechanism requires the commitment scheme to be deterministic, the hiding property of the commitment is still satisfied thanks to the pseudorandomness of the committed plaintext $v = f_k(\text{pw})$ (and assuming no more than t corruptions).

Since our realizations of $\mathcal{F}_{\text{TOPRF}}$, protocol 2HashTDH, requires the Random Oracle Model (ROM) for hash functions in the security analysis, we implement this commitment simply with another hash function modeled as a random oracle. Finally, since the user needs to verify the master-secret v as well as to derive a key K from it, we implement both operation using a single hash function call, i.e. we set $[C|K]$ to $H(v)$ where H hashes onto strings of length 2ℓ.

The proof of the following theorem is in the full version of this paper [20].

Theorem 2. *The TOPPSS scheme of Fig. 5 UC-realizes the PPSS functionality $\mathcal{F}_{\text{PPSS}}$ assuming access to the T-OPRF functionality $\mathcal{F}_{\text{TOPRF}}$ and to the authenticated message delivery functionality $\mathcal{F}_{\text{AUTH}}$, and assuming that hash function H is a random oracle.*

5   Concrete Instantiation of TOPPSS Using 2HashTDH

For concreteness we show an instantiation of TOPPSS with the T-OPRF functionality realized by protocol 2HashTDH from Fig. 3 in Sect. 3. In this figure we realize the $\mathcal{F}_{\text{DKG}}$ subprotocol assuming an honest user U, because in the context of a PPSS protocol, we only care about security for PPSS instances which were initialized with an honest user. Hence we simply have U create the sharing of the T-OPRF key and distributing it among the servers in $\mathcal{SI}$ (see a note on DKG in

---

Parameters: Security parameter ℓ, threshold parameters t, n, cyclic group $\langle g \rangle$ of prime order m, hash functions H_1, H_2, H_3 with ranges $\langle g \rangle$, $\{0,1\}^{2\ell}$ and $\{0,1\}^{2\ell}$.
Communication Setting: Communication between U and S_i's in INIT goes over secure channels, e.g. TLS, communication in REC not necessarily (see text).

INIT for user U on input $(sid, \mathcal{SI}, \mathsf{pw})$ where $\mathcal{SI} = \{S_1, ..., S_n\}$:

1. Pick $k \leftarrow_{\mathrm{R}} \mathbb{Z}_m$, generate $(k_1, ..., k_n)$ as a (t, n)-Shamir's secret sharing of k over $\mathbb{Z}_m$, and send $(sid, \mathcal{SI}, i, k_i)$ to each $S_i \in \mathcal{SI}$.
2. After receiving ACK's from all servers in $\mathcal{SI}$, compute $v := H_2(\mathsf{pw}, H_1(\mathsf{pw})^k)$, parse $H_3(v)$ as $[C|K]$, send (sid, C) to all $S_i \in \mathcal{SI}$ and output K.

INIT for server S_i:

1. On $(sid, \mathcal{SI}, i, k_i)$ from U, abort if sid not unique or $S_i \notin \mathcal{SI}$, otherwise record $(sid, \mathcal{SI}, i, k_i)$ and send (sid, ACK) to U.
2. On (sid, C) from U, append C to tuple $(sid, \mathcal{SI}, i, k_i)$.

REC for user U on input $(sid, ssid, \mathcal{SR}, \mathsf{pw})$ where $\mathcal{SR} = \{S'_1, ..., S'_{t+1}\}$:

1. Pick $r \leftarrow_{\mathrm{R}} \mathbb{Z}_m$, set $a := H_1(x)^r$, send $((sid, ssid), (\mathcal{SR}, a))$ to all $S'_i \in \mathcal{SR}$.
2. After receiving message $((sid, ssid), b_i, C)$ from all servers in $\mathcal{SR}$, abort if not all C's are the same. Otherwise compute $b := \prod_{S'_i \in \mathcal{SR}} b_i$ and $v := H_2(x, b^{1/r})$.
3. Parse $H_3(v)$ as $[C'|K']$. If $C' = C$ then output K', otherwise output FAIL.

REC for server $S_i \in \mathcal{SR}$:

1. On message $((sid, ssid), (\mathcal{SR}, a))$ from U, recover stored tuple $(sid, \mathcal{SI}, i, k_i, C)$. (Abort if there is no such tuple or if $a \notin \langle g \rangle$.) Compute $b_i := a^{\lambda_i \cdot k_i}$ where λ_i is an interpolation coefficient for index i and the subset of $t+1$ indices defined by $\mathcal{SR} \subseteq \mathcal{SI}$, and send $((sid, ssid), b_i, C)$ to U.

Fig. 6. Concrete instantiation of TOPPSS based on 2HashTDH T-OPRF.

Sect. 3.1). Note that if we implement $\mathcal{F}_{\mathrm{DKG}}$ in this user-centric way then we do not have to execute T-OPRF evaluation for U to compute $v = f_k(\mathsf{pw})$ as part of the initialization: User U can just compute $v = f_k(\mathsf{pw})$ locally because U picked the TOPRF key k (Fig. 6).

On the Role of Secure Channels. The communication in such instantiation of TOPPSS must go over secure channels in the *initialization phase*, which in practice could be implemented using e.g. TLS.[2] In the *reconstruction phase*, the communication does not have to go over secure channels, because TOPPSS is

---

[2] Note that if the $\mathcal{F}_{\mathrm{DKG}}$ was instantiated with the distributed key generation then authenticated channels would suffice for the communication between the user and the servers because the TOPRF evaluation protocol does not need secure channels. However, the standard realization of $\mathcal{F}_{\mathrm{DKG}}$ [30] would require secure channels between the servers.

secure in the password-only, i.e. PKI-free, model. However using TLS would offer a security benefit against the network adversary as a *hedge* against any server-spoofing attacks due to which the user might be tricked to run the PPSS reconstruction with the wrong set of servers. To see the benefit of running a PPSS protocol over TLS channels, denote the set of server identities which U inputs in the reconstruction as $\mathcal{SR}$. In the case of running PPSS reconstruction over TLS these can be equated with the public keys the user would use in the TLS sessions with the $t+1$ servers in the reconstruction. Consider the following two cases, and refer to the specification of the UC PPSS functionality $\mathcal{F}_{\text{PPSS}}$ of [19], which we include in the full version of this paper [20].

Case I: Every server S' in set $\mathcal{SR}$ is either incorrect (i.e. $S' \notin \mathcal{SI}$) and w.l.o.g. represents a malicious entity, or it is correct (i.e. $S' \in \mathcal{SI}$) but it is corrupted. In this case, according to $\mathcal{F}_{\text{PPSS}}$ specifications (see line 3b of the reconstruction phase), the adversary can perform one on-line password guess on such session. In other words, if the user runs reconstruction with incorrect/corrupt servers, the security is as in a (password-only) PAKE, i.e. the adversary can attempt to authenticate to such user using a password guess pw^*, and test if $\mathsf{pw}^* = \mathsf{pw}$.

Case II: There are *some* servers S' in set $\mathcal{SR}$ which are both correct (i.e. $S' \in \mathcal{SI}$) and uncorrupted. In this case, according to $\mathcal{F}_{\text{PPSS}}$ specifications (lines 3a and 3b of $\mathcal{F}_{\text{PPSS}}$), the adversary cannot learn anything from such instance, and can only either let it execute (line 3a) in which case U reconstructs the (correct!) secret K, or interfere with the protocol (line 3c) and make U output FAIL.

In short, if PPSS reconstruction is executed over insecure channels then the man-in-the-middle adversary could make every reconstruction instance fall into Case I. By contrast, executing it over TLS forces the reconstruction instances to fall into Case II, unless the adversary tricks U to execute the reconstruction for the set of servers $\mathcal{SR}$ which includes *only* corrupt entities, in which case such reconstruction instance (and only such instance) falls back into Case I.

Note on sid/ssid Monikers. As we explain above, it is not essential for security of reconstruction that the user remembers the servers in the initialization set $\mathcal{SI}$. It might also be helpful to clarify the potential security implications of sid/ssid monikers which we assume are inputs in the initialization and the reconstruction phase. String *sid* (which stands for "session ID" in the AKE and UC terminology) in the context of a PPSS scheme can be equated with a "user ID", because it is a string which servers in $\mathcal{SI}$ will use to disambiguate between multiple PPSS instances which they can potentially service. It is therefore sensible to require that U remembers this user ID string *sid* in addition to her password pw. On the other hand, string *ssid* could be a nonce, or some application-determined identifier of a unique PPSS reconstruction session.

References

1. Russian hackers amass over a billion internet passwords. New York Times, 08 June 2014. http://goo.gl/aXzqj8

2. Abdalla, M., Chevassut, O., Fouque, P.-A., Pointcheval, D.: A simple threshold authenticated key exchange from short secrets. In: Roy, B. (ed.) ASIACRYPT 2005. LNCS, vol. 3788, pp. 566–584. Springer, Heidelberg (2005). doi:10.1007/11593447_31

3. Abdalla, M., Cornejo, M., Nitulescu, A., Pointcheval, D.: Robust password-protected secret sharing. In: Askoxylakis, I., Ioannidis, S., Katsikas, S., Meadows, C. (eds.) ESORICS 2016. LNCS, vol. 9879, pp. 61–79. Springer, Cham (2016). doi:10.1007/978-3-319-45741-3_4

4. Bagherzandi, A., Jarecki, S., Saxena, N., Lu, Y.: Password-protected secret sharing. In: Proceedings of the 18th ACM Conference on Computer and Communications Security, pp. 433–444. ACM (2011)

5. Bellare, M., Namprempre, C., Pointcheval, D., Semanko, M., et al.: The one-more-rsa-inversion problems and the security of chaum's blind signature scheme. J. Cryptol. 16(3), 185–215 (2003)

6. Bellare, M., Pointcheval, D., Rogaway, P.: Authenticated key exchange secure against dictionary attacks. In: Preneel, B. (ed.) EUROCRYPT 2000. LNCS, vol. 1807, pp. 139–155. Springer, Heidelberg (2000). doi:10.1007/3-540-45539-6_11

7. Blazy, O., Chevalier, C., Vergnaud, D.: Mitigating server breaches in password-based authentication: secure and efficient solutions. In: Sako, K. (ed.) CT-RSA 2016. LNCS, vol. 9610, pp. 3–18. Springer, Cham (2016). doi:10.1007/978-3-319-29485-8_1

8. Boyko, V., MacKenzie, P., Patel, S.: Provably secure password-authenticated key exchange using Diffie-Hellman. In: Preneel, B. (ed.) EUROCRYPT 2000. LNCS, vol. 1807, pp. 156–171. Springer, Heidelberg (2000). doi:10.1007/3-540-45539-6_12

9. Brainard, J., Juels, A., Kaliski, B., Szydlo, M.: Nightingale: a new two-server approach for authentication with short secrets. In: 12th USENIX Security Symposium, pp. 201–213. IEEE Computer Society (2003)

10. Camenisch, J., Lehmann, A., Lysyanskaya, A., Neven, G.: Memento: how to reconstruct your secrets from a single password in a hostile environment. In: Garay, J.A., Gennaro, R. (eds.) CRYPTO 2014. LNCS, vol. 8617, pp. 256–275. Springer, Heidelberg (2014). doi:10.1007/978-3-662-44381-1_15

11. Camenisch, J., Lehmann, A., Neven, G.: Optimal distributed password verification. In: Proceedings of the 22nd ACM SIGSAC Conference on Computer and Communications Security, pp. 182–194. ACM (2015)

12. Canetti, R.: Universally composable security: a new paradigm for cryptographic protocols. In: Proceedings of the 42nd IEEE Symposium on Foundations of Computer Science, 2001, pp. 136–145. IEEE (2001)

13. Cash, D., Jarecki, S., Jutla, C., Krawczyk, H., Roşu, M.-C., Steiner, M.: Highly-scalable searchable symmetric encryption with support for boolean queries. In: Canetti, R., Garay, J.A. (eds.) CRYPTO 2013. LNCS, vol. 8042, pp. 353–373. Springer, Heidelberg (2013). doi:10.1007/978-3-642-40041-4_20

14. Raimondo, M., Gennaro, R.: Provably secure threshold password-authenticated key exchange. In: Biham, E. (ed.) EUROCRYPT 2003. LNCS, vol. 2656, pp. 507–523. Springer, Heidelberg (2003). doi:10.1007/3-540-39200-9_32

15. Ford, W., Kaliski, B.S.: Server-assisted generation of a strong secret from a password. In: Proeedings of the IEEE 9th International Workshops on Enabling Technologies: Infrastructure for Collaborative Enterprises 2000, (WET ICE 2000), pp. 176–180. IEEE (2000)

16. Gennaro, R., Jarecki, S., Krawczyk, H., Rabin, T.: Secure distributed key generation for discrete-log based cryptosystems. J. Cryptol. 20(1), 51–83 (2007)

17. Jarecki, S., Jutla, C., Krawczyk, H., Rosu, M., Steiner, M.: Outsourced symmetric private information retrieval. In: Proceedings of the 2013 ACM SIGSAC Conference on Computer & Communications Security, pp. 875–888. ACM (2013)

18. Jarecki, S., Kiayias, A., Krawczyk, H.: Round-optimal password-protected secret sharing and T-PAKE in the password-only model. In: Sarkar, P., Iwata, T. (eds.) ASIACRYPT 2014. LNCS, vol. 8874, pp. 233–253. Springer, Heidelberg (2014). doi:10.1007/978-3-662-45608-8_13

19. Jarecki, S., Kiayias, A., Krawczyk, H., Xu, J.: Highly-efficient and composable password-protected secret sharing (or: how to protect your bitcoin wallet online). In: 2016 IEEE European Symposium on Security and Privacy (EuroS&P), pp. 276–291. IEEE (2016)

20. Jarecki, S., Kiayias, A., Krawczyk, H., Xu, J.: Threshold oblivious PRF and minimal-cost password-protected secret sharing. Cryptology ePrint Archive (2017). http://eprint.iacr.org/2017/[TBD]

21. Jarecki, S., Krawczyk, H., Shirvanian, M., Saxena, N.: Device-enhanced password protocols with optimal online-offline protection. In: Proceedings of the 11th ACM on Asia Conference on Computer and Communications Security, pp. 177–188. ACM (2016)

22. Jarecki, S., Liu, X.: Fast secure computation of set intersection. In: Garay, J.A., Prisco, R. (eds.) SCN 2010. LNCS, vol. 6280, pp. 418–435. Springer, Heidelberg (2010). doi:10.1007/978-3-642-15317-4_26

23. Katz, J., MacKenzie, P., Taban, G., Gligor, V.: Two-server password-only authenticated key exchange. In: Ioannidis, J., Keromytis, A., Yung, M. (eds.) ACNS 2005. LNCS, vol. 3531, pp. 1–16. Springer, Heidelberg (2005). doi:10.1007/11496137_1

24. Kiefer, F., Manulis, M.: Distributed smooth projective hashing and its application to two-server password authenticated key exchange. In: Boureanu, I., Owesarski, P., Vaudenay, S. (eds.) ACNS 2014. LNCS, vol. 8479, pp. 199–216. Springer, Cham (2014). doi:10.1007/978-3-319-07536-5_13

25. Kiefer, F., Manulis, M.: Universally composable two-server PAKE. In: Bishop, M., Nascimento, A. (eds.) ISC 2016. LNCS, vol. 9866, pp. 147–166. Springer, Cham (2016). doi:10.1007/978-3-319-45871-7_10

26. MacKenzie, P., Shrimpton, T., Jakobsson, M.: Threshold password-authenticated key exchange. In: Yung, M. (ed.) CRYPTO 2002. LNCS, vol. 2442, pp. 385–400. Springer, Heidelberg (2002). doi:10.1007/3-540-45708-9_25

27. Naor, M., Pinkas, B., Reingold, O.: Distributed pseudo-random functions and KDCs. In: Stern, J. (ed.) EUROCRYPT 1999. LNCS, vol. 1592, pp. 327–346. Springer, Heidelberg (1999). doi:10.1007/3-540-48910-X_23

28. Shoup, V.: Lower bounds for discrete logarithms and related problems. In: Fumy, W. (ed.) EUROCRYPT 1997. LNCS, vol. 1233, pp. 256–266. Springer, Heidelberg (1997). doi:10.1007/3-540-69053-0_18

29. Szydlo, M., Kaliski, B.: Proofs for two-server password authentication. In: Menezes, A. (ed.) CT-RSA 2005. LNCS, vol. 3376, pp. 227–244. Springer, Heidelberg (2005). doi:10.1007/978-3-540-30574-3_16

30. Wikström, D.: Universally composable DKG with linear number of exponentiations. In: Blundo, C., Cimato, S. (eds.) SCN 2004. LNCS, vol. 3352, pp. 263–277. Springer, Heidelberg (2005). doi:10.1007/978-3-540-30598-9_19

31. Yi, X., Hao, F., Chen, L., Liu, J.K.: Practical threshold password-authenticated secret sharing protocol. In: Pernul, G., Ryan, P.Y.A., Weippl, E. (eds.) ESORICS 2015. LNCS, vol. 9326, pp. 347–365. Springer, Cham (2015). doi:10.1007/978-3-319-24174-6_18

Secure and Efficient Pairing at 256-Bit Security Level

Yutaro Kiyomura[1]($\boxtimes$), Akiko Inoue[2], Yuto Kawahara[1], Masaya Yasuda[3], Tsuyoshi Takagi[3], and Tetsutaro Kobayashi[1]

[1] NTT Secure Platform Laboratories, Musashino, Japan
{kiyomura.yutaro,kawahara.yuto,kobayashi.tetsutaro}@lab.ntt.co.jp
[2] NEC Central Research Laboratories, Kawasaki, Japan
a-inoue@cj.jp.nec.com
[3] Kyushu University, Fukuoka, Japan
{yasuda,takagi}@imi.kyushu-u.ac.jp

Abstract. At CRYPTO 2016, Kim and Barbulescu proposed an efficient number field sieve (NFS) algorithm for the discrete logarithm problem (DLP) in a finite field. The security of pairing-based cryptography (PBC) is based on the difficulty in solving the DLP. Hence, it has become necessary to revise the bitlength that the DLP is computationally infeasible against the efficient NFS algorithms. The timing of the main operations of PBC (i.e. pairing, scalar multiplication on the elliptic curves, and exponentiation on the finite field) generally becomes slower as the bitlength becomes longer, so it has become increasingly important to compute the main operations of PBC more efficiently. To choose a suitable pairing-friendly curve from among various pairing-friendly curves is one of the factors that affect the efficiency of computing the main operations of PBC. We should implement the main operations of PBC and compare the timing among some pairing-friendly curves in order to choose the suitable pairing-friendly curve precisely. In this paper, we focus on the five candidate pairing-friendly curves from the Barreto-Lynn-Scott (BLS) and Kachisa-Schaefer-Scott (KSS) families as the 256-bit secure pairing-friendly curves and show the following two results; (1) the revised bitlength that the DLP is computationally infeasible against the efficient NFS algorithms for each candidate pairing-friendly curve, (2) the suitable pairing-friendly curve by comparing the timing of the main operations of PBC among the candidate pairing-friendly curves using the revised bitlength.

1 Introduction

Many pairing-based cryptography (PBC) have been proposed, e.g., ID-based encryption [8,40], attribute-based encryption [41], and functional encryption [38]. A pairing on the elliptic curve is a non-degenerate bilinear map

---

A. Inoue—This work was done while the author was the student of Kyushu University.

D. Gollmann et al. (Eds.): ACNS 2017, LNCS 10355, pp. 59–79, 2017.
DOI: 10.1007/978-3-319-61204-1_4

$e : \mathbb{G}_1 \times \mathbb{G}_2 \to \mathbb{G}_3$, where $\mathbb{G}_1$, $\mathbb{G}_2$, $\mathbb{G}_3$ are a group with order r respectively. The security of the pairing is based on the difficulty in solving the discrete logarithm problem (DLP) in $\mathbb{G}_1$, $\mathbb{G}_2$, $\mathbb{G}_3$. The group $\mathbb{G}_1$, $\mathbb{G}_2$ are a subgroup on the elliptic curve, and the DLP on an elliptic curve (ECDLP) in $\mathbb{G}_1$, $\mathbb{G}_2$ must be computationally infeasible against the rho algorithm [16,39]. Hence, we should choose r with a secure bitlength against the rho algorithm. The group $\mathbb{G}_3$ is a subgroup on finite field $\mathbb{F}_{p^k}$, where p is a prime and $k \geq 1$ is an embedding degree, and the DLP on a finite field (FFDLP) in $\mathbb{G}_3$ must be computationally infeasible against the number field sieve (NFS) algorithms. There are various NFS algorithms (e.g. Classical-NFS [25], tower NFS (TNFS) [10,45], and special NFS (SNFS) [10,26]). We should choose p and k with a secure bitlength of p^k against the NFS algorithms.

The recommended bitlength of p^k of the pairing was discussed in the 2013 report of ENISA [15, Table 3.6], where the pairing and RSA have the same recommended bitlength. This was in accordance with a general belief stated, for example, by Lenstra: "An RSA modulus n and a finite field $\mathbb{F}_{p^k}$ therefore offer about the same level of security if n and p^k are of the same order of magnitude" [32, Sect. 5.1]. The recommended bitlength of RSA was derived from the complexity of the NFS algorithm for integer factorization [33]. In other words, the bitlength of p^k of the pairing was estimated considering the complexity of this NFS algorithm.

At CRYPTO 2016, Kim and Barbulescu proposed an efficient NFS algorithm called the extended tower number field sieve (exTNFS) algorithm [28]. This NFS algorithm greatly impacted the security of the mainstream pairing such as optimal ate pairing [47]. The complexity of the exTNFS algorithm was reduced from that of previous NFS algorithms by using the trivial equation $\mathbb{F}_{p^k} = \mathbb{F}_{p^{\eta\kappa}}$, where $\gcd(\eta, \kappa) = 1$. Kim and Barbulescu concluded that the bitlength of the pairing should increase roughly twice [28]. Therefore, we should revise to estimate the secure bitlength of p^k against the exTNFS in detail. Note that Menezes et al. estimated the bitlength of p^k for the pairing considering the exTNFS algorithm at 128- and 192-bit security levels [37].

Generally, faster timing of the main operations of PBC (i.e. pairing, scalar multiplication on the elliptic curves, and exponentiation on the finite field) is preferred to implement the PBC. To choose a suitable pairing-friendly curve from among various pairing-friendly curves is one of the factors that affect the efficiency of computing the main operations of PBC. Among the studies conducted before the exTNFS algorithm was proposed, Scott [44] theoretically chose the suitable pairing-friendly curve at each security level based on the bitlength of r, p^k and ρ-value given in Freeman et al. 's taxonomy [16]. However, Aranha et al. [3] discussed the suitable pairing-friendly curve different from that chosen theoretically by comparing the timing of the pairing among several pairing-friendly curves at 192-bit security level. To choose a suitable pairing-friendly curve at a certain security level, it is important to not only choose theoretically but also compare the timing of the main operations of PBC.

Our Contributions. Our goal with this paper is to obtain a secure and efficient pairing at 256-bit security level. To achieve this, our contribution is to revise the estimation of the bitlength of p^k due to the efficient NFS algorithms (e.g. Special exTNFS, Special TNFS) and choose the suitable pairing-friendly curve for efficiently computing the main operations of PBC. We focus on the Barreto-Lynn-Scott (BLS) [11] and Kachisa-Schaefer-Scott (KSS) [29] families that have high embedding degree and are easy to implement the pairing. We specifically choose the following five candidate pairing-friendly curves at the 256-bit security level; the BLS-k with $k = 24, 42, 48$ and KSS-k with $k = 32, 36$. For these curves, we estimate the secure bitlength p^k in detail against the efficient NFS algorithms by comparing the upper bound of norms of these algorithms using the Kim and Barbulescu's estimation method [28]. Furthermore, based on the revised bitlength of p^k, we search for the specific parameter of each candidate pairing-friendly curve to implement the main operations of PBC, and then compare the timing of these operations among the five candidate pairing-friendly curves. Finally, we show the suitable pairing-friendly curve at 256-bit security level.

2 Overview of Pairing

2.1 Definition and Properties

Let p be a prime and E be an elliptic curve defined over the finite field $\mathbb{F}_p$. Let r be a prime with $\gcd(p, r) = 1$. An embedding degree k is the smallest positive integer with $r \mid p^k - 1$. Let $\mathbb{G}_1, \mathbb{G}_2$ be a subgroup on the elliptic curve with order r and $\mathbb{G}_3$ be a subgroup on the finite field $\mathbb{F}_{p^k}$ with order r. A pairing e is defined by $e : \mathbb{G}_1 \times \mathbb{G}_2 \longrightarrow \mathbb{G}_3$; $(P, Q) \longmapsto f_{r,P}(Q)^{(p^k-1)/r}$, where the rational function $f_{r,P}$ satisfies $\operatorname{div}(f_{r,P}) = r(P) - r(\mathcal{O})$ for the point at infinity $\mathcal{O}$. For $P \in \mathbb{G}_1$, $Q \in \mathbb{G}_2$ and $a \in \mathbb{Z}$, a pairing e has the following properties;

- bilinearity: $e(aP, Q) = e(P, aQ) = e(P, Q)^a$,
- non-degeneracy: for all $P \in \mathbb{G}_1$, $e(P, Q) = 1$ then $Q = \mathcal{O}$ and for all $Q \in \mathbb{G}_2$, $e(P, Q) = 1$ then $P = \mathcal{O}$,
- efficiently computable: $e(P, Q)$ can be efficiently computed.

2.2 Optimal Ate Pairing

The optimal ate pairing proposed by Vercauteren [47] is the most efficient method of computing the pairing e. There are many implementation results of the optimal ate pairing [3,9,13,36,44]. Let m be an integer such that $r \nmid m$. Let $\lambda = mr$ and write $\lambda = \sum_{i=0}^{\omega} \alpha_i p^i$ where $\omega = \lfloor \log_p \lambda \rfloor$. Let $E[r]$ be an r-torsion subgroup. Define $\mathbb{G}_1 = E[r] \cap \operatorname{Ker}(\pi_p - [1]) = E(\mathbb{F}_p)[r]$, $\hat{\mathbb{G}}_2 = E[r] \cap \operatorname{Ker}(\pi_p - [p]) \subseteq E(\mathbb{F}_{p^k})[r]$ as the subgroup with r. Let E' be a twist of degree d of E with $\psi : E' \to E$ defined over $\mathbb{F}_{p^d}$, and define $\mathbb{G}_2 = \psi^{-1}(\hat{\mathbb{G}}_2)$. Note that d

depends on the pairing-friendly curve and is in $\{2, 3, 4, 6\}$ [24]. An optimal ate pairing a_k is defined by

$$a_k : \mathbb{G}_1 \times \mathbb{G}_2 \longrightarrow \mathbb{G}_3, \ (P, Q) \longmapsto \left(\prod_{i=0}^{\omega} f_{\alpha_i, Q}^{p^i}(P) \cdot \prod_{i=0}^{\omega-1} \frac{\ell_{[\beta_{i+1}]Q, [\alpha_i p^i]Q}(P)}{v_{[\beta_i]Q}(P)} \right)^{\frac{p^k - 1}{r}} \quad (1)$$

where $\beta_i = \sum_{j=i}^{\omega} \alpha_j p^j$, $\ell_{T,T'}$ is the line through T and T', and v_T is the vertical line through T, where T and T' are points on the elliptic curve.

3 Candidate Pairing-Friendly Curves at 256-Bit Security Level

In this section, we choose the five candidate pairing-friendly curves satisfying the security and efficiency from the BLS [11] and KSS [29] families to choose the suitable pairing-friendly curve at 256-bit security level. In this paper, we define $\text{len}(x)$ as the bitlength of x.

3.1 How to Choose Candidate Pairing-Friendly Curves

We show the security against the ECDLP and FFDLP and the efficiency for implementation of the main operations of PBC. An embedding degree k is determined by the chosen pairing-friendly curve, and the primes r and p are represented by the polynomial of a positive integer x.

Security. The parameters r, p, and k should satisfy the complexity of solving the DLP in $\mathbb{G}_1$, $\mathbb{G}_2$ and $\mathbb{G}_3$ to achieve the $\mathcal{K}$-bit security level. The definition of ECDLP in $\mathbb{G}_1$ and $\mathbb{G}_2$ is as follows. Given points $G, Y \in \mathbb{G}_1$ (or $\mathbb{G}_2$), find $x \in \mathbb{Z}$ such that $Y = xG$. An efficient algorithm for solving the ECDLP is the rho algorithm [16,39], which has the complexity of $\mathcal{O}(\sqrt{r})$. Therefore, we should choose the bitlength of r with $\text{len}(r) \geq 2\mathcal{K}$. The definition of the FFDLP in $\mathbb{G}_3$ is as follows. Given points $g, y \in \mathbb{G}_3$, find $x \in \mathbb{Z}$ such that $y = g^x$. An efficient algorithm for solving the FFDLP are the STNFS [10,26] and SexTNFS [28] algorithms. We give the bitlength of p^k which the FFDLP is computationally infeasible against these NFS algorithms in Sect. 5.

Efficiency. To efficiently compute the main operations of PBC (i.e. pairing, scalar multiplication in $\mathbb{G}_1$ and $\mathbb{G}_2$, and exponentiation in $\mathbb{G}_3$) with the above security, we consider the following conditions as affecting the efficiency of these operations.

- $\text{len}(r)$ and $\text{len}(p^k)$ are as small as possible.
- The ρ-value is approximately 1 ($\rho = \log p / \log r$).
- Parameter x in polynomials (e.g. $p(x)$, $r(x)$) has a low Hamming weight.
- The embedding degree k has the form $k = 2^i \cdot 3^j$ ($i \in \mathbb{Z}_{\geq 1}, j \in \mathbb{Z}_{\geq 0}$).
- The twist of degree d is 6 ($d = 6$ is maximum of degree).

These conditions are theoretically efficient ones, then the effect of each condition is uncertain in the implementation.

Table 1. Parameters for the five candidate pairing-friendly curves

BLS-24 [16, Construction 6.6]	$k = 24$, $\rho = 1.250$, $\deg(p(x)) = 10$, $\varphi(k) = 8$, $p(x) = (x - 1)^2(x^8 - x^4 + 1)/3 + x$, $r(x) = x^8 - x^4 + 1$, $t(x) = x + 1$
KSS-32 [29, Example 4.4]	$k = 32$, $\rho = 1.063$, $\deg(p(x)) = 18$, $\varphi(k) = 16$, $p(x) = (x^{18} - 6x^{17} + 13x^{16} + 57120x^{10} - 344632x^9 + 742560x^8$ $+ 815730721x^2 - 4948305594x + 10604499373)/2970292$, $r(x) = (x^{16} + 57120x^8 + 815730721)/(2 \cdot 13^8 \cdot 239^2)$, $t(x) = (-2x^9 - 56403x + 3107)/3107$
KSS-36 [29, Example 4.5]	$k = 36$, $\rho = 1.167$, $\deg(p(x)) = 14$, $\varphi(k) = 12$, $p(x) = (x^{14} - 4x^{13} + 7x^{12} + 683x^8 - 2510x^7 + 4781x^6 + 117649x^2$ $- 386569x + 823543)/28749$, $r(x) = (x^{12} + 683x^6 + 117649)/(7^6 \cdot 37^2)$, $t(x) = (2x^7 + 757x + 259)/259$
BLS-42 [16, Construction 6.6]	$k = 42$, $\rho = 1.333$, $\deg(p(x)) = 16$, $\varphi(k) = 12$, $p(x) = (x - 1)^2(x^{14} - x^7 + 1)/3 + x$, $r(x) = x^{12} + x^{11} - x^9 - x^8 + x^6 - x^4 - x^3 + x + 1$, $t(x) = x + 1$
BLS-48 [16, Construction 6.6]	$k = 48$, $\rho = 1.125$, $\deg(p(x)) = 18$, $\varphi(k) = 16$, $p(x) = (x - 1)^2(x^{16} - x^8 + 1)/3 + x$, $r(x) = x^{16} - x^8 + 1$, $t(x) = x + 1$

* deg(): degree of polynomial, φ(): Euler function

3.2   Selection of Candidate Pairing-Friendly Curves

In this subsection, we decide the candidate pairing-friendly curves from the BLS [11] and KSS [29] families to choose the suitable pairing-friendly curve at 256-bit security level based on Sect. 3.1. We focus on the BLS and KSS families that have high embedding degree and can be easy to construct the pairing. We specifically choose the following five candidate pairing-friendly curves at the 256-bit security level; the BLS-k with $k = 24, 42, 48$ and the KSS-k with $k = 32, 36$. In the case of the BLS-k, the small len(r) and len(p^k) in the BLS-42 and BLS-48 can be choose because these curve have high embedding degree k, and the implementation results in the BLS-24 exists [9,44]. In the case of the KSS-k, the KSS-36 has the small len(r), the KSS-32 has small ρ-value and simple tower construction for $\mathbb{F}_{p^k}$ since $k = 32 = 2^5$.

The detail of the five candidate pairing-friendly curves are as follows; the curves with $6 \mid k$ are defined by $E/\mathbb{F}_p : y^2 = x^3 + b$, and has the complex multiplication discriminant $D = 3$ and $d = 6$, the curves with $4 \mid k$ is defined by $E/\mathbb{F}_p : y^2 = x^3 + ax$, and has $D = 1$ and $d = 4$. Table 1 shows the parameters $p(x)$, $r(x)$, $t(x)$, k, ρ-value, $\deg(p(x))$, and Euler function $\varphi(k)$ for each curves. The parameters $n(x)$ and $f(x)$ satisfy $n(x) = p(x) + 1 - t(x)$ and $4p(x) - t(x)^2 = Df(x)^2$, respectively.

4   Overview of Number Field Sieve and Its Variants

In this section, we give an overview of the NFS algorithm and its variants to revise the bitlength of the five candidate pairing-friendly curves introduced in Sect. 3.2.

The FFDLP is classified into three cases by size of p: small, medium, or large. In medium and large cases, the NFS algorithms is the most efficient algorithm for solving the FFDLP. To accurately classify p, let $p = L_{p^k}(l_p, c_p)$, where $L_{p^k}(l_p, c_p) = \exp((c_p + o(1))(\log p^k)^{l_p}(\log \log p^k)^{1 - l_p})$. $o(1)$ becomes 0 when

$p^k \to \infty$. The prime p is called medium if $1/3 < l_p < 2/3$, large if $2/3 < l_p < 1$, boundary if $l_p = 2/3$.

Note that the above L_{p^k}-notation is just an asymptotic value. If we fix the value of p^k, the L_{p^k}-notation has a constant value c such that $c \times \exp((c_p + o(1))(\log p^k)^{l_p}(\log \log p^k)^{1-l_p})$ and $o(1) \neq 0$. Therefore, when we substitute the concrete value for p^k in L_{p^k}-notation, it is important to evaluate c and $o(1)$.

The NFS algorithms for solving the FFDLP are classified into three types: Classical-NFS, TNFS, and exTNFS, according to their mathematical constructions. The Classical-NFS algorithm was proposed in 2006, and the complexities are $L_{p^k}(1/3, (128/9)^{1/3})$ and $L_{p^k}(1/3, (64/9)^{1/3})$ in the medium and large cases, respectively. The TNFS algorithm was proposed in 1999 and later applied to the large case in 2015, where the complexity of the TNFS algorithm is also $L_{p^k}(1/3, (64/9)^{1/3})$ in the large case. Finally, the exTNFS algorithm proposed in 2015 is the generalization of combining the Classical-NFS and TNFS algorithms, and its complexities in medium and large cases are $L_{p^k}(1/3, (64/9)^{1/3})$.

4.1   Extended TNFS and Special-NFS Algorithms

In this subsection, we explain the exTNFS algorithm [28], which is effective for solving the FFDLP. We then give an overview of the Special-NFS algorithm. The NFS algorithms (not specified for exTNFS) are divided into the following four steps: 1. polynomial selection, 2. relation collection, 3. linear algebra, and 4. individual logarithm.

exTNFS Algorithm. We can use the exTNFS algorithm when the extension degree k is composite. Let $k = \eta \kappa$. We select an irreducible polynomial $h(t) \in \mathbb{Z}[t]$ over $\mathbb{Q}$ and $\mathbb{F}_p$ whose degree is η. We construct $\mathbb{Q}(\iota) = \mathbb{Q}[t]/h(t)$ and put $R = \mathbb{Z}[t]/h(t) \subset \mathbb{Q}(\iota)$.

Note that the Classical-NFS algorithm [25] is the case in which $R = \mathbb{Z}$ in the exTNFS algorithm, and the TNFS algorithm [45] is the case in which $\deg h = n$ in the exTNFS algorithm.

Polynomial Selection. We select polynomials f_1 and $f_2 \in R[X]$ that satisfy the condition that $f_1 \mod p$ and $f_2 \mod p$ have a common factor $\varphi(X)$ of degree κ, which is irreducible over $\mathbb{F}_{p^\eta} = R/pR$. In this section, $i \in \{1, 2\}$. Let K_i be the number fields defined by f_i above the fraction field of R and $\mathcal{O}_i$ be the integer ring of K_i. We denote the roots of f_i in $\mathbb{C}$ by θ_i and the degree of f_i by d_i. We then obtain two maps from $R[X]$ to $(R/pR)[X]/\varphi(X) \cong \mathbb{F}_{p^k}$.

Relation Collection. We select smoothness bound $B \in \mathbb{N}$ and define factor base $\mathcal{F}_i$ as follows: $\mathcal{F}_i = \{(\mathfrak{q}, \theta_i - \gamma) : \mathfrak{q} : \text{prime in } \mathbb{Q}(\iota) \text{ lying over a prime } p \leq B, f_i(\gamma) \equiv 0 \mod \mathfrak{q}\}$. We then obtain $a - bX \in R[X]$ by selecting $(a(t), b(t)) \in R^2$. The coefficients of $a(t)$ and $b(t)$ are bounded by A. Let $E = A^\eta$ be the sieve parameter. The norm of $a - b\theta_i$ in K_i is expressed as follows.

$$\mathcal{N}_{K_i/\mathbb{Q}}(a - b\theta_i) = \left| \text{Res}\left(h(t), \sum_{j \in [0,d_i]} f_{i,j} a(t)^j b(t)^{d_i - j} \right) \right|,$$

where $f_{i,j}$ is the coefficient of polynomial $f_i = \sum_{j=0}^{d_i} f_{i,j} X^j$. When $\mathcal{N}_{K_1/\mathbb{Q}}$ $(a - b\theta_1)$ and $\mathcal{N}_{K_2/\mathbb{Q}}(a - b\theta_2)$ are B-smooth, the $(a(t), b(t))$ pair is called a double smooth pair (an integer is B-smooth if the largest prime factor is less than B). When $(a(t), b(t))$ is a double smooth pair, $(a - b\theta_1)$ and $(a - b\theta_2)$ can be factored into the prime ideal in $\mathcal{O}_1$ and $\mathcal{O}_2$ using only the elements of $\mathcal{F}_1$ and $\mathcal{F}_2$, respectively. Therefore, we obtain the following notation: $(a - b\theta_i) = \prod_{\mathfrak{p} \in \mathcal{F}_i} \mathfrak{p}^{\mu_\mathfrak{p}}$ and the following relation up to units.

$$\phi_1((a - b\theta_1)) = \phi_2((a - b\theta_2)) \text{ in } \mathbb{F}_{p^k} \iff \phi_1\left(\prod_{\mathfrak{p} \in \mathcal{F}_1} \mathfrak{p}^{\mu_\mathfrak{p}} \right) = \phi_2\left(\prod_{\mathfrak{q} \in \mathcal{F}_2} \mathfrak{q}^{\mu_\mathfrak{q}} \right)$$

$$\iff \prod_{\mathfrak{p} \in \mathcal{F}_1} \phi_1(\mathfrak{p})^{\mu_\mathfrak{p}} = \prod_{\mathfrak{q} \in \mathcal{F}_2} \phi_2(\mathfrak{q})^{\mu_\mathfrak{q}}$$

Thus, this leads to

$$\sum_{\mathfrak{p} \in \mathcal{F}_1} \mu_\mathfrak{p} \log \phi_1(\mathfrak{p}) + \sum_j \lambda_{1,j} \log \Lambda_{1,j}$$
$$= \sum_{\mathfrak{q} \in \mathcal{F}_2} \mu_\mathfrak{q} \log \phi_2(\mathfrak{q}) + \sum_j \lambda_{2,j} \log \Lambda_{2,j} \mod p^k - 1, \tag{2}$$

where $\log \phi_1(\mathfrak{p})$, $\log \phi_2(\mathfrak{q})$, $\log \Lambda_{1,j}$ and $\log \Lambda_{2,j}$ are the unknowns called virtual logarithms [46], and $\lambda_{1,j}$ and $\lambda_{2,j}$ are computable values called character maps to distinguish the difference in units. Let N_λ be the number of character maps. When we collect more than $|\mathcal{F}_1| + |\mathcal{F}_2| + N_\lambda$ double smooth pairs $(a(t), b(t))$, we obtain the relations of (2).

In this section, we collect double smooth pairs (a, b), but it is possible to collect double smooth tupples $(a_1, a_2, \cdots, a_\tau)$. We call parameter τ a sieve dimension.

Linear Algebra. In collecting adequate relations in the previous step, we can construct and solve the simultaneous congruence. We obtain the values of the virtual logarithms $\log \phi_1(\mathfrak{p})$, $\log \phi_2(\mathfrak{q})$, $\log \Lambda_{1,j}$, and $\log \Lambda_{2,j}$.

Individual Logarithm. Finally, we compute the target logarithm x from the values of the virtual logarithms.

Special-NFS Algorithms. We collectively call three NFS algorithms (exTNFS, Classical-NFS, and TNFS) as the General NFS (GNFS) algorithms. The GNFS algorithms can be applied for special polynomial selection when p has a special form. The special cases of the GNFS algorithms are called Special-NFS (SNFS) algorithms. We consider the SNFS algorithms for solving the FFDLP to estimate the security of the pairing where p has a special form.

4.2   Larger Norm Implies Higher Complexity

In this section, we give an overview on the complexity of the NFS algorithms. The main steps to evaluate this complexity are as follows.

- We evaluate the upper bound of norms and probabilities in which the norms are B-smooth.

The smaller the upper bound of norms is, the higher the probability the norms are B-smooth. Therefore, we have to select polynomials so that the upper bounds of norms become small.

- We set the parameters appropriately so that we can collect adequate double smooth pairs from the sieve region.

The sieve region is the region to collect relations. When the sieve degree is τ, the sieve region is E^τ. We set the appropriate parameters to satisfy the inequality that $E^\tau \times$ (the probability of B-smooth of norm's upper bound) $\leq$ (the number of double smooth pairs we collect) $= B^{1+o(1)}$.

- Relation collection and linear algebra have the same complexities.

The whole complexity of NFS algorithms is the sum of the complexities of following two steps: relation collection and linear algebra. In the exTNFS algorithm, the complexity of relation collection is $O(E^2)$. We need to evaluate sizes of parameters because of the trade-off between relation collection and linear algebra.

When the norm is small, the probability that norms are B-smooth becomes high. We can obtain relations with a few trials. The complexity of relation collection becomes small, and the whole complexity becomes small. That is, the decrease in norms implies the reduction in the security of cryptography, which is based on the difficulty of the FFDLP. Therefore, we can estimate bitlengths by comparing the sizes of norms.

4.3   Comparing Norms of NFS Algorithms by Using Kim and Barbulescu's Estimation Method

We refer to the method of comparing norms in [28] to estimate and compare the norms of various NFS algorithms. The norms of each GNFS algorithm are listed in Table 2, and the norms of each SNFS algorithm are listed in Table 3 (part of Table 2 is omitted).

In Tables 2 and 3, E_G is the sieve parameter of the GNFS algorithms and E_S is the sieve parameter of the SNFS algorithms. In addition, d is the degree of polynomial selected in the step of polynomial selection. Note that d in SNFS algorithm is equal to the degree of p. The τ is the sieve dimension, and others are parameters used in each NFS algorithm. Sieve parameter E depends on the implementations. Kim and Barbulescu [28] used the formula

$$E_G = c_G L_{p^k}\left(\frac{1}{3}, \left(\frac{8}{9}\right)^{1/3}\right), E_S = c_S L_{p^k}\left(\frac{1}{3}, \left(\frac{4}{9}\right)^{1/3}\right).$$

Table 2. Norm sizes of GNFS algorithms **Table 3.** Norm sizes of SNFS algorithms

Algorithm	Norm product
NFS-JLSV$_1$	$E_G^{\frac{4k}{\tau}}(p^k)^{\frac{\tau-1}{k}}$
TNFS	$E_G^{\frac{2(d+1)}{\tau}}(p^k)^{\frac{2(\tau-1)}{d+1}}$
exTNFS-Conj	$E_G^{\frac{6\kappa}{\tau}}(p^k)^{\frac{\tau-1}{2\kappa}}$

Algorithm	Norm product
STNFS	$E_S^{\frac{2(d+1)}{\tau}}(p^k)^{\frac{\tau-1}{d}}$
SNFS-JP	$E_S^{\frac{2k(d+1)}{\tau}}(p^k)^{\frac{\tau-1}{kd}}$
SexTNFS	$E_S^{\frac{2\kappa(d+1)}{\tau}}(p^k)^{\frac{\tau-1}{\kappa d}}$

They determined $\log_2 c_G \approx -4.30$ using the results of three implementations [6,7,14]. Similarly, they determined $\log_2 c_S \approx -4.27$ using the results of an implementation [1]. After the values of E_G and E_S are determined, other parameters must be determined. The parameters, except τ, are computed using the theoretical optimal values. Then τ is determined as the best value in their bitsize of the norm.

5 Revise the Bitlength for Candidate Pairing-Friendly Curves

In this section, we revise to estimate the bitlengths for the five pairing-friendly curves (i.e. the BLS-k with $k = 24, 42, 48$ and KSS-k with $k = 32, 36$) at 256-bit security level by using the norms of NFS algorithms in the previous section.

5.1 Revised Estimation of Bitlength for BLS-48

In this subsection, we revised to estimate the bitlength for the BLS-48. We compare the norms of NFS algorithms based on the constants c_G and c_S and estimate the bitlength based on the initial norm of the GNFS algorithms at the 256-bit security level. The estimations for other pairing-friendly curves (i.e. the BLS-24, KSS-32, KSS-36 and BLS-42) are described in the Appendix A.

Determining Constants c_G and c_S. Before plotting norms, the constant values of c_G and c_S must be evaluated. As previously mentioned, c_G and c_S are evaluated from the implementation results. We discuss these values by adding new implementation results. In Kim and Barbulescu's study [28], $\log_2 c_G \approx -4.30$ and $\log_2 c_S \approx -4.27$; however, we evaluate c_G and c_S by adding to new results. First, we evaluate c_G using the result from Kleinjung [31] who solved the DLP in $\mathbb{F}_p$. We extrapolate from the pair ($\log_2 p^k = 768$, $\log_2 E_G \approx 35$) Kleinjung used [31] and obtain $\log_2 c_G \approx -3.26$. The sieve parameter E_G using Kleinjung's result is larger than that Kim and Barbulescu evaluated. Because E_G by Kim and Barbulescu [28] is evaluated more strictly, we plot the norms of the GNFS algorithms using $\log_2 c_G \approx -4.30$ they used. Next, we evaluate c_S using the results by Fried et al. [17] and Guillevic et al. [22]. We extrapolate from the pair ($\log_2 p^k = 1024$, $\log_2 E_S \approx 31$) used by Fried et al. [17] and obtain $\log_2 c_S \approx -3.43$. We also extrapolate from the pair $(\log_2 p^k = 510, \log_2 E_S \approx 26)$ used by Guillevic *et al.* [22] and obtain $\log_2 c_S \approx 0.67$. The sieve parameters E_S using the results from Fried *et al.* and Guillevic *et al.* are larger than those Kim and Barbulescu evaluated. As with E_G, because E_S from Kim and Barbulescu [28] is evaluated more strictly, we plot the norms of the SNFS algorithms using $\log_2 c_S \approx -4.27$ they used.

Initial Norm of General NFS Algorithms at 256-Bit Security Level. We define the initial norm as the norm of the GNFS algorithm for integer factorization, which corresponds to the bitlength at 256-bit security level. Let N be a composite number. The norm of this GNFS algorithm is $E_G^{d+1} N^{2/d+1}$, where d is the degree of the polynomial selected in the step of polynomial selection. The recommended bitlength of RSA at 256-bit security level is 15360-bit [5]. When N is 15360-bit, the optimal value of d is 15. We substitute the value of c_G, which we evaluated in the previous section, for the norm, and the initial norm is about 4006-bit.

Fastest Variant of NFS Algorithms for BLS-48. We concretely estimate the bitlengths of the PBC. We give details of the BLS-48, and the other curves are mentioned in the Appendix A. We fix the extension degree to $k = 48$ and examine the fastest NFS algorithm that solves the DLP in $\mathbb{F}_{p^{48}}$, where p is expressed by the BLS-48.

First, we plot the norms of three GNFS algorithms (i.e. NFS-JLSV$_1$, TNFS, and exTNFS-Conj) in Fig. 1. The E_G is as follows;

$$E_G = 2^{-4.3} L_{p^k}\left(\frac{1}{3}, \left(\frac{8}{9}\right)^{\frac{1}{3}}\right).$$

- NFS-JLSV$_1$. The norm of the NFS-JLSV$_1$ algorithm is expressed as $E_G^{\frac{4k}{\tau}}(p^k)^{\frac{\tau-1}{k}}$. The optimal value of τ is 9 when the norm is $E_G^{\frac{192}{\tau}}(p^k)^{\frac{\tau-1}{48}}$.
- TNFS. The norm of the TNFS algorithm is expressed as $E_G^{\frac{2(d+1)}{\tau}}(p^k)^{\frac{2(\tau-1)}{d+1}}$. The d is 15 using the formula $d = \sqrt[3]{3}\left(\log p^k / \log\log p^k\right)^{1/3}$. The optimal value of τ is 2 when the norm is $E_G^{\frac{32}{\tau}}(p^k)^{\frac{2(\tau-1)}{16}}$
- exTNFS-Conj. The norm of the exTNFS-Conj algorithm is expressed as $E_G^{\frac{6\kappa}{\tau}}(p^k)^{\frac{\tau-1}{2\kappa}}$. Kim and Barbulescu [28] used the case of $\gcd(\eta,\kappa) = 1$. However, Kim and Jeong proposed an algorithm allowing the choosing of η and κ freely from the co-primality condition, and their algorithm has the same complexities as when η and κ are co-prime [30]. Therefore, we use all cases of (η, κ). When $n = 48$, we can consider the cases of $(\eta, \kappa) = (2, 24), (3, 16),$ $(4, 12), (6, 8), (8, 6), (12, 4), (16, 3), (24, 2)$. When $(\eta, \kappa) = (2, 24), (3, 16),$ $(4, 12), (6, 8), (8, 6), (12, 4), (16, 3), (24, 2)$, the optimal value of τ is 7, 5, 4, 3, 2, 2, 2, 2, respectively.

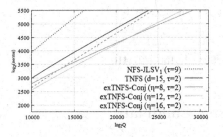 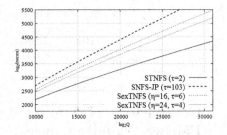

Fig. 1. Norms of GNFS algorithms in $\mathbb{F}_{p^{48}}$ **Fig. 2.** Norms of SNFS algorithms in $\mathbb{F}_{p^{48}}$

Next, we plot the norms of three SNFS algorithms (i.e. STNFS, SNFS-JP, and SexTNFS) in Fig. 2. In the BLS-48, $\deg(p(x)) = 18$. The E_S is as follows:

$$E_S = 2^{-4.3} L_{p^k}\left(\frac{1}{3}, \left(\frac{4}{9}\right)^{\frac{1}{3}}\right)$$

- STNFS. The norm of the STNFS algorithm is expressed as
  $E_G^{\frac{2(d+1)}{\tau}}(p^k)^{\frac{\tau-1}{d}}$. The optimal value of τ is 2 when the norm is $E_G^{\frac{38}{\tau}}(p^k)^{\frac{\tau-1}{18}}$.
- SNFS-JP. The norm of the SNFS-JP algorithm is expressed as
  $E_G^{\frac{2k(d+1)}{\tau}}(p^k)^{\frac{\tau-1}{kd}}$. The optimal value of τ is 103 when the norm is
  $E_G^{\frac{1824}{\tau}}(p^k)^{\frac{\tau-1}{864}}$.
- SexTNFS. The norm of the SexTNFS algorithm is expressed as
  $E_G^{\frac{2\kappa(d+1)}{\tau}}(p^k)^{\frac{\tau-1}{\kappa d}}$. We also use the all case of (η, κ). When $(\eta, \kappa) = (2, 24)$, $(3, 16)$, $(4, 12)$, $(6, 8)$, $(8, 6)$, $(12, 4)$, $(16, 3)$, $(24, 2)$, the optimal value of τ is 51, 34, 26, 17, 13, 9, 6, 4, respectively.

In Figs. 1 and 2, the STFNS algorithm has the smallest norm. When the norm is the initial norm of 4006-bit, the bitlength of the STNFS algorithm is 27410-bit. Therefore, we can estimate that the bitlength at 256-bit security level is 27410-bit.

5.2 Revised Bitlength at 256-Bit Security Level

We revise to estimate the bitlength for the five candidate pairing-friendly curves (i.e. the BLS-k with $k = 24, 42, 48$ and KSS-k with $k = 32, 36$) at 256-bit security level based on the Kim and Barbulescu's method [28]. The results are listed in Table 4. According to ENISA [15, Table 3.6], the bitlength of the pairing requires more than 15360-bit to achieve the 256-bit security level [16]. However, our revised estimation shows that it is necessary to increase the bitlength by more than 10000-bit to achieve the 256-bit security level. In other word, it is necessary to approximately multiply the bitlength by 1.7 times to achieve the 256-bit security level.

Table 4. Revised bitlength at 256-bit security level

	BLS-24	KSS-32	KSS-36	BLS-42	BLS-48
$len(p^k)$	25,990	27,410	28,280	28,150	27,410

6   Comparison of Timing Among Candidate Pairing-Friendly Curves

In this section, we measure and compare the timing of the main operations of PBC among the five candidate pairing-friendly curves using the revised bitlength to show a suitable pairing-friendly curve at 256-bit security level. Our implementation uses the efficient algorithms for computing the main operations of PBC.

6.1   Specific Parameter for Implementation

The specific parameter x_0 for each candidate pairing-friendly curve is required to decide the parameters of each curve in Table 1 and implement the main operations of PBC. The x_0 for the BLS-24 showed in [13] satisfies the revised bitlength of p^k in Table 4, but there are no documents showed the parameter satisfying the revised bitlength of p^k in Table 4 for other curves. Therefore, we should search for x_0 for each KSS-32, KSS-36, BLS-42, and BLS-48.

To efficiently compute the pairing, we search for the specific parameter x_0 with a low Hamming weight and $len(r) \geq 512$ and $len(p^k)$ always more than the bitlength in Table 4. Table 6 shows x_0, the bitlength of parameters, and Hamming weight of x_0 for the five candidate pairing-friendly curves. Note that the $\mathbb{G}_1$, $\mathbb{G}_2$, $\mathbb{G}_3$ are better to satisfy subgroup security [12] in order to resist against small-subgroup attacks as an optional security requirement. The information of implementation (i.e. the tower construction, elliptic curve E, twist E' of E, and $\ell_{T,T'}(P)$) are showed in Table 5.

6.2   Our Implemented Algorithms

In this subsection, we give an overview of the implemented efficient algorithms for computing the main operations of PBC. We implement the base field $\mathbb{F}_p$ arithmetic by using the GMP library [18]. Additionally, in the arithmetic of the tower field, we use the lazy reduction technique [42] which can reduce the number of the modulo operations of $\mathbb{F}_p$.

Pairing. The formulas of optimal ate pairing for each candidate pairing-friendly curve are given in Table 7. Note that these formulas can be produced from Eq. (1) and [47, Eq. (9)]. There are two steps involved in computing the optimal ate pairing; the miller loop (ML) $f' = f_{x,Q}(P) \cdot g$, where g is the part other than $f_{x,Q}(P)$ in Table 7, and final exponentiation (FE) $f'^{(p^k-1)/r}$.

Table 5. Information of implementation for the five candidate pairing-friendly curves

BLS-24	Fields	$\mathbb{F}_p \xrightarrow{u^2+1} \mathbb{F}_{p^2} \xrightarrow{v^2+u+1} \mathbb{F}_{p^4} \xrightarrow{w^3+v} \mathbb{F}_{p^{12}} \xrightarrow{z^2+w} \mathbb{F}_{p^{24}}$
	E, E'	$E/\mathbb{F}_p : y^2 = x^3 + 1,\ E'/\mathbb{F}_{p^4} : y^2 = x^3 - 1/v$
	$\ell_{T,T'}(P)$	$\big[\ \underbrace{y_P}_{1} : \underbrace{0}_{w} : \underbrace{0}_{w^2}\ \big|\ \underbrace{(-\lambda \cdot x_P)u}_{1} : \underbrace{cu}_{w} : \underbrace{0}_{w^2}\ \big]$ (top: 1 … z)
KSS-32	Fields	$\mathbb{F}_p \xrightarrow{u^2+2} \mathbb{F}_{p^2} \xrightarrow{v^2-u} \mathbb{F}_{p^4} \xrightarrow{w^2-v} \mathbb{F}_{p^8} \xrightarrow{z^2-w} \mathbb{F}_{p^{16}} \xrightarrow{s^2-z} \mathbb{F}_{p^{32}}$
	E, E'	$E/\mathbb{F}_p : y^2 = x^3 + 2x,\ E'/\mathbb{F}_{p^8} : y^2 = x^3 + 2x/w$
	$\ell_{T,T'}(P)$	$\big[\ \underbrace{y_P}_{1} : \underbrace{0}_{z}\ \big|\ \underbrace{-\lambda \cdot x_P}_{1} : \underbrace{c}_{z}\ \big]$ (top: 1 … s)
KSS-36	Fields	$\mathbb{F}_p \xrightarrow{u^2+1} \mathbb{F}_{p^2} \xrightarrow{v^3+u+1} \mathbb{F}_{p^6} \xrightarrow{w^3+v} \mathbb{F}_{p^{18}} \xrightarrow{z^2+w} \mathbb{F}_{p^{36}}$
	E, E'	$E/\mathbb{F}_p : y^2 = x^3 + 2,\ E'/\mathbb{F}_{p^6} : y^2 = x^3 - 2/v$
	$\ell_{T,T'}(P)$	$\big[\ \underbrace{y_P}_{1} : \underbrace{0}_{w} : \underbrace{0}_{w^2}\ \big|\ \underbrace{(-\lambda \cdot x_P)u}_{1} : \underbrace{cu}_{w} : \underbrace{0}_{w^2}\ \big]$ (top: 1 … z)
BLS-42	Fields	$\mathbb{F}_p \xrightarrow{u^7+2} \mathbb{F}_{p^7} \xrightarrow{v^3+u-1} \mathbb{F}_{p^{21}} \xrightarrow{w^2-v} \mathbb{F}_{p^{42}}$
	E, E'	$E/\mathbb{F}_p : y^2 = x^3 + 1,\ E'/\mathbb{F}_{p^6} : y^2 = x^3 + 1/(1-u)$
	$\ell_{T,T'}(P)$	$\big[\ \underbrace{y_P}_{1} : \underbrace{0}_{v} : \underbrace{0}_{v^2}\ \big|\ \underbrace{-\lambda \cdot x_P}_{1} : \underbrace{c}_{v} : \underbrace{0}_{v^2}\ \big]$ (top: 1 … w)
BLS-48	Fields	$\mathbb{F}_p \xrightarrow{u^2+1} \mathbb{F}_{p^2} \xrightarrow{v^2+u+1} \mathbb{F}_{p^4} \xrightarrow{w^2+v} \mathbb{F}_{p^8} \xrightarrow{z^3+w} \mathbb{F}_{p^{24}} \xrightarrow{s^2+z} \mathbb{F}_{p^{48}}$
	E, E'	$E/\mathbb{F}_p : y^2 = x^3 + 1,\ E'/\mathbb{F}_{p^8} : y^2 = x^3 - 1/w$
	$\ell_{T,T'}(P)$	$\big[\ \underbrace{y_P}_{1} : \underbrace{0}_{z} : \underbrace{0}_{z^2}\ \big|\ \underbrace{(-\lambda \cdot x_P)u}_{1} : \underbrace{cu}_{z} : \underbrace{0}_{z^2}\ \big]$ (top: 1 … s)

In the ML, the rational function $f_{x,Q}(P)$ can be computed using Miller's algorithm [34]. The computational cost of the ML is affected by bitlength x_0 and the Hamming weight of x_0. We can reduce the computational cost of the multiplication on $\mathbb{F}_{p^k}$ by using the sparse multiplication technique [35]. We use the affine pairing [2] since the computation of the inversion in $\mathbb{G}_2$ is fast.

In the FE, the equation $f'^{(p^k-1)/r}$ can be broken down into three components by using cyclotomic polynomial Φ_k as follows [43].

$$(p^k - 1)/r = \underbrace{[(p^{k/2} - 1)] \cdot [(p^{k/2} + 1)/\Phi_k(p)]}_{\text{easy part}} \cdot \underbrace{[\Phi_k(p)/r]}_{\text{hard part}}.$$

The computation of the easy part $m = f'^{(p^{k/2}-1)\cdot((p^{k/2}+1)/\Phi_k(p))}$ requires one conjugation, one inversion, some Frobenius operations and some multiplications on $\mathbb{F}_{p^k}$, so the computational cost of the easy part hardly affects that of the whole FE. The hard part can be computed by using the base p representation of $\Phi_k(p)/r$ as

Table 6. x_0, bitlength of parameters and Hamming weight of x_0

	x_0	len(x_0)	HW(x_0)	len(p^k)	len($p^{k/d}$)	len(p)	len(r)
BLS-24	$-1 + 2^{65} - 2^{75} + 2^{109}$	109	4	26122	4354	1089	872
KSS-32	$-1 - 2^2 - 2^{12} + 2^{14} + 2^{18} - 2^{30} + 2^{49}$	49	7	27536	6884	861	738
KSS-36	$2^5 + 2^{34} + 2^{40} + 2^{45} - 2^{58}$	58	5	28699	4784	798	669
BLS-42	$-1 + 2^2 - 2^8 + 2^{43}$	43	4	28830	4805	687	516
BLS-48	$-1 + 2^7 - 2^{10} - 2^{30} - 2^{32}$	33	5	27851	4642	581	518

* HW() : Hamming weight

Table 7. Formulas for computing optimal ate pairing $a_k(P,Q)$

BLS-k with $k = 24, 42, 48$	$(f_{x,Q}(P))^{(p^k-1)/r}$
KSS-32	$\left(f_{x,Q}(P) \cdot f_{2,Q}^{p^9}(P) \cdot \left(\overline{f_{3,Q}(P)} \right)^p \cdot \ell_{xQ,-3pQ}(P) \right)^{(p^{32}-1)/r}$
KSS-36	$\left(f_{x,Q}(P) \cdot f_{2,Q}^{p^7}(P) \cdot \left(\overline{f_{3,Q}(P)} \right)^p \cdot \ell_{xQ,-3pQ}(P) \right)^{(p^{36}-1)/r}$

* $\overline{t}$: conjugation of t in $\mathbb{F}_{p^k}$

$$m^{\Phi_k(p)/r} = (m^{\lambda_0}) \cdot (m^{\lambda_1})^p \cdots (m^{\lambda_{s-1}})^{p^{s-1}} \cdot (m^{\lambda_s})^{p^s}, \tag{3}$$

where λ_i is the polynomial by x and $s = \varphi(k) - 1$. For computing the Eq. (3), in the BLS-k, we can compute it with essentially just exponentiation by x since the BLS-k has a very convenient way to compute the each m^{λ_i} [13]. In the KSS-k, we apply the addition chain technique with Ghammam et al.'s λ_i-representation to compute the each m^{λ_i} efficiently since the coefficients of λ_i are dozens bits [23]. Additionally, to efficiently compute the exponentiation by x, we use the Karabina squaring technique [27] in the case of the curve with $d = 6$, and Granger-Scott squaring technique [19] in the case of the curve with $d = 4$ respectively. Hence, the computation of the hard part requires exponentiations by x $\deg(p(x)) - 1$ times, Frobenius operations s times, and squarings/multiplications on $\mathbb{F}_{p^k}$.

Scalar Mult. in $\mathbb{G}_1$ *and* $\mathbb{G}_2$. To efficiently compute the scalar multiplication in $\mathbb{G}_1$ and $\mathbb{G}_2$, we use the Gallant-Lambert-Vanstone (GLV) [20] and Galbraith-Lin-Scott (GLS) [21] which are the scalar decomposition methods By using the GLV/GLS, for given a scalar u and $P \in \mathbb{G}_1$ (or $\mathbb{G}_2$), the scalar u is decomposed into t scalars $u_1, u_2, \ldots, u_t$ with roughly the size len(u)$/t$, then we convert the multi-scalar multiplication $uP = u_1 P + u_2 \psi(P) + \cdots + u_t \psi^{t-1}(P)$ by using an efficient endomorphism ψ [9], where $t = 2$ in $\mathbb{G}_1$ and $t = \varphi(k)$ in $\mathbb{G}_2$ respectively. The number of the doubling in $\mathbb{G}_1$ and $\mathbb{G}_2$ can reduce to roughly $1/t$. Moreover, by using the width-w non adjacent form (w-NAF) [45], the number of the addition in $\mathbb{G}_1$ and $\mathbb{G}_2$ can be reduced in the computing the each scalar multiplication $u_i \psi^{i-1}(P)$. Note that we chose the optimal window size w for the scalars u_i. Let I_i and M_i be the cost of inversion and multiplication in $\mathbb{F}_{p^i}$ respectively. We use the Jacobian coordinates in $\mathbb{G}_1$ since $I_1 \approx 17.7M_1$. We use the affine coordinates in $\mathbb{G}_2$ since $I_{k/d} \approx 3.3M_{k/d}$ for BLS-24, KSS-32, KSS-36, BLS-48, and the Jacobian coordinates in $\mathbb{G}_2$ since $I_7 \approx 17.6M_7$ for BLS-42.

Table 8. Timing of computing pairing (ML, FE), scalar multiplication in $\mathbb{G}_1$, $\mathbb{G}_2$ and exponentiation in $\mathbb{G}_3$ (M clk: million clocks)

		BLS-24	KSS-32	KSS-36	BLS-42	BLS-48
Pairing	ML	53.80	32.04	37.61	36.36	**20.48**
	FE	**89.84**	197.26	147.49	100.95	96.36
	Total	143.64	229.30	185.10	137.31	**116.84**
Scalar Mult. in $\mathbb{G}_1$		12.00	8.31	6.13	3.94	**3.56**
Scalar Mult. in $\mathbb{G}_2$		38.87	53.32	35.01	49.43	**25.18**
Exp. in $\mathbb{G}_3$		71.00	62.88	77.30	**56.00**	63.46

* Scalar Mult. in $\mathbb{G}_2$ of BLS-42 only used the Jacobian coordinates because of $I_7 \approx 17.6 M_7$.

Exp. in $\mathbb{G}_3$. To efficiently compute the exponentiation in $\mathbb{G}_3$, we can use the GLS method and w-NAF since $\mathbb{G}_3 \subseteq \mathbb{G}_{\Phi_k(p)} = \{\alpha \in \mathbb{F}_{p^k} \mid \alpha^{\Phi_k(p)} = 1\}$ and the inversion in $\mathbb{G}_3$ can be efficiently computed by the conjugation [4].

6.3   Timing and Comparison

In this subsection, we show the timing of the main operations of PBC and compare those among the five candidate pairing-friendly curves.

Environment. We implement in C language, and its compiler is gcc 6.2.0 with -O3 option. We also measure on an Intel Core i7-6700 @ 3.4 GHz, RAM: 32 GB and OS: Ubuntu 16.04 (64-bit).

Results. Table 8 shows the timing of computing the pairing (ML, FE), scalar multiplication in $\mathbb{G}_1$, $\mathbb{G}_2$, and exponentiation in $\mathbb{G}_3$ for each candidate pairing-friendly curve. We discuss the timing of these operations among the five candidate pairing-friendly curves.

Pairing. The computational cost of the ML is affected by the bitlength and Hamming weight of x_0. The effect on the timing of the ML by the Hamming weight of x_0 is small since the Hamming weight of x_0 of each five candidate pairing-friendly curves is sufficiently small and much the same. As the embedding degree k increases, the bitlength of x_0 decrease, so the timing of the ML in the BLS-48 can be computed most efficiently.

The computational cost of the FE is affected by the coefficient of λ_i of Eq. (3), $\varphi(k)$, degree of $p(x)$, the bitlength of x_0 and p^k. The timings of the FE in the KSS-k are slower than that of the BLS-k since the computational cost of addition chain is required in addition to the exponentiation by x to compute the each m^{λ_i} in Eq. (3) of KSS-k. In the BLS-k, the coefficient of each λ_i of Eq. (3) is a few bits. Hence, the cost to compute all m^{λ_i} in the BLS-k is affected by the bitlength of p^k. The timing of the FE in the BLS-24 is faster than other BLS-k since the bitlength of x_0 of the BLS-24 is smaller.

Consequently, the timing of the pairing in the BLS-48 is faster than other pairing-friendly curves. The multi-pairing requires computing multiple MLs and one FE. Hence, the pairing-friendly curve with fast calculation of the ML has a significant effect on the efficiency of the computing the multi-pairing.

Scalar Mult. in $\mathbb{G}_1$. The input of scalar multiplication in $\mathbb{G}_1$ is a random element $P \in \mathbb{G}_1$ and a random scalar value of less than r. The computational cost is affected by the bitlength of r and the point addition/doubling in $\mathbb{G}_1$ affected by the bitlength of p. As k increases, the bitlength of p and r decrease. Hence, the timing of the scalar multiplication in $\mathbb{G}_1$ in the BLS-48 is faster than other pairing-friendly curves.

Scalar Mult. in $\mathbb{G}_2$. The input of scalar multiplication in $\mathbb{G}_2$ is a random element of $P \in \mathbb{G}_2$ and a random scalar value of less than r. Its computational cost is affected by the degree of twist d, and the bitlength r and $p^{k/d}$. The group $\mathbb{G}_2$ is a subgroup on $E'(\mathbb{F}_{p^{k/d}})$, and the bitlength of $p^{k/d}$ can be small when d is large. The bitlength of $p^{k/d}$ in the KSS-32 with $d = 4$ is about 2000-bit larger than that in the BLS-24, KSS-32, KSS-36, and BLS-48 with $d = 6$. Hence, the timing of the scalar multiplication in $\mathbb{G}_2$ in the KSS-32 is slower than other curves. Among the BLS-24, KSS-36 and BLS-48, as k increases, the bitlength of r decrease. Hence, the timing of the scalar multiplication in $\mathbb{G}_2$ in the BLS-48 is faster than other curves.

Exp. in $\mathbb{G}_3$. The input of exponentiation in $\mathbb{G}_3$ is a random element of $g \in \mathbb{G}_3$ and a random scalar value of less than r. The group $\mathbb{G}_3$ is a subgroup on finite field $\mathbb{F}_{p^k}$, and then the computational cost of the exponentiation in $\mathbb{G}_3$ is affected by the bitlength of r and the multiplication/squaring in $\mathbb{F}_{p^k}$. The BLS-42 and BLS-48 are theoretically better to compute the exponentiation in $\mathbb{G}_3$ efficiently since the bitlength of r is small rather than other curves. The number of the squaring in $\mathbb{F}_{p^k}$ in the BLS-48 is less than that in the BLS-42 because of the decomposition size. To compute the multi-exponentiation after the decomposition, the number of the multiplication in $\mathbb{F}_{p^k}$ in the BLS-42 is more reduced rather than in the BLS-48 because the bigger window size can be use in BLS-42 by w-NAF. As the result, the timing of the exponentiation in $\mathbb{G}_3$ in the BLS-42 is fastest in all candidate curves.

6.4   Impact on Timing by Revised Bitlength

In this subsection, we show the impact on the timing of the main operations of PBC by revised bitlength of p^k at 256-bit security level by comparing between our and previous implementations. Note that it is difficult to directly compare the timing of the main operations of PBC between our and previous implementations since the implemented algorithms and techniques are different.

In previous implementations, Scott [44] showed that the timing of the pairing is 88.8M clk, and Bos *et al.* [9] showed that the timing of the scalar multiplication in $\mathbb{G}_1$ and $\mathbb{G}_2$, and exponentiation in $\mathbb{G}_3$ are 5.2, 27.6, 47.1M clk respectively. These implemented in BLS-24 with about 15000-bit p^k and 500-bit r.

We compare the timing between our BLS-48 implementation and the previous BLS-24 implementations. Our BLS-48 implementation of the scalar multiplication in $\mathbb{G}_1$ is approximately 1.5 times faster than the previous BLS-24 implementations of that because the bitlength of r is the same between these implementations. Then, our BLS-48 implementation of other operations is approximately 1.0–1.3 times slower than the previous BLS-24 implementations of that due to the effect of the efficient NFS algorithms.

7 Conclusion

We give for the first time the revised bitlength which the DLP is computationally infeasible against the efficient NFS algorithms (e.g. SexTNFS, STNFS), and the timing of the main operations of PBC for the five candidate pairing-friendly curves (i.e. the BLS-k with $k = 24, 42, 48$, KSS-k with $k = 32, 36$) at 256-bit security level. On the security side, we show that it is necessary to increase bitlengths by more than 10000-bit from the previous estimation to achieve the 256-bit security level. On the implementation side, we show that the BLS-48 curve is the suitable curve at the 256-bit security level by comparing the timing of the main operations of PBC among the five candidate pairing-friendly curves with revised bitlengths. For more speeding up, we should implement $\mathbb{F}_p$-arithmetic in assembly, apply other efficient algorithms, etc.

A Norm Plots of BLS-24, KSS-32, KSS-36 and BLS-42

In this appendix, we show the norm plots of the GNFS and SNFS algorithms for the BLS-24, KSS-32, KSS-36, and BLS-42 in order to revise the bitlength of the these curves using the same method discussed in Sect. 5 (Figs. 3, 4, 5, 6, 7, 8, 9 and 10).

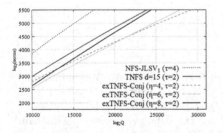 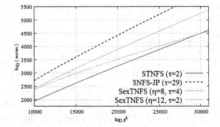

Fig. 3. Norms of GNFS algorithm in $\mathbb{F}_{p^{24}}$ **Fig. 4.** Norms of SNFS algorithm in $\mathbb{F}_{p^{24}}$

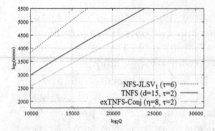

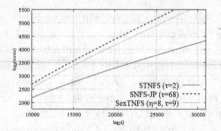

Fig. 5. Norms of GNFS algorithm in $\mathbb{F}_{p^{32}}$ **Fig. 6.** Norms of SNFS algorithm in $\mathbb{F}_{p^{32}}$

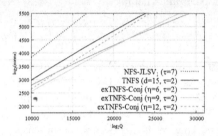

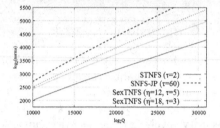

Fig. 7. Norms of GNFS algorithm in $\mathbb{F}_{p^{36}}$ **Fig. 8.** Norms of the SNFS algorithm in $\mathbb{F}_{p^{36}}$

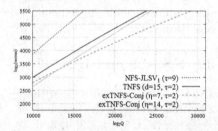

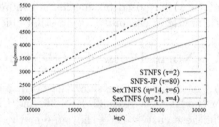

Fig. 9. Norms of GNFS algorithm in $\mathbb{F}_{p^{42}}$ **Fig. 10.** Norms of SNFS algorithm in $\mathbb{F}_{p^{42}}$

References

1. Aoki, K., Franke, J., Kleinjung, T., Lenstra, A.K., Osvik, D.A.: A kilobit special number field sieve factorization. In: Kurosawa, K. (ed.) ASIACRYPT 2007. LNCS, vol. 4833, pp. 1–12. Springer, Heidelberg (2007). doi:10.1007/978-3-540-76900-2_1
2. Acar, T., Lauter, K., Naehrig, M., Shumow, D.: Affine pairings on ARM. In: Abdalla, M., Lange, T. (eds.) Pairing 2012. LNCS, vol. 7708, pp. 203–209. Springer, Heidelberg (2013). doi:10.1007/978-3-642-36334-4_13

3. Aranha, D.F., Fuentes-Castañeda, L., Knapp, E., Menezes, A., Rodríguez-Henríquez, F.: Implementing pairings at the 192-bit security level. In: Abdalla, M., Lange, T. (eds.) Pairing 2012. LNCS, vol. 7708, pp. 177–195. Springer, Heidelberg (2013). doi:10.1007/978-3-642-36334-4_11
4. Aranha, D.F., Karabina, K., Longa, P., Gebotys, C.H., López, J.: Faster explicit formulas for computing pairings over ordinary curves. In: Paterson, K.G. (ed.) EUROCRYPT 2011. LNCS, vol. 6632, pp. 48–68. Springer, Heidelberg (2011). doi:10.1007/978-3-642-20465-4_5
5. Barker, E.B., Barker, W.C., Burr, W.E., Polk, W.T., Smid, M.E.: Recommendation for key management - part 1: General (Revision 4). NIST SP 800-57 (2016)
6. Barbulescu, R., Gaudry, P., Guillevic, A., Morain, F.: Improving NFS for the discrete logarithm problem in non-prime finite fields. In: Oswald, E., Fischlin, M. (eds.) EUROCRYPT 2015. LNCS, vol. 9056, pp. 129–155. Springer, Heidelberg (2015). doi:10.1007/978-3-662-46800-5_6
7. Bouvier, C., Gaudry, P., Imbert, L., Jeljeli, H., Thom, E.: Discrete logarithms in GF(p) — 180 digits. Announcement available at the NMBRTHRY archives, item 004703 (2014)
8. Boneh, D., Franklin, M.: Identity-based encryption from the weil pairing. In: Kilian, J. (ed.) CRYPTO 2001. LNCS, vol. 2139, pp. 213–229. Springer, Heidelberg (2001). doi:10.1007/3-540-44647-8_13
9. Bos, J.W., Costello, C., Naehrig, M.: Exponentiating in pairing groups. In: Lange, T., Lauter, K., Lisoněk, P. (eds.) SAC 2013. LNCS, vol. 8282, pp. 438–455. Springer, Heidelberg (2014). doi:10.1007/978-3-662-43414-7_22
10. Barbulescu, R., Gaudry, P., Kleinjung, T.: The tower number field sieve. In: Iwata, T., Cheon, J.H. (eds.) ASIACRYPT 2015. LNCS, vol. 9453, pp. 31–55. Springer, Heidelberg (2015). doi:10.1007/978-3-662-48800-3_2
11. Barreto, P.S.L.M., Lynn, B., Scott, M.: Constructing elliptic curves with prescribed embedding degrees. In: Cimato, S., Persiano, G., Galdi, C. (eds.) SCN 2002. LNCS, vol. 2576, pp. 257–267. Springer, Heidelberg (2003). doi:10.1007/3-540-36413-7_19
12. Barreto, P.S.L.M., Costello, C., Misoczki, R., Naehrig, M., Pereira, G.C.C.F., Zanon, G.: Subgroup security in pairing-based cryptography. In: Lauter, K., Rodríguez-Henríquez, F. (eds.) LATINCRYPT 2015. LNCS, vol. 9230, pp. 245–265. Springer, Cham (2015). doi:10.1007/978-3-319-22174-8_14
13. Costello, C., Lauter, K., Naehrig, M.: Attractive subfamilies of BLS curves for implementing high-security pairings. In: Bernstein, D.J., Chatterjee, S. (eds.) INDOCRYPT 2011. LNCS, vol. 7107, pp. 320–342. Springer, Heidelberg (2011). doi:10.1007/978-3-642-25578-6_23
14. Danilov, S.A., Popovyan, I.A.: Factorization of RSA-180, Cryptology ePrint Archive, Report 2010/270 (2010)
15. European Union Agency of Network and Information Security (ENISA): Algorithms, key sizes and parameters report, 2013 recommandations, version 1.0, October 2013
16. Freeman, D., Scott, M., Teske, E.: A taxonomy of pairing-friendly elliptic curves. J. Cryptol. 23, 224–280 (2010)
17. Fried, J., Gaudry, P., Heninger, N., Thomé, E.: A kilobit hidden SNFS discrete logarithm computation. In: Coron, J.-S., Nielsen, J.B. (eds.) EUROCRYPT 2017. LNCS, vol. 10210, pp. 202–231. Springer, Cham (2017). doi:10.1007/978-3-319-56620-7_8
18. The GNU Multiple Precision Arithmetic Library. https://gmplib.org/

19. Granger, R., Scott, M.: Faster squaring in the cyclotomic subgroup of sixth degree extensions. In: Nguyen, P.Q., Pointcheval, D. (eds.) PKC 2010. LNCS, vol. 6056, pp. 209–223. Springer, Heidelberg (2010). doi:10.1007/978-3-642-13013-7_13

20. Gallant, R.P., Lambert, R.J., Vanstone, S.A.: Faster point multiplication on elliptic curves with efficient endomorphisms. In: Kilian, J. (ed.) CRYPTO 2001. LNCS, vol. 2139, pp. 190–200. Springer, Heidelberg (2001). doi:10.1007/3-540-44647-8_11

21. Galbraith, S.D., Lin, X., Scott, M.: Endomorphisms for faster elliptic curve cryptography on a large class of curves. J. Crypto **24**, 446–469 (2011)

22. Guillevic, A., Morain, F., Thomé, E.: Solving discrete logarithms on a 170-bit MNT curve by pairing reduction, arXiv preprint arXiv:1605.07746 (2016)

23. Ghammam, L., Fouotsa, E.: Adequate elliptic curves for computing the product of n pairings. In: Duquesne, S., Petkova-Nikova, S. (eds.) WAIFI 2016. LNCS, vol. 10064, pp. 36–53. Springer, Cham (2016). doi:10.1007/978-3-319-55227-9_3

24. Hess, F., Smart, N., Vercauteren, F.: The eta pairing revisited. IEEE Trans. Inf. Theory **52**(10), 4595–4602 (2006)

25. Joux, A., Lercier, R., Smart, N., Vercauteren, F.: The number field sieve in the medium prime case. In: Dwork, C. (ed.) CRYPTO 2006. LNCS, vol. 4117, pp. 326–344. Springer, Heidelberg (2006). doi:10.1007/11818175_19

26. Joux, A., Pierrot, C.: The special number field sieve in $\mathbb{F}_{p^n}$. In: Cao, Z., Zhang, F. (eds.) Pairing 2013. LNCS, vol. 8365, pp. 45–61. Springer, Cham (2014). doi:10.1007/978-3-319-04873-4_3

27. Karabina, K.: Squaring in cyclotomic subgroups. Math. Comput. **82**, 555–579 (2013)

28. Kim, T., Barbulescu, R.: Extended tower number field sieve: a new complexity for the medium prime case. In: Robshaw, M., Katz, J. (eds.) CRYPTO 2016. LNCS, vol. 9814, pp. 543–571. Springer, Heidelberg (2016). doi:10.1007/978-3-662-53018-4_20

29. Kachisa, E.J., Schaefer, E.F., Scott, M.: Constructing brezing-weng pairing-friendly elliptic curves using elements in the cyclotomic field. In: Galbraith, S.D., Paterson, K.G. (eds.) Pairing 2008. LNCS, vol. 5209, pp. 126–135. Springer, Heidelberg (2008). doi:10.1007/978-3-540-85538-5_9

30. Kim, T., Jeong, J.: Extended tower number field sieve with application to finite fields of arbitrary composite extension degree. In: Fehr, S. (ed.) PKC 2017. LNCS, vol. 10174, pp. 388–408. Springer, Heidelberg (2017). doi:10.1007/978-3-662-54365-8_16

31. Kleinjung, T.: Discrete Logarithms in GF(p) – 768 bits. Announcement available at the NMBRTHRY archives, item **004917** (2016)

32. Lenstra, A.K.: Unbelievable security *matching AES security using public key systems*. In: Boyd, C. (ed.) ASIACRYPT 2001. LNCS, vol. 2248, pp. 67–86. Springer, Heidelberg (2001). doi:10.1007/3-540-45682-1_5

33. Lenstra, A.K., Lenstra, H.W. (eds.): The Development of the Number Field Sieve. LNM, vol. 1554. Springer, Heidelberg (1993). doi:10.1007/BFb0091534

34. Miller, V.S.: The weil pairing, and its efficient calculation. J. Cryptol. **17**, 235–261 (2004)

35. Mori, Y., Akagi, S., Nogami, Y., Shirase, M.: Pseudo 8–sparse multiplication for efficient ate–based pairing on barreto–naehrig curve. In: Cao, Z., Zhang, F. (eds.) Pairing 2013. LNCS, vol. 8365, pp. 186–198. Springer, Cham (2014). doi:10.1007/978-3-319-04873-4_11

36. Mitsunari, S.: A fast implementation of the optimal ate pairing over BN curve on intel haswell processor, Cryptology ePrint Archive, Report 2013/362 (2013)

37. Menezes, A., Sarker, P., Singh, S.: Challenges with assessing the impact of NFS advances on the security of pairing-based cryptography, Cryptology ePrint Archive, Report 2016/1102 (2016)
38. Okamoto, T., Takashima, K.: Fully secure functional encryption with general relations from the decisional linear assumption. In: Rabin, T. (ed.) CRYPTO 2010. LNCS, vol. 6223, pp. 191–208. Springer, Heidelberg (2010). doi:10.1007/978-3-642-14623-7_11
39. Pollard, J.: Monte Carlo methods for index computation (mod p). Math. Comput. **32**(143), 918–924 (1978)
40. Sakai, R., Ohgishi, K., Kasahara, M.: Cryptosystems based on pairing. In: SCIS 2000, C-20, pp. 26–28 (2000)
41. Sahai, A., Waters, B.: Fuzzy identity-based encryption. In: Cramer, R. (ed.) EUROCRYPT 2005. LNCS, vol. 3494, pp. 457–473. Springer, Heidelberg (2005). doi:10.1007/11426639_27
42. Devegili, A.J., Scott, M., Dahab, R.: Implementing cryptographic pairings over barreto-naehrig curves. In: Takagi, T., Okamoto, T., Okamoto, E., Okamoto, T. (eds.) Pairing 2007. LNCS, vol. 4575, pp. 197–207. Springer, Heidelberg (2007). doi:10.1007/978-3-540-73489-5_10
43. Scott, M., Benger, N., Charlemagne, M., Dominguez Perez, L.J., Kachisa, E.J.: On the final exponentiation for calculating pairings on ordinary elliptic curves. In: Shacham, H., Waters, B. (eds.) Pairing 2009. LNCS, vol. 5671, pp. 78–88. Springer, Heidelberg (2009). doi:10.1007/978-3-642-03298-1_6
44. Scott, M.: On the efficient implementation of pairing-based protocols. In: Chen, L. (ed.) IMACC 2011. LNCS, vol. 7089, pp. 296–308. Springer, Heidelberg (2011). doi:10.1007/978-3-642-25516-8_18
45. Schirokauer, O.: Using number fields to compute logarithms in finite fields. Math. Comp. **69**, 1267–1283 (2000)
46. Schirokauer, O.: Virtual logarithms. J. Algorithms **57**, 140–147 (2005)
47. Vercauteren, F.: Optimal pairings. IEEE Trans. Inf. Theory **56**(1), 455–461 (2010)

Data Protection and Mobile Security

No Free Charge Theorem: A Covert Channel via USB Charging Cable on Mobile Devices

Riccardo Spolaor[1](✉), Laila Abudahi[2], Veelasha Moonsamy[3], Mauro Conti[1], and Radha Poovendran[2]

[1] University of Padua, Padua, Italy
{rspolaor,conti}@math.unipd.it
[2] University of Washington, Seattle, USA
{abudahil,rp3}@uw.edu
[3] Radboud University, Nijmegen, The Netherlands
email@veelasha.org

Abstract. More and more people are regularly using mobile and battery-powered handsets, such as smartphones and tablets. At the same time, thanks to the technological innovation and to the high user demand, those devices are integrating extensive battery-draining functionalities, which results in a surge of energy consumption of these devices. This scenario leads many people to often look for opportunities to charge their devices at public charging stations: the presence of such stations is already prominent around public areas such as hotels, shopping malls, airports, gyms and museums, and is expected to significantly grow in the future. While most of the times the power comes for free, there is no guarantee that the charging station is nòt maliciously controlled by an adversary, with the intention to exfiltrate data from the devices that are connected to it.

In this paper, we illustrate for the first time how an adversary could leverage a maliciously controlled charging station to exfiltrate data from the smartphone via a USB charging cable (i.e., without using the data transfer functionality), controlling a simple app running on the device—and without requiring any permission to be granted by the user to send data out of the device. We show the feasibility of the proposed attack through a prototype implementation in Android, which is able to send out potentially sensitive information, such as IMEI and contacts' phone number.

1 Introduction

Market studies predicted that in 2011 smartphone sales would surpassed that of desktop PCs [31]. To this date, smartphones remain the most used handheld devices. This is partly due to the fact that these devices are more powerful and provide more functionalities than the traditional feature phones. As a result, users can perform a variety of tasks on an actual smartphone device, which in the past would have been possible only on a desktop PC. In order to carry out such tasks, the smartphone platform offers its users a plethora of applications (apps).

© Springer International Publishing AG 2017
D. Gollmann et al. (Eds.): ACNS 2017, LNCS 10355, pp. 83–102, 2017.
DOI: 10.1007/978-3-319-61204-1_5

Moreover, as users are constantly using apps (e.g., the gaming app, Pokémon Go) and would eventually require to recharge their smartphones, the demand for public charging stations have increased significantly in the last decade. Such stations can be seen in public areas such as airports, shopping malls, gyms and museums, where users can recharge their devices for free. In fact, this trend is also giving rise to a special type of business[1], which allows shop owners to install charging stations in their stores so as to boost their sales by providing free phone recharge to shoppers.

As the phone recharging is usually for free, however, at the same time one cannot be sure that the public charging stations are not maliciously controlled by an adversary. The Snowden revelations gave us proof that civilians are constantly under surveillance and nations are competing against each other by deploying smart technologies for collecting sensitive information en mass. In our work, we consider an adversary (e.g., manufacturers of public charging stations, Government agencies) whose aim is to take control over the public charging station and whose motive is to exfiltrate data from the user's smartphone once the device is plugged into the station.

In this paper, we demonstrate the feasibility of using power consumption (in the form of power bursts) to send out data over a Universal Serial Bus (USB) charging cable, which acts as a covert channel, to the public charging station. We implemented a proof-of-concept app, *PowerSnitch*, that can send out bits of data in the form of power bursts by manipulating the power consumption of the device's CPU. Interestingly, PowerSnitch does not require any special permission from the user at install-time (nor at run-time) to exfiltrate data out of the smartphone over the USB cable. On the adversary's side, we designed and implemented a decoder to retrieve the bits that have been transmitted via power bursts. Our empirical results show that we can successfully decode a payload of 512 bits with a 0% Bit Error Ratio (BER). In addition, we stress that the goal of this paper is to assess for the first time the feasibility of data transmission on such a covert channel and not to optimize its performance, which we will tackle as future work.

We focus primarily on Android, as it is currently the leading platform and has a large user base. However, we believe that this attack can be deployed on any other smartphone operating systems, as long as the device is connected to a power source at the public charging station.

Our contributions are as follows:

1. To our knowledge, we are the first to demonstrate the practicality of using the power feature of a USB charging cable as a covert channel to exfiltrate data, in the form of power bursts, from a device while it is connected to a power supplier. The attack works in Airplane mode as well.
2. We implemented a prototype of the attack, i.e., we designed and implemented its two components: (i) We built a proof-of-concept app, *PowerSnitch*, which does not require any permission granted by the user to communicate bits

---

[1] chargeitspot.com, chargetech.com.

of information in the form of power bursts back to the adversary; (ii) The decoder is deployed on the adversary side, i.e., public charging station to retrieve the binary information embedded in the power bursts.
3. We are able with our prototype to actually send out data using power bursts. Our prototype demonstrate the practical feasibility of the attack.

The rest of the paper is organized as follows. In Sect. 2, we present a brief literature overview of covert channel and data exfiltration techniques on smartphones. In Sect. 3, we include some background knowledge on Android operating system, and signal transmission and processing. In Sect. 4, we provide a description of our covert channel and decoder design, followed by the experimental results in Sect. 5 and discussion in Sect. 6. We conclude the paper in Sect. 7.

2 Related Work

In this section, we survey the existing work in the area of covert channels on mobile devices. We also present other non-conventional attack vectors, such as side channel information leakage via embedded sensors which can be used for data exfiltration.

Covert Channels – A covert channel can be considered as a secret channel used to exfiltrate information from a secured environment in an undetected manner. Chandra et al. [8] investigated the existence of different covert channels that can be used to communicate between two malicious applications. They examined the common resources (such as battery) shared between two malicious applications and how they could be exploited for covert communication. Similar studies presented in [14,18,21,26] exploited unknown covert channels in malicious and clean applications to leak out private information.

As demonstrated by Aloraini et al. [1], the adversary is further empowered as smartphones continue to have more computational power and extensive functionalities. The authors empirically showed that speech-like data can be sent over a cellular voice channel. The attack was successfully carried out with the help of a custom-built rootkit installed on Android devices. In [10], Do et al. demonstrated the feasibility of covertly exfiltrating data via SMS and inaudible audio transmission, without the user's knowledge, to other mobile devices including laptops.

In our work, we present a novel covert channel which exploits the USB charging cable by leaking information from a smartphone via power bursts. Our proposed method is non-invasive and can be deployed on non-rooted Android devices. We explain the attack in more detail in Sect. 4.1.

Power Consumption by Smartphones – In order to prolong the longevity of the smartphone's battery, it is crucial to understand how apps consume energy during execution and how to optimize such consumption. To this end, several works [4,6,23,33] have been proposed. Furthermore, the authors from [13,17] studied apps' power consumption to detect anomalous behavior on smartphones, thus leading to detection of malware.

Since existing work focus on energy consumption on the device, our attack would therefore go undetected as the smartphone's CPU sends small chunks of encoded data, which are translated into power bursts, back to the public charging station. Additionally, state-of-the-art attacks that have been performed while the smartphone is charging [15,19] exploit vulnerabilities of USB interface rather than actual energy consumption.

Attack Vectors using Side Channel Leaks – Modern smartphones are embedded with a plethora of sensors that allow users to interact seamlessly with the apps on their smartphones. However, these sensors have access to an abundance of information stored on the device that can get exfiltrated. These data leaks can be used as a side channel to infer, otherwise undisclosed, sensitive information about the user or device [2,16,32].

The authors from [3,22] demonstrated how accelerometer readings can be used to infer tap-, gesture- and keyboard-based input from users to unlock their smartphones. Similarly, Spreitzer [27] showed that the ambient-light sensor can be exploited to infer users' PIN input. Moreover, considering network traffic as a side-channel, it is possible to identify the set of apps installed on a victim's mobile device [28,29], and even infer the actions the victim is performing with a specific app [9].

As pointed out in the aforementioned existing work, the adversarial model did not require any special privileges to exploit side channel leaks to recover data exfiltrated via sensors. In this paper, we show that our custom app, PowerSnitch, does not require any special permissions to be granted by the user in order to communicate information (in terms of power bursts) to the adversary. Furthermore, we stress that while the INTERNET permission is one approach of data exfiltration, our proposed work is different as we show the feasibility and practicability of using a USB cable to exfiltrate data. In particular, our attack still works even when the phone is switched to Airplane mode and defeats existing USB charging protection dongles, as in [7], since we only require the USB power pins to exfiltrate data.

3   Background Knowledge

In this section, we briefly recall several concepts that we use in our paper about Android operating system in Sect. 3.1, and signal transmission and processing in Sect. 3.2.

3.1   Android System and Permissions

In the Android Operating System (OS), apps are distributed as APK files. These files are simple archives which contain bytecode, resources and metadata. A user can install or uninstall an app (thus the APK file) by directly interacting with the smartphone. When an Android app is running, its code is executed in a sandbox. In practice, an app runs isolated from the rest of the system, and it cannot directly access other apps' memory. The only way an app could gain memory access is via the mediation of inter-process communication techniques made available by Android. These measures are in place to prevent the access of malicious apps to other apps' data, which could potentially be privacy-sensitive.

Since Android apps run in a sandbox, they not only have restriction in shared memory usage, but also to most system resources. Instead, the Android OS provides an extensive set of Accessible Programming Interfaces (APIs), which allows access to system resources and services. In particular, the APIs that give access to potentially privacy-violating services (e.g., camera, microphone) or sensitive data (e.g., contacts) are protected by the Android Permission System [11]. An app that wants access to protected data or service must declare in the form of permission (identified by a string) in its manifest file. The list of permissions needed by an app is shown to the user when installing the app, and cannot be changed while an app is installed on the device. With the introduction of Android M (i.e., 6.0), permissions can be dynamically granted (by users) during an app's execution.

The permission system has also the goal of reducing the damage in case of a successful attack that manages to take control of an app, by limiting the resources that app's process has access to. Unfortunately, permission over-provisioning is a common malpractice, so much so that research efforts have been spent in trying to detect this problem [5]. Moreover, an app asking for permissions not related to its purpose (or functionality) can hide malicious behaviors (i.e., spyware or malware apps) [20].

3.2   Signal Transmission and Processing

In this section, we provide some background information on bit transmission, and signal processing and decoding used in our proposed decoder (see Sect. 4.4).

Bit Transmission – To enable bit transmission over our channel, an understanding of basic digital communication systems is essential. For proof-of-concept purposes, the design of our bit transmission system was inspired by amplitude-based modulation in the digital communication literature.

Amplitude-Shift Keying (ASK) is a form of digital modulation where digital bits are represented by variations in the amplitude of a carrier signal. To send bits over our channel, we used On-Off Signaling (OOS), which is the simplest form of ASK where digital data is represented by the presence and absence of some pulse $p(t)$ for a specific period of time. Figure 1a shows the difference between a Return-to-Zero (RZ) and a Non-Return-to-Zero (NRZ) on-off encoding. In NRZ encoding, bits are represented by a sufficient condition (a pulse) that occupies the entire bit period T_b while RZ encoding represents bits as pulses for a duration of $T_b/2$ before it returns to zero for the following $T_b/2$ period.

On the other hand, Fig. 1b shows the difference between a unipolar and a polar RZ on-off signaling. In a polar RZ encoding, two different conditions, different-sign pulses are used to encode different bits(zeros/ones) while the presence and absence of a single pulse, a positive one in our case, are used to encode different bits.

For the sake of our channel design, it is safe to assume that we can only increase the power consumption of a phone at certain times and hence, are able to generate only positive (high) bursts. Thus, a unipolar encoding seems more relative and applicable for our channel. Moreover, successive peaks, such as the first two zeros in Fig. 1a, are easier to identify, and thus decode, in the RZ-encoded signal than in the NRZ one. This advantage of RZ over NRZ becomes especially apparent in cases where the bit period is expected not to be restrictively fixed in the received signal whether it is due to expected high channel noises or lack of full control of the phone's CPU. Therefore, unipolar RZ on-off signaling was used to encode leaked bits over our covert channel.

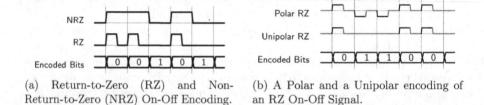

(a) Return-to-Zero (RZ) and Non-Return-to-Zero (NRZ) On-Off Encoding.

(b) A Polar and a Unipolar encoding of an RZ On-Off Signal.

Fig. 1. A comparison between bit encoding methods

Signal Processing and Decoding – After choosing the appropriate encoding method to transmit bits, it is also essential to think about the optimal receiver design and how to process the received signal and decode bits with minimum error probability at the receiver side of the channel. As known in the digital communication literature, matched filters are the optimal receivers for Additive White Gaussian Noise (AWGN) channels. We refer the reader to Sect. 4.2 of [24] for a detailed proof.

Matched Filters are obtained by correlating the received signal $R(t)$ with the known pulse that was first used to encode a transmitted bit, in this case $P(t)$ with period T_b. After correlation, the resulted signal is then sampled at time T_b, which means that the sampling rate equals to $1/T_b$ samples/seconds. This way, each bit is guaranteed to be represented by only one sample. The decoding decision will then be made based on that one sample value; if the sample value is more than a given threshold, this indicates the presence of $P(t)$; and hence a zero in our case, while a sample value below the threshold indicates the absence of $P(t)$ and hence a one is decoded.

However and most importantly, for matched filters to work as expected, it is essential to have fixed bit period T_b throughout the entire received signal. If the periods of the received bits were varying, the matched filter samples taken with the $1/T_b$ sampling rate will not be as optimal and representative of the bit data as expected and synchronization will be lost.

Since there exist infrequent phone-specific, OS-enforced conditions that can affect the power consumption of a phone, the noises on our channel are expected to be more complex to fit in an AWGN model. Hence, a matched filter receiver is most likely not the optimal receiver for our channel. More creative decoder design decisions are needed to maximize the throughput of our channel and minimize the error probability.

4   Covert Channel Using Mobile Device Energy Consumption

In this section, we elaborate on the components that make up our covert channel attack. We begin by giving an overview of the attack in Sect. 4.1. We then define the terms and parameters for transmission in Sect. 4.2, followed by a description of each component of the attack: PowerSnitch app in Sect. 4.3 and the energy traces decoder in Sect. 4.4.

4.1   Overview of Attack

As illustrated in Fig. 2, the attack scenario considers two components: the victim's Android mobile device (sender) and an accomplice's power supplier (receiver). Victim's mobile device is connected to a power supplier (controlled by the adversary) through a USB cable.

The left side of Fig. 2 depicts what happens after the victim has installed our proof-of-concept app, *PowerSnitch*. The app is able to exfiltrate victim's private information, which gets encoded as CPU bursts with a specific timing. Indeed, as the CPU is one of the most energy consuming resources in a device, a CPU burst can be directly measured as a "peak" based on the amount of energy absorbed by a mobile device. The right side of Fig. 2 illustrates how the energy supplier is able to measure (with a given sampling rate) the electric current provided to the mobile device connected to the public charging station. Then, such electric measurement, which is considered as a *signal*, is given as input to a decoder. It should be noted that the adversary, i.e., the public charging station, has control of the power supplier, and thus is able to control the amount of current provided to the device – even if it has the "fast charge" capability.

In our proposed covert channel attack, we consider situations in which users connect their mobile devices for more than 20 min. There are several scenarios that fulfill such time requirements. Examples are: (i) recharging a device overnight in a hotel room; (ii) making use of locked boxes in shopping malls for charging mobile phones; (iii) recharging devices on planes, in trains and cars.

In addition, we argue that those time requirements are more than reasonable since generally, 72% of users leave their phones on charging for more than 30 min, with an average time of 3 h and 54 min, as reported in [12]. This means that: (i) the mobile device is in stand-by mode; (ii) CPU and the use of other energy consuming resources (e.g., Wi-Fi or 3/4g data connection) usage is limited only to the OS and background apps. Moreover, since there is no user interaction, it is reasonable to assume that the phone screen, which has a relevant impact on energy consumption, will stay off for the aforementioned period of time.

Moreover, it is also worth noting that the attack is still feasible if there is no data connection between the victim's device and the power supplier, such as Media Transfer Protocol (MTP), Photo Transfer Protocol (PTP), Musical Instrument Digital Interface (MIDI). This is possible as our methodology only requires power consumption to send out the power bursts. Moreover, from Android version 6.0, when a device is connected via USB, it is set by default to "Charging" mode (i.e., just charge the device), thus no data connection is allowed unless the user switches on data connection manually. This improvement in security feature does not impact our proposed attack as we do not make use of data connection to transfer the power bursts.

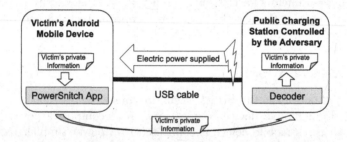

Fig. 2. The schema of the components involved in the attack.

4.2   Terminology and Transmission Parameters

In this section, we define the necessary terminology to identify concepts used in the rest of the paper:

- *Payload* is the information that has to be sent from the device to the receiver.
- *Transmission* is the whole sequence of bits transmitted in which the payload is encoded.

In order to obtain a successful communication, the sender and the receiver need to agree on the parameters of the transmission.

- *Period* is the time interval during which a bit is transmitted.
- *Duty cycle* is the ratio between burst and rest time in a period T_b. For example: if a burst lasts for $T_b/2$, the duty cycle will be 50%.
- *Preamble* is the sequence of bit used to synchronize the transmission. Usually a preamble is used at the beginning of a transmission, but it can also be used within a transmission in order to recover the synchronization in case of error. In our case, we used a preamble composed of 8 bits.

4.3 PowerSnitch App: Implementing the Attack on Android

The first component of our covert channel we discuss is the proof-of-concept which we called *PowerSnitch app*. This app, used for the covert channel exploit, has been designed as a service in order to be installed as a standalone app or a library in a repackaged app. Henceforth, we refer to both these variants simply with the term "app".

PowerSnitch app requires only the WAKE_LOCK permission and does not need root access to work. Such permission allows PowerSnitch app to wake and force execute the CPU while the device is in sleep mode, so that it can start to transmit the payload. We stress that since it is running as a background service, PowerSnitch app still works even when user authentication mechanisms (e.g., PIN, password) are in place. Moreover, since it does not use any conventional communication technology (e.g., Wi-Fi, Bluetooth, NFC), PowerSnitch app can exfiltrate information even if the device is in airplane mode. It is worth mentioning that Android M (i.e., 6.0) introduced the Doze mode [30], a battery power-saving optimization which reduces the apps activity when the device is inactive and running on battery for extended periods of time. When it is in place, Doze mode stops background CPU and network activity (ignoring wakelocks, job scheduler, Wi-Fi scan, etc.). Then on periodic time intervals (i.e., maintenance windows), the system runs all pending jobs, synchronization and alarms. However, such optimization is not active when a device is connected to a power source or when the screen is on. This means that Doze does not affect our proposal since we need the wakelock function but also the device to be plugged to a power source. Moreover, since our proposed attack needs also the status of the battery, it does not need any permission in order to obtain such information: in fact, it is sufficient to only register at run-time (not even in the manifest) a specific broadcast receiver (i.e., ACTION_BATTERY_CHANGED).

In Fig. 3, we illustrate the modules of PowerSnitch app. It is composed of three modules: *Payload encoder*, *Transmission controller* and *Bursts generator*. *Payload encoder* takes the payload as input and outputs an array of bits. The payload can be any element that can be serialized into an array of bits. We use strings as payloads, they are first decomposed into an array of characters and then, using the ASCII code of each character, into an array of bits. *Payload Encoder* can also add to its output array synchronization bits (e.g., the preamble), and error checking codes (e.g., CRC).

Transmission controller is in charge of monitoring the status of the device with the purpose of understanding when it is feasible to transmit through the covert channel. Indeed, in order to not be detected by the user, it checks whether all the following conditions are satisfied: (i) the USB cable is connected; (ii) the screen is off; and (iii) the battery is sufficiently charged (see Sect. 6). If our app receives a broadcast intent from the Android OS that invalidates one of the aforementioned conditions, *Transmission controller* module will interrupt the transmission. It is worth noticing that to obtain all this information, PowerSnitch app does not need any additional permission. From the GUI app used in

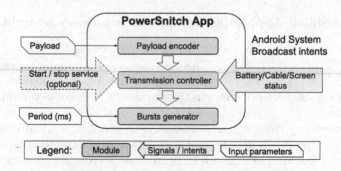

Fig. 3. The modules involved in the PowerSnitch app.

our experiments, we are also able to start or stop PowerSnitch app (represented in Fig. 3 with a dotted arrow).

The last component of PowerSnitch app is *Bursts generator*. The task of this component is to convert the encoded payload into bursts of energy consumption. These bursts will generate a signal that can be measured at the other end of the USB cable (i.e., the power supplier). In order to obtain these bursts of energy consumption, *Bursts generator* module can use a power consuming resource of the mobile device such as CPU, screen or flashlight. Our proof-of-concept, *Bursts generator* uses the CPU: a CPU burst is generated from a simple floating point operation repeated in a loop for a precise amount of time (given by transmission parameters).

4.4 Analysis of Energy Traces

To make better decoder design decisions, several channel traces were observed, collected and then used to calculate channel estimations and implement different simulations of the channel performance and behavior. A standard on-off signaling decoder needs to know the exact period of bits in the received signal in order to be able to decode them. However, a channel built based on a phone's power consumption is expected to have hard-to-model noises that, after examining the collected channel data traces, are actually affecting not only the peak periods but also the peak amplitudes. The amount of external power consumed by a phone can be largely affected by dominant OS-enforced, manufacturer-specific factors. For instance, different sudden drop patterns in power consumption especially when the phone is almost or completely charged, lack of control over the OS scheduler; when, how often and for how long do some heavy power-consuming OS background services run, as well as the precision and sampling rate of the power monitor on the receiver side of the channel.

Figure 4 shows a portion of the channel data captured after a transmission of ten successive bits (ten Zeros, therefore ten peaks) was initiated by our app on a Nexus 6 phone. It should be noted that the data was passed through a low-pass filter to get rid of harsh, high frequency noises in order to make the signal looks smoother. As a result, based on a threshold of 100 mA, ten peaks are successfully detected. Moreover, the width of each peak, and hence the period of each bit, is varying sufficiently. The first bit, for example, has a period of 300 ms while the eighth one has a period of only 195 ms. Although the intended bit period generated and transmitted by the app was 500 ms, the average period of the received bits was actually 311 ms, which the receiver has no way to predict in advance. Such variations in the received signal are expected to affect the performance of any decoder. An ideal matched filter receiver will have hard time decoding such inconsistent signal and synchronization will be lost very quickly. We elaborate further on this issue in the remaining sections.

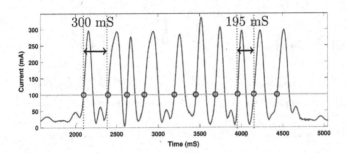

Fig. 4. A portion of a received signal showing the variations in peak widths and amplitudes.

Decoder Design. In this section, we provide additional explanation about the different processing stages that our decoder is taking the received signal through in order to overcome the channel inconsistencies and decode the sent bits with the minimum Bit Error Ratio (BER). In signal processing, the quality of a communication channel can be measured in terms of BER (represented as a percentage), which is the number of bit errors divided by the total number of transmitted bits over the channel. Channels affected by interference, distortion, noise, or synchronization errors have a high BER.

Figure 5 summarizes the different processing stages which will be discussed in the order they take place in, along with some background information and algorithm justifications, where applicable.

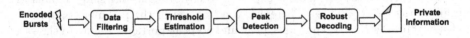

Fig. 5. Different phases of our decoder.

Data Filtering. First, the received signal is passed through a low-pass filter to get rid of the harsh high-frequency noises. For instance, Fig. 6 shows the same portion of a received signal before and after applying the low-pass filter. The low-pass filter helps not only to make the signal looks smoother, but also to make the threshold-based detection of real peaks easier by eliminating narrow-peak noises that can be falsely identified as real peaks or bits. Additionally, the low-pass filter used in our decoder adjusts its pass and stop frequencies based on the intended bit period generated by the phone in order to make sure that we do not over-filter or over-attenuate the signal.

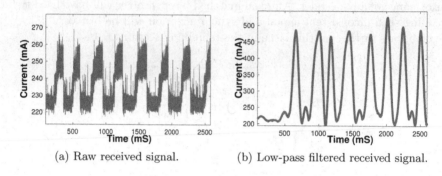

(a) Raw received signal.   (b) Low-pass filtered received signal.

Fig. 6. A portion of a received signal before and after applying the low-pass filter.

Threshold Estimation. The decoder detects peaks by decoding unipolar RZ on-off encoded bits. The presence or absence of a peak (a 0 or a 1 in our case, respectively) at a certain time and for a specific period is then translated to the corresponding bit. Peak detection is usually done by setting an appropriate threshold; anything above the threshold is a peak and anything below is just noise. However, deciding which threshold to use is not a trivial process especially with the unpredictable noise in our channel and the variations in width and amplitude of the received peaks.

The threshold value used by the decoder is highly critical to peaks detection, the resulted width of detected peaks and the decoder performance. Hence, we primarily use a known preamble data sent prior to the actual packet to estimate the threshold. The preamble consists of eight known bits (eight zeros in our case) at the start of the transmission, which means that the decoder is expecting eight peaks at the start. Since a unipolar RZ on-off encoded zero has a pulse for half of the bit period, the preamble is expected to have roughly the same number of peak and no-peak samples. Therefore, a histogram of the preamble samples is expected to split into two portions; peak and no-peak portions. Figure 7a shows a histogram of the preamble samples shown in Fig. 7b. As observed, the histogram has two distinguishable densities; each of them look like the probability density function of a Gaussian distribution.

Estimating the parameters (mean and variance) of two Gaussians that are believed to exist in one overall distribution is a complicated statistical problem. However, the Gaussian Mixture Model (GMM), introduced and explained in [25], is a probabilistic model commonly used to address this type of problem and to statistically estimate the parameters of existing Gaussian populations. To estimate the threshold, as shown in Fig. 8, the decoder uses the GMM to fit two Gaussians to the two histogram portions, find the mean of each one of them and then compute the threshold as the middle point between the two means. As a result, our decoder is able to estimate the threshold independently and without any previous knowledge of the expected amplitudes of the received bits. After that, each sample is converted to either a peak sample or no-peak sample based on whether the sample value is above or below the estimated threshold.

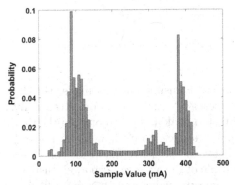
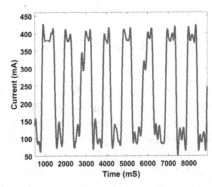

(a) A histogram of the preamble samples.   (b) A received preamble signal.

Fig. 7. A histogram of the preamble samples shows a mixture of two Gaussian-like densities.

Robust Decoding. Generally, the way a decoder translates the peak and no-peak samples to zeros and ones is highly time-sensitive. For instance, if the bit period is fixed and equals to T_b, the decoder simply checks the presence or absence of the peak in each T_b period. Since this decoding decision is made based on a very strict timing manner, the slightest error in the received bit periods will cause a quick loss of synchronization. As mentioned in the previous section, the received peak widths (and hence bit periods) over our channel are changing with a high variation around their mean. Therefore, our decoding decision cannot rely on an accurate notion of time. Instead, our decoder needs to assume a sufficient amount of error in the period of each received bit and to search for the peaks in a wider range instead of a strict period of time.

To address this level of time-insensitivity and achieve robustness to synchronization errors, our decoding decision was made based on the time difference between each two successive peaks. As an example, assume that two successive zeros were sent and hence two peaks were received. The difference between the start time of each peak should be rounded to the average bit period. It should be noted that the decoder computes the average bit period based on the received preamble data. However, if a zero-one-zero transmission was made, the time difference of the start of the two received peaks should be rounded to double of

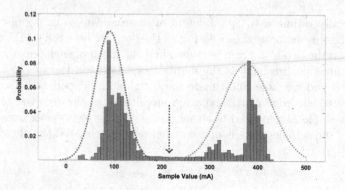

Fig. 8. Using the Gaussian Mixture Model to estimate the threshold.

the average bit period. If a zero-one-one-zero transmission was made, the difference should be rounded to triple the average period and so on. Eventually, synchronization is regained with every detected peak and based only on the time difference between peaks, the decoder makes a decision on how many no-peak bits (ones in our case) are transmitted between the zeros. The time difference does not have to be exactly equal to a multiple of the average bit period. Instead, a range of values can be rounded up to the same value and thus more flexible time-insensitive decoding decision is made.

5   Experimental Evaluation

In this section, we first describe the devices used in our experiments and the values for transmission parameters. We then report the results of the transmission evaluation.

5.1   Experiment Settings

In our experiments, we programmed the PowerSnitch app using Android Studio with API. The device used to measure the energy provided to the device via USB cable is Monsoon Power Monitor[2] in USB mode with 4.55 V in output. The decoder used to process signal was implemented in MATLAB. In order to evaluate the performance of the transmission, we send out a payload comprised of letters and numbers of ASCII code for a total of 512 bits. The values of period used range from 500 ms to 1000 ms with increments of 100 ms. It is worth mentioning that bits sent over our channel were not packeted and no error detection or correction techniques were used. For each phone and bit period, BER was computed after sending 512 bits at once and then number of bits that were incorrectly decoded was calculated.

---

[2] www.msoon.com/LabEquipment/PowerMonitor.

We evaluate the performance of our proposal on the following devices running Android OS: Nexus 4 with Android 5.1.1 (API 22), Nexus 5 with Android 6.0 (API 23), Nexus 6 with Android 6.0 (API 23) and Samsung S5 with Android 5.1.1 (API 22). We underline that the devices used in our experiments are actual personal devices, kindly lent by some users without any money reward. In order to replicate an actual real world scenario, we did not uninstall any app, nor stopped any app running in background. The only intervention we made on those devices is the installation of our PowerSnitch app.

5.2   Results

In Table 1, we report the performance of the decoder for processing the received power bursts on different mobile devices. The results presented in the table are in terms of BER in the transmission of the payload; the lower the BER, the better is the quality of the transmission. For Nexus devices (i.e., Nexus 4, 5 and 6), we achieve a zero or low BER of periods of 800 ms and 900 ms (i.e., 1.25 and 1.11 bits per seconds, respectively). While for Nexus 4 and 6, the BER remains under 20% and, for Nexus 5, it increases to 37% and 40% with periods 700 ms and 600 ms, respectively. For Samsung S5, the transmission BER is at 12.5% with a period of 1 s, and it slowly increases to around 21% with a period of half a second.

The higher BER for Nexus 5 (i.e., periods 700 ms and 600 ms in Table 1) are due to de-synchronization of the signal that the decoder was not able to recover. To cope with this problem, we can divide the payload into packets, where a packet header will be the preamble in order to recover the synchronization. A quick overview of the communication literature can show how a BER of 30% can be recovered using a simple Forward Error Correction (FEC) technique where the transmitter encodes the data using an Error Correction Code (ECC) prior to transmission; for example bits redundancy or parity checks.

Table 1. Results in terms of Bit Error Ratio (BER) as percentage.

Device		Period (milliseconds)					
Model	Operating system version	1000	900	800	700	600	500
Samsung S5	Android 5.1.1 (API 22)	12.5	13.5	13.31	16.33	17.9	21.42
Nexus 4	Android 5.1.1 (API 22)	13.5	0.78	0.0	0.0	13.33	16.21
Nexus 5	Android 6.0 (API 23)	21.0	0.0	0.95	36.82	40.35	13.4
Nexus 6	Android 6.0 (API 23)	1.07	0.0	0.21	0.0	4.05	7.42

6   Discussion and Optimizations

In this section, we elaborate further on the results obtained in the experimental evaluation of our proposed attack (Sect. 5). In particular, we discuss on interesting observation made during our experiments. We also present the optimizations that were implemented in the framework in order to make our proposed attack more robust.

An interesting phenomenon to notice is that, as observed in our experiments, the level of battery affects the quality of the transmission signal. In Fig. 9, we present the amount of electric current provided by the power supplier to a Nexus 6 during recharge (i.e., the first 35 min) and full battery states (i.e., after 35 min). Indeed, when the level for the battery is low (i.e., 0% to around 40%) the device consumes a high amount of energy, and almost all of it is used to recharge the battery.

When attempting to transmit data in the aforementioned conditions, we discover that the bursts were not easily distinguishable. In fact, the difference in terms of energy consumption between burst and rest was so small that it cannot be distinguished from noise; thus, they can be filtered out during the signal processing. Additionally, when the level of the battery is increased, the amount of energy consumed to recharge the battery gradually decreases. We observed that when the battery level is higher than 50%, the power bursts become more and more distinguishable. However the best condition under which the bursts are clear is when the battery is fully charged. Indeed, as we can notice from Fig. 9, the current drops down after the battery level reaches 100%, because there is no need to provide energy to the battery anymore - except to keep the device running.

The percentages mentioned above also depends from the power supplier used to provide energy to the device. In our experiments, we used Monsoon power monitor which provides as output at most 4.55 V. Due to the limitation of such power monitor, during the recharge of devices with fast charge technology (e.g., Samsung S5, Nexus 6 and 6P), which are able to work with 5.3 V and 2 mA, the energy consumed is almost constant until the battery is almost fully charged. Thus, we cannot decode any signal from the energy consumption.

In order to avoid to transmit when the receiver is not able to decode the signal, PowerSnitch checks whether the battery level is among a certain threshold ω. Such threshold ω can be obtained by PowerSnitch itself, simply knowing the model in which it is running. This information can be easily obtained without any permission (android.os.Build.MODEL and MANUFACTURER).

Optimizations. In what follows, we elaborate on the optimizations that were implemented in order to not be detected or make the victim suspicious. The first optimization is to keep a duty cycle (i.e., the time of burst in a period) under 50%. During an attack, if such optimization is not taken into account (i.e., a duty cycle greater than 75%), the victim may be alerted by two possible effects:

- the temperature of the device could increase significantly, in a way that could be perceived by touching it.
- if the attack takes place during the battery charge phase, the battery will take more time to recharge due to the high amount of energy used by CPU.

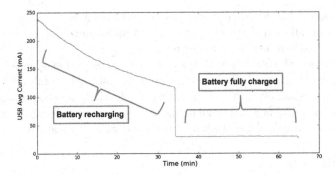

Fig. 9. Electric current provided to a Nexus 6 during recharge phase and battery fully charged.

However, as previously explained in Sect. 3.2, the duty cycle should be 50% of period (i.e., $T_b/2$) in order to achieve a RZ. Thus, the above effects are already taken care of in our proposed attack.

Another optimization involves the Android Debug Bridge (ADB) tool. It is possible to monitor CPU consumption of an Android device via ADB. Hence, one may use such debug tool to detect that something strange is happening on the device (i.e., a transmission on the covert channel using CPU bursts). Fortunately, PowerSnitch app could easily detect whether ADB setting is active through `Settings.Global.ADB_ENABLED`, once again provided by an Android API.

Another optimization to PowerSnitch app would be the ability to detect if the power supplier is an accomplice of the attack. The accomplice has to let PowerSnitch app know that it is listening to the covert channel by communicating something equivalent to a "hello message". In order to do so, we can rely on the information about the amount of electric current provided to recharge the battery. Such information is made available through `BatteryManager` object, provided by Android API. In particular, `BATTERY_PROPERTY_CURRENT_NOW` data field (available from API 21 and on devices with power gauge, such as Nexus series) of `BatteryManager` records an integer that represents the current entering the battery in terms of mA.

On one hand, the power supplier can then variate the current in output above and below a certain threshold θ with a precise timing. As a practical and non-limiting example, at a point in time during the recharging, the power supplier can output current with the following behavior: (i) below θ for t seconds, (ii) above θ for t seconds, (iii) again below θ for t seconds and finally (iv) above θ for good. On the other hand, since PowerSnitch app monitors `BATTERY_PROPERTY_CURRENT_NOW` and knows the aforementioned behavior (along with both θ and t), it will be able to understand that at the other end of the USB cable there is an accomplice power supplier ready to receive a transmission. This optimization is significant for reducing the chance to remain undetected, since PowerSnitch app will transmit data if and only if it is sure that an accomplice power supplier is listening. With such optimization, we will obtain a half-duplex communication channel, since the communication is bidirectional but only one participant (i.e., the device or the power source) is allowed to transmit at a time. This optimization is not currently implemented and will be considered as future work.

To summarize, the conditions under which the transmission of data is optimal and the chance of being detected is lowest are as follows: the mobile device has to be charged more than 50%, the screen has to be off, ADB tool should be switched off (which is true by default) and the phone must to be plugged with a USB charging cable to a public charging station which is controlled by the adversary.

7   Conclusion

In this paper, we demonstrate for the first time the practicality of using a (power-only) USB charging cable as a covert channel to exfiltrate data from a smartphone, which is connected to a charging station. Since there are no visible signs of the existence of a covert channel while the battery is recharging, the user is oblivious that data is being leaked from the device. Moreover, our proposed covert channel defeats existing USB charging protection dongles, as described in [7] because it requires only the USB power pins to exfiltrate data in the form of CPU power bursts.

To show the feasibility and practicality of our proposed covert channel, we implemented an app, *PowerSnitch*, which does not require the user to grant access to permissions at install-time (nor at run-time) on a non-rooted Android phone. Once the device is plugged in a compromised public charging station, the app encodes sensitive information and transmits it via power bursts back to the station. Our empirical results show that we are able to exfiltrate a payload encoded in power bursts at 1.25 bits per seconds with a BER under 1% on the Nexus 4-6 devices and a BER of around 13% for Samsung S5. As future work, we plan to investigate malicious power banks and how they can be exploited using our covert channel to exfiltrate data from smart devices. We will also work on the transmitter and decoder by extending the framework to include error correction algorithms and synchronization recover mechanisms to lower down the BER of data transmission—as this was not the main goal of this paper.

Acknowledgments. This work is supported by ONR grants N00014-14-1-0029 and N00014-16-1-2710, ARO grant W911NF-16-1-0485 and NSF grant CNS-1446866.

Veelasha Moonsamy is supported by the Technology Foundation STW (project 13499 - TYPHOON & ASPASIA) from the Dutch government.

Mauro Conti is supported by a Marie Curie Fellowship funded by the European Commission (agreement PCIG11-GA-2012-321980). This work is also partially supported by the EU TagItSmart! Project (agreement H2020-ICT30-2015-688061), the EU-India REACH Project (agreement ICI+/2014/342-896), "Physical-Layer Security for Wireless Communication", and "Content Centric Networking: Security and Privacy Issues" funded by the University of Padua. This work is partially supported by the grant n. 2017-166478 (3696) from Cisco University Research Program Fund and Silicon Valley Community Foundation.

We would like to thank Elia Dal Santo and Moreno Ambrosin for their insightful comments.

References

1. Aloraini, B., Johnson, D., Stackpole, B., Mishra, S.: A new covert channel over cellular voice channel in smartphones. Technical report (2015). arXiv preprint arXiv:1504.05647

2. Aviv, A.J., Gibson, K., Mossop, E., Blaze, M., Smith, J.M.: Smudge attacks on smartphone touch screens. In: Proceedings of USENIX WOOT (2010)

3. Aviv, A.J., Sapp, B., Blaze, M., Smith, J.M.: Practicality of accelerometer side channels on smartphones. In: Proceedings of USENIX ACSAC (2012)

4. Baghel, S., Keshav, K., Manepalli, V.: An investigation into traffic analysis for diverse data applications on smartphones. In: Proceedings of NCC (2012)

5. Bartel, A., Klein, J., Le Traon, Y., Monperrus, M.: Automatically securing permission-based software by reducing the attack surface: an application to android. In: Proceedings of ACM ASE (2012)

6. Carroll, A., Heiser, G.: An analysis of power consumption in a smartphone. In: Proceedings of USENIX ATC (2010)

7. Chacos, B.: USB condom promises to protect your dongle from infected ports. PC World, August 2014. http://tinyurl.com/hvlqkrt

8. Chandra, S., Lin, Z., Kundu, A., Khan, L.: Towards a systematic study of the covert channel attacks in smartphones. In: Tian, J., Jing, J., Srivatsa, M. (eds.) SecureComm 2014. LNICSSITE, vol. 152, pp. 427–435. Springer, Cham (2015). doi:10.1007/978-3-319-23829-6_29

9. Conti, M., Mancini, L.V., Spolaor, R., Verde, N.V.: Analyzing android encrypted network traffic to identify user actions. IEEE TIFS 11(1), 114–125 (2016)

10. Do, Q., Martini, B., Choo, K.K.R.: Exfiltrating data from android devices. Comput. Secur. 48, 74–91 (2015)

11. Felt, A.P., Chin, E., Hanna, S., Song, D., Wagner, D.: Android permissions demystified. In: Proceedings of ACM CCS (2011)

12. Ferreira, D., Dey, A.K., Kostakos, V.: Understanding human-smartphone concerns: a study of battery life. In: Proceedings of PerCom (2011)

13. Kim, H., Smith, J., Shin, K.G.: Detecting energy-greedy anomalies and mobile malware variants. In: Proceedings of ACM MobiSys (2008)

14. Lalande, J.-F., Wendzel, S.: Hiding privacy leaks in android applications using low-attention raising covert channels. In: Proceedings of ARES (2013)

15. Lau, B., Jang, Y., Song, C., Wang, T., Chung, P.H., Royal, P.: Mactans: injecting malware into IOS devices via malicious chargers. Black Hat, USA (2013)

16. Lin, L., Kasper, M., Güneysu, T., Paar, C., Burleson, W.: Trojan side-channels: lightweight hardware trojans through side-channel engineering. In: Clavier, C., Gaj, K. (eds.) CHES 2009. LNCS, vol. 5747, pp. 382–395. Springer, Heidelberg (2009). doi:10.1007/978-3-642-04138-9_27

17. Liu, L., Yan, G., Zhang, X., Chen, S.: VirusMeter: preventing your cellphone from spies. In: Kirda, E., Jha, S., Balzarotti, D. (eds.) RAID 2009. LNCS, vol. 5758, pp. 244–264. Springer, Heidelberg (2009). doi:10.1007/978-3-642-04342-0_13

18. Marforio, C., Ritzdorf, H., Francillon, A., Capkun, S.: Analysis of the communication between colluding applications on modern smartphones. In: Proceedings of USENIX ACSAC (2012)

19. Meng, W., Lee, W.H., Murali, S., Krishnan, S.: Charging me and i know your secrets!: towards juice filming attacks on smartphones. In: Proceedings of ACM CPS-SEC (2015)

20. Moonsamy, V., Rong, J., Liu, S.: Mining permission patterns for contrasting clean and malicious android applications. J. Future Gener. Comput. Syst. **36**, 122–132 (2013)
21. Novak, E., Tang, Y., Hao, Z., Li, Q., Zhang, Y.: Physical media covert channels on smart mobile devices. In: Proceedings of ACM UbiComp (2015)
22. Owusu, E., Han, J., Das, S., Perrig, A., Zhang, J.: ACCessory: password inference using accelerometers on smartphones. In: Proceedings of ACM HotMobile (2012)
23. Pathak, A., Charlie Hu, Y., Zhang, M.: Where is the energy spent inside my app?: Fine grained energy accounting on smartphones with Eprof. In: Proceedings of ACM EuroSys (2012)
24. Proakis, J.G.: Intersymbol Interference in Digital Communication Systems. Wiley, Hoboken (2003)
25. Reynolds, D.: Gaussian mixture models. Encycl. Biom., 827–832 (2015)
26. Schlegel, R., Zhang, K., Zhou, X.Y., Intwala, M., Kapadia, A., Wang, X.: Sound-comber: a stealthy and context-aware sound trojan for smartphones. In: Proceedings of NDSS (2011)
27. Spreitzer, R.: Pin skimming: exploiting the ambient-light sensor in mobile devices. In: Proceedings of ACM CCS SPSM (2014)
28. Stöber, T., Frank, M., Schmitt, J., Martinovic, I.: Who do you sync you are?: Smartphone fingerprinting via application behaviour. In: Proceedings of ACM WiSec (2013)
29. Taylor, V.F., Spolaor, R., Conti, M., Martinovic, I.: Appscanner: automatic fingerprinting of smartphone apps from encrypted network traffic. In: Proceedings of IEEE EuroS&P (2016)
30. Android Developers. Optimizing for Doze and App Standby. http://tinyurl.com/zvphw46
31. Business Insider. The Smartphone Market Is Now Bigger Than The PC Market (2011). http://goo.gl/XkM8XM
32. Yan, L., Guo, Y., Chen, X., Mei, H.: A study on power side channels on mobile devices. In: Proceedings of ACM Internetware (2015)
33. Yoon, C., Kim, D., Jung, W., Kang, C., Cha, H.: AppScope: application Energy metering framework for android smartphone using kernel activity monitoring. In: Proceedings of ATC (2012)

Are You Lying: Validating the Time-Location of Outdoor Images

Xiaopeng Li, Wenyuan Xu$^{(\boxtimes)}$, Song Wang, and Xianshan Qu

Department of CSE, University of South Carolina, Columbia, SC, USA
{xl4,wyxu,songwang,xqu}@cec.sc.edu

Abstract. Photos have been commonly used in our society to convey information, and the associated contextual information (i.e., the capture time and location) is a key part of what a photo conveys. However, the contextual information can be easily tampered or falsely claimed by forgers to achieve malicious goals, e.g., creating fear among the general public or distorting public opinions. Thus, this paper aims at verifying the capture time and location using the content of the photos only. Motivated by how the ancients estimate the time of the day by shadows, we designed algorithms based on projective geometry to estimate the sun position by leveraging shadows in the image. Meanwhile, we compute the sun position by applying astronomical algorithms according to the claimed capture time and location. By comparing the two estimations of the sun position, we are able to validate the consistency of the capture time and location, and hence the time-location of the photos. Experimental results show that our algorithms can estimate sun position and detect the inconsistency caused by falsified time, date, and latitude of location. By choosing the thresholds to be 9.2° and 4.8° for the sun position distance and altitude angle distance respectively, our framework can correctly identify 91.1% of the positive samples, with 7.7% error in identifying the negative samples. Note that we assume that the photos contain at least one vertical object and its shadow. Nevertheless, we believe this work serves as the first and important attempt in verifying the consistency of the contextual information only using the content of the photos.

Keywords: Capture time and location · Sun position · Shadows · Consistency · Projective geometry

1 Introduction

Benefiting from the development of digital technologies and internet, photos become increasingly common in our society. A huge number of photos are shared through social media platforms. People use photos to convey information and express emotions, and even employ them to illustrate news stories [14]. Meanwhile, people are exposed to fake photos that had been used for malicious purposes: fooling the world and creating chaos as well as panic [3,7]. For example,

© Springer International Publishing AG 2017
D. Gollmann et al. (Eds.): ACNS 2017, LNCS 10355, pp. 103–123, 2017.
DOI: 10.1007/978-3-319-61204-1_6

Fig. 1. A photo that was taken in Sept. 2013 was used for a news event happened in Jan. 2017.

the Hurricane Sandy hit the northeastern U. S. in 2012: numerious fake disaster photos and rumors were spread through social networks and caused panic and fear among the general public [10]. Therefore, the U.S. Federal Emergency Management Agency had to set up a "rumor control" section to defend against misinformation including fake photos on social networks [1]. In addition, fake photos have been used to distort public opinions. For instance, the fake refugee photos were shared online in the Europe's refugee crisis in 2015 and used to twist public opinion on asylum seekers [5].

Previous studies mainly focused on devising forensic techniques to detect photo tampering and manipulation. For example, researchers have proposed approaches to demonstrate copy-move manipulation [2,8,23] and leveraged shadows and lighting to determine photo tampering [4,20]. However, in addition to manipulating the content of a photo itself, the contextual information (i.e., the capture time and location) can also be falsified. For instance, the photo in Fig. 1 was claimed to be taken in January 2017, and was used on social medias to illustrate the news that a fleet of bikers were on the way to Washington D.C for President Trump's inauguration. However, the photo was actually published in 2013 for the anniversary of 9/11[1]. Thus, it is promising if we can validate the capture time and locations immediately purely using the photos themselves.

Determining whether the capture time or the location of an image is real is promising yet challenging. Although most images have timestamps and GPS information enclosed, these can be altered without traces once the format is known. Deciding whether a picture is taken at a place simply by experiences is infeasible since the image scenes may appear to be similar in various places, such as public lawns, parking lots, beaches and roadsides. Even if the capture location is true, the capture time can be falsified without any traces. Finding evidences from the content of an image to verify the time is difficult. Objects that reveal time directly (e.g., clocks and watches) are rarely seen in images.

---

[1] https://www.buzzfeed.com/tasneemnashrulla/bikers-for-trump-inauguration-fake-pictures.

Objects such as clothing, or colors of trees may indicate the capture time, but these indicators can only reveal a relatively long time span (e.g., a T-shirt is suitable from April through October in many places). So far, limited research has addressed this problem. Garg *et al.* [9] demonstrated the feasibility of using the Electric Network Frequency signal as a natural timestamp for video data in an indoor enviroment. Junejo and Foroosh [17] and Wu and Cao [27] used shadow trajectories to estimate the geo-location of stationary cameras from multiple outdoor images. Tsai *et al.* [26] and Kakar and Sudha [18] developed approaches that leverage the geolocation of images and the sun information to estimate the capture time for outdoor images. However, to the best of our knowledge, none has been done to validate both the capture time and location. In this paper, we study how to validate whether the image's capture time and location are true from a single outdoor image that has at least one shadow. Although we require a shadow in an image, we believe our work serves as the first attempt towards a full-fledged solution.

The basic idea is that the position of the sun is determined by time and location and can be utilized to check time-location consistency of outdoor images. Specifically, we estimate the sun position from two sets of information: (1) utilize vertical objects and their shadows in images to estimate sun position, and (2) use the claimed capture time and location in the metadata of images for estimation. Finally, we compare these two values and decide whether the claimed capture time and location are true.

In summary, we outline our three main contributions as below:

- We propose a framework that is called AYL for validating time-location consistency of outdoor images. We show that the variances of sun position correlate with the time and location, and the correlation can be used to determine whether the capture time and location of images are consistent.
- We demonstrate that the sun position can be acquired from shadows and design algorithms to estimate the sun position from one vertical object and its shadow in the image. The results show that the algorithms are effective.
- We implement the proposed framework and evaluate it using photos collected in 15 cities across the U.S. and China, which proves AYL to be effective.

2   Overview

We specify the threat model, overview the framework of AYL, and summarize the research challenges in this section.

2.1   Threat Model

We assume that an attacker modifies the capture time and location of an image for malicious purposes, but doesn't tamper or manipulate the image itself. Note that even if she modifies the image, we can detect it utilizing the prior works [2,4,8,20,23]. Below we describe how an attacker can modify the metadata.

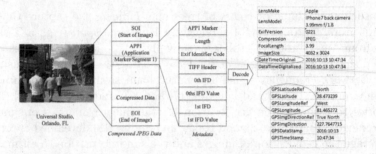

Fig. 2. The basic structure of JPEG compressed image files.

An image file contains not only the image itself but also the metadata that describes who, when, where, and how an image was taken [21,22]. Exchangeable image file format (Exif) is a popular standard that specifies the formats of images. The specification uses the existing file formats (e.g., JPEG) with the addition of specific metadata tags. Figure 2 shows the basic structure of JPEG compressed image files [16], and the application marker segment I (APPI) contains contextual information of images, e.g., the capture time, the image size, compression format, and details of cameras (focal length, camera maker) [16], etc. In particular, `DateTimeOriginal` records the capture time. `GPSLatitude` and `GPSLongitude` contain the GPS location (i.e., latitude and longitude) of where the image was taken. `GPSimgDirection` represents the direction measured by the magnetometer (i.e., the direction in which the camera faces). Modifying the capture time and the GPS information enclosed in metadata can be easily accomplished by using metadata editing tools such as ExifTool [12]. For example, the photo shown in Fig. 2 was taken in Orlando, FL, on 13th October 2016, at 10:47 a.m. An attacker can claim that the photo was taken in May 2016 in Los Angeles by changing `DateTimeOriginal` and other related GPS fields.

2.2 Overview of AYL

Our goal is to validate whether the claimed capture time and location are true. The capture time indicates the date and time when the photo was taken, and the capture location reveals where it was taken.

Basic Idea. Although an attacker can modify the metadata and claim that a photo was taken at time X and location Y, she won't be able to change the "time" and "location" information that is embedded in the photo. Thus, our framework works as follows. On one hand, we utilize the contents in outdoor images—vertical objects and resulting shadows—to extract the sun position that reflects when and where the image was taken. On the other hand, we utilize the metadata information—the claimed capture time and location—to obtain a second estimation of the sun position by applying astronomical algorithms. If these two estimations are close enough, we consider the capture time and location to be true with a high probability. Otherwise, they are considered to be falsified.

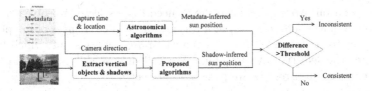

Fig. 3. The work flow of the proposed AYL framework.

Assumption. Without loss of generality, we assume that photos are taken using smartphone's rear cameras and the smartphone is held in such way that the camera looks at the front horizontally or vertically, and it is perpendicular to the ground. We further assume that the photographer stands on ground and the ground that interested shadows lie on is approximated to water level. Finally we assume that at least one vertical object and its shadow can be seen in the image. The objects can be human beings, road signs, lampposts, tree trunks and so on.

Workflow. Figure 3 shows the work flow of proposed approach. For convenience of description, we use the term *shadow-inferred sun position* to refer to the sun position estimated from shadows in the image, and use the term *metadata-inferred sun position* to refer to the sun position calculated from claimed capture time and location. In this paper, *capture time* denotes the date of year and the time of day unless otherwise indicated.

2.3 Research Challenges

Shadow-Inferred Sun Position. The first challenge is how to obtain sun position from a single image. Although shadows can be viewed in images, we still need to know the length ratio of objects and their shadows and the orientation of shadows to determine the sun position. However, the relative position of two objects in real world is no longer preserved when they are projected to a 2-d image. How to measure the actual length ratio and angles can be a challenging problem. Although single view reconstruction has been extensively studied, there is no generalized way to recover the relative positions of objects from one single image. To address above challenges, algorithms based on projective geometry are proposed in Sect. 4.

Validation. Once the *shadow-inferred sun position* is obtained, the next challenge would be how to validate that the capture time and location are true. To estimate the true capture time and location directly from the *shadow-inferred sun position* is difficult since a specific sun position can be viewed at various places and times. Conversely, the claimed capture time and location can determine a unique value of sun position that should be close enough to the *shadow-inferred sun position*. Then, we convert previous problem into a new problem: how to determine the two estimations—the *shadow-inferred sun position* and the *metadata-inferred sun position*—are close enough indicating the same sun position. Appropriate thresholds need to be selected to solve this problem. The sun moves across the sky at a varying speed. The changes of sun position with respect to time and location on the earth are not constant, which further complicates the selection of thresholds. We will discuss this problem in Sect. 5 and our experimental results are presented in Sect. 6.

3   Background

We discuss the basics on how the sun changes its position in the section, which serves as the foundation of our algorithm.

3.1   Sun Position Definition

The position of the sun in the sky is defined by an azimuth angle and an altitude angle. An azimuth angle describes the direction of the sun, whereas an altitude angle defines the height of the sun [24]. As shown in Fig. 4, the sun azimuth angle A is measured clockwise in the horizontal plane, from the north to the direction of the sun. Its value varies from 0° (north) through 90° (east), 180° (south), 270° (west), and up to 360° (north again). The altitude angle h is measured from the horizontal to the sun and it thus ranges from −90° (at the nadir) through 0° (on the horizon), up to 90° (at the zenith). For instance, when the sun crosses the meridian, its azimuth is 180° and altitude is at its largest value in a day.

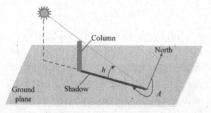

h = altitude angle, measured up from horizon
A = azimuth angle, measured from the north

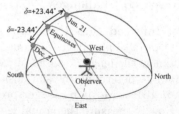

δ is the declination of the sun.

Fig. 4. An illustration of the altitude and azimuth angles of the sun.

Fig. 5. The path of the sun across the sky as observed on various dates in the northern hemisphere.

3.2   How Does the Sun Move

Observed from any location on the earth, the sun moves continuously across the sky throughout days and years. The relative position change is mainly caused by two types of motions of the earth: the rotation around its axis, and the revolution around the sun [24]. It takes about 24 h for the earth to finish one rotation around the earth's axis and about 365 days to complete one revolution around the sun. For an observer on the earth, the first motion contributes to the alternation of day and night, and the second motion leads to the alternation of seasons.

Daily Sun Path. Because of the earth's daily rotation, the sun appears to move along with the celestial sphere every day. It makes a 360° journey around the celestial sphere every 24 h. To an observer on the earth, the sun rises somewhere along the eastern horizon, and goes up to the highest point (zenith) around the noon, then goes down until it sets along the western horizon. Figure 5 shows three of the sun's daily paths viewed on the earth. Accordingly, the cast shadows of any objects move oppositely from somewhere along the west to somewhere along the east. The shadows' lengths vary with the sun's altitude angle. They become shorter and shorter since sunrise and reach the shortest when the sun is at its zenith. Then they become longer over time till sunset. Thus, the shadow that a camera takes at the same day and location but different times of the day will be totally different.

Yearly Sun Path. The sun's daily path across the sky also changes throughout the year. This is because the earth does not rotate on a stationary axis and the tilt in the axis varies each day with respect to the earth's orbit plane. To an observer on the earth, the sun looks higher in the summer than it looks in the winter at the same time in the day. As shown in Fig. 5, the sun follows different circles at different days in one year: most northerly on June 21st and most southerly on December 21st. The sun's motion along the north-south axis over a year is known as the declination of the sun, denoted by δ. Thus, the sun position inferred from photos taken at the same location and time but different days in a year will be different due to the sun's declination.

Sun Path at Different Latitudes. As the sun travels across the sky, the observed altitude angle varies based on the latitude of the observer. The further north or south we go from the equator, the lower the sun's altitude becomes. Figure 6 shows the sun's altitude angle versus the azimuth angle observed at 25° north latitude and 40° north latitude respectively. The sun's altitude angle observed at 25° north latitude is higher than the altitude angle observed at 40° north latitude at the same time. Thus, the sun position inferred from photos taken at the same time but different latitudes will be different.

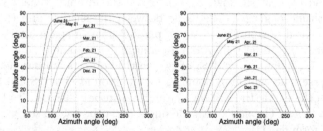

(a) at 25° north latitude in the U.S. (b) at 40° north latitude in the U.S.

Fig. 6. The same path of the sun observed at two latitudes.

4 Shadow-Inferred Sun Position

The framework AYL uses both the azimuth angle and altitude angle to determine the position of the sun in the sky. As shown in Fig. 4, the sun's altitude angle equals the angle between the shadow and the sun ray, and the sun's azimuth angle equals the angle measured across the shadow point of the top of the column, clockwise from the north to the direction of the shadow. In this section, we provide algorithms to estimate the altitude and azimuth angles of the sun from shadows in a photo. We study two scenarios and two corresponding algorithms to estimate the altitude angle. We also design an algorithm to measure the azimuth angle. For both algorithms, their sensitivities are analyzed.

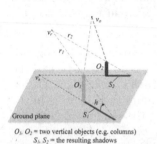

O_1, O_2 = two vertical objects (e.g. columns)
S_1, S_2 = the resulting shadows

Fig. 7. Estimate the sun's altitude angle with two shadows.

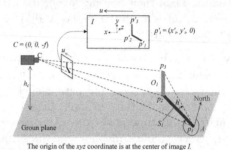

The origin of the xyz coordinate is at the center of image I.

Fig. 8. Estimate the sun's altitude and azimuth angle with one shadow.

4.1 Estimate Altitude

We consider two scenarios for estimating the sun's altitude angles: (a) photos that contain two vertical objects and their shadows, and (b) photos that contain only one vertical object and its shadow. Vertical objects refer to the ones that are perpendicular to the ground plane.

Two-Shadow Estimation. Figure 7 illustrates the first scenario, where two objects O_1 and O_2 cast shadows S_1 and S_2 on the ground plane, respectively. The sun's altitude angle h is the angle between the shadow and the sun ray. From the graphical perspective, a set of parallel lines in space intersect at one point when they are projected onto a 2-d image. This point is called vanishing point. In Fig. 7, shadows S_1 and S_2 of two vertical objects are parallel in space, and they intersect at vanishing point v_s on the ground plane. Since the sun is far away from the earth, the sun rays r_1 and r_2 can be considered to be parallel and intersect at v_r. The sun's altitude angle h can be calculated according to the following formula [11]:

$$h = \arccos\left(\frac{v_r{}^T \omega v_s}{\sqrt{v_r{}^T \omega v_r}\sqrt{v_s{}^T \omega v_s}}\right), \qquad (1)$$

Algorithm 1. Estimating the altitude angle h

Input: I: an image, f: the camera's focal length
Output: h
1: $G \leftarrow$ find the equation of the ground plane;
2: find the equations of lines $\{p'_1p_1, p'_2p_2, p'_3p_3\}$;
3: $\{(x_1, y_1, z_1), (x_2, y_2, z_2)\} \leftarrow$ compute the coordinates of the points $\{p_1, p_2\}$ by
   solving a set of equations G and p'_ip_i accordingly;
4: find the equation of line p_2p_3
5: $(x_3, y_3, z_3) \leftarrow$ compute the coordinates of p_3 by solving a set of equations p_2p_3 and
   p'_3p_3;
6: $\{\overrightarrow{p_1p_2}, \overrightarrow{p_1p_3}\} \leftarrow \{(x_2 - x_1, y_2 - y_1, z_2 - z_1), (x_3 - x_1, y_3 - y_1, z_3 - z_1)\}$
7: $h \leftarrow$ compute the angle between $\overrightarrow{p_1p_2}$ and $\overrightarrow{p_1p_3}$ using Eq. 9
8: **return** h

---

where ω is called the image of the absolute conic and given by the expression
[11,27]:

$$\omega \sim \begin{bmatrix} 1 & 0 & -u_0 \\ 0 & 1 & -v_0 \\ -u_0 & -v_0 & f^2 + u_0^2 + v_0^2 \end{bmatrix}. \tag{2}$$

This expression assumes that the camera has zero skew, the intersection of the
optical axis and the image plane is at the center of the image, and the pixels
are square. Such assumptions are true for current camera technologies [11,27].
In Eq. 2, (u_0, v_0) denotes the coordinates of the center point of the image, and
f denotes the camera's focal length. f is either included in the metadata of the
image or can be calculated by the following constraint on ω with respect to f:

$$v_s{}^T \omega v_o = 0, \tag{3}$$

where v_o is the vanishing points of the two vertical objects O_1 and O_2. When
the objects and their shadows are at perpendicular directions, v_o and v_s will
satisfy Eq. 3 [11]. Once we have the coordinates of v_o and v_s, we can obtain f
by solving Eq. 3.

One-Shadow Estimation. In this scenario, only one vertical object and its
shadow are visible in the image. Figure 8 illustrates this scenario where C denotes
the camera and I is the image. We assume that the image plane is perpendicular
to the ground plane and the direction $\overrightarrow{u}$ is parallel to the ground plane. So the
angle between the image plane and the ground plane is 90°. Let's denote the
image plane to be $z = 0$, and the coordinate frame is shown in Fig. 8. The center
of the image is the origin point $(0, 0, 0)$.

Algorithm 1 describes the steps to measure the altitude angle h given a vertical object and its shadow. Firstly, to find the equation of the ground plane
G, we define the distance between the camera and the ground plane to be h_c.
As we know G is perpendicular to the XY plane of the coordinate system, the
equation of G can be written as:

$$y = -h_c. \tag{4}$$

Next, we compute the equations of the lines p'_1p_1, p'_2p_2 and p'_3p_3. Since the line p'_ip_i passes through the point C $(0,0,f)$ and the point p'_i whose coordinates can be obtained from the image, it can be described by the two points as:

$$\frac{x}{x'_i} = \frac{y}{y'_i} = \frac{z+f}{f}, \tag{5}$$

where $(x'_i, y'_i, 0)$ are the coordinates of p'_i for $i = 1, 2, 3$.

Lines p'_1p_1 and p'_2p_2 intersect with plane G at points p_1 and p_2 respectively. By solving the Eqs. 4 and 5, the coordinates of p_1 and p_2 can be computed as follows:

$$p_i = (x'_i t_i, -h_c, f(t_i - 1)), \tag{6}$$

where $t_i = -\frac{h_c}{y'_i}$ for $i = 1, 2$. Then we have vector $\overrightarrow{p_1p_2} = (x'_2 t_2 - x'_1 t_1, 0, f(t_2 - t_1))$.

Now, we determine the coordinates of p_3 which is the intersection point of lines p'_3p_3 and p_2p_3. The equation of line p_2p_3 is given by:

$$x = x'_2 t_2, \quad z = f(t_2 - 1). \tag{7}$$

By solving the equations of p'_3p_3 and p_2p_3, we can obtain the coordinates of the point p_3:

$$p_3 = (x'_2 t_2, y'_3 t_2, f(t_2 - 1)). \tag{8}$$

Using the coordinates of p_2 and p_3, we have vector $\overrightarrow{p_1p_3} = (x'_2 t_2 - x'_1 t_1, h_c + y'_3 t_2, f(t_2 - t_1))$. The angle between $\overrightarrow{p_1p_3}$ and $\overrightarrow{p_1p_2}$ is the altitude angle and can be computed as follows:

$$\begin{aligned} h &= \arccos \frac{(\overrightarrow{p_1p_3})^T \overrightarrow{p_1p_2}}{\sqrt{(\overrightarrow{p_1p_3})^T \overrightarrow{p_1p_3}} \sqrt{(\overrightarrow{p_1p_2})^T \overrightarrow{p_1p_2}}}, \\ &= \arccos \frac{m}{\sqrt{m + (y'_3/y'_2 - 1)^2} \sqrt{m}}. \end{aligned} \tag{9}$$

where the intermediate variable $m = (\frac{x'_2}{y'_2} - \frac{x'_1}{y'_1})^2 + f^2(\frac{1}{y'_2} - \frac{1}{y'_1})^2$.

4.2 Estimate Azimuth

To estimate the sun's azimuth angle A from one shadow in an image, we design the following algorithm. The scenario is illustrated in Fig. 8. In particular, the point p_3 is not necessary to be visible for estimating the azimuth angle. Let C be the camera and the unit vector $\overrightarrow{u} = (1, 0, 0)$. The true north N is set to be the reference direction in our algorithm. The orientation of $\overrightarrow{u}$ with respect to N can be obtained by subtracting 90° from the image direction which is included in the metadata of the image.

The sun azimuth angle A equals the angle measured clockwise around point p_1 from due north to the shadow. We calculate A as follows:

$$A = \angle(N, \overrightarrow{u}) + \angle(\overrightarrow{u}, \overrightarrow{p_1p_2}), \tag{10}$$

where $\angle(\vec{u}, \overrightarrow{p_1p_2})$ denotes the angle measured clockwise from $\vec{u}$ to $\overrightarrow{p_1p_2}$, and $\angle(N, \vec{u})$ is the angle measured clockwise from N to $\vec{u}$, which is the orientation of $\vec{u}$. $\angle(\vec{u}, \overrightarrow{p_1p_2})$ is the only unknown variable in Eq. 10.

Next, we define the angle between $\vec{u}$ and $\overrightarrow{p_1p_2}$ to be α. $\angle(\vec{u}, \overrightarrow{p_1p_2})$ equals α if it is an acute angle. Otherwise, $\angle(\vec{u}, \overrightarrow{p_1p_2})$ is equal to $(360° - \alpha)$. The angle α can be calculated as:

$$\alpha = \arccos \frac{\vec{u}^T \overrightarrow{p_1p_2}}{\sqrt{\vec{u}^T \vec{u}} \sqrt{\overrightarrow{p_1p_2}^T \overrightarrow{p_1p_2}}}, \tag{11}$$

where $\overrightarrow{p_1p_2}$ has been calculated in Algorithm 1: $\overrightarrow{p_1p_2} = (x_2't_2 - x_1't_1, 0, f(t_2 - t_1))$ and $\vec{u} = (1, 0, 0)$. Then, we replace $\vec{u}$ and $\overrightarrow{p_1p_2}$ in Eq. 11 and compute it as:

$$\alpha = \arccos\left(\frac{\left(\frac{x_1'}{y_1'} - \frac{x_2'}{y_2'}\right)}{\sqrt{\left(\frac{x_1'}{y_1'} - \frac{x_2'}{y_2'}\right)^2 + f^2\left(\frac{1}{y_1'} - \frac{1}{y_2'}\right)^2}}\right). \tag{12}$$

4.3 Sensitivity Analysis

In this section, we quantify the estimation errors in the computing of the altitude angles and azimuth angles.

Errors of the Altitude Angle Inferred from Two Shadows. The estimation errors of altitude angles stem from the following factors: camera distortion and the detection errors of the objects and shadows. For a well designed camera, the systematic errors (e.g. camera distortion) are constant and can be calibrated if necessary. The detection errors of the shadows and objects in the image can be modeled as random variables. Consequently, the detection errors will result in random errors in the calculation of v_r and v_s. Without loss of generality, we consider the errors of v_r and v_s to be linear to the detection errors and define them to be Δv_r and Δv_s, respectively.

From the graphical perspective, the sun altitude angle h derived from vanishing points v_r and v_s has the geometric meaning as described in Fig. 9. Let C be the camera. The lines Cv_r and Cv_s are parallel to the shadow and the sun ray respectively. h represents the sun altitude angle and equals the angle formed by v_r, C and v_s. The error range of each vanishing point is a circle centered at

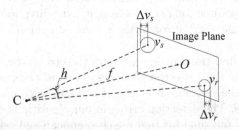

Fig. 9. Errors in the estimated altitude angle with two shadows.

the vanishing point with a radius of the maximum random error. Since Δv_r and Δv_s are small enough compared to the length of $|Cv_r|$ and $|Cv_s|$, they can be considered as two arcs with the center at C. In the worst case, the error Δh of the altitude angle can be calculated as below:

$$\Delta h = (\frac{\Delta v_r}{|Cv_r|} + \frac{\Delta v_s}{|Cv_s|})\frac{180°}{\pi}, \tag{13}$$

where $|Cv_r|$ and $|Cv_s|$ are the lengths between the camera and the two corresponding vanishing points: v_r and v_s. Thus, the error Δh depends on the random errors Δv_r and Δv_s.

Errors of the Altitude Angle Inferred from One Shadow. The sources of random errors in estimating the altitude angle from one shadow include the slope of the ground and the detection errors of interested object and its shadow. If the ground where the shadow located is not flat and has an error of ΔG with respect to the horizontal plane, ΔG will propagate as the altitude angle is estimated. In addition, the detection errors of the vertical object and its shadow can cause estimation errors. The detection errors can affect the angle estimation depending on the distances from the camera to the object and its shadow. The farther the distance, the larger the uncertainty of the estimated altitude angle. The errors in the altitude angle can be linear to the detection errors.

Errors of the Azimuth Angle. The sources of random errors in estimating the azimuth angle include the camera's orientation errors and the detection errors of shadows. To understand how a camera's orientation affects the estimation error of the angle between $\vec{u}$ and the shadow S_1, we define θ to be the angle between the image plane and the horizontal ground plane, and γ to be the angle between the camera and the horizontal plane. Assume $\theta = 90°$ and $\gamma = 0°$. And the estimated camera orientation is $\theta = 90° + \Delta\theta$ and $\gamma = 0° + \Delta\gamma$, where $\Delta\theta$ and $\Delta\gamma$ are random errors.

Figure 10 shows the impact of $\Delta\theta$ and $\Delta\gamma$ on the estimated direction of the shadow. First, the error $\Delta\theta$ will be propagated as we estimate the ground plane according to the camera's orientation. And due to this error, the estimated shadow direction will deviate from the true direction of the shadow. The deviation will be $\Delta\theta$ in the worst case. Second, the error $\Delta\gamma$ will also be propagated to the estimated ground plane. And this error will lead to a deviation in the estimated shadow direction as well, which is $\Delta\gamma$ in the worst case. In summary, the estimated shadow direction deviates from its true direction at most $\Delta\theta + \Delta\gamma$, which can produce $\Delta\theta + \Delta\gamma$ error in the estimated azimuth angle in the worst case.

In summary, we find three main sources of the errors: the detection errors of the objects and their shadows, the ground slope, and the camera's orientation errors. In general, the estimation errors of the sun position are linear to the three types of errors. The detection errors in our algorithms can be reduced by choosing the objects and shadows that are clear enough and using effective image detection algorithms. The errors caused by the slope of the ground will not be

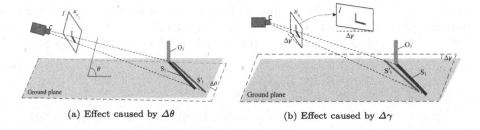

(a) Effect caused by $\Delta\theta$ (b) Effect caused by $\Delta\gamma$

Fig. 10. Effects caused by random errors in camera's orientation

greater than the slope angle and can be reduced greatly by measuring this angle. In addition, the camera's orientation errors can be reduced using inertial sensors to obtain the camera orientation.

5 Metadata-Inferred Sun Position and Validation

In this section, we describe the process to validate the consistency of a photo's capture time and location. The key idea is the following: we calculate the sun position using the capture time and location in the metadata of images. If the capture time and location are true, the sun position will match the one we estimated from shadows.

5.1 Metadata-Inferred Sun Position

As mentioned in Sect. 3, the position of the sun depends on the time of day, the date and the location of the observer. Its movement across the sky obeys the rules that have been studied in astronomy. In this section, we discuss the astronomical algorithms that are used to calculate *metadata-inferred sun position*, given the time and location.

We refer the time of day as the local time based on the standard time offsets of Coordinated Universal Time (UTC). However, the local standard time doesn't provide an intuitive connection with the sun position. In astronomy, the solar time is often used to discuss the sun position. It works because the sun finishes a 360° rotation around the celestial sphere every 24 h. The completed journey is divided into 24 h, and one solar hour means that the sun travels a 15° arc [19]. The instant when the sun is due south in the sky or the shadow points to exactly north is called solar noon, which is 12:00 for solar time. Every 15° arc the sun travels, one hour is added to 12:00 under the 24-h clock system, and the angle distance that the sun passes on the celestial sphere is defined as the hour angle H [19]. It is measured from the sun's solar noon position, and ranges from 0° to $+180°$ westwards and from 0° to $-180°$ eastwards. The conversion between the local standard time t_l to the solar time t_s is as follows [13,24]:

$$t_s = t_l + ET + \frac{4\ min}{deg}(\lambda_{std} - \lambda_l), \qquad (14)$$

where λ_l denotes the local longitude, and λ_{std} is the local longitude of standard time meridian, and ET stands for the equation of time, which describes the difference of the true solar time and the mean solar time [13]. The sun's hour angle is calculated as follows:

$$H = 15°(t_s - 12). \tag{15}$$

Using the observer's local horizon as a reference plane, the azimuth and altitude angles of the sun can be calculated as follows [24]:

$$\tan(A) = \frac{\sin H}{\sin \varphi \cos H - \cos \varphi \tan \delta}, \tag{16}$$

$$\sin(h) = \sin \delta \sin \varphi + \cos \varphi \cos \delta \cos H, \tag{17}$$

where φ is the latitude of the observer's location, and δ is the sun's declination angle and it can be calculated as below [15,24]:

$$\delta = -23.44° \cos\left(\frac{360°(N + 10)}{365°}\right), \tag{18}$$

where N is the number of days since January 1st. Note that the azimuth angle A calculated in Eq. 16 uses south as a reference. We can derive the azimuth angle according to its definition in Sect. 3.

5.2   Consistency Validation

Once obtaining the *shadow-inferred sun position* and *metadata-inferred sun position*, we check the difference between these two estimations by comparing their altitude angles and azimuth angles respectively. However, since there exists random and systemic errors in the *shadow-inferred sun position*, the estimation may not equal the "true" sun position. Thus, we have to select a threshold that is large enough to tolerate the errors yet small enough to detect the inconsistency between the *shadow-inferred sun position* and *metadata-inferred sun position*. Intuitively, the closer these two sun positions are to each other, the more likely the capture time and location are true.

We define the altitude angles of *shadow-inferred sun position* and *metadata-inferred sun position* to be h_s and h_m respectively, and the corresponding azimuth angles to be A_s and A_m. Then the distance of the two altitude angles is $d_h = |h_s - h_m|$, and the distance of the two azimuth angles is computed as $d_A = |A_s - A_m|$. The likelihood of the consistency is inversely proportional to d_h and d_A. However, the effects on d_h and d_A caused by fake capture time and/or location are different. For example, modifying the capture time from 12:00 p.m. to 13:00 p.m. may lead to 10° in d_A but only 2° in d_h. So two different thresholds for d_h and d_A have to be selected. The capture time and location are considered to be true only when both d_h and d_A are within the thresholds. Besides, the sun position can be described by a pair of azimuth angle and altitude angle: (A, h).

We can also use the sun position distance that is computed as $d_p = \sqrt{d_A^2 + d_h^2}$ to distinguish the two estimations of the sun position. Our goal is to choose appropriate variables and thresholds that can increase the probability of correct validation for inconsistent images and decrease the probability of false validation for consistent images. Section 6 details the selection of thresholds in the validation experiment.

6   Evaluation

This section presents the results of our experiments. To evaluate the performance of the sun position estimation algorithms, we conducted an experiment on November 8, 2016 in the U.S. and collected 60 photos. To validate the effectiveness of the framework AYL, we gathered 124 photos in China and the U.S in the span of four months, and examined whether we can detect the modifications of capture time, date and location.

6.1   Sun Position Estimation

To evaluate the accuracy of our sun position estimation algorithms, we collected 60 photos using the rear camera of an iPhone 7 from 9:30 a.m. to 14:30 p.m. at an interval of about 5 min on November 8, 2016 in Columbia, SC. As shown in Fig. 11(a) we set up the experiment in a relatively ideal situation: we place two columns (the red one and the grey one) on the ground vertically, and fixed the iPhone 7 on another vertical stick to take photos of these two columns and their shadows. Figure 11(b) shows the estimated altitude angles by applying the **two-shadow estimation** and **one-shadow estimation** algorithms to the photos. The ground truth sun positions are calculated using the astronomical algorithms in Sect. 5.1 according to the real time, latitude and longitude. The ground truth altitude angles are labeled with red and denoted as "Altitude". The other two

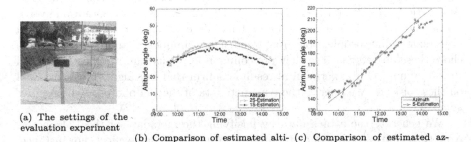

(a) The settings of the evaluation experiment

(b) Comparison of estimated altitude angle to ground truth.

(c) Comparison of estimated azimuth angle to ground truth.

Fig. 11. The experiment setting is shown in (a). And (b) presents the comparison of estimated altitude angle to ground truth altitude angle. (c) shows the comparison of estimated azimuth angle to ground truth azimuth angle.

curves in Fig. 11(b) represent the estimated altitude angles inferred from shadows. "2S-Estimation" is obtained by applying the **two-shadow estimation** algorithm, while "1S-Estimation" is plotted by applying the **one-shadow estimation** algorithm. We find that the average error in "2S-Estimation" is 1.43°, while it is 2.98° in "1S-Estimation". Figure 11(c) presents the estimated azimuth angles versus the ground truth azimuth angles. The curve in red is plotted using the ground truth azimuth angles, while the curve in blue is plotted using the data of estimated azimuth angles. The average error is approximately 4.3°.

The estimation errors of sun positions are mainly contributed by three factors. First, due to the ground slope and the camera's orientation, the image plane may not be precisely perpendicular to the ground plane, which causes errors. The second type of errors is random one that is introduced when extracting objects and shadows from the photos. Finally, errors can be created by the measurement drift of the compass over time. Due to the nature of the two algorithms, these types of errors will have different levels of impact on them. Figure 11(b) indicates that the **two-shadow estimation** algorithm outperforms the **one-shadow estimation** algorithm. It is partly because the **one-shadow estimation** algorithm is more sensitive to the ground slope. We believe that given the measurement of the slope, we shall be able to reduce the error. In summary, the algorithms in Sect. 4 are able to infer the sun position, either from two vertical objects and their shadows or from one object and its shadow.

6.2   Consistency Validation

To evaluate the performance of AYL and to understand threshold selection, we conducted a set of experiments.

Dataset. The data in this experiment was captured at 15 cities around the USA and China since September 2016. Our dataset consists of 124 photographs taken by 10 iPhones, including iPhone 5s, 6, 6 plus, 6s, 6s plus and 7. 61 out of the 124 photos were taken in China. Each photo encloses the metadata that includes the real capture time and location. 72 out of the 124 photos contain at least two vertical objects and their shadows, while 52 photos only contain one vertical object and its shadow. Our dataset mainly contains three types of vertical objects: standing people, poles (e.g. road signs, lampposts) and tree trunks. We chose these objects because they are common in reality and are mostly vertical to the ground. Our experimental results confirm that our algorithms work well on these objects. We refer to the true metadata of the 124 photos as the *positive samples.*

We generate the attack data by falsifying the metadata of the 124 photos. Note that multiple types of metadata may result in the same effect. For instance, modifying longitude one degree more to the west has the same effect on the sun position as changing the local time forward by four minutes. Thus, falsifying either longitude or the local time is equivalent. To simplify the analysis yet without loss of generality, we focus on three types of attacks that modify the following metadata:

- The falsified time of day, and true date and location.
- The falsified date, and true time and location.
- The falsified latitude of location, and true time and date.

We refer to the attack metadata as the *negative samples*. We have 124 negative samples for each type of attack metadata, The "fake" times of day are randomly generated in the range from 8:00 a.m. to 17:00 p.m. when the sun is likely to be seen. The "fake" dates are randomly generated from the range within one year. The "fake" latitudes of location are randomly generated in the range of 25° and 50° of the Northern Hemisphere where most of the U.S. and China locate. Here we didn't consider the attack data with falsified longitude. Because the result produced by only falsifying longitude can be equivalent to the result caused by falsifying the time of day accordingly.

Metric. We use ROC curves to evaluate the performance of AYL by varying thresholds for our system. An ROC curve represents Receiver Operating Characteristic curve and is created by plotting true positive rate (TPR) against false positive rate (FPR), as the threshold varies [6]. The true positive rate and false positive rate are defined as below.

$$TPR = \frac{\# \text{ of true positives}}{\# \text{ of (true positives + false negatives)}} = \frac{\# \text{ of true positives}}{\# \text{ of positives}}$$

$$FPR = \frac{\# \text{ of false positives}}{\# \text{ of (true negatives + false positives)}} = \frac{\# \text{ of false positives}}{\# \text{ of negatives}}$$

where a true positive denotes the result that a positive sample is correctly identified as such, and a false positive is the one that a negative sample is identified as a positive sample by mistakes. The point $(0, 1)$ on the ROC curve denotes 0 FPR and 100% TPR, which indicates an ideal system that can correctly identify all genuine photos and reject all falsified photos [25]. In our experiment, we select the optimal threshold as the one that yields the minimum distance from the corresponding point on the ROC curve to the ideal point $(0, 1)$. Another indicator that we use to evaluate the average performance of the validation is the area under the ROC curve (AUC). The closer it is to 1, the better the average performance is [6].

Performance and Threshold Selection. Based on the framework AYL, we performed consistency validation using the three types of attack metadata. To understand how the altitude angle and azimuth angle influence the performance of the validation, we examine three distances separately: the distance of the altitude angles d_h, the distance of the azimuth angles d_A, and the distance of the sun positions d_p. Here, the sun position is defined to be (A, h), in which A refers to the azimuth angle and h refers to the altitude angle. To decide the best distance variable which can yield the maximum AUC and the optimal threshold of the variable, we analyze the ROC curves that are plotted by varying the threshold of each type of distance.

The results are presented in the set of ROC curves shown in Fig. 12. Each ROC curve with distinct color is plotted by varying the threshold of one type

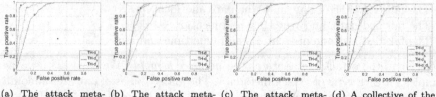

(a) The attack meta-
data with falsified time
of day.

(b) The attack meta-
data with falsified
date.

(c) The attack meta-
data with falsified lat-
itude.

(d) A collective of the
three types of attack
metadata.

Fig. 12. ROC curves based on different distance variables and different types of attack metadata.

of the three distances. "TH $-$ d$_p$" denotes varying the threshold of the sun position distance d_p. "TH $-$ d$_h$" and "TH $-$ d$_A$" denote varying the threshold of the altitude angle distance d_h and the azimuth angle distance d_A respectively. For each type of attack metadata, the randomly generating of 124 negative samples is repeated 5 times. Each false positive rate on the ROC curve is averaged over these repeated attack metadata. Figure 12(a) indicates that the detection based on d_A slightly outperforms the one based on d_h, to detect the attacks that falsify the photo's time of day. However, the d_h based detection achieves better performance in detecting the other types of attacks as shown in Fig. 12(b–c), especially in detecting falsified latitude. The result implies that d_h is more important in distinguishing different positions of the sun compared to d_A in general. Such a conclusion confirms with the result reported in Sect. 6.1, i.e., the average estimation error of the altitude angles is smaller than that of the azimuth angles. If only d_h is used for consistency validation, Fig. 12(d) guides us to choose the optimal threshold of d_h to be $3°$ and it achieves combined (TPR, FPR) values of $(89.5\%, 22\%)$, which means that 89.5% of positive samples can be correctly validated but 22% of negative samples will be mistakenly identified. In addition, Fig. 12(a–c) shows that the d_p based detection achieves the best performance in detecting falsified time, and has almost the same performance as the d_h based detection in detecting the other types of attacks. Once we only use d_p for consistency validation as shown in Fig. 12(d), we choose the optimal threshold of d_p to be $9.2°$, which achieves combined (TPR, FPR) values of $(92.7\%, 18.6\%)$ for all attacks.

To improve the performance further, we examine both the d_p and d_h to validate the consistency of time and location. That is, a sample has to satisfy both the thresholds of d_h and d_p to be accepted by AYL. Plotting the ROC curves and finding the global optimal thresholds by varying two thresholds can be tricky. Thus, we chose the local optimal threshold for one variable and varied the other threshold to plot the ROC curve. This approach may not generate the global optimal thresholds for the two variables, but it strikes a balance between the optimum and the computational cost. We chose the threshold of d_p to be $9.2°$ and varied the threshold of d_h. The resulting curve illustrates an improved performance than the one of using a single threshold as shown in Fig. 12(d).

Note that we cannot plot an integral ROC curve when the threshold of d_p is fixed since the highest true positive rate will be decided by the fixed threshold, which is 92.7%. The curve "TH $-$ $d_p d_h$" in Fig. 12(d) indicates that choosing the optimal threshold of d_h to be 4.8° can correctly identify 91.1% positive samples, and cannot identify 7.7% of negative samples.

Attacks Against AYL. Based on the above results, we analyze the robustness of the framework AYL when falsifying one of the three parameters—time of day, date and latitude of location, and falsifying more than one parameters. AYL cannot detect the falsifications that do not cause violations of both the thresholds of the altitude angle distance and the sun position distance. If an attacker modifies both the time and location of a photo such that the altitude angle and the sun position are within the thresholds, then the modification can fool AYL. Luckily. the motivation of falsifying the metadata of a photo is to use it for a chose event and the attacker may not be guaranteed to find such a combination.

Our framework can detect that the image shown in Fig. 1 was not taken at the claimed date and location. Although we do not have the required metadata (e.g., the time of day, the image direction and the camera orientation) and cannot estimate the azimuth angle as well as the exact sun position, we can estimate the altitude angle from the image. Given that the photo was claimed to be taken on or before January 16th in Florida, we can calculate the possible maximum altitude angle between January 1st and 16th to be 41°. Based on the image, we estimate the focal length to be 1287 pixels and the altitude angle to be 47.6°. The distance between the two estimates will be 6.6° which is larger than the threshold 4.8° in our experiments. Thus, we conclude that the date and location of this image were spoofed.

6.3 Discussions and Limitations

When estimating the altitude angle using the **one-shadow estimation** algorithm, an integrated vertical object and its shadow are required. However, vertical objects in the real world may not be absolutely vertical. By examining the scenario, we find that the direction of the sun ray is determined by a point on the object and the resulting point on the shadow. Even if the object is not exactly vertical, these two points are still able to decide the path of the sun ray and the altitude angle can be obtained from the sun ray and the shadow. Thus, we believe that our algorithm can eliminate this requirement, and AYL may not require to have the entire object in the photo if there exists a distinct point on the object.

In this paper, we assume that the camera is perpendicular to the ground and looks front horizontally or vertically when taking photos. Such assumption is used to simplify the algorithms for estimating the sun position. In fact, most smartphotnes are equipped with inertial sensors that have been widely used to estimate the orientation of the smartphone. If the sensor data is enclosed in the metadata, the device orientation can be obtained and used to determine the relationship between the device and the ground as well as the shadows.

A direction of future work is to estimate the sun position regardless of how cameras are oriented when taking photos.

7 Conclusion

We presented a new framework AYL which uses two estimations of sun position—*shadow-inferred sun position* and *metadata-inferred sun position*—to check whether the capture time and location of an outdoor image are true. Our framework exploits the relationship between the sun position in the sky and the time and location of an observer. We designed algorithms to obtain *shadow-inferred sun position* using only one vertical object and its shadow in the image. Our experiments show that the algorithms can estimate the sun position from shadows in the image with satisfactory accuracy. AYL utilizes both the altitude angle and azimuth angle for the consistency validation. The evaluation results guide us to choose the thresholds of altitude angle distance and sun position distance to be $4.8°$ and $9.2°$ respectively, which achieves combined (TPR, FPR) values of $(91.1\%, 7.7\%)$ for the consistency validation. We believe that our results illustrate the potential of using sun position to validate the consistency of the capture time and location. Our work raises an open question that whether other image contents can be leveraged for validating the consistency of image's contextual information.

References

1. Fema now has a rumor control section for misinformation. https://twitter.com/fema/status/264800761119113216
2. Bayram, S., Sencar, H.T., Memon, N.: An efficient and robust method for detecting copy-move forgery. In: IEEE ICASSP 2009, pp. 1053–1056. IEEE (2009)
3. Boididou, C., Papadopoulos, S., Kompatsiaris, Y., Schifferes, S., Newman, N.: Challenges of computational verification in social multimedia. In: ACM WWW 2014, pp. 743–748. ACM (2014)
4. Carvalho, T., Farid, H., Kee, E.: Exposing photo manipulation from user-guided 3D lighting analysis. In: SPIE/IS&T Electronic Imaging, p. 940902. International Society for Optics and Photonics (2015)
5. Dearden, L.: The fake refugee images that are being used to distort public opinion on asylum seekers, September 2015. http://www.independent.co.uk/news/world/europe/the-fake-refugee-images-that-are-being-used-to-distort-public-opinion-on-asylum-seekers-10503703.html
6. Fawcett, T.: An introduction to ROC analysis. Pattern Recogn. Lett. **27**(8), 861–874 (2006)
7. Ferrara, E.: Manipulation and abuse on social media by Emilio Ferrara with Chingman Au Yeung as coordinator. ACM SIGWEB Newslett. (Spring), Article No. 4 (2015)
8. Fridrich, J., Soukal, D., Lukas, J.: Detection of copy move forgery in digital images. In: Digital Forensic Research Workshop, August 2003

9. Garg, R., Varna, A.L., Wu, M.: Seeing ENF: natural time stamp for digital video via optical sensing and signal processing. In: ACM MM 2011, pp. 23–32. ACM (2011)
10. Gupta, A., Lamba, H., Kumaraguru, P., Joshi, A.: Faking sandy: characterizing and identifying fake images on twitter during hurricane sandy. In: ACM WWW 2013, pp. 729–736. ACM (2013)
11. Hartley, R., Zisserman, A.: Multiple View Geometry in Computer Vision, 2nd edn. Cambridge University Press, Cambridge (2004)
12. Harvey, P.: Exiftool. http://www.sno.phy.queensu.ca/~phil/exiftool/
13. Holbert, K.E., Srinivasan, D.: Solar energy calculations. In: Handbook Of Renewable Energy Technology, pp. 189–204 (2011)
14. Imran, M., Elbassuoni, S.M., Castillo, C., Diaz, F., Meier, P.: Extracting information nuggets from disaster-related messages in social media. In: Proceedings of ISCRAM, Baden-Baden, Germany (2013)
15. Iqbal, M.: An Introduction to Solar Radiation. Elsevier, Amsterdam (2012)
16. Japan Electronics and Information Technology Industries Association: Exchangeable image file format for digital still cameras: Exif Version 2.2, April 2002
17. Junejo, I.N., Foroosh, H.: Estimating geo-temporal location of stationary cameras using shadow trajectories. In: Forsyth, D., Torr, P., Zisserman, A. (eds.) ECCV 2008. LNCS, vol. 5302, pp. 318–331. Springer, Heidelberg (2008). doi:10.1007/978-3-540-88682-2_25
18. Kakar, P., Sudha, N.: Verifying temporal data in geotagged images via sun azimuth estimation. IEEE Trans. Inf. Forensics and Secur. **7**(3), 1029–1039 (2012)
19. Karttunen, H., Kroger, P., Oja, H., Poutanen, M., Donner, K.J.: Fundamental Astronomy, 5th edn. Springer, Heidelberg (2007)
20. Kee, E., O'brien, J.F., Farid, H.: Exposing photo manipulation from shading and shadows. ACM Trans. Graph. **33**(5), 165:1–165:21 (2014)
21. Metadata Working Group: Guidelines For Handling Image Metadata v2.0, November 2010
22. National Information Standards Organization: Understanding Metadata (2004)
23. Pan, X., Lyu, S.: Detecting image region duplication using sift features. In: IEEE ICASSP 2010, pp. 1706–1709. IEEE (2010)
24. Savoie, D.: Sundials: Design, Construction, and Use. Praxis Publishing, Chichester (2009)
25. Tian, J., Qu, C., Xu, W., Wang, S.: Kinwrite: handwriting-based authentication using kinect. In: NDSS 2013 (2013)
26. Tsai, T.H., Jhou, W.C., Cheng, W.H., Hu, M.C., Shen, I.C., Lim, T., Hua, K.L., Ghoneim, A., Hossain, M.A., Hidayati, S.C.: Photo sundial: estimating the time of capture in consumer photos. Neurocomputing **177**, 529–542 (2016)
27. Wu, L., Cao, X.: Geo-location estimation from two shadow trajectories. In: IEEE CVPR 2010, pp. 585–590 (2010)

Lights, Camera, Action! Exploring Effects of Visual Distractions on Completion of Security Tasks

Bruce Berg, Tyler Kaczmarek[✉], Alfred Kobsa, and Gene Tsudik

University of California Irvine, Irvine, CA, USA
{bgberg,tkaczmar,kobsa}@uci.edu, gts@ics.uci.edu

Abstract. Human errors in performing security-critical tasks are typically blamed on the complexity of those tasks. However, such errors can also occur because of (possibly unexpected) sensory distractions. A sensory distraction that produces negative effects can be abused by the adversary that controls the environment. Meanwhile, a distraction with positive effects can be artificially introduced to improve user performance.

The goal of this work is to explore the effects of visual stimuli on the performance of security-critical tasks. To this end, we experimented with a large number of subjects who were exposed to a range of unexpected visual stimuli while attempting to perform Bluetooth Pairing. Our results clearly demonstrate substantially increased task completion times and markedly lower task success rates. These negative effects are noteworthy, especially, when contrasted with prior results on audio distractions which had positive effects on performance of similar tasks. Experiments were conducted in a novel (fully automated and completely unattended) experimental environment. This yielded more uniform experiments, better scalability and significantly lower financial and logistical burdens. We discuss this experience, including benefits and limitations of the unattended automated experiment paradigm.

1 Introduction

It is widely believed that the human user is the weakest link in the security chain. Nonetheless, human participation is unavoidable in many security protocols. Such protocols require extensive usability testing, since users are unlikely to perform well when faced with overly difficult or intricate tasks. Typically, security-related usability testing entails evaluating human performance in a "best-case" scenario. In other words, testing is usually conducted in sterile lab-like environments.

At the same time, security protocols involving human users have become more commonplace. Examples include activities, such as: (1) using a personal device for verification of transaction amounts, (2) entering a PIN or a password and (3) solving a CAPTCHA, (4) comparing PINs when pairing Bluetooth devices, and (5) answering personal security questions.

© Springer International Publishing AG 2017
D. Gollmann et al. (Eds.): ACNS 2017, LNCS 10355, pp. 124–144, 2017.
DOI: 10.1007/978-3-319-61204-1_7

Since overall security of these tasks is determined by the human user (as the weakest link), extensive usability studies have been conducted. They aimed to assess users' ability to perform security tasks correctly and without undue delays, while providing an acceptable level of security [5,9,11,17].

However, the focus on maximizing successful protocol completion led developers to evaluate usability under contrived and unrealistic settings. In practice, security tasks can take place in noisy environments. In real-world settings, users are often exposed to various sensory stimuli. The impact of such stimuli on performance and completion of security tasks has not been well studied. A particular stimulus (e.g., a fire alarm or flickering lights) can be unintentional or hostile, i.e., introduced by the adversary that controls the physical environment. Furthermore, recent emergence of Internet of Things (IoT) devices (such as smart speakers and light fixtures) in home and office settings creates environments where compromised (malware-infected) devices can expose users to a variety of visual and audio stimuli.

There has been just one prior study that studied the effects of stimuli on the completion of security-critical tasks. It showed that introduction of unexpected audio stimuli during Bluetooth pairing actually improved subject performance [8]. This initial result, though interesting, motivates a more thorough study in order to fully understand the effects of a range of unexpected (and potentially malicious) stimuli.

Since modern user-aided security protocols focus on maximizing successful outcomes in an ideal environment, human errors are quite rare. For example, Uzun et al. [22] assume that:

> "...[A]ny non-zero fatal error rate in the sample size of 40 is unacceptable for security applications."

Consequently, numerous trials with many subjects are needed to gather data sufficient for making claims about human error rates. The scale is further exacerbated by the need to test multiple modalities, each with a distinct set of subjects. (This is because a given subject is less likely to make a similar mistake twice, even under different conditions.) Therefore, the number of required participants can quickly grow into hundreds, which presents a logistical challenge. To ease the burden of conducting a large-scale study, we designed and employed an entirely unattended and automated experimental setup, wherein subjects receive recorded instructions from a life-sized projection of a video-recorded experimenter ("avatar"), instead of a live experimenter.

We extensively experimented with subjects attempting to pair two Bluetooth devices (one of which was the subject's own device) in the presence of various unexpected visual stimuli. We tested a total of 169 subjects in the fully unattended experiment setting.[1] We initially hypothesized that visual stimuli would have beneficial or facilitatory effects on subject task completion, as was recently experienced with its audio counterpart [8]. Surprisingly, we discovered a marked

---

[1] All experiments described in this paper were fully authorized by the Institutional Review Board (IRB).

slowdown in task completion times across the board, and lower task success rates under certain stimuli.

The rest of the paper is organized as follows: The next section overviews related work and background material. Then, we present the design and setup of our experiments, followed by the presentation of our experimental results. Next, we derive conclusions and summarize lessons learned. The paper concludes with the discussion of limitations of our approach and directions for future work. Appendix 1 presents and analyzes performance of subjects arriving in groups. Appendix 2 contains the description of color spaces used to generate our stimuli. Details on the unattended experiment setup are in Appendix 3.

2 Background and Related Work

This section overviews related work in automated experiments, and human-assisted security methods. We also provide background information in psychology, particularly effects of sensory arousal on task performance, as well as effects of visual stimuli on arousal level and emotive state.

2.1 Automated Experiments

Other than recent results describing effects of audio distractions [8], we are unaware of any prior usability studies utilizing a fully automated and unattended physical environment.

However, some prior work reinforces validity of virtually-attended remote experiments and unattended online surveys, in contrast with same efforts in a traditional lab-based setting. Ollesch et al. [18] collected psychometric data in: (1) a physically attended experimental lab setting and (2) its virtually attended remote counterpart. No significant differences were found. This is further reinforced by Riva et al. [21] who compared data collected from (1) unattended online, and (2) attended offline, questionnaires. Finally, Lazem and Gracanin [14] replicated two classical social psychology experiments where both the participants and the experimenter were represented by avatars in Second Life[2], instead of being physically co-present. Here too, no significant differences were observed.

2.2 User Studies of Secure Device Pairing

Secure device pairing (mostly, but not only, via Bluetooth) has been extensively researched by experts in both security and usability. While initially pairing, the two devices have no prior knowledge of one another, i.e., there is no prior security context. Also, they can not rely on either a Trusted Third Party (TTP) or a Public Key Infrastructure (PKI) to facilitate the protocol. This makes device pairing especially vulnerable to man-in-the-middle (MiTM) attacks. This prompted the

---

[2] See secondlife.com.

design of numerous protocols requiring human involvement (integrity verification) over some out-of-band (OOB) channel, e.g., visual or audio comparison or copying/entering numbers.

For example, Short Authenticated String (SAS) protocols ask the user to compare two strings of about 20 bits each [13].

Uzun et al. [22] performed the first usability study of Bluetooth pairing techniques using SAS. It determined that the "compare-and-confirm" method – which involves the user comparing two 4-to-6-digit decimal numbers and indicating a match or lack thereof – was the most accurate and usable approach.

Kobsa et al. [11] compiled a comprehensive comparative usability study of eleven major secure device pairing methods. They measured task performance times, completion times, completion rates, perceived usability and perceived security. This led to the identification of most problematic as well as most effective pairing methods, for various device configurations.

Goodrich et al. [5] proposed an authentication protocol that used "Mad-Lib" style SAS. Each device in this protocol creates a nonsensical phrase based on the protocol outcome, and the user then determine if the two phrases match. This approach was found to be easier for non-specialist users.

Kainda et al. [9] examined usability of device pairing in a group setting. In this setting, up to 6 users tried to connect their devices to one another by participating in a SAS protocol. It was found that group effort decreased the expected rate of security and non-security failures. However, if a single individual was shown a SAS different from that of all others participants, the former often lied about the SAS in order to fit in with the group, demonstrating so-called "insecurity of conformity."

Gallego et al. [4] discovered that subject's performance in secure device pairing could be improved if it were to be scored. In other words, notifying subjects about their performance score resulted in fewer errors.

2.3 Effects of Sensory Stimulation

Sensory stimulation has variable impact on task performance. This is due to many contributing factors, including the subject's current level of arousal. The Yerkes-Dodson Law stipulates an inverse quadratic relationship between arousal and task performance [2]. It implies that, across all contributing stimulants, subjects who are either at a very low, or very high, level of arousal are not likely to perform well, and there exists an optimal level of arousal for correct task completion.

An extension to this law is the notion that completion of less complex tasks that produce lower levels of initial arousal in subjects benefits from inclusion of external stimuli. At the same time, completion of complex tasks that produce a high level of initial arousal suffers from the inclusion of external stimuli. Hockey [7] and Benignus et al. [1] classified this causal relationship by defining the complexity of a task as a function of the task's event rate (i.e., how many subtasks must be completed in a given time-frame) and the number of sources that originate these subtasks. External stimulation can serve to sharpen the focus of a subject at a low arousal level, improving task performance [19]. Conversely, it can overload subjects that are already at a high level of arousal, and induce errors in task completion [6].

O'Malley and Poplawsky [20] argued that sensory noise affects behavioral selectivity. Specifically, while a consistent positive or negative effect on task completion may not occur, a consistent negative effect was observed for tasks that require subjects to react to signals on their periphery. Meanwhile, a consistent positive effect on task completion was observed for tasks that require subjects to react to signals in the center of their field of attention. This leads to the claim that sensory stimulation has the effect of narrowing the subject's area of attention.

2.4   Unique Effects of Visual Stimuli

In addition to being general external stimuli that serve to raise arousal level, visual stimuli, particularly colors, have social and emotional implications. Naz and Epps [15] surveyed 98 college students about their emotional responses to five principal hues (red, blue, purple, green and yellow), five intermediate hues (yellow-red, green-yellow, blue-green, and red-purple) as well as three achromatic colors (white, gray, and black.) They found that principal hues are more likely to foster positive emotive responses. Furthermore, different colors within each group induce differing levels of arousal: some (red or green-yellow) increase arousal, while others (blue and green) are perceived as relaxing.

Moreover, visual stimuli were found to be dominating in multi-sensory contexts. Eimer [3] showed that in experiments with tactile, visual, and audio stimuli, subjects overwhelmingly utilized visual queues to localize tactile and auditory events.

3   Methodology

This section describes our experimental setup, procedures and subject parameters.

3.1   Apparatus

The experimental setting was designed to facilitate fully automated experiments with a wide range of sensory inputs. We located the experiment in a public, but low-traffic alcove at the top floor of the Computer Science Department building in a large public university. Figure 1 shows our setup from the subject's perspective (front view), and Fig. 2 depicts it from the side. More photos can be found in Appendix 2. The setup is comprised of readily available off-the-shelf components:

– A 60"-by-45" touch-sensitive interactive Smartboard (See footnote 2) whiteboard with a Hitachi CP-A300N short-throw projector (See footnote 2). The Smartboard acts as both an input and a display device. It reacts to tactile input, i.e., the user touches its surface, similar to a large touch-screen.

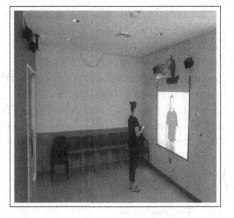

Fig. 1. Experimental environment: subject's perspective

Fig. 2. Experimental environment: side view

- A Logitech C920 HD Webcam (See footnote 2).
- Two pairs of BIC America RtR V44-2 speakers (See footnote 2): one alongside the smartboard, and the other – on the opposite wall. Their arrangement is such that the subject is typically standing in the center of the four speakers.
- Four programmable wirelessly controllable Phillips Hue A19 LED lightbulbs[3] to deliver the visual stimuli.

device. All prospective subjects were explicitly informed, during recruitment, that they would need to use their own personal device that supports Bluetooth communication. We could have instead provided a device to the subjects, which might have fostered a more uniform subject experience. However, there would have been some drawbacks:

- We wanted to avoid accidental errors due to the use of an unfamiliar device that might have a different user interface from that of the subject's own device. Mitigating this unfamiliarity would have required some training, which is incompatible with the unattended experiment setting.
- Virtually all current Bluetooth pairing scenarios involve at least one of the devices being owned by the person performing the pairing. Forcing subjects to use our device would have resulted in a more contrived or synthetic experience.
- From a purely practical perspective, an unattended portable device provided by us would have been more prone to damage or theft than other components, which are bulky and attached to walls and/or ceilings.

Not surprisingly, the majority of subjects' devices (152 out of 169) were smartphones. Tablets (13) and laptops (4) accounted for the rest.

---

[3] See: meethue.com for Hue Bulbs, smarttech.com for the Smartboard, logitech.com for the Webcam, bicamerica.com for speakers, and hitachi.com for the projector.

Bluetooth pairing is not as common as other security-critical tasks, such as password entry or CAPTCHA solving. However, we believe that Bluetooth pairing is the ideal security-critical task for the unattended experiment setup. It is preferred to passwords and PINs since it does not require subjects to reveal existing, or to select new, secrets. The security task at the core of Bluetooth pairing involves the user comparing two 6-digit decimal numbers – one displayed by each device being paired – and pressing a single button. This is a much more discrete and uniform activity than solving CAPTCHA-s, which vary widely in terms of difficulty and require higher-resolution displays as well as more extensive user input. These factors, even without external stimuli, would yield large variations in error rates and completion times.

3.2 Procedures

As mentioned earlier, instead of a live experimenter, we used a life-size video/audio recording of a experimenter giving instructions. This avatar is the subjects' only source of information about the experiment. Actual experimenter involvement is limited to strictly off-line activities, such as infrequent recalibration of avatar video volume and visual effects, as well as occasional repair of some components that suffered minor wear-and-tear damage throughout the study. This unattended setup allows the experiment to run without interruption 24/7 over a 5 month period.

Recall that the central goal of the experiment is to measure performance of subjects who attempt to pair their personal Bluetooth device to our Bluetooth device – an iMAC that uses the SmartBoard as an external display. This iMAC is hidden from the subject's view; it is situated directly on the other side of the SmartBoard wall in a separate office. During the pairing process, each subject is exposed to one randomly selected (from a fixed set) visual stimulus. This is done by rapid change in the ambient lighting of the room's four overhead lightbulbs to the chosen stimulus condition.

The experiment runs in four phases:

1. Initial: the subject walks in, presses a button on the wall which activates the experiment. Duration: instant.
2. Instruction: the avatar delivers instructions via Smartboard display and speakers. Duration: 45 s.
3. Pairing: the subject attempts to pair personal device with SmartBoard which represents the hidden iMAC desktop. In this phase, the subject is exposed to one (randomly selected out of 7) visual distraction stimulus. Duration: up to 3 min.
4. Final: the subject is prompted, on the SmartBoard, to enter some basic demographic information, as well as an email address to deliver the reward – an Amazon discount coupon. The information is entered directly into the Smart-Board, acting as a touch-screen input device. Duration: up to 6 min.

The total duration of the experiment ranged between 5 and 10 min.

In order to mitigate any disparities in task completion times between subjects that already had Bluetooth Discovery enabled and those who did not, the avatar informs subjects in the first 15 s of the instruction dialog that they will need to perform Bluetooth pairing with their personal device. This gives subjects over 30 s to enable Bluetooth Discovery Mode on thier device, if it is not enabled already.

We selected 6 visual effects that differed across two dimensions: color and intensity. In terms of color, we picked 3 values in the CIE chromatic space: Red, Blue, and Yellow-Green. Each is either *Solid*, i.e., shown at constant maximum intensity for the duration of the effect, or *Flickering*, i.e., its intensity grows and shrinks from the minimum to the maximum and back, completing one full cycle every second. In all settings, the maximum saturation was used. Color and intensity parameters for the 4 Phillips Hue bulbs under each condition are as follows (CCV stands for CIE Chromatic Value) [23]:

1. Red, CCV: $X = 0.674$, $Y = 0.322$
2. Blue, CCV: $X = 0.168$, $Y = 0.041$
3. Yellow-Green, CCV: $X = 0.408$, $Y = 0.517$
4. Solid intensity lumen output: 600 lm
5. Flickering intensity lumen range: 6 lm–600 lm

These color conditions were picked based on capabilities of programmable bulbs as well as background knowledge about emotive effects of color. Phillips Hue is an LED system based on creating white light. It can not create a blacklight effect or any achromatic light, which limits color selection to the subspace of the CIE color space [23] that Hue supports. (See Appendix 2 for more information).

With that restriction, we looked to the state-of-the-art about emotive reception and sensory effects of various colors in the Munsell color space [16]. (See Appendix 2 for more information). It has been shown that *principal hues* – Red, Yellow, Purple, Blue, and Green – are typically positively received. In contrast, *intermediate hues*, i.e., mixtures of any two principal hues, are more often negatively associated. Also, various colors have been shown to have either an arousing or a relaxing effect on subjects exposed to them. Based on this information, we chose three colors that differ as much as possible [15]:

- Red: Principal hue with positive emotional connotations, high associated arousal levels
- Blue: Principal hue with positive emotional connotations, low associated arousal levels
- Yellow-Green: Intermediate hue with negative emotional connotation, high associated arousal levels

Furthermore, we chose to have multiple modalities of light intensity for each color, with the expectation that a more complex modality would be more arousing and have a greater effect than its simple counterpart [12]. Not having found any previous work on the impact of exposure to colored light on performance of security-critical tasks, we include *Solid* light – the simplest modality of exposure that corresponds to the base level of stimulation. As a more complex modality, we included *Flickering* light.

Clearly, these two modalities were not the only possible choices. For example, it might have been intuitive to include even a more complex and startling *Strobing* light modality, achievable through rapid modulation of light intensity. It would have probably engendered a more profound impact on the subjects. However, ethical considerations coupled with the unattended nature of the experiment preclude using any modality that could endanger subjects with certain sensitivity conditions, such as photosensitive epilepsy. This led us to select a safe flickering frequency of 1 Hz.

We also found that all three light colors (under both intensity modalities) do not interfere with readability of a backlit personal wireless device or the image projected on the Smartboard. All experimenters, including one who used corrective lenses, could *correctly* read the screens of their personal devices, under all color conditions and intensity modalities.

3.3 Prior Results with a Similar Setup

A very similar setup was used in a previous study that assessed effects of unexpected audio distractions on 147 subjects performing Bluetooth pairing. As reported in [8], introduction of audio stimuli *significantly increased subject success rates* for every stimulus used. There was no significant impact on task completion time for any stimulus condition. This phenomenon was likely due to increased sensory arousal, as discussed in [8]. Our expectations for the impact of unexpected visual stimuli are rooted in these prior results.

3.4 Initial Hypotheses

We started out by hypothesizing that introduction of unexpected visual distractions during the process of human-aided pairing of two Bluetooth devices would have similar effects to those observed in prior experiments with audio distractions. Specifically, we expected two outcomes, as compared to a distraction-free setting:

[H1]: Lower error rates, and
[H2]: No effect on task completion times

3.5 Recruitment

The main challenge we encountered in the recruitment process is the scale of the experiments. Prior studies of usability of human-aided pairing protocols [5,9,17], demonstrated that 20–25 subjects per tested condition represents acceptable size for obtaining statistically significant findings. Our experiment has one condition for each of the six visual distraction variations, plus the control condition with no distractions. Therefore, collecting a meaningful amount of data requires at least 140 iterations of the experiment.

We used a four-pronged strategy to recruit subjects:

1. Email announcements sent to both graduate and undergraduate Computer Science students.
2. Posters placed (as signboards) near the entrance, and in the lobby, of a large campus building which housed the experimental setup.
3. Several instructors promoted participation in the experiment in their lectures.
4. Printed fliers handed out at various campus locations during daily peak pedestrian traffic times.

Recruitment efforts yielded 169 subjects in total, of whom 125 were male and 44 – female, corresponding to a 74%–26% gender split. This is expected, given that the location of our experimental setup was in the Computer Science and Engineering part of campus. Most subjects (161) were of college age (18–24 years), while 8 were in the 30+ group. This distribution is not surprising given the university population and the fact that older subjects generally correspond to researchers, faculty and staff, all of whom are much less likely to be attracted to being a subject in an experiment.

As follows from the above, our subjects' demographic was dominated by young, tech-savvy male undergraduate students.

4   Results

This section discusses the results, starting with data cleaning and proceeding to subject task completion effects.

4.1   Data Cleaning

We had to discard subject data for three reasons.

First, although instructions (in fliers, announcements and signs near the setup) specifically stated that subjects were to arrive alone, and perform the experiment without anyone else present, 37 groups (2 or more) of subjects participated. We found that the initial participant from each group performed in a manner consistent with individual subjects. However, subsequent group members who tried the experiment were (not surprisingly) significantly faster and more accurate in their task completion. Consequently, we discarded data of every subject who arrived in a group and was not the initial participant. We discuss this issue in more detail in Appendix 1.

The second reason for discarding data would have been due to subject auditory and/or visual impairment. A subject with an auditory impairment would have difficulties understanding the avatar's spoken instructions. A visually impaired subject would have difficulties with using the Smartboard and with the pairing process which relies on reading and comparing numbers. After carefully reviewing all subject video records, we could not identify any obvious visual or auditory impairment in any subject.

Some subjects successfully completed the experiment several times, perhaps hoping to receive multiple participation rewards. This occurred despite explicit instructions to the contrary. The system automatically rejected any repeated pairing attempts from devices already paired with the system, and any repeated attempts with different devices were discovered by visual inspection of subject trials. Every such repeated instance was discarded.

4.2 Task Failure Rate

Table 1 shows the number of subjects who, respectively, succeeded and failed at Bluetooth device pairing under each stimulus condition. It also details the failure rate for each condition.

Table 2 shows results from Barnard's exact test applied pairwise to the subject failure rate of the control condition and each stimulus. It demonstrates that differences between failure rates are statistically significant at the $\alpha = 0.05$ level with respect to all *Flickering* conditions: *Flickering Red*, *Flickering Blue*, and *Flickering Yellow-Green*. This even holds if we apply a conservative Bonferroni correction to account for three pairwise comparisons. This leads us to the mixed rejection of the initial hypothesis **H1**, as the failure rate increases significantly with the introduction of certain kinds of visual distractions, and remains unaffected by others. The next section discusses this further.

Table 3 shows odds ratios and 95% confidence interval for the failure rates under each stimulus, as compared to the control condition's failure rate. Interestingly, under this analysis, only the confidence intervals of *Flickering Blue* and *Flickering Yellow-Green* do not include a possible odds ratio of 1.0. Therefore – under this method of analysis – they are the only statistically significant stimuli at the $\alpha = 0.05$ level. The confidence interval defined for the *Flickering Red* condition challenges the claim of statistical significance at the $\alpha = 0.05$ level, as established by Barnard's exact test.

Table 1. Subject failure statistics

Stimulus	#Successful subjects	# failed subjects	Failure rate
None (control)	32	15	0.32
Solid Red	11	9	0.45
Flickering Red	9	11	0.55
Solid Blue	14	6	0.30
Flickering Blue	8	12	0.60
Solid Yellow-Green	10	12	0.54
Flickering Yellow-Green	7	13	0.65
Total	91	78	0.46

Table 2. Barnard's exact test on failure rates

Stimulus	Total pairings	Failure rate	Wald statistic	Nuisance parameter	p
None (control)	47	0.32	–	–	–
Solid Red	20	0.45	1.02	0.88	0.17
Flickering Red	20	0.55	1.77	0.86	0.04
Solid Blue	20	0.30	0.15	0.05	0.49
Flickering Blue	20	0.60	2.14	0.96	0.03
Solid Yellow-Green	22	0.54	1.79	0.94	0.06
Flickering Yellow-Green	20	0.65	2.51	0.91	0.01

Table 3. Subject failure rate by gender

Stimulus	Odds ratio wrt control	95% Confidence interval wrt control
None (control)	-	–
Solid Red	1.70	0.60-5.11
Flickering Red	2.61	0.89-7.63
Solid Blue	0.91	0.29-2.85
Flickering Blue	3.20	1.08–9.47
Solid Yellow-Green	1.79	0.91–7.24
Flickering Yellow-Green	3.96	1.31–11.6

Table 4. Subject failure rate by gender

Gender	# Successful subjects	# Unsuccessful subjects	Failure rate
Male	65	59	0.48
Female	25	20	0.44

We also examined subject failure rates by gender. As shown by Table 4 there is no statistically significant difference in failure rates between male and female participants; Wald statistic $= 0.36$, nuisance parameter $= 0.01$, $p = 0.46$.

4.3 Task Completion Times

Table 5[4] [5] shows average completion times in successful trials under each stimulus. After applying a conservative Bonferroni correction to account for six pairwise comparisons between individual stimulus conditions and the control condition, every stimulus condition shows an overwhelmingly large, statistically significant departure from the control condition. This results in rejection of hypothesis **H2**. The following section examines possible causes of this slowdown, as well as its implications.

Table 5. Avg times (sec) for successful pairing.

Stimulus	Mean time	Std Dev	DF wrt control	t-value wrt control	p
None	34.50	11.93	–	–	–
Solid Red	87.81	24.56	41	9.56	<0.001
Flickering Red	90.44	15.62	39	11.59	<0.001
Solid Blue	106.36	17.39	44	16.32	<0.001
Flickering Blue	91.25	24.11	38	9.61	<0.001
Solid Yellow-Green	90.30	19.08	40	11.1	<0.001
Flickering Yellow-Green	90.29	19.06	37	10.01	<0.001

Table 6. Cohen's d on completion times wrt Control

Stimulus	Cohen's d wrt control
None (control)	-
Solid Red	−3.42
Flickering Red	−4.49
Solid Blue	−5.33
Flickering Blue	−3.90
Solid Yellow-Green	−4.12
Flickering Yellow-Green	−4.29

---

[4] Std Dev = Standard Deviation.
[5] DF = Degrees of Freedom.

Table 6 shows Cohen's d for completion times under each stimulus when compared to the control condition. $|d| > 1.0$ in all cases, which means that every stimulus condition shows an overwhelmingly large, statistically significant departure from the control condition for the evaluation of Cohen's d. This result is statistically significant: it indicates that, with convincing probability, the mean completion time observed under the control is representative of a different distribution than that observed under each stimulus condition. This supports rejection of hypothesis **H2**.

Next, we looked into subject completion times for successful completion attempts by gender. Results are displayed in Table 7. A pairwise t-test shows that observed differences are not statistically significant; $t(84) = 0.04$, $p = 0.96$.

Table 7. Avg times (sec) by gender

Gender	Mean time	Standard deviation
Male	75.27	22.31
Female	75.20	24.10

Table 8. One-Way ANOVA test

	Sum of squares	DF	Mean square	F	p
Between groups	2964.28	5	592.86	1.466	0.217
Within groups	21440.33	53	404.535		
Total	24404.61	58			

Finally, we preformed Bartlett's test for homogeneity of variances as well as a One-Way analysis of variance (ANOVA) test between average task completion times of all stimulus conditions, excluding the control. Bartlett's test failed to reject the null hypothesis that all stimulus conditions share the same variance ($\chi^2 = 2.80$, $p = 0.731$). Furthermore, the one-way ANOVA test indicated no significant difference between any sample distributions (F = 1.466, p = 0.217.) Table 8 shows the results; their implications are discussed in the following section.

5   Discussion of Observed Effects

Several types of visual stimuli appear to have a negative effect on the subjects' successful completion of the Bluetooth Pairing task. However, collected data shows that this is not consistent across all stimuli. Instead, the negative effect may be tied to certain features of the particular stimulus. Instances of significant degradation in subject success rates were linked to the *Flickering* modality, for all color stimuli. This result implies that emotional perception of the stimulus may not be as much of a contributing factor to the overall increase of subject arousal as the presence of a dynamic visual stimulus. Also, in contrast with a previous study of audio distractions that observed positive effects [8], we noted no benefit to subject success rates under **any** visual stimulus.

These negative and neutral responses to static and dynamic light stimuli, respectively, are reinforced by the psychological concept of attentional selectivity. This concept assumes that the capture of an individual's attention by an aversive stimulus is likely to be momentary, occurring primarily when the stimulus is first introduced. In cognitive science, attention is conceptualized as a limited resource. For good evolutionary reason, the greatest demand on attention is in response to any change in one's environment. Once an assessment of the stimulus is made, and determined not to require additional action, attentional devotion to that stimulus fades quickly. This means that – while a static, adverse lighting change may remain adverse throughout its duration – its capacity to interfere with subject performance will fade rapidly after its onset. Instead, dynamically changing stimuli can more effectively capture subject attention and impair their performance, since many assessments are needed for many environmental changes occurring throughout the stimulus's duration.

Negative impact on subject task completion rates prompts a new attack vector for the adversary who controls ambient lighting. By taking advantage of color effects with shifting intensity levels, the adversary could force a user into failing Bluetooth pairing as a denial-of-service (DoS) attack. Moreover, the adversary might induce failure by using positively perceived colors of varying intensity. These colors may not even register as malicious in the user's mind, as they are innately associated with beneficial or pleasant emotions.

However, a much greater effect was observed in terms of average completion time. During review of subject trials, we noted that, upon exposure to the stimulus, subjects often take their gaze off their personal device (or the avatar) and focus their attention to the colorful, and possibly flickering, lights. The resulting delay frequently caused the subject's device to exit the Bluetooth pairing menu due to a time-out, and re-initiate the pairing protocol, resulting in much longer completion times overall.

Furthermore, as shown by Table 8, the introduced delay in subject task completion time was not based on the particular stimulus. Instead, the mere presence of a visual stimulus was enough to slow down successful subjects. Similar to the result in inducing user failure, the adversary is not forced to rely on an overtly malicious stimulus in order to cause substantial slowdown in task completion. However, the adversary has even more choices in stimulus selection, since all stimuli (including those with static intensity levels) were shown to impact task completion times the same way.

This effect shows further power for the adversary in control of ambient lighting. One possibility is that the adversary's goal is a denial-of-service attack by frustrating user's pairing attempts. In a more sinister scenario the adversary could try to "buy time" by introducing its own malicious device(s) alongside changes to ambient lighting and then leverage the user's lapse in focus (when being exposed to new sensory stimuli) to trick the user into pairing with that device. In the worst case, the adversary might take advantage of the user's inattentiveness while their gaze shifts away from their device and trick them into accepting a non-matching authenticator.

6   Unattended Setup: Limitations

Based on our earlier discussion of Data Cleaning, some subjects' data had to be removed from the dataset because they did not conduct the experiment alone. This occurred even though all recruitment materials (and means) as well as the avatar's instructions stated that subjects were to perform the task alone. This illustrates a basic limitation of the unattended setup: no one is present to enforce the rules in real time.[6]

We did not manage to capture fine-grained data about the subjects' awareness of a distraction. We have some anecdotal evidence from video recordings showing that some subjects noticed the distraction in obvious ways, e.g., verbal remarks or turning their heads. However, we have no evidence of subjects who failed to notice the stimulus. Information about subjects noticing a change in the environment is very important to the development of a realistic adversary model for future studies.

7   Study Shortcomings

In this section we discuss some shortcomings of our study.

7.1   Homogeneous Subjects

Our subject group was dominated by young, tech-savvy male college students. This is a consequence of the experiment's location. Replication of our experiment in a non-academic setting would be useful. However, recruiting a really diverse group of subjects is hard. Ideal venues might be stadiums, concert halls, fairgrounds or shopping malls. Unfortunately, deployment of our unattended setup in such public locations is logistically infeasible. Since these public areas already have many sensory stimuli, reliable adjustment of our subjects' arousal level in a consistent manner would be very hard. Furthermore, it would be very difficult to secure specialized and expensive experimental equipment.

In addition to being tech-savvy, young subjects are in general more apt to quickly recover from changes in the lighting of their surroundings than older adults [10]. It is possible that unexpected visual stimuli would have a different effect on an older (less technologically adept) population.

7.2   Sufficiently Diverse Stimuli

We selected six conditions to obtain as many diverse stimuli types as we could rigorously test, in addition to control. We first varied them by changing the regularity of the stimulus, expecting that a varying signal would have greater

---

[6] However, it would have been possible (though quite difficult in practice) to instrument our recording of the experiment to abort upon detecting simultaneous presence of multiple subjects.

impact on subjects' arousal than a steady signal. We then varied the colors, with the expectation that using colors that evoked different emotive responses and general arousal levels would impact task performance differently.

An ideal experiment would have included a stimulus with negative emotional connotation and low arousal levels. However, between three colors, two intensity conditions, and the control, we had seven total conditions to test. Furthermore, due to the nature of our experiment, we could only reasonably expect each subject to be tested under a single condition, since prior knowledge about the experiment would clearly bias the results. Adding just one additional stimulus (for both intensity modalities) would have required at least 40 more subjects. This would have placed a heavy logistical burden for our already nearly-depleted subject pool.

We also note that variance in intensity of our flickering modality did not approach the technical limit of Philips Hue bulbs. Instead, we deliberately limited the frequency of intensity fluctuations to 1Hz in order to avoid any possible negative reaction from light-sensitive subjects. This ethical issue does not reflect real-world conditions where an adversary (with no ethical qualms) could create a very fast strobing effect, possibly causing physical harm.

7.3   Synthetic Environment

Our unattended setup, while a step closer to an everyday setting than a sterile and highly controlled lab, is still quite synthetic. First, our choice to place it in a low-traffic area makes it quieter than many common settings. Second, our choice to situate it indoors makes it free of temperature fluctuations, air flow, and exposure to sunlight. Finally, our equipment (such as the Smartboard projector system) is not commonly encountered by most subjects.

7.4   Ideal Setting

Drawing upon aforementioned shortcomings, the ideal setting for our experiment would be one where:

- Subject demographics are more varied
- Subjects are not aware of the nature of the experiment until they are debriefed after task completion
- The environment is more commonplace
- The task is more security-critical

All of these criteria could be trivially met if, for example, we conducted the experiment at a busy bank ATM. The task at hand would be the obviously security-critical entry of the subject's PIN. A modern ATM comes standard with all of the features needed for our experiment: it has a keypad, a screen, a speaker (for visually impaired users), a video camera, and are in areas that are artificially lit. Similarly, a busy gas station would fit our needs, as each fuel pump typically includes a keypad for PIN entry, speakers, a screen, artificial lighting, and a video camera recording the transaction. However, despite their attractive qualities, there would be serious ethical and logistical obstacles to setting up an unattended automated experiment in one these location examples.

8   Conclusions and Future Work

As human participation in security-critical tasks becomes more commonplace, so does the incidence of users performing these tasks while subject to accidental or malicious distractors. This strongly motivates exploring user error rates and their reactions to various external stimuli. Our efforts described in this paper shed some light on understanding human errors in security-critical tasks by studying the effects of visual stimuli on users attempting to pair two Bluetooth devices.

We feel that this unattended experiment paradigm is a valuable approach that deserves further study. The development of standardized unattended and automated experimental setups could greatly lower the logistical and financial burdens associated with conducting large-scale user studies.

Given the observed negative effect on subject completion times, one interesting next step would be to conduct a similar experiment, where, instead of measuring subjects' ability to pair Bluetooth devices, we would examine the rates of incorrect pairing when the subjects are shown mis-matched numbers during the pairing process. This could help us determine whether (and how) visual distractions make users more likely to pair their device to some other (perhaps adversary-controlled) device.

Another direction is investigating effects of hybrid (e.g., audio/visual) distractions. Finally, we plan to conduct a study of subjects performing security-critical tasks, while being exposed to *multiple visual stimuli* lasting longer than 3 min. This might allow us to learn whether subjects' sensory arousal is the result of the surprise (due to the sudden visual stimulus), or an unavoidable psychophysical reaction.

Acknowledgments. This research has been supported by NSF Grant CNS-1544373.

Appendix 1: Analysis of Group Initiators

We considered potential differences in failure rates between subjects who performed the task alone, and those who did it as part of a group. As mentioned in the discussion of Data Cleaning, for each group, we only consider the initial participating group member, referred to as the Group Initiator. As Table 9 shows, there is no significant difference between failure rates of individual subjects and Group Initiators; Wald Statistic $= 0.34$, Nuisance parameter $= 0.01$, $p = 0.51$. Furthermore, as Table 10 shows, a pairwise t-test of completion times for individuals – compared to group initiators – shows that observed differences are not statistically significant; $t(84) = 0.09$, $p = 0.93$.

Table 9. Failure rates: initiators vs. individuals

Participant type	#Successful subjects	#Unsuccessful subjects	Failure rate
Group initiator	19	18	0.49
Individual	72	60	0.45

Table 10. Avg times (sec): initiators vs. individuals

Participant type	Mean time	Standard deviation
Group initiator	76.63	23.00
Individual	76.20	17.93

Appendix 2: A Few Colorful Words

Munsell Color System

The Munsell Color System is used for creating and describing colors. In it, all colors are grouped into two categories: primary and intermediate hues. Primary hues include: Red, Yellow, Purple, Blue, and Green, arranged in a circular shape as in Fig. 3. Intermediate hues are mixtures of two adjacent primary hues, such as Yellow-Green or Purple-Blue. Colors are defined on three dimensions: hue, lightness, and color purity. The Munsell system is based on human perception which makes it useful for rigorously defining human reaction to specific color forms. However basing the system on human perception makes the Munsell system a poor tool for direct conversion of light described by its physical wavelength into human-perceptible color.

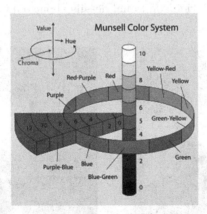

Fig. 3. Munsell color space (Image best viewed in color)

CIE Color Space

The Phillips Hue bulbs use the CIE color space. In CIE, colors are defined as a 2-dimensional space with X and Y values moving along a roughly triangular curve that corresponds to the translation of wavelengths of light to their human perception in the visible spectrum. The exact color range of the Philips Hue bulb is shown in Fig. 4.

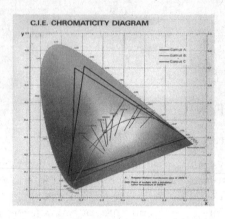

Fig. 4. Phillips Hue CIE color space (Image best viewed in color)

Appendix 3: Unattended Experiment Setup

Figures 5, 6, 7 and 8 provide additional details about our experimental setup.

Fig. 5. The experiment environment during the Solid Blue condition (Image best viewed in color)

Fig. 6. The subject's perspective during the Solid Red condition (Image best viewed in color.)

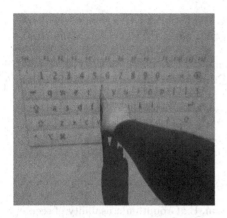 

Fig. 7. Subject entering email address on Smartboard

Fig. 8. Post-experimental review of video recordings (separate office)

References

1. Benignus, V.A., Otto, D.A., Knelson, J.H.: Effect of low-frequency random noises on performance of a numeric monitoring task. Percept. Mot. Skills **40**(1), 231–239 (1975)
2. Cohen, R.A.: Yerkes-Dodson law. In: Kreutzer, J.S., DeLuca, J., Caplan, B. (eds.) Encyclopedia of Clinical Neuropsychology, pp. 2737–2738. Springer, New York (2011)
3. Eimer, M.: Multisensory integration: how visual experience shapes spatial perception. Curr. Biol. **14**(3), R115–R117 (2004)
4. Gallego, A., Saxena, N., Voris, J.: Exploring extrinsic motivation for better security: a usability study of scoring-enhanced device pairing. In: Sadeghi, A.-R. (ed.) FC 2013. LNCS, vol. 7859, pp. 60–68. Springer, Heidelberg (2013). doi:10.1007/978-3-642-39884-1_6
5. Goodrich, M.T., Sirivianos, M., Solis, J., Soriente, C., Tsudik, G., Uzun, E.: Using audio in secure device pairing. Int. J. Secure. Netw. **4**(1), 57–68 (2009)
6. Harris, W., Stress, P.: The effects of intense noise stimulation and noxious stimulation upon perceptual performance. Ph.D. thesis, University of Southern California (1960)
7. Hockey, G.R.J.: Effect of loud noise on attentional selectivity. Q. J. Exp. Psychol. **22**(1), 28–36 (1970)
8. Kaczmarek, T., Kobsa, A., Sy, R., Tsudik, G.: An unattended study of users performing security critical tasks under adversarial noise. In: Proceedings of the NDSS Workshop on Useable Security, pp. 14:1–14:12 (2015)
9. Kainda, R., Flechais, I., Roscoe, A.W.: Usability and security of out-of-band channels in secure device pairing protocols. In: Proceedings of the 5th Symposium on Usable Privacy and Security, pp. 11:1–11:12 (2009). ACM ID: 1572547
10. Kline, D.W., Schieber F.: Vision and aging (1985)
11. Kobsa, A., Sonawalla, R., Tsudik, G., Uzun, E., Wang, Y.: Serial hook-ups: a comparative usability study of secure device pairing methods. In: Proceedings of the 5th Symposium on Usable Privacy and Security, pp. 10:1–10:12 (2009). ACM ID: 1572546

12. Koelega, H.S., Brinkman, J.-A., Zwep, B., Verbaten, M.N.: Dynamic vs static stimuli in their effect on visual vigilance performance. Percept. Mot. Skills **70**(3), 823–831 (1990)
13. Laur, S., Asokan, N., Nyberg, K.: Efficient mutual data authentication using manually authenticated strings. Cryptology ePrint Archive, report 2005/424 (2005). http://eprint.iacr.org/
14. Lazem, S., Gracanin, D.: Social traps in second life. In: 2010 Second International Conference on Games and Virtual Worlds for Serious Applications (VS-GAMES), pp. 133–140, March 2010
15. Naz, K., Epps, H.: Relationship between color and emotion: a study of college students. Coll. Stud. J. **38**(3), 396 (2004)
16. Nickerson, D.: History of the Munsell color system and its scientific application. J. Opt. Soc., 575–586 (1940)
17. Nithyanand, R., Saxena, N., Tsudik, G., Uzun, E.: Groupthink: usability of secure group association for wireless devices. In: Proceedings of the 12th ACM International Conference on Ubiquitous Computing, pp. 331–340 (2010). ACM ID: 1864399
18. Ollesch, H., Heineken, E., Schulte, F.P.: Physical or virtual presence of the experimenter: psychological online-experiments in different settings. Int. J. Internet Sci. **1**(1), 71–81 (2006)
19. Olmedo, E.L., Kirk, R.E.: Maintenance of vigilance by non-task-related stimulation in the monitoring environment. Percept. Mot. Skills **44**(3), 715–723 (1977)
20. O'Malley, J.J., Poplawsky, A.: Noise-induced arousal and breadth of attention. Percept. Mot. Skills **33**(3), 887–890 (1971)
21. Riva, G., Teruzzi, T., Anolli, L.: The use of the internet in psychological research: comparison of online and offline questionnaires. CyberPsychol. Behav. **6**(1), 73–80 (2003)
22. Uzun, E., Karvonen, K., Asokan, N.: Usability analysis of secure pairing methods. In: Dietrich, S., Dhamija, R. (eds.) FC 2007. LNCS, vol. 4886, pp. 307–324. Springer, Heidelberg (2007). doi:10.1007/978-3-540-77366-5_29
23. Wyszecki, G., Stiles, W.S.: Color Science, vol. 8. Wiley, New York (1982)

A Pilot Study of Multiple Password Interference Between Text and Map-Based Passwords

Weizhi Meng[1(✉)], Wenjuan Li[2], Wang Hao Lee[3], Lijun Jiang[2],
and Jianying Zhou[4]

[1] Department of Applied Mathematics and Computer Science,
Technical University of Denmark, Kongens Lyngby, Denmark
weme@dtu.dk

[2] Department of Computer Science, City University of Hong Kong,
Kowloon Tong, Hong Kong

[3] Infocomm Security Department, Institute for Infocomm Research,
Singapore, Singapore

[4] Singapore University of Technology and Design, Singapore, Singapore
jianying_zhou@sutd.edu.sg

Abstract. Today's computer users have to remember several passwords for each of their accounts. It is easily noticed that people may have difficulty in remembering multiple passwords, which result in a weak password selection. Previous studies have shown that recall success rates are not statistically dissimilar between textual passwords and graphical passwords. With the advent of map-based graphical passwords, this paper focuses on multiple password interference and presents a pilot study consisting of 60 participants to study the recall of multiple passwords between text passwords and map-based passwords under various account scenarios. Each participant has to create six distinct passwords for different account scenarios. It is found that participants in the map-based graphical password scheme could perform better than the textual password scheme in both short-term (one-hour session) and long term (after two weeks) password memorability tests (i.e., they made higher success rates). Our effort attempts to complement existing studies and stimulate more research on this issue.

Keywords: User authentication · Graphical passwords · Usable security · Multiple password interference · HCI

1 Introduction

Over the past few decades, text-based passwords are the most widely adopted method for user authentication. However, users may suffer from many issues when using text or pattern in the aspects of security and usability [6,35,38]. As an example, users find it difficult to remember their textual information for a long time due to the long-term memory (LTM) limitation [37]. As a result, they are likely to choose and use weak textual passwords. To mitigate these

© Springer International Publishing AG 2017
D. Gollmann et al. (Eds.): ACNS 2017, LNCS 10355, pp. 145–162, 2017.
DOI: 10.1007/978-3-319-61204-1_8

drawbacks, graphical passwords (GPs) have been proposed as an alternative in the literature, where users are able to interact with images during the registration and authentication [29]. Generally, graphical passwords can be categorized as recognition-based, pure recall-based and cued recall passwords.

Many research studies have shown that graphical passwords can indeed help users in remembering passwords (see Sect. 2). However, nowadays users have to remember not just one password, but many, which would add a significant burden on users' memory. As a result, the concern of *multiple password interference* is raised, where for users, remembering a password for an account (e.g., email account) affects their memory of other accounts' passwords (e.g., Facebook account). In the literature, this issue has not been widely investigated, whilst one previous research had paid attention to this issue. They made a study between text-based passwords and click-based graphical passwords (*PassPoints*) [8]. In particular, the click-based GP system requires users to click on several points on an image rather than enter textual information [36]. They found that the click-based GPs were significantly less susceptible to multiple password interference in the short-term, while the results were not statistically different from textual passwords in the long-term (i.e., after two weeks).

Our current work was motivated by two observations. (1) In the literature, there have been few studies investigating this important issue of multiple password memory in the area of graphical passwords. (2) Map-based graphical passwords have been recently developed, where users can click on several locations on a world map as their secrets. A map is believed to provide better recall for users, but has not been investigated for multiple password memory. Based on this claim, our main goal of this paper is to investigate the multiple password interference between text-based passwords and map-based graphical passwords. More specifically, we develop a prototype of map-based GP in our study, which allows a sequence of three click-points on a digital world map. Furthermore, we follow the steps in the most relevant work [8], in order to facilitate the comparison and validation of the collected results.

More formally, *memory interference* can be described as "the impaired ability to remember an item when it is similar to other items stored in memory" [3]. In this work, our primary goal is to investigate the multiple password interference in text passwords and map-based GPs. The contributions of our paper are summarized below:

- A user study consisting of 60 participants was conducted to investigate the multiple password interference issue between textual passwords and map-based graphical passwords; where 50 of them successfully return after two weeks to test recall of those passwords they had created. They were assigned to use either text password system or map-based GP system. Each participant should create six distinct passwords for different account scenarios.
- It is found that participants can perform better in recalling multiple passwords using the map-based graphical passwords, not only in the short-term (e.g., a one-hour session) but also in the long-term (namely two weeks). In comparison with the results from former study (e.g., PassPoints), our results

indicate that map-based graphical passwords can offer a better and positive impact on recalling multiple passwords in both short and long term.

It is worth emphasizing that our study aims to complement existing research and reveals that recall of multiple passwords between text and map-based graphical password schemes has a statistically significant difference in both short and long term. The results also encourage research studies to develop proper map-based schemes to improve the issue of multiple password interference.

The remainder of the paper is organized as follows. Section 2 describes related work about graphical password classification, map-based graphical passwords and multiple password interference. Section 3 presents our developed map-based GP system and introduces our study methodology including procedure and steps. Section 4 analyzes the collected results. We provide a discussion in Sect. 5 and present some limitations in Sect. 6. Finally, we conclude our paper in Sect. 7.

2   Related Work

2.1   Graphical Passwords

Graphical password schemes can be classified into three broad categories [5,7]: namely, recognition-based, pure recall-based and cued recall-based.

- *Recognition-based GPs.* This kind of scheme requires users to select images from a large gallery. For example, PassFaces [10,25] lets users identify a set of human faces during the authentication phase. To create a password, the user chooses four images of human faces from a fixed portfolio of faces. The Story scheme [10] lets users identify a few images with a theme (e.g., people and food) from an image gallery.
- *Pure recall-based GPs.* Such kind of scheme lets users draw a secret on an image. An example is the DAS proposed by Jermyn et al. [17], where users draw the password on a grid. Tao and Adams [30] proposed Pass-Go that lets users select intersections within a grid as a way to enter their password. Based on Pass-Go, Android unlock patterns are developed on Android phones, which are a tuned application requiring users to unlock their phones by inputting correct patterns. Then, Dunphy and Yan [12] explored whether a background image can improve the performance of graphical passwords.
- *Cued recall-based GPs.* This kind of scheme requires users to click on a sequence of points to construct their passwords. *PassPoints* by Wiedenbeck et al. [36] belongs to this category. In *PassPoints*, users recall a sequence of five selected points by clicking on them. For authentication, users have to click close to the chosen click points within some (adjustable) tolerance distance. Later, Chiasson et al. [9] proposed *Persuasive Cued Click-Points (PCCP)*, which lets users click a point on each of a sequence of background images.

The current graphical password schemes are designed around the actions of click, choice or draw. Combinations of these schemes have also been studied. For instance, Meng [20] developed a click-draw based graphical password scheme (CD-GPS) with the purpose of improving the image-based authentication in both security and usability by combining the actions of clicking, selecting and drawing. Several new GP systems can be referred to [14,18,19,23]. There are many studies focusing on security aspect of graphical passwords [4,16,17,26,32, 33] and relevant user behaviors [1,11,21,27,37,39].

2.2 Map-Based Graphical Passwords

Map-based GPs can be categorized as a kind of cued recall-based graphical password, in which users can recall their selected places on a world map. Despite some early research work, map-based GPs start becoming popular in recent years. In 2012, Georgakakis et al. [15] proposed a map-based graphical password scheme called *NAVI*, where the credentials of the user are his username and password formulated by drawing a route on a predefined Google map. In particular, a user selects the starting and the ending point, so that the route is calculated by the provider of the route planning service.

During the same time, Sun et al. [28] extended the above idea and suggested to use an extremely large image as the password space. They proposed a world map based graphical password authentication system called *PassMap*, in which a password consists of a sequence of two click-points on a large world map. To construct secrets, users can shift the world map to selected areas, and zoom the map to the desired zoom-level. An example of *PassMap* is shown in Fig. 1. Their collected results showed that after a week, the accuracy of login is 92.59%, and claimed that the performance was better than *PassPoints*. Later, Thorpe et al. [31] designed *GeoPass*, an interface for digital map-based authentication, where users can choose a single place as his or her password. For this scheme, users only have to remember the final location, rather than their method of navigating there. They included 35 participants in their study and found that nearly 97% of users could correctly remember their location-based-passwords over 8–9 days with very few failed login attempts. It is worth noting that *PassMap* and *GeoPass* are very similar in that secrets are constructed by clicking one or two places on a world map (e.g., Google map).

Fig. 1. An example of *PassMap*. The first click-point with red (Left) and the second click with red (Right). (Color figure online)

Recently, Meng [22] proposed *RouteMap* for better multiple password memory, which allows users to draw a route on a map as their password. They found that users could achieve better performance using *RouteMap* in terms of multiple password memory. From these studies, it is noticeable that users may have a better memorability regarding map-based graphical passwords.

2.3   Multiple Password Interference

Today, users have to remember more passwords rather than before in the course of their daily lives such as social networking accounts, personal email accounts, commercial email accounts and so on. In these cases, *multiple password interference* may become an important issue.

In the text password scheme, Vu et al. [34] had explored the memorability of multiple textual passwords for different accounts. The study discovered that users who were given five passwords have difficulty remembering them compared to those who only had to remember three. It is also revealed that users often selected passwords that have direct contextual relationship with the account(s) in question.

For the graphical password scheme, Everitt et al. [13] presented an early study on how frequency of access and interference affects recognition-based graphical passwords such as frequency of access to a graphical password. They employed *PassFaces* and discovered that many factors can noticeably impact authenticating with multiple facial graphical passwords. For example, participants who accessed four different graphical passwords per week were ten times more likely to completely fail to authenticate than participants who accessed a single password once per week.

Chiasson et al. [8] has conducted a study on comparing the recall of multiple text passwords against that of multiple click-based graphical passwords (namely *PassPoints*). Six account scenarios were simulated and they found that in the short-term, participants in the graphical password scheme performed significantly better than participants using the textual password scheme. They made fewer errors when recalling their graphical passwords. However, they found that recall success rates were not statistically different from each other after two weeks. More recently, Al-Ameen and Wright [2] showed that mental stories could be used to improve the interference issue in *GeoPass*. Meng et al. [24] presented a study on the memorability of multiple passwords between textual passwords and unlock patterns. The study also explored the difficulty of remembering those patterns for a prolonged period of time.

It is worth noting that recent research studies had showed that the performance of map-based GPs could be better than *PassPoints*; thus our current work is timely for the investigation of multiple password interference between textual passwords and map-based GPs. Our work is different from [2,24], as our used map-based scheme is not the same (i.e., *GeoPass* only allows one location to be selected). Thus, our study can complement existing research results.

3   User Study

Following the methodology in the literature [8], we conducted a lab-based user study involving two sessions: Session-1 and Session-2. All participants were students from a university who had no prior information security training, as well as no prior experience with graphical passwords.

The first session, Session-1 lasted for approximately an hour, recruiting a total of 60 participants ($M = 25.6$ years; $SD = 3$; 32 females). During the study, we randomly selected 30 participants to the text password scheme (MText), while the others were assigned to the graphical password scheme (MMP). For Session-2, after two weeks, 50 participants were successfully returned to the lab and recalled their created passwords. Also, we provide a feedback form for each session asking for users' feedback and attitude.

3.1   Implementation of Map-Based Graphical Passwords

As the source of existing map-based graphical passwords was not released, we developed a prototype of map-based graphical passwords in our lab computers with a 17-inch screen (Intel R, CPU 2.6 GHz). It is worth noting that our system provides a common platform that is able to realize several existing map-based GPs such as *PassMap* and *GeoPass*.

To obtain a world map, we utilized the JavaScript based Google Maps API, which provides an extensive move-by-dragging, zooming and search functions. When users zoom in/out on the map, our system reports the zoom-levels. The search function allows users to shift to a specific part of the map quickly and further zoom in to locate a specific area. Similar to [28], our system embedded a 640×420 pixel frame block for displaying the world map in a web page and road map instead of satellite-type map is used by default. The tolerance areas are 21×21 pixels.

We describe the registration and login phase of our developed map-based GP in Fig. 2 and Fig. 3, respectively.

- *Registration.* Figure 2 depicts that users are able to choose three locations on a world map in constructing their passwords. All the information like coordinates will be forwarded and stored in a database.
- *Login.* Figure 3 presents that users have to submit their names and three locations to the system for authentication. The system will compare the inputs with the stored information (i.e., checking whether the inputs are within tolerance). A user is successfully verified only if both credentials are correct.

It is worth noting that *PassPoints* requires users to select five points on an image as their secrets, *GeoPass* requires users to select one place while *PassMap* needs users to click two places. Intuitively, we consider the memory of one place in a map should be quite easy. To make a better comparison with *PassPoints*, our map-based GP system increases the memory burden a bit and thus demands users to choose three locations on a world map (as a study).

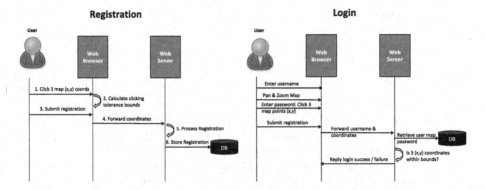

Fig. 2. Our developed system of map-based GP: registration phase.

Fig. 3. Our developed system of map-based GP: login phase.

3.2 Session-1: Procedure and Steps

Session-1 consists of three phases: password practice, password creation and password retention. During the first phase, participants must complete two trials as practice to get them familiarized with the graphical password creation. In this phase, participants are not required to remember their practice passwords.

In the second phase (password generation), each participant has to complete six trials. Each password is associated with a different account scenario. A total of six account scenarios are constructed: email, bank, instant messenger, library, online dating and work (commercial). The accounts were identified by distinct banners at the top of the system interface. For map-based GP system, Fig. 4 depicts an example of registration interface with library account. It is worth noting that the banners are the same for text and map-based passwords.

In Fig. 5, we detail the process of registration, where Fig. 5(a) shows the interface that requires users to input their names and Fig. 5(b) allows users to select locations on a Google world map to create their secrets. After selecting locations, the click points including X and Y coordinates and zoom-level will be

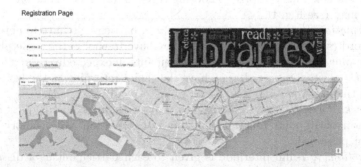

Fig. 4. The interface of map-based password creation.

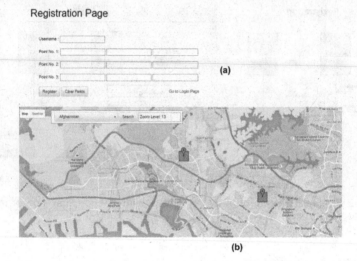

Fig. 5. Registration process: (a) inputting user name and (b) selecting locations.

shown in Fig. 5(a). To facilitate location search, our system provides a search bar in the map (see Fig. 5(b)) as well as zoom-level selection. Besides, users can switch the map between road and satellite map background.

In this stage, all accounts were presented to participants in the same order. In addition, participants were asked to construct their passwords which they could remember but would be difficult for attackers to guess. All participants were told that they would have to remember their created passwords for later use and each password was connected to a specific account scenario. Participants perform the steps below for each account:

- *Creation.* Participants have to create a password for each account. The textual password must consist of at least eight characters (this number is selected based on [8]) and entered twice. The password text was made visible during the creation process. For the graphical password, participants follow the steps in our system. This comprises choosing three different locations on a Google world map (see Fig. 4).
- *Confirmation.* In the text password scenario, participants have to enter their password (shown as asterisks), while they have to choose the same places in the graphical password scheme. If participants unexpectedly cannot remember their created password, they have the choice returning to the last step to create the password again.
- *Feedback.* Participants were invited to provide feedback to our questions (in the form of 10-point Likert-scale) about the creation and memory of the current passwords.
- *Distraction.* A 2-min interlude is given to each participant as a distraction task. First of all, participants counted backwards by 3 s from a random 4-digit number, and then participants were given a Mental Rotations Test puzzle

(MRT). Similar to [8], the purpose of these tasks was to clear participants' visual and textual working memory.

- *Login.* Participants were then invited to login with their created passwords. In the event that the password is mis-entered, they are allowed to re-enter their passwords multiple times with no pre-defined limit. If their created secrets are forgotten, they are allowed to return to the creation step.

Before the start of retention phase, participants were given a 10-min break where they were given a demographics questionnaire and could leave the experiment venue for the remaining break time. Upon returning from the break, account scenarios were presented to participants in a re-shuffled order with the help of a Latin square, where each row is connected with a participant and each column is connected with an account. All participants are required to recall each of their six textual or graphical passwords.

After this phase, participants were given another questionnaire to solicit feedback about their attitudes towards system usage and password memory.

3.3 Session-2: Procedure and Steps

Up to 50 participants returned to our lab after two weeks, among which 22 of them were previously assigned to the text password scheme. This session consists mainly of the 2-week retention phase, which follows the similar steps as per the initial retention phase during Session-1. The account scenarios were presented in the same order.

Similarly, participants are allowed to try multiple times with no limit to recall their created passwords until they gave up trying to recall. After the end of the session, a questionnaire was given to collect their feedback about password recall in these two schemes.

4   Results and Analysis

The Chi-square (χ^2) tests are mainly used to analyze data from the user studies. In all cases, we regard a value of $\rho < 0.05$ indicating that the results are statistically different between two schemes. In particular, we employ success rates as one of the major measures to evaluate participants' performance. The success rates for each step are presented in Table 1.

Table 1. The success rates of login, recall-1, recall-2 and password-confirmation. MText: multiple text passwords. MMP: multiple map-based passwords.

	Confirmation		Login		Recall-1			Recall-2		
	MText	MMP	MText	MMP	MText	MMP	χ^2	MText	MMP	χ^2
First attempt	97%	97%	94%	99%	62%	96%	$\chi^2 = 43.2,\ \rho < 0.05$	43%	72%	$\chi^2 = 30.1,\ \rho < 0.05$
Three attempts	99%	100%	98%	100%	86%	98%	$\chi^2 = 27.4,\ \rho < 0.05$	55%	79%	$\chi^2 = 23.3,\ \rho < 0.05$
More attempts	100%	100%	99%	100%	89%	99%	$\chi^2 = 17.5,\ \rho < 0.05$	64%	82%	$\chi^2 = 19.5,\ \rho < 0.05$

During password confirmation and login, it is found that there was no statistically significant difference in success rates between these two schemes. The observations are in line with the results from the previous research work [8]. In the next parts, we discuss success rates in retention and recall errors.

4.1 Recall Success Rates

Recall-1. In this phase (short term), it is found that the success-rate differences between the two password schemes are statistically significant. For correct-first-time attempts, participants in the text password scheme only achieved a success rate of 62%, but those in the map password scheme could reach a rate of 96%. With correct-on-multiple attempts, the textual password scheme yielded a success rate of up to 89%, indicating that 11% participants eventually gave up. The map password scheme in contrast yielded a rate of 99%. In other words, for 11% trials in the text password scheme, participants cannot remember their passwords, but there are only 1% unsuccessful trials for the map-based GP. These results indicated that participants in the graphical password scheme were more likely to successfully recall their created passwords than those who were under the text password scheme.

Recall-2. During this phase, we found that participants in the text password scheme had more difficulty in remembering their passwords, while those in the graphical password scheme could perform better. As an example, for the first attempt, the former could only achieve a success rate of 43%, but the latter could reach 72%. With multiple attempts, the text password scheme eventually yielded a rate of 64%, while a rate of 82% could be reached by the latter. Differing from the previous study [8], our existing results reveal that success rates have statistically significant differences between these two password schemes.

In Table 2, we present the success rates of male and female participants during Recall-2. For the graphical password scheme, 80% males can correctly enter map-based passwords within three attempts as compared to 78% female participants. Finally, male participants can reach 84% while females can reach 80%.

Table 2. Success rates of male and female for the Recall-2.

	MText: male	MText: female	χ^2
First attempt	42%	44%	$\chi^2 = 4.3,\ \rho > 0.05$
Within three attempts	53%	47%	$\chi^2 = 4.7,\ \rho > 0.05$
Multiple attempts	68%	60%	$\chi^2 = 5.4,\ \rho > 0.05$
	MMP: male	MMP: female	χ^2
First attempt	78%	76%	$\chi^2 = 4.1,\ \rho > 0.05$
Within three attempts	80%	78%	$\chi^2 = 4.5,\ \rho > 0.05$
Multiple attempts	84%	80%	$\chi^2 = 4.8,\ \rho > 0.05$

The results are similar under the text password scheme, where male participants could achieve 53% within three attempts and eventually 68% with multiple attempts. In contrast, female participants could achieve 47% and 60%, respectively. However, the results have no statistically significant differences ($\rho > 0.05$). This indicates that there is no apparent difference between male and female.

In [8], the results showed that males tend to perform better than females when using *PassPoints*, as males are good at visual-spatial tasks. By contrast, our system reduces such gap between males and females. Considering only statistics, participants in our study were performed generally better than the results reported by [8]. This implies that map-based GP can enable users to have a better memory of multiple passwords, as compared to *PassPoints*.

To understand the different observations, we further analyzed the collected data and informally interviewed the returned participants, where 90% of them gave their feedback. It is found that participants could connect their created map-based passwords with their past experience such as tours, visits, conference locations and so on. That is, when they have to input passwords in different account scenarios, they can link the account scenarios to their previous memory. For example, one participant said that *"When I have to input my map password to a work account, I can quickly remind of several locations where I have been"*. During the interview, we found that such relationships can generally enhance the memory accuracy and strength in the long-term.

4.2 Recall Errors

Recall-1. As shown in Table 3, participants made 183 errors in the text password scheme, whilst they made 15 errors in the graphical password scheme. Since we allow participants to try many times for an account, the number of errors is bigger than the number of trials.

To look closer to the types of errors, participants are vulnerable to multiple password interference under the text password scheme. It is found that 21 out of 30 participants made recall errors.

- *Wrong account.* Many participants tried a wrong password for an account. For example, some of them tried a password from instant message on an online dating account. Some of them made mistakes between bank account and work account.
- *Wrong account variant.* This error means that participants entered some variant of the correct password for another account.
- *Misspelled variant.* Up to 20 errors were made because of wrong spells. For example, some participants entered "lxyy1987" instead of "Lxyy1987".
- *Unknown errors.* These errors were not belonging to any types above. Some participants reflected that they may try a password other than one they created during the study.

The errors made in the text password condition and the graphical password condition are summarized in Table 3 and Table 4, respectively. The main observations regarding the graphical password condition are discussed as below.

Table 3. Classification of recall errors for the text password scheme.

Type of error	Recall-1	Recall-2
Wrong account	90	88
Wrong account variant	42	24
Misspelled variant	20	73
Unknown	31	40
Total number of errors	183	225

Table 4. Classification of recall errors for the map-based password scheme.

Type of error	Recall-1	Recall-2
Outside tolerance	5	15
Incorrect click order	2	7
Forgotten locations	3	8
Incorrect zoom level	5	46
Total number of errors	15	76

- *Outside tolerance.* There are up to five errors made for this sake, where participants could remember the general location, but may choose a place a bit far from the correct location. For example, some participants selected a place of car park, but clicked a "house" near the car park during the recall.
- *Incorrect click order.* Not many, but still two errors were made since participants selected their locations in a wrong order.
- *Forgotten locations.* Some participants may forget their locations, or tried a location from another account. All these errors belong to this type.
- *Incorrect zoom level.* Zoom level selection is a feature of map-based graphical passwords, which can help enlarge the password space. However, we find that a few participants may make errors by selecting a wrong zoom level, or even forget the correct level.

Recall-2. Overall, participants made significantly more errors when inputting their passwords after two-weeks, under either the text password scheme, or the graphical password scheme. There are totally 225 errors made related to text passwords, while 76 errors made for map-based passwords. The recall results have statistically significant differences in these two schemes: MText: ($t = 3.73$, $\rho < 0.05$) and MMP ($t = 4.81$, $\rho < 0.05$).

For the text password scheme, most errors were made because of "wrong account" and "misspelled variant", which cover up to 67% of the total errors. Some participants may cycle through their passwords or variations of their passwords when they forget the correct secret for a specific account. These observations are in line with [8].

For the graphical password scheme, most errors were caused by tolerance and zoom level, which claim a percentage of nearly 80% total errors. It is found that participants are more likely to click a place out of the tolerance or forget to check the zoom levels. For example, some participants could choose the correct location, but ignore the zoom level, resulting in a wrong trial.

5 Discussion

5.1 User Feedback

As stated earlier, after each session, we provided each participant some questions to collect their feedback and attitude towards system usage and multiple password interference. The major questions along with the scores are shown in Table 5. In particular, 10-point Likert scales were used in each feedback question ranging from 1 to 10, where 1-score indicates strong disagreement and 10-score indicates strong agreement.

- *System usage.* According to the first two questions, most participants considered that both system interfaces are easy to use (i.e., both scores are above 9 on average).
- *Recall-1.* Regarding the third and fourth question, it can be observed that most participants can remember multiple passwords within a short time. The score in the graphical password scheme is a bit higher than that in the text password scheme (8.9 versus 8.4).
- *Recall-2.* Most participants indicated that they had difficulty in remembering multiple textual passwords after two weeks, in which the score is only 4.5. In contrast, the score in the password scheme reached 7.6, which is much better. The feedback shows that participants can have a better capability of remembering multiple map-based passwords.
- *Preference.* It is found that participants gave a positive score (8.2 on average) to the map-based GP system, presenting their interests in using map-based passwords in practice. In comparison, the text password scheme only received a score of 5.3.

Table 5. Main questions and relevant scores from users' feedback.

Questions	Score (average)
1. The text password interface is easy to use	9.1
2. The map-based password interface is easy to use	9.0
3. I easily remembered the created map passwords after one hour	8.9
4. I easily remembered the created text passwords after one hour	8.4
5. I easily remembered the created map passwords after two weeks	7.6
6. I easily remembered the created text passwords after two weeks	4.5
7. I prefer using the map-based passwords instead of text passwords in practice	8.2
8. I prefer using the text passwords instead of map-based passwords in practice	5.3
9. I am able to manage multiple map-based passwords	8.3
10. I am able to manage multiple text passwords	6.3

- *Password management.* The last two questions have a score of 8.3 and 6.3, respectively. These indicate that most participants believed that they can handle multiple passwords in the graphical password scheme. Most of them stated that map locations are more easily to recall.

It is worth noting that these scores are subjective, but they reflect participants' confidence in their password generation and management. We also interviewed up to 80% (48 out of 60) participants to explore why they provide such scores. Our interests are mainly focusing on participants' feedback in relation to Recall-1, Recall-2 and password management.

- For Recall-1, most participants (40 out of 48) stated that they could remember both types of passwords in the short term. Some of them reflected that they had some techniques in remembering text passwords, or could employ some strategies based on their own experience.
- For Recall-2, most participants (43 out of 48) supported that they can remember map-based graphical passwords much better, as the map provides many cues for them to recall their selected locations. As an example, one participant revealed an example of the created map-based secret as: 'A school', 'B park' and 'C mall'. In real life, this participant will go to 'A school', through 'B park' and 'C mall' sequentially. Thus, it is easily to recall the password as long as he can remember this route for a specific account.

For password management, most participants (38 out of 48) advocated that they could have a better memory of multiple passwords in the graphical password scheme. For instance, one participant said that he could simply use different routes to create map-based passwords. Some participants said that their map-based passwords were generated based on their past experience in tourist and conference; thus, it is very convenient for them to remember the passwords for a long time.

5.2   Comparison with Map-Based GPs and PassPoints

Both map-based GPs and *PassPoints* belong to the kind of cued recall-based graphical password, where users are able to select a set of points on an image to construct the secrets. By considering the results from [8], both graphical passwords can be easier to recall than text passwords in the short term. These results indicate that most participants can manage multiple graphical passwords with different accounts.

By contrast, after two weeks, Chiasson et al. [8] showed there was no significant difference in recalling multiple passwords between *PassPoints* and text passwords. However, our study reveals that participants can still cope with multiple map-based passwords better than text passwords in the long term (after two weeks). There are two major reasons:

- *Background selection.* For map-based GPs, users can choose their own map background, i.e., they can zoom in or zoom out the map to a particular area and choose their preferred locations. In contrast, they have limited selections in *PassPoints*, with an image pool including several pre-loaded images.

– *Map locations.* Users can choose locations on a world map in the map-based GP. In *PassPoints*, they should click several points on an image. We find that locations are more useful for users to remember, as these places are usually familiar to them. At least, they have a general understanding of the selected location. Therefore, they can link the locations to facilitate their long-term memory. In comparison, five clicks in *PassPoints* may not provide much information for users to link these clicks to a particular account. For instance, users should click some objects to improve their memory; however, there may no strong relation between clicks, causing users unable to remember their clicks after a long period.

On the whole, our discussion revealed that users are empowered with better recall capability of multiple graphical passwords with the map-based graphical passwords rather than *PassPoints*.

6   Limitations

Lab Study. Our primary goal in this study is to examine the multiple password interference between textual and map-based graphical passwords. We followed the established methodology in the literature for study steps and working memory clearance. However, we acknowledge that the lab study may not reflect password usage in real lives. For instance, participants have to construct six new passwords one after another and recall them in a short time, which seldom occur in practice. In the lab environment, we found that participants may create passwords with medium strength. For example, participants in the text password scheme may generate some weak passwords, where easy patterns could be identified. By contrast, participants in the graphical password scheme can generate better secrets. Still, our study results provide useful information and complement existing research in this area.

Participants. Current participants in our study were mainly students, but this does not dimmish the results of our work. In future work, diverse participants can be considered to validate our results. Besides, all participants did not have any prior experience in utilizing any map-based graphical password system, but they have much experience in constructing text passwords. In this case, participants should have an advantage when recalling text passwords. Interestingly, our results showed that there was a significant difference in recalling text and map-based passwords. It is found that participants could handle multiple map-based passwords better than text passwords, since map can provide cues for participants when they recalled their secrets.

User Study. In the previous work with *PassPoints*, it is found that the connection between account and clicks is not clear. Participants were likely to select click-points in simple patterns such as a straight line or C-shape [8]. These indicate that participants tried to connect their passwords with their accounts for better recall. From our existing data, we reveal that there is a potential of enhancing multiple password memory through designing graphical passwords in an even proper way. That is, users can have a better recall capability of multiple graphical passwords when connecting passwords to their past experience. From this view, map provides many locations for users to select in terms of their own experience. Our implication could be verified in even larger studies.

Despite these limitations, our study provides interesting results to understand the effects of password interference, which can complement existing research.

7   Conclusion

Motivated by the recent development of map-based graphical password schemes, we conducted a two-phase user study with 60 participants to investigate the issue of multiple password interference between text passwords and map-based passwords. After two weeks, up to 50 participants successfully returned. In the study, each participant has to create and remember district passwords for six different account scenarios. Overall, we find that participants in the graphical password scheme can perform better than those in the text password scheme, not only in the short-term (one-hour session), but also in the long-term (after two weeks). Our current results reveal that participants can cope better with map when recalling multiple passwords. In particular, it is found that participants indeed made fewer recall errors in the map-based graphical password scheme than those in the text password scheme. Our work aims to complement existing work and stimulate more research in this area.

Acknowledgments. We would like to thank all participants for their hard work and collaboration in the user studies (e.g., data collection), and thank all anonymous reviewers for their helpful comments. This work was partially supported by SUTD start-up research grant SRG-ISTD-2017-124.

References

1. Alt, F., Schneegass, S., Shirazi, A.S., Hassib, M., Bulling, A.: Graphical passwords in the wild - understanding how users choose pictures and passwords in image-based authentication schemes. In: Proceedings of the 17th International Conference on Human-Computer Interaction with Mobile Devices and Services (MobileHCI), pp. 316–322 (2015)
2. Al-Ameen, M.N., Wright, M.: Multiple-password interference in the GeoPass user authentication scheme. In: Proceedings of NDSS Workshop on Usable Security (USEC), pp. 1–6 (2015)
3. Anderson, M.C., Neely, J.H.: Interference and inhibition in memory retrieval. In: Memory. Handbook of Perception and Cognition, chap. 8, 2nd edn, pp. 237–313. Academic Press (1996)
4. Bianchi, A., Oakley, I., Kim, H.: PassBYOP: bring your own picture for securing graphical passwords. IEEE Trans. Hum.-Mach. Syst. **46**(3), 380–389 (2015)
5. Biddle, R., Chiasson, S., Van Oorschot, P.C.: Graphical passwords: learning from the first twelve years. ACM Comput. Surv. **44**(4), 19 (2012)
6. Bonneau, J.: The science of guessing: analyzing an anonymized corpus of 70 million passwords. In: Proceedings of the 2012 IEEE Symposium on Security and Privacy, pp. 538–552 (2012)

7. Chiasson, S., Biddle, R., van Oorschot, P.C.: A second look at the usability of click-based graphical passwords. In: Proceedings of the 3rd Symposium on Usable Privacy and Security (SOUPS), pp. 1–12 (2007)

8. Chiasson, S., Forget, A., Stobert, E., van Oorschot, P.C., Biddle, R.: Multiple password interference in text passwords and click-based graphical passwords. In: Proceedings of the 2009 ACM Conference on Computer and Communications Security (CCS), pp. 500–511 (2009)

9. Chiasson, S., Stobert, E., Forget, A., Biddle, R.: Persuasive cued click-points: design, implementation, and evaluation of a knowledge-based authentication mechanism. IEEE Trans. Dependable Secure Comput. **9**(2), 222–235 (2012)

10. Davis, D., Monrose, F., Reiter, M.K.: On user choice in graphical password schemes. In: Proceedings of the 13th Conference on USENIX Security Symposium, pp. 1–11. USENIX Association (2004)

11. Dirik, A.E., Memon, N., Birget, J.C.: Modeling user choice in the PassPoints graphical password scheme. In: Proceedings of the 3rd Symposium on Usable Privacy and Security (SOUPS), pp. 20–28 (2007)

12. Dunphy, P., Yan, J.: Do background images improve "draw a secret" graphical passwords? In: Proceedings of the 14th ACM Conference on Computer and Communications Security (CCS), pp. 36–47 (2007)

13. Everitt, K.M., Bragin, T., Fogarty, J., Kohno, T.: A comprehensive study of frequency, interference, and training of multiple graphical passwords. In: Proceedings of the 27th International Conference on Human Factors in Computing Systems (CHI), pp. 889–898 (2009)

14. Gao, H., Liu, X.: A new graphical password scheme against spyware by using CAPTCHA. In: Proceedings of the 5th Symposium on Usable Privacy and Security (SOUPS), Article No. 21 (2009)

15. Georgakakis, E., Komninos, N., Douligeris, C.: NAVI: novel authentication with visual information. In: Proceedings of the 2012 IEEE Symposium on Computers and Communications (ISCC), pp. 588–595 (2012)

16. Gołofit, K.: Click passwords under investigation. In: Biskup, J., López, J. (eds.) ESORICS 2007. LNCS, vol. 4734, pp. 343–358. Springer, Heidelberg (2007). doi:10.1007/978-3-540-74835-9_23

17. Jermyn, I., Mayer, A., Monrose, F., Reiter, M.K., Rubin, A.D.: The design and analysis of graphical passwords. In: Proceedings of the 8th Conference on USENIX Security Symposium, pp. 1–14. USENIX Association, Berkeley (1999)

18. Liu, C.-L., Tsai, C.-J., Chang, T.-Y., Tsai, W.-J., Zhong, P.-K.: Implementing multiple biometric features for a recall-based graphical keystroke dynamics authentication system on a smart phone. J. Netw. Comput. Appl. **53**, 128–139 (2015)

19. Lopez, N., Rodriguez, M., Fellegi, C., Long, D., Schwarz, T.: Even or odd: a simple graphical authentication system. IEEE Lat. Am. Trans. **13**(3), 804–809 (2015)

20. Meng, Y.: Designing click-draw based graphical password scheme for better authentication. In: Proceedings of the 7th IEEE International Conference on Networking, Architecture, and Storage (NAS), pp. 39–48 (2012)

21. Meng, Y., Li, W.: Evaluating the effect of tolerance on click-draw based graphical password scheme. In: Chim, T.W., Yuen, T.H. (eds.) ICICS 2012. LNCS, vol. 7618, pp. 349–356. Springer, Heidelberg (2012). doi:10.1007/978-3-642-34129-8_32

22. Meng, W.: RouteMap: a route and map based graphical password scheme for better multiple password memory. Network and System Security. LNCS, vol. 9408, pp. 147–161. Springer, Cham (2015). doi:10.1007/978-3-319-25645-0_10

23. Meng, W., Wong, D.S., Furnell, S., Zhou, J.: Surveying the development of biometric user authentication on mobile phones. IEEE Commun. Surv. Tutor. **17**(3), 1268–1293 (2015)

24. Meng, W., Li, W., Jiang, L., Meng, L.: On multiple password interference of touch screen patterns and text passwords. In: Proceedings of the 2016 CHI Conference on Human Factors in Computing Systems (CHI), pp. 4818–4822 (2016)

25. Passfaces. http://www.realuser.com/

26. Pinkas, B., Sander, T.: Securing passwords against dictionary attacks. In: Proceedings of the 9th ACM Conference on Computer and communications security (CCS), pp. 161–170 (2002)

27. Shin, J., Kancharlapalli, S., Farcasin, M., Chan-Tin, E.: SmartPass: a smarter geolocation-based authentication scheme. Secur. Commun. Netw. **8**(18), 3927–3938 (2015)

28. Sun, H.-M., Chen, Y.-H., Fang, C.-C., Chang, S.-Y.: PassMap: a map based graphical-password authentication system. In: Proceedings of the 7th ACM Symposium on Information, Compuer and Communications Security (ASIACCS), pp. 99–100 (2012)

29. Suo, X., Zhu, Y., Owen, G.S.: Graphical passwords: a survey. In: Proceedings of the 21st Annual Computer Security Applications Conference (ACSAC), pp. 463–472 (2005)

30. Tao, H., Adams, C.: Pass-go: a proposal to improve the usability of graphical passwords. Int. J. Netw. Secur. **2**(7), 273–292 (2008)

31. Thorpe, J., MacRae, B., Salehi-Abari, A.: Usability and security evaluation of GeoPass: a geographic location-password scheme. In: Proceedings of the 2013 Symposium on Usable Privacy and Security (SOUPS), pp. 1–14 (2013)

32. van Oorschot, P.C., Stubblebine, S.: On countering online dictionary attacks with login histories and humans-in-the-loop. ACM Trans. Inf. Syst. Secur. **9**(3), 235–258 (2006)

33. van Oorschot, P.C., Salehi-Abari, A., Thorpe, J.: Purely automated attacks on passpoints-style graphical passwords. IEEE Trans. Inf. Forensics Secur. **5**(3), 393–405 (2010)

34. Vu, K.P.L., Proctor, R.W., Bhargav-Spantzel, A., Tai, B.-L., Cook, J., Schultz, E.E.: Improving password security and memorability to protect personal and organizational information. Int. J. Hum. Comput. Stud. **65**(8), 744–757 (2007)

35. Weir, M., Aggarwal, S., Collins, M., Stern, H.: Testing metrics for password creation policies by attacking large sets of revealed passwords. In: Proceedings of the 17th ACM Conference on Computer and Communications Security (CCS), pp. 162–175 (2010)

36. Wiedenbeck, S., Waters, J., Birget, J.-C., Brodskiy, A., Memon, N.: PassPoints: design and longitudinal evaluation of a graphical password system. Int. J. Hum. Comput. Stud. **63**(1–2), 102–127 (2005)

37. Wiedenbeck, S., Waters, J., Birget, J.-C., Brodskiy, A., Memon, N.: Authentication using graphical passwords: effects of tolerance and image choice. In: Proceedings of the 1st Symposium on Usable Privacy and Security (SOUPS) (2005)

38. Yan, J., Blackwell, A., Anderson, R., Grant, A.: Password memorability and security: empirical results. IEEE Secur. Priv. **2**, 25–31 (2004)

39. Zhu, B.B., Yan, J., Bao, G., Yang, M., Xu, N.: Captcha as graphical passwords - a new security primitive based on hard AI problems. IEEE Trans. Inf. Forensics Secur. **9**(6), 891–904 (2014)

Security Analysis

Hierarchical Key Assignment with Dynamic Read-Write Privilege Enforcement and Extended KI-Security

Yi-Ruei Chen[(✉)] and Wen-Guey Tzeng

Department of Computer Science, National Chiao Tung University,
Hsinchu 30010, Taiwan
yrchen.cs98g@nctu.edu.tw, wgtzeng@cs.nctu.edu.tw

Abstract. This paper addresses the problem of key assignment for controlling access of encrypted data in access hierarchies. We propose a hierarchical key assignment (HKA) scheme RW-HKA that supports dynamic reading and writing privilege enforcement simultaneously. It not only provides typical confidentiality guarantee in data encryption, but also allows users to verify the integrity of encrypted data. It can be applied to cloud-based systems for providing flexible access control on encrypted data in the clouds. For security, we define the extended key indistinguishable (EKI) security for RW-HKA schemes. An EKI-secure RW-HKA scheme is resistant to collusion such that no subset of users can conspire to distinguish a data decryption key, that is not legally accessible, from random strings. In this paper, we provide a generic construction of EKI-secure RW-HKA schemes based on sID-CPA secure identity-based broadcast encryption (IBBE) and strong one-time signature schemes. Furthermore, we provide a new IBBE scheme that is suitable in constructing an efficient RW-HKA scheme with a constant number of user private keys, constant size of encrypted data, and constant computation cost of a user in deriving a key for decryption. It is the first HKA scheme that achieves the aforementioned performance while supporting dynamic reading and writing privilege enforcement simultaneously.

Keywords: Hierarchical key assignment · Access control · Data outsourcing

1 Introduction

We address the problem of key assignment in access hierarchies to ensure that only authorized users are allowed to access certain data. Users are organized as a hierarchy, which is a poset of n disjoint security classes, according to the roles and clearances of users. A user with access privilege of a class is granted to access to the data in the descendant classes. A hierarchical key assignment

The research was supported in part by project 104-2221-E-009-112-MY3, Taiwan.

© Springer International Publishing AG 2017
D. Gollmann et al. (Eds.): ACNS 2017, LNCS 10355, pp. 165–183, 2017.
DOI: 10.1007/978-3-319-61204-1_9

(HKA) scheme is a method of assigning cryptographic keys to each class of the hierarchy. These keys are used to encrypt data and allow a user in a class to compute the data decryption keys of the descendant classes [1,18].

After the pioneer work of Akl and Taylor [1], the HKA problem has been extensively studied during the past 3 decades [1,12,18,38]. However, most of these schemes enforce dynamic reading privileges only. They cannot be directly applicable to enforce dynamic writing privileges simultaneously. An HKA scheme with dynamic reading and writing privileges enforcement can be used to provide flexible access control in cloud-based systems. For example, a private corporation can realize mobile office by outsourcing its encrypted data to the cloud and controlling the access right of its employees in a hierarchical structure [38]. This kind of HKA scheme is also useful in developing e-Health systems [33] and network-based computer systems in mobile ad hoc networks [28].

At the first glance, we can assign each class an asymmetric key pair (ek, dk) for data encryption and decryption. The writing and reading privileges can be separately assigned to the users by giving away ek and dk, respectively. Unfortunately, not all the asymmetric cryptosystems can be applicable to provide such a privilege separation successfully. For example, in the ElGamal encryption scheme, a user who knows dk can compute ek by itself. Given dk for the reading privilege would also give away ek for the writing privilege at the same time. Therefore, the separation of writing and reading privileges fails.

Contribution. In this paper, we propose a generic HKA scheme called *RW-HKA*, which supports dynamic reading and writing privilege enforcement simultaneously. The enforcement of privileges follows the known "no read-up" and "no write-down" principles. An authorized user in a class can only read the data in the descendant classes and write a datum into the ancestor classes in access hierarchies. Our RW-HKA scheme not only provides typical confidentiality guarantee in data encryption, but also allows users to verify the integrity of encrypted data. For security, we define the *extended key indistinguishable* (EKI) security for RW-HKA schemes. An EKI-secure RW-HKA scheme is resistant to collusion such that no subset of users can conspire to distinguish a data decryption key, that is not legally accessible, from random strings.

The construction of our generic RW-HKA scheme is based on sID-CPA secure identity-based broadcast encryption (IBBE) and strong one-time signature schemes. There are some suitable IBBE and signature schemes for implementing an RW-HKA scheme. In this paper, we provide a new IBBE scheme for constructing an efficient RW-HKA scheme in which the number of user private keys and the size of encrypted data are only small constants. In particular, the computation cost of a user in deriving the key of a descendant class for decryption is also a constant. This is the first HKA scheme that achieves the aforementioned performance while supporting dynamic reading and writing privilege enforcement.

Related Work. Akl and Taylor [1] first addressed the problem of assigning cryptographic keys in access hierarchies and proposed the first HKA scheme based on the hardness of factor problem. Later, Mackinnon et al. [31,32] improved the efficiency of Akl and Taylor's result by proposing an optimal algorithm for key generation and key derivation. Sandhu [34] addressed the rekeying issue for inserting or deleting classes and proposed a solution based on one-way functions in a specific situation that the security classes are organized as a rooted tree. Similar results were then proposed by Harn and Lin [26], Shen and Chen [37], Yang and Li [40], and Das et al. [21] for enhancing the cost for dynamic updates of access hierarchies. However, the above HKA schemes and some of their alternative schemes have two main common issues: inefficient rekeying and no formal security proofs. The cost of updating a cryptographic key is as the same as regenerating all the keys in the hierarchy. On the other hand, most of these schemes lacked rigorous security proofs. A relative large number of these schemes were shown to be vulnerable to collusion attacks [21,37,40].

To deal with the above issues, Atallah et al. [2,3] formalized two security notion of HKA schemes: key recovery (KR-security) and key indistinguishability (KI-security). KR-security ensures that a set of collusive users cannot obtain the keys for data decryption. In KI-security, the collusive users should not even be able to distinguish a data decryption key from random strings. Atallah et al. [2,3] also provided two dynamic and efficient approaches based on pseudorandom functions and symmetric key encryption schemes. The schemes are provably secure against KR-security and KI-security, respectively. After Atallah et al.'s work, many different constructions with the same security notions have been proposed [19,20,23,35,36]. De Santis et al. [35,36] provided two efficient constructions based on symmetric encryption schemes and public-key broadcast encryption schemes. D'Arco et al. [19,20] analyzed Akl and Taylor's scheme and considered different key assignment strategies to get the variations which are secure against KR-security. They also showed how to turn Akl and Taylor's scheme to be a KI-secure one. Freire and Paterson [23] provided a construction that is KI-secure based on the hardness of factoring Blum integers.

In 2013, Freire et al. [24] proposed a new security notion for HKA schemes called strong key indistinguishability (SKI-security). Comparing wit the KI-security, this notion provides an adversary with additional compromise ability to obtain data encryption keys. Freire et al. also proposed two SKI-secure HKA schemes based on pseudorandom functions and forward-secure pseudorandom generators. In 2014, Castiglione et al. [7] proved that the SKI-security notion is not stronger than the KI-security notion, i.e., these two security notions are equivalent. Cafaro et al. [6] showed a similar result in the unconditionally secure setting.

Recently, some of the approaches have been proposed to deal with the HKA problem in cloud computing [9,12,38]. Chen et al. [12] provided the first cloud-based HKA scheme called CloudHKA, which is proved to be KI-secure. CloudHKA first addressed the issue of revoking a user from accessing old data ciphers that were entitled to the user. In user revocation, the scheme outsources the cipher re-encryption to semi-trusted clouds by giving away some public re-encryption keys only. Tang et al. [38] proposed an KI-secure HKA scheme based on linear geometry. The scheme belongs to the directed derivation HKA schemes such as the schemes in [11,17,30], which only need to perform one (or constant) time of operation in key derivation. When comparing with other directed scheme, the scheme in Tang et al. [38] is the most efficient one in key derivation. Castiglione et al. [9] extended the conventional hierarchical access control model as a general one for certain additional sets of qualified users. They also proposed two KI-secure HKA schemes in the new model. The first construction is based on symmetric encryption schemes and secret sharing, and the second one is based on public-key threshold broadcast encryption schemes. Castiglione et al. [8] defined the concept of HKA schemes supporting dynamic updates and formalizing the relative security model. They also proposed a corresponding KI-secure HKA scheme with efficient key derivation and updating procedures, while requiring each user to store one private key only. Castiglione et al. [10] analyzed the Akl-Taylor scheme in the dynamic setting characterizing storage clouds. They considered different key assignment strategies and provided the corresponding schemes with KR-security.

2   Preliminaries

In this section, we introduce hierarchical access control (HAC) policy and two main building blocks of our generic RW-HKA scheme. The first one is identity-based broadcast encryption (IBBE) scheme and the second one is strong one-time signature scheme.

2.1   Hierarchical Access Control (HAC) Policy

A hierarchical access control (HAC) policy $\mathcal{P}$ is a 5-tuple $(\mathcal{SC}, \preceq, \mathcal{U}, \mathcal{D}, \lambda)$, where $\mathcal{SC} = \{SC_i : 1 \leq i \leq n\}$ is a set of security classes, $\preceq$ is a binary relation on $\mathcal{SC} \times \mathcal{SC}$, $\mathcal{U}$ is a set of users, $\mathcal{D}$ is a set of data, and $\lambda : \mathcal{U} \cup \mathcal{D} \to \mathcal{SC}$ is a function that associates each user and datum with a security class. $(\mathcal{SC}, \preceq)$ is a partial order set (poset) and $SC_j \preceq SC_i$ means that the security level of class SC_i is higher than or equal to that of SC_j. $\mathcal{P}$ requires the following two properties that indicate the "no read-up" and "no write-down" principles.

(1) *Simple security property*: A user $U \in \mathcal{U}$ can read a datum $D \in \mathcal{D}$ only if $\lambda(D) \preceq \lambda(U)$.
(2) *$\star$-property*: A user $U \in \mathcal{U}$ can write a datum $D \in \mathcal{D}$ only if $\lambda(U) \preceq \lambda(D)$.

The poset $(\mathcal{SC}, \preceq)$ is represented as a directed graph (access hierarchy) G. Each class SC_i is a node and there is a path of directed edges from SC_i to SC_j if and only if $SC_j \preceq SC_i$. G can be simplified by eliminating the edges that are implied by the transitive closure property. For example, Fig. 1 is an access hierarchy G with the nodes $SC_1, SC_2, ..., SC_6$ and directed edges (SC_1, SC_2), (SC_1, SC_3), (SC_2, SC_4), (SC_2, SC_5), (SC_3, SC_5), and (SC_3, SC_6).

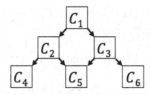

Fig. 1. An access hierarchy with 6 nodes and 6 directed edges.

2.2 Identity-Based Broadcast Encryption (IBBE) Scheme

An identity-based broadcast encryption (IBBE) scheme Ψ is a 4-tuple of poly-time algorithms $\Psi = (\mathbf{Gen}, \mathbf{Ext}, \mathbf{Enc}, \mathbf{Dec})$, where

- $\mathbf{Gen}(\tau, n) \rightarrow (msk, mpk)$. On input a security pkameter τ and the maximal size n of the set of receivers for each encryption, this key generation algorithm outputs a master secret key msk and public key mpk.
- $\mathbf{Ext}(mpk, msk, id_i) \rightarrow dk_i$. On input mpk, msk, and a receiver's identity id_i, this key extraction algorithm outputs a private key dk_i.
- $\mathbf{Enc}(mpk, \mathcal{S}, k) \rightarrow \mathsf{Hdr}_\mathcal{S}$. On input mpk, a set of receivers' identities $\mathcal{S}$, and a symmetric encryption key k, this encryption algorithm outputs a header cipher $\mathsf{Hdr}_\mathcal{S}$. The header $\mathsf{Hdr}_\mathcal{S}$ is for the receivers in $\mathcal{S}$ to derive k and k is used to encrypt a message m as a cipher broadcasted to $\mathcal{S}$.
- $\mathbf{Dec}(mpk, \mathsf{Hdr}_\mathcal{S}, dk_i) \rightarrow k$ or $\perp$. On input mpk, $\mathsf{Hdr}_\mathcal{S}$, and dk_i, this decryption algorithm outputs k for decryption if $id_i \in \mathcal{S}$, and outputs the stop symbol $\perp$ otherwise.

For correctness, we require that for all key pairs (msk, mpk) generated by $\mathbf{Gen}$, all secret keys dk_i output by $\mathbf{Ext}$, and all encryption key k,

$$\mathbf{Dec}(mpk, \mathbf{Enc}(mpk, \mathcal{S}, k), dk_i) = k$$

A standard security notion for IBBE schemes is sID-CPA security [4]. This notion of security is defined by the following game between an adversary $\mathcal{A}$ and a challenger $\mathcal{C}$. In the beginning, $\mathcal{A}$ chooses a target identity set $\mathcal{S}^*$ that it wants to attack.

Setup phase: $\mathcal{C}$ runs $\mathbf{Gen}(\tau, n)$ to obtain (msk, mpk), and gives mpk to $\mathcal{A}$.

Query phase 1: In this phase, $\mathcal{A}$ is allowed to issue the key extraction oracle $\mathcal{O}_{ext}$. For a query of identity $id_i \notin \mathcal{S}^*$, $\mathcal{O}_{ext}(id_i)$ returns dk_i.

Challenge phase: After $\mathcal{A}$ decides that query phase 1 is over, $\mathcal{C}$ runs $\mathbf{Enc}(mpk, \mathcal{S}^*, k)$ to obtain $\mathsf{Hdr}_{\mathcal{S}^*}$. Then $\mathcal{C}$ flips a random coin $b \in \{0, 1\}$ and sets $k_b \leftarrow k$ and $k_{1-b} \leftarrow \$$, where $\$$ is a random string. $\mathcal{C}$ returns $(\mathsf{Hdr}_{\mathcal{S}^*}, k_0, k_1)$ to $\mathcal{A}$.

Query phase 2: In this phase, $\mathcal{A}$ is also allowed to issue the oracle queries as in query phase 1.

Guess phase: $\mathcal{A}$ outputs its guess $b' \in \{0, 1\}$ for b.

$\mathcal{A}$ wins the sID-CPA security game if $b' = b$. The advantage of $\mathcal{A}$ in winning the sID-CPA security game is defined as:

$$\mathsf{Adv}_{\mathrm{IBBE},\mathcal{A}}^{\mathrm{sID\text{-}CPA}}(\tau) := |\Pr[\mathcal{A} \text{ outputs } 0|b = 0] - \Pr[\mathcal{A} \text{ outputs } 0|b = 1]|$$

We say that an IBBE scheme is sID-CPA secure if $\mathsf{Adv}_{\mathrm{IBBE},\mathcal{A}}^{\mathrm{sID\text{-}CPA}}(\tau)$ is negligible in τ for all probabilistic and poly-time adversary $\mathcal{A}$.

2.3 Strong One-Time Signature Scheme

A signature scheme Ω is a triple poly-time algorithms $\Omega = (\mathbf{KG}, \mathbf{Sig}, \mathbf{Ver})$, where

- $\mathbf{KG}(\kappa) \rightarrow (vk, sk)$. On input a security parameter κ, this key generation algorithm outputs a verification key vk and a signing key sk.
- $\mathbf{Sig}(sk, m) \rightarrow \sigma$. On input sk and a message m, this signing algorithm outputs a signature σ of m.
- $\mathbf{Ver}(vk, m, \sigma) \rightarrow b$. On input vk and a pair of (m, σ), this verification algorithm outputs a bit $b \in \{0, 1\}$, where $b = 1$ signifies "acceptance" and $b = 0$ signifies "rejection".

For correctness, we require that for all key pairs (vk, sk) generated by $\mathbf{KG}$ and all message m,

$$\mathbf{Ver}(vk, m, \mathbf{Sig}(sk, m)) = 1$$

In the standard security of one-time signature schemes, an adversary should be unable to forge a valid message and signature pair (m, σ) for all m chosen by the adversary. In the strong security case, it should be infeasible for an adversary to generate different signatures on the same message m. The security is defined by the following game between an adversary $\mathcal{A}$ and a challenger $\mathcal{C}$.

Setup phase. $\mathcal{C}$ runs $\mathbf{KG}(\kappa)$ to obtain (vk, sk) and gives vk to $\mathcal{A}$.

Attack phase. In this phase, $\mathcal{A}$ does one of the following two attacks. Firstly, $\mathcal{A}$ outputs (m^*, σ^*) for undefined m^* and σ^*. Secondly, $\mathcal{A}$ choses a message m and obtain its signature $\sigma^* \leftarrow \mathbf{Sig}(sk, m)$, then $\mathcal{A}$ outputs (m^*, σ^*) for $m^* \neq m$.

$\mathcal{A}$ wins the security game if $\mathbf{Ver}(vk, m^*, \sigma^*) = 1$. We say that an signature scheme is strong one-time secure if the probability of $\mathcal{A}$ wins the security game is negligible in κ for all probabilistic and poly-time adversary $\mathcal{A}$.

3 The Proposed RW-HKA Scheme

In this section, we first give an overview of our generic RW-HKA scheme and an intuition of its security. Then we provide a detailed description of our RW-HKA scheme, the corresponding EKI-security notion, and related analyzes in the rest sections.

3.1   An Overview

Our RW-HKA consists of the initialization algorithm **Initial** and key derivation algorithm **Derive**. In the beginning, the initializer runs **Initial** to generate all system parameters according to a given HAC policy $\mathcal{P}$, and assigns the keys for reading and writing access to authorized users accordingly. A user who has writing privilege of a class can encrypt-and-sign data into the corresponding classes such that an authorized user with appropriate reading privilege can run **Derive** for decryption.

More precisely, the initializer uses the signature scheme Ω to generate a pair (vk_i, sk_i) of verification and signing keys for each class SC_i, and regards vk_i as the public "identity" of SC_i. The decryption key dk_j w.r.t. the identity vk_j is generated with the underlying IBBE scheme Ψ. In key assignment, a user with reading privilege of SC_i is assigned dk_i for decrypting the data ciphers of SC_i and its descendant classes in the hierarchy. This is the simple security property (no read-up). A user with writing privilege of a class SC_i is assigned sk_z for generating a valid signature of Hdr_z in SC_z for $SC_i \preceq SC_z$. This is the $\star$-property (no write-down). The reading and writing privileges are separated in our RW-HKA since the generation of dk_i requires a master secret in Ψ and the generation of sk_i is independent of the generation of dk_i.

To encrypt a datum into SC_i, an authorized user takes the ancestor class set $\mathbb{A}_i = \{vk_z : SC_i \preceq SC_z\}$ as the broadcast set $\mathcal{S}$ and encrypts a data encryption key k with the underlying IBBE scheme Ψ. The resulting Hdr_i is then signed by using sk_i to obtain σ_i. The final cipher consists of the verification key vk_i, the IBBE cipher Hdr_i, and the signature σ. To decrypt a cipher $c_i = (vk_i, \mathsf{Hdr}_i, \sigma_i)$ in class SC_i, an authorized user first verifies the signature of Hdr_i w.r.t. vk_i and uses dk_j to decrypt Hdr_i if the verification has passed. The final decryption is successful if $SC_i \preceq SC_j$ since the "identity" vk_j is in the target broadcast set $\mathbb{A}_i$ for encrypting k.

In defining the EKI-security for RW-HKA, we model the collusion of authorized users as an adversary $\mathcal{A}$ for distinguishing data encryption keys from random strings in non-corrupted classes. $\mathcal{A}$ is allowed to access the oracles for corrupting (dk_i, sk_i)'s and decrypting c_i's for certain chosen classes. Consider a cipher $c = (vk, \mathsf{Hdr}, \sigma)$ that is not supposed to be decryptable by $\mathcal{A}$. The security of Ω ensures that any valid cipher $c' = (vk', \mathsf{Hdr}', \sigma')$ submitted by $\mathcal{A}$ to the decryption oracle have to satisfy $vk' \neq vk$. Simultaneously, the security of Ψ ensures that decryption of c' does not give $\mathcal{A}$ any further advantage in decrypting c.

3.2   Notion of RW-HKA Scheme with EKI-Security

In this section, we give a formal description of RW-HKA scheme and its corresponding EKI-security notion. The RW-HKA scheme Π is a pair of poly-time algorithms (**Initial, Derive**), where

- **Initial**$(\lambda, \tau, \mathcal{P}) \rightarrow (Pub, Sec)$. On input two security parameters λ, τ and an HAC policy $\mathcal{P}$, this initialization algorithm generates a pair (msk, mpk) of master secret key and public key. For each SC_i of $\mathcal{P}$, **Initial** generates a pair (ek_i, dk_i) of write-key and read-key and assigns them to the authorized users with writing and reading privileges accordingly. The public and secret information (Pub, Sec) are set as:

$$Pub = \{\mathcal{P}, mpk\}$$
$$Sec = \{msk\} \cup \{ek_i, dk_i : \forall SC_i\}$$

An authorized user can use ek_i to encrypt data encryption key k_i as c_i and use dk_i to decrypt c_i.

- **Derive**$(\mathcal{P}, mpk, c_i, dk_j) \to k_i$ or $\perp$. On input $\mathcal{P}$, mpk, a cipher c_i, and dk_j, this derivation algorithm outputs the stop symbol $\perp$ if $SC_i \npreceq SC_j$. Otherwise, **Derive** outputs the data decryption key k_i.

Correctness. We require that given an HAC policy $\mathcal{P}$, for all system information (Pub, Sec) generated by **Initial**, for all c_i of SC_i, and for all SC_j with $SC_i \preceq SC_j$, we have

$$\textbf{Derive}(\mathcal{P}, mpk, c_i, dk_j) = k_i$$

To separate the assignment of read and write access privileges, we require that the write-key ek_i and read-key dk_i cannot be derived from each other.

The security notion for an RW-HKA scheme is the extended key indistinguishability (EKI). $\mathcal{A}$ is allowed to access all public information and all secret information associated with a number of classes of its choice, and chooses a non-corrupted class SC_{i^*} that it want to attack. Formally, this notion of security is defined by the following EKI-security game between an adversary $\mathcal{A}$ and a challenger $\mathcal{C}$.

Setup phase. Given an HAC policy $\mathcal{P}$, $\mathcal{C}$ runs **Initial**$(\lambda, \tau, \mathcal{P})$ to generate (Pub, Sec) and gives Pub to $\mathcal{A}$.

Query phase 1. In this phase, $\mathcal{A}$ is allowed to issue the read-key corruption oracle $\mathcal{O}_{rc}$, the write-key corruption oracle $\mathcal{O}_{wc}$, and derivation oracle $\mathcal{O}_{der}$. For the query of a class index i and a valid access cipher c_i, $\mathcal{O}_{rc}(i)$, $\mathcal{O}_{wc}(i)$, and $\mathcal{O}_{der}(c_i)$ return the decryption key dk_i of SC_i, the encryption key ek_i of SC_i, and the data encryption key k_i of SC_i, respectively.

Challenge phase. After $\mathcal{A}$ decides that query phase 1 is over, $\mathcal{A}$ specifies a class index i^*, subject to $SC_{i^*} \npreceq SC_i$ for any corrupted dk_i in query phase 1. $\mathcal{C}$ flips a random coin $b \in \{0, 1\}$ and gives k_{i^*} of c_{i^*} to $\mathcal{A}$ if $b = 0$. Otherwise, it gives a random string \$ (with the same length of k_{i^*}) to $\mathcal{A}$.

Query phase 2. In this phase, $\mathcal{A}$ is also allowed to issue the oracle queries as in query phase 1, excepts that $\mathcal{O}_{rc}(i)$ for $SC_{i^*} \preceq SC_i$ and $\mathcal{O}_{der}(i^*)$.

Guess phase. $\mathcal{A}$ outputs its guess $b' \in \{0, 1\}$ for b.

$\mathcal{A}$ wins the EKI-security game if $b' = b$. The advantage of $\mathcal{A}$ winning the EKI-security game is defined as

$$\mathsf{Adv}_{\mathcal{A}}^{\mathrm{EKI}}(\lambda) := |\Pr[b' = b] - \frac{1}{2}|$$

We say that an RW-HKA scheme Π is EKI-secure if $\mathsf{Adv}_{\mathcal{A}}^{\mathrm{EKI}}(\lambda)$ is negligible in λ for all probabilistic and poly-time adversary $\mathcal{A}$.

3.3 A Generic Construction of RW-HKA Scheme

In this section, we provide our generic RW-HKA based on IBBE and signature schemes. Let $\mathbb{A}_i$ be the identity set of the ancestors of the class SC_i in $\mathcal{P}$, i.e., $\mathbb{A}_i = \{id_z : SC_i \preceq SC_z\}$. Let $\Psi = (\Psi.\mathbf{Gen}, \Psi.\mathbf{Ext}, \Psi.\mathbf{Enc}, \Psi.\mathbf{Dec})$ be an sID-CPA secure IBBE scheme, and $\Omega = (\Omega.\mathbf{KG}, \Omega.\mathbf{Sig}, \Omega.\mathbf{Ver})$ be a strong one-time signature scheme. Our RW-HKA scheme $\Pi = (\mathbf{Initial}, \mathbf{Derive})$ is as follows.

- **Initial**$(\lambda, \tau, \mathcal{P}) \rightarrow (Pub, Sec)$. On input two security parameters λ, τ and an HAC policy $\mathcal{P}$ with n security classes, this initialization algorithm first runs $\Psi.\mathbf{Gen}(\lambda, n)$ to generate a pair (msk, mpk) of system master secret key and public key. For each class SC_i of $\mathcal{P}$, this algorithm generates verification and signing key pair $(vk_i, sk_i) \leftarrow \Omega.\mathbf{KG}(\lambda)$, sets the identity id_i as the verification key vk_i, and assigns sk_i as the private part of encryption key ek_i. To generate the decryption key dk_i of SC_i, this algorithm runs $\Psi.\mathbf{Ext}$ to obtain $dk_i \leftarrow \Psi.\mathbf{Ext}(mpk, msk, id_i)$. The public and secret information (Pub, Sec) are set as:

$$Pub = \{\mathcal{P}, mpk\} \cup \{vk_i : \forall SC_i\}$$
$$Sec = \{msk\} \cup \{sk_i, dk_i : \forall SC_i\}$$

The initializer keeps Sec as secret and publishes Pub. To assign access privileges, the initializer gives dk_i to the users who has reading privilege of SC_i, and gives sk_i to the users who has write privilege of SC_i though secure channels. To encrypt data, an authorized user randomly picks a data encryption key k_i and encrypts it as

$$c_i = (vk_i, \mathsf{Hdr}_i, \sigma_i),$$

where $\mathsf{Hdr}_i \leftarrow \Psi.\mathbf{Enc}(mpk, \mathbb{A}_i, k_i)$ and $\sigma_i \leftarrow \Omega.\mathbf{Sig}(sk_i, \mathsf{Hdr}_i)$.
- **Derive**$(\mathcal{P}, mpk, c_i, dk_j) \rightarrow k_i$ or $\perp$. On input $\mathcal{P}, mpk, c_i = (vk_i, \mathsf{Hdr}_i, \sigma_i)$, and dk_j, if $SC_j \prec SC_i$, this algorithm outputs $\perp$. Otherwise, this algorithm outputs k_i by computing

$$k_i \leftarrow \Psi.\mathbf{Dec}(mpk, \mathsf{Hdr}_i, dk_j)$$

if $\Omega.\mathbf{Ver}(vk_i, \mathsf{Hdr}_{\mathbb{A}_i}, \sigma_i) = 1$ and outputs $\perp$ otherwise.

Correctness. Given an HAC policy $\mathcal{P}$, for all (Pub, Sec) generated by **Initial**, and for all SC_j with $SC_i \preceq SC_j$, we have $\mathbf{Derive}(\mathcal{P}, mpk, c_i, dk_j) = k_i$ since c_i can pass the verification of the signature scheme Ω and c_i is an encryption of k_i under the IBBE scheme Ψ with the broadcast set $\mathbb{A}_i$ that contains id_j. In addition, our RW-HKA scheme ensures that the write-key ek_i and read-key dk_i of a security class SC_i cannot be derived from each other since msk is kept secret and the generation of sk_i is independent of the generation of dk_i.

3.4   Supporting Dynamic Access Hierarchies and User Privileges

In this section, we describe the procedures of dealing with dynamic access hierarchies and user privileges. They are relation insertion, relation deletion, class insertion, and class deletion, as well as the procedures of revoking user privileges.

Relation Insertion. After inserting a new relation (SC_i, SC_j) into $\mathcal{P}$, the data owner updates $\mathcal{P}$ as $\mathcal{P}'$. The ciphers c_z of SC_z, $SC_z \preceq SC_i$, have to be updated since $\mathbb{A}_z$ is changed as $\mathbb{A}'_z \leftarrow \mathbb{A}_z \cup \mathbb{A}_i$. The initializer updates $c_z = (vk_z, \mathsf{Hdr}_z, \sigma_z)$ as $c'_z = (vk_z, \mathsf{Hdr}'_z, \sigma'_z)$, where $\mathsf{Hdr}'_i \leftarrow \varPsi.\mathbf{Enc}(mpk, \mathbb{A}'_z, k_z)$ and $\sigma'_i \leftarrow \varOmega.\mathbf{Sig}(sk_i, \mathsf{Hdr}'_z)$.

Relation Deletion. To delete a relation (SC_i, SC_j) from $\mathcal{P}$, the corresponding procedure is as the same as that in relation insertion with new $\mathbb{A}'_z \leftarrow \mathbb{A}_z \setminus \mathbb{A}_i$ for all $SC_z \prec SC_i$.

Class Insertion. To insert a new class SC_i into $\mathcal{P}$, the initializer associates SC_i with the class secrets as in our RW-HKA scheme. Then the initializer runs relation insertion to insert the incoming and outgoing relations of SC_i.

Class Deletion. To delete a class SC_i from $\mathcal{P}$, the initializer removes the associated information of SC_i, and deletes each of the incoming and outgoing relations of SC_i by using relation deletion.

Access Right Revocation. In our RW-HKA, an authorized user has writing or reading privilege of some classes. To revoke the writing privilege of SC_i from a user, the initializer updates (vk_i, sk_i) to $(vk'_i, sk'_i) \leftarrow \varOmega.\mathbf{KG}(\lambda)$ and updates id_i and ek_i accordingly. Then the initializer redistributes the new sk'_i to the other users who has writing privilege of SC_i through secure channels. Afterthen, the revoked user cannot generate a new c_i since it cannot compute a valid signature that can pass the verification with vk'_i.

To revoke the reading privilege of SC_i from a user, the initializer needs to remove ability of decrypting the ciphers associated with the descendant classes of SC_i. For each class $SC_z \preceq SC_i$, the initializer updates (vk_z, sk_z) to $(vk'_z, sk'_z) \leftarrow \varOmega.\mathbf{KG}(\lambda)$, and updates id_i and ek_i accordingly. The initializer also needs to update dk_z to $dk'_z \leftarrow \varPsi.\mathbf{Ext}(mpk, msk, id'_z)$ and c_i to c'_i with newly generated k'_z. Finally, the initializer redistributes the new ek'_z and dk'_z, $SC_z \preceq SC_i$, to the corresponding users through secure channels.

Local Rekeying. In the end of revoking the access privileges of a user from SC_i, the key redistribution is required in each of the descendant or ancestor classes of SC_i. It consumes a large bandwidth for transmission if there are many users in the affected classes. Similar to the results in [2,12,36], our RW-HKA can also support "local rekeying" with a slight modification. With the local rekeying property, the key redistribution is only required locally, i.e. only in SC_i, and do not "propagate" to the descendant or ancestor classes in the hierarchy. We modify our RW-HKA for supporting local rekeying property as follow.

In each class SC_z, the initializer prepares two extra symmetric keys rk_z and wk_z to encrypt dk_z and sk_z as the public access-tokens rt_z and wt_z, respectively. Rather than assigning dk_z and sk_z, the initializer assigns the private

rk_z and wk_z to the authorized users with writing and reading privileges, respectively. A user can obtain the keys dk_z and sk_z by using rk_z and wk_z to decrypt rt_z and wt_z, respectively.

Now, for the revocation of user privileges in SC_i, the updates of dk_z and sk_z in the other classes can be done by updating the public tokens rt_z and wt_z without changing the user private keys rk_z and wk_z. The key redistribution is only occurred in the local class SC_i since only rk_i and wk_i are required to be updated. Furthermore, we can use the group key management (GKM) schemes such as LKH [39] to maintain the keys rk_i and wk_i in each class SC_i. Here we recommend to use the semi-stateful GKM schemes [13–16] for practical application scenarios.

3.5   Security Analysis

In this section, we proof the security of our generic RW-HKA scheme. The following theorem shows that our RW-HKA scheme Π is EKI-secure based on the sID-CPA security of Ψ and the strong one-time security of Ω.

Theorem 1 (EKI-secure RW-HKA). *If Ψ is a sID-CPA secure IBBE scheme and Ω is a strong one-time signature scheme, our generic RW-HKA scheme Π is secure against extended key indistinguishability.*

Proof. We show how to turn an adversary $\mathcal{A}$ against our RW-HKA scheme Π into a forger against the signature scheme Ω and an attacker against the IBBE scheme Ψ. Assume that $\mathcal{A}$ specifies the class SC_{i*} to attack.

Let $SC_1, SC_2, \ldots, SC_m = SC_{i*}$ be a topological ordering of the ancestor classes of SC_{i*} in $\mathcal{P}$. We define a sequence of computational indistinguishable games $\mathbf{G}_0, \mathbf{G}_1, \ldots, \mathbf{G}_m$. Game $\mathbf{G}_0$ is identical to the original EKI-security game of Π. From $\mathbf{G}_0$ to $\mathbf{G}_m$, the later one incrementally makes a slight modification to the previous one while maintaining the indistinguishability among these two games on $\mathcal{A}$'s view. In each game $\mathbf{G}_i$, $0 \leq i \leq m$, the goal of $\mathcal{A}$ is to output a correct guess b' of b. More precisely, game $\mathbf{G}_i$ is defined as follows.

Game $\mathbf{G}_i, 1 \leq i \leq m$. This game is identical to game $\mathbf{G}_{i-1}$, except that the **Initial** algorithm is modified in such a way that the data encryption key k_i is substituted with a random string $\$$.

Let E_i be the event that $b' = b$ in Game $\mathbf{G}_i$. Let F_i be the event that $\mathcal{A}$ makes a valid forger of the challenged access cipher c_{i*}, i.e., $\mathcal{A}$ submits a valid cipher $c'_{i*} = (vk_{i*}, \mathsf{Hdr}'_{i*}, \sigma'_{i*})$ with $(\mathsf{Hdr}'_{i*}, \sigma'_{i*}) \neq (\mathsf{Hdr}_{i*}, \sigma_{i*})$ to the derivation oracle $\mathcal{O}_{der}$. We have the following two lemmas:

Lemma 1. $\Pr[F_i]$ is negligible.
Lemma 2. $|\Pr[E_{i-1} \wedge \overline{F}_{i-1}] - \Pr[E_i \wedge \overline{F}_i]|$ is negligible.

Lemma 1 ensures that $\mathcal{A}$ cannot forge a signature of the challenged cipher Hdr_{i^*}, i.e., it cannot obtain k_{i^*} by feeding Hdr_{i^*} to the derivation oracle $\mathcal{O}_{der}$. Besides the case of forgery, Lemma 2 ensures that $\mathcal{A}$ cannot distinguish the difference between game $\mathbf{G}_{i-1}$ and game $\mathbf{G}_i$.

To see that these two claims imply the theorem, we have:

$$\mathsf{Adv}_{\mathcal{A}}^{\mathrm{EKI}}(\lambda) = |\Pr[b' = b] - \frac{1}{2}|$$

$$\leq \sum_{i=1}^{m} |\Pr[E_{i-1}] - \Pr[E_i]|$$

$$\leq \sum_{i=1}^{m} |\Pr[F_{i-1}] - \Pr[F_i]| + \sum_{i=1}^{m} |\Pr[E_{i-1} \wedge \overline{F}_{i-1}] - \Pr[E_i \wedge \overline{F}_i]|,$$

since $\Pr[E_i]$ can be expressed as

$$\Pr[E_i] \leq |\Pr[E_i \wedge F_i] - \frac{1}{2}\Pr[F_i]| + |\Pr[E_i \wedge \overline{F}_i] + \frac{1}{2}\Pr[F_i]|$$

$$\leq \frac{1}{2}\Pr[F_i] + \Pr[E_i \wedge \overline{F}_i] + \frac{1}{2}\Pr[F_i]$$

$$= \Pr[F_i] + \Pr[E_i \wedge \overline{F}_i]$$

Note that in the last game $\mathbf{G}_m$, there is no information about the key $k_m = k_{i^*}$ is presented on $\mathcal{A}$'s view. It follows that the probability of a correct guess of b by $\mathcal{A}$ in game $\mathbf{G}_m$ is $\frac{1}{2}$, i.e., $\Pr[T_m] = \frac{1}{2}$.

Thus, $\mathsf{Adv}_{\mathcal{A}}^{\mathrm{EKI}}(\lambda)$ is negligible given the two lemmas. This concludes the proof of the theorem. $\qquad\square$

Lemma 1. $\Pr[F_i]$ *is negligible.*

Proof. We construct a probabilistic poly-time forger $\mathcal{F}$ that forges the signature σ_{i^*} w.r.t. the signature scheme Ω with probability $\Pr[F_i]$ as follows.

Given an HAC policy $\mathcal{P}$, $\mathcal{F}$ runs $\mathbf{Initial}(\lambda, \tau, \mathcal{P})$ to generate (Pub, Sec) and gives Pub to $\mathcal{A}$. $\mathcal{F}$ answers the oracle queries of $\mathcal{A}$ as in the EKI-security game. If $\mathcal{A}$ submits a valid access cipher $c_{i^*} = (vk_{i^*}, \mathsf{Hdr}'_{i^*}, \sigma'_{i^*})$ with $(\mathsf{Hdr}'_{i^*}, \sigma'_{i^*}) \neq (\mathsf{Hdr}_{i^*}, \sigma_{i^*})$ to the derivation oracle $\mathcal{O}_{der}$ in the query phase, $\mathcal{F}$ outputs $(\mathsf{Hdr}'_{i^*}, \sigma'_{i^*})$ as its forgery and stops. It is easy to see that the success probability of $\mathcal{F}$ to forge the signature at SC_{i^*} is exactly $\Pr[F_i]$. Thus, the security of the signature scheme Ω implies this lemma. $\qquad\square$

Lemma 2. $|\Pr[E_{i-1} \wedge \overline{F}_{i-1}] - \Pr[E_i \wedge \overline{F}_i]|$ *is negligible.*

Proof. Assume that $\mathcal{A}$ is able to distinguish game $\mathbf{G}_i$ and game $\mathbf{G}_{i-1}$. We use $\mathcal{A}$ to construct a probabilistic poly-time adversary $\mathcal{B}$ which breaks the sID-CPA security of Ψ as follows.

1. For a given HAC policy $\mathcal{P}$, the challenger $\mathcal{C}_{\Psi}$ generates $(msk, mpk) \leftarrow \mathbf{Gen}(\tau, n)$ and gives pk to $\mathcal{B}$.

2. $\mathcal{B}$ runs the following modified $\mathbf{Initial}(\lambda, \tau, \mathcal{P})$ to generates (Pub, Sec) and gives Pub to $\mathcal{A}$. In the beginning, $\mathcal{B}$ generates $(msk, mpk) \leftarrow \Psi.\mathbf{Gen}(\lambda, n)$. For each SC_z of $\mathcal{P}$, $\mathcal{B}$ generates (vk_z, sk_z), and k_z, and sets id_z and ek_z as in our RW-HKA scheme. To generate dk_z of each SC_z, $SC_{i^*} \not\preceq SC_z$, $\mathcal{B}$ makes the key extraction oracle query to obtain $dk_z \leftarrow \mathcal{O}_{ext}(id_z)$. For each $SC_z \neq SC_i$, $\mathcal{B}$ generates the cipher $c_z \leftarrow (vk_z, \mathsf{Hdr}_z, \sigma_z)$ as in our RW-HKA scheme. To generate c_i, $\mathcal{B}$ asks $\mathcal{C}_\Psi$ for the challenged message $(\mathsf{Hdr}_i, k_0, k_1)$, where $(k_b, k_{1-b}) \leftarrow (k, \$)$ with a random bit b chosen by $\mathcal{C}_\Psi$, and $\mathsf{Hdr}_i \leftarrow \mathbf{Enc}(ek_i, \mathbb{A}_i, k)$. $\mathcal{B}$ sets $c_i \leftarrow (vk_i, \mathsf{Hdr}_i, \sigma_i)$ and replaces k_i as k_0.

3. $\mathcal{B}$ answers the oracle queries of $\mathcal{A}$ as follows. For a read-key corruption oracle query $\mathcal{O}_{rc}(z)$, $\mathcal{B}$ returns dk_z if $SC_{i^*} \not\preceq SC_z$ and $\perp$ otherwise. For a write-key corruption oracle query $\mathcal{O}_{wc}(z)$, $\mathcal{B}$ returns sk_z. For a derivation oracle query $\mathcal{O}_{der}(z)$, $\mathcal{B}$ returns k_z if $SC_z \neq SC_{i^*}$ and $\perp$ otherwise.

4. At some point, $\mathcal{B}$ starts the challenge phase with $\mathcal{A}$ as in the EKI-security game. Afterthen, $\mathcal{A}$ can make more oracle queries as the above and returns its best answer in the end. Finally, $\mathcal{B}$ outputs 0 if $\mathcal{A}$ wins the game and outputs 1 otherwise.

In the above construction, $\mathcal{B}$ simulates the environment of $\mathcal{A}$ through interpolating between game $\mathbf{G}_{i-1}$ and game $\mathbf{G}_i$. In the generation c_i, $\mathcal{B}$ embeds the given Hdr_i into c_i and sets k_i as k_0. This is equivalent to game $\mathbf{G}_0$ if $b = 0$ since c_i is the cipher of k_0, and equivalent to game $\mathbf{G}_1$ otherwise. Eventually, $\mathcal{A}$ outputs its best answer. If $\mathcal{A}$ wins the game, $\mathcal{B}$ outputs 0, guessing for that $\mathcal{C}_\Psi$ encrypts k_0 as $\mathsf{Hdr}_{\mathbb{A}_i}$, and outputs 1 otherwise. Now we have

$$\mathsf{Adv}_{\Psi,\mathcal{B}}^{\mathrm{sID\text{-}CPA}}(\lambda) = |\Pr[\mathcal{B} \text{ outputs } 0|b = 0] - \Pr[\mathcal{B} \text{ outputs } 0|b = 1]|$$
$$= |\Pr[\mathcal{A} \text{ wins }|\mathbf{G}_0] - \Pr[\mathcal{A} \text{ wins }|\mathbf{G}_1]|$$
$$= |\Pr[E_{i-1} \wedge \overline{F}_{i-1}] - \Pr[E_i \wedge \overline{F}_i]|$$

Thus, the security of the IBBE scheme Ψ implies this lemma. □

3.6 Efficiency Analysis

In this section, we illustrate the efficiency of our generic RW-HKA scheme. We also demonstrate the efficiency of the result scheme when applying the existing and our newly constructed IBBE schemes.

Efficiency of Generic RW-HKA Scheme. For storage cost, the size of private information Sec is $O(|msk| + n)$ and the the size of public information Pub is $O(|\mathcal{P}| + |mpk| + n)$. For a user who has reading privilege of a class, it only needs to store one private key. A user who has writing privilege of a class SC_i is assigned $O(|\mathbb{A}_i|)$ keys for encrypting data into the ancestor classes of SC_i in the hierarchy. Each encrypted datum requires $O(|\mathsf{Hdr}_i|)$ of spaces for its header cipher.

For computation cost, the initializer runs one time of $\Psi.\mathbf{Gen}$, n times of $\Omega.\mathbf{KG}$, and n times of $\Psi.\mathbf{Ext}$ in generating system parameters. A user runs one time of $\Psi.\mathbf{Enc}$ and one time of $\Omega.\mathbf{Sig}$ for encrypting a datum. To derive a data encryption key, a user only needs to run one time of $\Omega.\mathbf{Ver}$ and one time of $\Psi.\mathbf{Dec}$.

Efficiency of Concrete RW-HKA Schemes. The requirement of the IBBE scheme for our RW-HKA scheme is the sID-CPA security. Most of the existing IBBE schemes satisfy this basic security notion. Table 1 shows the efficiency of our RW-HKA scheme when applying the concrete IBBE schemes proposed by Baek et al. [4], Delerablée [22], Gentry and Waters [25], Boneh et al. [5], Kim et al. [29], He et al. [27], and our newly constructed IBBE scheme proposed in Sect. 4, respectively.

Table 1. Efficiency of our RW-HKA scheme when applying concrete IBBE schemes

Applied IBBE	$|mpk|$	$|dk_i|$	$|\mathsf{Hdr}_i|$	der. cost	Security
Baek et al. [4]	$O(1)$	$O(1)$	$O(|\mathbb{A}_i|)$	$O(1)$	sID-CCA
Delerablée [22]	$O(n)$	$O(1)$	$O(1)$	$O(|\mathbb{A}_i|)$	sID-CPA
Gentry and Waters [25]	$O(n)$	$O(1)$	$O(\sqrt{n})$	$O(\sqrt{n})$	sID-CPA
Boneh et al. [5]	$O(\log n)$	$O(1)$	$O(1)$	$O(|\mathbb{A}_i|)$	sID-CPA
Kim et al. [29]	$O(n)$	$O(n)$	$O(1)$	$O(|\mathbb{A}_i|)$	sID-CCA
He et al. [27]	$O(1)$	$O(1)$	$O(|\mathbb{A}_i|)$	$O(1)$	sID-CCA
Ours	$O(n^2)$	$O(1)$	$O(1)$	$O(1)$	sID-CPA

† n: the number of security classes in hierarchy.
† $\mathbb{A}_i$: the set of ancestor classes of SC_i in hierarchy.

We consider the size of public information $|mpk|$, the number of private key of a user $|dk_i|$, the size of header cipher $|\mathsf{Hdr}_i|$ of each encrypted datum, and the computation cost of a user in deriving the decryption key of a datum. Here n is the number of security classes in the hierarchy and $\mathbb{A}_i$ is the set of the ancestor classes of SC_i in the hierarchy.

In Baek et al. [4] scheme, the size of public keys, user private keys, and decryption cost are constants. Thus, when applying this scheme for our RW-HKA, $|mpk|$, $|dk_i|$ and decryption cost are only constants. But the size of each cipher in Baek et al. [4] scheme is linear to the number of receivers so that $|\mathsf{Hdr}_i|$ of each cipher is linear to $|\mathbb{A}_i|$ in the result RW-HKA. It is not suitable in storing a large number of encrypted data since the total size of encrypted data is $O(\#(\text{data}) \cdot |\mathbb{A}_i|)$. In Delerablée [22], Boneh et al. [5], and Kim et al. [29] schemes, the size of each cipher remains a constant. When applying these schemes to our RW-HKA, it is efficient in storing many encrypted data since $|\mathsf{Hdr}_i|$ remains a constant. However, the decryption cost in these three schemes are linear to the number of receivers. It causes that each user needs $O(|\mathbb{A}_i|)$ of computation cost for key derivation in our RW-HKA. From Table 1, we see that only our IBBE scheme can provide constant size of $|dk_i|$ and $|\mathsf{Hdr}_i|$, and constant key derivation cost for RW-HKA simultaneously. Although the size $|mpk|$ of public information in our IBBE scheme is $O(n^2)$, a user of class SC_i only needs to take $|\mathbb{A}_i|$ of them in each time of encryption and $O(1)$ of them in each time of decryption.

4   Our Concrete Construction of IBBE Scheme

In this section, we propose a new IBBE scheme Ψ, which is proved to be sID-CPA secure. The scheme is modified from the IBBE scheme in Appendix A proposed by Delerablée [22]. The construction of our IBBE scheme $\Psi = (\textbf{Gen}, \textbf{Ext}, \textbf{Enc}, \textbf{Dec})$ is demonstrated as follows.

- $\textbf{Gen}(\tau, n) \rightarrow (msk, mpk)$. Given a security parameter τ and an integer n, this algorithm constructs a bilinear map $\hat{e} : \mathbb{G} \times \mathbb{G} \rightarrow \mathbb{G}_T$, where $\mathbb{G}$ and $\mathbb{G}_T$ are two multiplicative groups with prime order p and $|p| = \tau$. This algorithm randomly picks two generators $g \in \mathbb{G}$ and $h \in \mathbb{G}_T$, a value $\gamma \in \mathbb{Z}_p^*$, and a cryptographic hash function $\mathcal{H} : \{0,1\}^* \rightarrow \mathbb{Z}_p^*$. This algorithm then computes $w = g^\gamma$, $v = \hat{e}(g, h)$, an n-vector $\mathbf{t} = (t_i)_{\forall SC_i}$ with $t_i = \prod_{z \in \mathbb{A}_i} \mathcal{H}(id_z)$, an $(n+1)$-vector $\mathbf{h} = (h, h^\gamma, \ldots, h^{\gamma^n})$ and an $n \times n$ matrix $\mathbf{D} = [d_{i,j}]_{n \times n}$ with entries

$$d_{i,j} = \begin{cases} h^{\gamma^{-1} \cdot \Delta_{i,j}} & \text{if } SC_i \preceq SC_j \\ 0 & \text{otherwise} \end{cases} \text{, where}$$

$$\Delta_{i,j} = \prod_{z \in \mathbb{A}_i \setminus \{j\}} (\gamma + \mathcal{H}(id_z)) - t_i$$

The master secret key msk and public key mpk are defined as

$$msk = (g, \gamma)$$
$$mpk = (p, \mathbb{G}, \mathbb{G}_T, \hat{e}, w, v, \mathbf{t}, \mathbf{h}, \mathbf{D}, \mathcal{H})$$

- $\textbf{Ext}(mpk, msk, id_i) \rightarrow dk_i$. Given mpk, msk, and an identity id_i, this algorithm outputs the private decryption key dk_i as

$$dk_i = g^{\frac{1}{\gamma + \mathcal{H}(id_i)}}$$

- $\textbf{Enc}(mpk, \mathbb{A}_i, k_i) \rightarrow \mathsf{Hdr}_i$. Given mpk, $\mathbb{A}_i$, and a symmetric encryption key k_i, this algorithm randomly picks a value $r \in \mathbb{Z}_p^*$, and computes $\mathsf{Hdr}_i = (c_0, c_1, c_2)$, where

$$c_0 = k \cdot v^r$$
$$c_1 = w^{-r}$$
$$c_2 = h^{r \prod_{z \in \mathbb{A}_i} (\gamma + \mathcal{H}(id_z))}$$

- $\textbf{Dec}(mpk, \mathsf{Hdr}_i, dk_j) \rightarrow k$ or $\perp$. Given mpk, $\mathsf{Hdr}_i = (c_0, c_1, c_2)$, and dk_j, if $SC_i \not\preceq SC_j$, this algorithm outputs $\perp$. Otherwise, this algorithm computes

$$x = (\hat{e}(c_1, d_{i,j}) \cdot \hat{e}(dk_j, c_2))^{1/t_i}$$
$$k_i = c_0/x$$

We verify the correctness of the decryption process as follows.

$$
\begin{aligned}
x &= (\hat{e}(c_1, d_{i,j}) \cdot \hat{e}(dk_j, c_2))^{1/t_i} \\
&= (\hat{e}(w^{-r}, h^{\gamma^{-1} \cdot \Delta_{i,j}}) \cdot \hat{e}(g^{\frac{1}{\gamma + \mathcal{H}(id_j)}}, h^{r \cdot (\Delta_{i,j} + t_i) \cdot (\gamma + \mathcal{H}(id_j))}))^{1/t_i} \\
&= (\hat{e}(g, h)^{-r \cdot \Delta_{i,j}} \cdot \hat{e}(g, h)^{r \cdot (\Delta_{i,j} + t_i)})^{1/t_i} \\
&= (\hat{e}(g, h)^{r \cdot t_i})^{1/t_i} = v^r
\end{aligned}
$$

Thus, we have

$$
c_0 / x = (k_i \cdot v^r) / v^r = k_i
$$

The security of our IBBE scheme Ψ is stated with the following theorem. The key extraction and encryption algorithms are as the same with the scheme proposed by Delerablée [22]. In decryption, we move most of the computation to be pre-computed the key generation algorithm so that the decryptor only needs a constant time of operations in the result scheme. Notably, the pre-computational tasks are accomplished by using public information only. Thus, our IBBE scheme can be shown to be sID-CPA security with a standard reduction argument from the IBBE scheme proposed by Delerablée [22].

Theorem 2 (sID-CPA Security of Our IBBE Scheme). *Our IBBE scheme* Ψ *is secure against sID-CPA security.*

5 Conclusions

In this paper, we propose a generic HKA scheme called RW-HKA, which supports dynamic reading and writing privilege enforcement simultaneously. It not only provides typical confidentiality guarantee in data encryption, but also allows users to verify the integrity of encrypted data. The construction is based on sID-CPA secure IBBE and strong one-time signature schemes. It is proved to be secure against EKI-security, which is resistant to collusion of users for illegal accesses. We also provide a new IBBE scheme for constructing an efficient RW-HKA scheme with a constant number of user private keys, constant size of encrypted data, and constant computation cost of a user in deriving a key for decryption. It is the first HKA scheme that achieves such performance while supporting reading and writing privilege enforcement simultaneously.

A D07's IBBE Scheme [22]

- **Gen**$(\tau, n) \rightarrow (msk, pk)$. Given a security pkameter τ and an integer n, this algorithm constructs a bilinear map $\hat{e} : \mathbb{G} \times \mathbb{G} \rightarrow \mathbb{G}_T$, where $\mathbb{G}$ and $\mathbb{G}_T$ are two multiplicative groups with prime order p and $|p| = \tau$. This algorithm randomly picks two generators $g \in \mathbb{G}$ and $h \in \mathbb{G}_T$, a value $\gamma \in \mathbb{Z}_p^*$, and a cryptographic hash function $\mathcal{H} : \{0,1\}^* \to \mathbb{Z}_p^*$. The master secret key msk and the public key pk are defined as

$$msk = (g, \gamma), pk = (p, \mathbb{G}, \mathbb{G}_T, \hat{e}, w, v, h, h^\gamma, \dots, h^{\gamma^n}, \mathcal{H}),$$

where $w = g^\gamma$ and $v = \hat{e}(g, h)$.

- $\mathbf{Ext}(pk, msk, id_i) \to dk_i$. Given pk, msk, and an identity id_i, this algorithm outputs the private key dk_i as

$$dk_i = g^{\frac{1}{\gamma + \mathcal{H}(id_i)}}$$

- $\mathbf{Enc}(pk, \mathcal{S}, k) \to \mathsf{Hdr}_{\mathcal{S}}$. Given pk, a set $\mathcal{S}$ of some identities, and a symmetric encryption key k, this algorithm randomly picks a value $r \xleftarrow{\$} \mathbb{Z}_p^*$, and computes $\mathsf{Hdr}_{\mathcal{S}} = (c_0, c_1, c_2)$, where

$$c_0 = k \cdot v^r, c_1 = w^{-r}, c_2 = h^{r \prod_{id_j \in \mathcal{S}}(\gamma + \mathcal{H}(id_j))}$$

- $\mathbf{Dec}(pk, \mathsf{Hdr}_{\mathcal{S}}, dk_i) \to k$ or $\perp$. Given pk, $\mathsf{Hdr}_{\mathcal{S}} = (c_0, c_1, c_2)$, and dk_i, if $id_i \notin \mathcal{S}$, this algorithm outputs $\perp$. Otherwise, this algorithm computes

$$k = (\hat{e}(c_1, h^{\Delta_\gamma(id_i, \mathcal{S})}) \cdot \hat{e}(dk_i, c_2))^{\frac{1}{\prod_{id_j \mathcal{S} \wedge id_j \neq id_i} \mathcal{H}(id_j)}},$$

where

$$\Delta_\gamma(id_i, \mathcal{S}) = \gamma^{-1}(\prod_{id_j \mathcal{S} \wedge id_j \neq id_i}(\gamma + \mathcal{H}(id_j)) - \prod_{id_j \mathcal{S} \wedge id_j \neq id_i} \mathcal{H}(id_j))$$

References

1. Akl, S.G., Taylor, P.D.: Cryptographic solution to a problem of access control in a hierarchy. ACM Trans. Comput. Syst. (TOCS) 1(3), 239–248 (1983)
2. Atallah, M.J., Blanton, M., Fazio, N., Frikken, K.B.: Dynamic and efficient key management for access hierarchies. ACM Trans. Inf. Syst. Secur. 12(3), 18:1–18:43 (2009)
3. Mikhail, J.A., Keith, B.F., Marina, B.: Dynamic and efficient key management for access hierarchies. In: Proceedings of the 12th ACM Conference on Computer and Communications Security, CCS 2005, Alexandria, VA, USA, 7–11 November 2005, pp. 190–202 (2005)
4. Baek, J., Safavi-Naini, R., Susilo, W.: Efficient multi-receiver identity-based encryption and its application to broadcast encryption. In: Vaudenay, S. (ed.) PKC 2005. LNCS, vol. 3386, pp. 380–397. Springer, Heidelberg (2005). doi:10. 1007/978-3-540-30580-4_26
5. Boneh, D., Waters, B., Zhandry, M.: Low overhead broadcast encryption from multilinear maps. In: Garay, J.A., Gennaro, R. (eds.) CRYPTO 2014. LNCS, vol. 8616, pp. 206–223. Springer, Heidelberg (2014). doi:10.1007/978-3-662-44371-2_12
6. Cafaro, M., Civino, R., Masucci, B.: On the equivalence of two security notions for hierarchical key assignment schemes in the unconditional setting. IEEE Trans. Dependable Secure Comput. 12(4), 485–490 (2015)

7. Castiglione, A., De Santis, A., Masucci, B.: Key indistinguishability versus strong key indistinguishability for hierarchical key assignment schemes. IEEE Trans. Dependable Secure Comput. **13**(4), 451–460 (2016)
8. Castiglione, A., De Santis, A., Masucci, B., Palmieri, F., Castiglione, A., Huang, X.: Cryptographic hierarchical access control for dynamic structures. IEEE Trans. Inf. Forensics Secur. **11**(10), 2349–2364 (2016)
9. Castiglione, A., De Santis, A., Masucci, B., Palmieri, F., Castiglione, A., Li, J., Huang, X.: Hierarchical and shared access control. IEEE Trans. Inf. Forensics Secur. **11**(4), 850–865 (2016)
10. Castiglione, A., De Santis, A., Masucci, B., Palmieri, F., Huang, X., Castiglione, A.: Supporting dynamic updates in storage clouds with the Akl-Taylor scheme. Inf. Sci. **387**, 56–74 (2017)
11. Chen, T.-S., Chung, Y.-F.: Hierarchical access control based on chinese remainder theorem and symmetric algorithm. Comput. Secur. **21**(6), 565–570 (2002)
12. Chen, Y.-R., Chu, C.-K., Tzeng, W.-G., Zhou, J.: CloudHKA: a cryptographic approach for hierarchical access control in cloud computing. In: Jacobson, M., Locasto, M., Mohassel, P., Safavi-Naini, R. (eds.) ACNS 2013. LNCS, vol. 7954, pp. 37–52. Springer, Heidelberg (2013). doi:10.1007/978-3-642-38980-1_3
13. Chen, Y.-R., Tygar, J.D., Tzeng, W.-G.: Secure group key management using uni-directional proxy re-encryption schemes. In: 30th IEEE International Conference on Computer Communications, Joint Conference of the IEEE Computer and Communications Societies, INFOCOM 2011, 10–15 April 2011, Shanghai, China, pp. 1952–1960 (2011)
14. Chen, Y.-R., Tzeng, W.-G.: Efficient and provably-secure group key management scheme using key derivation. In: 11th IEEE International Conference on Trust, Security and Privacy in Computing and Communications, TrustCom 2012, Liverpool, United Kingdom, 25–27 June 2012, pp. 295–302 (2012)
15. Chen, Y.-R., Tzeng, W.-G.: Group key management with efficient rekey mechanism: a semi-stateful approach for out-of-synchronized members. Comput. Commun. **98**, 31–42 (2017)
16. Chou, K.-Y., Chen, Y.-R., Tzeng, W.-G.: An efficient and secure group key management scheme supporting frequent key updates on pay-tv systems. In: 13th Asia-Pacific Network Operations and Management Symposium, APNOMS 2011, Taipei, Taiwan, 21–23 September 2011, pp. 1–8 (2011)
17. Chung, Y.-F., Lee, H.-H., Lai, F., Chen, T.-S.: Access control in user hierarchy based on elliptic curve cryptosystem. Inf. Sci. **178**(1), 230–243 (2008)
18. Crampton, J., Martin, K.M., Wild, P.R.: On key assignment for hierarchical access control. In: 19th IEEE Computer Security Foundations Workshop, (CSFW-19 2006), Venice, Italy, 5–7 July 2006, pp. 98–111 (2006)
19. D'Arco, P., Santis, A., Ferrara, A.L., Masucci, B.: Security and tradeoffs of the Akl-Taylor scheme and its variants. In: Královič, R., Niwiński, D. (eds.) MFCS 2009. LNCS, vol. 5734, pp. 247–257. Springer, Heidelberg (2009). doi:10.1007/978-3-642-03816-7_22
20. D'Arco, P., De Santis, A., Ferrara, A.L., Masucci, B.: Variations on a theme by Akl and Taylor: security and tradeoffs. Theor. Comput. Sci. **411**(1), 213–227 (2010)
21. Das, M.L., Saxena, A., Gulati, V.P., Phatak, D.B.: Hierarchical key management scheme using polynomial interpolation. Oper. Syst. Rev. **39**(1), 40–47 (2005)
22. Delerablée, C.: Identity-based broadcast encryption with constant size ciphertexts and private keys. In: Kurosawa, K. (ed.) ASIACRYPT 2007. LNCS, vol. 4833, pp. 200–215. Springer, Heidelberg (2007). doi:10.1007/978-3-540-76900-2_12

23. Freire, E.S.V., Paterson, K.G.: Provably secure key assignment schemes from factoring. In: Parampalli, U., Hawkes, P. (eds.) ACISP 2011. LNCS, vol. 6812, pp. 292–309. Springer, Heidelberg (2011). doi:10.1007/978-3-642-22497-3_19

24. Freire, E.S.V., Paterson, K.G., Poettering, B.: Simple, efficient and strongly KI-secure hierarchical key assignment schemes. In: Dawson, E. (ed.) CT-RSA 2013. LNCS, vol. 7779, pp. 101–114. Springer, Heidelberg (2013). doi:10.1007/978-3-642-36095-4_7

25. Gentry, C., Waters, B.: Adaptive security in broadcast encryption systems (with short ciphertexts). In: Joux, A. (ed.) EUROCRYPT 2009. LNCS, vol. 5479, pp. 171–188. Springer, Heidelberg (2009). doi:10.1007/978-3-642-01001-9_10

26. Harn, L., Lin, H.-Y.: A cryptographic key generation scheme for multilevel data security. Comput. Secur. **9**(6), 539–546 (1990)

27. He, K., Weng, J., Liu, J., Liu, J.K., Liu, W., Deng, R.H.: Anonymous identity-based broadcast encryption with chosen-ciphertext security. In: Proceedings of the 11th ACM on Asia Conference on Computer and Communications Security, AsiaCCS 2016, Xi'an, China, 30 May–3 June 2016, pp. 247–255, 2016

28. Huang, D., Medhi, D.: A secure group key management scheme for hierarchical mobile ad hoc networks. Ad Hoc Netw. **6**(4), 560–577 (2008)

29. Kim, J., Susilo, W., Au, M.H., Seberry, J.: Adaptively secure identity-based broadcast encryption with a constant-sized ciphertext. IEEE Trans. Inf. Forensics Secur. **10**(3), 679–693 (2015)

30. Lin, Y.-L., Hsu, C.-L.: Secure key management scheme for dynamic hierarchical access control based on ECC. J. Syst. Softw. **84**(4), 679–685 (2011)

31. MacKinnon, S.J., Akl, S.G.: New key generation algorithms for multilevel security. In: Proceedings of the 1983 IEEE Symposium on Security and Privacy, Oakland, California, USA, 25–27 April 983, pp. 72–78 (1983)

32. MacKinnon, S.J., Taylor, P.D., Meijer, H., Akl, S.G.: An optimal algorithm for assigning cryptographic keys to control access in a hierarchy. IEEE Trans. Comput. **34**(9), 797–802 (1985)

33. Odelu, V., Das, A.K., Goswami, A.: An effective and secure key-management scheme for hierarchical access control in e-medicine system. J. Med. Syst. **37**(2), 9920 (2013)

34. Sandhu, R.S.: Cryptographic implementation of a tree hierarchy for access control. Inf. Process. Lett. **27**(2), 95–98 (1988)

35. De Santis, A., Ferrara, A.L., Masucci, B.: Efficient provably-secure hierarchical key assignment schemes. In: Kučera, L., Kučera, A. (eds.) MFCS 2007. LNCS, vol. 4708, pp. 371–382. Springer, Heidelberg (2007). doi:10.1007/978-3-540-74456-6_34

36. De Santis, A., Ferrara, A.L., Masucci, B.: Efficient provably-secure hierarchical key assignment schemes. Theoret. Comput. Sci. **412**(41), 5684–5699 (2011)

37. Shen, V.R.L., Chen, T.-S.: A novel key management scheme based on discrete logarithms and polynomial interpolations. Comput. Secur. **21**(2), 164–171 (2002)

38. Tang, S., Li, X., Huang, X., Xiang, Y., Lingling, X.: Achieving simple, secure and efficient hierarchical access control in cloud computing. IEEE Trans. Comput. **65**(7), 2325–2331 (2016)

39. Wong, C.K., Gouda, M.G., Lam, S.S.: Secure group communications using key graphs. IEEE/ACM Trans. Netw. **8**(1), 16–30 (2000)

40. Yang, C., Li, C.: Access control in a hierarchy using one-way hash functions. Comput. Secur. **23**(8), 659–664 (2004)

A Novel GPU-Based Implementation
of the Cube Attack
Preliminary Results Against Trivium

Marco Cianfriglia[1,2]([✉]), Stefano Guarino[2,3]([✉]), Massimo Bernaschi[2],
Flavio Lombardi[2], and Marco Pedicini[1,2]

[1] Roma Tre University, Rome, Italy
{cianfriglia,pedicini}@mat.uniroma3.it
[2] Istituto per le Applicazioni del Calcolo (IAC - CNR), Rome, Italy
{s.guarino,m.bernaschi,f.lombardi}@iac.cnr.it
[3] Sapienza University of Rome, Rome, Italy

Abstract. With black-box access to the cipher being its unique requirement, Dinur and Shamir's cube attack is a flexible cryptanalysis technique which can be applied to virtually any cipher. However, gaining a precise understanding of the characteristics that make a cipher vulnerable to the attack is still an open problem, and no implementation of the cube attack so far succeeded in breaking a real-world strong cipher. In this paper, we present a complete implementation of the cube attack on a GPU/CPU cluster able to improve state-of-the-art results against the Trivium cipher. In particular, our attack allows full key recovery up to 781 initialization rounds without brute-force, and yields the first ever maxterm after 800 initialization rounds. The proposed attack leverages a careful tuning of the available resources, based on an accurate analysis of the offline phase, that has been tailored to the characteristics of GPU computing. We discuss all design choices, detailing their respective advantages and drawbacks. Other than providing remarkable results, this paper shows how the cube attack can significantly benefit from accelerators like GPUs, paving the way for future work in the area.

Keywords: Cube attack · Trivium · GPU

1 Introduction

The security of a stream cipher relies on its ability to mimic the properties of the perfectly secure One Time Pad (OTP): predicting future keystream bits (*e.g.*, by recovering its inner state) must be computationally infeasible. In fact, as highlighted by *algebraic* and *correlation* attacks, any statistical correlation between output bits and linear combinations of input bits is a potential security breach for the cipher. Cryptographers are therefore caught in between implementation requirements, which suggest the use of efficient primitives such as *Feedback Shift Registers* (FSRs) or *Finite State Machines* (FSMs), and security requirements,

© Springer International Publishing AG 2017
D. Gollmann et al. (Eds.): ACNS 2017, LNCS 10355, pp. 184–207, 2017.
DOI: 10.1007/978-3-319-61204-1_10

which demand for solutions able to disguise the dependence of keystream-bits on the inner state of the registers. Many recent stream ciphers therefore rely upon irregular clocks, mutual clock control, non-linear and/or mutual feedback among different registers, or combinations of these solutions.

The cube attack, proposed by Dinur and Shamir [10], can be classified as an algebraic known-plaintext attack. Assuming that a chunk of keystream can be recovered from a known plaintext-ciphertext pair, the attack allows determining a set of linear equations binding key-bits. However, cube attacks significantly deviate from traditional algebraic attacks in that the equations are not recovered symbolically, but rather extracted through exhaustive searches over selected public/IV bits – the edges of the *cubes* the attack is named after. The possibility that a cube yields a linear equation depends on both its size and on the algebraic properties of the cipher. Since the Algebraic Normal Form (ANF) of the cipher (that is, its representation as a binary polynomial) is generally unknown beforehand, in practice the attack usually runs without clear prior insights into a convenient strategy for selecting the cubes – an approach made possible by the fact that the attack only requires black-box access to the attacked cipher. Exploring cubes of different (possibly large) size, trying many different sets of indices, and varying the binary assignment of the public bits not belonging to the tested cube are all promising solutions, but they all come at an exponential cost. In a sense, cube attacks can be therefore assimilated to *Time-Memory-Data Trade-Off* (TMDTO) attacks, as their success rate strongly depends on the extensiveness of the pre-computation stage, on the memory available to store the results of that stage, and on the amount of data usable to implement it. Consequently, identifying the most favourable design choices is the main pillar of a possibly successful cube attack.

Contributions. The present paper motivates and discusses in depth an implementation for Graphics Processing Unit (GPU) of the cube attack. The target cipher is Trivium [8,22], already considered in the literature to test the viability of the cube attack [10,14]. Our contributions can be summarized as follows: (i) We tailor the design and implementation of the cube attack to the characteristics of GPUs, in order to fully exploit parallelization while coping with limited memory. Our framework is extremely flexible and can be adapted to any other cipher at no more cost than some fine (performance) tuning, mostly related to memory allocation. (ii) We show the performance gain with respect to a CPU implementation, including results obtained on latest generation GPU cards. (iii) Our implementation allows for exhaustively assigning values to (subsets of) public variables with negligible additional costs. This means extending the quest for superpolys to a dimension never explored in previous works, and, by not being tied to a very small set of IV combinations, potentially weakening one of the basic requirements of the cube attack, that is, the assumption of a completely tweakable IV. (iv) Even though we run the attack with only a few preliminary sets of cubes – specifically selected to both validate our code and compare our results with the literature – our findings improve on the state-of-the-art for attacks against reduced-round versions of Trivium.

Roadmap. This paper is organized as follows: Sect. 2 introduces the cube attack and the targeted cipher Trivium; our implementation of the attack is described in Sect. 3, whereas experimental results are reported and discussed in Sect. 4; Sect. 5 gives an overview of related works; finally, Sect. 6, draws conclusions and suggests possible directions for future work.

2   Preliminaries

In this section, we first describe the theoretical implant of the cube attack, and we then briefly introduce Trivium. More details about Trivium are reported in Appendix A.

The Cube Attack. Let z denote a generic keystream bit produced by a stream cipher $\mathcal{E}$. z is the result of a function $E : \mathbb{F}_2^{n+k} \to \mathbb{F}_2$, computed over the $n + k$ input bits obtained from an Initial Vector IV of length n and a secret key K of length k. It is well known that z can be expressed as $z = p(\mathbf{x}, \mathbf{y})$, where p is the polynomial representation of E, $\mathbf{x} = (x_1, \ldots, x_n)$ is the vector of public variables (IV), $\mathbf{y} = (y_1, \ldots, y_k)$ is the vector of secret variables (K), and all variables in p appear with degree 1, at most. The cube attack relies on extracting from p a set of linear equations binding private variables in $\mathbf{y}$, through a suitable offline pre-computation phase involving public variables in $\mathbf{x}$.

Let $I = \{i_1, \ldots, i_m\} \subset \{1, \ldots, n\}$ and let us introduce the complement $\overline{I} = \{1, \ldots, n\} \backslash I$ of the set I. With a slight abuse of notation, let us consider variables in $\mathbf{x}$ as partitioned by I: $\mathbf{x} = (\mathbf{x}_I, \mathbf{x}_{\overline{I}})$, i.e., we tell apart the variables $\mathbf{x}_I$ indexed by I from those $\mathbf{x}_{\overline{I}}$ indexed by its complement $\overline{I}$. Let $t_I = x_{i_1} \cdots x_{i_m}$ be the monomial induced by I, that is, the product of all variables in $\mathbf{x}_I$. By writing $t_I(\mathbf{x}_I)$ we want to stress that t_I contains only variables in $\mathbf{x}_I$. If we factor $t_I(\mathbf{x}_I)$ out of $p(\mathbf{x}, \mathbf{y})$ we obtain

$$p(\mathbf{x}, \mathbf{y}) = t_I(\mathbf{x}_I) \cdot p_{S(I)}(\mathbf{x}, \mathbf{y}) + q(\mathbf{x}, \mathbf{y})$$

where the quotient $p_{S(I)}(\mathbf{x}, \mathbf{y})$ of the division is called the *superpoly* of I in p, whereas $q(\mathbf{x}, \mathbf{y})$ is the remainder of the division.

Now, for any binary vector $\mathbf{v}_{\overline{I}}$, we consider a fixed assignment for variables $\mathbf{x}_{\overline{I}}$[1], and let $C_I(\mathbf{v}_{\overline{I}})$ denote the *cube* induced by I and $\mathbf{v}_{\overline{I}}$, that is, the set of all 2^m possible binary assignments to $\mathbf{x}$ in which variables $\mathbf{x}_{\overline{I}}$ assume values specified by the binary vector $\mathbf{v}_{\overline{I}}$ and the remaining variables in $\mathbf{x}_I$ take all the possible combinations. It is easy to verify that all monomials in $p_{S(I)}$ do not contain any of the variables $\mathbf{x}_I$ (i.e., $p_{S(I)}(\mathbf{x}, \mathbf{y}) = p_{S(I)}(\mathbf{x}_{\overline{I}}, \mathbf{y})$), whereas all monomials in q do not contain *at least* one of the variables in $\mathbf{x}_I$. For this reason, regardless of $\mathbf{y}$, the sum of $p(\mathbf{x}, \mathbf{y})$ over all elements $\mathbf{v}$ of $C_I(\mathbf{v}_{\overline{I}})$ yields [10]

$$\sum_{\mathbf{v} \in C_I(\mathbf{v}_{\overline{I}})} p(\mathbf{v}, \mathbf{y}) = p_{S(I)}(\mathbf{v}_{\overline{I}}, \mathbf{y}) \tag{1}$$

which obviously does not depend on variables $\mathbf{x}_I$ anymore.

---

[1] The standard assumption is $\mathbf{v}_{\overline{I}} = \mathbf{0}$, but this is not actually required.

If $p_{S(I)}(\mathbf{v}_{\overline{I}}, \mathbf{y})$ is *linear*, the monomial $t_I(\mathbf{x}_I)$ is called a *maxterm for p with the assignment* $\mathbf{v}_{\overline{I}}$. If we can identify maxterms and find the symbolic expression of their superpolys, we obtain a system of linear equations that can be used to recover the secret key.

As $\mathbf{v}_{\overline{I}}$ is always clear from the context, to improve readability in the following we simply denote $C_I(\mathbf{v}_{\overline{I}})$ and $p_{S(I)}(\mathbf{v}_{\overline{I}}, \mathbf{y})$ as C_I and $p_{S(I)}(\mathbf{y})$, respectively.

Trivium. Trivium [8] is a stream cipher conceived by Christophe De Cannière and Bart Preneel, part of the eSTREAM portfolio. It generates up to 2^{64} bits of output from an 80-bit key K and an 80-bit Initial Vector IV, and it shows remarkable resistance to cryptanalysis despite its simplicity and its excellent performance. Trivium is composed by a 288-bit internal state consisting of three shift registers of length 93, 84 and 111, respectively. The feedback to each of these registers and the output bit of the cipher are obtained through non-linear combinations involving in total 15 out of the 288 internal state bits. To initialize the cipher, K and IV are written into two of the shift registers, with a fixed pattern filling the remaining bits. 1152 initialization rounds guarantee that the output begins to be produced only after all key-bits and IV-bits have been sufficiently mixed together to define the internal state of the registers.

3   The Proposed GPU Implementation of the Attack

In this section, we present, detail, and discuss our attack, designed to run on a cluster equipped with Graphics Processing Units (GPU). As previously mentioned, the success of a cube attack is highly dependent on suitable implementation choices. In order to better explain our own approach, we start with an analysis of the cube attack from a more implementative perspective.

3.1   Practical Cube Attack

At a high level, any practical implementation of the cube attack requires performing the following steps:

S1 Find as many maxterms as possible;
S2 For each maxterm, find the corresponding linear equation(s);
S3 Solve the obtained linear system.

Step S1. This is the core of the attack, where cubes that yield linear equations are identified. Choosing candidate maxterms (*i.e.*, cubes) is non-trivial. Intuitively, the degree of most maxterms lies in a specific range that depends on the (unknown) degree distribution of the monomials of the polynomial p. If the degree of t_I is too small, then $p_{S(I)}$ is most likely non-linear, but if the degree of t_I is too large, then $p_{S(I)}$ will probably be constant (*e.g.*, null). Moreover, since the complexity of the offline phase scales exponentially with $|I|$, the degree of tested potential maxterms is strongly influenced by practical limitations.

In [10], the authors propose a *random walk* to explore a maximal cube $C_{I_{\max}}$, *i.e.*, starting from a random subset $I \subset I_{\max}$ and iteratively testing the superpoly $p_{S(I)}$ to decide whether the degree of t_I should be increased or decreased. The underlying idea is to use a probabilistic approach to identify the optimal size $|I|$. In [14], the authors evaluate the cipher upon all vertices of a maximal cube $C_{I_{\max}}$, store the results in a table T of size $|T| = 2^{|I_{\max}|}$, and then apply the *Moebius transform* to the entire table T, thus computing at once the sums over $\binom{|I_{\max}|}{d}$ sub-cubes of $C_{I_{\max}}$ of degree d, for $d = 0, \ldots, |I_{\max}|$. These cubes are all possible sub-cubes of $C_{I_{\max}}$ in which the variables outside the cube have been set equal to 0. In this case the rationale is minimizing processing cost by reusing partial computations as much as possible. Interestingly, the authors of [14] show that specific cubes perform better than others, at least for reduced-round variants of Trivium, and use their findings to select the most promising maximal set $I_{\max}$.

None of these two strategies is suitable for GPUs. The stochastic nature of the random walk prevents the sequence of steps from being determined *a priori*, since the computation is performed only when (and if) needed. On the other hand, the Moebius transform requires a rigid schema of calculations and a large number of alternating read and write operations in memory that must be synchronized. Both approaches are conceived for implementations in which computational power is a constraint (while memory is not), and all advantages of using the Moebius Transform are lost in case of parallel processing. We rather perform an exhaustive search over a portion of a maximal cube, a solution that is highly parallelizable and feasible with our computational resources.

For each candidate maxterm t_I, we need to verify whether the superpoly $p_{S(I)}$ is linear. The goal being recovering key bits, any fixed assignment of variables $\mathbf{x}_{\bar{I}}$ with the bit vector $\mathbf{v}_{\bar{I}}$ can be used to get rid of variables. In order to guarantee that the degree of each superpoly is reduced to the bare minimum, the assignment $\mathbf{v}_{\bar{I}}$ to $\mathbf{0}$ is usually preferred, but we argue that this is not necessarily the best choice, as motivated later in Sect. 4.2. In any case, at this stage the superpoly $p_{S(I)}$ only depends on $\mathbf{y}$. In principle, assessing the linearity of $p_{S(I)}(\mathbf{y})$ requires finding all of its coefficients, but efficient *probabilistic* linearity tests [7, 21] can safely replace deterministic ones in most practical settings. Probabilistic tests involve verifying if

$$p_{S(I)}(\mathbf{u}_1 + \mathbf{u}_2) = p_{S(I)}(\mathbf{u}_1) + p_{S(I)}(\mathbf{u}_2) + p_{S(I)}(\mathbf{0}) \qquad (2)$$

holds for random pairs of vectors $\mathbf{u}_1, \mathbf{u}_2$. Practically, this means evaluating *numerically* four sums: $\sum_{\mathbf{v} \in C_I} E(\mathbf{v}, \mathbf{0})$, $\sum_{\mathbf{v} \in C_I} E(\mathbf{v}, \mathbf{u}_1)$, $\sum_{\mathbf{v} \in C_I} E(\mathbf{v}, \mathbf{u}_2)$, and $\sum_{\mathbf{v} \in C_I} E(\mathbf{v}, \mathbf{u}_1 + \mathbf{u}_2)$.

Probabilistic tests rely on the fact that (2) must be true for all $\mathbf{u}_1, \mathbf{u}_2$ if $p_{S(I)}$ is linear, whereas, in general, it holds with probability $\frac{1}{2}$. In particular, as done for previous cube attacks [10, 14], we will resort to a *complete-graph test* [21], which guarantees a slightly lesser accuracy than the (truly-random) BLR test [7] with far fewer evaluations of $p_{S(I)}$. Let us remark that what ultimately matters in the envisaged scenario is identifying "far-from-linear" superpolys [20]. To clarify, let us consider the superpoly $p_{S(I)}$

$$p_{S(I)}(\mathbf{y}) = l(\mathbf{y}) + \prod_{i=1}^{k} y_i$$

formed by a sum $l(\mathbf{y})$ of linear terms, plus one nonlinear term given by the product of all variables in $\mathbf{y}$. Despite the equality $p_{S(I)}(\mathbf{y}) = l(\mathbf{y})$ is *formally* wrong (the degree of $p_{S(I)}$ is as large as k), $p_{S(I)}(\mathbf{u}) = l(\mathbf{u})$ is *numerically* correct for all $\mathbf{u} \in \mathbb{F}_2^k$, except $\mathbf{u} = (1, 1, \ldots, 1)$. In other words, mistaking $p_{S(I)}$ for linear has practical consequences only if $\mathbf{u} = (1, 1, \ldots, 1)$.

Steps S2 and S3. Step **S2** consists in finding the *symbolic* expression of the superpoly of all identified maxterms, and the free term of the corresponding equation. Again, this turns into a set of *numerical* evaluations: the free term of $p_{S(I)}(\mathbf{y})$ is

$$p_{S(I)}(\mathbf{0}) = \sum_{\mathbf{v} \in C_I} E(\mathbf{v}, \mathbf{0})$$

whereas the coefficient of each variable y_i is

$$p_{S(I)}(\mathbf{e}_i) + p_{S(I)}(\mathbf{0}) = \sum_{\mathbf{v} \in C_I} E(\mathbf{v}, \mathbf{e}_i) + \sum_{\mathbf{v} \in C_I} E(\mathbf{v}, \mathbf{0})$$

where $\mathbf{e}_i$ is the unit vector with all null coordinates except $y_i = 1$. Once the polynomial $p_{S(I)}(\mathbf{y})$ is found, the attack assumes the availability of the 2^m keystream bits produced in correspondence to a fixed (unknown) assignment to the variables $\mathbf{y}$, as the variables $\mathbf{x}$ take all possible assignments in C_I. This produces the linear equation

$$p_{S(I)}(\mathbf{y}) = \sum_{\mathbf{v} \in C_I} E(\mathbf{v}, \mathbf{y})$$

whose left side is a linear combination of the key variables $\mathbf{y}$ with coefficients found *offline*, whereas the right side is a number found *online*, and whose solution is the sought unknown assignment to $\mathbf{y}$.

Finally, Step **S3** just requires solving the obtained linear system with any suitable technique described in the literature.

3.2   The Setting

Generally speaking, GPUs are processing units characterized by the following advantages and limitations:

Computing: Each unit features a large number (*i.e.*, *thousands*) of simple cores, that make possible running a much higher number of parallel threads compared to a standard CPU. More precisely, the GPU's basic processing unit is the *warp* consisting of 32 threads each. Threads are designed to work on 32-bit words, and the performance is maximized if all threads belonging to the same warp execute exactly the same operations at the same time on different but contiguous data.

Memory: The so-called *global* memory available on a GPU is limited, typically between 4 and 12 GB. Each thread can independently access data (random access is fully supported, but costly performance-wise). However, when threads in a warp access consecutive 32-bit words, the cost is equivalent to a single memory operation. Concurrent readings and writings by different threads to the same resources, which require some level of synchronization, should be avoided to prevent serialization that defeats parallelism.

The basic step of the attack is the sum of $E(\mathbf{v}, \mathbf{y})$ over all elements $\mathbf{v}$ of a cube C_I. Each time we sum over a cube, the key variables $\mathbf{y}$ are fixed, either to a random $\mathbf{u}_j$ for the linearity tests, or to $\mathbf{0}$ and to versors $\mathbf{e}_i$ for determining the superpoly. In both cases exactly the same sum $\sum_{\mathbf{v} \in C_I} E(\mathbf{v}, \mathbf{u}_j)$ must be performed for all elements of a set of keys $\{\mathbf{u}_1, \ldots, \mathbf{u}_M\}$.

We define the following strategy for carrying out the sums over a cube with the goal of maximizing the parallelization and fully exploiting at its best the computational power offered by GPUs:

- Assigning to all the threads within a warp the computation of the same cube C_I but with a different key $\mathbf{u}_j$. This choice guarantees that all threads perform the same operation at the same time for the entire computation.
- Leveraging the GPU computational power to calculate all the elements of a cube C_I, providing to the threads just a bit-mask representing the set I. With this approach we can exploit all available GPU memory to store the cubes evaluations and minimize, at the same time, the number of memory access operations.
- Defining a keystream generator function $E(\mathbf{x}, \mathbf{y})$ which outputs a 32-bit word, and letting each thread work on the whole word, fully leveraging the GPU computing model. This approach offers two remarkable benefits: (i) considering 32 keystream bits altogether is equivalent to concurrently attacking 32 different polynomials, and (ii) working on 32-bit integers fits much better with the GPUs features, whereas forcing the threads to work on single bits would critically affect the performance of the attack. Therefore, attacking 32 keystream bits altogether reduces (of a factor 32) the memory needed for storing the cubes' evaluation, thus imposing some limitations on the size of the cubes to be tested, as we will clarify later.
- Choosing the number M of keys to be a multiple of the warp size in order to perform the probabilistic linearity test on 32 keystream bits at the same time and for all M keys.

3.3   The Attack

A severe constraint in any GPU implementation is represented by the amount of memory $|T|$ currently available on GPUs. Moreover, for each cube, we need to consider M different keys in order to run the linearity test, thus reducing the amount of available memory even further to $|T|/M$. Storing single evaluations of the cipher in T means testing only sub-cubes of a maximal cube of size $|I_{\max}| = \log_2(|T|/M)$. With the memory available in current GPUs, $\log_2(|T|/M)$ is not large enough for any reasonably strong cipher. The new approach we propose is highly parallelizable, it can fully exploit the computational resources offered by GPU, and it is able to exploit GPU memory to test high order maximal cubes.

The proposed design of the attack relies on the following rationale: exploring only a portion of the maximal cube $C_{I_{\max}}$, considering only subsets $I \subseteq I_{\max}$ characterized by a non-empty *minimal* intersection $I_{\min}$. Quite naturally, a similar design leads to two distinct CUDA[2] kernels, respectively responsible for: (1) computing many variants of the cube $C_{I_{\min}}$, one for each of the possible combinations of the indices in $I_{\max} \setminus I_{\min}$, and writing the results in memory; (2) combining the stored results to test all cubes C_I such that $I_{\min} \subseteq I \subseteq I_{\max}$. Following this approach, the size of the explored $I_{\max}$ can be raised to $|I_{\max}| = |I_{\min}| + \log_2(|T|/M)$, with read and write memory operations carried out by different kernels.

According to the notation introduced in Sect. 2, the public variables are $\mathbf{x} = (x_1, \ldots, x_n)$. Now, let us distinguish these n public variables into three sets $\mathbf{x}_{\text{fix}} = (x_{i_1}, \ldots, x_{i_{d_{\text{fix}}}})$, $\mathbf{x}_{\text{free}} = (x_{j_1}, \ldots, x_{j_{d_{\text{free}}}})$, and $\mathbf{x}^*$, of size d_{fix}, d_{free}, and $n - d$, respectively, where $d = d_{\text{fix}} - d_{\text{free}}$. The variables $\mathbf{x}_{\text{fix}}$ correspond to the *fixed* components of $C_{I_{\max}}$ identified by $I_{\min}$, *i.e.*, $I_{\min} = \{i_1, \ldots, i_{d_{\text{fix}}}\}$, whereas the variables $\mathbf{x}_{\text{free}}$ correspond to the remaining *free* components of $C_{I_{\max}}$, *i.e.*, $I_{\max} \setminus I_{\min} = \{j_1, \ldots, j_{d_{\text{free}}}\}$ and $|I_{\max}| = d$. The variables $\mathbf{u}^*$ are the remaining public variables that fall outside $I_{\max}$.

The two kernels of our attack can be described as follows:

Kernel 1: It uses $2^{d_{\text{free}}}$ warps. Since, as described before, the 32 threads belonging to the same warp perform exactly the same operations but for different keys, in the following we simply consider a representative thread per warp and ignore the private variables $\mathbf{y}$.[3] For $t = 0, \ldots, 2^{d_{\text{free}}} - 1$, thread (*i.e.*, warp) s sums $E(\mathbf{u}, \mathbf{y})$ over each vertex of the cube $C_{I_{\min}}^s$ of size d_{fix} determined by the assignment of the d_{free}-bit representation $\mathbf{u}_{\text{free}}$ of integer s to the variables $\mathbf{x}_{\text{free}}$ and of $\mathbf{0}$ to the variable $\mathbf{u}^*$. Finally, thread s writes the sum in the sth entry of table T, so that, at the end of the execution of the kernel, each entry of T contains the sum over a cube of size d_{fix}. These evaluations allow for testing the monomial $t_{I_{\min}}$ with all the aforementioned assignments to the other $n - d_{\text{fix}}$ variables.

---

[2] CUDA is the software framework used for programming Nvidia GPUs.
[3] The work that here is assigned to a single thread can be actually split among any number of threads, reassembling the results at the end. We will not consider this possibility here for the sake of clarity.

Kernel 2: By simply combining the values stored in T at the end of Kernel 1, it is now possible to explore cubes of potentially any size $d_{\text{fix}} + \delta$, with $0 \leq \delta \leq d_{\text{free}}$. Although the exploration can potentially follow many other approaches (*e.g.*, a random walk as in [10]), the large computing power of our platform suggests to test cubes exhaustively. Moreover, we extend the exhaustive search to an area never reached, to the best of our knowledge, in the literature. For all I such that $I_{\min} \subseteq I \subseteq I_{\max}$, this kernel considers all variants of cube C_I obtained assigning all possible combinations of values to the variables in $I_{\max} \setminus I$. More precisely, for each possible choice of $\delta \in [0, d_{\text{free}}]$, there are exactly $\binom{d_{\text{free}}}{\delta} 2^{d_{\text{free}} - \delta}$ distinct cubes of size $d_{\text{fix}} + \delta$ available. In fact, we can choose δ free variables (the additional dimensions of the cube) in $\binom{d_{\text{free}}}{\delta}$ different ways, and we can choose the fixed assignment to the remaining $d_{\text{free}} - \delta$ variables in any of the $2^{d_{\text{free}} - \delta}$ possible combinations.

As a matter of fact, the number of cubes considered in [14] is $\sum_{\delta=0}^{d_{\text{free}}} \binom{d_{\text{free}}}{\delta} = 2^{d_{\text{free}}}$, whereas the number of cubes tested by our approach is significantly larger, namely, $\sum_{\delta=0}^{d_{\text{free}}} 2^{d_{\text{free}} - \delta} \binom{d_{\text{free}}}{\delta} = 3^{d_{\text{free}}}$. We would like to highlight that Kernel 2 is computationally dominated by Kernel 1, so the cost of our exhaustive search is negligible. Therefore, our design entails considering any possible assignment to variables outside the cube, to finally address the common conjecture (never proved in the literature), that assigning **0** is the best possible solution.

Let us underline that, in order to validate our implementation of the cube attack, we symbolically evaluated the polynomial p of Trivium up to 400 initialization rounds, and used p to identify all possible maxterms and their superpoly. We then ran the attack to find all maxterms whose variables belonged to selected sets I. Our experimental findings matched the symbolical findings. Further experimental validation of our code is reported in Sect. 4.

3.4 Performance Analysis

To evaluate the performance of our GPU based solution, we developed both a CPU and a GPU version of the cube attack. The cluster we used for the experiments is composed by 3 nodes, each equipped with 4 Tesla K80 with 12 GB of *global* memory and 4 Intel Xeon CPU E5-2640 with 128 GB of RAM. The CPU experiments were conducted on a parallel version based on OpenMP that exploits 32 cores of the four Intel(R) Xeon(R) CPU E5-2640. Each performance test was executed 5 times and the average time is reported. It is worth noticing that all versions rely on the same base functions to implement Trivium.

In Fig. 1a, we report the speed-up gained by the GPU version with respect to the parallel CPU version. We evaluated the two solutions over growing size maximal cubes $C_{I_{\max}}$, in which we anchor the size of $I_{\min}$, consequently causing the size of the set $I_{\max} \setminus I_{\min}$ to exponentially increase. Overall, the experiments show that the benefit of using the GPU version grows with the number of free variables d_{free} considered, reaching a speed-up up to $70\times$ when $d_{\text{free}} = 13$. The rationale is that the execution time of the CPU version increases almost linearly with d_{free} from the very beginning, whereas a similar trend can be observed for

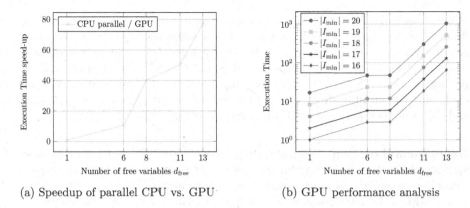

(a) Speedup of parallel CPU vs. GPU   (b) GPU performance analysis

Fig. 1. Performance experiments

the GPU version only when the number of blocks in use gets larger than the number of Streaming Multiprocessors (SMs) of the GPU, which happens when $d_{\text{free}} \geq 9$ in our case. Of course, slight fluctuations are possible, mostly due to the complex interactions among the multiple cache levels of a modern CPU. Moreover, we evaluated how the GPU solution scales when d_{free} increases. As reported in Fig. 1b, our solution scales linearly with the size of the problem, *i.e.*, exponentially with the size of the sub-cubes $C_{I_{\min}}$, thus paving the way for future works in the area.

Finally, we ran the attack under the control of the Nvidia profiler in order to measure the ALU occupancy achieved by our kernels. Kernel 1 is invoked just once per run to fill the whole table T, with an occupancy consistently over 95% when $d_{\text{free}} \geq 10$. Kernel 2 is instead invoked once per each $\delta \in [0, d_{\text{free}}]$, to compute all available cubes of size $d_{\text{fix}} + \delta$. The maximum occupancy exceeds 95% as soon as $d_{\text{free}} \geq 12$, with an average of approximately 50%. In either case the impact of d_{fix}, which determines the load of each thread, is negligible. Considering that d_{free} should be maximized to improve the attack success rate, our kernels guarantee an excellent use of resources in any realistic application. For instance, in our experiments discussed in Sect. 4 we set $d_{\text{free}} = 16$, which guarantees an occupancy above 99% for Kernel 1, and a maximum occupancy above 98% for Kernel 2.

4 Results

Finally, this section reports the results obtained by our GPU implementation of the cube attack against reduced-round Trivium. We recall that the attack ran on a cluster composed by 3 nodes, each equipped with 4 Tesla K80 with 12 GB of *global* memory and 4 Intel Xeon CPU E5-2640 with 128 GB of RAM.

As mentioned in Sect. 3.3, we performed a formal evaluation of our implementation, by checking our experimental results against Trivium's polynomials, explicitly computed up to 400 initialization rounds. In the following, the number of initialization rounds instead matches (and slightly overtakes) the best results from the literature, thus reaching a point where a symbolic evaluation would be prohibitive. Still, the results we exhibit are obtained from experiments specifically designed to reproduce tests carried out in the recent past [14], so as to provide, at the same time: (i) a direct comparison of our results with the state-of-the-art; (ii) an immediate means to assess the advantages of our approach, and (iii) a further validation of the correctness of our code.

Experimental Setting. In our attack, we consider two different reduced-round variants of Trivium, corresponding to 768 and 800 initialization rounds, respectively. As explained and motivated in Sect. 3.2, in our scheme, each call to Trivium produces 32 key-stream bits, which we use in our concurrent search for superpolys. The most significant practical consequence of a similar construction is the ability to devise attacks to Trivium reduced to any number of initialization rounds ranging from 768 to 831, at the cost of just two attacks, although the number of available superpolys decreases with the number of rounds. As a matter of fact, the jth output bit after 768 rounds can also be interpreted as the $(j-i)^{th}$ bit of output after $768 + i$ initialization rounds, for any $j \geq i$. In other words, an attack to Trivium reduced to $768 + i$ initialization rounds can count upon all superpolys found in correspondence of the jth output bit after 768 rounds, for all $j \geq i$.

For each of the two attacks (768 and 800 initialization rounds), we ran a set of independent runs, each using a different choice for the pair of sets of variables $I_{\min}, I_{\max}$ (with $I_{\min} \subset I_{\max}$) that define the minimal and maximal tested cubes $C_{I_{\min}}$ and $C_{I_{\max}}$. The size of $I_{\min}$ and $I_{\max} \setminus I_{\min}$ is $d_{\text{fix}} = 25$ and $d_{\text{free}} = 16$, respectively, for all runs, so that all maximal cubes have size $d = d_{\text{fix}} + d_{\text{free}} = 41$. Peculiarly to our implementation, when we test the monomial composed of all variables in some set $I_{\min} \subseteq I \subseteq I_{\max}$, we exhaustively assign values to all public variables in $I_{\max} \setminus I$, thus concurrently testing the linearity of $2^{41-|I|}$ possibly different superpolys. This feature of our attack – a possibility overlooked in the literature, but almost free-of-charge in our framework – provides primary benefits, as described in Sect. 4.2.

In all the reported experiments, we use a complete-graph linearity test based on combining 10 randomly sampled keys.

4.1   Summary of Results

As mentioned before, we implemented two attacks, against Trivium reduced to 768 (Trivium-768 in the following) and 800 (Trivium-800) initialization rounds, respectively. In both cases, our setting allows obtaining superpolys corresponding to 32 output bits altogether, at the cost of a single attack.

Results Against Trivium-768. For the attack against Trivium-768, we took inspiration from [14]: we launched 12 runs based on 12 different pairs $I_{\min}, I_{\max}$, chosen so as to guarantee that each of the 12 linearly independent superpolys found in [14] after 799 initialization rounds was to be found by one of our runs. The rationale of reproducing results from [14] was to both test the correctness of our implementation, and provide a better understanding of the advantages of our implementation with respect to the state-of-the-art. In this sense, let us highlight that a single run of ours cannot be directly compared with all results presented in [14], because each of our runs only explores the limited portion of the maximal cube $C_{I_{\max}}$ composed by all super-cubes of $C_{I_{\min}}$.

To better describe our results, let us introduce the binary matrix A whose element $A(i,j)$ is the coefficient of variable y_j in the ith available superpoly. The rank of A, denoted rk(A), clearly determines the number of key bits that can be recovered in the online phase of the attack based on the available superpolys, before recurring to brute-force.

As described before, the superpolys yielded by the ith output bit after round 768 are usable to attack Trivium for any number of initialization rounds between 768 and $768+i$. It is possible to define 32 different matrices $A_{768}, \ldots, A_{799}$: A_{768} includes all superpolys found, while each matrix A_{768+i} is obtained by incrementally removing the superpolys yielded by output bits $0, \ldots, i-1$. Figure 2a shows rk(A_i) as a function of i, comparing our findings with those of [14].

Overall, our results extend the state-of-the-art in a remarkable way, especially if we consider that our quest for maxterms was circumscribed to multiples of 12 *base* monomials of degree 25. In particular, let us highlight a few aspects that emerge from Fig. 2a:

- Since our runs were designed to include all 12 maxterms found in [14] after 799 initialization rounds, it is not surprising that rk(A_{799}) is at least 12. Yet, it is indeed larger: we found 3 more linearly independent superpolys, reaching rk$(A_{799}) = 15$.
- Although we did not force our tested cube to include the maxterms found in [14] after 784 rounds, we have rk$(A_{784}) = 59$, compared with rank 42 found in [14].
- Finally, and probably most importantly, our attack allows a full key recovery up to 781 initialization rounds.

Selected superpolys that guarantee the above ranks are reported in Appendix B, together with the corresponding maxterms. Very interesting is also *how* novel superpolys were found, a point that is better described in the following.

Results Against Trivium-800. To provide a further test of the quality of our attack, we launched a preliminary attack against Trivium-800. We kept unvaried all the parameters of the attack ($d_{\text{fix}} = 25$, $d_{\text{free}} = 16$, 32 output bits attacked altogether), but this time we only launched 4 runs, and we chose the sets $I_{\min}, I_{\max}$ at random. In total, we were able to find a single maxterm corresponding to 800 rounds, and no maxterms afterwards. This maxterm and the corresponding superpoly are also reported in Appendix B. Although our findings only allow to cut in half the complexity of a brute force attack, this is the first ever superpoly found considering more than 799 initialization rounds. We

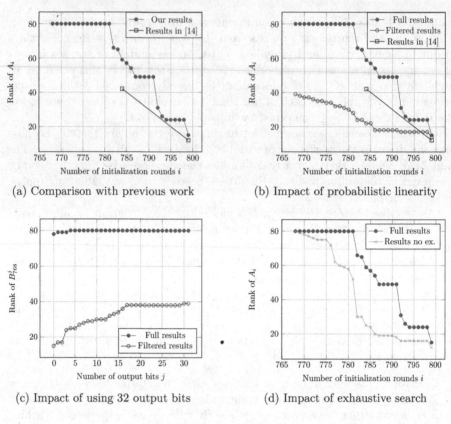

(a) Comparison with previous work   (b) Impact of probabilistic linearity

(c) Impact of using 32 output bits   (d) Impact of exhaustive search

Fig. 2. Our results

recall that our limited results should not appear as surprising: as previous work suggests [10,14], when the number of initialization rounds grows, a cube attack should increase the average degree of candidate maxterms and/or implement specific strategies for the selection of the index sets [14].

4.2 Further Discussion

Hereafter, we provide a more detailed analysis and a further discussion of our findings, considering two aspects in particular: the reliability of commonly used linearity tests, and the peculiar advantages of our attack design. Unless otherwise specified, in the following we always focus on Trivium-768.

On Probabilistic Linearity. A common practice in the cube attack related literature consists in using a probabilistic linearity test, meaning that a (small) chance exists that the superpolys found by an attack are not actually linear. In particular, the best results obtained with the cube attack against Trivium use a complete-graph test, which, with respect to the standard BLR test, trades-off accuracy for efficiency. The viability of a similar choice is supported by previous work [12,21], showing that the complete-graph test behaves essentially as a BLR test in testing a randomly chosen function f, with the quality of the former being especially high if the nonlinearity (minimum distance from any affine function) of f is large, that is, when the result of the test is particularly relevant.

Following the trend, we chose to implement a complete-graph test based on a set of 10 randomly chosen keys, exactly as done in [14]. However, while increasing the number of tests done *during* the attack was costly for us (it impacts on memory usage), implementing further test on the superpolys found *at the end* of the attack was not. We therefore decided to put our superpolys through additional tests involving other 15 keys chosen uniformly at random. Figure 2b compares $rk(A_i)$ as a function of i, for our full results and our filtered results, in which all superpolys that failed at least one of the additional tests have been removed. Let us stress once more that these two sets of results cannot be defined as wrong and correct, but they rather correspond to two different levels of trust in the found superpolys. In a sense, choosing between the two sets is equivalent to selecting the desired trade-off between efficiency and reliability of the attack: our full results permit a faster attack, which however may fail for a subset of all possible keys. Of course, many middle ways/intermediate approaches are possible. Investigating whether the reason of these failing tests is related to any of our design choices is left to future work.

On Using 32 Output Bits. A significant novelty of our implementation consists in the ability to concurrently attack 32 different polynomials, which describe 32 consecutive output bits of the target cipher. This choice is induced by GPUs features – as discussed in Sect. 3.2 – yet it is natural to assess what benefits it introduces. In Sect. 4.1 we showed that looking at 32 output bits altogether can be considered a way to concurrently attack 32 different reduced-round variants of Trivium. However, aiming to extend the attack to the full version of the cipher, our implementation can be used to check whether the same set of monomials yield different superpolys, hopefully involving different key variables, when we focus on different output bits. To this end, let us introduce a new set of matrices $B_{768}^0, \ldots, B_{768}^{31}$, where each B_{768}^j is obtained considering only the superpolys yielded by output bits $0, \ldots, j$ after 768 initialization rounds (*i.e.*, $A_{768} = B_{768}^{31}$). Figure 2c shows $rk(B_{768}^j)$ as a function of j, for both our full results and our filtered results. What the figure highlights is that considering several output bits altogether for the same version of the cipher, albeit possibly causing issues related to memory usage, does introduce the expected benefit, indeed a remarkable benefit if the matrix rank is initially (*i.e.*, when $j = 0$) low. This is the first ever result showing that considering a larger set of output bits is a viable alternative to exploring a larger cube.

On the Advantages of the Exhaustive Search. As described before, our implementation allows to find significantly more linearly independent superpolys than previous attempts from the literature. One of the reasons of our findings is the parallelization that makes possible to carry out, at a negligible cost, an exhaustive search over all public variables in $I_{\max} \setminus I$ when the cube C_I is under test. Figure 2d, again focusing on $\mathrm{rk}(A_i)$, compares our full results with results obtained without exhaustive search (shortened "no ex."), *i.e.*, setting all variables in $I_{\max} \setminus I$ to 0, as usually done in related work. What emerges is that through an exhaustive search it is indeed possible to remarkably increase $\mathrm{rk}(A_i)$. Significantly, the exhaustive search is what allows us to improve on the state-of-the-art for $i = 799$, which, among other things, suggests that the benefits of the exhaustive search are particularly relevant when increasing the number of tested cubes would be difficult otherwise (*e.g.*, by considering other monomials).

Another consequence of implementing an exhaustive search is that we found many redundant superpolys, *i.e.*, superpolys that are identical or just linearly dependent with the ones composing the maximal rank matrix $\tilde{A}$. A similar finding is extremely interesting, because we expect it to provide a wide choice of different IV combinations yielding superpolys that compose a maximal rank sub-matrix $\tilde{A}$, thus weakening the standard assumption that cube attacks require a completely tweakable IV.

5   Related Work

The cube attack is a widely applicable method of cryptanalysis introduced by Dinur and Shamir [10]. The underlying idea, similar to Vielhaber's AIDA [24], can be extended, *e.g.*, by assigning a *dynamic* value to IV bits not belonging to the tested cube [3,11], or by replacing cubes with generic subspaces of the IV space [25], and it is used in so-called *cube testers* to detect nonrandom behaviour rather than performing key extraction [4,5]. Despite the cube attack and its variants have shown promising results against several ciphers (*e.g.*, Trivium [10], Grain [11], Hummingbird-2 [13], Katan and Simon [3], Quavium [26]), Bernstein [6] expressed harsh criticism to the feasibility and convenience of cube attacks. Indeed, a general trend for cube attacks is to focus on reduced-round variants of a cipher, without any evidence that the full version can be equally attacked. However, while Bernstein suggests that the cube attack only works if the ANF of the cipher has low degree, Fouque and Vannet [14] argue (and, to some extent, experimentally show) that effective cube attacks can be run not aiming at the maximum degree of the ANF, but rather exploiting a nonrandom ANF by searching for maxterms of significantly lower degree. Along this line, O'Neil [18] suggests that even the full version of Trivium exhibits limited randomness, thus indicating the potential vulnerability of this cipher to cube attacks.

In recent years, several implementations of the cube attack attempted at breaking Trivium, our target cipher described in Sect. A. Quedenfeld and Wolf [19] found cubes for Trivium up to round 446. Srinivasan et al. [23] introduces a sufficient condition for testing a superpoly for linearity in $\mathbb{F}_2$ with a time complexity $O(2^{c+1}(k^2 + k))$, yielding 69 extremely sparse linearly independent superpolys for Trivium reduced to 576 rounds. In their seminal paper [10], Dinur and Shamir found 63, 53, and 35 linearly independent superpolys after, respectively, 672, 735, and 767 rounds. Fouque and Vannet [14] even improve over Dinur and Shamir, by obtaining 42 linearly independent superpolys after 784 rounds, and 12 linearly independent superpolys (plus 6 quadratic superpolys) after 799 rounds. To the best of our knowledge, these are the best results against Trivium to date, making our attack comparable to (or better than) the state-of-the-art.

Distributed computing and/or parallel processing have been explored in the literature to render attacks to crypto systems computationally or storage-wise feasible/practical. Smart et al. [15] develop a new methodology to assess cryptographic key strength using cloud computing. Marks et al. [16] provide numerical evidence of the potential of mixed GPU(AMD, Nvidia) and CPU technology to data encryption and decryption algorithms. Focusing on GPU, Milo et al. [17] leverage GPUs to quickly test passphrases used to protect private keyrings of OpenPGP cryptosystems, showing that the time complexity of the attack can be reduced up to three-orders of magnitude with respect to a standard procedure, and up to ten times with respect to a highly tuned CPU implementation. A relevant result is obtained by Agostini [2] leveraging GPUs to speed up Dictionary Attacks to the BitLocker technology commonly used in Windows OSes to encrypt disks. Finally, and most closely related to the present work, Fan and Gong [13] make use of GPUs to perform side channel cube attacks on Hummingbird-2. They describe an efficient term-by-term quadraticity test for extracting simple quadratic equations, leveraging the cube attack. Just like us, Fan and Gong speed-up the implementation of the proposed term-by-term quadraticity test by leveraging GPUs and finally recovering 48 out of 128 key bits of the Hummingbird-2 with a data complexity of about 2^{18} chosen plaintexts. However, we present a complete implementation of the cube attack thoroughly designed and optimized for GPUs. Our flexible construction allows an exhaustive exploration of subsets of IV bits, thus overcoming the limitations of dynamic cube attacks, which try to find the most suitable assignment to those bits by analyzing the target cipher.

6   Conclusions and Future Work

This work has discussed in depth an advanced GPU implementation of the cube attack aimed at breaking a reduced-round version of Trivium. The implemented attack allows extending the quest for superpolys to a dimension never explored in previous works, and weakens the previous cube attack assumption of a completely tweakable IV. An extensive experimental campaign is discussed and results validate the approach and improve over the state-of-the-art attacks against reduced-round versions of Trivium.

The tool, that we expect to release into the public domain, opens new perspectives by allowing a more comprehensive and hopefully exhaustive analysis of stream-ciphers security. For instance, along the line proposed in [1], we envisage developing our implementation to test the effectiveness of the generalized cube attack over $\mathbb{F}_n$.

A   Trivium Specifications

Trivium [8] is a synchronous stream cipher conceived by Christophe De Cannière and Bart Preneel, not patented, and specified as an International Standard under ISO/IEC 29192-3. Trivium combines a flexible trade-off between speed and gate count in hardware, and a reasonably efficient software implementation. Citing [9]: "Trivium is a hardware oriented design focussed on flexibility. It aims to be compact in environments with restrictions on the gate count, power-efficient on platforms with limited power resources, and fast in applications that require high-speed encryption". Particularly interesting is the fact that any state bit stays unused for at least 64 iterations after it has been modified. This means that up to 64 iterations can be parallelized and computed at once, allowing for a factor 64 reduction in the clock frequency without affecting the throughput.

Trivium generates up to 2^{64} bits of output from an 80-bit key K and an 80-bit Initial Vector IV, and it shows remarkable resistance to cryptanalysis despite its simplicity and its excellent performance. The 80-bit key K and the 80-bit IV, are used in Trivium to initialize three FSRs of length 93, 84 and 111, respectively. The internal states of the three registers are denoted $(s_1, \ldots, s_{93})$, $(s_{94}, \ldots, s_{177})$ and $(s_{178}, \ldots, s_{288})$ respectively. Fifteen out the 288 internal state bits are used at each round to compute the feedbacks to the three registers and the output bit of the cipher. However, to obtain a better mixing of the seed and to guarantee that each output bit is a complex non-linear function of all key-bits and IV-bits, the cipher undergoes 1152 initialization rounds without producing any output. In detail, the initial seed of the three registers is defined as follows:

$$(s_1, \ldots, s_{93}) \leftarrow (K_1, \ldots, K_{80}, 0, \ldots, 0)$$
$$(s_{94}, \ldots, s_{177}) \leftarrow (IV_1, \ldots, IV_{80}, 0, \ldots, 0)$$
$$(s_{178}, \ldots, s_{288}) \leftarrow (0, \ldots, 0, 1, 1, 1)$$

For each $t \geq 1$, the internal state of the cipher is updated as follows[4]:

$$(s_1, s_2, \ldots, s_{93}) \leftarrow (s_{243} + s_{286} \cdot s_{287} + s_{288} + s_{69}, s_1, \ldots, s_{92})$$
$$(s_{94}, s_{95}, \ldots, s_{177}) \leftarrow (s_{66} + s_{91} \cdot s_{92} + s_{93} + s_{171}, s_{94}, \ldots, s_{176})$$
$$(s_{178}, s_{179}, \ldots, s_{288}) \leftarrow (s_{162} + s_{175} \cdot s_{176} + s_{177} + s_{264}, s_{178}, \ldots, s_{287})$$

Finally, for each $t > 1152$, the output bit z_t is computed as:

$$z_t \leftarrow s_{66} + s_{93} + s_{162} + s_{177} + s_{243} + s_{288}$$

---

[4] Symbols $+$ and $\cdot$ denote sum and product over $\mathbb{F}_2$, *i.e.*, bitwise XOR and AND.

B   Tables of Maxterms and Superpolys

Trivium-781

maxterm bits	superpoly	round
3, 6, 8, 10, 12, 14, 18, 19, 20, 23, 25, 27, 31, 33, 38, 40, 43, 45, 48, 53, 54, 56, 58, 60, 62, 63, 69, 75, 77, 79, 80	x_{55}	781
1, 5, 7, 8, 10, 15, 16, 18, 20, 23, 25, 27, 32, 33, 36, 38, 40, 41, 43, 47, 49, 52, 53, 54, 56, 58, 63, 69, 71, 75, 77, 80	x_{69}	781
1, 6, 7, 8, 10, 12, 16, 19, 21, 24, 25, 27, 31, 33, 36, 38, 40, 41, 43, 47, 49, 52, 53, 56, 58, 63, 67, 69, 71, 73, 77, 80	x_{60}	781
1, 2, 3, 5, 6, 7, 8, 12, 14, 15, 16, 19, 21, 23, 25, 27, 36, 38, 40, 43, 45, 47, 49, 54, 56, 58, 60, 62, 69, 71, 73, 74, 80	$x_{51} + 1$	781
1, 2, 5, 6, 7, 8, 10, 12, 14, 15, 16, 19, 23, 25, 27, 36, 38, 40, 43, 45, 47, 52, 54, 56, 58, 60, 62, 69, 71, 73, 76, 79, 80	x_{45}	781
1, 2, 5, 6, 7, 8, 12, 14, 15, 16, 19, 21, 23, 25, 27, 33, 36, 38, 43, 45, 47, 49, 52, 54, 56, 58, 62, 65, 69, 71, 73, 76, 80	$x_{43} + x_{58}$	781
1, 3, 5, 6, 7, 8, 10, 12, 14, 15, 16, 19, 23, 27, 33, 36, 38, 40, 43, 45, 47, 52, 54, 56, 58, 60, 62, 69, 71, 73, 74, 79, 80	x_{23}	781
1, 3, 6, 7, 8, 10, 12, 14, 16, 19, 21, 23, 25, 33, 38, 40, 43, 45, 47, 49, 53, 54, 56, 58, 62, 63, 65, 69, 71, 75, 77, 79, 80	$x_8 + x_{35} + x_{64}$	781
1, 5, 7, 8, 10, 12, 14, 15, 16, 18, 20, 23, 24, 25, 27, 32, 33, 36, 40, 41, 43, 47, 49, 52, 53, 56, 58, 63, 69, 71, 75, 77, 80	$x_{67} + 1$	781
3, 5, 6, 8, 10, 12, 14, 16, 18, 19, 20, 23, 24, 25, 27, 31, 33, 38, 43, 45, 48, 53, 54, 56, 58, 60, 62, 63, 69, 75, 77, 79, 80	x_2	781
6, 8, 10, 12, 14, 15, 16, 18, 19, 20, 21, 23, 25, 27, 29, 31, 33, 36, 38, 40, 41, 42, 45, 49, 54, 56, 60, 62, 63, 69, 73, 75, 80	x_{58}	781
1, 2, 5, 6, 7, 8, 10, 12, 14, 15, 16, 19, 23, 25, 27, 36, 38, 40, 43, 45, 47, 49, 52, 54, 56, 58, 60, 62, 69, 71, 73, 74, 76, 79, 80	$x_{62} + 1$	781
1, 3, 5, 6, 8, 10, 12, 14, 15, 16, 19, 21, 25, 30, 31, 32, 33, 36, 38, 40, 41, 43, 45, 47, 49, 53, 54, 56, 63, 69, 71, 73, 75, 80	$x_3 + x_{25} + x_{39} + x_{40} + x_{51} + x_{66} + x_{67} + x_{78} + 1$	781
1, 3, 5, 6, 8, 12, 14, 15, 16, 19, 21, 25, 27, 31, 33, 38, 40, 41, 43, 45, 47, 48, 49, 50, 53, 54, 56, 60, 63, 69, 71, 73, 75, 80	$x_{10} + x_{13} + x_{14} + x_{19} + x_{25} + x_{28} + x_{29} + x_{31} + x_{37} + x_{40} + x_{46} + x_{52} + x_{53} + x_{55} + x_{56} + x_{57} + x_{60} + x_{61} + x_{62} + x_{64} + x_{66} + x_{68} + x_{69} + 1$	781
1, 3, 5, 7, 12, 14, 15, 16, 18, 19, 20, 21, 24, 25, 27, 31, 33, 36, 40, 41, 45, 49, 54, 56, 58, 60, 62, 63, 66, 71, 73, 75, 77, 80	x_{57}	781
1, 3, 6, 7, 8, 10, 12, 14, 16, 19, 21, 23, 25, 33, 36, 40, 41, 43, 45, 47, 49, 53, 54, 56, 58, 61, 63, 65, 69, 71, 75, 77, 79, 80	$x_{43} + x_{58} + x_{64} + x_{66} + x_{70}$	781
1, 3, 5, 6, 7, 8, 12, 13, 14, 15, 16, 19, 21, 23, 25, 27, 36, 38, 40, 43, 45, 47, 49, 52, 54, 56, 58, 62, 65, 69, 71, 73, 76, 79, 80	x_{65}	781
1, 3, 5, 6, 8, 10, 12, 14, 15, 16, 18, 19, 24, 25, 31, 32, 33, 36, 38, 40, 41, 43, 45, 47, 49, 50, 53, 54, 56, 63, 69, 71, 73, 75, 80	$x_{23} + x_{39} + x_{50} + x_{66} + x_{67} + x_{79} + 1$	781
1, 3, 5, 6, 8, 10, 12, 14, 15, 16, 19, 25, 30, 31, 32, 33, 38, 40, 41, 43, 45, 47, 49, 50, 53, 54, 56, 60, 63, 69, 71, 73, 75, 80	$x_9 + x_{18} + x_{24} + x_{26} + x_{32} + x_{33} + x_{34} + x_{42} + x_{51} + x_{53} + x_{54} + x_{58} + x_{59} + x_{64} + x_{66} + x_{68} + x_{69} + x_{80} + 1$	781
1, 3, 5, 6, 8, 10, 12, 14, 15, 16, 19, 21, 24, 25, 31, 32, 33, 36, 38, 40, 41, 43, 45, 47, 49, 50, 53, 54, 56, 63, 69, 71, 73, 75, 80	$x_{52} + x_{66} + x_{67} + x_{79}$	781
1, 3, 5, 6, 8, 12, 14, 15, 16, 18, 19, 21, 25, 27, 31, 33, 38, 40, 41, 43, 45, 47, 48, 49, 50, 53, 54, 56, 60, 63, 69, 71, 73, 75, 80	$x_{13} + x_{14} + x_{19} + x_{25} + x_{27} + x_{28} + x_{29} + x_{31} + x_{39} + x_{41} + x_{42} + x_{46} + x_{51} + x_{52} + x_{54} + x_{55} + x_{56} + x_{57} + x_{61} + x_{62} + x_{64} + x_{65} + x_{66} + x_{69} + x_{78}$	781
1, 3, 5, 6, 8, 12, 14, 15, 16, 18, 19, 24, 25, 27, 31, 32, 33, 36, 38, 40, 41, 45, 47, 48, 49, 53, 54, 56, 60, 63, 69, 71, 73, 75, 80	$x_{16} + x_{26} + x_{27} + x_{38} + x_{43} + x_{53} + x_{54} + x_{56} + x_{65} + x_{67} + x_{80}$	781
1, 3, 5, 6, 8, 12, 14, 15, 16, 19, 21, 24, 25, 31, 32, 33, 36, 38, 40, 41, 45, 47, 49, 50, 53, 54, 56, 60, 63, 69, 71, 73, 75, 80	$x_{25} + x_{27} + x_{30} + x_{54} + x_{57}$	781
1, 2, 3, 5, 6, 7, 8, 12, 14, 15, 16, 19, 23, 25, 27, 33, 36, 38, 40, 43, 45, 47, 52, 54, 56, 58, 60, 62, 65, 69, 70, 71, 73, 74, 76, 80	x_{42}	781
1, 3, 5, 6, 8, 10, 12, 14, 15, 16, 18, 19, 21, 25, 27, 30, 31, 32, 33, 38, 40, 41, 43, 45, 47, 48, 49, 53, 54, 56, 63, 69, 71, 73, 75, 80	$x_{14} + x_{29} + x_{41} + x_{55} + x_{61} + x_{62}$	781
1, 3, 5, 6, 8, 10, 12, 14, 15, 16, 18, 19, 24, 25, 31, 32, 33, 36, 38, 40, 41, 43, 45, 47, 48, 49, 50, 53, 54, 56, 63, 69, 71, 73, 75, 80	$x_{39} + x_{66}$	781
1, 3, 5, 6, 8, 10, 12, 14, 15, 16, 18, 19, 25, 30, 31, 32, 33, 38, 40, 41, 43, 45, 47, 48, 49, 50, 53, 54, 56, 60, 63, 69, 71, 73, 75, 80	$x_{24} + x_{55} + x_{61} + x_{66} + x_{67} + 1$	781
1, 3, 5, 6, 8, 10, 12, 14, 15, 16, 19, 21, 24, 25, 30, 31, 33, 38, 40, 41, 43, 45, 47, 49, 50, 53, 54, 56, 60, 63, 69, 70, 71, 73, 75, 80	$x_{12} + x_{27} + x_{32} + x_{33} + x_{40} + x_{42} + x_{51} + x_{53} + x_{57} + x_{58} + x_{60} + x_{64} + x_{80} + 1$	781
1, 3, 5, 6, 8, 10, 12, 14, 15, 16, 19, 21, 24, 25, 30, 31, 33, 38, 40, 41, 43, 45, 47, 49, 50, 53, 54, 56, 60, 63, 69, 70, 71, 73, 75, 80	$x_{27} + x_{32} + x_{42} + x_{53} + x_{58} + x_{60} + x_{64} + x_{78} + x_{80} + 1$	781
1, 3, 5, 6, 8, 12, 14, 15, 16, 18, 19, 24, 25, 27, 30, 31, 32, 33, 38, 40, 41, 45, 47, 49, 50, 53, 54, 56, 60, 63, 69, 70, 71, 73, 75, 80	$x_{11} + x_{24} + x_{25} + x_{29} + x_{30} + x_{31} + x_{40} + x_{41} + x_{45} + x_{50} + x_{52} + x_{53} + x_{54} + x_{56} + x_{58} + x_{61} + x_{65} + x_{66} + x_{67} + x_{68} + x_{77} + x_{79} + x_{80} + 1$	781

1, 3, 5, 6, 8, 12, 14, 15, 16, 19, 21, 24, 25, 30, 31, 32, 33, 36, 38, 40, 41, 43, 45, 47, 49, 50, 53, 54, 56, 60, 63, 69, 71, 73, 75, 80	$x_{14} + x_{16} + x_{27} + x_{29} + x_{30} + x_{31} + x_{40} + x_{41} + x_{42} + x_{43} + x_{54} + x_{55} + x_{56} + x_{57} + x_{58} + x_{64} + x_{79} + x_{80} + 1$	781
1, 3, 5, 6, 8, 12, 14, 15, 16, 19, 21, 24, 25, 30, 31, 32, 33, 36, 38, 40, 41, 43, 45, 47, 53, 54, 56, 60, 63, 69, 70, 71, 73, 75, 80	$x_{28} + x_{29} + x_{32} + x_{33} + x_{40} + x_{41} + x_{42} + x_{44} + x_{50} + x_{51} + x_{55} + x_{56} + x_{57} + x_{59} + x_{61} + x_{62} + x_{64} + x_{66} + x_{67} + x_{68} + x_{70} + x_{78} + 1$	781
1, 3, 5, 6, 8, 12, 14, 15, 16, 19, 21, 24, 25, 30, 31, 33, 36, 38, 40, 41, 43, 45, 47, 49, 50, 53, 54, 56, 60, 63, 69, 70, 71, 73, 75, 80	$x_{38} + x_{39} + x_{41} + x_{44} + x_{45} + x_{50} + x_{51} + x_{52} + x_{53} + x_{55} + x_{57} + x_{58} + x_{60} + x_{66} + x_{68} + x_{72} + x_{78} + x_{79} + 1$	781
1, 3, 5, 6, 8, 12, 14, 15, 16, 19, 21, 24, 25, 31, 33, 36, 38, 40, 41, 43, 45, 47, 48, 49, 50, 53, 54, 56, 60, 63, 69, 70, 71, 73, 75, 80	$x_{43} + x_{50} + x_{52} + x_{55} + x_{58} + x_{66} + x_{70} + x_{77} + 1$	781
1, 5, 6, 8, 12, 14, 15, 16, 18, 19, 24, 25, 27, 30, 31, 32, 33, 38, 40, 41, 43, 45, 47, 49, 50, 53, 54, 56, 60, 63, 69, 70, 71, 73, 75, 80	$x_{41} + x_{53} + x_{55} + x_{58} + x_{61} + x_{68}$	781
1, 5, 6, 8, 12, 14, 15, 16, 19, 24, 25, 27, 30, 31, 32, 33, 38, 40, 41, 43, 45, 47, 49, 50, 53, 54, 56, 60, 63, 69, 70, 71, 73, 75, 80	$x_{29} + x_{41} + x_{42} + x_{53} + x_{55} + x_{56} + x_{58} + x_{61} + x_{64} + x_{66} + x_{67} + x_{68} + x_{69}$	781
1, 5, 6, 8, 12, 14, 15, 16, 18, 19, 24, 25, 27, 30, 31, 32, 33, 38, 40, 41, 43, 45, 47, 49, 50, 53, 54, 56, 60, 63, 69, 70, 71, 73, 75, 80	$x_{14} + x_{55} + x_{58} + x_{61} + x_{64} + x_{66} + x_{68} + x_{80}$	781
1, 5, 6, 10, 12, 14, 15, 18, 19, 20, 21, 22, 23, 25, 27, 28, 29, 31, 33, 36, 38, 40, 41, 42, 45, 48, 49, 54, 60, 62, 63, 69, 73, 75, 77, 80	x_{64}	781
5, 6, 8, 10, 12, 14, 15, 16, 18, 19, 20, 21, 22, 23, 25, 27, 29, 31, 33, 36, 38, 40, 41, 42, 45, 48, 49, 54, 60, 62, 63, 69, 73, 75, 77, 80	$x_{66} + 1$	781
1, 2, 3, 5, 6, 7, 8, 12, 13, 14, 15, 16, 19, 21, 23, 27, 33, 36, 38, 40, 43, 45, 47, 52, 54, 56, 58, 60, 62, 69, 70, 71, 73, 74, 76, 79, 80	x_{56}	781
1, 3, 5, 6, 7, 8, 10, 12, 14, 15, 16, 19, 21, 23, 25, 33, 36, 40, 43, 45, 47, 49, 53, 54, 56, 58, 62, 63, 65, 69, 71, 74, 75, 77, 78, 79, 80	$x_{21} + x_{36} + x_{48} + x_{58} + x_{63} + 1$	781
1, 3, 5, 6, 7, 8, 10, 12, 14, 15, 16, 19, 21, 23, 25, 33, 38, 40, 43, 45, 47, 49, 53, 54, 56, 58, 62, 63, 65, 69, 71, 74, 75, 77, 78, 79, 80	$x_{19} + x_{27} + x_{45} + x_{54} + x_{64} + x_{66} + x_{72} + 1$	781
1, 3, 5, 6, 8, 12, 14, 15, 16, 18, 19, 24, 25, 30, 31, 32, 33, 36, 38, 40, 41, 43, 45, 47, 49, 50, 53, 54, 56, 60, 63, 69, 70, 71, 73, 75, 80	$x_5 + x_8 + x_{24} + x_{26} + x_{32} + x_{33} + x_{39} + x_{40} + x_{41} + x_{42} + x_{44} + x_{47} + x_{51} + x_{54} + x_{57} + x_{59} + x_{60} + x_{65} + x_{66} + x_{68} + x_{69} + x_{78} + x_{79}$	781
1, 3, 5, 6, 8, 12, 14, 15, 16, 19, 21, 25, 27, 31, 32, 33, 36, 38, 40, 41, 45, 47, 49, 50, 53, 54, 56, 60, 63, 69, 71, 73, 77, 80	$x_{25} + x_{52} + 1$	782
1, 5, 6, 7, 8, 10, 12, 14, 15, 19, 21, 24, 25, 27, 31, 36, 38, 39, 40, 41, 45, 47, 49, 53, 55, 56, 58, 62, 63, 66, 69, 71, 73, 77, 80	x_{40}	782
1, 3, 5, 6, 7, 10, 12, 15, 16, 19, 21, 23, 25, 27, 33, 36, 38, 40, 43, 45, 47, 49, 53, 56, 58, 61, 62, 63, 65, 69, 71, 74, 75, 77, 78	$x_{25} + 1$	782
1, 3, 5, 6, 10, 12, 14, 15, 16, 18, 19, 21, 25, 27, 31, 32, 33, 38, 40, 41, 43, 45, 47, 48, 49, 50, 53, 54, 56, 60, 63, 69, 71, 73, 75, 80	$x_{13} + x_{16} + x_{19} + x_{25} + x_{29} + x_{33} + x_{35} + x_{36} + x_{37} + x_{38} + x_{39} + x_{40} + x_{42} + x_{45} + x_{51} + x_{52} + x_{53} + x_{54} + x_{55} + x_{62} + x_{63} + x_{64} + x_{65} + x_{67} + x_{69} + x_{70} + x_{71} + x_{73} + x_{79} + x_{80} + 1$	782
5, 7, 8, 10, 12, 14, 15, 16, 19, 21, 23, 25, 27, 33, 34, 36, 38, 40, 45, 47, 49, 53, 54, 56, 60, 62, 63, 69, 71, 74, 80	$x_{38} + 1$	783
3, 5, 6, 7, 8, 10, 14, 15, 16, 21, 23, 25, 27, 33, 34, 36, 38, 40, 45, 47, 48, 49, 53, 54, 55, 56, 58, 61, 62, 63, 69, 71, 74, 80	$x_{27} + 1$	783
1, 5, 6, 8, 10, 12, 14, 15, 16, 18, 19, 21, 24, 25, 27, 30, 31, 32, 33, 38, 40, 41, 45, 47, 48, 49, 50, 53, 54, 56, 60, 63, 69, 71, 73, 75, 80	$x_{32} + x_{49} + x_{52} + x_{56} + x_{59} + x_{61} + x_{62} + x_{79} + 1$	783
1, 5, 6, 8, 10, 12, 14, 15, 16, 19, 21, 24, 25, 31, 33, 36, 38, 40, 41, 43, 45, 47, 48, 49, 50, 53, 54, 56, 60, 63, 69, 71, 73, 75, 80	$x_7 + x_{16} + x_{40} + x_{43} + x_{49} + x_{52} + x_{58} + x_{62} + x_{70} + x_{79} + 1$	783
1, 3, 5, 6, 8, 10, 12, 14, 15, 16, 18, 19, 24, 25, 30, 31, 32, 33, 38, 40, 41, 43, 45, 47, 49, 50, 53, 54, 56, 60, 63, 69, 71, 73, 75, 80	$x_{26} + x_{66} + x_{68} + 1$	783
1, 3, 6, 7, 10, 12, 15, 16, 19, 21, 23, 25, 33, 36, 38, 40, 43, 45, 47, 49, 53, 54, 56, 58, 61, 63, 65, 69, 71, 75, 77, 80	x_4	784
1, 5, 6, 7, 8, 10, 12, 14, 15, 16, 21, 23, 25, 27, 33, 36, 38, 40, 43, 45, 47, 49, 53, 54, 55, 56, 58, 60, 62, 63, 67, 69, 71, 80	$x_{53} + 1$	784
1, 6, 8, 10, 12, 14, 15, 16, 18, 19, 20, 21, 22, 23, 25, 27, 29, 33, 36, 38, 40, 41, 42, 45, 48, 49, 54, 60, 62, 63, 69, 73, 75, 80	x_{37}	784
3, 5, 6, 7, 8, 10, 12, 14, 15, 16, 19, 21, 23, 25, 27, 33, 36, 38, 40, 43, 45, 47, 49, 53, 54, 56, 58, 60, 62, 63, 69, 71, 74, 80	x_{36}	784
1, 5, 6, 7, 8, 10, 12, 14, 15, 16, 19, 21, 23, 25, 27, 29, 31, 33, 36, 38, 40, 41, 43, 45, 47, 49, 53, 54, 56, 58, 75, 76, 80	$x_{12} + 1$	785
1, 3, 6, 7, 10, 12, 14, 16, 19, 21, 23, 25, 27, 33, 36, 38, 40, 43, 45, 47, 49, 53, 54, 56, 58, 63, 65, 69, 71, 74, 75, 77, 78, 80	x_{34}	785
1, 5, 6, 7, 8, 12, 13, 14, 15, 16, 19, 21, 23, 25, 27, 33, 36, 38, 40, 43, 45, 47, 49, 52, 54, 56, 58, 60, 62, 69, 71, 73, 79, 80	x_{54}	785
1, 3, 5, 6, 7, 10, 12, 14, 16, 19, 21, 23, 25, 27, 33, 36, 38, 40, 43, 45, 47, 49, 53, 54, 56, 58, 61, 62, 63, 69, 71, 74, 75, 77, 78, 80	$x_{13} + x_{55} + x_{60} + x_{64}$	785
1, 3, 5, 6, 7, 10, 12, 14, 16, 19, 21, 23, 25, 27, 33, 36, 40, 41, 43, 45, 47, 49, 53, 54, 56, 58, 61, 62, 63, 65, 69, 71, 74, 75, 77, 78, 80	$x_{22} + x_{49} + x_{64}$	786
1, 3, 6, 7, 10, 12, 14, 16, 19, 21, 23, 25, 27, 33, 36, 40, 41, 43, 45, 47, 49, 53, 54, 56, 58, 61, 62, 63, 65, 69, 71, 74, 75, 77, 78, 80	$x_{14} + x_{23} + x_{41} + x_{47} + x_{49} + x_{50} + x_{58} + x_{64}$	786
1, 3, 5, 6, 7, 8, 10, 12, 14, 16, 19, 21, 23, 25, 27, 33, 36, 40, 43, 45, 47, 49, 53, 54, 56, 58, 61, 62, 63, 65, 69, 71, 74, 75, 77, 78, 80	$x_3 + x_4 + x_{20} + x_{22} + x_{30} + x_{34} + x_{38} + x_{40} + x_{42} + x_{45} + x_{49} + x_{51} + x_{58} + x_{61} + x_{65} + x_{67} + x_{69} + x_{72} + x_{78}$	786
1, 3, 6, 7, 8, 10, 12, 14, 15, 16, 19, 21, 23, 25, 27, 33, 38, 40, 43, 45, 47, 49, 53, 54, 56, 58, 62, 63, 69, 71, 75, 77, 79	$x_9 + x_{29} + x_{30} + x_{32} + x_{42} + x_{43} + x_{49} + x_{51} + x_{57} + x_{58} + x_{59} + x_{60} + x_{62} + x_{64} + x_{66} + x_{67} + x_{68} + x_{69} + x_{70} + x_{72} + x_{76}$	791
1, 3, 6, 7, 8, 10, 12, 15, 16, 19, 21, 23, 25, 27, 33, 36, 40, 41, 43, 45, 47, 49, 53, 56, 58, 62, 63, 69, 71, 75, 77, 80	$x_{17} + x_{26} + x_{30} + x_{32} + x_{41} + x_{43} + x_{47} + x_{57} + x_{62} + x_{65} + x_{66} + x_{70} + x_{72} + x_{74} + 1$	791

maxterm bits	superpoly	round
1, 3, 6, 7, 8, 10, 12, 15, 16, 19, 21, 23, 25, 27, 33, 36, 40, 41, 43, 45, 47, 49, 53, 56, 58, 62, 63, 69, 71, 75, 77, 80	$x_{14} + x_{17} + x_{26} + x_{30} + x_{43} + x_{47} + x_{50} + x_{57} + x_{58} + x_{59} + x_{65} + x_{70} + x_{72} + x_{74} + x_{77} + 1$	791
1, 3, 5, 6, 7, 8, 10, 12, 14, 16, 19, 21, 23, 25, 27, 33, 36, 40, 41, 43, 45, 47, 49, 53, 56, 58, 63, 65, 69, 71, 75, 77, 79	$x_{12} + x_{26} + x_{30} + x_{39} + x_{41} + x_{45} + x_{47} + x_{57} + x_{58} + x_{59} + x_{62} + x_{64} + x_{74} + x_{76} + 1$	791
1, 3, 5, 6, 7, 8, 10, 12, 15, 16, 19, 21, 23, 25, 27, 33, 36, 38, 40, 41, 43, 45, 47, 49, 53, 56, 58, 62, 63, 65, 69, 71, 75, 77	$x_5 + x_{18} + x_{20} + x_{26} + x_{28} + x_{29} + x_{30} + x_{31} + x_{32} + x_{41} + x_{42} + x_{44} + x_{50} + x_{51} + x_{56} + x_{57} + x_{62} + x_{64} + x_{67} + x_{69} + x_{70} + x_{71} + x_{74} + x_{77} + x_{78} + 1$	791
1, 3, 5, 6, 7, 8, 10, 12, 14, 15, 16, 19, 21, 23, 25, 27, 33, 38, 40, 41, 43, 45, 47, 49, 53, 56, 58, 61, 62, 63, 65, 69, 71, 75, 77	$x_1 + x_{28} + x_{32} + x_{47} + x_{58} + x_{59} + x_{62} + x_{64} + x_{74}$	791
1, 3, 5, 6, 7, 8, 10, 12, 14, 15, 16, 19, 21, 23, 25, 27, 33, 40, 41, 43, 45, 47, 49, 53, 56, 58, 61, 62, 63, 65, 69, 71, 75, 77, 80	$x_{10} + x_{11} + x_{12} + x_{13} + x_{15} + x_{17} + x_{19} + x_{20} + x_{29} + x_{31} + x_{32} + x_{33} + x_{37} + x_{39} + x_{40} + x_{41} + x_{42} + x_{44} + x_{46} + x_{48} + x_{49} + x_{50} + x_{53} + x_{57} + x_{60} + x_{67} + x_{70} + x_{71} + x_{76} + x_{78} + x_{79}$	791
1, 5, 6, 7, 8, 10, 12, 14, 15, 16, 19, 21, 23, 25, 27, 29, 31, 33, 36, 38, 40, 41, 42, 45, 47, 49, 53, 58, 63, 69, 71, 72, 76, 79, 80	x_{61}	791
1, 3, 6, 7, 8, 10, 12, 14, 15, 16, 19, 21, 23, 25, 33, 36, 38, 40, 41, 43, 45, 47, 49, 53, 54, 56, 58, 62, 63, 69, 71, 74, 75, 77, 78, 80	$x_{43} + x_{47} + x_{58} + x_{70} + x_{74} + 1$	791
1, 3, 5, 6, 7, 8, 10, 12, 14, 15, 16, 19, 21, 23, 25, 27, 33, 38, 40, 41, 43, 45, 47, 49, 53, 56, 58, 62, 63, 69, 71, 75, 77, 80	$x_{12} + x_{17} + x_{26} + x_{27} + x_{29} + x_{30} + x_{32} + x_{40} + x_{43} + x_{45} + x_{46} + x_{49} + x_{53} + x_{54} + x_{56} + x_{59} + x_{62} + x_{64} + x_{65} + x_{67} + x_{69} + x_{72} + x_{74} + x_{75}$	792
1, 3, 5, 6, 7, 8, 10, 12, 14, 15, 16, 19, 21, 23, 25, 27, 33, 36, 38, 40, 41, 43, 45, 47, 49, 53, 56, 58, 61, 63, 69, 71, 75, 77, 78, 79, 80	$x_{12} + x_{14} + x_{26} + x_{30} + x_{40} + x_{41} + x_{47} + x_{48} + x_{56} + x_{66} + x_{67} + x_{68} + x_{74} + x_{75} + 1$	792
1, 3, 5, 6, 7, 8, 10, 12, 14, 15, 16, 19, 21, 23, 25, 27, 33, 36, 38, 40, 41, 43, 45, 47, 49, 53, 54, 56, 58, 63, 69, 71, 74, 75, 77, 78, 79, 80	$x_{16} + x_{43} + x_{56}$	792
1, 3, 5, 6, 7, 8, 10, 12, 14, 15, 16, 19, 21, 23, 25, 27, 33, 36, 38, 40, 43, 45, 47, 49, 53, 54, 56, 58, 61, 63, 65, 69, 71, 75, 77, 78, 79, 80	$x_{14} + x_{16} + x_{26} + x_{29} + x_{30} + x_{41} + x_{45} + x_{55} + x_{56} + x_{59} + x_{62} + x_{64} + x_{66} + x_{68} + x_{70} + x_{71} + x_{72} + 1$	792
1, 3, 6, 7, 8, 10, 12, 14, 16, 19, 21, 23, 25, 27, 33, 36, 38, 40, 41, 43, 45, 47, 49, 53, 54, 56, 58, 61, 63, 69, 71, 75, 77, 80	$x_{45} + x_{72}$	793
1, 6, 7, 8, 10, 12, 14, 16, 19, 21, 23, 25, 27, 30, 31, 33, 38, 40, 43, 45, 47, 49, 51, 52, 56, 58, 63, 67, 69, 71, 73, 77, 80	$x_{10} + x_{55}$	798
1, 3, 5, 6, 7, 8, 10, 12, 15, 16, 19, 21, 23, 25, 27, 33, 38, 40, 43, 45, 47, 49, 53, 54, 56, 58, 62, 63, 65, 69, 71, 74, 75, 77, 80	$x_{36} + x_{52} + x_{60} + x_{63}$	798
1, 3, 5, 6, 7, 8, 10, 12, 15, 16, 19, 21, 23, 25, 27, 33, 38, 40, 43, 45, 47, 49, 53, 54, 56, 58, 61, 62, 63, 65, 69, 71, 74, 75, 77, 80	$x_6 + x_{11} + x_{25} + x_{33} + x_{36} + x_{53} + x_{60} + x_{62} + x_{63} + x_{64} + x_{79}$	798

Trivium-784

maxterm bits	superpoly	round
1, 3, 6, 7, 10, 12, 15, 16, 19, 21, 23, 25, 33, 36, 38, 40, 43, 45, 47, 49, 53, 54, 56, 58, 61, 63, 65, 69, 71, 75, 77, 80	x_4	784
1, 5, 6, 8, 10, 12, 14, 15, 16, 18, 19, 20, 21, 23, 25, 27, 29, 31, 33, 36, 38, 40, 41, 42, 45, 49, 54, 60, 62, 69, 73, 75, 80	x_{60}	784
5, 6, 7, 8, 10, 12, 14, 15, 16, 19, 21, 23, 25, 27, 33, 36, 38, 40, 43, 45, 47, 49, 53, 54, 56, 58, 61, 62, 63, 67, 69, 71, 80	$x_{56} + 1$	784
1, 3, 5, 6, 7, 10, 12, 15, 16, 19, 21, 23, 25, 27, 33, 36, 38, 40, 43, 45, 47, 49, 53, 54, 56, 58, 61, 63, 65, 69, 71, 75, 77, 78	$x_2 + x_9 + x_{13} + x_{14} + x_{22} + x_{23} + x_{30} + x_{36} + x_{38} + x_{39} + x_{40} + x_{42} + x_{47} + x_{48} + x_{51} + x_{56} + x_{65} + x_{67} + x_{68} + x_{69} + x_{74} + x_{75}$	784
1, 3, 5, 6, 8, 10, 12, 14, 15, 16, 19, 21, 25, 30, 31, 32, 33, 38, 40, 41, 43, 45, 47, 48, 49, 50, 53, 54, 56, 63, 69, 71, 75, 80	x_{38}	784
1, 3, 6, 7, 10, 12, 15, 16, 19, 21, 23, 25, 27, 33, 36, 38, 40, 41, 43, 45, 47, 49, 53, 54, 56, 58, 61, 63, 69, 71, 75, 77, 78, 79	$x_{38} + x_{47} + x_{74}$	784
1, 5, 6, 7, 8, 10, 12, 14, 15, 16, 21, 23, 25, 27, 33, 36, 38, 40, 43, 45, 47, 49, 53, 54, 55, 56, 58, 60, 62, 63, 67, 69, 71, 80	$x_{53} + 1$	784
1, 5, 6, 7, 8, 10, 12, 14, 15, 16, 21, 23, 25, 27, 33, 36, 38, 40, 43, 45, 47, 49, 53, 54, 56, 58, 61, 62, 63, 67, 69, 71, 80	x_{58}	784
1, 6, 8, 10, 12, 14, 15, 16, 18, 19, 20, 21, 22, 23, 25, 27, 29, 33, 36, 38, 40, 41, 42, 45, 48, 49, 54, 60, 62, 63, 69, 73, 75, 80	x_{37}	784
1, 6, 8, 10, 12, 14, 15, 16, 18, 19, 20, 21, 23, 25, 27, 29, 33, 36, 38, 40, 41, 45, 48, 49, 54, 56, 60, 62, 69, 73, 75, 77, 80	x_{64}	784
3, 5, 6, 7, 8, 10, 12, 14, 15, 16, 19, 21, 23, 25, 33, 36, 38, 40, 43, 45, 47, 49, 53, 54, 56, 58, 60, 62, 63, 69, 71, 74, 80	x_{36}	784
6, 8, 10, 12, 14, 15, 16, 18, 19, 20, 21, 22, 23, 25, 27, 29, 33, 36, 38, 40, 41, 45, 48, 49, 54, 56, 60, 62, 63, 69, 73, 75, 77, 80	x_{66}	784
5, 6, 7, 8, 10, 12, 14, 15, 16, 19, 21, 23, 25, 27, 33, 34, 36, 38, 40, 43, 45, 47, 49, 53, 54, 55, 56, 58, 60, 62, 63, 69, 71, 74, 80	$x_{67} + 1$	784
5, 6, 8, 10, 12, 14, 15, 18, 19, 20, 21, 22, 23, 25, 27, 28, 29, 31, 33, 36, 38, 40, 41, 42, 45, 49, 54, 60, 62, 63, 69, 73, 75, 80	x_{62}	784
1, 5, 7, 8, 10, 12, 14, 15, 16, 19, 21, 23, 25, 27, 33, 34, 36, 38, 40, 43, 45, 47, 48, 49, 53, 54, 56, 58, 60, 61, 62, 63, 69, 71, 74, 80	$x_{69} + 1$	784
3, 5, 7, 8, 10, 12, 14, 15, 16, 19, 21, 23, 25, 27, 33, 36, 38, 40, 43, 45, 47, 48, 49, 53, 54, 56, 58, 60, 61, 62, 63, 67, 69, 71, 74, 80	x_{40}	784
1, 5, 6, 7, 8, 10, 12, 14, 15, 16, 19, 21, 23, 25, 27, 29, 31, 33, 36, 38, 40, 41, 45, 47, 49, 53, 58, 60, 63, 71, 75, 76, 80	$x_{12} + 1$	785
1, 5, 6, 7, 10, 12, 16, 19, 21, 23, 24, 25, 27, 31, 33, 36, 38, 40, 41, 43, 47, 49, 52, 56, 58, 60, 63, 67, 69, 71, 73, 77, 80	x_{42}	785

1, 5, 6, 10, 12, 14, 15, 16, 19, 21, 25, 27, 30, 31, 33, 36, 38, 40, 41, 43, 45, 47, 48, 49, 53, 54, 56, 63, 69, 71, 73, 75, 80	x_{55}	785
1, 3, 6, 7, 10, 12, 14, 16, 19, 21, 23, 25, 27, 33, 36, 38, 40, 43, 45, 47, 49, 53, 54, 56, 58, 63, 65, 69, 71, 74, 75, 77, 78, 80	x_{34}	785
1, 5, 6, 7, 8, 12, 13, 14, 15, 16, 19, 21, 23, 25, 27, 33, 36, 38, 40, 43, 45, 47, 49, 52, 54, 56, 58, 60, 62, 69, 71, 73, 79, 80	x_{54}	785
1, 3, 5, 6, 7, 10, 12, 14, 16, 19, 21, 23, 25, 27, 33, 36, 40, 43, 45, 47, 49, 53, 54, 56, 58, 61, 62, 63, 69, 71, 74, 75, 77, 78, 80	$x_{13} + x_{55} + x_{60} + x_{64}$	785
1, 3, 5, 6, 7, 10, 12, 14, 16, 19, 21, 23, 25, 27, 33, 36, 40, 41, 43, 45, 47, 49, 53, 54, 56, 58, 62, 63, 65, 69, 71, 74, 75, 77, 78, 80	$x_{22} + x_{49} + x_{64}$	786
1, 3, 6, 7, 8, 10, 12, 14, 16, 19, 21, 23, 25, 27, 33, 36, 40, 43, 45, 47, 49, 53, 54, 56, 58, 61, 62, 63, 65, 69, 71, 74, 75, 77, 78, 80	$x_3 + x_4 + x_7 + x_{12} + x_{20} + x_{22} + x_{30} + x_{36} + x_{39} + x_{42} + x_{43} + x_{45} + x_{47} + x_{51} + x_{58} + x_{63} + x_{69} + x_{70} + x_{72} + x_{78} + 1$	786
1, 3, 6, 7, 10, 12, 14, 16, 19, 21, 23, 25, 27, 33, 36, 40, 41, 43, 45, 47, 49, 53, 54, 56, 58, 61, 62, 63, 65, 69, 71, 74, 75, 77, 78, 80	$x_{14} + x_{23} + x_{41} + x_{47} + x_{49} + x_{50} + x_{58} + x_{64}$	786
1, 3, 5, 6, 7, 8, 10, 12, 14, 16, 19, 21, 23, 25, 27, 33, 36, 40, 43, 45, 47, 49, 53, 54, 56, 58, 61, 62, 63, 65, 69, 71, 74, 75, 77, 78, 80	$x_3 + x_4 + x_{14} + x_{22} + x_{30} + x_{34} + x_{38} + x_{41} + x_{42} + x_{45} + x_{47} + x_{49} + x_{51} + x_{58} + x_{61} + x_{65} + x_{69} + x_{72} + x_{78}$	786
1, 3, 5, 6, 7, 8, 10, 12, 14, 16, 19, 21, 23, 25, 27, 33, 36, 40, 43, 45, 47, 49, 53, 54, 56, 58, 61, 62, 63, 65, 69, 71, 74, 75, 77, 78, 80	$x_3 + x_4 + x_{20} + x_{22} + x_{30} + x_{34} + x_{38} + x_{40} + x_{42} + x_{45} + x_{49} + x_{51} + x_{58} + x_{61} + x_{65} + x_{67} + x_{69} + x_{72} + x_{78}$	786
1, 3, 6, 7, 8, 10, 12, 14, 15, 16, 19, 21, 23, 25, 27, 33, 38, 40, 43, 45, 47, 49, 53, 56, 58, 62, 63, 69, 71, 75, 77, 79	$x_9 + x_{29} + x_{30} + x_{32} + x_{42} + x_{43} + x_{49} + x_{51} + x_{57} + x_{58} + x_{59} + x_{60} + x_{62} + x_{64} + x_{66} + x_{67} + x_{68} + x_{69} + x_{70} + x_{72} + x_{76}$	791
1, 3, 6, 7, 8, 10, 12, 15, 16, 19, 21, 23, 25, 27, 33, 36, 40, 41, 43, 45, 47, 49, 53, 56, 58, 62, 63, 69, 71, 75, 77, 80	$x_{17} + x_{26} + x_{30} + x_{32} + x_{41} + x_{43} + x_{47} + x_{57} + x_{62} + x_{65} + x_{66} + x_{70} + x_{72} + x_{74} + 1$	791
1, 3, 6, 7, 8, 10, 12, 15, 16, 19, 21, 23, 25, 27, 33, 36, 40, 41, 43, 45, 47, 49, 53, 56, 58, 62, 63, 69, 71, 75, 77, 80	$x_1 + x_{17} + x_{26} + x_{28} + x_{41} + x_{43} + x_{47} + x_{49} + x_{59} + x_{62} + x_{64} + x_{65} + x_{66} + x_{70} + x_{72} + x_{74} + x_{76} + 1$	791
1, 3, 6, 7, 8, 10, 12, 15, 16, 19, 21, 23, 25, 27, 33, 36, 40, 41, 43, 45, 47, 49, 53, 56, 58, 62, 63, 69, 71, 75, 77, 80	$x_{14} + x_{17} + x_{26} + x_{30} + x_{43} + x_{47} + x_{50} + x_{57} + x_{58} + x_{59} + x_{65} + x_{70} + x_{72} + x_{74} + x_{77} + 1$	791
1, 3, 5, 6, 7, 8, 10, 12, 14, 16, 19, 21, 23, 25, 27, 33, 36, 40, 41, 43, 45, 47, 49, 53, 56, 58, 63, 65, 69, 71, 75, 77, 79	$x_{12} + x_{26} + x_{30} + x_{39} + x_{41} + x_{45} + x_{47} + x_{57} + x_{58} + x_{59} + x_{62} + x_{64} + x_{74} + x_{76} + 1$	791
1, 3, 5, 6, 7, 8, 10, 12, 15, 16, 19, 21, 23, 25, 27, 33, 36, 38, 40, 41, 43, 45, 47, 49, 53, 56, 58, 62, 63, 65, 69, 71, 75, 77	$x_5 + x_{18} + x_{20} + x_{26} + x_{28} + x_{29} + x_{30} + x_{31} + x_{32} + x_{41} + x_{42} + x_{44} + x_{50} + x_{51} + x_{56} + x_{57} + x_{62} + x_{64} + x_{67} + x_{69} + x_{70} + x_{71} + x_{74} + x_{77} + x_{78} + 1$	791
1, 3, 6, 7, 8, 10, 12, 14, 15, 16, 19, 21, 23, 25, 27, 33, 40, 41, 43, 45, 47, 49, 53, 56, 58, 61, 62, 63, 65, 69, 71, 75, 77, 80	$x_7 + x_9 + x_{10} + x_{11} + x_{12} + x_{13} + x_{15} + x_{17} + x_{19} + x_{20} + x_{26} + x_{30} + x_{31} + x_{33} + x_{37} + x_{39} + x_{40} + x_{41} + x_{42} + x_{44} + x_{45} + x_{46} + x_{47} + x_{48} + x_{50} + x_{51} + x_{53} + x_{56} + x_{59} + x_{60} + x_{64} + x_{67} + x_{68} + x_{70} + x_{71} + x_{72} + x_{74} + x_{78} + x_{79} + 1$	791
1, 3, 5, 6, 7, 8, 10, 12, 14, 15, 16, 19, 21, 23, 25, 27, 33, 38, 40, 41, 43, 45, 47, 49, 53, 56, 58, 61, 62, 63, 65, 69, 71, 75, 77	$x_1 + x_{28} + x_{32} + x_{47} + x_{58} + x_{59} + x_{62} + x_{64} + x_{74}$	791
1, 3, 5, 6, 7, 8, 10, 12, 14, 15, 16, 19, 21, 23, 25, 27, 33, 40, 41, 43, 45, 47, 49, 53, 56, 58, 61, 62, 63, 65, 69, 71, 75, 77, 80	$x_3 + x_4 + x_6 + x_7 + x_9 + x_{10} + x_{13} + x_{15} + x_{19} + x_{22} + x_{28} + x_{30} + x_{33} + x_{34} + x_{35} + x_{38} + x_{39} + x_{40} + x_{41} + x_{43} + x_{44} + x_{47} + x_{48} + x_{49} + x_{50} + x_{53} + x_{54} + x_{55} + x_{56} + x_{58} + x_{61} + x_{62} + x_{65} + x_{66} + x_{67} + x_{68} + x_{69} + x_{71} + x_{72} + x_{76} + x_{77} + x_{78}$	791
1, 3, 5, 6, 7, 8, 10, 12, 14, 15, 16, 19, 21, 23, 25, 27, 33, 40, 41, 43, 45, 47, 49, 53, 56, 58, 61, 62, 63, 65, 69, 71, 75, 77, 80	$x_{10} + x_{11} + x_{12} + x_{13} + x_{15} + x_{17} + x_{19} + x_{20} + x_{29} + x_{31} + x_{32} + x_{33} + x_{37} + x_{39} + x_{40} + x_{41} + x_{42} + x_{44} + x_{46} + x_{48} + x_{49} + x_{50} + x_{53} + x_{57} + x_{60} + x_{67} + x_{70} + x_{71} + x_{76} + x_{78} + x_{79}$	791
1, 3, 5, 6, 7, 8, 10, 12, 14, 15, 16, 19, 21, 23, 25, 27, 33, 40, 41, 43, 45, 47, 49, 53, 56, 58, 61, 62, 63, 65, 69, 71, 75, 77, 80	$x_{10} + x_{11} + x_{12} + x_{13} + x_{14} + x_{15} + x_{17} + x_{19} + x_{20} + x_{26} + x_{29} + x_{31} + x_{33} + x_{37} + x_{39} + x_{40} + x_{42} + x_{44} + x_{46} + x_{48} + x_{53} + x_{57} + x_{58} + x_{59} + x_{60} + x_{66} + x_{67} + x_{68} + x_{70} + x_{71} + x_{72} + x_{77} + x_{78} + x_{79} + 1$	791
1, 3, 6, 7, 8, 10, 12, 14, 15, 16, 19, 21, 23, 25, 27, 33, 40, 41, 43, 45, 47, 49, 53, 54, 56, 58, 61, 62, 63, 65, 69, 71, 75, 77, 79	$x_3 + x_4 + x_6 + x_{11} + x_{15} + x_{17} + x_{19} + x_{20} + x_{22} + x_{30} + x_{34} + x_{35} + x_{37} + x_{38} + x_{43} + x_{47} + x_{51} + x_{54} + x_{57} + x_{58} + x_{60} + x_{61} + x_{64} + x_{65} + x_{67} + x_{68} + x_{70} + x_{72} + x_{74} + x_{77} + x_{79} + 1$	791
1, 3, 6, 7, 8, 10, 12, 14, 15, 16, 19, 21, 23, 25, 27, 33, 40, 41, 43, 45, 47, 49, 53, 54, 56, 58, 61, 62, 63, 65, 69, 71, 75, 77, 79	$x_3 + x_4 + x_6 + x_{11} + x_{12} + x_{15} + x_{17} + x_{18} + x_{19} + x_{20} + x_{22} + x_{30} + x_{34} + x_{35} + x_{37} + x_{38} + x_{39} + x_{42} + x_{43} + x_{45} + x_{47} + x_{50} + x_{54} + x_{56} + x_{57} + x_{58} + x_{61} + x_{65} + x_{69} + x_{70} + x_{74} + x_{78} + x_{79} + 1$	791
1, 3, 6, 7, 8, 10, 12, 14, 15, 16, 19, 21, 23, 25, 33, 36, 38, 40, 43, 45, 47, 49, 53, 54, 56, 58, 62, 63, 69, 71, 74, 75, 77, 78, 80	$x_1 + x_5 + x_9 + x_{14} + x_{18} + x_{20} + x_{26} + x_{28} + x_{32} + x_{41} + x_{42} + x_{43} + x_{45} + x_{47} + x_{49} + x_{66} + x_{67} + x_{69} + x_{70} + x_{76} + x_{78}$	791
1, 5, 6, 7, 8, 10, 12, 14, 15, 16, 19, 21, 23, 25, 27, 29, 31, 33, 36, 38, 40, 41, 42, 45, 47, 49, 53, 58, 63, 69, 71, 72, 76, 79, 80	x_{61}	791
1, 3, 5, 6, 7, 8, 10, 12, 14, 15, 16, 19, 21, 23, 25, 27, 33, 40, 41, 43, 45, 47, 49, 53, 54, 56, 58, 61, 62, 63, 65, 69, 71, 75, 77, 79	$x_3 + x_4 + x_6 + x_7 + x_9 + x_{11} + x_{17} + x_{18} + x_{20} + x_{22} + x_{26} + x_{28} + x_{29} + x_{31} + x_{34} + x_{35} + x_{37} + x_{38} + x_{39} + x_{41} + x_{46} + x_{50} + x_{51} + x_{54} + x_{55} + x_{56} + x_{58} + x_{61} + x_{65} + x_{66} + x_{67} + x_{74} + x_{76} + x_{78} + x_{79} + 1$	791
1, 3, 6, 7, 8, 10, 12, 14, 15, 16, 19, 21, 23, 25, 33, 36, 38, 40, 41, 43, 45, 47, 49, 53, 54, 56, 58, 62, 63, 69, 71, 74, 75, 77, 78, 80	$x_{43} + x_{47} + x_{58} + x_{70} + x_{74} + 1$	791

maxterm bits	superpoly	round
1, 3, 5, 6, 7, 8, 10, 12, 14, 15, 16, 19, 21, 23, 25, 27, 33, 38, 40, 41, 43, 45, 47, 49, 53, 56, 58, 62, 63, 69, 71, 75, 77, 80	$x_{12} + x_{17} + x_{26} + x_{27} + x_{29} + x_{30} + x_{32} + x_{40} + x_{43} + x_{45} + x_{46} + x_{49} + x_{53} + x_{54} + x_{56} + x_{59} + x_{62} + x_{64} + x_{65} + x_{67} + x_{69} + x_{72} + x_{74} + x_{75}$	792
1, 3, 5, 6, 7, 8, 10, 12, 14, 15, 16, 19, 21, 23, 25, 27, 33, 36, 38, 40, 41, 43, 45, 47, 49, 53, 54, 56, 58, 63, 69, 71, 75, 77, 78, 79, 80	$x_{12} + x_{26} + x_{39} + x_{56} + x_{68} + 1$	792
1, 3, 5, 6, 7, 8, 10, 12, 14, 15, 16, 19, 21, 23, 25, 27, 33, 36, 38, 40, 41, 43, 45, 47, 49, 53, 56, 58, 61, 63, 69, 71, 75, 77, 78, 79, 80	$x_{12} + x_{14} + x_{26} + x_{30} + x_{40} + x_{41} + x_{47} + x_{48} + x_{56} + x_{66} + x_{67} + x_{68} + x_{74} + x_{75} + 1$	792
1, 3, 5, 6, 7, 8, 10, 12, 14, 15, 16, 19, 21, 23, 25, 27, 33, 36, 38, 40, 41, 43, 45, 47, 49, 53, 54, 56, 58, 63, 69, 71, 74, 75, 77, 78, 79, 80	$x_{16} + x_{43} + x_{56}$	792
1, 3, 5, 6, 7, 8, 10, 12, 14, 15, 16, 19, 21, 23, 25, 27, 33, 36, 38, 40, 43, 45, 47, 49, 53, 54, 56, 58, 61, 63, 65, 69, 71, 75, 77, 78, 79, 80	$x_{14} + x_{16} + x_{26} + x_{29} + x_{30} + x_{41} + x_{45} + x_{56} + x_{59} + x_{62} + x_{64} + x_{66} + x_{68} + x_{70} + x_{71} + x_{72} + 1$	792
1, 3, 6, 7, 8, 10, 12, 14, 16, 19, 21, 23, 25, 27, 33, 36, 38, 40, 41, 43, 45, 47, 49, 53, 54, 56, 58, 61, 63, 69, 71, 75, 77, 80	$x_{45} + x_{72}$	793
1, 6, 7, 8, 10, 12, 14, 16, 19, 21, 23, 25, 27, 30, 31, 33, 38, 40, 43, 45, 47, 49, 51, 52, 56, 58, 63, 67, 69, 71, 73, 77, 80	$x_{10} + x_{55}$	798
1, 3, 5, 6, 7, 8, 10, 12, 15, 16, 19, 21, 23, 25, 27, 33, 38, 40, 43, 45, 47, 49, 53, 54, 56, 58, 62, 63, 65, 69, 71, 74, 75, 77, 80	$x_{36} + x_{52} + x_{60} + x_{63}$	798
1, 3, 5, 6, 7, 8, 10, 12, 15, 16, 19, 21, 23, 25, 27, 33, 38, 40, 43, 45, 47, 49, 53, 54, 56, 58, 62, 63, 65, 69, 71, 74, 75, 77, 80	$x_{10} + x_{17} + x_{27} + x_{36} + x_{37} + x_{40} + x_{52} + x_{59} + x_{60} + x_{63} + x_{66} + x_{67}$	798
1, 3, 5, 6, 7, 8, 10, 12, 14, 16, 19, 21, 23, 25, 33, 36, 38, 40, 43, 45, 47, 49, 53, 56, 58, 61, 62, 63, 65, 69, 71, 74, 75, 77, 78, 79	$x_{27} + x_{54} + x_{60}$	798
1, 3, 5, 6, 7, 8, 10, 12, 15, 16, 19, 21, 23, 25, 27, 33, 38, 40, 43, 45, 47, 49, 53, 54, 56, 58, 62, 63, 65, 69, 71, 74, 75, 77, 80	$x_6 + x_{11} + x_{25} + x_{33} + x_{36} + x_{53} + x_{60} + x_{62} + x_{63} + x_{64} + x_{79}$	798
1, 3, 5, 6, 7, 8, 10, 12, 15, 16, 19, 21, 23, 25, 27, 33, 38, 40, 43, 45, 47, 49, 53, 54, 56, 58, 61, 62, 63, 65, 69, 71, 74, 75, 77, 80	$x_6 + x_{11} + x_{25} + x_{33} + x_{36} + x_{52} + x_{53} + x_{60} + x_{62} + x_{63} + x_{64} + x_{79}$	798
5, 6, 7, 8, 10, 12, 14, 15, 16, 19, 21, 23, 25, 27, 33, 34, 36, 38, 40, 43, 45, 47, 49, 53, 54, 56, 58, 60, 61, 62, 63, 67, 69, 71, 74, 80	$x_{65} + x_{66} + x_{67} + 1$	798
1, 3, 5, 6, 7, 8, 10, 12, 14, 15, 16, 19, 21, 23, 25, 27, 33, 36, 38, 40, 43, 45, 47, 49, 54, 56, 58, 62, 69, 71, 73, 80	$x_{25} + 1$	799
1, 3, 5, 6, 7, 8, 10, 12, 14, 15, 16, 19, 21, 23, 25, 27, 33, 36, 40, 43, 45, 47, 49, 53, 56, 58, 63, 69, 71, 75, 77, 80	$x_{12} + x_{38} + x_{39} + x_{40}$	799

Trivium-799

maxterm bits	superpoly	round
1, 6, 8, 10, 12, 14, 18, 20, 23, 25, 27, 31, 33, 36, 38, 40, 43, 45, 47, 49, 53, 54, 56, 58, 62, 63, 69, 73, 75, 77, 80	x_{60}	799
1, 3, 5, 6, 7, 8, 10, 12, 14, 15, 16, 19, 21, 23, 25, 27, 33, 36, 38, 40, 43, 45, 47, 49, 53, 54, 56, 58, 62, 69, 71, 73, 80	$x_{25} + 1$	799
1, 3, 5, 6, 7, 8, 10, 12, 14, 15, 16, 19, 21, 23, 25, 27, 33, 36, 40, 43, 45, 47, 49, 53, 56, 58, 63, 69, 71, 75, 77, 80	$x_{25} + x_{40}$	799
1, 3, 5, 6, 7, 8, 10, 12, 14, 15, 16, 19, 21, 23, 25, 27, 33, 36, 40, 43, 45, 47, 49, 53, 56, 58, 63, 69, 71, 75, 77, 80	$x_{12} + x_{38} + x_{39} + x_{40}$	799
1, 5, 6, 7, 8, 10, 12, 14, 15, 16, 18, 20, 23, 25, 27, 33, 36, 38, 40, 41, 43, 47, 49, 53, 56, 58, 63, 69, 71, 75, 77, 80	$x_{67} + 1$	799
1, 5, 6, 7, 8, 10, 12, 14, 15, 16, 19, 21, 23, 25, 27, 31, 33, 36, 38, 40, 41, 45, 47, 49, 53, 56, 58, 63, 69, 71, 75, 80	x_{42}	799
1, 5, 6, 7, 8, 10, 12, 14, 15, 19, 21, 23, 25, 27, 31, 36, 38, 40, 41, 45, 47, 49, 53, 56, 58, 63, 69, 71, 73, 75, 77, 80	x_{53}	799
1, 5, 6, 8, 10, 12, 14, 15, 16, 18, 19, 20, 21, 23, 25, 27, 31, 33, 36, 38, 40, 41, 45, 49, 54, 56, 62, 69, 73, 75, 77, 80	x_{64}	799
1, 3, 5, 6, 7, 8, 10, 12, 14, 15, 16, 19, 21, 23, 25, 27, 33, 36, 38, 40, 43, 45, 47, 49, 53, 54, 56, 58, 62, 63, 69, 71, 80	$x_{36} + 1$	799
1, 3, 5, 6, 7, 8, 10, 12, 14, 15, 16, 19, 21, 23, 25, 27, 33, 40, 41, 43, 45, 47, 49, 53, 56, 58, 63, 65, 69, 71, 75, 77, 80	x_{38}	799
1, 5, 6, 7, 8, 10, 12, 14, 15, 16, 19, 21, 23, 25, 27, 33, 36, 38, 40, 43, 45, 47, 54, 56, 58, 60, 62, 65, 69, 70, 71, 73, 80	x_{56}	799
1, 3, 5, 6, 7, 8, 10, 12, 14, 15, 16, 19, 21, 23, 25, 33, 34, 36, 38, 40, 43, 45, 47, 49, 53, 54, 56, 58, 62, 63, 69, 71, 74, 80	$x_{69} + 1$	799
1, 6, 7, 8, 10, 12, 14, 16, 19, 21, 25, 27, 30, 31, 33, 36, 38, 40, 41, 43, 45, 47, 49, 51, 52, 56, 58, 63, 67, 69, 71, 73, 77, 80	$x_{66} + 1$	799
1, 3, 5, 6, 7, 8, 10, 12, 14, 15, 16, 19, 21, 23, 25, 27, 33, 34, 36, 38, 40, 43, 45, 47, 49, 53, 54, 56, 58, 62, 63, 67, 69, 71, 74, 80	x_{58}	799
1, 5, 6, 7, 8, 10, 12, 14, 15, 16, 19, 21, 23, 25, 27, 33, 36, 38, 40, 43, 45, 47, 49, 53, 54, 55, 56, 58, 62, 63, 67, 69, 71, 74, 80	x_{37}	799

Trivium-800

maxterm bits	superpoly	round
0, 5, 6, 7, 9, 11, 13, 17, 19, 20, 22, 24, 26, 30, 32, 33, 35, 37, 39, 42, 44, 46, 48, 52, 55, 57, 61, 62, 66, 68, 72, 74, 76, 79	x_{63}	800

References

1. Agnesse, A., Pedicini, M.: Cube attack in finite fields of higher order. In: Proceedings of 9th Australasian Information Security Conference, AISC 2011, pp. 9–14. ACS, Inc. (2011)

2. Agostini, E.: Bitlocker dictionary attack using GPUs. In: University of Cambridge Passwords 2015 Conference (2015). https://www.cl.cam.ac.uk/events/passwords2015/preproceedings.pdf

3. Ahmadian, Z., Rasoolzadeh, S., Salmasizadeh, M., Aref, M.R.: Automated dynamic cube attack on block ciphers: cryptanalysis of SIMON and KATAN. IACR Cryptology ePrint Archive **2015**, 40 (2015)

4. Aumasson, J.-P., Dinur, I., Meier, W., Shamir, A.: Cube testers and key recovery attacks on reduced-round MD6 and trivium. In: Dunkelman, O. (ed.) FSE 2009. LNCS, vol. 5665, pp. 1–22. Springer, Heidelberg (2009). doi:10.1007/978-3-642-03317-9_1

5. Baksi, A., Maitra, S., Sarkar, S.: New distinguishers for reduced round trivium and trivia-SC using cube testers. In: WCC2015-9th International Workshop on Coding and Cryptography 2015 (2015)

6. Bernstein, D.J.: Why haven't cube attacks broken anything? https://cr.yp.to/cubeattacks.html. Accessed 11 Nov 2016

7. Blum, M., Luby, M., Rubinfeld, R.: Self-testing/correcting with applications to numerical problems. In: ACM Symposium on Theory of Computing, pp. 73–83. ACM (1990)

8. De Cannière, C.: TRIVIUM: a stream cipher construction inspired by block cipher design principles. In: Katsikas, S.K., López, J., Backes, M., Gritzalis, S., Preneel, B. (eds.) ISC 2006. LNCS, vol. 4176, pp. 171–186. Springer, Heidelberg (2006). doi:10.1007/11836810_13

9. De Canniere, C., Preneel, B.: Trivium-specifications. eSTREAM, ECRYPT stream cipher project, report 2005/030 (2005)

10. Dinur, I., Shamir, A.: Cube attacks on tweakable black box polynomials. In: Joux, A. (ed.) EUROCRYPT 2009. LNCS, vol. 5479, pp. 278–299. Springer, Heidelberg (2009). doi:10.1007/978-3-642-01001-9_16

11. Dinur, I., Shamir, A.: Breaking grain-128 with dynamic cube attacks. In: Joux, A. (ed.) FSE 2011. LNCS, vol. 6733, pp. 167–187. Springer, Heidelberg (2011). doi:10.1007/978-3-642-21702-9_10

12. Dinur, I., Shamir, A.: Applying cube attacks to stream ciphers in realistic scenarios. Cryptogr. Commun. **4**(3–4), 217–232 (2012)

13. Fan, X., Gong, G.: On the security of Hummingbird-2 against side channel cube attacks. In: Armknecht, F., Lucks, S. (eds.) WEWoRC 2011. LNCS, vol. 7242, pp. 18–29. Springer, Heidelberg (2012). doi:10.1007/978-3-642-34159-5_2

14. Fouque, P.-A., Vannet, T.: Improving key recovery to 784 and 799 rounds of trivium using optimized cube attacks. In: Moriai, S. (ed.) FSE 2013. LNCS, vol. 8424, pp. 502–517. Springer, Heidelberg (2014). doi:10.1007/978-3-662-43933-3_26

15. Kleinjung, T., Lenstra, A.K., Page, D., Smart, N.P.: Using the cloud to determine key strengths. In: Galbraith, S., Nandi, M. (eds.) INDOCRYPT 2012. LNCS, vol. 7668, pp. 17–39. Springer, Heidelberg (2012). doi:10.1007/978-3-642-34931-7_3

16. Marks, M., Jantura, J., Niewiadomska-Szynkiewicz, E., Strzelczyk, P., Góźdź, K.: Heterogeneous GPU&CPU cluster for high performance computing in cryptography. Comput. Sci. **13**(2), 63–79 (2012)

17. Milo, F., Bernaschi, M., Bisson, M.: A fast, GPU based, dictionary attack to OpenPGP secret keyrings. J. Syst. Softw. **84**(12), 2088–2096 (2011)
18. O'Neil, S.: Algebraic structure defectoscopy (2007). Tools for Cryptanalysis 2007 Workshop. http://eprint.iacr.org/2007/378
19. Quedenfeld, F.M., Wolf, C.: Algebraic properties of the cube attack. IACR Cryptology ePrint Archive **2013**, 800 (2013)
20. Samorodnitsky, A.: Low-degree tests at large distances. In: Proceedings of 39th ACM symposium on Theory of Computing, pp. 506–515. ACM (2007)
21. Samorodnitsky, A., Trevisan, L.: A PCP characterization of NP with optimal amortized query complexity. In: Proceedings ACM Symposium on ToC, pp. 191–199. ACM (2000)
22. Shanmugam, D., Annadurai, S.: Secure implementation of stream cipher: trivium. In: Bica, I., Naccache, D., Simion, E. (eds.) SECITC 2015. LNCS, vol. 9522, pp. 253–266. Springer, Cham (2015). doi:10.1007/978-3-319-27179-8_18
23. Srinivasan, C., Pillai, U.U., Lakshmy, K., Sethumadhavan, M.: Cube attack on stream ciphers using a modified linearity test. J. Discret. Math. Sci. Cryptogr. **18**(3), 301–311 (2015)
24. Vielhaber, M.: Breaking ONE.FIVIUM by AIDA an algebraic IV differential attack (2007). http://eprint.iacr.org/2007/413
25. Winter, R., Salagean, A., Phan, R.C.-W.: Comparison of cube attacks over different vector spaces. In: Groth, J. (ed.) IMACC 2015. LNCS, vol. 9496, pp. 225–238. Springer, Cham (2015). doi:10.1007/978-3-319-27239-9_14
26. Zhang, S., Chen, G., Li, J.: Cube attack on reduced-round Quavium. ICMII-15 Adv. Comput. Sci. Res. (2015). doi:10.2991/icmii-15.2015.25

Related-Key Impossible-Differential Attack on Reduced-Round SKINNY

Ralph Ankele[1]([✉]), Subhadeep Banik[2], Avik Chakraborti[3], Eik List[4], Florian Mendel[5], Siang Meng Sim[2], and Gaoli Wang[6]

[1] Royal Holloway University of London, Egham, UK
ralph.ankele.2015@rhul.ac.uk
[2] Nanyang Technological University, Singapore, Singapore
bsubhadeep@ntu.edu.sg, ssim011@e.ntu.edu.sg
[3] NTT Secure Platform Laboratories, Tokyo, Japan
chakraborti.avik@lab.ntt.co.jp
[4] Bauhaus-Universität Weimar, Weimar, Germany
eik.list@uni-weimar.de
[5] Graz University of Technology, Graz, Austria
florian.mendel@iaik.tugraz.at
[6] East China Normal University, Shanghai, China
glwang@sei.ecnu.edu.cn

Abstract. At CRYPTO'16, Beierle et al. presented SKINNY, a family of lightweight tweakable block ciphers intended to compete with the NSA designs SIMON and SPECK. SKINNY can be implemented efficiently in both soft- and hardware and supports block sizes of 64 and 128 bits as well as tweakey sizes of 64, 128, 192 and 128, 256, 384 bits respectively. This paper presents a related-tweakey impossible-differential attack on up to 23 (out of 36) rounds of SKINNY-64/128 for different tweak sizes. All our attacks can be trivially extended to SKINNY-128/128.

Keywords: Symmetric Cryptography · Cryptanalysis · Tweakable block cipher · Impossible differential · Lightweight cryptography

1 Introduction

SKINNY is a family of lightweight tweakable block ciphers recently proposed at CRYPTO 2016 by Beierle et al. [3]. Its goal was to design a cipher that could be implemented highly efficiently on both soft- and hardware platforms, with performance comparable or better than the SIMON and SPECK families of block ciphers [1]. Like the NSA designs SIMON and SPECK, SKINNY supports a wide range of block sizes and tweak/key sizes – however, in contrast to the And-RX and Add-RX based NSA proposals, SKINNY is based on the better understood Substitution-Permutation-Network approach.

SKINNY offers a large security margin within the number of rounds for each member of the SKINNY family. The designers show that the currently best known attacks approach close to half of the number of rounds of the

© Springer International Publishing AG 2017
D. Gollmann et al. (Eds.): ACNS 2017, LNCS 10355, pp. 208–228, 2017.
DOI: 10.1007/978-3-319-61204-1_11

cipher. To motivate third-party cryptanalysis, the designers of SKINNY recently announced a cryptanalysis competition [2] for SKINNY-64/128 and SKINNY-128/128 with the obvious challenge of attacking more rounds than the preliminary analysis, concerning both the single- and related-key models.

Table 1. Summary of our attacks and comparison to existing cryptanalysis of SKINNY-64/128.

Instance	Rounds	Attack type	Time	Data	Memory	Ref
SKINNY-64/128	20	Impossible	$2^{121.1}$	$2^{47.7}$	$2^{74.7}$	[9]
SKINNY-64/128	21	Rectangle	$2^{87.9}$	$2^{54.0}$	$2^{54.0}$	[7]
SKINNY-64/128	21	Impossible	$2^{71.4}$	$2^{71.4\ b}$	$2^{68.0}$	Sect. 3.1
SKINNY-64/128	22	Rectangle	$2^{109.9}$	$2^{63.0}$	$2^{63.0}$	[7]
SKINNY-64/128	22	Impossible^a	$2^{71.6}$	$2^{71.4\ b}$	$2^{64.0}$	Sect. 3.2
SKINNY-64/128	23	Impossible	$2^{124.2}$	$2^{62.5}$	2^{124}	[7]
SKINNY-64/128	23	Impossible^a	2^{79}	$2^{71.4\ b}$	$2^{64.0}$	Sect. 3.3

^{a}The data complexity of our 21-round attack is beyond codebook. Our attack is more efficient than a full codebook attack in the case where SKINNY is used in a tweak-updating mode (i.e. where the tweak changes every time, but the key stays the same). This does not effect the 22/23 round attack as 48 bits of the tweakey are public (i.e. data complexity for full codebook would be 2^{64} from the state + 2^{48} from the tweak).
^{b}Our attack on 22/23 rounds uses the tweak against the recommendation of the SKINNY designers but still conform to the specification in [3].

Related Work. Recently and independent of our analysis Liu *et al.* [7] analyzed SKINNY in the related-tweakey model, showing impossible-differential and rectangle attacks on 19, 23, and 27 rounds of SKINNY-n/n, SKINNY-n/2n and SKINNY-n/3n, respectively. In [9], Tolba *et al.* showed impossible-differential attacks for 18, 20, 22 rounds of SKINNY-n/n, SKINNY-n/2n and SKINNY-n/3n, respectively. Additionally, Sadeghi *et al.* [8] studied related-tweakey impossible- differential and zero-correlation linear characteristics. In comparison to the other attacks, our 23-round related-tweakey impossible-differential attack on SKINNY-64/128 has the lowest time complexity so far. Table 1 summarizes our attacks and compares them to existing attacks on SKINNY-64/128.

Contributions and Outline. In this paper, we propose an impossible-differential attack on SKINNY-64/128 reduced to 23 rounds in the related-key model. The attack uses an 11-round impossible differential trail, to which six and four rounds can be added for obtaining a 21-round attack. Later, we show that another round can be appended leading to a 22-round attack, and even a 23-round attack.

The paper is organized as follows. In Sect. 2, we give a brief introduction to the SKINNY family of block ciphers. In Sect. 3, we detail the attack on SKINNY and provide time and memory complexities. Finally, Sect. 4 concludes the paper.

2   Description of SKINNY

Each round of SKINNY consists of the operations SUBCELLS, ADDROUNDCON-STANTS, ADDROUNDTWEAKEY, SHIFTROWS, and MIXCOLUMNS. The round operations are schematically illustrated in Fig. 1. A cell represents a 4-bit value in SKINNY-64/* and an 8-bit value in SKINNY-128/*.

We concentrate on SKINNY-64/128, which has a 64-bit block size and a 128-bit tweakey size. The data is arranged nibble-by-nibble in a row-wise fashion in a 4×4-matrix. SKINNY-64/128 recommends 36 rounds.

SUBCELLS (SC) substitutes each nibble x by $S(x)$, which is given below.

x	0	1	2	3	4	5	6	7	8	9	a	b	c	d	e	f
$S(x)$	c	6	9	0	1	a	2	b	3	8	5	d	4	e	7	f

ADDROUNDCONSTANTS (AC) adds LFSR-based round constants to Cells 0, 4, and 8 of the state.

ADDROUNDTWEAKEY (ART) adds the round tweakey to the first two state rows.

SHIFTROWS (SR) rotates the ith row, for $0 \leq i \leq 3$, by i positions to the right.

MIXCOLUMNS (MC) multiplies each column of the state by a matrix M:

$$M = \begin{bmatrix} 1 & 0 & 1 & 1 \\ 1 & 0 & 0 & 0 \\ 0 & 1 & 1 & 0 \\ 1 & 0 & 1 & 0 \end{bmatrix}$$

Tweakey Schedule. The tweakey schedule of SKINNY, as illustrated in Fig. 2, follows the TWEAKEY framework [5]. In contrast to the previous TWEAKEY designs DEOXYS-BC and JOLTIK-BC, SKINNY employs a significantly more lightweight strategy. In each round, only the two topmost rows of each tweakey word are extracted and XORed to the state. An additional round-dependent constant is also XORed to the state to prevent attacks from symmetry.

The 128-bit tweakey is arranged in two 64-bit tweakey words, represented by TK_1 and TK_2. In each round, the tweakey words are updated by a cell permutation P_T that ensures that the two bottom rows of a tweakey word in a

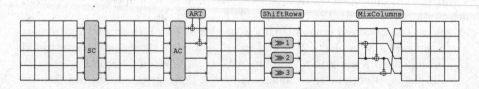

Fig. 1. Round function of SKINNY.

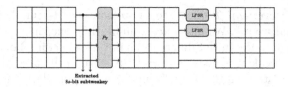

Fig. 2. Tweakey schedule of SKINNY.

certain round are exchanged with the two top rows in the tweakey word in the subsequent round. The permutation is given as:

$$P_T = \{9, 15, 8, 13, 10, 14, 12, 11, 0, 1, 2, 3, 4, 5, 6, 7\}$$

The permutation P_T has a period of 16, as visualized in Fig. 7 in the appendix. Moreover, each individual cell in the two topmost rows of TK_2 is transformed by a 4-bit LFSR to minimise the cancellation of differences from TK_1 and TK_2; TK_1 employs no LFSR transformation. The LFSR transformation L is given by

$$L(x_3, x_2, x_1, x_0) := (x_2, x_1, x_0, x_3 \oplus x_2),$$

where x_3, x_2, x_1, x_0 represent the individual bits of every tweakey nibble.

3   Related-Key Impossible-Differential Attack

Impossible-differential attacks were introduced independently by Biham *et al.* [4] and Knudsen [6]. They are widely used as an important cryptanalytic technique. The attack starts with finding an input difference that can never result in an output difference. By adding rounds before and/or after the impossible differential, one can collect pairs with certain plaintext and ciphertext differences. If there exists a pair that meets the input and output values of the impossible differential under some subkey, these subkeys must be wrong. In this way, we filter as many wrong keys as possible and exhaustively search the rest of the keys.

Notations. Let us state a few notations that are used in the attack description:

K^r represents the rth round key. This is equal to $TK_1^r \oplus TK_2^r$. Similarly, $k^r[i] = tk_1^r[i] \oplus tk_2^r[i]$ represents the individual i^{th} tweakey nibble in round r.
A^r represents the internal state before SC in round r.
B^r represents the internal state after SC in round r.
C^r represents the internal state after AT in round r.
D^r represents the internal state after SR in round r.
E^r represents the internal state after MC in round r. Furthermore, $E^r = A^{r+1}$.
L^t represents the t-times composition of LFSR function L.
$\overline{X}$ represents the corresponding variable X in the related-key setting.
$X[i]$ represents the ith nibble of the corresponding variable X.

Impossible-Differential Trail. Figure 3 presents the 11-round related-key differential trail that we use. We introduce a nibble difference in Cell 8 of the combined tweakey. Since the initial difference is in Cell 8, *i.e.* in one of the bottom two rows in the tweakey, it does not affect the state in the first round, and will be added to the state from the second round onwards. Similarly in the backward trail, the difference in the 11th round-tweakey appears in Cell 11 (in a bottom row), due to which we get an extra round in the backward direction.

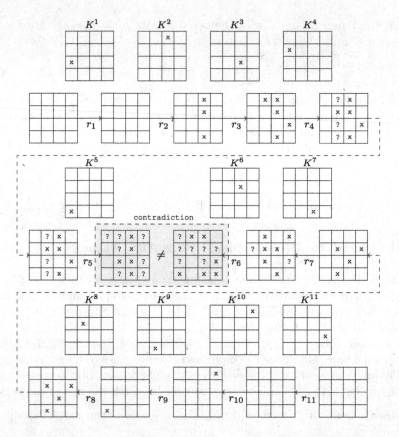

Fig. 3. Related-key impossible-differential trail over 11 rounds of SKINNY-64/128.

Lemma 1. *The equation $S(x \oplus \Delta_i) \oplus S(x) = \Delta_o$ has one solution x on average for $\Delta_i, \Delta_o \neq 0$. Similar result holds for the inverse S-Box S^{-1}.*

Proof. The above fact can be deduced by analyzing the Differential-Distribution Table (DDT) of the S-box S as illustrated in Table 2 in the appendix. The average can be calculated as $\frac{1}{225} \cdot \sum_{\Delta_i, \Delta_o \neq 0} DDT(\Delta_i, \Delta_o) \approx 1$. A similar exercise can be done for the inverse S-box yielding the same result.

Lemma 2. *For random values of x and $\Delta_i, \Delta_o \neq 0$, the equation $S(x \oplus \Delta_i) \oplus S(x) = \Delta_o$ holds with probability around 2^{-4}.*

Proof. The above fact can also be deduced by analyzing the Differential-Distribution Table (DDT) of the S-box S as illustrated in Table 2 in the appendix. The probability can be calculated as (let $\mathsf{Pr}[(x, \delta_i, \delta_o)$ denote the probability that the equation is satisfied for the triplet $x, \delta_i, \delta_o)$

$$\mathsf{Pr}[(x, \Delta_i, \Delta_o)] = \sum_{\delta_i, \delta_o \neq 0} \mathsf{Pr}[(x, \delta_i, \delta_o)|\Delta_i = \delta_i, \Delta_o = \delta_o]\mathsf{Pr}[\Delta_i = \delta_i, \Delta_o = \delta_o]$$

$$= \frac{1}{225} \cdot \sum_{\Delta_i, \Delta_o \neq 0} DDT(\Delta_i, \Delta_o) \cdot 2^{-4} \approx 2^{-4}$$

Attack on 21 Rounds. The impossible differential trail described in Fig. 3 can be extended by six and four rounds in backward and forward direction as will be explained in the following two lemmas.

Lemma 3. *It is possible to find plaintext pairs $P, \overline{P}$ and related-tweakey pairs $K, \overline{K}$ such that if the tweakey pairs differ only in nibble position 11, then there is no difference in the internal state after executing six rounds of SKINNY-64/128 with the plaintext-tweakey pairs (P, K) and $(\overline{P}, \overline{K})$.*

Proof. We will show how the required plaintext and tweakey pairs are generated. We choose the nibble at Position 11 to introduce the initial difference because after completing six rounds, the difference is shuffled to Cell 8 of the round key, which coincides with the beginning of the impossible- differential trail, shown in Fig. 3. It can be seen that the ADDROUNDTWEAKEY in the first round can be pushed behind the MIXCOLUMNS operation by changing the first round key to $\mathsf{Lin}(K_1)$ where $\mathsf{Lin} = \mathrm{MC} \circ \mathrm{SR}$ represents the linear layer (refer to Fig. 4).

$$\mathsf{Lin}(K^1) = \begin{bmatrix} k^1[0] & k^1[1] & k^1[2] & k^1[3] \\ k^1[0] & k^1[1] & k^1[2] & k^1[3] \\ k^1[7] & k^1[4] & k^1[5] & k^1[6] \\ k^1[0] & k^1[1] & k^1[2] & k^1[3] \end{bmatrix}$$

Furthermore, the initial difference between $K = TK_1^1 \oplus TK_2^1$ and $\overline{K} = \overline{TK_1^1} \oplus \overline{TK_2^1}$ can be selected in a specific form, so that in Round 6, the tweakey difference is zero. Let us denote $\delta_1 = tk_1^1[11] \oplus \overline{tk_1^1}[11]$ and $\delta_2 = tk_2^1[11] \oplus \overline{tk_2^1}[11]$. In Round 6, the difference will appear in Cell 0 of the round key and so we want:

$$k^6[0] \oplus \overline{k^6}[0] = tk_1^6[0] \oplus \overline{tk_1^6}[0] + tk_2^6[0] \oplus \overline{tk_2^6}[0]$$

$$= tk_1^1[11] \oplus \overline{tk_1^1}[11] \oplus L^3\left(tk_2^1[11]\right) \oplus L^3\left(\overline{tk_2^1}[11]\right)$$

$$= \delta_1 \oplus L^3(\delta_2) = 0$$

So, if the attacker chooses δ_1, δ_2 satisfying the equation $\delta_1 \oplus L^3(\delta_2) = 0$, then there is no difference introduced via the round-key addition in Round 6. The attacker should therefore follow the steps:

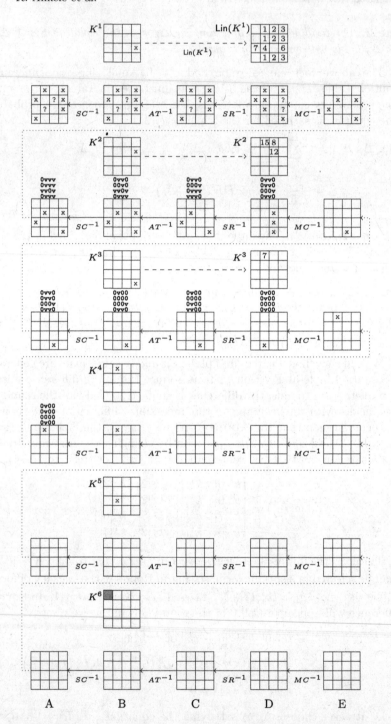

Fig. 4. Trail for the six forward rounds (the values of active nibbles in red are functions of δ_1, δ_2, the dark gray cell visualises the tweakey cancelation). (Color figure online)

1. Take any Plaintext P and compute the state after the first round MIX-COLUMNS, *i.e.* E^1.
2. Take any three-nibble difference $\Delta_1, \Delta_3, \Delta_4$ to construct $\overline{E^1}$ such that

$$
E^1 \oplus \overline{E^1} =
\begin{bmatrix}
0 & 0 & 0 & 0 \\
0 & \Delta_1 & 0 & \Delta_2 \\
\Delta_3 & 0 & 0 & 0 \\
0 & 0 & 0 & \Delta_4
\end{bmatrix}
$$

The value of Δ_2 will be determined shortly. The attacker can recover $\overline{P}$ by inverting the MC, SR, AC and SC layers on $\overline{E^1}$.
3. The attacker chooses the difference α in Cell 14 of E^2. She calculates then $k^1[1]$, $k^1[3]$, $k^1[7]$ so that

$$
B^2 \oplus \overline{B^2} = \mathsf{Lin}^{-1}(E^2) \oplus \mathsf{Lin}^{-1}(\overline{E^2}) =
\begin{bmatrix}
0 & 0 & 0 & 0 \\
0 & \alpha & 0 & \beta \\
\alpha & 0 & 0 & 0 \\
0 & 0 & 0 & \alpha
\end{bmatrix}.
$$

For example, $k^1[1]$ is a solution of the equation:

$$
S\left(E^1[5] \oplus k^1[1]\right) \oplus S\left(E^1[5] \oplus \Delta_1 \oplus k^1[1]\right) = \alpha.
$$

Lemma 1 ensures that the equation above has one solution on average.
4. β needs to be equal to $k^2[7] \oplus \overline{k^2}[7] = tk_1^2[7] \oplus tk_2^2[7] \oplus \overline{tk_1^2}[7] \oplus \overline{tk_2^2}[7]$. This is equal to $tk_1^1[11] \oplus L(tk_2^1[11]) \oplus \overline{tk_1^1}[11] \oplus L(\overline{tk_2^1}[11]) = \delta_1 \oplus L(\delta_2)$. So, the attacker chooses δ_1 and δ_2 satisfying $\delta_1 \oplus L^3(\delta_2) = 0$ and calculates $\beta = \delta_1 \oplus L(\delta_2)$. Δ_2 can then be determined as a solution of the equation:

$$
S\left(E^1[7] \oplus k^1[3]\right) \oplus S\left(E^1[7] \oplus \Delta_2 \oplus k^1[3]\right) = \beta \tag{1}
$$

The attacker now has the values of Δ_1, Δ_2, Δ_3, Δ_4 and so, he can compute $E^1, \overline{E^1}$ and hence $P, \overline{P}$.
5. However, the attacker still needs that in Round 4, the active nibble in $B^4[1]$ is equal to $\delta_1 \oplus L^2(\delta_2)$ to make all the state cells inactive in C^4, D^4, and E^4.
6. The attacker needs to guess three roundkey values in Round 1 (*i.e.* $k^1[2]$, $k^1[4]$, $k^1[6]$) and three roundkey values in Round 2 (*i.e.* $k^2[1] = tk_1^1[15] \oplus L(tk_2^1[15])$, $k^2[2] = tk_1^1[8] \oplus L(tk_2^1[8])$, $k^2[6] = tk_1^1[12] \oplus L(tk_2^1[12])$).
   If the attacker can guess these values, then he knows the actual values (marked with v) of the state cells for the plaintext pair $P, \overline{P}$ as opposed to only differences (marked by 0) in both Figs. 4 and 5.
7. Guessing the tweakey nibbles mentioned above enables the attacker to calculate the value of $B^3[1]$. Then, she calculates $k^3[1] = tk_1^1[7] \oplus L(tk_2^1[7])$ as follows. Since $D^3[1] = B^3[1] \oplus k^3[1]$ holds, we have:

$$
S\left(D^3[1] \oplus D^3[9] \oplus D^3[13]\right) \oplus S\left(D^3[1] \oplus D^3[9] \oplus \overline{D^3}[13]\right) = \delta_1 \oplus L^2(\delta_2).
$$

Since the knowledge of the guessed key nibbles already allows the attacker to calculate $D^3[9]$, $D^3[13]$, and $\overline{D^3}[13]$, $k^3[1] = tk_1^1[7] \oplus L(tk_2^1[7])$ is the solution to the equation above. Again, Lemma 1 guarantees one solution on average. Since the attacker has already determined $k^1[7] = tk_1^1[7] \oplus tk_2^1[7]$, this also determines the values of $tk_1^1[7]$ and $tk_2^1[7]$.

8. This guarantees that there are no more active nibbles after Round 4. The key difference does not add to the state in Round 5, and due to the fact that $\delta_1 \oplus L^3(\delta_2) = 0$, the tweak difference becomes 0 in Round 6.

Thus, by guessing six and calculating three key nibbles, we can construct $P, \overline{P}$ and $K, \overline{K}$ so that the internal state after six rounds has no active nibbles.

Lemma 4. *Given $C, \overline{C}$ as the two output ciphertexts after querying plaintext-tweakey pairs (P, K) and $(\overline{P}, \overline{K})$ to a 21-round SKINNY-64/128 encryption oracle. Then for a fraction 2^{-40} of the ciphertext pairs, it is possible to construct a backward trail for round 21 to round 18 by guessing intermediate tweakey nibbles so that there are no active nibbles in the internal state at the end of round 17.*

Proof. The attacker starts working backward from the ciphertext pairs $C, \overline{C}$ and proceeds as follows (illustrated in Fig. 5):

1. The attacker rejects ciphertext pairs which do not have seven inactive cells in Cells 3, 4, 5, 8, 9, 11, and 14) after peeling off the final MixColumns layer (*i.e.* D^{21}). Thus, a fraction of 2^{-28} pairs are filtered after this stage.
2. Furthermore, the attacker rejects ciphertext pairs which do not have the difference $\delta_1 \oplus L^{10}(\delta_2)$ in Cell 13 of A^{21}, *i.e.* reject if $A^{21}[13] \oplus \overline{A^{21}}[13] \neq \delta_1 \oplus L^{10}(\delta_2)$. Since calculating this cell does not require any key guess, she can do this filtering instantly leaving a fraction of 2^{-4} pairs after this stage.
3. Since the two bottommost rows of the state are not affected by the tweakey addition, and since $tk_1^1[7], tk_2^1[7]$ are already known, the attacker can calculate the actual values in Cells 0, 8, and 12 in A^{21} for the ciphertext pairs. These have to be equal since they are the output of the 20th-round MixColumns operation on the leftmost column which had only one active nibble in its input. If the active Cells 8 and 12 are different, the attacker can reject the pair. This adds another filter with probability 2^{-4}.
4. Since the actual values in Cell 0 in A^{21} for the ciphertext pairs were already calculated in the previous step, the attacker checks if the value of the active Cell 0 is equal to that of Cells 8 and 12, and rejects the pair otherwise. This adds another filter of probability 2^{-4}.
5. The attacker determines $k^{21}[5] = tk_1^1[4] \oplus L^{10}(tk_2^1[4])$ so that the active nibble in Cell 5 of A^{21} is $\delta_1 \oplus L^{10}(\delta_2)$. Since $A^{21}[5] = S^{-1}\left(k^{21}[5] \oplus C^{21}[5]\right)$, $k^{21}[5]$ is a solution to the equation below:

$$S^{-1}\left(k^{21}[5] \oplus C^{21}[5]\right) \oplus S^{-1}\left(k^{21}[5] \oplus \overline{C^{21}}[5]\right) = \delta_1 \oplus L^{10}(\delta_2).$$

6. The attacker determines $k^{21}[2] = tk_1^1[1] \oplus L^{10}(tk_2^1[1])$ and $k^{21}[6] = tk_1^1[2] \oplus L^{10}(tk_2^1[2])$ so that the active nibble in Cell 2 and 6 of A^{21} are equal to the

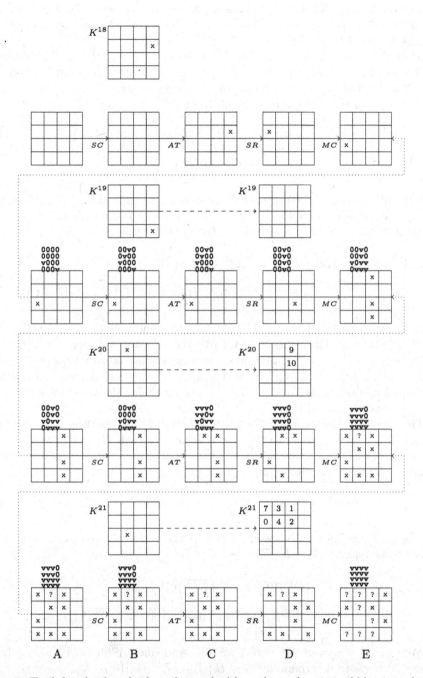

Fig. 5. Trail for the four backward rounds (the values of active nibbles in red are functions of δ_1 and δ_2). (Color figure online)

active nibble in Cell 14. Again, this works since those cells are output of the 20th-round MixColumns operation on Column 2 which had only one active nibble in its input.

7. Additionally, the attacker guesses $k^{21}[4] = tk_1^1[0] \oplus L^{10}(tk_2^1[0])$. This enables the attacker to compute the actual values for the entire leftmost column of A^{21} and hence to compute the leftmost column of D^{20}.

8. The value of the active nibble in cell 10 of A^{20} is given as:

$$A^{20}[10] \oplus \overline{A^{20}}[10] = S^{-1}\left(B^{20}[10]\right) \oplus S^{-1}\left(\overline{B^{20}}[10]\right)$$

$$= S^{-1}\left(D^{20}[8]\right) \oplus S^{-1}\left(\overline{D^{20}}[8]\right) = \eta. \tag{2}$$

Since the leftmost column of D^{20} is known, the attacker can calculate η, which must be equal to Cell 14 of A^{20} since they are output of the 19th-round MixColumns operation with one active input nibble.

$$A^{20}[14] \oplus \overline{A^{20}}[14] = S^{-1}\left(D^{20}[13]\right) \oplus S^{-1}\left(\overline{D^{20}}[13]\right)$$

$$= S^{-1}\left(A^{21}[1] \oplus A^{21}[13]\right) \oplus S^{-1}\left(\overline{A^{21}}[1] \oplus \overline{A^{21}}[13]\right). \tag{3}$$

It holds that $A^{21}[1] = S^{-1}\left(C^{21}[1] \oplus k^{21}[1]\right)$ and $\overline{A^{21}}[1] = S^{-1}(\overline{C^{21}}[1] \oplus k^{21}[1])$. By calculating Eqs. (2) and (3), the attacker can solve for $k^{21}[1] = tk_1^1[3] \oplus L^{10}(tk_2^1[3])$. One solution on average is guaranteed by Lemma 1.

9. The values $tk_1^1[i] \oplus tk_2^1[i]$, for $i = 1, 2, 3, 4$, were already determined during the calculation of the forward trail. So, using their values, the attacker can determine the actual values $tk_1^1[i]$, $tk_2^1[i]$ for $i = 1, 2, 3, 4$.

10. The attacker calculates $k^{20}[2] = tk_1^1[9] \oplus L^{10}(tk_2^1[9])$ so that the active nibble in Cell 2 in A^{20} is equal to the active value η in Cells 10 and 14 since they are output of the 19th-round MixColumns operation with one active input nibble. This is done by solving

$$\eta = A^{20}[2] \oplus \overline{A^{20}}[2] = S^{-1}\left(C^{20}[2] \oplus k^{20}[2]\right) \oplus S^{-1}\left(\overline{C^{20}}[2] \oplus k^{20}[2]\right). \tag{4}$$

11. The final condition to be satisfied is that the active nibble in Cell 8 of A^{19} has to be equal to $\delta_1 \oplus L^9(\delta_2) = \gamma$.

$$\gamma = S^{-1}\left(D^{19}[10]\right) \oplus S^{-1}\left(\overline{D^{19}}[10]\right)$$

$$= S^{-1}\left(A^{20}[6] \oplus A^{20}[14]\right) \oplus S^{-1}\left(\overline{A^{20}}[6] \oplus \overline{A^{20}}[14]\right). \tag{5}$$

Note that $A^{20}[6] = S^{-1}(C^{20}[6] \oplus k^{20}[6])$. And since $\overline{A^{20}}[6] = A^{20}[6]$, solving Eq. (5) helps to determine $k^{20}[6] = tk_1^1[10] \oplus L^{10}(tk_2^1[10])$.

The result follows since in the Steps 1–4, a total of $2^{-28-4-4-4} = 2^{-40}$ ciphertext pairs are filtered.

3.1   First Attack

Now, we put together the findings of Lemmas 3 and 4 into an attack procedure (see Fig. 8 in the appendix for details):

1. The attacker chooses the nibble values of the random base variable E^1 in all locations except Cells 5, 7, 8, and 15.
2. She chooses fixed differences δ_1, δ_2 satisfying $\delta_1 = L^3(\delta_2)$.
3. For each choice of $(E^1[5], E^1[7], E^1[8], E^1[15])$ (2^{16} choices):
   - Calculate P by inverting the first round.
   - Query the 21-round encryption oracle for P, K and $P, \overline{K}$.

So, for every choice of the base variable E^1, we have 2^{17} encryption calls. We can pair related plaintext and tweakey pairs in the following way: For every plaintext P_i, choose a plaintext P_j so that E^1 for P_i and P_j have a non-zero difference in all Cells 5, 7, 8, and 15. For every P_i, there exist $(2^4 - 1)^4 \approx 2^{15.6}$ such values of P_j, and so $2^{16+15.6} = 2^{31.6}$ pairs to work with. The attack now proceeds as follows. For each choice of P_i, P_j ($2^{31.6}$ choices):

- Denote $P = P_i$ and $\overline{P} = P_j$.
- The attacker can choose α and proceed with the steps of the above attack with one exception: She can no longer choose Δ_2 as in Step 4 of Lemma 3 since she has already chosen $P, \overline{P}, K, \overline{K}$.
- With probability 2^{-4} (as per Lemma 2), the plaintext pair satisfies Eq. (1) in Step 4 of Lemma 3 and proceeds; otherwise, she aborts.
- Request the ciphertext $\overline{C}$ for $(\overline{P}, \overline{K})$ and the ciphertext C for (P, K).
- If $C \oplus \overline{C}$ does not pass the 2^{-36} filter (Steps 1, 2, and 3 in Lemma 4), then abort and start again.
- If they pass the filter, the attacker can guess seven tweakey cells (2^{28} guesses) and calculate 17 key/tweak cells as follows:

#	Guessed	Rnd	Calculated	Rnd
1	$tk_1^1[i] \oplus tk_2^1[i]$ for $i = 2, 4, 6$	1		
2	$tk_1^1[i] \oplus L(tk_2^1[i])$ for $i = 8, 12, 15$	2		
3	$tk_1^1[i] \oplus L^{10}(tk_2^1[i])$ for $i = 0$	21		
4			$tk_1^1[i], tk_2^1[i]$ for $i = 7$	3
5			$tk_1^1[i], tk_2^1[i]$ for $i = 1, 2, 3, 4$	21
6			$tk_1^1[i] \oplus L^{10}(tk_2^1[i])$ for $i = 9, 10$	20

The 17 tweakey nibbles used for elimination are therefore:

(a) $tk_1^1[i], tk_2^1[i]$ for $i = 1, 2, 3, 4, 7$ (d) $tk_1^1[i] \oplus L(tk_2^1[i])$ for $i = 8, 12, 15$

(b) $tk_1^1[i] \oplus L^{10}(tk_2^1[i])$ for $i = 9, 10$ (e) $tk_1^1[i] \oplus tk_2^1[i]$ for $i = 6$

(c) $tk_1^1[i] \oplus L^{10}(tk_2^1[i])$ for $i = 0$

- A fraction of 2^{-4} tweakeys fulfills the condition required in Step 4 of Lemma 4.
- Therefore, the attacker has a set of $2^{28-4} = 2^{24}$ wrong key candidates.

The above procedure is repeated with 2^x chosen plaintexts until a single key solution remains for the 17 nibbles of the tweakey.

Complexity. For every base value of E^1, the attacker makes 2^{17} encryption calls. Out of these, she has $2^{31.6}$ pairs to work with. For each pair, the attacker can then choose α in $2^4 - 1$ ways, which gives her around $2^{35.6}$ initial guesses for the forward key nibbles $k^1[1]$, $k^1[3]$, and $k^1[7]$, of which a fraction of 2^{-4} passes the filter in Eq. (1). So, she has $2^{31.6}$ pairs to work with. In fact, for every pair (P_i, P_j) there is only one choice of α going forward on average.

Time complexity $= \max \left\{ 2^{x+17} \text{ encryptions}, 2^{x-4.4+24} \text{ guesses} \right\} = 2^{x+19.6}$.

The attacker gets wrong solutions for $2^{x-4.4+24} = 2^{x+19.6}$ incorrect solutions for 17 nibbles. To reduce the keyspace to 1 we need:

$$2^{17 \times 4} \cdot \left(1 - 2^{-17 \times 4} \right)^{2^{x+19.6}} \approx 2^{17 \times 4} e^{-2^{x-48.4}} = 1.$$

For this, we need $x = 55$. So, the total number of encryption calls to 21-round SKINNY-64/128 is $2^{55+17} = 2^{72}$ and the total number of guesses is $2^{74.6}$. We also need 2^{68} memory accesses, which are negligible in the total complexity. The memory complexity is upper bounded by storing one bit per key candidate which is therefore 2^{68} bits. The memory for storing the approximately $2 \cdot 2^{17}$ plaintexts and corresponding ciphertexts of a structure at each time is negligible.

3.2   22-Round Attack Under Partially Known Tweak

The attack above can be extended to 22-round SKINNY-64/128 under the assumption that 48 of the 128 bits in the tweakey are publicly known tweak (see Fig. 9 in the appendix for details). In particular, we assume that $tk_1^1[i], tk_2^1[i]$ for $i = 8, 11, 12, 13, 14, 15$ are reserved for the tweak. The remaining 80 bit constitute the secret key.

In this case, the attacker can add a round at the end (see Fig. 6 for details). Knowing six out of eight cells in the lower half of the tweakey blocks helps in the following way. From the ciphertext (i.e. E^{22}), one can revert the final round to compute E^{21} if we guess $k^{22}[4, 5]$, i.e. $tk_1^1[9, 10] \oplus L^{11}(tk_2^1[9, 10])$. The attack is almost the same as the previous attack, except that the tweakey indices $i = 8, 11, 12, 13, 14, 15$ and their functions are known and need not be guessed.

1. Generate $2^{31.6}$ plaintext/ciphertext pairs from every base choice of E^1 and 2^{17} encryption calls.
2. For each choice of P_i, P_j ($2^{31.6}$ choices):
   - Denote $P = P_i$ and $\overline{P} = P_j$.
   - The attacker can choose α and calculate $k^1[1]$, $k^1[3]$, and $k^1[7]$ as per Step 3 of Lemma 3.
   - She can no longer choose Δ_2 as in Step 4 of Lemma 3 since she has already chosen $P, \overline{P}, K, \overline{K}$.
   - With probability 2^{-4}, the plaintext pair satisfies Eq. (1) in Step 4 of Lemma 3 and proceeds; otherwise, she aborts.

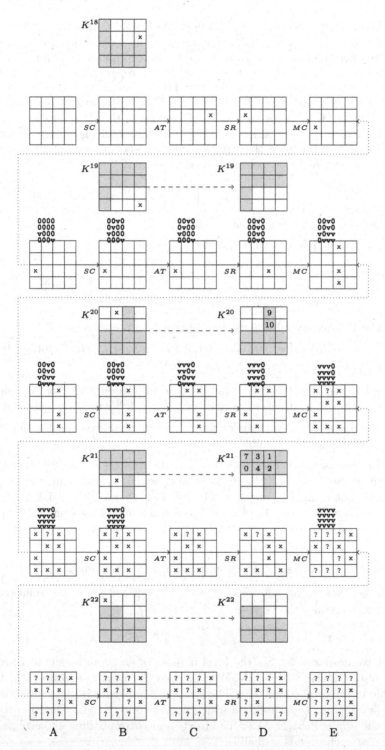

Fig. 6. Trail for the five backward rounds (the values of active nibbles in red are functions of δ_1, δ_2, grey cells are the key, white cells are the tweak). (Color figure online)

- The attacker doesn't need to guess the Round 2 tweakey nibbles since these are in the lower half of the tweakey blocks and therefore known.
- Retrieve the ciphertext $\overline{C}$ for $(\overline{P}, \overline{K})$ and the ciphertext C for (P, K).
- Guess $k^{22}[4,5] = tk_1^1[9,10] \oplus L^{11}(tk_2^1[9,10])$ to get E_{21}.
- If $E_{21} \oplus \overline{E}_{21}$ does not pass the 2^{-36} filter, then abort and restart.
- After determining $k^{20}[2] = tk_1^1[9] \oplus L^{10}(tk_2^1[9])$ and $k^{20}[6] = tk_1^1[10] \oplus L^{10}(tk_2^1[10])$ in Steps 10 and 11 of Lemma 4, the attacker can uniquely determine $tk_1^1[9,10]$ since $tk_1^1[9,10] \oplus L^{11}(tk_2^1[9,10])$ is already guessed.
- If they pass the filter, the attacker can guess six tweakey cells (2^{24} guesses) and calculate 16 key cells as follows:

#	Guessed	Rnd	Calculated	Rnd
1	$tk_1^1[i] \oplus tk_2^1[i]$ for $i = 2, 4, 6$	1		
2	$tk_1^1[i] \oplus L^{10}(tk_2^1[i])$ for $i = 0$	21		
3	$tk_1^1[i] \oplus L^{11}(tk_2^1[i])$ for $i = 9, 10$	22		
4			$tk_1^1[i], tk_2^1[i]$ for $i = 7$	3
5			$tk_1^1[i], tk_2^1[i]$ for $i = 1, 2, 3, 4$	21
6			$tk_1^1[i], tk_2^1[i]$ for $i = 9, 10$	20

The 16 tweakey nibbles used for elimination are therefore:

(a) $tk_1^1[i], tk_2^1[i]$ for $i = 1, 2, 3, 4, 7, 9, 10$.   (c) $tk_1^1[i] \oplus tk_2^1[i]$ for $i = 6$.

(b) $tk_1^1[i] \oplus L^{10}(tk_2^1[i])$ for $i = 0$.

- A fraction of 2^{-4} tweakeys fulfills the condition in Step 4 of Lemma 4.
- Therefore, the attacker has a set of $2^{24-4} = 2^{20}$ wrong key candidates.

The procedure above is repeated with 2^x chosen plaintexts until a single key solution remains for the 16 nibbles of the tweakey.

Complexity. For every base value of E^1, the attacker makes 2^{17} encryption calls. Out of these, she has $2^{31.6}$ pairs to work with. For each pair, the attacker can choose then α in $2^4 - 1$ ways, which gives her around $2^{35.6}$ initial guesses for the forward key nibbles $k^1[1], k^1[3], k^1[7]$, of which only a fraction of 2^{-4} passes the filter in Eq. (1). So, the attacker has $2^{31.6}$ pairs to work with. In effect, for every pair (P_i, P_j) there is only once choice of α going forward on average.

Time complexity $= \max\{2^{x+17} \text{ encryptions}, 2^{x-4.4+20} \text{ guesses}\} = 2^{x+17}$.

The attacker gets wrong solutions for $2^{x-4.4+20} = 2^{x+15.6}$ incorrect solutions for 16 nibbles. To reduce the keyspace to 1 we need:

$$2^{16 \times 4} \cdot \left(1 - 2^{-16 \times 4}\right)^{2^{x+15.6}} \approx 2^{16 \times 4} e^{-2^{x-48.4}} = 1.$$

For this, we need $x = 54$. So, the total number of encryption calls to 22-round SKINNY-64/128 is $2^{54+17} = 2^{71}$. We also need 2^{64} memory accesses, which are negligible in the total complexity. The memory complexity is upper bounded by storing one bit per key candidate which is therefore 2^{64} bits. The memory for storing the approximately $2 \cdot 2^{17}$ plaintexts and corresponding ciphertexts of a structure at each time is negligible.

3.3   23-Round Attack Under Partially Known Tweak

We can extend the 22 round attack to a 23 round attack by prepending one round at the beginning. In order to not disturb the notation, we denote the additonal round prepended at the beginning as the 0-th round. That is, the 23 rounds are labelled as rounds 0 to 22, and the variables A^0, B^0 etc. are defined as above. The plaintext is denoted by A^0 and the ciphertext by E^{22}. Note that, from the base value of E^1, the plaintext can be calculated if we guess $k^0[9, 10]$.

There are two principal differences to the 22-round attack.

1. When the attacker guesses $k^{22}[4, 5]$ which is $tk_1^1[9, 10] \oplus L^{11}(tk_2^1[9, 10])$ to invert the final round to get E_{21}, he uniquely determines $tk_1^1[9, 10]$ and $tk_2^1[9, 10]$. This is because at the beginning of the outer loop $k^0[9, 10]$ has already been guessed by the attacker to invert the initial round.
2. As the attacker can no longer determine $k^{20}[2] = tk_1^1[9] \oplus L^{10}(tk_2^1[9])$ and $k^{20}[6] = tk_1^1[10] \oplus L^{10}(tk_2^1[10])$ using Steps 10 and 11 of Lemma 4. The probability that with the given values of $tk_1^1[9, 10]$ and $tk_2^1[9, 10]$, Eqs. (4) and (5) are satisfied is 2^{-8}. This decreases the probability of ciphertext filter from 2^{-36} to 2^{-44}.

For each initial guess of $k^0[9, 10]$, the guessed and calculated key bytes are:

#	Guessed	Rnd	Calculated	Rnd
1	$tk_1^1[i] \oplus tk_2^1[i]$ for $i = 2, 4, 6$	1		
2	$tk_1^1[i] \oplus L^{10}(tk_2^1[i])$ for $i = 0$	21		
3	$tk_1^1[i] \oplus L^{11}(tk_2^1[i])$ for $i = 9, 10$	22		
4			$tk_1^1[i], tk_2^1[i]$ for $i = 7$	3
5			$tk_1^1[i], tk_2^1[i]$ for $i = 1, 2, 3, 4$	21

The 14 tweakey nibbles used for elimination are therefore:

(a) $tk_1^1[i], tk_2^1[i]$ for $i = 1, 2, 3, 4, 7$.   (c) $tk_1^1[i] \oplus tk_2^1[i]$ for $i = 6$.

(b) $tk_1^1[i] \oplus L^{10}(tk_2^1[i])$ for $i = 0$.   (d) $tk_1^1[i] \oplus L^{11}(tk_2^1[i])$ for $i = 9, 10$

As before, a fraction of 2^{-4} tweakeys fulfills the condition in Step 4 of Lemma 4. Therefore, the attacker has a set of $2^{24-4} = 2^{20}$ wrong key candidates.

Complexity. For each iteration of the outer loop, the complexity is calculated as follows: For every base value of E^1, the attacker makes 2^{17} encryption calls. Out of those, she has $2^{31.6}$ pairs to work with. For each pair, the attacker can choose then α in $2^4 - 1$ ways, which gives her around $2^{35.6}$ initial guesses for the forward key nibbles $k^1[1], k^1[3], k^1[7]$, of which only a fraction of 2^{-4} passes the filter in Eq. (1). In effect, for every pair (P_i, P_j) there is only one choice of α going forward on average.

Time complexity $= \max \left\{ 2^{x+17} \text{ encryptions}, 2^{x+31.6-44+20} \text{ guesses} \right\} = 2^{x+17}$.

The attacker gets $2^{x+31.6-44+20} = 2^{x+7.6}$ incorrect solutions for 14 nibbles. To reduce the keyspace to 1 we need:

$$2^{14\times 4} \cdot \left(1 - 2^{-14\times 4}\right)^{2^{x+7.6}} \approx 2^{14\times 4} e^{-2^{x-48.4}} = 1.$$

We need $x = 54$ leaving the total number of encryption calls to 22-round SKINNY-64/128 with $2^{54+17} = 2^{71}$. Multiplying this by 2^8 for the outer loop gives a total complexity of $2^{71+8} = 2^{79}$ which is just short of exhaustive search for the 80-bit key. We also need $2^{56+8} = 2^{64}$ memory accesses, which are negligible in the total complexity. The memory complexity is upper bounded by storing one bit per key candidate which is therefore 2^{64} bits. The memory for storing the approximately $2 \cdot 2^{17}$ plaintexts and ciphertexts of a structure is negligible.

4   Conclusion

In this paper, we outline related-key impossible-differential attacks against 21-round SKINNY-64/128 as well as attacks on 22 and 23 rounds under the assumption of having 48 of the 128-bit tweakey as public tweak. Our attacks are based on an 11-round impossible differential trail, to which we prepend six and append five rounds before and after the trail, respectively, to obtain an attack on 22 rounds. Finally, we can prepend a 23-rd round under similar assumptions.

Acknowledgements. This work was initiated during the group sessions of the 6th Asian Workshop on Symmetric Cryptography (ASK 2016) held in Nagoya, Japan. Ralph Ankele is supported by the European Union's Horizon 2020 research and innovation programme under grant agreement No. H2020-MSCA-ITN-2014-643161 ECRYPT-NET. Gaoli Wang is supported by National Natural Science Foundation of China (Grant Nos. 61572125, 61373142), Shanghai High-Tech Field Project (Grant No. 16511101400). Siang Meng Sim is supported by the Singapore National Research Foundation Fellowship 2012 (NRF-NRFF2012-06). This work has been supported in part by the Austrian Science Fund (project P26494-N15).

Appendix

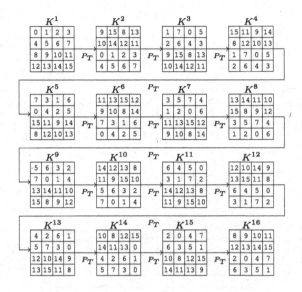

Fig. 7. The permutation P_T in the tweakey schedule has a period of 16.

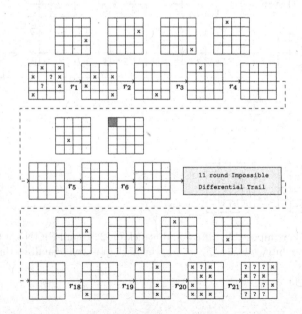

Fig. 8. Related-key impossible differential attack on 21-round SKINNY 64/128 (the dark gray cell visualises the cancelation of the tweakeys).

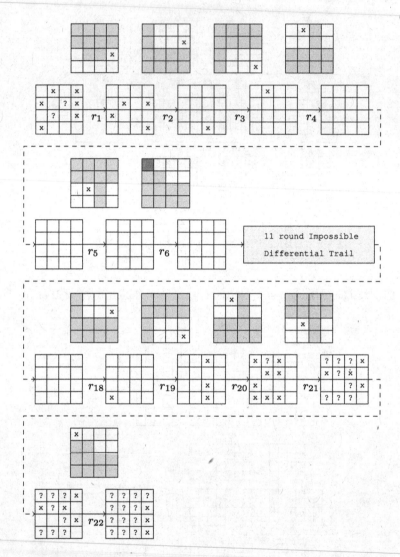

Fig. 9. Related-key impossible differential attack on 22 round SKINNY 64/128 (grey cells are the key, white cells are the tweak, the dark gray cell visualises the cancelation of the tweakeys)

Table 2. Difference-distribution table

	0	1	2	3	4	5	6	7	8	9	a	b	c	d	e	f
0	16	.	.	.	.	.	.	.	.	.	.	.	.	.	.	.
1	.	.	.	.	.	.	.	.	4	4	4	4	.	.	.	.
2	.	4	.	4	.	4	4	.	.	.	.	.	.	.	.	.
3	.	.	.	.	.	.	.	.	2	2	2	2	2	2	2	2
4	.	.	4	.	.	.	2	2	.	.	.	4	2	2	.	.
5	.	.	4	.	.	.	2	2	.	.	4	.	2	2	.	.
6	.	2	.	2	2	.	.	2	2	.	2	.	.	2	2	.
7	.	2	.	2	2	.	.	2	.	2	.	2	2	.	.	2
8	.	.	.	.	4	4	.	.	.	.	.	.	2	2	2	2
9	.	.	.	.	4	4	.	.	.	.	.	.	2	2	2	2
a	.	.	.	.	.	4	4	.	2	2	2	2	.	.	.	.
b	.	4	.	4	.	.	.	.	.	.	.	.	2	2	2	2
c	.	.	4	.	.	.	2	2	4	.	.	.	.	.	2	2
d	.	.	4	.	.	.	2	2	.	4	.	.	.	.	2	2
e	.	2	.	2	2	.	.	2	.	2	.	2	.	2	2	.
f	.	2	.	2	2	.	.	2	2	.	2	.	2	.	.	2

References

1. Beaulieu, R., Shors, D., Smith, J., Treatman-Clark, S., Weeks, B., Wingers, L.: The SIMON and SPECK Families of Lightweight Block Ciphers (2013). Cryptology ePrint Archive, Report 2013/404. http://eprint.iacr.org/
2. Beierle, C., Jean, J., Kölbl, S., Leander, G., Moradi, A., Peyrin, T., Sasaki, Y., Sasdrich, P., Sim, S.M.: Cryptanalysis competition (2016). https://sites.google.com/site/skinnycipher/cryptanalysis-competition
3. Beierle, C., Jean, J., Kölbl, S., Leander, G., Moradi, A., Peyrin, T., Sasaki, Y., Sasdrich, P., Sim, S.M.: The SKINNY family of block ciphers and its low-latency variant MANTIS. In: Robshaw, M., Katz, J. (eds.) CRYPTO 2016. LNCS, vol. 9815, pp. 123–153. Springer, Heidelberg (2016). doi:10.1007/978-3-662-53008-5_5
4. Biham, E., Biryukov, A., Shamir, A.: Cryptanalysis of skipjack reduced to 31 rounds using impossible differentials. In: Stern, J. (ed.) EUROCRYPT 1999. LNCS, vol. 1592, pp. 12–23. Springer, Heidelberg (1999). doi:10.1007/3-540-48910-X_2
5. Jean, J., Nikolić, I., Peyrin, T.: Tweaks and keys for block ciphers: the TWEAKEY framework. In: Sarkar, P., Iwata, T. (eds.) ASIACRYPT 2014. LNCS, vol. 8874, pp. 274–288. Springer, Heidelberg (2014). doi:10.1007/978-3-662-45608-8_15
6. Knudsen, L.: DEAL - A 128-bit Block Cipher. In: NIST AES Proposal (1998)
7. Liu, G., Ghosh, M., Ling, S.: Security Analysis of SKINNY under Related-Tweakey Settings (2016). Cryptology ePrint Archive, Report 2016/1108. http://eprint.iacr.org/2016/1108

8. Sadeghi, S., Mohammadi, T., Bagheri, N.: Cryptanalysis of Reduced round SKINNY Block Cipher (2016). Cryptology ePrint Archive, Report 2016/1120. http://eprint.iacr.org/2016/1120

9. Tolba, M., Abdelkhalek, A., Youssef, A.M.: Impossible differential cryptanalysis of reduced-round SKINNY. In: Joye, M., Nitaj, A. (eds.) AFRICACRYPT 2017. LNCS, vol. 10239, pp. 117–134. Springer, Cham (2017). doi:10.1007/978-3-319-57339-7_7

Faster Secure Multi-party Computation of AES and DES Using Lookup Tables

Marcel Keller, Emmanuela Orsini, Dragos Rotaru$^{(\boxtimes)}$, Peter Scholl, Eduardo Soria-Vazquez, and Srinivas Vivek

Department of Computer Science, University of Bristol, Bristol, UK
{m.keller,emmanuela.orsini,dragos.rotaru,peter.scholl,
eduardo.soria-vazquez,sv.venkatesh}@bristol.ac.uk

Abstract. We present an actively secure protocol for secure multi-party computation based on lookup tables, by extending the recent, two-party 'TinyTable' protocol of Damgård et al. (ePrint 2016). Like TinyTable, an attractive feature of our protocol is a very fast and simple online evaluation phase. We also give a new method for efficiently implementing the preprocessing material required for the online phase using arithmetic circuits over characteristic two fields. This improves over the suggested method from TinyTable by at least a factor of 50.

As an application of our protocol, we consider secure computation of the Triple DES and the AES block ciphers, computing the S-boxes via lookup tables. Additionally, we adapt a technique for evaluating (Triple) DES based on a polynomial representation of its S-boxes that was recently proposed in the side-channel countermeasures community. We compare the above two approaches with an implementation. The table lookup method leads to a very fast online time of over 230,000 blocks per second for AES and 45,000 for Triple DES. The preprocessing cost is not much more than previous methods that have a much slower online time.

Keywords: Multi-party computation · Block cipher · Implementation

1 Introduction

Secure multi-party computation (MPC) protocols allow useful computations to be performed on private data, without the data owners having to reveal their

This work has been partially supported by EPSRC via grant EP/N021940/1; by the Defense Advanced Research Projects Agency (DARPA) and Space and Naval Warfare Systems Center, Pacific (SSC Pacific) under contract No. N66001-15-C-4070; by EPSRC via grant EP/M016803; and by the European Union's H2020 Programme under grant agreement number ICT-644209 (HEAT) and the Marie Skłodowska-Curie grant agreement No. 643161 (ECRYPT-NET).

© Springer International Publishing AG 2017
D. Gollmann et al. (Eds.): ACNS 2017, LNCS 10355, pp. 229–249, 2017.
DOI: 10.1007/978-3-319-61204-1_12

inputs. The last decade has seen an enormous amount of progress in the practicality of MPC, with many works designing more efficient protocols and implementations. There has also been a growing interest in exploring the possible applications of MPC, with a number of works targeting specific computations such as auctions, statistics and stable matching [6, 7, 23, 29].

One promising application area that has recently emerged is the use of secure computation to protect long-term secret keys, for instance, in authentication servers or to protect company secrets [4]. Here, the secret key, sk, is split up into n pieces, or shares, such that certain subsets of the n shares are needed to reconstruct sk, and each share is stored on a different server (possibly in a different location and/or managed by a separate entity). When the key is needed by an application, say for a user logging in, the servers run an MPC protocol to authenticate the user, without ever revealing sk. Typically, the type of computation required here can be performed using a symmetric primitive such as a block cipher or hash function.

Several previous works in secure computation have studied the above type of application, and the AES function is even considered a standard benchmark for new protocols [5, 16, 35, 37, 39]. A recent line of works has even looked at special-purpose symmetric primitives, designed to have low complexity when evaluated in MPC [1, 2, 27]. However, in industries such as banking and the wider financial sector, strict regulations and legacy systems mean that switching to new primitives can be very expensive, or even impossible. Indeed, most banking systems today are using AES or Triple DES (3DES) to secure their data [24], but may still benefit greatly from MPC technologies to prevent theft and data breaches.

1.1   Our Contributions

In this work, we focus on the task of secure multi-party computation of the AES and the (Triple) DES block ciphers, in the setting of active security against any number of corrupted parties. We present a new technique for the preprocessing phase of efficient, secure computation of *table lookup* (with a secret index), and apply this to evaluating the S-boxes of AES and DES. In addition, we describe a new method of secure MPC evaluation of the DES S-boxes based on evaluating polynomials over binary finite fields, which reduces the number of non-linear field multiplications.

Our protocol for secure table lookup builds upon the recent 'TinyTable' protocol for secure two-party computation by Damgård et al. [18]. This protocol requires a preprocessing phase, which is independent of the inputs, where randomly masked (or 'scrambled') lookup tables on random data are created. In the online phase, where the function is securely evaluated, each (one-time) masked table can be used to perform a single table lookup on a private index in the MPC protocol. The online phase of TinyTable is *very efficient*, as each party only needs to send $\log_2 N$ bits over the network, for a table of size N.

However, the suggested technique for creating the masked tables is far less efficient: for secure computation of AES, it would take at least *256 times longer* to create the masked lookup tables, compared with using standard methods with a slower online time.

We extend and improve upon the TinyTable approach in two ways. Firstly, we show that the technique can easily be generalized to the multi-party setting and used in any SPDZ-like MPC protocol based on secret-sharing and information-theoretic MACs. Secondly, we describe a new, general approach for creating the masked tables using finite field arithmetic, which significantly improves the preprocessing cost of the protocol. Concretely, for a lookup table of size N, we can create the masked table using an arithmetic circuit over $\mathbb{F}_{2^k}$ with fewer than $N/k + \log N$ multiplications. This provides a range of possible instantiations with either binary or arithmetic circuit-based protocols. When using binary circuits, we only require $N - 2$ multiplications. For arithmetic circuits over $\mathbb{F}_{2^8}$, an AES S-box can be preprocessed with 33 multiplications, improving on the method in [18], which takes 1792 multiplications, by more than 50 times. With current practical protocols, it turns out to be even more efficient to work over $\mathbb{F}_{2^{40}}$, with only 11 multiplications. We remark that standard methods for computing AES based on polynomials or Boolean circuits can obtain better overall running times, but with a much slower online phase. The main goal of this work is to reduce the preprocessing cost whilst preserving the very fast online phase of TinyTable.

We also consider a new method for secure multi-party computation of DES based on a masking side-channel countermeasure technique. The DES S-box can be viewed as a lookup table mapping 6 bits to 4 bits, or as a polynomial over $\mathbb{F}_{2^6}$. A naïve method requires 62 field multiplications to evaluate a DES S-box polynomial over $\mathbb{F}_{2^6}$. There were many recent works that reduced the number of non-linear multiplications required to evaluate polynomials over binary finite fields, including the DES S-box polynomials [10,13,14,38,41]. A recent proposal by Pulkus and Vivek [38] showed that the DES S-boxes, when represented over a different field, $\mathbb{F}_{2^8}$, can be evaluated with only 3 non-linear multiplications. This is better than the best-known circuit over $\mathbb{F}_{2^6}$, which needs 4 non-linear multiplications. Applying the Pulkus–Vivek method in our context, we show how 1 round of the DES block cipher can be computed with just 24 multiplications over $\mathbb{F}_{2^8}$. This compares favorably with previous methods based on evaluating polynomials over $\mathbb{F}_{2^6}$ and boolean circuits.

Analogous to the MPC protocols based on table lookups, there are also masking side-channel countermeasures based on random-table lookups [11,12]. This analogy should not come as a surprise since the masking technique is also based on secret-sharing. The state-of-the-art for (higher-order) masking seems to suggest that the schemes based on evaluation of S-box polynomials usually outperform table-lookups based schemes in terms of time, RAM memory and randomness. We perform a similar comparison in the MPC context too. To this end, we evaluate the complexity of the various methods for secure computation of AES and 3DES, and present some implementation results. We implemented the protocols using the online phase of the SPDZ [16,19] MPC protocol. The preprocessing additionally requires some random multiplication triples and shared bits, for which we estimated costs using MASCOT [30] for arithmetic circuits, and based on the recent optimized TinyOT protocol [35,43] for binary circuits.

Our experiments show that the fastest online evaluation is achieved using lookup tables. The preprocessing for this method costs much less when using arithmetic circuits over larger fields, compared with a binary circuit protocol such as TinyOT [35,43], despite the quadratic (in the field bit length) communication cost of [30]. The polynomial-based methods for AES and DES still perform slightly better in the preprocessing phase, but for applications where a low online latency is desired, the lookup table approach is definitely preferred. If an application is mainly concerned with the total running time, then the polynomial-based methods actually lead to runtimes for AES that are comparable with the fastest recent 2-PC implementations using garbled circuits.

Related Work. A recent, independent work by Dessouky et al. [22] presented two different protocols for lookup table-based secure two-party computation in the semi-honest security model. The first protocol, OP-LUT, offers an online phase very similar to ours (and [18]), while the preprocessing stage, that is implemented using 1-out-of-N oblivious transfer, is incomparable to ours as we must work much harder to achieve active security.

The second protocol, SP-LUT, proposes a more efficient preprocessing phase, which only requires random 1-out-of-N oblivious transfer computation, but a slower online evaluation; however this protocol has a much lower overall communication compared to the previous one. These two protocols are also compared with the OTTT (One-Time Truth-Table) protocol by Ishai et al. [28] with parallel circuit based preprocessing [20]. More detailed comparisons with our protocols are provided in Sect. 5.2.

This work also provides an FPGA-based synthesis tool that transforms a high level function representation to multi-input/multi-output table-lookup representation, which could also be used with our protocol.

2   Preliminaries

We denote by λ the computational security parameter and κ the statistical security parameter. We consider the sets $\{0,1\}$ and $\mathbb{F}_2^k$ endowed with the structure of the fields $\mathbb{F}_2$ and $\mathbb{F}_{2^k}$, respectively. We denote by $\mathbb{F} = \mathbb{F}_{2^k}$ any finite field of characteristic two. Finally, we use $a \xleftarrow{\$} A$ as notation for a uniformly random sampling of a from a set A.

Note that by linearity we always mean $\mathbb{F}_2$-linearity, as we only consider fields of characteristic 2.

2.1   MPC Computation Model

Our protocol builds upon the *arithmetic black-box* model for MPC, represented by the functionality $\mathcal{F}_{\mathsf{ABB}}$ (shown in the full version). This functionality permits the parties to input and output secret-shared values and evaluate arbitrary binary circuits performing basic operations. This abstracts away the underlying details of secret sharing and MPC. Other than the standard **Add** and **Mult** commands, $\mathcal{F}_{ABB}$ also has a **BitDec** command for generating the bit decomposition of a given secret-shared value, two commands **Random** and **RandomBit** for generating random values according to different distributions and an **Open** command which allows the parties and the adversary to output values. **BitDec** can be implemented in a standard manner by opening and then bit-decomposing $x + r$, where r is obtained using k secret random bits.

We use the notation $[\![x]\!]$ to denote an authenticated and secret-shared value x, which is stored by $\mathcal{F}_{ABB}$. More precisely, this can be implemented with active security using the SPDZ protocol [16,19] based on additive secret sharing and unconditionally secure MACs. We also use the $+$ and $\cdot$ operators to denote calls to **Add** and **Mul** with the appropriate shared values in $\mathcal{F}_{ABB}$.

More concretely, our protocols are in the so called *preprocessing model* and consist of two different phases: an *online* computation, where the actual evaluation takes place, and a *preprocessing* phase that is independent of the parties' inputs. During the online evaluation, linear operations only require local computations thanks to the linearity of the secret sharing scheme and MAC. Multiplications and bit decompositions require random preprocessed data and interactions. More generally, the main task of the preprocessing step is to produce enough random secret data for the parties to use during the online computation: other than multiplication triples, which allow parties to compute products, it also provides random shared values. The preprocessing phase can be efficiently implemented using OT-based protocols for binary circuits [8,25,43] and arithmetic circuits [30].

Security Model. We describe our protocols in the universal composition (UC) framework of Canetti [9], and assume familiarity with this. Our protocols work with n parties from the set $\mathcal{P} = \{P_1, \ldots, P_n\}$, and we consider security against malicious, static adversaries, i.e. corruption may only take place before the protocols start, corrupting up to $n - 1$ parties.

3 Evaluating AES and DES S-box Polynomials

In this section, we recollect some of the previously known methods that aim to reduce the number of non-linear operations to evaluate univariate polynomials over binary finite fields, particularly, the AES and the DES S-boxes represented in this form. Note here that, by a non-linear multiplication, we mean those multiplications of polynomials that are neither multiplication by constants nor squaring operations. Since squaring is a linear operation in binary fields, once a monomial is computed, it can be repeatedly squared to generate as many more monomials as possible without costing any non-linear multiplication.

Due to limited space, a more detailed discussion can found in the full version.

3.1   AES S-box

The AES S-box evaluation on a given input (as an element of $\mathbb{F}_{2^8}$) consists of first computing its multiplicative inverse in $\mathbb{F}_{2^8}$ (mapping zero to zero), and then applying a bijective affine transformation. For the inverse S-box, the inverse affine transformation is applied first and then the multiplicative inverse. Note that the polynomial representation of the inverse function in $\mathbb{F}_{2^8}$ is X^{254}.

BitDecompostion Method. This approach, described by Damgård et al. [15], computes the squares X^{2^i}, for $i \in [7]$, and then multiplies them to get X^{254}. This method needs 6 non-linear multiplications.

Rivain–Prouff Method. This method, as presented in Gentry et al. [26], is a variant of the method of Rivain–Prouff [40] to evaluate the AES S-box polynomial using only 4 non-linear multiplications in $\mathbb{F}_{2^8}[X]$: $\{X, X^2\} \xrightarrow{\times} \{X^3, X^{12}\} \xrightarrow{\times} \{X^{14}\} \xrightarrow{\times} \{X^{15}, X^{240}\} \xrightarrow{\times} X^{254}$.

3.2   Des S-boxes

Cyclotomic Class Method. Recall that DES has eight 6-to-4-bit S-boxes. In this naïve method given by Carlet et al. [10], the DES S-boxes are represented as univariate polynomials over $\mathbb{F}_{2^6}$. In particular, the 4-bit S-box outputs are padded with zeros in the most significant bits and then identified with the elements of $\mathbb{F}_{2^6}$. It turns out that these polynomials have degree at most 62 [41].

Over $\mathbb{F}_{2^m}[X]$, define $C_i^m := \left\{ X^{i \cdot 2^j} : j = 0, 1, \ldots, m - 1 \right\}$ for $0 < i < 2^m$. Now we need to compute $C_0^6, C_1^6, C_3^6, C_5^6, C_7^6, C_9^6, C_{11}^6, C_{13}^6, C_{15}^6, C_{21}^6, C_{23}^6, C_{27}^6, C_{31}^6$, to cover all monomials up to degree 62, and this needs at most 11 non-linear multiplications. The target polynomial is then simply obtained as a linear combination of the computed monomials.

Pulkus–Vivek Method. This generic method to evaluate arbitrary polynomials over binary finite fields was proposed recently by Pulkus and Vivek [38] as an improvement over the method of Coron–Roy–Vivek [13,14]. In the PV method, the DES S-boxes are represented as polynomials over $\mathbb{F}_{2^8}$ instead of $\mathbb{F}_{2^6}$. The 6-bit input strings of the DES S-boxes are padded with zeroes in the two most significant positions and then naturally identified with the elements of $\mathbb{F}_{2^8}$. The four most significant coefficient bits of the polynomial outputs are *discarded* to obtain the desired 4-bit S-box output.

Firstly, a set of monomials $L = C_1^8 \cup C_3^8 \cup C_7^8$ in $\mathbb{F}_{2^8}[X]$ is computed. Then a polynomial, say $P(X)$, representing the given S-box is sought as $P(X) = p_1(X) \cdot q_1(X) + p_2(X)$, where $p_1(X)$, $q_1(X)$, and $p_2(X)$ have monomials only from the set L. In total, the PV method needs 3 non-linear multiplications in $\mathbb{F}_{2^8}[X]$ to evaluate each of the S-box polynomial.

3.3   MPC Evaluation of AES and DES S-box Polynomials

Here we detail the MPC evaluation of AES and DES S-boxes using the techniques described above. We recall that since the S-boxes, in both the ciphers we are considering, are the only non-linear components, they represent the only parts which actually need interactions in an MPC evaluation.

AES Evaluation. As we mention before in Sect. 3.1, the straightforward way to compute the S-box is using the BitDecomposition method, which requires 6 multiplications in $4 + 1$ rounds. We are considering the case of active security, so the AES evaluation is done in the field $\mathbb{F}_{2^{40}}$ instead of $\mathbb{F}_{2^8}$, via the embedding $\mathbb{F}_{2^8} \hookrightarrow \mathbb{F}_{2^{40}}$. This follows from the fact that we are using the SPDZ protocol which requires a field size of at least 2^κ, where κ is the statistical security parameter. This permits to have only one MAC per data item [15].

The evaluation proceeds as follow: first X is bit-decomposed so that all the squarings can be locally evaluated, and then X^{254} is obtained as described in [15]:

$$X^{254} = ((X^2 \cdot X^4) \cdot (X^8 \cdot X^{16})) \cdot ((X^{32} \cdot X^{64}) \cdot X^{128}).$$

This requires 4 rounds, out of which one is a call to BitDec. We also need an extra round for computing the inverse of the field embedding $\mathbb{F}_{2^8} \hookrightarrow \mathbb{F}_{2^{40}}$ to evaluate the S-box linear layer. We denote this method by AES-BD.

We denote by AES-RP the AES S-box evaluation that uses the Rivain–Prouff method (cf. Sect. 3.1). It requires $6 + 1$ rounds to compute the four powers $X^3, X^{14}, X^{15}, X^{254}$. Furthermore, this can be done with three calls to BitDec and four non-linear multiplications, but some of the openings can be done in parallel which yields to a depth-6 circuit. As before, we need an extra round to call BitDec and compute the S-box linear layer.

DES Evaluation. We denote by DES-PV the DES S-box evaluation using the Pulkus–Vivek method. Note that, although in side-channel world computing the squares is for free, since it is an $\mathbb{F}_2$-linear operation, in a secret-shared based MPC with MACs this is no longer true and we need to bit-decompose.

The squares from C_1^8, C_3^8, C_7^8, are obtained locally after X, X^3, X^7 are bit-decomposed. Here we need two multiplications, since $X^3 = X \cdot X^2$ and $X^7 = X^3 \cdot X^4$. The third multiplication occurs when computing the product $p_1(X) \cdot q_1(X)$, resulting in an S-box cost of only 3 triples, 24 bits and 5 communication rounds.

The number of rounds is due to 3 calls to BitDec (on X^3, X^7 and $p_1(X) \cdot q_1(X) + p_2(X)$) and 3 non-linear multiplications. Although at a first glance there seems to be six rounds, we have that $\mathsf{BitDec}(X^7)$ is independent of the $\mathsf{BitDec}(X^3)$, as we can compute X^7 without the call $\mathsf{BitDec}(X^3)$, resulting in only five rounds.

4 MPC Evaluation of Boolean Circuits Using Lookup Tables

In this section we describe an efficient MPC protocol for securely evaluating circuits over extension fields of $\mathbb{F}_2$ (including boolean circuits) containing additional 'lookup table' gates. This protocol is in the preprocessing model and follows the same approach proposed in [20], evaluating lookup table gates using preprocessed, masked lookup tables.

The functionality that we implement is $\mathcal{F}_{\text{ABB-LUT}}$ (Fig. 1), which augments the standard $\mathcal{F}_{\text{ABB}}$ functionality with a table lookup command. The concrete online cost of each table lookup is just $\log_2 N$ bits of communication per party, where N is the size of the table. Note that the functionality $\mathcal{F}_{\text{ABB-LUT}}$ works over a finite field $\mathbb{F}_{2^k}$, and has been simplified by assuming that the size of the range and domain of the lookup table T is not more than 2^k. However, our protocol actually works for general table sizes, and $\mathcal{F}_{\text{ABB-LUT}}$ can easily be extended to model this by representing a table lookup result with several field elements instead of one.

We now show how Protocol 1 implements the **Table Lookup** command of $\mathcal{F}_{\text{ABB-LUT}}$, given the right preprocessing material. For any non-linear function T, with ℓ input and m output bits, it is well known that it can be implemented as a lookup table of 2^ℓ components of m bits each. To evaluate $\mathsf{T}(\cdot)$ on a secret authenticated value $[\![x]\!], x \in \mathbb{F}_{2^\ell}$, the parties use a random authenticated T

Functionality $\mathcal{F}_{\text{ABB-LUT}}$

This functionality has all the features of $\mathcal{F}_{\text{ABB}}$, operating over $\mathbb{F}_{2^k}$, plus the following command.

Table Lookup: On command $(\mathsf{T}, \mathsf{id}_1, \mathsf{id}_2)$ from all parties, where $\mathsf{T} : \{0,1\}^\ell \to \{0,1\}^m$, for $\ell, m \leq k$, and id_1 is present in memory, retrieve (id_1, x) and store $(\mathsf{id}_2, \mathsf{T}(x))$.

Fig. 1. The ideal functionality for MPC using lookup tables

Functionality $\mathcal{F}_{\text{Prep-LUT}}$

This functionality has all of the same features as $\mathcal{F}_{\text{ABB}}$, with the following additional command.

Masked Table: On input $(\mathsf{MaskedTable}, \mathsf{T}, \mathsf{id})$ from all parties, where $\mathsf{T} : \{0,1\}^\ell \to \{0,1\}^m$ for $\ell, m \leq k$, sample a random value s, set $(\mathsf{Val}[\mathsf{id}_s], \mathsf{Val}[\mathsf{id}_{\mathsf{T}(s)}], \ldots, \mathsf{Val}[\mathsf{id}_{\mathsf{T}(s \oplus (2^\ell - 1))}]) \leftarrow (s, \mathsf{T}(s), \ldots, \mathsf{T}(s \oplus (2^\ell - 1)))$, and return $(\mathsf{id}_s, (\mathsf{id}_{\mathsf{T}(s)}, \ldots, \mathsf{id}_{\mathsf{T}(s \oplus (2^\ell - 1))}))$.

Fig. 2. Ideal functionality for the preprocessing of masked lookup tables.

Protocol 1. Secure online evaluation of SBox using lookup tables

Table Lookup: On input $[\![x]\!]$ compute $[\![\mathsf{T}(x)]\!]$ as follows.

1. Call $\mathcal{F}_{\mathsf{Prep-LUT}}$ on input $(\mathsf{MaskedTable}, \mathsf{T})$, and obtain a precomputed masked table $([\![s]\!], [\![\mathsf{Table}(s)]\!])$.
2. The parties open the value $h = x \oplus s$.
3. Locally compute $[\![\mathsf{T}(x)]\!] = [\![\mathsf{Table}(s)]\!][h]$, where $[\![\mathsf{Table}(s)]\!][h]$ is the hth component of $[\![\mathsf{Table}(s)]\!]$.

---

evaluation from $\mathcal{F}_{\mathsf{Prep-LUT}}$ (Fig. 2). More precisely, we would like the preprocessing to output values $([\![s]\!], [\![\mathsf{Table}(s)]\!])$, where $[\![s]\!]$ is a random authenticated value unknown to the parties and $[\![\mathsf{Table}(s)]\!]$) is the table

$$[\![\mathsf{Table}(s)]\!] = \left([\![\mathsf{T}(s)]\!], [\![\mathsf{T}(s \oplus 1)]\!], \ldots, [\![\mathsf{T}(s \oplus (2^\ell - 1))]\!] \right),$$

so that $[\![\mathsf{Table}(s)]\!][j], 0 \leq j \leq 2^\ell - 1$, denotes the element $[\![\mathsf{T}(s \oplus j)]\!]$. Given such a table, evaluating $[\![\mathsf{T}(x)]\!]$ is straightforward: first the parties open the value $h = x \oplus s$ and then they locally retrieve the value $[\![\mathsf{Table}(s)]\!][h] = [\![\mathsf{T}(s \oplus h)]\!] = [\![\mathsf{T}(s \oplus s \oplus x)]\!] = [\![\mathsf{T}(x)]\!]$.

Correctness easily follows from the linearity of the $[\![\cdot]\!]$-representation and the discussion above. Privacy follows from the fact that the value s used in **Table Lookup** is randomly chosen and is used only once, thus it perfectly blinds the secret value x.

4.1 The Preprocessing Phase: Securely Generating Masked Lookup Tables

In this section we describe how to securely implement $\mathcal{F}_{\mathsf{Prep-LUT}}$ (see Fig. 2), and in particular how to generate masked lookup tables which can be used for the online phase evaluation.

Recall that the goal is to obtain the shared values:

$$[\![\mathsf{Table}(s)]\!] = ([\![\mathsf{T}(s)]\!], [\![\mathsf{T}(s \oplus 1)]\!], \ldots, [\![\mathsf{T}(s \oplus (2^\ell - 1))]\!]).$$

Protocol 2 begins by taking a secret, random ℓ-bit mask $[\![s]\!] = ([\![s_0]\!], \ldots, [\![s_{\ell-1}]\!])$. Then, the parties expand s into a secret-shared bit vector $(s'_0, \ldots, s'_{2^\ell-1})$ which has a 1 in the s-th entry and is 0 elsewhere. We denote this procedure—the most expensive part of the protocol—by Demux, and describe how to perform it in the next section.

Once this is done, the parties can obtain the i-th entry of the masked lookup table by computing:

$$\mathsf{T}(i) \cdot [\![s'_0]\!] + \mathsf{T}(i \oplus 1) \cdot [\![s'_1]\!] + \cdots + \mathsf{T}(i \oplus (2^\ell - 1)) \cdot [\![s'_{2^\ell-1}]\!],$$

which is clearly $[\![\mathsf{T}(i \oplus s)]\!]$ as required. Note that since the S-box is public, this is a local computation for the parties. In the following we give an efficient protocol for computing Demux.

4.2   Computing Demux with Finite Field Multiplications

We now present a general method for computing Demux using fewer than $N/k +$ $\log N$ multiplications over $\mathbb{F}_{2^k}$, when k is any power of 2 and $N = 2^\ell$ is the table size. Launchbury et al. [32] previously described a protocol with $O(N)$ multiplications in $\mathbb{F}_2$, but our protocol has fewer multiplications than theirs for all choices of k.

As said before, Demux maps a binary representation $(s_0, \ldots, s_{\ell-1})$ of an integer $s = \sum_{i=0}^{\ell-1} s_i \cdot 2^i$ into a unary representation of fixed length 2^ℓ that contains a one in the position s and zeros elsewhere. A straightforward way to compute Demux is by computing, over $\mathbb{F}_{2^N}$ [1]:

$$[\![s']\!] = \prod_{i=0}^{\ell-1} ([\![s_i]\!] \cdot X^{2^i} + (1 - [\![s_i]\!])).$$

Notice that if $s_i = 1$ then the i-th term of the product equals X^{2^i}, whereas the term equals 1 if $s_i = 0$. This means the entire product evaluates to $s' = X^s$, where s is the integer representation of the bits $(s_0, \ldots, s_{\ell-1})$. Bit decomposing s' obtains the demuxed output as required. Unfortunately, this approach does not scale well with N, the table size, as we must exponentially increase the size of the field.

We now show how to compute this more generally, using operations over $\mathbb{F}_{2^k}$, where k is a power of two. We will only ever perform multiplications between elements of $\mathbb{F}_2$ and $\mathbb{F}_{2^k}$, and will consider elements of $\mathbb{F}_{2^k}$ as vectors over $\mathbb{F}_2$. Define the partial products, for $j = 1, \ldots, \ell$:

$$p_j(X) = \prod_{i=0}^{j-1} (s_i \cdot X^{2^i} + (1 - s_i)) \in \mathbb{F}_{2^N}$$

and note that $p_{j+1}(X) = p_j(X) \cdot (s_j \cdot X^{2^j} + (1 - s_j))$, for $j < \ell$.

Note also that the polynomial $p_j(X)$ has degree $< 2^j$, so $p_j(X)$ can be represented as a vector in $\mathbb{F}_2^{2^j}$ containing its coefficients, and more generally, a

---

Protocol 2. Protocol to generate secret shared table lookup

Table: On input (Table, P_i) from all the parties, do the following:

    1: Take ℓ random authenticated bits $[\![s_0]\!], \ldots, [\![s_{\ell-1}]\!]$, where each s_i is unknown to all the parties.

    2: Compute $([\![s'_0]\!], \ldots, [\![s'_{2^\ell-1}]\!]) \leftarrow \mathsf{Demux}([\![s_0]\!], \ldots, [\![s_{\ell-1}]\!])$

    3: $\forall i = 0, \ldots, 2^\ell - 1$, locally compute

$$[\![\mathsf{T}(i \oplus s)]\!] = \mathsf{T}(i) \cdot [\![s'_0]\!] + \mathsf{T}(i \oplus 1) \cdot [\![s'_1]\!] + \cdots + \mathsf{T}((2^\ell - 1) \oplus i) \cdot [\![s'_{2^\ell-1}]\!]$$

---

[1] A similar trick was used by Aliasgari et al. [3] for binary to unary conversion over prime fields.

Protocol 3. $(\llbracket s'_0 \rrbracket, \dots, \llbracket s'_{N-1} \rrbracket) \leftarrow \mathsf{Demux}(k, \llbracket s_0 \rrbracket, \dots, \llbracket s_{\ell-1} \rrbracket)$

Require: k a power of two, $u = N/k$, $\ell = \log_2 N$
Input: Bit decomposition of $s \in \{0, \dots, N-1\}$, with LSB first
Output: Satisfies $s'_s = 1$ and $s'_i = 0$ for all $i \neq s$

1: $\llbracket p \rrbracket = (1 - \llbracket s_0 \rrbracket, \llbracket s_0 \rrbracket)$ $\qquad$ // p starts in $\mathbb{F}_2^2$
2: **for** $j = 1$ to $\ell - 1$ **do**
3: $\quad \llbracket t \rrbracket = \llbracket s_j \rrbracket \cdot \llbracket p \rrbracket$ $\qquad$ // $\mathbb{F}_2 \times \mathbb{F}_2^{2^j}$ multiplication, 1 round
4: $\quad \llbracket p \rrbracket = (0^{2^j} \| \llbracket t \rrbracket) + (\llbracket p \rrbracket - \llbracket t \rrbracket \| 0^{2^j})$ $\qquad$ // p now in $\mathbb{F}_2^{2^{j+1}}$
5: Write $\llbracket p \rrbracket = (\llbracket b_0 \rrbracket, \dots, \llbracket b_{u-1} \rrbracket)$ $\qquad$ // $b_i \in \mathbb{F}_2^k$
6: **for** $i = 0$ to $u - 1$ **do**
7: $\quad (\llbracket s'_{ki} \rrbracket, \dots, \llbracket s'_{ki+k-1} \rrbracket) = \mathsf{BitDec}(\llbracket b_i \rrbracket)$ $\qquad$ // 1 round
8: **return** $(\llbracket s'_0 \rrbracket, \dots, \llbracket s'_{N-1} \rrbracket)$

---

vector p_j containing $\lceil 2^j/k \rceil$ elements of $\mathbb{F}_2^k$. This is the main observation that allows us to emulate the computation of s' using only $\mathbb{F}_{2^k}$ arithmetic.

Given a sharing of p_j represented in this way, a sharing of $p_j(X) \cdot X^{2^j}$ can be seen as the vector (increasing the powers of X from left to right):

$$(0^{2^j} \| p_j) \in \mathbb{F}_2^{2^{j+1}}$$

and a vector representation of $p_{j+1}(X)$ is:

$$\left((0^{2^j} \| s_j \cdot p_j) + ((1 - s_j) \cdot p_j \| 0^{2^j}) \right) \in \mathbb{F}_2^{2^{j+1}}.$$

Thus, given $\llbracket p_j \rrbracket$ represented as $\lceil 2^j/k \rceil$ shared elements of $\mathbb{F}_{2^k}$, we can compute $\llbracket p_{j+1} \rrbracket$ in MPC with $\lceil 2^j/k \rceil$ multiplications between $\llbracket s_j \rrbracket$ and a shared $\mathbb{F}_{2^k}$ element, plus some local additions.

Starting with $p_1(X) = s_0 \cdot X + (1 - s_0)$ we can iteratively apply the above method to compute $p_\ell = s'$, as shown in Protocol 3. The overall complexity of this protocol is given by

$$\sum_{j=1}^{\ell-1} \lceil 2^j/k \rceil < N/k + \ell$$

multiplications between bits and $\mathbb{F}_{2^k}$ elements.

Table 1 illustrates this trade-off between the field size and number of multiplications for some example parameters. We note that the main factor affecting the best choice of k is the cost of performing a multiplication in $\mathbb{F}_{2^k}$ in the underlying MPC protocol, and this may change as new protocols are developed. However, we compare costs of some current protocols in Sect. 5.

4.3 MPC Evaluation of AES and DES Using Lookup Tables

We now show how to use the lookup table MPC protocol described above to evaluate AES and DES.

Table 1. Number of $\mathbb{F}_2 \times \mathbb{F}_{2^k}$ multiplications for creating a masked lookup table of size N, for varying k.

N	$k = 1$	8	40	64	128
64	62	9	5	5	5
256	254	33	11	8	7
1024	1022	129	31	20	13

AES Evaluation. We require an MPC protocol which performs operations in $\mathbb{F}_{2^8}$. In practice, we actually embed $\mathbb{F}_{2^8}$ in $\mathbb{F}_{2^{40}}$, since we use the SPDZ protocol which requires a field size of at least 2^κ, for statistical security parameter κ. We implement the AES S-box using the table lookup method from Protocol 2 combined with Demux (Protocol 3) over $\mathbb{F}_{2^{40}}$, since this yields a lower communication cost (see Table 4). Notice that the data sent is highly dependent on the number of bits, triples and the field size.

In a naive implementation of this approach, we would have call BitDec on $[\![\mathsf{Table}(s)]\!]$, in order to perform the embedding $\mathbb{F}_{2^8} \hookrightarrow \mathbb{F}_{2^{40}}$. This is required since the table output is not embedded, but the MixColumns step needs this to perform multiplication by $X \in \mathbb{F}_{2^8}$ on each state.

With a more careful analysis we can avoid the BitDec calls by locally embedding the bit shares inside Protocol 2. We store the masked S-box table in bit decomposed form and then its bits are multiplied (in the clear) with Demux's output (secret-shared). This trick reduces the online communication by a factor of 8, halves the number of rounds required to evaluate AES and gives a very efficient online phase with only 10 rounds and 160 openings in $\mathbb{F}_{2^{40}}$.

DES Evaluation. Using the fact that DES S-boxes have size 64, we chose to use the Demux Protocol 3 with multiplications in $\mathbb{F}_{2^{40}}$, based on the costs in Table 4. Like AES, we try to isolate the input-dependent phase as much as possible with no extra cost.

Every DES round performs only bitwise addition and no embedding is necessary here. The masked table can be bit-decomposed without interaction, exactly as described above for AES, by multiplying clear bits with secret shared values. This yields a low number of openings, one per S-box look-up, so the total online cost for 3DES is 46 rounds with 384 openings.

5   Performance Evaluation

This section presents timings for 3DES and AES using the methods presented in previous sections. We also discuss trade-offs and different optimizations which turn out to be crucial for our running-times. The setup we have considered is that both the key and message used in the cipher are secret shared across two parties. We consider the input format for each block cipher as already embedded into

$\mathbb{F}_{2^{40}}$ for AES, or as a list of shared bits for DES. We implemented the protocols using the SPDZ software,[2] and estimated times for computing the multiplication triples and random bits needed based on the costs of MASCOT [30].

The results, shown in Tables 2 and 3, give measurements in terms of *latency* and *throughput*. Latency indicates the online phase time required to evaluate one block cipher, whereas throughput (which we consider for both online and offline phases) shows the maximum number of blocks per second which can be evaluated in parallel during one execution. We also measure the number of rounds of interaction of the protocols, and the number of *openings*, which is the total number of secret-shared field elements opened during the online evaluation.

Benchmarking Environment. The experiments were ran across two machines each with Intel i7-4790 CPUs running at 3.60 GHz, 16 GB of RAM connected over a 1 GBps LAN with an average ping of 0.3 ms (roundtrip). For experiments with 3–5 parties, we used three additional machines with i7-3770 CPUs at 3.1 GHz. In order to get accurate timings each experiment was averaged over 5 executions, each with at least 1000 cipher calls.

Security Parameters and Field Sizes. Secret-sharing based MPC can be usually split into 2 phases—preprocessing and online. In SPDZ-like systems, the preprocessing phase depends on a computational security parameter, and the online phase a statistical security parameter which depends on the field size. In our experiments the computational security parameter is $\lambda = 128$. The statistical security κ is 40 for every cipher except for 3DES-Raw which requires an embedding into a 42 bit field.

Results. The theoretical costs and practical results are shown in Tables 2 and 3, respectively. Timings are taken only for the encryption calls, excluding the key schedule mechanism.

AES-BD is implemented by embedding each block into $\mathbb{F}_{2^{40}}$, and then squaring the shares locally after the inputs are bit-decomposed. In this manner, each S-box computation costs 5 communication rounds and 6 multiplications. This method was described in [15].

3DES-Raw represents the 3DES cipher with the S-box evaluated as a polynomial of degree 62 over the field $\mathbb{F}_{2^6} = \mathbb{F}_2[x]/(x^6 + x^4 + x^3 + x + 1)$. To make the comparisons relevant with other ciphers in terms of active security we chose to embed the S-box input in $\mathbb{F}_{2^{42}}$, via the embedding $\mathbb{F}_{2^6} \hookrightarrow \mathbb{F}_{2^{42}}$, where $\mathbb{F}_{2^{42}} = \mathbb{F}_2[y]/(y^{42} + y^{21} + 1)$ and $y = x^7 + 1$. The S-boxes used for interpolating are taken from the PyCrypto library [34]. 3DES-Raw is implemented only for benchmarking purposes and it has no added optimizations. One S-box has a cost of 62 multiplications and 62 rounds.

3DES-PV is 3DES implemented with the Pulkus-Vivek method from Section 3.2. Since it has only a few multiplications in $\mathbb{F}_{2^{40}}$, the amount of preprocessing data required is very small, close to AES-BD. It suffers in terms of both latency and throughput due to the high number of communication rounds (needed for bit decomposition to perform the squarings).

---

[2] https://github.com/bristolcrypto/SPDZ-2.

Table 2. Communication cost for AES and 3DES in MPC.

Cipher	Online cost			Preprocessing cost			
	Rounds	Openings	Field	Triples	Bits	Field	Comm. (MB)
AES-BD	50	2240	$\mathbb{F}_{2^{40}}$	960	2560	$\mathbb{F}_{2^{40}}$	4.3
AES-RP	70	1920	$\mathbb{F}_{2^{40}}$	640	5120	$\mathbb{F}_{2^{40}}$	2.9
AES-LT	10	160	$\mathbb{F}_{2^{40}}$	1760	42240	$\mathbb{F}_{2^{40}}$	8.4
3DES-Raw	2979	48024	$\mathbb{F}_{2^{42}}$	23808	2688	$\mathbb{F}_{2^{42}}$	112
3DES-PV	230	3456	$\mathbb{F}_{2^{40}}$	1152	9216	$\mathbb{F}_{2^{40}}$	5.2
3DES-LT	46	384	$\mathbb{F}_{2^{40}}$	1920	26880	$\mathbb{F}_{2^{40}}$	8.8

Table 3. 1 GBps LAN timings for evaluating AES and 3DES in MPC.

Cipher	Online (single-thread)			Online (multi-thread)			Preprocessing[a]
	Latency (ms)	Batch size	ops/s	Batch size	ops/s	Threads	ops/s
AES-BD	5.20	64	758	1024	3164	16	30.7
AES-RP	7.19	1024	940	64	3872	16	**46.1**
AES-LT	**0.928**	2048	53918	512	**236191**	32	16.79
3DES-Raw	270	512	130	-	-	-	1.24
3DES-PV	36.98	512	86	512	366	32	**25.6**
3DES-LT	**4.254**	1024	10883	512	**45869**	16	15.3

[a]Extrapolated from timings for a 128-bit field

Surprisingly, AES-RP (the polynomial-based method from Sect. 3.1) has a better throughput than AES-BD although it requires 20 more rounds and 2 times more shared bits to evaluate. The explanation for this is that in AES-RP there are fewer openings, thus less data sent between parties.

AES-LT and 3DES-LT are the ciphers obtained with the lookup table protocol from Sect. 4. AES-LT achieves the lowest latency and the highest throughput in the online phase. The communication in the preprocessing phase is roughly twice the cost of the previous method, AES-BD.

Packing Optimization. We notice that in the online phase of AES-LT each opening requires to send 8 bit values embedded in $\mathbb{F}_{2^{40}}$. Instead of sending 40 bits across the network we select only the relevant bits, which for AES-LT are 8 bits. This reduces the communication by a factor of 5 and gives a throughput of 236k AES/second over LAN and a multi-threaded MPC engine.

The same packing technique is applied for 3DES-LT since during the protocol we only open 6 bit values from Protocol 1. These bits are packed into a byte and sent to the other party. Here the multi-threaded version of 3DES-LT improves the throughput only by a factor of 4.2x (vs AES-LT 4.4x) due to the higher number of rounds and openings along with the loss of 2 bits from packing.

General Costs of the Table Lookup Protocol. In Table 4, we estimate the communication cost for creating preprocessed, masked tables for a range of table sizes, using our protocol from Sect. 4.1. This requires multiplication triples over $\mathbb{F}_{2^k}$, where k is a parameter of the protocol. When $k = 1$, we give figures using a recent optimized variant [43] of the two-party TinyOT protocol [35]. For larger choices of k, the costs are based on the MASCOT protocol [30]. We note that even though MASCOT has a communication complexity in $O(k^2)$, it still gives the lowest costs (with $k = 40$) for all the table sizes we considered.

Table 4. Total communication cost (kBytes) of the $\mathbb{F}_2 \times \mathbb{F}_{2^k}$ multiplications needed in creating a masked lookup table of size N, with two parties. The $k = 1$ estimates are based on TinyOT [43], the others on MASCOT [30].

N	$k = 1$	40	64	128
64	35.01	21.8	43.52	112.64
256	143.45	47.96	69.63	157.7
1024	577.17	135.16	174.08	292.86

5.1 Multiparty Setting

We also ran the AES-LT protocol with different numbers of parties and measured the throughput of the preprocessing and online phases. Figure 3 indicates that the preprocessing gets more expensive as the number of parties increases, whereas the online phase throughput does not decrease by much. This is likely to be because the bottleneck for the preprocessing is in terms of communication (which is $O(n^2)$ in total), whereas the online phase is more limited by the local computation done by each party.

5.2 Comparison with Other Works

We now compare the performance of our protocols with other implementations in similar settings. Table 5 gives an overview of the most relevant previous works. We see that our AES-LT protocol comes very close to the best online throughput of TinyTable, whilst having a far more competitive offline cost.[3] Our AES-RP variant has a slower online phase, but is comparable to the best garbled circuit protocols overall.

TinyTable Protocol. The original, 2-party TinyTable protocol [18] presented implementations of the online phase only, with two different variants. The fastest variant is based on table lookup and obtains a throughput of around 340 thousand AES blocks per second over a 1Gbps LAN, which is 1.51x faster than our

---

[3] The reason for the very large preprocessing cost of TinyTable is due to the need to evaluate the S-box 256 times per table lookup.

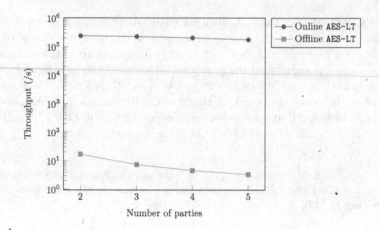

Fig. 3. Table lookup-based AES throughput for multiple parties.

Table 5. Performance comparison with other 2-PC protocols for evaluating AES in a LAN setting.

Protocol	Online		Comms. (total)	Notes
	Latency (ms)	Throughput (/s)		
TinyTable (binary) [18]	4.18	24500	3.07 MB	
TinyTable (optim.) [18]	1.02	339000	786.4 MB	
Wang et al. [43]	0.93	1075	2.57 MB	10 GBps
Rindal-Rosulek [39]	1.0	1000	1.6 MB	10 GBps
OP-LUT [22]	5	41670	0.103 MB	Passive
SP-LUT [22]	6	2208	0.044 MB	Passive
AES-LT	0.93	236200	8.4 MB	
AES-RP	7.19	940	2.9 MB	

online throughput. The latency (for sequential operations) is around 1ms, the same as ours. We attribute the difference in throughput to the additional local computation in our implementation, since we need to compute on MACs for every linear operation.

TinyTable does not report figures for the preprocessing phase. However, we estimate that using TinyOT and the naive method suggested in the paper would need would need over 1.3 million TinyOT triples for AES (34 ANDs for each S-box, repeated 256 times to create one masked table, for 16 S-boxes in 10 rounds). In contrast, our table lookup method uses around 160 thousand TinyOT triples, or just 2080 triples over $\mathbb{F}_{2^{40}}$ (cf. Table 1), per AES block.

Garbled Circuits. There are many implementations of AES for actively secure 2-PC using garbled circuits [33, 36, 39, 42, 43]. When measuring online throughput in a LAN setting, using garbled circuits gives much worse performance than methods based on table lookup, because evaluating a garbled circuit is much more expensive computationally. For example, out of all these works the lowest reported online time (even over a 10 GBps LAN) is 0.93 ms [43], and this does not improve in the amortized setting.

Some recent garbled circuit implementations, however, improve upon our performance in the preprocessing phase, where communication is typically the bottleneck. Wang et al. [43] require 2.57 MB of communication when 1024 circuits are being garbled at once, while Rindal and Rosulek [39] need only 1.6 MB. The runtime for both of these preprocessing phases is around 5 ms over a 10 GBps LAN; this would likely increase to at least 15–20 ms in a 1 GBps network, whereas our table lookup preprocessing takes around 60 ms using MASCOT. If a very fast online time is not required, our implementation of the Rivain–Prouff method would be more competitive, since this has a total amortized time of only 23 ms per AES block.

Secret-Sharing Based MPC. Other actively implementations of AES/DES using secret-sharing and dishonest majority based on secret sharing include those using SPDZ [15, 31] and MiniMAC [17, 21]. Our **AES-BD** method is the same as [15] and obtains faster performance than both SPDZ implementations. For DES, our TinyTable approach improves upon the times of the binary circuit implementation from [31] (which are for single-DES, so must be multiplied by 3) by over 100 times. Regarding MiniMAC, the implementation of [17] obtains slower online phase times than our work and TinyTable, and it is not known how to do the preprocessing with concrete efficiency.

OP-LUT and SP-LUT. The proposed 2-party protocols by Dessouky et al. [22] only offer security in the semi-honest setting. The preprocessing phase for both the protocols are based on 1-out-of-N oblivious transfer. In particular, the cost of the OP-LUT setup is essentially that of 1-out-of-N OT, while the cost of SP-LUT is the cost of 1-out-of-N *random* OT, which is much more efficient in terms of communication.

The online communication cost of OP-LUT is essentially the same as our online phase, since both protocols require each party to send $\log_2 N$ bits for a table of size N. However, we incur some additional local computation costs and a MAC check (at the end of the function evaluation) to achieve active security. The online phase of SP-LUT is less efficient, but the overall communication of this protocol is very low, only 0.055 MB for a single AES evaluation over a LAN setting with 1 GB network.

The work [22] reports figures for both preprocessing and online phase: using OP-LUT gives a latency of around 5 ms for 1 AES block in the LAN setting, and a throughput of 42000 blocks/s. These are both slower than our online phase figures using **AES-LT**. The preprocessing runtimes of both OP-LUT and

SP-LUT are much better than ours, however, achieving over 1000 blocks per second (roughly 80 times faster than AES-LT). This shows that we require a large overhead to obtain active security in the preprocessing, but the online phase cost is the same, or better.

References

1. Albrecht, M., Grassi, L., Rechberger, C., Roy, A., Tiessen, T.: MiMC: efficient encryption and cryptographic hashing with minimal multiplicative complexity. In: Cheon, J.H., Takagi, T. (eds.) ASIACRYPT 2016. LNCS, vol. 10031, pp. 191–219. Springer, Heidelberg (2016). doi:10.1007/978-3-662-53887-6_7
2. Albrecht, M.R., Rechberger, C., Schneider, T., Tiessen, T., Zohner, M.: Ciphers for MPC and FHE. In: Oswald, E., Fischlin, M. (eds.) EUROCRYPT 2015. LNCS, vol. 9056, pp. 430–454. Springer, Heidelberg (2015). doi:10.1007/978-3-662-46800-5_17
3. Aliasgari, M., Blanton, M., Zhang, Y., Steele, A.: Secure computation on floating point numbers. In: NDSS 2013. The Internet Society, February 2013
4. Araki, T., Furukawa, J., Lindell, Y., Nof, A., Ohara, K.: High-throughput semi-honest secure three-party computation with an honest majority. In: Weippl, E.R., Katzenbeisser, S., Kruegel, C., Myers, A.C., Halevi, S. (eds.) ACM CCS 2016, pp. 805–817. ACM Press, October 2016
5. Ben-Efraim, A., Lindell, Y., Omri, E.: Optimizing semi-honest secure multiparty computation for the internet. In: Weippl, E.R., Katzenbeisser, S., Kruegel, C., Myers, A.C., Halevi, S. (eds.) ACM CCS 16, pp. 578–590. ACM Press, October 2016
6. Bogdanov, D., Laur, S., Willemson, J.: Sharemind: a framework for fast privacy-preserving computations. In: Jajodia, S., Lopez, J. (eds.) ESORICS 2008. LNCS, vol. 5283, pp. 192–206. Springer, Heidelberg (2008). doi:10.1007/978-3-540-88313-5_13
7. Bogetoft, P., Damgård, I., Jakobsen, T., Nielsen, K., Pagter, J., Toft, T.: A practical implementation of secure auctions based on multiparty integer computation. In: Crescenzo, G., Rubin, A. (eds.) FC 2006. LNCS, vol. 4107, pp. 142–147. Springer, Heidelberg (2006). doi:10.1007/11889663_10
8. Burra, S.S., Larraia, E., Nielsen, J.B., Nordholt, P.S., Orlandi, C., Orsini, E., Scholl, P., Smart, N.P.: High performance multi-party computation for binary circuits based on oblivious transfer. Cryptology ePrint Archive, Report 2015/472 (2015). http://eprint.iacr.org/2015/472
9. Canetti, R.: Universally composable security: a new paradigm for cryptographic protocols. In: 42nd FOCS, pp. 136–145. IEEE Computer Society Press, October 2001
10. Carlet, C., Goubin, L., Prouff, E., Quisquater, M., Rivain, M.: Higher-order masking schemes for S-boxes. In: Canteaut, A. (ed.) FSE 2012. LNCS, vol. 7549, pp. 366–384. Springer, Heidelberg (2012). doi:10.1007/978-3-642-34047-5_21
11. Chari, S., Jutla, C.S., Rao, J.R., Rohatgi, P.: Towards sound approaches to counteract power-analysis attacks. In: Wiener, M. (ed.) CRYPTO 1999. LNCS, vol. 1666, pp. 398–412. Springer, Heidelberg (1999). doi:10.1007/3-540-48405-1_26
12. Coron, J.-S.: Higher order masking of look-up tables. In: Nguyen, P.Q., Oswald, E. (eds.) EUROCRYPT 2014. LNCS, vol. 8441, pp. 441–458. Springer, Heidelberg (2014). doi:10.1007/978-3-642-55220-5_25

13. Coron, J.-S., Roy, A., Vivek, S.: Fast evaluation of polynomials over binary finite fields and application to side-channel countermeasures. In: Batina, L., Robshaw, M. (eds.) CHES 2014. LNCS, vol. 8731, pp. 170–187. Springer, Heidelberg (2014). doi:10.1007/978-3-662-44709-3_10

14. Coron, J., Roy, A., Vivek, S.: Fast evaluation of polynomials over binary finite fields and application to side-channel countermeasures. J. Cryptogr. Eng. 5(2), 73–83 (2015). http://dx.doi.org/10.1007/s13389-015-0099-9

15. Damgård, I., Keller, M., Larraia, E., Miles, C., Smart, N.P.: Implementing AES via an actively/covertly secure dishonest-majority MPC protocol. In: Visconti, I., Prisco, R. (eds.) SCN 2012. LNCS, vol. 7485, pp. 241–263. Springer, Heidelberg (2012). doi:10.1007/978-3-642-32928-9_14

16. Damgård, I., Keller, M., Larraia, E., Pastro, V., Scholl, P., Smart, N.P.: Practical covertly secure MPC for dishonest majority – or: breaking the SPDZ limits. In: Crampton, J., Jajodia, S., Mayes, K. (eds.) ESORICS 2013. LNCS, vol. 8134, pp. 1–18. Springer, Heidelberg (2013). doi:10.1007/978-3-642-40203-6_1

17. Damgård, I., Lauritsen, R., Toft, T.: An empirical study and some improvements of the MiniMac protocol for secure computation. In: Abdalla, M., Prisco, R. (eds.) SCN 2014. LNCS, vol. 8642, pp. 398–415. Springer, Cham (2014). doi:10.1007/978-3-319-10879-7_23

18. Damgård, I., Nielsen, J.B., Nielsen, M., Ranellucci, S.: Gate-scrambling revisited - or: the TinyTable protocol for 2-party secure computation. Cryptology ePrint Archive, Report 2016/695 (2016). http://eprint.iacr.org/2016/695

19. Damgård, I., Pastro, V., Smart, N., Zakarias, S.: Multiparty computation from somewhat homomorphic encryption. In: Safavi-Naini, R., Canetti, R. (eds.) CRYPTO 2012. LNCS, vol. 7417, pp. 643–662. Springer, Heidelberg (2012). doi:10.1007/978-3-642-32009-5_38

20. Damgård, I., Zakarias, R.W.: Fast oblivious AES a dedicated application of the MiniMac protocol. In: Progress in Cryptology - AFRICACRYPT 2016–Proceedings of 8th International Conference on Cryptology in Africa, Fes, Morocco, 13–15 April 2016, pp. 245–264 (2016). http://dx.doi.org/10.1007/978-3-319-31517-1_13

21. Damgård, I., Zakarias, S.: Constant-overhead secure computation of Boolean circuits using preprocessing. In: Sahai, A. (ed.) TCC 2013. LNCS, vol. 7785, pp. 621–641. Springer, Heidelberg (2013). doi:10.1007/978-3-642-36594-2_35

22. Dessouky, G., Koushanfar, F., Sadeghi, A.R., Schneider, T., Zeitouni, S., Zohner, M.: Pushing the communication barrier in secure computation using lookup tables. In: 24th Annual Network and Distributed System Security Symposium (NDSS 2017). The Internet Society, 26 February–1 March 2017 (to appear). http://thomaschneider.de/papers/DKSSZZ17.pdf

23. Doerner, J., Evans, D., Shelat, A.: Secure stable matching at scale. In: Weippl, E.R., Katzenbeisser, S., Kruegel, C., Myers, A.C., Halevi, S. (eds.) ACM CCS 2016, pp. 1602–1613. ACM Press, October 2016

24. EMVCo: EMVCo Security QA (2017). https://www.emvco.com/faq.aspx?id=38. Accessed Feb 2017

25. Frederiksen, T.K., Keller, M., Orsini, E., Scholl, P.: A unified approach to MPC with preprocessing using OT. In: Iwata, T., Cheon, J.H. (eds.) ASIACRYPT 2015. LNCS, vol. 9452, pp. 711–735. Springer, Heidelberg (2015). doi:10.1007/978-3-662-48797-6_29

26. Gentry, C., Halevi, S., Smart, N.P.: Homomorphic evaluation of the AES circuit. In: Safavi-Naini, R., Canetti, R. (eds.) CRYPTO 2012. LNCS, vol. 7417, pp. 850–867. Springer, Heidelberg (2012). doi:10.1007/978-3-642-32009-5_49

27. Grassi, L., Rechberger, C., Rotaru, D., Scholl, P., Smart, N.P.: MPC-friendly symmetric key primitives. In: Weippl, E.R., Katzenbeisser, S., Kruegel, C., Myers, A.C., Halevi, S. (eds.) ACM CCS 2016, pp. 430–443. ACM Press, October 2016
28. Ishai, Y., Kushilevitz, E., Meldgaard, S., Orlandi, C., Paskin-Cherniavsky, A.: On the power of correlated randomness in secure computation. In: Sahai, A. (ed.) TCC 2013. LNCS, vol. 7785, pp. 600–620. Springer, Heidelberg (2013). doi:10.1007/978-3-642-36594-2_34
29. Jha, S., Kruger, L., Shmatikov, V.: Towards practical privacy for genomic computation. In: 2008 IEEE Symposium on Security and Privacy, pp. 216–230. IEEE Computer Society Press, May 2008
30. Keller, M., Orsini, E., Scholl, P.: MASCOT: Faster malicious arithmetic secure computation with oblivious transfer. In: Weippl, E.R., Katzenbeisser, S., Kruegel, C., Myers, A.C., Halevi, S. (eds.) ACM CCS 2016. pp. 830–842. ACM Press, October 2016
31. Keller, M., Scholl, P., Smart, N.P.: An architecture for practical actively secure MPC with dishonest majority. In: Sadeghi, A.R., Gligor, V.D., Yung, M. (eds.) ACM CCS 2013, pp. 549–560. ACM Press, November 2013
32. Launchbury, J., Diatchki, I.S., DuBuisson, T., Adams-Moran, A.: Efficient lookup-table protocol in secure multiparty computation. In: ACM SIGPLAN International Conference on Functional Programming, ICFP 2012, Copenhagen, Denmark, 9–15 September 2012, pp. 189–200 (2012). http://doi.acm.org/10.1145/2364527.2364556
33. Lindell, Y., Riva, B.: Blazing fast 2PC in the offline/online setting with security for malicious adversaries. In: Ray, I., Li, N., Kruegel, C. (eds.) ACM CCS 2015, pp. 579–590. ACM Press, October 2015
34. Litzenberger, D.C.: Pycrypto - the Python cryptography toolkit (2017). https://www.dlitz.net/software/pycrypto
35. Nielsen, J.B., Nordholt, P.S., Orlandi, C., Burra, S.S.: A new approach to practical active-secure two-party computation. In: Safavi-Naini, R., Canetti, R. (eds.) CRYPTO 2012. LNCS, vol. 7417, pp. 681–700. Springer, Heidelberg (2012). doi:10.1007/978-3-642-32009-5_40
36. Nielsen, J.B., Schneider, T., Trifiletti, R.: Constant round maliciously secure 2PC with function-independent preprocessing using LEGO. In: 24th NDSS Symposium. The Internet Society (2017). http://eprint.iacr.org/2016/1069
37. Pinkas, B., Schneider, T., Smart, N.P., Williams, S.C.: Secure two-party computation is practical. In: Matsui, M. (ed.) ASIACRYPT 2009. LNCS, vol. 5912, pp. 250–267. Springer, Heidelberg (2009). doi:10.1007/978-3-642-10366-7_15
38. Pulkus, J., Vivek, S.: Reducing the number of non-linear multiplications in masking schemes. In: Gierlichs, B., Poschmann, A.Y. (eds.) CHES 2016. LNCS, vol. 9813, pp. 479–497. Springer, Heidelberg (2016). doi:10.1007/978-3-662-53140-2_23
39. Rindal, P., Rosulek, M.: Faster malicious 2-party secure computation with online/offline dual execution. In: 25th USENIX Security Symposium, USENIX Security 2016, Austin, TX, USA, 10–12 August 2016, pp. 297–314 (2016). https://www.usenix.org/conference/usenixsecurity16/technical-sessions/presentation/rindal
40. Rivain, M., Prouff, E.: Provably secure higher-order masking of AES. In: Mangard, S., Standaert, F.-X. (eds.) CHES 2010. LNCS, vol. 6225, pp. 413–427. Springer, Heidelberg (2010). doi:10.1007/978-3-642-15031-9_28
41. Roy, A., Vivek, S.: Analysis and improvement of the generic higher-order masking scheme of FSE 2012. In: Bertoni, G., Coron, J.-S. (eds.) CHES 2013. LNCS, vol. 8086, pp. 417–434. Springer, Heidelberg (2013). doi:10.1007/978-3-642-40349-1_24

42. Wang, X., Malozemoff, A.J., Katz, J.: Faster two-party computation secure against malicious adversaries in the single-execution setting. In: EUROCRYPT 2017 Proceedings (2017)
43. Wang, X., Ranellucci, S., Katz, J.: Authenticated garbling and communication-efficient, constant-round, secure two-party computation. IACR Cryptology ePrint Archive 2017, 30 (2017). http://eprint.iacr.org/2017/030

Cryptographic Primitives

An Experimental Study of the BDD Approach for the Search LWE Problem

Rui Xu[1(✉)], Sze Ling Yeo[2], Kazuhide Fukushima[1], Tsuyoshi Takagi[3],
Hwajung Seo[2], Shinsaku Kiyomoto[1], and Matt Henricksen[4]

[1] KDDi Research Inc., Fujimino, Japan
ru-xu@kddi-research.jp
[2] Institute for Infocomm Research (I2R), Singapore, Singapore
[3] Institute of Mathematics for Industry (IMI), Kyushu University, Fukuoka, Japan
[4] Huawei Technologies, Singapore, Singapore

Abstract. The proved hardness of the Learning With Errors (LWE) problem, assuming the worst case intractability of classic lattice problems, has made it a standard building block in the recent design of lattice based cryptosystems. Nonetheless, a thorough understanding of the security of these schemes from the perspective of existing attacks remains an open problem. In this manuscript, we report our implementation of the Bounded Distance Decoding (BDD) approach for solving the search LWE problem. We implement a parallel version of the pruned enumeration method of the BDD strategy proposed by Liu and Nguyen. In our implementation we use the embarrassingly parallel design so that the power of multi-cores can be fully utilized. We let each thread take a randomized basis and perform independent enumerations to find the solution instead of parallelizing the enumeration algorithm itself. Other optimizations include fine-tuning the BKZ block size, the enumeration bound and the pruning coefficients and the optimal dimension of the LWE problem. Experiments are done using the TU Darmstadt LWE challenge. Finally we compare our implementation with a recent parallel BDD implementation by Kirshanova et al. [18] and show that our implementation is more efficient.

Keywords: Learning With Errors · Lattice based cryptography · Security evaluation · Bounded Distance Decoding

1 Introduction

Decades of development in the area of lattice-based cryptography have identified two important primitive hard problems, namely, the Shortest Integer Solution (SIS) problem [1] and the Learning With Errors (LWE) problem [24], to be standard building blocks of modern lattice-based cryptosystems.

In this work, we focus on the LWE problem proposed by Regev [24]. LWE has attracted more and more attention since its proposal. Initially LWE problem was reduced to the GAPSVP (the decision version of the shortest vector problem) or

© Springer International Publishing AG 2017
D. Gollmann et al. (Eds.): ACNS 2017, LNCS 10355, pp. 253–272, 2017.
DOI: 10.1007/978-3-319-61204-1_13

SIVP (Shortest Independent Vector Problem) under the quantum setting. This means that LWE is considered hard if there are no algorithms to efficiently solve the GAPSVP or SIVP using a quantum computer. Subsequently, the hardness reduction as been sharpened to accept a classic reduction to these standard lattice problems [8]. As such, LWE-based schemes are widely studied as potential primitives in the post-quantum era.

LWE. Let n be a positive integer, denoting the dimension of the lattice related with the LWE problem, q an odd prime, and let $\mathcal{D}$ be an error distribution over the integer ring modulo q, $\mathbb{Z}_q$. Denote by $\mathbf{s}$ a fixed secret vector in $\mathbb{Z}_q^n$ (in this manuscript we adopt the row vector convention to be consistent with software implementation) selected according to the uniform distribution on its support. Let $\mathcal{L}_{n,q,\mathcal{D}}$ be the probability distribution on $\mathbb{Z}_q^n \times \mathbb{Z}_q$ generated by choosing $\mathbf{a} \in \mathbb{Z}_q^n$ uniformly at random, choosing an error e according to $\mathcal{D}$ and returning

$$(\mathbf{a}, c) = (\mathbf{a}, \langle \mathbf{a}, \mathbf{s} \rangle + e)$$

in $\mathbb{Z}_q^n \times \mathbb{Z}_q$, where $\langle \cdot, \cdot \rangle$ denotes the inner product of two vectors in $\mathbb{Z}_q^n$. The search LWE problem is to find the secret vector $\mathbf{s}$ given a fixed number of m samples from $\mathcal{L}_{n,q,\mathcal{D}}$.

Although the intractability of LWE is well established by the reduction proofs, its concrete hardness is far from clear. In this work we follow the approach of Liu and Nguyen [21] to evaluate the performance of the BDD approach for solving LWE problem.

1.1   Our Contribution

In this manuscript, our main contributions include:

- We implement a parallel version of the BDD approach for solving the LWE problem. The implementation features an embarrassingly parallel design where each thread takes a randomized basis and performs an independent enumeration. The advantage of this design is that the power of multi-cores can be fully utilized.
- We give heuristic analysis on how to choose the optimal sub-dimension of the LWE instance. We use the Gaussian heuristic to estimate the cost of a pruned enumeration tree to find better sub-dimension which can reduce the time to solve an LWE instance.
- We compare our implementation with that of Kirshanova et al. [18] and show the advantages of our implementation. Specifically we show that the performance of our parallelization strategy is not limited by Amdahl's Law and the extreme pruning in our implementation brings huge speedup compared with the linear pruning used in the implementation of [18].
- We demonstrate that our implementation solves a couple of instances from the TU Darmstadt LWE challenge.

2   Preliminaries

2.1   Discrete Gaussian Distribution

We first describe the error distribution $\mathcal{D}$ in the LWE problem. In the general situation, any error distribution with small variance is fine for the LWE problem to be hard. However, in this work, similar to many other previous works regarding LWE, we focus on the discrete Gaussian distribution over the ring $\mathbb{Z}_q$ as the error distribution. Let $x \in \mathbb{Z}$. The discrete Gaussian distribution over $\mathbb{Z}$ with mean 0 and width parameter σ, denoted by $D_{\mathbb{Z},\sigma}$ assigns to each $x \in \mathbb{Z}$ the probability proportional to $\exp(-x^2/2\sigma^2)$. The error distribution we consider for the LWE problem is the discrete Gaussian distribution over $\mathbb{Z}_q$, denoted by $D_{\mathbb{Z}_q,\sigma}$, by accumulating the values of the probability mass function over all integers in each residue class mod q. In the original proposal of Regev, the width parameter associated with the moduli q is $\sigma = \frac{\alpha q}{\sqrt{2\pi}}$, where α is the relative error rate. With a slight abuse of notation, we also denote the discrete Gaussian distribution as $\mathcal{D}_{\mathbb{Z}_q,\alpha q}$. When the error distribution of an LWE instance is $\mathcal{D}_{\mathbb{Z}_q,\alpha q}$, we express the LWE instance as $\mathcal{L}_{n,q,\alpha}$.

2.2   Lattice

A lattice in $\mathbb{R}^m$ is a discrete additive subgroup generated by a (non-unique) basis $\mathbf{B} = (\mathbf{b}_1, \ldots, \mathbf{b}_m)^T$. Equivalently, the lattice $\Lambda(\mathbf{B})$ generated by $\mathbf{B}$ is given by $\Lambda(\mathbf{B}) = \{x | x = \sum_{i=1}^{m} z_i \mathbf{b}_i\}$, where z_i's are integers. Note that by our convention, the vector $\mathbf{b}_i$ in the basis matrix $\mathbf{B}$ is its row vector. The rank of the lattice $\Lambda(\mathbf{B})$ is defined as the rank of the basis matrix $\mathbf{B}$. If the rank of $\Lambda(\mathbf{B})$ equals m, we say that the lattice is full rank. A fundamental notion that lies in various lattice problems is the successive minimal $\lambda_k(\Lambda)$ which is defined to be the smallest real number r such that the lattice contains k linearly independent nonzero vectors of Euclidean length at most r. Specifically, $\lambda_1(\Lambda)$ is the length of the shortest nonzero vector of the lattice Λ.

The lattices we are interested in are a special type of lattices called q-ary lattices which are lattices satisfying $q\mathbb{Z}^m \subset \Lambda \subset \mathbb{Z}^m$. Fix positive integers $n \leq m \leq q$, where n serves as the main security parameter, and q is an odd prime. For any matrix $\mathbf{A} \in \mathbb{Z}^{m \times n}$, define the following two lattices.

$$\Lambda_q^{\perp}(\mathbf{A}) = \{\mathbf{x} \in \mathbb{Z}^m : \mathbf{x}\mathbf{A} = 0 \mod q\},$$

$$\Lambda_q(\mathbf{A}) = \{\mathbf{x} \in \mathbb{Z}^m : \mathbf{x} = \mathbf{A}\mathbf{s} \mod q \text{ for some } \mathbf{s} \in \mathbb{Z}_q^n\}.$$

It is easy to check that both $\Lambda_q^{\perp}(\mathbf{A})$ and $\Lambda_q(\mathbf{A})$ are q-ary lattices [23].

2.3   Lattice Reduction

As we have noticed in Sect. 2.2 that a lattice can be generated from different bases, the property of the basis plays a central role in the difficulty of various hard lattice problems. Informally, the more orthogonal the basis is, the easier the corresponding lattice problems are. As such, many attempts to solve hard lattice problems try to alter (often called reduce in the literature) the given basis in order to get basis which generates the same lattice while at the same time achieves the highest orthogonality possible. We adopt the convention that the first vector $\mathbf{b}_1$ in a reduced basis has the smallest length among the (reduced) basis vectors. After the lattice reduction algorithm, we can use the vector $\mathbf{b}_1$ as an approximation of the shortest vector. Since the determinate of a lattice is invariant under lattice reduction, when the basis is reduced, the length of each basis vector decreases. The common measurement of the quality of a lattice basis is called Hermite factor δ^m defined as: $||\mathbf{b}_1|| = \delta^m \text{vol}(\Lambda)^{1/m}$. We also refer to δ as the root-Hermite factor. A smaller root-Hermite factor typically implies a reduced basis with higher quality.

Lattice reduction algorithms can be viewed as a hierarchy of BKZ [26] based on the parameter blocksize β. The case when $\beta = 2$ is called LLL reduction, which was invented by Lenstra et al. [20]. LLL reduction is proven to run in polynomial time in the lattice dimension and outputs a short vector which is within an exponential factor of the minimal length of a lattice Λ, i.e., $\lambda_1(\Lambda)$. When $\beta = m$, i.e., the full size of the basis, the output basis is HKZ reduced [17] which implies solving the SVP. The situation when k lies in between 2 and m is known as the BKZ-β reduction which is the most referenced reduction algorithm in practice. Chen and Nguyen observed that the running time of BKZ reduction is mainly dominated by the root-Hermite factor δ and is less affected by the dimension m. See Chen and Nguyen [11] for a detailed analysis and their improvements over the standard BKZ as a collection of optimization known as BKZ 2.0. See also Albrecht et al. [2] for a thorough comparison of different estimations of the complexity of BKZ.

2.4 Pruned Enumeration

Gram-Schmidt Orthogonalization. Given a lattice basis $\mathbf{B} = (\mathbf{b}_1, \ldots, \mathbf{b}_m)^T$, the Gram-Schmidt Orthogonalization of $\mathbf{B}$ is denoted as $\mathbf{B}^* = (\mathbf{b}_1^*, \ldots, \mathbf{b}_m^*)$, where $\mathbf{b}_i^*$ is computed as $\mathbf{b}_i^* = \mathbf{b}_i - \sigma_{j=1}^{i-1}\mu_{i,j}\mathbf{b}_j^*$ for $i = 1, \ldots, m$, with $\mu_{i,j} = <\mathbf{b}_i, \mathbf{b}_j^* > /||\mathbf{b}_j^*||^2$ for all $1 \leq j \leq i \leq m$. Denote by $\pi_i(\cdot)$ the orthogonal projection onto $(\mathbf{b}_1, \mathbf{b}_2, \ldots, \mathbf{b}_{i-1})^\perp$. Then $\mathbf{b}_i^* = \pi_i(\mathbf{b}_i)$. Also $\pi_i(\Lambda(\mathbf{B}))$ is an $(m + 1 - i)$-dimensional lattice generated by the basis $(\pi_i(b_1), \pi_i(b_2), \ldots, \pi_i(b_{i-1}))^T$.

Lattice Enumeration. Given a target vector $\mathbf{t}$, a lattice basis $\mathbf{B} = (\mathbf{b}_1, \cdots, \mathbf{b}_m)^T$ and a radius R, lattice enumeration algorithm enumerates over all lattice vectors $\mathbf{v} \in \mathcal{L}$ such that $||\mathbf{v} - \mathbf{t}|| \leq R$ and finds the closest one. The enumeration algorithm enumerates over a search tree leveled by the enumeration depth $k \in [m]$. The root of the search is levelled using $k = 0$ and it represents the target vector. And $k = m$ corresponds to the leaves. The nodes at level k of the search tree consist of all vectors $\mathbf{v} \in \Lambda(\mathbf{B})$ such that $||\pi_{m+1-k}(\mathbf{t} - \mathbf{v})|| \leq R$. Gama et al. [14] use Gaussian heuristic to approximate the number of nodes in the k-th level of the enumeration tree as

$$H_k = \frac{V_k(R)}{\prod_{i=m+1-k}^{n}||\mathbf{b}_i^*||^2}, \tag{1}$$

where $V_k(R)$ denotes the volume of a k-dimensional ball of radius R. Then the total number of nodes in the enumeration tree is $N = \sum_{k=1}^{m} H_k$.

Gamma et al. also suggest to use extreme pruning to accelerate the enumeration algorithm. The idea of extreme pruning is by deliberately setting the probability that the solution vector is in the tree after pruning to be very small to cut lots of branches in the enumeration tree. Though the success probability of finding the desired solution becomes quite low, it is compensated by the huge reduction of the enumeration time. Their experiments show an exponential speed up over full enumeration. Formally, pruned enumeration bounds the enumeration tree by limiting the k-level nodes to those vectors $\mathbf{v} \in \Lambda(\mathbf{B})$ such that $\|\pi_{m+1-k}(\mathbf{v} - \mathbf{t})\| \leq R_k$ with R_k denoting the pruned radius and $R_1 \leq R_2 \leq \ldots \leq R_m = R$.

3   Related Work

We consider the search version of the LWE problem in this work. There are mainly three ways to solve the search LWE problem.

1. BKW approach: Blum, Kalai and Wasserman proposed BKW algorithm for the LPN (learning with parity noise) problem. Since LWE can be viewed as a generalization of LPN problem, BKW was also adapted to solve LWE by Albrecht et al. [3].
2. Algebraic approach: Arora-Ge [6] proposed to set up a system of algebraic equations over integers to describe the LWE problem and solve the search problem by solving the equation system. Later, this method was improved by using Gröbner basis techniques [4].
3. BDD approach: This approach views the search LWE problem as a decoding problem in a lattice. We will explain this idea in more details in the following.

Bounded Distance Decoding (**BDD**): Given m samples $(\mathbf{a_i}, c_i)$ following the given LWE distribution $\mathcal{L}_{n,q,\mathcal{D}}$, we organize the input into a matrix $\mathbf{A} \in \mathbb{Z}_q^{m \times n}$ whose rows constitute the m samples of the vector $\mathbf{a}_i$, and a vector $\mathbf{c} \in \mathbb{Z}_q^m$ whose i-th element is c_i from the i-th sample. Note that $\mathbf{c} = \mathbf{As} + \mathbf{e}$, where $\mathbf{e}$ is the error vector which follows the distribution $\mathcal{D}^n$. When the error distribution of the LWE problem is the discrete Gaussian distribution $\mathcal{D}_{\mathbb{Z}_q, \alpha q}$, we observe that the length of $\mathbf{e}$ is relatively small since each of its entries is distributed according to the discrete Gaussian. Consider the q-ary lattice

$$\Lambda_q(\mathbf{A}) = \{\mathbf{x} \in \mathbb{Z}^m : \mathbf{x} = \mathbf{As} \mod q \text{ for some } \mathbf{s} \in \mathbb{Z}_q^n\},$$

induced by $\mathbf{A}$. Then the vector $\mathbf{c}$ is bounded in distance from a vector $\mathbf{v} \in \Lambda_q(\mathbf{A})$. Finding the vector $\mathbf{v}$ from the q-ary lattice is called the **BDD** problem.

One approach to solve the BDD problem is to reduce the BDD problem to unique SVP problem by the embedding technique of Kannan [17] and solve the corresponding SVP problem. Another more common approach is to adapt the well-established Babai's algorithm to solve the BDD problem directly. In this regard, Lindner and Peikert [22] proposed a variant of Babai's nearest plane algorithm to solve the BDD problem. Bischof et al. [10] and Kirshanova et al. [18] have implemented parallel versions of this algorithm and investigated its practical performance. Liu and Nguyen [21] observed that Lindner and Peickert nearest plane algorithm can be viewed as a form of pruned enumeration where in the former, the pruning strategy bounds the coefficients instead of the usual way of bounding the projection lengths. Further, they propose to use the lattice enumeration with GNR extreme pruning strategy to accelerate the speed of finding the closest vector. This will be the approach we use in our experimental study. We refer the readers to the excellent survey by Albrecht et al. [2] for a comprehensive exploration of the concrete hardness of LWE.

4   Our Implementation

We choose to implement the Liu and Nguyen algorithm to study its practical performance. This algorithm uses enumeration with extreme pruning to solve the BDD problem. Given M samples $\{(\mathbf{a}_i, c_i)\}_{i=1,2,...,M}$ from the LWE distribution $\mathcal{L}_{n,q,\alpha}$, we use the matrix representation to express the LWE problem as $\mathbf{As}+\mathbf{e} = \mathbf{c}$. We outline the algorithm steps as follows:

Algorithm 1. LWE solver using BDD approach

Require: Inputs are $n, M, \alpha, q, \mathbf{A} \in \mathcal{Z}_q^{M \times n}, \mathbf{c} \in \mathcal{Z}_q^M$. The inputs satisfy $\mathbf{As} + \mathbf{e} = \mathbf{c}$ mod q for some hidden secret vector $\mathbf{s} \in \mathcal{Z}_q^n$ and error vector $\mathbf{e} \in \mathcal{Z}^M$ with its coefficients chosen independently according to the discrete Gaussian distribution $\mathcal{D}_{\mathbb{Z}_q, \alpha q}$.

Ensure: Output the hidden secret vector $\mathbf{s} \in \mathcal{Z}_q^n$ with high probability.

1: Decide an appropriate sub-dimension m for the given LWE instance.
2: Choose m random rows from the matrix $\mathbf{A}$ as $\mathbf{A}' \in \mathcal{Z}_q^{m \times n}$ and the corresponding m elements from the vector $\mathbf{c}' \in \mathcal{Z}_q^m$.
3: Compute the basis of the q-ary lattice $\Lambda_q(\mathbf{A}')$ generated by $\mathbf{A}'$, denote it by $\mathbf{B}$.
4: Choose an appropriate block size β and use BKZ-β to reduce the lattice basis $\mathbf{B}$.
5: Choose an appropriate enumeration radius R.
6: Compute a set of pruning coefficients and use pruned enumeration to find the closest vector $\mathbf{v}$ to the target vector $\mathbf{c}'$ using the enumeration radius R.
7: If the closest vector $\mathbf{v}$ can be found, then compute the secret vector $\mathbf{s}$ by solving the equation $\mathbf{A}'\mathbf{s} = \mathbf{c}' - \mathbf{v}$ mod q and return $\mathbf{s}$. Otherwise, goto Step 2.

The red and underlined parameters sub-dimension m, BKZ block size β and enumeration radius R in Algorithm 1 need to be optimized to achieve better performance for our LWE solver. However, it is easy to see that these parameters affect the running time and success probability of our LWE solver in an entangled way. Thus it is a multi-object optimization problem. In general, it is not trivial to solve such a multi-object optimization problem.

We give detailed explanation of the choice we make regarding each step of the algorithm in subsequent subsections.

4.1   Compute the Basis

This is easy linear algebra. Given a matrix $\mathbf{A}$ of size $m \times n$, we want to find the matrix $\mathbf{B}$ which is the basis of the q-ary lattice $\Lambda_q(\mathbf{A}) = \{\mathbf{x} \in \mathbb{Z}^m : \mathbf{x} = \mathbf{A}\mathbf{s} \mod q \text{ for some } \mathbf{s} \in \mathbb{Z}_q^n\}$. Recall that we use the convention of row vectors, so a lattice vector generated by the basis $\mathbf{B}$ can be represented by $\mathbf{z}\mathbf{B}$ where $\mathbf{z}$ is a row vector. If we forget for a moment that we are working on the modular ring $\mathbb{Z}_q$, then the basis for $\Lambda_q(\mathbf{A})$ is simply $\mathbf{A}^T$, the transpose of $\mathbf{A}$ since all lattice vectors except for those in $q\mathbb{Z}^m$ can be expressed by an integer linear combination of rows in matrix $\mathbf{A}^T$. To include $q\mathbb{Z}^m$ in order to make it q-ary lattice, we further compute the Hermite normal form of $\left[\begin{smallmatrix}\mathbf{A}^T \\ q\mathbf{I}_m\end{smallmatrix}\right]$ to get the basis of the q-ary lattice $\Lambda_q(\mathbf{A})$. In other words, $\mathbf{B} = \text{HNF}(\left[\begin{smallmatrix}\mathbf{A}^T \\ q\mathbf{I}_m\end{smallmatrix}\right])$, where $\text{HNF}(\mathbf{A})$ denotes the row Hermite normal form of a matrix $\mathbf{A}$ removing all zero rows. To see this, first note that $q\mathbb{Z}^m$ itself can be viewed as a lattice with the (m dimensional) identity matrix scaled by q as its basis. Thus, we get $\Lambda_q(\mathbf{A}) = \Lambda(\mathbf{A}^T) \cup \Lambda(q\mathbf{I}_m)$. The last step of our computation relies on the following fact:

Fact 1. Given two lattices $\Lambda(\mathbf{B}_1)$ and $\Lambda(\mathbf{B}_2)$ of the same dimension, the basis for the lattice generated by the union of $\Lambda(\mathbf{B}_1)$ and $\Lambda(\mathbf{B}_2)$ is $\text{HNF}(\left[\begin{smallmatrix}\mathbf{B}_1 \\ \mathbf{B}_2\end{smallmatrix}\right])$.

4.2   Enumeration Radius and Basis Randomization

The enumeration radius only affects the running time and success probability of the enumeration part. Consider an LWE instance $\mathcal{L}_{n,q,\alpha}$. Fix m samples from it to get the equation $\mathbf{A}'\mathbf{s} + \mathbf{e}' = \mathbf{c}' \mod q$. The exact BDD radius is the length of the error vector $\mathbf{e}' \in \mathcal{Z}^m$. Though we do not know the exact value of $||\mathbf{e}'||^2$, we know that its coefficients are generated from the discrete Gaussian distribution $\mathcal{D}_{\mathcal{Z}_q, \alpha q}$. According to the acceptance criteria of the LWE challenge, the requirement is that $||\mathbf{e}'|| \leq 2\sqrt{m}\alpha q$ for an LWE instance $\mathcal{L}_{n,q,\alpha}$ with m samples. Thus one option is to take $R = 2\sqrt{m}\alpha q$ as the **BDD** bound. More generally, we set squared **BDD** bound R^2 as $c \cdot m\alpha^2 q^2$ for some fixed constant c. To approximate the **BDD** bound R, we sampled error vectors according to the discrete Gaussian distribution and record the squared length of the error vector $\mathbf{e}'$. See Fig. 1 for our experiment results. The figure shows the histograms of the distribution of the squared norm of error vector $\mathbf{e}'$ with the bins set as the multiplier between $||\mathbf{e}'||^2$ and $m\alpha^2 q^2$ (i.e., the scalar c). From Fig. 1 we can see that $c = 1.3$ is an appropriate choice for making sure that the closest vector can be found with overwhelming probability and the distribution of the scalar c does not depend on the parameters n, α and m. However, if one chooses $c = 1$ so as to reduce the running time of the enumeration algorithm, then with probability about half, the closest vector can not be found within this radius.

Typical applications of pruned enumeration will first randomize the lattice basis by multiplying the basis matrix with a random unimodular matrix and then apply pruned enumeration to find the desired shortest vector or closest vector. If we adopt this method in our LWE solver, two problems arise.

1. The randomized basis usually has larger entries than the initial basis, thereby adding some burden to the lattice reduction algorithm.
2. If we want to choose a smaller enumeration radius such as setting the scalar c less than 1.3, we might miss the opportunity of finding the closest vector. This is primarily due to the fact that we are working with different bases of a fixed lattice and hence, the error norm is fixed.

Our randomization is natural and effective. We do not bother to do randomization over a fixed lattice but instead we choose a different lattice each time. In most instances, the number of samples M is larger than the LWE dimension n. It follows that after deciding on a sub-dimension m with $n < m < M$, we can randomly choose m samples from the total M samples to form a different lattice each time. This simple trick solves the two problems discussed above. First the entries of the generated basis are all less than or equal than q. Second, since we randomize over the different m-combinations of the samples, the error vector $\mathbf{e}'$ changes every time. Then we can choose a lower enumeration bound R and be confident that a fixed portion of the trials contribute to error vectors within the bound. For example, according to Fig. 1, if we choose $R^2 = m\alpha^2 q^2$ then the closest vector could be found within this bound with probability about 50 %.

Denote by p_{enum} the success probability of an enumeration algorithm given that the length of the error vector $||\mathbf{e}'||$ is indeed within the enumeration bound R. For simplicity we first consider $||\mathbf{e}'|| \leq 1.3 \cdot m\alpha^2 q^2$. Let $T_{enum}(c)$ be the time of the enumeration algorithm when setting the enumeration radius to be $c \cdot m\alpha^2 q^2$. We can estimate the total time of enumeration to find the closest vector (when setting $c = 1.3$) as $T(1.3) = T_{enum}(1.3)/p_{enum}$. Further assuming that changing the enumeration radius does not affect p_{enum}, we can approximate the success probability of the enumeration algorithm using different enumeration scalars c. For example if we choose $c = 1$ the probability that the error vector is within $1 \cdot m\alpha^2 q^2$ is about 0.5 so we get the total enumeration time of solving the BDD problem as $T(1) = 2 * T_{enum}(1)/p_{enum}$. In particular, choosing $c = 1$ may lead to a faster algorithm if $T_{enum}(1) < T_{enum}(1.3)/2$. Thus, by analyzing the impact of the enumeration radius on the running time of enumeration algorithm, we can choose nearly optimal enumeration radius. Finally, we can use the Gaussian heuristic Eq. (1) to approximate $T_{enum}(c)$ (see below).

4.3   Choose Sub-dimension

In the typical setting of LWE problem, the number of total samples M is bounded by a polynomial of the LWE dimension n. When treating LWE as a lattice problem, an important decision concerns a suitable choice for the dimension of the lattice. The dimension of the lattice equals the number of samples we choose. How many of the total M samples do we use to form the generating matrix $\mathbf{A}'$?

First, we show that if the sub-dimension were chosen too small, the sub LWE problem may not have a unique solution. Consider the following equation $\mathbf{A}'\mathbf{s} + \mathbf{e}' = \mathbf{c}'$. For any choice of $\mathbf{s}$, we can find an error vector $\mathbf{e}'$ satisfying the above equation. However, the LWE problem restricts the length of $\mathbf{e}'$. More

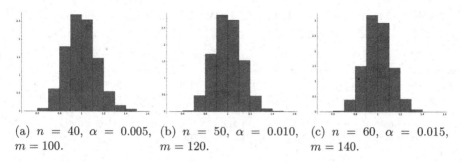

(a) $n = 40$, $\alpha = 0.005$,   (b) $n = 50$, $\alpha = 0.010$,   (c) $n = 60$, $\alpha = 0.015$,
$m = 100$. $m = 120$. $m = 140$.

Fig. 1. Histograms of square length of $\mathbf{e}'$ for different parameters.

precisely, each element of $\mathbf{e}'$ is chosen from the discrete Gaussian distribution with small variance and thus, it can not be too large. TU Darmstat University has held an LWE challenge website[1] similar to the famous SVP challenge. According to Buchmann et al. [9], the acceptance criteria for the correct answer of the LWE problem $\mathcal{L}_{n,q,\alpha}$ with M samples is that $||\mathbf{e}|| \leq 2\sqrt{M}\alpha q$. Based on this criteria, when we choose the sub-dimension to be m, we would also expect to find a secret vector $\mathbf{s}$ such that it leads to a error vector of length less than $2\sqrt{m}\alpha q$. Following the argument of Buchmann et al. [9], we calculate the probability that the sub LWE problem has more than one solution. For a chosen matrix $\mathbf{A}'$ of size $m \times n$, let $\Lambda_q(\mathbf{A}')$ denote the q-ary lattice generated by $\mathbf{A}'$. Recall that $\lambda_1(\Lambda_q(\mathbf{A}'))$ is the norm of the shortest non zero vector in $\Lambda_q(\mathbf{A})$. Assume that we have two solutions for the secret vector $\mathbf{s}_1$ and $\mathbf{s}_2$ satisfying the criteria $\mathbf{A}'\mathbf{s}_1 + \mathbf{e}'_1 = \mathbf{c}' = \mathbf{A}'\mathbf{s}_2 + \mathbf{e}'_2$ and $\mathbf{e}'_i \leq 2\sqrt{m}\alpha q$. Then by the triangle inequality, we have $||\mathbf{A}'(\mathbf{s}_1 - \mathbf{s}_2)|| \leq 4\sqrt{m}\alpha q$. Since $\mathbf{A}'(\mathbf{s}_1 - \mathbf{s}_2)$ is actually a vector in the q-ary lattice $\Lambda_q(\mathbf{A}')$, the fact that the sub LWE problem has more than one solution implies that $\lambda_1(\Lambda_q(\mathbf{A}')) \leq 4\sqrt{m}\alpha q$. On the other hand, Gaussian heuristic tells us that the expected length of the shortest vector of $\Lambda_q(\mathbf{A}')$ is $q^{1-\frac{n}{m}}\sqrt{\frac{m}{2\pi e}}$. In view of this, in our implementation we choose the sub-dimension m such that the corresponding Gaussian heuristic $q^{1-\frac{n}{m}}\sqrt{\frac{m}{2\pi e}}$ is larger than $4\sqrt{m}\alpha q$ so that the expected number of solutions is small.

The next question concerns how large the lattice sub-dimension m should be. Note that a large dimension invariably increases the time for the basis reduction. In [15], the authors experimentally showed that for a random input lattice, the root Hermite factor δ after a BKZ-beta reduction is independent of the lattice dimension. The following table shows the root Hermite factor obtained from various m random samples taken from the LWE challenge with $\alpha = 0.005$ under a BKZ-20 reduction averaged over 20 experiments for each pair of m and n (Table 1).

One sees that for each n, the value of δ is approximately 0.128 for the first values of m but deviates from this value for larger values of m. A closer examination reveals that in the latter case, the shortest vectors produced by the

---

[1] https://www.latticechallenge.org/lwe_challenge/challenge.php.

Table 1. Average root hermite factor for LWE instances with sub-dimension m

n	m										
	100	110	120	130	140	150	160	170	180	190	200
40	1.01272	1.01312	1.01291	1.01289	1.01308	1.01296	1.01160	1.01026	1.00915	1.00821	1.00741
45	1.01290	1.01276	1.01291	1.01301	1.01290	1.01298	1.01309	1.01193	1.01063	1.00954	1.00860
50	1.01282	1.01290	1.01282	1.01291	1.01301	1.01289	1.01309	1.01298	1.01215	1.01090	1.00983
55	1.01285	1.01298	1.01291	1.01291	1.01296	1.01310	1.01296	1.01296	1.01311	1.01229	1.01109
60	1.01281	1.01279	1.01287	1.01291	1.01296	1.01304	1.01301	1.01305	1.01309	1.01312	1.01236
65	1.01286	1.01280	1.01300	1.01304	1.01297	1.01303	1.01298	1.01312	1.01304	1.01322	1.01300

reduction algorithm are the unit vectors scaled by q. In general, we will like to have the lattice reduction to produce vectors with length less than q which tends to suggest that the input vectors are more well-mixed by the reduction algorithm to produce short vectors. In view of this, for a given BKZ-beta reduction, we will select the sub-dimension m such that the predicted shortest vector has length less than q, namely, $\delta^m q^{1-n/m} \leq q$ or, $m \leq \sqrt{n \log q / \log \delta}$, where δ is the expected root Hermite factor. For $\delta = 1.0128$, one checks that this gives the pairs (n, m) to be $(40, 152)$, $(45, 164)$, $(50, 175)$, $(55, 186)$, $(60, 196)$ and $(65, 206)$.

Apart from the lattice reduction, the size of m also affects the enumeration cost. Here, we examine the impact of m for the full enumeration tree. We propose to use the Gaussian heuristic to estimate the enumeration cost, i.e., Eq. (1) to estimate the (full) enumeration cost and to decide the optimal sub-dimension. The total cost is $N = \sum_{k=1}^{m} H_k$. We however do not know how to systematically solve the equation to find an optimal m which minimizes the total cost N. Instead we use numerical calculation to determine the optimal sub-dimension m for fixed BKZ block size β. We plot the total cost $N = \sum_{k=1}^{m} H_k$ for an LWE instance by varying the sub-dimension m. Figure 2 shows the estimated (logarithm of) full enumeration cost for different parameters. We deploy a conversion that for an LWE instance $\mathcal{L}_{n,q,\alpha}$, q is set to be the next prime of n^2 which follows the parameter setting of the LWE Challenge. Comparing Fig. 2a and b, we see that for fixed n and BKZ block size β, the optimal sub-dimension m does not depend on the relative error rate α. Figure 2a and c show the impact of BKZ block size β on the optimal sub-dimension. As we can see, by increasing β, the optimal sub-dimension m also increases and the full enumeration cost decreases for larger β. However, the larger β requires more BKZ reduction time. There is still a need for a trade-off between the BKZ reduction time and enumeration time by setting an appropriate block size β. We discuss this in the next subsection. Finally combining Fig. 2b and d, we can further get the impression that for fixed relative error rate α and BKZ block size β, the larger the dimension of the LWE instance n is, the larger sub-dimension we need to get an optimal performance. One problem of the numerical method to decide on the optimal sub-dimension is that we did not consider the BKZ reduction part. In practice we need to consider the running time of BKZ reduction algorithm, so the actual optimal sub-dimension is usually less than that viewed from the plot. However, the plot can still act as a rough guide to find the optimal sub-dimension. Due to page

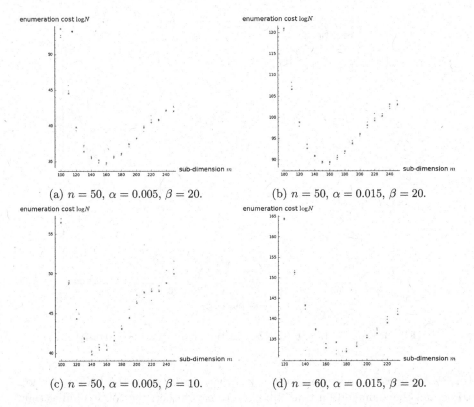

(a) $n = 50$, $\alpha = 0.005$, $\beta = 20$.

(b) $n = 50$, $\alpha = 0.015$, $\beta = 20$.

(c) $n = 50$, $\alpha = 0.005$, $\beta = 10$.

(d) $n = 60$, $\alpha = 0.015$, $\beta = 20$.

Fig. 2. Semi-log graph for full cost N. The parameters have the following reference: n is the LWE dimension, α is the relative error rate of the LWE instance and β is the block size for BKZ reduction algorithm used. Different colors stand for different trials. (Color figure online)

constraints, we defer the details of the estimation of enumeration cost and the relation between cost of full enumeration and that of enumeration with extreme pruning in Appendix A.

4.4   Balancing Reduction and Enumeration

Since we use enumeration to solve the BDD problem, we want to first reduce the lattice basis before applying enumeration. BKZ is now the de-facto standard of lattice reduction algorithm in cryptanalysis. We use the BKZ implementation in FPLLL [5] library to perform BKZ reduction.

The quality of the reduced basis and the running time of BKZ reduction algorithm highly depend on the block size β. The choice of an appropriate block size β affects the total running time of our LWE solver. Generally speaking, a larger block size β leads to longer running time of BKZ reduction algorithm but the highly reduced basis will decrease the running time of enumeration. So the folklore is that the optimal block size β should balance the running time

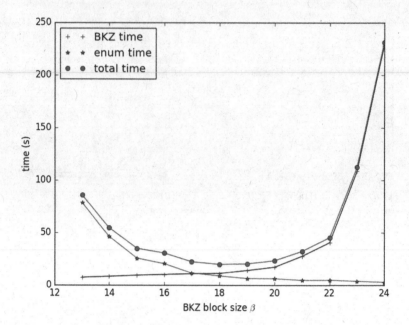

Fig. 3. Running time for BKZ reduction and pruned enumeration.

of BKZ reduction and enumeration. In other words, when the BKZ reduction time and the enumeration time are close to each other, the total running time is minimized. See Fig. 3 for an example. We plot the actual running time of BKZ reduction algorithm and pruned enumeration time for the LWE instance $\mathcal{L}_{40,1601,0.015}$ using sub-dimension $m = 120$. The enumeration radius is set to be $R^2 = 0.8m\alpha^2q^2$. The figure confirms the folklore that the optimal block size β should balance roughly the running time of BKZ reduction and enumeration. However, finding such optimal block size is not easy, especially for extreme pruning. In our experiments we manually tune the block size β by measuring the running time of the BKZ reduction part and pruned enumeration part.

4.5   Parallelization

Parallelism is ubiquitous in today's program design. We have multi-core CPUs even in our laptops. It is natural to implement the LWE solver algorithm in parallel. One option is to use parallel implementation of enumeration algorithm and parallel implementation of lattice reduction algorithm. Alternatively, One can use a sequential implementation of lattice reduction algorithm and enumeration algorithm but instead launch several threads to solve the BDD problem with different randomized basis. We choose the latter approach for its simplicity and its embarrassing parallelism. Although there are parallel implementations of lattice enumeration algorithms [12,13,16,19], we do not know any public available parallel implementation of BKZ reduction algorithm. Thus if we want to use parallel implementation of enumeration we might have to use a sequential implementation of BKZ reduction. Amdahl's Law sets a bound on the potential program speedup defined by the fraction of code (p) that can be parallelized as $speedup = \frac{1}{1-p}$. In using the combination of BKZ reduction and enumeration to solve the SVP or CVP problem, it is common knowledge that when the running time of BKZ reduction part and the enumeration part are roughly equal, the total running time is minimized (refer to the previous section and Fig. 3). If we want optimal performance then the fraction of parallelizable code would be about 1/2. It follows that regardless of how many threads are used, the speedup can be at most 2. We can circumvent this by using a small block size for the BKZ reduction, or plugging in the parallel enumeration into the BKZ reduction, but those methods are either complicated or do not achieve optimal performance gain.

In our implementation we use the embarrassingly parallel design to let each thread work on a different randomized basis, and thus there is no load balance issue. In order to achieve best performance we carefully choose the BKZ block size so that the BKZ reduction time is comparable to the enumeration time.

5   Experimental Results

Our implementation is written in C++, using the library FPLLL for BKZ reduction and lattice enumeration. Our program is compiled using gcc 5.4.0 on a desktop running Ubuntu 14.04 LTS. We test our LWE solver using the instances from the LWE challenge website. We use extreme pruning [14] for lattice enumeration as suggested by Liu and Nguyen [21]. Gamma et al. [14] suggest using numerical approximation to generate optimal pruning coefficients by fixing the successful probability and seeking for minimum overhead. Aono [7] also describes how to compute the optimal pruning coefficients. We follow Aono's approach to compute the optimal pruning coefficients in our implementation. We are preparing to release the source code after further optimization. At this moment, it is available upon request.

5.1   LWE Challenge

TU Darmstadt held a LWE challenge project. The challenge provides LWE instances with different parameters. The LWE challenge instance is identified by two parameters: the LWE dimension n and the relative error rate α. The other parameters of an LWE instance are set as follows:

- Moduli q is set as the next prime of n^2;
- Number of samples is set as $M = n^2$;
- Error distribution is set as the discrete Gaussian distribution with width parameter $\sigma = \alpha q$, i.e., the distribution $\mathcal{D}_{\mathbb{Z}_q, \sigma}$.

Using our implementation described in the preceding section, we solved several instances from the LWE Challenge website. Please refer to Table 2 for the

Table 2. Results on solving some instances from the LWE Challenge website

LWE parameters			BKZ reduction		Enumeration		#Trials	Total time
n	α	m	β	t	c	t		
40	0.005	100	10	-	1.3	-	2	4 s
40	0.01	120	10	4 s	1.3	4 s	2	16 s
40	0.015	120	18	12 s	0.8	10 s	819	18403 s
40	0.02	140	32	10.9 d	1.3	27.1 d	-	38 d
45	0.005	120	5	3 s	1.3	1 s	6	23 s
50	0.005	120	15	5 s	1.0	2 s	5	35 s
60	0.005	140	28	27 h	0.8	24 h	-	51 h

Note that the instances $\mathcal{L}_{40,0.02}$ and $\mathcal{L}_{60,0.005}$ take quite long time thus we use parallelized version of the solving algorithm, and we do not record the number of trials for these two instances.

detailed recording of the LWE parameters, the block size we used for BKZ and the running time for solving these instances. All the instances except two are run using a single thread on a desktop with a 3.60 GHz Intel Core i7 processor with eight cores and 32 GB 1600 MHz DDR3 memory. The instance $(n = 60, \alpha = 0.005)$ was run on a cluster consisting of 20 c4.8xlarge instances, each having 36 cores with a 60 GB memory (720 threads in total), on the Amazon EC2 platform. The instance $(n = 40, \alpha = 0.02)$ was solved on a cluster consisting of 8 desktops with a 3.60 GHz Intel Core i7 processor with eight cores and 32 GB 1600 MHz DDR3 memory (64 threads in total).

In the experiments we carefully choose the BKZ block size β to ensure the BKZ reduction time is comparable with the enumeration time so as to achieve the reduction on overall running time. Our experiments indeed confirm the folklore that when BKZ reduction time roughly equals that of enumeration time the total running time achieves the minimal. The squared BDD bound R^2 was set as $c \cdot m\alpha^2 q^2$. The successful probability in our pruning strategy is set to be 0.01. From the results of our experiments we find that the relative error rate α plays an important role in the hardness of the LWE problem.

5.2   Comparison with Other Implementations

Recently Kirshanova et al. [18] report a parallel implementation of BDD enumeration for solving LWE. They implement both the Lindner-Peikert [22] nearest planes algorithm and the pruned enumeration method[2] proposed by Liu and Nguyen [21]. They directly implement a pruned parallel enumeration algorithm. Their experiments show that the enumeration algorithm can be nicely parallelized. For example, they achieve a linear speedup by increasing the number of threads even until 10. However, the BKZ reduction is not parallelized. We can observe the impact of Amdahl's Law from their experimental results. For

---

[2] Kirshanova et al. use linear pruning instead of extreme pruning.

example in order to solve the LWE instance $\mathcal{L}_{80,4093,5}$ their serial implementation needs $4.3 + 13 = 17.3$ h. Their parallel implementation using 10 threads can reduce the enumeration time from 13 h to 1.5 h. Then the total running time is $4.3 + 1.5 = 5.8$ h. That is a 3x speedup by using 10 threads. Even when they increase the number of threads to 20, the total running time is $4.3 + 0.8 = 5.1$ h, which gives a 3.4x speedup by using 20 threads. Although one can circumvent this by using a very small block size for the BKZ reduction part, as we discussed in Sect. 4.4 this choice would increase the total running time of BKZ reduction and pruned enumeration.

On the contrary, our strategy to use extreme pruning and to use many threads working on different basis can scale quite well with respect to the number of threads. Moreover, using extreme pruning can highly reduce the time used by enumeration and thus reduce the total time needed for solving the LWE instance. We compare the running time of our implementation and that of Kirshanova et al. in Table 3. In the table, the time t for BKZ and enumeration stands for the total BKZ time and enumeration time for solving the corresponding LWE instance. In Kirshanova et al.'s setting they fixed the number of samples for the LWE instance and all their experiments use the fixed dimension. We try a different setting where the number of LWE samples are a polynomial of n, say n^2 so that we can use the optimal sub-dimension to reduce the difficulty of the LWE instance.

Table 3. Comparison between Kirshoanova et al. and ours results.

LWE			Kirshanova et al.					Ours implementation					#Trials
			Sub-dim	BKZ		Enum		Sub-dim	BKZ		Enum		
n	q	s	m	β	t	#Treads	t	m	β	t	c	t	
70	4093	6	140	20	65 min	1	44 min	140	20	19 min	1.0	36 s	18
			140	20	65 min	10	5 min	140	15	551 s	1.0	182 s	32
80	4093	5	150	25	4.3 h	1	13 h	150	15	2.8 h	0.8	2.4 h	347
			150	25	1.3 h	10	1.5 h	180	8	6 min	1.0	64 min	12
			150	25	1.3 h	20	50 min	180	10	417 s	1.0	117s	12

The first row of Table 3 shows that extreme pruning can indeed speedup the LWE solver. Kirshanova et al. need 109 min to solve the instance $\mathcal{L}_{70,4093,6}$ on a single thread, while our implementation can solve the instance using the same sub-dimension and block size $\beta = 20$ within 20 min. Further more, their implementation uses 70 min to solve the instance on 10 threads. Since our implementation uses 18 trials to solve the instance we can solve the instance within the time of two rounds if given 10 threads. Basically, we only need less than 4 min to solve the same instance given 10 threads. We note that the block size $\beta = 20$ is not optimal for our implementation. By changing β to 15, we solve the instance $\mathcal{L}_{70,4093,6}$ in 12 min on a single thread.

To further demonstrate the effectiveness of extreme pruning, we compare the performance of both our implementations for the instance $\mathcal{L}_{80,4093,5}$. When we use the same sub-dimension as $m = 150$, the running time of our implementation on a single thread is 5.2 h which is much smaller than 17.3 h of Kirshanova et al. But the advantage of our implementation lies also in another factor. Notice that the algorithm uses 347 trials to find the correct solution, which means a single trial uses on average only 1 min. Kirshanova et al. solve the instance in more than 2 h using 20 threads. Our implementation is expected to solve the instance in $347/20 = 18$ min. If we have more than 400 threads, we can solve the instance within 1 min. Moreover, if we apply the optimal sub-dimension trick we do not need so many threads to achieve the speedup. For example when we use 180 samples and BKZ block size $\beta = 10$, the total of 12 trials take $417 + 117 = 534$ s. Then with 12 threads our implementation is expected to solve the instance $\mathcal{L}_{80,4093,5}$ using 45 s. On the contrary, the BKZ reduction of Kirshanova et al.'s implementation alone takes 1.3 h.

6   Conclusion and Future Work

This current work described our choice of strategy to solve the BDD problem, namely the details of our implementation and our experimental results on several LWE challenge instances. Our implementation features a embarrassingly parallel design and the use of extreme pruning shows advantages over existing implementations. Potential future work include:

- We choose the optimal BKZ block size β manually in our experiments. This would be impossible for LWE instances with large dimension and/or large relative error rate. Thus it would be useful to explore the relation between the BKZ reduction time and (pruned) enumeration time and use some heuristics to decide the optimal BKZ block size.
- The success probabilities from our experiments seem to be higher than those estimated by Aono's algorithm, thereby resulting in fewer threads. Since we are using parallel implementations of the LWE solver, we have more room for a lower success probability. Lower successful probability can reduce the running time while we can simply add more threads to compensate the low probability of success. In fact our current environment of 20 c4.8xlarge Amazon EC2 instances contains in total more than 700 threads. We can deal with this problem in two ways: first, we can reduce the successful probability for the pruning strategy; second, we can deploy a two-level parallelization by using the first level to run the LWE solver in parallel and using the second level to run the parallel enumeration algorithm.

A   Estimate the Cost of BDD Enumeration

Recall that Gaussian heuristic suggests that the enumeration cost at level k of the full enumeration tree is

$$H_k = \frac{V_k(R)}{\prod_{i=m+1-k}^{m} ||\mathbf{b}_i^*||^2}$$

and the total cost is $N = \sum_{k=1}^{m} H_k$.

In the case of BDD enumeration, we set the enumeration bound $R = \sqrt{m}\alpha q$. As for the Gram-Schmidt vector $\mathbf{b}_i^*$, we use the GSA (Geometric Series Assumption) to approximate their lengths.

Geometric Series Assumption (GSA): Schnorr [25] introduced GSA which states that the Gram-Schmidt lengths $||\mathbf{b}_i^*||$ in a BKZ-reduced basis decrease geometrically with quotient r for some constant r related to the reduction algorithm. Specifically $||\mathbf{b}_{i+1}^*||/||\mathbf{b}_i^*|| = r$. Using Gaussian heuristic for the length of the shortest vector we approximate $||\mathbf{b}_1^*||$ as $\delta^m q^{\frac{m-n}{m}}$ where δ is the root Hermite factor achieved by a BKZ reduction algorithm.

Combining Gaussian heuristic and GSA together, we can reformulate Eq. (1) as

$$H_k = \frac{(\sqrt{m}\alpha q)^k \pi^{k/2}}{r^{(2m-k-1)k/2}\delta^{mk}q^{(m-n)k/m}\Gamma(k/2+1)}, \tag{2}$$

where r is the GSA constant, δ is the root Hermite factor achieved by the reduction algorithm and $\Gamma(\cdot)$ is the Gamma function. The total cost (number of nodes in the enumeration tree) is then calculated as

$$N = \sum_{k=1}^m H_k. \tag{3}$$

In the BDD case the enumeration cost behaves quite differently from that of the SVP enumeration. In the SVP case, the enumeration radius R is set roughly as $||\mathbf{b}_1^*||$ and the terms H_k reaches maximum when $k = m/2$. But for BDD enumeration, the enumeration radius[3] (bound) is set as $R = \sqrt{m}\alpha q$. This difference actually results in the fact that the cost may decrease as the sub-dimension m increases.

We numerically evaluated the leveled cost H_k and the results are shown in Fig. 4. Figure 4 displays the level cost H_k for different index k for an LWE instance $\mathcal{L}_{50,2503,0.005}$. The basis are reduced by the BKZ-20 reduction algorithm, while the enumeration radius is set as $R = \sqrt{m}\alpha q$. Figure 4a uses sub-dimension $m = 120$ and Fig. 4b uses sub-dimension $m = 140$. From the figures we can observe that the index which maximize H_k is no longer $m/2$ and for the smaller index the value of H_k may be less than 1. We next plot the total cost $N = \sum_{k=1}^m H_k$ for an LWE instance by varying the sub-dimension m.

In our experiments, we employ extreme pruning instead of the full enumeration tree. Here, we verify that extreme pruning does not affect the shape of the enumeration cost so that the estimation using full enumeration works effectively for choosing the optimal sub-dimension. Comparing Fig. 5 with Fig. 2, we can see that in the case of enumeration cost with extreme pruning the optimal sub-dimension m is usually a little bigger than that for full enumeration cost. However, the estimated costs do not differ too much. Moreover, when using a larger sub-dimension one has to take the cost of BKZ reduction into consideration. Our

---

[3] In our implementation we set $R^2 = cm(\alpha q)^2$ for some bound scalar c ranging from 0.8 to 1.3.

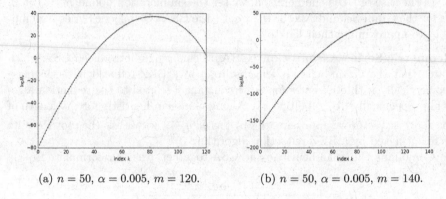

(a) $n = 50$, $\alpha = 0.005$, $m = 120$.   (b) $n = 50$, $\alpha = 0.005$, $m = 140$.

Fig. 4. Semi-log graph for level cost H_k.

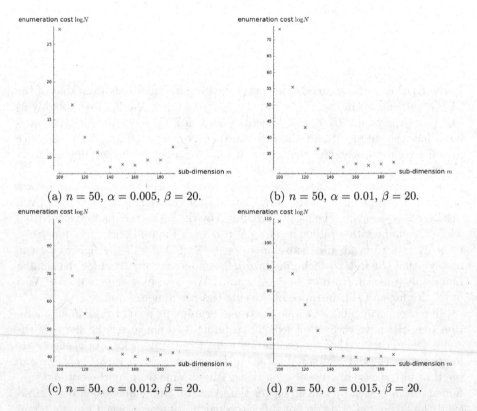

(a) $n = 50$, $\alpha = 0.005$, $\beta = 20$.   (b) $n = 50$, $\alpha = 0.01$, $\beta = 20$.

(c) $n = 50$, $\alpha = 0.012$, $\beta = 20$.   (d) $n = 50$, $\alpha = 0.015$, $\beta = 20$.

Fig. 5. Semi-log graph for cost N of pruned enumeration. The parameters have the following reference: n is the LWE dimension, α is the relative error rate of the LWE instance and β is the block size for BKZ reduction algorithm used.

experiments in solving the LWE challenge confirms that the estimation using full enumeration cost well serves our purpose to reduce the running time of the whole solving process.

References

1. Ajtai, M.: Generating hard instances of lattice problems. In: Proceedings of 28th Annual ACM Symposium on Theory of Computing, pp. 99–108. ACM (1996)
2. Albrecht, M.R., Player, R., Scott, S.: On the concrete hardness of learning with errors. Cryptology ePrint Archive, Report 2015/046 (2015)
3. Albrecht, M.R., Cid, C., Faugere, J., Fitzpatrick, R., Perret, L.: On the complexity of the BKW algorithm on LWE. Des. Codes Cryptogr. $74(2)$, 325–354 (2015)
4. Albrecht, M.R., Cid, C., Faugere, J., Perret, L.: Algebraic algorithms for LWE. Cryptology ePrint Archive, Report 2014/1018 (2014)
5. Albrecht, M.R., Cadé, D., Pujol, X., Stehlé, D.: fplll-4.0, a floating-point LLL implementation. http://perso.ens-lyon.fr/damien.stehle
6. Arora, S., Ge, R.: New algorithms for learning in presence of errors. In: Aceto, L., Henzinger, M., Sgall, J. (eds.) ICALP 2011. LNCS, vol. 6755, pp. 403–415. Springer, Heidelberg (2011). doi:10.1007/978-3-642-22006-7_34
7. Aono, Y.: A faster method for computing Gama-Nguyen-Regev's extreme pruning coefficients (2014). arXiv preprint arXiv:1406.0342
8. Brakerski, Z., Langlois, A., Peikert, C., Regev, O., Stehlé, D.: Classical hardness of learning with errors. In: Proceedings of 45th Annual ACM Symposium on Theory of Computing, STOC 2013, pp. 575–584. ACM, New York (2013)
9. Buchmann, J., Büscher, N., Göpfert, F., Katzenbeisser, S., Krämer, J., Micciancio, D., Siim, S., van Vredendaal, C., Walter, M.: Computation, creating cryptographic challenges using multi-party: the LWE challenge. In: Proceedings of 3rd ACM International Workshop on ASIA Public-Key Cryptography (AsiaCCS 2016), pp. 11–20 (2016)
10. Bischof, C., Buchmann, J., Dagdelen, Ö., Fitzpatrick, R., Göpfert, F., Mariano, A.: Nearest planes in practice. In: Ors, B., Preneel, B. (eds.) BalkanCryptSec 2014. LNCS, vol. 9024, pp. 203–215. Springer, Cham (2015). doi:10.1007/978-3-319-21356-9_14
11. Chen, Y., Nguyen, P.Q.: BKZ 2.0: better lattice security estimates. In: Lee, D.H., Wang, X. (eds.) ASIACRYPT 2011. LNCS, vol. 7073, pp. 1–20. Springer, Heidelberg (2011). doi:10.1007/978-3-642-25385-0_1
12. Detrey, J., Hanrot, G., Pujol, X., Stehlé, D.: Accelerating lattice reduction with FPGAs. In: Abdalla, M., Barreto, P.S.L.M. (eds.) LATINCRYPT 2010. LNCS, vol. 6212, pp. 124–143. Springer, Heidelberg (2010). doi:10.1007/978-3-642-14712-8_8
13. Dagdelen, Ö., Schneider, M.: Parallel enumeration of shortest lattice vectors. In: D'Ambra, P., Guarracino, M., Talia, D. (eds.) Euro-Par 2010. LNCS, vol. 6272, pp. 211–222. Springer, Heidelberg (2010). doi:10.1007/978-3-642-15291-7_21
14. Gama, N., Nguyen, P.Q., Regev, O.: Lattice enumeration using extreme pruning. In: Gilbert, H. (ed.) EUROCRYPT 2010. LNCS, vol. 6110, pp. 257–278. Springer, Heidelberg (2010). doi:10.1007/978-3-642-13190-5_13
15. Gama, N., Nguyen, P.Q.: Predicting Lattice Reduction. In: Smart, N. (ed.) EUROCRYPT 2008. LNCS, vol. 4965, pp. 31–51. Springer, Heidelberg (2008). doi:10.1007/978-3-540-78967-3_3

16. Hermans, J., Schneider, M., Buchmann, J., Vercauteren, F., Preneel, B.: Parallel shortest lattice vector enumeration on graphics cards. In: Bernstein, D.J., Lange, T. (eds.) AFRICACRYPT 2010. LNCS, vol. 6055, pp. 52–68. Springer, Heidelberg (2010). doi:10.1007/978-3-642-12678-9_4

17. Kannan, R.: Minkowski's convex body theorem and integer programming. Math. Oper. Res. **12**(3), 415–440 (1987)

18. Kirshanova, E., May, A., Wiemer, F.: Parallel implementation of BDD enumeration for LWE. In: Manulis, M., Sadeghi, A.-R., Schneider, S. (eds.) ACNS 2016. LNCS, vol. 9696, pp. 580–591. Springer, Cham (2016). doi:10.1007/978-3-319-39555-5_31

19. Kuo, P.-C., Schneider, M., Dagdelen, Ö., Reichelt, J., Buchmann, J., Cheng, C.-M., Yang, B.-Y.: Extreme enumeration on GPU and in clouds. In: Preneel, B., Takagi, T. (eds.) CHES 2011. LNCS, vol. 6917, pp. 176–191. Springer, Heidelberg (2011). doi:10.1007/978-3-642-23951-9_12

20. Lenstra, A., Lenstra, H.W., Lovász, L.: Factoring polynomials with rational coefficients. Math. Ann. **261**(4), 515–534 (1982)

21. Liu, M., Nguyen, P.Q.: Solving BDD by enumeration: an update. In: Dawson, E. (ed.) CT-RSA 2013. LNCS, vol. 7779, pp. 293–309. Springer, Heidelberg (2013). doi:10.1007/978-3-642-36095-4_19

22. Lindner, R., Peikert, C.: Better key sizes (and attacks) for LWE-based encryption. In: Kiayias, A. (ed.) CT-RSA 2011. LNCS, vol. 6558, pp. 319–339. Springer, Heidelberg (2011). doi:10.1007/978-3-642-19074-2_21

23. Micciancio, D., Regev, O.: Lattice-based cryptography. In: Post-Quantum Cryptography, p. 147 (2009)

24. Regev, O.: On lattices, learning with errors, random linear codes, cryptography. J. ACM (JACM) **56**(6), 34:1–34:40 (2009)

25. Schnorr, C.P.: Lattice reduction by random sampling and birthday methods. In: Alt, H., Habib, M. (eds.) STACS 2003. LNCS, vol. 2607, pp. 145–156. Springer, Heidelberg (2003). doi:10.1007/3-540-36494-3_14

26. Schnorr, C.P., Euchner, M.: Lattice basis reduction: improved practical algorithms and solving subset sum problems. Math. Program. **66**(1–3), 181–199 (1994)

Efficiently Obfuscating Re-Encryption Program Under DDH Assumption

Akshayaram Srinivasan[1(✉)] and Chandrasekaran Pandu Rangan[2]

[1] University of California, Berkeley, USA
akshayaram@berkeley.edu, prangan55@gmail.com
[2] Indian Institute of Technology, Madras, India

Abstract. The re-encryption functionality transforms a ciphertext encrypted under a public key pk_1 to a ciphertext of the same message encrypted under a different public key pk_2. Hohenberger et al. (TCC 2007) proposed a pairing-based obfuscator for the family of circuits implementing the re-encryption functionality under a new notion of obfuscation called as *average-case secure obfuscation*. Chandran et al. (PKC 2014) proposed a lattice-based construction for the same.

The construction given by Hohenberger et al. could only support polynomial sized message space and the proof of security relies on strong assumptions on bilinear groups. Chandran et al.'s construction could only satisfy a relaxed notion of correctness.

In this work, we propose a *simple* and *efficient* obfuscator for the re-encryption functionality that satisfies the strongest notion of correctness, supports encryption of messages from an exponential sized domain and relies on the *standard* DDH-assumption. This is the *first* construction that does not rely on pairings or lattices. All our proofs are in the standard model.

Keywords: Re-encryption functionality · Average-case secure obfuscation · DDH assumption · Standard model

1 Introduction

The goal of program obfuscation [2,22] is to make a program "unintelligible" while preserving its functionality. This security goal was formalized in the seminal work of Barak et al. [2]. The strongest and the most intuitive notion of

A. Srinivasan—Work partially done while author was a student at IIT-Madras. Research supported in part from 2017 AFOSR YIP Award, DARPA/ARL SAFEWARE Award W911NF15C0210, AFOSR Award FA9550-15-1-0274, NSF CRII Award 1464397, and research grants by the Okawa Foundation, Visa Inc., and Center for Long-Term Cybersecurity (CLTC, UC Berkeley) of Prof. Sanjam Garg. The views expressed are those of the author and do not reflect the official policy or position of the funding agencies.

C. Pandu Rangan—Research partially supported by Information Security Education and Awareness Program of Ministry of Information Technology, Government of India and Microsoft Research, Bangalore, India.

© Springer International Publishing AG 2017
D. Gollmann et al. (Eds.): ACNS 2017, LNCS 10355, pp. 273–292, 2017.
DOI: 10.1007/978-3-319-61204-1_14

security considered in their work is known as the *predicate black-box* (also known as virtual black-box) obfuscation. This definition requires any predicate that is efficiently computable given access to the obfuscated program must also be efficiently computable given black-box access to the same program. Unfortunately, Barak et al. showed that a *general* program obfuscator satisfying the predicate black-box property does not exist. In spite of the general impossibility result, program obfuscators satisfying this strong security notion have been constructed for simple function families in [10, 12, 29].

Average-Case Secure Obfuscation. Hohenberger et al. [24] noted that the predicate black-box property does not give meaningful security guarantees when the obfuscated functionality (like re-encryption) is a part of a larger cryptographic system (like the underlying encryption scheme). They discussed a scenario where having access to the obfuscated program could compromise the security of the larger cryptographic system even if the obfuscator satisfied the predicate black-box property.

To address this issue, Hohenberger et al. proposed a new definition of obfuscation which they termed as *average-case secure obfuscation*. This definition guarantees that any adversary against a cryptographic scheme having access to an obfuscated program can be transformed into an adversary with only black box access to the program given that the cryptographic scheme has *distinguishable attack property*. Informally, a cryptographic scheme is said to have distinguishable attack property if there exists a distinguisher that can "test" if a given algorithm can break the security of the scheme. Hohenberger et al. in fact showed that several natural cryptographic functionalities like semantically secure encryption and re-encryption have this property.

1.1   Prior Work

A cryptographic functionality that has been shown to be obfuscatable under the average-case security notion is the re-encryption function. The re-encryption functionality transforms a ciphertext encrypted under a public key pk_1 to a ciphertext of the same message encrypted under a different public key pk_2. Hohenberger et al. [24] designed an average case secure obfuscator for the re-encryption functionality under Decision Linear and a variant of 3-party Decisional Diffie-Hellman assumptions. However, this construction could only *support a message space of polynomial size*. The decryption algorithm in their work performs an exhaustive search on the message space by computing a pairing operation on each message and tests the output of the pairing against a specific value. Thus, it has to *compute a polynomial number of pairing operations in the worst case*. Moreover, the security of their construction is based on a strong assumption, namely, Strong 3-party DDH.

Remark 1. We note that it is possible to extend the system of [24] to a message space of arbitrary size by using their construction for the boolean space $\{0,1\}$. For an arbitrary message space $\mathcal{M}$, one can encrypt each message bit by bit and thus incurring a $O(\log(|\mathcal{M}|))$ overhead on encryption and decryption. For an exponential (in the security parameter λ) sized message space, the overhead on encryption and decryption algorithms would be $\texttt{poly}(\lambda)$. But it is desirable to have a system that performs constant number of operations (i.e., have a constant overhead) in every algorithm.

Chandran et al. [13] designed average case secure obfuscators for the re-encryption circuit assuming interactability of certain lattice problems. Their construction could only satisfy certain relaxed notions of correctness. In particular, they considered three relaxations of the correctness property. The first relaxation guarantees that the output of the original circuit and the obfuscated circuit are statistically close only on a subset of the actual inputs. The next relaxation guarantees that the output of the obfuscated program on a subset of inputs is correct with respect to some algorithm (like decryption). The final relaxation guarantees that the output of the obfuscated circuit and the original circuit are computationally indistinguishable. A natural question that arises from the prior work is:

> *Is there an efficient obfuscator for the re-encryption program under milder assumptions that satisfies the strongest notion of correctness, has a constant overhead in every algorithm and supports an exponential sized message space?*

1.2   Our Contributions

We highlight the main contributions of this work.

Main Result. In this work, we propose a new encryption - re-encryption system that supports encryption of messages from an exponential (in security parameter) space, involves a constant number of group exponentiation operations in all algorithms. We also design an average-case secure obfuscator for the re-encryption program which achieves the strongest notion of correctness (as in [23,24]). We prove the average case secure obfuscator property of our obfuscator and the security of our encryption - re-encryption system under the standard DDH assumption. All our proofs are in the *standard model*. Informally, the main result in this work is:

Informal Theorem 1. *Under the DDH-assumption, there exists an average-case secure obfuscator for the family of circuits implementing the re-encryption functionality.*

Remark 2. We observe that our construction of obfuscator for the re-encryption program is not secure when the holder of sk_2 has access to the obfuscated circuit. This is the case with all prior constructions of average-case secure obfuscators for re-encryption [13,24] as well as encrypted signatures [23]. The construction is secure as long as the obfuscated circuit is run by some (possibly malicious) party other than delegatee. It would be interesting to investigate the possibility of constructing average-case secure obfuscators which have "insider security." That is, they remain secure even when the delegatee has access to the obfuscated circuit. We leave this as an open problem.

Strengthening the Black-Box Security Model. Recall that in the average-case obfuscation paradigm, we must first design an encryption - re-encryption system that is secure against adversaries that have black box access to the re-encryption program. The security model considered by Hohenberger et al. [24] for this purpose is as follows: the challenger samples two public key-secret key pairs (pk_1, sk_1) and (pk_2, sk_2) and then provides pk_1, pk_2 to the adversary. The adversary (with oracle access to re-encryption program from pk_1 to pk_2) chooses two messages m_0 and m_1 and also gives information about the public key as well as the ciphertext level on which it wishes to be challenged. More precisely, the adversary can choose either to be challenged on a first level ciphertext[1] under pk_1 or a first level ciphertext under pk_2 or a second level ciphertext[2] under pk_2. The scheme is secure if the adversary is unable to distinguish between the corresponding encryptions of m_0 and m_1.

We observe that the above security model is insufficient in capturing the full security notion of encryption - re-encryption system (See Remark 4). In particular, the above security model allows the following trivial but insecure encryption - re-encryption system to be secure. Consider any semantically secure encryption scheme $\Pi = (\mathsf{KeyGen}, \mathsf{Encrypt}, \mathsf{Decrypt})$. To obtain a first level ciphertext of a message m under a public key pk, run the $\mathsf{Encrypt}$ algorithm on m and pk. The re-encryption program from pk_1 to pk_2 has sk_1, pk_1 and pk_2 hardwired into its description. When it is run with a first level ciphertext $c \leftarrow \mathsf{Encrypt}(m, pk_1)$, it decrypts the ciphertext using sk_1 and outputs $(\mathsf{Encrypt}(m, pk_1) || \mathsf{Encrypt}(sk_1, pk_2))$ where $||$ denotes concatenation. In order to decrypt a second level ciphertext, one can first decrypt the second component using sk_2 to obtain sk_1 and then decrypt the first component to obtain m. This system has an obvious drawback as it reveals sk_1 to the user with secret key sk_2. But one can prove that this system is secure under the security model considered in [24]. We also observe that it is possible to construct an average-case secure obfuscator for the above re-encryption program when one instantiates Π with a semantically secure encryption system which allows re-randomization of ciphertexts (e.g., the standard El-Gamal encryption).

To address this issue, we strengthen the security model for encryption - re-encryption system as follows. We consider the security of the system under two different security games.

1. The first game called as *Original Ciphertext Security* proceeds exactly as in [24] but the adversary is either challenged on a first level ciphertext under pk_1 or a first level ciphertext under pk_2.

---

[1] A first level ciphertext is the one which has not been re-encrypted. In other words, a first level ciphertext is given as input to the re-encryption program.

[2] A second level ciphertext under pk_2 is the output of re-encryption program from pk_1 to pk_2 on a first level ciphertext under pk_1.

2. In the second game called as the *Transformed Ciphertext Security*, in addition to (pk_1, pk_2) the adversary is also provided with sk_1. In the challenge phase, the adversary obtains a second level ciphertext under pk_2 as the challenge ciphertext. The goal of the adversary in both the games is to distinguish between encryptions of messages from encryptions of junk values.

Additionally, we require the encryption - re-encryption system to satisfy a special property called as *statistical independence*. Statistical independence requires that the output distribution of the re-encryption program (i.e., the distribution of the second level ciphertext) to be *statistically independent* of sk_1.

Note that the above trivial encryption - re-encryption system is not transformed ciphertext secure as the adversary with access to sk_1 can directly decrypt the first component of the challenge ciphertext to obtain the hidden message. We also stress that statistical independence property guarantees that the second level ciphertext cannot "leak" any information (in an information theoretic sense) regarding sk_1. This in particular, disallows other contrived examples which may reveal the secret key sk_1 to Bob but possibly is still transformed ciphertext secure.[3]

Remark 3. Though this security model was not explicitly considered, all prior works [13, 14, 24] satisfy this security notion.

Remark 4. We consider a stronger model for encryption - re-encryption security since the output of the re-encryption program (which we obfuscate in this work) does not have the same probability distribution as a fresh encryption of the message m under pk_2 (i.e., as an output of another encryption algorithm Encrypt2(m, pk_2) as in [24]). If the output was distributed identically to a fresh encryption under pk_2 then the security model given by Hohenberger et al. is sufficient for our purposes.[4] The above discussion regarding the issues with the security model is for the generalized case where the output distribution of the re-encryption program and distribution of a freshly encrypted ciphertext under pk_2 are not identical.

1.3 Related Work

Proxy Re-Encryption. A paradigm in cryptography which is closely related to re-encryption is *proxy re-encryption*. In a proxy re-encryption system, a semi-trusted proxy transforms ciphertexts intended for Alice (delegator) to a ciphertext of the same message for Bob (delegatee). Specifically, Alice provides the

---

[3] For example, we could consider the output of the re-encryption functionality to be (Encrypt(C, pk_2)||Encrypt(sk_1, pk_2)) where C is the input ciphertext. This is transformed ciphertext secure. However, the output of the re-encryption functionally is dependent on sk_1.

[4] The counter example discussed above will not work since the output of the re-encryption program is not identically distributed to a freshly generated ciphertext under pk_2.

proxy with a re-key $RK_{A \to B}$ which is a function of her secret key sk_1 and Bob's public key pk_2. The proxy runs a specific algorithm (called as the *re-encryption algorithm* in literature) which takes the ciphertext encrypted under Alice's public key and the re-key and outputs a ciphertext under Bob's public key. A (non-exhaustive) list of proxy re-encryption schemes under different notions of security can be found in [1,5,11,15,25]. The security goal is that given the re-encryption key, the proxy cannot learn any information about the underlying message from the first level ciphertext. This is formalized as a game between the challenger and the proxy. This must be contrasted with the simulation based security guarantee provided by obfuscation of re-encryption program. In particular, obfuscation of re-encryption circuit guarantees that no *non-black-box* information about the re-encryption circuit is "leaked" to the proxy. On the other hand, it is not directly evident if such guarantees can be given from proxy re-encryption systems as it may be possible that the re-key could leak some non-black-box information. For example, the re-key could reveal a function of the secret key that still does not contradict the semantic security of the encryption scheme.

FHE. Fully Homomorphic Encryption (FHE) [26] allows arbitrary computations to be performed on a ciphertext. The first construction of FHE was provided in the breakthrough work by Gentry [20]. Following this result, there have been several works constructing FHE from worst-case intractability of several lattice problems [7–9]. We note that using re-randomizable (or circuit private) FHE, it is possible to obfuscate the re-encryption program. However, all known constructions of FHE are from specific assumptions on lattices and there is still someway to go before they become "truly" practical. In contrast, we propose an obfuscation of re-encryption program from the standard DDH assumption and our construction is very efficient as it involves only a (small) constant number of group exponentiation operations.

Indistinguishability Obfuscation. Since the strongest notion of program obfuscation (namely, predicate black-box obfuscation) was shown to be impossible, Barak et al. [2] proposed a weakened notion of obfuscation called as *indistinguishability obfuscation* or iO. Indistinguishability obfuscation guarantees that for any two functionally equivalent circuits having the same size, obfuscations of the circuits are indistinguishable. The first candidate construction of iO was given in the recent breakthrough work of Garg et al. [17]. Subsequently, several cryptographic primitives like functional encryption [17], deniable encryption [27], non-interactive key exchange without a trusted setup [6], two-round multiparty computation protocols [16], hard instances of the complexity class PPAD [3,18] and trapdoor permutations [4,19] (to name a few) were constructed from iO and other assumptions like one-way functions. We note that indistinguishability guarantee provided by iO is strictly weaker than the security guarantee needed in this work. In addition, our goal is to obfuscate a specific functionality namely, the re-encryption functionality.

2   Preliminaries

A function $\mu(\cdot) : \mathbb{N} \to \mathbb{R}^+$ is said to be **negligible**, if for every positive polynomial $p(\cdot)$, there exists an N such that for all $n \geq N$, $\mu(n) < 1/p(n)$. Given a probability distribution D on a universe U, we denote $x \leftarrow D$ as the process of sampling an element x from U according to the distribution D. Given a finite set X, we use the notation $x \xleftarrow{\$} X$ for denoting the process of sampling x from the set X uniformly. If two probability distributions D and D' defined on a set X are identical, we denote it by $D \approx D'$. We use notation similar to [23] to denote the following randomized process: given n probability distributions $D_1, \cdots, D_n$, let $\{x_1 \leftarrow D_1; \cdots ; x_n \leftarrow D_n : f(x_1, \ldots, x_n)\}$ be the probability distribution of a (possibly randomized) function f. PPT machines refer to Probabilistic Polynomial Time Turing machines. All PPT machines run in time polynomial in the security parameter denoted by λ. We consider the non-uniform model of computation to model the adversaries. These machines take an additional auxiliary input z of length polynomial in λ. If p is a prime number then let $\mathbb{Z}_p^*$ denote the set $\{1, 2, \ldots, p-1\}$.

We assume familiarity with the notion of computational indistinguishability and statistical distance (a.k.a. variation distance) and skip the standard definitions. We state the following simple lemma regarding statistical distance. The proof can be derived directly from the definition.

Lemma 1. *For all distributions X_n and Y_n, for all PPT distinguishers D that output a single bit and for all $z \in \{0, 1\}^{poly(n)}$ we have,*

$$\Delta(D(X_n, z), D(Y_n, z)) = \left| Pr[b \leftarrow D(X_n, z) : b = 1] - Pr[b \leftarrow D(Y_n, z) : b = 1] \right|$$

The above lemma implies that $\{X_n\}_n \stackrel{c}{\approx} \{Y_n\}_n$ if and only if $D(X_n, z)$ and $D(Y_n, z)$ are statistically close for all PPT distinguishers D and for all auxiliary input z.

We now recall the Decisional Diffie-Hellman (DDH) assumption on prime order groups. Let Gen be an algorithm which takes 1^λ as input and randomly generates the parameters $(p, \mathbb{G}, g)$ where p is a λ bit prime, $\mathbb{G}$ is a multiplicative group of order p and g is a generator for $\mathbb{G}$.

Definition 1 (DDH assumption). *The DDH assumption states that the following distribution ensembles are computationally indistinguishable:*

$$\{(p, \mathbb{G}, g) \leftarrow Gen(1^\lambda); a, b \xleftarrow{\$} \mathbb{Z}_p^* : (g, g^a, g^b, g^{ab})\}_\lambda \stackrel{c}{\approx}$$

$$\{(p, \mathbb{G}, g) \leftarrow Gen(1^\lambda); a, b, c \xleftarrow{\$} \mathbb{Z}_p^* : (g, g^a, g^b, g^c)\}_\lambda$$

We assume the reader's familiarity with the syntax and security notion (multi message security) for a Public Key Encryption (PKE) system. We also assume familiarity with the El-Gamal encryption system. We recall the theorem regarding the multi-message security of El-Gamal encryption system and its variant.

Theorem 1 (Multi message security). *Assuming the DDH-assumption holds in the group* $\mathbb{G}$, *the El-Gamal encryption system is multi-message secure.*

We assume familiarity with of the concept of Pseudo Random Generator (PRG) and refer the reader to [21] for a formal definition.

2.1   Average-Case Secure Obfuscation

Let $\mathcal{C} = \{C_\lambda\}_{\lambda \in \mathbb{N}}$ be a family of polynomial sized circuits. For a length parameter λ, let C_λ be the set of circuits in $\mathcal{C}$ with input length $p_{in}(\lambda)$ and output length $p_{out}(\lambda)$ where $p_{in}(\cdot)$ and $p_{out}(\cdot)$ are polynomials. The circuit family $\mathcal{C}$ has an associated sampling algorithm Samp which takes 1^λ as input and outputs a circuit C chosen uniformly at random from C_λ. We also assume that there exists efficient (Encode, Decode) algorithms which encodes and decodes a given circuit C into binary strings when used as input/output of Turing machines. We make implicit use of such encoding and decoding algorithms and do not mention them explicitly.

We use similar notations (with some minor changes) as in [23] to denote probabilistic circuits. A probabilistic circuit $C(x; r)$ takes two inputs. The first input is called as the *regular input* and the second input is termed as the *random input*. The output of a probabilistic circuit on a regular input (denoted by $C(x; \cdot)$) can be viewed as a probability distribution where the randomness in the distribution comes from the random choice of r. We say that a machine $\mathcal{A}$ has *oracle access* to a probabilistic circuit C (denoted by $\mathcal{A}^{\mathcal{O}(C)}$) if during the oracle queries, $\mathcal{A}$ can only specify the regular input x to the circuit and the random input r is chosen uniformly at random from the corresponding sample space by the oracle $\mathcal{O}$. The output of a probabilistic machine $\mathcal{A}$ having oracle access to a probabilistic circuit C (denoted by $\mathcal{A}^{\mathcal{O}(C)}(x_1, \cdots, x_n)$) is a probability distribution where the randomness in the distribution comes from the random coins used by $\mathcal{A}$ as well as the random coins used by $\mathcal{O}$ in answering $\mathcal{A}$'s oracle queries. We say that $\mathcal{B}$ *evaluates* a probabilistic circuit C (or in other words, $\mathcal{B}$ is an evaluator of C) on regular input x, if $\mathcal{B}$ supplies the regular input as well as the random input r chosen uniformly at random from the corresponding sample space and outputs $C(x; r)$. We use $|C|$ to denote the size of a circuit C.

We recall the notion of *average case secure obfuscation* given in [24].

Definition 2 [23,24]. *A PPT machine* Obf *that takes as input a (probabilistic) circuit and outputs a new (probabilistic) circuit is an average-case secure obfuscator for the circuit family* $\mathcal{C} = \{C_\lambda\}_{\lambda \in \mathbb{N}}$ *with an associated sampling algorithm* Samp *if it satisfies the following properties:*

1. *Preserving Functionality: For all length parameter* $\lambda \in \mathbb{N}$ *and for all* $C \in C_\lambda$:

$$Pr[C' \leftarrow \mathsf{Obf}(C) : \exists x \in \{0,1\}^{p_{in}(\lambda)}, \Delta\big(C'(x; \cdot), C(x; \cdot)\big) \neq 0] = 0$$

2. *Polynomial Slowdown: There exists a polynomial* $p(\cdot)$ *such that for sufficiently large length parameters* λ, *for any* $C \in C_\lambda$, *we have*

$$Pr[C' \leftarrow \mathsf{Obf}(C) : |C'| \leq p(|C|)] = 1$$

3. *Average-case Secure Virtual Black Box: There exists a PPT machine (simulator)* Sim *such that for every PPT distinguisher D, there exists a negligible function* neg$(\cdot)$ *such that for every length parameter* λ *and for every* $z \in \{0,1\}^{poly(\lambda)}$:

$$|Pr[C \leftarrow \mathsf{Samp}(1^{\lambda}); C' \leftarrow \mathsf{Obf}(C); b \leftarrow D^{\mathcal{O}(C)}(C', z) : b = 1]-$$

$$Pr[C \leftarrow \mathsf{Samp}(1^{\lambda}); C' \leftarrow \mathsf{Sim}^{\mathcal{O}(C)}(1^{\lambda}, z); b \leftarrow D^{\mathcal{O}(C)}(C', z) : b = 1]| \leq \mathsf{neg}(\lambda)$$

Remark 5. The definition given in [24] considers a relaxed notion of correctness. Specifically, it allows a the output distribution of the obfuscated circuit and the original circuit to have a negligible statistical distance with a negligible probability. Here, as given in [23], we consider a stronger notion of correctness where we require that the output distribution of the original circuit and the obfuscated circuit to be identical.

3   Obfuscator for Re-Encryption Functionality

In this section, we describe our new encryption system, the re-encryption functionality which is to be obfuscated and finally the construction of an average case secure obfuscator for the functionality.

New Encryption Scheme. The new encryption system under consideration is a variant of the El-Gamal system.

New Encryption Scheme

- Setup(1^{λ}) : Let $(p, \mathbb{G}, g) \leftarrow \mathsf{Gen}(1^{\lambda})$. Let H be a pseudo random generator which takes as input an element from $\mathbb{G}$ and outputs an element in $\mathbb{Z}_p^*$.[a] Output the public parameters as $params = (p, g, \mathbb{G}, H)$ with message space $\mathcal{M} = \mathbb{G}$.
- KeyGen$(1^{\lambda}, params)$: Choose $x \xleftarrow{\$} \mathbb{Z}_p^*$ and set the public key pk to be (g, g^x) and the secret key $sk = x$.
- Encrypt1(m, pk) : Parse pk as (g, g^x). Choose a random $r \xleftarrow{\$} \mathbb{Z}_p^*$ and output $(m \cdot g^r, (g^x)^r)$.
- Decrypt1$(sk, [C_1, C_2])$: Parse the secret key sk as x. Output $m = (C_1) \cdot ((C_2)^{1/x})^{-1}$.

---

[a] The standard definition of pseudo random generator assumes the domain and the range to be bit strings. We note that it can be extended to any domain and range assuming efficient encoding and decoding functions from the domain to bit strings and from bit strings to range. The expansion factor of H depends on the actual encoding and decoding schemes used.

Re-encryption Functionality. Let (pk_1, sk_1) and (pk_2, sk_2) be two key-pairs which are obtained by running the KeyGen algorithm with independent random tapes. Let $h \xleftarrow{\$} \mathbb{G}$ be an element chosen uniformly and independently at random from the group $\mathbb{G}$. The PPT algorithm performing re-encryption from pk_1 to pk_2 (denoted by $\mathsf{Re-Enc}_{1 \rightarrow 2}$) is described below.

$$\mathsf{Re-Enc}_{1\to 2}$$

Input: $c_1 = [Y_1, Y_2]$ or special symbol denoted by keys^a

Constants:[b] $sk_1 = x$, $pk_1 = (g, g^x)$, $pk_2 = (g, g^y)$ and h

1. If $input = \mathsf{keys}$, output (pk_1, pk_2).
2. Else,
   - Compute $m = \mathsf{Decrypt1}(sk_1, c_1)$.
   - Choose $r', v, s \xleftarrow{\$} \mathbb{Z}_p^*$.
   - Output $[X_1, X_2, X_3, X_4, X_5] = [m \cdot g^{r'}, (g^{H(h)})^{r'} \cdot (g^y)^s, h \cdot (g^y)^v, g^v, g^s]$.

---

[a] $\mathsf{keys} \notin \mathbb{G} \times \mathbb{G}$. We need this for a technical part in the proof.

[b] Constants in a program denotes those values which are hardcoded in the program description.

Re-encryption Circuit Family. Let $C_{sk_1, pk_1, pk_2, h}$ be the description of a probabilistic circuit implementing the program $\mathsf{Re-Enc}_{1\to 2}$. We note that the constants in the above program are hardwired in the circuit description. These constants can be extracted when given the description of the circuit. Formally, the class of circuits implementing the re-encryption functionality for a given length parameter λ is,

$$C_\lambda = \{ C_{sk_1, pk_1, pk_2, h} : (pk_1, sk_1) \leftarrow \mathsf{KeyGen}(1^\lambda), (pk_2, sk_2) \leftarrow \mathsf{KeyGen}(1^\lambda), h \xleftarrow{\$} \mathbb{G} \}$$

The circuit family implementing the re-encryption functionality is given by $C = \{C_\lambda\}_{\lambda \in \mathbb{N}}$. The associated sampling algorithm Samp proceeds by choosing $(p, \mathbb{G}, g, H) \leftarrow \mathsf{Setup}(1^\lambda)$. It then samples $(pk_1, sk_1) \leftarrow \mathsf{KeyGen}(1^\lambda)$, $(pk_2, sk_2) \leftarrow \mathsf{KeyGen}(1^\lambda)$ and $h \xleftarrow{\$} \mathbb{G}$. It finally outputs the circuit description of $C_{sk_1, pk_1, pk_2, h}$.

The evaluator of the circuit $C_{sk_1, pk_1, pk_2, h}$ supplies the regular input which is either the ciphertext $c_1 = [C_1, C_2]$ or the special symbol keys and also supplies the random input rand chosen uniformly at random from $\{0, 1\}^{3\lambda}$ to the circuit for sampling r', v, s uniformly from $\mathbb{Z}_p^*$.

Decrypting the Circuit Output. The output of $\mathsf{Re-Enc}_{1\to 2}$ can be decrypted using the following algorithm $\mathsf{Decrypt2}$:

$$\mathsf{Decrypt2}$$

Input: $sk_2, [X_1, X_2, X_3, X_4, X_5]$:

1. Parse sk_2 as y.
2. Compute $h = X_3 \cdot (X_4{}^y)^{-1}$.
3. Compute $X_2' = X_2 \cdot ((X_5)^y)^{-1}$
4. Output $m = (X_1) \cdot ((X_2')^{1/(H(h))})^{-1}$.

Correctness. We note that correctness of $\mathsf{Decrypt1}$ algorithm directly follows from the correctness of El-Gamal encryption scheme and correctness of $\mathsf{Decrypt2}$ follows from inspection.

Obfuscator Construction. We now present the construction of an average-case secure obfuscator (denoted by Obf) for the re-encryption circuit family defined in Sect. 3.

Obf

Input: $C_{sk_1,pk_1,pk_2,h}$

1. Read $sk_1 = x$, $pk_1 = (g, g^x)$, $pk_2 = (g, g^y)$ and h from the description of the circuit $C_{sk_1,pk_1,pk_2,h}$.
2. Select $v \xleftarrow{\$} \mathbb{Z}_p^*$.
3. Compute $(Z_1, Z_2, Z_3) = (h \cdot (g^y)^v, g^v, H(h)/x)$.
4. Output the description of a circuit implementing the program Re $-$ Enc$'_{1\to2}$ described below with $pk_1, pk_2, Z_1, Z_2, Z_3$ as the constants in the program.

Re $-$ Enc$'_{1\to2}$

Input: $c_1 = [Y_1, Y_2]$ or special symbol denoted by keys.
Constants: $pk_1, pk_2, Z_1, Z_2, Z_3$.

1. If *input* = keys, output (pk_1, pk_2).
2. Else,
   - Choose two re-randomization values $r', v' \xleftarrow{\$} \mathbb{Z}_p^*$.
   - Re-randomize the input as $X_1 = Y_1 \cdot g^{r'}$, $\overline{X_2} = (Y_2 \cdot (g^x)^{r'})$ and the hardwired values as $X_3 = Z_1 \cdot (g^y)^{v'}$, $X_4 = Z_2 \cdot g^{v'}$.
   - Compute $\overline{\overline{X_2}} = (\overline{X_2})^{Z_3}$.
   - Choose $s \xleftarrow{\$} \mathbb{Z}_p^*$.
   - Compute $X_2 = \overline{\overline{X_2}} \cdot (g^y)^s$ and $X_5 = g^s$
   - Output $[X_1, X_2, X_3, X_4, X_5]$.

Let C' denote the circuit implementing Re $-$ Enc$'_{1\to2}$. The evaluator for the circuit C' provides either $c_1 = [C_1, C_2]$ or special symbol keys as the regular input and rand $\xleftarrow{\$} \{0,1\}^{3\lambda}$ as the random input for sampling r', v', s uniformly from $\mathbb{Z}_p^*$.

Remark 6. The obfuscated circuit C' is generated by the owner of sk_1 but can be evaluated by anyone. We assume (as described in Remark 2) that the evaluator of C' and the owner of sk_2 do not collude.

4   Security of New Encryption Scheme

We now describe the security model for semantic security of the encryption scheme when the adversary is given black box access to re-encryption functionality. In view of discussion presented in Sect. 1.2, we modify the security model given in [24] as follows.

4.1 Security Model

Let $C \leftarrow \mathsf{Samp}(1^\lambda)^5$ be the re-encryption circuit from pk_1 to pk_2.

Original Ciphertext Security. Let $\mathcal{A} = (\mathcal{A}_1, \mathcal{A}_2)$ be an adversary against the original ciphertext security.

Definition 3. *Let Π be an encryption scheme and let $IND_{b,ori}(\Pi, \mathcal{A} = (\mathcal{A}_1, \mathcal{A}_2), \lambda, i)$ where $b \in \{0, 1\}$ and $i \in \{1, 2\}$, denote the following experiment:*

$$IND_{b,ori}(\Pi, \mathcal{A} = (\mathcal{A}_1, \mathcal{A}_2), \lambda, i)$$

1. *params $\leftarrow$ Setup(1^λ). $(pk_1, sk_1) \leftarrow$ KeyGen(params) and $(pk_2, sk_2) \leftarrow$ KeyGen(params). Choose $h \xleftarrow{\$} \mathbb{G}$. Set $C = C_{sk_1, pk_1, pk_2, h}$ [a].*
2. *$(m_0, m_1, state) \leftarrow \mathcal{A}_1^{\mathcal{O}(C)}(pk_1, pk_2, params)$.*
3. *$C^* \leftarrow$ Encrypt1$(m_b, pk_i, params)$.*
4. *$b' \leftarrow \mathcal{A}_2^{\mathcal{O}(C)}(C^*, state)$. Output b'*

---

  [a] Note that setting C in this way is equivalent to sampling C using the Samp algorithm.

---

*The scheme Π is said to be **original ciphertext secure** with respect to the oracle access to C if for all PPT adversaries $\mathcal{A} = (\mathcal{A}_1, \mathcal{A}_2)$ and for all $i \in \{1, 2\}$, there exists a negligible function $\mu(\cdot)$ such that for all $\lambda \in \mathbb{N}$,*

$$\Delta\big(IND_{0,ori}(\Pi, \mathcal{A}, \lambda, i), IND_{1,ori}(\Pi, \mathcal{A}, \lambda, i)\big) \leq \mu(\lambda)$$

Transformed Ciphertext Security. Let $\mathcal{A} = (\mathcal{A}_1, \mathcal{A}_2)$ be an adversary against the transformed ciphertext security.

Definition 4. *Let Π be an encryption scheme and let $IND_{b,tran}(\Pi, \mathcal{A} = (\mathcal{A}_1, \mathcal{A}_2), \lambda)$ where $b \in \{0, 1\}$ denote the following experiment:*

$$IND_{b,tran}(\Pi, \mathcal{A} = (\mathcal{A}_1, \mathcal{A}_2), \lambda)$$

1. *params $\leftarrow$ Setup(1^λ). $(pk_1, sk_1) \leftarrow$ KeyGen(params) and $(pk_2, sk_2) \leftarrow$ KeyGen(params). Choose $h \xleftarrow{\$} \mathbb{G}$. Set $C = C_{sk_1, pk_1, pk_2, h}$.*
2. *$(m_0, m_1, state) \leftarrow \mathcal{A}_1^{\mathcal{O}(C)}(params, pk_1, pk_2, sk_1)$.*
3. *rand $\xleftarrow{\$} \{0, 1\}^{3\lambda}$. Compute $C^* \leftarrow C(\mathsf{Encrypt1}(m_b, pk_1, params); rand)$.*
4. *$b' \leftarrow \mathcal{A}_2^{\mathcal{O}(C)}(C^*, state)$. Output b'*

---

*The scheme Π is said to be **transformed ciphertext secure** with respect to the oracle access to C if for all PPT adversaries $\mathcal{A} = (\mathcal{A}_1, \mathcal{A}_2)$, there exists a negligible function $\mu(\cdot)$ such that for all $\lambda \in \mathbb{N}$,*

$$\Delta\big(IND_{0,tran}(\Pi, \mathcal{A}, \lambda), IND_{1,tran}(\Pi, \mathcal{A}, \lambda)\big) \leq \mu(\lambda)$$

---

[5] For the ease of exposition, we drop the subscripts sk_1, pk_1, pk_2, h.

Statistical Independence. Let us consider the following experiment.

$$\mathsf{Stat}(\Pi, \lambda, m)$$

1. $params \leftarrow \mathsf{Setup}(1^\lambda)$. $(pk_1, sk_1) \leftarrow \mathsf{KeyGen}(params)$ and $(pk_2, sk_2) \leftarrow \mathsf{KeyGen}(params)$. Choose $h \xleftarrow{\$} \mathbb{G}$. Set $C = C_{sk_1, pk_1, pk_2, h}$.
2. $\mathsf{rand} \xleftarrow{\$} \{0,1\}^{3\lambda}$. Compute $C^* \leftarrow C(\mathsf{Encrypt1}(m, pk_1, params); \mathsf{rand})$.
3. Output C^*

We require the output of $\mathsf{Stat}(\Pi, \lambda, m)$ to be statistically independent of sk_1.

4.2 Security Proof

We now show that the New Encryption Scheme is *original ciphertext secure* (in Theorem 2), *transformed ciphertext secure* (in Theorem 3) and has statistical independence property (in Lemma 2).

Theorem 2. *The New Encryption Scheme is **original ciphertext secure** with respect to the oracle $C_{sk_1, pk_1, pk_2, h}$ under the DDH-assumption.*

Proof. We give the proof of this theorem in the full version of our paper [28].

We now show the transformed ciphertext security of our construction.

Theorem 3. *The New Encryption Scheme is **transformed ciphertext secure** with respect to the oracle $C_{sk_1, pk_1, pk_2, h}$ under the multi-message security (2 messages) of El-Gamal encryption system (Theorem 1).*

Proof. The proof of this theorem appears in the full version of the paper [28].

We note that the statistical independence property of the re-encryption functionality directly follows from inspection of the output distribution of the re-encryption circuit. We record the following lemma.

Lemma 2. *The output distribution of $C_{sk_1, pk_1, pk_2, h}$ where $(pk_1, sk_1) \leftarrow \mathsf{KeyGen}(params)$, $(pk_2, sk_2) \leftarrow \mathsf{KeyGen}(params)$ and $h \xleftarrow{\$} \mathbb{G}$ is statistically independent of sk_1.*

5 Average-Case Virtual Black Box Property

We note that obfuscator construction preserves functionality (the formal proof appears in the full version). We note that the polynomial slowdown property of our construction can be easily verified. It is interesting to note that the obfuscated circuit computes seven exponentiations whereas the original circuit computes eight exponentiations.

We now show that Obf satisfies the average-case virtual black box property.

Lemma 3. Obf *satisfies the average case secure virtual black-box property.*

Proof. The proof techniques used here are similar to that of Hohenberger et al. in [24] and the details follow.

Let $C \leftarrow \mathsf{Samp}(1^\lambda)$ be a circuit chosen randomly from the set C_λ using the Samp algorithm. Let D be any distinguisher with oracle access to C.

We first describe our simulator Sim which has oracle access to the circuit C and takes as input the security parameter in unary form and auxiliary information string denoted by z.

$$\mathsf{Sim}^{\mathcal{O}(C)}$$

Input: $1^\lambda, z$

1. Query the oracle $\mathcal{O}(C)$ with the special symbol *keys* and obtain pk_1 and pk_2.
2. Parse pk_2 as (g, g^y).
3. Choose $h \xleftarrow{\$} \mathbb{G}$, $v \xleftarrow{\$} \mathbb{Z}_p^*$ and compute $(Z_1', Z_2') = (h \cdot (g^y)^v, g^v)$. Choose Z_3' uniformly at random from $\mathbb{Z}_p^*$.
4. Construct a circuit C' implementing the program $\mathsf{Re - Enc}'_{1 \to 2}$ with values $(pk_1, pk_2, Z_1', Z_2', Z_3')$ hardcoded in the program description.
5. Output the circuit description of C'.

It remains to show that the output distribution of the simulator is computationally indistinguishable to the output distribution of Obf even to distinguishers having oracle access to C.

We define two distributions $\mathsf{Nice}(D^{\mathcal{O}(C)}, \lambda, z)$ and $\mathsf{Junk}(D^{\mathcal{O}(C)}, \lambda, z)$ as follows:

$$\mathsf{Nice}(D^{\mathcal{O}(C)}, \lambda, z)$$

- $(p, g, \mathbb{G}, H) \leftarrow \mathsf{Setup}(1^\lambda)$. Choose $x, y \xleftarrow{\$} \mathbb{Z}_p^*$. Set $pk_1 = (g, g^x)$ and $pk_2 = (g, g^y)$.
- Choose $h \xleftarrow{\$} \mathbb{G}$ and $v \xleftarrow{\$} \mathbb{Z}_p^*$.
- Compute $Z_1 = h \cdot (g^y)^v$, $Z_2 = g^v$ and $Z_3 = H(h)/x$.
- Output $D^{\mathcal{O}(C)}(pk_1, pk_2, Z_1, Z_2, Z_3, z)$.

$$\mathsf{Junk}(D^{\mathcal{O}(C)}, \lambda, z)$$

- $(p, g, \mathbb{G}, H) \leftarrow \mathsf{Setup}(1^\lambda)$. Choose $x, y \xleftarrow{\$} \mathbb{Z}_p^*$. Set $pk_1 = (g, g^x)$ and $pk_2 = (g, g^y)$.
- Choose $h \xleftarrow{\$} \mathbb{G}$ and $v \xleftarrow{\$} \mathbb{Z}_p^*$.
- Compute $Z_1' = h \cdot (g^y)^v$, $Z_2' = g^v$ and $Z_3' \xleftarrow{\$} \mathbb{Z}_p^*$.
- Output $D^{\mathcal{O}(C)}(pk_1, pk_2, Z_1', Z_2', Z_3', z)$.

We first observe that for all $z \in \{0,1\}^{poly(\lambda)}$ and for all distinguishers D,

$$\{C \to \mathsf{Samp}(1^\lambda); C' \leftarrow \mathsf{Obf}(C) : D^{\mathcal{O}(C)}(C', z)\} \approx \mathsf{Nice}(D^{\mathcal{O}(C)}, \lambda, z)$$

$$\{C \to \mathsf{Samp}(1^\lambda); C' \leftarrow \mathsf{Sim}^{\mathcal{O}(C)}(1^\lambda, z) : D^{\mathcal{O}(C)}(C', z)\} \approx \mathsf{Junk}(D^{\mathcal{O}(C)}, \lambda, z)$$

In order to show that Obf satisfies the average case virtual black box property it is enough to show that (from Lemma 1), for all PPT distinguishers D, there exists a negligible function $\mu(\cdot)$ such that for all $z \in \{0,1\}^{poly(\lambda)}$,

$$\Delta\left(\mathsf{Nice}(D^{\mathcal{O}(C)}, \lambda, z)\}, \mathsf{Junk}(D^{\mathcal{O}(C)}, \lambda, z)\right) \leq \mu(\lambda)$$

We show that for all PPT distinguishers D, there exists a negligible function $\mu(\cdot)$ such that for all $z \in \{0,1\}^{poly(\lambda)}$,

$$\Delta\left(\mathsf{Nice}(D^{\mathcal{O}(C)}, \lambda, z)\}, \mathsf{Junk}(D^{\mathcal{O}(C)}, \lambda, z)\right) \leq \mu(\lambda)$$

We start with an useful lemma.

Lemma 4.

$$
\left\{
\begin{aligned}
&(p, g, \mathbb{G}, H) \leftarrow \mathsf{Setup}(1^\lambda); \\
&x, y \xleftarrow{\$} \mathbb{Z}_p^*; \\
&pk_1 = (g, g^x); \\
&pk_2 = (g, g^y); \\
&h \xleftarrow{\$} \mathbb{G}; \\
&Z_3', v \xleftarrow{\$} \mathbb{Z}_p^*: \\
&(pk_1, pk_2, h \cdot (g^y)^v, g^v, Z_3')
\end{aligned}
\right\}_\lambda
\overset{c}{\approx}
\left\{
\begin{aligned}
&(p, g, \mathbb{G}, H) \leftarrow \mathsf{Setup}(1^\lambda); \\
&x, y \xleftarrow{\$} \mathbb{Z}_p^*; \\
&pk_1 = (g, g^x); \\
&pk_2 = (g, g^y); \\
&h \xleftarrow{\$} \mathbb{G}; \\
&v \xleftarrow{\$} \mathbb{Z}_p^*: \\
&(pk_1, pk_2, h \cdot (g^y)^v, g^v, H(h)/x)
\end{aligned}
\right\}_\lambda
$$

Proof. The proof of this lemma appears in the full version of the paper [28].

First, we consider two distributions which are similar to Nice and Junk except that they consider a "dummy" distinguisher D^* which outputs whatever is given as input.

Proposition 1. $\{\mathsf{Nice}(D^*, \lambda, z)\}_\lambda \overset{c}{\approx} \{\mathsf{Junk}(D^*, \lambda, z)\}_\lambda$

Proof. The proof for the this proposition follows directly from the proof of Lemma 4. We note that left distribution in the lemma statement is identically distributed to $\mathsf{Junk}(D^*, \lambda, z)$ and the right distribution is identically distributed to $\mathsf{Nice}(D^*, \lambda, z)$. Hence,

$$\mathsf{Nice}(D^*, \lambda, z) \overset{c}{\approx} \mathsf{Junk}(D^*, \lambda, z)$$

□

We now consider two more distributions which proceed as Nice and Junk except that they consider distinguishers $D^{\mathcal{O}(R)}$ where R is a probabilistic circuit which on any input $[C_1, C_2]$, first checks if C_1, C_2 belong to $\mathbb{G}$ and if yes, outputs $[A, B, C, D, E]$ where A, B, C, D, E are chosen uniformly and independently from $\mathbb{G}$. Otherwise, it outputs $\bot$.

Note that input to $D^{\mathcal{O}(R)}$ is identically distributed to $\mathsf{Nice}(D^*, \lambda, z)$ in $\mathsf{Nice}(D^{\mathcal{O}(R)}, \lambda, z)$ and its input is identically distributed to $\mathsf{Junk}(D^*, \lambda, z)$ in $\mathsf{Junk}(D^{\mathcal{O}(R)}, \lambda, z)$. The following proposition is a direct consequence of Proposition 1.

288   A. Srinivasan and C. Pandu Rangan

Proposition 2. *For all PPT distinguihsers D, there exists a negligible function $\mu(\cdot)$ such that for all $z \in \{0,1\}^{poly(\lambda)}$ and for all $\lambda \in \mathbb{N}$, we have*

$$\Delta\big(\mathsf{Nice}(D^{\mathcal{O}(R)}, \lambda, z), \mathsf{Junk}(D^{\mathcal{O}(R)}, \lambda, z)\big) \leq \mu(\lambda)$$

Proof. Assume for the sake of contradiction that there exists a distinguisher $D^{\mathcal{O}(R)}$ which can distinguish between $\mathsf{Nice}(D^*, \lambda, z)$ and $\mathsf{Junk}(D^*, \lambda, z)$ with non-negligible advantage. We construct an distinguisher between D' (without the oracle access to R) which distinguishes between $\mathsf{Nice}(D^*, \lambda, z)$ and $\mathsf{Junk}(D^*, \lambda, z)$ with the same advantage.

D' runs D internally by giving its own input as input to D. When D requests an oracle access to R, D' can simulate the responses on its own (It will choose five independent random elements from the group and return as the response for any oracle query after checking whether the input belongs to $\mathbb{G} \times \mathbb{G}$). D' finally outputs what D outputs.

It is easy to see that $\dot{D}'$ as the same distinguishing advantage that D has and hence we have arrived at a contradiction to Proposition 1. $\qquad\square$

Consider any distinguisher D. Let us define,

$$\alpha(\lambda, z) = \Delta\big(\mathsf{Nice}(D^{\mathcal{O}(C)}, \lambda, z), \mathsf{Junk}(D^{\mathcal{O}(C)}, \lambda, z)\big)$$

$$\beta(\lambda, z) = \Delta\big(\mathsf{Nice}(D^{\mathcal{O}(R)}, \lambda, z), \mathsf{Junk}(D^{\mathcal{O}(R)}, \lambda, z)\big)$$

Let q_D be the number of oracle queries that D makes during its execution. Since D runs in polynomial time, q_D is polynomial in λ.

Proposition 3. *There exists an algorithm $\mathcal{B}$ against the multi-message ($2q_D$ messages) security of El-Gamal encryption scheme with an advantage $|\alpha(\lambda, z) - \beta(\lambda, z)|/2$.*

Proof. We prove the proposition by constructing an adversary $\mathcal{B}$ against the El-Gamal challenger with advantage $|\alpha(\lambda, z) - \beta(\lambda, z)|/2$.

$\mathcal{B}$ receives the public key g^y from the El-Gamal challenger. It chooses $x \xleftarrow{\$} \mathbb{Z}_p^*$ and sets $pk_1 = (g, g^x)$ and $pk_2 = (g, g^y)$. It chooses two message vectors $\boldsymbol{M_0}$ and $\boldsymbol{M_1}$ each of length $2q_D$ as follows. It sets $\boldsymbol{M_0} = \{1, 1, \ldots, 1\}$ (of length $2q_D$) and $\boldsymbol{M_1} = \{m_1, m_2, \ldots, m_{2q_D}\}$ where $m_1, \ldots, m_{2q_D}$ are chosen uniformly and independently at random from $\mathbb{G}$. It then receives the challenge ciphertext vector $\boldsymbol{C^*} = \{(g^{r_1}, Q_1), (g^{r_2}, Q_2), \ldots, (g^{r_{q_D}}, Q_{2q_D})\}$ and auxiliary information z. Note that for all $i \in \{1, 2, \ldots 2q_D\}$, r_i is a random element in $\mathbb{Z}_p^*$ and $Q_i = g^{yr_i}$ or an uniformly chosen element depending on whether $\boldsymbol{M_0}$ was encrypted or $\boldsymbol{M_1}$ was encrypted (due to the random choice of $m_1, \ldots, m_{2q_D}$).

$\mathcal{B}$ now uses D to determine whether the challenge ciphertext vector is an encryption of $\boldsymbol{M_0}$ or $\boldsymbol{M_1}$. It first generates the tuples which are distributed exactly as $\mathsf{Nice}(D^*, \lambda, z)$ and $\mathsf{Junk}(D^*, \lambda, z)$. It tosses a random coin c and runs D with input $\mathsf{Nice}(D^*, \lambda, z)$ if $c = 0$ and with input $\mathsf{Junk}(D^*, \lambda, z)$ if $c = 1$. $\mathcal{B}$ needs to answer the re-encryption oracle queries made by D. It uses the challenge ciphertext to answer those oracle queries. We show that if the challenge ciphertext is an encryption of M_0, then the oracle responses given by $\mathcal{B}$ are identically distributed to the output of the re-encryption circuit C. If the challenge ciphertext was an encryption of M_1, we show that the oracle responses are identically distributed to the output of R. The exact details follow.

$\mathcal{B}$ chooses $h \xleftarrow{\$} \mathbb{G}$ and $v \xleftarrow{\$} \mathbb{Z}_p^*$ and computes $Z_1 = h \cdot (g^y)^v$ and $Z_2 = g^v$. It then chooses $Z_3 = H(h)/x$ and $Z_3' \xleftarrow{\$} \mathbb{Z}_p^*$. It tosses a random coin and chooses $c \xleftarrow{\$} \{0, 1\}$. If $c = 0$, it runs $D^{\mathcal{O}(X)}(pk_1, pk_2, Z_1, Z_2, Z_3, z)$. Else it runs, $D^{\mathcal{O}(X)}(pk_1, pk_2, Z_1, Z_2, Z_3', z)$ where X is the circuit description of the program $\mathsf{Re-Enc}''_{1 \to 2}$ described below. Note that if $c = 0$, input to D is identical to $\mathsf{Nice}(D^*, \lambda, z)$. Else, it is identical to $\mathsf{Junk}(D^*, \lambda, z)$.

When D makes i^{th} oracle query $[C_1, C_2]$, $\mathcal{B}$ runs the following program and returns the output of the program to D as the response.

$$\mathsf{Re-Enc}''_{1 \to 2}$$

Constants: $pk_1, pk_2, Z_1, Z_2, Z_3$
Input: $[C_1, C_2], i$

1. Choose $r' \xleftarrow{\$} \mathbb{Z}_p^*$.
2. Compute $C_1' = C_1 \cdot g^{r'}$, $C_2' = C_2 \cdot (g^x)^{r'}$.
3. Compute $Z_1' = Z_1 \cdot (Q_i)$, $Z_2' = Z_2 \cdot (g^{r_i})$.
4. Compute $C_2'' = C_2'^{Z_3}$.
5. Compute $D_2 = C_2'' \cdot Q_{i+q_D}$.
6. Output $[C_1', D_2, Z_1', Z_2', g^{r_i+q_D}]$.

D finally outputs its guess. Let c' denote the output of D. If $c = c'$, $\mathcal{B}$ outputs 1. Else, it outputs 0.

We now prove the following two claims regarding the output of $\mathcal{B}$.

Claim. If M_0 was encrypted, the probability that $\mathcal{B}$ outputs 1 is given by $1/2 + \alpha(\lambda, z)/2$.

Proof. We claim that if M_0 was encrypted, then $\mathcal{B}$ perfectly simulates $D^{\mathcal{O}(C)}(pk_1, pk_2, h \cdot (g^y)^r, g^r, H(h)/x, z)$ or $D^{\mathcal{O}(C)}(pk_1, pk_2, h \cdot (g^y)^r, g^r, Z_3', z)$ depending upon the bit c. We already noted that the input to D is identically distributed to $\mathsf{Nice}(D^*, \lambda, z)$ or $\mathsf{Junk}(D^*, \lambda, z)$. It is enough to show that X simulates the circuit C perfectly. Since $Q_i = (g^y)^{r_i}$ for all $i \in [1, 2q_D]$ and Z_1, Z_2, Z_3 are properly generated as per the Obf algorithm, the output of X is given by,

$$(m \cdot g^{r+r'}, (g^{r+r'})^{H(h)} \cdot (g^y)^{r_{i+q_D}}, h \cdot (g^y)^{v+r_i}, g^{v+r_i}, g^{r_i+q_D})$$

which is identically distributed as the output of the re-encryption circuit since r', r_i, r_{i+q_D} are chosen uniformly at random from $\mathbb{Z}_p^*$. Hence, the probability that $\mathcal{B}$ outputs 1 in this case is same as the probability that $D^{\mathcal{O}(C)}$ outputs $c = c'$ which is same as $1/2 + \alpha(\lambda, z)/2$. $\qquad\square$

Claim. If M_1 was encrypted, the probability that $\mathcal{B}$ outputs 1 is given by $1/2 + \beta(\lambda, z)/2$.

Proof. We already noted that the input to D are perfectly generated according to either $\mathsf{Nice}(D^*, \lambda, z)$ or $\mathsf{Junk}(D^*, \lambda, z)$. We claim that the response given by $\mathcal{B}$ are same as the one given by R. The output of $\mathcal{B}$ is given by

$$(m \cdot g^{r+r'}, (g^{r+r'})^{H(h)} \cdot Q_{i+q_D}, h \cdot (g^y)^v \cdot Q_i, g^{v+r_i}, g^{r_i+q_D})$$

Since Q_i and Q_{i+q_D} are uniformly chosen random elements in $\mathbb{G}$ if M_1 was encrypted and r', r_i, r_{i+q_D} are chosen uniformly at random from $\mathbb{Z}_p^*$, we can easily see that all elements in the above distribution are random and independent for every invocation of the oracle.

Hence, in this case $\mathcal{B}$ perfectly simulates $D^{\mathcal{O}(R)}(pk_1, pk_2, h \cdot (g^y)^r, g^r, H(h)/x, z)$ or $D^{\mathcal{O}(R)}(pk_1, pk_2, h \cdot (g^y)^r, g^r, Z_3', z)$ depending on the bit c. Thus, the probability that $\mathcal{B}$ outputs 1 in this case is same as the probability that D^R outputs $c = c'$ which is given by $\beta(\lambda, z)/2 + 1/2$.   □

Hence the advantage of $\mathcal{B}$ in the multi message security game of the El-Gamal Encryption scheme is given by $|\alpha(\lambda, z) - \beta(\lambda, z)|/2$.   □

We know from Proposition 2 that $\beta(\lambda, z)$ is negligible. Hence from Proposition 3 we can infer that $\alpha(\lambda, z)$ is also negligible.

Hence, Obf satisfies the average case secure virtual black box property and this concludes the proof of Lemma.

    □

Since Obf satisfies the three requirements given in Definition 2, we conclude that Obf is an average-case secure obfuscator.

References

1. Ateniese, G., Kevin, F., Green, M., Hohenberger, S.: Improved proxy re-encryption schemes with applications to secure distributed storage. ACM Trans. Inf. Syst. Secur. (TISSEC) **9**(1), 1–30 (2006)
2. Barak, B., Goldreich, O., Impagliazzo, R., Rudich, S., Sahai, A., Vadhan, S., Yang, K.: On the (im)possibility of obfuscating programs. In: Kilian, J. (ed.) CRYPTO 2001. LNCS, vol. 2139, pp. 1–18. Springer, Heidelberg (2001). doi:10.1007/3-540-44647-8_1
3. Bitansky, N., Paneth, O., Rosen, A.: On the cryptographic hardness of finding a nash equilibrium. In: IEEE 56th Annual Symposium on Foundations of Computer Science, FOCS 2015, Berkeley, CA, USA, 17–20 October 2015, pp. 1480–1498 (2015)
4. Bitansky, N., Paneth, O., Wichs, D.: Perfect structure on the edge of chaos. In: Kushilevitz, E., Malkin, T. (eds.) TCC 2016. LNCS, vol. 9562, pp. 474–502. Springer, Heidelberg (2016). doi:10.1007/978-3-662-49096-9_20
5. Blaze, M., Bleumer, G., Strauss, M.: Divertible protocols and atomic proxy cryptography. In: Nyberg, K. (ed.) EUROCRYPT 1998. LNCS, vol. 1403, pp. 127–144. Springer, Heidelberg (1998). doi:10.1007/BFb0054122

6. Boneh, D., Zhandry, M.: Multiparty key exchange, efficient traitor tracing, and more from indistinguishability obfuscation. In: Garay, J.A., Gennaro, R. (eds.) CRYPTO 2014. LNCS, vol. 8616, pp. 480–499. Springer, Heidelberg (2014). doi:10.1007/978-3-662-44371-2_27

7. Brakerski, Z.: Fully homomorphic encryption without modulus switching from classical GapSVP. In: Safavi-Naini, R., Canetti, R. (eds.) CRYPTO 2012. LNCS, vol. 7417, pp. 868–886. Springer, Heidelberg (2012). doi:10.1007/978-3-642-32009-5_50 ·

8. Brakerski, Z., Vaikuntanathan, V.: Efficient fully homomorphic encryption from (standard) LWE. In: IEEE 52nd Annual Symposium on Foundations of Computer Science, FOCS 2011, Palm Springs, CA, USA, 22–25 October 2011, pp. 97–106 (2011)

9. Brakerski, Z., Vaikuntanathan, V.: Lattice-based FHE as secure as PKE. In: Innovations in Theoretical Computer Science, ITCS 2014, Princeton, NJ, USA, 12–14 January 2014, pp. 1–12 (2014)

10. Canetti, R., Dakdouk, R.R.: Obfuscating point functions with multibit output. In: Smart, N. (ed.) EUROCRYPT 2008. LNCS, vol. 4965, pp. 489–508. Springer, Heidelberg (2008). doi:10.1007/978-3-540-78967-3_28

11. Canetti, R., Hohenberger, S.: Chosen-ciphertext secure proxy re-encryption. In: Proceedings of 14th ACM Conference on Computer and Communications Security, pp. 185–194. ACM (2007)

12. Canetti, R., Rothblum, G.N., Varia, M.: Obfuscation of hyperplane membership. In: Micciancio, D. (ed.) TCC 2010. LNCS, vol. 5978, pp. 72–89. Springer, Heidelberg (2010). doi:10.1007/978-3-642-11799-2_5

13. Chandran, N., Chase, M., Liu, F.-H., Nishimaki, R., Xagawa, K.: Re-encryption, functional re-encryption, and multi-hop re-encryption: a framework for achieving obfuscation-based security and instantiations from lattices. In: Krawczyk, H. (ed.) PKC 2014. LNCS, vol. 8383, pp. 95–112. Springer, Heidelberg (2014). doi:10.1007/978-3-642-54631-0_6

14. Chandran, N., Chase, M., Vaikuntanathan, V.: Functional re-encryption and collusion-resistant obfuscation. In: Cramer, R. (ed.) TCC 2012. LNCS, vol. 7194, pp. 404–421. Springer, Heidelberg (2012). doi:10.1007/978-3-642-28914-9_23

15. Chow, S.S.M., Weng, J., Yang, Y., Deng, R.H.: Efficient unidirectional proxy re-encryption. In: Bernstein, D.J., Lange, T. (eds.) AFRICACRYPT 2010. LNCS, vol. 6055, pp. 316–332. Springer, Heidelberg (2010). doi:10.1007/978-3-642-12678-9_19

16. Garg, S., Gentry, C., Halevi, S., Raykova, M.: Two-round secure MPC from indistinguishability obfuscation. In: Lindell, Y. (ed.) TCC 2014. LNCS, vol. 8349, pp. 74–94. Springer, Heidelberg (2014). doi:10.1007/978-3-642-54242-8_4

17. Garg, S., Gentry, C., Halevi, S., Raykova, M., Sahai, A., Waters, B.: Candidate indistinguishability obfuscation and functional encryption for all circuits. In: 2013 IEEE 54th Annual Symposium on Foundations of Computer Science (FOCS), pp. 40–49. IEEE (2013)

18. Garg, S., Pandey, O., Srinivasan, A.: Revisiting the cryptographic hardness of finding a nash equilibrium. In: Robshaw, M., Katz, J. (eds.) CRYPTO 2016. LNCS, vol. 9815, pp. 579–604. Springer, Heidelberg (2016). doi:10.1007/978-3-662-53008-5_20

19. Garg, S., Pandey, O., Srinivasan, A., Zhandry, M.: Breaking the sub-exponential barrier in obfustopia. In: Coron, J.-S., Nielsen, J.B. (eds.) EUROCRYPT 2017. LNCS, vol. 10212, pp. 156–181. Springer, Cham (2017). doi:10.1007/978-3-319-56617-7_6

20. Gentry, C.: Fully homomorphic encryption using ideal lattices. In: Proceedings of 41st Annual ACM Symposium on Theory of Computing, STOC 2009, Bethesda, MD, USA, 31 May–2 June 2009, pp. 169–178 (2009)

21. Goldreich, O.: The Foundations of Cryptography - Basic Techniques, vol. 1. Cambridge University Press, Cambridge (2001)
22. Hada, S.: Zero-knowledge and code obfuscation. In: Okamoto, T. (ed.) ASIACRYPT 2000. LNCS, vol. 1976, pp. 443–457. Springer, Heidelberg (2000). doi:10.1007/3-540-44448-3_34
23. Hada, S.: Secure obfuscation for encrypted signatures. In: Gilbert, H. (ed.) EUROCRYPT 2010. LNCS, vol. 6110, pp. 92–112. Springer, Heidelberg (2010). doi:10.1007/978-3-642-13190-5_5
24. Hohenberger, S., Rothblum, G.N., Shelat, A., Vaikuntanathan, V.: Securely obfuscating re-encryption. In: Vadhan, S.P. (ed.) TCC 2007. LNCS, vol. 4392, pp. 233–252. Springer, Heidelberg (2007). doi:10.1007/978-3-540-70936-7_13
25. Libert, B., Vergnaud, D.: Unidirectional chosen-ciphertext secure proxy re-encryption. In: Cramer, R. (ed.) PKC 2008. LNCS, vol. 4939, pp. 360–379. Springer, Heidelberg (2008). doi:10.1007/978-3-540-78440-1_21
26. Rivest, R.L., Adleman, L., Dertouzos, M.L.: On data banks and privacy homomorphisms. Found. Secure Comput. 4(11), 169–180 (1978)
27. Sahai, A., Waters, B.: How to use indistinguishability obfuscation: deniable encryption, and more. In: Symposium on Theory of Computing, STOC 2014, New York, NY, USA, 31 May–03 June 2014, pp. 475–484 (2014)
28. Srinivasan, A., Rangan, C.P.: Efficiently obfuscating re-encryption program under DDH assumption. IACR Cryptology ePrint Archive 2015:822 (2015)
29. Wee, H.: On obfuscating point functions. In: Proceedings of 37th Annual ACM Symposium on Theory of Computing, pp. 523–532. ACM (2005)

Lattice-Based Group Signatures: Achieving Full Dynamicity with Ease

San Ling, Khoa Nguyen$^{(\boxtimes)}$, Huaxiong Wang, and Yanhong Xu$^{(\boxtimes)}$

Division of Mathematical Sciences, School of Physical and Mathematical Sciences,
Nanyang Technological University, Singapore, Singapore
{lingsan,khoantt,hxwang,xu0014ng}@ntu.edu.sg

Abstract. Lattice-based group signature is an active research topic in recent years. Since the pioneering work by Gordon et al. (Asiacrypt 2010), eight other schemes have been proposed, providing various improvements in terms of security, efficiency and functionality. However, most of the existing constructions work only in the static setting where the group population is fixed at the setup phase. The only two exceptions are the schemes by Langlois et al. (PKC 2014) that handles user revocations (but new users cannot join), and by Libert et al. (Asiacrypt 2016) which addresses the orthogonal problem of dynamic user enrollments (but users cannot be revoked).

In this work, we provide the first lattice-based group signature that offers full dynamicity (i.e., users have the flexibility in joining and leaving the group), and thus, resolve a prominent open problem posed by previous works. Moreover, we achieve this non-trivial feat in a relatively simple manner. Starting with Libert et al.'s fully static construction (Eurocrypt 2016) - which is arguably the most efficient lattice-based group signature to date, we introduce simple-but-insightful tweaks that allow to upgrade it directly into the fully dynamic setting. More startlingly, our scheme even produces slightly shorter signatures than the former. The scheme satisfies the strong security requirements of Bootle et al.'s model (ACNS 2016), under the Short Integer Solution (SIS) and the Learning With Errors (LWE) assumptions.

Keywords: Lattice-based group signatures · Full dynamicity · Updatable Merkle trees · Stern-like zero-knowledge protocols

1 Introduction

Group signature, introduced by Chaum and van Heyst [14], is a fundamental anonymity primitive which allows members of a group to sign messages on behalf of the whole group. Yet, users are kept accountable for the signatures they issue since a tracing authority can identify them should the need arise. There have been numerous works on group signatures in the last quarter-century.

Ateniese et al. [2] proposed the first scalable instantiation meeting the security properties that can be intuitively expected from the primitive, although

© Springer International Publishing AG 2017
D. Gollmann et al. (Eds.): ACNS 2017, LNCS 10355, pp. 293–312, 2017.
DOI: 10.1007/978-3-319-61204-1_15

clean security notions were not available yet at that time. Bellare et al. [4] filled this gap by providing strong security notions for static groups, in which the group population is fixed at the setup phase. Subsequently, Kiayias and Yung [23] and Bellare et al. [5] established the models capturing the partially dynamic setting, where users are able to join the group at any time, but once they have done so, they cannot leave the group. Sakai et al. [44] strengthened these models by suggesting the notion of opening soundness, guaranteeing that a valid signature only traces to one user. Efficient schemes satisfying these models have been proposed in the random oracle model [16,40] and in the standard model [20,32].

One essential functionality of group signatures is the support for membership revocation. Enabling this feature in an efficient manner is quite challenging, since one has to ensure that revoked users are no longer able to sign messages and the workloads of other parties (managers, non-revoked users, verifiers) do not significantly increase in the meantime. Several different approaches have been suggested [6,10,11,39,46] to address this problem, and notable pairing-based constructions supporting both dynamic joining and efficient revocation were given in [30,31,37]. Very recently, Bootle et al. [7] pointed out a few shortcomings of previous models, and put forward stringent security notions for fully dynamic group signatures. They also demonstrated a construction satisfying these notions based on the decisional Diffie-Hellman (DDH) assumption, following a generic transformation from a secure accountable ring signature scheme [8].

For the time being, existing schemes offering full dynamicity all rely on number-theoretic assumptions which are vulnerable to quantum attacks. To avoid putting all eggs in one basket, it is thus encouraging to consider instantiations based on alternative, post-quantum foundations, e.g., lattice assumptions. In view of this, let us now look at the topic of lattice-based group signatures.

LATTICE-BASED GROUP SIGNATURES. Lattice-based cryptography has been an exciting research area since the seminal works of Regev [42] and Gentry et al. [18]. Along with other primitives, lattice-based group signature has received noticeable attention in recent years. The first scheme was introduced by Gordon et al. [19] whose solution produced signature size linear in the number of group users N. Camenisch et al. [12] then extended [19] to achieve anonymity in the strongest sense. Later, Laguillaumie *et al.* [24] put forward the first scheme with the signature size logarithmic in N, at the cost of relatively large parameters. Simpler and more efficient solutions with $\mathcal{O}(\log N)$ signature size were subsequently given by Nguyen et al. [41] and Ling et al. [34]. Libert et al. [28] obtained substantial efficiency improvements via a construction based on Merkle trees which eliminates the need for GPV trapdoors [18]. More recently, a scheme supporting message-dependent opening (MDO) feature [43] was proposed in [29]. All the schemes mentioned above are designed for static groups.

The only two known lattice-based group signatures that have certain dynamic features were proposed by Langlois et al. [25] and Libert et al. [26]. The former is a scheme with verifier-local revocation (VLR) [6], which means that only the verifiers need to download the up-to-date group information. The latter addresses the orthogonal problem of dynamic user enrollments (but users cannot be revoked).

To achieve those partially dynamic functionalities, both of the proposals have to incorporate relatively complicated mechanisms[1] and both heavily rely on lattice trapdoors.

The discussed above situation is somewhat unsatisfactory, given that the full dynamicity feature is highly desirable in most applications of group signatures (e.g., protecting the privacy of commuters in public transportation), and it has been achieved based on number-theoretic assumptions. This motivates us to work on fully dynamic group signatures from lattices. Furthermore, considering that the journey to partial dynamicity in previous works [25,26] was shown not easy, we ask ourselves an inspiring question: Can we achieve full dynamicity with ease? At the end of the day, it is good to solve an open research question, but it would be even better and more exciting to do this in a simple way. To make it possible, we will likely need some new and insightful ideas.

OUR RESULTS AND TECHNIQUES. We introduce the first fully dynamic group signature from lattices. The scheme satisfies the strong security requirements put forward by Bootle et al. [7], under the Short Integer Solution (SIS) and the Learning With Errors (LWE) assumptions. As in all previous lattice-based group signatures, our scheme is analyzed in the random oracle model.

For a security parameter λ and maximum expected number of group users N, our scheme features signature size $\widetilde{\mathcal{O}}(\lambda \cdot \log N)$ and group public key size $\widetilde{\mathcal{O}}(\lambda^2 + \lambda \cdot \log N)$. The user's secret key has bit-size $\widetilde{\mathcal{O}}(\lambda) + \log N$. At each epoch when the group information is updated, the verifiers only need to download an extra $\widetilde{\mathcal{O}}(\lambda)$ bits in order to perform verification of signatures[2], while each active signer only has to download $\widetilde{\mathcal{O}}(\lambda \cdot \log N)$ bits. In Table 1, we give a detailed comparison of our scheme with known lattice-based group signatures, in terms of efficiency and functionality. The full dynamicity feature is achieved with a very reasonable cost and without having to rely on lattice trapdoors. Somewhat surprisingly, our scheme even produces shorter signatures than the scheme from [28] - which is arguably the most efficient lattice-based group signature known to date. Furthermore, these results are obtained in a relatively simple manner, thanks to three main ideas/techniques discussed below.

Our starting point is the scheme [28], which works in the static setting. Instead of relying on trapdoor-based ordinary signatures as in prior works, the LLNW scheme employs on a SIS-based Merkle tree accumulator. For a group of $N = 2^\ell$ users, the manager chooses uniformly random vectors $\mathbf{x}_0, \ldots, \mathbf{x}_{N-1}$; hashes them to $\mathbf{p}_0, \ldots, \mathbf{p}_{N-1}$, respectively; builds a tree on top of these hash values; and publishes the tree root $\mathbf{u}$. The signing key of user i consists of $\mathbf{x}_i$

---

[1] Langlois et al. considered users' "tokens" as functions of Bonsai signatures [13] and associated them with a sophisticated revocation technique, while Libert et al. used a variant of Boyen's signature [9] to sign users' public keys. Both underlying signature schemes require long keys and lattice trapdoors.

[2] We remark that in the DDH-based instantiation from [7] which relies on the accountable ring signature from [8], the verifiers have to periodically download public keys of active signers. Our scheme overcomes this issue, thanks to the use of an updatable accumulator constructed in Sect. 3.

Table 1. Comparison of known lattice-based group signatures, in terms of efficiency and functionality. The comparison is done based on two governing parameters: security parameter λ and the maximum expected number of group users $N = 2^\ell$. As for the scheme from [25], R denotes the number of revoked users at the epoch in question.

Scheme	Sig. size	Group PK size	Signer's SK size	Trap-door?	Model	Extra info per epoch
GKV [19]	$\widetilde{\mathcal{O}}(\lambda^2 \cdot N)$	$\widetilde{\mathcal{O}}(\lambda^2 \cdot N)$	$\widetilde{\mathcal{O}}(\lambda^2)$	yes	static	NA
CNR [12]	$\widetilde{\mathcal{O}}(\lambda^2 \cdot N)$	$\widetilde{\mathcal{O}}(\lambda^2)$	$\widetilde{\mathcal{O}}(\lambda^2)$	yes	static	NA
LLLS [24]	$\widetilde{\mathcal{O}}(\lambda \cdot \ell)$	$\mathcal{O}(\lambda^2 \cdot \ell)$	$\widetilde{\mathcal{O}}(\lambda^2)$	yes	static	NA
LLNW [25]	$\widetilde{\mathcal{O}}(\lambda \cdot \ell)$	$\widetilde{\mathcal{O}}(\lambda^2 \cdot \ell)$	$\widetilde{\mathcal{O}}(\lambda \cdot \ell)$	yes	VLR	Sign: **no** Ver: $\widetilde{\mathcal{O}}(\lambda) \cdot R$
NZZ [41]	$\widetilde{\mathcal{O}}(\lambda + \ell^2)$	$\widetilde{\mathcal{O}}(\lambda^2 \cdot \ell^2)$	$\widetilde{\mathcal{O}}(\lambda^2)$	yes	static	NA
LNW [34]	$\widetilde{\mathcal{O}}(\lambda \cdot \ell)$	$\widetilde{\mathcal{O}}(\lambda^2 \cdot \ell)$	$\widetilde{\mathcal{O}}(\lambda)$	yes	static	NA
LLNW [28]	$\widetilde{\mathcal{O}}(\lambda \cdot \ell)$	$\widetilde{\mathcal{O}}(\lambda^2 + \lambda \cdot \ell)$	$\widetilde{\mathcal{O}}(\lambda \cdot \ell)$	**FREE**	static	NA
LLM+ [26]	$\widetilde{\mathcal{O}}(\lambda \cdot \ell)$	$\widetilde{\mathcal{O}}(\lambda^2 \cdot \ell)$	$\widetilde{\mathcal{O}}(\lambda)$	yes	partially dynamic	NA
LMN [29]	$\widetilde{\mathcal{O}}(\lambda \cdot \ell)$	$\widetilde{\mathcal{O}}(\lambda^2 \cdot \ell)$	$\widetilde{\mathcal{O}}(\lambda)$	yes	MDO	NA
Ours	$\widetilde{\mathcal{O}}(\lambda \cdot \ell)$	$\widetilde{\mathcal{O}}(\lambda^2 + \lambda \cdot \ell)$	$\widetilde{\mathcal{O}}(\lambda) + \ell$	**FREE**	fully dynamic	Sign: $\frac{\widetilde{\mathcal{O}}(\lambda \cdot \ell)}{\text{Ver: } \widetilde{\mathcal{O}}(\lambda)}$

and the witness for the fact that $\mathbf{p}_i$ was accumulated in $\mathbf{u}$. When issuing signatures, the user proves knowledge of a valid pair $(\mathbf{x}_i, \mathbf{p}_i)$ and of the tree path from $\mathbf{p}_i$ to $\mathbf{u}$. The user also has to encrypt the binary representation $\mathrm{bin}(i)$ of his identity i, and prove that the ciphertext is well-formed. The encryption layer is also lattice-trapdoor-free, since it utilizes the Naor-Yung double-encryption paradigm [38] with Regev's LWE-based encryption scheme. To upgrade the LLNW scheme directly into a fully dynamic group signature, we now let the user compute the pair $(\mathbf{x}_i, \mathbf{p}_i)$ on his own (for enabling non-frameability), and we employ the following three ideas/techniques.

First, we add a dynamic ingredient into the static Merkle tree accumulator from [28]. To this end, we equip it with an efficient updating algorithm with complexity $\mathcal{O}(\log N)$: to change an accumulated value, we simply update the values at the corresponding leaf and along its path to the root.

Second, we create a simple rule to handle user enrollment and revocation efficiently (i.e., without resetting the whole tree). Specifically, we use the updating algorithm to set up the system so that: (i) If a user has not joined the group or has been revoked, the value at the leaf associated with him is set as $\mathbf{0}$; (ii) When a user joins the group, that value is set as his public key $\mathbf{p}_i$. Our setup guarantees that only active users (i.e., who has joined and has not been revoked at the given epoch) have their *non-zero* public keys accumulated into the updated root. This rule effectively separates active users who can sign from those who cannot: when signing messages, the user proceeds as in the LLNW scheme, and is asked to additionally prove in zero-knowledge that $\mathbf{p}_i \neq \mathbf{0}$. In other words, the seemingly big gap between being fully static and being fully dynamic has been reduced to a small difference!

Third, the arising question now is how to additionally prove the inequality $\mathbf{p}_i \neq \mathbf{0}$ in the framework of the Stern-like [45] protocol from [28]. One would naturally expect that this extra job could be done without losing too much in terms of efficiency. Here, the surprising and somewhat unexpected fact is that we can actually do it while *gaining* efficiency, thanks to the following simple idea. Recall that, in [28], to prove knowledge of $\mathbf{p}_i \in \{0,1\}^{nk}$, an extension technique from [33] is employed, in which $\mathbf{p}_i$ is extended into a vector of dimension $2nk$. We note that, the authors of [33] also suggested a slightly modified version of their technique, that allows to simultaneously prove that $\mathbf{p}_i \in \{0,1\}^{nk}$ and $\mathbf{p}_i$ is non-zero while working only with dimension $2nk - 1$. This intriguing tweak enables us to obtain a zero-knowledge protocol with slightly lower communication cost, and hence, group signatures with slightly smaller size than in [28].

To summarize, we solve a prominent open question in the field of lattice-based group signatures. Moreover, our solution is simple and comparatively efficient. Our results, while not yielding a truly practical scheme, would certainly help to bring the field one step closer to practice.

ORGANIZATION. In Sect. 2, we recall some background on fully dynamic group signatures and lattice-based cryptography. Section 3 develops an updatable Merkle tree accumulator. Our main scheme is constructed and analyzed in Sect. 4.

2 Preliminaries

2.1 Fully Dynamic Group Signatures

We recall the definition and security notions of fully dynamic group signatures (FDGS) presented by Bootle et al. [7]. A FDGS scheme involves the following entities: a group manager GM that determines who can join the group, a tracing manager TM who can open signatures, and a set of users who are potential group members. Users can join/leave the group at the discretion of GM. We assume GM will publish some information info_τ regularly associated with a distinct index τ (referred as epoch hereafter). Without loss of generality, assume there is one-to-one correspondence between information and the associated epoch. The information describes changes of the group, e.g., current group members or members that are excluded from the group. We assume the published group information is authentic. By comparing current group information with previous one, it allows any party to identify revoked users at current epoch. For simplicity assume $\tau_1 < \tau_2$ if info_{τ_1} is published before info_{τ_2}, i.e., the epoch preserves the order in which the corresponding group information was published. In existing models, the keys of authorities are supposed to be generated honestly; while in [7], Bootle et al. consider a stronger model where the keys of authorities can be maliciously generated by the adversary.

Syntax of Fully Dynamic Group Signatures. A FDGS scheme is a tuple of following polynomial-time algorithms.

GSetup(λ) $\rightarrow$ pp. On input security parameter λ, this algorithm generates public parameters pp.

$\langle$GKgen$_{\mathsf{GM}}$(pp), GKgen$_{\mathsf{TM}}$(pp)$\rangle$. This is an interactive protocol between the group manager GM and the tracing manager TM. If it completes successfully, algorithm GKgen$_{\mathsf{GM}}$ outputs a manager key pair (mpk, msk). Meanwhile, GM initializes the group information info and the registration table **reg**. The algorithm GKgen$_{\mathsf{TM}}$ outputs a tracing key pair (tpk, tsk). Set group public key gpk = (pp, mpk, tpk).

UKgen(pp) $\rightarrow$ (upk, usk). Given public parameters pp, this algorithm generates a user key pair (upk, usk).

$\langle$Join(info$_{\tau_{\mathrm{current}}}$, gpk, upk, usk); Issue(info$_{\tau_{\mathrm{current}}}$, msk, upk)$\rangle$. This is an interactive algorithm run by a user and GM. If it completes successfully, this user becomes a group member with an identifier uid and the Join algorithm stores secret group signing key gsk[uid] while Issue algorithm stores registration information in the table **reg** with index uid.

GUpdate(gpk, msk, info$_{\tau_{\mathrm{current}}}$, $\mathcal{S}$, **reg**) $\rightarrow$ info$_{\tau_{\mathrm{new}}}$. This is an algorithm run by GM to update group information while advancing the epoch. Given gpk,msk, info$_{\tau_{\mathrm{current}}}$, registration table **reg**, a set $\mathcal{S}$ of active users to be removed from the group, GM computes new group information info$_{\tau_{\mathrm{new}}}$ and may update the table **reg**. If there is no change to the group, GM outputs $\perp$.

Sign(gpk, gsk[uid], info$_\tau$, M) $\rightarrow$ Σ. This algorithm outputs a group signature Σ on message M by user uid. It outputs $\perp$ if this user is inactive at epoch τ.

Verify(gpk, info$_\tau$, M, Σ) $\rightarrow$ 0/1. This algorithm checks the validity of the signature Σ on message M at epoch τ.

Trace(gpk, tsk, info$_\tau$, **reg**, M, Σ) $\rightarrow$ (uid, Π_{trace}). This is an algorithm run by TM. Given the inputs, TM returns an identity uid of a group member who signed the message and a proof indicating this tracing result or $\perp$ if it fails to trace to a group member.

Judge(gpk, uid, info$_\tau$, Π_{trace}, M, Σ) $\rightarrow$ 0/1. This algorithm checks the validity of Π_{trace} outputted by the Trace algorithm.

Correctness and Security of Fully Dynamic Group Signatures. As put forward by Bootle et al. [7], a FDGS scheme must satisfy *correctness, anonymity, non-frameability, traceability* and *tracing soundness*.

Correctness demands that a signature generated by an honest and active user is always accepted by algorithm Verify, and that algorithm Trace can always identify that user, as well as produces a proof accepted by algorithm Judge.

Anonymity requires that it is infeasible for any PPT adversary to distinguish signatures generated by two active users of its choice at the chosen epoch, even if it can corrupt any user, can choose the key of GM, and is given access to the Trace oracle.

Non-Frameability makes sure that the adversary cannot generate a valid signature that traces to an honest user even if it can corrupt all other users, and can choose keys of both managers.

Traceability ensures that the adversary cannot produce a valid signature that cannot be traced to an active user at the chosen epoch, even if it can corrupt any user and can choose the key of TM.

Tracing Soundness requires that it is infeasible to produce a valid signature that traces to two different users, even if all group users and both managers are fully controlled by the adversary.

Formal definitions of correctness and security requirements are available in the full version.

2.2 Background on Lattices

We recall the average-case lattice problems SIS and LWE, together with their hardness results.

Definition 1 [1,18]. *The* $\mathsf{SIS}^\infty_{n,m,q,\beta}$ *problem is as follows: Given uniformly random matrix* $\mathbf{A} \in \mathbb{Z}_q^{n \times m}$, *find a non-zero vector* $\mathbf{x} \in \mathbb{Z}^m$ *such that* $\|\mathbf{x}\|_\infty \leq \beta$ *and* $\mathbf{A} \cdot \mathbf{x} = \mathbf{0} \bmod q$.

If $m, \beta = \mathsf{poly}(n)$, and $q > \beta \cdot \widetilde{\mathcal{O}}(\sqrt{n})$, then the $\mathsf{SIS}^\infty_{n,m,q,\beta}$ problem is at least as hard as worst-case lattice problem SIVP_γ for some $\gamma = \beta \cdot \widetilde{\mathcal{O}}(\sqrt{nm})$ (see, e.g., [18,36]). Specifically, when $\beta = 1$, $q = \widetilde{\mathcal{O}}(n)$, $m = 2n\lceil \log q \rceil$, the $\mathsf{SIS}^\infty_{n,m,q,1}$ problem is at least as hard as SIVP_γ with $\gamma = \widetilde{\mathcal{O}}(n)$.

In the last decade, numerous SIS-based cryptographic primitives have been proposed. In this work, we will extensively employ 2 such constructions:

- Our group signature scheme is based on the Merkle tree accumulator from [28], which is built upon a specific family of collision-resistant hash functions.
- Our zero-knowledge argument systems use the statistically hiding and computationally binding string commitment scheme from [22].

For appropriate setting of parameters, the security of the above two constructions can be based on the worst-case hardness of $\mathsf{SIVP}_{\widetilde{\mathcal{O}}(n)}$.

In the group signature in Sect. 4, we will employ the multi-bit version of Regev's encryption scheme [42], presented in [21]. The scheme is based on the hardness of the LWE problem.

Definition 2 [42]. *Let* $n, m \geq 1$, $q \geq 2$, *and let* χ *be a probability distribution on* $\mathbb{Z}$. *For* $\mathbf{s} \in \mathbb{Z}_q^n$, *let* $\mathcal{A}_{\mathbf{s},\chi}$ *be the distribution obtained by sampling* $\mathbf{a} \xleftarrow{\$} \mathbb{Z}_q^n$ *and* $e \hookleftarrow \chi$, *and outputting* $(\mathbf{a}, \mathbf{s}^\top \cdot \mathbf{a} + e) \in \mathbb{Z}_q^n \times \mathbb{Z}_q$. *The* $\mathsf{LWE}_{n,q,\chi}$ *problem asks to distinguish* m *samples chosen according to* $\mathcal{A}_{\mathbf{s},\chi}$ *(for* $\mathbf{s} \xleftarrow{\$} \mathbb{Z}_q^n$*) and* m *samples chosen according to the uniform distribution over* $\mathbb{Z}_q^n \times \mathbb{Z}_q$.

If q is a prime power, χ is the discrete Gaussian distribution $D_{\mathbb{Z},\alpha q}$, where $\alpha q \geq 2\sqrt{n}$, then $\mathsf{LWE}_{n,q,\chi}$ is as least as hard as $\mathsf{SIVP}_{\widetilde{\mathcal{O}}(n/\alpha)}$ (see [35, 36, 42]).

2.3 Stern-Like Protocols for Lattice-Based Relations

The zero-knowledge (ZK) argument systems appearing in this paper operate in the framework of Stern's protocol [45]. Let us now recall some background. This protocol was originally proposed in the context of code-based cryptography, and was later adapted into the lattice setting by Kawachi et al. [22]. Subsequently, it was empowered by Ling et al. [33] to handle the matrix-vector relations associated with the SIS and LWE problems, and further developed to design several lattice-based schemes: group signatures [25, 26, 28, 29, 34], policy-based signatures [15] and group encryption [27].

Stern-like protocols are quite useful in the context of lattice-based privacy-preserving systems, when one typically works with modular linear equations of the form $\sum_i \mathbf{M}_i \cdot \mathbf{x}_i = \mathbf{v} \bmod q$ - where $\{\mathbf{M}_i\}_i$, $\mathbf{v}$ are public, and one wants to prove in ZK that secret vectors $\{\mathbf{x}_i\}_i$ satisfy certain constraints, e.g., they have small norms and/or have coordinates arranged in a special way. The high-level ideas can be summarized as follows. If the given constraints are invariant under certain type of permutations of coordinates, then one readily uses uniformly random permutations to prove those constraints. Otherwise, one performs some pre-processings with $\{\mathbf{x}_i\}_i$ to reduce to the former case. Meanwhile, to prove that the modular linear equation holds, one makes use of a standard masking technique.

The basic protocol consists of 3 moves: commitment, challenge, response. If the statistically hiding and computationally binding string commitment scheme from [22] is employed in the first move, then one obtains a statistical zero-knowledge argument of knowledge (ZKAoK) with perfect completeness, constant soundness error $2/3$, and communication cost $\mathcal{O}(|w| \cdot \log q)$, where $|w|$ denotes the total bit-size of the secret vectors. In many applications, the protocol is repeated $\kappa = \omega(\log \lambda)$ times, for security parameter λ, to achieve negligible soundness error, and then made non-interactive via the Fiat-Shamir heuristic [17]. In the random oracle model, this results in a non-interactive zero-knowledge argument of knowledge (NIZKAoK) with bit-size $\mathcal{O}(|w| \cdot \log q) \cdot \omega(\log \lambda)$.

3 Updatable Lattice-Based Merkle Hash Trees

We first recall the lattice-based Merkle-tree accumulator from [28], and then, we equip it with a simple updating algorithm which allows to change an accumulated value in time logarithmic in the size of the accumulated set. This updatable hash tree will serve as the building block of our construction in Sect. 4.

3.1 Cryptographic Accumulators

An *accumulator scheme* is a tuple of polynomial-time algorithms defined below.

TSetup(λ). On input security parameter λ, output the public parameter pp.

TAcc$_{pp}$. On input a set $R = \{\mathbf{d}_0, \ldots, \mathbf{d}_{N-1}\}$ of N data values, output an accumulator value $\mathbf{u}$.

TWitness$_{pp}$. On input a data set R and a value $\mathbf{d}$, output $\perp$ if $\mathbf{d} \notin R$; otherwise output a witness w for the fact that $\mathbf{d}$ is accumulated in TAcc(R). (Typically, the size of w should be short (*e.g.*, constant or logarithmic in N) to be useful.)

TVerify$_{pp}$. On input accumulator value $\mathbf{u}$ and a value-witness pair $(\mathbf{d}, w)$, output 1 (which indicates that $(\mathbf{d}, w)$ is valid for the accumulator $\mathbf{u}$) or 0.

An accumulator scheme is called correct if for all pp $\leftarrow$ TSetup(λ), we have TVerify$_{pp}$(TAcc$_{pp}$(R), $\mathbf{d}$, TWitness$_{pp}$($R, \mathbf{d}$)) = 1 for all $\mathbf{d} \in R$.

A natural security requirement for accumulators, as considered in [3,11,28], says that it is infeasible to prove that a value $\mathbf{d}^*$ was accumulated in a value $\mathbf{u}$ if it was not. This property is formalized as follows.

Definition 3 [28]. *An accumulator scheme* (TSetup, TAcc, TWitness, TVerify) *is called secure if for all PPT adversaries $\mathcal{A}$:*

$$\Pr\big[\text{pp} \leftarrow \text{TSetup}(\lambda); (R, \mathbf{d}^*, w^*) \leftarrow \mathcal{A}(\text{pp}) :$$
$$\mathbf{d}^* \notin R \wedge \text{TVerify}_{pp}(\text{TAcc}_{pp}(R), \mathbf{d}^*, w^*) = 1\big] = \text{negl}(\lambda).$$

3.2 The LLNW Merkle-Tree Accumulator

NOTATIONS. Hereunder we will use the notation $x \xleftarrow{\$} S$ to indicate that x is chosen uniformly at random from finite set S. For bit $b \in \{0, 1\}$, we let $\bar{b} = 1 - b$. The Merkle-tree accumulator scheme from [28] works with parameters $n = \mathcal{O}(\lambda)$, $q = \widetilde{\mathcal{O}}(n^{1.5})$, $k = \lceil \log_2 q \rceil$, and $m = 2nk$. The set $\mathbb{Z}_q$ is identified by $\{0, \ldots, q-1\}$. Define the "powers-of-2" matrix

$$\mathbf{G} = \begin{bmatrix} 1\,2\,4\,\ldots\,2^{k-1} & & \\ & \ldots & \\ & & 1\,2\,4\,\ldots\,2^{k-1} \end{bmatrix} \in \mathbb{Z}_q^{n \times nk}.$$

Note that for every $\mathbf{v} \in \mathbb{Z}_q^n$, we have $\mathbf{v} = \mathbf{G} \cdot \text{bin}(\mathbf{v})$, where $\text{bin}(\mathbf{v}) \in \{0, 1\}^{nk}$ denotes the binary representation of $\mathbf{v}$. The scheme is built upon the following family of SIS-based collision-resistant hash functions.

Definition 4. *The function family $\mathcal{H}$ mapping $\{0, 1\}^{nk} \times \{0, 1\}^{nk}$ to $\{0, 1\}^{nk}$ is defined as $\mathcal{H} = \{h_{\mathbf{A}} \mid \mathbf{A} \in \mathbb{Z}_q^{n \times m}\}$, where for $\mathbf{A} = [\mathbf{A}_0 | \mathbf{A}_1]$ with $\mathbf{A}_0, \mathbf{A}_1 \in \mathbb{Z}_q^{n \times nk}$, and for any $(\mathbf{u}_0, \mathbf{u}_1) \in \{0, 1\}^{nk} \times \{0, 1\}^{nk}$, we have:*

$$h_{\mathbf{A}}(\mathbf{u}_0, \mathbf{u}_1) = \text{bin}\big(\mathbf{A}_0 \cdot \mathbf{u}_0 + \mathbf{A}_1 \cdot \mathbf{u}_1 \bmod q\big) \in \{0, 1\}^{nk}.$$

Note that $h_{\mathbf{A}}(\mathbf{u}_0, \mathbf{u}_1) = \mathbf{u} \Leftrightarrow \mathbf{A}_0 \cdot \mathbf{u}_0 + \mathbf{A}_1 \cdot \mathbf{u}_1 = \mathbf{G} \cdot \mathbf{u} \bmod q$.

A Merkle tree with $N = 2^\ell$ leaves, where ℓ is a positive integer, then can be constructed based on the function family $\mathcal{H}$ as follows.

TSetup(λ). Sample $\mathbf{A} \xleftarrow{\$} \mathbb{Z}_q^{n \times m}$, and output pp = $\mathbf{A}$.

$\mathsf{TAcc_A}(R = \{\mathbf{d}_0 \in \{0,1\}^{nk}, \ldots, \mathbf{d}_{N-1} \in \{0,1\}^{nk}\})$. For every $j \in [0, N-1]$, let $\mathrm{bin}(j) = (j_1, \ldots, j_\ell) \in \{0,1\}^\ell$ be the binary representation of j, and let $\mathbf{d}_j = \mathbf{u}_{j_1, \ldots, j_\ell}$. Form the tree of depth $\ell = \log N$ based on the N leaves $\mathbf{u}_{0,0,\ldots,0}, \ldots, \mathbf{u}_{1,1,\ldots,1}$ as follows:

1. At depth $i \in [\ell]$, the node $\mathbf{u}_{b_1,\ldots,b_i} \in \{0,1\}^{nk}$, for all $(b_1, \ldots, b_i) \in \{0,1\}^i$, is defined as $h_\mathbf{A}(\mathbf{u}_{b_1,\ldots,b_i,0}, \mathbf{u}_{b_1,\ldots,b_i,1})$.
2. At depth 0: The root $\mathbf{u} \in \{0,1\}^{nk}$ is defined as $h_\mathbf{A}(\mathbf{u}_0, \mathbf{u}_1)$.

The algorithm outputs the accumulator value $\mathbf{u}$.

$\mathsf{TWitness_A}(R, \mathbf{d})$. If $\mathbf{d} \notin R$, return $\perp$. Otherwise, $\mathbf{d} = \mathbf{d}_j$ for some $j \in [0, N-1]$ with binary representation $(j_1, \ldots, j_\ell)$. Output the witness w defined as:

$$ w = \left((j_1, \ldots, j_\ell), (\mathbf{u}_{j_1,\ldots,j_{\ell-1},\bar{j}_\ell}, \ldots, \mathbf{u}_{j_1,\bar{j}_2}, \mathbf{u}_{\bar{j}_1})\right) \in \{0,1\}^\ell \times \left(\{0,1\}^{nk}\right)^\ell, $$

for $\mathbf{u}_{j_1,\ldots,j_{\ell-1},\bar{j}_\ell}, \ldots, \mathbf{u}_{j_1,\bar{j}_2}, \mathbf{u}_{\bar{j}_1}$ computed by algorithm $\mathsf{TAcc_A}(R)$.

$\mathsf{TVerify_A}(\mathbf{u}, \mathbf{d}, w)$. Let the given witness w be of the form:

$$ w = \left((j_1, \ldots, j_\ell), (\mathbf{w}_\ell, \ldots, \mathbf{w}_1)\right) \in \{0,1\}^\ell \times \left(\{0,1\}^{nk}\right)^\ell. $$

The algorithm recursively computes the path $\mathbf{v}_\ell, \mathbf{v}_{\ell-1}, \ldots, \mathbf{v}_1, \mathbf{v}_0 \in \{0,1\}^{nk}$ as follows: $\mathbf{v}_\ell = \mathbf{d}$ and

$$ \forall i \in \{\ell-1, \ldots, 1, 0\} : \mathbf{v}_i = \begin{cases} h_\mathbf{A}(\mathbf{v}_{i+1}, \mathbf{w}_{i+1}), & \text{if } j_{i+1} = 0; \\ h_\mathbf{A}(\mathbf{w}_{i+1}, \mathbf{v}_{i+1}), & \text{if } j_{i+1} = 1. \end{cases} \tag{1} $$

Then it returns 1 if $\mathbf{v}_0 = \mathbf{u}$. Otherwise, it returns 0.

The following lemma states the correctness and security of the above Merkle tree accumulator.

Lemma 1 [28]. *The given accumulator scheme is correct and is secure in the sense of Definition 3, assuming the hardness of the $\mathsf{SIS}_{n,m,q,1}^\infty$ problem.*

3.3   An Efficient Updating Algorithm

Unlike the static group signature scheme from [28], our fully dynamic construction of Sect. 4 requires to regularly edit the accumulated values without having to reconstruct the whole tree. To this end, we equip the Merkle tree accumulator from [28] with a simple, yet efficient, updating algorithm: to change the value at a given leaf, we simply modify all values in the path from that leaf to the root. The algorithm, which takes as input a bit string $\mathrm{bin}(j) = (j_1, j_2, \ldots, j_\ell)$ and a value $\mathbf{d}^* \in \{0,1\}^{nk}$, is formally described below.

Given the tree in Sect. 3.2, algorithm $\mathsf{TUpdate_A}((j_1, j_2, \ldots, j_\ell), \mathbf{d}^*)$ performs the following steps:

1. Let $\mathbf{d}_j$ be the current value at the leaf of position determined by $\mathrm{bin}(j)$, and let $((j_1, \ldots, j_\ell), (\mathbf{w}_{j,\ell}, \ldots, \mathbf{w}_{j,1}))$ be its associated witness.

2. Set $\mathbf{v}_\ell := \mathbf{d}^*$ and recursively compute the path $\mathbf{v}_\ell, \mathbf{v}_{\ell-1}, \ldots, \mathbf{v}_1, \mathbf{v}_0 \in \{0,1\}^{nk}$ as in (1).

4. Set $\mathbf{u} := \mathbf{v}_0$; $\mathbf{u}_{j_1} := \mathbf{v}_1; \ldots; \mathbf{u}_{j_1,j_2,\ldots,j_{\ell-1}} := \mathbf{v}_{\ell-1}; \mathbf{u}_{j_1,j_2,\ldots,j_\ell} := \mathbf{v}_\ell = \mathbf{d}^*$.

It can be seen that the provided algorithm runs in time $\mathcal{O}(\ell) = \mathcal{O}(\log N)$. In Fig. 1, we give an illustrative example of a tree with $2^3 = 8$ leaves.

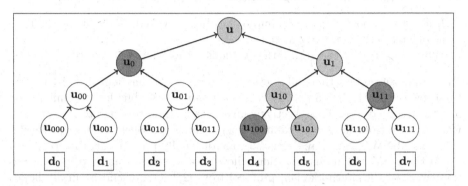

Fig. 1. A Merkle tree with $2^3 = 8$ leaves, which accumulates the data blocks $\mathbf{d}_0, \ldots, \mathbf{d}_7$ into the value $\mathbf{u}$ at the root. The bit string (101) and the pink nodes form a witness to the fact that $\mathbf{d}_5$ is accumulated in $\mathbf{u}$. If we replace $\mathbf{d}_5$ by a new value $\mathbf{d}^*$, we only need to update the yellow nodes. (Color figure online)

4 Our Fully Dynamic Group Signatures from Lattices

In this section, we construct our lattice-based fully dynamic group signature and prove its security in Bootle et al.'s model [7]. We start with the LLNW scheme [28], which works in the static setting.

While other constructions of lattice-based group signatures employ trapdoor-based ordinary signature schemes (e.g., [9,13]) to certify users, the LLNW scheme relies on a SIS-based Merkle tree accumulator which we recalled in Sect. 3.2. The GM, who manages a group of $N = 2^\ell$ users, chooses uniformly random vectors $\mathbf{x}_0, \ldots, \mathbf{x}_{N-1} \in \{0,1\}^m$; hashes them to $\mathbf{p}_0, \ldots, \mathbf{p}_{N-1} \in \{0,1\}^{nk}$, respectively; builds a tree on top of these hash values; and lets the tree root $\mathbf{u} \in \{0,1\}^{nk}$ be part of the group public key. The signing key of user i consists of $\mathbf{x}_i$ and the witness for the fact that $\mathbf{p}_i$ was accumulated in $\mathbf{u}$. When generating group signatures, the user proves knowledge of a valid pair $(\mathbf{x}_i, \mathbf{p}_i)$ and of the tree path from $\mathbf{p}_i$ to $\mathbf{u}$. The user also has to encrypt the binary representation $\text{bin}(i)$ of his identity i, and prove that the ciphertext is well-formed. The encryption layer utilizes the Naor-Yung double-encryption paradigm [38] with Regev's LWE-based cryptosystem, and thus, it is also lattice-trapdoor-free.

To upgrade the LLNW scheme directly into a fully dynamic group signature, some tweaks and new ideas are needed. First, to enable the non-frameability feature, we let the user compute the pair $(\mathbf{x}_i, \mathbf{p}_i)$ on his own. The second problem we have to consider is that Merkle hash trees seem to be a static primitive. To this end, we equip the accumulator with an efficient updating algorithm (see Sect. 3.3). Now, the challenging question is how to handle user enrollment and revocation in a simple manner (i.e., without having to reset the whole tree). To tackle these issues, we associate each of the N potential users with a leaf in the tree, and then use the updating algorithm to set up the system so that:

1. If a user has not joined the group or has been revoked, the value at the leaf associated with him is set as $\mathbf{0}$;
2. When a user joins the group, that value is set as his public key $\mathbf{p}_i$.

Our setup guarantees that only active users (i.e., who has joined and has not been revoked at the given epoch) have their *non-zero* public keys accumulated into the updated root. This effectively gives us a method to separate active users who can sign from those who cannot: when signing messages, the user proceeds as in the LLNW scheme, and is asked to additionally prove in ZK that $\mathbf{p}_i \neq \mathbf{0}$.

At this point, the arising question is how to prove the inequality $\mathbf{p}_i \neq \mathbf{0}$ in the framework of the Stern-like [45] protocol from [28]. One would naturally hope that this extra job could be done without losing too much in terms of efficiency. Here, the surprising and somewhat unexpected fact is that we can actually do it while *gaining* efficiency, thanks to a technique originally proposed in [33].

To begin with, let B_t^L denote the set of all vectors in $\{0,1\}^L$ having Hamming weight exactly t. In Stern-like protocols (see Sect. 2.3), a common technique for proving in ZK the possession of $\mathbf{p} \in \{0,1\}^{nk}$ consists of appending nk "dummy" entries to it to obtain $\mathbf{p}^* \in \mathsf{B}_{nk}^{2nk}$, and demonstrating to the verifier that a random permutation of $\mathbf{p}^*$ belongs to the "target set" B_{nk}^{2nk}. This suffices to convince the verifier that the original vector $\mathbf{p}$ belongs to $\{0,1\}^{nk}$, while the latter cannot learn any additional information about $\mathbf{p}$, thanks to the randomness of the permutation. This extending-then-permuting technique was first proposed in [33], and was extensively used in the underlying protocol of the LLNW scheme. Now, to address our question, we will employ a modified version of this technique, which was also initially suggested in [33]. Let us think of another "target set", so that it is possible to extend $\mathbf{p} \in \{0,1\}^{nk}$ to an element of that set if and only if $\mathbf{p}$ is non-zero. That set is B_{nk}^{2nk-1}. Indeed, the extended vector $\mathbf{p}^*$ belongs to B_{nk}^{2nk-1} if and only if the original vector has Hamming weight at least $nk - (nk - 1) = 1$, which means that it cannot be a zero-vector. When combining with the permuting step, this modification allows us to additionally prove the given inequality while working with smaller dimension. As a result, our fully dynamic scheme produces slightly shorter signatures than the original static scheme.

Finally, we remark that the fully dynamic setting requires a proof of correct opening, which boils down to proving correct decryption for Regev's encryption scheme. It involves modular linear equations with bounded-norm secrets, and can be easily handled using Stern-like techniques from [26,33].

4.1 Description of the Scheme

Our scheme is described as follows.

$\mathsf{GSetup}(\lambda)$. On input security parameter λ, this algorithm specifies the following:

- An expected number of potential users $N = 2^{\ell} = \mathsf{poly}(\lambda)$.
- Dimension $n = \mathcal{O}(\lambda)$, prime modulus $q = \widetilde{\mathcal{O}}(n^{1.5})$, and $k = \lceil \log_2 q \rceil$. These parameters implicitly determine the "powers-of-2" matrix $\mathbf{G} \in \mathbb{Z}_q^{n \times nk}$, as defined in Sect. 3.
- Matrix dimensions $m = 2nk$ for the hashing layer, and $m_E = 2(n+\ell)k$ for the encryption layer.
- An integer $\beta = \sqrt{n} \cdot \omega(\log n)$, and a β-bounded noise distribution χ.
- A hash function $\mathcal{H}_{\mathsf{FS}} : \{0,1\}^* \to \{1,2,3\}^{\kappa}$, where $\kappa = \omega(\log \lambda)$, to be modelled as a random oracle in the Fiat-Shamir transformations [17].
- Let $\mathsf{COM} : \{0,1\}^* \times \{0,1\}^m \to \mathbb{Z}_q^n$ be the string commitment scheme from [22], to be used in our zero-knowledge argument systems.
- Uniformly random matrix $\mathbf{A} \in \mathbb{Z}_q^{n \times m}$.

The algorithm outputs public parameters

$$\mathsf{pp} = \{\lambda, N, n, q, k, m, m_E, \ell, \beta, \chi, \kappa, \mathcal{H}_{\mathsf{FS}}, \mathsf{COM}, \mathbf{A}\}.$$

$\langle \mathsf{GKgen}_{\mathsf{GM}}(\mathsf{pp}), \mathsf{GKgen}_{\mathsf{TM}}(\mathsf{pp}) \rangle$. The group manager GM and the tracing manager TM initialize their keys and the public group information as follows.

- $\mathsf{GKgen}_{\mathsf{GM}}(\mathsf{pp})$. This algorithm samples $\mathsf{msk} \xleftarrow{\$} \{0,1\}^m$ and computes $\mathsf{mpk} = \mathbf{A} \cdot \mathsf{msk} \bmod q$, and outputs $(\mathsf{mpk}, \mathsf{msk})$. Here, we consider mpk as an identifier of the group managed by GM who has mpk as his public key. Furthermore, as in [7, Sect. 3.3, full version], we assume that the group information board is visible to everyone, but can only be edited by a party knowing msk.

- $\mathsf{GKgen}_{\mathsf{TM}}(\mathsf{pp})$. This algorithm initializes the Naor-Yung double-encryption mechanism with the ℓ-bit version Regev encryption scheme. It first chooses $\mathbf{B} \xleftarrow{\$} \mathbb{Z}_q^{n \times m_E}$. For each $i \in \{1,2\}$, it samples $\mathbf{S}_i \xleftarrow{\$} \chi^{n \times \ell}$, $\mathbf{E}_i \hookleftarrow \chi^{\ell \times m_E}$, and computes $\mathbf{P}_i = \mathbf{S}_i^{\top} \cdot \mathbf{B} + \mathbf{E}_i \in \mathbb{Z}_q^{\ell \times m_E}$. Then, TM sets $\mathsf{tsk} = (\mathbf{S}_1, \mathbf{E}_1)$, and $\mathsf{tpk} = (\mathbf{B}, \mathbf{P}_1, \mathbf{P}_2)$.

- TM sends tpk to GM who initializes the following:
  - Table $\mathbf{reg} := (\mathbf{reg}[0][1], \mathbf{reg}[0][2], \ldots, \mathbf{reg}[N-1][1], \mathbf{reg}[N-1][2])$, where for each $i \in [0, N-1]$: $\mathbf{reg}[i][1] = \mathbf{0}^{nk}$ and $\mathbf{reg}[i][2] = 0$. Looking ahead, $\mathbf{reg}[i][1]$ will be used to record the public key of a registered user, while $\mathbf{reg}[i][2]$ stores the epoch at which the user joins.
  - The Merkle tree $\mathcal{T}$ built on top of $\mathbf{reg}[0][1], \ldots, \mathbf{reg}[N-1][1]$. (Note that $\mathcal{T}$ is an all-zero tree at this stage, but it will be modified when a new user joins the group, or when GM computes the updated group information.)
  - Counter of registered users $c := 0$.

  Then, GM outputs $\mathsf{gpk} = (\mathsf{pp}, \mathsf{mpk}, \mathsf{tpk})$ and announces the initial group information $\mathsf{info} = \emptyset$. He keeps $\mathcal{T}$ and c for himself.

UKgen(pp). Each potential group user samples $\mathsf{usk} = \mathbf{x} \xleftarrow{\$} \{0,1\}^m$, and computes $\mathsf{upk} = \mathbf{p} = \mathsf{bin}(\mathbf{A} \cdot \mathbf{x}) \bmod q \in \{0,1\}^{nk}$.

Without loss of generality, we assume that every honestly generated upk is a non-zero vector. (Otherwise, the user would either pick $\mathbf{x} = \mathbf{0}^m$ or accidentally find a solution to the $\mathsf{SIS}^\infty_{n,m,q,1}$ problem associated with matrix $\mathbf{A}$ - both happen with negligible probability.)

⟨Join, Issue⟩. If the user with key pair $(\mathsf{upk}, \mathsf{usk}) = (\mathbf{p}, \mathbf{x})$ requests to join the group at epoch τ, he sends $\mathbf{p}$ to GM. If the latter accepts the request, then the two parties proceed as follows.

1. GM issues a member identifier for the user as $\mathsf{uid} = \mathsf{bin}(c) \in \{0,1\}^\ell$. The user then sets his long-term signing key as $\mathsf{gsk}[c] = (\mathsf{bin}(c), \mathbf{p}, \mathbf{x})$.
2. GM performs the following updates:
   - Update $\mathcal{T}$ by running algorithm $\mathsf{TUpdate}_{\mathbf{A}}(\mathsf{bin}(c), \mathbf{p})$.
   - Register the user to table $\mathbf{reg}$ as $\mathbf{reg}[c][1] := \mathbf{p}$; $\mathbf{reg}[c][2] := \tau$.
   - Increase the counter $c := c + 1$.

GUpdate(gpk, msk, info$_{\tau_{\text{current}}}$, $\mathcal{S}$, reg). This algorithm is run by GM to update the group information while also advancing the epoch. It works as follows.

1. Let the set $\mathcal{S}$ contain the public keys of registered users to be revoked. If $\mathcal{S} = \emptyset$, then go to Step 2.
   Otherwise, $\mathcal{S} = \{\mathbf{reg}[i_1][1], \ldots, \mathbf{reg}[i_r][1]\}$, for some $r \in [1, N]$ and some $i_1, \ldots, i_r \in [0, N-1]$. Then, for all $t \in [r]$, GM runs $\mathsf{TUpdate}_{\mathbf{A}}(\mathsf{bin}(i_t), \mathbf{0}^{nk})$ to update the tree $\mathcal{T}$.
2. At this point, by construction, each of the zero leaves in the tree $\mathcal{T}$ corresponds to either a revoked user or a potential user who has not yet registered. In other words, only active users who are allowed to sign in the new epoch τ_{new} have their non-zero public keys, denoted by $\{\mathbf{p}_j\}_j$, accumulated in the root $\mathbf{u}_{\tau_{\text{new}}}$ of the updated tree.
   For each j, let $w_j \in \{0,1\}^\ell \times (\{0,1\}^{nk})^\ell$ be the witness for the fact that $\mathbf{p}_j$ is accumulated in $\mathbf{u}_{\tau_{\text{new}}}$. Then GM publishes the group information of the new epoch as:
   $$\mathsf{info}_{\tau_{\text{new}}} = (\mathbf{u}_{\tau_{\text{new}}}, \{w_j\}_j).$$

We remark that the info$_\tau$ outputted at each epoch by GM is technically not part of the verification key. Indeed, as we will describe below, in order to verify signatures bound to epoch τ, the verifiers only need to download the first component $\mathbf{u}_\tau$ of size $\tilde{\mathcal{O}}(\lambda)$ bits. Meanwhile, each active signer only has to download the respective witness of size $\tilde{\mathcal{O}}(\lambda) \cdot \ell$.

Sign(gpk, gsk[j], info$_\tau$, M). To sign message M using the group information info$_\tau$ at epoch τ, the user possessing $\mathsf{gsk}[j] = (\mathsf{bin}(j), \mathbf{p}, \mathbf{x})$ first checks if info$_\tau$ includes a witness containing $\mathsf{bin}(j)$. If this is not the case, return $\bot$. Otherwise, the user downloads $\mathbf{u}_\tau$ and the witness of the form $(\mathsf{bin}(j), (\mathbf{w}_\ell, \ldots, \mathbf{w}_1))$ from info$_\tau$, and proceeds as follows.

1. Encrypt vector $\mathsf{bin}(j) \in \{0,1\}^\ell$ twice using Regev's encryption scheme. Namely, for each $i \in \{1, 2\}$, sample $\mathbf{r}_i \xleftarrow{\$} \{0,1\}^{m_E}$ and compute
$$\begin{aligned} \mathbf{c}_i &= (\mathbf{c}_{i,1}, \mathbf{c}_{i,2}) \\ &= \left(\mathbf{B} \cdot \mathbf{r}_i \bmod q, \; \mathbf{P}_i \cdot \mathbf{r}_i + \left\lceil \frac{q}{2} \right\rfloor \cdot \mathsf{bin}(j) \bmod q \right) \in \mathbb{Z}_q^n \times \mathbb{Z}_q^\ell. \end{aligned}$$

2. Generate a NIZKAoK Π_{gs} to demonstrate the possession of a valid tuple

$$\zeta = (\mathbf{x}, \mathbf{p}, \mathsf{bin}(j), \mathbf{w}_\ell, \ldots, \mathbf{w}_1, \mathbf{r}_1, \mathbf{r}_2) \tag{2}$$

such that:
   (i) $\mathsf{TVerify}_\mathbf{A}\left(\mathbf{u}_\tau, \mathbf{p}, \left(\mathsf{bin}(j), (\mathbf{w}_\ell, \ldots, \mathbf{w}_1)\right)\right) = 1$ and $\mathbf{A} \cdot \mathbf{x} = \mathbf{G} \cdot \mathbf{p} \bmod q$;
   (ii) $\mathbf{c}_1$ and $\mathbf{c}_2$ are both correct encryptions of $\mathsf{bin}(j)$ with randomness $\mathbf{r}_1$ and $\mathbf{r}_2$, respectively;
   (iii) $\mathbf{p} \neq \mathbf{0}^{nk}$.
   Note that statements (i) and (ii) were covered by the LLNW protocol [28]. Meanwhile, statement (iii) is handled using the technique described at the beginning of this Section. We thus obtain a Stern-like interactive zero-knowledge argument system which is a slight modification of the one from [28]. Due to space restriction, the details are presented in the full version.
   The protocol is repeated $\kappa = \omega(\log \lambda)$ times to achieve negligible soundness error and made non-interactive via the Fiat-Shamir heuristic as a triple $\Pi_{\mathsf{gs}} = (\{\mathrm{CMT}_i\}_{i=1}^\kappa, \mathrm{CH}, \{\mathrm{RSP}\}_{i=1}^\kappa)$, where

$$\mathrm{CH} = \mathcal{H}_{\mathsf{FS}}\left(M, (\{\mathrm{CMT}_i\}_{i=1}^\kappa, \mathbf{A}, \mathbf{u}_\tau, \mathbf{B}, \mathbf{P}_1, \mathbf{P}_2, \mathbf{c}_1, \mathbf{c}_2\right) \in \{1, 2, 3\}^\kappa.$$

3. Output the group signature

$$\Sigma = (\Pi_{\mathsf{gs}}, \mathbf{c}_1, \mathbf{c}_2). \tag{3}$$

$\mathsf{Verify}(\mathsf{gpk}, \mathsf{info}_\tau, M, \Sigma)$. This algorithm proceeds as follows:
   1. Download $\mathbf{u}_\tau \in \{0, 1\}^{nk}$ from info_τ.
   2. Parse Σ as $\Sigma = \left(\{\mathrm{CMT}_i\}_{i=1}^\kappa, (Ch_1, \ldots, Ch_\kappa), \{\mathrm{RSP}\}_{i=1}^\kappa, \mathbf{c}_1, \mathbf{c}_2\right)$. If $(Ch_1, \ldots, Ch_\kappa) \neq \mathcal{H}_{\mathsf{FS}}\left(M, (\{\mathrm{CMT}_i\}_{i=1}^\kappa, \mathbf{A}, \mathbf{u}_\tau, \mathbf{B}, \mathbf{P}_1, \mathbf{P}_2, \mathbf{c}_1, \mathbf{c}_2\right)$, then return 0.
   3. For each $i = 1$ to κ, run the verification phase of the protocol presented in the full version to check the validity of RSP_i with respect to CMT_i and Ch_i. If any of the conditions does not hold, then return 0.
   4. Return 1.

$\mathsf{Trace}(\mathsf{gpk}, \mathsf{tsk}, \mathsf{info}_\tau, \mathbf{reg}, M, \Sigma)$. This algorithm parses tsk as $(\mathbf{S}_1, \mathbf{E}_1)$, parses Σ as in (3), and performs the following steps.
   1. Use $\mathbf{S}_1$ to decrypt $\mathbf{c}_1 = (\mathbf{c}_{1,1}, \mathbf{c}_{1,2})$ to obtain a string $\mathbf{b}' \in \{0, 1\}^\ell$ (i.e., by computing $\lfloor (\mathbf{c}_{1,2} - \mathbf{S}_1^\top \cdot \mathbf{c}_{1,1})/(q/2) \rceil$.
   2. If info_τ does not include a witness containing $\mathbf{b}'$, then return $\perp$.
   3. Let $j' \in [0, N-1]$ be the integer having binary representation $\mathbf{b}'$. If the record $\mathbf{reg}[j'][1]$ in table $\mathbf{reg}$ is $\mathbf{0}^{nk}$, then return $\perp$.
   4. Generate a NIZKAoK Π_{trace} to demonstrate the possession of $\mathbf{S}_1 \in \mathbb{Z}^{n \times \ell}$, $\mathbf{E}_1 \in \mathbb{Z}^{\ell \times m_E}$, and $\mathbf{y} \in \mathbb{Z}^\ell$, such that:

$$\begin{cases} \|\mathbf{S}_1\|_\infty \leq \beta; \ \|\mathbf{E}_1\|_\infty \leq \beta; \ \|\mathbf{y}\|_\infty \leq \lceil q/5 \rceil; \\ \mathbf{S}_1^\top \cdot \mathbf{B} + \mathbf{E}_1 = \mathbf{P}_1 \bmod q; \\ \mathbf{c}_{1,2} - \mathbf{S}_1^\top \cdot \mathbf{c}_{1,1} = \mathbf{y} + \lfloor q/2 \rfloor \cdot \mathbf{b}' \bmod q. \end{cases} \tag{4}$$

As the statement involves modular linear equations with bounded-norm secrets, we can obtain a statistical zero-knowledge argument by employing the Stern-like interactive protocol from [26]. The protocol is repeated $\kappa = \omega(\log \lambda)$ times to achieve negligible soundness error and made non-interactive via the Fiat-Shamir heuristic as a triple $\Pi_{\text{trace}} = (\{\text{CMT}_i\}_{i=1}^\kappa, \text{CH}, \{\text{RSP}\}_{i=1}^\kappa)$, where

$$\text{CH} = \mathcal{H}_{\text{FS}}((\{\text{CMT}_i\}_{i=1}^\kappa, \text{gpk}, \text{info}_\tau, M, \Sigma, \mathbf{b}') \in \{1, 2, 3\}^\kappa. \qquad (5)$$

5. Set $\text{uid} = \mathbf{b}'$ and output $(\text{uid}, \Pi_{\text{trace}})$.

Judge$(\text{gpk}, \text{uid}, \text{info}_\tau, \Pi_{\text{trace}}, M, \Sigma)$. This algorithm consists of verifying the argument Π_{trace} w.r.t. common input $(\text{gpk}, \text{info}_\tau, M, \Sigma, \text{uid})$, in a similar manner as in algorithm Verify.

If Π_{trace} does not verify, return 0. Otherwise, return 1.

4.2   Analysis of the Scheme

EFFICIENCY. We first analyze the efficiency of the scheme described in Sect. 4.1, with respect to security parameter λ and parameter $\ell = \log N$.

- The public key gpk contains several matrices, and has bit-size $\widetilde{\mathcal{O}}(\lambda^2 + \lambda \cdot \ell)$.
- For each $j \in [0, N-1]$, the signing key gsk[j] has bit-size $\ell + nk + m = \widetilde{\mathcal{O}}(\lambda) + \ell$.
- At each epoch, the signature verifiers downloads $nk = \widetilde{\mathcal{O}}(\lambda)$ bits, while each active signer downloads $\widetilde{\mathcal{O}}(\lambda \cdot \ell)$ bits.
- The size of signature Σ is dominated by that of the Stern-like NIZKAoK Π_{gs}, which is $\mathcal{O}(|\zeta| \cdot \log q) \cdot \omega(\log \lambda)$, where $|\zeta|$ denotes the bit-size of the witness-tuple ζ in (2). Overall, Σ has bit-size $\widetilde{\mathcal{O}}(\lambda \cdot \ell)$.
- The Stern-like NIZKAoK Π_{trace} has bit-size $\widetilde{\mathcal{O}}(\ell^2 + \lambda \cdot \ell)$.

CORRECTNESS. We now demonstrate that the scheme is correct with overwhelming probability, based on the perfect completeness of Stern-like protocols, and the correctness of Regev's encryption scheme.

First, note that a signature $\Sigma = (\Pi_{\text{gs}}, \mathbf{c}_1, \mathbf{c}_2)$ generated by an active and honest user j is always accepted by algorithm Verify. Indeed, such a user can always compute a tuple $\zeta = (\mathbf{x}, \mathbf{p}, \text{bin}(j), \mathbf{w}_\ell, \ldots, \mathbf{w}_1, \mathbf{r}_1, \mathbf{r}_2)$ satisfying conditions (i),(ii) and (iii) in the Sign algorithm. The completeness of the underlying argument system then guarantees that Σ is always accepted by algorithm Verify.

Next, we show that algorithm Trace outputs $\text{bin}(j)$ with overwhelming probability, and produces a proof Π_{trace} accepted by algorithm Judge. Observe that, the decryption algorithm essentially computes

$$\mathbf{e} = \mathbf{c}_{1,2} - \mathbf{S}_1^T \mathbf{c}_{1,1} = \mathbf{E}_1 \cdot \mathbf{r}_1 + \lfloor q/2 \rceil \cdot \text{bin}(j) \mod q,$$

and sets the j-th bit of $\mathbf{b}'$ to be 0 if j-th entry of $\mathbf{e}$ is closer to 0 than to $\lfloor q/2 \rceil$ and 1 otherwise. Note that our parameters are set so that $\|\mathbf{E}_1 \cdot \mathbf{r}_1\|_\infty < q/5$, for $\mathbf{E}_1 \hookleftarrow \chi^{\ell \times m_E}$ and $\mathbf{r}_1 \overset{\$}{\leftarrow} \{0, 1\}^{m_E}$. This ensures that $\mathbf{b}' = \text{bin}(j)$ with overwhelming probability.

Further, as the user is active, info_τ must contain $w = (\mathsf{bin}(j), \mathbf{w}_\ell, \dots, \mathbf{w}_1)$ and $\mathbf{reg}[j][1]$ in table $\mathbf{reg}$ is not $\mathbf{0}^{nk}$. Therefore, algorithm Trace will move to the 4-th step, where it can always obtain the tuple $(\mathbf{S}_1, \mathbf{E}_1, \mathbf{y})$ satisfying the conditions (4). By the completeness of the argument system, Π_{trace} will be accepted by the algorithm Judge.

SECURITY. In Theorem 1, we prove that our scheme satisfies the security requirements of the Bootle et al.'s model [7].

Theorem 1. *Assume that the Stern-like argument systems used in Sect. 4.1 are simulation-sound. Then, in the random oracle model, the given fully dynamic group signature satisfies the anonymity, traceability, non-frameability and tracing soundness requirements under the* $\mathsf{LWE}_{n,q,\chi}$ *and* $\mathsf{SIS}^{\infty}_{n,m,q,1}$ *assumptions.*

In the random oracle model, the proof of Theorem 1 relies on the following facts:

1. The Stern-like zero-knowledge argument systems being used are simulation-sound;
2. The underlying encryption scheme, which is obtained from Regev cryptosystem [42] via the Naor-Yung transformation [38], is IND-CCA2 secure;
3. The Merkle tree accumulator we employ is secure in the sense of Definition 3;
4. For a properly generated key-pair $(\mathbf{x}, \mathbf{p})$, it is infeasible to find $\mathbf{x}' \in \{0,1\}^m$ such that $\mathbf{x}' \neq \mathbf{x}$ and $\mathsf{bin}(\mathbf{A} \cdot \mathbf{x}' \bmod q) = \mathbf{p}$.

Due to space restriction, details of the proof of Theorem 1 are provided in the full version of the paper.

Acknowledgements. The authors would like to thank Benoît Libert and the anonymous reviewers for helpful comments and discussions. The research was supported by Research Grant TL-9014101684-01 and the Singapore Ministry of Education under Research Grant MOE2013-T2-1-041.

References

1. Ajtai, M.: Generating hard instances of lattice problems (extended abstract). In: STOC 1996, pp. 99–108. ACM (1996)
2. Ateniese, G., Camenisch, J., Joye, M., Tsudik, G.: A practical and provably secure coalition-resistant group signature scheme. In: Bellare, M. (ed.) CRYPTO 2000. LNCS, vol. 1880, pp. 255–270. Springer, Heidelberg (2000). doi:10.1007/3-540-44598-6_16
3. Barić, N., Pfitzmann, B.: Collision-free accumulators and fail-stop signature schemes without trees. In: Fumy, W. (ed.) EUROCRYPT 1997. LNCS, vol. 1233, pp. 480–494. Springer, Heidelberg (1997). doi:10.1007/3-540-69053-0_33
4. Bellare, M., Micciancio, D., Warinschi, B.: Foundations of group signatures: formal definitions, simplified requirements, and a construction based on general assumptions. In: Biham, E. (ed.) EUROCRYPT 2003. LNCS, vol. 2656, pp. 614–629. Springer, Heidelberg (2003). doi:10.1007/3-540-39200-9_38

5. Bellare, M., Shi, H., Zhang, C.: Foundations of group signatures: the case of dynamic groups. In: Menezes, A. (ed.) CT-RSA 2005. LNCS, vol. 3376, pp. 136–153. Springer, Heidelberg (2005). doi:10.1007/978-3-540-30574-3_11

6. Boneh, D., Shacham, H.: Group signatures with verifier-local revocation. In: ACM CCCS 2004, pp. 168–177. ACM (2004)

7. Bootle, J., Cerulli, A., Chaidos, P., Ghadafi, E., Groth, J.: Foundations of fully dynamic group signatures. In: Manulis, M., Sadeghi, A.-R., Schneider, S. (eds.) ACNS 2016. LNCS, vol. 9696, pp. 117–136. Springer, Cham (2016). doi:10.1007/978-3-319-39555-5_7. Full version: https://eprint.iacr.org/2016/368.pdf

8. Bootle, J., Cerulli, A., Chaidos, P., Ghadafi, E., Groth, J., Petit, C.: Short accountable ring signatures based on DDH. In: Pernul, G., Ryan, P.Y.A., Weippl, E. (eds.) ESORICS 2015. LNCS, vol. 9326, pp. 243–265. Springer, Cham (2015). doi:10.1007/978-3-319-24174-6_13

9. Boyen, X.: Lattice mixing and vanishing trapdoors: a framework for fully secure short signatures and more. In: Nguyen, P.Q., Pointcheval, D. (eds.) PKC 2010. LNCS, vol. 6056, pp. 499–517. Springer, Heidelberg (2010). doi:10.1007/978-3-642-13013-7_29

10. Bresson, E., Stern, J.: Efficient revocation in group signatures. In: Kim, K. (ed.) PKC 2001. LNCS, vol. 1992, pp. 190–206. Springer, Heidelberg (2001). doi:10.1007/3-540-44586-2_15

11. Camenisch, J., Lysyanskaya, A.: Dynamic accumulators and application to efficient revocation of anonymous credentials. In: Yung, M. (ed.) CRYPTO 2002. LNCS, vol. 2442, pp. 61–76. Springer, Heidelberg (2002). doi:10.1007/3-540-45708-9_5

12. Camenisch, J., Neven, G., Rückert, M.: Fully anonymous attribute tokens from lattices. In: Visconti, I., Prisco, R. (eds.) SCN 2012. LNCS, vol. 7485, pp. 57–75. Springer, Heidelberg (2012). doi:10.1007/978-3-642-32928-9_4

13. Cash, D., Hofheinz, D., Kiltz, E., Peikert, C.: Bonsai trees, or how to delegate a lattice basis. In: Gilbert, H. (ed.) EUROCRYPT 2010. LNCS, vol. 6110, pp. 523–552. Springer, Heidelberg (2010). doi:10.1007/978-3-642-13190-5_27

14. Chaum, D., Heyst, E.: Group signatures. In: Davies, D.W. (ed.) EUROCRYPT 1991. LNCS, vol. 547, pp. 257–265. Springer, Heidelberg (1991). doi:10.1007/3-540-46416-6_22

15. Cheng, S., Nguyen, K., Wang, H.: Policy-based signature scheme from lattices. Des. Codes Cryptogr. **81**(1), 43–74 (2016)

16. Delerablée, C., Pointcheval, D.: Dynamic fully anonymous short group signatures. In: Nguyen, P.Q. (ed.) VIETCRYPT 2006. LNCS, vol. 4341, pp. 193–210. Springer, Heidelberg (2006). doi:10.1007/11958239_13

17. Fiat, A., Shamir, A.: How To prove yourself: practical solutions to identification and signature problems. In: Odlyzko, A.M. (ed.) CRYPTO 1986. LNCS, vol. 263, pp. 186–194. Springer, Heidelberg (1987). doi:10.1007/3-540-47721-7_12

18. Gentry, C., Peikert, C., Vaikuntanathan, V.: Trapdoors for hard lattices and new cryptographic constructions. In: STOC 2008, pp. 197–206. ACM (2008)

19. Gordon, S.D., Katz, J., Vaikuntanathan, V.: A group signature scheme from lattice assumptions. In: Abe, M. (ed.) ASIACRYPT 2010. LNCS, vol. 6477, pp. 395–412. Springer, Heidelberg (2010). doi:10.1007/978-3-642-17373-8_23

20. Groth, J.: Fully anonymous group signatures without random oracles. In: Kurosawa, K. (ed.) ASIACRYPT 2007. LNCS, vol. 4833, pp. 164–180. Springer, Heidelberg (2007). doi:10.1007/978-3-540-76900-2_10

21. Kawachi, A., Tanaka, K., Xagawa, K.: Multi-bit cryptosystems based on lattice problems. In: Okamoto, T., Wang, X. (eds.) PKC 2007. LNCS, vol. 4450, pp. 315–329. Springer, Heidelberg (2007). doi:10.1007/978-3-540-71677-8_21

22. Kawachi, A., Tanaka, K., Xagawa, K.: Concurrently secure identification schemes based on the worst-case hardness of lattice problems. In: Pieprzyk, J. (ed.) ASI-ACRYPT 2008. LNCS, vol. 5350, pp. 372–389. Springer, Heidelberg (2008). doi:10.1007/978-3-540-89255-7_23

23. Kiayias, A., Yung, M.: Secure scalable group signature with dynamic joins and separable authorities. Int. J. Secur. Netw. 1(1), 24–45 (2006)

24. Laguillaumie, F., Langlois, A., Libert, B., Stehlé, D.: Lattice-based group signatures with logarithmic signature size. In: Sako, K., Sarkar, P. (eds.) ASI-ACRYPT 2013. LNCS, vol. 8270, pp. 41–61. Springer, Heidelberg (2013). doi:10.1007/978-3-642-42045-0_3

25. Langlois, A., Ling, S., Nguyen, K., Wang, H.: Lattice-based group signature scheme with verifier-local revocation. In: Krawczyk, H. (ed.) PKC 2014. LNCS, vol. 8383, pp. 345–361. Springer, Heidelberg (2014). doi:10.1007/978-3-642-54631-0_20

26. Libert, B., Ling, S., Mouhartem, F., Nguyen, K., Wang, H.: Signature schemes with efficient protocols and dynamic group signatures from lattice assumptions. In: Cheon, J.H., Takagi, T. (eds.) ASIACRYPT 2016. LNCS, vol. 10032, pp. 373–403. Springer, Heidelberg (2016). doi:10.1007/978-3-662-53890-6_13

27. Libert, B., Ling, S., Mouhartem, F., Nguyen, K., Wang, H.: Zero-knowledge arguments for matrix-vector relations and lattice-based group encryption. In: Cheon, J.H., Takagi, T. (eds.) ASIACRYPT 2016. LNCS, vol. 10032, pp. 101–131. Springer, Heidelberg (2016). doi:10.1007/978-3-662-53890-6_4

28. Libert, B., Ling, S., Nguyen, K., Wang, H.: Zero-knowledge arguments for lattice-based accumulators: logarithmic-size ring signatures and group signatures without trapdoors. In: Fischlin, M., Coron, J.-S. (eds.) EUROCRYPT 2016. LNCS, vol. 9666, pp. 1–31. Springer, Heidelberg (2016). doi:10.1007/978-3-662-49896-5_1

29. Libert, B., Mouhartem, F., Nguyen, K.: A lattice-based group signature scheme with message-dependent opening. In: Manulis, M., Sadeghi, A.-R., Schneider, S. (eds.) ACNS 2016. LNCS, vol. 9696, pp. 137–155. Springer, Cham (2016). doi:10.1007/978-3-319-39555-5_8

30. Libert, B., Peters, T., Yung, M.: Group signatures with almost-for-free revocation. In: Safavi-Naini, R., Canetti, R. (eds.) CRYPTO 2012. LNCS, vol. 7417, pp. 571–589. Springer, Heidelberg (2012). doi:10.1007/978-3-642-32009-5_34

31. Libert, B., Peters, T., Yung, M.: Scalable group signatures with revocation. In: Pointcheval, D., Johansson, T. (eds.) EUROCRYPT 2012. LNCS, vol. 7237, pp. 609–627. Springer, Heidelberg (2012). doi:10.1007/978-3-642-29011-4_36

32. Libert, B., Peters, T., Yung, M.: Short group signatures via structure-preserving signatures: standard model security from simple assumptions. In: Gennaro, R., Robshaw, M. (eds.) CRYPTO 2015. LNCS, vol. 9216, pp. 296–316. Springer, Heidelberg (2015). doi:10.1007/978-3-662-48000-7_15

33. Ling, S., Nguyen, K., Stehlé, D., Wang, H.: Improved zero-knowledge proofs of knowledge for the ISIS problem, and applications. In: Kurosawa, K., Hanaoka, G. (eds.) PKC 2013. LNCS, vol. 7778, pp. 107–124. Springer, Heidelberg (2013). doi:10.1007/978-3-642-36362-7_8

34. Ling, S., Nguyen, K., Wang, H.: Group signatures from lattices: simpler, tighter, shorter, ring-based. In: Katz, J. (ed.) PKC 2015. LNCS, vol. 9020, pp. 427–449. Springer, Heidelberg (2015). doi:10.1007/978-3-662-46447-2_19

35. Micciancio, D., Mol, P.: Pseudorandom knapsacks and the sample complexity of LWE search-to-decision reductions. In: Rogaway, P. (ed.) CRYPTO 2011. LNCS, vol. 6841, pp. 465–484. Springer, Heidelberg (2011). doi:10.1007/978-3-642-22792-9_26

36. Micciancio, D., Peikert, C.: Hardness of SIS and LWE with small parameters. In: Canetti, R., Garay, J.A. (eds.) CRYPTO 2013. LNCS, vol. 8042, pp. 21–39. Springer, Heidelberg (2013). doi:10.1007/978-3-642-40041-4_2
37. Nakanishi, T., Fujii, H., Hira, Y., Funabiki, N.: Revocable group signature schemes with constant costs for signing and verifying. In: Jarecki, S., Tsudik, G. (eds.) PKC 2009. LNCS, vol. 5443, pp. 463–480. Springer, Heidelberg (2009). doi:10.1007/978-3-642-00468-1_26
38. Naor, M., Yung, M.: Public-key cryptosystems provably secure against chosen ciphertext attacks. In: STOC 1990, pp. 427–437. ACM (1990)
39. Nguyen, L.: Accumulators from bilinear pairings and applications. In: Menezes, A. (ed.) CT-RSA 2005. LNCS, vol. 3376, pp. 275–292. Springer, Heidelberg (2005). doi:10.1007/978-3-540-30574-3_19
40. Nguyen, L., Safavi-Naini, R.: Efficient and provably secure trapdoor-free group signature schemes from bilinear pairings. In: Lee, P.J. (ed.) ASIACRYPT 2004. LNCS, vol. 3329, pp. 372–386. Springer, Heidelberg (2004). doi:10.1007/978-3-540-30539-2_26
41. Nguyen, P.Q., Zhang, J., Zhang, Z.: Simpler efficient group signatures from lattices. In: Katz, J. (ed.) PKC 2015. LNCS, vol. 9020, pp. 401–426. Springer, Heidelberg (2015). doi:10.1007/978-3-662-46447-2_18
42. Regev, O.: On lattices, learning with errors, random linear codes, and cryptography. In: STOC 2005, pp. 84–93. ACM (2005)
43. Sakai, Y., Emura, K., Hanaoka, G., Kawai, Y., Matsuda, T., Omote, K.: Group signatures with message-dependent opening. In: Abdalla, M., Lange, T. (eds.) Pairing 2012. LNCS, vol. 7708, pp. 270–294. Springer, Heidelberg (2013). doi:10.1007/978-3-642-36334-4_18
44. Sakai, Y., Schuldt, J.C.N., Emura, K., Hanaoka, G., Ohta, K.: On the security of dynamic group signatures: preventing signature hijacking. In: Fischlin, M., Buchmann, J., Manulis, M. (eds.) PKC 2012. LNCS, vol. 7293, pp. 715–732. Springer, Heidelberg (2012). doi:10.1007/978-3-642-30057-8_42
45. Stern, J.: A new paradigm for public key identification. IEEE Trans. Inf. Theory 42(6), 1757–1768 (1996)
46. Tsudik, G., Xu, S.: Accumulating composites and improved group signing. In: Laih, C.-S. (ed.) ASIACRYPT 2003. LNCS, vol. 2894, pp. 269–286. Springer, Heidelberg (2003). doi:10.1007/978-3-540-40061-5_16

Breaking and Fixing Mobile App Authentication with OAuth2.0-based Protocols

Ronghai Yang[✉], Wing Cheong Lau, and Shangcheng Shi

Department of Information Engineering, The Chinese University of Hong Kong,
Hong Kong, China
{yr013,wclau,ss016}@ie.cuhk.edu.hk

Abstract. Although the OAuth2.0 protocol was originally designed to serve the authorization need for websites, mainstream identity providers like Google and Facebook have made significant changes on this protocol to support authentication for mobile apps. Prior research mainly focuses on how the features of mobile operating systems can affect the OAuth security. However, little has been done to analyze whether these significant modifications of the protocol call-flow can be well understood and implemented by app developers. Towards this end, we report a field-study on the Android OAuth2.0-based single-sign-on systems. In particular, we perform an in-depth static code analysis on three identity provider apps including Facebook, Google and Sina as well as their official SDKs to understand their OAuth-related transactions. We then dynamically test 600 top-ranked US and Chinese Android apps. Apart from various types of existing vulnerabilities, we also discover three previously unknown security flaws among these first-tier identity providers and a large number of popular 3rd-party apps. For example, 41% apps under study are susceptible to a newly discovered profile attack, which unlike prior works, enables remote account hijacking without any need to trick or interact with the victim. The prevalence of vulnerabilities further motivates us to propose/implement an alternative, fool-proof OAuth SDK for one of the affected IdPs to automatically prevent from these vulnerabilities. To facilitate the adoption of our proposed fixes, our solution requires minimal code changes by the 3rd-party-developers of the affected mobile apps.

Keywords: OAuth2.0 · OpenID Connect · Mobile app authentication

1 Introduction

The OAuth2.0 protocol was originally designed to serve the authorization need for 3rd-party websites. However, many major Identity Providers (IdPs) such as Facebook, Google and Sina, have recently adapted the OAuth2.0-based protocols to support Single-Sign-On (SSO) services for 3rd-party mobile apps (which take the role of Relying Party under the context of OAuth2.0). When OAuth2.0 is used as a SSO scheme, a user can log into the mobile Relying Party (RP), *e.g.,*

© Springer International Publishing AG 2017
D. Gollmann et al. (Eds.): ACNS 2017, LNCS 10355, pp. 313–335, 2017.
DOI: 10.1007/978-3-319-61204-1_16

IMDB, via the IdP without sharing the identity credential with the RP. In this paper, we will focus on OAuth2.0 and OpenID Connect (which is built on top of OAuth2.0), since they are the *de facto* SSO standards.

To support SSO services with 3rd-party mobile apps, the security of the adapted OAuth2.0 protocol has been evaluated by the literature. Chen *et al.* [12] first point out that some operating-system-provided components (*e.g.*, Intent/WebView for Android) are required to implement OAuth2.0 for mobile platforms. As further shown by Chen, the features of these components, if not well understood by mobile app developers, can be leveraged to compromise mobile SSO systems. Following this work, Ye *et al.* [34] apply model checking method to theoretically evaluate this modified protocol and Wang *et al.* [28,29] summarize those known vulnerabilities among 15 Chinese IdPs over different platforms.

Prior studies mainly focus on how the differences of mobile systems (*i.e.*, vulnerabilities in the system-provided components) can compromise mobile SSO systems. However, the security implications resulting from the major protocol changes, when adapting the OAuth2.0-based protocols to mobile platforms, are often left out. For example, the standard OAuth2.0 implicit flow has shown to be insecure for authentication and thus a revised version, *i.e.*, a variant of OpenID Connect (OIDC), is recommended. Nevertheless, they do not consider that the SSO results, even in the revised version, are passed through the user device. Consequently, these security-critical results are subject to tampering and further enable an adversary to infer the program logic of the RP server.

Towards this end, the goal of this research is to further the understanding of (1) how the changes of OAuth[1] protocol flow, if not well implemented, can lead to nontrivial security flaws, (2) the overall security quality of mobile SSO systems by checking whether existing vulnerabilities have been fixed or not. Our work consists of two pillars: (1) We perform a standard static code analysis of three first-tier Android IdP apps (Facebook, Google and Sina) and their corresponding SDKs widely used by the RP to understand their client-side program logic. (2) We develop a tool to dynamically test Top-600 US and Chinese Android apps to see how well the RP/IdP servers perform the OAuth transactions. From these studies, we have made the following technical contributions in this paper:

- We have identified the security-critical changes of the OAuth protocol call-flow.
- We have examined the implementations of 3 first-tier IdPs and 600 top-ranked US/Chinese Android Apps. In addition to different types of existing vulnerabilities, we have discovered three previously unknown vulnerabilities resulting from the incorrect implementations of the protocol call-flow modification.
- We have designed and implemented a foolproof solution which prevents future 3rd party app developers from committing the same mistakes. To ease the transition, our solution only requires minimal changes in the 3rd-party developed codes of a vulnerable mobile app.

---

[1] We use OAuth to denote OAuth2.0 and OpenID Connect, if not specified otherwise.

2 Background

In the OAuth ecosystem, four parties are involved to support SSO for 3rd-party mobile apps, namely, the backend server of the 3rd-party mobile app (RP server, for short), the backend server of the IdP (IdP server), the 3rd-party client-side mobile app (RP app) and the client-side mobile app of the IdP (IdP app). For ease of presentation, in the rest of this paper, we use notations in the parentheses to denote these four parties and use OAuth to denote OAuth2.0 as well as OIDC, if not specified otherwise.

The ultimate goal of SSO is for the IdP server to issue an identity proof, *e.g.*, the access token for OAuth2.0 and the id_token for OIDC, to the RP server. With this identity proof, the RP server can determine the user's identity and then log the user in.

2.1   The Implicit Flow of OAuth 2.0 for Mobile Platforms

OAuth2.0 [18] defines four types of authorization flows, out of which the implicit flow and authorization code flow[2] are widely used by the mobile platform and the website, respectively. Thanks to the demystification by Chen *et al.* [12], the standard implicit flow has proven to be insecure for authentication under mobile platforms. After revising the standard protocol, one believed-to-be-secure realization is illustrated in Fig. 1, although neither the RFC nor the IdP provides a complete call-flow diagram.

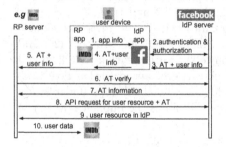

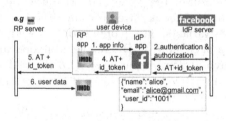

Fig. 1. The implicit flow of OAuth2.0 for mobile platforms

Fig. 2. The implicit flow of OpenID Connect

1. A user attempts to log into the RP with the IdP. The RP app sends its app information (*e.g.*, requested permissions) to the IdP app via a secure channel provided by the mobile operating system, for example the *Intent* channel in Android.

---

[2] In fact, the authorization code flow can also be securely used for mobile apps, but with the cost of worse performance.

2. Thanks to the secure channel, the IdP app can verify whether the RP app information is correct or not. If so, the IdP app then sends such information, as the authorization request, to the IdP server.
3. The IdP server compares information of the authorization request and that pre-registered by the RP developer. If it matches, the IdP server would believe the validity of the authorization request and thus issue an access token (AT) together with optional user information (*e.g.*, uid) to its own client-side app.
4. The IdP app returns the access token to the RP app via the secure channel maintained by the operating system.
5. The RP app sends the access token and user information to the RP server.
6. The RP server should call the security-focused SSO-API provided by the IdP to verify the access token.
7. If the access token is valid, the IdP server should respond the RP server with authorization information including which RP this access token is issued to.
8. Only if the access token belongs to this RP can the RP server accept it. Thereafter, the RP server can retrieve the user data with this verified access token.
9. The IdP server returns the user data associated with the access token.
10. The RP server can then identify the user and return his sensitive data to RP app.

2.2   The OpenID Connect Protocol

Since OAuth2.0 was originally designed to support authorization, to adapt it for authentication, it involves multiple high-latency round trips, *i.e.*, Step 6–Step 9 of Fig. 1. To support authentication more efficiently, IdPs like Google and Facebook have developed the OpenID Connect (OIDC) protocol [25] and its variants, on top of OAuth2.0. In addition to the authorization code flow and implicit flow, OIDC further supports another type of authorization flow, *i.e.*, hybrid flow. Regardless, *ALL* the real-world OIDC-enabled apps under study only implement the implicit flow. As such, we will focus the implicit flow in the rest of this paper.

Regarding the implicit flow, OIDC is backward compatible with OAuth2.0. The only difference is that, apart from the access token, IdPs also introduce a new parameter, *i.e.*, *id_token*, which is digitally signed by the IdP server. As illustrated in Fig. 2, the *id_token*, which consists of the user profile, is then sent to the RP server along with the original access token. Since the signature cannot be tampered/forged by an attacker, the RP server can now directly identify the user by extracting the user profile from the *id_token* without the trouble to retrieve the user profile from the IdP server.

2.3   Threat Model

The goal of an adversary is to break the mobile app authentication, *i.e.*, log into the RP app as the victim. Here, we trust the mobile operating system and assume it cannot be compromised. We assume the adversary can install the RP app and IdP app in her own device so as to communicate with the RP server and IdP server. The adversary, Eve, can act as a normal user and monitor/tamper the network traffic through her own device. In addition, we consider that an attacker can trick a user into installing a malicious app on her device. This app does not have any permission considered to be dangerous.

3   Major Protocol Changes that Affect Mobile OAuth Security

Although the above OAuth protocol seems simple, some security-related changes (when adapting the protocol from the website) are often overlooked by mobile app developers.

3.1   Untrusted Identity Proof

Websites typically employ the authorization code flow to support SSO services. In this case, the identity proof is transmitted via a secure HTTPS channel between the RP server and IdP server. On the contrary, mobile developers advocate the implicit flow, which passes the identity proof through the user device. Since the mobile device is untrustworthy (the attacker has full control of her own device), the identity proof is subject to tampering. Therefore, such an identity proof can only be accepted by the RP server for two reasons:

1. The RP server makes a direct server-to-server call to the IdP server to verify the identity proof (*i.e.*, Step 6–Step 9 in Fig. 1);
2. The identity proof is signed by the IdP server (*i.e.*, *id_token* in Fig. 2).

In other words, the RP server should establish a direct trust relationship with the IdP server to correctly process the identity proof.

Furthermore, there are different types of the so-called identity proof including the access token, the id_token, the user profile (returned at Step 9 of Fig. 1) and even the authorization code (if the authorization code flow is used). Since this notion is neither covered by the protocol specification nor research studies, RP developers need to determine which identity proof to use in which way, based on their own understanding of the protocol.

3.2   Heavy Client-Side Logic

Another major difference is about the user-agent. In the web-based SSO services, the user operates on the end-user's browser. By contrast, in the mobile SSO systems, the user interacts with the RP app and IdP app (the browser is split into two parts). Since mobile apps are more powerful, the RP app and IdP app are often responsible for more message exchange which instead is managed by the backend server in the case of website. This seems reasonable at the first sight. But some messages can only be processed at the server side, such as the user resource retrieved at Step 9 of Fig. 1. Otherwise, it can lead to the user-profile attack, as presented in Sect. 5.

Meanwhile, both the IdP app and RP app store much more data on the client side. For example, the IdP app keeps the user's identity information and the RP app stores the authorization information (e.g., the RP's name), which is displayed for the user to check/grant the permissions. Note that these security-critical data need to be retrieved from the server during SSO transactions. However, with such information in the client side, it may be tempting for an app developer to retrieve it from the phone directly. This can lead to the so-called user-profile attack and inconsistent RP app identity as presented in Sect. 5.

4   Our Approach

To evaluate the security implications resulting from the above protocol changes, we first perform dynamic testing on every RP/IdP app. The testing helps to understand the program logic on the server side. Secondly, to better understand the security practices on the client side, we conduct an in-depth static code analysis on the IdP apps (i.e., Facebook, Google and Sina) and their corresponding SDKs (used by RP apps). Given the limited number of IdP apps and SDKs, we can afford for the manual code examination.

4.1   Dynamic Testing

We design a tool to automatically fuzz every OAuth-related message. As shown in Fig. 3, we first set up a man-in-the-middle (MITM) proxy so that we can observe and tamper the network traffic going into and leaving our phone. We then manually operate the phone as a normal user and mimic the SSO process to generate a series of OAuth requests. For every request, our tool replaces each parameter with that from a different RP and user. Finally, the tool sends the fuzzed request to the receiver and checks whether the response is normal or not. Note that since the network traffic is typically protected by HTTPS, we employ the SSL-enabled proxy like *mitmproxy* [3].

The analysis of the communication between the IdP app and IdP server is straightforward, since such interactions share the same format for the same IdP. Unfortunately, fuzzing the messages between the RP app and RP server, i.e., Step 5 (ii) in Fig. 3, is more challenging than expected. Firstly, there are numerous interactions between the user mobile device and the RP server. It is therefore difficult to identify which request is used by the RP server to authenticate the user. Secondly, besides being protected by HTTPS, the message exchanges between the RP app and its backend server are often further encrypted or signed by the RP developer. Although it is possible to extract the cryptographic key from the Android app, such a practice may not be scalable. Therefore, it is usually easier to tamper the response from the IdP server to the IdP app, i.e., Step 3 (ii) instead. After all, all OAuth-related information received by the RP server can only be derived from the IdP server[3].

---

[3] One exception is Google's Android account management, as presented in Sect. 5.1.

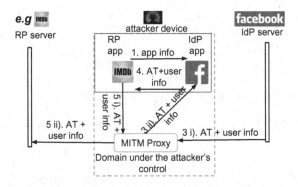

Fig. 3. The platform to analyze the implementations on the IdP/RP server side

4.2 Static Code Analysis

To understand the client-side logic, we decompile the binary code of the latest IdP app, and the official SDK (if it is compiled) widely used by the RP app. Although the IdP app and SDK usually are heavily obfuscated, the names of special activities and system APIs (*e.g.*, startActivity, getCallingPackage, *etc.*) are not changed. As such, we can identify the entry point of SSO services and then build a partial call flow graph. This provides an opportunity for us to only focus on the relatively small number of SSO-related security-critical activities. We then manually examine these activities to identify potential vulnerabilities. For example, we find that the SSO entry of Sina (*i.e.*, SSOActivity) by default verifies the received information, except when the information is from Sina itself. This practice is seemingly sound at first. However, it may be leveraged by a malicious RP app to bypass the security checking, if there is a "next Intent" [31] in Sina app. Therefore, during the code examination process, we will try to check the existence of the "next Intent". To confirm these vulnerabilities, we then build a toy RP app with the official SDK and launch the corresponding attacks on this toy app. By this way, we have identified the problem of inconsistent RP identity, as presented in Sect. 5.2.

5 Vulnerability Analysis

In addition to various existing vulnerabilities, the above method also helps to discover three unknown security flaws resulting from the inaccurate understanding/implementations of the protocol changes.

5.1 Profile Vulnerability

The identity proof is often incorrectly processed by the RP server and has led to the profile vulnerability, which enables an attacker to log into a susceptible RP as the victim by leveraging the victim's public user profile only. Note that all the

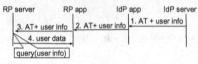 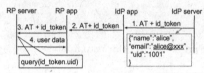

(a) RP app returns user information obtained via Android account management system

(b) The RP app returns user information obtained via API

Fig. 4. The RP app does not return correct identity proof of the user to the RP server

(a) No verification of the access token by the RP server

(b) No verification of the signature by the RP server

Fig. 5. The RP server does not verify the identity proof of the user

prior findings [12,29,34] require different types of interactions with the victim. In contrast, this newly discovered vulnerability can be exploited remotely and solely by the attacker without any need to trick or interact with the victim, for example via phishing attacks.

Different Types of Incorrect Implementations. We present two types of common but widespread mistakes (but the real-world misuses do not limited to these two types).

Return Incorrect Identity Proof to the RP Server. Some RP apps can directly retrieve the user information from the mobile device it is running on, regardless of the OAuth access token obtained from the IdP. Without the access token, the RP app only sends the user profile (*e.g.*, uid, email) to the RP server as the identity proof. As a consequence, the RP server has no way to correctly identify the user.

One interesting misuse is caused by Android Account Management System (AMS) [2,15] when using Google as the identity provider. Android AMS provides a centralized database (*i.e.*, */data/system/users/0/accounts.db*) for storing user accounts. While the main goal of AMS is to support seamless access user data via background synchronization, Google has integrated it to support SSO service. Specifically, when a user logs into his/her Google account, Google Login Service (*i.e.*, IdP app) will store the user's Google account information in the *accounts* table as shown below.

```
INSERT INTO "accounts" VALUES(1, 'alice@gmail.com', 'com.
    google', 'password', NULL);
```

With the Android permission of *GET_ACCOUNTS*, an app can easily get the user's Google account (*i.e.*, email address) by calling *getAccounts()* method. As shown in Fig. 4(a), some RP developers (*e.g.*, one antivirus app Psafe with 50+ million installs) directly return this email address (retrieved from the database) to the RP server as the identity proof, regardless of the access token (or id_token). As such, an adversary can insert a forged entry, *i.e.*, *victim's* email address, into the database on the *adversary's* mobile device.

Another typical example is shown in Fig. 4(b) where an RP app immediately retrieves the user data by calling the IdP API with the access token. However, this RP app only sends the user profile to its server as the identity proof. Although such a proof is protected by encryption/signature techniques in addition to TLS, an attacker can simply feed incorrect user information at Step 4 of Fig. 4(b).

The RP Server Does Not Verify the Identity Proof. As shown in Fig. 5(a), when the IdP servers return the user identity information (*e.g.*, user id/email address) along with the access token via OAuth, many RP servers (*e.g.*, Sohu with 80+ million monthly-active-user *etc.*) simply and incorrectly return sensitive user information to its own client-side app based on the received user-id WITHOUT verifying whether the received user-id is indeed bound to the issued access token (*i.e.*, lack of Step 6–Step 9 in Fig. 1).

Figure 5(b) shows another case where Facebook and Google adopt the OIDC-like protocol and digitally sign the user identity information. However, most RP servers just ignore this signature and insist on the traditional OAuth protocol. Worse still, some RP servers (*e.g.*, a free call/text app DingTong with 10+ million installs) even do not verify the signature but simply extract the user-id from the payload of the signature and accept the user-id as is the way without any authentication/validation.

Exploiting the Profile Vulnerability. Leveraging the same system setup of Fig. 3, an attacker can log into a susceptible app as the victim by exploiting the victim's profile with the following steps[4]:

1. The attacker setups a SSL-enabled MITM proxy for her own mobile device to monitor and tamper network traffic going into and leaving from her device.
2. The attacker installs the vulnerable RP app in her own mobile device.
3. The attacker signs into the vulnerable RP app with the attacker's own IdP login name and password.
4. When the IdP server returns the user profile to its client-side app, *i.e.*, Step 3 (ii) of Fig. 3, the attacker substitutes her own user-id (public user id for the case of Google+ and Sina users or guessable email address) with the victim's one using the MITM proxy. Although Facebook has started to issue private per-app user-id for each RP since May 2014, for backward compatibility reasons, to-date, Facebook still uses the public user-id to identify early adopters

---

[4] For the case of Google Android AMS, an attacker just needs to insert a forged entry in her own device and then follows the normal steps to complete the SSO procedure.

of a RP app. As such, a user of a vulnerable RP of Facebook is still suscepti-ble to our attack as long as he/she has signed into the RP app via Facebook before May 2014.

5. Since the RP server directly uses the user identity information returned by its client-side app to identify the user WITHOUT further validation, the attacker can therefore successfully sign into the RP as the victim.

Additional Challenges for the Exploit. The above exploit involves additional chal-lenges when the IdP client-side app, e.g., the one by Facebook, applies the certifi-cate pinning. In this case, the message sent by the IdP server (and then tampered by the attacker's MITM proxy) to its client-side app will not be accepted by the latter. As a workaround, the attacker can simply uninstall the IdP app so that the IdP SDK (widely used by RP apps) would automatically downgrade to carry out OAuth authentication via the built-in WebView browser. Being a general built-in browser, the WebView does not support the certificate pinning for a specific IdP.

But some IdPs do not support WebView. In this case, the attacker can-not just use off-the-shelf tools like Xposed SSLUnpinning [6] module, since the IdPs often use customized methods (instead of the native Android framework) to implement the certificate pinning. For such IdPs, the attacker has to reverse engi-neer the IdP app and manually remove the certificate pinning. To demonstrate the feasibility of this approach, we have successfully implemented a proof-of-concept hack on the Facebook app by reverse engineering the apk. But the RP app (more precisely, the IdP SDK) will also compare the certificate of Facebook with a previously hard-coded one (*i.e.*, the real certificate of Facebook). As such, we also need to bypass the certificate comparison function.

Vendor Responses. All of three IdPs under study acknowledged the security issue and pledged to help to notify the affected third-party app developers. In particular, Sina already sent a specific notification to ALL RP developers on its platform to inform them about the problem. The company also granted us the maximum amount of reward credits allowed by their bug-bounty program. It has also updated the Single-Sign-On section of its programming guide accordingly. Google has acknowledged our finding via their Google Security Hall of Fame and indicated that they will modify the corresponding documentation for their 3rd party app developers. Facebook has informed us that they are seeking a way to make their RP developers aware of this problem.

5.2   Inconsistent RP App Identity for the User

After authenticating the user at Step 2 of Fig. 1, the IdP app should retrieve the authorization information including the RP name, the requested permissions from its backend server. From the perspective of the user, the IdP app will pop up a dialog to indicate which RP requests what permissions from which user. Here, the name of the RP represents the identity of the RP which is verified by the IdP.

Since this authorization information is also stored by the RP app on the mobile device, we find that Google immediately retrieves such information from the device. Specifically, Google presents the name of the RP according to the value of *android:label* in the *AndroidManifest* file, when an RP app adopts the Intent[5] scheme (which is the default method used by Google's official SDK) to support SSO services. On one hand, Google in fact can correctly learn the identity of the RP. On the other hand, the *AndroidManifest* file can be arbitrarily defined by the RP app. As such, the displayed authorization page is inconsistent with Google's understanding.

Taking the advantage of this inconsistence, a malicious RP app can convince the user that the interacting RP, as verified by Google, is a benign and privileged app like IMDB. Due to the great trust placed on Google, the user is willing to grant the permissions if Google verifies the RP app as IMDB rather than some random app. As such, it would be easier for the malicious RP to obtain an access token. This access token enables an attacker to retrieve the victim's data hosted by Google. When combined with other attacks, *e.g.*, token hijacking, the attacker can also log into the victim's account on other Google-enabled benign apps.

Vendor Response. Google acknowledged this security bug. But as Google security team claims, this security issue was found independently by another concurrent work (but this issue is not fixed yet). Unfortunately, only the first report is in the scope of Google vulnerability reward program.

5.3   Treat the IdP App as a Special RP

Some IdPs including Google and Sina treat their client-side app as a special RP. If the user can successfully authenticate with the IdP, the IdP server would issue an access token to its client-side app. Note that this access token has higher privileges. It enables the IdP app to make sensitive transactions on behalf of the user, for example, to issue access token for any other RPs, *etc.* Since the adversary (acting as a normal user) can also obtain this access token, the adversary can easily launch application impersonation attack [20]. For example, the adversary can utilize this highly privileged access token to invoke sensitive APIs (*e.g.*, query the user information in a batch) with higher API quota. Note that such APIs/quotas originally are not allowed by the adversary or users.

Worse still, this privileged access token is not protected by HTTPS for Sina app. Therefore, an adversary can easily obtain it via eavesdropping. Using this special access token, the attacker can pretend to be the victim and make any transactions, for example, signing into any RP app on the Sina platform as the victim.

Vendor Response. We have reported this issue to Sina. Sina directly acknowledged this problem and have applied the corresponding fixes for their Android app.

---

[5] When an RP app chooses to use the WebView scheme, Google can correctly get the RP name from its own server.

6 Empirical Evaluation

We have studied the Android IdP apps and OAuth SDKs provided by three top-tier IdPs, *i.e.*, Facebook, Google and Sina. The number of registered users in these IdPs ranges from more than 800 million to over 2.5 billion as depicted in Table 1. We then comprehensively test the implementations of 600 top-ranked Android applications in US and China. Since more Chinese RPs support OAuth, we only select top 100 RPs (in overall category) and another top 100 apps in different categories for Sina from one major Chinese app store [4]. By contrast, we select 300 top-ranked RPs (in overall category) and 100 top-ranked RPs (in different categories) for Facebook and Google from Google Play. The top 100 apps in different categories is selected as follows: top 30 free apps in social, top 30 free apps in travel and local, top 30 free apps in fitness, top 10 free apps in communication. Out of these 600 apps, we identify that 182 apps use OAuth authentication service provided by one or more of the 3 IdPs mentioned above.

Table 1. Statistics for the usage of the protocol

IdPs (# of third-party RP)	# of IdP users (in Millions)	OAuth2.0		OpenID Connect		
		Insecure usage	Correct usage	Ignore id_token	Not verify id_token	Correct usage
Facebook (59)	>1,500	9	10	35 (20*)	2	3
Google (40)	>2,500	24	14	0+	0	2
Sina (83)	>8,00	78	5	N.A	N.A	N.A

– *: 20 out of 35 RPs are incorrectly implemented.
– +: Google customizes the OIDC protocol where typically only the id_token is issued to the RP. Therefore, this id_token cannot be ignored. Otherwise, there is not a valid identity proof.

In addition to different types of vulnerabilities, our studies also present first-hand information regarding the adoption rate as well as the misuse rate of OAuth2.0 and OIDC. As shown in Table 1, all of three IdPs support the OAuth2.0 protocol. Facebook and Google additionally develop and advocate OIDC-like protocols for SSO services. Since OIDC by default is supported by Facebook SDK, 68% apps of Facebook, as opposed to only 2 RP apps of Google, employ the OIDC protocol. Regardless, there are two types of misuses for these OIDC-enabled RPs:

– Ignore *id_token*: Some RPs ignore the *id_token*, in which case these RPs revert to the OAuth2.0 protocol and rely on the access token to authenticate the user. As such, these RPs (20 out of 35) share the same security issues of OAuth2.0 as illustrated in Table 2.
– Not verify *id_token*: Some RPs indeed rely on the *id_token* to verify the user. But 29% of them do not check whether the *id_token* is well signed.

Table 2. Statistics of the vulnerabilities for OAuth

IdPs (# of 3rd-party RP)	Profile attack	Token hijacking	Improper user agent	Access token disclosure	App secret disclosure	# of vulnerable RPs[a]
Facebook (59)	9 (15%)	27	1	2	0	31 (53%)
Google (40)	8 (20%)	20	1	1	0	24 (60%)
Sina (83)	58 (70%)	15	7	13	4	78 (94%)
Summary (182)	75 (41%)	62	9	16	4	133 (73%)

– [a]: One RP may be susceptible to multiple vulnerabilities, *e.g.*, the profile attack and token hijacking, at the same time.

These observations show that OAuth2.0 is still the most popular SSO protocol. Unfortunately, the implementations of OAuth2.0 (including the ignore *id_token* case of OIDC) are also more susceptible: 75% OAuth2.0-enabled RPs have at least one vulnerability whereas 29% OIDC-supported RPs are vulnerable. Different OAuth security vulnerabilities are summarized in Table 2. Below we first discuss the implication of the profile attack and then demonstrate the pervasiveness of existing vulnerabilities.

6.1   The Implication of the Profile Attack

As illustrated in Table 2, 41% of the RP under test are found to be vulnerable to the newly discovered profile attack. Table 3 depicts a partial list of the vulnerable mobile apps we have identified so far. Notice that the total number of downloads for this incomplete list of popular but vulnerable apps already exceeds 2.4 billion. Based on the SSO-user-adoption-rate of 51% according to the recent survey by Janrain [7], we conservatively estimate that more than one billion of different types of mobile app accounts are susceptible to the profile attack as of this writing.

After signing into the victim's vulnerable RP app account using our exploit, the attacker will have, in many cases, full access to the victim's sensitive and private information which is hosted by the vulnerable RP server. Just for the vulnerable apps listed in Table 3 alone, a massive amount of extremely sensitive personal information is wide-open for grab: this includes detailed travel itineraries, personal/intimate communication archives, family/private photos, personal finance records, as well as the viewing or shopping history of the victim. For some RPs, the online-currency/service credits associated with the victim's account are also at the disposal of the attacker.

Although our current attack is demonstrated over the Android platform, the exploit itself is platform-agnostic: any iOS or Android user of the vulnerable mobile app is affected as long as he/she has used the OAuth SSO service with the app before. As a proof of concept, we have conducted the same attack on two vulnerable iOS apps.

6.2   Re-Discover Known Vulnerabilities

Table 2 shows that our testing also rediscovers different types of existing security issues.

Table 3. A Partial list of vulnerable apps and the sensitive information exposed

Type of apps	IdP supported	# of app downloads (in Millions)	Type of private/sensitive information exposed	Feasible transactions by the attacker
Travel plan app	Sina	>270	Travel itineraries	-
Hotel booking app	Facebook, Google	>5	Lodging history	Pay for room bookings
Private chat app	Sina	>10	Private message/album	Send forged messages
Dating app	Google, Sina	>5	Dating history, preferences	Purchase gifts
Finance app1	Sina	>25	Personal income/expenses	-
Finance app2	Sina	>50	Stock list of interest	-
Call app	Facebook	>10	Contact list and call history	Call for free
Live video app	Sina	>15	The host the victim likes	Purchase gifts
Download app	Sina	>60	Download history	Enjoy VIP speed
Shopping apps	Facebook, Google	>100	Shopping history	-
Browser	Sina	>40	Browsing history	-
Video apps	Sina	>700	Video watching history	Purchase videos
Music apps	Google, Sina	>800	Playlist	Purchase sound-tracks
News apps	Sina	>350	News-reading history	-

1. Token hijacking [1]: At Step 6–Step 7 of Fig. 1, the RP server must check that the received access token is granted to the same RP. Unfortunately, 34% RPs fail to do so, which enables an adversary to sign onto a victim's benign RP account by leveraging an access token issued to a malicious RP.
2. Improper user agent [12]: In addition to the infeasibility to identify the RP app, the WebView, as a custom webkit browser embedded in the app, is also untrustworthy to be a OAuth user-agent. Since the WebView is under the control of the RP app, a malicious RP app is capable of stealing any information submitted by the user in the WebView (e.g., the user's IdP password) and modifying authorization information displayed by the WebView. Unfortunately, most RP apps support WebView, and worse still, 5% RP apps only support this problematic scheme.
3. Access token disclosure [27]: An access token should be transmitted securely. Thanks to the higher adoption rate of TLS, only 9% mobile RPs disclose their access token whereas 32% 3rd-party websites made this mistake [27].
4. App secret disclosure [29]: The confidential app secret should only be shared between the RP server and IdP server. There are 15 mobile RP apps deploy the authorization code flow rather than the implicit flow. Unfortunately, 27% of these apps inadvertently pass this secret through the user device. This allows an attacker to make operations on behalf of the RP, e.g., changing the RP's security setting.

7 Plausible Root Causes

Surprised by the prevalence of various incorrect implementations, we also try to analyze the underlying reason by examining the SDKs and the OAuth APIs provided by IdPs.

7.1 Unclear Developer Documentation

We found that many authentication-related security issues are caused by the lack of clear guidelines. Since OAuth2.0 (RFC 6749 [18]) was not designed for mobile app authentication, various IdPs have developed different home-brewed extensions of OAuth2.0-based APIs and SDKs to support SSO for mobile apps. Unfortunately, the implicit security assumptions and operational requirements of such home-brewed adaptations are often not clearly documented or well-understood by RP developers. For example, when Sina returns the user profile to the RP app at Step 4 of Fig. 1, the only purpose is to allow the RP app to display the user info (*e.g.*, user name, avatar, etc.). Despite of this specific intention, Sina makes the following confusing claim[6] in its programming guide [5]:

> For the convenience of app developers, the returned user information can avoid calling the user-profile API, *i.e.*, *users/show*.

As pointed out in Sect. 3, a server-to-server call is inevitable for the standard OAuth2.0 protocol to verify the untrusted identity proof. However, the above claim can mislead RP servers not to make the user-profile API call to the IdP server at Step 8 and Step 9. Instead, the RP server may directly use the returned user information from its client-side app as the identity proof and thus be vulnerable to the profile attack.

For another example, Google assumes the RP developers would adopt the OIDC protocol, and thus only shows how an RP app can authenticate with its backend server using the *id_token*. Unfortunately, the majority of apps use the OAuth2.0 protocol instead. For these apps, Google does not define the interactions between the RP app and RP server. Therefore, the RP developers without adequate security expertise have to implement the error-prone authentication services by themselves.

After reporting the security issues to the three IdPs, all of them recognize the need to improve their documentation by explicitly pointing out the implicit security requirements. For example, Sina now updates its claim [5] as follows:

> The third-party apps should not use *uid* to identify the logged-in user. Note that the access token is the only valid identity proof.

---

[6] Since Sina developers are not native English speakers, the statements are translated by the author.

7.2   Poor API Design of Sina

Since more Sina apps are vulnerable to the profile attack, we further examine its SDK and API design. We find that the poor API design of Sina may compound this problem. As an open social network, Sina by design allows any RP in possession of a valid access token, no matter whom the access token is issued to, to retrieve *any* user's basic profile via the *users/show* API: https://api.weibo.com/2/users/show.json?access_token=x&uid=x. Even if the RP server does not trust the user id and would like to use the access token to identify the user, it may incorrectly use the above API. In this case, Sina server would return the victim's user profile only based on the value of *uid* in the URL. Without realizing the exact semantic of the returned information, the RP developer may incorrectly interpret that the returned user profile is bound to the access token, and thus, log the user in.

In fact, Sina also provides a correct API (*i.e.*, *account/get_uid*) to get the user-id of the one, whom the access token is bound to. But Sina never specifies which API to use[7]. Due to the richer information provided by the inappropriate *users/show* API[8], we believe RP developers prefer the incorrect API. By contrast, Facebook and Google use the *people/me* API, which is more self-explained, and more importantly, is typically handled by IdP SDKs. For example, the PHP SDK of Facebook hard-encodes the user id *me*, and as such the *uid* fed by the attacker would be ignored by the SDK.

8   Defense

The community has proposed the following Current Best Practices (CBPs):

1. IdPs should provide clearer, and more security-focused developer guidelines.
2. The RP server should not trust any information even if it is signed by its own app. Trust should be anchored on the IdP server directly.
3. To implement OAuth2.0 for mobile apps, RP developers should use the authorization code flow instead of implicit flow by strictly following [14].
4. Use OIDC for authentication whenever possible.

If these CBPs can be correctly followed, then most vulnerabilities will not exist. Unfortunately, most RP developers never adhere these CBPs at all. Towards this end, the crux of the defense is to enforce the defined security checks, which cannot simply rely on the RP developers, but instead must be strongly enforced by the IdPs, since the latter has adequate security expertise. Therefore, we develop a general and foolproof solution, from the perspective of the IdP, to help the RP developers to automatically handle the error-prone SSO services. Our solution should achieve three goals:

---

[7] Worse still, Sina seems to recommend the incorrect one in their unclear developer documentation.

[8] The other API only returns the *uid* of the user.

1. All the identified vulnerabilities (including existing ones) can be prevented.
2. To ease the transition, the code changes by the RP developers should be as few as possible.
3. The solution should be backward compatible in a sense that the RP server can still support those users who use the older version of the RP app. After all, it is difficult for every user to upgrade his/her RP app.

With these goals in mind, we first review the architecture of existing SSO systems. As shown in Fig. 6, both the RP server and RP app can be split into two modules: the SDKs provided by the IdP serve the interactions with the IdP app/server and the upper layer codes implemented by the RP developer manage its own business logic. Currently, the identity proof is also (incorrectly) handled by the upper layer code. To prevent from SSO errors, we migrate such functionality to the lower layer SDKs, with the belief that the SDKs provided by IdPs can correctly process this security-critical identity proof. Specifically, upon the reception of the identity proof at Step 4 of Fig. 1, the client-side SDK can send it to the server-side SDK. The latter then performs the real authentication task by utilizing this identity proof. Below we will discuss more details.

8.1   Defense on the Client-Side SDK

As shown in Fig. 4, when the RP apps only return user information, there is no way for the RP servers to correctly identify the user. Therefore, the enforced version of the client-side SDK should automatically send required information including the access token (or the id_token for OIDC) to its backend server. With due consideration to the ease of transition, below we first discuss the design of the existing client-side SDK.

API Design of the Existing Client-Side SDK. We only take Sina, [9] one OAuth2.0 IdP, as the example. When authorization succeeds, the client-side SDK (used by the RP app) will utilize Android API *onActivityResult* to receive an access token along with a user id from Sina app. Before forwarding this result to the upper layer, *authorizeCallBack* API is first invoked to check whether the access token is in the correct format.

```
protected void onActivityResult(...){
    mSsoHandler.authorizeCallBack(requestCode,resultCode,result);
}
```

Initial Attempt Using Cookie-Based Scheme. Without affecting the upper layer behavior, we manage to authenticate the user in the SDK. Referring to the scheme in websites, one natural attempt, as shown in Fig. 6, is to use *cookie*.

---

[9] Our solution is applicable to OIDC protocol or other IdPs since they follow the same flavor.

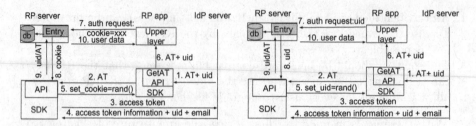

Fig. 6. The cookie-based defense   **Fig. 7.** The refined defense

1. After the client-side SDK (*i.e.*, *authorizeCallBack*) checks the format of the access token, instead of forwarding it to the upper layer, the client-side SDK first delivers the access token to the server-side SDK.
2. The server-side SDK exactly follows Step 3–Step 4 in Fig. 6 (corresponding to Step 6–Step 9 in Fig. 1) to identify the user via the access token.
3. If the verification succeeds, the server-side SDK then sets a *cookie* to the client-side app with a random nonce at Step 5 of Fig. 6. Only then would the client-side SDK forward the access token and the user profile to the upper layer code.
4. From the view of the upper layer, everything remains the same. Thus it can follow its original logics to interact with the RP server. The only difference is that a cookie would be automatically attached to the authentication request at Step 7 of Fig. 6.
5. Regardless of other information sent by the RP app, the server-side SDK only relies on the *cookie* to identify the user.

Unfortunately, this cookie-based solution is not applicable to every RP app: Unlike websites where a central browser can help to manage cookies, the Android app developers need to manage the cookies by themselves, for example using the *CookieManager*. Therefore, the *cookie* set by the underlying SDK may not be automatically used by the upper layer, if the latter adopts a customized *CookieStore*.

Refined Defense. Nevertheless, the cookie-based scheme still provides great insights. Note that the cookie is used to bind the requests of Step 2 and Step 7 in Fig. 6. Towards this end, the client-side SDK and its upper layer code must share some information like the *cookie*. Furthermore, the shared information must be a *secret*. Otherwise, an adversary can easily guess/compute this information and pretend to be anyone else.

Given these requirements, we revise the cookie-based solution. Referring to the practice of Facebook which issues private user-id on a per-app basis, we will randomly generate a one-time user id, instead of the *cookie*, as the secret to bind these two requests. More specifically, if authentication succeeds, the server-side SDK would return a randomly generated *uid* to its client-side SDK at Step 5 of Fig. 7. This *uid* plays the same role as *cookie*. Note that this *uid* can only be used for once so that an adversary cannot guess/compute its value. The old *uid* will be deleted once expired or used.

8.2   Defense on the Server-Side SDK

At Step 2 of Fig. 7, the server-side SDK can follow Step 3 and Step 4 in Fig. 7 (*i.e.*, correspond to Step 6–Step 9 in Fig. 1) to identify the user. Once authentication succeeds, the server-side SDK randomly generates a one-time *uid* with a specific prefix and maintains the mapping of $< uid, uid_{real} >$. At Step 7 of Fig. 7, the server-side upper layer should forward the authentication request to its SDK. Our newly added function in the SDK can then handle this request according to its content.

1. If the request only contains the user id, *i.e.*, *uid*, we check whether this *uid* starts with a specific prefix. If so, we check the mapping of $< uid, uid_{real} >$ and then get the user information uid_{real}. Otherwise, just abort the request.
2. If the request only contains the access token, we follow Step 3 and Step 4 in Fig. 7 to identify the user.
3. If the request contains the access token and user id, we first follow Case (1) to process the user id. If it fails, we then follow Case (2) to process the access token.

Remark. The correctness of our defense is based on two assumptions. Firstly, a correct implementation of OAuth, as shown in Figs. 1 and 2, is automatically immune to all the existing attacks. In fact, a formal proof is presented by [34]. More precisely, this work utilizes the model checking method to analyze the implementation-level protocol of Facebook and can only discover unauthorized storage access if the malicious app has root privileges. However, such a strong threat model is not considered in this paper. Secondly, the IdPs (*i.e.*, SDKs) with enough security expertise can accurately implement the protocol (as opposed to the RP developers who often make mistakes). Therefore, the crux of our solution is to enforce the IdP developers to correctly implement the OAuth protocol for the RP developers.

   Although the design of the defense seems complicated, we only add two more requests (*i.e.*, Step 2 and Step 5 at Fig. 7) into the current system. Note that the other steps (*e.g.*, Step 6, Step 8, *etc.*) already reside in the existing systems. For the ease of presentation, we intentionally hide these detailed (SDK-level) steps when discussing the protocol flow in Fig. 1. Note also that the proposed remedies does not affect the authorization code flow, although we can use the same idea to improve its security.

8.3   Implementation and Evaluation of the Proposed Defense

To demonstrate the feasibility of the proposed remedies, we implement the solution on a sample app provided by Sina. Like examples from the other IdPs, the sample app is built on top of the Android SDK. Note that this SDK is in the form of executable Jar file. To modify it, we thereby need to decompile the Jar file into Java code. While there are off-the-shelf Java decompiler tools like CFR or JAD, the extracted source code is not well structured and contains various errors, which requires non-trivial manual resolve.

We then follow the common practices to build a backend server on top of the official PHP SDK. The only change in the server-side SDK is to add a new function which performs the real authentication task. Meanwhile, we only add 5 lines of code in the server-side upper layer, which demonstrates the least programming efforts from the RP developers.

Table 4. The average running time under three different settings

Two types of app	User info (in ms)	Access token (in ms)	Access token & user info (in ms)
Vulnerable app	4490	7604	5092
Fixed app	6414	9667	6009

Evaluation. To demonstrate the effectiveness, we have launched different attacks listed in Table 2 on the sample app. It turns out that our solution can prevent from (or alert for insecure transmission, *e.g.*, token disclosure) all these attacks. For example, the profile vulnerability becomes impossible since an adversary cannot guess the randomly-generated one-time *uid*. Take token hijacking as another example. Since our defense exactly follows Fig. 1, our server-side SDK will verify whether this access token is issued to itself or not at Step 3 of Fig. 7 (corresponds to Step 6 of Fig. 1). Thus an access token of another RP will not be accepted.

To show the efficiency of our solution, we measure the running time of a complete SSO process. Specifically, we enumerate all three possible cases of the authentication request, as mentioned in Sect. 8.2. For each case, we operate the sample app to complete the OAuth2.0 process for 20 times under the same phone and network environment. The average time is presented in Table 4. It shows that the fixed app has only a small impact on the user performance.

The State-of-the-Art OAuth Defense. The Current Best Practices (CBP) [14] suggests the usage of authorization code flow to prevent from many existing problems. However, a correct implementation of these CBPs is still challenging for the RP developers (as is the case for the revised implicit flow). Furthermore, it requires lots of efforts to migrate from implicit flow (the current *de facto* standard) to authorization code flow. Xing *et al.* [32] develop invariants of HTTP parameters to protect third-party web service integrations like OAuth2.0. However, it cannot discover many attacks such as the profile attack. Another defense [11] uses program verifier to check whether the sequence of method calls satisfies the defined predicate. Despite its power, this method requires significant programming efforts for *all* involved parties. Compared to the state of the art, our defense not only can prevent from all the existing attacks, but also requires the minimal programming efforts from the RP developers.

9   Related Work

Security Analyses from the Protocol Perspective. IETF has presented comprehensive security considerations and threat models in RFC6749 [18] and RFC6819 [22] for OAuth2.0 protocol from the initial design. Additionally, the authorization code flow has proven to be secure cryptographically [10] as long as the TLS is properly used. Hu *et al.* [20] present the App Impersonation attack. These works mainly prove the authorization security for the web from the protocol design standpoint. However, we consider whether the protocol is securely implemented on mobile platforms for authentication.

Formal Security Proof for OAuth. The model checking method is extensively used to analyze the protocol specification [18] by numerous works including [8,9,16,24], just to name a few. Researchers also attempt to use the same method to reason about the protocol implementations by modeling the complex runtime platform, *e.g.*, browser [16,17]. All of these works assume a correct implementation of the protocol call-flow. However, we show that protocols are often implemented incorrectly.

Real-World Study on the Web-Based SSO Systems. More efforts have been contributed to the vulnerability detection of the website-based real-world systems, including [13,23,27,30]. Another research direction is to conduct large-scale testing on the web SSO systems, including SSOScan [26,35], OAuthT-ester [21,33]. Nevertheless, all the works mentioned so far only study the OAu-th/OIDC specification/implementations on the website. In contrast, we focus on the mobile platforms.

Analyses of Mobile OAuth SSO Systems. There are relatively few security analyses on OAuth under mobile environment. Chen *et al.* [12] shows how real-world OAuth systems can fall into the common pitfalls when leveraging the operating-system-provided components (*e.g.*, Intent, WebView, *etc.*). Ye *et al.* [34] utilize the model checking method to evaluate the OIDC-like protocol implemented by Facebook on Android platform. Summarizing these works, Wang *et al.* [19,29] collect the statistics of these known vulnerabilities. Prior works mainly focus on how the classic vulnerabilities (or features) of mobile systems can be leveraged to compromise SSO systems. In contrast, we analyze how the modified protocol call-flow itself can be incorrectly implemented.

10   Conclusion

In this paper, we report a field-study of mobile OAuth2.0-based SSO systems. We perform an in-depth static code analysis on three first-tier IdP apps and their official SDKs as well as dynamic testing on Top-600 Android apps in China and US. Besides the discovery of three previously unknown security flaws, we also demonstrate the prevalence of different types of existing vulnerabilities. The pervasiveness of these loopholes motivates us to design and implement a foolproof defense for those susceptible RPs with the aim of minimizing the programming efforts of RP developers. Our discoveries show that it is urgent for the various parties to re-examine their OAuth implementations and apply the fixes accordingly.

Acknowledgements. This project is supported in part by the Innovation and Technology Commission of Hong Kong (project no. ITS/216/15) and NSFC Grant (No. 61572415).

References

1. Access token hijacking. https://developers.facebook.com/docs/facebook-login/security#tokenhijacking
2. Android account manager. http://developer.android.com/reference/android/accounts/AccountManager.html
3. Man in the middle proxy. https://mitmproxy.org/
4. One major Chinese App store. http://sj.qq.com/myapp/category.htm
5. Sina access token API. http://open.weibo.com/wiki/OAuth2/access_token
6. SSL unpinning. https://github.com/ac-pm/SSLUnpinning_Xposed
7. Social login continues strong adoption (2014). http://janrain.com/blog/social-login-continues-strong-adoption/
8. Bai, G., Lei, J., Meng, G., Venkatraman, S.S., Saxena, P., Sun, J., Liu, Y., Dong, J.S.: AUTHSCAN: automatic extraction of web authentication protocols from implementations. In: NDSS (2013)
9. Bansal, C., Bhargavan, K., Maffeis, S.: Discovering concrete attacks on website authorization by formal analysis. In: IEEE CSF (2012)
10. Chari, S., Jutla, C.S., Roy, A.: Universally composable security analysis of OAuth v2.0. Cryptology ePrint Archive, Report 2011/526 (2011)
11. Chen, E.Y., Chen, S., Qadeer, S., Wang, R.: Securing multiparty online services via certification of symbolic transactions. In: IEEE S&P (2015)
12. Chen, E.Y., Pei, Y., Chen, S., Tian, Y., Kotcher, R., Tague, P.: OAuth demystified for mobile application developers. In: ACM CCS (2014)
13. Mainka, C., Vladislav Mladenov, J.S., Wich, T.: SoK: Single Sign-On security- an evaluation of OpenID Connect. In: IEEE EuroS&P (2017)
14. Denniss, W., Bradley, J.: OAuth 2.0 for native apps (2016)
15. Elenkov, N.: Android Security Internals: An In-Depth Guide to Android's Security Architecture. No Starch Press, San Francisco (2014)
16. Fett, D., Küsters, R., Schmitz, G.: An expressive model for the web infrastructure: definition and application to the browser ID SSO system. In: IEEE S&P (2014)
17. Fett, D., Küsters, R., Schmitz, G.: A comprehensive formal security analysis of OAuth 2.0. In: ACM CCS (2016)
18. Hardt, D.: The OAuth 2.0 authorization framework (2012)
19. Homakov, E.: The Achilles Heel of OAuth or Why Facebook Adds Special Fragment (2013)

20. Hu, P., Yang, R., Li, Y., Lau, W.C.: Application impersonation: problems of OAuth and API design in online social networks. In: ACM Conference on Online Social Networks, COSN (2014)
21. Li, W., Mitchell, C.J.: Analysing the security of Google's implementation of OpenID Connect. In: SIG SIDAR Conference on Detection of Intrusions and Malware & Vulnerability Assessment, DIMVA (2016)
22. Lodderstedt, T., McGloin, M., Hunt, P.: OAuth 2.0 threat model and security considerations (2013)
23. Mladenov, V., Mainka, C., Krautwald, J., Feldmann, F., Schwenk, J.: On the security of modern Single Sign-On protocols: OpenID Connect 1.0. CoRR abs/1508.04324 (2015)
24. Pai, S., Sharma, Y., Kumar, S., Pai, R.M., Singh, S.: Formal verification of OAuth 2.0 using Alloy framework. In: IEEE International Conference on Communication Systems and Network Technologies, CSNT (2011)
25. Sakimura, N., Bradley, J., Jones, M., de Medeiros, B., Mortimore, C.: OpenID Connect core 1.0. The OpenID Foundation (2014)
26. Shernan, E., Carter, H., Tian, D., Traynor, P., Butler, K.: More guidelines than rules: CSRF vulnerabilities from noncompliant OAuth 2.0 implementations. In: Almgren, M., Gulisano, V., Maggi, F. (eds.) DIMVA 2015. LNCS, vol. 9148, pp. 239–260. Springer, Cham (2015). doi:10.1007/978-3-319-20550-2_13
27. Sun, S., Beznosov, K.: The devil is in the (implementation) details: an empirical analysis of OAuth SSO systems. In: ACM CCS (2012)
28. Wang, H., Zhang, Y., Li, J., Gu, D.: The achilles heel of OAuth: a multi-platform study of OAuth-based authentication. In: ACM ACSAC (2016)
29. Wang, H., Zhang, Y., Li, J., Liu, H., Yang, W., Li, B., Gu, D.: Vulnerability assessment of OAuth implementations in Android applications. In: ACM ACSAC (2015)
30. Wang, R., Chen, S., Wang, X.: Signing me onto your accounts through Facebook and Google: a traffic-guided security study of commercially deployed Single-Sign-On web services. In: IEEE S&P (2012)
31. Wang, R., Xing, L., Wang, X., Chen, S.: Unauthorized origin crossing on mobile platforms: threats and mitigation. In: ACM CCS (2013)
32. Xing, L., Chen, Y., Wang, X., Chen, S.: InteGuard: toward automatic protection of third-party web service integrations. In: NDSS (2013)
33. Yang, R., Lee, G., Lau, W.C., Zhang, K., Hu, P.: Model-based security testing: an empirical study on OAuth 2.0 implementations. In: ACM ASIACCS (2016)
34. Ye, Q., Bai, G., Wang, K., Dong, J.S.: Formal analysis of a Single Sign-On protocol implementation for Android. In: International Conference on Engineering of Complex Computer Systems, ICECCS (2015)
35. Zhou, Y., Evans, D.: SSOScan: automated testing of web applications for Single Sign-On vulnerabilities. In: USENIX (2014)

Adaptive Proofs Have Straightline Extractors (in the Random Oracle Model)

David Bernhard, Ngoc Khanh Nguyen$^{(\boxtimes)}$, and Bogdan Warinschi

Computer Science Department, University of Bristol, Bristol, England
{csxdb,nn14160,csxbw}@bristol.ac.uk

Abstract. The concept of *adaptive security* for proofs of knowledge was recently studied by Bernhard et al. They formalised adaptive security in the ROM and showed that the non-interactive version of the Schnorr protocol obtained using the Fiat-Shamir transformation is not adaptively secure unless the one-more discrete logarithm problem is *easy*. Their only construction for adaptively secure protocols used the Fischlin transformation [11] which yields protocols with *straight-line extractors*. In this paper we provide two further key insights. Our main result shows that any adaptively secure protocol must have a straight-line extractor: even the most clever rewinding strategies cannot offer any benefits against adaptive provers. Then, we show that any Fiat-Shamir transformed Σ-protocol is not adaptively secure unless a related problem which we call the Σ-one-wayness problem is *easy*. This assumption concerns not just Schnorr but applies to a whole class of Σ-protocols including e.g. Chaum-Pedersen and representation proofs. We also prove that Σ-one-wayness is hard in an extension of the generic group model which, on its own is a contribution of independent interest. Taken together, these results suggest that the highly efficient proofs based on the popular Fiat-Shamir transformed Σ-protocols should be used with care in settings where adaptive security of such proofs is important.

1 Introduction

Noninteractive zero knowledge proofs are a useful tool widely deployed in modern cryptographic constructions. They allow a prover (e.g. the creator of a ciphertext) to send a single message which will convince a verifier of the veracity of a certain statement (e.g. that the plaintext underlying a ciphertext respects certain constraints), without revealing any further information. A particularly useful variant are the so called "proofs of knowledge" where there exists an extractor who can efficiently recover from the prover a witness that the statement is true. Over parties in the system (who should obtain no information from the witness), the extractor benefits from setup assumptions like the common reference string model or the random oracle mode [3]. Our focus is on the latter model which yields by far the most efficient noninteractive proofs of knowledge known to date. In a nutshell, in this paper we study different extraction strategies afforded by

© Springer International Publishing AG 2017
D. Gollmann et al. (Eds.): ACNS 2017, LNCS 10355, pp. 336–353, 2017.
DOI: 10.1007/978-3-319-61204-1_17

the random oracle model and identify fundamental efficiency limitations that such proofs need to face in practically relevant scenarios.

Background. Recall that in the random oracle setting the extractor learns the RO answers/queries that the prover makes. Here, we distinguish between several extraction strategies. [1] An extractor is *straight-line* if it only sees a single execution of the prover. An extractor is *rewinding* if it is allowed to launch and interact with further copies of the prover (with the same coins used to produce the statement) before returning a witness.

The distinction between straightline and rewinding extractors may be crucial in applications since for a rewinding extractor it is not clear how many times does it have to rewind to extract all witnesses from a prover who makes a sequence of n proofs. Shoup and Gennaro [20] first encountered this problem in the context of proving CCA security of a particular public-key encryption scheme; the "obvious" approach ends up rewinding 2^n times which leads to an inefficient reduction. Clearly, this problem disappears for a straight-line extractor.

The notion of *adaptive* proofs which lie somewhere between proofs with inefficient rewinding strategies and straight-line PoKs has been recently proposed by Bernhard et al. [5]. A proof scheme is adaptively secure if there is an extractor that can rewind, but must efficiently extract even from provers who make sequences of proofs. The notion is called adaptive because the extractor must return a witness for the first proof to the prover before the prover makes the second one, and so on.

As an application, Bernhard et al. study the Fiat-Shamir-Schnorr proofs: a non-interactive proof of knowledge obtained by applying the popular Fiat-Shamir [10] transform to the Schnorr [17] protocol for proving knowledge of discrete logarithm. The main theorem of [5] shows the Fiat-Shamir-Schnorr proof scheme is *not* an adaptive proof unless the one-more discrete logarithm problem is easy. This result essentially separates the usual PoK notion from adaptive proofs, but the separation has several shortcommings: (i) it relies on an inefficient interactive assumption, (ii) it is specific to a proof system for a specific problem (discrete logarithm), and (iii) it is specific to a class of proofs (those obtained from Σ-protocols via the Fiat-Shamir transform). That is, it does not pinpoint precisely the source of the difficulty and, in particular, it leaves open the question whether any adaptive proofs exist that are not also straight-line[2].

Our Contribution. In this paper we obtain a full characterization of adaptive proofs in the random oracle model and we leverage this result to provide more general results regarding the limitations of the Fiat-Shamir transform. Below we outline our contributions.

ADAPTIVE PROOFS $\equiv$ PROOFS WITH STRAIGHTLINE EXTRACTORS. Our main contribution is a negative answer to Bernhard et al.'s open question. It holds for

---

[1] Recall that in the random oracle model hash functions can only be computed by calling an oracle available to all parties (and which on a fresh input returns a truly random output).

[2] Straight-line proofs are trivially adaptively secure.

all non-interactive proof schemes in the ROM, whether or not they are derived from Σ-protocols:

Theorem 1 (Informal). *Consider an arbitrary non-interactive proof of knowledge system in the ROM. If the proof system has an efficient adaptive extractor against adaptive provers then it also has a straight-line extractor.*

The immediate consequence of this theorem is that when designing PoKs for an adaptive setting, one cannot rely on rewinding and instead one should ensure the existence of a straightline extractor. While a general strategy is to employ Fischlin's transformation [11], one may still want to rely on the more efficient construction that uses the Fiat-Shamir transformation whenever possible – the impossibility result of Bernhard et al. [5] only applies to Fiat-Shamir-Schnorr proofs.

LIMITATIONS OF THE FIAT-SHAMIR TRANSFORM. We show that the Fiat-Shamir transformation has intrinsic limitations. In particular, we generalize the results of [5] in two distinct directions. On the one hand, we show that it holds for arbitrary Σ-protocols for proving knowledge of preimages of linear functions (including Schnorr, Chaum–Pedersen and representation proofs). More interestingly, we weaken the condition under which these proofs are not adaptively secure from one-more discrete logarithm (resp. one-more one-wayness [1]) to the following assumption: a dishonest verifier in a single execution of the Σ-protocol cannot extract the witness. We call this assumption Σ-*one-wayness*. Our result thus improves from a "q-type" assumption, which does not admit an efficient game, to an efficient game with only three rounds.

This theorem answers the main open question of [5] and hints at a shortcoming of proofs based on the Fiat–Shamir transform: if used in a setting where the prover gets to adaptively chosen statements, an extractor would have to be (more or less) straight-line. However, if this is the case, the proof may not be that interesting anyway as the underlying witness is not well-protected:

Theorem 2 (Informal). *Suppose there is a straight-line extractor for a Fiat-Shamir transformed Σ-protocol Σ (to prove knowledge of a preimage of some linear function f). Then a dishonest verifier can extract a witness (a preimage under f) in a single run of Σ.*

Taken together, these results imply that Fiat-Shamir transformed Σ-protocols are not adaptively secure in any setting in which they might be useful. Take the Schnorr protocol as an example: if Fiat-Shamir-Schnorr is adaptively secure then either discrete logarithms are easy in the relevant group, in which case the Schnorr protocol is redundant, or the Schnorr protocol provably helps a dishonest verifier to extract the discrete log of the statement — in which case Schnorr is certainly not zero-knowledge.

The theorem of Bernhard et al. [5] contains an adaptive prover who makes a sequence of n proofs such that each one depends on all previous ones. The straightforward rewinding strategy — rewind on every proof to extract — ends up rewinding 2^n times since the rewound provers make new proofs which again have to be rewound. A combinatorial argument then shows that any strategy that rewinds fewer than 2^n times must have taken a discrete logarithm to find out the witness for one of the proofs output by the prover. The problem is that we do not know where, and also if we inject a challenge in one proof then we end up having to simulate all other proofs in the experiment. So far the solution to this problem was to reduce to the one-more discrete logarithm problem.

Our proof technique for Theorem 2 (formally, Theorem 14) is to take any prover and turn it into an adaptive prover who makes a chain of n proofs, together with some bookkeeping. We then show that any adaptive extractor against this prover must either take exponential time or reduce to a straight-line extractor against the original prover. Applying this theorem to the honest prover, we get a reduction to Σ-one-wayness. We summarize these results in the following table (FSS = Fiat-Shamir-Schnorr, DLOG = discrete logarithm, OMDL = one-more discrete logarithm).

Property	Breaks if FSS has
One-way (e.g. DLOG)	Straight line extractor [18]
Σ-one-way	Adaptive extractor (new)
One-more one-way (e.g. OMDL)	Adaptive extractor [5]

In conclusion, we suggest that adaptive proofs are not a new class of proofs but rather another description of the class of straight-line extractable proofs; and Fiat-Shamir transformed Σ-protocols for "useful" functions are *not* in this class.

WEAKER ASSUMPTIONS. The obvious next question is whether one can improve our results even further and only rely on one-wayness (e.g. in the case of Schnorr, DLOG) rather than Σ-one-wayness. We show using a meta-metareduction that no algebraic metareduction [16] from a programming extractor to one-wayness (e.g. DLOG) can exist, unless one-wayness is already easy. All previous metareductions in this area [5,18] including ours to Σ-one-wayness are algebraic: the only operations they perform on elements of the target group are group operations.

A GENERALIZATION OF THE GENERIC GROUP MODEL (GGM). To strengthen trust in the Σ-one-wayness hypothesis on which our impossibility relies we provide a justification in the generic group model. Interestingly, the first problem that one needs to face here is that the existing approaches to formalizing and using GGM is not suitable: in brief, an adversary that interacts with the Σ protocol for some problem gets to see not only group elements but also some information related to the exponents of these group elements – this ability is not considered the standard GGM formalizations. We suggest one approach to deal with this issue and use the resulting model to formally justify Σ-one-wayness.

Related Work. One-more type assumptions were introduced by Bellare et al. [1]. Their value in proving schemes secure is subject to some debate, as explained by Koblitz and Menezes [13] who also gave the first "weakened" one-more assumption. Problems with forking-based proofs were first noted by Shoup and Gennaro [20]; Paillier and Vergnaud [16] developed separation results for Schnorr-based signatures using metareductions that formed the first formal proof of a limitation of Schnorr-based techniques. Both Brown [8] and Bresson et al. [9] concurrently applied separation techniques to one-more problems. Fischlin and Fleischhacker [12] were the first to consider limitations of metareductions via meta-metareductions. The most recent results that motivated this paper are Seurin and Treger [18] who gave a very simple metareduction from a non-programming extractor for Schnorr proofs to discrete log; and Bernhard et al. [5] who introduced adaptive proofs.

2   Preliminaries

NOTATION. $f : A \to B$ is a function with domain A and range B; $\mathcal{A} : A \twoheadrightarrow B$ is a randomised algorithm on the same domain and range. We write security games in a language based on Bellare and Rogaway's code-based game-playing [2]. $y \leftarrow f(x)$ is assignment, $x \twoheadleftarrow R$ is uniform random sampling. $T[i]$ is the element at index i of table T.

An interactive, randomised algorithm $\mathcal{A}$ has access to a random string r and an input/output interface. It maintains its state between calls. A security game is such an algorithm that may at some point output "win" or "lose", which terminates the entire execution. We say that a security property is given by a game, to mean that the property holds if no efficient adversary can cause the game to output "win" with more than negligible probability in some underlying security parameter.

Σ-PROTOCOLS — Let k be a field. Let $\mathcal{W}, \mathcal{X}$ be k-vector spaces and let $\phi : \mathcal{W} \to \mathcal{X}$ be a k-linear map, (i.e. linear map w.r.t. the field k). Suppose further that one can sample uniformly from k and $\mathcal{W}$.

Definition 3. *The Σ-protocol Σ_ϕ is the following interactive protocol for a prover P to prove knowledge of a preimage under ϕ to a verifier V.*

$$
\begin{array}{lcl}
P(w \in \mathcal{W}, x = \phi(w)) & & V \\
r \twoheadleftarrow \mathcal{W}; a \leftarrow \phi(r) & \xrightarrow{\;(x,a)\;} & \\
 & \xleftarrow{\;c\;} & c \twoheadleftarrow k \\
s \leftarrow r + c \cdot w & \xrightarrow{\;s\;} & \phi(s) \overset{?}{=} a + c \cdot x
\end{array}
$$

We choose to let P transmit the statement x to V as part of the first round of the proof — of course, V may also know x in advance. The verifier accepts if the equation $\phi(s) = a + c \cdot x$ holds in $\mathcal{X}$, in which case we call (x, a, c, s) an accepting transcript. Instances of this template protocol include:

- Schnorr: $\mathcal{W} = k = GF(p)$, $\mathcal{X}$ is some group G of order p with a generator g (e.g. over an elliptic curve) and $\phi(w) = g^w$.

- Chaum-Pedersen: $\mathcal{W} = k$, $\mathcal{X} = G \times G$ for a group as above and $\phi(w) = (g^w, h^w)$ for two different generators g, h of G.
- Representation: $\mathcal{W} = k^n$, $\mathcal{X} = G$ and $\phi(w_1, \ldots, w_n) = \prod_{i=1}^{n} g_i{}^{w_i}$ for some known set of generators $\{g_i\}_{i \in I}$ of G.

Σ-protocols according to our definition automatically satisfy:

- Special soundness: if (x, a, c, s) and (x, a, c', s') are accepting transcripts with $c \neq c'$ then $1/(c - c') \cdot (s - s')$ is a preimage[3] of x under ϕ.
- Soundness: if $x' \in \mathcal{X} \setminus \operatorname{Im}[\phi]$, then a cheating prover gets a verifier to accept with probability at most $1/|k|$.
- Honest-verifier zero-knowledge: a verifier who chooses c as prescribed (at least, independently of a) gains no information from the protocol beyond the fact that the prover knows a preimage of x under ϕ.

PROOF SCHEMES — A (non-interactive) proof scheme for a relation ρ on sets X, W consists of a proof space Π and a pair of algorithms $\mathtt{prove} : X \times W \twoheadrightarrow \Pi$ and $\mathtt{verify} : X \times \Pi \to \{0, 1\}$ (i.e. $\mathtt{verify}$ is deterministic). An element $\pi \in \Pi$ satisfying $\mathtt{verify}(x, \pi) = 1$ is called a valid proof for x. For any (x, w) satisfying ρ, if $\pi \leftarrow \mathtt{prove}(x, w)$ then we require $\mathtt{verify}(x, \pi) = 1$. We further assume that there is an algorithm $\mathtt{sample} : \twoheadrightarrow X \times W$ that produces elements uniformly distributed in ρ (as a subset of $X \times W$). In the random oracle model (ROM), both $\mathtt{prove}$ and $\mathtt{verify}$ may call a function H that is modelled as a random oracle in security proofs. The relation ρ itself does not use H.

FIAT-SHAMIR — The Fiat-Shamir transformation turns Σ-protocols into non-interactive proof schemes that are full zero-knowledge proofs of knowledge in the random oracle model. The idea is simply to replace the verifier's challenge c by a hash over the statement x and the commitment a.

Definition 4. *Let $\phi : \mathcal{W} \to \mathcal{X}$ be a k-linear function where $\mathcal{W}$ and k are efficiently sampleable. Suppose that H is a function with domain (including) $\mathcal{X} \times \mathcal{X}$ and range k. Then the Fiat-Shamir transformed Σ-protocol $\mathcal{F}_\phi$ is the following proof scheme for sets $(\mathcal{X}, \mathcal{W})$ and relation $\rho(x, w) = 1 \iff \phi(w) = x$. The proof space is $\Pi = \mathcal{X} \times \mathcal{W}$ and the algorithms are*

- $\mathtt{sample}()$: *pick* $w \leftarrow \mathcal{W}$, *set* $x \leftarrow \phi(w)$ *and return* (x, w).
- $\mathtt{prove}(x, w)$: *pick* $r \leftarrow \mathcal{W}$, *set* $a \leftarrow \phi(r)$, $c \leftarrow H(x, a)$ *and* $s \leftarrow r + cw$. *The proof is* $\pi = (a, s)$.
- $\mathtt{verify}(x, (a, s))$: *check that* $\phi(s) = a + H(x, a) \cdot x$.

---

[3] The inversion $1/(c - c')$ is in the field k where it exists due to $c \neq c'$; the dot in this formula is field-vector multiplication.

3   Variations on the Theme of One-Wayness

Σ-protocols are only useful when the function ϕ is hard to invert: otherwise, they are trivially zero-knowledge proofs of knowledge, but so is the protocol in which the prover just sends the statement x to the verifier. For the same reasons, if ϕ is easy to invert then $\mathcal{F}_\phi$ is adaptively secure too. This shows that we cannot hope for a theorem of the form "Fiat-Shamir Schnorr is not adaptively secure", since it is adaptively secure e.g. in the group $(\mathbb{Z}_p, +)$ where taking discrete logarithms is easy. Consequently, limitation theorems take the form "if Schnorr is adaptively secure then some property (e.g. OMDL) is easy to break".

We discuss some possible security properties of the function ϕ and introduce Σ-one-wayness. Later on we will show that Fiat-Shamir proofs cannot be adaptively secure unless Σ-one-wayness of ϕ is easy to break (in which case, their use within protocols would be already questionable).

Σ-ONE-WAYNESS. One-wayness is the first obvious candidate property. Recall that a property defined by a game means that the property (in this case one-wayness of ϕ) holds if it is hard to make the game output "win". The game gives the adversary a uniformly chosen image x; the adversary wins by recovering any preimage w' s.t. $\phi(w') = x$.

Definition 5. *The one-wayness property for a function $\phi : \mathcal{W} \to \mathcal{X}$ is given by the following game.*

```
1 | w ← W; x ← φ(w)
2 | output x; input w' ∈ W
3 | if φ(w') = x then return "win" else return "lose" end
```

We propose a new security notion that we call Σ-one-wayness (for linear functions). This says that even a dishonest verifier (who chooses c arbitrarily, maybe depending on x and a) cannot extract w from a single run of the protocol. We do not claim that Σ-one-wayness is a sufficient security notion for Σ protocols (it says nothing about extracting partial information on w) but we postulate that it is only deployed if this condition is satisfied. Consequently, we are not proposing a new scheme that is secure under the Σ-one-wayness assumption, but we will claim that if the Fiat-Shamir proof scheme for a function ϕ is adaptively secure then the Σ-one-wayness property for ϕ is easy to break too. Σ-one-wayness is clearly stronger than one-wayness of the function ϕ, but it will turn out to be weaker than one-more one-wayness (which we define in a moment).

Definition 6. *The Σ-one-wayness property for a linear function $\phi : \mathcal{W} \to \mathcal{X}$ is defined by the following game.*

```
1 | w ← W; x ← φ(w)
2 | r ← W; a ← φ(r)
3 | output (x, a); input c ∈ k
4 | s ← r + c · w
5 | output s; input w' ∈ W
6 | if φ(w') = x then return "win" else return "lose" end
```

If Σ-one-wayness of a function ϕ is easy to break but one-wayness is hard to break, then running the protocol Σ_ϕ could leak a preimage to a dishonest verifier, that said verifier could otherwise not compute by herself. In this case we would discourage the use of the protocol Σ_ϕ (among other things it is certainly not zero-knowledge). If both Σ-one-wayness and one-wayness of ϕ are easy to break then Σ_ϕ is both harmless and useless. So, we think that Σ-one-wayness of ϕ is a necessary condition for the protocol Σ_ϕ to be deployed. Under this condition, we will show that the Fiat-Shamir transformed Σ_ϕ is not adaptively secure.

WEAK ONE-MORE ONE-WAYNESS. For Schnorr, the one-wayness property of the function $\phi(x) = g^x$ is the discrete logarithm property. Bernhard et al. [5] use a stronger assumption known as one-more discrete logarithm, which generalises to one-more one-wayness [1,9].

One-more one-wayness assumes an invertible function ϕ. The adversary is given two oracles which she can call in any order: a sampling oracle that picks a random preimage $w_i \in \mathcal{W}$ and reveals $x_i = \phi(w_i)$ and an opening oracle that on input x outputs $\phi^{-1}(x)$. To win the game, the adversary must recover the preimages of all samples x_i with fewer calls to the opening oracle than to the sampling oracle.

One-more one-wayness was first discussed by Bellare et al. [1] although the name first appears in a later paper [9]. Unlike the other properties here, it does not admit an efficient security game: the game itself needs to be able to invert ϕ on arbitrary inputs. Koblitz and Menezes [13] discussed a variant that only allows the adversary to open challenges themselves; this is insufficient for our applications. Instead we propose a new property: weak one-more one-wayness fixes this problem by restricting the adversary to asking linear combinations of the sampled challenges. Your task is still to recover all w_i with fewer queries to the linear combination oracle than to the sampling oracle. The requirement for ϕ to be invertible can also be dropped again.

Definition 7. *The weak and normal one-more one-wayness properties for a bijection $\phi : \mathcal{W} \rightarrow \mathcal{X}$ are given by the following games. In each game the adversary can call the* sample *and* open *oracles many times, in any order. The adversary wins the game if it can provide preimages under ϕ for all samples that it had obtained and yet it made fewer opening queries than sample queries.*

weak one-more one-wayness:

```
1  sample():
2      n ← n + 1
3      w[n] ← W
4      return φ(w[n])
5
6  open(c₁,...,cₙ ∈ kⁿ):
7      return ∑ⁿᵢ₌₁ cᵢ · wᵢ
```

one-more one-wayness:

```
1  sample():
2      (same as weak version)
3
4
5
6  open(x ∈ X):
7      return φ⁻¹(x)
```

The strong problem clearly reduces to the weak one. A weak adversary can still obtain a preimage of a particular sample by submitting the vector with 1 at the appropriate position and 0 elsewhere.

The point of the weak one-more one-wayness property is that the theorem by Bernhard et al. [5] can trivially be strengthened to show that Fiat-Shamir-Schnorr is not adaptively secure even under the weak one-more discrete logarithm property: their reduction only ever makes opening queries on elements that are linear combinations of samples with known coefficients. This is not surprising since their reduction is trying to "simulate" Schnorr proofs on sample elements.

Σ-one-wayness reduces to weak one-more one-wayness, even to a weaker version where the number of samples is additionally bounded at 2 and only a single linear combination query is allowed. Thus, we end up with a hierarchy of one-wayness/Σ-one-wayness/weak one-more one-wayness/one-more one-wayness, in order of increasing strength.

4 Adaptive Proofs

In this section we recall the notion of adaptive proofs and introduce templates for a couple of provers that form the basis of the results we prove in the next section.

Definition 8. *A prover is an algorithm $\mathcal{P}$ that outputs a statement/proof pair (x, π); in the ROM a prover may make random oracle calls. We assume that there is a uniformly sampleable space of random strings R associated to each prover and we write $\mathcal{P}(r)$ to mean running prover $\mathcal{P}$ on random string $r \in R$. In the ROM, we write $(x, \pi, l) \leftarrow \mathcal{P}(r)$ to mean that we also return the list l of random oracle queries made by this execution of the prover on random string r.*

A proof scheme is sound with error ε if for any prover $\mathcal{P}$, the probability of producing a pair (x, π) such that $\mathtt{verify}(x, \pi) = 1$ but no w exists making $\rho(x, w)$ hold, is at most ε.

A proof scheme in the random oracle model is straight-line extractable with error ε if there is an extractor $\mathcal{K}$ as follows. For any prover $\mathcal{P}$, pick $r \leftarrow R$ and execute $(x, \pi, l) \leftarrow \mathcal{P}(r)$. If $\mathtt{verify}(x, \pi) = 1$ w.r.t.[4] l then with probability at least $1 - \varepsilon$, $\mathcal{K}(x, \pi, l)$ returns a w such that $\rho(x, w)$ holds. It follows immediately that an extractable proof scheme with error ε is also sound with at most the same error.

A straight-line extractor, has black-box access to further copies of the prover in the following sense: it may start these copies and control all interaction with them, including picking the random string and answering all random oracle queries. As motivation for this (established) notion of straightline extractors, consider an extractor who is trying to extract from an honest prover. The code

---

[4] We say $\mathtt{verify}(x, \pi) = 1$ w.r.t. l if all elements on which $\mathtt{verify}$ queries the oracle on input (x, π) are contained in l, and $\mathtt{verify}$ outputs 1 on these inputs if given the appropriate responses in l.

of the honest prover is known, so the extractor can always simulate the honest prover on inputs of its choice. The extractor cannot see the random coins of the "main" copy of the honest prover from which it is trying to extract, however.

A proof scheme in the ROM has a *programming* straight-line extractor with error ε if there is a straight-line extractor $\mathcal{K}$ as follows. Let any prover $\mathcal{P}$ interact with $\mathcal{K}$ in the sense that $\mathcal{K}$ answers $\mathcal{P}$'s random oracle queries. If $\mathcal{P}$ outputs (x, π) such that $\texttt{verify}(x, \pi) = 1$ w.r.t. the oracle queries made by $\mathcal{P}$, then with probability at least $1 - \varepsilon$, $\mathcal{K}$ outputs a w such that $\rho(x, w)$ holds.

A rewinding extractor can, in addition to the capabilities of a straight-line extractor, launch further copies of the prover $\mathcal{P}$ with the *same random string* as the one that the extractor is trying to extract from, and answer their random oracle queries. The difference between straight-line and rewinding extractors is thus that rewinding extractors can run further copies of the prover that behave identically to the "main" one, as long as they receive the same inputs and outputs, and thus "fork" the prover instance from which they are trying to extract.

ADAPTIVE PROOFS. We present the adaptive proof game of Bernhard et al. [5]. The game is an interactive algorithm with two interfaces for an adaptive prover and an extractor. An adaptive prover for a proof scheme $(\Pi, \texttt{prove}, \texttt{verify})$ w.r.t. (X, W, ρ) is an algorithm that can repeatedly output pairs $(x, \pi) \in X \times \Pi$ and expect witnesses $w \in W$ in return. After a number of such interactions, the adaptive prover halts. In the random oracle model, an adaptive prover may also ask random oracle queries.

An adaptive extractor is an algorithm $\mathcal{K}$ that can interact with an adaptive prover: it repeatedly takes pairs $(x, \pi) \in X \times \Pi$ such that $\texttt{verify}(x, \pi) = 1$ as input and returns witnesses $w \in W$ such that $\rho(x, w) = 1$. In addition, an adaptive extractor may be rewinding, i.e. launch further copies of the adaptive prover that run on the same random string as the main one (managed by the game) and answer all their queries. The adaptive proof game does not check the correctness of witnesses for the other copies of the prover.

Definition 9. *A proof scheme $(\Pi, \texttt{prove}, \texttt{verify})$ is an n-proof (with error ε) in the random oracle model if there is an adaptive extractor $\mathcal{K}$ such that for any adaptive prover $\mathcal{P}$, the extractor wins the game in Fig. 1 (with probability at least $1 - \varepsilon$). The adaptive extractor $\mathcal{K}$ may launch and interact with further copies of the adaptive prover $\mathcal{P}$ on the same random string r as the main one, without the game mediating between them.*

In the n-proof game in Fig. 1, the prover $\mathcal{P}$ is trying to find a claim (x, π) that verifies, but from which the extractor $\mathcal{K}$ cannot extract a witness w. The extractor is trying to extract witnesses from all claims made by the prover.

The game uses three global variables. K stores the number of witnesses that the extractor has found so far. If this counter reaches n, the extractor wins and the scheme is an n-proof. Q stores a list of all the prover's random oracle queries so far. These are provided[5] to the extractor along with each of the prover's

---

[5] The original definition gave the extractor an extra $\texttt{list}$ oracle to query the prover's random oracle list. Our presentation is equivalent.

1	*initialise*:	18	$\mathcal{K}$ asks $\mathbf{ro}(x)$:
2	$Q \leftarrow [\,]$	19	$y \leftarrow \mathbf{ro}(x)$
3	$K \leftarrow 0$	20	send y to $\mathcal{K}$
4	$r \twoheadleftarrow R$	21	
5	run $\mathcal{P}(r)$	22	$\mathcal{K}$ outputs w:
6		23	if $\rho(\varXi, w)$ then
7	$\mathcal{P}$ asks $\mathbf{ro}(x)$:	24	$K \leftarrow K + 1$
8	$y \leftarrow \mathbf{ro}(x)$	25	if $K = n$ then
9	$Q \leftarrow Q :: (x, y)$	26	$\mathcal{K}$ wins; halt.
10	send y to $\mathcal{P}$	27	end
11		28	send w to $\mathcal{P}$
12	$\mathcal{P}$ outputs (x, π):	29	else
13	if not $\mathbf{verify}(x, \pi)$ then	30	$\mathcal{P}$ wins; halt.
14	$\mathcal{K}$ wins; halt.	31	end
15	end	32	$\mathcal{P}$ halts:
16	$\varXi \leftarrow x$	33	$\mathcal{K}$ wins; halt.
17	send (x, π, Q) to $\mathcal{K}$		

Fig. 1. The adaptive proof game. The extractor also has access to further copies of $\mathcal{P}(r)$ – to avoid cluttered notation we do not show this access explicitly.

claims. $\varXi$ stores the last statement that appeared in one of the prover's claims; it is used to check the validity of a witness returned by the extractor.

A n-proof for $n = 1$ is simply a proof of knowledge in the ROM: the prover makes a single claim (a pair containing a statement x and a proof π) and the extractor wins if it can obtain a witness. A proof scheme is an adaptive proof if there is an adaptive extractor that works for any polynomially bounded parameter n.

CANONICAL PROVERS. The canonical prover $\mathcal{P}_C$ samples a statement/witness pair and creates a proof. We write R_C for the randomness space of the canonical prover and $\mathcal{P}_C(r)$ to denote running the canonical prover on the random string $r \in R_C$.

Definition 10 (canonical prover). *Let $(\Pi, \mathbf{prove}, \mathbf{verify})$ be a proof scheme for (X, W, ρ) where r is uniformly sampleable via an algorithm* sample. *The canonical prover $\mathcal{P}_C$ for this scheme is the following algorithm.*

$$(x, w) \leftarrow \mathbf{sample}(); \ \pi \leftarrow \mathbf{prove}(x, w); \ \text{return} \ (x, \pi)$$

If required, the canonical prover can also return the list l of all random oracle queries made during its execution (by prove*).*

Since an extractor is supposed to work against any (efficient) prover, to argue that an extractor cannot exist it is enough to show that one cannot extract from the canonical prover. To deal with adaptive extractors, we propose the following construction of an adaptive chain prover $\mathcal{P}^n$ from any prover $\mathcal{P}$. It follows the idea of Shoup and Gennaro [20] in making a chain of "challenges" (in this case proofs) where each challenge depends on all previous ones and then asking queries on them *in reverse order*. This way, the obvious rewinding extractor using special soundness will take exponential time.

To make each challenge depend on previous ones, we use a function F to update the random string for each of the n contained copies of $\mathcal{P}$ based on the (random oracle) state of the previous copy. The final parameter l returned by $\mathcal{P}$ is the list of all random oracle queries made by this copy. Recall that $\mathcal{P}(r)$ means run prover $\mathcal{P}$ on random string $r \in R$ where R is the randomness space for this prover.

Definition 11 (adaptive chain prover). *Let* $(\Pi, \mathtt{prove}, \mathtt{verify})$ *be a proof scheme for* (X, W, ρ). *Let* $\mathcal{P}$ *be any prover (in the ROM) and let* R *be its randomness space. Let* L *be the space of possible random oracle input/output transcripts. Let* $F : R \times L \to R$ *be a function (which does not depend on the random oracle). The adaptive chain prover* $\mathcal{P}^n$ *of order* n *w.r.t. function* F *is the algorithm in Fig. 2, taking an* $r \in R$ *as input.*

```
                                          5   for i = n, 1, -1 do
        P^n(r):                           6       output (x_i, π_i)
                                          7       input w'
    1  for i = 1, n do                    8       if not ρ(x_i, w') then
    2      (x_i, π_i, l) ← P(r)           9          halt
    3      r ← F(r, l)                    10      end
    4  end                                11  end
```

Fig. 2. The adaptive chain prover $\mathcal{P}^n$. The provder depends on function F which here we assumed fixed. When F is a pseudorandom function, a key for F is sampled at the beginning of the execution of $\mathcal{P}^n$.

Later on, we will take F to be a (pseudo-)random function. This has the effect that two copies of $\mathcal{P}^n$ that get identical answers to their random oracle queries will behave identically, but two copies of $\mathcal{P}^n$ that "fork" will behave as copies of $\mathcal{P}$ with *independent* random strings from the forking point onwards.

The intuition behind this construction is that having access to copies of $\mathcal{P}$ on some uniformly random string r' cannot help you extract from a copy $\mathcal{P}(r)$, as long as r and r' are independent — certainly, an extractor could always simulate such copies herself if the code of $\mathcal{P}$ is known. We will use this idea to show that forking a copy of $\mathcal{P}^n$ is no help in extracting from the proofs made later on by another copy.

5   Limitations of the Fiat-Shamir Transformation

We recall the hierarchy of security definitions for functions (the last is the strongest): one-way/Σ-one-way/weak one-more one-way/one-more one-way.

KNOWN LIMITATIONS. Seurin and Treger [18] proved that Fiat-Shamir-Schnorr cannot have a non-programming straight-line extractor unless the underlying function is not one-way (i.e. one can take discrete logarithms). The following theorem generalizes this result to the case of arbitrary Σ-protocols.

Theorem 12. *Suppose there is a non-programming straight-line extractor K for the proof scheme $\mathcal{F}_\phi$. Then ϕ is not one-way. Specifically, there is an algorithm breaking one-wayness with approximately the same running time and success probability as the extractor K has against the canonical prover $\mathcal{P}_C$.*

Proof. Let x be a challenge from the one-way game for ϕ; we need to find a w' such that $\phi(w') = x$. We simulate a proof: pick $s \leftarrow \mathcal{W}$, $c \leftarrow k$ and $a \leftarrow \phi(s) - c \cdot x$. Then we give the extractor the statement x, proof (a, s) and a list of random oracle queries consisting of the entry $RO(x, a) = c$. These elements are identically distributed to what the extractor would see in an execution with the canonical prover $\mathcal{P}_C$ for $\mathcal{F}_\phi$. We pass any witness w' returned by the extractor on to the challenger to win with the same success probability.   □

Bernhard et al. [5] showed that substituting an adaptive extractor for a straight-line one gets a similar result for the one-more one-wayness assumption on the function ϕ — the proof of this theorem is nontrivial however. While their original proof only concerned Fiat-Shamir-Schnorr, a close inspection of the proof shows that it works for any Σ protocol and that the weak one-more assumption is sufficient too. In summary:

Theorem 13. *Suppose that there is an efficient adaptive extractor for $\mathcal{F}_\phi$. Then ϕ is not weak one-more one-way.*

We will not re-prove the theorem here. As to running time, as long as the extractor makes fewer than 2^n queries when running against a particular prover then the reduction to weak one-more one-wayness runs in the same time as the extractor, but its success probability is the inverse of the number of copies of the prover that the extractor causes to be invoked. Although Bernhard et al. [5] only prove the theorem for the case of the Fiat-Shamir-Schnorr protocol, their reduction is "black box" in the sense that it only needs to be able to sample and open instances of the underlying Σ-protocol. The exact same proof will work for our generalisation. We can write the prover in the cited theorem as $(\mathcal{P}_C)^n$, the adaptive chain prover derived from the canonical prover. This will allow us to conclude that any extractor against the chain prover implies a straight-line extractor against the canonical prover $\mathcal{P}_C$.

NEW RESULTS. If we switch to a programming straight-line extractor, we can show a separation result under the Σ-one-wayness assumption.

Theorem 14. *Suppose there is a programming straight-line extractor K for $\mathcal{F}_\phi$. Then ϕ is not Σ-one-way. Specifically, there exists a reduction with approximately the same running time and the same success probability as the extractor K against the canonical prover $\mathcal{P}_C$.*

Proof. We simulate the canonical prover towards the extractor. Receive x, a from the Σ-one-way challenger and ask the random oracle query $c \leftarrow RO(x, a)$ (which the extractor answers). Then send c to the Σ-one-way challenger to get s and send (x, a, s) to the extractor. Again, whenever the extractor provides the correct w', we win against the challenger. $\qquad\square$

This result is new, but not surprising — Fiat-Shamir transformed Σ-protocols are not supposed to be straight-line extractable and the Σ-one-wayness property is constructed exactly to make this reduction work. The value of said property is that we can also use it for adaptive extractors.

Our main contribution in this paper is a new theorem that says all adaptive proofs in the ROM admit a straight-line extractor.

Theorem 15. *Consider any non-interactive ROM proof scheme with an adaptive extractor $\mathcal{K}$. Suppose that, running against any n-prover $\widehat{\mathcal{P}}$, the extractor $\mathcal{K}$ causes at most $f(n) < 2^n$ copies of the prover to be run in the experiment and answers all extraction queries of the main run correctly with probability at least $p(n) > 0$. Then there is a programming straight-line extractor against any non-adaptive prover $\mathcal{P}$ with success probability $p(n)/(n \cdot f(n))$.*

Remark 16. The proof, which we provide in the full version, is information theoretic. It relies only on the number of copies of P instantiated by $\mathcal{K}$ and, in particular, it makes no assumptions on the efficiency of $\mathcal{P}$ and $\mathcal{K}$. It does not establish a relationship between the running time of $\mathcal{K}$ and the success probability; it is simply an application of the pigeonhole principle to derive a contradiction that whenever $f(n) < 2^n$ then $\mathcal{K}$ must essentially have "guessed correctly" rather than computed the preimage through interacting with $\mathcal{P}$. For example, if launching a new copy of $\mathcal{P}$ costs $\mathcal{K}$ one unit of time then the theorem provides negative results even for subexponential-time extractors.

Applying this theorem to protocols obtained via the Fiat-Shamir transform from Σ-protocols yields the following insight.

Corollary 17. *Suppose that the Fiat-Shamir transformed Σ-protocol $\mathcal{F}_\phi$ is adaptively secure. Then ϕ is not Σ-one-way secure.*

For the corollary, note that we have a programming straight-line extractor against the canonical prover $\mathcal{P}_C$ by applying Theorem 15 to the prover $(\mathcal{P}_C)^n$. The result then follows from Theorem 14.

We sketch the proof here and provide the full argument in the full version of this paper [7]. Let $\mathcal{P}$ be a non-adaptive prover. We construct a simulator $\mathcal{S}^n$ that is indistinguishable from the black box providing "rewinding" access for multiple copies of $\mathcal{P}^n$. The point of the simulator is that it shares state "between the copies". We then guess which instance of the prover (specifically, which proof) the extractor is going to answer without "forking" and inject the Σ-one-wayness challenge into it. The same combinatorial argument as in the proof of Bernhard et al. [5] shows that such an instance must exist if the extractor launches fewer than 2^n copies of the prover.

The core of such a simulation argument is to keep track of a history of each instance of the prover, since two copies of the prover with identical histories must behave identically towards the extractor. In $(\mathcal{P}_C)^n$, this history is implicitly tracked in the randomness r used for each copy of $\mathcal{P}_C$, however a collision in the random oracle could lead to two copies with different histories "merging". Our simulator computes an explicit history instead, namely the list of all random oracle queries so far.

As in Bernhard et al. [5] we define an event E that occurs whenever a copy of the prover gets its extraction query answered without having a "partner" (another copy, from which the witness was extracted by forking and special soundness). The novelty in our proof is that because we have cast the prover as a chain $(\mathcal{P}_C)^n$ with suitable state tracking, we can show that event E implies not only a "break" of the chain prover but also of one of the contained canonical provers $\mathcal{P}_C$. We then show that, if the simulator guessed correctly, event E implies that the simulator can solve its Σ-one-wayness challenge. This is a much weaker assumption than one-more one-wayness.

6   Generic Hardness of Σ-One-Wayness

In this section we show that Fiat-Shamir transformed Σ-protocol $\mathcal{F}_\phi$ is not adaptively secure in the generic setting. Thus if we want to build a protocol where we need an adaptively secure proof (such as to get CCA encryption), we would not use $\mathcal{F}_\phi$. By Corollary 17 we just need to prove that ϕ is Σ−one-way secure in the generic group model (GGM). Again, let $\mathcal{X}, \mathcal{W}$ be vector spaces over k and $\phi : \mathcal{W} \to \mathcal{X}$ be a k-linear map. We assume that k is a finite field and $\mathcal{X}, \mathcal{W}$ are finite dimensional, since sampling uniformly from an infinite set does not make much sense. Let $b_1, b_2, ..., b_n$ be basis vectors of $\mathcal{X}$, where $\dim(\mathcal{X}) = n$ and $\{b_1, ..., b_s\}$ be a basis for $\text{Im}(\phi)$. For each $1 \le i \le s$ denote a_i to be an element of $\mathcal{W}$ so that $\phi(a_i) = b_i$.

The generic group model is a model which analyses success of algorithms against representations of groups which do not reveal any information to adversary. There are many ways to formalise this idea [14,15,19]. We will follow the definition provided by Shoup [19]. Here, adversary is given access to images of elements of a group under a random injective map $\sigma : \mathcal{X} \to S \subset \{0,1\}^*$, called an *encoding function*. Group operations can be computed by making oracle queries. The adversary is given access to two oracles ADD and INV:

$$ADD(\sigma(x), \sigma(y)) = \sigma(x+y), \ INV(\sigma(x)) = \sigma(-x).$$

Note that the adversary cannot get any information from the representation $\sigma(x)$ of element x. A *generic algorithm* $\mathcal{A}$ for $\mathcal{X}$ on S is a probabilistic algorithm that takes as input an encoding list $(\sigma(x_1), \sigma(x_2), ..., \sigma(x_l))$ where each $x_i \in \mathcal{X}$ and σ is an encoding function of $\mathcal{X}$ on S. As the algorithm executes, it makes queries to ADD or INV oracles and then appends outputs of the queries to the encoding list. The output of $\mathcal{A}$ is a bit string denoted as $\mathcal{A}(\sigma; x_1, ..., x_l)$. We also want to extend the interface of the model and introduce the $FSS_{r,w}$ oracle:

$FSS_{r,w}(c) = r + cw$ for some $r, w \in W$ and $c \in k$. We may assume that when oracles encounter some $\sigma_1 \notin Im(\sigma)$, they return an error message.

The next theorem establishes generic hardness of Σ−one-wayness; we provide the proof in the full version of this paper [7].

Theorem 18. *Let w, r be random elements of W and A be a generic algorithm for X on $S \subset \{0, 1\}^*$ that makes at most m queries to ADD and INV oracles and exactly one query to $FSS_{r,w}$ oracle. Then the probability that $\phi(A(\sigma; b_1, ..., b_n, x, y)) = x$ is $O((m + n)^2/|X| + |ker(\phi)|/|W|)$, where $x = \phi(w)$ and $y = \phi(r)$.*

Note that Theorem 18 implies that every generic algorithm A, which wins Σ−one-wayness with high probability, must perform at least $\Omega(\alpha\sqrt{|X|})$ group operations, where $\alpha = \sqrt{1 - |\ker(\phi)|/|W|}$. In particular, if W is large then we get the lower bound $\Omega(\sqrt{|X|})$ for IES (described in [6]) by choosing $\phi(w) = g^w$.

7   Reducing to DLOG?

Given a non-programming straight line extractor in the ROM for a Fiat-Shamir transformed Σ-protocol $\mathcal{F}_\phi$ we can break one-wayness of ϕ; for a programming extractor or an adaptive extractor we can break Σ-one-wayness. This raises the question, can we break one-wayness given a programming extractor? Our answer is negative. We give the argument for the case of Schnorr proofs where one-wayness is the discrete logarithm (DLOG) problem; this also implies that there can be no generic metareduction to one-wayness for any Σ-protocol.

The metareductions in the theorems of Seurin and Treger [18], Bernhard et al. [5] and this paper are all algebraic (in the sense of Paillier and Vergnaud) [16] over the vector space[6] X, the range of the function ϕ. We therefore consider it a meaningful result to show that no algebraic metareduction to DLOG can exist (unless DLOG is already easy).

Theorem 19. *If there is an algebraic metareduction from a programming straight-line extractor for Fiat-Shamir-Schnorr proofs to the DLOG problem then there is also a meta-metareduction breaking the DLOG problem directly with approximately the same success probability.*

The proof is in [7]. The idea is that a metareduction M gets to see two bases in the group: the generator g and the challenge h from its DLOG challenger. Since we assumed a programming extractor, M must ask its random oracle queries to its extractor interface where our meta-metareduction will answer them. Any statement output by M (to the extractor) therefore has the form $(g^a h^b)$ for some (a, b) which are available to our meta-metareduction by use of the algebraic model. Proofs of the form $(a, 0)$ are independent of the challenge h; intuitively they should not help to compute the discrete logarithm of h so we just return

---

[6] Paillier and Vergnaud defined the algebraic model for groups; one can interpret a $GF(p)$ vector space as an Abelian group to use their definition of the algebraic model.

the witness a. The first time M outputs a proof with a statement of the form (a, b) with $b \neq 0$, we fork M on the relevant random oracle query and use special soundness to find the discrete logarithm of h to basis g.

8   Conclusions

Bernhard et al. introduced adaptive proofs, setting up a hierarchy of (1) proofs of knowledge (2) adaptive proofs and (3) straight-line extractable proofs, with a separation between (1) and (2). While useful for proving limitations of Σ-protocols, we have showed that adaptive proofs are not a new class of proof after all: all adaptively secure proofs admit a straight-line extractor against the canonical prover.

Along the way we have generalised previous results from Schnorr's protocol to Σ-protocols. In addition, we have weakened the counter-assumption from one-more one-wayness, which is a "q-type" interactive assumption (adversary gets an unbounded number of sample queries) and is not efficiently realisable to Σ-one-wayness, which both has a constant number of steps and an efficient security game.

Our result shows that the Fiat-Shamir transformation and Σ-protocols in general may be even weaker than previously thought. Namely, the non-interactive proof scheme $\mathcal{F}_\phi$ only achieves adaptive security if a single execution of the interactive protocol Σ_ϕ against a dishonest verifier already leaks the secret witness with non-negligible probability.

In essence, this shows that using proofs derived from the Fiat-Shamir scheme for some problem ϕ in a setting where adaptive security of such proofs is necessary requires care: these should be replaced with proofs that have straightline extractors. From a practice-oriented perspective, our results show that improving the efficiency of proofs that admit straightline extraction is an important line of future research.

References

1. Bellare, M., Namprempre, C., Pointcheval, D., Semanko, M.: The one-more-RSA-inversion problems and the security of Chaum's blind signature scheme. eprint 2001/002. Originally appeared as "The power of RSA inversion oracles and the security of Chaum's RSA-based blind signature scheme. In: Financial Cryptography. LNCS, vol. 2339, pp. 319–338. Springer, Heidelberg (2001)
2. Bellare, M., Rogaway, P.: The security of triple encryption and a framework for code-based game-playing proofs. In: Vaudenay, S. (ed.) EUROCRYPT 2006. LNCS, vol. 4004, pp. 409–426. Springer, Heidelberg (2006). doi:10.1007/11761679_25. The title cited is from the latest version on eprint at http://eprint.iacr.org/2004/331
3. Bellare, M., Rogaway, P.: Random oracles are practical: a paradigm for designing efficient protocols. In: Proceedings of the 1st ACM Conference on Computer and Communications Security, CCS 1993, pp. 62–73 (1993)
4. Bernhard, D.: Zero-knowledge proofs in theory and practice. Ph.D. thesis, University of Bristol (2014). www.cs.bris.ac.uk/bernhard/papers.html

5. Bernhard, D., Fischlin, M., Warinschi, B.: Adaptive proofs of knowledge in the random oracle model. In: Katz, J. (ed.) PKC 2015. LNCS, vol. 9020, pp. 629–649. Springer, Heidelberg (2015). doi:10.1007/978-3-662-46447-2_28

6. Bernhard, D., Fischlin, M., Warinschi, B.: On the hardness of proving CCA-security of signed ElGamal. In: Cheng, C.-M., Chung, K.-M., Persiano, G., Yang, B.-Y. (eds.) PKC 2016. LNCS, vol. 9614, pp. 47–69. Springer, Heidelberg (2016). doi:10.1007/978-3-662-49384-7_3

7. Bernhard, D., Nguyen, N.K., Warinschi, B.: Adaptive Proofs have Straightline Extractors (in the Random Oracle Model). Full version on eprint 2015/712

8. Brown, D.: Irreducibility to the One-More Evaluation Problems: More May Be Less. eprint 2007/435

9. Bresson, E., Monnerat, J., Vergnaud, D.: Separation results on the "one-more" computational problems. In: Malkin, T. (ed.) CT-RSA 2008. LNCS, vol. 4964, pp. 71–87. Springer, Heidelberg (2008). doi:10.1007/978-3-540-79263-5_5

10. Fiat, A., Shamir, A.: How to prove yourself: practical solutions to identification and signature problems. In: Odlyzko, A.M. (ed.) CRYPTO 1986. LNCS, vol. 263, pp. 186–194. Springer, Heidelberg (1987). doi:10.1007/3-540-47721-7_12

11. Fischlin, M.: Communication-efficient non-interactive proofs of knowledge with online extractors. In: Shoup, V. (ed.) CRYPTO 2005. LNCS, vol. 3621, pp. 152–168. Springer, Heidelberg (2005). doi:10.1007/11535218_10

12. Fischlin, M., Fleischhacker, N.: Limitations of the meta-reduction technique: the case of schnorr signatures. In: Johansson, T., Nguyen, P.Q. (eds.) EUROCRYPT 2013. LNCS, vol. 7881, pp. 444–460. Springer, Heidelberg (2013). doi:10.1007/978-3-642-38348-9_27. eprint 2013/140

13. Koblitz, N., Menezes, A.: Another look at non-standard discrete log and Diffie-Hellman problems. J. Math. Cryptol. 2(4), 311–326 (2008). eprint 2007/442

14. Maurer, U.: Abstract models of computation in cryptography. In: Smart, N.P. (ed.) Cryptography and Coding 2005. LNCS, vol. 3796, pp. 1–12. Springer, Heidelberg (2005). doi:10.1007/11586821_1

15. Nechaev, V.I.: Complexity of a determinate algorithm for the discrete logarithm. Math. Notes 55(2), 165–172 (1994)

16. Paillier, P., Vergnaud, D.: Discrete-log-based signatures may not be equivalent to discrete log. In: Roy, B. (ed.) ASIACRYPT 2005. LNCS, vol. 3788, pp. 1–20. Springer, Heidelberg (2005). doi:10.1007/11593447_1

17. Schnorr, C.P.: Efficient signature generation for smart cards. J. Cryptol. 4, 161–174 (1991). Springer

18. Seurin, Y., Treger, J.: A robust and plaintext-aware variant of signed ElGamal encryption. In: Dawson, E. (ed.) CT-RSA 2013. LNCS, vol. 7779, pp. 68–83. Springer, Heidelberg (2013). doi:10.1007/978-3-642-36095-4_5

19. Shoup, V.: Lower bounds for discrete logarithms and related problems. In: Fumy, W. (ed.) EUROCRYPT 1997. LNCS, vol. 1233, pp. 256–266. Springer, Heidelberg (1997). doi:10.1007/3-540-69053-0_18

20. Shoup, V., Gennaro, R.: Securing threshold cryptosystems against chosen ciphertext attack. In: Nyberg, K. (ed.) EUROCRYPT 1998. LNCS, vol. 1403, pp. 1–16. Springer, Heidelberg (1998). doi:10.1007/BFb0054113

More Efficient Construction of Bounded KDM Secure Encryption

Kaoru Kurosawa$^{(\boxtimes)}$ and Rie Habuka

Ibaraki University, Hitachi, Japan
kaoru.kurosawa.kk@vc.ibaraki.ac.jp

Abstract. Let sk_i be the secret-key of user i for $i = 1, \ldots, \ell$, and pk_j be the public-key of user $j \in \{1, \ldots, \ell\}$. A bounded Key Dependent Message (KDM) secure encryption scheme $\mathcal{E}_{\mathsf{b-KDM}}$ provides security even when one encrypts $f(sk_1, \ldots, sk_\ell)$ under pk_j for any function f which has arbitrarily fixed circuit size. An $\mathcal{E}_{\mathsf{b-KDM}}$ is known to be constructed from projection KDM seucrity. In this paper, we first show that it can be obtained from much weaker KDM security than the projection KDM security. We next present more efficient $\mathcal{E}_{\mathsf{b-KDM}}$ than before under various assumptions.

Keywords: KDM · Key dependent message · Encryption · Garbling scheme

1 Introduction

1.1 Background

Key dependent message (KDM) security has been studied by many researchers recently. A KDM-secure encryption scheme provides security even when one encrypts the secret-key sk under the corresponding public-key pk. More generally, a KDM-secure encryption scheme with respect to a set of functions $\mathcal{F}$ provides security even when one encrypts $f(sk_1, \ldots, sk_\ell)$ under pk_j for any function $f \in \mathcal{F}$, where sk_i is the secret-key of user i for $i = 1, \ldots, \ell$, and pk_j is the public-key of user $j \in \{1, \ldots, \ell\}$.

Boneh et al. [5] showed the first KDM-secure public-key encryption scheme in the standard model which is called BHHO encryption scheme. It is KDM-secure with respect to the set of all "affine in the exponent" functions under the DDH assumption.

Then a natural question is "can we construct a KDM-secure encryption scheme with respect to the set of all functions ?" Barak et al. [4] affirmatively solved this problem by showing that there exists a KDM-secure encryption scheme with respect to the set of all functions which have arbitrarily fixed circuit size. Such encryption schemes are called *bounded* KDM secure. Barak et al. constructed a bounded KDM secure encryption scheme under either the DDH assumption or the LWE assumption.

© Springer International Publishing AG 2017
D. Gollmann et al. (Eds.): ACNS 2017, LNCS 10355, pp. 354–372, 2017.
DOI: 10.1007/978-3-319-61204-1_18

On the other hand, a function is called projection if each output bit depends on at most one input bit. An encryption scheme is called *projection* KDM secure if it is KDM secure with respect to the set of all projection functions. Applebaum [1] showed that a bounded KDM secure encryption scheme can be obtained from a projection KDM secure encryption scheme.

Bellare et al. [6] then showed a clean method to construct a bounded KDM secure encryption scheme from a garbling scheme Ga and a projection KDM secure encryption scheme $\mathcal{E}$. Their scheme is more efficient than the schemes of [1,4]. (See [6, Fig. 20] for comparison.)[1] However, the size of ciphertexts is still very large.

To summarize, it is known that bounded KDM security can be obtained from projection KDM security. However, the size of ciphertexts is very large.

1.2 Our Contribution

In this paper, we first show that bounded KDM security can be obtained from much weaker KDM security than the projection KDM security. Based on this, we next present a more efficient construction of bounded KDM secure encryption schemes than Bellare et al. [6] under various assumptions.

In the scheme of Bellare et al. [6], a ciphertext of $f(sk_1, \ldots, sk_\ell)$ consists of (F, Y), where F is a garbled circuit and Y is a ciphertext of a projection KDM secure encryption scheme $\mathcal{E}$. Our scheme has the same structure, but $\mathcal{E}$ is required to be more weakly KDM secure. (More precisely, our weak KDM security is related to the underlying garbling scheme Ga.)

As a result, the size of our Y is k times smaller than that of Bellare et al. [6], where k is the security parameter. Further suppose that the number of gates of $f(sk_1, \ldots, sk_\ell)$ is $O(k\ell)$. Then the total size of our ciphertexts is $O(k^3\ell)$ under the DDH assumption while it is $O(k^4\ell)$ in the scheme of Bellare et al. [6], where BHHO encryption scheme is used in both schemes.

A projection KDM secure encryption scheme can be constructed from any KDM secure encryption scheme with respect to the set of affine (in the exponent) functions, where the secret-key is viewed as a bit string.[2] We can also construct our weakly KDM secure encryption scheme from such encryption schemes.

- As stated above, BHHO encryption scheme is KDM-secure with respect to the set of "affine in the exponent functions" under the DDH assumption.
- Brakerski and Goldwasser [3] showed a BHHO-like encryption scheme under the subgroup indistinguishablity assumptions. In particular, they presented a KDM secure encryption scheme with respect to the set of affine functions under the Paillier's decisional composite residuosity (DCR) assumption.
- Applebaum et al. [2] constructed a *symmetric-key* encryption scheme which is KDM-secure with respect to the set of affine functions under the LPN assumption.

---

[1] In [6], Bellare et al. mainly formalized Yao's garbled circuits as garbling schemes.

[2] Therefore we cannot use the KDM secure encryption schemes of [2, Sect. 3] [7].

Hence, we can construct a more efficient bounded KDM secure encryption scheme than Bellare et al. [6] under the DDH assumption, the DCR assumption and the LPN assumption respectively.

We also point out that the security proof of Bellare et al. [6] is not complete, and show how to fix it.

2   Preliminaries

PPT means probabilistic polynomial time, and PT means polynomial time. If A is a PPT algorithm, then $y \leftarrow A(x_1, \ldots, x_n; r)$ represents the act of running the algorithm A with inputs $x_1, \ldots, x_n$ and coins r to get an output y, and $y \leftarrow A(x_1, \ldots, x_n)$ represents the act of picking r at random and letting $y \leftarrow A(x_1, \ldots, x_n; r)$.

If X is a set, then $x \xleftarrow{\$} X$ represents the act of choosing x randomly from X. $|X|$ denotes the cardinality of X. If X is a string, then $|X|$ denotes the bit length of X, and $lsb(X)$ denotes the least significant bit of X. If X and Y are bit strings, $X\|Y$ denotes the concatenation.

Let k be a security parameter.

2.1   DDH Assumption

Let $\mathbb{G}$ be a group of prime order p. The DDH assumption (on $\mathbb{G}$) is that the distributions (g, g^x, g^y, g^{xy}) and (g, g^x, g^y, g^z) are computationally indistinguishable, where g is a random generator for $\mathbb{G}$ and $x, y, z \xleftarrow{\$} Z_q$.

2.2   KDM Security

Let $\mathcal{E} = (K, E, D)$ be a public key encryption scheme, $\mathcal{S} = \{S_k\}$ be the space of secret keys and $\mathcal{M} = \{M_k\}$ be the space of messages. For an integer $\ell > 0$, define

$$ALL_k^{(\ell)} = \{f \mid f : (S_k)^\ell \to M_k\}.$$

We then define the KDM$^{(\ell)}$ attack game between a challenger and an adversary $\mathcal{A}$ with respect to a function class $\mathcal{F} = \{\mathcal{F}_k\}$ such that $\mathcal{F}_k \subseteq ALL_k^{(\ell)}$ as follows.

Initialize. The challenger chooses $b \xleftarrow{\$} \{0,1\}$ and generates key pairs $(sk_i, pk_i) \leftarrow K(1^k)$ for all $i \in \{1, \ldots, \ell\}$. The challenger then sends $(pk_1, \ldots, pk_\ell)$ to the adversary $\mathcal{A}$.

Query. The adversary $\mathcal{A}$ makes adaptive queries of the form $(j, f) \in \{1, \ldots, \ell\} \times \mathcal{F}_k$. For each query, the challenger computes $x = f(sk_1, \ldots, sk_\ell)$ and sends the following ciphertext to the adversary $\mathcal{A}$.

$$c = \begin{cases} E_{pk_j}(x) & if\ b = 0 \\ E_{pk_j}(0^{|x|}) & if\ b = 1 \end{cases}$$

Finish. The adversary $\mathcal{A}$ outputs a guess $b' \in \{0, 1\}$.

Define $\mathsf{Adv}^{\mathrm{kdm}}_{\mathcal{E},\ell}(\mathcal{A}) = |2\Pr[b' = b] - 1|$. We say that $\mathcal{E}$ is KDM$^{(\ell)}$ secure with respect to $\mathcal{F}$ if $\mathsf{Adv}^{\mathrm{kdm}}_{\mathcal{E},\ell}(\mathcal{A})$ is negligible for any PPT adversary $\mathcal{A}$.

The KDM security of symmetric-key encryption schemes is defined similarly.

2.3 BHHO Encryption Scheme

Boneh et al. showed a KDM-secure public-key encryption scheme w.r.t. a class of affine in the exponent functions under the DDH assumption [5]. Let $\mathbb{G}$ be a group of prime order p and g be a generator of $\mathbb{G}$.

Key Generation. Let $t = \lceil 3 \log_2 p \rceil$. Choose $(g_1, \ldots, g_t) \overset{\$}{\leftarrow} \mathbb{G}^t, (s_1, \cdots, s_t) \overset{\$}{\leftarrow} \{0, 1\}^t$. Let $h \leftarrow (g_1^{s_1} \cdots g_t^{s_t})^{-1}$ and define the public and secret keys as

$$pk = (g_1, \ldots, g_t, h) \text{ and } sk = (s_1, \ldots, s_t).$$

Encryption. For a plaintext $m \in \mathbb{G}$, choose $r \overset{\$}{\leftarrow} Z_p$ and output a ciphertext

$$(g_1^r, \ldots, g_t^r, h^r \cdot m).$$

Decryption. For a ciphertext $(c_1, \ldots, c_t, d)$, output $m = d \cdot (c_1^{s_1} \cdots c_t^{s_t})$.

We say that f is an affine in the exponent function if $f(x_1, \ldots, x_n) = g^{a_0 + \sum_{i=1}^n a_i x_i}$, where $x_i \in \{0, 1\}^n$ and $a_i \in Z_p$. Define $\mathcal{F}_{\mathrm{affine}} = \{\mathcal{F}_k\}$ in such a way that $\mathcal{F}_k = ALL_k^{(\ell)} \cap \{f \mid f \text{ is an affine in the exponent function}\}$.

Proposition 1 [5]. *The above encryption scheme is KDM$^{(\ell)}$ secure w.r.t. $\mathcal{F}_{\mathrm{affine}}$ under the DDH assumption.*

2.4 Bounded-KDM Security

Let $q(k)$ be a polynomial in k. A function f is $q(k)$-bounded if it can be expressed as a boolean circuit such that the number of gates is $q(k)$. Define $\mathcal{F}_{\mathrm{q-gates}} = \{\mathcal{F}_k\}$, where $\mathcal{F}_k = ALL_k^{(\ell)} \cap \{f \mid f \text{ is } q(k)\text{-bounded}\}$. Then an encryption scheme $\mathcal{E}$ is q-bounded KDM$^{(\ell)}$ secure if it is KDM$^{(\ell)}$ secure w.r.t. $\mathcal{F}_{\mathrm{q-gates}}$.

2.5 Projection KDM Security

A function f is called a *projection* if each output bit depends on at most one input bit. Define $\mathcal{F}_{\mathrm{proj}} = \{\mathcal{F}_k\}$ in such a way that $\mathcal{F}_k = ALL_k^{(\ell)} \cap \{f \mid f \text{ is a projection}\}$. We then say that an encryption scheme $\mathcal{E}$ is projection KDM$^{(\ell)}$ secure if it is KDM$^{(\ell)}$ secure w.r.t. $\mathcal{F}_{\mathrm{proj}}$.

3   Garbling Scheme [6]

3.1   Circuits

A boolean circuit is a 5-tuple $f = (n, m, q, A, B, G)$. Here $n \geq 2$ is the number of inputs, $m \geq 1$ is the number of outputs, and $q \geq 1$ is the number of gates. We let Inputs $= \{1, ..., n\}$, Gates $= \{n+1, ..., n+q\}$, Wires $= \{1, ..., n+q\}$ and Outputs $= \{n + q - m + 1, ..., n + q\}$. Then A: Gates $\rightarrow$ Wires\Outputs is a function to identify each gate's first incoming wire, and B : Gates $\rightarrow$ Wires\Outputs is a function to identify each gate's second incoming wire. We require $A(g) < B(g) < g$ for each gate $g \in$ Gates. Finally G : Gates $\times \{0, 1\}^2 \rightarrow \{0, 1\}$ is a function that determines the functionality of each gate.

Each gate has two inputs, one output and arbitrary functionality. The wires are numbered 1 to $n+q$. The ith bit of the input is presented along wire i. Every non-input wire is the outgoing wire of some gate. The outgoing wire of each gate serves as the name of that gate. We denote the output of f on input $x \in \{0, 1\}^n$ by $y = f(x)$. See Appendix A for an example.

Let

$$\Phi_{topo}(f) = (n, m, q, A, B), \; \Phi_{size}(f) = (n, m, q)$$

We say that $\Phi_{topo}(f)$ and $\Phi_{size}(f)$ are side information functions.

3.2   Garbling Scheme

A garbling scheme is a three-tuple of algorithms Ga $=$ (Gb, En, Ev) such as follows.

- $(F, e) \leftarrow$ Gb$(1^k, f)$, where $f = (n, m, q, A, B, G)$ is a boolean circuit, F is its garbled circuit and e is a key of En.
- $X \leftarrow$ En(e, x), where $x \in \{0, 1\}^n$ is a real input and X is its garbled input.
- $y \leftarrow$ Ev(F, X), where y is the output of F on input X.

The correctness condition requires that if $(F, e) \leftarrow$ Gb$(1^k, f)$, then

$$\mathsf{Ev}[F, \mathsf{En}(e, x)] = f(x)$$

for any $x \in \{0, 1\}^n$.

3.3   Security of Garbling Schemes

For a garbling scheme Ga $=$ (Gb, En, Ev) and a side information function Φ, we consider a game between a challenger and an adversary $\mathcal{A}$ as follows.

1. $\mathcal{A}$ chooses (f_0, x_0) and (f_1, x_1) such that $f_0(x_0) = f_1(x_1)$ and $\Phi(f_0) = \Phi(f_1)$, and sends them to the challenger.
2. The challenger chooses $b \xleftarrow{\$} \{0, 1\}$, and computes $(F, e) \leftarrow$ Gb$(1^k, f_b)$ and $X \leftarrow$ En(e, x_b). He then sends (F, X) to $\mathcal{A}$.
3. $\mathcal{A}$ outputs a bit b'.

Define the advantage as $\text{Adv}_{\text{Ga},\Phi}^{garble}(\mathcal{A}) = |2\Pr(b' = b) - 1|$. We say that Ga is Φ-secure if $\text{Adv}_{\text{Ga},\Phi}^{garble}(\mathcal{A})$ is negligible for every PPT adversary $\mathcal{A}$.

Bellare et al. [6] showed a Φ_{topo}-secure garbling scheme $\text{Ga} = (\text{Gb}, \text{En}, \text{Ev})$ which is called Garble1. We illustrate it in Figs. 11 and 12 of Appendix B. We also show how to construct Φ_{size}-secure garbling schemes in Appendix C.

4 Bounded KDM Secure Encryption by Bellare et al.

4.1 Generic Construction

Bellare et al. [6] constructed a q-bounded $\text{KDM}^{(\ell)}$ secure encryption scheme $\mathcal{E}' = (K', E', D')$ from

- a Φ_{size}-secure garbling scheme $\text{Ga} = (\text{Gb}, \text{En}, \text{Ev})$ and
- a projection $\text{KDM}^{(\ell)}$ secure encryption scheme $\mathcal{E} = (K, E, D)$.

A ciphertext of $f(sk_1, \ldots, sk_\ell)$ under pk_j is given by $(F, E_{pk_j}(X))$, where

- F is a garbled circuit of an identity circuit ID, and
- X is a garbled input of a real input $f(sk_1, \ldots, sk_\ell)\|0^{\ell|sk|-L}$, where $L = |f(sk_1, \ldots, sk_\ell)|$.

The decryption is given by

$$F(X) = ID(f(sk_1, \ldots, sk_\ell)\|0^{\ell|sk|-L}) = f(sk_1, \ldots, sk_\ell)\|0^{\ell|sk|-L}.$$

More precisely, the key generation algorithm K' is the same as K, and it outputs $(pk, sk) \leftarrow K(1^k)$. The encryption algorithm E' and the decryption algorithm D' are given in Fig. 1. There ID is a circuit such that the number of gates is $q + n$, and $ID(z) = z$ for any $z \in \{0,1\}^n$, where $n = \max(|sk| \cdot \ell, |x|)$.

```
00  proc E'_{pk}(x)
01    n ← max(|sk| · ℓ, |x|)
02    z ← x‖0^{n−|x|}                          100  proc ConstructID(n)
03    ID←ConstructID(n)                        101  for i ∈ {n + 1, ..., n + q + 1} do
04    (F, e) ← Gb(1^k, ID)                     102    A(i) ← 1, B(i) ← i − 1,
05    X ← En(e, z)                                     G_i(a, b) ← { return  a}
06    Y ← E_{pk}(X)                            103  for i ∈ {n + q + 2, ..., q + 2n} do
07  return c = (F, Y)                          104    A(i) ← 1, B(i) ← i − n − q,
                                                       G_i(a, b) ← { return  b}
21  proc D'_{sk}(c)                            105  ID ← (n, n, q + n, A, B, G)
22    (F, Y) ← c                               106  return ID
23    X ← D_{sk}(Y)
24    y ← Ev(F, X)
25  return the first |x| bits of z
```

Fig. 1. Generic construction of q-bounded $\text{KDM}^{(\ell)}$ secure encryption scheme.

Proposition 2 [6]. *Suppose that*

– $\mathsf{Ga} = (\mathsf{Gb}, \mathsf{En}, \mathsf{Ev})$ *is* Φ_{size}-*secure and* $\mathsf{En}(e, \cdot)$ *is a projection function.*
– $\mathcal{E} = (K, E, D)$ *is projection* $KDM^{(\ell)}$ *secure.*

Then the above encryption scheme $\mathcal{E}'$ *is* q-*bounded* $KDM^{(\ell)}$ *secure. (See Fig. 2.)*

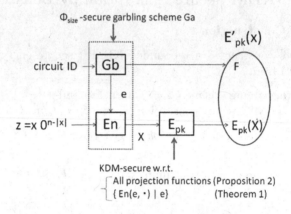

Fig. 2. q-bounded KDM secure $\mathcal{E}'(\mathsf{Ga}, \mathcal{E})$.

4.2 Instantiation Under DDH

Bellare et al. [6, Sect. 7.2] instantiated Proposition 2 by using Garble1 [6, Sect. 5][3] and BHHO encryption scheme [5]. (See Fig. 3.)

$$\left. \begin{array}{l} (\Phi_{topo}\text{-secure}) \ \mathsf{Garble1} \ \rightarrow \ (\Phi_{size}\text{-secure}) \ \widehat{\mathsf{Garble1}} \\ \mathsf{BHHO} \qquad\qquad\quad \rightarrow \ \text{projection KDM secure } \mathcal{E} \end{array} \right\} \rightarrow \mathcal{E}'$$

Fig. 3. q-bounded $KDM^{(\ell)}$ secure public-key encryption scheme $\mathcal{E}'$ of Bellare et al.

First construct a Φ_{size}-secure garbling scheme $\widehat{\mathsf{Garble1}} = (\mathsf{Gb}, \mathsf{En}, \mathsf{Ev})$ from (Φ_{topo}-secure) Garble1$=(\mathsf{Gb1}, \mathsf{En1}, \mathsf{Ev1})$ by using the method of Appendix C. (See the comments of Fig. 20 of [6].) We need to show that $\mathsf{En}(e, \cdot)$ is a projection. In Garble1,

$$e = (X_1^0, X_1^1, \ldots, X_n^0, X_n^1), \ X = \mathsf{En1}(e, x) = (X_1^{x_1}, \ldots, X_n^{x_n}),$$

---

[3] See Appendix B for Garble1.

where $X_i^j \in \{0,1\}^k$, $x = (x_1, \ldots, x_n)$ and $x_i \in \{0,1\}$. Therefore $\mathsf{En1}(e, \cdot)$ is a projection. Hence $\mathsf{En}(e, \cdot)$ is also a projection because $\mathsf{En} = \mathsf{En1}$ the method of Appendix C.

Note that

$$|X| = |X_1^{x_1}| + \ldots + |X_n^{x_n}| = kn. \tag{1}$$

Next construct a projection $\mathrm{KDM}^{(\ell)}$ secure encryption scheme $\mathcal{E} = (K, E, D)$ as follows. To compute $Y = E_{pk}(X)$, encrypt each bit of X by using BHHO encryption scheme. For more details, the key generation algorithm K is the same as that of BHHO encryption scheme, and for $X = (\epsilon_1, \ldots, \epsilon_{kn})$, let

$$E_{pk}(X) = (\mathrm{BHHO}_{pk}(g^{\epsilon_1}), \ldots, \mathrm{BHHO}_{pk}(g^{\epsilon_{kn}})), \tag{2}$$

where $\epsilon_i \in \{0,1\}$ and $\mathrm{BHHO}_{pk}(g^{\epsilon_i})$ denotes a BHHO ciphertext of g^{ϵ_i} for $i = 1, \ldots, kn$. This $\mathcal{E}$ is projection $\mathrm{KDM}^{(\ell)}$ secure under the DDH assumption.

The obtained encryption scheme $\mathcal{E}'$ from $\widehat{Garble}1$ and the above $\mathcal{E}$ is q-bounded $\mathrm{KDM}^{(\ell)}$ secure under the DDH assumption from Proposition 2.

Now suppose that we want to encrypt $f(sk_1, \ldots, sk_\ell)$. In Fig. 1, $X = \mathsf{En}(e, z)$ at line 05 has bit length kn from Eq. (1), where

$$n = \max(|sk| \cdot \ell, |f(sk_1, \ldots, sk_\ell)|) \tag{3}$$

Therefore $Y = E_{pk}(X)$ at line 06 consists of kn ciphertexts of BHHO encryption scheme. The total ciphertext consists of a garbled circuit F of ID and kn ciphertexts of BHHO encryption scheme. (See the third column of Fig. 20 of [6].)

5 How to Prove Proposition 2

In this section, we first point out that the proof of Proposition 2 by Bellare et al. [6] is not complete. We next show how to fix it.

5.1 Proof by Bellare et al.

Bellare et al. [6] proved Proposition 2 as follows. Let $\mathcal{A}$ be an adversary attacking $\mathcal{E}'$. To simplify the exposition, they first consider the case $\mathcal{A}$ makes only a single query. Then they sketch how to extend it to the general case.

SINGLE QUERY CASE. Suppose that $\mathcal{A}$ makes a single query (j, f). Let C be a circuit of n input wires, n output wires, and $q + n$ gates such that

$$C(x) = f(\text{the first } |sk|\ell \text{ bits of } x)\|0^{n-|f(sk_1, \cdots, sk_\ell)|},$$

where $n = \max(|sk| \cdot \ell, |f(sk_1, \ldots, sk_\ell)|)$. Then we have

$$C(\mathbf{K}) = f(sk_1\|\cdots\|sk_\ell)\|0^{n-|f(sk_1, \cdots, sk_\ell)|} \tag{4}$$

for $\mathbf{K} = sk_1\|\cdots\|sk_\ell\|0^{n-|sk|\cdot\ell}$.

We consider a series of Games $G_0, \ldots, G_4$ as shown in Fig. 4, where G_0 is equivalent to the KDM attack game with $b = 0$ because $C(\mathbf{K})$ is given by Eq. (4). Each game changes lines 04–05 of Fig. 1 step by step. Let $p_i = \Pr(b' = 1 \text{ in Game } G_i)$.

- $|p_0 - p_1|$ is negligible because $\mathrm{ID}(C(\mathbf{K})) = C(\mathbf{K})$, $\Phi_{size}(\mathrm{ID}) = \Phi_{size}(C)$ and the garbling scheme is Φ_{size}-secure.
- $|p_1 - p_2|$ is negligible because $\mathrm{En}(e, \cdot)$ is a projection function, and $\mathcal{E}$ is $\mathrm{KDM}^{(\ell)}$ secure w.r.t. {all projection functions}.
- $p_2 = p_3$ because $(F, E_{pk_j}(X))$ is the same in each game.
- $|p_3 - p_4|$ is negligible because $\mathrm{ID}(C(0^n)) = C(0^n)$, $\Phi_{size}(\mathrm{ID}) \doteq \Phi_{size}(C)$ and the garbling scheme is Φ_{size}-secure.

Their proof stops at this point. However, Game G_4 is not the KDM attack game with $b = 1$.

Game G_0
$(F, e) \leftarrow \mathsf{Gb}(1^k, \mathrm{ID})$,
$X \leftarrow \mathsf{En}(e, C(\mathbf{K}))$
return $(F, E_{pk_j}(X))$

Game G_1
$(F, e) \leftarrow \mathsf{Gb}(1^k, C)$,
$X \leftarrow \mathsf{En}(e, \mathbf{K})$
return $(F, E_{pk_j}(X))$

Game G_2
$(F, e) \leftarrow \mathsf{Gb}(1^k, C)$,
$X \leftarrow \mathsf{En}(e, \mathbf{K})$
return $(F, E_{pk_j}(0^{|X|}))$

Game G_3
$(F, e) \leftarrow \mathsf{Gb}(1^k, C)$,
$X \leftarrow \mathsf{En}(e, 0^n)$
return $(F, E_{pk_j}(0^{|X|}))$

Game G_4
$(F, e) \leftarrow \mathsf{Gb}(1^k, \mathrm{ID})$,
$X \leftarrow \mathsf{En}(e, C(0^n))$
return $(F, E_{pk_j}(0^{|X|}))$

Fig. 4. Proof of Proposition 2.

5.2   How to Fix the Proof

To complete the proof, we use the following proposition.

Proposition 3 [5]. *Suppose that the function class $\mathcal{F}$ contains all constant functions. Then the KDM security w.r.t. $\mathcal{F}$ implies IND-CPA security.*

Any constant function is a projection function. Therefore the set of all projection functions contain all constant functions.

Now we add Game G_5 and Game G_6 which are shown in Fig. 5. Then

Game G_5
$(F, e) \leftarrow \mathsf{Gb}(1^k, \mathrm{ID})$,
$X \leftarrow \mathsf{En}(e, 0^n)$
return $(F, E_{pk_j}(0^{|X|}))$

Game G_6
$(F, e) \leftarrow \mathsf{Gb}(1^k, \mathrm{ID})$,
$X \leftarrow \mathsf{En}(e, 0^n)$
return $(F, E_{pk_j}(X))$

Fig. 5. Game G_5 and Game G_6.

- $p_4 = p_5$ because $(F, E_{pk_j}(0^{|X|}))$ is the same in each game.
- $|p_5 - p_6|$ is negligible bacause $\mathcal{E}$ is IND-CPA secure from Proposition 3.

Game G_6 is the KDM attack game with $b = 1$ because the output of Game G_6 is $\mathcal{E}'_{pk_j}(0^{|f(sk_1\|\cdots\|sk_\ell)|})$. Finally $|p_0 - p_6|$ is negligible form the above discussion. This completes the proof.

6   Our Main Theorem

As shown above, bounded KDM security can be obtained from projection KDM security. However, the size of ciphertexts is still very large. In this section, we show that it can be obtained from a much weaker notion than the projection KDM security.

The encryption scheme $\mathcal{E}' = (K', E', D')$ given in Fig. 1 consists of

- a garbling scheme $\mathsf{Ga} = (\mathsf{Gb}, \mathsf{En}, \mathsf{Ev})$ and
- an encryption scheme $\mathcal{E} = (K, E, D)$ such that $K' = K$.

We denote such $\mathcal{E}'$ by $\mathcal{E}'(\mathsf{Ga}, \mathcal{E})$.

For a garbling scheme $\mathsf{Ga} = (\mathsf{Gb}, \mathsf{En}, \mathsf{Ev})$, first define $\widehat{\mathsf{En}}$ as

$$\widehat{\mathsf{En}}_{n,e}(x) = \mathsf{En}(e, x \| 0^{n-|x|}). \tag{5}$$

Next define the set of functions $\mathcal{F}(\mathsf{Ga}) = \{\mathcal{F}_k\}$ as follows.

$$\mathcal{F}_k = ALL_k^{(\ell)} \cap \{\widehat{\mathsf{En}}_{n,e} \mid (F, e) \leftarrow \mathsf{Gb}(1^k, C), \text{where } \Phi_{size}(C) = (n, n, n + q)\}.$$

Then our main Theorem is as follows. (See Fig. 2.)

Theorem 1. $\mathcal{E}'(\mathsf{Ga}, \mathcal{E})$ is q-bounded $KDM^{(\ell)}$ secure if Ga is Φ_{size}-secure, and $\mathcal{E}$ is IND-CPA secure and $KDM^{(\ell)}$-secure w.r.t. $\mathcal{F}(\mathsf{Ga})$.

The proof is the same as that of Proposition 2 which we fixed. In particular,

- $|p_1 - p_2|$ is negligible because $\mathcal{E}$ is $KDM^{(\ell)}$ secure w.r.t. $\mathcal{F}(\mathsf{Ga})$.
- $|p_5 - p_6|$ is negligible because $\mathcal{E}$ is IND-CPA secure.

Let $\widehat{\mathsf{Ga}}$ denote the Φ_{size}-secure garbling scheme which is obtained from a Φ_{topo}-secure garbling scheme Ga by using the method of Appendix C. Then we have the following corollary.

Corollary 1. $\mathcal{E}'(\widehat{\mathsf{Ga}}, \mathcal{E})$ is q-bounded $KDM^{(\ell)}$ secure if Ga is Φ_{topo}-secure, and $\mathcal{E}$ is IND-CPA secure and $KDM^{(\ell)}$-secure w.r.t. $\mathcal{F}(\mathsf{Ga})$.

Proof. Let $\widehat{\mathsf{Ga}} = (\mathsf{Gb}', \mathsf{En}', \mathsf{Ev}')$ and $\mathsf{Ga} = (\mathsf{Gb}, \mathsf{En}, \mathsf{Ev})$. Then $\mathsf{En}' = \mathsf{En}$ in the method of Appendix C. Hence we obtain this corollary from the definition of $\mathcal{F}(\mathsf{Ga})$. □

Let's compare Theorem 1 with Proposition 2.

- In Proposition 2, $\mathcal{E}$ needs to be $KDM^{(\ell)}$ secure w.r.t. the set of "all projection functions". Therefore Bellare et al. had to encrypt each bit of $X = \mathsf{En}(e, z)$ by using BHHO encryption scheme.
- In our theorem, on the other hand, $\mathcal{E}$ needs to be $KDM^{(\ell)}$ secure w.r.t. $\mathcal{F}(\mathsf{Ga})$ only. Therefore we can construct a more efficient $\mathcal{E}$ with an appropriate garbling scheme Ga as shown in the following sections.

Another difference is that in Theorem 1, $\mathcal{E}$ needs to be IND-CPA secure. However, it is not a matter because any encryption scheme must be IND-CPA secure anyway. See Table 1.

Table 1. Underlying primitive for bounded KDM security.

Scheme	$\mathcal{E}$ must be KDM secure w.r.t	En of Ga
BHR [6]	The set of all projection functions	Projection
Proposed	$\mathcal{F}(\mathsf{Ga})$	--

7 Our Instantiation Under DDH

In this section, we show a more efficient bounded KDM secure encryption scheme $\mathcal{E}'(\widehat{\mathsf{Ga}}, \mathcal{E})$ than Bellare et al. [6] under the DDH assumption based on Corollary 1. The encryption scheme $\mathcal{E}$ is based on BHHO encryption scheme [5], and the garbling scheme Ga is a variant of Garble1 [6] which we call Garble1+.

Let $\mathbb{G}$ be a group of prime order p and g be a generator of $\mathbb{G}$ as required in BHHO encryption scheme.

7.1 Garble1+

We define $(\mathsf{Gb}, \mathsf{Ev})$ of Garble1+ as shown in Figs. 6 and 7. Namely

$$e = (u_1^0, u_1^1, \dots, u_n^0, u_n^1), \tag{6}$$

$$\mathsf{En}(e, x) = (g^{u_1^{x_1}}, \dots, g^{u_n^{x_n}}), \tag{7}$$

where $u_i^j \in Z_p$, and

$$X_i^0 \leftarrow H(g^{u_i^0}), \; X_i^1 \leftarrow H(g^{u_i^1}) \tag{8}$$

for $i = 1, \dots, n$ in Gb, where $H : \mathbb{G} \rightarrow \{0,1\}^k$ is a hash function such that $H(v)$ is uniformly distributed over $\{0,1\}^k$ when $v \xleftarrow{\$} \mathbb{G}$.

The rest is the same as Garble1 [6] which is shown Appendix B. Then Garble1+ is Φ_{topo}-secure similarly to Garble1.

---

10 **proc** $\mathsf{Gb}(1^k, f)$
11 $(n, m, 1, A, B, G) \leftarrow f$
12 **for** $i \in$ Inputs $= \{1, \dots, n\}$ **do**
13    $(u_i^0, u_i^1) \xleftarrow{\$} (Z_p)^2$ such that $lsb(H(g^{u_i^0})) \neq lsb(H(g^{u_i^1}))$
14    $X_i^0 \leftarrow H(g^{u_i^0}), X_i^1 \leftarrow H(g^{u_i^1})$.
15 $e \leftarrow (u_1^0, u_1^1, \dots, u_n^0, u_n^1)$
16 **for** $i \in$ Wires/Outputs **do**
17    $(X_i^0, X_i^1) \xleftarrow{\$} (\{0,1\}^k)^2$ such that $lsb(X_i^0) \neq lsb(X_i^1)$.
The rest is the same as Garble1.

Fig. 6. Garble1+ (1).

```
20  proc Ev(F, X)                          30  proc En(e, x)
21  (P, Φ_topo(f)) ← F                      31  (u_1^0, u_1^1, ..., u_n^0, u_n^1) ← e
22  (n, m, q, A, B) ← Φ_topo(f)             32  (x_1, ..., x_n) ← x
23  (g^{u_1}, ..., g^{u_n}) ← X             33  X ← (g^{u_1^{x_1}}, ..., g^{u_n^{x_n}})
24  for i ∈ Inputs = {1, ..., n} do         34  return X
25     X_i ← H(g^{u_i}).
The rest is the same as Garble1.
```

Fig. 7. Garble1+ (2).

7.2 KDM-Secure Encryption w.r.t. $\mathcal{F}(Garble1+)$

We next show an efficient encryption scheme $\mathcal{E} = (K, E, D)$ which is KDM$^{(\ell)}$-secure w.r.t. $\mathcal{F}(Garble1+)$.

- The key generation algorithm K is the same as that of BHHO encryption scheme. Let $(sk, pk) ← K(1^k)$.
- For a message $(m_1, ..., m_n) \in \mathbb{G}^n$, let

$$E_{pk}(m_1, ..., m_n) = (\text{BHHO}_{pk}(m_1), ..., \text{BHHO}_{pk}(m_n)). \qquad (9)$$

We stress that $m_i \in \mathbb{G}^n$ in the above encryption scheme while $m_i \in \{1, g\}$ in that of Bellare et al. [6].

Theorem 2. *The above encryption scheme $\mathcal{E} = (K, E, D)$ is IND-CPA secure and KDM$^{(\ell)}$ secure w.r.t. $\mathcal{F}(Garble1+)$ under the DDH assumption.*

Proof. $\mathcal{E}$ is IND-CPA secure because BHHO encryption scheme is IND-CPA secure. We will prove that it is KDM$^{(\ell)}$ secure w.r.t. $\mathcal{F}(Garble1+)$.

In the KDM$^{(\ell)}$ attack game, the challenger generates key pairs $(sk_i, pk_i) ← K(1^k)$ for all $i \in \{1, ..., \ell\}$. Define $L = \ell|sk|$ and let

$$(sk_1, ..., sk_\ell) = (s_1, ..., s_L),$$

where $s_i \in \{0, 1\}$ for $i = 1, ..., L$.

Then for $e = (u_1^0, u_1^1, ..., u_n^0, u_n^1)$ of Eq. (6), we have

$$\widehat{\text{En}}_{n,e}(sk_1, ..., sk_\ell) = \widehat{\text{En}}_{n,e}(s_1, ..., s_L)$$
$$= (g^{u_1^{s_1}}, ..., g^{u_L^{s_L}}, g^{u_{L+1}^0}, ..., g^{u_n^0})$$

from Eqs. (7) and (5). Further from Eq. (9), we have

$$E_{pk}(\widehat{\text{En}}_{n,e}(sk_1, ..., sk_\ell)) = E_{pk}(g^{u_1^{s_1}}, ..., g^{u_L^{s_L}}, g^{u_{L+1}^0}, ..., g^{u_n^0})$$
$$= (\text{BHHO}_{pk}(g^{u_1^{s_1}}), ..., \text{BHHO}_{pk}(g^{u_L^{s_L}}),$$
$$\text{BHHO}_{pk}(g^{u_{L+1}^0}), ..., \text{BHHO}_{pk}(g^{u_n^0})).$$

366   K. Kurosawa and R. Habuka

We note that

$$u_i^0 + (u_i^1 - u_i^0) \times s_i = \begin{cases} u_i^0 \ if \ s_i = 0 \\ u_i^1 \ if \ s_i = 1 \end{cases}$$

Therefore we have

$$u_i^{s_i} = u_i^0 + (u_i^1 - u_i^0) \times s_i.$$

This means that $u_i^{s_i}$ is an affine function of $(s_1, \ldots, s_L)$ for fixed $e = (u_1^0, u_1^1, \ldots, u_n^0, u_n^1)$. More precisely, let $a_0 = u_i^0$ and

$$a_j = \begin{cases} u_i^1 - u_i^0 \ \text{if } j = i \\ 0 \qquad \text{otherwise} \end{cases}$$

Then $u_i^{s_i}$ is written as

$$u_i^{s_i} = a_0 + \sum_{j=1}^{L} a_j s_j.$$

Let

$$\mathsf{Exp}_e(s_1, \ldots, s_L)_i = g^{u_i^{s_i}} = g^{a_0 + \sum_{j=1}^{L} a_j s_j}. \tag{10}$$

Then $\mathsf{Exp}_e(s_1, \ldots, s_L)_i$ belongs to $\mathcal{F}_{\text{affine}}$ of Sect. 2.3 for $i = 1, \ldots, L$. Further for $i = L+1, \ldots, n$, $g^{u_i^0}$ is a constant function. Hence they also belong to $\mathcal{F}_{\text{affine}}$.

Now let $\mathcal{A}$ be an adversary who attacks $\mathcal{E}$ w.r.t. $\mathcal{F}(Garble1+)$. We construct an adversary $\mathcal{B}$ who attacks BHHO encryption scheme w.r.t. $\mathcal{F}_{\text{affine}}$ as follows.

Upon receiving $(pk_1, \ldots, pk_\ell)$ from the challenger, $\mathcal{B}$ sends them to $\mathcal{A}$. Suppose that $\mathcal{A}$ queries $(j, \widehat{\mathsf{En}}_{n,e})$ such that $e = (u_1^0, u_1^1, \ldots, u_n^0, u_n^1)$. Then $\mathcal{B}$ does the following.

1. For $i = 1, \ldots, L$, $\mathcal{B}$ queries $(j, \mathsf{Exp}_e(\cdot)_i)$ to the challenger, and receives a ciphertext c_i, where $\mathsf{Exp}_e(\cdot)_i$ is defined by Eq. (10).
2. For $i = L+1, \ldots, n$, $\mathcal{B}$ queries $(j, g^{u_i^0})$ to the challenger, and receives a ciphertext c_i.
3. $\mathcal{B}$ returns $C = (c_1, \ldots, c_n)$ to $\mathcal{A}$.

It is easy to see that C is a right challenge ciphertext for $(j, \widehat{\mathsf{En}}_{n,e})$. Finally $\mathcal{B}$ outputs whatever $\mathcal{A}$ does. Then we have $\mathsf{Adv}_{\mathcal{E},\ell}^{\text{kdm}}(\mathcal{A}) = \mathsf{Adv}_{\text{BHHO},\ell}^{\text{kdm}}(\mathcal{B})$. Therefore $\mathsf{Adv}_{\mathcal{E},\ell}^{\text{kdm}}(\mathcal{A})$ is negligible under the DDH assumption because $\mathsf{Adv}_{\text{BHHO},\ell}^{\text{kdm}}(\mathcal{B})$ is negligible under the DDH assumption. $\qquad\square$

7.3   Final Construction

Finally, our q-bounded KDM$^{(\ell)}$-secure encryption scheme $\mathcal{E}'$ is obtained by substituting Garble1+ and the encryption scheme $\mathcal{E}$ of Sect. 7.2 into Corollary 1 (Fig. 8).

Corollary 2. *Let $\mathcal{E}$ be the encryption scheme given by Sect. 7.2. Then $\mathcal{E}'(\widehat{Garble1+}, \mathcal{E})$ is q-bounded KDM$^{(\ell)}$-secure under the DDH assumption.*

Proof. From Corollary 1 and Theorem 2. $\qquad\square$

367 More Efficient Construction of Bounded KDM Secure Encryption

$$\left.\begin{array}{ll} (\varPhi_{topo}\text{-secure}) \text{ Garble1+} \rightarrow & (\varPhi_{size}\text{-secure}) \ \widehat{Garble1}+ \\ \text{BHHO} & \rightarrow \ \mathcal{F}(Garble1+)\text{-KDM secure } \mathcal{E} \end{array}\right\} \rightarrow \mathcal{E}'$$

Fig. 8. Proposed q-bounded KDM$^{(\ell)}$ secure public-key encryption scheme $\mathcal{E}'$.

8 Comparison

Let's compare two q-bounded KDM$^{(\ell)}$-secure encryption schemes under the DDH assumption (namely, the schemes in Sects. 4.2 and 7). In both schemes, $\mathcal{E}'_{pk}(x_1, \ldots, x_n)$ consists of (F, Y), where F is a garbled circuit of ID. We compare $Y = E_{pk}(X)$ in Tables 2 and 3.

Table 2. $Y = E_{pk}(X)$ (1).

Scheme	$\mathcal{E} = (K, E, D)$	X
BHR [6]	Projection KDM secure	$(X_1^{x_1}, \ldots, X_n^{x_n})$, where $X_i^{x_i} \xleftarrow{\$} \{0,1\}^k$
Proposed	$\mathcal{F}(Garble1+)$-KDM secure	$(g^{u_1^{x_1}}, \ldots, g^{u_n^{x_n}})$, where $u_i^{x_i} \xleftarrow{\$} Z_p$

Table 3. $Y = E_{pk}(X)$ (2).

Scheme	$Y = E_{pk}(X)$
BHR [6]	BHHO encryption of each bit of $X_i^{x_i} \in \{0,1\}^k$ for $i = 1, \ldots, n$
Proposed	BHHO encryption of $g^{u_i^{x_i}}$ for $i = 1, \ldots, n$

Our $\mathcal{E}$ is KDM secure w.r.t. only $\mathcal{F}(Garble1+)$ while it must be KDM secure w.r.t. the set of all projection functions in the scheme of Bellare et al. Therefore

- $E_{pk}(X)$ encrypts $g^{u_i^{x_i}}$ for $i = 1, \ldots, n$ in our scheme
- while it must encrypt each bit of $X_i^{x_i} \in \{0,1\}^k$ for $i = 1, \ldots, n$ in the scheme of Bellare et al.

The size of each ciphertext of BHHO encryption scheme is $O(k^2)$ from Sect. 2.3, where $\log p = O(k)$. Therefore

- $|Y| = O(k^2 n)$ in our scheme and
- $|Y| = O(k^3 n)$ in the scheme of Bellare et al.

Thus $|Y|$ is reduced to $1/k$ in our scheme.

The circuit ID has gate size $r = q + n$ from lines 101–105 of Fig. 1. Therefore the universal circuit which realizes ID has the gate size $O(r \log r)$ from Eq. (12). Hence $|F| = O(kr \log r)$ in both schemes because each gate of F has size $O(k)$. See Table 4 for the total size of ciphertexts $|F| + |Y|$.

Table 4. Size of ciphertexts (k is the security parameter.)

Scheme	Garbled circuit F	$Y = E_{pk}(X)$	Total size
BHR [6]	$O(kr \log r)$	$O(k^3 n)$	$O(kr \log r) + O(k^3 n)$
Proposed	$O(kr \log r)$	$O(k^2 n)$	$O(kr \log r) + O(k^2 n)$

When $f(sk_1, \ldots, sk_\ell)$ is encrypted, $n = \max(|sk| \cdot \ell, |f(sk_1, \ldots, sk_\ell)|)$. Suppose that $|f(sk_1, \ldots, sk_\ell)| = O(|sk| \cdot \ell)$ and $q = O(n)$. Then $n = O(|sk| \cdot \ell) = O(k\ell)$ and $r = q + n = O(n) = O(k\ell)$.

In this case, $O(k^2 n) = O(k^3 \ell)$, $O(k^3 n) = O(k^4 \ell)$ and $O(kr \log r) = O(k^2 \ell \log k)$ because $n = poly(k)$. Hence the total size of ciphertexts is $O(k^3 \ell)$ in our scheme, and $O(k^4 \ell)$ in the scheme of Bellare et al. Thus it is k times smaller in our scheme (Table 5).

Table 5. Size of ciphertexts for $q = O(|sk| \cdot \ell)$ and $|f(sk_1, \ldots, sk_\ell)| = O(|sk| \cdot \ell)$.

Scheme	Garbled circuit F	$Y = E_{pk}(X)$	Total size
BHR [6]	$O(k^2 \ell \log k)$	$O(k^4 \ell)$	$O(k^4 \ell)$
Proposed	$O(k^2 \ell \log k)$	$O(k^3 \ell)$	$O(k^3 \ell)$

9   Generalization

In general, we can construct

- a KDM secure encryption scheme w.r.t. $\mathcal{F}(Garble1)$ from any KDM secure encryption scheme w.r.t. the set of affine functions, and
- a KDM secure encryption scheme w.r.t. $\mathcal{F}(Garble1+)$ from any KDM secure encryption scheme w.r.t. the set of affine in the exponent functions,

where the secret-key is viewed as a bit string. By using each of them, we can construct a bounded KDM secure encryption scheme based on Corollary 1.

If the message space of our weakly KDM secure encryption scheme is large enough, then the size of our Y is k times smaller than that of Bellare et al. [6]. In this section, we show such examples.

9.1   Symmetric-Key Encryption Scheme

Proposition 2 holds for *symmetric-key* encryption schemes also as stated in [6]. Similarly, our Theorem 1 and Corollary 1 hold for symmetric-key encryption schemes too. The proofs are the same.

Applebaum et al. [2] showed a *symmetric-key* encryption scheme which is $KDM^{(\ell)}$ secure w.r.t. the set of affine functions (not in the exponent) under the LPN assumption. We call it ACPS encryption scheme. We can construct a bounded $KDM^{(\ell)}$ secure *symmetric-key* encryption scheme under the LPN assumption as follows. We use Garble1. Then

$$e = (X_1^0, X_1^1, \ldots, X_n^0, X_n^1), \ X = \mathsf{En}(e, x) = (X_1^{x_1}, \ldots, X_n^{x_n}),$$

where $X_i^j \in \{0,1\}^k$, $x = (x_1, \ldots, x_n)$ and $x_i \in \{0,1\}$. To compute $Y = E_{pk}(X)$ at line 06 of Fig. 1, encrypt each X_i^j by using ACPS encryption scheme. Namely

$$E(X) = (\mathrm{ACPS}(X_1^{x_1}), \ldots, \mathrm{ACPS}(X_n^{x_n})),$$

where $\mathrm{ACPS}(X_i^{x_i})$ denotes an ACPS ciphertext of $X_i^{x_i}$. Now $X_i^{x_i}$ is written as

$$X_i^{x_i} = X_i^0 + (X_i^1 - X_i^0) \times x_i. \tag{11}$$

This means that $X_i^{x_i}$ is an affine function of $(x_1, \ldots, x_n)$. Therefore we can show that E is $KDM^{(\ell)}$ secure w.r.t. $\mathcal{F}(Garble1)$. Hence $\mathcal{E}'(\widehat{Garble1}, \mathcal{E})$ is q-bounded $KDM^{(\ell)}$ secure under the LPN assumption from Corollary 1 (Fig. 9).

On the other hand, if we use Proposition 2, we must encrypt *each bit* of X_i^j by ACPS encryption scheme.

$$\left. \begin{array}{l} (\varPhi_{topo}\text{-secure}) \ \mathrm{Garble1} \ \rightarrow \quad (\varPhi_{size}\text{-secure}) \ \widehat{\mathrm{Garble1}} \\ \quad\quad \mathrm{ACPS} \quad\quad\quad \rightarrow \ \mathcal{F}(Garble1)\text{-KDM secure } \mathcal{E} \end{array} \right\} \rightarrow \mathcal{E}'$$

Fig. 9. Proposed q-bounded $KDM^{(\ell)}$ secure symmetric-key encryption scheme $\mathcal{E}'$.

9.2 Subgroup Indistinguishability Assumptions

Brakerski and Goldwasser [3] showed a BHHO-like encryption scheme under the subgroup indistinguishablity assumptions. In particular, they presented a KDM secure encryption scheme with respect to the set of affine functions under the Paillier's decisional composite residuosity (DCR) assumption. In this encryption scheme, the message space is Z_N, where $N = pq$.

Now we can construct a bounded $KDM^{(\ell)}$ secure public-key encryption scheme under the DCR assumption by applying our technique to this encryption scheme and Garble1.

A Example of Boolean Circuit

In an example of Fig. 10, $n = 4, m = 1, q = 3$, Inputs $= \{1, \ldots, 4\}$, Gates $= \{5, \ldots, 7\}$, Wires $= \{1, \ldots, 7\}$, Outputs $= \{7\}$, $G_5(x, y) = x \wedge y$ and

$$A(5) = 1, A(6) = 3, A(7) = 5,$$

$$B(5) = 2, B(6) = 4, B(7) = 6.$$

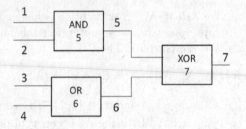

Fig. 10. A boolean circuit f.

B Garble1

Bellare et al. [6] showed a Φ_{topo}-secure garbling scheme $\mathsf{Ga} = (\mathsf{Gb}, \mathsf{En}, \mathsf{Ev})$ which is called Garble1.

For a k bit string A, let $A[1:k-1]$ denote the first $k-1$ bits of A. Define $B[1:k-1]$ similarly. For a pseudo-random function F, let

$$\mathbb{E}^T_{A,B}(M) = \mathsf{F}_{A[1:k-1]}(T) \oplus \mathsf{F}_{B[1:k-1]}(T) \oplus M,$$
$$\mathbb{D}^T_{A,B}(C) = \mathsf{F}_{A[1:k-1]}(T) \oplus \mathsf{F}_{B[1:k-1]}(T) \oplus C,$$

Then Garble1 is given in Figs. 11 and 12.

```
100  proc Gb(1^k, f)
101  (n, m, 1, A, B, G) ← f
102  for i ∈ Wires/Outputs do
103      Choose X_i^0 and X_i^1 such that lsb(X_i^0) ≠ lsb(X_i^1) randomly from {0,1}^k.
104  e ← (X_1^0, X_1^1, ..., X_n^0, X_n^1)
105  for i ∈ Outputs do
106      Choose X_i^0 and X_i^1 such that lsb(X_i^0) = 0 and lsb(X_i^1) = 1 randomly from {0,1}^k.
107  for (g, a, b) ∈ Gates × {0,1} × {0,1} do
108      i ← A(g), α ← lsb(X_i^a),
109      j ← B(g), β ← lsb(X_j^b),
110      T ← g ‖ α ‖ β,
111      P[g, α, β] ← E_{A,B}^T(X_g^{G_g(a,b)})
112  F ← (P, Φ_topo(f))
113  return (F, e)
```

Fig. 11. Garble1 (1).

C How to Achieve Φ_{size}-Security

In a topological circuit, only (n, m, q', A, B) is specified, but not the gate function G. A universal circuit $\mathsf{UC}_{(n,m,q)}$ is a topological circuit (n, m, q', A, B) which

```
120  proc Ev(F, X)
121  (P, Φ_topo(f)) ← F
122  (n, m, q, A, B) ← Φ_topo(f)
123  (X_1, ..., X_n) ← X
124  for g ← n + 1 to n + q do
125     i ← A(g), α ← lsb(X_i),
126     j ← B(g), β ← lsb(X_j),
127     T ← g ∥ α ∥ β,
128     X_g ← 𝔻^T_{A,B}(P[g, α, β])
129  for i ∈ Outputs do
130     y_g ← lsb(X_g)
131  return (y_{n+q-m+1}, ..., y_{n+q})
```

```
200  proc En(e, x)
201  (X_1^0, X_1^1, ..., X_n^0, X_n^1) ← e
202  (x_1, ..., x_n) ← x
203  X ← (X_1^{x_1}, ..., X_n^{x_n})
204  return X
```

Fig. 12. Garble1 (2).

can realize any boolean circuit f such that $\Phi_{size}(f) = (n, m, q)$ by using an appropriate gate function G. Valiant [8] constructed a universal circuit which has asymptotically optimal gate size

$$q' = O(q \log q). \tag{12}$$

Let $\mathsf{UC}_{(n,m,q)}(G)$ denote a boolean circuit which is obtained from $\mathsf{UC}_{(n,m,q)}$ and G.

A Φ_{size}-secure garbling scheme $\mathsf{Ga} = (\mathsf{Gb}, \mathsf{En}, \mathsf{Ev})$ can be obtained from a universal circuit and any Φ_{topo}-secure garbling scheme $(\mathsf{Gb1}, \mathsf{En1}, \mathsf{Ev1})$ as follows.

- $\mathsf{Gb}(1^k, f)$: Let $\Phi_{size}(f) = (n, m, q)$. Find a gate function G such that $\mathsf{UC}_{(n,m,q)}(G)$ realizes f. Then output $(F, e) \leftarrow \mathsf{Gb1}(1^k, \mathsf{UC}_{(n,m,q)}(G))$.
- $(\mathsf{En}, \mathsf{Ev}) = (\mathsf{En1}, \mathsf{Ev1})$.

References

1. Applebaum, B.: Key-dependent message security: generic amplification and completeness. In: Paterson, K.G. (ed.) EUROCRYPT 2011. LNCS, vol. 6632, pp. 527–546. Springer, Heidelberg (2011). doi:10.1007/978-3-642-20465-4_29
2. Applebaum, B., Cash, D., Peikert, C., Sahai, A.: Fast cryptographic primitives and circular-secure encryption based on hard learning problems. In: Halevi, S. (ed.) CRYPTO 2009. LNCS, vol. 5677, pp. 595–618. Springer, Heidelberg (2009). doi:10.1007/978-3-642-03356-8_35
3. Brakerski, Z., Goldwasser, S.: Circular and leakage resilient public-key encryption under subgroup indistinguishability. In: Rabin, T. (ed.) CRYPTO 2010. LNCS, vol. 6223, pp. 1–20. Springer, Heidelberg (2010). doi:10.1007/978-3-642-14623-7_1
4. Barak, B., Haitner, I., Hofheinz, D., Ishai, Y.: Bounded key-dependent message security. In: Gilbert, H. (ed.) EUROCRYPT 2010. LNCS, vol. 6110, pp. 423–444. Springer, Heidelberg (2010). doi:10.1007/978-3-642-13190-5_22
5. Boneh, D., Halevi, S., Hamburg, M., Ostrovsky, R.: Circular-secure encryption from decision Diffie-Hellman. In: Wagner, D. (ed.) CRYPTO 2008. LNCS, vol. 5157, pp. 108–125. Springer, Heidelberg (2008). doi:10.1007/978-3-540-85174-5_7

6. Bellare, M., Hoang, V.T., Rogaway, P.: Foundations of garbled circuits. In: ACM Conference on Computer and Communications Security, pp. 784–796 (2012). Cryptology ePrint Archive, Report 2012/265
7. Malkin, T., Teranishi, I., Yung, M.: Efficient circuit-size independent public key encryption with KDM security. In: Paterson, K.G. (ed.) EUROCRYPT 2011. LNCS, vol. 6632, pp. 507–526. Springer, Heidelberg (2011). doi:10.1007/978-3-642-20465-4_28
8. Valiant, L.G.: Universal circuits (preliminary report). In: STOC, pp. 196–203 (1976)

Signature Schemes with Randomized Verification

Cody Freitag[1], Rishab Goyal[1(✉)], Susan Hohenberger[2], Venkata Koppula[1], Eysa Lee[1], Tatsuaki Okamoto[3], Jordan Tran[4], and Brent Waters[1]

[1] University of Texas at Austin, Austin, USA
{cody.freitag,eysa.lee}@utexas.edu,
{rgoyal,kvenkata,bwaters}@cs.utexas.edu
[2] Johns Hopkins University, Baltimore, USA
susan@cs.jhu.edu
[3] NTT Secure Platform Laboratories, Musashino, Japan
okamoto.tatsuaki@lab.ntt.co.jp
[4] Princeton University, Princeton, USA

Abstract. A signature scheme consists of a setup, signing and verification algorithms. In most existing works, the verification algorithm is assumed to be deterministic. However, there could be signature schemes where the verification algorithm is randomized. In this work, we study signature schemes with randomized verification. Our results can be summarized as follows.

First, we present a security definition for signature schemes with randomized verification. The standard EUFCMA notion of security for signature schemes with deterministic verification is very restrictive when we consider randomized verification. Therefore, we propose a new security definition called χ-EUFCMA which captures a broad class of signature schemes with randomized verification.

Next, we analyse the security of Naor's transformation from Identity Based Encryption to signature schemes. Such a transformation results in a scheme with randomized verification. We show that this transformation can be proven χ-EUFCMA secure by choosing χ appropriately.

Finally, we show how a scheme with randomized verification can be generically transformed to one with deterministic verification.

1 Introduction

Digital signatures [6] are one of the central primitives of modern cryptography, both as an application, and as a building block for other primitives. A signature scheme consists of three algorithms: a setup algorithm that outputs a secret signing key and a public verification key, a signing algorithm that computes

S. Hohenberger—Supported by the National Science Foundation (NSF) CNS-1228443 and CNS-1414023, the Office of Naval Research under contract N00014-14-1-0333, and a Microsoft Faculty Fellowship.
B. Waters—Supported by NSF CNS-1228599 and CNS-1414082, DARPA SafeWare, Microsoft Faculty Fellowship, and Packard Foundation Fellowship.

© Springer International Publishing AG 2017
D. Gollmann et al. (Eds.): ACNS 2017, LNCS 10355, pp. 373–389, 2017.
DOI: 10.1007/978-3-319-61204-1_19

signatures on messages using the signing key, and a verification algorithm that uses the verification key to check if a signature corresponding to a message is valid. Clearly, the setup algorithm must be probabilistic. The signing algorithm can be transformed to a deterministic one by using a pseudorandom function (PRF). In such a transformation, a PRF key can be included as part of the secret key, and to sign a message, this PRF key can be used to generate the randomness for the signing algorithm. However, such a transformation cannot be applied to the verification algorithm, since verification is public.

Indeed, as noted by [8], "no generic transformation outside the random oracle model is known that makes the verification deterministic". As a result, when describing or using signatures, the verification algorithm is often implicitly or explicitly *assumed* to be deterministic. This is especially true for works which use signatures as part of a larger protocol. Moreover, in certain applications [8,9], it is important for the verification algorithm to be deterministic. This assumption regarding the determinism of the verification algorithm is often justified by the fact that most existing schemes have deterministic verification.

At the same time there are circumstances where a signature construction with a randomized verification will naturally arise. The most popular one is the transformation from identity based encryption (IBE) schemes to digital signatures suggested by Naor (sketched in [2]). This transformation, popularly referred to as Naor transformation, can be described as follows. The signing key consists of the IBE master secret key and the verification key is the IBE public key. To sign a message m, the signing algorithm generates an IBE secret key corresponding to the identity m. This signature can be verified by first choosing a random message x from the IBE message space, encrypting x for identity m and checking if the secret key can decrypt this ciphertext correctly. We would like to point out that the random coins used by verification algorithm consists of the IBE message x and the randomness required to encrypt x.[1]

The signature schemes derived from IBE schemes have some nice properties (for instance shorter signatures) as compared to the traditional signature schemes. Fortunately, for many IBE schemes, there are naturally derived signature schemes where the verification is deterministic. Boneh, Lynn and Shacham showed how to modify the Boneh-Franklin IBE scheme to get a signature scheme with deterministic verification [3]. Waters, in [10], showed an IBE scheme in the standard model, and then showed how it can be modified to get a signature scheme with deterministic verification. However, such signature scheme counterparts with deterministic verification do not exist for all IBE schemes. For example, the dual system based Waters IBE scheme [11] did not give an immediate signature with deterministic verification, and the generic Naor transformation was the only known approach. More generally, in future, there could be signature schemes with desirable properties, but where the verification algorithm is randomized.

---

[1] In this work we define take the "Naor transformation" to be how it was originally described in the Boneh-Franklin [2] paper. We empahsize that in this case there is a *single* message x that is encrypted and tested for verification.

Our Goals and Contributions. With this motivation, we explore a *broad* notion of signature schemes with randomized verification. We note that a few different works have considered randomized verification with varying degrees of formality (e.g., [7,8,11]). The main contributions of our work can be summarized as follows. First, we propose a formal security definition for signature schemes with randomized verification security. This security notion, called χ-Existential Unforgeability under Chosen Message Attack (χ-EUFCMA) security, captures a broad class of signature schemes with randomized verification. Next, we give a formal proof of Naor's IBE to signatures transformation. The proof of Naor transformation also illustrates the importance of our proposed security definition. Third, we show how to amplify the security of a χ-EUFCMA secure scheme. Finally, we show generic transformations from a signature scheme with randomized verification to one with deterministic verification. Each of these contributions is described in more detail below.

Defining a broad notion of security for signature schemes with randomized verification. Our first goal is to define a broad notion of security for signature schemes with randomized verification. Intuitively, a signature scheme is said to be secure if no polynomial time adversary can produce a forgery that gets verified with non-negligible probability. For deterministic verification, this can be easily captured via a security game [6] between a challenger and an adversary, where the adversary can query for polynomially many signatures, and eventually, must output a forgery σ^* on a fresh (unqueried) message m^*. The adversary wins if the verification algorithm accepts σ^* as a signature on message m^*, and we want that no polynomial time adversary wins with non-negligible probability.

Now one can achieve a standard notion of security for randomized verification if we just take the same game and make the attacker's success probability be over the coins of the challenger's verification algorithms (as well as the other coins in the game). While this is the most natural extension of the security to randomized verification, there are signature schemes that do not meet this definition, yet still offer some meaningful notion of security. Looking ahead, signature schemes that arise from the Naor transformation applied to IBE schemes with *small* message space does not satisfy this strong notion but still provide some type of security that we would like to be able to formally capture.

In order to capture a broad class of signature schemes with randomized verification, we introduce a weaker notion called χ-EUFCMA security. In this security game, the adversary receives a verification key from the challenger. It is allowed to make polynomially many signature queries, and must finally output a forgery σ^* on message m^* (not queried during the signature query phase). Informally, a signature scheme is said to be χ-EUFCMA secure if for any forgery σ^* on message m^* produced by a polynomially bounded adversary, the fraction of random coins that accept σ^*, m^* is at most χ. Ideally, we would want $\chi = 0$ (or negligible in the security parameter). However, as we will show below, this is not possible to achieve for certain Naor transformed signature schemes.

Naor Transformation for IBE schemes with small message spaces. For simplicity, let us consider the Naor transformation from IBE schemes with one bit message space (for example the Cocks IBE scheme [4]). Here, the adversary can send a malformed signature such that the adversary wins the game, but the signature-to-IBE reduction algorithm has 0 advantage. In more detail, consider an adversary that, with probability $1/2$, outputs a valid IBE key σ^* for ID m^*, and with probability $1/2$, outputs a malformed IBE key σ^* for m^*. This malformed key always outputs the incorrect decryption of the ciphertext. Such an adversary would break the signature scheme security, but if a reduction algorithm uses such a forgery directly to break the IBE security, then it has advantage 0. To get around this problem, we use a counting technique which is inspired by the artificial abort technique introduced by Waters [10]. The main idea behind Waters' artificial abort was to reduce the dependence between the event in which adversary wins and the one where simulation aborts. However, we require a technique which could efficiently test whether the forgery (i.e. IBE key) is malformed or not. To this end, the reduction algorithm runs the IBE decryption multiple times, each time on freshly generated ciphertext, and counts the number of correct decryptions. This helps the reduction in estimating whether the key is malformed or not. If the key is malformed, then the reduction algorithm simply aborts, otherwise it uses the forgery to break IBE security. The analysis is similar to that in [10].

The above issue with small message space IBE schemes was also noted by the work of Cui et al. [5]. They outlined a sketch of proof for large message spaces. However, even in the case of large message spaces, the issue of artificial abort needs to be addressed (although, unlike the small message case, the reduction algorithm does not suffer a loss in the winning advantage).

Amplifying the soundness from $\chi = 1 - 1/\mathsf{poly}(\cdot)$ *to* $\chi = 0$. Next, we show how to amplify the soundness from $\chi = 1 - 1/\mathsf{poly}(\lambda)$ to $\chi = 0$. Such a problem closely resembles that of transforming weak OWFs to strong OWFs. Therefore, the direct product amplification technique [12] could be used. So, if the verification algorithm is run sufficiently many times, each time with fresh randomness, and the signature is accepted only if all verifications succeed, then we could achieve $\chi = 0$-EUFCMA security.

Derandomizing the verification algorithm. Finally, we show how to transform any $\chi = 0$-EUFCMA secure signature scheme with randomized verification algorithm to a signature scheme with deterministic verification. We show two such transformations, one in the random oracle model (ROM), and another in the standard model. The transformation in the random oracle model is fairly straightforward. Here, the verification algorithm uses a hash function H which will be modeled as a random oracle in the proof. To verify a signature σ on message m, the verification algorithm computes $r = H(m, \sigma)$ and uses r as the randomness for the verification. To prove security, we first model the hash function H as a random oracle. Suppose there exists an attacker on this modified signature scheme. After the hash function is replaced with a random oracle, the attacker's forgery can be directly used as a forgery for the original $\chi = 0$-EUFCMA secure signature scheme. Here, note that it is crucial that the starting signature scheme with randomized verification is $\chi = 0$-EUFCMA secure, else the adversary's forgery may not be useful for the reduction as it will not translate to a forgery for the original signature scheme.

In our standard model transformation, the setup algorithm chooses a sufficiently large number of random strings $r_1, \ldots, r_\ell$ and includes them as part of the public verification key. The verification algorithm uses each of these strings as its randomness, and checks if m, σ verify in all cases. If so, it outputs 1, else it outputs 0. For security, suppose the adversary $\mathcal{A}$ sends a message m^* and a forgery σ^* such that the verification algorithm accepts m^*, σ^* when using $r_1, \ldots, r_\ell$ as the random strings. The reduction algorithm simply forwards m^*, σ^* as a forgery to the original signature scheme's challenger. To compute the winning probability of the reduction algorithm, we need to consider the case when the adversary $\mathcal{A}$ wins, but the reduction algorithm does not. This, in particular, can happen if the set of 'bad' random strings that result in (m^*, σ^*) being accepted is a negligible fraction of the total number of random strings. Suppose there exists a message m^* and forgery σ^* such that the number of 'bad' randomness strings that cause the verification algorithm to accept m^*, σ^* is a negligible fraction of the total number of randomness strings, and all strings $r_1, \ldots, r_\ell$ are in the bad set. We argue that since $r_1, \ldots, r_\ell$ are chosen uniformly at random, and since ℓ is a suitably large polynomial in the security parameter, this probability is negligible in the security parameter.

Conclusions. Many times in cryptography and security there is a concept that is thought to be well understood, but has never been rigorously proven or explored. Often when such an exploration takes place some new twists or facts will be discovered. In this case when we first decided to dig into the IBE to signature transformation we had an initial expectation that there would be a lossless connection from IBE to signatures. We actually were surprised that our initial attempts did not work out and that an artificial abort type mechanism was required for the small message spaces. Besides formally proving the IBE to signature transformation, we also show how to derandomize any signature scheme with randomized verification, both in the random oracle model as well as the standard model.

2 Definitions and Preliminaries

2.1 Signature Schemes with Deterministic Verification

A signature scheme $\mathcal{S} = (\mathsf{Setup_{Det}}, \mathsf{Sign_{Det}}, \mathsf{Verify_{Det}})$ with message space $\mathcal{M}$ consists of three algorithms, as follows:

1. $\mathsf{Setup_{Det}}(1^\lambda)$ is a randomized algorithm that takes security parameter λ as input and returns a pair of keys $(\mathsf{sk}, \mathsf{vk})$, where sk is the signing key and vk is the verification key.
2. $\mathsf{Sign_{Det}}(\mathsf{sk}, m)$ is a possibly randomized algorithm that takes as input the signing key sk, and a message m, and returns a signature σ.
3. $\mathsf{Verify_{Det}}(\mathsf{vk}, m, \sigma)$ is a determistic algorithm that takes as input the verification key vk, a message m, and a signature σ, and outputs 1 (accepts) if verification succeeds, and 0 (rejects) otherwise.

Definition 1. *A pair of polynomial time algorithms* $(Setup_{\mathrm{Det}}, Sign_{\mathrm{Det}}, Verify_{\mathrm{Det}})$ *is a EUFCMA secure signature scheme if it satisfies the following conditions*

1. *(Correctness) For all* $\lambda \in \mathbb{N}$, $m \in \mathcal{M}$,

$$
\Pr\left[Verify_{\mathrm{Det}}(\mathsf{vk}, m, \sigma) = 1 : \begin{array}{c} (\mathsf{sk}, \mathsf{vk}) \xleftarrow{\$} Setup_{\mathrm{Det}}(1^\lambda) \\ \sigma \xleftarrow{\$} Sign_{\mathrm{Det}}(\mathsf{sk}, m) \end{array} \right] = 1.
$$

2. *(Security) For every PPT attacker* $\mathcal{A}$ *there exists a negligible function* $\mathrm{negl}(\cdot)$ *such that for all* $\lambda \in \mathbb{N}$, $\mathrm{Adv}_{\mathcal{A}}^{\mathrm{EUFCMA}}(\lambda) \leq \mathrm{negl}(\lambda)$, *where advantage of* $\mathcal{A}$ *is defined as*

$$
\mathrm{Adv}_{\mathcal{A}}^{\mathrm{EUFCMA}}(\lambda) = \Pr\left[Verify_{\mathrm{Det}}(\mathsf{vk}, m^*, \sigma^*) = 1 : \begin{array}{c} (\mathsf{sk}, \mathsf{vk}) \xleftarrow{\$} Setup_{\mathrm{Det}}(1^\lambda) \\ (m^*, \sigma^*) = \mathcal{A}^{Sign_{\mathrm{Det}}(\mathsf{sk}, \cdot)}(1^\lambda, \mathsf{vk}) \end{array} \right],
$$

and $\mathcal{A}$ *should never have queried* m^* *to* $Sign_{\mathrm{Det}}$ *oracle.*

2.2 Identity-Based Encryption

An Identity-Based Encryption (IBE) scheme $\mathcal{E} = (Setup_{\mathrm{IBE}}, KeyGen_{\mathrm{IBE}}, Enc_{\mathrm{IBE}}, Dec_{\mathrm{IBE}})$ with message space $\mathcal{M}_{\mathrm{IBE}}$ and identity space $\mathcal{I}$ consists of the following polynomial time algorithms:

1. $Setup_{\mathrm{IBE}}(1^\lambda)$ is a randomized polynomial time algorithm that takes security parameter λ as input and returns $(\mathsf{pp}, \mathsf{msk})$, where pp are public parameters and msk is the master secret key.
2. $KeyGen_{\mathrm{IBE}}(\mathsf{pp}, \mathsf{msk}, \mathsf{ID})$ is a randomized polynomial time algorithm that takes as inputs the public parameters pp, the master secret key msk, and an identity ID and returns a secret key $\mathsf{sk}_{\mathsf{ID}}$ for ID.
3. $Enc_{\mathrm{IBE}}(\mathsf{pp}, m, \mathsf{ID})$ is a randomized polynomial time algorithm that takes as inputs the public parameters pp, a message m, and an identity ID and returns a ciphertext ct.
4. $Dec_{\mathrm{IBE}}(\mathsf{pp}, \mathsf{sk}_{\mathsf{ID}}, \mathsf{ct})$ is an polynomial time algorithm that takes as inputs the public parameters pp, a secret key $\mathsf{sk}_{\mathsf{ID}}$, and a ciphertext ct and returns a message m or $\bot$.

Definition 2. *A pair of polynomial time algorithms* $(Setup_{\mathrm{IBE}}, KeyGen_{\mathrm{IBE}}, Enc_{\mathrm{IBE}}, Dec_{\mathrm{IBE}})$ *is an adaptively-secure IBE scheme if it satisfies the following conditions*

1. *(Correctness) For all* $\lambda \in \mathbb{N}$, $m \in \mathcal{M}_{\mathrm{IBE}}$, $\mathsf{ID} \in \mathcal{I}$,

$$
\Pr\left[Dec_{\mathrm{IBE}}(\mathsf{pp}, \mathsf{sk}_{\mathsf{ID}}, \mathsf{ct}) = m : \begin{array}{c} (\mathsf{pp}, \mathsf{msk}) \xleftarrow{\$} Setup_{\mathrm{IBE}}(1^\lambda) \\ \mathsf{sk}_{\mathsf{ID}} \xleftarrow{\$} KeyGen_{\mathrm{IBE}}(\mathsf{pp}, \mathsf{msk}, \mathsf{ID}) \\ \mathsf{ct} \xleftarrow{\$} Enc_{\mathrm{IBE}}(\mathsf{pp}, m, \mathsf{ID}) \end{array} \right] = 1.
$$

2. *(Security)* For every PPT attacker $\mathcal{A} = (\mathcal{A}_0, \mathcal{A}_1)$ there exists a negligible function $negl(\cdot)$ such that for all $\lambda \in \mathbb{N}$, $\mathsf{Adv}_{\mathcal{A}}^{\mathrm{IBE}}(\lambda) \leq negl(\lambda)$, where advantage of $\mathcal{A}$ is defined as

$$\left| \Pr\left[\mathcal{A}_1^{\mathcal{O}_{\mathsf{ID}^*}}(\mathsf{st}, \mathsf{ct}_b) = b : \begin{array}{c} (\mathsf{pp}, \mathsf{msk}) \xleftarrow{\$} \mathit{Setup}_{\mathrm{IBE}}(1^\lambda); \quad b \xleftarrow{\$} \{0,1\} \\ (\mathsf{ID}^*, m_0^*, m_1^*, \mathsf{st}) = \mathcal{A}_0^{\mathcal{O}}(1^\lambda, \mathsf{pp}) \\ \mathsf{ct}_b \xleftarrow{\$} \mathit{Enc}_{\mathrm{IBE}}(\mathsf{pp}, m_b^*, \mathsf{ID}^*) \end{array} \right] - \frac{1}{2} \right|,$$

Here oracle $\mathcal{O}$ behaves same as $KeyGen_{\mathrm{IBE}}(\mathsf{pp}, \mathsf{msk}, \cdot)$, $\mathcal{A}$ should not have queried $\mathcal{O}$ for ID^* in the pre-challenge phase, and $\mathcal{O}_{\mathsf{ID}^*}$ behaves same as $\mathcal{O}$, except on input ID^* it outputs $\perp$.

3 Signatures Schemes with Randomized Verification

In traditional signature schemes (as defined in Sect. 2.1), the verification algorithm is assumed to be deterministic. However, in certain scenarios, it might be useful to consider verification algorithms that are randomized. In this section, we propose a EUFCMA equivalent definition that captures a broad class of signature schemes with randomized verification. Let χ be a fixed parameter. Intuitively, a signature scheme with randomized verification is said to be χ secure if any forgery produced by a PPT adversary is accepted by the verification algorithm with probability at most χ. For example, if $\chi = 1/2$, then in the security game, the adversary first gets the verification key, then queries for q signatures, and finally sends a forgery σ^* on message m^*. The scheme is χ secure if (σ^*, m^*) is accepted by the verification algorithm with probability at most $1/2$.

More formally, a randomized verification signature scheme $\mathcal{S}_{\mathrm{RV}} = (\mathsf{Setup}_{\mathrm{RV}}, \mathsf{Sign}_{\mathrm{RV}}, \mathsf{Verify}_{\mathrm{RV}})$ with message space $\mathcal{M}(\lambda)$ and verification coin space $\mathcal{R}(\lambda)$ consists of three algorithms, as follows:

1. $\mathsf{Setup}_{\mathrm{RV}}(1^\lambda)$ is a randomized polynomial time algorithm that takes security parameter λ as input and returns a pair of keys $(\mathsf{sk}, \mathsf{vk})$, where sk is the signing key and vk is the verification key.
2. $\mathsf{Sign}_{\mathrm{RV}}(\mathsf{sk}, m)$ is a possibly randomized polynomial time algorithm that takes as input the signing key sk, and a message m, and returns a signature σ.
3. $\mathsf{Verify}_{\mathrm{RV}}(\mathsf{vk}, m, \sigma; r)$ is a randomized polynomial time algorithm that takes as input the verification key vk, a message m, and a signature σ. It also uses random coins $r \in \mathcal{R}$, and outputs 1 (accepts) if verification succeeds, and 0 (rejects) otherwise.

Successful Verification Probability. We define a function over the tuple $(\lambda, \mathsf{vk}, m, \sigma)$ for a randomized verification signature scheme which captures the fraction of message-signature pairs (m, σ) which get verified under verification key vk as follows.

$$\mathrm{VerifyProb}(\lambda, \mathsf{vk}, m, \sigma) = \frac{|\{r \in \mathcal{R} : \mathsf{Verify}_{\mathrm{RV}}(\mathsf{vk}, m, \sigma; r) = 1\}|}{|\mathcal{R}|}.$$

In our security proofs, we will use the notation $\mathcal{R}_{m,\sigma}$ to denote the set $\{r \in \mathcal{R} : \mathsf{Verify}_{\mathrm{RV}}(\mathsf{vk}, m, \sigma; r) = 1\}$.

Definition 3. *A pair of polynomial time algorithms* $(\mathsf{Setup}_{\mathrm{RV}}, \mathsf{Sign}_{\mathrm{RV}}, \mathsf{Verify}_{\mathrm{RV}})$ *is a* $\chi(\lambda)$-*EUFCMA secure signature scheme with randomized verification if it satisfies the following conditions*

1. *(Correctness) For all* $\lambda \in \mathbb{N}$, $m \in \mathcal{M}$, $r \in \mathcal{R}$,

$$\Pr\left[\mathsf{Verify}_{\mathrm{RV}}(\mathsf{vk}, m, \sigma; r) = 1 : \begin{array}{c} (\mathsf{sk}, \mathsf{vk}) \xleftarrow{\$} \mathsf{Setup}_{\mathrm{RV}}(1^\lambda) \\ \sigma \xleftarrow{\$} \mathsf{Sign}_{\mathrm{RV}}(\mathsf{sk}, m) \end{array} \right] = 1.$$

2. *(Security) For every PPT attacker* $\mathcal{A}$ *there exists negligible functions* $negl_1(\cdot)$ *and* $negl_2(\cdot)$ *such that for all* $\lambda \in \mathbb{N}$, $\mathsf{Adv}_{\mathcal{A}}^{\mathrm{RV}}(\lambda) \leq negl_1(\lambda)$, *where advantage of* $\mathcal{A}$ *is defined as*

$$\Pr\left[\mathsf{VerifyProb}(\lambda, \mathsf{vk}, m^*, \sigma^*) > \chi(\lambda) + negl_2(\lambda) : \begin{array}{c} (\mathsf{sk}, \mathsf{vk}) \xleftarrow{\$} \mathsf{Setup}_{\mathrm{RV}}(1^\lambda) \\ (m^*, \sigma^*) = \mathcal{A}^{\mathsf{Sign}_{\mathrm{RV}}(\mathsf{sk}, \cdot)}(1^\lambda, \mathsf{vk}) \end{array} \right],$$

and $\mathcal{A}$ *should never have queried* m^* *to* $\mathsf{Sign}_{\mathrm{RV}}$ *oracle.*

4   Signature Scheme with Randomized Verification from IBE: Naor's Transformation

In this section, we construct signature schemes with randomized verification from IBE schemes using Naor's transformation. Naor suggested a generic method to construct signature schemes from IBE schemes, and a sketch of this transformation was given in [1]. At a high level, this transformation can be described as follows. The message space of the signature scheme is the identity space of the IBE scheme. To generate the signing key/verification key, the setup algorithm runs the IBE setup algorithm to compute an IBE public key/master secret key pair. The public key is set to be the verification key, while the master secret key is the signing key. To sign a message m, the signing algorithm generates a secret key corresponding to identity m. The verification algorithm chooses a uniformly random message x, encrypts it using the verification key for identity m. It then decrypts the encryption using the signature, and checks if the decryption is correct.

Intuitively, it appears that this signature scheme should be secure if the underlying IBE scheme is secure. However, there are certain subtleties, especially when the message space of the IBE scheme is small (see Sect. 1 for more details).

4.1   Construction

Let $\mathcal{E} = (\mathsf{Setup}_{\mathrm{IBE}}, \mathsf{KeyGen}_{\mathrm{IBE}}, \mathsf{Enc}_{\mathrm{IBE}}, \mathsf{Dec}_{\mathrm{IBE}})$ be an IBE scheme with message space $\mathcal{M}_{\mathrm{IBE}}$ and ID space $\mathcal{I}$. We define a randomized verification signature scheme $\mathcal{S}_{\mathrm{RV}} = (\mathsf{Setup}_{\mathrm{RV}}, \mathsf{Sign}_{\mathrm{RV}}, \mathsf{Verify}_{\mathrm{RV}})$ with message space $\mathcal{M} = \mathcal{I}$ as follows:

1. $\mathsf{Setup}_{RV}(1^\lambda)$: It computes $(\mathsf{pp}, \mathsf{msk}) \xleftarrow{\$} \mathsf{Setup}_{IBE}(1^\lambda)$ and returns $(\mathsf{sk}_{RV}, \mathsf{vk}_{RV})$ where $\mathsf{sk}_{RV} = \mathsf{msk}$ and $\mathsf{vk}_{RV} = \mathsf{pp}$.
2. $\mathsf{Sign}_{RV}(\mathsf{sk}_{RV}, m)$: Let $\mathsf{vk}_{RV} = \mathsf{pp}$, $\mathsf{sk}_{RV} = \mathsf{msk}$, and $m = \mathsf{ID}$. It computes $\mathsf{sk}_{ID} \xleftarrow{\$} \mathsf{KeyGen}_{IBE}(\mathsf{pp}, \mathsf{msk}, \mathsf{ID})$ and returns $\sigma = \mathsf{sk}_{ID}$.
3. $\mathsf{Verify}_{RV}(\mathsf{vk}_{RV}, m, \sigma; r)$: Let $\mathsf{vk}_{RV} = \mathsf{pp}$, $m = \mathsf{ID}$, $\sigma = \mathsf{sk}_{ID}$, and $r = d\|r'$. It computes $\mathsf{ct} = \mathsf{Enc}_{IBE}(\mathsf{pp}, d, \mathsf{ID})$ passing r' as the random coins to Enc_{IBE}. It outputs 1 if $\mathsf{Dec}_{IBE}(\mathsf{pp}, \mathsf{sk}_{ID}, \mathsf{ct}) = d$ and 0 otherwise.

4.2 Correctness and Security

Theorem 1. *If $\mathcal{E} = (Setup_{IBE}, KeyGen_{IBE}, Enc_{IBE}, Dec_{IBE})$ is an adaptively-secure IBE scheme (Definition 2) with message space $\mathcal{M}_{IBE}$ such that $|\mathcal{M}_{IBE}| = M$, then above construction forms a $1/M$-secure randomized verification signature scheme as per Definition 3.*

Proof. It is straightforward to verify the correctness as it follows directly from the correctness of IBE scheme. To prove $1/M$-EUFCMA security, consider a PPT adversary $\mathcal{A}$ that wins the $1/M$-EUFCMA security game described in Definition 3 with non-negligible advantage ϵ_1. This could be equivalently stated as that with probability ϵ_2, $\mathcal{A}$ outputs (m^*, σ^*) such that $\mathsf{VerifyProb}(\lambda, \mathsf{vk}, m^*, \sigma^*) > 1/M + \epsilon_2$ where ϵ_1 and ϵ_2 are non-negligible functions in λ. We construct a reduction algorithm $\mathcal{B}$ that acts as the challenger in the randomized verification signature security game and is the attacker in the adaptive IBE security game. We define algorithm $\mathcal{B}$ as follows:

1. The IBE challenger computes $(\mathsf{msk}, \mathsf{pp}) \xleftarrow{\$} \mathsf{Setup}_{IBE}(1^\lambda)$ and sends pp to the IBE attacker $\mathcal{B}$.
2. $\mathcal{B}$ sets $\mathsf{vk} = \mathsf{pp}$ and sends vk as the verification key to adversary $\mathcal{A}$.
3. Attacker $\mathcal{A}$ submits signature queries on a polynomial number of messages m_i. On each message, $\mathcal{B}$ sets $\mathsf{ID}_i = m_i$ and sends a secret key query for identity ID_i to the IBE challenger. $\mathcal{B}$ receives sk_{ID_i} and sends $\sigma_i = \mathsf{sk}_{ID_i}$ to attacker $\mathcal{A}$ as the signature for m_i.
4. Attacker $\mathcal{A}$ submits a forgery (m^*, σ^*) to $\mathcal{B}$ where $m^* \neq m_i$ for any queried m_i.
5. $\mathcal{B}$ sets $\mathsf{ID}^* = m^*$ and chooses $d_0^*, d_1^* \xleftarrow{\$} \mathcal{M}_{IBE}$ uniformly at random such that $d_0^* \neq d_1^*$. $\mathcal{B}$ sends ID^* as the challenge identity and d_0^*, d_1^* as the challenge messages to the IBE challenger. The IBE challenger chooses $b \xleftarrow{\$} \{0, 1\}$ and sends the challenge ciphertext $\mathsf{ct}^* \xleftarrow{\$} \mathsf{Enc}_{IBE}(\mathsf{pp}, d_b^*, \mathsf{ID}^*)$ to $\mathcal{B}$.
6. $\mathcal{B}$ runs $\mathsf{Verify}_{RV}(\mathsf{vk}, m^*, \sigma^*; r)$ q times, where $q = O(\lambda \cdot \epsilon_2^{-2})$ and $r \xleftarrow{\$} \mathcal{R}$ is freshly chosen for each run. Note that q is a polynomial when ϵ_2 is non-negligible. Let count be the number of times Verify_{RV} outputs 1. If $\mathsf{count} < q(1/M + \epsilon_2/2)$, $\mathcal{B}$ aborts and chooses $b' \xleftarrow{\$} \{0, 1\}$ at random as its guess. If $\mathsf{count} \geq q(1/M + \epsilon_2/2)$, $\mathcal{B}$ computes $d' = \mathsf{Dec}_{IBE}(\mathsf{pp}, \sigma^*, \mathsf{ct}^*)$. If $d' = d_0^*$, then $\mathcal{B}$ sets $b' = 0$. If $d' = d_1^*$, $\mathcal{B}$ sets $b' = 1$. Otherwise, d' is a non-challenge

message or $\perp$, so $\mathcal{B}$ guesses $b' \xleftarrow{\$} \{0,1\}$ uniformly at random and sends b' to IBE challenger as its guess.

We claim that $\mathcal{B}$ wins the IBE security game with non-negligible probability. To prove our claim, we define the following events:

- Abort: $\text{count} < q \left(\dfrac{1}{M} + \dfrac{\epsilon_2}{2} \right)$.
- Bad: $\text{VerifyProb}(\lambda, \text{vk}, m^*, \sigma^*) < \dfrac{1}{M}$
- Mid: $\dfrac{1}{M} \le \text{VerifyProb}(\lambda, \text{vk}, m^*, \sigma^*) < \dfrac{1}{M} + \epsilon_2$
- Good: $\text{VerifyProb}(\lambda, \text{vk}, m^*, \sigma^*) \ge \dfrac{1}{M} + \epsilon_2$.

We say that $\mathcal{B}$ aborts when event Abort occurs and it chooses a random bit $b' \xleftarrow{\$} \{0,1\}$ as its guess. Events Bad, Mid, and Good quantify the successful verification probability of the forgery (m^*, σ^*) that $\mathcal{A}$ produces. Event Bad signifies that (m^*, σ^*) verifies with probability no greater than $\chi = 1/M$. Event Good is the case when the forgery produced by $\mathcal{A}$ verifies with non-negligible advantage over $\chi = 1/M$. Then, event Mid represents the case in between Bad and Good such that Bad, Mid, and Good are mutually exclusive events.

The intuition behind running Verify_{RV} q times independently is to make sure that the reduction does not use the forgery whenever $\mathcal{A}$ provides a forgery that does worse than random guessing (of the encrypted message). This becomes necessary when we try to visualize the IBE message space to be very small. For instance, consider the scenario where $\mathcal{A}$ can win with advantage 1 in the randomized verification signature security game, but instead of always outputting a correct forgery it creates a bad forgery almost half the time. Therefore, if $\mathcal{B}$ directly uses the forgery to decrypt the challenge ciphertext which contains only one bit, then half the time it would be using a correct forgery and decrypting the challenge correctly, while at other times it would be decrypting incorrectly, and thus $\mathcal{B}$'s advantage in a direct reduction would be zero despite $\mathcal{A}$'s non-negligible advantage. Therefore, we want $\mathcal{B}$ to accept the forgery and not abort only if $\mathcal{A}$ produces a forgery such that verification succeeds with non-negligible probability over $1/M$. We provide a sketch of how these events relate to the probability that $\mathcal{B}$ wins as follows.

$$
\begin{aligned}
\Pr[\, \mathcal{B} \text{ wins} \,] &= \Pr[\, b = b' \,] \\
&= \Pr[\, b = b' \mid \text{Abort} \,] \cdot \Pr[\, \text{Abort} \,] + \Pr[\, b = b' \mid \overline{\text{Abort}} \,] \cdot \Pr[\, \overline{\text{Abort}} \,] \\
&\ge 1/2 \cdot \Pr[\, \text{Abort} \,] + \Pr[\, b = b' \mid \overline{\text{Abort}} \wedge \text{Bad} \,] \cdot \Pr[\, \overline{\text{Abort}} \wedge \text{Bad} \,] \\
&\quad + \Pr[\, b = b' \mid \overline{\text{Abort}} \wedge \text{Good} \,] \cdot \Pr[\, \overline{\text{Abort}} \wedge \text{Good} \,] \\
&\quad + \Pr[\, b = b' \mid \overline{\text{Abort}} \wedge \text{Mid} \,] \cdot \Pr[\, \overline{\text{Abort}} \wedge \text{Mid} \,]
\end{aligned}
$$

We know that $\Pr[\text{Good}] = \epsilon_1$. Using Chernoff bound, we can show that $\Pr[\text{Abort} \mid \text{Good}] \leq \text{negl}(\lambda)$, where $\text{negl}(\cdot)$ is a negligible function. Therefore, $\Pr[\overline{\text{Abort}} \wedge \text{Good}] \geq \epsilon_1 - \text{negl}(\lambda)$. Similarly, we can show that $\Pr[\overline{\text{Abort}} \wedge \text{Bad}] \leq \text{negl}(\lambda)$.

As described before, we want $\mathcal{B}$ to have non-negligible advantage in the IBE security game when $\mathcal{A}$ submits a forgery with high verification probability i.e. in event Good. Also we show that if the forgery submitted by $\mathcal{A}$ is verifies with at least $1/M$ probability i.e. event Mid, then $\mathcal{B}$ still wins the IBE security game with probability at least $1/2$. We consider the following three mutually exclusive events describing a particular run of Dec_{IBE} on challenge ciphertext ct^* given that $\mathcal{B}$ does not abort:

1. Win: Dec_{IBE} outputs the challenge message d_b^*.
2. Lose: Dec_{IBE} outputs the incorrect challenge message d_{1-b}^*.
3. Split: Dec_{IBE} outputs a non-challenge message or $\perp$

Note that $\mathcal{B}$ wins with probability 1 in event Win, probability 0 in event Lose, and probability $1/2$ in event Split. We use these different cases to analyze the probability the $\mathcal{B}$ wins in the event $\overline{\text{Abort}}$.

Claim 1. $\Pr[\, b = b' \mid \overline{\text{Abort}} \wedge \text{Good}\,] \geq \dfrac{\epsilon_2}{4} + \dfrac{1}{2}$

Proof. Let $p = \Pr[\text{Win} \mid \overline{\text{Abort}} \wedge \text{Good}] \geq 1/M + \epsilon_2/2$ since in event Good, $\text{VerifyProb}(\lambda, \text{vk}, m^*, \sigma^*) \geq 1/M + \epsilon_2/2$. Then, $\Pr[\text{Split} \mid \overline{\text{Abort}} \wedge \text{Good}] \geq (1 - p) \cdot (1 - 1/(M - 1))$ since the challenge messages were chosen uniformly at random from $\mathcal{M}_{\text{IBE}}$.

$$
\begin{aligned}
&\Pr[\, b = b' \mid \overline{\text{Abort}} \wedge \text{Good}\,] \\
&= \Pr[\, b = b' \mid \overline{\text{Abort}} \wedge \text{Good} \wedge \text{Win}\,] \cdot \Pr[\text{Win} \mid \overline{\text{Abort}} \wedge \text{Good}] \\
&\quad + \Pr[\, b = b' \mid \overline{\text{Abort}} \wedge \text{Good} \wedge \text{Lose}\,] \cdot \Pr[\text{Lose} \mid \overline{\text{Abort}} \wedge \text{Good}] \\
&\quad + \Pr[\, b = b' \mid \overline{\text{Abort}} \wedge \text{Good} \wedge \text{Split}\,] \cdot \Pr[\text{Split} \mid \overline{\text{Abort}} \wedge \text{Good}] \\
&\geq 1 \cdot p + 0 + \frac{1}{2} \cdot \left(1 - \frac{1}{M-1}\right) \cdot (1 - p) \\
&\geq \frac{\epsilon_2}{4} + \frac{1}{2}
\end{aligned}
$$

Claim 2. $\Pr[\, b = b' \mid \overline{\text{Abort}} \wedge \text{Mid}\,] \geq \dfrac{1}{2}$

Proof. The proof of the claim is the same as Claim 1 where $p = \Pr[\text{Win} \mid \overline{\text{Abort}} \wedge \text{Mid}] \geq \frac{1}{M}$. Therefore, we get that $\Pr[\, b = b' \mid \overline{\text{Abort}} \wedge \text{Mid}\,] \geq 1/2$.

Now we can complete our analysis of the winning probability of the reduction algorithm $\mathcal{B}$ in the IBE security game.

$$\Pr[\,\mathcal{B}\text{ wins}\,]$$

$$\geq \frac{1}{2} \cdot \Pr[\,\text{Abort}\,] + \Pr[\,b = b' \mid \overline{\text{Abort}} \wedge \text{Good}\,] \cdot \Pr[\,\overline{\text{Abort}} \wedge \text{Good}\,]$$

$$+ \Pr[\,b = b' \mid \overline{\text{Abort}} \wedge \text{Mid}\,] \cdot (\Pr[\,\overline{\text{Abort}}\,] - \Pr[\,\overline{\text{Abort}} \wedge \text{Bad}\,] - \Pr[\,\overline{\text{Abort}} \wedge \text{Good}\,])$$

$$\geq \frac{1}{2} \cdot \Pr[\,\text{Abort}\,] + \left(\frac{\epsilon_2}{4} + \frac{1}{2}\right) \cdot \Pr[\,\overline{\text{Abort}} \wedge \text{Good}\,]$$

$$+ \frac{1}{2} \cdot (\Pr[\,\overline{\text{Abort}}\,] - \Pr[\,\overline{\text{Abort}} \wedge \text{Bad}\,] - \Pr[\,\overline{\text{Abort}} \wedge \text{Good}\,])$$

$$= \frac{1}{2} + \frac{\epsilon_2}{4} \cdot \Pr[\,\overline{\text{Abort}} \wedge \text{Good}\,] - \frac{1}{2} \cdot \Pr[\,\overline{\text{Abort}} \wedge \text{Bad}\,]$$

$$= \frac{1}{2} + \frac{\epsilon_1 \epsilon_2}{4} - \text{negl}(\lambda)$$

Thus, $\text{Adv}_{\mathcal{B}}^{\text{IBE}} = \left| \Pr[\,\mathcal{B}\text{ wins}\,] - \frac{1}{2} \right| = \left| \frac{1}{2} + \frac{\epsilon_1 \epsilon_2}{4} - \text{negl}(\lambda) - \frac{1}{2} \right| = \frac{\epsilon_1 \epsilon_2}{4} - \text{negl}(\lambda)$, which is non-negligible if ϵ_1 and ϵ_2 are non-negligible.

We conclude that if $\mathcal{E}$ is an adaptively-secure IBE scheme, then our construction is a $\frac{1}{M}$-secure randomized verification signature scheme.

5   Amplifying Soundness

Ideally we would want from a randomized verification signature scheme that $\chi = 0$, which could be equivalently stated as that the forgery does not get verified with non-negligible probability. However not all constructions could be provably $\chi = 0$ secure, therefore we show a generic transformation from $\chi = 1 - 1/p$ to $\chi = 0$, where p is a polynomial in the security parameter λ.

5.1   Construction

Let $\mathcal{S}' = (\text{Setup}', \text{Sign}', \text{Verify}')$ be a $\chi = 1 - 1/p$ secure randomized verification signature scheme with message space $\mathcal{M}'$ and verification coin space $\mathcal{R}'$. We construct a $\chi = 0$ secure randomized verification signature scheme as follows:

1. $\text{Setup}_{\text{RV}}(1^{\lambda})$: It computes $(\text{sk}', \text{vk}') \xleftarrow{\$} \text{Setup}'(1^{\lambda})$ and returns (sk, vk) where $\text{sk} = \text{sk}'$ and $\text{vk} = \text{vk}'$.
2. $\text{Sign}_{\text{RV}}(\text{sk}, m)$: Let $\text{sk} = \text{sk}'$. It computes $\sigma \xleftarrow{\$} \text{Sign}'(\text{sk}', m)$ and outputs σ as the signature.
3. $\text{Verify}_{\text{RV}}(\text{vk}, m, \sigma; r)$: Let $\text{vk} = \text{vk}'$, $n = \Omega(p\lambda)$, and $r = (r_1, \ldots, r_n) \in (\mathcal{R}')^n$. For all $i \in [n]$, it runs $\text{Verify}'(\text{vk}', m, \sigma; r_i)$. It outputs 1 if all n verifications output 1, and 0 otherwise.

5.2   Correctness and Security

Theorem 2. *If $\mathcal{S}' = (\text{Setup}', \text{Sign}', \text{Verify}')$ be a $\chi = (1 - 1/p)$-EUFCMA secure signature scheme (Definition 3) with message space $\mathcal{M}'$ and verification coin* space $\mathcal{R}'$, then above construction forms a $\chi = 0$-EUFCMA secure signature scheme as per Definition 3 with message space $\mathcal{M}'$ and verification coin space $(\mathcal{R}')^n$, where $n = \Omega(p\lambda)$ and p is a polynomial in λ.

Proof. It is straightforward to verify the correctness as it follows directly from the correctness of underlying randomized verification signature scheme. The proof of soundness amplification is the standard direct product amplification [12]. For completeness, we sketch the argument below.

Let $\mathcal{R}^* = \{r \in \mathcal{R}' : \mathsf{Verify}'(\mathsf{vk}', m^*, \sigma^*; r) = 1\}$ denote the set of random coins on which verifier accepts the forgery (m^*, σ^*). Since, $\mathcal{S}'$ is $\chi = 1 - 1/p$ secure, we know that $\mathsf{VerifyProb}(\lambda, \mathsf{vk}', m^*, \sigma^*) \leq 1 - 1/p + \mu'$ with all but negligible probability, where μ' is a negligible function in λ. Therefore, we can write that with all but negligible probability $|\mathcal{R}^*| \leq (1 - 1/p + \mu')|\mathcal{R}'|$. After amplification, $\mathsf{Verify}_{\mathrm{RV}}$ runs Verify' on n uniformly random coins from $\mathcal{R}'$, thus set of random coins in $(\mathcal{R}')^n$ on which $\mathsf{Verify}_{\mathrm{RV}}$ accepts the forgery (m^*, σ^*) is $(\mathcal{R}^*)^n$. So, substituting $n = \Omega(p\lambda)$, we can write that with all but negligible probability,

$$\mathsf{VerifyProb}(\lambda, \mathsf{vk}, m^*, \sigma^*) = \frac{|\mathcal{R}^*|^n}{|\mathcal{R}'|^n} = \left(\frac{|\mathcal{R}^*|}{|\mathcal{R}'|}\right)^n \leq \left(1 - \frac{1}{p} + \mu'\right)^n \leq e^{-\frac{n}{p} + \mathrm{negl}} \leq e^{-\Omega(\lambda)}.$$

Therefore, we can conclude that if $\mathcal{S}'$ is a $\chi = (1 - 1/p)$-EUFCMA secure randomized verification signature scheme, then $\mathcal{S}$ is a $\chi = 0$-EUFCMA secure randomized verification signature scheme, where p is a polynomial in λ.

6 Derandomizing Verification in the Random Oracle Model

In this section, we show how to generically transform a signature scheme with randomized verification to a signature scheme with deterministic verification in the Random Oracle Model (ROM). Consider any $\chi = 0$-secure signature scheme with randomized verification.[2] Let ℓ denote the number of random bits used by the verification algorithm. To make the verification algorithm deterministic, we will use a hash function H that maps (message, signature) pairs to ℓ bit strings. The deterministic verification algorithm takes as input a signature σ, message m, computes $H(\sigma, m)$, and uses this as the randomness. For our security proof, the hash function H is modeled as a random oracle.

Let $\mathcal{S}_{\mathrm{RV}} = (\mathsf{Setup}_{\mathrm{RV}}, \mathsf{Sign}_{\mathrm{RV}}, \mathsf{Verify}_{\mathrm{RV}})$ be a $\chi = 0$ secure randomized verification signature scheme with message space $\mathcal{M}(\cdot)$, signature space $\Sigma(\cdot)$ and verification coin space $\mathcal{R}(\cdot)$. We construct a deterministic verification signature scheme with identical message and signature space:

1. $\mathsf{Setup}_{\mathrm{Det}}(1^\lambda)$: It computes $(\mathsf{sk}_{\mathrm{RV}}, \mathsf{vk}_{\mathrm{RV}}) \xleftarrow{\$} \mathsf{Setup}_{\mathrm{RV}}(1^\lambda)$ and returns $(\mathsf{sk}, \mathsf{vk}) = (\mathsf{sk}_{\mathrm{RV}}, \mathsf{vk}_{\mathrm{RV}})$.

---

[2] In the previous section, we showed how to amplify the security from $\chi = 1 - 1/\mathrm{poly}(\lambda)$ to $\chi = 0$. Therefore, our starting point in the transformation will be a randomized verification signature scheme with $\chi = 0$.

2. $\text{Sign}_{\text{Det}}(\text{sk}, m)$: Let $\text{sk} = \text{sk}_{\text{RV}}$. It computes $\sigma \xleftarrow{\$} \text{Sign}_{\text{RV}}(\text{sk}_{\text{RV}}, m)$ and outputs σ as the signature.
3. $\text{Verify}_{\text{Det}}(\text{vk}, m, \sigma)$: Let $\text{vk} = \text{vk}_{\text{RV}}$ and $r = H(m, \sigma)$. It outputs $\text{Verify}_{\text{RV}}(\text{vk}_{\text{RV}}, m, \sigma; r)$.

Theorem 3. *If $\mathcal{S}_{\text{RV}} = (\text{Setup}_{\text{RV}}, \text{Sign}_{\text{RV}}, \text{Verify}_{\text{RV}})$ be a $\chi = 0\text{-EUFCMA}$ secure randomized verification signature scheme (Definition 3) with message space $\mathcal{M}(\cdot)$, signature space $\Sigma(\cdot)$ and verification coin space $\mathcal{R}(\cdot)$, then above construction forms a EUFCMA secure deterministic verification signature scheme as per Definition 1 with message space $\mathcal{M}(\cdot)$ and signature space $\Sigma(\cdot)$.*

Proof. It is straightforward to verify the correctness as it follows directly from the correctness of underlying randomized verification signature scheme. For proving security in the random oracle model, we will assume the hash function H is modeled as a random oracle. Let q denote the number of queries made by the adversary to the random oracle. Without loss of generality, we can assume that the forgery (m^*, σ^*) is one of the q queries.

We can directly reduce an attack on our deterministic verification scheme construction to an attack on $\mathcal{S}_{\text{RV}}$. Our reduction algorithm $\mathcal{B}$ will receive a verification key vk_{RV} from $\mathcal{S}_{\text{RV}}$ challenger. It forwards vk_{RV} to $\mathcal{A}$. $\mathcal{B}$ relays all signature queries made by $\mathcal{A}$ to $\mathcal{S}_{\text{RV}}$ challenger. For answering hash queries $\mathcal{B}$ maintains a lookup table, and for each new query (m, σ) it chooses a uniformly random coin from $\mathcal{R}$. Finally, when $\mathcal{A}$ submits the forgery (m^*, σ^*), $\mathcal{B}$ forwards it to the challenger.

Suppose $\mathcal{A}$ wins with probability ϵ_1, and let $\mathcal{R}^* = \mathcal{R}_{m^*, \sigma^*}$.

$$\begin{aligned} \Pr[\mathcal{A} \text{ wins}] &= \Pr[\mathcal{A} \text{ outputs } (m^*, \sigma^*) \text{ and } H(m^*, \sigma^*) \in \mathcal{R}^*] \\ &\leq \Pr[\mathcal{A} \text{ outputs } (m^*, \sigma^*) \text{ and } H(m^*, \sigma^*) \in \mathcal{R}^* \text{ and } |\mathcal{R}^*|/|\mathcal{R}| \leq \text{negl}(\lambda)] \\ &\quad + \Pr[\mathcal{A} \text{ outputs } (m^*, \sigma^*) \text{ and } |\mathcal{R}^*|/|\mathcal{R}| \geq 1/\text{poly}(\lambda)] \end{aligned}$$

Now, note that $\Pr[\mathcal{A} \text{ outputs } (m^*, \sigma^*) \text{ and } H(m^*, \sigma^*) \in \mathcal{R}^* \text{ and } |\mathcal{R}^*|/|\mathcal{R}| \leq \text{negl}(\lambda)] \leq q \cdot \text{negl}(\lambda)$. This is because for each of the random oracle queries, the response is a uniformly random element in $\mathcal{R}$. As a result, using union bound, $\Pr[\exists \text{ RO query } (m_i, \sigma_i) \text{ such that } |\mathcal{R}_{m_i, \sigma_i}|/|\mathcal{R}| \leq \text{negl}(\lambda) \text{ and } H(m_i, \sigma_i) \in \mathcal{R}_{m_i, \sigma_i}] \leq q \cdot \text{negl}(\lambda)$.

Therefore, $\Pr[\mathcal{B} \text{ wins}] = \Pr[\mathcal{A} \text{ outputs } (m^*, \sigma^*) \text{ and } |\mathcal{R}^*|/|\mathcal{R}| \geq 1/\text{poly}(\lambda)] \geq \epsilon_1 - \text{negl}(\lambda)$.

7   Derandomizing Verification in the Standard Model

In this section, we present a transformation from a $\chi = 0\text{-EUFCMA}$ secure signature scheme with randomized verification to one that has a deterministic verification algorithm, and prove it EUFCMA secure in the standard model. At a high level, the transformed scheme can be described as follows. The verification key consists of n uniformly random strings $r_1, \ldots, r_n$ from $\mathcal{R}$, where n

is bounded by a sufficiently large polynomial in the security parameter. These random strings are then used by the verification algorithm to check if σ is a valid signature for message m. Note that the random strings $r_1, \ldots, r_n$ used during verification are chosen during Setup and kept as part of the verification key. At first sight, it might appear that, since the random strings are fixed during setup, the reduction to the underlying verification scheme will not go through because now the adversary's forgery may depend on the random coins used to verify. However, this is not the case because the underlying randomized verification signature scheme satisfies $\chi = 0$-EUFCMA security, therefore taking a union bound over all possible message-signature pairs we could rule out this case as if n is sufficiently large, then the probability that some bad message-signature pair gets verified using all $r_1, \ldots, r_n$ (independently) is negligible. Below we describe the transformation.

Let $\mathcal{S}_{\mathrm{RV}} = (\mathsf{Setup}_{\mathrm{RV}}, \mathsf{Sign}_{\mathrm{RV}}, \mathsf{Verify}_{\mathrm{RV}})$ be a $\chi = 0$-EUFCMA secure randomized verification signature scheme with message space $\mathcal{M}(\cdot)$, signature space $\Sigma(\cdot)$ and verification coin space $\mathcal{R}(\cdot)$. We construct a deterministic verification signature scheme with identical message and signature space:

1. $\mathsf{Setup}_{\mathrm{Det}}(1^\lambda)$: The setup algorithm first chooses $(\mathsf{sk}_{\mathrm{RV}}, \mathsf{vk}_{\mathrm{RV}}) \xleftarrow{\$} \mathsf{Setup}_{\mathrm{RV}}(1^\lambda)$. Next, it sets $\mathsf{sk} = \mathsf{sk}_{\mathrm{RV}}$, $n = \Omega(\log(|\mathcal{M}| \cdot |\Sigma|))$, and chooses n coins $\{r_1, \ldots, r_n\}$ where $r_i \xleftarrow{\$} \mathcal{R}$ for $i \le n$. The verification key vk consists of $(\mathsf{vk}_{\mathrm{RV}}, \{r_1, \ldots, r_n\})$.
2. $\mathsf{Sign}_{\mathrm{Det}}(\mathsf{sk}, m)$: Let $\mathsf{sk} = \mathsf{sk}_{\mathrm{RV}}$. The signing algorithm computes $\sigma \xleftarrow{\$} \mathsf{Sign}_{\mathrm{RV}}(\mathsf{sk}_{\mathrm{RV}}, m)$ and outputs σ as the signature.
3. $\mathsf{Verify}_{\mathrm{Det}}(\mathsf{vk}, m, \sigma)$: Let $\mathsf{vk} = (\mathsf{vk}_{\mathrm{RV}}, \{r_1, \ldots, r_n\})$. The verification algorithm checks if $\mathsf{Verify}_{\mathrm{RV}}(\mathsf{vk}_{\mathrm{RV}}, m, \sigma; r_i) = 1$ for all $i \le n$. If so, it outputs 1, else it outputs 0.

Theorem 4. *Let $\mathcal{S}_{\mathrm{RV}} = (Setup_{\mathrm{RV}}, Sign_{\mathrm{RV}}, Verify_{\mathrm{RV}})$ be a $\chi = 0$-EUFCMA secure randomized verification signature scheme (Definition 3) with message space $\mathcal{M}(\cdot)$, signature space $\Sigma(\cdot)$ and verification coin space $\mathcal{R}(\cdot)$. The above construction forms a EUFCMA secure deterministic verification signature scheme as per Definition 1 with message space $\mathcal{M}(\cdot)$ and signature space $\Sigma(\cdot)$.*

Proof. It is straightforward to verify the correctness as it follows directly from the correctness of underlying randomized verification signature scheme. To formally describe our reduction we distinguish between two types of forgers that an attacker $\mathcal{A}$ can emulate. Let $\mathcal{R}_{m^*,\sigma^*} = \{r \in \mathcal{R} : \mathsf{Verify}_{\mathrm{RV}}(\mathsf{vk}_{\mathrm{RV}}, m^*, \sigma^*; r) = 1\}$ denote the set of random coins on which verifier accepts the forgery (m^*, σ^*). We distinguish between two types of forgers as follows:

Type 1 forger: $\mathcal{A}$ with non-negligible probability, outputs a forgery (m^*, σ^*) such that $|\mathcal{R}_{m^*,\sigma^*}|/|\mathcal{R}|$ is *non-negligible* in λ.

Type 2 forger: $\mathcal{A}$ with non-negligible probability, outputs a forgery (m^*, σ^*) such that $|\mathcal{R}_{m^*,\sigma^*}|/|\mathcal{R}|$ is *negligible* in λ.

Claim 3. *If $\mathcal{A}$ emulates a Type 1 forger, then there exists a reduction algorithm $\mathcal{B}$ that breaks $\chi = 0$-EUFCMA security of randomized verification signature scheme $\mathcal{S}_{RV}$.*

Proof. In case of type 1 forgery, we can directly reduce an attack on our deterministic verification scheme construction to an attack on $\mathcal{S}_{RV}$. Our reduction algorithm $\mathcal{B}$ will receive a verification key vk_{RV} from $\mathcal{S}_{RV}$ challenger. It chooses n coins $\{r_1, \ldots, r_n\} \overset{\$}{\leftarrow} \mathcal{R}^n$ and forwards $(\mathsf{vk}_{RV}, \{r_1, \ldots, r_n\})$ to $\mathcal{A}$. $\mathcal{A}$ relays all signature queries made by $\mathcal{A}$ to $\mathcal{S}_{RV}$ challenger. By assumption, $|\mathcal{R}_{m^*, \sigma^*}|/|\mathcal{R}|$ is *non-negligible* for a type 1 forger, thus $\mathrm{VerifyProb}(\lambda, \mathsf{vk}_{RV}, m^*, \sigma^*) = |\mathcal{R}_{m^*, \sigma^*}|/|\mathcal{R}|$ is also non-negligible. Hence, if $\mathcal{A}$ emulates a type 1 forger, then it also breaks $\chi = 0$-EUFCMA security of randomized verification signature scheme $\mathcal{S}_{RV}$.

Claim 4. *If $\mathcal{A}$ emulates a Type 2 forger, then it does not break $\chi = 0$-EUFCMA security of randomized verification signature scheme $\mathcal{S}_{RV}$ with all but negligible probability.*

Proof. In case of type 2 forgery, we show that if we set $n = \Omega(\log(|\mathcal{M}| \cdot |\Sigma|))$, [3] then (with all but negligible probability) there does not exists any (m, σ) pair such that $\{r_1, \ldots, r_n\} \subseteq \mathcal{R}_{m,\sigma}$ and $|\mathcal{R}_{m,\sigma}|/|\mathcal{R}|$ is negligible. Let Bad denote the set of (m, σ) pairs such that $|\mathcal{R}_{m,\sigma}|/|\mathcal{R}|$ is negligible, and $R_{\max} = \max_{(m,\sigma) \in \mathsf{Bad}} |\mathcal{R}_{m,\sigma}|$. The proof follows from union bound over all such (m, σ) pairs in Bad. First, observe that

$$\forall (m, \sigma) \in \mathsf{Bad}, \quad \Pr[\mathrm{Verify}_{\mathrm{Det}}(\mathsf{vk}, m, \sigma) = 1] \leq \left(\frac{R_{\max}}{|\mathcal{R}|}\right)^n \leq (\mu')^n,$$

where μ' is a negligible function. Using union bound, we can write that

$$\Pr\left[\bigvee_{(m,\sigma) \in \mathsf{Bad}} \mathrm{Verify}_{\mathrm{Det}}(\mathsf{vk}, m, \sigma) = 1\right] \leq |\mathcal{M}| \cdot |\Sigma| \cdot (\mu')^n.$$

Since we set $n = \Omega(\log(|\mathcal{M}| \cdot |\Sigma|))$, the above expression is negligible in λ. Therefore, no type 2 forger can win with non-negligible probability over random choice of n verification coins.

Therefore, reduction algorithm $\mathcal{B}$ always guesses $\mathcal{A}$ to be a Type 1 forger and plays the standard reduction.

References

1. Boneh, D., Franklin, M.: Identity-based encryption from the weil pairing. In: Kilian, J. (ed.) CRYPTO 2001. LNCS, vol. 2139, pp. 213–229. Springer, Heidelberg (2001). doi:10.1007/3-540-44647-8_13

---

[3] Setting $n = \Omega(\log(|\mathcal{M}| \cdot |\Sigma|)/\log(1/\mu'))$ is sufficient where μ' is a negligible function that depends upon the parameters of the randomized verification scheme.

2. Boneh, D., Franklin, M.K.: Identity-based encryption from the weil pairing. SIAM J. Comput. **32**(3), 586–615 (2003)
3. Boneh, D., Lynn, B., Shacham, H.: Short signatures from the Weil pairing. J. Cryptol. **17**(4), 297–319 (2004)
4. Cocks, C.: An identity based encryption scheme based on quadratic residues. In: Honary, B. (ed.) Cryptography and Coding 2001. LNCS, vol. 2260, pp. 360–363. Springer, Heidelberg (2001). doi:10.1007/3-540-45325-3_32
5. Cui, Y., Fujisaki, E., Hanaoka, G., Imai, H., Zhang, R.: Formal security treatments for IBE-to-signature transformation: relations among security notions. IEICE Trans. **92–A**(1), 53–66 (2009)
6. Goldwasser, S., Micali, S., Rivest, R.L.: A digital signature scheme secure against adaptive chosen-message attacks. SIAM J. Comput. **17**(2), 281–308 (1988)
7. Naor, M.: On cryptographic assumptions and challenges. In: Boneh, D. (ed.) CRYPTO 2003. LNCS, vol. 2729, pp. 96–109. Springer, Heidelberg (2003). doi:10.1007/978-3-540-45146-4_6
8. Schröder, D., Unruh, D.: Security of blind signatures revisited. In: Fischlin, M., Buchmann, J., Manulis, M. (eds.) PKC 2012. LNCS, vol. 7293, pp. 662–679. Springer, Heidelberg (2012). doi:10.1007/978-3-642-30057-8_39
9. Unruh, D.: Everlasting multi-party computation. In: Canetti, R., Garay, J.A. (eds.) CRYPTO 2013. LNCS, vol. 8043, pp. 380–397. Springer, Heidelberg (2013). doi:10.1007/978-3-642-40084-1_22
10. Waters, B.: Efficient identity-based encryption without random oracles. In: Cramer, R. (ed.) EUROCRYPT 2005. LNCS, vol. 3494, pp. 114–127. Springer, Heidelberg (2005). doi:10.1007/11426639_7
11. Waters, B.: Dual system encryption: realizing fully secure IBE and HIBE under simple assumptions. In: Halevi, S. (ed.) CRYPTO 2009. LNCS, vol. 5677, pp. 619–636. Springer, Heidelberg (2009). doi:10.1007/978-3-642-03356-8_36
12. Yao, A.C.: Theory and application of trapdoor functions. In: Proceedings of the 23rd Annual Symposium on Foundations of Computer Science, SFCS 1982, pp. 80–91 (1982). http://dx.doi.org/10.1109/SFCS.1982.95

Side Channel Attack

Trade-Offs for S-Boxes: Cryptographic Properties and Side-Channel Resilience

Claude Carlet[1], Annelie Heuser[2], and Stjepan Picek[1,3,4(✉)]

[1] Universities of Paris VIII and Paris XIII, LAGA, UMR 7539, CNRS,
Saint-Denis, France
[2] CNRS/IRISA, Rennes, France
[3] Massachusetts Institute of Technology, CSAIL, Cambridge, USA
[4] Cyber Security Research Group, Delft University of Technology,
Mekelweg 2, Delft, The Netherlands
stjepan@computer.org

Abstract. When discussing how to improve side-channel resilience of a cipher, an obvious direction is to use various masking or hiding countermeasures. However, such schemes come with a cost, e.g. an increase in the area and/or reduction of the speed. When considering lightweight cryptography and various constrained environments, the situation becomes even more difficult due to numerous implementation restrictions. However, some options are possible like using S-boxes that are easier to mask or (more on a fundamental level), using S-boxes that possess higher inherent side-channel resilience. In this paper we investigate what properties should an S-box possess in order to be more resilient against side-channel attacks. Moreover, we find certain connections between those properties and cryptographic properties like nonlinearity and differential uniformity. Finally, to strengthen our theoretical findings, we give an extensive experimental validation of our results.

1 Introduction

When designing a block cipher, one needs to consider many possible cryptanalysis attacks and often give the best trade-off between the security, speed, ease of implementation, etc. Besides the two main directions in the form of linear [1] and differential [2] cryptanalyses, today the most prominent attacks come from the implementation attacks group where side-channel attacks (SCAs) play an important role. To protect against SCA, one common option is to use various countermeasures such as hiding or masking schemes [3] where one well known example is the threshold implementation [4]. However, such countermeasures come with a cost when implementing ciphers. If considering more resource constrained environments, one often does not have enough resources to implement standard ciphers like AES and therefore one needs to use lightweight cryptography. However, even lightweight ciphers can be too resource demanding especially when the cost of countermeasures is added. Therefore, although countermeasures

© Springer International Publishing AG 2017
D. Gollmann et al. (Eds.): ACNS 2017, LNCS 10355, pp. 393–414, 2017.
DOI: 10.1007/978-3-319-61204-1_20

represent the way how to go when considering SCA protection, there is no countermeasure (at least at the current state of the research) that offers sufficient protection against any attack while being cheap enough to be implemented in any environment.

In this paper, we consider how to improve SCA resilience of ciphers without imposing any extra cost. This is possible by considering the inherent resilience of ciphers. We particularly concentrate on block ciphers which utilize S-boxes and therefore study the resilience of S-boxes against side-channel attacks.

In the case of SCA concentrating only on 1-bit of the S-box output, a theoretical connection between the side-channel resistance and differential uniformity of S-boxes has been found in [5]. In particular, the authors showed that the higher the side-channel resistance, the smaller the differential resistance. However, as we show, this extension does not straightforwardly hold when considering more complex leakage models as the Hamming weight of the S-box output, which is the most prominent leakage model in side-channel analysis when considering Correlation Power Analysis (CPA) [6]. We therefore investigate S-box parameters which may influence the side-channel resistance while still having good or optimal cryptographic properties. The (almost) preservation of Hamming weight and a small Hamming distance between x and $F(x)$ are two properties each of which could strengthen the resistance to SCA from an intuitive perspective. Our theoretical and empirical findings show that notably in the case when exactly preserving the Hamming weight, the SCA resilience is improved. Moreover, we relax this assumption and investigate in S-boxes that almost preserve the Hamming weight. For our study, we employ the confusion coefficient [7] as a metric for side-channel resistance. Besides the signal-to-noise-ratio and the number of observed measurements, the confusion coefficient is the factor influencing the success rate of CPA and, moreover, it is the only factor that depends on the underlying considered algorithm and thus on the S-box. More precisely our main contributions are:

1. We calculate (resp. we bound above) the confusion coefficient value of a function F in the two scenarios where:
   (a) x and $F(x)$ have the same Hamming weight.
   (b) in average, $F(x)$ has a Hamming weight near that of x.
2. We observe that the S-boxes with no difference between the Hamming weights of their input and output have nonlinearity equal to 0; more generally, the same happens when the Hamming weight of x and the Hamming weight of $F(x)$ have always the same parity. Such functions are of course to be avoided from a cryptanalysis perspective. Furthermore, we show that more generally as well, for every S-box F, denoting by d_{w_H} the number of inputs x for which the Hamming weights of x and $F(x)$ have different parities, F has nonlinearity at most d_{w_H}. This implies that if the number of inputs x such that $w_H(x) \neq w_H(F(x))$ is at most d_{w_H}, the nonlinearity is at most d_{w_H}. We show in Example 2 that this does not make however the S-box necessarily weak. We emphasize that although these observations could be regarded trivial, they have practical consequences.

3. We show the connection between the number of fixed points in a function F and its nonlinearity.
4. We show that S-boxes such that $F(x)$ lies at a small Hamming distance from x (or more generally from an affine function of x) cannot have high nonlinearity although the obtainable values are not too bad for $n = 4, 8$.
5. In the practical part, we confirm our theoretical findings about the connection between (almost) preserving the Hamming weight and the confusion coefficient by investigating several S-boxes.
6. We investigate the relationship between the confusion coefficient of different key guesses and evaluate a number of S-boxes used in today's ciphers to show that their SCA resilience can significantly differ.

2 Preliminaries

2.1 Generalities on S-Boxes

Let n, m be positive integers, i.e., $n, m \in \mathbb{N}^+$. We denote by $\mathbb{F}_2^n$ the n-dimensional vector space over $\mathbb{F}_2$ and by $\mathbb{F}_{2^n}$ the finite field with 2^n elements. The set of all n-tuples of elements in the field $\mathbb{F}_2$ is denoted by $\mathbb{F}_2^n$, where $\mathbb{F}_2$ is the Galois field with two elements. Further, for any set S, we denote $S \backslash \{0\}$ by S^*. The usual inner product of a and b equals $a \cdot b = \bigoplus_{i=1}^{n} a_i b_i$ in $\mathbb{F}_2^n$.

The Hamming weight $w_H(a)$ of a vector a, where $a \in \mathbb{F}_2^n$, is the number of non-zero positions in the vector. An (n, m)-function is any mapping F from $\mathbb{F}_2^n$ to $\mathbb{F}_2^m$. An (n, m)-function F can be defined as a vector $F = (f_1, \cdots, f_m)$, where the Boolean functions $f_i : \mathbb{F}_2^n \to \mathbb{F}_2$ for $i \in \{1, \cdots, m\}$ are called the coordinate functions of F.

The component functions of an (n, m)-function F are all the linear combinations of the coordinate functions with non all-zero coefficients. Since for every n, there exists a field $\mathbb{F}_{2^n}$ of order 2^n, we can endow the vector space $\mathbb{F}_2^n$ with the structure of that field when convenient. If the vector space $\mathbb{F}_2^n$ is identified with the field $\mathbb{F}_{2^n}$ then we can take $a \cdot b = tr(ab)$ where $tr(x) = x + x^2 + \ldots + x^{2^{n-1}}$ is the trace function from $\mathbb{F}_{2^n}$ to $\mathbb{F}_2$. The addition of elements of the finite field $\mathbb{F}_{2^n}$ is denoted with "+", as usual in mathematics. Since, often, we identify $\mathbb{F}_2^n$ with $\mathbb{F}_{2^n}$ and if there is no ambiguity, we denote the addition of vectors of $\mathbb{F}_2^n$ when $n > 1$ with "+" as well.

An (n, m)-function F is balanced if it takes every value of $\mathbb{F}_2^m$ the same number 2^{n-m} of times.

The Walsh-Hadamard transform of an (n, m)-function F is (see e.g. [8]):

$$W_F(a, v) = \sum_{x \in \mathbb{F}_2^m} (-1)^{v \cdot F(x) + a \cdot x}, \quad a, v \in \mathbb{F}_2^m. \tag{1}$$

The *nonlinearity* nl of an (n, m)-function F equals the minimum nonlinearity of all its component functions $v \cdot F$, where $v \in \mathbb{F}_2^{m*}$ [8,9]:

$$nl = 2^{n-1} - \frac{1}{2} \max_{\substack{a \in \mathbb{F}_2^n \\ v \in \mathbb{F}_2^{m*}}} |W_F(a, v)|. \tag{2}$$

The nonlinearity of any (n, m) function F is bounded above by the so-called *covering radius bound*:

$$nl \leq 2^{n-1} - 2^{\frac{n}{2}-1}. \tag{3}$$

In the case $m = n$, a better bound exists. The nonlinearity of any (n, n) function F is bounded above by the so-called *Sidelnikov-Chabaud-Vaudenay bound* [10]:

$$nl \leq 2^{n-1} - 2^{\frac{n-1}{2}}. \tag{4}$$

Bound (4) is an equality if and only if F is an Almost Bent (AB) function, by definition of AB functions [8].

Let F be a function from $\mathbb{F}_2^n$ into $\mathbb{F}_2^m$ with $a \in \mathbb{F}_2^n$ and $b \in \mathbb{F}_2^m$. We denote:

$$D_F(a, b) = \{x \in \mathbb{F}_2^n : F(x) + F(x + a) = b\}. \tag{5}$$

The entry at the position (a, b) corresponds to the cardinality of the delta difference table $D_F(a, b)$ and is denoted as $\delta(a, b)$. The *differential uniformity* δ_F is then defined as [11]:

$$\delta_F = \max_{a \neq 0, b} \delta(a, b). \tag{6}$$

Functions that have differential uniformity equal to 2 are called the Almost Perfect Nonlinear (APN) functions. Every AB function is also APN, but the converse does not hold in general. AB functions exist only in an odd number of variables, while APN functions also exist for an even number of variables. When discussing the differential uniformity parameter for permutations, the best possible (and known) value is 2 for any odd n and also for $n = 6$. For n even and larger than 6, this is an open question. The differential uniformity value for the inverse function $F(x) = x^{2^n - 2}$ equals 4 when n is even and 2 when n is odd.

2.2   Side-Channel Resistance

Side-channel attacks analyze physical leakage that is unintentionally emitted during cryptographic operations in a device (e.g., through the power consumption [12] or electromagnetic emanation [13]). This side-channel leakage is statistically dependent on the intermediate processed values involving the secret key, which makes it possible to retrieve the secret from the measured data. In particular, as the attacker wants to retrieve the secret key, he makes predictions (hypotheses) on a small enumerable chunk (e.g., byte) of an intermediate state using all possible key values.

The side-channel resistance of implementations against Correlation Power Attack (CPA) [6] depends on three factors: the number of measurement traces, the signal-to-noise ratio (SNR) [14], and the confusion coefficient [7]. The relationship between the three factors is linear in case of low SNR [15]. The confusion coefficient measures the discrepancy between the hypothesis of an intermediate state using the correct (secret) key and any hypothesis made with a (wrong) key assumption. Therefore, as one compares possible intermediate processed values, the confusion coefficient depends on the underlying cryptographic algorithm and thus, if the attacker targets an S-box operation, on the side-channel resistance of that S-box. More precisely, let us assume the attacker exploits an intermediate processed value $F(k_c + t)$ during the first round that depends on the secret key $k_c \in \mathbb{F}_2^n$, an n-bit chunk of the plaintext $t \in \mathbb{F}_2^n$, and an S-box function F. Moreover, let us make the commonly accepted assumption that the device is leaking side-channel information as the Hamming weight (see e.g., [14]) of intermediate values with additive noise N:

$$w_H(F(k_c + t)) + N. \tag{7}$$

As the secret key k_c is unknown to the attacker, he computes for each key guess $k_g \in \mathbb{F}_2^n$ a hypothesis about the intermediate state:

$$y_{k_g,t} = y(k_g, t) = w_H(F(k_g + t)) \tag{8}$$

of the deterministic part of the leakage in Eq. (7). Interestingly, these hypotheses are not independent and their discrepancy is characterized by the confusion coefficient. Originally in [7] the confusion coefficient has been introduced for $(n, 1)$ Boolean functions:

$$\kappa(k_c, k_g) = Pr[(y(k_c, T)) \neq (y(k_g, T))], \tag{9}$$

with T being the random variable whose realization is t. In [5], the authors related $\kappa(k_c, k_g)$ in Eq. (9) to δ_F and showed that the higher the side-channel resistance, the smaller the differential resistance (that is, the higher δ_F). In fact, $\kappa(k_c, k_g)$ is represented as

$$\frac{1}{2^n} \sum_{t \in \mathbb{F}_2^n} (F(t + k_c) + F(t + k_g)), \tag{10}$$

which can then be straightforwardly connected to δ_F for 1-bit models.

In [16] the authors extend $\kappa(k_c, k_g)$ to the general multi-bit case for CPA and thus to (n, m)-functions F. In this paper, we use the definition given in [15] which is a standardized version of confusion coefficient given in [16] and thus a natural extension of Eq. (9):

$$\kappa(k_c, k_g) = \mathbb{E}\left\{ \left(\frac{1}{2}(y(k_c, T) - y(k_g, T)) \right)^2 \right\}, \tag{11}$$

where y is assumed to be standardized (i.e., $\mathbb{E}(y(\cdot, T)) = 0, Var(y(\cdot, T)) = 1$). More specifically, Eq. (11) enables us to compare confusion coefficients for different functions F. By substituting $y(*)$ with Eq. (8) and denoting $x = t \oplus k_c$ and $a = k_c + k_g$ we can write $\kappa(k_c, k_g)$ as

$$\mathbb{E}\left(\left(\frac{1}{2} \left(\frac{w_H(F(x))}{\sqrt{\frac{m}{4}}} - \frac{w_H(F(x+a))}{\sqrt{\frac{m}{4}}} \right) \right)^2 \right) = \mathbb{E}\left(\left(\frac{w_H(F(x)) - w_H(F(x+a))}{\sqrt{m}} \right)^2 \right). \tag{12}$$

Now, it is easy to see that from Eq. (12) we cannot straightforwardly derive a connection to δ_F for (n, m) functions. More precisely, for $m = 1$ the square is just 4 times the value of $F(t) + F(t+a)$ and then the confusion coefficient equals

$\delta(a, 1)$. For $m > 1$ we have the square of the difference between the weights of $F(t)$ and $F(t+a)$ which is not 4 times the weight of $b = F(t) + F(t+a)$ because the $1 - 0$ and the $0 - 1$ count with their signs in the sum. So there is no direct connection with δ_F anymore.

As a decisive criterion for comparison between confusion coefficients, the minimum value of $\kappa(k_c, k_g)$ was specified in [15] as it relates to the success rate when the SNR is low. Note that the higher is the minimum of the confusion coefficient, the lower is the side-channel resilience. This comes from the fact that the lower the confusion coefficient the smaller is the (Euclidean) distance between the correct key k_c and a key guess k_g and thus the harder it is for an attacker to distinguish if the leakage is arising due to a computation with k_c or k_g. A detailed discussion on this will be given in Subsect. 5.2. On the other hand, in [17] authors use $var(\kappa(k_c, k_g))$ as a criterion, where smaller values indicate lower side-channel resilience. Our experiments in Sect. 5 show that both metrics coincide with the empirical resilience using simulations.

In the case $\kappa(k_c, k_g) = 0$ or $\kappa(k_c, k_g) = 1$ for any $k_g \neq k_c$, CPA is not able to distinguish between k_c and this key guess k_g and will thus fail to reveal the secret key exclusively even if the number of measurements goes to infinity. More precisely, $\kappa(k_c, k_g) = 0$ means that for a key guess k_g one observes exactly the same intermediate values (see Eq. (8)) as for the correct key k_c. Contrary for $\kappa(k_c, k_g) = 1$ one observes the complementary value (can be seen from Eqs. (9) and (11)), however, as CPA takes the absolute value of correlation (due to hardware related properties [14]) an attacker again cannot distinguish between k_c and k_g in this case. In general, normalized confusion coefficient values close to 0.5 indicate that k_c and k_g can be easily distinguished (see Eq. (9)). We will show in Sect. 3 and empirically confirm in Sect. 5 that in case of preserving w_H there exists an key guess k_g such that $\kappa(k_c, k_g) = 1$.

3   S-Boxes (Almost) Preserving the Hamming Weight

3.1   Relation to the Confusion Coefficient

To obtain, for an (n, m)-function F, a connection between the confusion coefficient parameter and the Hamming weight preservation (i.e., the fact that, for every x, $F(x)$ has the same Hamming weight as x) or, more generally, a limited average Hamming weight modification, we start with Eq. (12). For any function F, we have:

$$\frac{1}{m}\mathbb{E}\left(\sum_{i=1}^{m} F_i(x) - \sum_{i=1}^{m} F_i(x + a)\right)^2$$

$$= \frac{1}{m}\mathbb{E}\left(\sum_{i=1}^{m}\left(\frac{1}{2} - \frac{1}{2}(-1)^{F_i(x)}\right) - \sum_{i=1}^{m}\left(\frac{1}{2} - \frac{1}{2}(-1)^{F_i(x+a)}\right)\right)^2$$

$$= \frac{1}{4m}\mathbb{E}\left(\sum_{i=1}^{m}\left((-1)^{F_i(x)} - (-1)^{F_i(x+a)}\right)\right)^2$$

$$= \frac{1}{4m} \sum_{1 \leq i,j \leq m} \mathbb{E}\left((-1)^{F_i(x)} - (-1)^{F_i(x+a)}\right)\left((-1)^{F_j(x)} - (-1)^{F_j(x+a)}\right)$$

$$= \frac{1}{4m} \sum_{1 \leq i,j \leq m} \mathbb{E}\left((-1)^{F_i(x)+F_j(x)} - (-1)^{F_i(x)+F_j(x+a)} - \right.$$

$$\left. (-1)^{F_i(x+a)+F_j(x)} + (-1)^{F_i(x+a)+F_j(x+a)}\right)$$

$$= \frac{1}{2m} \sum_{1 \leq i,j \leq m} \mathbb{E}\left((-1)^{F_i(x)+F_j(x)} - (-1)^{F_i(x)+F_j(x+a)}\right)$$

$$= \frac{1}{2m} \mathbb{E}\left(\left(\sum_{i=1}^{m}(-1)^{F_i(x)}\right)^2 - \left(\sum_{i=1}^{m}(-1)^{F_i(x)}\right)\left(\sum_{i=1}^{m}(-1)^{F_i(x+a)}\right)\right). \quad (13)$$

Lemma 1 addresses the case where F preserves the Hamming weight, whereas the scenario in which F modifies the Hamming weight in a limited way is described in Lemma 2. Note that the first scenario is a particular case of the second.

Lemma 1. *For an (n,n)-function such that, for every x, $F(x)$ has the same Hamming weight as x, the confusion coefficient equals $\frac{w_H(a)}{n}$.*

Proof. If F preserves the Hamming weight, that is, if $w_H(F(x)) = w_H(x)$ for every x (or more generally, if F is the composition of a function preserving the weight by an affine isomorphism on the right), then the confusion coefficient $\kappa(k_c, k_g) = \mathbb{E}\left(\left(\frac{w_H(F(x)) - w_H(F(x+a))}{\sqrt{n}}\right)^2\right)$, where $a = k_c + k_g$, becomes $\mathbb{E}\left(\left(\frac{w_H(x) - w_H(x+a)}{\sqrt{n}}\right)^2\right)$, and by applying Eq. (13) (which is valid for every F) to $F = Id$, we obtain:

$$\frac{1}{2n} \mathbb{E}\left(\left(\sum_{i=1}^{n}(-1)^{x_i}\right)^2 - \left(\sum_{i=1}^{n}(-1)^{x_i}\right)\left(\sum_{i=1}^{n}(-1)^{x_i+a_i}\right)\right) \quad (14)$$

$$= \frac{1}{2n} \mathbb{E}\left(\sum_{1 \leq i,j \leq m}(-1)^{x_i+x_j} - \sum_{1 \leq i,j \leq m}(-1)^{x_i+x_j+a_j}\right).$$

The expectations of all these sums for $i \neq j$ are null (since the character sums of nonzero linear functions are null), and we obtain:

$$\frac{1}{2n} \mathbb{E}\left(m - \sum_{1 \leq i \leq m}(-1)^{a_i}\right) = \frac{1}{n}\mathbb{E}\left(w_H(a)\right) = \frac{w_H(a)}{n}. \quad (15)$$

Example 1. *For $n = 4$, Lemma 1 gives $\min_{k_c \neq k_g} \kappa(k_c, k_g) = 0.25$ and for $w_H(a) = n$ we have $\kappa(k_c, k_g) = 1$, which means that the CPA distinguisher is not able to distinguish between these two hypotheses k_g and k_c (see Subsect. 2.2). Note that we give a more detailed discussion about the results and their ramifications in Sect. 5.*

Lemma 2. *For an (n,n)-function such that, on average, $F(x)$ has a Hamming weight near that of x, more precisely, where $\sum_x |w_H(F(x)) - w_H(x)| \leq d_{w_H}$, where d_{w_H} is some number, the standardized confusion coefficient is bounded above by $\frac{w_H(a)}{n} + \frac{4 d_{w_H}}{2^n}$.*

Proof. If $\mathbb{E}(|w_H(F(x)) - w_H(x)|) \leq \frac{d_{w_H}}{2^n}$, then according to Lemma 1 and its proof the confusion coefficient $\kappa(k_c, k_g) = \mathbb{E}\left(\left(\frac{w_H(F(x)) - w_H(F(x+a))}{\sqrt{n}} \right)^2 \right)$ is such that

$$
\begin{aligned}
\left| \kappa(k_c, k_g) - \frac{w_H(a)}{n} \right| &\leq \mathbb{E}\left(\left| \left(\frac{w_H(F(x)) - w_H(F(x+a))}{\sqrt{n}} \right)^2 - \left(\frac{w_H(x) - w_H(x+a)}{\sqrt{n}} \right)^2 \right| \right) \\
&= \mathbb{E}\left(\left| \left(\frac{w_H(F(x)) - w_H(F(x+a))}{\sqrt{n}} - \frac{w_H(x) - w_H(x+a)}{\sqrt{n}} \right) \right. \right. \\
&\qquad\qquad \left. \left. \left(\frac{w_H(F(x)) - w_H(F(x+a))}{\sqrt{n}} + \frac{w_H(x) - w_H(x+a)}{\sqrt{n}} \right) \right| \right) \\
&\leq \frac{2}{n} \left(\max_{x \in \mathbb{F}_2^n} w_H(F(x)) + \max_{x \in \mathbb{F}_2^n} w_H(x) \right) \mathbb{E}\left(|w_H(F(x)) - w_H(x)| \right) \\
&= \frac{4 d_{w_H}}{2^n}.
\end{aligned}
$$

3.2   Relation to Cryptographic Properties

We study the cryptographic consequences of the preservation of the Hamming weight. Again we first cover the specific case were the input and output of an S-box always have the same Hamming weight, and then the second case where the output has on average a Hamming weight close to that of the corresponding input (see Lemma 3).

If for every x, we have $w_H(F(x)) = w_H(x)$ then the sum (mod 2) of all coordinate functions of F equals the sum (mod 2) of all coordinates of x. This means that F has nonlinearity equal to zero since one of its component functions is linear. Of course, the same happens under the much weaker hypothesis that $w_H(F(x))$ and $w_H(x)$ have always the same parity. Therefore, an S-box function preserving the Hamming weight is cryptographically insecure.

However, if $\sum_x |w_H(F(x)) - w_H(x)| \leq d_{w_H}$, then we have $nl \leq d_{w_H}$. Indeed, this is a direct consequence of the following straightforward result, which has however much importance in our context:

Lemma 3. *If the Hamming weight of the Boolean function:*

$$
x \mapsto (w_H(F(x)) - w_H(x)) \ [mod\, 2],
$$

that is, $\sum_x ((w_H(F(x)) - w_H(x)) \ [mod\, 2])$, is at most d_{w_H}, then we have $nl \leq d_{w_H}$.

Indeed, the Hamming distance between the component function $\sum_i F_i \ [mod\, 2])$ and the linear function $\sum_i x_i \ [mod\, 2])$ is then at most d_{w_H}.

Example 2. *For a* $(4,4)$-*function* F *to have nonlinearity equal to* 4 *(optimal nonlinearity), it means that* d_{w_H} *must be at least* 4. *In order to construct functions with such properties, we ran a genetic algorithm as given by Picek et al.* [17]. *We use the same settings as there: 30 independent runs, population size equal to 50, 3-tournament selection, and mutation probability 0.3 per individual. The objective is the maximization of the following fitness function:*

$$fitness = nl + \Delta_{nl,4}(n \times 2^n - |w_H(F(x)) - w_H(x)|). \qquad (16)$$

Here, $\Delta_{nl,4}$ *represents the Kronecker delta function that equals 1 when nonlinearity is* 4 *and 0 otherwise. Notice we subtract the difference of the Hamming weights of the inputs and outputs of an S-box from the summed Hamming weight value for a* $(4,4)$-*function since we work with the maximization problem while that value should be minimized. Interestingly, we observed that finding S-boxes with those properties is a relatively easy task and that the obtained S-boxes never have more than 8 fixed points. We give examples of such S-boxes in Table 1, for instance,* S_5 *where nonlinearity equals* 4 *and* d_{w_H} *is* 4.

Next, inspired by our empirical results, we investigate whether it is theoretically possible to construct an S-box with even more fixed points while still having the maximal nonlinearity.

Lemma 4. *If an* (n,n)-*function has* k *fixed points then the maximal value of* $W_F(a,v)$ *when* $v \neq 0$ *is bounded below by* $(k-1)/(1-2^{-n})$. *If* nl *is the nonlinearity of an* (n,n)-*function, then its number* k *of fixed points is not larger than* $2^n - \lceil (2 - 2^{1-n}) \, nl \rceil$.

Proof. The number of fixed points k of an (n,n)-function F equals:

$$k = 2^{-n} \sum_{v \in \mathbb{F}_2^n} W_F(v,v) = 2^{-n} \sum_{x,v \in \mathbb{F}_2^n} (-1)^{v \cdot (x + F(x))}, \qquad (17)$$

which follows from Eq. (1) when $a = v$ and the property that $\sum_{v \in \mathbb{F}_2^n} (-1)^{v \cdot a}$ equals 2^n if $a = 0$ and is null otherwise. The value of $W_F(0,0)$ involved in Eq. (17) equals 2^n. We take it off and obtain:

$$k - 1 = 2^{-n} \sum_{v \in \mathbb{F}_2^{n*}} W_F(v,v). \qquad (18)$$

Then the arithmetic mean of $W_F(v,v)$ when $v \neq 0$ equals $(k-1)/(1-2^{-n})$. This implies that $\max_v W_F(v,v)$ is at least $(k-1)/(1-2^{-n})$ and the nonlinearity cannot be larger than $2^{n-1} - (k-1)/(2-2^{1-n})$. The inequality $nl \leq 2^{n-1} - (k-1)/(2-2^{1-n})$ is equivalent to $k \leq 2^n - \lceil (2-2^{1-n}) \, nl \rceil$.

4   S-Boxes Minimizing the Hamming Distance

4.1   Relation to the Confusion Coefficient

In real world applications, the device may not only leak in the Hamming weight, but also in the Hamming distance, therefore we now extend our study to the case were the leakage arises from the Hamming distance between x and $F(x)$. Again we first study the relation to the confusion coefficient and then give the connection to cryptographic properties.

By the triangular inequality, we have $|w_H(F(x)) - w_H(x)| \leq d_H(x, F(x))$. This implies that $\sum_x |w_H(F(x)) - w_H(x)| \leq \sum_x d_H(x, F(x))$.

Hence, if $\sum_x d_H(x, F(x)) \leq d_{d_H}$, we can use Lemma 2 and deduce that also in this scenario the confusion coefficient is bounded by $\frac{w_H(a)}{n} + \frac{4 d_{d_H}}{2^n}$.

4.2 Relation to Cryptographic Properties

From $\sum_x d_H(x, F(x)) \leq d_{d_H}$, up to adding a linear function (which does not change the nonlinearity nor the differential uniformity), considering S-boxes such that, for every x, $F(x)$ lies at a small distance from x corresponds to considering functions which take a too small number of values. We show that such functions have bad nonlinearity and bad differential uniformity.

Lemma 5. *Let F be any (n, m)-function such that $|F(\mathbb{F}_2^n)| \leq D$, then $\delta_F \geq \frac{2^n}{2^m - 1} \left(\frac{2^n}{D} - 1 \right)$ and $nl \leq 2^{n-1} - \frac{\frac{2^{n+m-1}}{D} - 2^{n-1}}{2^m - 1}$.*

Proof. By using the Cauchy-Schwartz inequality, we obtain $\sum_{a \in \mathbb{F}_2^{n*}} |D_a F^{-1}(0)|$
$= \sum_{b \in \mathbb{F}_2^m} |F^{-1}(b)|^2 - 2^n \geq \frac{(\sum_{b \in \mathbb{F}_2^m} |F^{-1}(b)|)^2}{D} - 2^n = \frac{2^{2n}}{D} - 2^n$, and there exists then $a \in \mathbb{F}_2^{n*}$ such that $|D_a F^{-1}(0)| \geq \frac{\frac{2^{2n}}{D} - 2^n}{2^m - 1}$. This proves the first assertion.

We have a partition of $\mathbb{F}_2^n$ into at most D parts by the preimages $F^{-1}(b)$, $b \in \mathbb{F}_2^m$, and there exists then $b \in \mathbb{F}_2^m$ such that $|F^{-1}(b)| \geq \frac{2^n}{D}$; for such b, we have $\sum_{x \in \mathbb{F}_2^n, v \in \mathbb{F}_2^m} (-1)^{v \cdot (F(x) + b)} \geq \frac{2^{n+m}}{D}$, which is equivalent to $\sum_{v \in \mathbb{F}_2^m, v \neq 0} (-1)^{v \cdot b} W_F(0, v) \geq \frac{2^{n+m}}{D} - 2^n$, and then there exists $v \neq 0$ such that $|W_F(0, v)| \geq \frac{\frac{2^{n+m}}{D} - 2^n}{2^m - 1}$, which implies that $nl \leq 2^{n-1} - \frac{\frac{2^{n+m-1}}{D} - 2^{n-1}}{2^m - 1}$. This proves the second assertion.

If D is small with respect to 2^m (so that 2^{n-1} is small with respect to $\frac{2^{n+m-1}}{D}$) and D is small with respect to $2^{n/2}$ (so that $\frac{2^n}{D}$ is large with respect to $2^{n/2}$), the nonlinearity is bad with respect to the covering radius bound $nl \leq 2^{n-1} - 2^{n/2-1}$. More precisely, if $D \leq \frac{2^m}{\lambda}$ with $\lambda > 1$, then $nl \leq 2^{n-1} - \frac{(\lambda-1)2^{n-1}}{2^m-1} < 2^{n-1} - (\lambda - 1)2^{n-m-1}$ and if $(\lambda - 1)2^{n-m}$ is significantly larger than $2^{n/2}$, the nonlinearity is bad with respect to the covering radius bound. We have also that if D is small with respect to 2^m then δ_F is large with respect to 2^{n-m} if $m < n$ and with 2 if $m = n$ (which are the smallest possible values of δ_F).

If F is an (n, n)-function and $x + F(x)$ has low weight for every x, say at most t_{d_H}, which is equivalent to saying that $d_H(x, F(x)) \leq t_{d_H}$ for every x, then its number of values is at most $D = \sum_{i=0}^{t_{d_H}} \binom{n}{i}$ and we can apply the result above to $x + F(x)$, which has the same nonlinearity and the same δ_F as F. As far as we know, these observations are new. Note that we also have the possibility of applying Lemma 3 and then we have that nonlinearity is bounded by t_{d_H}.

Remark 3. *Lemma 5 applies to the case when $d_H(x, F(x)) \leq t_{d_H}$ for every x where x equals $t \oplus k_g$. This represents a setting one would encounter when working for instance with software implementations. Now, if we consider a hardware setting (e.g., FPGA), then we are interested in the case $d_H(t, F(t \oplus k_g)) \leq t_{d_H}$ for every key. However, this case leads to the same observation as before but now with up to adding an affine function instead of up to adding a linear function as given in Lemma 5.*

5 Side-Channel Evaluation

5.1 Evaluation of S-Boxes with (Almost) w_H Preservation

As cryptographically non-optimal examples of S-boxes (almost) preserving w_H we consider five different functions F: the identity mapping (S_1), F not Id but preserving w_H (S_2), the identity mapping with an exchange of the images at position $x = 3$ and $x = 12$, i.e., $F(3) = 12$ and $F(12) = 3$, and as $w_H(3) = 2$ and $w_H(12) = 3$ we have $d_{w_H} = 2$ (see Lemma 1) (S_3), $F(x) = 2^n - x$ which gives the complementary Hamming weight (S_4). Finally, we investigate four S-box functions S_5 to S_8 with the smallest possible distance d_{w_H} that equals 4 and maximal possible nonlinearity equal to 4 (see Subsect. 3.2). S-box functions S_7 and S_8 have furthermore optimal differential uniformity (=4). The mappings are given in Table 1.

Table 1. Specifications of functions F, $(x)w_H(x)$

S-box	(0)0	(1)1	(2)1	(3)2	(4)1	(5)2	(6)2	(7)3	(8)1	(9)2	(10)2	(11)3	(12)2	(13)3	(14)3	(15)4
S_1	(0)0	(1)1	(2)1	(3)2	(4)1	(5)2	(6)2	(7)3	(8)1	(9)2	(10)2	(11)3	(12)2	(13)3	(14)3	(15)4
S_2	(0)0	(4)1	(8)1	(10)2	(1)1	(12)2	(3)2	(13)3	(2)1	(6)2	(9)2	(7)3	(5)2	(14)3	(11)3	(15)4
S_3	(0)0	(1)1	(2)1	(12)3	(4)1	(5)2	(6)2	(7)3	(8)1	(9)2	(10)2	(11)3	(3)1	(13)3	(14)3	(15)4
S_4	(15)4	(14)3	(13)3	(12)2	(11)3	(10)2	(9)2	(8)1	(7)3	(6)2	(5)2	(4)1	(3)2	(2)1	(1)1	(0)0
S_5	(0)0	(4)1	(8)1	(10)2	(1)1	(2)1	(9)2	(6)2	(3)2	(5)2	(7)3	(13)3	(12)2	(11)3	(14)3	(15)4
S_6	(0)0	(9)2	(2)1	(13)3	(8)1	(6)2	(5)2	(7)3	(1)1	(4)1	(12)2	(14)3	(3)2	(10)2	(11)3	(15)4
S_7	(0)0	(2)1	(5)2	(13)3	(8)1	(4)1	(6)2	(7)3	(1)1	(12)2	(10)2	(14)3	(3)2	(9)2	(11)3	(15)4
S_8	(0)0	(1)1	(10)2	(7)3	(2)1	(4)1	(5)2	(13)3	(8)1	(12)2	(3)2	(9)2	(6)2	(11)3	(14)3	(15)4

The confusion coefficients are illustrated in Fig. 1. Note that, the distribution of $\kappa(k_c, k_g)$ is independent on the particular choice of k_c (in the case there are no weak keys) and the values for $\kappa(k_c, k_g)$ are only permuted when choosing different value $k_c \in \mathbb{F}_2^n$. For our experiments we choose $k_c = 0$ and furthermore we order $\kappa(k_c, k_g)$ in an increasing order of magnitude for illustrative purpose. The minimum value of $\kappa(k_c, k_g)$ for $k_g \neq k_c$ is highlighted with a red cross as it is one indicator of the side-channel resistance. Moreover, we mark $\kappa(k_c, k_g) = 0$ or $\kappa(k_c, k_g) = 1$ with a red circle which points out that CPA is not able to distinguish between k_c and the marked k_g.

Figure 1a shows that, indeed, k_c is indistinguishable from one key hypothesis k_g if w_H is preserved. Or in other words, even if knowing t and observing $w_H(F(t + k_c)) + N$ with F equal to S_1 the attacker can not exclusively gain information about k_c even if the number of measurements $m \to \infty$. Moreover, it confirms Lemma 1. Note that in our example $a = k_g$, thus $\kappa(k_c, k_g) = \frac{w_H(k_g)}{4}$. Interestingly, when comparing our results to the study in [5], where the authors investigated $(n, 1)$-functions, we observe that the confusion coefficient takes different values which indeed confirms that the Hamming weight model is not a straightforward extension from 1-bit models. More precisely, in case of linear $(n, 1)$-function the authors observed that the confusion coefficient only takes values from $\{0,1\}$, whereas our examples illustrate (as well as our theoretical findings in Sect. 3) that the confusion coefficient is not restricted to only $\{0,1\}$, and is equal to 1 for only one particular k_g. Interestingly, for $d_{w_H} = 2$ (see in Fig. 1b) we also have that k_c is indistinguishable for one k_g. Moreover, apart from $\kappa(k_c, k_g) = 1$, only two different values are taken, each 7 times. This means that CPA is not able to distinguish between each of these 7 key guesses and in total only produces three different correlation values. When considering a complementary w_H preservation (e.g. $4 - w_H$) we achieve the same results as for w_H preservation (see also Fig. 1).

Note that, while being illustrative, these first four examples of F are not cryptographically optimal and thus are not suitable in practice. We therefore constructed four S-boxes (S_5 to S_8) with the smallest $d_{w_H}(= 4)$ while having optimal nonlinearity. Note that S_5, S_6 have suboptimal differential uniformity, while S_7, S_8 are cryptographically optimal (i.e. optimal nonlinearity and differential uniformity). Figures 1c to f show the confusion coefficient of S_5 to S_8. We can observe that all S-boxes have a very low minimum confusion coefficient that is even lower than for S_1 to S_4. Even more, as the previously investigated S-boxes, S_5 has $\kappa(k_c, k_g) = 1$. Therefore, we find an S-box with almost Hamming weight preserving for which even with an infinity amount of traces the secret key cannot exclusively be found. As the minimum value of the confusion coefficient of S_5 is low (=0.125) there exists additionally other key hypotheses which are harder to distinguish from the secret key. As a conclusion we can say that indeed exact w_H preserving results in a good side-channel resistance since we have $\kappa(k_c, k_g) = 1$. Moreover, when the w_H is almost preserved we present here S-boxes which have a very low minimum confusion coefficient.

5.2   A Closer Look at the Confusion Coefficient

To understand the exact reason why some (one or more) key guesses result in a smaller confusion coefficient than others and how this is related to F, we concentrate on the connection between k_c, k_g, F, and $\kappa(k_c, k_g)$. Loosely speaking, we are iterating on key guesses influencing the input of F while calculating the confusion coefficient on the measured output of F and being interested in the properties of F. To better address these connections, we split the problem into 2 individual problems.

First, we take a deeper look at the input of F, i.e., $t \oplus k_g$ where $\forall t, k_g \in \mathbb{F}_2^n$ (see Eq. (8)). Clearly, due to the $\oplus$ operation a particular permutation for different key guesses k_g is given. A 2-D representation for $t \oplus k_g$, where k_g is on the horizontal and t on the vertical axis, is given in Fig. 2, where again, for

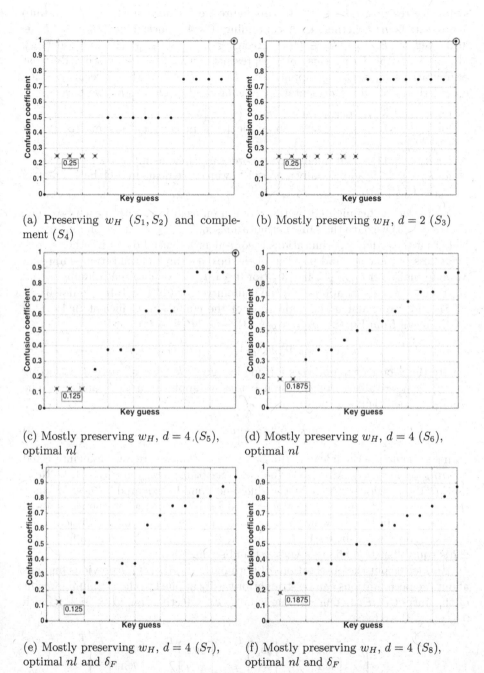

(a) Preserving w_H (S_1, S_2) and complement (S_4)

(b) Mostly preserving w_H, $d = 2$ (S_3)

(c) Mostly preserving w_H, $d = 4$ (S_5), optimal nl

(d) Mostly preserving w_H, $d = 4$ (S_6), optimal nl

(e) Mostly preserving w_H, $d = 4$ (S_7), optimal nl and δ_F

(f) Mostly preserving w_H, $d = 4$ (S_8), optimal nl and δ_F

Fig. 1. Confusion coefficients

simplicity reasons, $t, k_g \in \mathbb{F}_2^4$. In this figure we furthermore group $t \oplus k_g$ into 4 boxes $(n \times n)$ together, each containing 4×4 values: blue (B_0): $t \oplus k_g \in [0,3]$, yellow (B_1): $t \oplus k_g \in [4,7]$, green (B_2): $t \oplus k_g \in [8,11]$, and red (B_3): $t \oplus k_g \in [12,15]$. Using this color representation we can easily see 4 different permutations $\pi_0, \pi_1, \pi_2, \pi_3$ applied on $(B_0\ B_1\ B_2\ B_3)$. More precisely, when considering a column representation[1] among the key guesses k_g, we have:

- for $k_g \in [0,3]$: no permutation $(\pi_0 = \left(\begin{smallmatrix} 0 & 1 & 2 & 3 \\ 0 & 1 & 2 & 3 \end{smallmatrix}\right))$,
- for $k_g \in [4,7]$: pairwise swap of elements in each half of matrix $(\pi_1 = \left(\begin{smallmatrix} 0 & 1 & 2 & 3 \\ 1 & 0 & 3 & 2 \end{smallmatrix}\right))$,
- for $k_g \in [8,11]$: additionally reverse ordering of elements $(\pi_2 = \left(\begin{smallmatrix} 0 & 1 & 2 & 3 \\ 2 & 3 & 0 & 1 \end{smallmatrix}\right))$,
- for $k_g \in [12,15]$ additionally a pairwise swap of elements in each half of matrix $(\pi_3 = \left(\begin{smallmatrix} 0 & 1 & 2 & 3 \\ 3 & 2 & 1 & 0 \end{smallmatrix}\right))$.

Moreover, as highlighted by the zoom in on each box, within each box (i.e., $B_i, 0 \le i \le 3$) we have the same permutations $\pi_0, \ldots, \pi_3$ on the 4 column entries. Note that the order of permutations is equivalent for each box, or in other words, regardless of the color and position of the box the same permutation is applied. More formally, let $b_{ij} \in [4i, 4i+3]^4$ (for $0 \le i, j \le 3$) denote the columns within B_i, then b_{ij} equals π_j applied on the column vector $(4i\ 4i+1\ 4i+2\ 4i+3)$.

Second, we examine the expression of the confusion coefficient in Eq. (11) itself. Recall from Eq. (8), $y_{k_g,t} = y(k_g, t) = w_H(F(k_g + t))$. Let

$$y_{k_g} = (y(k_g, 0), y(k_g, 1), \ldots, y(k_g, 2^n - 1))$$

denote the vector of hypotheses for one key guess k_g over all texts t. Referring to Fig. 2, y_{k_g} relates to one column before its application to F and w_H. The confusion coefficient can be rewritten as

$$\kappa(k_c, k_g) = \frac{1}{4}\left\|y_{k_c} - y_{k_g}\right\|_2^2 \tag{19}$$

with $\|\cdot\|_2$ being the Euclidean norm. Let us recall that we are especially interested in $min_{k_g \ne k_c}\kappa(k_c, k_g)$. Moreover, the elements of $y_{k_c} - y_{k_g}$ are in $[-4, 4]$. Now, as Eq. (19) considers not only the difference but its squared values, we may conjecture that the minimum value is most likely reached when the elements of $y_{k_c} - y_{k_g}$ are in $[-1, 1]$, which is discussed in more detail and confirmed using several lightweight S-boxes in Appendix A. Roughly speaking, one difference of ± 2 is equivalent to 4 changes with ± 1 and so on.

Now let us put the observations of both parts together. Our previous findings about the permutations can be straightforwardly applied to the Hamming weight of the output of F. Let us assume w.l.o.g. $k_c = 0$, then for $k_g = 4i + j$ (with $0 \le i, j \le 3$) we have

$$y_{k_g} = \pi_i\left(\pi_j\begin{bmatrix} y_{0,0} \\ y_{0,1} \\ \vdots \\ y_{0,3} \end{bmatrix}^T\ \pi_j\begin{bmatrix} y_{0,4} \\ y_{0,5} \\ \vdots \\ y_{0,7} \end{bmatrix}^T\ \pi_j\begin{bmatrix} y_{0,8} \\ y_{0,9} \\ \vdots \\ y_{0,11} \end{bmatrix}^T\ \pi_j\begin{bmatrix} y_{0,12} \\ y_{0,13} \\ \vdots \\ y_{0,15} \end{bmatrix}^T\right)^T, \tag{20}$$

---

[1] Note that we also have the same permutations on the row entries, however, we are interested in particular in a column representation as they reflect the key hypotheses.

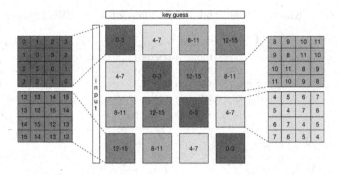

Fig. 2. Illustration of permutations of $t \oplus k_g$ $\forall t, k_g \in \mathbb{F}_2^4$ (input of F) (Color figure online)

with $y_0 = (y_{0,0}, y_{0,1}, \ldots, y_{0,15})$ and $(\cdot)^T$ denoting the transpose. Thus, we are looking for a function F such that the distance

$$
\left\| y_0 - \pi_i \left(\pi_j \begin{bmatrix} y_{0,0} \\ y_{0,1} \\ \vdots \\ y_{0,3} \end{bmatrix}^T \pi_j \begin{bmatrix} y_{0,4} \\ y_{0,5} \\ \vdots \\ y_{0,7} \end{bmatrix}^T \pi_j \begin{bmatrix} y_{0,8} \\ y_{0,9} \\ \vdots \\ y_{0,11} \end{bmatrix}^T \pi_j \begin{bmatrix} y_{0,12} \\ y_{0,13} \\ \vdots \\ y_{0,15} \end{bmatrix}^T \right)^T \right\|_2^2
\tag{21}
$$

is as small as possible for any $\pi_i, \pi_j \in \{\pi_0, \pi_1, \pi_2, \pi_3\}$.

This finding indicates that the order of the Hamming weight of the output of F plays a significant role. To be more precise, the minimum confusion coefficient may depend not only on the distribution of values along the 4 boxes (Example 4), but also on the order within each box (Example 5).

Example 4. *Note that the elements of y_{k_g} follow a binomial distribution due to the application of w_H. Therefore, 0 and 4 occur once, 1 and 3 occurs four times, and 2 six times. In order to reach a mininum squared Euclidean distance in Eq. (21) a natural strategy seems to be to distribute the values broadly among the 4 sets $[4i, 4i + 3]$ and to have a small difference between the values in one set. Let us consider the S-box of Midori [18] and Mysterion [19]. From Table 2 one can observe that for Midori we have the following sets: 2,2,3,2 − 3,3,4,3 − 1,2,1,2 − 0,1,1,2. So, the maximal distance between values is 2. Moreover, the first three sets only contain 2 different values and the last has 3. On the contrary, when looking at Mysterion (0,1,2,3 − 2,4,3,2 − 1,3,1,2 − 2,1,3,2), the structure looks less balanced. In particular, the maximal distance is 3 and we have always 3 different values within a set. When comparing the confusion coefficient in Fig. 3 we can observe that Midori has a much smaller minimum confusion coefficient and is thus more SCA resilient.*

Example 5. *Let us consider the S-box of KLEIN [20] and a small modification (S_9) in which we swap $F(1)$ with $F(3)$ (see Table 2). Note that both functions*

Table 2. Known S-boxes and one modification of KLEIN, $(x)w_H(x)$

S-box	(0)0	(1)1	(2)1	(3)2	(4)1	(5)2	(6)2	(7)3	(8)1	(9)2	(10)2	(11)3	(12)2	(13)3	(14)3	(15)4
Midori	(12)2	(10)2	(13)3	(3)2	(14)3	(11)3	(15)4	(7)3	(8)1	(9)2	(1)1	(5)2	(0)0	(2)1	(4)1	(6)2
Mysterion	(0)0	(1)1	(6)2	(11)3	(10)2	(15)4	(14)3	(3)2	(2)1	(7)3	(4)1	(12)2	(9)2	(8)1	(13)3	(5)2
KLEIN	(7)3	(4)1	(10)2	(9)2	(1)1	(15)4	(11)3	(0)0	(12)2	(3)2	(2)1	(6)2	(8)1	(14)3	(13)3	(5)2
S_9	(7)3	(9)2	(10)2	(4)1	(1)1	(15)4	(11)3	(0)0	(12)2	(3)2	(2)1	(6)2	(8)1	(14)3	(13)3	(5)2

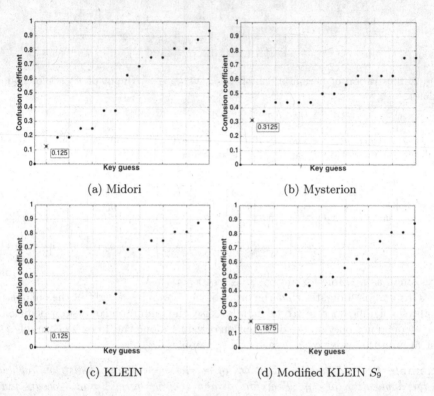

(a) Midori   (b) Mysterion

(c) KLEIN   (d) Modified KLEIN S_9

Fig. 3. Confusion coefficients of Midori, Mysterion, KLEIN, and KLEIN with a small modification

consist of the same values among the sets: 3,1,2,2 – 1,4,3,0 – 2,2,1,2 – 1,3,3,2. For both $min_{k_g \neq k_c} \kappa(k_c, k_g)$ is reached for $k_g = 11$, thus $\pi_1 = 2$ and $\pi_2 = 2$. However, as Fig. 3d shows, for KLEIN we have $min_{k_g \neq k_c} \kappa(k_c, k_g) = 0.125$, whereas $min_{k_g \neq k_c} \kappa(k_c, k_g) = 0.185$ for S_9, which relates to a squared Euclidean distance (see Eq. (21)) of 8 and 12, respectively.

Furthermore, in Appendix A we investigate several lightweight S-boxes in terms of minimum confusion coefficient and provide empirical evaluations. Note that, a preliminary study showing the difference of some lightweight S-boxes has been conducted in [21][2]. Our extended results in Appendix A theoretically and empirically confirm [21]. Moreover, the appendix provides details about the minimum Euclidean distance and the permutations π_i, π_j. Additionally, we take a deeper look at the expression of $y_{k_c} - y_{k_g}$ for the key hypothesis k_g that results in the smallest confusion coefficient (i.e., $\arg\min_{k_c \neq k_g} \kappa(k_c, k_g)$). We discover that for S_5 and the S-box proposed in [17], which has optimal properties of the confusion coefficient while holding optimal differential properties, the difference $\|y_{k_c} - y_{k_g}\|_2^2$ has a special particular structure, which is not observed for any other investigated 4-bit S-box.

Concluding, we derived specific criteria influencing the side-channel resistance (in particular in Eq. (21) and our findings in Appendix A) that could be exploited to optimize and find S-boxes in terms of side-channels resistance in future work – especially when adapted for $n > 4$.

6   Conclusions

In this paper, we prove a number of bounds between various cryptographic properties that can be related also with the side-channel resilience of a cipher. Our results confirm some well known intuitions that having an S-box more resilient against SCA will make it potentially more vulnerable against classical cryptanalyses. However, they also show that for the usual sizes of S-boxes, this weakening is moderate and trade-offs are then possible.

Since in this work we concentrated in our practical investigations on the Hamming weight model, in the future we plan to explore possible trade-offs for the Hamming distance model and to extend our (empirical) analysis to larger S-boxes using the theoretical findings in this paper.

Acknowledgments. This work has been supported in part by Croatian Science Foundation under the project IP-2014-09-4882. The parts of this work were done while the third author was affiliated with KU Leuven, Belgium.

A   Investigation of Known S-Boxes

We already described properties of the S-boxes of KLEIN, Midori, and Mysterion showing that Midori and KLEIN have both $\min_{k_g \neq k_c} \kappa(k_c, k_g) = 0.125$, whereas for Mysterion it equals 0.3125. Thus, the side-channel resistance of Mysterion is much smaller than that of KLEIN and Midori. Table 3 shows properties of several well-known S-boxes, where π_i and π_j indicate the permutations (see Eq. (21)) for the smallest squared Euclidean distance $(\min \| \cdot \|_2^2)$ and thus smallest confusion coefficient $(\min \kappa(k_c, k_g))$.

Note that the squared Euclidean distance should not serve as a new metric as it is in direct relation with the confusion coefficient, but its stated values

---

[2] Note that, in [22] the authors compared S-boxes regarding another (not normalized) version of the confusion coefficient and derived that their version is not aligned with their empirical results.

Table 3. S-box properties of known ciphers

Name	π_i	π_j	min $\| \cdot \|_2^2$	min $\kappa(k_c, k_g)$	$var(\kappa(k_c, k_g))$	Pw_H
KLEIN [20]	2	2	8	0.125	0.093	6
Midori [18]	0	1	8	0.125	0.096	4
Midori 2 [18]	2	3	16	0.250	0.059	5
Mysterion [19]	2	1	20	0.312	0.034	5
NOEKEON [23]	3	0	12	0.188	0.056	8
Piccolo [24]	1	0	24	0.375	0.028	4
PRESENT [25]	2	1	16	0.25	0.058	3
PRINCE [26]	1	3	12	0.188	0.059	3
RECTANGLE [27]	3	0	16	0.250	0.056	3
SKINNY [28]	1	0	16	0.250	0.037	8
S-box in [17]	0	1	8	0.125	0.104	4

should rather provide information how far y_{k_g} is apart from y_{k_c} in terms of the squared Hamming weight values. Further, as used in [17], we give $var(\kappa(k_c, k_g))$, where the higher the variance, the higher the side-channel resistance. Finally, we specify the Hamming weights preserved (Pw_H). One can observe that Piccolo has the highest minimum value of the confusion coefficient (and the highest minimum squared Euclidean norm) and thus its side-channel resistance is the lowest among the evaluated one. Next, there is Mysterion followed by SKINNY, RECTANGLE, PRESENT, and Midori 2 that all have the same minimum value of the confusion coefficient, but different variances thereof. Then, we have NOEKEON and PRINCE. The lowest minimum confusion coefficient is reached by KLEIN, Midori, and the S-box proposed in [17], which has been found under the constraint of optimal differential properties and the lowest confusion coefficient by using genetic algorithms. Interestingly for the latter one, Fig. 4 illustrates that for one key guess $\kappa(k_c, k_g) = 1$, which we do not observe for any other known S-boxes with optimal differential properties. Moreover, it corresponds to the confusion coefficient of S_5.

Additionally, we take a deeper look at the expression of $y_{k_c} - y_{k_g}$ for the key hypothesis k_g that results in the smallest confusion coefficient and we are interested if the elements in $|y_{k_c} - y_{k_g}|$ are in $[-1, 1]$ (see remark in Subsect. 5.2). Our investigations show that this does not hold for S-boxes with $\kappa(k_c, k_g) \geq 0.25$, but for the ones which are most side-channel resistant. In particular, Midori 2, Mysterion, PRESENT, RECTANGLE, and SKINNY contain two absolute difference of 2 (resulting in a Euclidean distance of 4), whereas Piccolo even has 4 absolute differences of 2. However, we could not observe any absolute difference greater than 2. On the contrary KLEIN, Midori, NOEKEON, PRINCE, and the S-box in [17] only contain absolute differences of one, which is thus equivalent to the Euclidean distance.

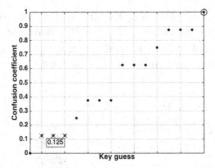

Fig. 4. Confusion coefficient of S-box in [17]

When considering the sum of differences among the 4 sets $[4s, 4s + 3]$ for $0 \leq s \leq 3$, we observed interesting distinctions. In particular, let us denote

$$\Delta_s = \left\| \begin{bmatrix} y_{0,4s} \\ y_{0,4s+1} \\ \vdots \\ y_{0,4s+3} \end{bmatrix} - \pi_i \left(\pi_j \begin{bmatrix} y_{0,4s} \\ y_{0,4s+1} \\ \vdots \\ y_{0,4s+3} \end{bmatrix} \right) \right\|_2^2, \tag{22}$$

with π_i and π_j being the permutation resulting in the minimum confusion coefficient.

Table 4 highlights that only for the S-box in [17] we have the same difference among all four sets. Note that, in future work this property may additionally help to detect and find S-boxes with better side-channel resistance for $n > 4$.

Table 4. Δ-property (Eq. (22)) of the most resilient known S-boxes

Name	Δ_0	Δ_1	Δ_2	Δ_3
KLEIN	1	3	3	1
Midori	2	2	0	4
NOEKEON	2	4	2	4
PRINCE	4	2	4	2
S-box in [17]	2	2	2	2

Finally, an empirical evaluation of the studied S-boxes is given in Fig. 5. To be reliable we conducted 5000 independent simulation experiments (SNR = 2) with random secret keys k_c and texts t. Figure 5a shows the first-order success rate (SR), i.e., the empirical probability that the correct secret key is exclusively found. As found due to the properties of the confusion coefficient, the S-box of Piccolo is the weakest, finding the correct key with a SR of 0.9 using

20 measurement traces, whereas KLEIN and Midori require around 35 and 40 traces to reach SR = 0.9. Since the S-box in [17] does not exclusively find the correct key and thus has a SR = 0, we additionally plot the guessing entropy [29] in Fig. 5b which confirms our findings that at least 2 key guesses have to be made.

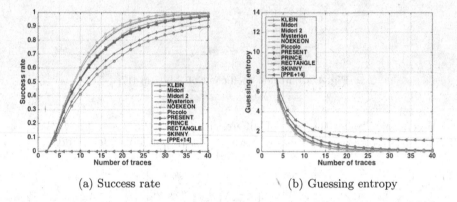

(a) Success rate   (b) Guessing entropy

Fig. 5. Empirical evaluation of known S-boxes

References

1. Matsui, M., Yamagishi, A.: A new method for known plaintext attack of FEAL cipher. In: Rueppel, R.A. (ed.) EUROCRYPT 1992. LNCS, vol. 658, pp. 81–91. Springer, Heidelberg (1993). doi:10.1007/3-540-47555-9_7

2. Biham, E., Shamir, A.: Differential cryptanalysis of DES-like cryptosystems. In: Menezes, A.J., Vanstone, S.A. (eds.) CRYPTO 1990. LNCS, vol. 537, pp. 2–21. Springer, Heidelberg (1991). doi:10.1007/3-540-38424-3_1

3. Mangard, S., Oswald, E., Popp, T.: Power Analysis Attacks: Revealing the Secrets of Smart Cards (Advances in Information Security). Springer-Verlag New York Inc., Secaucus (2007)

4. Nikova, S., Rechberger, C., Rijmen, V.: Threshold implementations against side-channel attacks and glitches. In: Ning, P., Qing, S., Li, N. (eds.) ICICS 2006. LNCS, vol. 4307, pp. 529–545. Springer, Heidelberg (2006). doi:10.1007/11935308_38

5. Heuser, A., Rioul, O., Guilley, S.: A theoretical study of kolmogorov-smirnov distinguishers. In: Prouff, E. (ed.) COSADE 2014. LNCS, vol. 8622, pp. 9–28. Springer, Cham (2014). doi:10.1007/978-3-319-10175-0_2

6. Brier, E., Clavier, C., Olivier, F.: Correlation power analysis with a leakage model. In: Joye, M., Quisquater, J.-J. (eds.) CHES 2004. LNCS, vol. 3156, pp. 16–29. Springer, Heidelberg (2004). doi:10.1007/978-3-540-28632-5_2

7. Fei, Y., Luo, Q., Ding, A.A.: A statistical model for DPA with novel algorithmic confusion analysis. In: Prouff, E., Schaumont, P. (eds.) CHES 2012. LNCS, vol. 7428, pp. 233–250. Springer, Heidelberg (2012). doi:10.1007/978-3-642-33027-8_14

8. Carlet, C.: Vectorial boolean functions for cryptography. In: Crama, Y., Hammer, P.L. (eds.) Boolean Models and Methods in Mathematics, Computer Science, and Engineering, 1st edn, pp. 398–469. Cambridge University Press, New York (2010)

9. Nyberg, K.: On the construction of highly nonlinear permutations. In: Rueppel, R.A. (ed.) EUROCRYPT 1992. LNCS, vol. 658, pp. 92–98. Springer, Heidelberg (1993). doi:10.1007/3-540-47555-9_8

10. Chabaud, F., Vaudenay, S.: Links between differential and linear cryptanalysis. In: Santis, A. (ed.) EUROCRYPT 1994. LNCS, vol. 950, pp. 356–365. Springer, Heidelberg (1995). doi:10.1007/BFb0053450

11. Nyberg, K.: Perfect nonlinear S-boxes. In: Davies, D.W. (ed.) EUROCRYPT 1991. LNCS, vol. 547, pp. 378–386. Springer, Heidelberg (1991). doi:10.1007/3-540-46416-6_32

12. Kocher, P., Jaffe, J., Jun, B.: Differential power analysis. In: Wiener, M. (ed.) CRYPTO 1999. LNCS, vol. 1666, pp. 388–397. Springer, Heidelberg (1999). doi:10.1007/3-540-48405-1_25

13. Gandolfi, K., Mourtel, C., Olivier, F.: Electromagnetic analysis: concrete results. In: Koç, Ç.K., Naccache, D., Paar, C. (eds.) CHES 2001. LNCS, vol. 2162, pp. 251–261. Springer, Heidelberg (2001). doi:10.1007/3-540-44709-1_21

14. Mangard, S., Oswald, E., Popp, T.: Power Analysis Attacks: Revealing the Secrets of Smart Cards. Springer, Heidelberg (2006). ISBN 0-387-30857-1. http://www.dpabook.org/

15. Guilley, S., Heuser, A., Rioul, O.: A key to success. In: Biryukov, A., Goyal, V. (eds.) INDOCRYPT 2015. LNCS, vol. 9462, pp. 270–290. Springer, Cham (2015). doi:10.1007/978-3-319-26617-6_15

16. Thillard, A., Prouff, E., Roche, T.: Success through confidence: evaluating the effectiveness of a side-channel attack. In: Bertoni, G., Coron, J.-S. (eds.) CHES 2013. LNCS, vol. 8086, pp. 21–36. Springer, Heidelberg (2013). doi:10.1007/978-3-642-40349-1_2

17. Picek, S., Papagiannopoulos, K., Ege, B., Batina, L., Jakobovic, D.: Confused by confusion: systematic evaluation of DPA resistance of various S-boxes. In: Meier, W., Mukhopadhyay, D. (eds.) INDOCRYPT 2014. LNCS, vol. 8885, pp. 374–390. Springer, Cham (2014). doi:10.1007/978-3-319-13039-2_22

18. Banik, S., Bogdanov, A., Isobe, T., Shibutani, K., Hiwatari, H., Akishita, T., Regazzoni, F.: Midori: a block cipher for low energy (extended version). Cryptology ePrint Archive, Report 2015/1142 (2015). http://eprint.iacr.org/

19. Journault, A., Standaert, F.X., Varici, K.: Improving the security and efficiency of block ciphers based on LS-designs. Codes Crypt. Des. **82**(1–2), 495–509 (2016)

20. Gong, Z., Nikova, S., Law, Y.W.: KLEIN: a new family of lightweight block ciphers. In: Juels, A., Paar, C. (eds.) RFIDSec 2011. LNCS, vol. 7055, pp. 1–18. Springer, Heidelberg (2012). doi:10.1007/978-3-642-25286-0_1

21. Heuser, A., Picek, S., Guilley, S., Mentens, N.: Side-channel analysis of lightweight ciphers: does lightweight equal easy? Cryptology ePrint Archive, Report 2017/261 (2017). http://eprint.iacr.org/2017/261

22. Lerman, L., Markowitch, O., Veshchikov, N.: Comparing Sboxes of ciphers from the perspective of side-channel attacks. IACR Cryptology ePrint Archive 2016/993 (2016)

23. Daemen, J., Peeters, M., Assche, G.V., Rijmen, V.: Nessie proposal: the block cipher Noekeon. Nessie submission (2000). http://gro.noekeon.org/

24. Shibutani, K., Isobe, T., Hiwatari, H., Mitsuda, A., Akishita, T., Shirai, T.: *Piccolo*: an ultra-lightweight blockcipher. In: Preneel, B., Takagi, T. (eds.) CHES 2011. LNCS, vol. 6917, pp. 342–357. Springer, Heidelberg (2011). doi:10.1007/978-3-642-23951-9_23

25. Bogdanov, A., Knudsen, L.R., Leander, G., Paar, C., Poschmann, A., Robshaw, M.J.B., Seurin, Y., Vikkelsoe, C.: PRESENT: an ultra-lightweight block cipher. In: Paillier, P., Verbauwhede, I. (eds.) CHES 2007. LNCS, vol. 4727, pp. 450–466. Springer, Heidelberg (2007). doi:10.1007/978-3-540-74735-2_31

26. Borghoff, J., et al.: PRINCE – a low-latency block cipher for pervasive computing applications. In: Wang, X., Sako, K. (eds.) ASIACRYPT 2012. LNCS, vol. 7658, pp. 208–225. Springer, Heidelberg (2012). doi:10.1007/978-3-642-34961-4_14

27. Zhang, W., Bao, Z., Lin, D., Rijmen, V., Yang, B., Verbauwhede, I.: RECTANGLE: a bit-slice lightweight block cipher suitable for multiple platforms. Sci. Chin. Inf. Sci. **58**(12), 1–15 (2015)

28. Beierle, C., Jean, J., Kölbl, S., Leander, G., Moradi, A., Peyrin, T., Sasaki, Y., Sasdrich, P., Sim, S.M.: The SKINNY family of block ciphers and its low-latency variant MANTIS. Cryptology ePrint Archive, Report 2016/660 (2016). http://eprint.iacr.org/2016/660

29. Standaert, F., Malkin, T., Yung, M.: A unified framework for the analysis of side-channel key recovery attacks (extended version). IACR Cryptology ePrint Archive 2006/139 (2006)

A Practical Chosen Message Power Analysis Approach Against Ciphers with the Key Whitening Layers

Chenyang Tu[1,2], Lingchen Zhang[1,2(✉)], Zeyi Liu[1,2,3], Neng Gao[1,2], and Yuan Ma[1,2]

[1] State Key Laboratory of Information Security,
Institute of Information Engineering, CAS, Beijing, China
{tuchenyang,zhanglingchen,liuzeyi,gaoneng,mayuan}@iie.ac.cn
[2] Data Assurance and Communication Security Research Center, CAS,
Beijing, China
[3] University of Chinese Academy of Sciences, Beijing, China

Abstract. The key whitening is a technique intended to enhance the strength of a block cipher. Although some research work involves DPA attacks against the key whitening layer in the compact architecture, there are no literatures dedicated in the influence of the key whitening layers in the loop architecture from the standpoint of DPA. In this paper, we propose a practical chosen message power analysis approach against the loop architecture of ciphers with the key whitening layers, thus proving that the key whitening technique does not enhance the security of ciphers regard to DPA. Our approach follows a reduction strategy: we recover the whitening key in the general cipher with the key whitening layer and reduce other complicated key whitening layers to the general case. In order to further manifest the validity of the new approach, we carry extensive experiments on two ISO standardized ciphers CLEFIA and Camellia implemented in loop architecture on FPGA, and the keys are recovered as expected.

Keywords: DPA · Key whitening · Chosen message · Loop architecture

1 Introduction

Key whitening is a technique intended to enhance the strength of a block cipher by adding key-relevant operations on plaintext and ciphertext without major changes to the algorithm [15,16]. The key whitening layer consists of steps that combine the data with portions of the key before the first round and after the last

---

C. Tu—This work is supported by the National Natural Science Foundation of China (Nos. 61402470 and 61602476) and National Basic Research Program of China (No. 2013CB338001). Besides, the work is also supported by Youth Innovation Promotion Association, CAS.

© Springer International Publishing AG 2017
D. Gollmann et al. (Eds.): ACNS 2017, LNCS 10355, pp. 415–434, 2017.
DOI: 10.1007/978-3-319-61204-1_21

round. The most common operation is XORing or modular adding the whitening key to the plaintext/ciphertext. The key whitening technique is adopted by many block ciphers, such as the ISO standardized Feistel-SP ciphers CLEFIA [2] and Camellia [3], and the lightweight ciphers DESL [15] and PRINCE [16].

Since first proposed by Kocher *et al.* in [1], DPA has proven to be a powerful method of side channel attack against many ciphers. In general, DPA can only deal with a small fraction of the long secret key (*e.g.* several round key bits) through a divide-and-conquer strategy, and its validity is highly dependent on the specific implementation. These traditional cryptographic implementations (*i.e.*, compact architecture [18]) are easily compromised by the conventional DPA, where the hypothesis space of the secret key fraction is only 2^8 or less [17]. For instance, DES, AES and many other ciphers under the compact architecture have shown to be vulnerable to DPA [1, 4, 7, 9, 10]. The ciphers with key whitening layers under the same architecture are also easily compromised by DPA, because key whitening layer is implemented independently of the substitution circuit [8].

With the advancement of circuit industry within recent years, a new architecture, *i.e.* loop architecture [18], is proposed, which computes a single round function in one clock cycle. Ciphers under the loop architecture are adopted in the higher computation speed and higher throughput application scenarios. Therefore, the capability of the loop implementation against the side channel attack has attracted researchers' great attention. It has been proved that the conventional DPA needs much more power traces to analyze ciphers under loop architecture with very high computational complexity [13]. In order to deal with the loop architecture efficiently and practically, the adversary usually launches the chosen message DPA [12] instead.

However, it is quite a challenge to launch the DPA methodology against the loop hardware implementations of ciphers with the key whitening layers. The key whitening layer is usually implemented within the first or last round in the loop architecture, which can increase the difficulty of the DPA methodology. In this case, the power consumption of the whitening operation is hard to be recognized from power traces, because the intermediate result of the whitening operation does not appear in registers or on bus as the case in [8]. Following the chosen message DPA [12], the adversary can only get the equivalent key (*i.e.* the value of "the round key add/xor the whitening key" as a unity) on the first/last round, but he would not be capable to directly determine either the whitening key or the round key.

Unless the adversary is able to directly obtain the whitening key or the round key in some way, the adversary has to peel off the first round with the equivalent key and perform DPA with an adaptive manner on the subsequent rounds, which means the adversary must solve one more round to deal with the key whitening layer. Generally, the cost of performing DPA in such way increases 50% to 100% than that without key whitening layer. In order to deal with the key whitening layer without increasing the DPA cost, the core issue is how to reveal either the whitening key or the round key directly in the first/last round. To the best of our knowledge, no research about this issue has been reported in the literature.

In this paper, we propose a practical chosen message DPA approach to recover the whitening key through a reduction strategy. First, by fully exploiting the relationship between the round key and the whitening key, we recover the whitening key in the general cipher with the key whitening layer. Then we successfully reduce other complicated key whitening layers to the general case. As a result, we show that the key whitening technique does not enhance the security of ciphers from the standpoint of DPA.

We take the Feistel-SP ciphers with the key whitening layers as an instance, due to the most comprehensive cases of the key whitening layers in these ciphers (*i.e.*, the key whitening operation on the left branch, on the right branch, and on both branches). According to the relationship in the round function, our approach can launch chosen message DPA to efficiently reveal the whitening key on the left branch. When the whitening key is on the right branch, we are able to reduce the recovery of the whitening key on the right branch to the left branch case through an adaptive chosen message manner. When the whitening keys are on both branches, we can reduce the recovery of the whitening keys to the left branch case and the right branch case respectively. Furthermore, we perform extensive experiments on two ISO standardized ciphers CLEFIA and Camellia with loop FPGA implementations. Experimental results show that all bits of the keys in both ciphers can be recovered as expected.

The remainder of this paper is organized as follows. In Sect. 2, the preliminaries are briefly described. Section 3 illustrates the practical chosen message power analysis method on Feistel-SP ciphers with the key whitening layers. Section 4 elaborates the practical attacks on two loop FPGA implementations of CLEFIA-128 and Camellia-128 in order to prove the effectiveness of our approach. We discuss and summarize the paper in Sect. 5.

2   Preliminaries

2.1   The Compact and Loop Architecture

The hardware implementations of ciphers usually follow two architectures, compact architecture and loop architecture [18], in order to adapt to different application scenarios. In the embedded application scenario, the size, power consumption, and cost of the cryptographic device are tightly restrained. On the other hand, in the higher computation speed and higher throughput application scenario, the performance and efficiency of the cryptographic device are the most important indicators. The compact architecture usually takes several clock cycles to accomplish a single round computation of the cryptographic algorithm, such as reusing the single substitution circuit (*e.g.* Sbox) several times as a subloop instead of using their duplications concurrently. Therefore, the compact architecture, which sacrifices performance to less circuit components, is often applied to the embedded cryptographic device, such as smart cards and wireless sensor nodes.

On the other hand, the loop architecture defines the round function of the cryptographic algorithm as several consecutive operations, which means that a single round is computed in one clock cycle. Compared to the compact architecture, the loop architecture, which has higher throughput or less calculation time, is usually applied to the higher computation speed and higher throughput application scenario, such as the hardware security module in the cloud computing environment, the instant messenger system, the network authentication system and the network routing device. The loop architecture is implemented in ASIC or FPGA chips in the hardware security module, with the advancement of circuit industry within recent years.

2.2 DPA on Key Whitening Layer

The key whitening layers can be compromised by DPA, as long as the adversary is able to locate the power consumption of the whitening operation on power traces. In the compact scenario, the power consumption of the whitening operation is convenient to be observed due to the independent implementation of the key whitening layer. Consequently, the adversary can reveal the whitening key through direct analysis of the power consumption of the whitening operation [8,11]. Unfortunately, the above strategy seems hard to apply in the loop scenario, because the whitening operations are implemented within the first/last round. It leads to much lower signal-to-noise of the slight power consumption which is too difficult to be detected. In this case, the adversary can only get the equivalent key (*i.e.* the value of "the round key adding the whitening key" as a unity) in DPA, but he would not be capable to directly determine either the whitening key or the round key. Therefore, how to reveal the whitening key or the round key in the loop architecture remains an open problem.

2.3 Feistel-SP Structure

The Feistel network is one of the most famous architectures in symmetric cryptography. According to the classifications of [5], the Feistel network has several derivatives, namely, the unbalanced Feistel network, the alternating Feistel network, the numeric Feistel network, and the famous type-1, type-2, and type-3 Feistel networks. Many practical block ciphers utilize the Feistel networks including DES (plain), Skipjack (unbalanced), BEAR/LION (alternating), CAST (type-1), CLEFIA(type-2) and MARS(type-3).

A typical SP type function often consists of three operations, *i.e.*, subkey addition, substitution, and permutation. In the subkey addition, a subkey is XORed to the state. The substitution is applied by Sbox-like non-linear bijection. In the permutation, a linear bijection (generally an MDS multiplication) is performed. Let $S_1, S_2, \cdots, S_t : \{0,1\}^s \longrightarrow \{0,1\}^s$ be non-linear bijections, $P : \{0,1\}^{st} \longrightarrow \{0,1\}^{st}$ be a linear bijection, $k = (k_1, k_2, \cdots, k_t)$ is the round key, then the round function $F : \{0,1\}^{st} \times \{0,1\}^{st} \longrightarrow \{0,1\}^{st}$ of SP type is defined by $F(x,k) = P(S_1(x_1 \oplus k_1), S_2(x_2 \oplus k_2), \cdots, S_t(x_t \oplus k_t))$. The notations s and t represent the size of the non-linear bijection and the number of the non-linear bijections, respectively.

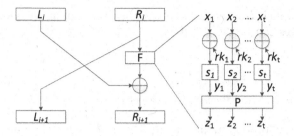

Fig. 1. The typical round function of the Feistel-SP structure

In the Feistel network, the core of the underlying round function is referred to as the F-function. In order to combine the advantages of SPN structures, the Feistel-SP ciphers use well-designed SPN functions as the F-functions. The typical round function of the Feistel-SP structure is shown in Fig. 1.

3   The Design of Our Approach

In this section, we describe the details of our practical chosen message DPA method on the loop architecture of Feistel-SP ciphers with the key whitening layers. Firstly, sharing the similar idea of chosen message DPA [12], we put forward the chosen message DPA which is also suitable for the case of Feistel-SP structure without the key whitening layer. Secondly, we analyze the difficulty to reveal the whitening key in the loop architecture directly. Then we propose an efficient approach to recover the whitening key through a reduction strategy. More precisely, we show how to recover the whitening key on the left branch, and reduce the recovery of the whitening key on the right branch to the left branch case through an adaptive chosen message manner. Combining these two strategies, we manage to perform practical DPA on loop architecture of Feistel-SP structure with the key whitening layer.

3.1   The Chosen Message DPA on the Loop Architecture of Feistel-SP

The typical loop implementation of Feistel-SP is shown in Fig. 1. Let L_i and R_i denote the left and right branches of the i-th round input respectively, and the size of the Sbox is 8-bit. The right branch R_i can be further split into t 8-bit cells, namely, $R_i = x_1||x_2||\cdots||x_t$. Each 8-bit cell x_j is first XORed with the corresponding 8-bit round key rk_j, then all t cells are processed with t parallel Sboxes $S_1, S_2, \cdots, S_t$[1]. Let $y_1||y_2||\cdots||y_t$ denote the output of the Sbox layer, the linear permutation P (normally multiplication with an MDS matrix) updates the state $y_1||y_2||\cdots||y_t$, and $z_1||z_2||\cdots||z_t$ is the output. The right branch of the i-th round output R_{i+1} is then calculated by XORing L_i and the output of P,

---

[1] The Sboxes can be identical or distinct.

while the left branch L_{i+1} is updated by directly assigning the value of R_i. The above procedures can be described as

$$
\begin{aligned}
L_{i+1} &= R_i, \\
R_{i+1} &= F(R_i, rk) \oplus L_i \\
&= P(S_1(x_1 \oplus rk_1), S_2(x_2 \oplus rk_2), \cdots, S_t(x_t \oplus rk_t)) \oplus L_i.
\end{aligned}
\tag{1}
$$

In the scenario of loop architecture, Feistel-SP treats the round function as several consecutive operations, thus no intermediate result except R_{i+1} is written into registers in the loop implementations. The power consumption of y_j is much less than R_{i+1} and hard to be observed. Therefore, we are only able to attack at this point when the data R_{i+1} is being written into registers as shown in Fig. 2.

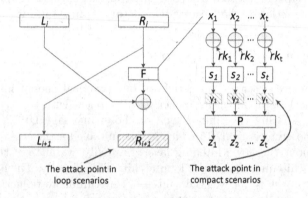

The attack point in loop scenarios

The attack point in compact scenarios

Fig. 2. Different DPA attack points of Feistel-SP

As shown in Fig. 2, regarding a specific cell y_j, the linear permutation P can be represented as follows:

$$
\begin{aligned}
P(y_1, y_2, \cdots, y_{j-1}, y_j, y_{j+1}, \cdots, y_t) &= P(0, 0, \cdots, 0, y_j, 0, \cdots, 0) \oplus \\
&\quad P(y_1, y_2, \cdots, y_{j-1}, 0, y_{j+1}, \cdots, y_t) \\
&= WP^j \oplus CP^j,
\end{aligned}
$$

where the superscript j indicates the function focusing on the cell y_j, and WP^j and CP^j denote the two components of the right half of the equation respectively. The above equation can be rewritten in byte-wise form as follows:

$$
P_{[1]}||P_{[2]}||\cdots||P_{[t]} = (WP^j_{[1]}||WP^j_{[2]}||\cdots||WP^j_{[t]}) \oplus (CP^j_{[1]}||CP^j_{[2]}||\cdots||CP^j_{[t]})
$$

where the subscript $[k]$ indicates the k-th byte output, and $WP^j_{[k]}$ and $CP^j_{[k]}$ denote the k-th bytes of WP^j and CP^j respectively. If we fix the values of all y_k where $k \neq j$, and keep y_j as a variable, then $WP^j_{[j]}$ can be seen as a function of

y_j and $CP^j_{[j]}$ is a constant. For convenience, we use $P'_{[j]}$ and $mask_{[j]}$ to represent $WP^j_{[j]}$ and $CP^j_{[j]}$ respectively.

Thus, the adversary chooses the specific byte of plaintext message, which corresponds the j-th byte of the target intermediate variable, while fixing other bytes of the plaintext message. The target byte is dependant on two unknown one-byte constant parameters, $i.e.$, the subkey rk_j and the $mask_{[j]}$ generated by P. Therefore, the size of guessed parameters from the whole round key is decreased to a pair of 8-bit values, $i.e.$, the hypothesis space of the secret value falls to 2^{16}, and the input space of the plaintext message is decreased to 2^8, which is suitable for practical DPA. Consequently, by alternately choosing the corresponding plaintext message byte for all possible positions, we can use the DPA attacking model shown in Fig. 3 to launch DPA and recover all t bytes of the round key (for the sake of simplicity, we assume the size of the round key is 32 bits).

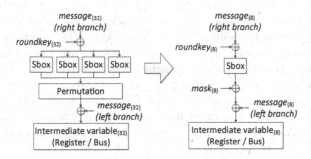

Fig. 3. DPA model of Feistel-SP in loop scenario

3.2 The Difficulty to Reveal the Whitening Key in the Loop Scenario

The whitening keys are generally used before the first round and after the last round. After the key whitening operations, the inputs (outputs) to the first (last) round are covered by the unknown whitening key from the plaintexts (ciphertexts). It seems that such operations would increase the size of unknown parameters, and raise the difficulty to launch a DPA attack. However, since the encryption and decryption of Feistel-SP ciphers follow similar procedures and both the pre-whitening and the post-whitening keys are almost equivalent from the perspective of DPA, we only discuss the encryption procedure in the first round.

Let ML (resp. MR) denote the left (resp. right) message branch, and wkL (resp. wkR) denote the left (resp. right) whitening key. As shown in Fig. 4, the whitening keys can be applied on the left branch, the right branch or both branches, which correspond to three types of whitening operations.

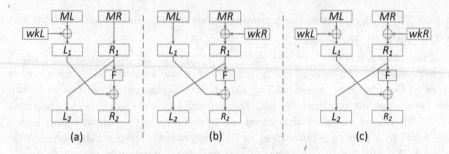

Fig. 4. The whitening key on (a) left branch (b) right branch (c) both branches

There are two main difficulties to apply DPA to reveal the whitening key in loop scenario. The first difficulty is that the power consumption of the whitening operation is hard to detect from power traces similar as Sect. 3.1. In the loop hardware implementation, the whitening operation and the round key addition operation are usually combined as one operation (*i.e. Message* $\oplus wk \oplus rk$), which is implemented by 3-input XOR gate or LUT. More precisely, there is no standalone whitening operation in the Feistel-SP computing procedure, thus the power consumption of the whitening operation is hard to detect. Therefore, the existing method against the whitening key as mentioned in [8] is not suitable.

Moreover, although we could choose R_{i+1} as the attack point, the whitening key is difficult to be separated from the round key and other intermediate variables by DPA. We assume that there are whitening keys on both branches. Due to the effect of whitening keys wkL and wkR, the DPA attacking model of Feistel-SP with whitening key in loop scenario is shown in Fig. 5(a). For the loop scenario, we use $rk \oplus wkR$ as the equivalent key and $mask \oplus wkL$ as the equivalent mask, thus the model is changed from Fig. 5(a) to (b). However, although we can recover the equivalent key and the equivalent mask by DPA, we are unable to directly determine the values of the whitening keys from the equivalent key and the equivalent mask.

3.3   Recovery of wkL

When we choose R_{i+1} as the attack point, the main drawback for DPA is the difficulty to separate the whitening key from the equivalent key and the equivalent mask. Luckily, we can achieve this goal through a reduction strategy. More precisely, we can efficiently recover the whitening key on the left branch wkL, and reduce the recovery of the whitening key on the right branch to the left branch case.

The recovery of wkL is based on the full exploitation of the complex relationship between the equivalent key and the equivalent mask with a chosen message DPA method. Hereafter we use subscript $[i]$ $(1 \leq i \leq t)$ to indicate the i-th byte of the corresponding variable or the output of one function, and use notation rk_j $(j \geq 1)$ to represent the round key in the j-th round of Feistel-SP. According to

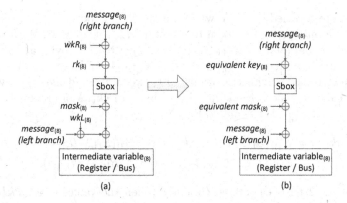

Fig. 5. Chosen message DPA model of Feistel-SP whitening key in the loop scenario

Fig. 4(a), two branches of the first round input $L_1\|R_1$ can be described by:

$$L_1\|R_1 = (ML \oplus wkL)\|MR. \tag{2}$$

Now, we will focus on the attack point R_2 as shown in Fig. 2. Equation 1 can be rewritten as:

$$\begin{aligned}
R_2 &= F(R_1, rk_1) \oplus L_1 \\
&= P(S_1(MR_{[1]} \oplus rk_{1,[1]}), S_2(MR_{[2]} \oplus rk_{1,[2]}), \\
&\quad \cdots, S_t(MR_{[t]} \oplus rk_{1,[t]})) \oplus wkL \oplus ML.
\end{aligned} \tag{3}$$

where $MR = MR_{[1]}\|MR_{[2]}\|\cdots\|MR_{[t]}$, and $rk_1 = rk_{1,[1]}\|rk_{1,[2]}\|\cdots\|rk_{1,[t]}$.

We focus on the first byte of R_2, and fix all $MR_{[j]}$ $(2 \le j \le t)$ to constants, that leads to the constant output of S_j. Thus, Eq. 3 can be rewritten in byte-wise form:

$$\begin{aligned}
R_{2,[1]} &= F_{[1]}(R_1, rk_1) \oplus L_{1,[1]} \\
&= P_{[1]}(S_1(MR_{[1]} \oplus rk_{1,[1]}), S_2(MR_{[2]} \oplus rk_{1,[2]}), \\
&\quad \cdots, S_t(MR_{[t]} \oplus rk_{1,[t]})) \oplus wkL_{[1]} \oplus ML_{[1]} \\
&= P'_{[1]}(S_1(MR_{[1]} \oplus rk_{1,[1]})) \oplus mask_{1,[1]} \oplus wkL_{[1]} \oplus ML_{[1]} \\
&= P'_{[1]}(S_1(MR_{[1]} \oplus rk_{1,[1]})) \oplus MASK_{1,[1]} \oplus ML_{[1]},
\end{aligned} \tag{4}$$

with $ML = ML_{[1]}\|ML_{[2]}\|\cdots\|ML_{[t]}$, $wkL = wkL_{[1]}\|wkL_{[2]}\|\cdots\|wkL_{[t]}$. Moreover, $mask_1 = mask_{1,[1]}\|mask_{1,[2]}\|\cdots\|mask_{1,[t]}$ is the intermediate variable which is generated in the first round. According to Sect. 3.1, $mask_{1,[j]}$ is a byte constant value if all MR_k $(k \ne j)$ are fixed since the round key rk_1 is pre-assigned, and the equivalent mask $MASK_1 = mask_1 \oplus wkL$.

At this time, $R_{2,[1]}$ is highly related to $rk_{1,[1]}$, and $R_{2,[2]}, R_{2,[3]}, \cdots, R_{2,[t]}$ will be treated as noise. Now, we can launch DPA against $R_{2,[1]}$ by enumerating 8-bit $MR_{[1]}$ while fixing other bits of MR and ML. Thus, both 8-bit $rk_{1,[1]}$ and 8-bit

$MASK_{1,[1]}$ are revealed by DPA, where the possible hypotheses space is 2^{16} and the possible input space of random test vector $MR_{[1]}$ is only 2^8. With the same approach, we could analyze $R_{2,[2]}, R_{2,[3]}, \cdots, R_{2,[t]}$ byte by byte, and reveal the values of $rk_{1,[2]}, rk_{1,[3]}, \cdots, rk_{1,[t]}$ and $MASK_{1,[2]}, MASK_{1,[3]}, \cdots, MASK_{1,[t]}$.

According to the complex relationship between rk_1 and $mask_{1,[j]}$, we could iteratively calculate each of $mask_{1,[j]}$ by all bytes of rk_1. More precisely, according to Eq. 4, $mask_{1,[1]}$ is calculated by:

$$mask_{1,[1]} = P'_{[1]}(S_1(MR_{[1]} \oplus rk_{1,[1]})) \oplus P_{[1]}(S_1(MR_{[1]} \oplus rk_{1,[1]}),$$
$$S_2(MR_{[2]} \oplus rk_{1,[2]}), \cdots, S_t(MR_{[t]} \oplus rk_{1,[t]})). \qquad (5)$$

$mask_{1,[j]}$ $(j \in [2,t])$ is calculated similarly. Then the values of all $wkL_{[k]}$ $(k \in [1,t])$ is iteratively calculated by $wkL_{[k]} = MASK_{1,[k]} \oplus mask_{1,[k]}$. Thus, the whitening key wkL is recovered, and our method is suitable for launching DPA to recover the whitening key on the left branch of Feistel-SP in loop architecture.

3.4   Recovery of wkR

According to Sect. 3.3, we reveal the round key rk_1 and the equivalent mask $MASK_1$ in the first round, and then successfully derive the left branch whitening key wkL from above two parameters. However, due to Figs. 4(b) and 5, the right branch whitening key wkR is hard to be distinguished from the equivalent key, because both wkR and rk_1 have exactly the same effects on the SP type F-function in the first round of Feistel-SP ciphers. More precisely, according to Eq. 4, the first byte of R_2 in this case would be rewritten as:

$$
\begin{aligned}
R_{2,[1]} &= F_{[1]}(R_1, rk_1) \oplus L_{1,[1]} \\
&= P_{[1]}(S_1(MR_{[1]} \oplus wkR_{[1]} \oplus rk_{1,[1]}), S_2(MR_{[2]} \oplus wkR_{[2]} \oplus rk_{1,[2]}), \\
&\quad \cdots, S_t(MR_{[t]} \oplus wkR_{[t]} \oplus rk_{1,[t]})) \oplus ML_{[1]} \\
&= P'_{[1]}(S_1(MR_{[1]} \oplus wkR_{[1]} \oplus rk_{1,[1]})) \oplus mask_{1,[1]} \oplus ML_{[1]} \\
&= P'_{[1]}(S_1(MR_{[1]} \oplus ek_{1,[1]})) \oplus mask_{1,[1]} \oplus ML_{[1]}. \qquad (6)
\end{aligned}
$$

with $wkR = wkR_{[1]} || wkR_{[2]} || \cdots || wkR_{[t]}$, and the equivalent key $ek_1 = rk_1 \oplus wkR$, and $mask_{1,[1]}$ is a byte constant value if $MR_{[2]}, MR_{[3]}, \cdots, MR_{[t]}$ are fixed since the equivalent key ek_1 is pre-assigned. In this scenario, we can only reveal each byte of ek_1 and $mask_1$ with the approach proposed in Sect. 3.3, and we are unable to split wkR from ek_1.

Therefore, we have to find relations between the case of the whitening key on the right branch and the case of the whitening key on the left branch in order to recover wkR. We focus on the second round of Feistel-SP. As shown in Fig. 6(a), the whitening key wkR is mixed up with $mask_2$ in the second round of Feistel-SP, and we choose R_3 as the attack point in the second round. Equations 3 and 4 can be rewritten as:

$$
\begin{aligned}
R_3 &= F(R_2, rk_2) \oplus L_2 \\
&= P(S_1(R_{2,[1]} \oplus rk_{2,[1]}), S_2(R_{2,[2]} \oplus rk_{2,[2]}), \\
&\quad \cdots, S_t(R_{2,[t]} \oplus rk_{2,[t]})) \oplus wkR \oplus MR.
\end{aligned} \tag{7}
$$

$$
\begin{aligned}
R_{3,[1]} &= F_{[1]}(R_2, rk_2) \oplus L_{2,[1]} \\
&= P_{[1]}(S_1(R_{2,[1]} \oplus rk_{2,[1]}), S_2(R_{2,[2]} \oplus rk_{2,[2]}), \\
&\quad \cdots, S_t(R_{2,[t]} \oplus rk_{2,[t]})) \oplus wkR_{[1]} \oplus MR_{[1]} \\
&= P'_{[1]}(S_1(R_{2,[1]} \oplus rk_{2,[1]})) \oplus mask_{2,[1]} \oplus wkR_{[1]} \oplus MR_{[1]} \\
&= P'_{[1]}(S_1(R_{2,[1]} \oplus rk_{2,[1]})) \oplus MASK_{2,[1]} \oplus MR_{[1]}.
\end{aligned} \tag{8}
$$

where $mask_{2,[1]}$ is a byte constant value if $R_{2,[2]}, R_{2,[3]}, \cdots, R_{2,[t]}$ are fixed since the round key rk_2 is pre-assigned. Therefore, as shown in Fig. 6(b), if we can control the input of the second round $L_2 \| R_2$ as the plaintext message $ML \| MR$, we will reduce the case of the whitening key on the right branch in the second round to the case of the whitening key on the left branch in the first round.

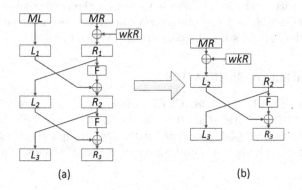

(a)   (b)

Fig. 6. The reduce from the right branch case to the left branch case

Fortunately, we can control the input of the second round $L_2 \| R_2$ in an adaptive chosen message manner. Firstly, we reveal ek_1 in the first round. Then based on the values of $L_2 \| R_2$ which meets the chosen message requirements, we calculate the corresponding plaintext message with ek_1 through the inverse transformation of the Feistel-SP round function. Finally, we could establish the chosen plaintext set which is suitable for the second round and reduce the case of wkR to the case of wkL, and reveal wkR for the second round of Feistel-SP.

On Recovering the Whitening Keys on both Branches. Our approach is also suitable for Feistel-SP ciphers with the key whitening layer on both branches, as shown in Fig. 4(c). The specific procedures of this scenario are as follows:

- **Step 1. Reveal the Whitening Key wkL in the First Round.** The first step of our method is to reveal wkL in the first round. We establish the chosen plaintext set which enumerates all possible values of the target bytes ($MR_{[1]}$), and fixes other bytes to constants. Then, we repeat the DPA attack against R_2 several times to recover all bytes of ek_1 and $MASK_1$. Finally, we calculate $mask_1$ by ek_1 and reveal wkL with the equation $wkL = MASK_1 \oplus mask_1$.
- **Step 2. Reveal the Whitening Key wkR in the Second Round.** The second step of our method is to establish the chosen plaintext set which is used for the second round, through an adaptive chosen message manner. According to Sect. 3.4, it can be done with similar strategy of Step 1 to recover wkR.

4   Applications

We apply these techniques to two typical Feistel-SP ciphers, CLEFIA-128 [2] and Camellia-128 [3], to verify the effectiveness of our method. We implement both ciphers with the loop architecture on a Virtex-5 Xilinx FPGA on SASEBO-GII board. Pearson Correlation Coefficient based Power Analysis (CPA) is applied during the security analysis. The aim is to recover the master keys in CLEFIA-128 and Camellia-128, and the master keys are recovered as expected in both experiments, thus manifesting the correctness of our approach.

4.1   Application to CLEFIA-128

Specification of CLEFIA-128. CLEFI-128 is a type-2 Feistel-SP cipher proposed at FSE 2007 by Shirai *et al.*. It is standardized by ISO [19] as a lightweight cipher. It encrypts a 128-bit plaintext into a 128-bit ciphertext with a 128-bit master key after applying the round function 18 times, as shown in Fig. 7.

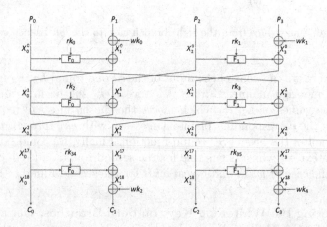

Fig. 7. CLEFIA encryption algorithm for 128-bit key

Let P and K be the 128-bit plaintext and the master key respectively. Thirty-six 32-bit round keys $rk_0, rk_1, \cdots, rk_{35}$ and four 32-bit whitening keys wk_0, wk_1, wk_2, wk_3 are generated from K. These whitening keys are defined as $wk_0 \| wk_1 \| wk_2 \| wk_3 = K$ according to the key schedule.

Let $X_0^i \| X_1^i \| X_2^i \| X_3^i (0 \le i \le 17)$ be an internal input state in each round. The plaintext is loaded into $P_0 \| P_1 \| P_2 \| P_3$. Next, X_1^0 and X_3^0 are updated by the pre-whitening layer, that is $(X_0^0, X_1^0, X_2^0, X_3^0) = (P_0, P_1 \oplus wk_0, P_2, P_3 \oplus wk_1)$. Then, the internal state is updated by the following computation up to the second last round (for $1 \le i \le 17$);

$$X_0^i = X_1^{i-1} \oplus F_0(X_0^{i-1}, rk_{2i-2}), X_1^i = X_2^{i-1},$$
$$X_2^i = X_3^{i-1} \oplus F_1(X_2^{i-1}, rk_{2i-1}), X_3^i = X_0^{i-1}.$$

Two SP type F-functions F_0 and F_1 consist of a 32-bit round key addition, an S-box transformation, and a multiplication by an MDS matrix, as shown in Fig. 8. Four parallel 8-bit Sboxes are applied, followed with an MDS multiplication in each of SP type F-functions. In addition, the MDS matrices for two SP type F-functions are different.

In the last round, $X_0^{18} \| X_1^{18} \| X_2^{18} \| X_3^{18}$ is computed by

$$X_0^{18} = X_0^{17}, X_1^{18} = X_1^{17} \oplus F_0(X_0^{17}, rk_{34}),$$
$$X_2^{18} = X_2^{17}, X_3^{18} = X_3^{17} \oplus F_1(X_2^{17}, rk_{35}).$$

Finally, C_1 and C_3 are updated by the post-whitening layer, i.e., $(C_0, C_1, C_2, C_3) = (X_0^{18}, X_1^{18} \oplus wk_2, X_2^{18}, X_3^{18} \oplus wk_3)$, and $C_0 \| C_1 \| C_2 \| C_3$ is the final ciphertext.

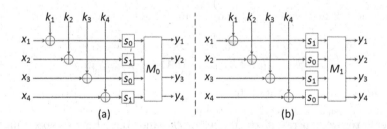

Fig. 8. CLEFIA SP type F-functions for (a) F_0 (b) F_1

Attack of CLEFIA-128. CLEFIA adopts 4-branch Type-2 Feistel network but uses two different diffusion matrices for the diffusion switching mechanism. It has the key whitening layer and only the left branch of the plaintext blocks is XORed with the whitening key. Our aim is to reveal the master keys through the recovery of the whitening keys.

Hereafter we use the notations P_j and C_j for CLEFIA-128 to represent the plaintext and ciphertext in the jth branch, X_j^i to represent the state in the jth

branch immediately after the round operation in round i, X_j^0 to represent the output of the key whitening layer before the first round. We use the notations F_0, F_1, S_0, S_1, M_0, and M_1 to further distinguish the different round functions of CLEFIA-128. The subscript $[n]$ indicates the nth byte of the state.

To reveal the master key K of CLEFIA-128, we focus on the whitening key. As shown in Fig. 7, the four 32-bit whitening keys wk_0, wk_1, wk_2, wk_3 are generated from K. These whitening keys are defined as $wk_0||wk_1||wk_2||wk_3 = K$. Thus, we do not need to calculate the inverse transformation of CLEFIA key schedule to reveal K, if we get the whitening key value. There are two key whitening layers in CLEFIA-128, the pre-whitening before the first round and the post-whitening after the last round. According to the Sect. 3.3, CLEFIA-128 belongs to the case of the whitening key on the left branch. Hence, our attack target is the first round and the last round of CLEFIA-128.

In order to recover both the pre-whitening and the post-whitening keys, the attack should be conducted in both the encryption and the decryption directions respectively. However, since the encryption and decryption of CLEFIA follow similar procedures and F_0 and F_1 are almost equivalent from the perspective of DPA, the attack procedures for recovering the whitening keys are almost identical. Thus we only describe the detailed process to recover wk_0.

The whitening key wk_0 is only XORed with 32-bit plaintext block P_1, and rk_0 is the round key. According to the specification of CLEFIA-128 and Eq. 1, the 32-bit output X_0^1 is described by

$$
\begin{aligned}
X_0^1 &= F_0(X_0^0, rk_0) \oplus X_1^0 \\
&= M_0(S_0(P_{0,[1]} \oplus rk_{0,[1]}), S_1(P_{0,[2]} \oplus rk_{0,[2]}), \\
&\quad \cdots, S_1(P_{0,[4]} \oplus rk_{0,[4]})) \oplus wk_0 \oplus P_1.
\end{aligned} \tag{9}
$$

Focusing on the first byte of X_0^1, Eq. 9 could be rewritten as

$$
\begin{aligned}
X_{0,[1]}^1 &= S_0(P_{0,[1]} \oplus rk_{0,[1]}) \oplus mask_{0,[1]} \oplus wk_{0,[1]} \oplus P_{1,[1]} \\
&= S_0(P_{0,[1]} \oplus rk_{0,[1]}) \oplus MASK_{0,[1]} \oplus P_{1,[1]}.
\end{aligned} \tag{10}
$$

where $mask_{0,[1]}$ is generated by $S_1(P_{0,[2]} \oplus rk_{0,[2]})$, $S_0(P_{0,[3]} \oplus rk_{0,[3]})$, and $S_1(P_{0,[4]} \oplus rk_{0,[4]})$, and $MASK_0$ is defined as $MASK_0 = mask_0 \oplus wk_0$.

Therefore, we can reveal the values of rk_0 and $MASK_0$ by chosen message DPA method as mentioned in Sect. 3.3. Then, we calculate $mask_0$ by rk_0 and reveal wk_0 with the equation $wk_0 = MASK_0 \oplus mask_0$. Thus, wk_1, wk_2, and wk_3 can be revealed by similar procedures. After deriving all the whitening keys, the master key K can be easily derived since $K = wk_0||wk_1||wk_2||wk_3$.

We preset the master key K as four consecutive 32-bit words denoted in hex form, FFEEDDCC-BBAA9988-77665544-33221100, in our FPGA implementation with the loop architecture. Thus, it is obvious that $wk_0 = $ FFEEDDCC and $rk_0 = $ F3E6CEF9. We use the attack result of the first Sbox (i.e., $rk_{0,[1]}$ and $MASK_{1,[1]}$) as an example. In Fig. 9(a), there are two max points F3D4 and F32B whose correlation coefficients are 0.0743 and −0.0743 respectively. According to the knowledge of the FPGA platform, the power consumption of the FPGA platform has a negative correlation with the Hamming distance power model. Thus, F32B is the attack result (*i.e.*, $rk_{0,[1]} = $ F3 and $MASK_{1,[1]} = $ 2B), which is revealed within 10000 power traces as shown in Fig. 9(b).

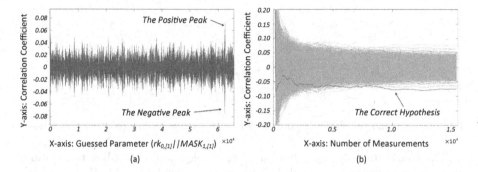

Fig. 9. Correlation traces in different hypothesis and different number of measurements on CLEFIA

We repeat the above procedure 3 times more, and all bytes of rk_0 and $MASK_1$ are revealed as $rk_0 = $ F3E6CEF9 and $MASK_1 = $ 2BF18258. Thus, the value of $mask_1$ is D41F5F94, which is calculated by the round key rk_0. And the wk_0 is FFEEDDCC, calculated by $mask_1 \oplus MASK_1$. The remaining parts of whitening keys are calculated by similar way. Now, we calculate K is FFEEDDCC-BBAA9988-77665544-33221100, which is identical with the preset master key.

In our method, we only attack 2 rounds, *i.e.*, the first round and the last round. Due to the equivalent key of the second round, the conventional DPA must be launched on the first 4 rounds for revealing enough consecutive round keys. Therefore, the cost of our method (*e.g.*, power traces or recovered round key fractions) is only half of the conventional method.

4.2 Application to Camellia-128

Specification of Camellia-128. Camellia-128, proposed at SAC 2000 by Aoki *et al.* [3], was jointly designed by NTT and Mitsubishi Electric Corporation. It is widely acknowledged and recommended by ISO [20], NESSIE [21], and CRYPTREC [22]. It encrypts a 128-bit plaintext into a 128-bit ciphertext with a 128-bit master key after applying the round function 18 times, as shown in Fig. 10.

Let P and K be a 128-bit plaintext and a secret key, respectively. Eighteen 64-bit round keys $rk_1, rk_2, \cdots, rk_{18}$ and four 64-bit whitening keys wk_1, wk_2, wk_3, wk_4 are generated from K. Let L_r and R_r ($0 \le r \le 18$) be left and right 64-bit of the internal state in each round. According to the key schedule, the pre-whitening keys are defined as $wk_1 \| wk_2 = K$. We omit the descriptions of the FL and FL^{-1} layers after the 6th and 12th rounds, since they have no impacts on our work.

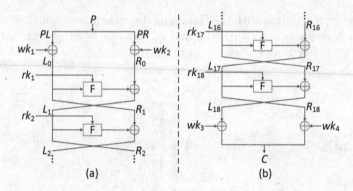

Fig. 10. Camellia encryption algorithm for (a) First two rounds (b) Last two rounds

The round function consists of a 64-bit subkey addition, Sbox transformation, and a diffusion layer called P-layer, as shown in Fig. 11. Eight parallel 8-bit Sboxes are applied, followed with the P-layer which operates on a 64-bit value $(z_1||z_2||\cdots||z_8)$. The corresponding output $(z_1'||z_2'||\cdots||z_8')$ is computed as follows.

$$z_1' = z[1,3,4,6,7,8], z_2' = z[1,2,4,5,7,8],$$
$$z_3' = z[1,2,3,5,6,8], z_4' = z[2,3,4,5,6,7],$$
$$z_5' = z[1,2,6,7,8], z_6' = z[2,3,5,7,8],$$
$$z_7' = z[3,4,5,6,8], z_8' = z[1,4,5,6,7].$$

Here, $z[s,t,u,\cdots]$ means $z_s \oplus z_t \oplus z_u \oplus \cdots$

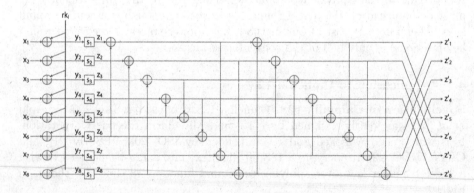

Fig. 11. Camellia SP type F-function

Attack of Camellia-128. Camellia is a Feistel-SP cipher but its P-layer does not satisfy the maximum branch number. It has the key whitening layer and the whole plaintext is XORed with the whitening key. Our aim is to reveal the master keys through the recovery of the whitening keys.

We use the notations L_i and R_i for Camellia-128 to represent the state immediately after the round operation in the ith round, especially L_0 and R_0 to represent the output of the whitening layer before the first round, respectively. We use the notations PL and PR to differentiate the left and right of plaintext, and S_1, S_2, S_3, S_4, and M to further distinguish the different Sboxes and diffusion operation of Camellia-128. The subscript $[n]$ indicates the nth byte of the state.

Camellia-128 belongs to the case of whitening keys on both branches. To reveal the master key K of Camellia-128, we focus on two 64-bit pre-whitening key wk_1 and wk_2, which are defined as $wk_1\|wk_2 = K$. The two whitening keys are XORed with two plaintext blocks PL and PR respectively, before the first round. Hence, our attack target is the first two rounds of Camellia-128. According to the attack procedure in the first round of Camellia-128 which is similar to that of CLEFIA-128, thus we only describe the detailed process to recover wk_1 in the second round.

Now, we show how to recover the whitening keys wk_1 by attacking the round function F of the second round. The whitening keys wk_1 and wk_2 are parallel XORed with two 64-bit plaintext blocks PL and PR respectively, and rk_1 and rk_2 is the round keys as shown in Fig. 10. The equivalent key ek_1 and the whitening key wk_2 are recovered in the first round by the attack process. Therefore, we focus on the second round.

According to the specification of Camellia-128 and Eq. 1, the 64-bit output L_2 is described by

$$
\begin{aligned}
L_2 = {} & F(L_1, rk_2) \oplus R_1 \\
= {} & M(S_1(L_{1,[1]} \oplus rk_{2,[1]}), S_2(L_{1,[2]} \oplus rk_{2,[2]}), \\
& \cdots, S_1(L_{1,[8]} \oplus rk_{1,[8]})) \oplus wk_1 \oplus PL.
\end{aligned}
\tag{11}
$$

Focusing on the first byte of L_2, Eq. 11 could be rewritten as

$$
\begin{aligned}
L_{2,[1]} = {} & S_1(L_{1,[1]} \oplus rk_{2,[1]}) \oplus mask_{2,[1]} \oplus wk_{1,[1]} \oplus PL_{[1]} \\
= {} & S_1(L_{1,[1]} \oplus rk_{2,[1]}) \oplus MASK_{2,[1]} \oplus PL_{[1]}.
\end{aligned}
\tag{12}
$$

where
$mask_{2,[1]}$ is generated by $S_2(L_{1,[2]} \oplus rk_{2,[2]}), S_3(L_{1,[3]} \oplus rk_{2,[3]}), \cdots, S_1(L_{1,[8]} \oplus rk_{2,[8]})$ and $MASK_2$ is defined as $MASK_2 = mask_2 \oplus wk_1$.

Before launch attack in the second round, we should calculate the corresponding plaintext message with ek_1 and wk_2 through the inverse transformation of Camellia round function, according to the value of L_1 which meets the chosen message requirements. Fortunately, we find that there is an one-to-one mapping relationship between PR and L_1. Thus, we can control L_2 by enumerating PR, and reveal the value wk_1 by the method as mentioned in Sect. 3.4. After deriving all the whitening keys, the master key K can be easily derived since $K = wk_1\|wk_2$.

We preset the master key K as four consecutive 32-bit words denoted in hex form, 01234567-89ABCDEF-FEDCBA98-76543210, in our FPGA implementation with the loop architecture. Thus, it is obvious that wk_1 = 01234567-89ABCDEF.

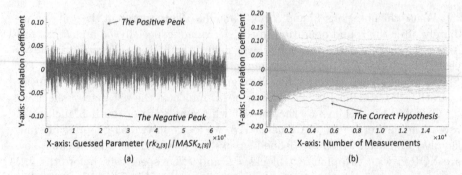

Fig. 12. Correlation traces in different (a) hypothesis and (b) number of measurements on Camellia

We use the attack result of the third Sbox (*i.e.*, $rk_{2,[3]}$ and $MASK_{2,[3]}$) as an example. In Fig. 12(a), there are two max points **40B8** and **4047** whose correlation coefficients are 0.107 and −0.107 respectively. According to the negative correlation between the power consumption of the FPGA platform and the Hamming distance power model, **4047** is the attack result (*i.e.*, $rk_{2,[3]} = 40$ and $MASK_{2,[3]} = 47$), which is revealed within 4000 power traces as shown in Fig. 12(b).

We repeat the above procedure 7 times more, and all bytes of rk_2, $MASK_2$, $mask_2$ and are revealed as similar as the case of CLEFIA-128. Thus, the wk_1 is **01234567-89ABCDEF**, calculated by $mask_2 \oplus MASK_2$. Now, we calculate K is **01234567-89ABCDEF-FEDCBA98-76543210**, which is identical with the preset master key.

In our method, we only attack the first two rounds. Due to the equivalent keys of the first two rounds, the conventional DPA must be launched on the first four rounds for revealing enough consecutive round keys. Therefore, the cost of our method is only half of the conventional method.

5 Conclusion and Discussion

In this paper, we propose a practical chosen message DPA approach to recover the whitening key through a reduction strategy, and show that the key whitening technique does not enhance the security of ciphers from the standpoint of DPA. Then we apply our method to CLEFIA-128 and Camellia-128, and the master keys are recovered as expected. Following the results presented in this work, several problems which are worth further investigations:

- **Optimizations.** The first natural question emerges is whether our method can be further optimized. One possible direction of improvement is taking advantage of more powerful DPA method, such as the adaptive strategy in [6,14] and the multiple CPA in [12], to discriminate the correct hypothesis from the key candidates in a more efficient way. Other potential optimizations are also possible directions for future research.

- **Countermeasures.** Next to optimality, another important question is to determine the countermeasures against such attack. Our method is suitable for the unprotected loop hardware implementations. Several common countermeasures on the compact architecture [17], can be considered to apply to the loop architecture in order to resist our approach, while the resource consumption will have a corresponding increase. Moreover, the countermeasures based on the mask methodology should be used with caution on the loop implementations of ciphers, because the mask countermeasures usually lead to slow computation speed. Considering the limitation of high computation speed and high throughput in the application scenario, the loop implementations of ciphers must keep the high performance, when the countermeasures against such attack are applied on these implementations. Thus, a trade-off between performance and security should be considered by the vendor.

References

1. Kocher, P., Jaffe, J., Jun, B.: Differential power analysis. In: Wiener, M. (ed.) CRYPTO 1999. LNCS, vol. 1666, pp. 388–397. Springer, Heidelberg (1999). doi:10.1007/3-540-48405-1_25

2. Shirai, T., Shibutani, K., Akishita, T., Moriai, S., Iwata, T.: The 128-bit blockcipher CLEFIA (extended abstract). In: Biryukov, A. (ed.) FSE 2007. LNCS, vol. 4593, pp. 181–195. Springer, Heidelberg (2007). doi:10.1007/978-3-540-74619-5_12

3. Aoki, K., Ichikawa, T., Kanda, M., Matsui, M., Moriai, S., Nakajima, J., Tokita, T.: *Camellia*: a 128-bit block cipher suitable for multiple platforms — design and analysis. In: Stinson, D.R., Tavares, S. (eds.) SAC 2000. LNCS, vol. 2012, pp. 41–54. Springer, Heidelberg (2001). doi:10.1007/3-540-44983-3_4

4. Brier, E., Clavier, C., Olivier, F.: Correlation power analysis with a leakage model. In: Joye, M., Quisquater, J.-J. (eds.) CHES 2004. LNCS, vol. 3156, pp. 16–29. Springer, Heidelberg (2004). doi:10.1007/978-3-540-28632-5_2

5. Hoang, V.T., Rogaway, P.: On generalized Feistel networks. In: Rabin, T. (ed.) CRYPTO 2010. LNCS, vol. 6223, pp. 613–630. Springer, Heidelberg (2010). doi:10.1007/978-3-642-14623-7_33

6. Kopf, B., Basin, D.A.: An information-theoretic model for adaptive side-channel attacks. In: Ning, P., De Capitani Vimercati di, S., Syverson, P.F. (eds.) Proceedings of the 14th ACM Conference on Computer and Communications Security, ACM-CCS 2007, pp. 286–296. ACM (2007)

7. Kim, Y., Ahn, J., Choi, H.: Power and electromagnetic analysis attack on a smart card implementation of CLEFIA. In: International Conference on Security and Management, SAM 2013 (2013)

8. Akkar, M.-L., Bevan, R., Dischamp, P., Moyart, D.: Power analysis, what is now possible. In: Okamoto, T. (ed.) ASIACRYPT 2000. LNCS, vol. 1976, pp. 489–502. Springer, Heidelberg (2000). doi:10.1007/3-540-44448-3_38

9. Lu, Y., O'Neill, M.P., McCanny, J.V.: Differential power analysis resistance of Camellia and countermeasure strategy on FPGAs. In: International Conference on Field-Programmable Technology, pp. 183–189 (2009)

10. Xiao, L., Heys, H.: A simple power analysis attack against the key schedule of the Camellia block cipher. Inf. Process. Lett. **95**, 409–412 (2005). Elsevier

11. Bayrak, A.G., Regazzoni, F., Novo, D., Ienne, P.: Sleuth: automated verification of software power analysis countermeasures. In: Bertoni, G., Coron, J.-S. (eds.) CHES 2013. LNCS, vol. 8086, pp. 293–310. Springer, Heidelberg (2013). doi:10.1007/978-3-642-40349-1_17

12. Moradi, A., Schneider, T.: Improved side-channel analysis attacks on xilinx bitstream encryption of 5, 6, and 7 series. In: Standaert, F.-X., Oswald, E. (eds.) COSADE 2016. LNCS, vol. 9689, pp. 71–87. Springer, Cham (2016). doi:10.1007/978-3-319-43283-0_5

13. Moradi, A., Kasper, M., Paar, C.: Black-box side-channel attacks highlight the importance of countermeasures. In: Dunkelman, O. (ed.) CT-RSA 2012. LNCS, vol. 7178, pp. 1–18. Springer, Heidelberg (2012). doi:10.1007/978-3-642-27954-6_1

14. Veyrat-Charvillon, N., Standaert, F.-X.: Adaptive chosen-message side-channel attacks. In: Zhou, J., Yung, M. (eds.) ACNS 2010. LNCS, vol. 6123, pp. 186–199. Springer, Heidelberg (2010). doi:10.1007/978-3-642-13708-2_12

15. Leander, G., Paar, C., Poschmann, A., Schramm, K.: New lightweight DES variants. In: Biryukov, A. (ed.) FSE 2007. LNCS, vol. 4593, pp. 196–210. Springer, Heidelberg (2007). doi:10.1007/978-3-540-74619-5_13

16. Borghoff, J., et al.: PRINCE – a low-latency block cipher for pervasive computing applications. In: Wang, X., Sako, K. (eds.) ASIACRYPT 2012. LNCS, vol. 7658, pp. 208–225. Springer, Heidelberg (2012). doi:10.1007/978-3-642-34961-4_14

17. Mangard, S., Oswald, E., Popp, T.: Power Analysis Attacks: Revealing the Secrets of Smart Cards. Springer, New York (2007). ISBN: 978-0-387-30857-9

18. Rodríguez-Henríquez, F., Saqib, N.A., Díaz-Pèrez, A., Koc, Ç.K.: Cryptographic Algorithms on Reconfigurable Hardware (Signals and Communication Technology). Springer, Heidelberg (2006)

19. ISO/IEC 29192-2:2011. Information technology - Security techniques - Lightweight cryptography-Part 2: Block ciphers (2011)

20. ISO/IEC 18033-3:2010. Information technology - Security techniques - Encryption Algorithms-Part 3: Block ciphers (2010)

21. New European Schemes for Signatures, Integrity, and Encryption (NESSIE). NESSIE Project Announces Final Selection of Crypto Algorithms (2003)

22. Cryptography Research and Evaluation Committees (CRYPTREC). e-Government recommended ciphers list (2003)

Side-Channel Attacks Meet Secure Network Protocols

Alex Biryukov, Daniel Dinu$^{(\boxtimes)}$, and Yann Le Corre

SnT, University of Luxembourg,
6, Avenue de la Fonte, 4364 Esch-sur-Alzette, Luxembourg
{alex.biryukov,dumitru-daniel.dinu,yann.lecorre}@uni.lu

Abstract. Side-channel attacks are powerful tools for breaking systems that implement cryptographic algorithms. The Advanced Encryption Standard (AES) is widely used to secure data, including the communication within various network protocols. Major cryptographic libraries such as OpenSSL or ARM mbed TLS include at least one implementation of the AES. In this paper, we show that most implementations of the AES present in popular open-source cryptographic libraries are vulnerable to side-channel attacks, even in a network protocol scenario when the attacker has limited control of the input. We present an algorithm for symbolic processing of the AES state for any input configuration where several input bytes are variable and known, while the rest are fixed and unknown as is the case in most secure network protocols. Then, we classify all possible inputs into 25 independent evaluation cases depending on the number of bytes controlled by attacker and the number of rounds that must be attacked to recover the master key. Finally, we describe an optimal algorithm that can be used to recover the master key using Correlation Power Analysis (CPA) attacks. Our experimental results raise awareness of the insecurity of unprotected implementations of the AES used in network protocol stacks.

Keywords: Side-channel attack · Secure network protocol · CPA · AES

1 Introduction

Side-channel attacks use observations made during the execution of an implementation of a cryptographic algorithm to recover secret information. From the multitude of side-channel attacks, Correlation Power Analysis (CPA) [5] stands out as a very efficient and reliable technique. Its success is augmented by the minimally invasive methods employed for the acquisition of the side-channel information. Some of the most frequently used sources of side-channel leakage are the power consumption or the electromagnetic (EM) emissions of a device under attack.

Nowadays, the AES [23] is the most popular symmetric cryptographic algorithm in use. It is widely deployed to secure data in transit or at rest. Various network protocols rely on the AES in different modes of operation to provide

© Springer International Publishing AG 2017
D. Gollmann et al. (Eds.): ACNS 2017, LNCS 10355, pp. 435–454, 2017.
DOI: 10.1007/978-3-319-61204-1_22

security services such as confidentiality and authenticity. The usage spectrum of the AES stretches from powerful servers and personal computers to resource constrained devices such as wireless sensor nodes. While the security of the algorithm and its implementations have been placed under scrutiny since it became the symmetric cryptographic standard, with a few notable exceptions, most of the previous work focused on the AES itself and less on the usage of the AES in complex systems.

By far, most of the experimental results reported in the side-channel literature are for implementations of the AES. They usually assume the attacker has full control of the AES input. This is not the case in a real world communication protocol, when often a major part of the input is fixed and only few bytes are variable. Moreover, sometimes the attacker cannot control these variable bytes and she has to passively observe executions of the targeted algorithm without being able to trigger encryptions of her own free will. With the notable exceptions of [16,24], the security of communication scenarios based on the AES against side-channel attacks has not been thoroughly analyzed so far. Thus, in this paper we analyze for the first time how much control of the AES input does an attacker need to recover the secret key of the cipher by performing a side-channel attack against a communication protocol.

Numerous standards for communication in the Internet of Things (IoT) such as IEEE 802.15.4 [15] and LoRaWAN [21] use the AES to encrypt and authenticate the Medium Access Control (MAC) layer frames. The 802.15.4 standard uses a variant of the AES-CCM [9,34], while LoRaWAN uses AES-CMAC [31]. The same CCM mode is used with the AES to encrypt the IPsec Encapsulating Security Payload (ESP) [14]. According to [29] the security architecture of IEEE 802.15.4 relies on four categories of security suites: none, AES-CTR, AES-CBC-MAC, and AES-CCM. A typical input for the AES-CTR and AES-CCM modes used in the IEEE 802.15.4 protocol is shown in Fig. 1. In this particular example, an attacker can manipulate up to 12 bytes of the input (Source Address and Frame Counter), while the other input bytes (Flags, Key Counter and Block Counter) are fixed. The attack on IEEE 802.15.4 wireless sensor nodes described in [24] assumes the control of only four input bytes (Frame Counter), while the remaining input bytes are constant. Thus the following question arises: *How many input bytes should an attacker change in the injected messages in order to fully recover the master key without triggering any network protection mechanism?*

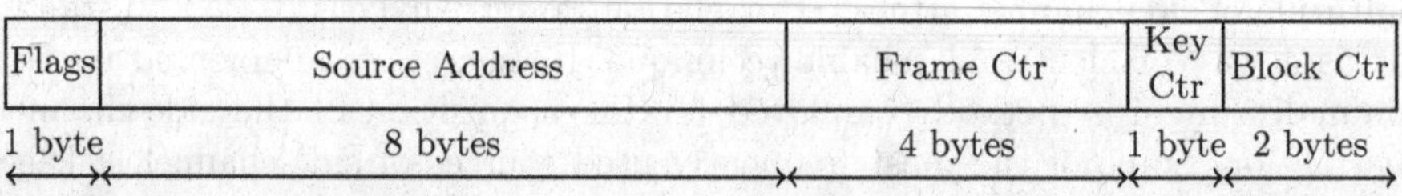

Flags	Source Address	Frame Ctr	Key Ctr	Block Ctr
1 byte	8 bytes	4 bytes	1 byte	2 bytes

Fig. 1. The first input block for the AES-CTR and AES-CCM modes used in IEEE 802.15.4.

While numerous network protocols use the AES to secure the communication between end nodes, major cryptographic libraries such as OpenSSL [25] and ARM mbed TLS [2] do not have a side-channel protected implementation of the AES for devices that do not support the AES-NI [13] instruction set as is the case with most IoT devices. Therefore, an elaborate analysis of the security of the unprotected implementations of the AES used in communication protocols is necessary. Only such a careful analysis can assess the impact of side-channel attacks on the security of real world systems using unprotected implementations of the AES.

In this work, we chose to focus on CPA attacks thanks to their efficiency and reliability. We opted for a non-invasive measurement setup and hence we selected the EM emissions of the target processor as source of side-channel leakage. The target is an ARM Cortex-M3 processor mounted on a STM32 Nucleo [32] board from STMicroelectronics. These processors are widely used for low-power applications and meet the requirements for use in the IoT.

The IoT will be a security nightmare if the whole ecosystem is not designed with security in mind. While many communication protocols for the IoT are in formative stages, the threat model of the IoT is less understood despite it is widely accepted that its attack surface is large. Although we focus on a particular side-channel attack (i.e. power/EM), other side-channel attacks such as timing, fault, cache or data remanence attacks might pose a similar or even a higher threat for the security of the IoT ecosystem. Attacks that do not exploit side-channel information, such as those used to compromise Internet-connected computers, should not be neglected since they have certain advantages over side-channel attacks. Thus, our work adds another piece to the security puzzle of the IoT by showing the need for side-channel countermeasures to prevent a somehow overlooked threat.

Research Contributions. This paper performs for the first time a thorough analysis of all possible attack scenarios against software implementations of the AES used to secure various communication protocols. Firstly, we present an algorithm for symbolic processing of a given input state of the AES. The algorithm outputs the number of rounds and the bytes that must be attacked to recover the secret key. Then, using this algorithm we perform a classification of all possible inputs depending on the number of rounds that must be attacked in order to recover the master key. The result is a set of 25 independent evaluation cases. Secondly, we describe an optimal algorithm that uses the above-mentioned symbolic representation to recover the master key of the AES using CPA attacks. The algorithm explores all possible combinations of input key bytes and discards the invalid key candidates, thus yielding only the correct master key if enough power traces with a good signal-to-noise ratio are provided. Afterwards, we evaluate the results of the attack algorithm in each of the 25 evaluation cases identified in the classification step using traces from an ARM Cortex-M3 processor.

Our results show that popular implementations of the AES found in well-known and widely used cryptographic libraries can be broken using CPA attacks.

The only requirement is that a part of the AES input is known and variable, while the rest is constant, which is a common scenario in communication protocols. Knowledge of the AES implementation strategy improves the attack results, but it is not crucial. All software tools presented in this paper are in the public domain[1] to support reproducibility of results and to maximize reusability.

2   Preliminaries

2.1   Description of the AES

We give a brief description of the AES [23] to recall relevant aspects of the algorithm and to introduce the notation used in this paper. For more details on the AES algorithm, we refer the reader to the official specifications.

The AES standard uses the 128-bit block length version of the Rijndael cipher [8] with three different key lengths: 128, 192, and 256 bits. The round function is applied to the 4×4 byte state matrix 10, 12, or 14 times depending on the key length. It comprises four transformations: SubBytes, ShiftRows, MixColumns, and AddRoundKey. The final round function does not include the MixColumns transformation.

Let $s_{i,j}$ be the state byte located at row i and column j ($0 \leq i, j \leq 3$), k_l the corresponding round key byte ($l = 16 \cdot r + i + 4 \cdot j$) and r the round number. After application of the AddRoundKey transformation, each byte of the state becomes $s'_{i,j} = s_{i,j} \oplus k_l$, where the "$\oplus$" symbol denotes bitwise exclusive or of two 8-bit values. The non-linear SubBytes operation transforms each byte of the state using an 8-bit S-box S as follows: $s'_{i,j} = S[s_{i,j}]$. The ShiftRows transformation performs a rotation of row i by i bytes to the left. In the MixColumns transformation, a polynomial multiplication over $GF(2^8)$ is applied to each column of the state matrix. The symbol "$\bullet$" is used for multiplication of two numbers in $GF(2^8)$, while $\{01\}$, $\{02\}$, and $\{03\}$ are 8-bit vectors representing elements from $GF(2^8)$.

The key schedule expands the master key into the 16-byte round keys. The round constant array Rcon contains the powers of $\{02\}$ in $GF(2^8)$ as described in the specifications.

2.2   Correlation Power Analysis

Correlation Power Analysis (CPA) [5] is a side-channel attack in which the attacker correlates the power model of a sensitive intermediate value of the target cryptographic algorithm with the measured power consumption or electromagnetic emission (EM) of the device running the target algorithm. Then, she chooses the key hypothesis that gives the maximum correlation coefficient as the most likely key. Compared to classical Differential Power Analysis (DPA) [17] attacks, CPA attacks have several advantages in terms of efficiency, robustness and number of experiments, but are more resource demanding. Agrawal et al. [1]

---

[1] https://github.com/cryptolu/aes-cpa.

introduced the electromagnetic emissions of a target device as a source of leakage for side-channel attacks.

A CPA attack can be split into two phases: acquisition and attack. In the acquisition phase, the attacker observes and records the leakage of the target device (power consumption or electromagnetic emission) for different inputs. While the acquisition of power consumption traces requires insertion of a resistor into the circuitry of the target device to measure the voltage across it, the observation of electromagnetic emission is non-invasive; it only requires an electromagnetic probe placed in the vicinity of the leaking spot. In the attack phase, the attacker correlates these observations with the modeled power consumption of the selection function to recover the secret key. A selection function combines a known input with the secret material to be recovered.

In this work we focus on the electromagnetic emissions of an ARM Cortex-M3 processor clocked at 8 MHz running various software implementations of the AES. The acquisition was performed from a spot above the chip using a Langer RF-K 7-4 H-field probe. The signal was amplified by 30dB and fed into a Teledine LeCroy WaveRunner 8254M-MS oscilloscope sampling at 500 MS/s. For more details on the measurement setup we refer the reader to the full version of this paper.

2.3 Attacking Temporary Key Bytes

To attack the AES in counter mode, Jaffe introduced a technique that propagates a DPA attack to later rounds. It can be used when just few key bytes of the AES input are known and variable, while the others are fixed (constant) and unknown [16]. Next we briefly describe how the unknown fixed bytes can be incorporated into a round key byte to recover a temporary key byte. Then, using these temporary key bytes the attack can be carried into later rounds until enough round key bytes are recovered to reverse the key schedule.

Using a CPA attack an adversary can recover only those key bytes that are XORed with variable and known state bytes in the AddRoundKey transformation. The gist of Jaffe's technique is that an attacker can still recover a temporary key byte when an input byte of the AddRoundKey transformation is the result of the MixColumns transformation applied to at least one known and variable input byte while the other input bytes are unknown and constant.

To better illustrate how this technique works, let us consider the first state byte $s'_{0,0}$ after performing the first round function:

$$s'_{0,0} = (\{02\} \bullet s_{0,0}) \oplus (\{03\} \bullet s_{1,1}) \oplus (\{01\} \bullet s_{2,2}) \oplus (\{01\} \bullet s_{3,3}) \oplus k_{16}$$

Suppose now that the input bytes $s_{0,0}$ and $s_{1,1}$ are known and variable (key bytes k_0 and k_5 were successfully recovered using a side-channel attack on the SubBytes transformation of the first round), while the other input bytes ($s_{2,2}$ and $s_{3,3}$) are unknown, but fixed. Thus $s'_{0,0}$ can be written as $(\{02\} \bullet s_{0,0}) \oplus (\{03\} \bullet s_{1,1}) \oplus k'_{16}$, where the constant part is included in the temporary key k'_{16} that will be recovered by attacking the SubBytes transformation of the second round; $k'_{16} = (\{01\} \bullet s_{2,2}) \oplus (\{01\} \bullet s_{3,3}) \oplus k_{16}$. The temporary key k'_{16} enables the computation of four state bytes in the following round. In this way, the attack is carried into the next rounds until all state bytes are known; consequently, the real key bytes can be recovered.

The technique works similarly when three input bytes are known and variable. Though, when only one input byte is known and variable, the attacker will recover the same two equally likely key candidates for two bytes of the same column of the cipher state. For example, when only $s_{3,3}$ is known and variable while the other input bytes are unknown and fixed, then $s'_{0,0} = (\{01\} \bullet s_{3,3}) \oplus k'_{16}$ and $s'_{1,0} = (\{01\} \bullet s_{3,3}) \oplus k'_{17}$. Thus attacking either of the two, an attacker will get two equally likely key bytes (k'_{16} and k'_{17}). If the state bytes are not processed in order by the SubBytes transformation, the attacker will not know which key byte corresponds to $s'_{0,0}$ and which key byte corresponds to $s'_{1,0}$.

2.4   Software Implementations of the AES

There are various ways to implement the AES in software depending on the execution time, code size and RAM consumption requirements. Other factors that influence the implementation strategy are the cipher mode of operation and the number of plaintext blocks to be encrypted. Schwabe and Stoffelen [30] identified four different strategies to implement the AES in software: traditional, T-tables, vector permute, and bit slicing. In this paper, we consider the following two implementation approaches for the AES that are relevant for a secure communication protocol:

- The *straightforward implementation* (**S-box** strategy) performs the four round transformations as described above. The substitution layer is implemented using a 256-byte lookup table based on S-box S. This implementation approach is suitable for 8-bit architectures.
- The *table based implementation* (**T-table** strategy) uses four lookup tables (T_0, T_1, T_2, and T_3) of 1024 bytes each to perform the SubBytes, ShiftRows, and MixColumns operations at the cost of 16 table lookups, 16 masks and 16 XORs per round, except for the final round. A low memory alternative uses just one T-table, but performs 12 additional rotations per round. This strategy was initially described by the designers of Rijndael [8]. It leads to very fast implementations on 32-bit platforms.

We did not analyze bit-sliced or vector permute implementations because such implementations are uncommon in cryptographic libraries due to the following limitations. The bit-sliced implementations process at least two blocks in parallel and thus they can be applied only to non-feedback modes of operation. The vector permute implementations require support for vector permute instructions, but most of the resource constrained microcontrollers for the IoT do not support such instructions.

An analysis of the existing AES implementations used by different open source cryptographic libraries is given in Table 1. The default implementations of the AES for platforms that do not support the AES-NI [13] instructions in popular cryptographic libraries such as OpenSSL [12,25] or mbed TLS [2,11] use the

Table 1. A summary of the existing AES implementations used by open source cryptographic libraries written in C/C++. All the T-table implementations are vulnerable to the attack described in this paper.

Library	Language	Version	Last update	AES-NI	T-table
Botan [27]	C++	2.1.0	Apr 2017	✓	✓
cryptlib [6]	C	3.4.3	Feb 2017	✓	✓
Crypto++ [7]	C++	5.6.5	Oct 2016	✓	✓
Libgcrypt [18]	C	1.7.6	Jan 2017	✓	✓
libtomcrypt [10]	C	1.17	Apr 2017	✗	✓
libsodium [19]	C	1.0.12	Mar 2017	✓	✗
mbed TLS [2]	C	2.4.2	Mar 2017	✓	✓
Nettle [22]	C	3.3	Oct 2016	✓	✓
OpenSSL [25]	C	1.1.0e	Feb 2017	✓	✓
wolfCrypt [35]	C	3.10.2	Feb 2017	✓	✓

T-table approach. Except for libsodium [19], all other cryptographic libraries analyzed have an implementation of the AES based on the T-table strategy. Moreover, these implementations are not protected against side-channel attacks such as DPA or cache attacks. It is well know that unprotected implementations of cryptographic algorithms are an easy target for DPA attacks. Recently, researchers from Rambus Cryptography Research Division have shown that even an unprotected software implementation based on AES-NI instructions can be attacked with DPA [28]. The T-table implementations of the AES are vulnerable to various cache attacks as shown in [20,26]. Although the unprotected T-table implementations are vulnerable to side channel attacks, nine out of the ten libraries considered in Table 1 have such an implementation of the AES.

3   Quantifying the Leakage

Biryukov *et al.* [4] introduced the correlation coefficient difference metric to analyze the leakage of different selection functions in the context of CPA. The correlation coefficient difference δ gives the difference between the correlation coefficient of the correct key and the correlation coefficient of the most likely key guess, where the most likely key is different from the correct key.

We use the correlation coefficient difference to quantify the leakages of two selection functions: φ_1 based on the AES S-box and φ_2 based on the AES T-table. The two selection functions are defined below:

$$\varphi_1 : \mathbb{F}_2^8 \mapsto \mathbb{F}_2^8, \quad \varphi_1(x \oplus k) = S(x \oplus k)$$

$$\varphi_2 : \mathbb{F}_2^8 \mapsto \mathbb{F}_2^{32}, \quad \varphi_2(x \oplus k) = T(x \oplus k)$$

Table 2. Correlation coefficient difference δ between the correlation of the correct key and the correlation of the most likely key [4], for different Hamming weights of the correct key; $\bar{\delta}$ and $\mathsf{SE}_{\bar{\delta}}$ are the mean and the standard error for a 95% confidence interval, respectively. The leakages are acquired from an ARM Cortex-M3 processor.

	Correct key									$\bar{\delta}$	$\mathsf{SE}_{\bar{\delta}}$
	0x00	0x01	0x03	0x07	0x0F	0x1F	0x3F	0x7F	0xFF		
φ_1	0.146	0.126	0.108	0.156	0.126	0.960	0.153	0.140	0.084	0.126	0.020
φ_2	0.104	0.072	0.143	0.074	0.070	0.126	0.078	0.044	0.028	0.082	0.028

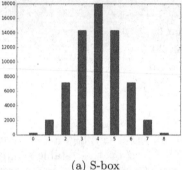
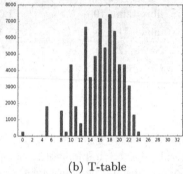

(a) S-box   (b) T-table

Fig. 2. Distribution of the Hamming weight of the output of the AES (a) S-box and (b) T-table for all possible input combinations.

When using simulated leakages, the values of the correlation coefficient difference are 0.813 and 0.7 for φ_1 and φ_2, respectively. These values are the same regardless of the correct key used. In the simulated environment, the leakages of the two selection functions are very high and the difference between them is about 14% of the first one. On the other hand, the mean correlation coefficient difference $\bar{\delta}$ for different values of the correct key using leakages acquired from an ARM Cortex-M3 processor is given in Table 2. The measurements were performed at a sampling rate of 500 MS/s using assembly implementations of the analyzed selection functions. Increasing the sampling rate to 1 GS/s does not significantly improve the results. The mean correlation coefficient difference $\bar{\delta}$ is positive for both selection functions, which means they leak enough information about the secret key such that an attacker can recover the key byte using only one key guess. In practice, the selection function based on the AES S-box leaks about 50% more than the selection function based on the AES T-table. This can be explained by the distribution of the Hamming weight of the two selection functions for all possible input combinations (See Fig. 2).

The reader can easily observe in Fig. 2a that the distribution of values in the case of the AES S-box follows a binomial distribution. On the other hand, the distribution of values in the case of the AES T-table shown Fig. 2b does not resemble a binomial distribution. Moreover, there are 14 out of 32 possible output values that never occur (i.e. 1, 2, 3, 4, 6, 7, 25, 26, 27, 28, 29, 30, 31, and 32). These differences between the distribution of the Hamming weights of the output of the two selection functions φ_1 and φ_2 explain why the leakage of φ_1 is greater than the leakage of φ_2 as quantified using the correlation coefficient difference. This means that a CPA attack against an implementation based on the T-table strategy requires more effort (i.e. power traces) compared to a CPA attack against an implementation based on the S-box strategy.

4 Generating the Evaluation Cases

In this section we describe the algorithm for symbolic processing of a given initial state to determine the number of rounds required to recover the master key of the AES. We used this algorithm to explore all possible attack cases and to choose the relevant evaluation cases for our scenario. The algorithm relies on the following symbolic representation of a byte situated at row i and column j of the AES state at the start of round r:

$$s_{i,j}^r = \begin{cases} 0, & \text{the corresponding key byte can not be recovered} \\ 1, & \text{the corresponding key byte can be recovered} \\ -n, & n \text{ temporary key bytes can be recovered} \end{cases}$$

Thus, the byte $s_{i,j}^r$ is variable if its symbolic representation is different from 0 and fixed (constant) when its symbolic representation is 0. Due to the MixColumns transformation, each column of the state at round $r + 1$ can be expressed as a function of four bytes of the state at round r. At the start of round $r + 1$ each byte of the state is updated using the following rules:

- if the number of variable input bytes is 0, then the symbolic representation of the output byte is set to 0;
- if the number of variable input bytes is 1, then the symbolic representation of the output byte is updated as follows:
  - if the variable input byte is multiplied by {01} in the MixColumns transformation, then the symbolic representation of the output byte is set to -2^{p+1}, where p is the number of independent input pairs. A new pair is added to the output byte;
  - else, the symbolic representation of the output byte is set to -2^p;
- if the number of variable input bytes is 2 or 3, then the symbolic representation of the output byte is set to -1;
- if the number of variable input bytes is 4, then the symbolic representation of the output byte is set to 1.

Besides updating the symbolic representation of the state, the algorithm keeps a list of key pairs for each byte of the state and carries this list into the next round. The algorithm stops when the symbolic representation of all bytes in a round is 1. It outputs the symbolic representation of the state and the associated key pairs. These can be used to compute the number of rounds

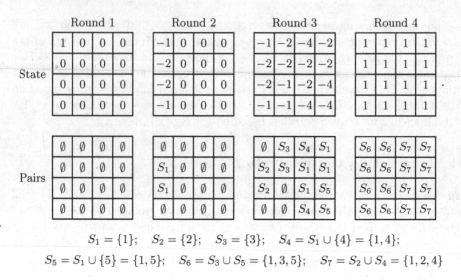

$$S_1 = \{1\}; \quad S_2 = \{2\}; \quad S_3 = \{3\}; \quad S_4 = S_1 \cup \{4\} = \{1,4\};$$
$$S_5 = S_1 \cup \{5\} = \{1,5\}; \quad S_6 = S_3 \cup S_5 = \{1,3,5\}; \quad S_7 = S_2 \cup S_4 = \{1,2,4\}$$

Fig. 3. Symbolic processing of an initial state.

required to recover the master key and the number of possible master keys. The pseudocode for the algorithm is given in the full version of this paper.

Figure 3 gives a graphical representation of how the algorithm works when only the first byte of the initial state is variable and known, while the other bytes are fixed and unknown. By attacking the result of the SubByte transformation applied to the first byte of the state in the first round, the key byte k_0 is recovered. This recovered key byte allows a carry of the attack into the second round where four key bytes $(k'_{16}, k'_{17}, k'_{18}, k'_{19})$ can be recovered by attacking the result of the SubBytes transformation. Because the attacker cannot distinguish between k'_{17} and k'_{18}, a new pair $S_1 = \{1\}$ is added to the corresponding state bytes. Then, the attacker targets the third round, where she can recover temporary key bytes for all state bytes. The pair S_1 from previous round affects all bytes of the third and fourth column of the state and thus the corresponding pairs are

Table 3. Possible attack outcomes for different number of bytes (**Bytes**) controlled by attacker. **Rnds** is the number of rounds that have to be attacked in order to recover the master key. **Prop. (%)** is the proportion of a given evaluation case with respect to all possible input configurations for a fixed number of bytes controlled by attacker.

Bytes	1	2	3	4	5	6	7	8	9	10	11	12	13	14	15	16
min(Rnds)	4	4	4	3	3	3	3	3	3	3	3	3	3	3	3	1
Prop. (%)	100	100	100	14.1	35.2	55.9	72.7	84.7	92.3	96.7	98.9	99.8	100	100	100	100
max(Rnds)	4	4	4	4	4	4	4	4	4	4	4	4	3	3	3	1
Prop. (%)	100	100	100	85.9	64.8	44.1	27.3	15.3	7.7	3.3	1.1	0.2	100	100	100	100
Trade-off	✗	✗	✗	✓	✓	✓	✓	✓	✓	✓	✓	✓	✗	✗	✗	✗

updated accordingly. In addition, new pairs are added when the attacker can not distinguish between key candidates as shown in Fig. 3. In the fourth round, the attacker is able to recover all round key bytes. Then, having all the round key bytes of the fourth round, she can reverse the AES key schedule to get the master key.

The attacker has to build 2^p possible round keys, where p is the number of independent pairs associated with the state bytes of the last attacked round. For the example in Fig. 3, the number of possible keys is 2^5 because $card(S) = card(S_6 \cup S_7) = card(\{1, 2, 3, 4, 5\}) = 5$. Thus, in addition to the number of rounds to attack, the algorithm for symbolic processing of an initial state gives the number of possible master keys to be recovered by an attacker. Though, the

Table 4. All evaluation cases with an example of a possible initial state for each evaluation case. **Bytes** gives the number of bytes controlled by attacker; **Rounds** gives the number of rounds that have to be attacked to recover the master key.

Case	Bytes	Rounds	Possible initial state
0	1	4	[1, 0, 0, 0, 0, 0, 0, 0, 0, 0, 0, 0, 0, 0, 0, 0]
1	2	4	[1, 1, 0, 0, 0, 0, 0, 0, 0, 0, 0, 0, 0, 0, 0, 0]
2	3	4	[1, 1, 1, 0, 0, 0, 0, 0, 0, 0, 0, 0, 0, 0, 0, 0]
3	4	3	[1, 1, 1, 1, 0, 0, 0, 0, 0, 0, 0, 0, 0, 0, 0, 0]
4	4	4	[1, 1, 1, 0, 0, 1, 0, 0, 0, 0, 0, 0, 0, 0, 0, 0]
5	5	3	[1, 1, 1, 1, 1, 0, 0, 0, 0, 0, 0, 0, 0, 0, 0, 0]
6	5	4	[1, 1, 1, 0, 0, 1, 1, 0, 0, 0, 0, 0, 0, 0, 0, 0]
7	6	3	[1, 1, 1, 1, 1, 1, 0, 0, 0, 0, 0, 0, 0, 0, 0, 0]
8	6	4	[1, 1, 1, 0, 0, 1, 1, 1, 0, 0, 0, 0, 0, 0, 0, 0]
9	7	3	[1, 1, 1, 1, 1, 1, 1, 0, 0, 0, 0, 0, 0, 0, 0, 0]
10	7	4	[1, 1, 1, 0, 0, 1, 1, 1, 1, 0, 0, 0, 0, 0, 0, 0]
11	8	3	[1, 1, 1, 1, 1, 1, 1, 1, 0, 0, 0, 0, 0, 0, 0, 0]
12	8	4	[1, 1, 1, 0, 0, 1, 1, 1, 1, 0, 1, 0, 0, 0, 0, 0]
13	9	3	[1, 1, 1, 1, 1, 1, 1, 1, 1, 0, 0, 0, 0, 0, 0, 0]
14	9	4	[1, 1, 1, 0, 0, 1, 1, 1, 1, 0, 1, 1, 0, 0, 0, 0]
15	10	3	[1, 1, 1, 1, 1, 1, 1, 1, 1, 1, 0, 0, 0, 0, 0, 0]
16	10	4	[1, 1, 1, 0, 0, 1, 1, 1, 1, 0, 1, 1, 1, 0, 0, 0]
17	11	3	[1, 1, 1, 1, 1, 1, 1, 1, 1, 1, 1, 0, 0, 0, 0, 0]
18	11	4	[1, 1, 1, 0, 0, 1, 1, 1, 1, 0, 1, 1, 1, 1, 0, 0]
19	12	3	[1, 1, 1, 1, 1, 1, 1, 1, 1, 1, 1, 1, 0, 0, 0, 0]
20	12	4	[1, 1, 1, 0, 0, 1, 1, 1, 1, 0, 1, 1, 1, 1, 0, 1]
21	13	3	[1, 1, 1, 1, 1, 1, 1, 1, 1, 1, 1, 1, 1, 0, 0, 0]
22	14	3	[1, 1, 1, 1, 1, 1, 1, 1, 1, 1, 1, 1, 1, 1, 0, 0]
23	15	3	[1, 1, 1, 1, 1, 1, 1, 1, 1, 1, 1, 1, 1, 1, 1, 0]
24	16	1	[1, 1, 1, 1, 1, 1, 1, 1, 1, 1, 1, 1, 1, 1, 1, 1]

attacker does not have to check all 2^p candidates to see which one is the correct one since she can discard the wrong candidates based on the difference between the correlation coefficients of the first two key candidates as we will show in Sect. 5.

Using the algorithm for symbolic processing of an initial state we evaluated all possible input combinations. More precisely, we considered all configurations of the initial state when the attacker controls i bytes of the input for $i \in [1, 16]$. When the attacker controls i bytes, there are $\binom{16}{i}$ possible input configurations. This results in $2^{16} - 1$ possible configurations of the initial state in total. Then, we classified these inputs into equivalence classes (evaluation cases) depending on the number of rounds that must be attacked in order to recover the master key. The results are summarized in Table 3. When the attacker controls between four and eleven bytes of the input, a trade-off between the input configuration and the number of rounds to be attacked is possible. When this is the case, the proportion of possible input configurations shows which evaluation case is more likely to appear if the initial state is chosen at random. Thus, when the attacker controls only four or five bytes of the input, it is crucial to carefully choose an input configuration from the limited set of possible input configurations that minimize the number of rounds to be attacked.

We give an example of a possible initial state for each of the 25 distinct evaluation cases identified after processing all possible input combinations in Table 4. Any possible input configuration for the AES encryption falls into one of these evaluation cases depending on the number of bytes controlled by attacker and the number of rounds that must be attacked in order to recover the master key.

5   The Attack

The attack we present in this section uses the symbolic representation of the AES state (described in Sect. 4) in conjunction with CPA attacks to recover individual bytes of the AES round keys. After executing Algorithm 1, the attacker has all round key bytes of round R. Thus, she is able to recover the master key of the cipher by reversing the key schedule.

The algorithm follows the symbolic representation of the state to infer which key bytes must be attacked and how many key candidates it should yield for each attacked key byte. By tracking the pairs associated with the recovered key bytes, the algorithm is able to discard all impossible round keys, thus saving computational resources. Indeed, the algorithm uses an optimal number of CPA attacks to recover the master key.

Initially, the set of known pairs is empty and all possible keys are considered valid. The algorithm keeps track of 2^p possible keys, where p is the total number of independent pairs in the symbolic representation of the state at round R.

The main loop of the algorithm runs through all rounds that must be attacked. At each round, the key bytes corresponding to variable state bytes are attacked to recover one or more temporary key bytes or a round key byte. Depending on the pairs associated with the byte to be attacked, there are three possible cases:

Algorithm 1. The attack

Require: *state* ▷ Initial state: 0 – fixed byte, 1 – variable byte
Require: $\lambda = (plaintexts, traces)$ ▷ Recorded in the acquisition phase
1: $state, pairs = \text{PROCESS}(state)$ ▷ Symbolic processing (Sect. 4)
2: $known\_pairs = \emptyset, \quad mapped\_pairs = \emptyset$
3: $keys[2^p] = \emptyset, \quad valid\_keys[2^p] = True$ ▷ p is the number of independent pairs
4: **for** $r = 1$ to R **do** ▷ R is the number of rounds to be attacked
5:   **for** $i = 0$ to 15 **do**
6:     **if** $state[r][i] \neq 0$ **then**
7:       **if** $pairs[r][i] == \emptyset$ **then** ▷ No pair
8:         $keys[0, \cdots, 2^p - 1][r][i] = \text{CPA}(\lambda, keys[0], r, i)$
9:       **else if** $pairs[r][i] \subseteq known\_pairs$ **then** ▷ Known pair(s)
10:        **if** $i \notin mapped\_pairs[pairs[r][i]]$ **then**
11:          $mask = 0, \quad temp\_keys = \emptyset, \quad \alpha_{max} = -1$
12:          **for** $pair \in pairs[r][i]$ **do**
13:            $mask = mask \vee 2^{pair-1}$
14:          **end for**
15:          **for** $j \in [0, 2^p - 1]$ **do**
16:            **if** $valid\_keys[j]$ and $temp\_keys[j \wedge mask] == \emptyset$ **then**
17:              $temp\_keys[j \wedge mask], \alpha = \text{CPA}(\lambda, keys[j], r, i)$
18:              **if** $\alpha > \alpha_{max}$ **then**
19:                $\alpha_{max} = \alpha$
20:              **end if**
21:            **end if**
22:            $valid\_keys[j][r][i] = temp\_keys[j \wedge mask]$
23:          **end for**
24:          **for** $j \in [0, 2^p - 1]$ **do**
25:            **if** $abs(state[r][i]) == 1$ and $\alpha + \beta < \alpha_{max}$ **then**
26:              $valid\_keys[j] = False$
27:            **end if**
28:          **end for**
29:        **end if**
30:      **else** ▷ New pair
31:        $mask = 2^{pairs[r][i]-new\_pair}, \quad k_1 = k_2 = \emptyset$
32:        **for** $j \in [0, 2^p - 1]$ **do**
33:          **if** $k_1[j \wedge mask] == \emptyset$ **then**
34:            $k_1[j \wedge mask], k_2[j \wedge mask] = \text{CPA}(\lambda, keys[j], r, i)$
35:          **end if**
36:          **if** $j \wedge 2^{new\_pair-1}$ **then**
37:            $keys[j][r][i] = k_1[j \wedge mask], \quad keys[j][r][i'] = k_2[j \wedge mask]$
38:          **else**
39:            $keys[j][r][i'] = k_2[j \wedge mask], \quad keys[j][r][i'] = k_1[j \wedge mask]$
40:          **end if**
41:        **end for**
42:        $known\_pairs = known\_pairs \cup new\_pair$
43:        Add (i, i') to $mapped\_pairs[new\_pair]$
44:      **end if**
45:    **end if**
46: **end for**
47: **end for**
48: **return** $keys[i]$, where $valid\_keys[i] == True$ for $i \in [0, 2^p - 1]$

- **No pair**: If the symbolic representation does not have a pair associated with the byte of the state to be used for the attack, then the algorithm will recover a single key byte which is distributed to all possible keys.
- **New pair**: If one of the pairs associated with the byte under attack is not present in the set of known pairs, then the algorithm will recover 2^u possible values for the corresponding key byte, where u is the number of known independent pairs associated with the byte under attack. The number of known pairs determines the number of CPA attacks to be performed. Using a mask based on the existing pairs and a mask for the new pair, the algorithm correctly distributes the recovered key byte values to all possible keys. The new pair is added to the set of known pairs and the two indexes of the state affected by the recovered temporary keys are mapped to this new pair. This mapping prevents the computation of the same temporary keys twice.
- **Known pairs(s)**: In the case where the t independent pairs associated with the key byte to be attacked are known but not mapped to the current state byte, the algorithm performs 2^t CPA attacks. Then, it distributes the attack results (the recovered key and the difference between the correlation coefficients of the first two most likely key candidates α) to the corresponding bytes of all possible keys. Afterwards, the possible keys for which the value of α is less than the maximum observed value α_{max} minus a threshold β are marked as invalid. In this way, only the combination of keys yielding the highest correlation peak is selected. At this moment, the input pairs are solved in the sense that the algorithm can uniquely assign each of the two temporary keys of a pair to the corresponding state bytes. As a consequence, the algorithm will not further process the possible keys marked as invalid. Thus, this optimization improves the algorithm efficiency by reducing the number of performed CPA attacks.

Finally, the algorithm returns all possible keys which are marked as valid. If the threshold β tends to zero, the algorithm will return only one possible key. When the quality of the side-channel acquisition is good (i.e. high signal-to-noise ratio) and there are enough power traces, the algorithm yields the correct key.

5.1   Optimality

We prove that our algorithm uses the minimum number of CPA attacks possible to recover the master key and thus is optimal. Hence, the lower bounds provided in Table 5 are optimal.

Theorem 1. *Algorithm 1 performs an optimal number of CPA attacks to recover the 16-byte master key of the AES.*

Proof. The only way an attacker can recover the 16-byte master key of the AES is to recover all key bytes of a round r and then to reverse the key schedule. Since the function that derives the round keys of round i from the round keys of round $i - 1$ is bijective, knowledge of all round key bytes of a round r leads to the knowledge of the master key.

Let us assume that Algorithm 1 uses n individual CPA attacks for a given initial state and it is not optimal. Thus, there exists at least an algorithm that is able to recover the master key using only m CPA attacks, with $m < n$. We show next that such an algorithm does not exist. If there exists an algorithm that uses less CPA attacks than Algorithm 1, then this algorithm attacks at least one key byte less. But if it does so, then the attack can not be carried into later rounds any more because the state byte corresponding to the unrecovered key yields unknown and variable state bytes after MixColumns transformation. These bytes can not be recovered using a CPA attack and thus the attack fails. As a consequence, there is no algorithm that uses less CPA attacks than Algorithm 1.   □

5.2   Choosing the Best Attack Strategy

For up to seven bytes controlled by attacker, our attack algorithm (Algorithm 1) is more efficient than the classic attack algorithm where all possible key bytes are attacked to recover the master key. The gain varies between 15% and 68% of the number of CPA attacks required by the classic attack. When an attacker has control of more than seven input bytes, our algorithm performs the same number of CPA attacks as the classic attack. At the same time, our algorithm gives a unique master key, provided that there are available enough traces with a high signal-to-noise ratio. This is not the case for a classic attack unless an additional mechanism to discard invalid keys, as the one in Algorithm 1, is employed.

An attacker willing to reduce the duration of the offline phase of the attack (without increasing the number of rounds that must be attacked) can use the results in Table 5 in corroboration with the data in Table 3 to adjust the attack accordingly. More precisely, if an attacker is able to control up to n bytes of the AES input, she can choose to control m ($m \leq n$) bytes of the input because m variable bytes minimize the number of CPA attacks required to recover the master key. This decision has to be made before performing the side-channel acquisition since it influences the chosen inputs. Another argument in favor of

Table 5. The number of individual CPA attacks required to recover the master key for different number of bytes (**Bytes**) controlled by attacker; **min(Rnds)/max(Rnds)** and **Bytes** precisely identify the evaluation case. **Classic attack** does not use the optimizations introduced in **Algorithm 1** to discard the invalid keys. **Gain** gives the number of CPA attacks saved by an attacker using **Algorithm 1** over an attacker using **Classic attack**.

Bytes		1	2	3	4	5	6	7	8	9	10	11	12	13	14	15	16
min(Rnds)	Clasic attack	150	104	188	80	66	52	46	40	41	42	43	44	45	46	47	16
	Algorithm 1	48	42	48	38	38	38	39	40	41	42	43	44	45	46	47	16
	Gain	102	62	140	42	28	14	7	0	0	0	0	0	0	0	0	0
max(Rnds)	Classic attack	150	104	188	110	72	50	51	52	53	54	55	56	45	46	47	16
	Algorithm 1	48	42	48	48	49	50	51	52	53	54	55	56	45	46	47	16
	Gain	102	62	140	62	23	0	0	0	0	0	0	0	0	0	0	0

using less variable input bytes is that the attack is much more difficult to detect if the injected packets have fewer variable bytes and mimic the appearance of a normal network traffic. For example, when $n = 12$, an attacker can choose $m = 4, 5,$ or 6 to reduce the complexity of the offline attack from 44 to 38 individual CPA attacks, while still attacking just three rounds. The result is an overall improvement of the attack efficiency by 14% over the classic attack.

An even better decision can be made with the help of experimental results for different configurations of the input from a similar target to the one to be attacked in addition to the results presented so far. For this reason, in the next section we determine experimentally the number of traces required to recover the master key for each evaluation case using EM leakages from an ARM Cortex-M3 processor.

6   Results

For the experimental evaluation, we considered two unprotected implementations of the AES written in ANSI C. The first implementation uses the S-box implementation strategy, while the second one uses the T-table implementation strategy. For each of the 25 evaluation cases we measured up to 2000 EM traces. The acquisition took about 90 min for an evaluation case. The samples were split into files corresponding to the AES round number. Then, we mounted the attack presented in Algorithm 1 using an increasing number of traces in the interval [100, 2000] with a step of 100 traces until the guessing entropy converged to zero.

For each implementation we considered two selection functions based on the AES S-box and T-table, respectively. The minimum number of traces for which the guessing entropy becomes zero and remains stable is pictorially shown in Fig. 4 for each evaluation case. All attacks recovered the full 16-byte master key with less than 1600 EM traces. In general, the master key was recovered with fewer traces when the selection function perfectly matched the implementation strategy. Though, our results show that full key recovery is possible even when the selection function does not perfectly match the attacked implementation. The attacks on the S-box implementation using the T-table selection function needed 204 more traces on average to recover the master key compared to the attacks on the same implementation using the S-box selection function. Similarly, using the S-box selection function instead of the T-table selection function to attack the implementation based on the T-table strategy required 354 more traces on average. For details on the exact number of traces required to recover the master key for each evaluation case and attack scenario we refer the reader to the full version of this paper.

Countermeasures. Our experimental results show that side-channel countermeasures such as masking must be employed in order to protect the AES implementations based on lookup tables (S-box and T-table implementation strategies) even in a communication protocol scenario, when the adversary has

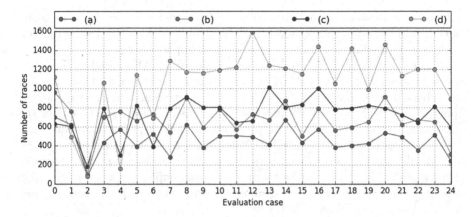

Fig. 4. The number of EM traces required to fully recover the master key. Scenarios: **(a)** S-box implementation, S-box selection function; **(b)** S-box implementation, T-table selection function; **(c)** T-table implementation, T-table selection function; **(d)** T-table implementation, S-box selection function.

a limited control of the input. Masking non-linear lookup tables is a challenging task since it adds a considerable penalty on execution time and memory usage [33].

Although not present in many cryptographic libraries due to their limitations (i.e. can not be used in a feedback mode of operation such as CCM), the bitsliced implementations have a lower CPA leakage than implementations using lookup tables [4], but they are still vulnerable to DPA attacks [3].

A lightweight primitive (block cipher or authenticated encryption), particularly one designed for efficient masking, is a good replacement for the AES-CCM when considering side-channel protection.

Other countermeasures, such as a key refreshing mechanism, can support a defense in depth approach. However, any additional countermeasure affects the overall efficiency of an IoT protocol and consequently the most effective ones (i.e. masking) must have priority given the resource constraints.

7   Conclusions

In this paper, we presented an extensive security analysis of AES software implementations against CPA attacks in the context of network protocols. In this scenario the attacker has control of several input bytes, while the remaining input bytes are fixed. To asses the resilience of AES implementations to all possible input combinations, we presented an algorithm for symbolic processing of the cipher state. Then, we classified all possible inputs into 25 independent evaluation cases depending on the number of input bytes controlled by attacker and the number of rounds that must be attacked to recover the master key. Finally, we described an optimal algorithm that recovers the master key by mounting the minimum number of CPA attacks possible. It makes clever decisions based on the set of key pairs that affects the key byte under attack and the correlation coefficient of possible key candidates to discard impossible keys.

We showed that unprotected implementations of the AES based on the S-box and T-table strategies can be broken even when the attacker controls only one input byte of the cipher with less than 1600 electromagnetic traces acquired from a 32-bit ARM Cortex-M3 processor in about one hour. Knowledge of the implementation strategy does not significantly improve the attack outcome, nor does it reduce the attack complexity. Thus, unprotected implementations of the AES should not be used to secure the communication between end devices in network protocols. Care must be taken when using implementations of the AES from popular open-source cryptographic libraries since most of them are not protected against side-channel attacks.

Acknowledgements. We would like to thank the ACNS 2017 reviewers for their valuable feedback and Johann Großschädl for proofreading the final version of this paper. The work of Daniel Dinu is supported by the CORE project ACRYPT (ID C12-15-4009992) funded by the Fonds National de la Recherche (FNR) Luxembourg.

References

1. Agrawal, D., Archambeault, B., Rao, J.R., Rohatgi, P.: The EM side-channel(s). In: Kaliski, B.S., Koç, K., Paar, C. (eds.) CHES 2002. LNCS, vol. 2523, pp. 29–45. Springer, Heidelberg (2003). doi:10.1007/3-540-36400-5_4
2. ARM. mbed TLS. https://tls.mbed.org/. Accessed Apr 2017
3. Balasch, J., Gierlichs, B., Reparaz, O., Verbauwhede, I.: DPA, bitslicing and masking at 1 GHz. In: Güneysu, T., Handschuh, H. (eds.) CHES 2015. LNCS, vol. 9293, pp. 599–619. Springer, Heidelberg (2015). doi:10.1007/978-3-662-48324-4_30
4. Biryukov, A., Dinu, D., Großschädl, J.: Correlation power analysis of lightweight block ciphers: from theory to practice. In: Manulis, M., Sadeghi, A.-R., Schneider, S. (eds.) ACNS 2016. LNCS, vol. 9696, pp. 537–557. Springer, Cham (2016). doi:10.1007/978-3-319-39555-5_29
5. Brier, E., Clavier, C., Olivier, F.: Correlation power analysis with a leakage model. In: Joye, M., Quisquater, J.-J. (eds.) CHES 2004. LNCS, vol. 3156, pp. 16–29. Springer, Heidelberg (2004). doi:10.1007/978-3-540-28632-5_2
6. cryptlib. The cryptlib Security Software Development Toolkit. http://www.cryptlib.com/. Accessed Apr 2017
7. Crypto++. Crypto++: a free C++ class library of cryptographic schemes. https://www.cryptopp.com/. Accessed Apr 2017
8. Daemen, J., Rijmen, V.: The Design of Rijndael: AES - The Advanced Encryption Standard. Springer, Heidelberg (2002)
9. Dworkin, M.J.: Recommendation for block cipher modes of operation: the CCM mode for authentication and confidentiality. NIST Special Publication 800-38C (2007)
10. GitHub. libtomcrypt: a fairly comprehensive, modular and portable cryptographic toolkit. https://github.com/libtom/libtomcrypt. Accessed Apr 2017
11. GitHub. mbed TLS - An open source, portable, easy to use, readable and flexible SSL library. https://github.com/ARMmbed/mbedtls/blob/development/library/aes.c. Accessed Apr 2017

12. GitHub. OpenSSL - TLS/SSL and crypto library. https://github.com/openssl/openssl/blob/master/crypto/aes/aes_core.c. Accessed Apr 2017
13. Hofemeier, G., Chesebrough, R.: Introduction to intel AES-NI and intel secure key instructions. Technical report. https://software.intel.com/sites/default/files/m/d/4/1/d/8/Introduction_to_Intel_Secure_Key_Instructions.pdf. Accessed Apr 2017
14. Housley, R.: Using Advanced Encryption Standard (AES) CCM Mode with IPsec Encapsulating Security Payload (ESP). RFC 4309, December 2005. https://tools.ietf.org/html/rfc4309
15. IEEE. IEEE Standard for Low-Rate Wireless Networks. https://standards.ieee.org/about/get/802/802.15.html
16. Jaffe, J.: A first-order DPA attack against AES in counter mode with unknown initial counter. In: Paillier, P., Verbauwhede, I. (eds.) CHES 2007. LNCS, vol. 4727, pp. 1–13. Springer, Heidelberg (2007). doi:10.1007/978-3-540-74735-2_1
17. Kocher, P., Jaffe, J., Jun, B.: Differential power analysis. In: Wiener, M. (ed.) CRYPTO 1999. LNCS, vol. 1666, pp. 388–397. Springer, Heidelberg (1999). doi:10.1007/3-540-48405-1_25
18. Libgcrypt. Libgcrypt: a general purpose cryptographic library based on the code from GnuPG. https://www.gnu.org/software/libgcrypt/. Accessed Apr 2017
19. libsodium. The Sodium crypto library (libsodium). https://download.libsodium.org/doc/. Accessed Apr 2017
20. Lipp, M., Gruss, D., Spreitzer, R., Maurice, C., Mangard, S.: Armageddon: cache attacks on mobile devices. In: Holz, T., Savage, S. (eds.) 25th USENIX Security Symposium, USENIX Security 16, Austin, TX, USA, 10–12 August 2016, pp. 549–564. USENIX Association (2016)
21. LoRa Alliance. Wide Area Networks for IoT. https://www.lora-alliance.org/. Accessed Apr 2017
22. Nettle. Nettle - a low-level cryptographic library. http://www.lysator.liu.se/nisse/nettle/. Accessed Apr 2017
23. NIST. Specification for the Advanced Encryption Standard (AES). Federal Information Processing Standards Publication 197 (2001)
24. O'Flynn, C., Chen, Z.: Power Analysis Attacks Against IEEE 802.15.4 Nodes. In: Standaert, F.-X., Oswald, E. (eds.) COSADE 2016. LNCS, vol. 9689, pp. 55–70. Springer, Cham (2016). doi:10.1007/978-3-319-43283-0_4
25. OpenSSL. Cryptography and SSL/TLS Toolkit. https://www.openssl.org/. Accessed Apr 2017
26. Osvik, D.A., Shamir, A., Tromer, E.: Cache attacks and countermeasures: the case of AES. In: Pointcheval, D. (ed.) CT-RSA 2006. LNCS, vol. 3860, pp. 1–20. Springer, Heidelberg (2006). doi:10.1007/11605805_1
27. Randombit. mbed TLS. https://botan.randombit.net/. Accessed Apr 2017
28. Saab, S., Rohatgi, P., Hampel, C.: Side-channel protections for cryptographic instruction set extensions. Cryptology ePrint Archive, Report 2016/700 (2016). http://eprint.iacr.org/2016/700
29. Sastry, N., Wagner, D.: Security considerations for IEEE 802.15.4 networks. In: Jakobsson, M., Perrig, A. (eds.) Proceedings of the 2004 ACM Workshop on Wireless Security, Philadelphia, PA, USA, 1 October 2004, pp. 32–42. ACM (2004)
30. Schwabe, P., Stoffelen, K.: All the AES you need on Cortex-M3 and M4. In: Selected Areas in Cryptography-SAC (2016)
31. Song, J., Poovendran, R., Lee, J., Iwata, T.: The AES-CMAC algorithm. RFC 4493, June 2006. https://tools.ietf.org/html/rfc4493
32. STMicroelectronics. STM32 MCU Nucleo. http://www.st.com/en/evaluation-tools/stm32-mcu-nucleo.html. Accessed Apr 2017

33. Vadnala, P.K.: Time-memory trade-offs for side-channel resistant implementations of block ciphers. In: Handschuh, H. (ed.) CT-RSA 2017. LNCS, vol. 10159, pp. 115–130. Springer, Cham (2017). doi:10.1007/978-3-319-52153-4_7
34. Whiting, D., Housley, R., and N. Ferguson. Counter with CBC-MAC (CCM). RFC 3610, September 2003. https://tools.ietf.org/html/rfc3610
35. wolfSSL. wolfCrypt Embedded Crypto Engine. https://www.wolfssl.com/wolfSSL/Products-wolfcrypt.html. Accessed Apr 2017

Cryptographic Protocol

Lattice-Based DAPS and Generalizations: Self-enforcement in Signature Schemes

Dan Boneh[✉], Sam Kim, and Valeria Nikolaenko

Stanford University, Stanford, USA
dabo@cs.stanford.edu

Abstract. Double authentication preventing signatures (DAPS) is a mechanism, due to Poettering and Stebila, for protecting certificate authorities (CAs) from coercion. We construct the first lattice-based DAPS signatures, thereby providing the first post-quantum DAPS system. We go further and generalize DAPS to a more general mechanism we call *predicate* authentication preventing signatures (PAPS). Here, for a given k-ary predicate ϕ, a PAPS system for ϕ is regular signature scheme. However, if the signer ever signs k messages $m_1, \ldots, m_k$ such that $\phi(m_1, \ldots, m_k)$ is true then these k signatures reveal the signer's secret key. This self-enforcement mechanism incentivizes the signer to never sign conflicting messages, namely messages that satisfy the predicate ϕ. The k conflicting messages can be signed at different times and the signatures may be generated independently of one another. We further generalize to the case when the signatures are generated by multiple signers. We motivate these primitives, give precise definitions, and provide several constructions. These primitives are challenging to construct and give rise to many new elegant open research questions.

1 Introduction

Suppose a web site such as facebook.com buys certificates for its domain from a certificate authority (CA) called XYZ. These certificates enable web browsers to establish a (one-sided) authenticated session with facebook.com. Sometime later, a law enforcement agency or a nation state that has jurisdiction over the CA compels XYZ to secretly issue a fresh certificate for facebook.com. The CA has no choice but to comply. The agency can then use this issued certificate in a man-in-the-middle attack on facebook.com. Web users have no way to detect that this is happening and that their traffic is being intercepted. We emphasize that the rogue certificate is issued by the same CA from which facebook.com normally buys its certificates. The only user-side signal is that a previously unseen public-key is being served in a facebook.com certificate, but this happens frequently under normal operation at a large site and would not generally look suspicious.

Some technologies, such as certificate transparency (CT) [LLK15] as well as CONIKS [MBB+15], are designed to detect situations where a CA such as XYZ issues a fake certificate for a domain. These technologies empower an origin domain, facebook.com in this case, to detect that a fake certificate was issued for its domain.

© Springer International Publishing AG 2017
D. Gollmann et al. (Eds.): ACNS 2017, LNCS 10355, pp. 457–477, 2017.
DOI: 10.1007/978-3-319-61204-1_23

Poettering and Stebila [PS14] proposed a very different defense against the scenario described above. Their idea, called double-authentication-preventing signatures, or DAPS for short, is as follows: suppose XYZ signs all its certificates using a signature scheme where the signing algorithm uses the secret signing key sk to sign a pair (subj, payload). Here subj is the domain-name to which the certificate is issued and payload is all other fields in the certificate. The resulting signature σ can be verified as a standard digital signature. The key property of DAPS is the following: suppose XYZ publishes two valid signatures σ_1 and σ_2 *for the same* subj *but for different payloads*, say one on (subj, payload$_1$) and another on (subj, payload$_2$). Then these two signatures enable anyone to expose XYZ's secret signing key sk. The point is that XYZ can argue that it should not be forced to issue the rogue certificate for facebook.com because that would expose its signing key thereby causing massive collateral damage to *all* of XYZ's customers. Whether this argument is effective remains to be seen, but the idea itself is interesting and, as we show below, leads to interesting and challenging cryptographic questions.

How to Use DAPS. There are many practical holes in the basic DAPS proposal described above that prevent it from being used as is, but with a bit of thought they can be addressed. However, our goal here is not to argue that DAPS will be deployed in practice, but rather to motivate this as an interesting cryptographic question. Towards this goal, we examine broader applications of DAPS as well as an elegant generalization.

Other Applications for DAPS. Beyond certificates, DAPS can be a useful "self-enforcement" security mechanism. For example, suppose Eve owns a certain patent and wants to sell the rights to the patent. Bob wants to buy the patent from Eve, but he is worried that Eve will sell the patent to multiple people. Using DAPS, Eve can use the patent number as the subject and use "owned by Bob" as the payload. If she tries to sell the same patent to two different people she will end up signing two pairs of messages with the same subject, but different payload. The resulting two signatures can be combined to expose Eve's private key. If this private key is of high value to Eve then this self-enforcement mechanism will prevent her from double-selling the same patent to two people. This way, Bob has some confidence in the exclusivity of the deal with Eve.

PAPS: DAPS for General Predicates. The previous paragraph motivates a more general elegant primitive which we call *predicate-authentication-preventing signatures*, or PAPS for short. Let $\mathcal{M}$ be a message space and let $\phi : \mathcal{M}^k \to \{0, 1\}$ be a predicate. A PAPS scheme for ϕ lets the signer sign any message $m \in \mathcal{M}$, just as in a regular signature scheme. However, if over the life of the secret key, the signer signs messages $m_1, \ldots, m_k \in \mathcal{M}$ such that $\phi(m_1, \ldots, m_k) = 1$ then these k signatures can be combined to expose the signer's secret signing key. This secret key extraction should work no matter how the k signatures are generated: as long as all k signatures are valid, it should be possible to extract the secret key. For security, as long as the predicate ϕ is never satisfied, the signature scheme should be existentially unforgeable under a chosen message attack, as for regular signatures.

Notice that DAPS is a special case of PAPS: the key space $\mathcal{M}$ is a set of pairs $\mathcal{M} = \mathcal{S} \times \mathcal{P}$ and the predicate ϕ_{DAPS} is simply the 2-ary predicate:

$$\phi_{\text{DAPS}}\big(\ (x_1, y_1),\ (x_2, y_2)\ \big) = 1 \quad \Leftrightarrow \quad x_1 = x_2 \text{ and } y_1 \neq y_2.$$

More general predicates come up naturally. For example, suppose a web site owns k machines and it wants to generate a different key-pair for each machine, necessitating a different certificate for each machine. The analogue of DAPS is a k-way DAPS where the message space $\mathcal{M}$ is again $\mathcal{M} = \mathcal{S} \times \mathcal{P}$, but now the predicate ϕ is the $(k+1)$-ary predicate

$$\phi\big(\ (x_1, y_1), \dots, (x_{k+1}, y_{k+1})\ \big) = 1 \quad \Leftrightarrow \quad \left\{ \begin{array}{l} x_1 = \cdots = x_{k+1} \text{ and} \\ y_1, \dots, y_{k+1} \text{ are all distinct.} \end{array} \right\}$$

This lets the site use a different certificate for each of its k machines, but if another certificate is issued then the CA's secret key is exposed. We give a construction for this predicate in Sect. 5 as well as for several other predicates.

Proving security of a PAPS construction is non trivial. For example, suppose the message space is $\mathcal{M} = \mathbb{F}_p$ for some prime p. Consider the 3-ary predicate ϕ defined as $\phi(x, y, z) = 1$ if and only if $x + y + z = 0$. That is, the secret key should leak if the signer ever signs three messages whose sum is zero. However, if the signer never signs three messages satisfying this condition, the signature scheme should be existentially unforgeable. To prove existential unforgeability, the simulator must interact with the adversary, answering all the adversary's adaptive signature queries, and using the adversary's existential forgery to solve a challenge problem. The problem is that, because the adversary's first two queries can be for arbitrary messages, the simulator must be prepared to provide a signature for *all* messages $m \in \mathcal{M}$. In particular, the simulator will know the signature on three messages x, y, z satisfying $x + y + z = 0$. But then the simulator can extract the secret signing key, and can produce any forgery by itself, meaning that the adversary is not helping the simulator. Nevertheless, in Sect. 5 we are able to prove security for several generalized predicates, though not for the 3-way summation predicate.

A Further Generalization: Multi-signer PAPS. We can further generalize the notion of PAPS to the setting of k signers where each signer has its own signing/public key-pair. As before, let ϕ be a k-ary predicate $\phi : \mathcal{M}^k \to \{0, 1\}$. Suppose that for $i = 1, \dots, k$ signer number i signs message $m_i \in \mathcal{M}$. Then, if $\phi(m_1, \dots, m_j) = 1$, then these k signatures (along with the k messages and k public keys) can be used to expose some secret s chosen at setup time.

Multi-signer PAPS come up naturally when considering certificates. Suppose that the agency, instead of asking XYZ to issue the rogue certificate for facebook.com, it asks a different CA to provide a certificate for facebook.com. We are now in a 2-signer scenario. If the predicate ϕ_{DAPS} can be made to work on the two signatures, despite them being from different CAs, then going to a different CA will not help the agency. We define multi-signer PAPS in Sect. 6 where we also give several constructions.

1.1   Contributions

In this work we build a lattice-based DAPS construction based on Short Integer Solutions (SIS) problem and the Learning with Errors (LWE) problem. Our construction builds upon the structure of the fully homomorphic signature scheme of Gorbunov et al. [GVW15]. In their construction, a signature consists of a preimage of a specially formed target matrix of a lattice trapdoor function. To make key leakage a feature rather than a form of insecurity, we carefully hash the messages to derive the target matrix such that two different message of the same subject leads to two matrices for which two preimage matrices leak a trapdoor. As in [PS14], we prove security in the random oracle model.

Also, as we discuss above, we extend DAPS to a more general primitive that we call predicate authentication preventing signatures (PAPS). In this setting, signatures of any messages that satisfy a certain predicate defined on these messages leak a signer's secret key. To motivate the notion, we show that for certain simple, but useful predicates, PAPS can already be constructed from DAPS.

Finally, we further extend PAPS to a multi-authority settings where signatures from different signers can also leak some shared private information. We give formal definitions in this setting and show that our lattice DAPS construction can be extended to this setting as well using the property that two short preimages of a specifically formed target matrix of lattice trapdoor functions can be merged to give a trapdoor of an extended lattice.

1.2   Related Work

Previous Works on DAPS. The notion of double-authentication-preventing signatures was introduced by Poettering and Stabila [PS14]. They provide a construction based on extractable trapdoor functions that can be constructed using the group of quadratic residues modulo a Blum integer. Subsequently, Bellare, Poettering, and Stebila [BPS15] gave a generic construction based on trapdoor identification schemes where the private randomness committed by the prover can be extracted using a trapdoor. We note that although lattice-based identification schemes have appeared in the literature [Lyu08,Lyu12], the construction from [PS14] does not directly give a lattice-based DAPS construction since lattice trapdoors are randomized with multiple preimages. Constructing DAPS from lattice-based assumptions is an interesting and important goal since they provide hardness even against quantum computers, a setting for which the previous two works do not provide security.

Delegating Restricted Signing Keys. A number of works in the literature have focused on schemes that allow an authority to delegate signing keys with some restricted functionalities (with function privacy). These include attribute-based signatures [MPR11], functional signatures [BGI14] and policy-based signature [BF14]. In these schemes, a signer is restricted to sign only certain messages that satisfy a predicate requirement. One difference between these notions and DAPS/PAPS is that in the former, the restriction is done by a central authority to restrict other signers, while in the latter, the authority restricts itself as a self-enforcement mechanism. Another major difference is that in the former, the restriction is determined with respect to each individual message while in the latter, the restriction is determined by all of the past messages that the signer signs, which is what makes DAPS and PAPS an interesting theoretical notion.

Ring/Group Signatures. A similar notion of double-signing preventing mechanism exists in the setting of group signatures [TX03] and ring signatures [RST01, BKM06] called Revocable-iff-Linked (RiffL) signatures [ALSY06]. In this setting, a signer can sign on behalf of a group; however, if it signs twice or more, then the identity of the signer is leaked. As in the discussion of the previous paragraph, one difference in the DAPS setting compared to RiffL is that DAPS is a self-enforcing mechanism, which means that the linkability is not enforced by another trusted authority of the system, but by itself. However, the more fundamental difference is that in DAPS, the act of double-signing immediately gives away the signing key or private data rather than simply leaking the information that it double signed. As was discussed in [PS14], there are instances where simply leaking the fact that a CA double signed may not be enough of a penalty (i.e. the 2011 Comodo incident[1]) and DAPS is designed to cope with these type of situations.

2   Preliminaries

Basic Notation. For an integer N, we write $[N]$ to denote the set $\{1, ..., N\}$. We use bold lowercase letters (e.g., $\mathbf{x}, \mathbf{w}$) to denote vectors and bold uppercase letters (e.g., $\mathbf{A}, \mathbf{G}$) to denote matrices. For a matrix $\mathbf{A}$, we use $\mathbf{A}^T$ to denote the trasnpose of $\mathbf{A}$ and for a vector $\mathbf{x}$, we use $\|\mathbf{x}\|$ to denote its Euclidean norm. In general, we write λ for the security parameters. We say a function $\epsilon(\lambda)$ is negligible in λ, if $\epsilon(\lambda) = o(1/\lambda^c)$ for every $c \in \mathbb{N}$, and we write $\mathsf{negl}(\lambda)$ to denote a negligible function in λ. We say that an event occurs with *negligible probability* if the probabilty of the event is $\mathsf{negl}(\lambda)$, and an event occurs with *overwhelming probability* if its complement occurs with negligible probability.

Entropy and Statistical Distance. The *statistical distance* between two random variables X and Y over a finite domain Ω is defined as

$$\mathsf{SD}(X, Y) = \frac{1}{2} \sum_{\omega \in \Omega} |\mathrm{Pr}[X = \omega] - \mathrm{Pr}[Y = \omega]|.$$

We say that two distribution ensembles $X = \{X_\lambda\}_{\lambda \in \mathbb{N}}$ and $Y = \{Y_\lambda\}_{\lambda \in \mathbb{N}}$ are *statistically indistinguishable*, denoted $\overset{\text{stat}}{\approx}$, if it holds that $\mathsf{SD}(X_\lambda, Y_\lambda)$ is negligible in λ. The *min-entropy* of a random variable X, denoted $\mathbf{H}_\infty(X|Y)$, is defined as

$$\mathbf{H}_\infty(X|Y) \overset{\text{def}}{=} -\log \left(\underset{y \leftarrow Y}{\mathbf{E}} \left[\max_x \mathrm{Pr}[X = x|Y = y] \right] \right)$$

---

[1] https://www.comodo.com/Comodo-Fraud-Incident-2011-03-23.html.

The optimal probability of an unbounded adversary guessing X given the correlated value Y is $2^{-\mathbf{H}_\infty(X|Y)}$.

2.1 Circular Security

In this section we briefly recall the notion of circular security. A public key encryption (PKE) consists of three algorithms $\Pi_{\text{pke}} = (\text{PKE.KeyGen}, \text{PKE.Encrypt}, \text{PKE.Decrypt})$ where PKE.KeyGen takes in a unary representation of the security parameter λ and outputs a public and secret key pair (pk, sk). The encryption algorithm takes in a public key pk and a message m and generates a ciphertext ct. The decryption algorithm takes in a secret key sk and a ciphertext ct and outputs a message m. For correctness, we require that for all $\lambda \in \mathbb{N}$, $(\text{pk}, \text{sk}) \leftarrow \text{PKE.KeyGen}(1^\lambda)$, we have that $\text{PKE.Decrypt}(\text{sk}, \text{PKE.KeyGen}(\text{pk}, m)) = m$ with overwhelming probability.

Definition 1 (Circular Security [CL01,BRS02]). *A public-key encryption scheme Π_{pke} is circular secure if for all efficient adversaries $\mathcal{A}$, there is a negligible function $\text{negl}(\lambda)$ such that*

$$\text{Adv}^{\text{circ}}_{\Pi_{\text{pke}},\mathcal{A}}(\lambda) \overset{\text{def}}{=} \left| \Pr[\text{Expt}^{(0)}_{\text{PKE},\mathcal{A}}(\lambda) = 1] - \Pr[\text{Expt}^{(1)}_{\text{PKE},\mathcal{A}}(\lambda) = 1] \right| \leq \text{negl}(\lambda),$$

where for each $b \in \{0,1\}$, and $\lambda \in \mathbb{N}$, the experient $\text{Expt}^{(b)}_{\text{PKE},\mathcal{A}}(\lambda)$ is defined as follows:

1. $(\text{pk}, \text{sk}) \leftarrow \text{PKE.KeyGen}(1^\lambda)$.
2. $\text{ct}_0 \leftarrow \text{PKE.Encrypt}(\text{pk}, 0^{|\text{sk}|})$.
3. $\text{ct}_1 \leftarrow \text{PKE.Encrypt}(\text{pk}, \text{sk})$.
4. $b' \leftarrow \mathcal{A}(\text{pk}, \text{ct}_b)$.
5. *Output* $b' \in \{0,1\}$.

Circular security assumption on lattice-based cryptosystems have been used extensively throughout the literature [Gen09,BV11,BV14,BGV12,GSW13] with some positive results showing that some common forms of lattice-based encryption schemes can be shown to be circular secure [ACPS09,ASP12].

2.2 Broadcast Encryption

A broadcast encryption scheme BE [FN93] consists of a tuple of algorithms $\Pi_{\text{be}} = (\text{BE.KeyGen}, \text{BE.Encrypt}, \text{BE.Decrypt})$ defined as follows:

1. $\text{BE.KeyGen}(1^\lambda, N) \to (\{\text{sk}_i\}_{i \in [N]}, \text{pk})$: On input the security parameter λ and a positive integer $N \in \mathbb{N}$, the key generation algorithm outputs a set of secret keys $\{\text{sk}_i\}_{i \in [N]}$ and a public key pk.
2. $\text{BE.Encrypt}(\text{pk}, \text{msg}, T) \to \text{ct}_T$: On input a public key pk, a message m, and a set of intended recipients $T \subseteq [N]$, the encryption algorithm outputs a ciphertext ct_T.
3. $\text{BE.Decrypt}(\text{sk}_i, \text{ct}) \to m'$: On input a secret key sk_i and a ciphertext ct, the decryption algorithm outputs a message m'.

Correctness. For correctness we require that for all $\lambda \in \mathbb{N}$, $N \in \mathbb{N}$, $T \subseteq [N]$, $(\{\mathsf{sk}_i\}_{i \in [N]}, \mathsf{pk}) \leftarrow \mathsf{BE.KeyGen}(1^\lambda, N)$, we have $\mathsf{BE.Decrypt}(\mathsf{sk}_i, \mathsf{BE.Encrypt}(\mathsf{pk}, \mathsf{ct}, T)) = m$ for all $i \in T$.

Security. For broadcast encryption scheme, we define the following security notion.

Definition 2. *A broadcast encryption scheme Π_{be} is secure if for all efficient adversaries $\mathcal{A}$, there is a negligible function $\mathsf{negl}(\lambda)$ such that*

$$\mathsf{Adv}_{\Pi_{\mathsf{be}}, \mathcal{A}}(\lambda) \overset{\text{def}}{=} \left| \Pr[\mathsf{Expt}^{(0)}_{\mathsf{BE}, \mathcal{A}}(\lambda) = 1] - \Pr[\mathsf{Expt}^{(1)}_{\mathsf{BE}, \mathcal{A}}(\lambda) = 1] \right| \leq \mathsf{negl}(\lambda),$$

where for each $b \in \{0, 1\}$, and $\lambda \in \mathbb{N}$, the experiment $\mathsf{Expt}^{(b)}_{\mathsf{BE}, \mathcal{A}}$ is defined as follows:

- $(\{\mathsf{sk}_i\}_{i \in [N]}, \mathsf{pk}) \leftarrow \mathsf{BE.KeyGen}(1^\lambda, N)$.
- $(T, m_0, m_1) \leftarrow \mathcal{A}_0(\mathsf{pk})$.
- $\mathsf{ct}_b \leftarrow \mathsf{BE.Encrypt}(\mathsf{pk}, m_b)$.
- $b' \leftarrow \mathcal{A}_1(\{\mathsf{sk}_i\}_{i \in [N] \setminus T}, \mathsf{ct}_b)$.
- *Output $b' \in \{0, 1\}$.*

A number of constructions for broadcast encryption schemes have been proposed in the literature [FN93, BGW05, BWZ14] with short ciphertext where the length of the ciphertext scales sublinearly in the number of users in the system. For linear length ciphertext, a broadcast encryption scheme can be constructed generically from a regular public key encryption scheme by concatenating the N instances of the public key encryption schemes.

2.3   Background on Lattices

In this section, we describe some of the results and notations for lattice-based cryptography that are used throughout the paper.

Lattice and Gaussians. Let n, q, m be positive integers. For a matrix $\mathbf{A} \in \mathbb{Z}_q^{n \times m}$, we let $\Lambda_q^\perp(\mathbf{A})$ denote the lattice $\{\mathbf{x} \in \mathbb{Z}^m : \mathbf{A} \cdot \mathbf{x} = \mathbf{0} \mod q\}$. More generally, for $\mathbf{u} \in \mathbb{Z}_q^n$, we let $\Lambda_q^{\mathbf{u}}(\mathbf{A})$ denote the shifted lattice $\{\mathbf{x} \in \mathbb{Z}^m : \mathbf{A} \cdot \mathbf{x} = \mathbf{u} \mod q\}$.

Regev [Reg09] defined a natural distribution on $\Lambda_q^{\mathbf{u}}(\mathbf{A})$ called a *discrete Gaussian* parameterized by a scalar $s > 0$. We use $\mathcal{D}_s(\Lambda_q^{\mathbf{u}}(\mathbf{A}))$ to denote this distribution. For a random matrix $\mathbf{A} \in \mathbb{Z}_q^{n \times m}$ and $s > \widetilde{O}(\sqrt{n})$, a vector $\mathbf{x}$ sampled from $\mathcal{D}_s(\Lambda_q^{\mathbf{u}}(\mathbf{A}))$ has Euclidean norm less than $s\sqrt{m}$ with overwhelming probability. For a matrix $\mathbf{U} = (\mathbf{u}_1 | \ldots | \mathbf{u}_k) \in \mathbb{Z}_q^{n \times k}$, we let $\mathcal{D}_s(\Lambda_q^{\mathbf{U}}(\mathbf{A}))$ be a distribution on matrices in $\mathbb{Z}^{m \times k}$ where the i-th column is sampled from $\mathcal{D}_s(\Lambda_q^{\mathbf{u}_i}(\mathbf{A}))$ for $i = 1, \ldots, k$.

The SIS Problem. Let n, m, q, β be positive integers. In the $\mathrm{SIS}(n, m, q, \beta)$ problem, the adversary is given a uniformly random matrix $\mathbf{A} \in \mathbb{Z}_q^{n \times m}$ and its goal is to find a vector $\mathbf{u} \in \mathbb{Z}_q^m$ with $\mathbf{u} \neq \mathbf{0}$ and $\|\mathbf{u}\| \leq \beta$ such that $\mathbf{A} \cdot \mathbf{u} = \mathbf{0}$.

The SIS problem is known to be as hard as certain worst-case lattice problems. In particular, for any $m = \mathsf{poly}(n)$, any $\beta > 0$, and any sufficiently large $q \geq \beta \cdot \mathsf{poly}(n)$, solving $\mathrm{SIS}(n, m, q, \beta)$ is at least as hard as approximating certain worst-case lattice problems such as the Shortest Vector Problem (GapSVP) and the Short Independent Vectors Problem (SIVP) on n-dimensional lattices to within $\beta \cdot \mathsf{poly}(n)$ factor [Ajt96, Mic04, MR07, MP13]. The hardness of SIS is also implied by the LWE problem.

The LWE Problem. Let n, m, q be positive integers and χ a noise distribution over $\mathbb{Z}_q$. In the $\mathrm{LWE}(n, m, q, \chi)$ problem, the adversary's goal is to distinguish between the two distributions:

$$(\mathbf{A}, \mathbf{A}^T \mathbf{s} + \mathbf{e}) \quad \text{and} \quad (\mathbf{A}, \mathbf{u})$$

where $\mathbf{A} \xleftarrow{\$} \mathbb{Z}_q^{n \times m}$, $\mathbf{s} \xleftarrow{\$} \mathbb{Z}_q^n$, $\mathbf{e} \leftarrow \chi^m$, and $\mathbf{u} \xleftarrow{\$} \mathbb{Z}_q^m$ are uniformly sampled.

We say that a noise distribution χ is B-bounded if its support is in $[-B, B]$. For any fixed $d > 0$, and sufficiently large q, taking χ as a certain q/n^d-bounded distribution, the $\mathrm{LWE}(n, m, q, \chi)$ problem is as hard as approximating certain worst-case lattice problems such as GapSVP and SIVP on n-dimensional lattices to within $\mathsf{poly}(n)$ factor [Reg09, Pei09, ACPS09, MM11, MP12, BLP+13].

Matrix Norms. For a matrix $\mathbf{R} \in \mathbb{Z}^{k \times m}$, we define the matrix norms:

- $\|\mathbf{R}\|$ denotes the ℓ_2 length of the longest column of $\mathbf{R}$.
- $\|\mathbf{R}\|_2$ is the operator norm of $\mathbf{R}$ defined as $\|\mathbf{R}\|_2 = \sup_{\|\mathbf{x}\|=1} \|\mathbf{Rx}\|$.

Note that always $\|\mathbf{R}\| \leq \|\mathbf{R}\|_2 \leq \sqrt{k}\|\mathbf{R}\|$ and that $\|\mathbf{R} \cdot \mathbf{S}\|_2 \leq \|\mathbf{R}\|_2 \cdot \|\mathbf{S}\|_2$.

Lattice Trapdoors. Here, we review the known results about lattice trapdoors which make the SIS and LWE problems easy to solve with knowledge such trapdoor. For this work, it is convenient to work with the notion of "gadget" based trapdoors as formalized in [MP12]. In such setting, there is a structured public gadget matrix $\mathbf{G} \in \mathbb{Z}_q^{n \times n\ell}$ for $\ell = \lceil \log q \rceil$. A trapdoor for a matrix $\mathbf{A} \in \mathbb{Z}_q^{n \times m}$ is an integer matrix $\mathbf{R} \in \mathbb{Z}_q^{n \times n\ell}$ such that $\mathbf{AR} = \mathbf{HG}$ for some invertible matrix $\mathbf{H} \in \mathbb{Z}_q^{n \times n}$. The quality of a trapdoor is measured by the operator norm of $\mathbf{R}$ where the smaller norm $\|\mathbf{R}\|_2$ means higher quality. We will often use the symbol $\mathbf{T_A}$ to denote the trapdoor matrix $\mathbf{R}$.

Since the exact constructions and algorithm details of such trapdoors are not needed for this work, we abstract out these details and summarize the relevant results in the lemma below.

Lemma 1 ([Ajt96, GPV08, AP11, MP12]). *There exist polynomial time algorithms* TrapGen, SamPre, Sam, Invert *such that the following holds. Given positive integers $n \geq 1, q \geq 2$, there exists $m^* = O(n \log q)$ such that for $k = \mathsf{poly}(n)$, we have:*

- TrapGen$(1^n, 1^m, q) \rightarrow (\mathbf{A}, \mathbf{T_A})$: A randomized algorithm that when $m \geq m^*$, outputs a full-rank matrix $\mathbf{A} \in \mathbb{Z}_q^{n \times m}$ and a trapdoor $\mathbf{T_A} \in \mathbb{Z}^{m \times n\ell}$ such that $\mathbf{A}$ is statistically close to uniform and $\|\mathbf{R}\|_2 = O(\sqrt{m})$.
- SamPre$(\mathbf{A}, \mathbf{T_A}, \mathbf{V}, s) \rightarrow \mathbf{U}$: A randomized algorithm that on input $\mathbf{A} \in \mathbb{Z}_q^{n \times m}$, a trapdoor $\mathbf{T_A}$ of $\mathbf{A}$, a matrix $\mathbf{V} \in \mathbb{Z}_q^{m \times k}$, and $s = \tilde{O}(\|\mathbf{T_A}\|_2)$, outputs a random sample $\mathbf{U} \in \mathbb{Z}^{m \times k}$ from the distribution $\mathcal{D}_s(\Lambda_q^{\mathbf{V}}(\mathbf{A}))$.
- Sam$(1^m, 1^k, q, s) \rightarrow \mathbf{U}$: A randomized algorithm that samples a matrix $\mathbf{U} \in \mathbb{Z}^{m \times k}$ such that each of its column is sampled from $\mathcal{D}_{\mathbb{Z}^m, s}$. We have that for $s \geq \omega(\sqrt{\log m})$ the matrix $\mathbf{V} = \mathbf{A} \cdot \mathbf{U}$ is statistically close to a uniform matrix in $\mathbb{Z}_q^{n \times k}$ and furthermore, the distribution of $\mathbf{U}$ given $\mathbf{V}$ is $\mathcal{D}_s(\Lambda_q^{\mathbf{V}}(\mathbf{A}))$.
- Invert$(\mathbf{A}, \mathbf{T_A}, \mathbf{b}) \rightarrow \mathbf{s}$: A deterministic algorithm that on input $\mathbf{A} \in \mathbb{Z}_q^{n \times m}$, a trapdoor $\mathbf{T_A}$ of $\mathbf{A}$, and an LWE vector $\mathbf{b} = \mathbf{A}^T \mathbf{s} + \mathbf{e}$ for $\|\mathbf{e}\| \leq q/O(\|\mathbf{T_A}\|_2)$, outputs the unique secret vector $\mathbf{s}$.

To simplify the notation, thoughout the paper, we will always assume that the gadget matrix $\mathbf{G}$ has the same width m as the matrix $\mathbf{A}$ output by the algorithm TrapGen.

2.4   FRD Encoding

In this section, we review an encoding function $H : \mathbb{Z}_q^n \rightarrow \mathbb{Z}_q^{n \times n}$ that maps vectors in $\mathbb{Z}_q^n$ to invertible matrices in $\mathbb{Z}_q^{n \times n}$ with the property that for any two distinct vectors $\mathbf{u}$ and $\mathbf{v}$, the difference between the outputs $H(\mathbf{u})$ and $H(\mathbf{v})$ is never singular, i.e., $\det(H(\mathbf{u}) - H(\mathbf{v})) \neq 0$.

Definition 3. *Let q be a prime and n a positive integer. We say that an efficiently computable function $H : \mathbb{Z}_q^n \rightarrow \mathbb{Z}_q^{n \times n}$ is a* full-rank difference *(FRD) encoding scheme if for all distinct $\mathbf{u}, \mathbf{v} \in \mathbb{Z}_q^n$, the matrix $H(\mathbf{u}) - H(\mathbf{v}) \in \mathbb{Z}_q^{n \times n}$ is full rank.*

This notion was formalized in [ABB10] and an injective encoding function satisfying the definition was explicitly constructed by generating an additive subgroup $\mathcal{G}_{\text{FRD}}$ of full-rank matrices by embedding ring multiplications into matrices (similar techniques were used in [CD09, PR06, LM06]). We do not explicitly provide the construction here and mainly use the result as a black box throughout this work.

3   Predicate-Authentication-Preventing Signatures

In this section, we formally define the notion of *predicate authentication preventing signatures* (PAPS) which generalizes the notion of *double authentication preventing signatures* (DAPS) that was introduced in [PS14].

3.1   PAPS Framework

The syntax for predicate-authentication-preventing signatures largely coincides with the syntax for standard signature schemes with an additional extraction algorithm that can extract some private information of the signer (i.e. the signing key) given the signatures of messages that satisfy a particular predicate.

Definition 4 (PAPS). *A predicate-authentication-preventing signature (PA PS) on a corresponding message space $\mathcal{M}$, and a predicate $f : \mathcal{M}^k \to \{0,1\}$ is a tuple of efficient algorithms $\Pi_{\mathsf{paps}} = (\mathsf{PAPS.KeyGen}, \mathsf{PAPS.Sign}, \mathsf{PAPS.Verify}, \mathsf{PAPS.Extract})$ defined as follows:*

- *$\mathsf{PAPS.KeyGen}(1^\lambda) \to (\mathsf{sk}, \mathsf{vk})$: On input a security parameter 1^λ, the key generation algorithm $\mathsf{PAPS.KeyGen}$ outputs a signing key sk and a verification key vk.*
- *$\mathsf{PAPS.Sign}(\mathsf{sk}, \mathsf{msg}) \to \sigma$: On input a signing key sk and a message $\mathsf{msg} \in \mathcal{M}$, the signing algorithm $\mathsf{PAPS.Sign}$ outputs a signature σ.*
- *$\mathsf{PAPS.Verify}(\mathsf{vk}, \mathsf{msg}, \sigma)$: On input a verification key vk, a message $\mathsf{msg} \in \mathcal{M}$, and a signature σ, the verification algorithm $\mathsf{PAPS.Verify}$ accepts/rejects.*
- *$\mathsf{PAPS.Extract}(\mathsf{vk}, \{(\mathsf{msg}_i, \sigma_i)\}_{i \in [k]}) \to \mathsf{sk}'$: On input a verification key vk, and a set of message and signature pairs, the extraction algorithm $\mathsf{PAPS.Extract}$ outputs the secret key sk'.*

Correctness. We say that a PAPS scheme is correct if, for all $\lambda \in \mathbb{N}$, $\mathsf{msg} \in \mathcal{M}$, and $\sigma \leftarrow \mathsf{PAPS.Sign}(\mathsf{sk}, \mathsf{msg})$, we have that $\mathsf{PAPS.Verify}(\mathsf{vk}, \mathsf{msg}, \sigma) = 1$.

3.2   Extraction

As in the case of [PS14], we can consider two notions of key extractability depending on whether the signer generates its keys at setup honestly or adversarially. We call the scenario for which the keys are always generated honestly as the *trusted setup model* and the scenario for which the keys can potentially be generated adversarially as the *untrusted setup model*. Before defining these two notions formally, we first define the notion of a *compromising set of signatures*.

Definition 5 (Compromising set of signatures). *Let $f : \mathcal{M}^k \to \{0,1\}$ be a predicate defined on k messages. Then, for a fixed verification key vk, a set of k message/signature pairs $\{(\mathsf{msg}_i, \sigma_i)\}_{i \in [k]}$ is f-compromising if each signature σ_i is a valid signature of msg_i and the k messages satisfy the predicate f; that is, if $\mathsf{PAPS.Verify}(\mathsf{vk}, \mathsf{msg}_i, \sigma_i) = 1$ for all $i = 1, ..., k$ and $f(\mathsf{msg}_1, ..., \mathsf{msg}_k) = 1$.*

We now define formally the two notions of key extractability.

Definition 6 (Extractability in trusted setup). *Fix a predicate $f : \mathcal{M}^k \to \{0,1\}$. We say that a PAPS scheme on a message space $\mathcal{M}$ is f-extractable in the* trusted setup model *if for all $\lambda \in \mathbb{N}$, secret $\in \mathcal{S}$, for all key pairs $(\mathsf{sk}, \mathsf{vk}) \leftarrow \mathsf{PAPS.KeyGen}(1^\lambda)$, and for all compromising set of signatures $S = \{(\mathsf{msg}_i, \sigma_i)\}_{i \in [k]}$, we have that $\mathsf{PAPS.Extract}(\mathsf{vk}, S) = \mathsf{sk}$ with overwhelming probability.*

For the untrusted setup model, instead of running the honest key generation algorithm, we allow an adversary to generate the keys along with a set of compromising set of signatures and require that the extraction algorithm succeeds on recovering the signing key from these set of signatures. Formally, we define extractability in the untrusted setup as follows.

Definition 7 (Extractability in untrusted setup). *Fix a predicate* $f :$ $\mathcal{M}^k \to \{0,1\}$. *We say that a* PAPS *scheme on a message space* $\mathcal{M}$ *is* f-*extractable if for all efficient adversary* $\mathcal{A}$, *we have that*

$$\Pr\left(\begin{array}{cc} (\mathsf{vk}, S = \{(\mathsf{msg}_i, \sigma_i)\}) \leftarrow \mathcal{A}(1^\lambda) & : & S\ f\text{-compromising} \\ \mathsf{sk}' \leftarrow \mathsf{PAPS.Extract}(\mathsf{vk}, S) & & \land\ \mathsf{sk}' = \mathsf{sk} \end{array} \right) = 1 - \mathsf{negl}(\lambda)$$

Double-Authentication-Preventing Signatures. The notion of double-authentication-preventing signatures is a special case of PAPS. Specifically, in DAPS, the data to be signed $\mathcal{M}_{\mathsf{DAPS}}$ is split into two parts: a *subject* and a *payload*. Then, we consider the following 2-ary predicate in this message space

$$F_{\mathsf{DAPS}}((\mathsf{subj}_0, \mathsf{payload}_0), (\mathsf{subj}_1, \mathsf{payload}_1)) = \begin{cases} 1 \text{ if } \mathsf{subj}_0 = \mathsf{subj}_1, \mathsf{payload}_0 \neq \mathsf{payload}_1 \\ 0 \text{ otherwise.} \end{cases}$$

The predicate is designed specifically for the certificate authority setting where a CA that signs two different messages pertaining to the same subject is penalized by leaking the CA's signing key. In this work, we will mainly focus on constructing PAPS for this particular predicate function.

Remark. An alternative formulation of extractability is allowing some private information of the signer to be extracted instead of the signing key. In this case, the key generation algorithm can take in some secret information to be leaked by a compromising set of signatures and generate the keys accordingly. This formulation generalizes the notion above where extraction always leaks the signing key and can be more befitting for certain applications (see Sect. 7).

3.3   Unforgeability

The security game for the unforgeability notion for PAPS is similar to the standard unforgeability notion for digital signatures where the adversary has access to a signing oracle and wins if it forges a new signature. However, since PAPS is designed precisely to leak the signing key on a compromising set of signatures, we require that the adversary's queries to the signing oracle is limited to non-compromising sets of signatures.

Definition 8 (Unforgeability). *An* f-*extractable* PAPS *scheme* $\Pi_{\mathsf{paps}} =$ (PAPS.KeyGen, PAPS.Sign, PAPS.Verify, PAPS.Extract) *is unforgeable if for all efficient adversary* $\mathcal{A}$, *we have that*

$$\mathsf{Adv}^{\mathsf{uf}}_{\Pi_{\mathsf{paps}}, \mathcal{A}}(\lambda) \stackrel{\mathsf{def}}{=} \Pr[\mathsf{Expt}_{\Pi_{\mathsf{paps}}, \mathcal{A}, \mathsf{uf}} = 1] \leq \mathsf{negl}(\lambda)$$

where the experiment $\mathsf{Expt}_{\Pi_{\mathsf{paps}}, \mathcal{A}, \mathsf{uf}}$ *is defined as follows:*

1. $(\mathsf{sk}, \mathsf{vk}) \leftarrow \mathsf{PAPS.KeyGen}(1^\lambda)$.
2. $(\mathsf{msg}^*, \sigma^*) \leftarrow \mathcal{A}_1^{\mathcal{O}_{\mathsf{Sign}}(\cdot)}(\mathsf{vk})$.
3. $\mathcal{A}$ *wins if* $\mathsf{Verify}^*(\mathsf{msg}^*, \sigma^*)$ *accepts*.

where the signing oracle $\mathcal{O}_{\mathsf{Sign}}(\cdot)$ *and* $\mathsf{Verify}^*(\cdot, \cdot)$ *are defined as follows:*

- *Oracle* $\mathcal{O}_{\mathsf{Sign}}(\cdot)$ *maintains a list* SignedList *of all the previous valid queries made by* $\mathcal{A}$. *For a query* msg, *that the adversary makes,* $\mathcal{O}_{\mathsf{Sign}}(\cdot)$ *checks whether there exists a compromising set of signatures in* SignedList $\cup$ $\{\mathsf{msg}\}$. *If this is the case, then* $\mathcal{O}_{\mathsf{Sign}}(\cdot)$ *outputs* $\perp$. *Otherwise, it outputs* $\mathsf{PAPS.Sign}(\mathsf{sk}, \mathsf{msg})$.
- *Verifier* $\mathsf{Verify}^*(\mathsf{msg}, \sigma)$ *accepts if* msg $\notin$ SignedList *and* $\mathsf{PAPS.Verify}(\mathsf{vk}, \mathsf{msg}, \sigma)$ *accepts*.

4 DAPS from Lattices

In this section, we describe our DAPS construction. For clarity of exposition, we first describe a variant of the dual-Regev encryption scheme [GPV08] which we will use in our construction as a blackbox. This way, we can abstract out the details of the encryption component and present our DAPS construction in a simpler and more intuitive way. After presenting our DAPS construction, we prove its extractable properties and also show that its security can be based on the security of the encryption scheme and the SIS hardness assumption.

4.1 Trapdoor Dual-Regev Encryption

In this section, we describe a simple variant of the dual-Regev encryption scheme [GPV08] that we will use as a blackbox for our DAPS construction. We present the encryption scheme here mainly for clarity of exposition and the only difference between the encryption scheme presented here and the original dual-Regev encryption scheme is the use of full trapdoors as the secret key of the scheme as opposed to a short preimage vector as is the case in [GPV08].

We use n as the security parameter λ. Let m, q be the trapdoor parameters dependent on n (Lemma 1). Let χ be a B-bounded noise distribution where $B = O(q/m)$. We construct a public key encryption scheme $\Pi_{\mathsf{pke}} = (\mathsf{PKE.KeyGen}, \mathsf{PKE.Encrypt}, \mathsf{PKE.Decrypt})$ as follows:

- $\mathsf{PKE.KeyGen}(1^n)$: On input the security parameters 1^n, the key generation algorithm generates trapdoor $(\mathbf{A}, \mathbf{T_A}) \leftarrow \mathsf{TrapGen}(1^n, 1^m, q)$ where $\mathbf{A} \in \mathbb{Z}_q^{n \times m}$. It sets the public key $\mathsf{pk} = \mathbf{A}$ and $\mathsf{sk} = \mathbf{T_A}$.
- $\mathsf{PKE.Encrypt}(\mathsf{pk}, m)$: On input the public key pk and a message $m \in \{0, 1\}$, the encryption algorithm samples uniformly random vectors $\mathbf{d}, \mathbf{s} \xleftarrow{\$} \mathbb{Z}_q^n$, and an error vector from the noise distribution $\mathbf{e} \leftarrow \chi^{m+1}$. It computes the vector

$$\mathbf{b} = [\mathbf{A} \mid \mathbf{d}]^T \mathbf{s} + \mathbf{e} + [\mathbf{0} \mid \lceil q/2 \rceil \cdot m].$$

It outputs the ciphertext $\mathsf{ct} = (\mathbf{d}, \mathbf{b}) \in \mathbb{Z}_q^n \times \mathbb{Z}_q^{m+1}$.

- PKE.Decrypt(sk, ct): On input the secret key sk and a ciphertext ct = $(\mathbf{d}, \mathbf{b})$, the decryption algorithm parses $\mathbf{b} = (b_0, b_1)$. It then computes $\mathbf{s} \leftarrow \mathsf{Invert}(\mathbf{A}, \mathbf{T_A}, \mathbf{b})$ and $m' = b_1 - \mathbf{d}^T \mathbf{s}$. It outputs $m \in \{0, 1\}$ such that m' is close to $\lceil q/2 \rceil \cdot m$.

Correctness. Let $\mathbf{b} = (\mathbf{b}_0 = \mathbf{A}^T \mathbf{s} + \mathbf{e}_0, b_1 = \mathbf{d}^T \mathbf{s} + e_1)$. The TrapGen algorithm outputs a trapdoor $\mathbf{T_A}$ such that $\|\mathbf{T_A}\|_2 = O(\sqrt{m})$. This means that for $B = O(q/\|\mathbf{T_A}\|\sqrt{m}) = O(q/m)$, the algorithm $\mathsf{Invert}(\mathbf{A}, \mathbf{T_A}, \mathbf{b})$ for $\mathbf{b} = \mathbf{A}^T \mathbf{s} + \mathbf{e}$ correctly outputs the secret vector $\mathbf{s}$. Then, $b_1 - \mathbf{d}^T \mathbf{s} = (\mathbf{d}^T \mathbf{s} + e_1 + \lceil q/2 \rceil \cdot m) - \mathbf{d}^T \mathbf{s} = \lceil q/2 \rceil \cdot m + \cdot e_1$ which is correctly decoded for given bound on the error distribution χ.

Remark. We note that the correctness still holds with a different trapdoor $\mathbf{T_A}'$ for which $\mathbf{T_A} \neq \mathbf{T_A}'$ as long as $\mathbf{T_A}'$ is of sufficient quality. In fact, we can flexibly adjust the parameter B for the noise distribution χ of the scheme to allow for correct decryption by a slightly lower quality trapdoor. For instance, if we take the parameter to be $B = O(q/m^{3/2})$, then a trapdoor $\mathbf{T_A}'$ of quality $\|\mathbf{T_A}'\|_2 \leq O(m)$ can correctly decrypt the ciphertext. This property will be used for the extractability of our DAPS construction.

Security. The security reduction easily follows from the security proof of the original dual-Regev encryption scheme and we do not reproduce it here.

Theorem 1. *The PKE scheme above is IND-CPA secure assuming that the $LWE(n, m, q, \chi)$ problem is hard.*

For our DAPS construction, we will also assume that the dual-Regev PKE scheme above is circular secure.

4.2 DAPS Construction

We present our construction for DAPS from lattices. Fix a security parameter $n \in \mathbb{N}$ and let m, q, s be the corresponding trapdoor parameters. We use a hash function $\mathsf{H}_{\mathsf{msg}} : \{0, 1\}^* \times \{0, 1\}^* \to \mathbb{Z}_q^{n \times n}$ that maps arbitrary messages to a full rank matrix in $\mathcal{G}_{\mathsf{FRD}}$ and a hash function $\mathsf{H}_{\mathsf{subj}} : \{0, 1\}^* \to \mathbb{Z}_q^{n \times m}$ that maps a subject to an SIS matrix for the scheme. We also use the dual-Regev public key encryption scheme $\Pi_{\mathsf{pke}} = (\mathsf{PKE.KeyGen}, \mathsf{PKE.Encrypt}, \mathsf{PKE.Decrypt})$ as decribed in Sect. 4.1 with a B-bounded noise distribution where $B = O(q/m^{3/2})$. We construct $\Pi_{\mathsf{daps}} = (\mathsf{DAPS.KeyGen}, \mathsf{DAPS.Sign}, \mathsf{DAPS.Verify}, \mathsf{DAPS.Extract})$ as follows:

- DAPS.KeyGen(1^n): On the security parameter 1^n, the key generation algorithm first runs $(\mathbf{A}, \mathbf{T_A}) \leftarrow \mathsf{PKE.KeyGen}(1^n)$. It then encrypts ct $\leftarrow$ PKE.Encrypt($\mathbf{A}, \mathbf{T_A}$) and set sk $= \mathbf{T_A}$ and vk $= (\mathbf{A}, \mathsf{ct})$.

- DAPS.Sign(sk, subj, payload): On input the signing key sk, a subject subj, and a message payload, the signing algorithm hashes the message $\mathbf{H}_{msg} \leftarrow \mathsf{H}_{msg}(subj, payload)$ and also hashes the subject $\mathbf{B}_{subj} \leftarrow \mathsf{H}_{subj}(subj)$. Then, it computes

$$\mathbf{U} \leftarrow \mathsf{SamPre}(\mathbf{A}, \mathbf{T_A}, \mathbf{B}_{subj} + \mathbf{H}_{msg} \cdot \mathbf{G}, s)$$

where $\mathbf{G}$ is the publicly known gadget matrix. Finally, it outputs $\mathbf{U}$ as the signature.
- DAPS.Verify(vk, subj, payload, σ): On input the verification key vk, a subject subj, a message payload, and a signature $\sigma = \mathbf{U}$, the algorithm hashes the message $\mathbf{H}_{msg} \leftarrow \mathsf{H}_{msg}(subj, payload)$ and also derives the matrix $\mathbf{B}_{subj} \leftarrow \mathsf{H}_{subj}(subj)$. Then, the algorithm verifies that

$$\mathbf{A} \cdot \mathbf{U} = \mathbf{B}_{subj} + \mathbf{H}_{msg} \cdot \mathbf{G}$$

and that $\|\mathbf{U}\| \leq s \cdot \sqrt{m}$.
- DAPS.Extract(vk, (subj$_1$, payload$_1$, σ_1), (subj$_2$, payload$_2$, σ_2)): On input a verification key vk = $(\mathbf{A}, \mathsf{ct})$, and two subject/message pairs (subj$_1$, payload$_1$), (subj$_2$, payload$_2$) and their signatures $\sigma_1 = \mathbf{U}_1, \sigma_2 = \mathbf{U}_2$, the extraction algorithm runs sk$' \leftarrow$ PKE.Decrypt($\mathbf{U}_1 - \mathbf{U}_2$) and outputs sk$'$.

Signing Correctness. The correctness follows easily from the correctness of the trapdoor algorithm SamPre (Lemma 1) and the tail bounds of discrete Gaussian.

Security and Extrability. We now state the security and extractability of the construction above.

Theorem 2. *The* DAPS *construction above is unforgeable assuming the hardness of* $\mathrm{SIS}(n, m, q, \beta)$ *for* $\beta = O(s)$ *and circular security of* Π_{pke} *modeling the hash functions* H_{msg} *and* H_{subj} *as random oracles.*

Theorem 3. *Assuming* H_{msg} *as a collision-resistant hash function, the* DAPS *construction above is* F_{DAPS}*-extractable.*

5   Extensions of DAPS to Other Predicates

In this work, we provide sample PAPS constructions for a number of simple predicates. Due to space limitations, we describe and prove our constructions in the full version.

6   Multi-authority Setting

For many practical situations, there is not a single authority signer, but multiple authorities who sign messages. For these type of situations, it is useful to extend the PAPS framework to the multi-authority setting where any compromising set of signatures by the different authorities reveals some private information of the signers.

We note that in this scenario, allowing a compromising set of signatures to reveal a secret key of a signer is not well-formulated in that any signer can compute a compromising set of signatures itself using its own signing key and extract another signer's secret key. Therefore, for the multi-authority setting, we let the extraction algorithm to reveal some predefined private data and define an additional algorithm PAPS.CommGen that takes in this private information that we denote by secret of a subset of the signers and generates a commitment comm pertaining to secret. For correctness, we require that a compromising set of signatures from these subset of signers along with the commitment comm allow anyone to reveal the private information secret.

Definition 9 (Multi-authority PAPS). *A multi-authority predicate-authen-tication-preventing signature (MAPAPS) on a corresponding message space $\mathcal{M}$, secret space $\mathcal{S}$, and a predicate $f : \mathcal{M}^k \rightarrow \{0,1\}$ is a tuple of efficient algorithms consisting of the* PAPS *algorithms* $\Pi_{\mathsf{paps}} = $ (PAPS.KeyGen, PAPS.Sign, PAPS.Verify, PAPS.Extract) *with two additional algorithms* (PAPS.CommGen, PAPS.CommExtract) *defined as follows:*

- PAPS.CommGen($\{\mathsf{vk}_i\}$, secret) $\rightarrow$ comm$_T$: *On input a set of verification keys $T = \{\mathsf{vk}_i\}$ and some private data secret $\in \mathcal{S}$, the commitment generation algorithm generates a public commitment* comm$_T$.
- PAPS.CommExtract(comm$_T$, $\{(\mathsf{msg}_i, \sigma_i)\}_{i \in [k]}$) $\rightarrow$ secret': *On input a public commitment* comm$_T$, *and a set of message and signature pairs, the extraction algorithm outputs private date* secret'.

Informally speaking, the PAPS.CommGen takes in a set of verification keys of the signers and generates a hiding commitment of some private data that belongs to these signers. On a compromising set of signatures produced by any of these signers allows anyone to extract this private information using the PAPS.CommExtract algorithm.

Correctness. As in the regular PAPS setting, we require that the algorithms PAPS.KeyGen, PAPS.Sign, PAPS.Verify satisfies the signing correctness require-ment as in Sect. 3.

6.1 Extractability

In addition to the extractability property of PAPS.Extract in the single authority setting as in Sect. 3, we require an additional extractability in the multi-authority case where it is required that PAPS.CommExtract extract private information from the public commitment comm.

Definition 10. *Let $f : \mathcal{M}^k \rightarrow \{0,1\}$ be a predicate defined on k messages and $T = \{\mathsf{vk}_i\}$ be a set of verification keys. Then, a set of k message/signature pairs*

$\{(\mathsf{msg}_j,\sigma_j)\}_{j\in[k]}$ *is f-compromising with respect to T if each signature σ_j is a valid signature of msg_j by some verification key $\mathsf{vk}_i \in T$, and the k messages satisfy the predicate f; that is, if for all $j \in [k]$, $\mathsf{PAPS.Verify}(\mathsf{vk}_i, \mathsf{msg}_j, \sigma_j) = 1$ for some $\mathsf{vk}_i \in T$ and $f(\mathsf{msg}_1, ..., \mathsf{msg}_k) = 1$.*

We define the standard extractability condition as follows.

Definition 11. *Fix a predicate $f : \mathcal{M}^k \rightarrow \{0,1\}$. We say that a MAPAPS scheme on a message space $\mathcal{M}$ is f-commitment-extractable if for all $\lambda \in N$, secret $\in \mathcal{S}$, a set of verification keys $T = \{\mathsf{vk}_i\}$, $\mathsf{comm}_T \leftarrow \mathsf{PAPS.CommGen}(T, \mathsf{secret})$ and for all compromising set of signatures $S = \{(\mathsf{msg}_j, \sigma_j)\}_{j\in[k]}$ with respect to T, we have that $\mathsf{PAPS.CommExtract}(\mathsf{comm}_T, S) = \mathsf{secret}$ with overwhelming probability.*

6.2 Security

Commitment Privacy. To prevent the extractability notion above from being satisfied trivially, we require a privacy requirement on the secret data of PAPS. Specifically, we require that an adversary with access limited to non-compromising sets of signatures do not learn information about the secret data. Let N be the number of authorities in the system.

Definition 12 (Privacy). *An f-extractable PAPS scheme $\Pi_{\mathsf{paps}} = (\mathsf{PAPS.KeyGen}, \mathsf{PAPS.CommGen}, \mathsf{PAPS.Sign}, \mathsf{PAPS.Verify}, \mathsf{PAPS.CommExtract})$ is data-hiding if for any efficient adversary $\mathcal{A}$, we have that*

$$\mathsf{Adv}^{\mathsf{dh}}_{\Pi_{\mathsf{paps}},\mathcal{A}}(\lambda) \overset{\mathrm{def}}{=} \left| \Pr[\mathsf{Expt}^{(0)}_{\Pi_{\mathsf{paps}},\mathcal{A},\mathsf{dh}}(\lambda) = 1] - \Pr[\mathsf{Expt}^{(1)}_{\Pi_{\mathsf{paps}},\mathcal{A},\mathsf{dh}}(\lambda) = 1] \right| \leq \mathsf{negl}(\lambda)$$

where $\mathsf{Expt}^{(b)}_{\Pi_{\mathsf{paps}},\mathcal{A},\mathsf{dh}}$ is defined as follows:

1. *$(\mathsf{sk}_i, \mathsf{vk}_i) \leftarrow \mathsf{PAPS.KeyGen}(1^\lambda)$ for $i = 1, ..., N$.*
2. *$(T, \mathsf{secret}_0, \mathsf{secret}_1) \leftarrow \mathcal{A}_0(\{\mathsf{vk}\}_{i\in[N]})$.*
3. *$\mathsf{comm}_b \leftarrow \mathsf{PAPS.CommGen}(\{\mathsf{vk}_i\}_{t\in T}, \mathsf{secret}_b)$.*
4. *Output $\mathcal{A}_1^{\mathcal{O}_{\mathsf{Sign}}(\cdot,\cdot)}(\{\mathsf{sk}_i\}_{i\in[N]\setminus T}, \mathsf{comm}_b)$.*

where the signing oracle $\mathcal{O}_{\mathsf{Sign}}(\cdot,\cdot)$ is defined as follows:

- *Oracle $\mathcal{O}_{\mathsf{Sign}}(\cdot,\cdot)$ maintains a list SignedList of all the previous valid queries made by $\mathcal{A}$. For a query msg and an index $i \in T$, that the adversary makes $\mathcal{O}_{\mathsf{Sign}}(\cdot,\cdot)$ checks whether there exists a compromising set of signatures with respect to T in SignedList $\cup \{\mathsf{msg}\}$. If this is the case, the $\mathcal{O}_{\mathsf{Sign}}(\cdot,\cdot)$ outputs $\bot$. Otherwise, it outputs $\mathsf{PAPS.Sign}(\mathsf{sk}, \mathsf{msg})$.*

7 Multi-authority DAPS

In this section we describe our construction of DAPS in the setting of multi-authority. We describe the scheme for the DAPS predicate, but the extensions in Sect. 5 translate directly to the multi authority setting since it uses the DAPS predicate as a black box. As in Sect. 4, we first describe a broadcast encryption scheme from the variant of the dual-Regev encryption scheme and then present our MADAPS construction using the broadcast encryption scheme. We prove that our scheme satisfies both the extractability and security requirements.

7.1 Broadcast Encryption

In this section, we present the broadcast encryption scheme. For clarity, we slightly alter the syntax of broadcast encryption where the global public key pk is divided into a number of public keys pk_i for each user in the system and the encryption algorithm takes in a subset of these public keys. Fix a security parameter n and let m, q, and χ be the corresponding parameter as in Sect. 4.1. We construct $\Pi_{\mathsf{be}} = (\mathsf{BE.KeyGen}, \mathsf{BE.Encrypt}, \mathsf{BE.Decrypt})$ as follows:

- BE.KeyGen($1^n, N$): On input the security parameter 1^n, the key generation algorithm generates trapdoors $(\mathbf{A}_i, \mathbf{T}_{\mathbf{A}i}) \leftarrow \mathsf{TrapGen}(1^n, 1^m, q)$ for $i = 1, ..., N$. It outputs $\{(\mathbf{A}_i, \mathbf{T}_{\mathbf{A}i})\}_{i \in [N]}$.
- BE.Encrypt($\{\mathsf{pk}_i\}_{i \in T}, m$): On input a set of public keys $\{\mathsf{pk}_i\}_{i \in T} = \{\mathbf{A}_i\}_{i \in T}$ and a message $m \in \{0, 1\}$, the encryption algorithm first samples uniformly random vectors $\mathbf{d}, \mathbf{s} \xleftarrow{\$} \mathbb{Z}_q^n$ and error vector $\mathbf{e}_0 \leftarrow \chi^m$ and then computes

$$\mathsf{ct}_0 = \mathbf{d}^T\mathbf{s} + \mathbf{e}_0 + \lceil q/2 \rceil \cdot m$$

Then for $t \in T$, it generates a fresh error vector from the noise distribution $\mathbf{e}_t \leftarrow \chi^m$ and computes

$$\mathsf{ct}_t = \mathbf{A}_t^T\mathbf{s} + \mathbf{e}_t$$

It outputs $\mathsf{ct}_T = (\mathsf{ct}_0, \{\mathsf{ct}_t\}_{t \in T})$.
- BE.Decrypt($\mathsf{sk}_i, \mathsf{ct}_T$): On input the secret key sk_i for $i \in T$ and a ciphertext $\mathsf{ct}_T = (\mathsf{ct}_0, \{\mathsf{ct}_t\}_{t \in T})$, the decryption algorithm computes $\mathbf{s} \leftarrow \mathsf{Invert}(\mathbf{A}, \mathbf{T}_{\mathbf{A}}, \mathsf{ct}_i)$. It then computes $m' = \mathsf{ct}_0 - \mathbf{d}^T\mathbf{s}$ and output $m \in \{0, 1\}$ such that m' is closest to $\lceil q/2 \rceil \cdot m$.

We note that each ciphertext component $(\mathsf{ct}_0, \mathsf{ct}_i)$ makes up a regular ciphertext of the dual-Regev encryption scheme. Therefore, the correctness of the BE scheme above follows directly from the correctness of the dual-Regev scheme. Also, the construction of the scheme above is a generic concatenation of the regular public key encryption scheme and therefore, the security follows directly from the security of the public key encryption scheme.

As in the case of the dual-Regev public key encryption scheme, one can correctly decrypt the ciphertext even with a different trapdoor for the public matrices $\mathbf{A}_i$ as long as it is of sufficient quality. Furthermore, any trapdoor of a concatenated matrix $[\mathbf{A}_i \mid \mathbf{A}_j]$ is sufficient to recover the encryption randomness used in the encryption and therefore decrypt a ciphertext intended for users i and j. More formally, there exists a decryption algorithm as follows:

- BE.Decrypt$'$(sk$_{i,j}$, ct$_T$): On input a trapdoor sk$_{i,j}$ = $\mathbf{T}_{[\mathbf{A}_i \mid \mathbf{A}_j]}$ for $i,j \in T$, and a ciphertext ct$_T$ = (ct$_0$, {ct$_t$}$_{t \in T}$), the decryption algorithm computes $\mathbf{s} \leftarrow$ Invert($[\mathbf{A}_i|\mathbf{A}_j]$, $\mathbf{T}_\mathbf{A}$, [ct$_i$|ct$_j$]). It then computes $m' = \text{ct}_0 - \mathbf{d}^T\mathbf{s}$ and outputs $m \in \{0,1\}$ such that m' is closest to $\lceil q/2 \rceil \cdot m$.

7.2 Multi-authority DAPS Construction

In this section, we extend the DAPS construction from Sect. 4.2 to the multi-authority setting (MADAPS). In addition to the algorithms in Π_{daps}, we define two additional algorithms DAPS.CommGen and DAPS.CommExtract as defined in Sect. 6. Let Π_{be} = (BE.KeyGen, BE.Encrypt, BE.Decrypt) be the broadcast encryption scheme as defined above with B-bounded noise distribution χ where $B = O(q/m^{3/2})$. Fix a security parameter $n \in \mathbb{N}$ and let m, q, s be the corresponding trapdoor parameters. We define the algorithms DAPS.CommGen and DAPS.CommExtract as follows:

- DAPS.CommGen({vk$_i$}, secret): On input a set of verification keys $T = \{\text{vk}_i\} = \{\mathbf{A}_i\}$ and some private data secret $\in \mathcal{S}$, the commitment generation algorithm computes ct$_T \leftarrow$ BE.Encrypt(T, secret) and sets comm = ct$_T$.
- DAPS.CommExtract(comm$_T$, (subj$_0$, payload$_0$, σ_0), (subj$_1$, payload$_1$, σ_1)): On input the parameters comm$_T$ and a pair of compromising pair of signatures $\sigma_0 = \mathbf{U}_i$ and $\sigma_1 = \mathbf{U}_j$ corresponding to vk$_i$ and vk$_j$ respectively, the extraction algorithm concatenates the keys $\tilde{\mathbf{U}} = \begin{bmatrix} \mathbf{U}_i \\ -\mathbf{U}_j \end{bmatrix}$. It runs secret$' \leftarrow$ BE.Decrypt$'(\tilde{\mathbf{U}}$, comm$_T$) and outputs the private data secret$'$.

Extractability and Privacy. We now state the extractability and privacy properties of the MADAPS construction above. We provide the proofs of the following theorems in the full version.

Theorem 4. *Assuming* H$_{\text{msg}}$ *is a collision-resistant hash function, the* MADAPS *construction above is* F$_{\text{DAPS}}$*-extractable.*

Theorem 5. *The* MADAPS *construction above is data-hiding assuming that the broadcast encryption scheme* Π_{be} *is secure.*

Acknowledgements. This work is supported by NSF, DARPA, the Simons foundation, and a grant from ONR. Opinions, findings and conclusions or recommendations expressed in this material are those of the author(s) and do not necessarily reflect the views of DARPA.

References

[ABB10] Agrawal, S., Boneh, D., Boyen, X.: Efficient Lattice (H)IBE in the standard model. In: Gilbert, H. (ed.) EUROCRYPT 2010. LNCS, vol. 6110, pp. 553–572. Springer, Heidelberg (2010). doi:10.1007/978-3-642-13190-5_28

[ACPS09] Applebaum, B., Cash, D., Peikert, C., Sahai, A.: Fast cryptographic primitives and circular-secure encryption based on hard learning problems. In: Halevi, S. (ed.) CRYPTO 2009. LNCS, vol. 5677, pp. 595–618. Springer, Heidelberg (2009). doi:10.1007/978-3-642-03356-8_35

[Ajt96] Ajtai, M.: Generating hard instances of lattice problems. In: STOC (1996)

[ALSY06] Au, M.H., Liu, J.K., Susilo, W., Yuen, T.H.: Constant-size ID-based linkable and revocable-iff-linked ring signature. In: Barua, R., Lange, T. (eds.) INDOCRYPT 2006. LNCS, vol. 4329, pp. 364–378. Springer, Heidelberg (2006). doi:10.1007/11941378_26

[AP11] Alwen, J., Peikert, C.: Generating shorter bases for hard random lattices. Theory Comput. Syst. 48(3), 535–553 (2011)

[ASP12] Alperin-Sheriff, J., Peikert, C.: Circular and KDM security for identity-based encryption. In: Fischlin, M., Buchmann, J., Manulis, M. (eds.) PKC 2012. LNCS, vol. 7293, pp. 334–352. Springer, Heidelberg (2012). doi:10.1007/978-3-642-30057-8_20

[BF14] Bellare, M., Fuchsbauer, G.: Policy-based signatures. In: Krawczyk, H. (ed.) PKC 2014. LNCS, vol. 8383, pp. 520–537. Springer, Heidelberg (2014). doi:10.1007/978-3-642-54631-0_30

[BGI14] Boyle, E., Goldwasser, S., Ivan, I.: Functional signatures and pseudorandom functions. In: Krawczyk, H. (ed.) PKC 2014. LNCS, vol. 8383, pp. 501–519. Springer, Heidelberg (2014). doi:10.1007/978-3-642-54631-0_29

[BGV12] Brakerski, Z., Gentry, C., Vaikuntanathan, V.: (Leveled) fully homomorphic encryption without bootstrapping. In: ITCS. ACM (2012)

[BGW05] Boneh, D., Gentry, C., Waters, B.: Collusion resistant broadcast encryption with short ciphertexts and private keys. In: Shoup, V. (ed.) CRYPTO 2005. LNCS, vol. 3621, pp. 258–275. Springer, Heidelberg (2005). doi:10.1007/11535218_16

[BKM06] Bender, A., Katz, J., Morselli, R.: Ring signatures: stronger definitions, and constructions without random oracles. In: Halevi, S., Rabin, T. (eds.) TCC 2006. LNCS, vol. 3876, pp. 60–79. Springer, Heidelberg (2006). doi:10.1007/11681878_4

[BLP+13] Brakerski, Z., Langlois, A., Peikert, C., Regev, O., Stehlé, D.: Classical hardness of learning with errors. In: STOC. ACM (2013)

[BPS15] Bellare, M., Poettering, B., Stebila, D.: Double-authentication-preventing signatures and trapdoor identification. Cryptology ePrint Archive, Report 2015/1157 (2015). http://eprint.iacr.org/

[BRS02] Black, J., Rogaway, P., Shrimpton, T.: Encryption-scheme security in the presence of key-dependent messages. In: Nyberg, K., Heys, H. (eds.) SAC 2002. LNCS, vol. 2595, pp. 62–75. Springer, Heidelberg (2003). doi:10.1007/3-540-36492-7_6

[BV11] Brakerski, Z., Vaikuntanathan, V.: Fully homomorphic encryption from ring-LWE and security for key dependent messages. In: Rogaway, P. (ed.) CRYPTO 2011. LNCS, vol. 6841, pp. 505–524. Springer, Heidelberg (2011). doi:10.1007/978-3-642-22792-9_29

[BV14]   Brakerski, Z., Vaikuntanathan, V.: Efficient fully homomorphic encryption from (standard) LWE. SIAM J. Comput. **43**(2), 831–871 (2014)

[BWZ14]   Boneh, D., Waters, B., Zhandry, M.: Low overhead broadcast encryption from multilinear maps. In: Garay, J.A., Gennaro, R. (eds.) CRYPTO 2014. LNCS, vol. 8616, pp. 206–223. Springer, Heidelberg (2014). doi:10.1007/978-3-662-44371-2_12

[CD09]   Cramer, R., Damgård, I.: On the amortized complexity of zero-knowledge protocols. In: Halevi, S. (ed.) CRYPTO 2009. LNCS, vol. 5677, pp. 177–191. Springer, Heidelberg (2009). doi:10.1007/978-3-642-03356-8_11

[CL01]   Camenisch, J., Lysyanskaya, A.: An Efficient System for Non-transferable Anonymous Credentials with Optional Anonymity Revocation. In: Pfitzmann, B. (ed.) EUROCRYPT 2001. LNCS, vol. 2045, pp. 93–118. Springer, Heidelberg (2001). doi:10.1007/3-540-44987-6_7

[FN93]   Fiat, A., Naor, M.: Broadcast encryption. In: Stinson, D.R. (ed.) CRYPTO 1993. LNCS, vol. 773, pp. 480–491. Springer, Heidelberg (1994). doi:10.1007/3-540-48329-2_40

[Gen09]   Gentry, C.: Fully homomorphic encryption using ideal lattices. STOC **9**, 169–178 (2009)

[GPV08]   Gentry, C., Peikert, C., Vaikuntanathan, V.: Trapdoors for hard lattices and new cryptographic constructions. In: STOC. ACM (2008)

[GSW13]   Gentry, C., Sahai, A., Waters, B.: Homomorphic encryption from learning with errors: conceptually-simpler, asymptotically-faster, attribute-based. In: Canetti, R., Garay, J.A. (eds.) CRYPTO 2013. LNCS, vol. 8042, pp. 75–92. Springer, Heidelberg (2013). doi:10.1007/978-3-642-40041-4_5

[GVW15]   Gorbunov, S., Vaikuntanathan, V., Wichs, D.: Leveled fully homomorphic signatures from standard lattices. In: STOC. ACM (2015)

[LLK15]   Laurie, B., Langley, A., Kasper, E.: Certificate Transparency. RFC 6962, (October 2015)

[LM06]   Lyubashevsky, V., Micciancio, D.: Generalized compact knapsacks are collision resistant. In: Bugliesi, M., Preneel, B., Sassone, V., Wegener, I. (eds.) ICALP 2006. LNCS, vol. 4052, pp. 144–155. Springer, Heidelberg (2006). doi:10.1007/11787006_13

[Lyu08]   Lyubashevsky, V.: Lattice-based identification schemes secure under active attacks. In: Cramer, R. (ed.) PKC 2008. LNCS, vol. 4939, pp. 162–179. Springer, Heidelberg (2008). doi:10.1007/978-3-540-78440-1_10

[Lyu12]   Lyubashevsky, V.: Lattice signatures without trapdoors. In: Pointcheval, D., Johansson, T. (eds.) EUROCRYPT 2012. LNCS, vol. 7237, pp. 738–755. Springer, Heidelberg (2012). doi:10.1007/978-3-642-29011-4_43

[MBB+15] Melara, M.S., Blankstein, A., Bonneau, J., Felten, E.W., Freedman, M.J.: CONIKS: bringing key transparency to end users. USENIX Secur. **15**, 383–398 (2015)

[Mic04]   Micciancio, D.: Almost perfect lattices, the covering radius problem, and applications to Ajtai's connection factor. SIAM J. Comput. **34**(1), 118–169 (2004)

[MM11]   Micciancio, D., Mol, P.: Pseudorandom knapsacks and the sample complexity of LWE search-to-decision reductions. In: Rogaway, P. (ed.) CRYPTO 2011. LNCS, vol. 6841, pp. 465–484. Springer, Heidelberg (2011). doi:10.1007/978-3-642-22792-9_26

[MP12]   Micciancio, D., Peikert, C.: Trapdoors for lattices: simpler, tighter, faster, smaller. In: Pointcheval, D., Johansson, T. (eds.) EUROCRYPT 2012.

LNCS, vol. 7237, pp. 700–718. Springer, Heidelberg (2012). doi:10.1007/978-3-642-29011-4_41

[MP13] Micciancio, D., Peikert, C.: Hardness of SIS and LWE with small parameters. In: Canetti, R., Garay, J.A. (eds.) CRYPTO 2013. LNCS, vol. 8042, pp. 21–39. Springer, Heidelberg (2013). doi:10.1007/978-3-642-40041-4_2

[MPR11] Maji, H.K., Prabhakaran, M., Rosulek, M.: Attribute-based signatures. In: Kiayias, A. (ed.) CT-RSA 2011. LNCS, vol. 6558, pp. 376–392. Springer, Heidelberg (2011). doi:10.1007/978-3-642-19074-2_24

[MR07] Micciancio, D., Regev, O.: Worst-case to average-case reductions based on Gaussian measures. SIAM J. Comput. **37**(1), 267–302 (2007)

[Pei09] Peikert, C.: Public-key cryptosystems from the worst-case shortest vector problem. In: STOC. ACM (2009)

[PR06] Peikert, C., Rosen, A.: Efficient collision-resistant hashing from worst-case assumptions on cyclic lattices. In: Halevi, S., Rabin, T. (eds.) TCC 2006. LNCS, vol. 3876, pp. 145–166. Springer, Heidelberg (2006). doi:10.1007/11681878_8

[PS14] Poettering, B., Stebila, D.: Double-authentication-preventing signatures. In: Kutyłowski, M., Vaidya, J. (eds.) ESORICS 2014. LNCS, vol. 8712, pp. 436–453. Springer, Cham (2014). doi:10.1007/978-3-319-11203-9_25

[Reg09] Regev, O.: On lattices, learning with errors, random linear codes, and cryptography. J. ACM (JACM) **56**(6), 34 (2009)

[RST01] Rivest, R.L., Shamir, A., Tauman, Y.: How to leak a secret. In: Boyd, C. (ed.) ASIACRYPT 2001. LNCS, vol. 2248, pp. 552–565. Springer, Heidelberg (2001). doi:10.1007/3-540-45682-1_32

[TX03] Tsudik, G., Xu, S.: Accumulating composites and improved group signing. In: Laih, C.-S. (ed.) ASIACRYPT 2003. LNCS, vol. 2894, pp. 269–286. Springer, Heidelberg (2003). doi:10.1007/978-3-540-40061-5_16

Forward-Secure Searchable Encryption on Labeled Bipartite Graphs

Russell W.F. Lai$^{(\boxtimes)}$ and Sherman S.M. Chow

Department of Information Engineering,
The Chinese University of Hong Kong, Shatin, N.T., Hong Kong
{russell,sherman}@ie.cuhk.edu.hk

Abstract. Forward privacy is a trending security notion of dynamic searchable symmetric encryption (DSSE). It guarantees the privacy of newly added data against the server who has knowledge of previous queries. The notion was very recently formalized by Bost (CCS '16) independently, yet the definition given is imprecise to capture how forward secure a scheme is. We further the study of forward privacy by proposing a generalized definition parametrized by a set of updates and restrictions on them. We then construct two forward private DSSE schemes over labeled bipartite graphs, as a generalization of those supporting keyword search over text files. The first is a generic construction from any DSSE, and the other is a concrete construction from scratch. For the latter, we designed a novel data structure called cascaded triangles, in which traversals can be performed in parallel while updates only affect the local regions around the updated nodes. Besides neighbor queries, our schemes support flexible edge additions and intelligent node deletions: The server can delete all edges connected to a given node, without having the client specify all the edges.

1 Introduction

In searchable symmetric encryption (SSE), an encrypted database can be queried with minimal leakage of information about the plaintext database to the hosting server. The client is additionally allowed to update the encrypted database in dynamic SSE (DSSE) without reencrypting from scratch. Since its introduction [16], many SSE schemes with different trade-offs between efficiency, security, and query expressiveness have been proposed [1]. Most earlier schemes were not dynamic. The first sublinear dynamic SSE scheme was proposed by Kamara *et al.* [11], but the query and update operations are inherently sequential. Some later schemes [8,10,17] feature parallelizable algorithms for queries and updates. Parallelism made DSSE an attractive solution for outsourcing data to cloud platform which fully leverages the multiprocessors.

Sherman Chow is supported in part by General Research Fund (Grant No. 14201914) and the Early Career Award from Research Grants Council, Hong Kong; and Huawei Innovation Research Program (HIRP) 2015 (Project No. YB2015110147).

© Springer International Publishing AG 2017
D. Gollmann et al. (Eds.): ACNS 2017, LNCS 10355, pp. 478–497, 2017.
DOI: 10.1007/978-3-319-61204-1_24

1.1 Security and Forward Privacy of SSE Schemes

Ideally, the knowledge of an encrypted database, together with a sequence of adaptively issued queries and updates, should not reveal any information about the plaintext database and the query results to the server. Although this can be achieved theoretically through techniques involving obfuscation [4] or oblivious RAM [7], the resulting solutions are not particularly efficient. Typically, a practical DSSE scheme tolerates the leakage of search and access patterns during queries, and some internal structure of the encrypted database during updates. Formally, the security is parametrized by a set of leakage functions describing these leakages. While some leakages seem to pose no harm, some have been exploited in attacks [9,18].

Forward privacy, advocated by Stefanov et al. [17], requires that newly added data remains private against the server, who has knowledge about previous queries. The property is arguably essential to all DSSE schemes, for otherwise, the ability to update in DSSE is somewhat useless as future data are less protected. Indeed, one of the recent attacks by Zhang et al. [18] exploits the leakage during updates in non-forward-private schemes.

Only a limited number of solutions [2,7,15,17] in the literature claimed to have forward privacy. The notion is not well understood in the earlier works [7, 15,17], and is only formally defined recently by Bost [2]. However, we argue that this definition cannot precisely describe in what sense a DSSE scheme is forward private. (See discussion in Sect. 3.2 for details.)

1.2 Our Formulation

We consider DSSE over labeled bipartite graphs, where nodes can be partitioned into two disjoint subsets X and Y, such that edges never connect two nodes from the same partition, and each edge is labeled with data from the set W. A neighbor query on node $x \in X$ (or $y \in Y$), returns a sequence of $(x, y, w) \in X \times Y \times W$ tuples if (x, y) is an edge on the graph labeled with w.

This abstract setting captures typical DSSE queries such as keyword searches over files (considering X and Y as the sets of keywords and files respectively), and labeled subgraphs queries over general graphs (considering X and Y as sets of nodes with outgoing and incoming edges respectively, and W as the set of edge labels). To generalize, we also consider neighbor queries over the entire bipartite graphs, i.e., both X and Y, which enables interesting bi-directional searching applications. Bi-directional search is useful to efficiently support update in DSSE [11] (since deleting a file in a DSSE supporting keyword searches implicitly requires finding all keywords the file contains). It also opens possibilities of interesting new queries such as related keyword search (which first searches for documents containing the queried keyword, then collects other keywords which are also contained in many of the matching documents).

1.3   Update Functionalities of DSSE Schemes

While the query functionality of SSE for keyword search is somewhat standard, the supported update types have large variations. Updates include additions and deletions, and can be edge-based or node-based. Schemes supporting only node additions are reasonable for some data type: *e.g.*, x as keywords and y as text files. Yet, edge updates allow fine-grained modification of existing data. In particular, schemes supporting edge additions are superior to those supporting only node additions, as the latter can be simulated by the former.

The benefits of supporting only edge deletions are however questionable, as they require the client to know about the edge to be deleted. It is unrealistic for the motivating application of SSE for keyword search: The client needs to know all keywords of a given file to completely remove the file from the server. It is desirable for a DSSE scheme to support node deletions: upon provided a node y from the client, the server can intelligently remove all edges connecting y.

To the best of our knowledge, most existing schemes only support either edge-based updates or node-based updates[1]. Supporting edge additions and node deletions simultaneously, while confining leakage, poses some technical challenges.

1.4   SSE as a Data Structure Problem

With edge additions and node deletions in mind, it is not an easy task to devise a parallel and dynamic (let alone forward private) SSE scheme. Intuitively, for data structures supporting parallel traversal, maintaining the traversal efficiency after an update often requires some global adjustment of the data structure. Consider a balanced binary search tree. A series of deletions can degenerate the tree into a linked list which requires sequential access; and balancing the tree may require the rotation of multiple tree nodes. (That may explain why the first parallel DSSE [10] utilizing a red-black tree which only stores all the files in the leaf level, resulting in an efficiency loss when compared with storing some of them in internal nodes.) In the context of DSSE, delegating the maintenance work to the server often implies excessive leakage of the internal data structure.

A notable approach for (non-dynamic) SSE schemes is the invert-index used by Curtmola *et al.* [5,6], in which the encrypted database consists of an index mapping hashed keywords to sets of files containing the keywords. This inverted-index allows the server to search in time linear in the number of matching files, which is optimal. Many subsequent works follow this framework (explicitly or implicitly), which utilize some data structure to represent the sets of data, pointed by the (hashed) queries in the index. The efficiency of queries and updates correspond to the efficiency of traversing and updating the sets respectively. On the other hand, the leakage of the internal structure of the encrypted database during updates corresponds to the amount of information required or changed to update the data structure storing the sets. Most efforts for designing (D)SSE schemes is dedicated to choosing or designing this data structure.

---

[1] A few exceptions include Lai-Chow [13] and a modified version of the one by Kamara *et al.* [11]. However, these schemes leak substantial information during updates.

1.5 Our Results

This work furthers the study of forward privacy of DSSE schemes over labeled bipartite graphs. We present three technical results. First, we give another formal, generalized definition of forward privacy. Specifically, our definition is parametrized by a set of updates and a restriction function on these updates. It generalizes the only existing one by Bost [2], by increasing the number of classes of leakage functions allowed, yet making each class more specific. Our definition still captures the essence of forward privacy even though the leakage in different classes might vary substantially. Since different existing SSE schemes implicitly assumed different flavors of forward privacy, we believe that our generalized, parameterized definition of forward privacy is of particular interest.

Second, we propose a simple generic construction of forward private DSSE from any DSSE, which preserves the efficiency of the base scheme. The forward privacy obtained is for edge additions, such that the addition of an edge (x, y) does not leak both x and y, hence hiding the edge. This generic transformation provides insights of what constitutes forward privacy in DSSE. Since the result applies on any DSSE, again we believe it is of independent interest.

Lastly, we construct a DSSE scheme from scratch which achieves a stronger forward privacy for edge additions, such that the addition of an edge (x, y) does not leak either x or y. Our construction utilizes a specially crafted data structure named cascaded triangles[2], which supports parallel queries and updates, and has the property that adding or deleting data only affects a constant amount of existing data. Thanks to cascaded triangles, our construction features minimal leakage, and optimal query and update complexity in terms of both computation and communication up to a constant factor.

Both of our constructions support flexible edge additions and intelligent node deletions: The server can delete all edges connected to a given node, without having the client specify all the edges. It is one of a few in the literature(See footnote 1).

2 Definitions

We present the necessary definitions for data representation and DSSE. For more detailed explanations, we refer the readers to the full version.

2.1 Data Representation

Let $\mathcal{X}$, $\mathcal{Y}$, and $\mathcal{W}$ be sets, where $\mathcal{X}$ and $\mathcal{Y}$ are disjoint, i.e., $\mathcal{X} \cap \mathcal{Y} = \phi$. We denote by $\mathcal{G} = \mathcal{G}(\mathcal{X}, \mathcal{Y}, \mathcal{W})$ a set of labeled bipartite graphs specified by these sets. For a labeled bipartite graph $G \in \mathcal{G}$, its edges are labeled with $w \in W \subseteq \mathcal{W}$, and

---

[2] While the design of cascaded triangles is original, we do not rule out the possibility that there are similar data structures outside the literature of SSE. To the best of our knowledge, we are unaware of any common similar data structure. There are false relatives such as fractional cascading which solves totally different problems.

are running across $X \subseteq \mathcal{X}$ and $Y \subseteq \mathcal{Y}$. Each edge can be uniquely represented by the tuple $(x, y, w) \in \mathcal{X} \times \mathcal{Y} \times \mathcal{W}$. The neighbor query function Qry maps a node $q = x$ (or $q = y$) and a graph G to a set of all edges connecting node x, denoted by $(x, *, *)$ (or a set of all edges connecting node y, denoted by $(*, y, *)$). Similarly, we use $(x, y, *)$ to denote the set of edges in the form of $(x, y, \cdot)$, which should be a singleton. The update function Udt maps an update u and a graph G to a new graph G'. The update $u = (\mathsf{Op}, \cdot, \cdot, \cdot)$ where $\mathsf{Op} = \mathsf{Add}$ or $\mathsf{Op} = \mathsf{Del}$ takes one of the following forms:

1. Edge Addition: $u = (\mathsf{Add}, x, y, w)$ adds the edge (x, y, w) to G.
2. Node Deletion: $u = (\mathsf{Del}, q)$ deletes the set of edges $(q, *, *)$ (or $(*, q, *)$) from G. Since $\mathcal{X}$ and $\mathcal{Y}$ are disjoint, there is no ambiguity.

2.2 Dynamic Searchable Symmetric Encryption (DSSE)

We present a definition of DSSE for the labeled bipartite graphs defined above, and briefly describe its security against adaptive chosen query attack (CQA2).

Definition 1. *A dynamic symmetric searchable encryption (DSSE) scheme for the space of labeled bipartite graphs specified by $\mathcal{G}$ is a tuple of algorithms and interactive protocols DSSE.$(\mathsf{Setup}, \mathsf{Qry}_e, \mathsf{Udt}_e)$ such that:*

- $(K, \mathsf{EDB}) \leftarrow \mathsf{Setup}(1^\lambda)$: *In the setup algorithm, the user inputs the security parameter λ. It outputs a secret key K, and an (initially empty) encrypted database EDB to be outsourced to the server. Alternatively, one can define a setup algorithm which takes as input the security parameter λ, and a graph G. In this case, the algorithm outputs a key K and an encrypted database EDB encrypting G.*
- $((K', R), (\mathsf{EDB}', R)) \leftarrow \mathsf{Qry}_e((K, q), \mathsf{EDB})$: *In the query protocol, the user inputs a secret key K and a query $q \in \mathcal{X} \cup \mathcal{Y}$. The server inputs the encrypted database EDB. The user outputs a possibly updated key K', while the server outputs a possibly updated encrypted database EDB'. Both the user and the server output a sequence of responses R. For non-interactive schemes, the user first runs $(K', \tau_q) \leftarrow \mathsf{QryTkn}(K, q)$ to generate a query token τ_q. The server then runs $(\mathsf{EDB}', R) \leftarrow \mathsf{Qry}_e(\tau_q, \mathsf{EDB})$ and outputs the query results R.*
- $(K', \mathsf{EDB}') \leftarrow \mathsf{Udt}_e((K, u), \mathsf{EDB})$: *In the update protocol, the user inputs a secret key K and an update $u \in \{\mathsf{Add}, \mathsf{Del}\} \times (\mathcal{X} \cup \mathcal{Y})$. The server inputs the encrypted database EDB. The user outputs a possibly updated key K'. The server outputs an updated encrypted database EDB'. For non-interactive schemes, the user first runs $(K', \tau_u) \leftarrow \mathsf{UdtTkn}(K, u)$ to generate an update token τ_u. The server then runs $\mathsf{EDB}' \leftarrow \mathsf{Udt}_e(\tau_u, \mathsf{EDB})$ and outputs the updated database EDB'.*

A DSSE scheme for the space $\mathcal{G}$ is said to be correct if, for all $\lambda \in \mathbb{N}$, all K and EDB output by $\mathsf{Setup}(1^\lambda)$, and all sequences of queries and updates, the responses to the plaintext queries equal those to the corresponding encrypted queries.

Let $\mathcal{L}_e$, $\mathcal{L}_q$, and $\mathcal{L}_u$ be stateful leakage algorithms. A DSSE scheme is said to be $(\mathcal{L}_e, \mathcal{L}_q, \mathcal{L}_u)$-secure against adaptive dynamic chosen-query attacks (CQA2), if there exists a simulator which, when given the leakages specified by the leakage algorithms, indistinguishably simulates the encrypted database, the query results, and the updates. The formal definition can be found in the full version.

3 Forward Privacy in DSSE

Intuitively, forward privacy ensures that newly added data remains hidden to the server who might have learned some secrets during previous queries, until it must be revealed by a later query. To formalize, instead of tinkering with the CQA2-security definition of DSSE, we define forward privacy based on the property of the leakage function $\mathcal{L}_u$. This is more convenient since the information leaked by $\mathcal{L}_u$ is sufficient for the simulator in the CQA2-security definition to simulate the updates. Similar to the semantic security of encryption schemes, we require that the leakages $\mathcal{L}_u$ on a pair of updates are indistinguishable, capturing the idea that not even a single bit about the update is leaked to the server. Note that the definition does not limit the types of the update u. Indeed, we can consider forward privacy for not only additions, but also deletions. In layman terms, suppose that the server learns during the query protocol about the association of q with some data, which is then deleted by the client. The server should not notice that the data are deleted until q is queried again: By that time the server can compare the query results and discover the deletion.

3.1 Our Definition

We first give a general definition of forward privacy parametrized by a set of updates and a restriction function, then discuss useful ways to parameterize it.

Definition 2 (Forward Privacy). *Let* DSSE *be a* $(\mathcal{L}_e, \mathcal{L}_q, \mathcal{L}_u)$-*CQA2 secure DSSE scheme for labeled bipartite graphs specified by* $\mathcal{G}$. *Let* $\mathcal{U}$ *be a set of updates and restriction* $p : \mathcal{U}^2 \rightarrow \{0, 1\}$ *be a predicate function. We say that* DSSE *is:* $(\mathcal{U}, p)$-*forward private, if for any* $u_b \in \mathcal{U}$ *where* $b \in \{0, 1\}$ *such that* $p(u_0, u_1) = 1$, *and any PPT distinguisher* $\mathcal{D}$, *it holds that*

$$|\Pr[\mathcal{D}(\mathcal{L}_u(u_0)) = 1] - \Pr[\mathcal{D}(\mathcal{L}_u(u_1)) = 1]| \leq \mathsf{negl}(\lambda).$$

Table 1 lists some useful combinations of $\mathcal{U}$ and p, denoted by $\mathcal{U}_i$ and p_i respectively for $i \in [6]$. One may also consider a set $\mathcal{U}$ which is a union of some of the (disjoint) $\mathcal{U}_i$'s, and a restriction p which is a composition of the corresponding p_i's: For $p_i : \mathcal{U}_i^2 \rightarrow \{0, 1\}$, define $p = p_i + p_j : (\mathcal{U}_i \cup \mathcal{U}_j)^2 \rightarrow \{0, 1\}$ such that $(p_i + p_j)(u) = 1$ if $u \in \mathcal{U}_i$ and $p_i(u) = 1$, or $u \in \mathcal{U}_j$ and $p_j(u) = 1$.

Note that there might exist schemes which are $(\mathcal{U}_i, p_i)$- and $(\mathcal{U}_j, p_j)$-forward private but not $(\mathcal{U}_i \cup \mathcal{U}_j, p_i + p_j)$-forward private. For example, if $\mathcal{U}_i$ and $\mathcal{U}_j$ are sets of additions and deletions respectively, while the distinguisher cannot tell which addition or deletion is chosen, it can separate additions from deletions.

Table 1. Useful combinations of $\mathcal{U}$ and p as parameters for forward privacy. (See Sect. 2.1 and the full version for details about the update operations.)

i	Type	Sets of updates $\mathcal{U}_i$	Updates $u_b \in \mathcal{U}_i$ $b \in \{0,1\}$	Possible conditions such that $p_i = 1$ (If there are no restrictions, $p_i \equiv 1$)
1	Edge	$\{(\mathsf{Add}, x, y, w)\}$	$(\mathsf{Add}, x_b, y_b, w_b)$	$x_0 = x_1$ or $y_0 = y_1$
2	Node	$\{(\mathsf{Add}, \mathbf{x}, y, \mathbf{w})\}$	$(\mathsf{Add}, \mathbf{x}_b, y_b, \mathbf{w}_b)$	$y_0 = y_1$
3	Node	$\{(\mathsf{Add}, x, \mathbf{y}, \mathbf{w})\}$	$(\mathsf{Add}, x_b, \mathbf{y}_b, \mathbf{w}_b)$	$x_0 = x_1$
4	Edge	$\{(\mathsf{Del}, x, y)\}$	(Del, x_b, y_b)	$x_0 = x_1$ or $y_0 = y_1$
5	Node	$\{(\mathsf{Del}, x)\}$	(Del, x_b)	$|(x_0, *, *)| = |(x_1, *, *)|$
6	Node	$\{(\mathsf{Del}, y)\}$	(Del, y_b)	$|(*, y_0, *)| = |(*, y_1, *)|$

3.2 Bost's Definition

The forward privacy definition of Bost [2] requires that the leakage function of $u = (\mathsf{Op}, x, y, w)$ can be written as a function of the operation Op, the node y, and $|(*, y, *)|$, i.e., the number of edges connected to y, where Op can be addition or deletion. We argue that the range of leakage functions which satisfy this requirement is so wide, such that the definition does not precisely describe in what sense a DSSE scheme is forward secure. At one extreme, consider a scheme of which the leakage function is given by $\mathcal{L}_u(\mathsf{Op}, x, y, w) = (\mathsf{Op}, y, |(*, y, *)|)$. If $\mathcal{L}_u$ accurately (not overly) captures the leakage, then adding or removing edges to or from a node y leaks the identity of the node y itself. Such scheme is vulnerable to frequency attacks: The attacker keeps a table mapping each y to the number of times y is updated. With the aid of external information, it can possibly extract information about y, such as its importance. Similar attacks have been demonstrated using search patterns [14]. At another extreme, Bost's construction [2] leaks nothing during updates, i.e., $\mathcal{L}_u(\mathsf{Op}, x, y, w) = \phi$. On the other hand, Bost's definition is also not general enough. There are other types of leakage functions, e.g., $\mathcal{L}_u(\mathsf{Op}, x, y, w) = (\mathsf{Op}, x, |(x, *, *)|)$, which intuitively capture forward privacy, but are not covered by this definition. In contrast, our definition of $(\mathcal{U}, p)$-forward privacy classified different types of update in a more fine-grained manner.

In the perspective of our model, Bost's definition can be regarded as special cases of $(\mathcal{U}_1, p_1)$ - (achieved by our generic construction in Sect. 4) and $(\mathcal{U}_4, p_4)$-forward privacy. Intuitively, the restrictions p_1 and p_4 mean that an update $u = (\mathsf{Op}, x, y, w)$ is protected by hiding one end of the connection between x and y. Therefore, similar to the above, it might be the case that the updates u_0 and u_1, where $u_b = (\mathsf{Add}, x_b, y, w)$ for $b \in \{0, 1\}$, are linkable as they correspond to the same node y, making the scheme vulnerable to the same frequency attack. On the other hand, $(\mathcal{U}_1, 1)$ - (achieved by our concrete construction in Sect. 5) and $(\mathcal{U}_4, 1)$-forward privacy completely hides the relation (x, y), making the addition and deletion of an edge (x, y, w) oblivious respectively (since w can be hidden simply by symmetric key encryption). Whether or not the stronger forward privacy is needed depends on the specific application scenarios.

3.3   Forward Privacy for Deletions

In the rest of this work, we focus on $\mathcal{U} = \mathcal{U}_1 = \{(\mathsf{Add}, x, y, w)\}$, with and without restrictions, *i.e.*, $p = p_1$ and $p \equiv 1$, respectively. In other words, we do not consider forward privacy for deletions. To argue for this design decision, we observe that while the scheme of Bost [2] performs "lazy edge deletion" (adding "deleted" edges rather than actually deleting), ours perform actual node deletion. The former increases the size of the encrypted database (by one edge), hence it is possible to make (edge) additions and deletions indistinguishable. The latter allows immediate space reclamation. This makes edge additions (which usually increase the size of the encrypted database) and node deletions easily distinguishable. Furthermore, since actual deletions of different nodes may result in a shrink of the database size in varying degrees, they are also easily distinguishable. In effect, we trade "forward privacy for (lazy) deletions" for efficiency. If forward privacy for deletions is a concern and lazy deletions are acceptable, using a similar technique of maintaining another instance of encrypted database for lazy deletions [2], schemes which are forward private for edge additions can be generically transformed to provide forward privacy for both edge additions and node deletions (simultaneously): To delete a node $y \in Y$, the client adds an edge connecting y to a special "deleted" node in X. We leave the details to the full version of this paper. In this sense, forward privacy for additions is a key property. We also believe that it is sufficient for practical applications by itself.

4   Forward Privacy from Any DSSE

In this section, we will show that a DSSE scheme with $(\mathcal{U}_1, p_1)$-forward privacy can be constructed from any DSSE scheme $\mathcal{E}$, where $\mathcal{U}_1$ and p_1 are defined in Table 1. For simplicity of our description below, we assume the base scheme to be non-interactive, so that the resulting scheme is also non-interactive[3]. Our transformation can be easily adapted to interactive schemes.

Our construction is inspired by that of Rizomiliotis and Gritzalis [15] which uses fresh keys for newly added data. The main idea is to locally maintain a table γ of pseudorandom function (PRF) keys K_x and counters c_x for each query $q = x$, so that adding an edge (x, y, w) is translated to adding another edge $(F(K_x, c_x), y, w)$. Our scheme also adopts the technique of Hahn and Kerschbaum [8], who observe that when the set $(x, *, *)$ is leaked upon querying on x, there is no need to protect the set by encryption any longer. To speed up subsequent queries, the server should thus transfer the set encrypted in the scheme to a plaintext bipartite graph $\hat{G}$. We assume an efficient data structure for representing the graph $\hat{G}$, so that neighbor queries, edge additions, and node deletions in $\hat{G}$ are parallelizable and have time complexity linear in the number of affected nodes only. There might be many ways to construct such a data structure. Cascaded triangles introduced in Sect. 5 is one example.

---

[3] Candidate base schemes include [13] and a modified version of [11].

4.1   Our Construction

Let $\mathcal{E}$ be a DSSE scheme for $\mathcal{G} = \mathcal{G}(\{0,1\}^\lambda, \mathcal{Y}, \mathcal{W})$, and $F : \{0,1\}^\lambda \times \mathcal{X} \to \{0,1\}^\lambda$ be a pseudorandom function (PRF). We construct a DSSE scheme for $\mathcal{G}' = \mathcal{G}'(\mathcal{X}, \mathcal{Y}, \mathcal{W})$. The resulting scheme supports queries over $\mathcal{X}$, assuming the base scheme supports queries over $\{0,1\}^\lambda$. Figures 1 and 2 formally describe the construction.

The setup algorithm initializes the base scheme $\mathcal{E}$, which yields a secret key $\tilde{K}$ and encrypted database $\tilde{\mathsf{EDB}}$. It also initializes an empty dictionary γ and an empty bipartite graph $\hat{G} \in \mathcal{G}'$. The new secret key K consists of $\tilde{K}$ and γ, while $\tilde{\mathsf{EDB}}$ and $\hat{G}$ are outsourced to the server. The dictionary γ maps a query $q = x \in \mathcal{X}$ to a PRF key K_x and a counter c_x.

To perform an update $u = (\mathsf{Add}, x, y, w)$, the client increments $c_x \leftarrow c_x + 1$, and transforms the update into $\tilde{u} = (\mathsf{Add}, F(K_x, c_x), y, w)$ of the base scheme. To perform an update $u = (\mathsf{Del}, x)$, the client removes the x-th row of γ, and sends to the server the update tokens for $(\mathsf{Del}, F(K_x, i))$ for $i \in [c_x]$. The update $u = (\mathsf{Del}, y)$ is processed as in the base scheme.

Finally, to query $q = x \in \mathcal{X}$, the client removes the x-th row of γ, and sends to the server the query tokens of $F(K_x, i)$ for $i \in [c_x]$. Given these tokens, the server retrieves the intended response R. Additionally, the client also sends the update tokens for $(\mathsf{Del}, F(K_x, i))$ for $i \in [c_x]$. The server uses these tokens to collect and remove the set of edges $R = (x, *, *)$ from the base scheme, and merge R to the plaintext graph $\hat{G}$.

The correctness follows directly from that of the base scheme $\mathcal{E}$.

$(K, \mathsf{EDB}) \leftarrow \mathsf{Setup}(1^\lambda)$

$(\tilde{K}, \tilde{\mathsf{EDB}}) \leftarrow \mathcal{E}.\mathsf{Setup}(1^\lambda)$
$\gamma = \phi, \hat{G} = \phi$
return $K = (\tilde{K}, \gamma), \mathsf{EDB} = (\tilde{\mathsf{EDB}}, \hat{G})$

$(K', \tau_q) \leftarrow \mathsf{QryTkn}(K, q)$

if $q = x \in \mathcal{X} \land \gamma[x] \neq \bot$ **then**
  $(K_x, c_x) \leftarrow \gamma[x]$
  $\gamma[x] \leftarrow \bot$
  for $i = 1, \dots, c_x$ **do**
    $\tilde{\tau}_i \leftarrow \mathcal{E}.\mathsf{QryTkn}(\tilde{K}, F(K_x, i))$
    $\tilde{\tau}_i^- \leftarrow \mathcal{E}.\mathsf{UdtTkn}(\tilde{K}, (\mathsf{Del}, F(K_x, i)))$
  endfor
  $\tau_q \leftarrow (x, \{\tilde{\tau}_i, \tilde{\tau}_i^-\}_{i=1}^{c_x})$
else
  $(q = x \in \mathcal{X}) ? \ \tau_q \leftarrow x \ : \ \tau_q \leftarrow \bot$
endif
return (K, τ_q)

$(\mathsf{EDB}, R) \leftarrow \mathsf{Qry}_e(\tau_q, \mathsf{EDB})$

Parse τ_q as $(x, \{\tilde{\tau}_i, \tilde{\tau}_i^-\}_{i=1}^{c_x})$
$R \leftarrow \mathsf{Qry}(x, \hat{G})$
for $i = 1, \dots, c_x$ **do**
  $\tilde{R} \leftarrow \mathcal{E}.\mathsf{Qry}_e(\tilde{\tau}_i, \tilde{\mathsf{EDB}})$
  $\tilde{\mathsf{EDB}} \leftarrow \mathcal{E}.\mathsf{Udt}_e(\tilde{\tau}_i^-, \tilde{\mathsf{EDB}})$
  $R \leftarrow R \cup \tilde{R}$
endfor
foreach $(\tilde{x}, y, w) \in R$ **do**
  $R \leftarrow R \setminus \{(\tilde{x}, y, w)\} \cup \{(x, y, w)\}$
  $\hat{G} \leftarrow \mathsf{Udt}((\mathsf{Add}, x, y, w), \hat{G})$
endfor
return (EDB, R)

$\mathsf{EDB}' \leftarrow \mathsf{Udt}_e(\tau_u^+, \mathsf{EDB}), u = (\mathsf{Add}, \dots)$

$\tilde{\mathsf{EDB}} \leftarrow \mathcal{E}.\mathsf{Udt}(\tilde{\tau}_u^+, \tilde{\mathsf{EDB}})$
return EDB

Fig. 1. Algorithms of generic construction for forward privacy

---

$(K', \tau_u^+) \leftarrow \mathsf{UdtTkn}(K, u = (\mathsf{Add}, \dots))$

parse u **as** (Add, x, y, w)
if $\gamma[x] = \bot$ **then**
   $K_x \leftarrow \{0,1\}^\lambda, c_x \leftarrow 1$
else
   $(K_x, c_x) \leftarrow \gamma[x], c_x \leftarrow c_x + 1$
endif
$\gamma[x] \leftarrow (K_x, c_x)$
$\tau_u^+ \leftarrow \mathcal{E}.\mathsf{UdtTkn}(\tilde{K}, (\mathsf{Add}, F(K_x, c_x), y, w))$
return (K, τ_u^+)

$\mathsf{EDB}' \leftarrow \mathsf{Udt}_e(\tau_u^-, \mathsf{EDB}),\ u = (\mathsf{Del}, \cdot)$

if $\tau_u^- = (x \in \mathcal{X}, \{\tilde{\tau}_i^-\}_{i=1}^{c_x})$ **then**
   $\hat{G} \leftarrow \mathsf{Udt}((\mathsf{Del}, x), \hat{G})$
   for $i = 1, \dots, c_x$ **do**
      $\widetilde{\mathsf{EDB}} \leftarrow \mathcal{E}.\mathsf{Udt}_e(\tau_i^-, \widetilde{\mathsf{EDB}})$
   endfor
elseif $\tau_u^- = (y \in \mathcal{Y}, \tilde{\tau}^-)$ **then**
   $\hat{G} \leftarrow \mathsf{Udt}((\mathsf{Del}, y), \hat{G})$
   $\widetilde{\mathsf{EDB}} \leftarrow \mathcal{E}.\mathsf{Udt}_e(\tau^-, \widetilde{\mathsf{EDB}})$
endif
return EDB

$(K', \tau_u^-) \leftarrow \mathsf{UdtTkn}(K, u = (\mathsf{Del}, \cdot))$

if $u = (\mathsf{Del}, x)$ **then**
   if $\gamma[x] \neq \bot$ **then**
      $(K_x, c_x) \leftarrow \gamma[x]$
      $\gamma[x] \leftarrow \bot$
      for $i = 1, \dots, c_x$ **do**
         $r_i \leftarrow F(K_x, i)$
         $\tilde{\tau}_i^- \leftarrow \mathcal{E}.\mathsf{UdtTkn}(\tilde{K}, (\mathsf{Del}, r_i))$
      endfor
      $\tau_u^- \leftarrow (x, \{\tilde{\tau}_i^-\}_{i=1}^{c_x})$
   else
      $\tau_u^- \leftarrow x$
   endif
elseif $u = (\mathsf{Del}, y)$ **then**
   $\tilde{\tau}^- \leftarrow \mathcal{E}.\mathsf{UdtTkn}(\tilde{K}, (\mathsf{Del}, y))$
   $\tau_u^- \leftarrow (y, \tilde{\tau}^-)$
else
   $\tau_u^- \leftarrow \bot$
endif
return τ_u^-

Fig. 2. Algorithms of generic construction for forward privacy (cont.)

4.2 Analysis

Efficiency. Our generic transformation almost preserves the efficiency of the underlying DSSE scheme. For most algorithms, the preservation is apparent. We highlight the slightly more complicated cases, namely, the query on x and the update (Del, x). In the former, c_x queries on $F(K_x, i)$ for $i \in [c_x]$ are executed, while in both cases c_x deletions $(\mathsf{Del}, F(K_x, i))$ for $i \in [c_x]$ are required. We analyze their efficiency assuming the following operations of the underlying DSSE scheme each takes constant time: the computation of the query token for each $F(K_x, i)$; the server computation for querying on each $F(K_x, i)$ (since by the pseudorandomness of F, the query should only return a single edge); the computation of each delete token; and the server computation for deleting each set $(F(K_x, i), *, *)$ (since each set is actually a singleton). Overall, the resulting scheme incurs $O(c_x)$ computation and communication costs for both the client and the server where c_x is the number of newly matched data item. These are extra costs on top of the costs for retrieving the previously matched data (in plaintext), *i.e.*, constant computation cost of the client, sublinear computation cost of the server, and sublinear communication cost of both. Since c_x is reset to zero whenever x is queried, the amortized extra costs of both the client and the server are low.

Security. Let $\mathcal{E}$ be an $(\tilde{\mathcal{L}}_e, \tilde{\mathcal{L}}_q, \tilde{\mathcal{L}}_u)$-CQA2 secure DSSE scheme for labeled bipartite graphs specified by $\mathcal{G}$; and $F : \{0,1\}^\lambda \times \mathbb{N} \to \{0,1\}^\lambda$ be a pseudorandom function. Our construction is CQA2-secure and forward private with respect to the following leakage functions. The proof can be found in the full version.

- $\mathcal{L}_e(\mathcal{G}') = \tilde{\mathcal{L}}_e(\mathcal{G})$,
- $\mathcal{L}_u(\mathsf{Add}, x, y, w) = (\tilde{x}, \tilde{\mathcal{L}}_u(\mathsf{Add}, \tilde{x}, y, w))$ for dummy node $\tilde{x} \leftarrow \{0,1\}^\lambda$,
- $\mathcal{L}_u(\mathsf{Del}, x) = (x, \{\tilde{\mathcal{L}}_u(\mathsf{Del}, \tilde{x}_i)\}_{i=1}^{c_x})$,
- $\mathcal{L}_u(\mathsf{Del}, y) = (y, \tilde{\mathcal{L}}_u(\mathsf{Del}, y))$,
- $\mathcal{L}_q(x) = (x, \mathsf{AP}_t(x), \{\tilde{\mathcal{L}}_q(\tilde{x}_i), \tilde{\mathcal{L}}_u(\mathsf{Del}, \tilde{x}_i)\}_{i=1}^{c_x})$, for dummy nodes $\tilde{x}_i$ defined by $\mathcal{L}_u(\mathsf{Add}, x, y, \cdot)$ for updates after the previous query on x.

Theorem 1. *Assume that $\mathcal{E}$ is $(\tilde{\mathcal{L}}_e, \tilde{\mathcal{L}}_q, \tilde{\mathcal{L}}_u)$-CQA2 secure, and F is pseudorandom, then the above construction is $(\mathcal{L}_e, \mathcal{L}_q, \mathcal{L}_u)$-CQA2 secure. Furthermore, let p'_1 be a function such that $p'_1(u_0, u_1) = 1$ if and only if $y_0 = y_1$ and $w_0 = w_1$[4]. Then the construction is $(\mathcal{U}_1, p'_1)$-forward private, where $\mathcal{U}_1$ is defined in Table 1.*

5   Forward Privacy from Scratch

We next construct (interactive) DSSE which achieves forward privacy directly. First, a new data structure, named *cascaded triangles*, is designed to represent labeled bipartite graphs which supports neighbor queries, edge additions, and node deletions efficiently. We then transform it into its encrypted version.

The construction of cascaded triangles is motivated by the following. Since the neighbor queries and node deletions require traversing the sets $(x, *, *)$ and $(*, y, *)$, their data structure representations are critical for the efficiency, and later the security of the resulting DSSE scheme. In particular, they determine whether the desired operations can be executed in parallel, and how much information has to be leaked to the server for performing such operations. For example, linked list [11] traversal and updates are inherently sequential, yet updating only has a local effect (on the previous and next nodes). However, random binary search tree [12] exhibits parallel traversal and updates, yet updating affects (or leaks) the subtree rooted from the altered node. Thus, cascaded triangles is designed to support parallel traversal and local updates simultaneously.

5.1   Warm Up: Plaintext Cascaded Triangles

Overview. Our goal is to store the bipartite graph G so that neighbor queries and deletions over X or Y can be executed in sublinear time. We can do so by pre-computing the set of edges connected to each x (and y), and storing the set by a data structure which allows efficient traversal. For any x, consider

---

[4] We can drop the restriction $w_0 = w_1$ by simply encrypting w during additions.

the set of edges connecting x. We pack this set into multiple perfect binary trees, called triangles, by first forming the largest triangle possible, subtracting the edges which are already packed, then continuing to form the next largest triangle. The resulting triangles thus have strictly decreasing (cascading) heights, except for the last two which may have equal heights. This invariant is to be maintained in any later updates. To add an edge connecting x, we check if the two shortest triangles have the same height. If so, we add a new node representing the new edge on top of the two triangles, merging them into one larger triangle. Otherwise, the new node is added as a new triangle of height 1. To delete an edge, we delete the node representing this edge by replacing it with the root of the shortest triangle, splitting the latter into two smaller triangles. We can see that the invariant is still maintained after each addition and each deletion. Finally, to traverse the data structure, one may use any (parallel) tree traversal algorithms.

Setup. Concretely, cascaded triangles consists of dictionaries γ, δ, and η. We can think of γ as local states stored at the client side, while δ and η are outsourced to the server. The dictionary η is the one to store the actual data. It maps an address addr to a tuple (a, b), where $b = (\mathsf{chd}_0, \mathsf{chd}_1)$ specifies the addresses of the left and right child respectively. b is maintained so that nodes in η form perfect binary trees (triangles). This means either both addresses $(\mathsf{chd}_0, \mathsf{chd}_1)$ are empty ($\perp$) or both are valid addresses occupied in η. To store an edge (x, y, w) into η, the edge is copied twice into $a^\triangle = a^\triangledown = (x, y, w)$. The tuples $(a^\triangle, b^\triangle)$ and $(a^\triangledown, b^\triangledown)$ are stored at random addresses $\mathsf{addr}^\triangle$ and $\mathsf{addr}^\triangledown$ in η respectively. The addresses $\mathsf{addr}^\triangle$ and $\mathsf{addr}^\triangledown$ are said to be *duals* of each other, and are registered in δ, i.e., $\delta[\mathsf{addr}^\triangle] = \mathsf{addr}^\triangledown$ and $\delta[\mathsf{addr}^\triangledown] = \mathsf{addr}^\triangle$.

Globally, we describe how the nodes in η are connected to each other via the addresses stored in b. We collect all $\eta[\mathsf{addr}] = (a, b)$ corresponding to the edges in $(q, *, *)$ (or $(*, q, *)$). Let $n_q = |(q, *, *)|$ (or $|(*, q, *)|$). We pack these tuples into triangles of cascading heights $h_1 \leq h_2 < h_3 < \ldots < h_k$, where $k \leq \lceil \lg n_q \rceil + 1$. Note the possible equality between h_1 and h_2 but not the others. The ordering of the nodes is implicitly determined by the update algorithms, but does not matter here. Using the procedures described in the overview, given the size n_q, the heights $h_1, \ldots, h_k$ are uniquely determined. It is possible to represent the heights compactly by a trinary string $h \in \{0, 1, 2\}^{\lceil \lg n_q \rceil}$, such that the i-th trit (trinary digit) is set to t, if there are t triangles of height i. Due to the constraints on the heights, only the least significant non-zero trit can be set to 2. Finally, the addresses of the roots and the heights of these triangles are stored in $\gamma[q] = (\mathsf{addr}_1, \ldots, \mathsf{addr}_k, h)$.

Queries and Traversal. Traversing the sets $(q, *, *)$ and $(*, q, *)$ are straightforward with the above structure: First, retrieve the roots of the triangles from $\gamma[q] = (\mathsf{addr}_1, \ldots, \mathsf{addr}_k, h)$. Then, use parallel tree-traversal algorithms to traverse the trees from the root starting at each of these addresses. Notice that the neighbor query function Qry on x and y are supported by traversing the sets $\gamma[x] = (x, *, *)$ and $\gamma[y] = (*, y, *)$ respectively.

Add. To add a new edge (x, y, w), we first retrieve $\gamma[x] = (\mathsf{addr}_1^{\triangle}, \ldots, \mathsf{addr}_k^{\triangle}, h^{\triangle})$ and check whether the triangles rooted at $\mathsf{addr}_1^{\triangle}$ and $\mathsf{addr}_2^{\triangle}$ have the same height.

To do so, we take a detour to describe the $+1$ *operation* in $h + 1$. Recall that only the least significant non-zero trit, say the i-th trit, in h can be set to 2. The operation $h + 1$ adds 1 to the i-th trit (instead of the least significant trit as in normal addition), which sets the i-th trit to 0 and carries 1 to the $(i + 1)$-trit. Denote this event by $\mathsf{Carry}(h + 1) = 1$. Otherwise, $h + 1$ simply adds 1 to the least significant trit as in normal addition, denoted by $\mathsf{Carry}(h + 1) = 0$. Later, for deletion, we would need the -1 *operation* which is the "reverse" of $+1$. Concretely, $h - 1$ subtracts 1 from the least significant non-zero trit, say the i-th trit, and set the $(i - 1)$-th trit to 2 if $i > 1$.

With the above procedures, checking the heights of the first two triangles can be done by simply checking whether the least significant non-zero trit in $h^{\triangle}$ equals 2, or equivalently whether $\mathsf{Carry}(h^{\triangle} + 1) = 1$. If that is the case, we add $(a^{\triangle}, b^{\triangle})$ where $a^{\triangle} = (x, y, w)$ and $b^{\triangle} = (\mathsf{addr}_1^{\triangle}, \mathsf{addr}_2^{\triangle})$ to a random address $\mathsf{addr}^{\triangle}$ in η. We then update $\gamma[x] \leftarrow (\mathsf{addr}^{\triangle}, \mathsf{addr}_3^{\triangle}, \ldots, \mathsf{addr}_k^{\triangle}, h^{\triangle} + 1)$, where the two addresses $\mathsf{addr}_1^{\triangle}$ and $\mathsf{addr}_2^{\triangle}$ are replaced by the new $\mathsf{addr}^{\triangle}$. Otherwise, we add $(a^{\triangle}, b^{\triangle})$ where $a^{\triangle} = (x, y, w)$ and $b^{\triangle} = (\bot, \bot)$ to a random address $\mathsf{addr}^{\triangle}$ in η, and update $\gamma[x] \leftarrow (\mathsf{addr}^{\triangle}, \mathsf{addr}_1^{\triangle}, \ldots, \mathsf{addr}_k^{\triangle}, h^{\triangle} + 1)$. The difference is highlighted in red.

Similarly, we retrieve $(\mathsf{addr}_1^{\triangledown}, \ldots, \mathsf{addr}_k^{\triangledown}, h^{\triangledown}) \leftarrow \gamma[y]$ and check whether $\mathsf{Carry}(h^{\triangledown} + 1) = 1$. If so, we add $((x, y, w), (\mathsf{addr}_1^{\triangledown}, \mathsf{addr}_2^{\triangledown}))$ to a random address $\mathsf{addr}^{\triangledown}$ in η, and update $\gamma[y] \leftarrow (\mathsf{addr}^{\triangledown}, \mathsf{addr}_3^{\triangledown}, \ldots, \mathsf{addr}_k^{\triangledown}, h^{\triangledown} + 1)$. Otherwise, we add $((x, y, w), (\bot, \bot))$ to $\mathsf{addr}^{\triangledown}$, and update $\gamma[y] \leftarrow (\mathsf{addr}^{\triangledown}, \mathsf{addr}_1^{\triangledown}, \ldots, \mathsf{addr}_k^{\triangledown}, h^{\triangledown} + 1)$.

Delete. To delete node x, or equivalently the set of edges $(x, *, *)$, we first traverse the set $(x, *, *)$ using the above traversal algorithm. We delete all the traversed nodes in η as well as the row $\gamma[x]$. It remains to delete the dual nodes of the traversed nodes. To do so, for each traversed address $\mathsf{addr}^{\triangle}$ with $a = (x, y, w)$, look up $\delta[\mathsf{addr}^{\triangle}] = \mathsf{addr}^{\triangledown}$ and $\gamma[y] = (\mathsf{addr}_1^{\triangledown}, \ldots, \mathsf{addr}_k^{\triangledown}, h^{\triangledown})$. We wish to replace the content of $\eta[\mathsf{addr}^{\triangledown}] = (a^{\triangledown}, b^{\triangledown})$ located in the middle of some triangle by the content of $\eta[\mathsf{addr}_1^{\triangledown}]$, the root of the smallest triangle, which splits the smallest triangle into two smaller ones. In this way, the heights of the resulting triangles still satisfy the required constraints.

Concretely, we perform the following steps. (1) Look up $\delta[\mathsf{addr}_1^{\triangledown}] = \mathsf{addr}_1^{\triangle}$. (2) Delete $\delta[\mathsf{addr}^{\triangle}]$ and $\delta[\mathsf{addr}_1^{\triangledown}]$. (3) Update $\delta[\mathsf{addr}_1^{\triangle}] \leftarrow \mathsf{addr}^{\triangledown}$ and $\delta[\mathsf{addr}^{\triangledown}] \leftarrow \mathsf{addr}_1^{\triangle}$. (4) Look up $\eta[\mathsf{addr}_1^{\triangledown}] = (a_1^{\triangledown}, b_1^{\triangledown})$, where $b_1^{\triangledown} = (\mathsf{addr}_0^{\triangledown}, \mathsf{addr}_1^{\triangledown})$. (5) Update $\eta[\mathsf{addr}^{\triangledown}] \leftarrow (a_1^{\triangledown}, b_1^{\triangledown})$ and delete $\eta[\mathsf{addr}_1^{\triangledown}]$. (6) Update $\gamma[y] = (\mathsf{addr}_0^{\triangledown}, \mathsf{addr}_1^{\triangledown}, \mathsf{addr}_2^{\triangledown}, \ldots, \mathsf{addr}_k^{\triangledown}, h^{\triangledown} - 1)$, where $h^{\triangledown} - 1$ is the reverse of $h^{\triangledown} + 1$. We omit the deletion of $(*, y, *)$ which is similar to the above.

Efficiency. The storage cost of cascaded triangles is $O(|X| + |Y| + |G|) = O(|G|)$. The complexity of querying (or deleting) x and y are $O(|(x, *, *)|)$ and $O(|(*, y, *)|)$ respectively. Addition of an edge can be computed in constant time.

5.2 Our Construction: Encrypted Cascaded Triangles

We now transform the plaintext cascaded triangles into its encrypted version. Recall that the goal of the client is to encrypt a labeled bipartite graph G into an encrypted database EDB which still supports neighbor queries, edge additions, and node deletions. To do so, the client represents G using cascaded triangles, stores γ locally, and outsources δ and the encrypted η to the server. The encryption should be *non-committing*, such that there exists a simulator which can simulate the ciphertexts for new data without any leakage. When some data is to be returned upon queries or is deleted, the simulator is given enough leakage so that it can "explain" the dummy ciphertexts. Furthermore, when the sets $(x, *, *)$ and $(*, y, *)$ are leaked upon querying x and y respectively, there is no need to protect the sets by encryption any longer [8]. To speed up subsequent queries, the server should transfer the set from the encrypted η to a plaintext labeled bipartite graph $\hat{G}$. Thus, we can conceptually split G into two disjoint subgraphs $G = \tilde{G} \cup \hat{G}$, where $\tilde{G}$ is encrypted and $\hat{G}$ is in plaintext.

Encrypted cascaded triangles are similar to the plaintext counterparts. We highlight the differences and omit the identical parts.

Setup and Overview. Let NCE.(KGen, Enc, Dec) be a symmetric-key non-committing encryption scheme[5]. The correctness of our scheme will follow directly from that of NCE. For each edge (x, y, w) in the encrypted subgraph $\tilde{G}$, $\eta[\mathsf{addr}^\triangle]$ stores the tuple $(c_a^\triangle, c_b^\triangle)$, where $c_a^\triangle$ and $c_b^\triangle$ are non-committing ciphertexts of $a^\triangle = (x, y, w)$ and $b^\triangle = (\mathsf{chd}_0, \mathsf{chd}_1)$ under the keys $\mathsf{ak}^\triangle$ and $\mathsf{bk}^\triangle$ respectively. The keys $\mathsf{ak}^\triangle$ and $\mathsf{bk}^\triangle$ are independently generated for each x. Similarly, $\eta[\mathsf{addr}^\triangledown]$ stores the tuple $(c_a^\triangledown, c_b^\triangledown)$ encrypted under $\mathsf{ak}^\triangledown$ and $\mathsf{bk}^\triangledown$ respectively. The keys $\mathsf{ak}^\triangledown$ and $\mathsf{bk}^\triangledown$ are independently generated for each y. For each x, $\gamma[x]$ additionally stores secret keys $\mathsf{ak}^\triangle$ and $\mathsf{bk}^\triangle$ associated to x. Similarly, for each y, $\gamma[y]$ additionally stores secret keys $\mathsf{ak}^\triangledown$ and $\mathsf{bk}^\triangledown$ associated to y. Formally, the setup protocol in Fig. 3 shows the initialization of these dictionaries.

Queries. Queries are similar to the plaintext case. Apart from the addresses, the client also sends ak and bk retrieved from $\gamma[q]$ to the server. Using bk, the server decrypts all $b = (\mathsf{chd}_0, \mathsf{chd}_1)$ and traverses the sub-trees. Using ak, it decrypts all $a = (x, y, w)$ which are returned as query results. The server also returns the previous query results $\hat{R}$ stored in $\hat{G}$.

---

[5] For example, a ciphertext for message m with randomness r can be computed as $c = (r, \mathsf{PRF}(K, r) \oplus m)$, where PRF, modeling a random oracle, is a pseudorandom function with secret key K. In practice, one may substitute PRF with an HMAC.

$(K, \text{EDB}) \leftarrow \text{Setup}(1^\lambda, |\mathcal{X}|, |\mathcal{Y}|, |\mathcal{W}|)$

$\gamma = \phi, \delta = \phi, \eta = \phi, \hat{G} = \phi$

return $K = \gamma, \text{EDB} = (\delta, \eta, \hat{G})$

$\text{Trav}(\text{EDB}, \{\text{addr}_j\}_{j=1}^k, \text{ak}, \text{bk})$

if $k > 1$ **then**

  $R = \phi, D = \phi$

  for $j = 1, \ldots, k$ **do**

    $(R', D') \leftarrow \text{Trav}(\text{EDB}, \text{addr}_j, \text{ak}, \text{bk})$

    $R = R \cup R', D = D \cup D'$

  endfor

else

  $(c_a, c_b) \leftarrow \eta[\text{addr}_1], \eta[\text{addr}_1] \leftarrow \perp$

  $\overline{\text{addr}_1} \leftarrow \delta[\text{addr}_1], \delta[\text{addr}_1] \leftarrow \perp$

  if $\text{ak} \neq \perp$ **then**

$\text{Trav}(\text{EDB}, \{\text{addr}_j\}_{j=1}^k, \text{ak}, \text{bk})$ (cont.)

    $(x, y, w) \leftarrow \text{NCE.Dec}(\text{ak}, c_a)$

  else

    $(x, y, w) \leftarrow c_a$

  endif

  $(\text{chd}_0, \text{chd}_1) \leftarrow \text{NCE.Dec}(\text{bk}, c_b)$

  if $\text{chd}_0 \neq \perp (\vee \text{chd}_1 \neq \perp)$ **then**

    $(R', D') \leftarrow \text{Trav}(\text{EDB}, \text{chd}_0, \text{ak}, \text{bk})$

    $(R'', D'') \leftarrow \text{Trav}(\text{EDB}, \text{chd}_1, \text{ak}, \text{bk})$

    $R = R' \cup R'' \cup \{(x, y, w)\}$

    $D = D' \cup D'' \cup \{\overline{\text{addr}_1}\}$

  else

    $R = \{(x, y, w)\}, D = \{\overline{\text{addr}_1}\}$

  endif

  endif

return (R, D)

Fig. 3. Setup and traverse algorithm of encrypted cascaded triangles

---

$((K', R), (\text{EDB}', R)) \leftarrow \text{Qry}_e((K, q), \text{EDB})$

Client

$R = \phi$

    $\xrightarrow{\quad q \quad}$

if $\gamma[q] \neq \perp$ **then**

    $\xleftarrow{\quad \hat{R} \quad}$

  $(\text{ak}, \text{bk}, \{\text{addr}_j\}_{j=1}^k, h) \leftarrow \gamma[q]$ $\xrightarrow{\text{ak}, \text{bk}, \{\text{addr}_j\}_{j=1}^k}$

  $\gamma[q] \leftarrow \perp$

    $\xleftarrow{\quad R \quad}$

  if $q = x \in \mathcal{X}$ **then**

    parse R **as** $\{(x, z_j, w_j)\}_j$

  else $/\!/ \; q = y \in \mathcal{Y}$

    parse R **as** $\{(z_j, y, w_j)\}_j$

  endif

    $(K, \text{EDB}) \leftarrow \text{DelDual}((K, \{z_j\}_j, (\text{EDB}, \{\overline{\text{addr}_j}\}_j))$

endif

$R \leftarrow R \cup \hat{R}$

return (K, R)

Server

$\hat{R} \leftarrow \text{Qry}(q, \hat{G})$

$(R, D) \leftarrow \text{Trav}(\text{EDB}, \{\text{addr}_j\}_{j=1}^k, \text{ak}, \text{bk})$

parse D **as** $(\overline{\text{addr}_j})_j$

$R \leftarrow R \cup \hat{R}$

foreach $(x, y, w) \in \hat{R}$ **do**

  $\hat{G} \leftarrow \text{Udt}((\text{Add}, x, y, w), \hat{G})$

endfor

return (EDB, R)

Fig. 4. Query protocol of encrypted cascaded triangles

$(K', \mathsf{EDB}') \leftarrow \mathsf{DelDual}((K, \{q_j\}_j), (\mathsf{EDB}, \{\overline{\mathsf{addr}_j}\}_j))$

Client		**Server**
for $j = 1, \ldots, n$ **do**		
$(\mathsf{ak}_j, \mathsf{bk}_j, \{\mathsf{addr}_{j,i}\}_{i=1}^{\ell_j}, h_j) \leftarrow \gamma[q_j]$	$\xrightarrow{\;\mathsf{addr}_j := \mathsf{addr}_{j,1}\;}$	$\mathsf{addr}_j^\bullet \leftarrow \delta(\mathsf{addr}_j)$
$(\mathsf{chd}_0, \mathsf{chd}_1) \leftarrow \mathsf{NCE.Dec}(\mathsf{bk}_j, \eta[\mathsf{addr}_j].c_b)$	$\xleftarrow{\;\eta[\mathsf{addr}_j].c_b\;}$	$\delta(\mathsf{addr}_j) \leftarrow \bot$
if $\mathsf{chd}_0 \neq \bot$ ($\lor \mathsf{chd}_1 \neq \bot$) **then**		$\delta(\mathsf{addr}_j^\bullet) \leftarrow \overline{\mathsf{addr}_j}$
$\mathsf{addr}_{j,0} \leftarrow \mathsf{chd}_0, \mathsf{addr}_{j,1} \leftarrow \mathsf{chd}_1$		$\delta(\overline{\mathsf{addr}_j}) \leftarrow \mathsf{addr}_j^\bullet$
$\gamma[q_j] \leftarrow (\mathsf{ak}_j, \mathsf{bk}_j, \{\mathsf{addr}_{j,i}\}_{i=0}^{\ell_j}), h_j - 1)$		$\eta[\overline{\mathsf{addr}_j}].c_a = \eta[\mathsf{addr}_j].c_a$
else		$\eta[\mathsf{addr}_j] \leftarrow \bot$
$\gamma[q_j] \leftarrow (\mathsf{ak}_j, \mathsf{bk}_j, \{\mathsf{addr}_{j,i}\}_{i=2}^{\ell_j}), h_j - 1)$		
endif		
endfor		

Fig. 5. Dual deletion protocol of encrypted cascaded triangles

As mentioned before, for the efficiency of subsequent queries, the server should remove revealed entries from η and add them to the plaintext subgraph $\hat{G}$. Thus, the client and the server cooperates to delete the set $(q, *, *)$ or $(*, q, *)$ from η. This conceptually removes the set from the encrypted subgraph $\tilde{G}$. Finally, the server adds the set to $\hat{G}$. Formally, the query protocol is shown in Fig. 4, which utilizes the subroutine Trav and subprotocol DelDual shown in Figs. 3 and 5 respectively.

Add. Instead of sending $(a^\triangle, b^\triangle)$ in the clear, the client sends their ciphertexts $(c_a^\triangle, c_b^\triangle)$, encrypted under $\mathsf{ak}^\triangle$ and $\mathsf{bk}^\triangle$ retrieved from $\gamma[x]$, to the server respectively. In the case where $\gamma[x] = \bot$, the client generates new secret keys $\mathsf{ak}^\triangle$ and $\mathsf{bk}^\triangle$ using the key generation algorithm of the non-committing encryption scheme. Sending $(a^\triangledown, b^\triangledown)$ requires a similar treatment. Figure 6 formally describes the addition protocol.

Delete. Deletion of $(x, *, *)$ is almost identical to querying x, except that the client does not send out $\mathsf{ak}^\triangle$. Instead, the server returns all $c_a^\triangle$ so that the client can decrypt them locally. After obtaining all the edges (x, y, w), the client and the server cooperate to delete $(x, *, *)$ from η as in the query algorithm. This conceptually removes $(x, *, *)$ from the encrypted subgraph $\tilde{G}$. Finally, the server also removes $(x, *, *)$ from the plaintext subgraph $\hat{G}$. Deletion of $(*, y, *)$ is done similarly. Figure 7 shows the deletion protocol, which also utilizes the subroutines Trav and DelDual in Figs. 3 and 5 respectively.

$(\cdot, \mathsf{EDB}') \leftarrow \mathsf{Udt}_e(\cdot, \mathsf{EDB})$

Receive $(\mathsf{addr}^\triangle, \mathsf{addr}^\triangledown, c_a^\triangle, c_b^\triangle, c_a^\triangledown, c_b^\triangledown)$

$\delta[\mathsf{addr}^\triangle] \leftarrow \mathsf{addr}^\triangledown, \quad \delta[\mathsf{addr}^\triangledown] \leftarrow \mathsf{addr}^\triangle$

$\eta[\mathsf{addr}^\triangle] \leftarrow (c_a^\triangle, c_b^\triangle), \quad \eta[\mathsf{addr}^\triangledown] \leftarrow (c_a^\triangledown, c_b^\triangledown)$

return EDB

$(K', \cdot) \leftarrow \mathsf{Udt}_e((K, u = (\mathsf{Add}, x, y, w)), \cdot)$

$\mathsf{addr}^\triangle, \mathsf{addr}^\triangledown \leftarrow \{0, 1\}^*$

for $(q, \Diamond) \in \{(x, \triangle), (y, \triangledown)\}$ **do**

  if $\gamma[q] = \bot$ **then**

    $\mathsf{ak}, \mathsf{bk} \leftarrow \mathsf{NCE.KGen}(1^\lambda)$

    $\mathsf{addr}_1, \mathsf{addr}_2 \leftarrow \bot$

    $\gamma[q] \leftarrow (\mathsf{ak}, \mathsf{bk}, \{\mathsf{addr}\}, 1)$

(cont.)

else

  $(\mathsf{ak}, \mathsf{bk}, \{\mathsf{addr}_j\}_{j=1}^k, h) \leftarrow \gamma[q]$

  if $\mathsf{Carry}(h + 1)$ **then**

    $\gamma[q] \leftarrow (\mathsf{ak}, \mathsf{bk}, \{\mathsf{addr}, \mathsf{addr}_j\}_{j=3}^k, h + 1)$

  else

    $\gamma[q] \leftarrow (\mathsf{ak}, \mathsf{bk}, \{\mathsf{addr}, \mathsf{addr}_j\}_{j=1}^k, h + 1)$

  endif

endif

$c_a^\Diamond \leftarrow \mathsf{NCE.Enc}(\mathsf{ak}, (x, y, w))$

$c_b^\Diamond \leftarrow \mathsf{NCE.Enc}(\mathsf{bk}, (\mathsf{addr}_1, \mathsf{addr}_2))$

endfor

Send $(\mathsf{addr}^\triangle, \mathsf{addr}^\triangledown, c_a^\triangle, c_b^\triangle, c_a^\triangledown, c_b^\triangledown)$

return K

Fig. 6. Addition protocol of encrypted cascaded triangles

$(K', \mathsf{EDB}') \leftarrow \mathsf{Udt}_e((K, u = (\mathsf{Del}, q)), \mathsf{EDB})$

if $\gamma[q] \neq \bot$ **then** $\xrightarrow{\quad q \quad}$ $\hat{G} \leftarrow \mathsf{Udt}((\mathsf{Del}, q), \hat{G})$

  $(\mathsf{ak}, \mathsf{bk}, \{\mathsf{addr}_i\}_i, h) \leftarrow \gamma[q]$ $\xrightarrow{\mathsf{bk}, \{\mathsf{addr}_i\}_i}$ $(R, D) \leftarrow \mathsf{Trav}(\mathsf{EDB}, \{\mathsf{addr}_i\}_i, \bot, \mathsf{bk})$

  $\gamma[q] \leftarrow \bot$ $\xleftarrow{\quad R \quad}$ **parse** D **as** $(\overline{\mathsf{addr}}_j)_j$

  if $q \in \mathcal{X}$ **then**

    $\{(q, z_j, w_j)\}_j \leftarrow \mathsf{NCE.Dec}(\mathsf{ak}, R)$

  else $/\!\!/ \; q \in \mathcal{Y}$

    $\{(z_j, q, w_j)\}_j \leftarrow \mathsf{NCE.Dec}(\mathsf{ak}, R)$

  endif

      $(K, \mathsf{EDB}) \leftarrow \mathsf{DelDual}((K, \{z_j\}_j), (\mathsf{EDB}, \{\overline{\mathsf{addr}}_j\}_j))$

endif

return K **return** EDB

Fig. 7. Deletion protocol of encrypted cascaded triangles

5.3 Analysis

Storage Cost. For each x, if $n_x = |(x, *, *) \cap \tilde{G}| > 0$, the client stores two λ-bit keys of non-committing encryption, one $\lceil \lg n_x \rceil$-trit string, and at most $\lceil \lg n_x \rceil + 1$ λ-bit addresses. Similar storage is required for each y. In the extreme case where the client adds all possible data and never queries or deletes, the storage cost is $O(\mathsf{poly} \cdot (|\mathcal{X}| \lg |\mathcal{Y}| + |\mathcal{Y}| \lg |\mathcal{X}|))$. However, querying and deleting x (or y) removes

$(x, *, *)$ (or $(*, y, *)$) from $\tilde{G}$, which waives the client local storage for x (or y). Thus, the storage of a reasonable client would be much smaller.

The storage cost of the server is linear in the number of edges (x, y, w) added to the server, which is optimal. Furthermore, if an edge (x, y, w) is revealed due to a previous query, it is stored in plaintext instead of ciphertext, where the former is much better in terms of locality [3].

Computation and Communication Cost. Both the client and the server perform essentially no work during setup: They just initialize empty dictionaries.

During a query on q, the client first looks up its dictionary $\gamma[q]$, which consists of two λ-bit keys of NCE, one $\lceil \lg n_q \rceil$-trit string, and at most $(\lceil \lg n_q \rceil + 1)$ λ-bit addresses. The query q, the keys, and the addresses are sent to the server. The server traverses the set $(x, *, *)$ if $q = x \in \mathcal{X}$, or the set $(*, y, *)$ if $q = y \in \mathcal{Y}$. In the former, the server needs to perform $O(|(x, *, *) \cap \tilde{G}|)$ decryption, and execute $\mathsf{Qry}(q, \hat{G})$ which takes time $|(x, *, *) \cap \hat{G}|$. It then returns the query result of size $|(x, *, *)| = |(x, *, *) \cap \tilde{G}| + |(x, *, *) \cap \hat{G}|$ to the client. The client looks up and sends $|(x, *, *) \cap \tilde{G}|$ addresses to the server, which performs the same amount of I/O tasks, and returns $O(|(x, *, *) \cap \tilde{G}|)$ ciphertexts to the client. The client finalizes by performing $O(|(x, *, *) \cap \tilde{G}|)$ decryption and the same amount of I/O tasks. Overall, both computation and communication complexities for both the client and the server are in the order of $O(\mathsf{poly} \cdot |(x, *, *)|)$, which is optimal for the server. In contrast to the presentation in the formal construction, the number of rounds can be compressed into 4.

Since deletion is almost identical to querying except for local decryption, their overall computation and communication complexities are identical.

For addition, the client performs a constant amount of I/O tasks to update γ, while sending 2 λ-bit addresses and 4 ciphertexts to the server. The server simply writes the ciphertexts to the specified addresses. Therefore, the overall computation, round, and communication complexities for both the client and the server are constant, which is again optimal.

Note that our scheme supports batch operations (querying, addition, and deletion) straightforwardly. In such case, the computation and communication complexities increase linearly while the round complexity remains unchanged.

Security. In the full version, we prove that our scheme is secure against adaptive chosen query attack with very minimal leakage. In particular, our scheme achieves $(\mathcal{U}_1, 1)$-forward privacy, where $\mathcal{U}_1$ is defined in Table 1. We begin by defining the leakage functions. For setup, $\mathcal{L}_u$ only leaks the sizes of the spaces, i.e., $|\mathcal{X}|$, $|\mathcal{Y}|$, and $|\mathcal{W}|$. For addition, $\mathcal{L}_u$ only leaks the update type and the time $\mathsf{Time}(x, y)$ of the addition as the new addresses are truly random and the ciphertexts can be simulated by the simulator of the non-committing encryption scheme. For deletion of $(x, *, *)$, $\mathcal{L}_u$ leaks the update type, x, and the time $\mathsf{Time}(x, y)$ when each of the edges $(x, y, w) \in (x, *, *)$ is added. It also leaks, for each y such that $(x, y, w) \in (x, *, *)$, the time $\mathsf{Time}(*, y)$ when the last edge in $(*, y, *)$ is added. Leakage for deleting $(*, y, *)$ is defined similarly. Finally, for queries on q, $\mathcal{L}_q$ leaks all information leaked by $\mathcal{L}_u$ upon deletion of $(x, *, *)$

if $q = x \in \mathcal{X}$, or $(*, y, *)$ if $q = y \in \mathcal{Y}$, and the access patterns $\mathsf{AP}_t(q)$ of q, assuming it is sorted by the time each response is added. Formally, we define:

- $\mathcal{L}_e(\mathcal{G}) = (|\mathcal{X}|, |\mathcal{Y}|, |\mathcal{W}|)$
- $\mathcal{L}_u(\mathsf{Add}, x, y, w) = (\mathsf{Add}, \mathsf{Time}(x, y))$
- $\mathcal{L}_u(\mathsf{Del}, x) = (\mathsf{Del}, x, \{(\mathsf{Time}(x, y), \mathsf{Time}(*, y)) : (x, y, \cdot) \in (x, *, *)\})$
- $\mathcal{L}_u(\mathsf{Del}, y) = (\mathsf{Del}, y, \{(\mathsf{Time}(x, y), \mathsf{Time}(x, *)) : (x, y, \cdot) \in (*, y, *)\})$
- $\mathcal{L}_q(x) = (x, \mathsf{AP}_t(x), \{(\mathsf{Time}(x, y), \mathsf{Time}(*, y)) : (x, y, \cdot) \in (x, *, *)\})$
- $\mathcal{L}_q(y) = (y, \mathsf{AP}_t(y), \{(\mathsf{Time}(x, y), \mathsf{Time}(x, *)) : (x, y, \cdot) \in (*, y, *)\})$

The simulation is sketched as follows. The simulator first initializes empty dictionaries δ and η, and the empty plaintext graph $\hat{G}$. It maintains a table T mapping time t to time-address tuples $((t_0, \mathsf{addr}), (t_1, \mathsf{addr}^\triangledown))$.

For addition at time t, it samples random addresses addr and $\mathsf{addr}^\triangledown$, and registers them in δ. It sets $T[t] \leftarrow ((t, \mathsf{addr}), (t, \mathsf{addr}^\triangledown))$. It simulates the ciphertexts in $\eta[\mathsf{addr}]$ and $\eta[\mathsf{addr}^\triangledown]$ using the simulator of NCE.

For deletion of $(x, *, *)$, it is given x, which allows it to delete $(x, *, *)$ from $\hat{G}$. It is also given the time $\mathsf{Time}(x, y)$ when each of the edges $(x, y, w) \in (x, *, *)$ is added, and for each y such that $(x, y, w) \in (x, *, *)$ the time $\mathsf{Time}(*, y)$ when the last edge in $(*, y, *)$ is added. It recalls from the table T all addr and $\mathsf{addr}^\triangledown$ pairs which are created at time $\mathsf{Time}(x, y)$ and $\mathsf{Time}(*, y)$. With the knowledge of these addresses, the simulator can maintain δ and η as in the real scheme. It must also output simulated bk and explain the ciphertexts c_b encrypting the addresses. To achieve this, the simulator passes the ciphertexts and the corresponding addresses to the simulator of the non-committing encryption scheme, where the latter outputs the simulated bk. Deletion of $(*, y, *)$ is simulated similarly.

Queries are simulated almost identically as in the simulation of deletions, except that the simulator must now also output simulated ak and explain the ciphertexts c_a encrypting the query result. Similar to the above, this can be done by calling the simulator of the non-committing encryption scheme.

Theorem 2. *Assume that* NCE.(KGen, Enc, Dec) *is a symmetric-key non-committing encryption scheme with message space* $\{0, 1\}^{\max(\lg |\mathcal{X}| + \lg |\mathcal{Y}| + \lg |\mathcal{W}|, 2\lambda)}$, *the above construction is* $(\mathcal{L}_e, \mathcal{L}_q, \mathcal{L}_u)$-CQA2 *secure. Furthermore, the above construction is* $(\mathcal{U}_1, 1)$-*forward private, where* $\mathcal{U}_1$ *is defined in Table 1.*

References

1. Bösch, C., Hartel, P., Jonker, W., Peter, A.: A survey of provably secure searchable encryption. ACM Comput. Surv. **47**(2), 18:1–18:51 (2014)
2. Bost, R.: $\sum o \varphi o \varsigma$: forward secure searchable encryption. In: Proceedings of the 2016 ACM SIGSAC Conference on Computer and Communications Security, Vienna, Austria, 24–28 October 2016, pp. 1143–1154 (2016)
3. Cash, D., Tessaro, S.: The locality of searchable symmetric encryption. In: Nguyen, P.Q., Oswald, E. (eds.) EUROCRYPT 2014. LNCS, vol. 8441, pp. 351–368. Springer, Heidelberg (2014). doi:10.1007/978-3-642-55220-5_20

4. Chen, Y.-C., Chow, S.S.M., Chung, K.-M., Lai, R.W.F., Lin, W.-K., Zhou, H.-S.: Cryptography for parallel RAM from indistinguishability obfuscation. In: Sudan, M. (ed.) ITCS 2016, Cambridge, MA, USA, 14–16 January 2016, pp. 179–190. ACM (2016)

5. Curtmola, R., Garay, J.A., Kamara, S., Ostrovsky, R.: Searchable symmetric encryption: improved definitions and efficient constructions. In: Juels, A., Wright, R.N., Sabrina De Capitani di Vimercati, (eds.) ACM CCS 2006, Alexandria, Virginia, USA, 30 October–3 November 2006, pp. 79–88. ACM Press (2006)

6. Curtmola, R., Garay, J.A., Kamara, S., Ostrovsky, R.: Searchable symmetric encryption: Improved definitions and efficient constructions. J. Comput. Secur. **19**(5), 895–934 (2011)

7. Garg, S., Mohassel, P., Papamanthou, C.: TWORAM: efficient oblivious RAM in two rounds with applications to searchable encryption. In: Robshaw, M., Katz, J. (eds.) CRYPTO 2016. LNCS, vol. 9816, pp. 563–592. Springer, Heidelberg (2016). doi:10.1007/978-3-662-53015-3_20

8. Hahn, F., Kerschbaum, F.: Searchable encryption with secure and efficient updates. In: Ahn, G.-J., Yung, M., Li, N. (eds.) ACM CCS 2014, Scottsdale, AZ, USA, 3–7 November 2014, pp. 310–320. ACM Press (2014)

9. Islam, S.M., Kuzu, M., Kantarcioglu, M.: Access pattern disclosure on searchable encryption: ramification, attack and mitigation. In: NDSS 2012, San Diego, CA, USA, 5–8 February 2012. The Internet Society (2012)

10. Kamara, S., Papamanthou, C.: Parallel and dynamic searchable symmetric encryption. In: Sadeghi, A.-R. (ed.) FC 2013. LNCS, vol. 7859, pp. 258–274. Springer, Heidelberg (2013). doi:10.1007/978-3-642-39884-1_22

11. Kamara, S., Papamanthou, C., Roeder, T.: Dynamic searchable symmetric encryption. In: Yu, T., Danezis, G., Gligor, V.D. (eds.) ACM CCS 2012, Raleigh, NC, USA, 16–18 October 2012, pp. 965–976. ACM Press (2012)

12. Lai, R.W.F., Chow, S.S.M.: Structured encryption with non-interactive updates and parallel traversal. In: 35th IEEE International Conference on Distributed Computing Systems, ICDCS 2015, Columbus, OH, USA, 29 June–2 July 2015, pp. 776–777 (2015)

13. Lai, R.W.F., Chow, S.S.M.: Parallel and dynamic structured encryption. In: SECURECOMM 2016 (2016, to appear)

14. Liu, C., Zhu, L., Wang, M., Tan, Y.: Search pattern leakage in searchable encryption: attacks and new construction. Inf. Sci. **265**, 176–188 (2014)

15. Rizomiliotis, P., Gritzalis, S.: ORAM based forward privacy preserving dynamic searchable symmetric encryption schemes. In: Proceedings of the 2015 ACM Workshop on Cloud Computing Security Workshop, CCSW 2015, Denver, Colorado, USA, 16 October 2015, pp. 65–76 (2015)

16. Song, D.X., Wagner, D., Perrig, A.: Practical techniques for searches on encrypted data. In: 2000 IEEE Symposium on Security and Privacy, Oakland, CA, USA, pp. 44–55. IEEE Computer Society Press, May 2000

17. Stefanov, E., Papamanthou, C., Shi, E.: Practical dynamic searchable encryption with small leakage. In: NDSS 2014, San Diego, CA, USA, 23–26 February 2014. The Internet Society (2014)

18. Zhang, Y., Katz, J., Papamanthou, C.: All your queries are belong to us: the power of file-injection attacks on searchable encryption. In: 25th USENIX Security Symposium, USENIX Security 16, Austin, TX, USA, 10–12 August 2016, pp. 707–720 (2016)

Bounds in Various Generalized Settings of the Discrete Logarithm Problem

Jason H.M. Ying$^{(\boxtimes)}$ and Noboru Kunihiro

The University of Tokyo, Tokyo, Japan
jason_ying@it.k.u-tokyo.ac.jp

Abstract. This paper examines the generic hardness of the generalized multiple discrete logarithm problem, where the solver has to solve k out of n instances for various settings of the discrete logarithm problem. For generic k and n, we introduce two techniques to establish the lower bounds for this computational complexity. One method can be shown to achieve asymptotically tight bounds for small inputs in the classical setting. The other method achieves bounds for larger inputs as well as being able to adapt for applications in other discrete logarithm settings. In the latter, we obtain the generalized lower bounds by applying partitions of n and furthermore show that our chosen method of partition achieves the best bounds. This work can be regarded as a generalization and extension on the hardness of the multiple discrete logarithm problem analyzed by Yun (EUROCRYPT '15). Some explicit bounds for various n with respect to k are also computed.

Keywords: Discrete logarithm · Generalized multiple discrete logarithm · Chebyshev's inequality · Optimization · Gaussian elimination

1 Introduction

Discrete logarithms arise in many aspects of cryptography. The hardness of the discrete logarithm problem is central in many cryptographic schemes; for instance in signatures, key exchange protocols and encryption schemes. In fact, many variants of the discrete logarithm problem have evolved over the years. Some of these include the Bilinear Diffie-Hellman Exponent Problem [6], the Bilinear Diffie-Hellman Inversion Problem [4], the Weak Diffie-Hellman Problem [14] and the Strong Diffie-Hellman Problem [5]. We first describe the classical discrete logarithm problem and its generalizations. Formal descriptions of the statements are as follow.

Let G be a cyclic group such that $|G| = p$ where p a prime and denote g to be a generator of G so that $G = \langle g \rangle$.

The discrete logarithm problem (DLP) is defined as follows: Given G, p and any h selected uniformly at random from G, find $x \in \mathbb{Z}_p$ satisfying $g^x = h$.

The k-Multiple Discrete Logarithm (k-MDL) is defined as follows:

© Springer International Publishing AG 2017
D. Gollmann et al. (Eds.): ACNS 2017, LNCS 10355, pp. 498–517, 2017.
DOI: 10.1007/978-3-319-61204-1_25

Definition 1 (MDL). *Given G, p and k elements h_1, h_2, ..., h_k selected uniformly at random from G, find non-negative integers x_1, x_2, ..., x_k satisfying $g^{x_i} = h_i \; \forall \; 1 \le i \le k$.*

In particular when $k = 1$, the 1-MDL is equivalent to DLP.

For $k \le n$, we define the (k,n)-Generalized Multiple Discrete Logarithm $((k,n)$-GMDL) as follows:

Definition 2 (GMDL). *Given G, p and n elements h_1, h_2, ..., h_n selected uniformly at random from G, find k pairs (i, x_i) satisfying $g^{x_i} = h_i$ where $i \in S$ and where S is a k-subset of $\{1, ..., n\}$.*

As the definition suggests, (k,n)-GMDL can be viewed as a generalization of k-MDL. In particular when $n = k$, the (k,k)-GMDL is equivalent to the k-MDL.

Cryptographic constructions based on DLP are applied extensively. For instance, an early application of the DLP in cryptography came in the form of the Diffie-Hellman key exchange protocol [8] for which the security is dependent on the hardness of the DLP. Among some of the others include the ElGamal encryption and signature schemes [9] as well as Schnorr's signature scheme and identification protocol [16]. The multiple discrete logarithm problem mainly arises from elliptic curve cryptography. NIST recommended a small set of fixed (or standard) curves for use in cryptographic schemes [1] to eliminate the computational cost of generating random secure elliptic curves. The implications for the security of standard elliptic curves over random elliptic curves were analysed based on the efficiency of solving multiple discrete logarithm problems [11].

In a generic group, no special properties which are exhibited by any specific groups or their elements are assumed. Algorithms for a generic group are termed as generic algorithms. There are a number of results pertaining to generic algorithms for DLP and k-MDL. Shoup showed that any generic algorithm for solving the DLP must perform $\Omega(\sqrt{p})$ group operations [17]. There are a few methods for computing discrete logarithm in approximately $\sqrt{p}$ operations. For example, Shanks Baby-Step-Giant-Step method computes the DLP in $\tilde{O}(\sqrt{p})$ operations. One other method is the Pollard's Rho Algorithm which can be achieved in $O(\sqrt{p})$ operations [15]. Since then, further practical improvements to the Pollard's Rho Algorithm have been proposed in [3,7,18] but the computational complexity remains the same. There exist index calculus methods which solve the DLP in subexponential time. However, such index calculus methods are not relevant in our context since they are not applicable for a generic group.

An extension of Pollard's Rho algorithm was proposed in [13] which solves k-MDL in $O(\sqrt{kp})$ group operations if $k \le O(p^{1/4})$. It was subsequently shown in [10] that $O(\sqrt{kp})$ can in fact be achieved without the imposed condition on k. The former's method of finding discrete logarithm is sequential in the sense that they are found one after another. However in the latter's method, all the discrete logarithms can only be obtained towards the end. Finally, it was presented in [19] that any generic algorithm solving k-MDL must require at least $\Omega(\sqrt{kp})$ group operations if $k = o(p)$.

Our Contributions. In the context of our work, suppose an adversary has knowledge or access to many instances of the discrete logarithm problem either from a generic underlying algebraic group or from a standard curve recommended by NIST. Our work investigates how difficult it is for such an adversary to solve subcollections of those instances. One of our result outcomes in this work shows that an adversary gaining access to additional instances of the DLP provides no advantage in solving some subcollection of them when k is small and for corresponding small classes of n. Our techniques are also applicable to other standard non-NIST based curves. For instance, the results in this work are relevant to Curve25519 [2] which has garnered considerable interest in recent years. Furthermore, we also establish formal lower bounds for the generic hardness of solving the GMDL problem for larger k values. As a corollary, these results provide the lower bounds of solving the GMDL problem for the full possible range of inputs k. Part of this work can be viewed as a generalization of the results in [19].

More specifically, we introduce two techniques to solve such generalized multiple discrete logarithm problems. The first we refer to as the matrix method which is also shown to achieve asymptotically tight bounds when the inputs are small. From this, we obtain the result that the GMDL problem is as hard as the MDL problem for $k = o\left(\frac{p^{1/3}}{\log^{2/3} p}\right)$ and $kn^2 = o\left(\frac{p}{\log^2 p}\right)$. This strictly improves the result of [13] where the equivalence is achieved for a smaller range of inputs satisfying $k = o(p^{\frac{1}{4}})$ and $k^2 n^2 = o(p)$. The second technique is referred as the block method which can be applied for larger inputs. We also show that the block partitioning in this method is optimized. Moreover, when n is relatively small with respect to k, the bounds that are obtained in this way are also asymptotically tight. Furthermore, we demonstrate that the block method can be adapted and applied to generalized versions of other discrete logarithm settings introduced in [12] to also obtain generic hardness bounds for such problems. For instance, part of this work also shows that solving one out of n instances of the Discrete Logarithm Problem with Auxiliary Inputs is as hard as solving a single given instance when n is not too large. In addition, we also explain why the matrix method cannot be extended to solve these problems.

2   Preliminaries

For generic groups of large prime order p, denote T_k, $T_{k,n}$ to be the expected workload (in group operations) of an optimal algorithm solving the k-MDL problem and (k, n)-GMDL problem respectively.

Lemma 1 and Corollary 1 for a special case $k = 1$ are attributed to [13].

Lemma 1. $T_1 \leq T_{1,n} + 2n \log_2 p$

Proof. Given an arbitrary $h \in G = \langle g \rangle$, obtaining x such that $g^x = h$ can be achieved in time T_1. For all i, $1 \leq i \leq n$, select integers r_i uniformly at random from the set $\{0, \ldots, p - 1\}$ and define $h_i := g^{r_i} h = g^{x + r_i}$. All the h_i are random since all the r_i and h are random. Apply a generic algorithm with inputs of $(h_1, h_2, \ldots, h_n)$ that solves the $(1, n)$-GMDL problem in time $T_{1,n}$. The resulting algorithm outputs (j, y) such that $h_j = g^y$, $1 \leq j \leq n$. Therefore, $x \equiv y - r_j \bmod p$, thus solving the 1-MDL problem within $T_{1,n} + 2n \log_2 p$ group multiplications. □

Corollary 1. *For all* $n = o\left(\frac{\sqrt{p}}{\log p}\right)$, $T_{1,n} = \Omega(\sqrt{p})$.

Proof. Since $T_1 = \Omega(\sqrt{p})$ [17], when $n = o\left(\frac{\sqrt{p}}{\log p}\right)$, it follows directly from Lemma 1 that $T_{1,n} = \Omega(\sqrt{p})$. □

It was also obtained in [13] that the GMDL problem is as hard as the MDL problem if $kn \ll \sqrt{p}$. Since $k \leq n$, this equivalence is valid for $k = o(p^{\frac{1}{4}})$ and $k^2 n^2 = o(p)$.

3   Generalized Bounds of $T_{k,n}$ for Small k

The first method we introduce is to obtain an improved lower bound of $T_{k,n}$ for small k. We refer to this as the Matrix technique.

We seek to obtain an upper bound of T_k based on $T_{k,n}$. Given $g^{x_i} = h_i$ $\forall$ $1 \leq i \leq k$, T_k represents the time to solve all such x_i. For all $1 \leq i \leq n$, denote y_i by the following[1]:

$$
\begin{pmatrix} y_1 \\ y_2 \\ \vdots \\ y_n \end{pmatrix} = \begin{pmatrix} 1 & \alpha_1 & \alpha_1^2 & \cdots & \alpha_1^{k-1} \\ 1 & \alpha_2 & \alpha_2^2 & \cdots & \alpha_2^{k-1} \\ \vdots & \vdots & \vdots & \ddots & \vdots \\ 1 & \alpha_n & \alpha_n^2 & \cdots & \alpha_n^{k-1} \end{pmatrix} \begin{pmatrix} x_1 \\ x_2 \\ \vdots \\ x_k \end{pmatrix}
$$

Next, multiply each g^{y_i} with a corresponding random element g^{r_i} where $0 \leq r_i \leq p - 1$. By considering these randomized $g^{y_i + r_i}$ as inputs to a (k, n)-GMDL solver, this solver outputs solutions to k out of n of such discrete logarithms. These solutions are of the form $y_i + r_i$. As such, a total of k values of y_i can be obtained by simply subtracting from their corresponding r_i. We claim that any k collections of y_i is sufficient to recover all of $x_1, x_2, \ldots, x_k$. Indeed, it suffices to show that any k-by-k submatrices of

$$
V = \begin{pmatrix} 1 & \alpha_1 & \alpha_1^2 & \cdots & \alpha_1^{k-1} \\ 1 & \alpha_2 & \alpha_2^2 & \cdots & \alpha_2^{k-1} \\ \vdots & \vdots & \vdots & \ddots & \vdots \\ 1 & \alpha_n & \alpha_n^2 & \cdots & \alpha_n^{k-1} \end{pmatrix}
$$

has non zero determinant. This can be satisfied by simply letting $\alpha_i = i$ since V is a Vandermonde matrix.

---

[1] This representation of y_i came about from a suggestion by Phong Q. Nguyen.

In this case, recovering $x_1, x_2, \ldots, x_k$ from k number of y_i requires solving a k-by-k system of linear equations. This can be achieved in $O(k^3)$ arithmetic operations using Gaussian elimination. Crucially, this does not involve any group operations. On the other hand, group operations are incurred from the computations of all $g^{y_i} = g^{x_1 + \alpha_i x_2 + \cdots + \alpha_i^{k-1} x_k}$. Since $\alpha_i = i$, this process requires the computations of each $(g^{x_j})^{i^{j-1}} \ \forall \ 1 \leq i \leq n, \ 1 \leq j \leq k$. Denote $a_{i,j} = (g^{x_j})^{i^{j-1}}$. By noting that $a_{i+1,j} = a_{i,j}^{(\frac{i+1}{i})^{j-1}}$, it can be concluded that computing $a_{i+1,j}$ given $a_{i,j}$ requires at most $2(j-1) \log_2 \frac{i+1}{i}$ group multiplications. Moreover, $a_{1,j} = g^{x_j}$ is already known. Hence the total number of groups multiplications required to compute all $(g^{x_j})^{i^{j-1}}$ is at most

$$\sum_{j=1}^{k} \sum_{i=1}^{n-1} 2(j-1) \log_2 \left(\frac{i+1}{i} \right) = k(k-1) \log_2 n.$$

Furthermore, each addition in the exponent of $g^{x_1 + \alpha_i x_2 + \cdots + \alpha_i^{k-1} x_k + r_i}$ constitutes a group multiplication. Therefore, kn group multiplications are necessary in this step. Thus, the total number of group multiplications required to compute all of $g^{y_i + r_i}$ is at most $kn + k(k-1) \log_2 n$. Since $k \leq n < p$, the above expression can be bounded from above by $2kn \log_2 p$ and so it follows that

$$T_k \leq T_{k,n} + 2kn \log_2 p$$

Since $T_k = \Omega(\sqrt{kp})$ [19], $T_{k,n}$ is asymptotically as large as T_k if $nk \log p \ll \sqrt{kp}$. Hence, $T_{k,n} = \Omega(\sqrt{kp})$ if $k = o\left(\frac{p^{1/3}}{\log^{2/3} p} \right)$ and $n\sqrt{k} = o\left(\frac{\sqrt{p}}{\log p} \right)$. Moreover, this bound is asymptotically tight since there exists an algorithm which solves k-MDL in $O(\sqrt{kp})$.

4   Generalized Bounds of $T_{k,n}$ for Larger k

The matrix technique has been shown to provide asymptotically tight bounds required to solve k out of n of the multiple discrete logarithm in the classical setting when the inputs are small. One main limitation of this technique is that it is only applicable to the classical DLP and cannot be extended for other variants or other settings of the DLP. This will be shown in further details in the subsequent sections. Moreover, in light of the fact that the bound of T_k can be achieved for large inputs extending to $o(p)$, the matrix method is not sufficient to obtain analogous bounds for larger such k values. In this section, we address these issues by introducing the block method to evaluate lower bounds of $T_{k,n}$ for general k, including large k values. Moreover, the block technique can also be applied to other variants or other settings in the MDLP. This will be also described in the later sections.

Proposition 1. *Suppose that* $n \geq k^2$. *Then,*

$$T_k \leq k T_{k,n} + 2n H_k \log_2 p$$

where H_k denotes the kth harmonic number, $H_k = \sum_{i=1}^{k} \frac{1}{i}$.

Proof. Given arbitraries $h_1, h_2, \ldots, h_k \in G = \langle g \rangle$, obtaining $x_1, x_2, \ldots, x_k$ such that $g^{x_i} = h_i \ \forall \ 1 \le i \le k$ can be achieved in time T_k. Consider n elements partitioned into k blocks each of size approximately s_k, where $s_k = \frac{n}{k}$. Each block is labelled i where i ranges from 1 to k. For each i, $1 \le i \le k$, select about s_k integers $r_{i,j}$ uniformly at random from $\mathbb{Z}_p$ and define $h_{i,j} := g^{r_{i,j}} h_i = g^{x_i + r_{i,j}}$. This generates n of $h_{i,j}$ which when applied to a generic algorithm for the (k, n)-GMDL problem, outputs k pseudo solutions. Computing each $h_{i,j}$ requires at most $2 \log_2 p$ group multiplications. Since each block is about size $\frac{n}{k} \ge k$, these k pseudo solutions might be derived from the same block. In which case, the algorithm outputs k of $((i', j), y_{i',j})$ such that $h_{i',j} = g^{y_{i',j}}$ for some i'. As a result, one can obtain $x_{i'} \equiv y_{i',j} - r_{i',j} \bmod p$ but derive no other information of other values of x_i. This invokes at most $T_{k,n} + 2n \log_2 p$ group operations. Figure 1 illustrates an overview of the first phase.

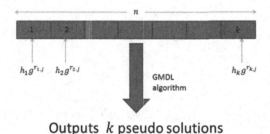

Outputs k pseudo solutions

Fig. 1. Overview of the first phase

The second phase proceeds as follows. Since 1 out of k discrete logarithms has been obtained, discard the block for which that determined discrete logarithm is contained previously. For each of $i \in \{1, \ldots, k\} \setminus \{i'\}$, select about $\frac{s_k}{k-1}$ integers $r_{i,j}$ uniformly at random from $\mathbb{Z}_p$ and define $h_{i,j} := g^{r_{i,j}} h_i = g^{x_i + r_{i,j}}$. Incorporate these new values of $h_{i,j}$ into the remaining $k-1$ blocks. Hence, each of the remaining $k-1$ unsolved discrete logarithms are contained separately in $k-1$ blocks each of size approximately $\frac{n}{k-1}$. This generates n of $h_{i,j}$ which when applied to a generic algorithm for the (k, n)-GMDL problem, outputs k pseudo solutions. Since each block is about size $\frac{n}{k-1} \ge k$, these k pseudo solutions might once again be derived from the same block. In which case, the algorithm outputs k of $((i'', j), y_{i'',j})$ such that $h_{i'',j} = g^{y_{i'',j}}$ for some i''. As a result, one can obtain $x_{i''} \equiv y_{i'',j} - r_{i'',j} \bmod p$ but derive no other information for the other remaining values of x_i. This second phase incurs at most $T_{k,n} + 2s_k \log_2 p$ group operations.

The third phase executes in a similar manner to the second phase as follows. For each of $i \in \{1, \ldots, k\} \setminus \{i', i''\}$, select about $\frac{s_k-1}{k-2}$ integers $r_{i,j}$ uniformly at random from $\mathbb{Z}_p$ and define $h_{i,j} := g^{r_{i,j}} h_i = g^{x_i + r_{i,j}}$. Incorporate these new values of $h_{i,j}$ into the remaining $k-2$ blocks. Hence, each of the remaining $k-2$ unsolved discrete logarithms are contained separately in $k-2$ blocks each of size approximately $\frac{n}{k-2}$. This generates n of $h_{i,j}$ which when applied to a generic

J.H.M. Ying and N. Kunihiro

algorithm for the (k,n)-GMDL problem, outputs k pseudo solutions. Since each block is about size $\frac{n}{k-2} \geq k$, these k pseudo solutions might be derived from the same block. In which case, the algorithm outputs k of $((i^{(3)}, j), y_{i^{(3)},j})$ such that $h_{i^{(3)},j} = g^{y_{i^{(3)},j}}$ for some $i^{(3)}$. As a result, one can obtain $x_{i^{(3)}} \equiv y_{i^{(3)},j} - r_{i^{(3)},j}$ mod p but derive no other information for the other remaining values of x_i. This second phase incurs at most $T_{k,n} + 2s_{k-1} \log_2 p$ group operations.

During each phase of the process, a generic algorithm for the (k,n)-GMDL problem can never guarantee outputs deriving from different blocks since $n \geq k^2$ implies $\frac{n}{k-i+1} \geq k$, $\forall 1 \leq i \leq k$. In general for $i \geq 2$, the maximum number of group operations required in the ith phase is given by $T_{k,n} + 2s_{k+2-i} \log_2 p$. The process terminates when all k discrete logarithms have been obtained. Since each phase outputs exactly 1 out of k discrete logarithms, all k discrete logarithms can be determined after k phases. Therefore, the following inequality can be obtained.

$$T_k \leq T_{k,n} + 2n\log_2 p + \sum_{i=2}^{k}(T_{k,n} + 2s_{k+2-i}\log_2 p)$$

Since $s_k = \frac{n}{k}$,

$$\sum_{i=2}^{k}(T_{k,n} + 2s_{k+2-i}\log_2 p) = (k-1)T_{k,n} + (2\log_2 p)\sum_{i=2}^{k}s_i = (k-1)T_{k,n} + (2\log_2 p)\sum_{i=2}^{k}\frac{n}{i}.$$

Hence, it follows that

$$T_k \leq kT_{k,n} + 2n(1 + \sum_{i=2}^{k}\frac{1}{i})\log_2 p = kT_{k,n} + 2nH_k\log_2 p.$$

This completes the proof. □

Remark 1. When $k = 1$, Proposition 1 corresponds to Lemma 1.

By regarding $k = k(p)$ as a function of p such that $\lim_{p\to+\infty} k(p) = +\infty$, the asymptotic bounds of $T_{k,n}$ can be obtained.

Theorem 1. *Suppose $k^3 \log^2 k = o\left(\frac{p}{\log^2 p}\right)$, then $T_{k,n} = \Omega\left(\sqrt{\frac{p}{k}}\right)$ for all n satisfying $n = k^2 + \Omega(1)$ and $n = o\left(\frac{\sqrt{kp}}{(\log k)(\log p)}\right)$.*

Proof. Clearly $k^3 \log^2 k = o\left(\frac{p}{\log^2 p}\right)$ implies that $k = o(p)$. Hence from [19], $T_k = \Omega(\sqrt{kp})$. It follows from Proposition 1 that if $nH_k\log_2 p \ll T_k$, then $T_{k,n} = \Omega(\frac{1}{k}\sqrt{kp}) = \Omega\left(\sqrt{\frac{p}{k}}\right)$. Moreover, since

$$\lim_{k\to+\infty}(H_k - \log k) = \lim_{k\to+\infty}[(\sum_{i=1}^{k}\frac{1}{i}) - \log k] = \gamma$$

where γ is the Euler-Mascheroni constant, the condition $nH_k \log_2 p \ll T_k$ implies that $n = o\left(\frac{\sqrt{kp}}{(\log k)(\log p)}\right)$. The lower bound $n = k^2 + \Omega(1)$ is obtained by noting that n has to be of size at least k^2 from the condition of Proposition 1. Finally, since the lower bound for n cannot be asymptotically greater than its upper bound, $k(p)$ has to satisfy $k^3 \log^2 k = o\left(\frac{p}{\log^2 p}\right)$. This completes the proof of Theorem 1.   □

Remark 2. Although Theorem 1 holds for a wide asymptotic range of n as given, the $T_{k,n}$ bound becomes sharper as n approaches $\frac{\sqrt{kp}}{(\log k)(\log p)}$. In essence, Theorem 1 does not yield interesting bounds but is a prelude to the more essential Theorem 2 which requires Proposition 1 and is hence included.

Proposition 2. *Suppose that $k < n < k^2$. Then,*

$$T_k \le (r + \frac{n}{k})T_{k,n} + 2rk \log_2 p + 2nH_{\lceil \frac{n}{k} \rceil} \log_2 p$$

where $r = \left\lceil \frac{\log(\frac{k^2}{n})}{\log(\frac{n}{n-k})} \right\rceil$.

Proof. The proof comprises two main phases. The former consists of the initializing phase followed by subsequent subphases. It utilizes the extended pigeon hole principle to obtain more than one solution during each of the initial subphases. The latter phase takes place after some point where the number of remaining unknown discrete logarithms is small enough such that each subphase can only recover one discrete logarithm. After this point, the method of determining all other discrete logarithms essentially mirrors that of the method described in the proof of Proposition 1. The formal proof and details are given as follows.

The initializing phase proceeds as follows. Given arbitraries $h_1, h_2, \ldots, h_k \in G = \langle g \rangle$, obtaining $x_1, x_2, \ldots, x_k$ such that $g^{x_i} = h_i \ \forall \ 1 \le i \le k$ can be achieved in time T_k. Consider n elements partitioned into k blocks each of size approximately s_k, where $s_k = \frac{n}{k}$. Each block is labelled i where i ranges from 1 to k. For each i, $1 \le i \le k$, select about s_k integers $r_{i,j}$ uniformly at random from $\mathbb{Z}_p$ and define $h_{i,j} := g^{r_{i,j}} h_i = g^{x_i + r_{i,j}}$. This generates n of $h_{i,j}$ which when applied to a generic algorithm for the (k, n)-GMDL problem, outputs k pseudo solutions. Computing each $h_{i,j}$ requires at most $2 \log_2 p$ group multiplications. Since each block is about size $\frac{n}{k} < k$, by the extended pigeon hole principle, at least $\frac{k^2}{n}$ out of these k solutions must be derived from distinct blocks. In other words, at least $\frac{k^2}{n}$ correspond to distinct i values and as a result, $\frac{k^2}{n}$ discrete logarithms out of k discrete logarithms can be obtained during this initializing phase. This invokes at most $T_{k,n} + 2n \log_2 p$ group operations.

The first subphase proceeds as follows. Since $\frac{k^2}{n}$ out of k discrete logarithms have been obtained, discard all the blocks for which those determined discrete logarithms are contained previously. Thus, about $k - \frac{k^2}{n} = \frac{k(n-k)}{n}$ blocks remain, each of size approximately $\frac{n}{k}$. For each of the remaining blocks i, select about k integers $r_{i,j}$ uniformly distributed across each i and uniformly at random from

$\mathbb{Z}_p$. Define $h_{i,j} := g^{r_{i,j}} h_i = g^{x_i + r_{i,j}}$. Incorporate these new values of $h_{i,j}$ into the remaining $\frac{k(n-k)}{n}$ blocks. Hence, each of the remaining $\frac{k(n-k)}{n}$ unsolved discrete logarithms are contained separately in $\frac{k(n-k)}{n}$ blocks each of size approximately $\frac{n^2}{k(n-k)}$. This generates n of $h_{i,j}$ which when applied to a generic algorithm for the (k,n)-GMDL problem, outputs k pseudo solutions. This incurs a maximum of $T_{k,n} + 2k \log_2 p$ group operations.

If $k \le \frac{n^2}{k(n-k)}$, these k pseudo solutions might be derived from the same block and hence phase two begins in which the method described in Proposition 1 can then be applied on this new set of blocks.

If $k > \frac{n^2}{k(n-k)}$, the extended pigeon hole principle ensures that at least $\frac{k^2(n-k)}{n^2}$ of the k pseudo solutions correspond to distinct i values and a result, about $\frac{k^2(n-k)}{n^2}$ discrete logarithms can be obtained in the first subphase.

Subsequent subphases are similar to the first subphase. In general for the rth subphase, since about $\frac{k^2(n-k)^{r-1}}{n^r}$ solutions will have been obtained in the $(r-1)^{th}$ subphase, discard all the blocks for which those determined discrete logarithms are contained previously. By a simple induction, it can be shown that the number of remaining blocks in the rth subphase is about $\frac{k(n-k)^r}{n^r}$. The induction proceeds as follows. The base case has already been verified in the first subphase. Suppose the result holds for $r = m-1$ for some $m \ge 2$. By the inductive hypothesis, the number of blocks remaining in the $(m-1)^{th}$ subphase is about $\frac{k(n-k)^{m-1}}{n^{m-1}}$. During the mth subphase, since about $\frac{k^2(n-k)^{m-1}}{n^m}$ solutions have already been obtained previously and are thus discarded, the number of remaining blocks is given by

$$\frac{k(n-k)^{m-1}}{n^{m-1}} - \frac{k^2(n-k)^{m-1}}{n^m} = \frac{k(n-k)^m}{n^m}.$$

This completes the induction. For each of the remaining blocks i, select about k integers $r_{i,j}$ uniformly distributed across each i and uniformly at random from $\mathbb{Z}_p$. Define $h_{i,j} := g^{r_{i,j}} h_i = g^{x_i + r_{i,j}}$. Incorporate these new values of $h_{i,j}$ into the remaining $\frac{k(n-k)^r}{n^r}$ blocks. Hence, each of the remaining $\frac{k(n-k)^r}{n^r}$ unsolved discrete logarithms are contained separately in $\frac{k(n-k)^r}{n^r}$ blocks each of size approximately $\frac{n^{r+1}}{k(n-k)^r}$. This generates n of $h_{i,j}$ which when applied to a generic algorithm for the (k,n)-GMDL problem, outputs k pseudo solutions. Each of the rth subphase requires at most $T_{k,n} + 2k \log_2 p$ group operations.

When $k \le \frac{n^{r+1}}{k(n-k)^r}$, the k outputs can only guarantee one solution. Hence, as soon as r satisfies the above inequality, the first main phase terminates at the end of the rth subphase and the second main phase commences. That is

$$k \le \frac{n^{r+1}}{k(n-k)^r} \implies r = \left\lceil \frac{\log(\frac{k^2}{n})}{\log(\frac{n}{n-k})} \right\rceil.$$

At the beginning of second main phase, there are a total of about $\frac{k(n-k)^r}{n^r} - 1$ unresolved discrete logarithms. The rest of the procedure follows starting from the second phase of Proposition 1 until the end. Hence, it can immediately be derived from the proof of Proposition 1 that the number of group operations required to solve them all is at most $2[(\frac{k(n-k)^r}{n^r} - 1)T_{k,n} + nH_{\lceil \frac{k(n-k)^r}{n^r} \rceil} - n]\log_2 p$. Since

$$k \leq \frac{n^{r+1}}{k(n-k)^r} \implies \frac{k(n-k)^r}{n^r} \leq \frac{n}{k},$$

the maximum number of group operations required in the second main phase is given by $(\frac{n}{k} - 1)T_{k,n} + 2(nH_{\lceil \frac{n}{k} \rceil} - n)\log_2 p$. The number of group operations required during the first main phase is the sum of the number required for the initializing phase and the number required for all the subphases. Therefore, the maximum number of group operations required in the first main phase is given by

$$T_{k,n} + 2n\log_2 p + r(T_{k,n} + 2k\log_2 p) = (r+1)T_{k,n} + 2n\log_2 p + 2rk\log_2 p.$$

Thus the maximum number of group operations required to execute both the first main phase and the second main phase is given by

$$(r+1)T_{k,n} + 2(n+rk)\log_2 p + (\frac{n}{k} - 1)T_{k,n} + 2(nH_{\lceil \frac{n}{k} \rceil} - n)\log_2 p.$$

Hence, $T_k \leq (r + \frac{n}{k})T_{k,n} + 2rk\log_2 p + 2nH_{\lceil \frac{n}{k} \rceil}\log_2 p$.   □

Remark 3. One other approach is to replace the (k, n)-GMDL solver with the $(1, n)$-GMDL solver during the second main phase. Both yield identical asymptotic results given in Sect. 7 as the computational bottleneck arises from the first main phase.

By regarding $k = k(p)$ as a function of p such that $\lim_{p \to +\infty} k(p) = +\infty$, the asymptotic bounds of $T_{k,n}$ can be obtained.

Theorem 2. *Suppose k, n satisfy the following conditions:*

1. $k = o(p)$ 2. $n = k^2 - \Omega(1)$
3. $\frac{n}{\sqrt{k}}\log(\frac{n}{k}) = o\left(\frac{\sqrt{p}}{\log p}\right)$ 4. $\frac{\log(\frac{k^2}{n})}{\log(\frac{n}{n-k})}\sqrt{k} = o\left(\frac{\sqrt{p}}{\log p}\right)$

Then,

$$T_{k,n} = \Omega\left(\frac{\sqrt{k}}{\frac{n}{k} + r}\sqrt{p}\right)$$

where $r = r(k, n) \geq 1$ is any function of k and n satisfying $r(k, n) = \Omega\left(\frac{\log(\frac{k^2}{n})}{\log(\frac{n}{n-k})}\right)$.

Proof. Condition 1 is necessary to utilize results in [19] for a lower bound of T_k. Condition 2 is required in order to apply the result of Proposition 2. Conditions 3 and 4 can be obtained by requiring $nH_{\lceil \frac{n}{k} \rceil} \log p \ll T_k$ and $rk \log p \ll T_k$ respectively and noting that $T_k = \Omega(\sqrt{kp})$. Hence from Proposition 2 and under these conditions, $T_{k,n} = \Omega\left(\frac{\sqrt{k}}{\frac{n}{k}+r}\sqrt{p}\right)$. □

It should be mentioned that $r(k,n)$ can be taken to be any function satisfying $r(k,n) = \Omega\left(\frac{\log(\frac{k^2}{k})}{\log(\frac{n}{n-k})}\right)$. However, it is clear that $T_{k,n}$ achieves sharper bounds for asymptotically smaller choices of r. One other point of note is that when $n = k$ where $k = o(p)$, all the 4 conditions are satisfied and $r(k,n)$ can be taken to be 1. In this case, $T_k = T_{k,k} = \Omega(\sqrt{kp})$. which indeed corresponds to the bound obtained in [19].

5   Optimizing the Partition of n

From the methods described in the proofs of Propositions 1 and 2, n is partitioned into blocks of approximately equal size at each phase. There are many ways to perform such partitions of n. The running time of each phase is partially determined by the number of uniformly randomly chosen $r_{i,j}$, which invariably depends on the partition of n. In this section, we show that the method of partition described in the proofs of the earlier Propositions minimizes the expected number of chosen $r_{i,j}$ required and hence results in the fastest running time among all other possible partitions. We first consider the case where a (k,n)-GMDL solver output solutions derived from the same block so only one discrete logarithm can be determined at each phase. We follow this up by considering general scenarios where a (k,n)-GMDL solver outputs solutions derived from multiple blocks.

5.1   Pseudo Solutions Deriving from the Same Block

Denote s_i to be the size of block i, $k \le s_i$, $1 \le i \le k$. So that $\sum_{i=1}^{k} s_i = n$. Let $p_{i,k}$ be the conditional probability that the k output solutions derive from block i given that the k output solutions derive from the same block. Then, $p_{i,k}$ can be expressed by the following.

$$p_{i,k} = \frac{\binom{s_i}{k}}{\sum_{j=1}^{k} \binom{s_j}{k}}$$

Suppose a solution is derived from block i. Upon discarding block i, s_i of $r_{i,j}$ have to be randomly chosen to fill the remaining blocks so that they sum back up to n. Let $E_k^{(1)}$ be the expected number of randomly chosen $r_{i,j}$ required. Then, $E_k^{(1)}$ can be expressed by the following.

$$E_k^{(1)} = \sum_{i=1}^{k} p_{i,k}s_i = \frac{\sum_{i=1}^{k} \binom{s_i}{k}s_i}{\sum_{i=1}^{k} \binom{s_i}{k}}$$

Our objective is therefore to minimize $E_k^{(1)}$ given $\sum_{i=1}^{k} s_i = n$. We expand the admissible values of s_i to the set of positive real numbers so that $s_i \in \mathbb{R}^+$. In this way, $\binom{s_i}{k}$ is defined as $\binom{s_i}{k} = \frac{s_i(s_i-1)...(s_i-k+1)}{k!}$. We prove the following result.

Theorem 3. *Given that* $\sum_{i=1}^{k} s_i = n$,

$$\frac{\sum_{i=1}^{k} \binom{s_i}{k} s_i}{\sum_{i=1}^{k} \binom{s_i}{k}} \geq \frac{n}{k}.$$

Proof. Without loss of generality, assume that $s_1 \leq s_2 \leq \cdots \leq s_k$. Thus, $\binom{s_1}{k} \leq \binom{s_2}{k} \leq \cdots \leq \binom{s_i}{k}$. By Chebyshev's sum inequality,

$$k \sum_{i=1}^{k} \binom{s_i}{k} s_i \geq \left(\sum_{i=1}^{k} s_i\right)\left(\sum_{i=1}^{k} \binom{s_i}{k}\right).$$

The result follows by replacing $\sum_{i=1}^{k} s_i$ with n in the above inequality.   □

Hence, $E_k^{(1)} \geq \frac{n}{k}$ and it is straightforward to verify that equality holds if $s_1 = s_2 = \cdots = s_k$. Therefore, the method of partitioning n into blocks of equal sizes at each phase as described in the proof of the Proposition 1 indeed minimizes the running time.

5.2   Pseudo Solutions Deriving from Multiple Blocks

Denote s_i to be the size of block i, $1 \leq i \leq k$. so that $\sum_{i=1}^{k} s_i = n$. Let $p_{i_1,i_2,...,i_m,k}$ be the conditional probability that the k output solutions derive from blocks $i_1, i_2, \ldots, i_m$ given that the k output solutions derive from m distinct blocks. Then, $p_{i_1,i_2,...,i_m,k}$ satisfies the following.

$$p_{i_1,i_2,...,i_m,k} \sum_{\{i_1...i_m\}\subseteq\{1,...,k\}} \sum_{k_1+\cdots+k_m=k} \binom{s_{i_1}}{k_1} \cdots \binom{s_{i_m}}{k_m} = \sum_{k_1+\cdots+k_m=k} \binom{s_{i_1}}{k_1} \cdots \binom{s_{i_m}}{k_m}$$

A more concise representation can be expressed as follows.

$$p_{i_1,i_2,...,i_m,k} \sum_{\{i_1...i_m\}\subseteq\{1,...,k\}} \sum_{k_1+\cdots+k_m=k} \prod_{t=1}^{m} \binom{s_{i_t}}{k_t} = \sum_{k_1+\cdots+k_m=k} \prod_{t=1}^{m} \binom{s_{i_t}}{k_t} \quad (1)$$

We can further simplify the above expression by the following lemma.

Lemma 2.

$$\sum_{k_1+\cdots+k_m=k} \prod_{t=1}^{m} \binom{s_{i_t}}{k_t} = \binom{s_{i_1} + \cdots + s_{i_m}}{k}$$

Proof. For brevity, denote $s = s_{i_1} + \cdots + s_{i_m}$.

Consider the polynomial $(1 + x)^s$. By the binomial theorem, $(1 + x)^s = \sum_{r=0}^{s} \binom{s_{i_1} + \cdots + s_{i_m}}{r} x^r$. On the other hand,

$$(1+x)^s = (1+x)^{s_{i_1}} \ldots (1+x)^{s_{i_m}} = \prod_{t=1}^{m} \sum_{r_t=0}^{s_{i_t}} \binom{s_{i_t}}{r_t} x^{r_t}.$$

In this instance, the coefficient of x^r is the sum of all products of binomial coefficients of the form $\binom{s_{i_t}}{r_t}$ where the r_t sum to r. Therefore,

$$\prod_{t=1}^{m} \sum_{r_t=0}^{s_{i_t}} \binom{s_{i_t}}{r_t} x^{r_t} = \sum_{r=0}^{s} \sum_{r_1+\ldots r_m=r} \prod_{t=1}^{m} \binom{s_{i_t}}{r_t} x^r.$$

Hence,

$$\sum_{r=0}^{s} \binom{s_{i_1} + \cdots + s_{i_m}}{r} x^r = \sum_{r=0}^{s} \sum_{r_1+\ldots r_m=r} \prod_{t=1}^{m} \binom{s_{i_t}}{r_t} x^r.$$

The result follows by equating the coefficients of x^r on both sides of the above equation. □

Corollary 2.

$$p_{i_1,i_2,\ldots,i_m,k} \sum_{\{i_1 \ldots i_m\} \subseteq \{1,\ldots,k\}} \binom{s_{i_1} + \cdots + s_{i_m}}{k} = \binom{s_{i_1} + \cdots + s_{i_m}}{k}$$

Proof. Follows directly from Eq. (1) and Lemma 2. □

Suppose m solutions are derived from blocks $i_1, \ldots, i_m$. Upon discarding blocks $i_1, \ldots, i_m$, $s_{i_1} + \cdots + s_{i_m}$ of $r_{i,j}$ have to be randomly chosen to fill the remaining blocks so that they sum back up to n. Denote $E_k^{(m)}$ to be the expected number of randomly chosen $r_{i,j}$ required. Then, a generalized form of $E_k^{(m)}$ can be expressed as follows.

$$E_k^{(m)} = \sum_{\{i_1 \ldots i_m\} \subseteq \{1,\ldots,k\}} p_{i_1,i_2,\ldots,i_m,k}(s_{i_1} + \cdots + s_{i_m})$$

From the result of Corollary 2, this implies that $E_k^{(m)}$ satisfies

$$E_k^{(m)} \sum_{\{i_1 \ldots i_m\} \subseteq \{1,\ldots,k\}} \binom{s_{i_1} + \cdots + s_{i_m}}{k} = \sum_{\{i_1 \ldots i_m\} \subseteq \{1,\ldots,k\}} \binom{s_{i_1} + \cdots + s_{i_m}}{k}(s_{i_1} + \cdots + s_{i_m}).$$

Once again, we seek to maximize $E_k^{(m)}$ given $\sum_{i=1}^{k} s_i = n$. As before, we expand the admissible values of s_i to the set of positive real numbers so that $s_i \in \mathbb{R}^+$. In this way, $\binom{s_i}{k}$ is defined as $\binom{s_i}{k} = \frac{s_i(s_i-1)\ldots(s_i-k+1)}{k!}$. We prove the following result.

Theorem 4. *Given that* $\sum_{i=1}^{k} s_i = n$,

$$\sum_{\{i_1 \ldots i_m\} \subseteq \{1, \ldots, k\}} \binom{s_{i_1} + \cdots + s_{i_m}}{k} (s_{i_1} + \cdots + s_{i_m}) \geq \frac{mn}{k} \sum_{\{i_1 \ldots i_m\} \subseteq \{1, \ldots, k\}} \binom{s_{i_1} + \cdots + s_{i_m}}{k}.$$

Proof. By Chebyshev's sum inequality,

$$\binom{k}{m} \sum_{\{i_1 \ldots i_m\} \subseteq \{1, \ldots, k\}} \binom{s_{i_1} + \cdots + s_{i_m}}{k} (s_{i_1} + \cdots + s_{i_m})$$

$$\geq \left(\sum_{\{i_1 \ldots i_m\} \subseteq \{1, \ldots, k\}} \binom{s_{i_1} + \cdots + s_{i_m}}{k} \right) \left(\sum_{\{i_1 \ldots i_m\} \subseteq \{1, \ldots, k\}} (s_{i_1} + \cdots + s_{i_m}) \right).$$

Since $\sum_{i=1}^{k} s_i = n$,

$$\sum_{\{i_1 \ldots i_m\} \subseteq \{1, \ldots, k\}} (s_{i_1} + \cdots + s_{i_m}) = \binom{k-1}{m-1} n.$$

Hence, we obtain

$$\binom{k}{m} \sum_{\{i_1 \ldots i_m\} \subseteq \{1, \ldots, k\}} \binom{s_{i_1} + \cdots + s_{i_m}}{k} (s_{i_1} + \cdots + s_{i_m})$$

$$\geq \binom{k-1}{m-1} n \sum_{\{i_1 \ldots i_m\} \subseteq \{1, \ldots, k\}} \binom{s_{i_1} + \cdots + s_{i_m}}{k}.$$

By elementary algebraic operations, it is straightforward to verify that $\frac{\binom{k-1}{m-1}}{\binom{k}{m}} = \frac{m}{k}$ from which the result follows. $\qquad \square$

Hence, $E_k^{(m)} \geq \frac{mn}{k}$ and it is straightforward to verify that equality holds if $s_1 = s_2 = \cdots = s_k$. Therefore, the method of partitioning n into blocks of equal sizes at each phase as described in the proof of the Proposition 2 indeed minimizes the running time.

6   Applications in Other MDLP Settings

We demonstrate how the block method can be adapted to obtain bounds in other generalized multiple discrete logarithm settings. We consider applications to the $(e_1, \ldots, e_d)$-Multiple Discrete Logarithm Problem with Auxiliary Inputs (MDLPwAI) as well as the $\mathbb{F}_p$-Multiple Discrete Logarithm Problem in the Exponent ($\mathbb{F}_p$-MDLPX). Let $G = \langle g \rangle$ be a cyclic group of large prime order p. Their formal definitions are as follow.

Definition 3 (MDLPwAI). *Given G, g, p, e_i and $g^{x_i{}^{e_1}}, g^{x_i{}^{e_2}}, \ldots g^{x_i{}^{e_d}} \ \forall \ 1 \leq i \leq k$, find non-negative integers $x_1, x_2, \ldots, x_k \in \mathbb{Z}_p$.*

Definition 4 ($\mathbb{F}_p$-MDLPX). *Let $\chi \in \mathbb{F}_p$ be an element of multiplicative order N. Given G, g, p, χ and $g^{\chi^{x_i}} \ \forall \ 1 \leq i \leq k$, find non-negative integers $x_1, x_2, \ldots, x_k \in \mathbb{Z}_N$.*

The computational complexity of MDLPwAI was analysed in [12]. In the same paper, the authors introduced the $\mathbb{F}_p$-MDLPX and also analysed its complexity.

Here, we define the Generalized Multiple Discrete Logarithm Problem (GMDLPwAI) with Auxiliary Inputs and the Generalized $\mathbb{F}_p$-Multiple Discrete Logarithm Problem in the Exponent ($\mathbb{F}_p$-GMDLPX) to be solving k out of n instances of the MDLPwAI and $\mathbb{F}_p$-MDLPX respectively. We provide the formal definitions below.

Definition 5 (GMDLPwAI). *Given G, g, p, e_i and $g^{x_i{}^{e_1}}, g^{x_i{}^{e_2}}, \ldots g^{x_i{}^{e_d}} \ \forall \ 1 \leq i \leq n$, find k pairs (i, x_i), $x_i \in \mathbb{Z}_p$, where $i \in S$ such that S is a k-subset of $\{1, \ldots, n\}$.*

Definition 6 ($\mathbb{F}_p$-GMDLPX). *Let $\chi \in \mathbb{F}_p$ be an element of multiplicative order N. Given G, g, p, χ and $g^{\chi^{x_i}} \ \forall \ 1 \leq i \leq n$, find k pairs (i, x_i), $x_i \in \mathbb{Z}_N$, where $i \in S$ such that S is a k-subset of $\{1, \ldots, n\}$.*

6.1   Block Based GMDLPwAI

The block method can be adapted to obtain bounds for the GMDLPwAI by randomizing the input elements in the following way. Given $g^{x_i{}^{e_1}}$, select random integers $r_{i,j} \in \mathbb{Z}_p^*$ and compute values of $(g^{x_i{}^{e_1}})^{r_{i,j}{}^{e_1}} = g^{(r_{i,j}x_i)^{e_1}}$ as inputs into the GMDLPwAI solver. For each $r_{i,j}$, reduce $r_{i,j}^{e_1}$ modulo p and then $g^{(r_{i,j}x_i)^{e_1}}$ can be computed within $2\log_2 p$ group operations. Repeat this procedure for all $e_2, \ldots, e_d$. We show how the x_i can be recovered. For instance, suppose the solver outputs solution (l, y) corresponding to some particular input $g^{(r_{i,j}x_i)^{e_1}}$. In which case, x_i can thus be obtained by solving $r_{i,j}x_i \equiv y \bmod p$. Such congruence equations are efficiently solvable since $\gcd(r_{i,j}, p) = 1$.

Let T_k' and $T_{k,n}'$ denote the time taken in group operations for an optimal algorithm to solve the MDLPwAI and GMDLPwAI problems respectively.

Suppose $k < n < k^2$. Then by adapting the block technique applied to the GMDL problem before, it can be shown that

$$T_k' \leq (r + \frac{n}{k})T_{k,n}' + 2d(rk + nH_{\lceil \frac{n}{k} \rceil})\log_2 p$$

where $r = \left\lceil \dfrac{\log(\frac{k^2}{n})}{\log(\frac{n}{n-k})} \right\rceil$.

It has been conjectured in [12] that $T'_k = \Omega(\sqrt{kp/d})$ for values of $e_i = i$. Assuming this conjecture, we can conclude from our results that for all polynomially bounded inputs with $d = O(p^{1/3-\epsilon})$, $\epsilon > 0$, $T'_{k,n}$ is bounded by

$$T'_{k,n} = \Omega\left(\frac{\sqrt{k}}{\frac{n}{k}+r}\sqrt{\frac{p}{d}}\right)$$

where $r = r(k,n) \geq 1$ is any function of k and n satisfying $r(k,n) = \Omega\left(\frac{\log(\frac{k^2}{n})}{\log(\frac{n}{n-k})}\right)$.

When $k = 1$, the above bound is not applicable since $n \geq k^2$ in this situation. Nevertheless, we show how an unconditional bound for $T'_{1,n}$ can still be obtained in this specific case without the assumption of any conjecture. Similar to the generalized case, randomize input elements of the form $g^{(r_{i,j}x_i)^{e_1}}$. One of the x_i can then be computed by solving $r_{i,j}x_i \equiv y \bmod p$ where y is a given known output. The process terminates here since one of the x_i has already been obtained. Hence, we have the following inequality: $T'_1 \leq T'_{1,n} + 2nd\log_2 p$. From the results of [5], $T'_1 = \Omega(\sqrt{p/d})$. It follows that for $n = o\left(\frac{\sqrt{p}}{d^{3/2}\log p}\right)$,

$$T'_{1,n} = \Omega(\sqrt{p/d}).$$

6.2   Matrix Based GMDLPwAI

In this section, our objective is to find values of e_i for which the matrix method can applied to solve the GMDLPwAI. We recall from Sect. 3 that the validity of the method for the k-MDL problem necessitates $g^{x_1+x_2+\cdots+x_k}$ to be efficiently computable from given values of $g^{x_1}, g^{x_2}, \ldots, g^{x_k}$. Moreover, the GMDLPwAI can be viewed as a generalization of the k-MDL problem (i.e. the GMDLPwAI reduces to the k-MDL problem when $d = 1$ and $e_1 = 1$). In a similar vein, it is required that $g^{f_i(x_1+x_2+\cdots+x_k)}$ to be efficiently computable, from the given known values of the GMDLPwAI. In the instance of GMDLPwAI, $f_i(x) = x^{e_i}$. In that regard, suppose $g^{f_{i0}(x_1+x_2+\cdots+x_k)} = g^{a_1 f_{i1}(x_1)}g^{a_2 f_{i2}(x_2)}\ldots g^{a_k f_{ik}(x_k)} \, \forall \, x_i$ and for some integer constants a_i so that it can be efficiently computed. We prove the following result.

Theorem 5. *Let $f_i(x) = x^{e_i}$, where $e_i \in \mathbb{Z}$ are not necessarily distinct. If for some integer constants a_i such that*

$$f_0(x_1 + x_2 + \cdots + x_k) \equiv a_1 f_1(x_1) + a_2 f_2(x_2) + \ldots a_k f_k(x_k) \mod p$$

for all odd primes p and for all $x_1, x_2, \ldots, x_k \in \mathbb{Z}_p$, then the only solutions are of the form $f_i = x^{c_i(p-1)+1}$ for some integer constants c_i.

Proof. For each $1 \leq i \leq k$, substitute $x_i = 1$ and $x_{i'} = 0$ for $i \neq i'$. We obtain $f_0(1) = a_i f_i(1) \, \forall \, i$. Hence, $a_i = 1 \, \forall \, i$. Upon establishing that all a_i values have to be 1, we proceed as follows.

514   J.H.M. Ying and N. Kunihiro

Let $x_2 = x_3 = \cdots = x_k = 0$. This implies that $f_0(x_1) \equiv f_1(x_1) \bmod p$ for all x_1. Similarly, let $x_1 = x_3 = \cdots = x_k = 0$. This implies that $f_0(x_2) \equiv f_2(x_2)$ $\bmod p$ for all x_2. Continuing in this fashion, it can be deduced that $f_0(x) \equiv f_1(x) \equiv \ldots f_k(x) \bmod p \; \forall \; x$. Next, let $x_3 = x_4 = \cdots = x_k = 0$. This implies $f_0(x_1+x_2) \equiv f_0(x_1)+f_0(x_2) \bmod p$. We claim that $f_0(x) \equiv xf_0(1) \bmod p \; \forall \; x$. It clear that the result holds true for $x = 0, 1$. By applying an inductive argument,

$$f_0(x+1) \equiv f_0(x) + f_0(1) \equiv xf_0(1) + f_0(1) \equiv (x+1)f_0(1) \bmod p$$

and the claim follows. Hence, p divides $x^{e_0} - x$ for all $x \in \mathbb{Z}_p$. Since $(\mathbb{Z}/p\mathbb{Z})^{\times} \cong C_{p-1}$, there exists a generator of the cyclic group $x_0 \in \mathbb{Z}_p$ such that if p divides $x_0^{e_0} - x_0$, then $e_0 \equiv 1 \bmod p-1$. Moreover, since we have earlier established that $f_0(x) \equiv f_1(x) \equiv \ldots f_k(x) \bmod p$, it can be concluded that $e_i \equiv 1 \bmod p-1$ for all i. Hence, $f_i = x^{c_i(p-1)+1}$ and it is straightforward to verify that these are indeed solutions to the original congruence equation.   □

From the result of Theorem 5, if e_i is of the form $e_i = c_i(p-1) + 1$, then $g^{f_i(x_1+x_2+\cdots+x_k)}$ be can efficiently computed. However, $g^{x^{c_i(p-1)+1}} = g^x$ so such e_i values reduces to the classical multiple discrete logarithm problem. Therefore, the matrix method is not applicable to solve the GMDLPwAI.

6.3   Block Based $\mathbb{F}_p$-GMDLPX

The block method can be also adapted to obtain bounds for the $\mathbb{F}_p$-GMDLPX by randomizing the input elements with the computations $(g^{x^{x_i}})^{x^{r_{i,j}}} = g^{x^{x_i+r_{i,j}}}$ where $r_{i,j} \in \mathbb{Z}_p$ are selected randomly. For each $r_{i,j}$, reduce $x^{r_{i,j}}$ modulo p and then $g^{x^{x_i+r_{i,j}}}$ can be computed within $2\log_2 p$ group operations.

Suppose the solver outputs solution (l, y) corresponding to some particular input $g^{x^{x_i+r_{i,j}}}$. In which case, x_i can thus be obtained by solving $r_{i,j} + x_i \equiv y$ $\bmod p$. The analysis and obtained bounds in this case is similar to the classical GMDL problem which was already discussed in Sect. 5 so we will omit the details here.

Let T_k'' and $T_{k,n}''$ denote the time taken in group operations for an optimal algorithm to solve the $\mathbb{F}_p$-MDLPX and $\mathbb{F}_p$-GMDLPX problems respectively. It has been shown in [12] that T_k'' can be achieved in $O(\sqrt{kN})$. If this is optimal, then our results show that

$$T_{k,n}'' = \Omega\left(\frac{\sqrt{k}}{\frac{n}{k}+r}\sqrt{N}\right)$$

where $r = r(k,n) \geq 1$ is any function of k and n satisfying $r(k,n) = \Omega\left(\frac{\log(\frac{k^2}{n})}{\log(\frac{n}{n-k})}\right)$, subject to the conditions given in Theorem 2.

6.4   Matrix Based $\mathbb{F}_p$-GMDLPX

The matrix method does not apply here since there is no known efficient method to compute $g^{x^{x_i+x_j}}$ given $g^{x^{x_i}}$ and $g^{x^{x_j}}$ if the Diffie-Hellman assumption holds.

7 Some Explicit Bounds of $T_{k,n}$

The conditions imposed in Theorem 2 might initially seem restrictive but in fact they are satisfied by large classes of k and n. In this section, we present some interesting explicit bounds of $T_{k,n}$ by varying n relative to k. For the remainder of this section, k can be taken to be any function satisfying $k = O(p^{1-\epsilon})$, for some $0 < \epsilon < 1$ and c is a constant. The proofs of Proposition 3 and 4 are straightforward applications of Theorem 2.

Proposition 3.

$$T_{k,k+c} = \Omega(\sqrt{kp})$$

where c is a positive constant.

Proposition 4.

$$T_{k,ck} = \Omega(\frac{\sqrt{k}}{\log k}\sqrt{p})$$

where c is a constant, $c > 1$.

Proposition 5.

$$T_{k,k\log^c k} = \Omega(\frac{\sqrt{k}}{\log^{1+c} k}\sqrt{p})$$

where c is a positive constant.

Proof.

$$\log \frac{n}{n-k} = \log \frac{\log^c k}{\log^c k - 1}.$$

Next, consider the function $\log \frac{\log^c k}{\log^c k - 1}$, for $k > e$ where e is the base of the natural logarithm. For brevity, denote $x = \log^c k$ so $x > 1$. By the Maclaurin series expansion, for $x > 1$,

$$\log \frac{x}{x-1} = -\log(1 - \frac{1}{x}) = \sum_{i=1}^{\infty} \frac{1}{ix^i} > \frac{1}{x}.$$

In particular, this proves that $\log \frac{\log^c k}{\log^c k - 1} > \frac{1}{\log^c k}$ when $k > e$. Hence for $k > e$,

$$\frac{\log(\frac{k^2}{n})}{\log(\frac{n}{n-k})} < (\log^c k) \log(\frac{k^2}{n}) = (\log^c k) \log(\frac{k}{\log^c k}) < \log^{1+c} k.$$

Thus $r(k,n)$ can be taken to be $\log^{1+c} k$ when $n = k \log^c k$. From Theorem 2, we obtain

$$T_{k,k\log^c k} = \Omega(\frac{\sqrt{k}}{\log^c k + \log^{1+c} k}\sqrt{p}) = \Omega(\frac{\sqrt{k}}{\log^{1+c} k}\sqrt{p}).$$

$\square$

Table 1 provides a summary of the results for the lower bounds of $T_{k,n}$ with different n relative to k.

Table 1. Some bounds of $T_{k,n}$

k	n	r	$T_{k,n}$
$o\left(\frac{p^{1/3}}{\log^{2/3}p}\right)$	$\{n : n\sqrt{k} = o\left(\frac{\sqrt{p}}{\log p}\right)\}$	N.A	$\Omega(\sqrt{kp})$
$\Omega\left(\frac{p^{1/3}}{\log^{2/3}p}\right)$	$k + c, c \geq 0$	constant	$\Omega(\sqrt{kp})$
$\Omega\left(\frac{p^{1/3}}{\log^{2/3}p}\right)$	$ck, c > 1$	$\log k$	$\Omega(\frac{\sqrt{k}}{\log k}\sqrt{p})$
$\Omega\left(\frac{p^{1/3}}{\log^{2/3}p}\right)$	$k\log^c k, c > 0$	$\log^{1+c} k$	$\Omega(\frac{\sqrt{k}}{\log^{1+c} k}\sqrt{p})$

8   Conclusion

In this paper, we established rigorous bounds for the generic hardness of the generalized multiple discrete logarithm problem which can be regarded a generalization of the multiple discrete logarithm problem. Some explicit bounds are also computed using both the matrix method and block method. Many instances of $T_{k,n}$ are shown to be in fact asymptotically optimal. The overall best bounds obtained here require the union of results from both of these techniques. Furthermore, we show that the block method can also be adapted to handle generalizations arising in other discrete logarithm problems. We similarly obtain bounds for these generalizations. For instance, a consequence of our result highlights that solving an instance of the MDLPwAI problem is as hard as solving the DLPwAI problem under certain conditions. We also demonstrated why the matrix method is not applicable to these and other variants of DLP.

Acknowledgement. The authors wish to thank Phong Nguyen for valuable discussions and all anonymous reviewers for their helpful comments. This research was partially supported by JST CREST Grant Number JPMJCR14D6, Japan and JSPS KAKENHI Grant Number 16H02780.

References

1. Digital signature standard (DSS). NIST (National Institute of Standards and Technology) FIPS, 186–4 (2013)
2. Bernstein, D.J.: Curve25519: new Diffie-Hellman speed records. In: Yung, M., Dodis, Y., Kiayias, A., Malkin, T. (eds.) PKC 2006. LNCS, vol. 3958, pp. 207–228. Springer, Heidelberg (2006). doi:10.1007/11745853_14
3. Bernstein, D.J., Lange, T., Schwabe, P.: On the correct use of the negation map in the Pollard Rho method. In: Public Key Cryptography, pp. 128–146 (2011)
4. Boneh, D., Boyen, X.: Efficient selective-ID secure identity-based encryption without random oracles. In: Cachin, C., Camenisch, J.L. (eds.) EUROCRYPT 2004. LNCS, vol. 3027, pp. 223–238. Springer, Heidelberg (2004). doi:10.1007/978-3-540-24676-3_14
5. Boneh, D., Boyen, X.: Short signatures without random oracles. In: Cachin, C., Camenisch, J.L. (eds.) EUROCRYPT 2004. LNCS, vol. 3027, pp. 56–73. Springer, Heidelberg (2004). doi:10.1007/978-3-540-24676-3_4

6. Boneh, D., Gentry, C., Waters, B.: Collusion resistant broadcast encryption with short ciphertexts and private keys. In: Shoup, V. (ed.) CRYPTO 2005. LNCS, vol. 3621, pp. 258–275. Springer, Heidelberg (2005). doi:10.1007/11535218_16
7. Cheon, J.H., Hong, J., Kim, M.: Accelerating Pollard's Rho algorithm on finite fields. J. Cryptol. **25**(2), 195–242 (2012)
8. Diffie, W., Hellman, M.E.: New directions in cryptography. IEEE Trans. Inf. Theory **22**(6), 644–654 (1976)
9. El Gamal, T.: A public key cryptosystem and a signature scheme based on discrete logarithm. IEEE Trans. Inf. Theory **31**(4), 469–472 (1985)
10. Fouque, P.-A., Joux, A., Mavromati, C.: Multi-user collisions: applications to discrete logarithm, even-Mansour and PRINCE. In: Sarkar, P., Iwata, T. (eds.) ASIACRYPT 2014. LNCS, vol. 8873, pp. 420–438. Springer, Heidelberg (2014). doi:10.1007/978-3-662-45611-8_22
11. Hitchcock, Y., Montague, P., Carter, G., Dawson, E.: The efficiency of solving multiple discrete logarithm problems and the implications for the security of fixed elliptic curves. Int. J. Inf. Secur. **3**(2), 86–98 (2004)
12. Kim, T.: Multiple discrete logarithm problems with auxiliary inputs. In: Iwata, T., Cheon, J.H. (eds.) ASIACRYPT 2015. LNCS, vol. 9452, pp. 174–188. Springer, Heidelberg (2015). doi:10.1007/978-3-662-48797-6_8
13. Kuhn, F., Struik, R.: Random walks revisited: extensions of Pollard's Rho algorithm for computing multiple discrete logarithms. In: Vaudenay, S., Youssef, A.M. (eds.) SAC 2001. LNCS, vol. 2259, pp. 212–229. Springer, Heidelberg (2001). doi:10.1007/3-540-45537-X_17
14. Mitsunari, S., Sakai, R., Kasahara, M.: A new traitor tracing. IEICE Trans. Fundam. Electron. Commun. Comput. Sci. **85**(2), 481–484 (2002)
15. Pollard, J.: Monte Carlo methods for index computations mod p. Math. Comput. **32**(143), 918–924 (1978)
16. Schnorr, C.P.: Efficient identification and signatures for smart cards. In: Brassard, G. (ed.) CRYPTO 1989. LNCS, vol. 435, pp. 239–252. Springer, New York (1990). doi:10.1007/0-387-34805-0_22
17. Shoup, V.: Lower bounds for discrete logarithms and related problems. In: Fumy, W. (ed.) EUROCRYPT 1997. LNCS, vol. 1233, pp. 256–266. Springer, Heidelberg (1997). doi:10.1007/3-540-69053-0_18
18. Teske, E.: On random walks for Pollard's Rho method. Math. Comput. **70**, 809–825 (2000)
19. Yun, A.: Generic hardness of the multiple discrete logarithm problem. In: Oswald, E., Fischlin, M. (eds.) EUROCRYPT 2015. LNCS, vol. 9057, pp. 817–836. Springer, Heidelberg (2015). doi:10.1007/978-3-662-46803-6_27

An Enhanced Binary Characteristic Set Algorithm and Its Applications to Algebraic Cryptanalysis

Sze Ling Yeo[1], Zhen Li[1(✉)], Khoongming Khoo[2], and Yu Bin Low[2]

[1] Infocomm Security Department, Institute for Infocomm Research,
1 Fusionopolis Way, #21-01 Connexis, Singapore 138632, Singapore
`slyeo@i2r.a-star.edu.sg, lizh0019@gmail.com`
[2] DSO National Laboratories, 20 Science Park Drive, Singapore 118230, Singapore
`{kkhoongm,lyubin}@dso.org.sg`

Abstract. Efficient methods to solve boolean polynomial systems underly the effectiveness of algebraic attacks on cryptographic ciphers and the security of multi-variate cryptosystems. Amongst various polynomial solving algorithms, the binary characteristic set algorithm was recently proposed to solve boolean polynomial systems including those arising from ciphers. In this paper, we propose some novel techniques to enhance the existing characteristic set solver. Specifically, we incorporate the ElimLin procedure and apply basic statistical learning techniques to improve the performance of the characteristic set algorithm. Our experiments show that our enhanced solver EBCSA performs better than existing algebraic methods on some ciphers, including CANFIL and PRESENT ciphers. We also perform the first algebraic cryptanalysis on the PRINCE cipher and an algebraic attack on Toyocrypt in a more practical/realistic setting as compared to previous attacks.

Keywords: Characteristic set algorithm · Algebraic cryptanalysis · ElimLin · Statistical learning

1 Introduction

1.1 Background

In the search for possible cryptographic algorithms in the post-quantum era, multi-variate cryptography has emerged as one of the potential candidates. This branch of cryptography comprises schemes based on the difficulty to solve multi-variate polynomials over finite fields. While solving generic multi-variate polynomial systems is known to be NP-hard, a number of polynomial systems arising from cryptographic constructions have been efficiently solved. A well-known example is the polynomial system from the HFE cryptosystem [1]. As such, it is important to have a better understanding of the existing methods to solve multi-variate polynomial systems constructed from cryptosystems.

© Springer International Publishing AG 2017
D. Gollmann et al. (Eds.): ACNS 2017, LNCS 10355, pp. 518–536, 2017.
DOI: 10.1007/978-3-319-61204-1_26

Various methods to solve multivariate polynomial systems exist. The most common approach to solve generic polynomial systems over finite fields is Gröbner basis algorithms [2–4]. Typically, a Gröbner basis with respect to the degree reverse lexicographical ordering is first computed via algorithms F_4 or F_5 [3,4]. It is then converted to a Gröbner basis with respect to the lexicographical ordering by algorithms such as the FGLM algorithm [5] which contains equations where variables are eliminated. This enables the variables to be solved one at a time. See [6] for more details.

Another approach to solve multivariate polynomial systems is the XL algorithm and its variants [7–10]. This class of algorithms performs well when the system under consideration is overdetermined, that is, the number of equations far exceeds the number of variables. Briefly, this method works by considering multiples of the generating polynomials bounded by some degree for which the number of independent equations exceeds the number of monomials. Common linear algebra techniques can then be applied to solve this resulting system.

In both the Gröbner basis and XL type approaches, new polynomials are gradually added with degrees larger than the degree of the generating polynomials. This often causes the number of monomials to increase very rapidly, thereby resulting in excessive memory requirements. The ElimLin algorithm was first proposed by Courtois to attack DES [11,12]. Essentially, the ElimLin algorithm seeks to successively find linear equations in the vector space generated by the original equations and subsequently, eliminating variables using these linear equations. Observe that his process constructs new polynomial systems with fewer variables without an increase in the degree. In fact, the "ElimLin" subroutine is incorporated into most Gröbner basis implementations.

All the above methods have been exploited in algebraic cryptanalysis of ciphers to solve for the unknown key variables in the polynomial system representing the ciphers. In general, block ciphers are less vulnerable to algebraic cryptanalysis as the polynomials usually have very high degrees or involve a large number of variables. Nonetheless, some block ciphers had been broken by algebraic attacks, including Keeloq [13]. As another example, ElimLin was used in [14] to solve 5 rounds of Present with half of the keybits fixed and 5 known plaintext/ciphertext pairs.

1.2 The Characteristic Set Algorithm

The characteristic set algorithm (or CSA for short) had been well-established to analyze algebraic properties of polynomial systems over algebraically-closed fields of characteristic 0, see [15,16]. In [17,18], the authors adapted this approach to solve polynomial systems over any finite field. Specifically, CSA seeks to decompose a polynomial system into monic triangular sets of polynomials with disjoint solution sets. Thus, the problem now boils down to solving a monic triangular set of polynomials which can be easily accomplished. In the case of a binary field, they further refined the decomposition algorithm so that the degrees of newly generated polynomials do not increase. The authors of [17] provided some experimental evidence which demonstrated that CSA seems to be more effective on sparse polynomial systems, particularly those exhibiting a block tri-angular structure. Examples of such systems include polynomials arising from linear feedback shift registers with nonlinear combining functions used in stream ciphers.

1.3   Our Contributions

In this paper, we propose some novel techniques to further enhance the binary CSA. The main features can be summarized as follows:

- we incorporate ElimLin into the binary CSA.
- we apply some statistical learning techniques to choose the next "splitting polynomial".
- we also adaptively choose the next polynomial set to process.
- we propose a preprocessing phase where the variables can be sorted.
- we allow for some flexibility in choosing the combination of features suitable for each polynomial system.

We call our enhanced version the enhanced binary characteristic set algorithm (EBCSA for short). In this paper, we present our experimental results on the block ciphers Present and Prince, as well as the stream ciphers Canfil and Toyocrypt. First, we benchmark EBCSA against prior versions of the characteristic set algorithms on Canfil ciphers. For Present cipher, we show that EBCSA outperforms previous algebraic methods as 5 rounds can be solved with fewer fixed keybits and known plaintext/ciphertext pairs. Moreover, we provide the first algebraic attack on 6 rounds of Present. As for the block cipher Prince, we carried out the first algebraic attack and our results showed that EBCSA is more effective than brute force search for 6 rounds of Prince-core (i.e. without whitening). Finally, we show that the stream cipher Toyocrypt can be solved by EBCSA in a more realistic setting compared to previous attacks, namely, using re-synchronization with sufficient random IV's for a small number of keystream bits.

1.4   Organization of This Paper

This paper is organized as follows. In Sect. 2, we set out the notations and terminologies. We then provide a general description of the binary characteristic set algorithm [17] and its variant [18]. In Sect. 3, we describe our new techniques and present our enhanced binary characteristic set algorithm. Our experimental results are then provided in Sect. 4. Finally, we wrap up with some concluding remarks and suggestions for future research.

2   A Description of the Characteristic Set Algorithm

2.1   Notations and the Set-Up

Let $F = \mathcal{F}_2$ denote the binary field. For a positive integer n, let $R = F[x_0, x_1, \ldots, x_{n-1}]$ be the multivariate polynomial ring in n variables. Since

$x^2 = x$ for all $x \in F$, we are interested in the quotient ring $\overline{R} = R/\{x_i^2 + x_i : i = 0, 1, \ldots, n-1\}$. In particular, every polynomial in R is equivalent to a representative $p \in \overline{R}$ such that the degree of each variable in p is at most 1. In the following, we will represent all the polynomials in this form.

Definition 1. *Fix σ to be a permutation on $\{0, 1, \ldots, n-1\}$. Let P be a polynomial in $\overline{R}$. We define the class of P with respect to σ to be the smallest i such that $x_{\sigma(i)}$ occurs in P. We denote the class of P by $\mathrm{cls}_\sigma(P)$. If σ is clear in the context, we will simply denote it by $\mathrm{cls}(P)$.*

Remark 1. – *For a constant polynomial $P = 0$ or 1, we define $\mathrm{cls}(P) = -\infty$.*
 – *Let σ be the identity permutation, that is, $\sigma(i) = i$ for $i = 0, 1, \ldots, n-1$. Then $\mathrm{cls}_\sigma(P)$ is the smallest i such that x_i occurs in P. On the other hand, suppose that $\sigma(i) = n-1-i$ for $i = 0, 1, \ldots, n-1$. Then $\mathrm{cls}_\sigma(P)$ is the largest i such that x_i occurs in P. This coincides with the definition in [17].*

For a polynomial $P \in \overline{R}$, let c be its class with respect to a permutation σ. Then P can be expressed as $P = I x_{\sigma(c)} + U$, where $x_{\sigma(c)}$ does not occur in both I and U. We term I and U as the left child and right child of P, respectively. By the definition of c, it follows that both $\mathrm{cls}_\sigma(I)$ and $\mathrm{cls}_\sigma(U)$ are strictly greater than c. Note that a monic polynomial is one in which $I = 1$.

Definition 2. *Let $\mathcal{A}$ be a finite set of polynomials in $\overline{R}$. $\mathcal{A}$ is called triangular if the classes of the polynomials in $\mathcal{A}$ are all distinct. Moreover, $\mathcal{A}$ is monic triangular if every polynomial in $\mathcal{A}$ is monic.*

The Main Problem: Let $\mathcal{P}$ be a finite set of boolean polynomials from $\overline{R}$. We wish to find the zero set of $\mathcal{P}$, that is, to find $Z(\mathcal{P}) = \{(x_0, x_1, \ldots, x_{n-1}) \in F^n : P(x_0, x_1, \ldots, x_{n-1}) = 0 \text{ for all } P \in \mathcal{P}\}$.

2.2 The Basic Structure of Binary CSA

In this section, we describe the main principles underlying binary CSA to help solve the main problem stated above. Our description is more generic than that in [17] in the sense that we allow for a variable ordering on the variables. Note that we will focus on the multiplication-free CSA or MFCS presented in [17] as it was already shown to result in the best performance for the binary case. As such, CSA will refer to MFCS from now on. Throughout this section, we fix a permutation σ and simply write $\mathrm{cls}(P)$ for the class of P. Essentially, the approach of binary CSA is a result of the following lemma.

Lemma 1. *Let P be a polynomial of $\overline{R}$ with class c and write $P = I x_{\sigma(c)} + U$. Then $Z(\{P\})$ can be split into $Z(\{P\}) = Z(\mathcal{P}_1) \cup Z(\mathcal{P}_2)$ such that: (1) $Z(\mathcal{P}_1)$ and $Z(\mathcal{P}_2)$ are disjoint; (2) One of $\mathcal{P}_1$ or $\mathcal{P}_2$ contains a monic polynomial with class c.*

Proof. As we are in the binary setting, the left child of P, namely I, can only take values 0 or 1 and $Z(\{I\})$ and $Z(\{I+1\})$ are disjoint. When $I = 0$, we have $U = P + Ix_{\sigma(c)} = 0$. On the other hand, if $I = 1$, we have $x_{\sigma(c)} + U = Ix_{\sigma(c)} + U = P = 0$. Let $\mathcal{P}_1 = \{I, U\}$ and $\mathcal{P}_2 = \{I + 1, x_{\sigma(c)} + U\}$. The lemma now follows.

Suppose that $\mathcal{A}$ is a monic triangular set of polynomials. Clearly, $\mathcal{A}$ cannot contain more than n polynomials. Suppose that $0 \leq c_1 < c_2 < \ldots < c_t \leq n - 1$ are the classes of the polynomials in $\mathcal{A}$. Then, $\mathcal{A}$ contains the polynomials $P_i = x_{\sigma(c_i)} + U_i$, $i = 1, 2, \cdots, t$, where $U_1, U_2, \ldots, U_t$ represents the right child of $P_1, \ldots, P_t$, respectively. Observe that since $c_1 < c_2 < \ldots < c_t$, U_t does not contain the variables $x_{\sigma(c_1)}, \ldots, x_{\sigma(c_t)}$. It follows that as long as the variables of U_t are fixed to some values, $x_{\sigma(c_t)}$ is determined. Similarly, U_{t-1} does not contain the variables $x_{\sigma(c_1)}, \ldots x_{\sigma(c_{t-1})}$ which allows one to compute $x_{\sigma(c_{t-1})}$. In this way, we can determine the zeroes of $\mathcal{A}$ by letting x_i, $i \notin \{\sigma(c_1), \ldots, \sigma(c_t)\}$ take values 0 or 1 and computing the corresponding values of $x_{\sigma(c_1)}, \ldots, x_{\sigma(c_t)}$ from the polynomials in $\mathcal{A}$.

In view of the above, the main goal of binary CSA is to decompose $Z(\mathcal{P})$ into disjoint sets based on Lemma 1. Specifically, we have the following theorem:

Theorem 1. [17, Theorems 4.1, 4.2] *Let $\mathcal{P}$ be a finite set of boolean polynomials of $\overline{R}$. The binary CSA decomposes $Z(\mathcal{P})$, in a finite number of steps, into a disjoint union of zero sets, namely,*

$$Z(\mathcal{P}) = \bigcup_{j=1}^{s} Z(\mathcal{A}_j),$$

where $\mathcal{A}_1, \ldots, \mathcal{A}_s$ are all monic triangular sets of boolean polynomials.

The main structure of CSA is described in Algorithm 1 in the Appendix. In order to achieve the goal of Theorem 1, the main function of binary CSA is the triset algorithm which takes in as input a finite set of boolean polynomials and outputs a monic triangular set $\mathcal{A}$ (possibly empty) as well as a set $\mathcal{P}^*$ of sets of boolean polynomials.

Remark 2. Beginning with the set $\mathcal{P}$, the whole CSA can be thought of as building a binary tree of sets $(\mathcal{P}^*)$ by adding $\mathcal{Q}^*$ as a branch to the corresponding input set to the triset function. The process stops when all the sets in the tree are processed and the result is a set $\mathcal{A}^*$ of monic triangular sets. Since the polynomial splits by letting $I = 0$ or 1, the binary CSA is essentially a generalization of brute force search where we split by letting the variables take 0 or 1.

Remark 3. Notice that we iteratively split polynomials to obtain sets with disjoint solutions. In this basic version of the triset function, we work on polynomials according to their class, namely, we pick the polynomial with the smallest class. In addition, the next input set to be fed into the triset function is generally taken to be the last generated set in $\mathcal{P}^*$.

2.3   BCSA

In [18], the authors proposed some additional features to improve the efficiency of the binary CSA in [17]. We call it BCSA and briefly describe the techniques here.

Linearization. The authors claimed that this technique was already implemented in the experiments of [17] but it was not explicitly mentioned. This technique applies to the triset function. Suppose that at a certain step, the polynomial $x_{\sigma(c)} + L$ is generated, where $\deg(L) \leq 1$. Then we substitute $x_{\sigma(c)}$ with L in the remaining polynomials and add $x_{\sigma(c)} + L$ to $\mathcal{A}$. Observe that this forms part of the ElimLin process (only the substitution part).

The Choose Function. As discussed in Remark 2, we continue to pick polynomials to split until a monic triangular set is obtained or an inconsistency is reached. In the triset function of CSA, we simply pick the next polynomial to be one with the smallest class in the set. To improve efficiency, it was suggested in [18] to sort the polynomials according to some complexity metric and to choose the "simplest" polynomial among the remaining polynomials to split. The rationale is to seek for more linear polynomials which will help to simplify other polynomials by linearization described above. The complexity metric proposed in BCSA orders the polynomials as follows: A polynomial P is simpler than a polynomial P' if $(\deg(P), \#I(P), \#U(P), n - 1 - \mathrm{cls}_\sigma(P)) < (\deg(P'), \#I(P'), \#U(P'), n - 1 - \mathrm{cls}_\sigma(P'))$ in a lexicographical manner, where $I(P)$ (resp. $U(P)$) and $I(P')$ (resp. $U(P')$) represent the left child (resp. right child) of P and P' respectively. Note that the description of the choose function in [18] with respect to the class of the polynomials is in a reverse order due to the difference in the definition of the class of a polynomial.

The Threshold Value. By Lemma (1), whenever we split a polynomial, we obtain a monic polynomial with the same class. In the triset function of CSA, we first split all the polynomials of the same class c to obtain all the monic polynomials with class c. We choose one of the monic polynomials P_0 to be in $\mathcal{A}$ and then add P_0 to all the monic polynomials to eliminate the class c from the remaining polynomials (the adding step).

Since BCSA splits polynomials in order of their complexity instead of their classes, it is not efficient to first obtain all the monic polynomials before adding them. As such, the authors of [18] introduced a threshold value t such that whenever t monic polynomials of the same class are obtained, the adding process will be executed to eliminate the class from $t - 1$ of these polynomials.

3   The Enhanced Binary CSA

In this paper, we propose additional techniques to enhance BCSA. Our version will be called EBCSA. Like BCSA, we keep the core of the structure, namely, EBCSA comprises the EBCSA-triset function and the main EBCS function, as shown in Fig. 1, where the new features introduced in EBCSA are marked in bold. Its basic structure is described in Algorithm 2 in the appendix. In the subsequent sections, we will describe the main features of EBCSA.

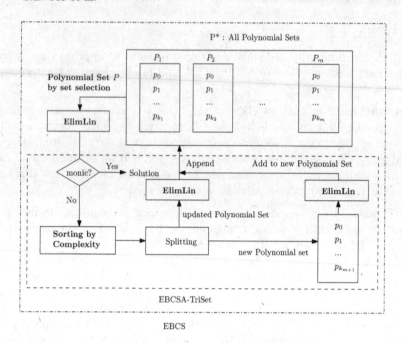

Fig. 1. Principles of the enhanced binary characteristic set algorithm

3.1   The ElimLin Technique

The first main feature of EBCSA is to integrate ElimLin into EBCSA-triset
[11,12]. Specifically, whenever a new polynomial is generated in $\mathcal{P}$, we perform
the following as long as the generated set contains a linear polynomial. First,
order the monomials in decreasing order of degrees (for example, the Grevlex
ordering). With respect to this monomial ordering, represent each polynomial
as a vector of coefficients. Next, perform Gaussian elimination on the matrix of
these vectors. If 1 is obtained, one can terminate EBCSA-triset with $\mathcal{A} = \emptyset$ and
$\mathcal{P}^*$. Otherwise, obtain all the linear polynomials $x_i + L_i$ and substitute x_i with
L_i in the remaining polynomials of $\mathcal{P}$ for all such i's.

A key merit of this process is that it allows us to detect all the linear polynomials in the vector space spanned by the polynomials [12], thereby simplifying
the remaining polynomials more quickly. At the same time, it detects an inconsistent set more rapidly. However, the main drawback is that it generally requires
more memory to construct the matrix and may not be feasible when our polynomial set contains too many polynomials.

3.2   An Improved Complexity Metric

In BCSA, a complexity metric was introduced in order to sort the polynomials.
With respect to this metric, the next simplest polynomial will be picked to
perform the splitting. Their metric compares the degree of the polynomials, the

number of terms in the left child and the right child of the polynomials as well
as the classes of the polynomials in a linear manner. We propose a better metric
based on the Sigmoid function.

For a child $\mathbf{p}$ of a polynomial $\mathbf{P}$, we define its complexity as

$$complex\,(x_1, x_2) = x_1 + Sigmoid\,(x_2, K) = x_1 + \frac{1 - e^{\frac{x_2}{K}}}{1 + e^{\frac{x_2}{K}}} \tag{1}$$

where $x_1 = \deg\,(\mathbf{p})$ determines the overall complexity and $x_2 = \#\,(\mathbf{p})$ counts the
number of terms in $\mathbf{p}$ and it is normalized by the Sigmoid function to the interval
$(0, 1)$. Here $K = -100$ is a scaling parameter which is optimized experimentally.
Then the complexity of the polynomial $\mathbf{P} = I \cdot x_{\sigma(c)} + U$ is defined as

$$Complex(\mathbf{P}) = Complex(\mathbf{I}) + \alpha\,(\mathbf{P}) \cdot Complex(\mathbf{U}) \tag{2}$$

and

$$\alpha\,(\mathbf{P}) = c^{a \cdot \deg(\mathbf{P}) - b} = 10^{3 \cdot \deg(\mathbf{P}) - 8} \tag{3}$$

where $Complex(\mathbf{p}) = complex(\deg\,(\mathbf{p}), \#\,(\mathbf{p}))$, and a, b and c are set to be 3,
8 and 10, respectively. $\mathbf{p}$ is taken as either $\mathbf{I}$ or $\mathbf{U}$. Note that as the degree of
the polynomial system increases, the complexity of the child $\mathbf{U}$ tends to be more
emphasized. Thus the coefficient $\alpha\,(\mathbf{P})$ is defined as an exponential function of
the degree. For instance, when $\deg\,(\mathbf{P})$ is 2, 3 and 4, the weighting coefficient
$\alpha\,(\mathbf{P})$ is 0.01, 10 and 10000, respectively.

In our model, the Sigmoid function is chosen as a candidate because it has
good properties, namely, it is a bounded differentiable real-valued function in
$(0, 1)$ that is defined for all real input values and has a positive derivative at
each value. The standard Sigmoid model is $Sigmoid\,(x) = \frac{1}{1 + e^{\frac{x - x_0}{K}}}$, and we are
using a modified version $Sigmoid\,(x) = \frac{1 - e^{\frac{x}{K}}}{1 + e^{\frac{x}{K}}}$. The complexity model in Eq. 2 is
a combination of linear model and the modified Sigmoid model, and it preserves
the good properties of the standard model. Similar logistic functions have been
effectively used in population growth modelling, artificial neural networks and
distance measure.

To select the best model mentioned above, we have tried many possible parameters for each model. Preliminary experimental results show that the modified
Sigmoid model $Sigmoid\,(x) = \frac{1 - e^{\frac{x}{K}}}{1 + e^{\frac{x}{K}}}$ exhibits the best overall performance on
our training sets. The training sets include 120 randomly selected polynomial
sets from more than 300 available polynomial sets generated from many different ciphers with degrees ranging from 2 to 4, and the rest of the sets are the
testing sets. To determine the optimal parameters of each model, a 10-fold cross-
validation [19] is applied. Hence, the parameters are dependent on the specific
cipher as well as the selected features (degree and number of terms).

In order to reduce the computational cost when selecting the optimal parameters, the parameters are confined to a limited range by preliminary observations.

Then the parameter setting (K, a, b, c) in Eqs. 1, 2 and 3 is determined by the cross-validation procedure as follows: we split the training sets into 10 equal sized parts; using 9 parts we fit the parameters, that is, record the parameter setting that produces the best performance; calculate its performance on the remaining 1 part as the validation set. We repeat these steps by using every part as the validation set. We repeat the whole procedure 3 times. Finally we get an average of the 3 solving times which corresponds to each parameter setting. We choose the setting which corresponds to the minimum solving time. For each of the models mentioned above, we repeat this procedure. As a result, the optimal parameter setting is determined as $(K, a, b, c) = (-100, 3, 8, 10)$. The parameters are not very sensitive, that is, when they are modified by around 10%, the solving time remains within a 10% change. This parameter setting is not guaranteed to be effective for all individual polynomial sets, but is optimal in the sense of overall performance.

By contrast, the choose function in BCSA uses a layered linear model with respect to the essential parameters. This model suffers from the following drawbacks. (1) There are no parameters to control, limiting the possibility to optimize the performance; (2) The "layered" comparison sometimes cannot get reasonable results, e.g. a polynomial with a higher degree but very few terms is not necessarily more complex than a polynomial with a lower degree but the number of terms is huge. As such, the proposed EBCSA complexity model has more flexibility to produce better performance by optimization over a number of polynomial sets. Indeed, we have carried out extensive experiments using different functions and the above function works well in general. We present our experimental results for solving Canfil functions as comparison between using the modified Sigmoid function and the choose function of BCSA in Subsect. 4.1.

3.3   The Set Metric

We introduce a set metric on the sets in the binary tree $\mathcal{P}^*$ to help us find a set with a solution more quickly. Hence, this feature is particularly useful if the system admits a known number of solutions, as well as the case where only a fixed number of solutions is required.

Observe that polynomial sets in the same branch may contain many identical polynomials. Roughly speaking, once many sets in a certain branch terminate without a solution, we like to choose sets from a different branch to increase the chances of a solution. In other words, we will look for a set which is most dissimilar to the previous terminated sets.

For the purpose of practical implementations, we propose the following statistical feature for a set $\mathcal{P}$ of polynomials: $S_{\mathcal{P}} = (m_1, s_1, m_2, s_2)$, where m_1, m_2 (respectively s_1, s_2) represent the mean (respectively the standard deviation) of the degrees and number of terms of the polynomials in $\mathcal{P}$ respectively. Suppose that $\mathcal{P}_1, \mathcal{P}_2, \ldots, \mathcal{P}_l$ are inconsistent, that is, terminate without any solution. Let $a(\mathcal{P}_1, \ldots, \mathcal{P}_l) = \frac{s_{\mathcal{P}_1} + \ldots + s_{\mathcal{P}_l}}{l}$, the average of the statistical features of the terminated sets. Then we choose a set $\mathcal{P}$ such that the Euclidean distance between $s_{\mathcal{P}}$ and $a(\mathcal{P}_1, \ldots, \mathcal{P}_l)$ is the largest on the fly.

3.4   Sorting the Variables

In our description of binary CSA, we have introduced the permutation σ to allow for the variables to be rearranged. In general, CSA tends to be more efficient if σ is chosen so that the variables $x_{\sigma(0)}, \ldots, x_{\sigma(x_{n-1})}$ occur in an increasing order of frequency in the polynomial system. The underlying reasons are as follows: recall that when performing linearization, we substitute a variable x_i with a linear polynomial L_i. While this reduces the number of variables, it may result in the polynomial system becoming more complex, particularly when L contains many terms. Hence, it is more advisable to substitute those x_i which do not occur too frequently. In terms of implementation, we simply rename the variables so that $x_0, x_1, \ldots, x_{n-1}$ occur in an ascending order of frequency in the polynomial system.

3.5   Combining the Various Features

We observe from our experiments that not all polynomial systems behave equally well with the same set of features. According to experimental results, some sets may be more efficient with linearization alone while others perform much better with the ElimLin technique. In addition, the set selection may not be necessary if the system has many solutions and we seek to find all of them. In view of the above, EBCSA allows for different combinations of features to yield the best performance. In our current implementation of EBCSA, the following choices are provided: (1) Linearization or ElimLin; (2) Set selection based on statistical feature or set index; and (3) Terminate with one solution or all solutions.

Remark 4. Observe that unlike BCSA, we do not need a threshold value here as whenever we obtain two monic polynomials with the same class, we add them and choose the simpler polynomial according to our complexity metric to be in the set $\mathcal{A}$. In addition, in case the ElimLin option is turned on, polynomials with the same leading term will automatically be added.

4   Experimental Results

We tested EBCSA with many different polynomial systems, specifically those from block and stream ciphers. Our experiments show significant improvements over existing algebraic methods. In this paper, we present three such experiments, namely, the block ciphers Present and Prince as well as the stream cipher Toyocrypt. Experiments on Canfil ciphers were also carried out to compare the efficiency of EBCSA against CSA and BCSA. Since our goal is to find the unknown key which is likely unique, we use the following configurations of EBCSA for all the experiments: Linearization/ElimLin, set selection, and termination when one solution is obtained.

The platform for testing is a personal computer (Ubuntu 14.04, Intel Xeon CPU E5640 2.67 GHz, 24 GB RAM). Only single-core CPU is used. In the following, we further classify EBCSA into EBCSA and EBCSA-GE (short for EBCSA with Gaussian Elimination), where the former refers to EBCSA with linearization and the latter representing EBCSA with the ElimLin feature switched on.

4.1 Experiments on Our Complexity and Set Metrics

In order to benchmark EBCSA against BCSA and CSA, we performed experiments on the Canfil functions similar to those carried out in [17,18]. The Canfil polynomial sets are boolean polynomial systems arising from linear feedback functions with nonlinear Canfil functions. In [17, Table 4], comparisons were made between CSA and the Gröbner basis algorithm for 8 such functions and the results showed that CSA outperformed the Gröbner basis method for these polynomial sets. In [18, Table 2], the authors recorded the number of splits (or branches) resulting from both CSA and BCSA for these same functions and concluded that BCSA was more effective on these sets.

In this paper, we carry out the same experiments using EBCSA and the number of splits (number of new non-terminating sets produced) for each of the 8 Canfil functions is recorded in Table 1. In addition, the timing on the Canfil functions using the approach of EBCSA with set selection is also listed.

Table 1. Comparison between EBCSA and BCSA in terms of the number of splits

#Splits\system	Canfil1	Canfil2	Canfil3	Canfil4	Canfil5	Canfil6	Canfil7	Canfil8
CSA	13719	23881	7251	1657	1086	3331	1551	180710
BCSA	11958	16074	3267	1316	574	671	1852	35547
EBCSA w.o set selection	5218	24616	5402	782	408	400	3970	90040
EBCSA w. set selection	5772	17299	3865	343	100	209	1046	35565
EBCSA time (s)	82.9	629.5	669.7	8.8	56.4	41.7	57.5	501.3

Observe that EBCSA performs better than BCSA in 5 out of the 8 Canfil experiments with significant improvements for Canfil1, Canfil4, Canfil5 and Canfil6. For Canfil7, CSA outperforms BCSA but is less effective than EBCSA (with set selection). Note that we have used a random key in our experiments and it was not stated if a random or a specific key was used in the experiments in [17,18].

The improvements of EBCSA suggest that our choice of a combined continuous Sigmoid function for polynomial complexity is more effective than the layered linear model used in BCSA. In addition, the use of set selection results in finding the unique solution more efficiently.

4.2 Experiments on Present Cipher

Present is a 31-round lightweight block cipher designed by Bogdanov et al. announced in CHES 2007 [20]. It has a block length of 64-bits and key lengths of either 80-bits or 128-bits. It is based on a permutation-substitution network with 16 4-bit S-boxes in each round. In [14], algebraic cryptanalysis was applied

to Present with the 80-bit master key. This involves representing the intermediate text by variables and each S-box by 21 quadratic equations. The authors
claimed that with 5 known plaintext/ciphertext (KP) pairs, 40 key bits can be
recovered by the method of ElimLin.

In the report of [21], results on algebraic attacks on Present with ElimLin
and the Polybori Gröbner basis implementation were presented. Even though
ElimLin was implemented as a sub-routine in Polybori implementation, it was
demonstrated that the stand-alone ElimLin implementation was more effective
to solve Present. In addition, it was reported that 6 rounds of Present remained
resistant to algebraic cryptanalysis using either ElimLin or Polybori.

As EBCSA-GE incorporates ElimLin into EBCSA, we perform experiments
on round-reduced Present with EBCSA-GE. Once again, we consider the implicit
representation of each s-box with 21 quadratic equations. Note that we place all
the text variables as the front variables while the 80 key variables are placed
behind. We perform our experiments with either 1 or 2 KP. We considered r
rounds of Present for $r = 5$ and $r = 6$. For experiments with one KP, we
did some pre-processing on the polynomial systems before running them on
EBCSA-GE. Specifically, we rename the variables so that the text variables
and the key variables are sorted in an ascending order of frequency. In each of
the experiments, we fix $x_{n-c-1}, x_{n-c}, \ldots, x_{n-1}$, (that is, c most frequent key
variables) and record the time taken as well as the number of splits as c varies.

Table 2. 5-round and 6-round present with 1 KP on EBCSA-GE

r	#FixedBits	64	60	56	54	52	50	48	44	40	38	36	34
5	Time (s)	<1	<1	<1	<1	<1	3.89	5.46	58.6	237.3	161.8	2148.7	>4 h
5	#Splits	0	0	0	0	0	3	16	174	710	556	6200	–
6	Time (s)	7.1	81.0	378.9	1979.2	1489.0	>4 h	>4 h	>4 h	>4 h	>4 h	>4 h	>4 h
6	#Splits	24	216	1234	6163	2172	–	–	–	–	–	–	–

Table 3. 5-round present with 2 KP on EBCSA-GE

#FixedBits	40	36	34	32	30	28	26
Time (s)	9.5	300.1	282.9	1562.9	888.4	2120.8	>4 h
#Splits	6	191	172	974	563	1362	–

One sees that EBCSA-GE outperforms both ElimLin and the Polybori implementation of [21] in the following sense. From Table 4 with 40 keybits fixed,
EBCSA-GE requires only 1 KP to solve in 4 min, which is of comparable magnitude to the timings of ElimLin and Polybori based on 5 KP's. With just 2
KP's, EBCSA-GE already outperforms ElimLin and Polybori, with a solving
time of 9.5s. Next with 36 keybits fixed, EBCSA-GE can solve the system in a
few minutes based on just 2 KP's while ElimLin and Polybori both take longer
to solve (3 to 4 h) and need more plaintexts (16 KP's). From Table 3 with 2

Table 4. A comparison of algebraic attacks on 5-round present

	Polybori [22]		Elimlin [22]		EBCSA-GE			
#FixedBits	40	36	40	36	40	40	36	36
#KP	5	16	5	16	1	2	1	2
Time (s)	241.2	>4 h	154.8	12.2	237.3	9.5	2148.7	300.1

KP, EBCSA-GE can solve up to 52 keybits (28 fixed keybits) for 5 rounds of Present, which is more keybits than what ElimLin and Polybori can solve from [21]. Finally from Table 2, EBCSA-GE can solve 6 rounds of Present with 1 KP and 52 fixed keybits, while the ElimLin and Polybori experiments from [21] are only conducted up to 5 rounds.

Recall that ElimLin essentially seeks to find linear polynomials in the space spanned by the polynomials and then eliminating their leading variables from the remaining polynomials via substitution. This method can only be successful if there are sufficient linear polynomials in the spanning set. This implies that more polynomials involving the key variables in the original set increases the chance of finding enough linear polynomials and hence, leading to a greater number of KPs as observed. In case there are insufficient linear polynomials, Gröbner basis may not be effective as an increase in the degree of the polynomials with a large number of variables will explode the memory. On the other hand, EBCSA-GE executes the splitting process without a rapid increase in memory and continues to find linear polynomials via ElimLin. By alternating between splitting and ElimLin processes, one sees that EBCSA-GE is more effective for Present cipher.

4.3 Experiments on Prince Cipher

Prince is another lightweight block cipher published in Asiacrypt 2012 [22]. It is a 12-round block cipher with a 64-bit block length and a 128-bit key length. It is optimized with respect to latency for hardware implementations. Its construction satisfies the reflection property and is symmetric about the middle layer. The cryptanalysis of Prince has gained much attention, especially after a challenge was posed by Ruhr-Universität Bochum [23] to find a practical attack for round-reduced Prince. To date, it was shown that round-reduced Prince is vulnerable to various attacks including Differential and Integral cryptanalysis as well as time-memory trade-off and meet-in-the-middle attacks [24–28]. To the best of our knowledge, no algebraic attack has been published on Prince.

We carry out the first algebraic attack on Prince. Specifically, we consider Prince-core, that is, without whitening. Hence, only the key K_1 is used. As in all block ciphers, we consider the implicit representation of the s-boxes, thereby resulting in 21 quadratic equations for each s-box. In addition, we introduce 64 intermediate variables for each round. Since whitening is not used, the variables for the first round and the last two rounds can be omitted.

Now, consider the 6-round Prince-core. This system has 384 variables (320 text variables and 64 key variables) and 2016 quadratic equations. We place the key variables behind $(x_{320}, \ldots, x_{383})$. We generate 2 different systems using different random keys. For each key, we fix some key bits and run the system with our EBCSA solver. First, Table 5 records the results for Prince with both EBCSA and EBCSA-GE.

Table 5. 6-round Prince-core: EBCSA vs EBCSA-GE

#Fixed bits	64	56	52	48	40
#Splits of EBCSA	53	533	11415	53147	–
EBCSA time (s)	4.3	54	1400	**5900**	>4 h
EBCSA-GE time (s)	**2**	**266**	**784**	>4 h	>4 h

Observe that when more bits are fixed, both EBCSA and EBCSA-GE should be executed in parallel to find the key as it is not certain which of the two versions will be more efficient. However, when fewer bits are fixed, EBCSA-GE tends to require much more memory to generate the large matrices, thereby slowing down the process significantly. In the following tables, we show the results of EBCSA for two random keys by varying the number of fixed keybits. By fitting the results with a suitable graph, we extrapolate the results to estimate the complexity without fixing any key bits. We conclude that the number of splits of Prince-core is generally better than the brute force $f = 2^x$, where x denotes the number of unknown keybits, as shown in Fig. 2. The results are also listed in Table 6, where the complexity of EBCSA is expected to provide an improvement of more than half the key size over the brute force search. We remark that this approach can be applied to test the complexity of other block ciphers as well.

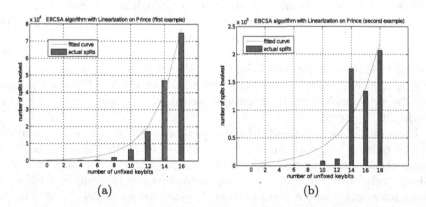

Fig. 2. Splits of Prince-core. (a) $f = 324.19 \times 2^{0.494x}$ (b) $f = 3287.44 \times 2^{0.337x}$

Table 6. Splits of Prince-core (a) $f = 324.19 \times 2^{0.494x}$ (b) $f = 3287.44 \times 2^{0.337x}$

#Fixed bits	64	62	60	58	56	54	52	50	48	46
(a)										
#Splits	45	105	167	301	1908	6415	17101	46994	74887	–
Linearization	3.51	6.89	11.66	21.44	138.27	471.83	1393.60	4010.22	5699.43	>4 h
(b)										
#Splits	49	106	121	433	888	8012	11893	173878	133736	207162
Linearization	3.52	6.51	8.03	29.97	67.09	599.15	956.47	12309.5	10288.5	15156.6

The comparison with the existing attack on 6-round Prince core from [24] is provided in Table 7. It is obvious that the proposed EBCSA algorithm is more practical in terms of required number of chosen plaintext.

Table 7. Comparison of existing attacks on 6-round Prince-core

Source	Data	Time	$D \times T$	Memory	Attacks
FSE13 [24]	2^{16}	2^{30}	2^{46}	2^{16}	Integral
Proposed EBCSA	1	$T \times K \approx 2^{60}$	2^{60}	2^{15}	First algebraic attack by EBCSA

Remark 5. **D**ata: number of chosen plaintext; **T**ime: equivalent Prince operations; complexity of attacks on Prince is measured by $D \times T < 2^{128}$; Memory: Number of Prince 64-bit WORDs.

4.4 Experiments on the Stream Cipher Toyocrypt

The Toyocrypt [29] is a stream cipher comprising a 128-bit Galois linear feedback shift register (GLFSR) passing through a nonlinear function f of degree 63. Obviously, a naive application of algebraic cryptanalysis will not work due to the high degree of f. In [30], the authors discovered that in fact, $(x_{23} + 1)f$ and $(x_{42} + 1)f$ are cubic polynomials. Further, in [31], the authors showed that with re-synchronization and 4 IV's, the degrees of the equations can be further reduced to 1. Note that in this case, the 4 IV's must span a 2-D vector space.

Several algebraic attacks on Toyocrypt have been proposed. Table 8 summarizes the assumptions and requirements needed to carry out existing algebraic attacks on Toyocrypt.

The attack of [30] requires 2^{18} keystream bits for one session, while the attack of [31] requires 128 keystream bits from 4 chosen IV resynchronizations. These attack scenarios may not be easy to achieve in practice. Here we consider a more realistic scenario where only a few keystream bits can be deduced from some known information about the plaintext. Besides, we allow the IV in different sessions to be selected randomly. Finally, we apply our attack to Toyocrypt with GLFSR as in the original specifications of the cipher unlike previous attacks which only considered Toyocrypt with LFSR as a good approximation.

Table 8. Comparison of assumptions on algebraic attacks on Toyocrypt

Source	Resynchronization needed?	No. of keystream-bits	Remarks on IV
EUROCRYPT 2003 [30], CRYPTO 2004 [32]	No	2^{18}	1 session with random IV
SAC 2004 [33]	Yes	2^{13}	Chosen IVs
CANS 2009 [31]	Yes	128	Chosen IVs
Proposed EBCSA	Yes	≤ 6	Random IVs

To be more precise, we fix a GLFSR with feedback taps $0, 1, 2, 7, 128$. For any 2 IV's,it follows from [31, Theorem 1] that the sum of $(x_{23} + 1)f$ (or $(x_{42} + 1)f$) at these two IV's result in equations of degree 2. Instead of choosing some fixed IV's, we randomly choose k IV's and construct the corresponding equations for $(x_{23} + 1)f + z(x_{23} + 1)$ and $(x_{42} + 1)f + z(x_{42} + 1)$, where z represents the keystream bit. Notice that these are degree 3 equations but the sum of any two of them will give a degree 2 equation.

We perform our experiments with at most 6 keystream bits using EBCSA-GE and record the results of our experiments in Table 9. In the preprocessing stage, we once again arranged the 128 variables ascendingly of their frequencies.

Table 9. Performance of Toyocrypt with EBCSA-GE

#Key stream	2	2	2	2	2	3	3	3	3	3	3	4	4	4	4	4	4	4	4	4	6
#IV	128	120	112	104	96	128	120	112	104	96	92	128	120	112	104	96	92	88	84	80	80
#Poly	512	480	448	416	384	768	720	672	624	576	552	1024	960	896	832	768	736	704	672	640	960
#Splits	0	1	1	1	1	0	1	1	1	1	1	0	1	1	1	1	1	2	2	2	2
#Elim Lin	6	9	8	7	11	4	8	8	6	8	7	4	6	5	8	5	9	9	13	13	11
Time (s)	7.1	41.0	318.5	24.1	32.7	4.7	19.4	17.1	178.7	59.4	516.4	7.4	134.0	413.0	76.1	79.7	71.1	97.0	542.1	669.3	187.4

5   Conclusions and Future Work

In this paper, we enhanced the binary characteristic set algorithm to work more effectively on both block ciphers and stream ciphers. Using EBCSA, we improve algebraic attacks on various well-known ciphers including Prince, Present and Toyocrypt. EBCSA uses CSA as its core and incorporates ElimLin and some statistical features to improve its performance. In our current version, we can turn ElimLin on or off for each input polynomial system. For future research, we will seek to identify some characteristics of the polynomial system to choose between EBCSA-GE or EBCSA. We can further integrate the criteria into the main algorithm so that ElimLin can be turned on or off for each individual set produced by EBCSA.

Acknowledgements. We are grateful to Dr. Matt Henricksen, Dr. Yap Wun She, Dr. Lee Hian Kiat and Ms. Ivana Thng for their helpful contributions throughout the project.

Appendix: Pseudocodes for CSA and EBCSA

Next, we present the main structure of EBCSA in Algorithm 2. Note that pre-processing of the input set can be carried out to sort the variables in a desired order, that is, we fix a permutation σ beforehand.

Algorithm 1. The CSA (or MFCS) algorithm [17]

- Set $\mathcal{P}^* = \emptyset$ and $\mathcal{A}^* = \emptyset$.
- CSA-triset function – Feed $\mathcal{P}$ into triset function and let $\mathcal{Q}^*$ and $\mathcal{A}$ be outputs:
  - Set $\mathcal{Q}^* = \emptyset$ and $\mathcal{A} = \emptyset$.
  - For $c = 0$ to $n - 1$ do:
    * If $1 \in \mathcal{P}$, return $\mathcal{A} = \emptyset$ and $\mathcal{Q}^*$.
    * Let $\mathcal{Q} = \{P \in \mathcal{P} : \mathrm{cls}(P) = c\}$, set $\mathcal{P} = \mathcal{P} \backslash \mathcal{Q}$, set $\mathcal{Q}_1 = \emptyset$.
    * While $\mathcal{Q} \neq \emptyset$ do
      · Pick $P \in \mathcal{Q}$ and set $\mathcal{Q} = \mathcal{Q} \backslash \{P\}$.
      Splitting step Write $P = I x_{\sigma(c)} + U$. Add $\mathcal{P} \cup \mathcal{Q} \cup \{I, U\}$ to $\mathcal{Q}^*$.
      · Add $I + 1$ to $\mathcal{P}$ and add $x_{\sigma(c)} + U$ to $\mathcal{Q}_1$.
    * Pick $P \in \mathcal{Q}_1$ and add P to $\mathcal{A}$.
  Adding step For every $P' \in \mathcal{Q}_1, P' \neq P$, add $P + P'$ to $\mathcal{P}$.
- Set $\mathcal{P}^* = \mathcal{P}^* \cup \mathcal{Q}^*$ and add $\mathcal{A}$ to $\mathcal{A}^*$.
- Repeat the whole procedure till $\mathcal{P}^*$ is empty.

Algorithm 2. Algorithm EBCSA

- **Input**: a set of polynomials $\mathbb{P} = \{p_1, \cdots, p_k\}$.
- **Outputs**: A monic triangular set $\mathbb{A}$ and a super set $\mathbb{P}^*$ for other potential solutions.
- **EBCSA-triset** function
  - Begin with the simplest P, e.g. $P_k = I x_{c_k} + U \in \mathbb{P}$.
  - Let $\mathbb{P}' = \mathbb{P} \backslash \{P_k = I x_{c_k} + U\}$. Split P_k: $\mathbb{P} \leftarrow \mathbb{P}' \cup \{I + 1\}$, $\mathbb{A} \leftarrow \mathbb{A} \cup \{x_{c_k} + U\}$ and $\mathbb{P}_{new} \leftarrow \mathbb{P}' \cup I \cup U$, append $\mathbb{P}_{new}$ to $\mathbb{P}^*$.
  - Process each new generated $\mathbb{P}_{new}$ or updated polynomial set $\mathbb{P}$, and terminate them as early as possible.
- Continue with every P until $\mathbb{P}$ is terminated or becomes the monic set $\mathbb{A}$.
- Select a new set $\mathbb{P}$ from $\mathbb{P}^*$ to apply the above algorithm.

References

1. Patarin, J.: Hidden fields equations (HFE) and isomorphisms of polynomials (IP): two new families of asymmetric algorithms. In: Maurer, U. (ed.) EURO-CRYPT 1996. LNCS, vol. 1070, pp. 33–48. Springer, Heidelberg (1996). doi:10.1007/3-540-68339-9_4
2. Buchberger, B.: Ein algorithmus zum auffinden der basiselemente des restklassenrings nach einem nulldimensionalen polynomideal. Universitat Innsbruck, Austria, Ph.D. thesis (1965)
3. Faugere, J.C.: A new efficient algorithm for computing gröbner bases (f4). J. Pure Appl. Algebra 139(1), 61–88 (1999)
4. Faugere, J.C.: A new efficient algorithm for computing Gröbner bases without reduction to zero (F5). In: Proceedings of ISSAC, ACM, pp. 75–83 (2002)
5. Faugere, J.C., Gianni, P., Lazard, D., Mora, T.: Efficient computation of zero-dimensional Gröbner bases by change of ordering. J. Symb. Comput. 16(4), 329–344 (1993)
6. Cox, D., Little, J., O'shea, D.: Ideals, Varieties, and Algorithms, vol. 3. Springer, New York (1992)
7. Kipnis, A., Shamir, A.: Cryptanalysis of the HFE public key cryptosystem by relinearization. In: Wiener, M. (ed.) CRYPTO 1999. LNCS, vol. 1666, pp. 19–30. Springer, Heidelberg (1999). doi:10.1007/3-540-48405-1_2
8. Courtois, N., Klimov, A., Patarin, J., Shamir, A.: Efficient algorithms for solving overdefined systems of multivariate polynomial equations. In: Preneel, B. (ed.) EUROCRYPT 2000. LNCS, vol. 1807, pp. 392–407. Springer, Heidelberg (2000). doi:10.1007/3-540-45539-6_27
9. Courtois, N.T., Pieprzyk, J.: Cryptanalysis of block ciphers with overdefined systems of equations. In: Zheng, Y. (ed.) ASIACRYPT 2002. LNCS, vol. 2501, pp. 267–287. Springer, Heidelberg (2002). doi:10.1007/3-540-36178-2_17
10. Courtois, N.T., Patarin, J.: About the XL algorithm over $GF(2)$. In: Joye, M. (ed.) CT-RSA 2003. LNCS, vol. 2612, pp. 141–157. Springer, Heidelberg (2003). doi:10.1007/3-540-36563-X_10
11. Courtois, N.T., Bard, G.V.: Algebraic cryptanalysis of the data encryption standard. In: Galbraith, S.D. (ed.) Cryptography and Coding 2007. LNCS, vol. 4887, pp. 152–169. Springer, Heidelberg (2007). doi:10.1007/978-3-540-77272-9_10
12. Courtois, N.T., Sepehrdad, P., Sušil, P., Vaudenay, S.: ElimLin algorithm revisited. In: Canteaut, A. (ed.) FSE 2012. LNCS, vol. 7549, pp. 306–325. Springer, Heidelberg (2012). doi:10.1007/978-3-642-34047-5_18
13. Indesteege, S., Keller, N., Dunkelman, O., Biham, E., Preneel, B.: A practical attack on KeeLoq. In: Smart, N. (ed.) EUROCRYPT 2008. LNCS, vol. 4965, pp. 1–18. Springer, Heidelberg (2008). doi:10.1007/978-3-540-78967-3_1
14. Nakahara Jr., J., Sepehrdad, P., Zhang, B., Wang, M.: Linear (Hull) and algebraic cryptanalysis of the block cipher PRESENT. In: Garay, J.A., Miyaji, A., Otsuka, A. (eds.) CANS 2009. LNCS, vol. 5888, pp. 58–75. Springer, Heidelberg (2009). doi:10.1007/978-3-642-10433-6_5
15. Aubry, P., Lazard, D., Maza, M.M.: On the theories of triangular sets. J. Symb. Comput. 28(1), 105–124 (1999)
16. Kalkbrener, M.: A generalized euclidean algorithm for computing triangular representations of algebraic varieties. J. Symb. Comput. 15(2), 143–167 (1993)
17. Fengjuan, C., Xiao-Shan, G., Chunming, Y.: A characteristic set method for solving boolean equations and applications in cryptanalysis of stream ciphers. J. Syst. Sci. Complex. 21(2), 191–208 (2008)

18. Huang, Z., Sun, Y., Lin, D.: On the efficiency of solving Boolean polynomial systems with the characteristic set method. arXiv preprint (2014). arXiv:1405.4596
19. Kohavi, R., et al.: A study of cross-validation and bootstrap for accuracy estimation and model selection. Int. Jt. Conf. Artif. Intell. 14(2), 1137–1145 (1995)
20. Bogdanov, A., Knudsen, L.R., Leander, G., Paar, C., Poschmann, A., Robshaw, M.J.B., Seurin, Y., Vikkelsoe, C.: PRESENT: an ultra-lightweight block cipher. In: Paillier, P., Verbauwhede, I. (eds.) CHES 2007. LNCS, vol. 4727, pp. 450–466. Springer, Heidelberg (2007). doi:10.1007/978-3-540-74735-2_31
21. Sepehrdad, P.: Statistical and algebraic cryptanalysis of lightweight and ultra-lightweight symmetric primitives. Ph.D. thesis, ÉCOLE POLYTECHNIQUE FÉDÉRALE DE LAUSANNE (2012)
22. Borghoff, J., Canteaut, A., Güneysu, T., Kavun, E.B., Knezevic, M., Knudsen, L.R., Leander, G., Nikov, V., Paar, C., Rechberger, C., Rombouts, P., Thomsen, S.S., Yalçın, T.: PRINCE – a low-latency block cipher for pervasive computing applications. In: Wang, X., Sako, K. (eds.) ASIACRYPT 2012. LNCS, vol. 7658, pp. 208–225. Springer, Heidelberg (2012). doi:10.1007/978-3-642-34961-4_14
23. Bochum, R.U.: The PRINCE Challenge. https://www.emsec.rub.de/research/research_startseite/prince-challenge/ (2014). Accessed 18 Jan 2017
24. Jean, J., Nikolić, I., Peyrin, T., Wang, L., Wu, S.: Security analysis of PRINCE. In: Moriai, S. (ed.) FSE 2013. LNCS, vol. 8424, pp. 92–111. Springer, Heidelberg (2014). doi:10.1007/978-3-662-43933-3_6
25. Dinur, I.: Cryptanalytic time-memory-data tradeoffs for FX-constructions with applications to PRINCE and PRIDE. In: Oswald, E., Fischlin, M. (eds.) EURO-CRYPT 2015. LNCS, vol. 9056, pp. 231–253. Springer, Heidelberg (2015). doi:10.1007/978-3-662-46800-5_10
26. Derbez, P., Perrin, L.: Meet-in-the-middle attacks and structural analysis of round-reduced PRINCE. In: Leander, G. (ed.) FSE 2015. LNCS, vol. 9054, pp. 190–216. Springer, Heidelberg (2015). doi:10.1007/978-3-662-48116-5_10
27. Canteaut, A., Fuhr, T., Gilbert, H., Naya-Plasencia, M., Reinhard, J.-R.: Multiple differential cryptanalysis of round-reduced PRINCE. In: Cid, C., Rechberger, C. (eds.) FSE 2014. LNCS, vol. 8540, pp. 591–610. Springer, Heidelberg (2015). doi:10.1007/978-3-662-46706-0_30
28. Canteaut, A., Naya-Plasencia, M., Vayssière, B.: Sieve-in-the-middle: improved MITM attacks. In: Canetti, R., Garay, J.A. (eds.) CRYPTO 2013. LNCS, vol. 8042, pp. 222–240. Springer, Heidelberg (2013). doi:10.1007/978-3-642-40041-4_13
29. Mihaljevic, M.J.: Cryptanalysis of toyocrypt-HS1 stream cipher. IEICE Trans. Fundam. Electron. Commun. Comput. Sci. 85(1), 66–73 (2002)
30. Courtois, N.T., Meier, W.: Algebraic attacks on stream ciphers with linear feedback. In: Biham, E. (ed.) EUROCRYPT 2003. LNCS, vol. 2656, pp. 345–359. Springer, Heidelberg (2003). doi:10.1007/3-540-39200-9_21
31. Zhang, A., Lim, C.-W., Khoo, K., Wei, L., Pieprzyk, J.: Extensions of the cube attack based on low degree annihilators. In: Garay, J.A., Miyaji, A., Otsuka, A. (eds.) CANS 2009. LNCS, vol. 5888, pp. 87–102. Springer, Heidelberg (2009). doi:10.1007/978-3-642-10433-6_7
32. Hawkes, P., Rose, G.G.: Rewriting variables: the complexity of fast algebraic attacks on stream ciphers. In: Franklin, M. (ed.) CRYPTO 2004. LNCS, vol. 3152, pp. 390–406. Springer, Heidelberg (2004). doi:10.1007/978-3-540-28628-8_24
33. Armknecht, F., Lano, J., Preneel, B.: Extending the resynchronization attack. In: Handschuh, H., Hasan, M.A. (eds.) SAC 2004. LNCS, vol. 3357, pp. 19–38. Springer, Heidelberg (2004). doi:10.1007/978-3-540-30564-4_2

SCRAPE: Scalable Randomness Attested by Public Entities

Ignacio Cascudo[1] and Bernardo David[2,3(✉)]

[1] Aalborg University, Aalborg, Denmark
[2] Aarhus University, Aarhus, Denmark
davidm.bernardo@gmail.com
[3] IOHK, Hong Kong, Hong Kong

Abstract. Uniform randomness beacons whose output can be publicly attested to be unbiased are required in several cryptographic protocols. A common approach to building such beacons is having a number parties run a coin tossing protocol with guaranteed output delivery (so that adversaries cannot simply keep honest parties from obtaining randomness, consequently halting protocols that rely on it). However, current constructions face serious scalability issues due to high computational and communication overheads. We present a coin tossing protocol for an honest majority that allows for any entity to verify that an output was honestly generated by observing publicly available information (even after the execution is complete), while achieving both guaranteed output delivery and scalability. The main building block of our construction is the first Publicly Verifiable Secret Sharing scheme for threshold access structures that requires only $O(n)$ exponentiations. Previous schemes required $O(nt)$ exponentiations (where t is the threshold) from each of the parties involved, making them unfit for scalable distributed randomness generation, which requires $t = n/2$ and thus $O(n^2)$ exponentiations.

1 Introduction

The problem of obtaining a reliable source of randomness has been studied since the early days of cryptography. Whereas individual parties can choose to trust locally available randomness sources, it has been shown that local randomness sources can be subverted [BLN16,DPSW16] and many applications require a common public randomness source that is guaranteed to be unbiased by a potential adversary. This necessity inspired the seminal work on Coin Tossing by Blum [Blu81], which allows two or more parties to generate an output that is guaranteed to be uniformly random as long as at least one of the parties is honest (and given that the protocol terminates).

I. Cascudo acknowledges support from the Danish Council for Independent Research, grant no. DFF-4002-00367.

B. David—This project has received funding from the European research Council (ERC) under the European Unions's Horizon 2020 research and innovation programme (grant agreement No. 669255).

D. Gollmann et al. (Eds.): ACNS 2017, LNCS 10355, pp. 537–556, 2017.
DOI: 10.1007/978-3-319-61204-1_27

The concept of a public *randomness beacon* that periodically issues fresh unpredictable and unbiased random values was proposed by Rabin [Rab83] in the context of contract signing and has found several other applications such as voting protocols [Adi08], generating public parameters for cryptographic schemes [BDF+15,LW15], privacy preserving instant messaging [WCGFJ12,vdHLZZ15], and anonymous browsing [DMS04,GRFJ14]. More recently, blockchain [Nak08,GKL15] applications such as smart contracts [KMS+16,B+14], sharding [CDE+16] and Proof-of-Stake based consensus protocols [KKR+16] have increased the need for randomness sources [BCG15].

Rabin's concept of randomness beacons fits the above applications very nicely but the proposed implementation in [Rab83] relies on a trusted third party. The goal of this paper is to construct a distributed randomness beacon guaranteeing output delivery and uniformly distributed randomness for the parties that use the beacon as long as a majority of these parties are honest. Moreover, in many of the aforementioned applications, parties that do not necessarily participate in randomness generation but wish to audit the protocol execution must be able to attest *a posteriori* that the randomness source is reliable and unbiased. Hence, we aim at constructing a publicly verifiable randomness beacon and not only a protocol that outputs randomness to the parties actively involved in its execution.

1.1   Related Works

A natural solution for obtaining randomness beacons consists in using a coin tossing protocol as proposed by Blum [Blu81] with its messages posted to a public bulletin board for later verification (or broadcast among the parties). However, it is known that in case half or more of the parties are corrupted, the adversary can bias the output of the protocol or even prevent the honest parties from obtaining any output at all by aborting protocol execution at a given point [Cle86]. Assuming that a majority of the players are honest, it is possible to guarantee output delivery [RBO89] through threshold verifiable secret sharing (VSS) [CGMA85] given that a broadcast channel is available. Basically, given that a majority of n parties are honest, each party can secret share its input into n shares such that $n/2$ are enough to reconstruct the secret, sending one share to each involved party before starting the coin tossing protocol. While the adversary cannot recover any input (since it has at most $n/2 - 1$ shares of each input), the honest parties collective know at least $n/2$ shares, which they can use to reconstruct the inputs of parties who abort and then finish the protocol.

While a coin tossing protocol with guaranteed output delivery (G.O.D.) with a honest majority based on VSS provides a reliable source of randomness, this approach still has two main issues: 1. most VSS schemes require interaction between the dealer and the other parties, which hinders scalability and 2. only parties who actively participate in the protocol can verify that it was executed correctly. While non-interactive VSS [Fel87] solves the interaction problem, it does not allow the protocol execution to be independently verified by entities that did not actively participate. A natural way to allow for any entity to verify that the outputs produced by such protocols is indeed honestly generated is to substitute traditional VSS by publicly verifiable secret sharing (PVSS) schemes [Sta96], which allow for anybody to verify the validity of shares and reconstructed secrets through information that can be made publicly available without requiring direct interaction between any of the parties. Variations of this approach have been proposed in [KKR+16, SJK+16].

While [KKR+16] instantiates a plain [RBO89]-style G.O.D. coin tossing protocol requiring communication among all parties (through a public ledger), [SJK+16] reduces the communication complexity by partitioning parties into committees that internally run a protocol with publicly verifiable outputs. Later on, a *client* that only communicates to the *leader* of each committee (instead of talking to all parties) can aggregate these outputs to obtain publicly verifiable randomness. However, while the vanilla approach of [KKR+16] achieves security assuming only an honest majority (meaning that the adversary corrupts less than half of all parties), the communication efficient approach of [SJK+16] only achieves security against an adversary that corrupts less than a third of all parties. Moreover, provided that there is an honest majority, the protocol of [KKR+16] guarantees that all parties get output regardless of which parties are corrupted, while in the protocols of [SJK+16], even if only the client is corrupted, it can abort and prevent all other parties from receiving randomness.

Even though coin tossing with G.O.D. built through PVSS can potentially achieve scalability and public verifiability, current PVSS constructions [Sta96, FO98, Sch99, BT99, RV05, HV09, Jha11, JVSN14] suffer from high computational overhead. In general, the parties are required to each compute $O(nt)$ exponentiations to verify n shares of a secret with threshold t, which translates into $O(n^2)$ exponentiations since $t = n/2$ in our randomness beacon application[1]. This computational overhead arises because the main idea behind these schemes is to commit to the coefficients of a polynomial used for a Shamir Secret Sharing [Sha79] and encrypt the shares, later using the commitments to the coefficients to independently compute commitments to the shares, which are proven in zero-knowledge to correspond to the encrypted shares. This approach was originally put forth in [Sch99], which uses the Fiat-Shamir heuristic (and consequently the random oracle model) to obtain the necessary non-interactive zero-knowledge proofs. Later on, variations of this protocol in the plain model were proposed in [RV05, JVSN14], which substitute the zero-knowledge proofs by checks based on Paillier Encryption [Pai99], and in [HV09, Jha11], which propose a pairing based method for checking share validity.

Other approaches for constructing public randomness beacons have been considered in [BCG15, BDF+15, LW15, BLMR14, BGM16]. Public verifiability (or

---

[1] In fact [Jha11] provides an alternative solution where only $O(n)$ exponentiations and a constant number of pairings are required for verification but $O(n)$ pairings are required for setup and $O(n^2)$ exponentiations in the target group of a bilinear map (more expensive than the other exponentiations performed in the source groups) are required for reconstruction.

auditability) in the context of general multiparty computation protocol has been previously considered in [BDO14,SV15].

1.2 Our Contributions

We introduce SCRAPE, a protocol that implements a publicly verifiable randomness beacon given an honest majority through a PVSS based guaranteed output delivery coin tossing protocol. Our main result lies at the core of SCRAPE: the *first* threshold PVSS scheme that only requires a *linear* number of exponentiations for sharing, verifying and reconstruction, whereas previous schemes only achieve quadratic complexity. This PVSS scheme can be instantiated both under the Decisional Diffie Hellman (DDH) assumption in the Random Oracle Model (ROM) and in the *plain model* under the Decisional Bilinear Square (DBS) assumption [HV09]. While improving on the computational complexity of previous schemes, our PVSS scheme retains a similarly low communication overhead, making it fit for applications with large amounts of users. We remark that our new PVSS schemes can also be used to improve the performance of [SJK+16].

Model: As in previous works [BDO14], we assume that the parties can use a "public bulletin board" to publish information that will be used for posterior verification. In fact, in the applications we are interested in, a *ledger* where messages can be posted for posterior verification is readily available, since the Bitcoin Backbone protocol itself implements such a mechanism (*i.e.* the *distributed ledger* analysed in [GKL15]). Nevertheless, our protocols are compatible with any public ledger, not only with that of [GKL15].

Our Techniques: We improve on Schoenmakers' PVSS scheme [Sch99] and its variants (which require $O(nt)$ exponentiations to verify n shares) by designing a share verification procedure that only requires $O(n)$ exponentiations (or pairings). Our procedure explores the fact that sharing a secret with Shamir Secret Sharing [Sha79] is equivalent to encoding the secret (plus randomness) with a Reed Solomon error correcting code, a fact which was first observed by McEliece and Sarwate in [MS81]. Since shares from Shamir Secret Sharing form a codeword of a Reed Solomon code, computing the inner product of a share vector with a codeword from the corresponding dual code should yield 0 if the shares are correctly computed. As in [Sch99], the dealer in our scheme shares the secret using Shamir Secret sharing, encrypts the shares $s_1, \ldots, s_n$ in ciphertexts of the form $h^{sk_i s_i}$ (where h^{sk_i} is a public key and sk_i is a secret key) but also commits to all shares by computing $v_i = g^{s_i}$, where g, h are two independently chosen generators of a group where the DLOG problem is assumed to be hard. The dealer also provides evidence that the shares in the ciphertexts are the same as the shares in the commitments. To verify the validity of the shares, anybody can sample a random codeword $c^\perp = (c_1^\perp, \ldots, c_n^\perp)$ of the dual code of the Reed Solomon code corresponding to the instance of Shamir Secret sharing that was used, compute the inner product of $c^\perp$ with the share vectors in the exponents of g (by computing $\prod_i v_i^{c_i^\perp} = g^{\sum_i s_i c_i^\perp}$) and check that it is equal to $g^0 = 1$. If the shares are not valid, this check fails with large probability. To prove that the shares in the ciphertexts and in the commitments are the same, the dealer can either use a non-interactive zero-knowledge (NIZK) proof constructed using the Fiat-Shamir heuristic as in [Sch99] (resulting in a construction in the ROM under the DDH assumption) or have the parties do pairing based checks as in [HV09] (resulting in a construction in the plain model under the DBS assumption).

Concrete Efficiency: In the DDH based construction in the ROM, the dealer is required to compute $5n$ exponentiations in the sharing phase, while verification and reconstruction respectively require $4n$ and $5t + 3$ exponentiations (given that all n shares are verified but only t shares are used in reconstruction). In the DBS based construction in the plain model, the dealer is required to compute $2n$ exponentiations in the sharing phase, while verification requires $2n$ pairings and reconstruction requires $2n$ pairings and $t + 1$ exponentiations (given that n decrypted shares are verified but only t shares are used in reconstruction). Previous results required nt extra exponentiations in the verification phase, resulting in $n^2/2$ extra exponentiations in the randomness beacon application, which requires $t = n/2$. In the random oracle model construction, extra NIZK data is needed, amounting to a total of $2n$ group elements and $n + 1$ ring elements published by the dealer. In the construction in the plain model, the dealer saves on the NIZK data and only posts $2n$ group elements, while requiring more expensive computation (*i.e.* pairings).

2   Preliminaries

In this section, we establish notation and introduce definitions that will be used throughout the paper. We denote uniformly sampling a random element x from a finite set $\mathcal{D}$ by $x \leftarrow \mathcal{D}$. We denote vectors as $\boldsymbol{x} = (x_1, \ldots, x_n)$. We denote the inner product of two vectors $\boldsymbol{x}, \boldsymbol{y}$ as $\langle \boldsymbol{x}, \boldsymbol{y} \rangle = \sum_{1 \leq i \leq n} x_i \cdot y_i$. For the sake of notation, the integer n will always be considered to be even, so that $n/2$ is an integer. In this paper q will always denote a prime number. We denote by $\mathbb{Z}_q$ the ring of integers modulo q and by $\mathbb{G}$ a finite multiplicative group of order q. Since q is prime, $\mathbb{Z}_q$ is a finite field and $\mathbb{G}$ is a cyclic group where every element $g \neq 1$ is a generator. We denote by $\mathbb{Z}_q[x]$ the ring of polynomials in one variable with coefficients in $\mathbb{Z}_q$. We denote by $log_g e$ the discrete logarithm of an element $e \in \mathbb{G}$ with respect to generator $g \in \mathbb{G}$.

2.1   Coding Theory

We define a $[n, k, d]$ code C to be a linear error correcting code over $\mathbb{Z}_q$ of length n, dimension k and minimum distance d. Its dual code $C^\perp$ is the vector space which consists of all vectors $\boldsymbol{c} \in \mathbb{Z}_q^n$ such that $\langle \boldsymbol{c}, \boldsymbol{c}^\perp \rangle = 0$ for all $\boldsymbol{c}$ in C. The dual code $C^\perp$ of an $[n, k, d]$ code C is an $[n, n - k, d^\perp]$ code (for some $d^\perp$). In this work, we will use the following basic linear algebra fact.

Lemma 1. *If* $v \in \mathbb{Z}_q^n \setminus C$, *and* $c^\perp$ *is chosen uniformly at random in* $C^\perp$ *then the probability that* $\langle v, c^\perp \rangle = 0$ *is exactly* $1/q$.

Proof. By linearity, a $c^\perp \in C^\perp$ is orthogonal to v if only if it is also orthogonal to every vector in the code D spanned by v and C, i.e., if and only if $c^\perp \in D^\perp$. Since $v \notin C$, then the dimension of D is $k + 1$ and hence the space $D^\perp$ has dimension $n - k - 1$. Therefore if $c^\perp$ is chosen uniformly at random in $C^\perp$ the probability that $\langle v, c^\perp \rangle = 0$ is

$$\frac{\#(D^\perp)}{\#(C^\perp)} = \frac{q^{n-k-1}}{q^{n-k}} = \frac{1}{q}.$$

Moreover, in this work we will always be under the assumption $n < q$ and we will use Reed-Solomon codes C of the following form

$$C = \{(p(1), p(2), \ldots, p(n)) : p(x) \in \mathbb{Z}_q[x], \deg p(x) \leq k - 1\}$$

where $p(x)$ ranges over all polynomials in $\mathbb{Z}_q[x]$ of degree at most $k - 1$. This is an $[n, k, n - k + 1]$-code. Its dual $C^\perp$ is an $[n, n - k, k + 1]$-code, which can be defined as follows

$$C^\perp = \{(v_1 f(1), v_2 f(2), \ldots, v_n f(n)) : f(x) \in \mathbb{Z}_q[x], \deg f(x) \leq n - k - 1\}$$

for the coefficients $v_i = \prod_{j=1, j \neq i}^n \frac{1}{i-j}$.

2.2 Shamir Secret Sharing

An (n, t) threshold secret sharing scheme allows a *dealer* D to split a secret s into n shares $S = (s_1, \ldots, s_n)$ distributed among n parties $P_1, \ldots, P_n$ such that it is possible to reconstruct the secret given t of the shares but no information at all is revealed if less shares are known. We refer to S as the *share vector* of the secret sharing scheme. The first threshold secret sharing scheme was introduced by Shamir in [Sha79]. In order to split a secret $s \in \mathbb{Z}_q$, the dealer samples $t - 1$ random coefficients $c_1, \ldots, c_{t-1} \leftarrow \mathbb{Z}_q$ and constructs a polynomial $p(x) = s + c_1 x + c_2 x^2 + \cdots + c_{t-1} x^{t-1}$. The shares are computed as $s_i = p(i)$ for $1 \leq i \leq n$. A party who possesses t shares can use Lagrange interpolation to recover the polynomial $p(x)$ and thus obtain s. On the other hand, a party who knows less than t shares has no information about the secret. McEliece *et al.* first observed that sharing a secret into n shares with Shamir Secret Sharing is equivalent to encoding the message $(x, c_1, \ldots, c_{t-1})$ under a $[n, t, n - t + 1]$ Reed Solomon code, implying that the share vector S is a codeword of such Reed Solomon code.

2.3 Assumptions

One of our constructions is proven in the Random Oracle Model [BR93], where it is assumed that the parties are given access to a function $H(x)$ that takes inputs of any size and returns unique uniformly random outputs of fixed size (returning the same output every time the input is the same). Such a function can be instantiated in practice by a cryptographic hash function. In this model we prove security of our protocols under the DDH assumption, that states given g, g^α, g^β it is hard for a PPT adversary to distinguish between $g^{\alpha\beta}$ from g^r, where g is a generator of a group $\mathbb{G}$ of order q and $\alpha, \beta, r \leftarrow \mathbb{Z}_q$.

We assume efficient non-degenerate bilinear groups described by $\Lambda :=$ $(q, \mathbb{G}, \mathbb{G}_T, e)$, where $\mathbb{G}$ and $\mathbb{G}_T$ are groups of order q and $\mathbb{G} \times \mathbb{G} \to \mathbb{G}_T$ is a bilinear map. We use symmetric pairings to describe the construction for the sake of clarity but remark that our construction can also be easily converted to asymmetric bilinear groups [AHO16], for which state-of-the-art pairing friendly curves [BN06] for which more efficient algorithms for computing pairings [AKL+11] are known. We prove our pairings based protocol secure under the Decisional Bilinear Square (DBS) assumption [HV09] that was shown in that paper to be equivalent to the Decisional Bilinear Quotient assumption [LV08] and related to the Decisional Bilinear Diffie Hellman assumption.

Assumption 1 *Decisional Bilinear Square (DBS) [HV09]*. *Let* $\Lambda :=$ $(q, \mathbb{G}, \mathbb{G}_T, e)$ *be a bilinear group. For a generator* $g \in \mathbb{G}$, *random values* $\mu, \nu, s \leftarrow \mathbb{Z}_q$ *and given* $u = g^\mu$ *and* $v = g^\nu$, *the following probability distributions are computationally indistinguishable:* $D_0 = (g, u, v, T_0 = e(u, u)^\nu)$ *and* $D_1 = (g, u, v, T_1 = e(u, u)^s)$.

Adversarial Model. We prove the security of our protocols in the stand alone setting against malicious adversaries, who may deviate from the protocol in any arbitrary way. We consider static adversaries, who have to choose which parties to corrupt before protocol execution begins.

Public Ledger and Broadcast Channel. We assume that the parties have access to a public ledger with *Liveness*, meaning that an adversary cannot prevent honest parties from adding information and agreeing on it, and *Persistence*, meaning that the information cannot be modified or removed a posteriori. It is known that such a ledger can be implemented by the Bitcoin backbone protocol assuming an honest majority, digital signatures and a Random Oracle [GKL15]. However, we remark that our protocols do not rely on any properties that are unique to the ledger of [GKL15], meaning that our constructions can also be instantiated over public ledgers in the plain model. Notice that access to a broadcast protocol is commonly assumed in multiparty protocols for an honest majority [RBO89] and that the same effect of broadcasting a messages can be achieved by writing it to the ledger. We also remark that the availability of a *public bulletin board* has been assumed in previous works on public verifiability for multiparty protocols [BDO14, SV15].

2.4   Publicly Verifiable Secret Sharing

We adopt the general model for PVSS schemes of [Sch99] and the security definitions of [RV05, HV09] (with some differences that we remark below). We consider a set of n parties $\mathcal{P} = \{P_1, \ldots, P_n\}$ and a dealer D who shares a secret among all the parties in $\mathcal{P}$. We will construct schemes for (n, t)-threshold access structures, meaning that the secret is split in n shares in such a way that knowing at most $t - 1$ shares reveals no information but a collection of t shares allows for secret reconstruction. Additionally, any external verifier V can check that the D is acting honest without learning any information about the shares or the secret. A PVSS protocol has four phases described below:

- **Setup:** The dealer D generates and publishes the parameters of the scheme. Every party P_i publishes a public key pk_i and withholds the corresponding secret key sk_i.
- **Distribution:** The dealer creates shares $s_1, \ldots, s_n$ for the secret s, encrypts share s_i with the public key pk_i for $i = 1, \ldots, n$ and publishes these encryptions $\hat{s}_i$, together with a proof $PROOF_D$ that these are indeed encryptions of a valid sharing of some secret.
- **Verification:** In this phase, any external V (not necessarily being a participant in the protocol) can verify non-interactively, given all the public information until this point, that the values $\hat{s}_i$ are encryptions of a valid sharing of some secret.
- **Reconstruction:** This phase is divided in two.
  Decryption of the shares: This phase can be carried out by any set $\mathcal{Q}$ of t or more parties. Every party P_i in $\mathcal{Q}$ decrypts the share s_i from the ciphertext $\hat{s}_i$ by using its secret key sk_i, and publishes s_i together with a (non-interactive) zero-knowledge proof $PROOF_i$ that this value is indeed a correct decryption of $\hat{s}_i$.
  Share pooling: Any external verifier V (not necessarily being a participant in the protocol) can now execute this phase. V first checks whether the proofs $PROOF_i$ are correct. If the check passes for less than t parties in $\mathcal{Q}$ then V aborts; otherwise V applies a reconstruction procedure to the set s_i of shares corresponding to parties P_i that passed the checks.

A PVSS scheme must provide three security guarantees: Correctness, Verifiability and IND1-Secrecy. These properties are defined below:

- **Correctness:** If the dealer and all players in $\mathcal{Q}$ are honest, then all verification checks in the verification and reconstruction phases pass and the secret can be reconstructed from the information published by the players in $\mathcal{Q}$ in the reconstruction phase.
- **Verifiability:** If the check in the verification step passes, then with high probability the values $\hat{s}_i$ are encryptions of a valid sharing of some secret. Furthermore if the check in the Reconstruction phase passes then the communicated values s_i are indeed the shares of the secret distributed by the dealer.
- **IND1-Secrecy:** Prior to the reconstruction phase, the public information together with the secret keys sk_i of any set of at most $t - 1$ players gives no information about the secret. Formally this is stated as in the following indistinguishability based definition adapted from [RV05, HV09]:

Definition 1 *Indistinguishability of secrets (IND1-secrecy).* *We say that the PVSS is IND1-secret if for any polynomial time adversary* $\mathcal{A}_{Priv}$ *corrupting at most* $t-1$ *parties,* $\mathcal{A}_{Priv}$ *has negligible advantage in the following game played against a challenger.*

1. The challenger runs the Setup phase of the PVSS as the dealer and sends all public information to $\mathcal{A}_{Priv}$. Moreover, it creates secret and public keys for all uncorrupted parties, and sends the corresponding public keys to $\mathcal{A}_{Priv}$.
2. $\mathcal{A}_{Priv}$ creates secret keys for the corrupted parties and sends the corresponding public keys to the challenger.
3. The challenger chooses values x_0 and x_1 at random in the space of secrets. Furthermore it chooses $b \leftarrow \{0,1\}$ uniformly at random. It runs the Distribution phase of the protocol with x_0 as secret. It sends $\mathcal{A}_{Priv}$ all public information generated in that phase, together with x_b.
4. $\mathcal{A}_{Priv}$ outputs a guess $b' \in \{0,1\}$.

The advantage of $\mathcal{A}_{Priv}$ *is defined as* $|\Pr[b = b'] - 1/2|$.

The IND1-secrecy definition is the one used in [RV05,HV09], except for the fact that we do not impose any privacy requirement after the Reconstruction phase. The difference stems from the fact that in [RV05,HV09] it was required that nobody learns the secret but the parties interacting during the reconstruction, while in our random beacon application the secret must be publicly reconstructed and published. We remark that our scheme can achieve both the relaxed definition required by the random beacon application and the stronger secrecy guarantees of [RV05,HV09] (through the use of private channels between parties or through the technique of [HV09] that requires extra data to be posted to the ledger). We also remark that our schemes can achieve the stronger secrecy notion formalized as IND2-secrecy in [RV05,HV09], which allows the adversary to choose arbitrary secrets. This is done by a black box transformation to the protocols that allows for sharing arbitrary secrets instead of random ones by using the random shared secret as a "one time pad" to encrypt an arbitrary secret, which is formally proven in [RV05,HV09].

2.5   Zero-Knowledge Proofs of Discrete Logarithm Knowledge

In our construction based on the DDH assumption in the random oracle model we will need a zero-knowledge proof of knowledge of a value $\alpha \in \mathbb{Z}_q$ such that $x = g^{\alpha}$ and $y = h^{\alpha}$ given g, x, h, y. We denote this proof by $DLEQ(g, x, h, y)$. Chaum and Pedersen constructed a sigma protocol to perform this proof in [CP93], their protocol works as follows:

1. The prover computes $a_1 = g^w$ and $a_2 = h^w$ where $w \leftarrow \mathbb{Z}_q$ and sends a_1, a_2 to the verifier.
2. The verifier sends a challenge $e \leftarrow \mathbb{Z}_q$ to the prover.
3. The prover sends a response $z = w - \alpha e$ to the verifier.

4. The verifier checks that $a_1 = g^z x^e$ and $a_2 = h^z y^e$ and accepts the proof if this holds.

This proof has the properties of completeness, soundness and zero-knowledge. In our proofs, we will specifically reference the soundness property, which means that a prover cannot convince a verifier of a fake statement except with a negligible *soundness error* ϵ. Notice that this sigma protocol can be transformed into a non-interactive zero-knowledge proof of knowledge of α in the random oracle model through the Fiat-Shamir heuristic [FS87,PS96]. We remark that, as in [Sch99], we need to compute this proof in parallel for n distinct pairs of values $(x_1, y_1), \ldots, (x_n, y_n)$. In this case, a single challenge e is computed by the prover as $e = H(x_1, y_1, \ldots, x_n, y_n, a_{1,1}, a_{2,1}, \ldots, a_{1,n}, a_{2,n})$, where the values $a_{1,i}, a_{2,i}$ are computed according to x_i, y_i as described above and $H(\cdot)$ is a random oracle (that can be of course substituted by a cryptographic hash function). The proof then consists of the challenge e along with responses z_i computed according to each x_i, y_i. The verifier can check the proof by computing $a'_{1,i} = g^{z_i} x_i^e$ and $a'_{2,i} = h^{z_i} y_i^e$, and verifying that $H(x_1, y_1, \ldots, x_n, y_n, a'_{1,1}, a'_{2,1}, \ldots, a'_{1,n}, a'_{2,n}) = e$.

3   PVSS Based on the DDH Assumption in the ROM

In this section, we construct a PVSS protocol secure under the DDH assumption in the Random Oracle Model. Our general approach is similar to that of Schoenmakers [Sch99] but differs significantly in the procedure used for share verification, which represents the main overhead in Schoenmakers' scheme.

3.1   Security Analysis

Notice that the Setup and Reconstruction phases are exactly equal to those of [Sch99], while our protocol differs in the Distribution and Verification phases, where we apply our new technique. The key observation is that maliciously generated encrypted shares $\hat{s}_1, \ldots, \hat{s}_n$ will only pass the verification procedure with probability $1/q$ plus the soundness error of the DLEQ proof, while $v_1, \ldots, v_n$ reveal no information about the secret h^s under the DDH assumption (by an argument similar to that of [Sch99]).

We formalize these observations below. First we consider IND1-secrecy. We remark that while we use our relaxed IND1-secrecy notion or our randomness beacon application (where no secrecy is preserved after reconstruction), Protocol π_{DDH} achieves the original stronger IND1-secrecy notion of [RV05,HV09] (where secrecy against parties outside the qualified set is guaranteed even after the reconstruction) if the reconstruction is carried out through private channels between the parties in the qualified set.

Theorem 1. *Under the decisional Diffie-Hellman assumption, the protocol π_{DDH} is IND1-secret against a static PPT adversary (Fig. 1).*

Protocol π_{DDH}

Let g and h be two independently chosen generators of a group $\mathbb{G}_q$ of order q. Let $H(\cdot)$ be a random oracle. Let C be the linear error correcting code corresponding to the (n, t)-threshold Shamir Secret Sharing scheme and let $C^{\perp}$ be its dual code. Protocol π_{DDH} is run between n parties $P_1, \ldots, P_n$, a dealer D and an external verifier V (in fact any number of external verifiers) who have access to a public ledger where they can post information for later verification. The protocol proceeds as follows:

1. **Setup:** Party P_i generates a secret key $sk_i \leftarrow \mathbb{Z}_q$, a public key $pk_i = h^{sk_i}$ and registers the public key pk_i by posting it to the public ledger, for $1 \leq i \leq n$.

2. **Distribution** The dealer D first samples $s \leftarrow \mathbb{Z}_q$. The secret is defined to be $S = h^s$. D chooses $t - 1$ coefficients $c_1, \ldots, c_{t-1} \leftarrow \mathbb{Z}_q$. D constructs a polynomial $p(x) = s + c_1 x + c_2 x^2 + \cdots + c_{t-1} x^{t-1}$ and computes shares $s_i = p(i)$ for $1 \leq i \leq n$. D encrypts the shares as $\hat{s}_i = pk_i^{s_i}$, computes commitments $v_i = g^{s_i}$ and computes $DLEQ(g, v_i, pk_i, \hat{s}_i)$, for $1 \leq i \leq n$ from a single challenge e as described in Section 2.5, obtaining $(e, z_1, \ldots, z_n)$. D publishes in the public ledger the encrypted shares $(\hat{s}_1, \ldots, \hat{s}_n)$ along with $PROOF_D = (v_1, \ldots, v_n, e, z_1, \ldots, z_n)$.

3. **Verification:** The verifier V first checks that the $DLEQ(g, v_i, pk_i, \hat{s}_i)$ provided by D is valid as described in Section 2.5. If the proof is valid, V samples a random codeword $\mathbf{c}^{\perp} = (c_1^{\perp}, \ldots, c_n^{\perp})$ of the dual code $C^{\perp}$ corresponding to the instance of Shamir's (n, t)-threshold secret sharing used by D and considers the shares valid if and only if the following expression is true:

$$\prod_{i=1}^{n} v_i^{c_i^{\perp}} = 1.$$

4. **Reconstruction:** If a set of t or more parties $\mathcal{Q}$ wishes to reconstruct the secret, each party $P_i \in \mathcal{Q}$ starts by publishing in the public ledger its decrypted share $\tilde{s}_i = \hat{s}_i^{\frac{1}{sk_i}} = h^{s_i}$ and $PROOF_i = DLEQ(h, pk_i, \tilde{s}_i, \hat{s}_i)$ (showing that the decrypted share $\tilde{s}_i$ corresponds to $\hat{s}_i$). Once every party in $\mathcal{Q}$ publishes their decrypted shares and $PROOF_i$, they first verify that the proofs are valid and, if this check succeeds, reconstruct the secret $S = h^s$ by Lagrange interpolation:

$$\prod_{P_i \in \mathcal{Q}} (\tilde{s}_i)^{\lambda_i} = \prod_{P_i \in \mathcal{Q}} h^{p(i)\lambda_i} = h^{p(0)} = h^s,$$

where $\lambda_i = \prod_{j \neq i} \frac{j}{j-i}$ are the Lagrange coefficients.

Fig. 1. Protocol π_{DDH}

Proof. We show that, if there exists an adversary $\mathcal{A}_{\mathrm{Priv}}$ which can break the IND1-secrecy property of protocol π_{DDH}, then there exists an adversary $\mathcal{A}_{\mathrm{DDH}}$ which can use $\mathcal{A}_{\mathrm{Priv}}$ to break the decisional Diffie-Hellman assumption with the same advantage. Without loss of generality we assume $\mathcal{A}_{\mathrm{Priv}}$ corrupts the $t - 1$ first parties.

Let $(g, g^\alpha, g^\beta, g^\gamma)$ be an instance of the DDH problem. Obviously if $\alpha = 0$ or $\beta = 0$ then the problem is trivial, so we assume these values are nonzero. Now $\mathcal{A}_{\mathrm{DDH}}$, using $\mathcal{A}_{\mathrm{Priv}}$, can simulate an IND1 game as follows:

1. The challenger sets $h = g^\alpha$ and runs the Setup phase of π_{DDH}. For $t \leq i \leq n$, $\mathcal{A}_{\mathrm{DDH}}$ selects uniformly random values $u_i \leftarrow \mathbb{Z}_p$ (these can be thought of implicitly defining sk_i as $sk_i = u_i/\alpha$) and sends the values $pk_i = g^{u_i}$ to $\mathcal{A}_{\mathrm{Priv}}$.
2. For $1 \leq i \leq t - 1$, $\mathcal{A}_{\mathrm{Priv}}$ chooses uniformly random values $sk_i \leftarrow \mathbb{Z}_q$ and sets $pk_i = h^{sk_i}$ and sends this to the challenger.
3. For $1 \leq i \leq t - 1$, the challenger chooses uniformly random values $s_i \leftarrow \mathbb{G}_q$ and sets $v_i = g^{s_i}$ and $\hat{s}_i = pk_i^{s_i}$.
   For $t \leq i \leq n$, it generates the values $v_i = g^{p(i)}$ where $p(x)$ is the unique polynomial of degree at most t determined by $p(0) = \beta$ and $p(i) = s_i$ for $i = 1, \ldots, t-1$. Note that $\mathcal{A}_{\mathrm{DDH}}$ does not know β, but it does know $g^\beta = g^{p(0)}$ and $g^{s_i} = g^{p(i)}$ for $1 \leq i \leq t - 1$, so it can use Lagrange interpolation in the exponent to compute the adequate v_i. It also creates the values $\hat{s}_i = v_i^{u_i}$. Note that then $\hat{s}_i = g^{u_i \cdot p(i)} = pk_i^{p(i)}$. From all the computed values, the challenger now creates the DLEQ proofs as the dealer does in the PVSS protocol. Finally it sends all this information together with the value g^γ (which plays the role of x_b in the IND game) to $\mathcal{A}_{\mathrm{Priv}}$.
4. $\mathcal{A}_{\mathrm{Priv}}$ makes a guess b'.

If $b' = 0$, $\mathcal{A}_{\mathrm{DDH}}$ guesses that $\gamma = \alpha \cdot \beta$. If $b' = 1$, $\mathcal{A}_{\mathrm{DDH}}$ guesses that γ is a random element in $\mathbb{Z}_p$.

The information that $\mathcal{A}_{\mathrm{Priv}}$ receives in step 3. is distributed exactly like a sharing of the value $h^\beta = g^{\alpha \cdot \beta}$ with the PVSS. Consequently, $\gamma = \alpha \cdot \beta$ if and only if the value g^γ sent to $\mathcal{A}_{\mathrm{Priv}}$ is the secret shared by the PVSS. It is now easy to see that the guessing advantage of $\mathcal{A}_{\mathrm{DDH}}$ is the same as the advantage of $\mathcal{A}_{\mathrm{Priv}}$.

The following two theorems guarantee the verifiability property of π_{DDH}.

Theorem 2. *If the dealer does not construct values $(v_i, \hat{s}_i)$ of the right form in the Distribution phase (i.e. either $log_g v_i \neq log_{pk_i} \hat{s}_i$ for some i, or $log_g v_i = log_{pk_i} \hat{s}_i = s_i$ for all i but the values s_i do not constitute a valid sharing of some secret in $\mathbb{Z}_q$ with the (n, t)-threshold Shamir secret sharing scheme), then this is detected in the verification step with probability at least $1 - \epsilon - 1/q$, where ϵ is the soundness error of the proof DLEQ.*

Proof. If the verification of DLEQ passes, then we have that, except with probability ϵ, for every $1 \leq i \leq n$, there exists s_i with $v_i = g^{s_i}$ and $\hat{s}_i = pk_i^{s_i}$. Now the values s_i are a valid sharing with the (n, t)-threshold Shamir secret sharing scheme if and only if the vector $\boldsymbol{v} = (s_1, \ldots, s_n) \in C$. Suppose that $(s_1, \ldots, s_n) \notin C$. Then by Lemma 1, since $\boldsymbol{c}^\perp$ is sampled uniformly at random then $\langle \boldsymbol{v}, \boldsymbol{c}^\perp \rangle \neq 0$ except with probability $1/q$. But then $\prod_{i=1}^n v_i^{c_i^\perp} = \prod_{i=1}^n g^{s_i \cdot c_i^\perp} = g^{\langle \boldsymbol{v}, \boldsymbol{c}^\perp \rangle} \neq 1$. Hence if the values s_i are not a valid Shamir sharing, then the check fails with probability $1 - 1/q$.

Theorem 3. *If a party in $\mathcal{Q}$ communicates an erroneous decryption share $\tilde{s}_i$ in the Reconstruction phase, then this is detected by the verifier with probability $1 - \epsilon$, where ϵ is the soundness error of the DLEQ proof.*

Proof. This is straightforward by definition since an adversary that succeeds in providing a DLEQ proof for an invalid decrypted share breaks DLEQ's soundness property.

4   PVSS Based on Pairings in the Plain Model

In this section, we construct a PVSS scheme based on the DBS assumption in the plain model (without requiring random oracles). This scheme uses the techniques of [HV09] to eliminate the need for the random oracle based NIZKs and instead use pairings to check that the encrypted shares $\hat{s}_i$ correspond to the committed shares v_i and, later on, check that the decrypted shares correspond to $\hat{s}_i$. We use the same information theoretical verification procedure as in the DDH based scheme. The protocol is described in Fig. 2.

4.1   Security Analysis

As in the DDH based protocol, notice that the Setup and Reconstruction phases are exactly equal to those of [HV09], while our protocol differs in the Distribution and Verification phases. Maliciously generated encrypted shares $\hat{s}_1, \ldots, \hat{s}_n$ will only pass the verification procedure with probability of $1/q$, while $v_1, \ldots, v_n$ again reveal no information about the secret $e(h, h)^s$ but this time under the BDS assumption. Again we remark that Protocol π_{DBS} achieves the original stronger IND1-secrecy notion of [RV05, HV09] (where secrecy against parties outside the qualified set is guaranteed even after the reconstruction) if the reconstruction is carried out through private channels between the parties in the qualified set. The proofs of Theorems 4 and 5 are similar to those of Theorems 1 and 2, respectively, and are presented in the full version of this work [CD17].

Theorem 4. *Under the DBS assumption, protocol π_{DBS} is IND1-secret against a static PPT adversary.*

Theorem 5. *If the dealer does not construct values $(v_i, \hat{s}_i)$ of the right form in the Distribution phase (i.e. either $\log_g v_i \neq \log_{pk_i} \hat{s}_i$ for some i, or $\log_g v_i = \log_{pk_i} \hat{s}_i = s_i$ for all $1 \leq i \leq n$ but the values s_i do not constitute a valid sharing of some secret in $\mathbb{Z}_q$ with the (n,t)-threshold Shamir secret sharing scheme), then this is detected in the verification step with probability at least $1 - 1/q$.*

Theorem 6. *If a party in $\mathcal{Q}$ communicates an erroneous decryption share $\tilde{s}_i$ in the Reconstruction phase, then this is detected by the verifier with probability 1.*

Proof. If $\tilde{s}_i = h^a$ with $a \neq s_i$ then $e(pk_i, \tilde{s}_i) = e(pk_i, h)^a \neq e(pk_i, h)^{s_i} = e(\hat{s}_i, h)$.

Protocol π_{DBS}

Let $\Lambda := (q, \mathbb{G}, \mathbb{G}_T, e)$ be a description of a bilinear group and g, h be two independently chosen generators of $\mathbb{G}$. Let C be the linear error correcting code equivalent to the (n, t)-threshold Shamir Secret Sharing scheme and $C^\perp$ its dual code. Protocol π_{DBS} is run between n parties $P_1, \ldots, P_n$, a dealer D and an external verifier V (in fact any number of external verifiers) who have access to a public ledger where they can post information for later verification. The protocol proceeds as follows:

1. **Setup:** Party P_i generates a secret key $sk_i \leftarrow \mathbb{Z}_q$, a public key $pk_i = h^{sk_i}$ and registers the public key pk_i by posting it to the public ledger, for $1 \le i \le n$. .

2. **Distribution** The dealer D first samples $s \leftarrow \mathbb{Z}_q$. The secret is defined to be $S = e(h, h)^s$. D chooses $t - 1$ coefficients $c_1, \ldots, c_{t-1} \leftarrow \mathbb{Z}_q$. D constructs a polynomial $p(x) = s + c_1 x + c_2 x^2 + \cdots + c_{t-1} x^{t-1}$ and computes shares $s_i = p(i)$ for $1 \le i \le n$. D encrypts the shares as $\hat{s}_i = pk_i^{s_i}$ and computes commitments $v_i = g^{s_i}$, for $1 \le i \le n$. D posts in the public ledger the encrypted shares $\hat{s}_1, \ldots, \hat{s}_n$ and $PROOF_D = (v_1, \ldots, v_n)$.

3. **Verification:** The verifier V first checks that $e(\hat{s}_i, g) = e(pk_i, v_i)$, for $1 \le i \le n$. If this check succeeds, V samples a random codeword $\mathbf{c}^\perp = (c_1^\perp, \ldots, c_n^\perp)$ of the dual code $C^\perp$ corresponding to the instance of Shamir's (n, t)-threshold secret sharing used by D and considers the shares valid if and only if the following expression is true:
$$\prod_{i=1}^{n} v_i^{c_i^\perp} = 1.$$

4. **Reconstruction:** If a set of t or more parties $\mathcal{Q}$ wishes to reconstruct the secret, each party $P_i \in \mathcal{Q}$ starts publishing in the public ledger its decrypted share $\tilde{s}_i = \hat{s}_i^{\frac{1}{sk_i}} = h^{s_i}$ (here $PROOF_i$ is an empty string). Once every party in $\mathcal{Q}$ publishes their decrypted shares, they first verify that $e(pk_i, \tilde{s}_i) = e(\hat{s}_i, h)$ for every $P_i \in \mathcal{Q}$. If this check succeeds, they reconstruct the value h^s by Lagrange interpolation:
$$\prod_{P_i \in \mathcal{Q}} (\tilde{s}_i)^{\lambda_i} = \prod_{P_i \in \mathcal{Q}} h^{p(i)\lambda_i} = h^{p(0)} = h^s,$$

where $\lambda_i = \prod_{j \ne i} \frac{j}{j-i}$ are the Lagrange coefficients. The secret is then computed as $S = e(h^s, h)$.

Fig. 2. Protocol π_{DBS}

5 Building the SCRAPE Randomness Beacon

Publicly verifiable secret sharing schemes have a multitude of applications as discussed in [Sch99], among them universally verifiable elections, threshold versions of El Gamal encryption and threshold software key escrow. However, we are specially interested in constructing SCRAPE, a protocol that implements a distributed randomness beacon that is guaranteed to be secure given an honest majority, a PVSS scheme and a public ledger. SCRAPE is basically a coin tossing protocol with guaranteed output delivery (G.O.D.), meaning that an adversary

---

Protocol π_{SCRAPE}

Protocol π_{DDH} is run between n parties $P_1, \ldots, P_n$ who have access to a public ledger where they can post information for later verification. A PVSS protocol is used as a sub protocol and it is assumed that the Setup phase is already done and the public keys pk_i of each party P_i are already registered in the ledger. The protocol proceeds as follows:

1. **Commit:** For $1 \leq j \leq n$, party P_j executes the Distribution phase of the PVSS sub protocol as the Dealer with threshold $t = \frac{n}{2}$, publishing the encrypted shares $\hat{s}_1^j, \ldots, \hat{s}_n^j$ and the verification information $PROOF_D^j$ on the public ledger, and also learning the random secret h^{s^j} and s^j. P_j also publishes a commitment to the secret exponent $\mathsf{Com}(s^j, r_j)$ (with fresh randomness $r_j \leftarrow \mathbb{Z}_q$), for $1 \leq j \leq n$.

2. **Reveal:** For every set of encrypted shares $\hat{s}_1^j, \ldots, \hat{s}_n^j$ and the verification information $PROOF_D^j$ published in the public ledger, all parties run the Verification phase of the PVSS sub protocol. Let $\mathcal{C}$ be the set of all parties who published commitments and valid shares. Once $\frac{n}{2}$ parties have posted their commitments and valid shares on the ledger, party P_j opens its commitment, posting $\mathsf{Open}(s^j, r_j)$ on the ledger, for $j \in \mathcal{C}$.

3. **Recovery:** For every party $P_a \in \mathcal{C}$ that does not publish $\mathsf{Open}(s^a, r_a)$ in the Reveal phase, party P_j runs the Reconstruction phase of the PVSS protocol posting $\tilde{s}_j^a$ and $PROOF_j^a$ to the public ledger, for $1 \leq j \leq n$. Once $\frac{n}{2}$ valid decrypted shares are published, every party reconstructs h^{s^a}.

The final randomness is computed as $\rho = \prod_{j \in \mathcal{C}} h^{s^j}$.

Fig. 3. Protocol π_{SCRAPE}

cannot prevent honest parties from obtaining a correct output (*e.g.* by aborting before the execution is finished). Moreover, SCRAPE is publicly verifiable, meaning that anybody can analyse past (and current) protocol transcripts to verify that the protocol is being correctly executed. The reason we aim at guaranteed output delivery is twofold: 1. Protecting against particularly adversarial behavior and 2. Tolerating non-byzantine failures in users after the commitment phase (*e.g.* power outage). When used to bootstrap blockchain based consensus protocols such as in [KKR+16], π_{SCRAPE} has to tolerate adversaries that can force a temporary loss of consensus making the users end up with conflicting random outputs or temporarily "disconnecting" users from the network or public ledger. We follow the general approach of [RBO89] to obtain guaranteed output delivery based on verifiable secret sharing, building on our PVSS schemes to achieve public verifiability for the final coin tossing protocol. More specifically, we use our PVSS protocols to instantiate the construction of [KKR+16], which proposed to combine a PVSS scheme with a public ledger to obtain publicly verifiable G.O.D. coin tossing. The protocol is described in Fig. 3. The security of π_{SCRAPE} follows from the security of the general construction of [KKR+16] and the security of our protocols that was proven in the previous sections. We refer the reader to [KKR+16] for a detailed discussion on the general protocol.

6   Concrete Complexity and Experiments

In this section, we discuss the concrete computational and communication complexity of our schemes, comparing them with previous work. We first start by discussing the computational complexity of our PVSS protocols, which is compared to that of the protocols of [Sch99, HV09] in terms of numbers of exponentiations and pairings required for each phase in Table 1. Notice that for our randomness beacon application we need $t = n/2$, which translates into an extra overhead of $n^2/2$ exponentiations required for the verification phase of previous protocols. Our protocols eliminate this quadratic overhead, resulting in much better scalability. For example, if 10000 users run SCRAPE based on [Sch99], 50004000 exponentiations are required in the verification phase, while our instantiation

Table 1. Concrete computational complexity in terms of numbers of exponentiations (Exp.) and pairings (Pair.) needed for each phase, considering that n shares are generated and t shares are used in reconstruction.

	Distribution	Verification		Reconstruction	
	Exp.	Exp.	Pair.	Exp.	Pair.
[HV09]	$n + t$	nt	$2n$	$t + 1$	$2t + 1$
Protocol π_{DBS}	$2n$	n	$2n$	$t + 1$	$2t + 1$
[Sch99]	$3n + t$	$nt + 4n$	-	$5t + 3$	-
Protocol π_{DDH}	$4n$	$5n$	-	$5t + 3$	-

Verification runtime of π_{DDH} vs. π_{DBS} vs. [Sch99]

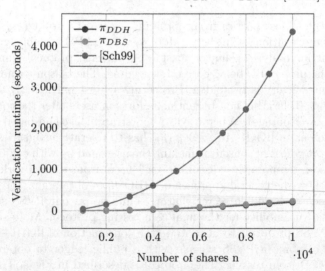

Fig. 4. Execution time of the verification phases of π_{DDH} vs. π_{DBS} vs. Schoenmakers' PVSS [Sch99] for a number of shares n from 1000 to 10000 and threshold $t = \frac{n}{2}$.

of SCRAPE would require only 50000 exponentiations, achieving a performance gain of over 100 times. In terms of communication complexity, our protocols match previous results [Sch99, HV09] in the reconstruction phase while requiring $2n$ group elements to be published in the distribution phase. In the randomness beacon application, this represents an overhead of only $0.5n$ in relation to [Sch99, HV09]. In order to evaluate the concrete performance of our proposed protocols, we have conducted experiments with implementations of π_{DDH}, π_{DBS} and the PVSS scheme of [Sch99], all written in Haskell. The experiments were run on a single core of a Linux machine with a Intel(R) Core(TM) i7-3770K CPU @ 3.50 GHz and 32 GB of RAM. Figure 4 compares the runtime of the verification phases of π_{DDH} and [Sch99], which represents the main improvement of π_{DDH}. Further discussion on concrete complexity and experimental results can be found in the full version of this work [CD17].

Acknowledgements. We thank Vincent Hanquez and Andrzej Rybczak for implementing π_{DDH} and π_{DBS}, respectively.

References

Adi08. Adida, B.: Helios: web-based open-audit voting. In: van Oorschot, P.C. (ed.) Proceedings of 17th USENIX Security Symposium, 28 July–1 August 2008, San Jose, CA, USA, pp. 335–348. USENIX Association (2008)

AHO16. Abe, M., Hoshino, F., Ohkubo, M.: Design in type-I, run in type-III: fast and scalable bilinear-type conversion using integer programming. In: Robshaw, M., Katz, J. (eds.) CRYPTO 2016. LNCS, vol. 9816, pp. 387–415. Springer, Heidelberg (2016). doi:10.1007/978-3-662-53015-3_14

AKL+11. Aranha, D.F., Karabina, K., Longa, P., Gebotys, C.H., López, J.: Faster explicit formulas for computing pairings over ordinary curves. In: Paterson, K.G. (ed.) EUROCRYPT 2011. LNCS, vol. 6632, pp. 48–68. Springer, Heidelberg (2011). doi:10.1007/978-3-642-20465-4_5

B+14. Buterin, V., et al.: A next-generation smart contract and decentralized application platform. White paper (2014)

BCG15. Bonneau, J., Clark, J., Goldfeder, S.: On bitcoin as a public randomness source. Cryptology ePrint Archive, Report 2015/1015 (2015). http://eprint.iacr.org/2015/1015

BDF+15. Baignères, T., Delerablée, C., Finiasz, M., Goubin, L., Lepoint, T., Rivain, M.: Trap me if you can - million dollar curve. Cryptology ePrint Archive, Report 2015/1249 (2015). http://eprint.iacr.org/2015/1249

BDO14. Baum, C., Damgård, I., Orlandi, C.: Publicly auditable secure multiparty computation. In: Abdalla, M., De Prisco, R. (eds.) SCN 2014. LNCS, vol. 8642, pp. 175–196. Springer, Cham (2014). doi:10.1007/978-3-319-10879-7_11

BGM16. Bentov, I., Gabizon, A., Mizrahi, A.: Cryptocurrencies without proof of work. In: Clark et al. [CMR+16], pp. 142–157 (2016)

BLMR14. Bentov, I., Lee, C., Mizrahi, A., Rosenfeld, M.: Proof of activity: extending bitcoin's proof of work via proof of stake [extended abstract]y. SIGMETRICS Perform. Eval. Rev. **42**(3), 34–37 (2014)

BLN16. Bernstein, D.J., Lange, T., Niederhagen, R.: Dual EC: a standardized back door. In: Ryan, P.Y.A., Naccache, D., Quisquater, J.-J. (eds.) The New Code-breakers. LNCS, vol. 9100, pp. 256–281. Springer, Heidelberg (2016). doi:10.1007/978-3-662-49301-4_17

Blu81. Blum, M.: Coin flipping by telephone. In: Gersho, A. (ed.) CRYPTO 1981, vol. ECE report 82-04, pp. 11–15. U.C. Santa Barbara, Department of Electrical and Computer Engineering (1981)

BN06. Barreto, P.S.L.M., Naehrig, M.: Pairing-friendly elliptic curves of prime order. In: Preneel, B., Tavares, S. (eds.) SAC 2005. LNCS, vol. 3897, pp. 319–331. Springer, Heidelberg (2006). doi:10.1007/11693383_22

BR93. Bellare, M., Rogaway, P.: Random oracles are practical: a paradigm for designing efficient protocols. In: Ashby, V. (ed.) ACM CCS 1993, pp. 62–73. ACM Press, November 1993

BT99. Boudot, F., Traoré, J.: Efficient publicly verifiable secret sharing schemes with fast or delayed recovery. In: Varadharajan, V., Mu, Y. (eds.) ICICS 1999. LNCS, vol. 1726, pp. 87–102. Springer, Heidelberg (1999). doi:10.1007/978-3-540-47942-0_8

CD17. Cascudo, I., David, B.: Scrape: scalable randomness attested by public entities. Cryptology ePrint Archive, Report 2017/216 (2017). http://eprint.iacr.org/2017/216

CDE+16. Croman, K., Decker, C., Eyal, I., Gencer, A.E., Juels, A., Kosba, A.E., Miller, A., Saxena, P., Shi, E., Sirer, E.G., Song, D., Wattenhofer, R.: On scaling decentralized blockchains - (a position paper). In: Clark et al. [CMR+16], pp. 106–125 (2016)

CGMA85. Chor, B., Goldwasser, S., Micali, S., Awerbuch, B.: Verifiable secret sharing and achieving simultaneity in the presence of faults (extended abstract). In: 26th FOCS, pp. 383–395. IEEE Computer Society Press, October 1985

Cle86. Cleve, R.: Limits on the security of coin flips when half the processors are faulty (extended abstract). In: Hartmanis, J. (ed.) Proceedings of 18th Annual ACM Symposium on Theory of Computing, 28–30 May 1986, Berkeley, California, USA, pp. 364–369. ACM (1986)

CMR+16. Clark, J., Meiklejohn, S., Ryan, P.Y.A., Wallach, D.S., Brenner, M., Rohloff, K. (eds.): FC 2016 Workshops. LNCS, vol. 9604. Springer, Heidelberg (2016)

CP93. Chaum, D., Pedersen, T.P.: Wallet databases with observers. In: Brickell, E.F. (ed.) CRYPTO 1992. LNCS, vol. 740, pp. 89–105. Springer, Heidelberg (1993). doi:10.1007/3-540-48071-4_7

DMS04. Dingledine, R., Mathewson, N., Syverson, P.: Tor: the second-generation onion router. In: Proceedings of 13th Conference on USENIX Security Symposium, SSYM 2004, Berkeley, CA, USA, vol. 13, p. 21. USENIX Association (2004)

DPSW16. Degabriele, J.P., Paterson, K.G., Schuldt, J.C.N., Woodage, J.: Backdoors in pseudorandom number generators: possibility and impossibility results. In: Robshaw, M., Katz, J. (eds.) CRYPTO 2016. LNCS, vol. 9814, pp. 403–432. Springer, Heidelberg (2016). doi:10.1007/978-3-662-53018-4_15

Fel87. Feldman, P.: A practical scheme for non-interactive verifiable secret sharing. In: 28th FOCS, pp. 427–437. IEEE Computer Society Press, October 1987

FO98. Fujisaki, E., Okamoto, T.: A practical and provably secure scheme for publicly verifiable secret sharing and its applications. In: Nyberg, K. (ed.) EUROCRYPT 1998. LNCS, vol. 1403, pp. 32–46. Springer, Heidelberg (1998). doi:10.1007/BFb0054115

FS87. Fiat, A., Shamir, A.: How to prove yourself: practical solutions to identification and signature problems. In: Odlyzko, A.M. (ed.) CRYPTO 1986. LNCS, vol. 263, pp. 186–194. Springer, Heidelberg (1987). doi:10.1007/3-540-47721-7_12

GKL15. Garay, J., Kiayias, A., Leonardos, N.: The bitcoin backbone protocol: analysis and applications. In: Oswald, E., Fischlin, M. (eds.) EUROCRYPT 2015. LNCS, vol. 9057, pp. 281–310. Springer, Heidelberg (2015). doi:10.1007/978-3-662-46803-6_10

GRFJ14. Ghosh, M., Richardson, M., Ford, B., Jansen, R.: A torpath to torcoin: proof-of-bandwidth altcoins for compensating relays. Technical report, DTIC Document (2014)

HV09. Heidarvand, S., Villar, J.L.: Public verifiability from pairings in secret sharing schemes. In: Avanzi, R.M., Keliher, L., Sica, F. (eds.) SAC 2008. LNCS, vol. 5381, pp. 294–308. Springer, Heidelberg (2009). doi:10.1007/978-3-642-04159-4_19

Jha11. Jhanwar, M.P.: A practical (non-interactive) publicly verifiable secret sharing scheme. In: Bao, F., Weng, J. (eds.) ISPEC 2011. LNCS, vol. 6672, pp. 273–287. Springer, Heidelberg (2011). doi:10.1007/978-3-642-21031-0_21

JVSN14. Jhanwar, M.P., Venkateswarlu, A., Safavi-Naini, R.: Ayineedi venkateswarlu, and reihaneh safavi-naini. paillier-based publicly verifiable (non-interactive) secret sharing. Des. Codes Crypt. 73(2), 529–546 (2014)

KKR+16. Kiayias, A., Konstantinou, I., Russell, A., David, B., Oliynykov, R.: A provably secure proof-of-stake blockchain protocol. Cryptology ePrint Archive, Report 2016/889 (2016). http://eprint.iacr.org/2016/889

KMS+16. Kosba, A.E., Miller, A., Shi, E., Wen, Z., Papamanthou, C.: Hawk: the blockchain model of cryptography and privacy-preserving smart contracts. In: 2016 IEEE Symposium on Security and Privacy, pp. 839–858. IEEE Computer Society Press, May 2016

LV08. Libert, B., Vergnaud, D.: Unidirectional chosen-ciphertext secure proxy re-encryption. In: Cramer, R. (ed.) PKC 2008. LNCS, vol. 4939, pp. 360–379. Springer, Heidelberg (2008). doi:10.1007/978-3-540-78440-1_21

LW15. Lenstra, A.K., Wesolowski, B.: A random zoo: sloth, unicorn, and trx. Cryptology ePrint Archive, Report 2015/366 (2015). http://eprint.iacr.org/2015/366

Mau96. Maurer, U.M. (ed.): EUROCRYPT 1996. LNCS, vol. 1070. Springer, Heidelberg (1996). doi:10.1007/3-540-68339-9

MS81. McEliece, R.J., Sarwate, D.V.: On sharing secrets and reed-solomon codes. Commun. ACM 24(9), 583–584 (1981)

Nak08. Nakamoto, S.: Bitcoin: A peer-to-peer electronic cash system (2008)

Pai99. Paillier, P.: Public-key cryptosystems based on composite degree residuosity classes. In: Stern, J. (ed.) EUROCRYPT 1999. LNCS, vol. 1592, pp. 223–238. Springer, Heidelberg (1999). doi:10.1007/3-540-48910-X_16

PS96. Pointcheval, D., Stern, J.: Security proofs for signature schemes. In: Maurer [Mau96], pp. 387–398 (1996)

Rab83. Rabin, M.O.: Transaction protection by beacons. J. Comput. Syst. Sci. 27(2), 256–267 (1983)

RBO89. Rabin, T., Ben-Or, M.: Verifiable secret sharing and multiparty protocols with honest majority (extended abstract). In: 21st ACM STOC, pp. 73–85. ACM Press. May 1989

RV05. Ruiz, A., Villar, J.L.: Publicly verifiable secret sharing from Paillier's cryptosystem. In: Wolf, C., Lucks, S., Yau, P.-W. (eds.) WEWoRC 2005 - Western European Workshop on Research in Cryptology. Leuven, Belgium, 5–7 July 2005. LNI, vol. 74, pp. 98–108. GI (2005)

Sch99. Schoenmakers, B.: A simple publicly verifiable secret sharing scheme and its application to electronic voting. In: Wiener, M. (ed.) CRYPTO 1999. LNCS, vol. 1666, pp. 148–164. Springer, Heidelberg (1999). doi:10.1007/3-540-48405-1_10

Sha79. Shamir, A.: How to share a secret. Commun. Assoc. Comput. Mach. **22**(11), 612–613 (1979)

SJK+16. Syta, E., Jovanovic, P., Kokoris Kogias, E., Gailly, N., Gasser, L., Khoffi, I., Fischer, M.J., Ford, B.: Scalable bias-resistant distributed randomness. Cryptology ePrint Archive, Report 2016/1067 (2016). (To appear at IEEE Security & Privacy 2017). http://eprint.iacr.org/2016/1067

Sta96. Stadler, M.: Publicly verifiable secret sharing. In: Maurer [Mau96], pp. 190–199 (1996)

SV15. Schoenmakers, B., Veeningen, M.: Universally verifiable multiparty computation from threshold homomorphic cryptosystems. In: Malkin, T., Kolesnikov, V., Lewko, A.B., Polychronakis, M. (eds.) ACNS 2015. LNCS, vol. 9092, pp. 3–22. Springer, Cham (2015). doi:10.1007/978-3-319-28166-7_1

vdHLZZ15. van den Hooff, J., Lazar, D., Zaharia, M., Zeldovich, N.: Vuvuzela: scalable private messaging resistant to traffic analysis. In: Proceedings of 25th Symposium on Operating Systems Principles, SOSP 2015, pp. 137–152. ACM, New York (2015)

WCGFJ12. Wolinsky, D.I., Corrigan-Gibbs, H., Ford, B., Johnson, A.: Dissent in numbers: making strong anonymity scale. In: Proceedings of 10th USENIX Conference on Operating Systems Design and Implementation, OSDI 2012, pp. 179–192. USENIX Association, Berkeley (2012)

cMix: Mixing with Minimal Real-Time Asymmetric Cryptographic Operations

David Chaum[1], Debajyoti Das[2(✉)], Farid Javani[3], Aniket Kate[2],
Anna Krasnova[4], Joeri De Ruiter[4], and Alan T. Sherman[3]

[1] Voting Systems Institute, Los Angeles, USA
das48@purdue.edu
[2] Purdue University, West Lafayette, USA
[3] Cyber Defense Lab, UMBC, Baltimore, USA
sherman@umbc.edu
[4] Radboud University, Nijmegen, Netherlands

Abstract. We introduce cMix, a new approach to anonymous communications. Through a precomputation, the core cMix protocol eliminates all expensive real-time public-key operations—at the senders, recipients and mixnodes—thereby decreasing real-time cryptographic latency and lowering computational costs for clients. The core real-time phase performs only a few fast modular multiplications.

In these times of surveillance and extensive profiling there is a great need for an anonymous communication system that resists global attackers. One widely recognized solution to the challenge of traffic analysis is a mixnet, which anonymizes a batch of messages by sending the batch through a fixed cascade of mixnodes. Mixnets can offer excellent privacy guarantees, including unlinkability of sender and receiver, and resistance to many traffic-analysis attacks that undermine many other approaches including onion routing. Existing mixnet designs, however, suffer from high latency in part because of the need for real-time public-key operations. Precomputation greatly improves the real-time performance of cMix, while its fixed cascade of mixnodes yields the strong anonymity guarantees of mixnets. cMix is unique in not requiring any real-time public-key operations by users. Consequently, cMix is the first mixing suitable for low latency chat for light-weight devices.

Our presentation includes a specification of cMix, security arguments, anonymity analysis, and a performance comparison with selected other approaches. We also give benchmarks from our prototype.

1 Introduction

Digital messaging has become a significant form of human communication, yet currently such systems do not provide basic protections of untraceability and unlinkability of messages. These protections are fundamental to freedom of inquiry, to freedom of expression, and increasingly to online privacy. Grave threats to privacy exist from global adversaries who construct traffic-analysis graphs detailing who communicates with whom.

© Springer International Publishing AG 2017
D. Gollmann et al. (Eds.): ACNS 2017, LNCS 10355, pp. 557–578, 2017.
DOI: 10.1007/978-3-319-61204-1_28

We introduce cMix, a new approach to anonymous communications. cMix is a new variant of fixed-cascade mixing networks (mixnets). cMix uses a pre-computation phase to avoid computationally intensive public-key cryptographic operations in its core real-time protocol. Senders/clients participate only in the real-time phase. Thus, senders never perform any public-key operations. cMix has drastically lower real-time cryptographic latency than do all other mixnets. Through its use of precomputation, and through its novel key management, cMix is markedly different from all previous mixnets. Due to its lack of public-key operations in its core real-time phase, it is very well suited for applications running on light-weight clients, including chat messaging systems running on smartphones, and for applications on low-power devices.

To provide anonymity online, an alternative approach to mixnets is onion routing, such as implemented in the widely used system TOR [46]. Onion-routing systems, however, have limitations on the level of anonymity achievable: most significantly, because they route different sessions of messages along different paths and they do not perform random permutations of messages, they are vulnerable to a variety of traffic-analysis attacks—for example [12,43], as well as intersection attacks [6,13,14].

By contrast, mixnets hold fundamentally greater promise to achieve higher levels of anonymity than can onion-routing systems because mixnets are resilient to traffic-analysis attacks. Specifically, since all mixnet messages travel through the same fixed cascade of mixnodes, observing the communication paths of messages within a mixnet is not useful to the adversary. Also, mixnets can process larger batches of messages than can onion-routing systems, which is important because the batch size is the size of the anonymity set. Using a fixed cascade achieves resilience against intersection attacks [6]. The main disadvantage of current mixnets is their performance, which is throttled by their use of real-time public-key cryptographic operations, which are much slower than symmetric-key operations.

cMix is a practical solution to the cryptographic latency problem. It also provides resistance to traffic analysis and intersection attacks, as do other fixed-route mixnet designs. cMix scales linearly in number of users. Our prototype implementation on Android clients demonstrates the practicality of cMix.

The main novel and significant contribution of cMix is that, by using pre-computation, cMix eliminates all expensive real-time public-key operations by sender, receiver, and mixnodes in its core protocol. In cMix, the clients never perform any public-key operations when sending messages. By splitting the computational load over a computationally intensive precomputation phase and a fast real-time phase, cMix's approach can also increase throughput, when demand is non-uniform. No other mixnet has achieved these performance characteristics. Thus, cMix greatly improves real-time performance (especially latency) over all existing traditional mixnets and all re-encryption mixnets, while enjoying their strong anonymity properties.

Our main contributions are the design, preliminary analysis, and prototype implementation of cMix, a new mixnet variant that, through precomputation, achieves lower real-time computational latency than do all existing mixnets (traditional and re-encryption), while still benefiting from the strong anonymity properties of mixnets over onion-routing systems.

In the rest of this paper we review related work, provide an overview of cMix, describe the core cMix protocol, explain some protocol enhancements, provide security arguments, compare cMix's performance with that of other mixnets, present benchmarks from our prototype implementation, discuss several issues raised by cMix, and present our conclusions.

2 Related Work

We briefly review selected background and related work on mixnets, onion routing, and precomputation for mixnets.

Mix Networks. In 1981, Chaum [9] introduced the concept of mixnets (often referred to as *decryption mixnets*) and gave basic cryptographic protocols whereby messages from a set of users are relayed by a sequence of trusted intermediaries, called *mixnodes* or *mixes*. A mixnode is simply a message relay (or proxy) that accepts a batch of encrypted messages, decrypts and randomly permutes them, and sends them on their way forward. The sender in a decryption mixnet must perform a number of public-key operations equal to the number of mixnodes. The length of the encrypted message is proportional to this number, and the length of the plaintext message is restricted for performance reasons.

Hybrid mixnets allow plaintext messages to have arbitrary length, by combining asymmetric and symmetric cryptography. First proposed in 1985 by Pfitzmann and Waidner [37], hybrid mixnets share a session key in the first message, and then use a stream cipher to encrypt further messages. Various proposals based on block ciphers followed [26,33]. The recent system called Riffle [31] provides both sender and receiver anonymity by using verifiable shuffles and private information retrieval. Periodocially, a client and all servers perform verifiable shuffles to exchange session keys, which are then used for several rounds in a manner similar to that performed by other hybrid networks.

In 1994, Park et al. [36] introduced *re-encryption mixnets*. Taking advantage of homomorphic properties of El-Gamal encryption, each mixnode re-encrypts the incoming message instead of decrypting it as in the original mixnet. Doing so reduces and fixes the size of the ciphertext message traveling through the mixnet. *Universal re-encryption mixnets* [23] do not require mixnodes to know public keys for re-encryption. Because the sender encrypts each message using the public key of the receiver, only the receiver is able to read the plaintext. Consequently, unlike other mixnets, universal re-encryption mixnets provide sender anonymity only against external observers, and not against message recipients.

Precomputation mixnets introduce a precomputation phase to decrease latency during message delivery. Jakobsson [25] introduced precompution to reduce the cost of node operations in re-encryption mixnets, though client costs remain the same. Adida and Wikström [1] considered an offline/online approach to mixing. Their protocol still requires several public-key operations in the online phase, and senders have to perform public-key operations. A notable distinction of the cMix precomputation mixnet is its shifting of *all* public-key operations in its core protocol to the precomputation phase. Moreover, only the mixnodes perform these public-key operations; no sender is involved.

Onion Routing. Higher latency of traditional mixnets can be unsatisfactory for several communication scenarios, such as web search or instant messaging. Over the past several years, a significant number of *low-latency* anonymity networks have been proposed [2,3,8,11,20,28,29,34,41], and some have been extensively employed in practice [15,46].

Common to many of them is *onion routing* [21,35], a technique whereby a message is wrapped in multiple layers of encryption, forming an *onion*. A common realization of an onion-routing system is to arrange a collection of onion routers (abbreviated ORs, also called hops or nodes) that relay traffic for users of the system. Users then randomly choose a path with few edges through the network of ORs and construct a circuit—a sequence of nodes that will route traffic. After the OR circuit is constructed, each of the nodes in the circuit shares a symmetric key with the anonymous user, which key is used to encrypt the layers of future onions. Upon receiving an onion, each node decrypts one of the layers, and forwards the message to the next node. Onion routing as it typically exists can be seen as a form of three-node mixing.

Low-latency anonymous communication networks based on onion routing [18,27,32,45], such as TOR [46], are susceptible to a variety of traffic-analysis attacks. By contrast, mixnet methodology ensures that the anonymity set of a user remains the same through the communication route and makes our protocol resistant to these network-level attacks.

There are similarities between our precomputation phase, which uses public-key operations, and the circuit-construction phase of onion routing. Similarly, there are similarities between our real-time phase, which uses symmetric-key operations, and the onion wrapping and unwrapping phases.

Unlike onion routing, however, our precomputation phase requires no participation from the senders. During enrollment, each of our senders establishes a separate shared secret with each mixnode, but this key establishment is performed infrequently. Furthermore, in contrast with onion routing, our senders do not perform anonymous key agreement [3,15,20,29] using a telescoping approach or layered public-key encryption; they can establish these keys using a Diffie-Hellman key exchange. These differences result in a significant reduction in the computation that the users need to perform and make our system more attractive to energy-constrained devices such as smartphones.

3   System Overview

Before defining cMix's core protocol, we first explain our architecture and communication model, adversarial model, and security goals.

3.1   Architecture and Communication Model

cMix is a new mixnet protocol that provides anonymous communication for its users (senders and receivers). The main goal is to ensure *unlinkability* of messages entering and leaving the system, though it is known which users are active at any given moment.

cMix has n mixnodes that compose a *fixed cascade*: all nodes are organized in a fixed order from the first node to the last. Within the cMix system this order can be systematically changed and rotated, without affecting users in any way. Any message sent by a user is forwarded through all n servers. As with any mixnet, cMix collects a certain number of messages in a batch before processing them. Section 8 discusses our strategy for assembling batches, though details may depend on the application.

To become a cMix user, one must first establish for each mixnode a shared key. Section 4.2 provides more details on how these keys can be established. *Round keys* derived from the shared keys are used in each *round* of communication. A round begins with the start of batch processing.

For each round, β messages are collected and randomly ordered. Each message must have the same length, and all messages in a batch are processed simultaneously. The other messages are not accepted and are sent in a subsequent round. To process messages quickly in real time, cMix performs precomputations that do not involve any user. The precomputations are performed in a separate phase during which cMix executes all public-key encryptions, enabling the real-time computations to be carried using only fast multiplications.

3.2   Adversarial Model

We assume authenticated communication channels between all mixnodes. Thus, an adversary can eavesdrop, forward, and delete messages between mixnodes, but not modify, replay, or inject new ones, without detection. For any communication not among mixnodes, we assume the adversary can eavesdrop, modify, or inject messages at any point of the network.

An adversary can also compromise users; however, we assume that at least two users are honest. Mixnodes can also be compromised, but at least one of them needs to be honest for the system to be secure. We assume compromised mixnodes to be malicious but cautious: they aim not to get caught violating the protocol.

The goal of the adversary is to compromise the anonymity of the communication initiator, or to link inputs and outputs of the system. As with other mixnets [1,9,23,25,31,36,37], we do not consider adversaries whose sole purpose is to launch denial-of-service (DoS) attacks.

We envision a deployment model in which there are dedicated trusted data centers serving as the mixnodes (perhaps competitively awarded). As such, they are incentivized not to be expelled. By contrast, some mixnets allow the mixnodes to enter and leave with low cost.

An implication of our deployment model is that it is sufficient to be able to detect a cheating node with sufficient probability at some point. By contrast, in more flexible deployment models, the nodes should prove more stringently that they have computed correctly before the output is opened. cMix does require that the exit node commit to its output prior to being able to read the system output, which protects against certain adaptive attacks.

3.3 Security Goals

cMix aims to satisfy each of the following two security properties:

- **Anonymity:** A protocol provides *anonymity* if the adversary cannot map any input message to the corresponding output message, with a probability significantly better than that of random guessing, even if the adversary compromises all but two users and all but one mixnode.
- **Integrity:** A protocol provides *integrity* if at the end of every run involving β honest users:
    1. either, the β messages from the honest users are delivered unaltered to the intended recipients,
    2. or, a malicious mixnode is detected with a non-negligible probability and no honest party is proven malicious.

4   The Core Protocol

We now present the core cMix protocol, beginning with some preliminary notations and concepts, followed by a detailed specification.

4.1 Preliminaries

We introduce the primitives and notations used to describe the protocol. There are n mixnodes that process β messages per batch. For simplicity we assume here that the system already knows for each sender what slot to use out of the β slots. When implementing the system this assignment can, for example, be achieved by including the sender's identity (possibly a pseudonym) when sending a message to the system.

All computations are performed in a prime-order cyclic group $\mathbf{G}$ satisfying the decision Diffie-Hellman (DDH) assumption. The order of the group is p, and g is a generator for this group. Let $\mathbf{G}^*$ be the set of non-identity elements of $\mathbf{G}$.

cMix uses a multi-party group-homomorphic cryptographic system. We make use of a system based on ElGamal, described by Benaloh [4], though any such system could be used. This system works as follows:

- $d_i \in \mathbb{Z}_p^*$: the secret share for mixnode i of the secret key d.
- e: the public key of the system, based on the mixnode shares of the secret key: $e = \prod_i g^{d_i}$.

- $\mathcal{E}_e(m) = (g^x, m \cdot e^x)$, $x \in_R \mathbb{Z}_p^*$: encryption of message m under the system's public key e. We call g^x the *random component* and $m \cdot e^x$ the *message component* of the ciphertext. When applying encryption on a vector of values, each value in the vector is encrypted individually—each with a fresh random value—and the result is a vector of ciphertexts.
- $\mathcal{D}_{d_i}(g^x) = (g^x)^{-d_i}$: the decryption share for mixnode i computed from the random component of a ciphertext using the mixnode's share of the secret key. As with encryption, applying this function on a vector of random values results in a vector of corresponding decryption shares.

To decrypt a ciphertext $(g^x, m \cdot e^x)$, all parties need to cooperate because the decryption shares for all mixnodes are required to retrieve the original message:

$$m \cdot e^x \cdot \prod_{i=1}^{n} \mathcal{D}_{d_i}(g^x) = m \cdot (\prod_{i=1}^{n} g^{d_i})^x \cdot \prod_{i=1}^{n} (g^x)^{-d_i} = m.$$

The cMix protocol uses the following values:

- $r_{i,a}, s_{i,a} \in \mathbf{G}^*$: random values (freshly generated for each round) of mixnode i for slot a. Thus, $\mathbf{r}_i = (r_{i,1}, r_{i,2}, \ldots, r_{i,\beta})$ is a vector of random values for the β slots in the message map at mixnode i. Similarly, $\mathbf{s}_i$ is also a vector of random values for mixnode i.
- π_i: a random permutation of the β slots used by mixnode i. The inverse of the permutation is denoted by π_i^{-1}.
- $k_{i,j} \in \mathbf{G}^*$: a group element shared between mixnode i and the sending user for slot j. These values are used as keys to blind messages.
- $M_j \in \mathbf{G}^*$: the message sent by user j. Like other values in the system, these values are group elements. They can be easily converted from, for example, an ASCII-encoded string. The group size determines the length of an individual message that can be sent.

For readability we introduce the following shorthand notations:

- $\mathbf{R}_i$: the product of all local random $\mathbf{r}$ values through mixnode i; i.e., $\mathbf{R}_i = \prod_{j=1}^{i} \mathbf{r}_j$.
- $\mathbf{S}_i$: the product and permutation of all local random s values:

$$\mathbf{S}_i = \begin{cases} \mathbf{s}_1 & i = 1 \\ \pi_i(\mathbf{S}_{i-1}) \times \mathbf{s}_i & 1 < i \leq n. \end{cases}$$

- $\mathbf{\Pi}_i(a)$: the permutation performed by cMix through mixnode i, i.e., the composition of all individual permutations:

$$\mathbf{\Pi}_i(a) = \begin{cases} \pi_1(a) & i = 1 \\ \pi_i(\mathbf{\Pi}_{i-1}(a)) & 1 < i \leq n. \end{cases}$$

- $\mathbf{k}_i$ and $\mathbf{k}_i^{-1}$: the vector of keys shared between mixnode i and the users for all β slots and their inverses, respectively; $\mathbf{k}_i = (k_{i,1}, k_{i,2}, \ldots, k_{i,\beta})$ and $\mathbf{k}_i^{-1} = (k_{i,1}^{-1}, k_{i,2}^{-1}, \ldots, k_{i,\beta}^{-1})$.

- K_j: the product of all shared keys of the sending user for slot j: $K_j = \prod_{i=1}^{n} k_{i,j}$.
- $\mathbf{K}$ is a vector of products of shared keys for the β slots; $\mathbf{K} = (K_1, K_2, \ldots, K_\beta)$.

4.2 Protocol Description

We now present the core protocol. In this explanation we focus on simplicity and clarity; see Sect. 5 and our extended paper [10] for a discussion of possible security issues and enhancements. We separately discuss each of the three protocol phases: setup, precomputation, and real time.

Setup. In the setup phase, the mixnodes establish their secret shares d_i and the shared public key e, which are used for the multi-party homomorphic encryption system.

The users also establish their keys $k_{i,j}$, which they share with all mixnodes. This can be done using any (offline) key distribution method. One way to derive these keys is using a Diffie-Hellman key exchange. The resulting key can be used as a seed to derive unique values for $k_{i,j}$ for every session. Depending on the chosen key distribution protocol, this would be the only time a user is possibly required to perform an asymmetric cryptographic operation. The key exchange must be performed once for each user, and this exchange can be carried during the user's enrollment into the system.

Precomputation. The precomputation phase is performed only by the mixnodes, without any involvement from the users. It is performed once for each real-time phase. The mixnodes establish shared values to circumvent the need for public-key operations during the real-time phase. The precomputation phase comprises three different steps given below. The goal of the precomputation phase is to compute the values $\boldsymbol{\Pi}_n(\mathbf{R}_n) \times \mathbf{S}_n$, which are used in the real-time phase. Figure 1 shows a schematic example of the first two steps of the precomputation phase.

Step 1 - Preprocessing. The mixnodes start by generating fresh $\mathbf{r}, \mathbf{s}, \boldsymbol{\pi}$ values. Then they collectively compute the product of all of their individual $\mathbf{r}$ values under encryption using the public key e of the system, which was computed during the setup phase. This computation takes place by each mixnode i sending the following message to the next mixnode:

$$\mathcal{E}_e(\mathbf{R}_i) = \begin{cases} \mathcal{E}_e(\mathbf{r}_1) & i = 1 \\ \mathcal{E}_e(\mathbf{R}_{i-1}) \times \mathcal{E}_e(\mathbf{r}_i) & 1 < i \leq n. \end{cases}$$

Each mixnode encrypts its own $\mathbf{r}$ values and uses the homomorphic property of the encryption system to compute the multiplication of this ciphertext with the input it receives from the previous mixnode. Eventually, the last mixnode sends the final values $\mathcal{E}_e(\mathbf{R}_n)$ to the first mixnode as input for the next step.

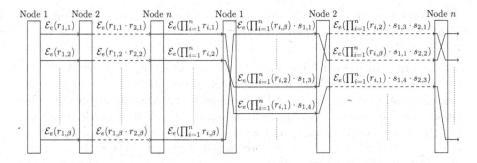

Fig. 1. A schematic example of the first two steps of the precomputation phase, which result in the values $\mathcal{E}_e(\boldsymbol{\varPi}_n(\mathbf{R}_n) \times \mathbf{S}_n)$.

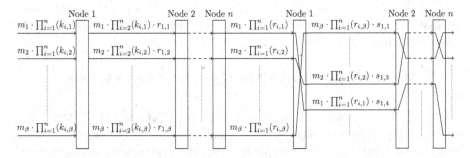

Fig. 2. A schematic example of the first two steps of the real-time phase, which result in the values $\boldsymbol{\varPi}_n(\mathbf{M} \times \mathbf{R}_n) \times \mathbf{S}_n$.

Step 2 - Mixing. In the second step, the mixnodes together mix the values and compute the results $\boldsymbol{\varPi}_n(\mathbf{R}_n) \times \mathbf{S}_n$, under encryption. The mixnodes perform this mixing by having each mixnode i send the following message to the next mixnode:

$$\mathcal{E}_e(\boldsymbol{\varPi}_i(\mathbf{R}_n) \times \mathbf{S}_i) = \begin{cases} \pi_1(\mathcal{E}_e(\mathbf{R}_n)) \times \mathcal{E}_e(\mathbf{s}_1) & i = 1 \\ \pi_i(\mathcal{E}_e(\boldsymbol{\varPi}_{i-1}(\mathbf{R}_n) \times \mathbf{S}_{i-1})) \times \mathcal{E}_e(\mathbf{s}_i) & 1 < i \leq n. \end{cases}$$

As with the first step, the last mixnode sends the final encrypted values $\mathcal{E}_e(\boldsymbol{\varPi}_n(\mathbf{R}_n) \times \mathbf{S}_n)$ to the first mixnode. These final values now must be decrypted together by all mixnodes, which happens in the last step of the precomputation.

Step 3 - Postprocessing. To complete the precomputation, the mixnodes decrypt the precomputed values. Each mixnode i computes its decryption shares $\mathcal{D}_{d_i}(\mathbf{x})$, where $(\mathbf{x}, \mathbf{c}) = \mathcal{E}_e(\boldsymbol{\varPi}_n(\mathbf{R}_n) \times \mathbf{S}_n))$. The message parts $\mathbf{c}$ are multiplied with all the decryption shares to retrieve the plaintext values $\boldsymbol{\varPi}_n(\mathbf{R}_n) \times \mathbf{S}_n$. This computation can be carried out either using another pass through the network (in which every mixnode multiplies in its own decryption share), or by having all mixnodes send their encryption shares to the last mixnode, which can then perform the multiplication. The last mixnode to be used in the real-time phase stores the decrypted precomputed values.

Real Time. For the real-time phase, each user constructs its input by taking its message m and multiplying it with its combined shared key k to compute the blinded message $m \times k$. This blinded message is then sent to the mixnet. One option would be to send the blinded messages to the first mixnode. Once the first mixnode receives enough blinded messages, it combines those messages to yield the vector $\mathbf{M} \times \mathbf{K}$. As in the precomputation phase, the real-time phase can again be split into three steps. Figure 2 gives a schematic example of the first two steps.

Step 1 - Preprocessing. During the preprocessing step, the mixnodes take out the keys $\mathbf{k}$ they share with the users and add in their $\mathbf{r}$ values to blind the original messages. This computation is performed by each mixnode i sending the following to the next mixnode:

$$\mathbf{M} \times \mathbf{K} \times (\prod_{j=1}^{i} \mathbf{k}_j^{-1} \times \mathbf{r}_j) = \mathbf{M} \times \mathbf{K} \times (\prod_{j=1}^{i-1} \mathbf{k}_j^{-1} \times \mathbf{r}_j) \times \mathbf{k}_i^{-1} \times \mathbf{r}_i.$$

The last mixnode sends the final values $\mathbf{M} \times \mathbf{R}_n = \mathbf{M} \times \mathbf{K} \times \prod_{j=1}^{n} (\mathbf{k}_i^{-1} \times \mathbf{r}_i)$, which are the blinded versions of the original messages, to the first mixnode as input for the next step. Now the user-specific keys $\mathbf{k}$ are taken out and replaced by the user-independent values $\mathbf{r}$.

Step 2 - Mixing. The second step performs the mixing to hide the association between sender and receiver. The $\mathbf{s}$ values are added in to hide which input message corresponds to which output message. Each mixnode i (except the last mixnode) sends the following message to the next mixnode:

$$\mathbf{\Pi}_i(\mathbf{M} \times \mathbf{R}_n) \times \mathbf{S}_i = \begin{cases} \pi_1(\mathbf{M} \times \mathbf{R}_n) \times \mathbf{s}_1 & i = 1 \\ \pi_i(\mathbf{\Pi}_{i-1}(\mathbf{M} \times \mathbf{R}_n) \times \mathbf{S}_{i-1}) \times \mathbf{s}_i & 1 < i < n. \end{cases}$$

Finally, the last mixnode computes:

$$\mathbf{\Pi}_n(\mathbf{M} \times \mathbf{R}_n) \times \mathbf{S}_n = \pi_n(\mathbf{\Pi}_{n-1}(\mathbf{M} \times \mathbf{R}_n) \times \mathbf{S}_{n-1}) \times \mathbf{s}_n.$$

Now every mixnode has performed its mixing, destroying the associations between senders and receivers.

Step 3 - Postprocessing. The last mixnode can perform the final step. This mixnode retrieves the locally stored precomputed values $\mathbf{\Pi}_n(\mathbf{R}_n) \times \mathbf{S}_n$. To retrieve the permuted messages it now needs only to perform the following computation, using the result from the previous mixing step:

$$\mathbf{\Pi}_n(\mathbf{M}) = \mathbf{\Pi}_n(\mathbf{M} \times \mathbf{R}_n) \times \mathbf{S}_n \times (\mathbf{\Pi}_n(\mathbf{R}_n) \times \mathbf{S}_n)^{-1}.$$

How the messages are then delivered to the recipients depends on the application and is independent from cMix. This step concludes the real-time phase, in which no public-key operations are performed.

5 Protocol Integrity

The cryptographic construction presented in Sect. 4 expects the mixnodes and users to be honest but curious. As for most mixnet protocols [9,26,33,36], cMix requires additional measures to ensure that mixnodes cannot tamper with the messages nor with the flow without detection. In this section, we augment the protocol to protect its anonymity and integrity against malicious attacks by users and by compromised mixnodes, which we shall call "adversarial mixnodes." The overall strategy relies on the assumption that compromised mixnodes are malicious but cautious.

5.1 Integrity of Values and Messages

To enable honest mixnodes to verifiably detect any malicious mixnodes that employ incorrect values or permutations in the precomputation or real-time phase, cMix augments communications with proofs. To this end, all messages exchanged between mixnodes are signed using a digital signature system with existential unforgeability under an adaptive chosen-message attack [22]. In addition, during precomputation, each mixnode commits to the permutation π_j it applies to the incoming slot j using a perfectly hiding commitment (Commit) scheme [24] and signs that commitment. Each node broadcasts its signed commitment using a reliable broadcast (Broadcast) protocol [17,44]. Doing so makes it possible to reconstruct and verify all individual values $r_{i,j}$, $s_{i,j}$ and $\pi_{i,j}$ that mixnode j applies to slot i.

In the real-time phase, the users become involved, making the process more complicated because they do not perform any public-key operations during the real-time phase. The values that we need to verify that depend on the users are the $k_{i,j}$ values (shared between a mixnode i and the user in slot j), and the $m_j \cdot K_j$ values (the blinded message that a user sends in slot j).

Because the k values are shared between a mixnode and a user, we need them to agree on the commitments to these values. If we detect an anomaly involving the k values, and a mixnode and user disagree on the value used, the commitment needs to provide proof of who misbehaved.

For this task, the following procedure can be followed in case the k values are derived from a seed that is agreed upon by a mixnode and user during the setup phase. After establishing this seed, the mixnode will compute a commitment of the seed and provide this commitment to the user. The mixnode generates the commitment in a way that enables the user to reveal the commitment and prove to other parties that the mixnode generated the commitment. The user must verify the commitment during the setup phase. This verification requires an additional public-key operation, though it will be performed only once provided the protocol runs without any disturbance.

Message Integrity at Entry. Messages at entry to cMix need to arrive at cMix and pass through the non-permuted part of the protocol without any undetected modification. Providing integrity of messages at this point is a challenge if one wishes to keep clients free from using any asymmetric cryptography during the real-time phase. We propose a construction where message authentication codes (*MAC*s) are generated over the input message to the cMix system. A shared-key *MAC* is sufficient here because the communicants (sender, mixnode) do not need to convince any third party.

To accomplish this goal, we introduce additional key values $l_{i,j}$, shared between mixnode i and the user for slot j, established and committed to in a similar way as for the k values. When the user sends the blinded message $m_j \cdot K_j$ to the system, the user also sends the following messages:

$$h_j = \text{Hash}(m_j \cdot K_j) \text{ and } (MAC_{l_{1,j}}(h_j), \dots, MAC_{l_{n,j}}(h_j)).$$

During the real-time phase, the first mixnode starts by checking its corresponding *MAC* value in the list. If it is incorrect, the mixnode informs the other mixnodes and does not forward the computed value for this slot. If the *MAC* value is correct, it forwards the h values and the *MAC* values for the other mixnodes, together with its basic computed values.

Each subsequent mixnode follows the same procedure. At the end of the first step of the real-time phase, all mixnodes have checked their *MAC* values on the received h values, or the value is not processed any further.

5.2   Message Tagging Detection

A message-tagging attack is an attack where the adversary can mark a message at some point during the process, such that it is recognizable when it is output, compromising unlinkability between the inputs and outputs [39]. To perform a tagging attack unnoticed, the tag should also be removed before the messages are output. Tagging attacks are a threat to all mixnets that use some form of malleable encryption, such as homomorphic encryption or group multiplications, where valid messages can be recognized when output by the mixnet. For example, Pfitzmann [38] presents such an attack on re-encryption mixnets.

A simple example of a tagging attack is the following: The last mixnode multiplies the blinded message in one of the slots j with an additional factor t in the first step of the real-time phase. Now the blinded message in slot j will be $m_j \cdot \prod_{i=1}^{n} r_{i,j} \cdot t$ at the end of the first step, whereas the values in the other slots stay the same. When the last mixnode performs Step 3, it will see the final messages before it outputs them. If the messages are recognizable as valid outputs, it will observe that one of the messages does not seem to be valid. If this invalid message becomes valid when multiplied with t^{-1}, this message is likely the tagged one. The mixnode can now link the message, and possibly the recipient, to the sender. The mixnode can remove the tag and output the original messages, making it unobservable to the users and other mixnodes that a tagging attack took place.

Detection. To protect against these kinds of attacks and make them detectable, only small changes are needed to the protocol:

- **Precomputation Phase - Step 3:** The mixnodes no longer send out their decryption shares to retrieve the precomputed values. Instead, they keep their shares secret and publish only a commitment to them. The last mixnode also publishes a commitment to the message component of the ciphertext. The commitments can be computed, for example, using only one signature per mixnode for the decryption shares for all slots simultaneously. The plaintext results of the precomputation phase are thus no longer retrieved in that phase itself.
- **Real-Time Phase - Step 3:** The output of the mixing step $\Pi_n(\mathbf{M} \times \mathbf{R}_n) \times \mathbf{S}_n$ is published by the last mixnode. Afterwards, all mixnodes release their decryption shares $\mathcal{D}_{d_i}(\mathbf{x})$ and the message component of the cipher-text $\mathbf{c}$. The output messages are then computed as follows, where $(\mathbf{x}, \mathbf{c}) = \mathcal{E}_e(\Pi_n(\mathbf{R}_n) \times \mathbf{S}_n)$:

$$\Pi_n(\mathbf{M} \times \mathbf{R}_n) \times \mathbf{S}_n \times \mathbf{c} \times \prod_{i=1}^{n} \mathcal{D}_{d_i}(\mathbf{x}).$$

Because the mixnodes committed to all of the values necessary to retrieve the precomputed value, they cannot change these values to take out a possible tag anymore. The output of the mixing step does not reveal anything yet about the messages, because the r and s values are still included. The precomputed value to take out these values is retrieved only after the output of the mixing step is made public. Therefore, the last mixnode would not know from which output slot a possible tag should be removed before the output of the mixing step is made public. Once the output is made public, the tag cannot be removed because all computations for the third step are fixed and can be verified by anyone.

5.3 Sending and Verifying Trap Messages

To ensure the protocol is functioning correctly, we let users send trap messages with a certain probability, and request to open message paths of these traps. Opening of a path includes verification of all messages exchanged between mixnodes, the incoming message from the user, as well as intermediate values and permutations.

Sending Trap Messages and Requesting to Open Them: To send a trap, a user starts by forming a string x with a round ID, user ID, and a statement that this is the dummy message. She calculates a MAC of x with every key $l_{i,j}$ shared between mixnode i and the user for slot j. The trap message is then the string together with its MAC values, i.e., $m = (x, MAC_{l_{1,j}}(x), \ldots, MAC_{l_{n,j}}(x))$. Note that these MAC values are different from the MAC values mentioned in Sect. 5.1, even though we are using the same keys in both places.

After the round is over, the trap message appears at the output. Each mixnode verifies the correctness of her associated user ID, round ID, and $MAC_{l_{i,j}}$

values. If any mixnode i detects that any of these values is incorrect, she stops the round. Otherwise, the mixnode sends an authenticated message to user j notifying that the dummy message is received correctly, including an incremented counter of all dummies received from this user.

If a user actually sent a dummy message, but she does not receive a notification message from each mixnode, she initiates the verification procedure. The verification procedure consists of the following steps:

(i) User j sends a request, tagged with MAC values, to open the path with her dummy message to all mixnodes.

(ii) A mixnode can either agree to participate in the path opening, or can dispute the MAC of the path-opening request.

(iii) If mixnode i disputes the MAC intended for her, both the user and mixnode open their local $l_{i,j}$ values to validate the MAC.

(iv) If both the user and the mixnode are using the same $l_{i,j}$ value, then the $MAC_{l_{i,j}}()$ values computed by them should be same, and the correctness of the request sent by the user can be easily validated. If the user has sent an invalid request, the request is dropped.

(v) If the user and the mixnode dispute over the $l_{i,j}$ value, then the seed of $l_{i,j}$ and the commitment to that seed (made by the mixnode and verified by the user during setup) are opened. The user and mixnode each recomputes $l_{i,j}$ from the seed. Two possibilities follow:

   (a) If the recomputed $l_{i,j}$ value is not equal to what the user claims, the path-opening request is dropped.

   (b) If the recomputed $l_{i,j}$ value is not equal to what the mixnode claims, she must either agree to participate in the path-opening, or be considered malicious.

(vi) Once all the mixnodes agree to participate in the path-opening, they first open the path of the precomputation phase, and then the real-time phase (both are explained below).

Path Opening for the Precomputation Phase: For the input slot j that we want to verify, the mixnodes have to decrypt exchanged messages for this slot and compute the r values in the non-permuted part. For the permuted part, mixnodes subsequently reveal the corresponding permutations and decrypt exchanged messages to obtain the s values. Doing so we can follow the computation through the mixnet and verify whether the precomputed value that was output was correctly computed.

To check, for example, whether mixnode i performed its computation correctly, that mixnode needs to present the signature from the previous mixnode on the values it received. The next mixnode will also have to present the signature it received from mixnode i to obtain proof what values mixnode i output. Once all information about the input, output, and values used in the expected computation are known, due to the signatures, commitments, and threshold decryption, it can be verified whether mixnode i performed the computation as expected or not.

Path Opening for the Real-Time Phase: Malicious mixnodes can also employ incorrect messages, values, and permutations in the real-time phase. We wish to make the mixnodes accountable for their behavior. One challenge is that malicious users may try to victimize some honest mixnodes by providing inconsistent inputs and later deny having done so.

First, the mixing step is verified. This check is performed in a similar fashion as for the precomputation phase. For the preprocessing step, a comparable approach is followed. To verify whether the correct input from the user to the system is used, all the keys l for the MAC values for the suspicious slot are output. The purpose is to detect if a mixnode or user changed the input. Because the values were processed in the mixing step, during the first step of the real-time phase, all mixnodes accepted the MAC values and thus should be able to provide a correct key. The mixnodes also reveal their r and k values for the corresponding slot, and the mixnodes check whether they performed their computations correctly. Although the mixnodes committed to the r values, they have not committed to individual k values. Therefore the sender must be involved in this process. The sender will also release all the k values it used for this message.

If a k value released by the sender does not match the one released by the corresponding mixnode, either the mixnode or the sender is misbehaving. The sender might do this to blame a mixnode of misbehaving to cause it to be removed from the system. This dispute needs to be resolved by having the user reveal the commitment by the mixnode on the shared keys. Doing so might also reveal k values used in previous sessions, but these values are only one of the components of K and therefore do not leak the original messages. There are two possibilities: either the user was malicious, in which case we do not care about his or her previous messages, or the mixnode was malicious, in which case we consider the k values to be compromised already. This procedure will reveal who acted maliciously and modified the output message.

5.4   Security Analysis

We now briefly highlight some of the security properties of cMix. For details, see the extended version of our paper [10].

As described in Sect. 3.3, we intend our protocol to have the following property: if any of the messages are altered by any node, (a) no honest party can be proven malicious, and (b) at least one malicious mixnode is detected with non-negligible probability. In [10], we argue how the above described integrity measures (integrity of values and messages, message tagging detection, and trap messages) achieve the integrity property.

We also claim: if $\mathcal{E}$ is a CPA-secure group-homomorphic encryption system, and Commit is a perfectly-hiding non-interactive commitment scheme, and if the protocol maintains integrity, then cMix offers anonymity. For a detailed anonymity analysis, see [10].

In [10] we also explain how cMix resists well-known attacks on mixnets [6,9, 12–14,16,39,40,43].

Galteland et al. [19] propose a tagging attack and an insider attack against the cMix protocol, as described in the preliminary cMix eprint. But security mechanisms specified in this preliminary cMix eprint prevent both attacks, as do alternative integrity mechanisms (e.g., trap messages) specified in the current cMix paper and its extended version [10]. In particular, as presented in the preliminary cMix eprint, cryptographic commitments enforced by Random Partial Checking (RPC) [30] prevent both attacks. Thus, these purported "Norwegian Attacks" do not work.

6   Comparison with Other Mixnets

We compare cMix with well-known fixed-cascade mixnet approaches based on performance. Specifically, as summarized in Table 1, we compare the performance of the core cMix protocol with that of each of the following three competing approaches: original mixnet and hybrid mixnets, re-encryption mixnet, and re-encryption mixnet with precomputation.

For each approach, we compare the precomputation and real-time costs, further compared by number of single-party public-key operations, multi-party public-key operations, and multiplications (each by client and by mixnodes). Note that, when using ElGamal encryption, the multiplication of two ciphertexts requires two multiplications.

Table 1. Performance comparison of the core cMix protocol with three competing mixnet [9,36] approaches: number of multiplication (Mult) and public-key (PK) operations performed for a batch of β messages processed by an n-node mixnet. One multi-party public-key operation (ops) requires all nodes to participate.

	Precomputation (ops for all mixnodes)			Real time (ops per client; ops for all mixnodes)		
	PK ops	Multi-party PK ops	Mult	PK ops	Multi-party PK ops	Mult
cMix (core)	$2n\beta$	β	$4n - 2\beta$	0; 0	0; 0	n; $3n\beta$
Original mix	-	-	-	n; $n\beta$	0; 0	0; 0
Re-encryption mix	-	-	-	1; $n\beta$	0; β	0; $2n\beta$
Re-encryption mix (precomputation)	$n\beta$	0	0	1; 0	0; β	0; $2n\beta$

The original mixnet [9] requires each sender to perform n encryptions; each of the n mixnodes in the cascade performs one decryption per message and β decryptions per batch. In total, all mixnodes perform n decryptions per message and βn decryptions per batch. Hybrid mixnets [26,33] require the same amount of asymmetric encryptions, but on a smaller plaintext.

In re-encryption mixnets [36], each client performs one encryption of its message using the mixnet's shared public key. Each node re-randomizes every message, instead of decrypting each one as with original mixnets, resulting in one public-key operation and one multiplication of ciphertexts per message per node. In addition, the nodes need to decrypt the output of the mixnet in a multi-party computation.

Re-encryption mixnets can be improved in a straightforward way using precomputation to perform the public-key operations necessary for the re-randomization, similarly to cMix's strategy. As shown in Table 1, however, with regard to real-time computation, cMix outperforms re-encryption mixnets with precomputation.

Universal re-encryption mixnets perform much more slowly because they require senders to encrypt messages with public keys of their recipients.

7 Proof of Concept

We implemented a proof-of-concept protoype in Python, including tagging detection as discussed in Sect. 5.2. Towards reducing the communication latency, we have also introduced a network handler.

Introducing an untrusted *network handler* reduces latency. In Steps 1 and 3 of the precomputation and real-time phases, only products of values known by the individual nodes are computed (see Sect. 4.2). To compute these products, it is not necessary to make a full pass through the mixnet. Instead, each node can send its values to an untrusted third party, which we call the network handler, who can compute the products and return the results to the mixnet. Doing so reduces latency of the network significantly because each node can send its values simultaneously to the handler, instead of forwarding its local result to the next node sequentially. The network handler does not learn any secret value, and it computes only values that would anyway become public. The network handler, and each of the mixnodes, is a single point of failure. In the event of failure, however, because the handler performs only public operations, it can be easily replaced by another entity—for example, by one of the mixnodes.

Each mixnode includes a keyserver (to establish shared keys with the users) and a mixnet server (to carry out the precomputations and real-time computations). We use Ed25519 [5] signatures to implement authenticated channels between mixnodes. For the precomputation, each mixnode uses parallel processes for the computation of the encryptions and the decryption shares. In the real-time phase, all operations are performed in a single thread.

We performed experiments by running the protoype on Amazon Web Services (AWS) instances, with each node comprising a c3.large with two virtual processors and 3.75 GB of RAM. For all values, we used a prime-order group of 2048 bits.

On the AWS instances, each 2048-bit ElGamal encryption took approximately 10 ms on average, and the computation of a decryption share took approximately 5 ms. Multiplication of group elements took only a fraction of a millisecond.

For our experiments we performed 100 precomputation and real-time phases for selected batch sizes up to 1000 with five mixnodes. Table 2 gives observed

Table 2. Timings measured on the network handler from the start of the phases until the final values or responses are computed. Timings are in seconds (wall clock) for 100 runs of the precomputation and real-time phases, for various batch sizes using five mixnodes.

Batch	Precomputation		Real time	
Size	Mean	Std. dev.	Mean	Std. dev.
10	0.48	0.07	0.07	0.02
50	1.99	0.04	0.21	0.04
100	4.00	0.30	0.38	0.06
200	7.74	0.09	0.75	0.08
300	11.46	0.13	1.09	0.08
400	15.24	0.11	1.44	0.08
500	19.08	0.23	1.80	0.11
1000	37.94	0.19	3.58	0.12

Table 3. Mean timings in seconds (CPU and wall clock) for 100 runs of the real-time phase of the cMix protocol measured on the mixnodes and network handler, for various batch sizes using five mixnodes. For the mixnodes the mean time is taken over all mixnodes.

Batch	Mixnode		Network handler	
Size	CPU	Wall	CPU	Wall
10	0.01	0.04	0.01	0.07
50	0.04	0.16	0.02	0.21
100	0.07	0.30	0.02	0.38
200	0.14	0.60	0.04	0.75
300	0.22	0.87	0.06	1.09
400	0.29	1.15	0.07	1.44
500	0.36	1.45	0.09	1.80
1000	0.73	2.88	0.19	3.58

timings on the network handler for selected batch sizes using five mixnodes, without any enhanced security mechanisms mentioned in Sect. 5. We measured elapsed time on the network handler from the time it instructed the nodes to start until either the precomputation finished successfully, or until it computed the final responses to be sent to the users in the real-time phase. Table 3 gives timings for the real-time phase per node and for the network handler, in both CPU and wall clock time. These timings show the low computational load on the nodes during this phase.

These timings demonstrate the high performance of the system in the real-time phase. The precomputation can be easily accelerated by performing more computations in parallel. Additional processors would significantly improve the time it takes to compute all necessary encryptions and decryption shares. For the real-time phase, a network connection with low latency would improve the timings.

8   Extensions, Discussion, and Future Work

Here, we first describe how receivers can send immediate responses. We then briefly discuss how to arrange messages into batches and how to deal with node failures. We also outline some of our future plans.

Return Path. It is easy to extend cMix to enable a receiver to send an immediate response through the mixnet, for example, to acknowledge receiving a message. To accomplish this goal, the nodes generate additional random values

s' and compute the permuted products S' during the precomputation phase. The nodes and users also generate fresh keys k', which will be used to encrypt the return message.

For a return path, the mixnodes apply the inverse permutations π^{-1} so that the responses will arrive at the original senders. Unless the recipient who sends a response shares keys with the system, no fresh r' values are needed because the message would not enter the system blinded and hence Step 1 of the real-time phase could be skipped. In Step 2, the system applies the inverse permutations π^{-1} and fresh s' values. In Step 3, to encrypt the response to the original sender, instead of multiplying only with its decryption component, each node multiplies with the product of its decryption component and k' value.

Batch Strategy. cMix follows the "threshold and timed mixing strategy" from Serjantov et al. [42], where a new round is started every t seconds only if there are at least β messages in the buffer. We expect at least β users to be active at any given time. When a smaller number of users is active, this strategy can lead to increased latency or even disruption. At the cost of increased energy consumption, one design choice is to inject dummy messages when needed to ensure enough traffic to have β messages every t seconds. Alternatively, empty slots can be used to verify if the precomputation was performed correctly, by revealing all committed values for these slots, where empty slots to use should be chosen at random.

Node Failure. Because cMix uses a fixed cascade of nodes, it is important to consider what happens if a node fails. First, we consider a node failure to be a highly rare event because we expect each node to be a highly reliable computing service that is capable of seamlessly handling failures. Second, the system will detect node failure and notify the senders and the other nodes; senders will be instructed to resend using a new cascade (e.g., the old cascade without the failed node). Each node can detect failures by listening for periodic "pings" from the other nodes.

To minimize possible disruption caused by a single failure, at the cost of increasing the precomputations, the following option can be deployed: Each node can have a reserve of precomputations ready to use for certain alternative cascades. For example, this reserve can include each of the alternative cascades formed by removing any one node from the current cascade.

Future Steps. Tasks we plan to work on in the future include the following: First, we would like to deploy cMix, including implementing and refining different applications. We would also like to carry out more performance studies. Second, we plan to explore alternative and even more efficient approaches for enforcing integrity of the nodes, to ensure that they cannot modify any message without detection. Third, currently, message length is restricted by the group modulus. We have begun to work out how to apply key-homomorphic pseudorandom functions [7] and an appropriate additive homomorphic encryption system to allow any length message. Fourth, we would like to explore possible ways of reusing a precomputation in a secure way.

9 Conclusion

cMix offers a promising new approach to anonymous communications, building on the strong anonimity properties of mixnets, and improving real-time cryptographic performance by eliminating real-time public-key operations in its core protocol. By replacing real-time public-key operations with precomputations, and by avoiding the user's direct involvement with the construction of the path through the mixnodes, cMix scales well for deployment with large anonymity sets and large numbers of mixnodes. Even though the adversary may know all senders and receivers in each batch, she cannot link any sender and receiver unless all mixnodes are compromised. cMix offers the potential for real-time performance improvements over existing mixnets, without losing any of their security guarantees.

Acknowledgments. We thank the anonymous reviewers for their comments. We also thank the following people for helpful suggestions: David Delatte, Russell Fink, Bryan Ford, Moritz Neikes, and Dhananjay Phatak.

Sherman was supported in part by the National Science Foundation under SFS grant 1241576 and a subcontract of INSuRE grant 1344369, and by the Department of Defense under CAE-R grant H98230-15-10294. Krasnova conducted this research within the Privacy and Identity Lab (PI.lab, http://www.pilab.nl) funded by SIDN.nl (http://www.sidn.nl/).

References

1. Adida, B., Wikström, D.: Offline/online mixing. In: Arge, L., Cachin, C., Jurdziński, T., Tarlecki, A. (eds.) ICALP 2007. LNCS, vol. 4596, pp. 484–495. Springer, Heidelberg (2007). doi:10.1007/978-3-540-73420-8_43
2. Backes, M., Goldberg, I., Kate, A., Mohammadi, E.: Provably secure and practical onion routing. In: Proceeding of the 25th IEEE Computer Security Foundations Symposium (CSF), pp. 369–385 (2012)
3. Backes, M., Kate, A., Mohammadi, E.: Ace: an efficient key-exchange protocol for onion routing. In: Proceeding of the 11th ACM Workshop on Privacy in the Electronic Society (WPES), pp. 55–64 (2012)
4. Benaloh, J.: Simple verifiable elections. In: Proceeding of USENIX/Accurate Electronic Voting Technology Workshop (EVT), p. 5 (2006)
5. Bernstein, D.J., Duif, N., Lange, T., Schwabe, P., Yang, B.Y.: High-speed high-security signatures. J. Cryptogr. Eng. **2**(2), 77–89 (2012)
6. Berthold, O., Pfitzmann, A., Standtke, R.: The disadvantages of free MIX routes and how to overcome them. In: Federrath, H. (ed.) Designing Privacy Enhancing Technologies. LNCS, vol. 2009, pp. 30–45. Springer, Heidelberg (2001). doi:10.1007/3-540-44702-4_3
7. Boneh, D., Lewi, K., Montgomery, H., Raghunathan, A.: Key homomorphic PRFs and their applications. In: Canetti, R., Garay, J.A. (eds.) CRYPTO 2013. LNCS, vol. 8042, pp. 410–428. Springer, Heidelberg (2013). doi:10.1007/978-3-642-40041-4_23
8. Camenisch, J., Lysyanskaya, A.: A formal treatment of onion routing. In: Shoup, V. (ed.) CRYPTO 2005. LNCS, vol. 3621, pp. 169–187. Springer, Heidelberg (2005). doi:10.1007/11535218_11

9. Chaum, D.: Untraceable electronic mail, return addresses, and digital pseudonyms. Commun. ACM **4**(2), 84–88 (1981)
10. Chaum, D., Das, D., Javani, F., Kate, A., Krasnova, A., Ruiter, J.D., Sherman, A.T.: cMix: mixing with minimal real-time asymmetric cryptographic operations. Cryptology ePrint Archive, Report 2016/008 (2016). https://eprint.iacr.org/2016/008.pdf
11. Chen, C., Asoni, D.E., Barrera, D., Danezis, G., Perrig, A.: HORNET: high-speed onion routing at the network layer. In: Proceeding of the 22nd ACM Conference on Computer and Communications Security, pp. 1441–1454 (2015)
12. Danezis, G.: The traffic analysis of continuous-time mixes. In: Martin, D., Serjantov, A. (eds.) PET 2004. LNCS, vol. 3424, pp. 35–50. Springer, Heidelberg (2005). doi:10.1007/11423409_3
13. Danezis, G., Diaz, C., Troncoso, C.: Two-sided statistical disclosure attack. In: Borisov, N., Golle, P. (eds.) PET 2007. LNCS, vol. 4776, pp. 30–44. Springer, Heidelberg (2007). doi:10.1007/978-3-540-75551-7_3
14. Danezis, G., Serjantov, A.: Statistical disclosure or intersection attacks on anonymity systems. In: Fridrich, J. (ed.) IH 2004. LNCS, vol. 3200, pp. 293–308. Springer, Heidelberg (2004). doi:10.1007/978-3-540-30114-1_21
15. Dingledine, R., Mathewson, N., Syverson, P.: Tor: the second-generation onion router. In: Proceeding of the 13th USENIX Security Symposium, pp. 303–320 (2004)
16. Dingledine, R., Shmatikov, V., Syverson, P.: Synchronous batching: from cascades to free routes. In: Martin, D., Serjantov, A. (eds.) PET 2004. LNCS, vol. 3424, pp. 186–206. Springer, Heidelberg (2005). doi:10.1007/11423409_12
17. Dolev, D., Reischuk, R., Strong, H.R.: Early stopping in byzantine agreement. J. ACM **37**(4), 720–741 (1990)
18. Evans, N.S., Dingledine, R., Grothoff, C.: A practical congestion attack on tor using long paths. In: Proceeding of the 18th USENIX Security Symposium, pp. 33–50 (2009)
19. Galteland, H., Mjølsnes, S.F., Olimid, R.F.: Attacks on cMix - some small overlooked details. Cryptology ePrint Archive, Report 2016/729 (2016). http://eprint.iacr.org/2016/729
20. Goldberg, I., Stebila, D., Ustaoglu, B.: Anonymity and one-way authentication in key exchange protocols. Des. Codes Crypt. **67**(2), 245–269 (2013)
21. Goldschlag, D.M., Reed, M.G., Syverson, P.F.: Onion routing. Commun. ACM **42**(2), 39–41 (1999)
22. Goldwasser, S., Micali, S., Rivest, R.L.: A digital signature scheme secure against adaptive chosen-message attacks. SIAM J. Comput. **17**(2), 281–308 (1995)
23. Golle, P., Jakobsson, M., Juels, A., Syverson, P.: Universal re-encryption for mixnets. In: Okamoto, T. (ed.) CT-RSA 2004. LNCS, vol. 2964, pp. 163–178. Springer, Heidelberg (2004). doi:10.1007/978-3-540-24660-2_14
24. Halevi, S., Micali, S.: Practical and provably-secure commitment schemes from collision-free hashing. In: Koblitz, N. (ed.) CRYPTO 1996. LNCS, vol. 1109, pp. 201–215. Springer, Heidelberg (1996). doi:10.1007/3-540-68697-5_16
25. Jakobsson, M.: Flash mixing. In: Proceedings of 18th Annual ACM Symposium on Principles of Distributed Computing (PODC), pp. 83–89 (1999)
26. Jakobsson, M., Juels, A.: An optimally robust hybrid mix network. In: Proceedings of 20th Annual ACM Symposium on Principles of Distributed Computing, pp. 284–292 (2001)

27. Jansen, R., Tschorsch, F., Johnson, A., Scheuermann, B.: The sniper attack: anonymously deanonymizing and disabling the Tor network. In: Proceedings of Network and Distributed System Security Symposium (NDSS 2014) (2014)
28. Kate, A., Goldberg, I.: Using Sphinx to improve onion routing circuit construction. In: Sion, R. (ed.) FC 2010. LNCS, vol. 6052, pp. 359–366. Springer, Heidelberg (2010). doi:10.1007/978-3-642-14577-3_30
29. Kate, A., Zaverucha, G.M., Goldberg, I.: Pairing-based onion routing with improved forward secrecy. ACM Trans. Inf. Syst. Secur. 13(4), 29:1–29:32 (2010)
30. Khazaei, S., Wikström, D.: Randomized partial checking revisited. In: Dawson, E. (ed.) CT-RSA 2013. LNCS, vol. 7779, pp. 115–128. Springer, Heidelberg (2013). doi:10.1007/978-3-642-36095-4_8
31. Kwon, A., Lazar, D., Devadas, S., Ford, B.: Riffle: an efficient communication system with strong anonymity. PoPETs 2016(2), 115–134 (2016)
32. Murdoch, S.J., Danezis, G.: Low-cost traffic analysis of Tor. In: Proceedings of 26th IEEE Symposium on Security and Privacy, pp. 183–195 (2005)
33. Ohkubo, M., Abe, M.: A length-invariant hybrid mix. In: Okamoto, T. (ed.) ASIACRYPT 2000. LNCS, vol. 1976, pp. 178–191. Springer, Heidelberg (2000). doi:10.1007/3-540-44448-3_14
34. Øverlier, L., Syverson, P.: Improving efficiency and simplicity of tor circuit establishment and hidden services. In: Borisov, N., Golle, P. (eds.) PET 2007. LNCS, vol. 4776, pp. 134–152. Springer, Heidelberg (2007). doi:10.1007/978-3-540-75551-7_9
35. Øverlier, L., Syverson, P.F.: Locating hidden servers. In: Proceedings of 27th IEEE Symposium on Security and Privacy, pp. 100–114 (2006)
36. Park, C., Itoh, K., Kurosawa, K.: Efficient anonymous channel and all/nothing election scheme. In: Helleseth, T. (ed.) EUROCRYPT 1993. LNCS, vol. 765, pp. 248–259. Springer, Heidelberg (1994). doi:10.1007/3-540-48285-7_21
37. Pfitzmann, A., Waidner, M.: Networks without user observability – design options. In: Pichler, F. (ed.) EUROCRYPT 1985. LNCS, vol. 219, pp. 245–253. Springer, Heidelberg (1986). doi:10.1007/3-540-39805-8_29
38. Pfitzmann, B.: Breaking an efficient anonymous channel. In: Santis, A. (ed.) EUROCRYPT 1994. LNCS, vol. 950, pp. 332–340. Springer, Heidelberg (1995). doi:10.1007/BFb0053448
39. Raymond, J.-F.: Traffic analysis: protocols, attacks, design issues, and open problems. In: Federrath, H. (ed.) Designing Privacy Enhancing Technologies. LNCS, vol. 2009, pp. 10–29. Springer, Heidelberg (2001). doi:10.1007/3-540-44702-4_2
40. Reed, M., Syverson, P., Goldschlag, D.: Anonymous connections and onion routing. IEEE J-SAC 16(4), 482–494 (1998)
41. Ruffing, T., Moreno-Sanchez, P., Kate, A.: P2P mixing and unlinkable Bitcoin transactions. In: NDSS 2017 (2017)
42. Serjantov, A., Dingledine, R., Syverson, P.: From a trickle to a flood: active attacks on several mix types. In: Petitcolas, F.A.P. (ed.) IH 2002. LNCS, vol. 2578, pp. 36–52. Springer, Heidelberg (2003). doi:10.1007/3-540-36415-3_3
43. Serjantov, A., Sewell, P.: Passive attack analysis for connection-based anonymity systems. In: Snekkenes, E., Gollmann, D. (eds.) ESORICS 2003. LNCS, vol. 2808, pp. 116–131. Springer, Heidelberg (2003). doi:10.1007/978-3-540-39650-5_7
44. Srikanth, T.K., Toueg, S.: Simulating authenticated broadcasts to derive simple fault-tolerant algorithms. Distrib. Comput. 2(2), 80–94 (1987)
45. Sun, Y., Edmundson, A., Vanbever, L., Li, O., Rexford, J., Chiang, M., Mittal, P.: Raptor: routing attacks on privacy in Tor. In: Proceedings of 24th USENIX Security Symposium, pp. 271–286 (2015)
46. The Tor project (2003). https://www.torproject.org/. Accessed April 2017

Almost Optimal Oblivious Transfer
from QA-NIZK

Olivier Blazy[1]([⊠]), Céline Chevalier[2], and Paul Germouty[1]

[1] Université de Limoges, XLim, Limoges, France
{olivier.blazy,paul.germouty}@unilim.fr
[2] CRED, Université Panthéon-Assas Paris II, Paris, France
celine.chevalier@u-paris2.fr

Abstract. We show how to build a UC-Secure Oblivious Transfer in the presence of Adaptive Corruptions from Quasi-Adaptive Non-Interactive Zero-Knowledge proofs. Our result is based on the work of Jutla and Roy at Asiacrypt 2015, where the authors proposed a constant-size very efficient PAKE scheme. As a stepping stone, we first show how a two-flow PAKE scheme can be generically transformed in an optimized way, in order to achieve an efficient three-flow Oblivious-Transfer scheme. We then compare our generic transformations to existing OT constructions and see that we manage to gain at least a factor 2 to the best known constructions. To the best of our knowledge, our scheme is the first UC-secure Oblivious Transfer with a constant size flow from the receiver, and nearly optimal size for the server.

Keywords: OT · UC · QA-NIZK · Pairing-based cryptography

1 Introduction

Sharing data securely and efficiently is at the core of modern cryptography. The concept of Oblivious Transfer was introduced in 1981 by Rabin [41]. It allows to retrieve information from a database without privacy risk. Such protocols allow a receiver to ask and have access to a single line of a database. This exchange is oblivious in two ways. On the one hand, it allows to preserve the privacy of the receiver by preventing the sender from learning which part of the database has been asked. On the other hand, it allows the receiver to recover only one line, the one he previously asked for. The receiver is said to be oblivious to the values he is not allowed to get and the sender is oblivious to the information received. Several instantiations and optimizations of such protocols have appeared in the literature, like [21,38]. More recently, new instantiations have been proposed, trying to reach round-optimality [29], or low communication costs [40]. Recent schemes like [1,9,10] manage to achieve round-optimality while maintaining a small communication cost. Choi *et al.* [22] also propose a generic method and an efficient instantiation secure against adaptive corruptions in the CRS model with erasures, but it is only 1-out-of-2 and it does not scale to 1-out-of-n OT, for $n > 2$.

© Springer International Publishing AG 2017
D. Gollmann et al. (Eds.): ACNS 2017, LNCS 10355, pp. 579–598, 2017.
DOI: 10.1007/978-3-319-61204-1_29

In *Password-Authenticated Key Exchange* (PAKE) protocols, proposed in 1992 by Bellovin and Merritt [7], authentication is done using a simple password, possibly drawn from a small space subject to exhaustive search. Since then, many schemes have been proposed and studied in various models, and ultimately proven in the UC framework.

Smooth Projective Hash Functions [23,26] have found several applications in the context of UC Secure protocols [3,8,34], and also more particularly in the context of Password Authenticated Key Exchange and Oblivious-Transfer Proposals [1,9]. These functions are defined such that their output can be computed in two different ways if the input belongs to a particular subset (the *language*), either using a private hashing key or a public projection key along with a private witness ensuring that the input belongs to the language. The value obtained is indistinguishable from random when the input does not belong to the language (this is called the *smoothness*) and in case the input does belong to the language but no witness is known (this is called the *pseudo-randomness*).

[28] proposed a technique to achieve *Non-Interactive Zero-Knowledge Proofs of Knowledge*. Since then, a tremendous amount of work has been dedicated to improving those proofs [13,24,27]. And more recently, a trend studied Quasi-Adaptive NIZK allowing more efficient protocols [2,30–32,35,36]. In this last paper, the authors proposed one of the first PAKE UC-Secure even in the presence of adaptive corruptions, that manages to have flows independent of the password length.

Both Smooth Projective Hash Functions and Quasi-Adaptive Non-Interactive Zero-Knowledge allow to produce very efficient protocols, with a very similar structure, and thus security arguments.

Contributions: Our paper proposes a generic transformation from Password-Authenticated Key Exchange to Oblivious Transfer. As the reverse transformation was already studied in [18], this allows to study one protocol for the other. Our framework allows to transform a UC-secure PAKE into a UC-secure OT with the same level of security (for instance resistance to adaptive corruptions).

We choose to focus on the transformation of a 2-round PAKE to a 3-round OT since this kind of PAKE is the most commonly encountered and the most efficient one. Furthermore, this allows us to give a generic optimization of our transformation in this specific case, exploiting the fact that the server does not need to hide his "password", since his password is a (public) number of line in the database. Note that our generic transformation could allow to transform an m-round PAKE into an $(m + 1)$-round OT, but the optimization would then have to be tailored to the considered protocol.

After showing an application of our technique on an already existing PAKE scheme from [1], which allows us to immediately recover the associated OT scheme given in their article, we show that our transformation also applies to the PAKE scheme from [32], allowing us to give a nearly optimal OT scheme.

Note that our technique works with every possible PAKE scheme, be they elliptic-curve-based or not. Furthermore, also note that it also allows to transform a BPR-secure PAKE [6] into an OT secure in a game-based security model.

In order to illustrate these last two points, we apply our framework to a (non-UC) secure lattice-based PAKE scheme proposed by [33], giving us a (non-UC) secure lattice-based OT scheme (see the paper full version [12]).

Comparison with Other Existing OT Schemes: The table given in Fig. 1 shows the costs of various known OT schemes based on elliptic curves.

While Barreto-Naehrig curves [5] are considered less and less secure for efficient parameters, we consider for the asymptotic scaling that elements in $\mathbb{G}_2$ are smaller than those in $\mathbb{G}_1$, hence, we switched some elements to $\mathbb{G}_2$ for big enough values of n.

Thanks to the use of a QA-NIZK technique, we manage to get rid of the extra logarithm overhead. And we end up being approximately four times more efficient than existing solutions without this overhead.

Communication cost comparisons of various Elliptic Curve based OT schemes

Paper	Assumption	# Group elements	# Rounds
Static Security (1-out-of-2)			
[40] + [25]	SXDH	51	8
[22]	SXDH	$26 + 7\ \mathbb{Z}_p$	4
Adaptive Security (1-out-of-2)			
[1]	SXDH	$12\ \mathbb{G}_1 + 1\ \mathbb{G}_2 + 2\mathbb{Z}_p$	3
[9][1]	DDH	$15\ \mathbb{G} + 2\ \mathbb{Z}_p$	3
[10]	SXDH	$12\ \mathbb{G}_1 + 4\ \mathbb{G}_2 + 2\ \mathbb{Z}_p$	3
This paper	SXDH	$6\ \mathbb{G}_1 + 2\ \mathbb{G}_2 + 2\ \mathbb{Z}_p$	3
Adaptive Security (1-out-of-n)			
[1]	SXDH	$\log n\ \mathbb{G}_1 + (n + 8\log n)\ \mathbb{G}_2 + n\ \mathbb{Z}_p$	3
[9][1]	DDH	$(n + 9\log n + 4)\ \mathbb{G} + 2n\ \mathbb{Z}_p$	3
[10]	SXDH	$4\ \mathbb{G}_1 + (4n + 4)\ \mathbb{G}_2 + n\ \mathbb{Z}_p$	3
This paper	SXDH	$4\ \mathbb{G}_1 + (n + 2)\ \mathbb{G}_2 + n\ \mathbb{Z}_p$	3

Fig. 1. Comparison to existing Oblivious Transfer. [1]It should be noted that the [9] OT does not rely on pairings. Classical implementations suggest that the elements in one of the groups in our scheme will be at least twice as big as those from the non-pairing curve, which would give roughly the same efficiency between both schemes in the 1-out-of-2 case, with an edge for the [9] construction in term of computational requirements. However, the scaling in their paper is then worse when increasing the number of lines (see the 1-out-of-n case).

Organization of the Paper: To present our fast Oblivious Transfer scheme in 3 flows, we first recall basic definitions and security notions needed for the understanding of the paper. Then, we propose our generic transformation from a 2-round PAKE scheme to a 3-round OT scheme and show how it can be generically improved by some optimizations. We then apply our framework to the PAKE from [1], and show it directly leads to their existing OT. We then recall the PAKE scheme described in [32], and show how to construct from it our new nearly optimal OT by applying our framework.

Finally, we sketch in the paper full version [12] a brief example to show how our framework behave on a non-UC lattice-based PAKE and how we get a game-based OT scheme from the BPR-secure PAKE scheme presented in [33].

2   Preliminaries

2.1   Definitions and Notations

In the following, in asymmetric protocols the sender is going to be called $\mathcal{S}$ while the receiver will be noted $\mathcal{R}$. For symmetrical ones like PAKE, we will rather use P_i and P_j. We will note the security parameter $\mathfrak{K}$, and n the size of the database (number of lines) for Oblivious Transfer schemes.

We are going to work in an asymmetric bilinear setting denoted as $(p, \mathbb{G}_1, \mathbb{G}_2, \mathbb{G}_T, e, g_1, g_2)$, where $\mathbb{G}_1$, $\mathbb{G}_2$, $\mathbb{G}_T$ are cyclic groups of order p for a $\mathfrak{K}$-bit prime p, g_1 and g_2 are generators of $\mathbb{G}_1$ and $\mathbb{G}_2$, respectively, and $e : \mathbb{G}_1 \times \mathbb{G}_2 \to \mathbb{G}_T$ is an efficiently computable (non-degenerated) bilinear map. Define $g_T := e(g_1, g_2)$, which is a generator in $\mathbb{G}_T$.

Definition 1 (DDH assumption). *Assuming a multiplicative group $\mathbb{G}$ of prime order p and generator g, we say that the DDH assumption is hard in $\mathbb{G}$ if, given a tuple $(g^x, g^y, g^z) \in \mathbb{G}^3$, it is hard to decide whether $z = xy$ or not.*

Definition 2 (Symmetric External Diffie Hellman (SXDH [4]) assumption). *This variant of DDH, used mostly in bilinear groups $(\mathbb{G}_1, \mathbb{G}_2)$ in which no computationally efficient homomorphism exists from $\mathbb{G}_2$ to $\mathbb{G}_1$ or $\mathbb{G}_1$ to $\mathbb{G}_2$, states that the DDH assumption is hard in both $\mathbb{G}_1$ and $\mathbb{G}_2$.*

Definition 3 ((Labelled) Quasi Adaptive Non-Interactive Zero Knowledge Argument). *Assuming a randomness distribution $\mathcal{D}_{\mathsf{param}}$ (for a public set of parameters param) and a set of languages $\{\mathfrak{L}_\rho\}_\rho$ parameterized by a randomness $\rho \leftarrow \mathcal{D}_{\mathsf{param}}$ and associated with a relation $\mathcal{R}_\rho$ (meaning that a word x belongs to $\mathfrak{L}_\rho$ if and only if there exists a witness w such that $\mathcal{R}_\rho(x, w)$ holds), a QA-NIZK argument Π for this set of languages $\{\mathfrak{L}_\rho\}_\rho$ consists of five probabilistic polynomial time (PPT) algorithms $\Pi = (\mathsf{Gen}_{\mathsf{param}}, \mathsf{Gen}_{\mathsf{crs}}, \mathsf{Prove}, \mathsf{Sim}, \mathsf{Ver})$ defined as follows:*

- $\mathsf{Gen}_{\mathsf{param}}(\mathfrak{K})$ *returns the public parameters param.*
- $\mathsf{Gen}_{\mathsf{crs}}(\mathsf{param}, \rho)$ *returns a common reference string crs and a trapdoor tk. We assume that crs contain the parameters param, a description of the language $\mathfrak{L}_\rho$ and a description of a label space $\mathcal{L}$.*
- Prove *takes as input crs, a label $\ell \in \mathcal{L}$, a word x of the language $\mathfrak{L}_\rho$ and a witness w corresponding to this word. If $(x, w) \notin \mathcal{R}_\rho$, it outputs failure. Otherwise, it outputs a proof π.*
- Ver *takes as input crs, a label ℓ, a word x and a proof π. It outputs a bit b (either 1 if the proof is correct, or 0 otherwise).*

– Sim *takes as input* crs, *a trapdoor* tk, *a label* $\ell \in \mathcal{L}$ *and a word* x. *It outputs a proof* π *for* x *(not necessarily in* $\mathfrak{L}_\rho$*) with respect to the label* ℓ.

These algorithms must satisfy the following properties

Perfect Completeness: *for all PPT adversary* $\mathcal{A}$, *all security parameter* $\mathfrak{K}$, *all public parameters* param $\leftarrow$ Gen$_{\text{param}}(\mathfrak{K})$, *all randomness* $\rho \leftarrow \mathcal{D}_{\text{param}}$, *all label* $\ell \in \mathcal{L}$ *and all* $(x, w) \in \mathcal{R}_\rho$, *the following holds:*

$$\Pr[(\text{crs}, \text{tk}) \leftarrow \text{Gen}_{\text{crs}}(\text{param}, \rho); \pi \leftarrow \text{Prove}(\text{crs}, \ell, x, w) : \text{Ver}(\text{crs}, \ell, x, \pi) = 1] = 1$$

Computational Adaptive Soundness: *For a given security parameter* $\mathfrak{K}$, *a scheme achieves computational adaptive soundness if for all PPT adversary* $\mathcal{A}$, *the probability*

$$\Pr[\text{param} \leftarrow \text{Gen}_{\text{param}}(\mathfrak{K}); \rho \leftarrow \mathcal{D}_{\text{param}}; (\text{crs}, \text{tk}) \leftarrow \text{Gen}_{\text{crs}}(\text{param}, \rho);$$
$$(\ell, x, \pi) \leftarrow \mathcal{A}(\text{param}, \text{crs}, \rho) : x \notin \mathfrak{L}_\rho \wedge \text{Ver}(\text{crs}, \ell, x, \pi) = 1]$$

is negligible.

Perfect Zero-Knowledge: *for all* $\mathfrak{K}$, *all* param $\leftarrow$ Gen$_{\text{param}}(\mathfrak{K})$, *all* $\rho \leftarrow \mathcal{D}_{\text{param}}$, *all* (crs, tk) $\leftarrow$ Gen$_{\text{crs}}(\text{param}, \rho)$, *all label* $\ell \in \mathcal{L}$ *and all* $(x, w) \in \mathcal{R}_\rho$, *the distributions* Prove(crs, ℓ, x, w) *and* Sim(crs, tk, ℓ, x) *are the same.*

2.2   Security Models

UC Framework. The goal of the UC framework [16] is to ensure that UC-secure protocols will continue to behave in the ideal way even if executed in a concurrent way in arbitrary environments. It is a simulation-based model, relying on the indistinguishability between the real world and the ideal world. In the ideal world, the security is provided by an ideal functionality $\mathcal{F}$, capturing all the properties required for the protocol and all the means of the adversary. In order to prove that a protocol Π emulates $\mathcal{F}$, one has to construct, for any polynomial adversary $\mathcal{A}$ (which controls the communication between the players), a simulator $\mathcal{S}$ such that no polynomial environment $\mathcal{Z}$ can distinguish between the real world (with the real players interacting with themselves and $\mathcal{A}$ and executing the protocol π) and the ideal world (with dummy players interacting with $\mathcal{S}$ and $\mathcal{F}$) with a significant advantage. The adversary can be either *adaptive, i.e.* allowed to corrupt users whenever it likes to, or *static, i.e.* required to choose which users to corrupt prior to the execution of the session sid of the protocol. After corrupting a player, $\mathcal{A}$ has complete access to the internal state and private values of the player, takes its entire control, and plays on its behalf.

Simple UC Framework. Canetti et al. formalized a simpler variant in [17], that we use here. This simplifies the description of the functionalities for the following reasons (in a nutshell): All channels are automatically assumed to be authenticated (as if we worked in the $\mathcal{F}_{\text{AUTH}}$-hybrid model); There is no need for *public delayed outputs* (waiting for the adversary before delivering a message to a party), neither for an explicit description of the corruptions. We refer the interested reader to [17] for details.

Password Authenticated Key Exchange. Before the UC framework, the classical security model for PAKE protocols was the so-called BPR model [6]. The first ideal functionality for PAKE protocols in the UC framework [16,20] was proposed by Canetti *et al.* [19], who showed how a simple variant of the Gennaro-Lindell methodology [26] could lead to a secure protocol.

We present the PAKE ideal functionality $\mathcal{F}_{pwKE}$ on Fig. 2. It was described in [19].

The functionality $\mathcal{F}_{\mathrm{pwKE}}$ is parameterized by a security parameter k. It interacts with an adversary $\mathcal{S}$ and a set of parties $P_1,\ldots,P_n$ via the following queries:

- **Upon receiving a query** (NewSession, sid, ssid, P_i, P_j, pw) **from party P_i:**
  Send (NewSession, sid, ssid, P_i, P_j) to $\mathcal{S}$. If this is the first NewSession query, or if this is the second NewSession query and there is a record (sid, ssid, P_j, P_i, pw$'$), then record (sid, ssid, P_i, P_j, pw) and mark this record **fresh**.
- **Upon receiving a query** (TestPwd, sid, ssid, P_i, pw$'$) **from the adversary $\mathcal{S}$:**
  If there is a record of the form (P_i, P_j, pw) which is **fresh**, then do: If pw $=$ pw$'$, mark the record **compromised** and reply to $\mathcal{S}$ with "correct guess". If pw $\neq$ pw$'$, mark the record **interrupted** and reply with "wrong guess".
- **Upon receiving a query** (NewKey, sid, ssid, P_i, sk) **from the adversary $\mathcal{S}$:**
  If there is a record of the form (sid, ssid, P_i, P_j, pw), and this is the first NewKey query for P_i, then:
  - If this record is **compromised**, or either P_i or P_j is corrupted, then output (sid, ssid, sk) to player P_i.
  - If this record is **fresh**, and there is a record (P_j, P_i, pw') with pw$'$ $=$ pw, and a key sk$'$ was sent to P_j, and (P_j, P_i, pw) was **fresh** at the time, then output (sid, ssid, sk$'$) to P_i.
  - In any other case, pick a new random key sk$'$ of length $\mathfrak{K}$ and send (sid, ssid, sk') to P_i.
  Either way, mark the record (sid, ssid, P_i, P_j, pw) as **completed**.

Fig. 2. Ideal Functionality for PAKE $\mathcal{F}_{\mathrm{pwKE}}$

The main idea behind this functionality is as follows: If neither party is corrupted and the adversary does not attempt any password guess, then the two players both end up with either the same uniformly-distributed session key if the passwords are the same, or uniformly-distributed independent session keys if the passwords are distinct. In addition, the adversary does not know whether this is a success or not. However, if one party is corrupted, or if the adversary successfully guessed the player's password (the session is then marked as compromised), the adversary is granted the right to fully determine its session key. There is in fact nothing lost by allowing it to determine the key. In case of a wrong guess (the session is then marked as interrupted), the two players are given independently-chosen random keys. A session that is nor compromised nor interrupted is called fresh, which is its initial status.

Finally notice that the functionality is not in charge of providing the password(s) to the participants. The passwords are chosen by the environment which then hands them to the parties as inputs. This guarantees security even in the case where two honest players execute the protocol with two different passwords: This models, for instance, the case where a user mistypes its password. It also implies that the security is preserved for all password distributions (not necessarily the uniform one) and in all situations where the passwords are related passwords and/or used in different protocols. Also note that allowing the environment to choose the passwords guarantees forward secrecy.

In case of corruption, the adversary learns the password of the corrupted player; After the NewKey -query, it additionally learns the session key.

Oblivious Transfer. The ideal functionality of an Oblivious Transfer (OT) protocol was given in [1,16,22]. We recall it in simple UC in Fig. 3 using the functionality introduced in [11]. Note that the BPR model [6] given for PAKE protocols can be adapted to give a game-based security model for OT schemes but this is well beyond the scope of this paper.

The party P_i is the sender $\mathcal{S}$, while the party P_j is the receiver $\mathcal{R}$. The former is provided with a database consisting of a set of n lines $(L_1, \ldots, L_n)$, while the latter is querying a particular line L_s (with $s \in \{1, \ldots, n\}$). Since there is no communication between them (the functionality deals with everything), it automatically ensures the oblivious property on both sides (the sender does not learn which line was queried, while the receiver does not learn any line other than L_s).

3   Generic Construction of an Oblivious Transfer from a UC-Secure PAKE

There is a trend in proven cryptography that consists in finding the link between primitives and show which primitives can be used to build other ones. A well-known example is the transformation of an IBE encryption scheme into a signature scheme, which was proposed by Naor as recalled in [15], and its reverse transformation from an affine MAC or signature scheme into an IBE scheme given in [14].

A neat result, presented by Canetti et al. in [18], shows how a UC-secure Oblivious Transfer scheme can be transformed into a UC-secure PAKE scheme. In this section we are going to go the other way, and show how a UC-secure PAKE can be transformed into a UC-secure Oblivious Transfer scheme. While

---

The functionality $\mathcal{F}_{(1,n)\text{-OT}}$ is parametrized by a security parameter $\mathfrak{K}$. It interacts with an adversary $\mathscr{S}$ and a set of parties $P_1, \ldots, P_N$ via the following queries:

- **Upon receiving an input** (Send, sid, ssid, $P_i, P_j, (L_1, \ldots, L_n)$) **from party** $\mathcal{P}_i$, with $L_i \in \{0,1\}^{\mathfrak{K}}$ for all i: record the tuple (sid, ssid, $P_i, P_j, (L_1, \ldots, L_n)$) and reveal (Send, sid, ssid, P_i, P_j) to the adversary $\mathscr{S}$. Ignore further Send-message with the same ssid from P_i.
- **Upon receiving an input** (Receive, sid, ssid, P_i, P_j, s) **from party** P_j: ignore the message if (sid, ssid, $P_i, P_j, (L_1, \ldots, L_n)$) is not recorded. Otherwise, reveal (Receive, sid, ssid, P_i, P_j) to the adversary $\mathscr{S}$ and send (Received, sid, ssid, P_i, P_j, L_s) to P_j and ignore further Receive-message with the same ssid from P_j.

Fig. 3. Ideal functionality for 1-out-of-n oblivious transfer $\mathcal{F}_{(1,n)\text{-OT}}$

considering this direction may seem like transforming a very strong primitive into a weaker one, this will allow us to prepare a framework to propose the most efficient Oblivious Transfer scheme to date. This also echoes to the results from [39] on a simulation-based transform.

3.1 Description of a UC-Secure PAKE Scheme

As explained in the introduction, we are going to consider PAKE protocols in two flows, since there already exists a multitude of PAKE schemes satisfying this requirement, and since minimizing the number of flows is one of the most important issues in modern instantiations. This allows us to generically give an optimization, leading to more efficient protocols.

In order to present the framework or our transformation, we formally split a PAKE scheme into the following four algorithms[1].

- Setup: An algorithm that generates the initial crs.
- flow$_1$(pw$_i$; ρ_i, ρ_i'): The algorithm run by the user P_i, using some randomness ρ_i and ρ_i' and possessing the password pw$_i$, to generate the first flow of the protocol. We note ρ_i the part of the randomness involved in hiding the password, and ρ_i' the remainder.
- flow$_2$(flow$_1$, pw$_j$; ρ_j, ρ_j'): The algorithm run by the user P_j, using some randomness ρ_j and possessing the password pw$_j$, after receiving the flow flow$_1$, to generate the second flow of the protocol. This algorithm also produces P_j's view of the key: $K_i^{(j)}$.
- Decap(flow$_2$, pw$_i$, $f(\rho_i, \rho_i')$): The algorithm run by P_i when receiving the flow flow$_2$ from P_j, using his password pw$_i$ and a function $f(\rho_i, \rho_i')$ of the randomness initially used and the remainder (which function can be anything from the identity to the null function), to recover P_i's view of the key: $K_j^{'(i)}$.

---

[1] Since it was shown in [19] that UC-secure PAKE is impossible in the plain model, we focus on the CRS model for the ease of readability.

3.2   Generic Construction of a UC-Secure OT Scheme

We denote by $\mathsf{DB} = (L_1, \ldots, L_n)$ the database of the server containing n lines, and by s the line requested by the user in an oblivious way. We assume the existence of a CPA-Secure encryption scheme, whose public parameters are going to be included in the public common reference string crs generated by the setup algorithm of the PAKE scheme. Such an encryption scheme $\mathcal{E}$ is described through four algorithms $(\mathsf{Setup}_{\mathrm{cpa}}, \mathsf{KeyGen}_{\mathrm{cpa}}, \mathsf{Encrypt}_{\mathrm{cpa}}, \mathsf{Decrypt}_{\mathrm{cpa}})$:

- $\mathsf{Setup}_{\mathrm{cpa}}(1^{\mathfrak{K}})$, where $\mathfrak{K}$ is the security parameter, generates the global parameters $\mathsf{param}_{\mathrm{cpa}}$ of the scheme;
- $\mathsf{KeyGen}_{\mathrm{cpa}}(\mathsf{param}_{\mathrm{cpa}})$ outputs a pair of keys: a (public) encryption key pk and a (private) decryption key sk;
- $\mathsf{Encrypt}_{\mathrm{cpa}}(\mathsf{pk}, M; \rho)$ outputs a ciphertext $c = \mathcal{E}(M)$ on the message M, under the encryption key pk, using the randomness ρ;
- $\mathsf{Decrypt}_{\mathrm{cpa}}(\mathsf{sk}, c)$ outputs the plaintext M encrypted in the ciphertext c, or $\perp$.

We also assume the existence of a Pseudo-Random Generator (PRG) F with input size equal to the plaintext size, and output size equal to the size of the messages in the database.

Using the PAKE scheme, the encryption scheme and the PRG, one can now obtain an OT scheme from a PAKE scheme, as described in Fig. 4. The idea of the protocol is as follows:

- First, a setup phase enables to generate the CRS, both for the PAKE and encryption schemes.
- The pre-flow is a technical requirement for the UC proof.
- When querying for line s, the receiver proceeds in three steps. The first one consists in generating a masking value R (technical requirement for the UC proof), the second one consists in running the first flow of the PAKE scheme for password s to generate the first flow denoted[2] as flow_0, and the third one consists in erasing anything but the needed values, and sending flow_0 to the sender.
- When answering to this query, the sender also proceeds in three steps. The first one consists in recovering the value R and the second one consists in running, for each line $k \in \{1, \ldots, n\}$, the second flow of the PAKE scheme to generate the second flow denoted as flow_k (using k as the password) as well as $\mathcal{S}$'s view of the session key, denoted as $(K_k)_{\mathcal{R}}^{(\mathcal{S})}$. Finally, in the third step, the server computes a xor Q_k of the line L_k and this session key and the value R and sends the set (flow_k, Q_k) back to $\mathcal{R}$.
- When receiving this set of values, the receiver uses flow_s to run the decapsulation algorithm of the PAKE scheme (with password s) to recover his view of the session key, denoted as $(K_s)_{\mathcal{S}}^{(\mathcal{R})}$. He finally uses R, Q_s and $(K_s)_{\mathcal{S}}^{(\mathcal{R})}$ to recover the expected line L_s.

---

[2] Note that we now denote as flow_0 (and not flow_1) the flow generated by the PAKE algorithm flow_1, in order to avoid the confusion with the flow_1 message generated by the sender (for line 1) during the database answer phase.

The correctness easily follows from the correctness of the PAKE protocol, since the views of $\mathcal{R}$ and $\mathcal{S}$ of the session key for the PAKE scheme executed for password s are the same. The scheme is oblivious with respect to $\mathcal{S}$ since the first flow of the PAKE scheme executed by $\mathcal{R}$ does not reveal any information on s. And it is oblivious with respect to $\mathcal{R}$ since the correctness of the PAKE scheme gives him no information on the session keys $(K_k)_{\mathcal{R}}^{(\mathcal{S})}$ for $k \neq s$. The security is stated in Theorem 4.

CRS generation:
crs $\overset{\$}{\leftarrow}$ Setup$(1^{\mathfrak{K}})$, param$_{\mathrm{cpa}}$ $\overset{\$}{\leftarrow}$ Setup$_{\mathrm{cpa}}(1^{\mathfrak{K}})$.

Pre-Flow:

1. $\mathcal{S}$ generates a key pair (pk, sk) $\overset{\$}{\leftarrow}$ KeyGen$_{\mathrm{cpa}}$(param$_{\mathrm{cpa}}$) for $\mathcal{E}$, stores sk and completely erases the random coins used by KeyGen$_{\mathrm{cpa}}$.
2. $\mathcal{S}$ publishes pk.

Index query on s:

1. $\mathcal{R}$ picks a random J, computes $R \leftarrow F(J)$ and sets $c \overset{\$}{\leftarrow}$ Encrypt$_{\mathrm{cpa}}$(pk, J).
2. $\mathcal{R}$ picks a random (ρ, ρ'), and runs flow$_1(s; \rho, \rho')$ to obtain flow$_0$.
3. $\mathcal{R}$ stores $f(\rho, \rho')$ needed for the Decap algorithm and R and completely erases the rest, including the random coins used by Encrypt$_{\mathrm{cpa}}$ and sends (c, flow_0) to $\mathcal{S}$.

Database answer:

1. $\mathcal{S}$ decrypts $J \leftarrow$ Decrypt$_{\mathrm{cpa}}$(sk, c) and computes $R \leftarrow F(J)$.
2. For each line k:
   - $\mathcal{S}$ picks a random (ρ_k, ρ'_k), runs flow$_2(\text{flow}_0, k; \rho_k, \rho'_k)$ to generate flow$_k$ and $(K_k)_{\mathcal{R}}^{(\mathcal{S})}$.
   - $\mathcal{S}$ then computes $Q_k = L_k \oplus (K_k)_{\mathcal{R}}^{(\mathcal{S})} \oplus R$.
3. $\mathcal{S}$ erases everything except $(\text{flow}_k, Q_k)_{k \in [\![1,n]\!]}$ and sends it to $\mathcal{R}$.

Data recovery:
$\mathcal{R}$ then using $f(\rho, \bot)$ computes $(K_s)_{\mathcal{S}}^{(\mathcal{R})} =$ Decap(flow$_s$, s, $f(\rho, \rho')$), sets $L_s = Q_s \oplus (K_s)_{\mathcal{S}}^{(\mathcal{R})} \oplus R$, and erases everything else.

Fig. 4. UC-Secure 1-out-of-n OT from a UC-Secure PAKE

Theorem 4. *Under the* UC-*security[3] of the* PAKE *protocol, the existence of a Pseudo-Random Generator* (PRG) *F with input size equal to the plaintext size, and output size equal to the size of the messages in the database, and the* IND-CPA *security of the encryption scheme, the transformation presented*

---

[3] Note that the transformation also allows to obtain an OT scheme from a BPR-secure PAKE scheme in a game-based security model, but since this is of less interest, we do not formally prove it due to lack of space.

*in Fig. 4 achieves a UC-secure 1-out-of-n Oblivious Transfer scheme, with the
same handling of corruptions as in the PAKE protocol.*

The high-level idea of the proof is quite simple. Recall that in a UC proof,
one has to simulate one of the two players in front of a corrupted player. The
first flow allows the simulated server to extract the requested line. One will
then be able to send random noise on all the lines besides the correct one. The
PAKE functionality ensures that this is possible. Then again for honest users, the
simulator will replace the remaining line by a random noise. In case of corruptions
of the server, one relies on the security of the CPA encryption, to ensure that
everything was done honestly (the adversary is not able to test the validity of
the retrieved R, and so for the correct line as $L \oplus K \oplus R$ where L and K are the
honest values, and R is simply set as $Q \ominus (L \oplus K)$). In case of corruption of the
user, we rely on the PAKE handling of user corruption to adapt the memory.

A more detailed proof can be found in the paper full version [12].

3.3 Generic Optimization

It should be noted that in the previous transformation we never used the fact
that the first user in the PAKE scheme (here, the receiver) does not learn the
password from the second (here, the sender). This is a property required by
PAKE but not useful in the transformation (since the sender executes n parallel
PAKE schemes, using the (public) number k of the lines as the password). This
argument is the key of the optimization we now propose in Fig. 5. The difference
is that in the index query, the receiver does not pick the second randomness ρ'
and in the database answer, the randomness ρ_k is not used anymore. The proof
remains the same as the previous one.

4 Warm-Up: Applying the Framework on [1]

4.1 Notations

Since in [1], the authors proposed both a UC-secure PAKE and a UC-secure Oblivious Transfer constructions, this section serves as an example of application of
our framework. More precisely, we apply here our (optimized) transformation to
their UC-secure PAKE and show that we obtain exactly their UC-secure Oblivious
Transfer.

There protocols make use of Commitment schemes and Smooth Projective
Hash Functions. Rigorous definitions can be found in the paper full version [12]
but we give here the notations used in the following.

A commitment scheme is a two-party primitive (between a committer and a
receiver) divided into two phases. In a first *commit* phase, the committer gives
the receiver an analogue of a sealed envelope containing a value m, while in the
second *opening* phase, the committer reveals m in such a way that the receiver
can verify it was indeed m that was contained in the envelope. It is required
that a committer cannot change the committed value (*i.e.*, he should not be

---

CRS generation:
crs $\xleftarrow{\$}$ Setup($1^{\mathfrak{K}}$), param$_{\text{cpa}}$ $\xleftarrow{\$}$ Setup$_{\text{cpa}}$($1^{\mathfrak{K}}$).

Pre-Flow:

1. $\mathcal{S}$ generates a key pair (pk, sk) $\xleftarrow{\$}$ KeyGen$_{\text{cpa}}$(param$_{\text{cpa}}$) for $\mathcal{E}$, stores sk and completely erases the random coins used by KeyGen$_{\text{cpa}}$.
2. $\mathcal{S}$ publishes pk.

Index query on s:

1. $\mathcal{R}$ picks a random J, computes $R \leftarrow F(J)$ and sets $c \xleftarrow{\$}$ Encrypt$_{\text{cpa}}$(pk, J).
2. $\mathcal{R}$ picks a random ρ, and runs flow$_1(s; \rho, \bot)$ to obtain flow$_0$.
3. $\mathcal{R}$ stores $f(\rho, \bot)$ needed for the Decap algorithm and R and completely erases the rest, including the random coins used by Encrypt$_{\text{cpa}}$ and sends (c, flow$_0$) to $\mathcal{S}$.

Database answer:

1. $\mathcal{S}$ decrypts $J \leftarrow$ Decrypt$_{\text{cpa}}$(sk, c) and computes $R \leftarrow F(J)$.
2. For each line k:
   - $\mathcal{S}$ picks a random ρ'_k, and runs flow$_2$(flow$_0$, k; $\bot$, ρ'_k) to generate flow$_k$ and $(K_k)_{\mathcal{R}}^{(\mathcal{S})}$.
   - $\mathcal{S}$ then computes $Q_k = L_k \oplus (K_k)_{\mathcal{R}}^{(\mathcal{S})} \oplus R$.
3. $\mathcal{S}$ erases everything except (flow$_k$, Q_k)$_{k \in [\![1,n]\!]}$ and sends it to $\mathcal{R}$.

Data recovery: $\mathcal{R}$ then using $f(\rho, \bot)$ computes $(K_s)_{\mathcal{S}}^{(\mathcal{R})} =$ Decap(flow$_s$, s, $f(\rho, \bot)$), sets $L_s = Q_s \oplus (K_s)_{\mathcal{S}}^{(\mathcal{R})} \oplus R$, and erases everything else.

Fig. 5. Optimized UC-Secure 1-out-of-n OT from a UC-Secure PAKE

able to open to a value different from the one he committed to), this is called the *binding* property. It is also required that the receiver cannot learn anything about m before the opening phase, this is called the *hiding* property.

Formally, a *non-interactive labelled commitment scheme* $\mathcal{C}$ is defined by three algorithms:

- SetupCom($1^{\mathfrak{K}}$) takes as input the security parameter $\mathfrak{K}$ and outputs the global parameters, passed through the CRS ρ to all other algorithms;
- Com$^{\ell}(x; \rho)$ takes as input a label ℓ, a message x and a random ρ, and outputs a pair (C, δ), where C is the commitment of x for the label ℓ, and δ is the corresponding opening data (a.k.a. decommitment information).
- VerCom$^{\ell}(C, x, \delta)$ takes as input a commitment C, a label ℓ, a message x, and the opening data δ and outputs 1 (true) if δ is a valid opening data for C, x and ℓ. It always outputs 0 (false) on $x = \bot$.

Smooth projective hash functions are defined such as their value can be computed in two different ways if the input belongs to a particular subset (the *language*), either using a private hashing key or a public projection key along with a private witness ensuring that the input belongs to the language. The projection key is derived from the hashing key. The hash value obtained is indistinguishable from random in case the input does not belong to the language (property of *smoothness*) and in case the input does belong to the language but no witness is known (property of *pseudo-randomness*).

Formally, a Smooth Projective Hash Function over a language $\mathfrak{L} \subset X$, onto a set $\mathcal{G}$, is defined by five algorithms (Setup, HashKG, ProjKG, Hash, ProjHash):

- Setup($1^{\mathfrak{K}}$) where $\mathfrak{K}$ is the security parameter, generates the global parameters param of the scheme, and the description of an $\mathcal{NP}$ language $\mathfrak{L}$;
- HashKG($\mathfrak{L}$, param), outputs a hashing key hk for the language $\mathfrak{L}$;
- ProjKG(hk, ($\mathfrak{L}$, param), W), derives the projection key hp from the hashing key hk.
- Hash(hk, ($\mathfrak{L}$, param), W), outputs a hash value $v \in \mathcal{G}$, thanks to the hashing key hk and W.
- ProjHash(hp, ($\mathfrak{L}$, param), W, w), outputs the hash value $v' \in \mathcal{G}$, thanks to the projection key hp and the witness w that $W \in \mathfrak{L}$.

4.2 Description of Their UC-Secure PAKE Scheme

Description of the Scheme: Both parties (denoted as P_b and $P_{\bar{b}}$) want to share a key by using a common password pw. This will thus fix a language $\mathfrak{L}_{pw}$, which is the set of valid commitments of pw, for a smooth projective hash function. Both flows are done simultaneously. In flow$_b$, each party P_b will generate a pair of hash keys (hk$_b$, hp$_b$) and then commit to the password pw$_b$ in C_b. Note that P_b stores the witness δ_b that this value C_b is a valid commitment of pw$_b$ and sends (hp$_b$, C_b) to the other party.

In the decapsulation phase, each party P_b then receives the commitment and the projection key of the other party and computes two hash values: On the one hand, the projected hash value using his own commitment C_b along with the witness δ_b and the projected key hp$_{\bar{b}}$ sent by the other party, and on the other hand, the (regular) hash value using the received commitment $C_{\bar{b}}$ and the secret hash key hk$_b$. The shared session key is then defined as the product of these two values.

The commitment scheme used has specific properties explained in [1] well beyond the scope of the paper. It needs to be compatible with SPHF, and both extractable and equivocable.

Completeness: If the two parties used the same password pw, then at the end of the protocol they will share the same value. Thanks to the property of the SPHF, the hash value and the projected value (on the same commitment C_b) will be equal because the commitment belongs to the good language $\mathfrak{L}_{pw}$ with the good witness δ_b.

4.3 Applying the Framework to Obtain a UC-Secure OT Scheme

Applying our (optimized) framework, we manage to fall back directly to the Oblivious Transfer that was proposed simultaneously in [1] (Fig. 6).

- Setup: $\rho \xleftarrow{\$} \mathsf{SetupCom}(1^{\mathfrak{K}})$.
- $\mathsf{flow}_b(\mathsf{pw}_b; \rho_b)$:
  - Generates $\mathsf{hk}_b \xleftarrow{\$} \mathsf{HashKG}(\mathfrak{L}_{\mathsf{pw}_b})$, $\mathsf{hp}_b \leftarrow \mathsf{ProjKG}(\mathsf{hk}_b, \mathfrak{L}_{\mathsf{pw}_b}, \perp)$
  - picks additional ρ'_b and computes $(C_b, \delta_b) \xleftarrow{\$} \mathsf{Com}^{\ell_b}(\mathsf{pw}_b; \rho'_b)$ with $\ell_b = (\mathsf{sid}, \mathsf{ssid}, P_b, P_{\bar{b}}, \mathsf{hp}_b)$. Sets $\rho_b = (\mathsf{hk}_b, \rho'_b)$.
  - stores hk_b, δ_b, completely erases random coins used before and sends (hp_b, C_b).
- $\mathsf{Decap}(\mathsf{flow}_b, \mathsf{pw}_b, f(\rho_b) = \mathsf{hk}_b, \delta_b)$:
  - Computes $H'_b \leftarrow \mathsf{ProjHash}(\mathsf{hp}_{\bar{b}}, \mathfrak{L}_{\mathsf{pw}_b}, (\ell_b, C_b), \delta_b)$ and $H_{\bar{b}} \leftarrow \mathsf{Hash}(\mathsf{hk}_b, \mathfrak{L}_{\mathsf{pw}_b}, (\ell_{\bar{b}}, C_{\bar{b}}))$ with $\ell_{\bar{b}} = (\mathsf{sid}, \mathsf{ssid}, P_{\bar{b}}, P_b, \mathsf{hp}_{\bar{b}})$
  - Computes $\mathsf{sk}_b = H'_b \cdot H_{\bar{b}}$ and erases everything else, except pw_b.

Fig. 6. UC-Secure PAKE from [1]

CRS generation:

$\rho \xleftarrow{\$} \mathsf{SetupCom}(1^{\mathfrak{K}})$, $\mathsf{param}_{\mathrm{cpa}} \xleftarrow{\$} \mathsf{Setup}_{\mathrm{cpa}}(1^{\mathfrak{K}})$.

Pre-flow :

1. $\mathcal{S}$ generates a key pair $(\mathsf{pk}, \mathsf{sk}) \xleftarrow{\$} \mathsf{KeyGen}_{\mathrm{cpa}}(\mathsf{param}_{\mathrm{cpa}})$ for $\mathcal{E}$, stores sk and completely erases the random coins used by $\mathsf{KeyGen}_{\mathrm{cpa}}$.
2. $\mathcal{S}$ publishes pk.

Index query on s:

1. $\mathcal{R}$ chooses a random value J, computes $R \leftarrow F(J)$ and encrypts J under pk: $c \xleftarrow{\$} \mathsf{Encrypt}_{\mathrm{cpa}}(\mathsf{pk}, J)$
2. $\mathcal{R}$ computes $(C, \delta) \xleftarrow{\$} \mathsf{Com}^{\ell}(s)$ with $\ell = (\mathsf{sid}, \mathsf{ssid}, \mathcal{R}, \mathcal{S})$
3. $\mathcal{R}$ stores δ_s, R, completely erases the random coins used and sends C and c to $\mathcal{S}$

Database answer $(L_1, \ldots, L_n)$:

1. $\mathcal{S}$ decrypts $J \leftarrow \mathsf{Decrypt}_{\mathrm{cpa}}(\mathsf{sk}, c)$ and gets $R \leftarrow F(J)$
2. For each line $k = 1, \ldots, n$,
   - $\mathcal{S}$ computes $\mathsf{hk}_k \xleftarrow{\$} \mathsf{HashKG}(\mathfrak{L}_k)$, $\mathsf{hp}_k \leftarrow \mathsf{ProjKG}(\mathsf{hk}_k, \mathfrak{L}_k, (\ell, C))$, $K_k \leftarrow \mathsf{Hash}(\mathsf{hk}_k, \mathfrak{L}_k, (\ell, C))$,
   - $\mathcal{S}$ computes $Q_k \leftarrow R \oplus K_k \oplus L_k$
3. $\mathcal{S}$ erases everything except $(\mathsf{hp}_k, Q_k)_{k=1,\ldots,n}$ and sends it to $\mathcal{R}$.

Data recovery:

Upon receiving $(\mathsf{hp}_k, Q_k)_{k=1,\ldots,n}$, $\mathcal{R}$ computes $K_s \leftarrow \mathsf{ProjHash}(\mathsf{hp}_s, \mathfrak{L}_s, (\ell, C), \delta)$ and gets $L_s \leftarrow R \oplus K_s \oplus Q_s$.
Then $\mathcal{R}$ erases everything except L_s.

Fig. 7. UC-Secure 1-out-of-n OT from an SPHF-Friendly Commitment

Theorem 5. *As the* PAKE *was UC-Secure in presence of adaptive corruption under* SXDH, *the Oblivious Transfer scheme presented in Fig. 7 is a secure 1-out-of-n Oblivious Transfer, with the same handling of corruptions as the* PAKE *protocol under* SXDH *and the* IND-CPA *security of the encryption scheme.*

5 New Efficient Oblivious Transfer Based on QA-NIZK

5.1 Description of Their UC-Secure PAKE Scheme

We now consider the instantiation from [32] in which the UC-PAKE allows for each party to hide his own password. As before, we first recall their protocol and then show how to use it in order to instantiate a new UC-secure Oblivious Transfer. This OT scheme ends up being the most efficient to date, as we show by comparing to the recent [10] in Fig. 8.

Paper	Pre-Flow	Query	Answer
[9]	$2\,\mathbb{G}$	$9\log n + 2\,\mathbb{G}$	$n\,\mathbb{G} + 2n\,\mathbb{Z}_p$
[10]	$2\,\mathbb{G}_2$	$4\,\mathbb{G}_1 + 2\,\mathbb{G}_2$	$4n\,\mathbb{G}_2 + n\,\mathbb{Z}_p$
This paper	$2\,\mathbb{G}_2$	$4\,\mathbb{G}_1$	$n\,\mathbb{G}_2 + n\,\mathbb{Z}_p$

Fig. 8. Comparison to previous schemes

Interestingly this protocol does not use Smooth Projective Hash Function, but Quasi Adaptive Non-Interactive Zero Knowledge proofs. It is presented in Fig. 9.

CRS generation:
$g_1 \xleftarrow{\$} \mathbb{G}_1, g_2 \xleftarrow{\$} \mathbb{G}_2, a, c, o, d, e, u_1, u_2 \xleftarrow{\$} \mathbb{Z}_q, H \xleftarrow{\$} \mathcal{H}$
Set $A = g_1^a, D = g_1^d, E = g_1^e, U_1 = g_1^{u_1}, U_2 = g_1^{u_2} \in \mathbb{G}_1$
And $B = g_2^c, O = g_2^o, V_1 = g_2^{u_1 c - d - oa}, V_2 = g_2^{u_2 c - e} \in \mathbb{G}_2$
crs $:= (g_1, g_2, A, B, O, F, D, E, U_1, U_2, V_1, V_2, H)$

flow$_b$(pw; (r_b, s_b)):

1. P_b computes $R_b = g_1^{r_b}, S_b = \mathsf{pw} \cdot A^{r_b}, T_b = (D \cdot E^{h_b})^{r_b}, \hat{\rho}_b = B^{s_b}, W_b = (U_1 \cdot U_2^{h_b})^{r_b}$, where $h_b = H(\mathsf{sid}, \mathsf{ssid}, P_b, P_{\bar{b}}, R_1, S_1, \hat{\rho}_b)$
2. P_b erases r_b, sends $R_b, S_b, T_b, \hat{\rho}_b$ and keeps W_b.

Key computing:

1. P_b computes $h_{\bar{b}} = H(\mathsf{sid}, \mathsf{ssid}, P_{\bar{b}}, P_b, R_{\bar{b}}, S_{\bar{b}}, \hat{\rho}_{\bar{b}})$
   Checks that all elements are consistent, and computes:
   $\rho = g_2^{s_b}, \theta = O^{s_b}, \gamma = (V_1 \cdot V_2^{h_{\bar{b}}})^{s_b}$
   $\mathsf{sk} = e(T_{\bar{b}}, \rho) \cdot e(S_{\bar{b}}/\mathsf{pw}, \theta) \cdot e(R_{\bar{b}}, \gamma) \cdot e(W_b, \hat{\rho}_{\bar{b}})$

Fig. 9. UC secure PAKE from [32]

As explained in the original paper, the scheme is loosely-based on Quasi-Adaptive NIZK, in the sense that the flows use computation very close to what would be expected. In the Fig. 9, $\mathcal{H}$ is a family of hash functions.

5.2 Applying the Framework to Obtain a UC-Secure OT Scheme

Using the instantiation from [32] described in Fig. 9, we apply our framework and directly obtain the protocol described in Fig. 10.

CRS generation:

$g_1 \xleftarrow{\$} \mathbb{G}_1, g_2 \xleftarrow{\$} \mathbb{G}_2, a, c, o, d, e, u_1, u_2 \xleftarrow{\$} \mathbb{Z}_p, H \xleftarrow{\$} \mathcal{H}$

Sets $A = g_1^a, D = g_1^d, E = g_1^e, U_1 = g_1^{u_1}, U_2 = g_1^{u_2} \in \mathbb{G}_1$

And $B = g_2^c, O = g_2^o, V_1 = g_2^{u_1 c - d - oa}, V_2 = g_2^{u_2 c - e} \in \mathbb{G}_2$

$\mathsf{crs} := (g_1, g_2, A, B, O, F, D, E, U_1, U_2, V_1, V_2, H)$

Pre-Flow:

1. $\mathcal{S}$ generates $(\mathsf{pk} = f_1 = g_1^\alpha, \mathsf{sk} = \alpha \xleftarrow{\$} \mathbb{Z}_p)$
2. $\mathcal{S}$ erases random coins, keeps sk and sends pk.

Index query on i:

1. $\mathcal{R}$ picks $J \xleftarrow{\$} \mathbb{G}_1, s \xleftarrow{\$} \mathbb{Z}_p$, sets $\mathsf{mask} \leftarrow F(J)$ and $c = (f_1^s J, g_1^s)$
2. $\mathcal{R}$ picks $r \xleftarrow{\$} \mathbb{Z}_p$ computes $R = g_1^r, S = G(i) \cdot A^r, T = (D \cdot E^{h_{\mathcal{R},\mathcal{S}}})^r, W = (U_1 \cdot U_2^{h_{\mathcal{R},\mathcal{S}}})^r$, where $h_{\mathcal{R},\mathcal{S}} = H(\mathsf{sid}, \mathsf{ssid}, \mathcal{R}, \mathcal{S}, R, S, c)$
3. $\mathcal{R}$ erases r, s, J, sends $c, C = (R, S, T)$ and keeps W, mask.

Database answer:

1. $\mathcal{S}$ sets $J = c_1 / c_2^\alpha, \mathsf{mask} = F(J)$ and $h_{\mathcal{R},\mathcal{S}} = H(\mathsf{sid}, \mathsf{ssid}, \mathcal{R}, \mathcal{S}, R, S, c)$
2. For $k = 1, \ldots, n$:
   (a) Picks $s_k \xleftarrow{\$} \mathbb{Z}_p$
   (b) Sets $\rho_k = B^{s_k}, \rho_k' = g_2^{s_k}, \theta_k = O^{s_k}, \gamma_k = (V_1 \cdot V_2^{h_{\mathcal{R},\mathcal{S}}})^{s_k}$
   (c) Computes $K_{\mathcal{R},\mathcal{S},k} = e(T, \rho_k') \cdot e(S/G(k), \theta_k) \cdot e(R, \gamma_k)$
   (d) Sets $Q_k = L_k \oplus \mathcal{H}(K_{\mathcal{R},\mathcal{S},k} \oplus \mathsf{mask})$
3. $\mathcal{S}$ sends all $(Q_k, \rho_k)_{k \in [\![1,n]\!]}$ and erases everything else.

Data recovery:

$\mathcal{R}$ computes $K_{\mathcal{R},\mathcal{S},i} = e(W, \rho_i)$, and recovers $L_i = Q_i \oplus \mathcal{H}(K_{\mathcal{R},\mathcal{S},i} \oplus \mathsf{mask})$

Fig. 10. UC secure OT from QA-NIZK

The function G is assumed to map line numbers to group elements, and it should be collision-resistant and efficiently computable. One can use for example, the G function from Lindell in [37].

Theorem 6. *As the PAKE was UC-Secure in presence of adaptive corruption under SXDH, and ElGamal is IND-CPA under SXDH too, the Oblivious Transfer* *scheme presented in Fig. 10 is a secure 1-out-of-n Oblivious Transfer, with the same handling of corruptions as the* PAKE *protocol under* SXDH.

The PAKE scheme directly fits in the framework, hence the security is inherited from the initial PAKE proven secure in [32].

The same basic arguments apply: We first start by switching the crs so as to be able to simulate password openings, which will allow an honest receiver to commit to a dummy line value, and retroactively compute the opening value for any possible line s. On the other hand, an honest server will extract the committed line number, and provide dummy answers for all the other ones. If the receiver is honest, then the server will also send a dummy answer for the valid line, since the encrypted value S provides the extra degree of freedom to open to any line value L_k provided belatedly by the environment.

6 Conclusion

We have just presented a framework for generically transforming any PAKE protocol into an Oblivious Transfer protocol. While doing so, we achieved to instantiate the most-efficient Oblivious-Transfer UC-Secure on elliptic curves.

We also propose an Oblivious-Transfer scheme on lattices in the paper full version [12], showing that our framework is not limited to regular elliptic curve cryptography. What remains now, is to find a UC-Secure Lattice based PAKE scheme.

Acknowledgments. This work was supported in part by the French ANR EnBid (ANR-14-CE28-0003) and ID-FIX (ANR-16-CE39-0004) Projects.

References

1. Abdalla, M., Benhamouda, F., Blazy, O., Chevalier, C., Pointcheval, D.: SPHF-friendly non-interactive commitments. In: Sako, K., Sarkar, P. (eds.) ASIACRYPT 2013. LNCS, vol. 8269, pp. 214–234. Springer, Heidelberg (2013). doi:10.1007/978-3-642-42033-7_12
2. Abdalla, M., Benhamouda, F., Pointcheval, D.: Disjunctions for hash proof systems: new constructions and applications. In: Oswald, E., Fischlin, M. (eds.) EUROCRYPT 2015. LNCS, vol. 9057, pp. 69–100. Springer, Heidelberg (2015). doi:10.1007/978-3-662-46803-6_3
3. Abdalla, M., Chevalier, C., Pointcheval, D.: Smooth projective hashing for conditionally extractable commitments. In: Halevi, S. (ed.) CRYPTO 2009. LNCS, vol. 5677, pp. 671–689. Springer, Heidelberg (2009). doi:10.1007/978-3-642-03356-8_39
4. Ateniese, G., Camenisch, J., Hohenberger, S., de Medeiros, B.: Practical group signatures without random oracles. Cryptology ePrint Archive, Report 2005/385 (2005). http://eprint.iacr.org/2005/385
5. Barreto, P.S.L.M., Naehrig, M.: Pairing-friendly elliptic curves of prime order. In: Preneel, B., Tavares, S. (eds.) SAC 2005. LNCS, vol. 3897, pp. 319–331. Springer, Heidelberg (2006). doi:10.1007/11693383_22

6. Bellare, M., Pointcheval, D., Rogaway, P.: Authenticated key exchange secure against dictionary attacks. In: Preneel, B. (ed.) EUROCRYPT 2000. LNCS, vol. 1807, pp. 139–155. Springer, Heidelberg (2000). doi:10.1007/3-540-45539-6_11

7. Bellovin, S.M., Merritt, M.: Encrypted key exchange: password-based protocols secure against dictionary attacks. In: 1992 IEEE Symposium on Security and Privacy, pp. 72–84. IEEE Computer Society Press, May 1992

8. Benhamouda, F., Blazy, O., Chevalier, C., Pointcheval, D., Vergnaud, D.: New techniques for SPHFs and efficient one-round PAKE protocols. In: Canetti, R., Garay, J.A. (eds.) CRYPTO 2013. LNCS, vol. 8042, pp. 449–475. Springer, Heidelberg (2013). doi:10.1007/978-3-642-40041-4_25

9. Blazy, O., Chevalier, C.: Generic construction of UC-secure oblivious transfer. In: Malkin, T., Kolesnikov, V., Lewko, A.B., Polychronakis, M. (eds.) ACNS 2015. LNCS, vol. 9092, pp. 65–86. Springer, Cham (2015). doi:10.1007/978-3-319-28166-7_4

10. Blazy, O., Chevalier, C.: Structure-preserving smooth projective hashing. In: Cheon, J.H., Takagi, T. (eds.) ASIACRYPT 2016. LNCS, vol. 10032, pp. 339–369. Springer, Heidelberg (2016). doi:10.1007/978-3-662-53890-6_12

11. Blazy, O., Chevalier, C., Germouty, P.: Adaptive oblivious transfer and generalization. In: Cheon, J.H., Takagi, T. (eds.) ASIACRYPT 2016. LNCS, vol. 10032, pp. 217–247. Springer, Heidelberg (2016). doi:10.1007/978-3-662-53890-6_8

12. Blazy, O., Chevalier, C., Germouty, P.: Almost optimal oblivious transfer from QA-NIZK. Cryptology ePrint Archive, Report 2017/358 (2017). http://eprint.iacr.org/2017/358

13. Blazy, O., Fuchsbauer, G., Izabachène, M., Jambert, A., Sibert, H., Vergnaud, D.: Batch Groth–Sahai. In: Zhou, J., Yung, M. (eds.) ACNS 2010. LNCS, vol. 6123, pp. 218–235. Springer, Heidelberg (2010). doi:10.1007/978-3-642-13708-2_14

14. Blazy, O., Kiltz, E., Pan, J.: (Hierarchical) identity-based encryption from affine message authentication. In: Garay, J.A., Gennaro, R. (eds.) CRYPTO 2014. LNCS, vol. 8616, pp. 408–425. Springer, Heidelberg (2014). doi:10.1007/978-3-662-44371-2_23

15. Boneh, D., Franklin, M.: Identity-based encryption from the Weil pairing. In: Kilian, J. (ed.) CRYPTO 2001. LNCS, vol. 2139, pp. 213–229. Springer, Heidelberg (2001). doi:10.1007/3-540-44647-8_13

16. Canetti, R.: Universally composable security: a new paradigm for cryptographic protocols. In: 42nd FOCS, pp. 136–145. IEEE Computer Society Press, October 2001

17. Canetti, R., Cohen, A., Lindell, Y.: A simpler variant of universally composable security for standard multiparty computation. In: Gennaro, R., Robshaw, M. (eds.) CRYPTO 2015. LNCS, vol. 9216, pp. 3–22. Springer, Heidelberg (2015). doi:10.1007/978-3-662-48000-7_1

18. Canetti, R., Dachman-Soled, D., Vaikuntanathan, V., Wee, H.: Efficient password authenticated key exchange via oblivious transfer. In: Fischlin, M., Buchmann, J., Manulis, M. (eds.) PKC 2012. LNCS, vol. 7293, pp. 449–466. Springer, Heidelberg (2012). doi:10.1007/978-3-642-30057-8_27

19. Canetti, R., Halevi, S., Katz, J., Lindell, Y., MacKenzie, P.: Universally composable password-based key exchange. In: Cramer, R. (ed.) EUROCRYPT 2005. LNCS, vol. 3494, pp. 404–421. Springer, Heidelberg (2005). doi:10.1007/11426639_24

20. Canetti, R., Krawczyk, H.: Universally composable notions of key exchange and secure channels. In: Knudsen, L.R. (ed.) EUROCRYPT 2002. LNCS, vol. 2332, pp. 337–351. Springer, Heidelberg (2002). doi:10.1007/3-540-46035-7_22

21. Canetti, R., Lindell, Y., Ostrovsky, R., Sahai, A.: Universally composable two-party and multi-party secure computation. In: 34th ACM STOC, pp. 494–503. ACM Press, May 2002
22. Choi, S.G., Katz, J., Wee, H., Zhou, H.-S.: Efficient, adaptively secure, and composable oblivious transfer with a single, global CRS. In: Kurosawa, K., Hanaoka, G. (eds.) PKC 2013. LNCS, vol. 7778, pp. 73–88. Springer, Heidelberg (2013). doi:10.1007/978-3-642-36362-7_6
23. Cramer, R., Shoup, V.: Universal hash proofs and a paradigm for adaptive chosen ciphertext secure public-key encryption. In: Knudsen, L.R. (ed.) EUROCRYPT 2002. LNCS, vol. 2332, pp. 45–64. Springer, Heidelberg (2002). doi:10.1007/3-540-46035-7_4
24. Escala, A., Groth, J.: Fine-tuning Groth-Sahai proofs. In: Krawczyk, H. (ed.) PKC 2014. LNCS, vol. 8383, pp. 630–649. Springer, Heidelberg (2014). doi:10.1007/978-3-642-54631-0_36
25. Garay, J.A., Wichs, D., Zhou, H.-S.: Somewhat non-committing encryption and efficient adaptively secure oblivious transfer. In: Halevi, S. (ed.) CRYPTO 2009. LNCS, vol. 5677, pp. 505–523. Springer, Heidelberg (2009). doi:10.1007/978-3-642-03356-8_30
26. Gennaro, R., Lindell, Y.: A framework for password-based authenticated key exchange. In: Biham, E. (ed.) EUROCRYPT 2003. LNCS, vol. 2656, pp. 524–543. Springer, Heidelberg (2003). doi:10.1007/3-540-39200-9_33
27. Ghadafi, E., Smart, N.P., Warinschi, B.: Groth–Sahai proofs revisited. In: Nguyen, P.Q., Pointcheval, D. (eds.) PKC 2010. LNCS, vol. 6056, pp. 177–192. Springer, Heidelberg (2010). doi:10.1007/978-3-642-13013-7_11
28. Groth, J., Sahai, A.: Efficient non-interactive proof systems for bilinear groups. In: Smart, N. (ed.) EUROCRYPT 2008. LNCS, vol. 4965, pp. 415–432. Springer, Heidelberg (2008). doi:10.1007/978-3-540-78967-3_24
29. Horvitz, O., Katz, J.: Universally-composable two-party computation in two rounds. In: Menezes, A. (ed.) CRYPTO 2007. LNCS, vol. 4622, pp. 111–129. Springer, Heidelberg (2007). doi:10.1007/978-3-540-74143-5_7
30. Jutla, C.S., Roy, A.: Shorter quasi-adaptive NIZK proofs for linear subspaces. In: Sako, K., Sarkar, P. (eds.) ASIACRYPT 2013. LNCS, vol. 8269, pp. 1–20. Springer, Heidelberg (2013). doi:10.1007/978-3-642-42033-7_1
31. Jutla, C.S., Roy, A.: Switching lemma for bilinear tests and constant-size NIZK proofs for linear subspaces. In: Garay, J.A., Gennaro, R. (eds.) CRYPTO 2014. LNCS, vol. 8617, pp. 295–312. Springer, Heidelberg (2014). doi:10.1007/978-3-662-44381-1_17
32. Jutla, C.S., Roy, A.: Dual-system simulation-soundness with applications to UC-PAKE and more. In: Iwata, T., Cheon, J.H. (eds.) ASIACRYPT 2015. LNCS, vol. 9452, pp. 630–655. Springer, Heidelberg (2015). doi:10.1007/978-3-662-48797-6_26
33. Katz, J., Vaikuntanathan, V.: Smooth projective hashing and password-based authenticated key exchange from lattices. In: Matsui, M. (ed.) ASIACRYPT 2009. LNCS, vol. 5912, pp. 636–652. Springer, Heidelberg (2009). doi:10.1007/978-3-642-10366-7_37
34. Katz, J., Vaikuntanathan, V.: Round-optimal password-based authenticated key exchange. In: Ishai, Y. (ed.) TCC 2011. LNCS, vol. 6597, pp. 293–310. Springer, Heidelberg (2011). doi:10.1007/978-3-642-19571-6_18
35. Kiltz, E., Wee, H.: Quasi-adaptive NIZK for linear subspaces revisited. In: Oswald, E., Fischlin, M. (eds.) EUROCRYPT 2015. LNCS, vol. 9057, pp. 101–128. Springer, Heidelberg (2015). doi:10.1007/978-3-662-46803-6_4

36. Libert, B., Peters, T., Joye, M., Yung, M.: Non-malleability from malleability: simulation-sound quasi-adaptive NIZK proofs and CCA2-secure encryption from homomorphic signatures. In: Nguyen, P.Q., Oswald, E. (eds.) EUROCRYPT 2014. LNCS, vol. 8441, pp. 514–532. Springer, Heidelberg (2014). doi:10.1007/978-3-642-55220-5_29

37. Lindell, Y.: Highly-efficient universally-composable commitments based on the DDH assumption. In: Paterson, K.G. (ed.) EUROCRYPT 2011. LNCS, vol. 6632, pp. 446–466. Springer, Heidelberg (2011). doi:10.1007/978-3-642-20465-4_25

38. Naor, M., Pinkas, B.: Efficient oblivious transfer protocols. In: Kosaraju, S.R. (ed.) 12th SODA, pp. 448–457. ACM-SIAM, January 2001

39. Nguyen, M.-H.: The relationship between password-authenticated key exchange and other cryptographic primitives. In: Kilian, J. (ed.) TCC 2005. LNCS, vol. 3378, pp. 457–475. Springer, Heidelberg (2005). doi:10.1007/978-3-540-30576-7_25

40. Peikert, C., Vaikuntanathan, V., Waters, B.: A framework for efficient and composable oblivious transfer. In: Wagner, D. (ed.) CRYPTO 2008. LNCS, vol. 5157, pp. 554–571. Springer, Heidelberg (2008). doi:10.1007/978-3-540-85174-5_31

41. Rabin, M.O.: How to exchange secrets with oblivious transfer. Technical report TR81, Harvard University (1981)

OnionPIR: Effective Protection of Sensitive Metadata in Online Communication Networks

Daniel Demmler, Marco Holz$^{(\boxtimes)}$, and Thomas Schneider

Technische Universität Darmstadt, Darmstadt, Germany
{daniel.demmler,thmomas.schneider}@cysec.de, onionpir@marcoholz.de

Abstract. While great effort has been put into securing the *content* of messages transmitted over digital infrastructures, practical protection of *metadata* is still an open research problem. Scalable mechanisms for protecting users' anonymity and hiding their social graph are needed. One technique that we focus on in this work is private information retrieval (PIR), an active field of research that enables private querying of data from a public database without revealing which data has been requested and a fundamental building block for private communication. We introduce two significant improvements for the multi-server scheme RAID-PIR (ACM CCSW'14): precomputing queries using the *Method of four Russians* and optimizing the database layout for parallel queries. We then propose *OnionPIR*, an anonymous messaging service as example application for PIR combined with onion routing that prevents the leakage of communication meta-data. By providing and evaluating a prototype, we show that OnionPIR is usable in practice. Based on our results, we conclude that it is possible to build and deploy such a service today, while its operating expenses are within the order of magnitude of those of traditional messaging services that leak metadata.

Keywords: Private information retrieval · Tor · Privacy · Meta-data protection

1 Introduction

Communication has never been more important for our society than it is today. Large parts of our infrastructure depend on digital computer networks and would collapse when online communication suddenly were not possible anymore. With the rise of instant messengers and the availability of mobile devices, communication shifted from the offline world to online communication platforms. End-to-end encryption is nowadays often deployed to protect the content of exchanged messages such that it cannot be read by an adversary. However, these digital platforms produce large amounts of metadata, i.e., who communicates with whom at what time. Electronic mass surveillance programs strengthen the need for systems providing communication channels without leaking metadata, which has shown to be of tremendous value [Lan15,MMM16], e.g., for people living in

© Springer International Publishing AG 2017
D. Gollmann et al. (Eds.): ACNS 2017, LNCS 10355, pp. 599–619, 2017.
DOI: 10.1007/978-3-319-61204-1_30

suppressive regimes, for whom trying to contact government-critical organizations can be a tremendous risk. The right to remain anonymous is a fundamental right in our modern society.

Developing systems that protect the users' privacy and at the same time scale to a large number of users is challenging and a very active research field. Private information retrieval (PIR) can be used as a fundamental building block for those systems and also has many other applications, e.g., private querying of entries from patent registers or medical databases, or improving the scalability of Tor, as proposed in PIR-Tor [MOT+11].

Moreover, building blocks of private and untraceable communication services could potentially also be reused in other privacy-critical applications that are currently under active research, such as electronic voting systems [BV14] or privacy-preserving location-based services [MCA06, HCE11, DSZ14].

Outline and Our Contributions. In Sect. 2, we propose and implement two optimizations for the multi-server PIR scheme RAID-PIR [DHS14] that lead to significant performance improvements. The optimization reduces the server's computation time by using the Method of four Russians for precomputation (Sect. 2.2), thus increasing the overall throughput of the system. The second optimization targets the process of privately retrieving multiple database entries within a single PIR query by adjusting the database layout to increase the number of entries that can be queried in parallel (Sect. 2.3). After introducing multiple generic implementation enhancements (Sect. 2.4), we perform benchmarks to evaluate our optimizations and measure speedups of up to factor 7x over RAID-PIR for many cases when querying through a WAN connection and a performance close to the theoretical optimum when using a DSL connection for the queries (Sect. 2.5).

Based on our improved implementation of PIR, we propose *OnionPIR*, a novel anonymous communication system that combines PIR with onion routing to create a scalable way to privately exchange messages (Sect. 3). OnionPIR is designed to be ready for deployment and provides a way to establish a secure communication channel without the need of exchanging any type of cryptographic keys out-of-band. We built a proof-of-concept implementation and our evaluation demonstrates its practical performance. We will publish our implementations as open-source software upon acceptance of this paper.

Finally, we provide an overview and comparison with related work in Sect. 4.

2   Private Information Retrieval and Improvements

Private information retrieval (PIR) is an active research topic and enables the querying of information from one or multiple servers without disclosing which information was requested. Since this technique is a fundamental building block for higher-level protocols and applications, its performance is of prime importance. In this section, we briefly summarize RAID-PIR [DHS14] and introduce and implement two significant performance improvements.

2.1 PIR Background

PIR protocols can be grouped into information-theoretic schemes and computational schemes. RAID-PIR [DHS14], the work that we base our system on, is a multi-server scheme based on the original information-theoretic design by Chor et al. [CKGS95]. Its security guarantees rely on a non-collusion assumption between multiple servers. By using fast XOR operations, it achieves great performance. The alternatives are single-server PIR schemes that rely on computational assumptions, as first introduced in [CGN98] and soon followed by another approach [KO97]. These schemes are typically based on computationally expensive, partially homomorphic encryption. A recent scheme using lattice-based cryptography is presented in [AMBFK16]. Both approaches were combined in [DG14] in a design called hybrid PIR.

Notation. Throughout the paper we use the following notation: The database that may be queried by clients is divided into B blocks of size b bits each and distributed to k servers. When a client wants to query the n-th block from the database, it generates $k - 1$ random bitstrings, called queries, of length B and constructs the k-th bitstring in a way that the XOR of all k queries results in a bitstring where all bits except the n-th are zero. Each of the k queries is then sent to a different server which will XOR all blocks whose corresponding bit i is set in the query. Each server returns the resulting block of XORs of length b bits. To recover the n-th block, the clients computes the XOR of all received blocks.

$$
\begin{array}{c|c|c|c|c|}
q_1 & 11010 & 01010 & 11101 & 01001 \\
q_2 & 10110 & 01101 & 10101 & 00111 \\
q_3 & 01100 & 10001 & 01110 & 10110 \\
q_4 & 00100 & 10110 & 00110 & 11000 \\
\hline
e_3 & 00100 & 00000 & 00000 & 00000 \\
\end{array}
$$

Fig. 1. RAID-PIR Query. The four queries q_i that are sent to the $k = 4$ servers consist of one (orange) *flip chunk* and three (black) randomly generated chunks so that the XOR of all queries is the plaintext query e_3 that retrieves the third database block. (Color figure online)

RAID-PIR improves this scheme by splitting each query into k chunks, as shown in Fig. 1. The queries are constructed in a way that all chunks sent to the l-th server, except the l-th chunk are generated from a seed using a pseudo random generator (PRG). In addition, a redundancy parameter r is introduced as trade-off between security and performance and chosen such that $2 \leq r \leq k$. It allows reducing the downstream bandwidth and server-side computational costs. For that matter, each server will only handle r chunks of the database. This also reduces the number of servers that have to collude in order to retrieve the plaintext of a query from k to r.

Another optimization provides the ability to request multiple blocks of the database within a single query. This is achieved by XORing the requested blocks only within a chunk at the server side and returning one block per chunk instead of one block per query. However, this optimization has the limitation that the requested blocks have to be in different chunks of the database.

In the following sections, we introduce two additional improvements to RAID-PIR that further increase its efficiency.

The first one speeds up the generation of the PIR responses at the server-side by using precomputations (Sect. 2.2), while the second one achieves a significant speedup when querying multiple blocks in parallel by using a database layout where blocks are uniformly distributed (Sect. 2.3).

2.2   Method of Four Russians

The *Method of four Russians* [ADKF70], is a technique for matrix multiplication with a limited number of possible values per entry. Since RAID-PIR makes heavy usage of such kinds of multiplications, this method can be used to reduce the computation time of the RAID-PIR servers for generating responses for PIR queries. This leads to lower latency and higher throughput and comes at the (low) cost of a preprocessing phase for the servers and increased memory requirements.

The Method of four Russians bases on the assumption that the number of possible values for the cells of a matrix is finite. Here, it is assumed that all matrices in the multiplication $X \cdot Y = Z$ are done over the field with two elements $(\mathbb{F}_2)$, i.e. only binary elements are being multiplied. Note, that this property is not required for the algorithm (i.e., it is also applicable to numbers in $\mathbb{Z}$) but it reduces the complexity of the algorithm and is sufficient for our use case. In $\mathbb{F}_2$, the multiplicative operation is AND and the additive operation is XOR.

Crucial for understanding the idea behind the four Russians algorithm, is the fact that the indices of all non-zero elements in row r in the first matrix X indicate a subset of rows in the second matrix Y that have to be XORed in order to produce the r-th row of the output matrix Z. As a naïve approach, it would now be possible to precompute all possible XOR combinations of all rows in Y, hereby creating a lookup table (LUT) that returns the final results for the rows of Z. In other words, a lookup could be performed using each row in X as key to the LUT, returning the corresponding row in Z. This would reduce the computational complexity from $\mathcal{O}(n^3)$ to $\mathcal{O}(n)$ assuming the lookup is done in constant time for matrices of size $n \times n$. Unfortunately, this naïve approach is impractical due to the exponential computational and memory complexity of the precomputation of $\mathcal{O}(n \cdot 2^n)$, which is worse than the regular approach for $n \geq 4$.

The reason why only n XOR operations have to be performed for each of the 2^n possible combinations of rows of the matrix Y is, that it is possible to use a Gray code to reorder the possible combinations in a way that two consecutive combinations only differ by a single bit that corresponds to only one row of Y. For each of the 2^n combinations, it is now sufficient to XOR the last precomputed result with the row of Y that changed in the Gray code. The binary Gray code g of a number m can be calculated as $g = (m \oplus (m \gg 1))$. In Fig. 2 we depict the calculation of LUT entries for a given example matrix Y^*.

$$Y^* = \begin{bmatrix} 0\ 1\ 1\ 1 \\ 1\ 1\ 0\ 0 \\ 1\ 0\ 0\ 1 \\ 1\ 0\ 1\ 1 \end{bmatrix}$$

Gray code	decimal	result
0000	0	0000
0001	1	$1011 = 0000 \oplus Y^*[4,:]$
0011	3	$0010 = 1011 \oplus Y^*[3,:]$
0010	2	$1001 = 0010 \oplus Y^*[4,:]$
0110	6	$0101 = 1001 \oplus Y^*[2,:]$
...	...	...

Fig. 2. Example LUT precomputation for matrix Y^*. $Y^*[n,:]$ denotes the n-th row of matrix Y^*.

Instead of precomputing a LUT for the full matrix Y, it is divided into groups of $t \in \mathbb{N}$ rows each. The precomputation is then done within these n/t groups using the Gray code as described above, resulting in a computational complexity of $\mathcal{O}(2^t \cdot n^2/t)$. If $t = \log_2(n)$ the complexity simplifies to $\mathcal{O}(n^3/\log_2(n))$, resulting in a speedup of t, compared to traditional matrix multiplication, for all queries that will be answered after the precomputation phase.

While the matrix Y is divided into groups horizontally, the matrix X has to be divided into vertical groups of t columns. To multiply the two $n \times n$ matrices, X is now traversed column-wise, grouped by t columns forming a group each. For each group, the corresponding LUT can now be created by t rows of Y and a lookup can be performed for all t-bit parts of the n rows of X in this group. This procedure is depicted in Fig. 3. We note that this optimization is generic and can potentially be used in any PIR scheme in which queries can be expressed as matrix-matrix or vector-matrix multiplication, e.g., [CKGS95, LG15, Hen16].

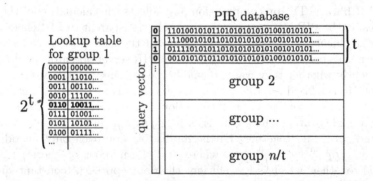

Fig. 3. Precomputation in the PIR database for $t = 4$. A query vector is one *row* in the matrix X. The first 4 bits of the query vector represent the index in the LUT for the first group of Y (the PIR database).

Implementation: An efficient implementation of the Method of four Russians is provided in [ABH10]. However, since it only offers full matrix-matrix multiplications and uses different data structures than RAID-PIR, we implemented our own version and directly integrate it into the RAID-PIR library[1] to avoid costly memory movements and exploit memory locality. While the traditional Method of four Russians assumes the multiplication of two full matrices, this is not the case in RAID-PIR. Often, only a single PIR request has to be answered, making X effectively a vector. Several PIR queries could be batched to create matrix X at the cost of increased latency of the individual queries, similar to [LG15].

In our implementation we do the LUT precomputation once while loading the database and store the LUTs in main memory. For every query we achieve a theoretical speedup of t. The downside of this approach is the increased memory requirement. Without loss of generality, we now assume that Y is a RAID-PIR database of size $B \times b$, i.e., B blocks of b bits each. While a larger t decreases the computation time for the generation of response vectors, it also excessively increases the memory needed to store the LUTs. When t blocks of the PIR database are put in a group, only B/t XOR operations have to be performed per query, resulting in a speedup of factor t. On the other hand, each of the B/t LUTs requires $2^t \cdot b$ bits of memory, where b is the database block size. The choice of $t = 4$ gives a good trade-off between theoretical speedup and memory requirements for larger databases (see Table 1) and is used in our implementation.

Table 1. Comparison of speedup and memory for the Method of four Russians

t (speedup)	2	3	4	5	6	8
$\frac{1}{t} \cdot 2^t$ (memory overhead)	2.00	2.66	4.00	6.40	10.67	32.00

The optimizations were implemented in C and integrated into the existing RAID-PIR library. The precomputation only affects the calculation of the XOR responses by the PIR servers and is completely transparent to PIR clients. This means that clients do not have to be aware of the changes and can send the same queries as before. Furthermore, different PIR server operators could decide independently whether they want to enable the precomputation or not.

In [LG15] a related approach to speedup server computation, based on the Strassen algorithm for matrix multiplication is introduced. This comes at the expense of higher latency for each individual query, as multiple queries are collected and processed together. Furthermore, this approach only introduces a speedup of $q/q^{0.80735} = q^{0.19265}$, where q is the number of queries that are processed together, which is less efficient than our approach (constant speedup t) when q is small (e.g., for $q < 1334$ and $t = 4$).

2.3 Uniform Distribution of Data Entries

In the following we present an optimization that is specific to so-called multi-block (MB) queries as proposed in RAID-PIR, that enable a client to privately request more than one block in a single query. The improvements of RAID-PIR,

---

[1] https://github.com/encryptogroup/RAID-PIR.

compared to the original CKGS scheme [CKGS95], base on the fact that the PIR blocks are grouped into k chunks that divide the database into equally sized parts. PIR servers receiving an MB query will reply with one block per chunk. The XOR of the corresponding blocks is calculated as before, but only within the boundaries of these chunks. Since only one block per chunk can be queried in one multi-block query, the blocks to retrieve have to be located in different chunks. The performance analysis of RAID-PIR showed that no speedup could be achieved for a large consecutive file residing in a single chunk. However, a large speedup could be measured when querying many smaller files that were distributed among the database. In RAID-PIR, files larger than the block size b are simply placed in adjacent blocks to build the database. While this does not affect the performance of a query when retrieving the blocks consecutively, querying multiple blocks in parallel is often not possible, as they most likely happen to be located in the same chunk.

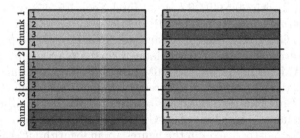

Fig. 4. Uniform distribution of the data entries in the PIR database. While the first file (green) filled up only the first chunk in the original RAID-PIR database layout (left), it is now uniformly distributed over the whole database (right). (Color figure online)

Implementation: To improve the performance of multi-block queries for files larger than the block size b, we change the layout of the database. Instead of placing files in consecutive blocks, they are now uniformly distributed over the whole database, as shown in Fig. 4. For each file, the number of blocks n needed to store its contents is calculated. The file is then placed such that between two of its blocks $B/n - 1$ blocks are used for other files, where B is the total number of blocks. If a chosen block is already assigned to another file, the next possible free block of the database is used instead. The first bytes of the next file will then fill the rest of the previous file's last block. It is guaranteed that all B blocks of the database contain data and no block, except the last, one is only partially filled. Since both the PIR servers and the clients have to know the new locations of the database entries, both client-side and server-side code is extended to support the new layout.

2.4   Generic Implementation Improvements

We also introduced several implementation improvements in RAID-PIR, that are included in all measurements in Sect. 2.5. We pipelined the communication on the client side and decoupled the sending of queries and reception of answers. Furthermore, only a single seed is sent from the client to the servers during a session and a state is kept while the connection is alive. We also introduced SSE2 intrinsics on very wide data types. This feature is available in all x86 CPUs since several years and leads to a more efficient bit-wise XOR computation, which is one of the most crucial operations in RAID-PIR for both clients and PIR servers.

2.5   PIR Benchmark Results

In this section, we evaluate the performance of our modified implementation of RAID-PIR [DHS14]. All measurements include the generic optimizations from Sect. 2.4. We show the influence of the optimizations from Sects. 2.2 and 2.3.

Benchmark Setting. Three PIR servers were deployed as `r3.xlarge` instances on Amazon EC2 with 30.5 GiB RAM, an Intel Xeon E5-2670 v2 processor and a 1 Gbit/s ethernet connection. The PIR queries were performed by a `t2.micro` instance with 1 GiB RAM, one core of an Intel Xeon E5-2676 v3 processor and a 250 Mbit/s ethernet connection (0.5 ms latency) in the WAN setup and a notebook with 8 GiB RAM, an Intel Core i3-3120M processor and a consumer grade DSL Internet connection (4.5 Mbit/s downstream, 400 kbit/s upstream, 30 ms latency) in the DSL setup. The used database consists of 964 files of Ubuntu security updates adding up to a total size of 3.8 GiB. All results are average runtimes of 5 iterations. Note that the y-axes in the figures are in logarithmic scale.

PIR Server Startup Duration. The time to load the database into the PIR servers' RAM is mostly independent of the block size and took ≈40 s for the database of 3.8 GiB. The four Russian precomputation took between 5 s and 7 s. The startup time scales linearly with the database size.

File Query via WAN. In the remainder of this section, the number of servers is set to $k = 3$ and the redundancy parameter is $r = 2$. The block size is varied from $b = 16$ kiB to $b = 4$ MiB.

Small File Query via WAN: The results for querying 10 small files adding up to 2.9 MiB in the WAN setup are depicted in the left part of Fig. 5. Single-block queries are abbreviated as SB, multi-block queries as MB, the Method of four Russians as 4R and the uniform distribution of the data entries as UD. Here, the multi-block queries are significantly faster than single-block queries even when no optimizations are applied since the requested files are already distributed among the whole database. Therefore, uniform distribution of the data entries does not have a significant impact on the overall performance. Four Russians precomputation improves the runtimes by factor 2 and does not introduce improvements any more when the cache size of the processor is not sufficient for larger block sizes. For a block size of $b = 16$ kiB, the runtime decreases from 10.1 s for the

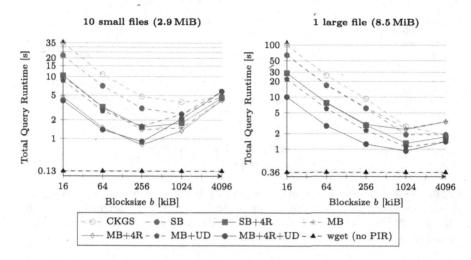

Fig. 5. WAN Benchmarks: Runtimes for varying block size b with $k = 3$ servers, redundancy parameter $r = 2$ for 10 small files (2.9 MiB, left) and a large file (8.5 MiB, right). CKGS: [CKGS95], SB: Single-Block scheme, MB: Multi-Block scheme, 4R: Four Russians precomputation, UD: Uniform distribution of data entries. DB size 3.8 GiB.

originally best performing multi-block queries to 4.1 s when all optimizations are applied. The best performance is obtained for a block size of $b = 256$ kiB where 7 queries are performed to retrieve 18 blocks in 0.8 s.

Large File Query via WAN: In the right part of Fig. 5, one large 8.5 MiB file is requested. As already observed in [DHS14], single-block and multi-block queries take similar time when querying one large file in the WAN setup, where bandwidth and latency typically are not the bottleneck. The four Russians precomputation leads to a significant speedup and effectively improves the runtime of most test cases by factor 2. This is a good result even though the theoretical speedup, that does not cover data locality and communication overhead, is 4. When the CPU cache size is too small to store the precomputed LUTs and all interim results for large block sizes, the Method of four Russians' speedup decreases and is not measurable any more for $b = 4$ MiB.

The speedup gained by uniformly distributing the data entries in the database is even greater and improves the overall runtime by approximately a factor of 3 compared to the original multi-block queries for small block sizes. While all blocks have to be queried from the first chunk for a non-uniformly distributed database, it is now possible to retrieve 3 blocks from 3 different chunks in one multi-block query in parallel. When both optimizations are combined, the runtime decreases from 64 s (or 100 s for the original CKGS scheme) to 10 s for block size $b = 16$ kiB and reaches its minimum for $b = 1$ MiB where the runtime decreases from 1.9 s for the originally best performing single-block queries to 0.9 s for multi-block queries using both optimizations.

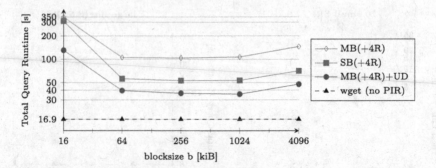

Fig. 6. Large File, DSL: Runtimes for varying block sizes b with $k = 3$ servers, redundancy parameter $r = 2$ and one large file (8.5 MiB). CKGS: original PIR scheme [CKGS95], SB: Single-Block scheme, MB: Multi-Block scheme, 4R: Four Russians precomputation, UD: Uniform distribution of data entries. DB size 3.8 GiB.

Large File Query via DSL. In Fig. 6, the results for querying one large file (8.5 MiB, as in the first test case) over the consumer-grade DSL connection are depicted. The first interesting observation is that the results for single- and multi-block queries with a block size of $b = 16$ kiB do not differ significantly. The reason for this is the low upstream bandwidth of the DSL connection. For small block sizes, the required upstream bandwidth, which depends on the number of blocks, has a larger impact than the downstream bandwidth, which depends on the block size. To query a database containing $B = 246360$ blocks of 16 kiB each, $B/3$ bits have to be transferred to each server (plus additional 16 Bytes for the PRG seeds) resulting in a total transmission time of $3 \cdot (B/3 \text{ bit} + 16 \text{ Byte})/(400 \text{ kbit/s}) \approx 602 \text{ ms}$. This phenomenon can also be observed in the benchmarks run in [DHS14].

Due to this high latency and communication overhead of the PIR queries, the server-side four Russians precomputation does not significantly affect the overall runtime of a client's request. As in the first test case, single-block queries provide better performance than multi-block queries without the uniform distribution. However, when the data entries are distributed uniformly, a significant speedup to multi-block queries can be observed and – as already expected – the multi-block queries supersede the single-block approach because the number of requests reduces by about a third. It was also shown in [DHS14] that larger block sizes have the disadvantage that large parts of some requested blocks contain no relevant data and therefore lead to a slower runtime.

For a redundancy parameter of $r = 2$, the data that needs to be sent to the client is about twice the size of the raw data. The best results are achieved for $b = 1$ MiB using both new optimizations. Retrieving the file without PIR, using wget over an unencrypted HTTP connection takes 16.9 s and is only 2.1 times faster, indicating that the runtime of 35.5 s for the PIR approach almost reached the theoretically optimal results for the DSL setup, as we need twice the communication. When the network is the bottleneck, computation has only a minor influence on total runtime. Our results also show that the previously assumed optimal block size of $b = B = \sqrt{\text{DB size}}$ does not lead to good performance, especially in the DSL setting with asymmetric up- and downstream bandwidth. For a database size of 3.8 GiB that would be ≈22 kiB block size, which our results show to be far worse than larger block sizes.

Query Time for Varying Database Size. To demonstrate how the size of the database influences the query runtime we performed the same queries as in Sect. 2.5 using multi block queries with four Russians precomputation and uniformly distributed database blocks. We varied the size of the database and depict our results in Fig. 7. The time that the clients needs to generate the queries to the servers is almost constant and always around or below 50 ms. The communication varies due to queries being sent over a public Internet connection and also remains mostly constant between 200 ms and 300 ms. This is due to the fact, that even though the database and thus the number of blocks increases, the query size only grows from 250 Bytes to 4 kiB, which is negligible compared to the block size $b = 256$ kiB, that the client receives. Server computation time and thus total time increase linearly with the database size.

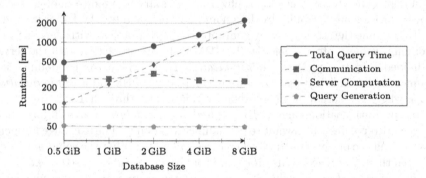

Fig. 7. Varying DB size: Runtimes for varying database size, $k = 3$ servers, redundancy parameter $r = 2$ for querying one large file (8.5 MiB) and block size $b = 256$ kiB. All queries are multi-block (MB) queries using four Russians precomputation and uniformly distributed database blocks. Note the logarithmic scale of the y-axis.

3   OnionPIR

As an application for our previously presented improvements, we introduce a private communication system called *OnionPIR* for anonymous communication. We use RAID-PIR with our optimizations as a building block for public key distribution, and onion routing to get a system efficient enough for practical use.

3.1   Motivation

When two parties register under pseudonymous identities to a classical messaging service and connect to the service by using Tor [DMS04], a malicious server can link those two pseudonyms together. Even if these pseudonyms do not reveal information about the users behind them, it is possible to build a social graph isomorphic to the one built with information from other sources. These two graphs can then be mapped together to reveal information about users. Therefore, Tor alone is not enough to provide protection against metadata leakage. Additionally, it has turned out that the assumption of users willing to establish a shared secret is problematic. In order to provide protection against mass surveillance, a system to establish private communication channels based on already existing contact information such as phone numbers or email addresses is needed.

3.2 System Model and Goals

Existing privacy-preserving communication systems do often not scale for large numbers of users or require the exchange of a shared secret out-of-band, which is error-prone and leads to usability issues most users are not willing to accept. The success of the popular messaging app Signal[2] and the adaption of its protocol in WhatsApp and Facebook Messenger are significantly founded on the combination of strong security and great usability. Since most users do not have an in-depth knowledge of cryptography, it is necessary to build technologies that provide privacy and usability by design and give reasonable defaults for its users.

To combine the strong privacy guarantees of PIR protocols with the efficiency of onion routing protocols, such as Tor [DMS04], the interaction with the server is divided into two phases: an *initialization phase* where the communication channels are established and keys are exchanged via PIR and a *communication phase* where the actual message exchange takes place through Tor.

In the initialization phase, PIR is used to *privately* exchange information between two parties which want to communicate securely. The exchange happens in a way that no one except these two parties will find out that the communication between them happened. This procedure could theoretically be used to exchange all kinds of data. However, when it comes to practical applicability, querying large amounts of information via PIR does not scale well for a larger number of users. Hence, in the communication phase no PIR techniques will be used.

Instead, the information retrieved from PIR is then used to place messages, encrypted using Authenticated Encryption with Associated Data (AEAD), in an anonymous inbox, called "dead drop". This dead drop can only be read and written at the OnionPIR control server in the communication phase by clients that know its identifier. By using onion routing to hide the identity of the communication partners, the server is not able to determine who sent and received messages. For real-time communication, the users could also establish direct connections using TCP streams or web sockets [FM11] through the server.

The OnionPIR system, depicted in Fig. 8, serves clients who want to communicate with each other and two types of servers. All honest clients form the anonymity set among which a user is anonymous. That means that a potential

---

[2] downloaded by 1–5 Mio. users according to the Google Play Store, https://play.google.com/store/apps/details?id=org.thoughtcrime.securesms.

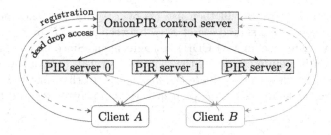

Fig. 8. OnionPIR system. The OnionPIR control server handles user registrations, distributes users' public keys to the PIR servers and holds the dead drops. Clients perform PIR queries to privately receive other users' public keys from the PIR servers. Clients connect to the OnionPIR control server directly or via Tor (dashed lines).

adversary cannot determine which users within this set communicate with each other. The central OnionPIR control server handles user registration and serves as a content provider for the PIR servers. It also acts as a database for the "dead drops" in the communication phase. The PIR servers are used only in the initialization phase to privately distribute the public keys of clients.

3.3 Protocol Description

In this section we describe the OnionPIR protocol. A (simplified) version is depicted in Fig. 9. When a client A registers for the service, it first runs through an account verification process and sends its public key to the OnionPIR control server which will later distribute it to the PIR servers. Another client B, that has an address book entry for A (e.g., a mobile number or e-mail address), will later *privately* query A's public key via PIR. In addition, each user periodically queries for his or her own public key to make sure the OnionPIR control server does not distribute bad keys to the PIR servers (see Sect. 3.4 for details). Note, that a single PIR server is never able to replace keys unless it colludes with other PIR servers because the responses of multiple servers will be combined to

Client A	OnionPIR	Client B
store PK_A →		← store PK_B
authenticated access		authenticated access
		query PK_A →
← query PK_B		via PIR
via PIR		
derive $ID_{A \to B}$	communication phase	.... derive $ID_{A \to B}$
write to dead drop $ID_{A \to B}$ →		
via Tor		read dead drop $ID_{A \to B}$ →
		via Tor

Fig. 9. Simplified OnionPIR protocol. The registration, dead drop and PIR servers were abstracted into one server. Key renewal and the answer of client B are not shown.

retrieve the public key. B will then use his own private key and the received public key to generate a shared secret K_{AB} between A and B by performing an Elliptic Curve Diffie-Hellman (ECDH) key agreement. Since no communication with the server is required to derive the shared secret, this type of key agreement protocols is often also called *private key agreement*. In the same way, A is also able to derive the shared secret K_{AB} using her own private key and the public key of B.

Symmetric Keys. The shared secret K_{AB} and both parties' public keys are used to derive identifiers of the dead drops and keys used to exchange messages. Client A generates two different symmetric keys derived from the initial shared secret K_{AB} and the respective public keys by a keyed-hash message authentication code (HMAC): $K_{A \to B} = \mathrm{HMAC}_{K_{AB}}(pk_B)$ for sending messages to B and $K_{B \to A} = \mathrm{HMAC}_{K_{AB}}(pk_A)$ for receiving messages from B. These secrets constantly get replaced by new ones, transmitted alongside every message to provide forward secrecy.

Dead Drop IDs. The identifiers of the dead drop for sending from A to B is built by computing an HMAC with the key $K_{A \to B}$ of a nonce $N_{A \to B}$ that increases after a given time period. Hence, the identifier $ID_{A \to B} = \mathrm{HMAC}_{K_{A \to B}}(N_{A \to B})$ changes in a fixed interval, even if no messages were exchanged at all. This prevents the server from identifying clients that disconnect for several days and would otherwise reconnect using the same identifiers. However, a fixed point in time at which the nonce changes (i.e. a unix timestamp rounded to the current day) is not a good choice. Assuming synchronized clocks often is error-prone[3]. If all clients would update their identifiers at a given point in time, e.g., at midnight, the server could detect a correlation between all identifiers of a user whose clock is out of sync. Thus, the nonces will be handled per contact and the secret key $K_{A \to B}$ is used to generate a point in time at which the nonce $N_{A \to B}$ will be increased. This leads to a different update time for all identifiers of dead drops.

Sending Messages. When a client A wants to send a message m to B, it encrypts the message using Authenticated Encryption with Associated Data (AEAD) using $K_{A \to B}$ so that only B can read it. Note, that the ciphertext of the message m does not reveal any information about the sender or the receiver as explained in [Ber09, Sect. 9]. A then stores the encrypted message in the dead drop $ID_{A \to B}$, which A accesses via Tor. B is now able to fetch and decrypt A's message from this dead drop. The shared secret will only be replaced by a new one transmitted alongside a message when B acknowledges the retrieval of the

---

[3] The need for secure time synchronization protocols lead to a number of secure time synchronization concepts such as ANTP [DSZ16], NTS (https://tools.ietf.org/html/draft-ietf-ntp-network-time-security-14) or Roughtime (https://roughtime.googlesource.com/roughtime).

new secret in a dead drop derived from the *existing* shared secret. Until then, A will not store any messages in a dead drop derived from the new secret and will retransmit messages until B sends the acknowledgment. This procedure will guarantee that no messages will be lost. Note, that $ID_{A \to B}$ may still change over time because of the used nonce $N_{A \to B}$ which is known by both parties.

Querying for new public keys using PIR is done in a fixed interval, e.g. once a day, to discover new users of the system. It is mandatory not to write to the dead drop specified by the shared secret immediately after receiving a new public key. This would allow an adversary to correlate a PIR request of a given user with (multiple) new requests to the dead drop database, even if the server does not know which public keys were queried. Instead, the first access to the dead drop database for new identifiers is delayed until a random point in time between the current query and the next one derived from the shared key $K_{A \to B}$. This ensures that the server cannot correlate the dead drop access to the PIR request because it could also have been initiated by any other clients that did a PIR request in the fixed interval. Note, that this delay is only necessary when a new contact is discovered. New users joining the service will therefore also have to delay their first interaction with other users.

Initial Contact. If B wants to contact A without A knowing about that (and thus not checking the associated dead drop at $ID_{B \to A}$), an anonymous signaling mechanism is required. We propose a per-user fixed and public dead drop that *all* other users write to, to establish contact. The fixed dead drop ID is $ID_A = \mathrm{HMAC}_0(pk_A)$. Messages into this dead drop are encrypted using hybrid encryption, similar to PGP, where A's public key pk_A is used to encrypt an ephemeral symmetric key, which encrypts the request message. This reveals how many contact requests A receives, which we consider as non-critical. However, this number could be obscured by sending dummy requests.

3.4   Analysis

Correctness. Correctness of RAID-PIR is shown in [DHS14], correctness of Tor is explained in [DMS04] and has already been well-proven in practice. Messages are acknowledged and retransmitted if identifiers change and the message has not been read yet. Therefore it is guaranteed that messages will reach their desired destination. Correctness of the ECDH key exchange is shown in [Ber09]. All operations involved in the private establishment of identifiers for the dead drops are deterministic and therefore result in the same identifiers for both parties.

Security. OnionPIR's security is based on the security guarantees of the underlying protocols and their assumptions. First, we assume that RAID-PIR is secure and does not leak any metadata about the queried information. A security argumentation for this is given in [DHS14]. In particular, it is important that the PIR servers are run by different non-colluding operators. Note, that the security guarantees are still fulfilled if less than r servers collude, where r is the redundancy parameter. A good choice for the operators would be a number of NGOs located in different legal territories. It is also mandatory that the different PIR servers are not in (physical) control of the same data center operator.

Next, we assume that Tor provides anonymity for the users that tunnel their connections through this anonymity network. This assumption implies that there is no global passive adversary which is able to monitor and analyze user traffic and colludes with the operator of the PIR servers or gains unauthorized access to them. Note, that it is necessary to at least de-anonymize two specific Tor connections in order to learn if two users are communicating with each other. Therefore, an attacker would have to be able to de-anonymize all users of the service to gain the full social graph of a given user. OnionPIR does not put any effort in hiding the fact that a user is using the service at all.

Anonymity is guaranteed among all honest users. A malicious user that announces its list of requests to the dead drops, would effectively remove himself from the anonymity set. Note, that no user is able to gain information about communication channels it is not participating in. In addition, no user is able to prove that a communication took place because the used AEAD guarantees repudiability.

OnionPIR relies on a Trust On First Use (TOFU) strategy to lessen the burden of manually exchanging keys. For the purpose of detecting the distribution of faulty keys, a client queries not only for the public keys of its contacts, but also for its own public key. This query can be performed at nearly no cost when querying together with other contacts using RAID-PIR's multi-block query. Since the PIR servers are not able to determine which PIR blocks are being requested, they are not able to manipulate the resulting response in a meaningful way, unless they collude. Of course, additional security can be achieved by adding out-of-band key verification, e.g., by announcing public keys on personal websites. Another interesting option are plausibility checks for the updates of the PIR database in the PIR servers and thereby extending the existing anytrust model.

Access to the dead drops is not protected against any type of manipulation by third parties since identifiers are only known to the involved parties. An adversary that is interested in deleting the messages for a specific client would have to brute-force the identifier of the dead drop which is impossible in practice. Protection against a server deleting messages or blocking access to the system is out of scope of this work and would require a federated or decentralized system.

Protection against malicious clients trying to flood the dead drops with large amounts of data could be achieved by making use of blind signatures [Cha83]. A client could encrypt a number of random tokens and authenticate against the server who will then blindly sign them. The tokens can then be decrypted by the client and sent to the server with each write access to the dead drops. While the server is able to determine that these tokens have a valid signature, it cannot identify the client who generated them. Since a server only signs a fixed number of tokens in a given time interval per client, this approach rate-limits write requests to the database.

Complexity and Efficiency. OnionPIR aims at providing efficient anonymous communication. Many existing systems (see Sect. 4) require a high communication overhead or high computational costs. The dead drop database is therefore combined with scalable onion routing and can be implemented as simple key-value storage. The servers needed in the communication phase can be deployed with low operating costs, comparable to traditional communication services.

The initialization phase in which the PIR requests are performed is crucial for the scalability of the system. As shown in Sect. 2.5, PIR is a valid choice that offers reasonable performance and allows users to detect malicious actions of the servers. The 3.8 GiB PIR database used for the benchmarks in Sect. 2.5 is sufficient to store public keys of 127 Mio. users, using 256 bit elliptic curve public keys.

The database size and the server's computation will grow linearly in the number of users. The PIR servers' ingress traffic for PIR queries depends on the number of blocks B in the database and therefore also scales linearly. Thanks to the PRG used in [DHS14], the size of a PIR query is $\lceil B/8 \rceil$ Bytes for all servers combined (excluding PRG seeds and overhead for lower level transport protocols). The egress bandwidth is constant since the size of the response only depends on the block size b (and the number of chunks for multi-block queries).

3.5 Implementation

We implemented a desktop application for secure messaging with metadata protection based on OnionPIR. Our implementation is written in Python and C, using our version of RAID-PIR including the optimizations from Sect. 2 for PIR, the Networking and Cryptography library (NaCl) [BLS12] for cryptographic operations, and Stem[4] as controller library for Tor. The system is divided into a client, the OnionPIR control server and PIR servers. The client offers a GUI as depicted in Fig. 10. Our open-source implementation is publicly available.[5]

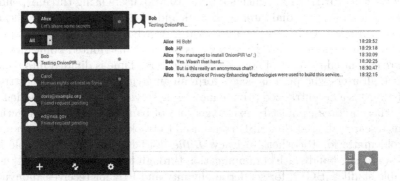

Fig. 10. Screenshot of the OnionPIR client GUI.

---

[4] https://stem.torproject.org/.
[5] https://github.com/encryptogroup/onionPIR.

4   Related Work

Nowadays, end-to-end encryption is available and deployed at large scale. In the past, these technologies were not accessible to a large user base because they required expert knowledge or were just not convenient, and thus only used by enthusiasts. For example, users of OpenPGP [CDF+07] have to manually build a web of trust and should be familiar with public key cryptography, while S/MIME [Ram99] requires certificate management. When the Signal Protocol was integrated into popular messaging services like WhatsApp[6] or Facebook Messenger[7] in 2016, private messaging became available to an extremely large user base. However, protecting not only communication content but also its metadata are still under active research. OnionPIR currently relies on RAID-PIR [DHS14] for public key distribution, but could also employ different PIR schemes. PIR is an active research area and there are other viable candidates [DG14, Hen16, AMBFK16]. Similarly, we rely on Tor to provide anonymity, which can also be achieved by using alternative techniques such as mixing networks [Cha81].

Next, we present system proposals that are related to OnionPIR. *Redphone* was one of the first applications that tackled metadata leakage using Bloom filters [Blo70] for private contact discovery. Redphone's encrypted call features were integrated into Textsecure/Signal, however, without anonymity due to scalability [Ope14]. With *DP5* [BDG15] users can anonymously exchange online presence information based on PIR. DP5 divides time in long-term and short-term epochs which are in the order of days and minutes respectively, which makes it impractical for real-time communication. Users must share symmetric keys before the protocol run, which can be hard to achieve in practice. Recently, *Alpenhorn* [LZ16] was proposed, which is based on identity-based encryption (IBE) for key distribution, a mix network [CJK+16, Cha81] for privacy and a so called keywheel construct for forward secrecy. Alpenhorn was integrated into Vuvuzela [vdHLZZ15], which supports 10 Mio. users using three Alpenhorn servers with an average dial latency of 150 s and a client bandwidth overhead of 3.7 KiB/s. *Riposte* [CGBM15] enables a large user base to privately post public messages to a message board. Its security is based on distributed point functions that are evaluated by a set of non-colluding servers. Time is divided into epochs, in which all users who post messages form an anonymity set. *Ricochet* (https://ricochet.im) is a decentralized private messaging service based on Tor hidden services, whose addresses must be exchanged out-of-band to establish connections. Recent research showed that HSDirs are used to track users [SN16], which might be problematic for Ricochet's privacy. *Riffle* [KLDF16] provides scalable low-latency and low-bandwidth communication through mix networks [Cha81] using verifiable shuffles [BG12] for sender anonymity and PIR for receiver anonymity. In Riffle, time is divided into epochs and each client sends and receives messages

---

[6] https://www.whatsapp.com/security/WhatsApp-Security-Whitepaper.pdf.

[7] https://fbnewsroomus.files.wordpress.com/2016/07/secret_conversations_whitepaper -1.pdf.

even if they do not communicate. *The Pynchon Gate* [SC05] is an anonymous mail system that guarantees only receiver anonymity by using PIR.

Acknowledgments. We thank the anonymous reviewers for their valuable feedback on our paper. This work has been co-funded by the German Federal Ministry of Education and Research (BMBF) and by the Hessen State Ministry for Higher Education, Research and the Arts (HMWK) within CRISP, and by the DFG as part of project S5 within the CRC 1119 CROSSING.

References

[ABH10] Albrecht, M., Bard, G., Hart, W.: Efficient multiplication of dense matrices over GF(2). ACM Trans. Math. Softw. **37**, 9:1–9:14 (2010)

[ADKF70] Arlazarov, V., Dinic, E., Kronrod, M., Faradzev, I.: On economical construction of the transitive closure of a directed graph. USSR Acad. Sci. **134**, 1209–1210 (1970)

[AMBFK16] Aguilar-Melchor, C., Barrier, J., Fousse, L., Killijian, M.-O.: XPIR: private information retrieval for everyone. In: Privacy Enhancing Technologies Symposium (PETS 2016), no. 2, pp. 155–174 (2016)

[BDG15] Borisov, N., Danezis, G., Goldberg, I.: DP5: a private presence service. In: Privacy Enhancing Technologies Symposium (PETS 2015), no. 2, pp. 4–24 (2015)

[Ber09] Bernstein, D.J.: Cryptography in NaCl (2009). https://cr.yp.to/highspeed/naclcrypto-20090310.pdf

[BG12] Bayer, S., Groth, J.: Efficient zero-knowledge argument for correctness of a shuffle. In: Pointcheval, D., Johansson, T. (eds.) EUROCRYPT 2012. LNCS, vol. 7237, pp. 263–280. Springer, Heidelberg (2012). doi:10.1007/978-3-642-29011-4_17

[Blo70] Bloom, B.H.: Space/time trade-offs in hash coding with allowable errors. Commun. ACM **13**(7), 422–426 (1970)

[BLS12] Bernstein, D.J., Lange, T., Schwabe, P.: The security impact of a new cryptographic library. In: Hevia, A., Neven, G. (eds.) LATINCRYPT 2012. LNCS, vol. 7533, pp. 159–176. Springer, Heidelberg (2012). doi:10.1007/978-3-642-33481-8_9

[BV14] Budurushi, J., Volkamer, M.: Feasibility analysis of various electronic voting systems for complex elections. In: International Conference for E-Democracy and Open Government 2014 (2014)

[CDF+07] Callas, J., Donnerhacke, L., Finney, H., Shaw, D., Thayer, R.: OpenPGP message format. RFC 4880, RFC Editor, November 2007. http://www.rfc-editor.org/rfc/rfc4880.txt

[CGBM15] Corrigan-Gibbs, H., Boneh, D., Mazières, D.: Riposte: an anonymous messaging system handling millions of users. In: IEEE Symposium on Security and Privacy (S&P 2015), pp. 321–338 (2015)

[CGN98] Chor, B., Gilboa, N., Naor, M.: Private information retrieval by keywords. IACR Cryptology ePrint Archive, Report 1998/003 (1998). http://eprint.iacr.org/1998/003

[Cha81] Chaum, D.L.: Untraceable electronic mail, return addresses, and digital pseudonyms. Commun. ACM **24**, 84–90 (1981)

[Cha83] Chaum, D.: Blind signature systems. In: Advances in Cryptology - CRYPTO 1983, p. 153 (1983)

[CJK+16] Chaum, D., Javani, F., Kate, A., Krasnova, A., de Ruiter, J., Sherman, A.T., Das, D.: cMix: anonymization by high-performance scalable mixing. IACR Cryptology ePrint Archive, Report 2016/008 (2016). http://eprint.iacr.org/2016/008

[CKGS95] Chor, B., Kushilevitz, E., Goldreich, O., Sudan, M.: Private information retrieval. In: Foundations of Computer Science (FOCS 1995), pp. 41–50 (1995)

[DG14] Devet, C., Goldberg, I.: The best of both worlds: combining information-theoretic and computational PIR for communication efficiency. In: Cristofaro, E., Murdoch, S.J. (eds.) PETS 2014. LNCS, vol. 8555, pp. 63–82. Springer, Cham (2014). doi:10.1007/978-3-319-08506-7_4

[DHS14] Demmler, D., Herzberg, A., Schneider, T.: RAID-PIR: practical multi-server PIR. In: ACM Cloud Computing Security Workshop (CCSW 2014), pp. 45–56 (2014)

[DMS04] Dingledine, R., Mathewson, N., Syverson, P.: Tor: the second-generation onion router. In: USENIX Security Symposium 2004, p. 21 (2004)

[DSZ14] Demmler, D., Schneider, T., Zohner, M.: Ad-hoc secure two-party computation on mobile devices using hardware tokens. In: USENIX Security Symposium 2014, pp. 893–908 (2014)

[DSZ16] Dowling, B., Stebila, D., Zaverucha, G.: Authenticated network time synchronization. In: USENIX Security Symposium 2016, pp. 823–840 (2016)

[FM11] Fette, I., Melnikov, A.: The websocket protocol. RFC 6455, RFC Editor, December 2011. http://www.rfc-editor.org/rfc/rfc6455.txt

[HCE11] Huang, Y., Chapman, P., Evans, D.: Privacy-preserving applications on smartphones. In: USENIX Workshop on Hot Topics in Security (HotSec 2011), p. 4 (2011)

[Hen16] Henry, R.: Polynomial batch codes for efficient IT-PIR. In: Privacy Enhancing Technologies Symposium (PETS 2016), pp. 202–218 (2016)

[KLDF16] Kwon, A., Lazar, D., Devadas, S., Ford, B.: Riffle: an efficient communication system with strong anonymity. In: Privacy Enhancing Technologies Symposium (PETS 2016), pp. 115–134 (2016)

[KO97] Kushilevitz, E., Ostrovsky, R.: Replication is not needed: single database, computationally-private information retrieval. In: Foundations of Computer Science (FOCS 1997), pp. 364–373 (1997)

[Lan15] Landau, S.: Mining the metadata: and its consequences. In: International Conference on Software Engineering (ICSE 2015), pp. 4–5 (2015)

[LG15] Lueks, W., Goldberg, I.: Sublinear scaling for multi-client private information retrieval. In: Böhme, R., Okamoto, T. (eds.) FC 2015. LNCS, vol. 8975, pp. 168–186. Springer, Heidelberg (2015). doi:10.1007/978-3-662-47854-7_10

[LZ16] Lazar, D., Zeldovich, N.: Alpenhorn: bootstrapping secure communication without leaking metadata. In: USENIX Symposium on Operating Systems Design and Implementation (OSDI 2016), pp. 571–586 (2016)

[MCA06] Mokbel, M.F., Chow, C.-Y., Aref, W.G.: The new Casper: query processing for location services without compromising privacy. In: International Conference on Very Large Data Bases (VLDB 2006), pp. 763–774 (2006)

[MMM16] Mayer, J., Mutchler, P., Mitchell, J.C.: Evaluating the privacy properties of telephone metadata. Natl. Acad. Sci. **113**(20), 5536–5541 (2016)

[MOT+11] Mittal, P., Olumofin, F., Troncoso, C., Borisov, N., Goldberg, I.: PIR-tor: scalable anonymous communication using private information retrieval. In: USENIX Security Symposium 2011, p. 31 (2011)

[Ope14]   Open Whisper Systems. The difficulty of private contact discovery (2014). https://whispersystems.org/blog/contact-discovery/

[Ram99]   Ramsdell, B.: S/MIME version 3 message specification. RFC 2633, RFC Editor, June 1999. http://www.rfc-editor.org/rfc/rfc2633.txt

[SC05]   Sassaman, L., Cohen, B., Gate, T.P.: A secure method of pseudonymous mail retrieval. In: Workshop on Privacy in the Electronic Society (WPES 2005), pp. 1–9 (2005)

[SN16]   Sanatinia, A., Noubir, G.: HOnions: towards detection and identification of misbehaving tor HSDirs. In: Hot Topics in Privacy Enhancing Technologies Symposium (HotPETS 2016) (2016)

[vdHLZZ15] van den Hooff, J., Lazar, D., Zaharia, M., Zeldovich, N.: Vuvuzela: scalable private messaging resistant to traffic analysis. In: Symposium on Operating Systems Principles (SOSP 2015), pp. 137–152 (2015)

Data and Server Security

Accountable Storage

Giuseppe Ateniese[1], Michael T. Goodrich[2], Vassilios Lekakis[3],
Charalampos Papamanthou[5], Evripidis Paraskevas[5(✉)],
and Roberto Tamassia[4]

[1] Department of Computer Science, Stevens Institute of Technology, Hoboken, USA
gatenies@stevens.edu
[2] Department of Computer Science, University of California, Irvine, USA
goodrich@uci.edu
[3] Department of Computer Science, University of Maryland, College Park, USA
lex@cs.umd.edu
[4] Department of Computer Science, Brown University, Providence, USA
rt@cs.brown.edu
[5] Department of Electrical and Computer Engineering, University of Maryland,
College Park, USA
cpap@umd.edu, evripar@terpmail.umd.edu

Abstract. We introduce *Accountable Storage* (AS), a framework enabling a client to outsource n file blocks to a server while being able (any time after outsourcing) to provably compute how many bits were discarded or corrupted by the server. Existing techniques (e.g., proofs of data possession or storage) can address the accountable storage problem, with linear server computation and bandwidth. Instead, our optimized protocols achieve $O(\delta \log n)$ complexity (where δ is the maximum number of corrupted blocks that can be tolerated) through the novel use of invertible Bloom filters and a new primitive called *proofs of partial storage*. With accountable storage, a client can be compensated with a dollar amount proportional to the number d of corrupted bits (that he can now provably compute). We integrate our protocol with Bitcoin, supporting automatic such compensations. Our implementation is open-source and shows our protocols perform well in practice.

1 Introduction

Cloud computing is revolutionizing our digital world, posing new security and privacy challenges. E.g., businesses and individuals are reluctant to outsource their databases for fear of having their data lost or damaged. Thus, they would benefit from technologies that would allow them to manage their risk of data loss, just like insurance allows them to manage their risk of physical or financial losses, e.g., from fire or liability.

As a first step, a client needs a mechanism for verifying that a cloud provider is storing her entire database intact, and fortunately, *Provable Data Possession* (PDP) [3,11,13] and *Proofs of Retrievability* (POR) [10,19,26–28], have been conceived as a solution to the integrity problem of remote databases. PDP and

© Springer International Publishing AG 2017
D. Gollmann et al. (Eds.): ACNS 2017, LNCS 10355, pp. 623–644, 2017.
DOI: 10.1007/978-3-319-61204-1_31

POR scheme can verify whether the server possesses the database originally uploaded by the client by having the server generate a proof in response to a challenge.

However, they leave unsettled several risk management issues. Arguably, an important question is

> *What happens if a PDP or POR scheme shows that a client's outsourced database has been damaged?*

The objective of this work is to design new efficient protocols for *Accountable Storage* (AS) that enable the client to reliably and quickly assess the damage and at the same time automatically get compensated using the Bitcoin protocol.

To be precise, suppose Alice outsources her file blocks $b_1, b_2, \ldots, b_n$ to a potentially malicious cloud storage provider, Bob. Since Alice does not trust Bob, she wishes, at any point in time, to be able to compute the amount of damage, if any, that her file blocks have undergone, by engaging in a simple challenge-response protocol with Bob. For instance, she wishes to *provably* compute the value of a *damage metric*, such as

$$d = \sum_{i=1}^{n} w_i \cdot ||b_i \oplus b_i'||, \tag{1}$$

where b_i' is the file currently stored by Bob at the time of the challenge, $||.||$ denotes Hamming distance and w_i is a weight corresponding to file b_i. If $d = 0$, Alice is entitled to no dollar credit. Bob can easily prove to Alice that this is the case through existing protocols, as noted above. If $d > 0$, however, then Alice should receive a compensation proportional to the damage d, which should be provided automatically.[1]

Naive Approaches for AS. A PDP protocol [3,4,11,13,29] enables a server to prove to a client that all of the client's data is stored intact. One could design an AS protocol by using a PDP protocol only for the portion of storage that the server possesses. This could determine the damage, d (e.g., when all weights w_i are equal to 1). However this approach requires using of PDP at the bit level, and in particular computing one 2048-bit tag for each bit of our file collection which is very storage-inefficient.

To overcome the above problem, one could use PDP at the block level, but at the same time keep some redundancy locally. Specifically, before outsourcing the n blocks at the server, the client could store δ extra check blocks locally (e.g., computed with an error-correcting code). The client could then verify through PDP that a set of at most δ blocks have gone missing and retrieve the lost blocks by executing the decoding algorithm on the remote intact $n - \delta$ data blocks and

---

[1] We highlight that such fine-grained compensation models, which work at the bit level as opposed to at the file block level, allow Alice to better manage her risk for damage even within the same file. For example, compensation for an unusable movie stored by Bob could be larger than that for a usable movie whose resolution has deteriorated by just 5%.

the δ local check blocks (then the recovered blocks can be used to compute d). This procedure has $O(n)$ communication, since the $n - \delta$ blocks at the server must be sent to the client. IRIS [28] is a system along these lines, requiring the whole file system be streamed over to the client for recovery.

Finally we note here that while PDP techniques combined with redundant blocks stored at the client can be used to solve the accountable storage problem (even inefficiently, as shown above), POR techniques cannot. This is because POR techniques (e.g., [26]) cannot provide proofs of retrievability for a certain portion of the file (as is the case with PDP), but only for the whole file—this is partly due to the fact that error-correcting codes are used on top of all the file blocks.

Our AS Protocol. Our protocol for assessing damage d from Relation 1 is based on *recovering the actual blocks $b_1, b_2, \ldots, b_\delta$ and XORing them with the corrupted blocks $b'_1, b'_2, \ldots, b'_\delta$ returned by the server.* For recovery, we use the invertible Bloom filter (IBF) data structure [12,15]. An IBF is an array of t cells and can store $O(t)$ elements. Unlike a Bloom filter [7], its elements can be enumerated with high probability.

Let $B = \{b_1, b_2, \ldots, b_n\}$ be the set of outsourced blocks and let δ be the maximum number of corrupted blocks that can be tolerated. In preprocessing, the client computes an IBF $\mathbf{T}_B$ with $O(\delta)$ cells, on the blocks $b_1, \ldots, b_n$. $\mathbf{T}_B$ is stored locally. Computing $\mathbf{T}_B$ is similar to computing a Bloom filter: every cell of $\mathbf{T}_B$ is mapped to a XOR over a set of at most n blocks, thus the local storage is $O(\delta)$. To outsource the blocks, the client computes homomorphic tags, $\mathbf{T}_i$ (as in [3]), for each block b_i. The client then stores $(b_i, \mathbf{T}_i)$ with the cloud and deletes $b_1, b_2, \ldots, b_n$ from local storage. In the challenge phase, the client asks the server to construct an IBF $\mathbf{T}_K$ of $O(\delta)$ cells on the set of blocks K the server currently has—this is the "proof" the server sends to the client. Then the client takes the "difference" $\mathbf{T}_L = \mathsf{subtract}(\mathbf{T}_B, \mathbf{T}_K)$ and recovers the elements of the difference $B - K$ (since $|B - K| \le \delta$ and $\mathbf{T}_L$ has $O(\delta)$ cells). Recovering blocks in $B - K$ enables the client to compute d using Relation 1. Clearly, the bandwidth of this protocol is proportional to δ (due to the size of the IBFs), and not to the total number of outsourced blocks n. Our optimized construction in Sect. 5 achieves sublinear server and client complexities as well.

Fairness Through Integration with Bitcoin. The above protocol assures that Bob (the server) cannot succeed in persuading Alice that the damage of her file blocks is $d' < d$. After Alice is persuaded, compensation proportional to d must be sent to her. But Bob could try to cheat again. Specifically, Bob could try to give Alice a smaller compensation or even worse, disappear. To deal with this problem, we develop a modified version of the recently-introduced timed commitment in Bitcoin [2]. At the beginning of the AS protocol, Bob deposits a large amount, A, of bitcoins, where A is contractually agreed on and is typically higher than the maximum possible damage to Alice's file blocks. The Bitcoin-integrated AS protocol of Sect. 6 ensures that unless Bob fully and timely compensates Alice for damage d, then A bitcoins are automatically and irrevocably transferred to Alice. At the same time, if Alice tries to cheat (e.g., by asking for compensation higher than the contracted amount), our protocol ensures that she gets no compensation at all while Bob gets back all A of his bitcoins.

Structure of the Paper. Section 2 presents background on IBFs and Bitcoin, Sect. 3 gives definitions, and Sects. 4 and 5 present our constructions. We present our Bitcoin protocol in Sect. 6, our evaluation in Sect. 7 and conclude in Sect. 8.

2 Preliminaries

Let τ denote the security parameter, δ denote an upper bound on the number of corrupted blocks that can be tolerated, n denote the number of file blocks, and $b_1, b_2, \ldots, b_n$ denote the file blocks. Each block b_i has λ bits. The first $\log n$ bits of each block b_i are used for storing the index i of the block, which can be retrieved through function index(). Namely $i = \text{index}(b_i)$. Let also $h_1, h_2, \ldots, h_k$ be k hash functions chosen at random from a universal family of functions $\mathcal{H}$ [9] such that $h_i : \{0,1\}^\lambda \to \{1, 2, \ldots, t\}$ for some parameter t.

Invertible Bloom Filters. An *Invertible Bloom Filter (IBF)* [12,15] can be used to compactly store a set of blocks $\{b_1, b_2, \ldots, b_n\}$: It uses a table (array) $\mathbf{T}$ of $t = (k+1)\delta$ cells. Each cell of the IBF's table $\mathbf{T}$ contains the following two fields[2]: (1) dataSum: XOR of blocks b_i mapped to this cell; (2) hashSum: XOR of cryptographic tags (to be defined later) T_i for all blocks b_i mapped to this cell. As in Bloom filters, we use functions $h_1, \ldots, h_k$ to decide which blocks map to which cells.

An IBF supports simple algorithms for insertion and deletion via algorithm update in Fig. 1. For $B \subseteq A$, one can also take the *difference* of IBFs $\mathbf{T}_A$ and $\mathbf{T}_B$, to produce an IBF $\mathbf{T}_D \leftarrow \text{subtract}(\mathbf{T}_A, \mathbf{T}_B)$ representing the difference set $D = A - B$. Finally, given $\mathbf{T}_D$, we can enumerate its contents by using algorithm listDiff from [12]:

Lemma 1 (Adjusted from Eppstein et al. [12]). *Let $B \subseteq A$ be two sets having $\leq \delta$ blocks in their difference $A - B$, let $\mathbf{T}_A$ and $\mathbf{T}_B$ be their IBFs constructed using k hash functions and let $\mathbf{T}_D \leftarrow \text{subtract}(\mathbf{T}_A, \mathbf{T}_B)$. All IBFs have $t = (k+1)\delta$ cells and their hashSum field is computed using a function mapping blocks to at least $k \log \delta$ bits. Then there is an algorithm $\text{listDiff}(\mathbf{T}_D)$ that recovers $A - B$ with probability $1 - O(\delta^{-k})$.*

Bitcoin Basics. Bitcoin [23] is a decentralized digital currency system where transactions are recorded on a public ledger (the blockchain) and are verified through the collective effort of *miners*. A bitcoin address is the hash of an ECDSA public key. Let A and B be two bitcoin addresses. A *standard* transaction contains a signature from A and mandates a certain amount of bitcoins be transferred from A to B. If A's signature is valid, the transaction is inserted into a block which is then stored in the blockchain.

---

[2] Note that we do not use the count field, as in [12,15].

Algorithm T ← update(b_i, T)

for each $j = 1, \ldots, k$ do
   Set $T[h_j(b_i)]$.dataSum $\oplus = b_i$;
   Set $T[h_j(b_i)]$.hashSum $\oplus = T_i$;
return T;

Algorithm T_D ← subtract(T_A, T_B)

for each $i = 1, \ldots, t$ do
   $T_D[i]$.dataSum $= T_A[i]$.dataSum $\oplus T_B[i]$.dataSum;
   $T_D[i]$.hashSum $= T_A[i]$.hashSum $\oplus T_B[i]$.hashSum;
return T_D;

Fig. 1. Update and subtraction algorithms in IBFs.

Bitcoin allows for more complicated transactions (as we are using here), whose validation requires more than just a signature. In particular, each transaction can specify a *locktime* containing a timestamp t at which the transaction is locked (before time t, even if a valid signature is provided, the transaction is not final). Slightly changing the notation from [2], a Bitcoin transaction T_x can be represented as the table below, where Prev is the transaction (say T_y) that T_x is redeeming, InputsToPrev are inputs that T_x is sending to T_y so that T_y's redeeming can take place, Conditions is a program written in the Bitcoin scripting language (outputting a boolean) controlling whether T_x can be redeemed or not (given inputs from another transaction), Amount is the value in bitcoins, and Locktime is the locktime. For standard transactions, InputsToPrev is a signature with the sender's secret key, and Conditions implements a signature verification with the recipient's public key. Also, standard transactions have locktime set to 0, meaning they are locked and final.

| Prev : |
| InputsToPrev : |
| Conditions : |
| Amount : |
| Locktime : |

3   Accountable Storage Definitions

We now define an AS scheme. An AS scheme does not allow the client to compute damage d directly. Instead, it allows the client to use the server's proof to retrieve the blocks $\mathcal{L}$ that are not stored by the server any more (or are stored corrupted). By having the server send the current blocks he stores in the position of blocks in $\mathcal{L}$ (in addition to the proof), computing the damage d is straightforward.

Definition 1 (δ-AS scheme). *A δ-AS scheme $\mathcal{P}$ is the collection of four PPT algorithms:*

1. $\{pk, sk, state, T_1, \ldots, T_n\} \leftarrow$ Setup$(b_1, \ldots, b_n, \delta, 1^\tau)$ *takes as inputs file blocks $b_1, \ldots, b_n$, a parameter δ and the security parameter τ and returns a public key pk, a secret key sk, tags $T_1, \ldots, T_n$ and a client state state.*
2. chal $\leftarrow$ GenChal(1^τ) *generates a challenge for the server;*
3. $\mathcal{V} \leftarrow$ GenProof$(pk, \beta_{i_1}, \ldots, \beta_{i_m}, T_{i_1}, \ldots, T_{i_m}, chal)$ *takes as inputs a public key pk, a collection of $m \leq n$ blocks and their corresponding tags. It returns a proof of accountability $\mathcal{V}$;*
4. $\{reject, \mathcal{L}\} \leftarrow$ CheckProof$(pk, state, \mathcal{V}, chal)$ *takes as inputs a public key pk and a proof of accountability $\mathcal{V}$. It returns a list of blocks $\mathcal{L}$ or reject.*

Relation to Proofs of Storage. A δ-AS scheme is a generalization of proof-of-storage (PoS) schemes, such as [3,19]. In particular, a 0-AS scheme (i.e., where we set $\delta = 0$) is equivalent to PoS protocols, where there is no tolerance for corrupted/lost blocks.

Definition 2 (δ-AS scheme correctness). *Let $\mathcal{P}$ be a δ-AS scheme. Let $\{\text{pk}, \text{sk}, \text{state}, T_1, \ldots, T_n\} \leftarrow \text{Setup}(b_1, \ldots, b_n, \delta, 1^\tau)$ for some set of blocks $B = \{b_1, \ldots, b_n\}$. Let now $\mathcal{L} \subseteq B$ such that $|\mathcal{L}| \leq \delta$, $\text{chal} \leftarrow \text{GenChal}(1^\tau)$ and $\mathcal{V} \leftarrow \text{GenProof}(\text{pk}, B - \mathcal{L}, T(B - \mathcal{L}), \text{chal})$, where $T(B - \mathcal{L})$ denotes the tags corresponding to the blocks in $B - \mathcal{L}$. A δ-AS scheme is correct if the probability that $\mathcal{L} \leftarrow \text{CheckProof}(\text{pk}, \text{state}, \mathcal{V}, \text{chal})$ is at least $1 - \text{neg}(\tau)$.*[3]

To define the security of a δ-AS scheme, the adversary adaptively asks for tags on a set of blocks $B = \{b_1, b_2, \ldots, b_n\}$ that he chooses. After the adversary gets access to the tags, his goal is to output a proof $\mathcal{V}$, so that if $\mathcal{L}$ is output by algorithm CheckProof, where $|\mathcal{L}| \leq \delta$, then (a) either $\mathcal{L}$ is *not* a subset of the original set of blocks B; (b) or the adversary does *not* store all remaining blocks in $B - \mathcal{L}$ intact.

Such a proof is invalid since it would allow the verifier to either recover the *wrong* set of blocks (e.g., a set of blocks whose Hamming distance from the corrupted blocks is a lot smaller) or to accept a corruption of more than δ file blocks.

Definition 3 (δ-AS security). *Let $\mathcal{P}$ be a δ-AS scheme as in Definition 1 and $\mathcal{A}$ be a PPT adversary. We define security using the following steps.*

1. **Setup.** *$\mathcal{A}$ chooses $\delta \in [0, n)$, blocks $B = \{b_1, b_2, \ldots, b_n\}$ and is given $T_1, \ldots, T_n$ and pk output by $\{\text{pk}, \text{sk}, \text{state}, T_1, \ldots, T_n\} \leftarrow \text{Setup}(b_1, \ldots, b_n, \delta, 1^\tau)$.*[4]
2. **Forge.** *$\mathcal{A}$ is given $\text{chal} \leftarrow \text{GenChal}(1^\tau)$ and outputs a proof of accountability $\mathcal{V}$.*

Suppose $\mathcal{L} \leftarrow \text{CheckProof}(\text{pk}, \text{state}, \mathcal{V}, \text{chal})$. We say that the δ-AS scheme $\mathcal{P}$ is secure if, with probability at least $1 - \text{neg}(\tau)$: (i) $\mathcal{L} \subseteq B$; and (ii) there exists a PPT knowledge extractor $\mathcal{E}$ that can extract all the remaining file blocks in $B - \mathcal{L}$.

Note here that if the set $\mathcal{L}$ is empty, then the above definition is equivalent to the original PDP security definition [3]. Also note that the notion of a knowledge extractor is similar to the standard one, introduced in the context of proofs of knowledge [5]. If the adversary can output an accepting proof, then he can execute GenProof repeatedly until it extracts the selected blocks.

---

[3] Function $\lambda : \mathbb{N} \to \mathbb{R}$ is $\text{neg}(\tau)$ iff $\forall$ nonzero polynomials $p(\tau)$ there exists N so that $\forall \tau > N$ it is $\lambda(\tau) < 1/p(\tau)$.

[4] $\mathcal{A}$ could also choose blocks adaptively, after seeing tags for already requested blocks. Our proof of security handles that.

4   Our Basic Construction

We now give an overview of our basic construction: On input blocks $B = \{b_1, \ldots, b_n\}$ in local storage, the client decides on a parameter δ (meaning that he can tolerate up to δ corrupted files) and computes the local state, tags, public and secret key by running $\{\mathsf{pk}, \mathsf{sk}, \mathsf{state}, \mathbf{T}_1, \ldots, \mathbf{T}_n\} \leftarrow \mathsf{Setup}(b_1, \ldots, b_n, \delta, 1^\tau)$. In our construction the tag $\mathbf{T}_i$ is set to $(h(i)g^{b_i})^d \mod N$, as in [3], where $h(.)$ is a collision-resistant hash function, N is an RSA modulus and (e, d) denote an RSA public/private key pair. The client then sends blocks $b_1, \ldots, b_n$ and tags $\mathbf{T}_1, \ldots, \mathbf{T}_n$ to the server and locally stores the state state, which is an IBF of the blocks $b_1, b_2, \ldots, b_n$.

At challenge phase, the client runs $\mathsf{chal} \leftarrow \mathsf{GenChal}(1^\tau)$ that picks a random challenge s and sends it to the server. To generate a proof of accountability (see Fig. 2-left) with $\mathsf{GenProof}$, the server computes an IBF $\mathbf{T}_K$ on the set of blocks that he (believes he) stores, along with a proof of data possession [3] on the same set of blocks. The indices of these blocks are stored in a set Kept. For the computation of the PDP proof, the server uses randomness derived from the challenge s.

To verify the proof, the client takes the difference $\mathbf{T}_L = \mathsf{subtract}(\mathbf{T}_B, \mathbf{T}_K)$ and executes algorithm $\mathsf{recover}$ from Fig. 2-right, which is a modified version of $\mathsf{listDiff}$ from [12]. Algorithm $\mathsf{recover}$ adds blocks whose tags verify to the set of lost blocks $\mathcal{L}$. Then it checks the PDP proof for those block indices corresponding to blocks that were not output by $\mathsf{recover}$. If this PDP proof does not reject, then the client is persuaded that the server stores everything except for blocks in $\mathcal{L}$. To make sure $\mathsf{recover}$ does not fail with a noticeable probability, our construction sets the parameters according to the following corollary. The detailed algorithms of our construction are in Fig. 3.

Corollary 1. *Let τ be the security parameter and B and K be two sets such that $K \subseteq B$ and $|B - K| \leq \delta$. Let $\mathbf{T}_B$ and $\mathbf{T}_K$ be IBFs constructed by algorithm*

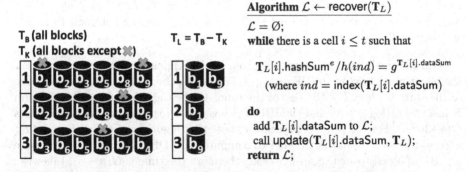

Fig. 2. (**Left**) On input $b_1, b_2, \ldots, b_9$, the client outputs an IBF $\mathbf{T}_B$ of three cells using two hash functions. The server loses blocks b_1 and b_9. $\mathbf{T}_K$ is computed on blocks $b_2, b_3, \ldots, b_8$ and $\mathbf{T}_L$ contains the lost blocks b_1 and b_9. (**Right**) The algorithm for recovering the lost blocks.

Alg. $\{\mathsf{pk}, \mathsf{sk}, \mathsf{state}, \mathrm{T}_1, \ldots, \mathrm{T}_n\} \leftarrow \mathsf{Setup}(b_1, \ldots, b_n, \delta, 1^\tau)$. Let

- $N = pq$ be an RSA modulus of 2τ bits.
- Let e be a random prime and d be a number such that $ed = 1 \mod \phi(N)$.
- $\mathcal{H}$ be a set of k random hash functions $\mathcal{H} = \{h_1, h_2, \ldots, h_k\}$ such that $h_i : \{0,1\}^* \to [t]$, where $t = (k+1)\delta$ and $k = \tau/\log \delta$.
- g be a generator of $\mathbb{QR}_N$.
- $\theta_s : \{0,1\}^\tau \times \{0,1\}^* \to \{0,1\}^\tau$ be a pseudorandom function.
- $h : \{0,1\}^* \to \mathbb{QR}_N$ be a collision-resistant hash function (modeled as random oracle).

Set $\mathsf{pk} = (N, e, \mathcal{H}, g, \theta_s, h)$ and $\mathsf{sk} = d$. For block b_i compute tag

$$\mathrm{T}_i = \left(h(i) g^{b_i} \right)^d \mod N.$$

Set state to be the IBF $\mathbf{T}_B$ on $b_1, \ldots, b_n$ computed using Algorithm update from Figure 1.

Alg. $\mathsf{chal} \leftarrow \mathsf{GenChal}(1^k)$. Pick a random $s \in \{0,1\}^\tau$ and output $\mathsf{chal} = s$.

Alg. $\mathcal{V} \leftarrow \mathsf{GenProof}(\mathsf{pk}, b_{i_1}, \ldots, b_{i_m}, \mathrm{T}_{i_1}, \ldots, \mathrm{T}_{i_m}, \mathsf{chal})$. Let $\{b_{i_1}, \ldots, b_{i_m}\} \subseteq \{b_1, \ldots, b_n\}$ be the blocks that the server stores. Set $\mathsf{Kept} = \{i_1, i_2, \ldots, i_m\}$.

1. (PDP proof) Compute a combined tag T along with a combined sum S as

$$\mathrm{T} = \prod_{i \in \mathsf{Kept}} \mathrm{T}_i^{a_i} \text{ and } S = \sum_{i \in \mathsf{Kept}} a_i b_i, \text{ where } a_i = \theta_{\mathsf{chal}}(i). \tag{2}$$

2. (IBF proof) Compute IBF $\mathbf{T}_K$ on the set of blocks $\{b_i : i \in \mathsf{Kept}\}$.

The proof $\mathcal{V}$ is the tuple $\{\mathrm{T}, S, \mathbf{T}_K\}$.

Alg. $\{\mathsf{reject}, \mathcal{L}\} \leftarrow \mathsf{CheckProof}(\mathsf{pk}, \mathsf{state}, \mathcal{V}, \mathsf{chal})$. Parse $\mathcal{V}$ as $\{\mathrm{T}, S, \mathbf{T}_K\}$, state as $\mathbf{T}_B$.

1. Set $\mathbf{T}_L \leftarrow \mathsf{subtract}(\mathbf{T}_B, \mathbf{T}_K)$
2. Set $\mathcal{L} \leftarrow \mathsf{recover}(\mathbf{T}_L)$ and $\mathsf{Kept} = [n]$.
3. For each $b \in \mathcal{L}$, set $\mathsf{Kept} = \mathsf{Kept} - \{\mathsf{index}(b)\}$.
4. Output reject if $\mathrm{T}^e / \prod_{i \in \mathsf{Kept}} h(i)^{a_i} \neq g^S$, where $a_i = \theta_{\mathsf{chal}}(i)$. Otherwise output $\mathcal{L}$.

Fig. 3. Our δ-AS scheme construction.

update *of Fig. 1 using* $\tau/\log \delta$ *hash functions. The IBFs* $\mathbf{T}_B$ *and* $\mathbf{T}_K$ *have* $t = (\tau/\log \delta + 1)\delta$ *cells and employ tags in the* $\mathsf{hashSum}$ *field that map blocks to* τ *bits. Then with probability at least* $1 - 2^{-\tau}$, *algorithm* $\mathsf{recover}(\mathsf{subtract}(\mathbf{T}_B, \mathbf{T}_K))$ *will output* $\mathcal{L} = B - K$.

Our detailed proof of security is given in the Appendix. The local state that the client must keep is an IBF of $t = (k+1)\delta$ cells, therefore the asymptotic size of the state is $O(\delta)$. For the size of the proof $\mathcal{V}$, the tag T has size $O(1)$, the sum S has size $O(\log n + \lambda)$ and the IBF $\mathbf{T}_K$ has size $O(\delta)$. Overall, the size of $\mathcal{V}$ is $O(\delta + \log n)$. For the proof computation, note that algorithm $\mathsf{GenProof}$ must first access at least $n - \delta$ blocks in order to compute the PDP proof and then compute an IBF of δ cells over the same blocks, therefore the time is $O(n + \delta)$. Likewise, the verification algorithm needs to verify a PDP proof for a linear number of blocks and to process a proof of size $O(\delta + \log n)$, thus its computation time is again $O(n + \delta)$.

Theorem 1 (δ-AS scheme). *Let n be the number of blocks. For all $\delta \leq n$, there exists a δ-AS scheme such that: (1) It is correct according to Definition 2; (2) It is secure in the random oracle model based on the RSA assumption and according to Definition 3; (3) The proof has size $O(\delta + \log n)$ and its computation at the server takes $O(n + \delta)$ time; (4) Verification at the client takes $O(n + \delta)$ time and requires local state of size $O(\delta)$; (5) The space at the server is $O(n)$.*

We now make two observations related to our construction. First, note that the server could potentially launch a DoS attack, by *pretending it does not store some of the blocks* so that the client is forced to spend cycles retrieving these blocks. This is not an issue, since as we will see later, the server will be penalized for that, so it is not in its best interest. Second, note that the tags that the client initially uploads are publicly verifiable so anyone can check their validity—therefore the client cannot upload bogus tags and blame the server later for that.

Streaming and Appending Blocks. Our construction assumes the client has all blocks available in the beginning. This is not necessary. Blocks b_i could come one at a time, and the client could easily update its local state with algorithm update$(b_i, \mathbf{T}, 1)$, compute the new tag $\mathbf{T}_i$ and send the pair $(b_i, \mathbf{T}_i)$ to the server for storage. This also means that our construction is partially-dynamic, supporting append-only updates. Modifying a block is not so straightforward due to replay attacks. However techniques from various fully-dynamic PDP schemes could be potentially used for this problem (e.g., [13]).

5   Sublinear Construction Using Proofs of Partial Storage

In the previous construction, the server and client run in $O(n + \delta)$ time. In this section we present optimizations that reduce the server and client performance to $O(\delta \log n)$. Recall that the proof generation in Fig. 3 has two distinct, linear-time parts: First, proving that a subset of blocks is kept intact (in particular the blocks with indices in Kept), and second, computing an IBF on this set of blocks. We show here how to execute both these tasks in sublinear time using (i) partial proofs of storage; (ii) a data structure based on segment trees that the client must prepare during preprocessing.

Proofs of Partial Storage. In our original construction, we prove that a subset of blocks is kept intact (in particular the blocks with indices in Kept) using a PDP-style proof, as originally introduced by Ateniese et al. [3]. In our new construction we will replace that part with a new primitive called *proofs of partial storage*. To motivate proofs of partial storage, let us recall how proofs of storage [26] work.

Proofs of storage provide the same guarantees with PDP-style proofs [3] but are much more practical in terms of proof construction time. In particular, one can construct a PoS proof in constant time as follows. Along with the original blocks $b_1, b_2, \ldots, b_n$ the client outsources an additional n redundant blocks $\beta_1, \beta_2, \ldots, \beta_n$ computed with an error-correcting code such as Reed-Solomon, such that any n out of the $2n$ blocks $b_1, b_2, \ldots, b_n, \beta_1, \beta_2, \ldots, \beta_n$ can be used to retrieve the original blocks $b_1, b_2, \ldots, b_n$. Also, the client outsources tags T_i (as computed in Algorithm Setup in Fig. 3) for all $2n$ blocks. Now, during the challenge phase, the client picks a constant-sized subset of random blocks to challenge (out of the $2n$ blocks), say $\tau = 128$ blocks. Because the subset is chosen at random every time, the server, with probability at least $1 - 2^{-\tau}$, will pass the challenge (i.e., provide verifying tags for the challenged blocks) only if he stores at least half of the blocks $b_1, b_2, \ldots, b_n, \beta_1, \beta_2, \ldots, \beta_n$—which means that the original blocks $b_1, b_2, \ldots, b_n$ are recoverable.

Unfortunately, we cannot use proofs of storage as described above directly, since we want to prove that a *subset of the blocks* is stored intact, and the above construction applies to the whole set of blocks. In the following we describe how to fix this problem using a segment-tree-like data structure.

Our New Construction: Using a Segment Tree. A segment tree T is a binary search tree that stores the set B of n key-value pairs (i, b_i) at the leaves of the tree (ordered by the key). Let v be an internal node of the tree T. Denote with cover(v) the set of blocks that are included in the leaves of the subtree rooted on node v. Let also $|v| = |\mathsf{cover}(v)|$. Every internal node v of T has a label label(v) that stores:

1. All blocks $b_1, b_2, \ldots, b_{|v|}$ contained in cover(v) along with respective tags T_i. The tags are computed as in Algorithm Setup in Fig. 3;
2. Another $|v|$ redundant blocks $\beta_1, \beta_2, \ldots, \beta_{|v|}$ computed using Reed-Solomon codes such that *any* $|v|$ out of the $2|v|$ blocks $b_1, b_2, \ldots, b_{|v|}, \beta_1, \beta_2, \ldots, \beta_{|v|}$ are enough to retrieve the original blocks $b_1, b_2, \ldots, b_{|v|}$. Along with every redundant block β_i, we also store its tag T_i.
3. An IBF T_v on the blocks contained in cover(v);

By using the segment tree, one can compute functions on *any subset* of $n - \delta$ blocks in $O(\delta \log n)$ time (instead of taking $O(n - \delta)$ time): For example, if $i_1, i_2, \ldots, i_\delta$ are the indices of the omitted δ blocks, the desired IBF T_K can be computed by combining (i.e., XORing the dataSum and hashSum fields and adding the count fields):

- The IBF T_1 corresponding to indices from 1 to $i_1 - 1$;
- The IBF T_2 corresponding to indices from $i_1 + 1$ to $i_2 - 1$;
- ...
- The IBF $T_{\delta+1}$ corresponding to indices from $i_\delta + 1$ to i_n.

Each one of the above IBFs can be computed in $O(\log n)$ time by combining a logarithmic number of IBFs stored at internal nodes of the segment tree and therefore the total complexity of computing the final IBF T_K is $O(\delta \log n)$. Similarly, a partial proof of storage for the lost blocks with indices $i_1, i_2, \ldots, i_\delta$ can be computed by returning.

- A proof of storage corresponding to indices from 1 to $i_1 - 1$;
- A proof of storage corresponding to indices from $i_1 + 1$ to $i_2 - 1$;
- ...
- A proof of storage corresponding to indices from $i_\delta + 1$ to i_n.

Again, each one of the above proofs of storage can be computed by returning $O(\log n)$ partial proofs of storage so in total, one needs to return $O(\delta \log n)$ proofs of storage. Note however that our segment tree increases our space to $O(n \log n)$ and also setting it up requires $O(n \log n)$ time. Therefore we have the following:

Theorem 2 (Sublinear δ-AS scheme). *Let n be the number of blocks. For all $\delta \leq n$, there exists a δ-AS scheme such that: (1) It is correct according to Definition 2; (2) It is secure in the random oracle model based on the RSA assumption and according to Definition 3; (3) The proof has size $O(\delta \log n)$ and its computation at the server takes $O(\delta \log n)$ time; (4) Verification at the client takes $O(\delta \log n)$ time and requires local state of size $O(\delta)$; (5) The space at the server is $O(n \log n)$.*

6 Bitcoin Integration

After the client computes the damage d using the AS protocol described in the previous section, we would like to enable automatic compensation by the server to the client in the amount of d bitcoins. The server initially makes a "security deposit" of A bitcoins by means of a special bitcoin transaction that automatically transfers A bitcoins to the client unless the server transfers d bitcoins to the client before a given deadline. Here, the amount A is a parameter that is contractually established by the client and server and is meant to be larger than the maximum damage that can be incurred by the server.[5]

We have designed a variation of the AS protocol integrated with Bitcoin that, upon termination, achieves one of the following outcomes within an established deadline:

1. If both the server and the client follow the protocol, the client gets exactly d bitcoins from the server and the server gets back his A bitcoins.
2. If the server does not follow the protocol (e.g., he tries to give fewer than d bitcoins to the client, fails to respond in a timely manner, or tries to forge an AS proof), the client gets A bitcoins from the server automatically.
3. If the client requests more than d bitcoins from the server by providing invalid evidence, the server receives all A deposited bitcoins back and the client receives nothing.

---

[5] Of course, this is just a simple setting, a proof of concept. Clearly other technical and financial instruments can be used to improve this approach if committing such a large amount of A bitcoins is too demanding.

Primitives Used in Our Protocol. Our protocol is using two primitives, which we describe informally in the following.

- A *trusted and tamper-resilient channel* between the client and the server, e.g., a bulletin board. This can be easily implemented by requesting all messages exchanged between the server and the client be posted on the blockchain (note that it is easy for a party P to post arbitrary data D on the blockchain by making a transaction to itself and by including D in the body of the transaction). From now on, we will assume that all messages are posted to the blockchain creating a history *hist*.
- A *trusted bitcoin arbitrator* BA. This is a trusted party that only intervenes in case of disputes. BA will always examine the history of transactions *hist* to assess the situation and determine whether to help the server. In all other cases it can remain offline.

Bitcoin Transactions. Our protocol uses three non-standard bitcoin transactions:

1. safeGuard(y): This transaction is posted by the server S and it effectively "freezes" A bitcoins to a hash output y. It can be redeemed by a transaction (called retBtcs) posted by the server S which provides the preimage x of $y = \mathsf{H}(x)$ or by a transaction (called fuse(t)) signed by both the client and the server. For the needs of our protocol, fuse(t) has a locktime t. The safeGuard transaction is the following:

Prev :	aTransaction
InputsToPrev :	$\mathsf{sig}_S([\mathsf{safeGuard}])$
Conditions :	$body, \sigma_1, \sigma_2, x$:
	$\mathsf{H}(x) = y \wedge \mathsf{ver}_S(body, \sigma_1)$
	$\vee$
	$\mathsf{ver}_S(body, \sigma_1) \wedge \mathsf{ver}_C(body, \sigma_2)$
Amount :	A ฿
Locktime :	0

We note here that transaction safeGuard(y) is based on the timed commitment over Bitcoin by Andrychowicz et al. [2], with an important difference: the committed value x (where $y = \mathsf{H}(x)$) is chosen by the verifier (client) and not by the committer (server). The server just uses y.

2. retBtcs: Once the server gets a hold of value x, it can post the following transaction to redeem safeGuard and retrieve his A bitcoins.

Prev :	safeGuard
InputsToPrev :	$[\text{retBtcs}], \text{sig}_S([\text{retBtcs}]), \perp, x$
Conditions :	$body, \sigma :$ $\text{ver}_S(body, \sigma)$
Amount :	A ฿
Locktime :	0

3. safeGuard can also be redeemed by fuse(t), as mentioned before:

Prev :	safeGuard
InputsToPrev :	$[\text{fuse}], \text{sig}_S([\text{fuse}]), \text{sig}_C([\text{fuse}]), \perp$
Conditions :	$body, \sigma :$ $\text{ver}_C(body, \sigma)$
Amount :	A ฿
Locktime :	t

Protocol Details. We now describe our protocol in detail, as depicted in Fig. 4. Let S denote the server and C the client. For each step $i = 1, \ldots, 10$, there is a deadline, t_i, to complete the step, where timelock t of the fuse(t) transaction is $\gg t_{10}$. We recall that, as mentioned in the beginning of this section, all messages exchanged between the client and the server are recorded on the blockchain, creating the history $hist$.

- **Step 1:** C picks a random secret x and sends the following items to S: (i) a hash $hash = \text{H}(\mathbf{T}_B)$ of the IBF of the original blocks he stores; (ii) an encryption $\text{Enc}_\text{P}(x)$ of x under BA's public key, P; (iii) a cryptographic hash of x, $y = \text{H}(x)$; and (iv) a zero-knowledge proof, ZKP_1, that $\text{H}(x)$ and $\text{Enc}_\text{P}(x)$ encode the same secret x. If ZKP_1 does not verify or is not sent within time t_1, S aborts the protocol.
- **Step 2:** S posts bitcoin transaction safeGuard(y) for A bitcoins. It also sends to the client a signature of the Fuse(t) transaction, $\sigma_S = \text{sig}_S([\text{fuse}])$. If this transaction is not posted within time t_2 or the signature is not valid or is not sent within time t_2, C aborts the protocol.
- **Step 3:** S and C run the AS protocol from the previous section. S returns proof $\mathcal{V} = \{\text{T}, S, \mathbf{T}_K\}$ and blocks $b'_1, b'_2, \ldots, b'_\delta$ for the current blocks he stores, in the position of the original blocks $b_1, b_2, \ldots, b_\delta$ that he lost. The client C computes now the damage d using CheckProof. If CheckProof rejects or S delays it past time t_3, C jumps to Step 9.
- **Step 4:** C notifies S that the damage is d and sends a zero-knowledge proof, ZKP_2, to S for that. If C fails to do so by t_4 or ZKP_2 fails to verify, S jumps to Step 6. We note here that ZKP_2 is for the statement $(hash, \mathcal{V}, b'_1, b'_2, \ldots, b'_\delta, d, \text{chal})$: $\exists$ $secret$ $\mathbf{T}_B$ such that

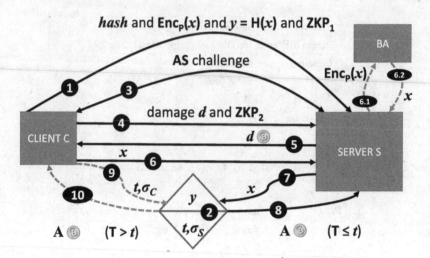

Fig. 4. Integration of the AS protocol with Bitcoin. The dotted lines indicate what happens when a party is trying to cheat. The rhomboid indicates the safeGuard(y) transaction (at the bottom of the rhomboid we show the server signature σ_S on the fuse(t) transaction) that is redeemed either by retBtcs (arrow 7) or by fuse (arrow 9), depending on the flow of the protocol.

$$hash = \mathsf{H}(\mathbf{T}_B) \,\&\, \{b_i\}_{i=1}^{\delta} \leftarrow \mathsf{CheckProof}(\mathsf{pk}, \mathbf{T}_B, \mathcal{V}, \mathsf{chal}) \,\&\, d = \sum_{i=1}^{\delta} ||b_i \oplus b_i'||.$$

The zero-knowledge proof ZKP_2 is needed here since the recovery algorithm takes as input the sensitive state of the client $\mathbf{T}_B$, which we want to hide from the server—otherwise the server can recover the original blocks himself and claim that there was no damage.

- **Step 5:** S sends d bitcoins to C. If S has not done so by time t_5, C jumps to Step 9.
- **Step 6:** C sends secret x to S. If S has not received x by t_6, S contacts BA and asks the BA to examine the history $hist$ up to that moment. BA checks $hist$ and if it is valid, BA sends x to S.
- **Step 7:** If S has secret x, S posts transaction retBtcs.
- **Step 8:** If transaction retBtcs is valid, S receives A bitcoins before timelock t.
- **Step 9:** C waits until time t, computes $\sigma_C = \mathsf{sig}_S([\mathsf{fuse}])$ and posts transaction fuse(t) using σ_C and σ_S.
- **Step 10:** If transaction fuse is valid, C receives A bitcoins.

It is easy to see that when the above protocol terminates, one of the three outcomes described in the beginning of this section is achieved. We emphasize that BA can determine whether C properly followed the protocol by analyzing $hist$. If C has not done so and S reports it, BA will reveal x to S at any point in time. Also, we note here that for the zero-knowledge proofs ZKP_1 and ZKP_2, we can use a SNARK with zero-knowledge [25], that was recently implemented and shown to be practical.

Global safeGuard. The protocol above protects the client at each AS challenge. But the cloud provider could stop interacting, simply disappear, and never be reachable by the client. Instead of the client aborting, we can use a *global* safeguard transaction at the time the client and the server initiate their business relationship (i.e., when the client uploads the original file blocks and they both sign the SLA). This global transaction is meant to protect the client if the server cannot be reached at all or refuses to collaborate, but creates a scalability problem given that the server has to escrow a large amount of bitcoins for every client/customer. We do not address this problem technically but we expect it can be mitigated through financial mechanisms (securities, commodities, credit, etc.) typically deployed for traditional escrow accounts.

Removing the Bitcoin Arbitrator. Even though BA is only involved in case of disputes, it is preferable to remove it completely. Unfortunately, this seems impossible to achieve *efficiently* given the limitations of the Bitcoin scripting language. We sketch in this section two possible approaches to remove the BA. These will be further explored in future work.

The first approach relies on a secure two-party computation protocol. In a secure two-party computation protocol (2PC), party A inputs x and party B inputs y and they want to compute $f_A(x, y)$ and $f_B(x, y)$ respectively, without learning each other's input other than what can be inferred from the output of the two functions. Yao's seminal result [31] showed that oblivious transfer implies 2PC secure against honest-but-curious adversaries. This result can be extended to generically deal with malicious adversaries through zero-knowledge proofs or more efficiently via the *cut-and-choose* method [20] or LEGO and MiniLEGO [14,24] (other efficient solutions were proposed in [18,30]).

To remove the BA, it is enough to create a symmetric version of our original scheme where both parties create a safeGuard transaction and then exchange the secrets of both commitments through a fair exchange protocol embedded into a 2PC. The secrets must be verifiable in the sense that the fair exchange must ensure the secrets open the initial commitments or fail (as in "committed 2PC" by Jarecki and Shmatikov [18]). Unfortunately, generic techniques for 2PC results in quite impractical schemes and this is the reason why we prefer a practical solution with an arbiter. An efficient 2PC protocol with Bitcoin is proposed in [22] but it does not provide fairness since the 2PC protocol can be interrupted at any time by one of the parties. In the end, since this generic approach is too expensive in practice, we will not elaborate on it any further in this paper.

Another promising approach to remove the BA is to adopt smart contracts. Smart contracts are digital contracts that run through a blockchain. Ethereum [1] is a new cryptocurrency system that provides a Turing-complete language to write such contracts, which is expected to enable several decentralized applications without trusted entities. Smart contracts will enable our protocol to be fully automated without any arbitrators or trusted parties in between. To use a smart contract to run our protocol, the contract is expected to receive a deposit from the server, inputs from both parties, and then it will decide the money flow accordingly based on the CheckProof result. The contract in that case will maintain some properties to ensure fair execution, for example both parties should be incentivized to follow the protocol, and if any party does not follow the protocol (by aborting for example), there should be a mechanism to end the protocol properly for the honest party. To make our protocol fit in the smart contract model, we will need to address the fact that the CheckProof computation would be too expensive to be performed by the contract, due to its overhead. In Ethereum for example, the participants must pay for the cost of running the contract, which is run by the miners.

In order to address the points above, zero knowledge SNARKs [6] could be employed to help reduce miners' overhead, and thus reduce the computational cost of the verification algorithm running on the network, while preserving the secrecy of the inputs.

7   Evaluation

We prototyped the proposed Accountable Storage (AS) scheme in Python 2.7.5. Our implementation is open-source[6] and consists of 4 K lines of source code. We use the pycrypto library 2.6.1 [21] and an RSA modulus N of size 1024 bits. We serialize the protocol messages using Google Protocol Buffers [16] and perform all the modulo exponentation operations using GMPY2 [17], which is a C-coded Python extension module that supports fast multiple precision arithmetic (the use of GMPY2 gave us 60% speedup in exponentiations in comparison with the regular python arithmetic library).

We divide the prototype in two major components. The first is responsible for data pre-processing, issuing proof challenges and verifying proofs (including recover process). The second produces proof every time it receives a challenge. Both modules utilize the IBF data structure to produce and verify proofs. Our prototype uses parallel computing via the Python multiprocessing module to carry out many of the heavy, but independent, cryptographic operations simultaneously. We used a *single-producer, many-consumers* approach to divide the available tasks in a pool of [8–12] processes-workers. The workers use message passing to coordinate and update the results of their computations. This approach significantly enhanced the performance of preprocessing as well as the proof generation and checking phase of the protocol. Our parallel implementation provides an approximate 5x speedup over a sequential implementation.

Finally note that since it can be easily estimated, we have not evaluated the Bitcoin part of our protocol which is dominated by the time it takes for transactions to be part of the blockchain. Nowadays this latency is approximately 10 min.

Experimental Setup: Our experimental setup involves two nodes, one implementing the server and another implementing the client functionality. The two

---

[6] https://github.com/vlekakis/delta-AccountableStorage.

nodes communicate through a Local Area Network (LAN). The two machines are equipped with an Intel 2.3 Ghz Core i7 processor and have 16 GB of RAM.

Our data are randomly generated filesystems. Every file-system includes different number of equally-sized blocks. The number of blocks ranges from 100 to 500000 and the different sizes of blocks used are 1 KB, 2 KB, 4 KB and 8 KB. The total filesystem size varies from 100 KB to 4.1 GB. Our experiments consist of 10 trials of challenge/proof exchanges between the client and the server for different filesystems. Throughout the evaluation we report the average values over these 10 trials. For some of the large filesystems consisting of 100000 and 500000 number of blocks, we have estimated the results based on the experiments in lower filesystems due to computational power limitations.

In our experiments, we select the tolerance parameter δ, which indicates the maximum amount of data blocks that can be lost, to be equal to $\log_2(n)$. One other possible choice of δ is to set it equal to $\sqrt{n}$. We select the logarithm of the number of blocks as δ, because this provides a harder condition on how many blocks can be lost or corrupted from the cloud server.

For the IBF construction, we have used the blocks and their generated tags. The selected number of hash functions used for the IBF construction is $k = 6$. This choice of hash functions leads to a very low probability of failure of the recovery algorithm, which depends on the values of k and δ.

Preprocessing Overheads: We first examine the memory overhead of the preprocessing phase, which is shown in Table 1. The first column describes the available number of blocks in a filesystem and the second represents the estimated total size of the tags needed. The preprocessing memory overhead is proportional to the number of blocks in a filesystem.

Figure 5 shows the CPU-time-related overheads of the preprocessing of the protocol. These overheads are divided to tag generation and the creation of the client state represented by the IBF T_B. The tag generation time (Fig. 5a) increases linearly with both the available number of blocks and the size of each block. While this cost is significant for large file systems, it is an operation that client performs only once at the setup phase. On the other hand, the cost of construction of the IBF (Fig. 5b) is estimated to be negligible; the IBF construction of our biggest filesystem is estimated to take around 42 s.

Table 1. Memory footprint of the AS scheme (KB)

n	Tag size (KB)	Proof size (KB)			
		$1KB$	$2KB$	4 KB	8 KB
10^2	32	353	692	1369	2722
10^3	236	528	1035	2048	4073
10^4	2644	762	1493	2593	5875
10^5	24895	937	1898	3526	7226
$5 * 10^5$	118326	1077	2182	4054	8308

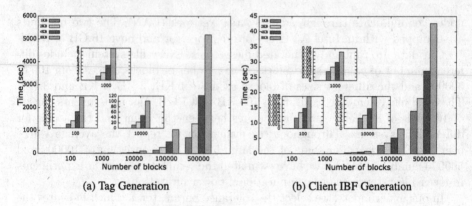

(a) Tag Generation   (b) Client IBF Generation

Fig. 5. Preprocessing overheads

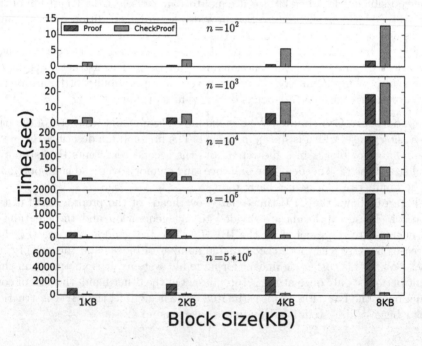

Fig. 6. Proof generation and proof check (including recover) time

Challenge-Proof Overheads: We now examine memory and CPU-related overheads for the challenge-proof exchange and the recovery phase. The last four columns of Table 1 show the proof sizes (in KB) for $\delta = \log_2(n)$, which increase proportionally to the block size.

Every subgraph of Fig. 6 shows how different block sizes affect the performance of the challenge-proof exchange for a given number of blocks. The left bar in the figure shows the proof generation time and the right bar the proof check along that includes the time recover the lost blocks. We notice that larger block sizes are estimated to increase the time-overhead of challenge-proof exchange. We also notice that the proof check and in particular the recover process introduces higher time overhead in comparison to proof generation for small size filesystems (100 and 1000 number of blocks). This is expected, because the tag verification (publicly verifiable) in the recover process (in Fig. 2) introduces significant time overhead compared to proof generation (in Fig. 3) for a small number of blocks. However, in higher size filesystems, we observe from Fig. 6 that the time overhead of proof generation increases exponentially and overcomes the proof check time (including recover process) overhead. This is also expected, because the number of blocks (denoted by Kept set) used in the proof generation process is much higher than the number of blocks used in the recover process.

8   Conclusions

In this paper we put forth the notion of accountability in cloud storage. Unlike existing work such as proof-of-storage schemes and verifiable computation, we design protocols that respond to a verification failure, enabling the client to assess the damage that has occurred in a storage repository. We also present a protocol that enables automatic compensation of the client, based on the amount of damage, and is implemented over Bitcoin. Our implementation shows that our system can be used in practice.

Acknowledgments. Research supported in part by an NSF CAREER award CNS-1652259, NSF grants CNS-1525044, CNS-1526950, CNS-1228639 and CNS-1526631, a NIST award and by the Defense Advanced Research Projects Agency (DARPA) under agreement no. AFRL FA8750-15-2-0092. The views expressed are those of the authors and do not reflect the official policy or position of the Department of Defense or the U.S. Government.

Appendix

RSA Assumption

Definition 4. *Let $N = pq$ be an RSA modulus, where p and q are τ-bit primes. Given N, e and g, where g is randomly chosen from $\mathbb{Z}_N^*$ and e is a prime of $\Theta(\tau)$ bits, there is no PPT algorithm that can output $y^{1/e} \bmod N$, except with probability $\mathsf{neg}(\tau)$.*

Proof of Security of Construction in Fig. 3.

Setup. $\mathcal{A}$ chooses parameter $\delta \in [0, n)$, blocks $B = \{b_1, b_2, \ldots, b_n\}$ and is given pk and $\mathsf{T}_1, \ldots, \mathsf{T}_n$ as output by $\{\mathsf{pk}, \mathsf{sk}, \mathsf{state}, \mathsf{T}_1, \ldots, \mathsf{T}_n\} \leftarrow \mathsf{Setup}(b_1, \ldots, b_n, \delta, 1^\tau)$. The random oracle is programmed so that it returns $r_i^e g^{-b_i}$ on input i for some random r_i, i.e., $h(i) = r_i^e g^{-b_i} \bmod N$.

Forge. $\mathcal{A}$ is given chal $\leftarrow$ GenChal(1^τ), computes proof of accountability $\mathcal{V}$ and returns $\mathcal{V}$. Suppose $\mathcal{L} \leftarrow$ CheckProof(pk, sk, state, $\mathcal{V}$, chal). We must show that with probability $\geq 1 - \mathsf{neg}(\tau)$ it is (i) $\mathcal{L} \subseteq B$; and (ii) there exists a PPT knowledge extractor $\mathcal{E}$ that can extract *all the remaining file blocks* in $B - \mathcal{L}$.

1. **Showing $\mathcal{L} \subseteq B$.** Note that all blocks in $\mathcal{L}$ are output by Algorithm recover of Fig. 2. In this algorithm a block b_i' can enter $\mathcal{L}$ only if its tag verifies. Suppose now $b_i' \notin B$ (namely $b_i' \neq b_i$) and $\mathsf{tag}^e/h(i) = g^{b_i'}$ for some arbitrary tag computed by the adversary. But since $h(i) = r_i^e g^{-b_i}$, this can be written as $\mathsf{tag}^e/r_i^e g^{-b_i} = g^{b_i'}$ which gives $g^{b_i'-b_i} = (\mathsf{tag}/r_i)^e$. Since e is a prime and $b_i - b_i' \neq 0$, there exist α and β such that $(b_i' - b_i) \times \alpha + e \times \beta = 1$, giving $g^{1/e} = g^{-\beta}(\mathsf{tag}/r_i)^{e \times \alpha}$, breaking the RSA assumption—see Definition 4.

2. **Showing there exists a PPT knowledge extractor $\mathcal{E}$ that can extract all the remaining file blocks in $B - \mathcal{L}$.** We now show how to build an extractor that, after $\ell = |B - \mathcal{L}|$ interactions with the adversary, he can extract the blocks $\{b_i : i \in B - \mathcal{L}\}$. The extractor will challenge the adversary exactly ℓ times, each time with different randomness. Let $S_1, S_2, \ldots, S_\ell, \mathrm{T}^{(1)}, \mathrm{T}^{(2)}, \ldots, \mathrm{T}^{(\ell)}$ be the sums and tags he receives by $\mathcal{A}$ during each challenge, as in Eq. (2). We have two cases:

   (a) $S_j = \sum_{i \in B-\mathcal{L}} a_{ij} b_i$, for $j \in B - \mathcal{L}$ (a_{ij} denotes the randomness of the j-th challenge corresponding to the i-th block). In this case, the extractor can solve a system of ℓ linear equations and retrieve the original blocks $\{b_i : i \in B - \mathcal{L}\}$.

   (b) Suppose there exists $j \in B - \mathcal{L}$ such that $S_j \neq \sum_{i \in B-\mathcal{L}} a_{ij} b_i = S$. For simplicity of notation, let's set $\mathrm{T}^{(j)} = \mathrm{T}$ and $S_j = \bar{S}$. Then by the CheckProof algorithm we have

$$\frac{\mathrm{T}^e}{\prod_{i \in B-\mathcal{L}} h(i)^{a_i}} = g^{\bar{S}}.$$

   But since $h(i) = r_i^e g^{-b_i}$ we have that

$$g^{\bar{S}-S} = \left(\frac{\mathrm{T}}{\prod_{i \in B-\mathcal{L}} r_i^{a_i}}\right)^e = \mathcal{Z}^e.$$

   Therefore we have $\mathcal{Z}^e = g^{\bar{S}-S}$. Again, since e is prime and $S \neq \bar{S}$ we can use the same trick as before, and break the RSA assumption. $\qquad\square$

References

1. Ethereum: A platform for decentralized applications. www.ethereum.org/
2. Andrychowicz, M., Dziembowski, S., Malinowski, D., Mazurek, L.: Secure multi-party computations on bitcoin. In: IEEE SSP (2014)
3. Ateniese, G., Burns, R., Curtmola, R., Herring, J., Kissner, L., Peterson, Z., Song, D.: Provable data possession at untrusted stores. In: ACM CCS (2007)
4. Ateniese, G., Di Pietro, R., Mancini, L.V., Tsudik, G.: Scalable and efficient provable data possession. In: SecureComm (2008)

5. Bellare, M., Goldreich, O.: On defining proofs of knowledge. In: Brickell, E.F. (ed.) CRYPTO 1992. LNCS, vol. 740, pp. 390–420. Springer, Heidelberg (1993). doi:10. 1007/3-540-48071-4_28

6. Ben-Sasson, E., Chiesa, A., Genkin, D., Tromer, E., Virza, M.: SNARKs for C: verifying program executions succinctly and in zero knowledge. In: Canetti, R., Garay, J.A. (eds.) CRYPTO 2013. LNCS, vol. 8043, pp. 90–108. Springer, Heidelberg (2013). doi:10.1007/978-3-642-40084-1_6

7. Bloom, B.H.: Space/time trade-offs in hash coding with allowable errors. Comm. ACM **13**, 422–426 (1970)

8. Boneh, D., Naor, M.: Timed commitments. In: Bellare, M. (ed.) CRYPTO 2000. LNCS, vol. 1880, pp. 236–254. Springer, Heidelberg (2000). doi:10.1007/3-540-44598-6_15

9. Carter, I.L., Wegman, M.N.: Universal classes of hash functions. In: ACM STOC (1977)

10. Cash, D., Küpçü, A., Wichs, D.: Dynamic proofs of retrievability via oblivious RAM. In: Johansson, T., Nguyen, P.Q. (eds.) EUROCRYPT 2013. LNCS, vol. 7881, pp. 279–295. Springer, Heidelberg (2013). doi:10.1007/978-3-642-38348-9_17

11. Curtmola, R., Khan, O., Burns, R.C., Ateniese, G.: MR-PDP: multiple-replica provable data possession. In: ICDCS (2008)

12. Eppstein, D., Goodrich, M.T., Uyeda, F., Varghese, G.: What's the difference? Efficient set reconciliation without prior context. In: SIGCOMM (2011)

13. Erway, C., Küpçü, A., Papamanthou, C., Tamassia, R.: Dynamic provable data possession. In: ACM CCS (2009)

14. Frederiksen, T.K., Jakobsen, T.P., Nielsen, J.B., Nordholt, P.S., Orlandi, C.: MiniLEGO: efficient secure two-party computation from general assumptions. In: Johansson, T., Nguyen, P.Q. (eds.) EUROCRYPT 2013. LNCS, vol. 7881, pp. 537–556. Springer, Heidelberg (2013). doi:10.1007/978-3-642-38348-9_32

15. Goodrich, M.T., Mitzenmacher, M.: Invertible Bloom Lookup Tables. ArXiv e-prints, January 2011

16. Google. Google protocol buffers. www.developers.google.com/protocol-buffers/

17. Van Horsen, C.: Gmpy2: Mupltiple-precision arithmetic for python. www.gmpy2. readthedocs.org/en/latest/intro.html/

18. Jarecki, S., Shmatikov, V.: Efficient two-party secure computation on committed inputs. In: Naor, M. (ed.) EUROCRYPT 2007. LNCS, vol. 4515, pp. 97–114. Springer, Heidelberg (2007). doi:10.1007/978-3-540-72540-4_6

19. Juels, A., Kaliski Jr. B.S.: PORs: proofs of retrievability for large files. In: ACM CCS (2007)

20. Lindell, Y., Pinkas, B.: An efficient protocol for secure two-party computation in the presence of malicious adversaries. In: Naor, M. (ed.) EUROCRYPT 2007. LNCS, vol. 4515, pp. 52–78. Springer, Heidelberg (2007). doi:10.1007/978-3-540-72540-4_4

21. Litzenberger, D.C.: Pycrypto - the python cryptography toolkit. www.dlitz.net/software/pycrypto/

22. Andrychowicz, M., Dziembowski, S., Malinowski, D., Mazurek, L.: Fair two-party computations via the bitcoin deposits. In: FC (2014)

23. Nakamoto, S.: Bitcoin: a peer-to-peer electronic cash system. www.bitcoin.org/bitcoin.pdf

24. Nielsen, J.B., Orlandi, C.: LEGO for two-party secure computation. In: Reingold, O. (ed.) TCC 2009. LNCS, vol. 5444, pp. 368–386. Springer, Heidelberg (2009). doi:10.1007/978-3-642-00457-5_22

25. Parno, B., Howell, J., Gentry, C., Raykova, M.: Pinocchio: nearly practical verifiable computation. In: IEEE SSP (2013)
26. Shacham, H., Waters, B.: Compact proofs of retrievability. In: Pieprzyk, J. (ed.) ASIACRYPT 2008. LNCS, vol. 5350, pp. 90–107. Springer, Heidelberg (2008). doi:10.1007/978-3-540-89255-7_7
27. Shi, E., Stefanov, E., Papamanthou, C.: Practical dynamic proofs of retrievability. In: ACM CCS (2013)
28. Stefanov, E., van Dijk, M., Oprea, A., Juels, A.: Iris: a scalable cloud file system with efficient integrity checks. In: ACSAC (2012)
29. Wang, Q., Wang, C., Li, J., Ren, K., Lou, W.: Enabling public verifiability and data dynamics for storage security in cloud computing. In: Backes, M., Ning, P. (eds.) ESORICS 2009. LNCS, vol. 5789, pp. 355–370. Springer, Heidelberg (2009). doi:10.1007/978-3-642-04444-1_22
30. Woodruff, D.P.: Revisiting the efficiency of malicious two-party computation. In: Naor, M. (ed.) EUROCRYPT 2007. LNCS, vol. 4515, pp. 79–96. Springer, Heidelberg (2007). doi:10.1007/978-3-540-72540-4_5
31. Yao, A.C.-C.: How to generate and exchange secrets. In: SFCS (1986)

Maliciously Secure Multi-Client ORAM

Matteo Maffei[1], Giulio Malavolta[2], Manuel Reinert[3(✉)],
and Dominique Schröder[2]

[1] TU Wien, Wien, Austria
matteo.maffei@tuwien.ac.at
[2] Friedrich-Alexander Universität Erlangen-Nürnberg, Nürnberg, Germany
{giulio.malavolta,dominique.schroeder}@fau.de
[3] CISPA, Saarland University, Saarbrücken, Germany
reinert@cs.uni-saarland.de

Abstract. Oblivious RAM (ORAM) has emerged as an enabling technology to secure cloud-based storage services. The goal of this cryptographic primitive is to conceal not only the data but also the access patterns from the server. While the early constructions focused on a single client scenario, a few recent works have focused on a setting where multiple clients may access the same data, which is crucial to support data sharing applications. All these works, however, either do not consider malicious clients or they significantly constrain the definition of obliviousness and the system's practicality. It is thus an open question whether a natural definition of obliviousness can be enforced in a malicious multi-client setting and, if so, what the communication and computational lower bounds are.

In this work, we formalize the notion of maliciously secure multi-client ORAM, we prove that the server-side computational complexity of any secure realization has to be $\Omega(n)$, and we present a cryptographic instantiation of this primitive based on private information retrieval techniques, which achieves an $O(\sqrt{N})$ communication complexity. We further devise an efficient access control mechanism, built upon a novel and generally applicable realization of plaintext equivalence proofs for ciphertext vectors. Finally, we demonstrate how our lower bound can be bypassed by leveraging a trusted proxy, obtaining logarithmic communication and server-side computational complexity. We implemented our scheme and conducted an experimental evaluation, demonstrating the feasibility of our approach.

1 Introduction

Oblivious RAM. Cloud storage has rapidly become a central component in the digital society, providing a seamless technology to save large amounts of data, to synchronize them across multiple devices, and to *share* them with other parties. Popular data-sharing, cloud-based applications are e.g., personal health record management systems (PHRs), like those employed in Austria [22] and Estonia [20], collaborative platforms (e.g., Google Docs), and credit score

© Springer International Publishing AG 2017
D. Gollmann et al. (Eds.): ACNS 2017, LNCS 10355, pp. 645–664, 2017.
DOI: 10.1007/978-3-319-61204-1_32

systems (e.g., Experian, Equifax, and TransUnion in US) are just a few popular data-sharing, cloud-based applications taking advantage of such features. A stringent and well-understood security requirement is *access control*: read and write access should be granted only to authorized clients.

While access control protects user's data from other clients, encryption on the server's side is needed to obtain privacy guarantees against cloud administrators. Encryption is, however, not enough: as shown in the literature [28, 39], the capability to observe which data are accessed by which users allows the cloud administrator to learn sensitive information: for instance, it has been shown that the access patterns to a DNA sequence allow for determining the patient's disease. The property of hiding data accesses is called *obliviousness* and the corresponding cryptographic construction *Oblivious RAM* (ORAM): while the first constructions were highly inefficient [23], recent groundbreaking research paved the way for a tremendous efficiency boost, exploiting ingenious tree-based constructions [2, 3, 8, 14, 15, 24, 32, 39, 44, 45, 47], server side computations [26, 35], and trusted hardware [5, 27, 31, 42, 48].

Except for a few recent noticeable exceptions, discussed below, a fundamental limitation of all these constructions is that they target a single-client architecture, where the data owner is the only party allowed to read outsourced data, which does not make them suitable for data sharing services. The fundamental challenge to solve is to enforce access control and obliviousness *simultaneously*. These properties are seemingly contradictory: can the server check the correctness of data accesses with respect to the access control policy at all, if it is not allowed to learn anything about them?

Multi-Client ORAM. A few recent constructions gave positive answers to this question, devising ORAM constructions in the multi-client setting, which specifically allow the data owner to share data with other clients while imposing fine-grained access control policies. Although, at a first glance, these constructions share the same high-level goal, they actually differ in a number of important aspects. Therefore we find it interesting to draw a systematic comparison among these approaches (cf. Table 1). First of all, obliviousness is normally defined against the server, but in a multi-client setting it is important to consider it against the clients too (**MC**), since they might be curious or, even worse, collude with the server. This latter aspect is important, since depending on the application, the cloud administrator might create fake clients or just have common interests with one of the legitimate clients. Some constructions allow multiple data owners to operate on the same ORAM (**MD**), while others require them to use disjoint ORAMs: the latter are much less efficient, since if the client does not want to reveal the owner of the accessed entry (e.g., to protect her anonymity, think for instance of the doctor accessing the patient's record), then the client has to perform a fake access to each other ORAM, thereby introducing a multiplicative factor of $O(m)$, where m is the number of data owners. Some constructions require the data owner to periodically access the dataset in order to validate previous accesses (**PI**), some others rely on server-side client synchronization, which can be achieved for instance by a shared log on the server, a gossiping protocol among clients, etc. (**CS**), while others assume a trusted proxy (**Pr**). Among these, gossiping is the mildest assumption since it can be realized directly on the server side as described by [30]. Another aspect to consider is the possibility for the data owner to specify fine-grained access control mechanisms (**AC**). Finally, some constructions enable concurrent accesses to the ORAM (**P**). The final three columns compare the asymptotic complexity of server-side and client-side computations as well as communication.

Table 1. Comparison of the related work supporting multiple clients to our constructions. The abbreviations mean: **MC**: oblivious against malicious clients, **MD**: supports multiple data owners sharing their data in one ORAM, **PI**: requires the periodic interaction with the data owner, **CS**: requires synchronization among clients, **AC**: access control, **Pr**: trusted proxy, **P**: parallel accesses, **S comp.**: server computation complexity, **C comp.**: client communication complexity, **Comm.**: communication complexity.

Work	MC	MD	PI	CS	Pr	AC	P	S comp.	C comp.	Comm.
Franz et al. [21]	✓	✗	✓	✗	✗	✓	✗	$O(\sqrt{n})$	$O(\sqrt{n})$	$O(\sqrt{n})$
GORAM [33]	✗	✗	✗	✓	✗	✓	✗	$O(\log(n))$	$O(\log(n))$	$O(\log(n))$
PIR-MCORAM (this work)	✓	✓	✗	✓	✗	✓	✗	$O(n)$	$O(\sqrt{n})$	$O(\sqrt{n})$
BCP-OPRAM [7]	✗	✗	✓	✓	✗	✗	✓	$\Omega(\log^3(n))$	$\Omega(\log^3(n))$	$\Omega(\log^3(n))$
CLT-OPRAM [10]	✗	✗	✓	✓	✗	✗	✓	$O(\log^2(n))$	$O(\log^2(n))$	$O(\log^2(n))$
PrivateFS [49]	✗	✗	✗	✓	✗	✗	✓	$O(\log^2(n))$	$O(1)$	$O(\log^2(n))$
Shroud [31]	✗	✗	✗	✗	✓	✗	✓	$O(\log^2(n))$	$O(1)$	$O(\log^2(n))$
TaoStore [42]	✗	✗	✗	✗	✓	✗	✓	$O(\log(n))$	$O(1)$	$O(\log(n))$
TAO-MCORAM (this work)	✓	✓	✗	✗	✓	✓	✓	$O(\log(n))$	$O(1)$	$O(\log(n))$

Franz et al. pioneered the line of work on multi-client ORAM, introducing the concept of delegated ORAM [21]. The idea of this construction, based on simple symmetric cryptography, is to let clients commit their changes to the server and to let the data owner periodically validate them according to the access control policy, finally transferring the valid entries into the actual database. Assuming periodic accesses from the data owner, however, constrains the applicability of this technique. Furthermore, this construction does not support multiple data owners. Finally, it guarantees the obliviousness of access patterns with respect to the server as well as malicious clients, excluding however the accesses on data readable by the adversary. While excluding write operations is necessary (an adversary can clearly notice that the data has changed), excluding read operations is in principle not necessary and limits the applicability of the obliviousness definition: for instance, we would like to hide the fact that an oncologist accessed the PHR of a certain patient even from parties with read access to the PHR (e.g., the pharmacy, which can read the prescription but not the diagnosis).

More recently, Maffei *et al.* [33] proposed the notion of group ORAM, in which the server performs access control by verifying client-provided zero-knowledge proofs: this approach enables direct client accesses without any interaction with the data owner and more generic access control policies. The scheme relies on a gossiping protocol, but malicious clients are considered only in the context of access control and, indeed, obliviousness does not hold against them.

Another line of work, summarized in the lower part of Table 1, focuses on the parallelization of client accesses, which is crucial to scale to a large number of clients, while retaining obliviousness guarantees. Most of them [5,31,42,48] assume a trusted proxy performing accesses on behalf of users, with TaoStore [42] being the most efficient and secure among them. These constructions do not formally consider obliviousness against malicious clients nor access control, although a contribution of this work is to prove that a simple variant of TaoStore [42] guarantees both. Finally, instead of a trusted proxy, BCP-OPRAM [7] and CLT-OPRAM [10] rely on a gossiping protocol while PrivateFS [49] assumes a client-maintained log on the server-side, but they do not achieve obliviousness against malicious clients nor access control. Moreover, PrivateFS guarantees concurrent client accesses only if the underlying ORAM already does so.

To summarize, the progress in the field does not answer a few foundational questions, which touch the core of the application of ORAM technologies in cloud-based data-sharing applications. First, is it possible at all to enforce the obliviousness of data accesses without constraining the security definition or placing severe system assumptions? If the answer is positive, it would be interesting to know at what computational cost.

Our Contributions. This work answers the questions above, providing a foundational framework for multi-client ORAM. In particular,

- We give for the first time a formal definition of *obliviousness against malicious clients in the multi-client setting*. Intuitively, none should be able to determine which entry is read by which client. However, write operations are oblivious only with respect to the server and to those clients who cannot read the modified entry, since clients with read access can obviously notice that the entry has changed.
- We establish an insightful *computational lower bound*: in a multi-client setting where clients have *direct access* to the database, the number of operations *on the server side* has to be linear in the database size. Intuitively, the reason is that if a client does not want to access all entries in a read operation, then it must know where the required entry is located in the database. Since malicious clients can share this information with the server, the server can determine for each read operation performed by an honest client, which among the entries the adversary has access to might be the subject of the read, and which certainly not.
- We present PIR-MCORAM, the first *cryptographic construction* that ensures the obliviousness of data accesses as well as access control in a malicious multi-client setting. Our construction relies on Private Information Retrieval (PIR) [11] to achieve obliviousness and uses new accumulation technique

based on an oblivious gossiping protocol to reduce the communication bandwidth in an amortized fashion. Moreover, it combines public-key cryptography and zero-knowledge proofs for access control.

- We present a novel technique based on universal pair-wise hash functions [9] in order to speed up the efficiency of Plaintext Equivalence Proofs, a computationally demanding cryptographic building block of PIR-MCORAM. This construction is generally applicable and we show that it improves solutions recently adopted in the multi-client ORAM literature [33] by one order of magnitude.
- To bypass the aforementioned lower bound, we consider the recently proposed *proxy-based Setting* [5,31,42,48,49], which assumes the presence of a trusted party mediating the accesses between clients and server. We prove, in particular, that a simple variant of TaoStore [42] guarantees obliviousness in the malicious setting as well as access control.
- We implement PIR-MCORAM and conduct an *experimental evaluation* of our schemes. PIR-MCORAM constitutes a practical solution for databases of modest size: for instance, DNA encoded in VCF files requires approximately 125MB [40]. Thus, an extended personal health record fits without problems in a 256MB database, for which a read or write operation in PIR-MCORAM takes approximately 14 seconds amortized. TaoStore offers much better performance as well as support for parallel accesses, but it assumes a trusted proxy.

2   A Lower Bound for Maliciously Secure Multi-Client ORAM

In this section, we study how much computational effort is necessary to securely realize ORAM in the malicious multi-client setting. Our result shows that *any* construction, regardless of the underlying computational assumptions, must access the entire memory (up to a constant factor) in every operation. Our lower bound can be seen as a generalization of the result on history independence of Roche et al. [41], in the sense that they consider a "catastrophic attack" where the complete state of the client is leaked to the adversary, whereas we allow only the corruption of a certain subset of clients. Note that, while the bound in [41] concerns the *communication* complexity, our result only bounds the *computation* complexity on the server side.

Before stating our lower bound, we formalize the notion of Multi-Client ORAM in the malicious setting. We follow the definitional framework introduced by Maffei *et al.* [33], refining the obliviousness definition in order to consider malicious clients possibly colluding with the server.

2.1   Multi-Client Oblivious RAM

In a Multi-Client ORAM scheme the parties consist of the data owner $\mathcal{O}$, several clients $\mathcal{C}_1, \ldots, \mathcal{C}_k$, and the server $\mathcal{S}$. The data owner outsources its database $\mathcal{DB}$

to S while granting access to the clients $C_1, \ldots, C_k$ in a selective manner. This is expressed by an access control matrix ACM which has an entry ACM(i, idx) for every client C_i and every entry idx in the database, characterizing which access right C_i has for entry idx: either no access ($\bot$), read-only access (R), or read-write access (RW). We treat ACM as a global variable for the data owner so as to ease the presentation. Moreover, ACM is only accessible to the data owner and not to any client. We write $o \leftarrow A(\ldots)$ to denote that algorithm A on some input generates output o. Likewise, we write $\langle o_C, o_S \rangle \leftarrow \langle A(\ldots), S_A(\ldots) \rangle$ to denote that the protocol A executed between the client and the server yields client output o_C and server output o_S.

Definition 1 (Multi-Client ORAM [33]). *A Multi-Client ORAM scheme Θ is composed of the following (interactive)* PPT *algorithms:*

$(cap_O, \mathcal{DB}) \leftarrow \text{gen}(1^\lambda, n)$. *The generation algorithm initializes a database $\mathcal{DB}$ of size n and an empty access control matrix* ACM. *Finally, the algorithm returns the data owner's capability cap_O.*

$cap_i \leftarrow \text{addCl}(cap_O, i)$. *The input of the add client algorithm is the data owner's capability cap_O and a client identifier i. It appends a row corresponding to i in* ACM *such that for all $j \in \{1, \ldots, n\}$:* ACM(i, j) $= \bot$. *The algorithm outputs the capability for client C_i.*

$\langle \bot, \mathcal{DB}' \rangle \leftarrow \langle \text{addE}(cap_O, \text{idx}, \text{data}), S_{\text{addE}}(\mathcal{DB}) \rangle$. *The add entry algorithm takes as input the data owner's capability cap_O, an index idx, and a data data in interaction with S that takes $\mathcal{DB}$ as input. It appends a column corresponding to idx in* ACM *such that for all $i \in \{1, \ldots, k\}$:* ACM(i, idx) $= \bot$, *writes data at position idx in $\mathcal{DB}$, and outputs the modified database $\mathcal{DB}'$ on S.*

$\langle \bot, \mathcal{DB}' \rangle \leftarrow \langle \text{chMode}(cap_O, \text{idx}, i, p), S_{\text{chMode}}(\mathcal{DB}) \rangle$. *The change mode algorithm takes as input the data owner's capability cap_O, some index idx, a client identifier i, and a permission $p \in \{\text{R}, \text{RW}, \bot\}$ in interaction with S that takes $\mathcal{DB}$ as input. It updates the entry* ACM(i, idx) *to p and returns the modified database $\mathcal{DB}'$ on S.*

$\langle \text{data}, \bot \rangle \leftarrow \langle \text{read}(\text{idx}, cap_i), S_{\text{read}}(\mathcal{DB}) \rangle$. *The read algorithm takes as input an index idx and a client capability cap_i on the client side and the database $\mathcal{DB}$ on S and returns a data data on the client and generates no output on the server.*

$\langle \text{data}', \mathcal{DB}' \rangle \leftarrow \langle \text{write}(\text{idx}, cap_i, \text{data}), S_{\text{write}}(\mathcal{DB}) \rangle$. *The write algorithm takes as input an index idx, a client capability cap_i, and a data data on the client side and the database $\mathcal{DB}$ on S. Let data$'$ be the data stored at idx in $\mathcal{DB}$. The protocol modifies $\mathcal{DB}$ at index idx to data. Finally, it returns data$'$ on the client side as well as the modified database $\mathcal{DB}'$ on S.*

Attacker Model. The data owner is assumed to be trusted, since she is interested to protect her data. We allow the server to be fully compromised and to corrupt an arbitrary subset of clients. As explained below, this attacker model is relaxed when it comes to the integrity of outsourced data, which can only be achieved by assuming an honest-but-curious server (while still allowing for client compromise), as discussed below.

Security. A Multi-Client ORAM has four fundamental security properties. The first three concern access control and are intuitively described below.

Secrecy: only users with at least read permissions on an entry can learn its content.

Integrity: only users with write permissions on an entry can change its content.

Tamper Resistance: only users with write permissions on an entry can change its content in a way that the updated entry is considered valid by honest clients.

The difference between integrity and tamper-resistance is that integrity prevents unauthorized changes and thus requires an honest-but-curious server to perform access control, while tamper resistance is a weaker property that allows clients to detect unauthorized changes a-posteriori and thus can in principle be achieved even if the server is malicious.

Obliviousness Against Malicious Clients. Intuitively, a Multi-Client ORAM is secure if the server and an arbitrary subset of clients cannot get any information about the access patterns of honest clients, other than what is trivially leaked by the entries that the corrupted clients have read access to. The original obliviousness definition [33] does not allow the server to corrupt honest clients: here we extend it to handle static corruption of the clients and, in order to avoid trivial attacks, we restrict the queries of the adversary to the write oracle to indices that the set of corrupted clients cannot read.

Definition 2 (Obliviousness against Malicious Clients). *A Multi-Client ORAM Θ is secure against malicious clients, if for all* PPT *adversaries $\mathcal{A}$ the success probability of $\mathcal{A}$ in the following experiment is negligibly close to $1/2$.*

1. *$\mathcal{A}$ commits to a set of client identifiers* ID.
2. *The challenger samples $b \in \{0,1\}$, executes $(\mathsf{ACM}, \mathcal{DB}) \leftarrow \mathsf{gen}(1^\lambda, n)$ and forwards $\mathcal{DB}$ to $\mathcal{A}$ and hands over the capabilities of all the clients $\in$ ID to $\mathcal{A}$.*
3. *The adversary has access to the following interfaces that he can query adaptively and in any order.*

   $\mathsf{addCl}_{cap_\mathcal{O}}(i)$: *The challenger adds an empty row entry to* ACM *corresponding to i.*

   $\mathsf{addE}_{cap_\mathcal{O}}(\mathsf{idx}, \mathsf{data})$: *The challenger runs $\langle \mathsf{addE}(\mathsf{idx}, \mathsf{data}, cap_\mathcal{O}), \mathcal{A}\rangle$ in interaction with $\mathcal{A}$.*

   $\mathsf{chMode}_{cap_\mathcal{O}}(\mathsf{idx}, i, \{\mathsf{R}, \mathsf{RW}, \perp\})$: *The challenger runs $\langle \mathsf{chMode}(cap_\mathcal{O}, \mathsf{idx}, i, \{\mathsf{R}, \mathsf{RW}, \perp\}), \mathcal{A}\rangle$ in interaction with $\mathcal{A}$.*

   $\mathsf{read}(\mathsf{idx}, i)$: *The challenger runs $\langle \mathsf{read}(\mathsf{idx}, cap_i), \mathcal{A}\rangle$ in interaction with $\mathcal{A}$.*

   $\mathsf{write}(\mathsf{idx}, i, \mathsf{data})$: *The challenger runs $\langle \mathsf{write}(\mathsf{idx}, cap_i, \mathsf{data}), \mathcal{A}\rangle$ in interaction with $\mathcal{A}$.*

   $\mathsf{query}((\mathsf{op}_0, \mathsf{op}_1), (\mathsf{idx}_0, \mathsf{idx}_1), (i_0, i_1), (\mathsf{data}_0, \mathsf{data}_1))$: *The challenger checks in case that $\mathsf{op}_0 = \mathsf{write}$ or $\mathsf{op}_1 = \mathsf{write}$ if there is an $i \in$ ID such that $\mathsf{ACM}(i, \mathsf{idx}_0) \neq \perp$ and $\mathsf{ACM}(i, \mathsf{idx}_1) \neq \perp$, if this is the case the challenger aborts. Otherwise it executes $\langle \mathsf{read}(\mathsf{idx}_b, cap_{i_b}), \mathcal{A}\rangle$ (or $\langle \mathsf{write}(\mathsf{idx}_b, cap_{i_b},$*

data$_b$), $\mathcal{A}$⟩, *depending on the operation) in interaction with* $\mathcal{A}$. *In case* op$_0$ = write *or* op$_1$ = write, *from this moment on, the queries of* $\mathcal{A}$ *to the interface* chMode *on any* $i \in$ ID *and* idx$_0$ *or* idx$_1$ *are forbidden.*

4. $\mathcal{A}$ *outputs a bit* b', *the challenger returns 1 if* $b' = b$.

2.2 Formal Result

In the following we state a formal lower bound on the computational complexity of any ORAM secure against malicious clients. We denote by physical addresses of a database the memory addresses associated with each storage cell of the memory. Intuitively, the lower bound says that the server has to access a constant fraction of the dataset for any read and write operation.

Theorem 1. *Let* n *be the number of entries in the database and* Θ *be a multi-client ORAM scheme. If* Θ *accesses on average* $o(n)$ *physical addresses for each read and write operation (over the random coins of the read or write operation, respectively),* Θ *is not secure against malicious clients (see Definition 2).*

We formally prove this theorem in our full version [34].

2.3 Discussion

Given the lower bound established in the previous section, we know that any multi-client ORAM scheme that is secure against malicious clients *must* read and write a constant fraction of the database on every access. However, the bound does not impose any restriction on the required communication bandwidth. In fact, it does not exclude constructions with sublinear communication complexity, where the server performs a significant amount of computation. In particular, the aforementioned lower bound calls for the deployment of private information retrieval (PIR) [11] technologies, which allow a client to read an entry from a database without the server learning which entry has been read.

The problem of private database modification is harder. A naïve approach would be to let the client change each entry in the database $\mathcal{DB}$ upon every access, which is however too expensive. Homomorphic encryption might be a natural candidate to outsource the computation to the server and to reduce the required bandwidth: unfortunately, Ostrovsky and Skeith III [37] showed that no private database modification (or PIR writing) scheme with sublinear communication (in the worst case) can be implemented using algebraic cryptographic constructions, such as linearly homomorphic encryption schemes. This result does not apply to schemes based on fully-homomorphic encryption, which is however hardly usable in practice due to the high computation cost associated with the currently known schemes.

The following sections describe our approach to bypass these two lower bounds. First we show how to integrate non-algebraic techniques, specifically out-of-band communication among clients, in order to achieve sublinear *amortized* communication complexity (Sect. 3). Second, we show how to leverage a trusted proxy performing the access to the server on behalf of clients in order to reach a logarithmic overhead in communication and server-side computation, with constant client-slide computation (Sect. 5).

3 PIR-MCORAM

In this section, we present a high-level description of PIR-MCORAM, a Multi-Client ORAM scheme based on PIR. The full details can be found in our full version [34]. Our construction is inspired by Franz et al. [21], who proposed to augment the database with a stack of modified entries, which is periodically flushed into the database by the data owner. In our construction, we let each client C_i maintain its own temporary stack of entries S_i that is stored on the server side in addition to the regular database DB. These stacks contain recent changes to entries in DB and to entries in other clients' stacks, which are not yet propagated to DB. In contrast to the approach by Franz et al. [21], clients themselves are responsible to flush their stack once it is filled (i.e., after $|S_i|$ many operations), *without requiring any intervention of the data owner*. An *oblivious gossiping protocol*, which can be realized using standard techniques [16, 30], allows clients to find the most up-to-date entry in the database, thereby obtaining a sublinear communication bandwith even for write operations and thus bypassing the impossibility result by Ostrovsky and Skeith III [37].

More precisely, when operating on index j, the client performs a PIR read on DB and on all stacks S_i, which can easily be realized since all stacks are stored on the server. Thanks to the oblivious gossiping protocol, the client knows which index is the most current one. At this point, the client appends either a dummy entry (read) or a real entry (write) to its personal stack. If the stack is full, the client flushes it. Flushing means to apply all changes in the personal stack to the database. To be oblivious, the client has to ensure that *all* entries in DB change. Moreover, for guaranteeing correctness, the client has to ensure that it does not overwrite entries which are more recent than those in its stack.

After explaining how to achieve obliviousness, we also need to discuss how to realize access control and how to protect the clients against the server. Data secrecy (i.e., read access control) is obtained via public-key encryption. Tamper-resistance (i.e., a-posteriori detection of illegal changes) is achieved by letting each client sign the modified entry so that others can check that this entry was produced by a client with write access. Data integrity (i.e., write access control) is achieved by further letting each client prove to the server that it is eligible to write the entry. As previously mentioned, data integrity is stronger than tamper-resistance, but assumes an honest-but-curious server: a malicious server may collude with malicious clients and thus store arbitrary information without checking integrity proofs.

3.1 Analysis

We elaborate on the communication complexity of our solution. We assume that $|DB| = N$, that there are M clients, and we set the stack length $len_S = \sqrt{N}$

for every client. The worst case for an operation, hence, happens every $\sqrt{N}$-th operation for a client C_i, meaning that besides extracting the data from the database and adding an entry to the personal stack, C_i has also to flush the stack. We analyze the four algorithms independently: extracting data requires two PIR reads, one on DB and the other on the concatenation of all stacks. Thus, the overall cost is $\mathsf{PIR}(N) + \mathsf{PIR}(M\sqrt{N})$. Adding an entry to the personal stack always requires to upload one entry, independently of whether this replacement is real or dummy.

Our flushing algorithm assumes that C_i holds $\sqrt{N}$ entries and then down-and-uploads every entry of DB. Thus, the overall complexity is $2N + \sqrt{N}$. A similar analysis shows that if the client holds only $O(1)$ many entries, then C_i down-and-uploads DB but additionally performs a PIR step for every downloaded entry in its own stack to retrieve a potential replacement, resulting in a complexity of $2N + N \cdot \mathsf{PIR}(\sqrt{N})$.

To conclude, the construction achieves a worst-case complexity of $O(\mathsf{PIR}(N) + \mathsf{PIR}(M\sqrt{N}) + N)$ and $O(\mathsf{PIR}(N) + \mathsf{PIR}(M\sqrt{N}) + N\mathsf{PIR}(\sqrt{N}))$ for $O(\sqrt{N})$ and $O(1)$ client-side memory, respectively. By amortizing the flush step over $\sqrt{N}$ many operations, we achieve an amortized complexity of $O(\mathsf{PIR}(N) + \mathsf{PIR}(M\sqrt{N}) + \sqrt{N})$ or $O(\mathsf{PIR}(N) + \mathsf{PIR}(M\sqrt{N}) + \sqrt{N}\mathsf{PIR}(\sqrt{N}))$, respectively. Since our construction is parametric over the specific PIR protocol, we can leverage the progress in this field: at present, the best $\mathsf{PIR}(N)$ is $O(\log\log(N))$ [17] and, hence, the amortized cost becomes $O(\log\log(M\sqrt{N}) + \sqrt{N})$ or $O(\log\log(M\sqrt{N}) + \sqrt{N}\log\log(N))$, respectively. Since, in most scenarios, $M\sqrt{N} < 2^{2^{N/2}}$, we get $O(\sqrt{N})$ and $O(\sqrt{N}\log\log(N))$.

3.2   Discussion

The construction presented in this section leverages PIR for reading entries and an accumulated PIR writing technique to replace old entries with newer ones. Due to the nature of PIR, one advantage of the construction is its possibility to allow multiple clients to concurrently read from the database and to append single entries to their stacks. This is no longer possible when a client flushes her personal stack since the database is entirely updated, which might lead to inconsistent results when reading from the database. To overcome this drawback, we present a fully concurrent, maliciously secure Multi-Client ORAM in Sect. 5. Another drawback of the flush algorithm is the cost of the integrity (zero-knowledge) proofs. Since we have to use public-key encryption as the top-layer encryption scheme for every entry to allow for proving properties about the underlying plaintexts, the number of proofs to be computed, naïvely implemented, is proportional to the block size. Varying block sizes require us to split an entry into chunks and encrypt every chunk separately since the message space of public-key encryption is a constant amount of bits. The zero-knowledge proof has then to be computed on every of these encrypted chunks. To overcome this linear dependency, we present a new proof paradigm to make the number of computed zero-knowledge proofs independent of the block size in Sect. 4.

4   Integrity Proof Revised: The Hash-and-Proof Paradigm

In this section, we focus on the integrity proofs employed in our construction, presenting a novel and generally applicable cryptographic technique to boost their efficiency.

Plaintext Equivalence Proofs. To guarantee the integrity of the database, our construction requires extensive use of proofs showing that the ciphertexts were correctly rerandomized. In the literature, these proofs are called plaintext-equivalence-proofs (PEPs) and are the main efficiency bottleneck of our writing algorithm. Since the block size of an entry is in general *much larger* than the message space of the encryption scheme, we have to compute zero-knowledge proofs over vectors of ciphertexts. In this case, the integrity proof shows *for each* of these ciphertext vectors that they have been correctly rerandomized. The computational cost for these proofs scales linearly with the block size, which is clearly an undesirable dependency. In fact, this problem is not unique to our setting but affects any system deploying PEPs over long entries, among others verifiable secret shuffling [4,25,36] and mix networks [29].

In the following, we put forward a general technique to improve the computational efficiency of PEPs over ciphertext vectors. Our approach is fully black-box, non-interactive, and its proof size is *independent* of the number of ciphertexts of each entry. Thus, our technique can be used to boost the efficiency of not only PIR-MCORAM, but also any system based on PEPs. The basic idea behind our solution is to homomorphically compute a pairwise independent hash function [9] over the plaintexts of the two vectors and a PEP over the two resulting ciphertexts. Intuitively, a pairwise independent hash function is a collection of compressing functions such that the probability of two inputs to yield the same output is negligibly small in the size of the output domain (over the random choice of the function). This property ensures that the soundness of the proof is preserved.

General Problem Description. Let $\Pi_{\mathsf{PKE}} = (\mathsf{Gen}_{\mathsf{PKE}}, \mathsf{E}, \mathsf{D}, \mathsf{Rnd})$ be a randomizable, additively homomorphic public-key encryption scheme and $(\mathcal{P}, \mathcal{V})$ be a zero-knowledge proof system (ZKP) that takes as input two instances of ciphertexts $(c, b) \in \mathsf{E}(ek, m)^2$ for some $m \in \mathcal{M}$ and outputs a proof π for the statement $\exists r : b = \mathsf{Rnd}(ek, c, r)$. Construct a zero-knowledge proof system $(\mathcal{P}^*, \mathcal{V}^*)$ that takes as input two vectors of ciphertexts of length n, $(\mathbf{c}, \mathbf{b}) \in \mathsf{E}(ek, \mathbf{m})^{n \times 2}$ for some vector $\mathbf{m}$ and a vector $\mathbf{r}$ of randomnesses of the same length and outputs a proof π^* of the following statement: for all $i \in \{1, \ldots, n\}$ there exists a value r_i such that $b_i = \mathsf{Rnd}(ek, c_i, r_i)$. The efficiency goal is to make the size of the proof as well as the invocations of $(\mathcal{P}, \mathcal{V})$ independent of n. Knowing the decryption key dk, this statement is equivalent to the following one: for all $i \in \{1, \ldots, n\}$ we have $\mathsf{D}(dk, b_i) = \mathsf{D}(dk, c_i)$.

Our Solution. Let $\mathcal{M} = \mathbb{F}_p$ be the message space of Π_{PKE} for some field $\mathbb{F}_p$, such as the ElGamal or the Paillier encryption scheme [18,38]. We describe our solution as an honest-verifier Σ-protocol which can be made non-interactive and resilient against any malicious verifier by applying the Fiat-Shamir heuristic [19]. In the following, $\mathsf{E}(ek, z_0; z_1)$ denotes the encryption of z_0 with key ek and randomness z_1.

(1) $\mathcal{P}^*$ sends the vectors $(\mathbf{c}, \mathbf{b})$ to $\mathcal{V}^*$.
(2) $\mathcal{V}^*$ samples a vector $\mathbf{z} \in \mathbb{F}_p^{n+2}$ uniformly at random and sends it to $\mathcal{P}^*$.
(3) $\mathcal{P}^*$ computes $c' \leftarrow \mathsf{E}(ek, z_0; z_1) \bigotimes_{i=1}^{n} z_{i+1} \cdot c_i$ and $b' \leftarrow \mathsf{E}(ek, z_0; z_1) \bigotimes_{i=1}^{n} z_{i+1} \cdot b_i$ and runs $\mathcal{P}$ on inputs (c', b') to obtain π; $\mathcal{P}^*$ sends π to $\mathcal{V}^*$, who can recompute (c', b') and run $\mathcal{V}$ on $((c', b'), \pi)$. $\mathcal{V}^*$ returns the output of $\mathcal{V}$.

Security Analysis. In the following we state the formal guarantees of our techniques.

Theorem 2 (Hash-and-Proof). *Let Π_{PKE} be an additively homomorphic CPA-secure public-key encryption scheme and let $(\mathcal{P}, \mathcal{V})$ be a ZKP for PEPs over Π_{PKE}. Then $(\mathcal{P}^*, \mathcal{V}^*)$ is a ZKP for PEPs over Π_{PKE}.*

Proof. The correctness of Π_{PKE} and of the ZKP $(\mathcal{P}, \mathcal{V})$ imply the *correctness* of the protocol described above. The *zero-knowledge* of the protocol follows from the zero-knowledge of $(\mathcal{P}, \mathcal{V})$. Arguing about the *soundness* requires a more accurate analysis: we define as $\mathsf{cheat}_{(\mathcal{P}^*, \mathcal{V}^*)}$ the event where a malicious $\mathcal{P}^*$ fools $\mathcal{V}^*$ into accepting a proof over a false statement. This event happens with probability

$$
\begin{aligned}
\Pr\left[\mathsf{cheat}_{(\mathcal{P}^*, \mathcal{V}^*)}\right] = & \Pr\left[\mathsf{cheat}_{(\mathcal{P}, \mathcal{V})} \mid \mathsf{D}(dk, c') = \mathsf{D}(dk, b')\right] \\
& \cdot \Pr\left[\mathsf{D}(dk, c') = \mathsf{D}(dk, b')\right] + \\
& \Pr\left[\mathsf{cheat}_{(\mathcal{P}, \mathcal{V})} \mid \mathsf{D}(dk, c') \neq \mathsf{D}(dk, b')\right] \\
& \cdot \Pr\left[\mathsf{D}(dk, c') \neq \mathsf{D}(dk, b')\right]
\end{aligned}
$$

where the probabilities are taken over the random coins of $\mathcal{P}^*$ and $\mathcal{V}^*$. By the soundness of $(\mathcal{P}, \mathcal{V})$ we get

$$
\begin{aligned}
\Pr\left[\mathsf{cheat}_{(\mathcal{P}^*, \mathcal{V}^*)}\right] & \leq 1 \cdot \Pr\left[\mathsf{D}(dk, c') = \mathsf{D}(dk, b')\right] + \mu \cdot \Pr\left[\mathsf{D}(dk, c') \neq \mathsf{D}(dk, b')\right] \\
& \leq \mu + \Pr\left[\mathsf{D}(dk, c') = \mathsf{D}(dk, b')\right]
\end{aligned}
$$

where μ is a negligible function in the security parameter. Therefore, to prove soundness, it is sufficient to show that when $\mathsf{cheat}_{(\mathcal{P}^*, \mathcal{V}^*)}$ happens, then the probability $\Pr\left[\mathsf{D}(dk, c') = \mathsf{D}(dk, b')\right]$ is a negligible function in the security parameter. We shall note that, due to the homomorphic properties of Π_{PKE}, the resulting plaintext of c' and b' are $z_0 + \sum_{i=1}^{n} z_{i+1} \mathsf{D}(dk, c_i) \in \mathbb{F}_p$, and $z_0 + \sum_{i=1}^{n} z_{i+1} \mathsf{D}(dk, b_i) \in \mathbb{F}_p$, respectively. It is easy to see that this corresponds to the computation of the *universal pair-wise hash function* $\mathsf{h}_{(\mathbf{z})}$ as described by Carter and Wegman in [9] (Proposition 8). It follows that for all $\mathbf{c} \neq \mathbf{b}$ the resulting plaintexts of c' and b' are uniformly distributed over $\mathbb{F}_p$, thus $\Pr\left[\mathsf{D}(dk, c') = \mathsf{D}(dk, b')\right] = p^{-2}$, which is a negligible function in the security parameter. This concludes our proof. $\square$

5 Proxy-Based Realization

Driven by the goal of building an efficient and scaleable Multi-Client ORAM that is secure against malicious users, we explore the usage of a trusted proxy mediating accesses between clients and the server, an approach advocated in recent parallel ORAM constructions [5,42,48]. In contrast to previous works, we are not only interested in parallel accesses, but also in handling access control and providing obliviousness against multiple, possibly malicious, clients.

TaoStore [42]. In a nutshell, trusted proxy-based ORAM constructions implement a single-client ORAM which is run by the trusted entity on behalf of clients, which connect to it with read and write requests in a parallel fashion. We leverage the state of the art, TaoStore [42], which implements a variant of a Path-ORAM [46] client on the proxy and allows for retrieving multiple paths from the server concurrently. More specifically, the proxy consists of the *processor* and the *sequencer*. The processor performs read and write requests to the untrusted server: this is the most complex part of TaoStore and we leave it untouched. The sequencer is triggered by client requests and forwards them to the processor which executes them in a concurrent fashion.

Our Modifications. Since the proxy is trusted, it can enforce access control. In particular, we can change the sequencer so as to let it know the access control matrix and check for every client's read and write requests whether they are eligible or not. As already envisioned by Sahin *et al.* [42], the underlying ORAM construction can be further refined in order to make it secure against a malicious server, either by following the approach based on Merkle-trees proposed by Stefanov *et al.* [46] or by using authenticated encryption as suggested by Sahin *et al.* [42]. In the rest of the paper, we call the system TAO-MCORAM.

6 Security and Privacy Results

In the following we report the security results for PIR-MCORAM: those for TAO-MCORAM follow along the same lines. For the formal definition of the security properties we refer to [33]. Note that in our proofs we consider the adaptive version of each definition where the attacker is allowed to spawn and corrupt clients without restrictions. As a consequence, our instantiation requires us to fix in advance the number of clients M supported by the construction. Alternatively, one could consider the selective versions of the security definitions where the attacker is required to commit in advance to the client subset that he wants to corrupt. We postpone full proofs to our full version [34].

Theorem 3 (Secrecy). *Let Π_{PKE} be a CPA-secure encryption scheme, then PIR-MCORAM achieves secrecy.*

Theorem 4 (Integrity). *Let Π_{DS} be an existentially unforgeable digital signature scheme, ZKP be a zero-knowledge proof of knowledge protocol, and Π_{PKE} be a CCA-secure encryption scheme, then PIR-MCORAM achieves integrity.*

Theorem 5 (Tamper Resistance). *Let Π_{DS} be an existentially unforgeable digital signature scheme and let Π_{PKE} be a CCA-secure encryption scheme, then PIR-MCORAM achieves tamper resistance.*

Theorem 6 (Obliviousness against mal. clients). *Let PIR be a private information retrieval scheme, let Π_{PKE} be a CPA-secure encryption scheme, let Π_{DS} be an existentially unforgeable digital signature scheme, and let ZKP be a zero-knowledge proof of knowledge protocol, then PIR-MCORAM is secure against malicious clients.*

7   Evaluation

In this section, we describe our implementation and report on the experimental results. We start by reviewing the cryptographic schemes that we deploy: all of them are instantiated with a security parameter of 128 bits [6].

Cryptographic Instantiations. We deploy ElGamal encryption [18] in a hybrid fashion to construct an entry in the database. Using the hybrid technique, we decrease the entry size from $O(MB)$ to $O(M + B)$ since the data is encrypted only once and the corresponding secret key is encrypted for all clients with read access. In contrast, we encrypt the signing keys of the Schnorr signature scheme [43] using the Cramer-Shoup encryption scheme [13]. We use XPIR [1], the state of the art in computational PIR.

Finally, in order to construct integrity proofs, we use an OR-proof [12] over a conjunction of plaintext-equivalence proofs [29] (PEP) on the ElGamal ciphertexts forming one entry and a standard discrete logarithm proof [43] showing that the client knows the signing key corresponding to the authenticated verification key. In the homomorphic hash version, the conjunction of PEPs reduces to the computation of the homomorphic hash plus one PEP. As a matter of fact, since the public components necessary to verify a proof (the new and old ciphertexts and the verification key) and the secret components necessary to compute the proof (the randomness used for rerandomization or the signing key) are independent of the number of clients, all deployed proofs solely depend on the block size.

Implementation and Experiments. We implemented the cryptographic components of PIR-MCORAM in Java and we use a wrapper to GMP to speed up computations.

We used an Intel Xeon E5-4650L with 2.60 GHz, 8 cores, and 20 MB cache for the client and server experiments. We performed micro-benchmarks for PIR-MCORAM while varying the storage size from 32 MB to 2 GB and the block size from 4 KB to 1 MB, both for the solution with and without the homomorphic hash computation. We measured partial computation times as well as the end-to-end access time where we assume a network with 100 Mbit/s downstream and 10 Mbit/s upstream. In order to show the efficiency of our homomorphic hash construction and demonstrate its generic applicability, we also compare GORAM [33] with batched shuffle proofs, as originally presented, with a variant thereof where we replace the batched shuffle proofs with our homomorphic hash plus one shuffle proof.

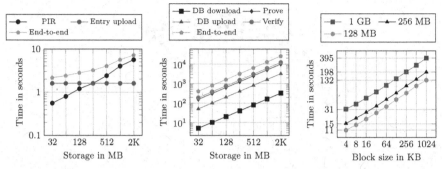

(a) Access without flush, 1 MB block size. (b) Flush, 1 MB block size. (c) Amortized time, varying block size.

Fig. 1. The end-to-end running time of an operation in PIR-MCORAM.

Discussion. Figures 1 and 2 report the results for PIR-MCORAM. Figure 1a shows the end-to-end and partial running times of an access to the ORAM when the flush algorithm is not executed, whereas Fig. 1b depicts the worst case running time (i.e., with flush operation). For the example of the medical record which usually fits into 128 MB (resp. 256 MB for additional files such as X-ray images), the amortized times per access range from 11 (resp. 15) seconds for 4 KB up to 131 (resp. 198) seconds for 1 MB sized entries (see Fig. 1c).

Figure 2 shows the improvement as we compare the combined proof computation and proof verification time in the flush algorithm of PIR-MCORAM, first as described in Sect. 3 and then with the integrity proof based on the universal homomorphic hash (see Sect. 4). We observe that our expectations are fulfilled: the larger the block size, the more effect has the universal hash computation since the number of proofs to compute decreases. Concretely, with 1 MB block size we gain a speed-up of about 4% for flush operations with respect to the construction without homomorphic hash.

To demonstrate its general applicability, we instantiate our proof technique into GORAM [33], which uses so-called batched shuffle proofs, achieving much better results. In GORAM, clients have to compute integrity proofs, which are proofs of shuffle correctness–a much more expensive primitive than PEPs. To overcome the efficiency problem, the authors have developed batched shuffle proofs: the idea is to homomorphically sum up half of the columns of the database matrix at random and to perform a shuffle proof on the resulting list of ciphertexts. To achieve soundness, the protocol has to be repeated $k = 128$ times. We observe that we can replace batched shuffle proofs by our protocol: clients compute the homomorphic hash on the old and the new ciphertexts and then one shuffle proof on the resulting lists of ciphertexts. As shown in Table 2, this modification speeds up GORAM by one order of magnitude (14x on the client and 10.8x on the server).

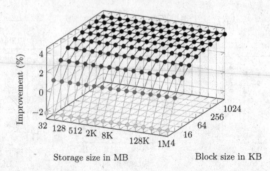

Fig. 2. The improvement in percent when comparing the combined proof computation time on the client and proof verification time on the server for varying storage and block sizes, once without and once with the universal homomorphic hash.

Table 2. Comparison of GORAM [33] with batched shuffle proofs and GORAM instantiated with our homomorphic hash (HH) variant for 10 users, 1 GB storage, and 8 KB block size.

Construction	Client time	Server time
GORAM with $k = 128$	91.315 s	39.213 s
GORAM with HH	5.980 s	3.384 s
Improvement	**14x**	**10.8x**

Finally, our solution TAO-MCORAM only adds access control to the actual computation of TaoStore's trusted proxy [42]. Interestingly enough, TaoStore's bottleneck is not computation, but communication. Hence, our modifications do not cause any noticeable slowdown on the throughput of TaoStore. Consequently, we end up with a throughput of about 40 operations per second when considering an actual deployment of TAO-MCORAM in a cloud-based setting [42].

8 Conclusion

This work studies the problem of obliviousness in multi-client outsourced storage. We establish a lower bound on the server-side computational complexity, showing that any secure realization has to involve at least $\Omega(n)$ computation steps. We further present a novel cryptographic instantiation, which achieves an amortized communication overhead of $O(\sqrt{n})$ by combining private information retrieval technologies, a new accumulation technique, and an oblivious gossiping protocol. Access control is enforced by efficient integrity proofs, which leverage a new construction for Plaintext Equivalence Proofs based on a homomorphic universal pair-wise hash function. Finally, we showed how to bypass our lower bound by leveraging a trusted proxy [42], thereby achieving logarithmic communication and server side computational complexity.

This work opens up a number of interesting research directions. Among those, it would be interesting to prove a lower bound on the communication complexity. Furthermore, we would like to relax the obliviousness property in order to bypass the lower bound established in this paper, coming up with more efficient constructions and quantifying the associated privacy loss.

Acknowledgements. This research is based upon work supported by the German research foundation (DFG) through the collaborative research center 1223, by the German Federal Ministry of Education and Research (BMBF) through the Center for IT-Security, Privacy and Accountability (CISPA), and by the state of Bavaria at the Nuremberg Campus of Technology (NCT). NCT is a research cooperation between the Friedrich-Alexander-Universität Erlangen-Nürnberg (FAU) and the Technische Hochschule Nürnberg Georg Simon Ohm (THN). Dominique Schröder is supported by the German Federal Ministry of Education and Research (BMBF) through funding for the project PROMISE. Finally, we thank the reviewers for their helpful comments.

References

1. Aguilar-Melchor, C., Barrier, J., Fousse, L., Killijian, M.O.: XPIR : private information retrieval for everyone. In: Proceedings of the Privacy Enhancing Technologies Symposium (PETS 2016), pp. 155–174. De Gruyter (2016)
2. Ajtai, M.: Oblivious RAMs without cryptographic assumptions. In: Proceedings of ACM Symposium on Theory of Computing (STOC 2010), pp. 181–190. ACM (2010)
3. Apon, D., Katz, J., Shi, E., Thiruvengadam, A.: Verifiable oblivious storage. In: Krawczyk, H. (ed.) PKC 2014. LNCS, vol. 8383, pp. 131–148. Springer, Heidelberg (2014). doi:10.1007/978-3-642-54631-0_8
4. Bayer, S., Groth, J.: Efficient zero-knowledge argument for correctness of a shuffle. In: Pointcheval, D., Johansson, T. (eds.) EUROCRYPT 2012. LNCS, vol. 7237, pp. 263–280. Springer, Heidelberg (2012). doi:10.1007/978-3-642-29011-4_17
5. Bindschaedler, V., Naveed, M., Pan, X., Wang, X., Huang, Y.: Practicing oblivious access on cloud storage: the gap, the fallacy, and the new way forward. In: Proceedings of the Conference on Computer and Communications Security (CCS 2015), pp. 837–849. ACM (2015)
6. BlueKrypt: Cryptograhpic Key Length Recommendation. www.keylength.com
7. Boyle, E., Chung, K.-M., Pass, R.: Oblivious parallel RAM and applications. In: Kushilevitz, E., Malkin, T. (eds.) TCC 2016 Part II. LNCS, vol. 9563, pp. 175–204. Springer, Heidelberg (2016). doi:10.1007/978-3-662-49099-0_7
8. Carbunar, B., Sion, R.: Regulatory compliant oblivious RAM. In: Zhou, J., Yung, M. (eds.) ACNS 2010. LNCS, vol. 6123, pp. 456–474. Springer, Heidelberg (2010). doi:10.1007/978-3-642-13708-2_27
9. Carter, J.L., Wegman, M.N.: Universal classes of hash functions (extended abstract). In: Proceedings of the ACM Symposium on Theory of Computing (STOC 1977), pp. 106–112. ACM (1977)
10. Chen, B., Lin, H., Tessaro, S.: Oblivious parallel RAM: improved efficiency and generic constructions. In: Kushilevitz, E., Malkin, T. (eds.) TCC 2016. LNCS, vol. 9563, pp. 205–234. Springer, Heidelberg (2016). doi:10.1007/978-3-662-49099-0_8

11. Chor, B., Kushilevitz, E., Goldreich, O., Sudan, M.: Private information retrieval. J. ACM **45**(6), 965–981 (1998)
12. Cramer, R., Damgård, I., Schoenmakers, B.: Proofs of partial knowledge and simplified design of witness hiding protocols. In: Desmedt, Y.G. (ed.) CRYPTO 1994. LNCS, vol. 839, pp. 174–187. Springer, Heidelberg (1994). doi:10.1007/3-540-48658-5_19
13. Cramer, R., Shoup, V.: A practical public key cryptosystem provably secure against adaptive chosen ciphertext attack. In: Krawczyk, H. (ed.) CRYPTO 1998. LNCS, vol. 1462, pp. 13–25. Springer, Heidelberg (1998). doi:10.1007/BFb0055717
14. Damgård, I., Meldgaard, S., Nielsen, J.B.: Perfectly secure oblivious RAM without random oracles. In: Ishai, Y. (ed.) TCC 2011. LNCS, vol. 6597, pp. 144–163. Springer, Heidelberg (2011). doi:10.1007/978-3-642-19571-6_10
15. Dautrich, J., Stefanov, E., Shi, E.: Burst ORAM: minimizing ORAM response times for bursty access patterns. In: Proceedings of the USENIX Security Symposium (USENIX 2014), pp. 749–764. USENIX Association (2014)
16. Demers, A., Greene, D., Hauser, C., Irish, W., Larson, J., Shenker, S., Sturgis, H., Swinehart, D., Terry, D.: Epidemic algorithms for replicated database maintenance. In: Proceedings of the Symposium on Principles of Distributed Computing (PODC 1987), pp. 1–12. ACM (1987)
17. Dong, C., Chen, L.: A fast single server private information retrieval protocol with low communication cost. In: Kutyłowski, M., Vaidya, J. (eds.) ESORICS 2014 Part I. LNCS, vol. 8712, pp. 380–399. Springer, Cham (2014). doi:10.1007/978-3-319-11203-9_22
18. ElGamal, T.: A public key cryptosystem and a signature scheme based on discrete logarithms. In: Blakley, G.R., Chaum, D. (eds.) CRYPTO 1984. LNCS, vol. 196, pp. 10–18. Springer, Heidelberg (1985). doi:10.1007/3-540-39568-7_2
19. Fiat, A., Shamir, A.: How to prove yourself: practical solutions to identification and signature problems. In: Odlyzko, A.M. (ed.) CRYPTO 1986. LNCS, vol. 263, pp. 186–194. Springer, Heidelberg (1987). doi:10.1007/3-540-47721-7_12
20. The EH Foundation. http://www.e-tervis.ee
21. Franz, M., Williams, P., Carbunar, B., Katzenbeisser, S., Peter, A., Sion, ·R., Sotakova, M.: Oblivious outsourced storage with delegation. In: Danezis, G. (ed.) FC 2011. LNCS, vol. 7035, pp. 127–140. Springer, Heidelberg (2012). doi:10.1007/978-3-642-27576-0_11
22. GmbH, E.: ELGA. https://www.elga.gv.at
23. Goldreich, O., Ostrovsky, R.: Software protection and simulation on oblivious RAMs. J. ACM **43**(3), 431–473 (1996)
24. Goodrich, M.T., Mitzenmacher, M.: Privacy-preserving access of outsourced data via oblivious RAM simulation. In: Aceto, L., Henzinger, M., Sgall, J. (eds.) ICALP 2011. LNCS, vol. 6756, pp. 576–587. Springer, Heidelberg (2011). doi:10.1007/978-3-642-22012-8_46
25. Groth, J.: A verifiable secret shuffle of homomorphic encryptions. In: Desmedt, Y.G. (ed.) PKC 2003. LNCS, vol. 2567, pp. 145–160. Springer, Heidelberg (2003). doi:10.1007/3-540-36288-6_11
26. Huang, Y., Goldberg, I.: Outsourced private information retrieval with pricing and access control. In: Proceedings of the Annual ACM Workshop on Privacy in the Electronic Society (WPES 2013). ACM (2013)
27. Iliev, A., Smith, S.W.: Protecting client privacy with trusted computing at the server. IEEE Secur. Priv. **3**(2), 20–28 (2005)

28. Islam, M., Kuzu, M., Kantarcioglu, M.: Access pattern disclosure on searchable encryption: ramification, attack and mitigation. In: Proceedings of the Annual Network & Distributed System Security Symposium (NDSS 2012). Internet Society (2012)
29. Jakobsson, M., Juels, A.: Millimix: mixing in small batches. Technical report, pp. 99–33. DIMACS (1999)
30. Kim, B.H., Lie, D.: Caelus: verifying the consistency of cloud services with battery-powered devices. In: Proceedings of the IEEE Symposium on Security & Privacy (S&P 2015), pp. 880–896. IEEE Press (2015)
31. Lorch, J.R., Parno, B., Mickens, J., Raykova, M., Schiffman, J.: Shroud: ensuring private access to large-scale data in the data center. In: Proceedings of the USENIX Conference on File and Storage Technologies (FAST 2013), pp. 199–214. USENIX Association (2013)
32. Maas, M., Love, E., Stefanov, E., Tiwari, M., Shi, E., Asanovic, K., Kubiatowicz, J., Song, D.: PHANTOM: practical oblivious computation in a secure processor. In: Proceedings of the Conference on Computer and Communications Security (CCS 2013), pp. 311–324. ACM (2013)
33. Maffei, M., Malavolta, G., Reinert, M., Schröder, D.: Privacy and access control for outsourced personal records. In: Proceedings of the IEEE Symposium on Security & Privacy (S&P 2015). IEEE Press (2015)
34. Maffei, M., Malavolta, G., Reinert, M., Schröder, D.: Maliciously Secure Multi-Client ORAM. Cryptology ePrint Archive, Report 2017/329 (2017). eprint.iacr.org
35. Mayberry, T., Blass, E.O., Chan, A.H.: Efficient private file retrieval by combining ORAM and PIR. In: Proceedings of the Annual Network & Distributed System Security Symposium (NDSS 2014). Internet Society (2013)
36. Neff, C.A.: A verifiable secret shuffle and its application to e-voting. In: Proceedings of Conference on Computer and Communications Security (CCS 2001), pp. 116–125. ACM (2001)
37. Ostrovsky, R., III, W.E.S.: Algebraic Lower Bounds for Computing on Encrypted Data. In: Electronic Colloquium on Computational Complexity (ECCC), vol. 14, no. 022 (2007)
38. Paillier, P.: Public-key cryptosystems based on composite degree residuosity classes. In: Stern, J. (ed.) EUROCRYPT 1999. LNCS, vol. 1592, pp. 223–238. Springer, Heidelberg (1999). doi:10.1007/3-540-48910-X_16
39. Pinkas, B., Reinman, T.: Oblivious RAM revisited. In: Rabin, T. (ed.) CRYPTO 2010. LNCS, vol. 6223, pp. 502–519. Springer, Heidelberg (2010). doi:10.1007/978-3-642-14623-7_27
40. Robinson, R.J.: How big is the human genome? https://medium.com/precision-medicine/how-big-is-the-human-genome-e90caa3409b0
41. Roche, D.S., Aviv, A., Choi, S.G.: A practical oblivious map data structure with secure deletion and history independence. In: Proceedings of the IEEE Symposium on Security & Privacy (S&P 2016). IEEE Press (2016)
42. Sahin, C., Zakhary, V., Abbadi, A.E., Lin, H.R., Tessaro, S.: TaoStore: overcoming asynchronicity in oblivious data storage. In: Proceedings of the IEEE Symposium on Security & Privacy (S&P 2016). IEEE Press (2016)
43. Schnorr, C.P.: Efficient identification and signatures for smart cards. In: Quisquater, J.-J., Vandewalle, J. (eds.) EUROCRYPT 1989. LNCS, vol. 434, pp. 688–689. Springer, Heidelberg (1990). doi:10.1007/3-540-46885-4_68
44. Shi, E., Chan, T.-H.H., Stefanov, E., Li, M.: Oblivious RAM with $O((\log N)^3)$ worst-case cost. In: Lee, D.H., Wang, X. (eds.) ASIACRYPT 2011. LNCS, vol. 7073, pp. 197–214. Springer, Heidelberg (2011). doi:10.1007/978-3-642-25385-0_11

45. Stefanov, E., Shi, E., Song, D.: Towards practical oblivious RAM. In: Proceedings of the Annual Network & Distributed System Security Symposium (NDSS 2012). Internet Society (2012)

46. Stefanov, E., van Dijk, M., Shi, E., Fletcher, C., Ren, L., Yu, X., Devadas, S.: Path ORAM: an extremely simple oblivious RAM protocol. In: Proceedings of the Conference on Computer and Communications Security (CCS 2013). ACM (2013)

47. Stefanov, E., Shi, E.: Multi-cloud oblivious storage. In: Proceedings of the Conference on Computer and Communications Security (CCS 2013), pp. 247–258. ACM (2013)

48. Stefanov, E., Shi, E.: ObliviStore: high performance oblivious cloud storage. In: Proceedings of the IEEE Symposium on Security & Privacy (S&P 2013), pp. 253–267. IEEE Press (2013)

49. Williams, P., Sion, R., Tomescu, A.: PrivateFS: a parallel oblivious file system. In: Proceedings of the Conference on Computer and Communications Security (CCS 2012), pp. 977–988. ACM (2012)

Legacy-Compliant Data Authentication for Industrial Control System Traffic

John Henry Castellanos[✉], Daniele Antonioli, Nils Ole Tippenhauer, and Martín Ochoa

Singapore University of Technology and Design, Singapore, Singapore
{john_castellanos,daniele_antonioli}@mymail.sutd.edu.sg,
{nils_tippenhauer,martin_ochoa}@sutd.edu.sg

Abstract. Industrial Control Systems (ICS) commonly rely on unencrypted and unauthenticated communication between devices such as Programmable Logic Controllers, Human-Machine-Interfaces, sensors, and actuators. In this work, we discuss solutions to extend such environments with established cryptographic authentication schemes. In particular, we consider schemes that are legacy compliant in the sense that authentication data is embedded as additional payload for domain specific protocols, for example the industrial EtherNet/IP protocol. To that end, we propose a selective protocol (that signs every critical packet sent) and a protocol that aggregates groups of packets based on real-time requirements and the available throughput, for various realistic hardware configurations. We evaluate our analysis by implementing an authenticated channel in a realistic Water Treatment testbed.

Keywords: Industrial Control Systems · Authentication · Network security

1 Introduction

Industrial Control Systems (ICS) commonly rely on unencrypted and unauthenticated communication between the industrial devices such as Programmable Logic Controllers (PLC), Human-Machine-Interfaces (HMI), sensors, and actuators. The use of cryptographic schemes in such devices is often hindered by their long lifetime, compatibility issues, low processing power of the embedded devices, and real-time requirements in the communication [9]. Most common industrial communication protocols do not feature any built-in capabilities for authentication (e.g., Modbus/TCP, EtherNet/IP). In the past, critical infrastructure control networks' were isolated from the office and corporate networks, and thus malware and other advanced attacks did not pose a realistic threat to such control networks.

Nowadays, with the increased connectivity to general IT infrastructures such as a LAN, and the Internet itself, attacks such as the well-known Man-in-the-Middle between ICS devices are more realistic [29]. For ICS communications,

D. Gollmann et al. (Eds.): ACNS 2017, LNCS 10355, pp. 665–685, 2017.
DOI: 10.1007/978-3-319-61204-1_33

666   J.H. Castellanos et al.

message *integrity* is paramount, while *confidentiality* of messages exchanged is less important. In particular, if an attacker can alter the values of the sensors or the commands being sent to the actuators, he could effectively alter the control of the ICS and potentially cause the malfunctioning of the physical system.

A number of works (e.g., [2,3,5,23,26,31] to name a few) highlight the importance of securing the networks of ICS. Most of them also remark that by using cryptography a non negligible overhead can be introduced, and that such systems usually have strict timing constraints. However, to the best of our knowledge, no detailed analysis on cryptography-enabled authentication has been reported so far for ICS under realistic constraints. In particular, data in ICS is also often passing through intermediate gateways and other industrial network appliances. For that reason, the solution must be legacy-compliant (which can prevent solutions such as TLS encapsulation).

In this paper, we explore the application of well-known cryptographic primitives to verify authenticity of communication between devices in modern ICS, embedded as payload in legacy industrial protocols. In particular, we are interested in solutions that would either be feasible to implement in constrained legacy devices, or could be provided by low-cost additional devices. The focus of the paper is *to detect* manipulations of legitimate traffic by the attacker, and the presence of new messages introduced by the attacker. If confidentiality is required, an additional suitable encryption scheme should be used. In addition, the discussion of consequences of successful detection of message manipulation attacks is out of scope of this work. The mitigation to such an attack is a subject on its own, since in general one cannot simply drop incorrectly signed packets: on one hand, even a single missing packet could have unforeseen consequences in controlling physical processes, on the other hand if packets are dropped attackers could easily implement denial of service attacks [15].

In order to maximize the efficiency of legacy-compliant authentication while preserving security, we propose a basic authentication protocol that signs a subset of *critical* packets. This is in contrast to closely related work, that usually focuses on authenticating *all* of the communication between nodes in ICS [18,25,28,33]. In addition, we propose a second protocol that aggregates groups of packets based on the real-time requirements, network throughput, and the processor capacity for various realistic hardware configurations. We evaluate our analysis experimentally in an industrial Water Treatment testbed [11] (Secure Water Treatment, SWaT) by means of additional network components. However our protocols can be deployed by controller manufacturers as a firmware modification of existing Ethernet modules.

We summarize our contributions as follows. In this work we: **(a)** discuss design options for legacy-compliant basic authentication protocols in the context of ICS networks; **(b)** perform an experimental performance evaluation of a number of cryptographic primitives on several hardware platforms; **(c)** propose a novel aggregated authentication protocol, adapted to requirements of ICS; **(d)** empirically evaluate the proposed protocol in a Water Treatment testbed.

2   Preliminaries

In the following we introduce some fundamental notions on Industrial Control Systems, and introduce the attacker model and the expected security guarantees of our solution.

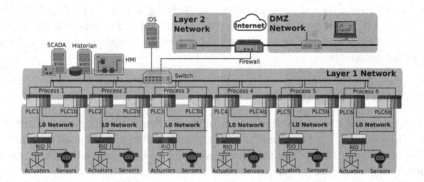

Fig. 1. SWaT network architecture.

2.1   Industrial Control Systems

ICSs are a subset of Cyber-Physical Systems that monitor industrial infrastructures such as Water Treatment systems, nuclear power plants, smart grids, and electric power distribution systems [22]. Securing a safety-critical ICS against malicious attackers is crucial to avoid catastrophic events that may result in natural disasters, economical crises, and loss of human life. The threat model of ICS often assumes a strong attacker, the system under attack provides a large attack surface because an ICS may be vulnerable both to *cyber-attacks* and *physical attacks*. Unfortunately, a combination of cyber and physical attacks can result in a severe damage of the system even without physical access to the system, e.g., [2,32].

Figure 1 shows an example of the architecture of an ICS network. The network is layered to logically separate devices and monitored processes. Compared to a standard corporate network, an ICS network includes a wider range of devices. The ICS has to connect several older legacy hardware and new hardware, and it has to manage software with different capabilities and interfaces. In addition, a traditional ICS network is expected to have a long life time (e.g., twenty years), and many of its components are unlikely to change or be upgraded over the years. That is why the protocols we propose in the remainder of the paper target *both* high-end ICS devices (able to tolerate the computation overhead), and low-end ICS devices (with the introduction of an external module that is able to tap into the ICS network).

Different industrial protocols have been used in ICS. They evolved from serial communication networks (e.g., RS-485, RS-232) to bus systems (e.g.,

Fieldbus), and then to Ethernet-based communications such as EtherNet/IP (ENIP) [6]. ENIP is a modern, object-oriented application layer industrial protocol, that implements the Common Industrial Protocol (CIP) specifications [24] over the TCP/IP protocol stack. ENIP can be extended to support custom commands and device profiles, and it provides a native compatibility with traditional TCP/IP based IT corporate network. It is important to notice that our authentication scheme *does not* depend on the underlying industrial protocol. We use ENIP as an example protocol as we have a local Water Treatment testbed that uses ENIP. However, the same scheme should easily translate to other modern industrial protocols.

2.2   Security Guarantees

In this work, we discuss solutions to provide authenticity of network traffic for ICS. The solution is not designed to provide confidentiality, and does not prevent denial of service attacks. The attacker is assumed to be able to eavesdrop, insert, drop, and manipulate messages on the network. The attacker does not have access to pre-shared keys, and he is constrained by polynomial computational power on the size of the key, where the computational hardness assumption are simular to the ones proposed in [27].

Moreover, one key observation of this work is that not all packets being transmitted by nodes in ICS need to be authenticated: in the following we will discuss (using ENIP as an example) how to select a subset of critical packets that need to be protected, whereas we argue that other packets are less critical and do not necessarily need to be authenticated since their manipulation does not pose a threat to ICS.

3   Traffic Authentication for ICS: The SPA Protocol

In this section we introduce a first efficient protocol to guarantee authenticity of critical communication in ICS. In addition to providing authenticity, the solution also needs to integrate well into existing systems. In particular, the solution needs to fulfill *real-time* requirements and *legacy-compliance* integration. Note that as stated in the previous section, we will use ENIP as a running example to illustrate our protocol, since it is the protocol used in our evaluation setting. Our idea can be applied to other industrial network protocols (i.e. Modbus TCP [19] and PROFINET [10]), however the implementation details will vary, essentially because a signature must be appended to certain packets, which is easily accomplished in ENIP as we will discuss later.

Our solution must be computationally efficient enough to allow resource-constrained devices (such as PLCs) to sign and verify packets fast enough. In particular, the solution should be able to handle high volume traffic loads, without introducing high queuing, and processing delay. We start by proposing a first solution we called SPA (Selective Packet Authentication). The SPA protocol relies on a simple algorithm: we assume a setting where two devices, that want to communicate, possess a common pre-shared key, and we propose to sign only selected outgoing packets using well-known cryptographic algorithms, such as authenticated signatures schemes [12,14].

This solution is conceptually simple, however it potentially conflicts with the real-time constraints outlined earlier. In Sect. 4, we discuss our refinement of this idea, we present a mathematical analysis able to capture the constraints, in order to guarantee the normal operation of the ICS.

3.1 Legacy-Compliance

ICS often integrate devices that cannot easily be replaced or updated. As a result, a number of legacy industrial protocols are established that are widely supported, but do not feature any security capabilities by design. In general, such protocols allow the reading of distinct memory locations (e.g., in Modbus/TCP) or *tags* (in EtherNet/IP) that represent sensor values or similar. We argue that an upgrade of all such devices in an ICS is costly; therefore an authentication system needs to impact the existing system as little as possible. In particular, the use of TLS tunnels to transmit data would not only incur computational overhead, but could also fail to pass through industrial network appliances or intermediate gateways. Therefore, we propose to embed the authentication data as additional payload in the existing industrial protocols. This will ensure that receivers that are not aware of the authentication scheme can at least process the normal payload without benefiting from the authentication data. Intermediate devices that are unaware of the authentication scheme could also just pass along the authentication data as normal payload. As result, our authentication solution can be integrated in legacy systems, e.g., by introducing external modules to data sources, or through firmware modification of Network modules.

3.2 SPA Protocol Description

The intuition behind the SPA protocol is that (a) only a subset of transmitted messages is security relevant, and (b) selective signing of that subset is more efficient than signing all messages in the stream. For simplicity, we will refer to sign as the process of generating a Message Authentication Code using a pre-shared key for the case of symmetric key cryptography or, a signature using the counterpart's public key for the asymmetric case.

Let p be a critical data packet (we discuss how to identify them later in the context of ENIP). The SPA protocol, shown in Fig. 2, calculates a signature $Sig_k\{p\}$ of the packet p, generates a new packet $p' = (p, Sig_k\{p\})$, and sends it to B. By using the inverse function Ver_K (where $K = k$ in the symmetric case or the corresponding public key otherwise), B verifies the authenticity of the message as $p = Ver_k\{Sig_k\{p\}\}$. Note that the signature scheme Sig guarantees that a computationally constrained adversary cannot forge a signature (with high probability).

We note that to prevent replay attacks, the payload should contain a timestamp or a counter to identify the ENIP session or packet. Therefore, if an active adversary replays the message in a future ENIP session, the application layer will check the nonce and mark it as invalid. On the other hand, since we do not sign non-critical packets in the data stream (nor e.g., TCP synchronisation packets) an attacker could manipulate the content of those messages. Such an attack is easily detected by orthogonal means [7]. By design, ICS systems will trigger alarms in case of problems in network connectivity. In particular, we cannot prevent against denial of service attacks by means of authenticity and such attacks are out of scope.

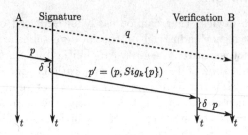

Fig. 2. SPA protocol overview: p is a critical message that is signed by the signature module. q is a non-critical message, which is simply forwarded; δ is the delay introduced by authentication and verification.

Real-Time Requirements and Backlogs. ICS operate under very strict real-time operation conditions, with maximal critical response time, very high availability requirements, and low tolerance for high delay or jitter conditions. As described before, in order to guarantee the authenticity of messages, we authenticate outgoing packets using a *Signature* module, and verify incoming packets using a *Verification* module respectively. We define $g(\Delta)$ as the number of packets generated by a device in an interval of time Δ, and $s(\Delta)$ as the rate at which packets are being signed. As an example, $g(\Delta)$ can be 1000 packets per second if $\Delta = 1$ second, while $s(\Delta)$ generally depends on the processing speed, e.g., 300 packets per second. Similarly we denote $r(\Delta)$ the number of packets received by a node and $v(\Delta)$ the number of verifications performed.

In order to be compliant with the real-time requirements of a given system, essentially one needs to authenticate/verify packets at least as fast at the rate they are being produced/received, that is $g \leq s$ and $r \leq v$ (see Appendix A).

3.3   Application to ENIP-CIP

We now discuss the application of the SPA protocol to Common Industrial Protocol (CIP) traffic to ensure legacy compliance, leveraging its extension possibilities. SPA payload is given by $p = (payload, ENIPsequencenr, ENIPsessionid)$ based on a critical ENIP packet from A to B. The ENIP session ID is a randomly generated 32-bit integer and will serve as a counter as discussed in Sect. 3.2. Note that although 32-bit is a relatively small search space, in this context we do not rely on it for security, but merely to prevent replay attacks. An attacker still needs to forge a secure MAC or cryptographic signature to bypass verification, as we will illustrate in Sect. 5.

When replacing the original p with p' of SPA, we increase the length of any packet p that we sign. The packet extension must conform to the ENIP standard. The structure is defined as follows: *Type ID*: $0x00c1$. *Length*: size of the signature specified in Bytes. *Data*: the signature $Sig_k\{p\}$.

The device receiving p' will search for the Type ID $0x00c1$, verify the content of the payload, and then remove the signature. In the case of a mismatch, i.e., $p \neq Ver_k\{Sig_k\{p\}\}$ the device will raise an alarm.

Identification of Critical Data. The integrity of messages exchanged in an ICS system can be protected in different layers of the OSI network stack. We identified critical data from the pool of CIP services observed in the traffic captures. In particular, protection is required for data that can affect the normal operation of the control of a physical process.

The identified critical services are: *Read Data* (Service 0x4C), *Write Data* (Service 0x4D), and *Read Tag Fragmented Data* (Service 0x52). In Sect. 5, we discuss in more detail this choice in the context of other CIP services observed in our Water Treatment testbed. By authenticating critical packets only, we increase security and minimise the computation overhead.

3.4 Ad-Hoc Protocols vs TLS

We have chosen to focus on ad-hoc protocols at the application layer, rather than of using TLS [4], for various reasons. Firstly, we want to reduce the computation overhead, by only authenticating packets that contain critical payloads (such as commands to actuators and values from sensors). In comparison, while TLS would sign every packet in a ENIP connection, SPA protocol would only sign and verify a comparatively small amount of those ignoring packets such as TCP handshake, EtherNet/IP communication control messages and CIP non critical service messages, as shown in Fig. 3. We believe that integrity attacks on TCP handshake and ACK messages can be detected by orthogonal methods. On the other hand, we want to be backwards compatible with devices that do not support message authentication, a feature that we achieve by using the extension capabilities of ENIP. Such a feature would not be achievable by using TLS.

We now discuss performance of SPA vs TLS. Let v the speed in packets per second that a given cryptographic signature can provide for a given average packet size. For simplicity we assume that both TCP packets and ENIP packets are about the same size, although in practice TCP will be slightly bigger. Then if the number of critical packets is c, then the actual throughput on the total of generated packets g will be higher, since we only need to sign $c \cdot g$. Thus, effectively we will increase the tolerance on the generated packets g by a factor of c, allowing a maximum of $\frac{v}{c}$ packets per second. This is illustrated in Fig. 3.

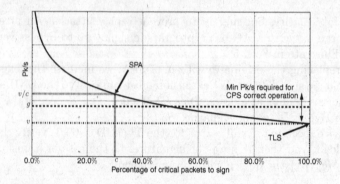

Fig. 3. The y-axis represents the tolerance to a network constraint's in packets/second. The x-axis represents the percentage of critical packets authenticated.

Example. In practice, the amount of sent and received packets is symmetric, as we will discuss in the following sections. In addition, the signature and verification time are similar. Let $g = r$ be 10^3 packets per second, and let $c = 0.4$. Let $s = v$ be 10^5 packets per second. In this case, cryptography-enabled authentication does not create backlogs neither for TLS (signing every packet) nor for SPA since $g + r = 2 \cdot 10^3 < s + v = 2 \cdot 10^5$. This example is based in the real-time communication constraints of SWaT and the signature and verification rates for SHA-256 based HMAC on a virtualized ARM similar to the one used in the SWaT PLCs. A deeper analysis and discussion is presented in Sect. 5.

In sum, the proposed SPA protocol has both advantages and drawbacks in terms of the authentication goals. We briefly summarized them as:

Advantages. The protocol is conceptually simple and easy to implement and to integrate into legacy systems by means of an external component as we will discuss in Sect. 5. Moreover, it gives fine-granularity detection capabilities, since it can be pointed out which packet was altered in transit with almost instantaneous detection times. Additionally, by using the extension capabilities of ENIP, the protocol is backwards compatible with devices that do not implement a verification module.

Disadvantages. The main disadvantage of the SPA protocol is that although it is designed to sign critical packets only, the overheads might still be unacceptable for devices and/or algorithms with low signature/verification rates. This is the main motivation for proposing an extended protocol, as we will discuss in the following section.

4   Extensions: The ASPA Protocol

The SPA protocol presented in the previous section and the constraint analysis, give us guidelines to determine whether the scheme is feasible, depending on the packet generation rate, packet size, and signature algorithm performance on links that have a reliable connectivity. However, there exist two limitations to the practical application of this protocol: first, we would like the protocol to be used even in scenarios where (for legacy or cost reasons) the underlying hardware is not as fast as necessary (i.e. in the sense of the signature rate s vs generation rate g as discussed in the previous section), and second, even faster hardware might be insufficient for stronger signature algorithms, that are computationally more expensive.

In particular, strong signature algorithms would enable the use of asymmetric cryptography, that has several advantages in terms of key management. For example, the use of public/private keys in asymmetric cryptography allows dynamic addition of new devices to the system if centrally signed certificates are used. In addition, the compromise of a single device will only expose a single private key, instead of exposing a secret key shared between many devices.

Our protocol can be extended to deal with more expensive cryptographic algorithms and slower CPUs. The intuition behind this extension, called *ASPA* (Aggregated Selective Packet Authentication) is the following: typically, authenticated signature schemes are relatively inefficient for short amount of data, but they get more efficient for large amounts of data. Thus, on average, it is usually faster to perform an aggregate signature over multiple packets instead of signing them individually, and as a result the aggregate signature increases the signing rate s.

4.1   The ASPA Protocol

Let $P(T)$ be the sequence of outgoing packets in the time interval T. Let n be the expected number of packets in this time $\mathbf{E}[\|P(T)\|] = n$ such that the typical set is $P(T) = \{p_1, \ldots, p_n\}$. In the ASPA protocol, shown in Fig. 4, the *Signature* module simply forwards packets p_i to B, and in addition accumulates their payload in a queue. After time T has passed, it signs the accumulated queue, and sends the *aggregated signature* together with the sequence number of the first and the last element of the queue. The *Verification* module will forward all received messages to the original destination (without immediately validating a signature), and in addition store the received messages in a queue

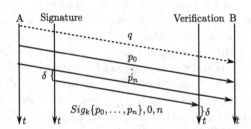

Fig. 4. ASPA overview. p_i are critical messages which are aggregated. q is a non-critical message, thus not authenticated. δ is the delay introduced by authentication.

until the aggregated signature arrives. Then, the verification module can check whether the signature matches the content of the respective packets received. If the Verification module cannot validate the signature, an attack was detected. For example values of T and $P(T)$, we refer to Sect. 5. We note that the ASPA protocol relies on lower layers (i.e., TCP) to ensure that no message gets lost during transmission. That will ensure that the verification and signature modules always have the same view of exchanged messages.

Security Trade-Off. In the first proposed protocol, an attack can be detected as soon as a spoofed packet with an invalid signature is received. In the ASPA protocol, we can only detect an attack after T time has passed, and an invalid signature is received. Depending on the value of T, and the particular ICS system, this could be problematic (or not). This is related with the problem of ICS resilience and reaction time in case of failure, and therefore we see it as orthogonal. For most practical applications (see Sect. 5), the values of T will be small and thus the reaction time will not differ much from the reaction time of the first protocol.

4.2 Performance Advantage

Symmetric Authentication. Let $\delta(size, cpu, alg)$ be the time needed to sign a packet of size *size* with algorithm *alg* on CPU *cpu*. The signature scheme is typically based on HMAC and an underlying hash function (such as SHA-256). In that case, there is a constant b such that $\delta(b', cpu, alg) \approx \delta(b, cpu, alg)$ for $b' \leq b$. However, for $B > b$: $\delta(B, cpu, alg) = \delta(b, cpu, alg) + \left\lceil \frac{B-b}{b} \right\rceil \cdot c$ where $c < \delta(b, cpu, alg)$. Therefore, given an expected packet number of a certain expected size, it is more efficient to sign multiple packets instead of just one, in terms of the rate per interval s. As we will discuss in the next section, the optimization value converges after a certain number of packets, as is to be expected.

Asymmetric Authentication. In the case of signatures based on asymmetric cryptography (such as ECDSA [12]), typically the payload is first hashed and then an operation involving the private key is performed on the digest (such as decryption). Since the cryptographic operation is orders of magnitude slower than the hashing operation (i.e., hundreds of ms vs. μs in some cases), signing multiple packets takes almost as long as signing a single packet.

The above described effects on the signing rate s can be observed in the plotted values for different algorithms and values of n of Fig. 6. Note that a similar discussion about the signing rate s and the packet generation rate g, presented in the previous section, also applies for the ASPA protocol. In practise, a device might have active connections (TCP streams) to m devices at the same time. As result, there will be m queues $Q_{\Delta,i}^S$, possibly signed with different keys. In that case, the constraint becomes $\forall_i\ g_i < s_i$, where $g = \sum_{i=1}^{m} g_i$. In the case of ICS such as SWaT, devices typically communicate only with one or two devices (e.g., $m = 2$). For the sake of simplicity we assume a single queue in the following.

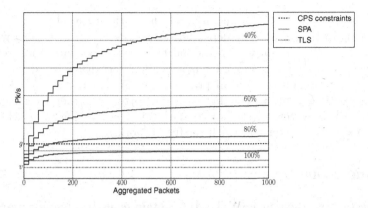

Fig. 5. The x-axis represents chunks of packets, y-axis tolerance in terms of packets per second. The step functions represent different percentages of critical packets, segmented lines various CPS communication constraints. CPU power and packet size are constant.

ASPA vs TLS. In Fig. 5, we illustrate the general case comparison against TLS. Let $v = \frac{1}{\delta}$ the number of packets per second an ideal implementation of TLS could sign. By using ASPA to aggregate n packets we can tolerate approximately: $v' = \frac{n}{\delta + (n-1)\cdot\delta'}$ packets per second, which will be faster than v and will converge to a constant since: $\lim_{n\to\infty} \frac{n}{\delta + (n-1)\cdot\delta'} = \frac{1}{\delta'}$. We will discuss empirically this phenomenon in the evaluation section for various hardware configurations.

Example. Let $p = r$ be 10^3 packets/second as in the previous example (as observed by us in SWaT). Let $s = v$ be 300 packets/second using the SPA protocol of the previous section and ECDSA and a Raspberry Pi3 (for more details see Sect. 5). In that case, cryptography-enabled authentication clearly causes backlogs since $g + r = 2 \cdot 10^3 > s + v = 600$. As discussed above, signing one packet takes almost as much time as signing multiple packets, and then we can implement the ASPA protocol by accumulating chunks of 50 packets (for $T \approx 30$ ms), thus augmenting s to about $1,5 \cdot 10^4$ s packets per second. This implies a minimum reaction time to attacks of about 60 ms.

In sum, the ASPA protocol improves the signing and verification rates per packet (s and v), while at the same time providing good reaction time.

Advantages. The ASPA protocol is useful in situations where signing each packet individually is not feasible due to slow hardware or constraints of the ICS network. In particular, it offers a significant advantage when signing multiple packets with ECC based authentication. The ASPA protocol can be used selectively for critical messages (as in the SPA protocol). Alternatively, it could be used to provide delayed authentication for all messages, as the amount of data included in the aggregated signature has a negligible impact on the overall time of creating the signature.

Disadvantages. The coarse granularity of the authentication potentially can delay reaction times to attacks, depending on the size T of the signing window. However in our evaluation we have found that the ideal window for authentication is a few dozens of packets for most signing algorithms/hardware, which allows for a fast reaction time in practice. In our experience, actual attacks require a relatively long interaction with the system in order to influence its physical state. However, we stress that the subject of reaction to attacks is out of the scope of this paper and is left for future work.

5   Validation

In this section, we use a real SWaT [11] to obtain realistic examples of real-time constraints. We investigate the amount of critical traffic shared between devices in the network, in order to approximate the expected number of packets per time interval, and the expected size of those packets. Furthermore, we benchmark symmetric and asymmetric signature algorithms for a variety of hardware platforms, and discuss the feasibility of their implementation with respect to the constraints derived in the previous sections. We use standard algorithms such as SHA (instead of light-weight algorithms specifically designed for embedded systems) to allow better comparison against TLS. When using light-weight algorithms, the presented numbers are expected to improve.

5.1   SWaT

SWaT [11] (see Fig. 1) consists of an ICS with a process made up of six stages. In a nutshell, the process begins by collecting the incoming water in a tank, then it performs a chemical treatment stage, it filters the treated water through an Ultrafiltration (UF) system. Afterwards, the water is de-chlorinated using a combination of chemicals and Ultraviolet lamps, and then fed to a Reverse Osmosis (RO) stage. A backwash process cleans the membranes in UF using the water produced by RO. The cyber portion of SWaT consists of a layered communications network, PLCs, HMIs, SCADA, and a Historian.

The validation focuses on benchmarking integrity, and authenticity controls in the plant network, that connects the PLCs, the HMI and the SCADA system. The network is using the EtherNet/IP industrial protocols on top of an (Ethernet based) TCP/IP network. We performed several network capture by setting up a mirroring port on the plant network industrial switch. From those captures we identified ENIP-CIP communications among 21 hosts, through implicit and explicit messages. Implicit messages (UDP/2222) are used in our plant for keep-alive signals, while explicit messages (TCP/44818) are used for configuring, monitoring and controlling the plant stages. The plant performs its communication to a rate of 16.000 ENIP-CIP messages per second on average over all stages. About 14,3% of ENIP-CIP connections belong to UDP/2222, and the rest 85,7% to TCP/44818. We can split TCP connections between TCP session traffic (42,7%), and CIP explicit messages (42,9%).

We focused on CIP explicit messages, and we tried to extract a subset of CIP services that deals with critical data. Manipulation of those services could affect the state of the controlled physical process. We selected the following services as critical: **Read Data** (Service 0x4C); **Write Data** (Service 0x4D); **Read Tag Fragmented Data** (Service 0x52).

Read Data and Read Tag Fragmented Data. CIP services are classified as critical data since an attacker might rise a fake alarm in the SCADA system or he might hide a safety-related event modifying its data on the fly. The *Write Data* CIP service is classified as critical because an attacker might directly modify the behavior of actuators pushing data into PLCs. By selecting CIP services with critical data, our proposal only signs around 42% of SWaT's traffic (including TCP and UDP), avoiding processing and bandwidth overheads for non-critical data CIP services such as the *Get all attributes* service.

We performed an in-depth analysis of the capture file in order to estimate the frequency, and the size of the packets received per second, by an arbitrary testbed's device. Table 1 shows a summary of the results targeting PLC2. We decided to use PLC2 as an upper bound because we estimated that it is the "busiest" device in our testbed. As you can see from the table, PLC2 sends 1127 packets per second on average, and it receives 1168 packets per second on average.

Table 1. Frequency and size of critical packets shared by host PLC2 to others.

		Sent	Received
ENIP message	Request	561 Pk/s	607 Pk/s
		$\mu = 63B$, $\sigma = 3.36$	$\mu = 69B$, $\sigma = 5.32$
	Response	566 Pk/s	561 Pk/s
		$\mu = 75B$, $\sigma = 58.16$	$\mu = 86B$, $\sigma = 9.42$
Total Pk/s		1127	1168

5.2 Hardware Benchmark

In order to evaluate the efficiency of the underlying primitives, and therefore the packet signing rate $s(t, cpu, alg)$, we used different types of hardware platforms.

Controllino is an open source Hardware PLC, based on Arduino Mega 2560 board, with an ATmega2560 CPU (16 MHz), 256 KB of flash memory, an Ethernet connector, and two serial interfaces. For our experiments, we used the spaniakos cryptographic library [8]. **ARM** We used the QEMU emulator with the following settings: ARM926EJ-S rev 5 processors family at 530MHz, 256MB of RAM, Debian 3.2.51 32 bit Operating System, and the libgcrypt-1.6.5 cryptographic library. **Raspberry Pi** is a single-board computer of credit card-size. It was initially developed for educational purposes, but because of its low energy

J.H. Castellanos et al.

consumption and low cost, it has become into a popular multipurpose hardware. We choose it as a possible hardware to implement the authentication and integrity mechanism. The characteristics of the Raspberry Pi model 2 (RPi2) are: Quad-core ARM Cortex-A7 processor at 900 MHz, 1 GB of RAM, 4 USB ports, 40 GPIO pins and an Ethernet port. The cryptographic library used was again libgcrypt-1.6.5 [17]. **PC** We used a workstation with the following specifications: Intel Core i5-5300U processor at 2.30GHz, 3 GB of RAM, Xubuntu 15.10 64 bit OS, and libgcrypt-1.6.5 as cryptographic library.

Table 2. Benchmark of HMAC-SHA256 and ECDSA signature process for different packet sizes over 5 types of hardware (rounded values). Times are in μs.

	HMAC - Time				
Size	Controllino	ARM	RPi2	RPi3	PC
64 B	$2.2 \cdot 10^4$	76	53	15	2
128 B	$3.3 \cdot 10^4$	78	58	16	2
256 B	$5.5 \cdot 10^4$	84	69	18	3
512 B	$1 \cdot 10^5$	117	89	32	4
1 KB	$1.8 \cdot 10^5$	171	130	35	6
2 KB	$3.6 \cdot 10^5$	252	211	58	10
4 KB	$7 \cdot 10^5$	474	374	104	18
	ECDSA - Time				
4 KB	N/A	$1.5 \cdot 10^5$	$1 \cdot 10^5$	$3.2 \cdot 10^4$	$3.1 \cdot 10^3$

Table 3. Benchmark of performance of HMAC-SHA256 and ECDSA authentication for different packet sizes and hardware. Average size per packet 73 Bytes from Table 1.

	HMAC - Average Pkt/s				
Size	Contr.	ARM	RPi2	RPi3	PC
64 B	40	$1.1 \cdot 10^4$	$1.6 \cdot 10^4$	$5.8 \cdot 10^4$	$4.4 \cdot 10^5$
128 B	53	$2.2 \cdot 10^4$	$3 \cdot 10^4$	$1.1 \cdot 10^5$	$8.8 \cdot 10^5$
256 B	64	$4.2 \cdot 10^4$	$5 \cdot 10^4$	$1.9 \cdot 10^5$	$1.2 \cdot 10^6$
512 B	70	$6 \cdot 10^4$	$7.9 \cdot 10^4$	$2.2 \cdot 10^5$	$1.7 \cdot 10^6$
1 KB	78	$8.2 \cdot 10^4$	$1.1 \cdot 10^5$	$4 \cdot 10^5$	$2.3 \cdot 10^6$
2 KB	78	$1.1 \cdot 10^5$	$1.3 \cdot 10^5$	$4.8 \cdot 10^5$	$2.8 \cdot 10^6$
4 KB	80	$1.2 \cdot 10^5$	$1.5 \cdot 10^5$	$5.4 \cdot 10^5$	$3 \cdot 10^6$
	ECDSA - Average Pkt/s				
4 KB	N/A	$3.7 \cdot 10^2$	$5.6 \cdot 10^2$	$1.7 \cdot 10^3$	$1.8 \cdot 10^4$

We also benchmarked the elliptic curve based ECDSA signing standard. As discussed previously, the cryptographic operation dominates the execution time of the hashing component of the algorithm. This makes the times for signing payloads of up to 4KB almost identical to small payloads of 32B. The results are reported in the last row of Table 2. The resulting s for ASPA for aggregated signatures of about 58 packets (4KB) is reported in Table 3.

5.3 Discussion

We now provide a summary of our empirical results using different hardware platforms and cryptographic algorithms combination over our SWaT. In Table 4, we compare the SPA protocol (first two columns) with the ASPA protocol for $n = 58$ packets. As we can see, with a processor such as the one in the Controllino (16MHz), it is infeasible to cope with the real-time requirements of the SWaT networks for all algorithms considered. As shown in Fig. 6, and taking as reference our plant constraints, a symmetric signature is supported by most hardware. On the other hand, ECC signatures are possible in Raspberry Pi2, Pi3 and PCs thanks to our extension.

From a communications cost perspective, a signature in HMAC would add an overhead of 28% in size for an average ENIP packet, while ECDSA would add about 57%. Since we are only signing critical data, which corresponds to 42% of total traffic, our overhead in bandwidth will be 12% and 24% for HMAC and ECDSA respectively.

In sum, we have shown that in some cases, there is an advantage on using aggregation and selectiveness over traditional authenticated tunnels in terms of efficiency. Although we have shown an example for a concrete testbed and some hardware configurations, the possible configurations in practice could vary

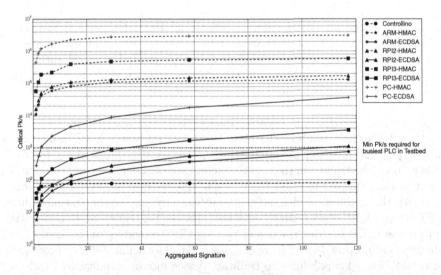

Fig. 6. ASPA performance on various hardware. PLC2, as the busiest, requires 1000 Pk/s for implementing a multi-link authentication process, y-axis is in log scale.

Table 4. Feasibility of algorithms vs. hardware in the SWaT. Slow hardware does not tolerate the minimum required signing rate even for ASPA. However, thanks to ASPA we can use RPi2, RPi3 and PCs to implement ECC based authentication.

	Sequential				Parallel			
	HMAC	ECC	HMAC-ASPA	ECC-ASPA	HMAC	ECC	HMAC-ASPA	ECC-ASPA
Controllino	×	×	×	×	×	×	×	×
ARM	✓	×	✓	×	✓	×	✓	×
RPi2	✓	×	✓	✓	✓	×	✓	✓
RPi3	✓	×	✓	✓	✓	×	✓	✓
PC	✓	×	✓	✓	✓	×	✓	✓

highly. For instance, in a full scale water plant the number of sensors and actuators could be one order of magnitude higher ($g = 10^4$), whereas the available hardware could range between the Controllino (16 MhZ) and the ARM processor (500 MhZ). Moreover, the underlying cryptographic algorithm could also vary, thus affecting the concrete performance.

Implementation in SWaT. We have successfully tested the proposed approach in a link between two PLCs of SWaT using Raspberry PIs and summarize our implementation efforts in Appendix B.

6   Related Work

In [23], the authors analyze different attacks on PLCs and several protection techniques. The authors state that the bulk of the security issues in ICS is the lack of security in the network communications. A public key based authentication protocol is proposed. However, the authors mention that their techniques could degrade the performance of the system, and they do not provide any further analysis.

A detailed analysis on cryptography-enabled authentication is reported in [31]. The authors analyze the security of electric power grids and identify authenticity and integrity as the most important security properties after availability. Some constraints on the message delays are considered during the analysis based on the expected message frequency as discussed in industrial standards.

In [1], the authors discuss integrity, authenticity, and authorization policies of ENIP and CIP. Two security profiles (providing integrity and confidentiality) for the ENIP are described and two for CIP (providing authorization and integrity) are proposed as a future extension. Authors define security profile as a set of well-defined capabilities to facilitate device interoperability, and end-user selection of devices with the appropriate security capability. In particular, the

CIP authorization profile will provide secure communications between CIP end-points, to ensure device and user authenticity. Authors describe ENIP over TLS (for TCP-based communication) and DTLS (for UDP).

Authors propose to implement confidentiality and integrity checks between paired devices called SCADA Cryptographic Modules (SCM) [33]. They are featured with two ports (plaintext and ciphertext). While it receives plaintext messages from a device through the plaintext port, it encrypts the message using AES and send it to its remote SCM through the ciphertext port. The message includes a MAC code (HMAC-SHA-1 or CBC-MAC) for integrity verification. In [28], authors propose a bump-in-the-wire solution to add security features such as Data privacy and authenticity to ICS communication. The solution suggests adding two devices (each per peer) into the communication channel (YASIR transmitter and YASIR receiver). YASIR Transmitter encrypts the message using AES-CTR, and attach an HMAC-SHA-1 signature. A YASIR Receiver decrypts and checks the message's integrity. In the case of integrity violation, YASIR Receiver adds an error byte to the frame; it guarantees that destination device discards the final message.

In IP-based communications, authors at [18] introduce a DNP protocol modification. They propose to assure confidentiality using DES cryptographic primitive and Integrity-Authenticity with HMAC-MD5-96 or HMAC- SHA-1-96. Their work suggests a change in the DNP's data structure. It could thwart their adoption by industry. Recently, researchers create a new MODBUS security development module (NMSDM) [25] that replaces MODBUS message. They propose to use cryptographic algorithms such as RSA, AES and SHA-2 to guarantee security properties as confidentiality, integrity, authentication and non-repudiation. Their proposal is to add a *cryptographic buffer* on top of PDU, as part of the TCP payload. Like ours, it allows adding security properties to the communication without modifying the protocol by itself. Both works [18,25] search to protect every message in an ICS, and computing overhead of cryptographic implementation are not taken into account.

In automotive applications, authors at [21,30] present authentication techniques called CANAuth and LeiA for CAN bus protocol. Similarly to ICS, CAN bus systems relies on hard real-time constraints. The authors shows that state of the art authentication mechanisms for broadcast networks cannot be used due to time (and storage) constraints. LeiA and CANAuth, however, are specifically designed for CAN bus systems. A low encryption overhead is also of paramount importance in Wireless Sensor Network (WSN), e.g., [13,20]. This is mainly due to bandwidth, latency and energy cost introduced by encryption. Solutions such as [13] provide authentication, but are not analyzed in the context of real-time constraints.

To the best of our knowledge, we are the first to propose the aggregated selective packet authentication protocol in the context of ICS. The idea is inspired by a delayed authentication scheme for GPS proposed in [16].

7   Conclusions

In this work, we discussed the introduction of efficient legacy-compliant authentication in ICS networks. By design, our protocols are backward compatible with devices not implementing them, as they transmit authentication data as payload in legacy industrial protocols.

We have shown that with the advent of inexpensive and fast hardware (such as the Raspberry Pi), it is feasible to enhance legacy plants with constraints similar to the ones of SWaT with authentic channels for strong signature algorithms with simple protocols. However, introducing state of the art asymmetric algorithms (which are advantageous from the key management point of view), or implementing our solution in slower hardware requires careful performance trade-offs. In future work, we plan to compare the real-time constraints of SWaT with constraints found in other ICS test-beds (i.e. smartgrid).

A   Appendix: Real-Time Requirements and Backlogs

The delay produced by the signature module can be represented as a queue of packets pending to be signed after a time interval Δ: $Q_\Delta^S = g(\Delta) - s(\Delta)$. We assume an empty queue at the beginning of the time interval Δ. To avoid an avalanche effect on the packet queue Q_Δ^S the queue must remain below a certain constant threshold C: $\forall \Delta.\ Q_\Delta^S < C$. In particular, since Δ is arbitrary $C \approx 0$ (i.e., $\forall \Delta.\ Q_\Delta^S < 0$) and thus: $\forall \Delta.\ g(\Delta) < s(\Delta)$. Similarly, if $r(\Delta)$ is the number of expected packets in a time interval Δ, then the number v of verified packets in this interval must be $v(\Delta, cpu, alg) > r(\Delta)$ to avoid backlogs of the verification queue Q_Δ^V.

$s(\Delta)$ depends not only on the time but also on the algorithm used, the CPU capacity, and the size of the packets. In a fraction of time Δ, the device will produce not only packets at different rates, but will also nondeterministically produce packets of different sizes. To simplify our analysis, and without loss of generality, we thus set $\Delta = 1$ s, and $s(cpu, alg)$ the rate at which packets of a certain constant expected size can be signed using a certain cpu and a given signature algorithm alg. We use s, v, g and r to abbreviate the rates per second of the signing, verification, outgoing (or *generated*) and incoming packets respectively.

Parallel versus Sequential Signature and Verification. We are assuming an architecture where network communications are handled by dedicated hardware. This avoids sharing the main device CPU (such as the main PLC computing unit), which is busy computing other control and communication related tasks. In practice this is usually the case, since communications are often handled by a dedicated network module. For a parallel implementation of the protocol, where the signature and verification components have at least a single core each, then we have effectively two queues that we can consider separate, Q_Δ^S and Q_Δ^V. Otherwise, for a sequential implementation of the protocol, an incoming packet that needs to be verified has to potentially wait until an outgoing packet is signed; and vice versa, an outgoing packet has to wait until a verification is finished. We have thus the constraints: $\forall \, \Delta. \, Q^S_\Delta < C \wedge Q^V_\Delta < C'$ which we can simplify by considering a single queue $Q_\Delta = Q^S_\Delta \cup Q^V_\Delta = (g(\Delta) - s(\Delta)) + (r(\Delta) - v(\Delta)) < C + C'$. As above, $C + C' \approx 0$ and thus: $g(\Delta) + r(\Delta) < s(\Delta) + v(\Delta)$. In other words, the rate of signature and verification in a time interval has to be faster than the amount of packets sent and received in that same time interval.

B   Appendix: Authenticated Link Using Raspberry Pi

We used two (paired) Raspberry Pi3, (as shown in Fig. 7) each with two Ethernet adapters. We choose a link between two PLCs in our SWaT to perform the test. The devices were configured as Ethernet bridges and placed as physical Man-in-the-Middle over the link. This choice was motivated because the source code of the Network Adapter in the SWaT was unavailable to us (as such proprietary code is confidential), and we could not directly implement our solution without additional hardware. However, a vendor could easily deploy our solution.

Once connected, the devices passively listen to packets from and to the PLCs. When a critical-data packet is identified, it is captured and the ENIP payload is signed with HMAC-SHA256 algorithm using a pre-shared key. The concatenation of the captured packet and its signature is injected back into the communication channel.

Similarly, the remote Raspberry Pi3 is placed as a verification module in front of the destination PLC. Once the verification module identifies a packet coming from its counter-

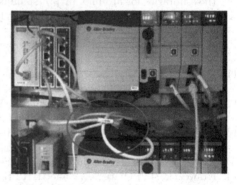

Fig. 7. Raspberry Pi3 connected between PLCs.

part, the packet is analysed looking for an attached signature, the signature is extracted and verified against the ENIP payload using HMAC-SHA256 algorithm with the pre-shared key. The packet is converted back to its original version and it is delivered to its destination.

We configured four additional variables in the PLCs to store information about the authentication processes: *Signed messages*: Number of signed messages by its cryptographic module; *Checked messages*: Number of signed message correctly verified by its cryptographic module; *Wrong-signature messages*: Number of signed messages which signature does not correspond to its payload detected by its cryptographic module; *No-signed messages*: Messages with critical data from a peered host with no-attached signature detected by its cryptographic module.

References

1. Batke, B., Wiberg, J., Dubè, D.: CIP security phase 1 secure transport for Ethernet/IP. In: ODVA Industry Conference (2015)
2. Cárdenas, A.A., Amin, S.M., Sinopoli, B., Giani, A., Perrig, A., Sastry, S.S.: Challenges for securing cyber physical systems. In: Workshop on Future Directions in Cyber-physical Systems Security, DHS, July 2009
3. Cárdenas, A.A., Baras, J.S.: Evaluation of classifiers: practical considerations for security applications. In: AAAI Workshop on Evaluation Methods for Machine Learning (2006)
4. Dierks, T.: The transport layer security (TLS) protocol version 1.2 (2008). https://www.ietf.org/rfc/rfc5246.txt
5. Fletcher, K.K., Liu, X.: Security requirements analysis, specification, prioritization and policy development in cyber-physical systems. In: Secure Software Integration Reliability Improvement Companion (SSIRI-C), pp. 106–113 (2011)
6. Galloway, B., Hancke, G.: Introduction to industrial control networks. Commun. Surv. Tutor. 15(2), 860–880 (2013). IEEE
7. Gomes, N., Mattos, L.: Attacks detection based on IP and TCP protocols violation. Int. J. Forensic Comput. Sci. 1, 49–56 (2006)
8. Hash libraries for arduino. http://spaniakos.github.io/Cryptosuite/
9. Igure, V.M., Laughter, S.A., Williams, R.D.: Security issues in scada networks. Comput. Secur. 25(7), 498–506 (2006)
10. P. Inc.: Profinet and it. Technical report, PROFIBUS Nutzerorganisation e.V. (2008)
11. iTrust: Center for Research in Cyber Security. Secure water treatment test-bed. http://itrust.sutd.edu.sg/research/testbeds/secure-water-treatment-swat/
12. Johnson, D., Menezes, A., Vanstone, S.: The elliptic curve digital signature algorithm (ECDSA). Int. J. Inf. Secur. 1(1), 36–63 (2001)
13. Karlof, C., Sastry, N., Wagner, D.: TinySec: a link layer security architecture for wireless sensor networks. In: Proceedings of the International Conference on Embedded Networked Sensor Systems, SenSys 04, pp. 162–175. ACM (2004)
14. Krawczyk, H., Bellare, M., Canetti, R.: HMAC: keyed-hashing for message authentication (1997). https://www.ietf.org/rfc/rfc2104.txt
15. Krotofil, M., Cárdenas, A.A., Manning, B., Larsen, J.: CPS: driving cyber-physical systems to unsafe operating conditions by timing DoS attacks on sensor signals. In: Proceedings of the Computer Security Applications Conference (ACSAC), pp. 146–155. ACM (2014)
16. Kuhn, M.G.: An asymmetric security mechanism for navigation signals. In: Fridrich, J. (ed.) IH 2004. LNCS, vol. 3200, pp. 239–252. Springer, Heidelberg (2004). doi:10.1007/978-3-540-30114-1_17
17. Gnu cryptographic library. https://www.gnu.org/software/libgcrypt/
18. Majdalawieh, M., Parisi-Presicce, F., Wijesekera, D.: DNPSec: distributed network protocol version 3 (DNP3) security framework. In: Elleithy, K., Sobh, T., Mahmood, A., Iskander, M., Karim, M. (eds.) Advances in Computer, Information, and Systems Sciences, and Engineering, vol. 3, pp. 227–234. Springer, Dordrecht (2007). doi:10.1007/1-4020-5261-8_36
19. Modbus-IDA. Modbus messaging on tcp/ip implementation guide v1.0b. Technical report, Modbus Organization (2006)

20. Nie, P., Vähä-Herttua, J., Aura, T., Gurtov, A.: Performance analysis of HIP diet exchange for wsn security establishment. In: Proceedings of the ACM Symposium on QoS and Security for Wireless and Mobile Networks, Q2SWinet 11, pp. 51–56. ACM (2011)
21. Radu, A.I., Garcia, F.D.: LeiA: a lightweight authentication protocol for CAN. In: Askoxylakis, I., Ioannidis, S., Katsikas, S., Meadows, C. (eds.) ESORICS 2016. LNCS, vol. 9879, pp. 283–300. Springer, Cham (2016). doi:10.1007/978-3-319-45741-3_15
22. Rajkumar, R., Lee, I., Sha, L., Stankovic, J.: Cyber-physical systems: the next computing revolution. In: 2010 47th ACM/IEEE on Design Automation Conference (DAC), pp. 731–736, June 2010
23. Sandaruwan, G., Ranaweera, P., Oleshchuk, V.A.: PLC security and critical infrastructure protection. In: Industrial and Information Systems (ICIIS), pp. 81–85. IEEE (2013)
24. Schiffer, V., Vangompel, D., Voss, R.: The common industrial protocol (CIP) and the family of CIP networks. ODVA, Ann Arbor (2006)
25. Shahzad, A., Lee, M., Lee, Y.-K.K., Kim, S., Xiong, N., Choi, J.-Y.Y., Cho, Y.: Real time MODBUS transmissions and cryptography security designs and enhancements of protocol sensitive information. Symmetry 7(3), 1176–1210 (2015)
26. Slay, J., Miller, M.: Lessons learned from the maroochy water breach. In: Goetz, E., Shenoi, S. (eds.) ICCIP 2007. IIFIP, vol. 253, pp. 73–82. Springer, Boston, MA (2008). doi:10.1007/978-0-387-75462-8_6
27. Smart, N., Babbage, S., Catalano, D., Cid, C., Weger, B. d., Dunkelman, O., Ward, M.: Ecrypt ii yearly report on algorithms and keysizes (2011–2012). In: European Network of Excellence in Cryptology (ECRYPT II) (2012)
28. Tsang, P.P., Smith, S.W.: YASIR: a low-latency, high-integrity security retrofit for legacy SCADA systems. In: Jajodia, S., Samarati, P., Cimato, S. (eds.) SEC 2008. ITIFIP, vol. 278, pp. 445–459. Springer, Boston, MA (2008). doi:10.1007/978-0-387-09699-5_29
29. Urbina, D., Giraldo, J., Tippenhauer, N.O., Cárdenas, A.: Attacking fieldbus communications in ICS: applications to the SWaT testbed. In: Proceedings of Singapore Cyber Security Conference (SG-CRC), January 2016
30. Van Herrewege, A., Singelee, D., Verbauwhede, I.: CANAuth-a simple, backward compatible broadcast authentication protocol for CAN bus. In: ECRYPT Workshop on Lightweight Cryptography, vol. 2011 (2011)
31. Wang, W., Lu, Z.: Cyber security in the smart grid: survey and challenges. Comput. Netw. 57(5), 1344–1371 (2013)
32. Weinberger, S.: Computer security: is this the start of cyberwarfare? Nature 174, 142–145 (2011)
33. Wright, A.K., Kinast, J.A., McCarty, J.: Low-latency cryptographic protection for SCADA communications. Acns 3089, 263–277 (2004)

Multi-client Oblivious RAM Secure Against Malicious Servers

Erik-Oliver Blass[1], Travis Mayberry[2(✉)], and Guevara Noubir[3]

[1] Airbus Group Innovations, Munich, Germany
erik-oliver.blass@airbus.com
[2] US Naval Academy, Annapolis, MD, USA
mayberry@usna.edu
[3] Northeastern University, Boston, MA, USA
noubir@ccs.neu.edu

Abstract. This paper tackles the open problem whether an Oblivious RAM can be shared among multiple clients in the presence of a fully malicious server. Current ORAM constructions rely on clients knowing the ORAM state to not reveal information about their access pattern. With multiple clients, a straightforward approach requires clients exchanging updated state to maintain security. However, clients on the internet usually cannot directly communicate with each other due to NAT and firewall settings. Storing state on the server is the only option, but a malicious server can arbitrarily tamper with that information.

We first extend the classical square-root ORAM by Goldreich and the hierarchical one by Goldreich and Ostrovsky to add multi-client security. We accomplish this by separating the *critical* portions of the access, which depend on the state of the ORAM, from the non-critical parts (cache access) that can be executed securely in any state. Our second contribution is a secure multi-client variant of Path ORAM. To enable secure meta-data update during evictions in Path ORAM, we employ our first result, small multi-client secure classical ORAMs, as a building block. Depending on the block size, the communication complexity of our multi-client secure construction reaches a low $O(\log N)$ communication complexity per client, similar to state-of-the-art single-client ORAMs.

1 Introduction

The main metric of an ORAM's performance, communication overhead, has improved by orders of magnitude over the last few years. However, at least one significant hurdle to actual adoption remains: security of modern ORAMs relies on there being only a single client at all times. While there already exist multi-client ORAMs secure in the face of a *semi-honest* server [18], existence of an ORAM secure against a fully *malicious* server is still an open question. The main challenge here stems from the fact that today's ORAMs modify some of the data on the server after every access. If a malicious server "rewinds" the data and presents an old version to a client, further interactions may reveal details about the access pattern. In single client scenarios, this is typically solved by

© Springer International Publishing AG 2017
D. Gollmann et al. (Eds.): ACNS 2017, LNCS 10355, pp. 686–707, 2017.
DOI: 10.1007/978-3-319-61204-1_34

storing a small token on the client, such as the root of a hash tree [22]. This token authenticates and verifies freshness of all data retrieved from the server, ensuring that no such rewind attack is possible.

In this paper, we address the fundamental problem how multiple clients can share data stored in a single ORAM. With multiple clients, an authentication token is not sufficient. Data may not pass one client's authentication, simply because it has been modified by one of the other clients. If clients could communicate with each other using a secure out-of-band channel, then it becomes trivial to continually exchange and update each other with the most recent token. However, existence of secure out-of-band communication is often not a reasonable assumption for modern devices. As we will see, it is the absence of out-of-band-communication which makes multi-client ORAM technically challenging.

Current solutions for multi-client ORAM work only in the presence of semi-honest (honest-but-curious) adversaries, which cannot perform rewind attacks on the clients. Often, this is not a very satisfying model, since rewind attacks are very easy to execute for real-world adversaries and would be difficult to detect. Consequently, we only address fully malicious servers. Goodrich et al. [12], in their paper examining multi-client ORAM, recently proposed as an open question whether one could be secure for multiple clients against a malicious server.

Technical Highlights. We introduce the first construction for a multi-client ORAM and prove access pattern indistinguishability, even if the server is fully malicious. Our contribution is twofold, specifically:

- We start by focusing on two ORAM constructions that follow a "classical" approach, the *square-root* ORAM by Goldreich [8] and the *hierarchical* ORAM by Goldreich and Ostrovsky [9]. We adapt these ORAMs for multi-client security. Our approach is to separate client accesses into two parts: non-critical portions, which can be performed securely in the presence of a malicious server and critical portions, which cannot but contains efficient integrity checks which will reveal any malicious behavior and allow the client to terminate the protocol.
- The "classical" ORAM constructions have been largely overshadowed by more recent tree-based ORAMs [23,25]. Consequently, we go on to demonstrate how a multi-client secure Path ORAM [25] can be constructed. We solve the key challenge of realizing a multi-client secure version of the *read* protocol by storing Path ORAM's metadata using small "classical" ORAMs as building blocks. For block sizes in $\Omega(\log^4 N)$, this results in a multi-client ORAM which has overall communication complexity of $O(\phi \cdot \log N)$.

Table 1 summarizes asymptotic behavior for our new multi-client ORAMs and compares them to their corresponding single-client ORAMs.

2   Motivation: Multi-client ORAM

Instead of a single ORAM accessed by a single client, we envision multiple clients securely exchanging or sharing data stored in a single ORAM. For example, imag-

Table 1. Communication and storage worst-case complexity for existing single-client ORAMs and our new multi-client versions. ϕ is the number of different clients supported by the ORAM. $\hat{O}$ denotes amortized complexity.

	Communication		Storage	
	Single client	Multi client	Single client	Multi client
Square-root [8]	$\hat{O}(\sqrt{N})$	$\hat{O}(\phi \cdot \sqrt{N})$	$O(N)$	$O(\phi \cdot N)$
(Deamortized) Hierarchical [9, 15]	$O(\log^3 N)$	$O(\phi \cdot \log^3 N)$	$O(N \cdot \log N)$	$O(\phi \cdot N \cdot \log N)$
Tree-based [25]	$O(\log N)$	$O(\phi \cdot \log N)$	$O(N \cdot \log N)$	$O(\phi \cdot N \cdot \log N)$

ine multiple employees of a company that read from and write into the same database stored at an untrusted server. Similar to standard ORAM security, sharing data and jointly working on the database should not leak the employees' access patterns to the server. Alternatively, we can also envision a single person with multiple different devices (laptop, tablet, smartphone) accessing the same data hosted at an untrusted server (e.g., Dropbox). Again, working on the same data should not reveal access patterns. Throughout this paper, we consider the terms "multi-client" and "multi-user" to be equivalent. As suggested by Goodrich et al. [12], we assume that clients all trust each other and leave expansion of our results for more fine-grained access control as future work. In the multi-device scenarios above, it is reasonable that clients trust each other since they belong to a single user.

To provide security, ORAM protocols are *stateful*. Hiding client accesses to a certain data block is typically achieved by performing shuffling or reordering of blocks, such that two accesses are not recognizable as being the same. An obvious attack that a malicious server can do is to undo or "rewind" that shuffling after the first access and present the same, original view (state) of the data to the client when they make the second access. If the client was to blindly execute their access, and it was the same block of data as the first access, it would result in the same pattern of interactions with the server that the first access did. The server would immediately have broken the security of the ORAM scheme. This is a straightforward attack that can be easily defeated in case of a single client: as an internal state, the client stores and updates a token for authentication and freshness, see Ren et al. [22].

However, with two or more clients sharing data in an ORAM, this attack becomes a new challenge. After watching one client retrieve some data, the adversary rewinds the ORAM's state and present the original view to the second client. If the second client accesses the same data (or not) that the first client did, the server will recognize it, therefore violating security. Without having some secure side-channel to exchange authentication tokens after every access, it is difficult for clients to detect such an attack.

2.1 Technical Challenges

A multi-client ORAM has to overcome a new technical challenge. Roughly speaking, the server is fully malicious and can present different ORAM states to different clients, i.e., different devices of the same user. As different clients do not have a direct communication channel to synchronize *their* state, it is difficult for them to synchronize on an ORAM state. We expand on this challenge below.

Adversary Model: This paper tackles the scenario of ϕ trusted clients sharing storage on a fully malicious server (the adversary). Other works such as Maffei et al. [18] have addressed the problem against a semi-honest server, but in many scenarios that may not be sufficient. Real-world attacks that clients need to defend against include, e.g., insider attacks from a cloud provider hosting the server and outside hackers compromising the server. Such attacks would allow for malicious adversarial behavior. In general, there is no strong line between the two adversarial models that suggests that one is more reasonable to defend against. To cope with all possible adversaries, it is therefore important to protect against malicious adversaries, too.

No Out-of-Band Communication: We assume that beyond a single cryptographic key (possibly derived from a password) the clients do not share any long-term secrets and cannot communicate with each other except through the malicious server. This matches with existing cloud settings, since most consumer devices are behind NAT and cannot be directly contacted from the Internet. Major real-world cryptographic applications, for instance WhatsApp [26] and Semaphor [24], have all messages between clients relayed through the server for this reason.

We emphasize that in the malicious setting the server always has the option to simply stop responding or send purposefully wrong data as a denial-of-service attack. This cannot be avoided, but is also not a significant problem since it will be easily detected by clients. In contrast, the attacks we focus on in this paper are those where the server tries to compromise security without being detected.

2.2 Other Applications

A major application of this work is in supporting private cloud storage that is accessible from multiple devices. In reality, this is one of the most compelling use cases for cloud storage a la Dropbox, Google Drive, iCloud, etc. and is a good target for a privacy-preserving solution. Beyond simple storage, Oblivious RAM is used as a subroutine in other constructions, such as dynamic proofs of retrievability [3]. Any construction which uses ORAM and wishes to support multiple clients must rely on an ORAM that is secure for multiple clients.

Finally, an interesting application comes from the release of Intel's new SGX-enabled processors. SGX enables a trusted enclave to run protected code which cannot be examined from the outside, even by the operating system or hypervisor running on the machine. The major remaining channel for leakage in this system is in the pattern of accesses that the enclave code makes to the untrusted RAM lying outside the processor. It has already been noted that Oblivious RAM may be a valuable tool to eliminate that leakage [5]. Furthermore, it is expected that systems may have multiple enclaves running at the same time which wish to share information through the untrusted RAM. This scenario corresponds exactly with our multi-client ORAM setting, so the solution here could be used to securely offer that functionality.

2.3 Related Work

Most existing work on Oblivious RAM assumes only a single client. Franz et al. [7] proposed an solution for multiple clients using the original square-root ORAM, but relies on a semi-honest server. Goodrich et al. [12] extend that work to more modern tree-based ORAMs, still relying on a semi-honest server. Recent work by Maffei et al. [18] supports multiple clients with read/write access control, but again requires a semi-honest server. Other efficient solutions are possible with an even weaker security model by using trusted hardware on the server side [13,17].

There is also a concurrent line of work in Parallel ORAMs which target ORAMs running on multi-core or multi-processor systems [2,4,20]. These schemes either do not target malicious adversaries or require constant and continuous communication between clients to synchronize their state. As stated above, this is not a viable solution for clients which are not always on or may be across different networks.

Currently, no solution exists that allows for multiple clients interacting with a malicious server and without direct client-to-client communication, or constant polling to create the same effect.

3 Security Definition

We briefly recall standard ORAM concepts. An ORAM provides an interface to read from and write to blocks of a RAM (an array of storage blocks). It supports $\mathsf{Read}(x)$, to read from the block at address x, and $\mathsf{Write}(x, v)$ to write value v to block x. The ORAM allows storage of N blocks, each of size B.

To securely realize this functionality, an ORAM interacts with a malicious storage device. Below, we use Σ to represent the *interface* between a client and the actual storage device. A client with access to Σ issues Read and Write requests as needed. We stress that we make no assumptions about how the storage device responds to these requests, and allow for arbitrary malicious behavior. For instance, an adversary could respond to a Read request with old or corrupt data, refuse to actually perform a Write correctly etc. As related work, the (untrusted) storage device is part of a *server* as in envisioned applications for a (multi-client) ORAM such as outsourced cloud storage.

Definition 1 (ORAM Operation OP**).** *An operation* OP *is defined as* OP $=$ (o, x, v), *where* $o = \{\text{Read}, \text{Write}\}$, x *is the virtual address of the block to be accessed and* v *is the value to write to that block.* $v = \bot$ *when* $o = $ Read.

We now present our multi-client ORAM security definition which slightly augments the standard, single-client ORAM definition. We emphasize that clients only interact with the server by read or write operations to a memory location; there are no other messages sent. Therefore the "protocol" is fully defined by these patterns of accesses.

Definition 2 (Multi-client ORAM Π**).** *A multi-client ORAM* Π $=$ (Init, ClientInit, Access) *comprises the following three algorithms.*

1. Init(λ, N, B, ϕ) *initializes* Π. *It takes as input security parameter* λ, *total number of blocks* N, *block size* B, *and number of clients* ϕ. Init *initializes the storage device represented by* Σ *and outputs a key* κ.
2. ClientInit$(\lambda, N, B, j, \kappa)$ *uses security parameter* λ, *number of blocks* N, *block size* B, *and a secret key* κ *to initialize client* u_j. *It outputs a client state* st_{u_j}.
3. Access$(\text{OP}, \Sigma, st_{u_j})$ *performs operation* OP *on the ORAM using client* u_j's *state* st_{u_j} *and interface* Σ. Access *outputs a new state* st_{u_j} *for client* u_j.

In contrast to single-client ORAM, a multi-client ORAM introduces the notion of clients. This is modeled by different per-client states, st_{u_i} for client u_i. After initializing the multi-client ORAM with Init, Algorithm ClientInit is run by each client (with no communication between them) separately and outputs their initial local states st_{u_i}. The ClientInit function only requires that each client have a shared key κ, which could be derived from a password. Whenever client u_i executes Access on the multi-client ORAM, they can attempt to update the multi-client ORAM represented by Σ, and update their own local state st_{u_i}, but not the other clients' local states.

Finally, we define the security of a multi-client ORAM against malicious servers. Consider the game-based experiment $\text{Sec}^{\text{ORAM}}_{\mathcal{A},\Pi}(\lambda)$ below. In this game, $\mathcal{A}$ has complete control over the storage device and how it responds to client requests. For ease of exposition, we model this as $\mathcal{A}$ outputting their own compromised version $\Sigma_{\mathcal{A}}$ of an interface to the storage device. It is this malicious interface $\Sigma_{\mathcal{A}}$ that clients will subsequently use for their Access operations. Interface $\Sigma_{\mathcal{A}}$ is controlled by the adversary and updates state $st_{\mathcal{A}}$. That is, $\mathcal{A}$ learns all clients' calls to the interface, therewith the clients' requested access pattern, and can adaptively react to clients' interface requests. To initialize the ORAM, ϕ time steps are used to do the setup for each client, then the game continues for poly(λ) additional steps where the adversary interactively specifies operations and maliciously modifies the storage.

Experiment 1 (Experiment $\mathsf{Sec}^{\mathsf{ORAM}}_{\mathcal{A},\Pi}(\lambda)$)

1 $b \xleftarrow{\$} \{0,1\}$
2 $(\kappa, \Sigma) \leftarrow \mathsf{Init}(\lambda, B, N, \phi)$
3 **for** $i = 1$ **to** ϕ **do**
4      $st_{u_i} \leftarrow \mathsf{ClientInit}(\lambda, N, B, \kappa, i)$
5 **end**
6 $(\mathsf{OP}_{\phi+1,0}, \mathsf{OP}_{\phi+1,1}, i, st_{\mathcal{A}}, \Sigma_{\mathcal{A}}) \leftarrow \mathcal{A}(\lambda, N, B, \phi, \Sigma)$
7 **for** $j = \phi + 1$ **to** $poly(\lambda)$ **do**
8      $st_{u_i} \leftarrow \mathsf{Access}(\mathsf{OP}_{j,b}, \Sigma_{\mathcal{A}}, st_{u_i})$
9      $(\mathsf{OP}_{j+1,0}, \mathsf{OP}_{j+1,1}, i, st_{\mathcal{A}}) \leftarrow \mathcal{A}(P, st_{\mathcal{A}})$
10 **end**
11 $b' \leftarrow \mathcal{A}(st_{\mathcal{A}})$
12 **output** 1 **iff** $b = b'$

In summary, $\mathcal{A}$ gets oracle access to the ORAM and can adaptively query it during $poly(\lambda)$ rounds. In each round, $\mathcal{A}$ selects a client u_i and determines two operations $\mathsf{OP}_{j,0}$ and $\mathsf{OP}_{j,1}$. The oracle performs operation OP_b as client u_i with state st_{u_i}, interacting with the adversary-controlled $\Sigma_{\mathcal{A}}$ using protocol Π. Eventually, $\mathcal{A}$ guesses b.

Definition 3 (Multi-client ORAM security). *An ORAM $\Pi = (\mathsf{Init}, \mathsf{Access})$ is multi-client secure iff for all PPT adversaries $\mathcal{A}$, there exists a function $\epsilon(\lambda)$ negligible in security parameter λ such that*

$$Pr[\mathsf{Sec}^{\mathsf{ORAM}}_{\mathcal{A},\Pi}(\lambda) = 1] < \frac{1}{2} + \epsilon(\lambda).$$

Our game-based definition is equivalent to ORAM's standard security definition with two exceptions: we allow the adversary to arbitrarily change the state of the on-server storage Σ, and we split the ORAM algorithm into ϕ different "pieces" which cannot share state among themselves.

As discussed above, this work assumes that all clients trust each other and do not conspire. For ease of exposition, we assume that all client share key κ used for encryption and MAC computations that we will introduce later.

Consistency: An orthogonal concern to security for multi-client schemes is *consistency*, whether the clients each see the same version of the database when they access it. Because the clients in our model do not have any way of communicating except through the malicious adversary, it is possible for $\mathcal{A}$ to "desynchronize" the clients so that their updates are not propagated to each other. Our multi-client ORAM guarantees that in this case the clients still have complete security and access pattern privacy, but consistency cannot be guaranteed. This is a well known problem with the only solution being *fork* consistency [16], which we achieve.

4   Multi-client Security for Classical ORAMs

We start by transforming two classical ORAM constructions, the original square-root solution by Goldreich [8] and the hierarchical one by Goldreich and

Ostrovsky [9], into multi-client secure versions, retaining the same communication complexity per client. Our exposition focuses in the beginning on details for the multi-client square-root ORAM, as the hierarchical ORAM is borrowing from the same ideas.

Recall that the square-root ORAM algorithm works by dividing the server storage into two parts: the main memory and the cache, which is of size $O(\sqrt{N})$. The main memory is shuffled by a pseudo-random permutation π. Every access reads the entire cache, plus one block in the main memory. If the block the client wants is found in the cache, a "dummy" location is read in the main memory, otherwise the actual location of the target block is read and it is inserted into the cache for later accesses. After $\sqrt{N}$ accesses, the cache is full and the client must download the entire ORAM and reshuffle it into a fresh state.

Specific Challenge: When considering a multi-client scenario, it becomes easy for a malicious server to break security of the square-root ORAM. For example, client u_1 can access a block x that is not in the cache, requiring u_1 to read $\pi(x)$ from main memory and insert it into the cache. The malicious server now restores the cache to the state it was in before u_1's access added block x. If a second client u_2 also attempts to access block x, the server will now observe that both clients read from the same location $\pi(x)$ in main memory and know that u_1 and u_2 have accessed the same block (or not). Without the clients having a way to communicate directly with each other and pass information that allows them to verify the changes to the cache, the server can always "rewind" the cache back to a previous state. This will eventually force one client to leak information about their accesses.

Rationale: Our approach for multi-client security is based on the observation that the *cache update* part of the square-root solution is secure by itself. Updating the cache only involves downloading the cache, changing one element in it, re-encrypting, and finally storing it back on the server. Downloading and later uploading the cache implies always "touching" the same $\sqrt{N}$ blocks. This is independent of what the malicious server presents to a client as Σ and also independent of the block being updated by the client. Changing values inside the cache cannot leak any information to the server, as its content is always newly IND-CPA encrypted. Succinctly, being similar to a trivial ORAM, updating a cache is automatically multi-client secure.

However, reading can leak information. Reading from the main ORAM is conditional on what the client finds in the cache. We call this part the *critical* part of the access, and the cache update correspondingly *non-critical*. To counteract this leakage, we implement the following changes to enable multiple clients for the square-root ORAM:

1. **Separate ORAMs:** Instead of a single ORAM, we use a sequence of ORAMs, $\Sigma = \mathsf{ORAM}_1, \mathsf{ORAM}_2, \ldots, \mathsf{ORAM}_\phi$, one for each client. Client u_i will perform the critical part of their access only on ORAM_i's main memory

Input: Storage interface Σ, Security parameter λ, number of blocks in each
       ORAM N, block size B, client number i, key κ
Output: Client state st_{u_i}
1 Generate permutation $\pi_{i,0}$ from key κ;
2 Initialize $\sqrt{N} + N$ main memory blocks (size B, shuffled with $\pi_{i,0}$) and $\sqrt{N}$
  cache blocks;
3 Set cache counter $\chi_i = 0$; Set epoch counter $\gamma_i = 0$;
4 $data_i = \mathsf{Enc}_\kappa(\text{main memory})\|\mathsf{Enc}_\kappa(\text{cache}\|\chi_i\|\gamma_i)$;
5 $mac_i = \mathsf{MAC}_\kappa(\mathsf{Enc}_\kappa(\text{cache}\|\chi_i\|\gamma_i))$;
6 $\mathsf{ORAM}_i = data_i\|mac_i$;
7 On Σ, replace the ith ORAM by ORAM_i;
8 **output** $st_{u_i} = \{\kappa, \chi_i\}$;

Algorithm 1. $\mathsf{ClientInit}(\Sigma, \lambda, N, B, i, \kappa)$, initialize client square-root ORAM

and cache. Thus, each client can guarantee they will not read the same address
from their ORAM's main memory twice. However, any change to the cache
as part of ORAM $\mathsf{Read}(x)$ or $\mathsf{Write}(x, v)$ operations will be written to every
ORAM's cache. Updating the cache on any ORAM is already guaranteed to
be multi-client secure and does not leak information.

2. **Authenticated Caches:** For each client u_i to guarantee that they will not
   repeat access to the main memory of ORAM_i, the cache is stored together
   with an encrypted *access counter* χ on the server. Each client stores locally
   a MAC over both the cache and the encrypted access counter χ of their
   own ORAM. Every access to their own cache increments the counter and
   updates the MAC. Since clients read only from their own ORAMs, and they
   can always verify the counter value for the last time that they performed a
   read, the server cannot roll back beyond that point. Two reads will never be
   performed with the cache in the same state.

4.1 Details

We detail the above ideas in two algorithms: Algorithm 1 shows the per client
initialization procedure $\mathsf{ClientInit}$, and Algorithm 2 describes the way a client
performs an Access with our multi-client secure square-root ORAM. The Init
algorithm is trivial in our case, as it initializes Σ to ϕ empty arrays ORAM_j.
Each array is of size $N + 2 \cdot \sqrt{N}$ blocks, each block has size B bits.

Before explaining $\mathsf{ClientAccess}$, we first introduce the notion of an *epoch*. In
general, after $\sqrt{N}$ accesses to a square-root ORAM, its cache is "full", and the
whole ORAM needs to be re-shuffled. Re-shuffling requires computing a new
permutation π. Per ORAM, a permutation can be used for $\sqrt{N}$ operations,
i.e., one *epoch*. The next $\sqrt{N}$ operations, i.e., the next epoch, will use another
permutation and so on. In the two algorithms, we use an epoch counter γ_i.
Therewith, π_{i,γ_i} denotes the permutation of client u_i in ORAM_i's epoch γ_i. For

Input: Mult-client ORAM Σ, address x, new value v, client u_i, $st_{u_i} = \{\kappa, \chi_i\}$
Output: Value of block x, new state st_{u_i}

1 From ORAM$_i$ in Σ: read $c_i = \mathsf{Enc}_\kappa(\mathsf{cache}||\chi_i||\gamma_i)$ and mac_i;
2 $mac_i' = \mathsf{MAC}_\kappa(c_i)$;
3 if $mac_i' \neq mac_i$ then output Abort;
4 Decrypt c_i to get cache and counter χ_i';
5 if $\chi_i' < \chi_i$ then output Abort;
6 if $block\ x \notin cache$ then
7   Read and decrypt block $\pi_{i,\gamma_i}(x)$ from ORAM$_i$'s main memory;
8 else
9   Read next dummy block from ORAM$_i$'s main memory;
10 if $v = \bot$ then // operation is a Read
11      $\nu \leftarrow$ existing value of block x;
12 else // operation is a Write
13      $\nu \leftarrow v$;
14 Append block (x, ν) to cache;
15 if $cache\ is\ full$ then
16      $\gamma_i = \gamma_i + 1$; Compute new permutation π_{i,γ_i};
17   Read and decrypt ORAM$_i$'s main memory;
18   Shuffle cache and main memory using $\pi_{i,\gamma}$;
19   Update ORAM$_i$ with $\mathsf{Enc}_\kappa(\mathsf{main\ memory})$;
20 $\chi_i = \chi_i' + 1$; $mac_i = \mathsf{MAC}_\kappa(\mathsf{Enc}_\kappa(\mathsf{cache}||\chi_i||\gamma_i))$;
21 Update ORAM$_i$ with $\mathsf{Enc}_\kappa(\mathsf{cache}||\chi_i||\gamma_i)$ and mac_i;
22 for $j \neq i$ do // for all ORAM$_j \neq$ ORAM$_i$
23   Read and decrypt cache and χ_j from ORAM$_j$; Read and verify mac_i from ORAM$_j$;
24   Append block (x, ν) to cache;
25   if $cache\ is\ full$ then
26          $\gamma_j = \gamma_j + 1$; Compute new permutation π_{j,γ_j};
27   Read and decrypt ORAM$_j$'s main memory;
28   Shuffle cache and main memory using π_{j,γ_j};
29   Update ORAM$_j$ with $\mathsf{Enc}_\kappa(\mathsf{main\ memory})$;
30      $mac_j = \mathsf{MAC}_\kappa(\mathsf{Enc}_\kappa(\mathsf{cache}||\chi_j||\gamma_j))$;
31   Update ORAM$_j$ with $\mathsf{Enc}_\kappa(\mathsf{cache}||\chi_j||\gamma_j)$ and mac_j;
32 end
33 output $(\nu, st_{u_i} = \{\kappa, \chi_i\})$;

Algorithm 2. Access(OP, Σ, st_{u_i}), Read, Write for multi-client square-root ORAM

any client, to be able to know the current epoch of ORAM$_i$, we store γ_i together with the ORAM's cache on the server.

On a side note, we point out that there are various ways to generate pseudo-random permutations π_{i,γ_i} on N elements in a deterministic fashion. For example, in the a cloud context, one can use $\mathsf{PRF}_\kappa(i||\gamma_i)$ as the seed in a PRG and therewith perform Knuth's Algorithm P (Fisher-Yates shuffle) [14]. Alternatively, one can use the idea of random tags followed by oblivious sorting by Goldreich and Ostrovsky [9].

In addition to the epoch counter, we also introduce a per client *cache counter* χ_i. Using χ_i, client u_i counts the number of accesses of u_i to the main memory and cache of their own ORAM_i. After each access to ORAM_i by client u_i, χ_i is incremented. Each client u_i keeps a local copy of χ_i and therewith verifies freshness of data presented by the server. As we will see below, this method ensures multi-client ORAM security. Note in Algorithm 2 that a client u_j never increases χ_i of another client u_i. Only u_i ever updates χ_i.

In our algorithms, Enc_κ is an IND-CPA encryption such as AES-CBC. For convenience, we only write Enc_κ(main memory), although the main memory needs to be encrypted block by block to allow for the retrieval of specific blocks. Also, for the encryption of main memory blocks, Enc_κ offers authenticated encryption such as encrypt-then-MAC.

A client can determine whether a *cache is full* in Algorithm 2 by the convention that empty blocks in the cache decrypt to $\perp$. As long as there are blocks in the cache remaining with value $\perp$, the cache is not full.

ClientInit: Each client runs the ClientInit algorithm to initialize their ORAM. The server stores the ORAMs (with MACs) computed with a single key κ. Each client receives their state from the ClientInit algorithm, comprising the cache counter. Note that although not captured in the security definition, our scheme also allows for dynamic adding and removing of clients. Removing is as simple as just asking the server to delete one of the ORAMs, and adding could be done by running ClientInit, but instead of initializing the blocks to be empty, the client first downloads a copy of another client's ORAM to get the most recent version of the database.

Access: After verifying the MAC for ORAM_i and whether its cache is not from before u_i's last access, u_i performs a standard Read or Write operation for block x on ORAM_i. If the cache is full, u_i re-shuffles ORAM_i updating π. In addition, u_i also adds block x to all other clients' ORAMs. Note that for this, u_i does not read from the other ORAMs, but only completely downloads and re-encrypts their cache.

Our scheme is effectively running ϕ traditional square-root ORAMs in parallel, making the overall complexity $O(\phi\sqrt{N})$. Due to limited space, see the full version of this paper [1] for a detailed security analysis.

Fork Consistency: When a client makes an access, they add an element to the cache for all ϕ clients. Therefore, at any given timestep, if the server is not maliciously changing caches, all caches will have the same number of elements in them. Since each cache is verified by a MAC, the server cannot remove individual elements from a cache. The only viable attack is to present an old view of a cache which was at one point valid, but does not contain new updates that have been added by other clients. If the server chooses to do this, he creates a *fork* between the views of the clients which have seen the update and those that have not. Since the server can never "merge" caches together, but only present entire caches that have been verified with a MAC by a legitimate client, there is no

way to reconcile two forks that were created without a client finding out. This
achieves fork consistency for our scheme.

Complexity: Making the square-root solution multi-client secure does not
induce any additional asymptotic complexity, per client. Each access requires
downloading the cache of size $\sqrt{N}$ and accessing one block from the main memory. Every $\sqrt{N}$ accesses, the main memory and cache must be shuffled, requiring
N communication if the client has enough storage to temporarily hold the database locally. If not, then Goldreich and Ostrovsky [9] noticed that one can use
a Batcher sorting network to obliviously shuffle the database with complexity
$O(N \log^2 N)$, or the AKS algorithm with complexity $O(N \log N)$. One can also
reduce the hidden constant using a more efficient Zig-Zag sort [10]. In the first
scenario, the amortized overall complexity is then $O(\phi\sqrt{N})$, while the second is
$O(\phi\sqrt{N} \log N)$.

Goodrich et al. [11] also propose a way to deamortize the classical square-root
ORAM such that it obtains a worst-case overhead factor of $\sqrt{N} \cdot \log^2(N)$. Their
method involves dividing the work of shuffling over the $\sqrt{N}$ operations during an
epoch such that when the cache is full there is a newly shuffled main memory to
swap in right away. Since the shuffling is completely oblivious (does not depend
on any pattern of data accesses) and memoryless (the clients only need to know
what step of the shuffle they are on in order to continue the shuffle), it can be
considered a "non-critical" portion of the algorithm and no special protections
need to be added for malicious security.

Note on Computational Complexity: While Algorithm 1 returns the whole
updated state Σ, in practice a client only needs to update the other clients'
caches (up to $\sqrt{N}$ times). In addition to the communication complexity involved,
there is also computation the client must perform in our scheme. Fortunately,
the computation is exactly proportional to the communication and easily quantifiable. Every block of data retrieved from the server has a MAC that must be
verified and a layer of encryption that must be removed. Since modern ciphers
and hash functions are very efficient, and can even be done in hardware on many
computers, communication is the clear bottleneck. For comparison, encryption
and MACs are common on almost every secure network protocol, so we consider
only the communication overhead in our analysis.

Unified Cache: A natural optimization to this scheme is to have one single
shared cache instead of a separate one for each user. If the server behaves honestly, then all caches will contain the same blocks and be in the same state
anyway, so a single cache can save some communication and storage. To still
protect against a malicious server, one must be careful in this case to store ϕ different counters with the cache and have each client only increment their counter
when they do an access. This ensures that if a client inserts a block into the cache
it cannot be "rolled back" past the point that they inserted that block without

them noticing. Since the cost to reshuffle the ORAMs dominates complexity of our scheme, this optimization does not change asymptotic performance. Resulting in only a small constant improvement and making the presentation and proof unnecessary difficult, we omit full discussion of this technique.

4.2   Hierarchical Construction

In addition to the square-root ORAM, Goldreich and Ostrovsky [9] also propose a generalization which achieves poly-log overhead. In order to do this, it has a hierarchical series of caches instead of a single cache. Each cache has 2^j slots in it, for j from 1 to $\log N$, where each slot is a bucket holding $O(\log N)$ blocks. At the bottom of the hierarchy is the main memory which has $2 \cdot N$ buckets.

The reader is encouraged to refer to the original paper [9] for full details, but the main idea is that each level of the cache is structured as a hash table. Up to 2^{j-1} blocks can be stored in cache level j, half the space is reserved for dummies like in the previous construction. After accessing 2^{j-1} blocks, the entire level is retrieved and shuffled into the next level. Shuffling involves generating a new hash function and rehashing all the blocks into their new locations in level $j+1$, until the shuffling percolates all the way to the bottom, and the client must shuffle main memory to start again. Level j must be shuffled after 2^{j-1} accesses, resulting in an amortized poly-logarithmic cost.

To actually access a block, a client queries the caches in order using the unique hash function at each level. When the block is found, the remainder of the queries will be on dummy blocks to hide that the block was already found. After reading, and potentially changing the value of the block, it is added back into the first level of the cache and the cache is shuffled as necessary.

Multi-client Security: As this scheme is a generalization of the square-root one, our modifications extend naturally to provide multi-client security. Again, each client should have their own ORAM which they read from. Writing to other clients' ORAMs is done by inserting the block into the top level of their cache and then shuffling as necessary. The only difference this time is that each level of the cache must be independently authenticated. Since the cache levels are now hash tables, and computing a MAC over every level for each access would require downloading the whole data structure, we can instead use a Merkle tree [19]. This allows for efficient verification and updating of pieces of the cache without having access to the entire thing, and it maintains poly-logarithmic communication complexity. The root of the Merkle tree will contain the counter that is incremented by the ORAM owner when they perform an access.

Deamortizing: Other authors have proposed deamortized versions of the hierarchical construction that achieve worst-case poly-logarithmic complexity, such as Kushilevitz et al. [15] and Ostrovsky and Shoup [21]. We will use as an example the "warm-up" construction from Kushilevitz et al. [15], Sect. 6.1. This is a direct deamortization of the original hierarchical scheme described above. They deamortize by using three separate hash tables at each level of the ORAM, labelled "active", "inactive", and "output". Instead of shuffling all at one time after 2^{j-1} accesses (which would lead to worst case $O(N)$ complexity), their approach is now different. When the cache fills up at level j, it is marked "inactive", and the old "inactive" buffer is cleared and marked "active". The idea will be that the "inactive" buffer is shuffled over time with each ORAM access, so that no worst-case $O(N)$ operations are required. As it is shuffled, the contents are copied into the "output" buffer. Accesses can continue while the "inactive" buffer is being shuffled, as long as a read operation searches both the "active" and "inactive" buffers (since a block could be in either one). ·

When the shuffle completes, the "output" buffer will contain the newly shuffled contents that go into level $j + 1$. This buffer is marked as "active" for level $j + 1$, the "active" buffer on level j is marked "inactive" and the "inactive" buffer is cleared and marked "active", restarting the whole process. Since the shuffle is spread out over 2^{j-1} accesses, and the shuffling was the only part that was worst-case $O(N)$, this makes a full construction that now has worst-case $O(\log^3 N)$ communication complexity.

In terms of multi-client security, the only important aspects of this process is that no elements be removed from "active" or "inactive" buffers that the owner of the ORAM has put there – until a shuffle is complete, starting a new epoch. The shuffling itself is automatically data oblivious and therewith "non-critical", in the terms we have established in this paper. Using a Merkle tree and counters, as described in the amortized version, will assure that the server cannot roll back the cache to any state prior to the last access by the owner, guaranteeing security.

Kushilevitz et al. [15] also propose an improved hierarchical scheme that achieves $O(\log^2 N/\log\log N)$ complexity, which is substantially more involving. As the deamortized hierarchical ORAM as described above is sufficient for our main contribution in Sect. 5, we leave it to future research to adapt Kushilevitz et al. [15]'s scheme for multi-client security.

5   Tree-Based Construction

While pioneering the research, classical ORAMs have been outperformed by newer tree-based ORAMs which achieve better average and worst-case complexity and low constants in practice. We now proceed to show how these constructions can be modified to also support multiple clients. Our strategy will be similar to before, but with one major twist: in order to avoid linear worst case complexity, tree-based ORAMs do only small local "shuffling," which turns out to make separating a client access into critical and non-critical parts much more difficult. When writing, one must not only add a new version of the block to the ORAM, but also explicitly mark the old version as obsolete, requiring a conditional access. This is in contrast with our previous construction where old versions of a block would simply be discarded during the shuffle.

5.1 Overview

For this section, we will use Path ORAM [25] as the basis for our multi-client scheme, but the concepts apply similarly to other tree-based schemes.

Although the interface exposed to the client by Path ORAM is the same as other ORAM protocols, it is easiest to understand the Access operation as being broken down into three parts: ReadAndRemove, Add, and Evict [23]. ReadAndRemove, as the name suggests, reads a block from the ORAM and removes it, while Add adds it back to the ORAM, potentially with a different value. These two operations used together form the basis of the Access operation, but it begins to illustrate the difficulty we have making this scheme multi-client secure: changing the value of a block implicitly requires reading it, meaning that both reading and writing are equally *critical* and not easily separated as our previous construction. The third operation, Evict, is a partial shuffling that is done after each access in order to maintain the integrity of the tree.

The RAM in Path ORAM is structured as a tree with N leaf nodes. Each node in the tree holds up to Z blocks, where Z is a small constant. Each block is tagged with a value uniform in the range $[0, N)$. As an invariant, blocks will always be located on the path from the root of the tree to the leaf node corresponding to their tag. Over the lifecycle of the tree, blocks will enter at the root and filter their way down toward the leaves, making room for new blocks to in turn enter at the root. The client has a map storing for every block which leaf node the block is tagged for.

ReadAndRemove: To retrieve block x, the client looks up in the map which leaf node it is tagged for and retrieves all nodes from the root to that leaf node, denoted $\mathcal{P}(x)$. By the tree invariant, block x will be found somewhere on the path $\mathcal{P}(x)$. The client then removes block x from the node it was found in, reencrypts all the nodes and puts them back in the RAM.

Add: To put a block back in the ORAM, the client simply retrieves the root node and inserts the block into one of its free slots, reencrypting and writing the node back afterwards. The map is updated with a new random tag for this block in the interval $[0, N)$. If there is not enough room in the root node, the client keeps the block locally in a "stash" of size $Y = O(\log N)$, waiting for a later opportunity to insert the block into the tree.

Evict: So that the stash does not become too large, after every operation the client also performs an eviction which moves blocks down the tree to free up space. Eviction consists of picking a path in the tree (using reverse lexicographic order [6]) and moving blocks on that path as far down the tree as they can go, without violating the invariant. Additionally, the client inserts any matching block from the stash into the path.

Recursive Map: Typically, the client's map, which stores the tag for each block, has size $O(N \cdot \log N)$ bit and is often too large to store locally. Yet, if block size B is at least $2 \cdot \log N$ bit, the map can itself be stored *recursively* in an ORAM on the server, inducing a total communication complexity of $O(\log^2 N)$ blocks.

Additionally, Stefanov et al. [25] show that if $B = \Omega(\log^2 N)$ bit, communication complexity can be reduced to $O(\log N)$ blocks.

Integrity: Because of its tree structure, it is straightforward to ensure integrity in Path ORAM. Similar to a Merkle tree, the client can store a MAC in every node of the tree that is computed over the contents of that node and the respective MACs of its two children. Since the client accesses entire paths in the tree at once, verifying and updating the MAC values when an access is done incurs minimal overhead. This is a common strategy with tree-based ORAMs, which we will make integral use of in our scheme. We will also include client u_i's counter χ_u in the root MAC as before, to prevent rollback attacks (see below).

Challenge. Looking at Path ORAM, there exist several additional challenges when trying to add multi-client capabilities with our previous strategy. First, recursively storing the map into ORAMs imposes a problem. To resolve a tag, each path accessed in the recursive ORAMs has to be different for each client. If we separate the map into ϕ separate ORAMs (which we will do), the standard recursive lookup results in a large blowup in communication costs. At the top level of the recursion, we would have ϕ ORAMs, one for each client. Yet, each of those will fan out to ϕ ORAMs to obliviously support the next level of recursion, each of which will have ϕ more, going down $\log n$ levels. The overall communication complexity for the tag lookup would be $\phi^{\log N} \in \Omega(N)$.

Second, an **Add** in Path ORAM cannot be performed without **ReadAndRemove**, so we cannot easily split the access into *critical* and *non-critical* parts like before.

Rationale. To remedy these problems, we institute major changes to Path ORAM:

1. **Unified Tagging:** Instead of separately tagging blocks in each of the ORAMs and storing the tags recursively, we will have a unified tagging system where the tag for a block can be computed for any of the separate ORAMs from a common "base tag." This is crucial to avoiding the $O(N)$ communication overhead that would otherwise be induced by the recursive map as described above. For a block x, the map will resolve to a base tag value t. This same tag value is stored in every client's recursive ORAM. Let h be a PRF mapping from $[0, 2^\lambda) \times [1, \phi]$ to $[0, N)$. The idea is now that the leaf that block x will be percolating to in the recursive ORAM tree differs for every ORAM of every client u_i and is pseudo-randomly determined by value $h(t, i)$. This way, (1) the paths accessed in all recursive map ORAMs for all clients differ for the same block x, and (2) only one lookup is necessary at each level of the recursive map to get the leaf node tag for all ϕ ORAMs.

2. **Secure Block Removal:** The central problem with **ReadAndRemove** is that it is required before every **Add** so that the tree will not fill up with old, obsolete blocks which cannot be removed. Unlike the square-root ORAM, the shuffling process (eviction) happens locally and cannot know about other versions of

a block which exist on different paths. We solve this problem by including metadata on each bucket. For every node in the tree, we include an encrypted array which indicates the ID of every block in that node. Removing a block from the tree can then be performed by simply changing the metadata to indicate that the slot is empty. It will be overwritten by the eviction routine with a real block if that slot is ever needed. If B is large, this metadata is substantially smaller than the real blocks. We can then store it in a less efficient classical ORAM described above which is itself multi-client secure. This allows us to take advantage of the better complexity provided by tree-based ORAMs for the majority of the data, while falling back on a simpler ORAM for the metadata which is independent of B.

We also note that Path ORAM's stash concept cannot be used in a multi-client setting. Since the clients do not have a way of communicating with each other out of band, all shared state (which includes the stash) must be stored in the RAM. This has already been noted by Goodrich et al. [12], and since the size of the stash does not exceed $\log N$, storing it in the RAM (encrypted and integrity protected) does not affect the overall complexity. As before, we also introduce an eviction counter e for each ORAM. Client u_i will verify whether, for each of their recursive ORAMs, this eviction counter is fresh.

5.2   Details

To initialize the multi-client ORAM (Algorithm 3), ϕ separate ORAMs are created and the initial states (containing the shared key) are distributed to each client. For each client u_i, the ORAM takes the form of a series of trees $T_{j,i}$. The first tree stores the data blocks, while the remaining trees recursively store the map which relates block addresses to leaf nodes. In addition to this, as described above, each tree has its own sub-ORAM to keep track of block metadata. The stash of each (sub-)ORAM is called $S_{0,i}$, and the metadata (classical) ORAM $M_{j,i}$.

To avoid confusion between different ORAM initialization functions, MInit is a reference to Algorithm 1, i.e., initialization of a multi-client secure classical ORAM.

For simplicity, we assume that Enc_κ encrypts each node of a tree separately, therewith allowing individual node access. Also, we assume authenticated encryption, using the per node integrity protection previously mentioned.

As noted above, the functions (ReadAndRemove, Add) can be used to implement (Read, Write), which in turn can implement a simple interface (Access). Because our construction introduces dependencies between ReadAndRemove and Add, in Algorithm 4 we illustrate a unified Access function for our scheme. The client starts with the root block and traverses the recursive map upwards, finds the address of block x, and finally retrieves it from the main tree. For each recursive tree, it retrieves a tag value t allowing to locate the correct block in the next tree. After retrieving a block in each tree, the client marks that block as free in the metadata ORAM so that it can be overwritten during a future

Input: Security parameter λ, number of blocks in each ORAM N, block size B,
   number of clients ϕ, initialization sub-routine for multi-client classical
   ORAM MInit
Output: Initial ORAM state Σ_{init}, initial per client states $\{st_{u_1}, \dots, st_{u_\phi}\}$

1 $\kappa \xleftarrow{\$} \{0,1\}^\lambda$;
2 **for** $j = 1$ *to* ϕ **do**
3      $i = 0$;
4      $N_0 = N$;
5   **while** $N_i > 1$ **do**
6   Initialize a tree $T_{j,i}$ with N_i leaf nodes;
7   Set eviction counter $e_{j,i} = 0$;
           // The stash must also be stored on the server
8   Create array $S_{j,i}$ with Y blocks;
           // Use a sub-ORAM to hold block metadata
9          $M_{j,i} = \text{MInit}(\lambda, 2n_i \cdot Z, Z \cdot \log n_i, \phi)$;
10         $N_{i+1} = N_i \cdot \lceil \log N_i / B \rceil$;
11         $i = i + 1$;
12   **end**
       // Let m be number of recursive trees made in previous loop
13     $m = i$;
14   Create a root block $\mathcal{R}_j$;
15   Set ORAM counter $\chi_j = 0$;
16     $\text{ORAM}_j = \text{Enc}_\kappa((T_{j,0}, M_{j,0}, S_{j,0}, e_{j,0}) \| \dots \| (T_{j,m}, M_{j,m}, S_{j,m}, e_{j,m}) \| \chi_j \| \mathcal{R}_j)$;
17   Send $st_{u_j} = \{\kappa, \chi_j, e_{j,0}, \dots, e_{j_m}\}$ to client u_j;
18 **end**
19 Initialize Σ to hold $\{\text{ORAM}_1, \dots, \text{ORAM}_\phi\}$;

Algorithm 3. $\text{Init}(\lambda, N, B, \phi)$, initialize multi-client tree-based ORAM

eviction. This is necessary to maintain the integrity of the tree and ensure that it does not overflow. At the same time, the client also marks that block free in the metadata of each other client and inserts the new block value into the root of their trees. This is analogous to the previous scheme where a client reads from their own ORAM and writes back to the ORAMs of the other clients. We use a straightforward MAC technique for paths MACPath, which we present in the extended version of this paper [1].

Again, we avoid confusion between different ORAM access operations by referring to the multi-client secure classical ORAM access operation of Algorithm 2 as MAccess.

Algorithm 5 illustrates the eviction procedure. Since eviction does not take as input any client access, it is non-critical. The client simply downloads a path in the tree which is specified by eviction counter e and retrieves it in its entirety. The only modification that we make from the original Path ORAM scheme is that we read block metadata from the sub-ORAM that indicates which blocks in the path are free and can be overwritten by new blocks being pushed down the tree.

Input: Address x, client u_i, $st_{u_i} = \{\kappa, \chi_i\}$, sub-routine for multi-client classical ORAM MAccess

Output: The value of block x

// Let m be recursion depth, n_j the number of blocks in tree j

1 Retrieve root block $\mathcal{R}$;
2 $pos = x/n$; $x_m = \lfloor pos \cdot (B/\lambda) \rfloor$; $t_m = \mathcal{R}[x_m]$// Find tag t_m of address x;
3 $t'_m \xleftarrow{\$} [0, 2^\lambda)$// Compute new tag t'_m for x;
4 **for** $j = m$ to 0 **do**
5      $leaf_j = h(t_j, u_i)$// Compute leaf of client u_i's ORAM$_i$;
6   Read path $\mathcal{P}(leaf_j)$ and $S_{i,j}$ from $T_{i,j}$, locating block x_j;
7   Retrieve MAC values for $\mathcal{P}(leaf_j)$ as V and the stored counter as χ'_i;
8   **if** $V \neq \mathsf{MACPath}(\Sigma, st_{u_i}, \mathcal{P}(leaf_j), S, \chi'_i) \vee \chi'_i \neq \chi_i$ **then** Abort ;
9   Re-encrypt and write back $\mathcal{P}(leaf_j)$ and $S_{i,j}$ to $T_{i,j}$;
        // Let (a, b) be the node and slot that x_j was found at
10     $\mathsf{MAccess}(M_j, (\mathsf{write}, a \cdot Z + b, \bot), u_i)$;
11   **if** $j \neq 0$ **then**
12         $t'_j \xleftarrow{\$} [0, 2^\lambda)$ // Sample a new value for t;
           // Block x_j contains multiple t values
13   Extract t_{j-1} from block x_j;
14   Update block x_j with new value t'_{j-1} and new leaf tag t'_j;
15   **else**
16   Set v to the value of block x_j;
17   If OP is a write, update x_j with new value;
18   Insert block x_j into the stash $S_{i,j}$;
19     $\chi_i = \chi_i + 1$;
20   Update MAC of stash to $\mathsf{MAC}_\kappa(S_{i,j}, \text{MAC of root bucket}, \chi_i, e_{i,j})$;
       // Update the block in other client's ORAMs
21   **for** $p \neq i$ **do**
22   Retrieve path $\mathcal{P}(h(t_j, u_p))$ from $T_{p,j}$ and update metadata so block x_j is removed;
23   Insert block x_j into the stash $S_{p,j}$ of $T_{p,j}$;
24   Update MAC of root bucket in $T_{p,j}$;
25   **end**
26   **output** $(v, st_{u_i} = \{\kappa, \chi_i, e_{i,0}, \ldots, e_{i,m}\})$;
27 **end**

Algorithm 4. $\mathsf{Access}(\mathsf{OP}, \Sigma, st_{u_i})$, Read or Write on multi-client tree-based ORAM

The overhead from the additional metadata ORAMs that we have in our construction is fortunately not dependent on the block size B. Therefore, if B is large enough, we can achieve as low as the overhead of single-client Path ORAM, $O(\log N)$, or a total complexity of $O(\phi \log N)$ for ϕ users. However, this only applies if the block size B is sufficiently large, at least $\Omega(\log^4 N)$. Otherwise, for smaller B, the complexity can be up to $O(\phi \log^5 N)$. Due to limited space, we refer to the extended version of this paper [1] for a detailed security analysis.

Input: Address x, new value v, client u_i, $st_{u_i} = \{k, \chi_i\}$, the number of
      recursive trees m
Output: The value of block x
1 **for** $j = 1$ to ϕ **do**
2   **for** $r = 1$ to m **do**
3   Retrieve eviction counter $e_{j,r}$ for $T_{j,r}$;
4   Retrieve path $\mathcal{P}(e_{j,r})$, $S_{j,r}$ and MAC chain V;
           // Verify integrity of the path and eviction counter
5   **if** $V \neq \mathsf{MACPath}(\Sigma, st_{u_i}, \mathcal{P}(\text{leaf}_j), S_{j,r}, \chi'_i, e_{j,r})$ **then** Abort ;
6   Read metadata for path from $M_{j,r}$;
7   Move blocks out of the stash and down the path as far as possible;
8   Reencrypt $\mathcal{P}(e_{j,r})$ and $S_{j,r}$ and write back to server;
9   Update metadata for path $M_{j,r}$;
10         $e_{j,r} = e_{j,r} + 1$;
11   **end**
12 **end**

Algorithm 5. $\mathsf{Evict}(\Sigma, st_{u_i})$ – Perform Evict on multi-client tree-based ORAM

Complexity: The complexity of our scheme is dominated by the cost of an
eviction. For a client to read a path in each of $O(\log N)$ recursive trees, for each
of the ϕ different ORAMs, it takes $O(\phi \cdot B \cdot \log^2 N)$ bits of communication.
Additionally, the client must make $O(\phi \cdot \log^2 N)$ accesses to a metadata ORAM.
If $\mu(N, B)$ denotes the cost of a single access in such a sub-ORAM, the overall
communication complexity is then $O(\phi \cdot \log^2 N \cdot [B + \mu(N, \log N)])$ bit. The
deamortized hierarchical ORAM by Kushilevitz et al. [15] has $O(\log^3 N)$ blocks
communication complexity, where each block is of size $\log N$ bit (the meta-data
we need for our construction). Taking this hierarchical ORAM as a sub-ORAM,
the total communication complexity computes to $O(\phi \cdot \log^2 N [B + \log^4 N])$ bits.
If $B \in \Omega(\log^4 N)$ then the communication complexity, in terms of blocks, is
$O(\phi \log^2 N)$, otherwise it is at most $O(\phi \log^5 N)$, i.e., with the assumption $B \in$
$\Omega(\log N)$ (the minimal possible block size for Path ORAM to work).

Additionally, if we use the recursive optimization trick from Stefanov et al.
[25] to reduce the overhead from the Path ORAM part of the construction from
$O(\log^2 N)$ to $O(\log N)$, we can achieve a total complexity of $O(\log N)$ for blocks
of size $\Omega(\log^4 N)$.

Although a complexity linear in ϕ may seem at first to be expensive, we
stress that this is a substantial improvement over naive solutions which achieve
the same level of security. The only straightforward way to have multi-client
security against malicious servers is for each client to append their updates to a
master list, and for clients to scan this list to find the most updated version of
a block during reads. This is not only linear in the size of the database, but in
the number of operations performed over the entire life of the ORAM.

One notable difference in parameters from basic Path ORAM is that we
require a block size of at least $c \cdot \lambda$, where $c \geq 2$. Path ORAM only needs $c \cdot \log n$,
and for security parameter λ, $\lambda > \log N$ holds. In our scheme, the map trees do
not directly hold addresses, but t values which are of size λ. In order for the map

recursion to terminate in $O(\log N)$ steps, blocks must be big enough to hold at least two t values of size λ. If the block size is $\Omega(\lambda^2)$, we can also take advantage of the asymmetric block optimization from Stefanov et al. [25] to reduce the complexity to $O(\phi \cdot (\log^6 n + B \cdot \log N)$. Then, if additionally $B \in \Omega(\log^5 N)$, the total complexity is reduced to $O(\log N)$ per client.

6   Conclusion

We have presented the first techniques that allow multi-client ORAM, specifically secure against *fully malicious* servers. Our multi-client ORAMs are reasonably efficient with communication complexities as low as $O(\log N)$ per client. Future work will focus on efficiency improvements, including reducing worst-case complexity to sublinear in ϕ. Additionally, the question of whether tree-based constructions are more efficient than classical ones is not as clear in the multi-client setting as it is for a single client. Although tree ORAMs are more efficient for a number of parameter choices, they incur substantial overhead from using sub-ORAMs to hold tree metadata. This is not required for the classical constructions. Future research may focus on achieving a "pure" tree-based construction which does not depend on another ORAM as a subroutine. Finally, it may be interesting to investigate whether multiple clients can be supported with a more fine-grained access control, secure against fully malicious servers.

References

1. Blass, E.-O., Mayberry, T., Noubir, G.: Multi-client oblivious ram secure against malicious servers. Cryptology ePrint Archive, Report 2015/121 (2015). http://eprint.iacr.org/2015/121
2. Boyle, E., Chung, K.-M., Pass, R.: Oblivious parallel RAM and applications. In: Kushilevitz, E., Malkin, T. (eds.) TCC 2016. LNCS, vol. 9563, pp. 175–204. Springer, Heidelberg (2016). doi:10.1007/978-3-662-49099-0_7
3. Cash, D., Küpçü, A., Wichs, D.: Dynamic proofs of retrievability via oblivious RAM. In: Johansson, T., Nguyen, P.Q. (eds.) EUROCRYPT 2013. LNCS, vol. 7881, pp. 279–295. Springer, Heidelberg (2013). doi:10.1007/978-3-642-38348-9_17
4. Chan, T.-H.H., Shi, E.: Circuit OPRAM: a (somewhat) tight oblivious parallel RAM. Cryptology ePrint Archive, Report 2016/1084 (2016). http://eprint.iacr.org/2016/1084
5. Costan, V., Devadas, S.: Intel SGX explained. Technical report, Cryptology ePrint Archive, Report 2016/086 (2016). https://eprint.iacr.org/2016/086
6. Fletcher, C.W., Ren, L., Kwon, A., Van Dijk, M., Stefanov, E., Devadas, S., Tiny, O.: A Low-Latency, Low-Area Hardware ORAM Controller. Cryptology ePrint Archive, Report 2014/431 (2014). http://eprint.iacr.org/
7. Franz, M., Williams, P., Carbunar, B., Katzenbeisser, S., Peter, A., Sion, R., Sotakova, M.: Oblivious outsourced storage with delegation. In: Danezis, G. (ed.) FC 2011. LNCS, vol. 7035, pp. 127–140. Springer, Heidelberg (2012). doi:10.1007/978-3-642-27576-0_11
8. Goldreich, O.: Towards a theory of software protection and simulation by oblivious RAMs. In: Symposium on Theory of Computing, pp. 182–194, New York, USA (1987)

9. Goldreich, O., Ostrovsky, R.: Software protection, simulation on oblivious RAMs. J. ACM **43**(3), 431–473 (1996). ISSN 0004–5411
10. Goodrich, M.T.: Zig-zag sort: a simple deterministic data-oblivious sorting algorithm running in o(n log n) time. In: Proceedings of the 46th Annual ACM Symposium on Theory of Computing, STOC 2014, pp. 684–693 (2014). ISBN 978-1-4503-2710-7
11. Goodrich, M.T., Mitzenmacher, M., Ohrimenko, O., Tamassia, R.: Oblivious RAM simulation with efficient worst-case access overhead. In: Proceedings of Workshop on Cloud Computing Security Workshop, pp. 95–100, Chicago, USA (2011)
12. Goodrich, M.T., Mitzenmacher, M., Ohrimenko, O., Tamassia, R.: Privacy-preserving group data access via stateless oblivious RAM simulation. In: Proceedings of the Twenty-Third Annual ACM-SIAM Symposium on Discrete Algorithms, pp. 157–167 (2012)
13. Iliev, A., Smith, S.W.: Protecting client privacy with trusted computing at the server. IEEE Secur. Priv. **3**(2), 20–28 (2005)
14. Knuth, D.E.: The Art of Computer Programming, Seminumerical Algorithms, 2nd edn., vol. 2, chap. 3.4.2, pp. 139–140. Addison Wesley 1981. ISBN 978-0201896848
15. Kushilevitz, E., Lu, S., Ostrovsky, R.: On the (in) security of hash-based oblivious RAM and a new balancing scheme. In: Proceedings of Symposium on Discrete Algorithms, pp. 143–156, Kyoto, Japan (2012)
16. Li, J., Krohn, M.N., Mazières, D., Shasha, D.: Secure untrusted data repository (SUNDR). In: Proceedings of Operating System Design and Implementation, pp. 121–136, San Francisco, USA (2004)
17. Lorch, J.R., Parno, B., Mickens, J., Raykova, M., Schiffman, J.: Shroud: ensuring private access to large-scale data in the data center. In: USENIX Conference on File and Storage Technologies, pp. 199–213 (2013)
18. Maffei, M., Malavolta, G., Reinert, M., Schröder, D.: Privacy and access control for outsourced personal records. In: IEEE Symposium on Security and Privacy, pp. 341–358. IEEE (2015)
19. Merkle, R.C.: A digital signature based on a conventional encryption function. In: Pomerance, C. (ed.) CRYPTO 1987. LNCS, vol. 293, pp. 369–378. Springer, Heidelberg (1988). doi:10.1007/3-540-48184-2_32
20. Nayak, K., Katz, J.: An oblivious parallel RAM with $o(\log^2 n)$ parallel runtime blowup. Cryptology ePrint Archive, Report 2016/1141 (2016). http://eprint.iacr.org/2016/1141
21. Ostrovsky, R., Shoup, V.: Private information storage. In: Proceedings of Symposium on Theory of Computing, pp. 294–303. ACM (1997)
22. Ren, L., Fletcher, C.W., Yu, X., van Dijk, M., Devadas, S.: Integrity verification for path oblivious-RAM. In: Proceedings of High Performance Extreme Computing Conference, pp. 1–6, Waltham, USA (2013)
23. Shi, E., Chan, T.-H.H., Stefanov, E., Li, M.: Oblivious RAM with $O((\log N)^3)$ worst-case cost. In: Lee, D.H., Wang, X. (eds.) ASIACRYPT 2011. LNCS, vol. 7073, pp. 197–214. Springer, Heidelberg (2011). doi:10.1007/978-3-642-25385-0_11
24. SpiderOak. Semaphor (2016). https://spideroak.com/solutions/semaphor
25. Stefanov, E., van Dijk, M., Shi, E., Fletcher, C., Ren, L., Yu, X., Devadas, S., Path, O.: An extremely simple oblivious RAM protocol. In: Proceedings of Conference on Computer & Communications Security, pp. 299–310, Berlin, Germany (2013). ISBN 978-1-4503-2477-9
26. WhatsApp. Whatsapp encryption overview (2016). https://www.whatsapp.com/security/WhatsApp-Security-Whitepaper.pdf

Author Index

710   Author Index

Printed in the United States
by Publishers, Inc.

Printed in the United States
By Bookmasters